SECOND EDITION

# CALCULUS
## EARLY TRANSCENDENTALS

MICHAEL SULLIVAN

Chicago State University

KATHLEEN MIRANDA

State University of New York,
Old Westbury

w.h.freeman
Macmillan Learning
New York

*Vice President, STEM:* Daryl Fox
*Program Director:* Andrew Dunaway
*Program Manager:* Nikki Miller Dworsky
*Development Editor:* Andrew Sylvester
*Associate Editor:* Andy Newton
*Editorial Assistant:* Justin Jones
*Senior Marketing Manager:* Nancy Bradshaw
*Marketing Assistant:* Savannah DiMarco
*Executive Media Editor:* Catriona Kaplan
*Assistant Media Editor:* Doug Newman
*Director of Digital Production:* Keri deManigold
*Senior Media Project Manager:* Alison Lorber
*Photo Editor:* Christine Buese
*Photo Research:* Donna Ranieri
*Director of Content Management Enhancement:* Tracey Kuehn
*Managing Editor:* Lisa Kinne
*Director of Design, Content Management:* Diana Blume
*Cover Design:* Lumina Datamatics
*Text Design:* Diana Blume
*Senior Content Project Manager:* Edgar Doolan
*Senior Workflow Manager:* Paul Rohloff
*Illustration Coordinator:* Janice Donnola
*Illustrations:* Network Graphics
*Composition:* Lumina Datamatics Ltd.
*Printing and Binding:* LSC Communications
*Cover and Title Page Image:* Background All/Shutterstock

Library of Congress Control Number: 2018935580

ISBN-13: 978-1-319-01835-1
ISBN-10: 1-319-01835-1

1  2  3  4  5  6       23  22  21  20  19  18

W. H. Freeman and Company
One New York Plaza
Suite 4500
New York, NY 10004-1562
www.macmillanlearning.com

# What Is Calculus?

Calculus is a part of mathematics that evolved much later than other subjects. Algebra, geometry, and trigonometry were developed in ancient times, but calculus as we know it did not appear until the seventeenth century.

The first evidence of calculus has its roots in ancient mathematics. For example, In his book *A History of* π, Petr Beckmann explains that Greek mathematician Archimedes (287–212 BCE) "took the step from the concept of 'equal to' to the concept of 'arbitrarily close to' or 'as closely as desired'…and reached the threshold of the differential calculus, just as his method of squaring the parabola reached the threshold of the integral calculus."* But it was not until Sir Isaac Newton and Gottfried Wilhelm Leibniz, each working independently, expanded, organized, and applied these early ideas, that the subject we now know as calculus was born.

Although we attribute the birth of calculus to Newton and Leibniz, many other mathematicians, particularly those in the eighteenth and nineteenth centuries, contributed greatly to the body and rigor of calculus. You will encounter many of their names and contributions as you pursue your study of calculus.

But what is calculus?

The simple answer is: calculus models change. Since the world and most things in it are constantly changing, mathematics that explains change becomes immensely useful.

Calculus has two major branches, differential calculus and integral calculus. Let's take a peek at what calculus is by looking at two problems that prompted the development of calculus.

## The Tangent Problem—The Basis of Differential Calculus

Suppose we want to find the slope of the line tangent to the graph of a function at some point $P = (x_1, y_1)$. See Figure 1(a). Since the tangent line necessarily contains the point $P$, it remains only to find the slope to identify the tangent line. Suppose we repeatedly zoom in on the graph of the function at the point $P$. See Figure 1(b). If we can zoom in close enough, then the graph of the function will look approximately linear, and we can choose a point $Q$, on the graph of the function different from the point $P$, and use the formula for slope.

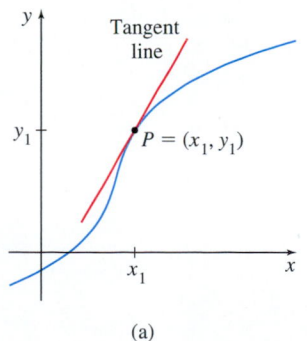

(a)

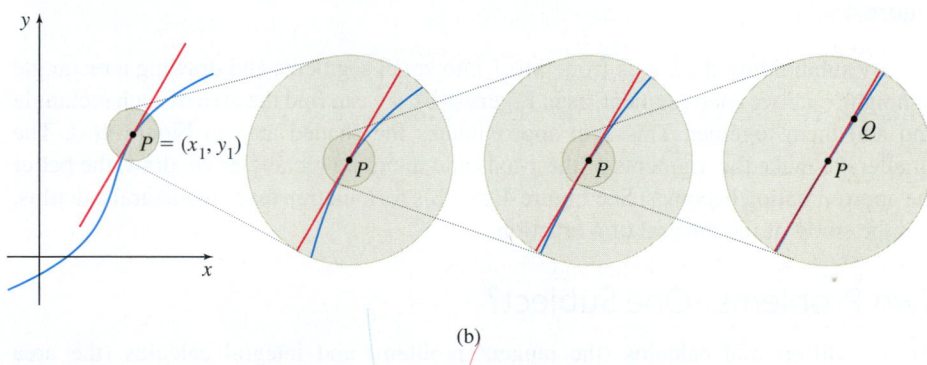

(b)

**Figure 1**

Repeatedly zooming in on the point $P$ is equivalent to choosing a point $Q$ closer and closer to the point $P$. Notice that as we zoom in on $P$, the line connecting the points $P$ and $Q$, called a secant line, begins to look more and more like the tangent line to the graph of the function at the point $P$. If the point $Q$ can be made as close as we please to the point $P$, without equaling the point $P$, then the slope of the tangent line $m_{\tan}$ can be found. This formulation leads to differential calculus, the study of the derivative of a function.

The derivative gives us information about how a function changes at a given instant and can be used to solve problems involving velocity and acceleration; marginal cost and profit; and the rate of change of a chemical reaction. Derivatives are the subjects of Chapters 2 through 4.

## The Area Problem—The Basis of Integral Calculus

If we want to find the area of a rectangle or the area of a circle, formulas are available. (See Figure 2.) But what if the figure is curvy, but not circular as in Figure 3? How do we find this area?

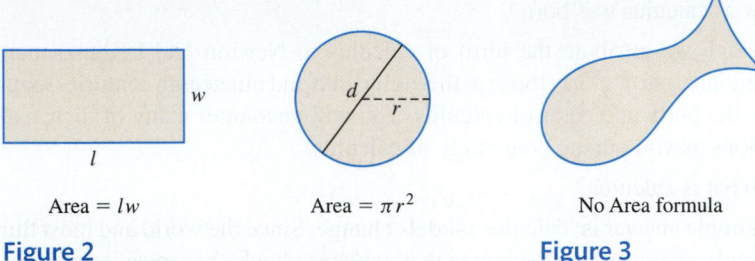

Area = $lw$          Area = $\pi r^2$          No Area formula

**Figure 2**                              **Figure 3**

Calculus provides a way. Look at Figure 4(a). It shows the graph of $y = x^2$ from $x = 0$ to $x = 1$. Suppose we want to find the area of the shaded region.

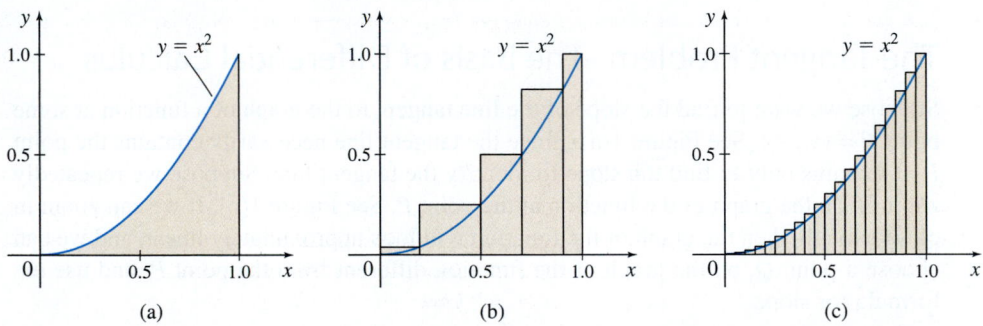

(a)                    (b)                    (c)

**Figure 4**

By subdividing the $x$-axis from 0 to 1 into small segments and drawing a rectangle of height $x^2$ above each segment, as in Figure 4(b), we can find the area of each rectangle and add them together. This sum approximates the shaded area in Figure 4(a). The smaller we make the segments of the $x$-axis and the more rectangles we draw, the better the approximation becomes. See Figure 4(c). This formulation leads to integral calculus, and the study of the integral of a function.

## Two Problems—One Subject?

At first differential calculus (the tangent problem) and integral calculus (the area problem) appear to be different, so why call both of them calculus? The Fundamental Theorem of Calculus establishes that the derivative and the integral are related. In fact, one of Newton's teachers, Isaac Barrow, recognized that the tangent problem and the area problem are closely related, and that derivatives and integrals are inverses of each other. Both Newton and Leibniz formalized this relationship between derivatives and integrals in the *Fundamental Theorem of Calculus*.

---

* Beckmann, P. (1976). *A History of π* (3rd. ed., p. 64). New York: St. Martin's Press.

# Contents | *Calculus,* *Early Transcendentals,* Second Edition

# Preface

## To the Student

As you begin your calculus course, you may feel anxious about the large number of theorems, definitions, problems, and pages that you see in the book. Don't worry, your concerns are normal. As educators we understand the challenges you face in this course, and as authors we have crafted a text that gives you the tools to meet them. If you pay attention in class, keep up with your homework, and read and study what is in this book, you will build the knowledge and skills you need to be successful. Here are some ways that you can use the book to your benefit.

- **Read actively.** When you are busy, it is tempting to skip reading and go right to the problems. Don't! This book has examples with clear explanations to help you learn how to logically solve mathematics problems. Reading the book actively, with a pencil in hand, and doing the worked examples along with us will help you develop a clearer understanding than you will get from simple memorization.

- **Read before class, not after.** If you do, the classroom lecture will sound familiar, and you can ask questions about material you don't understand. You'll be amazed at how much more you get out of class if you read first.

- **Use the examples.** Mastering calculus requires a deep understanding of the concepts. Read and work the examples, and do the suggested NOW WORK Problems listed after them. Also do all the problems your instructor assigns. Be sure to seek help as soon as you feel you need it.

- **Use the book's features.** In the classroom, we use many methods to communicate. These methods are reflected in various features that we have created for this book. These features, described on the following pages, are designed to make it easier for you to understand, review, and practice. Taking advantage of the features will help you to master calculus.

- **Engage with the applications.** Students come to this course with a wide variety of educational goals. You may be majoring in life sciences or social sciences, business or economics, engineering or mathematics. This book has many applied exercises—some written by students like you—related to courses in a wide variety of majors.

Please do not hesitate to contact us with any questions, suggestions, or comments that might improve this text. We look forward to hearing from you, and we hope that you enjoy learning calculus and do well—not only in calculus but in all your studies.

Best Wishes,

*Mike*    *Kathleen*

Michael Sullivan and Kathleen Miranda

## To the Instructor

The challenges facing calculus instructors are daunting. Diversity among students, both in their mathematical preparedness for learning calculus and in their ultimate educational and career goals, is vast and growing. Increased emphasis by universities and administrators on structured assessment and its related data analytics puts additional demands on instructors.

There is not enough classroom time to teach every topic in the syllabus, to answer every student's questions, or to delve into the rich examples and applications that showcase the beauty and utility of calculus. As mathematics instructors, we share these frustrations. As authors, our goal is to craft a calculus textbook that supports your teaching philosophy, promotes student understanding, and identifies measurable learning outcomes.

*Calculus, Early Transcendentals,* Second Edition, is a mathematically precise calculus book that embraces proven pedagogical strategies to increase both student and instructor success.

- **The language of the text is simple, clear, and precise.** Definitions and theorems are carefully stated, and titled when appropriate. Numbering of definitions, equations, and theorems is kept to a minimum.

- **The mathematics is rigorous.** Proofs of theorems are presented throughout the text in one of three ways: (1) following the statement of a theorem; (2) preceding a theorem followed by "We have proved the following theorem"; or (3) in Appendix B, if the proof is more involved. (When the proof of a theorem is beyond the scope of this book, we note that it may be found in advanced calculus or numerical analysis texts.)

- **The text is adaptable to your class and your teaching style.** Proofs can be included or omitted and exercises are paired, grouped, and graduated from easy to difficult. By dividing exercises into **Skill Building, Applications and Extensions,** and **Challenge Problems,** we make it easier for you to tailor assignments to the specific needs of your students based on their ability and preparedness.

- **The structural pedagogical features,** described on the pages that follow, can be used to design an assessment tool for your students. Learning objectives are listed at the start of each section; the structure of the solved examples and/or the proofs can be used to write rubrics for evaluation; and assessment questions can be drawn from the Now Work Problems, end of section exercises, and chapter review problems. Additional media and supplement resources available to you and your students are described on page xiii.

Let us know what you think of the text and do not hesitate to contact us with any suggestions or comments that might improve it. We look forward to hearing from you.

Best Wishes,

*Mike*    *Kathleen*

Michael Sullivan and Kathleen Miranda

For more information about the book and the authors' explanation of its organization and presentation,
go to macmillanlearning.com/exploresullivan2e

# Pedagogical Features Promote Student Success

**Just In Time Review** Throughout the text there are margin notes labeled **NEED TO REVIEW?** followed by a topic and page references. These references point to a previous presentation of a concept related to the discussion at hand.

**RECALL** margin notes provide a quick refresher of key results that are being used in theorems, definitions, and examples.

**NEED TO REVIEW?** Summation properties are discussed in Section P.8, pp. 72–73.

**RECALL** $\sum_{i=1}^{n} i = \frac{n(n+1)}{2}$ (p. 72)

Using summation properties, we get

$$S_n = \sum_{i=1}^{n} \frac{300}{n^2} i = \frac{300}{n^2} \sum_{i=1}^{n} i = \frac{300}{n^2} \cdot \frac{n(n+1)}{2} = 150\left(\frac{n+1}{n}\right) = 150\left(1 + \frac{1}{n}\right)$$

**IN WORDS** The average value $\bar{y}$ of a function $f$ equals the value $f(u)$ in the Mean Value Theorem for Integrals.

**DEFINITION  Average Value of a Function over an Interval**

Let $f$ be a function that is continuous on the closed interval $[a, b]$. The **average value $\bar{y}$ of $f$ over $[a, b]$** is

$$\bar{y} = \frac{1}{b-a} \int_a^b f(x)\, dx \qquad (8)$$

**IN WORDS** These notes translate complex formulas, theorems, proofs, rules, and definitions using plain language that provide students with an alternate way to learn the concepts.

**NOTE** In Section 2.3 we proved the Simple Power Rule, $\frac{d}{dx} x^n = nx^{n-1}$ where $n$ is a positive integer. Here we have extended the Simple Power Rule from positive integers to all integers. In Chapter 3 we extend the result to include all real numbers.

**CAUTIONS** and **NOTES** provide supporting details or warnings about common pitfalls. **ORIGINS** give biographical information or historical details about key figures and discoveries in the development of calculus.

**CAUTION** In writing an indefinite integral $\int f(x)\, dx$, remember to include the "$dx$."

**ORIGINS** Willard F. Libby (1908–1980) grew up in California and went to college and graduate school at UC Berkeley. Libby was a physical chemist who taught at the University of Chicago and later at UCLA. While at Chicago, he developed the methods for using natural carbon-14 to date archaeological artifacts. Libby won the Nobel Prize in Chemistry in 1960 for this work.

**Effective Use of Color** The text contains an abundance of graphs and illustrations that carefully utilize color to make concepts easier to visualize.

**Dynamic Figures** The text includes many pieces of art that students can interact with through the online e-Book. These dynamic figures, indicated by the icon **DF** next to the figure label, illustrate select principles of calculus, including limits, rates of change, solids of revolution, convergence, and divergence.

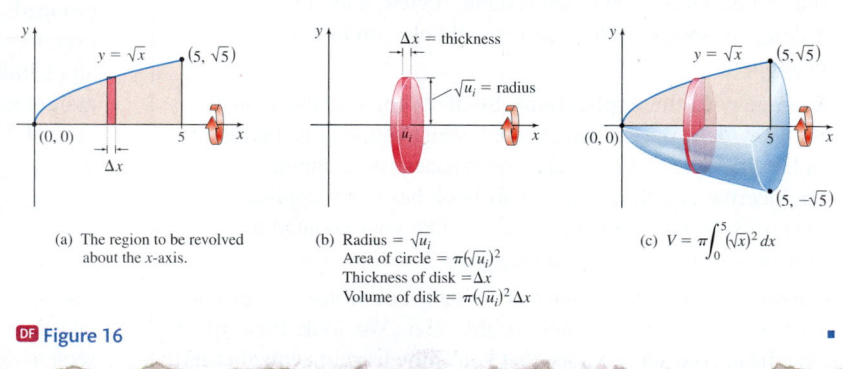

(a) The region to be revolved about the $x$-axis.

(b) Radius $= \sqrt{u_i}$
Area of circle $= \pi(\sqrt{u_i})^2$
Thickness of disk $= \Delta x$
Volume of disk $= \pi(\sqrt{u_i})^2 \Delta x$

(c) $V = \pi \int_0^5 (\sqrt{x})^2\, dx$

**DF Figure 16**

## 5.3 The Fundamental Theorem of Calculus

**OBJECTIVES** *When you finish this section, you should be able to:*

1 Use Part 1 of the Fundamental Theorem of Calculus (p. 378)
2 Use Part 2 of the Fundamental Theorem of Calculus (p. 379)
3 Interpret the integral of a rate of change (p. 381)
4 Interpret the integral as an accumulation function (p. 382)

**Learning Objectives** Each section begins with a set of **Objectives** that serve as a broad outline of the section. The objectives help students study effectively by focusing attention on the concepts being covered. Each learning objective is supported by appropriate definitions, theorems, and proofs. One or more carefully chosen examples enhance the learning objective, and where appropriate, an application example is also included. Learning objectives help instructors prepare a syllabus that includes the important topics of calculus, and concentrate instruction on mastery of these topics.

**Examples with Detailed and Annotated Solutions**  Examples are named according to their purpose. The text includes hundreds of examples with detailed step-by-step solutions. Additional annotations provide the formula used, or the reasoning needed, to perform a particular step of the solution. Where procedural steps for solving a type of problem are given in the text, these steps are followed in the solved examples. Often a figure or a graph is included to complement the solution.

**Immediate Reinforcement of Skills**  Following most examples there is the statement, NOW WORK . This callout directs students to a related exercise from the section's problem set. Doing a related problem immediately after working through a solved example encourages active learning, enhances understanding, and strengthens a student's NOW WORK Problem 51. ability to use the objective. This practice also serves as a confidence-builder for students.

---

| EXAMPLE 4 | **Using Implicit Differentiation with the Chain Rule** |
| --- | --- |

Find $y'$ if $\tan^2 y = e^x$. Express $y'$ as a function of $x$ alone.

**Solution** Assuming $y$ is a differentiable function of $x$, differentiate both sides with respect to $x$.

$$\frac{d}{dx}\tan^2 y = \frac{d}{dx}e^x$$

$$2\tan y \sec^2 y \frac{dy}{dx} = e^x$$

Use the Chain Rule on the left.
$$\frac{d}{dx}\tan y = \sec^2 y\frac{dy}{dx}$$

$$y' = \frac{dy}{dx} = \frac{e^x}{2\tan y \sec^2 y}$$

Solve for $y' = \dfrac{dy}{dx}$.

$$= \frac{e^x}{2e^{x/2}(1+e^x)} = \frac{e^{x/2}}{2(1+e^x)}$$

$\tan y = e^{x/2}$
$\sec^2 y = 1 + \tan^2 y = 1 + e^x$

NOW WORK Problem 25. ∎

---

**CHAPTER 5 PROJECT**  **Managing the Klamath River**

There is a gauge on the Klamath River, just downstream from the dam at Keno, Oregon. The U.S. Geological Survey has posted flow rates for this gauge every month since 1930. The averages of these monthly measurements since 1930 are given in Table 1. Notice that the data in Table 1 measure the rate of change of the volume $V$ in cubic feet of water each second over one year; that is, the table gives $\dfrac{dV}{dt} = V'(t)$ in cubic feet per second, where $t$ is in months.

**Chapter Projects**  The case studies that open each chapter demonstrate how major concepts apply to recognizable and often contemporary situations in biology, environmental studies, astronomy, engineering, technology, and other fields. At the end of the chapter, there is an extended project with questions that guide students through a solution to the situation.

## 4.2 Assess Your Understanding

### Concepts and Vocabulary

1. *True or False*  Any function $f$ that is defined on a closed interval $[a, b]$ will have both an absolute maximum value and an absolute minimum value.

2. *Multiple Choice*  A number $c$ in the domain of a function $f$ is called a(n) [(**a**) extreme value (**b**) critical number (**c**) local number] of $f$ if either $f'(c) = 0$ or $f'(c)$ does not exist.

3. *True or False*  At a critical number, there is a local extreme value.

---

## The Exercises are Divided into Four Categories

**Concepts and Vocabulary** provide a selection of quick fill-in-the-blank, multiple choice, and true/false questions. These problems assess a student's comprehension of the main points of the section.

---

### Skill Building

*In Problems 5–10, find the derivative of each function f at any real number c. Use Form (1) on page 163.*

5. $f(x) = 10$  6. $f(x) = -4$  7. $f(x) = 2x + 3$

8. $f(x) = 3x - 5$  9. $f(x) = 2 - x^2$  10. $f(x) = 2x^2 + 4$

*In Problems 11–16, differentiate each function f and determine the domain of f'. Use Form (3) on page 164.*

11. $f(x) = 5$  12. $f(x) = -2$

13. $f(x) = 3x^2 + x + 5$  14. $f(x) = 2x^2 - x - 7$

15. $f(x) = 5\sqrt{x-1}$  16. $f(x) = 4\sqrt{x+3}$

*In Problems 45–48, each function f is continuous for all real numbers, and the graph of $y = f'(x)$ is given.*

(a) Does the graph of f have any horizontal tangent lines? If yes, explain why and identify where they occur.

(b) Does the graph of f have any vertical tangent lines? If yes, explain why, identify where they occur, and determine whether the point is a cusp of f.

(c) Does the graph of f have any corners? If yes, explain why and identify where they occur.

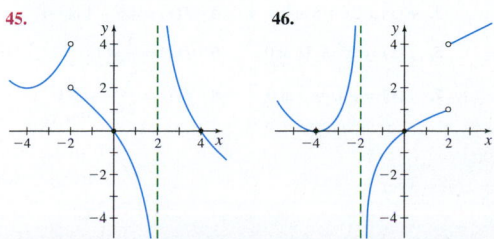

45.

46.

**Skill Building** problems are grouped into subsets, usually corresponding to the objectives of the section. Problem numbers in red are the NOW WORK Problems that are correlated to the solved examples. The skill building problems are paired by odd/even and usually progress from easy to difficult.

**Applications and Extensions** consist of applied problems, questions that require proof, and problems that extend the concepts of the section. (See the Applications Index for a full list of the applied examples and problems in the text.)

**Applications and Extensions**

**21. Work to Lift an Elevator** How much work is required if six cables, each weighing 0.36 lb/in., lift a 10,000-lb elevator 400 ft? Assume the cables work together and equally share the weight of the elevator.

**22. Work by Gravity** A cable with a uniform linear mass density of 9 kg/m is being unwound from a cylindrical drum. If 15 m are already unwound, what is the work done by gravity in unwinding another 60 m of the cable?

**23. Work to Lift a Bucket and Chain** A uniform chain 10 m long and with mass 20 kg is hanging from the top of a building 10 m high. If a bucket filled with cement of mass 75 kg is attached to the end of the chain, how much work is required to pull the bucket and chain to the top of the building?

**Challenge Problems**

**86.** Explain why the function $f(x) = x^n + ax + b$, where $n$ is a positive even integer, has at most two distinct real zeros.

**88.** Explain why the function $f(x) = x^n + ax^2 + b$, where $n$ is a positive odd integer, has at most three distinct real zeros.

**92.** Show that $x \leq \ln(1+x)^{1+x}$ for all $x > -1$.
   [*Hint:* Consider $f(x) = -x + \ln(1+x)^{1+x}$.]

**Challenge Problems** provide more difficult extensions of the section's material. They are often combined with concepts learned in previous chapters. Challenge problems are intended to be thought-provoking and require some ingenuity to solve—they are an excellent addition to assignments for exceptionally strong students or for use in group work.

# The Chapter Review is in Three Parts

**Things to Know** is a detailed list of important definitions, formulas, and theorems contained in the chapter, and page references to where they can be found.

## Chapter Review

### THINGS TO KNOW

**2.1 Rates of Change and the Derivative**
 • **Definition** Derivative of a function at a number
$$f'(c) = \lim_{x \to c} \frac{f(x) - f(c)}{x - c}$$
provided the limit exists. (p. 159)

*Three Interpretations of the Derivative*
 • *Geometric* If $y = f(x)$, the derivative $f'(c)$ is the slope of the tangent line to the graph of $f$ at the point $(c, f(c))$. (p. 154)
 • *Rate of change of a function* If $y = f(x)$, the derivative $f'(c)$ is the rate of change of $f$ with respect to $x$ at $c$. (p. 156)
 • *Physical* If the signed distance $s$ from the origin at time $t$ of an object in rectilinear motion is given by the position function $s = f(t)$, the derivative $f'(t_0)$ is the velocity of the object at time $t_0$. (p. 158)

### OBJECTIVES

| Section | You should be able to ... | Examples | Review Exercises |
|---|---|---|---|
| 2.1 | 1 Find equations for the tangent line and the normal line to the graph of a function (p. 155) | 1 | 67–70 |
| | 2 Find the rate of change of a function (p. 156) | 2, 3 | 1, 2, 73 (a) |
| | 3 Find average velocity and instantaneous velocity (p. 157) | 4, 5 | 71(a), (b); 72(a), (b) |
| | 4 Find the derivative of a function at a number (p. 159) | 6–8 | 3–8, 75 |

**Objectives Table** is a section-by-section list of each learning objective of the chapter (with page references), as well as references to the solved examples and review problems related to that learning objective.

**Review Problems** provide exercises related to the key concepts of the chapter and are matched to the learning objectives of each section of the chapter.

### REVIEW EXERCISES

*In Problems 1 and 2, use a definition of the derivative to find the rate of change of f at the indicated numbers.*

**1.** $f(x) = \sqrt{x}$ at **(a)** $c = 1$  **(b)** $c = 4$ 1, 2
   **(c)** $c$ any positive real number

**2.** $f(x) = \dfrac{2}{x-1}$ at **(a)** $c = 0$  **(b)** $c = 2$
   **(c)** $c$ any real number, $c \neq 1$

*In Problems 3–8, use a definition of the derivative to find the derivative of each function at the given number.*

**3.** $F(x) = 2x + 5$ at 2

**4.** $f(x) = 4x^2 + 1$ at $-1$

**5.** $f(x) = 3x^2 + 5x$ at 0

**6.** $f(x) = \dfrac{3}{x}$ at 1

**7.** $f(x) = \sqrt{4x+1}$ at 0

**8.** $f(x) = \dfrac{x+1}{2x-3}$ at 1

**71. Rectilinear Motion** As an object in rectilinear motion moves, its signed distance $s$ (in meters) from the origin at time $t$ (in seconds) is given by the position function
$$s = f(t) = t^2 - 6t$$
   **(a)** Find the average velocity of the object from 0 to 5 s.
   **(b)** Find the velocity at $t = 0$, at $t = 5$, and at any time $t$.
   **(c)** Find the acceleration at any time $t$.

**73. Business** The price $p$ in dollars per pound when $x$ pounds of a commodity are demanded is modeled by the function
$$p(x) = \frac{10,000}{5x + 100} - 5, \text{ when between 0 and 90 lb are demanded}$$
   (purchased).
   **(a)** Find the rate of change of price with respect to demand.
   **(b)** What is the revenue function $R$? (Recall, revenue $R$ equals price times amount purchased.)
   **(c)** What is the marginal revenue $R'$ at $x = 10$ and at $x = 40$ lb?

**75.** If $f(x) = 2 + |x - 3|$ for all $x$, determine whether the derivative $f'$ exists at $x = 3$.

# New to the Second Edition

The Second Edition of *Calculus, Early Transcendentals,* is the culmination of more than 100 reviews of the first edition, more than 20 focus groups, and 4 years of work. Contributions came from a variety of sources: students who studied from the first edition, as well as professors at community colleges, 4-year universities, and research universities, some of whom taught from the first edition and some of whom did not. We also commissioned a team of reputable mathematicians to review every aspect of the revised presentation. Using this invaluable input, as well as insights from our own experiences teaching calculus, we have revised and modified as needed, while preserving the writing style and presentation that students using the first edition praise.

We acknowledge the crucial importance of clarity and rigor, particularly considering the investment an instructor makes when adopting a calculus textbook. In preparing for this revision, we analyzed every component of the text: content, organization, examples, exercises, rigor, mathematical accuracy, depth and breadth of coverage, accessibility, and readability. Reviewers suggested how and where to revise, cut, deepen, or expand the coverage. Many offered suggestions for new exercises or nuances to enhance a student's learning experience. Wherever new Objectives or Examples were added, corresponding problems in the exercise sets were also added.

A brief description of the major changes to the content and organization in the second edition is provided below. It is by no means comprehensive and does not reflect the more subtle, but no less important, changes we made in our quest to increase clarity and deepen conceptual understanding.

## Chapter P Preparing for Calculus
- Sequences; Summation Notation; the Binomial Theorem has been moved from Appendix A to Chapter P, and now includes a full exercise set.

## Chapter 2 The Derivative
- Section 2.2 includes NEW EXAMPLES on obtaining information about a function from the graph of its derivative function and determining whether a (piecewise-defined) function has a derivative at a number.

## Chapter 3 More About Derivatives
- Implicit Differentiation and Derivatives of Inverse Trigonometric Functions are now treated in separate sections.
- The discussion of Taylor Polynomials (previously Section 3.5) is now in Chapter 8 (Infinite Series).

## Chapter 4 Applications of the Derivative
- Section 4.3 includes a NEW EXAMPLE on obtaining information about the graph of a function from the graph of its derivative.
- Section 4.4 includes NEW EXAMPLES on obtaining information about the graph of a function $f$ from the graphs of $f'$ and $f''$.

## Chapter 5 The Integral
- Section 5.2 The Definite Integral has been rewritten and expanded to accommodate the following objectives:
  - Form Riemann sums
  - Define a definite integral as a limit of Riemann sums
  - Approximate a definite integral using Riemann sums
  - Know conditions that guarantee a definite integral exists
  - Find a definite integral using the limit of Riemann sums
  - Form Riemann sums from a table
- To improve conceptual understanding, three new interpretations of the definite integral have been included:
  - Section 5.3: Objective 3: Interpret the integral as a rate of change, accompanied by two new examples, and Objective 4: Interpret the integral as an accumulation function, accompanied by a new example.
  - Section 5.4: Objective 4: Interpreting integrals involving rectilinear motion.
- Section 5.5: The Indefinite Integral and Substitution are now treated in one section.
- Section 5.6 now covers separable first-order differential equations and growth and decay models (both uninhibited and inhibited).

## Chapter 6 Applications of the Integral
- Section 6.5: NEW Objective 3: Find the surface area of a solid of revolution.

## Chapter 7 Techniques of Integration
- Section 7.1: NEW Objective 2: Find a definite integral using integration by parts.
- Section 7.2: NEW Summary tables are now included.
- Section 7.5: NEW Objective 5: Solve the logistic differential equation.
- Section 7.6: NEW Objective 2: Approximate an integral using the Midpoint Rule.
- NEW Section 7.9: Mixed Practice provides exercises that "mix up" all the integrals.

## Chapter 8 Infinite Series
- Section 8.3: NEW Objective 5: Approximating the sum of a convergent series.
- Section 8.4 now includes an expanded form of the Limit Comparison Test.
- Section 8.10: NEW Objective 1: Approximate a function and its graph using a Taylor Polynomial.

## Chapter 9 Parametric Equations; Polar Equations
- Arc length, formerly in Section 9.2, now appears in Section 9.3.

## Chapter 11 Vector Functions
- Section 11.5 has been expanded to include more examples of both indefinite and definite integrals of vector functions.

## Chapter 12 Functions of Several Variables
- Section 12.3 now includes an example that uses the definition to find the partial derivatives of a function on two variables, and an example that uses the Chain Rule to find partial derivatives has been added.

## Chapter 13 Directional Derivatives, Gradients, and Extrema
- Section 13.1 has an increased emphasis on the geometric interpretation of the gradient.
- Section 13.2: NEW Objective 3: Find an equation of a tangent plane to a surface defined explicitly.

## Chapter 14 Multiple Integrals
- The presentation now uses double and triple Riemann sums to define double and triple integrals and to develop applications.

## Chapter 15 Vector Calculus
- The presentation in Sections 15.2, 15.3, and 15.4 now distinguishes between line integrals of scalar functions and line integrals of vector functions.

## Chapter 16 Differential Equations
- Section 16.2 has two new topics
  - Using a Slope Field.
  - Euler's Method to Approximate Solutions to a First-order Differential Equation.

## WebAssign Premium

Macmillan's **WebAssign Premium** integrates Sullivan/Miranda's *Calculus* into a powerful online system with an interactive e-book, powerful answer evaluator, and easy-to-assign exercise bank that includes algorithmically generated homework and quizzes. Combined with Macmillan's esteemed content including "CalcTools" (described below), WebAssign Premium offers course and assignment customization to extend and enhance the classroom experience for instructors and students.

Student resources available through WebAssign Premium include,

- Macmillan's CalcTools
  - **Dynamic Figures:** Around 100 figures from the text have been recreated in an interactive format for students to manipulate and explore as they master key concepts. Tutorial videos explain how to use the figures.
  - **CalcClips:** These whiteboard tutorial videos provide a step-by-step walk-through illustrating key concepts from examples adapted from the book.
  - **LearningCurve** *macmillan learning* This powerful, self-paced formative assessment tool provides instant feedback tied to specific sections of the text. Question difficulty level and topic selection adapt based on the individual student's performance.
- **Tutorial questions** that break up questions into segments to help students work through learning a concept.
- A full, interactive, and easily navigated **e-book** with highlighting and notetaking features.
- **Additional student resources,** including the text's student solutions manual, Maple™ Manual, and Mathematica® Manual.

And for instructors,

- Over **7,000 exercises culled directly from the end-of-chapter sections of the text,** with detailed solutions available to students at your discretion.
- Ready-to-use **Course Pack Assignments** created and curated from the full exercise bank greatly decreases your preparation time.
- A suite of **Instructor Resources,** including iClicker questions, Instructor's Manual, PowerPoint lecture slides, a printable test bank, and more.

## WeBWorK

**WeBWorK** webwork.maa.org
Macmillan Learning offers hundreds of algorithmically generated questions (with full solutions) authored directly from the book through this free, open-source online homework system created at the University of Rochester. Adopters also have access to a shared national library test bank with thousands of additional questions, including questions tagged to the book's topics.

## Additional Resources

For Instructors
*Instructor's Solutions Manual*
Contains worked-out solutions to all exercises in the text.
ISBN: (complete) 978-1-319-10828-1; (SV) 978-1-319-10836-6; (MV) 978-1-319-10833-5

*Test Bank*
Includes a comprehensive set of multiple-choice test items.

*Instructor's Resource Manual*
Provides sample course outlines, suggested class time, key points, lecture materials, discussion topics, class activities, worksheets, projects, and questions to accompany the dynamic figures.
ISBN: 978-1-319-10828-1

**iClicker** *iClicker* is a two-way radio-frequency classroom response solution developed by educators for educators. Each step of iClicker's development has been informed by teaching and learning. To learn more about packaging iClicker with this textbook, please contact your local sales representative or visit www.iclicker.com.

*Lecture Slides* offer a customizable, detailed lecture presentation of concepts covered in each chapter.

*Image Slides* contain all textbook figures and tables.

For Students
*Student Solutions Manual*
Contains worked-out solutions to all odd-numbered exercises in the text.
ISBN: (SV) 978-1-319-06756-4; (MV) 978-1-319-06754-0

*Software Manuals*
Maple™ and Mathematica® software manuals serve as basic introductions to popular mathematical software options.

# Acknowledgments

Ensuring the mathematical rigor, accuracy, and precision, as well as the complete coverage and clear language that we have strived for with this text, requires not only a great deal of time and effort on our part, but also on the part of an entire team of exceptional reviewers. The following people have also provided immeasurable support reading drafts of chapters and offering insight and advice critical to the success of the second edition:

Kent Aeschliman, *Oakland Community College*

Anthony Aidoo, *Eastern Connecticut State University*

Martha Allen, *Georgia College & State University*

Jason Aran, *Drexel University*

Alyssa Armstrong, *Wittenberg University*

Mark Ashbrook, *Arizona State University*

Beyza Aslan, *University of North Florida*

Dr. Mathai Augustine, *Cleveland State Community College*

C. L. Bandy, *Texas State University*

Scott Barnett, *Henry Ford College*

William C. Bauldry, *Appalachian State University*

Nick Belloit, *Florida State College at Jacksonville*

Robert W. Benim, *University of Georgia*

Daniel Birmajer, *Nazareth College*

Dr. Benkam B. Bobga, *University of North Georgia, Gainesville Campus*

Laurie Boudreaux, *Nicholls State University*

Alain Bourget, *California State University, Fullerton*

David W. Boyd, *Valdosta State University*

James R. Bozeman, Ph.D., *Lyndon State College*

Naala Brewer, *Arizona State University, Tempe Campus*

Light Bryant, *Arizona Western College*

Meghan Burke, *Kennesaw State University*

James R. Bush, *Waynesburg University*

Sam Butler, *University of Nebraska, Omaha*

Joseph A. Capalbo, *Bryant University*

Debra Carney, *Colorado School of Mines*

Stephen N. Chai, *Miles College*

David Chan, *Virginia Commonwealth University*

Youn-Sha Chan, *University of Houston—Downtown*

E. William Chapin, Jr., *University of Maryland Eastern Shore*

Kevin Charlwood, *Washburn University, Topeka*

Jeffrey Cohen, *El Camino College*

Hugh Cornell, *University of North Florida*

Ana-Maria Croicu, *Kennesaw State University*

Charles Curtis, *Missouri Southern State University*

Seth Daugherty, *St. Louis Community College*

Christopher B. Davis, *Tennessee Technological University*

Shirley Davis, *South Plains College*

John C. D. Diamantopoulos, *Northeastern State University*

Roberto Diaz, *Fullerton College*

Tim Doyle, *DePaul University*

Deborah A. Eckhardt, *St. Johns River State College*

Steven Edwards, *Kennesaw State University*

Amy H. Erickson, *Georgia Gwinnett College*

Keith Erickson, *Georgia Gwinnett College*

Karen Ernst, *Hawkeye Community College*

Nancy Eschen, *Florida State College at Jacksonville*

Kevin Farrell, *Lyndon State College*

Ruth Feigenbaum, Ph.D., *Bergen Community College*

Mohammad Ganjizadeh, *Tarrant County College*

Dennis Garity, *Oregon State University*

Tom Geil, *Milwaukee Area Technical College*

Bekki George, *University of Houston*

Dr. Jeff Gervasi, *Porterville College*

Aaron Gibbs, *Johnson County Community College*

William Griffiths IV, *Kennesaw State University*

Mark Grinshpon, *Georgia State University*

Gary Grohs, *Elgin Community College*

Boyko Gyurov, *Georgia Gwinnett College*

Teresa Hales Peacock, *Nash Community College*

Sami M. Hamid, *University of North Florida*

Craig Hardesty, *Hillsborough Community College*

Alan T. Hayashi, *Oxnard College*

Mary Beth Headlee, *The State College of Florida—Venice Campus*

Beata Hebda, *University of North Georgia*

Piotr Hebda, *University of North Georgia*

Shahryar Heydari, *Piedmont College*

Kaat Higham, *Bergen Community College*

Syed Hussain, *Orange Coast College*

Mohamed Jamaloodeen, *Georgia Gwinnett College*

Dr. Christopher C. Jett, *University of West Georgia*

Paul W. Jones II, *University of South Alabama*

Victor Kaftal, *University of Cincinnati*

Dr. Robin S. Kalder, *Central Connecticut State University*

John Khoury, *Eastern Florida State College*

Minsu Kim, *University of North Georgia*

Wei Kitto, *Florida State College at Jacksonville*

Ashok Kumar, *Valdosta State University*

Hong-Jian Lai, *West Virginia University*

Carmen Latterell, *University of Minnesota Duluth*

Barbara A. Lawrence, *Borough of Manhattan Community College*

Richard C. Le Borne, *Tennessee Technological University*

Glenn Ledder, *University of Nebraska, Lincoln*

Jeffrey Ledford, *Virginia Commonwealth University*

Namyong Lee, *Minnesota State University, Mankato*

Rowan Lindley, *Westchester Community College*

Yung-Way Liu, *Tennessee Technological University*

Doron Lubinsky, *Georgia Institute of Technology*

Frank Lynch, *Eastern Washington University*

Jeffrey W. Lyons, *Nova Southeastern University*

Filix Maisch, *Oregon State University*

Mark Marino, *Erie Community College*

Jeff McGowan, *Central Connecticut State University*

John R. Metcalf, *St. Johns River State College*

Ashod Minasian, *El Camino College*

Matthew Mitchell, *Florida State College at Jacksonville*

Val Mohanakumar, *Hillsborough Community College*

Ronald H. Moore, *Florida State College at Jacksonville*

Tom Morley, *Georgia Institute of Technology*
Catherine Moushon, *Elgin Community College*
Keith Nabb, *Moraine Valley Community College*
Jeffrey Neugebauer, *Eastern Kentucky University*
Shai Neumann, *Eastern Florida State College*
Mike Nicholas, *Colorado School of Mines*
Jon Oaks, *Macomb Community College*
C. Altay Özgener, *State College of Florida, Manatee—Sarasota*
Joshua Palmatier, *State University of New York at Oneonta*
Sam Pearsall, *Los Angeles Pierce College*
Stan Perrine, *Georgia Gwinnett College*
Davidson B. Pierre, *State College of Florida, Manatee—Sarasota*
Cynthia Piez, *University of Idaho*
Daniel Pinzon, *Georgia Gwinnett College*
Katherine Pinzon, *Georgia Gwinnett College*

Mihaela Poplicher, *University of Cincinnati*
Elise Price, *Tarrant County College*
Brooke P. Quinlan, *Hillsborough Community College*
Mahbubur Rahman, *University of North Florida*
Joel Rappoport, *Florida State College at Jacksonville*
William Rogge, *University of Nebraska— Lincoln*
Richard Rossi, *Montana Tech of the University of Montana*
Bernard Rothman, *Ramapo College*
Daniel Russow, *Arizona Western College*
Kristen R. Schreck, *Moraine Valley Community College*
Randy Scott, *Santiago Canyon College*
Daniel Seabold, *Hofstra University*
Vicki Sealey, *West Virginia University*
Plamen Simeonov, *University of Houston– Downtown*
Derrick Thacker, *Northeast State Community College*
Jen Tyne, *The University of Maine*

Jossy Uvah, *University of West Florida*
William Veczko, *St. Johns River State College*
H. Joseph Vorder Bruegge, *Hillsborough Community College*
M. Vorderbruegge, *Hillsborough Community College*
Kathy Vranicar, *University of Nebraska, Omaha*
Kathryn C. Waddel, *Waynesburg University*
Martha Ellen Waggoner, *Simpson College*
Qing Wang, *Shepherd University*
Yajni Warnapala, *Roger Williams University*
Mike Weimerskirch, *University of Minnesota*
Benjamin Wiles, *Purdue University*
Catalina Yang, *Oxnard College*
Dr. Michael A. Zeitzew, *El Camino College*
Hong Zhang, *University of Wisconsin Oshkosh*

Additional applied exercises, as well as many of the case study and chapter projects that open and close each chapter of this text were contributed by a team of creative individuals. We would like to thank all of our exercise contributors for their work on this vital and exciting component of the text:

Wayne Anderson, *Sacramento City College* (Physics)
Allison Arnold, *University of Georgia* (Mathematics)
Kevin Cooper, *Washington State University* (Mathematics)
Adam Graham-Squire, *High Point University* (Mathematics)

Sergiy Klymchuk, *Auckland University of Technology* (Mathematics)
Eric Mandell, *Bowling Green State University* (Physics and Astronomy)
Eric Martell, *Millikin University* (Physics and Astronomy)
Barry McQuarrie, *University of Minnesota, Morris* (Mathematics)

Rachel Renee Miller, *Colorado School of Mines* (Physics)
Kanwal Singh, *Sarah Lawrence College* (Physics)
John Travis, *Mississippi College* (Mathematics)
Gordon Van Citters, *National Science Foundation* (Astronomy)

We would like to thank the dozens of instructors who provided invaluable input throughout the development of the first edition of this text.

Marwan A. Abu-Sawwa, *University of North Florida*
Jeongho Ahn, *Arkansas State University*
Weam M. Al-Tameemi, *Texas A&M International University*
Martha Allen, *Georgia College & State University*
Roger C. Alperin, *San Jose State University*
Robin Anderson, *Southwestern Illinois College*
Allison W. Arnold, *University of Georgia*
Mathai Augustine, *Cleveland State Community College*
Carroll Bandy, *Texas State University*
Scott E. Barnett, *Henry Ford Community College*
Emmanuel N. Barron, *Loyola University Chicago*
Abby Baumgardner, *Blinn College, Bryan*
Thomas Beatty, *Florida Gulf State University*

Robert W. Bell, *Michigan State University*
Nicholas G. Belloit, *Florida State College at Jacksonville*
Mary Beth Headlee, *State College of Florida, Manatee-Sarasota*
Daniel Birmajer, *Nazareth College of Rochester*
Justin Bost, *Rowan-Cabarrus Community College-South Campus*
Laurie Boudreaux, *Nicholls State University*
Alain Bourget, *California State University at Fullerton*
Jennifer Bowen, *The College of Wooster*
David Boyd, *Valdosta State University*
Brian Bradie, *Christopher Newport University*
James Brandt, *University of California, Davis*
Jim Brandt, *Southern Utah University*
Light R. Bryant, *Arizona Western College*

Kirby Bunas, *Santa Rosa Junior College*
Dietrich Burbulla, *University of Toronto*
Joni Burnette Pirnot, *State College of Florida, Manatee-Sarasota*
James Bush, *Waynesburg University*
Shawna M. Bynum, *Napa Valley College*
Joe Capalbo, *Bryant University*
Mindy Capaldi, *Valparaiso University*
Luca Capogna, *University of Arkansas*
Deborah Carney, *Colorado School of Mines*
Jenna P. Carpenter, *Louisiana Tech University*
Nathan Carter, *Bentley University*
Vincent Castellana, *Craven Community College*
Stephen Chai, *Miles College*
Julie Clark, *Hollins University*
Adam Coffman, *Indiana University-Purdue University, Fort Wayne*
William Cook, *Appalachian State University*

Sandy Cooper, *Washington State University*
David A. Cox, *Amherst College*
Mark Crawford, *Waubonsee Community College*
Charles N. Curtis, *Missouri Southern State University*
Larry W. Cusick, *California State University, Fresno*
Alain D'Amour, *Southern Connecticut State University*
Rajendra Dahal, *Coastal Carolina University*
John Davis, *Baylor University*
Ernesto Diaz, *Dominican University of California*
Robert Diaz, *Fullerton College*
Geoffrey D. Dietz, *Gannon University*
Della Duncan-Schnell, *California State University, Fresno*
Deborah A. Eckhardt, *St. Johns River State College*
Karen Ernst, *Hawkeye Community College*
Mark Farag, *Fairleigh Dickinson University*
Kevin Farrell, *Lyndon State College*
Judy Fethe, *Pellissippi State Community College*
Md Firozzaman, *Arizona State University*
Tim Flaherty, *Carnegie Mellon University*
Walden Freedman, *Humboldt State University*
Kseniya Fuhrman, *Milwaukee School of Engineering*
Melanie Fulton
Douglas R. Furman, *SUNY Ulster Community College*
Rosa Garcia Seyfried, *Harrisburg Area Community College*
Tom Geil, *Milwaukee Area Technical College*
Jeff Gervasi, *Porterville College*
William T. Girton, *Florida Institute of Technology*
Darren Glass, *Gettysburg College*
Jerome A. Goldstein, *University of Memphis*
Giséle Goldstein, *University of Memphis*
Lourdes M. Gonzalez, *Miami Dade College*
Pavel Grinfield, *Drexel University*
Mark Grinshpon, *Georgia State University*
Jeffery Groah, *Lone Star College*
Gary Grohs, *Elgin Community College*
Paul Gunnells, *University of Massachusetts, Amherst*
Semion Gutman, *University of Oklahoma*
Teresa Hales Peacock, *Nash Community College*
Christopher Hammond, *Connecticut College*
James Handley, *Montana Tech*
Alexander L. Hanhart, *New York University*
Gregory Hartman, *Virginia Military Institute*

Karl Havlak, *Angelo State University*
LaDawn Haws, *California State University, Chico*
Janice Hector, *DeAnza College*
Anders O.F. Hendrickson, *St. Norbert College*
Shahryar Heydari, *Piedmont College*
Max Hibbs, *Blinn College*
Rita Hibschweiler, *University of New Hampshire*
David Hobby, *SUNY at New Paltz*
Michael Holtfrerich, *Glendale Community College*
Keith E. Howard, *Mercer University*
Tracey Hoy, *College of Lake County*
Syed I. Hussain. Orange Coast College
Maiko Ishii, *Dawson College*
Nena Kabranski, *Tarrent County College*
William H. Kazez, *The University of Georgia*
Michael Keller, *University of Tulsa*
Steven L. Kent, *Youngstown State University*
Eric S. Key, *University of Wisconsin, Milwaukee*
Michael Kirby, *Colorado State University*
Stephen Kokoska, *Bloomsburg University of Pennsylvania*
Alex Kolesnik, *Ventura College*
Natalia Kouzniak, *Simon Fraser University*
Sunil Kumar Chebolu, *Illinois State University*
Ashok Kumar, *Valdosta State University*
Geoffrey Laforte, *University of West Florida*
Tamara J. Lakins, *Allegheny College*
Justin Lambright, *Anderson University*
Peter Lampe, *University of Wisconsin-Whitewater*
Donald Larson, *Penn State Altoona*
Carmen Latterell, *University of Minnesota, Duluth*
Glenn W. Ledder, *University of Nebraska, Lincoln*
Namyong Lee, *Minnesota State University, Mankato*
Denise LeGrand, *University of Arkansas, Little Rock*
Serhiy Levkov and his team
Benjamin Levy, *Wentworth Institute of Technology*
James Li-Ming Wang, *University of Alabama*
Aihua Li, *Montclair State University*
Amy H. Lin Erickson, *Georgia Gwinnett College*
Rowan Lindley, *Westchester Community College*
Roger Lipsett, *Brandeis University*
Joanne Lubben, *Dakota Weslyan University*
Matthew Macauley, *Clemson University*
Filix Maisch, *Oregon State University*
Heath M. Martin, *University of Central Florida*

Vania Mascioni, *Ball State University*
Betsy McCall, *Columbus State Community College*
Philip McCartney, *Northern Kentucky University*
Kate McGivney, *Shippensburg University*
Douglas B. Meade, *University of South Carolina*
Jie Miao, *Arkansas State University*
John Mitchell, *Clark College*
Val Mohanakumar, *Hillsborough Community College*
Catherine Moushon, *Elgin Community College*
Suzanne Mozdy, *Salt Lake Community College*
Gerald Mueller, *Columbus State Community College*
Will Murray, *California State University, Long Beach*
Kevin Nabb, *Moraine Valley Community College*
Vivek Narayanan, *Rochester Institute of Technology*
Larry Narici, *St. John's University*
Raja Nicolas Khoury, *Collin College*
Bogdan G. Nita, *Montclair State University*
Charles Odion, *Houston Community College*
Giray Ökten, *Florida State University*
Nicholas Ormes, *University of Denver*
Chihiro Oshima, *Santa Fe College*
Kurt Overhiser, *Valencia College*
Edward W. Packel, *Lake Forest College*
Joshua Palmatier, *SUNY College at Oneonta*
Chad Pierson, *University of Minnesota, Duluth*
Cynthia Piez, *University of Idaho*
Jeffrey L. Poet, *Missouri Western State University*
Shirley Pomeranz, *The University of Tulsa*
Vadim Ponomarenko, *San Diego State University*
Elise Price, *Tarrant County College, Southeast Campus*
Harald Proppe, *Concordia University*
Frank Purcell, *Twin Prime Editorial*
Michael Radin, *Rochester Institute of Technology*
Jayakumar Ramanathan, *Eastern Michigan University*
Joel Rappaport, *Florida State College at Jacksonville*
Marc Renault, *Shippensburg University*
Suellen Robinson, *North Shore Community College*
William Rogge, *University of Nebraska, Lincoln*
Yongwu Rong, *George Washington University*
Amber Rosin, *California State Polytechnic University*

Richard J. Rossi, *Montana Tech of the University of Montana*

Bernard Rothman, *Ramapo College of New Jersey*

Dan Russow, *Arizona Western College*

Adnan H. Sabuwala, *California State University, Fresno*

Alan Saleski, *Loyola University Chicago*

John Samons, *Florida State College at Jacksonville*

Brandon Samples, *Georgia College & State University*

Jorge Sarmiento, *County College of Morris*

Ned Schillow, *Lehigh Carbon Community College*

Kristen R. Schreck, *Moraine Valley Community College*

Randy Scott, *Santiago Canyon College*

George F. Seelinger, *Illinois State University*

Andrew Shulman, *University of Illinois at Chicago*

Mark Smith, *Miami University*

John Sumner, *University of Tampa*

Geraldine Taiani, *Pace University*

Barry A. Tesman, *Dickinson College*

Derrick Thacker, *Northeast State Community College*

Millicent P. Thomas, *Northwest University*

Tim Trenary, *Regis University*

Kiryl Tsishchanka and his team

Pamela Turner, *Hutchinson Community College*

Jen Tyne, *University of Maine*

David Unger, *Southern Illinois University-Edwardsville*

Marie Vanisko, *Carroll College*

William Veczko, *St. Johns River State College*

James Vicknair, *California State University, San Bernardino*

Robert Vilardi, *Troy University, Montgomery*

David Vinson, *Pellissippi Community College*

Klaus Volpert, *Villanova University*

Bryan Wai-Kei, *University of South Carolina, Salkehatchie*

David Walnut, *George Mason University*

Lianwen Wang, *University of Central Missouri*

Qing Wang, *Shepherd University*

E. William Chapin, Jr., *University of Maryland, Eastern Shore*

Rebecca Wong, *West Valley College*

Kerry Wyckoff, *Brigham Young University*

Jeffrey Xavier Watt, *Indiana University-Purdue University Indianapolis*

Carolyn Yackel, *Mercer University*

Catalina Yang, *Oxnard College*

Yvonne Yaz, *Milwaukee State College of Engineering*

Hong Zhang, *University of Wisconsin, Oshkosh*

Qiao Zhang, *Texas Christian University*

Qing Zhang, *University of Georgia*

Because the student is the ultimate beneficiary of this material, we would be remiss to neglect their input in its creation. We would like to thank students from the following schools who have provided us with feedback and suggestions for exercises throughout the text, and have written application exercises that appear in many of the chapters:

*Barry University*
*Boston University*
*Bronx Community College*
*California State University—Bakersfield*
*Catholic University*
*Colorado State University*
*Idaho State University*
*Lamar University*
*Lander University*

*University of Maryland*
*Minnesota State University*
*Millikin University*
*University of Missouri—St. Louis*
*Murray State University*
*University of North Georgia*
*North Park University*
*University of North Texas*
*University of South Florida*

*Southern Connecticut State University*
*St. Norbert College*
*Texas State University—San Marcos*
*Towson State University*
*Trine University*
*State University of West Georgia*
*University of Wisconsin—River Falls*

Finally, we would like to thank the editorial, production, and marketing teams at W. H. Freeman, whose support, knowledge, and hard work were instrumental in the publication of this text: Ruth Baruth and Terri Ward, who initially brought our text to Macmillan; Nikki Miller Dworsky, Andrew Sylvester, Katharine Munz, Justin Jones and Andy Newton, who formed the Editorial team; Diana Blume, Edgar Doolan, Paul Rohloff, Christine Buese, and Janice Donnola on the Project Management, Design, and Production teams; and Nancy Bradshaw and Leslie Allen on the Marketing and Market Development teams. Thanks as well to the diligent group at Lumina, lead by Sakthivel Sivalingam and Misbah Ansari, for their expert composition, and to Ron Weickart at Network Graphics for his skill and creativity in executing the art program.

*Mike*
Michael Sullivan

*Kathleen*
Kathleen Miranda

# Applications Index

Note: *Italics* indicates Example.

# P Preparing for Calculus

Nasa/Getty Images

Until now, the mathematics you have encountered has centered mainly on algebra, geometry, and trigonometry. These subjects have a long history, well over 2000 years. But calculus is relatively new; it was developed less than 400 years ago. In the time since, it has profoundly influenced every facet of human activity from biology to medicine to economics and physics, to name just a few. The year 2019 marks the 50th anniversary of the Apollo 11 mission, which put the first humans on the Moon, an achievement made possible by mathematics, in particular calculus.

Calculus deals with change and how the change in one quantity affects other quantities. Fundamental to these ideas are functions and their properties.

In this chapter, we discuss many of the functions used in calculus. We also provide a review of techniques from precalculus used to obtain the graphs of functions and to transform known functions into new functions.

Your instructor may choose to cover all or part of this chapter. Regardless, throughout the text, you will see the **NEED TO REVIEW?** marginal notes. They reference specific topics, often discussed in Chapter P.

# P.1 Functions and Their Graphs

**OBJECTIVES** *When you finish this section, you should be able to:*

1  Evaluate a function (p. 3)
2  Find the difference quotient of a function (p. 4)
3  Find the domain of a function (p. 5)
4  Identify the graph of a function (p. 6)
5  Analyze a piecewise-defined function (p. 8)
6  Obtain information from or about the graph of a function (p. 8)
7  Use properties of functions (p. 10)
8  Find the average rate of change of a function (p. 12)

Often there are situations where one variable is somehow linked to another variable. For example, the price of a gallon of gas is linked to the price of a barrel of oil. A person can be associated to her telephone number(s). The volume $V$ of a sphere depends on its radius $R$. The force $F$ exerted by an object corresponds to its acceleration $a$. These are examples of a **relation**, a correspondence between two sets called the **domain** and the **range**. If $x$ is an element of the domain and $y$ is an element of the range, and if a relation exists from $x$ to $y$, then we say that $y$ **corresponds** to $x$ or that $y$ **depends on** $x$, and we write $x \rightarrow y$. It is often helpful to think of $x$ as the **input** and $y$ as the **output** of the relation. See Figure 1.

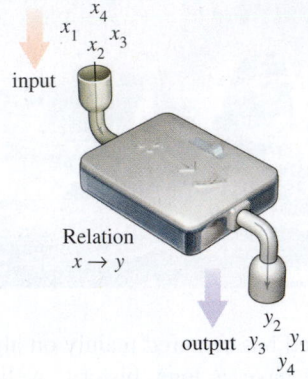

**Figure 1**

Suppose an astronaut standing on the Moon throws a rock 20 meters up and starts a stopwatch as the rock begins to fall back down. If $x$ represents the number of seconds on the stopwatch and if $y$ represents the altitude of the rock at that time, then there is a relation between time $x$ and altitude $y$. If the altitude of the rock is measured at $x = 1, 2, 2.5, 3, 4,$ and $5$ seconds, then the altitude is approximately $y = 19.2, 16.8, 15, 12.8, 7.2,$ and $0$ meters, respectively. This is an example of a relation expressed **verbally**.

The astronaut could also express this relation **numerically**, **graphically**, or **algebraically**. The relation can be expressed by a table of numbers (see Table 1) or by the set of ordered pairs $\{(0, 20), (1, 19.2), (2, 16.8), (2.5, 15), (3, 12.8), (4, 7.2), (5, 0)\}$, where the first element of each pair denotes the time $x$ and the second element denotes the altitude $y$. The relation also can be expressed visually, using either a graph, as in Figure 2, or a map, as in Figure 3. Finally, the relation can be expressed algebraically using the formula

$$y = 20 - 0.8x^2$$

**TABLE 1**

| Time, $x$ (in seconds) | Altitude, $y$ (in meters) |
|---|---|
| 0 | 20 |
| 1 | 19.2 |
| 2 | 16.8 |
| 2.5 | 15 |
| 3 | 12.8 |
| 4 | 7.2 |
| 5 | 0 |

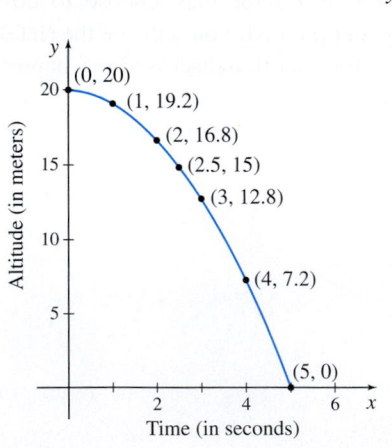

**Figure 2**

**Figure 3**

**NOTE** Not every relation is a function. If any element $x$ in the set $X$ corresponds to more than one element $y$ in the set $Y$, then the relation is not a function.

In this example, notice that if $X$ is the set of times from 0 to 5 seconds and $Y$ is the set of altitudes from 0 to 20 meters, then each element of $X$ corresponds to one and only one element of $Y$. Each given time value yields a **unique**, that is, exactly one, altitude value. Any relation with this property is called a *function from X into Y*.

**DEFINITION** Function

Let $X$ and $Y$ be two nonempty sets.* A **function** $f$ from $X$ into $Y$ is a relation that associates with each element of $X$ exactly one element of $Y$.

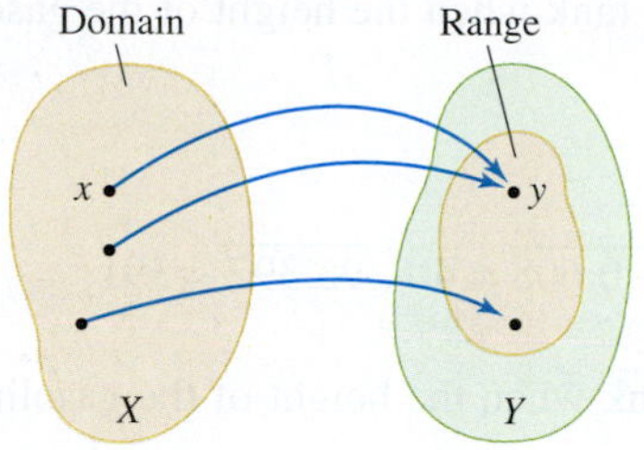

Domain    Range

$X$    $Y$

**Figure 4**

The set $X$ is called the **domain** of the function. For each element $x$ in $X$, the corresponding element $y$ in $Y$ is called the **value** of the function at $x$, or the **image** of $x$. The set of all the images of the elements in the domain is called the **range** of the function. Since there may be elements in $Y$ that are not images of any $x$ in $X$, the range of a function is a subset of $Y$. See Figure 4.

## 1 Evaluate a Function

Functions are often denoted by letters such as $f$, $F$, $g$, and so on. If $f$ is a function, then for each element $x$ in the domain, the corresponding image in the range is denoted by the symbol $f(x)$, read "$f$ of $x$." $f(x)$ is called the **value of** $f$ **at** $x$. The variable $x$ is called the **independent variable** or the **argument** because it can be assigned any element from the domain, while the variable $y$ is called the **dependent variable** because its value depends on $x$.

**EXAMPLE 1** Evaluating a Function

For the function $f$ defined by $f(x) = 2x^2 - 3x$, find:

**(a)** $f(5)$     **(b)** $f(x+h)$     **(c)** $f(x+h) - f(x)$

**Solution (a)** $f(5) = 2 \cdot 5^2 - 3 \cdot 5 = 50 - 15 = 35$

**(b)** The function $f(x) = 2x^2 - 3x$ gives us a rule to follow. To find $f(x+h)$, expand $(x+h)^2$, multiply the result by 2, and then subtract the product of 3 and $(x+h)$.

$$f(x+h) = 2(x+h)^2 - 3(x+h) = 2(x^2 + 2hx + h^2) - 3x - 3h$$

In $f(x)$ replace $x$ by $x+h$

$$= 2x^2 + 4hx + 2h^2 - 3x - 3h$$

**(c)** $f(x+h) - f(x) = [2x^2 + 4hx + 2h^2 - 3x - 3h] - [2x^2 - 3x] = 4hx + 2h^2 - 3h$ ∎

**NOW WORK** Problem **13**.

**EXAMPLE 2** Finding the Amount of Gasoline in a Tank

A Shell station stores its gasoline in an underground tank that is a right circular cylinder lying on its side. The volume $V$ of gasoline in the tank (in gallons) is given by the formula

$$V(h) = 40h^2 \sqrt{\frac{96}{h} - 0.608}$$

where $h$ is the height (in inches) of the gasoline as measured on a depth stick. See Figure 5.

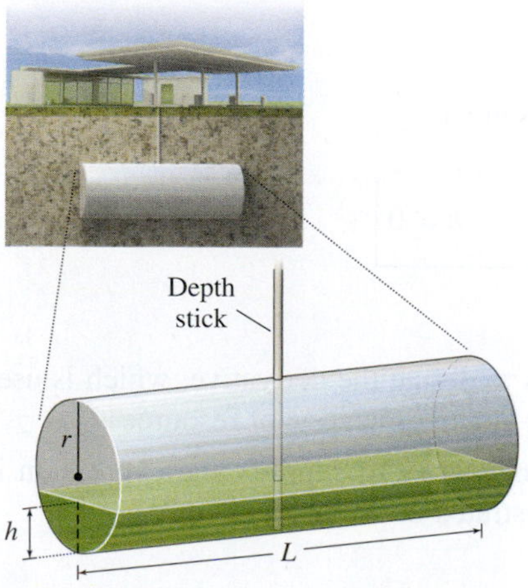

**(a)** If $h = 12$ inches, how many gallons of gasoline are in the tank?

**(b)** If $h = 1$ inch, how many gallons of gasoline are in the tank?

Depth stick

$r$

$h$    $L$

**Figure 5**

---

*The sets $X$ and $Y$ will usually be sets of real numbers, defining a **real function**. The two sets could also be sets of complex numbers, defining a **complex function**, or $X$ could be a set of real numbers and $Y$ a set of vectors, defining a **vector-valued function**. In the broad definition, $X$ and $Y$ can be any two sets.

**Solution (a)** We evaluate $V$ when $h = 12$.

$$V(12) = 40 \cdot 12^2 \sqrt{\frac{96}{12} - 0.608} = 40 \cdot 144\sqrt{8 - 0.608} = 5760\sqrt{7.392} \approx 15{,}660$$

There are about 15,660 gallons of gasoline in the tank when the height of the gasoline in the tank is 12 inches.

**(b)** Evaluate $V$ when $h = 1$.

$$V(1) = 40 \cdot 1^2 \sqrt{\frac{96}{1} - 0.608} = 40\sqrt{96 - 0.608} = 40\sqrt{95.392} \approx 391$$

There are about 391 gallons of gasoline in the tank when the height of the gasoline in the tank is 1 inch. ∎

## Implicit Form of a Function

In general, a function $f$ defined by an equation in $x$ and $y$ is given **implicitly**. If it is possible to solve the equation for $y$ in terms of $x$, then we write $y = f(x)$ and say the function is given **explicitly**. For example,

| Implicit Form | Explicit Form |
|---|---|
| $x^2 - y = 6$ | $y = f(x) = x^2 - 6$ |
| $xy = 4$ | $y = g(x) = \dfrac{4}{x}$ |

In calculus we sometimes deal with functions that are defined implicitly and that cannot be expressed in explicit form. For example, if the function with independent variable $x$ and dependent variable $y$ is defined by $\sin(xy) = xy - 2y + \cos x - \sin y$, there is no method to solve for $y$ and express the function explicitly.

## 2 Find the Difference Quotient of a Function

An important concept in calculus involves working with a certain quotient. For a given function $y = f(x)$, the inputs $x$ and $x + h$, $h \neq 0$, result in the images $f(x)$ and $f(x + h)$. The quotient of their differences

$$\frac{f(x+h) - f(x)}{(x+h) - x} = \frac{f(x+h) - f(x)}{h}$$

with $h \neq 0$, is called the *difference quotient of $f$ at $x$.*

**DEFINITION Difference Quotient**

The **difference quotient** of a function $f$ at $x$ is given by

$$\boxed{\frac{f(x+h) - f(x)}{h} \qquad h \neq 0}$$

The difference quotient is used in calculus to define the derivative, which is used in applications such as the velocity of an object and optimization of resources.

When finding a difference quotient, it is necessary to simplify the expression in order to cancel the $h$ in the denominator, as illustrated in the next example.

**EXAMPLE 3  Finding the Difference Quotient of a Function**

Find the difference quotient of each function.

**(a)** $f(x) = 2x^2 - 3x$     **(b)** $f(x) = \sqrt{x}$

**Solution**

**(a)** $\dfrac{f(x+h)-f(x)}{h} = \dfrac{[2(x+h)^2 - 3(x+h)] - [2x^2 - 3x]}{h}$

$\uparrow$

$f(x+h) = 2(x+h)^2 - 3(x+h)$

$= \dfrac{2(x^2 + 2xh + h^2) - 3x - 3h - 2x^2 + 3x}{h}$     Simplify.

$= \dfrac{2x^2 + 4xh + 2h^2 - 3x - 3h - 2x^2 + 3x}{h}$     Distribute.

$= \dfrac{4xh + 2h^2 - 3h}{h}$     Combine like terms.

$= \dfrac{h(4x + 2h - 3)}{h}$     Factor out $h$.

$= 4x + 2h - 3$     Divide out the factor $h$.

**(b)** $\dfrac{f(x+h)-f(x)}{h} = \dfrac{\sqrt{x+h}-\sqrt{x}}{h}$     $f(x+h) = \sqrt{x+h}$

$= \dfrac{\sqrt{x+h}-\sqrt{x}}{h} \cdot \dfrac{\sqrt{x+h}+\sqrt{x}}{\sqrt{x+h}+\sqrt{x}}$     Rationalize the numerator.

$= \dfrac{(\sqrt{x+h})^2 - (\sqrt{x})^2}{h(\sqrt{x+h}+\sqrt{x})}$     $(A-B)(A+B) = A^2 - B^2$

$= \dfrac{h}{h(\sqrt{x+h}+\sqrt{x})}$     $(\sqrt{x+h})^2 - (\sqrt{x})^2$
$= x+h-x = h$

$= \dfrac{1}{\sqrt{x+h}+\sqrt{x}}$     Divide out the factor $h$.     ∎

**NOW WORK** Problem 23.

## 3 Find the Domain of a Function

In applications, the domain of a function is sometimes specified. For example, we might be interested in the population of a city from 1990 to 2019. The domain of the function is time, in years, and is restricted to the interval [1990, 2019]. Other times the domain is restricted by the context of the function itself. For example, the volume $V$ of a sphere, given by the function $V = \dfrac{4}{3}\pi R^3$, makes sense only if the radius $R$ is greater than 0. But often the domain of a function $f$ is not specified; only the formula defining the function is given. In such cases, the **domain** of $f$ is the largest set of real numbers for which the value $f(x)$ is defined and is a real number.

### EXAMPLE 4  Finding the Domain of a Function

Find the domain of each of the following functions:

**(a)** $f(x) = x^2 + 5x$     **(b)** $g(x) = \dfrac{3x}{x^2 - 4}$

**(c)** $h(t) = \sqrt{4 - 3t}$     **(d)** $F(u) = \dfrac{5u}{\sqrt{u^2 - 1}}$

**Solution (a)** Since $f(x) = x^2 + 5x$ is defined for any real number $x$, the domain of $f$ is the set of all real numbers.

**(b)** Since division by zero is not defined, $x^2 - 4$ cannot be 0, that is, $x \neq -2$ and $x \neq 2$.

The function $g(x) = \dfrac{3x}{x^2 - 4}$ is defined for any real number except $x = -2$ and $x = 2$.

So, the domain of $g$ is the set of real numbers $\{x \,|\, x \neq -2, x \neq 2\}$.

**NEED TO REVIEW?** Solving inequalities is discussed in Appendix A.1, pp. A-6 to A-8.

**NEED TO REVIEW?** Interval notation is discussed in Appendix A.1, p. A-5.

(c) Since the square root of a negative number is not a real number, the value of $4 - 3t$ must be nonnegative. The solution of the inequality $4 - 3t \geq 0$ is $t \leq \dfrac{4}{3}$, so the domain of $h$ is the set of real numbers $\left\{ t \mid t \leq \dfrac{4}{3} \right\}$ or the interval $\left( -\infty, \dfrac{4}{3} \right]$.

(d) Since the square root is in the denominator, the value of $u^2 - 1$ must be not only nonnegative, it also cannot equal zero. That is, $u^2 - 1 > 0$. The solution of the inequality $u^2 - 1 > 0$ is the set of real numbers $\{u \mid u < -1\} \cup \{u \mid u > 1\}$ or the set $(-\infty, -1) \cup (1, \infty)$. ∎

If $x$ is in the domain of a function $f$, we say that $f$ **is defined at** $x$, or $f(x)$ **exists**. If $x$ is not in the domain of $f$, we say that $f$ **is not defined at** $x$, or $f(x)$ **does not exist**. The domain of a function is expressed using inequalities, interval notation, set notation, or words, whichever is most convenient. Notice the various ways the domain was expressed in the solution to Example 4.

NOW WORK Problem **17**.

## 4 Identify the Graph of a Function

In applications, often a graph reveals the relationship between two variables more clearly than an equation. For example, Table 2 shows the average price of gasoline at a particular gas station in the United States (for the years 1985–2016 adjusted for inflation, based on 2014 dollars). If we plot these data using year as the independent variable and price as the dependent variable, and then connect the points (year, price) we obtain Figure 6.

**TABLE 2** Average Retail Price of Gasoline (2014 dollars)

| Year | Price | Year | Price | Year | Price |
|------|-------|------|-------|------|-------|
| 1985 | 2.55 | 1996 | 1.80 | 2007 | 3.19 |
| 1986 | 1.90 | 1997 | 1.76 | 2008 | 3.56 |
| 1987 | 1.89 | 1998 | 1.49 | 2009 | 2.58 |
| 1988 | 1.81 | 1999 | 1.61 | 2010 | 3.00 |
| 1989 | 1.87 | 2000 | 2.03 | 2011 | 3.69 |
| 1990 | 2.03 | 2001 | 1.90 | 2012 | 3.72 |
| 1991 | 1.91 | 2002 | 1.76 | 2013 | 3.54 |
| 1992 | 1.82 | 2003 | 1.99 | 2014 | 3.43 |
| 1993 | 1.74 | 2004 | 2.31 | 2015 | 2.74 |
| 1994 | 1.71 | 2005 | 2.74 | 2016 | 2.24 |
| 1995 | 1.72 | 2006 | 3.01 | | |

*Source:* U.S. Energy Information Administration

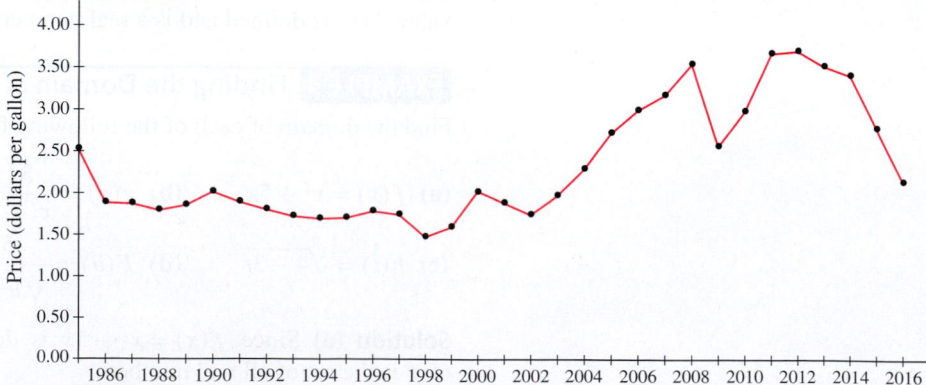

**Figure 6** Graph of the average retail price of gasoline (2014 dollars).

The graph shows that for each date on the horizontal axis there is only one price on the vertical axis. So, the graph represents a function, although the rule for determining the price from the year is not given.

NEED TO REVIEW? The graph of an equation is discussed in Appendix A.3, pp. A-16 to A-18.

When a function is defined by an equation in $x$ and $y$, the **graph of the function** is the set of points $(x, y)$ in the $xy$-plane that satisfy the equation.

But not every collection of points in the $xy$-plane represents the graph of a function. Recall that a relation is a function only if each element $x$ in the domain corresponds to exactly one image $y$ in the range. This means the graph of a function never contains two points with the same $x$-coordinate and different $y$-coordinates. Compare the graphs in Figures 7 and 8. In Figure 7, every number $x$ is associated with exactly one number $y$, but in Figure 8 some numbers $x$ are associated with three numbers $y$. Figure 7 shows the graph of a function; Figure 8 shows a graph that is not the graph of a function.

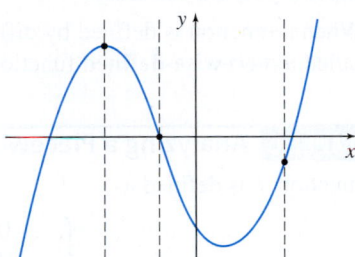

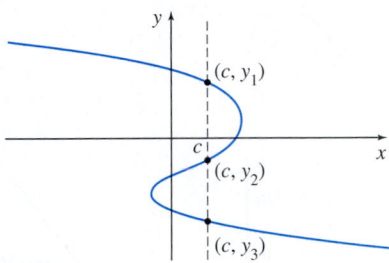

**Figure 7** Function: Exactly one $y$ for each $x$. Every vertical line intersects the graph in at most one point.

**Figure 8** Not a function: $x = c$ has 3 $y$'s associated with it. The vertical line $x = c$ intersects the graph in three points.

For a graph to be a graph of a function, it must satisfy the *Vertical-line Test*.

**THEOREM  Vertical-line Test**

**NOTE** The phrase "if and only if" means the statements on each side of the phrase are equivalent. That is, they have the same meaning.

A set of points in the $xy$-plane is the graph of a function if and only if every vertical line intersects the graph in at most one point.

**EXAMPLE 5  Identifying the Graph of a Function**

Which graphs in Figure 9 represent the graph of a function?

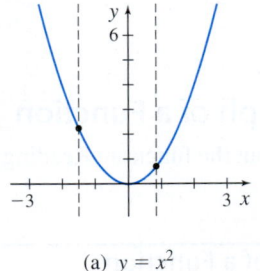

(a) $y = x^2$

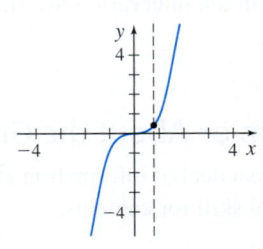

(b) $y = x^3$

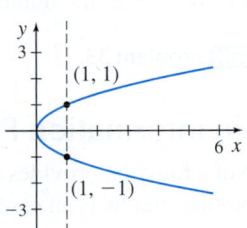

(c) $x = y^2$

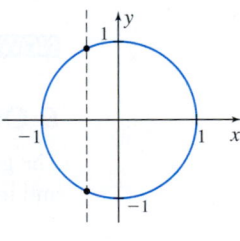

(d) $x^2 + y^2 = 1$

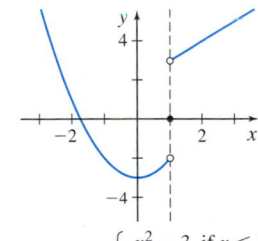

(e) $f(x) = \begin{cases} x^2 - 3 & \text{if } x < 1 \\ 0 & \text{if } x = 1 \\ x + 2 & \text{if } x > 1 \end{cases}$

**Figure 9**

**Solution** The graphs in Figure 9(a), 9(b), and 9(e) are graphs of functions because every vertical line intersects each graph in at most one point. The graphs in Figure 9(c) and 9(d) are not graphs of functions because there is a vertical line that intersects each graph in more than one point. ∎

**NOW WORK** Problems 31(a) and (b).

Notice that although the graph in Figure 9(e) represents a function, it looks different from the graphs in (a) and (b). The graph consists of two pieces plus a point, and they are not connected. Also notice that different equations describe different pieces of the graph. Functions with graphs similar to the one in Figure 9(e) are called *piecewise-defined functions*.

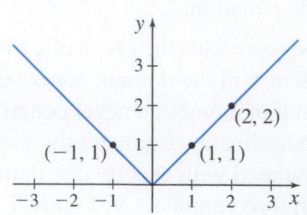

**Figure 10** $f(x)=|x|$

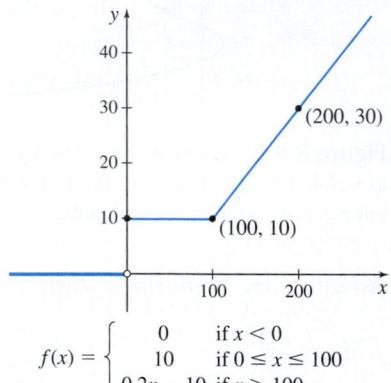

$$f(x) = \begin{cases} 0 & \text{if } x < 0 \\ 10 & \text{if } 0 \le x \le 100 \\ 0.2x - 10 & \text{if } x > 100 \end{cases}$$

**Figure 11**

**RECALL** $x$-intercepts are numbers on the $x$-axis at which a graph touches or crosses the $x$-axis.

## 5 Analyze a Piecewise-Defined Function

Sometimes a function is defined differently on different parts of its domain. For example, the *absolute value function* $f(x)=|x|$ is actually defined by two equations: $f(x)=x$ if $x \ge 0$ and $f(x)=-x$ if $x<0$. These equations are usually combined into one expression as

$$f(x) = |x| = \begin{cases} x & \text{if } x \ge 0 \\ -x & \text{if } x < 0 \end{cases}$$

Figure 10 shows the graph of the absolute value function. Notice that the graph of $f$ satisfies the Vertical-line Test.

When a function is defined by different equations on different parts of its domain, it is called a **piecewise-defined** function.

### EXAMPLE 6  Analyzing a Piecewise-Defined Function

The function $f$ is defined as

$$f(x) = \begin{cases} 0 & \text{if } x < 0 \\ 10 & \text{if } 0 \le x \le 100 \\ 0.2x - 10 & \text{if } x > 100 \end{cases}$$

(a) Evaluate $f(-1)$, $f(100)$, and $f(200)$.

(b) Graph $f$.

(c) Find the domain, range, and the $x$- and $y$-intercepts of $f$.

**Solution** (a) $f(-1)=0$; $f(100)=10$; $f(200)=0.2(200)-10=30$

(b) The graph of $f$ consists of three pieces corresponding to each equation in the definition. The graph is the horizontal line $y=0$ on the interval $(-\infty, 0)$, the horizontal line $y=10$ on the interval $[0, 100]$, and the line $y=0.2x-10$ on the interval $(100, \infty)$, as shown in Figure 11.

(c) $f$ is a piecewise-defined function. Look at the values that $x$ can take on: $x<0, 0 \le x \le 100, x > 100$. We conclude the domain of $f$ is all real numbers. The range of $f$ is the number 0 and all real numbers greater than or equal to 10. The $x$-intercepts are all the numbers in the interval $(-\infty, 0)$; the $y$-intercept is 10. ∎

**NOW WORK** Problem **33**.

## 6 Obtain Information From or About the Graph of a Function

The graph of a function provides a great deal of information about the function. Reading and interpreting graphs is an essential skill for calculus.

### EXAMPLE 7  Obtaining Information from the Graph of a Function

The graph of $y=f(x)$ is given in Figure 12. ($x$ might represent time and $y$ might represent the position of an object attached to a spring that has been compressed 4 units above its at rest position and is then released. The object oscillates between 4 units above and 4 units below its at rest position.)

(a) What are $f(0)$, $f\left(\dfrac{3\pi}{2}\right)$, and $f(3\pi)$?

(b) What is the domain of $f$?

(c) What is the range of $f$?

(d) List the intercepts of the graph.

(e) How many times does the line $y=2$ intersect the graph of $f$?

(f) For what values of $x$ does $f(x)=-4$?

(g) For what values of $x$ is $f(x)>0$?

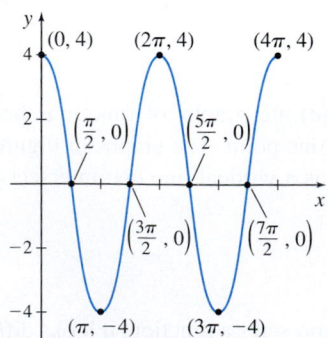

**Figure 12**

**Solution** (a) Since the point $(0, 4)$ is on the graph of $f$, the $y$-coordinate 4 is the value of $f$ at 0; that is, $f(0) = 4$. Similarly, when $x = \dfrac{3\pi}{2}$, then $y = 0$, so $f\left(\dfrac{3\pi}{2}\right) = 0$, and when $x = 3\pi$, then $y = -4$, so $f(3\pi) = -4$.

(b) The points on the graph of $f$ have $x$-coordinates between 0 and $4\pi$ inclusive. The domain of $f$ is $\{x \mid 0 \le x \le 4\pi\}$ or the closed interval $[0, 4\pi]$.

(c) Every point on the graph of $f$ has a $y$-coordinate between $-4$ and 4 inclusive. The range of $f$ is $\{y \mid -4 \le y \le 4\}$ or the closed interval $[-4, 4]$.

(d) The intercepts of the graph of $f$ are $(0, 4)$, $\left(\dfrac{\pi}{2}, 0\right)$, $\left(\dfrac{3\pi}{2}, 0\right)$, $\left(\dfrac{5\pi}{2}, 0\right)$, and $\left(\dfrac{7\pi}{2}, 0\right)$.

> **RECALL** Intercepts are points at which a graph crosses or touches a coordinate axis.

(e) Draw the graph of the line $y = 2$ on the same set of coordinate axes as the graph of $f$. The line intersects the graph of $f$ four times.

(f) Find points on the graph of $f$ for which $y = f(x) = -4$; there are two such points: $(\pi, -4)$ and $(3\pi, -4)$. So $f(x) = -4$ when $x = \pi$ and when $x = 3\pi$.

(g) $f(x) > 0$ when the $y$-coordinate of a point $(x, y)$ on the graph of $f$ is positive. This occurs when $x$ is in the set $\left[0, \dfrac{\pi}{2}\right) \cup \left(\dfrac{3\pi}{2}, \dfrac{5\pi}{2}\right) \cup \left(\dfrac{7\pi}{2}, 4\pi\right]$. ∎

**NOW WORK** Problems **37, 39, 41, 43, 45, 47,** and **49.**

---

**EXAMPLE 8** **Obtaining Information About the Graph of a Function**

Consider the function $f(x) = \dfrac{x+1}{x+2}$.

(a) What is the domain of $f$?

(b) Is the point $\left(1, \dfrac{1}{2}\right)$ on the graph of $f$?

(c) If $x = 2$, what is $f(x)$? What is the corresponding point on the graph of $f$?

(d) If $f(x) = 2$, what is $x$? What is the corresponding point on the graph of $f$?

(e) What are the $x$-intercepts of the graph of $f$ (if any)? What point(s) on the graph of $f$ correspond(s) to the $x$-intercept(s)?

**Solution** (a) The domain of $f$ consists of all real numbers except $-2$; that is, the set $\{x \mid x \ne -2\}$. The function is not defined at $-2$.

(b) When $x = 1$, then $f(1) = \underset{\underset{x=1}{\uparrow}}{\dfrac{1+1}{1+2}} = \dfrac{2}{3}$. The point $\left(1, \dfrac{2}{3}\right)$ is on the graph of $f$; the

point $\left(1, \dfrac{1}{2}\right)$ is not on the graph of $f$.

(c) If $x = 2$, then $f(2) = \underset{\underset{x=2}{\uparrow}}{\dfrac{2+1}{2+2}} = \dfrac{3}{4}$. The point $\left(2, \dfrac{3}{4}\right)$ is on the graph of $f$.

(d) If $f(x) = 2$, then $\dfrac{x+1}{x+2} = 2$. Solving for $x$, we find

$$x + 1 = 2(x + 2)$$
$$x + 1 = 2x + 4$$
$$x = -3$$

The point $(-3, 2)$ is on the graph of $f$.

(e) The $x$-intercepts of the graph of $f$ occur when $y = 0$. That is, they are the solutions of the equation $f(x) = 0$. The $x$-intercepts are also called the **real zeros** or **roots** of the function $f$.

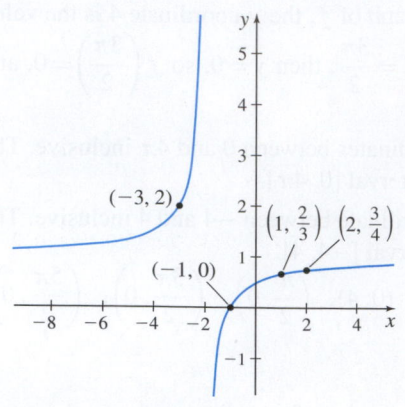

**Figure 13** $f(x) = \dfrac{x+1}{x+2}$

**NEED TO REVIEW?** Symmetry of graphs is discussed in Appendix A.3, pp. A-17 to A-18.

The real zeros of the function $f(x) = \dfrac{x+1}{x+2}$ satisfy the equation $x + 1 = 0$, so $x = -1$. The only $x$-intercept is $-1$, so the point $(-1, 0)$ is on the graph of $f$. ∎

Figure 13 shows the graph of $f$.

**NOW WORK** Problems 55, 57, and 59.

## 7 Use Properties of Functions

One of the goals of calculus is to develop techniques for graphing functions. Here we review some properties of functions that help obtain the graph of a function.

> **DEFINITION Even and Odd Functions**
>
> A function $f$ is **even** if, for every number $x$ in its domain, the number $-x$ is also in the domain and
>
> $$\boxed{f(-x) = f(x)}$$
>
> A function $f$ is **odd** if, for every number $x$ in its domain, the number $-x$ is also in the domain and
>
> $$\boxed{f(-x) = -f(x)}$$

For example, $f(x) = x^2$ is an even function since

$$f(-x) = (-x)^2 = x^2 = f(x)$$

Also, $g(x) = x^3$ is an odd function since

$$g(-x) = (-x)^3 = -x^3 = -g(x)$$

See Figure 14 for the graph of $f(x) = x^2$ and Figure 15 for the graph of $g(x) = x^3$. Notice that the graph of the even function $f(x) = x^2$ is symmetric with respect to the $y$-axis, and the graph of the odd function $g(x) = x^3$ is symmetric with respect to the origin.

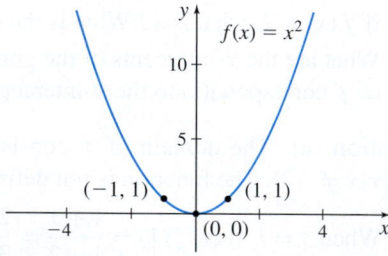

**Figure 14** The function $f(x) = x^2$ is even. The graph of $f$ is symmetric with respect to the $y$-axis.

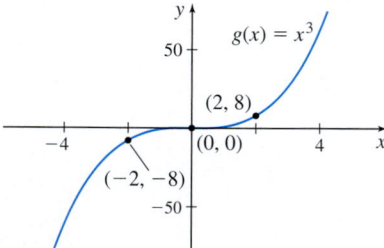

**Figure 15** The function $g(x) = x^3$ is odd. The graph of $g$ is symmetric with respect to the origin.

> **THEOREM Graphs of Even and Odd Functions**
>
> - A function is even if and only if its graph is symmetric with respect to the $y$-axis.
> - A function is odd if and only if its graph is symmetric with respect to the origin.

**EXAMPLE 9  Identifying Even and Odd Functions**

Determine whether each of the following functions is even, odd, or neither. Then determine whether its graph is symmetric with respect to the $y$-axis, the origin, or neither.

(a) $f(x) = x^2 - 5$   (b) $g(x) = \dfrac{4x}{x^2 - 5}$   (c) $h(x) = \sqrt[3]{5x^3 - 1}$

(d) $F(x) = |x|$   (e) $H(x) = \dfrac{x^2 + 2x - 1}{(x-5)^2}$

**Solution** **(a)** The domain of $f$ is $(-\infty, \infty)$, so for every number $x$ in its domain, $-x$ is also in the domain. Replace $x$ by $-x$ and simplify.

$$f(-x) = (-x)^2 - 5 = x^2 - 5 = f(x)$$

Since $f(-x) = f(x)$, the function $f$ is even. So the graph of $f$ is symmetric with respect to the $y$-axis.

**(b)** The domain of $g$ is $\{x \mid x \neq \pm\sqrt{5}\}$, so for every number $x$ in its domain, $-x$ is also in the domain. Replace $x$ by $-x$ and simplify.

$$g(-x) = \frac{4(-x)}{(-x)^2 - 5} = \frac{-4x}{x^2 - 5} = -g(x)$$

Since $g(-x) = -g(x)$, the function $g$ is odd. So the graph of $g$ is symmetric with respect to the origin.

**(c)** The domain of $h$ is $(-\infty, \infty)$, so for every number $x$ in its domain, $-x$ is also in the domain. Replace $x$ by $-x$ and simplify.

$$h(-x) = \sqrt[3]{5(-x)^3 - 1} = \sqrt[3]{-5x^3 - 1} = \sqrt[3]{-(5x^3 + 1)} = -\sqrt[3]{5x^3 + 1}$$

Since $h(-x) \neq h(x)$ and $h(-x) \neq -h(x)$, the function $h$ is neither even nor odd. The graph of $h$ is not symmetric with respect to the $y$-axis and not symmetric with respect to the origin.

**(d)** The domain of $F$ is $(-\infty, \infty)$, so for every number $x$ in its domain, $-x$ is also in the domain. Replace $x$ by $-x$ and simplify.

$$F(-x) = |-x| = |-1| \cdot |x| = |x| = F(x)$$

The function $F$ is even. So the graph of $F$ is symmetric with respect to the $y$-axis.

**(e)** The domain of $H$ is $\{x \mid x \neq 5\}$. The number $x = -5$ is in the domain of $H$, but $x = 5$ is not in the domain. So the function $H$ is neither even nor odd, and the graph of $H$ is not symmetric with respect to the $y$-axis or the origin. ∎

**NOW WORK** Problem **61.**

Another important property of a function is to know where it is increasing or decreasing.

**DEFINITION**

A function $f$ is **increasing** on an interval $I$ if, for any choice of $x_1$ and $x_2$ in $I$, with $x_1 < x_2$, then $f(x_1) < f(x_2)$.

A function $f$ is **decreasing** on an interval $I$ if, for any choice of $x_1$ and $x_2$ in $I$, with $x_1 < x_2$, then $f(x_1) > f(x_2)$.

A function $f$ is **constant** on an interval $I$ if, for all choices of $x$ in $I$, the values of $f(x)$ are equal.

Notice in the definition for an increasing (decreasing) function $f$, the value $f(x_1)$ is *strictly* less than (*strictly* greater than) the value $f(x_2)$. If a nonstrict inequality is used, we obtain the definitions for *nondecreasing* and *nonincreasing* functions.

**IN WORDS** From left to right, the graph of an increasing function goes up, the graph of a decreasing function goes down, and the graph of a constant function remains at a fixed height. From left to right, the graph of a nondecreasing function never goes down, and the graph of a nonincreasing function never goes up. See Figure 16 on page 12.

**DEFINITION**

A function $f$ is **nondecreasing** on an interval $I$ if, for any choice of $x_1$ and $x_2$ in $I$, with $x_1 < x_2$, then $f(x_1) \leq f(x_2)$.

A function $f$ is **nonincreasing** on an interval $I$ if, for any choice of $x_1$ and $x_2$ in $I$, with $x_1 < x_2$, then $f(x_1) \geq f(x_2)$.

Figure 16 illustrates the definitions. In Chapter 4 we use calculus to find where a function is increasing or decreasing or is constant.

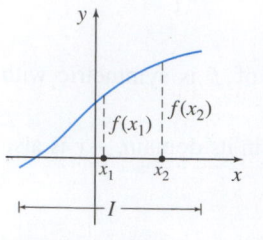
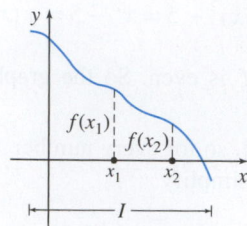
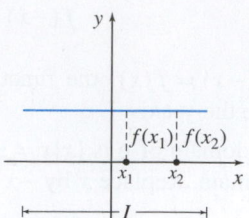
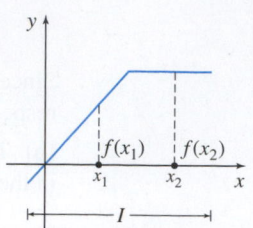
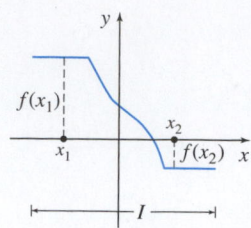

(a) For $x_1 < x_2$ in $I$, $f(x_1) < f(x_2)$; $f$ is increasing on $I$.

(b) For $x_1 < x_2$ in $I$, $f(x_1) > f(x_2)$; $f$ is decreasing on $I$.

(c) For all $x$ in $I$, the values of $f$ are equal; $f$ is constant on $I$.

(d) For $x_1 < x_2$ in $I$, $f(x_1) \leq f(x_2)$; $f$ is nondecreasing on $I$.

(e) For $x_1 < x_2$ in $I$, $f(x_1) \geq f(x_2)$; $f$ is nonincreasing on $I$.

**Figure 16**

NOW WORK Problems 51 and 53.

## 8 Find the Average Rate of Change of a Function

The average rate of change of a function plays an important role in calculus. It provides information about how a change in the independent variable $x$ of a function $y = f(x)$ causes a change in the dependent variable $y$. We use the symbol $\Delta x$, read "delta $x$," to represent the change in $x$ and $\Delta y$ to represent the change in $y$. Then by forming the quotient $\dfrac{\Delta y}{\Delta x}$, we arrive at an *average rate of change*.

**DEFINITION** Average Rate of Change

If $a$ and $b$, where $a \neq b$, are in the domain of a function $y = f(x)$, the **average rate of change of** $f$ from $a$ to $b$ is defined as

$$\boxed{\frac{\Delta y}{\Delta x} = \frac{f(b) - f(a)}{b - a} \qquad a \neq b}$$

**NOTE** $\dfrac{\Delta y}{\Delta x}$ represents the change in $y$ with respect to $x$.

EXAMPLE 10  **Finding the Average Rate of Change**

Find the average rate of change of $f(x) = 3x^2$:

**(a)** From 1 to 3    **(b)** From 1 to $x$, $x \neq 1$

**Solution (a)** The average rate of change of $f(x) = 3x^2$ from 1 to 3 is

$$\frac{\Delta y}{\Delta x} = \frac{f(3) - f(1)}{3 - 1} = \frac{27 - 3}{3 - 1} = \frac{24}{2} = 12$$

See Figure 17.

Notice that the average rate of change of $f(x) = 3x^2$ from 1 to 3 is the slope of the line containing the points $(1, 3)$ and $(3, 27)$.

**(b)** The average rate of change of $f(x) = 3x^2$ from 1 to $x$ is

$$\frac{\Delta y}{\Delta x} = \frac{f(x) - f(1)}{x - 1} = \frac{3x^2 - 3}{x - 1} = \frac{3(x^2 - 1)}{x - 1}$$

$$= \frac{3(x - 1)(x + 1)}{x - 1} = 3(x + 1) = 3x + 3$$

provided $x \neq 1$. ∎

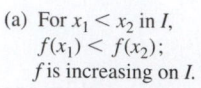

**Figure 17** $f(x) = 3x^2$

NOW WORK Problem 65.

CALC
CLIP

## EXAMPLE 11  Analyzing a Cost Function

The weekly cost $C$, in dollars, of manufacturing $x$ lightbulbs is

$$C(x) = 7500 + \sqrt{125x}$$

(a) Find the average rate of change of the weekly cost $C$ of manufacturing from 100 to 101 lightbulbs.

(b) Find the average rate of change of the weekly cost $C$ of manufacturing from 1000 to 1001 lightbulbs.

(c) Interpret the results from parts (a) and (b).

**Solution (a)**  The weekly cost of manufacturing 100 lightbulbs is

$$C(100) = 7500 + \sqrt{125 \cdot 100} = 7500 + \sqrt{12{,}500} \approx \$7611.80$$

The weekly cost of manufacturing 101 lightbulbs is

$$C(101) = 7500 + \sqrt{125 \cdot 101} = 7500 + \sqrt{12{,}625} \approx \$7612.36$$

The average rate of change of the weekly cost $C$ from 100 to 101 is

$$\frac{\Delta C}{\Delta x} = \frac{C(101) - C(100)}{101 - 100} \approx \frac{7612.36 - 7611.80}{1} = \$0.56$$

**(b)**  The weekly cost of manufacturing 1000 lightbulbs is

$$C(1000) = 7500 + \sqrt{125 \cdot 1000} = 7500 + \sqrt{125{,}000} \approx \$7853.55$$

The weekly cost of manufacturing 1001 lightbulbs is

$$C(1001) = 7500 + \sqrt{125 \cdot 1001} = 7500 + \sqrt{125{,}125} \approx \$7853.73$$

The average rate of change of the weekly cost $C$ from 1000 to 1001 is

$$\frac{\Delta C}{\Delta x} = \frac{C(1001) - C(1000)}{1001 - 1000} \approx \frac{7853.73 - 7853.55}{1} = \$0.18$$

**(c)**  Part (a) tells us that the unit cost of manufacturing the 101st lightbulb is \$0.56. From (b) we learn that the unit cost of manufacturing the 1001st lightbulb is only \$0.18. The unit cost of manufacturing the 1001st lightbulb is less than the unit cost of manufacturing the 101st lightbulb. ∎

NOW WORK **Problem 73.**

## The Secant Line

The average rate of change of a function has an important geometric interpretation. Look at the graph of $y = f(x)$ in Figure 18 on page 14. Two points are labeled on the graph: $(a, f(a))$ and $(b, f(b))$. The line containing these two points is called a **secant line**. The slope of a secant line is

$$m_{\text{sec}} = \frac{f(b) - f(a)}{b - a} \underset{\substack{\uparrow \\ h = b - a}}{=} \frac{f(a+h) - f(a)}{h}$$

Notice that the slope of a secant line equals the average rate of change of $f$ from $a$ to $b$ and also equals the difference quotient of $f$ at $a$.

**THEOREM  Slope of a Secant Line**

The average rate of change of a function $f$ from $a$ to $b$ equals the slope of the secant line containing the two points $(a, f(a))$ and $(b, f(b))$ on the graph of $f$.

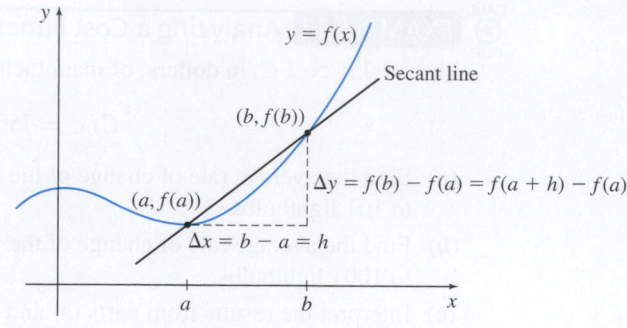

**Figure 18** A secant line of the function $y = f(x)$.

## P.1 Assess Your Understanding

### Concepts and Vocabulary

1. If $f$ is a function defined by $y = f(x)$, then $x$ is called the _____ variable, and $y$ is the _____ variable.

2. *True or False*   The independent variable is sometimes referred to as the argument of the function.

3. *True or False*   If no domain is specified for a function $f$, then the domain of $f$ is taken to be the set of all real numbers.

4. *True or False*   The domain of the function $f(x) = \dfrac{3(x^2 - 1)}{x - 1}$ is $\{x \mid x \neq \pm 1\}$.

5. *True or False*   A function can have more than one $y$-intercept.

6. A set of points in the $xy$-plane is the graph of a function if and only if every _____ line intersects the graph in at most one point.

7. If the point $(5, -3)$ is on the graph of $f$, then $f(\_\_\_) = \_\_\_$.

8. Find $a$ so that the point $(-1, 2)$ is on the graph of $f(x) = ax^2 + 4$.

9. *Multiple Choice*   A function $f$ is [(a) increasing (b) decreasing (c) nonincreasing (d) nondecreasing (e) constant] on an interval $I$ if, for any choice of $x_1$ and $x_2$ in $I$, with $x_1 < x_2$, then $f(x_1) < f(x_2)$.

10. *Multiple Choice*   A function $f$ is [(a) even (b) odd (c) neither even nor odd] if for every number $x$ in its domain, the number $-x$ is also in the domain and $f(-x) = f(x)$. A function $f$ is [(a) even (b) odd (c) neither even nor odd] if for every number $x$ in its domain, the number $-x$ is also in the domain and $f(-x) = -f(x)$.

11. *True or False*   Even functions have graphs that are symmetric with respect to the origin.

12. The average rate of change of $f(x) = 2x^3 - 3$ from 0 to 2 is _____.

### Practice Problems

*In Problems 13–16, for each function find:*

(a) $f(0)$        (b) $f(-x)$        (c) $-f(x)$

(d) $f(x + 1)$        (e) $f(x + h)$

13. $f(x) = 3x^2 + 2x - 4$        14. $f(x) = \dfrac{x}{x^2 + 1}$

15. $f(x) = |x| + 4$        16. $f(x) = \sqrt{3 - x}$

*In Problems 17–22, find the domain of each function.*

17. $f(x) = x^3 - 1$        18. $f(x) = \dfrac{x}{x^2 + 1}$

19. $v(t) = \sqrt{t^2 - 9}$        20. $g(x) = \sqrt{\dfrac{2}{x - 1}}$

21. $h(x) = \dfrac{x + 2}{x^3 - 4x}$        22. $s(t) = \dfrac{\sqrt{t + 1}}{t - 5}$

*In Problems 23–28, find the difference quotient of f. That is, find*
$$\frac{f(x + h) - f(x)}{h}, \quad h \neq 0$$

23. $f(x) = -3x + 1$        24. $f(x) = \dfrac{1}{x + 3}$

25. $f(x) = \sqrt{x + 7}$        26. $f(x) = \dfrac{2}{\sqrt{x + 7}}$

27. $f(x) = x^2 + 2x$        28. $f(x) = (2x + 3)^2$

*In Problems 29–32, determine whether the graph is that of a function by using the Vertical-line Test. If it is, use the graph to find*

(a) *the domain and range*

(b) *the intercepts, if any*

(c) *any symmetry with respect to the x-axis, y-axis, or the origin.*

29.

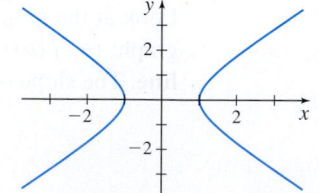

30.

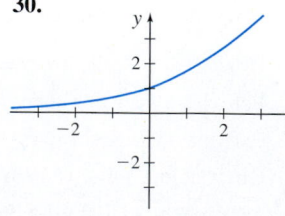

31.

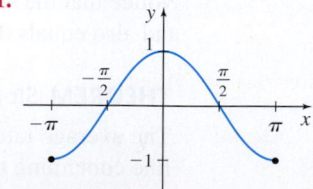

32.

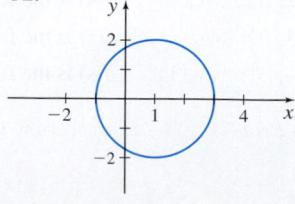

---

**1.** = NOW WORK problem        = Graphing technology recommended        **CAS** = Computer Algebra System recommended

*In Problems 33–36, for each piecewise-defined function:*

*(a)* *Find $f(-1)$, $f(0)$, $f(1)$, and $f(8)$.*

*(b)* *Graph $f$.*

*(c)* *Find the domain, range, and intercepts of $f$.*

**33.** $f(x) = \begin{cases} x+3 & \text{if } -2 \le x < 1 \\ 5 & \text{if } x = 1 \\ -x+2 & \text{if } x > 1 \end{cases}$

**34.** $f(x) = \begin{cases} 2x+5 & \text{if } -3 \le x < 0 \\ -3 & \text{if } x = 0 \\ -5x & \text{if } x > 0 \end{cases}$

**35.** $f(x) = \begin{cases} 1+x & \text{if } x < 0 \\ x^2 & \text{if } x \ge 0 \end{cases}$

**36.** $f(x) = \begin{cases} \dfrac{1}{x} & \text{if } x < 0 \\ \sqrt[3]{x} & \text{if } x \ge 0 \end{cases}$

*In Problems 37–54, use the graph of the function $f$ to answer the questions.*

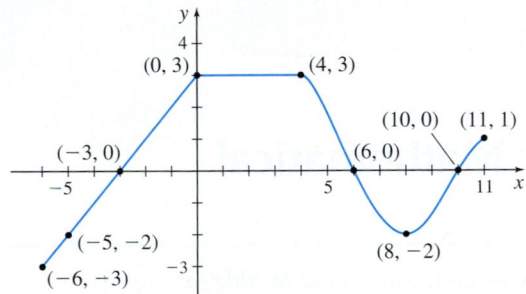

**37.** Find $f(0)$ and $f(-6)$.

**38.** Is $f(3)$ positive or negative?

**39.** Is $f(-4)$ positive or negative?

**40.** For what values of $x$ is $f(x) = 0$?

**41.** For what values of $x$ is $f(x) > 0$?

**42.** What is the domain of $f$?

**43.** What is the range of $f$?

**44.** What are the $x$-intercepts?

**45.** What is the $y$-intercept?

**46.** How often does the line $y = \dfrac{1}{2}$ intersect the graph?

**47.** How often does the line $x = 5$ intersect the graph?

**48.** For what values of $x$ does $f(x) = 3$?

**49.** For what values of $x$ does $f(x) = -2$?

**50.** On what interval(s) is the function $f$ increasing?

**51.** On what interval(s) is the function $f$ decreasing?

**52.** On what interval(s) is the function $f$ constant?

**53.** On what interval(s) is the function $f$ nonincreasing?

**54.** On what interval(s) is the function $f$ nondecreasing?

*In Problems 55–60, answer the questions about the function*

$$g(x) = \frac{x+2}{x-6}$$

**55.** What is the domain of $g$?

**56.** Is the point $(3, 14)$ on the graph of $g$?

**57.** If $x = 4$, what is $g(x)$? What is the corresponding point on the graph of $g$?

**58.** If $g(x) = 2$, what is $x$? What is (are) the corresponding point(s) on the graph of $g$?

**59.** List the $x$-intercepts, if any, of the graph of $g$.

**60.** What is the $y$-intercept, if there is one, of the graph of $g$?

*In Problems 61–64, determine whether the function is even, odd, or neither. Then determine whether its graph is symmetric with respect to the $y$-axis, the origin, or neither.*

**61.** $h(x) = \dfrac{x}{x^2 - 1}$     **62.** $f(x) = \sqrt[3]{3x^2 + 1}$

**63.** $G(x) = \sqrt{x}$         **64.** $F(x) = \dfrac{2x}{|x|}$

**65.** Find the average rate of change of $f(x) = -2x^2 + 4$:

  **(a)** From 1 to 2    **(b)** From 1 to 3

  **(c)** From 1 to 4    **(d)** From 1 to $x$, $x \ne 1$

**66.** Find the average rate of change of $s(t) = 20 - 0.8t^2$:

  **(a)** From 1 to 4    **(b)** From 1 to 3

  **(c)** From 1 to 2    **(d)** From 1 to $t$, $t \ne 1$

*In Problems 67–72, the graph of a piecewise-defined function is given. Write a definition for each piecewise-defined function. Then state the domain and the range of the function.*

**67.**

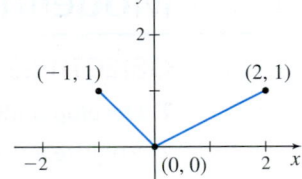

**68.**

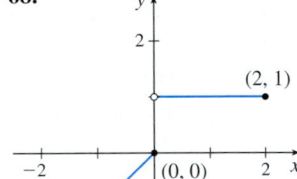

**69.**

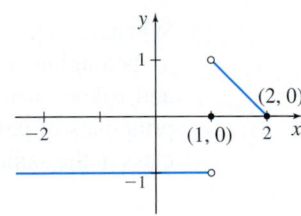

**70.**

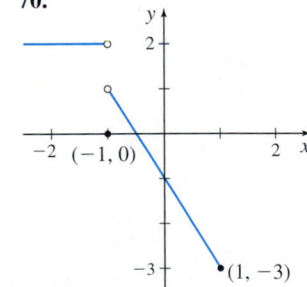

**71.**

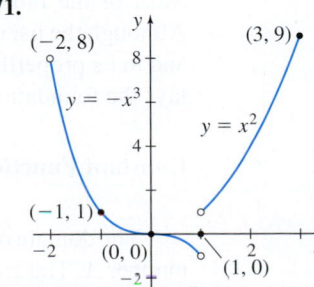

**72.**

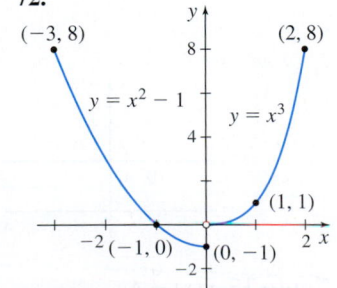

**73.** The monthly cost $C$, in dollars, of manufacturing $x$ road bikes is given by the function

$$C(x) = 0.004x^3 - 0.6x^2 + 250x + 100,500$$

(a) Find the average rate of change of the cost $C$ of manufacturing from 100 to 101 road bikes.

(b) Find the average rate of change of the cost $C$ of manufacturing from 500 to 501 road bikes.

(c) Interpret the results from parts (a) and (b) in the context of the problem.

**74.** The graph in the right column represents the distance $d$ (in miles) that Kevin was from home as a function of time $t$ (in hours). Answer the questions by referring to the graph. In parts (a)–(g), how many hours elapsed and how far was Kevin from home during this time?

(a) From $t = 0$ to $t = 2$      (b) From $t = 2$ to $t = 2.5$

(c) From $t = 2.5$ to $t = 2.8$   (d) From $t = 2.8$ to $t = 3$

(e) From $t = 3$ to $t = 3.9$     (f) From $t = 3.9$ to $t = 4.2$

(g) From $t = 4.2$ to $t = 5.3$

(h) What is the farthest distance that Kevin was from home?

(i) How many times did Kevin return home?

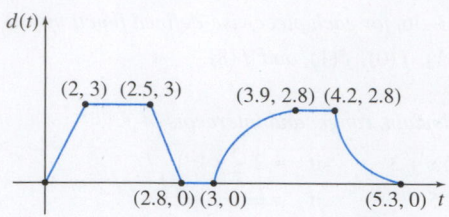

**75. Population as a Function of Age** The function

$$P(a) = 0.014a^2 - 5.073a + 327.287$$

represents the population $P$ (in millions) of Americans that were $a$ years of age or older in 2012.

(a) Identify the independent and dependent variables.

(b) Evaluate $P(20)$. Interpret $P(20)$ in the context of this problem.

(c) Evaluate $P(0)$. Interpret $P(0)$ in the context of this problem.

*Source:* U.S. Census Bureau

# P.2 Library of Functions; Mathematical Modeling

**OBJECTIVES** *When you finish this section, you should be able to:*

**1** Develop a library of functions (p. 16)

**2** Analyze a polynomial function and its graph (p. 19)

**3** Find the domain and the intercepts of a rational function (p. 22)

**4** Construct a mathematical model (p. 23)

When a collection of functions have common properties, they can be "grouped together" as belonging to a **class of functions**. Polynomial functions, exponential functions, and trigonometric functions are examples of classes of functions. As we investigate principles of calculus, we will find that often a principle applies to all functions in a class in the same way.

## 1 Develop a Library of Functions

Most of the functions in this section will be familiar to you; several might be new. Although the list may seem familiar, pay special attention to the domain of each function and to its properties, particularly to the shape of each graph. Knowledge of these graphs lays the foundation for later graphing techniques.

**Constant Function**          $\boxed{f(x) = A \qquad A \text{ is a real number}}$

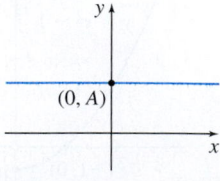

**Figure 19** $f(x) = A$

The domain of a **constant function** is the set of all real numbers; its range is the single number $A$. The graph of a constant function is a horizontal line whose $y$-intercept is $A$; it has no $x$-intercept if $A \neq 0$. A constant function is an even function. See Figure 19.

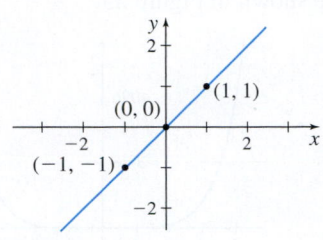

**Figure 20** $f(x) = x$

**NEED TO REVIEW?** Equations of lines are discussed in Appendix A.3, pp. A-18 to A-20.

**Identity Function** $\boxed{f(x) = x}$

The domain and the range of the **identity function** are the set of all real numbers. Its graph is the line through the origin whose slope is $m = 1$. Its only intercept is $(0, 0)$. The identity function is an odd function; its graph is symmetric with respect to the origin. It is increasing over its domain. See Figure 20.

Notice that the graph of $f(x) = x$ bisects quadrants I and III.

The graphs of both the constant function and the identity function are straight lines. These functions belong to the class of *linear functions*:

**Linear Functions** $\boxed{f(x) = mx + b \quad m \text{ and } b \text{ are real numbers}}$

The domain of a **linear function** is the set of all real numbers. The graph of a linear function is a line with slope $m$ and $y$-intercept $b$. If $m > 0$, $f$ is an increasing function, and if $m < 0$, $f$ is a decreasing function. If $m = 0$, then $f$ is a constant function, and its graph is the horizontal line, $y = b$. See Figure 21.

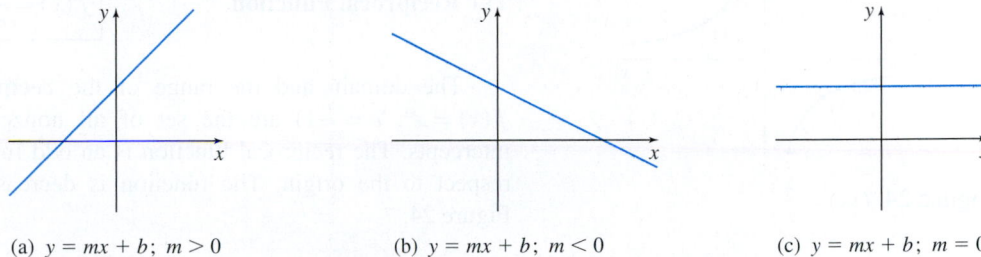

(a) $y = mx + b$; $m > 0$     (b) $y = mx + b$; $m < 0$     (c) $y = mx + b$; $m = 0$

**Figure 21**

In a *power function*, the independent variable $x$ is raised to a power.

**Power Functions** $\boxed{f(x) = x^a \quad a \text{ is a real number}}$

Below we examine power functions $f(x) = x^n$, where $n \geq 1$ is a positive integer. The domain of these power functions is the set of all real numbers. The only intercept of their graph is the point $(0, 0)$.

If $f$ is a power function and $n$ is a positive odd integer, then $f$ is an odd function whose range is the set of all real numbers and the graph of $f$ is symmetric with respect to the origin. The points $(-1, -1)$, $(0, 0)$, and $(1, 1)$ are on the graph of $f$. As $x$ becomes unbounded in the negative direction, $f$ also becomes unbounded in the negative direction. Similarly, as $x$ becomes unbounded in the positive direction, $f$ also becomes unbounded in the positive direction. The function $f$ is increasing over its domain.

The graphs of several odd power functions are given in Figure 22.

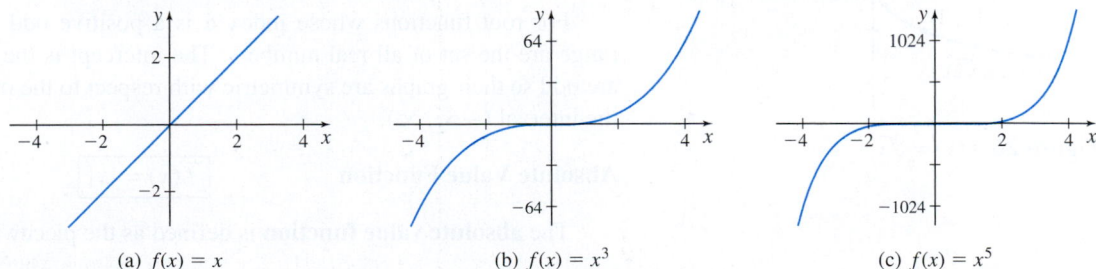

(a) $f(x) = x$     (b) $f(x) = x^3$     (c) $f(x) = x^5$

**Figure 22**

If $f$ is a power function and $n$ is a positive even integer, then $f$ is an even function whose range is $\{y | y \geq 0\}$. The graph of $f$ is symmetric with respect to the $y$-axis. The points $(-1, 1)$, $(0, 0)$, and $(1, 1)$ are on the graph of $f$. As $x$ becomes unbounded in either the negative direction or the positive direction, $f$ becomes unbounded in the positive direction. The function is decreasing on the interval $(-\infty, 0]$ and is increasing on the interval $[0, \infty)$.

The graphs of several even power functions are shown in Figure 23.

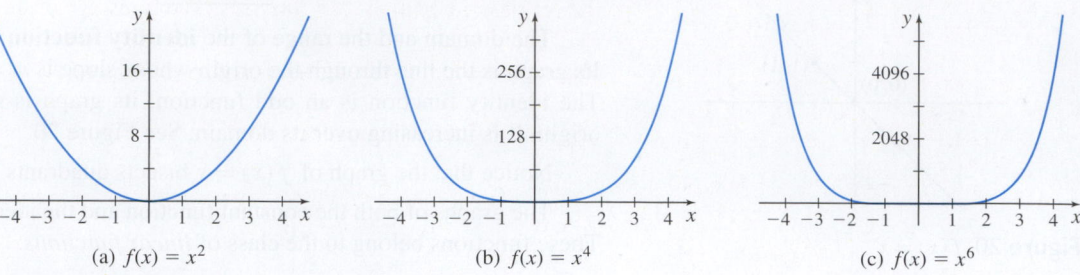

(a) $f(x) = x^2$    (b) $f(x) = x^4$    (c) $f(x) = x^6$

**Figure 23**

Look closely at Figures 22 and 23. As the integer exponent $n$ increases, the graph of $f$ is flatter (closer to the $x$-axis) when $x$ is in the interval $(-1, 1)$ and steeper when $x$ is in interval $(-\infty, -1)$ or $(1, \infty)$.

**The Reciprocal Function**   $\boxed{f(x) = \dfrac{1}{x}}$

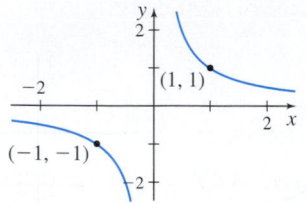

**Figure 24** $f(x) = \dfrac{1}{x}$

The domain and the range of the **reciprocal function** (the power function $f(x) = x^a$, $a = -1$) are the set of all nonzero real numbers. The graph has no intercepts. The reciprocal function is an odd function so the graph is symmetric with respect to the origin. The function is decreasing on $(-\infty, 0)$ and on $(0, \infty)$. See Figure 24.

**Root Functions**   $\boxed{f(x) = x^{1/n} = \sqrt[n]{x} \qquad n \geq 2 \text{ is a positive integer}}$

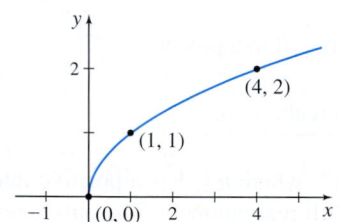

**Figure 25** $f(x) = \sqrt{x}$

Root functions are also power functions. If $n = 2$, $f(x) = x^{1/2} = \sqrt{x}$ is the **square root function**. The domain and the range of the square root function are the set of nonnegative real numbers. The intercept is the point $(0, 0)$. The square root function is neither even nor odd; it is increasing on the interval $[0, \infty)$. See Figure 25.

For root functions whose index $n$ is a positive even integer, the domain and the range are the set of nonnegative real numbers. The intercept is the point $(0, 0)$. Such functions are neither even nor odd; they are increasing on their domain, the interval $[0, \infty)$.

If $n = 3$, $f(x) = x^{1/3} = \sqrt[3]{x}$ is the **cube root function**. The domain and range of the cube root function are all real numbers. The intercept of the graph of the cube root function is the point $(0, 0)$.

Because $f(-x) = \sqrt[3]{-x} = -\sqrt[3]{x} = -f(x)$, the cube root function is odd. Its graph is symmetric with respect to the origin. The cube root function is increasing on the interval $(-\infty, \infty)$. See Figure 26.

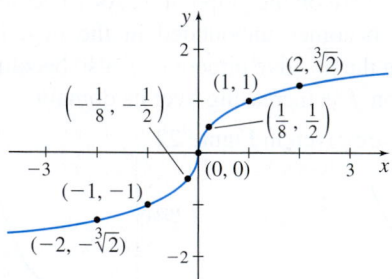

**Figure 26** $f(x) = \sqrt[3]{x}$

For root functions whose index $n$ is a positive odd integer, the domain and the range are the set of all real numbers. The intercept is the point $(0, 0)$. Such functions are odd so their graphs are symmetric with respect to the origin. They are increasing on the interval $(-\infty, \infty)$.

**Absolute Value Function**   $\boxed{f(x) = |x|}$

The **absolute value function** is defined as the piecewise-defined function

$$f(x) = |x| = \begin{cases} x & \text{if } x \geq 0 \\ -x & \text{if } x < 0 \end{cases}$$

or as the function

$$f(x) = |x| = \sqrt{x^2}$$

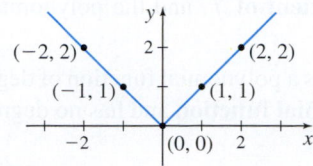

**Figure 27** $f(x) = |x|$

The domain of the absolute value function is the set of all real numbers. The range is the set of nonnegative real numbers. The intercept of the graph of $f$ is the point $(0, 0)$. Because $f(-x) = |-x| = |x| = f(x)$, the absolute value function is even. Its graph is symmetric with respect to the $y$-axis. See Figure 27.

**NOW WORK** Problems **11** and **15**.

The **floor function**, also known as the **greatest integer function**, is defined as the largest integer less than or equal to $x$:

$$f(x) = \lfloor x \rfloor = \text{largest integer less than or equal to } x$$

The domain of the floor function $\lfloor x \rfloor$ is the set of all real numbers; the range is the set of all integers. The $y$-intercept of $\lfloor x \rfloor$ is 0, and the $x$-intercepts are the numbers in the interval $[0, 1)$. The floor function is constant on every interval of the form $[k, k+1)$, where $k$ is an integer, and is nondecreasing on its domain. See Figure 28.

The **ceiling function** is defined as the smallest integer greater than or equal to $x$:

$$f(x) = \lceil x \rceil = \text{smallest integer greater than or equal to } x$$

**IN WORDS** The floor function can be thought of as the "rounding down" function. The ceiling function can be thought of as the "rounding up" function.

The domain of the ceiling function $\lceil x \rceil$ is the set of all real numbers; the range is the set of integers. The $y$-intercept of $\lceil x \rceil$ is 0, and the $x$-intercepts are the numbers in the interval $(-1, 0]$. The ceiling function is constant on every interval of the form $(k, k+1]$, where $k$ is an integer, and is nondecreasing on its domain. See Figure 29.

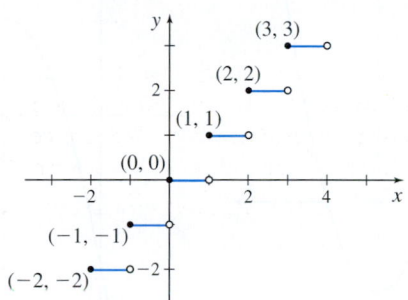

**Figure 28** $f(x) = \lfloor x \rfloor$

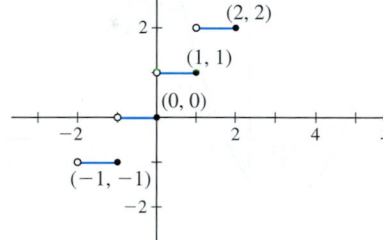

**Figure 29** $f(x) = \lceil x \rceil$

For example, for the floor function, $\lfloor 3 \rfloor = 3$ and $\lfloor 2.9 \rfloor = 2$, but for the ceiling function, $\lceil 3 \rceil = 3$ and $\lceil 2.9 \rceil = 3$.

The floor and ceiling functions are examples of **step functions**. At each integer the function has a *discontinuity*. That is, at integers the function jumps from one value to another without taking on any of the intermediate values.

## 2 Analyze a Polynomial Function and Its Graph

A **monomial** is a function of the form $y = ax^n$, where $a \neq 0$ is a real number and $n \geq 0$ is an integer. *Polynomial functions* are formed by adding a finite number of monomials.

**DEFINITION** Polynomial Function

A **polynomial function** is a function of the form

$$f(x) = a_n x^n + a_{n-1} x^{n-1} + \cdots + a_1 x + a_0$$

where $a_n, a_{n-1}, \ldots, a_1, a_0$ are real numbers and $n$ is a nonnegative integer. The domain of a polynomial function is the set of all real numbers.

If $a_n \neq 0$, then $a_n$ is called the **leading coefficient of** $f$, and the polynomial has **degree** $n$.

The constant function $f(x) = A$, where $A \neq 0$, is a polynomial function of degree 0. The constant function $f(x) = 0$ is the **zero polynomial function** and has no degree. Its graph is a horizontal line containing the point $(0, 0)$.

If the degree of a polynomial function is 1, then it is a linear function of the form $f(x) = mx + b$, where $m \neq 0$. The graph of a linear function is a straight line with slope $m$ and $y$-intercept $b$.

Any polynomial function $f$ of degree 2 can be written in the form

$$f(x) = ax^2 + bx + c$$

where $a$, $b$, and $c$ are constants and $a \neq 0$. The square function $f(x) = x^2$ is a polynomial function of degree 2. Polynomial functions of degree 2 are also called **quadratic functions**. The graph of a quadratic function is known as a **parabola** and is symmetric about its **axis of symmetry**, the vertical line $x = -\dfrac{b}{2a}$. Figure 30 shows the graphs of typical parabolas and their axes of symmetry. The $x$-intercepts, if any, of a quadratic function satisfy the quadratic equation $ax^2 + bx + c = 0$.

**NEED TO REVIEW?** Quadratic equations and the discriminant are discussed in Appendix A.1, pp. A-3 to A-4.

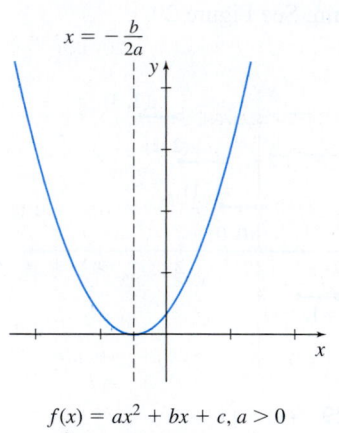

$f(x) = ax^2 + bx + c, a > 0$
$b^2 - 4ac = 0$
One $x$-intercept

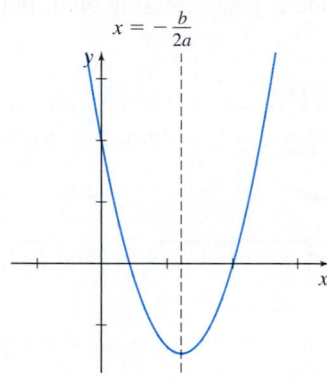

$f(x) = ax^2 + bx + c, a > 0$
$b^2 - 4ac > 0$
Two $x$-intercepts

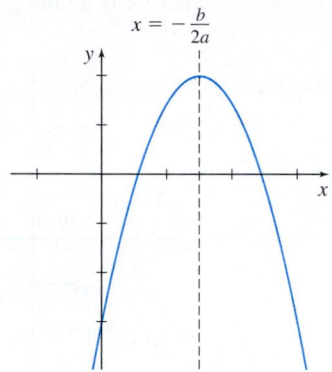

$f(x) = ax^2 + bx + c; a < 0$
$b^2 - 4ac > 0$
Two $x$-intercepts

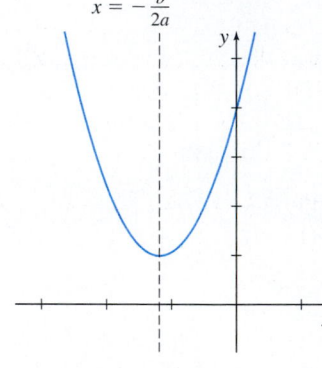

$f(x) = ax^2 + bx + c; a > 0$
$b^2 - 4ac < 0$
No $x$-intercepts

**Figure 30**

Graphs of polynomial functions have many properties in common.

The zeros of a polynomial function $f$ give insight into the graph of $f$. If $r$ is a real zero of a polynomial function $f$, then $r$ is an $x$-intercept of the graph of $f$, and $(x - r)$ is a factor of $f$.

**NOTE** Finding the zeros of a polynomial function $f$ is easy if $f$ is linear, quadratic, or in factored form. Otherwise, finding the zeros can be difficult.

If $(x - r)$ occurs more than once in the factored form of $f$, then $r$ is called a **repeated** or **multiple zero of** $f$. In particular, if $(x - r)^m$ is a factor of $f$, but $(x - r)^{m+1}$, where $m \geq 1$ is an integer, is not a factor of $f$, then $r$ is called a **zero of multiplicity** $m$ **of** $f$. When the multiplicity of a zero $r$ is an even integer, the graph of $f$ will touch (but not cross) the $x$-axis at $r$; when the multiplicity of $r$ is an odd integer, then the graph of $f$ will cross the $x$-axis at $r$. As Figure 31 illustrates, when the graph crosses the $x$-axis, the sign of $f(x)$ changes, and when the graph touches the $x$-axis, the sign of $f(x)$ remains the same.

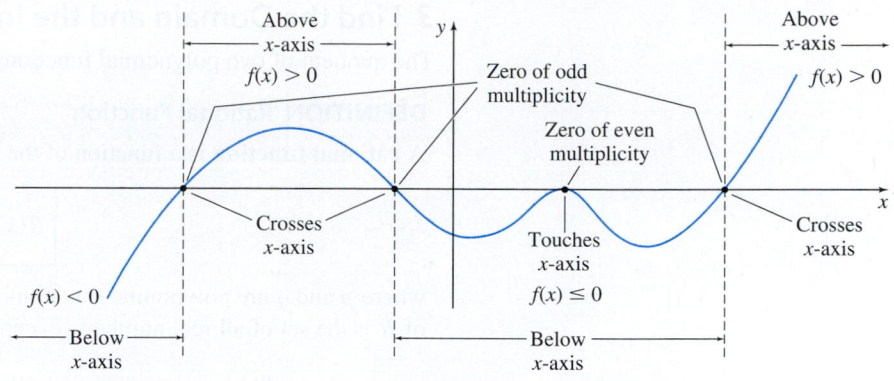

**Figure 31** The graph of a polynomial function.

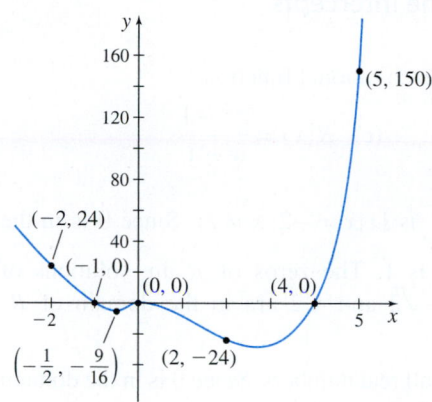

**Figure 32** $f(x) = x^2(x-4)(x+1)$
$= x^4 - 3x^3 - 4x^2$

**TABLE 3**

| $x$ | $f(x)$ | $y = -2x^3$ |
|---|---|---|
| 10 | $-1494$ | $-2000$ |
| 100 | $-1,949,904$ | $-2,000,000$ |
| 500 | $-248,749,504$ | $-250,000,000$ |
| 1000 | $-1,994,999,004$ | $-2,000,000,000$ |

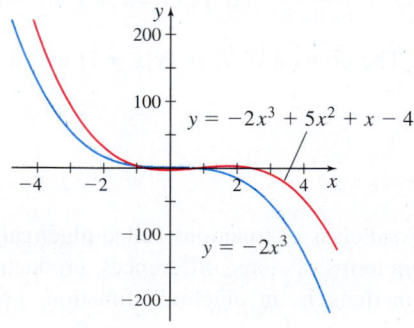

**Figure 33**

## EXAMPLE 1  Analyzing the Graph of a Polynomial Function

For the polynomial function $f(x) = x^2(x-4)(x+1)$:

**(a)** Find the $x$- and $y$-intercepts of the graph of $f$.

**(b)** Determine whether the graph crosses or touches the $x$-axis at each $x$-intercept.

**(c)** Plot at least one point to the left and right of each $x$-intercept and connect the points to obtain the graph.

**Solution** **(a)** The $y$-intercept is $f(0) = 0$. The $x$-intercepts are the zeros of the function: 0, 4, and $-1$.

**(b)** 0 is a zero of multiplicity 2; the graph of $f$ will touch the $x$-axis at 0. The numbers 4 and $-1$ are zeros of multiplicity 1; the graph of $f$ will cross the $x$-axis at 4 and $-1$.

**(c)** Since $f(-2) = 24$, $f\left(-\dfrac{1}{2}\right) = -\dfrac{9}{16}$, $f(2) = -24$, and $f(5) = 150$, the points $(-2, 24)$, $\left(-\dfrac{1}{2}, -\dfrac{9}{16}\right)$, $(2, -24)$, and $(5, 150)$ are on the graph. See Figure 32. ∎

The behavior of the graph of a function for large values of $x$, either positive or negative, is referred to as its **end behavior**.

For large values of $x$, either positive or negative, the graph of the polynomial function

$$f(x) = a_n x^n + a_{n-1} x^{n-1} + \cdots + a_1 x + a_0 \quad a_n \neq 0$$

resembles the graph of the power function

$$y = a_n x^n$$

For example, the end behavior of the graph of the polynomial function $f(x) = x^4 - 3x^3 - 4x^2$, shown in Figure 32, resembles the graph of $y = x^4$.

As another example, if $f(x) = -2x^3 + 5x^2 + x - 4$, then the graph of $f$ will behave like the graph of $y = -2x^3$ for very large values of $x$, either positive or negative. We can see that the graphs of $f$ and $y = -2x^3$ "behave" the same by considering Table 3 and Figure 33.

**NOW WORK** Problem 21.

## 3 Find the Domain and the Intercepts of a Rational Function

The quotient of two polynomial functions $p$ and $q$ is called a *rational function*.

> **DEFINITION** Rational Function
>
> A **rational function** is a function of the form
>
> $$R(x) = \frac{p(x)}{q(x)}$$
>
> where $p$ and $q$ are polynomial functions and $q$ is not the zero polynomial. The domain of $R$ is the set of all real numbers, except those for which the denominator $q$ is 0.

If $R(x) = \dfrac{p(x)}{q(x)}$ is a rational function, the real zeros, if any, of the numerator, which are also in the domain of $R$, are the $x$-intercepts of the graph of $R$.

CALC
▶ CLIP

**EXAMPLE 2** Finding the Domain and the Intercepts of a Rational Function

Find the domain and the intercepts (if any) of each rational function:

**(a)** $R(x) = \dfrac{2x^2 - 4}{x^2 - 4}$ **(b)** $R(x) = \dfrac{x}{x^2 + 1}$ **(c)** $R(x) = \dfrac{x^2 - 1}{x - 1}$

**Solution (a)** The domain of $R(x) = \dfrac{2x^2 - 4}{x^2 - 4}$ is $\{x \mid x \neq -2; x \neq 2\}$. Since 0 is in the domain of $R$ and $R(0) = 1$, the $y$-intercept is 1. The zeros of $R$ are solutions of the equation $2x^2 - 4 = 0$ or $x^2 = 2$. Since $-\sqrt{2}$ and $\sqrt{2}$ are in the domain of $R$, the $x$-intercepts are $-\sqrt{2}$ and $\sqrt{2}$.

**(b)** The domain of $R(x) = \dfrac{x}{x^2 + 1}$ is the set of all real numbers. Since 0 is in the domain of $R$ and $R(0) = 0$, the $y$-intercept is 0, and the $x$-intercept is also 0.

**(c)** The domain of $R(x) = \dfrac{x^2 - 1}{x - 1}$ is $\{x \mid x \neq 1\}$. Since 0 is in the domain of $R$ and $R(0) = 1$, the $y$-intercept is 1. The $x$-intercept(s), if any, satisfy the equation

$$x^2 - 1 = 0$$
$$x^2 = 1$$
$$x = -1 \quad \text{or} \quad x = 1$$

Since 1 is not in the domain of $R$, the only $x$-intercept is $-1$. ∎

Figure 34 shows the graphs of $R(x) = \dfrac{x^2 - 1}{x - 1}$ and $f(x) = x + 1$. Notice the "hole" in the graph of $R$ at the point $(1, 2)$. Also, $R(x) = \dfrac{x^2 - 1}{x - 1}$ and $f(x) = x + 1$ are not the same function: their domains are different. The domain of $R$ is $\{x \mid x \neq 1\}$ and the domain of $f$ is the set of all real numbers.

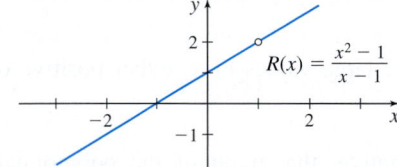

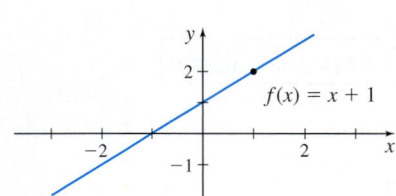

**Figure 34**

NOW WORK Problem 27.

## Algebraic Functions

Every function discussed so far belongs to a broad class of functions called **algebraic functions**. A function $f$ that can be expressed in terms of sums, differences, products, quotients, powers, or roots of polynomial functions is an algebraic function. For example, the function $f$ defined by

$$f(x) = \frac{3x^3 - x^2(x + 1)^{4/3}}{\sqrt{x^4 + 2}}$$

is an algebraic function. Functions that are not algebraic are called **transcendental functions**. Examples of transcendental functions include the trigonometric functions, exponential functions, and logarithmic functions, which are discussed later.

## 4 Construct a Mathematical Model

Problems in engineering and the sciences often can by solved using mathematical models that involve functions. To build a model, verbal descriptions must be translated into the language of mathematics by assigning symbols to represent the independent and dependent variables and then finding a function that relates the variables.

**EXAMPLE 3**  **Constructing a Model from a Verbal Description**

A liquid is poured into a container in the shape of a right circular cone with radius 4 meters and height 16 meters, as shown in Figure 35. Express the volume $V$ of the liquid as a function of the height $h$ of the liquid.

**Solution** The formula for the volume of a right circular cone of radius $r$ and height $h$ is

$$V = \frac{1}{3}\pi r^2 h$$

The volume depends on two variables, $r$ and $h$. To express $V$ as a function of $h$ only, we use the fact that a cross section of the cone and the liquid form two similar triangles. See Figure 36.

Corresponding sides of similar triangles are in proportion. Since the cone's radius is 4 meters and its height is 16 meters, we have

$$\frac{r}{h} = \frac{4}{16} = \frac{1}{4}$$

$$r = \frac{1}{4}h$$

Then

$$V = \frac{1}{3}\pi r^2 h = \underset{\underset{r = \frac{1}{4}h}{\uparrow}}{\frac{1}{3}\pi \left(\frac{1}{4}h\right)^2 h} = \frac{1}{48}\pi h^3$$

So $V = V(h) = \dfrac{1}{48}\pi h^3$ expresses the volume $V$ as a function of the height of the liquid. Since $h$ is measured in meters, $V$ will be expressed in cubic meters. ∎

**NOW WORK** Problem 29.

In many applications, data are collected and are used to build the mathematical model. If the data involve two variables, the first step is to plot ordered pairs using rectangular coordinates. The resulting graph is called a **scatter plot**.

Scatter plots are used to help suggest the type of relation that exists between the two variables. Once the general shape of the relation is recognized, a function can be chosen whose graph closely resembles the shape in the scatter plot.

**EXAMPLE 4**  **Identifying the Shape of a Scatter Plot**

Determine whether you would model the relation between the two variables shown in each scatter plot in Figure 37 with a linear function or a quadratic function.

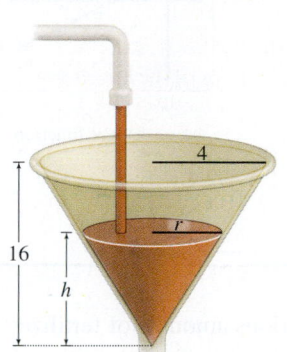

**Figure 35**

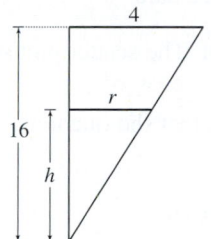

**Figure 36**

**NEED TO REVIEW?** Similar triangles and geometry formulas are discussed in Appendix A.2, pp. A-13 to A-15.

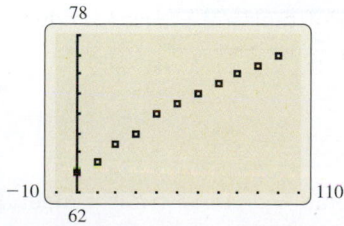
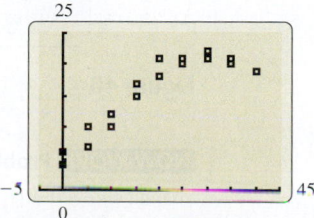
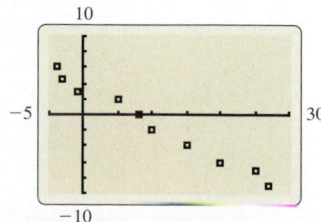

**Figure 37**

**Solution** For each scatter plot, we choose a function whose graph closely resembles the shape of the scatter plot. See Figure 38.

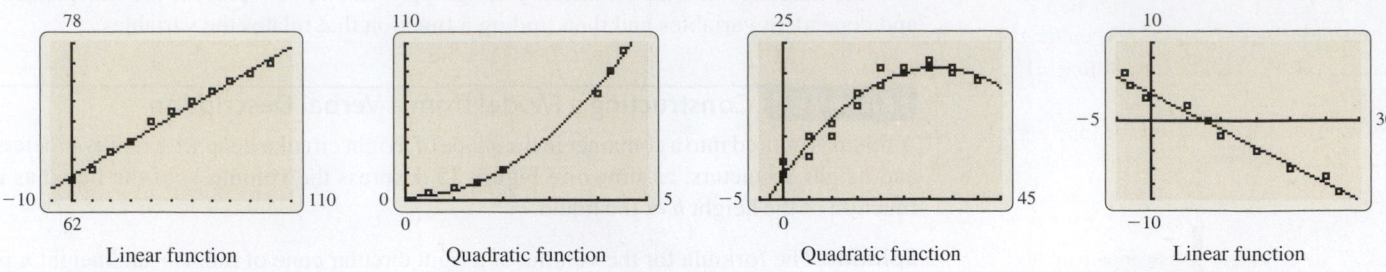

Linear function          Quadratic function          Quadratic function          Linear function

**Figure 38**

■ **EXAMPLE 5** **Building a Function from Data**

The data shown in Table 4 measure crop yield for various amounts of fertilizer:

(a) Draw a scatter plot of the data and determine a possible type of relation that may exist between the two variables.

(b) Use technology to find the function of best fit to these data.

**Solution** **(a)** Figure 39 shows the data and a scatter plot. The scatter plot suggests the graph of a quadratic function.

**(b)** The graphing calculator screen in Figure 40 shows that the quadratic function of best fit is

$$Y(x) = -0.017x^2 + 1.077x + 3.894$$

where $x$ represents the amount of fertilizer used and $Y$ represents crop yield. The graph of the quadratic model is illustrated in Figure 41.

**TABLE 4**

| Plot | Fertilizer, $x$ (Pounds/100 ft²) | Yield (Bushels) |
|------|----------------------------------|-----------------|
| 1    | 0                                | 4               |
| 2    | 0                                | 6               |
| 3    | 5                                | 10              |
| 4    | 5                                | 7               |
| 5    | 10                               | 12              |
| 6    | 10                               | 10              |
| 7    | 15                               | 15              |
| 8    | 15                               | 17              |
| 9    | 20                               | 18              |
| 10   | 20                               | 21              |
| 11   | 25                               | 20              |
| 12   | 25                               | 21              |
| 13   | 30                               | 21              |
| 14   | 30                               | 22              |
| 15   | 35                               | 21              |
| 16   | 35                               | 20              |
| 17   | 40                               | 19              |
| 18   | 40                               | 19              |

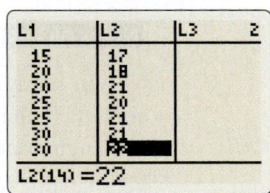

**Figure 39**

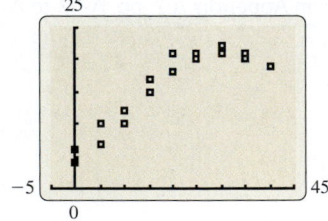

**Figure 40**          **Figure 41**

■

NOW WORK **Problem 31.**

# P.2 Assess Your Understanding

## Concepts and Vocabulary

1. *Multiple Choice* The function $f(x) = x^2$ is [(a) increasing (b) decreasing (c) neither] on the interval $[0, \infty)$.

2. *True or False* The floor function $f(x) = \lfloor x \rfloor$ is an example of a step function.

3. *True or False* The cube function is odd and is increasing on the interval $(-\infty, \infty)$.

4. *True or False* The cube root function is decreasing on the interval $(-\infty, \infty)$.

5. *True or False* The domain and the range of the reciprocal function are all real numbers.

6. A number $r$ for which $f(r) = 0$ is called a(n) _____ of the function $f$.

7. *Multiple Choice* If $r$ is a zero of even multiplicity of a function $f$, the graph of $f$ [(a) crosses (b) touches (c) doesn't intersect] the $x$-axis at $r$.

8. *True or False* The $x$-intercepts of the graph of a polynomial function are called real zeros of the function.

9. *True or False* The function $f(x) = \left[x + \sqrt[5]{x^2 - \pi}\right]^{2/3}$ is an algebraic function.

10. *True or False* The domain of every rational function is the set of all real numbers.

## Practice Problems

*In Problems 11–18, match each graph to its function:*

A. *Constant function*  
B. *Identity function*  
C. *Square function*  
D. *Cube function*  
E. *Square root function*  
F. *Reciprocal function*  
G. *Absolute value function*  
H. *Cube root function*

11.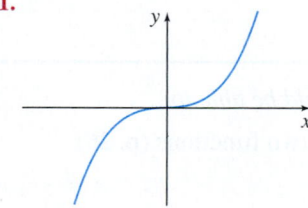

12.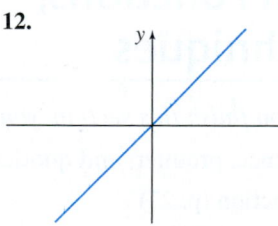

13.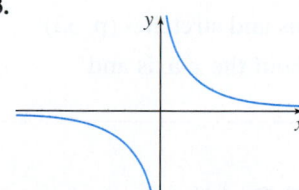

14.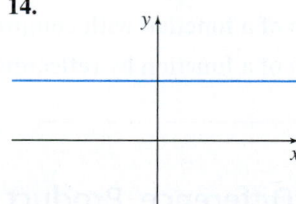

15.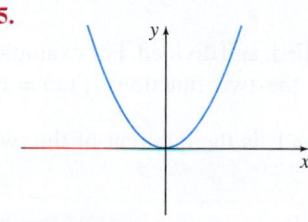

16.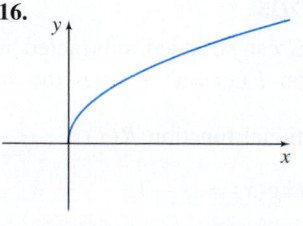

17.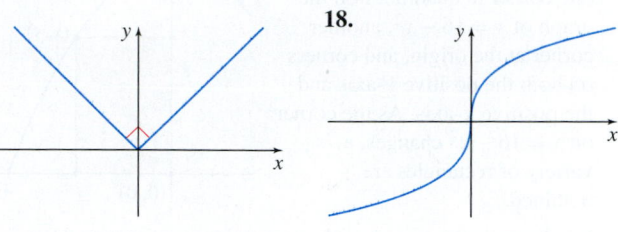
18.

19. If $f(x) = \lfloor 2x \rfloor$, find:

    (a) $f(1.2)$      (b) $f(1.6)$      (c) $f(-1.8)$

20. If $f(x) = \left\lceil \dfrac{x}{2} \right\rceil$, find:

    (a) $f(1.2)$      (b) $f(1.6)$      (c) $f(-1.8)$

*In Problems 21 and 22, for each polynomial function $f$:*

(a) *List each real zero and its multiplicity.*

(b) *Find the $x$- and $y$-intercepts of the graph of $f$.*

(c) *Determine whether the graph of $f$ crosses or touches the $x$-axis at each $x$-intercept.*

(d) *Determine the end behavior of the graph of $f$.*

21. $f(x) = 3(x - 7)(x + 4)^3$      22. $f(x) = 4x(x^2 + 1)(x - 2)^3$

*In Problems 23 and 24, decide which of the polynomial functions in the list might have the given graph. (More than one answer is possible.)*

23. (a) $f(x) = -4x(x - 1)(x - 2)$

    (b) $f(x) = x^2(x - 1)^2(x - 2)$

    (c) $f(x) = 3x(x - 1)(x - 2)$

    (d) $f(x) = x(x - 1)^2(x - 2)^2$

    (e) $f(x) = x^3(x - 1)(x - 2)$

    (f) $f(x) = -x(1 - x)(x - 2)$

24. (a) $f(x) = 2x^3(x - 1)(x - 2)^2$

    (b) $f(x) = x^2(x - 1)(x - 2)$

    (c) $f(x) = x^3(x - 1)^2(x - 2)$

    (d) $f(x) = x^2(x - 1)^2(x - 2)^2$

    (e) $f(x) = 5x(x - 1)^2(x - 2)$

    (f) $f(x) = -2x(x - 1)^2(2 - x)$

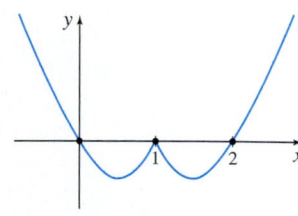

*In Problems 25–28, find the domain and the intercepts of each rational function.*

25. $R(x) = \dfrac{5x^2}{x + 3}$      26. $H(x) = \dfrac{-4x^2}{(x - 2)(x + 4)}$

27. $R(x) = \dfrac{3x^2 - x}{x^2 + 4}$      28. $R(x) = \dfrac{3(x^2 - x - 6)}{4(x^2 - 9)}$

---

**1.** = NOW WORK problem      〔∿〕 = Graphing technology recommended      〔CAS〕 = Computer Algebra System recommended

**29. Constructing a Model** The rectangle shown in the figure has one corner in quadrant I on the graph of $y = 16 - x^2$, another corner at the origin, and corners on both the positive $y$-axis and the positive $x$-axis. As the corner on $y = 16 - x^2$ changes, a variety of rectangles are obtained.

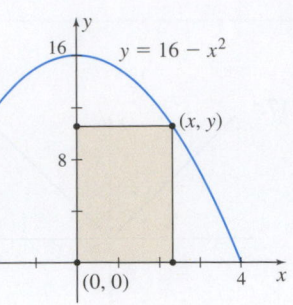

**(a)** Express the area $A$ of the rectangles as a function of $x$.

**(b)** What is the domain of $A$?

**30. Constructing a Model** The rectangle shown in the figure is inscribed in a semicircle of radius 2. Let $P = (x, y)$ be the point in quadrant I that is a vertex of the rectangle and is on the circle. As the point $(x, y)$ on the circle changes, a variety of rectangles are obtained.

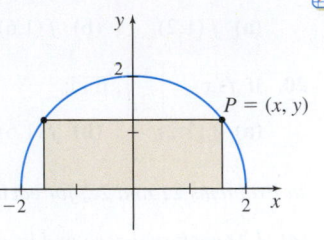

**(a)** Express the area $A$ of the rectangles as a function of $x$.

**(b)** Express the perimeter $p$ of the rectangles as a function of $x$.

**31. Height of a Ball** A ballplayer throws a ball at an inclination of $45°$ to the horizontal. The following data represent the height $h$ (in feet) of the ball at the instant that it has traveled $x$ feet horizontally:

| Distance, $x$ | 20 | 40 | 60 | 80 | 100 | 120 | 140 | 160 | 180 | 200 |
|---|---|---|---|---|---|---|---|---|---|---|
| Height, $h$ | 25 | 40 | 55 | 65 | 71 | 77 | 77 | 75 | 71 | 64 |

**(a)** Draw a scatter plot of the data with $x$ as the independent variable. Comment on the type of relation that may exist between the two variables.

**(b)** Use technology to verify that the quadratic function of best fit to these data is

$$h(x) = -0.0037x^2 + 1.03x + 5.7$$

Use this function to determine the horizontal distance the ball travels before it hits the ground.

**(c)** Approximate the height of the ball when it has traveled 10 feet.

**32. Educational Attainment** The following data represent the percentage of the U.S. population whose age is $x$ (in years) who did not have a high school diploma as of January 2016:

| Age, $x$ | 30 | 40 | 50 | 60 | 70 | 80 |
|---|---|---|---|---|---|---|
| Percentage Without a High School Diploma, $P$ | 11.6 | 11.7 | 10.4 | 10.4 | 17.0 | 24.6 |

*Source:* U.S. Census Bureau.

**(a)** Draw a scatter plot of the data, treating age as the independent variable. Comment on the type of relation that may exist between the two variables.

**(b)** Use technology to verify that the cubic function of best fit to these data is

$$P(x) = 0.00026x^3 - 0.0303x^2 + 1.0877x - 0.5071$$

**(c)** Use this model to predict the percentage of 35-year-olds who do not have a high school diploma.

# P.3 Operations on Functions; Graphing Techniques

**OBJECTIVES** *When you finish this section, you should be able to:*

1 Form the sum, difference, product, and quotient of two functions (p. 26)

2 Form a composite function (p. 27)

3 Transform the graph of a function with vertical and horizontal shifts (p. 30)

4 Transform the graph of a function with compressions and stretches (p. 32)

5 Transform the graph of a function by reflecting it about the $x$-axis and the $y$-axis (p. 33)

## 1 Form the Sum, Difference, Product, and Quotient of Two Functions

Functions, like numbers, can be added, subtracted, multiplied, and divided. For example, the polynomial function $F(x) = x^2 + 4x$ is the sum of the two functions $f(x) = x^2$ and $g(x) = 4x$. The rational function $R(x) = \dfrac{x^2}{x^3 - 1}$, $x \neq 1$, is the quotient of the two functions $f(x) = x^2$ and $g(x) = x^3 - 1$.

**DEFINITION  Operations on Functions**

If $f$ and $g$ are two functions, their sum, $f + g$; their difference, $f - g$; their product, $f \cdot g$; and their quotient, $\dfrac{f}{g}$, are defined by

- Sum:    $(f + g)(x) = f(x) + g(x)$ • Difference: $(f - g)(x) = f(x) - g(x)$

- Product:  $(f \cdot g)(x) = f(x) \cdot g(x)$  • Quotient:  $\left(\dfrac{f}{g}\right)(x) = \dfrac{f(x)}{g(x)}$, $g(x) \neq 0$

For every operation except division, the domain of the resulting function consists of the intersection of the domains of $f$ and $g$. The domain of a quotient $\dfrac{f}{g}$ consists of the numbers $x$ that are common to the domains of both $f$ and $g$ but excludes the numbers $x$ for which $g(x) = 0$.

---

**EXAMPLE 1**  **Forming the Sum, Difference, Product, and Quotient of Two Functions**

Let $f$ and $g$ be two functions defined as

$$f(x) = \sqrt{x - 1} \qquad \text{and} \qquad g(x) = \sqrt{4 - x}$$

Find the following functions and determine their domain:

**(a)** $(f + g)(x)$    **(b)** $(f - g)(x)$    **(c)** $(f \cdot g)(x)$    **(d)** $\left(\dfrac{f}{g}\right)(x)$

**Solution**  The domain of $f$ is $\{x \mid x \geq 1\}$, and the domain of $g$ is $\{x \mid x \leq 4\}$.

**(a)** $(f + g)(x) = f(x) + g(x) = \sqrt{x - 1} + \sqrt{4 - x}$. The domain of $(f + g)(x)$ is the closed interval $[1, 4]$.

**(b)** $(f - g)(x) = f(x) - g(x) = \sqrt{x - 1} - \sqrt{4 - x}$. The domain of $(f - g)(x)$ is the closed interval $[1, 4]$.

**(c)** $(f \cdot g)(x) = f(x) \cdot g(x) = (\sqrt{x - 1})(\sqrt{4 - x}) = \sqrt{-x^2 + 5x - 4}$. The domain of $(f \cdot g)(x)$ is the closed interval $[1, 4]$.

**(d)** $\left(\dfrac{f}{g}\right)(x) = \dfrac{f(x)}{g(x)} = \dfrac{\sqrt{x - 1}}{\sqrt{4 - x}} = \dfrac{\sqrt{-x^2 + 5x - 4}}{4 - x}$. The domain of $\left(\dfrac{f}{g}\right)(x)$ is the half-open interval $[1, 4)$. ∎

**NOW WORK**  Problem **11**.

## 2 Form a Composite Function

Suppose an oil tanker is leaking, and your job requires you to find the area of the circular oil spill surrounding the tanker. You determine that the radius of the spill is increasing at a rate of 3 meters per minute. That is, the radius $r$ of the spill is a function of the time $t$ in minutes since the leak began and can be written as $r(t) = 3t$.

For example, after 20 minutes the radius of the spill is $r(20) = 3 \cdot 20 = 60$ meters. Recall that the area $A$ of a circle is a function of its radius $r$; that is, $A(r) = \pi r^2$. So, the area of the oil spill after 20 minutes is $A(60) = \pi(60^2) = 3600\pi$ square meters. Notice that the argument $r$ of the function $A$ is itself a function and that the area $A$ of the oil spill is found at any time $t$ by evaluating the function $A = A(r(t))$.

Functions such as $A = A(r(t))$ are called *composite functions*.

Another example of a composite function is $y = (2x+3)^2$. If $y = f(u) = u^2$ and $u = g(x) = 2x+3$, then substitute $g(x) = 2x+3$ for $u$ to get the original function:

$$y = f(u) = f(g(x)) = (2x+3)^2$$
$$\underset{u=g(x)}{\uparrow} \quad \underset{g(x)=2x+3}{\uparrow}$$

In general, suppose that $f$ and $g$ are two functions and that $x$ is a number in the domain of $g$. By evaluating $g$ at $x$, we obtain $g(x)$. Now, if $g(x)$ is in the domain of the function $f$, we can evaluate $f$ at $g(x)$, obtaining $f(g(x))$. The correspondence from $x$ to $f(g(x))$ is called *composition*.

> **DEFINITION  Composite Function**
>
> Given two functions $f$ and $g$, the **composite function**, denoted by $f \circ g$ (read "$f$ composed with $g$") is defined by
>
> $$\boxed{(f \circ g)(x) = f(g(x))}$$
>
> The domain of $f \circ g$ is the set of all numbers $x$ in the domain of $g$ for which $g(x)$ is in the domain of $f$.

Figure 42 illustrates the definition. Only those numbers $x$ in the domain of $g$ for which $g(x)$ is in the domain of $f$ are in the domain of $f \circ g$. As a result, the domain of $f \circ g$ is a subset of the domain of $g$, and the range of $f \circ g$ is a subset of the range of $f$.

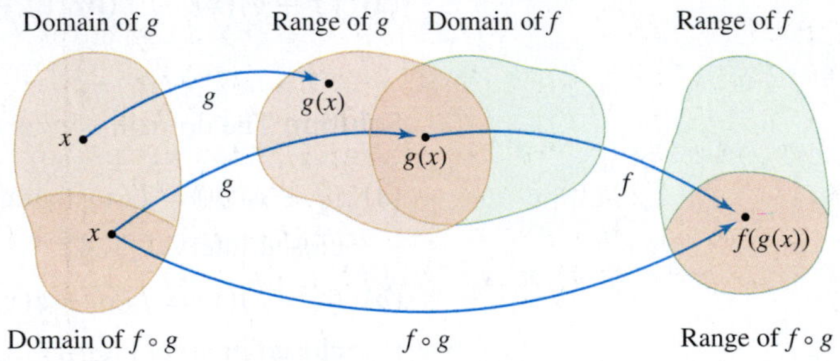

Domain of $g$     Range of $g$    Domain of $f$     Range of $f$

Domain of $f \circ g$            $f \circ g$            Range of $f \circ g$

**Figure 42**

---

**EXAMPLE 2   Evaluating a Composite Function**

Suppose that $f(x) = \dfrac{1}{x+2}$ and $g(x) = \dfrac{4}{x-1}$.

NOTE  The "inside" function $g$, in $f(g(x))$, is evaluated first.

**(a)**  $(f \circ g)(0) = f(g(0)) = f(-4) = \dfrac{1}{-4+2} = -\dfrac{1}{2}$ $\qquad g(x) = \dfrac{4}{x-1}; \; g(0) = -4$

**(b)**  $(g \circ f)(1) = g(f(1)) = g\left(\dfrac{1}{3}\right) = \dfrac{4}{\dfrac{1}{3}-1} = \dfrac{4}{-\dfrac{2}{3}} = -6$
$\qquad \underset{\uparrow}{\;}$
$f(x) = \dfrac{1}{x+2}; \; f(1) = \dfrac{1}{3}$

**(c)**  $(f \circ f)(1) = f(f(1)) = f\left(\dfrac{1}{3}\right) = \dfrac{1}{\dfrac{1}{3}+2} = \dfrac{1}{\dfrac{7}{3}} = \dfrac{3}{7}$

**(d)**  $(g \circ g)(-3) = g(g(-3)) = g(-1) = \dfrac{4}{-1-1} = -2$
$\qquad \underset{\uparrow}{\;}$
$g(x) = \dfrac{4}{x-1}; \; g(-3) = \dfrac{4}{-3-1} = -1$    ∎

**NOW WORK** Problem **15.**

### EXAMPLE 3   Finding the Domain of a Composite Function

Suppose that $f(x) = \dfrac{1}{x+2}$ and $g(x) = \dfrac{4}{x-1}$. Find $f \circ g$ and its domain.

#### Solution

$$(f \circ g)(x) = f(g(x)) = \frac{1}{g(x)+2} = \frac{1}{\dfrac{4}{x-1}+2} = \frac{x-1}{4+2(x-1)} = \frac{x-1}{2x+2}$$

To find the domain of $f \circ g$, first note that the domain of $g$ is $\{x \mid x \neq 1\}$, so we exclude 1 from the domain of $f \circ g$. Next note that the domain of $f$ is $\{x \mid x \neq -2\}$, which means $g(x)$ cannot equal $-2$. To determine what additional values of $x$ to exclude, we solve the equation $g(x) = -2$:

$$\frac{4}{x-1} = -2 \qquad\qquad g(x) = -2$$
$$4 = -2(x-1)$$
$$4 = -2x + 2$$
$$2x = -2$$
$$x = -1$$

We also exclude $-1$ from the domain of $f \circ g$.

The domain of $f \circ g$ is $\{x \mid x \neq -1, x \neq 1\}$. ∎

We could also find the domain of $f \circ g$ by first finding the domain of $g$: $\{x \mid x \neq 1\}$. So, exclude 1 from the domain of $f \circ g$. From $(f \circ g)(x) = \dfrac{x-1}{2x+2} = \dfrac{x-1}{2(x+1)}$, notice that $x \neq -1$, so we exclude $-1$ from the domain of $f \circ g$. Therefore, the domain of $f \circ g$ is $\{x \mid x \neq -1, x \neq 1\}$.

NOW WORK **Problem 23.**

In general, the composition of two functions $f$ and $g$ is not commutative. That is, $f \circ g$ almost never equals $g \circ f$. For example, in Example 3,

$$(g \circ f)(x) = g(f(x)) = \frac{4}{f(x)-1} = \frac{4}{\dfrac{1}{x+2}-1} = \frac{4(x+2)}{1-(x+2)} = -\frac{4(x+2)}{x+1}$$

Functions $f$ and $g$ for which $f \circ g = g \circ f$ will be discussed in the next section.

Some techniques in calculus require us to "decompose" a composite function. For example, the function $H(x) = \sqrt{x+1}$ is the composition $f \circ g$ of the functions $f(x) = \sqrt{x}$ and $g(x) = x+1$.

CALC
▶
CLIP

### EXAMPLE 4   Decomposing a Composite Function

Find functions $f$ and $g$ so that $f \circ g = F$ when:

**(a)** $F(x) = \dfrac{1}{x+1}$   **(b)** $F(x) = (x^3 - 4x - 1)^{100}$   **(c)** $F(t) = \sqrt{2-t}$

#### Solution **(a)** If we let $f(x) = \dfrac{1}{x}$ and $g(x) = x+1$, then

$$(f \circ g)(x) = f(g(x)) = \frac{1}{g(x)} = \frac{1}{x+1} = F(x)$$

**(b)** If we let $f(x) = x^{100}$ and $g(x) = x^3 - 4x - 1$, then

$$(f \circ g)(x) = f(g(x)) = f(x^3 - 4x - 1) = (x^3 - 4x - 1)^{100} = F(x)$$

**(c)** If we let $f(t) = \sqrt{t}$ and $g(t) = 2 - t$, then

$$(f \circ g)(t) = f(g(t)) = f(2 - t) = \sqrt{2 - t} = F(t) \qquad \blacksquare$$

Although the functions $f$ and $g$ chosen in Example 4 are not unique, there is usually a "natural" selection for $f$ and $g$ that first comes to mind. When decomposing a composite function, the "natural" selection for $g$ is often an expression inside parentheses, in a denominator, or under a radical.

NOW WORK Problem 27.

## 3 Transform the Graph of a Function with Vertical and Horizontal Shifts

At times we need to graph a function that is very similar to a function with a known graph. Often techniques, called **transformations**, can be used to draw the new graph.

First we consider *translations*. **Translations** shift the graph from one position to another without changing its shape, size, or direction.

For example, let $f$ be a function with a known graph, say, $f(x) = x^2$. If $k$ is a positive number, then adding $k$ to $f$ adds $k$ to each $y$-coordinate, causing the graph of $f$ to **shift vertically up** $k$ units. On the other hand, subtracting $k$ from $f$ subtracts $k$ from each $y$-coordinate, causing the graph of $f$ to **shift vertically down** $k$ units. See Figure 43.

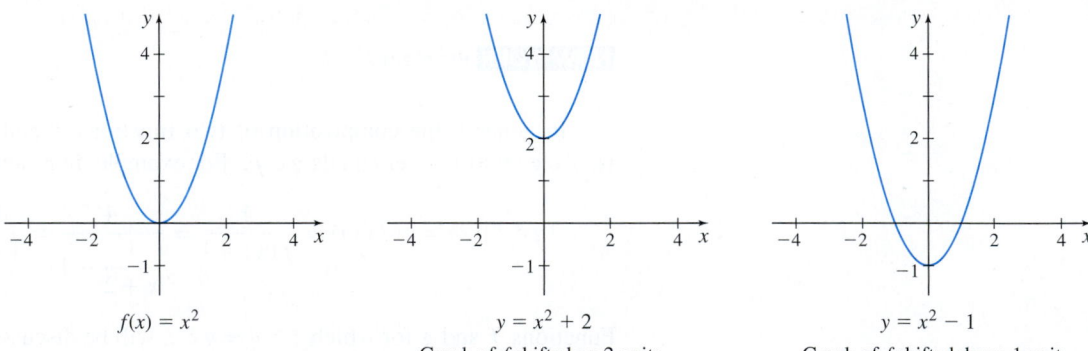

$f(x) = x^2$

$y = x^2 + 2$
Graph of $f$ shifted up 2 units

$y = x^2 - 1$
Graph of $f$ shifted down 1 unit

DF **Figure 43**

So, adding (or subtracting) a positive constant to a function shifts the graph of the original function vertically up (or down). Now we investigate how to shift the graph of a function right or left.

Again, let $f$ be a function with a known graph, say, $f(x) = x^2$, and let $h$ be a positive number. To shift the graph to the right $h$ units, subtract $h$ from the argument of $f$. In other words, replace the argument $x$ of a function $f$ by $x - h$, $h > 0$. The graph of the new function $y = f(x - h)$ is the graph of $f$ **shifted horizontally right** $h$ units. On the other hand, if we replace the argument $x$ of a function $f$ by $x + h$, $h > 0$, the graph of the new function $y = f(x + h)$ is the graph of $f$ **shifted horizontally left** $h$ units. See Figure 44.

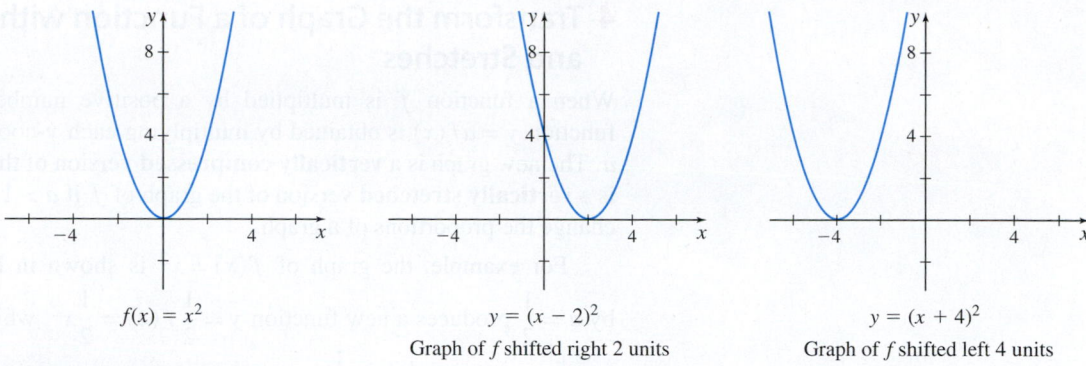

$f(x) = x^2$

$y = (x - 2)^2$
Graph of $f$ shifted right 2 units

$y = (x + 4)^2$
Graph of $f$ shifted left 4 units

**DF** **Figure 44**

The graph of a function $f$ can be moved anywhere in the plane by combining vertical and horizontal shifts.

---

**EXAMPLE 5** **Combining Vertical and Horizontal Shifts**

Use transformations to graph the function $f(x) = (x + 3)^2 - 5$.

**Solution** Graph $f$ in steps:

- Observe that $f$ is basically a square function, so begin by graphing $y = x^2$ in Figure 45(a).
- Replace the argument $x$ with $x + 3$ to obtain $y = (x + 3)^2$. This shifts the graph of $f$ horizontally to the left 3 units, as shown in Figure 45(b).
- Finally, subtract 5 from each $y$-coordinate, which shifts the graph in Figure 45(b) vertically down 5 units and results in the graph of $f(x) = (x + 3)^2 - 5$ shown in Figure 45(c).

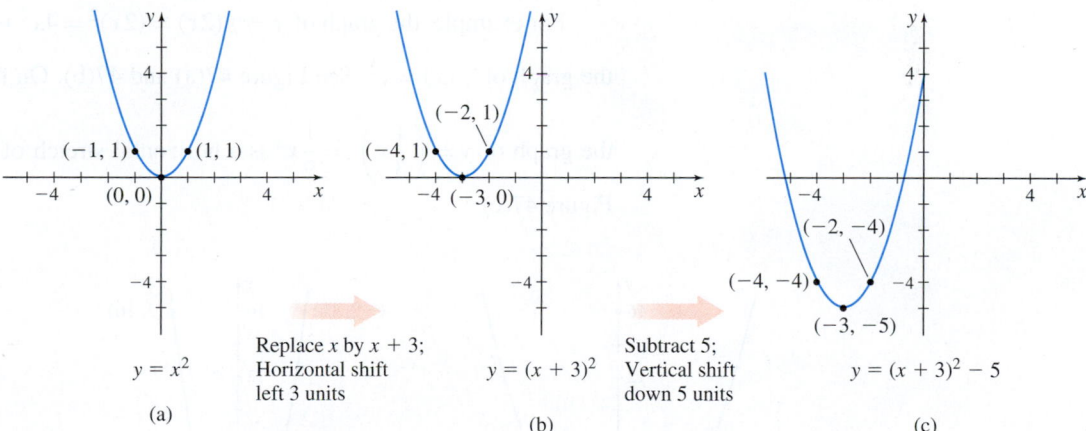

$y = x^2$
Replace $x$ by $x + 3$;
Horizontal shift
left 3 units
(a)

$y = (x + 3)^2$
Subtract 5;
Vertical shift
down 5 units
(b)

$y = (x + 3)^2 - 5$
(c)

**Figure 45**

Notice the points plotted in Figure 45. Using key points can be helpful in keeping track of the transformation that has taken place.

**NOW WORK** Problem **39.**

## 4 Transform the Graph of a Function with Compressions and Stretches

When a function $f$ is multiplied by a positive number $a$, the graph of the new function $y = af(x)$ is obtained by multiplying each $y$-coordinate on the graph of $f$ by $a$. The new graph is a **vertically compressed** version of the graph of $f$ if $0 < a < 1$ and is a **vertically stretched** version of the graph of $f$ if $a > 1$. Compressions and stretches change the proportions of a graph.

For example, the graph of $f(x) = x^2$ is shown in Figure 46(a). Multiplying $f$ by $a = \dfrac{1}{2}$ produces a new function $y = \dfrac{1}{2} f(x) = \dfrac{1}{2} x^2$, which vertically compresses the graph of $f$ by a factor of $\dfrac{1}{2}$, as shown in Figure 46(b). On the other hand, if $a = 3$, then multiplying $f$ by 3 produces a new function $y = 3f(x) = 3x^2$, and the graph of $f$ is vertically stretched by a factor of 3, as shown in Figure 46(c).

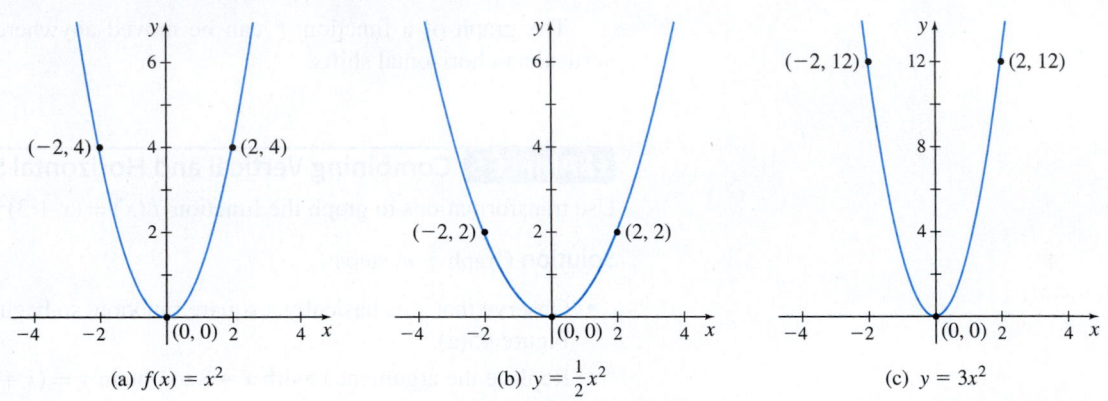

(a) $f(x) = x^2$

(b) $y = \dfrac{1}{2} x^2$

Graph of $f$ is vertically compressed

(c) $y = 3x^2$

Graph of $f$ is vertically stretched

**Figure 46**

If the argument $x$ of a function $f$ is multiplied by a positive number $a$, the graph of the new function $y = f(ax)$ is a **horizontal compression** of the graph of $f$ when $a > 1$ and a **horizontal stretch** of the graph of $f$ when $0 < a < 1$.

For example, the graph of $y = f(2x) = (2x)^2 = 4x^2$ is a horizontal compression of the graph of $f(x) = x^2$. See Figure 47(a) and 47(b). On the other hand, if $a = \dfrac{1}{3}$, then the graph of $y = \left(\dfrac{1}{3} x\right)^2 = \dfrac{1}{9} x^2$ is a horizontal stretch of the graph of $f(x) = x^2$. See Figure 47(c).

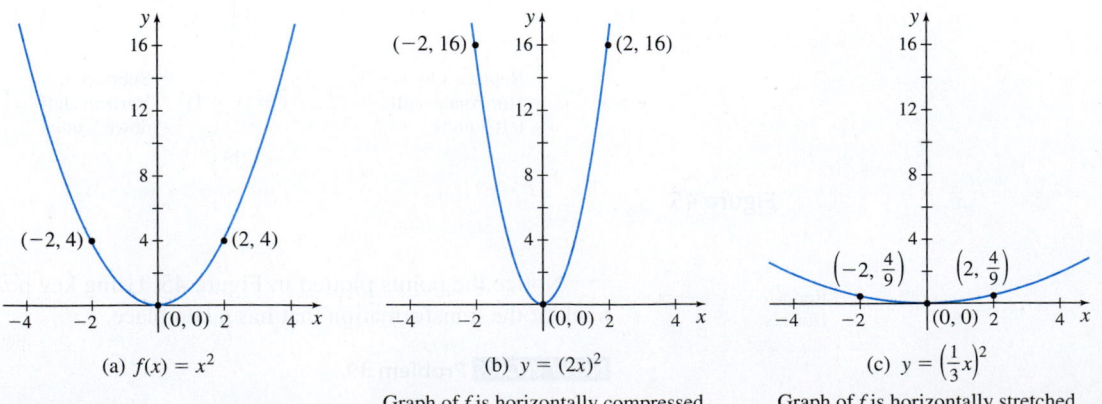

(a) $f(x) = x^2$

(b) $y = (2x)^2$

Graph of $f$ is horizontally compressed

(c) $y = \left(\dfrac{1}{3} x\right)^2$

Graph of $f$ is horizontally stretched

**Figure 47**

## 5 Transform the Graph of a Function by Reflecting It About the $x$-axis or the $y$-axis

The third type of transformation, reflection about the $x$- or $y$-axis, changes the orientation of the graph of the function $f$ but keeps the shape and the size of the graph intact.

When a function $f$ is multiplied by $-1$, the graph of the new function $y = -f(x)$ is the **reflection about the $x$-axis** of the graph of $f$. For example, if $f(x) = \sqrt{x}$, then the graph of the new function $y = -f(x) = -\sqrt{x}$ is the reflection of the graph of $f$ about the $x$-axis. See Figure 48(a) and 48(b).

If the argument $x$ of a function $f$ is multiplied by $-1$, then the graph of the new function $y = f(-x)$ is the **reflection about the $y$-axis** of the graph of $f$. For example, if $f(x) = \sqrt{x}$, then the graph of the new function $y = f(-x) = \sqrt{-x}$ is the reflection of the graph of $f$ about the $y$-axis. See Figure 48(a) and 48(c). Notice in this example that the domain of $y = \sqrt{-x}$ is all real numbers for which $-x \geq 0$, or equivalently, $x \leq 0$.

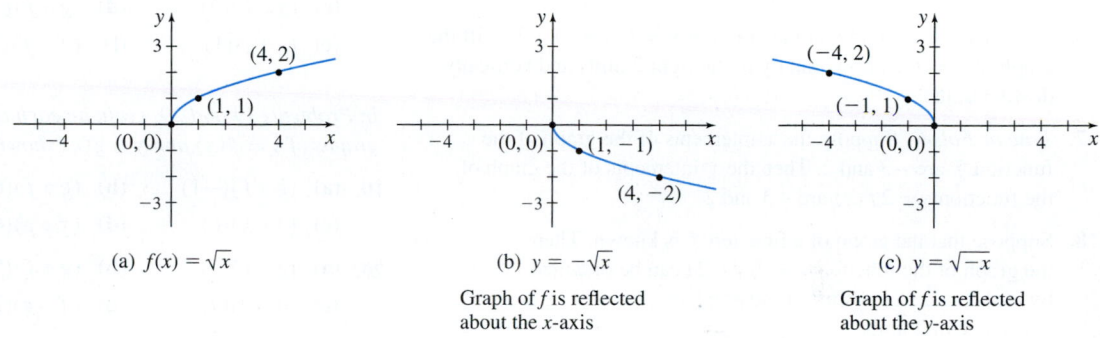

(a) $f(x) = \sqrt{x}$

(b) $y = -\sqrt{x}$
Graph of $f$ is reflected about the $x$-axis

(c) $y = \sqrt{-x}$
Graph of $f$ is reflected about the $y$-axis

**DF Figure 48**

### EXAMPLE 6  Combining Transformations

Use transformations to graph the function $f(x) = \sqrt{1-x} + 2$.

**Solution** We graph $f$ in steps:

- Observe that $f$ is basically a square root function, so we begin by graphing $y = \sqrt{x}$. See Figure 49(a).
- Now replace the argument $x$ with $x + 1$ to obtain $y = \sqrt{x+1}$, which shifts the graph of $y = \sqrt{x}$ horizontally to the left 1 unit, as shown in Figure 49(b).
- Then replace $x$ with $-x$ to obtain $y = \sqrt{-x+1} = \sqrt{1-x}$, which reflects the graph about the $y$-axis. See Figure 49(c).
- Finally, add 2 to each $y$-coordinate, which shifts the graph vertically up 2 units and results in the graph of $f(x) = \sqrt{1-x} + 2$ shown in Figure 49(d).

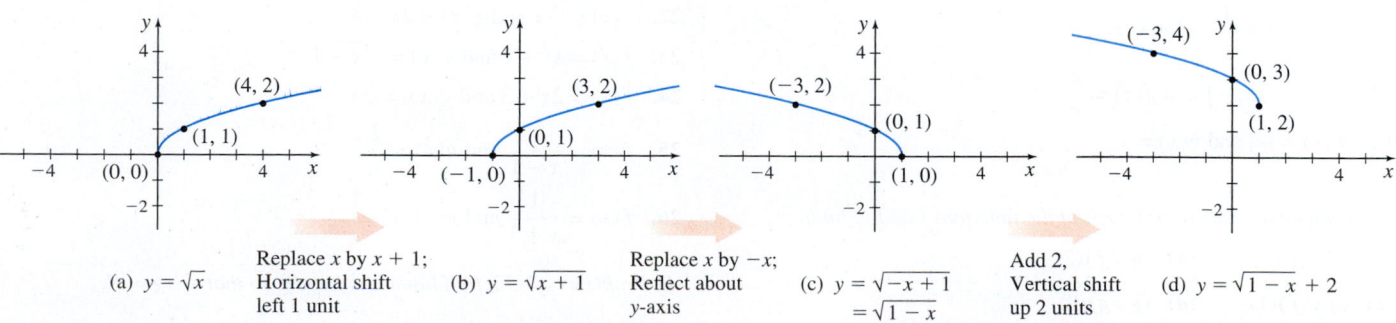

(a) $y = \sqrt{x}$

Replace $x$ by $x + 1$;
Horizontal shift
left 1 unit

(b) $y = \sqrt{x+1}$

Replace $x$ by $-x$;
Reflect about
$y$-axis

(c) $y = \sqrt{-x+1}$
$= \sqrt{1-x}$

Add 2,
Vertical shift
up 2 units

(d) $y = \sqrt{1-x} + 2$

**Figure 49**

**NOW WORK** Problem 43.

## P.3 Assess Your Understanding

### Concepts and Vocabulary

**1.** If the domain of a function $f$ is $\{x \mid 0 \le x \le 7\}$ and the domain of a function $g$ is $\{x \mid -2 \le x \le 5\}$, then the domain of the sum function $f + g$ is _____.

**2.** *True or False*   If $f$ and $g$ are functions, then the domain of $\dfrac{f}{g}$ consists of all numbers $x$ that are in the domains of both $f$ and $g$.

**3.** *True or False*   The domain of $f \cdot g$ consists of the numbers $x$ that are in the domains of both $f$ and $g$.

**4.** *True or False*   The domain of the composite function $f \circ g$ is the same as the domain of $g(x)$.

**5.** *True or False*   The graph of $y = -f(x)$ is the reflection of the graph of $y = f(x)$ about the $x$-axis.

**6.** *True or False*   To obtain the graph of $y = f(x + 2) - 3$, shift the graph of $y = f(x)$ horizontally to the right 2 units and vertically down 3 units.

**7.** *True or False*   Suppose the $x$-intercepts of the graph of the function $f$ are $-3$ and 2. Then the $x$-intercepts of the graph of the function $y = 2f(x)$ are $-3$ and 2.

**8.** Suppose that the graph of a function $f$ is known. Then the graph of the function $y = f(x - 2)$ can be obtained by a(n)_____ shift of the graph of $f$ to the _____ a distance of 2 units.

**9.** Suppose that the graph of a function $f$ is known. Then the graph of the function $y = f(-x)$ can be obtained by a reflection about the _____-axis of the graph of $f$.

**10.** Suppose the $x$-intercepts of the graph of the function $f$ are $-2$, 1, and 5. The $x$-intercepts of $y = f(x + 3)$ are _____, _____, and _____.

### Practice Problems

*In Problems 11–14, the functions f and g are given. Find each of the following functions and determine their domain:*

**(a)** $(f + g)(x)$    **(b)** $(f - g)(x)$

**(c)** $(f \cdot g)(x)$    **(d)** $\left(\dfrac{f}{g}\right)(x)$

**11.** $f(x) = 3x + 4$ and $g(x) = 2x - 3$

**12.** $f(x) = 1 + \dfrac{1}{x}$ and $g(x) = \dfrac{1}{x}$

**13.** $f(x) = \sqrt{x + 1}$ and $g(x) = \dfrac{2}{x}$

**14.** $f(x) = |x|$ and $g(x) = x$

*In Problems 15 and 16, for each of the functions f and g, find:*

**(a)** $(f \circ g)(4)$    **(b)** $(g \circ f)(2)$

**(c)** $(f \circ f)(1)$    **(d)** $(g \circ g)(0)$

**15.** $f(x) = 2x$ and $g(x) = 3x^2 + 1$

**16.** $f(x) = \dfrac{3}{x + 1}$ and $g(x) = \sqrt{x}$

*In Problems 17 and 18, evaluate each expression using the values given in the table.*

**17.**

| $x$ | $-3$ | $-2$ | $-1$ | 0 | 1 | 2 | 3 |
|-----|------|------|------|---|---|---|---|
| $f(x)$ | $-7$ | $-5$ | $-3$ | $-1$ | 3 | 5 | 7 |
| $g(x)$ | 8 | 3 | 0 | $-1$ | 0 | 3 | 8 |

**(a)** $(f \circ g)(1)$    **(b)** $(f \circ g)(-1)$
**(c)** $(g \circ f)(-1)$    **(d)** $(g \circ f)(1)$
**(e)** $(g \circ g)(-2)$    **(f)** $(f \circ f)(-1)$

**18.**

| $x$ | $-3$ | $-2$ | $-1$ | 0 | 1 | 2 | 3 |
|-----|------|------|------|---|---|---|---|
| $f(x)$ | 11 | 9 | 7 | 5 | 3 | 1 | $-1$ |
| $g(x)$ | $-8$ | $-3$ | 0 | 1 | 0 | $-3$ | $-8$ |

**(a)** $(f \circ g)(1)$    **(b)** $(f \circ g)(2)$
**(c)** $(g \circ f)(2)$    **(d)** $(g \circ f)(3)$
**(e)** $(g \circ g)(1)$    **(f)** $(f \circ f)(3)$

*In Problems 19 and 20, evaluate each composite function using the graphs of $y = f(x)$ and $y = g(x)$ shown in the figure below.*

**19. (a)** $(g \circ f)(-1)$    **(b)** $(g \circ f)(6)$
   **(c)** $(f \circ g)(6)$    **(d)** $(f \circ g)(4)$

**20. (a)** $(g \circ f)(1)$    **(b)** $(g \circ f)(5)$
   **(c)** $(f \circ g)(7)$    **(d)** $(f \circ g)(2)$

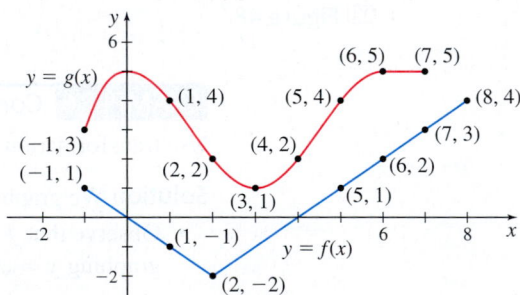

*In Problems 21–26, for the given functions f and g, find:*

**(a)** $f \circ g$    **(b)** $g \circ f$    **(c)** $f \circ f$    **(d)** $g \circ g$

*State the domain of each composite function.*

**21.** $f(x) = 3x + 1$ and $g(x) = 8x$

**22.** $f(x) = -x$ and $g(x) = 2x - 4$

**23.** $f(x) = x^2 + 1$ and $g(x) = \sqrt{x - 1}$

**24.** $f(x) = 2x + 3$ and $g(x) = \sqrt{x}$

**25.** $f(x) = \dfrac{x}{x - 1}$ and $g(x) = \dfrac{2}{x}$

**26.** $f(x) = \dfrac{1}{x + 3}$ and $g(x) = -\dfrac{2}{x}$

*In Problems 27–32, find functions f and g so that $f \circ g = F$.*

**27.** $F(x) = (2x + 3)^4$    **28.** $F(x) = (1 + x^2)^3$

**29.** $F(x) = \sqrt{x^2 + 1}$    **30.** $F(x) = \sqrt{1 - x^2}$

**31.** $F(x) = |2x + 1|$    **32.** $F(x) = |2x^2 + 3|$

---

**1.** = NOW WORK problem    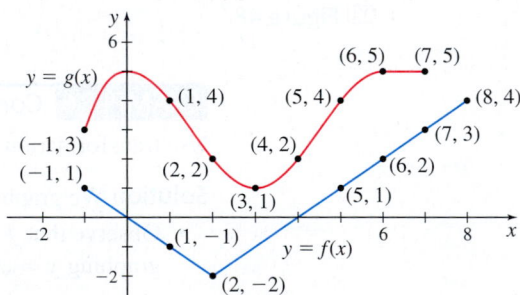 = Graphing technology recommended    CAS = Computer Algebra System recommended

*In Problems 33–46, graph each function using the graphing techniques of shifting, compressing, stretching, and/or reflecting. Begin with the graph of a basic function and show all stages.*

**33.** $f(x) = x^3 + 2$

**34.** $g(x) = x^3 - 1$

**35.** $h(x) = \sqrt{x - 2}$

**36.** $f(x) = \sqrt{x + 1}$

**37.** $g(x) = 4\sqrt{x}$

**38.** $f(x) = \frac{1}{2}\sqrt{x}$

**39.** $f(x) = (x - 1)^3 + 2$

**40.** $g(x) = 3(x - 2)^2 + 1$

**41.** $h(x) = \dfrac{1}{2x}$

**42.** $f(x) = \dfrac{4}{x} + 2$

**43.** $G(x) = 2|1 - x|$

**44.** $g(x) = -(x + 1)^3 - 1$

**45.** $g(x) = -4\sqrt{x - 1}$

**46.** $f(x) = 4\sqrt{2 - x}$

*In Problems 47 and 48, the graph of a function f is illustrated. Use the graph of f as the first step in graphing each of the following functions:*

**(a)** $F(x) = f(x) + 3$

**(b)** $G(x) = f(x + 2)$

**(c)** $P(x) = -f(x)$

**(d)** $H(x) = f(x + 1) - 2$

**(e)** $Q(x) = \dfrac{1}{2}f(x)$

**(f)** $g(x) = f(-x)$

**(g)** $h(x) = f(2x)$

**47.**

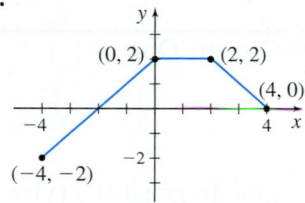

**48.**

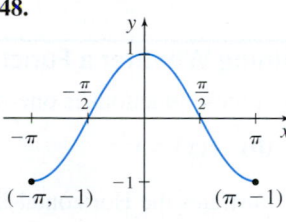

**49. Period of a Pendulum**   The period $T$ (in seconds) of a simple pendulum is a function of its length $l$ (in meters) defined by the equation

$$T = T(l) = 2\pi \sqrt{\frac{l}{g}}$$

where $g \approx 9.8$ meters/second$^2$ is the acceleration due to gravity.

**(a)** Use technology to graph the function $T = T(l)$.

**(b)** Now graph the functions $T = T(l + 1)$, $T = T(l + 2)$, and $T = T(l + 3)$.

**(c)** Discuss how adding to the length $l$ changes the period $T$.

**(d)** Now graph the functions $T = T(2l)$, $T = T(3l)$, and $T = T(4l)$.

**(e)** Discuss how multiplying the length $l$ by 2, 3, and 4 changes the period $T$.

**50.** Suppose $(1, 3)$ is a point on the graph of $y = g(x)$.

**(a)** What point is on the graph of $y = g(x + 3) - 5$?

**(b)** What point is on the graph of $y = -2g(x - 2) + 1$?

**(c)** What point is on the graph of $y = g(2x + 3)$?

# P.4 Inverse Functions

**OBJECTIVES** *When you finish this section, you should be able to:*

**1** Determine whether a function is one-to-one (p. 35)

**2** Determine the inverse of a function defined by a set of ordered pairs (p. 37)

**3** Obtain the graph of the inverse function from the graph of a one-to-one function (p. 38)

**4** Find the inverse of a one-to-one function defined by an equation (p. 38)

## 1 Determine Whether a Function Is One-to-One

By definition, for a function $y = f(x)$, if $x$ is in the domain of $f$, then $x$ has one, and only one, image $y$ in the range. If a function $f$ also has the property that no $y$ in the range of $f$ is the image of more than one $x$ in the domain, then the function is called a *one-to-one function*.

**DEFINITION One-to-One Function**

A function $f$ is a **one-to-one function** if any two different inputs in the domain correspond to two different outputs in the range. That is, if $x_1 \neq x_2$, then $f(x_1) \neq f(x_2)$.

**IN WORDS** A function is not one-to-one if there are two different inputs in the domain corresponding to the same output.

Figure 50 illustrates the distinction among one-to-one functions, functions that are not one-to-one, and relations that are not functions.

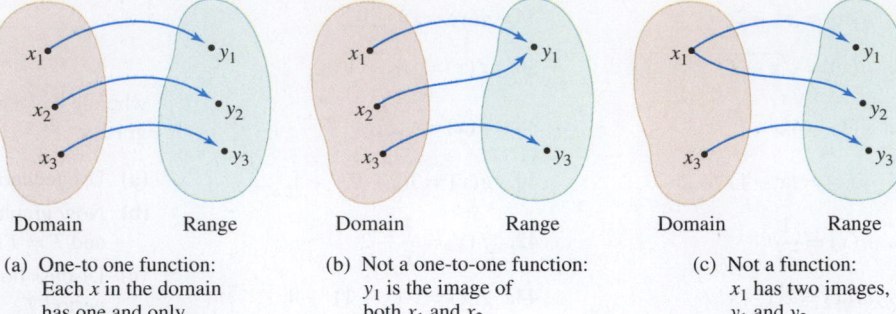

Domain    Range          Domain    Range          Domain    Range

(a) One-to one function:    (b) Not a one-to-one function:    (c) Not a function:
    Each $x$ in the domain          $y_1$ is the image of              $x_1$ has two images,
    has one and only               both $x_1$ and $x_2$.              $y_1$ and $y_2$.
    image in the range.

**Figure 50**

If the graph of a function $f$ is known, there is a simple test called the *Horizontal-line Test,* to determine whether $f$ is a one-to-one function.

> **THEOREM  Horizontal-line Test**
>
> The graph of a function in the $xy$-plane is the graph of a one-to-one function if and only if every horizontal line intersects the graph in at most one point.

**EXAMPLE 1    Determining Whether a Function Is One-to-One**

Determine whether each of these functions is one-to-one:

**(a)** $f(x) = x^2$        **(b)** $g(x) = x^3$

**Solution (a)** Figure 51 illustrates the Horizontal-line Test for the graph of $f(x) = x^2$. The horizontal line $y = 1$ intersects the graph of $f$ twice, at $(1, 1)$ and at $(-1, 1)$, so $f$ is not one-to-one.

**(b)** Figure 52 illustrates the Horizontal-line Test for the graph of $g(x) = x^3$. Because every horizontal line intersects the graph of $g$ exactly once, it follows that $g$ is one-to-one.

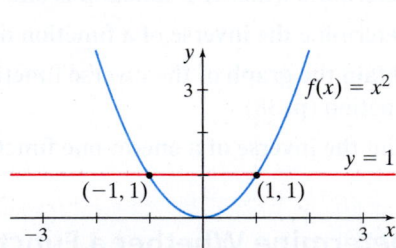

A horizontal line intersects the graph twice; $f$ is not one-to-one.

**Figure 51**

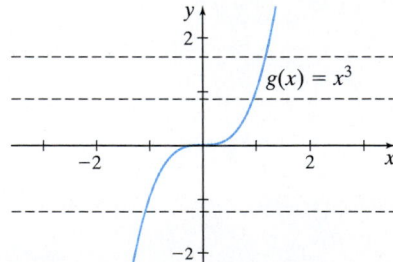

Every horizontal line intersects the graph exactly once; $g$ is one-to-one.

**Figure 52**

**NOW WORK** Problem 9.

Notice that the one-to-one function $g(x) = x^3$ also is an increasing function on its domain. Because an increasing (or decreasing) function will always have different $y$-values for different $x$-values, a function that is increasing (or decreasing) on an interval is also a one-to-one function on that interval.

**THEOREM** One-to-One Function

- A function that is increasing on an interval $I$ is a one-to-one function on $I$.
- A function that is decreasing on an interval $I$ is a one-to-one function on $I$.

Suppose that $f$ is a one-to-one function. Then to each $x$ in the domain of $f$, there is exactly one image $y$ in the range (because $f$ is a function); and to each $y$ in the range of $f$, there is exactly one $x$ in the domain (because $f$ is one-to-one). The correspondence from the range of $f$ back to the domain of $f$ is also a function, called the *inverse function of $f$*. The symbol $f^{-1}$ is used to denote the inverse of $f$.

**DEFINITION** Inverse Function

**NOTE** $f^{-1}$ is not the reciprocal function. That is, $f^{-1}(x) \neq \dfrac{1}{f(x)}$. The reciprocal function $\dfrac{1}{f(x)}$ is written $[f(x)]^{-1}$.

Let $f$ be a one-to-one function. The **inverse of** $f$, denoted by $f^{-1}$, is the function defined on the range of $f$ for which

$$\boxed{x = f^{-1}(y) \quad \text{if and only if} \quad y = f(x)}$$

We will discuss how to find inverses for three representations of functions: (1) sets of ordered pairs, (2) graphs, and (3) equations. We begin with finding the inverse of a function represented by a set of ordered pairs.

## 2 Determine the Inverse of a Function Defined by a Set of Ordered Pairs

If the function $f$ is a set of ordered pairs $(x, y)$, then the inverse of $f$, denoted $f^{-1}$, is the set of ordered pairs $(y, x)$.

**EXAMPLE 2** Finding the Inverse of a Function Defined by a Set of Ordered Pairs

Find the inverse of the one-to-one function:

$$\{(-3, -5), (-1, 1), (0, 2), (1, 3)\}$$

State the domain and the range of the function and its inverse.

**Solution** The inverse of the function is found by interchanging the entries in each ordered pair. The inverse is

$$\{(-5, -3), (1, -1), (2, 0), (3, 1)\}$$

The domain of the function is $\{-3, -1, 0, 1\}$; the range of the function is $\{-5, 1, 2, 3\}$. The domain of the inverse function is $\{-5, 1, 2, 3\}$; the range of the inverse function is $\{-3, -1, 0, 1\}$. ∎

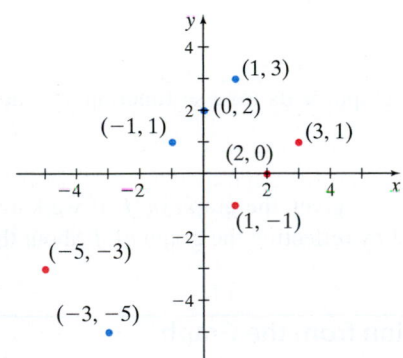

**Figure 53**

Figure 53 shows the one-to-one function (in blue) and its inverse (in red).

NOW WORK Problem 19.

Remember, if $f$ is a one-to-one function, it has an inverse $f^{-1}$. See Figure 54. Based on the results of Example 2 and Figure 53, two properties of a one-to-one function $f$ and its inverse function $f^{-1}$ become apparent:

$$\text{Domain of } f = \text{Range of } f^{-1} \qquad \text{Range of } f = \text{Domain of } f^{-1}$$

The next theorem provides a means for verifying that two functions are inverses of one another.

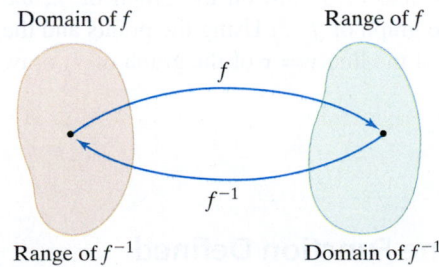

**Figure 54**

**THEOREM**

Given a one-to-one function $f$ and its inverse function $f^{-1}$, then

- $(f^{-1} \circ f)(x) = f^{-1}(f(x)) = x$    where $x$ is in the domain of $f$
- $(f \circ f^{-1})(x) = f(f^{-1}(x)) = x$    where $x$ is in the domain of $f^{-1}$

EXAMPLE 3 **Verifying Inverse Functions**

Verify that the inverse of $f(x) = \dfrac{1}{x-1}$ is $f^{-1}(x) = \dfrac{1}{x} + 1$. For what values of $x$ is $f^{-1}(f(x)) = x$? For what values of $x$ is $f(f^{-1}(x)) = x$?

**Solution** The domain of $f$ is $\{x \mid x \neq 1\}$ and the domain of $f^{-1}$ is $\{x \mid x \neq 0\}$. Now

$$f^{-1}(f(x)) = f^{-1}\left(\frac{1}{x-1}\right) = \frac{1}{\left(\dfrac{1}{x-1}\right)} + 1 = x - 1 + 1 = x \quad \text{provided } x \neq 1$$

$$f(f^{-1}(x)) = f\left(\frac{1}{x} + 1\right) = \frac{1}{\left(\dfrac{1}{x} + 1\right) - 1} = \frac{1}{\dfrac{1}{x}} = x \quad \text{provided } x \neq 0$$

NOW WORK **Problem 15.**

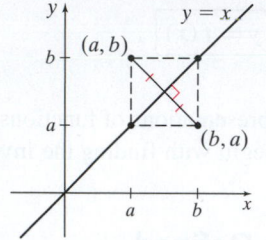

**Figure 55**

## 3 Obtain the Graph of the Inverse Function from the Graph of a One-to-One Function

Suppose $(a, b)$ is a point on the graph of a one-to-one function $f$ defined by $y = f(x)$. Then $b = f(a)$. This means that $a = f^{-1}(b)$, so $(b, a)$ is a point on the graph of the inverse function $f^{-1}$. Figure 55 shows the relationship between the point $(a, b)$ on the graph of $f$ and the point $(b, a)$ on the graph of $f^{-1}$. The line segment containing $(a, b)$ and $(b, a)$ is perpendicular to the line $y = x$ and is bisected by the line $y = x$. (Do you see why?) The point $(b, a)$ on the graph of $f^{-1}$ is the reflection about the line $y = x$ of the point $(a, b)$ on the graph of $f$.

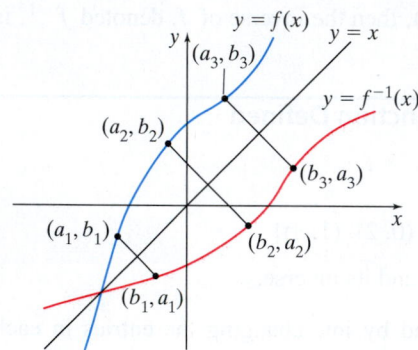

**Figure 56**

**THEOREM  Symmetry of Inverse Functions**

The graph of a one-to-one function $f$ and the graph of its inverse function $f^{-1}$ are symmetric with respect to the line $y = x$.

We can use this result to find the graph of $f^{-1}$ given the graph of $f$. If we know the graph of $f$, then the graph of $f^{-1}$ is obtained by reflecting the graph of $f$ about the line $y = x$. See Figure 56.

EXAMPLE 4 **Graphing the Inverse Function from the Graph of a Function**

The graph in Figure 57 is that of a one-to-one function $y = f(x)$. Draw the graph of its inverse function.

**Solution** Since the points $(-2, -1)$, $(-1, 0)$, and $(2, 1)$ are on the graph of $f$, the points $(-1, -2)$, $(0, -1)$, and $(1, 2)$ are on the graph of $f^{-1}$. Using the points and the fact that the graph of $f^{-1}$ is the reflection about the line $y = x$ of the graph of $f$, draw the graph of $f^{-1}$, as shown in Figure 58. ∎

NOW WORK **Problem 27.**

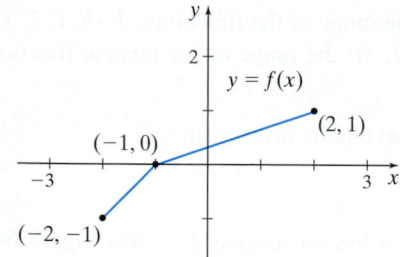

**Figure 57**

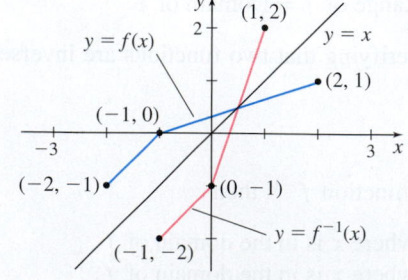

**Figure 58**

## 4 Find the Inverse of a One-to-One Function Defined by an Equation

Since the graphs of a one-to-one function $f$ and its inverse function $f^{-1}$ are symmetric with respect to the line $y = x$, the inverse function $f^{-1}$ can be obtained by interchanging the roles of $x$ and $y$ in $f$. If $f$ is defined by the equation

$$y = f(x)$$

then $f^{-1}$ is defined by the equation

$$x = f(y) \qquad \text{Interchange } x \text{ and } y.$$

The equation $x = f(y)$ defines $f^{-1}$ *implicitly*. If the implicit equation can be solved for $y$, we will have the *explicit* form of $f^{-1}$, that is,

$$y = f^{-1}(x)$$

### Steps for Finding the Inverse of a One-to-One Function

*Step 1*  Write $f$ in the form $y = f(x)$.

*Step 2*  Interchange the variables $x$ and $y$ to obtain $x = f(y)$. This equation defines the inverse function $f^{-1}$ implicitly.

*Step 3*  If possible, solve the implicit equation for $y$ in terms of $x$ to obtain the explicit form of $f^{-1}$: $y = f^{-1}(x)$.

*Step 4*  Check the result by showing that $f^{-1}(f(x)) = x$ and $f(f^{-1}(x)) = x$.

CALC
▶ **EXAMPLE 5**  **Finding the Inverse Function**
CLIP

The function $f(x) = 2x^3 - 1$ is one-to-one. Find its inverse.

**Solution**  We follow the steps given above.

*Step 1*  Write $f$ as $y = 2x^3 - 1$.

*Step 2*  Interchange the variables $x$ and $y$.

$$x = 2y^3 - 1$$

This equation defines $f^{-1}$ implicitly.

*Step 3*  Solve the implicit form of the inverse function for $y$.

$$x + 1 = 2y^3$$

$$y^3 = \frac{x+1}{2}$$

$$y = \sqrt[3]{\frac{x+1}{2}} = f^{-1}(x)$$

*Step 4*  Check the result.

$$f^{-1}(f(x)) = f^{-1}(2x^3 - 1) = \sqrt[3]{\frac{(2x^3 - 1) + 1}{2}} = \sqrt[3]{\frac{2x^3}{2}} = \sqrt[3]{x^3} = x$$

$$f(f^{-1}(x)) = f\left(\sqrt[3]{\frac{x+1}{2}}\right) = 2\left(\sqrt[3]{\frac{x+1}{2}}\right)^3 - 1 = 2\left(\frac{x+1}{2}\right) - 1$$

$$= x + 1 - 1 = x$$

See Figure 59 for the graphs of $f$ and $f^{-1}$.

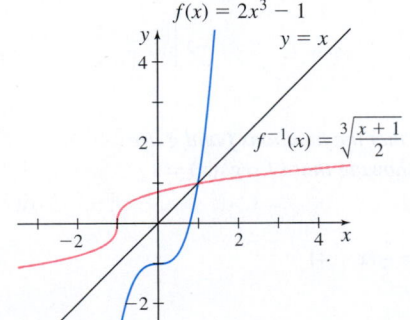

$f(x) = 2x^3 - 1$

$f^{-1}(x) = \sqrt[3]{\dfrac{x+1}{2}}$

**Figure 59**

NOW WORK  Problem 31.

If a function $f$ is not one-to-one, it has no inverse function. But sometimes we can restrict the domain of such a function so that it is a one-to-one function. Then on the restricted domain the new function has an inverse function.

## EXAMPLE 6    Finding the Inverse of a Domain-Restricted Function

Find the inverse of $f(x) = x^2$ if $x \geq 0$.

**NEED TO REVIEW?** Principal roots are discussed in Appendix A.1, p. A-9.

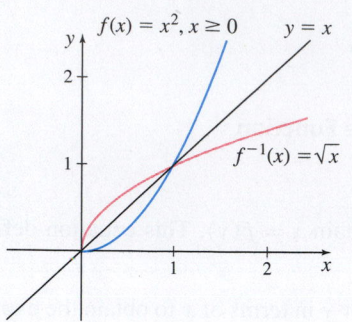

**Figure 60**

**Solution** The function $f(x) = x^2$ is not one-to-one (see Example 1(a)). However, by restricting the domain of $f$ to $x \geq 0$, the new function $f$ is one-to-one, so $f^{-1}$ exists. To find $f^{-1}$, follow the steps.

**Step 1**  $y = x^2$, where $x \geq 0$.

**Step 2**  Interchange the variables $x$ and $y$: $x = y^2$, where $y \geq 0$. This is the inverse function written implicitly.

**Step 3**  Solve for $y$: $y = \sqrt{x} = f^{-1}(x)$. (Since $y \geq 0$, only the principal square root is obtained.)

**Step 4**  Check that $f^{-1}(x) = \sqrt{x}$ is the inverse function of $f$.

$$f^{-1}(f(x)) = \sqrt{f(x)} = \sqrt{x^2} = |x| = x \qquad \text{where } x \geq 0$$
$$f(f^{-1}(x)) = [f^{-1}(x)]^2 = [\sqrt{x}]^2 = x \qquad \text{where } x \geq 0$$

The graphs of $f(x) = x^2$, $x \geq 0$, and $f^{-1}(x) = \sqrt{x}$ are shown in Figure 60.

**NOW WORK** Problem 37.

# P.4 Assess Your Understanding

## Concepts and Vocabulary

1. *True or False*  If every vertical line intersects the graph of a function $f$ at no more than one point, $f$ is a one-to-one function.

2. If the domain of a one-to-one function $f$ is $[4, \infty)$, the range of its inverse function $f^{-1}$ is _____.

3. *True or False*  If $f$ and $g$ are inverse functions, the domain of $f$ is the same as the domain of $g$.

4. *True or False*  If $f$ and $g$ are inverse functions, their graphs are symmetric with respect to the line $y = x$.

5. *True or False*  If $f$ and $g$ are inverse functions, then $(f \circ g)(x) = f(x) \cdot g(x)$.

6. *True or False*  If a function $f$ is one-to-one, then $f(f^{-1}(x)) = x$, where $x$ is in the domain of $f$.

7. Given a collection of points $(x, y)$, explain how you would determine if it represents a one-to-one function $y = f(x)$.

8. Given the graph of a one-to-one function $y = f(x)$, explain how you would graph the inverse function $f^{-1}$.

## Practice Problems

*In Problems 9–14, the graph of a function f is given. Use the Horizontal-line Test to determine whether f is one-to one.*

**9.**

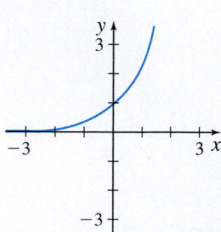

**10.**

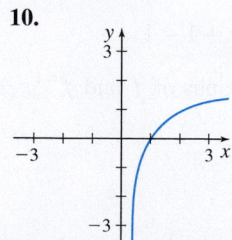

**11.**

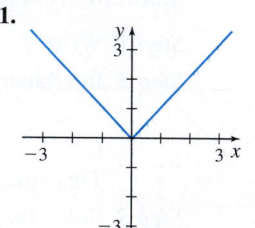

**12.**

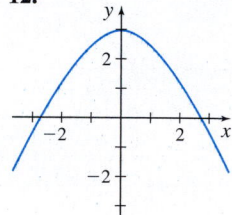

**13.**

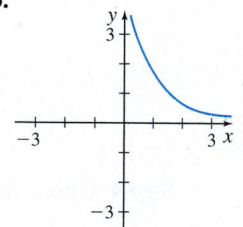

**14.**

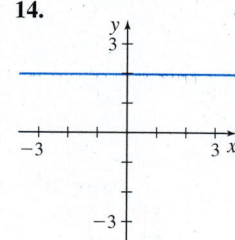

*In Problems 15–18, verify that the functions f and g are inverses of each other by showing that $(f \circ g)(x) = x$ and $(g \circ f)(x) = x$.*

**15.**  $f(x) = 3x + 4$; $g(x) = \dfrac{1}{3}(x - 4)$

**16.**  $f(x) = x^3 - 8$; $g(x) = \sqrt[3]{x + 8}$

**17.**  $f(x) = \dfrac{1}{x}$; $g(x) = \dfrac{1}{x}$

**18.**  $f(x) = \dfrac{2x + 3}{x + 4}$; $g(x) = \dfrac{4x - 3}{2 - x}$

---

**1.** = NOW WORK problem        ⟋⟍ = Graphing technology recommended        CAS = Computer Algebra System recommended

In Problems 19–22, **(a)** determine whether the function is one-to-one. If it is one-to-one, **(b)** find the inverse of each function. **(c)** State the domain and the range of the function and its inverse.

**19.** $\{(-3, 5), (-2, 9), (-1, 2), (0, 11), (1, -5)\}$

**20.** $\{(-2, 2), (-1, 6), (0, 8), (1, -3), (2, 8)\}$

**21.** $\{(-2, 1), (-3, 2), (-10, 0), (1, 9), (2, 1)\}$

**22.** $\{(-2, -8), (-1, -1), (0, 0), (1, 1), (2, 8)\}$

In Problems 23–28, the graph of a one-to-one function $f$ is given. Draw the graph of the inverse function. For convenience, the graph of $y = x$ is also given.

**23.** 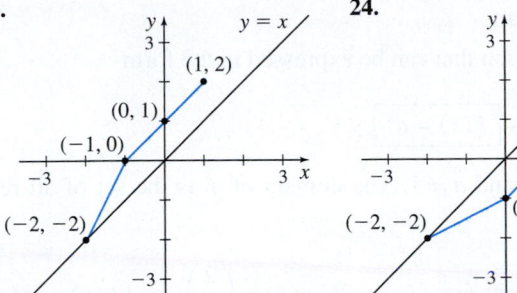 **24.**

**27.** **28.**
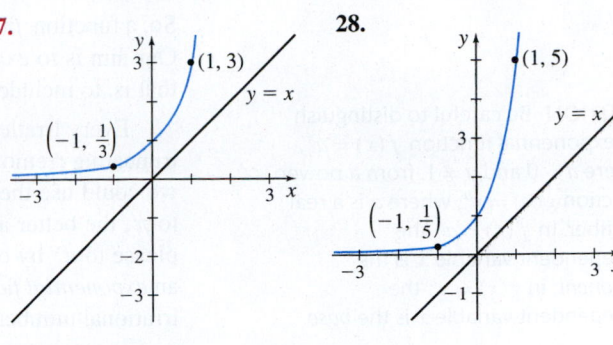

**25.** 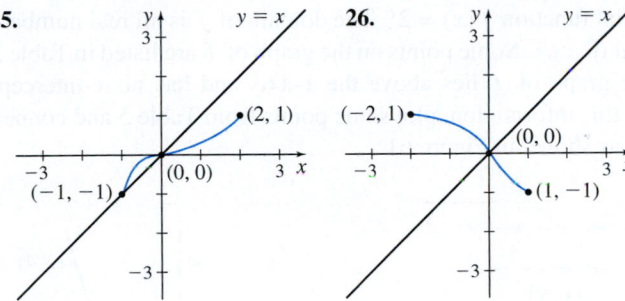 **26.**

In Problems 29–38, the function $f$ is one-to-one.

**(a)** Find its inverse and check the result.

**(b)** Find the domain and the range of $f$ and the domain and the range of $f^{-1}$.

**29.** $f(x) = 4x + 2$

**30.** $f(x) = 1 - 3x$

**31.** $f(x) = \sqrt[3]{x + 10}$

**32.** $f(x) = 2x^3 + 4$

**33.** $f(x) = \dfrac{1}{x - 2}$

**34.** $f(x) = \dfrac{2x}{3x - 1}$

**35.** $f(x) = \dfrac{2x + 3}{x + 2}$

**36.** $f(x) = \dfrac{-3x - 4}{x - 2}$

**37.** $f(x) = x^2 + 4, \ x \geq 0$

**38.** $f(x) = (x - 2)^2 + 4, \ x \leq 2$

# P.5 Exponential and Logarithmic Functions

**OBJECTIVES** *When you finish this section, you should be able to:*

1 Analyze an exponential function (p. 41)

2 Define the number $e$ (p. 44)

3 Analyze a logarithmic function (p. 45)

4 Solve exponential equations and logarithmic equations (p. 48)

We begin our study of transcendental functions with the exponential and logarithmic functions. In Sections P.6 and P.7, we investigate the trigonometric functions and their inverse functions.

## 1 Analyze an Exponential Function

The expression $a^r$, where $a > 0$ is a fixed real number and $r = \dfrac{m}{n}$ is a rational number, in lowest terms with $n \geq 2$, is defined as

$$a^r = a^{m/n} = (a^{1/n})^m = (\sqrt[n]{a})^m$$

So, a function $f(x) = a^x$ can be defined so that its domain is the set of rational numbers. Our aim is to expand the domain of $f$ to include both rational and irrational numbers, that is, to include all real numbers.

> **CAUTION** Be careful to distinguish an exponential function $f(x) = a^x$, where $a > 0$ and $a \neq 1$, from a power function $g(x) = x^a$, where $a$ is a real number. In $f(x) = a^x$ the independent variable $x$ is the *exponent*; in $g(x) = x^a$ the independent variable $x$ is the *base*.

Every irrational number $x$ can be approximated by a rational number $r$ formed by truncating (removing) all but a finite number of digits from $x$. For example, for $x = \pi$, we could use the rational numbers $r = 3.14$ or $r = 3.14159$, and so on. The closer $r$ is to $\pi$, the better approximation $a^r$ is to $a^\pi$. In general, we can make $a^r$ as close as we please to $a^x$ by choosing $r$ sufficiently close to $x$. Using this argument, we can define an *exponential function* $f(x) = a^x$, where $x$ includes all the rational numbers and all the irrational numbers.

### DEFINITION Exponential Function

An **exponential function** is a function that can be expressed in the form

$$f(x) = a^x$$

where $a$ is a positive real number and $a \neq 1$. The domain of $f$ is the set of all real numbers.

> **NOTE** The base $a = 1$ is excluded from the definition of an exponential function because $f(x) = 1^x = 1$ (a constant function). Bases that are negative are excluded because $a^{1/n}$, where $a < 0$ and $n$ is an even integer, is not defined.

Examples of exponential functions are $f(x) = 2^x$, $g(x) = \left(\dfrac{2}{3}\right)^x$, and $h(x) = \pi^x$.

Consider the exponential function $f(x) = 2^x$. The domain of $f$ is all real numbers; the range of $f$ is the interval $(0, \infty)$. Some points on the graph of $f$ are listed in Table 5. Since $2^x > 0$ for all $x$, the graph of $f$ lies above the $x$-axis and has no $x$-intercept. The $y$-intercept is 1. Using this information, plot some points from Table 5 and connect them with a smooth curve, as shown in Figure 61.

**TABLE 5**

| $x$ | $f(x) = 2^x$ | $(x, y)$ |
|---|---|---|
| $-10$ | $2^{-10} \approx 0.00098$ | $(-10, 0.00098)$ |
| $-3$ | $2^{-3} = \dfrac{1}{8}$ | $\left(-3, \dfrac{1}{8}\right)$ |
| $-2$ | $2^{-2} = \dfrac{1}{4}$ | $\left(-2, \dfrac{1}{4}\right)$ |
| $-1$ | $2^{-1} = \dfrac{1}{2}$ | $\left(-1, \dfrac{1}{2}\right)$ |
| $0$ | $2^0 = 1$ | $(0, 1)$ |
| $1$ | $2^1 = 2$ | $(1, 2)$ |
| $2$ | $2^2 = 4$ | $(2, 4)$ |
| $3$ | $2^3 = 8$ | $(3, 8)$ |
| $10$ | $2^{10} = 1024$ | $(10, 1024)$ |

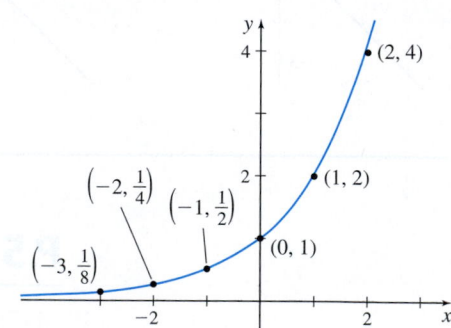

**Figure 61** $f(x) = 2^x$

The graph of $f(x) = 2^x$ is typical of all exponential functions of the form $f(x) = a^x$ with $a > 1$, a few of which are graphed in Figure 62.

Notice that every graph in Figure 62 lies above the $x$-axis, passes through the point $(0, 1)$, and is increasing. Also notice that the graphs with larger bases are steeper when $x > 0$, but when $x < 0$, the graphs with larger bases are closer to the $x$-axis.

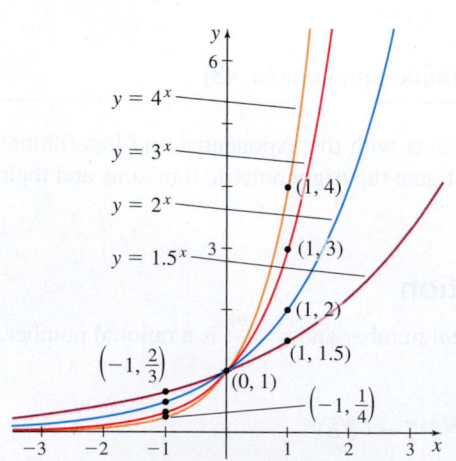

**Figure 62** $f(x) = a^x; a > 1$

All functions of the type $f(x) = a^x$, $a > 1$, have the following properties:

## Properties of an Exponential Function $f(x) = a^x$, $a > 1$

- The domain is the set of all real numbers; the range is the set of positive real numbers.
- There are no $x$-intercepts; the $y$-intercept is 1.
- The exponential function $f$ is increasing on the interval $(-\infty, \infty)$.
- The graph of $f$ contains the points $\left(-1, \dfrac{1}{a}\right)$, $(0, 1)$, and $(1, a)$.
- $\dfrac{f(x+1)}{f(x)} = a$
- Because $f(x) = a^x$ is a function, if $u = v$, then $a^u = a^v$.
- Because $f(x) = a^x$ is a one-to-one function, if $a^u = a^v$, then $u = v$.

## THEOREM  Laws of Exponents

If $u$, $v$, $a$, and $b$ are real numbers with $a > 0$ and $b > 0$, then

$$a^u \cdot a^v = a^{u+v} \qquad \frac{a^u}{a^v} = a^{u-v} \qquad (a^u)^v = a^{uv} \qquad (ab)^u = a^u \cdot b^u \qquad \left(\frac{a}{b}\right)^u = \frac{a^u}{b^u}$$

**NEED TO REVIEW?** The Laws of Exponents are discussed in Appendix A.1, pp. A-8 to A-9.

For example, we can use the Laws of Exponents to show the following property of an exponential function:

$$\frac{f(x+1)}{f(x)} = \frac{a^{x+1}}{a^x} = a^{(x+1)-x} = a^1 = a$$

## EXAMPLE 1  Graphing an Exponential Function

Graph the exponential function $g(x) = \left(\dfrac{1}{2}\right)^x$.

**Solution** We begin by writing $\dfrac{1}{2}$ as $2^{-1}$. Then

$$g(x) = \left(\frac{1}{2}\right)^x = (2^{-1})^x = 2^{-x}$$

Now use the graph of $f(x) = 2^x$ shown in Figure 61 and reflect the graph about the $y$-axis to obtain the graph of $g(x) = 2^{-x}$. See Figure 63.

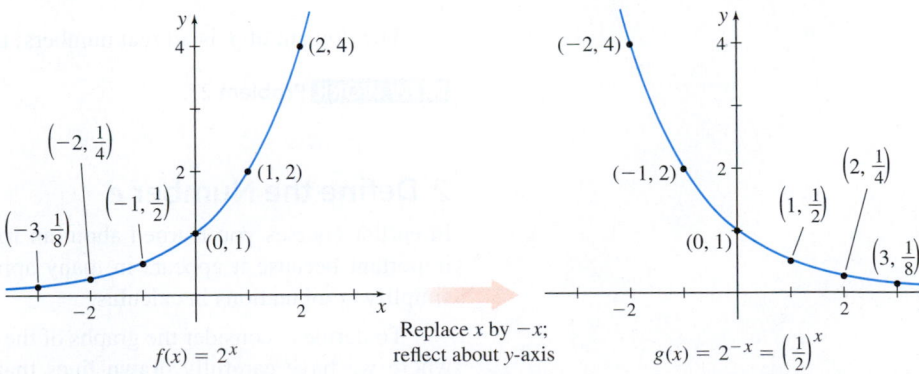

Replace $x$ by $-x$;
reflect about $y$-axis

$f(x) = 2^x$ 

$g(x) = 2^{-x} = \left(\dfrac{1}{2}\right)^x$

**Figure 63**

NOW WORK **Problem 19.**

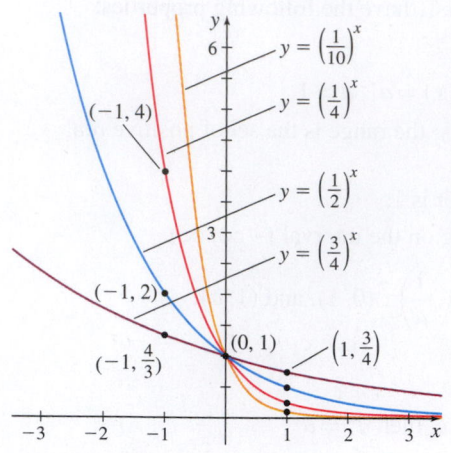

**Figure 64** $f(x) = a^x, 0 < a < 1$

The graph of $g(x) = \left(\dfrac{1}{2}\right)^x$ in Figure 63 is typical of all exponential functions that have a base between 0 and 1. Figure 64 illustrates the graphs of several more exponential functions whose bases are between 0 and 1. Notice that the graphs with smaller bases are steeper when $x < 0$, but when $x > 0$, these graphs are closer to the $x$-axis.

**Properties of an Exponential Function** $f(x) = a^x, 0 < a < 1$

- The domain is the set of all real numbers; the range is the set of positive real numbers.
- There are no $x$-intercepts; the $y$-intercept is 1.
- The exponential function $f$ is decreasing on the interval $(-\infty, \infty)$.
- The graph of $f$ contains the points $\left(-1, \dfrac{1}{a}\right)$, $(0, 1)$, and $(1, a)$.
- $\dfrac{f(x + 1)}{f(x)} = a$
- If $u = v$, then $a^u = a^v$.
- If $a^u = a^v$, then $u = v$.

**EXAMPLE 2  Graphing an Exponential Function Using Transformations**

Graph $f(x) = 3^{-x} - 2$ and determine the domain and range of $f$.

**Solution** We begin with the graph of $y = 3^x$. Figure 65 shows the steps.

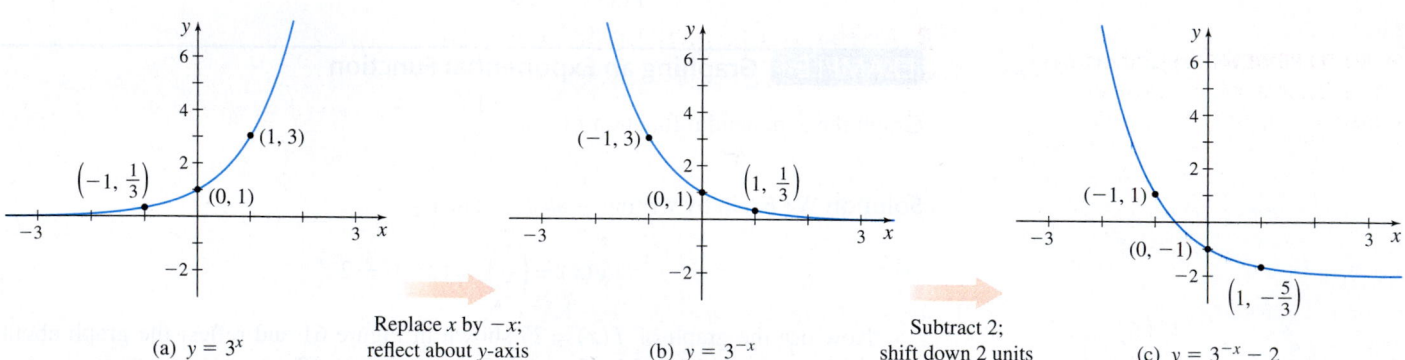

(a) $y = 3^x$  → Replace $x$ by $-x$; reflect about $y$-axis → (b) $y = 3^{-x}$  → Subtract 2; shift down 2 units → (c) $y = 3^{-x} - 2$

**Figure 65**

The domain of $f$ is all real numbers; the range of $f$ is the interval $(-2, \infty)$. ∎

**NOW WORK** Problem 27.

## 2 Define the Number $e$

In earlier courses you learned about an irrational number, called $e$. The number $e$ is important because it appears in many applications and because it has properties that simplify computations in calculus.

To define $e$, consider the graphs of the functions $y = 2^x$ and $y = 3^x$ in Figure 66(a), where we have carefully drawn lines that just touch each graph at the point $(0, 1)$. (These lines are *tangent lines*, which we discuss in Chapter 1.) Notice that the slope of the tangent line to $y = 3^x$ is greater than 1 (approximately 1.10) and that the slope of the tangent line to the graph of $y = 2^x$ is less than 1 (approximately 0.69). Between these graphs there is an exponential function $y = a^x$, whose base is between 2 and 3, and whose tangent line to the graph at the point $(0, 1)$ has a slope of exactly 1, as shown in Figure 66(b). The base of this exponential function is the number $e$.

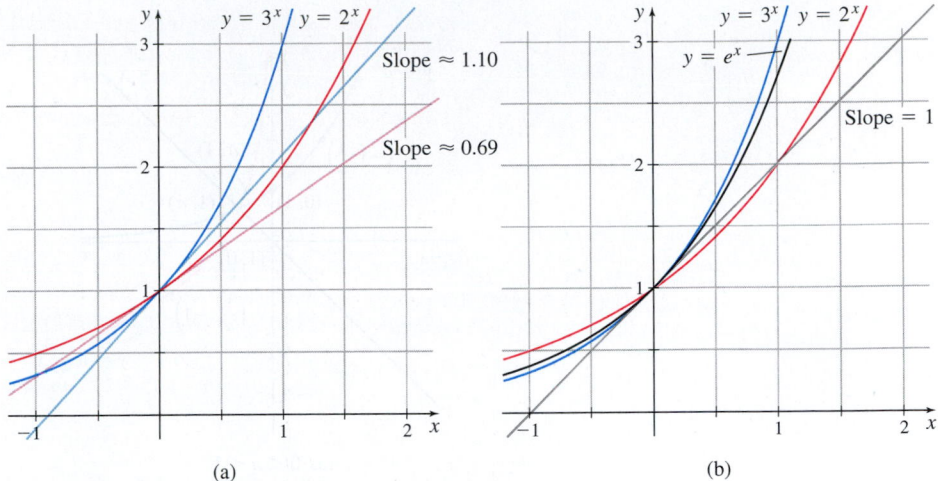

(a)                                              (b)

**Figure 66**

imageBROKER/Alamy

**ORIGINS** The irrational number we call $e$ was named to honor the Swiss mathematician Leonhard Euler (1707–1783), who is considered one of the greatest mathematicians of all time. Euler was encouraged to study mathematics by his father. Considered the most prolific mathematician who ever lived, he made major contributions to many areas of mathematics. One of Euler's greatest gifts was his ability to explain difficult mathematical concepts in simple language.

**DEFINITION The number $e$**

The **number** $e$ is defined as the base of the exponential function whose tangent line to the graph at the point $(0, 1)$ has slope 1. The function $f(x) = e^x$ occurs with such frequency that it is usually referred to as *the* **exponential function**.

The number $e$ is an irrational number, and in Chapter 3, we will show $e \approx 2.71828$.

## 3 Analyze a Logarithmic Function

**NEED TO REVIEW?** Logarithms and their properties are discussed in Appendix A.1, pp. A-10 to A-11.

Recall that a one-to-one function $y = f(x)$ has an inverse function that is defined implicitly by the equation $x = f(y)$. Since an exponential function $y = f(x) = a^x$, where $a > 0$ and $a \neq 1$, is a one-to-one function, it has an inverse function, called a *logarithmic function*, that is defined implicitly by the equation

$$x = a^y \qquad \text{where } a > 0 \quad \text{and} \quad a \neq 1$$

**DEFINITION Logarithmic Function**

The **logarithmic function with base** $a$, where $a > 0$ and $a \neq 1$, is denoted by $y = \log_a x$ and is defined by

$$\boxed{y = \log_a x \quad \text{if and only if} \quad x = a^y}$$

The domain of the logarithmic function $y = \log_a x$ is $x > 0$.

**IN WORDS** A logarithm is an exponent. That is, if $y = \log_a x$, then $y$ is the exponent in $x = a^y$.

In other words, if $f(x) = a^x$, where $a > 0$ and $a \neq 1$, its inverse function is $f^{-1}(x) = \log_a x$.

If the base of a logarithmic function is the number $e$, then it is called the **natural logarithmic function**, and it is given a special symbol, **ln** (from the Latin *logarithmus naturalis*). That is,

$$\boxed{y = \ln x \quad \text{if and only if} \quad x = e^y}$$

Since the exponential function $y = a^x$, $a > 0, a \neq 1$ and the logarithmic function $y = \log_a x$ are inverse functions, the following properties hold:

- $\log_a(a^x) = x$ for all real numbers $x$
- $a^{\log_a x} = x$ for all $x > 0$

Because exponential functions and logarithmic functions are inverses of each other, the graph of the logarithmic function $y = \log_a x$, $a > 0$ and $a \neq 1$, is the reflection of the graph of the exponential function $y = a^x$, about the line $y = x$, as shown in Figure 67.

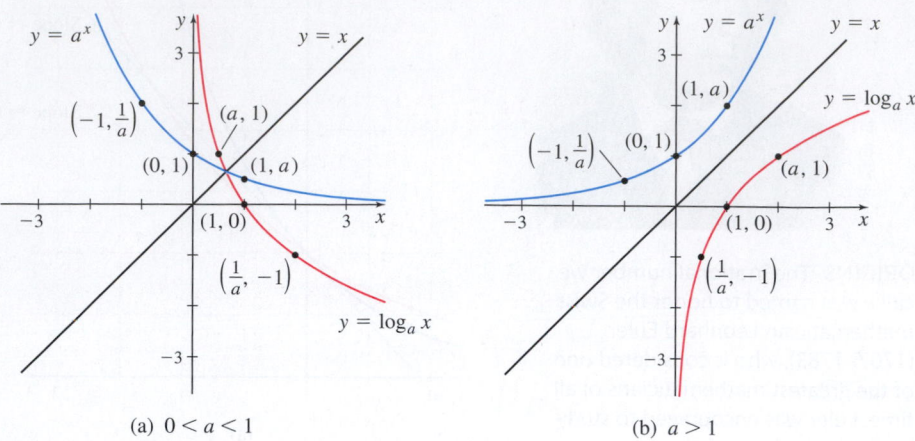

(a) $0 < a < 1$                                  (b) $a > 1$

**Figure 67**

Based on the graphs in Figure 67, we see that

$$\log_a 1 = 0 \qquad \log_a a = 1 \qquad a > 0 \quad \text{and} \quad a \neq 1$$

**EXAMPLE 3  Graphing a Logarithmic Function**

Graph:

**(a)** $f(x) = \log_2 x$       **(b)** $g(x) = \log_{1/3} x$       **(c)** $F(x) = \ln x$

**Solution** **(a)** To graph $f(x) = \log_2 x$, graph $y = 2^x$ and reflect it about the line $y = x$. See Figure 68(a).

**(b)** To graph $g(x) = \log_{1/3} x$, graph $y = \left(\dfrac{1}{3}\right)^x$ and reflect it about the line $y = x$. See Figure 68(b).

**(c)** To graph $F(x) = \ln x$, graph $y = e^x$ and reflect it about the line $y = x$. See Figure 68(c).

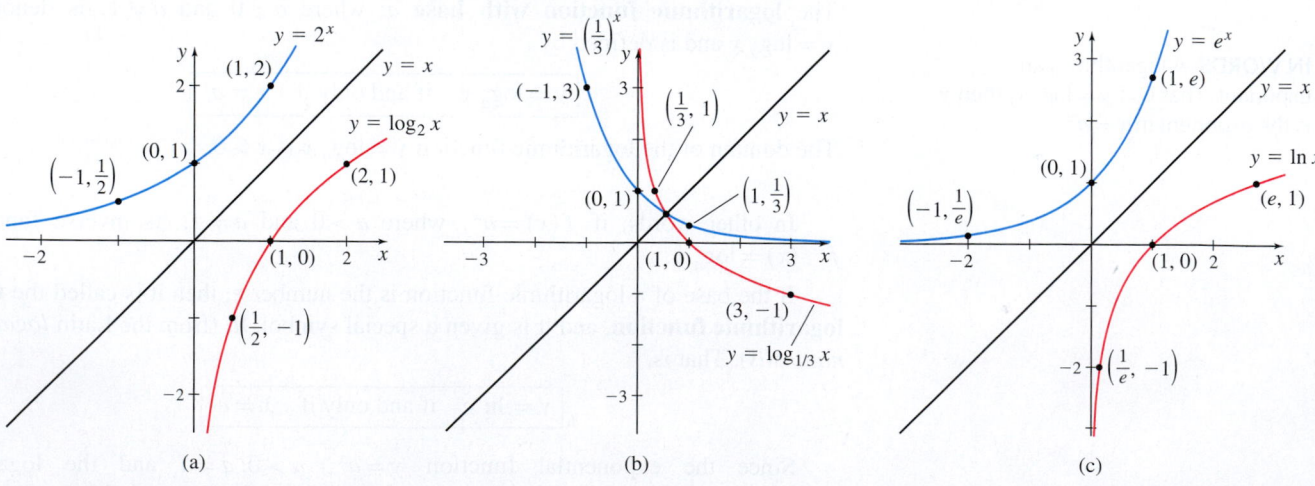

(a)                                  (b)                                  (c)

**Figure 68**

NOW WORK Problem 39.

Since a logarithmic function is the inverse of an exponential function, it follows that

- Domain of the logarithmic function = Range of the exponential function
  $$= (0, \infty)$$
- Range of the logarithmic function = Domain of the exponential function
  $$= (-\infty, \infty)$$

The domain of a logarithmic function is the set of *positive* real numbers, so the argument of a logarithmic function must be greater than zero.

### EXAMPLE 4  Finding the Domain of a Logarithmic Function

Find the domain of each function:

**(a)** $F(x) = \log_2(x + 3)$    **(b)** $g(x) = \ln\left(\dfrac{1+x}{1-x}\right)$    **(c)** $h(x) = \log_{1/2}|x|$

**NEED TO REVIEW?** Solving inequalities is discussed in Appendix A.1, pp. A-5 to A-8.

**Solution (a)** The argument of a logarithm must be positive. So to find the domain of $F(x) = \log_2(x + 3)$, we solve the inequality $x + 3 > 0$. The domain of $F$ is $\{x \mid x > -3\}$.

**(b)** Since $\ln\left(\dfrac{1+x}{1-x}\right)$ requires $\dfrac{1+x}{1-x} > 0$, we find the domain of $g$ by solving the inequality $\dfrac{1+x}{1-x} > 0$. Since $\dfrac{1+x}{1-x}$ is not defined for $x = 1$, and the solution to the equation $\dfrac{1+x}{1-x} = 0$ is $x = -1$, we use $-1$ and $1$ to separate the real number line into three intervals $(-\infty, -1)$, $(-1, 1)$, and $(1, \infty)$. Then we choose a test number in each interval and evaluate the rational expression $\dfrac{1+x}{1-x}$ at these numbers to determine if the expression is positive or negative. For example, we chose the numbers $-2$, $0$, and $2$ and found that $\dfrac{1+x}{1-x} > 0$ on the interval $(-1, 1)$. See the table on the left. So the domain of $g(x) = \ln\left(\dfrac{1+x}{1-x}\right)$ is $\{x \mid -1 < x < 1\}$.

| Interval | Test Number | Sign of $\dfrac{1+x}{1-x}$ |
|---|---|---|
| $(-\infty, -1)$ | $-2$ | Negative |
| $(-1, 1)$ | $0$ | Positive |
| $(1, \infty)$ | $2$ | Negative |

**(c)** $\log_{1/2}|x|$ requires $|x| > 0$. So the domain of $h(x) = \log_{1/2}|x|$ is $\{x \mid x \neq 0\}$. ∎

**NOW WORK** Problem 31.

---

**Properties of a Logarithmic Function** $f(x) = \log_a x$, $a > 0$ and $a \neq 1$

- The domain of $f$ is the set of all positive real numbers; the range is the set of all real numbers.
- The $x$-intercept of the graph of $f$ is 1. There is no $y$-intercept.
- A logarithmic function is decreasing on the interval $(0, \infty)$ if $0 < a < 1$ and increasing on the interval $(0, \infty)$ if $a > 1$.
- The graph of $f$ contains the points $(1, 0)$, $(a, 1)$, and $\left(\dfrac{1}{a}, -1\right)$.
- Because $f(x) = \log_a x$ is a function, if $u = v$, then $\log_a u = \log_a v$.
- Because $f(x) = \log_a x$ is a one-to-one function, if $\log_a u = \log_a v$, then $u = v$.

See Figure 69 on page 48 for typical graphs with base $a$, $a > 1$; see Figure 70 for graphs with base $a$, $0 < a < 1$.

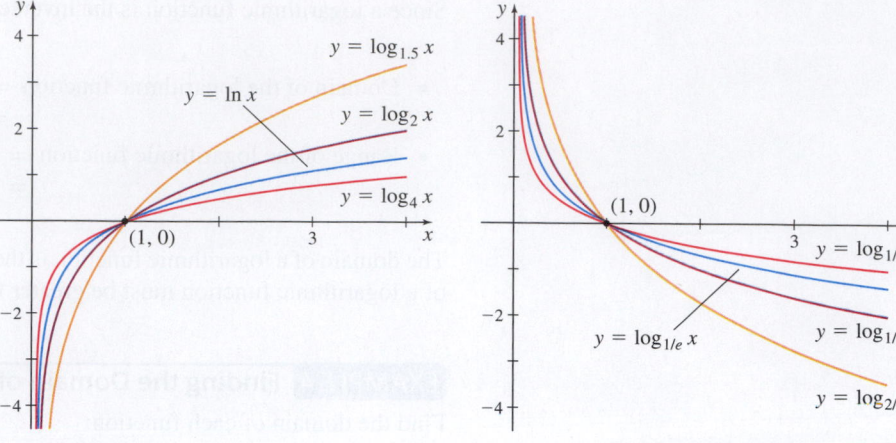

**Figure 69** $y = \log_a x, a > 1$    **Figure 70** $y = \log_a x, 0 < a < 1$

## 4 Solve Exponential Equations and Logarithmic Equations

Equations that involve terms of the form $a^x$, $a > 0$, $a \ne 1$, are referred to as **exponential equations**. For example, the equation $5^{2x+3} = 5^x$ is an exponential equation.

**IN WORDS** In an exponential equation, if the bases are equal, then the exponents are equal.

The one-to-one property of exponential functions

$$\boxed{\text{if } a^u = a^v \quad \text{then } u = v} \tag{1}$$

can be used to solve certain kinds of exponential equations.

---

### EXAMPLE 5   Solving Exponential Equations

Solve each exponential equation:

**(a)** $4^{2x-1} = 8^{x+3}$    **(b)** $e^{-x^2} = (e^x)^2 \cdot \dfrac{1}{e^3}$

**Solution (a)** We begin by expressing both sides of the equation with the same base so that we can use the one-to-one property (1).

$$
\begin{aligned}
4^{2x-1} &= 8^{x+3} \\
(2^2)^{2x-1} &= (2^3)^{x+3} && 4 = 2^2,\ 8 = 2^3 \\
2^{2(2x-1)} &= 2^{3(x+3)} && (a^r)^s = a^{rs} \\
2(2x-1) &= 3(x+3) && \text{If } a^u = a^v, \text{ then } u = v. \\
4x - 2 &= 3x + 9 && \text{Simplify.} \\
x &= 11 && \text{Solve.}
\end{aligned}
$$

The solution is 11.

**(b)** We use the Laws of Exponents to obtain the base $e$ on the right side.

$$(e^x)^2 \cdot \frac{1}{e^3} = e^{2x} \cdot e^{-3} = e^{2x-3}$$

As a result,

$$
\begin{aligned}
e^{-x^2} &= e^{2x-3} \\
-x^2 &= 2x - 3 && \text{If } a^u = a^v, \text{ then } u = v. \\
x^2 + 2x - 3 &= 0 \\
(x+3)(x-1) &= 0 \\
x = -3 \quad &\text{or} \quad x = 1
\end{aligned}
$$

The solution set is $\{-3, 1\}$. ∎

**NOW WORK** Problem 49.

To use the one-to-one property of exponential functions, each side of the equation must be written with the same base. Since for many exponential equations it is not possible to write each side with the same base, we need a different strategy to solve such equations.

---

**EXAMPLE 6** **Solving Exponential Equations**

Solve the exponential equations:

**(a)** $10^{2x} = 50$        **(b)** $8 \cdot 3^x = 5$

**Solution** **(a)** Since 10 and 50 cannot be written with the same base, we write the exponential equation as a logarithm.

$$10^{2x} = 50 \quad \text{if and only if} \quad \log 50 = 2x$$

**RECALL** Logarithms to the base 10 are called **common logarithms** and are written without a subscript. That is, $x = \log_{10} y$ is written $x = \log y$.

Then $x = \dfrac{\log 50}{2}$ is an exact solution of the equation. Using a calculator, an approximate

solution is $x = \dfrac{\log 50}{2} \approx 0.849$.

**(b)** It is impossible to write 8 and 5 as a power of 3, so we write the exponential equation as a logarithm.

$$8 \cdot 3^x = 5$$

$$3^x = \frac{5}{8}$$

$$x = \log_3 \frac{5}{8}$$

**NEED TO REVIEW?** The Change-of-Base Formula,

$\log_a u = \dfrac{\log_b u}{\log_b a}, a \neq 1, b \neq 1,$

and $u$ positive real numbers, is discussed in Appendix A.1, p. A-11.

The Change-of-Base Formula can be used with a calculator to obtain an approximate solution.

$$x = \log_3 \frac{5}{8} = \frac{\ln \dfrac{5}{8}}{\ln 3} \approx -0.428 \qquad \blacksquare$$

Alternatively, we could have solved each of the equations in Example 6 by taking the natural logarithm (or the common logarithm) of each side. For example,

$$8 \cdot 3^x = 5$$

$$3^x = \frac{5}{8}$$

$$\ln 3^x = \ln \frac{5}{8} \qquad \text{If } u = v, \text{ then } \log_a u = \log_a v.$$

$$x \ln 3 = \ln \frac{5}{8} \qquad \log_a u^r = r \log_a u$$

$$x = \frac{\ln \dfrac{5}{8}}{\ln 3}$$

**NOW WORK** Problem 51.

Equations that contain logarithms are called **logarithmic equations**. Care must be taken when solving logarithmic equations algebraically. In the expression $\log_a y$, remember that $a$ and $y$ are positive and $a \neq 1$. Be sure to check each apparent solution in the original equation and to discard any solutions that are extraneous.

Some logarithmic equations can be solved by changing the logarithmic equation to an exponential equation using the fact that $x = \log_a y$ if and only if $y = a^x$.

EXAMPLE 7  **Solving Logarithmic Equations**

Solve each equation:

**(a)** $\log_3(4x - 7) = 2$        **(b)** $\log_x 64 = 2$

**Solution** **(a)** Change the logarithmic equation to an exponential equation.

$$\log_3(4x - 7) = 2$$
$$4x - 7 = 3^2 \qquad \text{Change to an exponential equation.}$$
$$4x - 7 = 9$$
$$4x = 16$$
$$x = 4$$

*Check:* For $x = 4$, $\log_3(4x - 7) = \log_3(4 \cdot 4 - 7) = \log_3 9 = 2$ since $3^2 = 9$.
The solution is 4.

**(b)** Change the logarithmic equation to an exponential equation.

$$\log_x 64 = 2$$
$$x^2 = 64 \qquad \text{Change to an exponential equation.}$$
$$x = 8 \quad \text{or} \quad x = -8 \qquad \text{Solve.}$$

The base of a logarithm is always positive. As a result, $-8$ is an extraneous solution, so we discard it. Now we check the solution 8.

*Check:* For $x = 8$, $\log_8 64 = 2$, since $8^2 = 64$.
The solution is 8. ∎

NOW WORK **Problem 57.**

The properties of logarithms that result from the fact that a logarithmic function is one-to-one can be used to solve some equations that contain two logarithms with the same base.

CALC

CLIP  EXAMPLE 8  **Solving a Logarithmic Equation**

Solve the logarithmic equation $2 \ln x = \ln 9$.

**Solution** Each logarithm has the same base, so

$$2 \ln x = \ln 9$$
$$\ln x^2 = \ln 9 \qquad \qquad r \log_a u = \log_a u^r$$
$$x^2 = 9 \qquad \qquad \text{If } \log_a u = \log_a v, \text{ then } u = v.$$
$$x = 3 \quad \text{or} \quad x = -3$$

We discard the solution $x = -3$ since $-3$ is not in the domain of $f(x) = \ln x$. The solution is 3. ∎

NOW WORK **Problem 61.**

Although, each of these equations was relatively easy to solve, this is not generally the case. Many solutions to exponential and logarithmic equations need to be approximated using technology.

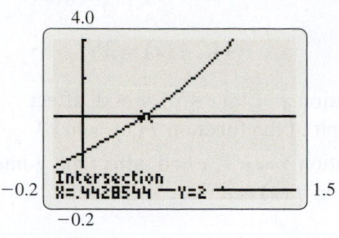

**Figure 71**

**EXAMPLE 9** **Approximating the Solution to an Exponential Equation**

Solve $x + e^x = 2$. Express the solution rounded to three decimal places.

**Solution** We can approximate the solution to the equation by graphing the two functions $Y_1 = x + e^x$ and $Y_2 = 2$. Then we use graphing technology to approximate the intersection of the graphs. Since the function $Y_1$ is increasing (do you know why?) and the function $Y_2$ is constant, there will be only one point of intersection. Figure 71 shows the graphs of the two functions and their intersection. They intersect when $x \approx 0.4428544$, so the solution of the equation is 0.443 rounded to three decimal places. ∎

**NOW WORK** Problem 65.

# P.5 Assess Your Understanding

## Concepts and Vocabulary

1. The graph of every exponential function $f(x) = a^x$, $a > 0$ and $a \neq 1$, passes through three points: _____, _____, and _____.

2. *True or False* The graph of the exponential function $f(x) = \left(\dfrac{3}{2}\right)^x$ is decreasing.

3. If $3^x = 3^4$, then $x = $ _____.

4. If $4^x = 8^2$ then $x = $ _____.

5. *True or False* The graphs of $y = 3^x$ and $y = \left(\dfrac{1}{3}\right)^x$ are symmetric with respect to the line $y = x$.

6. *True or False* The range of the exponential function $f(x) = a^x$, $a > 0$ and $a \neq 1$, is the set of all real numbers.

7. The number $e$ is defined as the base of the exponential function $f$ whose tangent line to the graph of $f$ at the point (0, 1) has slope _____.

8. The domain of the logarithmic function $f(x) = \log_a x$ is _____.

9. The graph of every logarithmic function $f(x) = \log_a x$, $a > 0$ and $a \neq 1$, passes through three points: _____, _____, and _____.

10. *Multiple Choice* The graph of $f(x) = \log_2 x$ is [(a) increasing (b) decreasing (c) neither].

11. *True or False* If $y = \log_a x$, then $y = a^x$.

12. *True or False* The graph of $f(x) = \log_a x$, $a > 0$ and $a \neq 1$, has an $x$-intercept equal to 1 and no $y$-intercept.

13. *True or False* $\ln e^x = x$ for all real numbers.

14. $\ln e = $ _____.

15. Explain what the number $e$ is.

16. What is the $x$-intercept of the function $h(x) = \ln(x + 1)$?

## Practice Problems

17. Suppose that $g(x) = 4^x + 2$.

    (a) What is $g(-1)$? What is the corresponding point on the graph of $g$?

    (b) If $g(x) = 66$, what is $x$? What is the corresponding point on the graph of $g$?

18. Suppose that $g(x) = 5^x - 3$.

    (a) What is $g(-1)$? What is the corresponding point on the graph of $g$?

    (b) If $g(x) = 122$, what is $x$? What is the corresponding point on the graph of $g$?

*In Problems 19–24, the graph of an exponential function is given. Match each graph to one of the following functions:*

(a) $y = 3^{-x}$     (b) $y = -3^x$     (c) $y = -3^{-x}$

(d) $y = 3^x - 1$     (e) $y = 3^{x-1}$     (f) $y = 1 - 3^x$

**19.**

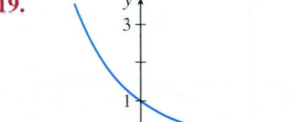

**20.**

**21.**

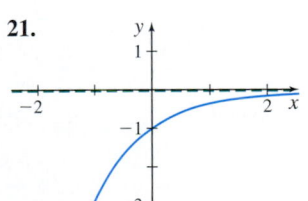

**22.**

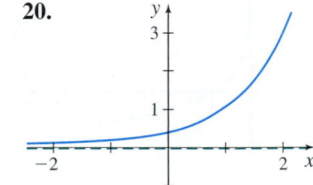

**23.**

**24.**

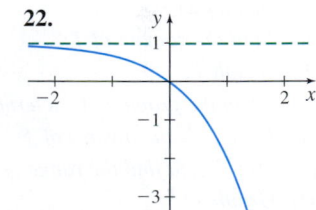

*In Problems 25–30, use transformations to graph each function. Find the domain and range.*

25. $f(x) = 2^{x+2}$

26. $f(x) = 1 - 2^{-x/3}$

27. $f(x) = 4\left(\dfrac{1}{3}\right)^x$

28. $f(x) = \left(\dfrac{1}{2}\right)^{-x} + 1$

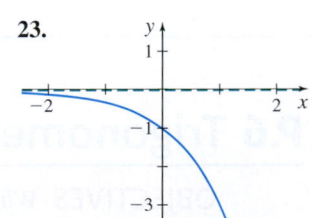

29. $f(x) = e^{-x}$

30. $f(x) = 5 - e^x$

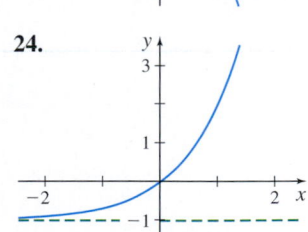

---

**1.** = NOW WORK problem     📈 = Graphing technology recommended     CAS = Computer Algebra System recommended

*In Problems 31–34, find the domain of each function.*

**31.** $F(x) = \log_2 x^2$

**32.** $g(x) = 8 + 5\ln(2x + 3)$

**33.** $f(x) = \ln(x - 1)$

**34.** $g(x) = \sqrt{\ln x}$

*In Problems 35–40, the graph of a logarithmic function is given. Match each graph to one of the following functions:*

**(a)** $y = \log_3 x$

**(b)** $y = \log_3(-x)$

**(c)** $y = -\log_3 x$

**(d)** $y = \log_3 x - 1$

**(e)** $y = \log_3(x - 1)$

**(f)** $y = 1 - \log_3 x$

**35.**
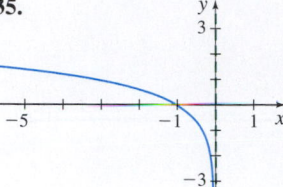

**36.**
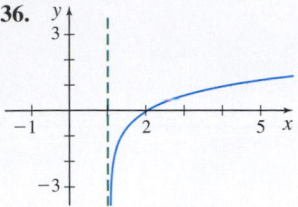

**37.**
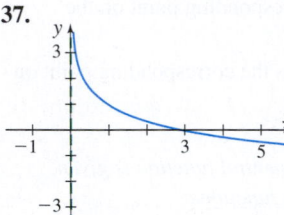

**38.**
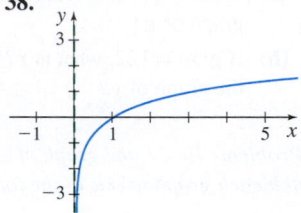

**39.**
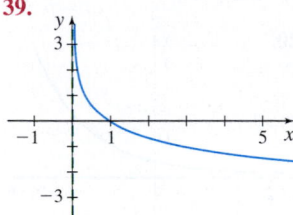

**40.**
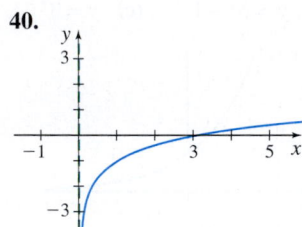

*In Problems 41–44,*
**(a)** *Find the domain of* $f$.
**(b)** *Graph* $f$.
**(c)** *From the graph of* $f$, *determine the range of* $f$.
**(d)** *Find* $f^{-1}$, *the inverse of* $f$.
**(e)** *Use* $f^{-1}$ *to find the range of* $f$.
**(f)** *Graph* $f^{-1}$.

**41.** $f(x) = \ln(x + 4)$

**42.** $f(x) = \frac{1}{2}\log(2x)$

**43.** $f(x) = 3e^x + 2$

**44.** $f(x) = 2^{x/3} + 4$

**45.** How does the transformation $y = \ln(x + c)$, $c > 0$, affect the $x$-intercept of the graph of the function $f(x) = \ln x$?

**46.** How does the transformation $y = e^{cx}$, $c > 0$, affect the $y$-intercept of the graph of the function $f(x) = e^x$?

*In Problems 47–62, solve each equation.*

**47.** $3^{x^2} = 9^x$

**48.** $5^{x^2 + 8} = 125^{2x}$

**49.** $e^{3x} = \dfrac{e^2}{e^x}$

**50.** $e^{4x} \cdot e^{x^2} = e^{12}$

**51.** $e^{1 - 2x} = 4$

**52.** $e^{1 - x} = 5$

**53.** $5(2^{3x}) = 9$

**54.** $0.3(4^{0.2x}) = 0.2$

**55.** $3^{1 - 2x} = 4^x$

**56.** $2^{x + 1} = 5^{1 - 2x}$

**57.** $\log_2(2x + 1) = 3$

**58.** $\log_3(3x - 2) = 2$

**59.** $\log_x\left(\dfrac{1}{8}\right) = 3$

**60.** $\log_x 64 = -3$

**61.** $\ln(2x + 3) = 2\ln 3$

**62.** $\dfrac{1}{2}\log_3 x = 2\log_3 2$

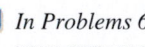 *In Problems 63–66, use technology to solve each equation. Express your answer rounded to three decimal places.*

**63.** $\log_5(x + 1) - \log_4(x - 2) = 1$

**64.** $\ln x = x$

**65.** $e^x + \ln x = 4$

**66.** $e^x = x^2$

**67.** **(a)** If $f(x) = \ln(x + 4)$ and $g(x) = \ln(3x + 1)$, graph $f$ and $g$ on the same set of axes.
**(b)** Find the point(s) of intersection of the graphs of $f$ and $g$ by solving $f(x) = g(x)$.
**(c)** Based on the graph, solve $f(x) > g(x)$.

**68.** **(a)** If $f(x) = 3^{x + 1}$ and $g(x) = 2^{x + 2}$, graph $f$ and $g$ on the same set of axes.
**(b)** Find the point(s) of intersection of the graphs of $f$ and $g$ by solving $f(x) = g(x)$. Round answers to three decimal places.
**(c)** Based on the graph, solve $f(x) > g(x)$.

---

# P.6 Trigonometric Functions

**OBJECTIVES** *When you finish this section, you should be able to:*

**1** Work with properties of trigonometric functions (p. 52)

**2** Graph the trigonometric functions (p. 53)

**NEED TO REVIEW?** Trigonometric functions are discussed in Appendix A.4, pp. A-27 to A-32.

## 1 Work with Properties of Trigonometric Functions

Table 6 lists the six trigonometric functions and the domain and range of each function.

**TABLE 6**

| Function | Symbol | Domain | Range |
|---|---|---|---|
| sine | $y = \sin x$ | All real numbers | $\{y \mid -1 \le y \le 1\}$ |
| cosine | $y = \cos x$ | All real numbers | $\{y \mid -1 \le y \le 1\}$ |
| tangent | $y = \tan x$ | $\left\{x \mid x \ne \text{ odd integer multiples of } \dfrac{\pi}{2}\right\}$ | All real numbers |
| cosecant | $y = \csc x$ | $\{x \mid x \ne \text{ integer multiples of } \pi\}$ | $\{y \mid y \le -1 \text{ or } y \ge 1\}$ |
| secant | $y = \sec x$ | $\left\{x \mid x \ne \text{ odd integer multiples of } \dfrac{\pi}{2}\right\}$ | $\{y \mid y \le -1 \text{ or } y \ge 1\}$ |
| cotangent | $y = \cot x$ | $\{x \mid x \ne \text{ integer multiples of } \pi\}$ | All real numbers |

**NOTE** In calculus, radians are generally used to measure angles, unless degrees are specifically mentioned.

An important property common to all trigonometric functions is that they are *periodic*.

**DEFINITION** Periodic Function

A function $f$ is called **periodic** if there is a positive number $p$ with the property that whenever $x$ is in the domain of $f$, so is $x + p$, and

$$\boxed{f(x + p) = f(x)}$$

If there is a smallest number $p$ with this property, it is called the (**fundamental**) **period** of $f$.

The sine, cosine, cosecant, and secant functions are periodic with period $2\pi$; the tangent and cotangent functions are periodic with period $\pi$.

**THEOREM** Period of Trigonometric Functions

$$\begin{array}{lll} \sin(x + 2\pi) = \sin x & \cos(x + 2\pi) = \cos x & \tan(x + \pi) = \tan x \\ \csc(x + 2\pi) = \csc x & \sec(x + 2\pi) = \sec x & \cot(x + \pi) = \cot x \end{array}$$

Because the trigonometric functions are periodic, once the values of the function over one period are known, the values over the entire domain are known. This property is useful for graphing trigonometric functions.

The next result, also useful for graphing the trigonometric functions, is a consequence of the even-odd identities, namely, $\sin(-x) = -\sin x$ and $\cos(-x) = \cos x$. From these, we have

$$\tan(-x) = \frac{\sin(-x)}{\cos(-x)} = \frac{-\sin x}{\cos x} = -\tan x \qquad \sec(-x) = \frac{1}{\cos(-x)} = \frac{1}{\cos x} = \sec x$$

$$\cot(-x) = \frac{1}{\tan(-x)} = \frac{1}{-\tan x} = -\cot x \qquad \csc(-x) = \frac{1}{\sin(-x)} = \frac{1}{-\sin x} = -\csc x$$

**THEOREM** Even-Odd Properties of the Trigonometric Functions

The sine, tangent, cosecant, and cotangent functions are odd, so their graphs are symmetric with respect to the origin.

The cosine and secant functions are even, so their graphs are symmetric with respect to the $y$-axis.

## 2 Graph the Trigonometric Functions

**NEED TO REVIEW?** The values of the trigonometric functions for select numbers are discussed in Appendix A.4, pp. A-29 and A-31.

To graph $y = \sin x$, we use Table 7 on page 54 to obtain points on the graph. Then we plot some of these points and connect them with a smooth curve. Since the sine function has a period of $2\pi$, continue the graph to the left of 0 and to the right of $2\pi$. See Figure 72.

**TABLE 7**

| $x$ | $y = \sin x$ | $(x, y)$ |
|---|---|---|
| $0$ | $0$ | $(0, 0)$ |
| $\dfrac{\pi}{6}$ | $\dfrac{1}{2}$ | $\left(\dfrac{\pi}{6}, \dfrac{1}{2}\right)$ |
| $\dfrac{\pi}{2}$ | $1$ | $\left(\dfrac{\pi}{2}, 1\right)$ |
| $\dfrac{5\pi}{6}$ | $\dfrac{1}{2}$ | $\left(\dfrac{5\pi}{6}, \dfrac{1}{2}\right)$ |
| $\pi$ | $0$ | $(\pi, 0)$ |
| $\dfrac{7\pi}{6}$ | $-\dfrac{1}{2}$ | $\left(\dfrac{7\pi}{6}, -\dfrac{1}{2}\right)$ |
| $\dfrac{3\pi}{2}$ | $-1$ | $\left(\dfrac{3\pi}{2}, -1\right)$ |
| $\dfrac{11\pi}{6}$ | $-\dfrac{1}{2}$ | $\left(\dfrac{11\pi}{6}, -\dfrac{1}{2}\right)$ |
| $2\pi$ | $0$ | $(2\pi, 0)$ |

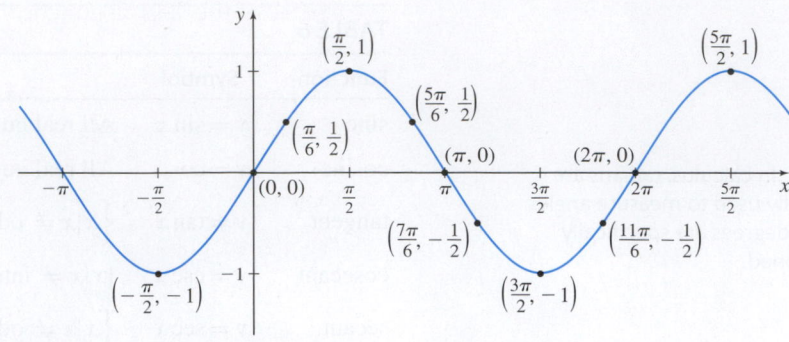

**Figure 72** $f(x) = \sin x$

Notice the symmetry of the graph with respect to the origin. This is a consequence of $f$ being an odd function.

The graph of $y = \sin x$ illustrates some facts about the sine function.

**Properties of the Sine Function** $f(x) = \sin x$

- The domain of $f$ is the set of all real numbers.
- The range of $f$ consists of all real numbers in the closed interval $[-1, 1]$.
- The sine function is an odd function, so its graph is symmetric with respect to the origin.
- The sine function has a period of $2\pi$.
- The $x$-intercepts of $f$ are $\ldots, -2\pi, -\pi, 0, \pi, 2\pi, 3\pi, \ldots$; the $y$-intercept is $0$.
- The maximum value of $f$ is $1$ and occurs at $x = \ldots, -\dfrac{3\pi}{2}, \dfrac{\pi}{2}, \dfrac{5\pi}{2}, \dfrac{9\pi}{2}, \ldots$; the minimum value of $f$ is $-1$ and occurs at $x = \ldots, -\dfrac{\pi}{2}, \dfrac{3\pi}{2}, \dfrac{7\pi}{2}, \dfrac{11\pi}{2}, \ldots$.

The graph of the cosine function is obtained in a similar way. Locate points on the graph of the cosine function $f(x) = \cos x$ for $0 \leq x \leq 2\pi$. Then connect the points with a smooth curve and continue the graph to the left of $0$ and to the right of $2\pi$ to obtain the graph of $y = \cos x$. See Figure 73.

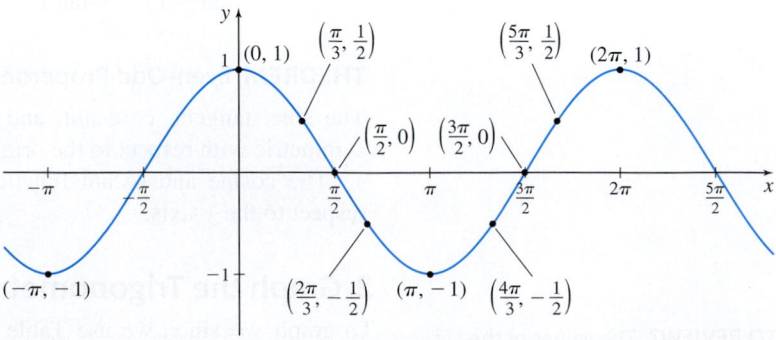

**Figure 73** $f(x) = \cos x$

The graph of $y = \cos x$ illustrates some facts about the cosine function.

**Properties of the Cosine Function** $f(x) = \cos x$

- The domain of $f$ is the set of all real numbers.
- The range of $f$ consists of all real numbers in the closed interval $[-1, 1]$.
- The cosine function is an even function, so its graph is symmetric with respect to the $y$-axis.
- The cosine function has a period of $2\pi$.
- The $x$-intercepts of $f$ are $\ldots, -\dfrac{3\pi}{2}, -\dfrac{\pi}{2}, \dfrac{\pi}{2}, \dfrac{3\pi}{2}, \dfrac{5\pi}{2}, \ldots$;

  the $y$-intercept is 1.
- The maximum value of $f$ is 1 and occurs at $x = \ldots, -2\pi, 0, 2\pi, 4\pi, 6\pi, \ldots$; the minimum value of $f$ is $-1$ and occurs at $x = \ldots, -\pi, \pi, 3\pi, 5\pi, \ldots$.

Many variations of the sine and cosine functions can be graphed using transformations.

EXAMPLE 1  **Graphing Variations of $f(x) = \sin x$ Using Transformations**

Use the graph of $f(x) = \sin x$ to graph $g(x) = 2 \sin x$.

**Solution** Notice that $g(x) = 2f(x)$, so the graph of $g$ is a vertical stretch of the graph of $f(x) = \sin x$. Figure 74 illustrates the transformation.

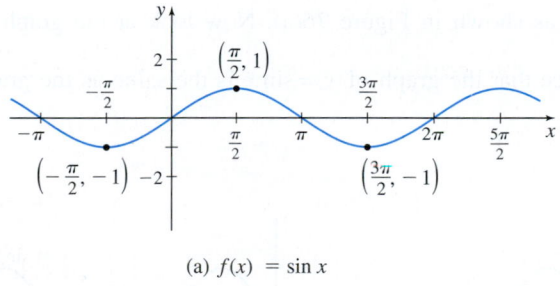

(a) $f(x) = \sin x$

Multiply by 2

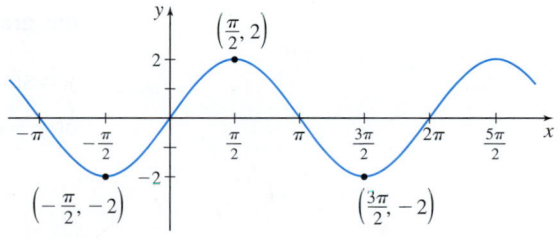

(b) $g(x) = 2 \sin x$

**DF Figure 74**

Notice that the values of $g(x) = 2 \sin x$ lie between $-2$ and 2, inclusive.

In general, the values of the functions $f(x) = A \sin x$ and $g(x) = A \cos x$, where $A \neq 0$, will satisfy the inequalities

$$-|A| \leq A \sin x \leq |A| \qquad \text{and} \qquad -|A| \leq A \cos x \leq |A|$$

respectively. The number $|A|$ is called the **amplitude** of $f(x) = A \sin x$ and of $g(x) = A \cos x$.

EXAMPLE 2  **Graphing Variations of $f(x) = \cos x$ Using Transformations**

Use the graph of $f(x) = \cos x$ to graph $g(x) = \cos(3x)$.

**Solution** The graph of $g(x) = \cos(3x)$ is a horizontal compression of the graph of $f(x) = \cos x$. Figure 75 on page 56 shows the transformation.

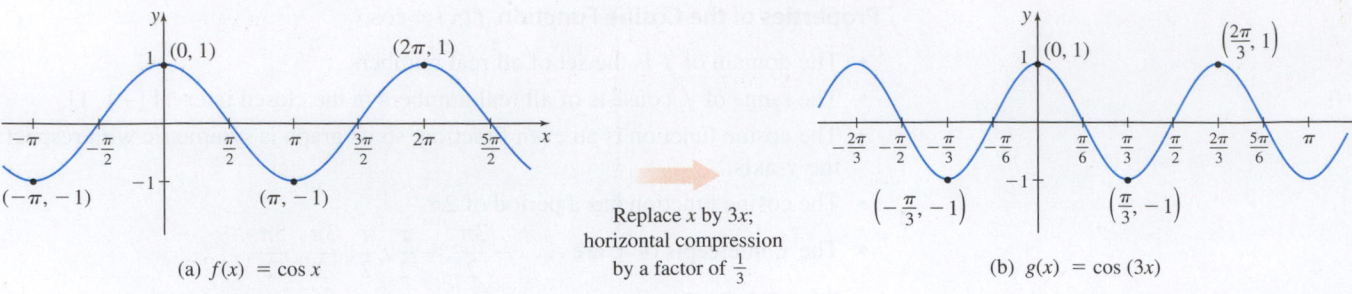

(a) $f(x) = \cos x$

Replace $x$ by $3x$;
horizontal compression
by a factor of $\frac{1}{3}$

(b) $g(x) = \cos(3x)$

**DF** **Figure 75**

From the graph, we notice that the period of $g(x) = \cos(3x)$ is $\dfrac{2\pi}{3}$.

**NOW WORK** Problems **27** and **29**.

In general, if $\omega > 0$, the functions $f(x) = \sin(\omega x)$ and $g(x) = \cos(\omega x)$ have period $T = \dfrac{2\pi}{\omega}$. If $\omega > 1$, the graphs of $f(x) = \sin(\omega x)$ and $g(x) = \cos(\omega x)$ are horizontally compressed, and the period of the functions is less than $2\pi$. If $0 < \omega < 1$, the graphs of $f(x) = \sin(\omega x)$ and $g(x) = \cos(\omega x)$ are horizontally stretched, and the period of the functions is greater than $2\pi$.

One period of the graph of $f(x) = \sin(\omega x)$ or $g(x) = \cos(\omega x)$ is called a **cycle**.

### Sinusoidal Graphs

If we shift the graph of the function $y = \cos x$ to the right $\dfrac{\pi}{2}$ units, we obtain the graph of $y = \cos\left(x - \dfrac{\pi}{2}\right)$, as shown in Figure 76(a). Now look at the graph of $y = \sin x$ in Figure 76(b). Notice that the graph of $y = \sin x$ is the same as the graph of $y = \cos\left(x - \dfrac{\pi}{2}\right)$.

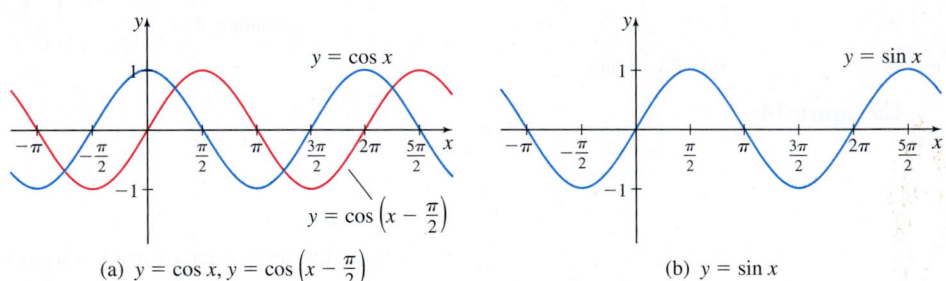

(a) $y = \cos x,\ y = \cos\left(x - \dfrac{\pi}{2}\right)$

(b) $y = \sin x$

**DF** **Figure 76**

Figure 76 suggests that

$$\boxed{\sin x = \cos\left(x - \dfrac{\pi}{2}\right)} \qquad (1)$$

**NEED TO REVIEW?** Trigonometric identities are discussed in Appendix A.4, pp. A-32 to A-35.

**Proof**  To prove this identity, we use the difference formula for $\cos(A - B)$ with $A = x$ and $B = \dfrac{\pi}{2}$.

$$\cos\left(x - \dfrac{\pi}{2}\right) = \cos x \cos\dfrac{\pi}{2} + \sin x \sin\dfrac{\pi}{2} = \cos x \cdot 0 + \sin x \cdot 1 = \sin x$$

Because of this relationship, the graphs of $y = A\sin(\omega x)$ or $y = A\cos(\omega x)$ are referred to as **sinusoidal graphs,** and the functions and their variations are called **sinusoidal functions.**

**THEOREM** Amplitude and Period

For the graphs of $y = A\sin(\omega x)$ and $y = A\cos(\omega x)$, where $A \neq 0$ and $\omega > 0$,

$$\text{Amplitude} = |A| \qquad \text{Period} = T = \frac{2\pi}{\omega}$$

> CALC
> ▶ CLIP

**EXAMPLE 3** Finding the Amplitude and Period of a Sinusoidal Function

Determine the amplitude and period of $y = -3\sin(4\pi x)$.

**Solution** Comparing $y = -3\sin(4\pi x)$ to $y = A\sin(\omega x)$, we find that $A = -3$ and $\omega = 4\pi$. Then,

$$\text{Amplitude} = |A| = |-3| = 3 \qquad \text{Period} = T = \frac{2\pi}{\omega} = \frac{2\pi}{4\pi} = \frac{1}{2} \qquad\blacksquare$$

**NOW WORK** Problem 33.

**TABLE 8**

| $x$ | $y = \tan x$ | $(x, y)$ |
|-----|--------------|----------|
| $0$ | $0$ | $(0, 0)$ |
| $\dfrac{\pi}{6}$ | $\dfrac{\sqrt{3}}{3}$ | $\left(\dfrac{\pi}{6}, \dfrac{\sqrt{3}}{3}\right)$ |
| $\dfrac{\pi}{4}$ | $1$ | $\left(\dfrac{\pi}{4}, 1\right)$ |
| $\dfrac{\pi}{3}$ | $\sqrt{3}$ | $\left(\dfrac{\pi}{3}, \sqrt{3}\right)$ |

The function $y = \tan x$ is an odd function with period $\pi$. It is not defined at odd multiples of $\dfrac{\pi}{2}$. Do you see why? So, we construct Table 8 for $0 \leq x \leq \dfrac{\pi}{3}$. Then we plot the points from Table 8, connect them with a smooth curve, and reflect the graph about the origin, as shown in Figure 77. To investigate the behavior of $\tan x$ near $\dfrac{\pi}{2}$, we use the identity $\tan x = \dfrac{\sin x}{\cos x}$. When $x$ is close to, but less than, $\dfrac{\pi}{2}$, $\sin x$ is close to 1 and $\cos x$ is a positive number close to 0, so the ratio $\dfrac{\sin x}{\cos x}$ is a large, positive number. The closer $x$ gets to $\dfrac{\pi}{2}$, the larger $\tan x$ becomes. Figure 78 shows the graph of $y = \tan x$, $-\dfrac{\pi}{2} < x < \dfrac{\pi}{2}$. The graph of $y = \tan x$ is obtained by repeating the graph in Figure 78. See Figure 79.

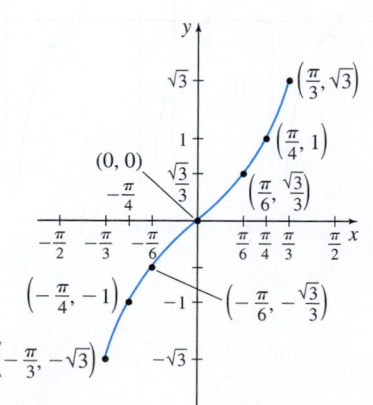

**Figure 77** $y = \tan x$, $-\dfrac{\pi}{3} \leq x \leq \dfrac{\pi}{3}$

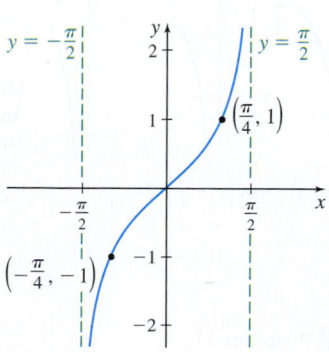

**Figure 78** $y = \tan x$, $-\dfrac{\pi}{2} < x < \dfrac{\pi}{2}$

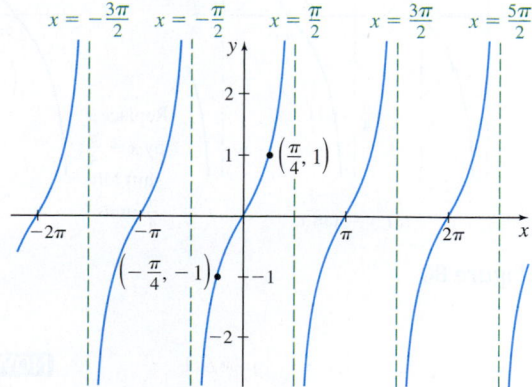

**Figure 79** $y = \tan x$, $-\infty < x < \infty$, $x \neq$ odd multiples of $\dfrac{\pi}{2}$

The graph of $y = \tan x$ illustrates the following properties of the tangent function:

---

**Properties of the Tangent Function** $f(x) = \tan x$

- The domain of $f$ is the set of all real numbers, except odd multiples of $\dfrac{\pi}{2}$.

- The range of $f$ consists of all real numbers.

- The tangent function is an odd function, so its graph is symmetric with respect to the origin.

- The tangent function is periodic with period $\pi$.

- The $x$-intercepts of $f$ are $\ldots, -2\pi, -\pi, 0, \pi, 2\pi, 3\pi, \ldots$; the $y$-intercept is 0.

---

**EXAMPLE 4   Graphing Variations of $f(x) = \tan x$ Using Transformations**

Use the graph of $f(x) = \tan x$ to graph $g(x) = -\tan\left(x + \dfrac{\pi}{4}\right)$.

**Solution** Figure 80 illustrates the steps used in graphing $g(x) = -\tan\left(x + \dfrac{\pi}{4}\right)$.

- Begin by graphing $f(x) = \tan x$. See Figure 80(a).

- Replace the argument $x$ by $x + \dfrac{\pi}{4}$ to obtain $y = \tan\left(x + \dfrac{\pi}{4}\right)$, which shifts the graph horizontally to the left $\dfrac{\pi}{4}$ unit, as shown in Figure 80(b).

- Multiply $\tan\left(x + \dfrac{\pi}{4}\right)$ by $-1$, which reflects the graph about the $x$-axis and results in the graph of $y = -\tan\left(x + \dfrac{\pi}{4}\right)$, as shown in Figure 80(c).

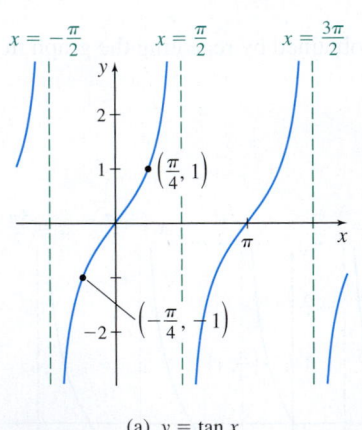
(a) $y = \tan x$

Replace $x$ by $x + \dfrac{\pi}{4}$; shift left $\dfrac{\pi}{4}$ units

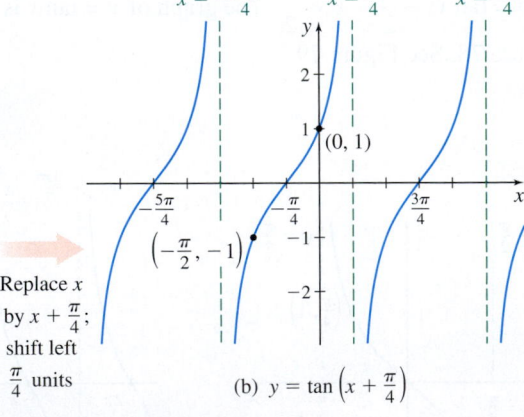
(b) $y = \tan\left(x + \dfrac{\pi}{4}\right)$

Multiply by $-1$; reflect about $x$-axis

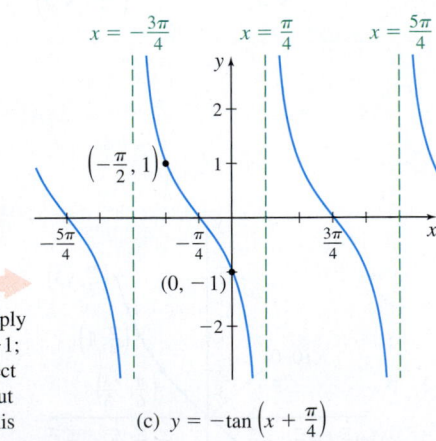
(c) $y = -\tan\left(x + \dfrac{\pi}{4}\right)$

**Figure 80**

■

**NOW WORK** Problem **31**.

The graph of the cotangent function is obtained similarly. $y = \cot x$ is an odd function, with period $\pi$. Because $\cot x$ is not defined at multiples of $\pi$, graph $y$ on the interval $(0, \pi)$ and then repeat the graph, as shown in Figure 81.

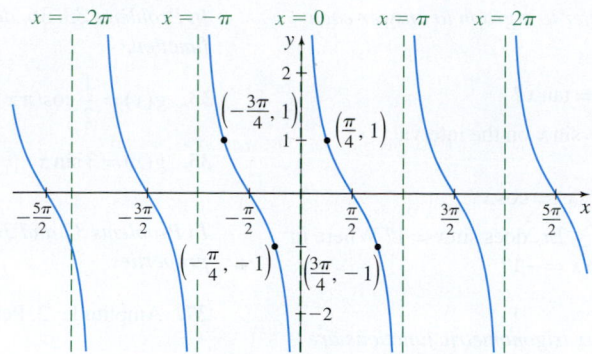

**Figure 81** $f(x) = \cot x$, $-\infty < x < \infty$, $x \neq$ multiples of $\pi$

The cosecant and secant functions, sometimes referred to as **reciprocal functions**, are graphed by using the reciprocal identities

$$\csc x = \frac{1}{\sin x} \quad \text{and} \quad \sec x = \frac{1}{\cos x}$$

The graphs of $y = \csc x$ and $y = \sec x$ are shown in Figures 82 and 83, respectively.

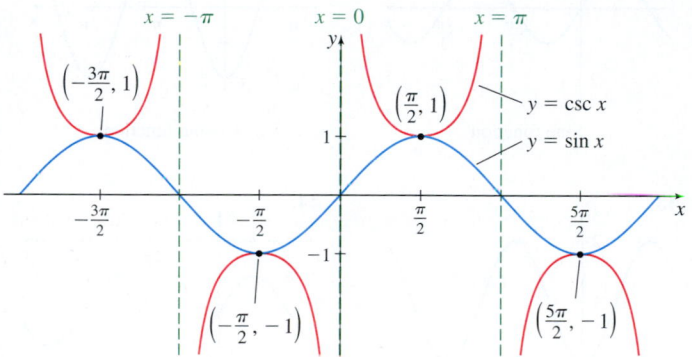

**Figure 82**

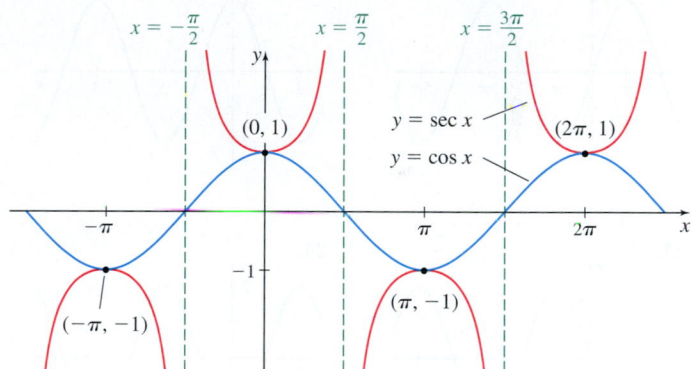

**Figure 83**

# P.6 Assess Your Understanding

## Concepts and Vocabulary

1. The sine, cosine, cosecant, and secant functions have period _____; the tangent and cotangent functions have period _____.

2. The domain of the tangent function $f(x) = \tan x$ is _____.

3. The range of the sine function $f(x) = \sin x$ is _____.

4. Explain why $\tan\left(\frac{\pi}{4} + 2\pi\right) = \tan\frac{\pi}{4}$.

5. *True or False* The range of the secant function is the set of all positive real numbers.

6. The function $f(x) = 3\cos(6x)$ has amplitude _____ and period _____.

7. *True or False* The graphs of $y = \sin x$ and $y = \cos x$ are identical except for a horizontal shift.

8. *True or False* The amplitude of the function $f(x) = 2\sin(\pi x)$ is 2 and its period is $\frac{\pi}{2}$.

9. *True or False* The graph of the sine function has infinitely many $x$-intercepts.

10. The graph of $y = \tan x$ is symmetric with respect to the _____.

11. The graph of $y = \sec x$ is symmetric with respect to the _____.

12. Explain, in your own words, what it means for a function to be periodic.

## Practice Problems

*In Problems 13–16, use the even-odd properties to find the exact value of each expression.*

13. $\tan\left(-\dfrac{\pi}{4}\right)$

14. $\sin\left(-\dfrac{3\pi}{2}\right)$

15. $\csc\left(-\dfrac{\pi}{3}\right)$

16. $\cos\left(-\dfrac{\pi}{6}\right)$

---

**1.** = NOW WORK problem      = Graphing technology recommended      CAS = Computer Algebra System recommended

In Problems 17–20, if necessary, refer to a graph to answer each question.

**17.** What is the $y$-intercept of $f(x) = \tan x$?

**18.** Find the $x$-intercepts of $f(x) = \sin x$ on the interval $[-2\pi, 2\pi]$.

**19.** What is the smallest value of $f(x) = \cos x$?

**20.** For what numbers $x$, $-2\pi \le x \le 2\pi$, does $\sin x = 1$? Where in the interval $[-2\pi, 2\pi]$ does $\sin x = -1$?

In Problems 21–26, the graphs of six trigonometric functions are given. Match each graph to one of the following functions:

**(a)** $y = 2\sin\left(\dfrac{\pi}{2}x\right)$

**(b)** $y = 2\cos\left(\dfrac{\pi}{2}x\right)$

**(c)** $y = 3\cos(2x)$

**(d)** $y = -3\sin(2x)$

**(e)** $y = -2\cos\left(\dfrac{\pi}{2}x\right)$

**(f)** $y = -2\sin\left(\dfrac{1}{2}x\right)$

**21.**
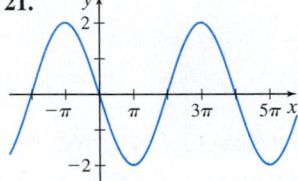

**22.**
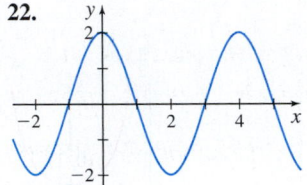

**23.**
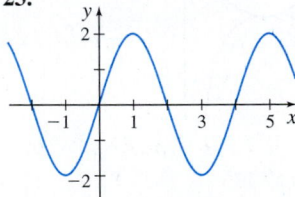

**24.**
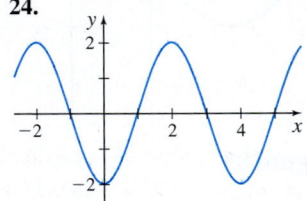

**25.**
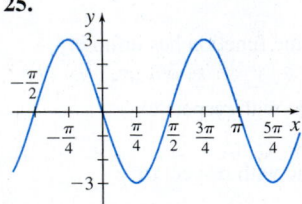

**26.**
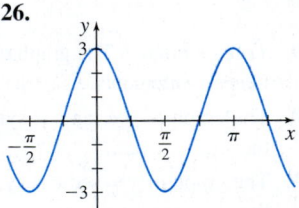

In Problems 27–32, graph each function using transformations. Be sure to label key points and show at least two periods.

**27.** $f(x) = 4\sin(\pi x)$

**28.** $f(x) = -3\cos x$

**29.** $f(x) = 3\cos(2x) - 4$

**30.** $f(x) = 4\sin(2x) + 2$

**31.** $f(x) = \tan\left(\dfrac{\pi}{2}x\right)$

**32.** $f(x) = 4\sec\left(\dfrac{1}{2}x\right)$

In Problems 33–36, determine the amplitude and period of each function.

**33.** $g(x) = \dfrac{1}{2}\cos(\pi x)$

**34.** $f(x) = \sin(2x)$

**35.** $g(x) = 3\sin x$

**36.** $f(x) = -2\cos\left(\dfrac{3}{2}x\right)$

In Problems 37 and 38, write the sine function that has the given properties.

**37.** Amplitude: 2, Period: $\pi$

**38.** Amplitude: $\dfrac{1}{3}$, Period: 2

In Problems 39 and 40, write the cosine function that has the given properties.

**39.** Amplitude: $\dfrac{1}{2}$, Period: $\pi$

**40.** Amplitude: 3, Period: $4\pi$

In Problems 41–48, for each graph, find an equation involving the indicated trigonometric function.

**41.**
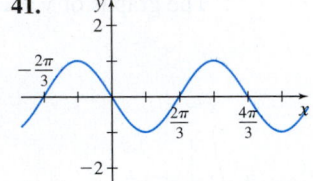
Sine function

**42.**
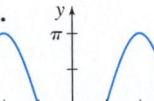
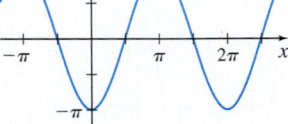
Cosine function

**43.**
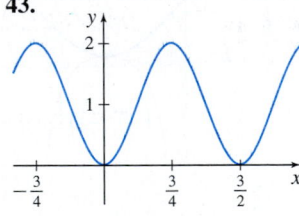
Cosine function

**44.**
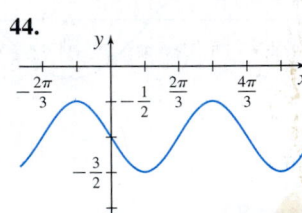
Sine function

**45.**
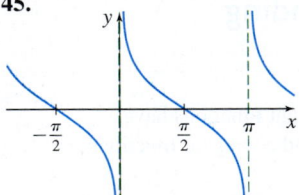
Cotangent function

**46.**
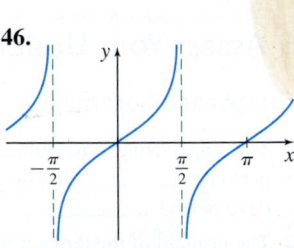
Tangent function

**47.**
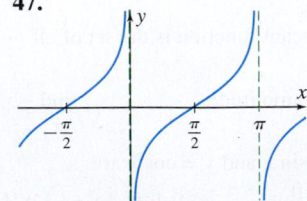
Tangent function

**48.**
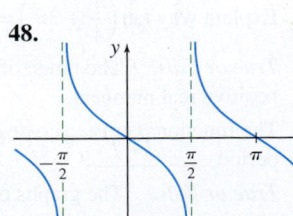
Cotangent function

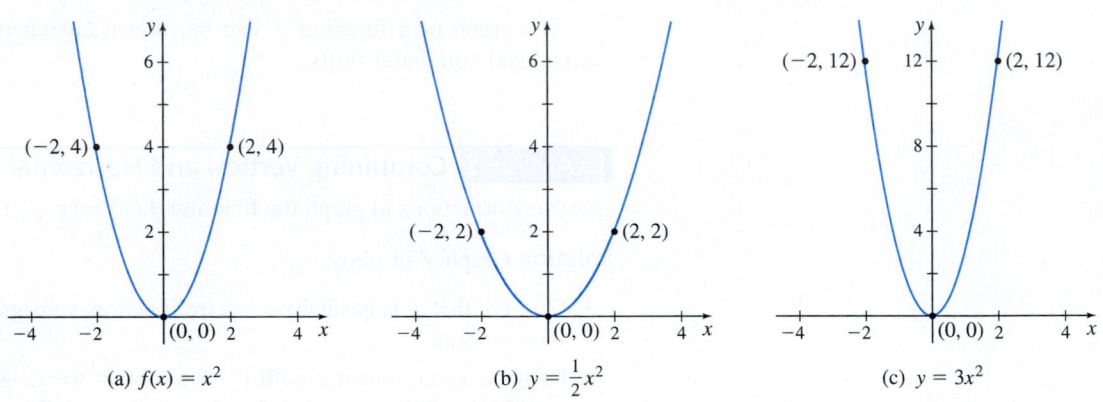

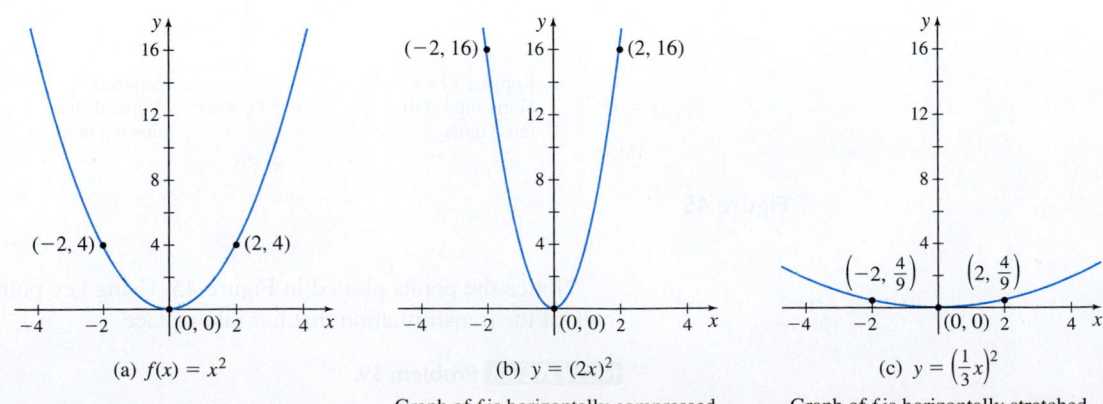

## 5 Transform the Graph of a Function by Reflecting It About the x-axis or the y-axis

The third type of transformation, reflection about the $x$- or $y$-axis, changes the orientation of the graph of the function $f$ but keeps the shape and the size of the graph intact.

When a function $f$ is multiplied by $-1$, the graph of the new function $y = -f(x)$ is the **reflection about the x-axis** of the graph of $f$. For example, if $f(x) = \sqrt{x}$, then the graph of the new function $y = -f(x) = -\sqrt{x}$ is the reflection of the graph of $f$ about the $x$-axis. See Figure 48(a) and 48(b).

If the argument $x$ of a function $f$ is multiplied by $-1$, then the graph of the new function $y = f(-x)$ is the **reflection about the y-axis** of the graph of $f$. For example, if $f(x) = \sqrt{x}$, then the graph of the new function $y = f(-x) = \sqrt{-x}$ is the reflection of the graph of $f$ about the $y$-axis. See Figure 48(a) and 48(c). Notice in this example that the domain of $y = \sqrt{-x}$ is all real numbers for which $-x \geq 0$, or equivalently, $x \leq 0$.

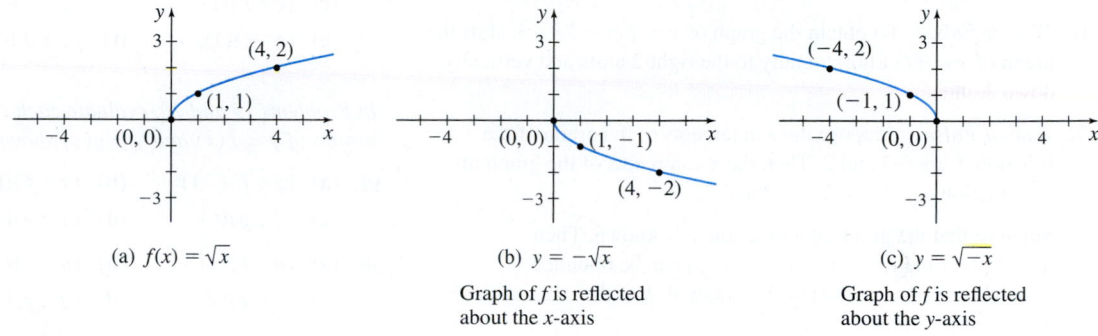

(a) $f(x) = \sqrt{x}$

(b) $y = -\sqrt{x}$
Graph of $f$ is reflected about the $x$-axis

(c) $y = \sqrt{-x}$
Graph of $f$ is reflected about the $y$-axis

**DF Figure 48**

### EXAMPLE 6   Combining Transformations

Use transformations to graph the function $f(x) = \sqrt{1 - x} + 2$.

**Solution** We graph $f$ in steps:

- Observe that $f$ is basically a square root function, so we begin by graphing $y = \sqrt{x}$. See Figure 49(a).
- Now replace the argument $x$ with $x + 1$ to obtain $y = \sqrt{x + 1}$, which shifts the graph of $y = \sqrt{x}$ horizontally to the left 1 unit, as shown in Figure 49(b).
- Then replace $x$ with $-x$ to obtain $y = \sqrt{-x + 1} = \sqrt{1 - x}$, which reflects the graph about the $y$-axis. See Figure 49(c).
- Finally, add 2 to each $y$-coordinate, which shifts the graph vertically up 2 units and results in the graph of $f(x) = \sqrt{1 - x} + 2$ shown in Figure 49(d).

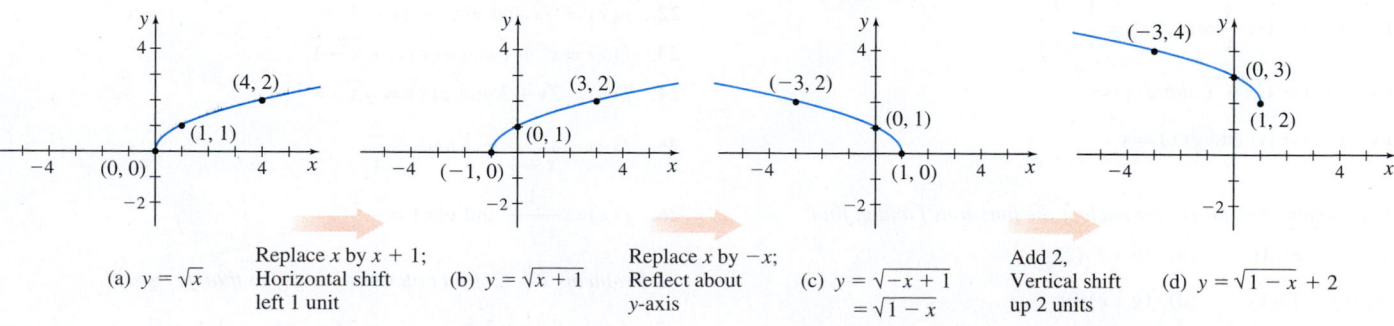

(a) $y = \sqrt{x}$

Replace $x$ by $x + 1$;
Horizontal shift
left 1 unit

(b) $y = \sqrt{x + 1}$

Replace $x$ by $-x$;
Reflect about
$y$-axis

(c) $y = \sqrt{-x + 1}$
$\quad = \sqrt{1 - x}$

Add 2,
Vertical shift
up 2 units

(d) $y = \sqrt{1 - x} + 2$

**Figure 49**

NOW WORK Problem 43.

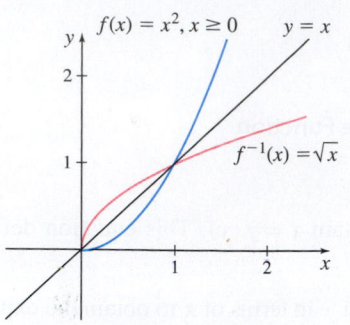

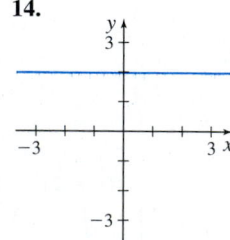

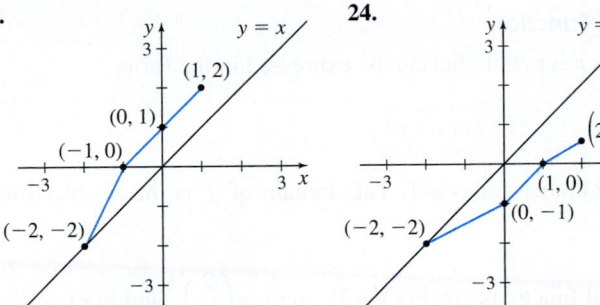

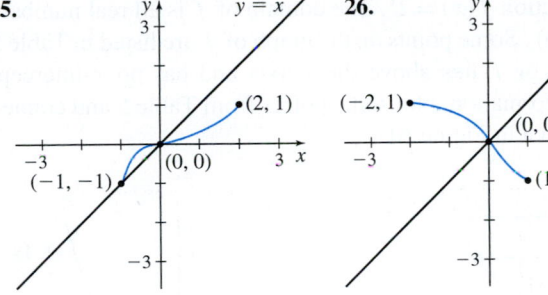

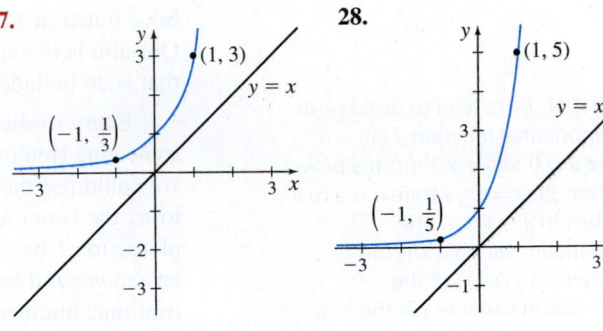

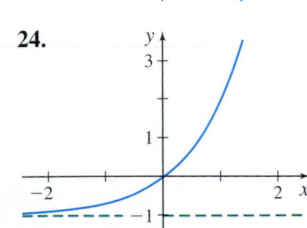

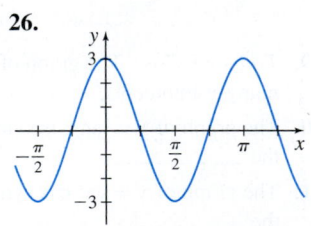

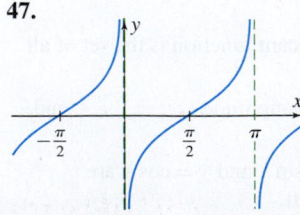
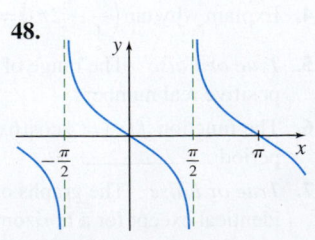

# P.7 Inverse Trigonometric Functions

**OBJECTIVES** *When you finish this section, you should be able to:*

1 Define the inverse trigonometric functions (p. 61)
2 Use the inverse trigonometric functions (p. 65)
3 Solve trigonometric equations (p. 65)

In Section P.4 we discussed inverse functions, and we concluded that if a function is one-to-one, it has an inverse function. We also observed that if a function is not one-to-one, it may be possible to restrict its domain so that the restricted function is one-to-one. For example, the function $y = x^2$ is not one-to-one. However, if we restrict the domain to $x \geq 0$, the function is one-to-one.

Other properties of a one-to-one function $f$ and its inverse function $f^{-1}$ that we discussed in Section P.4 are summarized below.

- $f^{-1}(f(x)) = x$ for every $x$ in the domain of $f$.
- $f(f^{-1}(x)) = x$ for every $x$ in the domain of $f^{-1}$.
- Domain of $f$ = range of $f^{-1}$ and range of $f$ = domain of $f^{-1}$.
- The graph of $f$ and the graph of $f^{-1}$ are reflections of one another about the line $y = x$.
- If a function $y = f(x)$ has an inverse function, the implicit equation of the inverse function is $x = f(y)$. If we solve this equation for $y$, we obtain the explicit form $y = f^{-1}(x)$.

We now investigate the inverse trigonometric functions. In calculus the inverses of the sine function, the tangent function, and the secant function are used most often. So we define these inverse functions first.

## 1 Define the Inverse Trigonometric Functions

Although the trigonometric functions are not one-to-one, we can restrict the domains of the trigonometric functions so that each new function will be one-to-one on its restricted domain. Then an inverse function can be defined on each restricted domain.

Figure 84 shows the graph of $y = \sin x$. Because every horizontal line $y = b$, where $b$ is between $-1$ and $1$, inclusive, intersects the graph of $y = \sin x$ infinitely many times, it follows from the Horizontal-line Test that the function $y = \sin x$ is not one-to-one.

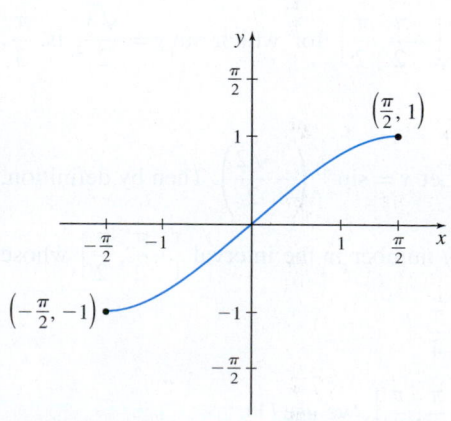

$y = \sin x$, $-\frac{\pi}{2} \leq x \leq \frac{\pi}{2}$, $-1 \leq y \leq 1$

**Figure 85**

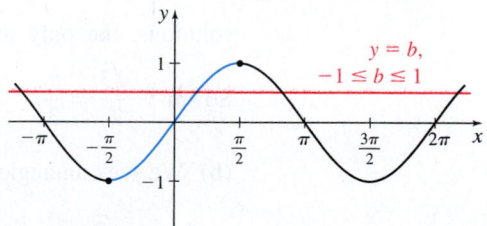

**Figure 84** $y = \sin x$, $-\infty < x < \infty$, $-1 \leq y \leq 1$

### The Inverse Sine Function

However, if the domain of the function $f(x) = \sin x$ is restricted to the interval $\left[-\frac{\pi}{2}, \frac{\pi}{2}\right]$, the function $y = \sin x$, $-\frac{\pi}{2} \leq x \leq \frac{\pi}{2}$, is one-to-one and has an inverse function. See Figure 85.

**IN WORDS** We read $y = \sin^{-1} x$ as "$y$ is the number (or angle) whose sine equals $x$."

To obtain an equation for the inverse of $y = \sin x$ on $-\dfrac{\pi}{2} \le x \le \dfrac{\pi}{2}$, we interchange $x$ and $y$. Then the implicit form of the inverse is $x = \sin y$, $-\dfrac{\pi}{2} \le y \le \dfrac{\pi}{2}$. The explicit form of the inverse function is called the *inverse sine* of $x$ and is written $y = \sin^{-1} x$.

**DEFINITION  Inverse Sine Function**

The **inverse sine function**, denoted by $y = \sin^{-1} x$, or $y = \arcsin x$, is defined as

$$y = \sin^{-1} x \quad \text{if and only if} \quad x = \sin y$$
$$\text{where } -1 \le x \le 1 \quad \text{and} \quad -\frac{\pi}{2} \le y \le \frac{\pi}{2}$$

**CAUTION** The superscript $^{-1}$ that appears in $y = \sin^{-1} x$ is not an exponent but the symbol used to denote the inverse function. (To avoid confusion, $y = \sin^{-1} x$ is sometimes written $y = \arcsin x$.)

The domain of $y = \sin^{-1} x$ is the closed interval $[-1, 1]$, and the range is the closed interval $\left[-\dfrac{\pi}{2}, \dfrac{\pi}{2}\right]$. The graph of $y = \sin^{-1} x$ is the reflection of the restricted portion of the graph of $f(x) = \sin x$ about the line $y = x$. See Figure 86.

Because $y = \sin x$ defined on the closed interval $\left[-\dfrac{\pi}{2}, \dfrac{\pi}{2}\right]$ and $y = \sin^{-1} x$ are inverse functions,

$$\sin^{-1}(\sin x) = x \quad \text{if } x \text{ is in the closed interval } \left[-\frac{\pi}{2}, \frac{\pi}{2}\right] \tag{1}$$

$$\sin(\sin^{-1} x) = x \quad \text{if } x \text{ is in the closed interval } [-1, 1] \tag{2}$$

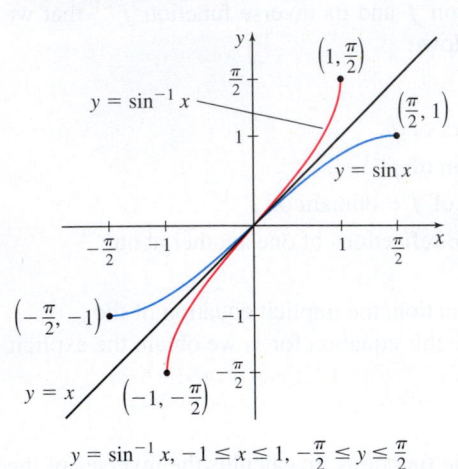

$y = \sin^{-1} x, \ -1 \le x \le 1, \ -\dfrac{\pi}{2} \le y \le \dfrac{\pi}{2}$

**Figure 86**

**EXAMPLE 1  Finding the Values of an Inverse Sine Function**

Find the exact value of:

**(a)** $\sin^{-1} \dfrac{\sqrt{3}}{2}$    **(b)** $\sin^{-1}\left(-\dfrac{\sqrt{2}}{2}\right)$    **(c)** $\sin^{-1}\left(\sin \dfrac{\pi}{8}\right)$    **(d)** $\sin^{-1}\left(\sin \dfrac{5\pi}{8}\right)$

**Solution  (a)** We seek an angle whose sine is $\dfrac{\sqrt{3}}{2}$. Let $y = \sin^{-1} \dfrac{\sqrt{3}}{2}$. Then by definition, $\sin y = \dfrac{\sqrt{3}}{2}$, where $-\dfrac{\pi}{2} \le y \le \dfrac{\pi}{2}$. Although $\sin y = \dfrac{\sqrt{3}}{2}$ has infinitely many solutions, the only number in the interval $\left[-\dfrac{\pi}{2}, \dfrac{\pi}{2}\right]$ for which $\sin y = \dfrac{\sqrt{3}}{2}$, is $\dfrac{\pi}{3}$.

So $\sin^{-1} \dfrac{\sqrt{3}}{2} = \dfrac{\pi}{3}$.

**(b)** We seek an angle whose sine is $-\dfrac{\sqrt{2}}{2}$. Let $y = \sin^{-1}\left(-\dfrac{\sqrt{2}}{2}\right)$. Then by definition, $\sin y = -\dfrac{\sqrt{2}}{2}$, where $-\dfrac{\pi}{2} \le y \le \dfrac{\pi}{2}$. The only number in the interval $\left[-\dfrac{\pi}{2}, \dfrac{\pi}{2}\right]$ whose sine is $-\dfrac{\sqrt{2}}{2}$, is $-\dfrac{\pi}{4}$. So $\sin^{-1}\left(-\dfrac{\sqrt{2}}{2}\right) = -\dfrac{\pi}{4}$.

**(c)** Since the number $\dfrac{\pi}{8}$ is in the interval $\left[-\dfrac{\pi}{2}, \dfrac{\pi}{2}\right]$, we use (1).

$$\sin^{-1}\left(\sin \frac{\pi}{8}\right) = \frac{\pi}{8}$$

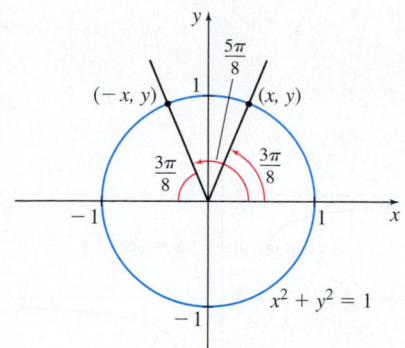

**Figure 87**

**(d)** Since the number $\dfrac{5\pi}{8}$ is not in the closed interval $\left[-\dfrac{\pi}{2}, \dfrac{\pi}{2}\right]$, we cannot use (1).

Instead, we find a number $x$ in the interval $\left[-\dfrac{\pi}{2}, \dfrac{\pi}{2}\right]$ for which $\sin x = \sin \dfrac{5\pi}{8}$.

Using Figure 87, we see that $\sin \dfrac{5\pi}{8} = y = \sin \dfrac{3\pi}{8}$. The number $\dfrac{3\pi}{8}$ is in the

interval $\left[-\dfrac{\pi}{2}, \dfrac{\pi}{2}\right]$, so

$$\sin^{-1}\left(\sin \frac{5\pi}{8}\right) = \sin^{-1}\left(\sin \frac{3\pi}{8}\right) = \frac{3\pi}{8} \qquad \blacksquare$$

NOW WORK Problems **9** and **21**.

### The Inverse Tangent Function

The tangent function $y = \tan x$ is not one-to-one. To define the *inverse tangent function*, we restrict the domain of $y = \tan x$ to the open interval $\left(-\dfrac{\pi}{2}, \dfrac{\pi}{2}\right)$. On this interval, $y = \tan x$ is one-to-one. See Figure 88.

Now use the steps for finding the inverse of a one-to-one function, listed on page 39:

***Step 1*** $y = \tan x,\ -\dfrac{\pi}{2} < x < \dfrac{\pi}{2}$.

***Step 2*** Interchange $x$ and $y$ to obtain $x = \tan y,\ -\dfrac{\pi}{2} < y < \dfrac{\pi}{2}$, the implicit form of the inverse tangent function.

***Step 3*** The explicit form is called the *inverse tangent* of $x$ and is denoted by $y = \tan^{-1} x$.

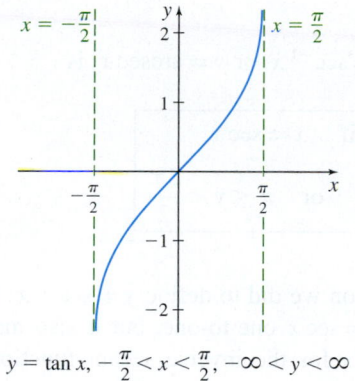

$y = \tan x,\ -\dfrac{\pi}{2} < x < \dfrac{\pi}{2},\ -\infty < y < \infty$

**Figure 88**

> **DEFINITION** Inverse Tangent Function
>
> The **inverse tangent function**, symbolized by $y = \tan^{-1} x$, or $y = \arctan x$, is
>
> | $y = \tan^{-1} x$ | if and only if | $x = \tan y$ |
> |---|---|---|
> | where $-\infty < x < \infty$ | and | $-\dfrac{\pi}{2} < y < \dfrac{\pi}{2}$ |

The domain of the function $y = \tan^{-1} x$ is the set of all real numbers, and its range is $-\dfrac{\pi}{2} < y < \dfrac{\pi}{2}$. To graph $y = \tan^{-1} x$, reflect the graph of $y = \tan x,\ -\dfrac{\pi}{2} < x < \dfrac{\pi}{2}$, about the line $y = x$, as shown in Figure 89.

NOW WORK Problem **13**.

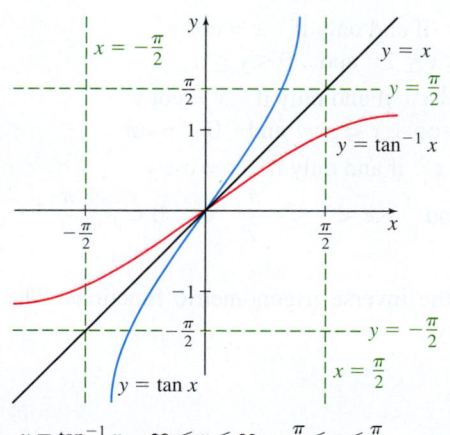

$y = \tan^{-1} x,\ -\infty < x < \infty,\ -\dfrac{\pi}{2} < y < \dfrac{\pi}{2}$

**Figure 89**

### The Inverse Secant Function

To define the *inverse secant function*, we restrict the domain of the function $y = \sec x$ to the set $\left[0, \dfrac{\pi}{2}\right) \cup \left[\pi, \dfrac{3\pi}{2}\right)$. See Figure 90(a) on page 64. An equation for the inverse function is obtained by following the steps for finding the inverse of a one-to-one function. By interchanging $x$ and $y$, we obtain the implicit form of the inverse function: $x = \sec y$, where $0 \le y < \dfrac{\pi}{2}$ or $\pi \le y < \dfrac{3\pi}{2}$. The explicit form is called the *inverse secant* of $x$ and is written $y = \sec^{-1} x$. The graph of $y = \sec^{-1} x$ is given in Figure 90(b).

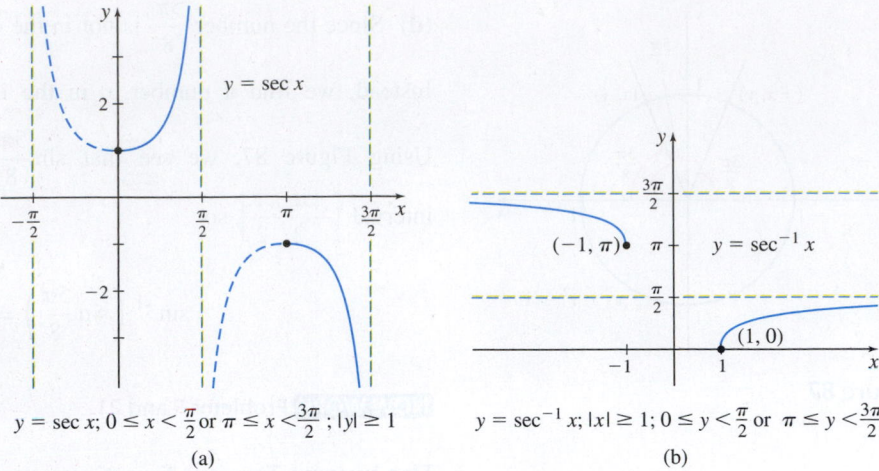

$y = \sec x;\ 0 \le x < \dfrac{\pi}{2}\text{ or }\pi \le x < \dfrac{3\pi}{2};\ |y| \ge 1$

(a)

$y = \sec^{-1} x;\ |x| \ge 1;\ 0 \le y < \dfrac{\pi}{2}\text{ or }\pi \le y < \dfrac{3\pi}{2}$

(b)

**Figure 90**

---

**DEFINITION** Inverse Secant Function

The **inverse secant function**, symbolized by $y = \sec^{-1} x$, or $y = \operatorname{arcsec} x$, is

$$y = \sec^{-1} x \quad \text{if and only if} \quad x = \sec y$$

$$\text{where } |x| \ge 1 \quad \text{and} \quad 0 \le y < \frac{\pi}{2} \quad \text{or} \quad \pi \le y < \frac{3\pi}{2}$$

You may wonder why we chose the restriction we did to define $y = \sec^{-1} x$. The reason is that this restriction not only makes $y = \sec x$ one-to-one, but it also makes $\tan x \ge 0$. With $\tan x \ge 0$, the derivative formula for the inverse secant function is simpler, as we will see in Chapter 3.

The remaining three inverse trigonometric functions are defined as follows.

---

**DEFINITION**

- **Inverse Cosine Function**  $y = \cos^{-1} x$  if and only if  $x = \cos y$
  where $-1 \le x \le 1$  and  $0 \le y \le \pi$
- **Inverse Cotangent Function**  $y = \cot^{-1} x$  if and only if  $x = \cot y$
  where $-\infty < x < \infty$  and  $0 < y < \pi$
- **Inverse Cosecant Function**  $y = \csc^{-1} x$  if and only if  $x = \csc y$
  where $|x| \ge 1$  and  $-\pi < y \le -\dfrac{\pi}{2}$  or  $0 < y \le \dfrac{\pi}{2}$

---

The following three identities involve the inverse trigonometric functions. The proofs may be found in books on trigonometry.

---

**THEOREM**

- $\cos^{-1} x = \dfrac{\pi}{2} - \sin^{-1} x,\quad -1 \le x \le 1$
- $\cot^{-1} x = \dfrac{\pi}{2} - \tan^{-1} x,\quad -\infty < x < \infty$
- $\csc^{-1} x = \dfrac{\pi}{2} - \sec^{-1} x,\quad |x| \ge 1$

---

The identities above are used whenever the inverse cosine, inverse cotangent, and inverse cosecant functions are needed. The graphs of the inverse cosine, inverse cotangent, and inverse cosecant functions can also be obtained by using these identities to transform the graphs of $y = \sin^{-1} x$, $y = \tan^{-1} x$, and $y = \sec^{-1} x$, respectively.

## 2 Use the Inverse Trigonometric Functions

 **EXAMPLE 2** **Writing a Trigonometric Expression as an Algebraic Expression**

Write $\sin(\tan^{-1} u)$ as an algebraic expression containing $u$.

**Solution** Let $\theta = \tan^{-1} u$ so that $\tan \theta = u$, where $-\dfrac{\pi}{2} < \theta < \dfrac{\pi}{2}$. We note that in the interval $\left(-\dfrac{\pi}{2}, \dfrac{\pi}{2}\right)$, $\sec \theta > 0$. Then

$$\sin(\tan^{-1} u) = \sin \theta = \sin \theta \cdot \frac{\cos \theta}{\cos \theta} = \tan \theta \cos \theta$$

$$= \frac{\tan \theta}{\sec \theta} \underset{\substack{\uparrow \\ \sec^2 \theta = 1 + \tan^2 \theta;\ \sec \theta > 0}}{=} \frac{\tan \theta}{\sqrt{1 + \tan^2 \theta}} = \frac{u}{\sqrt{1 + u^2}}$$

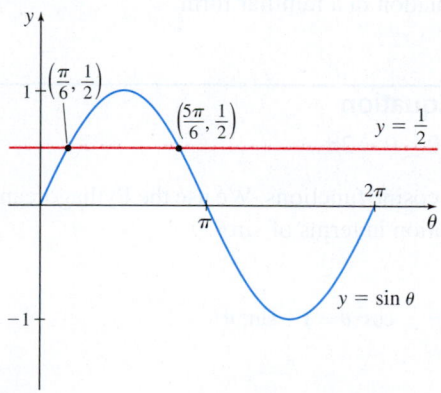

**Figure 91** $\tan \theta = u;\ -\dfrac{\pi}{2} < \theta < \dfrac{\pi}{2}$

An alternate method of obtaining the solution to Example 2 uses right triangles. Let $\theta = \tan^{-1} u$ so that $\tan \theta = u$, $-\dfrac{\pi}{2} < \theta < \dfrac{\pi}{2}$, and label the right triangles drawn in Figure 91. Using the Pythagorean Theorem, the hypotenuse of each triangle is $\sqrt{1 + u^2}$. Then $\sin(\tan^{-1} u) = \sin \theta = \dfrac{u}{\sqrt{1 + u^2}}$. ∎

**NOW WORK** Problem 67.

## 3 Solve Trigonometric Equations

Inverse trigonometric functions can be used to solve trigonometric equations.

**EXAMPLE 3** **Solving Trigonometric Equations**

Solve the equations:

**(a)** $\sin \theta = \dfrac{1}{2}$      **(b)** $\cos \theta = 0.4$

Give a general formula for all the solutions and list all the solutions in the interval $[-2\pi, 2\pi]$.

**Solution** **(a)** Use the inverse sine function $y = \sin^{-1} x$, $-\dfrac{\pi}{2} \le y \le \dfrac{\pi}{2}$.

$$\sin \theta = \frac{1}{2}$$

$$\theta = \sin^{-1} \frac{1}{2} \qquad -\frac{\pi}{2} \le \theta \le \frac{\pi}{2}$$

$$\theta = \frac{\pi}{6}$$

**Figure 92**

Over the interval $[0, 2\pi]$, one period, there are two angles $\theta$ for which $\sin \theta = \dfrac{1}{2}$: $\dfrac{\pi}{6}$ and $\dfrac{5\pi}{6}$. See Figure 92. Using the period $2\pi$, the solutions of $\sin \theta = \dfrac{1}{2}$ are given by the general formula

$$\theta = \frac{\pi}{6} + 2k\pi \quad \text{or} \quad \theta = \frac{5\pi}{6} + 2k\pi \qquad \text{where } k \text{ is any integer}$$

The solutions in the interval $[-2\pi, 2\pi]$ are

$$\left\{ -\frac{11\pi}{6}, -\frac{7\pi}{6}, \frac{\pi}{6}, \frac{5\pi}{6} \right\}$$

**(b)** A calculator must be used to solve $\cos \theta = 0.4$. With your calculator in radian mode,

$$\theta = \cos^{-1}(0.4) \approx 1.159279 \qquad 0 \le \theta \le \pi$$

Rounded to three decimal places, $\theta = \cos^{-1} 0.4 = 1.159$ radians. But there is another angle $\theta$ in the interval $[0, 2\pi]$ for which $\cos \theta = 0.4$, namely, $\theta \approx 2\pi - 1.159 \approx 5.124$ radians.

Because the cosine function has period $2\pi$, all the solutions of $\cos \theta = 0.4$ are given by the general formulas

$$\theta \approx 1.159 + 2k\pi \quad \text{or} \quad \theta \approx 5.124 + 2k\pi \qquad \text{where } k \text{ is any integer}$$

The solutions in the interval $[-2\pi, 2\pi]$ are $\{-5.124, -1.159, 1.159, 5.124\}$. ∎

**NOW WORK** Problem 63.

---

**EXAMPLE 4** Solving a Trigonometric Equation

Solve the equation $\sin(2\theta) = \dfrac{1}{2}$, where $0 \le \theta < 2\pi$.

**Solution** In the interval $[0, 2\pi)$, the sine function has the value $\dfrac{1}{2}$ at $\theta = \dfrac{\pi}{6}$ and at $\theta = \dfrac{5\pi}{6}$, as shown in Figure 93. Since the period of the sine function is $2\pi$ and the argument in the equation $\sin(2\theta) = \dfrac{1}{2}$ is $2\theta$, we write the general formula for all the solutions.

$$2\theta = \frac{\pi}{6} + 2k\pi \quad \text{or} \quad 2\theta = \frac{5\pi}{6} + 2k\pi \qquad \text{where } k \text{ is any integer}$$

$$\theta = \frac{\pi}{12} + k\pi \qquad\qquad \theta = \frac{5\pi}{12} + k\pi$$

The solutions of $\sin(2\theta) = \dfrac{1}{2}$, $0 \le \theta < 2\pi$, are $\left\{ \dfrac{\pi}{12}, \dfrac{5\pi}{12}, \dfrac{13\pi}{12}, \dfrac{17\pi}{12} \right\}$. ∎

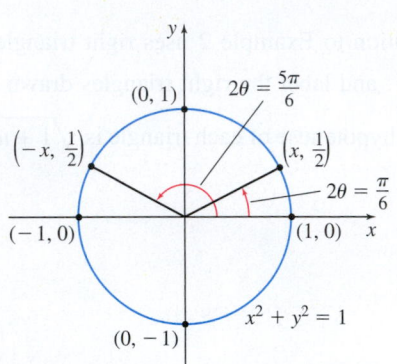

**Figure 93**

**NEED TO REVIEW?** Trigonometric identities are discussed in Appendix Section A.4, pp. A-32 to A-35.

**NOW WORK** Problem 49.

Many trigonometric equations can be solved by applying algebra techniques such as factoring or using the quadratic formula. It is often necessary to begin by using a trigonometric identity to express the given equation in a familiar form.

---

**EXAMPLE 5** Solving a Trigonometric Equation

Solve the equation $3\sin\theta - \cos^2\theta = 3$, where $0 \le \theta < 2\pi$.

**Solution** The equation involves both sine and cosine functions. We use the Pythagorean identity $\sin^2\theta + \cos^2\theta = 1$ to rewrite the equation in terms of $\sin\theta$.

$$3\sin\theta - \cos^2\theta = 3$$
$$3\sin\theta - (1 - \sin^2\theta) = 3 \qquad \color{blue}{\cos^2\theta = 1 - \sin^2\theta}$$
$$\sin^2\theta + 3\sin\theta - 4 = 0$$

This is a quadratic equation in $\sin\theta$. Factor the left side and solve for $\sin\theta$.

$$(\sin\theta + 4)(\sin\theta - 1) = 0$$
$$\sin\theta + 4 = 0 \qquad \text{or} \qquad \sin\theta - 1 = 0$$
$$\sin\theta = -4 \qquad \text{or} \qquad \sin\theta = 1$$

The range of the sine function is $-1 \le y \le 1$, so $\sin\theta = -4$ has no solution. Solving $\sin\theta = 1$, we obtain

$$\theta = \sin^{-1} 1 = \frac{\pi}{2}$$

The only solution in the interval $[0, 2\pi)$ is $\dfrac{\pi}{2}$. ∎

**NOW WORK** Problem 53.

# P.7 Assess Your Understanding

## Concepts and Vocabulary

1. $y = \sin^{-1} x$ if and only if $x =$ _____, where $-1 \le x \le 1$ and $-\dfrac{\pi}{2} \le y \le \dfrac{\pi}{2}$.

2. *True or False*   $\sin^{-1}(\sin x) = x$, where $-1 \le x \le 1$.

3. *True or False*   The domain of $y = \sin^{-1} x$ is $-\dfrac{\pi}{2} \le x \le \dfrac{\pi}{2}$.

4. *True or False*   $\sin(\sin^{-1} 0) = 0$.

5. *True or False*   $y = \tan^{-1} x$ means $x = \tan y$, where $-\infty < x < \infty$ and $-\dfrac{\pi}{2} \le y \le \dfrac{\pi}{2}$.

6. *True or False*   The domain of the inverse tangent function is the set of all real numbers.

7. *True or False*   $\sec^{-1} 0.5$ is not defined.

8. *True or False*   Trigonometric equations can have multiple solutions.

## Practice Problems

*In Problems 9–20, find the exact value of each expression. Do not use a calculator.*

9. $\sin^{-1} 0$

10. $\tan^{-1} 1$

11. $\sec^{-1}(-1)$

12. $\sec^{-1} \dfrac{2\sqrt{3}}{3}$

13. $\tan^{-1}(-1)$

14. $\tan^{-1} 0$

15. $\sin^{-1} \dfrac{\sqrt{2}}{2}$

16. $\tan^{-1} \dfrac{\sqrt{3}}{3}$

17. $\tan^{-1} \sqrt{3}$

18. $\sin^{-1}\left(-\dfrac{\sqrt{3}}{2}\right)$

19. $\sec^{-1}(-\sqrt{2})$

20. $\sin^{-1}\left(-\dfrac{\sqrt{2}}{2}\right)$

*In Problems 21–28, find the exact value of each expression. Do not use a calculator.*

21. $\sin^{-1}\left(\sin \dfrac{\pi}{5}\right)$

22. $\sin^{-1}\left[\sin\left(-\dfrac{\pi}{10}\right)\right]$

23. $\tan^{-1}\left[\tan\left(-\dfrac{3\pi}{8}\right)\right]$

24. $\sin^{-1}\left(\sin \dfrac{3\pi}{7}\right)$

25. $\sin^{-1}\left(\sin \dfrac{9\pi}{8}\right)$

26. $\sin^{-1}\left[\sin\left(-\dfrac{5\pi}{3}\right)\right]$

27. $\tan^{-1}\left(\tan \dfrac{4\pi}{5}\right)$

28. $\tan^{-1}\left[\tan\left(-\dfrac{2\pi}{3}\right)\right]$

*In Problems 29–44, find the exact value, if any, of each composite function. If there is no value, write, "It is not defined." Do not use a calculator.*

29. $\sin\left(\sin^{-1} \dfrac{1}{4}\right)$

30. $\sin\left[\sin^{-1}\left(-\dfrac{2}{3}\right)\right]$

31. $\tan(\tan^{-1} 4)$

32. $\tan[\tan^{-1}(-2)]$

33. $\sin(\sin^{-1} 1.2)$

34. $\sin[\sin^{-1}(-2)]$

35. $\tan(\tan^{-1} \pi)$

36. $\sin[\sin^{-1}(-1.5)]$

37. $\cos(\sin^{-1} 0.3)$

38. $\sin(\tan^{-1} 2)$

39. $\tan(\sec^{-1} 2)$

40. $\cos(\tan^{-1} 3)$

41. $\sin(\sin^{-1} 0.2 + \tan^{-1} 2)$

42. $\cos(\sec^{-1} 2 + \sin^{-1} 0.1)$

43. $\tan(2 \sin^{-1} 0.4)$

44. $\cos(2 \tan^{-1} 5)$

*In Problems 45–62, solve each equation on the interval $0 \le \theta < 2\pi$.*

45. $\tan \theta = -\dfrac{\sqrt{3}}{3}$

46. $\sec \dfrac{3\theta}{2} = -2$

47. $2 \sin \theta + 3 = 2$

48. $1 - \cos \theta = \dfrac{1}{2}$

49. $\sin(3\theta) = -1$

50. $\cos(2\theta) = \dfrac{1}{2}$

51. $4 \cos^2 \theta = 1$

52. $2 \sin^2 \theta - 1 = 0$

53. $2 \sin^2 \theta - 5 \sin \theta + 3 = 0$

54. $2 \cos^2 \theta - 7 \cos \theta - 4 = 0$

55. $1 + \sin \theta = 2 \cos^2 \theta$

56. $\sec^2 \theta + \tan \theta = 0$

57. $\sin \theta + \cos \theta = 0$

58. $\tan \theta = \cot \theta$

59. $\cos(2\theta) + 6 \sin^2 \theta = 4$

60. $\cos(2\theta) = \cos \theta$

61. $\sin(2\theta) + \sin(4\theta) = 0$

62. $\cos(4\theta) - \cos(6\theta) = 0$

*In Problems 63–66, use a calculator to solve each equation on the interval $0 \le \theta < 2\pi$. Round answers to three decimal places.*

63. $\tan \theta = 5$

64. $\cos \theta = 0.6$

65. $2 + 3 \sin \theta = 0$

66. $4 + \sec \theta = 0$

67. Write $\cos(\sin^{-1} u)$ as an algebraic expression containing $u$, where $|u| \le 1$.

68. Write $\tan(\sin^{-1} u)$ as an algebraic expression containing $u$, where $|u| \le 1$.

69. Show that $y = \sin^{-1} x$ is an odd function. That is, show $\sin^{-1}(-x) = -\sin^{-1} x$.

70. Show that $y = \tan^{-1} x$ is an odd function. That is, show $\tan^{-1}(-x) = -\tan^{-1} x$.

71. (a) On the same set of axes, graph $f(x) = 3 \sin(2x) + 2$ and $g(x) = \dfrac{7}{2}$ on the interval $[0, \pi]$.

    (b) Solve $f(x) = g(x)$ on the interval $[0, \pi]$ and label the points of intersection on the graph drawn in (a).

    (c) Shade the region bounded by $f(x) = 3 \sin(2x) + 2$ and $g(x) = \dfrac{7}{2}$ between the points found in (b) on the graph drawn in (a).

    (d) Solve $f(x) > g(x)$ on the interval $[0, \pi]$.

72. (a) On the same set of axes, graph $f(x) = -4 \cos x$ and $g(x) = 2 \cos x + 3$ on the interval $[0, 2\pi]$.

    (b) Solve $f(x) = g(x)$ on the interval $[0, 2\pi]$ and label the points of intersection on the graph drawn in (a).

---

**1.** = NOW WORK problem       📐 = Graphing technology recommended       [CAS] = Computer Algebra System recommended

(c) Shade the region bounded by $f(x) = -4\cos x$ and $g(x) = 2\cos x + 3$ between the points found in (b) on the graph drawn in (a).

(d) Solve $f(x) > g(x)$ on the interval $[0, 2\pi]$.

73. **Area Under a Graph** The area $A$ under the graph of $y = \dfrac{1}{1 + x^2}$ and above the $x$-axis between $x = a$ and $x = b$ is given by

$$A = \tan^{-1} b - \tan^{-1} a$$

See the figure.

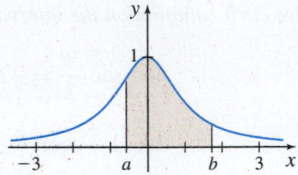

(a) Find the exact area under the graph of $y = \dfrac{1}{1 + x^2}$ and above the $x$-axis between $x = 0$ and $x = \sqrt{3}$.

(b) Find the exact area under the graph of $y = \dfrac{1}{1 + x^2}$ and above the $x$-axis between $x = -\dfrac{\sqrt{3}}{3}$ and $x = 1$.

74. **Area Under a Graph** The area $A$ under the graph of $y = \dfrac{1}{\sqrt{1 - x^2}}$ and above the $x$-axis between $x = a$ and $x = b$ is given by

$$A = \sin^{-1} b - \sin^{-1} a$$

See the figure.

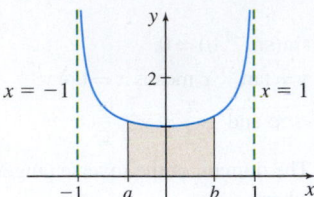

(a) Find the exact area under the graph of $y = \dfrac{1}{\sqrt{1 - x^2}}$ and above the $x$-axis between $x = 0$ and $x = \dfrac{\sqrt{3}}{2}$.

(b) Find the exact area under the graph of $y = \dfrac{1}{\sqrt{1 - x^2}}$ and above the $x$-axis between $x = -\dfrac{1}{2}$ and $x = \dfrac{1}{2}$.

# P.8 Sequences; Summation Notation; the Binomial Theorem

**OBJECTIVES** *When you finish this section, you should be able to:*

1 Write the first several terms of a sequence (p. 69)
2 Write the terms of a recursively defined sequence (p. 70)
3 Use summation notation (p. 71)
4 Find the sum of the first $n$ terms of a sequence (p. 72)
5 Use the Binomial Theorem (p. 73)

When you hear the word *sequence* as used in "a sequence of events," you probably think of something that happens first, then second, and so on. In mathematics, the word *sequence* also refers to outcomes that are first, second, and so on.

**DEFINITION** Sequence

A **sequence** is a function whose domain is the set of positive integers.

In a sequence the inputs are $1, 2, 3, \ldots$. Because a sequence is a function, it has a graph. Figure 94(a) shows the graph of the function $f(x) = \dfrac{1}{x}, x > 0$. If all the points on this graph are removed except those whose $x$-coordinates are positive integers, that is, if all points are removed except $(1, 1)$, $\left(2, \dfrac{1}{2}\right)$, $\left(3, \dfrac{1}{3}\right)$, and so on, the remaining points would be the graph of the sequence $f(n) = \dfrac{1}{n}$, as shown in Figure 94(b). Notice that we use $n$ to represent the independent variable in a sequence. This is to remind us that $n$ is a positive integer.

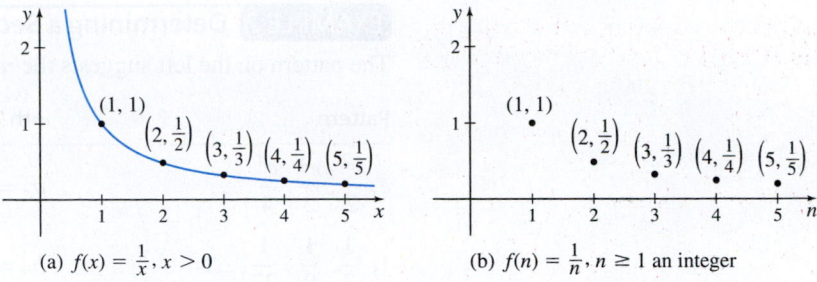

(a) $f(x) = \frac{1}{x}, x > 0$          (b) $f(n) = \frac{1}{n}, n \geq 1$ an integer

**Figure 94**

## 1  Write the First Several Terms of a Sequence

A sequence is usually represented by listing its values in order. For example, the sequence whose graph is given in Figure 94(b) might be represented as

$$f(1), f(2), f(3), f(4), \ldots \quad \text{or} \quad 1, \frac{1}{2}, \frac{1}{3}, \frac{1}{4}, \ldots$$

The list never ends, as the dots (**ellipses**) indicate. The numbers in this ordered list are called the **terms** of the sequence.

In dealing with sequences, we usually use subscripted letters such as $a_1$ to represent the first term, $a_2$ for the second term, $a_3$ for the third term, and so on.

For the sequence $f(n) = \frac{1}{n}$, we write

$$\underbrace{a_1 = f(1) = 1}_{\text{First term}} \quad \underbrace{a_2 = f(2) = \frac{1}{2}}_{\text{Second term}} \quad \underbrace{a_3 = f(3) = \frac{1}{3}}_{\text{Third term}} \quad \underbrace{a_4 = f(4) = \frac{1}{4}}_{\text{Fourth term}} \cdots \quad \underbrace{a_n = f(n) = \frac{1}{n}}_{n\text{th term}} \cdots$$

In other words, we usually do not use the traditional function notation $f(n)$ for sequences. For this particular sequence, we have a rule for the $n$th term, which is $a_n = \frac{1}{n}$, so it is easy to find any term of the sequence.

When a formula for the $n$th term (sometimes called the **general term**) of a sequence is known, we usually represent the entire sequence by placing braces around the formula for the $n$th term. For example, the sequence whose $n$th term is $b_n = \left(\frac{1}{2}\right)^n$ can be represented either as

$$b_1 = \frac{1}{2}, \quad b_2 = \frac{1}{4}, \quad b_3 = \frac{1}{8}, \quad \ldots, \quad b_n = \frac{1}{2^n}, \quad \ldots,$$

or as

$$\{b_n\} = \left\{\left(\frac{1}{2}\right)^n\right\}$$

**EXAMPLE 1**   **Writing the Terms of a Sequence**

Write the first six terms of the sequence: $\{S_n\} = \left\{(-1)^{n-1} \dfrac{2}{n}\right\}$

**Solution** $S_1 = (-1)^0 \dfrac{2}{1} = 2$          $S_2 = (-1)^1 \dfrac{2}{2} = -1$          $S_3 = (-1)^2 \dfrac{2}{3} = \dfrac{2}{3}$

$S_4 = -\dfrac{1}{2}$          $S_5 = \dfrac{2}{5}$          $S_6 = -\dfrac{1}{3}$

The first six terms of the sequence are $2, -1, \dfrac{2}{3}, -\dfrac{1}{2}, \dfrac{2}{5}, -\dfrac{1}{3}$. ∎

**NOW WORK** Problem **17**.

EXAMPLE 2 **Determining a Sequence from a Pattern**

The pattern on the left suggests the *n*th term of the sequence on the right.

| Pattern | *n*th Term | Sequence |
|---|---|---|
| $e, \dfrac{e^2}{2}, \dfrac{e^3}{3}, \dfrac{e^4}{4}, \ldots$ | $a_n = \dfrac{e^n}{n}$ | $\{a_n\} = \left\{\dfrac{e^n}{n}\right\}$ |
| $1, \dfrac{1}{3}, \dfrac{1}{9}, \dfrac{1}{27}, \ldots$ | $b_n = \dfrac{1}{3^{n-1}}$ | $\{b_n\} = \left\{\dfrac{1}{3^{n-1}}\right\}$ |
| $1, 3, 5, 7, \ldots$ | $c_n = 2n - 1$ | $\{c_n\} = \{2n - 1\}$ |
| $1, 4, 9, 16, 25, \ldots$ | $d_n = n^2$ | $\{d_n\} = \{n^2\}$ |
| $1, -\dfrac{1}{2}, \dfrac{1}{3}, -\dfrac{1}{4}, \dfrac{1}{5}, \ldots$ | $e_n = (-1)^{n+1}\left(\dfrac{1}{n}\right)$ | $\{e_n\} = \left\{(-1)^{n+1}\left(\dfrac{1}{n}\right)\right\}$ |

■

NOW WORK Problem 21.

## 2 Write the Terms of a Recursively Defined Sequence

A second way of defining a sequence is to assign a value to the first (or the first few) term(s) and specify the *n*th term by a formula or an equation that involves one or more of the terms preceding it. Such a sequence is defined **recursively**, and the rule or formula is called a **recursive formula**.

EXAMPLE 3 **Writing the Terms of a Recursively Defined Sequence**

Write the first five terms of the recursively defined sequence.

$$s_1 = 1 \qquad s_n = 4s_{n-1} \qquad n \geq 2$$

**Solution** The first term is given as $s_1 = 1$. To get the second term, we use $n = 2$ in the formula to obtain $s_2 = 4s_1 = 4 \cdot 1 = 4$. To obtain the third term, we use $n = 3$ in the formula, getting $s_3 = 4s_2 = 4 \cdot 4 = 16$. Each new term requires that we know the value of the preceding term. The first five terms are

$$s_1 = 1$$
$$s_2 = 4s_{2-1} = 4s_1 = 4 \cdot 1 = 4$$
$$s_3 = 4s_{3-1} = 4s_2 = 4 \cdot 4 = 16$$
$$s_4 = 4s_{4-1} = 4s_3 = 4 \cdot 16 = 64$$
$$s_5 = 4s_{5-1} = 4s_4 = 4 \cdot 64 = 256$$

■

NOW WORK Problem 29.

EXAMPLE 4 **Writing the Terms of a Recursively Defined Sequence**

Write the first five terms of the recursively defined sequence.

$$u_1 = 1 \qquad u_2 = 1 \qquad u_n = u_{n-2} + u_{n-1} \qquad n \geq 3$$

**Solution** We are given the first two terms. Obtaining the third term requires that we know each of the previous two terms.

$$u_1 = 1$$
$$u_2 = 1$$
$$u_3 = u_1 + u_2 = 1 + 1 = 2$$
$$u_4 = u_2 + u_3 = 1 + 2 = 3$$
$$u_5 = u_3 + u_4 = 2 + 3 = 5$$

■

The sequence defined in Example 4 is called a **Fibonacci sequence**; its terms are called **Fibonacci numbers**. These numbers appear in a wide variety of applications.

**NOW WORK** Problem 35.

## 3 Use Summation Notation

It is often important to be able to find the sum of the first $n$ terms of a sequence $\{a_n\}$, namely,

$$a_1 + a_2 + a_3 + \cdots + a_n$$

**NOTE** Summation notation is called **sigma notation** in some books.

Rather than write down all these terms, we introduce **summation notation**. Using summation notation, the sum is written

$$a_1 + a_2 + a_3 + \cdots + a_n = \sum_{k=1}^{n} a_k$$

The symbol $\sum$ (the uppercase Greek letter sigma, which is an $S$ in our alphabet) is simply an instruction to sum, or add up, the terms. The integer $k$ is called the **index of summation**; it tells us where to start the sum ($k = 1$) and where to end it ($k = n$). The expression $\sum_{k=1}^{n} a_k$ is read, "The sum of the terms $a_k$ from $k = 1$ to $k = n$."

**EXAMPLE 5    Expanding Summation Notation**

Write out each sum:

**NOTE** If $n \geq 0$ is an integer, the **factorial symbol** ! means $0! = 1$, $1! = 1$, and $n! = 1 \cdot 2 \cdot 3 \cdot \cdots \cdot (n-1) \cdot n$, where $n > 1$.

**(a)** $\displaystyle\sum_{k=1}^{n} \frac{1}{k}$        **(b)** $\displaystyle\sum_{k=1}^{n} k!$

**Solution (a)** $\displaystyle\sum_{k=1}^{n} \frac{1}{k} = 1 + \frac{1}{2} + \frac{1}{3} + \cdots + \frac{1}{n}$    **(b)** $\displaystyle\sum_{k=1}^{n} k! = 1! + 2! + \cdots + n!$  ∎

**NOW WORK** Problem 41.

**EXAMPLE 6    Writing a Sum in Summation Notation**

Express each sum using summation notation:

**(a)** $1^2 + 2^2 + 3^2 + \cdots + 9^2$        **(b)** $1 + \frac{1}{2} + \frac{1}{4} + \frac{1}{8} + \cdots + \frac{1}{2^{n-1}}$

**Solution (a)** The sum $1^2 + 2^2 + 3^2 + \cdots + 9^2$ has 9 terms, each of the form $k^2$; it starts at $k = 1$ and ends at $k = 9$:

$$1^2 + 2^2 + 3^2 + \cdots + 9^2 = \sum_{k=1}^{9} k^2$$

**(b)** The sum

$$1 + \frac{1}{2} + \frac{1}{4} + \frac{1}{8} + \cdots + \frac{1}{2^{n-1}}$$

has $n$ terms, each of the form $\dfrac{1}{2^{k-1}}$. It starts at $k = 1$ and ends at $k = n$.

$$1 + \frac{1}{2} + \frac{1}{4} + \frac{1}{8} + \cdots + \frac{1}{2^{n-1}} = \sum_{k=1}^{n} \frac{1}{2^{k-1}}$$  ∎

The index of summation does not need to begin at 1 or end at $n$. For example, the sum in Example 6(b) could be expressed as

$$\sum_{k=0}^{n-1} \frac{1}{2^k} = 1 + \frac{1}{2} + \frac{1}{4} + \frac{1}{8} + \cdots + \frac{1}{2^{n-1}}$$

Letters other than $k$ can be used as the index. For example,

$$\sum_{j=1}^{n} j! \qquad \text{and} \qquad \sum_{i=1}^{n} i!$$

each represent the same sum as Example 5(b).

**NOW WORK** Problem 51.

## 4 Find the Sum of the First $n$ Terms of a Sequence

When working with summation notation, the following properties are useful for finding the sum of the first $n$ terms of a sequence.

**THEOREM** Properties Involving Summation Notation

If $\{a_n\}$ and $\{b_n\}$ are two sequences and $c$ is a real number, then:

$$\bullet \quad \sum_{k=1}^{n} (ca_k) = c\sum_{k=1}^{n} a_k \tag{1}$$

$$\bullet \quad \sum_{k=1}^{n} (a_k + b_k) = \sum_{k=1}^{n} a_k + \sum_{k=1}^{n} b_k \tag{2}$$

$$\bullet \quad \sum_{k=1}^{n} (a_k - b_k) = \sum_{k=1}^{n} a_k - \sum_{k=1}^{n} b_k \tag{3}$$

$$\bullet \quad \sum_{k=j+1}^{n} a_k = \sum_{k=1}^{n} a_k - \sum_{k=1}^{j} a_k \quad \text{where } 0 < j < n \tag{4}$$

The proof of property (1) follows from the distributive property of real numbers. The proofs of properties (2) and (3) are based on the commutative and associative properties of real numbers. Property (4) states that the sum from $j + 1$ to $n$ equals the sum from 1 to $n$ minus the sum from 1 to $j$. It can be helpful to use this property when the index of summation begins at a number larger than 1.

**THEOREM** Formulas for Sums of the First $n$ Terms of a Sequence

$$\bullet \quad \sum_{k=1}^{n} 1 = \underbrace{1 + 1 + 1 + \cdots + 1}_{n \text{ terms}} = n \tag{5}$$

$$\bullet \quad \sum_{k=1}^{n} k = 1 + 2 + 3 + \cdots + n = \frac{n(n+1)}{2} \tag{6}$$

$$\bullet \quad \sum_{k=1}^{n} k^2 = 1^2 + 2^2 + 3^2 + \cdots + n^2 = \frac{n(n+1)(2n+1)}{6} \tag{7}$$

$$\bullet \quad \sum_{k=1}^{n} k^3 = 1^3 + 2^3 + 3^3 + \cdots + n^3 = \left[\frac{n(n+1)}{2}\right]^2 \tag{8}$$

Notice the difference between formulas (5) and (6). In formula (5), the constant 1 is being summed from 1 to $n$, while in (6), the index of summation $k$ is being summed from 1 to $n$.

**EXAMPLE 7** **Finding Sums**

Find each sum:

(a) $\displaystyle\sum_{k=1}^{100} \frac{1}{2}$   (b) $\displaystyle\sum_{k=1}^{5}(3k)$   (c) $\displaystyle\sum_{k=1}^{10}(k^3+1)$   (d) $\displaystyle\sum_{k=6}^{20}(4k^2)$

**Solution** (a) $\displaystyle\sum_{k=1}^{100} \frac{1}{2} = \underset{(1)}{\frac{1}{2}} \sum_{k=1}^{100} \underset{(5)}{1} = \frac{1}{2} \cdot 100 = 50$

(b) $\displaystyle\sum_{k=1}^{5}(3k) = \underset{(1)}{3} \sum_{k=1}^{5} \underset{(6)}{k} = 3\left(\frac{5(5+1)}{2}\right) = 3 \cdot 15 = 45$

(c) $\displaystyle\sum_{k=1}^{10}(k^3+1) = \underset{(2)}{\sum_{k=1}^{10}k^3} + \underset{(8) \text{ and } (5)}{\sum_{k=1}^{10}1} = \left[\frac{10(10+1)}{2}\right]^2 + 10 = 3025 + 10 = 3035$

(d) Notice that the index of summation starts at 6.

$$\sum_{k=6}^{20}(4k^2) = \underset{(1)}{4} \sum_{k=6}^{20}k^2 = \underset{(4)}{4}\left[\sum_{k=1}^{20}k^2 - \sum_{k=1}^{5}k^2\right]$$

$$= \underset{(7)}{4}\left[\frac{20 \cdot 21 \cdot 41}{6} - \frac{5 \cdot 6 \cdot 11}{6}\right] = 4[2870 - 55] = 11,260 \quad\blacksquare$$

**NOW WORK** Problem 61.

## 5 Use the Binomial Theorem

You already know formulas for expanding $(x+a)^n$ for $n=2$ and $n=3$. The *Binomial Theorem* is a formula for the expansion of $(x+a)^n$ for any positive integer $n$.

If $n=1, 2, 3$, and $4$, the expansion of $(x+a)^n$ is straightforward.

- $(x+a)^1 = x+a$      Two terms, beginning with $x^1$ and ending with $a^1$

- $(x+a)^2 = x^2 + 2ax + a^2$      Three terms, beginning with $x^2$ and ending with $a^2$

- $(x+a)^3 = x^3 + 3ax^2 + 3a^2x + a^3$      Four terms, beginning with $x^3$ and ending with $a^3$

- $(x+a)^4 = x^4 + 4ax^3 + 6a^2x^2 + 4a^3x + a^4$      Five terms, beginning with $x^4$ and ending with $a^4$

Notice that each expansion of $(x+a)^n$ begins with $x^n$, ends with $a^n$, and has $n+1$ terms. As you look at the terms from left to right, the powers of $x$ are decreasing by one, while the powers of $a$ are increasing by one. As a result, we might conjecture that the expansion of $(x+a)^n$ would look like this:

$$(x+a)^n = x^n + \underline{\quad} ax^{n-1} + \underline{\quad} a^2x^{n-2} + \cdots + \underline{\quad} a^{n-1}x + a^n$$

where the blanks are numbers to be found. This, in fact, is the case.

Before filling in the blanks, we introduce the symbol $\dbinom{n}{j}$.

**DEFINITION**

If $j$ and $n$ are integers with $0 \le j \le n$, the symbol $\binom{n}{j}$ is defined as

$$\binom{n}{j} = \frac{n!}{j!\,(n-j)!}$$

**EXAMPLE 8** Evaluating $\binom{n}{j}$

**(a)** $\binom{3}{1} = \dfrac{3!}{1!\,(3-1)!} = \dfrac{3!}{1!\,2!} = \dfrac{3 \cdot 2 \cdot 1}{1(2 \cdot 1)} = \dfrac{6}{2} = 3$

**(b)** $\binom{4}{2} = \dfrac{4!}{2!\,(4-2)!} = \dfrac{4!}{2!\,2!} = \dfrac{4 \cdot 3 \cdot 2 \cdot 1}{(2 \cdot 1)(2 \cdot 1)} = \dfrac{24}{4} = 6$

**(c)** $\binom{8}{7} = \dfrac{8!}{7!\,(8-7)!} = \dfrac{8!}{7!\,1!} = \underset{\substack{\uparrow \\ 8! = 8 \cdot 7!}}{\dfrac{8 \cdot 7!}{7! \cdot 1!}} = \dfrac{8}{1} = 8$

**RECALL** $0! = 1$

**(d)** $\binom{n}{0} = \dfrac{n!}{0!\,(n-0)!} = \dfrac{n!}{0!\,n!} = \dfrac{n!}{1 \cdot n!} = 1$

**(e)** $\binom{n}{n} = \dfrac{n!}{n!\,(n-n)!} = \dfrac{n!}{n!\,0!} = \dfrac{n!}{n! \cdot 1} = 1$ ∎

**NOW WORK** Problem 67.

**THEOREM** Binomial Theorem

Let $x$ and $a$ be real numbers. For any positive integer $n$,

$$(x+a)^n = \binom{n}{0}x^n + \binom{n}{1}ax^{n-1} + \binom{n}{2}a^2x^{n-2} + \cdots + \binom{n}{j}a^j x^{n-j}$$

$$+ \cdots + \binom{n}{n}a^n = \sum_{j=0}^{n} \binom{n}{j}a^j x^{n-j}$$

Because of its appearance in the Binomial Theorem, the symbol $\binom{n}{j}$ is called a **binomial coefficient**.

**EXAMPLE 9** Expanding a Binomial

Use the Binomial Theorem to expand $(x+2)^5$.

**Solution** We use the Binomial Theorem with $a = 2$ and $n = 5$. Then

$$(x+2)^5 = \binom{5}{0}x^5 + \binom{5}{1}2x^4 + \binom{5}{2}2^2x^3 + \binom{5}{3}2^3x^2 + \binom{5}{4}2^4x + \binom{5}{5}2^5$$

$$= 1 \cdot x^5 + 5 \cdot 2x^4 + 10 \cdot 4x^3 + 10 \cdot 8x^2 + 5 \cdot 16x + 1 \cdot 32$$

$$= x^5 + 10x^4 + 40x^3 + 80x^2 + 80x + 32$$ ∎

**NOW WORK** Problem 77.

# P.8 Assess Your Understanding

## Concepts and Vocabulary

1. A(n) _____ is a function whose domain is the set of positive integers.

2. *True or False* $3! = 6$

3. If $n \geq 0$ is an integer, then $n! = $ _____ when $n \geq 2$.

4. The sequence $a_1 = 5$, $a_n = 3a_{n-1}$ is an example of a(n) _____ sequence.

5. The notation $a_1 + a_2 + a_3 + \cdots + a_n = \displaystyle\sum_{k=1}^{n} a_k$ is an example of _____ notation.

6. *True or False* $\displaystyle\sum_{k=1}^{n} k = 1 + 2 + 3 + \cdots + n = \frac{n(n+1)}{2}$

7. If $s_1 = 1$ and $s_n = 4s_{n-1}$, then $s_3 = $ _____ .

8. $\dbinom{n}{0} = $ _____ ; $\dbinom{n}{1} = $ _____

9. *True or False* $\dbinom{n}{j} = \dfrac{j!}{(n-j)! \, n!}$

10. The _____ can be used to expand expressions like $(2x + 3)^6$.

## Practice Problems

*In Problems 11–18, write the first five terms of each sequence.*

11. $\{s_n\} = \{2n\}$

12. $\{s_n\} = \{n^2\}$

13. $\{a_n\} = \left\{ \dfrac{n}{n+1} \right\}$

14. $\{a_n\} = \left\{ \dfrac{n+1}{2n} \right\}$

15. $\{b_n\} = \{(-1)^n n^2\}$

16. $\{b_n\} = \left\{ (-1)^{n-1} \dfrac{n}{n+1} \right\}$

17. $\{s_n\} = \left\{ (-1)^{n-1} \dfrac{2^n}{3^n + 1} \right\}$

18. $\{s_n\} = \left\{ (-1)^{n-1} \left( \dfrac{4}{3} \right)^n \right\}$

*In Problems 19–26, the given pattern continues. Find the nth term of a sequence suggested by the pattern.*

19. $2, \dfrac{3}{2}, \dfrac{4}{3}, \dfrac{5}{4}, \ldots$

20. $\dfrac{1}{2 \cdot 3}, \dfrac{1}{3 \cdot 4}, \dfrac{1}{4 \cdot 5}, \dfrac{1}{5 \cdot 6}, \ldots$

21. $1, \dfrac{1}{4}, \dfrac{1}{9}, \dfrac{1}{16}, \dfrac{1}{25}, \ldots$

22. $\dfrac{1}{3}, \dfrac{2}{5}, \dfrac{3}{7}, \dfrac{4}{9}, \dfrac{5}{11}, \ldots$

23. $2, -4, 6, -8, 10, \ldots$

24. $2, -2, 2, -2, 2, \ldots$

25. $2, -3, 4, -5, 6, \ldots$

26. $1, \dfrac{1}{2}, 3, \dfrac{1}{4}, 5, \dfrac{1}{6}, 7, \dfrac{1}{8}, \ldots$

*In Problems 27–38, a sequence is defined recursively. Write the first five terms of the sequence.*

27. $a_1 = 3$; $a_n = 2 + a_{n-1}$

28. $a_1 = 1$; $a_n = 4 + a_{n-1}$

29. $a_1 = 2$; $a_n = n + a_{n-1}$

30. $a_1 = -1$; $a_n = n + a_{n-1}$

31. $a_1 = 1$; $a_n = 3a_{n-1}$

32. $a_1 = 1$; $a_n = -a_{n-1}$

33. $a_1 = 2$; $a_n = \dfrac{a_{n-1}}{n}$

34. $a_1 = -3$; $a_n = n + a_{n-1}$

35. $a_1 = 1$; $a_2 = -1$; $a_n = a_{n-1} \cdot a_{n-2}$

36. $a_1 = -1$; $a_2 = 2$; $a_n = a_{n-2} + n a_{n-1}$

37. $a_1 = A$; $a_n = a_{n-1} + d$

38. $a_1 = A$; $a_n = r a_{n-1}$, $r \neq 0$

*In Problems 39–48, write out each sum.*

39. $\displaystyle\sum_{k=1}^{n} (k - 2)$

40. $\displaystyle\sum_{k=1}^{n} (2k)$

41. $\displaystyle\sum_{k=1}^{n} \dfrac{k^2}{k+1}$

42. $\displaystyle\sum_{k=1}^{n} \dfrac{(k+1)^2}{k}$

43. $\displaystyle\sum_{k=0}^{n} \dfrac{1}{2^k}$

44. $\displaystyle\sum_{k=0}^{n} \left( \dfrac{2}{3} \right)^k$

45. $\displaystyle\sum_{k=0}^{n} \dfrac{k+1}{2^{k+1}}$

46. $\displaystyle\sum_{k=1}^{n} \dfrac{2k+2}{k}$

47. $\displaystyle\sum_{k=2}^{n} (-1)^k 2^k$

48. $\displaystyle\sum_{k=1}^{n} (-1)^{k+1} \ln k$

*In Problems 49–56, express each sum using summation notation.*

49. $2 + 4 + 6 + \cdots + 100$

50. $\dfrac{1}{2} + \dfrac{2^3}{2} + \dfrac{3^3}{2} + \cdots + \dfrac{7^3}{2}$

51. $2 + \dfrac{3}{2} + \dfrac{4}{3} + \cdots + \dfrac{35+1}{35}$

52. $\dfrac{3}{2} + \dfrac{5}{2} + \dfrac{7}{2} + \dfrac{9}{2} + \cdots + \dfrac{2(12)+1}{2}$

53. $3 - \dfrac{3}{10} + \dfrac{3}{100} - \dfrac{3}{1000} + \cdots + (-1)^{10} \dfrac{3}{10^{10}}$

54. $-\dfrac{2}{3} + \dfrac{2}{9} - \dfrac{2}{27} + \cdots + (-1)^8 \dfrac{2}{3^8}$

55. $\dfrac{1}{e} - \dfrac{2}{e^2} + \dfrac{3}{e^3} - \cdots + (-1)^{n+1} \dfrac{n}{e^n}$

56. $1 - \dfrac{5}{2} + \dfrac{5^2}{3} - \dfrac{5^3}{4} + \cdots + (-1)^{n-1} \dfrac{5^{n-1}}{n}$

*In Problems 57–66, find the sum of each sequence.*

57. $\displaystyle\sum_{k=1}^{25} \dfrac{4}{5}$

58. $\displaystyle\sum_{k=1}^{50} 6$

59. $\displaystyle\sum_{k=1}^{30} (3k)$

60. $\displaystyle\sum_{k=1}^{20} \left( -\dfrac{2k}{3} \right)$

61. $\displaystyle\sum_{k=1}^{100} (2k - 1)$

62. $\displaystyle\sum_{k=1}^{18} (4k + 3)$

63. $\displaystyle\sum_{k=1}^{20} (k^2 - 2)$

64. $\displaystyle\sum_{k=0}^{10} (k^2 + 1)$

65. $\displaystyle\sum_{k=6}^{50} \left( \dfrac{k}{2} \right)$

66. $\displaystyle\sum_{k=3}^{20} (3k)$

*In Problems 67–72, evaluate each expression.*

67. $\dbinom{5}{2}$

68. $\dbinom{7}{4}$

69. $\dbinom{9}{3}$

70. $\dbinom{8}{4}$

71. $\dbinom{50}{1}$

72. $\dbinom{100}{2}$

---

**1.** = NOW WORK problem    〖�⁄∖〗 = Graphing technology recommended    CAS = Computer Algebra System recommended

*In Problems 73–80, expand each expression using the Binomial Theorem.*

**73.** $(x+1)^4$

**74.** $(x-1)^6$

**75.** $(x-3)^5$

**76.** $(x+2)^4$

**77.** $(2x+3)^4$

**78.** $(2x-1)^5$

**79.** $(x^2-y^2)^3$

**80.** $(x^2+y^2)^4$

**81. Credit Card Debt** John has a balance of $3000 on his Discover card that charges 1% interest per month on any unpaid balance. John can afford to pay $100 toward the balance each month. His balance each month after making a $100 payment is given by the recursively defined sequence

$$B_0 = \$3000 \qquad B_n = 1.01 B_{n-1} - 100$$

Determine John's balance after making the first payment. That is, determine $B_1$.

 **82. Approximating $f(x) = e^x$** In calculus it can be shown that

$$f(x) = e^x = \sum_{k=0}^{\infty} \frac{x^k}{k!}$$

We can approximate the value of $f(x) = e^x$ for any $x$ using the sum

$$f(x) = e^x \approx \sum_{k=0}^{n} \frac{x^k}{k!}$$

for some $n$.

**(a)** Approximate $f(1.3)$ with $n=4$.

**(b)** Approximate $f(1.3)$ with $n=7$.

**(c)** Use a calculator to approximate $f(1.3)$.

**(d)** Use trial and error along with a graphing calculator's SEQuence mode to determine the value of $n$ required to approximate $f(1.3)$ correct to eight decimal places.

# 1 Limits and Continuity

**CHAPTER 1 PROJECT** The Chapter Project on page 152 looks at a hypothetical situation involving pollution in a lake and explores some legal arguments that might be made.

Joe Raedle/Getty Images

## Oil Spills and Dispersant Chemicals

On April 20, 2010, the Deepwater Horizon drilling rig exploded and initiated the worst marine oil spill in recent history. Oil gushed from the well for three months and released millions of gallons of crude oil into the Gulf of Mexico. One technique used to help clean up during and after the spill was the use of the chemical dispersant Corexit, pictured above in the waters off the coast of Louisiana. Oil dispersants allow the oil particles to spread more freely in the water, thus allowing the oil to biodegrade more quickly. Their use is debated, however, because some of their ingredients are carcinogens. Further, the use of oil dispersants can increase toxic hydrocarbon levels affecting sea life. Over time, the pollution caused by the oil spill and the dispersants will eventually diminish and sea life will return, more or less, to its previous condition. In the short term, however, pollution raises serious questions about the health of the local sea life and the safety of fish and shellfish for human consumption.

The concept of a limit is central to calculus. To understand calculus, it is essential to know what it means for a function to have a limit, and then how to find a limit of a function. Chapter 1 explains what a limit is, shows how to find a limit of a function, and demonstrates how to prove that limits exist using the definition of limit.

We begin the chapter using numerical and graphical approaches to explore the idea of a limit. Although these methods seem to work well, there are instances in which they fail to identify the correct limit.

In Section 1.2, we use analytic techniques to find limits. Some of the proofs of these techniques are found in Section 1.6, others in Appendix B. A limit found by correctly applying these analytic techniques is precise; there is no doubt that it is correct.

In Sections 1.3–1.5, we continue to study limits and some ways that they are used. For example, we use limits to define *continuity*, an important property of a function.

Section 1.6 provides a precise definition of *limit*, the so-called $\varepsilon$-$\delta$ (epsilon-delta) definition, which we use to show when a limit does, and does not, exist.

# 1.1 Limits of Functions Using Numerical and Graphical Techniques

**OBJECTIVES** *When you finish this section, you should be able to:*

1 Discuss the idea of a limit (p. 80)
2 Investigate a limit using a table (p. 80)
3 Investigate a limit using a graph (p. 82)

Calculus can be used to solve certain fundamental questions in geometry. Two of these questions are:

- Given a function $f$ and a point $P$ on its graph, what is the slope of the tangent line to the graph of $f$ at $P$? See Figure 1.
- Given a nonnegative function $f$ whose domain is the closed interval $[a, b]$, what is the area of the region enclosed by the graph of $f$, the $x$-axis, and the vertical lines $x = a$ and $x = b$? See Figure 2.

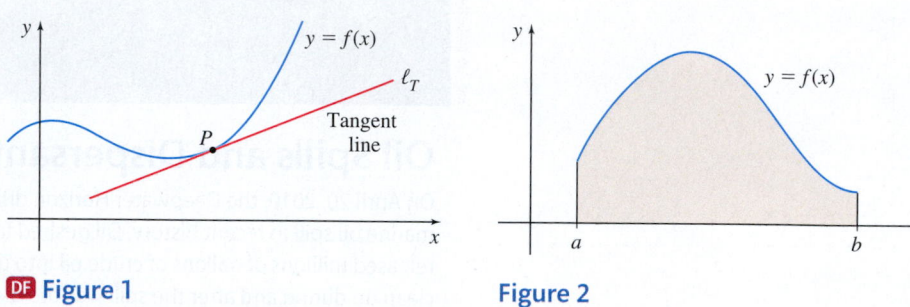

**DF Figure 1**          **Figure 2**

These questions, traditionally called the **tangent problem** and the **area problem**, were solved by Gottfried Wilhelm von Leibniz and Sir Isaac Newton during the late seventeenth and early eighteenth centuries. The solutions to the two seemingly different problems are both based on the idea of a limit. Their solutions not only are related to each other but are also applicable to many other problems in science and geometry. Here we begin to discuss the tangent problem. The discussion of the area problem begins in Chapter 5.

## The Slope of the Tangent Line to a Graph

Notice that the line $\ell_T$ in Figure 1 just touches the graph of $f$ at the point $P$. This unique line is the *tangent line* to the graph of $f$ at $P$. But how is the tangent line defined?

In plane geometry, a tangent line to a circle is defined as a line having exactly one point in common with the circle, as shown in Figure 3. However, this definition does not work for graphs in general. For example, in Figure 4, the three lines $\ell_1$, $\ell_2$, and $\ell_3$ contain the point $P$ and have exactly one point in common with the graph of $f$, but they do not meet the requirement of just touching the graph at $P$. On the other hand, the line $\ell_T$ just touches the graph of $f$ at $P$, but it intersects the graph at another point. It is the slope of the tangent line $\ell_T$ that distinguishes it from all other lines containing $P$.

So before defining a tangent line, we investigate its slope, which we denote by $m_{\tan}$. We begin with the graph of a function $f$, a point $P$ on its graph, and the tangent line $\ell_T$ to $f$ at $P$, as shown in Figure 5.

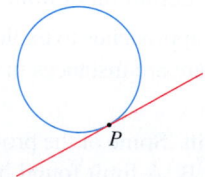

**Figure 3** Tangent line to a circle at the point $P$.

**NEED TO REVIEW?** The slope of a line is discussed in Appendix A.3, p. A-18.

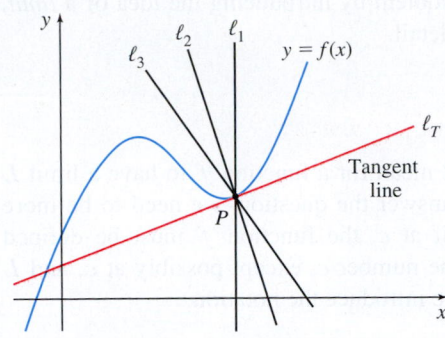

**Figure 4**

The tangent line $\ell_T$ to the graph of $f$ at $P$ must contain the point $P$. We denote the coordinates of $P$ by $(c, f(c))$. Since finding a slope requires two points, and we have only one point on the tangent line $\ell_T$, we proceed as follows.

Suppose we choose any point $Q = (x, f(x))$, other than $P$, on the graph of $f$, as shown in Figure 6. ($Q$ can be to the left or to the right of $P$; we chose $Q$ to be to the right of $P$.) The line containing the points $P = (c, f(c))$ and $Q = (x, f(x))$ is called a **secant line** of the graph of $f$. The slope $m_{\text{sec}}$ of this secant line is

$$m_{\text{sec}} = \frac{f(x) - f(c)}{x - c} \tag{1}$$

Figure 7 shows three different points $Q_1$, $Q_2$, and $Q_3$ on the graph of $f$ that are successively closer to the point $P$, and three associated secant lines $\ell_1$, $\ell_2$, and $\ell_3$. The closer the point $Q$ is to the point $P$, the closer the secant line is to the tangent line $\ell_T$. The line $\ell_T$, the *limiting position* of these secant lines, is the *tangent line to the graph of $f$ at $P$.*

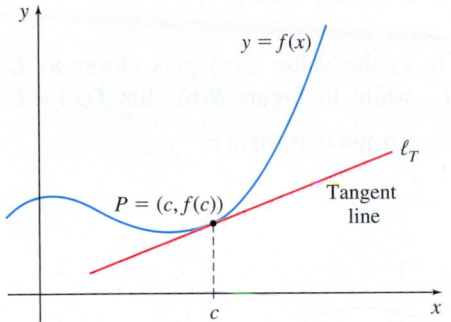

**Figure 5** $m_{\text{tan}}$ = slope of the tangent line.

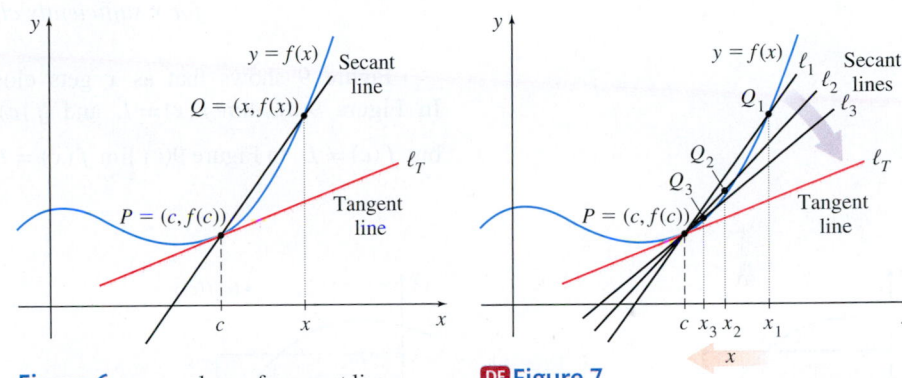

**Figure 6** $m_{\text{sec}}$ = slope of a secant line.    **DF Figure 7**

If the limiting position of the secant lines is the tangent line, then the limit of the slopes of the secant lines should equal the slope of the tangent line. Notice in Figure 7 that as the point $Q$ moves closer to the point $P$, the numbers $x$ get closer to $c$. So, equation (1) suggests that

$$m_{\text{tan}} = \text{Slope of the tangent line to } f \text{ at } P$$

$$= \text{Limit of } \frac{f(x) - f(c)}{x - c} \text{ as } x \text{ gets closer to } c$$

In symbols, we write

$$m_{\text{tan}} = \lim_{x \to c} \frac{f(x) - f(c)}{x - c}$$

The notation $\lim_{x \to c}$ is read, "the limit as $x$ approaches $c$."

The **tangent line** to the graph of a function $f$ at a point $P = (c, f(c))$ is the line containing the point $P$ whose slope is

$$m_{\text{tan}} = \lim_{x \to c} \frac{f(x) - f(c)}{x - c}$$

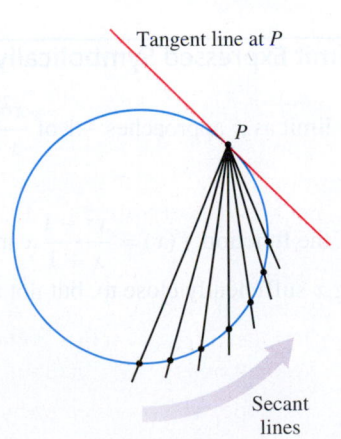

**Figure 8**

provided the limit exists.

As Figure 8 illustrates, this new idea of a tangent line is consistent with the traditional definition of a tangent line to a circle.

**NOW WORK** Problem 2.

We have begun to answer the tangent problem by introducing the idea of a *limit*. Now we describe the idea of a limit in more detail.

## 1 Discuss the Idea of a Limit

We begin by asking a question: What does it mean for a function $f$ to have a limit $L$ as $x$ approaches some fixed number $c$? To answer the question, we need to be more precise about $f$, $L$, and $c$. To have a limit at $c$, the function $f$ must be defined everywhere in an open interval containing the number $c$, except possibly at $c$, and $L$ must be a number. Using these restrictions, we introduce the notation

$$\lim_{x \to c} f(x) = L$$

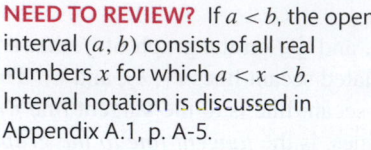

**NEED TO REVIEW?** If $a < b$, the open interval $(a, b)$ consists of all real numbers $x$ for which $a < x < b$. Interval notation is discussed in Appendix A.1, p. A-5.

which is read, "the limit as $x$ approaches $c$ of $f(x)$ is equal to the number $L$." The notation $\lim_{x \to c} f(x) = L$ can be described as

*The value $f(x)$ can be made as close as we please to $L$, for $x$ sufficiently close to $c$, but not equal to $c$.*

Figure 9 shows that as $x$ gets closer to $c$, the value $f(x)$ gets closer to $L$. In Figure 9(a), $\lim_{x \to c} f(x) = L$ and $f(c) = L$, while in Figure 9(b), $\lim_{x \to c} f(x) = L$, but $f(c) \neq L$. In Figure 9(c) $\lim_{x \to c} f(x) = L$, but $f$ is not defined at $c$.

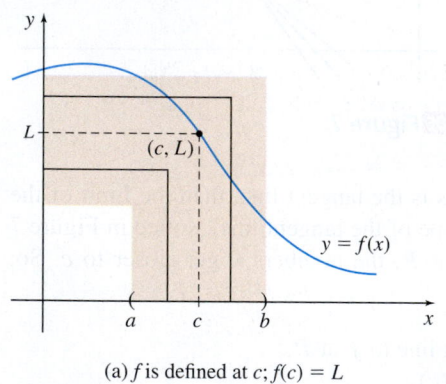

(a) $f$ is defined at $c$; $f(c) = L$

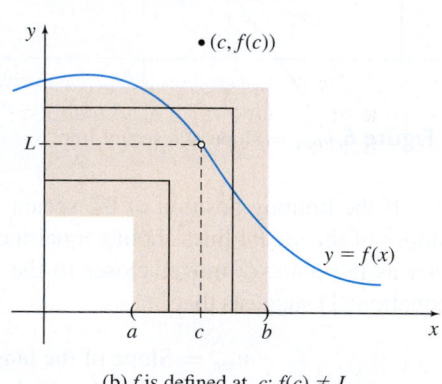

(b) $f$ is defined at $c$; $f(c) \neq L$

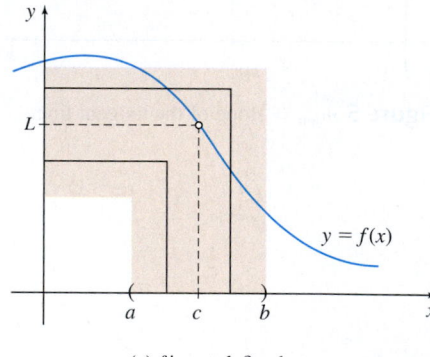

(c) $f$ is not defined at $c$

**Figure 9** $\lim_{x \to c} f(x) = L$

---

**EXAMPLE 1** Reading and Interpreting a Limit Expressed Symbolically

• In words, $\lim_{x \to -1} \dfrac{x^2 - 1}{x + 1} = -2$, is read as, "The limit as $x$ approaches $-1$ of $\dfrac{x^2 - 1}{x + 1}$ is equal to the number $-2$."

• The limit may be interpreted as "The value of the function $f(x) = \dfrac{x^2 - 1}{x + 1}$ can be made as close as we please to $-2$ by choosing $x$ sufficiently close to, but not equal to, $-1$." ∎

**NOW WORK** Problem 1.

## 2 Investigate a Limit Using a Table

We can use a table to better understand what it means for a function to have a limit as $x$ approaches a number $c$.

**EXAMPLE 2** **Investigating a Limit Using a Table**

Investigate $\lim_{x \to 2}(2x + 5)$ using a table.

**Solution** We create Table 1 by evaluating $f(x) = 2x + 5$ at values of $x$ near 2, choosing numbers $x$ slightly less than 2 and numbers $x$ slightly greater than 2.

**TABLE 1**

| $x$ | numbers $x$ slightly less than 2 | | | | | numbers $x$ slightly greater than 2 | | | |
|---|---|---|---|---|---|---|---|---|---|
| $x$ | 1.99 | 1.999 | 1.9999 | 1.99999 | $\to 2 \leftarrow$ | 2.00001 | 2.0001 | 2.001 | 2.01 |
| $f(x) = 2x + 5$ | 8.98 | 8.998 | 8.9998 | 8.99998 | $f(x)$ approaches 9 | 9.00002 | 9.0002 | 9.002 | 9.02 |

Table 1 suggests that the value of $f(x) = 2x + 5$ can be made "as close as we please" to 9 by choosing $x$ "sufficiently close" to 2. This suggests that $\lim_{x \to 2}(2x + 5) = 9$. ∎

**NOW WORK** Problem 9.

In creating Table 1, first we used numbers $x$ close to 2 but less than 2, and then we used numbers $x$ close to 2 but greater than 2. When $x < 2$, we say, "$x$ is approaching 2 from the left," and the number 9 is called the **left-hand limit**. When $x > 2$, we say, "$x$ is approaching 2 from the right," and the number 9 is called the **right-hand limit**. Together, these are called the **one-sided limits** of $f$ as $x$ approaches 2.

One-sided limits are symbolized as follows. The left-hand limit, written

$$\lim_{x \to c^-} f(x) = L_{\text{left}}$$

is read, "The limit as $x$ approaches $c$ from the left of $f(x)$ equals $L_{\text{left}}$." It means that the value of $f$ can be made as close as we please to the number $L_{\text{left}}$ by choosing $x < c$ and sufficiently close to $c$.

Similarly, the right-hand limit, written

$$\lim_{x \to c^+} f(x) = L_{\text{right}}$$

is read, "The limit as $x$ approaches $c$ from the right of $f(x)$ equals $L_{\text{right}}$." It means that the value of $f$ can be made as close as we please to the number $L_{\text{right}}$ by choosing $x > c$ and sufficiently close to $c$.

**EXAMPLE 3** **Investigating a Limit Using a Calculator to Set Up a Table**

Investigate $\lim_{x \to 0} \dfrac{e^x - 1}{x}$ using a table.

**Solution** The domain of $f(x) = \dfrac{e^x - 1}{x}$ is $\{x \mid x \neq 0\}$. So, $f$ is defined everywhere in an open interval containing the number 0, except for 0.

Using a calculator, set up Table 2 to investigate the left-hand limit $\lim_{x \to 0^-} \dfrac{e^x - 1}{x}$ and the right-hand limit $\lim_{x \to 0^+} \dfrac{e^x - 1}{x}$. First, we evaluate $f$ at numbers less than 0, but close to zero, and then at numbers greater than 0, but close to zero.

**TABLE 2**

| $x$ | $x$ approaches 0 from the left | | | | | $x$ approaches 0 from the right | | | |
|---|---|---|---|---|---|---|---|---|---|
| $x$ | $-0.01$ | $-0.001$ | $-0.0001$ | $-0.00001$ | $\to 0 \leftarrow$ | 0.00001 | 0.0001 | 0.001 | 0.01 |
| $f(x) = \dfrac{e^x - 1}{x}$ | 0.995 | 0.9995 | 0.99995 | 0.999995 | $f(x)$ approaches 1 | 1.000005 | 1.00005 | 1.0005 | 1.005 |

Table 2 suggests that $\lim\limits_{x \to 0^-} \dfrac{e^x - 1}{x} = 1$ and $\lim\limits_{x \to 0^+} \dfrac{e^x - 1}{x} = 1$.

This suggests $\lim\limits_{x \to 0} \dfrac{e^x - 1}{x} = 1$. ∎

**NOW WORK** Problem 13.

**EXAMPLE 4**  **Investigating a Limit Using Technology to Set Up a Table**

Investigate $\lim\limits_{x \to 0} \dfrac{\sin x}{x}$ using a table.

**Solution** The domain of the function $f(x) = \dfrac{\sin x}{x}$ is $\{x \mid x \neq 0\}$. So, $f$ is defined everywhere in an open interval containing 0, except for 0.

We investigate the one-sided limits of $\dfrac{\sin x}{x}$ as $x$ approaches 0 using a graphing calculator to set up a table. When making the table, we choose numbers $x$ (in radians) slightly less than 0 and numbers slightly greater than 0. See Figure 10.

The table in Figure 10 suggests that $\lim\limits_{x \to 0^-} f(x) = 1$ and $\lim\limits_{x \to 0^+} f(x) = 1$. This suggests that $\lim\limits_{x \to 0} \dfrac{\sin x}{x} = 1$. ∎

**NOW WORK** Problem 15.

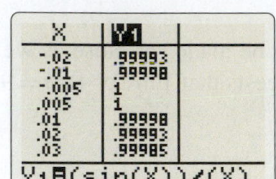

**Figure 10**

**NOTE** $f(x) = \dfrac{\sin x}{x}$ is an even function, so the $Y_1$ values in Figure 10 are symmetric about $x = 0$.

## 3 Investigate a Limit Using a Graph

The graph of a function can also help us investigate limits. Figure 11 shows the graphs of three different functions $f$, $g$, and $h$. Observe that in each function, *as $x$ gets closer to $c$, whether from the left or from the right, the value of the function gets closer to the number $L$.* This is the key idea of a limit.

Notice in Figure 11(b) that the value of $g$ at $c$ does not affect the limit. Notice in Figure 11(c) that $h$ is not defined at $c$, but the value of $h$ gets closer to the number $L$ for $x$ sufficiently close to $c$. This suggests that the limit of a function as $x$ approaches $c$ does not depend on the value, if it exists, of the function at $c$.

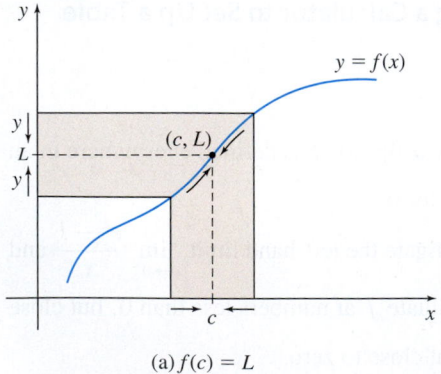

(a) $f(c) = L$

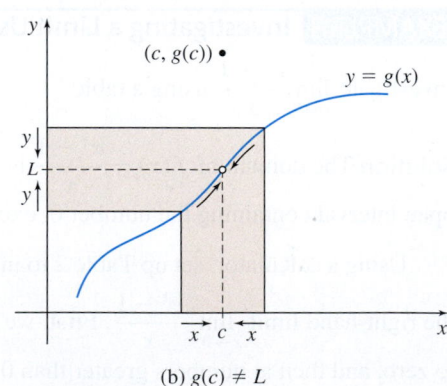

(b) $g(c) \neq L$

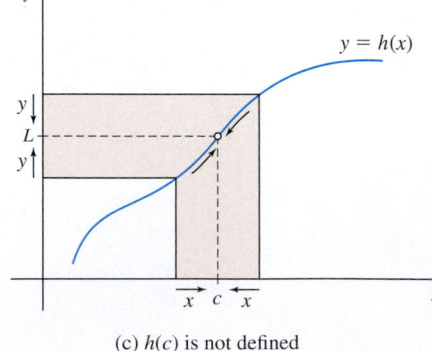

(c) $h(c)$ is not defined

**Figure 11**

**NEED TO REVIEW?**
Piecewise-defined functions are discussed in Section P.1, p. 8.

**EXAMPLE 5**  **Investigating a Limit Using a Graph**

Use a graph to investigate $\lim\limits_{x \to 2} f(x)$ if $f(x) = \begin{cases} 3x + 1 & \text{if } x \neq 2 \\ 10 & \text{if } x = 2 \end{cases}$.

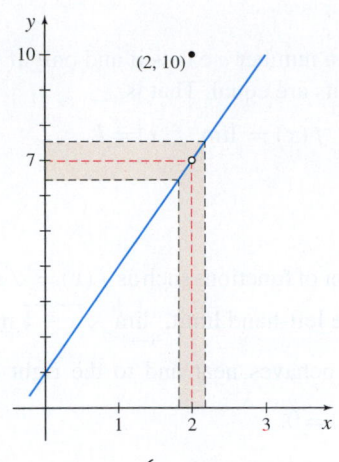

**Figure 12** $f(x) = \begin{cases} 3x+1 & \text{if } x \neq 2 \\ 10 & \text{if } x = 2 \end{cases}$

**Solution** The function $f$ is a piecewise-defined function. Its graph is shown in Figure 12. Observe that as $x$ approaches 2 from the left, the value of $f$ is close to 7, and as $x$ approaches 2 from the right, the value of $f$ is close to 7. In fact, we can make the value of $f$ as close as we please to 7 by choosing $x$ sufficiently close to 2 but not equal to 2. This suggests $\lim\limits_{x \to 2} f(x) = 7$. ∎

If we use a table to investigate $\lim\limits_{x \to 2} f(x)$, the result is the same. See Table 3.

**TABLE 3**

| | $x$ approaches 2 from the left | | $x$ approaches 2 from the right |
|---|---|---|---|
| $x$ | 1.99  1.999  1.9999  1.99999 | → 2 ← | 2.00001  2.0001  2.001  2.01 |
| $f(x)$ | 6.97  6.997  6.9997  6.99997 | $f(x)$ approaches 7 | 7.00003  7.0003  7.003  7.03 |

Figure 12 shows that $f(2) = 10$ but that this value has no impact on the limit as $x$ approaches 2. In fact, $f(2)$ can equal any number, and it would have no effect on the limit as $x$ approaches 2.

We make the following observations:

- The limit $L$ of a function $y = f(x)$ as $x$ approaches a number $c$ does not depend on the value of $f$ at $c$.
- The limit $L$ of a function $y = f(x)$ as $x$ approaches a number $c$ is unique; that is, a function cannot have more than one limit as $x$ approaches $c$. (A proof of this property is given in Appendix B.)
- If there is *no single number* that the value of $f$ approaches as $x$ gets close to $c$, we say that $f$ has no limit as $x$ approaches $c$, or more simply, that the *limit of $f$ does not exist at $c$*.

Examples 6 and 7 illustrate situations in which a limit does not exist.

CALC
CLIP

**EXAMPLE 6  Investigating a Limit Using a Graph**

Use a graph to investigate $\lim\limits_{x \to 0} f(x)$ if $f(x) = \begin{cases} x & \text{if } x < 0 \\ 1 & \text{if } x > 0 \end{cases}$.

**Solution** Figure 13 shows the graph of $f$. We first investigate the one-sided limits. The graph suggests that, as $x$ approaches 0 from the left,

$$\lim_{x \to 0^-} f(x) = 0$$

and as $x$ approaches 0 from the right,

$$\lim_{x \to 0^+} f(x) = 1$$

Since there is no single number that the values of $f$ approach when $x$ is close to 0, we conclude that $\lim\limits_{x \to 0} f(x)$ does not exist. ∎

Table 4 uses a numerical approach to support the conclusion that $\lim\limits_{x \to 0} f(x)$ does not exist.

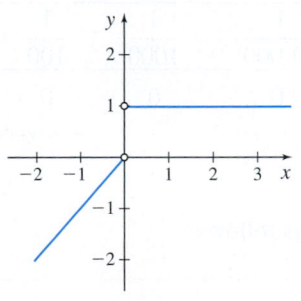

**Figure 13** $f(x) = \begin{cases} x & \text{if } x < 0 \\ 1 & \text{if } x > 0 \end{cases}$

**TABLE 4**

| | $x$ approaches 0 from the left | | $x$ approaches 0 from the right |
|---|---|---|---|
| $x$ | −0.01  −0.001  −0.0001 | → 0 ← | 0.0001  0.001  0.01 |
| $f(x)$ | −0.01  −0.001  −0.0001 | no single number | 1  1  1 |

Examples 5 and 6 lead to the following result.

**THEOREM**

The limit of a function $y = f(x)$ as $x$ approaches a number $c$ exists if and only if both one-sided limits exist at $c$ and both one-sided limits are equal. That is,

$$\lim_{x \to c} f(x) = L \quad \text{if and only if} \quad \lim_{x \to c^-} f(x) = \lim_{x \to c^+} f(x) = L$$

**NOW WORK** Problems **25** and **31**.

A one-sided limit is used to describe the behavior of functions such as $f(x) = \sqrt{x - 1}$ near $x = 1$. Since the domain of $f$ is $\{x \mid x \geq 1\}$, the left-hand limit, $\lim\limits_{x \to 1^-} \sqrt{x - 1}$ makes no sense. But $\lim\limits_{x \to 1^+} \sqrt{x - 1} = 0$ suggests how $f$ behaves near and to the right of 1. See Figure 14 and Table 5. They suggest $\lim\limits_{x \to 1^+} f(x) = 0$.

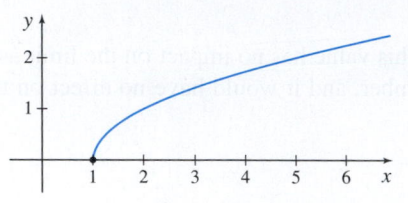

**Figure 14** $f(x) = \sqrt{x - 1}$

**TABLE 5**

| | x approaches 1 from the right | | | | | |
|---|---|---|---|---|---|---|
| $x$ | 1.009 | 1.0009 | 1.00009 | 1.000009 | 1.0000009 | $x \to 1$ |
| $f(x) = \sqrt{x-1}$ | 0.0949 | 0.03 | 0.00949 | 0.003 | 0.000949 | $f(x)$ approaches 0 |

Using numeric tables and/or graphs gives us an idea of what a limit might be. That is, these methods suggest a limit, but there are dangers in using these methods, as the following example illustrates.

**EXAMPLE 7** **Investigating a Limit**

Investigate $\lim\limits_{x \to 0} \sin \dfrac{\pi}{x^2}$.

**Solution** The domain of the function $f(x) = \sin \dfrac{\pi}{x^2}$ is $\{x \mid x \neq 0\}$.

Suppose we let $x$ approach zero in the following way:

**TABLE 6**

| | x approaches 0 from the left | | | | | x approaches 0 from the right | | | |
|---|---|---|---|---|---|---|---|---|---|
| $x$ | $-\dfrac{1}{10}$ | $-\dfrac{1}{100}$ | $-\dfrac{1}{1000}$ | $-\dfrac{1}{10{,}000}$ | $\to 0 \leftarrow$ | $\dfrac{1}{10{,}000}$ | $\dfrac{1}{1000}$ | $\dfrac{1}{100}$ | $\dfrac{1}{10}$ |
| $f(x) = \sin \dfrac{\pi}{x^2}$ | 0 | 0 | 0 | 0 | $f(x)$ approaches 0 | 0 | 0 | 0 | 0 |

Table 6 suggests that $\lim\limits_{x \to 0} \sin \dfrac{\pi}{x^2} = 0$.

Alternatively, suppose we let $x$ approach zero as follows:

**TABLE 7**

| | x approaches 0 from the left | | | | | | x approaches 0 from the right | | | |
|---|---|---|---|---|---|---|---|---|---|---|
| $x$ | $-\dfrac{2}{3}$ | $-\dfrac{2}{5}$ | $-\dfrac{2}{7}$ | $-\dfrac{2}{9}$ | $-\dfrac{2}{11}$ | $\to 0 \leftarrow$ | $\dfrac{2}{11}$ | $\dfrac{2}{9}$ | $\dfrac{2}{7}$ | $\dfrac{2}{5}$ | $\dfrac{2}{3}$ |
| $f(x) = \sin \dfrac{\pi}{x^2}$ | 0.707 | 0.707 | 0.707 | 0.707 | 0.707 | $f(x)$ approaches 0.707 | 0.707 | 0.707 | 0.707 | 0.707 | 0.707 |

Table 7 suggests that $\lim\limits_{x \to 0} \sin \dfrac{\pi}{x^2} = \dfrac{\sqrt{2}}{2} \approx 0.707$.

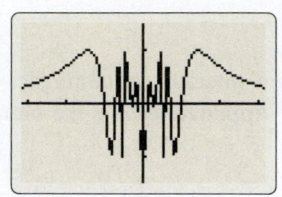

**Figure 15** $f(x) = \sin \dfrac{\pi}{x^2}, \; -\pi \le x \le \pi$

In fact, by carefully selecting $x$, we can make $f$ appear to approach any number in the interval $[-1, 1]$.

Now look at Figure 15, which illustrates that the graph of $f(x) = \sin \dfrac{\pi}{x^2}$ oscillates rapidly as $x$ approaches 0. This suggests that the value of $f$ does not approach a single number and that $\lim\limits_{x \to 0} \sin \dfrac{\pi}{x^2}$ does not exist. ∎

NOW WORK **Problem 55.**

So, how do we find a limit with certainty? The answer lies in giving a very precise definition of limit. The next example helps explain the definition.

---

**EXAMPLE 8** **Analyzing a Limit**

In Example 2, we claimed that $\lim\limits_{x \to 2} (2x + 5) = 9$.

**(a)** How close must $x$ be to 2, so that $f(x) = 2x + 5$ differs from 9 by less than 0.1?
**(b)** How close must $x$ be to 2, so that $f(x) = 2x + 5$ differs from 9 by less than 0.05?

> **RECALL** On the number line, the distance between two points with coordinates $a$ and $b$ is $|a - b|$.

> **NEED TO REVIEW?** Inequalities involving absolute values are discussed in Appendix A.1, p. A-7.

**Solution (a)** The function $f(x) = 2x + 5$ differs from 9 by less than 0.1 if the distance between $f(x)$ and 9 is less than 0.1 unit. That is, if $|f(x) - 9| < 0.1$.

$$|(2x + 5) - 9| < 0.1$$
$$|2x - 4| < 0.1$$
$$|2(x - 2)| < 0.1$$
$$|x - 2| < \frac{0.1}{2} = 0.05$$
$$-0.05 < x - 2 < 0.05$$
$$1.95 < x < 2.05$$

So, if $1.95 < x < 2.05$, then $f(x)$ differs from 9 by less than 0.1.

**(b)** The function $f(x) = 2x + 5$ differs from 9 by less than 0.05 if $|f(x) - 9| < 0.05$. That is, if

$$|(2x + 5) - 9| < 0.05$$
$$|2x - 4| < 0.05$$
$$|x - 2| < \frac{0.05}{2} = 0.025$$

So, if $1.975 < x < 2.025$, then $f(x)$ differs from 9 by less than 0.05. ∎

Notice that the closer we require $f$ to be to the limit 9, the narrower the interval for $x$ becomes. See Figure 16.

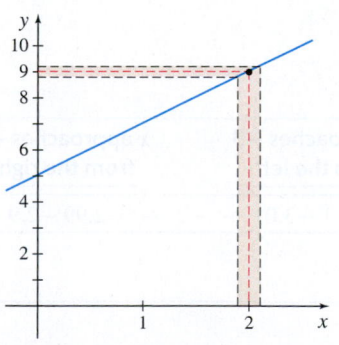

**DF Figure 16** $f(x) = 2x + 5$

NOW WORK **Problem 57.**

The discussion in Example 8 forms the basis of the definition of a limit. We state the definition here but postpone the details until Section 1.6. It is customary to use the Greek letters $\varepsilon$ (epsilon) and $\delta$ (delta) in the definition, so we call it the *$\varepsilon$-$\delta$ definition of a limit*.

> **DEFINITION** $\varepsilon$-$\delta$ Definition of a Limit
>
> Let $f$ be a function defined everywhere in an open interval containing $c$, except possibly at $c$. Then the **limit of the function** $f$ as $x$ **approaches** $c$ is the number $L$, written
>
> $$\lim_{x \to c} f(x) = L$$
>
> if, given any number $\varepsilon > 0$, there is a number $\delta > 0$ so that
>
> $$\text{whenever } 0 < |x - c| < \delta \quad \text{then } |f(x) - L| < \varepsilon$$

Notice in the definition that $f$ is defined everywhere in an open interval containing $c$ except possibly at $c$.

## 1.1 Assess Your Understanding

### Concepts and Vocabulary

1. *Multiple Choice* The limit as $x$ approaches $c$ of a function $f$ is written symbolically as [(a) $\lim_{c \to x} f(x)$ (b) $\lim_{x \to c} f(x)$ (c) $\lim_{x \to c} f(x)$]

2. *True or False* The tangent line to the graph of $f$ at a point $P = (c, f(c))$ is the limiting position of the secant lines passing through $P$ and a point $(x, f(x))$, $x \neq c$, as $x$ moves closer to $c$.

3. *True or False* If $f$ is not defined at $x = c$, then $\lim_{x \to c} f(x)$ does not exist.

4. *True or False* The limit $L$ of a function $y = f(x)$ as $x$ approaches the number $c$ depends on the value of $f$ at $c$.

5. *True or False* If $f(c)$ is defined, this suggests that $\lim_{x \to c} f(x)$ exists.

6. *True or False* The limit of a function $y = f(x)$ as $x$ approaches a number $c$ equals $L$ if at least one of the one-sided limits as $x$ approaches $c$ equals $L$.

### Skill Building

*In Problems 7–12, complete each table and investigate the limit.*

7. $\lim_{x \to 1} 2x$

| | $x$ approaches 1 from the left $\longrightarrow$ | | | | $\longleftarrow$ $x$ approaches 1 from the right | | |
|---|---|---|---|---|---|---|---|
| $x$ | 0.9 | 0.99 | 0.999 | $\to 1 \leftarrow$ | 1.001 | 1.01 | 1.1 |
| $f(x) = 2x$ | | | | | | | |

8. $\lim_{x \to 2} (x + 3)$

| | $x$ approaches 2 from the left $\longrightarrow$ | | | | $\longleftarrow$ $x$ approaches 2 from the right | | |
|---|---|---|---|---|---|---|---|
| $x$ | 1.9 | 1.99 | 1.999 | $\to 2 \leftarrow$ | 2.001 | 2.01 | 2.1 |
| $f(x) = x + 3$ | | | | | | | |

9. $\lim_{x \to 0} (x^2 + 2)$

| | $x$ approaches 0 from the left $\longrightarrow$ | | | | $\longleftarrow$ $x$ approaches 0 from the right | | |
|---|---|---|---|---|---|---|---|
| $x$ | $-0.1$ | $-0.01$ | $-0.001$ | $\to 0 \leftarrow$ | 0.001 | 0.01 | 0.1 |
| $f(x) = x^2 + 2$ | | | | | | | |

10. $\lim_{x \to -1} (x^2 - 2)$

| | $x$ approaches $-1$ from the left $\longrightarrow$ | | | | $\longleftarrow$ $x$ approaches $-1$ from the right | | |
|---|---|---|---|---|---|---|---|
| $x$ | $-1.1$ | $-1.01$ | $-1.001$ | $\to -1 \leftarrow$ | $-0.999$ | $-0.99$ | $-0.9$ |
| $f(x) = x^2 - 2$ | | | | | | | |

11. $\lim_{x \to -3} \dfrac{x^2 - 9}{x + 3}$

| | $x$ approaches $-3$ from the left $\longrightarrow$ | | | | $\longleftarrow$ $x$ approaches $-3$ from the right | | |
|---|---|---|---|---|---|---|---|
| $x$ | $-3.5$ | $-3.1$ | $-3.01$ | $\to -3 \leftarrow$ | $-2.99$ | $-2.9$ | $-2.5$ |
| $f(x) = \dfrac{x^2 - 9}{x + 3}$ | | | | | | | |

12. $\lim_{x \to -1} \dfrac{x^3 + 1}{x + 1}$

| | $x$ approaches $-1$ from the left $\longrightarrow$ | | | | $\longleftarrow$ $x$ approaches $-1$ from the right | | |
|---|---|---|---|---|---|---|---|
| $x$ | $-1.1$ | $-1.01$ | $-1.001$ | $\to -1 \leftarrow$ | $-0.999$ | $-0.99$ | $-0.9$ |
| $f(x) = \dfrac{x^3 + 1}{x + 1}$ | | | | | | | |

---

**1.** = NOW WORK problem    = Graphing technology recommended    CAS = Computer Algebra System recommended

*In Problems 13–16, use technology to complete the table and investigate the limit.*

**13.** $\lim\limits_{x \to 0} \dfrac{2 - 2e^x}{x}$

| | x approaches 0 from the left | | | | | x approaches 0 from the right | | |
|---|---|---|---|---|---|---|---|---|
| $x$ | −0.2 −0.1 −0.01 | → | 0 | ← | 0.01 0.1 0.2 | | | |
| $f(x) = \dfrac{2 - 2e^x}{x}$ | | | | | | | | |

**14.** $\lim\limits_{x \to 1} \dfrac{\ln x}{x - 1}$

| | x approaches 1 from the left | | | | | x approaches 1 from the right | | |
|---|---|---|---|---|---|---|---|---|
| $x$ | 0.9 0.99 0.999 | → | 1 | ← | 1.001 1.01 1.1 | | | |
| $f(x) = \dfrac{\ln x}{x - 1}$ | | | | | | | | |

**15.** $\lim\limits_{x \to 0} \dfrac{1 - \cos x}{x}$, where $x$ is measured in radians

| | x approaches 0 from the left | | | | | x approaches 0 from the right | | |
|---|---|---|---|---|---|---|---|---|
| $x$ (in radians) | −0.2 −0.1 −0.01 | → | 0 | ← | 0.01 0.1 0.2 | | | |
| $f(x) = \dfrac{1 - \cos x}{x}$ | | | | | | | | |

**16.** $\lim\limits_{x \to 0} \dfrac{\sin x}{1 + \tan x}$, where $x$ is measured in radians

| | x approaches 0 from the left | | | | | x approaches 0 from the right | | |
|---|---|---|---|---|---|---|---|---|
| $x$ (in radians) | −0.2 −0.1 −0.01 | → | 0 | ← | 0.01 0.1 0.2 | | | |
| $f(x) = \dfrac{\sin x}{1 + \tan x}$ | | | | | | | | |

*In Problems 17–20, use the graph to investigate*

**(a)** $\lim\limits_{x \to 2^-} f(x)$　**(b)** $\lim\limits_{x \to 2^+} f(x)$　**(c)** $\lim\limits_{x \to 2} f(x)$.

**17.**

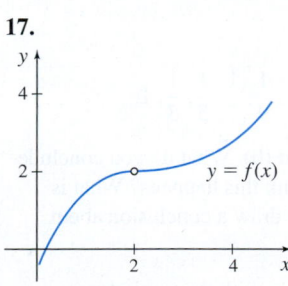

**18.**

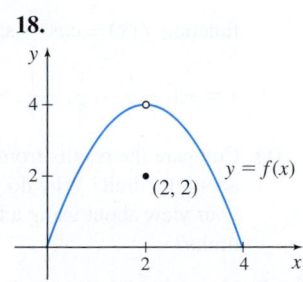

**19.**

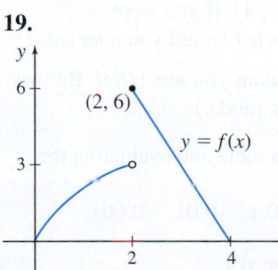

**20.**

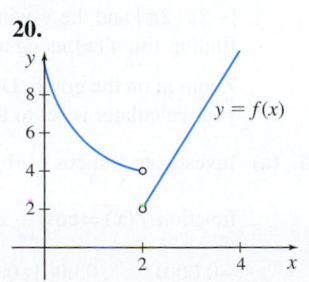

*In Problems 21–28, use the graph to investigate $\lim\limits_{x \to c} f(x)$. If the limit does not exist, explain why.*

**21.**

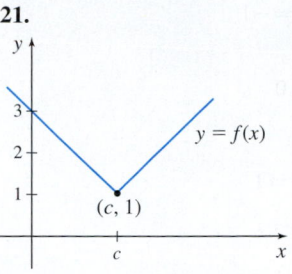

**22.**

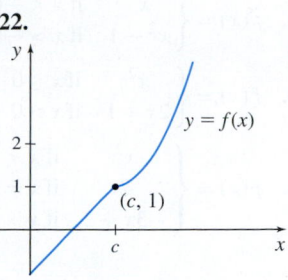

**23.**

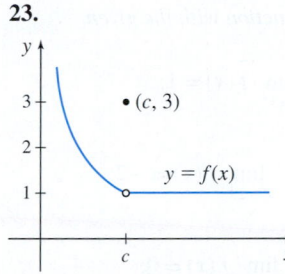

**24.**

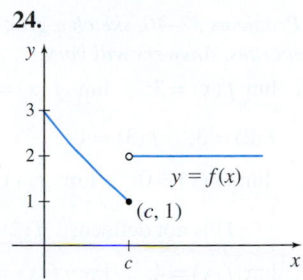

**25.**

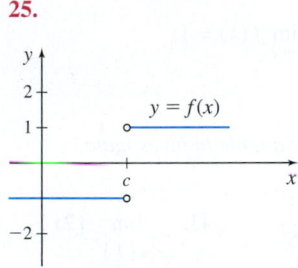

**26.**

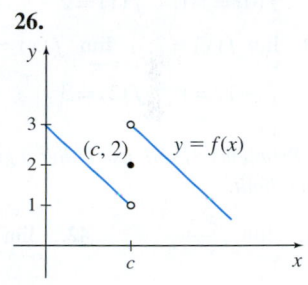

**27.**

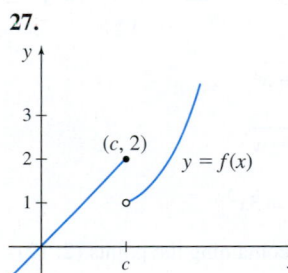

**28.**

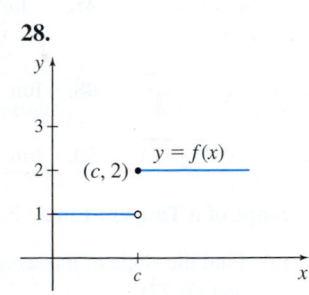

*In Problems 29–36, use a graph to investigate $\lim\limits_{x \to c} f(x)$ at the number $c$.*

**29.** $f(x) = \begin{cases} 2x + 5 & \text{if } x \le 2 \\ 4x + 1 & \text{if } x > 2 \end{cases}$ at $c = 2$

**30.** $f(x) = \begin{cases} 2x + 1 & \text{if } x \le 0 \\ 2x & \text{if } x > 0 \end{cases}$ at $c = 0$

**31.** $f(x) = \begin{cases} 3x - 1 & \text{if } x < 1 \\ 4 & \text{if } x = 1 \\ 4x & \text{if } x > 1 \end{cases}$ at $c = 1$

**32.** $f(x) = \begin{cases} x + 2 & \text{if } x < 2 \\ 4 & \text{if } x = 2 \\ x^2 & \text{if } x > 2 \end{cases}$ at $c = 2$

**33.** $f(x) = \begin{cases} 2x^2 & \text{if } x < 1 \\ 3x^2 - 1 & \text{if } x > 1 \end{cases}$ at $c = 1$

**34.** $f(x) = \begin{cases} x^3 & \text{if } x < -1 \\ x^2 - 1 & \text{if } x > -1 \end{cases}$ at $c = -1$

**35.** $f(x) = \begin{cases} x^2 & \text{if } x \leq 0 \\ 2x + 1 & \text{if } x > 0 \end{cases}$ at $c = 0$

**36.** $f(x) = \begin{cases} x^2 & \text{if } x < 1 \\ 2 & \text{if } x = 1 \\ -3x + 2 & \text{if } x > 1 \end{cases}$ at $c = 1$

## Applications and Extensions

*In Problems 37–40, sketch a graph of a function with the given properties. Answers will vary.*

**37.** $\lim\limits_{x \to 2} f(x) = 3$;  $\lim\limits_{x \to 3^-} f(x) = 3$;  $\lim\limits_{x \to 3^+} f(x) = 1$;

$f(2) = 3$;  $f(3) = 1$

**38.** $\lim\limits_{x \to -1} f(x) = 0$;  $\lim\limits_{x \to 2^-} f(x) = -2$;  $\lim\limits_{x \to 2^+} f(x) = -2$;

$f(-1)$ is not defined;  $f(2) = -2$

**39.** $\lim\limits_{x \to 1} f(x) = 4$;  $\lim\limits_{x \to 0^-} f(x) = -1$;  $\lim\limits_{x \to 0^+} f(x) = 0$;

$f(0) = -1$;  $f(1) = 2$

**40.** $\lim\limits_{x \to 2} f(x) = 2$;  $\lim\limits_{x \to -1} f(x) = 0$;  $\lim\limits_{x \to 1} f(x) = 1$;

$f(-1) = 1$;  $f(2) = 3$

*In Problems 41–50, use either a graph or a table to investigate each limit.*

**41.** $\lim\limits_{x \to 5^+} \dfrac{|x - 5|}{x - 5}$

**42.** $\lim\limits_{x \to 5^-} \dfrac{|x - 5|}{x - 5}$

**43.** $\lim\limits_{x \to \left(\frac{1}{2}\right)^-} \lfloor 2x \rfloor$

**44.** $\lim\limits_{x \to \left(\frac{1}{2}\right)^+} \lfloor 2x \rfloor$

**45.** $\lim\limits_{x \to \left(\frac{2}{3}\right)^-} \lfloor 2x \rfloor$

**46.** $\lim\limits_{x \to \left(\frac{2}{3}\right)^+} \lfloor 2x \rfloor$

**47.** $\lim\limits_{x \to 2^+} \sqrt{|x| - x}$

**48.** $\lim\limits_{x \to 2^-} \sqrt{|x| - x}$

**49.** $\lim\limits_{x \to 2^+} \sqrt[3]{\lfloor x \rfloor - x}$

**50.** $\lim\limits_{x \to 2^-} \sqrt[3]{\lfloor x \rfloor - x}$

**51. Slope of a Tangent Line**  For $f(x) = 3x^2$:

(a) Find the slope of the secant line containing the points $(2, 12)$ and $(3, 27)$.

(b) Find the slope of the secant line containing the points $(2, 12)$ and $(x, f(x))$, $x \neq 2$.

(c) Create a table to investigate the slope of the tangent line to the graph of $f$ at 2 using the result from (b).

(d) On the same set of axes, graph $f$, the tangent line to the graph of $f$ at the point $(2, 12)$, and the secant line from (a).

**52. Slope of a Tangent Line**  For $f(x) = x^3$:

(a) Find the slope of the secant line containing the points $(2, 8)$ and $(3, 27)$.

(b) Find the slope of the secant line containing the points $(2, 8)$ and $(x, f(x))$, $x \neq 2$.

(c) Create a table to investigate the slope of the tangent line to the graph of $f$ at 2 using the result from (b).

(d) On the same set of axes, graph $f$, the tangent line to the graph of $f$ at the point $(2, 8)$, and the secant line from (a).

**53. Slope of a Tangent Line**  For $f(x) = \dfrac{1}{2}x^2 - 1$:

(a) Find the slope $m_{\text{sec}}$ of the secant line containing the points $P = (2, f(2))$ and $Q = (2 + h, f(2 + h))$.

(b) Use the result from (a) to complete the following table:

| $h$ | $-0.5$ | $-0.1$ | $-0.001$ | $0.001$ | $0.1$ | $0.5$ |
|---|---|---|---|---|---|---|
| $m_{\text{sec}}$ | | | | | | |

(c) Investigate the limit of the slope of the secant line found in (a) as $h \to 0$.

(d) What is the slope of the tangent line to the graph of $f$ at the point $P = (2, f(2))$?

(e) On the same set of axes, graph $f$ and the tangent line to $f$ at $P = (2, f(2))$.

**54. Slope of a Tangent Line**  For $f(x) = x^2 - 1$:

(a) Find the slope $m_{\text{sec}}$ of the secant line containing the points $P = (-1, f(-1))$ and $Q = (-1 + h, f(-1 + h))$.

(b) Use the result from (a) to complete the following table:

| $h$ | $-0.1$ | $-0.01$ | $-0.001$ | $-0.0001$ | $0.0001$ | $0.001$ | $0.01$ | $0.1$ |
|---|---|---|---|---|---|---|---|---|
| $m_{\text{sec}}$ | | | | | | | | |

(c) Investigate the limit of the slope of the secant line found in (a) as $h \to 0$.

(d) What is the slope of the tangent line to the graph of $f$ at the point $P = (-1, f(-1))$?

(e) On the same set of axes, graph $f$ and the tangent line to $f$ at $P = (-1, f(-1))$.

**55.** (a) Investigate $\lim\limits_{x \to 0} \cos \dfrac{\pi}{x}$ by using a table and evaluating the function $f(x) = \cos \dfrac{\pi}{x}$ at

$$x = -\frac{1}{2}, -\frac{1}{4}, -\frac{1}{8}, -\frac{1}{10}, -\frac{1}{12}, \ldots, \frac{1}{12}, \frac{1}{10}, \frac{1}{8}, \frac{1}{4}, \frac{1}{2}.$$

(b) Investigate $\lim\limits_{x \to 0} \cos \dfrac{\pi}{x}$ by using a table and evaluating the function $f(x) = \cos \dfrac{\pi}{x}$ at

$$x = -1, -\frac{1}{3}, -\frac{1}{5}, -\frac{1}{7}, -\frac{1}{9}, \ldots, \frac{1}{9}, \frac{1}{7}, \frac{1}{5}, \frac{1}{3}, 1.$$

(c) Compare the results from (a) and (b). What do you conclude about the limit? Why do you think this happens? What is your view about using a table to draw a conclusion about limits?

(d) Use technology to graph $f$. Begin with the $x$-window $[-2\pi, 2\pi]$ and the $y$-window $[-1, 1]$. If you were finding $\lim\limits_{x \to 0} f(x)$ using a graph, what would you conclude? Zoom in on the graph. Describe what you see. (*Hint:* Be sure your calculator is set to the radian mode.)

**56.** (a) Investigate $\lim\limits_{x \to 0} \cos \dfrac{\pi}{x^2}$ by using a table and evaluating the function $f(x) = \cos \dfrac{\pi}{x^2}$ at $x = -0.1, -0.01, -0.001,$

$-0.0001, \ldots, 0.0001, 0.001, 0.01, 0.1.$

(b) Investigate $\lim\limits_{x \to 0} \cos \dfrac{\pi}{x^2}$ by using a table and evaluating the

function $f(x) = \cos \dfrac{\pi}{x^2}$ at

$x = -\dfrac{2}{3}, -\dfrac{2}{5}, -\dfrac{2}{7}, -\dfrac{2}{9}, \ldots, \dfrac{2}{9}, \dfrac{2}{7}, \dfrac{2}{5}, \dfrac{2}{3}.$

(c) Compare the results from (a) and (b). What do you conclude about the limit? Why do you think this happens? What is your view about using a table to draw a conclusion about limits?

(d) Use technology to graph $f$. Begin with the $x$-window $[-2\pi, 2\pi]$ and the $y$-window $[-1, 1]$. If you were finding $\lim\limits_{x \to 0} f(x)$ using a graph, what would you conclude?

Zoom in on the graph. Describe what you see.

(*Hint*: Be sure your calculator is set to the radian mode.)

**57.** (a) Use a table to investigate $\lim\limits_{x \to 2} \dfrac{x-8}{2}$.

(b) How close must $x$ be to 2, so that $f(x)$ is within 0.1 of the limit?

(c) How close must $x$ be to 2, so that $f(x)$ is within 0.01 of the limit?

**58.** (a) Use a table to investigate $\lim\limits_{x \to 2} (5 - 2x)$.

(b) How close must $x$ be to 2, so that $f(x)$ is within 0.1 of the limit?

(c) How close must $x$ be to 2, so that $f(x)$ is within 0.01 of the limit?

**59. First-Class Mail** As of January 1, 2017, the U.S. Postal Service charged $0.47 postage for first-class letters weighing up to and including 1 ounce, plus a flat fee of $0.21 for each additional or partial ounce up to and including 3.5 ounces. First-class letter rates do not apply to letters weighing more than 3.5 ounces.

*Source*: U.S. Postal Service Notice 123

(a) Find a function $C$ that models the first-class postage charged, in dollars, for a letter weighing $w$ ounces. Assume $w > 0$.

(b) What is the domain of $C$?

(c) Graph the function $C$.

(d) Use the graph to investigate $\lim\limits_{w \to 2^-} C(w)$ and $\lim\limits_{w \to 2^+} C(w)$.

Do these suggest that $\lim\limits_{w \to 2} C(w)$ exists?

(e) Use the graph to investigate $\lim\limits_{w \to 0^+} C(w)$.

(f) Use the graph to investigate $\lim\limits_{w \to 3.5^-} C(w)$.

**60. First-Class Mail** As of January 1, 2017, the U.S. Postal Service charged $0.94 postage for first-class large envelopes weighing up to and including 1 ounce, plus a flat fee of $0.21 for each additional or partial ounce up to and including 13 ounces. First-class rates do not apply to large envelopes weighing more than 13 ounces.

*Source*: U.S. Postal Service Notice 123

(a) Find a function $C$ that models the first-class postage charged, in dollars, for a large envelope weighing $w$ ounces. Assume $w > 0$.

(b) What is the domain of $C$?

(c) Graph the function $C$.

(d) Use the graph to investigate $\lim\limits_{w \to 1^-} C(w)$ and $\lim\limits_{w \to 1^+} C(w)$.

Do these suggest that $\lim\limits_{w \to 1} C(w)$ exists?

(e) Use the graph to investigate $\lim\limits_{w \to 12^-} C(w)$ and $\lim\limits_{w \to 12^+} C(w)$.

Do these suggest that $\lim\limits_{w \to 12} C(w)$ exists?

(f) Use the graph to investigate $\lim\limits_{w \to 0^+} C(w)$.

(g) Use the graph to investigate $\lim\limits_{w \to 13^-} C(w)$.

**61. Correlating Student Success to Study Time** Professor Smith claims that a student's final exam score is a function of the time $t$ (in hours) that the student studies. He claims that the closer to seven hours one studies, the closer to 100% the student scores on the final. He claims that studying significantly less than seven hours may cause one to be underprepared for the test, while studying significantly more than seven hours may cause "burnout."

(a) Write Professor Smith's claim symbolically as a limit.

(b) Write Professor Smith's claim using the $\varepsilon$-$\delta$ definition of limit.

*Source*: Submitted by the students of Millikin University.

**62.** The definition of the slope of the tangent line to the graph of $y = f(x)$ at the point $(c, f(c))$ is $m_{\tan} = \lim\limits_{x \to c} \dfrac{f(x) - f(c)}{x - c}$.

Another way to express this slope is to define a new variable $h = x - c$. Rewrite the slope of the tangent line $m_{\tan}$ using $h$ and $c$.

**63.** If $f(2) = 6$, can you conclude anything about $\lim\limits_{x \to 2} f(x)$? Explain your reasoning.

**64.** If $\lim\limits_{x \to 2} f(x) = 6$, can you conclude anything about $f(2)$? Explain your reasoning.

**65.** The graph of $f(x) = \dfrac{x-3}{3-x}$ is a straight line with a point punched out.

(a) What straight line and what point?

(b) Use the graph of $f$ to investigate the one-sided limits of $f$ as $x$ approaches 3.

(c) Does the graph suggest that $\lim\limits_{x \to 3} f(x)$ exists? If so, what is it?

**66.** (a) Use a table to investigate $\lim\limits_{x \to 0} (1 + x)^{1/x}$.

(b) Use graphing technology to graph $g(x) = (1 + x)^{1/x}$.

(c) What do (a) and (b) suggest about $\lim\limits_{x \to 0} (1 + x)^{1/x}$?

(d) Find $\lim\limits_{x \to 0} (1 + x)^{1/x}$.

## Challenge Problems

*For Problems 67–70, investigate each of the following limits.*

$$f(x) = \begin{cases} 1 & \text{if } x \text{ is an integer} \\ 0 & \text{if } x \text{ is not an integer} \end{cases}$$

**67.** $\lim\limits_{x \to 2} f(x)$  **68.** $\lim\limits_{x \to 1/2} f(x)$  **69.** $\lim\limits_{x \to 3} f(x)$  **70.** $\lim\limits_{x \to 0} f(x)$

# 1.2 Limits of Functions Using Properties of Limits

**OBJECTIVES** *When you finish this section, you should be able to:*

1 Find the limit of a sum, a difference, and a product (p. 91)
2 Find the limit of a power and the limit of a root (p. 93)
3 Find the limit of a polynomial (p. 95)
4 Find the limit of a quotient (p. 96)
5 Find the limit of an average rate of change (p. 98)
6 Find the limit of a difference quotient (p. 98)

In Section 1.1, we used a numerical approach (tables) and a graphical approach to investigate limits. We saw that these approaches are not always reliable. The only way to be sure a limit is correct is to use the $\varepsilon$-$\delta$ definition of a limit. In this section, we state without proof results based on the $\varepsilon$-$\delta$ definition. Some of the results are proved in Section 1.6 and others in Appendix B.

We begin with two basic limits.

## THEOREM  The Limit of a Constant

If $f(x) = A$, where $A$ is a constant, then for any real number $c$,

$$\lim_{x \to c} f(x) = \lim_{x \to c} A = A \qquad (1)$$

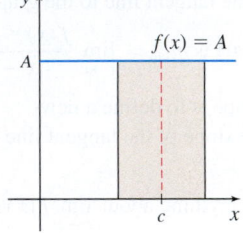

**Figure 17** For $x$ close to $c$, the value of $f$ remains at $A$; $\lim_{x \to c} A = A$.

The theorem is proved in Section 1.6. See Figure 17 and Table 8 for graphical and numerical support of $\lim_{x \to c} A = A$.

**TABLE 8**

|  | $x$ approaches $c$ from the left | | | | $x$ approaches $c$ from the right | | |
|---|---|---|---|---|---|---|---|
| $x$ | $c - 0.01$ | $c - 0.001$ | $c - 0.0001$ | $\to c \leftarrow$ | $c + 0.0001$ | $c + 0.001$ | $c + 0.01$ |
| $f(x) = A$ | $A$ | $A$ | $A$ | $f(x)$ remains at $A$ | $A$ | $A$ | $A$ |

For example,

$$\lim_{x \to 5} 2 = 2 \qquad \lim_{x \to \sqrt{2}} \frac{1}{3} = \frac{1}{3} \qquad \lim_{x \to 5} (-\pi) = -\pi$$

## THEOREM  The Limit of the Identity Function

If $f(x) = x$, then for any real number $c$,

$$\lim_{x \to c} f(x) = \lim_{x \to c} x = c \qquad (2)$$

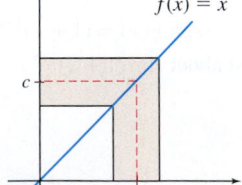

**Figure 18** For $x$ close to $c$, the value of $f$ is just as close to $c$; $\lim_{x \to c} x = c$.

This theorem is proved in Section 1.6. See Figure 18 and Table 9 for graphical and numerical support of $\lim_{x \to c} x = c$.

**TABLE 9**

|  | $x$ approaches $c$ from the left | | | | $x$ approaches $c$ from the right | | |
|---|---|---|---|---|---|---|---|
| $x$ | $c - 0.01$ | $c - 0.001$ | $c - 0.0001$ | $\to c \leftarrow$ | $c + 0.0001$ | $c + 0.001$ | $c + 0.01$ |
| $f(x) = x$ | $c - 0.01$ | $c - 0.001$ | $c - 0.0001$ | $f(x)$ approaches $c$ | $c + 0.0001$ | $c + 0.001$ | $c + 0.01$ |

For example,

$$\lim_{x \to -5} x = -5 \qquad \lim_{x \to \sqrt{3}} x = \sqrt{3} \qquad \lim_{x \to 0} x = 0$$

## 1 Find the Limit of a Sum, a Difference, and a Product

Many functions are combinations of sums, differences, and products of a constant function and the identity function. The following properties can be used to find the limit of such functions.

> **THEOREM  Limit of a Sum**
>
> If $f$ and $g$ are functions for which $\lim\limits_{x \to c} f(x)$ and $\lim\limits_{x \to c} g(x)$ both exist, then $\lim\limits_{x \to c}[f(x) + g(x)]$ exists and
>
> $$\boxed{\lim_{x \to c}[f(x) + g(x)] = \lim_{x \to c} f(x) + \lim_{x \to c} g(x)}$$

**IN WORDS**  The limit of the sum of two functions equals the sum of their limits.

A proof is given in Appendix B.

**EXAMPLE 1  Finding the Limit of a Sum**

Find $\lim\limits_{x \to -3}(x + 4)$.

**Solution** $F(x) = x + 4$ is the sum of two functions $f(x) = x$ and $g(x) = 4$.

From the limits given in (1) and (2), we have

$$\lim_{x \to -3} f(x) = \lim_{x \to -3} x = -3 \qquad \text{and} \qquad \lim_{x \to -3} g(x) = \lim_{x \to -3} 4 = 4$$

Then using the Limit of a Sum, we have

$$\lim_{x \to -3}(x + 4) = \lim_{x \to -3} x + \lim_{x \to -3} 4 = -3 + 4 = 1 \qquad \blacksquare$$

> **THEOREM  Limit of a Difference**
>
> If $f$ and $g$ are functions for which $\lim\limits_{x \to c} f(x)$ and $\lim\limits_{x \to c} g(x)$ both exist, then $\lim\limits_{x \to c}[f(x) - g(x)]$ exists and
>
> $$\boxed{\lim_{x \to c}[f(x) - g(x)] = \lim_{x \to c} f(x) - \lim_{x \to c} g(x)}$$

**IN WORDS**  The limit of the difference of two functions equals the difference of their limits.

**EXAMPLE 2  Finding the Limit of a Difference**

Find $\lim\limits_{x \to 4}(6 - x)$.

**Solution** $F(x) = 6 - x$ is the difference of two functions $f(x) = 6$ and $g(x) = x$.

$$\lim_{x \to 4} f(x) = \lim_{x \to 4} 6 = 6 \qquad \text{and} \qquad \lim_{x \to 4} g(x) = \lim_{x \to 4} x = 4$$

Then using the Limit of a Difference, we have

$$\lim_{x \to 4}(6 - x) = \lim_{x \to 4} 6 - \lim_{x \to 4} x = 6 - 4 = 2 \qquad \blacksquare$$

> **THEOREM  Limit of a Product**
>
> If $f$ and $g$ are functions for which $\lim\limits_{x \to c} f(x)$ and $\lim\limits_{x \to c} g(x)$ both exist, then $\lim\limits_{x \to c}[f(x) \cdot g(x)]$ exists and
>
> $$\boxed{\lim_{x \to c}[f(x) \cdot g(x)] = \lim_{x \to c} f(x) \cdot \lim_{x \to c} g(x)}$$

**IN WORDS**  The limit of the product of two functions equals the product of their limits.

A proof is given in Appendix B.

EXAMPLE 3  **Finding the Limit of a Product**

Find:

(a) $\lim\limits_{x\to3} x^2$        (b) $\lim\limits_{x\to-5} (-4x)$

**Solution** (a) $F(x)=x^2$ is the product of two functions, $f(x)=x$ and $g(x)=x$. Then using the Limit of a Product, we have

$$\lim\limits_{x\to3} x^2 = \lim\limits_{x\to3} x \cdot \lim\limits_{x\to3} x = 3\cdot 3 = 9$$

(b) $F(x)=-4x$ is the product of two functions, $f(x)=-4$ and $g(x)=x$. Then using the Limit of a Product, we have

$$\lim\limits_{x\to-5} (-4x) = \lim\limits_{x\to-5}(-4) \cdot \lim\limits_{x\to-5} x = (-4)(-5) = 20 \quad \blacksquare$$

A *corollary*\* of the Limit of a Product Theorem is the special case when $f(x)=k$ is a constant function.

> **COROLLARY  Limit of a Constant Times a Function**
>
> If $g$ is a function for which $\lim\limits_{x\to c} g(x)$ exists and if $k$ is any real number, then $\lim\limits_{x\to c}[kg(x)]$ exists and
>
> $$\boxed{\lim\limits_{x\to c}[kg(x)] = k \lim\limits_{x\to c} g(x)}$$

**IN WORDS** The limit of a constant times a function equals the constant times the limit of the function.

You are asked to prove this corollary in Problem 103.

Limit properties often are used in combination.

EXAMPLE 4  **Finding a Limit**

Find:

(a) $\lim\limits_{x\to1}[2x(x+4)]$        (b) $\lim\limits_{x\to2^+}[4x(2-x)]$

**Solution** (a)

$$\lim\limits_{x\to1}[(2x)(x+4)] = \left[\lim\limits_{x\to1}(2x)\right]\left[\lim\limits_{x\to1}(x+4)\right] \qquad \text{Limit of a Product}$$

$$= \left[2\cdot \lim\limits_{x\to1} x\right]\cdot\left[\lim\limits_{x\to1} x + \lim\limits_{x\to1} 4\right] \qquad \begin{array}{l}\text{Limit of a Constant Times a}\\ \text{Function, Limit of a Sum}\end{array}$$

$$= (2\cdot 1)\cdot(1+4) = 10 \qquad \text{Use (2) and (1), and simplify.}$$

**NOTE** The limit properties are also true for one-sided limits.

(b) We use properties of limits to find the one-sided limit.

$$\lim\limits_{x\to2^+}[4x(2-x)] = 4\lim\limits_{x\to2^+}[x(2-x)] = 4\left[\lim\limits_{x\to2^+} x\right]\left[\lim\limits_{x\to2^+}(2-x)\right]$$

$$= 4\cdot 2\left[\lim\limits_{x\to2^+} 2 - \lim\limits_{x\to2^+} x\right] = 4\cdot 2\cdot(2-2) = 0 \quad \blacksquare$$

NOW WORK **Problem 13.**

To find the limit of piecewise-defined functions at numbers where the defining equation changes requires the use of one-sided limits.

---

\*A **corollary** is a theorem that follows directly from a previously proved theorem.

**RECALL** The limit $L$ of a function $y = f(x)$ as $x$ approaches a number $c$ exists if and only if

$$\lim_{x \to c^-} f(x) = \lim_{x \to c^+} f(x) = L.$$

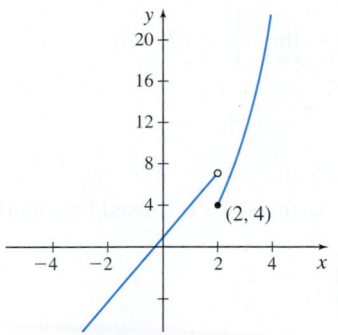

**Figure 19** $f(x) = \begin{cases} 3x + 1 & \text{if} \quad x < 2 \\ 2x(x-1) & \text{if} \quad x \geq 2 \end{cases}$

**ORIGINS  Oliver Heaviside** (1850–1925) was a self-taught mathematician and electrical engineer. He developed a branch of mathematics called **operational calculus** in which differential equations are solved by converting them to algebraic equations. Heaviside applied vector calculus to electrical engineering and, perhaps most significantly, he simplified *Maxwell's equations* to the form used by electrical engineers to this day. In 1902 Heaviside claimed there is a layer surrounding Earth from which radio signals bounce, allowing the signals to travel around the Earth. Heaviside's claim was proved true in 1923. The layer, contained in the ionosphere, is named the **Heaviside layer**. The function we discuss here is one of his minor contributions to mathematics and electrical engineering.

**EXAMPLE 5  Finding a Limit for a Piecewise-defined Function**

Find $\lim\limits_{x \to 2} f(x)$, if it exists.

$$f(x) = \begin{cases} 3x + 1 & \text{if} \quad x < 2 \\ 2x(x-1) & \text{if} \quad x \geq 2 \end{cases}$$

**Solution** Since the rule for $f$ changes at 2, we need to find the one-sided limits of $f$ as $x$ approaches 2.

For $x < 2$, we use the left-hand limit. Also, because $x < 2$, $f(x) = 3x + 1$.

$$\lim_{x \to 2^-} f(x) = \lim_{x \to 2^-} (3x + 1) = \lim_{x \to 2^-} (3x) + \lim_{x \to 2^-} 1 = 3 \lim_{x \to 2^-} x + 1 = 3 \cdot 2 + 1 = 7$$

For $x \geq 2$, we use the right-hand limit. Also, because $x \geq 2$, $f(x) = 2x(x-1)$.

$$\lim_{x \to 2^+} f(x) = \lim_{x \to 2^+} [2x(x-1)] = \lim_{x \to 2^+} (2x) \cdot \lim_{x \to 2^+} (x-1)$$

$$= 2 \lim_{x \to 2^+} x \cdot \left[ \lim_{x \to 2^+} x - \lim_{x \to 2^+} 1 \right] = 2 \cdot 2 [2 - 1] = 4$$

Since $\lim\limits_{x \to 2^-} f(x) = 7 \neq \lim\limits_{x \to 2^+} f(x) = 4$, $\lim\limits_{x \to 2} f(x)$ does not exist. ∎

See Figure 19.

**NOW WORK** Problem 73.

The **Heaviside function**, $u_c(t) = \begin{cases} 0 & \text{if} \quad t < c \\ 1 & \text{if} \quad t \geq c \end{cases}$, is a function that is used in electrical engineering to model a switch. The switch is off if $t < c$, and it is on if $t \geq c$.

**EXAMPLE 6  Finding a Limit of the Heaviside Function**

Find $\lim\limits_{t \to 0} u_0(t)$, where $u_0(t) = \begin{cases} 0 & \text{if} \quad t < 0 \\ 1 & \text{if} \quad t \geq 0 \end{cases}$

**Solution** Since the Heaviside function changes rules at $t = 0$, we find the one-sided limits as $t$ approaches 0.

For $t < 0$, $\lim\limits_{t \to 0^-} u_0(t) = \lim\limits_{t \to 0^-} 0 = 0$    and    for $t \geq 0$, $\lim\limits_{t \to 0^+} u_0(t) = \lim\limits_{t \to 0^+} 1 = 1$

Since the one-sided limits as $t$ approaches 0 are not equal, $\lim\limits_{t \to 0} u_0(t)$ does not exist. ∎

**NOW WORK** Problem 81.

## 2 Find the Limit of a Power and the Limit of a Root

Using the Limit of a Product, if $\lim\limits_{x \to c} f(x)$ exists, then

$$\lim_{x \to c} [f(x)]^2 = \lim_{x \to c} [f(x) \cdot f(x)] = \lim_{x \to c} f(x) \cdot \lim_{x \to c} f(x) = \left[ \lim_{x \to c} f(x) \right]^2$$

Repeated use of this property produces the next corollary.

**COROLLARY  Limit of a Power**

If $\lim\limits_{x \to c} f(x)$ exists and if $n$ is a positive integer, then

$$\boxed{\lim_{x \to c} [f(x)]^n = \left[ \lim_{x \to c} f(x) \right]^n}$$

## EXAMPLE 7   Finding the Limit of a Power

Find:

(a)  $\lim\limits_{x \to 2} x^5$     (b)  $\lim\limits_{x \to 1} (2x - 3)^3$     (c)  $\lim\limits_{x \to c} x^n$     $n$ a positive integer

**Solution (a)**  $\lim\limits_{x \to 2} x^5 = \left( \lim\limits_{x \to 2} x \right)^5 = 2^5 = 32$

(b)  $\lim\limits_{x \to 1} (2x - 3)^3 = \left[ \lim\limits_{x \to 1} (2x - 3) \right]^3 = \left[ \lim\limits_{x \to 1} (2x) - \lim\limits_{x \to 1} 3 \right]^3 = (2 - 3)^3 = -1$

(c)  $\lim\limits_{x \to c} x^n = \left[ \lim\limits_{x \to c} x \right]^n = c^n$  ∎

The result from Example 7(c) is worth remembering since it is used frequently:

$$\lim\limits_{x \to c} x^n = c^n$$

where $c$ is a number and $n$ is a positive integer.

**NOW WORK** Problem 15.

### THEOREM   Limit of a Root

If $\lim\limits_{x \to c} f(x)$ exists and if $n \geq 2$ is an integer, then

$$\lim\limits_{x \to c} \sqrt[n]{f(x)} = \sqrt[n]{\lim\limits_{x \to c} f(x)}$$

provided $f(x) > 0$ if $n$ is even.

## EXAMPLE 8   Finding the Limit of a Root

Find $\lim\limits_{x \to 4} \sqrt[3]{x^2 + 11}$.

**Solution**

$$\lim\limits_{x \to 4} \sqrt[3]{x^2 + 11} \underset{\substack{\uparrow \\ \text{Limit of a Root}}}{=} \sqrt[3]{\lim\limits_{x \to 4} (x^2 + 11)} = \sqrt[3]{\lim\limits_{x \to 4} x^2 + \lim\limits_{x \to 4} 11}$$

$$= \sqrt[3]{4^2 + 11} = \sqrt[3]{27} = 3$$  ∎

**NOW WORK** Problem 19.

The Limit of a Power and the Limit of a Root are used together to find the limit of a function with a rational exponent.

### THEOREM   Limit of a Fractional Power $[f(x)]^{m/n}$

If $f$ is a function for which $\lim\limits_{x \to c} f(x)$ exists and if $[f(x)]^{m/n}$ is defined for positive integers $m$ and $n$, then

$$\lim\limits_{x \to c} [f(x)]^{m/n} = \left[ \lim\limits_{x \to c} f(x) \right]^{m/n}$$

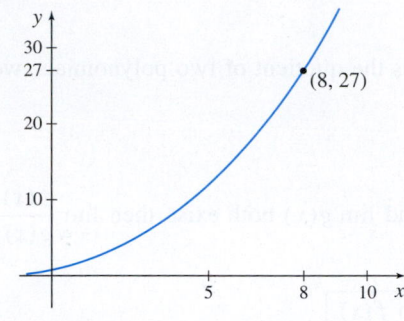

**Figure 20** $f(x) = (x+1)^{3/2}$

**EXAMPLE 9** **Finding the Limit of a Fractional Power $[f(x)]^{m/n}$**

Find $\lim\limits_{x \to 8}(x+1)^{3/2}$.

**Solution** Let $f(x) = x+1$. Near 8, $x+1 > 0$, so $(x+1)^{3/2}$ is defined. Then

$$\lim_{x \to 8}[f(x)]^{3/2} = \lim_{x \to 8}(x+1)^{3/2} = \left[\lim_{x \to 8}(x+1)\right]^{3/2} = [8+1]^{3/2} = 9^{3/2} = 27$$

$$\uparrow$$
$$\lim_{x \to c}[f(x)]^{m/n} = \left[\lim_{x \to c}f(x)\right]^{m/n}$$

See Figure 20.

**NOW WORK** Problem **23**.

## 3 Find the Limit of a Polynomial

Sometimes $\lim\limits_{x \to c} f(x)$ can be found by substituting $c$ for $x$ in $f(x)$. For example,

$$\lim_{x \to 2}(5x^2) = 5\lim_{x \to 2}x^2 = 5 \cdot 2^2 = 20$$

Since $\lim\limits_{x \to c} x^n = c^n$ if $n$ is a positive integer, we can use the Limit of a Constant Times a Function to obtain a formula for the limit of a monomial $f(x) = ax^n$.

$$\boxed{\lim_{x \to c}(ax^n) = ac^n}$$

where $a$ is any number.

Since a polynomial is the sum of monomials and the limit of a sum is the sum of the limits, we have the following result.

> **THEOREM** **Limit of a Polynomial Function**
>
> If $P$ is a polynomial function, then
>
> $$\boxed{\lim_{x \to c}P(x) = P(c)}$$
>
> for any number $c$.

**IN WORDS** To find the limit of a polynomial as $x$ approaches $c$, evaluate the polynomial at $c$.

**Proof** If $P$ is a polynomial function, that is, if

$$P(x) = a_n x^n + a_{n-1}x^{n-1} + \cdots + a_1 x + a_0$$

where $n$ is a nonnegative integer, then

$$\lim_{x \to c}P(x) = \lim_{x \to c}\left(a_n x^n + a_{n-1}x^{n-1} + \cdots + a_1 x + a_0\right)$$

$$= \lim_{x \to c}\left(a_n x^n\right) + \lim_{x \to c}\left(a_{n-1}x^{n-1}\right) + \cdots + \lim_{x \to c}(a_1 x) + \lim_{x \to c}a_0$$

$$= a_n c^n + a_{n-1}c^{n-1} + \cdots + a_1 c + a_0 \qquad \text{Limit of a Monomial}$$

$$= P(c)$$

**EXAMPLE 10** **Finding the Limit of a Polynomial**

Find the limit of each polynomial:

**(a)** $\lim\limits_{x \to 3}(4x^2 - x + 2) = 4 \cdot 3^2 - 3 + 2 = 35$

**(b)** $\lim\limits_{x \to -1}(7x^5 + 4x^3 - 2x^2) = 7(-1)^5 + 4(-1)^3 - 2(-1)^2 = -13$

**(c)** $\lim\limits_{x \to 0}(10x^6 - 4x^5 - 8x + 5) = 10 \cdot 0^6 - 4 \cdot 0^5 - 8 \cdot 0 + 5 = 5$

**NOW WORK** Problem **29**.

## 4 Find the Limit of a Quotient

To find the limit of a rational function, which is the quotient of two polynomials, we use the following result.

### THEOREM Limit of a Quotient

If $f$ and $g$ are functions for which $\lim\limits_{x \to c} f(x)$ and $\lim\limits_{x \to c} g(x)$ both exist, then $\lim\limits_{x \to c} \dfrac{f(x)}{g(x)}$ exists and

$$\lim_{x \to c} \frac{f(x)}{g(x)} = \frac{\lim\limits_{x \to c} f(x)}{\lim\limits_{x \to c} g(x)}$$

provided $\lim\limits_{x \to c} g(x) \neq 0$.

**IN WORDS** The limit of the quotient of two functions equals the quotient of their limits, provided that the limit of the denominator is not zero.

**NEED TO REVIEW?** Rational functions are discussed in Section P.2, p. 22.

### COROLLARY Limit of a Rational Function

If the number $c$ is in the domain of a rational function $R(x) = \dfrac{p(x)}{q(x)}$, then

$$\lim_{x \to c} R(x) = R(c) \tag{3}$$

You are asked to prove this corollary in Problem 104.

### EXAMPLE 11 Finding the Limit of a Rational Function

Find:

(a) $\displaystyle\lim_{x \to 1} \frac{3x^3 - 2x + 1}{4x^2 + 5}$     (b) $\displaystyle\lim_{x \to -2} \frac{2x + 4}{3x^2 - 1}$

**Solution** (a) Since 1 is in the domain of the rational function $R(x) = \dfrac{3x^3 - 2x + 1}{4x^2 + 5}$,

$$\lim_{x \to 1} R(x) \underset{\substack{\uparrow \\ \text{Use (3)}}}{=} R(1) = \frac{3 - 2 + 1}{4 + 5} = \frac{2}{9}$$

(b) Since $-2$ is in the domain of the rational function $H(x) = \dfrac{2x + 4}{3x^2 - 1}$,

$$\lim_{x \to -2} H(x) \underset{\substack{\uparrow \\ \text{Use (3)}}}{=} H(-2) = \frac{-4 + 4}{12 - 1} = \frac{0}{11} = 0$$

∎

**NOW WORK** Problem 33.

### EXAMPLE 12 Finding the Limit of a Quotient

Find $\displaystyle\lim_{x \to 4} \frac{\sqrt{3x^2 + 1}}{x - 1}$.

**Solution** We seek the limit of the quotient of two functions. Since the limit of the denominator $\lim\limits_{x \to 4}(x - 1) \neq 0$, we use the Limit of a Quotient.

$$\lim_{x \to 4} \frac{\sqrt{3x^2 + 1}}{x - 1} \underset{\substack{\uparrow \\ \text{Limit of a Quotient}}}{=} \frac{\lim\limits_{x \to 4} \sqrt{3x^2 + 1}}{\lim\limits_{x \to 4}(x - 1)} \underset{\substack{\uparrow \\ \text{Limit of a Root}}}{=} \frac{\sqrt{\lim\limits_{x \to 4}(3x^2 + 1)}}{\lim\limits_{x \to 4}(x - 1)} = \frac{\sqrt{3 \cdot 4^2 + 1}}{4 - 1} = \frac{\sqrt{49}}{3} = \frac{7}{3}$$

∎

**NOW WORK** Problem 31.

Based on these examples, you might be tempted to conclude that finding a limit as $x$ approaches $c$ is simply a matter of substituting the number $c$ into the function. The next few examples show that substitution cannot always be used and other strategies need to be employed.

The limit as $x$ approaches $c$ of a rational function can be found using substitution, provided the number $c$ is in the domain of the rational function. The next example shows a strategy that can be tried when $c$ is not in the domain.

**EXAMPLE 13**  **Finding the Limit of a Rational Function**

Find $\lim\limits_{x \to -2} \dfrac{x^2 + 5x + 6}{x^2 - 4}$.

**Solution** Since $-2$ is not in the domain of the rational function, substitution cannot be used. But this does not mean that the limit does not exist! Factoring the numerator and the denominator, we find

$$\frac{x^2 + 5x + 6}{x^2 - 4} = \frac{(x + 2)(x + 3)}{(x + 2)(x - 2)}$$

Since $x \neq -2$, and we are interested in the limit as $x$ approaches $-2$, the factor $x + 2$ can be divided out. Then

$$\lim_{x \to -2} \frac{x^2 + 5x + 6}{x^2 - 4} = \lim_{x \to -2} \frac{(x + 2)(x + 3)}{(x + 2)(x - 2)} = \lim_{x \to -2} \frac{x + 3}{x - 2} = \frac{-2 + 3}{-2 - 2} = -\frac{1}{4}$$

$$\underset{\text{Factor}}{\uparrow} \qquad \underset{\substack{x \neq -2 \\ \text{Divide out } (x + 2)}}{\uparrow} \qquad \underset{\substack{\text{Use the Limit of a} \\ \text{Rational Function}}}{\uparrow}$$

∎

**NOW WORK** Problem 35.

The Limit of a Quotient property can be used only when the limit of the denominator of the function is not zero. The next example illustrates a strategy to try if radicals are present.

**EXAMPLE 14**  **Finding the Limit of a Quotient**

Find $\lim\limits_{x \to 5} \dfrac{\sqrt{x} - \sqrt{5}}{x - 5}$.

**Solution** The domain of $h(x) = \dfrac{\sqrt{x} - \sqrt{5}}{x - 5}$ is $\{x \mid x \geq 0, x \neq 5\}$. Since the limit of the denominator is

$$\lim_{x \to 5} (x - 5) = 0$$

we cannot use the Limit of a Quotient property. A different strategy is necessary. We rationalize the numerator of the quotient.

$$\frac{\sqrt{x} - \sqrt{5}}{x - 5} = \frac{(\sqrt{x} - \sqrt{5})}{(x - 5)} \cdot \frac{(\sqrt{x} + \sqrt{5})}{(\sqrt{x} + \sqrt{5})} = \frac{x - 5}{(x - 5)(\sqrt{x} + \sqrt{5})} = \frac{1}{\sqrt{x} + \sqrt{5}}$$

$$\underset{x \neq 5}{\uparrow}$$

Do you see why rationalizing the numerator works? It causes the term $x - 5$ to appear in the numerator, and since $x \neq 5$, the factor $x - 5$ can be divided out. Then

**NOTE** When finding a limit, remember to include "$\lim\limits_{x \to c}$" at each step until you let $x \to c$.

$$\lim_{x \to 5} \frac{\sqrt{x} - \sqrt{5}}{x - 5} = \lim_{x \to 5} \frac{1}{\sqrt{x} + \sqrt{5}} = \frac{\lim\limits_{x \to 5} 1}{\lim\limits_{x \to 5} (\sqrt{x} + \sqrt{5})} = \frac{1}{\sqrt{5} + \sqrt{5}} = \frac{1}{2\sqrt{5}} = \frac{\sqrt{5}}{10}$$

$$\underset{\text{Use the Limit of a Quotient}}{\uparrow}$$

∎

**NOW WORK** Problem 41.

## 5 Find the Limit of an Average Rate of Change

The next two examples illustrate limits that we encounter in Chapter 2.

NEED TO REVIEW? Average rate of change is discussed in Section P.1, p. 12.

If $c$ and $x$, where $x \neq c$, are in the domain of a function $y = f(x)$, the average rate of change of $f$ from $c$ to $x$ is

$$\frac{\Delta y}{\Delta x} = \frac{f(x) - f(c)}{x - c} \qquad x \neq c$$

### EXAMPLE 15  Finding the Limit of an Average Rate of Change

(a)  Find the average rate of change of $f(x) = x^2 + 3x$ from 2 to $x$, $x \neq 2$.

(b)  Find the limit as $x$ approaches 2 of the average rate of change of $f(x) = x^2 + 3x$ from 2 to $x$.

**Solution** (a)  The average rate of change of $f$ from 2 to $x$ is

$$\frac{\Delta y}{\Delta x} = \frac{f(x) - f(2)}{x - 2} = \frac{(x^2 + 3x) - [2^2 + 3 \cdot 2]}{x - 2} = \frac{x^2 + 3x - 10}{x - 2} = \frac{(x + 5)(x - 2)}{x - 2}$$

(b)  The limit of the average rate of change is

$$\lim_{x \to 2} \frac{f(x) - f(2)}{x - 2} = \lim_{x \to 2} \frac{(x + 5)(x - 2)}{x - 2} = \lim_{x \to 2} (x + 5) = 7 \qquad \blacksquare$$

NOW WORK Problem 63.

## 6 Find the Limit of a Difference Quotient

In Section P.1, we defined the difference quotient of a function $f$ at $x$ as

$$\frac{f(x + h) - f(x)}{h} \qquad h \neq 0$$

CALC
CLIP

### EXAMPLE 16  Finding the Limit of a Difference Quotient

(a)  For $f(x) = 2x^2 - 3x + 1$, find the difference quotient $\dfrac{f(x + h) - f(x)}{h}$, $h \neq 0$.

(b)  Find the limit as $h$ approaches 0 of the difference quotient of $f(x) = 2x^2 - 3x + 1$.

**Solution** (a)  To find the difference quotient of $f$, we begin with $f(x + h)$.

$$f(x + h) = 2(x + h)^2 - 3(x + h) + 1 = 2(x^2 + 2xh + h^2) - 3x - 3h + 1$$

$$= 2x^2 + 4xh + 2h^2 - 3x - 3h + 1$$

Now

$$f(x + h) - f(x) = (2x^2 + 4xh + 2h^2 - 3x - 3h + 1) - (2x^2 - 3x + 1) = 4xh + 2h^2 - 3h$$

Then the difference quotient is

$$\frac{f(x + h) - f(x)}{h} = \frac{4xh + 2h^2 - 3h}{h} = \frac{h(4x + 2h - 3)}{h} = 4x + 2h - 3, \quad h \neq 0$$

(b)  $\displaystyle\lim_{h \to 0} \frac{f(x + h) - f(x)}{h} = \lim_{h \to 0} (4x + 2h - 3) = 4x + 0 - 3 = 4x - 3 \qquad \blacksquare$

NOW WORK Problem 69.

## Summary

### Two Basic Limits

- $\lim_{x \to c} A = A$, where $A$ is a constant.
- $\lim_{x \to c} x = c$

### Properties of Limits

If $f$ and $g$ are functions for which $\lim_{x \to c} f(x)$ and $\lim_{x \to c} g(x)$ both exist, and $k$ is a constant, then

- **Limit of a Sum or a Difference:**
  $\lim_{x \to c}[f(x) \pm g(x)] = \lim_{x \to c} f(x) \pm \lim_{x \to c} g(x)$

- **Limit of a Product:** $\lim_{x \to c}[f(x) \cdot g(x)] = \lim_{x \to c} f(x) \cdot \lim_{x \to c} g(x)$

- **Limit of a Constant Times a Function:**
  $\lim_{x \to c}[kg(x)] = k \lim_{x \to c} g(x)$

- **Limit of a Power:** $\lim_{x \to c}[f(x)]^n = \left[\lim_{x \to c} f(x)\right]^n$
  where $n$ is a positive integer

- **Limit of a Root:** $\lim_{x \to c} \sqrt[n]{f(x)} = \sqrt[n]{\lim_{x \to c} f(x)}$
  where $n \geq 2$ is an integer, provided $f(x) > 0$ if $n$ is even

- **Limit of $[f(x)]^{m/n}$:** $\lim_{x \to c}[f(x)]^{m/n} = \left[\lim_{x \to c} f(x)\right]^{m/n}$
  provided $[f(x)]^{m/n}$ is defined for positive integers $m$ and $n$

- **Limit of a Quotient:** $\lim_{x \to c} \dfrac{f(x)}{g(x)} = \dfrac{\lim\limits_{x \to c} f(x)}{\lim\limits_{x \to c} g(x)}$
  provided $\lim_{x \to c} g(x) \neq 0$

- **Limit of a Polynomial Function:** $\lim_{x \to c} P(x) = P(c)$

- **Limit of a Rational Function:** $\lim_{x \to c} R(x) = R(c)$
  if $c$ is in the domain of $R$

## 1.2 Assess Your Understanding

### Concepts and Vocabulary

1. (a) $\lim_{x \to 4} (-3) = $ _____; (b) $\lim_{x \to 0} \pi = $ _____

2. If $\lim_{x \to c} f(x) = 3$, then $\lim_{x \to c}[f(x)]^5 = $ _____.

3. If $\lim_{x \to c} f(x) = 64$, then $\lim_{x \to c} \sqrt[3]{f(x)} = $ _____.

4. (a) $\lim_{x \to -1} x = $ _____; (b) $\lim_{x \to e} x = $ _____

5. (a) $\lim_{x \to 0} (x - 2) = $ ____; (b) $\lim_{x \to 1/2} (3 + x) = $ _____

6. (a) $\lim_{x \to 2} (-3x) = $ _____; (b) $\lim_{x \to 0} (3x) = $ _____

7. *True or False*   If $p$ is a polynomial function, then $\lim_{x \to 5} p(x) = p(5)$.

8. If the domain of a rational function $R$ is $\{x \mid x \neq 0\}$, then $\lim_{x \to 2} R(x) = R(\underline{\quad})$.

9. *True or False*   Properties of limits cannot be used for one-sided limits.

10. *True or False*   If $f(x) = \dfrac{(x+1)(x+2)}{x+1}$ and $g(x) = x + 2$, then $\lim_{x \to -1} f(x) = \lim_{x \to -1} g(x)$.

### Skill Building

*In Problems 11–44, find each limit using properties of limits.*

11. $\lim_{x \to 3} [2(x + 4)]$

12. $\lim_{x \to -2} [3(x + 1)]$

13. $\lim_{x \to -2} [x(3x - 1)(x + 2)]$

14. $\lim_{x \to -1} [x(x - 1)(x + 10)]$

15. $\lim_{t \to 1} (3t - 2)^3$

16. $\lim_{x \to 0} (-3x + 1)^2$

17. $\lim_{x \to 4} (3\sqrt{x})$

18. $\lim_{x \to 8} \left(\dfrac{1}{4}\sqrt[3]{x}\right)$

19. $\lim_{x \to 3} \sqrt{5x - 4}$

20. $\lim_{t \to 2} \sqrt{3t + 4}$

21. $\lim_{t \to 2} \left[t\sqrt{(5t + 3)(t + 4)}\right]$

22. $\lim_{t \to -1} \left[t\sqrt[3]{(t + 1)(2t - 1)}\right]$

23. $\lim_{x \to 9} \left(\sqrt{x} + x + 4\right)^{1/2}$

24. $\lim_{t \to 2} \left(t\sqrt{2t} + 4\right)^{1/3}$

25. $\lim_{t \to -1} [4t(t + 1)]^{2/3}$

26. $\lim_{x \to 0} (x^2 - 2x)^{3/5}$

27. $\lim_{t \to 1} (3t^2 - 2t + 4)$

28. $\lim_{x \to 0} (-3x^4 + 2x + 1)$

29. $\lim_{x \to \frac{1}{2}} (2x^4 - 8x^3 + 4x - 5)$

30. $\lim_{x \to -\frac{1}{3}} (27x^3 + 9x + 1)$

31. $\lim_{x \to 4} \dfrac{x^2 + 4}{\sqrt{x}}$

32. $\lim_{x \to 3} \dfrac{x^2 + 5}{\sqrt{3x}}$

33. $\lim_{x \to -2} \dfrac{2x^3 + 5x}{3x - 2}$

34. $\lim_{x \to 1} \dfrac{2x^4 - 1}{3x^3 + 2}$

35. $\lim_{x \to 2} \dfrac{x^2 - 4}{x - 2}$

36. $\lim_{x \to -2} \dfrac{x + 2}{x^2 - 4}$

37. $\lim_{x \to -1} \dfrac{x^3 - x}{x + 1}$

38. $\lim_{x \to -1} \dfrac{x^3 + x^2}{x^2 - 1}$

39. $\lim_{x \to -8} \left(\dfrac{2x}{x + 8} + \dfrac{16}{x + 8}\right)$

40. $\lim_{x \to 2} \left(\dfrac{3x}{x - 2} - \dfrac{6}{x - 2}\right)$

41. $\lim_{x \to 2} \dfrac{\sqrt{x} - \sqrt{2}}{x - 2}$

42. $\lim_{x \to 3} \dfrac{\sqrt{x} - \sqrt{3}}{x - 3}$

43. $\lim_{x \to 4} \dfrac{\sqrt{x + 5} - 3}{(x - 4)(x + 1)}$

44. $\lim_{x \to 3} \dfrac{\sqrt{x + 1} - 2}{x(x - 3)}$

---

**1.** = NOW WORK problem        〰 = Graphing technology recommended        CAS = Computer Algebra System recommended

*In Problems 45–50, find each one-sided limit using properties of limits.*

**45.** $\lim\limits_{x \to 3^-} (x^2 - 4)$

**46.** $\lim\limits_{x \to 2^+} (3x^2 + x)$

**47.** $\lim\limits_{x \to 3^-} \dfrac{x^2 - 9}{x - 3}$

**48.** $\lim\limits_{x \to 3^+} \dfrac{x^2 - 9}{x - 3}$

**49.** $\lim\limits_{x \to 3^-} \left( \sqrt{9 - x^2} + x \right)^2$

**50.** $\lim\limits_{x \to 2^+} \left( 2\sqrt{x^2 - 4} + 3x \right)$

*In Problems 51–58, use the information below to find each limit.*

$$\lim_{x \to c} f(x) = 5 \qquad \lim_{x \to c} g(x) = 2 \qquad \lim_{x \to c} h(x) = 0$$

**51.** $\lim\limits_{x \to c} [f(x) - 3g(x)]$

**52.** $\lim\limits_{x \to c} [5f(x)]$

**53.** $\lim\limits_{x \to c} [g(x)]^3$

**54.** $\lim\limits_{x \to c} \dfrac{f(x)}{g(x) - h(x)}$

**55.** $\lim\limits_{x \to c} \dfrac{h(x)}{g(x)}$

**56.** $\lim\limits_{x \to c} [4f(x) \cdot g(x)]$

**57.** $\lim\limits_{x \to c} \left[ \dfrac{1}{g(x)} \right]^2$

**58.** $\lim\limits_{x \to c} \sqrt[3]{5g(x) - 3}$

*In Problems 59 and 60, use the graphs of the functions and properties of limits to find each limit, if it exists. If the limit does not exist, write, "The limit does not exist," and explain why.*

**59. (a)** $\lim\limits_{x \to 4} [f(x) + g(x)]$

**(b)** $\lim\limits_{x \to 4} \{f(x)[g(x) - h(x)]\}$

**(c)** $\lim\limits_{x \to 4} [f(x) \cdot g(x)]$

**(d)** $\lim\limits_{x \to 4} [2h(x)]$

**(e)** $\lim\limits_{x \to 4} \dfrac{g(x)}{f(x)}$

**(f)** $\lim\limits_{x \to 4} \dfrac{h(x)}{f(x)}$

**60. (a)** $\lim\limits_{x \to 3} \{2[f(x) + h(x)]\}$

**(b)** $\lim\limits_{x \to 3^-} [g(x) + h(x)]$

**(c)** $\lim\limits_{x \to 3} \sqrt[3]{h(x)}$

**(d)** $\lim\limits_{x \to 3} \dfrac{f(x)}{h(x)}$

**(e)** $\lim\limits_{x \to 3} [h(x)]^3$

**(f)** $\lim\limits_{x \to 3} [f(x) - 2h(x)]^{3/2}$

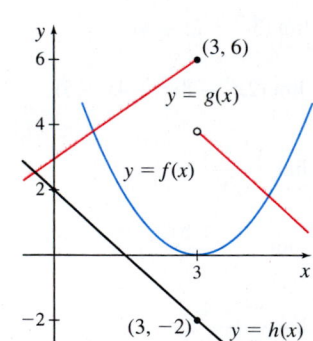

*In Problems 61–66, for each function $f$, find the limit as $x$ approaches $c$ of the average rate of change of $f$ from $c$ to $x$. That is, find*

$$\lim_{x \to c} \frac{f(x) - f(c)}{x - c}$$

**61.** $f(x) = 3x^2, \quad c = 1$

**62.** $f(x) = 8x^3, \quad c = 2$

**63.** $f(x) = -2x^2 + 4, \quad c = 1$

**64.** $f(x) = 20 - 0.8x^2, \quad c = 3$

**65.** $f(x) = \sqrt{x}, \quad c = 1$

**66.** $f(x) = \sqrt{2x}, \quad c = 5$

*In Problems 67–72, find the limit of the difference quotient for each function $f$. That is, find $\lim\limits_{h \to 0} \dfrac{f(x + h) - f(x)}{h}$.*

**67.** $f(x) = 4x - 3$

**68.** $f(x) = 3x + 5$

**69.** $f(x) = 3x^2 + 4x + 1$

**70.** $f(x) = 2x^2 + x$

**71.** $f(x) = \dfrac{2}{x}$

**72.** $f(x) = \dfrac{3}{x^2}$

*In Problems 73–80, find $\lim\limits_{x \to c^-} f(x)$ and $\lim\limits_{x \to c^+} f(x)$ for the given number $c$. Based on the results, determine whether $\lim\limits_{x \to c} f(x)$ exists.*

**73.** $f(x) = \begin{cases} 2x - 3 & \text{if } x \le 1 \\ 3 - x & \text{if } x > 1 \end{cases}$ at $c = 1$

**74.** $f(x) = \begin{cases} 5x + 2 & \text{if } x < 2 \\ 1 + 3x & \text{if } x \ge 2 \end{cases}$ at $c = 2$

**75.** $f(x) = \begin{cases} 3x - 1 & \text{if } x < 1 \\ 4 & \text{if } x = 1 \\ 2x & \text{if } x > 1 \end{cases}$ at $c = 1$

**76.** $f(x) = \begin{cases} 3x - 1 & \text{if } x < 1 \\ 2 & \text{if } x = 1 \\ 2x & \text{if } x > 1 \end{cases}$ at $c = 1$

**77.** $f(x) = \begin{cases} x - 1 & \text{if } x < 1 \\ \sqrt{x - 1} & \text{if } x > 1 \end{cases}$ at $c = 1$

**78.** $f(x) = \begin{cases} \sqrt{9 - x^2} & \text{if } 0 < x < 3 \\ \sqrt{x^2 - 9} & \text{if } x > 3 \end{cases}$ at $c = 3$

**79.** $f(x) = \begin{cases} \dfrac{x^2 - 9}{x - 3} & \text{if } x \ne 3 \\ 6 & \text{if } x = 3 \end{cases}$ at $c = 3$

**80.** $f(x) = \begin{cases} \dfrac{x - 2}{x^2 - 4} & \text{if } x \ne 2 \\ 1 & \text{if } x = 2 \end{cases}$ at $c = 2$

### Applications and Extensions

**Heaviside Functions**  *In Problems 81 and 82, find the limit, if it exists, of the given Heaviside function at $c$.*

**81.** $u_1(t) = \begin{cases} 0 & \text{if } t < 1 \\ 1 & \text{if } t \ge 1 \end{cases}$ at $c = 1$

**82.** $u_3(t) = \begin{cases} 0 & \text{if } t < 3 \\ 1 & \text{if } t \ge 3 \end{cases}$ at $c = 3$

*In Problems 83–92, find each limit.*

**83.** $\lim\limits_{h \to 0} \dfrac{(x + h)^2 - x^2}{h}$

**84.** $\lim\limits_{h \to 0} \dfrac{\sqrt{x + h} - \sqrt{x}}{h}$

**85.** $\lim\limits_{h \to 0} \dfrac{\dfrac{1}{x + h} - \dfrac{1}{x}}{h}$

**86.** $\lim\limits_{h \to 0} \dfrac{\dfrac{1}{(x + h)^3} - \dfrac{1}{x^3}}{h}$

87. $\lim\limits_{x\to 0}\left[\dfrac{1}{x}\left(\dfrac{1}{4+x}-\dfrac{1}{4}\right)\right]$    88. $\lim\limits_{x\to -1}\left[\dfrac{2}{x+1}\left(\dfrac{1}{3}-\dfrac{1}{x+4}\right)\right]$

89. $\lim\limits_{x\to 7}\dfrac{x-7}{\sqrt{x+2}-3}$    90. $\lim\limits_{x\to 2}\dfrac{x-2}{\sqrt{x+2}-2}$

91. $\lim\limits_{x\to 1}\dfrac{x^3-3x^2+3x-1}{x^2-2x+1}$    92. $\lim\limits_{x\to -3}\dfrac{x^3+7x^2+15x+9}{x^2+6x+9}$

93. **Cost of Water**  The Jericho Water District determines quarterly water costs, in dollars, using the following rate schedule:

| Water used (in thousands of gallons) | Cost |
|---|---|
| $0\le x\le 10$ | $9.00 |
| $10<x\le 30$ | $9.00 + 0.95 for each thousand gallons in excess of 10,000 gallons |
| $30<x\le 100$ | $28.00 + 1.65 for each thousand gallons in excess of 30,000 gallons |
| $x>100$ | $143.50 + 2.20 for each thousand gallons in excess of 100,000 gallons |

*Source*: Jericho Water District, Syosset, NY.

(a) Find a function $C$ that models the quarterly cost, in dollars, of using $x$ thousand gallons of water.

(b) What is the domain of the function $C$?

(c) Find each of the following limits. If the limit does not exist, explain why.

$$\lim\limits_{x\to 5}C(x)\quad \lim\limits_{x\to 10}C(x)\quad \lim\limits_{x\to 30}C(x)\quad \lim\limits_{x\to 100}C(x)$$

(d) What is $\lim\limits_{x\to 0^+}C(x)$?

(e) Graph the function $C$.

94. **Cost of Electricity**  In December 2016, Florida Power and Light had the following monthly rate schedule for electric usage in single-family residences:

| Monthly customer charge | $7.87 |
|---|---|
| Fuel charge | |
| $\le 1000$ kWH | $0.02173 per kWH |
| $>1000$ kWH | $21.73 + 0.03173 for each kWH in excess of 1000 |

*Source*: Florida Power and Light, Miami, FL.

(a) Find a function $C$ that models the monthly cost, in dollars, of using $x$ kWH of electricity.

(b) What is the domain of the function $C$?

(c) Find $\lim\limits_{x\to 1000}C(x)$, if it exists. If the limit does not exist, explain why.

(d) What is $\lim\limits_{x\to 0^+}C(x)$?

(e) Graph the function $C$.

95. **Low-Temperature Physics**  In thermodynamics, the average molecular kinetic energy (energy of motion) of a gas having molecules of mass $m$ is directly proportional to its temperature $T$ on the absolute (or Kelvin) scale. This can be expressed as $\dfrac{1}{2}mv^2=\dfrac{3}{2}kT$, where $v=v(T)$ is the speed of a typical molecule at time $t$ and $k$ is a constant, known as the **Boltzmann constant**.

(a) What limit does the molecular speed $v$ approach as the gas temperature $T$ approaches absolute zero (0 K or $-273°$C or $-469°$F)?

(b) What does this limit suggest about the behavior of a gas as its temperature approaches absolute zero?

96. For the function $f(x)=\begin{cases}3x+5 & \text{if } x\le 2\\ 13-x & \text{if } x>2\end{cases}$, find

(a) $\lim\limits_{h\to 0^-}\dfrac{f(2+h)-f(2)}{h}$

(b) $\lim\limits_{h\to 0^+}\dfrac{f(2+h)-f(2)}{h}$

(c) Does $\lim\limits_{h\to 0}\dfrac{f(2+h)-f(2)}{h}$ exist?

97. Use the fact that $|x|=\begin{cases}x & \text{if } x\ge 0\\ -x & \text{if } x<0\end{cases}$ to show that $\lim\limits_{x\to 0}|x|=0$.

98. Use the fact that $|x|=\sqrt{x^2}$ to show that $\lim\limits_{x\to 0}|x|=0$.

99. Find functions $f$ and $g$ for which $\lim\limits_{x\to c}[f(x)+g(x)]$ may exist even though $\lim\limits_{x\to c}f(x)$ and $\lim\limits_{x\to c}g(x)$ do not exist.

100. Find functions $f$ and $g$ for which $\lim\limits_{x\to c}[f(x)g(x)]$ may exist even though $\lim\limits_{x\to c}f(x)$ and $\lim\limits_{x\to c}g(x)$ do not exist.

101. Find functions $f$ and $g$ for which $\lim\limits_{x\to c}\dfrac{f(x)}{g(x)}$ may exist even though $\lim\limits_{x\to c}f(x)$ and $\lim\limits_{x\to c}g(x)$ do not exist.

102. Find a function $f$ for which $\lim\limits_{x\to c}|f(x)|$ may exist even though $\lim\limits_{x\to c}f(x)$ does not exist.

103. Prove that if $g$ is a function for which $\lim\limits_{x\to c}g(x)$ exists and if $k$ is any real number, then $\lim\limits_{x\to c}[kg(x)]$ exists and $\lim\limits_{x\to c}[kg(x)]=k\lim\limits_{x\to c}g(x)$.

104. Prove that if the number $c$ is in the domain of a rational function $R(x)=\dfrac{p(x)}{q(x)}$, then $\lim\limits_{x\to c}R(x)=R(c)$.

**Challenge Problems**

105. Find $\lim\limits_{x\to a}\dfrac{x^n-a^n}{x-a}$, $n$ a positive integer.

106. Find $\lim\limits_{x\to -a}\dfrac{x^n+a^n}{x+a}$, $n$ a positive integer.

107. Find $\lim\limits_{x\to 1}\dfrac{x^m-1}{x^n-1}$, $m$, $n$ positive integers.

108. Find $\lim\limits_{x\to 0}\dfrac{\sqrt[3]{1+x}-1}{x}$.

109. Find $\lim\limits_{x\to 0}\dfrac{\sqrt{(1+ax)(1+bx)}-1}{x}$.

110. Find $\lim\limits_{x\to 0}\dfrac{\sqrt{(1+a_1x)(1+a_2x)\cdots(1+a_nx)}-1}{x}$.

111. Find $\lim\limits_{h\to 0}\dfrac{f(h)-f(0)}{h}$ if $f(x)=x|x|$.

# 1.3 Continuity

**OBJECTIVES** *When you finish this section, you should be able to:*

**1** Determine whether a function is continuous at a number (p. 102)

**2** Determine intervals on which a function is continuous (p. 105)

**3** Use properties of continuity (p. 107)

**4** Use the Intermediate Value Theorem (p. 109)

Sometimes $\lim\limits_{x \to c} f(x)$ equals $f(c)$ and sometimes it does not. In fact, $f(c)$ may not even be defined and yet $\lim\limits_{x \to c} f(x)$ may exist. In this section, we investigate the relationship between $\lim\limits_{x \to c} f(x)$ and $f(c)$. Figure 21 shows some possibilities.

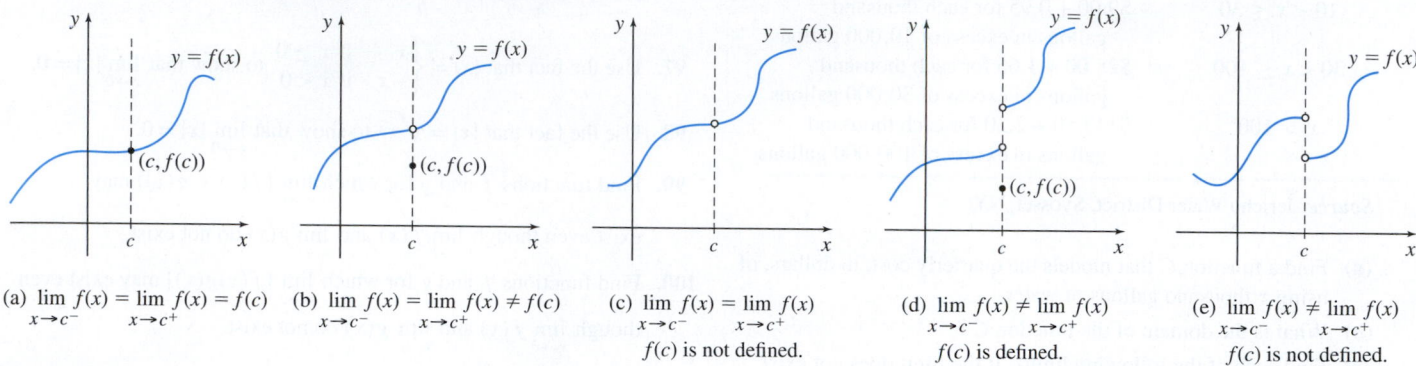

(a) $\lim\limits_{x \to c^-} f(x) = \lim\limits_{x \to c^+} f(x) = f(c)$    (b) $\lim\limits_{x \to c^-} f(x) = \lim\limits_{x \to c^+} f(x) \neq f(c)$    (c) $\lim\limits_{x \to c^-} f(x) = \lim\limits_{x \to c^+} f(x)$

$f(c)$ is not defined.

(d) $\lim\limits_{x \to c^-} f(x) \neq \lim\limits_{x \to c^+} f(x)$    (e) $\lim\limits_{x \to c^-} f(x) \neq \lim\limits_{x \to c^+} f(x)$

$f(c)$ is defined.      $f(c)$ is not defined.

**Figure 21**

Of these five graphs, the "nicest" one is Figure 21(a). There, $\lim\limits_{x \to c} f(x)$ exists and is equal to $f(c)$. Functions that have this property are said to be *continuous at the number c*. This agrees with the intuitive notion that a function is continuous if its graph can be drawn without lifting the pencil. The functions in Figures 21(b)–(e) are not continuous at $c$, since each has a break in the graph at $c$. This leads to the definition of *continuity at a number*.

> **DEFINITION** Continuity at a Number
>
> A function $f$ is **continuous at a number** $c$ if the following three conditions are met:
>
> - $f(c)$ is defined (that is, $c$ is in the domain of $f$)
> - $\lim\limits_{x \to c} f(x)$ exists
> - $\lim\limits_{x \to c} f(x) = f(c)$

If a function $f$ is defined on an open interval containing $c$, except possibly at $c$, then the function is **discontinuous at** $c$, whenever any one of the above three conditions is not satisfied.

## 1 Determine Whether a Function Is Continuous at a Number

**EXAMPLE 1** Determining Whether a Function Is Continuous at a Number

**(a)** Determine whether $f(x) = 3x^2 - 5x + 4$ is continuous at 1.

**(b)** Determine whether $g(x) = \dfrac{x^2 + 9}{x^2 - 4}$ is continuous at 2.

**Solution (a)** We begin by checking the conditions for continuity. First, 1 is in the domain of $f$ and $f(1) = 2$. Second, $\lim\limits_{x \to 1} f(x) = \lim\limits_{x \to 1}(3x^2 - 5x + 4) = 2$, so $\lim\limits_{x \to 1} f(x)$ exists. Third, $\lim\limits_{x \to 1} f(x) = f(1)$. Since the three conditions are met, $f$ is continuous at 1.

**(b)** Since 2 is not in the domain of $g$, the function $g$ is discontinuous at 2. ∎

Figure 22 shows the graphs of $f$ and $g$ from Example 1. Notice that $f$ is continuous at 1, and its graph is drawn without lifting the pencil. But the function $g$ is discontinuous at 2, and to draw its graph, you must lift your pencil at $x = 2$.

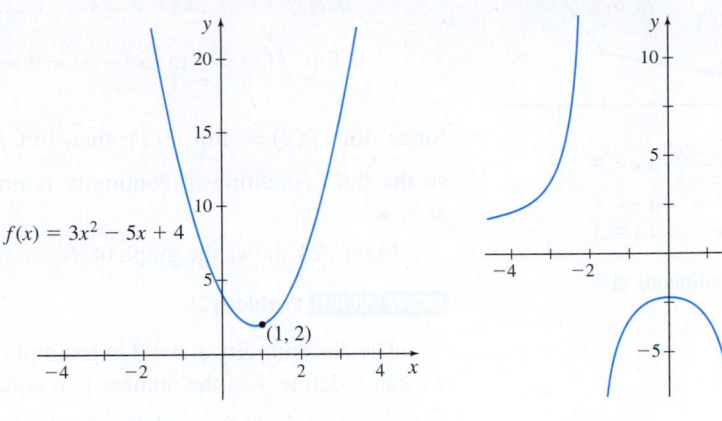

(a) $f$ is continuous at 1.     (b) $g$ is discontinuous at 2.

**Figure 22**

NOW WORK Problem 19.

---

**EXAMPLE 2**  **Determining Whether a Function Is Continuous at a Number**

Determine whether $f(x) = \dfrac{\sqrt{x^2 + 2}}{x^2 - 4}$ is continuous at the numbers $-2$, $0$, and $2$.

**Solution** The domain of $f$ is $\{x \mid x \neq -2,\ x \neq 2\}$. Since $f$ is not defined at $-2$ and $2$, the function $f$ is not continuous at $-2$ and at $2$. The number $0$ is in the domain of $f$.

That is, $f$ is defined at $0$, and $f(0) = -\dfrac{\sqrt{2}}{4}$. Also,

$$\lim_{x \to 0} f(x) = \lim_{x \to 0} \frac{\sqrt{x^2 + 2}}{x^2 - 4} = \frac{\lim_{x \to 0} \sqrt{x^2 + 2}}{\lim_{x \to 0}(x^2 - 4)} = \frac{\sqrt{\lim_{x \to 0}(x^2 + 2)}}{\lim_{x \to 0} x^2 - \lim_{x \to 0} 4}$$

$$= \frac{\sqrt{0 + 2}}{0 - 4} = -\frac{\sqrt{2}}{4} = f(0)$$

The three conditions of continuity at a number are met. So, the function $f$ is continuous at $0$. ■

Figure 23 shows the graph of $f$.

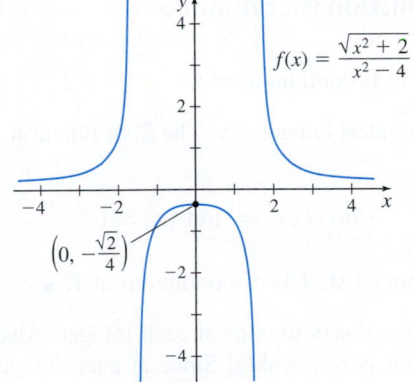

**Figure 23** $f$ is continuous at 0; $f$ is discontinuous at $-2$ and $2$.

NOW WORK Problem 21.

---

CALC
CLIP

**EXAMPLE 3**  **Determining Whether a Piecewise-Defined Function Is Continuous**

Determine whether the function

$$f(x) = \begin{cases} \dfrac{x^2 - 9}{x - 3} & \text{if } x < 3 \\[2mm] 9 & \text{if } x = 3 \\[2mm] x^2 - 3 & \text{if } x > 3 \end{cases}$$

is continuous at 3.

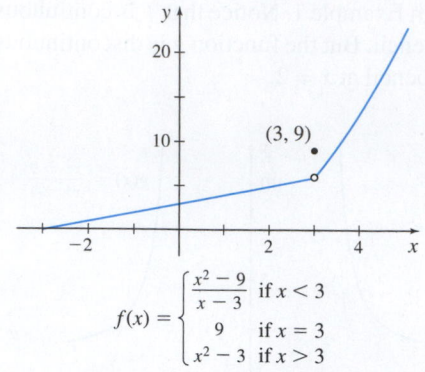

$$f(x) = \begin{cases} \dfrac{x^2 - 9}{x - 3} & \text{if } x < 3 \\ 9 & \text{if } x = 3 \\ x^2 - 3 & \text{if } x > 3 \end{cases}$$

**Figure 24** $f$ is discontinuous at 3.

**Solution** Since $f(3) = 9$, the function $f$ is defined at 3. To check the second condition, we investigate the one-sided limits.

$$\lim_{x \to 3^-} f(x) = \lim_{x \to 3^-} \frac{x^2 - 9}{x - 3} = \lim_{x \to 3^-} \frac{(x - 3)(x + 3)}{x - 3} \underset{\substack{\uparrow \\ \text{Divide out } x - 3}}{=} \lim_{x \to 3^-} (x + 3) = 6$$

$$\lim_{x \to 3^+} f(x) = \lim_{x \to 3^+} (x^2 - 3) = 9 - 3 = 6$$

Since $\lim\limits_{x \to 3^-} f(x) = \lim\limits_{x \to 3^+} f(x)$, then $\lim\limits_{x \to 3} f(x)$ exists. But, $\lim\limits_{x \to 3} f(x) = 6$ and $f(3) = 9$, so the third condition of continuity is not satisfied. The function $f$ is discontinuous at 3. ■

Figure 24 shows the graph of $f$.

**NOW WORK** Problem 25.

The discontinuity at $c = 3$ in Example 3 is called a *removable discontinuity* because we can redefine $f$ at the number $c$ to equal $\lim\limits_{x \to c} f(x)$ and make $f$ continuous at $c$. So, in Example 3, if $f(3)$ is redefined to be 6, then $f$ would be continuous at 3.

**IN WORDS** Visually, the graph of a function $f$ with a removable discontinuity has a hole at $(c, \lim\limits_{x \to c} f(x))$. The discontinuity is removable because we can fill in the hole by appropriately defining $f$. The function will then be continuous at $x = c$.

**DEFINITION** Removable Discontinuity

Let $f$ be a function that is defined everywhere in an open interval containing $c$, except possibly at $c$. The number $c$ is called a **removable discontinuity** of $f$ if the function is discontinuous at $c$ but $\lim\limits_{x \to c} f(x)$ exists. The discontinuity is removed by defining (or redefining) the value of $f$ at $c$ to be $\lim\limits_{x \to c} f(x)$.

**NOW WORK** Problems 13 and 35.

**EXAMPLE 4** Determining Whether a Function Is Continuous at a Number

Determine whether the floor function $f(x) = \lfloor x \rfloor$ is continuous at 1.

**NEED TO REVIEW?** The floor function is discussed in Section P.2, p. 19.

**Solution** The floor function $f(x) = \lfloor x \rfloor = $ the greatest integer $\leq x$. The floor function $f$ is defined at 1 and $f(1) = 1$. But

$$\lim_{x \to 1^-} f(x) = \lim_{x \to 1^-} \lfloor x \rfloor = 0 \qquad \text{and} \qquad \lim_{x \to 1^+} f(x) = \lim_{x \to 1^+} \lfloor x \rfloor = 1$$

So, $\lim\limits_{x \to 1} \lfloor x \rfloor$ does not exist. Since $\lim\limits_{x \to 1} \lfloor x \rfloor$ does not exist, $f$ is discontinuous at 1. ■

Figure 25 illustrates that the floor function is discontinuous at each integer. Also, none of the discontinuities of the floor function is removable. Since at each integer the value of the floor function "jumps" to the next integer, without taking on any intermediate values, the discontinuity at integer values is called a **jump discontinuity**.

**NOW WORK** Problem 53.

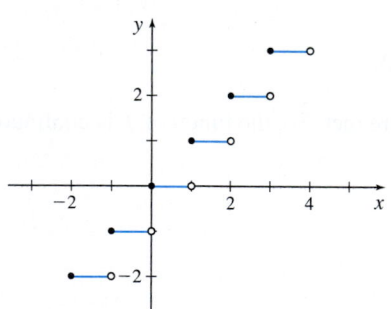

**Figure 25** $f(x) = \lfloor x \rfloor$

We have defined what it means for a function $f$ to be *continuous* at a number. Now we define *one-sided continuity* at a number.

**DEFINITION** One-Sided Continuity at a Number

Let $f$ be a function defined on the interval $(a, c]$. Then $f$ is **continuous from the left at the number $c$** if

$$\lim_{x \to c^-} f(x) = f(c)$$

Let $f$ be a function defined on the interval $[c, b]$. Then $f$ is **continuous from the right at the number $c$** if

$$\lim_{x \to c^+} f(x) = f(c)$$

In Example 4, we showed that the floor function $f(x) = \lfloor x \rfloor$ is discontinuous at $x = 1$. But since

$$f(1) = \lfloor 1 \rfloor = 1 \qquad \text{and} \qquad \lim_{x \to 1^+} f(x) = \lfloor x \rfloor = 1$$

the floor function is continuous from the right at 1. In fact, the floor function is discontinuous at each integer $n$, but it is continuous from the right at every integer $n$. (Do you see why?)

## 2 Determine Intervals on Which a Function Is Continuous

So far, we have considered only continuity at a number $c$. Now we use one-sided continuity to define continuity on an interval.

**DEFINITION  Continuity on an Interval**

- A function $f$ is **continuous on an open interval** $(a, b)$ if $f$ is continuous at every number in $(a, b)$.
- A function $f$ is **continuous on an interval** $[a, b)$ if $f$ is continuous on the open interval $(a, b)$ and continuous from the right at the number $a$.
- A function $f$ is **continuous on an interval** $(a, b]$ if $f$ is continuous on the open interval $(a, b)$ and continuous from the left at the number $b$.
- A function $f$ is **continuous on a closed interval** $[a, b]$ if $f$ is continuous on the open interval $(a, b)$, continuous from the right at $a$, and continuous from the left at $b$.

Figure 26 gives examples of graphs over different types of intervals.

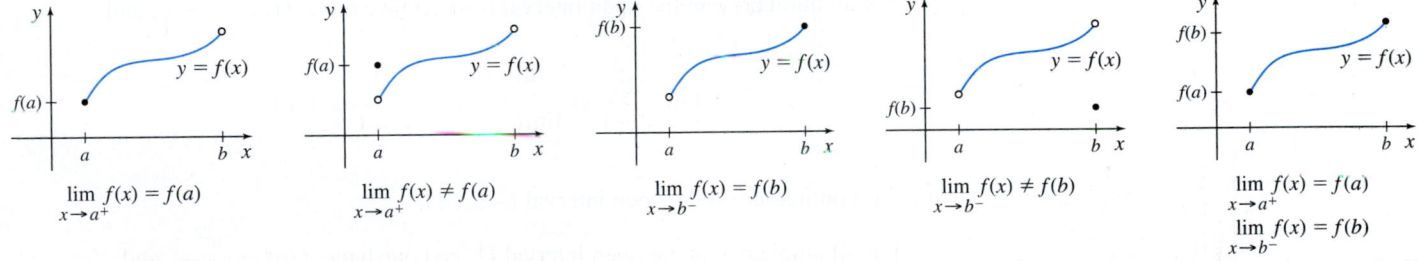

(a) $f$ is continuous on $[a, b)$.  (b) $f$ is continuous on $(a, b)$.  (c) $f$ is continuous on $(a, b]$.  (d) $f$ is continuous on $(a, b)$.  (e) $f$ is continuous on $[a, b]$.

**Figure 26**

For example, the graph of the floor function $f(x) = \lfloor x \rfloor$ in Figure 25 illustrates that $f$ is continuous on every interval $[n, n+1)$, $n$ an integer. In each interval, $f$ is continuous from the right at the left endpoint $n$ and is continuous at every number in the open interval $(n, n+1)$.

**EXAMPLE 5  Determining Whether a Function Is Continuous on a Closed Interval**

Is the function $f(x) = \sqrt{4 - x^2}$ continuous on the closed interval $[-2, 2]$?

**Solution** The domain of $f$ is $\{x \mid -2 \le x \le 2\}$. So, $f$ is defined for every number in the closed interval $[-2, 2]$.

For any number $c$ in the open interval $(-2, 2)$,

$$\lim_{x \to c} f(x) = \lim_{x \to c} \sqrt{4 - x^2} = \sqrt{\lim_{x \to c} (4 - x^2)} = \sqrt{4 - c^2} = f(c)$$

So, $f$ is continuous on the open interval $(-2, 2)$.

To determine whether $f$ is continuous on $[-2, 2]$, we investigate the limit from the right at $-2$ and the limit from the left at 2. Then

$$\lim_{x \to -2^+} f(x) = \lim_{x \to -2^+} \sqrt{4 - x^2} = 0 = f(-2)$$

So, $f$ is continuous from the right at $-2$. Similarly,

$$\lim_{x \to 2^-} f(x) = \lim_{x \to 2^-} \sqrt{4 - x^2} = 0 = f(2)$$

So, $f$ is continuous from the left at 2. We conclude that $f$ is continuous on the closed interval $[-2, 2]$. ∎

Figure 27 shows the graph of $f$.

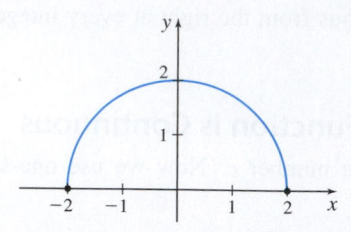

**Figure 27** $f(x) = \sqrt{4 - x^2}, -2 \le x \le 2$

**NOW WORK** Problems **37** and **77**.

---

**EXAMPLE 6**    **Determining Where $f(x) = \dfrac{x^2}{x - 1}$ Is Continuous**

Determine where the function $f(x) = \dfrac{x^2}{x - 1}$ is continuous.

**Solution**  The domain of $f(x) = \dfrac{x^2}{x - 1}$ is the set $\{x \mid x \ne 1\}$, which is the same as the set $(-\infty, 1) \cup (1, \infty)$.

Since $f$ is not defined at the number 1, $f$ is discontinuous at 1.

We examine $f$ on the intervals $(-\infty, 1)$ and $(1, \infty)$.

For all numbers $c$ in the open interval $(-\infty, 1)$ we have $f(c) = \dfrac{c^2}{c - 1}$, and

$$\lim_{x \to c} \frac{x^2}{x - 1} = \frac{\lim\limits_{x \to c} x^2}{\lim\limits_{x \to c} (x - 1)} = \frac{c^2}{c - 1} = f(c)$$

So, $f$ is continuous on the open interval $(-\infty, 1)$.

For all numbers $c$ in the open interval $(1, \infty)$, we have $f(c) = \dfrac{c^2}{c - 1}$, and

$$\lim_{x \to c} \frac{x^2}{x - 1} = \frac{\lim\limits_{x \to c} x^2}{\lim\limits_{x \to c} (x - 1)} = \frac{c^2}{c - 1} = f(c)$$

So, $f$ is continuous on the open interval $(1, \infty)$.

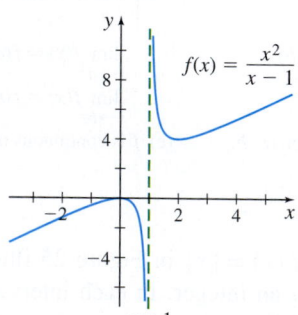

**Figure 28** $f(x) = \dfrac{x^2}{x - 1}$ is continuous on $(-\infty, 1) \cup (1, \infty)$.

The function $f(x) = \dfrac{x^2}{x - 1}$ is continuous on $(-\infty, 1) \cup (1, \infty)$. ∎

Figure 28 shows the graph of $f$.

When listing the properties of a function in Chapter P, we included the function's domain, its symmetry, and its zeros. Now we add continuity to the list by asking, "Where is the function continuous?" We answer this question here for two important classes of functions: polynomial functions and rational functions.

---

**THEOREM**

- A polynomial function is continuous on its domain, all real numbers.
- A rational function is continuous on its domain.

**Proof** If $P$ is a polynomial function, its domain is the set of real numbers. For a polynomial function,

$$\lim_{x \to c} P(x) = P(c)$$

for any number $c$. That is, a polynomial function is continuous at every real number.

If $R(x) = \dfrac{p(x)}{q(x)}$ is a rational function, then $p(x)$ and $q(x)$ are polynomials and the domain of $R$ is $\{x \mid q(x) \neq 0\}$. The Limit of a Rational Function (p. 96) states that for all $c$ in the domain of a rational function,

$$\lim_{x \to c} R(x) = R(c)$$

So a rational function is continuous at every number in its domain. ∎

Rational functions deserve special attention. For example, the rational function $R(x) = \dfrac{x^2 - 2x + 1}{x - 1}$ is continuous for $\{x \mid x \neq 1\}$. But notice in Figure 29 that its graph has a hole at $(1, 0)$. The rational function $f(x) = \dfrac{1}{x}$ is continuous for $\{x \mid x \neq 0\}$. But notice in Figure 30 the behavior of the graph as $x$ goes from negative numbers to positive numbers.

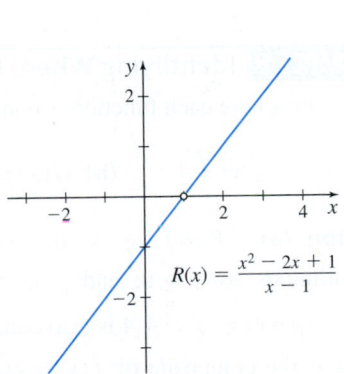

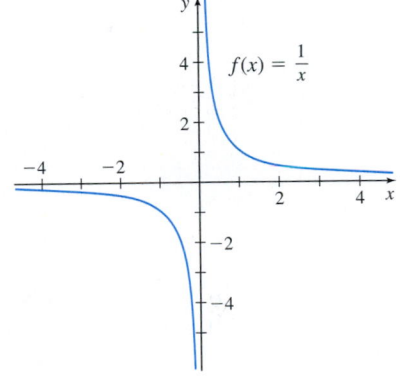

**Figure 29** The graph of $R$ has a hole at $(1, 0)$.     **Figure 30** $f$ is not defined at 0.

## 3 Use Properties of Continuity

Once we know where functions are continuous, we can build other continuous functions.

> **THEOREM Continuity of a Sum, Difference, Product, and Quotient**
>
> If the functions $f$ and $g$ are continuous at a number $c$, and if $k$ is a real number, then the functions $f + g$, $f - g$, $f \cdot g$, and $kf$ are also continuous at $c$. If $g(c) \neq 0$, the function $\dfrac{f}{g}$ is continuous at $c$.

The proofs of these properties are based on properties of limits. For example, the proof of the continuity of $f + g$ is based on the Limit of a Sum property. That is, if $\lim\limits_{x \to c} f(x)$ and $\lim\limits_{x \to c} g(x)$ exist, then $\lim\limits_{x \to c} [f(x) + g(x)] = \lim\limits_{x \to c} f(x) + \lim\limits_{x \to c} g(x)$.

**EXAMPLE 7** Identifying Where Functions Are Continuous

Determine where each function is continuous:

**(a)** $F(x) = x^2 + 5 - \dfrac{x}{x^2 + 4}$     **(b)** $G(x) = x^3 + 2x + \dfrac{x^2}{x^2 - 1}$

**Solution**

**(a)** $F$ is the difference of the two functions $f(x) = x^2 + 5$ and $g(x) = \dfrac{x}{x^2 + 4}$, each of which is continuous for all real numbers. The function $F = f - g$ is also continuous for all real numbers.

**(b)** $G$ is the sum of the two functions $f(x) = x^3 + 2x$, which is continuous for all real numbers, and $g(x) = \dfrac{x^2}{x^2 - 1}$, which is continuous for $\{x \mid x \neq -1, \ x \neq 1\}$. Since $G = f + g$, $G$ is continuous for, $\{x \mid x \neq -1, x \neq 1\}$. ∎

**NOW WORK** Problem 45.

**NEED TO REVIEW?** Composite functions are discussed in Section P.3, pp. 27–30.

The continuity of a composite function depends on the continuity of its components.

**THEOREM** Continuity of a Composite Function

If a function $g$ is continuous at $c$ and a function $f$ is continuous at $g(c)$, then the composite function $(f \circ g)(x) = f(g(x))$ is continuous at $c$. That is,

$$\lim_{x \to c}(f \circ g)(x) = \lim_{x \to c} f(g(x)) = f[\lim_{x \to c} g(x)] = f(g(c))$$

**EXAMPLE 8** Identifying Where Functions Are Continuous

Determine where each function is continuous:

**(a)** $F(x) = \sqrt{x^2 + 4}$     **(b)** $G(x) = \sqrt{x^2 - 1}$     **(c)** $H(x) = \dfrac{x^2 - 1}{x^2 - 4} + \sqrt{x - 1}$

**Solution** **(a)** $F = f \circ g$ is the composite of $f(x) = \sqrt{x}$ and $g(x) = x^2 + 4$; $f$ is continuous for $x \geq 0$; and $g$ is continuous for all real numbers. The function $F = (f \circ g)(x) = \sqrt{x^2 + 4}$ is also continuous for all real numbers.

**(b)** $G$ is the composite of $f(x) = \sqrt{x}$ and $g(x) = x^2 - 1$; $f$ is continuous for $x \geq 0$; and $g$ is continuous for all real numbers. The function $G = (f \circ g)(x) = \sqrt{x^2 - 1}$ is continuous for $x^2 - 1 \geq 0$, the set $\{x \mid x \leq -1\} \cup \{x \mid x \geq 1\}$.

**(c)** $H$ is the sum of $f(x) = \dfrac{x^2 - 1}{x^2 - 4}$ and the function $g(x) = \sqrt{x - 1}$; $f$ is continuous on $\{x \mid x \neq -2, x \neq 2\}$; and $g$ is continuous on $\{x \mid x \geq 1\}$. The function $H$ is continuous on $[1, 2) \cup (2, \infty)$. ∎

**NOW WORK** Problem 47.

**NEED TO REVIEW?** Inverse functions are discussed in Section P.4, pp. 37–40.

Recall that for any function $f$ that is one-to-one over its domain, its inverse $f^{-1}$ is also a function, and the graphs of $f$ and $f^{-1}$ are symmetric with respect to the line $y = x$. It is intuitive that if $f$ is continuous, then so is $f^{-1}$. See Figure 31. The following theorem, whose proof is given in Appendix B, confirms this.

**THEOREM** Continuity of an Inverse Function

If $f$ is a one-to-one function that is continuous on its domain, then its inverse function $f^{-1}$ is also continuous on its domain.

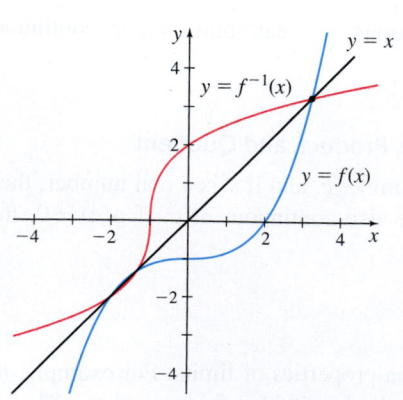

**Figure 31**

## 4 Use the Intermediate Value Theorem

Functions that are continuous on a closed interval have many important properties. One of them is stated in the *Intermediate Value Theorem*. The proof of the Intermediate Value Theorem may be found in most books on advanced calculus.

> **THEOREM  The Intermediate Value Theorem**
>
> Let $f$ be a function that is continuous on a closed interval $[a, b]$ and $f(a) \neq f(b)$. If $N$ is any number between $f(a)$ and $f(b)$, then there is at least one number $c$ in the open interval $(a, b)$ for which $f(c) = N$.

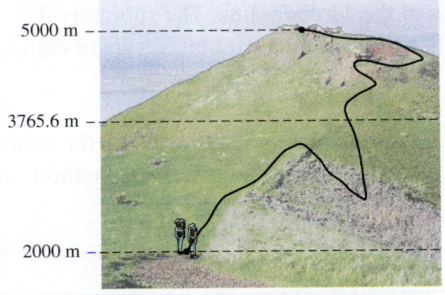

5000 m

3765.6 m

2000 m

To get a better idea of this result, suppose you climb a mountain, starting at an elevation of 2000 meters and ending at an elevation of 5000 meters. No matter how many ups and downs you take as you climb, at some time your altitude must be 3765.6 meters, or any other number between 2000 and 5000.

In other words, a function $f$ that is continuous on a closed interval $[a, b]$ with $f(a) \neq f(b)$, must take on all values between $f(a)$ and $f(b)$. Figure 32 illustrates this. Figure 33 shows why the continuity of the function is crucial. Notice in Figure 33 that there is a hole in the graph of $f$ at the point $(c, N)$. Because of the discontinuity at $c$, there is no number $c$ in the open interval $(a, b)$ for which $f(c) = N$.

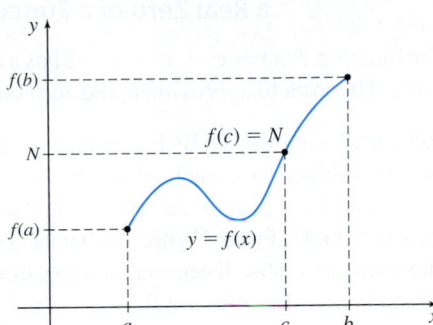

**Figure 32** $f$ takes on every value between $f(a)$ and $f(b)$.

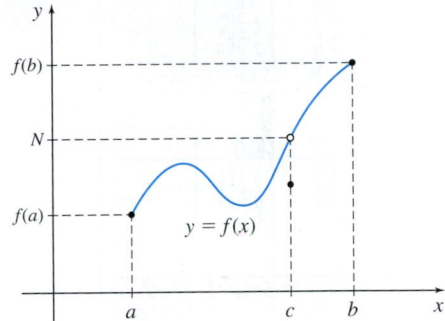

**Figure 33** The discontinuity at $c$ results in no number $c$ in $(a, b)$ for which $f(c) = N$.

The Intermediate Value Theorem is an existence theorem. It states that there is at least one number $c$ for which $f(c) = N$, but it does not tell us how to find $c$. However, we can use the Intermediate Value Theorem to locate an open interval $(a, b)$ that contains $c$.

An immediate application of the Intermediate Value Theorem involves locating the zeros of a function. Suppose a function $f$ is continuous on the closed interval $[a, b]$ and $f(a)$ and $f(b)$ have opposite signs. Then by the Intermediate Value Theorem, there is at least one number $c$ between $a$ and $b$ for which $f(c) = 0$. That is, $f$ has at least one zero between $a$ and $b$.

**EXAMPLE 9  Using the Intermediate Value Theorem**

Use the Intermediate Value Theorem to show that

$$f(x) = x^3 + x^2 - x - 2$$

has a zero between 1 and 2.

**Solution** Since $f$ is a polynomial, it is continuous on the closed interval $[1, 2]$. Because $f(1) = -1$ and $f(2) = 8$ have opposite signs, the Intermediate Value Theorem states that $f(c) = 0$ for at least one number $c$ in the interval $(1, 2)$. That is, $f$ has at least one zero between 1 and 2. Figure 34 shows the graph of $f$ on a graphing calculator. ∎

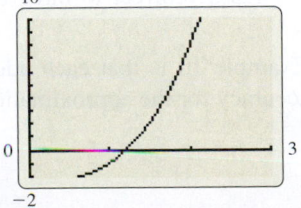

**Figure 34** $f(x) = x^3 + x^2 - x - 2$

NOW WORK Problem **59**.

The Intermediate Value Theorem can be used to approximate a zero in the interval $(a, b)$ by dividing the interval $[a, b]$ into smaller subintervals. There are two popular methods of subdividing the interval $[a, b]$.

- The **bisection method** bisects $[a, b]$, that is, divides $[a, b]$ into two equal subintervals and compares the sign of $f\left(\dfrac{a+b}{2}\right)$ to the signs of the previously computed values $f(a)$ and $f(b)$. The subinterval whose endpoints have opposite signs is then bisected, and the process is repeated.

- The second method divides $[a, b]$ into 10 subintervals of equal length and compares the signs of $f$ evaluated at each of the 11 endpoints. The subinterval whose endpoints have opposite signs is then divided into 10 subintervals of equal length and the process is repeated.

We choose to use the second method because it lends itself well to the table feature of a graphing calculator. You are asked to use the bisection method in Problems 107–114.

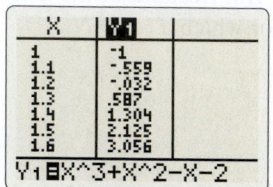

**Figure 35**

🔃 **EXAMPLE 10**   Using the Intermediate Value Theorem to Approximate a Real Zero of a Function

The function $f(x) = x^3 + x^2 - x - 2$ has a zero in the interval $(1, 2)$. Use the Intermediate Value Theorem to approximate the zero correct to three decimal places.

**Solution**  Using the TABLE feature on a graphing utility, subdivide the interval $[1, 2]$ into 10 subintervals, each of length 0.1. Then find the subinterval whose endpoints have opposite signs, or the endpoint whose value equals 0 (in which case, the exact zero is found). From Figure 35, since $f(1.2) = -0.032$ and $f(1.3) = 0.587$, by the Intermediate Value Theorem, a zero lies in the interval $(1.2, 1.3)$. Correct to one decimal place, the zero is 1.2.

Repeat the process by subdividing the interval $[1.2, 1.3]$ into 10 subintervals, each of length 0.01. See Figure 36. We conclude that the zero is in the interval $(1.20, 1.21)$, so correct to two decimal places, the zero is 1.20.

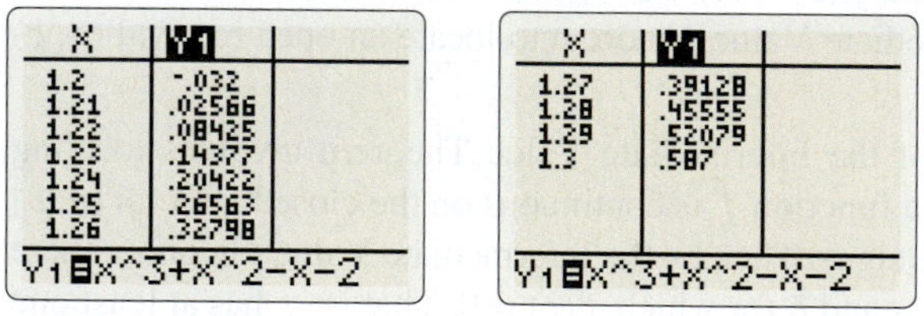

**Figure 36**

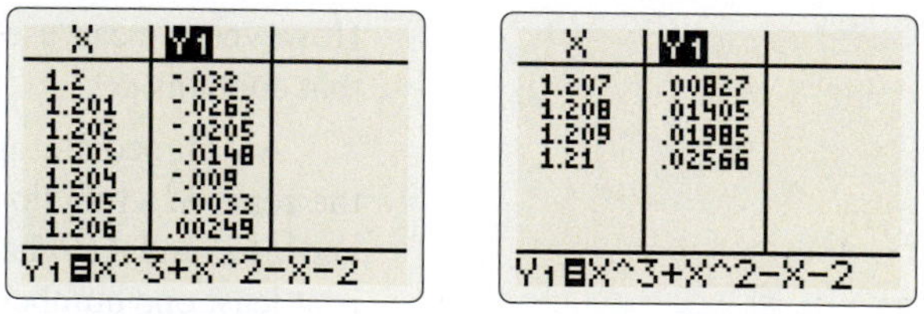

**Figure 37**

Now subdivide the interval $[1.20, 1.21]$ into 10 subintervals, each of length 0.001. See Figure 37.

We conclude that the zero of the function $f$ is 1.205, correct to three decimal places.  ∎

Notice that a benefit of the method used in Example 10 is that each additional iteration results in one additional decimal place of accuracy for the approximation.

**NOW WORK** Problem 65.

# 1.3 Assess Your Understanding

## Concepts and Vocabulary

1. *True or False* A polynomial function is continuous at every real number.

2. *True or False* Piecewise-defined functions are never continuous at numbers where the function changes equations.

3. The three conditions necessary for a function $f$ to be continuous at a number $c$ are _____, _____, and _____.

4. *True or False* If $f$ is continuous at 0, then $g(x) = \dfrac{1}{4}f(x)$ is continuous at 0.

5. *True or False* If $f$ is a function defined everywhere in an open interval containing $c$, except possibly at $c$, then the number $c$ is called a removable discontinuity of $f$ if the function $f$ is not continuous at $c$.

6. *True or False* If a function $f$ is discontinuous at a number $c$, then $\lim\limits_{x \to c} f(x)$ does not exist.

7. *True or False* If a function $f$ is continuous on an open interval $(a, b)$, then it is continuous on the closed interval $[a, b]$.

8. *True or False* If a function $f$ is continuous on the closed interval $[a, b]$, then $f$ is continuous on the open interval $(a, b)$.

*In Problems 9 and 10, explain whether each function is continuous or discontinuous.*

9. The velocity of a ball thrown up into the air as a function of time, if the ball lands 5 seconds after it is thrown and stops.

10. The temperature of an oven used to bake a potato as a function of time.

11. *True or False* If a function $f$ is continuous on a closed interval $[a, b]$, and $f(a) \neq f(b)$, then the Intermediate Value Theorem guarantees that the function takes on every value between $f(a)$ and $f(b)$.

12. *True or False* If a function $f$ is continuous on a closed interval $[a, b]$ and $f(a) \neq f(b)$, but both $f(a) > 0$ and $f(b) > 0$, then according to the Intermediate Value Theorem, $f$ does not have a zero on the open interval $(a, b)$.

## Skill Building

*In Problems 13–18, use the graph of $y = f(x)$ (below).*

(a) Determine whether $f$ is continuous at $c$.

(b) If $f$ is discontinuous at $c$, state which condition(s) of the definition of continuity is (are) not satisfied.

(c) If $f$ is discontinuous at $c$, determine if the discontinuity is removable.

(d) If the discontinuity is removable, define (or redefine) $f$ at $c$ to make $f$ continuous at $c$.

13. $c = -3$    14. $c = 0$

15. $c = 2$    16. $c = 3$

17. $c = 4$    18. $c = 5$

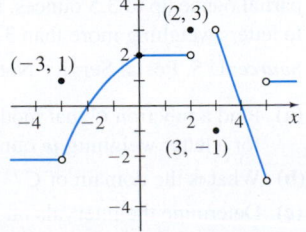

*In Problems 19–32, determine whether the function $f$ is continuous at $c$.*

19. $f(x) = x^2 + 1$  at $c = -1$

20. $f(x) = x^3 - 5$  at $c = 5$

21. $f(x) = \dfrac{x}{x^2 + 4}$  at $c = -2$

22. $f(x) = \dfrac{x}{x - 2}$  at $c = 2$

23. $f(x) = \begin{cases} 2x + 5 & \text{if } x \leq 2 \\ 4x + 1 & \text{if } x > 2 \end{cases}$  at $c = 2$

24. $f(x) = \begin{cases} 2x + 1 & \text{if } x \leq 0 \\ 2x & \text{if } x > 0 \end{cases}$  at $c = 0$

25. $f(x) = \begin{cases} 3x - 1 & \text{if } x < 1 \\ 4 & \text{if } x = 1 \\ 2x & \text{if } x > 1 \end{cases}$  at $c = 1$

26. $f(x) = \begin{cases} 3x - 1 & \text{if } x < 1 \\ 2 & \text{if } x = 1 \\ 2x & \text{if } x > 1 \end{cases}$  at $c = 1$

27. $f(x) = \begin{cases} 3x - 1 & \text{if } x < 1 \\ 2x & \text{if } x > 1 \end{cases}$  at $c = 1$

28. $f(x) = \begin{cases} 3x - 1 & \text{if } x < 1 \\ 2 & \text{if } x = 1 \\ 3x & \text{if } x > 1 \end{cases}$  at $c = 1$

29. $f(x) = \begin{cases} x^2 & \text{if } x \leq 0 \\ 2x & \text{if } x > 0 \end{cases}$  at $c = 0$

30. $f(x) = \begin{cases} x^2 & \text{if } x < -1 \\ 2 & \text{if } x = -1 \\ -3x + 2 & \text{if } x > -1 \end{cases}$  at $c = -1$

31. $f(x) = \begin{cases} 4 - 3x^2 & \text{if } x < 0 \\ 4 & \text{if } x = 0 \\ \sqrt{\dfrac{16 - x^2}{4 - x}} & \text{if } 0 < x < 4 \end{cases}$  at $c = 0$

32. $f(x) = \begin{cases} \sqrt{4 + x} & \text{if } -4 \leq x \leq 4 \\ \sqrt{\dfrac{x^2 - 3x - 4}{x - 4}} & \text{if } x > 4 \end{cases}$  at $c = 4$

*In Problems 33–36, each function $f$ has a removable discontinuity at $c$. Define $f(c)$ so that $f$ is continuous at $c$.*

33. $f(x) = \dfrac{x^2 - 4}{x - 2}$,  $c = 2$

34. $f(x) = \dfrac{x^2 + x - 12}{x - 3}$,  $c = 3$

35. $f(x) = \begin{cases} 1 + x & \text{if } x < 1 \\ 4 & \text{if } x = 1 \\ 2x & \text{if } x > 1 \end{cases}$  $c = 1$

**36.** $f(x) = \begin{cases} x^2 + 5x & \text{if } x < -1 \\ 0 & \text{if } x = -1 \quad c = -1 \\ x - 3 & \text{if } x > -1 \end{cases}$

*In Problems 37–40, determine whether each function $f$ is continuous on the given interval. If the answer is no, find the numbers in the interval, if any, at which $f$ is continuous.*

**37.** $f(x) = \dfrac{x^2 - 9}{x - 3}$ on the interval $[-3, 3]$

**38.** $f(x) = 1 + \dfrac{1}{x}$ on the interval $[-1, 0]$

**39.** $f(x) = \dfrac{1}{\sqrt{x^2 - 9}}$ on the interval $[-3, 3]$

**40.** $f(x) = \sqrt{9 - x^2}$ on the interval $[-3, 3]$

*In Problems 41–50, determine where each function $f$ is continuous by using properties of continuity.*

**41.** $f(x) = 2x^2 + 5x - \dfrac{1}{x}$    **42.** $f(x) = x + 1 + \dfrac{2x}{x^2 + 5}$

**43.** $f(x) = (x - 1)(x^2 + x + 1)$    **44.** $f(x) = \sqrt{x}(x^3 - 5)$

**45.** $f(x) = \dfrac{x - 9}{\sqrt{x} - 3}$    **46.** $f(x) = \dfrac{x - 4}{\sqrt{x} - 2}$

**47.** $f(x) = \sqrt{\dfrac{x^2 + 1}{2 - x}}$    **48.** $f(x) = \sqrt{\dfrac{4}{x^2 - 1}}$

**49.** $f(x) = (2x^2 + 5x - 3)^{2/3}$    **50.** $f(x) = (x + 2)^{1/2}$

*In Problems 51–56, use the function*

$$f(x) = \begin{cases} \sqrt{15 - 3x} & \text{if } x < 2 \\ \sqrt{5} & \text{if } x = 2 \\ 9 - x^2 & \text{if } 2 < x < 3 \\ \lfloor x - 2 \rfloor & \text{if } 3 \le x \end{cases}$$

**51.** Is $f$ continuous at 0? Why or why not?

**52.** Is $f$ continuous at 4? Why or why not?

**53.** Is $f$ continuous at 3? Why or why not?

**54.** Is $f$ continuous at 2? Why or why not?

**55.** Is $f$ continuous at 1? Why or why not?

**56.** Is $f$ continuous at 2.5? Why or why not?

*In Problems 57 and 58:*

**(a)** *Use technology to graph $f$ using a suitable scale on each axis.*

**(b)** *Based on the graph from (a), determine where $f$ is continuous.*

**(c)** *Use the definition of continuity to determine where $f$ is continuous.*

**(d)** *What advice would you give a fellow student about using technology to determine where a function is continuous?*

**57.** $f(x) = \dfrac{x^3 - 8}{x - 2}$    **58.** $f(x) = \dfrac{x^2 - 3x + 2}{3x - 6}$

*In Problems 59–64, use the Intermediate Value Theorem to determine which of the functions must have zeros in the given interval. Indicate those for which the theorem gives no information. Do not attempt to locate the zeros.*

**59.** $f(x) = x^3 - 3x$ on $[-2, 2]$

**60.** $f(x) = x^4 - 1$ on $[-2, 2]$

**61.** $f(x) = \dfrac{x}{(x + 1)^2} - 1$ on $[10, 20]$

**62.** $f(x) = x^3 - 2x^2 - x + 2$ on $[3, 4]$

**63.** $f(x) = \dfrac{x^3 - 1}{x - 1}$ on $[0, 2]$

**64.** $f(x) = \dfrac{x^2 + 3x + 2}{x^2 - 1}$ on $[-3, 0]$

*In Problems 65–72, verify that each function has a zero in the indicated interval. Then use the Intermediate Value Theorem to approximate the zero correct to three decimal places by repeatedly subdividing the interval containing the zero into 10 subintervals.*

**65.** $f(x) = x^3 + 3x - 5$; interval: $[1, 2]$

**66.** $f(x) = x^3 - 4x + 2$; interval: $[1, 2]$

**67.** $f(x) = 2x^3 + 3x^2 + 4x - 1$; interval: $[0, 1]$

**68.** $f(x) = x^3 - x^2 - 2x + 1$; interval: $[0, 1]$

**69.** $f(x) = x^3 - 6x - 12$; interval: $[3, 4]$

**70.** $f(x) = 3x^3 + 5x - 40$; interval: $[2, 3]$

**71.** $f(x) = x^4 - 2x^3 + 21x - 23$; interval: $[1, 2]$

**72.** $f(x) = x^4 - x^3 + x - 2$; interval: $[1, 2]$

*In Problems 73 and 74,*

**(a)** *Use the Intermediate Value Theorem to show that $f$ has a zero in the given interval.*

**(b)** *Use technology to find the zero rounded to three decimal places.*

**73.** $f(x) = \sqrt{x^2 + 4x} - 2$ in $[0, 1]$

**74.** $f(x) = x^3 - x + 2$ in $[-2, 0]$

## Applications and Extensions

*In Problems 75–78, determine whether each function is*

**(a)** *continuous from the left*    **(b)** *continuous from the right*

*at the numbers $c$ and $d$.*

**75.** $f(x) = \begin{cases} x^2 & \text{if } -1 < x < 1 \\ x - 1 & \text{if } |x| \ge 1 \end{cases}$    at $c = -1$ and $d = 1$

**76.** $f(x) = \begin{cases} x^2 - 1 & \text{if } -1 \le x \le 1 \\ |x + 1| & \text{if } |x| > 1 \end{cases}$    at $c = -1$ and $d = 1$

**77.** $f(x) = \sqrt{(x + 1)(x - 5)}$ at $c = -1$ and $d = 5$

**78.** $f(x) = \sqrt{(x - 1)(x - 2)}$ at $c = 1$ and $d = 2$

**79. First-Class Mail**  As of January 1, 2017, the U.S. Postal Service charged $0.47 postage for first-class letters weighing up to and including 1 ounce, plus a flat fee of $0.21 for each additional or partial ounce up to 3.5 ounces. First-class letter rates do not apply to letters weighing more than 3.5 ounces.

*Source*: U.S. Postal Service Notice 123.

**(a)** Find a function $C$ that models the first-class postage charged for a letter weighing $w$ ounces. Assume $w > 0$.

**(b)** What is the domain of $C$?

**(c)** Determine the intervals on which $C$ is continuous.

**(d)** At numbers where $C$ is not continuous (if any), what type of discontinuity does $C$ have?

**(e)** What are the practical implications of the answer to (d)?

**80. First-Class Mail** As of January 1, 2017, the U.S. Postal Service charged $0.94 postage for first-class large envelopes weighing up to and including 1 ounce, plus a flat fee of $0.21 for each additional or partial ounce up to 13 ounces. First-class rates do not apply to large envelopes weighing more than 13 ounces.
*Source*: U.S. Postal Service Notice 123.

(a) Find a function $C$ that models the first-class postage charged for a large envelope weighing $w$ ounces. Assume $w > 0$.

(b) What is the domain of $C$?

(c) Determine the intervals on which $C$ is continuous.

(d) At numbers where $C$ is not continuous (if any), what type of discontinuity does $C$ have?

(e) What are the practical implications of the answer to (d)?

**81. Cost of Electricity** In December 2016, Florida Power and Light had the following monthly rate schedule for electric usage in single-family residences:

| | |
|---|---|
| Monthly customer charge | $7.87 |
| Fuel charge | |
| ≤1000 kWH | 0.02173 per kWH |
| >1000 kWH | $21.73 + 0.03173 for each kWH in excess of 1000 |

*Source*: Florida Power and Light, Miami, FL.

(a) Find a function $C$ that models the monthly cost of using $x$ kWH of electricity.

(b) What is the domain of $C$?

(c) Determine the intervals on which $C$ is continuous.

(d) At numbers where $C$ is not continuous (if any), what type of discontinuity does $C$ have?

(e) What are the practical implications of the answer to (d)?

**82. Cost of Water** The Jericho Water District determines quarterly water costs, in dollars, using the following rate schedule:

| Water used (in thousands of gallons) | Cost |
|---|---|
| $0 \leq x \leq 10$ | $9.00 |
| $10 < x \leq 30$ | $9.00 + 0.95 for each thousand gallons in excess of 10,000 gallons |
| $30 < x \leq 100$ | $28.00 + 1.65 for each thousand gallons in excess of 30,000 gallons |
| $x > 100$ | $143.50 + 2.20 for each thousand gallons in excess of 100,000 gallons |

*Source*: Jericho Water District, Syosset, NY.

(a) Find a function $C$ that models the quarterly cost of using $x$ thousand gallons of water.

(b) What is the domain of $C$?

(c) Determine the intervals on which $C$ is continuous.

(d) At numbers where $C$ is not continuous (if any), what type of discontinuity does $C$ have?

(e) What are the practical implications of the answer to (d)?

**83. Gravity on Europa** Europa, one of the larger satellites of Jupiter, has an icy surface and appears to have oceans beneath the ice. This makes it a candidate for possible extraterrestrial life. Because Europa is much smaller than most planets, its gravity is weaker. If we think of Europa as a sphere with uniform internal density, then inside the sphere, the gravitational field $g$ is given by

NASA/JPL/DLR

$$g(r) = \frac{Gm}{R^3} r \quad 0 \leq r < R$$

where $R$ is the radius of the sphere, $r$ is the distance from the center of the sphere, and $G$ is the universal gravitation constant. Outside a uniform sphere of mass $m$, the gravitational field $g$ is given by

$$g(r) = \frac{Gm}{r^2} \quad R < r$$

(a) For the gravitational field of Europa to be continuous at its surface, what must $g(r)$ equal? [*Hint*: Investigate $\lim_{r \to R} g(r)$.]

(b) Determine the gravitational field at Europa's surface. This will indicate the type of gravity environment organisms will experience. Use the following measured values: Europa's mass is $4.8 \times 10^{22}$ kilograms, its radius is $1.569 \times 10^6$ meters, and $G = 6.67 \times 10^{-11}$.

(c) Compare the result found in (b) to the gravitational field on Earth's surface, which is 9.8 meter/second². Is the gravity on Europa less than or greater than that on Earth?

**84.** Find constants $A$ and $B$ so that the function below is continuous for all $x$. Graph the resulting function.

$$f(x) = \begin{cases} (x-1)^2 & \text{if } -\infty < x < 0 \\ (A-x)^2 & \text{if } 0 \leq x < 1 \\ x + B & \text{if } 1 \leq x < \infty \end{cases}$$

**85.** Find constants $A$ and $B$ so that the function below is continuous for all $x$. Graph the resulting function.

$$f(x) = \begin{cases} x + A & \text{if } -\infty < x < 4 \\ (x-1)^2 & \text{if } 4 \leq x \leq 9 \\ Bx + 1 & \text{if } 9 < x < \infty \end{cases}$$

**86.** For the function $f$ below, find $k$ so that $f$ is continuous at 2.

$$f(x) = \begin{cases} \dfrac{\sqrt{2x+5} - \sqrt{x+7}}{x-2} & \text{if } x \geq -\dfrac{5}{2}, x \neq 2 \\ k & \text{if } x = 2 \end{cases}$$

**87.** Suppose $f(x) = \dfrac{x^2 - 6x - 16}{(x^2 - 7x - 8)\sqrt{x^2 - 4}}$.

(a) For what numbers $x$ is $f$ defined?

(b) For what numbers $x$ is $f$ discontinuous?

(c) Which discontinuities, if any, found in (b) are removable?

**88. Intermediate Value Theorem**

(a) Use the Intermediate Value Theorem to show that the function $f(x) = \sin x + x - 3$ has a zero in the interval $[0, \pi]$.

(b) Approximate the zero rounded to three decimal places.

**89. Intermediate Value Theorem**

(a) Use the Intermediate Value Theorem to show that the function $f(x) = e^x + x - 2$ has a zero in the interval $[0, 2]$.

(b) Approximate the zero rounded to three decimal places.

*In Problems 90–93, verify that each function intersects the given line in the indicated interval. Then use the Intermediate Value Theorem to approximate the point of intersection correct to three decimal places by repeatedly subdividing the interval into 10 subintervals.*

**90.** $f(x) = x^3 - 2x^2 - 1$; line: $y = -1$; interval: $(1, 4)$

**91.** $g(x) = -x^4 + 3x^2 + 3$; line: $y = 3$; interval: $(1, 2)$

**92.** $h(x) = \dfrac{x^3 - 5}{x^2 + 1}$; line: $y = 1$; interval: $(1, 3)$

**93.** $r(x) = \dfrac{x - 6}{x^2 + 2}$; line: $y = -1$; interval: $(0, 3)$

**94.** Graph a function that is continuous on the closed interval $[5, 12]$, that is negative at both endpoints, and has exactly three distinct zeros in this interval. Does this contradict the Intermediate Value Theorem? Explain.

**95.** Graph a function that is continuous on the closed interval $[-1, 2]$, that is positive at both endpoints, and has exactly two zeros in this interval. Does this contradict the Intermediate Value Theorem? Explain.

**96.** Graph a function that is continuous on the closed interval $[-2, 3]$, is positive at $-2$ and negative at $3$, and has exactly two zeros in this interval. Is this possible? Does this contradict the Intermediate Value Theorem? Explain.

**97.** Graph a function that is continuous on the closed interval $[-5, 0]$, is negative at $-5$ and positive at $0$, and has exactly three zeros in the interval. Is this possible? Does this contradict the Intermediate Value Theorem? Explain.

**98. (a)** Explain why the Intermediate Value Theorem gives no information about the zeros of the function $f(x) = x^4 - 1$ on the interval $[-2, 2]$.

(b) Use technology to determine whether $f$ has a zero on the interval $[-2, 2]$.

**99. (a)** Explain why the Intermediate Value Theorem gives no information about the zeros of the function $f(x) = \ln(x^2 + 2)$ on the interval $[-2, 2]$.

(b) Use technology to determine whether $f$ has a zero on the interval $[-2, 2]$.

**100. Intermediate Value Theorem**

(a) Use the Intermediate Value Theorem to show that the functions $y = x^3$ and $y = 1 - x^2$ intersect somewhere between $x = 0$ and $x = 1$.

(b) Use technology to find the coordinates of the point of intersection rounded to three decimal places.

(c) Use technology to graph both functions on the same set of axes. Be sure the graph shows the point of intersection.

**101. Intermediate Value Theorem** An airplane is traveling at a speed of 620 miles per hour and then encounters a slight headwind that slows it to 608 miles per hour. After a few minutes, the headwind eases and the plane's speed increases to 614 miles per hour. Explain why the plane's speed is 610 miles per hour on at least two different occasions during the flight.

*Source:* Submitted by the students of Millikin University.

**102.** Suppose a function $f$ is defined and continuous on the closed interval $[a, b]$. Is the function $h(x) = \dfrac{1}{f(x)}$ also continuous on the closed interval $[a, b]$? Discuss the continuity of $h$ on $[a, b]$.

**103.** Given the two functions $f$ and $h$:

$$f(x) = x^3 - 3x^2 - 4x + 12 \qquad h(x) = \begin{cases} \dfrac{f(x)}{x - 3} & \text{if } x \neq 3 \\ p & \text{if } x = 3 \end{cases}$$

(a) Find all the zeros of the function $f$.

(b) Find the number $p$ so that the function $h$ is continuous at $x = 3$. Justify your answer.

(c) Determine whether $h$, with the number found in (b), is even, odd, or neither. Justify your answer.

**104.** The function $f(x) = \dfrac{|x|}{x}$ is not defined at 0. Explain why it is impossible to define $f(0)$ so that $f$ is continuous at 0.

**105.** Find two functions $f$ and $g$ that are each continuous at $c$, yet $\dfrac{f}{g}$ is not continuous at $c$.

**106.** Discuss the difference between a discontinuity that is removable and one that is nonremovable. Give an example of each.

**Bisection Method for Approximating Zeros of a Function** *Suppose the Intermediate Value Theorem indicates that a function $f$ has a zero in the interval $(a, b)$. The bisection method approximates the zero by evaluating $f$ at the midpoint $m_1$ of the interval $(a, b)$. If $f(m_1) = 0$, then $m_1$ is the zero we seek and the process ends. If $f(m_1) \neq 0$, then the sign of $f(m_1)$ is opposite that of either $f(a)$ or $f(b)$ (but not both), and the zero lies in that subinterval. Evaluate $f$ at the midpoint $m_2$ of this subinterval. Continue bisecting the subinterval containing the zero until the desired degree of accuracy is obtained.*
*In Problems 107–114, use the bisection method three times to approximate the zero of each function in the given interval.*

**107.** $f(x) = x^3 + 3x - 5$;   interval: $[1, 2]$

**108.** $f(x) = x^3 - 4x + 2$;   interval: $[1, 2]$

**109.** $f(x) = 2x^3 + 3x^2 + 4x - 1$;   interval: $[0, 1]$

**110.** $f(x) = x^3, -x^2 - 2x + 1$;   interval: $[0, 1]$

**111.** $f(x) = x^3 - 6x - 12$;   interval: $[3, 4]$

**112.** $f(x) = 3x^3 + 5x - 40$;   interval: $[2, 3]$

**113.** $f(x) = x^4 - 2x^3 + 21x - 23$;   interval $[1, 2]$

**114.** $f(x) = x^4 - x^3 + x - 2$;   interval: $[1, 2]$

**115. Intermediate Value Theorem** Use the Intermediate Value Theorem to show that the function $f(x) = \sqrt{x^2 + 4x} - 2$ has a zero in the interval $[0, 1]$. Then approximate the zero correct to one decimal place.

**116. Intermediate Value Theorem** Use the Intermediate Value Theorem to show that the function $f(x) = x^3 - x + 2$ has a zero in the interval $[-2, 0]$. Then approximate the zero correct to two decimal places.

**117. Continuity of a Sum** If $f$ and $g$ are each continuous at $c$, prove that $f + g$ is continuous at $c$.
(*Hint:* Use the Limit of a Sum Property.)

**118. Intermediate Value Theorem** Suppose that the functions $f$ and $g$ are continuous on the interval $[a, b]$. If $f(a) < g(a)$ and $f(b) > g(b)$, prove that the graphs of $y = f(x)$ and $y = g(x)$ intersect somewhere between $x = a$ and $x = b$.
[*Hint:* Define $h(x) = f(x) - g(x)$ and show $h(x) = 0$ for some $x$ between $a$ and $b$.]

**Challenge Problems**

**119. Intermediate Value Theorem** Let $f(x) = \dfrac{1}{x-1} + \dfrac{1}{x-2}$.

Use the Intermediate Value Theorem to prove that there is a real number $c$ between 1 and 2 for which $f(c) = 0$.

**120. Intermediate Value Theorem** Prove that there is a real number $c$ between 2.64 and 2.65 for which $c^2 = 7$.

**121.** Show that the existence of $\displaystyle\lim_{h \to 0} \frac{f(a+h) - f(a)}{h}$ implies $f$ is continuous at $x = a$.

**122.** Find constants $A$, $B$, $C$, and $D$ so that the function below is continuous for all $x$. Graph the resulting function.

$$f(x) = \begin{cases} \dfrac{x^2 + x - 2}{x - 1} & \text{if } -\infty < x < 1 \\ A & \text{if } \quad x = 1 \\ B(x - C)^2 & \text{if } \quad 1 < x < 4 \\ D & \text{if } \quad x = 4 \\ 2x - 8 & \text{if } \quad 4 < x < \infty \end{cases}$$

**123.** Let $f$ be a function for which $0 \leq f(x) \leq 1$ for all $x$ in $[0, 1]$. If $f$ is continuous on $[0, 1]$, show that there exists at least one number $c$ in $[0, 1]$ such that $f(c) = c$. [*Hint*: Let $g(x) = x - f(x)$.]

# 1.4 Limits and Continuity of Trigonometric, Exponential, and Logarithmic Functions

**OBJECTIVES** *When you finish this section, you should be able to:*

1 Use the Squeeze Theorem to find a limit (p. 115)
2 Find limits involving trigonometric functions (p. 116)
3 Determine where the trigonometric functions are continuous (p. 120)
4 Determine where an exponential or a logarithmic function is continuous (p. 121)

Until now we have found limits using the basic limits

$$\lim_{x \to c} A = A \qquad \lim_{x \to c} x = c$$

and properties of limits. But there are many limit problems that cannot be found by directly applying these techniques. To find such limits requires different results, such as the *Squeeze Theorem*,* or basic limits involving trigonometric and exponential functions.

## 1 Use the Squeeze Theorem to Find a Limit

To use the Squeeze Theorem to find $\displaystyle\lim_{x \to c} g(x)$, we need to know, or be able to find, two functions $f$ and $h$ that "sandwich" the function $g$ between them for all $x$ close to $c$. That is, in some interval containing $c$, the functions $f$, $g$, and $h$ satisfy the inequality

$$f(x) \leq g(x) \leq h(x)$$

Then if $f$ and $h$ have the same limit $L$ as $x$ approaches $c$, the function $g$ is "squeezed" to the same limit $L$ as $x$ approaches $c$. See Figure 38.

We state the Squeeze Theorem here. The proof is given in Appendix B.

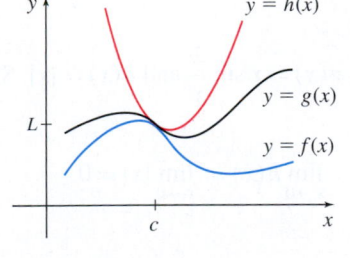

$\displaystyle\lim_{x \to c} f(x) = L,\ \lim_{x \to c} h(x) = L,\ \lim_{x \to c} g(x) = L$

**Figure 38**

**THEOREM Squeeze Theorem**

Suppose the functions $f$, $g$, and $h$ have the property that for all $x$ in an open interval containing $c$, except possibly at $c$,

$$f(x) \leq g(x) \leq h(x)$$

If

$$\lim_{x \to c} f(x) = \lim_{x \to c} h(x) = L$$

then

$$\lim_{x \to c} g(x) = L$$

*The Squeeze Theorem is also known as the Sandwich Theorem and the Pinching Theorem.

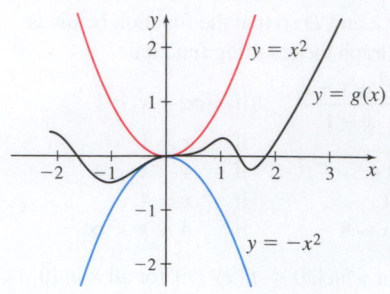

**Figure 39**

For example, suppose we wish to find $\lim\limits_{x\to 0} g(x)$, and we know that

$$-x^2 \le g(x) \le x^2$$

for all $x \ne 0$. Since $\lim\limits_{x\to 0}(-x^2)=0$ and $\lim\limits_{x\to 0} x^2=0$, the Squeeze Theorem tells us that $\lim\limits_{x\to 0} g(x)=0$. Figure 39 illustrates how $g$ is "squeezed" between $y=x^2$ and $y=-x^2$ near 0.

---

**EXAMPLE 1**   **Using the Squeeze Theorem to Find a Limit**

Use the Squeeze Theorem to find $\lim\limits_{x\to 0}\left(x\sin\dfrac{1}{x}\right)$.

**Solution** If $x \ne 0$, then $g(x)=x\sin\dfrac{1}{x}$ is defined. We seek two functions that "squeeze" $g(x)=x\sin\dfrac{1}{x}$ near 0. Since $-1 \le \sin\theta \le 1$ for all $\theta$, we begin with the inequality

$$\left|\sin\frac{1}{x}\right| \le 1 \qquad x \ne 0$$

Since $x \ne 0$ and we seek to squeeze $g(x)=x\sin\dfrac{1}{x}$, we multiply both sides of the inequality by $|x|$, $x \ne 0$. Since $|x|>0$, the direction of the inequality is preserved. [Note that if we multiply $\left|\sin\dfrac{1}{x}\right| \le 1$ by $x$, we would not know whether the inequality symbol would remain the same or be reversed since we do not know whether $x>0$ or $x<0$.]

$$|x|\left|\sin\frac{1}{x}\right| \le |x| \qquad \textbf{Multiply both sides by } |x|>0.$$

$$\left|x\sin\frac{1}{x}\right| \le |x| \qquad |a|\cdot|b|=|a\,b|.$$

$$-|x| \le x\sin\frac{1}{x} \le |x| \qquad |a|\le b \text{ is equivalent to } -b\le a\le b.$$

Now use the Squeeze Theorem with $f(x)=-|x|$, $g(x)=x\sin\dfrac{1}{x}$ and $h(x)=|x|$. Since $f(x) \le g(x) \le h(x)$ and

$$\lim_{x\to 0} f(x)= \lim_{x\to 0}(-|x|)=0 \qquad \text{and} \qquad \lim_{x\to 0} h(x)= \lim_{x\to 0}|x|=0$$

it follows that

$$\lim_{x\to 0} g(x)= \lim_{x\to 0}\left(x\sin\frac{1}{x}\right)=0 \qquad\blacksquare$$

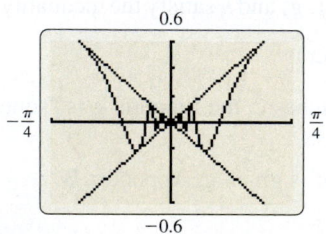

**Figure 40** $g(x)=x\sin\dfrac{1}{x}$ is squeezed between $f(x)=-|x|$ and $h(x)=|x|$.

Figure 40 shows how $g(x)=x\sin\dfrac{1}{x}$ is squeezed between $y=-|x|$ and $y=|x|$.

**NOW WORK**   Problems 5 and 47.

## 2 Find Limits Involving Trigonometric Functions

Knowing $\lim\limits_{x\to c} A=A$ and $\lim\limits_{x\to c} x=c$ helped us to find the limits of many algebraic functions. Knowing several basic trigonometric limits can help to find many limits involving trigonometric functions.

**THEOREM Two Basic Trigonometric Limits**

$$\lim_{x \to 0} \sin x = 0 \qquad \lim_{x \to 0} \cos x = 1$$

The graphs of $y = \sin x$ and $y = \cos x$ in Figure 41 suggest that $\lim_{x \to 0} \sin x = 0$ and $\lim_{x \to 0} \cos x = 1$. The proofs of these limits use the Squeeze Theorem. Problem 63 provides an outline of the proof that $\lim_{x \to 0} \sin x = 0$.

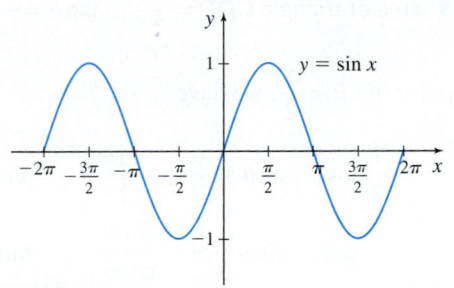
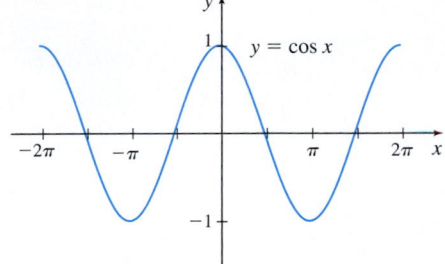

**Figure 41**

A third basic trigonometric limit, $\lim_{\theta \to 0} \dfrac{\sin \theta}{\theta} = 1$, is important in calculus. In Section 1.1, a table suggested that $\lim_{\theta \to 0} \dfrac{\sin \theta}{\theta} = 1$. The function $f(\theta) = \dfrac{\sin \theta}{\theta}$, whose graph is shown in Figure 42, is defined for all real numbers $\theta \neq 0$. The graph suggests $\lim_{\theta \to 0} \dfrac{\sin \theta}{\theta} = 1$.

**Figure 42** $f(\theta) = \dfrac{\sin \theta}{\theta}$

**THEOREM**

If $\theta$ is measured in radians, then

$$\lim_{\theta \to 0} \frac{\sin \theta}{\theta} = 1$$

**Proof** Although $\dfrac{\sin \theta}{\theta}$ is a quotient, we have no way to divide out $\theta$. To find $\lim_{\theta \to 0^+} \dfrac{\sin \theta}{\theta}$, we let $\theta$ be a positive acute central angle of a unit circle, as shown in Figure 43(a). Notice that $COP$ is a sector of the circle. Add the point $B = (\cos \theta, 0)$ to the graph and form triangle $BOP$. Next extend the terminal side of angle $\theta$ until it intersects the line $x = 1$ at the point $D$, forming a second triangle $COD$. The $x$-coordinate of $D$ is 1. Since the length of the line segment $\overline{OC}$ is $OC = 1$, then the $y$-coordinate of $D$ is $CD = \dfrac{CD}{OC} = \tan \theta$. So, $D = (1, \tan \theta)$. See Figure 43(b).

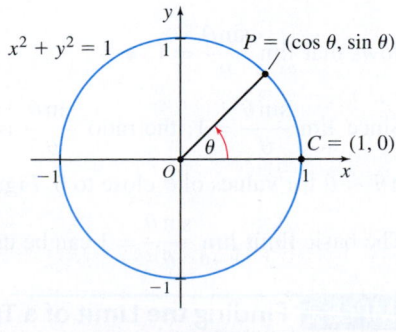

(a) $0 < \theta < \dfrac{\pi}{2}$    (b)

**Figure 43**

We see from Figure 43(b) that

$$\text{area of triangle } BOP \leq \text{area of sector } COP \leq \text{area of triangle } COD \qquad (1)$$

Each of these areas can be expressed in terms of $\theta$ as

- area of triangle $BOP = \dfrac{1}{2}\cos\theta \cdot \sin\theta$      **area of a triangle** $= \dfrac{1}{2}$ **base** × **height**

- area of sector $COP = \dfrac{\theta}{2}$      **area of a sector** $= \dfrac{1}{2}r^2\theta; r = 1$

- area of triangle $COD = \dfrac{1}{2}\cdot 1\cdot\tan\theta = \dfrac{\tan\theta}{2}$      **area of a triangle** $= \dfrac{1}{2}$ **base** × **height**

So, for $0 < \theta < \dfrac{\pi}{2}$, we have

$$\frac{1}{2}\cos\theta \cdot \sin\theta \leq \frac{\theta}{2} \leq \frac{\tan\theta}{2} \qquad \text{From (1)}$$

$$\cos\theta \cdot \sin\theta \leq \theta \leq \frac{\sin\theta}{\cos\theta} \qquad \text{Multiply all parts by 2; } \tan\theta = \frac{\sin\theta}{\cos\theta}.$$

$$\cos\theta \leq \frac{\theta}{\sin\theta} \leq \frac{1}{\cos\theta} \qquad \text{Divide by } \sin\theta; \text{ since } 0 < \theta < \frac{\pi}{2}, \sin\theta > 0.$$

$$\frac{1}{\cos\theta} \geq \frac{\sin\theta}{\theta} \geq \cos\theta \qquad \text{Invert each term in the inequality;}$$
$$\text{reverse the inequality signs.}$$

$$\cos\theta \leq \frac{\sin\theta}{\theta} \leq \frac{1}{\cos\theta} \qquad \text{Rewrite using } \leq.$$

Now use the Squeeze Theorem with $f(\theta) = \cos\theta$, $g(\theta) = \dfrac{\sin\theta}{\theta}$, and $h(\theta) = \dfrac{1}{\cos\theta}$. Since $f(\theta) \leq g(\theta) \leq h(\theta)$, and since

$$\lim_{\theta \to 0^+} f(\theta) = \lim_{\theta \to 0^+} \cos\theta = 1 \quad \text{and} \quad \lim_{\theta \to 0^+} h(\theta) = \lim_{\theta \to 0^+} \frac{1}{\cos\theta} = \frac{\lim\limits_{\theta \to 0^+} 1}{\lim\limits_{\theta \to 0^+} \cos\theta} = \frac{1}{1} = 1$$

then by the Squeeze Theorem, $\lim\limits_{\theta \to 0^+} g(\theta) = \lim\limits_{\theta \to 0^+} \dfrac{\sin\theta}{\theta} = 1$.

Now use the fact that $g(\theta) = \dfrac{\sin\theta}{\theta}$ is an even function, that is,

$$g(-\theta) = \frac{\sin(-\theta)}{-\theta} = \frac{-\sin\theta}{-\theta} = \frac{\sin\theta}{\theta} = g(\theta)$$

So,

$$\lim_{\theta \to 0^-} \frac{\sin\theta}{\theta} = \lim_{\theta \to 0^+} \frac{\sin(-\theta)}{-\theta} = \lim_{\theta \to 0^+} \frac{\sin\theta}{\theta} = 1$$

It follows that $\lim\limits_{\theta \to 0} \dfrac{\sin\theta}{\theta} = 1$. ∎

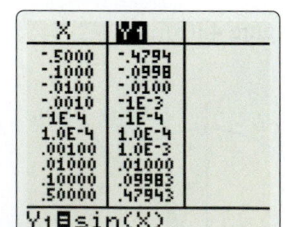

**Figure 44**

Since $\lim\limits_{\theta \to 0} \dfrac{\sin\theta}{\theta} = 1$, the ratio $\dfrac{\sin\theta}{\theta}$ is close to 1 for values of $\theta$ close to 0. That is, $\sin\theta \approx \theta$ for values of $\theta$ close to 0. Figure 44 illustrates this property.

The basic limit $\lim\limits_{\theta \to 0} \dfrac{\sin\theta}{\theta} = 1$ can be used to find the limits of similar expressions.

**EXAMPLE 2**   **Finding the Limit of a Trigonometric Function**

Find:

**(a)** $\lim\limits_{\theta \to 0} \dfrac{\sin(3\theta)}{\theta}$      **(b)** $\lim\limits_{\theta \to 0} \dfrac{\sin(5\theta)}{\sin(2\theta)}$

**Solution (a)** Since $\lim\limits_{\theta \to 0} \dfrac{\sin(3\theta)}{\theta}$ is not in the same form as $\lim\limits_{\theta \to 0} \dfrac{\sin \theta}{\theta}$, we multiply the numerator and the denominator by 3 and make the substitution $t = 3\theta$.

$$\lim_{\theta \to 0} \frac{\sin(3\theta)}{\theta} = \lim_{\theta \to 0} \frac{3\sin(3\theta)}{3\theta} = \lim_{\substack{t \to 0 \\ \uparrow \\ t = 3\theta \\ t \to 0 \text{ as } \theta \to 0}} \left(3\,\frac{\sin t}{t}\right) = 3 \lim_{t \to 0} \frac{\sin t}{t} = 3 \cdot 1 = 3$$

$$\lim_{t \to 0} \frac{\sin t}{t} = 1$$

**(b)** We begin by dividing the numerator and the denominator by $\theta$. Then

$$\frac{\sin(5\theta)}{\sin(2\theta)} = \frac{\dfrac{\sin(5\theta)}{\theta}}{\dfrac{\sin(2\theta)}{\theta}}$$

Now we follow the approach in (a) on the numerator and on the denominator.

$$\lim_{\theta \to 0} \frac{\sin(5\theta)}{\theta} = \lim_{\theta \to 0} \frac{5\sin(5\theta)}{5\theta} = \lim_{\substack{t \to 0 \\ \uparrow \\ t = 5\theta \\ t \to 0 \text{ as } \theta \to 0}} \frac{5\sin t}{t} = 5 \lim_{t \to 0} \frac{\sin t}{t} = 5$$

$$\lim_{\theta \to 0} \frac{\sin(2\theta)}{\theta} = \lim_{\theta \to 0} \frac{2\sin(2\theta)}{2\theta} = \lim_{\substack{t \to 0 \\ \uparrow \\ t = 2\theta \\ t \to 0 \text{ as } \theta \to 0}} \frac{2\sin t}{t} = 2 \lim_{t \to 0} \frac{\sin t}{t} = 2$$

$$\lim_{\theta \to 0} \frac{\sin(5\theta)}{\sin(2\theta)} = \frac{\displaystyle\lim_{\theta \to 0} \frac{\sin(5\theta)}{\theta}}{\displaystyle\lim_{\theta \to 0} \frac{\sin(2\theta)}{\theta}} = \frac{5}{2}$$ ∎

**NOW WORK** Problems 23 and 25.

Example 3 establishes an important limit used in Chapter 2.

---

**EXAMPLE 3** **Finding a Basic Trigonometric Limit**

Establish the formula

$$\boxed{\lim_{\theta \to 0} \frac{\cos \theta - 1}{\theta} = 0}$$

where $\theta$ is measured in radians.

**Solution** First we rewrite the expression $\dfrac{\cos \theta - 1}{\theta}$ as the product of two terms whose limits are known. For $\theta \neq 0$,

$$\frac{\cos \theta - 1}{\theta} = \left(\frac{\cos \theta - 1}{\theta}\right)\left(\frac{\cos \theta + 1}{\cos \theta + 1}\right)$$

$$= \frac{\cos^2 \theta - 1}{\theta(\cos \theta + 1)} \underset{\substack{\uparrow \\ \sin^2 \theta + \cos^2 \theta = 1}}{=} \frac{-\sin^2 \theta}{\theta(\cos \theta + 1)} = \left(\frac{\sin \theta}{\theta}\right)\frac{(-\sin \theta)}{\cos \theta + 1}$$

Now we find the limit.

$$\lim_{\theta \to 0} \frac{\cos \theta - 1}{\theta} = \lim_{\theta \to 0}\left[\left(\frac{\sin \theta}{\theta}\right)\left(\frac{(-\sin \theta)}{\cos \theta + 1}\right)\right] = \left[\lim_{\theta \to 0} \frac{\sin \theta}{\theta}\right]\left[\lim_{\theta \to 0} \frac{-\sin \theta}{\cos \theta + 1}\right]$$

$$= 1 \cdot \frac{\displaystyle\lim_{\theta \to 0}(-\sin \theta)}{\displaystyle\lim_{\theta \to 0}(\cos \theta + 1)} = \frac{0}{2} = 0$$ ∎

**NOW WORK** Problem 31.

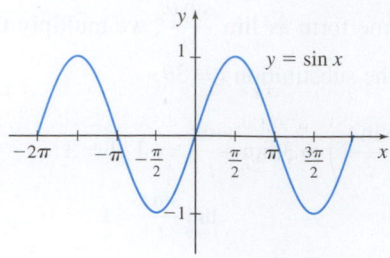

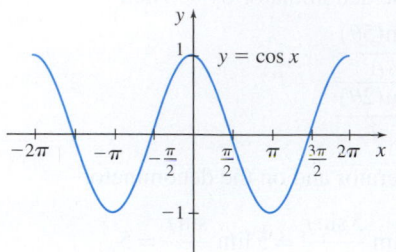

**Figure 45**

## 3 Determine Where the Trigonometric Functions Are Continuous

The graphs of $f(x) = \sin x$ and $g(x) = \cos x$ in Figure 45 suggest that $f$ and $g$ are continuous for all real numbers.

**EXAMPLE 4 Showing $f(x) = \sin x$ Is Continuous at 0**

- $f(0) = \sin 0 = 0$, so $f$ is defined at 0.

- $\lim\limits_{x \to 0} f(x) = \lim\limits_{x \to 0} \sin x = 0$, so the limit at 0 exists.

- $\lim\limits_{x \to 0} \sin x = \sin 0 = 0$.

Since all three conditions of continuity are satisfied, $f(x) = \sin x$ is continuous at 0. ∎

In a similar way, we can show that $g(x) = \cos x$ is continuous at 0. That is, $\lim\limits_{x \to 0} \cos x = \cos 0 = 1$.

You are asked to prove the following theorem in Problem 67.

**THEOREM**

- The sine function $y = \sin x$ is continuous for all real numbers.
- The cosine function $y = \cos x$ is continuous for all real numbers.

Based on this theorem the following two limits can be added to the list of basic limits.

$$\lim_{x \to c} \sin x = \sin c \qquad \text{for all real numbers } c$$

$$\lim_{x \to c} \cos x = \cos c \qquad \text{for all real numbers } c$$

Using the facts that the sine and cosine functions are continuous for all real numbers, we can use basic trigonometric identities to determine where the remaining four trigonometric functions are continuous:

- $y = \tan x$: Since $\tan x = \dfrac{\sin x}{\cos x}$, from the Continuity of a Quotient property, $y = \tan x$ is continuous at all real numbers except those for which $\cos x = 0$. That is, $y = \tan x$ is continuous for all real numbers except odd multiples of $\dfrac{\pi}{2}$.

- $y = \sec x$: Since $\sec x = \dfrac{1}{\cos x}$, from the Continuity of a Quotient property, $y = \sec x$ is continuous at all real numbers except those for which $\cos x = 0$. That is, $y = \sec x$ is continuous for all real numbers except odd multiples of $\dfrac{\pi}{2}$.

- $y = \cot x$: Since $\cot x = \dfrac{\cos x}{\sin x}$, from the Continuity of a Quotient property, $y = \cot x$ is continuous at all real numbers except those for which $\sin x = 0$. That is, $y = \cot x$ is continuous for all real numbers except integer multiples of $\pi$.

- $y = \csc x$: Since $\csc x = \dfrac{1}{\sin x}$, from the Continuity of a Quotient property, $y = \csc x$ is continuous at all real numbers except those for which $\sin x = 0$. That is, $y = \csc x$ is continuous for all real numbers except integer multiples of $\pi$.

Recall that a one-to-one function that is continuous on its domain has an inverse function that is continuous on its domain.

**NEED TO REVIEW?** Inverse trigonometric functions are discussed in Section P.7, pp. 61–66.

Since each of the six trigonometric functions is continuous on its domain, then each is continuous on the restricted domain used to define its inverse trigonometric function. This means the inverse trigonometric functions are continuous on their domains. These results are summarized in Table 10.

**TABLE 10** Continuity of the Trigonometric Functions and Their Inverses

| Name | Function | Domain | Properties |
|------|----------|--------|------------|
| sine | $f(x) = \sin x$ | all real numbers | continuous on the interval $(-\infty, \infty)$ |
| cosine | $f(x) = \cos x$ | all real numbers | continuous on the interval $(-\infty, \infty)$ |
| tangent | $f(x) = \tan x$ | $\left\{x \mid x \neq \text{odd multiples of } \dfrac{\pi}{2}\right\}$ | continuous at all real numbers except odd multiples of $\dfrac{\pi}{2}$ |
| cosecant | $f(x) = \csc x$ | $\{x \mid x \neq \text{multiples of } \pi\}$ | continuous at all real numbers except multiples of $\pi$ |
| secant | $f(x) = \sec x$ | $\left\{x \mid x \neq \text{odd multiples of } \dfrac{\pi}{2}\right\}$ | continuous at all real numbers except odd multiples of $\dfrac{\pi}{2}$ |
| cotangent | $f(x) = \cot x$ | $\{x \mid x \neq \text{multiples of } \pi\}$ | continuous at all real numbers except multiples of $\pi$ |
| inverse sine | $f(x) = \sin^{-1} x$ | $-1 \leq x \leq 1$ | continuous on the closed interval $[-1, 1]$ |
| inverse cosine | $f(x) = \cos^{-1} x$ | $-1 \leq x \leq 1$ | continuous on the closed interval $[-1, 1]$ |
| inverse tangent | $f(x) = \tan^{-1} x$ | all real numbers | continuous on the interval $(-\infty, \infty)$ |
| inverse cosecant | $f(x) = \csc^{-1} x$ | $|x| \geq 1$ | continuous on the set $(-\infty, -1] \cup [1, \infty)$ |
| inverse secant | $f(x) = \sec^{-1} x$ | $|x| \geq 1$ | continuous on the set $(-\infty, -1] \cup [1, \infty)$ |
| inverse cotangent | $f(x) = \cot^{-1} x$ | all real numbers | continuous on the interval $(-\infty, \infty)$ |

## 4 Determine Where an Exponential or a Logarithmic Function Is Continuous

The graphs of an exponential function $y = a^x$ and its inverse function $y = \log_a x$ are shown in Figure 46. The graphs suggest that an exponential function and a logarithmic function are continuous on their domains. We state the following theorem without proof.

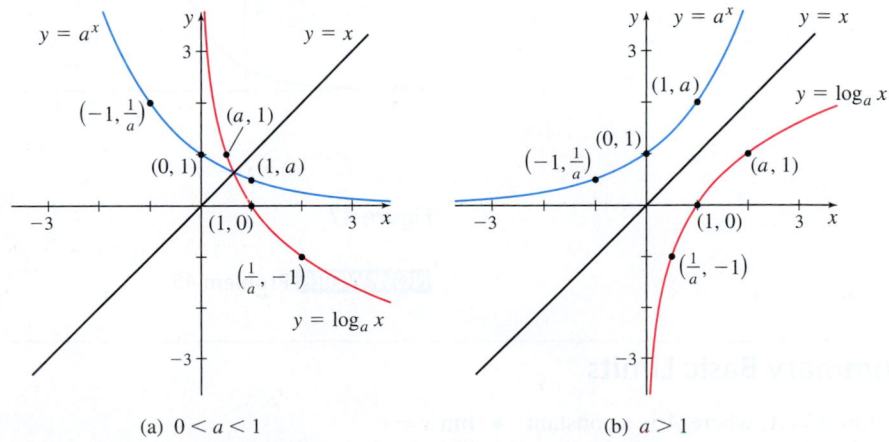

(a) $0 < a < 1$          (b) $a > 1$

**Figure 46**

> **THEOREM Continuity of Exponential and Logarithmic Functions**
>
> - An exponential function is continuous on its domain, all real numbers.
> - A logarithmic function is continuous on its domain, all positive real numbers.

Based on this theorem, the following two limits can be added to the list of basic limits.

**IN WORDS** The limit of an exponential function and a logarithmic function can be found using substitution.

$$\lim_{x \to c} a^x = a^c \qquad \text{for all real numbers } c, \ a > 0, \ a \neq 1$$

$$\lim_{x \to c} \log_a x = \log_a c \qquad \text{for any real number } c > 0, \ a > 0, \ a \neq 1$$

**CALC CLIP**

**EXAMPLE 5   Showing a Composite Function Is Continuous**

Show that

**(a)** $f(x) = e^{2x}$ is continuous for all real numbers.

**(b)** $F(x) = \sqrt[3]{\ln x}$ is continuous for $x > 0$.

**Solution (a)** The domain of the exponential function is the set of all real numbers, so $f$ is defined for any number $c$. That is, $f(c) = e^{2c}$. Also for any number $c$,

$$\lim_{x \to c} f(x) = \lim_{x \to c} e^{2x} = \lim_{x \to c} (e^x)^2 = \left[ \lim_{x \to c} e^x \right]^2 = (e^c)^2 = e^{2c} = f(c)$$

Since $\lim_{x \to c} f(x) = f(c)$ for any number $c$, then $f$ is continuous for all real numbers $c$.

**(b)** The logarithmic function $f(x) = \ln x$ is continuous on its domain, the set of all positive real numbers. The function $g(x) = \sqrt[3]{x}$ is continuous for all real numbers. The function $F(x) = (g \circ f)(x) = \sqrt[3]{\ln x}$ is continuous for all real numbers $x > 0$. ∎

Figures 47(a) and 47(b) illustrate the graphs of $f$ and $F$.

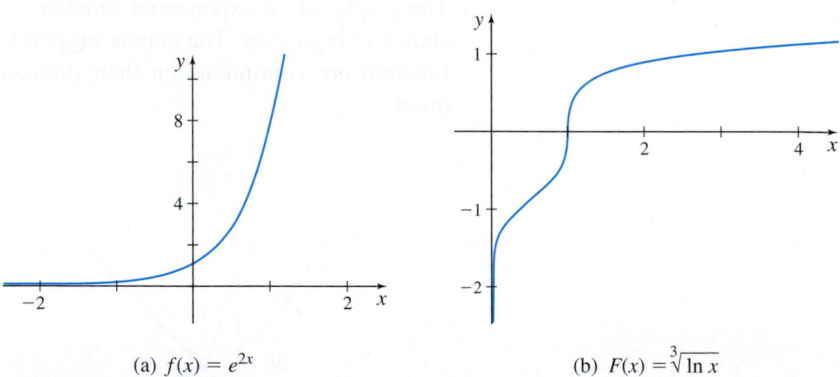

(a) $f(x) = e^{2x}$              (b) $F(x) = \sqrt[3]{\ln x}$

**Figure 47**

**NOW WORK** Problem 45.

## Summary Basic Limits

- $\lim_{x \to c} A = A$, where $A$ is a constant
- $\lim_{x \to c} x = c$
- $\lim_{x \to c} \sin x = \sin c$
- $\lim_{x \to c} \cos x = \cos c$
- $\lim_{\theta \to 0} \dfrac{\sin \theta}{\theta} = 1$
- $\lim_{\theta \to 0} \dfrac{\cos \theta - 1}{\theta} = 0$
- $\lim_{x \to c} a^x = a^c, \ a > 0, a \neq 1$
- $\lim_{x \to c} \log_a x = \log_a c, \ c > 0, a > 0, a \neq 1$

# 1.4 Assess Your Understanding

## Concepts and Vocabulary

1. $\lim\limits_{x \to 0} \sin x = $ _____

2. *True or False*    $\lim\limits_{x \to 0} \dfrac{\cos x - 1}{x} = 1$

3. The Squeeze Theorem states that if the functions $f$, $g$, and $h$ have the property $f(x) \le g(x) \le h(x)$ for all $x$ in an open interval containing $c$, except possibly at $c$, and if $\lim\limits_{x \to c} f(x) = \lim\limits_{x \to c} h(x) = L$, then $\lim\limits_{x \to c} g(x) = $ _____.

4. *True or False*    $f(x) = \csc x$ is continuous for all real numbers except $x = 0$.

## Skill Building

*In Problems 5–8, use the Squeeze Theorem to find each limit.*

5. Suppose $-x^2 + 1 \le g(x) \le x^2 + 1$ for all $x$ in an open interval containing 0. Find $\lim\limits_{x \to 0} g(x)$.

6. Suppose $-(x-2)^2 - 3 \le g(x) \le (x-2)^2 - 3$ for all $x$ in an open interval containing 2. Find $\lim\limits_{x \to 2} g(x)$.

7. Suppose $\cos x \le g(x) \le 1$ for all $x$ in an open interval containing 0. Find $\lim\limits_{x \to 0} g(x)$.

8. Suppose $-x^2 + 1 \le g(x) \le \sec x$ for all $x$ in an open interval containing 0. Find $\lim\limits_{x \to 0} g(x)$.

*In Problems 9–22, find each limit.*

9. $\lim\limits_{x \to 0} (x^3 + \sin x)$

10. $\lim\limits_{x \to 0} (x^2 - \cos x)$

11. $\lim\limits_{x \to \pi/3} (\cos x + \sin x)$

12. $\lim\limits_{x \to \pi/3} (\sin x - \cos x)$

13. $\lim\limits_{x \to 0} \dfrac{\cos x}{1 + \sin x}$

14. $\lim\limits_{x \to 0} \dfrac{\sin x}{1 + \cos x}$

15. $\lim\limits_{x \to 0} \dfrac{3}{1 + e^x}$

16. $\lim\limits_{x \to 0} \dfrac{e^x - 1}{1 + e^x}$

17. $\lim\limits_{x \to 0} (e^x \sin x)$

18. $\lim\limits_{x \to 0} (e^{-x} \tan x)$

19. $\lim\limits_{x \to 1} \ln \left( \dfrac{e^x}{x} \right)$

20. $\lim\limits_{x \to 1} \ln \left( \dfrac{x}{e^x} \right)$

21. $\lim\limits_{x \to 0} \dfrac{e^{2x}}{1 + e^x}$

22. $\lim\limits_{x \to 0} \dfrac{1 - e^x}{1 - e^{2x}}$

*In Problems 23–34, find each limit.*

23. $\lim\limits_{x \to 0} \dfrac{\sin(7x)}{x}$

24. $\lim\limits_{x \to 0} \dfrac{\sin \frac{x}{3}}{x}$

25. $\lim\limits_{\theta \to 0} \dfrac{\theta + 3 \sin \theta}{2\theta}$

26. $\lim\limits_{x \to 0} \dfrac{2x - 5 \sin(3x)}{x}$

27. $\lim\limits_{\theta \to 0} \dfrac{\sin \theta}{\theta + \tan \theta}$

28. $\lim\limits_{\theta \to 0} \dfrac{\tan \theta}{\theta}$

29. $\lim\limits_{\theta \to 0} \dfrac{5}{\theta \cdot \csc \theta}$

30. $\lim\limits_{\theta \to 0} \dfrac{\sin(3\theta)}{\sin(2\theta)}$

31. $\lim\limits_{\theta \to 0} \dfrac{1 - \cos^2 \theta}{\theta}$

32. $\lim\limits_{\theta \to 0} \dfrac{\cos(4\theta) - 1}{2\theta}$

33. $\lim\limits_{\theta \to 0} (\theta \cdot \cot \theta)$

34. $\lim\limits_{\theta \to 0} \left[ \sin \theta \left( \dfrac{\cot \theta - \csc \theta}{\theta} \right) \right]$

*In Problems 35–38, determine whether $f$ is continuous at the number $c$.*

35. $f(x) = \begin{cases} 3 \cos x & \text{if } x < 0 \\ 3 & \text{if } x = 0 \\ x + 3 & \text{if } x > 0 \end{cases}$    at $c = 0$

36. $f(x) = \begin{cases} \cos x & \text{if } x < 0 \\ 0 & \text{if } x = 0 \\ e^x & \text{if } x > 0 \end{cases}$    at $c = 0$

37. $f(\theta) = \begin{cases} \sin \theta & \text{if } \theta \le \dfrac{\pi}{4} \\ \cos \theta & \text{if } \theta > \dfrac{\pi}{4} \end{cases}$    at $c = \dfrac{\pi}{4}$

38. $f(x) = \begin{cases} \tan^{-1} x & \text{if } x < 1 \\ \ln x & \text{if } x \ge 1 \end{cases}$    at $c = 1$

*In Problems 39–46, determine where $f$ is continuous.*

39. $f(x) = \sin \left( \dfrac{x^2 - 4x}{x - 4} \right)$

40. $f(x) = \cos \left( \dfrac{x^2 - 5x + 1}{2x} \right)$

41. $f(\theta) = \dfrac{1}{1 + \sin \theta}$

42. $f(\theta) = \dfrac{1}{1 + \cos^2 \theta}$

43. $f(x) = \dfrac{\ln x}{x - 3}$

44. $f(x) = \ln (x^2 + 1)$

45. $f(x) = e^{-x} \sin x$

46. $f(x) = \dfrac{e^x}{1 + \sin^2 x}$

## Applications and Extensions

*In Problems 47–50, use the Squeeze Theorem to find each limit.*

47. $\lim\limits_{x \to 0} \left( x^2 \sin \dfrac{1}{x} \right)$

48. $\lim\limits_{x \to 0} \left[ x \left( 1 - \cos \dfrac{1}{x} \right) \right]$

49. $\lim\limits_{x \to 0} \left[ x^2 \left( 1 - \cos \dfrac{1}{x} \right) \right]$

50. $\lim\limits_{x \to 0} \left[ \sqrt{x^3 + 3x^2} \sin \left( \dfrac{1}{x} \right) \right]$

*In Problems 51–54, show that each statement is true.*

51. $\lim\limits_{x \to 0} \dfrac{\sin(ax)}{\sin(bx)} = \dfrac{a}{b}; \quad b \ne 0$

52. $\lim\limits_{x \to 0} \dfrac{\cos(ax)}{\cos(bx)} = 1$

53. $\lim\limits_{x \to 0} \dfrac{\sin(ax)}{bx} = \dfrac{a}{b}; \quad b \ne 0$

54. $\lim\limits_{x \to 0} \dfrac{1 - \cos(ax)}{bx} = 0; \quad a \ne 0, b \ne 0$

55. Show that $\lim\limits_{x \to 0} \dfrac{1 - \cos x}{x^2} = \dfrac{1}{2}$.

56. **Squeeze Theorem** If $0 \le f(x) \le 1$ for every number $x$, show that $\lim\limits_{x \to 0} [x^2 f(x)] = 0$.

---

1. = NOW WORK problem    📐 = Graphing technology recommended    CAS = Computer Algebra System recommended

**57. Squeeze Theorem** If $0 \le f(x) \le M$ for every $x$, show that $\lim\limits_{x \to 0} [x^2 f(x)] = 0$.

**58.** The function $f(x) = \dfrac{\sin(\pi x)}{x}$ is not defined at 0. Decide how to define $f(0)$ so that $f$ is continuous at 0.

**59.** Define $f(0)$ and $f(1)$ so that the function $f(x) = \dfrac{\sin(\pi x)}{x(1-x)}$ is continuous on the interval $[0, 1]$.

**60.** Is $f(x) = \begin{cases} \dfrac{\sin x}{x} & \text{if } x \ne 0 \\ 1 & \text{if } x = 0 \end{cases}$ continuous at 0?

**61.** Is $f(x) = \begin{cases} \dfrac{1 - \cos x}{x} & \text{if } x \ne 0 \\ 0 & \text{if } x = 0 \end{cases}$ continuous at 0?

**62. Squeeze Theorem** Show that $\lim\limits_{x \to 0} \left[ x^n \sin\left(\dfrac{1}{x}\right) \right] = 0$, where $n$ is a positive integer. (*Hint:* Look first at Problem 56.)

**63.** Prove $\lim\limits_{\theta \to 0} \sin \theta = 0$.

(*Hint:* Use a unit circle as shown in the figure, first assuming $0 < \theta < \dfrac{\pi}{2}$.

Then use the fact that $\sin \theta$ is less than the length of the arc $AP$, and the Squeeze Theorem, to show that $\lim\limits_{\theta \to 0^+} \sin \theta = 0$. Then use a similar argument with $-\dfrac{\pi}{2} < \theta < 0$ to show $\lim\limits_{\theta \to 0^-} \sin \theta = 0$.)

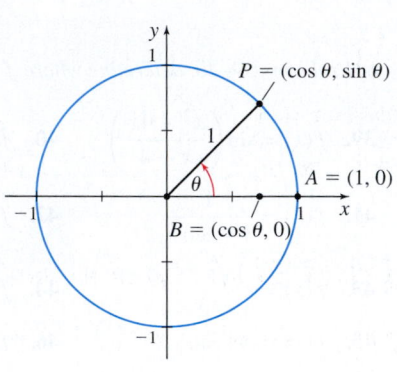

**64.** Prove $\lim\limits_{\theta \to 0} \cos \theta = 1$. Use either the proof outlined in Problem 63 or a proof using the result $\lim\limits_{\theta \to 0} \sin \theta = 0$ and a Pythagorean identity.

**65.** Without using limits, explain how you can decide where $f(x) = \cos(5x^3 + 2x^2 - 8x + 1)$ is continuous.

**66.** Explain the Squeeze Theorem. Draw a graph to illustrate your explanation.

**Challenge Problems**

**67.** Use the Sum Formulas $\sin(a+b) = \sin a \cos b + \cos a \sin b$ and $\cos(a+b) = \cos a \cos b - \sin a \sin b$ to show that the sine function and cosine function are continuous on their domains.

**68.** Find $\lim\limits_{x \to 0} \dfrac{\sin x^2}{x}$.

**69. Squeeze Theorem** If $f(x) = \begin{cases} 1 & \text{if } x \text{ is rational} \\ 0 & \text{if } x \text{ is irrational} \end{cases}$ show that $\lim\limits_{x \to 0} [x f(x)] = 0$.

**70.** Suppose points $A$ and $B$ with coordinates $(0, 0)$ and $(1, 0)$, respectively, are given. Let $n$ be a number greater than 0, and let $\theta$ be an angle with the property $0 < \theta < \dfrac{\pi}{1+n}$.

Construct a triangle $ABC$ where $\overline{AC}$ and $\overline{AB}$ form the angle $\theta$, and $\overline{BC}$ and $\overline{BA}$ form the angle $n\theta$ (see the figure below). Let $D$ be the point of intersection of $\overline{AB}$ with the perpendicular from $C$ to $\overline{AB}$. What is the limiting position of $D$ as $\theta$ approaches 0?

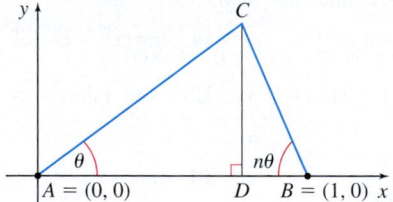

---

# 1.5 Infinite Limits; Limits at Infinity; Asymptotes

**OBJECTIVES** *When you finish this section, you should be able to:*

**1** Find infinite limits (p. 125)

**2** Find the vertical asymptotes of a graph (p. 128)

**3** Find limits at infinity (p. 128)

**4** Find the horizontal asymptotes of a graph (p. 135)

**5** Find the asymptotes of the graph of a rational function (p. 135)

**RECALL** The symbols $\infty$ (infinity) and $-\infty$ (negative infinity) are not numbers. The symbol $\infty$ expresses unboundedness in the positive direction, and $-\infty$ expresses unboundedness in the negative direction.

We have described $\lim\limits_{x \to c} f(x) = L$ by saying if a function $f$ is defined everywhere in an open interval containing $c$, except possibly at $c$, then the value $f(x)$ can be made as close as we please to $L$ by choosing numbers $x$ sufficiently close to $c$. Here $c$ and $L$ are real numbers. In this section, we extend the language of limits to allow $c$ to be $\infty$ or $-\infty$ (*limits at infinity*) and to allow $L$ to be $\infty$ or $-\infty$ (*infinite limits*). These limits are useful for locating *asymptotes* that aid in graphing some functions.

We begin with infinite limits.

## 1 Find Infinite Limits

Consider the function $f(x) = \dfrac{1}{x^2}$. Table 11 lists values of $f$ for selected numbers $x$ that are close to 0 and Figure 48 shows the graph of $f$.

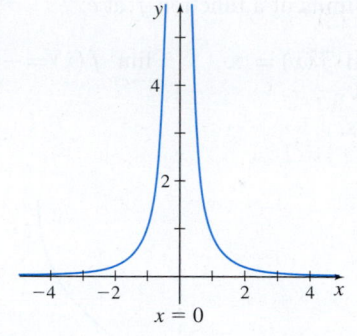

**Figure 48** $f(x) = \dfrac{1}{x^2}$

**TABLE 11**

| | x approaches 0 from the left | | | | x approaches 0 from the right | | |
|---|---|---|---|---|---|---|---|
| $x$ | −0.01 | −0.001 | −0.0001 | → 0 ← | 0.0001 | 0.001 | 0.01 |
| $f(x) = \dfrac{1}{x^2}$ | 10,000 | $10^6$ | $10^8$ | $f(x)$ becomes unbounded | $10^8$ | $10^6$ | 10,000 |

As $x$ approaches 0, the value of $\dfrac{1}{x^2}$ increases without bound. Since the value of $\dfrac{1}{x^2}$ is not approaching a single real number, the *limit* of $f(x)$ as $x$ approaches 0 *does not exist*. However, since the numbers are increasing without bound, we describe the behavior of the function near zero by writing

$$\lim_{x \to 0} \frac{1}{x^2} = \infty$$

and say that $f(x) = \dfrac{1}{x^2}$ has an **infinite limit** as $x$ approaches 0.

In other words, a function $f$ has an infinite limit at $c$ if $f$ is defined everywhere in an open interval containing $c$, except possibly at $c$, and $f(x)$ becomes unbounded when $x$ is sufficiently close to $c$.*

**EXAMPLE 1**  Investigating $\displaystyle\lim_{x \to 0^-} \frac{1}{x}$ and $\displaystyle\lim_{x \to 0^+} \frac{1}{x}$

Consider the function $f(x) = \dfrac{1}{x}$. Table 12 shows values of $f$ for selected numbers $x$ that are close to 0 and Figure 49 shows the graph of $f$.

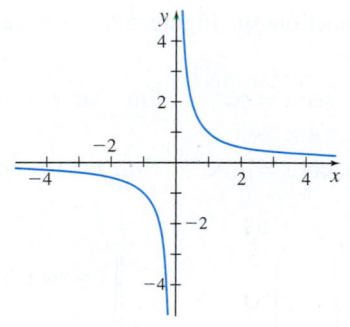

**Figure 49** $f(x) = \dfrac{1}{x}$

**TABLE 12**

| | x approaches 0 from the left | | | | x approaches 0 from the right | | | |
|---|---|---|---|---|---|---|---|---|
| $x$ | −0.01 | −0.001 | −0.0001 | → 0 ← | 0.0001 | 0.001 | 0.01 | 0.1 |
| $f(x) = \dfrac{1}{x}$ | −100 | −1000 | −10,000 | $f(x)$ becomes unbounded | 10,000 | 1000 | 100 | 10 |

Here as $x$ gets closer to 0 from the right, the value of $f(x) = \dfrac{1}{x}$ can be made as large as we please. That is, $\dfrac{1}{x}$ becomes unbounded in the positive direction. So,

$$\lim_{x \to 0^+} \frac{1}{x} = \infty$$

Similarly, the notation

$$\lim_{x \to 0^-} \frac{1}{x} = -\infty$$

is used to indicate that $\dfrac{1}{x}$ becomes unbounded in the negative direction as $x$ approaches 0 from the left. ∎

*A precise definition of an infinite limit is given in Section 1.6.

So, there are four possible one-sided infinite limits of a function $f$ at $c$:

$$\lim_{x \to c^-} f(x) = \infty, \qquad \lim_{x \to c^-} f(x) = -\infty, \qquad \lim_{x \to c^+} f(x) = \infty, \qquad \lim_{x \to c^+} f(x) = -\infty$$

See Figure 50 for illustrations of these possibilities.

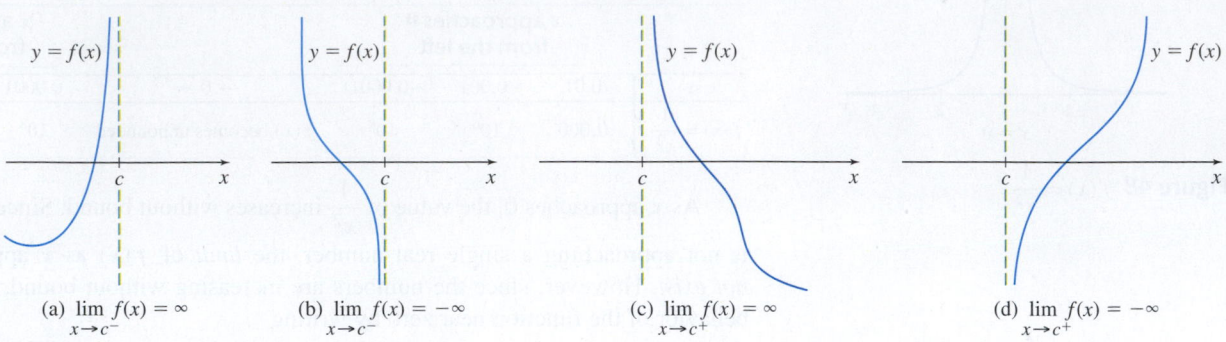

(a) $\lim_{x \to c^-} f(x) = \infty$    (b) $\lim_{x \to c^-} f(x) = -\infty$    (c) $\lim_{x \to c^+} f(x) = \infty$    (d) $\lim_{x \to c^+} f(x) = -\infty$

**Figure 50**

When we know the graph of a function, we can use it to investigate infinite limits.

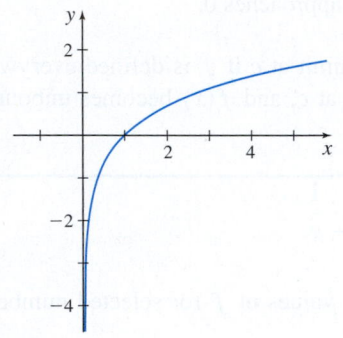

**Figure 51** $f(x) = \ln x$

### EXAMPLE 2  Investigating an Infinite Limit

Investigate $\lim_{x \to 0^+} \ln x$.

**Solution** The domain of $f(x) = \ln x$ is $\{x \mid x > 0\}$. Notice that the graph of $f(x) = \ln x$ in Figure 51 decreases without bound as $x$ approaches 0 from the right. The graph suggests that

$$\lim_{x \to 0^+} \ln x = -\infty$$

∎

**NOW WORK** Problem 11.

Based on the graphs of the trigonometric functions in Figure 52, we have the following infinite limits:

$$\lim_{x \to \pi/2^-} \tan x = \infty \qquad \lim_{x \to \pi/2^+} \tan x = -\infty \qquad \lim_{x \to \pi/2^-} \sec x = \infty \qquad \lim_{x \to \pi/2^+} \sec x = -\infty$$

$$\lim_{x \to 0^-} \csc x = -\infty \qquad \lim_{x \to 0^+} \csc x = \infty \qquad \lim_{x \to 0^-} \cot x = -\infty \qquad \lim_{x \to 0^+} \cot x = \infty$$

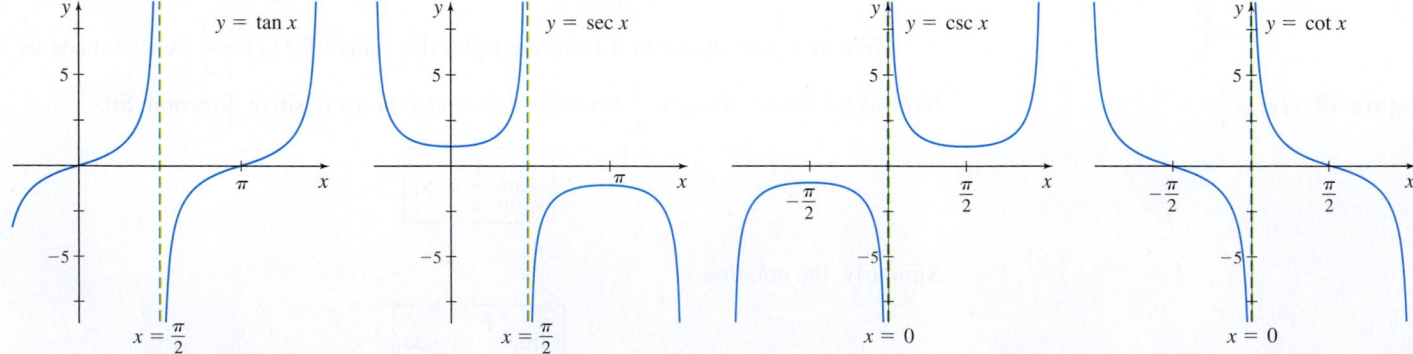

**Figure 52**

Limits of quotients, in which the limit of the numerator is a nonzero number and the limit of the denominator is 0, often result in infinite limits. (If the limits of both the numerator and the denominator are 0, the quotient is called an *indeterminate form*. Limits of indeterminate forms are discussed in Section 4.5.)

**EXAMPLE 3  Finding an Infinite Limit**

Find each limit: **(a)** $\lim\limits_{x \to 3^+} \dfrac{2x+1}{x-3}$    **(b)** $\lim\limits_{x \to 1^-} \dfrac{x^2}{\ln x}$

**Solution  (a)** $\lim\limits_{x \to 3^+} \dfrac{2x+1}{x-3}$ is the right-hand limit of the quotient of two functions, $f(x) = 2x+1$ and $g(x) = x-3$. Since $\lim\limits_{x \to 3^+} (x-3) = 0$, we cannot use the Limit of a Quotient. So we need a different strategy. As $x$ approaches $3^+$, we have

$$\lim_{x \to 3^+} (2x+1) = 7 \quad \text{and} \quad \lim_{x \to 3^+} (x-3) = 0$$

Here the limit of the numerator is 7. The denominator is positive and approaching 0. So the quotient is positive and becoming unbounded. We conclude that

$$\lim_{x \to 3^+} \frac{2x+1}{x-3} = \infty$$

**(b)**  As in (a), we cannot use the Limit of a Quotient because $\lim\limits_{x \to 1^-} \ln x = 0$. Here we have

$$\lim_{x \to 1^-} x^2 = 1 \qquad \lim_{x \to 1^-} \ln x = 0$$

We are seeking a left-hand limit, so for numbers close to 1 but less than 1, the denominator $\ln x < 0$. So the quotient is negative and becoming unbounded. We conclude that

$$\lim_{x \to 1^-} \frac{x^2}{\ln x} = -\infty$$ ∎

The discussion below may prove useful when finding $\lim\limits_{x \to c^-} \dfrac{f(x)}{g(x)}$, where $\lim\limits_{x \to c^-} f(x)$ is a nonzero number and $\lim\limits_{x \to c^-} g(x) = 0$.

---

**Analyzing $\lim\limits_{x \to c^-} \dfrac{f(x)}{g(x)}$ when $\lim\limits_{x \to c^-} f(x) = L$, $L \neq 0$, and $\lim\limits_{x \to c^-} g(x) = 0$.**

- If $L > 0$ and $g(x) < 0$ for numbers $x$ close to $c$, but less than $c$, then

$$\lim_{x \to c^-} \frac{f(x)}{g(x)} = -\infty$$

- If $L > 0$ and $g(x) > 0$ for numbers $x$ close to $c$, but less than $c$, then

$$\lim_{x \to c^-} \frac{f(x)}{g(x)} = \infty$$

- If $L < 0$ and $g(x) < 0$ for numbers $x$ close to $c$, but less than $c$, then

$$\lim_{x \to c^-} \frac{f(x)}{g(x)} = \infty$$

- If $L < 0$ and $g(x) > 0$ for numbers $x$ close to $c$, but less than $c$, then

$$\lim_{x \to c^-} \frac{f(x)}{g(x)} = -\infty$$

---

Similar arguments are valid for right-hand limits.

**NOW WORK** Problem **27**.

## 2 Find the Vertical Asymptotes of a Graph

Figure 53 illustrates the possibilities that can occur when a function has an infinite limit at $c$. In each case, notice that the graph of $f$ has a vertical asymptote $x = c$.

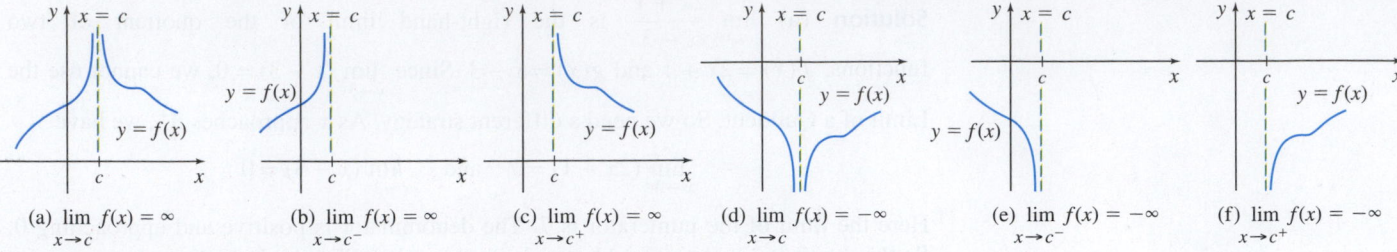

(a) $\lim\limits_{x \to c} f(x) = \infty$    (b) $\lim\limits_{x \to c^-} f(x) = \infty$    (c) $\lim\limits_{x \to c^+} f(x) = \infty$    (d) $\lim\limits_{x \to c} f(x) = -\infty$    (e) $\lim\limits_{x \to c^-} f(x) = -\infty$    (f) $\lim\limits_{x \to c^+} f(x) = -\infty$

**Figure 53**

---

**DEFINITION  Vertical Asymptote**

The line $x = c$ is a **vertical asymptote** of the graph of the function $f$ if any of the following is true:

$$\lim_{x \to c^-} f(x) = \infty \qquad \lim_{x \to c^+} f(x) = \infty \qquad \lim_{x \to c^-} f(x) = -\infty \qquad \lim_{x \to c^+} f(x) = -\infty$$

---

For rational functions, a vertical asymptote may occur where the denominator equals 0.

**EXAMPLE 4  Finding a Vertical Asymptote**

Find any vertical asymptote(s) of the graph of $f(x) = \dfrac{x}{(x-3)^2}$.

**Solution**  The domain of $f$ is $\{x \mid x \neq 3\}$. Since 3 is the only number for which the denominator of $f$ equals zero, we construct Table 13 and investigate the one-sided limits of $f$ as $x$ approaches 3. Table 13 suggests that

$$\lim_{x \to 3} \frac{x}{(x-3)^2} = \infty$$

So, $x = 3$ is a vertical asymptote of the graph of $f$.

**TABLE 13**

| | x approaches 3 from the left | | | | x approaches 3 from the right | | |
|---|---|---|---|---|---|---|---|
| $x$ | 2.9 | 2.99 | 2.999 | $\to 3 \leftarrow$ | 3.001 | 3.01 | 3.1 |
| $f(x) = \dfrac{x}{(x-3)^2}$ | 290 | 29,900 | 2,999,000 | $f(x)$ becomes unbounded | 3,001,000 | 30,100 | 310 |

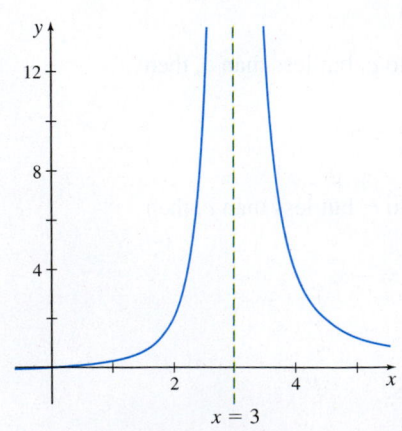

**Figure 54** $f(x) = \dfrac{x}{(x-3)^2}$

Figure 54 shows the graph of $f(x) = \dfrac{x}{(x-3)^2}$ and its vertical asymptote. ∎

**NOW WORK** Problems 15 and 63 (find any vertical asymptotes).

## 3 Find Limits at Infinity

Now we investigate what happens as $x$ becomes unbounded in either the positive direction or the negative direction. Suppose as $x$ becomes unbounded, the value of a function $f$ approaches some real number $L$. Then the number $L$ is called the *limit of $f$ at infinity*.

For example, the graph of $f(x) = \dfrac{1}{x}$ in Figure 55 suggests that as $x$ becomes unbounded in either the positive direction or the negative direction, the values $f(x)$ get closer to 0. Table 14 illustrates this for a few numbers $x$.

**TABLE 14**

| $x$ | $\pm 100$ | $\pm 1000$ | $\pm 10,000$ | $\pm 100,000$ | $x$ becomes unbounded |
|---|---|---|---|---|---|
| $f(x) = \dfrac{1}{x}$ | $\pm 0.01$ | $\pm 0.001$ | $\pm 0.0001$ | $\pm 0.00001$ | $f(x)$ approaches 0 |

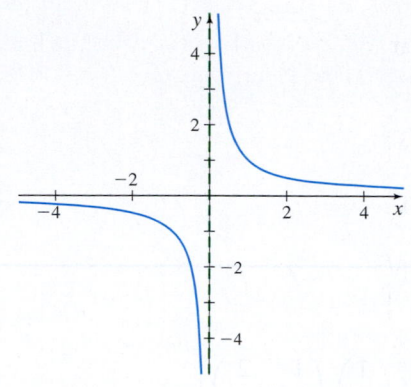

**Figure 55** $f(x) = \dfrac{1}{x}$

Since $f$ can be made as close as we please to 0 by choosing $x$ sufficiently large, we write

$$\lim_{x \to \infty} \frac{1}{x} = 0$$

and say that the limit as $x$ approaches infinity of $f$ is equal to 0. Similarly, as $x$ becomes unbounded in the negative direction, the function $f(x) = \dfrac{1}{x}$ can be made as close as we please to 0, and we write

$$\lim_{x \to -\infty} \frac{1}{x} = 0$$

Although we do not prove $\displaystyle\lim_{x \to -\infty} \frac{1}{x} = 0$ here, it can be proved using the $\varepsilon$-$\delta$ Definition of a Limit at Infinity and is left as an exercise in Section 1.6.

The limit properties discussed in Section 1.2 hold for infinite limits. Although these properties are stated below for limits as $x \to \infty$, they are also valid for limits as $x \to -\infty$.

> **THEOREM Properties of Limits at Infinity**
>
> If $k$ is a real number, $n \geq 2$ is an integer, and the functions $f$ and $g$ approach real numbers as $x \to \infty$, then the following properties are true:
>
> - $\displaystyle\lim_{x \to \infty} A = A$, where $A$ is a constant
> - $\displaystyle\lim_{x \to \infty} [kf(x)] = k \lim_{x \to \infty} f(x)$
> - $\displaystyle\lim_{x \to \infty} [f(x) \pm g(x)] = \lim_{x \to \infty} f(x) \pm \lim_{x \to \infty} g(x)$
> - $\displaystyle\lim_{x \to \infty} [f(x)g(x)] = \left[ \lim_{x \to \infty} f(x) \right] \left[ \lim_{x \to \infty} g(x) \right]$
> - $\displaystyle\lim_{x \to \infty} \frac{f(x)}{g(x)} = \frac{\displaystyle\lim_{x \to \infty} f(x)}{\displaystyle\lim_{x \to \infty} g(x)}$, provided $\displaystyle\lim_{x \to \infty} g(x) \neq 0$
> - $\displaystyle\lim_{x \to \infty} [f(x)]^n = \left[ \lim_{x \to \infty} f(x) \right]^n$
> - $\displaystyle\lim_{x \to \infty} \sqrt[n]{f(x)} = \sqrt[n]{\lim_{x \to \infty} f(x)}$, where $f(x) > 0$ if $n$ is even

Since $\displaystyle\lim_{x \to \infty} \frac{1}{x} = 0$ and $\displaystyle\lim_{x \to -\infty} \frac{1}{x} = 0$, we can use properties of limits at infinity to obtain two additional properties.

> If $p > 0$ is a rational number and $k$ is any real number, then
>
> $$\lim_{x \to \infty} \frac{k}{x^p} = 0 \quad \text{and} \quad \lim_{x \to -\infty} \frac{k}{x^p} = 0 \tag{1}$$

provided $x^p$ is defined when $x < 0$.

You are asked to prove (1) in Problem 92.

**EXAMPLE 5** Finding Limits at Infinity

Use property (1).

(a) $\lim\limits_{x \to \infty} \dfrac{5}{x^3} = 0$ 　　　　(b) $\lim\limits_{x \to \infty} \dfrac{1}{\sqrt{x}} = 0$

(c) $\lim\limits_{x \to -\infty} \dfrac{-6}{x^{1/3}} = 0$ 　　　(d) $\lim\limits_{x \to -\infty} \dfrac{4}{5x^{8/3}} = 0$ ■

**EXAMPLE 6** Finding Limits at Infinity

Find each limit.

(a) $\lim\limits_{x \to \infty} \left(2 + \dfrac{3}{x}\right)$ 　　　(b) $\lim\limits_{x \to \infty} \left[\left(2 + \dfrac{1}{x}\right)\left(\dfrac{1}{3} - \dfrac{2}{x^3}\right)\right]$

(c) $\lim\limits_{x \to \infty} \left[\dfrac{1}{4}\left(\dfrac{3x^2 + 5x - 1}{x^2}\right)\right]$

**Solution** We use properties of limits at infinity and property (1).

(a) $\lim\limits_{x \to \infty} \left(2 + \dfrac{3}{x}\right) = \lim\limits_{x \to \infty} 2 + \lim\limits_{x \to \infty} \dfrac{3}{x} = 2 + 0 = 2$

(b) $\lim\limits_{x \to \infty} \left[\left(2 + \dfrac{1}{x}\right)\left(\dfrac{1}{3} - \dfrac{2}{x^3}\right)\right] = \lim\limits_{x \to \infty}\left(2 + \dfrac{1}{x}\right) \cdot \lim\limits_{x \to \infty}\left(\dfrac{1}{3} - \dfrac{2}{x^3}\right)$

$$= \left(\lim\limits_{x \to \infty} 2 + \lim\limits_{x \to \infty} \dfrac{1}{x}\right) \cdot \left(\lim\limits_{x \to \infty} \dfrac{1}{3} - \lim\limits_{x \to \infty} \dfrac{2}{x^3}\right)$$

$$= (2 + 0)\left(\dfrac{1}{3} - 0\right) = \dfrac{2}{3}$$

(c) $\lim\limits_{x \to \infty} \left[\dfrac{1}{4}\left(\dfrac{3x^2 + 5x - 1}{x^2}\right)\right] = \dfrac{1}{4} \lim\limits_{x \to \infty} \dfrac{3x^2 + 5x - 1}{x^2} = \dfrac{1}{4} \lim\limits_{x \to \infty}\left(3 + \dfrac{5}{x} - \dfrac{1}{x^2}\right)$

$$= \dfrac{1}{4}\left(\lim\limits_{x \to \infty} 3 + \lim\limits_{x \to \infty} \dfrac{5}{x} - \lim\limits_{x \to \infty} \dfrac{1}{x^2}\right) = \dfrac{1}{4} \cdot 3 = \dfrac{3}{4}$$ ■

**NOW WORK** Problem 45.

For some algebraic functions involving a quotient, we find limits at infinity by dividing the numerator and the denominator by the highest power of $x$ that appears in the denominator.

**CALC CLIP**

**EXAMPLE 7** Finding a Limit at Infinity

Find each limit.

(a) $\lim\limits_{x \to \infty} \dfrac{3x^{3/2} + 5x^{1/2}}{x^3 - x^{2/3}}$ 　　　(b) $\lim\limits_{x \to \infty} \dfrac{\sqrt{4x^2 + 10}}{x - 5}$

**Solution (a)** Divide the numerator and the denominator by $x^3$, the highest power of $x$ that appears in the denominator.

$$\lim\limits_{x \to \infty} \dfrac{3x^{3/2} + 5x^{1/2}}{x^3 - x^{2/3}} \underset{\uparrow}{=} \lim\limits_{x \to \infty} \dfrac{\dfrac{3x^{3/2} + 5x^{1/2}}{x^3}}{\dfrac{x^3 - x^{2/3}}{x^3}} = \lim\limits_{x \to \infty} \dfrac{\dfrac{3x^{3/2}}{x^3} + \dfrac{5x^{1/2}}{x^3}}{1 - \dfrac{x^{2/3}}{x^3}}$$

Divide the numerator and<br>the denominator by $x^3$.

$$= \lim\limits_{x \to \infty} \dfrac{\dfrac{3}{x^{3/2}} + \dfrac{5}{x^{5/2}}}{1 - \dfrac{1}{x^{7/3}}} = \dfrac{\lim\limits_{x \to \infty}\left(\dfrac{3}{x^{3/2}} + \dfrac{5}{x^{5/2}}\right)}{\lim\limits_{x \to \infty}\left(1 - \dfrac{1}{x^{7/3}}\right)} = \dfrac{0 + 0}{1 - 0} = 0$$

**(b)** Divide the numerator and the denominator by $x$, the highest power of $x$ that appears in the denominator. But remember, since $\sqrt{x^2} = |x| = x$ when $x \geq 0$, in the numerator the divisor in the square root will be $x^2$.

$$\lim_{x \to \infty} \frac{\sqrt{4x^2 + 10}}{x - 5} = \lim_{x \to \infty} \frac{\sqrt{\dfrac{4x^2 + 10}{x^2}}}{\dfrac{x - 5}{x}} = \lim_{x \to \infty} \frac{\sqrt{4 + \dfrac{10}{x^2}}}{1 - \dfrac{5}{x}}$$

↑ Divide the numerator and the denominator by $x$.
Remember to divide the numerator by $\sqrt{x^2} = x$.

$$= \frac{\lim_{x \to \infty} \sqrt{4 + \dfrac{10}{x^2}}}{\lim_{x \to \infty}\left(1 - \dfrac{5}{x}\right)} = \frac{\sqrt{\lim_{x \to \infty}\left(4 + \dfrac{10}{x^2}\right)}}{1} = \sqrt{4} = 2$$

↑ Use properties of limits at infinity.

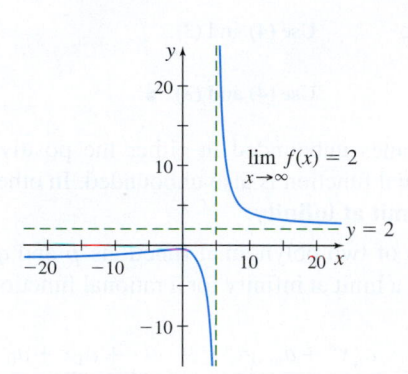

**Figure 56** $f(x) = \dfrac{\sqrt{4x^2 + 10}}{x - 5}$

The graph of $f(x) = \dfrac{\sqrt{4x^2 + 10}}{x - 5}$ is shown in Figure 56.

NOW WORK **Problems 47 and 57.**

For polynomial and rational functions, there is an easier approach to finding limits at infinity.

## Limits at Infinity of Polynomial and Rational Functions

We begin by investigating limits at infinity of power functions. For any power function $p(x) = x^n$, if $n$ is a positive integer, then $p$ increases without bound as $x \to \infty$. That is,

$$\lim_{x \to \infty} x^n = \infty \qquad n \text{ a positive integer} \tag{2}$$

To investigate $\lim_{x \to -\infty} x^n$, $n$ a positive integer, we need to consider whether $n$ is even or odd. Remember, as $x \to -\infty$, $x$ is unbounded in the negative direction, so $x < 0$.

If $n$ is positive and even, then $x^n > 0$ when $x < 0$. If $n$ is positive and odd, then $x^n < 0$ when $x < 0$. Since in both cases $x^n$ becomes unbounded as $x \to -\infty$, we have

$$\lim_{x \to -\infty} x^n = \infty \text{ if } n \text{ is a positive even integer}$$

$$\lim_{x \to -\infty} x^n = -\infty \text{ if } n \text{ is a positive odd integer} \tag{3}$$

Now consider a polynomial function, $p(x) = a_n x^n + a_{n-1} x^{n-1} + \cdots + a_1 x + a_0$, $a_n \neq 0$. To find $\lim_{x \to \pm\infty} p(x)$, factor out $x^n$. Then

$$\lim_{x \to \pm\infty} p(x) = \lim_{x \to \pm\infty} (a_n x^n + a_{n-1} x^{n-1} + \cdots + a_1 x + a_0)$$

$$= \lim_{x \to \pm\infty} \left[ x^n \left( a_n + \frac{a_{n-1}}{x} + \cdots + \frac{a_1}{x^{n-1}} + \frac{a_0}{x^n} \right) \right]$$

Since $\lim_{x \to \pm\infty} \left( a_n + \dfrac{a_{n-1}}{x} + \cdots + \dfrac{a_1}{x^{n-1}} + \dfrac{a_0}{x^n} \right) = a_n$, we have

$$\lim_{x \to \pm\infty} p(x) = a_n \lim_{x \to \pm\infty} x^n \tag{4}$$

Using (2), (3), and (4), we can find the limit at infinity of any polynomial function.

---

EXAMPLE 8 **Finding Limits at Infinity of a Polynomial Function**

**(a)** $\lim\limits_{x\to\infty}(3x^3-x^2+5x+1)=3\lim\limits_{x\to\infty}x^3=\infty$      Use (4) and (2).

**(b)** $\lim\limits_{x\to\infty}(x^5+2x)=\lim\limits_{x\to\infty}x^5=\infty$      Use (4) and (2).

**(c)** $\lim\limits_{x\to-\infty}(2x^4-x^2+2x+8)=2\lim\limits_{x\to-\infty}x^4=\infty$      Use (4) and (3).

**(d)** $\lim\limits_{x\to-\infty}(x^3+10x+7)=\lim\limits_{x\to-\infty}x^3=-\infty$      Use (4) and (3). ∎

So, for a polynomial function, as $x$ becomes unbounded in either the positive direction or the negative direction, the polynomial function is also unbounded. In other words, polynomial functions have an **infinite limit at infinity**.

Since a rational function $R$ is the quotient of two polynomial functions $p$ and $q$, where $q$ is not the zero polynomial, we can find a limit at infinity for a rational function using only the leading terms of $p$ and $q$.

Consider the rational function $R(x)=\dfrac{a_nx^n+a_{n-1}x^{n-1}+\cdots+a_1x+a_0}{b_mx^m+b_{m-1}x^{m-1}+\cdots+b_1x+b_0}$, where $a_n\neq 0$, $b_m\neq 0$.

Then we simplify $R$ as follows:

$$R(x)=\frac{x^n\left(a_n+\dfrac{a_{n-1}}{x}+\cdots+\dfrac{a_1}{x^{n-1}}+\dfrac{a_0}{x^n}\right)}{x^m\left(b_m+\dfrac{b_{m-1}}{x}+\cdots+\dfrac{b_1}{x^{m-1}}+\dfrac{b_0}{x^m}\right)}$$

Since $\lim\limits_{x\to\infty}\left(a_n+\dfrac{a_{n-1}}{x}+\cdots+\dfrac{a_1}{x^{n-1}}+\dfrac{a_0}{x^n}\right)=a_n$

and $\lim\limits_{x\to\infty}\left(b_m+\dfrac{b_{m-1}}{x}+\cdots+\dfrac{b_1}{x^{m-1}}+\dfrac{b_0}{x^m}\right)=b_m$, we have

$$\lim\limits_{x\to\infty}R(x)=\lim\limits_{x\to\infty}\frac{x^n}{x^m}\cdot\frac{\lim\limits_{x\to\infty}\left(a_n+\dfrac{a_{n-1}}{x}+\cdots+\dfrac{a_1}{x^{n-1}}+\dfrac{a_0}{x^n}\right)}{\lim\limits_{x\to\infty}\left(b_m+\dfrac{b_{m-1}}{x}+\cdots+\dfrac{b_1}{x^{m-1}}+\dfrac{b_0}{x^m}\right)}=\frac{a_n}{b_m}\lim\limits_{x\to\infty}\frac{x^n}{x^m}$$

Or more generally,

$$\boxed{\lim\limits_{x\to\pm\infty}R(x)=\lim\limits_{x\to\pm\infty}\frac{a_nx^n+a_{n-1}x^{n-1}+\cdots+a_1x+a_0}{b_mx^m+b_{m-1}x^{m-1}+\cdots+b_1x+b_0}=\frac{a_n}{b_m}\lim\limits_{x\to\pm\infty}\frac{x^n}{x^m}}\quad(5)$$

---

EXAMPLE 9 **Finding Limits at Infinity of a Rational Function**

Find each limit.

**(a)** $\lim\limits_{x\to\infty}\dfrac{3x^2-2x+8}{x^2+1}$      **(b)** $\lim\limits_{x\to\infty}\dfrac{4x^2-5x}{x^3+1}$      **(c)** $\lim\limits_{x\to\infty}\dfrac{5x^4-3x^2}{2x^2+1}$

**Solution** Each of these limits at infinity involve a rational function, so we use (5).

**(a)** $\lim\limits_{x\to\infty}\dfrac{3x^2-2x+8}{x^2+1}=3\lim\limits_{x\to\infty}\dfrac{x^2}{x^2}=3$

**(b)** $\lim\limits_{x\to\infty}\dfrac{4x^2-5x}{x^3+1}=4\lim\limits_{x\to\infty}\dfrac{x^2}{x^3}=4\lim\limits_{x\to\infty}\dfrac{1}{x}=0$

**(c)** $\lim\limits_{x\to\infty}\dfrac{5x^4-3x^2}{2x^2+1}=\dfrac{5}{2}\lim\limits_{x\to\infty}\dfrac{x^4}{x^2}=\dfrac{5}{2}\lim\limits_{x\to\infty}x^2=\infty$ ∎

The graphs of the functions $F(x)=\dfrac{3x^2-2x+8}{x^2+1}$ and $G(x)=\dfrac{4x^2-5x}{x^3+1}$ are shown in Figure 57(a) and (b), respectively. Notice how each graph behaves as $x$ becomes unbounded in the positive direction. The graph of the function $H(x) = \dfrac{5x^4 - 3x^2}{2x^2 + 1}$, which

has an infinite limit at infinity, is shown in Figure 57(c).

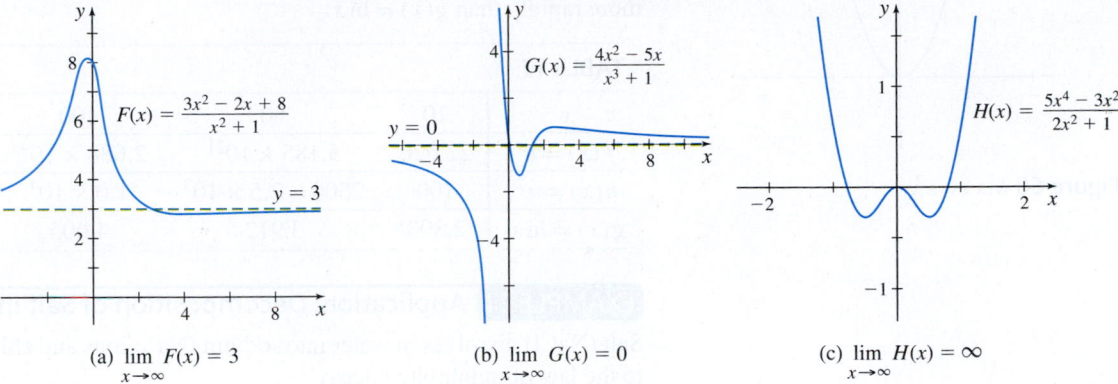

(a) $\lim\limits_{x \to \infty} F(x) = 3$        (b) $\lim\limits_{x \to \infty} G(x) = 0$        (c) $\lim\limits_{x \to \infty} H(x) = \infty$

**Figure 57**

**NOW WORK** Problems **49** and **59**.

## Other Infinite Limits at Infinity

We have seen that all polynomial functions have infinite limits at infinity, and some rational functions have infinite limits at infinity. Other functions also have an infinite limit at infinity.

For example, consider the function $f(x) = e^x$.

The graph of the exponential function, shown in Figure 58, suggests that

$$\lim_{x \to -\infty} e^x = 0 \qquad \lim_{x \to \infty} e^x = \infty$$

These limits are supported by the information in Table 15.

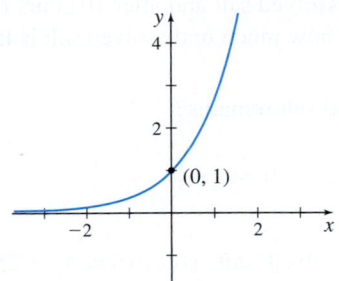

**Figure 58** $f(x) = e^x$

**TABLE 15**

| $x$ | $-1$ | $-5$ | $-10$ | $-20$ | $x$ approaches $-\infty$ |
|---|---|---|---|---|---|
| $f(x) = e^x$ | 0.36788 | 0.00674 | 0.00005 | $2 \times 10^{-9}$ | $f(x)$ approaches 0 |
| $x$ | 1 | 5 | 10 | 20 | $x$ approaches $\infty$ |
| $f(x) = e^x$ | $e \approx 2.71828$ | 148.41 | 22,026 | $4.85 \times 10^8$ | $f(x)$ becomes unbounded |

**EXAMPLE 10** **Finding the Limit at Infinity of $g(x) = \ln x$**

Find $\lim\limits_{x \to \infty} \ln x$.

**Solution** Table 16 and the graph of $g(x) = \ln x$ in Figure 59 suggest that $g(x) = \ln x$ has an infinite limit at infinity. That is,

$$\lim_{x \to \infty} \ln x = \infty$$

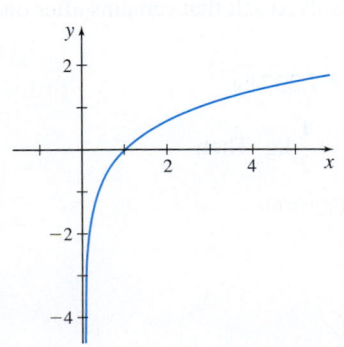

**Figure 59** $g(x) = \ln x$

**TABLE 16**

| $x$ | $e^{10}$ | $e^{100}$ | $e^{1000}$ | $e^{10,000}$ | $e^{100,000}$ | $\to x$ becomes unbounded |
|---|---|---|---|---|---|---|
| $g(x) = \ln x$ | 10 | 100 | 1000 | 10,000 | 100,000 | $\to g(x)$ becomes unbounded |

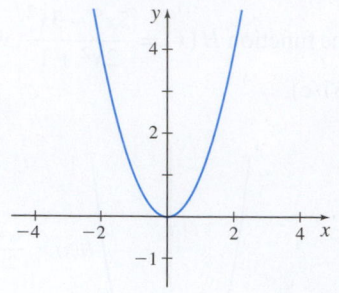

**Figure 60** $h(x) = x^2$

Now let's compare the graph of $f(x) = e^x$ in Figure 58, the graph of $g(x) = \ln x$ in Figure 59, and the graph of $h(x) = x^2$ in Figure 60. As $x$ becomes unbounded in the positive direction, all three of the functions increase without bound. But in Table 17, $f(x) = e^x$ approaches infinity more rapidly than $h(x) = x^2$, which approaches infinity more rapidly than $g(x) = \ln x$.

**TABLE 17**

| $x$ | 10 | 50 | 100 | 1000 | 10,000 |
|---|---|---|---|---|---|
| $f(x) = e^x$ | 22,026 | $5.185 \times 10^{21}$ | $2.688 \times 10^{43}$ | $e^{1000}$ | $e^{10,000}$ |
| $h(x) = x^2$ | 100 | $2500 = 2.5 \times 10^3$ | $1.0 \times 10^4$ | $1.0 \times 10^6$ | $1.0 \times 10^8$ |
| $g(x) = \ln x$ | 2.303 | 3.912 | 4.605 | 6.908 | 9.210 |

## EXAMPLE 11 Application: Decomposition of Salt in Water

Salt (NaCl) dissolves in water into sodium ($Na^+$) ions and chloride ($Cl^-$) ions according to the law of uninhibited decay

$$A(t) = A_0 e^{kt}$$

where $A = A(t)$ is the amount (in kilograms) of undissolved salt present at time $t$ (in hours), $A_0$ is the original amount of undissolved salt, and $k$ is a negative number that represents the rate of dissolution.

**(a)** If initially there are 25 kilograms (kg) of undissolved salt and after 10 hours (h) there are 15 kg of undissolved salt remaining, how much undissolved salt is left after one day?

**(b)** How long will it take until $\frac{1}{2}$ kg of undissolved salt remains?

**(c)** Find $\lim\limits_{t \to \infty} A(t)$.

**(d)** Interpret the answer found in (c).

**NEED TO REVIEW?** Solving exponential equations is discussed in Section P.5, pp. 48–49.

**Solution** **(a)** Initially, there are 25 kg of undissolved salt, so $A(0) = A_0 = 25$. To find the number $k$ in $A(t) = A_0 e^{kt}$, we use the fact that at $t = 10$, then $A(10) = 15$. That is,

$$A(10) = 15 = 25 e^{10k} \qquad A(t) = A_0 e^{kt}, \; A_0 = 25; \; A(10) = 15; \; t = 10$$

$$e^{10k} = \frac{3}{5}$$

$$10k = \ln \frac{3}{5}$$

$$k = \frac{1}{10} \ln 0.6$$

So, $A(t) = 25 e^{\left(\frac{1}{10} \ln 0.6\right)t}$. The amount of undissolved salt that remains after one day (24 h) is

$$A(24) = 25 e^{\left(\frac{1}{10} \ln 0.6\right)24} \approx 7.337 \, \text{kg}$$

**(b)** We want to find $t$ so that $A(t) = 25 e^{\left(\frac{1}{10} \ln 0.6\right)t} = \frac{1}{2}$ kg. Then

$$\frac{1}{2} = 25 e^{\left(\frac{1}{10} \ln 0.6\right)t}$$

$$e^{\left(\frac{1}{10} \ln 0.6\right)t} = \frac{1}{50}$$

$$\left(\frac{1}{10} \ln 0.6\right) t = \ln \frac{1}{50}$$

$$t \approx 76.582$$

After approximately 76.6 h, $\frac{1}{2}$ kg of undissolved salt will remain.

(c) Since $\dfrac{1}{10}\ln 0.6 \approx -0.051$, we have $\lim\limits_{t\to\infty} A(t) = \lim\limits_{t\to\infty}(25e^{-0.051t}) = \lim\limits_{t\to\infty}\dfrac{25}{e^{0.051t}} = 0$

(d) As $t$ becomes unbounded, the amount of undissolved salt approaches $0\,\text{kg}$. Eventually, there will be no undissolved salt in the water. ∎

**NOW WORK** Problem 79.

## 4 Find the Horizontal Asymptotes of a Graph

Limits at infinity have an important geometric interpretation. When $\lim\limits_{x\to\infty} f(x) = M$, it means that as $x$ becomes unbounded in the positive direction, the value of $f$ can be made as close as we please to a number $M$. That is, the graph of $y = f(x)$ is as close as we please to the horizontal line $y = M$ by choosing $x$ sufficiently large. Similarly, $\lim\limits_{x\to-\infty} f(x) = L$ means that the graph of $y = f(x)$ is as close as we please to the horizontal line $y = L$ for $x$ unbounded in the negative direction. These lines are *horizontal asymptotes* of the graph of $f$.

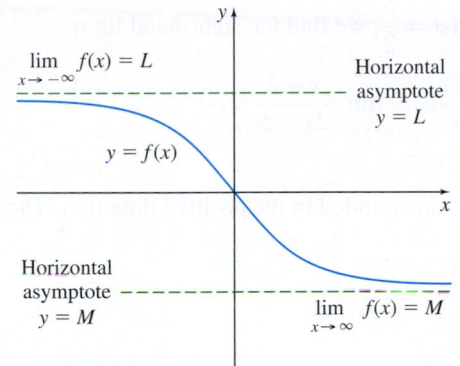

$$\lim_{x\to-\infty} f(x) = L$$

Horizontal asymptote $y = L$

$y = f(x)$

Horizontal asymptote $y = M$

$$\lim_{x\to\infty} f(x) = M$$

**Figure 61**

> **DEFINITION  Horizontal Asymptote**
>
> The line $y = L$ is a **horizontal asymptote** of the graph of a function $f$ for $x$ unbounded in the negative direction if $\lim\limits_{x\to-\infty} f(x) = L$.
>
> The line $y = M$ is a **horizontal asymptote** of the graph of a function $f$ for $x$ unbounded in the positive direction if $\lim\limits_{x\to\infty} f(x) = M$.

In Figure 61, $y = L$ is a horizontal asymptote as $x \to -\infty$ because $\lim\limits_{x\to-\infty} f(x) = L$. The line $y = M$ is a horizontal asymptote as $x \to \infty$ because $\lim\limits_{x\to\infty} f(x) = M$. To identify horizontal asymptotes, we find the limits of $f$ at infinity.

**EXAMPLE 12  Finding the Horizontal Asymptotes of a Graph**

Find the horizontal asymptotes, if any, of the graph of $f(x) = \dfrac{3x-2}{4x-1}$.

**Solution** We examine the two limits at infinity: $\lim\limits_{x\to-\infty}\dfrac{3x-2}{4x-1}$ and $\lim\limits_{x\to\infty}\dfrac{3x-2}{4x-1}$.

Since $\lim\limits_{x\to-\infty}\dfrac{3x-2}{4x-1} = \dfrac{3}{4}$, the line $y = \dfrac{3}{4}$ is a horizontal asymptote of the graph of $f$ for $x$ unbounded in the negative direction.

Since $\lim\limits_{x\to\infty}\dfrac{3x-2}{4x-1} = \dfrac{3}{4}$, the line $y = \dfrac{3}{4}$ is a horizontal asymptote of the graph of $f$ for $x$ unbounded in the positive direction. ∎

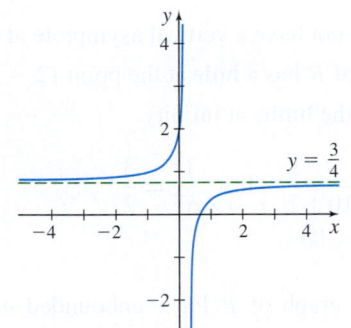

$y = \dfrac{3}{4}$

**Figure 62** $f(x) = \dfrac{3x-2}{4x-1}$

Figure 62 shows the graph of $f(x) = \dfrac{3x-2}{4x-1}$ and the horizontal asymptote $y = \dfrac{3}{4}$.

**NOW WORK** Problem 63 (find any horizontal asymptotes).

## 5 Find the Asymptotes of the Graph of a Rational Function

In the next example we find the horizontal asymptotes and vertical asymptotes, if any, of the graph of a rational function.

**EXAMPLE 13  Finding the Asymptotes of the Graph of a Rational Function**

Find any asymptotes of the graph of the rational function $R(x) = \dfrac{3x^2 - 12}{2x^2 - 9x + 10}$.

**Solution** We begin by factoring $R$.

$$R(x) = \frac{3x^2 - 12}{2x^2 - 9x + 10} = \frac{3(x-2)(x+2)}{(2x-5)(x-2)}$$

The domain of $R$ is $\left\{ x \,\middle|\, x \neq \frac{5}{2} \text{ and } x \neq 2 \right\}$. Since $R$ is a rational function, it is continuous on its domain, that is, all real numbers except $x = \frac{5}{2}$ and $x = 2$.

To check for vertical asymptotes, we find the limits as $x$ approaches $\frac{5}{2}$ and 2. First we consider $\lim\limits_{x \to \frac{5}{2}^-} R(x)$.

$$\lim_{x \to \frac{5}{2}^-} R(x) = \lim_{x \to \frac{5}{2}^-} \frac{3(x-2)(x+2)}{(2x-5)(x-2)} = \lim_{x \to \frac{5}{2}^-} \frac{3(x+2)}{2x-5} = 3 \lim_{x \to \frac{5}{2}^-} \frac{x+2}{2x-5} = -\infty$$

That is, as $x$ approaches $\frac{5}{2}$ from the left, $R$ becomes unbounded in the negative direction. The graph of $R$ has a vertical asymptote on the left at $x = \frac{5}{2}$.

To determine the behavior to the right of $x = \frac{5}{2}$, we find the right-hand limit.

$$\lim_{x \to \frac{5}{2}^+} R(x) = \lim_{x \to \frac{5}{2}^+} \frac{3(x+2)}{2x-5} = 3 \lim_{x \to \frac{5}{2}^+} \frac{x+2}{2x-5} = \infty$$

As $x$ approaches $\frac{5}{2}$ from the right, $R$ becomes unbounded in the positive direction. The graph of $R$ has a vertical asymptote on the right at $x = \frac{5}{2}$.

Next we consider $\lim\limits_{x \to 2} R(x)$.

$$\lim_{x \to 2} R(x) = \lim_{x \to 2} \frac{3(x-2)(x+2)}{(2x-5)(x-2)} = \lim_{x \to 2} \frac{3(x+2)}{2x-5} = \frac{3(2+2)}{2 \cdot 2 - 5} = \frac{12}{-1} = -12$$

Since the limit is not infinite, the function $R$ does not have a vertical asymptote at 2. Since 2 is not in the domain of $R$, the graph of $R$ has a **hole** at the point $(2, -12)$.

To check for horizontal asymptotes, we find the limits at infinity.

$$\lim_{x \to \pm\infty} R(x) = \lim_{x \to \pm\infty} \frac{3x^2 - 12}{2x^2 - 9x + 10} = \frac{3}{2} \cdot \lim_{x \to \pm\infty} \frac{x^2}{x^2} = \frac{3}{2} \quad (5)$$

The line $y = \frac{3}{2}$ is a horizontal asymptote of the graph of $R$ for $x$ unbounded in the negative direction and for $x$ unbounded in the positive direction. ∎

The graph of $R$ and its asymptotes are shown in Figure 63. Notice the hole in the graph at the point $(2, -12)$.

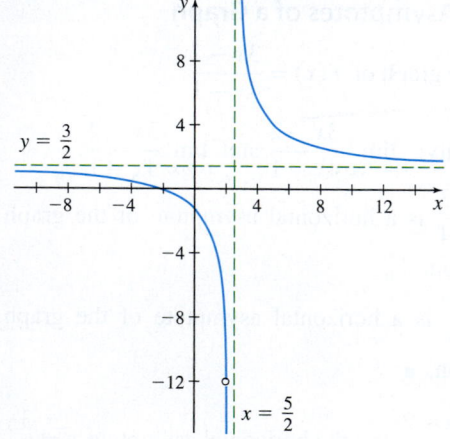

**Figure 63** $R(x) = \dfrac{3x^2 - 12}{2x^2 - 9x + 10}$

**NOW WORK** Problem 69.

## 1.5 Assess Your Understanding

### Concepts and Vocabulary

1. *True or False*  $\infty$ is a number.

2. (a) $\lim\limits_{x \to 0^-} \dfrac{1}{x} =$ _____ (b) $\lim\limits_{x \to 0^+} \dfrac{1}{x} =$ _____ (c) $\lim\limits_{x \to 0^+} \ln x =$ _____

3. *True or False*  The graph of a rational function has a vertical asymptote at every number $x$ at which the function is not defined.

4. If $\lim\limits_{x \to 4} f(x) = \infty$, then the line $x = 4$ is a(n) _____ asymptote of the graph of $f$.

---

**1.** = NOW WORK problem    = Graphing technology recommended    CAS = Computer Algebra System recommended

**5.** (a) $\lim\limits_{x\to\infty}\dfrac{1}{x}=$ ___ (b) $\lim\limits_{x\to\infty}\dfrac{1}{x^2}=$ ___ (c) $\lim\limits_{x\to\infty}\ln x=$ ___

**6.** *True or False*    $\lim\limits_{x\to-\infty}5=0$.

**7.** (a) $\lim\limits_{x\to-\infty}e^x=$ ___ (b) $\lim\limits_{x\to\infty}e^x=$ ___ (c) $\lim\limits_{x\to\infty}e^{-x}=$ ___

**8.** *True or False*    The graph of a function can have at most two horizontal asymptotes.

## Skill Building

*In Problems 9–16, use the accompanying graph of $y=f(x)$.*

**9.** Find $\lim\limits_{x\to\infty}f(x)$.

**10.** Find $\lim\limits_{x\to-\infty}f(x)$.

**11.** Find $\lim\limits_{x\to-1^-}f(x)$.

**12.** Find $\lim\limits_{x\to-1^+}f(x)$.

**13.** Find $\lim\limits_{x\to3^-}f(x)$.

**14.** Find $\lim\limits_{x\to3^+}f(x)$.

**15.** Identify all vertical asymptotes.

**16.** Identify all horizontal asymptotes.

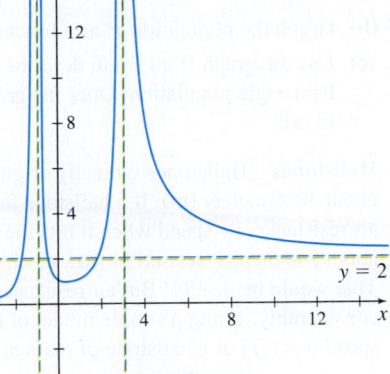

*In Problems 17–26, use the graph below of $y=f(x)$.*

**17.** Find $\lim\limits_{x\to\infty}f(x)$.

**18.** Find $\lim\limits_{x\to-\infty}f(x)$.

**19.** Find $\lim\limits_{x\to-3^-}f(x)$.

**20.** Find $\lim\limits_{x\to-3^+}f(x)$.

**21.** Find $\lim\limits_{x\to0^-}f(x)$.

**22.** Find $\lim\limits_{x\to0^+}f(x)$.

**23.** Find $\lim\limits_{x\to4^-}f(x)$.

**24.** Find $\lim\limits_{x\to4^+}f(x)$.

**25.** Identify all vertical asymptotes.

**26.** Identify all horizontal asymptotes.

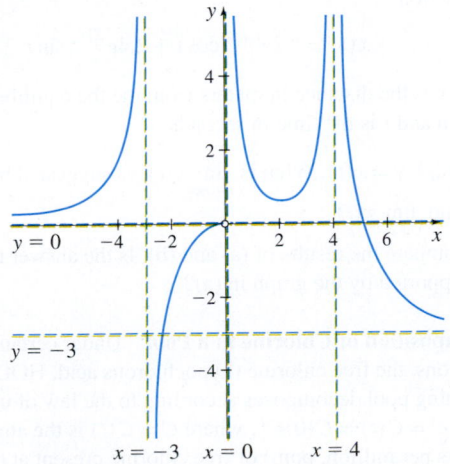

*In Problems 27–42, find each limit.*

**27.** $\lim\limits_{x\to2^-}\dfrac{3x}{x-2}$

**28.** $\lim\limits_{x\to-4^+}\dfrac{2x+1}{x+4}$

**29.** $\lim\limits_{x\to2^+}\dfrac{5}{x^2-4}$

**30.** $\lim\limits_{x\to1^-}\dfrac{2x}{x^3-1}$

**31.** $\lim\limits_{x\to-1^+}\dfrac{5x+3}{x(x+1)}$

**32.** $\lim\limits_{x\to0^-}\dfrac{5x+3}{5x(x-1)}$

**33.** $\lim\limits_{x\to-3^-}\dfrac{1}{x^2-9}$

**34.** $\lim\limits_{x\to2^+}\dfrac{x}{x^2-4}$

**35.** $\lim\limits_{x\to3}\dfrac{1-x}{(3-x)^2}$

**36.** $\lim\limits_{x\to-1}\dfrac{x+2}{(x+1)^2}$

**37.** $\lim\limits_{x\to\pi^-}\cot x$

**38.** $\lim\limits_{x\to-\pi/2^-}\tan x$

**39.** $\lim\limits_{x\to\pi/2^+}\csc(2x)$

**40.** $\lim\limits_{x\to-\pi/2^-}\sec x$

**41.** $\lim\limits_{x\to-1^+}\ln(x+1)$

**42.** $\lim\limits_{x\to1^+}\ln(x-1)$

*In Problems 43–60, find each limit.*

**43.** $\lim\limits_{x\to\infty}\dfrac{5}{x^2+4}$

**44.** $\lim\limits_{x\to-\infty}\dfrac{1}{x^2-9}$

**45.** $\lim\limits_{x\to\infty}\dfrac{2x+4}{5x}$

**46.** $\lim\limits_{x\to\infty}\dfrac{x+1}{x}$

**47.** $\lim\limits_{x\to\infty}\dfrac{x^{3/2}+2x-x^{1/2}}{5x-x^{1/2}}$

**48.** $\lim\limits_{x\to\infty}\dfrac{6x^2+x^{3/4}}{2x^2+x^{3/2}-2x^{1/2}}$

**49.** $\lim\limits_{x\to-\infty}\dfrac{x^2+1}{x^3-1}$

**50.** $\lim\limits_{x\to\infty}\dfrac{x^2-2x+1}{x^3+5x+4}$

**51.** $\lim\limits_{x\to\infty}\left[\dfrac{3x}{2x+5}-\dfrac{x^2+1}{4x^2+8}\right]$

**52.** $\lim\limits_{x\to\infty}\left[\dfrac{1}{x^2+x+4}-\dfrac{x+1}{3x-1}\right]$

**53.** $\lim\limits_{x\to-\infty}\left[2e^x\left(\dfrac{5x+1}{3x}\right)\right]$

**54.** $\lim\limits_{x\to-\infty}\left[e^x\left(\dfrac{x^2+x-3}{2x^3-x^2}\right)\right]$

**55.** $\lim\limits_{x\to\infty}\dfrac{\sqrt{x}+2}{3x-4}$

**56.** $\lim\limits_{x\to\infty}\dfrac{\sqrt{3x^3}+2}{x^2+6}$

**57.** $\lim\limits_{x\to\infty}\sqrt{\dfrac{3x^2-1}{x^2+4}}$

**58.** $\lim\limits_{x\to\infty}\left(\dfrac{16x^3+2x+1}{2x^3+3x}\right)^{2/3}$

**59.** $\lim\limits_{x\to-\infty}\dfrac{5x^3}{x^2+1}$

**60.** $\lim\limits_{x\to-\infty}\dfrac{x^4}{x-2}$

*In Problems 61–66, find any horizontal or vertical asymptotes of the graph of $f$.*

**61.** $f(x)=3+\dfrac{1}{x}$

**62.** $f(x)=2-\dfrac{1}{x^2}$

**63.** $f(x)=\dfrac{x^2}{x^2-1}$

**64.** $f(x)=\dfrac{2x^2-1}{x^2-1}$

**65.** $f(x)=\dfrac{\sqrt{2x^2-x+10}}{2x-3}$

**66.** $f(x)=\dfrac{\sqrt[3]{x^2+5x}}{x-6}$

*In Problems 67–72, for each rational function $R$:*

*(a) Find the domain of $R$.*

*(b) Find any horizontal asymptotes of $R$.*

*(c) Find any vertical asymptotes of $R$.*

*(d) Discuss the behavior of the graph at numbers where $R$ is not defined.*

**67.** $R(x)=\dfrac{-2x^2+1}{2x^3+4x^2}$

**68.** $R(x)=\dfrac{x^3}{x^4-1}$

**69.** $R(x)=\dfrac{x^2+3x-10}{2x^2-7x+6}$

**70.** $R(x)=\dfrac{x(x-1)^2}{(x+3)^3}$

**71.** $R(x)=\dfrac{x^3-1}{x-x^2}$

**72.** $R(x)=\dfrac{4x^5}{x^3-1}$

## Applications and Extensions

*In Problems 73 and 74:*

*(a) Sketch a graph of a function f that has the given properties.*

*(b) Define a function that describes the graph.*

**73.** $f(3) = 0$    $\lim\limits_{x \to \infty} f(x) = 1$    $\lim\limits_{x \to -\infty} f(x) = 1$

$\lim\limits_{x \to 1^-} f(x) = \infty$    $\lim\limits_{x \to 1^+} f(x) = -\infty$

**74.** $f(2) = 0$    $\lim\limits_{x \to \infty} f(x) = 0$    $\lim\limits_{x \to -\infty} f(x) = 0$    $\lim\limits_{x \to 0} f(x) = \infty$

$\lim\limits_{x \to 5^-} f(x) = -\infty$    $\lim\limits_{x \to 5^+} f(x) = \infty$

**75. Newton's Law of Cooling**  Suppose an object is heated to a temperature $u_0$. Then at time $t = 0$, the object is put into a medium with a constant lower temperature $T$ causing the object to cool. **Newton's Law of Cooling** states that the temperature $u$ of the object at time $t$ is given by $u = u(t) = (u_0 - T)e^{kt} + T$, where $k < 0$ is a constant.

(a) Find $\lim\limits_{t \to \infty} u(t)$. Is this the value you expected? Explain why or why not.

(b) Find $\lim\limits_{t \to 0^+} u(t)$. Is this the value you expected? Explain why or why not.

*Source*: Submitted by the students of Millikin University.

**76. Environment**  A utility company burns coal to generate electricity. The cost $C$, in dollars, of removing $p\%$ of the pollutants emitted into the air is

$$C = \frac{70,000p}{100 - p} \qquad 0 \le p < 100$$

Find the cost of removing:

(a) 45% of the pollutants.

(b) 90% of the pollutants.

(c) Find $\lim\limits_{p \to 100^-} C$.

(d) Interpret the answer found in (c).

**77. Pollution Control**  The cost $C$, in thousands of dollars, to remove a pollutant from a lake is

$$C(x) = \frac{5x}{100 - x} \qquad 0 \le x < 100$$

where $x$ is the percent of pollutant removed. Find $\lim\limits_{x \to 100^-} C(x)$. Interpret your answer.

**78. Population Model**  A rare species of insect was discovered in the Amazon Rain Forest. To protect the species, entomologists declared the insect endangered and transferred 25 insects to a protected area. The population $P$ of the new colony $t$ days after the transfer is

$$P(t) = \frac{50(1 + 0.5t)}{2 + 0.01t}$$

(a) What is the projected size of the colony after 1 year (365 days)?

(b) What is the largest population that the protected area can sustain? That is, find $\lim\limits_{t \to \infty} P(t)$.

(c) Graph the population $P$ as a function of time $t$.

(d) Use the graph from (c) to describe the regeneration of the insect population. Does the graph support the answer to (b)?

**79. Population of an Endangered Species**  Often environmentalists capture several members of an endangered species and transport them to a controlled environment where they can produce offspring and regenerate their population. Suppose six American bald eagles are captured, tagged, transported to Montana, and set free. Based on past experience, the environmentalists expect the population to grow according to the model

$$P(t) = \frac{500}{1 + 82.3e^{-0.162t}}$$

where $t$ is measured in years.

(a) If the model is correct, how many bald eagles can the environment sustain? That is, find $\lim\limits_{t \to \infty} P(t)$.

(b) Graph the population $P$ as a function of time $t$.

(c) Use the graph from (b) to describe the growth of the bald eagle population. Does the graph support the answer to (a)?

**80. Hailstones**  Hailstones typically originate at an altitude of about 3000 meters (m). If a hailstone falls from 3000 m with no air resistance, its speed when it hits the ground would be about 240 meters/second (m/s), which is 540 miles/hour (mi/h)! That would be deadly! But air resistance slows the hailstone considerably. Using a simple model of air resistance, the speed $v = v(t)$ of a hailstone of mass $m$ as a function of time $t$ is given by $v(t) = \dfrac{mg}{k}(1 - e^{-kt/m})$ m/s, where $g = 9.8$ m/s$^2$ and $k$ is a constant that depends on the size of the hailstone, its mass, and the conditions of the air. For a hailstone with a diameter $d = 1$ centimeter (cm) and mass m $= 4.8 \times 10^{-4}$ kg, $k$ has been measured to be $3.4 \times 10^{-4}$ kg/s.

(a) Determine the limiting speed of the hailstone by finding $\lim\limits_{t \to \infty} v(t)$. Express your answer in meters per second and miles per hour, using the fact that 1 mi/h $\approx 0.447$ m/s. This speed is called the **terminal speed** of the hailstone.

(b) Graph $v = v(t)$. Does the graph support the answer to (a)?

**81. Damped Harmonic Motion**  The motion of a spring is given by the function

$$x(t) = 1.2e^{-t/2} \cos t + 2.4e^{-t/2} \sin t$$

where $x$ is the distance in meters from the the equilibrium position and $t$ is the time in seconds.

(a) Graph $y = x(t)$. What is $\lim\limits_{t \to \infty} x(t)$, as suggested by the graph?

(b) Find $\lim\limits_{t \to \infty} x(t)$.

(c) Compare the results of (a) and (b). Is the answer to (b) supported by the graph in (a)?

**82. Decomposition of Chlorine in a Pool**  Under certain water conditions, the free chlorine (hypochlorous acid, HOCl) in a swimming pool decomposes according to the law of uninhibited decay, $C = C(t) = C(0)e^{kt}$, where $C = C(t)$ is the amount (in parts per million, ppm) of free chlorine present at time $t$ (in hours) and $k$ is a negative number that represents the rate of decomposition. After shocking his pool, Ben immediately tested the water and found the concentration of free chlorine to be $C_0 = C(0) = 2.5$ ppm. Twenty-four hours later, Ben tested the water again and found the amount of free chlorine to be 2.2 ppm.

**(a)** What amount of free chlorine will be left after 72 hours?

**(b)** When the free chlorine reaches 1.0 ppm, the pool should be shocked again. How long can Ben go before he must shock the pool again?

**(c)** Find $\lim_{t \to \infty} C(t)$.

**(d)** Interpret the answer found in (c).

**83. Decomposition of Sucrose**   Reacting with water in an acidic solution at 35°C, the amount $A$ of sucrose ($C_{12}H_{22}O_{11}$) decomposes into glucose ($C_6H_{12}O_6$) and fructose ($C_6H_{12}O_6$) according to the law of uninhibited decay $A = A(t) = A(0)e^{kt}$, where $A = A(t)$ is the amount (in moles) of sucrose present at time $t$ (in minutes) and $k$ is a negative number that represents the rate of decomposition. An initial amount $A_0 = A(0) = 0.40$ mole of sucrose decomposes to 0.36 mole in 30 minutes.

**(a)** How much sucrose will remain after 2 hours?

**(b)** How long will it take until 0.10 mole of sucrose remains?

**(c)** Find $\lim_{t \to \infty} A(t)$.

**(d)** Interpret the answer found in (c).

**84. Macrophotography**   A camera lens can be approximated by a thin lens. A thin lens of focal length $f$ obeys the thin-lens equation $\dfrac{1}{f} = \dfrac{1}{p} + \dfrac{1}{q}$, where $p > f$ is the distance from the lens to the object being photographed and $q$ is the distance from the lens to the image formed by the lens. See the figure below. To photograph an object, the object's image must be formed on the photo sensors of the camera, which can only occur if $q$ is positive.

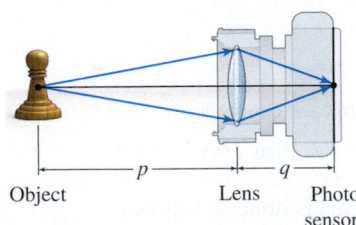

Object            Lens      Photo
                            sensors

**(a)** Is the distance $q$ of the image from the lens continuous as the distance of the object being photographed approaches the focal length $f$ of the lens? (*Hint*: First solve the thin-lens equation for $q$ and then find $\lim_{p \to f^+} q$.)

**(b)** Use the result from (a) to explain why a camera (or any lens) cannot focus on an object placed close to its focal length.

*In Problems 85 and 86, find conditions on $a$, $b$, $c$, and $d$ so that the graph of $f$ has no horizontal or vertical asymptotes.*

**85.** $f(x) = \dfrac{ax^3 + b}{cx^4 + d}$

**86.** $f(x) = \dfrac{ax + b}{cx + d}$

**87.** Explain why the following properties are true. Give an example of each.

**(a)** If $n$ is an even positive integer, then $\lim\limits_{x \to c} \dfrac{1}{(x-c)^n} = \infty$.

**(b)** If $n$ is an odd positive integer, then $\lim\limits_{x \to c^-} \dfrac{1}{(x-c)^n} = -\infty$.

**(c)** If $n$ is an odd positive integer, then $\lim\limits_{x \to c^+} \dfrac{1}{(x-c)^n} = \infty$.

**88.** Explain why a rational function, whose numerator and denominator have no common zeros, will have vertical asymptotes at each point of discontinuity.

**89.** Explain why a polynomial function of degree 1 or higher cannot have any asymptotes.

**90.** If $P$ and $Q$ are polynomials of degree $m$ and $n$, respectively, discuss $\lim\limits_{x \to \infty} \dfrac{P(x)}{Q(x)}$ when:

**(a)** $m > n$      **(b)** $m = n$      **(c)** $m < n$

**91.** ◩ **(a)** Use a table to investigate $\lim\limits_{x \to \infty} \left(1 + \dfrac{1}{x}\right)^x$.

[CAS] **(b)** Find $\lim\limits_{x \to \infty} \left(1 + \dfrac{1}{x}\right)^x$.

**(c)** Compare the results from (a) and (b). Explain the possible causes of any discrepancy.

**92.** Prove that $\lim\limits_{x \to \pm\infty} \dfrac{k}{x^p} = 0$, for any real number $k$, and $p > 0$, provided $x^p$ is defined when $x < 0$.

## Challenge Problems

**93. Kinetic Energy**   At low speeds the kinetic energy $K$, that is, the energy due to the motion of an object of mass $m$ and speed $v$, is given by the formula $K = K(v) = \dfrac{1}{2}mv^2$. But this formula is only an approximation to the general formula, and works only for speeds much less than the speed of light, $c$. The general formula, which holds for all speeds, is

$$K_{\text{gen}}(v) = mc^2 \left[ \frac{1}{\sqrt{1 - \dfrac{v^2}{c^2}}} - 1 \right]$$

**(a)** As an object is accelerated closer and closer to the speed of light, what does its kinetic energy $K_{\text{gen}}$ approach?

**(b)** What does the result suggest about the possibility of reaching the speed of light?

**94.** $\lim\limits_{x \to \infty} \left(1 + \dfrac{1}{x}\right) = 1$, but $\lim\limits_{x \to \infty} \left(1 + \dfrac{1}{x}\right)^x > 1$. Discuss why the property $\lim\limits_{x \to \infty} [f(x)]^n = \left[\lim\limits_{x \to \infty} f(x)\right]^n$ cannot be used to find the second limit.

# 1.6 The $\varepsilon$-$\delta$ Definition of a Limit

**OBJECTIVE**   *When you finish this section, you should be able to:*

**1** Use the $\varepsilon$-$\delta$ definition of a limit (p. 141)

Throughout the chapter, we stated that we could be sure a limit was correct only if it was based on the $\varepsilon$-$\delta$ definition of a limit. In this section, we examine this definition and how to use it to prove a limit exists, to verify the value of a limit, and to show that a limit does not exist.

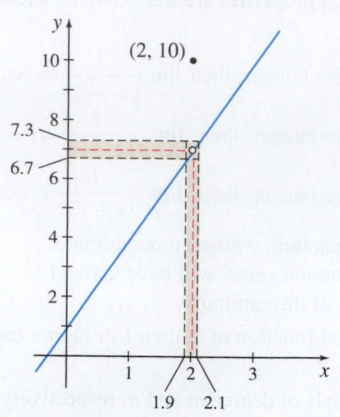

**Figure 64** $f(x) = \begin{cases} 3x + 1 & \text{if } x \neq 2 \\ 10 & \text{if } x = 2 \end{cases}$

Consider the function $f$ defined by

$$f(x) = \begin{cases} 3x + 1 & \text{if } x \neq 2 \\ 10 & \text{if } x = 2 \end{cases}$$

whose graph is given in Figure 64.

As $x$ gets closer to 2, but $x \neq 2$, the value $f(x)$ gets closer to 7. If in fact, by taking $x$ close enough to 2, we can make $f(x)$ as close to 7 as we please, then $\lim\limits_{x \to 2} f(x) = 7$.

For example, if we want $f(x)$ to differ from 7 by less than 0.3, then $|f(x) - 7| < 0.3$, and

$$-0.3 < f(x) - 7 < 0.3$$
$$-0.3 < (3x + 1) - 7 < 0.3$$
$$-0.3 < 3x - 6 < 0.3$$
$$-0.3 < 3(x - 2) < 0.3$$
$$\frac{-0.3}{3} < x - 2 < \frac{0.3}{3}$$
$$-0.1 < x - 2 < 0.1$$
$$|x - 2| < 0.1$$

That is, whenever $x \neq 2$ and $x$ differs from 2 by less than 0.1, then $f(x)$ differs from 7 by less than 0.3.

Now generalizing the question, we ask, for $x \neq 2$, how close must $x$ be to 2 to guarantee that $f(x)$ differs from 7 by less than any given positive number $\varepsilon$? ($\varepsilon$ might be extremely small.) The statement "$f(x)$ differs from 7 by less than $\varepsilon$" means

$$|f(x) - 7| < \varepsilon$$
$$-\varepsilon < f(x) - 7 < \varepsilon$$
$$7 - \varepsilon < f(x) < 7 + \varepsilon \qquad \text{Add 7 to each expression.}$$
$$7 - \varepsilon < 3x + 1 < 7 + \varepsilon \qquad \text{Since } x \neq 2, \text{ then } f(x) = 3x + 1.$$

Now we want the middle term to be $x - 2$. This is done as follows:

$$7 - \varepsilon < 3x + 1 < 7 + \varepsilon$$
$$6 - \varepsilon < 3x < 6 + \varepsilon \qquad \text{Subtract 1 from each expression.}$$
$$2 - \frac{\varepsilon}{3} < x < 2 + \frac{\varepsilon}{3} \qquad \text{Divide each expression by 3.}$$
$$-\frac{\varepsilon}{3} < x - 2 < \frac{\varepsilon}{3} \qquad \text{Subtract 2 from each expression.}$$
$$|x - 2| < \frac{\varepsilon}{3}$$

The answer to our question is

"$x$ must be within $\dfrac{\varepsilon}{3}$ of 2, but not equal to 2, to guarantee that $f(x)$ is within $\varepsilon$ of 7."

That is, whenever $x \neq 2$ and $x$ differs from 2 by less than $\dfrac{\varepsilon}{3}$, then $f(x)$ differs from 7 by less than $\varepsilon$, which can be restated as

$$\text{whenever} \quad \begin{bmatrix} x \neq 2 \text{ and } x \text{ differs from 2} \\ \text{by less than } \delta = \frac{\varepsilon}{3} \end{bmatrix} \quad \text{then} \quad \begin{bmatrix} f(x) \text{ differs from 7} \\ \text{by less than } \varepsilon \end{bmatrix}$$

We shorten the phrase "$\varepsilon$ is any given positive number" by writing $\varepsilon > 0$. Then the statement on the right is written

$$|f(x) - 7| < \varepsilon$$

Similarly, the statement "$x$ differs from 2 by less than $\delta$" is written $|x - 2| < \delta$. The statement "$x \neq 2$" is handled by writing

$$0 < |x - 2| < \delta$$

So for our example, we write:

Given any $\varepsilon > 0$, then there is a number $\delta > 0$ so that

whenever $0 < |x - 2| < \delta$, then $|f(x) - 7| < \varepsilon$

Since the number $\delta = \dfrac{\varepsilon}{3}$ satisfies the inequalities for any number $\varepsilon > 0$, we conclude that $\lim\limits_{x \to 2} f(x) = 7$.

**NOW WORK** Problem 41.

This discussion above explains the *definition of the limit of a function*.

---

**DEFINITION** Limit of a Function

Let $f$ be a function defined everywhere in an open interval containing $c$, except possibly at $c$. Then the **limit as $x$ approaches $c$ of $f(x)$ is $L$**, written

$$\boxed{\lim_{x \to c} f(x) = L}$$

if, given any number $\varepsilon > 0$, there is a number $\delta > 0$ so that

$$\boxed{\text{whenever } \; 0 < |x - c| < \delta \quad \text{then} \quad |f(x) - L| < \varepsilon}$$

---

This definition is commonly called the **$\varepsilon$-$\delta$ definition of a limit** of a function.

Figure 65 illustrates the definition for three choices of $\varepsilon$. Compare Figures 65(a) and 65(b). Notice that in Figure 65(b), the smaller $\varepsilon$ requires a smaller $\delta$. Figure 65(c) illustrates what happens if $\delta$ is too large for the choice of $\varepsilon$; here there are values of $f$, for example, at $x_1$ and $x_2$, for which $|f(x) - L| \not< \varepsilon$.

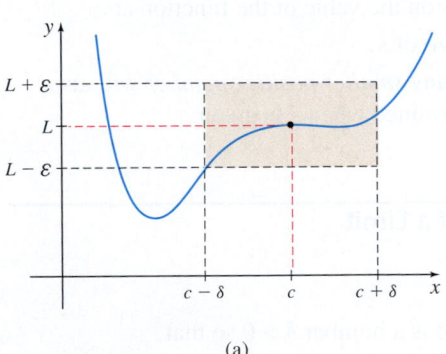

(a)

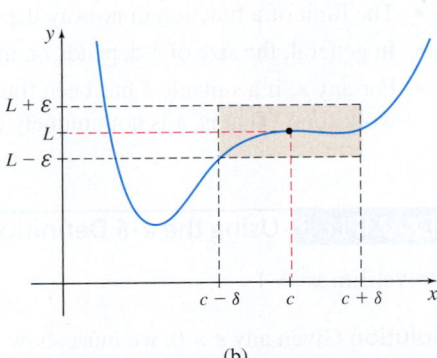

(b)

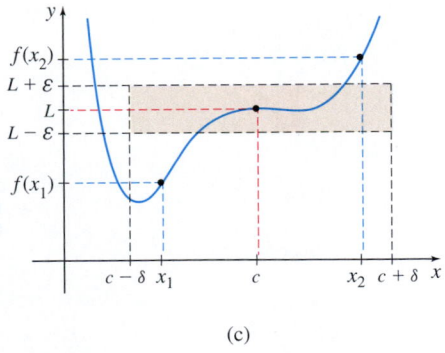

(c)

**Figure 65**

## 1 Use the $\varepsilon$-$\delta$ Definition of a Limit

**EXAMPLE 1** Using the $\varepsilon$-$\delta$ Definition of a Limit

Use the $\varepsilon$-$\delta$ definition of a limit to prove $\lim\limits_{x \to -1} (1 - 2x) = 3$.

**Solution** Given any $\varepsilon > 0$, we must show there is a number $\delta > 0$ so that

$$\text{whenever} \quad 0 < |x - (-1)| = |x + 1| < \delta \quad \text{then} \quad |(1 - 2x) - 3| < \varepsilon$$

The idea is to find a connection between $|x - (-1)| = |x + 1|$ and $|(1 - 2x) - 3|$. Since

$$|(1 - 2x) - 3| = |-2x - 2| = |-2(x + 1)| = |-2| \cdot |x + 1| = 2|x + 1|$$

we see that for any $\varepsilon > 0$,

$$\text{whenever} \quad |x - (-1)| = |x + 1| < \delta = \frac{\varepsilon}{2} \quad \text{then} \quad |(1 - 2x) - 3| = 2|x + 1| < 2\delta = \varepsilon$$

That is, given any $\varepsilon > 0$ there is a $\delta$, $\delta = \dfrac{\varepsilon}{2}$, so that whenever $0 < |x - (-1)| < \delta$, we have $|(1 - 2x) - 3| < \varepsilon$. This proves that $\lim\limits_{x \to -1} (1 - 2x) = 3$. ∎

**NOW WORK** Problem 21.

A geometric interpretation of the $\varepsilon$-$\delta$ definition is shown in Figure 66. We see that whenever $x$ on the horizontal axis is between $-1 - \delta$ and $-1 + \delta$, but not equal to $-1$, then $f(x)$ on the vertical axis is between the horizontal lines $y = 3 + \varepsilon$ and $y = 3 - \varepsilon$. So, $\lim\limits_{x \to -1} f(x) = 3$ describes the behavior of $f$ near $-1$.

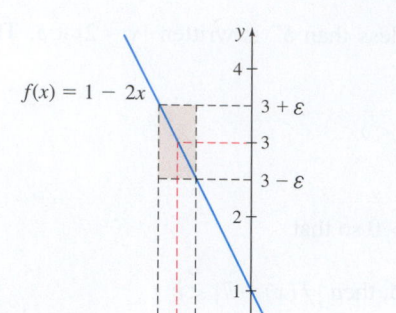

**Figure 66** $\lim\limits_{x \to -1} (1 - 2x) = 3$

### EXAMPLE 2    Using the $\varepsilon$-$\delta$ Definition of a Limit

Use the $\varepsilon$-$\delta$ definition of a limit to prove that

**(a)** $\lim\limits_{x \to c} A = A$, where $A$ and $c$ are real numbers

**(b)** $\lim\limits_{x \to c} x = c$, where $c$ is a real number

**Solution** **(a)** $f(x) = A$ is the constant function whose graph is a horizontal line. Given any $\varepsilon > 0$, we must find $\delta > 0$ so that whenever $0 < |x - c| < \delta$, then $|f(x) - A| < \varepsilon$.

Since $|A - A| = 0$, then $|f(x) - A| < \varepsilon$ no matter what positive number $\delta$ is used. That is, any choice of $\delta$ guarantees that whenever $0 < |x - c| < \delta$, then $|f(x) - A| < \varepsilon$.

**(b)** $f(x) = x$ is the identity function. Given any $\varepsilon > 0$, we must find $\delta$ so that whenever $0 < |x - c| < \delta$, then $|f(x) - c| = |x - c| < \varepsilon$. The easiest choice is to make $\delta = \varepsilon$. That is, whenever $0 < |x - c| < \delta = \varepsilon$, then $\underset{\underset{f(x)\,=\,x}{\uparrow}}{|f(x) - c|} = |x - c| < \varepsilon$. ∎

Some observations about the $\varepsilon$-$\delta$ definition of a limit are given below:

- The limit of a function in no way depends on the value of the function at $c$.
- In general, the size of $\delta$ depends on the size of $\varepsilon$.
- For any $\varepsilon$, if a suitable $\delta$ has been found, any *smaller positive number* will also work for $\delta$. That is, $\delta$ is not uniquely determined when $\varepsilon$ is given.

CALC
CLIP
### EXAMPLE 3    Using the $\varepsilon$-$\delta$ Definition of a Limit

Prove: $\lim\limits_{x \to 2} x^2 = 4$

**Solution** Given any $\varepsilon > 0$, we must show there is a number $\delta > 0$ so that

$$\text{whenever} \quad 0 < |x - 2| < \delta \quad \text{then} \quad |x^2 - 4| < \varepsilon$$

To establish a connection between $|x^2 - 4|$ and $|x - 2|$, we write $|x^2 - 4|$ as

$$|x^2 - 4| = |(x + 2)(x - 2)| = |x + 2| \cdot |x - 2|$$

Now if we can find a number $K$ for which $|x + 2| < K$, then we can choose $\delta = \dfrac{\varepsilon}{K}$.

To find $K$, we restrict $x$ to some interval centered at 2. For example, suppose the distance between $x$ and 2 is less than 1. Then

$$|x - 2| < 1$$

$$-1 < x - 2 < 1$$

$$1 < x < 3 \qquad \text{Simplify.}$$

$$1 + 2 < x + 2 < 3 + 2 \qquad \text{Add 2 to each part.}$$

$$3 < x + 2 < 5$$

In particular, we have $|x + 2| < 5$. It follows that whenever $|x - 2| < 1$,

$$|x^2 - 4| = |x + 2| \cdot |x - 2| < 5\,|x - 2|$$

If $|x - 2| < \delta = \dfrac{\varepsilon}{5}$, then $\left| x^2 - 4 \right| < 5\,|x - 2| < 5 \cdot \dfrac{\varepsilon}{5} = \varepsilon$, as desired.

But before choosing $\delta = \dfrac{\varepsilon}{5}$, we must remember that there are two constraints on $|x - 2|$. Namely,

$$|x - 2| < 1 \qquad \text{and} \qquad |x - 2| < \frac{\varepsilon}{5}$$

To ensure that both inequalities are satisfied, we select $\delta$ to be the smaller of the numbers 1 and $\dfrac{\varepsilon}{5}$, abbreviated as $\delta = \min\left\{ 1, \dfrac{\varepsilon}{5} \right\}$. Now

$$\text{whenever } |x - 2| < \delta = \min\left\{ 1, \frac{\varepsilon}{5} \right\} \text{ then } |x^2 - 4| < \varepsilon$$

proving $\displaystyle\lim_{x \to 2} x^2 = 4$. ∎

**NOW WORK** Problem 23.

In Example 3, the decision to restrict $x$ so that $|x - 2| < 1$ was completely arbitrary. However, since we are looking for $x$ close to 2, the interval chosen should be small. In Problem 40, you are asked to verify that if we had restricted $x$ so that $|x - 2| < \dfrac{1}{3}$, then the choice for $\delta$ would be less than or equal to the smaller of $\dfrac{1}{3}$ and $\dfrac{3\varepsilon}{13}$; that is, $\delta \leq \min\left\{ \dfrac{1}{3}, \dfrac{3\varepsilon}{13} \right\}$.

**EXAMPLE 4   Using the $\varepsilon$-$\delta$ Definition of a Limit**

Prove $\displaystyle\lim_{x \to c} \dfrac{1}{x} = \dfrac{1}{c}$, where $c > 0$.

**Solution** The domain of $f(x) = \dfrac{1}{x}$ is $\{x \mid x \neq 0\}$.

For any $\varepsilon > 0$, we need to find a positive number $\delta$ so that whenever $0 < |x - c| < \delta$, then $\left| \dfrac{1}{x} - \dfrac{1}{c} \right| < \varepsilon$. For $x \neq 0$, and $c > 0$, we have

$$\left| \frac{1}{x} - \frac{1}{c} \right| = \left| \frac{c - x}{xc} \right| = \frac{|c - x|}{|x| \cdot |c|} = \frac{|x - c|}{c\,|x|}$$

The idea is to find a connection between

$$|x - c| \qquad \text{and} \qquad \frac{|x - c|}{c\,|x|}$$

We proceed as in Example 3. Since we are interested in $x$ near $c$, we restrict $x$ to a small interval around $c$, say, $|x - c| < \dfrac{c}{2}$. Then

$$-\frac{c}{2} < x - c < \frac{c}{2}$$

$$\frac{c}{2} < x < \frac{3c}{2} \qquad \text{Add } c \text{ to each expression.}$$

Since $c > 0$, then $x > \dfrac{c}{2} > 0$, and $\dfrac{1}{x} < \dfrac{2}{c}$. Now

$$\left| \frac{1}{x} - \frac{1}{c} \right| = \frac{|x - c|}{c|x|} < \frac{2}{c^2} \cdot |x - c| \qquad \text{Substitute } \frac{1}{|x|} < \frac{2}{c}.$$

We can make $\left| \dfrac{1}{x} - \dfrac{1}{c} \right| < \varepsilon$ by choosing $\delta = \dfrac{c^2}{2}\varepsilon$. Then

whenever $|x - c| < \delta = \dfrac{c^2}{2}\varepsilon$, we have $\left| \dfrac{1}{x} - \dfrac{1}{c} \right| < \dfrac{2}{c^2} \cdot |x - c| < \dfrac{2}{c^2} \cdot \left( \dfrac{c^2}{2} \cdot \varepsilon \right) = \varepsilon$

But remember, there are two restrictions on $|x - c|$.

$$|x - c| < \frac{c}{2} \qquad \text{and} \qquad |x - c| < \frac{c^2}{2} \cdot \varepsilon$$

So, given any $\varepsilon > 0$, we choose $\delta = \min\left( \dfrac{c}{2}, \dfrac{c^2}{2} \cdot \varepsilon \right)$. Then whenever $0 < |x - c| < \delta$,

we have $\left| \dfrac{1}{x} - \dfrac{1}{c} \right| < \varepsilon$. This proves $\lim\limits_{x \to c} \dfrac{1}{x} = \dfrac{1}{c}, c > 0$. ∎

NOW WORK **Problem 25.**

The $\varepsilon$-$\delta$ definition of a limit can be used to show that a limit does not exist or that a limit is not equal to a specific number. Example 5 illustrates how the $\varepsilon$-$\delta$ definition of a limit is used to show that a limit is not equal to a specific number.

**EXAMPLE 5    Showing a Limit Is Not Equal to a Specific Number**

Use the $\varepsilon$-$\delta$ definition of a limit to prove the statement $\lim\limits_{x \to 3} (4x - 5) \neq 10$.

**NOTE** In a proof by contradiction, we assume that the conclusion is not true and then show this leads to a contradiction.

**Solution** We use a proof by contradiction. Assume $\lim\limits_{x \to 3} (4x - 5) = 10$ and choose $\varepsilon = 1$. (Any smaller positive number $\varepsilon$ will also work.) Then there is a number $\delta > 0$ so that

$$\text{whenever} \quad 0 < |x - 3| < \delta \quad \text{then} \quad |(4x - 5) - 10| < 1$$

We simplify the right inequality.

$$|(4x - 5) - 10| = |4x - 15| < 1$$

$$-1 < 4x - 15 < 1$$

$$14 < 4x < 16$$

$$3.5 < x < 4$$

**NOTE** For example, if $\delta = \dfrac{1}{4}$, then

$$3 - \frac{1}{4} < x < 3 + \frac{1}{4}$$

$$2.75 < x < 3.25$$

contradicting $3.5 < x < 4$.

According to our assumption, whenever $0 < |x - 3| < \delta$, then $3.5 < x < 4$. Regardless of the value of $\delta$, the inequality $0 < |x - 3| < \delta$ is satisfied by a number $x$ that is less than 3. This contradicts the fact that $3.5 < x < 4$. The contradiction means that $\lim\limits_{x \to 3} (4x - 5) \neq 10$. ∎

NOW WORK **Problem 33.**

**EXAMPLE 6** **Showing a Limit Does Not Exist**

The **Dirichlet function** is defined by

$$f(x) = \begin{cases} 1 & \text{if } x \text{ is rational} \\ 0 & \text{if } x \text{ is irrational} \end{cases}$$

Prove $\lim_{x \to c} f(x)$ does not exist for any $c$.

**Solution** We use a proof by contradiction. That is, we assume that $\lim_{x \to c} f(x)$ exists and show that this leads to a contradiction.

Assume $\lim_{x \to c} f(x) = L$ for some number $c$. Now if we are given $\varepsilon = \dfrac{1}{2}$ (or any smaller positive number), then there is a positive number $\delta$, so that

whenever    $0 < |x - c| < \delta$    then    $|f(x) - L| < \dfrac{1}{2}$

Suppose $x_1$ is a rational number satisfying $0 < |x_1 - c| < \delta$, and $x_2$ is an irrational number satisfying $0 < |x_2 - c| < \delta$. Then from the definition of the function $f$,

$$f(x_1) = 1 \qquad \text{and} \qquad f(x_2) = 0$$

Using these values in the inequality $|f(x) - L| < \varepsilon$, we get

$$|f(x_1) - L| = |1 - L| < \frac{1}{2} \qquad \text{and} \qquad |f(x_2) - L| = |0 - L| < \frac{1}{2}$$

$$-\frac{1}{2} < 1 - L < \frac{1}{2} \qquad\qquad -\frac{1}{2} < -L < \frac{1}{2}$$

$$-\frac{3}{2} < -L < -\frac{1}{2} \qquad\qquad \frac{1}{2} > L > -\frac{1}{2}$$

$$\frac{1}{2} < L < \frac{3}{2} \qquad\qquad -\frac{1}{2} < L < \frac{1}{2}$$

From the left inequality, we have $L > \dfrac{1}{2}$, and from the right inequality, we have $L < \dfrac{1}{2}$. Since it is impossible for both inequalities to be satisfied, we conclude that $\lim_{x \to c} f(x)$ does not exist. ∎

**EXAMPLE 7** **Using the $\varepsilon$-$\delta$ Definition of a Limit**

Prove that if $\lim_{x \to c} f(x) > 0$, then there is an open interval around $c$, for which $f(x) > 0$ everywhere in the interval, except possibly at $c$.

**Solution** Suppose $\lim_{x \to c} f(x) = L > 0$. Then given any $\varepsilon > 0$, there is a $\delta > 0$ so that

whenever    $0 < |x - c| < \delta$    then    $|f(x) - L| < \varepsilon$

If $\varepsilon = \dfrac{L}{2}$, then from the definition of limit, there is a $\delta > 0$ so that

whenever $0 < |x - c| < \delta$    then $|f(x) - L| < \dfrac{L}{2}$    or equivalently,    $\dfrac{L}{2} < f(x) < \dfrac{3L}{2}$

Since $\dfrac{L}{2} > 0$, the last statement proves our assertion that $f(x) > 0$ for all $x$ in the interval $(c - \delta, c + \delta)$, except possibly at $c$. ∎

In Problem 43, you are asked to prove the theorem stating that if $\lim\limits_{x \to c} f(x) < 0$, then there is an open interval around $c$, for which $f(x) < 0$ everywhere in the interval, except possibly at $c$.

We close this section with the $\varepsilon$-$\delta$ definitions of limits at infinity and infinite limits.

> **DEFINITION** Limit at Infinity
>
> Let $f$ be a function defined on an open interval $(b, \infty)$. Then $f$ has a **limit at infinity**
>
> $$\lim_{x \to \infty} f(x) = L$$
>
> where $L$ is a real number, if given any number $\varepsilon > 0$, there is a positive number $M$ so that whenever $x > M$, then $|f(x) - L| < \varepsilon$.
>
> If $f$ is a function defined on an open interval $(-\infty, a)$, then
>
> $$\lim_{x \to -\infty} f(x) = L$$
>
> if given any number $\varepsilon > 0$, there is a negative number $N$ so that whenever $x < N$, then $|f(x) - L| < \varepsilon$.

Figures 67 and 68 illustrate limits at infinity.

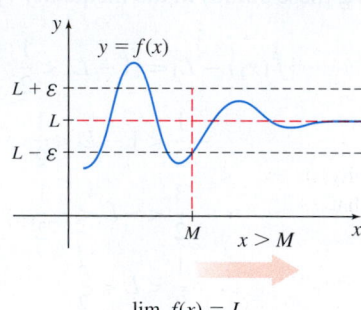

$$\lim_{x \to \infty} f(x) = L$$

**Figure 67** For any $\varepsilon > 0$, there is a positive number $M$ so that whenever $x > M$, then $|f(x) - L| < \varepsilon$.

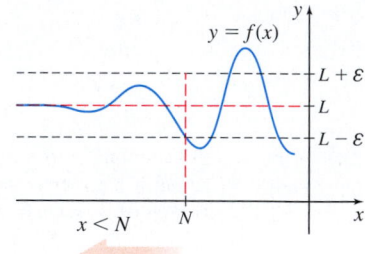

$$\lim_{x \to -\infty} f(x) = L$$

**Figure 68** For any $\varepsilon > 0$, there is a negative number $N$ so that whenever $x < N$, then $|f(x) - L| < \varepsilon$.

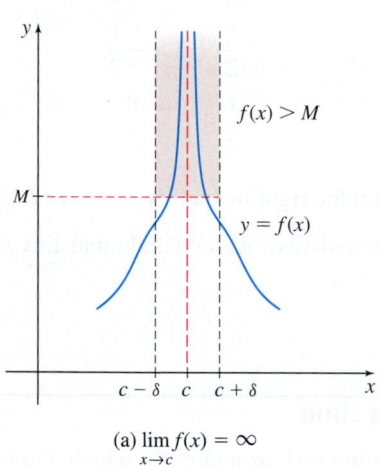

(a) $\lim\limits_{x \to c} f(x) = \infty$

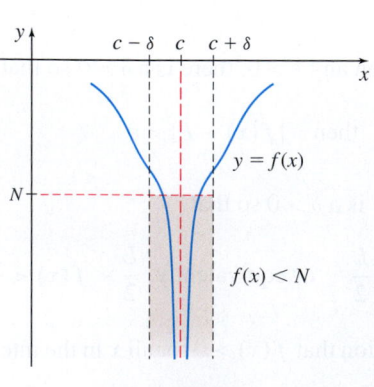

(b) $\lim\limits_{x \to c} f(x) = -\infty$

**Figure 69**

> **DEFINITION** Infinite Limit
>
> Let $f$ be a function defined everywhere on an open interval containing $c$, except possibly at $c$. Then $f(x)$ becomes unbounded in the positive direction (has an infinite limit) as $x$ approaches $c$, written
>
> $$\lim_{x \to c} f(x) = \infty$$
>
> if, for every positive number $M$, a positive number $\delta$ exists so that
>
> whenever $0 < |x - c| < \delta$ then $f(x) > M$.
>
> Similarly, $f(x)$ becomes unbounded in the negative direction (has an infinite limit) as $x$ approaches $c$, written
>
> $$\lim_{x \to c} f(x) = -\infty$$
>
> if, for every negative number $N$, a positive number $\delta$ exists so that
>
> whenever $0 < |x - c| < \delta$ then $f(x) < N$.

Figure 69 illustrates infinite limits.

> **DEFINITION  Infinite Limit at Infinity**
>
> Let $f$ be a function defined on an open interval $(b, \infty)$. Then $f$ has an **infinite limit at infinity**
>
> $$\lim_{x \to \infty} f(x) = \infty$$
>
> if for any positive number $M$, there is a corresponding positive number $N$ so that whenever $x > N$, then $f(x) > M$.

A similar definition applies for infinite limits at negative infinity.

## 1.6 Assess Your Understanding

### Concepts and Vocabulary

**1.** *True or False*   The limit of a function as $x$ approaches $c$ depends on the value of the function at $c$.

**2.** *True or False*   In the $\varepsilon$-$\delta$ definition of a limit, we require $0 < |x - c|$ to ensure that $x \neq c$.

**3.** *True or False*   In an $\varepsilon$-$\delta$ proof of a limit, the size of $\delta$ usually depends on the size of $\varepsilon$.

**4.** *True or False*   When proving $\lim\limits_{x \to c} f(x) = L$ using the $\varepsilon$-$\delta$ definition of a limit, we try to find a connection between $|f(x) - c|$ and $|x - c|$.

**5.** *True or False*   Given any $\varepsilon > 0$, suppose there is a $\delta > 0$ so that whenever $0 < |x - c| < \delta$, then $|f(x) - L| < \varepsilon$. Then $\lim\limits_{x \to c} f(x) = L$.

**6.** *True or False*   A function $f$ has a limit $L$ at infinity, if for any given $\varepsilon > 0$, there is a positive number $M$ so that whenever $x > M$, then $|f(x) - L| > \varepsilon$.

### Skill Building

*In Problems 7–12, for each limit, find the largest $\delta$ that "works" for the given $\varepsilon$.*

**7.** $\lim\limits_{x \to 1} (2x) = 2, \quad \varepsilon = 0.01$  **8.** $\lim\limits_{x \to 2} (-3x) = -6, \quad \varepsilon = 0.01$

**9.** $\lim\limits_{x \to 2} (6x - 1) = 11$   **10.** $\lim\limits_{x \to -3} (2 - 3x) = 11$

$\varepsilon = \dfrac{1}{2}$   $\varepsilon = \dfrac{1}{3}$

**11.** $\lim\limits_{x \to 2} \left(-\dfrac{1}{2}x + 5\right) = 4$   **12.** $\lim\limits_{x \to \frac{5}{6}} \left(3x + \dfrac{1}{2}\right) = 3$

$\varepsilon = 0.01$   $\varepsilon = 0.3$

**13.** For the function $f(x) = 4x - 1$, we have $\lim\limits_{x \to 3} f(x) = 11$.

For each $\varepsilon > 0$, find a $\delta > 0$ so that

whenever   $0 < |x - 3| < \delta$   then   $|(4x - 1) - 11| < \varepsilon$

(a) $\varepsilon = 0.1$       (b) $\varepsilon = 0.01$
(c) $\varepsilon = 0.001$    (d) $\varepsilon > 0$ is arbitrary

**14.** For the function $f(x) = 2 - 5x$, we have $\lim\limits_{x \to -2} f(x) = 12$.

For each $\varepsilon > 0$, find a $\delta > 0$ so that

whenever   $0 < |x + 2| < \delta$   then   $|(2 - 5x) - 12| < \varepsilon$

(a) $\varepsilon = 0.2$       (b) $\varepsilon = 0.02$
(c) $\varepsilon = 0.002$    (d) $\varepsilon > 0$ is arbitrary

**15.** For the function $f(x) = \dfrac{x^2 - 9}{x + 3}$, we have $\lim\limits_{x \to -3} f(x) = -6$.

For each $\varepsilon > 0$, find a $\delta > 0$ so that

whenever   $0 < |x + 3| < \delta$   then   $\left| \dfrac{x^2 - 9}{x + 3} - (-6) \right| < \varepsilon$

(a) $\varepsilon = 0.1$       (b) $\varepsilon = 0.01$       (c) $\varepsilon > 0$ is arbitrary

**16.** For the function $f(x) = \dfrac{x^2 - 4}{x - 2}$, we have $\lim\limits_{x \to 2} f(x) = 4$.

For each $\varepsilon > 0$, find a $\delta > 0$ so that

whenever   $0 < |x - 2| < \delta$   then   $\left| \dfrac{x^2 - 4}{x - 2} - 4 \right| < \varepsilon$

(a) $\varepsilon = 0.1$       (b) $\varepsilon = 0.01$       (c) $\varepsilon > 0$ is arbitrary

*In Problems 17–32, write a proof for each limit using the $\varepsilon$-$\delta$ definition of a limit.*

**17.** $\lim\limits_{x \to 2} (3x) = 6$   **18.** $\lim\limits_{x \to 3} (4x) = 12$

**19.** $\lim\limits_{x \to 0} (2x + 5) = 5$   **20.** $\lim\limits_{x \to -1} (2 - 3x) = 5$

**21.** $\lim\limits_{x \to -3} (-5x + 2) = 17$   **22.** $\lim\limits_{x \to 2} (2x - 3) = 1$

**23.** $\lim\limits_{x \to 2} (x^2 - 2x) = 0$   **24.** $\lim\limits_{x \to 0} (x^2 + 3x) = 0$

**25.** $\lim\limits_{x \to 1} \dfrac{1 + 2x}{3 - x} = \dfrac{3}{2}$   **26.** $\lim\limits_{x \to 2} \dfrac{2x}{4 + x} = \dfrac{2}{3}$

**27.** $\lim\limits_{x \to 0} \sqrt[3]{x} = 0$   **28.** $\lim\limits_{x \to 1} \sqrt{2 - x} = 1$

**29.** $\lim\limits_{x \to -1} x^2 = 1$   **30.** $\lim\limits_{x \to 2} x^3 = 8$

**31.** $\lim\limits_{x \to 3} \dfrac{1}{x} = \dfrac{1}{3}$   **32.** $\lim\limits_{x \to 2} \dfrac{1}{x^2} = \dfrac{1}{4}$

**33.** Use the $\varepsilon$-$\delta$ definition of a limit to show that the statement $\lim\limits_{x \to 3} (3x - 1) = 12$ is false.

**34.** Use the $\varepsilon$-$\delta$ definition of a limit to show that the statement $\lim\limits_{x \to -2} (4x) = -7$ is false.

---

**1.** = NOW WORK problem        📶 = Graphing technology recommended        [CAS] = Computer Algebra System recommended

## Applications and Extensions

**35.** Show that $\left| \dfrac{1}{x^2+9} - \dfrac{1}{18} \right| < \dfrac{7}{234} |x-3|$ if $2 < x < 4$.

Use this to show that $\lim\limits_{x \to 3} \dfrac{1}{x^2+9} = \dfrac{1}{18}$.

**36.** Show that $|(2+x)^2 - 4| \le 5\,|x|$ if $-1 < x < 1$.

Use this to show that $\lim\limits_{x \to 0} (2+x)^2 = 4$.

**37.** Show that $\left| \dfrac{1}{x^2+9} - \dfrac{1}{13} \right| \le \dfrac{1}{26}\,|x-2|$ if $1 < x < 3$.

Use this to show that $\lim\limits_{x \to 2} \dfrac{1}{x^2+9} = \dfrac{1}{13}$.

**38.** Use the $\varepsilon$-$\delta$ definition of a limit to show that $\lim\limits_{x \to 1} x^2 \ne 1.31$.
(*Hint:* Use $\varepsilon = 0.1$.)

**39.** If $m$ and $b$ are any constants, prove that
$$\lim_{x \to c} (mx+b) = mc+b$$

**40.** Verify that if $x$ is restricted so that $|x-2| < \dfrac{1}{3}$ in the proof of $\lim\limits_{x \to 2} x^2 = 4$, then the choice for $\delta$ would be less than or equal to the smaller of $\dfrac{1}{3}$ and $\dfrac{3\varepsilon}{13}$; that is, $\delta \le \min\left\{ \dfrac{1}{3}, \dfrac{3\varepsilon}{13} \right\}$.

**41.** For $x \ne 3$, how close to 3 must $x$ be to guarantee that $2x-1$ differs from 5 by less than 0.1?

**42.** For $x \ne 0$, how close to 0 must $x$ be to guarantee that $3^x$ differs from 1 by less than 0.1?

**43.** Prove that if $\lim\limits_{x \to c} f(x) < 0$, then there is an open interval around $c$ for which $f(x) < 0$ everywhere in the interval, except possibly at $c$.

**44.** Use the $\varepsilon$-$\delta$ definition of a limit at infinity to prove that
$$\lim_{x \to -\infty} \frac{1}{x} = 0$$

**45.** Use the $\varepsilon$-$\delta$ definition of a limit at infinity to prove that
$$\lim_{x \to \infty} \left( -\frac{1}{\sqrt{x}} \right) = 0$$

**46.** For $\lim\limits_{x \to -\infty} \dfrac{1}{x^2} = 0$, find a value of $N$ that satisfies the $\varepsilon$-$\delta$ definition of limits at infinity for $\varepsilon = 0.1$.

**47.** Use the $\varepsilon$-$\delta$ definition of limit to prove that no number $L$ exists so that $\lim\limits_{x \to 0} \dfrac{1}{x} = L$.

**48.** Explain why in the $\varepsilon$-$\delta$ definition of a limit, the inequality $0 < |x-c| < \delta$ has two strict inequality symbols.

**49.** The $\varepsilon$-$\delta$ definition of a limit states, in part, that $f$ is defined everywhere in an open interval containing $c$, except possibly at $c$. Discuss the purpose of including the phrase, *except possibly at $c$*, and why it is necessary.

**50.** In the $\varepsilon$-$\delta$ definition of a limit, what does $\varepsilon$ measure? What does $\delta$ measure? Give an example to support your explanation.

**51.** Discuss $\lim\limits_{x \to 0} f(x)$ and $\lim\limits_{x \to 1} f(x)$ if
$$f(x) = \begin{cases} x^2 & \text{if } x \text{ is rational} \\ 0 & \text{if } x \text{ is irrational} \end{cases}.$$

**52.** Discuss $\lim\limits_{x \to 0} f(x)$ if $f(x) = \begin{cases} x^2 & \text{if } x \text{ is rational} \\ \tan x & \text{if } x \text{ is irrational} \end{cases}.$

## Challenge Problems

**53.** Use the $\varepsilon$-$\delta$ definition of limit to prove that
$$\lim_{x \to 1} (4x^3 + 3x^2 - 24x + 22) = 5.$$

**54.** If $\lim\limits_{x \to c} f(x) = L$ and $\lim\limits_{x \to c} g(x) = M$, prove that $\lim\limits_{x \to c} [f(x) + g(x)] = L + M$. Use the $\varepsilon$-$\delta$ definition of limit.

**55.** For $\lim\limits_{x \to \infty} \dfrac{2-x}{\sqrt{5+4x^2}} = -\dfrac{1}{2}$, find a value of $M$ that satisfies the $\varepsilon$-$\delta$ definition of a limit at infinity for $\varepsilon = 0.01$.

**56.** Use the $\varepsilon$-$\delta$ definition of a limit to prove that the linear function $f(x) = ax + b$ is continuous for all real numbers.

**57.** Show that the function $f(x) = \begin{cases} 0 & \text{if } x \text{ is rational} \\ x & \text{if } x \text{ is irrational} \end{cases}$
is continuous only at $x = 0$.

**58.** Suppose that $f$ is defined on an interval $(a, b)$ and there is a number $K$ so that
$$|f(x) - f(c)| \le K|x-c|$$
for all $c$ in $(a, b)$ and $x$ in $(a, b)$. Such a constant $K$ is called a **Lipschitz constant**. Find a Lipschitz constant for $f(x) = x^3$ on the interval $(0, 2)$.

# Chapter Review

## THINGS TO KNOW

### 1.1 Limits of Functions Using Numerical and Graphical Techniques

- Slope of a secant line: $m_{\sec} = \dfrac{f(x) - f(c)}{x - c}$   (p. 79)

- Slope of a tangent line: $m_{\tan} = \lim\limits_{x \to c} \dfrac{f(x) - f(c)}{x - c}$   (p. 79)

- $\lim\limits_{x \to c} f(x) = L$: read, "The limit as $x$ approaches $c$ of $f(x)$ is equal to the number $L$." (p. 80)

- $\lim\limits_{x \to c} f(x) = L$: interpreted as, "The value $f(x)$ can be made as close as we please to $L$, for $x$ sufficiently close to $c$, but not equal to $c$." (p. 80)

- One-sided limits (p. 81)
- The limit $L$ of a function $y = f(x)$ as $x$ approaches a number $c$ does not depend on the value of $f$ at $c$. (p. 83)
- The limit $L$ of a function $y = f(x)$ as $x$ approaches a number $c$ is unique. A function cannot have more than one limit as $x$ approaches $c$.   (p. 83)

*Theorem*

The limit $L$ of a function $y = f(x)$ as $x$ approaches a number $c$ exists if and only if both one-sided limits exist at $c$ and both one-sided limits are equal. That is, $\lim\limits_{x \to c} f(x) = L$ if and only if
$$\lim_{x \to c^-} f(x) = \lim_{x \to c^+} f(x) = L. \quad \text{(p. 84)}$$

## 1.2 Limits of Functions Using Properties of Limits

### Basic Limits

- $\lim\limits_{x \to c} A = A$, $A$ a constant, for any real number $c$.    (p. 90)

- $\lim\limits_{x \to c} x = c$, for any real number $c$.    (p. 90)

### Properties of Limits    If $f$ and $g$ are functions for which $\lim\limits_{x \to c} f(x)$ and $\lim\limits_{x \to c} g(x)$ both exist and if $k$ is any real number, then:

- $\lim\limits_{x \to c}[f(x) \pm g(x)] = \lim\limits_{x \to c} f(x) \pm \lim\limits_{x \to c} g(x)$    (pp. 91)

- $\lim\limits_{x \to c}[f(x) \cdot g(x)] = \lim\limits_{x \to c} f(x) \cdot \lim\limits_{x \to c} g(x)$    (p. 91)

- $\lim\limits_{x \to c}[kg(x)] = k \lim\limits_{x \to c} g(x)$    (p. 92)

- $\lim\limits_{x \to c}[f(x)]^n = \left[\lim\limits_{x \to c} f(x)\right]^n$, $n$ is a positive integer    (p. 93)

- $\lim\limits_{x \to c} \sqrt[n]{f(x)} = \sqrt[n]{\lim\limits_{x \to c} f(x)}$, where $n \geq 2$ is an integer, provided $f(x) > 0$ if $n$ is even    (p. 94)

- $\lim\limits_{x \to c}[f(x)]^{m/n} = \left[\lim\limits_{x \to c} f(x)\right]^{m/n}$, provided $[f(x)]^{m/n}$ is defined for positive integers $m$ and $n$    (p. 94)

- $\lim\limits_{x \to c} \dfrac{f(x)}{g(x)} = \dfrac{\lim\limits_{x \to c} f(x)}{\lim\limits_{x \to c} g(x)}$, provided $\lim\limits_{x \to c} g(x) \neq 0$    (p. 96)

- If $P$ is a polynomial function, then $\lim\limits_{x \to c} P(x) = P(c)$.    (p. 95)

- If $R$ is a rational function and if $c$ is in the domain of $R$, then $\lim\limits_{x \to c} R(x) = R(c)$.    (p. 96)

## 1.3 Continuity

### Definitions

- Continuity at a number    (p. 102)
- Removable discontinuity    (p. 104)
- One-sided continuity at a number    (p. 104)
- Continuity on an interval    (p. 105)

### Properties of Continuity

- A polynomial function is continuous on its domain, all real numbers.    (p. 106)

- A rational function is continuous on its domain.    (p. 106)

- If the functions $f$ and $g$ are continuous at a number $c$, and if $k$ is a real number, then the functions $f + g$, $f - g$, $f \cdot g$, and $kf$ are also continuous at $c$. If $g(c) \neq 0$, the function $\dfrac{f}{g}$ is continuous at $c$.    (p. 107)

- If a function $g$ is continuous at $c$ and a function $f$ is continuous at $g(c)$, then the composite function $(f \circ g)(x) = f(g(x))$ is continuous at $c$.    (p. 108)

- If $f$ is a one-to-one function that is continuous on its domain, then its inverse function $f^{-1}$ is also continuous on its domain.    (p. 108)

### The Intermediate Value Theorem    Let $f$ be a function that is continuous on a closed interval $[a, b]$ and $f(a) \neq f(b)$. If $N$ is any number between $f(a)$ and $f(b)$, then there is at least one number $c$ in the open interval $(a, b)$ for which $f(c) = N$.    (p. 109)

## 1.4 Limits and Continuity of Trigonometric, Exponential, and Logarithmic Functions

### Basic Limits

- $\lim\limits_{\theta \to 0} \dfrac{\sin \theta}{\theta} = 1$    (p. 117)
- $\lim\limits_{\theta \to 0} \dfrac{\cos \theta - 1}{\theta} = 0$    (p. 119)
- $\lim\limits_{x \to c} \sin x = \sin c$    (p. 120)
- $\lim\limits_{x \to c} \cos x = \cos c$    (p. 120)
- $\lim\limits_{x \to c} a^x = a^c$;    $a > 0$,    $a \neq 1$    (p. 122)
- $\lim\limits_{x \to c} \log_a x = \log_a c$;    $c > 0$,    $a > 0$,    $a \neq 1$    (p. 122)

### Squeeze Theorem    If the functions $f$, $g$, and $h$ have the property that for all $x$ in an open interval containing $c$, except possibly at $c$, $f(x) \leq g(x) \leq h(x)$, and if $\lim\limits_{x \to c} f(x) = \lim\limits_{x \to c} h(x) = L$, then $\lim\limits_{x \to c} g(x) = L$.    (p. 115)

### Properties of Continuity

- The six trigonometric functions are continuous on their domains.    (p. 120)

- The six inverse trigonometric functions are continuous on their domains.    (p. 121)

- An exponential function is continuous for all real numbers.    (p. 122)

- A logarithmic function is continuous for all positive real numbers.    (p. 122)

## 1.5 Infinite Limits; Limits at Infinity; Asymptotes

### Basic Limits

- $\lim\limits_{x \to 0^-} \dfrac{1}{x} = -\infty$    (p. 125)
- $\lim\limits_{x \to 0^+} \dfrac{1}{x} = \infty$    (p. 125)
- $\lim\limits_{x \to 0} \dfrac{1}{x^2} = \infty$    (p. 125)
- $\lim\limits_{x \to 0^+} \ln x = -\infty$    (p. 126)
- $\lim\limits_{x \to \infty} \dfrac{1}{x} = 0$    (p. 129)
- $\lim\limits_{x \to -\infty} \dfrac{1}{x} = 0$    (p. 129)
- $\lim\limits_{x \to -\infty} e^x = 0$    (p. 133)
- $\lim\limits_{x \to \infty} e^x = \infty$    (p. 133)
- $\lim\limits_{x \to \infty} \ln x = \infty$    (p. 133)

### Definitions

- Vertical asymptote    (p. 128)
- Horizontal asymptote    (p. 135)

### Properties of Limits at Infinity    (p. 129): If $k$ is a real number, $n \geq 2$ is an integer, and the functions $f$ and $g$ approach real numbers as $x \to \infty$, then:

- $\lim\limits_{x \to \infty} A = A$, where $A$ is a constant

- $\lim\limits_{x \to \infty}[kf(x)] = k \lim\limits_{x \to \infty} f(x)$

- $\lim\limits_{x \to \infty}[f(x) \pm g(x)] = \lim\limits_{x \to \infty} f(x) \pm \lim\limits_{x \to \infty} g(x)$

- $\lim\limits_{x \to \infty}[f(x)g(x)] = \left[\lim\limits_{x \to \infty} f(x)\right]\left[\lim\limits_{x \to \infty} g(x)\right]$

$\bullet\ \lim\limits_{x\to\infty}\dfrac{f(x)}{g(x)}=\dfrac{\lim\limits_{x\to\infty}f(x)}{\lim\limits_{x\to\infty}g(x)}$ provided $\lim\limits_{x\to\infty}g(x)\neq 0$

$\bullet\ \lim\limits_{x\to\infty}[f(x)]^{n}=\left[\lim\limits_{x\to\infty}f(x)\right]^{n}$

$\bullet\ \lim\limits_{x\to\infty}\sqrt[n]{f(x)}=\sqrt[n]{\lim\limits_{x\to\infty}f(x)}$, where $f(x)>0$ if $n$ is even

$\bullet\ \lim\limits_{x\to\pm\infty}\dfrac{k}{x^{p}}=0$, $p>0$, provided $x^{p}$ is defined when $x<0$

$\bullet\ \lim\limits_{x\to\infty}x^{n}=\infty$, $n>0$, an integer (p. 131)

$\bullet\ \lim\limits_{x\to-\infty}x^{n}=\infty$ if $n$ is a positive, even integer

$\quad \lim\limits_{x\to-\infty}x^{n}=-\infty$ if $n$ is a positive, odd integer (p. 131)

$\bullet$ For a polynomial function
$p(x)=a_{n}x^{n}+\cdots+a_{0}, a_{n}\neq 0$

$\quad \lim\limits_{x\to\pm\infty}p(x)=\lim\limits_{x\to\pm\infty}a_{n}x^{n}$ (p. 131)

$\bullet$ The limit at infinity of $R(x)=\dfrac{p(x)}{q(x)}$ (p. 132)

## 1.6 The $\varepsilon$-$\delta$ Definition of a Limit

**Definitions**

- Limit of a Function   (p. 141)
- Limit at Infinity   (p. 146)
- Infinite Limit   (p. 146)
- Infinite Limit at Infinity   (p. 147)

**Properties of limits**

- If $\lim\limits_{x\to c}f(x)>0$, then there is an open interval around $c$, for which $f(x)>0$ everywhere in the interval, except possibly at $c$.   (p. 145)

- If $\lim\limits_{x\to c}f(x)<0$, then there is an open interval around $c$, for which $f(x)<0$ everywhere in the interval, except possibly at $c$.   (p. 146)

## OBJECTIVES

| Section | You should be able to ... | Examples | Review Exercises |
|---|---|---|---|
| **1.1** | **1** Discuss the idea of a limit (p. 80) | 1 | 4 |
| | **2** Investigate a limit using a table (p. 80) | 2–4 | 1 |
| | **3** Investigate a limit using a graph (p. 82) | 5–8 | 2, 3 |
| **1.2** | **1** Find the limit of a sum, a difference, and a product (p. 91) | 1–6 | 8, 10, 12, 14, 22, 26, 29, 30, 47, 48 |
| | **2** Find the limit of a power and the limit of a root (p. 93) | 7–9 | 11, 18, 28, 55 |
| | **3** Find the limit of a polynomial (p. 95) | 10 | 10, 22 |
| | **4** Find the limit of a quotient (p. 96) | 11–14 | 13–17, 19–21, 23–25, 27, 56 |
| | **5** Find the limit of an average rate of change (p. 98) | 15 | 37 |
| | **6** Find the limit of a difference quotient (p. 98) | 16 | 5, 6, 49 |
| **1.3** | **1** Determine whether a function is continuous at a number (p. 102) | 1–4 | 31–36 |
| | **2** Determine intervals on which a function is continuous (p. 105) | 5, 6 | 39–42 |
| | **3** Use properties of continuity (p. 107) | 7, 8 | 39–42 |
| | **4** Use the Intermediate Value Theorem (p. 109) | 9, 10 | 38, 44–46 |
| **1.4** | **1** Use the Squeeze Theorem to find a limit (p. 115) | 1 | 7, 69 |
| | **2** Find limits involving trigonometric functions (p. 116) | 2, 3 | 9, 51–55 |
| | **3** Determine where the trigonometric functions are continuous (p. 120) | 4 | 63–65 |
| | **4** Determine where an exponential or a logarithmic function is continuous (p. 121) | 5 | 43 |
| **1.5** | **1** Find infinite limits (p. 125) | 1–3 | 57, 58 |
| | **2** Find the vertical asymptotes of a graph (p. 128) | 4 | 61, 62 |
| | **3** Find limits at infinity (p. 128) | 5–11 | 59, 60 |
| | **4** Find the horizontal asymptotes of a graph (p. 135) | 12 | 61, 62 |
| | **5** Find the asymptotes of the graph of a rational function (p. 135) | 13 | 67, 68 |
| **1.6** | **1** Use the $\varepsilon$-$\delta$ definition of a limit (p. 141) | 1–7 | 50, 66 |

## REVIEW EXERCISES

**1.** Use a table of numbers to investigate $\lim\limits_{x\to 0}\dfrac{1-\cos x}{1+\cos x}$.

*In Problems 2 and 3, use a graph to investigate $\lim\limits_{x\to c}f(x)$.*

**2.** $f(x)=\begin{cases} 2x-5 & \text{if } x<1 \\ 6-9x & \text{if } x\geq 1 \end{cases}$   at $c=1$

**3.** $f(x)=\begin{cases} x^{2}+2 & \text{if } x<2 \\ 2x+1 & \text{if } x\geq 2 \end{cases}$   at $c=2$

4. Write the statement "$f(x)$ can be made as close as we please to 5 by choosing $x$ sufficiently close to, but not equal to, 3" symbolically.

In Problems 5 and 6, for each function find the limit of the difference quotient $\lim\limits_{h\to 0}\dfrac{f(x+h)-f(x)}{h}$.

5. $f(x)=\dfrac{3}{x}$

6. $f(x)=3x^2+2x$

7. Find $\lim\limits_{x\to 0}f(x)$ if $1+\sin x\le f(x)\le |x|+1$

In Problems 8–22, find each limit.

8. $\lim\limits_{x\to 2}\left(2x-\dfrac{1}{x}\right)$

9. $\lim\limits_{x\to\pi}(x\cos x)$

10. $\lim\limits_{x\to -1}(x^3+3x^2-x-1)$

11. $\lim\limits_{x\to 0}\sqrt[3]{x(x+2)^3}$

12. $\lim\limits_{x\to 0}[(2x+3)(x^5+5x)]$

13. $\lim\limits_{x\to 3}\dfrac{x^3-27}{x-3}$

14. $\lim\limits_{x\to 3}\left(\dfrac{x^2}{x-3}-\dfrac{3x}{x-3}\right)$

15. $\lim\limits_{x\to 2}\dfrac{x^2-4}{x-2}$

16. $\lim\limits_{x\to -1}\dfrac{x^2+3x+2}{x^2+4x+3}$

17. $\lim\limits_{x\to -2}\dfrac{x^3+5x^2+6x}{x^2+x-2}$

18. $\lim\limits_{x\to 1}\left(x^2-3x+\dfrac{1}{x}\right)^{15}$

19. $\lim\limits_{x\to 2}\dfrac{3-\sqrt{x^2+5}}{x^2-4}$

20. $\lim\limits_{x\to 0}\left\{\dfrac{1}{x}\left[\dfrac{1}{(2+x)^2}-\dfrac{1}{4}\right]\right\}$

21. $\lim\limits_{x\to 0}\dfrac{(x+3)^2-9}{x}$

22. $\lim\limits_{x\to 1}[(x^3-3x^2+3x-1)(x+1)^2]$

In Problems 23–28, find each one-sided limit, if it exists.

23. $\lim\limits_{x\to -2^+}\dfrac{x^2+5x+6}{x+2}$

24. $\lim\limits_{x\to 5^+}\dfrac{|x-5|}{x-5}$

25. $\lim\limits_{x\to 1^-}\dfrac{|x-1|}{x-1}$

26. $\lim\limits_{x\to 3/2^+}\lfloor 2x\rfloor$

27. $\lim\limits_{x\to 4^-}\dfrac{x^2-16}{x-4}$

28. $\lim\limits_{x\to 1^+}\sqrt{x-1}$

In Problems 29 and 30, find $\lim\limits_{x\to c^-}f(x)$ and $\lim\limits_{x\to c^+}f(x)$ for the given c. Determine whether $\lim\limits_{x\to c}f(x)$ exists.

29. $f(x)=\begin{cases}2x+3 & \text{if }x<2\\9-x & \text{if }x\ge 2\end{cases}$ at $c=2$

30. $f(x)=\begin{cases}3x+1 & \text{if }x<3\\10 & \text{if }x=3\\4x-2 & \text{if }x>3\end{cases}$ at $c=3$

In Problems 31–36, determine whether f is continuous at c.

31. $f(x)=\begin{cases}5x-2 & \text{if }x<1\\5 & \text{if }x=1\\2x+1 & \text{if }x>1\end{cases}$ at $c=1$

32. $f(x)=\begin{cases}x^2 & \text{if }x<-1\\2 & \text{if }x=-1\\-3x-2 & \text{if }x>-1\end{cases}$ at $c=-1$

33. $f(x)=\begin{cases}4-3x^2 & \text{if }x<0\\4 & \text{if }x=0\\\sqrt{16-x^2} & \text{if }0<x\le 4\end{cases}$ at $c=0$

34. $f(x)=\begin{cases}\sqrt{4+x} & \text{if }-4\le x\le 4\\\sqrt{\dfrac{x^2-16}{x-4}} & \text{if }x>4\end{cases}$ at $c=4$

35. $f(x)=\lfloor 2x\rfloor$ at $c=\dfrac{1}{2}$

36. $f(x)=|x-5|$ at $c=5$

37. (a) Find the average rate of change of $f(x)=2x^2-5x$ from 1 to x.

(b) Find the limit as x approaches 1 of the average rate of change found in (a).

38. A function f is defined on the interval $[-1, 1]$ with the following properties: f is continuous on $[-1, 1]$ except at 0, negative at $-1$, positive at 1, but with no zeros. Does this contradict the Intermediate Value Theorem?

In Problems 39–43 find all numbers x for which f is continuous.

39. $f(x)=\dfrac{x}{x^3-27}$

40. $f(x)=\dfrac{x^2-3}{x^2+5x+6}$

41. $f(x)=\dfrac{2x+1}{x^3+4x^2+4x}$

42. $f(x)=\sqrt{x-1}$

43. $f(x)=2^{-x}$

44. Use the Intermediate Value Theorem to determine whether $2x^3+3x^2-23x-42=0$ has a zero in the interval $[3, 4]$.

In Problems 45 and 46, use the Intermediate Value Theorem to approximate the zero correct to three decimal places.

45. $f(x)=8x^4-2x^2+5x-1$ on the interval $[0, 1]$.

46. $f(x)=3x^3-10x+9$; zero between $-3$ and $-2$.

47. Find $\lim\limits_{x\to 0^+}\dfrac{|x|}{x}(1-x)$ and $\lim\limits_{x\to 0^-}\dfrac{|x|}{x}(1-x)$.

What can you say about $\lim\limits_{x\to 0}\dfrac{|x|}{x}(1-x)$?

48. Find $\lim\limits_{x\to 2}\left(\dfrac{x^2}{x-2}-\dfrac{2x}{x-2}\right)$. Then comment on the statement that this limit is given by $\lim\limits_{x\to 2}\dfrac{x^2}{x-2}-\lim\limits_{x\to 2}\dfrac{2x}{x-2}$.

49. Find $\lim\limits_{h\to 0}\dfrac{f(x+h)-f(x)}{h}$ for $f(x)=\sqrt{x}$.

50. For $\lim\limits_{x\to 3}(2x+1)=7$, find the largest possible $\delta$ that "works" for $\varepsilon=0.01$.

In Problems 51–60, find each limit.

51. $\lim\limits_{x\to 0}\cos(\tan x)$

52. $\lim\limits_{x\to 0}\dfrac{\sin\dfrac{x}{4}}{x}$

53. $\lim\limits_{x\to 0}\dfrac{\tan(3x)}{\tan(4x)}$

54. $\lim\limits_{x\to 0}\dfrac{\cos\dfrac{x}{3}-1}{x}$

**55.** $\lim\limits_{x \to 0} \left( \dfrac{\cos x - 1}{x} \right)^{10}$

**56.** $\lim\limits_{x \to 0} \dfrac{e^{4x} - 1}{e^x - 1}$

**57.** $\lim\limits_{x \to \pi/2^+} \tan x$

**58.** $\lim\limits_{x \to -3} \dfrac{2+x}{(x+3)^2}$

**59.** $\lim\limits_{x \to \infty} \dfrac{3x^3 - 2x + 1}{x^3 - 8}$

**60.** $\lim\limits_{x \to \infty} \dfrac{3x^4 + x}{2x^2}$

*In Problems 61 and 62, find any vertical and horizontal asymptotes of $f$.*

**61.** $f(x) = \dfrac{4x - 2}{x + 3}$

**62.** $f(x) = \dfrac{2x}{x^2 - 4}$

**63.** Let $f(x) = \begin{cases} \dfrac{\tan x}{2x} & \text{if } x \neq 0 \\ \dfrac{1}{2} & \text{if } x = 0 \end{cases}$. Is $f$ continuous at 0?

**64.** Let $f(x) = \begin{cases} \dfrac{\sin(3x)}{x} & \text{if } x \neq 0 \\ 1 & \text{if } x = 0 \end{cases}$. Is $f$ continuous at 0?

**65.** The function $f(x) = \dfrac{\cos\left(\pi x + \dfrac{\pi}{2}\right)}{x}$ is not defined at 0. Decide how to define $f(0)$ so that $f$ is continuous at 0.

**66.** Use the $\varepsilon$-$\delta$ definition of a limit to prove $\lim\limits_{x \to -3} (x^2 - 9) \neq 18$.

**67.** **(a)** Sketch a graph of a function $f$ that has the following properties:
$$f(-1) = 0 \quad \lim\limits_{x \to \infty} f(x) = 2 \quad \lim\limits_{x \to -\infty} f(x) = 2$$
$$\lim\limits_{x \to 4^-} f(x) = -\infty \quad \lim\limits_{x \to 4^+} f(x) = \infty$$

**(b)** Define a function that describes your graph.

**68.** **(a)** Find the domain and the intercepts (if any) of
$$R(x) = \dfrac{2x^2 - 5x + 2}{5x^2 - x - 2}.$$

**(b)** Discuss the behavior of the graph of $R$ at numbers where $R$ is not defined.

**(c)** Find any vertical or horizontal asymptotes of the function $R$.

**69.** If $1 - x^2 \leq f(x) \leq \cos x$ for all $x$ in the interval $-\dfrac{\pi}{2} < x < \dfrac{\pi}{2}$, show that $\lim\limits_{x \to 0} f(x) = 1$.

---

## CHAPTER 1 PROJECT    Pollution in Clear Lake

The Toxic Waste Disposal Company (TWDC) specializes in the disposal of a particularly dangerous pollutant, Agent Yellow (AY). Unfortunately, instead of safely disposing of this pollutant, the company simply dumped AY in (formerly) Clear Lake. Fortunately, they have been caught and are now defending themselves in court.

The facts below are not in dispute. As a result of TWDC's activity, the current concentration of AY in Clear Lake is now 10 ppm (parts per million). Clear Lake is part of a chain of rivers and lakes. Fresh water flows into Clear Lake and the contaminated water flows downstream from it. The Department of Environmental Protection estimates that the level of contamination in Clear Lake will fall by 20% each year. These facts can be modeled as

$$p(0) = 10 \qquad p(t+1) = 0.80p(t)$$

where $p = p(t)$, measured in ppm, is the concentration of pollutants in the lake at time $t$, in years.

**1.** Explain how the above equations model the facts.

**2.** Create a table showing the values of $t$ for $t = 0, 1, 2, \ldots, 20$.

**3.** Show that $p(t) = 10(0.8)^t$.

**4.** Use technology to graph $p = p(t)$.

**5.** What is $\lim\limits_{t \to \infty} p(t)$?

Lawyers for TWDC looked at the results in 1–5 above and argued that their client has not done any real damage. They concluded that Clear Lake would eventually return to its former clear and unpolluted state. They even called in a mathematician, who wrote the following on a blackboard:

$$\lim\limits_{t \to \infty} p(t) = 0$$

and explained that this bit of mathematics means, descriptively, that after many years the concentration of AY will, indeed, be close to zero.

Concerned citizens booed the mathematician's testimony. Fortunately, one of them has taken calculus and knows a little bit about limits. She noted that, although "after many years the concentration of AY will approach zero," the townspeople like to swim in Clear Lake and state regulations prohibit swimming unless the concentration of AY is below 2 ppm. She proposed a fine of $100,000 per year for each full year that the lake is unsafe for swimming. She also questioned the mathematician, saying, "Your testimony was correct as far as it went, but I remember from studying calculus that talking about the eventual concentration of AY after many, many years is only a small part of the story. The more precise meaning of your statement $\lim\limits_{t \to \infty} p(t) = 0$ is that given some tolerance $T$ for the concentration of AY, there is some time $N$ (which may be very far in the future) so that for all $t > N$, $p(t) < T$."

**6.** Using a table or a graph for $p = p(t)$, find $N$ so that if $t > N$, then $p(t) < 2$.

**7.** How much is the fine?

Her words were greeted by applause. The town manager sprang to his feet and noted that although a tolerance of 2 ppm was fine for swimming, the town used Clear Lake for its drinking water and until the concentration of AY dropped below 0.5 ppm, the water would be unsafe for drinking. He proposed a fine of $200,000 per year for each full year the water was unfit for drinking.

**8.** Using a table or a graph for $p = p(t)$, find $N$ so that if $t > N$, then $p(t) < 0.5$.

**9.** How much is the fine?

**10.** How would you find if you were on the jury trying TWDC? If the jury found TWDC guilty, what fine would you recommend? Explain your answers.

# 2 The Derivative

**CHAPTER 2 PROJECT** In the Chapter Project on page 206 at the end of this chapter, we explore some of the physics at work that allowed engineers and pilots to successfully maneuver the Lunar Module to the Moon's surface.

Rolls Press/Popperfoto/Getty Images

## The Apollo Lunar Module
### *"One Giant Leap for Mankind"*

On May 25, 1961, in a special address to Congress, U.S. President John F. Kennedy proposed the goal "before this decade is out, of landing a man on the Moon and returning him safely to the Earth." Roughly eight years later, on July 16, 1969, a Saturn V rocket launched from the Kennedy Space Center in Florida, carrying the *Apollo 11* spacecraft and three astronauts—Neil Armstrong, Buzz Aldrin, and Michael Collins—bound for the Moon.

The *Apollo* spacecraft had three parts: the Command Module with a cabin for the three astronauts; the Service Module that supported the Command Module with propulsion, electrical power, oxygen, and water; and the Lunar Module for landing on the Moon. After its launch, the spacecraft traveled for three days until it entered into lunar orbit. Armstrong and Aldrin then moved into the Lunar Module, which they landed in the flat expanse of the Sea of Tranquility. After more than 21 hours, the first humans to touch the surface of the Moon crawled into the Lunar Module and lifted off to rejoin the Command Module, which Collins had been piloting in lunar orbit. The three astronauts then headed back to Earth, where they splashed down in the Pacific Ocean on July 24.

Chapter 2 opens by returning to the tangent problem to find an equation of the tangent line to the graph of a function $f$ at a point $P = (c, f(c))$. Remember in Section 1.1 we found that the slope of a tangent line was a limit,

$$m_{\tan} = \lim_{x \to c} \frac{f(x) - f(c)}{x - c}$$

This limit turns out to be one of the most significant ideas in calculus, the *derivative*.

In this chapter, we introduce interpretations of the derivative, treat the derivative as a function, and consider some properties of the derivative. By the end of the chapter, you will have a collection of basic derivative formulas and derivative rules that will be used throughout your study of calculus.

# 2.1 Rates of Change and the Derivative

**OBJECTIVES** *When you finish this section, you should be able to:*

1 Find equations for the tangent line and the normal line to the graph of a function (p. 155)
2 Find the rate of change of a function (p. 156)
3 Find average velocity and instantaneous velocity (p. 157)
4 Find the derivative of a function at a number (p. 159)

In Chapter 1 we discussed the tangent problem: *Given a function f and a point P on its graph, what is the slope of the tangent line to the graph of f at P?* See Figure 1, where $\ell_T$ is the tangent line to the graph of $f$ at the point $P = (c, f(c))$.

The tangent line $\ell_T$ to the graph of $f$ at $P$ must contain the point $P$. Since finding the slope requires two points, and we have only one point on the tangent line $\ell_T$, we reason as follows.

Suppose we choose any point $Q = (x, f(x))$, other than $P$, on the graph of $f$. ($Q$ can be to the left or to the right of $P$; we chose $Q$ to be to the right of $P$.) The line containing the points $P = (c, f(c))$ and $Q = (x, f(x))$ is a secant line of the graph of $f$. The slope $m_{\sec}$ of this secant line is

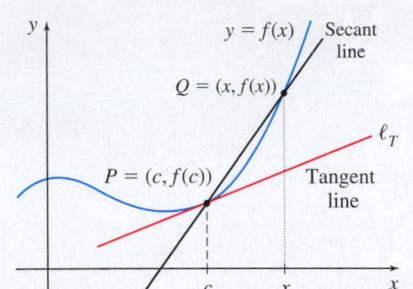

**Figure 1** $m_{\sec}$ = slope of the secant line.

$$m_{\sec} = \frac{f(x) - f(c)}{x - c} \tag{1}$$

Figure 2 shows three different points $Q_1$, $Q_2$, and $Q_3$ on the graph of $f$ that are successively closer to the point $P$, and three associated secant lines $\ell_1$, $\ell_2$, and $\ell_3$. The closer the points $Q$ are to the point $P$, the closer the secant lines are to the tangent line $\ell_T$. The line $\ell_T$, the *limiting position* of these secant lines, is the *tangent line to the graph of f at P*.

If the limiting position of the secant lines is the tangent line, then the limit of the slopes of the secant lines should equal the slope of the tangent line. Notice in Figure 2 that as the points $Q_1$, $Q_2$, and $Q_3$ move closer to the point $P$, the numbers $x$ get closer to $c$. So, equation (1) suggests that

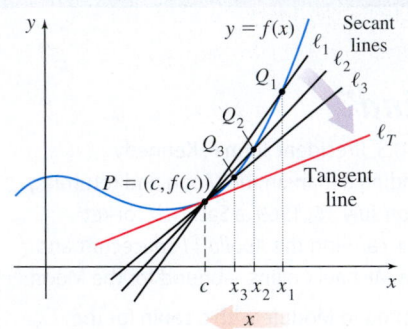

**DF Figure 2** $m_{\tan} = \lim\limits_{x \to c} \frac{f(x) - f(c)}{x - c}$

$$m_{\tan} = \text{Slope of the tangent line to } f \text{ at } P$$
$$= \text{Limit of } \frac{f(x) - f(c)}{x - c} \text{ as } x \text{ gets closer to } c$$
$$= \lim_{x \to c} \frac{f(x) - f(c)}{x - c}$$

provided the limit exists.

**DEFINITION  Tangent Line**

The **tangent line** to the graph of $f$ at a point $P$ is the line containing the point $P = (c, f(c))$ and having the slope

$$m_{\tan} = \lim_{x \to c} \frac{f(x) - f(c)}{x - c} \tag{2}$$

provided the limit exists.

**NOTE** It is possible for the limit in (2) not to exist. The geometric significance of this is discussed in the next section.

The limit in equation (2) that defines the slope of the tangent line occurs so frequently that it is given a special notation $f'(c)$, read, "$f$ prime of $c$," and called **prime notation**:

$$f'(c) = \lim_{x \to c} \frac{f(x) - f(c)}{x - c} \tag{3}$$

## 1 Find Equations for the Tangent Line and the Normal Line to the Graph of a Function

**THEOREM** Equation of a Tangent Line

If $m_{\tan} = f'(c)$ exists, then an equation of the tangent line to the graph of a function $y = f(x)$ at the point $P = (c, f(c))$ is

$$\boxed{y - f(c) = f'(c)(x - c)}$$

**RECALL** Two lines, neither of which are horizontal, with slopes $m_1$ and $m_2$, respectively, are perpendicular if and only if

$$m_1 = -\frac{1}{m_2}$$

The line perpendicular to the tangent line at a point $P$ on the graph of a function $f$ is called the **normal line** to the graph of $f$ at $P$.

An equation of the normal line to the graph of a function $y = f(x)$ at the point $P = (c, f(c))$ is

$$\boxed{y - f(c) = -\frac{1}{f'(c)}(x - c)}$$

provided $f'(c)$ exists and is not equal to zero. If $f'(c) = 0$, the tangent line is horizontal, the normal line is vertical, and the equation of the normal line is $x = c$.

**CALC CLIP**

**EXAMPLE 1** Finding Equations for the Tangent Line and the Normal Line

(a) Find the slope of the tangent line to the graph of $f(x) = x^2$ at $c = -2$.

(b) Use the result from (a) to find an equation of the tangent line when $c = -2$.

(c) Find an equation of the normal line to the graph of $f$ at the point $(-2, 4)$.

(d) Graph $f$, the tangent line to $f$ at $(-2, 4)$, and the normal line to $f$ at $(-2, 4)$ on the same set of axes.

**RECALL** One way to find the limit of a quotient when the limit of the denominator is 0 is to factor the numerator and divide out common factors.

**Solution (a)** At $c = -2$, the slope of the tangent line is

$$f'(-2) = \lim_{x \to -2} \frac{f(x) - f(-2)}{x - (-2)} = \lim_{x \to -2} \frac{x^2 - (-2)^2}{x + 2} = \lim_{x \to -2} \frac{x^2 - 4}{x + 2}$$

$$= \lim_{x \to -2} (x - 2) = -4$$

**NEED TO REVIEW?** The point-slope form of a line is discussed in Appendix A.3, p. A-19.

**(b)** We use the result from (a) and the point-slope form of an equation of a line to obtain an equation of the tangent line. An equation of the tangent line containing the point $(-2, f(-2)) = (-2, 4)$ is

$$y - f(-2) = f'(-2)[x - (-2)] \qquad \text{Point-slope form of an equation of the tangent line.}$$

$$y - 4 = -4 \cdot (x + 2) \qquad f(-2) = 4; \quad f'(-2) = -4$$

$$y = -4x - 4 \qquad \text{Simplify.}$$

**(c)** Since the slope of the tangent line to $f$ at $(-2, 4)$ is $-4$, the slope of the normal line to $f$ at $(-2, 4)$ is $\dfrac{1}{4}$.

Using the point-slope form of an equation of a line, an equation of the normal line is

$$y - 4 = \frac{1}{4}(x + 2)$$

$$y = \frac{1}{4}x + \frac{9}{2}$$

**(d)** The graphs of $f$, the tangent line to the graph of $f$ at the point $(-2, 4)$, and the normal line to the graph of $f$ at $(-2, 4)$ are shown in Figure 3. ∎

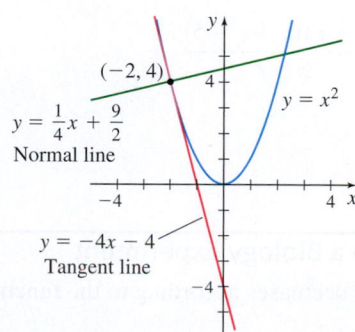

**Figure 3** $f(x) = x^2$

**NOW WORK** Problem 11.

## 2 Find the Rate of Change of a Function

Everything in nature changes. Examples include climate change, change in the phases of the Moon, and change in populations. To describe natural processes mathematically, the ideas of change and rate of change are often used.

Recall that the average rate of change of a function $y = f(x)$ over the interval from $c$ to $x$ is given by

$$\text{Average rate of change} = \frac{f(x) - f(c)}{x - c} \qquad x \neq c$$

### DEFINITION  Instantaneous Rate of Change

The **instantaneous rate of change** of $f$ at $c$ is the limit as $x$ approaches $c$ of the average rate of change. Symbolically, the instantaneous rate of change of $f$ at $c$ is

$$\lim_{x \to c} \frac{f(x) - f(c)}{x - c}$$

provided the limit exists.

The expression "instantaneous rate of change" is often shortened to *rate of change*.

Using prime notation, the **rate of change** of $f$ at $c$ is $f'(c) = \lim\limits_{x \to c} \dfrac{f(x) - f(c)}{x - c}$.

### EXAMPLE 2  Finding a Rate of Change

Find the rate of change of the function $f(x) = x^2 - 5x$ at:

(a) $c = 2$

(b) Any real number $c$

**Solution** (a) For $c = 2$,

$$f(x) = x^2 - 5x \qquad \text{and} \qquad f(2) = 2^2 - 5 \cdot 2 = -6$$

The rate of change of $f$ at $c = 2$ is

$$f'(2) = \lim_{x \to 2} \frac{f(x) - f(2)}{x - 2} = \lim_{x \to 2} \frac{(x^2 - 5x) - (-6)}{x - 2} = \lim_{x \to 2} \frac{x^2 - 5x + 6}{x - 2}$$

$$= \lim_{x \to 2} \frac{(x - 2)(x - 3)}{x - 2} = \lim_{x \to 2} (x - 3) = -1$$

(b) If $c$ is any real number, then $f(c) = c^2 - 5c$, and the rate of change of $f$ at $c$ is

$$f'(c) = \lim_{x \to c} \frac{f(x) - f(c)}{x - c} = \lim_{x \to c} \frac{(x^2 - 5x) - (c^2 - 5c)}{x - c} = \lim_{x \to c} \frac{(x^2 - c^2) - 5(x - c)}{x - c}$$

$$= \lim_{x \to c} \frac{(x - c)(x + c) - 5(x - c)}{x - c} = \lim_{x \to c} \frac{(x - c)(x + c - 5)}{x - c}$$

$$= \lim_{x \to c} (x + c - 5) = 2c - 5 \qquad \blacksquare$$

NOW WORK Problem 17.

### EXAMPLE 3  Finding the Rate of Change in a Biology Experiment

In a metabolic experiment, the mass $M$ of glucose decreases according to the function

$$M(t) = 4.5 - 0.03t^2$$

where $M$ is measured in grams (g) and $t$ is the time in hours (h). Find the reaction rate $M'(t)$ at $t = 1$ h.

**Solution** The reaction rate at $t = 1$ is $M'(1)$.

$$M'(1) = \lim_{t \to 1} \frac{M(t) - M(1)}{t - 1} = \lim_{t \to 1} \frac{(4.5 - 0.03t^2) - (4.5 - 0.03)}{t - 1}$$

$$= \lim_{t \to 1} \frac{-0.03t^2 + 0.03}{t - 1} = \lim_{t \to 1} \frac{(-0.03)(t^2 - 1)}{t - 1} = \lim_{t \to 1} \frac{(-0.03)(t - 1)(t + 1)}{t - 1}$$

$$= -0.03 \cdot 2 = -0.06$$

The reaction rate at $t = 1$ h is $-0.06$ g/h. That is, the mass $M$ of glucose at $t = 1$ h is decreasing at the rate of 0.06 g/h. ∎

NOW WORK Problem **43**.

## 3 Find Average Velocity and Instantaneous Velocity

*Average velocity* is a physical example of an average rate of change. For example, consider an object moving along a horizontal line with the positive direction to the right, or moving along a vertical line with the positive direction upward. Motion along a line is referred to as **rectilinear motion**. The object's location at time $t = 0$ is called its **initial position**. The initial position is usually marked as the origin $O$ on the line. See Figure 4. We assume the position $s$ at time $t$ of the object from the origin is given by a function $s = f(t)$. Here $s$ is the signed, or directed, distance (using some measure of distance such as centimeters, meters, feet, etc.) of the object from $O$ at time $t$ (in seconds or hours). The function $f$ is usually called the **position function** of the object.

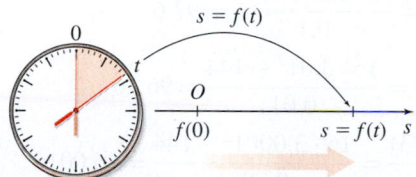

**Figure 4** $t$ is the travel time. $s$ is the signed distance of the object from the origin at time $t$.

### DEFINITION Average Velocity

The signed distance $s$ from the origin at time $t$ of an object in rectilinear motion is given by the position function $s = f(t)$. If at time $t_0$ the object is at $s_0 = f(t_0)$ and at time $t_1$ the object is at $s_1 = f(t_1)$, then the change in time is $\Delta t = t_1 - t_0$ and the change in position is $\Delta s = s_1 - s_0 = f(t_1) - f(t_0)$. The average rate of change of position with respect to time is

$$\boxed{\frac{\Delta s}{\Delta t} = \frac{f(t_1) - f(t_0)}{t_1 - t_0} \qquad t_1 \neq t_0}$$

and is called the **average velocity** of the object over the interval $[t_0, t_1]$. See Figure 5.

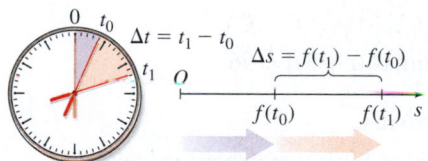

**Figure 5** The average velocity is $\dfrac{\Delta s}{\Delta t}$.

### EXAMPLE 4  Finding Average Velocity

The Mike O'Callaghan–Pat Tillman Memorial Bridge spanning the Colorado River opened on October 16, 2010. Having a span of 1900 ft, it is the longest arch bridge in the Western Hemisphere, and its roadway is 890 ft above the Colorado River.

If a rock falls from the roadway, the function $s = f(t) = 16t^2$ gives the distance $s$, in feet, that the rock falls after $t$ seconds for $0 \le t \le 7.458$. Here 7.458 s is the approximate time it takes the rock to fall 890 ft into the river. The average velocity of the rock during its fall is

$$\frac{\Delta s}{\Delta t} = \frac{f(7.458) - f(0)}{7.458 - 0} = \frac{890 - 0}{7.458} \approx 119.335 \text{ ft/s}$$ ∎

Scott Prokop/Shutterstock

**NOTE** Here the motion occurs along a vertical line with the positive direction downward.

The average velocity of the rock in Example 4 approximates the average velocity over the interval [0, 7.458]. But the average velocity does not tell us about the velocity at any particular instant of time. That is, it gives no information about the rock's *instantaneous velocity*.

We can investigate the instantaneous velocity of the rock, say, at $t = 3$ s, by computing average velocities for short intervals of time beginning at $t = 3$. First we compute the average velocity for the interval beginning at $t = 3$ and ending at $t = 3.5$. The corresponding distances the rock has fallen are

$$f(3) = 16 \cdot 3^2 = 144 \text{ ft} \qquad \text{and} \qquad f(3.5) = 16 \cdot 3.5^2 = 196 \text{ ft}$$

Then $\Delta t = 3.5 - 3.0 = 0.5$, and during this 0.5-s interval,

$$\text{Average velocity} = \frac{\Delta s}{\Delta t} = \frac{f(3.5) - f(3)}{3.5 - 3} = \frac{196 - 144}{0.5} = 104 \text{ ft/s}$$

Table 1 shows average velocities of the rock for smaller intervals of time.

**TABLE 1**

| Time interval | Start $t_0 = 3$ | End $t$ | $\Delta t$ | $\dfrac{\Delta s}{\Delta t} = \dfrac{f(t) - f(t_0)}{t - t_0} = \dfrac{16t^2 - 144}{t - 3}$ |
|---|---|---|---|---|
| [3, 3.1] | 3 | 3.1 | 0.1 | $\dfrac{\Delta s}{\Delta t} = \dfrac{f(3.1) - f(3)}{3.1 - 3} = \dfrac{16 \cdot 3.1^2 - 144}{0.1} = 97.6$ |
| [3, 3.01] | 3 | 3.01 | 0.01 | $\dfrac{\Delta s}{\Delta t} = \dfrac{f(3.01) - f(3)}{3.01 - 3} = \dfrac{16 \cdot 3.01^2 - 144}{0.01} = 96.16$ |
| [3, 3.0001] | 3 | 3.0001 | 0.0001 | $\dfrac{\Delta s}{\Delta t} = \dfrac{f(3.0001) - f(3)}{3.0001 - 3} = \dfrac{16 \cdot 3.0001^2 - 144}{0.0001} = 96.0016$ |

The average velocity of 96.0016 over the time interval $\Delta t = 0.0001$ s should be very close to the instantaneous velocity of the rock at $t = 3$ s. As $\Delta t$ gets closer to 0, the average velocity gets closer to the instantaneous velocity. So, to obtain the instantaneous velocity at $t = 3$ precisely, we use the limit of the average velocity as $\Delta t$ approaches 0 or, equivalently, as $t$ approaches 3.

$$\lim_{\Delta t \to 0} \frac{\Delta s}{\Delta t} = \lim_{t \to 3} \frac{f(t) - f(3)}{t - 3} = \lim_{t \to 3} \frac{16t^2 - 16 \cdot 3^2}{t - 3} = \lim_{t \to 3} \frac{16(t^2 - 9)}{t - 3}$$

$$= \lim_{t \to 3} \frac{16(t - 3)(t + 3)}{t - 3} = \lim_{t \to 3}[16(t + 3)] = 96$$

The rock's instantaneous velocity at $t = 3$ s is 96 ft/s.

We generalize this result to obtain a definition for instantaneous velocity.

**DEFINITION  Instantaneous Velocity**

If $s = f(t)$ is the position function of an object at time $t$, the **instantaneous velocity** $v$ of the object at time $t_0$ is defined as the limit of the average velocity $\dfrac{\Delta s}{\Delta t}$ as $\Delta t$ approaches 0. That is,

$$v = \lim_{\Delta t \to 0} \frac{\Delta s}{\Delta t} = \lim_{t \to t_0} \frac{f(t) - f(t_0)}{t - t_0}$$

provided the limit exists.

We usually shorten "instantaneous velocity" and just use the word "velocity."

**NOW WORK** Problem 31.

**EXAMPLE 5  Finding Velocity**

Find the velocity $v$ of the falling rock from Example 4 at:

**(a)** $t_0 = 1$ s after it begins to fall

**(b)** $t_0 = 7.4$ s, just before it hits the Colorado River

**(c)** At any time $t_0$.

**Solution** **(a)** Use the definition of instantaneous velocity with $f(t) = 16t^2$ and $t_0 = 1$.

$$v = \lim_{\Delta t \to 0} \frac{\Delta s}{\Delta t} = \lim_{t \to 1} \frac{f(t) - f(1)}{t - 1} = \lim_{t \to 1} \frac{16t^2 - 16}{t - 1} = \lim_{t \to 1} \frac{16(t^2 - 1)}{t - 1}$$

$$= \lim_{t \to 1} \frac{16(t - 1)(t + 1)}{t - 1} = \lim_{t \to 1} [16(t + 1)] = 32$$

At 1 s, the velocity of the rock is 32 ft/s.

**(b)** For $t_0 = 7.4$ s,

$$v = \lim_{\Delta t \to 0} \frac{\Delta s}{\Delta t} = \lim_{t \to 7.4} \frac{f(t) - f(7.4)}{t - 7.4} = \lim_{t \to 7.4} \frac{16t^2 - 16 \cdot (7.4)^2}{t - 7.4}$$

$$= \lim_{t \to 7.4} \frac{16[t^2 - (7.4)^2]}{t - 7.4} = \lim_{t \to 7.4} \frac{16(t - 7.4)(t + 7.4)}{t - 7.4}$$

$$= \lim_{t \to 7.4} [16(t + 7.4)] = 16(14.8) = 236.8$$

**NOTE** Did you know? 236.8 ft/s is more than 161 mi/h!

At 7.4 s, the velocity of the rock is 236.8 ft/s.

**(c)**

$$v = \lim_{t \to t_0} \frac{f(t) - f(t_0)}{t - t_0} = \lim_{t \to t_0} \frac{16t^2 - 16t_0^2}{t - t_0} = \lim_{t \to t_0} \frac{16(t - t_0)(t + t_0)}{t - t_0}$$

$$= 16 \lim_{t \to t_0} (t + t_0) = 32t_0$$

At $t_0$ seconds, the velocity of the rock is $32t_0$ ft/s. ∎

**NOW WORK** Problem 33.

## 4 Find the Derivative of a Function at a Number

Slope of a tangent line, rate of change of a function, and velocity are all found using the same limit,

$$f'(c) = \lim_{x \to c} \frac{f(x) - f(c)}{x - c}$$

The common underlying idea is the mathematical concept of *derivative*.

---

**DEFINITION  Derivative of a Function at a Number**

If $y = f(x)$ is a function and $c$ is in the domain of $f$, then the **derivative** of $f$ at $c$, denoted by $f'(c)$, is the number

$$\boxed{f'(c) = \lim_{x \to c} \frac{f(x) - f(c)}{x - c}}$$

provided this limit exists.

---

**EXAMPLE 6  Finding the Derivative of a Function at a Number**

Find the derivative of $f(x) = 2x^2 - 3x - 2$ at $x = 2$. That is, find $f'(2)$.

**Solution** Using the definition of the derivative, we have

$$f'(2) = \lim_{x \to 2} \frac{f(x) - f(2)}{x - 2} = \lim_{x \to 2} \frac{(2x^2 - 3x - 2) - 0}{x - 2} \qquad f(2) = 2 \cdot 4 - 3 \cdot 2 - 2 = 0$$

$$= \lim_{x \to 2} \frac{(x - 2)(2x + 1)}{x - 2}$$

$$= \lim_{x \to 2} (2x + 1) = 5$$   ∎

**NOW WORK** Problem 23.

So far we have given three interpretations of the derivative:

- *Geometric interpretation:* If $y = f(x)$, the derivative $f'(c)$ is the slope of the tangent line to the graph of $f$ at the point $(c, f(c))$.
- *Rate of change of a function interpretation:* If $y = f(x)$, the derivative $f'(c)$ is the rate of change of $f$ at $c$.
- *Physical interpretation:* If the signed distance $s$ from the origin at time $t$ of an object in rectilinear motion is given by the position function $s = f(t)$, the derivative $f'(t_0)$ is the velocity of the object at time $t_0$.

### EXAMPLE 7    Finding an Equation of a Tangent Line

(a)  Find the derivative of $f(x) = \sqrt{2x}$ at $x = 8$.

(b)  Use the derivative $f'(8)$ to find an equation of the tangent line to the graph of $f$ at the point $(8, 4)$.

**Solution** (a)  The derivative of $f$ at 8 is

$$f'(8) = \lim_{x \to 8} \frac{f(x) - f(8)}{x - 8} = \lim_{\substack{\uparrow \\ x \to 8}} \frac{\sqrt{2x} - 4}{x - 8} = \lim_{\substack{\uparrow \\ x \to 8}} \frac{\left(\sqrt{2x} - 4\right)\left(\sqrt{2x} + 4\right)}{(x - 8)\left(\sqrt{2x} + 4\right)}$$

$$f(8) = \sqrt{2 \cdot 8} = 4 \qquad \text{Rationalize the numerator.}$$

$$= \lim_{x \to 8} \frac{2x - 16}{(x - 8)\left(\sqrt{2x} + 4\right)} = \lim_{x \to 8} \frac{2(x - 8)}{(x - 8)\left(\sqrt{2x} + 4\right)} = \lim_{x \to 8} \frac{2}{\sqrt{2x} + 4} = \frac{1}{4}$$

(b)  The slope of the tangent line to the graph of $f$ at the point $(8, 4)$ is $f'(8) = \dfrac{1}{4}$. Using the point-slope form of a line, we get

$$y - 4 = f'(8)(x - 8) \qquad y - y_1 = m(x - x_1)$$

$$y - 4 = \frac{1}{4}(x - 8) \qquad f'(8) = \frac{1}{4}$$

$$y = \frac{1}{4}x + 2$$

The graphs of $f$ and the tangent line to the graph of $f$ at $(8, 4)$ are shown in Figure 6.

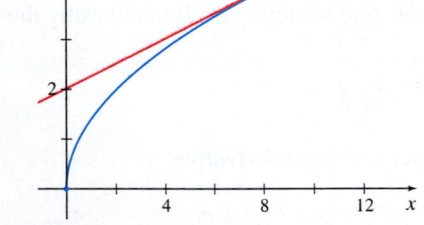

**Figure 6**

NOW WORK Problem 15.

### EXAMPLE 8    Approximating the Derivative of a Function Defined by a Table

The table below lists several values of a function $y = f(x)$ that is continuous on the interval $[-1, 5]$ and has a derivative at each number in the interval $(-1, 5)$. Approximate the derivative of $f$ at 2.

| $x$ | 0 | 1 | 2 | 3 | 4 |
|-----|---|---|---|----|----|
| $f(x)$ | 0 | 3 | 12 | 33 | 72 |

**Solution**  There are several ways to approximate the derivative of a function defined by a table. Each uses an average rate of change to approximate the rate of change of $f$ at 2, which is the derivative of $f$ at 2.

- Using the average rate of change from 2 to 3, we have

$$\frac{f(3) - f(2)}{3 - 2} = \frac{33 - 12}{1} = 21$$

With this choice, $f'(2)$ is approximately 21.

- Using the average rate of change from 1 to 2, we have

$$\frac{f(2) - f(1)}{2 - 1} = \frac{12 - 3}{1} = 9$$

With this choice, $f'(2)$ is approximately 9.

- A third approximation can be found by averaging the above two approximations.

Then $f'(2)$ is approximately $\dfrac{21 + 9}{2} = 15.$ ∎

**NOW WORK** Problem 51.

## 2.1 Assess Your Understanding

### Concepts and Vocabulary

1. *True or False*   The derivative is used to find instantaneous velocity.

2. *True or False*   The derivative can be used to find the rate of change of a function.

3. The notation $f'(c)$ is read $f$ _____ of $c$; $f'(c)$ represents the _____ of the tangent line to the graph of $f$ at the point _____ .

4. *True or False*   If it exists, $\lim\limits_{x \to 3} \dfrac{f(x) - f(3)}{x - 3}$ is the derivative of the function $f$ at 3.

5. If $f(x) = 6x - 3$, then $f'(3) = $ _____.

6. The velocity of an object, the slope of a tangent line, and the rate of change of a function are three different interpretations of the mathematical concept called the _____.

### Skill Building

*In Problems 7–16,*

   (a) *Find an equation for the tangent line to the graph of each function at the indicated point.*

   (b) *Find an equation of the normal line to each function at the indicated point.*

   (c) *Graph the function, the tangent line, and the normal line at the indicated point on the same set of coordinate axes.*

7. $f(x) = 3x^2$ at $(-2, 12)$      8. $f(x) = x^2 + 2$ at $(-1, 3)$

9. $f(x) = x^3$ at $(-2, -8)$     10. $f(x) = x^3 + 1$ at $(1, 2)$

11. $f(x) = \dfrac{1}{x}$ at $(1, 1)$      12. $f(x) = \sqrt{x}$ at $(4, 2)$

13. $f(x) = \dfrac{1}{x + 5}$ at $\left(1, \dfrac{1}{6}\right)$    14. $f(x) = \dfrac{2}{x + 4}$ at $\left(1, \dfrac{2}{5}\right)$

15. $f(x) = \dfrac{1}{\sqrt{x}}$ at $(1, 1)$     16. $f(x) = \dfrac{1}{x^2}$ at $(1, 1)$

*In Problems 17–20, find the rate of change of $f$ at the indicated numbers.*

17. $f(x) = 5x - 2$   at (a) $c = 0$,   (b) $c = 2$

18. $f(x) = x^2 - 1$   at (a) $c = -1$,   (b) $c = 1$

19. $f(x) = \dfrac{x^2}{x + 3}$   at (a) $c = 0$,   (b) $c = 1$

20. $f(x) = \dfrac{x}{x^2 - 1}$   at (a) $c = 0$,   (b) $c = 2$

*In Problems 21–30, find the derivative of each function at the given number.*

21. $f(x) = 2x + 3$ at 1      22. $f(x) = 3x - 5$ at 2

23. $f(x) = x^2 - 2$ at 0      24. $f(x) = 2x^2 + 4$ at 1

25. $f(x) = 3x^2 + x + 5$ at $-1$    26. $f(x) = 2x^2 - x - 7$ at $-1$

27. $f(x) = \sqrt{x}$ at 4      28. $f(x) = \dfrac{1}{x^2}$ at 2

29. $f(x) = \dfrac{2 - 5x}{1 + x}$ at 0    30. $f(x) = \dfrac{2 + 3x}{2 + x}$ at 1

31. **Approximating Velocity**   An object in rectilinear motion moves according to the position function $s(t) = 10t^2$ ($s$ in centimeters and $t$ in seconds). Approximate the velocity of the object at time $t_0 = 3$ s by letting $\Delta t$ first equal 0.1 s, then 0.01 s, and finally 0.001 s. What limit does the velocity appear to be approaching? Organize the results in a table.

32. **Approximating Velocity**   An object in rectilinear motion moves according to the position function $s(t) = 5 - t^2$ ($s$ in centimeters and $t$ in seconds). Approximate the velocity of the object at time $t_0 = 1$ by letting $\Delta t$ first equal 0.1, then 0.01, and finally 0.001. What limit does the velocity appear to be approaching? Organize the results in a table.

33. **Rectilinear Motion**   As an object in rectilinear motion moves, its signed distance $s$ (in meters) from the origin after $t$ seconds is given by the position function $s = f(t) = 3t^2 + 4t$. Find the velocity $v$ at $t_0 = 0$. At $t_0 = 2$. At any time $t_0$.

34. **Rectilinear Motion**   As an object in rectilinear motion moves, its signed distance $s$ (in meters) from the origin after $t$ seconds is given by the position function $s = f(t) = 2t^3 + 4$. Find the velocity $v$ at $t_0 = 0$. At $t_0 = 3$. At any time $t_0$.

35. **Rectilinear Motion**   As an object in rectilinear motion moves, its signed distance $s$ from the origin at time $t$ is given by the position function $s = s(t) = 3t^2 - \dfrac{1}{t}$, where $s$ is in centimeters and $t$ is in seconds. Find the velocity $v$ of the object at $t_0 = 1$ and $t_0 = 4$.

36. **Rectilinear Motion**   As an object in rectilinear motion moves, its signed distance $s$ from the origin at time $t$ is given by the position function $s = s(t) = \sqrt{4t}$, where $s$ is in centimeters and $t$ is in seconds. Find the velocity $v$ of the object at $t_0 = 1$ and $t_0 = 4$.

---

**1.** = NOW WORK problem     〔⋀〕 = Graphing technology recommended     〔CAS〕 = Computer Algebra System recommended

**37.** The Princeton Dinky is the shortest rail line in the country. It runs for 2.7 miles, connecting Princeton University to the Princeton Junction railroad station. The Dinky starts from the university and moves north toward Princeton Junction. Its distance from Princeton is shown in the graph below, where the time $t$ is in minutes and the distance $s$ of the Dinky from Princeton University is in miles.

MancusoMichael/AP Images

(a) When is the Dinky headed toward Princeton University?

(b) When is it headed toward Princeton Junction?

(c) When is the Dinky stopped?

(d) Find its average velocity on a trip from Princeton to Princeton Junction.

(e) Find its average velocity for the round trip shown in the graph, that is, from $t = 0$ to $t = 13$.

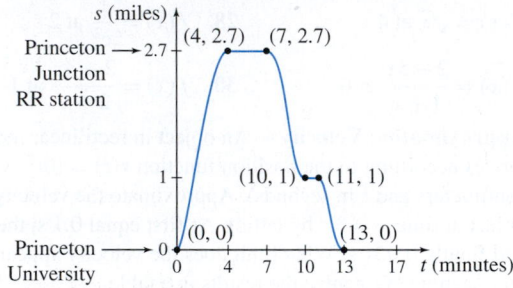

**38.** Barbara walks to the deli, which is six blocks east of her house. After walking two blocks, she realizes she left her phone on her desk, so she runs home. After getting the phone, and closing and locking the door, Barbara starts on her way again. At the deli, she waits in line to buy a bottle of **vitaminwater**™, and then she jogs home. The graph below represents Barbara's journey. The time $t$ is in minutes and $s$ is Barbara's distance, in blocks, from home.

(a) At what times is she headed toward the deli?

(b) At what times is she headed home?

(c) When is the graph horizontal? What does this indicate?

(d) Find Barbara's average velocity from home until she starts back to get her phone.

(e) Find Barbara's average velocity from home to the deli after getting her phone.

(f) Find her average velocity from the deli to home.

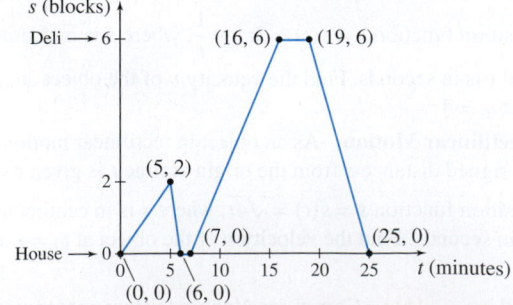

**Applications and Extensions**

**39. Slope of a Tangent Line**   An equation of the tangent line to the graph of a function $f$ at $(2, 6)$ is $y = -3x + 12$. What is $f'(2)$?

**40. Slope of a Tangent Line**   An equation of the tangent line of a function $f$ at $(3, 2)$ is $y = \frac{1}{3}x + 1$. What is $f'(3)$?

**41. Tangent Line**   Does the tangent line to the graph of $y = x^2$ at $(1, 1)$ pass through the point $(2, 5)$?

**42. Tangent Line**   Does the tangent line to the graph of $y = x^3$ at $(1, 1)$ pass through the point $(2, 5)$?

**43. Respiration Rate**   A human being's respiration rate $R$ (in breaths per minute) is given by $R = R(p) = 10.35 + 0.59p$, where $p$ is the partial pressure of carbon dioxide in the lungs. Find the rate of change in respiration when $p = 50$.

**44. Instantaneous Rate of Change**   The volume $V$ of the right circular cylinder of height 5 m and radius $r$ m shown in the figure is $V = V(r) = 5\pi r^2$. Find the instantaneous rate of change of the volume with respect to the radius when $r = 3$ m.

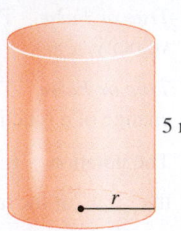

**45. Market Share**   During a month-long advertising campaign, the total sales $S$ of a magazine is modeled by the function $S(x) = 5x^2 + 100x + 10{,}000$, where $x$ represents the number of days since the campaign began, $0 \le x \le 30$.

(a) What is the average rate of change of sales from $x = 10$ to $x = 20$ days?

(b) What is the instantaneous rate of change of sales when $x = 10$ days?

**46. Demand Equation**   The demand equation for an item is $p = p(x) = 90 - 0.02x$, where $p$ is the price in dollars and $x$ is the number of units (in thousands) made.

(a) Assuming all units made can be sold, find the revenue function $R(x) = xp(x)$.

(b) **Marginal Revenue**   Marginal revenue is defined as the additional revenue earned by selling an additional unit. If we use $R'(x)$ to measure the marginal revenue, find the marginal revenue when 1 million units are sold.

**47. Gravity**   If a ball is dropped from the top of the Empire State Building, 1002 ft above the ground, the distance $s$ (in feet) it falls after $t$ seconds is $s(t) = 16t^2$.

(a) What is the average velocity of the ball for the first 2 s?

(b) How long does it take for the ball to hit the ground?

(c) What is the average velocity of the ball during the time it is falling?

(d) What is the velocity of the ball when it hits the ground?

**48. Velocity**   A ball is thrown upward. Its height $h$ in feet is given by $h(t) = 100t - 16t^2$, where $t$ is the time elapsed in seconds.

(a) What is the velocity $v$ of the ball at $t = 0$ s, $t = 1$ s, and $t = 4$ s?

(b) At what time $t$ does the ball strike the ground?

(c) At what time $t$ does the ball reach its highest point? (*Hint:* At the time the ball reaches its maximum height, it is stationary. So, its velocity $v = 0$.)

**49. Gravity** A rock is dropped from a height of 88.2 m and falls toward Earth in a straight line. In $t$ seconds the rock falls $4.9t^2$ m.

   **(a)** What is the average velocity of the rock for the first 2 s?

   **(b)** How long does it take for the rock to hit the ground?

   **(c)** What is the average velocity of the rock during its fall?

   **(d)** What is the velocity $v$ of the rock when it hits the ground?

**50. Velocity** At a certain instant, the speedometer of an automobile reads $V$ mi/h. During the next $\dfrac{1}{4}$ s the automobile travels 20 ft. Approximate $V$ from this information.

**51.** A tank is filled with 80 l of water at 7 a.m. ($t = 0$). Over the next 12 h the water is continuously used. The table below gives the amount of water $A(t)$ (in liters) remaining in the tank at selected times $t$, where $t$ measures the number of hours after 7 a.m.

| $t$ | 0 | 2 | 5 | 7 | 9 | 12 |
|-----|----|----|----|----|----|----|
| $A(t)$ | 80 | 71 | 66 | 60 | 54 | 50 |

   **(a)** Use the table to approximate $A'(5)$.

   **(b)** Using appropriate units, interpret $A'(5)$ in the context of the problem.

**52.** The table below lists the outside temperature $T$, in degrees Fahrenheit, in Naples, Florida, on a certain day in January, for selected times $x$, where $x$ is the number of hours since 12 a.m.

| $x$ | 5 | 7 | 9 | 11 | 12 | 13 | 14 | 16 | 17 |
|-----|----|----|----|----|----|----|----|----|----|
| $T(x)$ | 62 | 71 | 74 | 78 | 81 | 83 | 84 | 85 | 78 |

   **(a)** Use the table to approximate $T'(11)$.

   **(b)** Using appropriate units, interpret $T'(11)$ in the context of the problem.

**53. Rate of Change** Show that the rate of change of a linear function $f(x) = mx + b$ is the slope $m$ of the line $y = mx + b$.

**54. Rate of Change** Show that the rate of change of a quadratic function $f(x) = ax^2 + bx + c$ is a linear function of $x$.

**55. Agriculture** The graph represents the diameter $d$ (in centimeters) of a maturing peach as a function of the time $t$ (in days) it is on the tree.

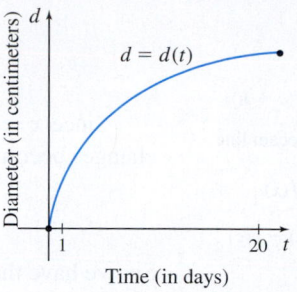

   **(a)** Interpret the derivative $d'(t)$ as a rate of change.

   **(b)** Which is larger, $d'(1)$ or $d'(20)$?

   **(c)** Interpret both $d'(1)$ and $d'(20)$.

**56. Business** The graph represents the demand $d$ (in gallons) for olive oil as a function of the cost $c$ (in dollars per gallon) of the oil.

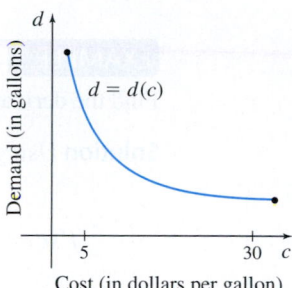

   **(a)** Interpret the derivative $d'(c)$.

   **(b)** Which is larger, $d'(5)$ or $d'(30)$? Give an interpretation to $d'(5)$ and $d'(30)$.

**57. Volume of a Cube** A metal cube with each edge of length $x$ centimeters is expanding uniformly as a consequence of being heated.

   **(a)** Find the average rate of change of the volume of the cube with respect to an edge as $x$ increases from 2.00 to 2.01 cm.

   **(b)** Find the instantaneous rate of change of the volume of the cube with respect to an edge at the instant when $x = 2$ cm.

---

# 2.2 The Derivative as a Function

**OBJECTIVES** *When you finish this section, you should be able to:*

**1** Define the derivative function (p. 163)

**2** Graph the derivative function (p. 165)

**3** Identify where a function has no derivative (p. 167)

## 1 Define the Derivative Function

The derivative of $f$ at a real number $c$ has been defined as the real number

$$f'(c) = \lim_{x \to c} \frac{f(x) - f(c)}{x - c} \tag{1}$$

provided the limit exists.

Another way to find the derivative of $f$ at any real number is obtained by rewriting the expression $\dfrac{f(x) - f(c)}{x - c}$ and letting $x = c + h$, $h \neq 0$. Then

$$\frac{f(x) - f(c)}{x - c} = \frac{f(c + h) - f(c)}{(c + h) - c} = \frac{f(c + h) - f(c)}{h}$$

Since $x = c + h$, then as $x$ approaches $c$, $h$ approaches 0. Equation (1) with these changes becomes

$$f'(c) = \lim_{x \to c} \frac{f(x) - f(c)}{x - c} = \lim_{h \to 0} \frac{f(c + h) - f(c)}{h}$$

So, we have the following alternate form for the derivative of $f$ at a real number $c$.

$$f'(c) = \lim_{h \to 0} \frac{f(c + h) - f(c)}{h} \qquad (2)$$

See Figure 7.

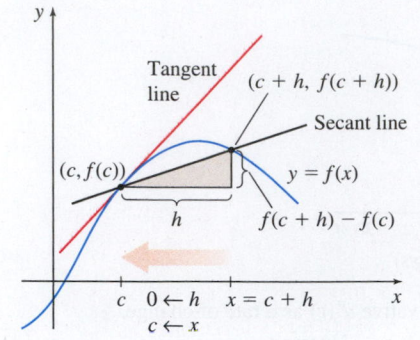

**Figure 7** The slope of the tangent line at $c$ is
$$f'(c) = \lim_{h \to 0} \frac{f(c + h) - f(c)}{h}$$

---

**EXAMPLE 1** **Finding the Derivative of a Function at a Number $c$**

Find the derivative of the function $f(x) = x^2 - 5x$ at any real number $c$ using Form (2).

**Solution** Using Form (2), we have

$$f'(c) = \lim_{h \to 0} \frac{f(c + h) - f(c)}{h} = \lim_{h \to 0} \frac{[(c + h)^2 - 5(c + h)] - (c^2 - 5c)}{h}$$

$$= \lim_{h \to 0} \frac{[(c^2 + 2ch + h^2) - 5c - 5h] - c^2 + 5c}{h} = \lim_{h \to 0} \frac{2ch + h^2 - 5h}{h}$$

$$= \lim_{h \to 0} \frac{h(2c + h - 5)}{h} = \lim_{h \to 0} (2c + h - 5) = 2c - 5 \qquad \blacksquare$$

**NOTE** Compare the solution and answer found in Example 1 to Example 2(b) on p. 156.

Based on Example 1, if $f(x) = x^2 - 5x$, then $f'(c) = 2c - 5$ for any choice of $c$. That is, the derivative $f'$ is a function and, using $x$ as the independent variable, we can write $f'(x) = 2x - 5$.

**DEFINITION** **The Derivative Function $f'$**

The **derivative function** $f'$ of a function $f$ is

$$f'(x) = \lim_{h \to 0} \frac{f(x + h) - f(x)}{h} \qquad (3)$$

**IN WORDS** In Form (3) the derivative is the limit of a difference quotient.

provided the limit exists. If $f$ has a derivative, then $f$ is said to be **differentiable**.

The domain of the function $f'$ is the set of real numbers in the domain of $f$ for which the limit (3) exists. So the domain of $f'$ is a subset of the domain of $f$.

We can use either Form (1) or Form (3) to find derivatives. However, if we want the derivative of $f$ at a specified number $c$, we usually use Form (1) to find $f'(c)$. If we want to find the derivative function of $f$, we usually use Form (3) to find $f'(x)$. In this section, we use the definitions of the derivative, Forms (1) and (3), to investigate derivatives. In the next section, we begin to develop formulas for finding the derivatives.

NOTE The instruction "differentiate $f$" means "find the derivative of $f$."

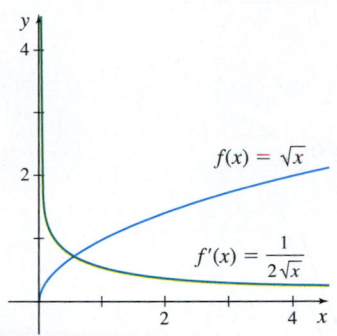

**Figure 8**

### EXAMPLE 2  Finding the Derivative Function

Differentiate $f(x) = \sqrt{x}$ and determine the domain of $f'$.

**Solution** The domain of $f$ is $\{x \mid x \geq 0\}$. To find the derivative of $f$, we use Form (3). Then

$$f'(x) = \lim_{h \to 0} \frac{f(x+h) - f(x)}{h} = \lim_{h \to 0} \frac{\sqrt{x+h} - \sqrt{x}}{h}$$

We rationalize the numerator to find the limit.

$$f'(x) = \lim_{h \to 0} \left[ \frac{\sqrt{x+h} - \sqrt{x}}{h} \cdot \frac{\sqrt{x+h} + \sqrt{x}}{\sqrt{x+h} + \sqrt{x}} \right] = \lim_{h \to 0} \frac{(x+h) - x}{h \left( \sqrt{x+h} + \sqrt{x} \right)}$$

$$= \lim_{h \to 0} \frac{h}{h \left( \sqrt{x+h} + \sqrt{x} \right)} = \lim_{h \to 0} \frac{1}{\sqrt{x+h} + \sqrt{x}} = \frac{1}{2\sqrt{x}}$$

The limit does not exist when $x = 0$. But for all other $x$ in the domain of $f$, the limit does exist. So, the domain of the derivative function $f'(x) = \dfrac{1}{2\sqrt{x}}$ is $\{x \mid x > 0\}$.  ∎

In Example 2, notice that the domain of the derivative function $f'$ is a proper subset of the domain of the function $f$. The graphs of both $f$ and $f'$ are shown in Figure 8.

**NOW WORK** Problem 15.

### EXAMPLE 3  Interpreting the Derivative as a Rate of Change

The surface area $S$ (in square meters) of a balloon is expanding as a function of time $t$ (in seconds) according to $S = S(t) = 5t^2$. Find the rate of change of the surface area of the balloon with respect to time. What are the units of $S'(t)$?

**Solution** An interpretation of the derivative function $S'(t)$ is the rate of change of $S = S(t)$.

$$S'(t) = \lim_{h \to 0} \frac{S(t+h) - S(t)}{h} = \lim_{h \to 0} \frac{5(t+h)^2 - 5t^2}{h} \qquad \text{Form (3)}$$

$$= \lim_{h \to 0} \frac{5(t^2 + 2th + h^2) - 5t^2}{h}$$

$$= \lim_{h \to 0} \frac{5t^2 + 10th + 5h^2 - 5t^2}{h} = \lim_{h \to 0} \frac{10th + 5h^2}{h}$$

$$= \lim_{h \to 0} \frac{(10t + 5h)h}{h} = \lim_{h \to 0} (10t + 5h) = 10t$$

Since $S'(t)$ is the limit of the quotient of a change in area divided by a change in time, the units of the rate of change are square meters per second (m²/s). The rate of change of the surface area $S$ of the balloon with respect to time is $10t$ m²/s.  ∎

**NOW WORK** Problem 67.

## 2 Graph the Derivative Function

There is a relationship between the graph of a function and the graph of its derivative.

### EXAMPLE 4  Graphing a Function and Its Derivative

Find $f'$ if $f(x) = x^3 - 1$. Then graph $y = f(x)$ and $y = f'(x)$ on the same set of coordinate axes.

**Solution** $f(x) = x^3 - 1$ so

$$f(x+h) = (x+h)^3 - 1 = x^3 + 3hx^2 + 3h^2x + h^3 - 1$$

Using Form (3), we find

$$f'(x) = \lim_{h \to 0} \frac{f(x+h) - f(x)}{h} = \lim_{h \to 0} \frac{(x^3 + 3hx^2 + 3h^2x + h^3 - 1) - (x^3 - 1)}{h}$$

$$= \lim_{h \to 0} \frac{3hx^2 + 3h^2x + h^3}{h} = \lim_{h \to 0} \frac{h(3x^2 + 3hx + h^2)}{h}$$

$$= \lim_{h \to 0} (3x^2 + 3hx + h^2) = 3x^2$$

The graphs of $f$ and $f'$ are shown below in Figure 9. ∎

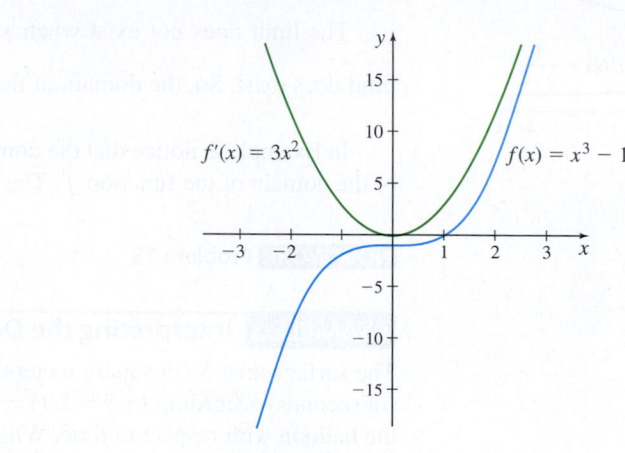

**Figure 9**

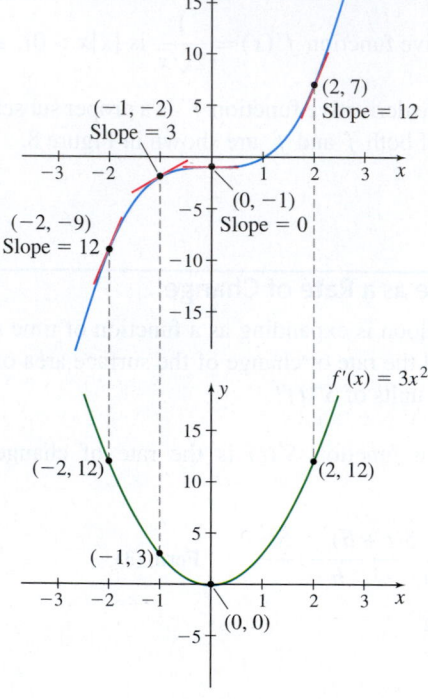

**Figure 10**

Figure 10 illustrates several tangent lines to the graph of $f(x) = x^3 - 1$. Observe that the tangent line to the graph of $f$ at $(0, -1)$ is horizontal, so its slope is 0. Then $f'(0) = 0$, so the graph of $f'$ contains the point $(0, 0)$. Also notice that every tangent line to the graph of $f$ has a nonnegative slope, so $f'(x) \geq 0$. That is, the range of the function $f'$ is $\{y | y \geq 0\}$. Finally, notice that the slope of each tangent line is the $y$-coordinate of the corresponding point on the graph of the derivative $f'$.

**NOW WORK** Problem 19.

With these ideas in mind, we can obtain a rough sketch of the derivative function $f'$, even if we know only the graph of the function $f$.

**EXAMPLE 5** Graphing the Derivative Function

Use the graph of the function $y = f(x)$, shown in Figure 11, to sketch the graph of the derivative function $y = f'(x)$.

**Solution** We begin by drawing tangent lines to the graph of $f$ at the points shown in Figure 11. See the graph at the top of Figure 12. At the points $(-2, 3)$ and $\left(\frac{3}{2}, -2\right)$ the tangent lines are horizontal, so their slopes are 0. This means $f'(-2) = 0$ and $f'\left(\frac{3}{2}\right) = 0$, so the points $(-2, 0)$ and $\left(\frac{3}{2}, 0\right)$ are on the graph of the derivative function at the bottom of Figure 12. Now we estimate the slope of the tangent lines at the other selected points.

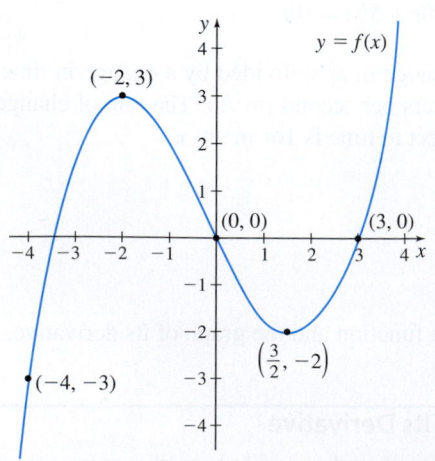

**Figure 11**

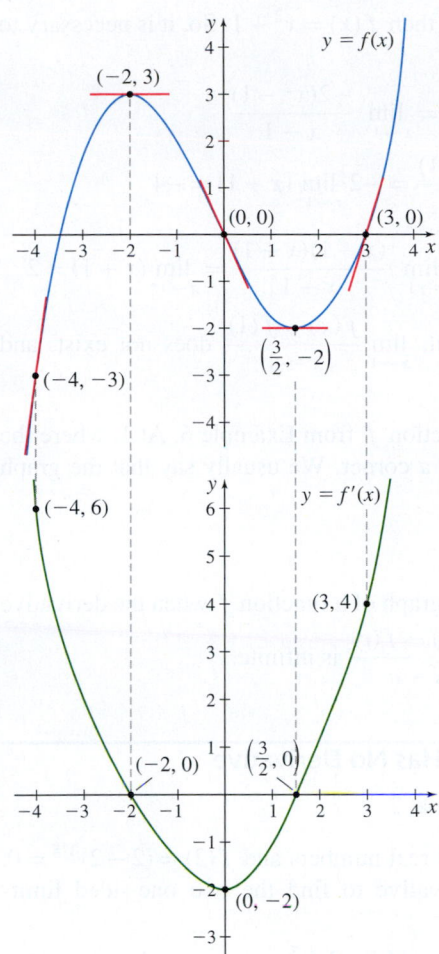

**DF** Figure 12

For example, at the point $(-4, -3)$, the slope of the tangent line is positive and the line is rather steep. We estimate the slope to be close to 6, and we plot the point $(-4, 6)$ on the bottom graph of Figure 12. Continue the process and then connect the points with a smooth curve. ∎

Notice in Figure 12 that at the points on the graph of $f$ where the tangent lines are horizontal, the graph of the derivative $f'$ intersects the $x$-axis. Also notice that wherever the graph of $f$ is increasing, the slopes of the tangent lines are positive, that is, $f'$ is positive, so the graph of $f'$ is above the $x$-axis. Similarly, wherever the graph of $f$ is decreasing, the slopes of the tangent lines are negative, so the graph of $f'$ is below the $x$-axis.

**NOW WORK** Problem **29.**

### 3 Identify Where a Function Has No Derivative

Suppose a function $f$ is continuous on an open interval containing the number $c$. The function $f$ has no derivative at the number $c$ if $\lim\limits_{x \to c} \dfrac{f(x) - f(c)}{x - c}$ does not exist. Three (of several) ways this can happen are:

- $\lim\limits_{x \to c^-} \dfrac{f(x) - f(c)}{x - c}$ exists and $\lim\limits_{x \to c^+} \dfrac{f(x) - f(c)}{x - c}$ exists, but they are not equal.*
  When this happens the graph of $f$ has a **corner** at $(c, f(c))$. For example, the absolute value function $f(x) = |x|$ has a corner at $(0, 0)$. See Figure 13.
- The one-sided limits are both infinite and both equal $\infty$ or both equal $-\infty$. When this happens, the graph of $f$ has a vertical tangent line at $(c, f(c))$. For example, the cube root function $f(x) = \sqrt[3]{x}$ has a vertical tangent at $(0, 0)$. See Figure 14.
- Both one-sided limits are infinite, but one equals $-\infty$ and the other equals $\infty$. When this happens, the graph of $f$ has a vertical tangent line at the point $(c, f(c))$, which is referred to as a **cusp**. For example, the function $f(x) = x^{2/3}$ has a cusp at $(0, 0)$. See Figure 15.

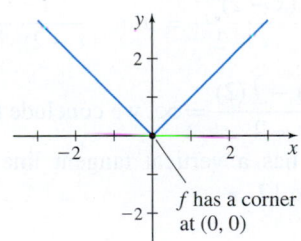

**Figure 13** $f(x) = |x|$;
$f'(0)$ does not exist.

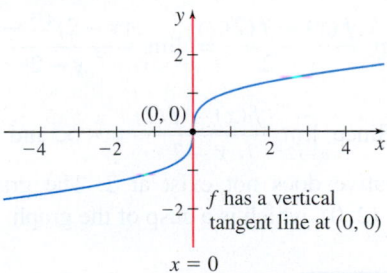

**Figure 14** $f(x) = \sqrt[3]{x}$;
$f'(0)$ does not exist.

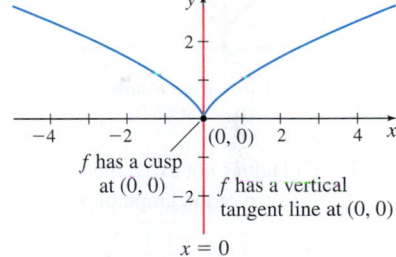

**Figure 15** $f(x) = x^{2/3}$;
$f'(0)$ does not exist.

**EXAMPLE 6**  **Identifying Where a Function Has No Derivative**

Given the piecewise defined function $f(x) = \begin{cases} -2x^2 + 4 & \text{if } x < 1 \\ x^2 + 1 & \text{if } x \geq 1 \end{cases}$,
determine whether $f'(1)$ exists.

**Solution** Use Form (1) of the definition of a derivative to determine whether $f'(1)$ exists.

$$\lim_{x \to 1} \frac{f(x) - f(1)}{x - 1} = \lim_{x \to 1} \frac{f(x) - 2}{x - 1} \qquad f(1) = 1^2 + 1 = 2$$

---

*The one-sided limits, $\lim\limits_{x \to c^-} \frac{f(x) - f(c)}{x - c}$ and $\lim\limits_{x \to c^+} \frac{f(x) - f(c)}{x - c}$ are called the **left-hand derivative** of $f$ at $c$, denoted $f'_-(c)$, and the **right-hand derivative** of $f$ at $c$, denoted $f'_+(c)$, respectively. Using properties of limits, the derivative $f'(c)$ exists if and only if $f'_-(c) = f'_+(c)$.

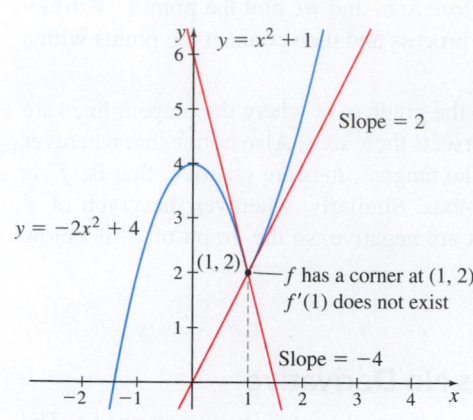

**Figure 16** $f$ has a corner at $(1, 2)$.

If $x < 1$, then $f(x) = -2x^2 + 4$; if $x \geq 1$, then $f(x) = x^2 + 1$. So, it is necessary to find the one-sided limits at 1.

$$\lim_{x \to 1^-} \frac{f(x) - f(1)}{x - 1} = \lim_{x \to 1^-} \frac{(-2x^2 + 4) - 2}{x - 1} = \lim_{x \to 1^-} \frac{-2(x^2 - 1)}{x - 1}$$
$$= -2 \lim_{x \to 1^-} \frac{(x - 1)(x + 1)}{x - 1} = -2 \lim_{x \to 1^-} (x + 1) = -4$$

$$\lim_{x \to 1^+} \frac{f(x) - f(1)}{x - 1} = \lim_{x \to 1^+} \frac{(x^2 + 1) - 2}{x - 1} = \lim_{x \to 1^+} \frac{(x - 1)(x + 1)}{x - 1} = \lim_{x \to 1^+} (x + 1) = 2$$

Since the one-sided limits are not equal, $\displaystyle\lim_{x \to 1} \frac{f(x) - f(1)}{x - 1}$ does not exist, and so $f'(1)$ does not exist. ■

Figure 16 illustrates the graph of the function $f$ from Example 6. At 1, where the derivative does not exist, the graph of $f$ has a corner. We usually say that the graph of $f$ is not *smooth* at a corner.

**NOW WORK** Problem 39.

Example 7 illustrates the behavior of the graph of a function $f$ when the derivative at a number $c$ does not exist because $\displaystyle\lim_{x \to c} \frac{f(x) - f(c)}{x - c}$ is infinite.

**CALC**
**CLIP**

**EXAMPLE 7    Showing That a Function Has No Derivative**

Show that $f(x) = (x - 2)^{4/5}$ has no derivative at 2.

**Solution** The function $f$ is continuous for all real numbers and $f(2) = (2 - 2)^{4/5} = 0$. Use Form (1) of the definition of the derivative to find the two one-sided limits at 2.

$$\lim_{x \to 2^-} \frac{f(x) - f(2)}{x - 2} = \lim_{x \to 2^-} \frac{(x - 2)^{4/5} - 0}{x - 2} = \lim_{x \to 2^-} \frac{(x - 2)^{4/5}}{x - 2} = \lim_{x \to 2^-} \frac{1}{(x - 2)^{1/5}} = -\infty$$

$$\lim_{x \to 2^+} \frac{f(x) - f(2)}{x - 2} = \lim_{x \to 2^+} \frac{(x - 2)^{4/5} - 0}{x - 2} = \lim_{x \to 2^+} \frac{(x - 2)^{4/5}}{x - 2} = \lim_{x \to 2^+} \frac{1}{(x - 2)^{1/5}} = \infty$$

Since $\displaystyle\lim_{x \to 2^-} \frac{f(x) - f(2)}{x - 2} = -\infty$ and $\displaystyle\lim_{x \to 2^+} \frac{f(x) - f(2)}{x - 2} = \infty$, we conclude that the derivative does not exist at 2. The graph of $f$ has a vertical tangent line at the point $(2, 0)$, which is a cusp of the graph. See Figure 17. ■

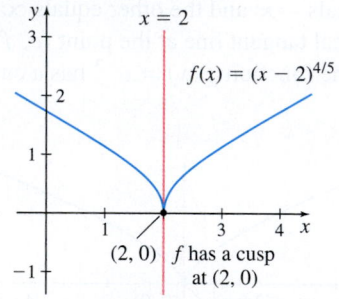

**Figure 17** $f'(2)$ does not exist; the point $(2, 0)$ is a cusp of the graph of $f$.

**NOW WORK** Problem 35.

**EXAMPLE 8    Obtaining Information About $y = f(x)$ from the Graph of Its Derivative Function**

Suppose $y = f(x)$ is continuous for all real numbers. Figure 18 shows the graph of its derivative function $f'$.

**(a)** Does the graph of $f$ have any horizontal tangent lines? If yes, explain why and identify where they occur.

**(b)** Does the graph of $f$ have any vertical tangent lines? If yes, explain why, identify where they occur, and determine whether the point is a cusp of $f$.

**(c)** Does the graph of $f$ have any corners? If yes, explain why and identify where they occur.

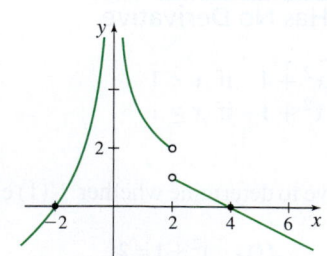

**Figure 18** $y = f'(x)$

**Solution (a)** Since the derivative $f'$ equals the slope of a tangent line, horizontal tangent lines occur where the derivative equals 0. Since $f'(x) = 0$ for $x = -2$ and $x = 4$, the graph of $f$ has two horizontal tangent lines, one at the point $(-2, f(-2))$ and the other at the point $(4, f(4))$.

**(b)** As $x$ approaches 0, the derivative function $f'$ becomes unbounded in the positive direction both for $x < 0$ and for $x > 0$. The graph of $f$ has a vertical tangent line at $x = 0$. The point $(0, f(0))$ is not a cusp because the unboundedness is in the same direction on each side of 0.

**(c)** The derivative is not defined at 2 but the one-sided derivatives have unequal finite limits as $x$ approaches 2. So the graph of $f$ has a corner at $(2, f(2))$. ∎

**NOW WORK** Problem **45.**

## Differentiability and Continuity

**NEED TO REVIEW?** Continuity is discussed in Section 1.3, pp. 102–108.

In Chapter 1, we investigated the continuity of a function. Here we have been investigating the differentiability of a function. An important connection exists between continuity and differentiability.

### THEOREM

If a function $f$ has a derivative at a number $c$, then $f$ is continuous at $c$.

**Proof** To show that $f$ is continuous at $c$, we need to verify that $\lim\limits_{x \to c} f(x) = f(c)$. We begin by observing that if $x \neq c$, then

$$f(x) - f(c) = \left[ \frac{f(x) - f(c)}{x - c} \right] (x - c)$$

We take the limit of both sides as $x \to c$, and use the fact that the limit of a product equals the product of the limits (we show later that each limit exists).

$$\lim_{x \to c} [f(x) - f(c)] = \lim_{x \to c} \left\{ \left[ \frac{f(x) - f(c)}{x - c} \right] (x - c) \right\} = \left[ \lim_{x \to c} \frac{f(x) - f(c)}{x - c} \right] \left[ \lim_{x \to c} (x - c) \right]$$

Since $f$ has a derivative at $c$, we know that

$$\lim_{x \to c} \frac{f(x) - f(c)}{x - c} = f'(c)$$

is a number. Also for any real number $c$, $\lim\limits_{x \to c} (x - c) = 0$. So

$$\lim_{x \to c} [f(x) - f(c)] = [f'(c)] \left[ \lim_{x \to c} (x - c) \right] = f'(c) \cdot 0 = 0$$

That is, $\lim\limits_{x \to c} f(x) = f(c)$, so $f$ is continuous at $c$. ∎

An equivalent statement of this theorem gives a condition under which a function has no derivative.

### COROLLARY

If a function $f$ is discontinuous at a number $c$, then $f$ has no derivative at $c$.

**IN WORDS** Differentiability implies continuity, but continuity does not imply differentiability.

Let's look at some of the possibilities. In Figure 19(a), the function $f$ is continuous at the number 1 and has a derivative at 1. The function $g$, graphed in Figure 19(b), is continuous at the number 0, but it has no derivative at 0. So continuity at a number $c$ provides no prediction about differentiability. On the other hand, the function $h$ graphed in Figure 19(c) illustrates the corollary: If $h$ is discontinuous at a number, it has no derivative at that number.

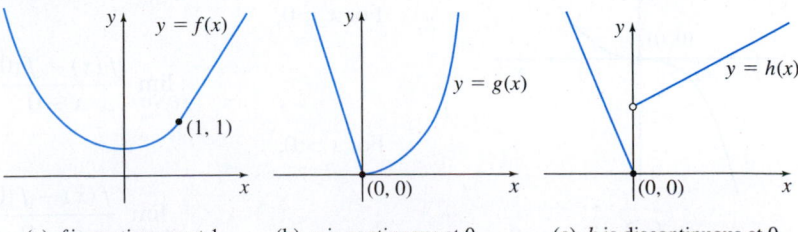

(a) $f$ is continuous at 1, and $f'(1)$ exists.

(b) $g$ is continuous at 0, but $g'(0)$ does not exist.

(c) $h$ is discontinuous at 0, so $h'(0)$ does not exist.

**Figure 19**

The corollary is useful if we are seeking the derivative of a function $f$ that we suspect is discontinuous at a number $c$. If we can show that $f$ is discontinuous at $c$, then the corollary affirms that the function $f$ has no derivative at $c$. For example, since the floor function $f(x) = \lfloor x \rfloor$ is discontinuous at every integer $c$, it has no derivative at an integer.

### EXAMPLE 9  Determining Whether a Function Has a Derivative at a Number

Determine whether the function

$$f(x) = \begin{cases} 2x + 2 & \text{if } x < 3 \\ 5 & \text{if } x = 3 \\ x^2 - 1 & \text{if } x > 3 \end{cases}$$

has a derivative at 3.

**Solution** Since $f$ is a piecewise-defined function, it may be discontinuous at 3 and therefore has no derivative at 3. So we begin by determining whether $f$ is continuous at 3.

Since $f(3) = 5$, the function $f$ is defined at 3. Use one-sided limits to check whether $\lim\limits_{x \to 3} f(x)$ exists.

$$\lim_{x \to 3^-} f(x) = \lim_{x \to 3^-} (2x + 2) = 8 \qquad \lim_{x \to 3^+} f(x) = \lim_{x \to 3^+} (x^2 - 1) = 8$$

Since $\lim\limits_{x \to 3^-} f(x) = \lim\limits_{x \to 3^+} f(x)$, then $\lim\limits_{x \to 3} f(x)$ exists. But $\lim\limits_{x \to 3} f(x) = 8$ and $f(3) = 5$, so $f$ is discontinuous at 3. From the corollary, since $f$ is discontinuous at 3, the function $f$ does not have a derivative at 3. ■

Figure 20 shows the graph of $f$.

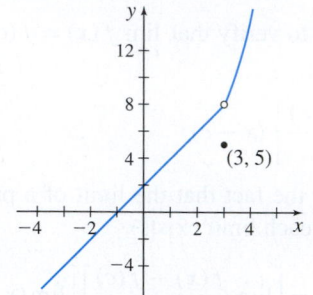

**Figure 20** $f(x) = \begin{cases} 2x + 2 & \text{if } x < 3 \\ 5 & \text{if } x = 3 \\ x^2 - 1 & \text{if } x > 3 \end{cases}$

**NOW WORK** Problem 43.

In Example 9, the function $f$ is discontinous at 3, so by the corollary, the derivative of $f$ at 3 does not exist. But when a function is continuous at a number $c$, then sometimes the derivative at $c$ exists and other times the derivative at $c$ does not exist.

### EXAMPLE 10  Determining Whether a Function Has a Derivative at a Number

Determine whether each piecewise-defined function has a derivative at $c$. If the function has a derivative at $c$, find it.

**(a)** $f(x) = \begin{cases} x^3 & \text{if } x < 0 \\ x^2 & \text{if } x \geq 0 \end{cases} \quad c = 0$ **(b)** $g(x) = \begin{cases} 1 - 2x & \text{if } x \leq 1 \\ x - 2 & \text{if } x > 1 \end{cases} \quad c = 1$

**Solution** **(a)** See Figure 21. The function $f$ is continuous at 0, which you should verify. To determine whether $f$ has a derivative at 0, we examine the one-sided limits at 0 using Form (1).

For $x < 0$,

$$\lim_{x \to 0^-} \frac{f(x) - f(0)}{x - 0} = \lim_{x \to 0^-} \frac{x^3 - 0}{x} = \lim_{x \to 0^-} x^2 = 0$$

For $x > 0$,

$$\lim_{x \to 0^+} \frac{f(x) - f(0)}{x - 0} = \lim_{x \to 0^+} \frac{x^2 - 0}{x} = \lim_{x \to 0^+} x = 0$$

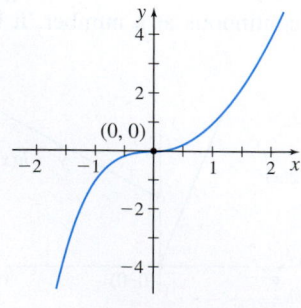

**Figure 21** $f(x) = \begin{cases} x^3 & \text{if } x < 0 \\ x^2 & \text{if } x \geq 0 \end{cases}$

Since both one-sided limits are equal, $f$ has a derivative at 0, and $f'(0) = 0$.

**(b)** See Figure 22. The function $g$ is continuous at 1, which you should verify. To determine whether $g$ has a derivative at 1, examine the one-sided limits at 1 using Form (1).

For $x < 1$,

$$\lim_{x \to 1^-} \frac{g(x) - g(1)}{x - 1} = \lim_{x \to 1^-} \frac{(1 - 2x) - (-1)}{x - 1} = \lim_{x \to 1^-} \frac{2 - 2x}{x - 1}$$

$$= \lim_{x \to 1^-} \frac{-2(x - 1)}{x - 1} = \lim_{x \to 1^-} (-2) = -2$$

For $x > 1$,

$$\lim_{x \to 1^+} \frac{g(x) - g(1)}{x - 1} = \lim_{x \to 1^+} \frac{(x - 2) - (-1)}{x - 1} = \lim_{x \to 1^+} \frac{x - 1}{x - 1} = 1$$

The one-sided limits are not equal, so $\lim_{x \to 1} \frac{g(x) - g(1)}{x - 1}$ does not exist. That is, $g$ has no derivative at 1. ∎

Notice in Figure 21 the tangent lines to the graph of $f$ turn smoothly around the origin. On the other hand, notice in Figure 22 the tangent lines to the graph of $g$ change abruptly at the point $(1, -1)$, where the graph of $g$ has a corner.

**NOW WORK** Problem 41.

**Figure 22** $g(x) = \begin{cases} 1 - 2x & \text{if } x \le 1 \\ x - 2 & \text{if } x > 1 \end{cases}$

---

## 2.2 Assess Your Understanding

### Concepts and Vocabulary

1. *True or False*  The domain of a function $f$ and the domain of its derivative function $f'$ are always equal.

2. *True or False*  If a function is continuous at a number $c$, then it is differentiable at $c$.

3. *Multiple Choice*  If $f$ is continuous at a number $c$ and if

   $\lim_{x \to c} \dfrac{f(x) - f(c)}{x - c}$ is infinite, then the graph of $f$ has

   [**(a)** a horizontal **(b)** a vertical **(c)** no] tangent line at $c$.

4. The instruction, "Differentiate $f$," means to find the ___ of $f$.

### Skill Building

*In Problems 5–10, find the derivative of each function f at any real number c. Use Form (1) on page 163.*

5. $f(x) = 10$      6. $f(x) = -4$      7. $f(x) = 2x + 3$

8. $f(x) = 3x - 5$    9. $f(x) = 2 - x^2$   10. $f(x) = 2x^2 + 4$

*In Problems 11–16, differentiate each function f and determine the domain of f'. Use Form (3) on page 164.*

11. $f(x) = 5$      12. $f(x) = -2$

13. $f(x) = 3x^2 + x + 5$    14. $f(x) = 2x^2 - x - 7$

15. $f(x) = 5\sqrt{x - 1}$    16. $f(x) = 4\sqrt{x + 3}$

*In Problems 17–22, differentiate each function f. Graph $y = f(x)$ and $y = f'(x)$ on the same set of coordinate axes.*

17. $f(x) = \dfrac{1}{3}x + 1$    18. $f(x) = -4x - 5$

19. $f(x) = 2x^2 - 5x$    20. $f(x) = -3x^2 + 2$

21. $f(x) = x^3 - 8x$    22. $f(x) = -x^3 - 8$

*In Problems 23–26, for each figure determine if the graphs represent a function f and its derivative f'. If they do, indicate which is the graph of f and which is the graph of f'.*

23.
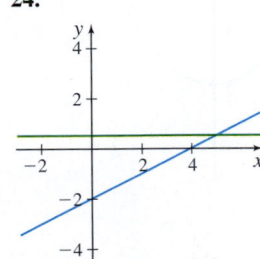

24.

25.
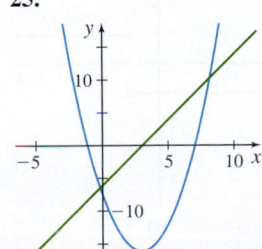

26.
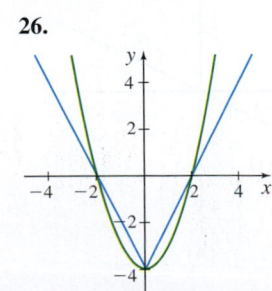

---

**1.** = NOW WORK problem      Ⓜ = Graphing technology recommended      **CAS** = Computer Algebra System recommended

*In Problems 27–30, use the graph of f to obtain the graph of f'.*

**27.**

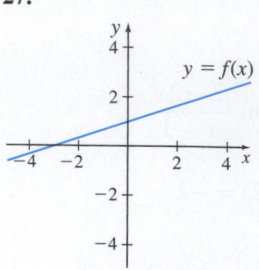

**28.**

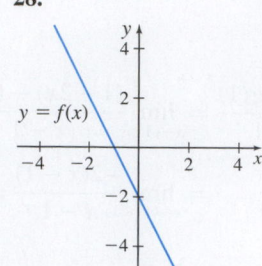

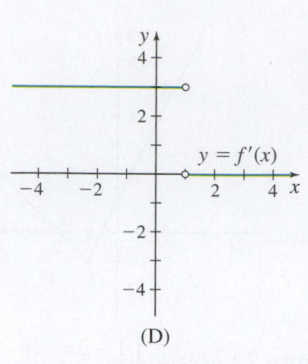

(C)                    (D)

**29.**

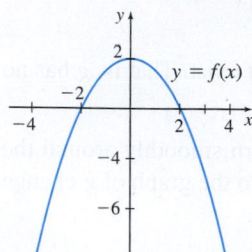

**30.**

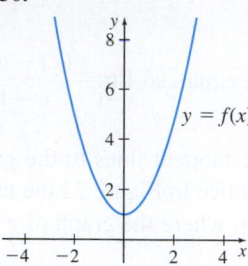

*In Problems 35–44, determine whether each function f has a derivative at c. If it does, what is $f'(c)$? If it does not, give the reason why.*

**35.** $f(x) = x^{2/3}$ at $c = -8$      **36.** $f(x) = 2x^{1/3}$ at $c = 0$

**37.** $f(x) = |x^2 - 4|$ at $c = 2$      **38.** $f(x) = |x^2 - 4|$ at $c = -2$

**39.** $f(x) = \begin{cases} 2x + 3 & \text{if } x < 1 \\ x^2 + 4 & \text{if } x \geq 1 \end{cases}$ at $c = 1$

**40.** $f(x) = \begin{cases} 3 - 4x & \text{if } x < -1 \\ 2x + 9 & \text{if } x \geq -1 \end{cases}$ at $c = -1$

**41.** $f(x) = \begin{cases} -4 + 2x & \text{if } x \leq \dfrac{1}{2} \\ 4x^2 - 4 & \text{if } x > \dfrac{1}{2} \end{cases}$ at $c = \dfrac{1}{2}$

**42.** $f(x) = \begin{cases} 2x^2 + 1 & \text{if } x < -1 \\ -1 - 4x & \text{if } x \geq -1 \end{cases}$ at $c = -1$

**43.** $f(x) = \begin{cases} 2x^2 + 1 & \text{if } x < -1 \\ 2 + 2x & \text{if } x \geq -1 \end{cases}$ at $c = -1$

**44.** $f(x) = \begin{cases} 5 - 2x & \text{if } x < 2 \\ x^2 & \text{if } x \geq 2 \end{cases}$ at $c = 2$

*In Problems 31–34, the graph of a function f is given. Match each graph to the graph of its derivative f' in A–D.*

**31.**

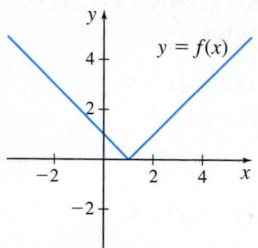

**32.**

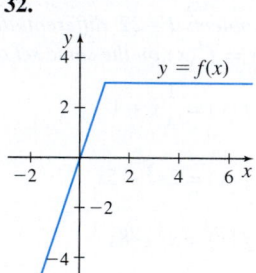

**33.**

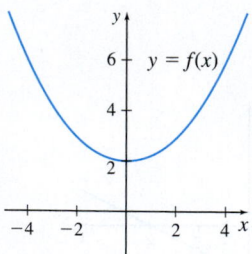

**34.**

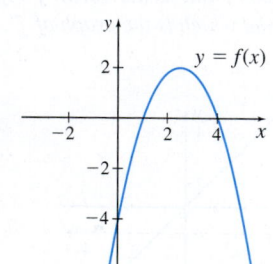

*In Problems 45–48, each function f is continuous for all real numbers, and the graph of $y = f'(x)$ is given.*

**(a)** *Does the graph of f have any horizontal tangent lines? If yes, explain why and identify where they occur.*

**(b)** *Does the graph of f have any vertical tangent lines? If yes, explain why, identify where they occur, and determine whether the point is a cusp of f.*

**(c)** *Does the graph of f have any corners? If yes, explain why and identify where they occur.*

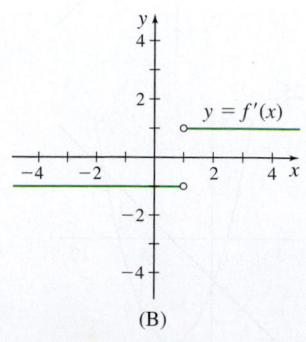

(A)                    (B)

**45.**

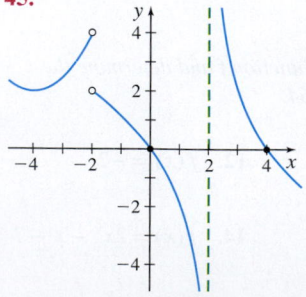

**46.**

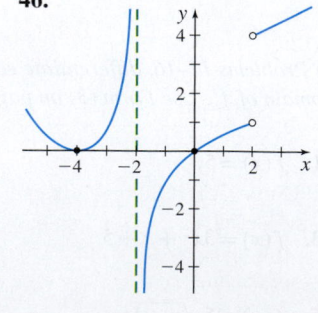

**47.**

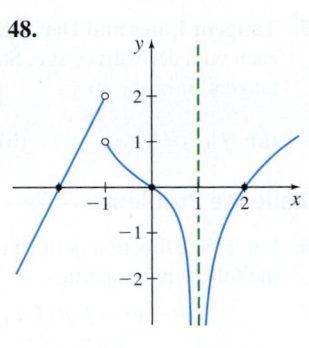

**48.**

*In Problems 49 and 50, use the given points* $(c, f(c))$ *on the graph of the function f.*

**(a)** For which numbers c does $\lim_{x \to c} f(x)$ exist but f is not continuous at c?

**(b)** For which numbers c is f continuous at c but not differentiable at c?

**49.**

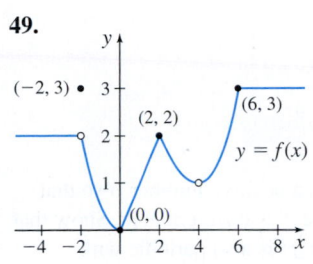

**50.**

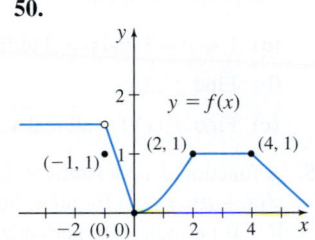

*In Problems 51–54, find the derivative of each function.*

**51.** $f(x) = mx + b$

**52.** $f(x) = ax^2 + bx + c$

**53.** $f(x) = \dfrac{1}{x^2}$

**54.** $f(x) = \dfrac{1}{\sqrt{x}}$

## Applications and Extensions

*In Problems 55–66, each limit represents the derivative of a function f at some number c. Determine f and c in each case.*

**55.** $\lim\limits_{h \to 0} \dfrac{(2+h)^2 - 4}{h}$

**56.** $\lim\limits_{h \to 0} \dfrac{(2+h)^3 - 8}{h}$

**57.** $\lim\limits_{x \to 1} \dfrac{x^2 - 1}{x - 1}$

**58.** $\lim\limits_{x \to 1} \dfrac{x^4 - 1}{x - 1}$

**59.** $\lim\limits_{h \to 0} \dfrac{\sqrt{9 + h} - 3}{h}$

**60.** $\lim\limits_{h \to 0} \dfrac{(8+h)^{1/3} - 2}{h}$

**61.** $\lim\limits_{x \to \pi/6} \dfrac{\sin x - \dfrac{1}{2}}{x - \dfrac{\pi}{6}}$

**62.** $\lim\limits_{x \to \pi/4} \dfrac{\cos x - \dfrac{\sqrt{2}}{2}}{x - \dfrac{\pi}{4}}$

**63.** $\lim\limits_{x \to 0} \dfrac{2(x+2)^2 - (x+2) - 6}{x}$

**64.** $\lim\limits_{x \to 0} \dfrac{3x^3 - 2x}{x}$

**65.** $\lim\limits_{h \to 0} \dfrac{(3+h)^2 + 2(3+h) - 15}{h}$

**66.** $\lim\limits_{h \to 0} \dfrac{3(h-1)^2 + h - 3}{h}$

**67. Units** The volume $V$ (in cubic feet) of a balloon is expanding according to $V = V(t) = 4t$, where $t$ is the time (in seconds). At what rate is the volume changing with respect to time? What are the units of $V'(t)$?

**68. Units** The area $A$ (in square miles) of a circular patch of oil is expanding according to $A = A(t) = 2t$, where $t$ is the time (in hours). At what rate is the area changing with respect to time? What are the units of $A'(t)$?

**69. Units** A manufacturer of precision digital switches has a daily cost $C$ (in dollars) of $C(x) = 10{,}000 + 3x$, where $x$ is the number of switches produced daily. What is the rate of change of cost with respect to $x$? What are the units of $C'(x)$?

**70. Units** A manufacturer of precision digital switches has daily revenue $R$ (in dollars) of $R(x) = 5x - \dfrac{x^2}{2000}$, where $x$ is the number of switches produced daily. What is the rate of change of revenue with respect to $x$? What are the units of $R'(x)$?

**71.** $f(x) = \begin{cases} x^3 & \text{if } x \le 0 \\ x^2 & \text{if } x > 0 \end{cases}$

**(a)** Determine whether $f$ is continuous at 0.

**(b)** Determine whether $f'(0)$ exists.

**(c)** Graph the function $f$ and its derivative $f'$.

**72.** For the function $f(x) = \begin{cases} 2x & \text{if } x \le 0 \\ x^2 & \text{if } x > 0 \end{cases}$

**(a)** Determine whether $f$ is continuous at 0.

**(b)** Determine whether $f'(0)$ exists.

**(c)** Graph the function $f$ and its derivative $f'$.

**73. Velocity** The distance $s$ (in feet) of an automobile from the origin at time $t$ (in seconds) is given by the position function

$$s = s(t) = \begin{cases} t^3 & \text{if } 0 \le t < 5 \\ 125 & \text{if } t \ge 5 \end{cases}$$

(This could represent a crash test in which a vehicle is accelerated until it hits a brick wall at $t = 5$ s.)

**(a)** Find the velocity just before impact (at $t = 4.99$ s) and just after impact (at $t = 5.01$ s).

**(b)** Is the velocity function $v = s'(t)$ continuous at $t = 5$?

**(c)** How do you interpret the answer to (b)?

**74. Population Growth** A simple model for population growth states that the rate of change of population size $P$ with respect to time $t$ is proportional to the population size. Express this statement as an equation involving a derivative.

**75. Atmospheric Pressure** Atmospheric pressure $p$ decreases as the distance $x$ from the surface of Earth increases, and the rate of change of pressure with respect to altitude is proportional to the pressure. Express this law as an equation involving a derivative.

**76. Electrical Current** Under certain conditions, an electric current $I$ will die out at a rate (with respect to time $t$) that is proportional to the current remaining. Express this law as an equation involving a derivative.

**77. Tangent Line** Let $f(x) = x^2 + 2$. Find all points on the graph of $f$ for which the tangent line passes through the origin.

**78. Tangent Line** Let $f(x) = x^2 - 2x + 1$. Find all points on the graph of $f$ for which the tangent line passes through the point $(1, -1)$.

**79. Area and Circumference of a Circle**  A circle of radius $r$ has area $A = \pi r^2$ and circumference $C = 2\pi r$. If the radius changes from $r$ to $r + \Delta r$, find the:

(a) Change in area.

(b) Change in circumference.

(c) Average rate of change of area with respect to radius.

(d) Average rate of change of circumference with respect to radius.

(e) Rate of change of circumference with respect to radius.

**80. Volume of a Sphere**  The volume $V$ of a sphere of radius $r$ is $V = \dfrac{4\pi r^3}{3}$. If the radius changes from $r$ to $r + \Delta r$, find the:

(a) Change in volume.

(b) Average rate of change of volume with respect to radius.

(c) Rate of change of volume with respect to radius.

**81.** Use the definition of the derivative to show that $f(x) = |x|$ has no derivative at 0.

**82.** Use the definition of the derivative to show that $f(x) = \sqrt[3]{x}$ has no derivative at 0.

**83.** If $f$ is an even function that is differentiable at $c$, show that its derivative function is odd. That is, show $f'(-c) = -f'(c)$.

**84.** If $f$ is an odd function that is differentiable at $c$, show that its derivative function is even. That is, show $f'(-c) = f'(c)$.

**85. Tangent Lines and Derivatives**  Let $f$ and $g$ be two functions, each with derivatives at $c$. State the relationship between their tangent lines at $c$ if:

(a) $f'(c) = g'(c)$    (b) $f'(c) = -\dfrac{1}{g'(c)}$    $g'(c) \neq 0$

**Challenge Problems**

**86.** Let $f$ be a function defined for all real numbers $x$. Suppose $f$ has the following properties:

$$f(u + v) = f(u)f(v) \qquad f(0) = 1 \qquad f'(0) \text{ exists}$$

(a) Show that $f'(x)$ exists for all real numbers $x$.

(b) Show that $f'(x) = f'(0)f(x)$.

**87.** A function $f$ is defined for all real numbers and has the following three properties:

$$f(1) = 5 \qquad f(3) = 21 \qquad f(a + b) - f(a) = kab + 2b^2$$

for all real numbers $a$ and $b$ where $k$ is a fixed real number independent of $a$ and $b$.

(a) Use $a = 1$ and $b = 2$ to find $k$.

(b) Find $f'(3)$.

(c) Find $f'(x)$ for all real $x$.

**88.** A function $f$ is **periodic** if there is a positive number $p$ so that $f(x + p) = f(x)$ for all $x$. Suppose $f$ is differentiable. Show that if $f$ is periodic with period $p$, then $f'$ is also periodic with period $p$.

# 2.3 The Derivative of a Polynomial Function; The Derivative of $y = e^x$

**OBJECTIVES**  *When you finish this section, you should be able to:*

1 Differentiate a constant function (p. 175)

2 Differentiate a power function (p. 175)

3 Differentiate the sum and the difference of two functions (p. 177)

4 Differentiate the exponential function $y = e^x$ (p. 180)

Finding the derivative of a function from the definition can become tedious, especially if the function $f$ is complicated. Just as we did for limits, we derive some basic derivative formulas and some properties of derivatives that make finding a derivative simpler.

Before getting started, we introduce other notations commonly used for the derivative $f'(x)$ of a function $y = f(x)$. The most common ones are

$$y' \qquad \frac{dy}{dx} \qquad Df(x)$$

**Leibniz notation** $\dfrac{dy}{dx}$ may be written in several equivalent ways as

$$\frac{dy}{dx} = \frac{d}{dx}y = \frac{d}{dx}f(x)$$

where $\dfrac{d}{dx}$ is an instruction to find the derivative (with respect to the independent variable $x$) of the function $y = f(x)$.

In **operator notation** $Df(x)$, $D$ is said to *operate* on the function, and the result is the derivative of $f$. To emphasize that the operation is performed with respect to the independent variable $x$, it is sometimes written $Df(x) = D_x f(x)$.

We use prime notation or Leibniz notation, or sometimes a mixture of the two, depending on which is more convenient. We do not use the notation $Df(x)$ in this book.

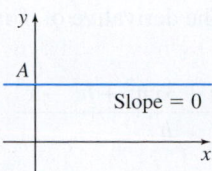

**Figure 23** $f(x) = A$

## 1 Differentiate a Constant Function

See Figure 23. Since the graph of a constant function $f(x) = A$ is a horizontal line, the tangent line to $f$ at any point is also a horizontal line, whose slope is 0. Since the derivative is the slope of the tangent line, the derivative of $f$ is 0.

**THEOREM Derivative of a Constant Function**

If $f$ is the constant function $f(x) = A$, then

$$\boxed{f'(x) = 0}$$

That is, if $A$ is a constant, then

$$\boxed{\frac{d}{dx}A = 0}$$

**Proof** If $f(x) = A$, then its derivative function is given by

$$f'(x) = \lim_{h \to 0} \frac{f(x+h) - f(x)}{h} = \lim_{h \to 0} \frac{A - A}{h} = 0$$

The definition of a     $f(x) = A$<br>derivative, Form (3)    $f(x+h) = A$

---

**EXAMPLE 1 Differentiating a Constant Function**

**(a)** If $f(x) = \sqrt{3}$, then $f'(x) = 0$    **(b)** If $f(x) = -\dfrac{1}{2}$, then $f'(x) = 0$

**(c)** If $f(x) = \pi$, then $\dfrac{d}{dx}\pi = 0$    **(d)** If $f(x) = 0$, then $\dfrac{d}{dx}0 = 0$

## 2 Differentiate a Power Function

Next we analyze the derivative of a power function $f(x) = x^n$, where $n \geq 1$ is an integer.

When $n = 1$, then $f(x) = x$ is the identity function and its graph is the line $y = x$, as shown in Figure 24.

The slope of the line $y = x$ is 1, so we would expect $f'(x) = 1$.

**Proof** $f'(x) = \dfrac{d}{dx}x = \lim_{h \to 0} \dfrac{f(x+h) - f(x)}{h} = \lim_{h \to 0} \dfrac{(x+h) - x}{h} = \lim_{h \to 0} \dfrac{h}{h} = 1$

$f(x) = x, \; f(x+h) = x + h$

**Figure 24** $f(x) = x$

**THEOREM Derivative of $f(x) = x$**

If $f(x) = x$, then

$$\boxed{f'(x) = \frac{d}{dx}x = 1}$$

When $n = 2$, then $f(x) = x^2$ is the square function. The derivative of $f$ is

$$f'(x) = \frac{d}{dx}x^2 = \lim_{h \to 0} \frac{(x+h)^2 - x^2}{h} = \lim_{h \to 0} \frac{x^2 + 2hx + h^2 - x^2}{h}$$

$$= \lim_{h \to 0} \frac{h(2x + h)}{h} = \lim_{h \to 0}(2x + h) = 2x$$

The slope of the tangent line to the graph of $f(x) = x^2$ is different for every number $x$. Figure 25 shows the graph of $f$ and several of its tangent lines. Notice that the slope of each tangent line drawn is twice the value of $x$.

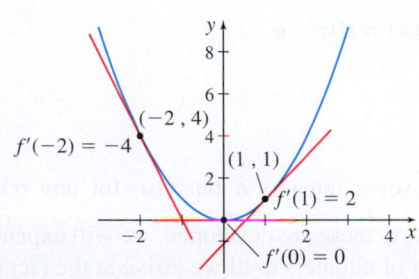

**DF Figure 25** $f(x) = x^2$

When $n = 3$, then $f(x) = x^3$ is the cube function. The derivative of $f$ is

$$f'(x) = \lim_{h \to 0} \frac{(x+h)^3 - x^3}{h} = \lim_{h \to 0} \frac{x^3 + 3x^2h + 3xh^2 + h^3 - x^3}{h}$$

$$= \lim_{h \to 0} \frac{h(3x^2 + 3xh + h^2)}{h} = \lim_{h \to 0} (3x^2 + 3xh + h^2) = 3x^2$$

**IN WORDS** The derivative of $x$ raised to an integer power $n \geq 1$ is $n$ times $x$ raised to the power $n - 1$.

Notice that the derivative of each of these power functions is another power function, whose degree is 1 less than the degree of the original function and whose coefficient is the degree of the original function. This rule holds for all power functions as the following theorem, called the *Simple Power Rule*, indicates.

**THEOREM  Simple Power Rule**

The derivative of the power function $y = x^n$, where $n \geq 1$ is an integer, is

$$\boxed{y' = \frac{d}{dx}x^n = nx^{n-1}}$$

**NEED TO REVIEW?** The Binomial Theorem is discussed in Section P.8, pp. 73–74.

**Proof** If $f(x) = x^n$ and $n$ is a positive integer, then $f(x+h) = (x+h)^n$. We use the Binomial Theorem to expand $(x+h)^n$. Then

$$f(x+h) = (x+h)^n = x^n + nx^{n-1}h + \frac{n(n-1)}{2}x^{n-2}h^2 + \frac{n(n-1)(n-2)}{6}x^{n-3}h^3 + \cdots + nxh^{n-1} + h^n$$

$$f'(x) = \lim_{h \to 0} \frac{f(x+h) - f(x)}{h}$$

$$= \lim_{h \to 0} \frac{\left[x^n + nx^{n-1}h + \frac{n(n-1)}{2}x^{n-2}h^2 + \frac{n(n-1)(n-2)}{6}x^{n-3}h^3 + \cdots + nxh^{n-1} + h^n\right] - x^n}{h}$$

$$= \lim_{h \to 0} \frac{nx^{n-1}h + \frac{n(n-1)}{2}x^{n-2}h^2 + \frac{n(n-1)(n-2)}{6}x^{n-3}h^3 + \cdots + nxh^{n-1} + h^n}{h} \qquad \text{Simplify.}$$

$$= \lim_{h \to 0} \frac{h\left[nx^{n-1} + \frac{n(n-1)}{2}x^{n-2}h + \frac{n(n-1)(n-2)}{6}x^{n-3}h^2 + \cdots + nxh^{n-2} + h^{n-1}\right]}{h} \qquad \text{Factor } h \text{ in the numerator.}$$

$$= \lim_{h \to 0} \left[nx^{n-1} + \frac{n(n-1)}{2}x^{n-2}h + \frac{n(n-1)(n-2)}{6}x^{n-3}h^2 + \cdots + nxh^{n-2} + h^{n-1}\right] \qquad \text{Divide out the common } h.$$

$$= nx^{n-1} \qquad \text{Take the limit. Only the first term remains.} \qquad \blacksquare$$

**EXAMPLE 2  Differentiating a Power Function**

**(a)** $\dfrac{d}{dx}x^5 = 5x^4$    **(b)** If $g(x) = x^{10}$, then $g'(x) = 10x^9$.  ∎

**NOW WORK** Problem 1.

*Note:* $\dfrac{d}{dx}x^n = nx^{n-1}$ is true not only for positive integers $n$ but also for any real number $n$. But the proof requires future results. As these are developed, we will expand the Power Rule to include an ever-widening set of numbers until we arrive at the fact it is true when $n$ is a real number.

But what if we want to find the derivative of the function $f(x) = ax^n$ when $a \neq 1$? The next theorem, called the *Constant Multiple Rule*, provides a way.

**THEOREM** Constant Multiple Rule

If a function $f$ is differentiable and $k$ is a constant, then $F(x) = kf(x)$ is a function that is differentiable and

$$\boxed{F'(x) = kf'(x)}$$

**IN WORDS** The derivative of a constant times a differentiable function $f$ equals the constant times the derivative of $f$.

**Proof** Use the definition of a derivative, Form (3).

$$F'(x) = \lim_{h \to 0} \frac{F(x+h) - F(x)}{h} = \lim_{h \to 0} \frac{kf(x+h) - kf(x)}{h}$$

$$= \lim_{h \to 0} \frac{k[f(x+h) - f(x)]}{h} = k \cdot \lim_{h \to 0} \frac{f(x+h) - f(x)}{h} = k \cdot f'(x) \qquad \blacksquare$$

Using Leibniz notation, the Constant Multiple Rule takes the form

$$\boxed{\frac{d}{dx}[kf(x)] = k\left[\frac{d}{dx} f(x)\right]}$$

A change in the symbol used for the independent variable does not affect the derivative formula. For example, $\dfrac{d}{dt}t^2 = 2t$ and $\dfrac{d}{du}u^5 = 5u^4$.

---

**EXAMPLE 3** **Differentiating a Constant Times a Power Function**

Find the derivative of each function:

**(a)** $f(x) = 5x^3$    **(b)** $g(u) = -\dfrac{1}{2}u^2$    **(c)** $u(x) = \pi^4 x^3$

**Solution** Notice that each of these functions involves the product of a constant and a power function. So, we use the Constant Multiple Rule followed by the Simple Power Rule.

**(a)** $f(x) = 5 \cdot x^3$, so $f'(x) = 5\left[\dfrac{d}{dx}x^3\right] = 5 \cdot 3x^2 = 15x^2$

**(b)** $g(u) = -\dfrac{1}{2} \cdot u^2$, so $g'(u) = -\dfrac{1}{2} \cdot \dfrac{d}{du}u^2 = -\dfrac{1}{2} \cdot 2u^1 = -u$

**(c)** $u(x) = \pi^4 x^3$, so $u'(x) = \pi^4 \cdot \dfrac{d}{dx}x^3 = \pi^4 \cdot 3x^2 = 3\pi^4 x^2 \qquad \blacksquare$

$\uparrow$
$\pi$ is a constant

**NOW WORK** Problem 31.

## 3 Differentiate the Sum and the Difference of Two Functions

We can find the derivative of a function that is the sum of two functions whose derivatives are known by adding the derivatives of each function.

**THEOREM** Sum Rule

**IN WORDS** The derivative of the sum of two differentiable functions equals the sum of their derivatives. That is, $(f + g)' = f' + g'$.

If two functions $f$ and $g$ are differentiable and if $F(x) = f(x) + g(x)$, then $F$ is differentiable and

$$\boxed{F'(x) = f'(x) + g'(x)}$$

**Proof** If $F(x) = f(x) + g(x)$, then

$$F(x+h) - F(x) = [f(x+h) + g(x+h)] - [f(x) + g(x)]$$
$$= [f(x+h) - f(x)] + [g(x+h) - g(x)]$$

So, the derivative of $F$ is

$$F'(x) = \lim_{h \to 0} \frac{[f(x+h) - f(x)] + [g(x+h) - g(x)]}{h}$$

$$= \lim_{h \to 0} \frac{f(x+h) - f(x)}{h} + \lim_{h \to 0} \frac{g(x+h) - g(x)}{h} \qquad \text{The limit of a sum is the sum of the limits.}$$

$$= f'(x) + g'(x) \qquad \blacksquare$$

In Leibniz notation, the Sum Rule takes the form

$$\boxed{\frac{d}{dx}[f(x) + g(x)] = \frac{d}{dx}f(x) + \frac{d}{dx}g(x)}$$

---

**EXAMPLE 4**  **Differentiating the Sum of Two Functions**

Find the derivative of $f(x) = 3x^2 + 8$.

**Solution** Here $f$ is the sum of $3x^2$ and 8. So, we begin by using the Sum Rule.

$$f'(x) = \underset{\uparrow}{\frac{d}{dx}}(3x^2 + 8) = \underset{\uparrow}{\frac{d}{dx}}(3x^2) + \underset{\uparrow}{\frac{d}{dx}}8 = 3\frac{d}{dx}x^2 + 0 = \underset{\uparrow}{3 \cdot 2x} = 6x \qquad \blacksquare$$

Sum Rule    Constant Multiple Rule    Simple Power Rule

**NOW WORK** Problem 7.

---

**THEOREM  Difference Rule**

If the functions $f$ and $g$ are differentiable and if $F(x) = f(x) - g(x)$, then $F$ is differentiable, and $F'(x) = f'(x) - g'(x)$. That is,

$$\boxed{\frac{d}{dx}[f(x) - g(x)] = \frac{d}{dx}f(x) - \frac{d}{dx}g(x)}$$

**IN WORDS** The derivative of the difference of two differentiable functions is the difference of their derivatives. That is, $(f - g)' = f' - g'$.

The proof of the Difference Rule is left as an exercise. (See Problem 78.)

The Sum and Difference Rules extend to sums (or differences) of more than two functions. That is, if the functions $f_1, f_2, \ldots, f_n$ are all differentiable, and $a_1, a_2, \ldots, a_n$ are constants, then

$$\boxed{\frac{d}{dx}[a_1 f_1(x) + a_2 f_2(x) + \cdots + a_n f_n(x)] = a_1 \frac{d}{dx}f_1(x) + a_2 \frac{d}{dx}f_2(x) + \cdots + a_n \frac{d}{dx}f_n(x)}$$

Combining the rules for finding the derivative of a constant, a power function, and a sum or difference allows us to differentiate any polynomial function.

---

**EXAMPLE 5**  **Differentiating a Polynomial Function**

(a)  Find the derivative of $f(x) = 2x^4 - 6x^2 + 2x - 3$.

(b)  What is $f'(2)$?

(c)  Find the slope of the tangent line to the graph of $f$ at the point $(1, -5)$.

(d)  Find an equation of the tangent line to the graph of $f$ at the point $(1, -5)$.

(e)  Find an equation of the normal line to the graph of $f$ at the point $(1, -5)$.

(f)  Use technology to graph $f$, the tangent line, and the normal line to the graph of $f$ at the point $(1, -5)$ on the same screen.

**Solution (a)**

$$f'(x) = \frac{d}{dx}(2x^4 - 6x^2 + 2x - 3) = \underset{\underset{\text{Sum/Difference Rules}}{\uparrow}}{\frac{d}{dx}(2x^4) - \frac{d}{dx}(6x^2) + \frac{d}{dx}(2x) - \frac{d}{dx}3}$$

$$= \underset{\underset{\text{Constant Multiple Rule}}{\uparrow}}{2 \cdot \frac{d}{dx}x^4 - 6 \cdot \frac{d}{dx}x^2 + 2 \cdot \frac{d}{dx}x - 0}$$

$$= \underset{\underset{\text{Simple Power Rule}}{\uparrow}}{2 \cdot 4x^3 - 6 \cdot 2x + 2 \cdot 1} = \underset{\underset{\text{Simplify}}{\uparrow}}{8x^3 - 12x + 2}$$

**(b)** $f'(2) = 8 \cdot 2^3 - 12 \cdot 2 + 2 = 64 - 24 + 2 = 42$.

**(c)** The slope of the tangent line at the point $(1, -5)$ equals $f'(1)$.

$$f'(1) = 8 \cdot 1^3 - 12 \cdot 1 + 2 = 8 - 12 + 2 = -2$$

**(d)** Use the point-slope form of an equation of a line to find an equation of the tangent line at $(1, -5)$.

$$y - (-5) = -2(x - 1)$$

$$y = -2(x - 1) - 5 = -2x + 2 - 5 = -2x - 3$$

The line $y = -2x - 3$ is tangent to the graph of $f(x) = 2x^4 - 6x^2 + 2x - 3$ at the point $(1, -5)$.

**(e)** Since the normal line and the tangent line at the point $(1, -5)$ on the graph of $f$ are perpendicular and the slope of the tangent line is $-2$, the slope of the normal line is $\frac{1}{2}$.

Use the point-slope form of an equation of a line to find an equation of the normal line.

$$y - (-5) = \frac{1}{2}(x - 1)$$

$$y = \frac{1}{2}(x - 1) - 5 = \frac{1}{2}x - \frac{1}{2} - 5 = \frac{1}{2}x - \frac{11}{2}$$

The line $y = \frac{1}{2}x - \frac{11}{2}$ is normal to the graph of $f$ at the point $(1, -5)$.

**(f)** The graphs of $f$, the tangent line, and the normal line to $f$ at $(1, -5)$ are shown in Figure 26. ∎

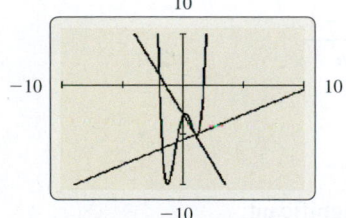

**Figure 26** $f(x) = 2x^4 - 6x^2 + 2x - 3$

**NOW WORK** Problem 33.

In some applications, we need to solve equations or inequalities involving the derivative of a function.

**EXAMPLE 6  Solving Equations and Inequalities Involving Derivatives**

**(a)** Find the points on the graph of $f(x) = 4x^3 - 12x^2 + 2$, where $f$ has a horizontal tangent line.

**(b)** Where is $f'(x) > 0$? Where is $f'(x) < 0$?

**Solution (a)** The slope of a horizontal tangent line is 0. Since the derivative of $f$ equals the slope of the tangent line, we need to find the numbers $x$ for which $f'(x) = 0$.

$$f'(x) = 12x^2 - 24x = 12x(x - 2)$$

$$12x(x - 2) = 0 \qquad\qquad\qquad f'(x) = 0$$

$$x = 0 \text{ or } x = 2 \qquad\qquad\qquad \text{Solve.}$$

At the points $(0, f(0)) = (0, 2)$ and $(2, f(2)) = (2, -14)$, the graph of the function $f(x) = 4x^3 - 12x^2 + 2$ has horizontal tangent lines.

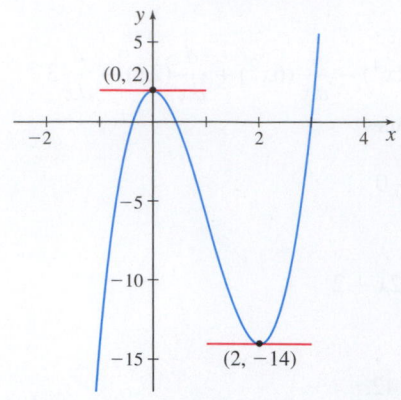

**Figure 27** $f(x) = 4x^3 - 12x^2 + 2$

**(b)** Since $f'(x) = 12x(x - 2)$ and we want to solve the inequalities $f'(x) > 0$ and $f'(x) < 0$, we use the zeros of $f'$, 0 and 2, and form a table using the intervals $(-\infty, 0)$, $(0, 2)$, and $(2, \infty)$. ∎

**TABLE 2**

| Interval | $(-\infty, 0)$ | $(0, 2)$ | $(2, \infty)$ |
|---|---|---|---|
| Sign of $f'(x) = 12x(x - 2)$ | Positive | Negative | Positive |

We conclude $f'(x) > 0$ on $(-\infty, 0) \cup (2, \infty)$ and $f'(x) < 0$ on $(0, 2)$. ∎

Figure 27 shows the graph of $f$ and the two horizontal tangent lines.

NOW WORK Problem 37.

## 4 Differentiate the Exponential Function $y = e^x$

None of the differentiation rules developed so far allows us to find the derivative of an exponential function. To differentiate $f(x) = a^x$, we return to the definition of a derivative.

NEED TO REVIEW? Exponential functions are discussed in Section P.5, pp. 41–44.

We begin by making some general observations about the derivative of $f(x) = a^x$, $a > 0$ and $a \neq 1$. We then use these observations to find the derivative of the exponential function $y = e^x$.

Suppose $f(x) = a^x$, where $a > 0$ and $a \neq 1$. The derivative of $f$ is

$$f'(x) = \lim_{h \to 0} \frac{f(x + h) - f(x)}{h} = \lim_{h \to 0} \frac{a^{x+h} - a^x}{h} = \lim_{h \to 0} \frac{a^x \cdot a^h - a^x}{h}$$
$$\uparrow \quad a^{x+h} = a^x \cdot a^h$$

$$= \lim_{h \to 0} \left[ a^x \cdot \frac{a^h - 1}{h} \right] = a^x \cdot \lim_{h \to 0} \frac{a^h - 1}{h}$$
$$\uparrow$$
Factor out $a^x$.

provided $\lim_{h \to 0} \dfrac{a^h - 1}{h}$ exists.

Three observations about the derivative are significant:

- $f'(0) = a^0 \lim_{h \to 0} \dfrac{a^h - 1}{h} = \lim_{h \to 0} \dfrac{a^h - 1}{h}$.

- $f'(x)$ is a multiple of $a^x$. In fact, $\dfrac{d}{dx} a^x = f'(0) \cdot a^x$.

- If $f'(0)$ exists, then $f'(x)$ exists, and the domain of $f'$ is the same as that of $f(x) = a^x$, all real numbers.

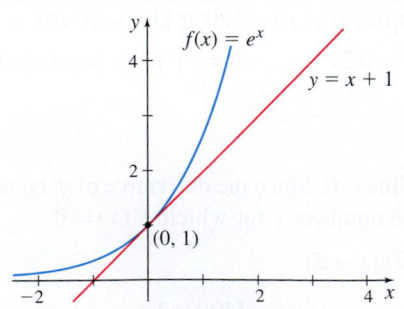

**Figure 28**

NEED TO REVIEW? The number $e$ is discussed in Section P.5, pp. 44–45.

The slope of the tangent line to the graph of $f(x) = a^x$ at the point $(0, 1)$ is $f'(0) = \lim_{h \to 0} \dfrac{a^h - 1}{h}$, and the value of this limit depends on the base $a$. In Section P.5, the number $e$ was defined as that number for which the slope of the tangent line to the graph of $y = a^x$ at the point $(0, 1)$ equals 1. That is, if $f(x) = e^x$, then $f'(0) = 1$ so that

$$\lim_{h \to 0} \frac{e^h - 1}{h} = 1$$

Figure 28 shows $f(x) = e^x$ and the tangent line $y = x + 1$ with slope 1 at the point $(0, 1)$.

Since $\dfrac{d}{dx}a^x = f'(0) \cdot a^x$, if $f(x) = e^x$, then $\dfrac{d}{dx}e^x = f'(0) \cdot e^x = 1 \cdot e^x = e^x$.

**THEOREM  Derivative of the Exponential Function $y = e^x$**

The derivative of the exponential function $y = e^x$ is

$$y' = \frac{d}{dx}e^x = e^x \qquad (1)$$

CALC
CLIP
**EXAMPLE 7  Differentiating an Expression Involving $y = e^x$**

Find the derivative of $f(x) = 4e^x + x^3$.

**Solution** The function $f$ is the sum of $4e^x$ and $x^3$. Then

$$f'(x) = \frac{d}{dx}(4e^x + x^3) = \frac{d}{dx}(4e^x) + \frac{d}{dx}x^3 = 4\frac{d}{dx}e^x + 3x^2 = 4e^x + 3x^2 \qquad \blacksquare$$

$\uparrow$ dx  $\qquad$  $\uparrow$ dx  $\quad$  $\uparrow$
Sum Rule  $\qquad$  Constant Multiple Rule;  Use (1).
$\qquad\qquad\qquad$ Simple Power Rule

**NOW WORK** Problem 25.

Now we know $\dfrac{d}{dx}e^x = e^x$. To find the derivative of $f(x) = a^x$, $a > 0$ and $a \neq 1$, we need more information. See Chapter 3.

## 2.3 Assess Your Understanding

**Concepts and Vocabulary**

1. $\dfrac{d}{dx}\pi^2 = $ _____ ;    $\dfrac{d}{dx}x^3 = $ _____ .

2. When $n$ is a positive integer, the Simple Power Rule states that $\dfrac{d}{dx}x^n = $ _____ .

3. *True or False*   The derivative of a power function of degree greater than 1 is also a power function.

4. If $k$ is a constant and $f$ is a differentiable function, then $\dfrac{d}{dx}[kf(x)] = $ _____ .

5. The derivative of $f(x) = e^x$ is _____ .

6. *True or False*   The derivative of an exponential function $f(x) = a^x$, where $a > 0$ and $a \neq 1$, is always a constant multiple of $a^x$.

**Skill Building**

In Problems 7–26, find the derivative of each function using the formulas of this section. (a, b, c, and d, when they appear, are constants.)

7. $f(x) = 3x + \sqrt{2}$

8. $f(x) = 5x - \pi$

9. $f(x) = x^2 + 3x + 4$

10. $f(x) = 4x^4 + 2x^2 - 2$

11. $f(u) = 8u^5 - 5u + 1$

12. $f(u) = 9u^3 - 2u^2 + 4u + 4$

13. $f(s) = as^3 + \dfrac{3}{2}s^2$

14. $f(s) = 4 - \pi s^2$

15. $f(t) = \dfrac{1}{3}(t^5 - 8)$

16. $f(x) = \dfrac{1}{5}(x^7 - 3x^2 + 2)$

17. $f(t) = \dfrac{t^3 + 2}{5}$

18. $f(x) = \dfrac{x^7 - 5x}{9}$

19. $f(x) = \dfrac{x^3 + 2x + 1}{7}$

20. $f(x) = \dfrac{1}{a}(ax^2 + bx + c), a \neq 0$

21. $f(x) = ax^2 + bx + c$

22. $f(x) = ax^3 + bx^2 + cx + d$

23. $f(x) = 4e^x$

24. $f(x) = -\dfrac{1}{2}e^x$

25. $f(u) = 5u^2 - 2e^u$

26. $f(u) = 3e^u + 10$

In Problems 27–32, find each derivative.

27. $\dfrac{d}{dt}\left(\sqrt{3}\,t + \dfrac{1}{2}\right)$

28. $\dfrac{d}{dt}\left(\dfrac{2t^4 - 5}{8}\right)$

29. $\dfrac{dA}{dR}$ if $A(R) = \pi R^2$

30. $\dfrac{dC}{dR}$ if $C = 2\pi R$

31. $\dfrac{dV}{dr}$ if $V = \dfrac{4}{3}\pi r^3$

32. $\dfrac{dP}{dT}$ if $P = 0.2T$

In Problems 33–36:

(a) Find the slope of the tangent line to the graph of each function $f$ at the indicated point.

(b) Find an equation of the tangent line at the point.

(c) Find an equation of the normal line at the point.

(d) Graph $f$ and the tangent line and normal line found in (b) and (c) on the same set of axes.

33. $f(x) = x^3 + 3x - 1$ at $(0, -1)$

34. $f(x) = x^4 + 2x - 1$ at $(1, 2)$

35. $f(x) = e^x + 5x$ at $(0, 1)$

36. $f(x) = 4 - e^x$ at $(0, 3)$

---

**1.** = NOW WORK problem    $\boxed{\wedge}$ = Graphing technology recommended    $\boxed{\text{CAS}}$ = Computer Algebra System recommended

*In Problems 37–42:*

**(a)** *Find the points, if any, at which the graph of each function f has a horizontal tangent line.*

**(b)** *Find an equation for each horizontal tangent line.*

**(c)** *Solve the inequality $f'(x) > 0$.*

**(d)** *Solve the inequality $f'(x) < 0$.*

**(e)** *Graph f and any horizontal lines found in (b) on the same set of axes.*

**(f)** *Describe the graph of f for the results obtained in parts (c) and (d).*

**37.** $f(x) = 3x^2 - 12x + 4$    **38.** $f(x) = x^2 + 4x - 3$

**39.** $f(x) = x + e^x$    **40.** $f(x) = 2e^x - 1$

**41.** $f(x) = x^3 - 3x + 2$    **42.** $f(x) = x^4 - 4x^3$

**43. Rectilinear Motion** At $t$ seconds, an object in rectilinear motion is $s$ meters from the origin, where $s(t) = t^3 - t + 1$. Find the velocity of the object at $t = 0$ and at $t = 5$.

**44. Rectilinear Motion** At $t$ seconds, an object in rectilinear motion is $s$ meters from the origin, where $s(t) = t^4 - t^3 + 1$. Find the velocity of the object at $t = 0$ and at $t = 1$.

**Rectilinear Motion** *In Problems 45 and 46, each position function gives the signed distance s from the origin at time t of an object in rectilinear motion:*

**(a)** *Find the velocity v of the object at any time t.*

**(b)** *When is the object at rest?*

**45.** $s(t) = 2 - 5t + t^2$    **46.** $s(t) = t^3 - \dfrac{9}{2}t^2 + 6t + 4$

*In Problems 47 and 48, use the graphs to find each derivative.*

**47.** Let $u(x) = f(x) + g(x)$ and $v(x) = f(x) - g(x)$.

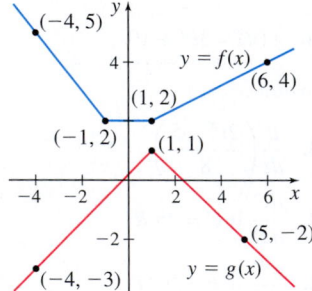

**(a)** $u'(0)$    **(b)** $u'(4)$
**(c)** $v'(-2)$    **(d)** $v'(6)$
**(e)** $3u'(5)$    **(f)** $-2v'(3)$

**48.** Let $F(t) = f(t) + g(t)$ and $G(t) = g(t) - f(t)$.

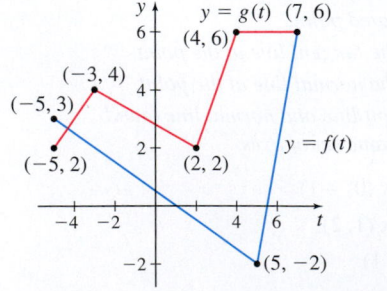

**(a)** $F'(0)$    **(b)** $F'(3)$
**(c)** $F'(-4)$    **(d)** $G'(-2)$
**(e)** $G'(-1)$    **(f)** $G'(6)$

*In Problems 49 and 50, for each function f:*

**(a)** *Find $f'(x)$ by expanding $f(x)$ and differentiating the polynomial.*

**CAS (b)** *Find $f'(x)$ using a CAS.*

**(c)** *Show that the results found in parts (a) and (b) are equivalent.*

**49.** $f(x) = (2x - 1)^3$    **50.** $f(x) = (x^2 + x)^4$

## Applications and Extensions

*In Problems 51–56, find each limit.*

**51.** $\displaystyle \lim_{h \to 0} \frac{5\left(\dfrac{1}{2} + h\right)^8 - 5\left(\dfrac{1}{2}\right)^8}{h}$

**52.** $\displaystyle \lim_{h \to 0} \frac{6(2+h)^5 - 6 \cdot 2^5}{h}$

**53.** $\displaystyle \lim_{h \to 0} \frac{\sqrt{3}(8+h)^5 - \sqrt{3} \cdot 8^5}{h}$

**54.** $\displaystyle \lim_{h \to 0} \frac{\pi(1+h)^{10} - \pi}{h}$

**55.** $\displaystyle \lim_{h \to 0} \frac{a(x+h)^3 - ax^3}{h}$

**56.** $\displaystyle \lim_{h \to 0} \frac{b(x+h)^n - bx^n}{h}$

*In Problems 57–62, find an equation of the tangent line(s) to the graph of the function f that is (are) parallel to the line L.*

**57.** $f(x) = 3x^2 - x$;  $L: y = 5x$

**58.** $f(x) = 2x^3 + 1$;  $L: y = 6x - 1$

**59.** $f(x) = e^x$;  $L: y - x - 5 = 0$

**60.** $f(x) = -2e^x$;  $L: y + 2x - 8 = 0$

**61.** $f(x) = \dfrac{1}{3}x^3 - x^2$;  $L: y = 3x - 2$

**62.** $f(x) = x^3 - x$;  $L: x + y = 0$

**63. Tangent Lines** Let $f(x) = 4x^3 - 3x - 1$.

**(a)** Find an equation of the tangent line to the graph of $f$ at $x = 2$.

**(b)** Find the coordinates of any points on the graph of $f$ where the tangent line is parallel to $y = x + 12$.

**(c)** Find an equation of the tangent line to the graph of $f$ at any points found in (b).

**(d)** Graph $f$, the tangent line found in (a), the line $y = x + 12$, and any tangent lines found in (c) on the same screen.

**64. Tangent Lines** Let $f(x) = x^3 + 2x^2 + x - 1$.

**(a)** Find an equation of the tangent line to the graph of $f$ at $x = 0$.

**(b)** Find the coordinates of any points on the graph of $f$ where the tangent line is parallel to $y = 3x - 2$.

**(c)** Find an equation of the tangent line to the graph of $f$ at any points found in (b).

**(d)** Graph $f$, the tangent line found in (a), the line $y = 3x - 2$, and any tangent lines found in (c) on the same screen.

**65. Tangent Line** Show that the line perpendicular to the $x$-axis and containing the point $(x, y)$ on the graph of $y = e^x$ and the tangent line to the graph of $y = e^x$ at the point $(x, y)$ intersect the $x$-axis 1 unit apart. See the figure.

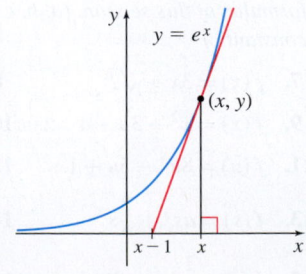

**66. Tangent Line** Show that the tangent line to the graph of $y = x^n$, $n \geq 2$ an integer, at $(1, 1)$ has $y$-intercept $1 - n$.

**67. Tangent Lines** If $n$ is an odd positive integer, show that the tangent lines to the graph of $y = x^n$ at $(1, 1)$ and at $(-1, -1)$ are parallel.

**68. Tangent Line** If the line $3x - 4y = 0$ is tangent to the graph of $y = x^3 + k$ in the first quadrant, find $k$.

**69. Tangent Line** Find the constants $a$, $b$, and $c$ so that the graph of $y = ax^2 + bx + c$ contains the point $(-1, 1)$ and is tangent to the line $y = 2x$ at $(0, 0)$.

**70. Tangent Line** Let $T$ be the tangent line to the graph of $y = x^3$ at the point $\left( \dfrac{1}{2}, \dfrac{1}{8} \right)$. At what other point $Q$ on the graph of $y = x^3$ does the line $T$ intersect the graph? What is the slope of the tangent line at $Q$?

**71. Military Tactics** A dive bomber is flying from right to left along the graph of $y = x^2$. When a rocket bomb is released, it follows a path that is approximately along the tangent line. Where should the pilot release the bomb if the target is at $(1, 0)$?

**72. Military Tactics** Answer the question in Problem 71 if the plane is flying from right to left along the graph of $y = x^3$.

**73. Fluid Dynamics** The velocity $v$ of a liquid flowing through a cylindrical tube is given by the **Hagen–Poiseuille equation** $v = k(R^2 - r^2)$, where $R$ is the radius of the tube, $k$ is a constant that depends on the length of the tube and the velocity of the liquid at its ends, and $r$ is the variable distance of the liquid from the center of the tube. See the figure below.

(a) Find the rate of change of $v$ with respect to $r$ at the center of the tube.

(b) What is the rate of change halfway from the center to the wall of the tube?

(c) What is the rate of change at the wall of the tube?

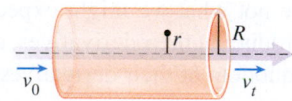

**74. Rate of Change** Water is leaking out of a swimming pool that measures 20 ft by 40 ft by 6 ft. The amount of water in the pool at a time $t$ is $W(t) = 35,000 - 20t^2$ gallons, where $t$ equals the number of hours since the pool was last filled. At what rate is the water leaking when $t = 2$ h?

**75. Luminosity of the Sun** The luminosity $L$ of a star is the rate at which it radiates energy. This rate depends on the temperature $T$ and surface area $A$ of the star's photosphere (the gaseous surface that emits the light). Luminosity is modeled by the equation $L = \sigma A T^4$, where $\sigma$ is a constant known as the **Stefan–Boltzmann constant**, and $T$ is expressed in the absolute (Kelvin) scale for which 0 K is absolute zero. As with most stars, the Sun's temperature has gradually increased over the 6 billion years of its existence, causing its luminosity to slowly increase.

(a) Find the rate at which the Sun's luminosity changes with respect to the temperature of its photosphere. Assume that the surface area $A$ remains constant.

(b) Find the rate of change at the present time. The temperature of the photosphere is presently 5800 K (10,000 °F), the radius of the photosphere is $r = 6.96 \times 10^8$ m, and $\sigma = 5.67 \times 10^{-8} \dfrac{W}{m^2 \, K^4}$.

(c) Assuming that the rate found in (b) remains constant, how much would the luminosity change if its photosphere temperature increased by 1 K (1 °C or 1.8 °F)? Compare this change to the present luminosity of the Sun.

**76. Medicine: Poiseuille's Equation** The French physician Poiseuille discovered that the volume $V$ of blood (in cubic centimeters per unit time) flowing through an artery with inner radius $R$ (in centimeters) can be modeled by

$$V(R) = kR^4$$

where $k = \dfrac{\pi}{8vl}$ is constant (here $v$ represents the viscosity of blood and $l$ is the length of the artery).

(a) Find the rate of change of the volume $V$ of blood flowing through the artery with respect to the radius $R$.

(b) Find the rate of change when $R = 0.03$ and when $R = 0.04$.

(c) If the radius of a partially clogged artery is increased from 0.03 to 0.04 cm, estimate the effect on the rate of change of the volume $V$ with respect to $R$ of the blood flowing through the enlarged artery.

(d) How do you interpret the results found in (b) and (c)?

**77. Derivative of an Area** Let $f(x) = mx$, $m > 0$. Let $F(x)$, $x > 0$, be defined as the area of the shaded region in the figure. Find $F'(x)$.

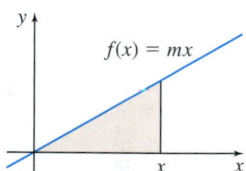

**78. The Difference Rule** Prove that if $f$ and $g$ are differentiable functions and if $F(x) = f(x) - g(x)$, then

$$F'(x) = f'(x) - g'(x)$$

**79. Simple Power Rule** Let $f(x) = x^n$, where $n$ is a positive integer. Use a factoring principle to show that

$$f'(c) = \lim_{x \to c} \frac{f(x) - f(c)}{x - c} = nc^{n-1}$$

**80. Normal Lines** For what nonnegative number $b$ is the line given by $y = -\dfrac{1}{3}x + b$ normal to the graph of $y = x^3$?

**81. Normal Lines** Let $N$ be the normal line to the graph of $y = x^2$ at the point $(-2, 4)$. At what other point $Q$ does $N$ meet the graph?

## Challenge Problems

**82. Tangent Line**   Find $a, b, c, d$ so that the tangent line to the graph of the cubic $y = ax^3 + bx^2 + cx + d$ at the point $(1, 0)$ is $y = 3x - 3$ and at the point $(2, 9)$ is $y = 18x - 27$.

**83. Tangent Line**   Find the fourth degree polynomial that contains the origin and to which the line $x + 2y = 14$ is tangent at both $x = 4$ and $x = -2$.

**84. Tangent Lines**   Find equations for all the lines containing the point $(1, 4)$ that are tangent to the graph of $y = x^3 - 10x^2 + 6x - 2$. At what points do each of the tangent lines touch the graph?

**85.** The line $x = c$, where $c > 0$, intersects the cubic $y = 2x^3 + 3x^2 - 9$ at the point $P$ and intersects the parabola $y = 4x^2 + 4x + 5$ at the point $Q$, as shown in the figure on the right.

(a) If the line tangent to the cubic at the point $P$ is parallel to the line tangent to the parabola at the point $Q$, find the number $c$.

(b) Write an equation for each of the two tangent lines described in (a).

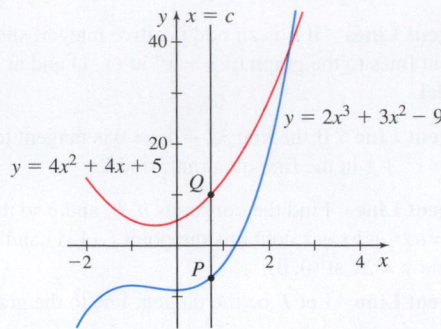

**86.** $f(x) = Ax^2 + B, A > 0$.

(a) Find $c, c > 0$, in terms of $A$ so that the tangent lines to the graph of $f$ at $(c, f(c))$ and $(-c, f(-c))$ are perpendicular.

(b) Find the slopes of the tangent lines in (a).

(c) Find the coordinates, in terms of $A$ and $B$, of the point of intersection of the tangent lines in (a).

# 2.4 Differentiating the Product and the Quotient of Two Functions; Higher-Order Derivatives

**OBJECTIVES**   *When you finish this section, you should be able to:*

**1** Differentiate the product of two functions (p. 184)

**2** Differentiate the quotient of two functions (p. 186)

**3** Find higher-order derivatives (p. 189)

**4** Work with acceleration (p. 190)

In this section, we obtain formulas for differentiating products and quotients of functions. As it turns out, the formulas are not what we might expect. The derivative of the product of two functions is *not* the product of their derivatives, and the derivative of the quotient of two functions is *not* the quotient of their derivatives.

## 1 Differentiate the Product of Two Functions

Consider the two functions $f(x) = 2x$ and $g(x) = x^3$. Both are differentiable, and their derivatives are $f'(x) = 2$ and $g'(x) = 3x^2$. Form the product

$$F(x) = f(x)g(x) = 2x \cdot x^3 = 2x^4$$

Now find $F'$ using the Constant Multiple Rule and the Simple Power Rule.

$$F'(x) = 2 \cdot 4x^3 = 8x^3$$

Notice that $f'(x)g'(x) = 2 \cdot 3x^2 = 6x^2$ is not equal to $F'(x) = \dfrac{d}{dx}[f(x)g(x)] = 8x^3$. We conclude that the derivative of a product of two functions is *not* the product of their derivatives.

To find the derivative of the product of two differentiable functions $f$ and $g$, we let $F(x) = f(x)g(x)$ and use the definition of a derivative, namely,

$$F'(x) = \lim_{h \to 0} \frac{[f(x+h)g(x+h)] - [f(x)g(x)]}{h}$$

We can express $F'$ in an equivalent form that contains the difference quotients for $f$ and $g$, by subtracting and adding $f(x+h)g(x)$ to the numerator.

$$F'(x) = \lim_{h \to 0} \frac{f(x+h)g(x+h) - f(x+h)g(x) + f(x+h)g(x) - f(x)g(x)}{h}$$

$$= \lim_{h \to 0} \frac{f(x+h)[g(x+h) - g(x)] + [f(x+h) - f(x)]g(x)}{h} \qquad \text{Group and factor.}$$

$$= \left[\lim_{h \to 0} f(x+h)\right] \left[\lim_{h \to 0} \frac{g(x+h) - g(x)}{h}\right] + \left[\lim_{h \to 0} \frac{f(x+h) - f(x)}{h}\right] \left[\lim_{h \to 0} g(x)\right] \qquad \text{Use properties of limits.}$$

$$= \left[\lim_{h \to 0} f(x+h)\right] g'(x) + f'(x) \left[\lim_{h \to 0} g(x)\right] \qquad \text{Definition of a derivative.}$$

$$= f(x)g'(x) + f'(x)g(x) \qquad \lim_{h \to 0} g(x) = g(x) \text{ since } h \text{ is not present.}$$

Since $f$ is differentiable, it is continuous, so $\lim_{h \to 0} f(x+h) = f(x)$.

We have proved the following theorem.

> **THEOREM** Product Rule
>
> If $f$ and $g$ are differentiable functions and if $F(x) = f(x)g(x)$, then $F$ is differentiable, and the derivative of the product $F$ is
>
> $$\boxed{F'(x) = [f(x)g(x)]' = f(x)g'(x) + f'(x)g(x)}$$
>
> In Leibniz notation, the Product Rule takes the form
>
> $$\boxed{\frac{d}{dx}F(x) = \frac{d}{dx}[f(x)g(x)] = f(x)\left[\frac{d}{dx}g(x)\right] + \left[\frac{d}{dx}f(x)\right]g(x)}$$

**IN WORDS** The derivative of the product of two differentiable functions equals the first function times the derivative of the second function plus the derivative of the first function times the second function. That is, $(fg)' = f(g') + (f')g$.

---

**EXAMPLE 1**  **Differentiating the Product of Two Functions**

Find $y'$ if $y = (1 + x^2)e^x$.

**Solution** The function $y$ is the product of two functions: a polynomial, $f(x) = 1 + x^2$, and the exponential function, $g(x) = e^x$. By the Product Rule,

$$y' = \frac{d}{dx}[(1+x^2)e^x] = (1+x^2)\left[\frac{d}{dx}e^x\right] + \left[\frac{d}{dx}(1+x^2)\right]e^x = (1+x^2)e^x + 2xe^x$$

$$\underset{\text{Product Rule}}{\uparrow}$$

At this point, we have found the derivative, but it is customary to simplify the answer. Then

$$y' = (1 + x^2 + 2x)e^x = (x+1)^2 e^x \qquad \blacksquare$$

$$\underset{\text{Factor out } e^x.}{\uparrow} \qquad\qquad \underset{\text{Factor.}}{\uparrow}$$

**NOW WORK** Problem 9.

Do not use the Product Rule unnecessarily! When one of the factors is a constant, use the Constant Multiple Rule. For example, it is easier to work

$$\frac{d}{dx}[5(x^2 + 1)] = 5\frac{d}{dx}(x^2 + 1) = 5 \cdot 2x = 10x$$

than it is to work

$$\frac{d}{dx}[5(x^2 + 1)] = 5\frac{d}{dx}(x^2 + 1) + \left[\frac{d}{dx}5\right](x^2 + 1) = 5 \cdot 2x + 0 \cdot (x^2 + 1) = 10x$$

Also, it is easier to simplify $f(x) = x^2(4x - 3)$ before finding the derivative. That is, it is easier to work

$$\frac{d}{dx}[x^2(4x-3)] = \frac{d}{dx}(4x^3 - 3x^2) = 12x^2 - 6x$$

than it is to use the Product Rule

$$\frac{d}{dx}[x^2(4x-3)] = x^2 \frac{d}{dx}(4x-3) + \left(\frac{d}{dx}x^2\right)(4x-3) = (x^2)(4) + (2x)(4x-3)$$

$$= 4x^2 + 8x^2 - 6x = 12x^2 - 6x$$

CALC
▶ CLIP  **EXAMPLE 2  Differentiating a Product in Two Ways**

Find the derivative of $F(v) = (5v^2 - v + 1)(v^3 - 1)$ in two ways:

**(a)** By using the Product Rule

**(b)** By multiplying the factors of the function before finding its derivative

**Solution (a)** $F$ is the product of the two functions $f(v) = 5v^2 - v + 1$ and $g(v) = v^3 - 1$. Using the Product Rule, we get

$$F'(v) = (5v^2 - v + 1)\left[\frac{d}{dv}(v^3 - 1)\right] + \left[\frac{d}{dv}(5v^2 - v + 1)\right](v^3 - 1)$$

$$= (5v^2 - v + 1)(3v^2) + (10v - 1)(v^3 - 1)$$

$$= 15v^4 - 3v^3 + 3v^2 + 10v^4 - 10v - v^3 + 1$$

$$= 25v^4 - 4v^3 + 3v^2 - 10v + 1$$

**(b)** Here we multiply the factors of $F$ before differentiating.

$$F(v) = (5v^2 - v + 1)(v^3 - 1) = 5v^5 - v^4 + v^3 - 5v^2 + v - 1$$

Then

$$F'(v) = 25v^4 - 4v^3 + 3v^2 - 10v + 1 \qquad \blacksquare$$

Notice that the derivative is the same whether you differentiate and then simplify, or whether you multiply the factors and then differentiate. Use the approach that you find easier.

NOW WORK  **Problem 13.**

## 2 Differentiate the Quotient of Two Functions

The derivative of the quotient of two functions is *not* equal to the quotient of their derivatives. Instead, the derivative of the quotient of two functions is found using the *Quotient Rule*.

**THEOREM  Quotient Rule**

If two functions $f$ and $g$ are differentiable and if $F(x) = \dfrac{f(x)}{g(x)}$, $g(x) \neq 0$, then $F$ is differentiable, and the derivative of the quotient $F$ is

$$F'(x) = \left[\frac{f(x)}{g(x)}\right]' = \frac{f'(x)g(x) - f(x)g'(x)}{[g(x)]^2}$$

**IN WORDS** The derivative of a quotient of two functions is the derivative of the numerator times the denominator, minus the numerator times the derivative of the denominator, all divided by the denominator squared. That is,

$$\left(\frac{f}{g}\right)' = \frac{f'g - fg'}{g^2}$$

In Leibniz notation, the Quotient Rule takes the form

$$\frac{d}{dx}F(x) = \frac{d}{dx}\left[\frac{f(x)}{g(x)}\right] = \frac{\left[\dfrac{d}{dx}f(x)\right]g(x) - f(x)\left[\dfrac{d}{dx}g(x)\right]}{[g(x)]^2}$$

**Proof** We use the definition of a derivative to find $F'(x)$.

$$F'(x) = \lim_{h \to 0} \frac{F(x+h) - F(x)}{h} = \lim_{h \to 0} \frac{\dfrac{f(x+h)}{g(x+h)} - \dfrac{f(x)}{g(x)}}{h}$$

$$\uparrow \quad F(x) = \frac{f(x)}{g(x)}$$

$$= \lim_{h \to 0} \frac{f(x+h)g(x) - f(x)g(x+h)}{h[g(x+h)g(x)]}$$

We write $F'$ in an equivalent form that contains the difference quotients for $f$ and $g$ by subtracting and adding $f(x)g(x)$ to the numerator.

$$F'(x) = \lim_{h \to 0} \frac{f(x+h)g(x) - f(x)g(x) + f(x)g(x) - f(x)g(x+h)}{h[g(x+h)g(x)]}$$

Now group and factor the numerator.

$$F'(x) = \lim_{h \to 0} \frac{[f(x+h) - f(x)]g(x) - f(x)[g(x+h) - g(x)]}{h[g(x+h)g(x)]}$$

$$= \lim_{h \to 0} \frac{\left[\dfrac{f(x+h) - f(x)}{h}\right]g(x) - f(x)\left[\dfrac{g(x+h) - g(x)}{h}\right]}{g(x+h)g(x)}$$

$$= \frac{\displaystyle\lim_{h \to 0}\left[\dfrac{f(x+h) - f(x)}{h}\right] \cdot \lim_{h \to 0} g(x) - \lim_{h \to 0} f(x) \cdot \lim_{h \to 0}\left[\dfrac{g(x+h) - g(x)}{h}\right]}{\displaystyle\lim_{h \to 0} g(x+h) \cdot \lim_{h \to 0} g(x)}$$

$$= \frac{f'(x)g(x) - f(x)g'(x)}{[g(x)]^2} \qquad \blacksquare$$

**RECALL** Since $g$ is differentiable, it is continuous; so, $\displaystyle\lim_{h \to 0} g(x+h) = g(x)$.

---

**EXAMPLE 3** **Differentiating the Quotient of Two Functions**

Find $y'$ if $y = \dfrac{x^2 + 1}{2x - 3}$.

**Solution** The function $y$ is the quotient of $f(x) = x^2 + 1$ and $g(x) = 2x - 3$. Using the Quotient Rule, we have

$$y' = \frac{d}{dx} \frac{x^2 + 1}{2x - 3} = \frac{\left[\dfrac{d}{dx}(x^2 + 1)\right](2x - 3) - (x^2 + 1)\left[\dfrac{d}{dx}(2x - 3)\right]}{(2x - 3)^2}$$

$$= \frac{(2x)(2x - 3) - (x^2 + 1)(2)}{(2x - 3)^2} = \frac{4x^2 - 6x - 2x^2 - 2}{(2x - 3)^2} = \frac{2x^2 - 6x - 2}{(2x - 3)^2}$$

provided $x \neq \dfrac{3}{2}$. $\blacksquare$

**NOW WORK** Problem **23**.

---

**COROLLARY** **Derivative of the Reciprocal of a Function**

If a function $g$ is differentiable, then

**IN WORDS** The derivative of the reciprocal of a function is the negative of the derivative of the denominator divided by the square of the denominator. That is, $\left(\dfrac{1}{g}\right)' = -\dfrac{g'}{g^2}$.

$$\boxed{\frac{d}{dx} \frac{1}{g(x)} = -\frac{\dfrac{d}{dx}g(x)}{[g(x)]^2} = -\frac{g'(x)}{[g(x)]^2}} \qquad (1)$$

provided $g(x) \neq 0$.

The proof of the corollary is left as an exercise. (See Problem 98.)

**EXAMPLE 4** Differentiating the Reciprocal of a Function

(a) $\dfrac{d}{dx}\dfrac{1}{x^2+x} \underset{\text{Use (1).}}{=} -\dfrac{\dfrac{d}{dx}(x^2+x)}{(x^2+x)^2} = -\dfrac{2x+1}{(x^2+x)^2}$

(b) $\dfrac{d}{dx}e^{-x} = \dfrac{d}{dx}\dfrac{1}{e^x} \underset{\text{Use (1).}}{=} -\dfrac{\dfrac{d}{dx}e^x}{(e^x)^2} = -\dfrac{e^x}{e^{2x}} = -\dfrac{1}{e^x} = -e^{-x}$ ∎

**NOW WORK** Problem 25.

Notice that the derivative of the reciprocal of a function $f$ is *not* the reciprocal of the derivative. That is,

$$\dfrac{d}{dx}\dfrac{1}{f(x)} \neq \dfrac{1}{f'(x)}$$

The rule for the derivative of the reciprocal of a function allows us to extend the Simple Power Rule to all integers. Here is the proof.

Suppose $n$ is a negative integer and $x \neq 0$. Then $m = -n$ is a positive integer, and

$$\dfrac{d}{dx}x^n = \dfrac{d}{dx}\dfrac{1}{x^m} \underset{\text{Use (1).}}{=} -\dfrac{\dfrac{d}{dx}x^m}{(x^m)^2} \underset{\text{Simple Power Rule}}{=} -\dfrac{mx^{m-1}}{x^{2m}} = -mx^{m-1-2m} = -mx^{-m-1} \underset{\text{Substitute } n=-m.}{=} nx^{n-1}$$

**NOTE** In Section 2.3 we proved the Simple Power Rule, $\dfrac{d}{dx}x^n = nx^{n-1}$ where $n$ is a positive integer. Here we have extended the Simple Power Rule from positive integers to all integers. In Chapter 3 we extend the result to include all real numbers.

**THEOREM** Power Rule

The derivative of $y = x^n$, where $n$ any integer, is

$$\boxed{y' = \dfrac{d}{dx}x^n = nx^{n-1}}$$

**EXAMPLE 5** Differentiating Using the Power Rule

(a) $\dfrac{d}{dx}x^{-1} = -x^{-2} = -\dfrac{1}{x^2}$

(b) $\dfrac{d}{du}\dfrac{1}{u^2} = \dfrac{d}{du}u^{-2} = -2u^{-3} = -\dfrac{2}{u^3}$

(c) $\dfrac{d}{ds}\dfrac{4}{s^5} = 4\dfrac{d}{ds}s^{-5} = 4 \cdot (-5)s^{-6} = -20s^{-6} = -\dfrac{20}{s^6}$ ∎

**NOW WORK** Problem 31.

**EXAMPLE 6** Using the Power Rule in Electrical Engineering

**Ohm's Law** states that the current $I$ running through a wire is inversely proportional to the resistance $R$ in the wire and can be written as $I = \dfrac{V}{R}$, where $V$ is the voltage. Find the rate of change of $I$ with respect to $R$ when $V = 12$ volts.

**Solution** The rate of change of $I$ with respect to $R$ is the derivative $\dfrac{dI}{dR}$. We write Ohm's Law with $V = 12$ as $I = \dfrac{V}{R} = 12R^{-1}$ and use the Power Rule.

$$\dfrac{dI}{dR} = \dfrac{d}{dR}(12R^{-1}) = 12 \cdot \dfrac{d}{dR}R^{-1} = 12(-1R^{-2}) = -\dfrac{12}{R^2}$$

The minus sign in $\dfrac{dI}{dR}$ indicates that the current $I$ decreases as the resistance $R$ in the wire increases. ∎

**NOW WORK** Problem 91.

## 3 Find Higher-Order Derivatives

Since the derivative $f'$ is a function, it makes sense to ask about the derivative of $f'$. The derivative (if there is one) of $f'$ is also a function called the **second derivative** of $f$ and denoted by $f''$, read "$f$ double prime."

By continuing in this fashion, we can find the **third derivative** of $f$, the **fourth derivative** of $f$, and so on, provided that these derivatives exist. Collectively, these are called **higher-order derivatives**.

Leibniz notation also can be used for higher-order derivatives. Table 3 summarizes the notation for higher-order derivatives.

**TABLE 3**

| | Prime Notation | | Leibniz Notation | |
|---|---|---|---|---|
| **First Derivative** | $y'$ | $f'(x)$ | $\dfrac{dy}{dx}$ | $\dfrac{d}{dx}f(x)$ |
| **Second Derivative** | $y''$ | $f''(x)$ | $\dfrac{d^2y}{dx^2}$ | $\dfrac{d^2}{dx^2}f(x)$ |
| **Third Derivative** | $y'''$ | $f'''(x)$ | $\dfrac{d^3y}{dx^3}$ | $\dfrac{d^3}{dx^3}f(x)$ |
| **Fourth Derivative** | $y^{(4)}$ | $f^{(4)}(x)$ | $\dfrac{d^4y}{dx^4}$ | $\dfrac{d^4}{dx^4}f(x)$ |
| $\vdots$ | | | | |
| **$n$th Derivative** | $y^{(n)}$ | $f^{(n)}(x)$ | $\dfrac{d^ny}{dx^n}$ | $\dfrac{d^n}{dx^n}f(x)$ |

---

**EXAMPLE 7**  **Finding Higher-Order Derivatives of a Power Function**

Find the second, third, and fourth derivatives of $y = 2x^3$.

**Solution** Use the Power Rule and the Constant Multiple Rule to find each derivative. The first derivative is

$$y' = \frac{d}{dx}(2x^3) = 2 \cdot \frac{d}{dx}x^3 = 2 \cdot 3x^2 = 6x^2$$

The next three derivatives are

$$y'' = \frac{d^2}{dx^2}(2x^3) = \frac{d}{dx}(6x^2) = 6 \cdot \frac{d}{dx}x^2 = 6 \cdot 2x = 12x$$

$$y''' = \frac{d^3}{dx^3}(2x^3) = \frac{d}{dx}(12x) = 12$$

$$y^{(4)} = \frac{d^4}{dx^4}(2x^3) = \frac{d}{dx}12 = 0 \qquad\blacksquare$$

All derivatives of this function $f$ of order 4 or more equal 0. This result can be generalized. For a power function $f$ of degree $n$, where $n$ is a positive integer,

$$f(x) = x^n$$
$$f'(x) = nx^{n-1}$$
$$f''(x) = n(n-1)x^{n-2}$$
$$\vdots$$
$$f^{(n)}(x) = n(n-1)(n-2) \cdot \ldots \cdot 3 \cdot 2 \cdot 1$$

**NOTE** If $n > 1$ is an integer, the product
$$n \cdot (n-1) \cdot (n-2) \cdot \ldots \cdot 3 \cdot 2 \cdot 1$$
is often written $n!$ and is read, "$n$ factorial." The **factorial symbol** $!$ means $0! = 1$, $1! = 1$, and $n! = 1 \cdot 2 \cdot 3 \cdot \ldots \cdot (n-1) \cdot n$, where $n > 1$.

The $n$th-order derivative of $f(x) = x^n$ is a constant, so all derivatives of order greater than $n$ equal 0.

It follows from this discussion that the $n$th derivative of a polynomial of degree $n$ is a constant and that all derivatives of order $n + 1$ and higher equal 0.

**NOW WORK** Problem 41.

## EXAMPLE 8  Finding Higher-Order Derivatives

Find the second and third derivatives of $y = (1 + x^2)e^x$.

**Solution** In Example 1, we found that $y' = (1 + x^2)e^x + 2xe^x = (x^2 + 2x + 1)e^x$. To find $y''$, use the Product Rule with $y'$.

$$y'' = \frac{d}{dx}[(x^2 + 2x + 1)e^x] = (x^2 + 2x + 1)\left(\frac{d}{dx}e^x\right) + \left[\frac{d}{dx}(x^2 + 2x + 1)\right]e^x$$

$$\underset{\text{Product Rule}}{\uparrow}$$

$$= (x^2 + 2x + 1)e^x + (2x + 2)e^x = (x^2 + 4x + 3)e^x$$

$$y''' = \frac{d}{dx}[(x^2 + 4x + 3)e^x] = (x^2 + 4x + 3)\frac{d}{dx}e^x + \left[\frac{d}{dx}(x^2 + 4x + 3)\right]e^x$$

$$\underset{\text{Product Rule}}{\uparrow}$$

$$= (x^2 + 4x + 3)e^x + (2x + 4)e^x = (x^2 + 6x + 7)e^x \qquad \blacksquare$$

**NOW WORK** Problem 45.

## 4 Work with Acceleration

For an object in rectilinear motion whose signed distance $s$ from the origin at time $t$ is the position function $s = s(t)$, the derivative $s'(t)$ has a physical interpretation as the velocity of the object. The second derivative $s''$, which is the rate of change of the velocity, is called *acceleration*.

**DEFINITION** Acceleration

For an object in rectilinear motion, its signed distance $s$ from the origin at time $t$ is given by a position function $s = s(t)$. The first derivative $\dfrac{ds}{dt}$ is the velocity $v = v(t)$ of the object at time $t$.

The **acceleration** $a = a(t)$ of an object at time $t$ is defined as the rate of change of velocity with respect to time. That is,

$$\boxed{a = a(t) = \frac{dv}{dt} = \frac{d}{dt}v = \frac{d}{dt}\left(\frac{ds}{dt}\right) = \frac{d^2s}{dt^2}}$$

**IN WORDS** Acceleration is the second derivative of a position function with respect to time.

## EXAMPLE 9  Analyzing Vertical Motion

A ball is propelled vertically upward from the ground with an initial velocity of 29.4 m/s. The height $s$ (in meters) of the ball above the ground is approximately $s = s(t) = -4.9t^2 + 29.4t$, where $t$ is the number of seconds that elapse from the moment the ball is released.

**(a)** What is the velocity of the ball at time $t$? What is its velocity at $t = 1$ s?

**(b)** When will the ball reach its maximum height?

**(c)** What is the maximum height the ball reaches?

**(d)** What is the acceleration of the ball at any time $t$?

**(e)** How long is the ball in the air?

**(f)** What is the velocity of the ball upon impact with the ground? What is its speed?

**(g)** What is the total distance traveled by the ball?

**Solution (a)** Since $s = s(t) = -4.9t^2 + 29.4t$, then

$$v = v(t) = \frac{ds}{dt} = -9.8t + 29.4$$

$$v(1) = -9.8 + 29.4 = 19.6$$

At $t = 1$ s, the velocity of the ball is 19.6 m/s.

**(b)** The ball reaches its maximum height when $v(t) = 0$.

$$v(t) = -9.8t + 29.4 = 0$$
$$9.8t = 29.4$$
$$t = 3$$

The ball reaches its maximum height after 3 s.

**(c)** The maximum height is

$$s = s(3) = -4.9 \cdot 3^2 + 29.4 \cdot 3 = 44.1$$

The maximum height of the ball is 44.1 m.

**(d)** The acceleration of the ball at any time $t$ is

$$a = a(t) = \frac{d^2s}{dt^2} = \frac{dv}{dt} = \frac{d}{dt}(-9.8t + 29.4) = -9.8 \text{ m/s}^2$$

**(e)** There are two ways to answer the question "How long is the ball in the air?" *First way:* Since it takes 3 s for the ball to reach its maximum height, it follows that it will take another 3 s to reach the ground, for a total time of 6 s in the air. *Second way:* When the ball reaches the ground, $s = s(t) = 0$. Solve for $t$:

$$s(t) = -4.9t^2 + 29.4t = 0$$
$$t(-4.9t + 29.4) = 0$$
$$t = 0 \quad \text{or} \quad t = \frac{29.4}{4.9} = 6$$

The ball is at ground level at $t = 0$ and at $t = 6$, so the ball is in the air for 6 s.

**(f)** Upon impact with the ground, $t = 6$ s. So the velocity is

$$v(6) = (-9.8)(6) + 29.4 = -29.4$$

Upon impact the direction of the ball is downward, and its speed is 29.4 m/s.

**(g)** The total distance traveled by the ball is

$$s(3) + s(3) = 2\,s(3) = 2(44.1) = 88.2 \text{ m} \qquad \blacksquare$$

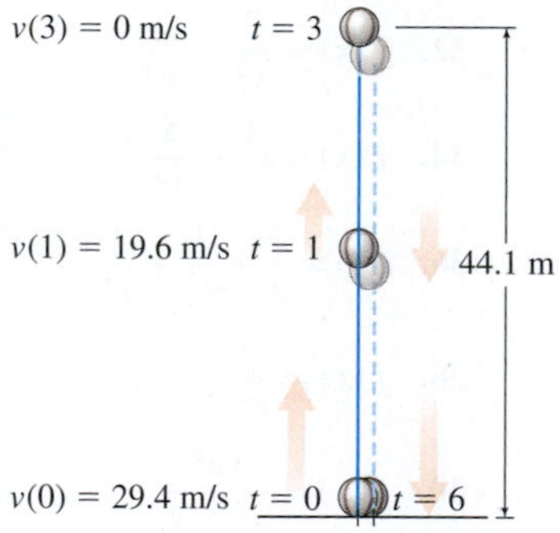

See Figure 29 for an illustration.

**NOTE** *Speed* and *velocity* are not the same. Speed measures how fast an object is moving and is defined as the absolute value of its velocity. Velocity measures both the speed and the direction of an object and may be a positive number or a negative number or zero.

$v(3) = 0$ m/s    $t = 3$

$v(1) = 19.6$ m/s    $t = 1$    44.1 m

$v(0) = 29.4$ m/s    $t = 0$    $t = 6$

**Figure 29**

**NOTE** The Earth is not perfectly round; it bulges slightly at the equator, and its mass is not distributed uniformly. As a result, the acceleration of a freely falling body varies slightly.

NOW WORK **Problem 83.**

In Example 9, the acceleration of the ball is constant. In fact, acceleration is the same for all falling objects at the same location, provided air resistance is not taken into account. In the sixteenth century, Galileo (1564–1642) discovered this by experimentation.* He also found that all falling bodies obey the law stating that the distance $s$ they fall when dropped is proportional to the square of the time $t$ it takes to fall that distance, and that the constant of proportionality $c$ is the same for all objects. That is,

$$s = -ct^2$$

---

*In a famous legend, Galileo dropped a feather and a rock from the top of the Leaning Tower of Pisa, to show that the acceleration due to gravity is constant. He expected them to fall together, but he failed to account for air resistance that slowed the feather. In July 1971, Apollo 15 astronaut David Scott repeated the experiment on the Moon, where there is no air resistance. He dropped a hammer and a falcon feather from his shoulder height. Both hit the Moon's surface at the same time. A video of this experiment may be found at the NASA website.

The velocity $v$ of the falling object is

$$v = \frac{ds}{dt} = \frac{d}{dt}(-ct^2) = -2ct$$

and its acceleration $a$ is

$$a = \frac{dv}{dt} = \frac{d^2s}{dt^2} = -2c$$

which is a constant. Usually, we denote the constant $2c$ by $g$ so $c = \frac{1}{2}g$. Then

$$a = -g \qquad v = -gt \qquad s = -\frac{1}{2}gt^2$$

The number $g$ is called the **acceleration due to gravity**. For our planet, $g$ is approximately 32 ft/s$^2$, or 9.8 m/s$^2$. On the planet Jupiter, $g \approx 26.0$ m/s$^2$, and on our moon, $g \approx 1.60$ m/s$^2$.

## 2.4 Assess Your Understanding

### Concepts and Vocabulary

1. *True or False* The derivative of a product is the product of the derivatives.

2. If $F(x) = f(x)g(x)$, then $F'(x) =$ _____ .

3. *True or False* $\dfrac{d}{dx}x^n = nx^{n+1}$, for any integer $n$.

4. If $f$ and $g \neq 0$ are two differentiable functions, then
$$\frac{d}{dx}\frac{f(x)}{g(x)} = \underline{\hspace{1.5cm}} .$$

5. *True or False* $f(x) = \dfrac{e^x}{x^2}$ can be differentiated using the

   Quotient Rule or by writing $f(x) = \dfrac{e^x}{x^2} = x^{-2}e^x$ and using the

   Product Rule.

6. If $g \neq 0$ is a differentiable function, then $\dfrac{d}{dx}\dfrac{1}{g(x)} =$ _____ .

7. If $f(x) = x$, then $f''(x) =$ _____ .

8. When an object in rectilinear motion is modeled by the position function $s = s(t)$, then the acceleration $a$ of the object at time $t$ is given by $a = a(t) =$ _____ .

### Skill Building

*In Problems 9–40, find the derivative of each function.*

9. $f(x) = xe^x$

10. $f(x) = x^2e^x$

11. $f(x) = x^2(x^3 - 1)$

12. $f(x) = x^4(x + 5)$

13. $f(x) = (3x^2 - 5)(2x + 1)$

14. $f(x) = (3x - 2)(4x + 5)$

15. $s(t) = (2t^5 - t)(t^3 - 2t + 1)$

16. $F(u) = (u^4 - 3u^2 + 1)(u^2 - u + 2)$

17. $f(x) = (x^3 + 1)(e^x + 1)$

18. $f(x) = (x^2 + 1)(e^x + x)$

19. $g(s) = \dfrac{2s}{s + 1}$

20. $F(z) = \dfrac{z + 1}{2z}$

21. $G(u) = \dfrac{1 - 2u}{1 + 2u}$

22. $f(w) = \dfrac{1 - w^2}{1 + w^2}$

23. $f(x) = \dfrac{4x^2 - 2}{3x + 4}$

24. $f(x) = \dfrac{-3x^3 - 1}{2x^2 + 1}$

25. $f(w) = \dfrac{1}{w^3 - 1}$

26. $g(v) = \dfrac{1}{v^2 + 5v - 1}$

27. $s(t) = t^{-3}$

28. $G(u) = u^{-4}$

29. $f(x) = -\dfrac{4}{e^x}$

30. $f(x) = \dfrac{3}{4e^x}$

31. $f(x) = \dfrac{10}{x^4} + \dfrac{3}{x^2}$

32. $f(x) = \dfrac{2}{x^5} - \dfrac{3}{x^3}$

33. $f(x) = 3x^3 - \dfrac{1}{3x^2}$

34. $f(x) = x^5 - \dfrac{5}{x^5}$

35. $s(t) = \dfrac{1}{t} - \dfrac{1}{t^2} + \dfrac{1}{t^3}$

36. $s(t) = \dfrac{1}{t} + \dfrac{1}{t^2} + \dfrac{1}{t^3}$

37. $f(x) = \dfrac{e^x}{x^2}$

38. $f(x) = \dfrac{x^2}{e^x}$

39. $f(x) = \dfrac{x^2 + 1}{xe^x}$

40. $f(x) = \dfrac{xe^x}{x^2 - x}$

*In Problems 41–54, find $f'$ and $f''$ for each function.*

41. $f(x) = 3x^2 + x - 2$

42. $f(x) = -5x^2 - 3x$

43. $f(x) = e^x - 3$

44. $f(x) = x - e^x$

45. $f(x) = (x + 5)e^x$

46. $f(x) = 3x^4e^x$

47. $f(x) = (2x + 1)(x^3 + 5)$

48. $f(x) = (3x - 5)(x^2 - 2)$

49. $f(x) = x + \dfrac{1}{x}$

50. $f(x) = x - \dfrac{1}{x}$

51. $f(t) = \dfrac{t^2 - 1}{t}$

52. $f(u) = \dfrac{u + 1}{u}$

53. $f(x) = \dfrac{e^x + x}{x}$

54. $f(x) = \dfrac{e^x}{x}$

55. Find $y'$ and $y''$ for **(a)** $y = \dfrac{1}{x}$ and **(b)** $y = \dfrac{2x - 5}{x}$.

56. Find $\dfrac{dy}{dx}$ and $\dfrac{d^2y}{dx^2}$ for **(a)** $y = \dfrac{5}{x^2}$ and **(b)** $y = \dfrac{2 - 3x}{x}$.

---

**1.** = NOW WORK problem    🔲 = Graphing technology recommended    CAS = Computer Algebra System recommended

**Rectilinear Motion** *In Problems 57–60, find the velocity v and acceleration a of an object in rectilinear motion whose signed distance s from the origin at time t is modeled by the position function s = s(t).*

**57.** $s(t) = 16t^2 + 20t$

**58.** $s(t) = 16t^2 + 10t + 1$

**59.** $s(t) = 4.9t^2 + 4t + 4$

**60.** $s(t) = 4.9t^2 + 5t$

*In Problems 61–68, find the indicated derivative.*

**61.** $f^{(4)}(x)$ if $f(x) = x^3 - 3x^2 + 2x - 5$

**62.** $f^{(5)}(x)$ if $f(x) = 4x^3 + x^2 - 1$

**63.** $\dfrac{d^8}{dt^8}\left(\dfrac{1}{8}t^8 - \dfrac{1}{7}t^7 + t^5 - t^3\right)$

**64.** $\dfrac{d^6}{dt^6}(t^6 + 5t^5 - 2t + 4)$

**65.** $\dfrac{d^7}{du^7}(e^u + u^2)$

**66.** $\dfrac{d^{10}}{du^{10}}(2e^u)$

**67.** $\dfrac{d^5}{dx^5}(-e^x)$

**68.** $\dfrac{d^8}{dx^8}(12x - e^x)$

*In Problems 69–72:*

**(a)** *Find the slope of the tangent line for each function f at the given point.*

**(b)** *Find an equation of the tangent line to the graph of each function f at the given point.*

**(c)** *Find the points, if any, where the graph of the function has a horizontal tangent line.*

**(d)** *Graph each function, the tangent line found in (b), and any tangent lines found in (c) on the same set of axes.*

**69.** $f(x) = \dfrac{x^2}{x - 1}$ at $\left(-1, -\dfrac{1}{2}\right)$

**70.** $f(x) = \dfrac{x}{x + 1}$ at $(0, 0)$

**71.** $f(x) = \dfrac{x^3}{x + 1}$ at $\left(1, \dfrac{1}{2}\right)$

**72.** $f(x) = \dfrac{x^2 + 1}{x}$ at $\left(2, \dfrac{5}{2}\right)$

*In Problems 73–80:*

**(a)** *Find the points, if any, at which the graph of each function f has a horizontal tangent line.*

**(b)** *Find an equation for each horizontal tangent line.*

**(c)** *Solve the inequality $f'(x) > 0$.*

**(d)** *Solve the inequality $f'(x) < 0$.*

**(e)** *Graph f and any horizontal lines found in (b) on the same set of axes.*

**(f)** *Describe the graph of f for the results obtained in (c) and (d).*

**73.** $f(x) = (x + 1)(x^2 - x - 11)$

**74.** $f(x) = (3x^2 - 2)(2x + 1)$

**75.** $f(x) = \dfrac{x^2}{x + 1}$

**76.** $f(x) = \dfrac{x^2 + 1}{x}$

**77.** $f(x) = xe^x$

**78.** $f(x) = x^2 e^x$

**79.** $f(x) = \dfrac{x^2 - 3}{e^x}$

**80.** $f(x) = \dfrac{e^x}{x^2 + 1}$

*In Problems 81 and 82, use the graphs to determine each derivative.*

**81.** Let $u(x) = f(x) \cdot g(x)$ and $v(x) = \dfrac{g(x)}{f(x)}$.

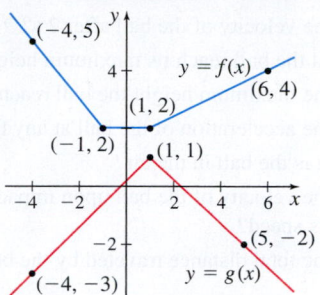

**(a)** $u'(0)$

**(b)** $u'(4)$

**(c)** $v'(-2)$

**(d)** $v'(6)$

**(e)** $\dfrac{d}{dx}\dfrac{1}{f(x)}$ at $x = -2$

**(f)** $\dfrac{d}{dx}\dfrac{1}{g(x)}$ at $x = 4$

**82.** Let $F(t) = f(t) \cdot g(t)$ and $G(t) = \dfrac{f(t)}{g(t)}$.

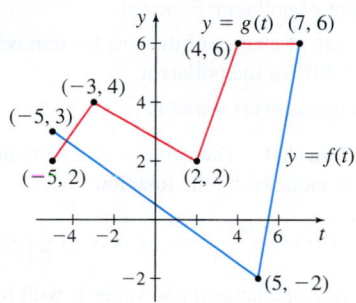

**(a)** $F'(0)$

**(b)** $F'(3)$

**(c)** $F'(-4)$

**(d)** $G'(-2)$

**(e)** $G'(-1)$

**(f)** $\dfrac{d}{dt}\dfrac{1}{f(t)}$ at $t = 3$

## Applications and Extensions

**83. Vertical Motion** An object is propelled vertically upward from the ground with an initial velocity of 39.2 m/s. The signed distance s (in meters) of the object from the ground after t seconds is given by the position function $s = s(t) = -4.9t^2 + 39.2t$.

**(a)** What is the velocity of the object at time $t$?

**(b)** When will the object reach its maximum height?

**(c)** What is the maximum height?

**(d)** What is the acceleration of the object at any time $t$?

**(e)** How long is the object in the air?

**(f)** What is the velocity of the object upon impact with the ground? What is its speed?

**(g)** What is the total distance traveled by the object?

**84. Vertical Motion** A ball is thrown vertically upward from a height of 6 ft with an initial velocity of 80 ft/s. The signed distance $s$ (in feet) of the ball from the ground after $t$ seconds is given by the position function $s = s(t) = 6 + 80t - 16t^2$.

(a) What is the velocity of the ball after 2 s?

(b) When will the ball reach its maximum height?

(c) What is the maximum height the ball reaches?

(d) What is the acceleration of the ball at any time $t$?

(e) How long is the ball in the air?

(f) What is the velocity of the ball upon impact with the ground? What is its speed?

(g) What is the total distance traveled by the ball?

**85. Environmental Cost** The cost $C$, in thousands of dollars, for the removal of a pollutant from a certain lake is given by the function $C(x) = \dfrac{5x}{110 - x}$, where $x$ is the percent of pollutant removed.

(a) What is the domain of $C$?

(b) Graph $C$.

(c) What is the cost to remove 80% of the pollutant?

(d) Find $C'(x)$, the rate of change of the cost $C$ with respect to the amount of pollutant removed.

(e) Find the rate of change of the cost for removing 40%, 60%, 80%, and 90% of the pollutant.

(f) Interpret the answers found in (e).

**86. Investing in Fine Art** The value $V$ of a painting $t$ years after it is purchased is modeled by the function

$$V(t) = \frac{100t^2 + 50}{t} + 400 \quad 1 \le t \le 5$$

(a) Find the rate of change in the value $V$ with respect to time.

(b) What is the rate of change in value after 2 years?

(c) What is the rate of change in value after 3 years?

(d) Interpret the answers in (b) and (c).

**87. Drug Concentration** The concentration of a drug in a patient's blood $t$ hours after injection is given by the function $f(t) = \dfrac{0.4t}{2t^2 + 1}$ (in milligrams per liter).

(a) Find the rate of change of the concentration with respect to time.

(b) What is the rate of change of the concentration after 10 min? After 30 min? After 1 hour?

(c) Interpret the answers found in (b).

(d) Graph $f$ for the first 5 hours after administering the drug.

(e) From the graph, approximate the time (in minutes) at which the concentration of the drug is highest. What is the highest concentration of the drug in the patient's blood?

**88. Population Growth** A population of 1000 bacteria is introduced into a culture and grows in number according to the formula $P(t) = 1000 \left( 1 + \dfrac{4t}{100 + t^2} \right)$, where $t$ is measured in hours.

(a) Find the rate of change in population with respect to time.

(b) What is the rate of change in population at $t = 1$, $t = 2$, $t = 3$, and $t = 4$?

(c) Interpret the answers found in (b).

(d) Graph $P = P(t)$, $0 \le t \le 20$.

(e) From the graph, approximate the time (in hours) when the population is the greatest. What is the maximum population of the bacteria in the culture?

**89. Economics** The price-demand function for a popular e-book is given by $D(p) = \dfrac{100,000}{p^2 + 10p + 50}$, $4 \le p \le 20$, where $D = D(p)$ is the quantity demanded at the price $p$ dollars.

(a) Find $D'(p)$, the rate of change of demand with respect to price.

(b) Find $D'(5)$, $D'(10)$, and $D'(15)$.

(c) Interpret the results found in (b).

**90. Intensity of Light** The intensity of illumination $I$ on a surface is inversely proportional to the square of the distance $r$ from the surface to the source of light. If the intensity is 1000 units when the distance is 1 m from the light, find the rate of change of the intensity with respect to the distance when the source is 10 meters from the surface.

**91. Ideal Gas Law** The Ideal Gas Law, used in chemistry and thermodynamics, relates the pressure $p$, the volume $V$, and the absolute temperature $T$ (in Kelvin) of a gas, using the equation $pV = nRT$, where $n$ is the amount of gas (in moles) and $R = 8.31$ is the ideal gas constant. In an experiment, a spherical gas container of radius $r$ meters is placed in a pressure chamber and is slowly compressed while keeping its temperature at 273 K.

(a) Find the rate of change of the pressure $p$ with respect to the radius $r$ of the chamber.

(*Hint:* The volume $V$ of a sphere is $V = \dfrac{4}{3}\pi r^3$.)

(b) Interpret the sign of the answer found in (a).

(c) If the sphere contains 1.0 mol of gas, find the rate of change of the pressure when $r = \dfrac{1}{4}$ m.

(*Note:* The metric unit of pressure is the pascal, Pa).

**92. Body Density** The density $\rho$ of an object is its mass $m$ divided by its volume $V$; that is, $\rho = \dfrac{m}{V}$. If a person dives below the surface of the ocean, the water pressure on the diver will steadily increase, compressing the diver and therefore increasing body density. Suppose the diver is modeled as a sphere of radius $r$.

(a) Find the rate of change of the diver's body density with respect to the radius $r$ of the sphere.

(*Hint:* The volume $V$ of a sphere is $V = \dfrac{4}{3}\pi r^3$.)

(b) Interpret the sign of the answer found in (a).

(c) Find the rate of change of the diver's body density when the radius is 45 cm and the mass is 80,000 g (80 kg).

***Jerk and Snap*** *Problems 93–96 use the following discussion:* Suppose that an object is moving in rectilinear motion so that its signed distance $s$ from the origin at time $t$ is given by the position function $s = s(t)$. The velocity $v = v(t)$ of the object at time $t$ is the rate of change of $s$ with respect to time, namely, $v = v(t) = \dfrac{ds}{dt}$. The acceleration $a = a(t)$

of the object at time $t$ is the rate of change of the velocity with respect to time,

$$a = a(t) = \frac{dv}{dt} = \frac{d}{dt}\left(\frac{ds}{dt}\right) = \frac{d^2s}{dt^2}$$

There are also physical interpretations of the third derivative and the fourth derivative of $s = s(t)$. The **jerk** $J = J(t)$ of the object at time $t$ is the rate of change of the acceleration $a$ with respect to time; that is,

$$J = J(t) = \frac{da}{dt} = \frac{d}{dt}\left(\frac{dv}{dt}\right) = \frac{d^2v}{dt^2} = \frac{d^3s}{dt^3}$$

The **snap** $S = S(t)$ of the object at time $t$ is the rate of change of the jerk $J$ with respect to time; that is,

$$S = S(t) = \frac{dJ}{dt} = \frac{d^2a}{dt^2} = \frac{d^3v}{dt^3} = \frac{d^4s}{dt^4}$$

Engineers take jerk into consideration when designing elevators, aircraft, and cars. In these cases, they try to minimize jerk, making for a smooth ride. But when designing thrill rides, such as roller coasters, the jerk is increased, making for an exciting experience.

**93. Rectilinear Motion**  As an object in rectilinear motion moves, its signed distance $s$ from the origin at time $t$ is given by the position function $s = s(t) = t^3 - t + 1$, where $s$ is in meters and $t$ is in seconds.

(a) Find the velocity $v$, acceleration $a$, jerk $J$, and snap $S$ of the object at time $t$.

(b) When is the velocity of the object 0 m/s?

(c) Find the acceleration of the object at $t = 2$ and at $t = 5$.

(d) Does the jerk of the object ever equal 0 m/s$^3$?

(e) How would you interpret the snap for this object in rectilinear motion?

**94. Rectilinear Motion**  As an object in rectilinear motion moves, its signed distance $s$ from the origin at time $t$ is given by the position function $s = s(t) = \frac{1}{6}t^4 - t^2 + \frac{1}{2}t + 4$, where $s$ is in meters and $t$ is in seconds.

(a) Find the velocity $v$, acceleration $a$, jerk $J$, and snap $S$ of the object at any time $t$.

(b) Find the velocity of the object at $t = 0$ and at $t = 3$.

(c) Find the acceleration of the object at $t = 0$. Interpret your answer.

(d) Is the jerk of the object constant? In your own words, explain what the jerk says about the acceleration of the object.

(e) How would you interpret the snap for this object in rectilinear motion?

**95. Elevator Ride Quality**  The ride quality of an elevator depends on several factors, two of which are acceleration and jerk. In a study of 367 persons riding in a 1600-kg elevator that moves at an average speed of 4 m/s, the majority of riders were comfortable in an elevator with vertical motion given by

$$s(t) = 4t + 0.8t^2 + 0.333t^3$$

(a) Find the acceleration that the riders found acceptable.

(b) Find the jerk that the riders found acceptable.

*Source*: Elevator Ride Quality, January 2007, http://www.lift-report.de/index.php/news/176/368/Elevator-Ride-Quality.

**96. Elevator Ride Quality**  In a hospital, the effects of high acceleration or jerk may be harmful to patients, so the acceleration and jerk need to be lower than in standard elevators. It has been determined that a 1600-kg elevator that is installed in a hospital and that moves at an average speed of 4 m/s should have vertical motion

$$s(t) = 4t + 0.55t^2 + 0.1167t^3$$

(a) Find the acceleration of a hospital elevator.

(b) Find the jerk of a hospital elevator.

*Source*: Elevator Ride Quality, January 2007, http://www.lift-report.de/index.php/news/176/368/Elevator-Ride-Quality.

**97. Current Density in a Wire**  The current density $J$ in a wire is a measure of how much an electrical current is compressed as it flows through a wire and is modeled by the function $J(A) = \dfrac{I}{A}$, where $I$ is the current (in amperes) and $A$ is the cross-sectional area of the wire. In practice, current density, rather than merely current, is often important. For example, superconductors lose their superconductivity if the current density is too high.

(a) As current flows through a wire, it heats the wire, causing it to expand in area $A$. If a constant current is maintained in a cylindrical wire, find the rate of change of the current density $J$ with respect to the radius $r$ of the wire.

(b) Interpret the sign of the answer found in (a).

(c) Find the rate of change of current density with respect to the radius $r$ when a current of 2.5 amps flows through a wire of radius $r = 0.50$ mm.

**98. Derivative of a Reciprocal, Function**  Prove that if a function $g$ is differentiable, then $\dfrac{d}{dx}\dfrac{1}{g(x)} = -\dfrac{g'(x)}{[g(x)]^2}$, provided $g(x) \neq 0$.

**99. Extended Product Rule**  Show that if $f$, $g$, and $h$ are differentiable functions, then

$$\frac{d}{dx}[f(x)g(x)h(x)] = f(x)g(x)h'(x) + f(x)g'(x)h(x) + f'(x)g(x)h(x)$$

From this, deduce that

$$\frac{d}{dx}[f(x)]^3 = 3[f(x)]^2 f'(x)$$

*In Problems 100–105, use the Extended Product Rule (Problem 99) to find $y'$.*

**100.** $y = (x^2 + 1)(x - 1)(x + 5)$

**101.** $y = (x - 1)(x^2 + 5)(x^3 - 1)$

**102.** $y = (x^4 + 1)^3$            **103.** $y = (x^3 + 1)^3$

**104.** $y = (3x + 1)\left(1 + \dfrac{1}{x}\right)(x^{-5} + 1)$

**105.** $y = \left(1 - \dfrac{1}{x}\right)\left(1 - \dfrac{1}{x^2}\right)\left(1 - \dfrac{1}{x^3}\right)$

**106. (Further) Extended Product Rule**  Write a formula for the derivative of the product of four differentiable functions. That is, find a formula for $\dfrac{d}{dx}[f_1(x)f_2(x)f_3(x)f_4(x)]$. Also find a formula for $\dfrac{d}{dx}[f(x)]^4$.

**107.** If $f$ and $g$ are differentiable functions, show that

if $F(x) = \dfrac{1}{f(x)g(x)}$, then

$$F'(x) = -F(x)\left[\frac{f'(x)}{f(x)} + \frac{g'(x)}{g(x)}\right]$$

provided $f(x) \neq 0$, $g(x) \neq 0$.

**108. Higher-Order Derivatives**  If $f(x) = \dfrac{1}{1-x}$, find a formula

for the $n$th derivative of $f$. That is, find $f^{(n)}(x)$.

**109.** Let $f(x) = \dfrac{x^6 - x^4 + x^2}{x^4 + 1}$. Rewrite $f$ in the form

$(x^4 + 1)f(x) = x^6 - x^4 + x^2$. Now find $f'(x)$ without using the quotient rule.

**110.** If $f$ and $g$ are differentiable functions with $f \neq -g$, find the

derivative of $\dfrac{fg}{f+g}$.

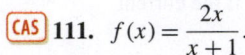 **111.** $f(x) = \dfrac{2x}{x+1}$.

(a) Use technology to find $f'(x)$.

(b) Simplify $f'$ to a single fraction using either algebra or a CAS.

(c) Use technology to find $f^{(5)}(x)$. (*Hint:* Your CAS may have a method for finding higher-order derivatives without finding other derivatives first.)

**Challenge Problems** —————————————

**112.** Suppose $f$ and $g$ have derivatives up to the fourth order. Find the first four derivatives of the product $fg$ and simplify the answers. In particular, show that the fourth derivative is

$$\frac{d^4}{dx^4}(fg) = f^{(4)}g + 4f^{(3)}g^{(1)} + 6f^{(2)}g^{(2)} + 4f^{(1)}g^{(3)} + fg^{(4)}$$

Identify a pattern for the higher-order derivatives of $fg$.

**113.** Suppose $f_1(x), \dots, f_n(x)$ are differentiable functions.

(a) Find $\dfrac{d}{dx}[f_1(x) \cdot \dots \cdot f_n(x)]$.

(b) Find $\dfrac{d}{dx}\dfrac{1}{f_1(x) \cdot \dots \cdot f_n(x)}$.

**114.** Let $a, b, c$, and $d$ be real numbers. Define

$$\begin{vmatrix} a & b \\ c & d \end{vmatrix} = ad - bc$$

This is called a $2 \times 2$ **determinant** and it arises in the study of linear equations. Let $f_1(x)$, $f_2(x)$, $f_3(x)$, and $f_4(x)$ be differentiable and let

$$D(x) = \begin{vmatrix} f_1(x) & f_2(x) \\ f_3(x) & f_4(x) \end{vmatrix}$$

Show that

$$D'(x) = \begin{vmatrix} f_1'(x) & f_2'(x) \\ f_3(x) & f_4(x) \end{vmatrix} + \begin{vmatrix} f_1(x) & f_2(x) \\ f_3'(x) & f_4'(x) \end{vmatrix}$$

**115.** Let $f_0(x) = x - 1$

$$f_1(x) = 1 + \frac{1}{x-1}$$

$$f_2(x) = 1 + \frac{1}{1 + \dfrac{1}{x-1}}$$

$$f_3(x) = 1 + \frac{1}{1 + \dfrac{1}{1 + \dfrac{1}{x-1}}}$$

(a) Write $f_1$, $f_2$, $f_3$, $f_4$, and $f_5$ in the form $\dfrac{ax+b}{cx+d}$.

(b) Using the results from (a), write the sequence of numbers representing the coefficients of $x$ in the numerator, beginning with $f_0(x) = x - 1$.

(c) Write the sequence in (b) as a recursive sequence. (*Hint:* Look at the sum of consecutive terms.)

(d) Find $f_0'$, $f_1'$, $f_2'$, $f_3'$, $f_4'$, and $f_5'$.

# 2.5 The Derivative of the Trigonometric Functions

**OBJECTIVE**  *When you finish this section, you should be able to:*

1  Differentiate trigonometric functions (p. 196)

## 1 Differentiate Trigonometric Functions

To find the derivatives of $y = \sin x$ and $y = \cos x$, we use the limits

$$\lim_{\theta \to 0} \frac{\sin \theta}{\theta} = 1 \qquad \text{and} \qquad \lim_{\theta \to 0} \frac{\cos \theta - 1}{\theta} = 0$$

that were established in Section 1.4.

**THEOREM**  Derivative of $y = \sin x$

The derivative of $y = \sin x$ is $y' = \cos x$. That is,

$$y' = \frac{d}{dx}\sin x = \cos x$$

**Proof**

$$y' = \lim_{h \to 0} \frac{\sin(x+h) - \sin x}{h} \qquad \text{The definition of a derivative}$$

$$= \lim_{h \to 0} \frac{\sin x \cos h + \sin h \cos x - \sin x}{h} \qquad \sin(A+B) = \sin A \cos B + \sin B \cos A$$

$$= \lim_{h \to 0} \left[ \frac{\sin x \cos h - \sin x}{h} + \frac{\sin h \cos x}{h} \right] \qquad \text{Rearrange terms.}$$

$$= \lim_{h \to 0} \left[ \sin x \cdot \frac{\cos h - 1}{h} + \frac{\sin h}{h} \cdot \cos x \right] \qquad \text{Factor.}$$

$$= \left[ \lim_{h \to 0} \sin x \right] \left[ \lim_{h \to 0} \frac{\cos h - 1}{h} \right] + \left[ \lim_{h \to 0} \cos x \right] \left[ \lim_{h \to 0} \frac{\sin h}{h} \right] \qquad \text{Use properties of limits.}$$

$$= \sin x \cdot 0 + \cos x \cdot 1 = \cos x \qquad \lim_{\theta \to 0} \frac{\cos \theta - 1}{\theta} = 0; \quad \lim_{\theta \to 0} \frac{\sin \theta}{\theta} = 1 \quad \blacksquare$$

**NEED TO REVIEW?** The trigonometric functions are discussed in Section P.6, pp. 52–59. Trigonometric identities are discussed in Appendix A.4, pp. A-32 to A-35.

The geometry of the derivative $\dfrac{d}{dx} \sin x = \cos x$ is shown in Figure 30. On the graph of $f(x) = \sin x$, the horizontal tangents are marked as well as the tangent lines that have slopes of 1 and $-1$. The derivative function is plotted on the second graph and those points are connected with a smooth curve.

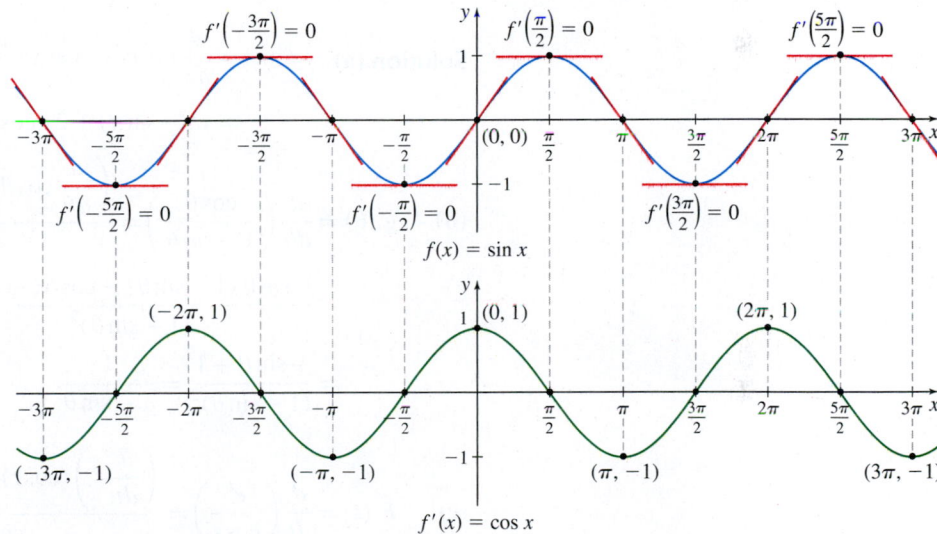

**Figure 30**

To find derivatives involving the trigonometric functions, use the sum, difference, product, and quotient rules and the derivative formulas from Sections 2.3 and 2.4.

---

**EXAMPLE 1  Differentiating the Sine Function**

Find $y'$ if:

**(a)** $y = x + 4 \sin x$    **(b)** $y = x^2 \sin x$    **(c)** $y = \dfrac{\sin x}{x}$    **(d)** $y = e^x \sin x$

**Solution (a)** Use the Sum Rule and the Constant Multiple Rule.

$$y' = \frac{d}{dx}(x + 4 \sin x) = \frac{d}{dx} x + \frac{d}{dx}(4 \sin x) = 1 + 4 \frac{d}{dx} \sin x = 1 + 4 \cos x$$

**(b)** Use the Product Rule.

$$y' = \frac{d}{dx}(x^2 \sin x) = x^2 \left[ \frac{d}{dx} \sin x \right] + \left[ \frac{d}{dx} x^2 \right] \sin x = x^2 \cos x + 2x \sin x$$

**(c)** Use the Quotient Rule.

$$y' = \frac{d}{dx}\left(\frac{\sin x}{x}\right) = \frac{\left[\dfrac{d}{dx}\sin x\right]\cdot x - \sin x \cdot \left[\dfrac{d}{dx}x\right]}{x^2} = \frac{x\cos x - \sin x}{x^2}$$

**(d)** Use the Product Rule.

$$y' = \frac{d}{dx}(e^x \sin x) = e^x \frac{d}{dx}\sin x + \left(\frac{d}{dx}e^x\right)\sin x$$

$$= e^x \cos x + e^x \sin x = e^x(\cos x + \sin x)$$

**NOW WORK** Problems 5, 29, and 45.

> **THEOREM** Derivative of $y = \cos x$
>
> The derivative of $y = \cos x$ is
>
> $$y' = \frac{d}{dx}\cos x = -\sin x$$

You are asked to prove this in Problem 75.

**CALC** **CLIP**

**EXAMPLE 2** Differentiating Trigonometric Functions

Find the derivative of each function:

**(a)** $f(x) = x^2 \cos x$ **(b)** $g(\theta) = \dfrac{\cos\theta}{1 - \sin\theta}$ **(c)** $F(t) = \dfrac{e^t}{\cos t}$

**Solution (a)** $\quad f'(x) = \dfrac{d}{dx}(x^2\cos x) = x^2\dfrac{d}{dx}\cos x + \left(\dfrac{d}{dx}x^2\right)(\cos x)$

$$= x^2(-\sin x) + 2x\cos x = 2x\cos x - x^2\sin x$$

**(b)** $\quad g'(\theta) = \dfrac{d}{d\theta}\left(\dfrac{\cos\theta}{1-\sin\theta}\right) = \dfrac{\left(\dfrac{d}{d\theta}\cos\theta\right)(1-\sin\theta) - (\cos\theta)\left[\dfrac{d}{d\theta}(1-\sin\theta)\right]}{(1-\sin\theta)^2}$

$$= \frac{-\sin\theta\,(1-\sin\theta) - \cos\theta(-\cos\theta)}{(1-\sin\theta)^2} = \frac{-\sin\theta + \sin^2\theta + \cos^2\theta}{(1-\sin\theta)^2}$$

$$= \frac{-\sin\theta + 1}{(1-\sin\theta)^2} = \frac{1}{1-\sin\theta}$$

**(c)** $\quad F'(t) = \dfrac{d}{dt}\left(\dfrac{e^t}{\cos t}\right) = \dfrac{\left(\dfrac{d}{dt}e^t\right)(\cos t) - e^t\left(\dfrac{d}{dt}\cos t\right)}{\cos^2 t} = \dfrac{e^t\cos t - e^t(-\sin t)}{\cos^2 t}$

$$= \frac{e^t(\cos t + \sin t)}{\cos^2 t}$$

**NOW WORK** Problem 13.

**EXAMPLE 3** Identifying Horizontal Tangent Lines

Find all points on the graph of $f(x) = x + \sin x$ where the tangent line is horizontal.

**Solution** Since tangent lines are horizontal at points on the graph of $f$ where $f'(x) = 0$, begin by finding $f'(x) = 1 + \cos x$.

Now solve the equation: $f'(x) = 1 + \cos x = 0$.

$$\cos x = -1$$
$$x = (2k+1)\pi \qquad \text{where } k \text{ is an integer}$$

At each of the points $((2k+1)\pi, (2k+1)\pi)$, where $k$ is an integer, the graph of $f$ has a horizontal tangent line. See Figure 31.

Notice that each of the points with a horizontal tangent line lies on the line $y = x$. ∎

**NOW WORK** Problem 57.

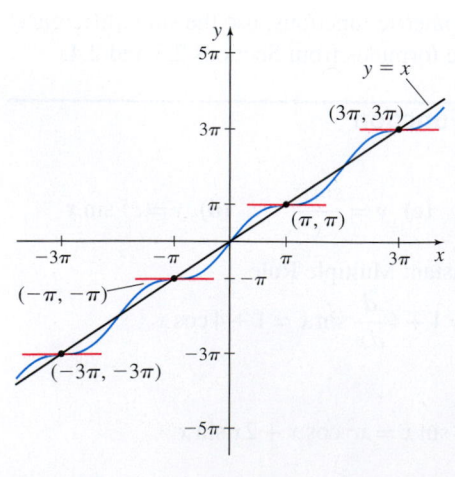

**Figure 31** $f(x) = x + \sin x$

The derivatives of the remaining four trigonometric functions are obtained using trigonometric identities and basic derivative rules. We establish the formula for the derivative of $y = \tan x$ in Example 4. You are asked to prove formulas for the derivative of the secant function, the cosecant function, and the cotangent function in the exercises. (See Problems 76–78.)

**EXAMPLE 4  Differentiating $y = \tan x$**

Show that the derivative of $y = \tan x$ is

$$y' = \frac{d}{dx} \tan x = \sec^2 x$$

**Solution**

$$y' = \frac{d}{dx} \tan x = \frac{d}{dx} \frac{\sin x}{\cos x} = \frac{\left[\dfrac{d}{dx} \sin x\right] \cos x - \sin x \left[\dfrac{d}{dx} \cos x\right]}{\cos^2 x}$$

$$\underset{\text{Identity}}{\uparrow} \qquad \underset{\text{Quotient Rule}}{\uparrow}$$

$$= \frac{\cos x \cdot \cos x - \sin x \cdot (-\sin x)}{\cos^2 x} = \frac{\cos^2 x + \sin^2 x}{\cos^2 x} = \frac{1}{\cos^2 x} = \sec^2 x \qquad ■$$

**NOW WORK Problem 15.**

Table 4 lists the derivatives of the six trigonometric functions along with the domain of each derivative.

**TABLE 4**

| Derivative Function | Domain of the Derivative Function |
|---|---|
| $\dfrac{d}{dx} \sin x = \cos x$ | $(-\infty, \infty)$ |
| $\dfrac{d}{dx} \cos x = -\sin x$ | $(-\infty, \infty)$ |
| $\dfrac{d}{dx} \tan x = \sec^2 x$ | $\left\{ x \mid x \neq \dfrac{2k+1}{2}\pi,\, k \text{ an integer} \right\}$ |
| $\dfrac{d}{dx} \cot x = -\csc^2 x$ | $\{ x \mid x \neq k\pi,\, k \text{ an integer} \}$ |
| $\dfrac{d}{dx} \csc x = -\csc x \cot x$ | $\{ x \mid x \neq k\pi,\, k \text{ an integer} \}$ |
| $\dfrac{d}{dx} \sec x = \sec x \tan x$ | $\left\{ x \mid x \neq \dfrac{2k+1}{2}\pi,\, k \text{ an integer} \right\}$ |

**NOTE** If the trigonometric function begins with the letter $c$, that is, cosine, cotangent, or cosecant, then its derivative has a minus sign.

**EXAMPLE 5  Finding the Second Derivative of a Trigonometric Function**

Find $f''\left(\dfrac{\pi}{4}\right)$ if $f(x) = \sec x$.

**Solution** If $f(x) = \sec x$, then $f'(x) = \sec x \tan x$ and

$$f''(x) = \frac{d}{dx}(\sec x \tan x) = \sec x \left(\frac{d}{dx} \tan x\right) + \left(\frac{d}{dx} \sec x\right) \tan x$$

$$\underset{\text{Use the Product Rule.}}{\uparrow}$$

$$= \sec x \cdot \sec^2 x + (\sec x \tan x) \tan x = \sec^3 x + \sec x \tan^2 x$$

$$f''\left(\frac{\pi}{4}\right) = \sec^3\left(\frac{\pi}{4}\right) + \sec\left(\frac{\pi}{4}\right)\tan^2\left(\frac{\pi}{4}\right) = \left(\sqrt{2}\right)^3 + \sqrt{2} \cdot 1^2 = 2\sqrt{2} + \sqrt{2} = 3\sqrt{2}$$

$$\underset{\sec\frac{\pi}{4} = \sqrt{2};\ \tan\frac{\pi}{4} = 1}{\uparrow}$$

$$\qquad ■$$

**NOW WORK Problem 35.**

## Application: Simple Harmonic Motion

**Simple harmonic motion** is a repetitive motion that can be modeled by a trigonometric function. A swinging pendulum and an oscillating spring are examples of simple harmonic motion.

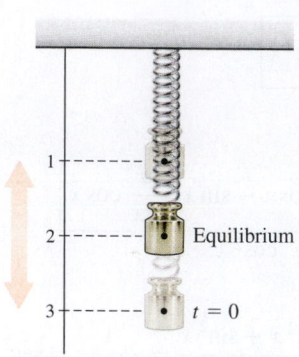

**DF Figure 32**

EXAMPLE 6   **Analyzing Simple Harmonic Motion**

An object hangs on a spring, making the spring 2 m long in its equilibrium position. See Figure 32. If the object is pulled down 1 m and released, it oscillates up and down. The length $l$ of the spring after $t$ seconds is modeled by the function $l(t) = 2 + \cos t$.

**(a)** How does the length of the spring vary?

**(b)** Find the velocity of the object.

**(c)** At what position is the speed of the object a maximum?

**(d)** Find the acceleration of the object.

**(e)** At what position is the acceleration equal to 0?

**Solution  (a)**   Since $l(t) = 2 + \cos t$ and $-1 \le \cos t \le 1$, the length of the spring oscillates between 1 and 3 m.

**(b)**  The velocity $v$ of the object is

$$v = l'(t) = \frac{d}{dt}(2 + \cos t) = -\sin t$$

**(c)**  Speed is the magnitude of velocity. Since $v = -\sin t$, the speed of the object is $|v| = |-\sin t| = |\sin t|$. Since $-1 \le \sin t \le 1$, the object moves the fastest when $|v| = |\sin t| = 1$. This occurs when $\sin t = \pm 1$ or, equivalently, when $\cos t = 0$. So, the speed is a maximum when $l(t) = 2$, that is, when the spring is at the equilibrium position.

**(d)**  The acceleration $a$ of the object is given by

$$a = l''(t) = \frac{d}{dt}l'(t) = \frac{d}{dt}(-\sin t) = -\cos t$$

**(e)**  Since $a = -\cos t$, the acceleration is zero when $\cos t = 0$. So, $a = 0$ when $l(t) = 2$, that is, when the spring is at the equilibrium position. This is the same time at which the speed is maximum. ∎

Figure 33 shows the graphs of the length of the spring $y = l(t)$, the velocity $y = v(t)$, and the acceleration $y = a(t)$.

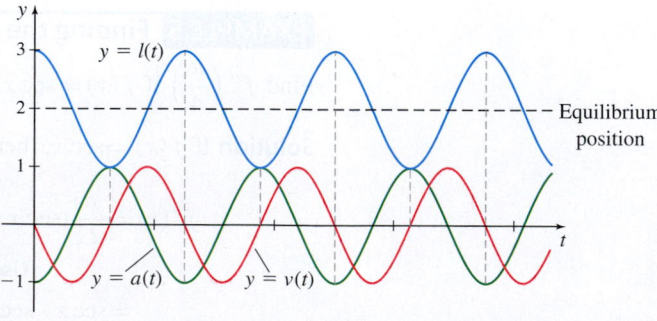

**Figure 33**

NOW WORK **Problem 65.**

# 2.5 Assess Your Understanding

## Concepts and Vocabulary

**1.** *True or False*  $\dfrac{d}{dx}\cos x = \sin x$

**2.** *True or False*  $\dfrac{d}{dx}\tan x = \cot x$

**3.** *True or False*  $\dfrac{d^2}{dx^2}\sin x = -\sin x$

**4.** *True or False*  $\dfrac{d}{dx}\sin \dfrac{\pi}{3} = \cos \dfrac{\pi}{3}$

## Skill Building

*In Problems 5–38, find $y'$.*

**5.** $y = x - \sin x$

**6.** $y = \cos x - x^2$

**7.** $y = \tan x + \cos x$

**8.** $y = \sin x - \tan x$

**9.** $y = 3\sin\theta - 2\cos\theta$

**10.** $y = 4\tan\theta + \sin\theta$

**11.** $y = \sin x \cos x$

**12.** $y = \cot x \tan x$

**13.** $y = t\cos t$

**14.** $y = t^2 \tan t$

**15.** $y = e^x \tan x$

**16.** $y = e^x \sec x$

**17.** $y = \pi \sec u \tan u$

**18.** $y = \pi u \tan u$

**19.** $y = \dfrac{\cot x}{x}$

**20.** $y = \dfrac{\csc x}{x}$

**21.** $y = x^2 \sin x$

**22.** $y = t^2 \tan t$

**23.** $y = t\tan t - \sqrt{3}\sec t$

**24.** $y = x\sec x + \sqrt{2}\cot x$

**25.** $y = \dfrac{\sin\theta}{1-\cos\theta}$

**26.** $y = \dfrac{x}{\cos x}$

**27.** $y = \dfrac{\sin t}{1+t}$

**28.** $y = \dfrac{\tan u}{1+u}$

**29.** $y = \dfrac{\sin x}{e^x}$

**30.** $y = \dfrac{\cos x}{e^x}$

**31.** $y = \dfrac{\sin\theta + \cos\theta}{\sin\theta - \cos\theta}$

**32.** $y = \dfrac{\sin\theta - \cos\theta}{\sin\theta + \cos\theta}$

**33.** $y = \dfrac{\sec t}{1+t\sin t}$

**34.** $y = \dfrac{\csc t}{1+t\cos t}$

**35.** $y = \csc\theta \cot\theta$

**36.** $y = \tan\theta \cos\theta$

**37.** $y = \dfrac{1+\tan x}{1-\tan x}$

**38.** $y = \dfrac{\csc x - \cot x}{\csc x + \cot x}$

*In Problems 39–50, find $y''$.*

**39.** $y = \sin x$

**40.** $y = \cos x$

**41.** $y = \tan\theta$

**42.** $y = \cot\theta$

**43.** $y = t\sin t$

**44.** $y = t\cos t$

**45.** $y = e^x \sin x$

**46.** $y = e^x \cos x$

**47.** $y = 2\sin u - 3\cos u$

**48.** $y = 3\sin u + 4\cos u$

**49.** $y = a\sin x + b\cos x$

**50.** $y = a\sec\theta + b\tan\theta$

*In Problems 51–56:*

(a) *Find an equation of the tangent line to the graph of $f$ at the indicated point.*

(b) *Graph the function and the tangent line.*

**51.** $f(x) = \sin x$ at $(0, 0)$

**52.** $f(x) = \cos x$ at $\left(\dfrac{\pi}{3}, \dfrac{1}{2}\right)$

**53.** $f(x) = \tan x$ at $(0, 0)$

**54.** $f(x) = \tan x$ at $\left(\dfrac{\pi}{4}, 1\right)$

**[N] 55.** $f(x) = \sin x + \cos x$ at $\left(\dfrac{\pi}{4}, \sqrt{2}\right)$

**[N] 56.** $f(x) = \sin x - \cos x$ at $\left(\dfrac{\pi}{4}, 0\right)$

*In Problems 57–60:*

(a) *Find all points on the graph of $f$ where the tangent line is horizontal.*

**[N]** (b) *Graph the function and the horizontal tangent lines on the interval $[-2\pi, 2\pi]$.*

**57.** $f(x) = 2\sin x + \cos x$

**58.** $f(x) = \cos x - \sin x$

**59.** $f(x) = \sec x$

**60.** $f(x) = \csc x$

## Applications and Extensions

*In Problems 61 and 62, find the nth derivative of each function.*

**61.** $f(x) = \sin x$

**62.** $f(\theta) = \cos\theta$

**63.** What is $\displaystyle\lim_{h\to 0} \dfrac{\cos\left(\dfrac{\pi}{2}+h\right) - \cos\dfrac{\pi}{2}}{h}$?

**64.** What is $\displaystyle\lim_{h\to 0} \dfrac{\sin(\pi + h) - \sin\pi}{h}$?

**65. Simple Harmonic Motion**  The signed distance $s$ (in meters) of an object from the origin at time $t$ (in seconds) is modeled by the position function $s(t) = \dfrac{1}{8}\cos t$.

(a) Find the velocity $v$ of the object.

(b) When is the speed of the object a maximum?

(c) Find the acceleration $a$ of the object.

(d) When is the acceleration equal to 0?

(e) Graph $s$, $v$, and $a$ on the same screen.

---

**1.** = NOW WORK problem      [N] = Graphing technology recommended      [CAS] = Computer Algebra System recommended

**66. Simple Harmonic Motion**
An object attached to a coiled spring is pulled down a distance $d = 5$ cm from its equilibrium position and then released as shown in the figure. The motion of the object at time $t$ seconds is simple harmonic and is modeled by $d(t) = -5 \cos t$.

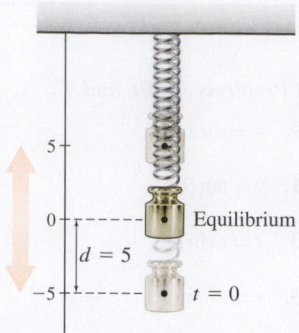

(a) As $t$ varies from 0 to $2\pi$, how does the length of the spring vary?

(b) Find the velocity $v$ of the object.

(c) When is the speed of the object a maximum?

(d) Find the acceleration $a$ of the object.

(e) When is the acceleration equal to 0?

(f) Graph $d$, $v$, and $a$ on the same screen.

**67. Rate of Change** A large, 8-ft high decorative mirror is placed on a wood floor and leaned against a wall. The weight of the mirror and the slickness of the floor cause the mirror to slip.

(a) If $\theta$ is the angle between the top of the mirror and the wall, and $y$ is the distance from the floor to the top of the mirror, what is the rate of change of $y$ with respect to $\theta$?

(b) In feet/radian, how fast is the top of the mirror slipping down the wall when $\theta = \dfrac{\pi}{4}$?

**68. Rate of Change** The sides of an isosceles triangle are sliding outward. See the figure.

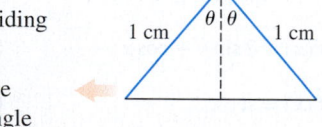

(a) Find the rate of change of the area of the triangle with respect to $\theta$.

(b) How fast is the area changing when $\theta = \dfrac{\pi}{6}$?

**69. Sea Waves** Waves in deep water tend to have the symmetric form of the function $f(x) = \sin x$. As they approach shore, however, the sea floor creates drag, which changes the shape of the wave. The trough of the wave widens and the height of the wave increases, so the top of the wave is no longer symmetric with the trough. This type of wave can be represented by a function such as

$$w(x) = \frac{4}{2 + \cos x}$$

(a) Graph $w = w(x)$ for $0 \le x \le 4\pi$.

(b) What is the maximum and the minimum value of $w$?

(c) Find the values of $x$, $0 < x < 4\pi$, at which $w'(x) = 0$.

(d) Evaluate $w'$ near the peak at $\pi$, using $x = \pi - 0.1$, and near the trough at $2\pi$, using $x = 2\pi - 0.1$.

(e) Explain how these values confirm a nonsymmetric wave shape.

**70. Swinging Pendulum** A simple pendulum is a small-sized ball swinging from a light string. As it swings, the supporting string makes an angle $\theta$ with the vertical. See the figure. At an angle $\theta$, the tension in the string is $T = \dfrac{W}{\cos \theta}$, where $W$ is the weight of the swinging ball.

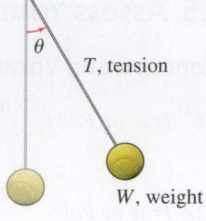

(a) Find the rate of change of the tension $T$ with respect to $\theta$ when the pendulum is at its highest point ($\theta = \theta_{\max}$).

(b) Find the rate of change of the tension $T$ with respect to $\theta$ when the pendulum is at its lowest point.

(c) What is the tension at the lowest point?

**71. Restaurant Sales** A restaurant in Naples, Florida is very busy during the winter months and extremely slow over the summer. But every year the restaurant grows its sales. Suppose over the next two years, the revenue $R$, in units of $10,000, is projected to follow the model

$$R = R(t) = \sin t + 0.3t + 1 \qquad 0 \le t \le 12$$

where $t = 0$ corresponds to November 1, 2018; $t = 1$ corresponds to January 1, 2019; $t = 2$ corresponds to March 1, 2019; and so on.

(a) What is the projected revenue for November 1, 2018; March 1, 2019; September 1, 2019; and January 1, 2020?

(b) What is the rate of change of revenue with respect to time?

(c) What is the rate of change of revenue with respect to time for January 1, 2020?

(d) Graph the revenue function and the derivative function $R' = R'(t)$.

(e) Does the graph of $R$ support the facts that every year the restaurant grows its sales and that sales are higher during the winter and lower during the summer? Explain.

**72. Polarizing Sunglasses**
Polarizing sunglasses are filters that transmit only light for which the electric field oscillations are in a specific direction. Light is polarized naturally by scattering off the molecules in the atmosphere and by reflecting off many (but not all) types of surfaces. If light of intensity $I_0$ is already polarized in a certain direction, and the transmission direction of the polarizing filter makes an angle with that direction, then the intensity $I$ of the light after passing through the filter is given by **Malus's Law**, $I(\theta) = I_0 \cos^2 \theta$.

REUTERS/Alamy

(a) As you rotate a polarizing filter, $\theta$ changes. Find the rate of change of the light intensity $I$ with respect to $\theta$.

(b) Find both the intensity $I(\theta)$ and the rate of change of the intensity with respect to $\theta$, for the angles $\theta = 0°$, $45°$, and $90°$. (Remember to use radians for $\theta$.)

**73.** If $y = \sin x$ and $y^{(n)}$ is the *n*th derivative of *y* with respect to *x*, find the smallest positive integer *n* for which $y^{(n)} = y$.

**74.** Use the identity $\sin A - \sin B = 2 \cos \dfrac{A+B}{2} \sin \dfrac{A-B}{2}$, with $A = x + h$ and $B = x$, to prove that

$$\frac{d}{dx} \sin x = \lim_{h \to 0} \frac{\sin(x+h) - \sin x}{h} = \cos x$$

**75.** Use the definition of a derivative to prove $\dfrac{d}{dx} \cos x = -\sin x$.

**76. Derivative of $y = \sec x$**   Use a derivative rule to show that

$$\frac{d}{dx} \sec x = \sec x \tan x$$

**77. Derivative of $y = \csc x$**   Use a derivative rule to show that

$$\frac{d}{dx} \csc x = -\csc x \cot x$$

**78. Derivative of $y = \cot x$**   Use a derivative rule to show that

$$\frac{d}{dx} \cot x = -\csc^2 x$$

**79.** Let $f(x) = \cos x$. Show that finding $f'(0)$ is the same as finding $\displaystyle\lim_{x \to 0} \frac{\cos x - 1}{x}$.

**80.** Let $f(x) = \sin x$. Show that finding $f'(0)$ is the same as finding $\displaystyle\lim_{x \to 0} \frac{\sin x}{x}$.

**81.** If $y = A \sin t + B \cos t$, where *A* and *B* are constants, show that $y'' + y = 0$.

**Challenge Problem** ———————————

**82.** For a differentiable function *f*, let $f^*$ be the function defined by

$$f^*(x) = \lim_{h \to 0} \frac{f(x+h) - f(x-h)}{h}$$

**(a)** Find $f^*(x)$ for $f(x) = x^2 + x$.

**(b)** Find $f^*(x)$ for $f(x) = \cos x$.

**(c)** Write an equation that expresses the relationship between the functions $f^*$ and $f'$, where $f'$ denotes the derivative of *f*. Justify your answer.

# Chapter Review

## THINGS TO KNOW

### 2.1 Rates of Change and the Derivative

- **Definition**   Derivative of a function at a number

$$f'(c) = \lim_{x \to c} \frac{f(x) - f(c)}{x - c}$$

provided the limit exists. (p. 159)

***Three Interpretations of the Derivative***

- *Geometric*   If $y = f(x)$, the derivative $f'(c)$ is the slope of the tangent line to the graph of *f* at the point $(c, f(c))$. (p. 154)
- *Rate of change of a function*   If $y = f(x)$, the derivative $f'(c)$ is the rate of change of *f* with respect to *x* at *c*. (p. 156)
- *Physical*   If the signed distance *s* from the origin at time *t* of an object in rectilinear motion is given by the position function $s = f(t)$, the derivative $f'(t_0)$ is the velocity of the object at time $t_0$. (p. 158)

### 2.2 The Derivative as a Function

- **Definition of a derivative function** (Form 3) (p. 164)

$$f'(x) = \lim_{h \to 0} \frac{f(x+h) - f(x)}{h}, \text{ provided the limit exists.}$$

- **Theorem**   If a function *f* has a derivative at a number *c*, then *f* is continuous at *c*. (p. 169)
- **Corollary**   If a function *f* is discontinuous at a number *c*, then *f* has no derivative at *c*. (p. 169)

### 2.3 The Derivative of a Polynomial Function; The Derivative of $y = e^x$

- **Leibniz notation**   $\dfrac{dy}{dx} = \dfrac{d}{dx} y = \dfrac{d}{dx} f(x)$ (p. 174)

- **Basic derivatives**

$$\frac{d}{dx} A = 0 \quad A \text{ is a constant}  \text{ (p. 175)} \qquad \frac{d}{dx} x = 1 \text{ (p. 175)}$$

$$\frac{d}{dx} e^x = e^x \text{ (p. 181)}$$

- *Simple Power Rule*   $\dfrac{d}{dx} x^n = n x^{n-1}, \quad n \geq 1$, an integer (p. 176)

***Properties of Derivatives***

- *Sum Rule*   $\dfrac{d}{dx}[f + g] = \dfrac{d}{dx} f + \dfrac{d}{dx} g$ (p. 177)

$$(f + g)' = f' + g'$$

- *Difference Rule*   $\dfrac{d}{dx}[f - g] = \dfrac{d}{dx} f - \dfrac{d}{dx} g$ (p. 178)

$$(f - g)' = f' - g'$$

- *Constant Multiple Rule*   If *k* is a constant, $\dfrac{d}{dx}[kf] = k \dfrac{d}{dx} f$ (p. 177)

$$(kf)' = k \cdot f'$$

### 2.4 Differentiating the Product and the Quotient of Two Functions; Higher-Order Derivatives

***Properties of Derivatives***

- *Product Rule*   $\dfrac{d}{dx}(fg) = f\left(\dfrac{d}{dx} g\right) + \left(\dfrac{d}{dx} f\right) g$ (p. 185)

$$(fg)' = fg' + f'g$$

- *Quotient Rule*   $\dfrac{d}{dx}\left(\dfrac{f}{g}\right) = \dfrac{\left(\dfrac{d}{dx} f\right) g - f\left(\dfrac{d}{dx} g\right)}{g^2}$ (p. 186)

$$\left(\frac{f}{g}\right)' = \frac{f'g - fg'}{g^2}$$

provided $g(x) \neq 0$

- *Reciprocal Rule*   $\dfrac{d}{dx}\left(\dfrac{1}{g}\right) = -\dfrac{\dfrac{d}{dx} g}{g^2}$ (p. 187)

$$\left(\frac{1}{g}\right)' = -\frac{g'}{g^2}$$

provided $g(x) \neq 0$

- **Power Rule**   $\dfrac{d}{dx}x^n = nx^{n-1}$, $n$ an integer (p. 188)

- **Higher-order derivatives**   See Table 3 (p. 189)

- **Position Function**   $s = s(t)$

- **Velocity**   $v = v(t) = \dfrac{ds}{dt}$

- **Acceleration**   $a = a(t) = \dfrac{dv}{dt} = \dfrac{d^2s}{dt^2}$ (p. 190)

## 2.5 The Derivative of the Trigonometric Functions
**Basic Derivatives**

$\dfrac{d}{dx}\sin x = \cos x$ (p. 196)      $\dfrac{d}{dx}\sec x = \sec x \tan x$ (p. 199)

$\dfrac{d}{dx}\cos x = -\sin x$ (p. 198)      $\dfrac{d}{dx}\csc x = -\csc x \cot x$ (p. 199)

$\dfrac{d}{dx}\tan x = \sec^2 x$ (p. 199)      $\dfrac{d}{dx}\cot x = -\csc^2 x$ (p. 199)

## OBJECTIVES

| Section | You should be able to … | Examples | Review Exercises |
|---|---|---|---|
| 2.1 | 1 Find equations for the tangent line and the normal line to the graph of a function (p. 155) | 1 | 67–70 |
|  | 2 Find the rate of change of a function (p. 156) | 2, 3 | 1, 2, 73 (a) |
|  | 3 Find average velocity and instantaneous velocity (p. 157) | 4, 5 | 71(a), (b); 72(a), (b) |
|  | 4 Find the derivative of a function at a number (p. 159) | 6–8 | 3–8, 75 |
| 2.2 | 1 Define the derivative function (p. 163) | 1–3 | 9–12, 77 |
|  | 2 Graph the derivative function (p. 165) | 4, 5 | 9–12, 15–18 |
|  | 3 Identify where a function has no derivative (p. 167) | 6–10 | 13, 14, 75 |
| 2.3 | 1 Differentiate a constant function (p. 175) | 1 |  |
|  | 2 Differentiate a power function (p. 175) | 2, 3 | 19–22 |
|  | 3 Differentiate the sum and the difference of two functions (p. 177) | 4–6 | 23–26, 33, 34, 40, 51, 52, 67 |
|  | 4 Differentiate the exponential function $y = e^x$ (p. 180) | 7 | 44, 45, 53, 54, 56, 59, 69 |
| 2.4 | 1 Differentiate the product of two functions (p. 184) | 1, 2 | 27, 28, 36, 46, 48–50, 53–56, 60 |
|  | 2 Differentiate the quotient of two functions (p. 186) | 3–6 | 29–35, 37–43, 47, 57–59, 68, 73, 74 |
|  | 3 Find higher-order derivatives (p. 189) | 7, 8 | 61–66, 71, 72, 76 |
|  | 4 Work with acceleration (p. 190) | 9 | 71, 72, 76 |
| 2.5 | 1 Differentiate trigonometric functions (p. 196) | 1–6 | 49–60, 70 |

## REVIEW EXERCISES

*In Problems 1 and 2, use a definition of the derivative to find the rate of change of f at the indicated numbers.*

**1.** $f(x) = \sqrt{x}$ at **(a)** $c = 1$   **(b)** $c = 4$   1, 2
   **(c)** $c$ any positive real number

**2.** $f(x) = \dfrac{2}{x-1}$ at **(a)** $c = 0$   **(b)** $c = 2$
   **(c)** $c$ any real number, $c \neq 1$

*In Problems 3–8, use a definition of the derivative to find the derivative of each function at the given number.*

**3.** $F(x) = 2x + 5$ at 2      **4.** $f(x) = 4x^2 + 1$ at $-1$

**5.** $f(x) = 3x^2 + 5x$ at 0      **6.** $f(x) = \dfrac{3}{x}$ at 1

**7.** $f(x) = \sqrt{4x+1}$ at 0      **8.** $f(x) = \dfrac{x+1}{2x-3}$ at 1

*In Problems 9–12, use a definition of the derivative to find the derivative of each function. Graph f and f' on the same set of axes.*

**9.** $f(x) = x - 6$      **10.** $f(x) = 7 - 3x^2$

**11.** $f(x) = \dfrac{1}{2x^3}$      **12.** $f(x) = \pi$

*In Problems 13 and 14, determine whether the function f has a derivative at c. If it does, find the derivative. If it does not, explain why. Graph each function.*

**13.** $f(x) = |x^3 - 1|$ at $c = 1$

**14.** $f(x) = \begin{cases} 4 - 3x^2 & \text{if } x \leq -1 \\ -x^3 & \text{if } x > -1 \end{cases}$ at $c = -1$

*In Problems 15 and 16, determine if the graphs represent a function f and its derivative f'. If they do, indicate which is the graph of f and which is the graph of f'.*

**15.**

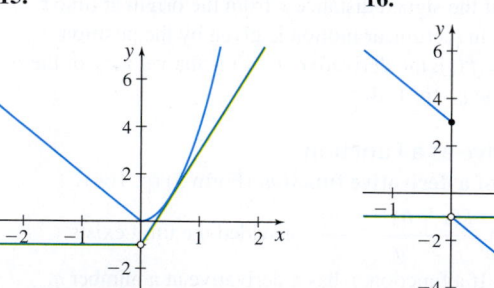

**16.**

**17.** Use the information in the graph of $y = f(x)$ to sketch the graph of $y = f'(x)$.

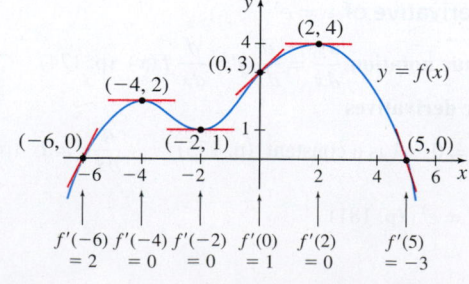

**18.** Match the graph of $y = f(x)$ with the graph of its derivative.

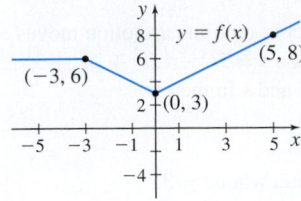

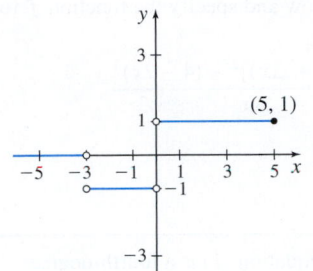

(A)

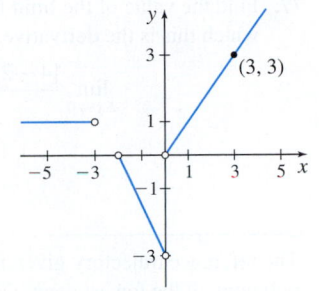

(B)

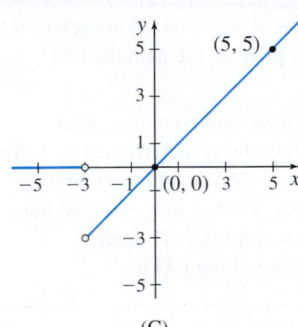

(C)

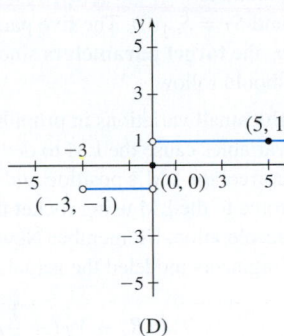

(D)

*In Problems 19–60, find the derivative of each function. Where a or b appears, it is a constant.*

**19.** $f(x) = x^5$

**20.** $f(x) = ax^3$

**21.** $f(x) = \dfrac{x^4}{4}$

**22.** $f(x) = -6x^2$

**23.** $f(x) = 2x^2 - 3x$

**24.** $f(x) = 3x^3 + \dfrac{2}{3}x^2 - 5x + 7$

**25.** $F(x) = 7(x^2 - 4)$

**26.** $F(x) = \dfrac{5(x+6)}{7}$

**27.** $f(x) = 5(x^2 - 3x)(x - 6)$

**28.** $f(x) = (2x^3 + x)(x^2 - 5)$

**29.** $f(x) = \dfrac{6x^4 - 9x^2}{3x^3}$

**30.** $f(x) = \dfrac{2x + 2}{5x - 3}$

**31.** $f(x) = \dfrac{7x}{x - 5}$

**32.** $f(x) = 2x^{-12}$

**33.** $f(x) = 2x^2 - 5x^{-2}$

**34.** $f(x) = 2 + \dfrac{3}{x} + \dfrac{4}{x^2}$

**35.** $f(x) = \dfrac{a}{x} - \dfrac{b}{x^3}$

**36.** $f(x) = (x^3 - 1)^2$

**37.** $f(x) = \dfrac{3}{(x^2 - 3x)^2}$

**38.** $f(x) = \dfrac{x^2}{x + 1}$

**39.** $s(t) = \dfrac{t^3}{t - 2}$

**40.** $f(x) = 3x^{-2} + 2x^{-1} + 1$

**41.** $F(z) = \dfrac{1}{z^2 + 1}$

**42.** $f(v) = \dfrac{v - 1}{v^2 + 1}$

**43.** $g(z) = \dfrac{1}{1 - z + z^2}$

**44.** $f(x) = 3e^x + x^2$

**45.** $s(t) = 1 - e^t$

**46.** $f(x) = ae^x(2x^2 + 7x)$

**47.** $f(x) = \dfrac{1 + x}{e^x}$

**48.** $f(x) = (2xe^x)^2$

**49.** $f(x) = x \sin x$

**50.** $s(t) = \cos^2 t$

**51.** $G(u) = \tan u + \sec u$

**52.** $g(v) = \sin v - \dfrac{1}{3} \cos v$

**53.** $f(x) = e^x \sin x$

**54.** $f(x) = e^x \csc x$

**55.** $f(x) = 2 \sin x \cos x$

**56.** $f(x) = (e^x + b) \cos x$

**57.** $f(x) = \dfrac{\sin x}{\csc x}$

**58.** $f(x) = \dfrac{1 - \cot x}{1 + \cot x}$

**59.** $f(\theta) = \dfrac{\cos \theta}{2e^\theta}$

**60.** $f(\theta) = 4\theta \cot \theta \tan \theta$

*In Problems 61–66, find the first derivative and the second derivative of each function.*

**61.** $f(x) = (5x + 3)^2$

**62.** $f(x) = xe^x$

**63.** $g(u) = \dfrac{u}{2u + 1}$

**64.** $F(x) = e^x(\sin x + 2 \cos x)$

**65.** $f(u) = \dfrac{\cos u}{e^u}$

**66.** $F(x) = \dfrac{\sin x}{x}$

*In Problems 67–70, for each function:*

**(a)** *Find an equation of the tangent line to the graph of the function at the indicated point.*

**(b)** *Find an equation of the normal line to the function at the indicated point.*

**(c)** *Graph the function, the tangent line, and the normal line on the same screen.*

**67.** $f(x) = 2x^2 - 3x + 7$ at $(-1, 12)$

**68.** $y = \dfrac{x^2 + 1}{2x - 1}$ at $\left(2, \dfrac{5}{3}\right)$

**69.** $f(x) = x^2 - e^x$ at $(0, -1)$

**70.** $s(t) = 1 + 2 \sin t$ at $(\pi, 1)$

**71. Rectilinear Motion**  As an object in rectilinear motion moves, its signed distance $s$ (in meters) from the origin at time $t$ (in seconds) is given by the position function

$$s = f(t) = t^2 - 6t$$

**(a)** Find the average velocity of the object from 0 to 5 s.

**(b)** Find the velocity at $t = 0$, at $t = 5$, and at any time $t$.

**(c)** Find the acceleration at any time $t$.

**72. Rectilinear Motion**  As an object in rectilinear motion moves, its signed distance $s$ from the origin at time $t$ is given by the position function $s(t) = t - t^2$, where $s$ is in centimeters and $t$ is in seconds.

**(a)** Find the average velocity of the object from 1 to 3 s.

**(b)** Find the velocity of the object at $t = 1$ s and $t = 3$ s.

**(c)** What is its acceleration at $t = 1$ and $t = 3$?

**73. Business**  The price $p$ in dollars per pound when $x$ pounds of a commodity are demanded is modeled by the function

$$p(x) = \frac{10,000}{5x + 100} - 5,$$ when between 0 and 90 lb are demanded (purchased).

   **(a)**  Find the rate of change of price with respect to demand.

   **(b)**  What is the revenue function $R$? (Recall, revenue $R$ equals price times amount purchased.)

   **(c)**  What is the marginal revenue $R'$ at $x = 10$ and at $x = 40$ lb?

**74.**  If $f(x) = \dfrac{x-1}{x+1}$ for all $x \neq -1$, find $f'(1)$.

**75.**  If $f(x) = 2 + |x - 3|$ for all $x$, determine whether the derivative $f'$ exists at $x = 3$.

**76. Rectilinear Motion**  An object in rectilinear motion moves according to the position function $s = 2t^3 - 15t^2 + 24t + 3$, where $t$ is measured in minutes and $s$ in meters.

   **(a)**  When is the object at rest?

   **(b)**  Find the object's acceleration when $t = 3$.

**77.**  Find the value of the limit below and specify the function $f$ for which this is the derivative.

$$\lim_{\Delta x \to 0} \frac{[4 - 2(x + \Delta x)]^2 - (4 - 2x)^2}{\Delta x}$$

---

## CHAPTER 2 PROJECT    The Lunar Module

Rolls Press/Popperfoto/ Getty Images

The Lunar Module (LM) was a small spacecraft that detached from the Apollo Command Module and was designed to land on the Moon. Fast and accurate computations were needed to bring the LM from an orbiting speed of about 5500 ft/s to a speed slow enough to land it within a few feet of a designated target on the Moon's surface. The LM carried a 70-lb computer to assist in guiding it successfully to its target. The approach to the target was split into three phases, each of which followed a reference trajectory specified by NASA engineers.[*] The position and velocity of the LM were monitored by sensors that tracked its deviation from the preassigned path at each moment. Whenever the LM strayed from the reference trajectory, control thrusters were fired to reposition it. In other words, the LM's position and velocity were adjusted by changing its acceleration.

The reference trajectory for each phase was specified by the engineers to have the form

$$r_{\text{ref}}(t) = R_T + V_T t + \frac{1}{2} A_T t^2 + \frac{1}{6} J_T t^3 + \frac{1}{24} S_T t^4 \tag{1}$$

The variable $r_{\text{ref}}$ represents the intended position of the LM at time $t$ before the end of the landing phase. The engineers specified the end of the landing phase to take place at $t = 0$, so that during the phase, $t$ was always negative. Note that the LM was landing in three dimensions, so there were actually three equations like (1). Since each of those equations had this same form, we will work in one dimension, assuming, for example, that $r$ represents the distance of the LM above the surface of the Moon.

**1.**  If the LM follows the reference trajectory, what is the reference velocity $v_{\text{ref}}(t)$?

**2.**  What is the reference acceleration $a_{\text{ref}}(t)$?

**3.**  The rate of change of acceleration is called **jerk**. Find the reference jerk $J_{\text{ref}}(t)$.

**4.**  The rate of change of jerk is called **snap**. Find the reference snap $S_{\text{ref}}(t)$.

**5.**  Evaluate $r_{\text{ref}}(t)$, $v_{\text{ref}}(t)$, $a_{\text{ref}}(t)$, $J_{\text{ref}}(t)$, and $S_{\text{ref}}(t)$ when $t = 0$.

The reference trajectory given in equation (1) is a fourth-degree polynomial, the lowest degree polynomial that has enough free parameters to satisfy all the mission criteria. Now we see that the parameters $R_T = r_{\text{ref}}(0)$, $V_T = v_{\text{ref}}(0)$, $A_T = a_{\text{ref}}(0)$, $J_T = J_{\text{ref}}(0)$, and $S_T = S_{\text{ref}}(0)$. The five parameters in equation (1) are referred to as the **target parameters** since they provide the path the LM should follow.

But small variations in propulsion, mass, and countless other variables cause the LM to deviate from the predetermined path. To correct the LM's position and velocity, NASA engineers apply a force to the LM using rocket thrusters. That is, they changed the acceleration. (Remember Newton's second law, $F = ma$.) Engineers modeled the actual trajectory of the LM by

$$r(t) = R_T + V_T t + \frac{1}{2} A_T t^2 + \frac{1}{6} J_A t^3 + \frac{1}{24} S_A t^4 \tag{2}$$

We know the target parameters for position, velocity, and acceleration. We need to find the actual parameters for jerk and snap to know the proper force (acceleration) to apply.

**6.**  Find the actual velocity $v = v(t)$ of the LM.

**7.**  Find the actual acceleration $a = a(t)$ of the LM.

**8.**  Use equation (2) and the actual velocity found in Problem 6 to express $J_A$ and $S_A$ in terms of $R_T$, $V_T$, $A_T$, $r(t)$, and $v(t)$.

**9.**  Use the results of Problems 7 and 8 to express the actual acceleration $a = a(t)$ in terms of $R_T$, $V_T$, $A_T$, $r(t)$, and $v(t)$.

The result found in Problem 9 provides the acceleration (force) required to keep the LM in its reference trajectory.

**10.**  When riding in an elevator, the sensation one feels just before the elevator stops at a floor is jerk. Would you want jerk to be small or large in an elevator? Explain. Would you want jerk to be small or large on a roller coaster ride? Explain. How would you explain snap?

---

[*] A. R. Klumpp, "Apollo Lunar-Descent Guidance," MIT Charles Stark Draper Laboratory, R-695, June 1971,
http://www.hq.nasa.gov/alsj/ApolloDescentGuidnce.pdf.

# 3

# More About Derivatives

**CHAPTER 3 PROJECT** In the Chapter Project on page 261, we explore a basic model to predict world population and examine its accuracy.

Steve Raymer/Getty Images

## World Population Growth

In the late 1700s Thomas Malthus predicted that population growth, if left unchecked, would outstrip food resources and lead to mass starvation. His prediction turned out to be incorrect, since in 1800 world population had not yet reached one billion and currently world population is in excess of seven billion. Malthus' most dire predictions did not come true largely because improvements in food production made his models inaccurate. On the other hand, his population growth model is still used today, and population growth remains an important issue in human progress. For example, many people are crowded into growing urban areas, such as India's Kolkata (pictured above), which has a population of over 14.3 million and a density of over 63,000 people per square mile.

In this chapter, we continue exploring properties of derivatives, beginning with the Chain Rule, which allows us to differentiate composite functions. The Chain Rule also provides the means to establish derivative formulas for exponential functions, logarithmic functions, and hyperbolic functions.

We also solve problems involving relative error in the measurements of two related variables and use the derivative to approximate the real zeros of a function.

# 3.1 The Chain Rule

**OBJECTIVES** *When you finish this section, you should be able to:*

1 Differentiate a composite function (p. 208)
2 Differentiate $y = a^x$, $a > 0$, $a \neq 1$ (p. 212)
3 Use the Power Rule for functions to find a derivative (p. 213)
4 Use the Chain Rule for multiple composite functions (p. 214)

Using the differentiation rules developed so far, it would be difficult to differentiate the function

$$F(x) = (x^3 - 4x + 1)^{100}$$

**NEED TO REVIEW?** Composite functions and their properties are discussed in Section P.3, pp. 27–29.

But $F$ is the composite function of $f(u) = u^{100}$ and $u = g(x) = x^3 - 4x + 1$, so $F(x) = f(g(x)) = (f \circ g)(x) = (x^3 - 4x + 1)^{100}$. In this section, we derive the *Chain Rule*, a result that enables us to find the derivative of a composite function. We use the Chain Rule to find the derivative in applications involving functions such as $A(t) = 102 - 90e^{-0.21t}$ (market penetration, Problem 101) and $v(t) = \dfrac{mg}{k}(1 - e^{-kt/m})$ (the terminal velocity of a falling object, Problem 103).

## 1 Differentiate a Composite Function

Suppose $y = (f \circ g)(x) = f(g(x))$ is a composite function, where $y = f(u)$ is a differentiable function of $u$, and $u = g(x)$ is a differentiable function of $x$. What is the derivative of $(f \circ g)(x)$? It turns out that the derivative of the composite function $f \circ g$ is the product of the derivatives $f'(u) = f'(g(x))$ and $g'(x)$.

**THEOREM Chain Rule**

If a function $g$ is differentiable at $x$ and a function $f$ is differentiable at $g(x)$, then the composite function $f \circ g$ is differentiable at $x$ and

$$\boxed{(f \circ g)'(x) = f'(g(x)) \cdot g'(x)}$$

**IN WORDS** If you think of the function $y = f(u)$ as the outside function and the function $u = g(x)$ as the inside function, then the derivative of $f \circ g$ is the derivative of the outside function, evaluated at the inside function, times the derivative of the inside function. That is,

$$\frac{d}{dx}(f \circ g)(x) = f'(g(x)) \cdot g'(x)$$

For differentiable functions $y = f(u)$ and $u = g(x)$, the Chain Rule, in Leibniz notation, takes the form

$$\boxed{\frac{dy}{dx} = \frac{dy}{du} \cdot \frac{du}{dx}}$$

where in $\dfrac{dy}{du}$ we substitute $u = g(x)$.

**Partial Proof** The Chain Rule is proved using the definition of a derivative. First we observe that if $x$ changes by a small amount $\Delta x$, the corresponding change in $u = g(x)$ is $\Delta u$ and $\Delta u$ depends on $\Delta x$. Also,

$$g'(x) = \frac{du}{dx} = \lim_{\Delta x \to 0} \frac{\Delta u}{\Delta x}$$

Since $y = f(u)$, the change $\Delta u$, which could equal 0, causes a change $\Delta y$. Suppose $\Delta u$ is never 0. Then

$$f'(u) = \frac{dy}{du} = \lim_{\Delta u \to 0} \frac{\Delta y}{\Delta u}$$

To find $\dfrac{dy}{dx}$, we write

$$\frac{dy}{dx} = \lim_{\Delta x \to 0} \frac{\Delta y}{\Delta x} = \lim_{\substack{\Delta x \to 0 \\ \Delta u \neq 0}} \left( \frac{\Delta y}{\Delta x} \cdot \frac{\Delta u}{\Delta u} \right) = \lim_{\Delta x \to 0} \left( \frac{\Delta y}{\Delta u} \cdot \frac{\Delta u}{\Delta x} \right)$$

$$= \left( \lim_{\Delta x \to 0} \frac{\Delta y}{\Delta u} \right) \left( \lim_{\Delta x \to 0} \frac{\Delta u}{\Delta x} \right)$$

Since the differentiable function $u = g(x)$ is continuous, $\Delta u \to 0$ as $\Delta x \to 0$, so in the first factor we can replace $\Delta x \to 0$ by $\Delta u \to 0$. Then

$$\frac{dy}{dx} = \left( \lim_{\Delta u \to 0} \frac{\Delta y}{\Delta u} \right) \left( \lim_{\Delta x \to 0} \frac{\Delta u}{\Delta x} \right) = \frac{dy}{du} \cdot \frac{du}{dx}$$

This proves the Chain Rule if $\Delta u$ is never 0. To complete the proof, we need to consider the case when $\Delta u$ may be 0. (This part of the proof is given in Appendix B.) ∎

---

**EXAMPLE 1  Differentiating a Composite Function**

Find the derivative of:

**(a)** $y = (x^3 - 4x + 1)^{100}$ **(b)** $y = \cos\left(3x - \dfrac{\pi}{4}\right)$

**Solution (a)** In the composite function $y = (x^3 - 4x + 1)^{100}$, let $u = x^3 - 4x + 1$. Then $y = u^{100}$. Now $\dfrac{dy}{du}$ and $\dfrac{du}{dx}$ are

$$\frac{dy}{du} = \frac{d}{du} u^{100} = 100u^{99} = 100 \underset{\substack{\uparrow \\ u = x^3 - 4x + 1}}{(x^3 - 4x + 1)^{99}}$$

and

$$\frac{du}{dx} = \frac{d}{dx}(x^3 - 4x + 1) = 3x^2 - 4$$

Now use the Chain Rule to find $\dfrac{dy}{dx}$.

$$\frac{dy}{dx} \underset{\substack{\uparrow \\ \text{Chain Rule}}}{=} \frac{dy}{du} \cdot \frac{du}{dx} = 100(x^3 - 4x + 1)^{99}(3x^2 - 4)$$

**(b)** In the composite function $y = \cos\left(3x - \dfrac{\pi}{4}\right)$, let $u = 3x - \dfrac{\pi}{4}$. Then $y = \cos u$ and

$$\frac{dy}{du} = \frac{d}{du} \cos u = -\sin u = \underset{\substack{\uparrow \\ u = 3x - \frac{\pi}{4}}}{-\sin\left(3x - \frac{\pi}{4}\right)} \quad \text{and} \quad \frac{du}{dx} = \frac{d}{dx}\left(3x - \frac{\pi}{4}\right) = 3$$

Now use the Chain Rule.

$$\frac{dy}{dx} \underset{\substack{\uparrow \\ \text{Chain Rule}}}{=} \frac{dy}{du} \cdot \frac{du}{dx} = -\sin\left(3x - \frac{\pi}{4}\right) \cdot 3 = -3\sin\left(3x - \frac{\pi}{4}\right)$$ ∎

**NOW WORK** Problems 9 and 37.

---

**EXAMPLE 2  Differentiating a Composite Function**

Find $y'$ if:

**(a)** $y = e^{x^2 - 4}$ **(b)** $y = \sin(4e^x)$

**Solution (a)** For $y = e^{x^2 - 4}$, let $u = x^2 - 4$. Then $y = e^u$ and

$$\frac{dy}{du} = \frac{d}{du} e^u = e^u = \underset{\substack{\uparrow \\ u = x^2 - 4}}{e^{x^2 - 4}} \quad \text{and} \quad \frac{du}{dx} = \frac{d}{dx}(x^2 - 4) = 2x$$

Using the Chain Rule, we get

$$y' = \frac{dy}{dx} = \frac{dy}{du} \cdot \frac{du}{dx} = e^{x^2 - 4} \cdot 2x = 2xe^{x^2 - 4}$$

**(b)** For $y = \sin(4e^x)$, let $u = 4e^x$. Then $y = \sin u$ and

$$\frac{dy}{du} = \frac{d}{du}\sin u = \cos u = \cos(4e^x) \quad \text{and} \quad \frac{du}{dx} = \frac{d}{dx}(4e^x) = 4e^x$$
$$\uparrow$$
$$u = 4e^x$$

Using the Chain Rule, we get

$$y' = \frac{dy}{dx} = \frac{dy}{du} \cdot \frac{du}{dx} = \cos(4e^x) \cdot 4e^x = 4e^x \cos(4e^x) \qquad \blacksquare$$

For composite functions $y = f(u(x))$, where $f$ is an exponential or trigonometric function, the Chain Rule simplifies finding $y'$. For example,

- If $y = e^{u(x)}$, where $u = u(x)$ is a differentiable function of $x$, then by the Chain Rule

$$\frac{dy}{dx} = \frac{dy}{du} \cdot \frac{du}{dx} = e^{u(x)}\frac{du}{dx}$$

That is,

$$\boxed{\frac{d}{dx}e^{u(x)} = e^{u(x)}\frac{du}{dx}}$$

- Similarly, if $u = u(x)$ is a differentiable function,

$$\frac{d}{dx}\sin u(x) = \cos u(x)\frac{du}{dx} \qquad \frac{d}{dx}\sec u(x) = \sec u(x)\tan u(x)\frac{du}{dx}$$

$$\frac{d}{dx}\cos u(x) = -\sin u(x)\frac{du}{dx} \qquad \frac{d}{dx}\csc u(x) = -\csc u(x)\cot u(x)\frac{du}{dx}$$

$$\frac{d}{dx}\tan u(x) = \sec^2 u(x)\frac{du}{dx} \qquad \frac{d}{dx}\cot u(x) = -\csc^2 u(x)\frac{du}{dx}$$

**NOW WORK** Problem 41.

---

**EXAMPLE 3**  **Finding an Equation of a Tangent Line**

Find an equation of the tangent line to the graph of $y = 5e^{4x}$ at the point $(0, 5)$.

**Solution** The slope of the tangent line to the graph of $y = f(x)$ at the point $(0, 5)$ is $f'(0)$.

$$f'(x) = \frac{d}{dx}(5e^{4x}) = 5\frac{d}{dx}e^{4x} = 5e^{4x} \cdot \frac{d}{dx}(4x) = 5e^{4x} \cdot 4 = 20e^{4x}$$
$$\uparrow \qquad\qquad \uparrow$$
$$\text{Constant} \qquad u = 4x; \ \frac{d}{dx}e^u = e^u\frac{du}{dx}$$
$$\text{Multiple Rule}$$

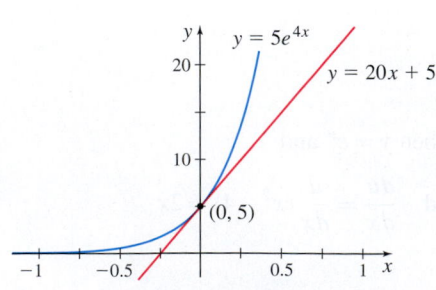

$y = 5e^{4x}$

$y = 20x + 5$

$(0, 5)$

**Figure 1**

$m_{\tan} = f'(0) = 20e^0 = 20$. Using the point-slope form of a line,

$$y - 5 = 20(x - 0) \qquad y - y_0 = m_{\tan}(x - x_0)$$
$$y = 20x + 5$$
$$\blacksquare$$

The graph of $y = 5e^{4x}$ and the line $y = 20x + 5$ are shown in Figure 1.

**NOW WORK** Problem 77.

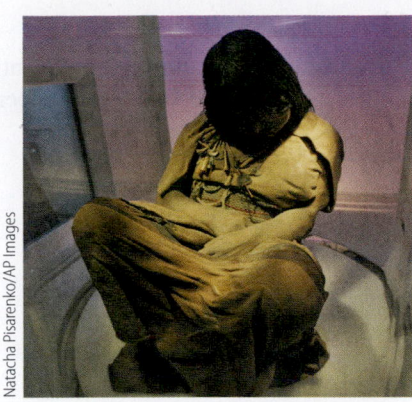

**EXAMPLE 4** **Application: Carbon-14 Dating**

All carbon on Earth contains some carbon-14, which is radioactive. When a living organism dies, the carbon-14 begins to decay at a fixed rate. The formula $P(t) = 100e^{-0.000121t}$ gives the percentage of carbon-14 present at time $t$ years. Notice that when $t = 0$, the percentage of carbon-14 present is 100%. When the preserved bodies of 15-year-old La Doncella and two younger children were found in Peru in 2005, 93.5% of the carbon-14 remained in their bodies, indicating that the three had died about 550 years earlier.

**(a)** What is the rate of change of the percentage of carbon-14 present in a 550-year-old fossil?

**(b)** What is the rate of change of the percentage of carbon-14 present in a 2000-year-old fossil?

**Solution (a)** The rate of change of $P$ is given by its derivative

$$P'(t) = \frac{d}{dt}\left(100e^{-0.000121t}\right) = 100\left(-0.000121e^{-0.000121t}\right) = -0.0121e^{-0.000121t}$$

$$\underset{\underset{\frac{d}{dx}e^{u(x)} = e^{u(x)}\frac{du}{dx}}{\uparrow}}{}$$

At $t = 550$ years,

$$P'(550) = -0.0121e^{-0.000121(550)} \approx -0.0113$$

The percentage of carbon-14 present in a 550-year-old fossil is decreasing at the rate of 1.13% per year.

**(b)** When $t = 2000$ years, the rate of change is

$$P'(2000) = -0.0121e^{-0.000121(2000)} \approx -0.0095$$

The percentage of carbon-14 present in a 2000-year-old fossil is decreasing at the rate of 0.95% per year. ∎

**NOW WORK** Problem **99.**

When we stated the Chain Rule, we expressed it two ways: using prime notation and using Leibniz notation. In each example so far, we have used Leibniz notation. But when solving numerical problems, using prime notation is often easier. In this form, the Chain Rule states that if a function $g$ is differentiable at $x_0$ and a function $f$ is differentiable at $g(x_0)$, then

$$\boxed{(f \circ g)'(x_0) = f'(g(x_0))g'(x_0)}$$

**EXAMPLE 5** **Differentiating a Composite Function**

Suppose $h = f \circ g$. Find $h'(1)$ given that:

$$f(1) = 2 \quad f'(1) = 1 \quad f'(2) = -4 \quad g(1) = 2 \quad g'(1) = -3 \quad g'(2) = 5$$

**Solution** Based on the Chain Rule using prime notation, we have

$$h'(x_0) = (f \circ g)'(x_0) = f'(g(x_0))\, g'(x_0)$$

When $x_0 = 1$,

$$h'(1) = f'(g(1)) \cdot g'(1) = f'(2) \cdot (-3) = (-4)(-3) = 12$$

$$\underset{g(1)=2;\ g'(1)=-3 \qquad f'(2)=-4}{\uparrow \qquad\qquad\qquad \uparrow}$$

∎

**NOW WORK** Problem **81.**

## 2 Differentiate $y = a^x$, $a > 0$, $a \neq 1$

NEED TO REVIEW? Properties of logarithms are discussed in Appendix A.1, pp. A-10 to A-11.

The Chain Rule allows us to establish a formula for differentiating an exponential function $y = a^x$ for any base $a > 0$ and $a \neq 1$. We start with the following property of logarithms:

$$a^x = e^{\ln a^x} = e^{x \ln a} \qquad a > 0, a \neq 1$$

Then

$$\frac{d}{dx} a^x = \frac{d}{dx} e^{x \ln a} = e^{x \ln a} \frac{d}{dx}(x \ln a) = e^{x \ln a} \ln a = a^x \ln a$$

$$\underset{\uparrow}{\frac{d}{dx} e^{u(x)} = e^{u(x)} \frac{du}{dx}}$$

---

**THEOREM  Derivative of $y = a^x$**

The derivative of an exponential function $y = a^x$, where $a > 0$ and $a \neq 1$, is

$$\boxed{y' = \frac{d}{dx} a^x = a^x \ln a}$$

If $u = u(x)$ is a differentiable function, then

**IN WORDS** The rate of change of any exponential function $y = a^x$ is proportional to the function $y$ itself.

$$\boxed{\frac{d}{dx} a^{u(x)} = a^{u(x)} \ln a \frac{du}{dx}}$$

---

The second result follows directly from the proof of $\dfrac{d}{dx} a^x$.

---

**EXAMPLE 6  Differentiating Exponential Functions**

Find the derivative of each function:

**(a)**  $f(x) = 2^x$     **(b)**  $F(x) = 3^{-x}$     **(c)**  $g(x) = \left(\dfrac{1}{2}\right)^{x^2+1}$

**Solution (a)**  $f$ is an exponential function with base $a = 2$.

$$f'(x) = \frac{d}{dx} 2^x = 2^x \ln 2 \qquad \frac{d}{dx} a^x = a^x \ln a$$

**(b)**  $F'(x) = \dfrac{d}{dx} 3^{-x} = 3^{-x} \ln 3 \dfrac{d}{dx}(-x) \qquad \dfrac{d}{dx} a^{u(x)} = a^{u(x)} \ln a \dfrac{du}{dx}$

$$= -3^{-x} \ln 3$$

**(c)**  $g'(x) = \dfrac{d}{dx} \left(\dfrac{1}{2}\right)^{x^2+1} = \left(\dfrac{1}{2}\right)^{x^2+1} \ln \dfrac{1}{2} \dfrac{d}{dx}(x^2+1) \qquad a^{u(x)} = a^{u(x)} \ln a \dfrac{du}{dx}$

$$= \left(\dfrac{1}{2}\right)^{x^2+1} \ln \dfrac{1}{2} \cdot (2x) \qquad \frac{d}{dx}(x^2+1) = 2x$$

$$= \frac{x}{2^{x^2}} \ln \frac{1}{2} \qquad \text{Simplify.} \qquad \blacksquare$$

**NOW WORK** Problem 47.

## 3 Use the Power Rule for Functions to Find a Derivative

We use the Chain Rule to establish other derivative formulas, such as a formula for the derivative of a function raised to a power.

**THEOREM  Power Rule for Functions**

If $g$ is a differentiable function and $n$ is an integer, then

$$\frac{d}{dx}[g(x)]^n = n[g(x)]^{n-1}g'(x)$$

**Proof**  If $y = [g(x)]^n$, let $y = u^n$ and $u = g(x)$. Then

$$\frac{dy}{du} = nu^{n-1} = n[g(x)]^{n-1} \qquad \text{and} \qquad \frac{du}{dx} = g'(x)$$

By the Chain Rule,

$$y' = \frac{d}{dx}[g(x)]^n = \frac{dy}{du} \cdot \frac{du}{dx} = n[g(x)]^{n-1}g'(x) \qquad \blacksquare$$

---

**EXAMPLE 7  Using the Power Rule for Functions to Find a Derivative**

**(a)**  If $f(x) = (3 - x^3)^{-5}$, then

$$f'(x) = \frac{d}{dx}(3 - x^3)^{-5} = -5(3 - x^3)^{-5-1} \cdot \frac{d}{dx}(3 - x^3) \qquad \text{Use the Power Rule for Functions.}$$

$$= -5(3 - x^3)^{-6} \cdot (-3x^2) = 15x^2(3 - x^3)^{-6} = \frac{15x^2}{(3 - x^3)^6}$$

**(b)**  If $f(\theta) = \cos^3 \theta$, then $f(\theta) = (\cos \theta)^3$, and

$$f'(\theta) = \frac{d}{d\theta}(\cos \theta)^3 = 3(\cos \theta)^{3-1} \cdot \frac{d}{d\theta}\cos \theta \qquad \text{Use the Power Rule for Functions.}$$

$$= 3\cos^2 \theta \cdot (-\sin \theta) = -3\cos^2 \theta \sin \theta \qquad \blacksquare$$

**NOW REWORK**  Example 7 using the Chain Rule and rework Example 1(a) using the Power Rule for Functions. Decide for yourself which method is easier.

**NOW WORK**  Problem 21.

Often other derivative rules are used along with the Power Rule for Functions.

CALC
CLIP

**EXAMPLE 8  Using the Power Rule for Functions with Other Derivative Rules**

Find the derivative of:

**(a)**  $f(x) = e^x(x^2 + 1)^3$    **(b)**  $g(x) = \left(\dfrac{3x + 2}{4x^2 - 5}\right)^5$

**Solution (a)** The function $f$ is the product of $e^x$ and $(x^2+1)^3$, so we first use the Product Rule.

$$f'(x) = e^x \left[ \frac{d}{dx}(x^2+1)^3 \right] + \left[ \frac{d}{dx} e^x \right] (x^2+1)^3 \qquad \text{Use the Product Rule.}$$

To complete the solution, we use the Power Rule for Functions to find $\dfrac{d}{dx}(x^2+1)^3$.

$$f'(x) = e^x \left[ 3(x^2+1)^2 \cdot \frac{d}{dx}(x^2+1) \right] + e^x (x^2+1)^3 \qquad \text{Use the Power Rule for Functions.}$$

$$= e^x[3(x^2+1)^2 \cdot 2x] + e^x(x^2+1)^3 = e^x[6x(x^2+1)^2 + (x^2+1)^3]$$

$$\underset{\underset{\text{Factor out } (x^2+1)^2}{\uparrow}}{=} e^x(x^2+1)^2[6x + (x^2+1)] = e^x(x^2+1)^2(x^2+6x+1)$$

**(b)** $g$ is a function raised to a power, so we begin with the Power Rule for Functions.

$$g'(x) = \frac{d}{dx} \left( \frac{3x+2}{4x^2-5} \right)^5 = 5 \left( \frac{3x+2}{4x^2-5} \right)^4 \left[ \frac{d}{dx} \frac{3x+2}{4x^2-5} \right] \qquad \begin{array}{l}\text{Use the Power Rule}\\\text{for Functions.}\end{array}$$

$$= 5 \left( \frac{3x+2}{4x^2-5} \right)^4 \left[ \frac{(3)(4x^2-5) - (3x+2)(8x)}{(4x^2-5)^2} \right] \qquad \text{Use the Quotient Rule.}$$

$$= \frac{5(3x+2)^4[(12x^2-15) - (24x^2+16x)]}{(4x^2-5)^6} \qquad \text{Distribute.}$$

$$= \frac{5(3x+2)^4[-12x^2-16x-15]}{(4x^2-5)^6} \qquad \text{Simplify.} \qquad \blacksquare$$

**NOW WORK** Problem 29.

## 4 Use the Chain Rule for Multiple Composite Functions

The Chain Rule can be extended to multiple composite functions. For example, if the functions

$$y = f(u) \qquad u = g(v) \qquad v = h(x)$$

are each differentiable functions of $u$, $v$, and $x$, respectively, then the composite function $y = (f \circ g \circ h)(x) = f(g(h(x)))$ is a differentiable function of $x$ and

$$\boxed{y' = \frac{dy}{dx} = \frac{dy}{du} \cdot \frac{du}{dv} \cdot \frac{dv}{dx}}$$

where $u = g(v) = g(h(x))$ and $v = h(x)$. This "chain" of factors is the basis for the name Chain Rule.

**EXAMPLE 9** Differentiating a Multiple Composite Function

Find $y'$ if:

**(a)** $y = 5\cos^2(3x+2)$      **(b)** $y = \sin^3 \left( \dfrac{\pi}{2}x \right)$

**Solution (a)** For $y = 5\cos^2(3x+2)$, use the Chain Rule with $y = 5u^2$, $u = \cos v$, and $v = 3x+2$. Then $y = 5u^2 = 5\cos^2 v = 5\cos^2(3x+2)$ and

$$\frac{dy}{du} = \frac{d}{du}(5u^2) = 10u = 10\cos(3x+2)$$
$$\underset{\underset{v=3x+2}{u=\cos v}}{\uparrow}$$

$$\frac{du}{dv} = \frac{d}{dv}\cos v = -\sin v = -\sin(3x+2)$$
$$\underset{v=3x+2}{\uparrow}$$

$$\frac{dv}{dx} = \frac{d}{dx}(3x+2) = 3$$

Then

$$y' = \frac{dy}{dx} \underset{\text{Chain Rule}}{\uparrow} = \frac{dy}{du} \cdot \frac{du}{dv} \cdot \frac{dv}{dx} = [10\cos(3x+2)][-\sin(3x+2)][3]$$

$$= -30\cos(3x+2)\sin(3x+2)$$

**(b)** For $y = \sin^3\left(\frac{\pi}{2}x\right)$, use the Chain Rule with $y = u^3$, $u = \sin v$, and $v = \frac{\pi}{2}x$.

Then $y = u^3 = (\sin v)^3 = \left[\sin\left(\frac{\pi}{2}x\right)\right]^3 = \sin^3\left(\frac{\pi}{2}x\right)$, and

$$\frac{dy}{du} = \frac{d}{du}u^3 = 3u^2 \underset{\substack{u=\sin v \\ v=\frac{\pi}{2}x}}{=} 3\left[\sin\left(\frac{\pi}{2}x\right)\right]^2 = 3\sin^2\left(\frac{\pi}{2}x\right)$$

$$\frac{du}{dv} = \frac{d}{dv}\sin v = \cos v \underset{v=\frac{\pi}{2}x}{=} \cos\left(\frac{\pi}{2}x\right)$$

$$\frac{dv}{dx} = \frac{d}{dx}\left(\frac{\pi}{2}x\right) = \frac{\pi}{2}$$

Then

$$y' = \frac{dy}{dx} \underset{\text{Chain Rule}}{\uparrow} = \frac{dy}{du} \cdot \frac{du}{dv} \cdot \frac{dv}{dx} = 3\sin^2\left(\frac{\pi}{2}x\right) \cdot \cos\left(\frac{\pi}{2}x\right) \cdot \frac{\pi}{2}$$

$$= \frac{3\pi}{2}\sin^2\left(\frac{\pi}{2}x\right)\cos\left(\frac{\pi}{2}x\right)$$  ∎

NOW WORK Problems 59 and 63.

# 3.1 Assess Your Understanding

## Concepts and Vocabulary

1. The derivative of a composite function $(f \circ g)(x)$ can be found using the _____ Rule.

2. *True or False* If $y = f(u)$ and $u = g(x)$ are differentiable functions, then $y = f(g(x))$ is differentiable.

3. *True or False* If $y = f(g(x))$ is a differentiable function, then $y' = f'(g'(x))$.

4. To find the derivative of $y = \tan(1 + \cos x)$, using the Chain Rule, begin with $y = $ _____ and $u = $ _____.

5. If $y = (x^3 + 4x + 1)^{100}$, then $y' = $ _____.

6. If $f(x) = e^{3x^2+5}$, then $f'(x) = $ _____.

7. *True or False* The Chain Rule can be applied to multiple composite functions.

8. $\dfrac{d}{dx}\sin x^2 = $ _____.

## Skill Building

*In Problems 9–14, write y as a function of x. Find $\dfrac{dy}{dx}$ using the Chain Rule.*

9. $y = u^5$, $u = x^3 + 1$      10. $y = u^3$, $u = 2x + 5$

11. $y = \dfrac{u}{u+1}$, $u = x^2 + 1$

12. $y = \dfrac{u-1}{u}$, $u = x^2 - 1$

13. $y = (u+1)^2$, $u = \dfrac{1}{x}$

14. $y = (u^2 - 1)^3$, $u = \dfrac{1}{x+2}$

*In Problems 15–32, find the derivative of each function using the Power Rule for Functions.*

15. $f(x) = (3x + 5)^2$

16. $f(x) = (2x - 5)^3$

17. $f(x) = (6x - 5)^{-3}$

18. $f(t) = (4t + 1)^{-2}$

19. $g(x) = (x^2 + 5)^4$

20. $F(x) = (x^3 - 2)^5$

21. $f(u) = \left(u - \dfrac{1}{u}\right)^3$

22. $f(x) = \left(x + \dfrac{1}{x}\right)^3$

23. $g(x) = (4x + e^x)^3$

24. $F(x) = (e^x - x^2)^2$

25. $f(x) = \tan^2 x$

26. $f(x) = \sec^3 x$

27. $f(z) = (\tan z + \cos z)^2$

28. $f(z) = (e^z + 2\sin z)^3$

29. $y = (x^2 + 4)^2(2x^3 - 1)^3$

30. $y = (x^2 - 2)^3(3x^4 + 1)^2$

31. $y = \left(\dfrac{\sin x}{x}\right)^2$

32. $y = \left(\dfrac{x + \cos x}{x}\right)^5$

---

**1.** = NOW WORK problem      ⊿ = Graphing technology recommended      CAS = Computer Algebra System recommended

*In Problems 33–54, find $y'$.*

**33.** $y = \sin(4x)$

**34.** $y = \cos(5x)$

**35.** $y = 2\sin(x^2 + 2x - 1)$

**36.** $y = \dfrac{1}{2}\cos(x^3 - 2x + 5)$

**37.** $y = \sin\dfrac{1}{x}$

**38.** $y = \sin\dfrac{3}{x}$

**39.** $y = \sec(4x)$

**40.** $y = \cot(5x)$

**41.** $y = e^{1/x}$

**42.** $y = e^{1/x^2}$

**43.** $y = \dfrac{1}{x^4 - 2x + 1}$

**44.** $y = \dfrac{3}{x^5 + 2x^2 - 3}$

**45.** $y = \dfrac{100}{1 + 99e^{-x}}$

**46.** $y = \dfrac{1}{1 + 2e^{-x}}$

**47.** $y = 2^{\sin x}$

**48.** $y = (\sqrt{3})^{\cos x}$

**49.** $y = 6^{\sec x}$

**50.** $y = 3^{\tan x}$

**51.** $y = 5xe^{3x}$

**52.** $y = x^3 e^{2x}$

**53.** $y = x^2 \sin(4x)$

**54.** $y = x^2 \cos(4x)$

*In Problems 55–58, find $y'$ (a and b are constants).*

**55.** $y = e^{-ax}\sin(bx)$

**56.** $y = e^{ax}\cos(-bx)$

**57.** $y = \dfrac{e^{ax} - 1}{e^{ax} + 1}$

**58.** $y = \dfrac{e^{-ax} + 1}{e^{bx} - 1}$

*In Problems 59–62, write y as a function of x. Find $\dfrac{dy}{dx}$ using the Chain Rule.*

**59.** $y = u^3,\ u = 3v^2 + 1,\ v = \dfrac{4}{x^2}$

**60.** $y = 3u,\ u = 3v^2 - 4,\ v = \dfrac{1}{x}$

**61.** $y = u^2 + 1,\ u = \dfrac{4}{v},\ v = x^2$

**62.** $y = u^3 - 1,\ u = -\dfrac{2}{v},\ v = x^3$

*In Problems 63–70, find $y'$.*

**63.** $y = e^{-2x}\cos(3x)$

**64.** $y = e^{\pi x}\tan(\pi x)$

**65.** $y = \cos(e^{x^2})$

**66.** $y = \tan(e^{x^2})$

**67.** $y = e^{\cos(4x)}$

**68.** $y = e^{\csc^2 x}$

**69.** $y = 4\sin^2(3x)$

**70.** $y = 2\cos^2(x^2)$

*In Problems 71 and 72, find the derivative of each function by:*
**(a)** *Using the Chain Rule.*
**(b)** *Using the Power Rule for Functions.*
**(c)** *Expanding and then differentiating.*
**(d)** *Verify the answers from parts (a)–(c) are equal.*

**71.** $y = (x^3 + 1)^2$

**72.** $y = (x^2 - 2)^3$

*In Problems 73–78:*
**(a)** *Find an equation of the tangent line to the graph of $f$ at the given point.*
**(b)** *Find an equation of the normal line to the graph of $f$ at the given point.*
**(c)** *Use technology to graph $f$, the tangent line, and the normal line on the same screen.*

**73.** $f(x) = (x^2 - 2x + 1)^5$ at $(1, 0)$

**74.** $f(x) = (x^3 - x^2 + x - 1)^{10}$ at $(0, 1)$

**75.** $f(x) = \dfrac{x}{(x^2 - 1)^3}$ at $\left(2, \dfrac{2}{27}\right)$

**76.** $f(x) = \dfrac{x^2}{(x^2 - 1)^2}$ at $\left(2, \dfrac{4}{9}\right)$

**77.** $f(x) = \sin(2x) + \cos\dfrac{x}{2}$ at $(0, 1)$

**78.** $f(x) = \sin^2 x + \cos^3 x$ at $\left(\dfrac{\pi}{2}, 1\right)$

*In Problems 79 and 80, find the indicated derivative.*

**79.** $\dfrac{d^2}{dx^2}\cos(x^5)$

**80.** $\dfrac{d^3}{dx^3}\sin^3 x$

**81.** Suppose $h = f \circ g$. Find $h'(1)$ if $f'(2) = 6$, $f(1) = 4$, $g(1) = 2$, and $g'(1) = -2$.

**82.** Suppose $h = f \circ g$. Find $h'(1)$ if $f'(3) = 4$, $f(1) = 1$, $g(1) = 3$, and $g'(1) = 3$.

**83.** Suppose $h = g \circ f$. Find $h'(0)$ if $f(0) = 3$, $f'(0) = -1$, $g(3) = 8$, and $g'(3) = 0$.

**84.** Suppose $h = g \circ f$. Find $h'(2)$ if $f(1) = 2$, $f'(1) = 4$, $f(2) = -3$, $f'(2) = 4$, $g(-3) = 1$, and $g'(-3) = 3$.

**85.** If $y = u^5 + u$ and $u = 4x^3 + x - 4$, find $\dfrac{dy}{dx}$ at $x = 1$.

**86.** If $y = e^u + 3u$ and $u = \cos x$, find $\dfrac{dy}{dx}$ at $x = 0$.

## Applications and Extensions

*In Problems 87–94, find the indicated derivative.*

**87.** $\dfrac{d}{dx}f(x^2 + 1)$ (*Hint:* Let $u = x^2 + 1$.)

**88.** $\dfrac{d}{dx}f(1 - x^2)$

**89.** $\dfrac{d}{dx}f\left(\dfrac{x+1}{x-1}\right)$

**90.** $\dfrac{d}{dx}f\left(\dfrac{1-x}{1+x}\right)$

**91.** $\dfrac{d}{dx}f(\sin x)$

**92.** $\dfrac{d}{dx}f(\tan x)$

**93.** $\dfrac{d^2}{dx^2}f(\cos x)$

**94.** $\dfrac{d^2}{dx^2}f(e^x)$

**95. Rectilinear Motion** An object is in rectilinear motion and its position $s$, in meters, from the origin at time $t$ seconds is given by $s = s(t) = A\cos(\omega t + \phi)$, where $A$, $\omega$, and $\phi$ are constants.

**(a)** Find the velocity $v$ of the object at time $t$.
**(b)** When is the velocity of the object 0?
**(c)** Find the acceleration $a$ of the object at time $t$.
**(d)** When is the acceleration of the object 0?

**96. Rectilinear Motion** A bullet is fired horizontally into a bale of paper. The distance $s$ (in meters) the bullet travels into the bale of paper in $t$ seconds is given by

$$s = s(t) = 8 - (2 - t)^3 \quad 0 \le t \le 2$$

**(a)** Find the velocity $v$ of the bullet at any time $t$.
**(b)** Find the velocity of the bullet at $t = 1$.
**(c)** Find the acceleration $a$ of the bullet at any time $t$.
**(d)** Find the acceleration of the bullet at $t = 1$.
**(e)** How far into the bale of paper did the bullet travel?

**97. Rectilinear Motion**  Find the acceleration $a$ of a car if the distance $s$, in feet, it has traveled along a highway at time $t \geq 0$ seconds is given by

$$s(t) = \frac{80}{3}\left[t + \frac{3}{\pi}\sin\left(\frac{\pi}{6}t\right)\right]$$

**98. Rectilinear Motion**  An object moves in rectilinear motion so that at time $t \geq 0$ seconds, its position from the origin is $s(t) = \sin e^t$, in feet.

(a) Find the velocity $v$ and acceleration $a$ of the object at any time $t$.

(b) At what time does the object first have zero velocity?

(c) What is the acceleration of the object at the time $t$ found in (b)?

**99. Resistance**  The resistance $R$ (measured in ohms) of an 80-meter-long electric wire of radius $x$ (in centimeters) is given by the formula $R = R(x) = \dfrac{0.0048}{x^2}$. The radius $x$ is given by $x = 0.1991 + 0.000003T$ where $T$ is the temperature in kelvin. How fast is $R$ changing with respect to $T$ when $T = 320$ K?

**100. Pendulum Motion in a Car**  The motion of a pendulum swinging in the direction of motion of a car moving at a low, constant speed, can be modeled by

$$s = s(t) = 0.05\sin(2t) + 3t \qquad 0 \leq t \leq \pi$$

where $s$ is the distance in meters and $t$ is the time in seconds.

(a) Find the velocity $v$ at $t = \dfrac{\pi}{8}$, $t = \dfrac{\pi}{4}$, and $t = \dfrac{\pi}{2}$.

(b) Find the acceleration $a$ at the times given in (a).

(c) Graph $s = s(t)$, $v = v(t)$, and $a = a(t)$ on the same screen.

*Source*: Mathematics students at Trine University, Angola, Indiana.

**101. Economics**  The function $A(t) = 102 - 90\,e^{-0.21t}$ represents the relationship between $A$, the percentage of the market penetrated by the latest generation smart phones, and $t$, the time in years, where $t = 0$ corresponds to the year 2018.

(a) Find $\lim\limits_{t\to\infty} A(t)$ and interpret the result.

(b) Graph the function $A = A(t)$, and explain how the graph supports the answer in (a).

(c) Find the rate of change of $A$ with respect to time.

(d) Evaluate $A'(5)$ and $A'(10)$ and interpret these results.

(e) Graph the function $A' = A'(t)$, and explain how the graph supports the answers in (d).

**102. Meteorology**  The atmospheric pressure at a height of $x$ meters above sea level is $P(x) = 10^4 e^{-0.00012x}$ kg/m². What is the rate of change of the pressure with respect to the height at $x = 500$ m? At $x = 750$ m?

**103. Hailstones**  Hailstones originate at an altitude of about 3000 m, although this varies. As they fall, air resistance slows down the hailstones considerably. In one model of air resistance, the speed of a hailstone of mass $m$ as a function of time $t$ is given by $v(t) = \dfrac{mg}{k}(1 - e^{-kt/m})$ m/s,

where $g = 9.8$ m/s² is the acceleration due to gravity and $k$ is a constant that depends on the size of the hailstone and the conditions of the air.

(a) Find the acceleration $a(t)$ of a hailstone as a function of time $t$.

(b) Find $\lim\limits_{t\to\infty} v(t)$. What does this limit say about the speed of the hailstone?

(c) Find $\lim\limits_{t\to\infty} a(t)$. What does this limit say about the acceleration of the hailstone?

**104. Mean Earnings**  The mean earnings $E$, in dollars, of workers 18 years and over are given in the table below:

| Year | 1975 | 1980 | 1985 | 1990 | 1995 | 2000 | 2005 | 2010 | 2015 |
|---|---|---|---|---|---|---|---|---|---|
| Mean Earnings | 8,552 | 12,665 | 17,181 | 21,793 | 26,792 | 32,604 | 41,231 | 49,733 | 48,000 |

*Source*: U.S. Bureau of the Census, Current Population Survey.

(a) Find the exponential function of best fit and show that it equals $E = E(t) = 9854(1.05)^t$, where $t$ is the number of years since 1974.

(b) Find the rate of change of $E$ with respect to $t$.

(c) Find the rate of change at $t = 26$ (year 2000).

(d) Find the rate of change at $t = 36$ (year 2010).

(e) Find the rate of change at $t = 41$ (year 2015).

(f) Compare the answers to (c), (d), and (e). Interpret each answer and explain the differences.

**105. Rectilinear Motion**  An object moves in rectilinear motion so that at time $t > 0$ its position $s$ from the origin is $s = s(t)$. The velocity $v$ of the object is $v = \dfrac{ds}{dt}$, and its acceleration is $a = \dfrac{dv}{dt} = \dfrac{d^2 s}{dt^2}$. If the velocity $v = v(s)$ is expressed as a function of $s$, show that the acceleration $a$ can be expressed as $a = v\dfrac{dv}{ds}$.

**106. Student Approval**  Professor Miller's student approval rating is modeled by the function $Q(t) = 21 + \dfrac{10\sin\dfrac{2\pi t}{7}}{\sqrt{t} - \sqrt{20}}$,

where $0 \leq t \leq 16$ is the number of weeks since the semester began.

(a) Find $Q'(t)$.

(b) Evaluate $Q'(1)$, $Q'(5)$, and $Q'(10)$.

(c) Interpret the results obtained in (b).

(d) Use technology to graph $Q(t)$ and $Q'(t)$.

(e) How would you explain the results in (d) to Professor Miller?

*Source*: Mathematics students at Millikin University, Decatur, Illinois.

**107. Angular Velocity**  If the disk in the figure is rotated about a vertical line through an angle $\theta$, torsion in the wire attempts to turn the disk in the opposite direction. The motion $\theta$ at time $t$ (assuming no friction or air resistance) obeys the equation

$$\theta(t) = \frac{\pi}{3}\cos\left(\frac{1}{2}\sqrt{\frac{2k}{5}}\,t\right)$$

where $k$ is the coefficient of torsion of the wire.

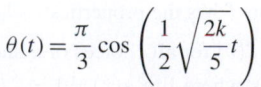

(a) Find the angular velocity $\omega = \dfrac{d\theta}{dt}$ of the disk at any time $t$.

(b) What is the angular velocity at $t = 3$?

**108. Harmonic Motion**   A weight hangs on a spring making it 2 m long when it is stretched out (see the figure). If the weight is pulled down and then released, the weight oscillates up and down, and the length $l$ of the spring after $t$ seconds is given by $l(t) = 2 + \cos(2\pi t)$.

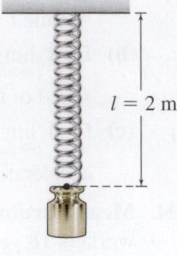

$l = 2$ m

(a) Find the length $l$ of the spring at the times $t = 0$, $\dfrac{1}{2}$, $1$, $\dfrac{3}{2}$, and $\dfrac{5}{8}$.

(b) Find the velocity $v$ of the weight at time $t = \dfrac{1}{4}$.

(c) Find the acceleration $a$ of the weight at time $t = \dfrac{1}{4}$.

**109.** Find $F'(1)$ if $f(x) = \sin x$ and $F(t) = f(t^2 - 1)$.

**110. Normal Line**   Find the point on the graph of $y = e^{-x}$ where the normal line to the graph passes through the origin.

**111.** Use the Chain Rule and the fact that $\cos x = \sin\left(\dfrac{\pi}{2} - x\right)$ to show that $\dfrac{d}{dx}\cos x = -\sin x$.

**112.** If $y = e^{2x}$, show that $y'' - 4y = 0$.

**113.** If $y = e^{-2x}$, show that $y'' - 4y = 0$.

**114.** If $y = Ae^{2x} + Be^{-2x}$, where $A$ and $B$ are constants, show that $y'' - 4y = 0$.

**115.** If $y = Ae^{ax} + Be^{-ax}$, where $A$, $B$, and $a$ are constants, show that $y'' - a^2 y = 0$.

**116.** If $y = Ae^{2x} + Be^{3x}$, where $A$ and $B$ are constants, show that $y'' - 5y' + 6y = 0$.

**117.** If $y = Ae^{-2x} + Be^{-x}$, where $A$ and $B$ are constants, show that $y'' + 3y' + 2y = 0$.

**118.** If $y = A\sin(\omega t) + B\cos(\omega t)$, where $A$, $B$, and $\omega$ are constants, show that $y'' + \omega^2 y = 0$.

**119.** Show that $\dfrac{d}{dx} f(h(x)) = 2xg(x^2)$, if $\dfrac{d}{dx} f(x) = g(x)$ and $h(x) = x^2$.

**120.** Find the $n$th derivative of $f(x) = (2x + 3)^n$.

**121.** Find a general formula for the $n$th derivative of $y$.
   (a) $y = e^{ax}$       (b) $y = e^{-ax}$

**122.** (a) What is $\dfrac{d^{10}}{dx^{10}} \sin(ax)$?

   (b) What is $\dfrac{d^{25}}{dx^{25}} \sin(ax)$?

   (c) Find the $n$th derivative of $f(x) = \sin(ax)$.

**123.** (a) What is $\dfrac{d^{11}}{dx^{11}} \cos(ax)$?

   (b) What is $\dfrac{d^{12}}{dx^{12}} \cos(ax)$?

   (c) Find the $n$th derivative of $f(x) = \cos(ax)$.

**124.** If $y = e^{-at}[A\sin(\omega t) + B\cos(\omega t)]$, where $A$, $B$, $a$, and $\omega$ are constants, find $y'$ and $y''$.

**125.** Show that if a function $f$ has the properties:
   - $f(u + v) = f(u)f(v)$ for all choices of $u$ and $v$
   - $f(x) = 1 + xg(x)$, where $\lim\limits_{x \to 0} g(x) = 1$,

   then $f' = f$.

**126.** Find the $n$th derivative of $f(x) = \dfrac{1}{3x - 4}$.

**127.** Let $f_1(x), \ldots, f_n(x)$ be $n$ differentiable functions. Find the derivative of $y = f_1(f_2(f_3(\ldots(f_n(x)\ldots))))$.

**128.** Let
$$f(x) = \begin{cases} x^2 \sin \dfrac{1}{x} & \text{if } x \neq 0 \\ 0 & \text{if } x = 0 \end{cases}$$

Show that $f'(0)$ exists, but that $f'$ is not continuous at 0.

**129.** Define $f$ by
$$f(x) = \begin{cases} e^{-1/x^2} & \text{if } x \neq 0 \\ 0 & \text{if } x = 0 \end{cases}$$

Show that $f$ is differentiable on $(-\infty, \infty)$ and find $f'$ for each value of $x$.
[*Hint:* To find $f'(0)$, use the definition of the derivative. Then show that $1 < x^2 e^{1/x^2}$ for $x \neq 0$.]

**130.** Suppose $f(x) = x^2$ and $g(x) = |x - 1|$. The functions $f$ and $g$ are continuous on their domains, the set of all real numbers.

   (a) Is $f$ differentiable at all real numbers? If not, where does $f'$ not exist?

   (b) Is $g$ differentiable at all real numbers? If not, where does $g'$ not exist?

   (c) Can the Chain Rule be used to differentiate the composite function $(f \circ g)(x)$ for all $x$? Explain.

   (d) Is the composite function $(f \circ g)(x)$ differentiable? If so, what is its derivative?

**131.** Suppose $f(x) = x^4$ and $g(x) = x^{1/3}$. The functions $f$ and $g$ are continuous on their domains, the set of all real numbers.

   (a) Is $f$ differentiable at all real numbers? If not, where does $f'$ not exist?

   (b) Is $g$ differentiable at all real numbers? If not, where does $g'$ not exist?

   (c) Can the Chain Rule be used to differentiate the composite function $(f \circ g)(x)$ for all $x$? Explain.

   (d) Is the composite function $(f \circ g)(x)$ differentiable? If so, what is its derivative?

**132.** The function $f(x) = e^x$ has the property $f'(x) = f(x)$. Give an example of another function $g(x)$ such that $g(x)$ is defined for all real $x$, $g'(x) = g(x)$, and $g(x) \neq f(x)$.

**133. Harmonic Motion**   The motion of the piston of an automobile engine is approximately simple harmonic. If the stroke of a piston (twice the amplitude) is 10 cm and the angular velocity $\omega$ is 60 revolutions per second, then the motion of the piston is given by $s(t) = 5\sin(120\pi t)$ cm.

   (a) Find the acceleration $a$ of the piston at the end of its stroke $\left(t = \dfrac{1}{240} \text{ second}\right)$.

   (b) If the mass of the piston is 1 kg, what resultant force must be exerted on it at this point?
   (*Hint:* Use Newton's Second Law, that is, $F = ma$.)

# 3.2 Implicit Differentiation

**OBJECTIVES** *When you finish this section, you should be able to:*

**1** Find a derivative using implicit differentiation (p. 219)
**2** Find higher-order derivatives using implicit differentiation (p. 222)
**3** Differentiate functions with rational exponents (p. 223)

So far we have differentiated only functions $y = f(x)$ where the dependent variable $y$ is expressed *explicitly* in terms of the independent variable $x$. There are functions that are not written in the form $y = f(x)$, but are written in the *implicit* form $F(x, y) = 0$. For example, $x$ and $y$ are related implicitly in the equations

$$xy - 4 = 0 \qquad y^2 + 3x^2 - 1 = 0 \qquad e^{x^2 - y^2} - \cos(xy) = 0$$

**NEED TO REVIEW?** The implicit form of a function is discussed in Section P.1, p. 4.

The implicit form $xy - 4 = 0$ can easily be written explicitly as the function $y = \dfrac{4}{x}$. Also, $y^2 + 3x^2 - 1 = 0$ can be written explicitly as the two functions $y_1 = \sqrt{1 - 3x^2}$ and $y_2 = -\sqrt{1 - 3x^2}$. In the equation $e^{x^2 - y^2} - \cos(xy) = 0$, it is impossible to express $y$ as an explicit function of $x$. To find the derivative of a function defined implicitly, we use the technique of *implicit differentiation*.

## 1 Find a Derivative Using Implicit Differentiation

The method used to differentiate an implicitly defined function is called **implicit differentiation**. It does not require rewriting the function explicitly, but it does require that the dependent variable $y$ is a differentiable function of the independent variable $x$. So throughout the section, we make the assumption that there is a differentiable function $y = f(x)$ defined by the implicit equation. This assumption made, the method consists of differentiating both sides of the implicitly defined function with respect to $x$ and then solving the resulting equation for $\dfrac{dy}{dx}$.

---

**EXAMPLE 1**  **Finding a Derivative Using Implicit Differentiation**

Find $\dfrac{dy}{dx}$ if $xy - 4 = 0$.

**(a)** Use implicit differentiation.
**(b)** Solve for $y$ and then differentiate.
**(c)** Verify the results of (a) and (b) are the same.

**Solution (a)** To differentiate implicitly, we assume $y$ is a differentiable function of $x$ and differentiate both sides with respect to $x$.

$$\frac{d}{dx}(xy - 4) = \frac{d}{dx}0 \qquad \text{\color{blue}Differentiate both sides with respect to } x.$$

$$\frac{d}{dx}(xy) - \frac{d}{dx}4 = 0 \qquad \text{\color{blue}Use the Difference Rule.}$$

$$x \cdot \frac{d}{dx}y + \left(\frac{d}{dx}x\right)y - 0 = 0 \qquad \text{\color{blue}Use the Product Rule.}$$

$$x\frac{dy}{dx} + y = 0 \qquad \text{\color{blue}Simplify.}$$

$$\frac{dy}{dx} = -\frac{y}{x} \qquad \text{\color{blue}Solve for } \frac{dy}{dx}. \tag{1}$$

**(b)** Solve $xy - 4 = 0$ for $y$, obtaining $y = \dfrac{4}{x} = 4x^{-1}$. Then

$$\frac{dy}{dx} = \frac{d}{dx}(4x^{-1}) = -4x^{-2} = -\frac{4}{x^2} \tag{2}$$

**(c)** At first glance, the results in (1) and (2) appear to be different. However, since $xy - 4 = 0$, or equivalently, $y = \dfrac{4}{x}$, the result from (1) becomes

$$\underset{\underset{(1)}{\uparrow}}{\frac{dy}{dx}} = \underset{\underset{y = \frac{4}{x}}{\uparrow}}{-\frac{y}{x}} = -\frac{\frac{4}{x}}{x} = -\frac{4}{x^2}$$

which is the same as (2). ∎

In most instances, we will not know the explicit form of the function (as we did in Example 1) and so we will leave the derivative $\dfrac{dy}{dx}$ expressed in terms of $x$ and $y$ [as in (1)].

**NOW WORK** Problem 17.

The Power Rule for Functions is

$$\frac{d}{dx}[f(x)]^n = n[f(x)]^{n-1} f'(x)$$

where $n$ is an integer. If $y = f(x)$, it takes the form

$$\boxed{\frac{d}{dx} y^n = n y^{n-1} \frac{dy}{dx}}$$

This is convenient notation to use with implicit differentiation when $y^n$ appears.

$$\underset{\underset{n=1}{\uparrow}}{\frac{d}{dx} y} = 1 \cdot \frac{dy}{dx} = \frac{dy}{dx} \qquad \underset{\underset{n=2}{\uparrow}}{\frac{d}{dx} y^2} = 2y \frac{dy}{dx} \qquad \underset{\underset{n=3}{\uparrow}}{\frac{d}{dx} y^3} = 3y^2 \frac{dy}{dx}$$

To differentiate an implicit function:

- Assume that $y$ is a differentiable function of $x$.
- Differentiate both sides of the equation with respect to $x$.
- Solve the resulting equation for $y' = \dfrac{dy}{dx}$.

CALC
▶ CLIP
**EXAMPLE 2   Using Implicit Differentiation to Find an Equation of a Tangent Line**

Find an equation of the tangent line to the graph of the ellipse $3x^2 + 4y^2 = 2x$ at the point $\left(\dfrac{1}{2}, -\dfrac{1}{4}\right)$. Graph the equation and the tangent line on the same set of coordinate axes.

**Solution** First use implicit differentiation to find the slope of the tangent line.

Assuming $y$ is a differentiable function of $x$, differentiate both sides with respect to $x$.

$$\frac{d}{dx}(3x^2+4y^2)=\frac{d}{dx}(2x)$$     Differentiate both sides with respect to $x$.

$$\frac{d}{dx}(3x^2)+\frac{d}{dx}(4y^2)=2$$     Use the Sum Rule.

$$3\frac{d}{dx}x^2+4\frac{d}{dx}y^2=2$$     Use the Constant Multiple Rule.

$$6x+4\left(2y\frac{dy}{dx}\right)=2$$     $\frac{d}{dx}y^2=2y\frac{dy}{dx}$

$$6x+8y\frac{dy}{dx}=2$$     Simplify.

$$\frac{dy}{dx}=\frac{2-6x}{8y}=\frac{1-3x}{4y}$$     Solve for $\frac{dy}{dx}$.

provided $y\neq 0$.

Now evaluate $\dfrac{dy}{dx}=\dfrac{1-3x}{4y}$ at $\left(\dfrac{1}{2},-\dfrac{1}{4}\right)$.

**NOTE** The notation $\dfrac{dy}{dx}\Big|_{(x_0,y_0)}$ is used when the derivative $\dfrac{dy}{dx}$ is to be evaluated at a point $(x_0,y_0)$.

$$\frac{dy}{dx}\bigg|_{(\frac{1}{2},-\frac{1}{4})}=\frac{1-3x}{4y}\bigg|_{(\frac{1}{2},-\frac{1}{4})}=\frac{1-3\cdot\frac{1}{2}}{4\cdot\left(-\frac{1}{4}\right)}=\frac{1}{2}$$

The slope of the tangent line to the graph of $3x^2+4y^2=2x$ at the point $\left(\dfrac{1}{2},-\dfrac{1}{4}\right)$ is $\dfrac{1}{2}$. An equation of the tangent line is

$$y+\frac{1}{4}=\frac{1}{2}\left(x-\frac{1}{2}\right)$$

$$y=\frac{1}{2}x-\frac{1}{2}$$

Figure 2 shows the graph of the ellipse $3x^2+4y^2=2x$ and the graph of the tangent line $y=\dfrac{1}{2}x-\dfrac{1}{2}$ at the point $\left(\dfrac{1}{2},-\dfrac{1}{4}\right)$. ∎

**DF** Figure 2

In figure: $y=\frac{1}{2}x-\frac{1}{2}$; $3x^2+4y^2=2x$; $\left(\frac{1}{2},-\frac{1}{4}\right)$

**NOW WORK** Problems 15 and 53.

**EXAMPLE 3** **Using Implicit Differentiation to Find an Equation of a Tangent Line**

Find an equation of the tangent line to the graph of $e^y\cos x=x+1$ at the point $(0,0)$.

**Solution** Assuming $y$ is a differentiable function of $x$, use implicit differentiation to find the slope of the tangent line.

$$\frac{d}{dx}(e^y\cos x)=\frac{d}{dx}(x+1)$$

$$e^y\cdot\frac{d}{dx}(\cos x)+\left(\frac{d}{dx}e^y\right)\cdot\cos x=1$$     Use the Product Rule on the left.

$$e^y(-\sin x)+e^y\frac{dy}{dx}\cdot\cos x=1$$     $\frac{d}{dx}e^y=e^y\frac{dy}{dx}$ using the Chain Rule.

$$(e^y\cos x)\frac{dy}{dx}=1+e^y\sin x$$

$$y'=\frac{1+e^y\sin x}{e^y\cos x}$$

provided $\cos x\neq 0$.

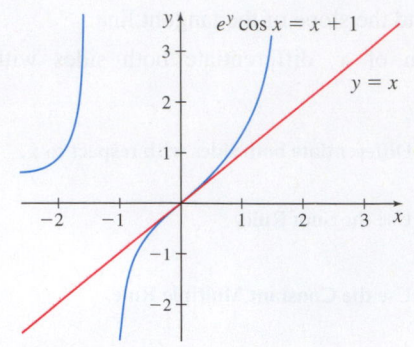

**Figure 3**

At the point $(0, 0)$, the derivative is

$$\left.\frac{dy}{dx}\right|_{(0,0)} = \frac{1 + e^0 \sin 0}{e^0 \cos 0} = \frac{1 + 1 \cdot 0}{1 \cdot 1} = 1$$

so the slope of the tangent line to the graph at $(0, 0)$ is 1. An equation of the tangent line to the graph at the point $(0, 0)$ is $y = x$. ∎

The graph of $e^y \cos x = x + 1$ and the tangent line to the graph at $(0, 0)$ are shown in Figure 3.

---

**EXAMPLE 4   Using Implicit Differentiation with the Chain Rule**

Find $y'$ if $\tan^2 y = e^x$. Express $y'$ as a function of $x$ alone.

**Solution** Assuming $y$ is a differentiable function of $x$, differentiate both sides with respect to $x$.

$$\frac{d}{dx}\tan^2 y = \frac{d}{dx}e^x$$

|  |  |
|---|---|
|  | Use the Chain Rule on the left. |
| $2\tan y \sec^2 y \dfrac{dy}{dx} = e^x$ | $\dfrac{d}{dx}\tan y = \sec^2 y \dfrac{dy}{dx}$ |
| $y' = \dfrac{dy}{dx} = \dfrac{e^x}{2\tan y \sec^2 y}$ | Solve for $y' = \dfrac{dy}{dx}$. |
| $= \dfrac{e^x}{2e^{x/2}(1 + e^x)} = \dfrac{e^{x/2}}{2(1 + e^x)}$ | $\tan y = e^{x/2}$ $\sec^2 y = 1 + \tan^2 y = 1 + e^x$ |

**NOW WORK** Problem 25.

## 2 Find Higher-Order Derivatives Using Implicit Differentiation

Implicit differentiation can be used to find higher-order derivatives.

---

**EXAMPLE 5   Finding Higher-Order Derivatives**

Use implicit differentiation to find $y'$ and $y''$ if $y^2 - x^2 = 5$. Express $y''$ in terms of $x$ and $y$.

**Solution** First, we assume there is a twice differentiable function $y = f(x)$ that satisfies $y^2 - x^2 = 5$. Now we find $y'$.

$$\frac{d}{dx}(y^2 - x^2) = \frac{d}{dx}5$$

$$\frac{d}{dx}y^2 - \frac{d}{dx}x^2 = 0$$

$$2yy' - 2x = 0 \qquad \frac{d}{dx}y^2 = 2y\frac{dy}{dx} = 2yy' \qquad (3)$$

$$y' = \frac{2x}{2y} = \frac{x}{y} \qquad \text{Solve for } y'. \qquad (4)$$

provided $y \neq 0$.

Equations (3) and (4) both involve $y'$. Either one can be used to find $y''$. We use (3) because it avoids differentiating a quotient.

$$\frac{d}{dx}(2yy' - 2x) = \frac{d}{dx}0$$

$$\frac{d}{dx}(yy') - \frac{d}{dx}x = 0 \qquad \text{Use the Difference Rule.}$$

$$y \cdot \frac{d}{dx}y' + \left(\frac{d}{dx}y\right)y' - 1 = 0 \qquad \text{Use the Product Rule.}$$

$$yy'' + (y')^2 - 1 = 0$$

$$y'' = \frac{1 - (y')^2}{y} \qquad \text{Solve for } y''. \qquad (5)$$

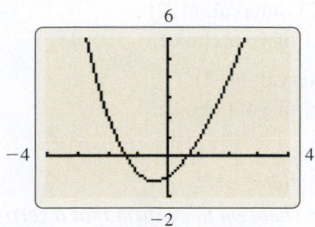

**Figure 14** $f(x) = \sin x + x^2 - 1$

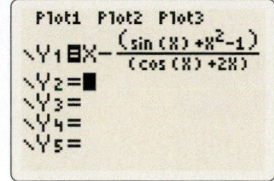

**Figure 15**

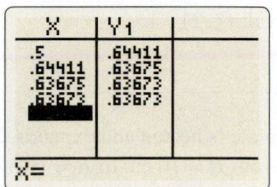

**Figure 16**

**EXAMPLE 6** Using Technology with Newton's Method

The graph of the function $f(x) = \sin x + x^2 - 1$ is shown in Figure 14.

(a) Use the Intermediate Value Theorem to confirm that $f$ has a zero in the interval $(0, 1)$.

(b) Use technology with Newton's Method and a first approximation of $c_1 = 0.5$ to find a fourth approximation to the zero.

**Solution** (a) The function $f(x) = \sin x + x^2 - 1$ is continuous on its domain, all real numbers, so it is continuous on the closed interval $[0, 1]$. Since $f(0) = -1$ and $f(1) = \sin 1 \approx 0.841$ have opposite signs, the Intermediate Value Theorem guarantees that $f$ has a zero in the interval $(0, 1)$.

(b) We begin by finding $f'(x) = \cos x + 2x$. To use Newton's Method with a graphing calculator, enter

$$x - \frac{\sin x + x^2 - 1}{\cos x + 2x} \qquad x - \frac{f(x)}{f'(x)}$$

into the $Y =$ editor, as shown in Figure 15. Then create a table (Figure 16) by entering the initial value $0.5$ in the $X$ column. The graphing calculator computes

$$0.5 - \frac{\sin 0.5 + 0.5^2 - 1}{\cos 0.5 + 2(0.5)} = 0.64410789$$

and displays $0.64411$ in column $Y_1$ next to $0.5$. The value $Y_1$ is the second approximation $c_2$ that we use in the next iteration. Now enter $0.64410789$ in the $X$ column of the next row, and the new entry in column $Y_1$ is the third approximation $c_3$. Repeat the process until the desired approximation is obtained. The fourth approximation to the zero of $f$ is $0.63673$. ∎

**NOW WORK** Problem 43.

## When Newton's Method Fails

You may wonder, does Newton's Method always work? The answer is no. The list below, while not exhaustive, gives some conditions under which Newton's Method fails.

- Newton's Method fails if the conditions of the theorem are not met

  (a) $f'(c_n) = 0$: Algebraically, division by 0 is not defined. Geometrically, the tangent line is parallel to the $x$-axis and so has no $x$-intercept.

  (b) $f'(c)$ is undefined: The process cannot be used.

- Newton's Method fails if the initial estimate $c_1$ is not "good enough:"

  (a) Choosing an initial estimate too far from the required zero could result in approximating a different zero of the function.

  (b) The convergence could approach the zero so slowly that hundreds of iterations are necessary.

- Newton's Method fails if the terms oscillate between two values and so never get closer to the zero.

Problems 76–79 illustrate some of these possibilities.

# 3.5 Assess Your Understanding

**Concepts and Vocabulary**

1. *Multiple Choice* If $y = f(x)$ is a differentiable function, the differential $dy =$
   [(a) $\Delta y$ (b) $\Delta x$ (c) $f(x)dx$ (d) $f'(x)dx$]

2. A linear approximation to a differentiable function $f$ near $x_0$ is given by the function $L(x) = $ _____.

3. *True or False* The difference $|\Delta y - dy|$ measures the departure of the graph of $y = f(x)$ from the graph of the tangent line to $f$.

4. If $Q$ is a quantity to be measured and $\Delta Q$ is the error made in measuring $Q$, then the relative error in the measurement at $x_0$ is given by the ratio _____.

---

**1.** = NOW WORK problem     ⚠ = Graphing technology recommended     CAS = Computer Algebra System recommended

**5.** *True or False*   Newton's Method uses tangent lines to the graph of $f$ to approximate the zeros of $f$.

**6.** *True or False*   Before using Newton's Method, we need a first approximation for the zero.

## Skill Building

*In Problems 7–18, find the differential $dy$ of each function.*

**7.** $y = x^3 - 2x + 1$

**8.** $y = e^x + 2x - 1$

**9.** $y = 4(x^2 + 1)^{3/2}$

**10.** $y = \sqrt{x^2 - 1}$

**11.** $y = 3\sin(2x) + x$

**12.** $y = \cos^2(3x) - x$

**13.** $y = e^{-x}$

**14.** $y = e^{\sin x}$

**15.** $y = xe^x$

**16.** $y = \dfrac{e^{-x}}{x}$

**17.** $y = \sin^{-1}(2x)$

**18.** $y = \tan^{-1} x^2$

*In Problems 19–24:*
*(a) Find the differential $dy$ for each function $f$.*
*(b) Evaluate $dy$ and $\Delta y$ at the given value of $x$ when*
    *(i) $\Delta x = 0.5$, (ii) $\Delta x = 0.1$, and (iii) $\Delta x = 0.01$.*
*(c) Find the error $|\Delta y - dy|$ for each choice of $dx = \Delta x$.*

**19.** $f(x) = e^x$ at $x = 1$

**20.** $f(x) = e^{-x}$ at $x = 1$

**21.** $f(x) = x^{2/3}$ at $x = 2$

**22.** $f(x) = x^{-1/2}$ at $x = 1$

**23.** $f(x) = \cos x$ at $x = \pi$

**24.** $f(x) = \tan x$ at $x = 0$

*In Problems 25–32:*
*(a) Find the linear approximation $L(x)$ to $f$ at $x_0$.*
*(b) Graph $f$ and $L$ on the same set of axes.*

**25.** $f(x) = (x + 1)^5$,   $x_0 = 2$

**26.** $f(x) = x^3 - 1$,   $x_0 = 0$

**27.** $f(x) = \sqrt{x}$,   $x_0 = 4$

**28.** $f(x) = x^{2/3}$,   $x_0 = 1$

**29.** $f(x) = \ln x$,   $x_0 = 1$

**30.** $f(x) = e^x$,   $x_0 = 1$

**31.** $f(x) = \cos x$,   $x_0 = \dfrac{\pi}{3}$

**32.** $f(x) = \sin x$,   $x_0 = \dfrac{\pi}{6}$

**33.** Use differentials to approximate the change in:

    **(a)** $y = f(x) = x^2$   as $x$ changes from 3 to 3.001.

    **(b)** $y = f(x) = \dfrac{1}{x + 2}$   as $x$ changes from 2 to 1.98.

**34.** Use differentials to approximate the change in:

    **(a)** $y = x^3$   as $x$ changes from 3 to 3.01.

    **(b)** $y = \dfrac{1}{x - 1}$   as $x$ changes from 2 to 1.98.

*In Problems 35–42, for each function:*
*(a) Use the Intermediate Value Theorem to confirm that a zero exists in the given interval.*
*(b) Use Newton's Method with the first approximation $c_1$ to find $c_3$, the third approximation to the real zero.*

**35.** $f(x) = x^3 + 3x - 5$, interval: $(1, 2)$. Let $c_1 = 1.5$.

**36.** $f(x) = x^3 - 4x + 2$, interval: $(1, 2)$. Let $c_1 = 1.5$.

**37.** $f(x) = 2x^3 + 3x^2 + 4x - 1$, interval: $(0, 1)$. Let $c_1 = 0.5$.

**38.** $f(x) = x^3 - x^2 - 2x + 1$, interval: $(0, 1)$. Let $c_1 = 0.5$

**39.** $f(x) = x^3 - 6x - 12$, interval: $(3, 4)$. Let $c_1 = 3.5$.

**40.** $f(x) = 3x^3 + 5x - 40$, interval: $(2, 3)$. Let $c_1 = 2.5$.

**41.** $f(x) = x^4 - 2x^3 + 21x - 23$, interval: $(1, 2)$. Use a first approximation $c_1$ of your choice.

**42.** $f(x) = x^4 - x^3 + x - 2$, interval: $(1, 2)$. Use a first approximation $c_1$ of your choice.

*In Problems 43–48, for each function:*
*(a) Use the Intermediate Value Theorem to confirm that a zero exists in the given interval.*
*(b) Use technology with Newton's Method to find $c_5$, the fifth approximation to the real zero. Use the midpoint of the interval for the first approximation $c_1$.*

**43.** $f(x) = x + e^x$, interval: $(-1, 0)$

**44.** $f(x) = x - e^{-x}$, interval: $(0, 1)$

**45.** $f(x) = x^3 + \cos^2 x$, interval: $(-1, 0)$

**46.** $f(x) = x^2 + 2\sin x - 0.5$, interval: $(0, 1)$

**47.** $f(x) = 5 - \sqrt{x^2 + 2}$, interval: $(4, 5)$

**48.** $f(x) = 2x^2 + x^{2/3} - 4$, interval: $(1, 2)$

## Applications and Extensions

**49. Area of a Disk**   A circular plate is heated and expands. If the radius of the plate increases from $R = 10$ cm to $R = 10.1$ cm, use differentials to approximate the increase in the area of the top surface.

**50. Volume of a Cylinder**   In a wooden block 3 cm thick, an existing circular hole with a radius of 2 cm is enlarged to a hole with a radius of 2.2 cm. Use differentials to approximate the volume of wood that is removed.

**51. Volume of a Balloon**   Use differentials to approximate the change in volume of a spherical balloon of radius 3 m as the balloon swells to a radius of 3.1 m.

**52. Volume of a Paper Cup**   A manufacturer produces paper cups in the shape of a right circular cone with a radius equal to one-fourth the height. Specifications call for the cups to have a top diameter of 4 cm. After production, it is discovered that the diameter measures only 3.8 cm. Use differentials to approximate the loss in capacity of the cup.

**53. Volume of a Sphere**

    **(a)** Use differentials to approximate the volume of material needed to manufacture a hollow sphere if its inner radius is 2 m and its outer radius is 2.1 m.

    **(b)** Is the approximation overestimating or underestimating the volume of material needed?

    **(c)** Discuss the importance of knowing the answer to (b) if the manufacturer receives an order for 10,000 spheres.

**54. Distance Traveled**   A bee flies around a circle traced on an equator of a ball with a radius of 7 cm at a constant distance of 2 cm from the ball. An ant travels along the same circle but on the ball.

    **(a)** Use differentials to approximate how many more centimeters the bee travels than the ant in one round-trip journey.

    **(b)** Does the linear approximation overestimate or underestimate the difference in the distances the bugs travel? Explain.

**55. Estimating Height**   To find the height of a building, the length of the shadow of a 3-m pole placed 9 m from the building is measured. See the figure. This measurement is found to be 1 m, with a percentage error of 1%. The height of the building is estimated to be 30 m. Use differentials to find the percentage error in the estimate.

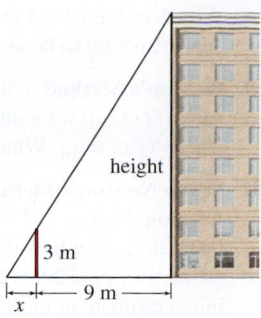

**56. Pendulum Length**   The period of the pendulum of a grandfather clock is $T = 2\pi\sqrt{\dfrac{l}{g}}$, where $l$ is the length (in meters) of the pendulum, $T$ is the period (in seconds), and $g$ is the acceleration due to gravity (9.8 m/s$^2$). Suppose an increase in temperature increases the length $l$ of the pendulum, a thin wire, by 1%. What is the corresponding percentage error in the period? How much time will the clock lose (or gain) each day?

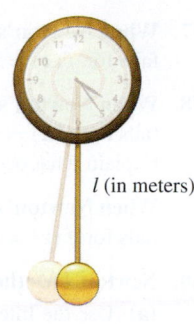

$l$ (in meters)

**57. Pendulum Length**   Refer to Problem 56. If the pendulum of a grandfather clock is normally 1 m long and the length is increased by 10 cm, use differentials to approximate the number of minutes the clock will lose (or gain) each day.

**58. Luminosity of the Sun**   The luminosity $L$ of a star is the rate at which it radiates energy. This rate depends on the temperature $T$ (in kelvin where 0 K is absolute zero) and the surface area $A$ of the star's photosphere (the gaseous surface that emits the light). Luminosity at time $t$ is given by the formula $L(t) = \sigma A T^4$, where $\sigma$ is a constant, known as the **Stefan–Boltzmann constant**.

As with most stars, the Sun's temperature has gradually increased over the 5 billion years of its existence, causing its luminosity to slowly increase. For this problem, we assume that increased luminosity $L$ is due only to an increase in temperature $T$. That is, we treat $A$ as a constant.

(a) Find the rate of change of the temperature $T$ of the Sun with respect to time $t$. Write the answer in terms of the rate of change of the Sun's luminosity $L$ with respect to time $t$.

(b) 4.5 billion years ago, the Sun's luminosity was only 70% of what it is now. If the rate of change of luminosity $L$ with respect to time $t$ is constant, then $\dfrac{\Delta L}{\Delta t} = \dfrac{0.3 L_c}{\Delta t} = \dfrac{0.3 L_c}{4.5}$, where $L_c$ is the current luminosity. Use differentials to approximate the current rate of change of the temperature $T$ of the Sun in degrees per century.

**59. Climbing a Mountain**   Weight $W$ is the force on an object due to the pull of gravity. On Earth, this force is given by Newton's Law of Universal Gravitation: $W = \dfrac{GmM}{r^2}$, where $m$ is the mass of the object, $M = 5.974 \times 10^{24}$ kg is the mass of Earth, $r$ is the distance of the object from the center of the Earth, and $G = 6.67 \times 10^{-11}$ m$^3$/(kg · s$^2$) is the universal gravitational constant. Suppose a person weighs 70 kg at sea level, that is, when $r = 6370$ km (the radius of Earth). Use differentials to approximate the person's change in weight at the top of Mount Everest, which is 8.8 km above sea level. (In this problem we treat kilograms as weight.)

**60. Body Mass Index**   The **body mass index (BMI)** is given by the formula

$$\text{BMI} = 703\,\frac{m}{h^2}$$

where $m$ is the person's weight in pounds and $h$ is the person's height in inches. A BMI between 18.5 and 25 indicates that weight is normal, whereas a BMI greater than 25 indicates that a person is overweight.

(a) Suppose a man who is 5 ft 6 in. weighs 142 lb in the morning when he first wakes up, but he weighs 148 lb in the afternoon. Calculate his BMI in the morning and use differentials to approximate the change in his BMI in the afternoon. Round both answers to three decimal places.

(b) Did this linear approximation overestimate or underestimate the man's afternoon weight? Explain.

(c) A woman who weighs 165 lb estimates her height at 68 in. with a possible error of ± 1.5 in. Calculate her BMI, assuming a height of 68 in. Then use differentials to approximate the possible error in her calculation of BMI. Round the answers to three decimal places.

(d) In the situations described in (a) and (c), how do you explain the classification of each person as normal or overweight?

**61. Error Estimation**   The radius of a spherical ball is found by measuring the volume of the sphere (by finding how much water it displaces). It is determined that the volume is 40 cubic centimeters (cm$^3$), with a tolerance of no more than 1%.

(a) Use differentials to approximate the relative error in measuring the radius.

(b) Find the corresponding percentage error in measuring the radius.

(c) Interpret the results in the context of the problem.

**62. Error Estimation**   The oil pan of a car has the shape of a hemisphere with a radius of 8 cm. The depth $h$ of the oil is measured at 3 cm, with a percentage error of no more than 10%. [*Hint:* The volume $V$ for a spherical segment is $V = \dfrac{1}{3}\pi h^2 (3R - h)$, where $R$ is the radius of the sphere.]

(a) Use differentials to approximate the relative error in measuring the volume.

(b) Find the corresponding percentage error in measuring volume.

(c) Interpret the results in the context of the problem.

**63. Error Estimation**   A closed container is filled with liquid. As the temperature $T$, in degrees Celsius, increases, the pressure $P$, in atmospheres, exerted on the container increases according to the ideal gas law, $P = 50T$.

(a) If the relative error in measuring the temperature is no more than 0.01, what is the relative error in measuring the pressure?

(b) Find the corresponding percentage error in measuring the pressure.

(c) Interpret the results in the context of the problem.

**64. Error Estimation**   A container filled with liquid is held at a constant pressure. As the temperature $T$, in degrees Celsius, decreases, the volume $V$, in cm$^3$, of the container decreases according to the ideal gas law, $V = 15T$.

(a) If the relative error in measuring the temperature is no more than 0.02, what is the relative error in measuring the volume?

(b) Find the corresponding percentage error in measuring the volume.

(c) Interpret the results in the context of the problem.

**65. Percentage Error**   If the percentage error in measuring the edge of a cube is 2%, what is the percentage error in computing its volume?

**66. Focal Length**   To photograph an object, a camera's lens forms an image of the object on the camera's photo sensors. A camera lens can be approximated by a thin lens, which obeys the thin-lens equation $\dfrac{1}{f} = \dfrac{1}{p} + \dfrac{1}{q}$, where $p$ is the distance from the lens to the object being photographed, $q$ is the distance from the lens to the image of the object, and $f$ is the focal length of the lens. A camera whose lens has a focal length of 50 mm is being used to photograph a dog. The dog is originally 15 m from the lens, but moves 0.33 m (about a foot) closer to the lens. Use differentials to approximate the distance the image of the dog moved.

**Using Newton's Method to Solve Equations**   *In Problems 67–70, use Newton's Method to solve each equation.*

**67.** $e^{-x} = \ln x$

**68.** $e^{-x} = x - 4$

**69.** $e^x = x^2$

**70.** $e^x = 2\cos x,\ x > 0$

**71. Approximating $e$**   Use Newton's Method to approximate the value of $e$ by finding the zero of the equation $\ln x - 1 = 0$. Use $c_1 = 3$ as the first approximation and find the fourth approximation to the zero. Compare the results from this approximation to the value of $e$ obtained with a calculator.

**72.** Show that the linear approximation of a function $f(x) = (1 + x)^k$, where $x$ is near 0 and $k$ is any number, is given by $y = 1 + kx$.

**73.** Does it seem reasonable that if a first degree polynomial approximates a differentiable function in an interval near $x_0$, a higher-degree polynomial should approximate the function over a wider interval? Explain your reasoning.

**74.** Why does a function need to be differentiable at $x_0$ for a linear approximation to be used?

**75. Newton's Method**   Suppose you use Newton's Method to solve $f(x) = 0$ for a differentiable function $f$, and you obtain $x_{n+1} = x_n$. What can you conclude?

**76. When Newton's Method Fails**   Verify that the function $f(x) = -x^3 + 6x^2 - 9x + 6$ has a zero in the interval $(2, 5)$. Show that Newton's Method fails if an initial estimate of $c_1 = 2.9$ is chosen. Repeat Newton's Method with an initial estimate of $c_1 = 3.0$. Explain what occurs for each of these two choices. (The zero is near $x = 4.2$.)

**77. When Newton's Method Fails**   Show that Newton's Method fails for $f(x) = x^3 - 2x + 2$ with an initial estimate of $c_1 = 0$.

**78. When Newton's Method Fails**   Show that Newton's Method fails for $f(x) = x^8 - 1$ if an initial estimate of $c_1 = 0.1$ is chosen. Explain what occurs.

**79. When Newton's Method Fails**   Show that Newton's Method fails for $f(x) = (x - 1)^{1/3}$ with an initial estimate of $c_1 = 2$.

**80. Newton's Method**

(a) Use the Intermediate Value Theorem to show that $f(x) = x^4 + 2x^3 - 2x - 2$ has a zero in the interval $(-2, -1)$.

(b) Use Newton's Method to find $c_3$, a third approximation to the zero from (a).

(c) Explain why the initial approximation, $c_1 = -1$, cannot be used in (b).

**81. Specific Gravity**   A solid wooden sphere of diameter $d$ and specific gravity $S$ sinks in water to a depth $h$, which is determined by the equation $2x^3 - 3x^2 - S = 0$, where $x = \dfrac{h}{d}$. Use Newton's Method to find a third approximation to $h$ for a maple ball of diameter 6 in. for which $S = 0.786$. Use $x = 1.6$ as the first approximation.

**Challenge Problem**

**82. Kepler's Equation**   The equation $x - p\sin x = M$, called **Kepler's equation**, occurs in astronomy. Use Newton's Method to find a second approximation to $x$ when $p = 0.2$ and $M = 0.85$. Use $c_1 = 1$ as your first approximation.

# 3.6 Hyperbolic Functions

**OBJECTIVES**   *When you finish this section, you should be able to:*

**1** Define the hyperbolic functions (p. 250)

**2** Establish identities for hyperbolic functions (p. 252)

**3** Differentiate hyperbolic functions (p. 253)

**4** Differentiate inverse hyperbolic functions (p. 254)

## 1 Define the Hyperbolic Functions

Functions involving certain combinations of $e^x$ and $e^{-x}$ occur so frequently in applied mathematics that they warrant special study. These functions, called *hyperbolic functions,* have properties similar to those of trigonometric functions. Because of this, they are named the *hyperbolic sine* (sinh), the *hyperbolic cosine* (cosh), the *hyperbolic tangent* (tanh), the *hyperbolic cotangent* (coth), the *hyperbolic cosecant* (csch), and the *hyperbolic secant* (sech).

**DEFINITION**

The **hyperbolic sine function** and **hyperbolic cosine function** are defined as

$$y = \sinh x = \frac{e^x - e^{-x}}{2} \qquad y = \cosh x = \frac{e^x + e^{-x}}{2}$$

Hyperbolic functions are related to a hyperbola in much the same way as the trigonometric functions (sometimes called *circular functions*) are related to the circle. Just as any point $P$ on the unit circle $x^2 + y^2 = 1$ has coordinates $(\cos t, \sin t)$, as shown in Figure 17, a point $P$ on the hyperbola $x^2 - y^2 = 1$ has coordinates $(\cosh t, \sinh t)$, as shown in Figure 18. Moreover, both the sector of the circle shown in Figure 17 and the shaded portion of Figure 18 have areas that each equal $\dfrac{t}{2}$.

**Figure 17**

**Figure 18**

The functions $y = \sinh x$ and $y = \cosh x$ are defined for all real numbers. The hyperbolic sine function, $y = \sinh x$, is an odd function, so its graph is symmetric with respect to the origin; the hyperbolic cosine function, $y = \cosh x$, is an even function, so its graph is symmetric with respect to the $y$-axis. Their graphs may be found by combining the graphs of $y = \dfrac{e^x}{2}$ and $y = \dfrac{e^{-x}}{2}$, as illustrated in Figures 19 and 20, respectively. The range of $y = \sinh x$ is all real numbers, and the range of $y = \cosh x$ is the interval $[1, \infty)$.

The remaining four hyperbolic functions are combinations of the hyperbolic cosine and hyperbolic sine functions, and their relationships are similar to those of the trigonometric functions.

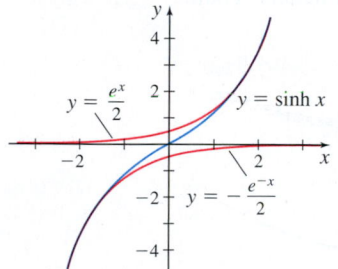

**Figure 19** $y = \sinh x = \dfrac{e^x - e^{-x}}{2}$

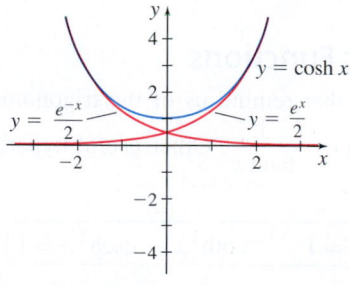

**Figure 20** $y = \cosh x = \dfrac{e^x + e^{-x}}{2}$

Hyperbolic tangent: $\qquad y = \tanh x = \dfrac{\sinh x}{\cosh x} = \dfrac{e^x - e^{-x}}{e^x + e^{-x}}$

Hyperbolic cotangent: $\qquad y = \coth x = \dfrac{\cosh x}{\sinh x} = \dfrac{e^x + e^{-x}}{e^x - e^{-x}}$

Hyperbolic secant: $\qquad y = \operatorname{sech} x = \dfrac{1}{\cosh x} = \dfrac{2}{e^x + e^{-x}}$

Hyperbolic cosecant: $\qquad y = \operatorname{csch} x = \dfrac{1}{\sinh x} = \dfrac{2}{e^x - e^{-x}}$

The graphs of these four hyperbolic functions are shown in Figures 21 through 24.

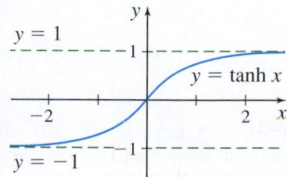

**Figure 21** $y = \tanh x$

**Figure 22** $y = \coth x$

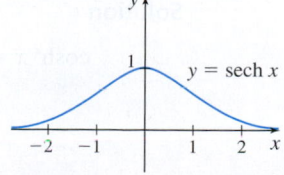

**Figure 23** $y = \operatorname{sech}$

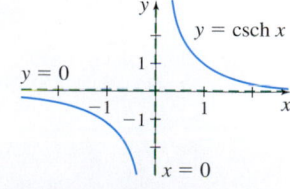

**Figure 24** $y = \operatorname{csch} x$

**EXAMPLE 1    Evaluating Hyperbolic Functions**

Find the exact value of:

**(a)** $\cosh 0$        **(b)** $\operatorname{sech} 0$        **(c)** $\tanh(\ln 2)$

**Solution (a)** $\cosh 0 = \dfrac{e^0 + e^0}{2} = \dfrac{2}{2} = 1$

**(b)** $\operatorname{sech} 0 = \dfrac{1}{\cosh 0} = \dfrac{1}{1} = 1$

**(c)** $\tanh(\ln 2) = \dfrac{e^{\ln 2} - e^{-\ln 2}}{e^{\ln 2} + e^{-\ln 2}} = \dfrac{2 - e^{\ln(1/2)}}{2 + e^{\ln(1/2)}} = \dfrac{2 - \dfrac{1}{2}}{2 + \dfrac{1}{2}} = \dfrac{3}{5}$    ∎

**NOW WORK** Problem 9.

The hyperbolic cosine function has an interesting physical interpretation. If a cable or chain of uniform density is suspended at its ends, such as with high-voltage lines, it will assume the shape of the graph of a hyperbolic cosine.

Suppose we fix our coordinate system, as in Figure 25, so that the cable, which is suspended from endpoints of equal height, lies in the $xy$-plane, with the lowest point of the cable at the point $(0, b)$. Then the shape of the cable will be modeled by the equation

$$y = a \cosh \frac{x}{a} + b - a$$

where $a$ is a constant that depends on the weight per unit length of the cable and on the tension or horizontal force holding the ends of the cable. The graph of this equation is called a **catenary**, from the Latin word *catena* that means "chain."

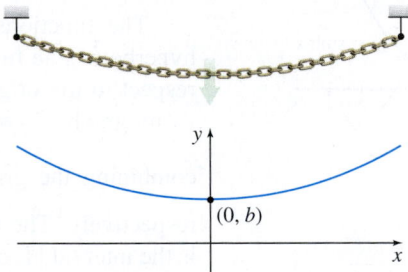

**Figure 25** $y = a \cosh \dfrac{x}{a} + b - a$

## 2 Establish Identities for Hyperbolic Functions

There are identities for the hyperbolic functions that remind us of the trigonometric identities. For example, $\tanh x = \dfrac{\sinh x}{\cosh x}$ and $\coth x = \dfrac{1}{\tanh x}$. Other useful hyperbolic identities include

$$\boxed{\cosh^2 x - \sinh^2 x = 1 \qquad \tanh^2 x + \operatorname{sech}^2 x = 1 \qquad \coth^2 x - \operatorname{csch}^2 x = 1}$$

**EXAMPLE 2    Establishing Identities for Hyperbolic Functions**

Establish the identity $\cosh^2 x - \sinh^2 x = 1$.

**Solution**

$$\cosh^2 x - \sinh^2 x = \left(\frac{e^x + e^{-x}}{2}\right)^2 - \left(\frac{e^x - e^{-x}}{2}\right)^2$$

$$= \frac{e^{2x} + 2e^0 + e^{-2x}}{4} - \frac{e^{2x} - 2e^0 + e^{-2x}}{4} = \frac{2 + 2}{4} = 1$$    ∎

Below we list other identities involving hyperbolic functions that can be established.

**Sum formulas:**

$$\sinh(A + B) = \sinh A \cosh B + \cosh A \sinh B$$
$$\cosh(A + B) = \cosh A \cosh B + \sinh A \sinh B$$

**Even/odd properties:**    $\sinh(-A) = -\sinh A$     $\cosh(-A) = \cosh A$

The derivations of these identities are left as exercises. (See Problems 17–20.)

NOW WORK Problem 19.

## 3 Differentiate Hyperbolic Functions

Since the hyperbolic functions are algebraic combinations of $e^x$ and $e^{-x}$, they are differentiable at all real numbers for which they are defined. For example,

$$\frac{d}{dx}\sinh x = \frac{d}{dx}\left(\frac{e^x - e^{-x}}{2}\right) = \frac{1}{2}\left[\frac{d}{dx}e^x - \frac{d}{dx}e^{-x}\right] = \frac{1}{2}(e^x + e^{-x}) = \cosh x$$

$$\frac{d}{dx}\cosh x = \frac{d}{dx}\left(\frac{e^x + e^{-x}}{2}\right) = \frac{1}{2}\left[\frac{d}{dx}e^x + \frac{d}{dx}e^{-x}\right] = \frac{1}{2}(e^x - e^{-x}) = \sinh x$$

$$\frac{d}{dx}\operatorname{csch} x = \frac{d}{dx}\left(\frac{1}{\sinh x}\right) = \frac{-\dfrac{d}{dx}\sinh x}{\sinh^2 x} = \frac{-\cosh x}{\sinh^2 x} = -\frac{1}{\sinh x} \cdot \frac{\cosh x}{\sinh x}$$

$$= -\operatorname{csch} x \coth x$$

The formulas for the derivatives of the hyperbolic functions are listed below.

| | | |
|---|---|---|
| $\dfrac{d}{dx}\sinh x = \cosh x$ | $\dfrac{d}{dx}\tanh x = \operatorname{sech}^2 x$ | $\dfrac{d}{dx}\operatorname{csch} x = -\operatorname{csch} x \coth x$ |
| $\dfrac{d}{dx}\cosh x = \sinh x$ | $\dfrac{d}{dx}\coth x = -\operatorname{csch}^2 x$ | $\dfrac{d}{dx}\operatorname{sech} x = -\operatorname{sech} x \tanh x$ |

CALC
⊙ **EXAMPLE 3**  **Differentiating Hyperbolic Functions**
CLIP

Find $y'$.

**(a)** $y = x^2 - 2\sinh x$     **(b)** $y = \cosh(x^2 + 1)$

**Solution (a)** Use the Difference Rule.

$$y' = \frac{d}{dx}(x^2 - 2\sinh x) = \frac{d}{dx}x^2 - 2\frac{d}{dx}\sinh x = 2x - 2\cosh x$$

**(b)** Use the Chain Rule with $y = \cosh u$ and $u = x^2 + 1$.

$$y' = \frac{dy}{dx} = \frac{dy}{du} \cdot \frac{du}{dx} = \sinh u \cdot 2x = 2x\sinh(x^2 + 1)$$ ∎

NOW WORK Problem 31.

### EXAMPLE 4   Finding the Angle Between a Catenary and Its Support

A cable is suspended between two poles of the same height that are 20 m apart, as shown in Figure 26(a). If the poles are placed at $(-10, 0)$ and $(10, 0)$, the equation that models the height of the cable is $y = 10 \cosh \dfrac{x}{10} + 15$. Find the angle $\theta$ at which the cable meets a pole.

**Solution**  The angle $\theta$ at which the cable meets the pole equals the angle between the tangent line to the catenary at the point $(10, 30.4)$ and the pole. The slope of the tangent line to the cable is given by

$$y' = \frac{d}{dx} \left( 10 \cosh \frac{x}{10} + 15 \right) = 10 \cdot \frac{1}{10} \sinh \frac{x}{10} = \sinh \frac{x}{10}$$

At $x = 10$, the slope $m_{\tan}$ of the tangent line is $m_{\tan} = \sinh \dfrac{10}{10} = \sinh 1$.

To find the angle $\theta$ between the cable and the pole, we form a right triangle using the tangent line and the pole, as shown in Figure 26(b).

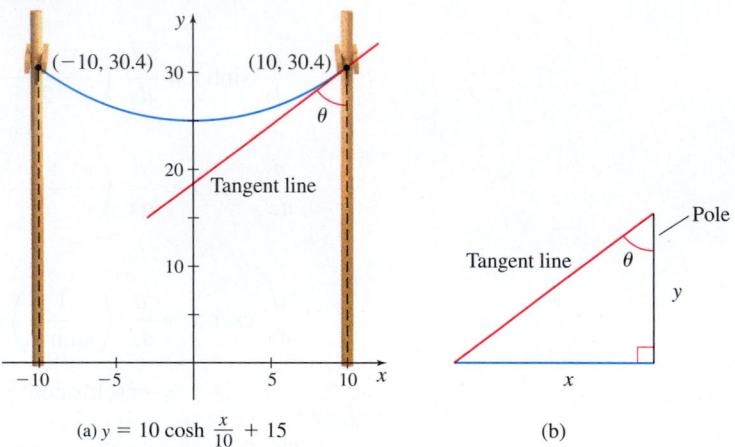

(a) $y = 10 \cosh \dfrac{x}{10} + 15$     (b)

**Figure 26**

From Figure 26(b), the slope of the tangent line is $m_{\tan} = \dfrac{y}{x} = \sinh 1$.

So $\tan \theta = \dfrac{x}{y} = \dfrac{1}{\sinh 1}$ and $\theta = \tan^{-1} \left( \dfrac{1}{\sinh 1} \right) \approx 0.7050$ radians $\approx 40.4°$.  ■

**NOW WORK** Problem 61.

## 4 Differentiate Inverse Hyperbolic Functions

The graph of $y = \sinh x$, shown in Figure 19 on p. 251, suggests that every horizontal line intersects the graph at exactly one point. In fact, the function $y = \sinh x$ is one-to-one, and so it has an inverse function. We denote the inverse by $y = \sinh^{-1} x$ and define it as

$$y = \sinh^{-1} x \qquad \text{if and only if} \qquad x = \sinh y$$

The domain of $y = \sinh^{-1} x$ is the set of real numbers, and the range is also the set of real numbers. See Figure 27(a).

The graph of $y = \cosh x$ (see Figure 20 on p. 251) shows that every horizontal line above $y = 1$ intersects the graph of $y = \cosh x$ at two points so $y = \cosh x$ is not one-to-one. However, if the domain of $y = \cosh x$ is restricted to the nonnegative values of $x$, we have a one-to-one function that has an inverse. We denote the inverse function by $y = \cosh^{-1} x$ and define it as

$$y = \cosh^{-1} x \qquad \text{if and only if} \qquad x = \cosh y \qquad y \geq 0$$

The domain of $y = \cosh^{-1} x$ is $x \geq 1$, and the range is $y \geq 0$. See Figure 27(b).

**NEED TO REVIEW?** One-to-one functions and inverse functions are discussed in Section P.4, pp. 35–40.

The other inverse hyperbolic functions are defined similarly. Their graphs are given in Figure 27(c)–(f).

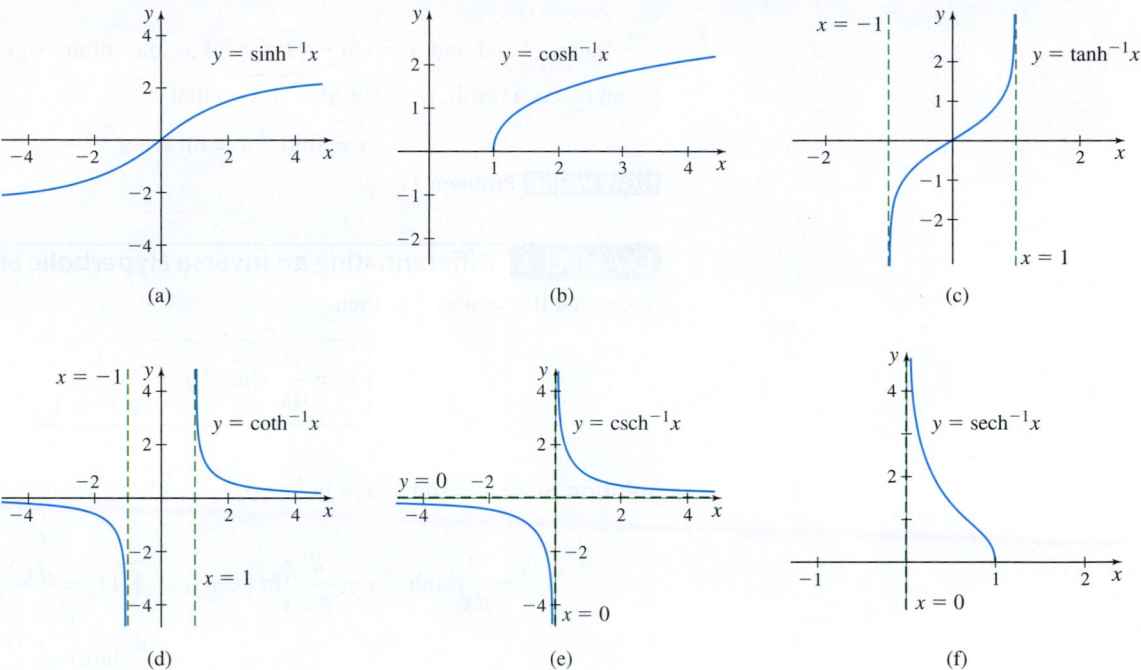

**Figure 27**

Since the hyperbolic functions are defined in terms of the exponential function, the inverse hyperbolic functions can be expressed in terms of natural logarithms.

$$y = \sinh^{-1} x = \ln\left(x + \sqrt{x^2 + 1}\right) \qquad \text{for all real } x$$

$$y = \cosh^{-1} x = \ln\left(x + \sqrt{x^2 - 1}\right) \qquad x \geq 1$$

$$y = \tanh^{-1} x = \frac{1}{2} \ln\left(\frac{1 + x}{1 - x}\right) \qquad |x| < 1$$

$$y = \coth^{-1} x = \frac{1}{2} \ln\left(\frac{x + 1}{x - 1}\right) \qquad |x| > 1$$

We show that $\sinh^{-1} x = \ln\left(x + \sqrt{x^2 + 1}\right)$ here, and leave the others as exercises. (See Problems 71–73.)

---

**EXAMPLE 5** **Expressing the Inverse Hyperbolic Sine Function as a Natural Logarithm**

Express $y = \sinh^{-1} x$ as a natural logarithm.

**Solution** Since $y = \sinh^{-1} x$, where $x = \sinh y$, we have

$$x = \frac{e^y - e^{-y}}{2}$$

$$2xe^y = (e^y)^2 - 1 \qquad \text{Multiply both sides by } 2e^y.$$

$$(e^y)^2 - 2xe^y - 1 = 0$$

**NEED TO REVIEW?** The quadratic formula is discussed in Appendix A.1, p. A-3.

This is a quadratic equation in $e^y$. Use the quadratic formula and solve for $e^y$.

$$e^y = \frac{2x \pm \sqrt{4x^2 + 4}}{2} = x \pm \sqrt{x^2 + 1}$$

Since $e^y > 0$ and $x < \sqrt{x^2 + 1}$ for all $x$, the minus sign on the right side is not possible. As a result, $e^y = x + \sqrt{x^2 + 1}$ so that

$$y = \sinh^{-1} x = \ln(x + \sqrt{x^2 + 1})$$    ∎

**NOW WORK** Problem 71.

**EXAMPLE 6**  **Differentiating an Inverse Hyperbolic Sine Function**

Show that if $y = \sinh^{-1} x$, then

$$y' = \frac{d}{dx} \sinh^{-1} x = \frac{1}{\sqrt{x^2 + 1}}.$$

**Solution** Since $y = \sinh^{-1} x = \ln\left(x + \sqrt{x^2 + 1}\right)$, we have

$$y' = \frac{d}{dx} \sinh^{-1} x = \frac{d}{dx}\left[\ln\left(x + \sqrt{x^2 + 1}\right)\right] = \frac{\frac{d}{dx}\left(x + \sqrt{x^2 + 1}\right)}{x + \sqrt{x^2 + 1}}$$

$$\underset{\frac{d}{dx}\ln(u) = \frac{u'(x)}{u(x)}}{\uparrow}$$

$$= \frac{1 + \frac{1}{2}(x^2 + 1)^{-1/2} \cdot 2x}{x + \sqrt{x^2 + 1}} = \frac{1 + \dfrac{x}{\sqrt{x^2 + 1}}}{x + \sqrt{x^2 + 1}} = \frac{\dfrac{\sqrt{x^2 + 1} + x}{\sqrt{x^2 + 1}}}{x + \sqrt{x^2 + 1}}$$

$$= \frac{x + \sqrt{x^2 + 1}}{(x + \sqrt{x^2 + 1})\sqrt{x^2 + 1}} = \frac{1}{\sqrt{x^2 + 1}}$$    ∎

**NOW WORK** Problem 43.

Similarly, we have the following derivative formulas. If $y = \cosh^{-1} x$, then

$$y' = \frac{d}{dx}\cosh^{-1} x = \frac{1}{\sqrt{x^2 - 1}} \qquad x > 1$$

If $y = \tanh^{-1} x$, then

$$y' = \frac{d}{dx}\tanh^{-1} x = \frac{1}{1 - x^2} \qquad |x| < 1$$

**EXAMPLE 7**  **Differentiating an Inverse Hyperbolic Tangent Function**

Find $y'$ if $y = \tanh^{-1}\sqrt{x}$.

**Solution** Use the Chain Rule with $y = \tanh^{-1} u$ and $u = \sqrt{x}$. Then

$$y' = \frac{dy}{dx} = \frac{dy}{du} \cdot \frac{du}{dx} = \frac{d}{du}\tanh^{-1} u \cdot \frac{d}{dx}\sqrt{x}$$

$$= \frac{1}{1 - u^2} \cdot \frac{1}{2\sqrt{x}} \underset{u = \sqrt{x}}{=} \frac{1}{1 - x} \cdot \frac{1}{2\sqrt{x}} = \frac{\sqrt{x}}{2x(1 - x)}$$    ∎

**NOW WORK** Problem 41.

# 3.6 Assess Your Understanding

## Concepts and Vocabulary

**1.** *True or False*   $\operatorname{csch} x = \dfrac{1}{\cosh x}$.

**2.** In terms of $\sinh x$ and $\cosh x$, $\tanh x = $ _____.

**3.** *True or False*   The domain of $y = \cosh x$ is $[1, \infty)$.

**4.** *Multiple Choice*   The function $y = \cosh x$ is
   [**(a)** even **(b)** odd **(c)** neither].

**5.** *True or False*   $\cosh^2 x + \sinh^2 x = 1$.

**6.** *True or False*   $\dfrac{d}{dx} \sinh x = \dfrac{1}{\sqrt{x^2 + 1}}$.

**7.** *True or False*   The function $y = \sinh^{-1} x$ is defined for all real numbers $x$.

**8.** *True or False*   The functions $f(x) = \tanh^{-1} x$ and $g(x) = \coth^{-1} x$ are identical except for their domains.

## Skill Building

*In Problems 9–14, find the exact value of each expression.*

**9.** $\operatorname{csch}(\ln 3)$

**10.** $\operatorname{sech}(\ln 2)$

**11.** $\cosh^2(5) - \sinh^2(5)$

**12.** $\cosh(-\ln 2)$

**13.** $\tanh 0$

**14.** $\sinh\left(\ln \dfrac{1}{2}\right)$

*In Problems 15–24, establish each identity.*

**15.** $\tanh^2 x + \operatorname{sech}^2 x = 1$

**16.** $\coth^2 x - \operatorname{csch}^2 x = 1$

**17.** $\sinh(-A) = -\sinh A$

**18.** $\cosh(-A) = \cosh A$

**19.** $\sinh(A + B) = \sinh A \cosh B + \cosh A \sinh B$

**20.** $\cosh(A + B) = \cosh A \cosh B + \sinh A \sinh B$

**21.** $\sinh(2x) = 2 \sinh x \cosh x$

**22.** $\cosh(2x) = \cosh^2 x + \sinh^2 x$

**23.** $\cosh(3x) = 4 \cosh^3 x - 3 \cosh x$

**24.** $\tanh(2x) = \dfrac{2 \tanh x}{1 + \tanh^2 x}$

*In Problems 25–48, find the derivative of each function.*

**25.** $f(x) = \sinh(3x)$

**26.** $f(x) = \cosh \dfrac{x}{2}$

**27.** $f(x) = \cosh(x^2 + 1)$

**28.** $f(x) = \cosh(2x^3 - 1)$

**29.** $g(x) = \coth \dfrac{1}{x}$

**30.** $g(x) = \tanh(x^2)$

**31.** $F(x) = \sinh x \cosh(4x)$

**32.** $F(x) = \sinh(2x) \cosh(-x)$

**33.** $s(t) = \cosh^2 t$

**34.** $s(t) = \tanh^2 t$

**35.** $G(x) = e^x \cosh x$

**36.** $G(x) = e^x(\cosh x + \sinh x)$

**37.** $f(x) = x^2 \operatorname{sech} x$

**38.** $f(x) = x^3 \tanh x$

**39.** $v(t) = \cosh^{-1}(4t)$

**40.** $v(t) = \sinh^{-1}(3t)$

**41.** $f(x) = \tanh^{-1}(x^2 - 1)$

**42.** $f(x) = \cosh^{-1}(2x + 1)$

**43.** $g(x) = x \sinh^{-1} x$

**44.** $g(x) = x^2 \cosh^{-1} x$

**45.** $s(t) = \tanh^{-1}(\tan t)$

**46.** $s(t) = \sinh^{-1}(\sin t)$

**47.** $f(x) = \cosh^{-1}\left(\sqrt{x^2 - 1}\right), x > \sqrt{2}$

**48.** $f(x) = \sinh^{-1}\left(\sqrt{x^2 + 1}\right)$

*In Problems 49–52, find $y'$.*

**49.** $\sinh x + \cosh y = x + y$

**50.** $\sinh y - \cosh x = x - y$

**51.** $e^x \cosh y = \ln x$

**52.** $e^y \sinh x = \ln y$

*In Problems 53–56, find $y''$.*

**53.** $y = (\sinh x)^2$

**54.** $y = (\cosh x)^3$

**55.** $y = \cosh x^3$

**56.** $y = \sinh x^2$

*In Problems 57–60, find an equation of the tangent line to the graph of f at the given point.*

**57.** $f(x) = \sinh(2x)$ at $(0, 0)$

**58.** $f(x) = \cosh x^2$ at $(0, 1)$

**59.** $f(x) = e^x \cosh x$ at $(0, 1)$

**60.** $f(x) = e^{-x} \sinh x$ at $(0, 0)$

## Applications and Extensions

**61. Catenary**   A cable is suspended between two supports of the same height that are 100 m apart. If the supports are placed at $(-50, 0)$ and $(50, 0)$, the equation that models the height of the cable is $y = 12 \cosh \dfrac{x}{12} + 20$. Find the angle $\theta$ at which the cable meets a support.

**62. Catenary**   A town hangs strings of holiday lights across the road between utility poles. Each set of poles is 12 m apart. The strings hang in catenaries modeled by $y = 15 \cosh \dfrac{x}{15} - 10$.

   **(a)** Find the slope of the tangent line where the lights meet the pole.

   **(b)** Find the angle at which the string of lights meets a pole.

Randy Mckown/Dreamstime.com

**63. Catenary**   The famous Gateway Arch to the West in St. Louis, Missouri, is constructed in the shape of a modified inverted catenary. (Modified because the weight is not evenly dispersed throughout the arch.) If $y$ is the height of the arch (in feet) and $x = 0$ corresponds to the center of the arch (its highest point), an equation for the arch is given by

$$y = -68.767 \cosh\left(\dfrac{0.711x}{68.767}\right) + 693.859 \text{ ft}$$

   **(a)** Find the maximum height of the arch.

   **(b)** Find the width of the arch at ground level.

   **(c)** What is the slope of the arch 50 ft from its center?

   **(d)** What is the slope of the arch 200 ft from its center?

   **(e)** Find the angle (in degrees) that the arch makes with the ground. (Assume the ground is level.)

   **(f)** Graph the equation that models the Gateway Arch, and explain how the graph supports the answers found in (a)–(e).

---

**1.** = NOW WORK problem      = Graphing technology recommended      CAS = Computer Algebra System recommended

**64.** Establish the identity $(\cosh x + \sinh x)^n = \cosh(nx) + \sinh(nx)$ for any real number $n$.

**65.** (a) Show that if $y = \cosh^{-1} x$, then $y' = \dfrac{1}{\sqrt{x^2 - 1}}$.

(b) What is the domain of $y'$?

**66.** (a) Show that if $y = \tanh^{-1} x$, then $y' = \dfrac{1}{1 - x^2}$.

(b) What is the domain of $y'$?

**67.** Show that $\dfrac{d}{dx} \tanh x = \operatorname{sech}^2 x$.

**68.** Show that $\dfrac{d}{dx} \coth x = -\operatorname{csch}^2 x$.

**69.** Show that $\dfrac{d}{dx} \operatorname{sech} x = -\operatorname{sech} x \tanh x$.

**70.** Show that $\dfrac{d}{dx} \operatorname{csch} x = -\operatorname{csch} x \coth x$.

**71.** Show that $\tanh^{-1} x = \dfrac{1}{2} \ln \dfrac{1+x}{1-x}$, $-1 < x < 1$.

**72.** Show that $\cosh^{-1} x = \ln(x + \sqrt{x^2 - 1})$, $x \geq 1$.

**73.** Show that $\coth^{-1} x = \dfrac{1}{2} \ln \dfrac{x+1}{x-1}$, $|x| > 1$.

**Challenge Problems**

*In Problems 74 and 75, find each limit.*

**74.** $\lim\limits_{x \to 0} \dfrac{\sinh x}{x}$

**75.** $\lim\limits_{x \to 0} \dfrac{\cosh x - 1}{x}$

**76.** (a) Sketch the graph of $y = f(x) = \dfrac{1}{2}(e^x + e^{-x})$.

(b) Let $R$ be a point on the graph and let $r$, $r \neq 0$, be the $x$-coordinate of $R$. The tangent line to the graph of $f$ at $R$ crosses the $x$-axis at the point $Q$. Find the coordinates of $Q$ in terms of $r$.

(c) If $P$ is the point $(r, 0)$, find the length of the line segment $PQ$ as a function of $r$ and the limiting value of this length as $r$ increases without bound.

**77.** What happens if you try to find the derivative of $f(x) = \sin^{-1}(\cosh x)$? Explain why this occurs.

**78.** Let $f(x) = x \sinh^{-1} x$.

(a) Show that $f$ is an even function.

(b) Find $f'(x)$ and $f''(x)$.

# Chapter Review

## THINGS TO KNOW

### 3.1 The Chain Rule

**Properties of Derivatives:**

- Chain Rule: $(f \circ g)'(x) = f'(g(x)) \cdot g'(x)$ (p. 208)    $\dfrac{dy}{dx} = \dfrac{dy}{du} \cdot \dfrac{du}{dx}$

- Power Rule for functions: $\dfrac{d}{dx}[g(x)]^n = n[g(x)]^{n-1} g'(x)$, where $n$ is an integer (p. 213)

**Basic Derivative Formulas:**

- $\dfrac{d}{dx} e^{u(x)} = e^{u(x)} \dfrac{du}{dx}$ (p. 210)
- Derivatives of trigonometric functions (p. 210)
- $\dfrac{d}{dx} a^x = a^x \ln a$   $a > 0$ and $a \neq 1$ (p. 212)

### 3.2 Implicit Differentiation

To differentiate an implicit function (p. 220):

- Assume $y$ is a differentiable function of $x$.
- Differentiate both sides of the equation with respect to $x$.
- Solve the resulting equation for $y' = \dfrac{dy}{dx}$.

**Properties of Derivatives:**

- Power Rule for rational exponents: $\dfrac{d}{dx} x^{p/q} = \dfrac{p}{q} \cdot x^{(p/q)-1}$, provided $x^{p/q}$ and $x^{p/q-1}$ are defined. (p. 223)

- Power Rule for functions: $\dfrac{d}{dx}[u(x)]^r = r[u(x)]^{r-1} u'(x)$, $r$ a rational number; provided $u^r$ and $u^{r-1}$ are defined. (p. 224)

### 3.3 Derivatives of the Inverse Trigonometric Functions

**Theorem:** The derivative of an inverse function (p. 227)

**Basic Derivative Formulas:**

- $\dfrac{d}{dx} \sin^{-1} x = \dfrac{1}{\sqrt{1 - x^2}}$   $-1 < x < 1$ (p. 229)
- $\dfrac{d}{dx} \tan^{-1} x = \dfrac{1}{1 + x^2}$ (p. 230)
- $\dfrac{d}{dx} \sec^{-1} x = \dfrac{1}{x\sqrt{x^2 - 1}}$   $|x| > 1$ (p. 230)

### 3.4 Derivatives of Logarithmic Functions

**Basic Derivative Formulas:**

- $\dfrac{d}{dx} \log_a x = \dfrac{1}{x \ln a}$, $x > 0, a > 0, a \neq 1$ (p. 233)
- $\dfrac{d}{dx} \ln x = \dfrac{1}{x}$, $x > 0$ (p. 233)
- $\dfrac{d}{dx} \ln u(x) = \dfrac{u'(x)}{u(x)}$ (p. 234)

**Steps for Using Logarithmic Differentiation** (p. 236):

- **Step 1** If the function $y = f(x)$ consists of products, quotients, and powers, take the natural logarithm of each side. Then simplify using properties of logarithms.
- **Step 2** Differentiate implicitly, and use $\dfrac{d}{dx} \ln y = \dfrac{y'}{y}$.
- **Step 3** Solve for $y'$, and replace $y$ with $f(x)$.

**Theorems:**

- **Power Rule** If $a$ is a real number, then $\dfrac{d}{dx} x^a = ax^{a-1}$ (p. 237)
- The number $e$ can be expressed as $\lim\limits_{h \to 0}(1 + h)^{1/h} = e$ or $\lim\limits_{n \to \infty}\left(1 + \dfrac{1}{n}\right)^n = e$. (p. 237)

## 3.5 Differentials; Linear Approximations; Newton's Method

- The differential $dx$ of $x$ is defined as $dx = \Delta x \neq 0$, where $\Delta x$ is the change in $x$. (p. 241)
- The differential $dy$ of $y = f(x)$ is defined as $dy = f'(x)dx$. (p. 241)
- The **linear approximation** to a differentiable function $f$ at $x = x_0$ is given by $L(x) = f(x_0) + f'(x_0)(x - x_0)$. (p. 243)
- **Relative error** at $x_0$ in $Q = \dfrac{|\Delta Q|}{Q(x_0)}$ (p. 244)
- **Newton's Method** for finding the zero of a function. (p. 246)

## 3.6 Hyperbolic Functions

*Definitions:*

- Hyperbolic sine: $y = \sinh x = \dfrac{e^x - e^{-x}}{2}$ (p. 251)

- Hyperbolic cosine: $y = \cosh x = \dfrac{e^x + e^{-x}}{2}$ (p. 251)

*Hyperbolic Identities* (pp. 251–253):

- $\tanh x = \dfrac{\sinh x}{\cosh x}$
- $\coth x = \dfrac{\cosh x}{\sinh x}$
- $\operatorname{sech} x = \dfrac{1}{\cosh x}$
- $\operatorname{csch} x = \dfrac{1}{\sinh x}$
- $\cosh^2 x - \sinh^2 x = 1$
- $\tanh^2 x + \operatorname{sech}^2 x = 1$
- $\coth^2 x - \operatorname{csch}^2 x = 1$
- Sum Formulas:

$$\sinh(A + B) = \sinh A \cosh B + \cosh A \sinh B$$
$$\cosh(A + B) = \cosh A \cosh B + \sinh A \sinh B$$

- Even/odd Properties:

$$\sinh(-A) = -\sinh A \qquad \cosh(-A) = \cosh A$$

*Inverse Hyperbolic Functions* (p. 255):

- $y = \sinh^{-1} x = \ln\left(x + \sqrt{x^2 + 1}\right)$ for all real $x$

- $y = \cosh^{-1} x = \ln\left(x + \sqrt{x^2 - 1}\right) \quad x \geq 1$

- $y = \tanh^{-1} x = \dfrac{1}{2} \ln \dfrac{1 + x}{1 - x} \quad |x| < 1$

- $y = \coth^{-1} x = \dfrac{1}{2} \ln \dfrac{x + 1}{x - 1} \quad |x| > 1$

*Basic Derivative Formulas* (pp. 253, 256):

- $\dfrac{d}{dx} \sinh x = \cosh x$
- $\dfrac{d}{dx} \operatorname{sech} x = -\operatorname{sech} x \tanh x$
- $\dfrac{d}{dx} \cosh x = \sinh x$
- $\dfrac{d}{dx} \operatorname{csch} x = -\operatorname{csch} x \coth x$
- $\dfrac{d}{dx} \tanh x = \operatorname{sech}^2 x$
- $\dfrac{d}{dx} \coth x = -\operatorname{csch}^2 x$
- $\dfrac{d}{dx} \sinh^{-1} x = \dfrac{1}{\sqrt{x^2 + 1}}$
- $\dfrac{d}{dx} \cosh^{-1} x = \dfrac{1}{\sqrt{x^2 - 1}} \quad x > 1$
- $\dfrac{d}{dx} \tanh^{-1} x = \dfrac{1}{1 - x^2} \quad |x| < 1$

## OBJECTIVES

| Section | You should be able to … | Examples | Review Exercises |
|---|---|---|---|
| 3.1 | 1 Differentiate a composite function (p. 208) | 1–5 | 1, 13, 24 |
| | 2 Differentiate $y = a^x$, $a > 0$, $a \neq 1$ (p. 212) | 6 | 19, 22 |
| | 3 Use the Power Rule for functions to find a derivative (p. 213) | 7, 8 | 1, 11, 12, 14, 17 |
| | 4 Use the Chain Rule for multiple composite functions (p. 214) | 9 | 15, 18, 61 |
| 3.2 | 1 Find a derivative using implicit differentiation (p. 219) | 1–4 | 43–52, 73, 77 |
| | 2 Find higher-order derivatives using implicit differentiation (p. 222) | 5 | 49–52 |
| | 3 Differentiate functions with rational exponents (p. 223) | 6, 7 | 2–8, 15, 16, 61–64 |
| 3.3 | 1 Find the derivative of an inverse function (p. 227) | 1, 2 | 53, 54 |
| | 2 Find the derivative of the inverse trigonometric functions (p. 228) | 3, 4 | 32–38 |
| 3.4 | 1 Differentiate logarithmic functions (p. 233) | 1–3 | 20, 21, 23, 25–30, 52, 72 |
| | 2 Use logarithmic differentiation (p. 235) | 4–7 | 9, 10, 31, 71 |
| | 3 Express $e$ as a limit (p. 237) | 8 | 55, 56 |
| 3.5 | 1 Find the differential of a function and interpret it geometrically (p. 241) | 1 | 65, 69, 70 |
| | 2 Find the linear approximation to a function (p. 243) | 2 | 67 |
| | 3 Use differentials in applications (p. 244) | 3, 4 | 66, 68 |
| | 4 Use Newton's Method to approximate a real zero of a function (p. 245) | 5, 6 | 74–76 |
| 3.6 | 1 Define the hyperbolic functions (p. 250) | 1 | 57, 58 |
| | 2 Establish identities for hyperbolic functions (p. 252) | 2 | 59, 60 |
| | 3 Differentiate hyperbolic functions (p. 253) | 3, 4 | 39–41 |
| | 4 Differentiate inverse hyperbolic functions (p. 254) | 5–7 | 42 |

## REVIEW EXERCISES

*In Problems 1–42, find the derivative of each function. When a, b, or n appear, they are nonzero constants.*

**1.** $y = (ax+b)^n$

**2.** $y = \sqrt{2ax}$

**3.** $y = x\sqrt{1-x}$

**4.** $y = \dfrac{1}{\sqrt{x^2+1}}$

**5.** $y = (x^2+4)^{3/2}$

**6.** $F(x) = \dfrac{x^2}{\sqrt{x^2-1}}$

**7.** $z = \dfrac{\sqrt{2ax-x^2}}{x}$

**8.** $y = \sqrt{x} + \sqrt[3]{x}$

**9.** $y = (e^x - x)^{5x}$

**10.** $\phi(x) = \dfrac{(x^2-a^2)^{3/2}}{\sqrt{x+a}}$

**11.** $f(x) = \dfrac{x^2}{(x-1)^2}$

**12.** $u = (b^{1/2} - x^{1/2})^2$

**13.** $y = x\sec(2x)$

**14.** $u = \cos^3 x$

**15.** $y = \sqrt{a^2 \sin \dfrac{x}{a}}$

**16.** $\phi(z) = \sqrt{1+\sin z}$

**17.** $u = \sin v - \dfrac{1}{3}\sin^3 v$

**18.** $y = \tan\sqrt{\dfrac{\pi}{x}}$

**19.** $y = 1.05^x$

**20.** $v = \ln(y^2+1)$

**21.** $z = \ln(\sqrt{u^2+25} - u)$

**22.** $y = x^2 + 2^x$

**23.** $y = \ln[\sin(2x)]$

**24.** $f(x) = e^{-x}\sin(2x+\pi)$

**25.** $g(x) = \ln(x^2 - 2x)$

**26.** $y = \ln\dfrac{x^2+1}{x^2-1}$

**27.** $y = e^{-x}\ln x$

**28.** $w = \ln\left(\sqrt{x+7} - \sqrt{x}\right)$

**29.** $y = \dfrac{1}{12}\ln\dfrac{x}{\sqrt{144-x^2}}$

**30.** $y = \ln(\tan^2 x)$

**31.** $f(x) = \dfrac{e^x(x^2+4)}{x-2}$

**32.** $y = \sin^{-1}(x-1) + \sqrt{2x-x^2}$

**33.** $y = 2\sqrt{x} - 2\tan^{-1}\sqrt{x}$

**34.** $y = 4\tan^{-1}\dfrac{x}{2} + x$

**35.** $y = \sin^{-1}(2x-1)$

**36.** $y = x^2\tan^{-1}\dfrac{1}{x}$

**37.** $y = x\tan^{-1} x - \ln\sqrt{1+x^2}$

**38.** $y = \sqrt{1-x^2}(\sin^{-1} x)$

**39.** $y = \tanh\dfrac{x}{2} + \dfrac{2x}{4+x^2}$

**40.** $y = x\sinh x$

**41.** $y = \sqrt{\sinh x}$

**42.** $y = \sinh^{-1} e^x$

*In Problems 43–48, find $y' = \dfrac{dy}{dx}$ using implicit differentiation.*

**43.** $x = y^5 + y$

**44.** $x = \cos^5 y + \cos y$

**45.** $\ln x + \ln y = x\cos y$

**46.** $\tan(xy) = x$

**47.** $y = x + \sin(xy)$

**48.** $x = \ln(\csc y + \cot y)$

*In Problems 49–52, find $y'$ and $y''$.*

**49.** $xy + 3y^2 = 10x$

**50.** $y^3 + y = x^2$

**51.** $xe^y = 4x^2$

**52.** $\ln(x+y) = 8x$

**53.** The function $f(x) = e^{2x}$ has an inverse function $g$. Find $g'(1)$.

**54.** The function $f(x) = \sin x$ defined on the restricted domain $\left[-\dfrac{\pi}{2}, \dfrac{\pi}{2}\right]$ has an inverse function $g$. Find $g'\left(\dfrac{1}{2}\right)$.

*In Problems 55–56, express each limit in terms of the number e.*

**55.** $\displaystyle\lim_{n\to\infty}\left(1+\dfrac{2}{5n}\right)^n$

**56.** $\displaystyle\lim_{h\to 0}(1+3h)^{2/h}$

*In Problems 57 and 58, find the exact value of each expression.*

**57.** $\sinh 0$

**58.** $\cosh(\ln 3)$

*In Problems 59 and 60, establish each identity.*

**59.** $\sinh x + \cosh x = e^x$

**60.** $\tanh(x+y) = \dfrac{\tanh x + \tanh y}{1 + \tanh x \tanh y}$

**61.** If $f(x) = \sqrt{1-\sin^2 x}$, find the domain of $f'$.

**62.** If $f(x) = x^{1/2}(x-2)^{3/2}$ for all $x \geq 2$, find the domain of $f'$.

**63.** Let $f$ be the function defined by $f(x) = \sqrt{1+6x}$.

    **(a)** What are the domain and the range of $f$?

    **(b)** Find the slope of the tangent line to the graph of $f$ at $x = 4$.

    **(c)** Find the $y$-intercept of the tangent line to the graph of $f$ at $x = 4$.

    **(d)** Give the coordinates of the point on the graph of $f$ where the tangent line is parallel to the line $y = x + 12$.

**64. Tangent and Normal Lines** Find equations of the tangent and normal lines to the graph of $y = x\sqrt{x} + (x-1)^2$ at the point $(2, 2\sqrt{3})$.

**65.** Find the differential $dy$ if $x^3 + 2y^2 = x^2 y$.

**66. Measurement Error** If $p$ is the period of a pendulum of length $L$, the acceleration due to gravity may be computed by the formula $g = \dfrac{4\pi^2 L}{p^2}$. If $L$ is measured with negligible error, but an error of no more than 2% may occur in the measurement of $p$, what is the approximate percentage error in the computation of $g$?

**67. Linear Approximation** Find the linear approximation to

$$y = x + \ln x \text{ at } x = 1$$

**68. Measurement Error** If the percentage error in measuring the edge of a cube is no more than 5%, what is the percentage error in computing its volume?

**69.** For the function $f(x) = \tan x$:

    **(a)** Find the differential $dy$ and $\Delta y$ when $x = 0$.

    **(b)** Compare $dy$ to $\Delta y$ when $x = 0$ and (i) $\Delta x = 0.5$, (ii) $\Delta x = 0.1$, and (iii) $\Delta x = 0.01$.

**70.** For the function $f(x) = \ln x$:

    **(a)** Find the differential $dy$ and $\Delta y$ when $x = 1$.

    **(b)** Compare $dy$ to $\Delta y$ when $x = 1$ and (i) $\Delta x = 0.5$, (ii) $\Delta x = 0.1$, and (iii) $\Delta x = 0.01$.

**71.** If $f(x) = (x^2+1)^{(2-3x)}$, find $f'(1)$.

**72.** Find $\lim\limits_{x \to 2} \dfrac{\ln x - \ln 2}{x - 2}$.

**73.** Find $y'$ at $x = \dfrac{\pi}{2}$ and $y = \pi$ if $x \sin y + y \cos x = 0$.

*In Problems 74 and 75, for each function:*

*(a) Use the Intermediate Value Theorem to confirm that a zero exists in the given interval.*

*(b) Use Newton's Method to find $c_3$, the third approximation to the real zero.*

**74.** $f(x) = 8x^4 - 2x^2 + 5x - 1$,  interval: $(0, 1)$.  Let $c_1 = 0.5$.

**75.** $f(x) = 2 - x + \sin x$,  interval: $\left(\dfrac{\pi}{2}, \pi\right)$. Let $c_1 = \dfrac{\pi}{2}$.

**76. (a)** Use the Intermediate Value Theorem to confirm that the function $f(x) = 2 \cos x - e^x$ has a zero in the interval $(0, 1)$.

**(b)** Use technology with Newton's Method to find $c_5$, the fifth approximation to the real zero. Use the midpoint of the interval for the first approximation $c_1$.

**77. Tangent Line** Find an equation of the tangent line to the graph of $4xy - y^2 = 3$ at the point $(1, 3)$.

---

## CHAPTER 3 PROJECT    World Population

The Law of Uninhibited Growth states that, under certain conditions, the rate of change of a population is proportional to the size of the population at that time. One consequence of this law is that the time it takes for a population to double remains constant. For example, suppose a certain bacteria obeys the Law of Uninhibited Growth. Then if the bacteria take five hours to double from 100 organisms to 200 organisms, in the next five hours they will double again from 200 to 400. We can model this mathematically using the formula

$$P(t) = P_0 \, 2^{t/D}$$

where $P(t)$ is the population at time $t$, $P_0$ is the population at time $t = 0$, and $D$ is the doubling time.

If we use this formula to model population growth, a few observations are in order. For example, the model is continuous, but actual population growth is discrete. That is, an actual population would change from 100 to 101 individuals in an instant, as opposed to a model that has a continuous flow from 100 to 101. The model also produces fractional answers, whereas an actual population is counted in whole numbers. For large populations, however, the growth is continuous enough for the model to match real-world conditions, at least for a short time. In general, as growth continues, there are obstacles to growth at which point the model will fail to be accurate. Situations that follow the model of the Law of Uninhibited Growth vary from the introduction of invasive species into a new environment to the spread of a deadly virus for which there is no immunization. Here, we investigate how accurately the model predicts world population.

1. The world population on July 1, 1959, was approximately $2.983435 \times 10^9$ persons and had a doubling time of $D = 40$ years. Use these data and the Law of Uninhibited Growth to write a formula for the world population $P = P(t)$. Use this model to solve Problems 2 through 4.

2. Find the rate of change of the world population $P = P(t)$ with respect to time $t$.

3. Find the rate of change on July 1, 2020 of the world population with respect to time. (Note that $t = 0$ is July 1, 1959.) Round the answer to the nearest whole number.

4. Approximate the world population on July 1, 2020. Round the answer to the nearest person.

5. According to the United Nations, the world population on July 1, 2015 was $7.349472 \times 10^9$. Use $t_0 = 2015$, $P_0 = 7.349472 \times 10^9$, and $D = 40$ and find a new formula to model the world population $P = P(t)$.

6. Use the new model from Problem 5 to find the rate of change of the world population on July 1, 2020.

7. Compare the results from Problems 3 and 6. Interpret and explain any discrepancy between the two rates of change.

8. Use the new model to approximate the world population on July 1, 2020. Round the answer to the nearest person.

9. Discuss possible reasons for the discrepancies in the approximations of the 2020 population. Do you think one set of data gives better results than the other?

*Source: UN World Population 2015* © 2016.

https://esa.un.org/unpd/wpp/Publications/Files/World_Population_2015

Bloomberg/Getty Images

**CHAPTER 4 PROJECT** The Chapter Project on page 356 examines two economic indicators, the Unemployment Rate and the GDP growth rate, and investigates relationships they might have to each other.

## The U.S. Economy

There are many ways economists collect and report data about the U.S. economy. Two such reports are the Unemployment Rate and the Gross Domestic Product (GDP). The Unemployment Rate gives the percentage of people over the age of 16 who are unemployed and are actively looking for a job. The GDP growth rate measures the rate of increase (or decrease) in the market value of all goods and services produced over a period of time. The graphs on the right show the Unemployment Rate and the GDP growth rate in the United States from 2007–2016.

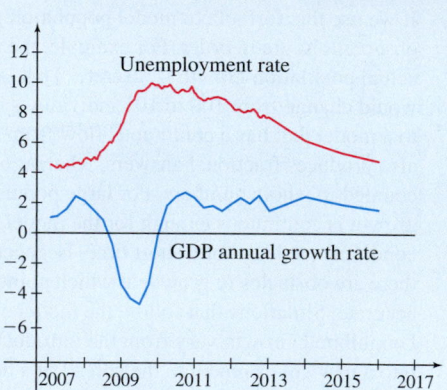

In Chapters 2 and 3, we developed a variety of formulas for finding derivatives. We also began to investigate how derivatives are used to analyze velocity and acceleration and to approximate the zeros of a function.

In this chapter, we continue to explore applications of the derivative. We use the derivative to solve problems involving rates of change of variables that are related, to find optimal (minimum or maximum) values of functions, and to investigate properties of the graph of a function.

# 4.1 Related Rates

**OBJECTIVE** *When you finish this section, you should be able to:*

1 Solve related rate problems (p. 263)

In the natural sciences and in many of the social and behavioral sciences, there are quantities that are related to each other, but that vary with time. For example, the pressure of an ideal gas of fixed volume is proportional to the temperature, yet each of these variables may change over time. Problems involving the rates of change of related variables are called **related rate problems**. In a related rate problem, we seek the rate at which one of the variables is changing at a certain time when the rates of change of the other variables are known.

## 1 Solve Related Rate Problems

We approach related rate problems by writing an equation involving time-dependent variables. This equation is often obtained by investigating the geometric and/or physical conditions imposed by the problem. We then differentiate the equation with respect to time and obtain a new equation that involves the variables and their rates of change with respect to time.

For example, suppose an object falling into still water causes a circular ripple, as illustrated in Figure 1. Both the radius and the area of the circle created by the ripple increase with time and their rates of growth are related. We use the formula for the area of a circle

$$A = \pi r^2$$

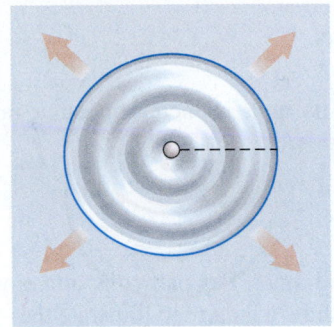

**Figure 1**

to relate the radius and the area. Both $A$ and $r$ are functions of time $t$, and so the area of the circle can be expressed as

$$A(t) = \pi [r(t)]^2$$

Now we differentiate both sides with respect to $t$, obtaining

$$\frac{dA}{dt} = 2\pi r \frac{dr}{dt}$$

The derivatives (rates of change) are related by this equation, so we call them **related rates**. We can solve for one of these rates if the value of the other rate and the values of the variables are known.

### EXAMPLE 1  Solving a Related Rate Problem

A golfer hits a ball into a pond of still water, causing a circular ripple as shown in Figure 2. If the radius of the circle increases at the constant rate of 0.5 m/s, how fast is the area of the circle increasing when the radius of the ripple is 2 m?

**Solution** The quantities that are changing, that is, the variables of the problem, are

$t =$ the time (in seconds) elapsed from the time when the ball hits the water

$r =$ the radius (in meters) of the ripple $t$ seconds after the ball hits the water

$A =$ the area (in square meters) of the circle formed by the ripple $t$ seconds after the ball hits the water

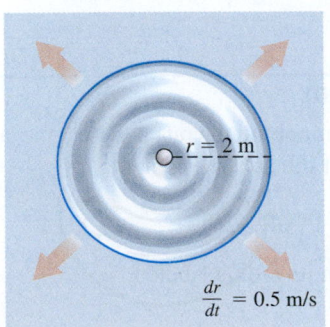

**Figure 2**

The rates of change with respect to time are

$\dfrac{dr}{dt} =$ the rate (in meters per second) at which the radius of the ripple is increasing

$\dfrac{dA}{dt} =$ the rate (in meters squared per second) at which the area of the circle is increasing

It is given that $\dfrac{dr}{dt} = 0.5$ m/s. We seek $\dfrac{dA}{dt}$ when $r = 2$ m.

The relationship between $A$ and $r$ is given by the formula for the area of a circle:

$$A = \pi r^2$$

Since $A$ and $r$ are functions of $t$, we differentiate with respect to $t$ to obtain

$$\frac{dA}{dt} = 2\pi r \frac{dr}{dt}$$

Since $\dfrac{dr}{dt} = 0.5 \text{ m/s}$,

$$\frac{dA}{dt} = 2\pi r (0.5) = \pi r$$

When $r = 2 \text{ m}$,

$$\frac{dA}{dt} = 2\pi \approx 6.283$$

When the radius of the circular ripple is $2 \text{ m}$, its area is increasing at a rate of approximately $6.283 \text{ m}^2/\text{s}$. ∎

### Steps for Solving a Related Rate Problem

**Step 1**  Read the problem carefully, perhaps several times. Pay particular attention to the rate you are asked to find. If possible, draw a picture to illustrate the problem.

**Step 2**  Identify the variables, assign symbols to them, and label the picture. Identify the rates of change as derivatives. Be sure to list the units for each variable and each rate of change. Indicate what is given and what is asked for.

**Step 3**  Write an equation that relates the variables. It may be necessary to use more than one equation.

**Step 4**  Differentiate both sides of the equation(s).

**Step 5**  Substitute numerical values for the variables and the derivatives. Solve for the unknown rate.

**CAUTION**  Numerical values cannot be substituted (Step 5) until after the equation has been differentiated (Step 4).

NOW WORK  Problem 7.

### EXAMPLE 2  Solving a Related Rate Problem

A spherical balloon is inflated at the rate of $10 \text{ m}^3/\text{min}$. Find the rate at which the surface area of the balloon is increasing when the radius of the sphere is $3 \text{ m}$.

**Solution**  We follow the steps for solving a related rate problem.

**Step 1**  Figure 3 shows a sketch of the balloon with its radius labeled.

**Step 2**  Identify the variables of the problem:

$t =$ the time (in minutes) measured from the moment the balloon begins inflating

$R =$ the radius (in meters) of the balloon at time $t$

$V =$ the volume (in meters cubed) of the balloon at time $t$

$S =$ the surface area (in meters squared) of the balloon at time $t$

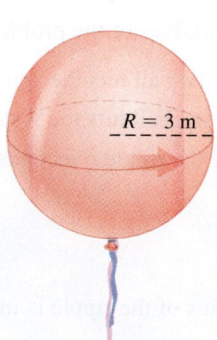

$R = 3 \text{ m}$

**Figure 3**

Identify the rates of change:

$\dfrac{dR}{dt}$ = the rate of change of the radius of the balloon (in meters per minute)

$\dfrac{dV}{dt}$ = the rate of change of the volume of the balloon (in meters cubed per minute)

$\dfrac{dS}{dt}$ = the rate of change of the surface area of the balloon (in meters squared per minute)

We are given $\dfrac{dV}{dt} = 10\,\text{m}^3/\text{min}$, and we seek $\dfrac{dS}{dt}$ when $R = 3\,\text{m}$.

**Step 3** Since both the volume $V$ of the balloon (a sphere) and its surface area $S$ can be expressed in terms of the radius $R$, we use two equations to relate the variables.

**NEED TO REVIEW?** Geometry formulas are discussed in Appendix A.2, p. A-15.

$$V = \frac{4}{3}\pi R^3 \quad \text{and} \quad S = 4\pi R^2 \qquad \text{where } V, S, \text{ and } R \text{ are functions of } t$$

**Step 4** Differentiate both sides of the equations with respect to time $t$.

$$\frac{dV}{dt} = 4\pi R^2 \frac{dR}{dt} \qquad \text{and} \qquad \frac{dS}{dt} = 8\pi R \frac{dR}{dt}$$

Combine the equations by solving for $\dfrac{dR}{dt}$ in the equation on the left and substituting the result into the equation for $\dfrac{dS}{dt}$ on the right. Then

$$\frac{dS}{dt} = 8\pi R \left( \frac{\dfrac{dV}{dt}}{4\pi R^2} \right) = \frac{2}{R}\frac{dV}{dt}$$

**Step 5** Substitute $R = 3\,\text{m}$ and $\dfrac{dV}{dt} = 10\,\text{m}^3/\text{min}$.

$$\frac{dS}{dt} = \frac{2}{3} \cdot 10 = \frac{20}{3} \approx 6.667$$

When the radius of the balloon is $3\,\text{m}$, its surface area is increasing at a rate of approximately $6.667\,\text{m}^2/\text{min}$. ∎

**NOW WORK** Problems 11 and 17.

---

CALC
CLIP

**EXAMPLE 3  Solving a Related Rate Problem**

A rectangular swimming pool 10 m long and 5 m wide has a depth of 3 m at one end and 1 m at the other end. See Figure 4. If water is pumped into the pool at the rate of 300 liters per minute (1/min), at what rate is the water level rising when it is 1.5 m deep at the deep end?

**Solution**

**Step 1** Draw a picture of the cross-sectional view of the pool, as shown in Figure 4.

**Step 2** The width of the pool is 5 m, the water level (measured at the deep end) is $h$ (in meters), the distance from the wall at the deep end to the edge of the water is $L$ (in meters), and the volume of water in the pool is $V$ (in cubic meters, $\text{m}^3$). Each of the variables $h$, $L$, and $V$ varies with respect to time $t$.

We are given $\dfrac{dV}{dt} = 300\ 1/\text{min}$ and are asked to find $\dfrac{dh}{dt}$ when $h = 1.5\,\text{m}$.

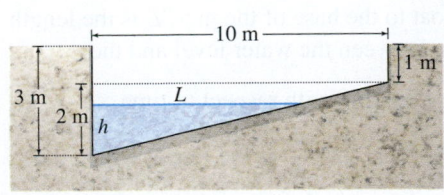

**Figure 4**

**Step 3** The volume $V$ is related to $L$ and $h$ by the formula

$$V = (\text{Cross-sectional triangular area})(\text{width}) = \left(\frac{1}{2}Lh\right)(5) = \frac{5}{2}Lh$$

NEED TO REVIEW? Similar triangles are discussed in Appendix A.2, pp. A-12 to A-14.

Using similar triangles, $L$ and $h$ are related by the equation

$$\frac{L}{h} = \frac{10}{2} \quad \text{so} \quad L = 5h$$

Now we can write $V$ as

$$V = \frac{5}{2}Lh = \frac{5}{2}\underset{\underset{L=5h}{\uparrow}}{(5h)}h = \frac{25}{2}h^2 \tag{1}$$

Both $V$ and $h$ vary with time $t$.

**Step 4** Differentiate both sides of equation (1) with respect to $t$.

$$\frac{dV}{dt} = 25h\frac{dh}{dt}$$

**NOTE** 1000 liters $= 1\,\text{m}^3$.

**Step 5** Substitute $h = 1.5$ m and $\dfrac{dV}{dt} = 300$ l/min $= \dfrac{300}{1000}$ m$^3$/min $= 0.3$ m$^3$/min. Then

$$0.3 = 25(1.5)\frac{dh}{dt} \qquad \qquad \frac{dV}{dt} = 25h\frac{dh}{dt}$$

$$\frac{dh}{dt} = \frac{0.3}{25(1.5)} = 0.008$$

When the height of the water at the deep end is 1.5 m, the water level is rising at a rate of 0.008 m/min. ∎

**NOW WORK** Problem 23.

---

### EXAMPLE 4    Solving a Related Rate Problem

A person standing at the end of a pier is docking a boat by pulling a rope at the rate of 2 m/s. The end of the rope is 3 m above water level. See Figure 5(a). How fast is the boat approaching the base of the pier when 5 m of rope are left to pull in? (Assume the rope never sags, and that the rope is attached to the boat at water level.)

#### Solution

**Step 1** Draw illustrations, like Figure 5, representing the problem. Label the sides of the triangle 3 and $x$, and label the hypotenuse of the triangle $L$.

**Step 2** $x$ is the distance (in meters) from the boat to the base of the pier, $L$ is the length of the rope (in meters), and the distance between the water level and the person's hand is 3 m. Both $x$ and $L$ are changing with respect to time, so $\dfrac{dx}{dt}$ is the rate at which the boat approaches the pier, and $\dfrac{dL}{dt}$ is the rate at which the rope is pulled in. Both $\dfrac{dx}{dt}$ and $\dfrac{dL}{dt}$ are measured in meters/second (m/s). We are given $\dfrac{dL}{dt} = 2$ m/s, and we seek $\dfrac{dx}{dt}$ when $L = 5$ m.

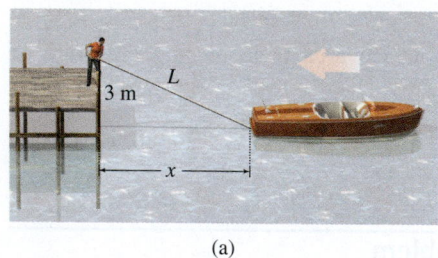

(a)

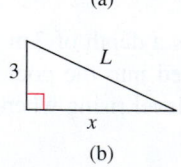

(b)

**Figure 5**

NEED TO REVIEW? The Pythagorean Theorem is discussed in Appendix A.2, pp. A-11 to A-12.

**Step 3** The lengths 3, $x$, and $L$ are the sides of a right triangle and, by the Pythagorean Theorem, are related by the equation

$$x^2 + 3^2 = L^2 \tag{2}$$

***Step 4*** Differentiate both sides of equation (2) with respect to $t$.

$$2x \frac{dx}{dt} = 2L \frac{dL}{dt}$$

$$\frac{dx}{dt} = \frac{L}{x} \frac{dL}{dt} \qquad \text{Solve for } \frac{dx}{dt}.$$

***Step 5*** When $L = 5$, we use equation (2) to find $x = 4$. Since $\dfrac{dL}{dt} = 2$,

$$\frac{dx}{dt} = \frac{5}{4} \cdot 2 \qquad L = 5, x = 4, \frac{dL}{dt} = 2$$

$$\frac{dx}{dt} = 2.5$$

When 5 m of rope are left to be pulled in, the boat is approaching the pier at a rate of 2.5 m/s. ∎

**NOW WORK** Problem 29.

---

**EXAMPLE 5**  **Solving a Related Rate Problem**

A revolving light, located 5 km from a straight shoreline, turns at a constant angular speed of 3 rad/min. With what speed is the spot of light moving along the shore when the beam makes an angle of 60° with the shoreline?

**RECALL** For motion that is circular, angular speed $\omega$ is defined as the rate of change of a central angle $\theta$ of the circle with respect to time. That is, $\omega = \dfrac{d\theta}{dt}$, where $\theta$ is measured in radians.

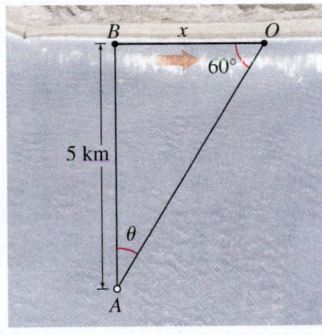

**Figure 6**

**Solution** Figure 6 illustrates the triangle that describes the problem.

$$x = \text{the distance (in kilometers) of the beam of light from the point } B$$
$$\theta = \text{the angle (in radians) the beam of light makes with } AB$$

Both variables $x$ and $\theta$ change with time $t$ (in minutes). The rates of change are

$$\frac{dx}{dt} = \text{the speed of the spot of light along the shore (in kilometers per minute)}$$

$$\frac{d\theta}{dt} = \text{the angular speed of the beam of light (in radians per minute)}$$

We are given $\dfrac{d\theta}{dt} = 3$ rad/min and we seek $\dfrac{dx}{dt}$ when the angle $AOB = 60°$.

From Figure 6,

$$\tan \theta = \frac{x}{5} \qquad \text{so} \qquad x = 5 \tan \theta$$

Then

$$\frac{dx}{dt} = 5 \sec^2 \theta \frac{d\theta}{dt}$$

When $AOB = 60°$, angle $\theta = 30° = \dfrac{\pi}{6}$ radian. Since $\dfrac{d\theta}{dt} = 3$ rad/min,

$$\frac{dx}{dt} = \frac{5}{\cos^2 \theta} \frac{d\theta}{dt} = \frac{5}{\left( \cos \dfrac{\pi}{6} \right)^2} \cdot 3 = \frac{15}{\dfrac{3}{4}} = 20$$

When $\theta = 30°$, the light is moving along the shore at a speed of 20 km/min. ∎

**NOW WORK** Problem 37.

## 4.1 Assess Your Understanding

### Concepts and Vocabulary

1. If a spherical balloon of volume $V$ is inflated at a rate of $10 \, \text{m}^3/\text{min}$, where $t$ is the time (in minutes), what is the rate of change of $V$ with respect to $t$?

2. For the balloon in Problem 1, if the radius $r$ is increasing at the rate of $0.5 \, \text{m/min}$, what is the rate of change of $r$ with respect to $t$?

*In Problems 3 and 4, $x$ and $y$ are differentiable functions of $t$. Find $\dfrac{dx}{dt}$ when $x = 3$, $y = 4$, and $\dfrac{dy}{dt} = 2$.*

3. $x^2 + y^2 = 25$    4. $x^3 y^2 = 432$

### Skill Building

5. Suppose $h$ is a differentiable function of $t$ and suppose that
$$\frac{dh}{dt} = \frac{5}{16}\pi \quad \text{when } h = 8.$$
Find $\dfrac{dV}{dt}$ when $h = 8$ if $V = \dfrac{1}{12}\pi h^3$.

6. Suppose $x$ and $y$ are differentiable functions of $t$ and suppose that when $t = 20$, $\dfrac{dx}{dt} = 5$, $\dfrac{dy}{dt} = 4$, $x = 150$, and $y = 80$. Find $\dfrac{ds}{dt}$ when $t = 20$ if $s^2 = x^2 + y^2$.

7. Suppose $h$ is a differentiable function of $t$ and suppose that when $h = 3$, $\dfrac{dh}{dt} = \dfrac{1}{12}$. Find $\dfrac{dV}{dt}$ when $h = 3$ if $V = 80h^2$.

8. Suppose $x$ is a differentiable function of $t$ and suppose that when $x = 15$, $\dfrac{dx}{dt} = 3$. Find $\dfrac{dy}{dt}$ when $x = 15$ if $y^2 = 625 - x^2$, $y \geq 0$.

9. **Volume of a Cube** If each edge of a cube is increasing at the constant rate of $3 \, \text{cm/s}$, how fast is the volume of the cube increasing when the length $x$ of an edge is $10 \, \text{cm}$?

10. **Volume of a Sphere** If the radius of a sphere is increasing at $1 \, \text{cm/s}$, find the rate of change of its volume when the radius is $6 \, \text{cm}$.

11. **Radius of a Sphere** If the surface area of a sphere is shrinking at the constant rate of $0.1 \, \text{cm}^2/\text{h}$, find the rate of change of its radius when the radius is $\dfrac{20}{\pi} \, \text{cm}$.

12. **Surface Area of a Sphere** If the radius of a sphere is increasing at the constant rate of $2 \, \text{cm/min}$, find the rate of change of its surface area when the radius is $100 \, \text{cm}$.

13. **Dimensions of a Triangle** Consider a right triangle with hypotenuse of (fixed) length $45 \, \text{cm}$ and variable legs of lengths $x$ and $y$, respectively. If the leg of length $x$ increases at the rate of $2 \, \text{cm/min}$, at what rate is $y$ changing when $x = 4 \, \text{cm}$?

14. **Change in Area** The fixed sides of an isosceles triangle are of length $1 \, \text{cm}$. (See the figure.) If the sides slide outward at a speed of $1 \, \text{cm/min}$, at what rate is the area enclosed by the triangle changing when $\theta = 30°$?

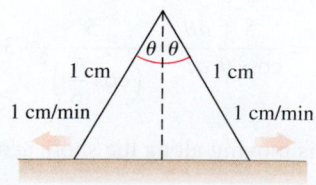

15. **Area of a Triangle** An isosceles triangle has equal sides $4 \, \text{cm}$ long and the included angle is $\theta$. If $\theta$ increases at the rate of $2°/\text{min}$, at what rate is the area of the triangle changing when $\theta$ is $30°$?

$$\left(\textit{Hint: } \frac{d\theta}{dt} = \frac{2\pi}{180} \text{ rad/min.}\right)$$

16. **Area of a Rectangle** In a rectangle with a diagonal $15 \, \text{cm}$ long, one side is increasing at the rate of $2\sqrt{5} \, \text{cm/s}$. Find the rate of change of the area when that side is $10 \, \text{cm}$ long.

17. **Change in Surface Area** A spherical balloon filled with gas has a leak that causes the gas to escape at a rate of $1.5 \, \text{m}^3/\text{min}$. At what rate is the surface area of the balloon shrinking when the radius is $4 \, \text{m}$?

18. **Frozen Snow Ball** Suppose that the volume of a spherical ball of ice decreases (by melting) at a rate proportional to its surface area. Show that its radius decreases at a constant rate.

### Applications and Extensions

19. **Change in Inclination** A ladder $5 \, \text{m}$ long is leaning against a wall. If the lower end of the ladder slides away from the wall at the rate of $0.5 \, \text{m/s}$, at what rate is the inclination $\theta$ of the ladder with respect to the ground changing when the lower end is $4 \, \text{m}$ from the wall?

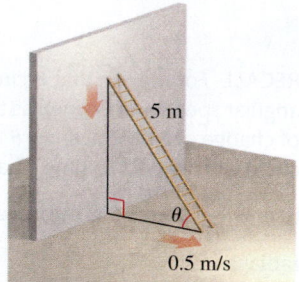

20. **Angle of Elevation** A man $2 \, \text{m}$ tall walks horizontally at a constant rate of $1 \, \text{m/s}$ toward the base of a tower $25 \, \text{m}$ tall. When the man is $10 \, \text{m}$ from the tower, at what rate is the angle of elevation changing if that angle is measured from the horizontal to the line joining the top of the man's head to the top of the tower?

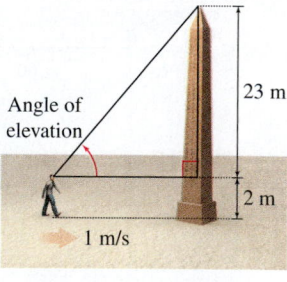

21. **Filling a Pool** A public swimming pool is $30 \, \text{m}$ long and $5 \, \text{m}$ wide. Its depth is $3 \, \text{m}$ at the deep end and $1 \, \text{m}$ at the shallow end. If water is pumped into the pool at the rate of $15 \, \text{m}^3/\text{min}$, how fast is the water level rising when it is $1 \, \text{m}$ deep at the deep end? Use Figure 4 as a guide.

22. **Filling a Tank** Water is flowing into a vertical cylindrical tank of diameter $6 \, \text{m}$ at the rate of $5 \, \text{m}^3/\text{min}$. Find the rate at which the depth of the water is rising.

23. **Fill Rate** A container in the form of a right circular cone (vertex down) has radius $4 \, \text{m}$ and height $16 \, \text{m}$. See the figure on the next page. If water is poured into the container at the constant rate of $16 \, \text{m}^3/\text{min}$, how fast is the water level rising when the water is $8 \, \text{m}$ deep?

(*Hint:* The volume $V$ of a cone of radius $r$ and height $h$ is $V = \dfrac{1}{3}\pi r^2 h$.)

---

**1.** = NOW WORK problem    📐 = Graphing technology recommended    [CAS] = Computer Algebra System recommended

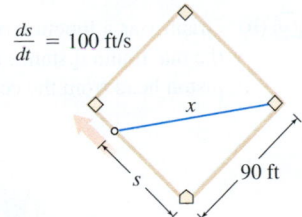

24. **Building a Sand Pile**  Sand is being poured onto the ground, forming a conical pile whose height equals one-fourth of the diameter of the base. If the sand is falling at the rate of 20 cm³/s, how fast is the height of the sand pile increasing when it is 3 cm high?

25. **Is There a Leak?**  A cistern in the shape of a cone 4 m deep and 2 m in diameter at the top is being filled with water at a constant rate of 3 m³/min.

    (a) At what rate is the water rising when the water in the tank is 3 m deep?

    (b) If, when the water is 3 m deep, it is observed that the water rises at a rate of 0.5 m/min, at what rate is water leaking from the tank?

26. **Change in Area**  The vertices of a rectangle are at $(0, 0)$, $(0, e^x)$, $(x, 0)$, and $(x, e^x)$, $x > 0$. If $x$ increases at the rate of 1 unit per second, at what rate is the area increasing when $x = 10$ units?

27. **Distance from the Origin**  An object is moving along the parabola $y^2 = 4(3 - x)$. When the object is at the point $(-1, 4)$, its $y$-coordinate is increasing at the rate of 3 units per second. How fast is the distance from the object to the origin changing at that instant?

28. **Funneling Liquid**  A conical funnel 15 cm in diameter and 15 cm deep is filled with a liquid that runs out at the rate of 5 cm³/min. At what rate is the depth of liquid changing when its depth is 8 cm?

29. **Baseball**  A baseball is hit along the third-base line with a speed of 100 ft/s. At what rate is the ball's distance from first base changing when it crosses third base? See the figure.

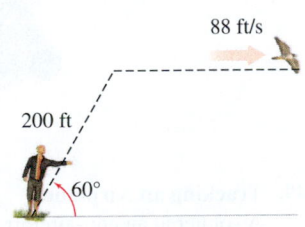

30. **Flight of a Falcon**  A peregrine falcon flies up from its trainer at an angle of 60° until it has flown 200 ft. It then levels off and continues to fly away. If the constant speed of the bird is 88 ft/s, how fast is the falcon moving away from the falconer after it is 6 s in flight. See the figure.

    *Source*: http://www.rspb.org.uk

31. **Boyle's Law**  A gas is said to be compressed adiabatically if there is no gain or loss of heat. When such a gas is diatomic (has two atoms per molecule), it satisfies the equation $PV^{1.4} = k$, where $k$ is a constant, $P$ is the pressure, and $V$ is the volume. At a given instant, the pressure is 20 kg/cm², the volume is 32 cm³, and the volume is decreasing at the rate of 2 cm³/min. At what rate is the pressure changing?

32. **Heating a Plate**  When a metal plate is heated, it expands. If the shape of the plate is circular and its radius is increasing at the rate of 0.02 cm/s, at what rate is the area of the top surface increasing when the radius is 3 cm?

33. **Pollution**  After a rupture, oil begins to escape from an underwater well. If, as the oil rises, it forms a circular slick whose radius increases at a rate of 0.42 ft/min, find the rate at which the area of the spill is increasing when the radius is 120 ft.

34. **Flying a Kite**  A girl flies a kite at a height 30 m above her hand. If the kite flies horizontally away from the girl at the rate of 2 m/s, at what rate is the string being let out when the length of the string released is 70 m? Assume that the string remains taut.

35. **Falling Ladder**  An 8-m ladder is leaning against a vertical wall. If a person pulls the base of the ladder away from the wall at the rate of 0.5 m/s, how fast is the top of the ladder moving down the wall when the base of the ladder is

    (a) 3 m from the wall?

    (b) 4 m from the wall?

    (c) 6 m from the wall?

36. **Beam from a Lighthouse**  A light in a lighthouse 2000 m from a straight shoreline is rotating at 2 revolutions per minute. How fast is the beam moving along the shore when it passes a point 500 m from the point on the shore opposite the lighthouse? (*Hint:* One revolution = $2\pi$ rad.)

37. **Moving Radar Beam**  A radar antenna, making one revolution every 5 s, is located on a ship that is 6 km from a straight shoreline. How fast is the radar beam moving along the shoreline when the beam makes an angle of 45° with the shore?

38. **Tracking a Rocket**  When a rocket is launched, it is tracked by a tracking dish on the ground located a distance $D$ from the point of launch. The dish points toward the rocket and adjusts its angle of elevation $\theta$ to the horizontal (ground level) as the rocket rises. Suppose a rocket rises vertically at a constant speed of 2.0 m/s, with the tracking dish located 150 m from the launch point. Find the rate of change of the angle $\theta$ of elevation of the tracking dish with respect to time $t$ (tracking rate) for each of the following:

    (a) Just after launch.

    (b) When the rocket is 100 m above the ground.

    (c) When the rocket is 1.0 km above the ground.

    (d) Use the results in (a)–(c) to describe the behavior of the tracking rate as the rocket climbs higher and higher. What limit does the tracking rate approach as the rocket gets extremely high?

39. **Lengthening Shadow**  A child, 1 m tall, is walking directly under a street lamp that is 6 m above the ground. If the child walks away from the light at the rate of 20 m/min, how fast is the child's shadow lengthening?

40. **Approaching a Pole**  A boy is walking toward the base of a pole 20 m high at the rate of 4 km/h. At what rate (in meters per second) is the distance from his feet to the top of the pole changing when he is 5 m from the pole?

**41. Riding a Ferris Wheel** A Ferris wheel is 50 ft in diameter and its center is located 30 ft above the ground. See the image. If the wheel rotates once every 2 min, how fast is a passenger rising when he is 42.5 ft above the ground? How fast is he moving horizontally?

Yee Kong Mau/Getty Images

**42. Approaching Cars** Two cars approach an intersection, one heading east at the rate of 30 km/h and the other heading south at the rate of 40 km/h. At what rate are the two cars approaching each other at the instant when the first car is 100 m from the intersection and the second car is 75 m from the intersection? Assume the cars maintain their respective speeds.

**43. Parting Ways** An elevator in a building is located on the fifth floor, which is 25 m above the ground. A delivery truck is positioned directly beneath the elevator at street level. If, simultaneously, the elevator goes down at a speed of 5 m/s and the truck pulls away at a speed of 8 m/s, how fast will the elevator and the truck be separating 1 s later? Assume the speeds remain constant at all times.

**44. Pulley** In order to lift a container 10 m, a rope is attached to the container and, with the help of a pulley, the container is hoisted. See the figure. If a person holds the end of the rope and walks away from beneath the pulley at the rate of 2 m/s, how fast is the container rising when the person is 5 m away? Assume the end of the rope in the person's hand was originally at the same height as the top of the container.

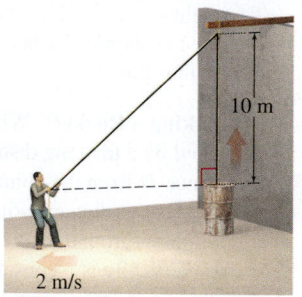

10 m

2 m/s

**45. Business** A manufacturer of precision digital switches has daily cost $C$ and revenue $R$, in dollars, of $C(x) = 10,000 + 3x$ and $R(x) = 5x - \dfrac{x^2}{2000}$, respectively, where $x$ is the daily production of switches. Production is increasing at the rate of 50 switches per day when production is 1000 switches.

(a) Find the rate of increase in cost when production is 1000 switches per day.

(b) Find the rate of increase in revenue when production is 1000 switches per day.

(c) Find the rate of increase in profit when production is 1000 switches per day. (*Hint:* Profit = Revenue − Cost)

**46. An Enormous Growing Black Hole** In December 2011 astronomers announced the discovery of the two most massive black holes identified to date. The holes appear to be quasar* remnants, each having a mass equal to 10 billion Suns. Huge black holes typically grow by swallowing nearby matter, including whole stars. In this way, the size of the **event horizon** (the distance from the center of the black hole to the position at which no light can escape) increases. From general relativity, the radius $R$ of the event horizon for a black hole of mass $m$ is $R = \dfrac{2Gm}{c^2}$, where $G$ is the gravitational constant and $c$ is the speed of light in a vacuum.

(a) If one of these huge black holes swallows one Sun-like star per year, at what rate (in kilometers per year) will its event horizon grow? The mass of the Sun is $1.99 \times 10^{30}$ kg, the speed of light in a vacuum is $c = 3.00 \times 10^8$ m/s, and $G = 6.67 \times 10^{-11}$ m³/(kg · s²).

(b) By what percent does the event horizon change per year?

**47. Weight in Space** An object that weighs $K$ lb on the surface of Earth weighs approximately

$$W(R) = K \left( \frac{3960}{3960 + R} \right)^2$$

pounds when it is a distance of $R$ mi from Earth's surface. Find the rate at which the weight of an object weighing 1000 lb on Earth's surface is changing when it is 50 mi above Earth's surface and is being lifted at the rate of 10 mi/s.

**48. Pistons** In a certain piston engine, the distance $x$ in meters between the center of the crank shaft and the head of the piston is given by $x = \cos\theta + \sqrt{16 - \sin^2\theta}$, where $\theta$ is the angle between the crank and the path of the piston head. See the figure below.

(a) If $\theta$ increases at the constant rate of 45 rad/s, what is the speed of the piston head when $\theta = \dfrac{\pi}{6}$?

(b) Graph $x$ as a function of $\theta$ on the interval $[0, \pi]$. Determine the maximum distance and the minimum distance of the piston head from the center of the crank shaft.

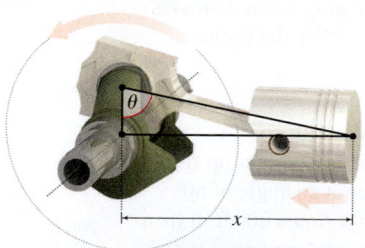

$\theta$

$x$

**49. Tracking an Airplane** A soldier at an anti-aircraft battery observes an airplane flying toward him at an altitude of 4500 ft. See the figure. When the angle of elevation of the battery is 30°, the soldier must increase the angle of elevation by 1°/s to keep the plane in sight. What is the ground speed of the plane?

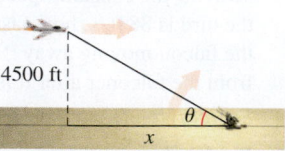

4500 ft

$\theta$

$x$

---

*A **quasar** is an astronomical object that emits massive amounts of electromagnetic radiation.

**50. Change in the Angle of Elevation** A hot air balloon is rising at a speed of 100 m/min. If an observer is standing 200 m from the lift-off point, what is the rate of change of the angle of elevation of the observer's line of sight when the balloon is 600 m high?

**51. Rate of Rotation** A searchlight is following a plane flying at an altitude of 3000 ft in a straight line over the light; the plane's velocity is 500 mi/h. At what rate is the searchlight turning when the distance between the light and the plane is 5000 ft? See the figure.

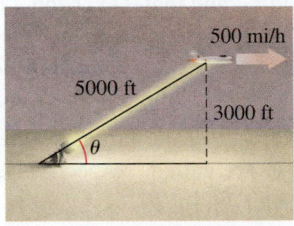

**52. Police Chase** A police car approaching an intersection at 80 ft/s spots a suspicious vehicle on the cross street. When the police car is 210 ft from the intersection, the policeman turns a spotlight on the vehicle, which is at that time just crossing the intersection at a constant rate of 60 ft/s. See the figure. How fast must the light beam be turning 2 s later in order to follow the suspicious vehicle?

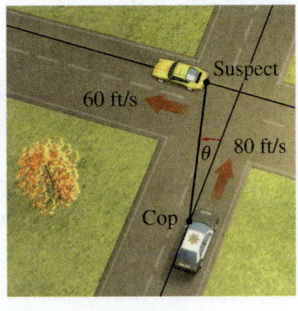

**53. Change in Volume** The height $h$ and width $x$ of an open box with a square base are related to its volume by the formula $V = hx^2$. Discuss how the volume changes

**(a)** if $h$ decreases with time, but $x$ remains constant.

**(b)** if both $h$ and $x$ change with time.

**54. Rate of Change** Let $y = 2e^{\cos x}$. If both $x$ and $y$ vary with time in such a way that $y$ increases at a steady rate of 5 units per second, at what rate is $x$ changing when $x = \dfrac{\pi}{2}$?

**Challenge Problems** ————————

**55. Moving Shadows** The dome of an observatory is a hemisphere 60 ft in diameter. A boy is playing near the observatory at sunset. He throws a ball upward so that its shadow climbs to the highest point on the dome.

How fast is the shadow moving along the dome $\dfrac{1}{2}$ s after the ball begins to fall? How did you use the fact that it was sunset in solving the problem? (*Note:* A ball falling from rest covers a distance $s = 16t^2$ ft in $t$ seconds.)

**56. Moving Shadows** A railroad train is moving at a speed of 15 mi/h past a station 800 ft long. The track has the shape of the parabola $y^2 = 600x$ as shown in the figure. If the sun is just rising in the east, find how fast the shadow $S$ of the locomotive $L$ is moving along the wall at the instant it reaches the end of the wall.

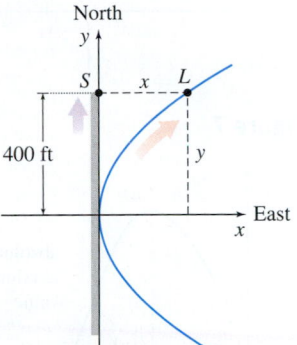

**57. Change in Area** The hands of a clock are 2 in and 3 in long. See the figure. As the hands move around the clock, they sweep out the triangle $OAB$. At what rate is the area of the triangle changing at 12:10 p.m.?

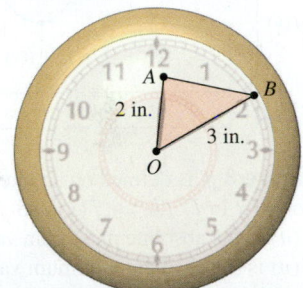

**58. Distance** A train is traveling northeast at a rate of 25 ft/s along a track that is elevated 20 ft above the ground. The track passes over the street below at an angle of 30°. See the figure. Five seconds after the train passes over the road, a car traveling east passes under the tracks going 40 ft/s. How fast are the train and the car separating after 3 s?

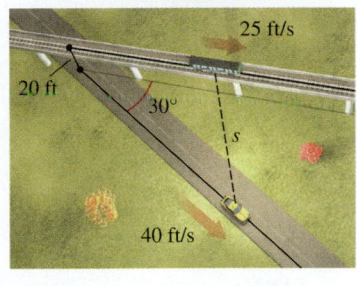

---

# 4.2 Maximum and Minimum Values; Critical Numbers

**OBJECTIVES** *When you finish this section, you should be able to:*

**1** Identify absolute maximum and minimum values and local extreme values of a function (p. 272)

**2** Find critical numbers (p. 275)

**3** Find absolute maximum and absolute minimum values (p. 276)

Often problems in engineering and economics seek to find an optimal, or best, solution to a problem. For example, state and local governments try to set tax rates to optimize revenue. If a problem like this can be modeled by a function, then finding the maximum or the minimum values of the function solves the problem.

# 1  Identify Absolute Maximum and Minimum Values and Local Extreme Values of a Function

Figure 7 illustrates the graph of a function $f$ defined on a closed interval $[a, b]$. Pay particular attention to the graph at the numbers $x_1$, $x_2$, and $x_3$. In small open intervals containing $x_1$ or containing $x_3$ the value of $f$ is greatest at these numbers. We say that $f$ has *local maxima* at $x_1$ and $x_3$, and that $f(x_1)$ and $f(x_3)$ are *local maximum values of $f$*. Similarly, in small open intervals containing $x_2$, the value of $f$ is least at $x_2$. We say $f$ has a *local minimum* at $x_2$ and $f(x_2)$ is a *local minimum value of $f$*. ("Maxima" is the plural of "maximum"; "minima" is the plural of "minimum.")

On the closed interval $[a, b]$, the largest value of $f$ is $f(x_3)$, while the smallest value of $f$ is $f(b)$. These are called, respectively, the *absolute maximum value* and *absolute minimum value* of $f$ on $[a, b]$.

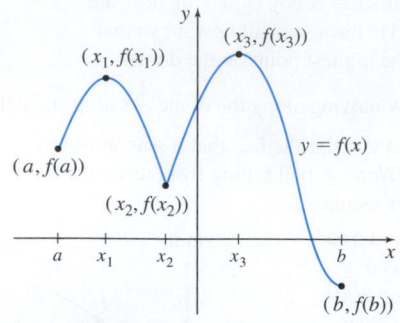

**Figure 7**

## DEFINITION  Absolute Extrema

Let $f$ be a function defined on an interval $I$. If there is a number $u$ in the interval for which $f(u) \geq f(x)$ for all $x$ in $I$, then $f$ has an **absolute maximum** at $u$ and the number $f(u)$ is called the **absolute maximum value** of $f$ on $I$.

If there is a number $v$ in $I$ for which $f(v) \leq f(x)$ for all $x$ in $I$, then $f$ has an **absolute minimum** at $v$ and the number $f(v)$ is the **absolute minimum value** of $f$ on $I$.

The values $f(u)$ and $f(v)$ are sometimes called **absolute extrema** or the **extreme values** of $f$ on $I$. ("Extrema" is the plural of the Latin noun *extremum*.)

The *absolute* maximum value and the *absolute* minimum value, if they exist, are the largest and smallest values, respectively, of a function $f$ on the interval $I$. The interval may be open, closed, or neither. See Figure 8. Contrast this idea with that of a *local* maximum value and a *local* minimum value. These are the largest and smallest values of $f$ in *some open interval* in $I$. The next definition makes this distinction precise.

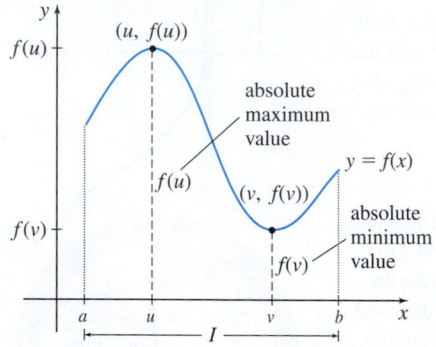

**Figure 8** $f$ is defined on an interval $I$. For all $x$ in $I$, $f(u) \geq f(x)$ and $f(v) \leq f(x)$. $f(u)$ is the absolute maximum value of $f$. $f(v)$ is the absolute minimum value of $f$.

## DEFINITION  Local Extrema

Let $f$ be a function defined on some interval $I$ and let $u$ and $v$ be numbers in $I$. If there is an open interval in $I$ containing $u$ so that $f(u) \geq f(x)$ for all $x$ in this open interval, then $f$ has a **local maximum** (or **relative maximum**) at $u$, and the number $f(u)$ is called a **local maximum value**.

Similarly, if there is an open interval in $I$ containing $v$ so that $f(v) \leq f(x)$ for all $x$ in this open interval, then $f$ has a **local minimum** (or a **relative minimum**) at $v$, and the number $f(v)$ is called a **local minimum value.**

The term **local extreme value** describes either a local maximum value of $f$ or a local minimum value of $f$.

Figure 9 illustrates the definition.

Notice in the definition of a local maximum that the interval that contains $u$ is required to be *open*. Notice also in the definition of a local maximum that the value $f(u)$ must be greater than or equal to *all* other values of $f$ in this open interval. The word "local" is used to emphasize that this condition holds on *some* open interval containing $u$. Similar remarks hold for a local minimum. This means that a local extremum cannot occur at the endpoint of an interval.

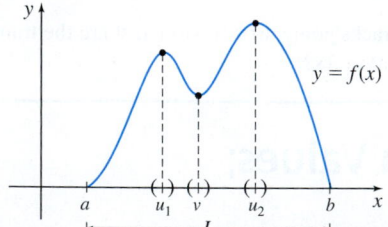

**Figure 9** $f$ has a local maximum at $u_1$ and at $u_2$. $f$ has a local minimum at $v$.

**NEED TO REVIEW?** Continuity is discussed in Section 1.3, pp. 102–110.

---

**EXAMPLE 1**  **Identifying Maximum and Minimum Values and Local Extreme Values from the Graph of a Function**

Figures 10–15, show the graphs of six different functions. For each function:

**(a)** Find the domain.

**(b)** Determine where the function is continuous.

**(c)** Identify the absolute maximum value and the absolute minimum value, if they exist.

**(d)** Identify any local extreme values.

**Solution**

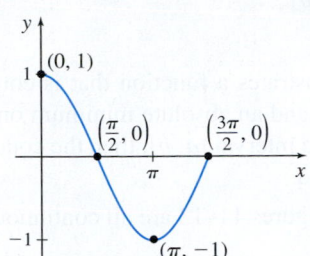

**Figure 10**

(a) Domain: $\left[0, \dfrac{3\pi}{2}\right]$

(b) Continuous on $\left[0, \dfrac{3\pi}{2}\right]$

(c) Absolute maximum value:
    $f(0) = 1$
    Absolute minimum value:
    $f(\pi) = -1$

(d) No local maximum value;
    local minimum value: $f(\pi) = -1$

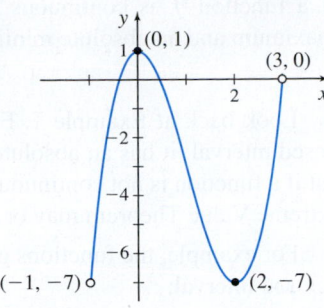

**Figure 11**

(a) Domain: $(-1, 3)$

(b) Continuous on $(-1, 3)$

(c) Absolute maximum value:
    $f(0) = 1$
    Absolute minimum value:
    $f(2) = -7$

(d) Local maximum value:
    $f(0) = 1$
    Local minimum value:
    $f(2) = -7$

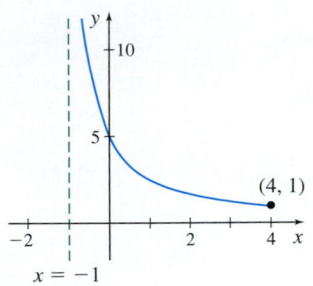

**Figure 12**

(a) Domain: $(-1, 4]$

(b) Continuous on $(-1, 4]$

(c) No absolute maximum value;
    absolute minimum value:
    $f(4) = 1$

(d) No local maximum value;
    no local minimum value

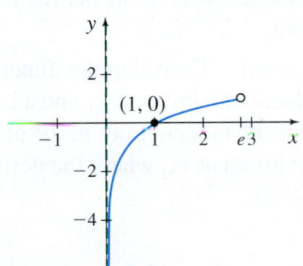

**Figure 13**

(a) Domain: $(0, e)$

(b) Continuous on $(0, e)$

(c) No absolute maximum value;
    no absolute minimum value

(d) No local maximum value;
    no local minimum value

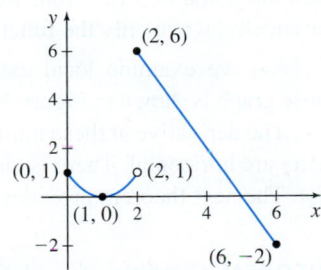

**Figure 14**

(a) Domain: $[0, 6]$

(b) Continuous on $[0, 6]$ except
    at $x = 2$

(c) Absolute maximum value:
    $f(2) = 6$
    Absolute minimum value:
    $f(6) = -2$

(d) Local maximum value: $f(2) = 6$
    Local minimum value: $f(1) = 0$

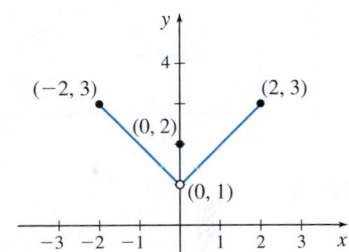

**Figure 15**

(a) Domain: $[-2, 2]$

(b) Continuous on $[-2, 2]$ except
    at $x = 0$

(c) Absolute maximum value:
    $f(-2) = 3$,   $f(2) = 3$
    No absolute minimum value

(d) Local maximum value: $f(0) = 2$
    No local minimum value

■

**NOW WORK** Problem 7.

Example 1 illustrates that a function $f$ can have both an absolute maximum value and an absolute minimum value, can have one but not the other, or can have neither an absolute maximum value nor an absolute minimum value. The next theorem provides conditions for which a function $f$ will always have absolute extrema. Although the theorem seems simple, the proof requires advanced topics and may be found in most advanced calculus books.

**THEOREM Extreme Value Theorem**

If a function $f$ is continuous on a closed interval $[a, b]$, then $f$ has an absolute maximum and an absolute minimum on $[a, b]$.

Look back at Example 1. Figure 10 illustrates a function that is continuous on a closed interval; it has an absolute maximum and an absolute minimum on the interval. But if a function is not continuous on a closed interval $[a, b]$, then the conclusion of the Extreme Value Theorem may or may not hold.

For example, the functions graphed in Figures 11–13 are all continuous, but not on a closed interval:

- In Figure 11, the function has both an absolute maximum and an absolute minimum.
- In Figure 12, the function has an absolute minimum but no absolute maximum.
- In Figure 13, the function has neither an absolute maximum nor an absolute minimum.

On the other hand, the functions graphed in Figures 14 and 15 are each defined on a closed interval, but neither is continuous on that interval:

- In Figure 14, the function has an absolute maximum and an absolute minimum.
- In Figure 15, the function has an absolute maximum but no absolute minimum.

**NOW WORK** Problem 9.

The Extreme Value Theorem is an *existence theorem*. It states that, if a function is continuous on a closed interval, then extreme values exist. It does not tell us how to find these extreme values. Although we can locate both absolute and local extrema given the graph of a function, we need tools that allow us to locate the extreme values analytically, when only the function $f$ is known.

Next we examine local extrema more closely. Consider the function $y = f(x)$ whose graph is shown in Figure 16. There is a local maximum at $x_1$ and a local minimum at $x_2$. The derivative at these numbers is 0, since the tangent lines to the graph of $f$ at $x_1$ and $x_2$ are horizontal. There is also a local maximum at $x_3$, where the derivative fails to exist. The next theorem provides the details.

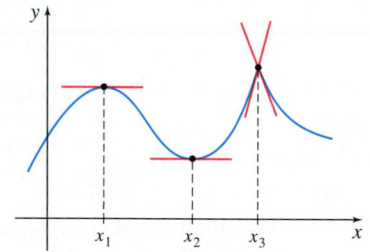

**Figure 16**

**THEOREM Condition for a Local Maximum or a Local Minimum**

If a function $f$ has a local maximum or a local minimum at the number $c$, then either $f'(c) = 0$ or $f'(c)$ does not exist.

**Proof** Suppose $f$ has a local maximum at $c$. Then, by definition,

$$f(c) \geq f(x)$$

for all $x$ in some open interval containing $c$. Equivalently,

$$f(x) - f(c) \leq 0$$

The derivative of $f$ at $c$ is

$$f'(c) = \lim_{x \to c} \frac{f(x) - f(c)}{x - c}$$

provided the limit exists. If this limit does not exist, then $f'(c)$ does not exist and there is nothing further to prove.

If this limit does exist, then

$$\lim_{x \to c^-} \frac{f(x) - f(c)}{x - c} = \lim_{x \to c^+} \frac{f(x) - f(c)}{x - c} \tag{1}$$

**ORIGINS** Pierre de Fermat (1601–1665), a lawyer whose contributions to mathematics were made in his spare time, ranks as one of the great "amateur" mathematicians. Although Fermat is often remembered for his famous "last theorem," he established many fundamental results in number theory and, with Pascal, cofounded the theory of probability. Since his work on calculus was the best done before Newton and Leibniz, he must be considered one of the principal founders of calculus.

In the limit on the left, $x < c$ and $f(x) - f(c) \le 0$, so $\dfrac{f(x) - f(c)}{x - c} \ge 0$ and

$$\lim_{x \to c^-} \frac{f(x) - f(c)}{x - c} \ge 0 \qquad (2)$$

[Do you see why? If the limit were negative, then there would be an open interval about $c$, $x < c$, on which $f(x) - f(c) < 0$; refer to Section 1.6, Example 7, p. 145.]

Similarly, in the limit on the right side of (1), $x > c$ and $f(x) - f(c) \le 0$, so $\dfrac{f(x) - f(c)}{x - c} \le 0$ and

$$\lim_{x \to c^+} \frac{f(x) - f(c)}{x - c} \le 0 \qquad (3)$$

Since the limits (2) and (3) must be equal, we have

$$f'(c) = \lim_{x \to c} \frac{f(x) - f(c)}{x - c} = 0$$

The proof when $f$ has a local minimum at $c$ is similar and is left as an exercise (Problem 92). ∎

For differentiable functions, the following theorem, often called **Fermat's Theorem**, is simpler.

> **Theorem** If a differentiable function $f$ has a local maximum or a local minimum at $c$, then $f'(c) = 0$.

In other words, for differentiable functions, local extreme values occur at points where the tangent line to the graph of $f$ is horizontal.

As the theorems show, the numbers at which a function $f'(x) = 0$ or at which $f'(x)$ does not exist provide a clue for locating where $f$ has local extrema. But unfortunately, knowing that $f'(c) = 0$ or that $f'$ does not exist at $c$ does not guarantee a local extremum occurs at $c$. For example, in Figure 17, $f'(x_3) = 0$, but $f$ has neither a local maximum nor a local minimum at $x_3$. Similarly, $f'(x_4)$ does not exist, but $f$ has neither a local maximum nor a local minimum at $x_4$.

Even though there may be no local extrema found at the numbers $c$ for which $f'(c) = 0$ or $f'(c)$ do not exist, the collection of all such numbers provides *all* the *possibilities* where $f$ *might* have local extreme values. For this reason, we call these numbers *critical numbers*.

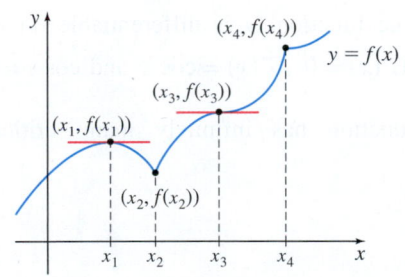

**Figure 17**

---

**DEFINITION** Critical Number

A **critical number** of a function $f$ is a number $c$ in the domain of $f$ for which either $f'(c) = 0$ or $f'(c)$ does not exist.

---

## 2 Find Critical Numbers

**EXAMPLE 2** Finding Critical Numbers

Find any critical numbers of the following functions:

**(a)** $f(x) = x^3 - 6x^2 + 9x + 2$ **(b)** $R(x) = \dfrac{1}{x - 2}$

**(c)** $g(x) = \dfrac{(x - 2)^{2/3}}{x}$ **(d)** $G(x) = \sin x$

**Solution (a)** Since $f$ is a polynomial, it is differentiable at every real number. So, the critical numbers occur where $f'(x) = 0$.

$$f'(x) = 3x^2 - 12x + 9 = 3(x-1)(x-3)$$

$f'(x) = 0$ at $x = 1$ and $x = 3$; the numbers 1 and 3 are the critical numbers of $f$.

**(b)** The domain of $R(x) = \dfrac{1}{x-2}$ is $\{x \mid x \neq 2\}$, and $R'(x) = -\dfrac{1}{(x-2)^2}$. $R'$ exists for all numbers $x$ in the domain of $R$ (remember, 2 is not in the domain of $R$). Since $R'$ is never 0, $R$ has no critical numbers.

**(c)** The domain of $g(x) = \dfrac{(x-2)^{2/3}}{x}$ is $\{x \mid x \neq 0\}$, and the derivative of $g$ is

$$g'(x) = \frac{x \cdot \left[\dfrac{2}{3}(x-2)^{-1/3}\right] - 1 \cdot (x-2)^{2/3}}{x^2} \underset{\uparrow}{=} \frac{2x - 3(x-2)}{3x^2(x-2)^{1/3}} = \frac{6-x}{3x^2(x-2)^{1/3}}$$

$$\text{Multiply by } \frac{3(x-2)^{1/3}}{3(x-2)^{1/3}}$$

Critical numbers occur where $g'(x) = 0$ or where $g'(x)$ does not exist. Since $g'(6) = 0$, $x = 6$ is a critical number. Next, $g'(x)$ does not exist where

$$3x^2(x-2)^{1/3} = 0$$
$$3x^2 = 0 \quad \text{or} \quad (x-2)^{1/3} = 0$$
$$x = 0 \quad \text{or} \quad x = 2$$

We ignore 0 since it is not in the domain of $g$. The critical numbers of $g$ are 6 and 2.

**(d)** The domain of $G$ is all real numbers. The function $G$ is differentiable on its domain, so the critical numbers occur where $G'(x) = 0$. $G'(x) = \cos x$ and $\cos x = 0$ at $x = \pm\dfrac{\pi}{2},\ \pm\dfrac{3\pi}{2},\ \pm\dfrac{5\pi}{2}, \ldots$ This function has infinitely many critical numbers. ∎

**NOW WORK** Problem 25.

We are not yet ready to give a procedure for determining whether a function $f$ has a local maximum, a local minimum, or neither at a critical number. This requires the *Mean Value Theorem*, which is the subject of the next section. However, the critical numbers do help us find the extreme values of a function $f$.

## 3 Find Absolute Maximum and Absolute Minimum Values

The following theorem provides a way to find the absolute extreme values of a function $f$ that is continuous on a closed interval $[a, b]$.

**THEOREM** Locating Absolute Extreme Values

Let $f$ be a function that is continuous on a closed interval $[a, b]$. Then the absolute maximum value and the absolute minimum value of $f$ are the largest and the smallest values, respectively, found among the following:

- The values of $f$ at the critical numbers in the open interval $(a, b)$
- $f(a)$ and $f(b)$, the values of $f$ at the endpoints $a$ and $b$

For any function $f$ satisfying the conditions of this theorem, the Extreme Value Theorem reveals that extreme values exist, and this theorem tells us how to find them.

> **Steps for Finding the Absolute Extreme Values of a Function $f$ That Is Continuous on a Closed Interval $[a, b]$**
>
> **Step 1** Locate all critical numbers in the open interval $(a, b)$.
>
> **Step 2** Evaluate $f$ at each critical number and at the endpoints $a$ and $b$.
>
> **Step 3** The largest value is the absolute maximum value; the smallest value is the absolute minimum value.

**EXAMPLE 3  Finding Absolute Maximum and Minimum Values**

Find the absolute maximum value and the absolute minimum value of each function:

**(a)** $f(x) = x^3 - 6x^2 + 9x + 2$ on $[0, 2]$    **(b)** $g(x) = \dfrac{(x-2)^{2/3}}{x}$ on $[1, 10]$

**Solution** **(a)** The function $f$, a polynomial function, is continuous on the closed interval $[0, 2]$, so the Extreme Value Theorem guarantees that $f$ has an absolute maximum value and an absolute minimum value on the interval. Follow the steps for finding extreme values.

**Step 1** From Example 2(a), the critical numbers of $f$ are 1 and 3. We exclude 3, since it is not in the interval $(0, 2)$.

**Step 2** Find the value of $f$ at the critical number 1 and at the endpoints 0 and 2:

$$f(1) = 6 \qquad f(0) = 2 \qquad f(2) = 4$$

**Step 3** The largest value 6 is the absolute maximum value of $f$; the smallest value 2 is the absolute minimum value of $f$.

**(b)** The function $g$ is continuous on the closed interval $[1, 10]$, so $g$ has an absolute maximum and an absolute minimum on the interval.

**Step 1** From Example 2(c), the critical numbers of $g$ are 2 and 6. Both critical numbers are in the interval $(1, 10)$.

**Step 2** Evaluate $g$ at the critical numbers 2 and 6 and at the endpoints 1 and 10:

| $x$ | $g(x) = \dfrac{(x-2)^{2/3}}{x}$ | |
| --- | --- | --- |
| 1 | $\dfrac{(1-2)^{2/3}}{1} = (-1)^{2/3} = 1$ | ← absolute maximum value |
| 2 | $\dfrac{(2-2)^{2/3}}{2} = 0$ | ← absolute minimum value |
| 6 | $\dfrac{(6-2)^{2/3}}{6} = \dfrac{4^{2/3}}{6} \approx 0.42$ | |
| 10 | $\dfrac{(10-2)^{2/3}}{10} = \dfrac{8^{2/3}}{10} = 0.4$ | |

**Step 3** The largest value 1 is the absolute maximum value; the smallest value 0 is the absolute minimum value. ∎

NOW WORK **Problem 49.**

For piecewise-defined functions $f$, we need to look carefully at the number(s) where the rules for the function change.

**EXAMPLE 4    Finding Absolute Maximum and Minimum Values**

Find the absolute maximum value and absolute minimum value of the function

$$f(x) = \begin{cases} 2x - 1 & \text{if } 0 \le x \le 2 \\ x^2 - 5x + 9 & \text{if } 2 < x \le 3 \end{cases}$$

**Solution** The function $f$ is continuous on the closed interval $[0, 3]$. (You should verify this.) To find the absolute maximum value and absolute minimum value, we follow the three-step procedure.

**Step 1** Find the critical numbers in the open interval $(0, 3)$:

- On the open interval $(0, 2)$: $f(x) = 2x - 1$ and $f'(x) = 2$. Since $f'(x) \ne 0$ on the interval $(0, 2)$, there are no critical numbers in $(0, 2)$.

- On the open interval $(2, 3)$: $f(x) = x^2 - 5x + 9$ and $f'(x) = 2x - 5$.

  Solving $f'(x) = 2x - 5 = 0$, we find $x = \dfrac{5}{2}$. Since $\dfrac{5}{2}$ is in the

  interval $(2, 3)$, $\dfrac{5}{2}$ is a critical number.

- At $x = 2$, the rule for $f$ changes, so we investigate the one-sided limits

  of $\dfrac{f(x) - f(2)}{x - 2}$.

  $$\lim_{x \to 2^-} \frac{f(x) - f(2)}{x - 2} = \lim_{x \to 2^-} \frac{(2x - 1) - 3}{x - 2} = \lim_{x \to 2^-} \frac{2x - 4}{x - 2}$$

  $$= \lim_{x \to 2^-} \frac{2(x - 2)}{x - 2} = 2$$

  $$\lim_{x \to 2^+} \frac{f(x) - f(2)}{x - 2} = \lim_{x \to 2^+} \frac{(x^2 - 5x + 9) - 3}{x - 2} = \lim_{x \to 2^+} \frac{x^2 - 5x + 6}{x - 2}$$

  $$= \lim_{x \to 2^+} \frac{(x - 2)(x - 3)}{x - 2} = \lim_{x \to 2^+} (x - 3) = -1$$

  The one-sided limits are not equal, so the derivative does not exist at 2; 2 is a critical number.

**Step 2** Evaluate $f$ at the critical numbers $\dfrac{5}{2}$ and 2 and at the endpoints 0 and 3.

| $x$ | $f(x)$ | |
|---|---|---|
| 0 | $2 \cdot 0 - 1 = -1$ | ← absolute minimum value |
| 2 | $2 \cdot 2 - 1 = 3$ | ← absolute maximum value |
| $\dfrac{5}{2}$ | $\left(\dfrac{5}{2}\right)^2 - 5 \cdot \dfrac{5}{2} + 9 = \dfrac{25}{4} - \dfrac{25}{2} + 9 = \dfrac{11}{4}$ | |
| 3 | $3^2 - 5 \cdot 3 + 9 = 9 - 15 + 9 = 3$ | ← absolute maximum value |

**Step 3** The largest value 3 is the absolute maximum value; the smallest value $-1$ is the absolute minimum value. ∎

The graph of $f$ is shown in Figure 18. Notice that the absolute maximum occurs at 2 and at 3.

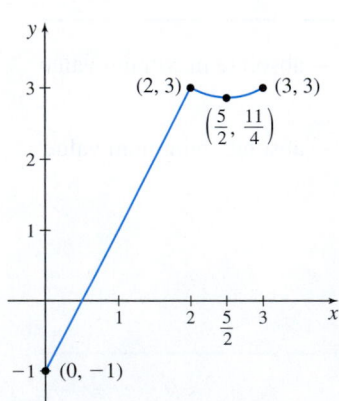

**Figure 18**

$$f(x) = \begin{cases} 2x - 1 & \text{if } 0 \le x \le 2 \\ x^2 - 5x + 9 & \text{if } 2 < x \le 3 \end{cases}$$

**NOW WORK** Problem 61.

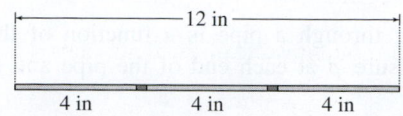

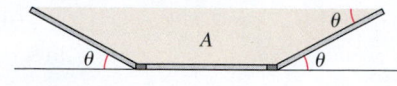

**DF Figure 19**

### EXAMPLE 5   Constructing a Rain Gutter

A rain gutter is to be constructed using a piece of aluminum 12 in. wide. After marking a length of 4 in. from each edge, the piece of aluminum is bent up at an angle $\theta$, as illustrated in Figure 19. The area $A$ of a cross section of the opening, expressed as a function of $\theta$, is

$$A(\theta) = 16\sin\theta(\cos\theta + 1) \qquad 0 \le \theta \le \frac{\pi}{2}$$

Find the angle $\theta$ that maximizes the area $A$. (This bend will allow the most water to flow through the gutter.)

**Solution**  The function $A = A(\theta)$ is continuous on the closed interval $\left[0, \frac{\pi}{2}\right]$. To find the angle $\theta$ that maximizes $A$, follow the three-step procedure.

**Step 1**  Locate all critical numbers in the open interval $\left(0, \frac{\pi}{2}\right)$.

$$\begin{aligned}
A'(\theta) &= 16\sin\theta(-\sin\theta) + 16\cos\theta(\cos\theta + 1) \\
&= 16[-\sin^2\theta + \cos^2\theta + \cos\theta] \\
&= 16[(\cos^2\theta - 1) + \cos^2\theta + \cos\theta] \\
&= 16[2\cos^2\theta + \cos\theta - 1] = 16(2\cos\theta - 1)(\cos\theta + 1)
\end{aligned}$$

The critical numbers satisfy the equation $A'(\theta) = 0, 0 < \theta < \frac{\pi}{2}$.

$$16(2\cos\theta - 1)(\cos\theta + 1) = 0$$

$$2\cos\theta - 1 = 0 \quad \text{or} \quad \cos\theta + 1 = 0$$

$$\cos\theta = \frac{1}{2} \qquad\qquad \cos\theta = -1$$

$$\theta = \frac{\pi}{3} \text{ or } \frac{5\pi}{3} \qquad\qquad \theta = \pi$$

Of these solutions, only $\frac{\pi}{3}$ is in the interval $\left(0, \frac{\pi}{2}\right)$. So, $\frac{\pi}{3}$ is the only critical number.

**Step 2**  Evaluate $A$ at the critical number $\frac{\pi}{3}$ and at the endpoints 0 and $\frac{\pi}{2}$.

| $\theta$ | $A(\theta) = 16\sin\theta(\cos\theta + 1)$ |
|---|---|
| $0$ | $16\sin 0(\cos 0 + 1) = 0$ |
| $\dfrac{\pi}{3}$ | $16\sin\dfrac{\pi}{3}\left(\cos\dfrac{\pi}{3} + 1\right) = 16\left(\dfrac{\sqrt{3}}{2}\right)\left(\dfrac{1}{2} + 1\right)$ $= 12\sqrt{3} \approx 20.785 \;\longleftarrow\; \text{absolute maximum value}$ |
| $\dfrac{\pi}{2}$ | $16\sin\dfrac{\pi}{2}\left(\cos\dfrac{\pi}{2} + 1\right) = 16(1)(0 + 1) = 16$ |

**Step 3**  If the aluminum is bent at an angle of $\frac{\pi}{3}$ (60°), the area of the opening is a maximum. The maximum area is approximately 20.785 in$^2$.  ∎

**NOW WORK** Problem 71.

## Application: Physics in Medicine

From physics, the volume $V$ of fluid flowing through a pipe is a function of the radius $r$ of the pipe and the difference in pressure $p$ at each end of the pipe and is given by

$$V = kpr^4$$

where $k$ is a constant.

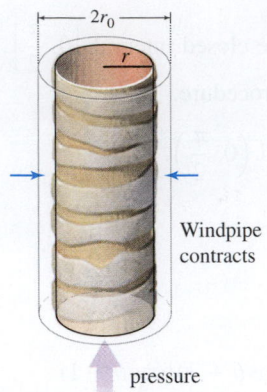

**— $2r_0$ —**

Windpipe contracts

pressure

**Figure 20**

### EXAMPLE 6 Analyzing a Cough

Coughing is caused by increased pressure in the lungs and is accompanied by a decrease in the diameter of the windpipe. See Figure 20. The radius $r$ of the windpipe decreases with increased pressure $p$ according to the formula $r_0 - r = cp$, where $r_0$ is the radius of the windpipe when there is no difference in pressure and $c$ is a positive constant. The volume $V$ of air flowing through the windpipe is

$$V = kpr^4$$

where $k$ is a constant. Find the radius $r$ that allows the most air to flow through the windpipe. Restrict $r$ so that $0 < \dfrac{r_0}{2} \leq r \leq r_0$.

**Solution** Since $p = \dfrac{r_0 - r}{c}$, we can express $V$ as a function of $r$:

$$V = V(r) = k\left(\frac{r_0 - r}{c}\right)r^4 = \frac{kr_0}{c}r^4 - \frac{k}{c}r^5 \qquad \frac{r_0}{2} \leq r \leq r_0$$

Now find the absolute maximum of $V$ on the interval $\left[\dfrac{r_0}{2}, r_0\right]$.

$$V'(r) = \frac{4k\,r_0}{c}r^3 - \frac{5k}{c}r^4 = \frac{k}{c}r^3(4r_0 - 5r)$$

The only critical number in the interval $\left(\dfrac{r_0}{2}, r_0\right)$ is $r = \dfrac{4r_0}{5}$.

Evaluate $V$ at the critical number and at the endpoints, $\dfrac{r_0}{2}$ and $r_0$.

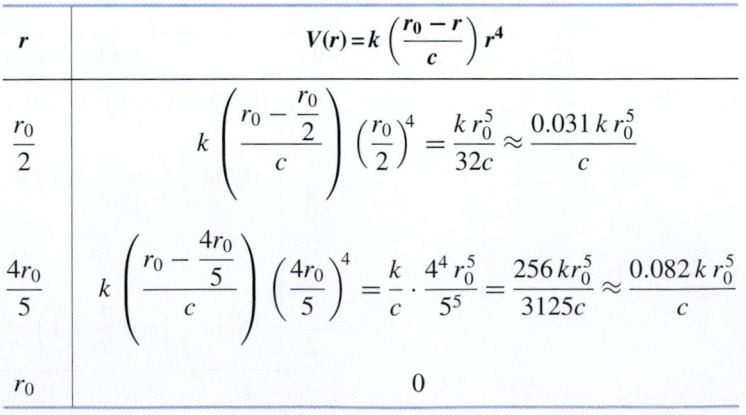

| $r$ | $V(r) = k\left(\dfrac{r_0 - r}{c}\right)r^4$ |
|---|---|
| $\dfrac{r_0}{2}$ | $k\left(\dfrac{r_0 - \dfrac{r_0}{2}}{c}\right)\left(\dfrac{r_0}{2}\right)^4 = \dfrac{k\,r_0^5}{32c} \approx \dfrac{0.031\,k\,r_0^5}{c}$ |
| $\dfrac{4r_0}{5}$ | $k\left(\dfrac{r_0 - \dfrac{4r_0}{5}}{c}\right)\left(\dfrac{4r_0}{5}\right)^4 = \dfrac{k}{c}\cdot\dfrac{4^4\,r_0^5}{5^5} = \dfrac{256\,kr_0^5}{3125c} \approx \dfrac{0.082\,k\,r_0^5}{c}$ |
| $r_0$ | $0$ |

The largest of these three values is $\dfrac{256\,kr_0^5}{3125c}$. So, the maximum air flow occurs when the radius of the windpipe is $\dfrac{4r_0}{5}$, that is, when the windpipe contracts by 20%. ∎

# 4.2 Assess Your Understanding

## Concepts and Vocabulary

**1.** *True or False*   Any function $f$ that is defined on a closed interval $[a, b]$ will have both an absolute maximum value and an absolute minimum value.

**2.** *Multiple Choice*   A number $c$ in the domain of a function $f$ is called a(n)
[(**a**) extreme value (**b**) critical number (**c**) local number]
of $f$ if either $f'(c) = 0$ or $f'(c)$ does not exist.

**3.** *True or False*   At a critical number, there is a local extreme value.

**4.** *True or False*   If a function $f$ is continuous on a closed interval $[a, b]$, then its absolute maximum value is found at a critical number.

**5.** *True or False*   The Extreme Value Theorem tells us where the absolute maximum and absolute minimum can be found.

**6.** *True or False*   If $f$ is differentiable on the interval $(0, 4)$ and $f'(2) = 0$, then $f$ has a local maximum or a local minimum at 2.

## Skill Building

*In Problems 7 and 8, use the graphs below to determine whether the function $f$ has an absolute extremum and/or a local extremum or neither at $x_1$, $x_2$, $x_3$, $x_4$, $x_5$, $x_6$, $x_7$, and $x_8$.*

**7.**

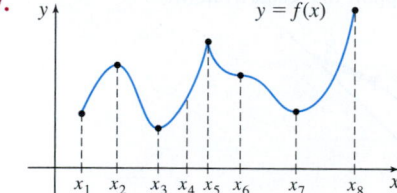

**8.**

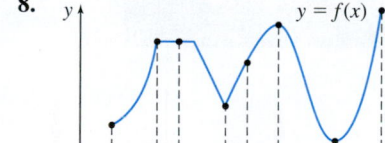

*In Problems 9–12, provide a graph of a continuous function $f$ that has the following properties:*

**9.** domain $[0, 8]$, absolute maximum at 0, absolute minimum at 3, local minimum at 7

**10.** domain $[-5, 5]$, absolute maximum at 3, absolute minimum at $-3$

**11.** domain $[3, 10]$ and has no local extreme points

**12.** no absolute extreme values, is differentiable at 4 and has a local minimum at 4, is not differentiable at 0, but has a local maximum at 0

*In Problems 13–36, find the critical numbers, if any, of each function.*

**13.** $f(x) = x^2 - 8x$

**14.** $f(x) = 1 - 6x + x^2$

**15.** $f(x) = x^3 - 3x^2$

**16.** $f(x) = x^3 - 6x$

**17.** $f(x) = x^4 - 2x^2 + 1$

**18.** $f(x) = 3x^4 - 4x^3$

**19.** $f(x) = x^{2/3}$

**20.** $f(x) = x^{1/3}$

**21.** $f(x) = 2\sqrt{x}$

**22.** $f(x) = 4 - \sqrt{x}$

**23.** $f(x) = x + \sin x, \ 0 \le x \le \pi$

**24.** $f(x) = x - \cos x, \quad -\dfrac{\pi}{2} \le x \le \dfrac{\pi}{2}$

**25.** $f(x) = x\sqrt{1 - x^2}$

**26.** $f(x) = x^2\sqrt{2 - x}$

**27.** $f(x) = \dfrac{x^2}{x - 1}$

**28.** $f(x) = \dfrac{x}{x^2 - 1}$

**29.** $f(x) = (x + 3)^2(x - 1)^{2/3}$

**30.** $f(x) = (x - 1)^2(x + 1)^{1/3}$

**31.** $f(x) = \dfrac{(x - 3)^{1/3}}{x - 1}$

**32.** $f(x) = \dfrac{(x + 3)^{2/3}}{x + 1}$

**33.** $f(x) = \dfrac{\sqrt[3]{x^2 - 9}}{x}$

**34.** $f(x) = \dfrac{\sqrt[3]{4 - x^2}}{x}$

**35.** $f(x) = \begin{cases} 3x & \text{if } 0 \le x < 1 \\ 4 - x & \text{if } 1 \le x \le 2 \end{cases}$

**36.** $f(x) = \begin{cases} x^2 & \text{if } 0 \le x < 1 \\ 1 - x^2 & \text{if } 1 \le x \le 2 \end{cases}$

*In Problems 37–64, find the absolute maximum value and absolute minimum value of each function on the indicated interval. Notice that the functions in Problems 37–58 are the same as those in Problems 13–34 above.*

**37.** $f(x) = x^2 - 8x$ on $[-1, 10]$

**38.** $f(x) = 1 - 6x + x^2$ on $[0, 4]$

**39.** $f(x) = x^3 - 3x^2$ on $[1, 4]$

**40.** $f(x) = x^3 - 6x$ on $[-1, 1]$

**41.** $f(x) = x^4 - 2x^2 + 1$ on $[0, 2]$

**42.** $f(x) = 3x^4 - 4x^3$ on $[-2, 0]$

**43.** $f(x) = x^{2/3}$ on $[-1, 1]$

**44.** $f(x) = x^{1/3}$ on $[-1, 1]$

**45.** $f(x) = 2\sqrt{x}$ on $[1, 4]$

**46.** $f(x) = 4 - \sqrt{x}$ on $[0, 4]$

**47.** $f(x) = x + \sin x$ on $[0, \pi]$

**48.** $f(x) = x - \cos x$ on $\left[-\dfrac{\pi}{2}, \dfrac{\pi}{2}\right]$

**49.** $f(x) = x\sqrt{1 - x^2}$ on $[-1, 1]$

**50.** $f(x) = x^2\sqrt{2 - x}$ on $[0, 2]$

**51.** $f(x) = \dfrac{x^2}{x - 1}$ on $\left[-1, \dfrac{1}{2}\right]$

**52.** $f(x) = \dfrac{x}{x^2 - 1}$ on $\left[-\dfrac{1}{2}, \dfrac{1}{2}\right]$

 **1.** = NOW WORK problem      = Graphing technology recommended      CAS = Computer Algebra System recommended

**53.** $f(x) = (x+3)^2(x-1)^{2/3}$ on $[-4, 5]$

**54.** $f(x) = (x-1)^2(x+1)^{1/3}$ on $[-2, 7]$

**55.** $f(x) = \dfrac{(x-3)^{1/3}}{x-1}$ on $[2, 11]$

**56.** $f(x) = \dfrac{(x+3)^{2/3}}{x+1}$ on $[-4, -2]$

**57.** $f(x) = \dfrac{\sqrt[3]{x^2-9}}{x}$ on $[3, 6]$

**58.** $f(x) = \dfrac{\sqrt[3]{4-x^2}}{x}$ on $[-4, -1]$

**59.** $f(x) = e^x - 3x$ on $[0, 1]$

**60.** $f(x) = e^{\cos x}$ on $[-\pi, 2\pi]$

**61.** $f(x) = \begin{cases} 2x+1 & \text{if } 0 \le x < 1 \\ 3x & \text{if } 1 \le x \le 3 \end{cases}$

**62.** $f(x) = \begin{cases} x+3 & \text{if } -1 \le x \le 2 \\ 2x+1 & \text{if } 2 < x \le 4 \end{cases}$

**63.** $f(x) = \begin{cases} x^2 & \text{if } -2 \le x < 1 \\ x^3 & \text{if } 1 \le x \le 2 \end{cases}$

**64.** $f(x) = \begin{cases} x+2 & \text{if } -1 \le x < 0 \\ 2-x & \text{if } 0 \le x \le 1 \end{cases}$

## Applications and Extensions

*In Problems 65–68, for each function $f$:*

*(a) Find the derivative $f'$.*

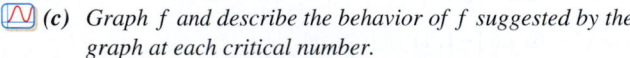 *(b) Use technology to find the critical numbers of $f$.*

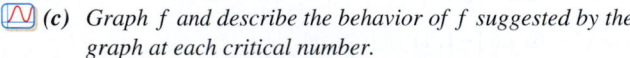 *(c) Graph $f$ and describe the behavior of $f$ suggested by the graph at each critical number.*

**65.** $f(x) = 3x^4 - 2x^3 - 21x^2 + 36x$

**66.** $f(x) = x^2 + 2x - \dfrac{2}{x}$

**67.** $f(x) = \dfrac{(x^2 - 5x + 2)\sqrt{x+5}}{\sqrt{x^2+2}}$

**68.** $f(x) = \dfrac{(x^2 - 9x + 16)\sqrt{x+3}}{\sqrt{x^2-4x+6}}$

*In Problems 69 and 70, for each function $f$:*

*(a) Find the derivative $f'$.*

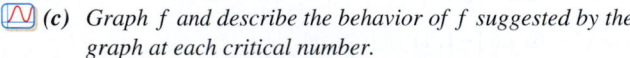 *(b) Use technology to find the absolute maximum value and the absolute minimum value of $f$ on the closed interval $[0, 5]$.*

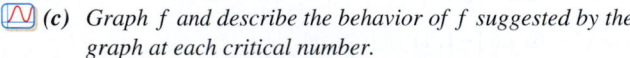 *(c) Graph $f$. Are the results from (b) supported by the graph?*

**69.** $f(x) = x^4 - 12.4x^3 + 49.24x^2 - 68.64x$

**70.** $f(x) = e^{-x}\sin(2x) + e^{-x/2}\cos(2x)$

**71. Cost of Fuel** A truck has a top speed of 75 mi/h, and when traveling at the rate of $x$ mi/h, it consumes fuel at the rate

of $\dfrac{1}{200}\left(\dfrac{2500}{x} + x\right)$ gallon per mile. If the price of fuel

is \$3.60/ gal, the cost $C$ (in dollars) of driving 200 mi is given by

$$C(x) = (3.60)\left(\dfrac{2500}{x} + x\right)$$

(a) What is the most economical speed for the truck to travel? Use the interval $[10, 75]$.

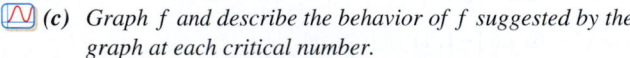 (b) Graph the cost function $C$.

**72. Trucking Costs** If the driver of the truck in Problem 71 is paid \$28.00 per hour and wages are added to the cost of fuel, what is the most economical speed for the truck to travel?

**73. Projectile Motion** An object is propelled upward at an angle $\theta$, $45° < \theta < 90°$, to the horizontal with an initial velocity of $v_0$ ft/s from the base of an inclined plane that makes an angle of $45°$ to the horizontal. See the illustration below. If air resistance is ignored, the distance $R$, in feet, that the object travels up the inclined plane is given by the function

$$R(\theta) = \dfrac{v_0^2\sqrt{2}}{16}\cos\theta\,(\sin\theta - \cos\theta)$$

(a) Find the angle $\theta$ that maximizes $R$. What is the maximum value of $R$?

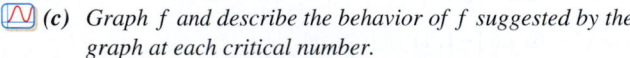 (b) Graph $R = R(\theta)$, $45° \le \theta \le 90°$, using $v_0 = 32$ ft/s.

(c) Does the graph support the result from (a)?

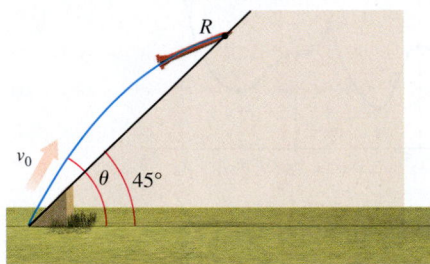

**74. Height of a Cable** An electric cable is suspended between two poles of equal height that are 20 m apart, as illustrated in the figure. The shape of the cable is modeled by the equation

$$y = 10\cosh\dfrac{x}{10} + 15$$

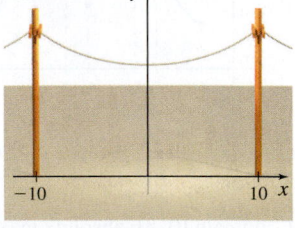

with the $x$-axis placed along the ground and the two poles at $(-10, 0)$ and $(10, 0)$. What is the height of the cable at its lowest point?

**75. Golf Ball Speed** The fastest speed of a golf ball on record is 97.05 m/s (217.1 mi/h). It was hit by Ryan Winther at the Orange County National Driving Range in Orlando, Florida on January 13, 2013. When a ball is hit at an angle $\theta$ to the horizontal, $0° \le \theta \le 90°$, and lands at the same level from which it was hit, the horizontal range $R$ of the ball is given

by $R = \dfrac{2v_0^2}{g}\sin\theta\cos\theta$, where $v_0$ is the initial speed of the ball

and $g = 9.8$ m/s$^2$ is the acceleration due to gravity.

(a) Show that the golf ball achieves its maximum range if the golfer hits it at an angle of $45°$.

(b) What is the maximum range that could be achieved by the record golf ball speed?

*Source*: http://www.guinnessworldrecords.com.

**76. Optics**   When light goes through a thin slit, it spreads out (diffracts). After passing through the slit, the intensity $I$ of the light on a distant screen is given by $I = I_0 \left( \dfrac{\sin \alpha}{\alpha} \right)^2$,

where $I_0$ is the original intensity of the light and $\alpha$ depends on the angle away from the center.

(a) What is the intensity of the light as $\alpha \rightarrow 0$?

(b) Show that the bright spots, that is, the places where the intensity has a local maximum, occur when $\tan \alpha = \alpha$.

(c) Is the intensity of the light the same at each bright spot?

**Economics**   *In Problems 77 and 78, use the following discussion:*

*In determining a tax rate on consumer goods, the government is always faced with the question, "What tax rate produces the largest tax revenue?" Imposing a tax may cause the price of the goods to increase and reduce the demand for the product. A very large tax may reduce the demand to zero, with the result that no tax is collected. On the other hand, if no tax is levied, there is no tax revenue at all. (Tax revenue R is the product of the tax rate t times the actual quantity q, in dollars, consumed.)*

**77.** The government has determined that the relationship between the quantity $q$ of a product consumed and the related tax rate $t$ is $t = \sqrt{27 - 3q^2}$. Find the tax rate that maximizes tax revenue. How much tax is generated by this rate?

**78.** On a particular product, government economists determine that the relationship between the tax rate $t$ and the quantity $q$ consumed is $t + 3q^2 = 18$. Find the tax rate that maximizes tax revenue and the revenue generated by the tax.

**79. Catenary**   A town hangs strings of holiday lights across the road between utility poles. Each set of poles is 12 m apart. The strings hang in catenaries modeled by $y = 15 \cosh \dfrac{x}{15} - 10$ with the poles at $(\pm 6, 0)$. What is the height of the string of lights at its lowest point?

**80. Harmonic Motion**   An object of mass 1 kg moves in simple harmonic motion, with an amplitude $A = 0.24$ m and a period of 4 s. The position $s$ of the object is given by $s(t) = A \cos(\omega t)$, where $t$ is the time in seconds.

(a) Find the position of the object at time $t$ and at time $t = 0.5$ s.

(b) Find the velocity $v = v(t)$ of the object.

(c) Find the velocity of the object when $t = 0.5$ s.

(d) Find the acceleration $a = a(t)$ of the object.

(e) Use Newton's Second Law of Motion, $F = ma$, to find the magnitude and direction of the force acting on the object when $t = 0.5$ s.

(f) Find the minimum time required for the object to move from its initial position to the point where $s = -0.12$ m.

(g) Find the velocity of the object when $s = -0.12$ m.

**81.** (a) Find the critical numbers of
$$f(x) = \frac{x^3}{3} - 0.055x^2 + 0.0028x - 4.$$

(b) Find the absolute extrema of $f$ on the interval $[-1, 1]$.

(c) Graph $f$ on the interval $[-1, 1]$.

**82. Extreme Value**

(a) Find the minimum value of $y = x - \cosh^{-1} x$.

(b) Graph $y = x - \cosh^{-1} x$.

**83. Locating Extreme Values**   Find the absolute maximum value and the absolute minimum value of $f(x) = \sqrt{1 + x^2} + |x - 2|$ on $[0, 3]$, and determine where each occurs.

**84.** The function $f(x) = Ax^2 + Bx + C$ has a local minimum at 0, and its graph contains the points $(0, 2)$ and $(1, 8)$. Find $A$, $B$, and $C$.

**85.** (a) Determine the domain of the function
$$f(x) = [(16 - x^2)(x^2 - 9)]^{1/2}.$$

(b) Find the absolute maximum value of $f$ on its domain.

**86. Absolute Extreme Values**   Without finding them, explain why the function $f(x) = \sqrt{x(2 - x)}$ must have an absolute maximum value and an absolute minimum value. Then find the absolute extreme values in two ways (one with and one without calculus).

**87. Put It Together**   If a function $f$ is continuous on the closed interval $[a, b]$, which of the following is necessarily true?

(a) $f$ is differentiable on the open interval $(a, b)$.

(b) If $f(u)$ is an absolute maximum value of $f$, then $f'(u) = 0$.

(c) $\lim\limits_{x \to c} f(x) = f(\lim\limits_{x \to c} x)$ for $a < c < b$.

(d) $f'(x) = 0$, for some $x$, $a \leq x \leq b$.

(e) $f$ has an absolute maximum value on $[a, b]$.

**88.** Write a paragraph that explains the similarities and differences between an absolute extreme value and a local extreme value.

**89.** Explain in your own words the method for finding the extreme values of a continuous function that is defined on a closed interval.

**90.** A function $f$ is defined and continuous on the closed interval $[a, b]$. Why can't $f(a)$ be a local extreme value on $[a, b]$?

**91.** Show that if $f$ has a local minimum at $c$, then $g(x) = -f(x)$ has a local maximum at $c$.

**92.** Show that if $f$ has a local minimum at $c$, then either $f'(c) = 0$ or $f'(c)$ does not exist.

**Challenge Problem** ─────────────────────

**93.** (a) Prove that a rational function of the form $f(x) = \dfrac{ax^{2n} + b}{cx^n + d}$, $n \geq 1$ an integer, has at most five critical numbers.

(b) Give an example of such a rational function with exactly five critical numbers.

# 4.3 The Mean Value Theorem

**ORIGINS** Michel Rolle (1652–1719) was a French mathematician whose formal education was limited to elementary school. At age 24, he moved to Paris and married. To support his family, Rolle worked as an accountant, but he also began to study algebra and became interested in the theory of equations. Although Rolle is primarily remembered for the theorem proved here, he was also the first to use the notation $\sqrt[n]{\ }$ for the $n$th root.

**OBJECTIVES** *When you finish this section, you should be able to:*

1 **Use Rolle's Theorem (p. 284)**
2 **Work with the Mean Value Theorem (p. 285)**
3 **Identify where a function is increasing and decreasing (p. 288)**

In this section, we prove several theorems. The most significant of these, the *Mean Value Theorem*, is used to develop tests for locating local extreme values. To prove the Mean Value Theorem, we need *Rolle's Theorem.*

## 1 Use Rolle's Theorem

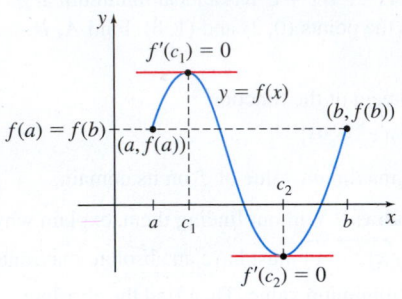

**Figure 21**

> **THEOREM** Rolle's Theorem
>
> Let $f$ be a function defined on a closed interval $[a, b]$. If:
>
> * $f$ is continuous on $[a, b]$,
> * $f$ is differentiable on $(a, b)$, and
> * $f(a) = f(b)$,
>
> then there is at least one number $c$ in the open interval $(a, b)$ for which $f'(c) = 0$.

Before we prove Rolle's Theorem, notice that the graph in Figure 21 meets the three conditions of Rolle's Theorem and has at least one number $c$ at which $f'(c) = 0$.

In contrast, the graphs in Figure 22 show that the conclusion of Rolle's Theorem may not hold when one or more of the three conditions are not met.

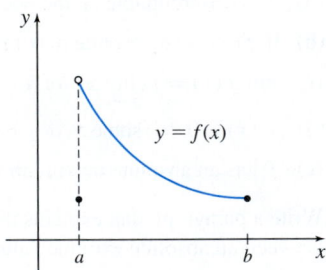

(a) $f$ is defined on $[a, b]$.
  $f$ is not continuous at $a$.
  $f$ is differentiable on $(a, b)$.
  $f(a) = f(b)$.
  No $c$ in $(a, b)$ at which $f'(c) = 0$.

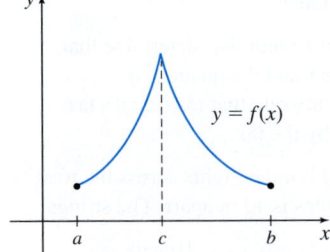

(b) $f$ is defined on $[a, b]$.
  $f$ is continuous on $[a, b]$.
  $f$ is not differentiable on $(a, b)$,
    no derivative at $c$.
  $f(a) = f(b)$.
  No $c$ in $(a, b)$ at which $f'(c) = 0$.

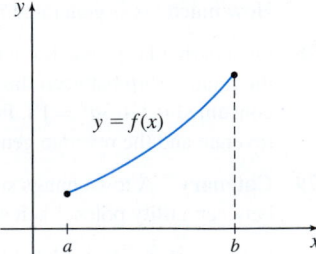

(c) $f$ is defined on $[a, b]$.
  $f$ is continuous on $[a, b]$.
  $f$ is differentiable on $(a, b)$.
  $f(a) \neq f(b)$.
  No $c$ in $(a, b)$ at which $f'(c) = 0$.

**Figure 22**

**Proof** Because $f$ is continuous on a closed interval $[a, b]$, the Extreme Value Theorem guarantees that $f$ has an absolute maximum value and an absolute minimum value on $[a, b]$. There are two possibilities:

1. If $f$ is a constant function on $[a, b]$, then $f'(x) = 0$ for all $x$ in $(a, b)$.

2. If $f$ is not a constant function on $[a, b]$, then, because $f(a) = f(b)$, either the absolute maximum or the absolute minimum occurs at some number $c$ in the open interval $(a, b)$. Then $f(c)$ is a local maximum value (or a local minimum value), and, since $f$ is differentiable on $(a, b)$, $f'(c) = 0$. ∎

EXAMPLE 1   **Using Rolle's Theorem**

Find the $x$-intercepts of $f(x) = x^2 - 5x + 6$, and show that $f'(c) = 0$ for some number $c$ in the open interval formed by the two $x$-intercepts. Find $c$.

**Solution**   At the $x$-intercepts, $f(x) = 0$.

$$f(x) = x^2 - 5x + 6 = (x - 2)(x - 3) = 0$$

So, $x = 2$ and $x = 3$ are the $x$-intercepts of the graph of $f$, and $f(2) = f(3) = 0$.

Since $f$ is a polynomial, it is continuous on the closed interval $[2, 3]$ formed by the $x$-intercepts and is differentiable on the open interval $(2, 3)$. The three conditions of Rolle's Theorem are satisfied, guaranteeing that there is a number $c$ in the open interval $(2, 3)$ for which $f'(c) = 0$. Since $f'(x) = 2x - 5$, the number $c$ for which $f'(x) = 0$ is $c = \dfrac{5}{2}$. ∎

See Figure 23 for the graph of $f$.

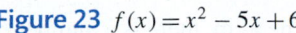

**Figure 23** $f(x) = x^2 - 5x + 6$

NOW WORK   **Problem 9.**

## 2 Work with the Mean Value Theorem

The importance of Rolle's Theorem is that it can be used to obtain other results, many of which have wide-ranging application. Perhaps the most important of these is the *Mean Value Theorem,* sometimes called the *Theorem of the Mean for Derivatives.*

A geometric example helps explain the Mean Value Theorem. Consider the graph of a function $f$ that is continuous on a closed interval $[a, b]$ and differentiable on the open interval $(a, b)$, as illustrated in Figure 24. Then there is at least one number $c$ between $a$ and $b$ at which the slope $f'(c)$ of the tangent line to the graph of $f$ equals the slope of the secant line joining the points $A = (a, f(a))$ and $B = (b, f(b))$. That is, there is at least one number $c$ in the open interval $(a, b)$ for which

$$f'(c) = \frac{f(b) - f(a)}{b - a}$$

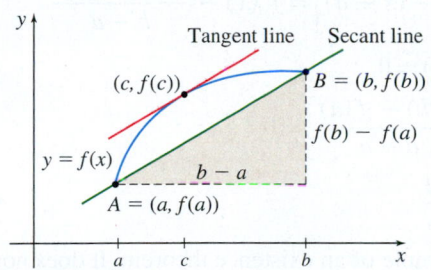

**DF** **Figure 24** The tangent line at $(c, f(c))$ is parallel to the secant line from $A$ to $B$.

---

**THEOREM   Mean Value Theorem**

Let $f$ be a function defined on a closed interval $[a, b]$. If

- $f$ is continuous on $[a, b]$
- $f$ is differentiable on $(a, b)$

then there is at least one number $c$ in the open interval $(a, b)$ for which

$$\boxed{f'(c) = \frac{f(b) - f(a)}{b - a}} \tag{1}$$

---

**Proof**   The slope $m_{\text{sec}}$ of the secant line containing the points $(a, f(a))$ and $(b, f(b))$ is

$$m_{\text{sec}} = \frac{f(b) - f(a)}{b - a}$$

Then using the point $(a, f(a))$ and the point-slope form of an equation of a line, an equation of this secant line is

$$y - f(a) = \frac{f(b) - f(a)}{b - a}(x - a)$$

$$y = f(a) + \frac{f(b) - f(a)}{b - a}(x - a)$$

**NOTE** The function $g$ has a geometric significance: Its value equals the vertical distance from the graph of $y = f(x)$ to the secant line joining $(a, f(a))$ to $(b, f(b))$. See Figure 25.

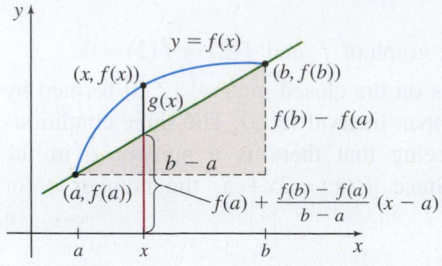

**Figure 25**

Now construct a function $g$ that satisfies the conditions of Rolle's Theorem, by defining $g$ as

$$g(x) = f(x) - \left[ f(a) + \underbrace{\frac{f(b) - f(a)}{b - a}(x - a)}_{} \right] \qquad (2)$$

$$\underbrace{\qquad\qquad\qquad\qquad}_{\substack{\text{Height of the secant line} \\ \text{containing } (a, f(a)) \text{ and } (b, f(b))}}$$

for all $x$ on $[a, b]$.

Since $f$ is continuous on $[a, b]$ and differentiable on $(a, b)$, it follows that $g$ is also continuous on $[a, b]$ and differentiable on $(a, b)$. Also,

$$g(a) = f(a) - \left[ f(a) + \frac{f(b) - f(a)}{b - a}(a - a) \right] = 0$$

$$g(b) = f(b) - \left[ f(a) + \frac{f(b) - f(a)}{b - a}(b - a) \right] = 0$$

Now, since $g$ satisfies the three conditions of Rolle's Theorem, there is a number $c$ in $(a, b)$ at which $g'(c) = 0$. Since $f(a)$ and $\dfrac{f(b) - f(a)}{b - a}$ are constants, the derivative of $g$, from Equation (2) above, is given by

$$g'(x) = f'(x) - \frac{d}{dx}\left[ f(a) + \frac{f(b) - f(a)}{b - a}(x - a) \right] = f'(x) - \frac{f(b) - f(a)}{b - a}$$

Now evaluate $g'$ at $c$ and use the fact that $g'(c) = 0$.

$$g'(c) = f'(c) - \left[ \frac{f(b) - f(a)}{b - a} \right] = 0$$

$$f'(c) = \frac{f(b) - f(a)}{b - a} \qquad \blacksquare$$

The Mean Value Theorem is another example of an existence theorem. It does not tell us how to find the number $c$; it merely states that at least one number $c$ exists. Often, the number $c$ can be found.

---

**CALC CLIP**

**EXAMPLE 2    Verifying the Mean Value Theorem**

**(a)** Verify that the function $f(x) = x^3 - 3x + 5$, $-1 \le x \le 1$ satisfies the conditions of the Mean Value Theorem.

**(b)** Find the number(s) $c$ guaranteed by the Mean Value Theorem.

**(c)** Interpret the number(s) geometrically.

**Solution** **(a)** Since $f$ is a polynomial function, $f$ is continuous on the closed interval $[-1, 1]$ and differentiable on the open interval $(-1, 1)$. The conditions of the Mean Value Theorem are met.

**(b)** Evaluate $f$ at the endpoints $-1$ and $1$ and then find $f'(x)$.

$$f(-1) = 7 \qquad f(1) = 3 \qquad \text{and} \qquad f'(x) = 3x^2 - 3$$

The number(s) $c$ in the open interval $(-1, 1)$ guaranteed by the Mean Value Theorem satisfy the equation

$$f'(c) = \frac{f(1) - f(-1)}{1 - (-1)}$$

$$3c^2 - 3 = \frac{3 - 7}{1 - (-1)} = \frac{-4}{2} = -2$$

$$3c^2 = 1$$

$$c = \sqrt{\frac{1}{3}} = \frac{\sqrt{3}}{3} \qquad \text{or} \qquad c = -\sqrt{\frac{1}{3}} = -\frac{\sqrt{3}}{3}$$

There are two numbers in the interval $(-1, 1)$ that satisfy the Mean Value Theorem.

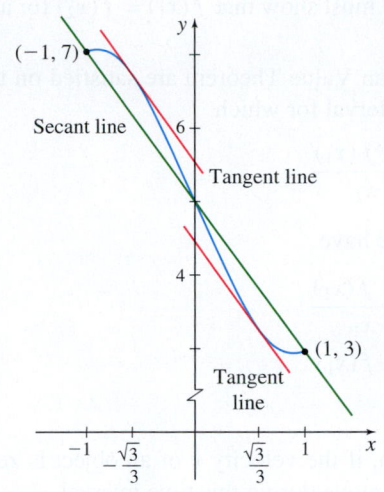

**Figure 26** $f(x) = x^3 - 3x + 5$,
$-1 \leq x \leq 1$

**(c)** At each number, $-\dfrac{\sqrt{3}}{3}$ and $\dfrac{\sqrt{3}}{3}$, the slope of the tangent line to the graph of $f$ equals the slope of the secant line connecting the points $(-1, 7)$ and $(1, 3)$. That is,

$$m_{\tan} = f'\left(-\frac{\sqrt{3}}{3}\right) = f'\left(\frac{\sqrt{3}}{3}\right) = -2 \quad \text{and} \quad m_{\sec} = \frac{f(1) - f(-1)}{1 - (-1)} = \frac{3 - 7}{2} = -2$$

The tangent lines and the secant line are parallel.

See Figure 26. ∎

**NOW WORK** Problem 25.

The Mean Value Theorem can be applied to rectilinear motion. Suppose the position function $s = f(t)$ models the signed distance $s$ of an object from the origin at time $t$. If $f$ is continuous on $[a, b]$ and differentiable on $(a, b)$, the Mean Value Theorem tells us that there is at least one time $t$ at which the velocity of the object equals its average velocity.

Let $t_1$ and $t_2$ be two distinct times in $[a, b]$. Then

$$\text{Average velocity over } [t_1, t_2] = \frac{f(t_2) - f(t_1)}{t_2 - t_1}$$

$$\text{Instantaneous velocity } v(t) = \frac{ds}{dt} = f'(t)$$

The Mean Value Theorem states there is a time $t_0$ in the interval $(t_1, t_2)$ for which

$$f'(t_0) = \frac{f(t_2) - f(t_1)}{t_2 - t_1}$$

That is, the velocity at time $t_0$ equals the average velocity over the interval $[t_1, t_2]$.

**EXAMPLE 3** **Applying the Mean Value Theorem to Rectilinear Motion**

Use the Mean Value Theorem to show that a car that travels 110 mi in 2 h must have had a speed of 55 miles per hour (mi/h) at least once during the 2 h.

**Solution** Let $s = f(t)$ represent the distance $s$ the car has traveled after $t$ hours. Its average velocity during the time period from 0 to 2 h is

$$\text{Average velocity} = \frac{f(2) - f(0)}{2 - 0} = \frac{110 - 0}{2 - 0} = 55 \text{ mi/h}$$

Using the Mean Value Theorem, there is a time $t_0$, $0 < t_0 < 2$, at which

$$f'(t_0) = \frac{f(2) - f(0)}{2 - 0} = 55$$

That is, the car had a velocity of 55 mi/h at least once during the 2 h period. ∎

**NOW WORK** Problem 57.

The following corollaries are consequences of the Mean Value Theorem.

**IN WORDS** If $f'(x) = 0$ for all numbers in $(a, b)$, then $f$ is a constant function on the interval $(a, b)$.

**COROLLARY**

If a function $f$ is continuous on the closed interval $[a, b]$ and is differentiable on the open interval $(a, b)$, and if $f'(x) = 0$ for all numbers $x$ in $(a, b)$, then $f$ is constant on $(a, b)$.

**Proof**  To show that $f$ is a constant function, we must show that $f(x_1) = f(x_2)$ for any two numbers $x_1$ and $x_2$ in the interval $(a, b)$.

Suppose $x_1 < x_2$. The conditions of the Mean Value Theorem are satisfied on the interval $[x_1, x_2]$, so there is a number $c$ in this interval for which

$$f'(c) = \frac{f(x_2) - f(x_1)}{x_2 - x_1}$$

Since $f'(c) = 0$ for all numbers in $(a, b)$, we have

$$0 = \frac{f(x_2) - f(x_1)}{x_2 - x_1}$$
$$0 = f(x_2) - f(x_1)$$
$$f(x_1) = f(x_2)$$  ∎

**IN WORDS**  If an object in rectilinear motion has velocity $v = 0$, then the object is at rest.

Applying the corollary to rectilinear motion, if the velocity $v$ of an object is zero over an interval $(t_1, t_2)$, then the object does not move during this time interval.

**COROLLARY**

If the functions $f$ and $g$ are differentiable on an open interval $(a, b)$ and if $f'(x) = g'(x)$ for all numbers $x$ in $(a, b)$, then there is a number $C$ for which $f(x) = g(x) + C$ on $(a, b)$.

**IN WORDS**  If two functions $f$ and $g$ have the same derivative, then the functions differ by a constant, and the graph of $f$ is a vertical shift of the graph of $g$.

**Proof**  We define the function $h$ so that $h(x) = f(x) - g(x)$. Then $h$ is differentiable since it is the difference of two differentiable functions, and

$$h'(x) = f'(x) - g'(x) = 0$$

Since $h'(x) = 0$ for all numbers $x$ in the interval $(a, b)$, $h$ is a constant function and we can write

$$h(x) = f(x) - g(x) = C \qquad \text{for some number } C$$

That is,

$$f(x) = g(x) + C$$  ∎

## 3 Identify Where a Function Is Increasing and Decreasing

A third corollary that follows from the Mean Value Theorem gives us a way to determine where a function is increasing and where a function is decreasing. Recall that functions are increasing or decreasing on *intervals,* either open or closed or half-open or half-closed.

**NEED TO REVIEW?**  Increasing and decreasing functions are discussed in Section P.1, pp. 11–12.

**COROLLARY**  Increasing/Decreasing Function Test

Let $f$ be a function that is differentiable on the open interval $(a, b)$:

- If $f'(x) > 0$ on $(a, b)$, then $f$ is increasing on $(a, b)$.

- If $f'(x) < 0$ on $(a, b)$, then $f$ is decreasing on $(a, b)$.

The proof of the first part of the corollary is given here. The proof of the second part is left as an exercise. See Problem 82.

**Proof**  Since $f$ is differentiable on $(a, b)$, it is continuous on $(a, b)$. To show that $f$ is increasing on $(a, b)$, choose two numbers $x_1$ and $x_2$ in $(a, b)$, with $x_1 < x_2$. We need to show that $f(x_1) < f(x_2)$.

Since $x_1$ and $x_2$ are in $(a, b)$, $f$ is continuous on the closed interval $[x_1, x_2]$ and differentiable on the open interval $(x_1, x_2)$. By the Mean Value Theorem, there is a number $c$ in the interval $(x_1, x_2)$ for which

$$f'(c) = \frac{f(x_2) - f(x_1)}{x_2 - x_1}$$

Since $f'(x) > 0$ for all $x$ in $(a, b)$, it follows that $f'(c) > 0$. Since $x_1 < x_2$, then $x_2 - x_1 > 0$. As a result,

$$f(x_2) - f(x_1) > 0$$

$$f(x_1) < f(x_2)$$

So, $f$ is increasing on $(a, b)$. ∎

There are a few important observations to make about the Increasing/Decreasing Function Test:

- In using the Increasing/Decreasing Function Test, we determine open intervals on which the derivative is positive or negative.
- Suppose $f$ is continuous on the closed interval $[a, b]$ and differentiable on the open interval $(a, b)$.
  - If $f'(x) > 0$ on $(a, b)$, then $f$ is increasing on the closed interval $[a, b]$.
  - If $f'(x) < 0$ on $(a, b)$, then $f$ is decreasing on the closed interval $[a, b]$.
- The Increasing/Decreasing Function Test is valid if the interval $(a, b)$ is $(-\infty, b)$ or $(a, \infty)$ or $(-\infty, \infty)$.

**EXAMPLE 4** **Identifying Where a Function Is Increasing and Decreasing**

Determine where the function $f(x) = 2x^3 - 9x^2 + 12x - 5$ is increasing and where it is decreasing.

**NEED TO REVIEW?** Solving inequalities is discussed in Appendix A.1, pp. A-5 to A-8.

**Solution** The function $f$ is a polynomial so $f$ is continuous and differentiable at every real number. We find $f'$.

$$f'(x) = 6x^2 - 18x + 12 = 6(x - 2)(x - 1)$$

The Increasing/Decreasing Function Test states that $f$ is increasing on intervals where $f'(x) > 0$ and that $f$ is decreasing on intervals where $f'(x) < 0$. We solve these inequalities by using the numbers 1 and 2 to form three intervals on the $x$-axis, as shown in Figure 27. Then we determine the sign of $f'(x)$ on each interval, as shown in Table 1.

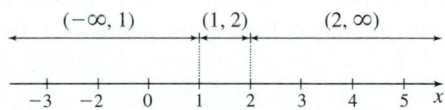

**Figure 27**

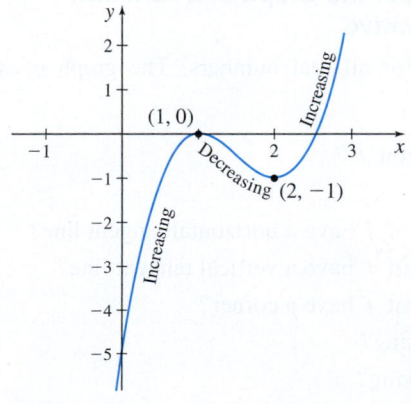

**DF** **Figure 28** $f(x) = 2x^3 - 9x^2 + 12x - 5$

**TABLE 1**

| Interval | Sign of $x - 1$ | Sign of $x - 2$ | Sign of $f'(x) = 6(x - 2)(x - 1)$ | Conclusion |
|---|---|---|---|---|
| $(-\infty, 1)$ | Negative $(-)$ | Negative $(-)$ | Positive $(+)$ | $f$ is increasing |
| $(1, 2)$ | Positive $(+)$ | Negative $(-)$ | Negative $(-)$ | $f$ is decreasing |
| $(2, \infty)$ | Positive $(+)$ | Positive $(+)$ | Positive $(+)$ | $f$ is increasing |

We conclude that $f$ is increasing on the intervals $(-\infty, 1)$ and $(2, \infty)$, and $f$ is decreasing on the interval $(1, 2)$. Since $f$ is continuous on its domain, $f$ is increasing on the intervals $(-\infty, 1]$ and $[2, \infty)$ and is decreasing on the interval $[1, 2]$. ∎

Figure 28 shows the graph of $f$. Notice that the graph of $f$ has horizontal tangent lines at the points $(1, 0)$ and $(2, -1)$.

**NOW WORK** Problem 31.

**EXAMPLE 5** **Identifying Where a Function Is Increasing and Decreasing**

Determine where the function $f(x) = (x^2 - 1)^{2/3}$ is increasing and where it is decreasing.

**Solution** $f$ is continuous for all real numbers $x$, and

$$f'(x) = \frac{2}{3}(x^2 - 1)^{-1/3}(2x) = \frac{4x}{3(x^2 - 1)^{1/3}}$$

The critical numbers of $f$ are $-1, 0, 1$.

The Increasing/Decreasing Function Test states that $f$ is increasing on intervals where $f'(x) > 0$ and decreasing on intervals where $f'(x) < 0$. We solve these inequalities by using the critical numbers $-1$, $0$, and $1$ to form four intervals on the $x$-axis: $(-\infty, -1)$, $(-1, 0)$, $(0, 1)$, and $(1, \infty)$ as shown in Table 2.

**TABLE 2**

| Interval | Sign of $4x$ | Sign of $(x^2 - 1)^{1/3}$ | Sign of $f'(x) = \dfrac{4x}{3(x^2 - 1)^{1/3}}$ | Conclusion |
|---|---|---|---|---|
| $(-\infty, -1)$ | Negative $(-)$ | Positive $(+)$ | Negative $(-)$ | $f$ is decreasing on $(-\infty, -1)$ |
| $(-1, 0)$ | Negative $(-)$ | Negative $(-)$ | Positive $(+)$ | $f$ is increasing on $(-1, 0)$ |
| $(0, 1)$ | Positive $(+)$ | Negative $(-)$ | Negative $(-)$ | $f$ is decreasing on $(0, 1)$ |
| $(1, \infty)$ | Positive $(+)$ | Positive $(+)$ | Positive $(+)$ | $f$ is increasing on $(1, \infty)$ |

Then we determine the sign of $f'$ in each interval. We conclude that $f$ is increasing on the intervals $(-1, 0)$ and $(1, \infty)$ and that $f$ is decreasing on the intervals $(-\infty, -1)$ and $(0, 1)$.

Since $f$ is continuous on its domain, $f$ is increasing on the intervals $[-1, 0]$ and $[1, \infty)$ and is decreasing on the intervals $(-\infty, -1]$ and $[0, 1]$. ∎

Figure 29 shows the graph of $f$. Since $f'(0) = 0$, the graph of $f$ has a horizontal tangent line at the point $(0, 1)$. Notice that $f'$ does not exist at $\pm 1$. Since $f'$ becomes unbounded at $-1$ and $1$, the graph of $f$ has vertical tangent lines at the points $(-1, 0)$ and $(1, 0)$, both of which are cusps.

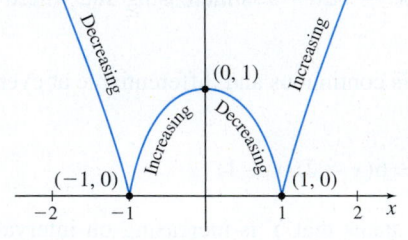

**Figure 29** $f(x) = (x^2 - 1)^{2/3}$

**NOW WORK** Problem 37.

---

**EXAMPLE 6**   **Obtaining Information About the Graph of a Function from the Graph of Its Derivative**

Suppose $f$ is a function that is continuous for all real numbers. The graph of its derivative function $f'$ is shown in Figure 30.

(a) What is the domain of the derivative function $f'$?
(b) List the critical numbers of $f$.
(c) At what numbers $x$, if any, does the graph of $f$ have a horizontal tangent line?
(d) At what numbers $x$, if any, does the graph of $f$ have a vertical tangent line?
(e) At what numbers $x$, if any, does the graph of $f$ have a corner?
(f) On what intervals is the graph of $f$ increasing?
(g) On what intervals is the graph of $f$ decreasing?

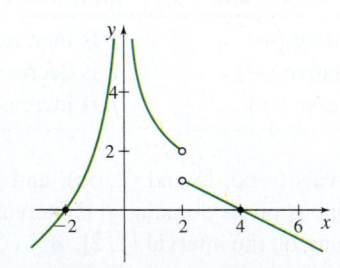

**Figure 30** The graph of $y = f'(x)$

**Solution (a)** The domain of $f'$ is the set of all real numbers except $0$ and $2$.

**(b)** A critical number of $f$ is a number $c$ in the domain of $f$ for which either $f'(c) = 0$ or $f'(c)$ doesn't exist.

- $f'(x) = 0$ where the graph of $f'$ crosses the $x$-axis, namely at $-2$ and $4$.
- $f'(x)$ does not exist for $x = 0$ and $x = 2$.

The critical numbers of $f$ are $-2, 0, 2$, and $4$.

**(c)** The graph of $f$ has a horizontal tangent line when its slope is zero, namely, at $x = -2$ and $x = 4$.

**(d)** The graph of $f$ has a vertical tangent line where the derivative $f'$ is unbounded. This occurs at $x = 0$.

**(e)** From the graph of $f'$ we see that the derivative has unequal one-sided limits at $x = 2$. As $x$ approaches 2 from the left, the derivative is 2, and as $x$ approaches 2 from the right, the derivative is 1. This means the graph of $f$ has a corner at $x = 2$.

**(f)** The graph of $f$ is increasing when $f'(x) > 0$; that is, when the graph of $f'$ is above the $x$-axis. This occurs on the intervals $(-2, 0)$ and $(0, 4)$. Since $f$ is continuous for all real numbers, $f$ is continuous on $[-2, 4]$.

**(g)** The graph of $f$ is decreasing when $f'(x) < 0$; that is, when the graph of $f'$ is below the $x$-axis. This occurs on the intervals $(-\infty, -2)$ and $(4, \infty)$. ■

**NOW WORK** Problem 47.

## Application: Agricultural Economics

**EXAMPLE 7   Determining Crop Yield\***

A variation of the von Liebig model states that the yield $f(x)$ of a plant, measured in bushels, responds to the amount $x$ of potassium in a fertilizer according to the following square root model:

$$f(x) = -0.057 - 0.417x + 0.852\sqrt{x}$$

For what amounts of potassium will the yield increase? For what amounts of potassium will the yield decrease?

**Solution** The yield is increasing when $f'(x) > 0$.

$$f'(x) = -0.417 + \frac{0.426}{\sqrt{x}} = \frac{-0.417\sqrt{x} + 0.426}{\sqrt{x}}$$

Now $f'(x) > 0$ when

$$-0.417\sqrt{x} + 0.426 > 0$$
$$0.417\sqrt{x} < 0.426$$
$$\sqrt{x} < 1.022$$
$$x < 1.044$$

The crop yield is increasing when the amount of potassium in the fertilizer is less than 1.044 and is decreasing when the amount of potassium in the fertilizer is greater than 1.044. ■

# 4.3 Assess Your Understanding

**Concepts and Vocabulary**

**1.** *True or False*  If a function $f$ is defined and continuous on a closed interval $[a, b]$, differentiable on the open interval $(a, b)$, and if $f(a) = f(b)$, then Rolle's Theorem guarantees that there is at least one number $c$ in the interval $(a, b)$ for which $f(c) = 0$.

**2.** In your own words, give a geometric interpretation of the Mean Value Theorem.

**3.** *True or False*  If two functions $f$ and $g$ are differentiable on an open interval $(a, b)$ and if $f'(x) = g'(x)$ for all numbers $x$ in $(a, b)$, then $f$ and $g$ differ by a constant on $(a, b)$.

**4.** *True or False*  When the derivative $f'$ is positive on an open interval $I$, then $f$ is positive on $I$.

**Skill Building**

*In Problems 5–16, verify that each function satisfies the three conditions of Rolle's Theorem on the given interval. Then find all numbers $c$ in $(a, b)$ guaranteed by Rolle's Theorem.*

**5.** $f(x) = x^2 - 3x$ on $[0, 3]$

**6.** $f(x) = x^2 + 2x$ on $[-2, 0]$

**7.** $g(x) = x^2 - 2x - 2$ on $[0, 2]$

**8.** $g(x) = x^2 + 1$ on $[-1, 1]$

**9.** $f(x) = x^3 - x$ on $[-1, 0]$

**10.** $f(x) = x^3 - 4x$ on $[-2, 2]$

**\*Source:** Quirino Paris. (1992). The von Liebig Hypothesis. *American Journal of Agricultural Economics*, 74(4), 1019–1028.

**1.** = NOW WORK problem          〰 = Graphing technology recommended          CAS = Computer Algebra System recommended

**11.** $f(t) = t^3 - t + 2$ on $[-1, 1]$   **12.** $f(t) = t^4 - 3$ on $[-2, 2]$

**13.** $s(t) = t^4 - 2t^2 + 1$ on $[-2, 2]$   **14.** $s(t) = t^4 + t^2$ on $[-2, 2]$

**15.** $f(x) = \sin(2x)$ on $[0, \pi]$

**16.** $f(x) = \sin x + \cos x$ on $[0, 2\pi]$

*In Problems 17–20, state why Rolle's Theorem cannot be applied to the function $f$ on the given interval.*

**17.** $f(x) = x^2 - 2x + 1$ on $[-2, 1]$   **18.** $f(x) = x^3 - 3x$ on $[2, 4]$

**19.** $f(x) = x^{1/3} - x$ on $[-1, 1]$   **20.** $f(x) = x^{2/5}$ on $[-1, 1]$

*In Problems 21–30,*

*(a) Verify that each function satisfies the conditions of the Mean Value Theorem on the indicated interval.*

*(b) Find the number(s) $c$ guaranteed by the Mean Value Theorem.*

*(c) Interpret the number(s) $c$ geometrically.*

**21.** $f(x) = x^2 + 1$ on $[0, 2]$

**22.** $f(x) = x + 2 + \dfrac{3}{x - 1}$ on $[2, 7]$

**23.** $f(x) = \ln \sqrt{x}$ on $[1, e]$   **24.** $f(x) = xe^x$ on $[0, 1]$

**25.** $f(x) = x^3 - 5x^2 + 4x - 2$ on $[1, 3]$

**26.** $f(x) = x^3 - 7x^2 + 5x$ on $[-2, 2]$

**27.** $f(x) = \dfrac{x + 1}{x}$ on $[1, 3]$   **28.** $f(x) = \dfrac{x^2}{x + 1}$ on $[0, 1]$

**29.** $f(x) = \sqrt[3]{x^2}$ on $[1, 8]$   **30.** $f(x) = \sqrt{x - 2}$ on $[2, 4]$

*In Problems 31–46, determine where each function is increasing and where each is decreasing.*

**31.** $f(x) = x^3 + 6x^2 + 12x + 1$   **32.** $f(x) = -x^3 + 3x^2 + 4$

**33.** $f(x) = x^3 - 3x + 1$   **34.** $f(x) = x^3 - 6x^2 - 3$

**35.** $f(x) = x^4 - 4x^2 + 1$   **36.** $f(x) = x^4 + 4x^3 - 2$

**37.** $f(x) = x^{2/3}(x^2 - 4)$   **38.** $f(x) = x^{1/3}(x^2 - 7)$

**39.** $f(x) = |x^3 + 3|$   **40.** $f(x) = |x^2 - 4|$

**41.** $f(x) = 3 \sin x$ on $[0, 2\pi]$

**42.** $f(x) = \cos(2x)$ on $[0, 2\pi]$

**43.** $f(x) = xe^x$   **44.** $g(x) = x + e^x$

**45.** $f(x) = e^x \sin x,\ 0 \le x \le 2\pi$

**46.** $f(x) = e^x \cos x,\ 0 \le x \le 2\pi$

*In Problems 47–50, the graph of the derivative function $f'$ of a function $f$ that is continuous for all real numbers is given.*

*(a) What is the domain of the derivative function $f'$?*

*(b) List the critical numbers of $f$.*

*(c) At what numbers $x$, if any, does the graph of $f$ have a horizontal tangent line?*

*(d) At what numbers $x$, if any, does the graph of $f$ have a vertical tangent line?*

*(e) At what numbers $x$, if any, does the graph of $f$ have a corner?*

*(f) On what intervals is the graph of $f$ increasing?*

*(g) On what intervals is the graph of $f$ decreasing?*

**47.**
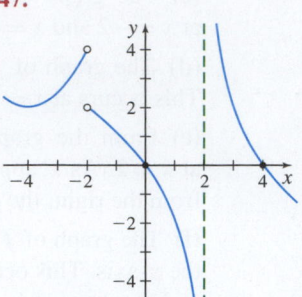

**48.**
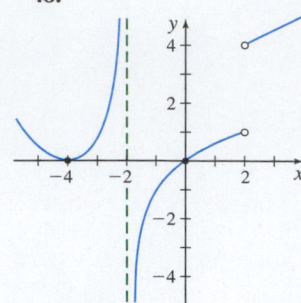

**49.**
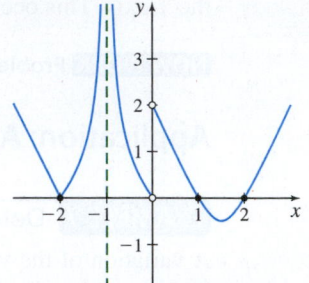

**50.**
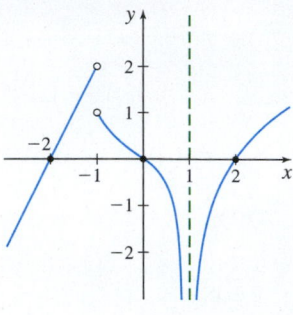

### Applications and Extensions

**51.** Show that the function $f(x) = 2x^3 - 6x^2 + 6x - 5$ is increasing for all $x$.

**52.** Show that the function $f(x) = x^3 - 3x^2 + 3x$ is increasing for all $x$.

**53.** Show that the function $f(x) = \dfrac{x}{x + 1}$ is increasing on any interval not containing $x = -1$.

**54.** Show that the function $f(x) = \dfrac{x + 1}{x}$ is decreasing on any interval not containing $x = 0$.

**55. Mean Value Theorem**   Draw the graph of a function $f$ that is continuous on $[a, b]$ but not differentiable on $(a, b)$, and for which the conclusion of the Mean Value Theorem does not hold.

**56. Mean Value Theorem**   Draw the graph of a function $f$ that is differentiable on $(a, b)$ but not continuous on $[a, b]$, and for which the conclusion of the Mean Value Theorem does not hold.

**57. Rectilinear Motion**   An automobile travels 20 mi down a straight road at an average velocity of 40 mi/h. Show that the automobile must have a velocity of exactly 40 mi/h at some time during the trip. (Assume that the position function is differentiable.)

**58. Rectilinear Motion**   Suppose a car is traveling on a highway. At 4:00 p.m., the car's speedometer reads 40 mi/h. At 4:12 p.m., it reads 60 mi/h. Show that at some time between 4:00 and 4:12 p.m., the acceleration was exactly 100 mi/h².

**59. Rectilinear Motion**   Two stock cars start a race at the same time and finish in a tie. If $f_1(t)$ is the position of one car at time $t$ and $f_2(t)$ is the position of the second car at time $t$, show that at some time during the race they have the same velocity. [*Hint:* Set $f(t) = f_2(t) - f_1(t)$.]

**60. Rectilinear Motion**   Suppose $s = f(t)$ is the position of an object from the origin at time $t$. If the object is at a specific location at $t = a$, and returns to that location at $t = b$, then $f(a) = f(b)$. Show that there is at least one time $t = c$, $a < c < b$ for which $f'(c) = 0$. That is, show that there is a time $c$ when the velocity of the object is 0.

**61. Loaded Beam**   The vertical deflection $d$ (in feet), of a particular 5-foot-long loaded beam can be approximated by

$$d = d(x) = -\frac{1}{192}x^4 + \frac{25}{384}x^3 - \frac{25}{128}x^2$$

where $x$ (in feet) is the distance from one end of the beam.

(a) Verify that the function $d = d(x)$ satisfies the conditions of Rolle's Theorem on the interval $[0, 5]$.

(b) What does the result in (a) say about the ends of the beam?

(c) Find all numbers $c$ in $(0, 5)$ that satisfy the conclusion of Rolle's Theorem. Then find the deflection $d$ at each number $c$.

(d) Graph the function $d$ on the interval $[0, 5]$.

**62.** For the function $f(x) = x^4 - 2x^3 - 4x^2 + 7x + 3$:

(a) Find the critical numbers of $f$ rounded to three decimal places.

(b) Find the intervals where $f$ is increasing and decreasing.

**63. Rolle's Theorem**   Use Rolle's Theorem with the function $f(x) = (x - 1) \sin x$ on $[0, 1]$ to show that the equation $\tan x + x = 1$ has a solution in the interval $(0, 1)$.

**64. Rolle's Theorem**   Use Rolle's Theorem to show that the function $f(x) = x^3 - 2$ has exactly one real zero.

**65. Rolle's Theorem**   Use Rolle's Theorem to show that the function $f(x) = (x - 8)^3$ has exactly one real zero.

**66. Rolle's Theorem**   Without finding the derivative, show that if $f(x) = (x^2 - 4x + 3)(x^2 + x + 1)$, then $f'(x) = 0$ for at least one number between 1 and 3. Check by finding the derivative and using the Intermediate Value Theorem.

**67. Rolle's Theorem**   Consider $f(x) = |x|$ on the interval $[-1, 1]$. Here $f(1) = f(-1) = 1$ but there is no $c$ in the interval $(-1, 1)$ at which $f'(c) = 0$. Explain why this does not contradict Rolle's Theorem.

**68. Mean Value Theorem**   Consider $f(x) = x^{2/3}$ on the interval $[-1, 1]$. Verify that there is no $c$ in $(-1, 1)$ for which

$$f'(c) = \frac{f(1) - f(-1)}{1 - (-1)}$$

Explain why this does not contradict the Mean Value Theorem.

**69. Mean Value Theorem**   The Mean Value Theorem guarantees that there is a real number $N$ in the interval $(0, 1)$ for which $f'(N) = f(1) - f(0)$ if $f$ is continuous on the interval $[0, 1]$ and differentiable on the interval $(0, 1)$. Find $N$ if $f(x) = \sin^{-1} x$.

**70. Mean Value Theorem**   Show that when the Mean Value Theorem is applied to the function $f(x) = Ax^2 + Bx + C$ in the interval $[a, b]$, the number $c$ referred to in the theorem is the midpoint of the interval.

**71.** (a) Apply the Increasing/Decreasing Function Test to the function $f(x) = \sqrt{x}$. What do you conclude?

(b) Is $f$ increasing on the interval $[0, \infty)$? Explain.

**72.** Explain why the function $f(x) = ax^4 + bx^3 + cx^2 + dx + e$ must have a zero between 0 and 1 if

$$\frac{a}{5} + \frac{b}{4} + \frac{c}{3} + \frac{d}{2} + e = 0$$

**73. Put It Together**   If $f'(x)$ and $g'(x)$ exist and $f'(x) > g'(x)$ for all real $x$, then which of the following statements must be true about the graph of $y = f(x)$ and the graph of $y = g(x)$?

(a) They intersect exactly once.

(b) They intersect no more than once.

(c) They do not intersect.

(d) They could intersect more than once.

(e) They have a common tangent at each point of intersection.

**74.** Prove that there is no $k$ for which the function

$$f(x) = x^3 - 3x + k$$

has two distinct zeros in the interval $[0, 1]$.

**75.** Show that $e^x > x^2$ for all $x > 0$. (*Hint:* Show that $f(x) = e^x - x^2$ is an increasing function for $x > 0$.)

**76.** Show that $e^x > 1 + x$ for all $x > 0$.

**77.** Show that $0 < \ln x < x$ for $x > 1$.

**78.** Show that $\tan \theta \geq \theta$ for all $\theta$ in the open interval $\left(0, \frac{\pi}{2}\right)$.

**79.** Establish the identity $\sin^{-1} x + \cos^{-1} x = \frac{\pi}{2}$ by showing that the derivative of $y = \sin^{-1} x + \cos^{-1} x$ is 0. Then use the fact that when $x = 0$, then $y = \frac{\pi}{2}$.

**80.** Establish the identity $\tan^{-1} x + \cot^{-1} x = \frac{\pi}{2}$ by showing that the derivative of $y = \tan^{-1} x + \cot^{-1} x$ is 0. Then use the fact that when $x = 1$, then $y = \frac{\pi}{2}$.

**81.** Let $f$ be a function that is continuous on the closed interval $[a, b]$ and differentiable on the open interval $(a, b)$. If $f(x) = 0$ for three different numbers $x$ in $(a, b)$, show that there must be at least two numbers in $(a, b)$ at which $f'(x) = 0$.

**82. Proof of the Increasing/Decreasing Function Test**   Let $f$ be a function that is differentiable on an open interval $(a, b)$. Show that if $f'(x) < 0$ for all numbers in $(a, b)$, then $f$ is a decreasing function on $(a, b)$. (See the Corollary on p. 288.)

**83.** Suppose that the domain of $f$ is an open interval $(a, b)$ and $f'(x) > 0$ for all $x$ in the interval. Show that $f$ cannot have an extreme value on $(a, b)$.

## Challenge Problems

**84.** Use Rolle's Theorem to show that between any two real zeros of a polynomial function $f$, there is a real zero of its derivative function $f'$.

**85.** Find where the general cubic $f(x) = ax^3 + bx^2 + cx + d$ is increasing and where it is decreasing by considering cases depending on the value of $b^2 - 3ac$. [*Hint:* $f'(x)$ is a quadratic function; examine its discriminant.]

**86.** Explain why the function $f(x) = x^n + ax + b$, where $n$ is a positive even integer, has at most two distinct real zeros.

**87.** Explain why the function $f(x) = x^n + ax + b$, where $n$ is a positive odd integer, has at most three distinct real zeros.

**88.** Explain why the function $f(x) = x^n + ax^2 + b$, where $n$ is a positive odd integer, has at most three distinct real zeros.

**89.** Explain why the function $f(x) = x^n + ax^2 + b$, where $n$ is a positive even integer, has at most four distinct real zeros.

**90. Mean Value Theorem**   Use the Mean Value Theorem to verify that

$$\frac{1}{9} < \sqrt{66} - 8 < \frac{1}{8}$$

(*Hint:* Consider $f(x) = \sqrt{x}$ on the interval $[64, 66]$.)

**91.** Given $f(x) = \dfrac{ax^n + b}{cx^n + d}$, where $n \geq 2$ is a positive integer and $ad - bc \neq 0$, find the critical numbers and the intervals on which $f$ is increasing and decreasing.

**92.** Show that $x \leq \ln(1+x)^{1+x}$ for all $x > -1$.
[*Hint:* Consider $f(x) = -x + \ln(1+x)^{1+x}$.]

**93.** Show that $a \ln \dfrac{b}{a} \leq b - a < b \ln \dfrac{b}{a}$ if $0 < a < b$.

**94.** Show that for any positive integer $n$, $\dfrac{n}{n+1} < \ln\left(1 + \dfrac{1}{n}\right)^n < 1$.

**95.** It can be shown that

$$\frac{d}{dx} \cot^{-1} x = \frac{d}{dx} \tan^{-1} \frac{1}{x}, \quad x \neq 0$$

(See Section 3.3, Assess Your Understanding Problem 58.) You might infer from this that $\cot^{-1} x = \tan^{-1} \dfrac{1}{x} + C$ for all $x \neq 0$, where $C$ is a constant. Show, however, that

$$\cot^{-1} x = \begin{cases} \tan^{-1} \dfrac{1}{x} & \text{if } x > 0 \\[2mm] \tan^{-1} \dfrac{1}{x} + \pi & \text{if } x < 0 \end{cases}$$

What is the explanation of the incorrect inference?

# 4.4 Local Extrema and Concavity

**OBJECTIVES**  *When you finish this section, you should be able to:*

**1** Use the First Derivative Test to find local extrema (p. 294)

**2** Use the First Derivative Test with rectilinear motion (p. 297)

**3** Determine the concavity of a function (p. 298)

**4** Find inflection points (p. 300)

**5** Use the Second Derivative Test to find local extrema (p. 302)

So far we know that if a function $f$, defined on a closed interval, has a local maximum or a local minimum at a number $c$ in the open interval, then $c$ is a critical number. We are now ready to see how the derivative is used to determine whether a function $f$ has a local maximum, a local minimum, or neither at a critical number.

## 1 Use the First Derivative Test to Find Local Extrema

All local extreme values of a function $f$ occur at critical numbers. While the value of $f$ at each critical number is a candidate for being a local extreme value for $f$, not every critical number gives rise to a local extreme value.

How do we distinguish critical numbers that give rise to local extreme values from those that do not? And then how do we determine if a local extreme value is a local maximum value or a local minimum value? Figure 31 provides a clue. If you look from left to right along the graph of $f$, you see that the graph of $f$ is increasing to the left of $x_1$, where a local maximum occurs, and is decreasing to its right. The function $f$ is decreasing to the left of $x_2$, where a local minimum occurs, and is increasing to its right. So, knowing where a function $f$ increases and decreases enables us to find local maximum values and local minimum values.

The next theorem is usually referred to as the *First Derivative Test*, since it relies on information obtained from the first derivative of a function.

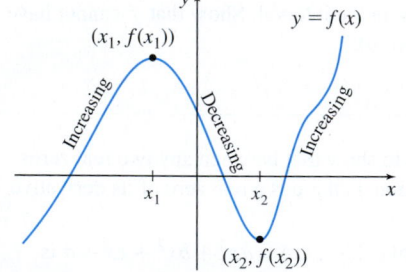

**Figure 31**

**THEOREM  First Derivative Test**

Let $f$ be a function that is continuous on an interval $I$. Suppose that $c$ is a critical number of $f$ and $(a, b)$ is an open interval in $I$ containing $c$:

- If $f'(x) > 0$ for $a < x < c$ and $f'(x) < 0$ for $c < x < b$, then $f(c)$ is a local maximum value.
- If $f'(x) < 0$ for $a < x < c$ and $f'(x) > 0$ for $c < x < b$, then $f(c)$ is a local minimum value.
- If $f'(x)$ has the same sign on both sides of $c$, then $f(c)$ is neither a local maximum value nor a local minimum value.

**IN WORDS**  If $c$ is a critical number of $f$ and if $f$ is increasing to the left of $c$ and decreasing to the right of $c$, then $f(c)$ is a local maximum value. If $f$ is decreasing to the left of $c$ and increasing to the right of $c$, then $f(c)$ is a local minimum value.

**Partial Proof** If $f'(x) > 0$ on the interval $(a, c)$, then $f$ is increasing on $(a, c)$. Also, if $f'(x) < 0$ on the interval $(c, b)$, then $f$ is decreasing on $(c, b)$. So, for all $x$ in $(a, b)$, $f(x) \leq f(c)$. That is, $f(c)$ is a local maximum value. See Figure 32(a). ∎

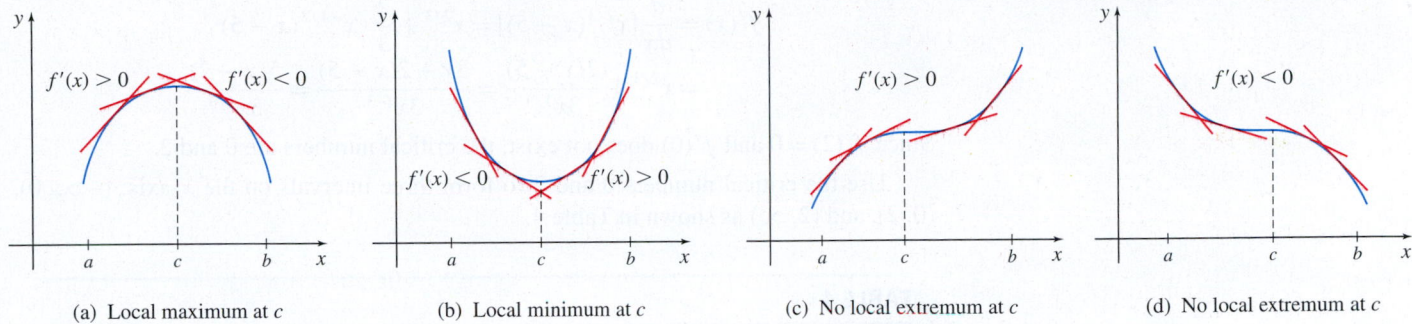

(a) Local maximum at $c$  (b) Local minimum at $c$  (c) No local extremum at $c$  (d) No local extremum at $c$

**Figure 32**

The proofs of the second and third bullets are left as exercises (see Problems 119 and 120). Figure 32(b) illustrates the behavior of $f'$ near a local minimum value. In Figures 32(c) and 32(d), $f'$ has the same sign on both sides of $c$, so $f$ has neither a local maximum nor a local minimum at $c$.

**EXAMPLE 1  Using the First Derivative Test to Find Local Extrema**

Find the local extrema of $f(x) = x^4 - 4x^3$.

**Solution** Since $f$ is a polynomial function, $f$ is continuous and differentiable at every real number. We begin by finding the critical numbers of $f$.

$$f'(x) = 4x^3 - 12x^2 = 4x^2(x - 3)$$

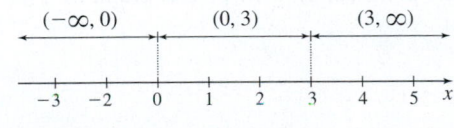

**Figure 33**

The critical numbers are 0 and 3. We use the critical numbers 0 and 3 to form three intervals on the $x$-axis, as shown in Figure 33. Then we determine where $f$ is increasing and where it is decreasing by determining the sign of $f'(x)$ in each interval. See Table 3.

**TABLE 3**

| Interval | Sign of $x^2$ | Sign of $x - 3$ | Sign of $f'(x) = 4x^2(x - 3)$ | Conclusion |
|---|---|---|---|---|
| $(-\infty, 0)$ | Positive (+) | Negative (−) | Negative (−) | $f$ is decreasing |
| $(0, 3)$ | Positive (+) | Negative (−) | Negative (−) | $f$ is decreasing |
| $(3, \infty)$ | Positive (+) | Positive (+) | Positive (+) | $f$ is increasing |

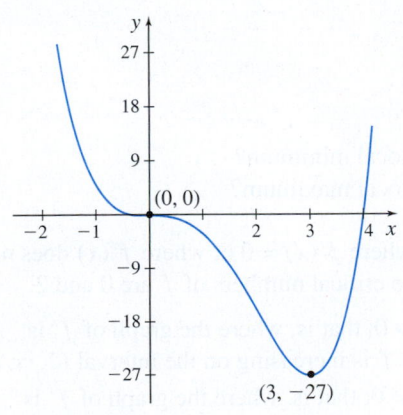

**Figure 34** $f(x) = x^4 - 4x^3$

Using the First Derivative Test, $f$ has neither a local maximum nor a local minimum at 0, and $f$ has a local minimum at 3. The local minimum value is $f(3) = -27$. ∎

The graph of $f$ is shown in Figure 34. Notice that the tangent line to the graph of $f$ is horizontal at the points $(0, 0)$ and $(3, -27)$.

**NOW WORK** Problem 13.

**EXAMPLE 2    Using the First Derivative Test to Find Local Extrema**

Find the local extrema of $f(x) = x^{2/3}(x - 5)$.

**Solution** The domain of $f$ is all real numbers and $f$ is continuous on its domain.

$$f'(x) = \frac{d}{dx}[x^{2/3}(x - 5)] = x^{2/3} + \frac{2}{3} \cdot x^{-1/3}(x - 5)$$

$$= x^{2/3} + \frac{2(x - 5)}{3x^{1/3}} = \frac{3x + 2(x - 5)}{3x^{1/3}} = \frac{5(x - 2)}{3x^{1/3}}$$

Since $f'(2) = 0$ and $f'(0)$ does not exist, the critical numbers are 0 and 2.

Use the critical numbers 0 and 2 to form three intervals on the $x$-axis: $(-\infty, 0)$, $(0, 2)$, and $(2, \infty)$ as shown in Table 4.

**TABLE 4**

| Interval | Sign of $x - 2$ | Sign of $x^{1/3}$ | Sign of $f'(x) = \dfrac{5(x - 2)}{3x^{1/3}}$ | Conclusion |
|---|---|---|---|---|
| $(-\infty, 0)$ | Negative ($-$) | Negative ($-$) | Positive ($+$) | $f$ is increasing |
| $(0, 2)$ | Negative ($-$) | Positive ($+$) | Negative ($-$) | $f$ is decreasing |
| $(2, \infty)$ | Positive ($+$) | Positive ($+$) | Positive ($+$) | $f$ is increasing |

The sign of $f'(x)$ on each interval tells us where $f$ is increasing and decreasing. By the First Derivative Test, $f$ has a local maximum at 0 and a local minimum at 2; $f(0) = 0$ is the local maximum value and $f(2) = -3\sqrt[3]{4}$ is the local minimum value. ∎

Since $f'(2) = 0$, the graph of $f$ has a horizontal tangent line at the point $(2, -3\sqrt[3]{4})$. Since $f'(0)$ does not exist and $\lim\limits_{x \to 0^-} f'(x) = \infty$ and $\lim\limits_{x \to 0^+} f'(x) = -\infty$, the graph of $f$ has a vertical tangent line at the point $(0, 0)$, a cusp. The graph of $f$ is shown in Figure 35.

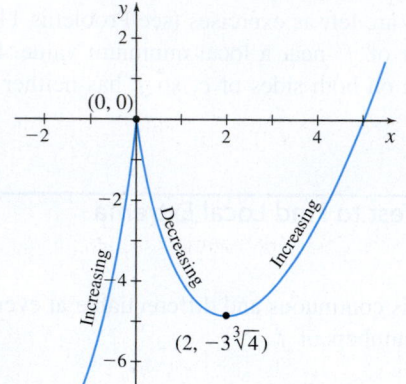

**Figure 35** $f(x) = x^{2/3}(x - 5)$

**NOW WORK** Problems **21** and **51(a)**.

**EXAMPLE 3    Obtaining Information About the Graph of $f$ from the Graph of Its Derivative**

A function $f$ is continuous for all real numbers. Figure 36 shows the graph of its derivative function $f'$.

**(a)** Determine the critical numbers of $f$.
**(b)** Where is $f$ increasing?
**(c)** Where is $f$ decreasing?
**(d)** At what numbers $x$, if any, does $f$ have a local minimum?
**(e)** At what numbers $x$, if any, does $f$ have a local maximum?

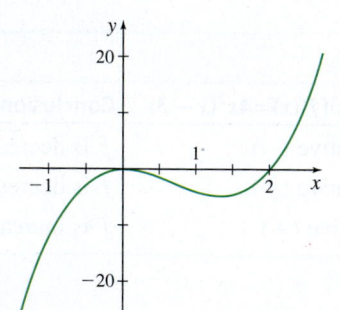

**Figure 36** $y = f'(x)$

**Solution (a)** The critical numbers of $f$ occur where $f'(x) = 0$ or where $f'(x)$ does not exist. Here $f'$ exists for all real numbers and the critical numbers of $f$ are 0 and 2.

**(b)** The function $f$ is increasing where $f'(x) > 0$; that is, where the graph of $f'$ is above the $x$-axis. This occurs for $x > 2$. So $f$ is increasing on the interval $(2, \infty)$.

**(c)** The function $f$ is decreasing where $f'(x) < 0$; that is, where the graph of $f'$ is below the $x$-axis. So $f$ is decreasing on the intervals $(-\infty, 0)$ and $(0, 2)$.

**(d)** $f$ is decreasing for $x < 2$ and increasing for $x > 2$, so by the First Derivative Test, $f$ has a local minimum at $x = 2$.

**(e)** $f$ has no local maximum. (Do you see why?) There is no point $(c, f(c))$ on the graph of $f$ where $f$ is increasing for $x < c$ and decreasing for $x > c$. ∎

**NOW WORK** Problem **35**.

## 2 Use the First Derivative Test with Rectilinear Motion

Increasing and decreasing functions can be used to investigate the motion of an object in rectilinear motion. Suppose the signed distance $s$ of an object from the origin at time $t$ is given by the position function $s = f(t)$. We assume the motion is along a horizontal line with the positive direction to the right. The velocity $v$ of the object is $v = \dfrac{ds}{dt}$.

- If $v = \dfrac{ds}{dt} > 0$, the position $s$ of the object is increasing with time $t$, and the object moves to the right.

- If $v = \dfrac{ds}{dt} < 0$, the position $s$ of the object is decreasing with time $t$, and the object moves to the left.

- If $v = \dfrac{ds}{dt} = 0$, the object is at rest.

This information, along with the First Derivative Test, can be used to find the local extreme values of $s = f(t)$ and to determine at what times $t$ the direction of the motion of the object changes.

Similarly, if the acceleration of the object, $a = \dfrac{dv}{dt} > 0$, then the velocity of the object is increasing, and if $a = \dfrac{dv}{dt} < 0$, then the velocity is decreasing. Again, this information, along with the First Derivative Test, is used to find the local extreme values of the velocity.

**EXAMPLE 4**  **Using the First Derivative Test with Rectilinear Motion**

An object moves along a horizontal line so that its signed distance $s$ from the origin at time $t \geq 0$, in seconds, is given by the position function

$$s = t^3 - 9t^2 + 15t + 3$$

**(a)** Determine the time intervals when the object is moving to the right.
**(b)** Determine the time intervals when the object is moving to the left.
**(c)** When does the object reverse direction?
**(d)** Draw a figure that illustrates the motion of the object.
**(e)** When is the velocity of the object increasing and when is it decreasing?
**(f)** Draw a figure that illustrates the velocity of the object.

**Solution**  To investigate the motion, we first find the velocity $v$.

$$v = \frac{ds}{dt} = \frac{d}{dt}(t^3 - 9t^2 + 15t + 3) = 3t^2 - 18t + 15 = 3(t^2 - 6t + 5) = 3(t - 1)(t - 5)$$

The critical numbers are 1 and 5. We use the critical numbers to form three intervals on the nonnegative $t$-axis: $(0, 1)$, $(1, 5)$, and $(5, \infty)$ as shown in Table 5.

**TABLE 5**

| Time Interval | Sign of $t - 1$ | Sign of $t - 5$ | Velocity, $v$ | Position $s$ | Motion of the Object |
|---|---|---|---|---|---|
| $(0, 1)$ | Negative $(-)$ | Negative $(-)$ | Positive $(+)$ | Increasing | To the right |
| $(1, 5)$ | Positive $(+)$ | Negative $(-)$ | Negative $(-)$ | Decreasing | To the left |
| $(5, \infty)$ | Positive $(+)$ | Positive $(+)$ | Positive $(+)$ | Increasing | To the right |

We use Table 5 to describe the motion of the object.

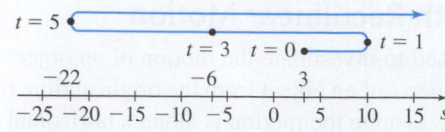

**Figure 37** $s(t) = t^3 - 9t^2 + 15t + 3$

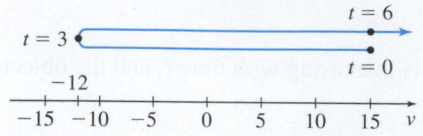

**Figure 38** $v(t) = 3(t-1)(t-5)$

(a) The object moves to the right for the first second and again after 5 s.

(b) The object moves to the left on the interval $(1, 5)$.

(c) The object reverses direction at $t = 1$ and $t = 5$.

(d) Figure 37 illustrates the motion of the object.

(e) To determine when the velocity increases or decreases, we find the acceleration.

$$a = \frac{dv}{dt} = \frac{d}{dt}(3t^2 - 18t + 15) = 6t - 18 = 6(t - 3)$$

Since $a < 0$ on the interval $(0, 3)$, the velocity $v$ decreases for the first 3 s. On the interval $(3, \infty)$, $a > 0$, so the velocity $v$ increases from 3 s onward.

(f) Figure 38 illustrates the velocity of the object. ∎

An interesting phenomenon occurs during the first three seconds. The velocity of the object is decreasing on the interval $(0, 3)$, as shown in Figure 38.

On the interval $(0, 3)$ the speed of the object is

$$|v(t)| = 3|t^2 - 6t + 5| = 3|(t - 1)(t - 5)|$$

• If $0 < t < 1$, $|v(t)| = 3(t^2 - 6t + 5)$ and $\dfrac{d|v|}{dt} = 6(t - 3) < 0$, so the speed is decreasing.

• If $1 < t < 3$, $|v(t)| = -3(t^2 - 6t + 5)$ and $\dfrac{d|v|}{dt} = -6(t - 3) > 0$, so the speed is increasing.

That is, the velocity of the object is decreasing from $t = 0$ to $t = 3$, but its speed is decreasing from $t = 0$ to $t = 1$ and is increasing from $t = 1$ to $t = 3$.

NOW WORK Problem 29.

## 3 Determine the Concavity of a Function

Figure 39 shows the graphs of two familiar functions: $y = x^2$, $x \geq 0$, and $y = \sqrt{x}$. Each graph starts at the origin, passes through the point $(1, 1)$, and is increasing. But there is a noticeable difference in their shapes. The graph of $y = x^2$ bends upward; it is *concave up*. The graph of $y = \sqrt{x}$ bends downward; it is *concave down*.

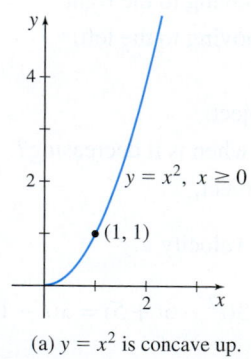

(a) $y = x^2$ is concave up.

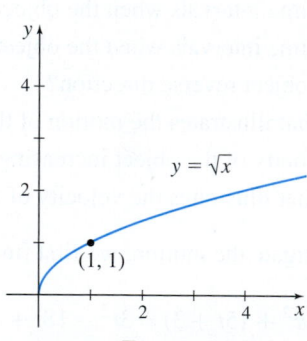

(b) $y = \sqrt{x}$ is concave down.

**Figure 39**

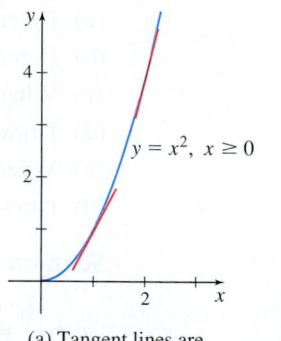

(a) Tangent lines are below the graph.

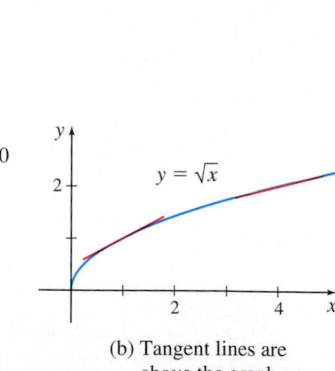

(b) Tangent lines are above the graph.

**Figure 40**

Suppose we draw tangent lines to the graphs of $y = x^2$ and $y = \sqrt{x}$, as shown in Figure 40. Notice that the graph of $y = x^2$ lies above all its tangent lines, and moving from left to right, the slopes of the tangent lines are increasing. That is, the derivative $y' = \dfrac{d}{dx}x^2$ is an increasing function.

On the other hand, the graph of $y = \sqrt{x}$ lies below all its tangent lines, and moving from left to right, the slopes of the tangent lines are decreasing. That is, the derivative $y' = \dfrac{d}{dx}\sqrt{x}$ is a decreasing function. This discussion leads to the following definition.

**DEFINITION** Concave Up; Concave Down

Let $f$ be a function that is continuous on a closed interval $[a, b]$ and differentiable on the open interval $(a, b)$.

- $f$ is **concave up** on $(a, b)$ if the graph of $f$ lies above each of its tangent lines throughout $(a, b)$.
- $f$ is **concave down** on $(a, b)$ if the graph of $f$ lies below each of its tangent lines throughout $(a, b)$.

We can formulate a test to determine where a function $f$ is concave up or concave down, provided $f''$ exists. Since $f''$ equals the rate of change of $f'$, it follows that if $f''(x) > 0$ on an open interval, then $f'$ is increasing on that interval, and if $f''(x) < 0$ on an open interval, then $f'$ is decreasing on that interval. As we observed in Figure 40, when $f'$ is increasing, the graph is concave up, and when $f'$ is decreasing, the graph is concave down. These observations lead to the *Test for Concavity*.

**THEOREM** Test for Concavity

Let $f$ be a function that is continuous on a closed interval $[a, b]$. Suppose $f'$ and $f''$ exist on the open interval $(a, b)$.

- If $f''(x) > 0$ on the interval $(a, b)$, then $f$ is concave up on $(a, b)$.
- If $f''(x) < 0$ on the interval $(a, b)$, then $f$ is concave down on $(a, b)$.

**NOTE** When $f'$ and $f''$ both exist on an open interval $(a, b)$, the function $f$ is called **twice differentiable** on $(a, b)$.

**Proof** Suppose $f''(x) > 0$ on the interval $(a, b)$, and $c$ is any fixed number in $(a, b)$. An equation of the tangent line to $f$ at the point $(c, f(c))$ is

$$y = f(c) + f'(c)(x - c)$$

We need to show that the graph of $f$ lies above each of its tangent lines for all $x$ in $(a, b)$. That is, we need to show that

$$f(x) \geq f(c) + f'(c)(x - c) \qquad \text{for all } x \text{ in } (a, b)$$

If $x = c$, then $f(x) = f(c)$ and we are finished.

If $x \neq c$, then by applying the Mean Value Theorem to the function $f$, there is a number $x_1$ between $c$ and $x$, for which

$$f'(x_1) = \frac{f(x) - f(c)}{x - c}$$

Now solve for $f(x)$:

$$f(x) = f(c) + f'(x_1)(x - c) \tag{1}$$

There are two possibilities: Either $c < x_1 < x$ or $x < x_1 < c$.

Suppose $c < x_1 < x$. Since $f''(x) > 0$ on the interval $(a, b)$, it follows that $f'$ is increasing on $(a, b)$. For $x_1 > c$, this means that $f'(x_1) > f'(c)$. As a result, from (1), we have

$$f(x) > f(c) + f'(c)(x - c)$$

That is, the graph of $f$ lies above each of its tangent lines to the right of $c$ in $(a, b)$.

Similarly, if $x < x_1 < c$, then $f(x) > f(c) + f'(c)(x - c)$.

In all cases, $f(x) \geq f(c) + f'(c)(x - c)$ so $f$ is concave up on $(a, b)$.

The proof that if $f''(x) < 0$, then $f$ is concave down is left as an exercise. See Problem 121. ∎

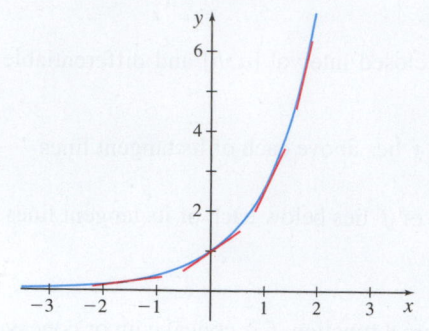

**Figure 41** $f(x) = e^x$ is concave up on its domain.

### EXAMPLE 5  Determining the Concavity of $f(x) = e^x$

Show that $f(x) = e^x$ is concave up on its domain.

**Solution** The domain of $f(x) = e^x$ is all real numbers. The first and second derivatives of $f$ are

$$f'(x) = e^x \qquad f''(x) = e^x$$

Since $f''(x) > 0$ for all real numbers, by the Test for Concavity, $f$ is concave up on its domain. ∎

Figure 41 shows the graph of $f(x) = e^x$ and a selection of tangent lines to the graph. Notice that for any $x$, the graph of $f$ lies above its tangent lines.

### CALC CLIP ▶ EXAMPLE 6  Finding Local Extrema and Determining Concavity

(a) Find any local extrema of the function $f(x) = x^3 - 6x^2 + 9x + 30$.

(b) Determine where $f(x) = x^3 - 6x^2 + 9x + 30$ is concave up and where it is concave down.

**Solution (a)** The first derivative of $f$ is

$$f'(x) = 3x^2 - 12x + 9 = 3(x - 1)(x - 3)$$

The function $f$ is a polynomial, so its critical numbers occur when $f'(x) = 0$. So, 1 and 3 are critical numbers of $f$. Now

- $f'(x) = 3(x - 1)(x - 3) > 0$ if $x < 1$ or $x > 3$.
- $f'(x) < 0$ if $1 < x < 3$.

So $f$ is increasing on $(-\infty, 1)$ and on $(3, \infty)$; $f$ is decreasing on $(1, 3)$.

At 1, $f$ has a local maximum, and at 3, $f$ has a local minimum. The local maximum value is $f(1) = 34$; the local minimum value is $f(3) = 30$.

(b) To determine concavity, we use the second derivative:

$$f''(x) = 6x - 12 = 6(x - 2)$$

Solve the inequalities $f''(x) < 0$ and $f''(x) > 0$ and use the Test for Concavity.

- $f''(x) < 0$     if $x < 2$,     so $f$ is concave down on $(-\infty, 2)$
- $f''(x) > 0$     if $x > 2$,     so $f$ is concave up on $(2, \infty)$ ∎

Figure 42 shows the graph of $f$. The point $(2, 32)$ where the concavity of $f$ changes is of special importance and is called an *inflection point*.

**DF** Figure 42 $f(x) = x^3 - 6x^2 + 9x + 30$

**NOW WORK** Problem 51(b).

## 4 Find Inflection Points

**DEFINITION** Inflection Point

Suppose $f$ is a function that is differentiable on an open interval $(a, b)$ containing $c$. If the concavity of $f$ changes at the point $(c, f(c))$, then $(c, f(c))$ is an **inflection point** of $f$.

If $(c, f(c))$ is an inflection point of $f$, then on one side of $c$ the slopes of the tangent lines are increasing (or decreasing), and on the other side of $c$ the slopes of the tangent lines are decreasing (or increasing). This means the derivative $f'$ must have a local maximum or a local minimum at $c$. In either case, it follows that $f''(c) = 0$ or $f''(c)$ does not exist.

### THEOREM  A Condition for an Inflection Point

Let $f$ denote a function that is differentiable on an open interval $(a, b)$ containing $c$. If $(c, f(c))$ is an inflection point of $f$, then either $f''(c) = 0$ or $f''$ does not exist at $c$.

Notice the wording in the theorem. If you *know* that $(c, f(c))$ is an inflection point of $f$, then the second derivative of $f$ at $c$ is 0 or does not exist. The converse is not necessarily true. In other words, a number at which $f''(x) = 0$ or at which $f''$ does not exist will not always identify an inflection point.

### Steps for Finding the Inflection Points of a Function $f$

*Step 1* Find all numbers in the domain of $f$ at which $f''(x) = 0$ or at which $f''$ does not exist.

*Step 2* Use the Test for Concavity to determine the concavity of $f$ on both sides of each of these numbers.

*Step 3* If the concavity changes, there is an inflection point; otherwise, there is no inflection point.

### EXAMPLE 7  Finding Inflection Points

Find the inflection points of $f(x) = x^{5/3}$.

**Solution** We follow the steps for finding an inflection point.

*Step 1* The domain of $f$ is all real numbers. The first and second derivatives of $f$ are

$$f'(x) = \frac{5}{3}x^{2/3} \qquad f''(x) = \frac{10}{9}x^{-1/3} = \frac{10}{9x^{1/3}}$$

The second derivative of $f$ does not exist when $x = 0$. So, $(0, 0)$ is a possible inflection point.

*Step 2* Now use the Test for Concavity.

• If $x < 0$  then  $f''(x) < 0$ so $f$ is concave down on $(-\infty, 0)$.
• If $x > 0$  then  $f''(x) > 0$ so $f$ is concave up on $(0, \infty)$.

*Step 3* Since the concavity of $f$ changes at 0, we conclude that $(0, 0)$ is an inflection point of $f$. ∎

Figure 43 shows the graph of $f$ with its inflection point $(0, 0)$ marked.

A change in concavity does not, of itself, guarantee an inflection point. For example, Figure 44 shows the graph of $f(x) = \frac{1}{x}$. The first and second derivatives are $f'(x) = -\frac{1}{x^2}$ and $f''(x) = \frac{2}{x^3}$. Then

$$f''(x) = \frac{2}{x^3} < 0 \quad \text{if } x < 0 \quad \text{and} \quad f''(x) = \frac{2}{x^3} > 0 \quad \text{if } x > 0$$

So, $f$ is concave down on $(-\infty, 0)$ and concave up on $(0, \infty)$, yet $f$ has no inflection point at $x = 0$ because 0 is not in the domain of $f$.

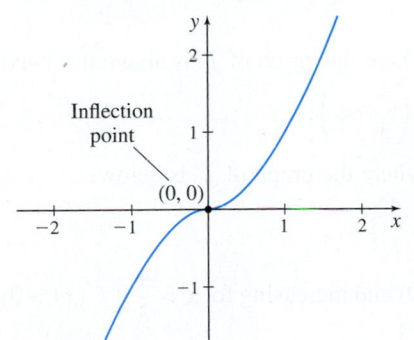

**Figure 43** $f(x) = x^{5/3}$

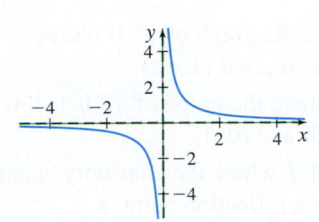

**Figure 44** $f(x) = \frac{1}{x}$

NOW WORK Problems 49 and 51(c).

**EXAMPLE 8** Obtaining Information About the Graph of $f$ from the Graphs of $f'$ and $f''$

A function $f$ is continuous for all real numbers. The graphs of $f'$ and $f''$ are given in Figure 45.

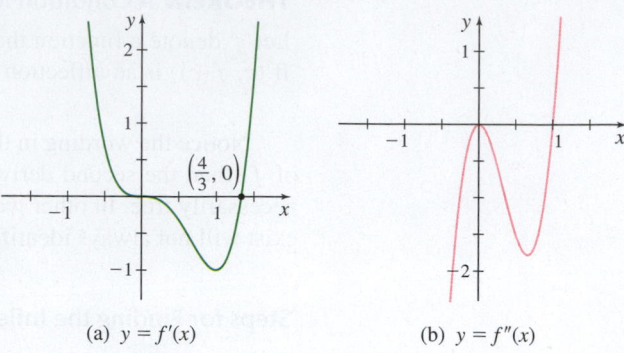

(a) $y = f'(x)$            (b) $y = f''(x)$

**Figure 45**

(a) Determine the critical numbers of $f$.

(b) Where is $f$ increasing?

(c) Where is $f$ decreasing?

(d) At what numbers $x$, if any, does $f$ have a local minimum?

(e) At what numbers $x$, if any, does $f$ have a local maximum?

(f) Where is $f$ concave up?

(g) Where is $f$ concave down?

(h) List any inflection points.

**Solution** (a) The critical numbers of $f$ occur where $f'(x) = 0$ or where $f'$ does not exist. See Figure 45(a). Since $f'$ exists for all real numbers and since $f'(0) = 0$ and $f'\left(\dfrac{4}{3}\right) = 0$, the only critical numbers are $0$ and $\dfrac{4}{3}$.

(b) $f$ is increasing where $f'(x) > 0$; that is, where the graph of $f'$ is above the $x$-axis. This occurs on the intervals $(-\infty, 0)$ and $\left(\dfrac{4}{3}, \infty\right)$.

(c) $f$ is decreasing where $f'(x) < 0$; that is, where the graph of $f'$ is below the $x$-axis. This occurs on the interval $\left(0, \dfrac{4}{3}\right)$.

(d) Since $f$ is decreasing for $x < \dfrac{4}{3}$ ($f'(x) < 0$) and increasing for $x > \dfrac{4}{3}$ ($f'(x) > 0$), the function $f$ has a local minimum at $x = \dfrac{4}{3}$.

(e) Since $f$ is increasing for $x < 0$ ($f'(x) > 0$) and decreasing for $x > 0$ ($f'(x) < 0$), the function $f$ has local maximum at $x = 0$.

(f) $f$ is concave up where $f''(x) > 0$; that is, where the graph of $f''$ is above the $x$-axis. See Figure 45(b). This occurs on the interval $(1, \infty)$.

(g) $f$ is concave down where $f''(x) < 0$; that is, where the graph of $f''$ is below the $x$-axis. This occurs on the intervals $(-\infty, 0)$ and $(0, 1)$.

(h) Inflection points occur at points on the graph of $f$ where the concavity changes. The concavity changes at $x = 1$ so $(1, f(1))$ is an inflection point. ∎

**NOW WORK** Problem 63.

## 5 Use the Second Derivative Test to Find Local Extrema

Suppose $c$ is a critical number of $f$ and $f'(c) = 0$. This means that the graph of $f$ has a horizontal tangent line at the point $(c, f(c))$. If $f''$ exists on an open interval containing $c$ and $f''(c) > 0$, then the graph of $f$ is concave up on the interval, as shown in Figure 46(a). Intuitively, it would seem that $f(c)$ is a local minimum value of $f$. If, on the other hand, $f''(c) < 0$, the graph of $f$ is concave down on the interval, and $f(c)$ would appear to be a local maximum value of $f$, as shown in Figure 46(b). The next theorem, known as the *Second Derivative Test*, confirms our intuition.

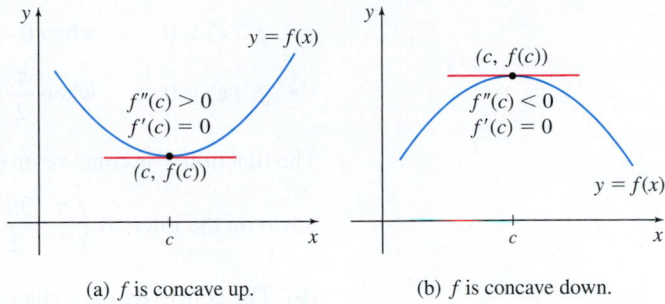

(a) $f$ is concave up.     (b) $f$ is concave down.

**Figure 46**

---

**THEOREM  Second Derivative Test**

Let $f$ be a function for which $f'$ and $f''$ exist on an open interval $(a, b)$. Suppose $c$ lies in $(a, b)$ and is a critical number of $f$:

- If $f''(c) < 0$, then $f(c)$ is a local maximum value.
- If $f''(c) > 0$, then $f(c)$ is a local minimum value.

---

**Proof**  Suppose $f''(c) < 0$. Then

$$f''(c) = \lim_{x \to c} \frac{f'(x) - f'(c)}{x - c} < 0$$

Now since $\lim_{x \to c} \dfrac{f'(x) - f'(c)}{x - c} < 0$, there is an open interval about $c$ for which

$$\frac{f'(x) - f'(c)}{x - c} < 0$$

everywhere in the interval, except possibly at $c$ itself (refer to Example 7, p. 145, in Section 1.6). Since $c$ is a critical number, $f'(c) = 0$, so

$$\frac{f'(x)}{x - c} < 0$$

For $x < c$ on this interval, $f'(x) > 0$, and for $x > c$ on this interval, $f'(x) < 0$. By the First Derivative Test, $f(c)$ is a local maximum value. ∎

In Problem 122, you are asked to prove that if $f''(c) > 0$, then $f(c)$ is a local minimum value.

---

**EXAMPLE 9  Using the Second Derivative Test to Identify Local Extrema**

(a) Determine where $f(x) = x - 2\cos x$, $0 \le x \le 2\pi$, is concave up and concave down.

(b) Find any inflection points.

(c) Use the Second Derivative Test to identify any local extreme values.

**Solution (a)** Since $f$ is continuous on the closed interval $[0, 2\pi]$ and $f'$ and $f''$ exist on the open interval $(0, 2\pi)$, we can use the Test for Concavity. The first and second derivatives of $f$ are

$$f'(x) = \frac{d}{dx}(x - 2\cos x) = 1 + 2\sin x \qquad \text{and} \qquad f''(x) = \frac{d}{dx}(1 + 2\sin x) = 2\cos x$$

To determine concavity, we solve the inequalities $f''(x) < 0$ and $f''(x) > 0$. Since $f''(x) = 2\cos x$, we have

- $f''(x) > 0$     when $0 < x < \dfrac{\pi}{2}$ and $\dfrac{3\pi}{2} < x < 2\pi$

- $f''(x) < 0$     when $\dfrac{\pi}{2} < x < \dfrac{3\pi}{2}$

The function $f$ is concave up on the intervals $\left(0, \dfrac{\pi}{2}\right)$ and $\left(\dfrac{3\pi}{2}, 2\pi\right)$, and $f$ is concave down on the interval $\left(\dfrac{\pi}{2}, \dfrac{3\pi}{2}\right)$.

**(b)** The concavity of $f$ changes at $\dfrac{\pi}{2}$ and $\dfrac{3\pi}{2}$, so the points $\left(\dfrac{\pi}{2}, \dfrac{\pi}{2}\right)$ and $\left(\dfrac{3\pi}{2}, \dfrac{3\pi}{2}\right)$ are inflection points of $f$.

**(c)** We find the critical numbers by solving the equation $f'(x) = 0$.

$$1 + 2\sin x = 0 \qquad 0 \le x \le 2\pi$$

$$\sin x = -\frac{1}{2}$$

$$x = \frac{7\pi}{6} \qquad \text{or} \qquad x = \frac{11\pi}{6}$$

Using the Second Derivative Test, we get

$$f''\left(\frac{7\pi}{6}\right) = 2\cos\left(\frac{7\pi}{6}\right) = -\sqrt{3} < 0$$

and

$$f''\left(\frac{11\pi}{6}\right) = 2\cos\left(\frac{11\pi}{6}\right) = \sqrt{3} > 0$$

So,

$$f\left(\frac{7\pi}{6}\right) = \frac{7\pi}{6} - 2\cos\frac{7\pi}{6} = \frac{7\pi}{6} + \sqrt{3} \approx 5.397$$

is a local maximum value, and

$$f\left(\frac{11\pi}{6}\right) = \frac{11\pi}{6} - 2\cos\frac{11\pi}{6} = \frac{11\pi}{6} - \sqrt{3} \approx 4.028$$

is a local minimum value. ∎

See Figure 47 for the graph of $f$.

Sometimes the Second Derivative Test cannot be used, and sometimes it gives no information.

- If the second derivative of the function does not exist at a critical number, the Second Derivative Test cannot be used.
- If the second derivative exists at a critical number, but equals 0 there, the Second Derivative Test gives no information.

In these situations, the First Derivative Test must be used to identify local extreme points.

An example is the function $f(x) = x^4$. Both $f'(x) = 4x^3$ and $f''(x) = 12x^2$ exist for all real numbers. Since $f'(0) = 0$, 0 is a critical number of $f$. But $f''(0) = 0$, so the Second Derivative Test gives no information about the behavior of $f$ at 0.

**NOW WORK** Problem 59.

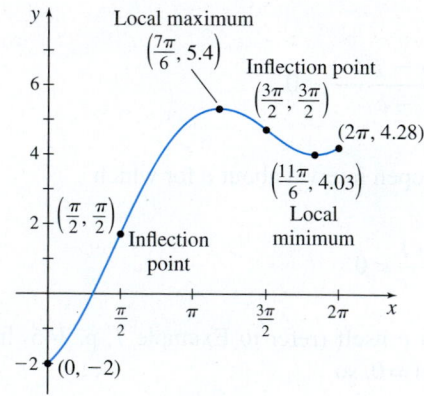

**Figure 47** $y = x - 2\cos x,\ 0 \le x \le 2\pi$

# Application: The Logistic Function in Business

**EXAMPLE 10** Analyzing Monthly Sales

Unit monthly sales $R$ of a new product over a period of time are expected to follow the logistic function

$$R = R(t) = \frac{20{,}000}{1 + 50e^{-t}} - \frac{20{,}000}{51} \qquad t \ge 0$$

where $t$ is measured in months.

(a) When are the monthly sales increasing? When are they decreasing?

(b) Find the rate of change of sales.

(c) When is the rate of change of sales $R'$ increasing? When is it decreasing?

(d) When is the rate of change of sales a maximum?

(e) Find any inflection points of $R$.

(f) Interpret the result found in (e) in the context of the problem.

**Solution** (a) We find $R'(t)$ and use the Increasing/Decreasing Function Test.

$$R'(t) = \frac{d}{dt}\left(\frac{20{,}000}{1 + 50e^{-t}} - \frac{20{,}000}{51}\right) = 20{,}000 \cdot \frac{50e^{-t}}{(1+50e^{-t})^2} = \frac{1{,}000{,}000e^{-t}}{(1+50e^{-t})^2}$$

Since $e^{-t} > 0$ for all $t \ge 0$, then $R'(t) > 0$ for $t \ge 0$. The sales function $R$ is an increasing function. So, monthly sales are always increasing.

(b) The rate of change of sales is given by the derivative $R'(t) = \dfrac{1{,}000{,}000e^{-t}}{(1+50e^{-t})^2}$, $t \ge 0$.

(c) Using the Increasing/Decreasing Function Test with $R'$, the rate of change of sales $R'$ is increasing when its derivative $R''(t) > 0$; $R'(t)$ is decreasing when $R''(t) < 0$.

$$R''(t) = \frac{d}{dt}R'(t) = 1{,}000{,}000\left[\frac{-e^{-t}(1+50e^{-t})^2 + 100e^{-2t}(1+50e^{-t})}{(1+50e^{-t})^4}\right]$$

$$= 1{,}000{,}000e^{-t}\left[\frac{-1 - 50e^{-t} + 100e^{-t}}{(1+50e^{-t})^3}\right] = \frac{1{,}000{,}000e^{-t}}{(1+50e^{-t})^3}(50e^{-t} - 1)$$

Since $e^{-t} > 0$ for all $t$, the sign of $R''$ depends on the sign of $50e^{-t} - 1$.

$$50e^{-t} - 1 > 0 \qquad\qquad 50e^{-t} - 1 < 0$$
$$50e^{-t} > 1 \qquad\qquad 50e^{-t} < 1$$
$$50 > e^t \qquad\qquad 50 < e^t$$
$$t < \ln 50 \qquad\qquad t > \ln 50$$

Since $R''(t) > 0$ for $t < \ln 50 \approx 3.912$ and $R''(t) < 0$ for $t > \ln 50 \approx 3.912$, the rate of change of sales is increasing for the first 3.9 months and is decreasing from 3.9 months on.

(d) The critical number of $R'$ is $\ln 50 \approx 3.912$. Using the First Derivative Test, the rate of change of sales is a maximum about 3.9 months after the product is introduced.

(e) Since $R''(t) > 0$ for $t < \ln 50$ and $R''(t) < 0$ for $t > \ln 50$, the point $(\ln 50, 9608)$ is the inflection point of $R$.

(f) The sales function $R$ is an increasing function, but at the inflection point $(\ln 50, 9608)$ the rate of change in sales begins to decrease. ∎

See Figure 48 for the graphs of $R$ and $R'$.

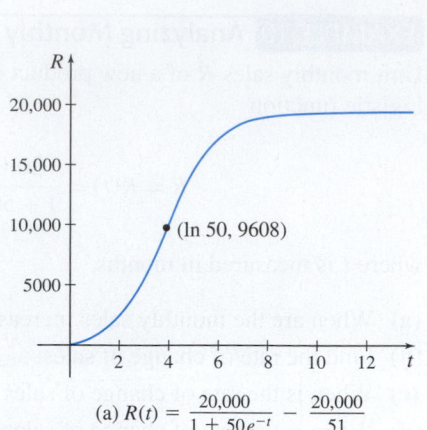

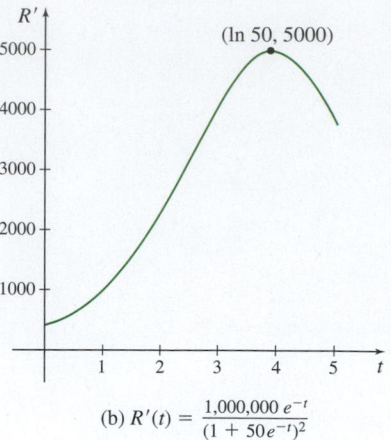

(a) $R(t) = \dfrac{20{,}000}{1 + 50e^{-t}} - \dfrac{20{,}000}{51}$

(b) $R'(t) = \dfrac{1{,}000{,}000\, e^{-t}}{(1 + 50e^{-t})^2}$

**Figure 48**

NOW WORK **Problem 93.**

## 4.4 Assess Your Understanding

### Concepts and Vocabulary

1. *True or False*  If a function $f$ is continuous on the interval $[a, b]$, differentiable on the interval $(a, b)$, and changes from an increasing function to a decreasing function at the point $(c, f(c))$, then $(c, f(c))$ is an inflection point of $f$.

2. *True or False*  Suppose $c$ is a critical number of $f$ and $(a, b)$ is an open interval containing $c$. If $f'(x)$ is positive on both sides of $c$, then $f(c)$ is a local maximum value.

3. *Multiple Choice*  Suppose a function $f$ is continuous on a closed interval $[a, b]$ and differentiable on the open interval $(a, b)$. If the graph of $f$ lies above each of its tangent lines on the interval $(a, b)$, then on $(a, b)$ $f$ is

   [(**a**) concave up (**b**) concave down (**c**) neither].

4. *Multiple Choice*  If the acceleration of an object in rectilinear motion is negative, then the velocity of the object is

   [(**a**) increasing (**b**) decreasing (**c**) neither].

5. *Multiple Choice*  Suppose $f$ is a function that is differentiable on an open interval containing $c$ and the concavity of $f$ changes at the point $(c, f(c))$. Then the point $(c, f(c))$ on the graph of $f$ is a(n)

   [(**a**) inflection point (**b**) critical point (**c**) both (**d**) neither].

6. *Multiple Choice*  Suppose a function $f$ is continuous on a closed interval $[a, b]$ and both $f'$ and $f''$ exist on the open interval $(a, b)$. If $f''(x) > 0$ on the interval $(a, b)$, then on $(a, b)$ $f$ is

   [(**a**) increasing (**b**) decreasing (**c**) concave up (**d**) concave down].

7. *True or False*  Suppose $f$ is a function for which $f'$ and $f''$ exist on an open interval $(a, b)$ and suppose $c$, $a < c < b$, is a critical number of $f$. If $f''(c) = 0$, then the Second Derivative Test cannot be used to determine if there is a local extremum at $c$.

8. *True or False*  Suppose a function $f$ is differentiable on the open interval $(a, b)$. If either $f''(c) = 0$ or $f''$ does not exist at the number $c$ in $(a, b)$, then $(c, f(c))$ is an inflection point of $f$.

### Skill Building

*In Problems 9–12, the graph of a function $f$ is given.*

(a) *Identify the points where each function has a local maximum value, a local minimum value, or an inflection point.*

(b) *Identify the intervals on which each function is increasing, decreasing, concave up, or concave down.*

**9.**

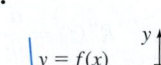
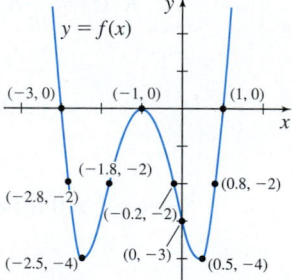

**10.**

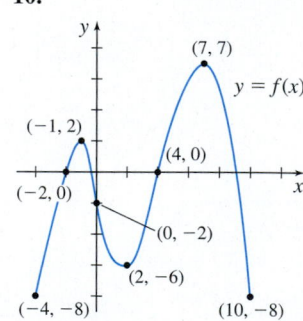

**11.**

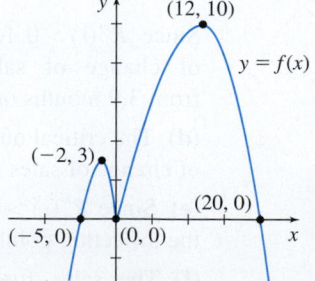

**12.**

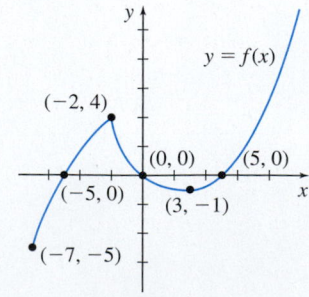

---

**1.** = NOW WORK problem          = Graphing technology recommended          **CAS** = Computer Algebra System recommended

*In Problems 13–26, for each function:*

*(a)*  *Find the critical numbers.*

*(b)*  *Use the First Derivative Test to find any local extrema.*

**13.**  $f(x) = x^3 - 6x^2 + 2$

**14.**  $f(x) = x^3 + 6x^2 + 12x + 1$

**15.**  $f(x) = 3x^4 - 4x^3$

**16.**  $h(x) = x^4 + 2x^3 - 3$

**17.**  $f(x) = (5 - 2x)e^x$

**18.**  $f(x) = (x - 8)e^x$

**19.**  $f(x) = x^{2/3} + x^{1/3}$

**20.**  $f(x) = \frac{1}{2}x^{2/3} - x^{1/3}$

**21.**  $g(x) = x^{2/3}(x^2 - 4)$

**22.**  $f(x) = x^{1/3}(x^2 - 9)$

**23.**  $f(x) = \dfrac{\ln x}{x^3}$

**24.**  $h(x) = \dfrac{\ln x}{\sqrt{x^3}}$

**25.**  $f(\theta) = \sin\theta - 2\cos\theta$

**26.**  $f(x) = x + 2\sin x$

*In Problems 27–34, an object in rectilinear motion moves along a horizontal line with the positive direction to the right. The position s of the object from the origin at time $t \geq 0$ (in seconds) is given by the function $s = s(t)$.*

*(a)*  *Determine the intervals during which the object moves to the right and the intervals during which it moves to the left.*

*(b)*  *When does the object reverse direction?*

*(c)*  *When is the velocity of the object increasing and when is it decreasing?*

*(d)*  *Draw a figure to illustrate the motion of the object.*

*(e)*  *Draw a figure to illustrate the velocity of the object.*

**27.**  $s = t^2 - 2t + 3$

**28.**  $s = 2t^2 + 8t - 7$

**29.**  $s = 2t^3 + 6t^2 - 18t + 1$

**30.**  $s = 3t^4 - 16t^3 + 24t^2$

**31.**  $s = 2t - \dfrac{6}{t}, \quad t > 0$

**32.**  $s = 3\sqrt{t} - \dfrac{1}{\sqrt{t}}, \quad t > 0$

**33.**  $s = 2\sin(3t), \quad 0 \leq t \leq \dfrac{2\pi}{3}$

**34.**  $s = 3\cos(\pi t), \quad 0 \leq t \leq 2$

*In Problems 35–38, the function f is continuous for all real numbers and the graph of its derivative function f′ is given.*

*(a)*  *Determine the critical numbers of f.*

*(b)*  *Where is f increasing?*

*(c)*  *Where is f decreasing?*

*(d)*  *At what numbers x, if any, does f have a local minimum?*

*(e)*  *At what numbers x, if any, does f have a local maximum?*

**35.**

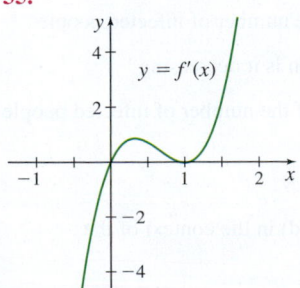

**36.**

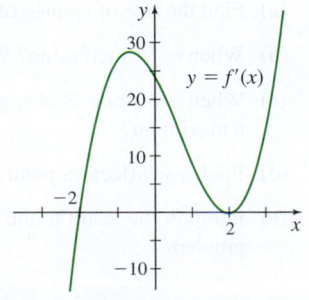

**37.**

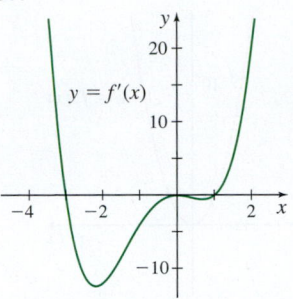

**38.**

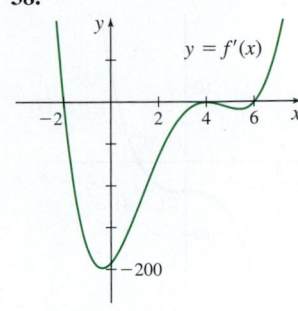

*In Problems 39–62:*

*(a)*  *Find the local extrema of f.*

*(b)*  *Determine the intervals on which f is concave up and on which it is concave down.*

*(c)*  *Find any points of inflection.*

**39.**  $f(x) = 2x^3 - 6x^2 + 6x - 3$

**40.**  $f(x) = 2x^3 + 9x^2 + 12x - 4$

**41.**  $f(x) = x^4 - 4x$

**42.**  $f(x) = x^4 + 4x$

**43.**  $f(x) = 5x^4 - x^5$

**44.**  $f(x) = 4x^6 + 6x^4$

**45.**  $f(x) = 3x^5 - 20x^3$

**46.**  $f(x) = 3x^5 + 5x^3$

**47.**  $f(x) = x^2 e^x$

**48.**  $f(x) = x^3 e^x$

**49.**  $f(x) = 6x^{4/3} - 3x^{1/3}$

**50.**  $f(x) = x^{2/3} - x^{1/3}$

**51.**  $f(x) = x^{2/3}(x^2 - 8)$

**52.**  $f(x) = x^{1/3}(x^2 - 2)$

**53.**  $f(x) = x^2 - \ln x$

**54.**  $f(x) = \ln x - x$

**55.**  $f(x) = \dfrac{x}{(1 + x^2)^{5/2}}$

**56.**  $f(x) = \dfrac{\sqrt{x}}{1 + x}$

**57.**  $f(x) = x^2\sqrt{1 - x^2}$

**58.**  $f(x) = x\sqrt{1 - x}$

**59.**  $f(x) = x - 2\sin x, \quad 0 \leq x \leq 2\pi$

**60.**  $f(x) = 2\cos^2 x - \sin^2 x, \quad 0 \leq x \leq 2\pi$

**61.**  $f(x) = \cosh x$

**62.**  $f(x) = \sinh x$

*In Problems 63–66, the function f is continuous for all real numbers and the graphs of f′ and f″ are given.*

*(a)*  *Determine the critical numbers of f.*

*(b)*  *Where is f increasing?*

*(c)*  *Where is f decreasing?*

*(d)*  *At what numbers x, if any, does f have a local minimum?*

*(e)*  *At what numbers x, if any, does f have a local maximum?*

*(f)*  *Where is f concave up?*

*(g)*  *Where is f concave down?*

*(h)*  *List any inflection points.*

**63.**

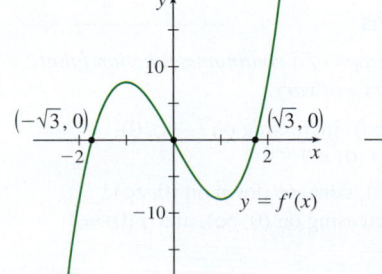
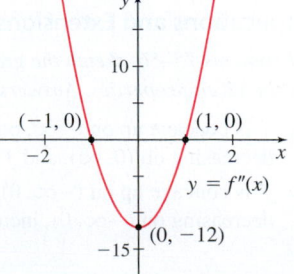

**64.**

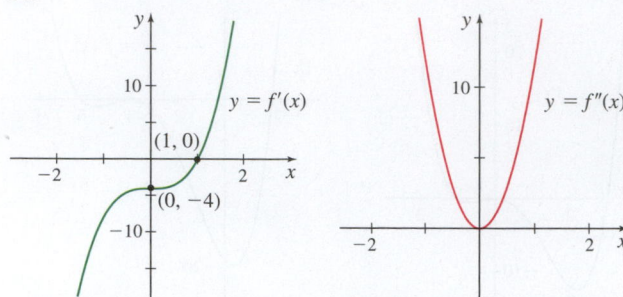

**65.**

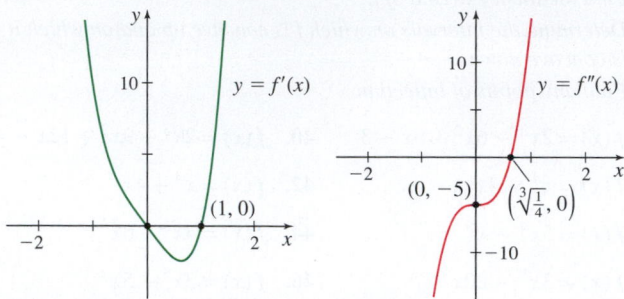

**66.**

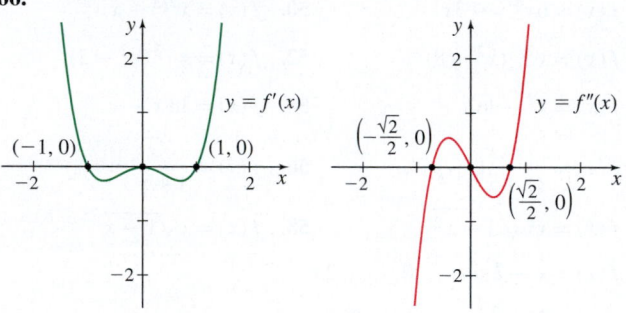

*In Problems 67–74, find the local extrema of each function f by:*

**(a)** *Using the First Derivative Test.*

**(b)** *Using the Second Derivative Test.*

**(c)** *Discuss which of the two tests you found easier.*

**67.** $f(x) = -2x^3 + 15x^2 - 36x + 7$

**68.** $f(x) = x^3 + 10x^2 + 25x - 25$

**69.** $f(x) = x^4 - 8x^2 - 5$    **70.** $f(x) = x^4 + 2x^2 + 2$

**71.** $f(x) = 3x^5 + 5x^4 + 1$    **72.** $f(x) = 60x^5 + 20x^3$

**73.** $f(x) = (x - 3)^2 e^x$    **74.** $f(x) = (x + 1)^2 e^{-x}$

## Applications and Extensions

*In Problems 75–86, sketch the graph of a continuous function f that has the given properties. Answers will vary.*

**75.** $f$ is concave up on $(-\infty, \infty)$, increasing on $(-\infty, 0)$, decreasing on $(0, \infty)$, and $f(0) = 1$.

**76.** $f$ is concave up on $(-\infty, 0)$, concave down on $(0, \infty)$, decreasing on $(-\infty, 0)$, increasing on $(0, \infty)$, and $f(0) = 1$.

**77.** $f$ is concave down on $(-\infty, 1)$, concave up on $(1, \infty)$, decreasing on $(-\infty, 0)$, increasing on $(0, \infty)$, $f(0) = 1$, and $f(1) = 2$.

**78.** $f$ is concave down on $(-\infty, 0)$, concave up on $(0, \infty)$, increasing on $(-\infty, \infty)$, and $f(0) = 1$ and $f(1) = 2$.

**79.** $f'(x) > 0$ if $x < 0$; $f'(x) < 0$ if $x > 0$; $f''(x) > 0$ if $x < 0$; $f''(x) > 0$ if $x > 0$ and $f(0) = 1$.

**80.** $f'(x) > 0$ if $x < 0$; $f'(x) < 0$ if $x > 0$; $f''(x) > 0$ if $x < 0$; $f''(x) < 0$ if $x > 0$ and $f(0) = 1$.

**81.** $f''(0) = 0$; $f'(0) = 0$; $f''(x) > 0$ if $x < 0$; $f''(x) > 0$ if $x > 0$ and $f(0) = 1$.

**82.** $f''(0) = 0$; $f'(x) > 0$ if $x \neq 0$; $f''(x) < 0$ if $x < 0$; $f''(x) > 0$ if $x > 0$ and $f(0) = 1$.

**83.** $f'(0) = 0$; $f'(x) < 0$ if $x \neq 0$; $f''(x) > 0$ if $x < 0$; $f''(x) < 0$ if $x > 0$; $f(0) = 1$.

**84.** $f''(0) = 0$; $f'(0) = \dfrac{1}{2}$; $f''(x) > 0$ if $x < 0$; $f''(x) < 0$ if $x > 0$ and $f(0) = 1$.

**85.** $f'(0)$ does not exist; $f''(x) > 0$ if $x < 0$; $f''(x) > 0$ if $x > 0$ and $f(0) = 1$.

**86.** $f'(0)$ does not exist; $f''(x) < 0$ if $x < 0$; $f''(x) > 0$ if $x > 0$ and $f(0) = 1$.

[CAS] *In Problems 87–90, for each function:*

**(a)** *Determine the intervals on which f is concave up and on which it is concave down.*

**(b)** *Find any points of inflection.*

**(c)** *Graph f and describe the behavior of f at each inflection point.*

**87.** $f(x) = e^{-(x-2)^2}$    **88.** $f(x) = x^2\sqrt{5 - x}$

**89.** $f(x) = \dfrac{2 - x}{2x^2 - 2x + 1}$    **90.** $f(x) = \dfrac{3x}{x^2 + 3x + 5}$

**91. Inflection Point** For the function $f(x) = ax^3 + bx^2$, find $a$ and $b$ so that the point $(1, 6)$ is an inflection point of $f$.

**92. Inflection Point** For the cubic polynomial function $f(x) = ax^3 + bx^2 + cx + d$, find $a$, $b$, $c$, and $d$ so that 0 is a critical number, $f(0) = 4$, and the point $(1, -2)$ is an inflection point of $f$.

**93. Public Health** In a town of 50,000 people, the number of people at time $t$ who have influenza is $N(t) = \dfrac{10{,}000}{1 + 9999e^{-t}}$, where $t$ is measured in days. Note that the flu is spread by the one person who has it at $t = 0$.

**(a)** Find the rate of change of the number of infected people.

**(b)** When is $N'$ increasing? When is it decreasing?

**(c)** When is the rate of change of the number of infected people a maximum?

**(d)** Find any inflection points of $N$.

**(e)** Interpret the result found in (d) in the context of the problem.

**94. Business**   The profit $P$ (in millions of dollars) generated by introducing a new technology is expected to follow the logistic function $P(t) = \dfrac{300}{1 + 50e^{-0.2t}}$, where $t$ is the number of years after its release.

(a)  When is annual profit increasing? When is it decreasing?

(b)  Find the rate of change in profit.

(c)  When is the rate of change in profit increasing? When is it decreasing?

(d)  When is the rate of change in profit a maximum?

(e)  Find any inflection points of $P$.

(f)  Interpret the result found in (e) in the context of the problem.

**95. Population Model**   The following data represent the population of the United States:

| Year | Population | Year | Population |
|------|------------|------|------------|
| 1900 | 76,212,168 | 1960 | 179,323,175 |
| 1910 | 92,228,486 | 1970 | 203,302,031 |
| 1920 | 106,021,537 | 1980 | 226,542,203 |
| 1930 | 123,202,624 | 1990 | 248,709,873 |
| 1940 | 132,164,569 | 2000 | 281,421,906 |
| 1950 | 151,325,798 | 2010 | 308,745,538 |

*Source*: U.S. Census Bureau.

An ecologist finds the data fit the logistic function

$$P(t) = \frac{762{,}176{,}717.8}{1 + 8.743e^{-0.0162t}}.$$

(a)  Draw a scatterplot of the data using the years since 1900 as the independent variable and population as the dependent variable.

(b)  Verify that $P$ is the logistic function of best fit.

(c)  Find the rate of change in population.

(d)  When is $P'$ increasing? When is it decreasing?

(e)  When is the rate of change in population a maximum?

(f)  Find any inflection points of $P$.

(g)  Interpret the result found in (f) in the context of the problem.

**96. Biology**   The amount of yeast biomass in a culture after $t$ hours is given in the table below.

| Time (in hours) | Yeast Biomass | Time (in hours) | Yeast Biomass | Time (in hours) | Yeast Biomass |
|-----------------|---------------|-----------------|---------------|-----------------|---------------|
| 0 | 9.6 | 7 | 257.3 | 13 | 629.4 |
| 1 | 18.3 | 8 | 350.7 | 14 | 640.8 |
| 2 | 29.0 | 9 | 441.0 | 15 | 651.1 |
| 3 | 47.2 | 10 | 513.3 | 16 | 655.9 |
| 4 | 71.1 | 11 | 559.7 | 17 | 659.6 |
| 5 | 119.1 | 12 | 594.8 | 18 | 661.8 |
| 6 | 174.6 | | | | |

*Source*: Tor Carlson, *Uber Geschwindigkeit und Grosse der Hefevermehrung in Wurrze, Biochemische Zeitschrift*, Bd. 57, 1913, pp. 313–334.

The logistic function $y = \dfrac{663.0}{1 + 71.6e^{-0.5470t}}$, where $t$ is time, models the data.

(a)  Draw a scatterplot of the data using time $t$ as the independent variable.

(b)  Verify that $y$ is the logistic function of best fit.

(c)  Find the rate of change in biomass.

(d)  When is $y'$ increasing? When is it decreasing?

(e)  When is the rate of change in the biomass a maximum?

(f)  Find any inflection points of $y$.

(g)  Interpret the result found in (f) in the context of the problem.

**97. U.S. Budget**   The United States budget documents the amount of money (revenue) the federal government takes in (through taxes, for example) and the amount (expenses) it pays out (for social programs, defense, etc.). When revenue exceeds expenses, we say there is a **budget surplus**, and when expenses exceed revenue, there is a **budget deficit**. The function

$$B = B(t) = -12.8t^3 + 163.4t^2 - 614.0t + 390.6$$

where $0 \le t \le 12$ approximates revenue minus expenses for the years 2000 to 2012, with $t = 0$ representing the year 2000 and $B$ in billions of dollars.

(a)  Find all the local extrema of $B$. (Round the answers to two decimal places.)

(b)  Do the local extreme values represent a budget surplus or a budget deficit?

(c)  Find the intervals on which $B$ is concave up or concave down. Identify any inflection points of $B$.

(d)  What does the concavity on either side of the inflection point(s) indicate about the rate of change of the budget? Is it increasing at an increasing rate? Increasing at a decreasing rate?

(e)  Graph the function $B$. Given that the budget for 2015 was 3.9 trillion dollars, does $B$ seem to be an accurate predictor for the budget for the years 2015 and beyond?

**98.** If $f(x) = ax^3 + bx^2 + cx + d$, $a \ne 0$, how does the quantity $b^2 - 3ac$ determine the number of potential local extrema?

**99.** If $f(x) = ax^3 + bx^2 + cx + d$, $a \ne 0$, find $a$, $b$, $c$, and $d$ so that $f$ has a local minimum at 0, a local maximum at 4, and the graph contains the points $(0, 5)$ and $(4, 33)$.

**100.** Find the local minimum of the function

$$f(x) = \frac{2}{x} + \frac{8}{1 - x}, \quad 0 < x < 1.$$

**101.** Find the local extrema and the inflection points of $y = \sqrt{3}\sin x + \cos x$, $0 \le x \le 2\pi$.

**102.** If $x > 0$ and $n > 1$, can the expression $x^n - n(x - 1) - 1$ ever be negative?

**103.** Why must the First Derivative Test be used to find the local extreme values of the function $f(x) = x^{2/3}$?

**104. Put It Together**   Which of the following is true of the function

$$f(x) = x^2 + e^{-2x}$$

(a) $f$ is increasing  (b) $f$ is decreasing  (c) $f$ is discontinuous at 0

(d) $f$ is concave up  (e) $f$ is concave down

**105. Put It Together** If a function $f$ is continuous for all $x$ and if $f$ has a local maximum at $(-1, 4)$ and a local minimum at $(3, -2)$, which of the following statements must be true?

(a) The graph of $f$ has an inflection point somewhere between $x = -1$ and $x = 3$.

(b) $f'(-1) = 0$.

(c) The graph of $f$ has a horizontal asymptote.

(d) The graph of $f$ has a horizontal tangent line at $x = 3$.

(e) The graph of $f$ intersects both axes.

**106. Vertex of a Parabola** If $f(x) = ax^2 + bx + c$, $a \neq 0$, prove that $f$ has a local maximum at $-\dfrac{b}{2a}$ if $a < 0$ and has a local minimum at $-\dfrac{b}{2a}$ if $a > 0$.

**107.** Show that $\sin x \leq x$, $0 \leq x \leq 2\pi$. (*Hint:* Let $f(x) = x - \sin x$.)

**108.** Show that $1 - \dfrac{x^2}{2} \leq \cos x$, $0 \leq x \leq 2\pi$.

(*Hint:* Use the result of Problem 107.)

**109.** Show that $2\sqrt{x} > 3 - \dfrac{1}{x}$, for $x > 1$.

**110.** Use calculus to show that $x^2 - 8x + 21 > 0$ for all $x$.

**111.** Use calculus to show that $3x^4 - 4x^3 - 12x^2 + 40 > 0$ for all $x$.

**112.** Show that the function $f(x) = ax^2 + bx + c$, $a \neq 0$, has no inflection points. For what values of $a$ is $f$ concave up? For what values of $a$ is $f$ concave down?

**113.** Show that every polynomial function of degree 3

$f(x) = ax^3 + bx^2 + cx + d$ has exactly one inflection point.

**114.** Prove that a polynomial of degree $n \geq 3$ has at most $(n - 1)$ critical numbers and at most $(n - 2)$ inflection points.

**115.** Show that the function $f(x) = (x - a)^n$, $a$ a constant, has exactly one inflection point if $n \geq 3$ is an odd integer.

**116.** Show that the function $f(x) = (x - a)^n$, $a$ a constant, has no inflection point if $n \geq 2$ is an even integer.

**117.** Show that the function $f(x) = \dfrac{ax + b}{ax + d}$ has no critical points and no inflection points.

**118. (a)** Show that the second derivative of the logistic function

$$f(t) = \frac{c}{1 + ae^{-bt}} \qquad a > 0, c > 0$$

is

$$f''(t) = ab^2 ce^{-bt}(ae^{-bt} - 1)$$

**(b)** Show that the logistic function $f$ has an inflection point at

$$\left( \frac{\ln a}{b}, \frac{c}{2} \right)$$

**119. First Derivative Test Proof** Let $f$ be a function that is continuous on an interval $I$. Suppose $c$ is a critical number of $f$ and $(a, b)$ is an open interval in $I$ containing $c$. Prove that if $f'(x) < 0$ for $a < x < c$ and $f'(x) > 0$ for $c < x < b$, then $f(c)$ is a local minimum value.

**120. First Derivative Test Proof** Let $f$ be a function that is continuous on an interval $I$. Suppose $c$ is a critical number of $f$ and $(a, b)$ is an open interval in $I$ containing $c$. Prove that if $f'(x)$ has the same sign on both sides of $c$, then $f(c)$ is neither a local maximum value nor a local minimum value.

**121. Test of Concavity Proof** Let $f$ be a function that is continuous on a closed interval $[a, b]$. Suppose $f'$ and $f''$ exist on the open interval $(a, b)$. Prove that if $f''(x) < 0$ on the interval $(a, b)$, then $f$ is concave down on $(a, b)$.

**122. Second Derivative Test Proof** Let $f$ be a function for which $f'$ and $f''$ exist on an open interval $(a, b)$. Suppose $c$ lies in $(a, b)$ and is a critical number of $f$. Prove that if $f''(c) > 0$, then $f(c)$ is a local minimum value.

## Challenge Problems

**123.** Find the inflection point of $y = (x + 1) \tan^{-1} x$.

**124. Bernoulli's Inequality** Prove

$(1 + x)^n > 1 + nx$ for $x > -1$, $x \neq 0$, and $n > 1$.

(*Hint:* Let $f(x) = (1 + x)^n - (1 + nx)$.)

# 4.5 Indeterminate Forms and L'Hôpital's Rule

**OBJECTIVES** *When you finish this section, you should be able to:*

**1** Identify indeterminate forms of the type $\dfrac{0}{0}$ and $\dfrac{\infty}{\infty}$ (p. 310)

**2** Use L'Hôpital's Rule to find a limit (p. 312)

**3** Find the limit of an indeterminate form of the type $0 \cdot \infty$, $\infty - \infty$, $0^0$, $1^\infty$, or $\infty^0$ (p. 315)

NEED TO REVIEW? The limit of a quotient is discussed in Section 1.2, pp. 96–98.

In this section, we reexamine the limit of a quotient and explore what we can do when previous strategies cannot be used. The discussion here will prove helpful in the next section.

## 1 Identify Indeterminate Forms of the Type $\dfrac{0}{0}$ and $\dfrac{\infty}{\infty}$

To find $\displaystyle\lim_{x \to c} \frac{f(x)}{g(x)}$ we usually first try to use the Limit of a Quotient:

$$\lim_{x \to c} \frac{f(x)}{g(x)} = \frac{\displaystyle\lim_{x \to c} f(x)}{\displaystyle\lim_{x \to c} g(x)} \tag{1}$$

But the quotient property cannot always be used. For example, to find $\lim\limits_{x\to 2}\dfrac{x^2-4}{x-2}$, we cannot use equation (1), since the numerator and the denominator each approach 0 (resulting in the form $\dfrac{0}{0}$). Instead, we use algebra and obtain

$$\lim_{x\to 2}\frac{x^2-4}{x-2}=\lim_{x\to 2}\frac{(x-2)(x+2)}{x-2}=\lim_{x\to 2}(x+2)=4$$

**NEED TO REVIEW?** Limits at infinity and infinite limits are discussed in Section 1.5, pp. 128–135.

To find $\lim\limits_{x\to\infty}\dfrac{3x-2}{x^2+5}$, we cannot use equation (1), since the numerator and the denominator each become unbounded (resulting in the form $\dfrac{\infty}{\infty}$). Instead, we use the end behavior of polynomial functions. Then

$$\lim_{x\to\infty}\frac{3x-2}{x^2+5}=\lim_{x\to\infty}\frac{3}{x}=0$$

As a third example, to find $\lim\limits_{x\to 0}\dfrac{\sin x}{x}$, we cannot use equation (1) since it leads to the form $\dfrac{0}{0}$. Instead, we use a geometric argument (the Squeeze Theorem) to show that

$$\lim_{x\to 0}\frac{\sin x}{x}=1$$

**NOTE** The word "indeterminate" conveys the idea that the limit cannot be found without additional work.

When $\lim\limits_{x\to c}\dfrac{f(x)}{g(x)}=\dfrac{\lim\limits_{x\to c}f(x)}{\lim\limits_{x\to c}g(x)}$ leads to the form $\dfrac{0}{0}$ or $\dfrac{\infty}{\infty}$, we say that $\dfrac{f}{g}$ is an *indeterminate form at c.* Then depending on the functions $f$ and $g$, $\lim\limits_{x\to c}\dfrac{f(x)}{g(x)}$, if it exists, could equal 0, $\infty$, or some nonzero real number.

> **DEFINITION** Indeterminate Form at $c$ of the Type $\dfrac{0}{0}$ or the Type $\dfrac{\infty}{\infty}$
>
> If the functions $f$ and $g$ are each defined in an open interval containing the number $c$, except possibly at $c$, then the quotient $\dfrac{f(x)}{g(x)}$ is called an **indeterminate form at $c$ of the type $\dfrac{0}{0}$** if
>
> $$\lim_{x\to c}f(x)=0 \qquad \text{and} \qquad \lim_{x\to c}g(x)=0$$
>
> and an **indeterminate form at $c$ of the type $\dfrac{\infty}{\infty}$** if
>
> $$\lim_{x\to c}f(x)=\pm\infty \qquad \text{and} \qquad \lim_{x\to c}g(x)=\pm\infty$$

**NOTE** $\dfrac{0}{0}$ and $\dfrac{\infty}{\infty}$ are symbols used to denote an indeterminate form.

These definitions also hold for limits at infinity.

---

**EXAMPLE 1**   Identifying an Indeterminate Form of the Type $\dfrac{0}{0}$ or $\dfrac{\infty}{\infty}$

(a) $\dfrac{\cos(3x)-1}{2x}$ is an indeterminate form at 0 of the type $\dfrac{0}{0}$ since

$$\lim_{x\to 0}[\cos(3x)-1]=0 \qquad \text{and} \qquad \lim_{x\to 0}(2x)=0$$

(b) $\dfrac{x-1}{x^2+2x-3}$ is an indeterminate form at 1 of the type $\dfrac{0}{0}$ since

$$\lim_{x\to 1}(x-1)=0 \qquad \text{and} \qquad \lim_{x\to 1}(x^2+2x-3)=0$$

**(c)** $\dfrac{x^2 - 2}{x - 3}$ is not an indeterminate form at 3 of the type $\dfrac{0}{0}$ since $\displaystyle\lim_{x \to 3}(x^2 - 2) \neq 0$.

**(d)** $\dfrac{x^2}{e^x}$ is an indeterminate form at $\infty$ of the type $\dfrac{\infty}{\infty}$ since

$$\lim_{x \to \infty} x^2 = \infty \qquad \text{and} \qquad \lim_{x \to \infty} e^x = \infty \qquad \blacksquare$$

**NOW WORK** Problem 7.

## 2 Use L'Hôpital's Rule to Find a Limit

A theorem, named after the French mathematician Guillaume François de L'Hôpital (pronounced "low-pee-tal"), provides a method for finding the limit of an indeterminate form.

> **THEOREM** L'Hôpital's Rule
>
> Suppose the functions $f$ and $g$ are differentiable on an open interval $I$ containing the number $c$, except possibly at $c$, and $g'(x) \neq 0$ for all $x \neq c$ in $I$. Let $L$ denote either a real number or $\pm\infty$, and suppose $\dfrac{f(x)}{g(x)}$ is an indeterminate form at $c$ of the type $\dfrac{0}{0}$ or $\dfrac{\infty}{\infty}$. If $\displaystyle\lim_{x \to c} \dfrac{f'(x)}{g'(x)} = L$, then $\displaystyle\lim_{x \to c} \dfrac{f(x)}{g(x)} = L$.

L'Hôpital's Rule is also valid for limits at infinity and one-sided limits. A partial proof of L'Hôpital's Rule is given in Appendix B. A limited proof is given here that assumes $f'$ and $g'$ are continuous at $c$ and $g'(c) \neq 0$.

**Proof** Suppose $\displaystyle\lim_{x \to c} f(x) = 0$ and $\displaystyle\lim_{x \to c} g(x) = 0$. Then $f(c) = 0$ and $g(c) = 0$. Since both $f'$ and $g'$ are continuous at $c$ and $g'(c) \neq 0$, we have

$$\lim_{x \to c} \frac{f'(x)}{g'(x)} \underset{\substack{\uparrow \\ \text{Quotient} \\ \text{Property}}}{=} \frac{\displaystyle\lim_{x \to c} f'(x)}{\displaystyle\lim_{x \to c} g'(x)} \underset{\substack{\uparrow \\ f',\, g' \\ \text{continuous}}}{=} \frac{f'(c)}{g'(c)} = \frac{\displaystyle\lim_{x \to c} \dfrac{f(x) - f(c)}{x - c}}{\displaystyle\lim_{x \to c} \dfrac{g(x) - g(c)}{x - c}}$$

$$= \lim_{x \to c} \frac{\dfrac{f(x) - f(c)}{x - c}}{\dfrac{g(x) - g(c)}{x - c}} = \lim_{x \to c} \frac{f(x) - f(c)}{g(x) - g(c)} \underset{\substack{\uparrow \\ f(c) = 0 \\ g(c) = 0}}{=} \lim_{x \to c} \frac{f(x)}{g(x)} \qquad \blacksquare$$

**ORIGINS** Guillaume François de L'Hôpital (1661–1704) was a French nobleman. When he was 30, he hired Johann Bernoulli to tutor him in calculus. Several years later, he entered into a deal with Bernoulli. L'Hôpital paid Bernoulli an annual sum for Bernoulli to share his mathematical discoveries with him but no one else. In 1696 L'Hôpital published the first textbook on differential calculus. It was immensely popular, the last edition being published in 1781 (seventy-seven years after L'Hôpital's death). The book included the rule we study here. After L'Hôpital's death, Johann Bernoulli made his deal with L'Hôpital public and claimed that L'Hôpital's textbook was his own material. His position was dismissed because he often made such claims. However, in 1921, the manuscripts of Bernoulli's lectures to L'Hôpital were found, showing L'Hôpital's calculus book was indeed largely Johann Bernoulli's work.

**IN WORDS** If $\displaystyle\lim_{x \to c} \dfrac{f(x)}{g(x)}$ leads to $\dfrac{0}{0}$ or $\dfrac{\infty}{\infty}$, L'Hôpital's Rule states that the limit of the quotient equals the limit of the quotient of their derivatives.

### Steps for Finding a Limit Using L'Hôpital's Rule

**Step 1** Check that $\dfrac{f}{g}$ is an indeterminate form at $c$ of the type $\dfrac{0}{0}$ or $\dfrac{\infty}{\infty}$.

If it is not, do not use L'Hôpital's Rule.

**Step 2** Differentiate $f$ and $g$ separately.

**Step 3** Find $\displaystyle\lim_{x \to c} \dfrac{f'(x)}{g'(x)}$. This limit is equal to $\displaystyle\lim_{x \to c} \dfrac{f(x)}{g(x)}$, provided the limit is a number or $\infty$ or $-\infty$.

**Step 4** If $\dfrac{f'}{g'}$ is an indeterminate form at $c$, repeat the process.

**EXAMPLE 2**  **Using L'Hôpital's Rule to Find a Limit**

Find $\lim\limits_{x \to 0} \dfrac{\tan x}{6x}$.

**Solution** We follow the steps for finding a limit using L'Hôpital's Rule.

***Step 1*** Since $\lim\limits_{x \to 0} \tan x = 0$ and $\lim\limits_{x \to 0} (6x) = 0$, the quotient $\dfrac{\tan x}{6x}$ is an indeterminate form at 0 of the type $\dfrac{0}{0}$.

***Step 2*** $\dfrac{d}{dx} \tan x = \sec^2 x$ and $\dfrac{d}{dx}(6x) = 6$.

***Step 3*** $\lim\limits_{x \to 0} \dfrac{\dfrac{d}{dx} \tan x}{\dfrac{d}{dx}(6x)} = \lim\limits_{x \to 0} \dfrac{\sec^2 x}{6} = \dfrac{1}{6}$.

It follows from L'Hôpital's Rule that $\lim\limits_{x \to 0} \dfrac{\tan x}{6x} = \dfrac{1}{6}$. ∎

**CAUTION** When using L'Hôpital's Rule, we find the derivative of the numerator and the derivative of the denominator separately. Be sure not to find the derivative of the quotient. Also, using L'Hôpital's Rule to find a limit of an expression that is not an indeterminate form may result in an incorrect answer.

In the solution of Example 2, we were careful to determine that the limit of the ratio of the derivatives, that is, $\lim\limits_{x \to 0} \dfrac{\sec^2 x}{6}$, existed or became infinite before using L'Hôpital's Rule. However, the usual practice is to combine Step 2 and Step 3 as follows:

$$\lim\limits_{x \to 0} \dfrac{\tan x}{6x} = \lim\limits_{x \to 0} \dfrac{\dfrac{d}{dx} \tan x}{\dfrac{d}{dx}(6x)} = \lim\limits_{x \to 0} \dfrac{\sec^2 x}{6} = \dfrac{1}{6}$$

**NOW WORK** Problem 29.

At times, it is necessary to use L'Hôpital's Rule more than once.

**EXAMPLE 3**  **Using L'Hôpital's Rule to Find a Limit**

Find $\lim\limits_{x \to 0} \dfrac{\sin x - x}{x^2}$.

**Solution** Use the steps for finding a limit using L'Hôpital's Rule.

***Step 1*** Since $\lim\limits_{x \to 0} (\sin x - x) = 0$ and $\lim\limits_{x \to 0} x^2 = 0$, the expression $\dfrac{\sin x - x}{x^2}$ is an indeterminate form at 0 of the type $\dfrac{0}{0}$.

***Steps 2 and 3*** Use L'Hôpital's Rule.

$$\lim\limits_{x \to 0} \dfrac{\sin x - x}{x^2} \underset{\substack{\uparrow \\ \text{L'Hôpital's} \\ \text{Rule}}}{=} \lim\limits_{x \to 0} \dfrac{\dfrac{d}{dx}(\sin x - x)}{\dfrac{d}{dx}x^2} = \lim\limits_{x \to 0} \dfrac{\cos x - 1}{2x} = \dfrac{1}{2} \lim\limits_{x \to 0} \dfrac{\cos x - 1}{x}$$

***Step 4*** Since $\lim\limits_{x \to 0} (\cos x - 1) = 0$ and $\lim\limits_{x \to 0} x = 0$, the expression $\dfrac{\cos x - 1}{x}$ is an indeterminate form at 0 of the type $\dfrac{0}{0}$. So, use L'Hôpital's Rule again.

$$\lim\limits_{x \to 0} \dfrac{\sin x - x}{x^2} = \dfrac{1}{2} \lim\limits_{x \to 0} \dfrac{\cos x - 1}{x} \underset{\substack{\uparrow \\ \text{L'Hôpital's} \\ \text{Rule}}}{=} \dfrac{1}{2} \lim\limits_{x \to 0} \dfrac{\dfrac{d}{dx}(\cos x - 1)}{\dfrac{d}{dx}x} = \dfrac{1}{2} \lim\limits_{x \to 0} \dfrac{-\sin x}{1} = 0$$

∎

**NOW WORK** Problem 39.

EXAMPLE 4 Using L'Hôpital's Rule to Find a Limit at Infinity

Find: **(a)** $\lim\limits_{x \to \infty} \dfrac{\ln x}{x}$ **(b)** $\lim\limits_{x \to \infty} \dfrac{x}{e^x}$ **(c)** $\lim\limits_{x \to \infty} \dfrac{e^x}{x}$

**Solution (a)** Since $\lim\limits_{x \to \infty} \ln x = \infty$ and $\lim\limits_{x \to \infty} x = \infty$, $\dfrac{\ln x}{x}$ is an indeterminate form at $\infty$ of the type $\dfrac{\infty}{\infty}$. Using L'Hôpital's Rule,

$$\lim_{x \to \infty} \frac{\ln x}{x} \underset{\substack{\uparrow \\ \text{L'Hôpital's} \\ \text{Rule}}}{=} \lim_{x \to \infty} \frac{\dfrac{d}{dx} \ln x}{\dfrac{d}{dx} x} = \lim_{x \to \infty} \frac{\dfrac{1}{x}}{1} = \lim_{x \to \infty} \frac{1}{x} = 0$$

**(b)** $\lim\limits_{x \to \infty} x = \infty$ and $\lim\limits_{x \to \infty} e^x = \infty$, so $\dfrac{x}{e^x}$ is an indeterminate form at $\infty$ of the type $\dfrac{\infty}{\infty}$. Using L'Hôpital's Rule,

$$\lim_{x \to \infty} \frac{x}{e^x} \underset{\substack{\uparrow \\ \text{L'Hôpital's} \\ \text{Rule}}}{=} \lim_{x \to \infty} \frac{\dfrac{d}{dx} x}{\dfrac{d}{dx} e^x} = \lim_{x \to \infty} \frac{1}{e^x} = 0$$

**(c)** From (b), we know that $\dfrac{e^x}{x}$ is an indeterminate form at $\infty$ of the type $\dfrac{\infty}{\infty}$. Using L'Hôpital's Rule,

$$\lim_{x \to \infty} \frac{e^x}{x} \underset{\substack{\uparrow \\ \text{L'Hôpital's} \\ \text{Rule}}}{=} \lim_{x \to \infty} \frac{\dfrac{d}{dx} e^x}{\dfrac{d}{dx} x} = \lim_{x \to \infty} \frac{e^x}{1} = \infty$$ ∎

NOW WORK Problem 35.

The results from Example 4 tell us that $y = x$ grows faster than $y = \ln x$, and that $y = e^x$ grows faster than $y = x$. In fact, these inequalities are true for all positive powers of $x$. That is,

$$\boxed{\lim_{x \to \infty} \frac{\ln x}{x^n} = 0 \qquad \lim_{x \to \infty} \frac{x^n}{e^x} = 0}$$

for $n \geq 1$ an integer. You are asked to verify these results in Problems 95 and 96.

Sometimes simplifying first reduces the effort needed to find the limit.

EXAMPLE 5 Using L'Hôpital's Rule to Find a Limit

Find $\lim\limits_{x \to 0} \dfrac{\tan x - \sin x}{x^2 \tan x}$.

**Solution** $\dfrac{\tan x - \sin x}{x^2 \tan x}$ is an indeterminate form at $0$ of the type $\dfrac{0}{0}$. We simplify the expression before using L'Hôpital's Rule. Then it is easier to find the limit.

$$\frac{\tan x - \sin x}{x^2 \tan x} = \frac{\dfrac{\sin x}{\cos x} - \sin x}{x^2 \cdot \dfrac{\sin x}{\cos x}} = \frac{\dfrac{\sin x - \sin x \cos x}{\cos x}}{\dfrac{x^2 \sin x}{\cos x}} = \frac{\sin x (1 - \cos x)}{x^2 \sin x} = \frac{1 - \cos x}{x^2}$$

**CAUTION** Be sure to verify that a simplified form is an indeterminate form at $c$ before using L'Hôpital's Rule.

Since $\dfrac{1-\cos x}{x^2}$ is an indeterminate form at 0 of the type $\dfrac{0}{0}$, we use L'Hôpital's Rule.

$$\lim_{x \to 0} \frac{\tan x - \sin x}{x^2 \tan x} = \lim_{x \to 0} \frac{1-\cos x}{x^2} \underset{\substack{\uparrow \\ \text{L'Hôpital's Rule}}}{=} \lim_{x \to 0} \frac{\dfrac{d}{dx}(1-\cos x)}{\dfrac{d}{dx}x^2} = \lim_{x \to 0} \frac{\sin x}{2x}$$

$$= \frac{1}{2}\lim_{x \to 0} \frac{\sin x}{x} \underset{\substack{\uparrow \\ \lim\limits_{x \to 0} \frac{\sin x}{x}=1}}{=} \frac{1}{2}$$

Compare this solution to one that uses L'Hôpital's Rule at the start.

In Example 6, we use L'Hôpital's Rule to find a one-sided limit.

**EXAMPLE 6   Using L'Hôpital's Rule to Find a One-Sided Limit**

Find $\lim\limits_{x \to 0^+} \dfrac{\cot x}{\ln x}$.

**Solution** Since $\lim\limits_{x \to 0^+} \cot x = \infty$ and $\lim\limits_{x \to 0^+} \ln x = -\infty$, $\dfrac{\cot x}{\ln x}$ is an indeterminate form at $0^+$ of the type $\dfrac{\infty}{\infty}$. Using L'Hôpital's Rule,

$$\lim_{x \to 0^+} \frac{\cot x}{\ln x} \underset{\substack{\uparrow \\ \text{L'Hôpital's Rule}}}{=} \lim_{x \to 0^+} \frac{\dfrac{d}{dx}\cot x}{\dfrac{d}{dx}\ln x} = \lim_{x \to 0^+} \frac{-\csc^2 x}{\dfrac{1}{x}} = -\lim_{x \to 0^+} \frac{x}{\sin^2 x} \underset{\substack{\uparrow \\ \text{L'Hôpital's Rule}}}{=} -\lim_{x \to 0^+} \frac{\dfrac{d}{dx}x}{\dfrac{d}{dx}\sin^2 x}$$

$$= -\lim_{x \to 0^+} \frac{1}{2\sin x \cos x} = \left(-\frac{1}{2}\right)\left(\lim_{x \to 0^+} \frac{1}{\sin x}\right)\left(\lim_{x \to 0^+} \frac{1}{\cos x}\right)$$

$$= \left(-\frac{1}{2}\right)\left(\lim_{x \to 0^+} \frac{1}{\sin x}\right)(1) = -\infty$$

**NOW WORK** Problem 61.

## 3 Find the Limit of an Indeterminate Form of the Type $0 \cdot \infty$, $\infty - \infty$, $0^0$, $1^\infty$, or $\infty^0$

L'Hôpital's Rule can only be used for indeterminate forms of the types $\dfrac{0}{0}$ and $\dfrac{\infty}{\infty}$. We need to rewrite indeterminate forms of the type $0 \cdot \infty$, $\infty - \infty$, $0^0$, $1^\infty$, or $\infty^0$ in the form $\dfrac{0}{0}$ or $\dfrac{\infty}{\infty}$ before using L'Hôpital's Rule to find the limit.

### Indeterminate Forms of the Type $0 \cdot \infty$

**NOTE** It might be tempting to argue that $0 \cdot \infty$ is 0, since "anything" times 0 is 0, but $0 \cdot \infty$ is not a product of numbers. That is, $0 \cdot \infty$ is not "zero times infinity"; it symbolizes "a quantity tending to zero" times "a quantity tending to infinity."

Suppose $\lim\limits_{x \to c} f(x) = 0$ and $\lim\limits_{x \to c} g(x) = \infty$. Depending on the functions $f$ and $g$, $\lim\limits_{x \to c}[f(x) \cdot g(x)]$, if it exists, could equal 0, $\infty$, or some nonzero number. The product $f \cdot g$ is called an **indeterminate form at $c$ of the type $0 \cdot \infty$.**

To find $\lim\limits_{x \to c}(f \cdot g)$, rewrite the product $f \cdot g$ as one of the following quotients:

$$f \cdot g = \frac{f}{\dfrac{1}{g}} \qquad \text{or} \qquad f \cdot g = \frac{g}{\dfrac{1}{f}}$$

The right side of the equation on the left is an indeterminate form at $c$ of the type $\dfrac{0}{0}$; the right side of the equation on the right is of the type $\dfrac{\infty}{\infty}$. Then use L'Hôpital's Rule with one of these quotients, choosing the one for which the derivatives are easier to find. If the first choice does not work, then try the other one.

**EXAMPLE 7  Finding the Limit of an Indeterminate Form of the Type $0 \cdot \infty$**

Find:

**(a)** $\displaystyle\lim_{x \to 0^+} (x \ln x)$    **(b)** $\displaystyle\lim_{x \to \infty} \left( x \sin \dfrac{1}{x} \right)$

**NOTE** We choose to use $\dfrac{\ln x}{\dfrac{1}{x}}$ rather than $\dfrac{x}{\dfrac{1}{\ln x}}$ because it is easier to find the derivatives of $\ln x$ and $\dfrac{1}{x}$ than it is to find the derivatives of $x$ and $\dfrac{1}{\ln x}$.

**Solution (a)** Since $\displaystyle\lim_{x \to 0^+} x = 0$ and $\displaystyle\lim_{x \to 0^+} \ln x = -\infty$, then $x \ln x$ is an indeterminate form at $0^+$ of the type $0 \cdot \infty$. Change $x \ln x$ to an indeterminate form of the type $\dfrac{\infty}{\infty}$ by writing $x \ln x = \dfrac{\ln x}{\dfrac{1}{x}}$. Then use L'Hôpital's Rule.

$$\lim_{x \to 0^+} (x \ln x) = \lim_{x \to 0^+} \frac{\ln x}{\dfrac{1}{x}} = \lim_{x \to 0^+} \frac{\dfrac{d}{dx} \ln x}{\dfrac{d}{dx} \dfrac{1}{x}} = \lim_{x \to 0^+} \frac{\dfrac{1}{x}}{-\dfrac{1}{x^2}} = \lim_{x \to 0^+} (-x) = 0$$

**(b)** Since $\displaystyle\lim_{x \to \infty} x = \infty$ and $\displaystyle\lim_{x \to \infty} \sin \dfrac{1}{x} = 0$, then $x \sin \dfrac{1}{x}$ is an indeterminate form at $\infty$ of the type $0 \cdot \infty$. Change $x \sin \dfrac{1}{x}$ to an indeterminate form of the type $\dfrac{0}{0}$ by writing

**NOTE** Notice in the solution to (b) that we did not need to use L'Hôpital's Rule.

$$\lim_{x \to \infty} x \sin \frac{1}{x} = \lim_{x \to \infty} \frac{\sin \dfrac{1}{x}}{\dfrac{1}{x}} \underset{\substack{\uparrow \\ \text{Let } t = \frac{1}{x}}}{=} \lim_{t \to 0^+} \frac{\sin t}{t} = 1 \qquad \blacksquare$$

**NOW WORK** Problem 45.

**NOTE** The indeterminate form of the type $\infty - \infty$ is a convenient notation for any of the following: $\infty - \infty, \; -\infty - (-\infty), \; \infty + (-\infty)$. Note that $\infty + \infty = \infty$ and $(-\infty) + (-\infty) = -\infty$ are not indeterminate forms because there is no confusion about these sums.

## Indeterminate Forms of the Type $\infty - \infty$

Suppose $\displaystyle\lim_{x \to c} f(x) = \infty$ and $\displaystyle\lim_{x \to c} g(x) = \infty$. Then depending on the functions $f$ and $g$, $\displaystyle\lim_{x \to c} [f(x) - g(x)]$, if it exists, could equal 0, $\infty$, or some nonzero number. In such instances the difference $f(x) - g(x)$ is called an **indeterminate form at $c$ of the type $\infty - \infty$.**

If the limit of a function results in the indeterminate form $\infty - \infty$, it is generally possible to rewrite the function as an indeterminate form of the type $\dfrac{0}{0}$ or $\dfrac{\infty}{\infty}$ by using algebra or trigonometry.

**EXAMPLE 8  Finding the Limit of an Indeterminate Form of the Type $\infty - \infty$**

Find $\displaystyle\lim_{x \to 0^+} \left( \dfrac{1}{x} - \dfrac{1}{\sin x} \right)$.

**Solution** Since $\displaystyle\lim_{x \to 0^+} \dfrac{1}{x} = \infty$ and $\displaystyle\lim_{x \to 0^+} \dfrac{1}{\sin x} = \infty$, then $\dfrac{1}{x} - \dfrac{1}{\sin x}$ is an indeterminate form at $0^+$ of the type $\infty - \infty$. Rewrite the difference as a single fraction.

$$\lim_{x \to 0^+} \left( \frac{1}{x} - \frac{1}{\sin x} \right) = \lim_{x \to 0^+} \frac{\sin x - x}{x \sin x}$$

Then $\dfrac{\sin x - x}{x \sin x}$ is an indeterminate form at $0^+$ of the type $\dfrac{0}{0}$. Now use L'Hôpital's Rule.

$$\lim_{x \to 0^+} \left( \frac{1}{x} - \frac{1}{\sin x} \right) = \lim_{x \to 0^+} \frac{\sin x - x}{x \sin x} \underset{\uparrow}{=} \lim_{x \to 0^+} \frac{\dfrac{d}{dx}(\sin x - x)}{\dfrac{d}{dx}(x \sin x)} = \lim_{x \to 0^+} \frac{\cos x - 1}{x \cos x + \sin x}$$

$$\text{L'Hôpital's Rule}$$

$$\underset{\uparrow}{=} \lim_{x \to 0^+} \frac{\dfrac{d}{dx}(\cos x - 1)}{\dfrac{d}{dx}(x \cos x + \sin x)} = \lim_{x \to 0^+} \frac{-\sin x}{(-x \sin x + \cos x) + \cos x}$$

$$\text{Type } \frac{0}{0}\text{; use L'Hôpital's Rule}$$

$$= \lim_{x \to 0^+} \frac{\sin x}{x \sin x - 2 \cos x} = \frac{0}{-2} = 0 \qquad \blacksquare$$

**NOW WORK** Problem 47.

## Indeterminate Forms of the Type $1^\infty$, $0^0$, or $\infty^0$

A function of the form $[f(x)]^{g(x)}$ may result in an indeterminate form of the type $1^\infty$, $0^0$, or $\infty^0$. To find the limit of such a function, we let $y = [f(x)]^{g(x)}$ and take the natural logarithm of each side.

**NEED TO REVIEW?** Properties of logarithms are discussed in Appendix A.1, pp. A-10 to A-11.

$$\ln y = \ln [f(x)]^{g(x)} = g(x) \ln f(x)$$

The expression on the right will then be an indeterminate form of the type $0 \cdot \infty$ and we find the limit using the method of Example 7.

> **Steps for Finding $\lim\limits_{x \to c}[f(x)]^{g(x)}$ When $[f(x)]^{g(x)}$ Is an Indeterminate Form at $c$ of the Type $1^\infty$, $0^0$, or $\infty^0$**
>
> **Step 1** Let $y = [f(x)]^{g(x)}$ and take the natural logarithm of each side, obtaining $\ln y = g(x) \ln f(x)$.
>
> **Step 2** Find $\lim\limits_{x \to c} \ln y$.
>
> **Step 3** If $\lim\limits_{x \to c} \ln y = L$, then $\lim\limits_{x \to c} y = e^L$.

These steps also can be used for limits at infinity and for one-sided limits.

**EXAMPLE 9** **Finding the Limit of an Indeterminate Form of the Type $0^0$**

Find $\lim\limits_{x \to 0^+} x^x$.

**Solution** The expression $x^x$ is an indeterminate form at $0^+$ of the type $0^0$. Follow the steps for finding $\lim\limits_{x \to c}[f(x)]^{g(x)}$.

**Step 1** Let $y = x^x$. Then $\ln y = x \ln x$.

**Step 2** $\lim\limits_{x \to 0^+} \ln y = \lim\limits_{x \to 0^+} (x \ln x) = 0$ [from Example 7(a)].

**CAUTION** Do not stop after finding $\lim\limits_{x \to c} \ln y \, (= L)$. Remember, we want to find $\lim\limits_{x \to c} y \, (= e^L)$.

**Step 3** Since $\lim\limits_{x \to 0^+} \ln y = 0$, $\lim\limits_{x \to 0^+} y = e^0 = 1$. $\qquad \blacksquare$

**NOW WORK** Problem 51.

EXAMPLE 10  **Finding the Limit of an Indeterminate Form of the Type $1^\infty$**

Find $\lim\limits_{x \to 0^+} (1+x)^{1/x}$.

**Solution** The expression $(1+x)^{1/x}$ is an indeterminate form at $0^+$ of the type $1^\infty$.

**Step 1** Let $y = (1+x)^{1/x}$. Then $\ln y = \dfrac{1}{x} \ln(1+x)$.

**Step 2** $\lim\limits_{x \to 0^+} \ln y = \lim\limits_{x \to 0^+} \dfrac{\ln(1+x)}{x} = \lim\limits_{x \to 0^+} \dfrac{\dfrac{d}{dx} \ln(1+x)}{\dfrac{d}{dx} x} = \lim\limits_{x \to 0^+} \dfrac{\dfrac{1}{1+x}}{1} = 1$

Type $\dfrac{0}{0}$; use L'Hôpital's Rule

**Step 3** Since $\lim\limits_{x \to 0^+} \ln y = 1$, $\lim\limits_{x \to 0^+} y = e^1 = e$.  ∎

NOW WORK Problem 85.

# 4.5 Assess Your Understanding

## Concepts and Vocabulary

**1. True or False** $\dfrac{f(x)}{g(x)}$ is an indeterminate form at $c$ of the type $\dfrac{0}{0}$ if $\lim\limits_{x \to c} \dfrac{f(x)}{g(x)}$ does not exist.

**2. True or False** If $\dfrac{f(x)}{g(x)}$ is an indeterminate form at $c$ of the type $\dfrac{0}{0}$, then L'Hôpital's Rule states that $\lim\limits_{x \to c} \dfrac{f(x)}{g(x)} = \lim\limits_{x \to c} \left[ \dfrac{d}{dx} \dfrac{f(x)}{g(x)} \right]$.

**3. True or False** $\dfrac{1}{x}$ is an indeterminate form at $0$.

**4. True or False** $x \ln x$ is not an indeterminate form at $0^+$ because $\lim\limits_{x \to 0^+} x = 0$ and $\lim\limits_{x \to 0^+} \ln x = -\infty$, and $0 \cdot -\infty = 0$.

**5.** In your own words, explain why $\infty - \infty$ is an indeterminate form, but $\infty + \infty$ is not an indeterminate form.

**6.** In your own words, explain why $0 \cdot \infty \neq 0$.

## Skill Building

In Problems 7–26:

(a) Determine whether each expression is an indeterminate form at $c$.

(b) If it is, identify the type. If it is not an indeterminate form, state why.

**7.** $\dfrac{1-e^x}{x}$,  $c=0$

**8.** $\dfrac{1-e^x}{x-1}$,  $c=0$

**9.** $\dfrac{e^x}{x}$,  $c=0$

**10.** $\dfrac{e^x}{x}$,  $c=\infty$

**11.** $\dfrac{\ln x}{x^2}$,  $c=\infty$

**12.** $\dfrac{\ln(x+1)}{e^x-1}$,  $c=0$

**13.** $\dfrac{\sec x}{x}$,  $c=0$

**14.** $\dfrac{x}{\sec x - 1}$,  $c=0$

**15.** $\dfrac{\sin x(1-\cos x)}{x^2}$,  $c=0$

**16.** $\dfrac{\sin x - 1}{\cos x}$,  $c=\dfrac{\pi}{2}$

**17.** $\dfrac{\tan x - 1}{\sin(4x-\pi)}$,  $c=\dfrac{\pi}{4}$

**18.** $\dfrac{e^x - e^{-x}}{1-\cos x}$,  $c=0$

**19.** $x^2 e^{-x}$,  $c=\infty$

**20.** $x \cot x$,  $c=0$

**21.** $\csc\dfrac{x}{2} - \cot\dfrac{x}{2}$,  $c=0$

**22.** $\dfrac{x}{x-1} + \dfrac{1}{\ln x}$,  $c=1$

**23.** $\left(\dfrac{1}{x^2}\right)^{\sin x}$,  $c=0$

**24.** $(e^x + x)^{1/x}$,  $c=0$

**25.** $(x^2-1)^x$,  $c=0$

**26.** $(\sin x)^x$,  $c=0$

In Problems 27–42, identify each quotient as an indeterminate form of the type $\dfrac{0}{0}$ or $\dfrac{\infty}{\infty}$. Then find the limit.

**27.** $\lim\limits_{x \to 2} \dfrac{x^2+x-6}{x^2-3x+2}$

**28.** $\lim\limits_{x \to 1} \dfrac{2x^3+5x^2-4x-3}{x^3+x^2-10x+8}$

**29.** $\lim\limits_{x \to 1} \dfrac{\ln x}{x^2-1}$

**30.** $\lim\limits_{x \to 0} \dfrac{\ln(1-x)}{e^x-1}$

**31.** $\lim\limits_{x \to 0} \dfrac{e^x - e^{-x}}{\sin x}$

**32.** $\lim\limits_{x \to 0} \dfrac{\tan(2x)}{\ln(1+x)}$

**33.** $\lim\limits_{x \to 1} \dfrac{\sin(\pi x)}{x-1}$

**34.** $\lim\limits_{x \to \pi} \dfrac{1+\cos x}{\sin(2x)}$

**35.** $\lim\limits_{x \to \infty} \dfrac{x^2}{e^x}$

**36.** $\lim\limits_{x \to \infty} \dfrac{e^x}{x^4}$

**37.** $\lim\limits_{x \to \infty} \dfrac{\ln x}{e^x}$

**38.** $\lim\limits_{x \to \infty} \dfrac{x+\ln x}{x \ln x}$

**39.** $\lim\limits_{x \to 0} \dfrac{e^x - 1 - \sin x}{1-\cos x}$

**40.** $\lim\limits_{x \to 0} \dfrac{e^x - e^{-x} - 2\sin x}{3x^3}$

**41.** $\lim\limits_{x \to 0} \dfrac{\sin x - x}{x^3}$

**42.** $\lim\limits_{x \to 0} \dfrac{x^3}{\cos x - 1}$

**1.** = NOW WORK problem        〰 = Graphing technology recommended        CAS = Computer Algebra System recommended

*In Problems 43–58, identify each expression as an indeterminate form of the type $0 \cdot \infty$, $\infty - \infty$, $0^0$, $1^\infty$, or $\infty^0$. Then find the limit.*

**43.** $\lim\limits_{x \to 0^+} (x^2 \ln x)$

**44.** $\lim\limits_{x \to \infty} (xe^{-x})$

**45.** $\lim\limits_{x \to \infty} [x(e^{1/x} - 1)]$

**46.** $\lim\limits_{x \to \pi/2} [(1 - \sin x) \tan x]$

**47.** $\lim\limits_{x \to \pi/2} (\sec x - \tan x)$

**48.** $\lim\limits_{x \to 0} \left( \cot x - \dfrac{1}{x} \right)$

**49.** $\lim\limits_{x \to 1} \left( \dfrac{1}{\ln x} - \dfrac{x}{\ln x} \right)$

**50.** $\lim\limits_{x \to 0} \left( \dfrac{1}{x} - \dfrac{1}{e^x - 1} \right)$

**51.** $\lim\limits_{x \to 0^+} (2x)^{3x}$

**52.** $\lim\limits_{x \to 0^+} x^{x^2}$

**53.** $\lim\limits_{x \to \infty} (x + 1)^{e^{-x}}$

**54.** $\lim\limits_{x \to \infty} (1 + x^2)^{1/x}$

**55.** $\lim\limits_{x \to 0^+} (\csc x)^{\sin x}$

**56.** $\lim\limits_{x \to \infty} x^{1/x}$

**57.** $\lim\limits_{x \to \pi/2^-} (\sin x)^{\tan x}$

**58.** $\lim\limits_{x \to 0} (\cos x)^{1/x}$

*In Problems 59–90, find each limit.*

**59.** $\lim\limits_{x \to 0^+} \dfrac{\cot x}{\cot(2x)}$

**60.** $\lim\limits_{x \to \infty} \dfrac{\ln(\ln x)}{\ln x}$

**61.** $\lim\limits_{x \to 1/2^-} \dfrac{\ln(1 - 2x)}{\tan(\pi x)}$

**62.** $\lim\limits_{x \to 1^-} \dfrac{\ln(1 - x)}{\cot(\pi x)}$

**63.** $\lim\limits_{x \to \infty} \dfrac{x^4 + x^3}{e^x + 1}$

**64.** $\lim\limits_{x \to \infty} \dfrac{x^2 + x - 1}{e^x + e^{-x}}$

**65.** $\lim\limits_{x \to 0} \dfrac{xe^{4x} - x}{1 - \cos(2x)}$

**66.** $\lim\limits_{x \to 0} \dfrac{x \tan x}{1 - \cos x}$

**67.** $\lim\limits_{x \to 0} \dfrac{\tan^{-1} x}{x}$

**68.** $\lim\limits_{x \to 0} \dfrac{\tan^{-1} x}{\sin^{-1} x}$

**69.** $\lim\limits_{x \to 0} \dfrac{\cos x - 1}{\cos(2x) - 1}$

**70.** $\lim\limits_{x \to 0} \dfrac{\tan x - \sin x}{x^3}$

**71.** $\lim\limits_{x \to 0^+} (x^{1/2} \ln x)$

**72.** $\lim\limits_{x \to \infty} [(x - 1)e^{-x^2}]$

**73.** $\lim\limits_{x \to \pi/2} [\tan x \ln(\sin x)]$

**74.** $\lim\limits_{x \to 0^+} [\sin x \ln(\sin x)]$

**75.** $\lim\limits_{x \to 0} [\csc x \ln(x + 1)]$

**76.** $\lim\limits_{x \to \pi/4} [(1 - \tan x) \sec(2x)]$

**77.** $\lim\limits_{x \to a} \left[ (a^2 - x^2) \tan \left( \dfrac{\pi x}{2a} \right) \right]$

**78.** $\lim\limits_{x \to 1^+} \left[ (1 - x) \tan \left( \dfrac{1}{2} \pi x \right) \right]$

**79.** $\lim\limits_{x \to 1} \left( \dfrac{1}{\ln x} - \dfrac{1}{x - 1} \right)$

**80.** $\lim\limits_{x \to 1} \left( \dfrac{x}{x - 1} - \dfrac{1}{\ln x} \right)$

**81.** $\lim\limits_{x \to \pi/2} \left( x \tan x - \dfrac{\pi}{2} \sec x \right)$

**82.** $\lim\limits_{x \to \pi} (\cot x - x \csc x)$

**83.** $\lim\limits_{x \to 1^-} (1 - x)^{\tan(\pi x)}$

**84.** $\lim\limits_{x \to 0^+} x^{\sqrt{x}}$

**85.** $\lim\limits_{x \to 0} \left( \dfrac{\sin x}{x} \right)^{1/x}$

**86.** $\lim\limits_{x \to \infty} \left( 1 + \dfrac{5}{x} + \dfrac{3}{x^2} \right)^x$

**87.** $\lim\limits_{x \to (\pi/2)^-} (\tan x)^{\cos x}$

**88.** $\lim\limits_{x \to 0^+} (x^2 + x)^{-\ln x}$

**89.** $\lim\limits_{x \to 0} (\cosh x)^{e^x}$

**90.** $\lim\limits_{x \to 0^+} (\sinh x)^x$

## Applications and Extensions

**91. Wolf Population**   In 2014 there were 229 wolves in Wyoming outside of Yellowstone National Park. Suppose the population $w$ of wolves in the region at time $t$ follows the logistic growth curve

$$w = w(t) = \dfrac{Ke^{rt}}{\dfrac{K}{40} + e^{rt} - 1}$$

where $K = 252$, $r = 0.283$, and $t = 0$ represents the population in the year 2000.

*Source*: Federal Wildlife Service

(a) Find $\lim\limits_{t \to \infty} w(t)$.

(b) Interpret the answer found in (a) in the context of the problem.

(c) Use technology to graph $w = w(t)$.

**92. Skydiving**   The downward velocity $v$ of a skydiver with nonlinear air resistance can be modeled by

$$v = v(t) = -A + RA \dfrac{e^{Bt+C} - 1}{e^{Bt+C} + 1}$$

where $t$ is the time in seconds, and $A$, $B$, $C$, and $R$ are positive constants with $R > 1$.

(a) Find $\lim\limits_{t \to \infty} v(t)$.

(b) Interpret the limit found in (a).

(c) If the velocity $v$ is measured in feet per second, reasonable values of the constants are $A = 108.6$, $B = 0.554$, $C = 0.804$, and $R = 2.62$. Graph the velocity of the skydiver with respect to time.

**93. Electricity**   The equation governing the amount of current $I$ (in amperes) in a simple $RL$ circuit consisting of a resistance $R$ (in ohms), an inductance $L$ (in henrys), and an electromotive force $E$ (in volts) is

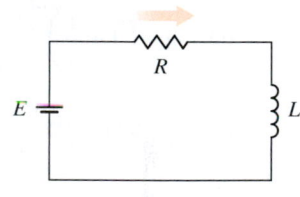

$$I = \dfrac{E}{R}(1 - e^{-Rt/L}).$$

(a) Find $\lim\limits_{t \to \infty} I(t)$ and $\lim\limits_{R \to 0^+} I(t)$.

(b) Interpret these limits.

**94.** Find $\lim\limits_{x \to 0} \dfrac{a^x - b^x}{x}$, where $a \neq 1$ and $b \neq 1$ are positive real numbers.

**95.** Show that $\lim\limits_{x \to \infty} \dfrac{\ln x}{x^n} = 0$, for $n \geq 1$ an integer.

**96.** Show that $\lim\limits_{x \to \infty} \dfrac{x^n}{e^x} = 0$ for $n \geq 1$ an integer.

**97.** Show that $\lim\limits_{x \to 0^+} (\cos x + 2 \sin x)^{\cot x} = e^2$.

**98.** Find $\lim\limits_{x \to \infty} \dfrac{P(x)}{e^x}$, where $P$ is a polynomial function.

**99.** Find $\lim\limits_{x \to \infty} [\ln(x + 1) - \ln(x - 1)]$.

**100.** Show that $\lim\limits_{x \to 0^+} \dfrac{e^{-1/x^2}}{x} = 0$.   *Hint:* Write $\dfrac{e^{-1/x^2}}{x} = \dfrac{\dfrac{1}{x}}{e^{1/x^2}}$.

**101.** If $n$ is an integer, show that $\displaystyle\lim_{x\to 0^+} \frac{e^{-1/x^2}}{x^n} = 0$.

**102.** Show that $\displaystyle\lim_{x\to\infty} \sqrt[x]{x} = 1$.

**103.** Show that $\displaystyle\lim_{x\to\infty} \left(1+\frac{a}{x}\right)^x = e^a$, $a$ any real number.

**104.** Show that $\displaystyle\lim_{x\to\infty} \left(\frac{x+a}{x-a}\right)^x = e^{2a}$, $a \neq 0$.

**105. (a)** Show that the function below has a derivative at 0. What is $f'(0)$?

$$f(x) = \begin{cases} e^{-1/x^2} & \text{if } x \neq 0 \\ 0 & \text{if } x = 0 \end{cases}$$

**(b)** Graph $f$.

**106.** If $a, b \neq 0$ and $c > 0$ are real numbers, show that

$$\lim_{x\to c} \frac{x^a - c^a}{x^b - c^b} = \frac{a}{b} c^{a-b}.$$

**107.** Prove L'Hôpital's rule when $\dfrac{f(x)}{g(x)}$ is an indeterminate form at $-\infty$ of the type $\dfrac{0}{0}$.

**Challenge Problems**

**108.** Explain why L'Hôpital's Rule does not apply to $\displaystyle\lim_{x\to 0} \frac{x^2 \sin\frac{1}{x}}{\sin x}$.

**109.** Find each limit:

**(a)** $\displaystyle\lim_{x\to\infty} \left(1+\frac{1}{x}\right)^{-x^2}$     **(b)** $\displaystyle\lim_{x\to\infty} \left(1+\frac{\ln a}{x}\right)^x$, $a > 1$

**(c)** $\displaystyle\lim_{x\to\infty} \left(1+\frac{1}{x}\right)^{x^2}$     **(d)** $\displaystyle\lim_{x\to\infty} \left(1+\frac{\sin x}{x}\right)^x$

**(e)** $\displaystyle\lim_{x\to\infty} (e^x)^{-1/\ln x}$     **(f)** $\displaystyle\lim_{x\to\infty} \left[\left(\frac{1}{a}\right)^x\right]^{-1/x}$, $0 < a < 1$

**(g)** $\displaystyle\lim_{x\to\infty} x^{1/x}$     **(h)** $\displaystyle\lim_{x\to\infty} (a^x)^{1/x}$, $a > 1$

**(i)** $\displaystyle\lim_{x\to\infty} [(2+\sin x)^x]^{1/x}$     **(j)** $\displaystyle\lim_{x\to 0^+} x^{-1/\ln x}$

**110.** Find constants $A$, $B$, $C$, and $D$ so that

$$\lim_{x\to 0} \frac{\sin(Ax) + Bx + Cx^2 + Dx^3}{x^5} = \frac{4}{15}.$$

**111.** A function $f$ has derivatives of all orders.

**(a)** Find $\displaystyle\lim_{h\to 0} \frac{f(x+2h) - 2f(x+h) + f(x)}{h^2}$.

**(b)** Find $\displaystyle\lim_{h\to 0} \frac{f(x+3h) - 3f(x+2h) + 3f(x+h) - f(x)}{h^3}$.

**(c)** Generalize parts (a) and (b).

**112.** The formulas in Problem 111 can be used to approximate derivatives. Approximate $f'(2)$, $f''(2)$, and $f'''(2)$ from the table. The data are for $f(x) = \ln x$. Compare the exact values with your approximations.

| $x$ | 2.0 | 2.1 | 2.2 | 2.3 | 2.4 |
|---|---|---|---|---|---|
| $f(x)$ | 0.6931 | 0.7419 | 0.7885 | 0.8329 | 0.8755 |

**113.** Consider the function $f(t, x) = \dfrac{x^{t+1} - 1}{t+1}$, where $x > 0$ and $t \neq -1$.

**(a)** For $x$ fixed at $x_0$, show that $\displaystyle\lim_{t\to -1} f(t, x_0) = \ln x_0$.

**(b)** For $x$ fixed, define a function $F(t, x)$, where $x > 0$, that is continuous so $F(t, x) = f(t, x)$ for all $t \neq -1$.

**(c)** For $t$ fixed, show that $\dfrac{d}{dx} F(t, x) = x^t$, for $x > 0$ and all $t$.

*Source*: Michael W. Ecker (2012, September), Unifying Results via L'Hôpital's Rule. *Journal of the American Mathematical Association of Two Year Colleges*, 4(1) pp. 9–10.

# 4.6 Using Calculus to Graph Functions

**OBJECTIVE** *When you finish this section, you should be able to:*

1 Graph a function using calculus (p. 320)

In precalculus, algebra was used to graph a function. In this section, we combine precalculus methods with differential calculus to obtain a more detailed graph.

## 1 Graph a Function Using Calculus

The following steps are used to graph a function by hand or to analyze a technology-generated graph of a function.

**Steps for Graphing a Function $y = f(x)$**

**Step 1** Find the domain of $f$ and any intercepts.

**Step 2** Identify any asymptotes, and examine the end behavior.

**Step 3** Find $y = f'(x)$ and use it to find any critical numbers of $f$. Determine the slope of the tangent line at the critical numbers. Also find $f''$.

**Step 4** Use the Increasing/Decreasing Function Test to identify intervals on which $f$ is increasing ($f'(x) > 0$) and the intervals on which $f$ is decreasing ($f'(x) < 0$).

**Step 5** Use either the First Derivative Test or the Second Derivative Test (if applicable) to identify local maximum values and local minimum values.

**Step 6** Use the Test for Concavity to determine intervals where the function is concave up ($f''(x) > 0$) and concave down ($f''(x) < 0$). Identify any inflection points.

**Step 7** Graph the function using the information gathered. Plot additional points as needed. It may be helpful to find the slope of the tangent line at such points.

CALC
CLIP
**EXAMPLE 1   Using Calculus to Graph a Polynomial Function**

Graph $f(x) = 3x^4 - 8x^3$.

**Solution** Follow the steps for graphing a function.

**Step 1** $f$ is a fourth-degree polynomial; its domain is all real numbers. $f(0) = 0$, so the $y$-intercept is 0. We find the $x$-intercepts by solving $f(x) = 0$.

$$3x^4 - 8x^3 = 0$$
$$x^3(3x - 8) = 0$$
$$x = 0 \quad \text{or} \quad x = \frac{8}{3}$$

There are two $x$-intercepts: 0 and $\frac{8}{3}$. Plot the intercepts $(0, 0)$ and $\left(\frac{8}{3}, 0\right)$.

**NOTE** When graphing a function, piece together a sketch of the graph as you go, to ensure that the information is consistent and makes sense. If it appears to be contradictory, check your work. You may have made an error.

**Step 2** Polynomials have no asymptotes, but the end behavior of $f$ resembles the power function $y = 3x^4$.

**Step 3** $f'(x) = 12x^3 - 24x^2 = 12x^2(x - 2)$     $f''(x) = 36x^2 - 48x = 12x(3x - 4)$

For polynomials, the critical numbers occur when $f'(x) = 0$.

$$12x^2(x - 2) = 0$$
$$x = 0 \quad \text{or} \quad x = 2$$

The critical numbers are 0 and 2. At the points $(0, 0)$ and $(2, -16)$, the tangent lines are horizontal. Plot these points.

**Step 4** Use the critical numbers 0 and 2 to form three intervals on the $x$-axis: $(-\infty, 0)$, $(0, 2)$, and $(2, \infty)$. Then determine the sign of $f'(x)$ on each interval.

| Interval | Sign of $f'$ | Conclusion |
|---|---|---|
| $(-\infty, 0)$ | negative | $f$ is decreasing on $(-\infty, 0)$ |
| $(0, 2)$ | negative | $f$ is decreasing on $(0, 2)$ |
| $(2, \infty)$ | positive | $f$ is increasing on $(2, \infty)$ |

**Step 5** Use the First Derivative Test. Based on the table, $f(0) = 0$ is neither a local maximum value nor a local minimum value, and $f(2) = -16$ is a local minimum value.

**Step 6** $f''(x) = 36x^2 - 48x = 12x(3x - 4)$; the zeros of $f''$ are $x = 0$ and $x = \dfrac{4}{3}$.

Use the numbers 0 and $\dfrac{4}{3}$ to form three intervals on the $x$-axis: $(-\infty, 0)$, $\left(0, \dfrac{4}{3}\right)$, and $\left(\dfrac{4}{3}, \infty\right)$. Then determine the sign of $f''(x)$ on each interval.

| Interval | Sign of $f''$ | Conclusion |
|---|---|---|
| $(-\infty, 0)$ | positive | $f$ is concave up on $(-\infty, 0)$ |
| $\left(0, \dfrac{4}{3}\right)$ | negative | $f$ is concave down on $\left(0, \dfrac{4}{3}\right)$ |
| $\left(\dfrac{4}{3}, \infty\right)$ | positive | $f$ is concave up on $\left(\dfrac{4}{3}, \infty\right)$ |

The concavity of $f$ changes at 0 and $\dfrac{4}{3}$, so the points $(0, 0)$ and $\left(\dfrac{4}{3}, -\dfrac{256}{27}\right)$ are inflection points. Plot the inflection points.

**Step 7** Using the points plotted and the information about the shape of the graph, graph the function. See Figure 49. ∎

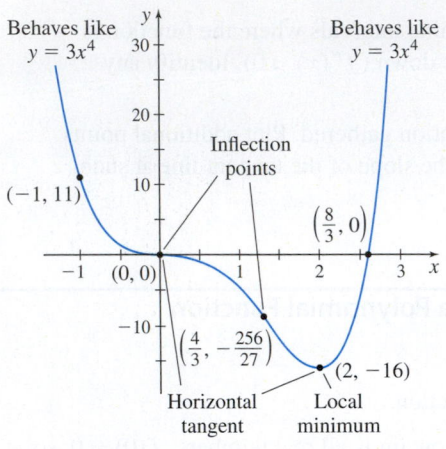

**DF** **Figure 49** $f(x) = 3x^4 - 8x^3$

Behaves like $y = 3x^4$

Behaves like $y = 3x^4$

$(-1, 11)$

Inflection points

$\left(\dfrac{8}{3}, 0\right)$

$(0, 0)$

$\left(\dfrac{4}{3}, -\dfrac{256}{27}\right)$

$(2, -16)$

Horizontal tangent

Local minimum

**NOW WORK** Problem 1.

---

**EXAMPLE 2** **Using Calculus to Graph a Rational Function**

Graph $f(x) = \dfrac{6x^2 - 6}{x^3}$.

**Solution** Follow the steps for graphing a function.

**Step 1** $f$ is a rational function; the domain of $f$ is $\{x \,|\, x \neq 0\}$. The $x$-intercepts are $-1$ and 1. There is no $y$-intercept. Plot the intercepts $(-1, 0)$ and $(1, 0)$.

**Step 2** The degree of the numerator is less than the degree of the denominator, so $f$ has a horizontal asymptote. We find the horizontal asymptotes by finding the limits at infinity.

$$\lim_{x \to \infty} \frac{6x^2 - 6}{x^3} = \lim_{x \to \infty} \frac{6x^2}{x^3} = \lim_{x \to \infty} \frac{6}{x} = 0$$

The line $y = 0$ is a horizontal asymptote as $x \to \infty$. It is also a horizontal asymptote as $x \to -\infty$.

We identify vertical asymptotes by checking for infinite limits. $[\lim_{x \to 0} f(x)$ is the only one to check—do you see why?]

$$\lim_{x \to 0^-} \frac{6x^2 - 6}{x^3} = \infty \qquad \lim_{x \to 0^+} \frac{6x^2 - 6}{x^3} = -\infty$$

The line $x = 0$ is a vertical asymptote. Draw the asymptotes on the graph.

**Step 3**   $f'(x) = \dfrac{d}{dx} \dfrac{6x^2 - 6}{x^3} = \dfrac{12x(x^3) - (6x^2 - 6)(3x^2)}{x^6}$    Use the Quotient Rule.

$$= \dfrac{-6x^4 + 18x^2}{x^6} = \dfrac{6x^2(3 - x^2)}{x^6} = \dfrac{6(3 - x^2)}{x^4}$$

$f''(x) = \dfrac{d}{dx} \dfrac{6(3 - x^2)}{x^4} = 6 \cdot \dfrac{(-2x)x^4 - (3 - x^2)(4x^3)}{x^8}$    Use the Quotient Rule.

$$= 6 \cdot \dfrac{2x^5 - 12x^3}{x^8} = \dfrac{12(x^2 - 6)}{x^5}$$

Critical numbers occur where $f'(x) = 0$ or where $f'(x)$ does not exist. Since $f'(x) = 0$ when $x = \pm\sqrt{3}$, there are two critical numbers, $\sqrt{3}$ and $-\sqrt{3}$. (The derivative does not exist at 0, but, since 0 is not in the domain of $f$, it is not a critical number.)

**Step 4**  Use the numbers $-\sqrt{3}$, 0, and $\sqrt{3}$ to form four intervals on the $x$-axis: $(-\infty, -\sqrt{3})$, $(-\sqrt{3}, 0)$, $(0, \sqrt{3})$ and $(\sqrt{3}, \infty)$. Then determine the sign of $f'(x)$ on each interval.

| Interval | Sign of $f'$ | Conclusion |
|---|---|---|
| $(-\infty, -\sqrt{3})$ | negative | $f$ is decreasing on $(-\infty, -\sqrt{3})$ |
| $(-\sqrt{3}, 0)$ | positive | $f$ is increasing on $(-\sqrt{3}, 0)$ |
| $(0, \sqrt{3})$ | positive | $f$ is increasing on $(0, \sqrt{3})$ |
| $(\sqrt{3}, \infty)$ | negative | $f$ is decreasing on $(\sqrt{3}, \infty)$ |

**Step 5**  Use the First Derivative Test. From the table in Step 4, we conclude

- $f(-\sqrt{3}) = -\dfrac{4}{\sqrt{3}} \approx -2.309$ is a local minimum value

- $f(\sqrt{3}) = \dfrac{4}{\sqrt{3}} \approx 2.309$ is a local maximum value

Plot the local extreme points.

**Step 6**  Use the Test for Concavity. $f''(x) = \dfrac{12(x^2 - 6)}{x^5} = 0$ when $x = \pm\sqrt{6}$. Now use the numbers $-\sqrt{6}$, 0, and $\sqrt{6}$ to form four intervals on the $x$-axis: $(-\infty, -\sqrt{6})$, $(-\sqrt{6}, 0)$, $(0, \sqrt{6})$ and $(\sqrt{6}, \infty)$. Then determine the sign of $f''(x)$ on each interval.

| Interval | Sign of $f''$ | Conclusion |
|---|---|---|
| $(-\infty, -\sqrt{6})$ | negative | $f$ is concave down on $(-\infty, -\sqrt{6})$ |
| $(-\sqrt{6}, 0)$ | positive | $f$ is concave up on $(-\sqrt{6}, 0)$ |
| $(0, \sqrt{6})$ | negative | $f$ is concave down on $(0, \sqrt{6})$ |
| $(\sqrt{6}, \infty)$ | positive | $f$ is concave up on $(\sqrt{6}, \infty)$ |

The concavity of the function $f$ changes at $-\sqrt{6}$, 0, and $\sqrt{6}$. So, the points $\left(-\sqrt{6}, -\dfrac{5\sqrt{6}}{6}\right)$ and $\left(\sqrt{6}, \dfrac{5\sqrt{6}}{6}\right)$ are inflection points. Since 0 is not in the domain of $f$, there is no inflection point at 0. Plot the inflection points.

**Step 7** Now use the information about where the function increases/decreases and the concavity to complete the graph. See Figure 50. Notice the apparent symmetry of the graph with respect to the origin. We can verify this symmetry by showing $f(-x) = -f(x)$ and concluding that $f$ is an odd function.

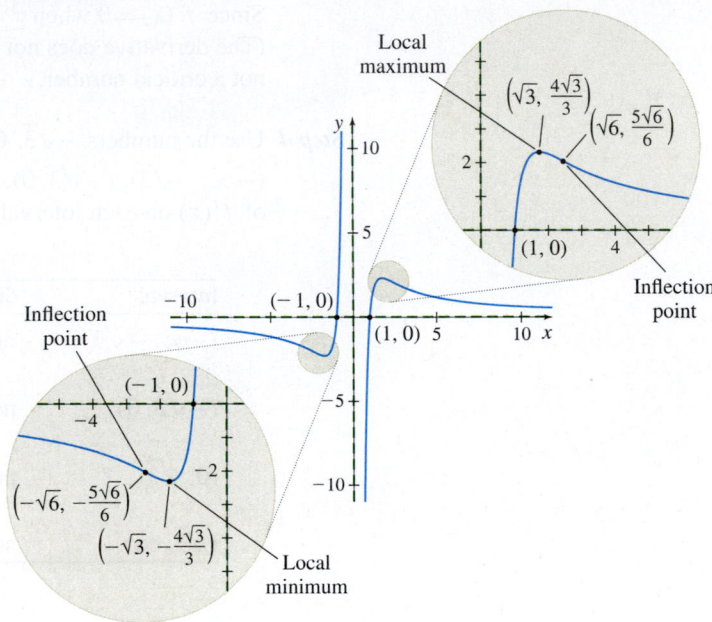

**Figure 50** $f(x) = \dfrac{6x^2 - 6}{x^3}$

**NOW WORK** Problem 11.

Some functions have a graph with an *oblique asymptote*. That is, the graph of the function approaches a line $y = mx + b$, $m \neq 0$, as $x$ becomes unbounded in either the positive or negative direction.

In general, for a function $y = f(x)$, the line $y = mx + b$, $m \neq 0$, is an **oblique asymptote** of the graph of $f$, if

$$\lim_{x \to -\infty} [f(x) - (mx + b)] = 0 \qquad \text{or} \qquad \lim_{x \to \infty} [f(x) - (mx + b)] = 0$$

In Chapter 1, we saw that rational functions can have vertical or horizontal asymptotes. Rational functions can also have oblique asymptotes, but cannot have both a horizontal asymptote and an oblique asymptote.

- Rational functions for which the degree of the numerator is 1 more than the degree of the denominator have an oblique asymptote.
- Rational functions for which the degree of the numerator is less than or equal to the degree of the denominator have a horizontal asymptote.

**EXAMPLE 3**  **Using Calculus to Graph a Rational Function**

Graph $f(x) = \dfrac{x^2 - 2x + 6}{x - 3}$.

**Solution**

***Step 1***  $f$ is a rational function; the domain of $f$ is $\{x \mid x \neq 3\}$. There are no $x$-intercepts, since $x^2 - 2x + 6 = 0$ has no real solutions. (Its discriminant is negative.) The $y$-intercept is $f(0) = -2$. Plot the intercept $(0, -2)$.

***Step 2***  Identify any vertical asymptotes by checking for infinite limits (3 is the only number to check). Since $\displaystyle\lim_{x \to 3^-} \dfrac{x^2 - 2x + 6}{x - 3} = -\infty$

and $\displaystyle\lim_{x \to 3^+} \dfrac{x^2 - 2x + 6}{x - 3} = \infty$, the line $x = 3$ is a vertical asymptote.

The degree of the numerator of $f$ is 1 more than the degree of the denominator, so $f$ will have an oblique asymptote. We divide $x^2 - 2x + 6$ by $x - 3$ to find the line $y = mx + b$.

$$f(x) = \frac{x^2 - 2x + 6}{x - 3} = x + 1 + \frac{9}{x - 3}$$

Since $\displaystyle\lim_{x \to \infty} [f(x) - (x + 1)] = \lim_{x \to \infty} \dfrac{9}{x - 3} = 0$, then $y = x + 1$ is an oblique asymptote of the graph of $f$. Draw the asymptotes on the graph.

***Step 3***
$$f'(x) = \frac{d}{dx} \frac{x^2 - 2x + 6}{x - 3} = \frac{(2x - 2)(x - 3) - (x^2 - 2x + 6)(1)}{(x - 3)^2}$$

$$= \frac{x^2 - 6x}{(x - 3)^2} = \frac{x(x - 6)}{(x - 3)^2}$$

$$f''(x) = \frac{d}{dx} \frac{x^2 - 6x}{(x - 3)^2} = \frac{(2x - 6)(x - 3)^2 - (x^2 - 6x)[2(x - 3)]}{(x - 3)^4}$$

$$= (x - 3)\frac{(2x^2 - 12x + 18) - (2x^2 - 12x)}{(x - 3)^4} = \frac{18}{(x - 3)^3}$$

$f'(x) = 0$ at $x = 0$ and at $x = 6$. So, 0 and 6 are critical numbers. The tangent lines are horizontal at 0 and at 6. Since 3 is not in the domain of $f$, 3 is not a critical number.

***Step 4***  Use the numbers 0, 3, and 6 to form four intervals on the $x$-axis: $(-\infty, 0)$, $(0, 3)$, $(3, 6)$ and $(6, \infty)$. Then determine the sign of $f'(x)$ on each interval.

| Interval | Sign of $f'(x)$ | Conclusion |
|---|---|---|
| $(-\infty, 0)$ | positive | $f$ is increasing on $(-\infty, 0)$ |
| $(0, 3)$ | negative | $f$ is decreasing on $(0, 3)$ |
| $(3, 6)$ | negative | $f$ is decreasing on $(3, 6)$ |
| $(6, \infty)$ | positive | $f$ is increasing on $(6, \infty)$ |

***Step 5***  Using the First Derivative Test, we find that $f(0) = -2$ is a local maximum value and $f(6) = 10$ is a local minimum value. Plot the points $(0, -2)$ and $(6, 10)$.

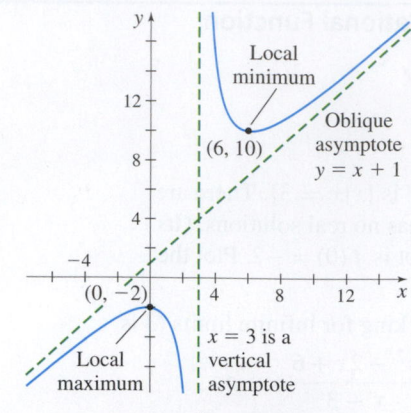

**Figure 51** $f(x) = \dfrac{x^2 - 2x + 6}{x - 3}$

**NEED TO REVIEW?** Properties of Limits at Infinity are discussed in Section 1.5, pp. 129–131.

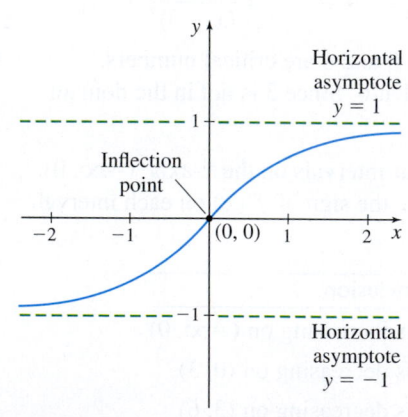

**Figure 52** $f(x) = \dfrac{x}{\sqrt{x^2 + 4}}$

**Step 6** $f''(x) = \dfrac{18}{(x - 3)^3}$. If $x < 3$, $f''(x) < 0$, and if $x > 3$, $f''(x) > 0$. From the Test for Concavity, we conclude $f$ is concave down on the interval $(-\infty, 3)$ and $f$ is concave up on the interval $(3, \infty)$. The concavity changes at 3, but 3 is not in the domain of $f$. So, the graph of $f$ has no inflection point.

**Step 7** Use the information to complete the graph of the function. See Figure 51. ∎

NOW WORK Problem 13.

---

**EXAMPLE 4    Using Calculus to Graph a Function**

Graph $f(x) = \dfrac{x}{\sqrt{x^2 + 4}}$.

**Solution**

**Step 1** The domain of $f$ is all real numbers. The only intercept is $(0, 0)$. Plot the point $(0, 0)$.

**Step 2** Check for horizontal asymptotes. Since

$$\lim_{x \to \infty} f(x) = \lim_{x \to \infty} \frac{x}{\sqrt{x^2 + 4}} = \sqrt{\lim_{x \to \infty} \frac{x^2}{x^2 + 4}} = 1$$

the line $y = 1$ is a horizontal asymptote as $x \to \infty$. Since $\lim\limits_{x \to -\infty} f(x) = -1$, the line $y = -1$ is also a horizontal asymptote as $x \to -\infty$. Draw the horizontal asymptotes on the graph.

Since $f$ is defined for all real numbers, there are no vertical asymptotes.

**Step 3**  $f'(x) = \dfrac{d}{dx} \dfrac{x}{\sqrt{x^2 + 4}} = \dfrac{1 \cdot \sqrt{x^2 + 4} - x \left[ \dfrac{1}{2}(x^2 + 4)^{-1/2} \cdot 2x \right]}{x^2 + 4}$

$= \dfrac{\sqrt{x^2 + 4} - \dfrac{x^2}{\sqrt{x^2 + 4}}}{x^2 + 4} = \dfrac{x^2 + 4 - x^2}{(x^2 + 4)\sqrt{x^2 + 4}} = \dfrac{4}{(x^2 + 4)^{3/2}}$

$f''(x) = \dfrac{d}{dx} \dfrac{4}{(x^2 + 4)^{3/2}} = 4 \left[ -\dfrac{3}{2}(x^2 + 4)^{-5/2} \right] \cdot 2x = -\dfrac{12x}{(x^2 + 4)^{5/2}}$

Since $f'(x)$ is never 0 and $f'$ exists for all real numbers, there are no critical numbers.

**Step 4** Since $f'(x) > 0$ for all $x$, $f$ is increasing on $(-\infty, \infty)$.

**Step 5** Because there are no critical numbers, there are no local extreme values.

**Step 6** Test for concavity. Since $f''(x) > 0$ on the interval $(-\infty, 0)$, $f$ is concave up on $(-\infty, 0)$; and since $f''(x) < 0$ on the interval $(0, \infty)$, $f$ is concave down on $(0, \infty)$. The concavity changes at 0, and 0 is in the domain of $f$, so the point $(0, 0)$ is an inflection point of $f$. Plot the inflection point.

**Step 7** Figure 52 shows the graph of $f$. ∎

Notice the apparent symmetry of the graph with respect to the origin. We can verify this by showing $f(-x) = -f(x)$, that is, by showing $f$ is an odd function.

NOW WORK Problem 25.

**EXAMPLE 5**  **Using Calculus to Graph a Function**

Graph $f(x) = 4x^{1/3} - x^{4/3}$.

**Solution**

**Step 1**  The domain of $f$ is all real numbers. Since $f(0) = 0$, the $y$-intercept is 0. Now $f(x) = 0$ when $4x^{1/3} - x^{4/3} = x^{1/3}(4-x) = 0$ or when $x = 0$ or $x = 4$. So, the $x$-intercepts are 0 and 4. Plot the intercepts $(0, 0)$ and $(4, 0)$.

**Step 2**  Since $\lim\limits_{x \to \pm\infty} f(x) = \lim\limits_{x \to \pm\infty} [x^{1/3}(4-x)] = -\infty$, there is no horizontal asymptote. Since the domain of $f$ is all real numbers, there is no vertical asymptote.

**Step 3**  $f'(x) = \dfrac{d}{dx}(4x^{1/3} - x^{4/3}) = \dfrac{4}{3}x^{-2/3} - \dfrac{4}{3}x^{1/3} = \dfrac{4}{3}\left(\dfrac{1}{x^{2/3}} - x^{1/3}\right)$

$\qquad = \dfrac{4}{3} \cdot \dfrac{1-x}{x^{2/3}}$

$f''(x) = \dfrac{d}{dx}\left(\dfrac{4}{3}x^{-2/3} - \dfrac{4}{3}x^{1/3}\right) = -\dfrac{8}{9}x^{-5/3} - \dfrac{4}{9}x^{-2/3} = -\dfrac{4}{9}\left(\dfrac{2}{x^{5/3}} + \dfrac{1}{x^{2/3}}\right)$

$\qquad = -\dfrac{4}{9} \cdot \dfrac{2+x}{x^{5/3}}$

Since $f'(x) = \dfrac{4}{3} \cdot \dfrac{1-x}{x^{2/3}} = 0$ when $x = 1$ and $f'(x)$ does not exist at $x = 0$, the critical numbers are 0 and 1. At the point $(1, 3)$, the tangent line to the graph is horizontal; at the point $(0, 0)$, the tangent line is vertical. Plot these points.

**Step 4**  Use the critical numbers 0 and 1 to form three intervals on the $x$-axis: $(-\infty, 0)$, $(0, 1)$ and $(1, \infty)$. Then determine the sign of $f'(x)$ on each interval.

| Interval | Sign of $f'$ | Conclusion |
|---|---|---|
| $(-\infty, 0)$ | positive | $f$ is increasing on $(-\infty, 0)$ |
| $(0, 1)$ | positive | $f$ is increasing on $(0, 1)$ |
| $(1, \infty)$ | negative | $f$ is decreasing on $(1, \infty)$ |

**Step 5**  By the First Derivative Test, $f(1) = 3$ is a local maximum value and $f(0) = 0$ is not a local extreme value.

**Step 6**  Now test for concavity by using the numbers $-2$ and 0 to form three intervals on the $x$-axis: $(-\infty, -2)$, $(-2, 0)$ and $(0, \infty)$. Then determine the sign of $f''(x)$ on each interval.

| Interval | Sign of $f''$ | Conclusion |
|---|---|---|
| $(-\infty, -2)$ | negative | $f$ is concave down on the interval $(-\infty, -2)$ |
| $(-2, 0)$ | positive | $f$ is concave up on the interval $(-2, 0)$ |
| $(0, \infty)$ | negative | $f$ is concave down on the interval $(0, \infty)$ |

The concavity changes at $-2$ and at 0. Since

$$f(-2) = 4(-2)^{1/3} - (-2)^{4/3} = 4\sqrt[3]{-2} - \sqrt[3]{16} \approx -7.560,$$

the inflection points are $(-2, -7.560)$ and $(0, 0)$. Plot the inflection points.

**Step 7**  The graph of $f$ is given in Figure 53. ∎

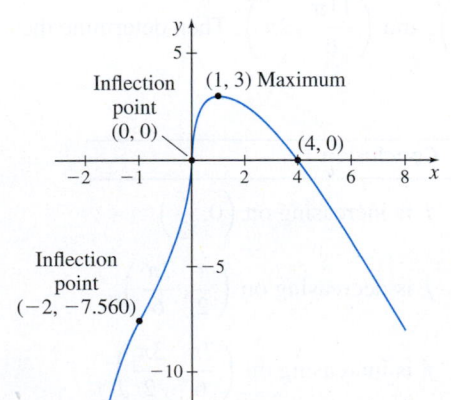

**Figure 53** $f(x) = 4x^{1/3} - x^{4/3}$

**NOW WORK** Problem 19.

EXAMPLE 6  **Using Calculus to Graph a Trigonometric Function**

Graph $f(x) = \sin x - \cos^2 x$, $0 \leq x \leq 2\pi$.

**Solution**

**Step 1**  The domain of $f$ is $\{x | 0 \leq x \leq 2\pi\}$. Since $f(0) = \sin 0 - \cos^2 0 = -1$, the $y$-intercept is $-1$. The $x$-intercepts satisfy the equation

$$\sin x - \cos^2 x = \sin x - (1 - \sin^2 x) = \sin^2 x + \sin x - 1 = 0$$

**NEED TO REVIEW?**  Trigonometric equations are discussed in Section P.7, pp. 65–66.

This trigonometric equation is quadratic in form, so we use the quadratic formula.

$$\sin x = \frac{-1 \pm \sqrt{1 - 4(1)(-1)}}{2} = \frac{-1 \pm \sqrt{5}}{2}$$

Since $\dfrac{-1 - \sqrt{5}}{2} < -1$ and $-1 \leq \sin x \leq 1$, the $x$-intercepts occur at

$$x = \sin^{-1}\left(\frac{-1 + \sqrt{5}}{2}\right) \approx 0.666 \text{ and at } x = \pi - \sin^{-1}\left(\frac{-1 + \sqrt{5}}{2}\right) \approx 2.475.$$

Plot the intercepts.

**Step 2**  The function $f$ has no asymptotes.

**Step 3**  $f'(x) = \dfrac{d}{dx}(\sin x - \cos^2 x) = \cos x + 2 \cos x \sin x = \cos x (1 + 2 \sin x)$

$$f''(x) = \frac{d}{dx}[\cos x (1 + 2 \sin x)] = 2 \cos^2 x - \sin x (1 + 2 \sin x)$$

$$= 2 \cos^2 x - \sin x - 2 \sin^2 x = -4 \sin^2 x - \sin x + 2$$

The critical numbers occur where $f'(x) = 0$. That is, where

$$\cos x = 0 \qquad \text{or} \qquad 1 + 2 \sin x = 0$$

In the interval $[0, 2\pi]$: $\cos x = 0$ if $x = \dfrac{\pi}{2}$ or if $x = \dfrac{3\pi}{2}$;

and $1 + 2 \sin x = 0$ when $\sin x = -\dfrac{1}{2}$, that is, when $x = \dfrac{7\pi}{6}$ or $x = \dfrac{11\pi}{6}$.

So, the critical numbers are $\dfrac{\pi}{2}, \dfrac{7\pi}{6}, \dfrac{3\pi}{2}$, and $\dfrac{11\pi}{6}$. At each of these numbers, the tangent lines are horizontal.

**Step 4**  Use the numbers $0, \dfrac{\pi}{2}, \dfrac{7\pi}{6}, \dfrac{3\pi}{2}, \dfrac{11\pi}{6}$, and $2\pi$ to form five intervals: $\left(0, \dfrac{\pi}{2}\right)$, $\left(\dfrac{\pi}{2}, \dfrac{7\pi}{6}\right), \left(\dfrac{7\pi}{6}, \dfrac{3\pi}{2}\right), \left(\dfrac{3\pi}{2}, \dfrac{11\pi}{6}\right)$, and $\left(\dfrac{11\pi}{6}, 2\pi\right)$. Then determine the sign of $f'(x)$ on each interval.

| Interval | Sign of $f'$ | Conclusion |
|---|---|---|
| $\left(0, \dfrac{\pi}{2}\right)$ | positive | $f$ is increasing on $\left(0, \dfrac{\pi}{2}\right)$ |
| $\left(\dfrac{\pi}{2}, \dfrac{7\pi}{6}\right)$ | negative | $f$ is decreasing on $\left(\dfrac{\pi}{2}, \dfrac{7\pi}{6}\right)$ |
| $\left(\dfrac{7\pi}{6}, \dfrac{3\pi}{2}\right)$ | positive | $f$ is increasing on $\left(\dfrac{7\pi}{6}, \dfrac{3\pi}{2}\right)$ |
| $\left(\dfrac{3\pi}{2}, \dfrac{11\pi}{6}\right)$ | negative | $f$ is decreasing on $\left(\dfrac{3\pi}{2}, \dfrac{11\pi}{6}\right)$ |
| $\left(\dfrac{11\pi}{6}, 2\pi\right)$ | positive | $f$ is increasing on $\left(\dfrac{11\pi}{6}, 2\pi\right)$ |

**Step 5** By the First Derivative Test, $f\left(\dfrac{\pi}{2}\right)=1$ and $f\left(\dfrac{3\pi}{2}\right)=-1$ are local maximum

values, and $f\left(\dfrac{7\pi}{6}\right)=-\dfrac{5}{4}$ and $f\left(\dfrac{11\pi}{6}\right)=-\dfrac{5}{4}$ are local minimum values.
Plot these points.

**Step 6** Use the Test for Concavity. To solve $f''(x)>0$ and $f''(x)<0$, we first solve
the equation $f''(x)=-4\sin^2 x-\sin x+2=0$, or equivalently,

$$4\sin^2 x+\sin x-2=0 \qquad 0\le x\le 2\pi$$

$$\sin x=\dfrac{-1\pm\sqrt{1+32}}{8}$$

$$\sin x\approx 0.593 \qquad\text{or}\qquad \sin x\approx -0.843$$
$$x\approx 0.63 \qquad x\approx 2.51 \qquad x\approx 4.14 \qquad x\approx 5.28$$

Use these numbers to form five subintervals of $[0,2\pi]$ and determine the
sign of $f''(x)$ on each interval.

| Interval | Sign of $f''$ | Conclusion |
|---|---|---|
| $(0,0.63)$ | positive | $f$ is concave up on the interval $(0,0.63)$ |
| $(0.63,2.51)$ | negative | $f$ is concave down on the interval $(0.63,2.51)$ |
| $(2.51,4.14)$ | positive | $f$ is concave up on the interval $(2.51,4.14)$ |
| $(4.14,5.28)$ | negative | $f$ is concave down on the interval $(4.14,5.28)$ |
| $(5.28,2\pi)$ | positive | $f$ is concave up on the interval $(5.28,2\pi)$ |

The inflection points are $(0.63,-0.06)$, $(2.51,-0.06)$, $(4.14,-1.13)$,
and $(5.28,-1.13)$. Plot the inflection points.

**Step 7** The graph of $f$ is given in Figure 54. ∎

**NOW WORK** Problem 33.

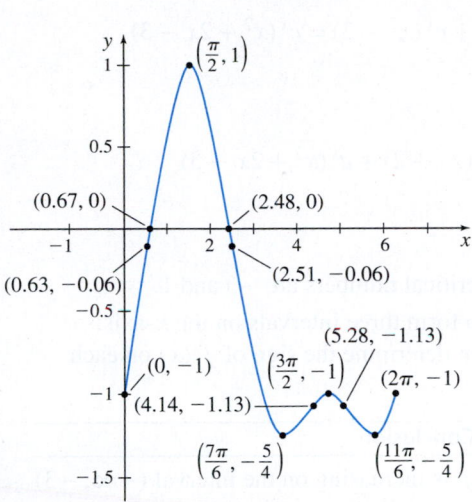

**Figure 54** $f(x)=\sin x-\cos^2 x$,
$0\le x\le 2\pi$

**NOTE** The function $f$ in Example 6
is periodic with period $2\pi$. To graph
this function over its unrestricted
domain, the set of real numbers,
repeat the graph in Figure 54 over
intervals of length $2\pi$.

---

**EXAMPLE 7    Using Calculus to Graph a Function**

Graph $f(x)=e^x(x^2-3)$.

**Solution**

**Step 1** The domain of $f$ is all real numbers. Since $f(0)=-3$, the $y$-intercept is $-3$.
The $x$-intercepts occur where

$$f(x)=e^x(x^2-3)=0$$
$$x=\sqrt{3}\qquad\text{or}\qquad x=-\sqrt{3}$$

Plot the intercepts.

**Step 2** Since the domain of $f$ is all real numbers, there are no vertical asymptotes. To
determine if there are any horizontal asymptotes, we find the limits at infinity.
First we find

$$\lim_{x\to-\infty}f(x)=\lim_{x\to-\infty}[e^x(x^2-3)]$$

Since $e^x(x^2-3)$ is an indeterminate form at $-\infty$ of the type $0\cdot\infty$, we

write $e^x(x^2-3)=\dfrac{x^2-3}{e^{-x}}$ and use L'Hôpital's Rule.

$$\lim_{x\to-\infty}f(x)=\lim_{x\to-\infty}[e^x(x^2-3)]=\lim_{x\to-\infty}\dfrac{x^2-3}{e^{-x}}$$
$$=\lim_{\underset{\substack{\uparrow\\ \text{Type }\frac{\infty}{\infty};\ \text{use}\\ \text{L'Hôpital's Rule}}}{x\to-\infty}}\dfrac{2x}{-e^{-x}}=\lim_{\underset{\substack{\uparrow\\ \text{Type }\frac{\infty}{\infty};\ \text{use}\\ \text{L'Hôpital's Rule}}}{x\to-\infty}}\dfrac{2}{e^{-x}}=0$$

The line $y=0$ is a horizontal asymptote as $x\to-\infty$.

Since $\lim\limits_{x\to\infty}[e^x(x^2-3)]=\infty$, there is no horizontal asymptote as $x\to\infty$.
Draw the asymptote on the graph.

**Step 3**  $f'(x)=\dfrac{d}{dx}[e^x(x^2-3)]=e^x(2x)+e^x(x^2-3)=e^x(x^2+2x-3)$

$$=e^x(x+3)(x-1)$$

$f''(x)=\dfrac{d}{dx}[e^x(x^2+2x-3)]=e^x(2x+2)+e^x(x^2+2x-3)$

$$=e^x(x^2+4x-1)$$

Solving $f'(x)=0$, we find that the critical numbers are $-3$ and $1$.

**Step 4**  Use the critical numbers $-3$ and $1$ to form three intervals on the $x$-axis: $(-\infty,-3)$, $(-3,1)$ and $(1,\infty)$. Then determine the sign of $f'(x)$ on each interval.

| Interval | Sign of $f'$ | Conclusion |
|---|---|---|
| $(-\infty,-3)$ | positive | $f$ is increasing on the interval $(-\infty,-3)$ |
| $(-3,1)$ | negative | $f$ is decreasing on the interval $(-3,1)$ |
| $(1,\infty)$ | positive | $f$ is increasing on the interval $(1,\infty)$ |

**Step 5**  Use the First Derivative Test to identify the local extrema. From the table in Step 4, there is a local maximum at $-3$ and a local minimum at $1$. Then $f(-3)=6e^{-3}\approx 0.30$ is a local maximum value and $f(1)=-2e\approx -5.44$ is a local minimum value. Plot the local extrema.

**Step 6**  To determine the concavity of $f$, first we solve $f''(x)=0$. We find $x=-2\pm\sqrt{5}$. Now use the numbers $-2-\sqrt{5}\approx -4.24$ and $-2+\sqrt{5}\approx 0.24$ to form three intervals on the $x$-axis: $(-\infty,-2-\sqrt{5})$, $(-2-\sqrt{5},-2+\sqrt{5})$ and $(-2+\sqrt{5},\infty)$. Then apply the Test for Concavity.

| Interval | Sign of $f''$ | Conclusion |
|---|---|---|
| $(-\infty,-2-\sqrt{5})$ | positive | $f$ is concave up on the interval $(-\infty,-2-\sqrt{5})$ |
| $(-2-\sqrt{5},-2+\sqrt{5})$ | negative | $f$ is concave down on the interval $(-2-\sqrt{5},-2+\sqrt{5})$ |
| $(-2+\sqrt{5},\infty)$ | positive | $f$ is concave up on the interval $(-2+\sqrt{5},\infty)$ |

The inflection points are $(-4.24,0.22)$ and $(0.24,-3.73)$. Plot the inflection points.

**Step 7**  The graph of $f$ is given in Figure 55. ∎

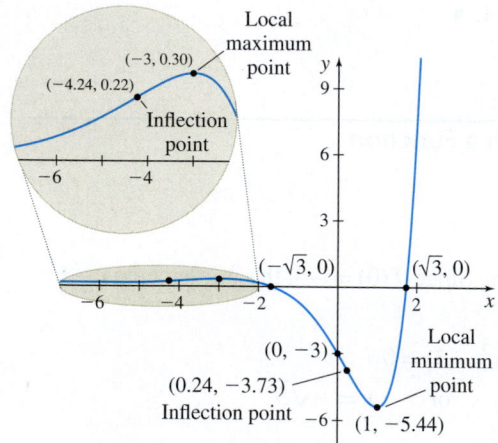

**Figure 55** $f(x)=e^x(x^2-3)$

NOW WORK Problem 39.

---

## 4.6 Assess Your Understanding

### Skill Building

*In Problems 1–42, use calculus to graph each function. Follow the steps for graphing a function on page 321.*

**1.** $f(x)=x^4-6x^2+10$

**2.** $f(x)=x^4-4x$

**3.** $f(x)=x^5-10x^2$

**4.** $f(x)=x^5-3x^3+4$

**5.** $f(x)=3x^5+5x^4$

**6.** $y=60x^5+20x^3$

**7.** $f(x)=\dfrac{2}{x^2-4}$

**8.** $f(x)=\dfrac{1}{x^2-1}$

**9.** $f(x)=\dfrac{2x-1}{x+1}$

**10.** $f(x)=\dfrac{x-2}{x}$

**11.** $f(x)=\dfrac{x}{x^2+1}$

**12.** $f(x)=\dfrac{2x}{x^2-4}$

**13.** $f(x)=\dfrac{x^2+1}{2x}$

**14.** $f(x)=\dfrac{x^2-1}{2x}$

---

**1.** = NOW WORK problem          📈 = Graphing technology recommended          CAS = Computer Algebra System recommended

**15.** $f(x) = \dfrac{x^2 - 1}{x^2 + 2x - 3}$

**16.** $f(x) = \dfrac{x^2 - x - 12}{x^2 - 9}$

**17.** $f(x) = \dfrac{x(x^3 + 1)}{(x^2 - 4)(x + 1)}$

**18.** $f(x) = \dfrac{x(x^3 - 1)}{(x + 1)^2(x - 1)}$

**19.** $f(x) = 1 + \dfrac{1}{x} + \dfrac{1}{x^2}$

**20.** $f(x) = \dfrac{2}{x} + \dfrac{1}{x^2}$

**21.** $f(x) = \sqrt{3 - x}$

**22.** $f(x) = x\sqrt{x + 2}$

**23.** $f(x) = x + \sqrt{x}$

**24.** $f(x) = \sqrt{x} - \sqrt{x + 1}$

**25.** $f(x) = \dfrac{x^2}{\sqrt{x + 1}}$

**26.** $f(x) = \dfrac{x}{\sqrt{x^2 + 2}}$

**27.** $f(x) = \dfrac{1}{(x + 1)(x - 2)}$

**28.** $f(x) = \dfrac{1}{(x - 1)(x + 3)}$

**29.** $f(x) = x^{2/3} + 3x^{1/3} + 2$

**30.** $f(x) = x^{5/3} - 5x^{2/3}$

**31.** $f(x) = \sin x - \cos x$

**32.** $f(x) = \sin x + \tan x$

**33.** $f(x) = \sin^2 x - \cos x$

**34.** $f(x) = \cos^2 x + \sin x$

**35.** $y = \ln x - x$

**36.** $y = x \ln x$

**37.** $f(x) = \ln(4 - x^2)$

**38.** $y = \ln(x^2 + 2)$

**39.** $f(x) = 3e^{3x}(5 - x)$

**40.** $f(x) = 3e^{-3x}(x - 4)$

**41.** $f(x) = e^{-x^2}$

**42.** $f(x) = e^{1/x}$

## Applications and Extensions

📐 *In Problems 43–52, for each function:*

   *(a) Graph the function.*

   *(b) Identify any asymptotes.*

   *(c) Use the graph to identify intervals on which the function increases or decreases and the intervals where the function is concave up or down.*

   *(d) Approximate the local extreme values using the graph.*

   *(e) Compare the approximate local maxima and local minima to the exact local extrema found by using calculus.*

   *(f) Approximate any inflection points using the graph.*

**43.** $f(x) = \dfrac{x^{2/3}}{x - 1}$

**44.** $f(x) = \dfrac{5 - x}{x^2 + 3x + 4}$

**45.** $f(x) = x + \sin(2x)$

**46.** $f(x) = x - \cos x$

**47.** $f(x) = \ln(x\sqrt{x - 1})$

**48.** $f(x) = \ln(\tan^2 x)$

**49.** $f(x) = \sqrt[3]{\sin x}$

**50.** $f(x) = e^{-x}\cos x$

**51.** $y^2 = x^2(6 - x),\ y \geq 0$

**52.** $y^2 = x^2(4 - x^2),\ y \geq 0$

*In Problems 53–56, graph a function $f$ that is continuous on the interval [1, 6] and satisfies the given conditions.*

**53.** $f'(2)$ does not exist

   $f'(3) = -1$

   $f''(3) = 0$

   $f'(5) = 0$

   $f''(x) < 0,\quad 2 < x < 3$

   $f''(x) > 0,\quad x > 3$

**54.** $f'(2) = 0$

   $f''(2) = 0$

   $f'(3)$ does not exist

   $f'(5) = 0$

   $f''(x) > 0,\quad 2 < x < 3$

   $f''(x) > 0,\quad x > 3$

**55.** $f'(2) = 0$

   $\lim\limits_{x \to 3^-} f'(x) = \infty$

   $\lim\limits_{x \to 3^+} f'(x) = \infty$

   $f'(5) = 0$

   $f''(x) > 0,\quad x < 3$

   $f''(x) < 0,\quad x > 3$

**56.** $f'(2) = 0$

   $\lim\limits_{x \to 3^-} f'(x) = -\infty$

   $\lim\limits_{x \to 3^+} f'(x) = -\infty$

   $f'(5) = 0$

   $f''(x) < 0,\quad x < 3$

   $f''(x) > 0,\quad x > 3$

**57.** Graph a function $f$ defined and continuous for $-1 \leq x \leq 2$ that satisfies the following conditions:

$$f(-1) = 1 \quad f(1) = 2 \quad f(2) = 3 \quad f(0) = 0 \quad f\!\left(\dfrac{1}{2}\right) = 3$$

$$\lim\limits_{x \to -1^+} f'(x) = -\infty \quad \lim\limits_{x \to 1^-} f'(x) = -1 \quad \lim\limits_{x \to 1^+} f'(x) = \infty$$

   $f$ has a local minimum at 0   $f$ has a local maximum at $\dfrac{1}{2}$

**58. Graph of a Function**   Which of the following is true about the graph of $y = \ln|x^2 - 1|$ in the interval $(-1, 1)$?

   **(a)** The graph is increasing.

   **(b)** The graph has a local minimum at $(0, 0)$.

   **(c)** The graph has a range of all real numbers.

   **(d)** The graph is concave down.

   **(e)** The graph has an asymptote $x = 0$.

**59. Properties of a Function**   Suppose $f(x) = \dfrac{1}{x} + \ln x$ is defined only on the closed interval $\dfrac{1}{e} \leq x \leq e$.

   **(a)** Determine the numbers $x$ at which $f$ has its absolute maximum and absolute minimum.

   **(b)** For what numbers $x$ is the graph concave up?

   **(c)** Graph $f$.

**60. Probability**   The function $f(x) = \dfrac{1}{\sqrt{2\pi}}e^{-x^2/2}$, encountered in probability theory, is called the **standard normal density function**. Determine where this function is increasing and decreasing, find all local maxima and local minima, find all inflection points, and determine the intervals where $f$ is concave up and concave down. Then graph the function.

*In Problems 61–64, graph each function. Use L'Hôpital's Rule to find any asymptotes.*

**61.** $f(x) = \dfrac{\sin(3x)}{x\sqrt{4 - x^2}}$

**62.** $f(x) = x\sqrt{x}$

**63.** $f(x) = x^{1/x}$

**64.** $y = \dfrac{1}{x}\tan x \quad -\dfrac{\pi}{2} < x < \dfrac{\pi}{2}$

# 4.7 Optimization

**OBJECTIVE** *When you finish this section, you should be able to:*

1 Solve optimization problems (p. 332)

Investigating maximum and/or minimum values of a function has been a recurring theme throughout most of this chapter. We continue this theme, using calculus to solve *optimization problems.*

**Optimization** is a process of determining the best solution to a problem. Often the best solution is one that maximizes or minimizes some quantity. For example, a business owner's goal usually involves minimizing cost while maximizing profit, or an engineer's goal may be to minimize the load on a beam. In this section, we model optimization problems using a function.

## 1 Solve Optimization Problems

Even though each individual problem has unique features, it is possible to outline a general method for obtaining an optimal solution.

### Steps for Solving Optimization Problems

**Step 1** Read the problem until you understand it and can identify the quantity for which a maximum or minimum value is to be found. Assign a symbol to represent it.

**Step 2** Assign symbols to represent the other variables in the problem. If possible, draw a picture and label it. Determine relationships among the variables.

**Step 3** Express the quantity to be maximized or minimized as a function of one of the variables, and determine a meaningful domain for the function.

**Step 4** Find the absolute maximum value or absolute minimum value of the function.

**NOTE** It is good practice to have in mind meaningful estimates of the answer.

---

**EXAMPLE 1**  **Maximizing an Area**

A farmer with 4000 m of fencing wants to enclose a rectangular plot that borders a straight river, as shown in Figure 56. If the farmer does not fence the side along the river, what is the largest rectangular area that can be enclosed?

**Solution**

**Step 1** The quantity to be maximized is the area; we denote it by $A$.

**Step 2** We denote the dimensions of the rectangle by $x$ and $y$ (both in meters), with the length of the side $y$ parallel to the river. The area is $A = xy$. Because there are 4000 m of fence available, the variables $x$ and $y$ are related by the equation

$$x + y + x = 4000$$

$$y = 4000 - 2x$$

**Step 3** Now express the area $A$ as a function of $x$.

$$A = A(x) = x(4000 - 2x) = 4000x - 2x^2 \qquad \color{blue}{A = xy, \quad y = 4000 - 2x}$$

The domain of $A$ is the closed interval $[0, 2000]$.

**NOTE** Since $A = A(x)$ is continuous on the closed interval $[0, 2000]$, it will have an absolute maximum.

**Step 4** To find the number $x$ that maximizes $A(x)$, differentiate $A$ with respect to $x$ and find the critical numbers:

$$A'(x) = 4000 - 4x = 0$$

**Figure 56** Area $A = xy$

The only critical number is $x = 1000$. The maximum value of $A$ occurs either at the critical number or at an endpoint of the interval $[0, 2000]$.

$$A(1000) = 2,000,000 \qquad A(0) = 0 \qquad A(2000) = 0$$

The maximum value is $A(1000) = 2,000,000 \, \text{m}^2$, which results from fencing a rectangular plot that measures 2000 m along the side parallel to the river and 1000 m along each of the other two sides. ∎

We could have reached the same conclusion by noting that $A''(x) = -4 < 0$ for all $x$, so $A$ is always concave down, and the local maximum at $x = 1000$ must be the absolute maximum of the function.

**NOW WORK** Problem 3.

CALC
CLIP

**EXAMPLE 2   Maximizing a Volume**

From each corner of a square piece of sheet metal 18 cm on a side, remove a small square and turn up the edges to form an open box. What are the dimensions of the box with the largest volume?

**Solution**

**Step 1** The quantity to be maximized is the volume of the box; denote it by $V$. Denote the length of each side of the small squares by $x$ and the length of each side after the small squares are removed by $y$, as shown in Figure 57. Both $x$ and $y$ are in centimeters.

**Step 2** Then $y = (18 - 2x)$ cm. The height of the box is $x$ cm, and the area of the base of the box is $y^2$ cm$^2$. So, the volume of the box is $V = xy^2$ cm$^3$.

**Step 3** To express $V$ as a function of one variable, substitute $y = 18 - 2x$ into the formula for $V$. Then the function to be maximized is

$$V = V(x) = x(18 - 2x)^2$$

Since both $x \geq 0$ and $18 - 2x \geq 0$, we find $x \leq 9$, meaning the domain of $V$ is the closed interval $[0, 9]$. (All other numbers make no physical sense—do you see why?)

**Step 4** To find the value of $x$ that maximizes $V$, differentiate $V$ and find the critical numbers.

$$V'(x) = \frac{d}{dx}[x(18 - 2x)^2] = \underset{\substack{\uparrow \\ \text{Product Rule}}}{2x(18 - 2x)(-2) + (18 - 2x)^2} = \underset{\substack{\uparrow \\ \text{Simplify and factor.}}}{(18 - 2x)(18 - 6x)}$$

Now solve $V'(x) = 0$ for $x$. The solutions are

$$x = 9 \qquad \text{or} \qquad x = 3$$

The only critical number in the open interval $(0, 9)$ is 3. Evaluate $V$ at 3 and at the endpoints 0 and 9.

$$V(0) = 0 \qquad V(3) = 3(18 - 6)^2 = 432 \qquad V(9) = 0$$

The maximum volume is 432 cm$^3$. The box with the maximum volume has a height of 3 cm. Since $y = 18 - 2 \cdot 3 = 12$ cm, the base of the box is square and measures 12 cm by 12 cm. ∎

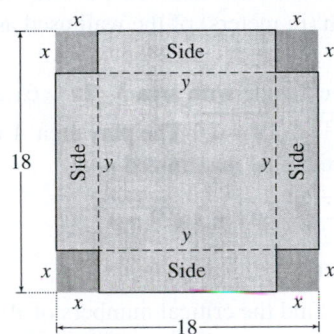

**Figure 57** $y = 18 - 2x$

**NOW WORK** Problems 7 and 9.

EXAMPLE 3   **Maximizing an Area**

A manufacturer makes a flexible square play yard (Figure 58a), with hinges at the four corners. It can be opened at one corner and attached at right angles to a wall or the side of a house, as shown in Figure 58(b). If each side is 3 m in length, the open configuration doubles the available area from 9 m² to 18 m². Is there a configuration that will more than double the play area?

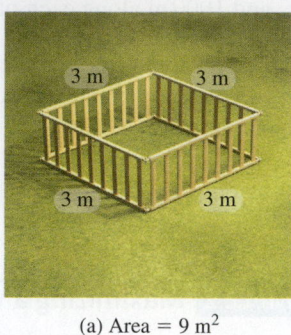

(a) Area = 9 m²                    (b) Area = 18 m²

**Figure 58**

**Solution**

**Step 1**  We want to maximize the play area $A$.

**Step 2**  Since the play yard must be attached at right angles to the wall, the possible configurations depend on the amount of wall used as a fifth side, as shown in Figure 59. Use $x$ to represent half the length (in meters) of the wall used as the fifth side.

**Step 3**  Partition the play area into two sections: a rectangle with area $3 \cdot 2x = 6x$ and a triangle with base $2x$ and altitude $\sqrt{3^2 - x^2} = \sqrt{9 - x^2}$. The play area $A$ is the sum of the areas of the two sections. The area to be maximized is

$$A = A(x) = 6x + \frac{1}{2}(2x)\sqrt{9 - x^2} = 6x + x\sqrt{9 - x^2}$$

The domain of $A$ is the closed interval $[0, 3]$.

**Step 4**  To find the maximum area of the play yard, find the critical numbers of $A$.

$$A'(x) = 6 + x\left[\frac{1}{2}(-2x)(9 - x^2)^{-1/2}\right] + \sqrt{9 - x^2} \quad \text{Use the Sum Rule and Product Rule.}$$

$$= 6 - \frac{x^2}{\sqrt{9 - x^2}} + \sqrt{9 - x^2} = \frac{6\sqrt{9 - x^2} - 2x^2 + 9}{\sqrt{9 - x^2}}$$

Critical numbers occur when $A'(x) = 0$ or where $A'(x)$ does not exist. $A'$ does not exist at $-3$ and $3$, and $A'(x) = 0$ at

$$6\sqrt{9 - x^2} - 2x^2 + 9 = 0 \qquad\qquad A'(x) = 0$$

$$\sqrt{9 - x^2} = \frac{2x^2 - 9}{6} \qquad\qquad \text{Isolate the radical on the left.}$$

$$9 - x^2 = \left(\frac{2x^2 - 9}{6}\right)^2 \qquad\qquad \text{Square both sides.}$$

$$9 - x^2 = \frac{4x^4 - 36x^2 + 81}{36} \qquad\qquad \text{Solve for } x.$$

$$324 - 36x^2 = 4x^4 - 36x^2 + 81$$

$$324 = 4x^4 + 81$$

$$x^4 = \frac{324 - 81}{4} = \frac{243}{4}$$

$$x = \sqrt[4]{\frac{243}{4}} \approx 2.792$$

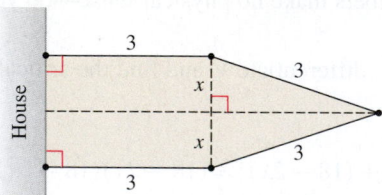

**DF Figure 59**

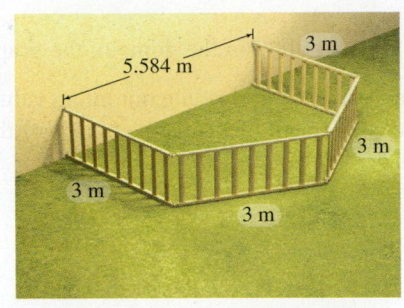

**Figure 60** Area ≈ 19.817 m²

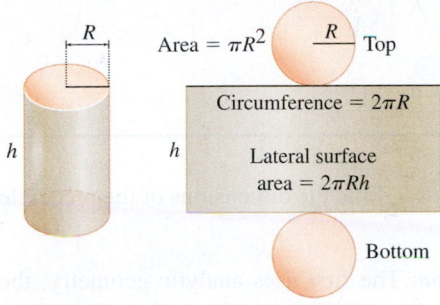

**Figure 61**

**NEED TO REVIEW?** Geometry
formulas are discussed in
Appendix A.2, p. A-15.

The only critical number in the open interval $(0, 3)$ is $\sqrt[4]{\dfrac{243}{4}} \approx 2.792$.

Now evaluate $A(x)$ at the endpoints 0 and 3 and at the critical number $x \approx 2.792$.

$$A(0) = 0 \qquad A(3) = 18 \qquad A(2.792) \approx 19.817$$

Using a wall of length $2x \approx 2(2.792) = 5.584$ m maximizes the area; the configuration shown in Figure 60 increases the play area by approximately 10% (from 18 to 19.817 m²). ∎

---

**EXAMPLE 4  Minimizing Cost**

A manufacturer needs to produce a cylindrical container with a capacity of 1000 cm³. The top and bottom of the container are made of material that costs \$0.05 per square centimeter, while the side of the container is made of material costing \$0.03 per square centimeter. Find the dimensions that will minimize the company's cost of producing the container.

**Solution** Figure 61 shows a cylindrical container and the area of its top, bottom, and lateral surface. As shown in the figure, let $h$ denote the height of the container and $R$ denote the radius. The total area of the bottom and top is $2\pi R^2$ cm². The area of the lateral surface of the can is $2\pi Rh$ cm².

The variables $h$ and $R$ are related. Since the capacity (volume) $V$ of the cylinder is 1000 cm³,

$$V = \pi R^2 h = 1000$$

$$h = \frac{1000}{\pi R^2}$$

The cost $C$, in dollars, of manufacturing the container is

$$C = (0.05)(2\pi R^2) + (0.03)(2\pi Rh) = 0.1\pi R^2 + 0.06\pi Rh$$

$\underset{\text{The cost of the top}}{\uparrow} \qquad \underset{\text{The cost of}}{\uparrow}$
The cost of the top        The cost of
and the bottom              the side

By substituting for $h$, $C$ can be expressed as a function of $R$.

$$C = C(R) = 0.1\pi R^2 + (0.06\pi R)\left(\frac{1000}{\pi R^2}\right) = 0.1\pi R^2 + \frac{60}{R}$$

This is the function to be minimized. The domain of $C$ is $\{R \mid R > 0\}$.

To find the minimum cost, differentiate $C$ with respect to $R$.

$$C'(R) = 0.2\pi R - \frac{60}{R^2} = \frac{0.2\pi R^3 - 60}{R^2}$$

Solve $C'(R) = 0$ to find the critical numbers.

$$0.2\pi R^3 - 60 = 0$$

$$R^3 = \frac{300}{\pi}$$

$$R = \sqrt[3]{\frac{300}{\pi}} \approx 4.571 \text{ cm}$$

Now find $C''(R)$ and use the Second Derivative Test.

$$C''(R) = \frac{d}{dR}\left(0.2\pi R - \frac{60}{R^2}\right) = 0.2\pi + \frac{120}{R^3}$$

$$C''\left(\sqrt[3]{\frac{300}{\pi}}\right) = 0.2\pi + \frac{120\pi}{300} = 0.6\pi > 0$$

$C$ has a local minimum at $\sqrt[3]{\dfrac{300}{\pi}}$. Since $C''(R) > 0$ for all $R$ in the domain, the graph of $C$ is always concave up, and the local minimum value is the absolute minimum value. The radius of the container that minimizes the cost is $R \approx 4.571$ cm. The height of the container that minimizes the cost of the material is

$$h = \frac{1000}{\pi R^2} \approx \frac{1000}{20.892\pi} \approx 15.236 \text{ cm}$$

The minimum cost of the container is

$$C\left(\sqrt[3]{\frac{300}{\pi}}\right) = 0.1\pi \left(\sqrt[3]{\frac{300}{\pi}}\right)^2 + \frac{60}{\sqrt[3]{\dfrac{300}{\pi}}} \approx \$19.69 \qquad \blacksquare$$

**NOTE** If the costs of the materials for the top, bottom, and lateral surfaces of a cylindrical container are all equal, then the minimum total cost occurs when the surface area is minimum. It can be shown (see Problem 39) that for any fixed volume, the minimum surface area of a cylindrical container is obtained when the height equals twice the radius.

NOW WORK Problem 11.

EXAMPLE 5 **Maximizing Area**

A rectangle is inscribed in a semicircle of radius 2. Find the dimensions of the rectangle that has the maximum area.

**Solution** We present two methods of solution: The first uses analytic geometry, the second uses trigonometry. To begin, place the semicircle with its diameter along the $x$-axis and center at the origin. Then inscribe a rectangle in the semicircle, as shown in Figure 62. The length of the inscribed rectangle is $2x$ and its height is $y$.

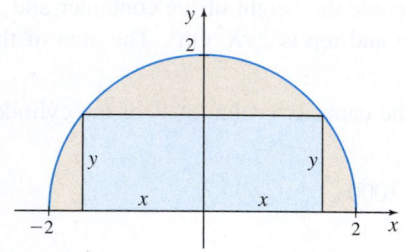

**Figure 62**

**Analytic Geometry Method** The area $A$ of the inscribed rectangle is $A = 2xy$ and the equation of the semicircle is $x^2 + y^2 = 4$, $y \geq 0$. We solve for $y$ in $x^2 + y^2 = 4$ and obtain $y = \sqrt{4 - x^2}$. Now substitute this expression for $y$ in the area formula for the rectangle to express $A$ as a function of $x$ alone.

$$A = A(x) = 2x\sqrt{4 - x^2} \qquad 0 \leq x \leq 2$$
$$\underset{A = 2xy;\ y = \sqrt{4 - x^2}}{\uparrow}$$

Then using the Product Rule,

$$A'(x) = 2\left[x \cdot \frac{1}{2}(4 - x^2)^{-1/2}(-2x) + \sqrt{4 - x^2}\right] = 2\left[\frac{-x^2}{\sqrt{4 - x^2}} + \sqrt{4 - x^2}\right]$$

$$= 2\left[\frac{-x^2 + 4 - x^2}{\sqrt{4 - x^2}}\right] = 2\left[\frac{-2(x^2 - 2)}{\sqrt{4 - x^2}}\right] = -\frac{4(x^2 - 2)}{\sqrt{4 - x^2}}$$

The only critical number in the open interval $(0, 2)$ is $\sqrt{2}$, where $A'(\sqrt{2}) = 0$. [The numbers $-\sqrt{2}$ and $-2$ are not in the domain of $A$, and $2$ is not in the open interval $(0, 2)$.] The values of $A$ at the endpoints $0$ and $2$ and at the critical number $\sqrt{2}$ are

$$A(0) = 0 \qquad A(\sqrt{2}) = 4 \qquad A(2) = 0$$

The maximum area of the inscribed rectangle is $4$, and it corresponds to the rectangle whose length is $2x = 2\sqrt{2}$ and whose height is $y = \sqrt{2}$.

**Trigonometry Method** Using Figure 62, draw the radius $r = 2$ from $O$ to the vertex of the rectangle and place the angle $\theta$ in the standard position, as shown in Figure 63. Then $x = 2\cos\theta$ and $y = 2\sin\theta$, $0 \leq \theta \leq \dfrac{\pi}{2}$. The area $A$ of the rectangle is

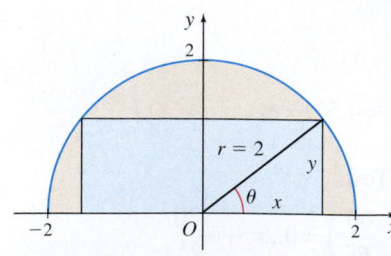

**Figure 63**

$$A = 2xy = 2(2\cos\theta)(2\sin\theta) = 8\cos\theta\sin\theta = 4\sin(2\theta)$$
$$\underset{\sin(2\theta) = 2\sin\theta\cos\theta}{\uparrow}$$

Since the area $A$ is a differentiable function of $\theta$, we obtain the critical numbers by finding $A'(\theta)$ and solving the equation $A'(\theta) = 0$.

$$A'(\theta) = 8\cos(2\theta) = 0 \quad A(\theta) = 4\sin(2\theta)$$

$$\cos(2\theta) = 0$$

$$2\theta = \cos^{-1} 0 = \frac{\pi}{2}$$

$$\theta = \frac{\pi}{4}$$

Now find $A''(\theta)$ and use the Second Derivative Test.

$$A''(\theta) = -16\sin(2\theta)$$

$$A''\left(\frac{\pi}{4}\right) = -16\sin\left(2 \cdot \frac{\pi}{4}\right) = -16\sin\frac{\pi}{2} = -16 < 0$$

So at $\theta = \dfrac{\pi}{4}$, the area $A$ is maximized and the maximum area of the inscribed rectangle is

$$A\left(\frac{\pi}{4}\right) = 4\sin\left(2 \cdot \frac{\pi}{4}\right) = 4 \text{ square units}$$

The rectangle with the maximum area has

$$\text{length } 2x = 4\cos\frac{\pi}{4} = 2\sqrt{2} \text{ and height } y = 2\sin\frac{\pi}{4} = \sqrt{2} \qquad ■$$

**NOW WORK** Problem 41.

---

**EXAMPLE 6**  **Snell's Law of Refraction**

**ORIGINS** In 1621 Willebrord Snell (c. 1580–1626) discovered the law of refraction, one of the basic principles of geometric optics. Pierre de Fermat (1601–1665) was able to prove the law mathematically using the principle that light follows the path that takes the least time.

Light travels at different speeds in different media (air, water, glass, etc.). Suppose that light travels from a point $A$ in one medium, where its speed is $c_1$, to a point $B$ in another medium, where its speed is $c_2$. See Figure 64. We use Fermat's principle that light always travels along the path that requires the least time to prove Snell's Law of Refraction.

$$\boxed{\frac{\sin\theta_1}{c_1} = \frac{\sin\theta_2}{c_2}}$$

**Solution** Position the coordinate system, as illustrated in Figure 65. The light passes from one medium to the other at the point $P$. Since the shortest distance between two points is a line, the path taken by the light is made up of two line segments—from $A = (0, a)$ to $P = (x, 0)$ and from $P = (x, 0)$ to $B = (k, -b)$, where $a$, $b$, and $k$ are positive constants.

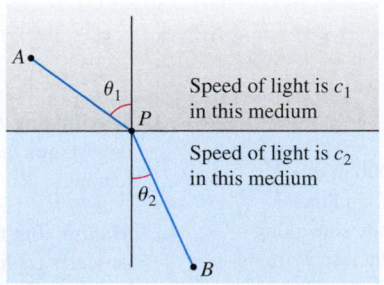

**Figure 64**

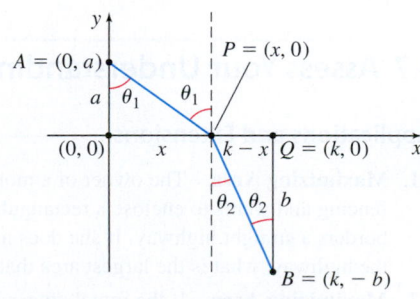

**Figure 65**

Since

$$\text{Time} = \frac{\text{Distance}}{\text{Speed}}$$

the travel time $t_1$ from $A = (0, a)$ to $P = (x, 0)$ is

$$t_1 = \frac{\sqrt{x^2 + a^2}}{c_1} \qquad c_1 \text{ is the speed of light in medium 1.}$$

and the travel time $t_2$ from $P = (x, 0)$ to $B = (k, -b)$ is

$$t_2 = \frac{\sqrt{(k - x)^2 + b^2}}{c_2} \qquad c_2 \text{ is the speed of light in medium 2.}$$

The total time $T = T(x)$ is given by

$$T(x) = t_1 + t_2 = \frac{\sqrt{x^2 + a^2}}{c_1} + \frac{\sqrt{(k - x)^2 + b^2}}{c_2}$$

To find the least time, find the critical numbers of $T$.

$$T'(x) = \frac{x}{c_1 \sqrt{x^2 + a^2}} - \frac{k - x}{c_2 \sqrt{(k - x)^2 + b^2}} = 0 \tag{1}$$

From Figure 65,

$$\frac{x}{\sqrt{x^2 + a^2}} = \sin \theta_1 \qquad \text{and} \qquad \frac{k - x}{\sqrt{(k - x)^2 + b^2}} = \sin \theta_2 \tag{2}$$

Using the result from (2) in equation (1), we have

$$T'(x) = \frac{\sin \theta_1}{c_1} - \frac{\sin \theta_2}{c_2} = 0$$

$$\frac{\sin \theta_1}{c_1} = \frac{\sin \theta_2}{c_2}$$

Now to be sure that the minimum value of $T$ occurs when $T'(x) = 0$, we show that $T''(x) > 0$. From (1),

$$T''(x) = \frac{d}{dx} \frac{x}{c_1 \sqrt{x^2 + a^2}} - \frac{d}{dx} \frac{k - x}{c_2 \sqrt{(k - x)^2 + b^2}}$$

$$= \frac{a^2}{c_1 (x^2 + a^2)^{3/2}} + \frac{b^2}{c_2 [(k - x)^2 + b^2]^{3/2}} > 0$$

Since $T''(x) > 0$ for all $x$, $T$ is concave up for all $x$, and the minimum value of $T$ occurs at the critical number. That is, $T$ is a minimum when $\dfrac{\sin \theta_1}{c_1} = \dfrac{\sin \theta_2}{c_2}$. ∎

## 4.7 Assess Your Understanding

### Applications and Extensions

1. **Maximizing Area**   The owner of a motel has 3000 m of fencing and wants to enclose a rectangular plot of land that borders a straight highway. If she does not fence the side along the highway, what is the largest area that can be enclosed?

2. **Maximizing Area**   If the motel owner in Problem 1 decides to also fence the side along the highway, except for 5 m to allow for access, what is the largest area that can be enclosed?

3. **Maximizing Area**   Find the dimensions of the rectangle with the largest area that can be enclosed on all sides by $L$ meters of fencing.

4. **Maximizing the Area of a Triangle**   An isosceles triangle has a perimeter of fixed length $L$. What should the dimensions of the triangle be if its area is to be a maximum? See the figure.

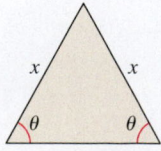

5. **Maximizing Area**   A gardener with 200 m of available fencing wishes to enclose a rectangular field and then divide it into two plots with a fence parallel to one of the sides, as shown in the figure. What is the largest area that can be enclosed?

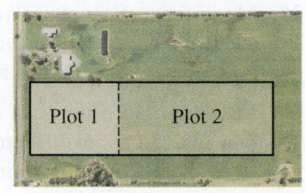

6. **Minimizing Fencing**   A realtor wishes to enclose 600 m$^2$ of land in a rectangular plot and then divide it into two plots with a fence parallel to one of the sides. What are the dimensions of the rectangular plot that require the least amount of fencing?

7. **Maximizing the Volume of a Box**   An open box with a square base is to be made from a square piece of cardboard that measures 12 cm on each side. A square will be cut out from each corner of the cardboard and the sides will be turned up to form the box. Find the dimensions that yield the maximum volume.

8. **Maximizing the Volume of a Box**   An open box with a rectangular base is to be made from a piece of tin measuring 30 cm by 45 cm by cutting out a square from each corner and turning up the sides. Find the dimensions that yield the maximum volume.

9. **Minimizing the Surface Area of a Box**   An open box with a square base is to have a volume of 2000 cm$^3$. What should the dimensions of the box be if the amount of material used is to be a minimum?

10. **Minimizing the Surface Area of a Box**   If the box in Problem 9 is to be closed on top, what should the dimensions of the box be if the amount of material used is to be a minimum?

11. **Minimizing the Cost to Make a Can**   A cylindrical container that has a capacity of 10 m$^3$ is to be produced. The top and bottom of the container are to be made of a material that costs $20 per square meter, while the side of the container is to be made of a material costing $15 per square meter. Find the dimensions that will minimize the cost of the material.

12. **Minimizing the Cost of Fencing**   A builder wishes to fence in 60,000 m$^2$ of land in a rectangular shape. For security reasons, the fence along the front part of the land will cost $20 per meter, while the fence for the other three sides will cost $10 per meter. How much of each type of fence should the builder buy to minimize the cost of the fence? What is the minimum cost?

13. **Maximizing Revenue**   A car rental agency has 24 cars (each an identical model). The owner of the agency finds that at a price of $18 per day, all the cars can be rented; however, for each $1 increase in rental cost, one of the cars is not rented. What should the agency charge to maximize income?

14. **Maximizing Revenue**   A charter flight club charges its members $200 per year. But for each new member in excess of 60, the charge for every member is reduced by $2. What number of members leads to a maximum revenue?

15. **Minimizing Distance**   Find the coordinates of the points on the graph of the parabola $y = x^2$ that are closest to the point $\left(2, \dfrac{1}{2}\right)$.

16. **Minimizing Distance**   Find the coordinates of the points on the graph of the parabola $y = 2x^2$ that are closest to the point $(1, 4)$.

17. **Minimizing Distance**   Find the coordinates of the points on the graph of the parabola $y = 4 - x^2$ that are closest to the point $(6, 2)$.

18. **Minimizing Distance**   Find the coordinates of the points on the graph of $y = \sqrt{x}$ that are closest to the point $(4, 0)$.

19. **Minimizing Transportation Cost**   A truck has a top speed of 75 mi/h and, when traveling at a speed of $x$ mi/h, consumes gasoline at the rate

of $\dfrac{1}{200}\left(\dfrac{1600}{x} + x\right)$

gallons per mile. The truck is to be taken on a 200-mi trip by a driver who is paid at the rate of $b$ per hour plus a commission of $c$. Since the time required for the trip at $x$ mi/h is $\dfrac{200}{x}$, the cost $C$ of the trip, when gasoline costs $a$ per gallon, is

$$C = C(x) = \left(\frac{1600}{x} + x\right)a + \frac{200}{x}b + c$$

Find the speed that minimizes the cost $C$ under each of the following conditions:

**(a)**  $a = \$3.50, b = 0, c = 0$

**(b)**  $a = \$3.50, b = \$10.00, c = \$500$

**(c)**  $a = \$4.00, b = \$20.00, c = 0$

20. **Optimal Placement of a Cable Box**   A telephone company is asked to provide cable service to a customer whose house is located 2 km away from the road along which the cable lines run. The nearest cable box is located 5 km down the road. As shown in the figure below, let $5 - x$ denote the distance from the box to the connection so that $x$ is the distance from this point to the point on the road closest to the house. If the cost to connect the cable line is $500 per kilometer along the road and $600 per kilometer away from the road, where along the road from the box should the company connect the cable line so as to minimize construction cost?

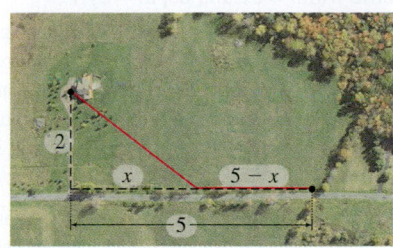

21. **Minimizing a Path**   Two houses $A$ and $B$ on the same side of a road are a distance $p$ apart, with distances $q$ and $r$, respectively, from the center of the road, as shown in the figure below. Find the length of the shortest path that goes from $A$ to the road and then on to the other house $B$.

**(a)**  Use calculus.

**(b)**  Use only elementary geometry.

(*Hint:* Reflect $B$ across the road to a point $C$ that is also a distance $r$ from the center of the road.)

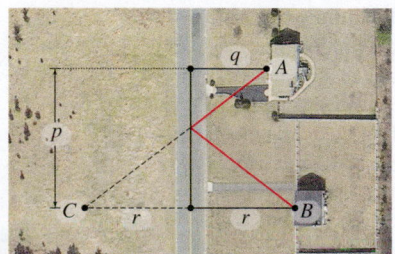

**22. Minimizing Travel Time**   A small island is 3 km from the nearest point $P$ on the straight shoreline of a large lake. A town is 12 km down the shore from $P$. See the figure. If a person on the island can row a boat 2.5 km/h and can walk 4 km/h, where should the boat be landed so that the person arrives in town in the shortest time?

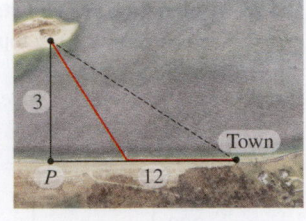

**23. Supporting a Wall**   Find the length of the shortest beam that can be used to brace a wall if the beam is to pass over a second wall 2 m high and 5 m from the first wall. See the figure. What is the angle of elevation of the beam?

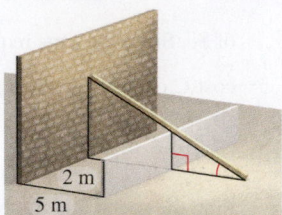

**24. Maximizing the Strength of a Beam**   The strength of a rectangular beam is proportional to the product of the width and the cube of its depth. Find the dimensions of the strongest beam that can be cut from a log whose cross section has the form of a circle of fixed radius $R$. See the figure.

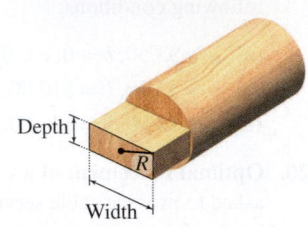

**25. Maximizing the Strength of a Beam**   If the strength of a rectangular beam is proportional to the product of its width and the square of its depth, find the dimensions of the strongest beam that can be cut from a log whose cross section has the form of the ellipse $10x^2 + 9y^2 = 90$.
(*Hint:* Choose width $= 2x$ and depth $= 2y$.)

**26. Maximizing the Strength of a Beam**   The strength of a beam made from a certain wood is proportional to the product of its width and the cube of its depth. Find the dimensions of the rectangular cross section of the beam with maximum strength that can be cut from a log whose original cross section is in the form of the ellipse $b^2x^2 + a^2y^2 = a^2b^2$, $a \geq b$.

**27. Pricing Wine**   A winemaker in Walla Walla, Washington, is producing the first vintage for her own label, and she needs to know how much to charge per case of wine. It costs her $132 per case to make the wine. She understands from industry research and an assessment of her marketing list that she can sell $x = 1430 - \dfrac{11}{6}p$ cases of wine, where $p$ is the price of a case in dollars. She can make at most 1100 cases of wine in her production facility. How many cases of wine should she produce, and what price $p$ should she charge per case to maximize her profit $P$? (*Hint:* Maximize the profit $P = xp - 132x$, where $x$ equals the number of cases.)

**28. Optimal Window Dimensions**   A Norman window has the shape of a rectangle surmounted by a semicircle of diameter equal to the width of the rectangle, as shown in the figure. If the perimeter of the window is 10 m, what dimensions will admit the most light?

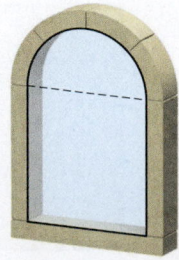

**29. Maximizing Volume**   The sides of a V-shaped trough are 28 cm wide. Find the angle between the sides of the trough that results in maximum capacity.

**30. Maximizing Volume**   A metal rain gutter is to have 10-cm sides and a 10-cm horizontal bottom, with the sides making equal angles with the bottom, as shown in the figure. How wide should the opening across the top be for maximum carrying capacity?

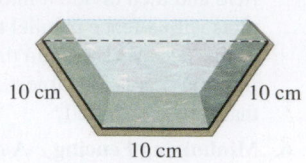

**31. Minimizing Construction Cost**   A proposed tunnel with a fixed cross-sectional area is to have a horizontal floor, vertical walls of equal height, and a ceiling that is a semicircular cylinder. If the ceiling costs three times as much per square meter to build as the vertical walls and the floor, find the most economical ratio of the diameter of the semicircular cylinder to the height of the vertical walls.

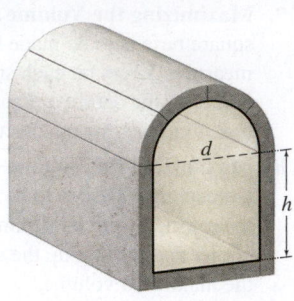

**32. Minimizing Construction Cost**   An observatory is to be constructed in the form of a right circular cylinder surmounted by a hemispherical dome. If the hemispherical dome costs three times as much per square meter as the cylindrical wall, what are the most economical dimensions for a given volume? Neglect the floor.

**33. Intensity of Light**   The intensity of illumination at a point varies inversely as the square of the distance between the point and the light source. Two lights, one having an intensity eight times that of the other, are 6 m apart. At what point between the two lights is the total illumination least?

**34. Drug Concentration**   The concentration of a drug in the bloodstream $t$ hours after injection into muscle tissue is given by $C(t) = \dfrac{2t}{16 + t^2}$. When is the concentration greatest?

**35. Optimal Wire Length**   A wire is to be cut into two pieces. One piece will be bent into a square, and the other piece will be bent into a circle. If the total area enclosed by the two pieces is to be 64 cm$^2$, what is the minimum length of wire that can be used? What is the maximum length of wire that can be used?

**36. Optimal Wire Length**   A wire is to be cut into two pieces. One piece will be bent into an equilateral triangle, and the other piece will be bent into a circle. If the total area enclosed by the two pieces is to be 64 cm$^2$, what is the minimum length of wire that can be used? What is the maximum length of wire that can be used?

**37. Optimal Area**   A wire 35 cm long is cut into two pieces. One piece is bent into the shape of a square, and the other piece is bent into the shape of a circle.
(a) How should the wire be cut so that the area enclosed is a minimum?
(b) How should the wire be cut so that the area enclosed is a maximum?
(c) Graph the area enclosed as a function of the length of the piece of wire used to make the square. Show that the graph confirms the results of (a) and (b).

**38. Optimal Area**   A wire 35 cm long is cut into two pieces. One piece is bent into the shape of an equilateral triangle, and the other piece is bent into the shape of a circle.

  (a) How should the wire be cut so that the area enclosed is a minimum?

  (b) How should the wire be cut so that the area enclosed is a maximum?

  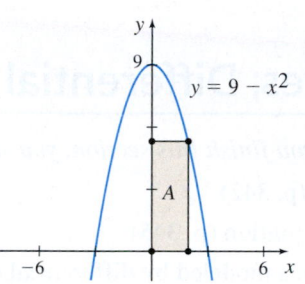 (c) Graph the area enclosed as a function of the length of the piece of wire used to make the triangle. Show that the graph confirms the results of (a) and (b).

**39. Optimal Dimensions for a Can**   Show that a cylindrical container of fixed volume $V$ requires the least material (minimum surface area) when its height is twice its radius.

**40. Maximizing Area**   Find the triangle of largest area that has two sides along the positive coordinate axes if its hypotenuse is tangent to the graph of $y = 3e^{-x}$.

**41. Maximizing Area**   Find the largest area of a rectangle with one vertex on the parabola $y = 9 - x^2$, another at the origin, and the remaining two on the positive $x$-axis and positive $y$-axis, respectively. See the figure.

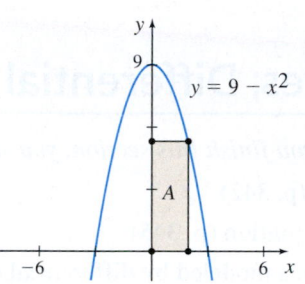

**42. Maximizing Volume**   Find the dimensions of the right circular cone of maximum volume having a slant height of 4 ft. See the figure.

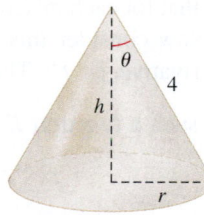

**43. Minimizing Distance**   Let $a$ and $b$ be two positive real numbers. Find the line through the point $(a, b)$ and connecting the points $(0, y_0)$ and $(x_0, 0)$ so that the distance from $(x_0, 0)$ to $(0, y_0)$ is a minimum. (In general, $x_0$ and $y_0$ will depend on the line.)

**44. Maximizing Velocity**   An object moves on the $x$-axis in such a way that its velocity at time $t$ seconds, $t \geq 1$, is given by $v = \dfrac{\ln t}{t}$ cm/s. At what time $t$ does the object attain its maximum velocity?

**45. Physics**   A heavy object of mass $m$ is to be dragged along a horizontal surface by a rope making an angle $\theta$ to the horizontal. The force $F$ required to move the object is given by the formula

$$F = \frac{cmg}{c \sin \theta + \cos \theta}$$

where $g$ is the acceleration due to gravity and $c$ is the **coefficient of friction** of the surface. Show that the force is least when $\tan \theta = c$.

**46. Chemistry**   A self-catalytic chemical reaction results in the formation of a product that causes its formation rate to increase. The reaction rate $V$ of many self-catalytic chemicals is given by

$$V = kx(a - x) \qquad 0 \leq x \leq a$$

where $k$ is a positive constant, $a$ is the initial amount of the chemical, and $x$ is the variable amount of the chemical. For what value of $x$ is the reaction rate a maximum?

**47. Optimal Viewing Angle**   A picture 4 m in height is hung on a wall with the lower edge 3 m above the level of an observer's eye. How far from the wall should the observer stand in order to obtain the most favorable view? (That is, the picture should subtend the maximum angle.)

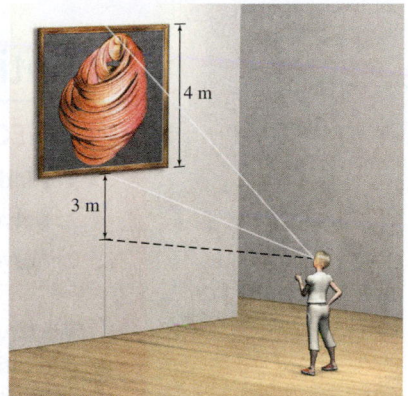

**48. Maximizing Transmission Speed**   Traditional telephone cable is made up of a core of copper wires covered by an insulating material. If $x$ is the ratio of the radius of the core to the thickness of the insulating material, the speed $v$ of signaling is $v = kx^2 \ln \dfrac{1}{x}$, where $k$ is a constant. Determine the ratio $x$ that results in maximum speed.

**49. Absolute Minimum**   If $a$, $b$, and $c$ are positive constants, show that the minimum value of $f(x) = ae^{cx} + be^{-cx}$ is $2\sqrt{ab}$.

## Challenge Problems

**50. Maximizing Length**   The figure shows two corridors meeting at a right angle. One has width 1 m, and the other, width 8 m. Find the length of the longest pipe that can be carried horizontally from one corridor, around the corner, and into the other corridor.

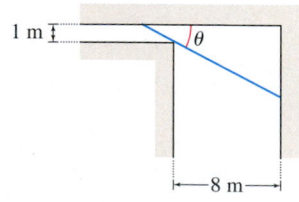

**51. Optimal Height of a Lamp**   In the figure, a circular area of radius 20 ft is surrounded by a walk. A light is placed above the center of the area. What height most strongly illuminates the walk? The intensity of illumination is given by $I = \dfrac{\sin \theta}{s}$, where $s$ is the distance from the source and $\theta$ is the angle at which the light strikes the surface.

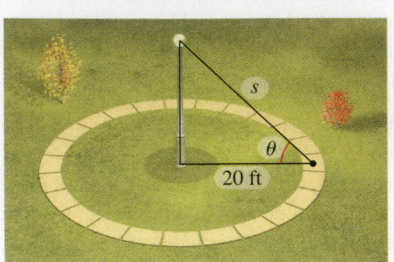

**52. Maximizing Area**   Show that the rectangle of largest area that can be inscribed under the graph of $y = e^{-x^2}$ has two of its vertices at the point of inflection of $y$. See the figure.

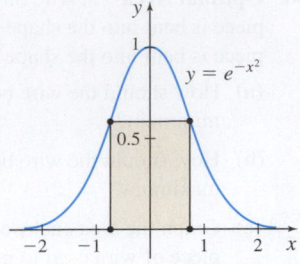

**CAS** **53. Minimizing Distance**   Find the point $(x, y)$ on the graph of $f(x) = e^{-x/2}$ that is closest to the point $(1, 8)$.

# 4.8 Antiderivatives; Differential Equations

**OBJECTIVES**  *When you finish this section, you should be able to:*

1  Find antiderivatives (p. 342)

2  Solve a differential equation (p. 345)

3  Solve applied problems modeled by differential equations (p. 347)

We have already learned that for each differentiable function $f$, there is a corresponding derivative function $f'$. Now consider this question: For a given function $f$, is there a function $F$ whose derivative is $f$? That is, is it possible to find a function $F$ so that $F' = \dfrac{dF}{dx} = f$? If such a function $F$ can be found, it is called an *antiderivative* of $f$.

> **DEFINITION**  Antiderivative
>
> A function $F$ is called an **antiderivative** of the function $f$ if $F'(x) = f(x)$ for all $x$ in the domain of $f$.

## 1  Find Antiderivatives

For example, an antiderivative of the function $f(x) = 2x$ is $F(x) = x^2$, since

$$F'(x) = \frac{d}{dx} x^2 = 2x$$

Another function $F$ whose derivative is $2x$ is $F(x) = x^2 + 3$, since

$$F'(x) = \frac{d}{dx}(x^2 + 3) = 2x$$

This leads us to suspect that the function $f(x) = 2x$ has many antiderivatives. Indeed, any of the functions $x^2$ or $x^2 + \dfrac{1}{2}$ or $x^2 + 2$ or $x^2 + \sqrt{5}$ or $x^2 - 1$ has the property that its derivative is $2x$. Any function $F(x) = x^2 + C$, where $C$ is a constant, is an antiderivative of $f(x) = 2x$.

Are there other antiderivatives of $2x$ that are not of the form $x^2 + C$? A corollary of the Mean Value Theorem (p. 288) tells us the answer is no.

**COROLLARY**

If $f$ and $g$ are differentiable functions and if $f'(x) = g'(x)$ for all numbers $x$ in an interval $(a, b)$, then there exists a number $C$ for which $f(x) = g(x) + C$ on $(a, b)$.

This result can be stated in the following way.

**THEOREM  The General Form of an Antiderivative**

If a function $F$ is an antiderivative of a function $f$ defined on an interval $I$, then any other antiderivative of $f$ has the form $F(x) + C$, where $C$ is an (arbitrary) constant.

All the antiderivatives of $f$ can be obtained from the expression $F(x) + C$ by letting $C$ range over all real numbers. For example, all the antiderivatives of $f(x) = x^5$ are of the form $F(x) = \dfrac{x^6}{6} + C$, where $C$ is a constant. Figure 66 shows the graphs of $F(x) = \dfrac{x^6}{6} + C$ for some numbers $C$. The antiderivatives of a function $f$ are a family of functions, each one a vertical translation of the others.

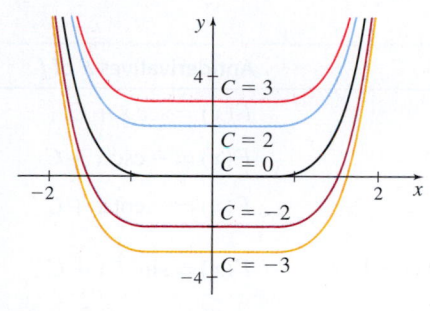

**Figure 66** $F(x) = \dfrac{x^6}{6} + C$

**EXAMPLE 1  Finding the Antiderivatives of a Function**

Find all the antiderivatives of:

**(a)** $f(x) = 0$      **(b)** $g(\theta) = -\sin\theta$      **(c)** $h(x) = x^{1/2}$

**Solution (a)** Since the derivative of a constant function is 0, all the antiderivatives of $f$ are of the form $F(x) = C$, where $C$ is a constant.

**(b)** Since $\dfrac{d}{d\theta}\cos\theta = -\sin\theta$, all the antiderivatives of $g(\theta) = -\sin\theta$ are of the form $G(\theta) = \cos\theta + C$, where $C$ is a constant.

**(c)** The derivative of $\dfrac{2}{3}x^{3/2}$ is $\dfrac{2}{3} \cdot \dfrac{3}{2}x^{3/2-1} = x^{1/2}$. So, all the antiderivatives of $h(x) = x^{1/2}$ are of the form $H(x) = \dfrac{2}{3}x^{3/2} + C$, where $C$ is a constant. ∎

In Example 1(c), you may wonder how we knew to choose $\dfrac{2}{3}x^{3/2}$. For any real number $a$, the Power Rule states $\dfrac{d}{dx}x^a = ax^{a-1}$. That is, differentiation reduces the exponent by 1. Antidifferentiation is the inverse process, so it suggests we increase the exponent by 1. This is how we obtain the factor $x^{3/2}$ of the antiderivative $\dfrac{2}{3}x^{3/2}$. The factor $\dfrac{2}{3}$ is needed so that when we differentiate $\dfrac{2}{3}x^{3/2}$, the result is $\dfrac{2}{3} \cdot \dfrac{3}{2}x^{3/2-1} = x^{1/2}$.

**NOW WORK** Problem 11.

Let $f(x) = x^a$, where $a$ is a real number and $a \neq -1$. (The case for which $a = -1$ requires special attention.) Then the function $F(x)$ defined by

$$F(x) = \frac{x^{a+1}}{a+1} \qquad a \neq -1$$

is an antiderivative of $f(x) = x^a$. That is, all the antiderivatives of $f(x) = x^a, a \neq -1$, are of the form $\dfrac{x^{a+1}}{a+1} + C$, where $C$ is a constant.

Now consider the case when $a = -1$. Then $f(x) = x^{-1} = \dfrac{1}{x}$. We know that $\dfrac{d}{dx} \ln|x| = \dfrac{1}{x} = x^{-1}$, for $x \neq 0$. So, all the antiderivatives of $f(x) = x^{-1}$ are of the form $\ln|x| + C$, provided $x \neq 0$.

**CAUTION** Be sure to include the absolute value bars when finding the antiderivative of $\dfrac{1}{x}$.

Table 6 includes these results along with the antiderivatives of some other common functions. Each listing is based on a derivative formula.

**TABLE 6**

| Function $f$ | Antiderivatives $F$ of $f$ | Function $f$ | Antiderivatives $F$ of $f$ |
|---|---|---|---|
| $f(x) = 0$ | $F(x) = C$ | $f(x) = \sec x \tan x$ | $F(x) = \sec x + C$ |
| $f(x) = 1$ | $F(x) = x + C$ | $f(x) = \csc x \cot x$ | $F(x) = -\csc x + C$ |
| $f(x) = x^a, \quad a \neq -1$ | $F(x) = \dfrac{x^{a+1}}{a+1} + C$ | $f(x) = \csc^2 x$ | $F(x) = -\cot x + C$ |
| $f(x) = x^{-1} = \dfrac{1}{x}$ | $F(x) = \ln|x| + C$ | $f(x) = \dfrac{1}{\sqrt{1-x^2}}, \quad |x| < 1$ | $F(x) = \sin^{-1} x + C$ |
| $f(x) = e^x$ | $F(x) = e^x + C$ | $f(x) = \dfrac{1}{1+x^2}$ | $F(x) = \tan^{-1} x + C$ |
| $f(x) = a^x$ | $F(x) = \dfrac{a^x}{\ln a} + C, \quad a > 0, a \neq 1$ | $f(x) = \dfrac{1}{x\sqrt{x^2-1}}, \quad |x| > 1$ | $F(x) = \sec^{-1} x + C$ |
| $f(x) = \sin x$ | $F(x) = -\cos x + C$ | $f(x) = \sinh x$ | $F(x) = \cosh x + C$ |
| $f(x) = \cos x$ | $F(x) = \sin x + C$ | $f(x) = \cosh x$ | $F(x) = \sinh x + C$ |
| $f(x) = \sec^2 x$ | $F(x) = \tan x + C$ | | |

The next two theorems are consequences of properties of derivatives and the relationship between derivatives and antiderivatives.

**THEOREM Sum Rule and Difference Rule**

If the functions $F_1$ and $F_2$ are antiderivatives of the functions $f_1$ and $f_2$, respectively, then $F_1 + F_2$ is an antiderivative of $f_1 + f_2$ and $F_1 - F_2$ is an antiderivative of $f_1 - f_2$.

The Sum Rule can be extended to any finite sum of functions.

**IN WORDS** An antiderivative of the sum of two functions equals the sum of the antiderivatives of the functions. An antiderivative of a number times a function equals the number times an antiderivative of the function.

**THEOREM Constant Multiple Rule**

If $k$ is a real number and if $F$ is an antiderivative of $f$, then $kF$ is an antiderivative of $kf$.

**EXAMPLE 2** Finding the Antiderivatives of a Function

Find all the antiderivatives of $f(x) = e^x + \dfrac{6}{x^2} - \sin x$.

**Solution** Since $f$ is the sum of three functions, we use the Sum Rule. That is, we find the antiderivatives of each function individually and then add.

**NOTE** Since the antiderivatives of $e^x$ are $e^x + C_1$, the antiderivatives of $\dfrac{6}{x^2}$ are $-\dfrac{6}{x} + C_2$, and the antiderivatives of $\sin x$ are $-\cos x + C_3$, the constant $C$ in Example 2 is $C_1 + C_2 + C_3$.

- An antiderivative of $e^x$ is $e^x$.
- An antiderivative of $6x^{-2}$ is $6 \cdot \dfrac{x^{-2+1}}{-2+1} = 6 \cdot \dfrac{x^{-1}}{-1} = -\dfrac{6}{x}$.
- Finally, an antiderivative of $\sin x$ is $-\cos x$.

Then all the antiderivatives of the function $f$ are given by

$$F(x) = e^x - \dfrac{6}{x} + \cos x + C$$

where $C$ is a constant. ∎

**NOW WORK** Problem **29**.

## 2 Solve a Differential Equation

In studies of physical, chemical, biological, and other phenomena, scientists attempt to find mathematical laws that describe and predict observed behavior. These laws often involve the derivatives of an unknown function $F$, which must be determined.

For example, suppose we seek all functions $y = F(x)$ for which

$$\frac{dy}{dx} = F'(x) = f(x)$$

An equation of the form $\dfrac{dy}{dx} = f(x)$ is an example of a *differential equation*.

Any function $y = F(x)$, for which $\dfrac{dy}{dx} = F'(x) = f(x)$, is a **solution** of the differential equation. The **general solution** of a differential equation $\dfrac{dy}{dx} = f(x)$ consists of all the antiderivatives of $f$.

For example, the general solution of the differential equation $\dfrac{dy}{dx} = 5x^2 + 2$ is

$$y = \frac{5}{3}x^3 + 2x + C$$

In the differential equation $\dfrac{dy}{dx} = 5x^2 + 2$, suppose the solution must satisfy the **boundary condition** when $x = 3$, then $y = 5$. We use the boundary condition with the general solution to find the constant $C$.

$$
\begin{aligned}
y &= \frac{5}{3}x^3 + 2x + C && \textcolor{blue}{\textbf{General Solution}} \\
5 &= \frac{5}{3} \cdot 3^3 + 2 \cdot 3 + C && \textcolor{blue}{\textbf{Boundary Condition: } x = 3, \quad y = 5} \\
5 &= 45 + 6 + C \\
C &= -46
\end{aligned}
$$

The **particular solution** of the differential equation $\dfrac{dy}{dx} = 5x^2 + 2$ satisfying the boundary condition when $x = 3$, then $y = 5$ is

$$y = \frac{5}{3}x^3 + 2x - 46$$

We can verify the solution of a differential equation by finding the derivative of the solution. For example, to verify the solution found above, we differentiate $y = \dfrac{5}{3}x^3 + 2x - 46$ obtaining

$$\frac{dy}{dx} = \frac{5}{3} \cdot 3x^2 + 2 = 5x^2 + 2$$

which is the original differential equation.

---

**CALC CLIP** ▶ **EXAMPLE 3  Finding the Particular Solution of a Differential Equation**

Find the particular solution of the differential equation $\dfrac{dy}{dx} = x^2 + 2x + 1$ with the boundary condition when $x = 3$, then $y = -1$. Verify the solution.

**Solution** We begin by finding the general solution of the differential equation, namely

$$y = \frac{x^3}{3} + x^2 + x + C$$

To determine the number $C$, use the boundary condition when $x = 3$, then $y = -1$.

$$-1 = \frac{3^3}{3} + 3^2 + 3 + C \qquad x = 3, \quad y = -1$$

$$C = -22$$

The particular solution of the differential equation with the boundary condition when $x = 3$, then $y = -1$, is

$$y = \frac{x^3}{3} + x^2 + x - 22$$

To verify this solution we find $\dfrac{dy}{dx}$. Then

$$\frac{dy}{dx} = \frac{3x^2}{3} + 2x + 1 = x^2 + 2x + 1$$

which is the original differential equation. ∎

**NOW WORK** Problem 41.

Differential equations often involve higher-order derivatives. For example, the equation $\dfrac{d^2y}{dx^2} = 12x^2$ is an example of a *second-order differential equation*. The **order** of a differential equation is the order of the highest-order derivative of $y$ appearing in the equation.

In solving higher-order differential equations, the number of arbitrary constants in the general solution equals the order of the differential equation. For particular solutions, a first-order differential equation requires one boundary condition; a second-order differential equation requires two boundary conditions; and so on.

**EXAMPLE 4**  **Finding the Particular Solution of a Second-Order Differential Equation**

Find the particular solution of the differential equation $\dfrac{d^2y}{dx^2} = 12x^2$ with the boundary conditions when $x = 0$, then $y = 1$ and when $x = 3$, then $y = 8$. Verify the solution.

**Solution** All the antiderivatives of $\dfrac{d^2y}{dx^2} = 12x^2$ are

$$\frac{dy}{dx} = 4x^3 + C_1$$

All the antiderivatives of $\dfrac{dy}{dx} = 4x^3 + C_1$ are

$$y = x^4 + C_1x + C_2$$

This is the general solution of the differential equation. To find $C_1$ and $C_2$ and the particular solution of the differential equation, use the boundary conditions.

- When $x = 0$  then  $y = 1$:     $1 = 0^4 + C_1 \cdot 0 + C_2$    so  $C_2 = 1$
- When $x = 3$  then  $y = 8$:     $8 = 3^4 + 3C_1 + 1$    so  $C_1 = -\dfrac{74}{3}$

The particular solution with the given boundary conditions is

$$y = x^4 - \frac{74}{3}x + 1$$

To verify this solution find the second derivative of $y$. Then

$$\frac{dy}{dx} = 4x^3 - \frac{74}{3}$$

$$\frac{d^2y}{dx^2} = 12x^2$$

which is the original differential equation. ■

**NOW WORK** Problem 45.

## 3 Solve Applied Problems Modeled by Differential Equations

### Rectilinear Motion

For an object in rectilinear motion, the functions $s = s(t)$, $v = v(t)$, and $a = a(t)$ represent the position $s$, velocity $v$, and acceleration $a$, respectively, of the object at time $t$. The three quantities $s$, $v$, and $a$ are related by the differential equations

$$\frac{ds}{dt} = v(t) \qquad \frac{dv}{dt} = a(t) \qquad \frac{d^2s}{dt^2} = a(t)$$

If the acceleration $a = a(t)$ is a known function of the time $t$, then the velocity can be found by solving the differential equation $\dfrac{dv}{dt} = a(t)$. Similarly, if the velocity $v = v(t)$ is a known function of $t$, then the position $s$ from the origin at time $t$ is the solution of the differential equation $\dfrac{ds}{dt} = v(t)$.

In physical problems, boundary conditions are often the values of the velocity $v$ and position $s$ at time $t = 0$. In such cases, $v(0) = v_0$ and $s(0) = s_0$ are referred to as **initial conditions**.

---

**EXAMPLE 5**  **Solving a Rectilinear Motion Problem**

Find the position function $s = s(t)$ of an object from the origin at time $t$ if its acceleration $a$ is

$$a(t) = 8t - 3$$

and the initial conditions are $v_0 = v(0) = 4$ and $s_0 = s(0) = 1$.

**Solution** First solve the differential equation $\dfrac{dv}{dt} = a(t) = 8t - 3$ and use the initial condition $v_0 = v(0) = 4$.

$$v(t) = 4t^2 - 3t + C_1$$

$$v_0 = v(0) = 4 \cdot 0^2 - 3 \cdot 0 + C_1 = 4 \qquad t = 0, \; v_0 = 4$$

$$C_1 = 4$$

The velocity of the object at time $t$ is $v(t) = 4t^2 - 3t + 4$.

The position function $s = s(t)$ of the object at time $t$ satisfies the differential equation

$$\frac{ds}{dt} = v(t) = 4t^2 - 3t + 4$$

Then

$$s(t) = \frac{4}{3}t^3 - \frac{3}{2}t^2 + 4t + C_2$$

Using the initial condition, $s_0 = s(0) = 1$, we have

$$s_0 = s(0) = 0 - 0 + 0 + C_2 = 1$$

$$C_2 = 1$$

The position function $s = s(t)$ of the object from the origin at any time $t$ is

$$s = s(t) = \frac{4}{3}t^3 - \frac{3}{2}t^2 + 4t + 1$$ ∎

**NOW WORK** Problem 47.

---

**EXAMPLE 6** **Solving a Rectilinear Motion Problem**

When the brakes of a car are applied, the car decelerates at a constant rate of $10 \, \text{m/s}^2$. If the car is to stop within $20 \, \text{m}$ after the brakes are applied, what is the maximum velocity the car could have been traveling? Express the answer in miles per hour.

**Solution** Let the position function $s = s(t)$ represent the distance $s$ in meters the car has traveled $t$ seconds after the brakes are applied. Let $v_0$ be the velocity of the car at the time the brakes are applied ($t = 0$). Since the car decelerates at the rate of $10 \, \text{m/s}^2$, its acceleration $a$, in meters per second squared, is

$$a(t) = \frac{dv}{dt} = -10$$

Solve the differential equation $\frac{dv}{dt} = -10$ for $v = v(t)$.

$$v(t) = -10t + C_1$$

When $t = 0$, $v(0) = v_0$, the velocity of the car when the brakes are applied, so $C_1 = v_0$. Then

$$v(t) = \frac{ds}{dt} = -10t + v_0$$

Now solve the differential equation $v(t) = \frac{ds}{dt}$ for $s = s(t)$.

$$s(t) = -5t^2 + v_0 t + C_2$$

Since the distance $s$ is measured from the point at which the brakes are applied, the second initial condition is $s(0) = 0$. Then $s(0) = -5 \cdot 0 + v_0 \cdot 0 + C_2 = 0$, so $C_2 = 0$. The distance $s$, in meters, the car travels $t$ seconds after applying the brakes is

$$s(t) = -5t^2 + v_0 t$$

The car stops completely when its velocity equals 0. That is, when

$$v(t) = -10t + v_0 = 0$$

$$t = \frac{v_0}{10}$$

This is the time it takes the car to come to rest. Substituting $\frac{v_0}{10}$ for $t$ in the position function $s = s(t)$, the distance the car has traveled is

$$s\left(\frac{v_0}{10}\right) = -5\left(\frac{v_0}{10}\right)^2 + v_0\left(\frac{v_0}{10}\right) = \frac{v_0^2}{20}$$

If the car is to stop within $20 \, \text{m}$, then $s \leq 20$; that is, $\frac{v_0^2}{20} \leq 20$ or equivalently $v_0^2 \leq 400$. The maximum possible velocity $v_0$ for the car is $v_0 = 20 \, \text{m/s}$.

To express 20 m/s in miles per hour, we proceed as follows:

$$v_0 = 20 \text{ m/s} = \left(\frac{20 \text{ m}}{\text{s}}\right)\left(\frac{1 \text{ km}}{1000 \text{ m}}\right)\left(\frac{3600 \text{ s}}{1 \text{ h}}\right)$$

$$= 72 \text{ km/h} \approx \left(\frac{72 \text{ km}}{\text{h}}\right)\left(\frac{1 \text{ mi}}{1.6 \text{ km}}\right) = 45 \text{ mi/h}$$

The car's maximum possible velocity to stop within 20 m is 45 mi/h. ∎

NOW WORK **Problem 61.**

## Freely Falling Objects

An object falling toward Earth is a common example of motion with (nearly) constant acceleration. In the absence of air resistance, all objects, regardless of size, weight, or composition, fall with the same acceleration when released from the same point above Earth's surface, and if the distance fallen is not too great, the acceleration remains constant throughout the fall. This ideal motion, in which air resistance and the small change in acceleration with altitude are neglected, is called **free fall**. The constant acceleration of a freely falling object is called the **acceleration due to gravity** and is denoted by the symbol $g$. Near Earth's surface, its magnitude is approximately 32 ft/s², or 9.8 m/s², and its direction is down toward the center of Earth.

---

**EXAMPLE 7**  **Solving a Problem Involving Free Fall**

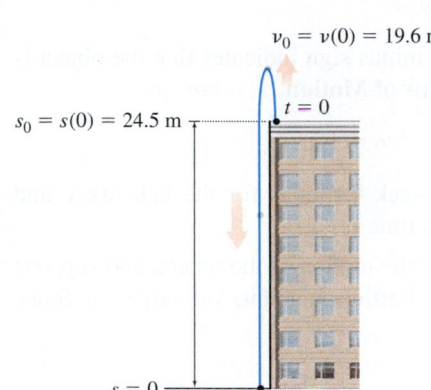

$v_0 = v(0) = 19.6$ m/s

$s_0 = s(0) = 24.5$ m

$t = 0$

$s = 0$

**Figure 67**

A rock is thrown straight up with an initial velocity of 19.6 m/s from the roof of a building 24.5 m above ground level, as shown in Figure 67.

**(a)** How long does it take the rock to reach its maximum altitude?

**(b)** What is the maximum altitude of the rock?

**(c)** If the rock misses the edge of the building on the way down and eventually strikes the ground, what is the total time the rock is in the air?

**Solution**  To answer the questions, we need to find the velocity $v = v(t)$ and the distance $s = s(t)$ of the rock as functions of time. We begin measuring time when the rock is released. If $s$ is the distance, in meters, of the rock from the ground, then since the rock is released at a height of 24.5 m, $s_0 = s(0) = 24.5$ m.

The initial velocity of the rock is given as $v_0 = v(0) = 19.6$ m/s. If air resistance is ignored, the only force acting on the rock is gravity. Since the acceleration due to gravity is $-9.8$ m/s², the acceleration $a$ of the rock is

$$a = \frac{dv}{dt} = -9.8$$

Solving the differential equation, we get

$$v(t) = -9.8t + v_0$$

Using the initial condition, $v_0 = v(0) = 19.6$ m/s, the velocity of the rock at any time $t$ is

$$v(t) = -9.8t + 19.6 \tag{1}$$

Now solve the differential equation $\dfrac{ds}{dt} = v(t) = -9.8t + 19.6$. Then

$$s(t) = -4.9t^2 + 19.6t + s_0$$

Using the initial condition, $s(0) = 24.5$ m, the distance $s$ of the rock from the ground at any time $t$ is

$$s(t) = -4.9t^2 + 19.6t + 24.5 \tag{2}$$

Now we can answer the questions.

(a) The rock reaches its maximum altitude when its velocity is 0.

$$v(t) = -9.8t + 19.6 = 0 \quad \text{From (1)}$$
$$t = 2$$

The rock reaches its maximum altitude at $t = 2$ s.

(b) To find the maximum altitude, evaluate $s(2)$. The maximum altitude of the rock is

$$s(2) = -4.9 \cdot 2^2 + 19.6 \cdot 2 + 24.5 = 44.1 \text{ m} \quad \text{From (2)}$$

(c) We find the total time the rock is in the air by solving $s(t) = 0$.

$$-4.9t^2 + 19.6t + 24.5 = 0$$
$$t^2 - 4t - 5 = 0$$
$$(t - 5)(t + 1) = 0$$
$$t = 5 \quad \text{or} \quad t = -1$$

The only meaningful solution is $t = 5$. The rock is in the air for 5 s. ∎

Now we examine the general problem of freely falling objects.

If $F$ is the weight of an object of mass $m$, then according to Galileo, assuming air resistance is negligible, a freely falling object obeys the equation

$$F = -mg$$

where $g$ is the acceleration due to gravity. The minus sign indicates that the object is falling. Also, according to **Newton's Second Law of Motion**, $F = ma$, so

$$ma = -mg \qquad \text{or} \qquad a = -g$$

where $a$ is the acceleration of the object. We seek formulas for the velocity $v$ and distance $s$ from Earth of a freely falling object at time $t$.

Let $t = 0$ be the instant we begin to measure the motion of the object, and suppose at this instant the object's vertical distance above Earth is $s_0$ and its velocity is $v_0$. Since the acceleration $a = \dfrac{dv}{dt}$ and $a = -g$, we have

$$\boxed{a = \frac{dv}{dt} = -g} \qquad (3)$$

Solving the differential equation (3) for $v$ and using the initial condition $v_0 = v(0)$, we obtain

$$\boxed{v(t) = -gt + v_0}$$

Now since $\dfrac{ds}{dt} = v(t)$, we have

$$\frac{ds}{dt} = -gt + v_0 \qquad (4)$$

Solving the differential equation (4) for $s$ and using the initial condition $s_0 = s(0)$, we obtain

$$\boxed{s(t) = -\frac{1}{2}gt^2 + v_0 t + s_0}$$

**NOW WORK** Problem **63.**

### Newton's First Law of Motion

We close this section with a special case of the *Law of Inertia*, which was originally stated by Galileo.

> **THEOREM** Newton's First Law of Motion
>
> If no force acts on a body, then a body at rest remains at rest and a body moving with constant velocity continues to do so.

**Proof** The force $F$ acting on a body of mass $m$ is given by Newton's Second Law of Motion $F = ma$, where $a$ is the acceleration of the body. If there is no force acting on the body, then $F = 0$. In this case, the acceleration $a$ must be 0. But $a = \dfrac{dv}{dt}$, where $v$ is the velocity of the body. So,

$$\frac{dv}{dt} = 0 \qquad \text{and} \qquad v = C, \text{ a constant}$$

That is, the body is at rest ($v = 0$) or else in a state of uniform motion ($v$ is a nonzero constant). ∎

## 4.8 Assess Your Understanding

### Concepts and Vocabulary

1. A function $F$ is called a(n) _____ of a function $f$ if $F' = f$.
2. *True or False* If $F$ is an antiderivative of $f$, then $F(x) + C$, where $C$ is a constant, is also an antiderivative of $f$.
3. All the antiderivatives of $y = x^{-1}$ are _____.
4. *True or False* An antiderivative of $\sin x$ is $-\cos x + \pi$.
5. *True or False* The general solution of a differential equation $\dfrac{dy}{dx} = f(x)$ consists of all the antiderivatives of $f'(x)$.
6. *True or False* Free fall is an example of motion with constant acceleration.
7. *True or False* To find a particular solution of a differential equation $\dfrac{dy}{dx} = f(x)$, we need a boundary condition.
8. *True or False* If $F_1$ and $F_2$ are both antiderivatives of a function $f$ on an interval $I$, then $F_1 - F_2 = C$, a constant, on $I$.

### Skill Building

*In Problems 9–36, find all the antiderivatives of each function.*

9. $f(x) = 2$
10. $f(x) = \dfrac{1}{2}$
11. $f(x) = 4x^5$
12. $f(x) = x$
13. $f(x) = 5x^{3/2}$
14. $f(x) = x^{5/2} + 2$
15. $f(x) = 2x^{-2}$
16. $f(x) = 3x^{-3}$
17. $f(x) = \sqrt{x}$
18. $f(x) = \dfrac{1}{\sqrt{x}}$
19. $f(x) = 4x^3 - 3x^2 + 1$
20. $f(x) = x^2 - x$
21. $f(x) = (2 - 3x)^2$
22. $f(x) = (3x - 1)^2$
23. $f(x) = \dfrac{3x - 2}{x}$
24. $f(x) = \dfrac{x^2 + 4}{x}$
25. $f(x) = \dfrac{3x^{1/2} - 4}{x}$
26. $f(x) = \dfrac{4x^{3/2} - 1}{x}$
27. $f(x) = 2x - 3\cos x$
28. $f(x) = 2\sin x - \cos x$
29. $f(x) = 4e^x + x$
30. $f(x) = e^{-x} + \sec^2 x$
31. $f(x) = \dfrac{7}{1 + x^2}$
32. $f(x) = x + \dfrac{10}{\sqrt{1 - x^2}}$
33. $f(x) = e^x + \dfrac{1}{x\sqrt{x^2 - 1}}$
34. $f(x) = \sin x + \dfrac{1}{1 + x^2}$
35. $f(x) = 3\sinh x$
36. $f(x) = 3 + \cosh x$

*In Problems 37–46, find the particular solution of each differential equation having the given boundary condition(s). Verify the solution.*

37. $\dfrac{dy}{dx} = 3x^2 - 2x + 1$, when $x = 2$, $y = 1$
38. $\dfrac{dv}{dt} = 3t^2 - 2t + 1$, when $t = 1$, $v = 5$
39. $\dfrac{dy}{dx} = x^{1/3} + x\sqrt{x} - 2$, when $x = 1$, $y = 2$
40. $s'(t) = t^4 + 4t^3 - 5$, $s(2) = 5$
41. $\dfrac{ds}{dt} = t^3 + \dfrac{1}{t^2}$, when $t = 1$, $s = 2$

---

**1.** = NOW WORK problem      ∿ = Graphing technology recommended      [CAS] = Computer Algebra System recommended

**42.** $\dfrac{dy}{dx} = \sqrt{x} - x\sqrt{x} + 1$,   when $x = 1$, $y = 0$

**43.** $f'(x) = x - 2\sin x$,   $f(\pi) = 0$

**44.** $\dfrac{dy}{dx} = x^2 - 2\sin x$,   when $x = \pi$, $y = 0$

**45.** $\dfrac{d^2 y}{dx^2} = e^x$,   when $x = 0$, $y = 2$,   when $x = 1$, $y = e$

**46.** $f''(\theta) = \sin\theta + \cos\theta$, $f'\left(\dfrac{\pi}{2}\right) = 2$ and $f(\pi) = 4$

*In Problems 47–52, the acceleration of an object in rectilinear motion is given. Find the position function $s = s(t)$ of the object under the given initial conditions.*

**47.** $a(t) = -32\,\text{ft/s}^2$,   $s(0) = 0\,\text{ft}$,   $v(0) = 128\,\text{ft/s}$

**48.** $a(t) = -980\,\text{cm/s}^2$,   $s(0) = 5\,\text{cm}$,   $v(0) = 1980\,\text{cm/s}$

**49.** $a(t) = 3t\,\text{m/s}^2$,   $s(0) = 2\,\text{m}$,   $v(0) = 18\,\text{m/s}$

**50.** $a(t) = 5t - 2\,\text{ft/s}^2$,   $s(0) = 0\,\text{ft}$,   $v(0) = 8\,\text{ft/s}$

**51.** $a(t) = \sin t\,\text{ft/s}^2$,   $s(0) = 0\,\text{ft}$, $v(0) = 5\,\text{ft/s}$

**52.** $a(t) = e^t\,\text{m/s}^2$,   $s(0) = 0\,\text{m}$, $v(0) = 4\,\text{m/s}$

## Applications and Extensions

*In Problems 53 and 54, find all the antiderivatives of each function.*

**53.** $f(u) = \dfrac{u^2 + 10u + 21}{3u + 9}$

**54.** $f(t) = \dfrac{t^3 - 5t + 8}{t^5}$

*In Problems 55 and 56, find the particular solution of each differential equation having the given boundary condition.*

**55.** $f'(t) = \dfrac{t^4 + 3t - 1}{t}$ if $f(1) = \dfrac{1}{4}$

**56.** $g'(x) = \dfrac{x^2 - 1}{x^4 - 1}$ if $g(0) = 0$

**57.** Use the fact that

$$\frac{d}{dx}(x\cos x + \sin x) = -x\sin x + 2\cos x$$

to find $F$ if

$$\frac{dF}{dx} = -x\sin x + 2\cos x \qquad \text{and} \qquad F(0) = 1$$

**58.** Use the fact that

$$\frac{d}{dx}\sin x^2 = 2x\cos x^2$$

to find $h$ if

$$\frac{dh}{dx} = x\cos x^2 \qquad \text{and} \qquad h(0) = 2$$

**59. Rectilinear Motion**   A car decelerates at a constant rate of $10\,\text{m/s}^2$ when its brakes are applied. If the car must stop within 15 m after applying the brakes, what is the maximum allowable velocity for the car? Express the answer in meters per second and in miles per hour.

**60. Rectilinear Motion**   A car can accelerate from 0 to 60 km/h in 10 seconds. If the acceleration is constant, how far does the car travel during this time?

**61. Rectilinear Motion**   A BMW 6 series can accelerate from 0 to 60 mi/h in 5 s. If the acceleration is constant, how far does the car travel during this time?
*Source*: BMW USA.

**62. Free Fall**   A ball is thrown straight up from ground level, with an initial velocity of 19.6 m/s. How high is the ball thrown? How long will it take the ball to return to ground level?

**63. Free Fall**   A child throws a ball straight up. If the ball is to reach a height of 9.8 m, what is the minimum initial velocity that must be imparted to the ball? Assume the initial height of the ball is 1 m.

**64. Free Fall**   A ball thrown directly down from a roof 49 m high reaches the ground in 3 s. What is the initial velocity of the ball?

**65. Inertia**   A constant force is applied to an object that is initially at rest. If the mass of the object is 4 kg and if its velocity after 6 s is 12 m/s, determine the force applied to it.

**66. Rectilinear Motion**   Starting from rest, with what constant acceleration must a car move to travel 2 km in 2 min? (Give your answer in centimeters per second squared.)

**67. Downhill Speed of a Skier**   The down slope acceleration $a$ of a skier is given by $a = a(t) = g\sin\theta$, where $t$ is time, in seconds, $g = 9.8\,\text{m/s}^2$ is the acceleration due to gravity, and $\theta$ is the angle of the slope. If the skier starts from rest at the lift, points his skis straight down a $20°$ slope, and does not turn, how fast is he going after 5 s?

**68. Free Fall**   A child on top of a building 24 m high drops a rock and then 1 s later throws another rock straight down. What initial velocity must the second rock be given so that the dropped rock and the thrown rock hit the ground at the same time?

**69. Free Fall**   The world's high jump record, set on July 27, 1993, by Cuban jumper Javier Sotomayor, is 2.45 m. If this event were held on the Moon, where the acceleration due to gravity is $1.6\,\text{m/s}^2$, what height would Sotomayor have attained? Assume that he propels himself with the same force on the Moon as on Earth.

**70. Free Fall**   The 2-m high jump is common today. If this event were held on the Moon, where the acceleration due to gravity is $1.6\,\text{m/s}^2$, what height would be attained? Assume that an athlete can propel him- or herself with the same force on the Moon as on Earth.

## Challenge Problems

*In Problems 71–72, find all the antiderivatives of $f$.*

**71.** $f(x) = \dfrac{2x^3 + 2x + 3}{1 + x^2}$

**72.** $f(x) = \dfrac{5 + 7\sin x}{2\cos^2 x}$

**73. Radiation**   Radiation, such as X-rays or the radiation from radioactivity, is absorbed as it passes through tissue or any other material. The rate of change in the intensity $I$ of the radiation with respect to the depth $x$ of tissue is directly proportional to the intensity $I$. This proportion can be expressed as an equation by introducing a positive constant of proportionality $k$, where $k$ depends on the properties of the tissue and the type of radiation.

**(a)** Show that $\dfrac{dI}{dx} = -kI$, $k > 0$.

**(b)** Explain why the minus sign is necessary.

**(c)** Solve the differential equation in (a) to find the intensity $I$ as a function of the depth $x$ in the tissue. The intensity of the radiation when it enters the tissue is $I(0) = I_0$.

**(d)** Find the value of $k$ if the intensity is reduced by 90% of its maximum value at a depth of 2.0 cm.

**74. Moving Shadows**   A lamp on a post 10 m high stands 25 m from a wall. A boy standing 5 m from the lamp and 20 m from the

wall throws a ball straight up with an initial velocity of 19.6 m/s. The acceleration due to gravity is $a = -9.8$ m/s². The ball is thrown up from an initial height of 1 m above ground.

**(a)** How fast is the shadow of the ball moving on the wall 3 s after the ball is released?

**(b)** Explain if the ball is moving up or down.

**(c)** How far is the ball above ground at $t = 3$ s?

# Chapter Review

## THINGS TO KNOW

### 4.1 Related Rates

Steps for solving a related rate problem (p. 264)

### 4.2 Maximum and Minimum Values; Critical Numbers

*Definitions:*

- Absolute maximum; absolute minimum (p. 272)
- Absolute maximum value; absolute minimum value (p. 272)
- Local maximum; local minimum (p. 272)
- Local maximum value; local minimum value (p. 272)
- Critical number (p. 275)

*Theorems:*

- **Extreme Value Theorem** If a function $f$ is continuous on a closed interval $[a, b]$, then $f$ has an absolute maximum and an absolute minimum on $[a, b]$. (p. 274)
- If a function $f$ has a local maximum or a local minimum at the number $c$, then either $f'(c) = 0$ or $f'(c)$ does not exist. (p. 274)

*Procedure:*

- Steps for finding the absolute extreme values of a function $f$ that is continuous on a closed interval $[a, b]$. (p. 277)

### 4.3 The Mean Value Theorem

- **Rolle's Theorem** (p. 284)
- **Mean Value Theorem** (p. 285)
- **Corollaries to the Mean Value Theorem**
  - If $f'(x) = 0$ for all numbers $x$ in $(a, b)$, then $f$ is constant on $(a, b)$. (p. 287)
  - If $f'(x) = g'(x)$ for all numbers $x$ in $(a, b)$, then there is a number $C$ for which $f(x) = g(x) + C$ on $(a, b)$. (p. 288)
- **Increasing/Decreasing Function Test** (p. 288)

### 4.4 Local Extrema and Concavity

*Definitions:*

- Concave up; concave down (p. 299)
- Inflection point (p. 300)

*Theorems:*

- First Derivative Test (p. 294)
- Test for concavity (p. 299)
- A condition for an inflection point (p. 301)
- Second Derivative Test (p. 303)

*Procedure:*

- Steps for finding the inflection points of a function (p. 301)

### 4.5 Indeterminate Forms and L'Hôpital's Rule

*Definitions:*

- Indeterminate form at $c$ of the type $\dfrac{0}{0}$ or the type $\dfrac{\infty}{\infty}$ (p. 311)
- Indeterminate form at $c$ of the type $0 \cdot \infty$ or $\infty - \infty$ (pp. 315–316)
- Indeterminate form at $c$ of the type $1^\infty, 0^0,$ or $\infty^0$ (p. 317)

*Theorem:*

- L'Hôpital's Rule (p. 312)

*Procedures:*

- Steps for finding a limit using L'Hôpital's Rule (p. 312)
- Steps for finding $\lim\limits_{x \to c}[f(x)]^{g(x)}$, where $[f(x)]^{g(x)}$ is an indeterminate form at $c$ of the type $1^\infty, 0^0,$ or $\infty^0$ (p. 317)

### 4.6 Using Calculus to Graph Functions

*Procedure:*

- Steps for graphing a function $y = f(x)$. (p. 321)

### 4.7 Optimization

*Procedure:*

- Steps for solving optimization problems (p. 332)

### 4.8 Antiderivatives; Differential Equations

*Definitions:*

- Antiderivative (p. 342)
- General solution of a differential equation (p. 345)
- Particular solution of a differential equation (p. 345)
- Boundary condition (p. 345)
- Initial condition (p. 347)

*Basic Antiderivatives*   See Table 6 (p. 344)

*Antidifferentiation Properties:*

- **Sum Rule and Difference Rule**: If functions $F_1$ and $F_2$ are antiderivatives of the functions $f_1$ and $f_2$, respectively, then $F_1 + F_2$ is an antiderivative of $f_1 + f_2$ and $F_1 - F_2$ is an antiderivative of $f_1 - f_2$. (p. 344)
- **Constant Multiple Rule**: If $k$ is a real number and if $F$ is an antiderivative of $f$, then $kF$ is an antiderivative of $kf$. (p. 344)

*Theorems:*

- If a function $F$ is an antiderivative of a function $f$ on an interval $I$, then any other antiderivative of $f$ has the form $F(x) + C$, where $C$ is an (arbitrary) constant. (p. 343)
- Newton's First Law of Motion (p. 351)
- Newton's Second Law of Motion (p. 350)

## OBJECTIVES

## REVIEW EXERCISES

**1. Related rates** A spherical snowball is melting at the rate of $2\ \text{cm}^3/\text{min}$. How fast is the surface area changing when the radius is 5 cm?

**2. Related rates** A lighthouse is 3 km from a straight shoreline. Its light makes one revolution every 8 s. How fast is the light moving along the shore when it makes an angle of $30°$ with the shoreline?

**3. Related rates** Two planes at the same altitude are approaching an airport, one from the north and one from the west. The plane from the north is flying at 250 mi/h and is 30 mi from the airport. The plane from the west is flying at 200 mi/h and is 20 mi from the airport. How fast are the planes approaching each other at that instant?

*In Problems 4 and 5, use the graphs below to determine whether each function has an absolute extremum and/or a local extremum or neither at the indicated points.*

**4.**

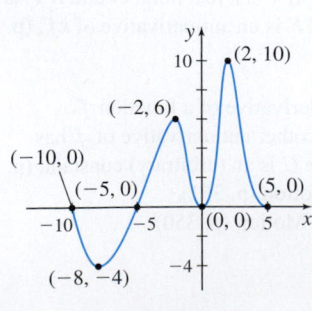

**5.**

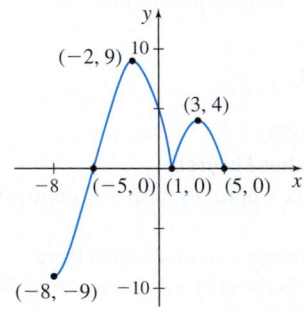

**6. Critical Numbers** $f(x) = \dfrac{x^2}{2x - 1}$

(a) Find all the critical numbers of $f$.

(b) Find the local extrema of $f$.

**7. Critical Numbers** Find all the critical numbers of $f(x) = \cos(2x)$ on the closed interval $[0, \pi]$.

*In Problems 8 and 9, find the absolute maximum value and absolute minimum value of each function on the indicated interval.*

**8.** $f(x) = x - \sin(2x)$ on $[0, 2\pi]$

**9.** $f(x) = \dfrac{3}{2}x^4 - 2x^3 - 6x^2 + 5$ on $[-2, 3]$

**10. Rolle's Theorem** Verify that the hypotheses for Rolle's Theorem are satisfied for the function $f(x) = x^3 - 4x^2 + 4x$ on the interval $[0, 2]$. Find the coordinates of the point(s) at which there is a horizontal tangent line to the graph of $f$.

**11. Mean Value Theorem** Verify that the hypotheses for the Mean Value Theorem are satisfied for the function $f(x) = \dfrac{2x - 1}{x}$ on the interval $[1, 4]$. Find a point on the graph of $f$ that has a tangent line whose slope equals the slope of the secant line joining $(1, 1)$ to $\left(4, \dfrac{7}{4}\right)$.

**12. Mean Value Theorem** Does the Mean Value Theorem apply to the function $f(x) = \sqrt{x}$ on the interval $[0, 9]$? If not, why not? If so, find the number $c$ referred to in the theorem.

*In Problems 13–15, find the local extrema of each function:*

**(a)** *Using the First Derivative Test.*

**(b)** *Using the Second Derivative Test, if possible. If the Second Derivative Test cannot be used, explain why.*

**13.** $f(x) = x^3 - x^2 - 8x + 1$     **14.** $f(x) = x^2 - 24x^{2/3}$

**15.** $f(x) = x^4 e^{-2x}$

**16. Rectilinear Motion** The position $s$ of an object from the origin at time $t$ is given by $s = s(t) = t^4 + 2t^3 - 36t^2$. Draw figures to illustrate the motion of the object and its velocity.

*In Problems 17–22, graph each function. Follow the steps given in Section 4.6.*

**17.** $f(x) = -x^3 - x^2 + 2x$     **18.** $f(x) = x^{1/3}(x^2 - 9)$

**19.** $f(x) = xe^x$     **20.** $f(x) = \dfrac{x - 3}{x^2 - 4}$

**21.** $f(x) = x\sqrt{x - 3}$     **22.** $f(x) = x^3 - 3\ln x$

*In Problems 23 and 24, for each function:*

**(a)** *Determine the intervals where each function is increasing and decreasing.*

**(b)** *Determine the intervals on which each function $f$ is concave up and concave down.*

**(c)** *Identify any inflection points.*

**23.** $f(x) = x^4 + 12x^2 + 36x - 11$

**24.** $f(x) = 3x^4 - 2x^3 - 24x^2 - 7x + 2$

**25.** If $y$ is a function and $y' > 0$ for all $x$ and $y'' < 0$ for all $x$, which of the following could be part of the graph of $y = f(x)$? See illustrations (A) through (D).

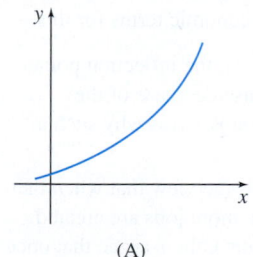

(A)

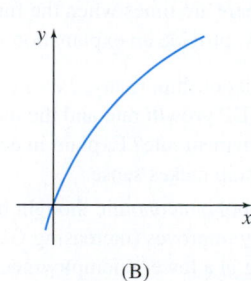

(B)

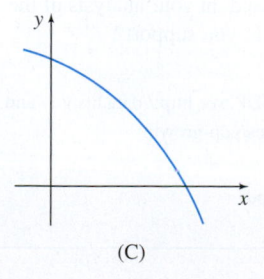

(C)

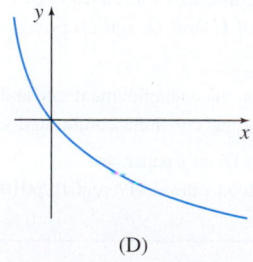

(D)

**26.** Graph a function $f$ that has the following properties:

$f(-3) = 2; \quad f(-1) = -5; \quad f(2) = -4;$

$f(6) = -1; \quad f'(-3) = f'(6) = 0;$

$\lim_{x \to 0^-} f(x) = -\infty; \quad \lim_{x \to 0^+} f(x) = \infty;$

$f''(x) > 0$ if $x < -3$ or $0 < x < 4;$

$f''(x) < 0$ if $-3 < x < 0$ or $4 < x$

**27.** Graph a function $f$ that has the following properties:

$f(-2) = 2; \quad f(5) = 1; \quad f(0) = 0;$

$f'(x) > 0$ if $x < -2$ or $5 < x;$

$f'(x) < 0$ if $-2 < x < 2$ or $2 < x < 5;$

$f''(x) > 0$ if $x < 0$ or $2 < x$ and $f''(x) < 0$ if $0 < x < 2;$

$\lim_{x \to 2^-} f(x) = -\infty; \quad \lim_{x \to 2^+} f(x) = \infty$

**28. Mean Value Theorem** For the function $f(x) = x\sqrt{x + 1}$, $0 \le x \le b$, the number $c$ satisfying the Mean Value Theorem is $c = 3$. Find $b$.

**29. Maximizing Volume** An open box is to be made from a piece of cardboard by cutting squares out of each corner and folding up the sides. If the size of the cardboard is 2 ft by 3 ft, what size squares (in inches) should be cut out to maximize the volume of the box?

**30. Minimizing Distance** Find the point on the graph of $2y = x^2$ nearest to the point $(4, 1)$.

*In Problems 31–38, find all the antiderivatives of each function.*

**31.** $f(x) = 0$     **32.** $f(x) = x^{1/2}$

**33.** $f(x) = \cos x$     **34.** $f(x) = \sec x \tan x$

**35.** $f(x) = \dfrac{2}{x}$     **36.** $f(x) = -2x^{-3}$

**37.** $f(x) = 4x^3 - 9x^2 + 10x - 3$     **38.** $f(x) = e^x + \dfrac{4}{x}$

**39. Velocity** A box moves down an inclined plane with an acceleration $a(t) = t^2(t - 3) \text{ cm/s}^2$. It covers a distance of 10 cm in 2 s. What was the original velocity of the box?

10 cm

**40. Free Fall** Two objects begin a free fall from rest at the same height 1 s apart. How long after the first object begins to fall will the two objects be 10 m apart?

*In Problems 41–44, determine if the expression is an indeterminate form at $0$. If it is, identify its type.*

**41.** $\dfrac{xe^{3x} - x}{1 - \cos(2x)}$     **42.** $\left(\dfrac{1}{x}\right)^{\tan x}$

**43.** $\dfrac{1}{x^2} - \dfrac{1}{x^2 \sec x}$     **44.** $\dfrac{\tan x - x}{x - \sin x}$

*In Problems 45–56, find each limit.*

**45.** $\lim\limits_{x \to \pi/2} \dfrac{\sec^2 x}{\sec^2 (3x)}$

**46.** $\lim\limits_{x \to 0} \left[ \dfrac{2}{\sin^2 x} - \dfrac{1}{1 - \cos x} \right]$

**47.** $\lim\limits_{x \to 0} \dfrac{e^x - e^{-x}}{\sin x}$

**48.** $\lim\limits_{x \to 0^-} x \cot (\pi x)$

**49.** $\lim\limits_{x \to 0} \dfrac{\tan x + \sec x - 1}{\tan x - \sec x + 1}$

**50.** $\lim\limits_{x \to a} \dfrac{ax - x^2}{a^4 - 2a^3 x + 2ax^3 - x^4}$

**51.** $\lim\limits_{x \to 0} \dfrac{x - \sin x}{x^3}$

**52.** $\lim\limits_{x \to 0} \dfrac{\tan x - \sin x}{\sin^3 x}$

**53.** $\lim\limits_{x \to \infty} (1 + 4x)^{2/x}$

**54.** $\lim\limits_{x \to 1} \left[ \dfrac{2}{x^2 - 1} - \dfrac{1}{x - 1} \right]$

**55.** $\lim\limits_{x \to 4} \dfrac{x^2 - 16}{x^2 + x - 20}$

**56.** $\lim\limits_{x \to 0^+} (\cot x)^x$

*In Problems 57–60, find the particular solution of each differential equation having the given boundary conditions. Verify the solution.*

**57.** $\dfrac{dy}{dx} = e^x$, when $x = 0$, $y = 2$

**58.** $\dfrac{dy}{dx} = \dfrac{1}{2} \sec x \tan x$, when $x = 0$, $y = 7$

**59.** $\dfrac{dy}{dx} = \dfrac{2}{x}$, when $x = 1$, then $y = 4$

**60.** $\dfrac{d^2 y}{dx^2} = x^2 - 4$, when $x = 3$, $y = 2$, when $x = 2$, $y = 2$

**61. Maximizing Profit**   A manufacturer has determined that the cost $C$ of producing $x$ items is given by

$$C(x) = 200 + 35x + 0.02x^2 \text{ dollars}$$

Each item produced can be sold for $78. How many items should she produce to maximize profit?

**62. Optimization**   The sales of a new sound system over a period of time are expected to follow the logistic curve

$$f(x) = \dfrac{5000}{1 + 5e^{-x}} \qquad x \geq 0$$

where $x$ is measured in years. In what year is the sales rate a maximum?

**63. Maximum Area**   Find the area of the rectangle of largest area in the fourth quadrant that has vertices at $(0, 0)$, $(x, 0)$, $x > 0$, and $(0, y)$, $y < 0$. The fourth vertex is on the graph of $y = \ln x$.

**64. Differential Equation**   A motorcycle accelerates at a constant rate from 0 to 72 km/h in 10 s. How far has it traveled in that time?

---

| CHAPTER 4 PROJECT | **The U.S. Economy** |

Economic data are quite variable and very complicated, making them difficult to model. The models we use here are rough approximations of the Unemployment Rate and Gross Domestic Product (GDP).

The unemployment rate can be modeled by

$$U = U(t) = \dfrac{5 \cos \dfrac{2t}{\pi}}{2 + \sin \dfrac{2t}{\pi}} + 6 \qquad 0 \leq t \leq 25$$

where $t$ is in years and $t = 0$ represents January 1, 1985.

The GDP growth rate can be modeled by

$$G = G(t) = 5 \cos \left( \dfrac{2t}{\pi} + 5 \right) + 2 \qquad 0 \leq t \leq 25$$

where $t$ is in years and $t = 0$ represents January 1, 1985. If the GDP growth rate is increasing, the economy is **expanding** and if it is decreasing, the economy is **contracting**. A **recession** occurs when GDP growth rate is negative.

1. During what years was the U.S. economy in recession?

2. Find the critical numbers of $U = U(t)$.

3. Determine the intervals on which $U$ is increasing and on which it is decreasing.

4. Find all the local extrema of $U$.

5. Find the critical numbers of $G = G(t)$.

6. Determine the intervals on which $G$ is increasing and on which it is decreasing.

7. Find all the local extrema of $G$.

8. (a) Find the inflection points of $G$.

   (b) Describe in economic terms what the inflection points of $G$ represent.

9. Graph $U = U(t)$ and $G = G(t)$ on the same set of coordinate axes.

10. Okun's Law states that an increase in the unemployment rate tends to coincide with a decrease in the GDP growth rate.

    (a) During years of recession, what is happening to the unemployment rate? Explain in economic terms why this makes sense.

    (b) Use your answer to part (a) to explain whether the functions $U$ and $G$ generally agree with Okun's Law.

    (c) If there are times when the functions do not satisfy Okun's Law, provide an explanation in economic terms for this.

11. What relationship, if any, exists between the inflection points of the GDP growth rate and the increase/decrease of the unemployment rate? Explain in economic terms why such a relationship makes sense.

12. One school of economic thought holds the view that when the economy improves (increasing GDP), more jobs are created, resulting in a lower unemployment rate. Others argue that once the unemployment rate improves (decreases), more people are working and this increases GDP. Based on your analysis of the graphs of $U$ and $G$, which position do you support?

---

For real data on the unemployment rate and GDP, see http://data.bls.gov and http://www.tradingeconomics.com/united-states/gdp-growth

To read Arthur Okun's paper, see http://cowles.econ.yale.edu/P/cp/p01b/p0190.pdf

# 5

# The Integral

JuliaNicole/Getty Images

**CHAPTER 5 PROJECT** The Chapter Project on page 426 examines ways to obtain the total flow of a river over any period of time from data that provide flow rates.

## Managing the Klamath River

The Klamath River starts in the eastern lava plateaus of Oregon, passes through a farming region of that state, and then crosses northern California. Because it is the largest river in the region, runs through a variety of terrain, and has different land uses. Historically, the Klamath supported large salmon runs. Now it is used to irrigate agricultural lands and to generate electric power for the region. Downstream, it runs through federal wild lands and is used for fishing, rafting, and kayaking.

If a river is to be well-managed for such a variety of uses, its flow must be understood. For that reason, the U.S. Geological Survey (USGS) maintains a number of gauges along the Klamath. These typically measure the depth of the river, which can then be expressed as a rate of flow in cubic feet per second ($ft^3/s$). In order to understand the river flow fully, we must be able to find the total amount of water that flows down the river over any period of time.

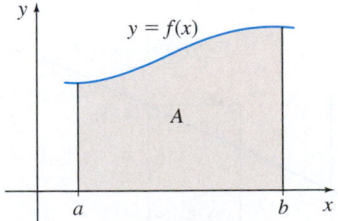

**Figure 1** $A$ is the area of the region enclosed by the graph of $f$, the $x$-axis, and the lines $x = a$ and $x = b$.

We began our study of calculus by asking two questions from geometry. The first, the tangent problem, "What is the slope of the tangent line to the graph of a function?" led to the derivative of a function.

The second question was the area problem, "Given a function $f$, defined and nonnegative on a closed interval $[a, b]$, what is the area of the region enclosed by the graph of $f$, the $x$-axis, and the vertical lines $x = a$ and $x = b$?" Figure 1 illustrates this region. Answering the area problem question leads to the *integral* of a function.

The first two sections of Chapter 5 show how the concept of the integral evolves from the area problem. At first glance, the area problem and the tangent problem look quite dissimilar. However, much of calculus is built on a surprising relationship between the two problems and their associated concepts. This relationship is the basis for the *Fundamental Theorem of Calculus*, discussed in Section 5.3.

# 5.1 Area

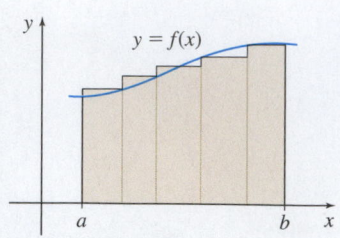

**Figure 2** The sum of the areas of the rectangles approximates the area of the region.

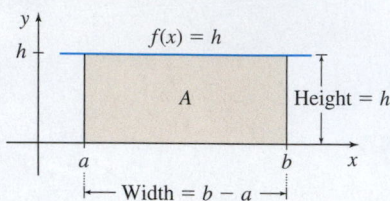

**Figure 3** $A = h(b - a)$.

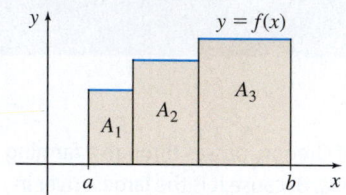

**Figure 4** $A = A_1 + A_2 + A_3$

**OBJECTIVES** *When you finish this section, you should be able to:*

**1** Approximate the area under the graph of a function (p. 358)

**2** Find the area under the graph of a function (p. 362)

The area problem asks, "Given a function $f$, defined and nonnegative on a closed interval $[a, b]$, what is the area $A$ of the region enclosed by the graph of $f$, the $x$-axis, and the vertical lines $x = a$ and $x = b$?" To answer the question posed in the area problem, we approximate the area $A$ of the region by adding the areas of a finite number of rectangles, as shown in Figure 2. This approximation of the area $A$ will lead to the *integral*, which gives the *exact* area $A$ of the region. We begin by reviewing the formula for the area of a rectangle.

The area $A$ of a rectangle with width $w$ and height $h$ is given by the geometry formula

$$A = hw$$

See Figure 3. The graph of a constant function $f(x) = h$, for some positive number $h$, is a horizontal line that lies above the $x$-axis. The region enclosed by this line, the $x$-axis, and the lines $x = a$ and $x = b$ is the rectangle whose area $A$ is the product of the width $(b - a)$ and the height $h$.

$$A = h(b - a)$$

If the graph of $y = f(x)$ consists of three horizontal lines, each of positive height as shown in Figure 4, the area $A$ enclosed by the graph of $f$, the $x$-axis, and the lines $x = a$ and $x = b$ is the sum of the rectangular areas $A_1$, $A_2$, and $A_3$.

## 1 Approximate the Area Under the Graph of a Function

**EXAMPLE 1** **Approximating the Area Under the Graph of a Function**

Approximate the area $A$ of the region enclosed by the graph of $f(x) = \dfrac{1}{2}x + 3$, the $x$-axis, and the lines $x = 2$ and $x = 4$.

**Solution** Figure 5 illustrates the area $A$ to be approximated.

We begin by drawing a rectangle of width $4 - 2 = 2$ and height $f(2) = 4$. The area of the rectangle, $2 \cdot 4 = 8$, approximates the area $A$, but it underestimates $A$, as seen in Figure 6(a).

Alternatively, $A$ can be approximated by a rectangle of width $4 - 2 = 2$ and height $f(4) = 5$. See Figure 6(b). This approximation of the area equals $2 \cdot 5 = 10$, but it overestimates $A$. We conclude that

$$8 < A < 10$$

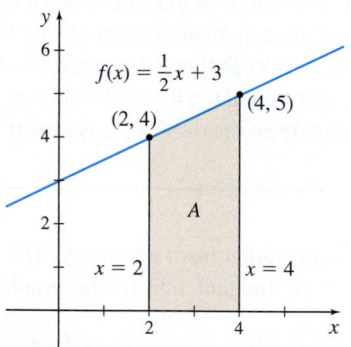

**Figure 5**

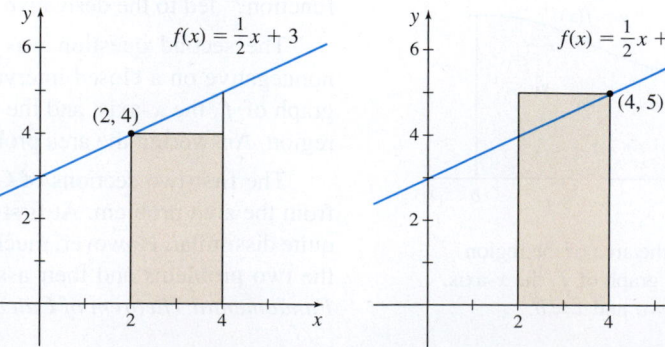

(a) The area $A$ is underestimated.  (b) The area $A$ is overestimated.

**Figure 6**

The approximation of the area $A$ of the region can be improved by dividing the closed interval $[2, 4]$ into two subintervals, $[2, 3]$ and $[3, 4]$. Now draw two rectangles: one rectangle with width $3 - 2 = 1$ and height $f(2) = \frac{1}{2} \cdot 2 + 3 = 4$; the other rectangle with width $4 - 3 = 1$ and height $f(3) = \frac{1}{2} \cdot 3 + 3 = \frac{9}{2}$. As Figure 7(a) illustrates, the sum of the areas of the two rectangles

$$1 \cdot 4 + 1 \cdot \frac{9}{2} = \frac{17}{2} = 8.5$$

underestimates the area.

Repeat this process by drawing two rectangles, one of width 1 and height $f(3) = \frac{9}{2}$; the other of width 1 and height $f(4) = \frac{1}{2} \cdot 4 + 3 = 5$. The sum of the areas of these two rectangles,

$$1 \cdot \frac{9}{2} + 1 \cdot 5 = \frac{19}{2} = 9.5$$

overestimates the area as shown in Figure 7(b). We conclude that

$$8.5 < A < 9.5$$

obtaining a better approximation to the area.

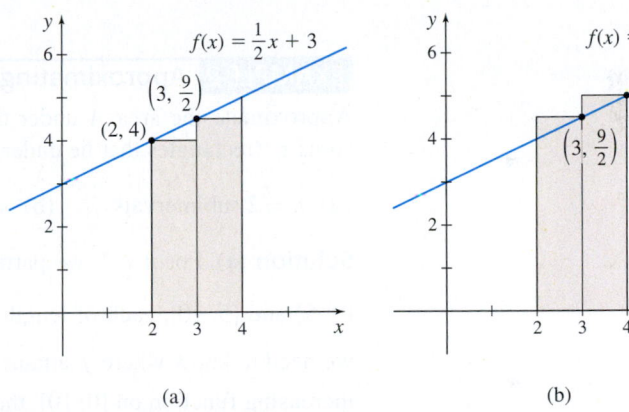

(a)                                             (b)

**DF** **Figure 7**                                                          ∎

Observe that as the number $n$ of subintervals of the interval $[2, 4]$ increases, the approximation of the area $A$ improves. For $n = 1$, the error in approximating $A$ is 1 square unit, but for $n = 2$, the error is only 0.5 square unit.

**NOW WORK** **Problem 5.**

In general, the procedure for approximating the area $A$ is based on the idea of summing the areas of rectangles. We shall refer to the area $A$ of the region enclosed by the graph of a function $y = f(x) \geq 0$, the $x$-axis, and the lines $x = a$ and $x = b$ as the **area under the graph of $f$ from $a$ to $b$**.

We make two assumptions about the function $f$:

- $f$ is continuous on the closed interval $[a, b]$.
- $f$ is nonnegative on the closed interval $[a, b]$.

We divide, or **partition**, the interval $[a, b]$ into $n$ nonoverlapping subintervals:

$$[x_0, x_1], [x_1, x_2], \ldots, [x_{i-1}, x_i], \ldots, [x_{n-1}, x_n] \qquad x_0 = a, x_n = b$$

each of the same length. See Figure 8. Since there are $n$ subintervals and the length of the interval $[a, b]$ is $b - a$, the common length $\Delta x$ of each subinterval is

$$\Delta x = \frac{b - a}{n}$$

**NOTE** The actual area in Figure 5 is 9 square units, obtained by using the formula for the area $A$ of a trapezoid with base $b$ and parallel heights $h_1$ and $h_2$:

$$A = \frac{1}{2}b(h_1 + h_2) = \frac{1}{2}(2)(4 + 5) = 9.$$

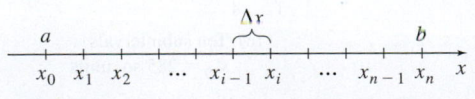

**Figure 8**

**NEED TO REVIEW?** The Extreme Value Theorem is discussed in Section 4.2, p. 274.

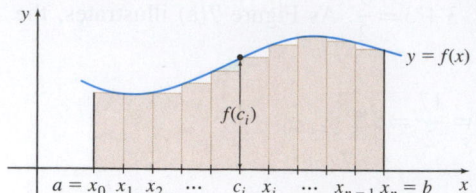

**Figure 9** $f(c_i)$ is the absolute minimum value of $f$ on $[x_{i-1}, x_i]$.

**NEED TO REVIEW?** Summation notation is discussed in Section P.8, pp. 71–72.

Since $f$ is continuous on the closed interval $[a, b]$, it is continuous on every subinterval $[x_{i-1}, x_i]$ of $[a, b]$. By the Extreme Value Theorem, there is a number in each subinterval where $f$ attains its absolute minimum. Label these numbers $c_1, c_2, c_3, \ldots, c_n$, so that $f(c_i)$ is the absolute minimum value of $f$ in the subinterval $[x_{i-1}, x_i]$. Now construct $n$ rectangles, each having $\Delta x$ as its base and $f(c_i)$ as its height, as illustrated in Figure 9. This produces $n$ narrow rectangles of uniform width $\Delta x = \dfrac{b-a}{n}$ and heights $f(c_1), f(c_2), \ldots, f(c_n)$, respectively. The areas of the $n$ rectangles are

$$\text{Area of the first rectangle} = f(c_1)\Delta x$$
$$\text{Area of the second rectangle} = f(c_2)\Delta x$$
$$\vdots$$
$$\text{Area of the } n\text{th (and last) rectangle} = f(c_n)\Delta x$$

The sum $s_n$ of the areas of the $n$ rectangles approximates the area $A$. That is,

$$A \approx s_n = f(c_1)\Delta x + f(c_2)\Delta x + \cdots + f(c_i)\Delta x + \cdots + f(c_n)\Delta x = \sum_{i=1}^{n} f(c_i)\Delta x$$

Since the rectangles used to approximate the area $A$ lie under the graph of $f$, the sum $s_n$, called a **lower sum**, *underestimates* $A$. That is, $s_n \le A$.

---

**EXAMPLE 2** **Approximating Area Using Lower Sums**

Approximate the area $A$ under the graph of $f(x) = x^2$ from 0 to 10 by using lower sums $s_n$ (rectangles that lie under the graph) for:

**(a)** $n = 2$ subintervals  **(b)** $n = 5$ subintervals  **(c)** $n = 10$ subintervals

**Solution** **(a)** For $n = 2$, we partition the closed interval $[0, 10]$ into two subintervals $[0, 5]$ and $[5, 10]$, each of length $\Delta x = \dfrac{10-0}{2} = 5$. See Figure 10(a). To compute $s_2$, we need to know where $f$ attains its minimum value in each subinterval. Since $f$ is an increasing function on $[0, 10]$, the absolute minimum is attained at the left endpoint of each subinterval. So, for $n = 2$, the minimum of $f$ on $[0, 5]$ occurs at 0 and the minimum of $f$ on $[5, 10]$ occurs at 5. The lower sum $s_2$ is

$$s_2 = \sum_{i=1}^{2} f(c_i)\Delta x = \Delta x \sum_{i=1}^{2} f(c_i) = 5[f(0) + f(5)] = 5(0 + 25) = 125$$

$$\underset{\substack{\uparrow \\ \Delta x = 5 \\ f(c_1) = f(0);\ f(c_2) = f(5)}}{} \quad \underset{\substack{\uparrow \\ f(0) = 0 \\ f(5) = 25}}{}$$

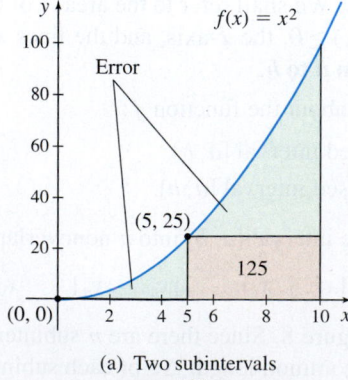

(a) Two subintervals
$s_2 = 125$ sq. units

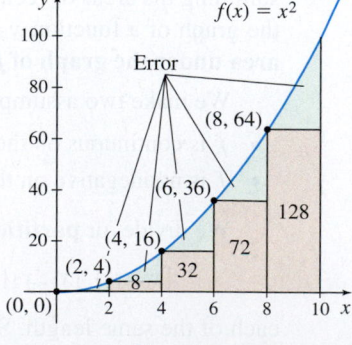

(b) Five subintervals
$s_5 = 240$ sq. units

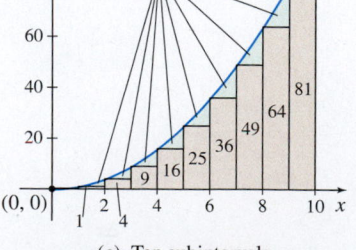

(c) Ten subintervals
$s_{10} = 285$ sq. units

**DF** **Figure 10**

**(b)** For $n = 5$, partition the interval $[0, 10]$ into five subintervals $[0, 2]$, $[2, 4]$, $[4, 6]$, $[6, 8]$, $[8, 10]$, each of length $\Delta x = \dfrac{10 - 0}{5} = 2$. See Figure 10(b). The lower sum $s_5$ is

$$s_5 = \sum_{i=1}^{5} f(c_i)\Delta x = \Delta x \sum_{i=1}^{5} f(c_i) = 2[f(0) + f(2) + f(4) + f(6) + f(8)]$$

$$= 2(0 + 4 + 16 + 36 + 64) = 240$$

**(c)** For $n = 10$, partition $[0, 10]$ into 10 subintervals, each of length $\Delta x = \dfrac{10 - 0}{10} = 1$. See Figure 10(c). The lower sum $s_{10}$ is

$$s_{10} = \sum_{i=1}^{10} f(c_i)\Delta x = \Delta x \sum_{i=1}^{10} f(c_i) = 1[f(0) + f(1) + f(2) + \cdots + f(9)]$$

$$= 0 + 1 + 4 + 9 + 16 + 25 + 36 + 49 + 64 + 81 = 285 \qquad \blacksquare$$

**NOW WORK** Problem **13(a)**.

In general, as Figure 11(a) illustrates, the error due to using lower sums $s_n$ (rectangles that lie below the graph of $f$) occurs because a portion of the region lies outside the rectangles. To improve the approximation of the area, we increase the number of subintervals. For example, in Figure 11(b), there are four subintervals and the error is reduced; in Figure 11(c), there are eight subintervals and the error is further reduced. By taking a finer and finer partition of the interval $[a, b]$, that is, by increasing the number $n$ of subintervals without bound, we can make the sum of the areas of the rectangles as close as we please to the actual area of the region. (A proof of this statement is usually found in books on advanced calculus.)

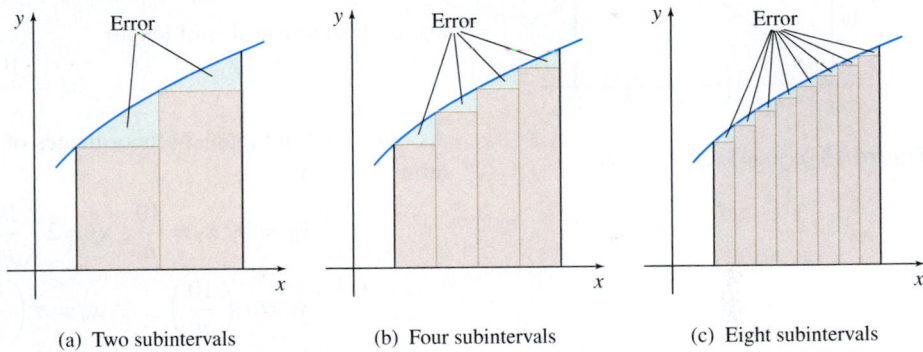

(a) Two subintervals        (b) Four subintervals        (c) Eight subintervals

**Figure 11**

For functions that are continuous on a closed interval, $\lim\limits_{n \to \infty} s_n$ always exists. With this in mind, we now define the area under the graph of a function $f$ from $a$ to $b$.

**DEFINITION  Area $A$ Under the Graph of a Function from $a$ to $b$**

Suppose a function $f$ is nonnegative and continuous on a closed interval $[a, b]$. Partition $[a, b]$ into $n$ subintervals $[x_0, x_1], [x_1, x_2], \ldots, [x_{i-1}, x_i], \ldots, [x_{n-1}, x_n]$, each of length

$$\Delta x = \frac{b - a}{n}$$

In each subinterval $[x_{i-1}, x_i]$, let $f(c_i)$ equal the absolute minimum value of $f$ on the subinterval. Form the lower sums

$$s_n = \sum_{i=1}^{n} f(c_i)\Delta x = f(c_1)\Delta x + \cdots + f(c_n)\Delta x$$

The **area $A$ under the graph of $f$ from $a$ to $b$** is the number

$$\boxed{A = \lim_{n \to \infty} s_n}$$

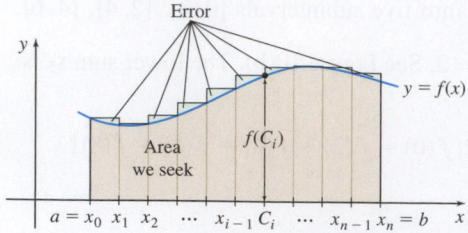

**Figure 12** $f(C_i)$ is the absolute maximum value of $f$ on $[x_{i-1}, x_i]$.

The area $A$ is defined using lower sums $s_n$ (rectangles that lie below the graph of $f$). By a parallel argument, we can choose values $C_1, C_2, \ldots, C_n$ so that the height $f(C_i)$ of the $i$th rectangle is the absolute maximum value of $f$ on the $i$th subinterval, as shown in Figure 12. The corresponding **upper sums** $S_n$ (rectangles that extend above the graph of $f$) *overestimate* the area $A$. So, $S_n \geq A$. It can be shown that as $n$ increases without bound, the limit of the upper sums $S_n$ equals the limit of the lower sums $s_n$. That is,

$$\lim_{n \to \infty} s_n = \lim_{n \to \infty} S_n = A$$

**NOW WORK** Problem 13(b).

## 2 Find the Area Under the Graph of a Function

In the next example, instead of using a specific number of rectangles to *approximate* area, we partition the interval $[a, b]$ into $n$ subintervals, obtaining $n$ rectangles. By letting $n \to \infty$, we find the *actual* area under the graph of $f$ from $a$ to $b$.

### EXAMPLE 3   Finding Area Using Upper Sums

Find the area $A$ under the graph of $f(x) = 3x$ from 0 to 10 using upper sums $S_n$ (rectangles that extend above the graph of $f$). That is, find $A = \lim_{n \to \infty} S_n$.

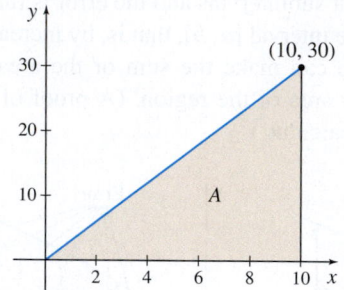

**Figure 13** $f(x) = 3x, 0 \leq x \leq 10$

**Solution** Figure 13 illustrates the area $A$. Partition the closed interval $[0, 10]$ into $n$ subintervals

$$[x_0, x_1], [x_1, x_2], \ldots, [x_{i-1}, x_i], \ldots, [x_{n-1}, x_n]$$

where

$$0 = x_0 < x_1 < x_2 < \cdots < x_i < \cdots < x_{n-1} < x_n = 10$$

and each subinterval is of length

$$\Delta x = \frac{10 - 0}{n} = \frac{10}{n}$$

As illustrated in Figure 14, coordinates of the endpoints of each subinterval, written in terms of $n$, are

$$x_0 = 0, \ x_1 = \frac{10}{n}, \ x_2 = 2\left(\frac{10}{n}\right), \ldots, \ x_{i-1} = (i-1)\left(\frac{10}{n}\right),$$

$$x_i = i\left(\frac{10}{n}\right), \ldots, \ x_n = n\left(\frac{10}{n}\right) = 10$$

$$\Delta x = \frac{10}{n}$$

**Figure 14**

To find $A$ using upper sums $S_n$ (rectangles that extend above the graph of $f$), we need the absolute maximum value of $f$ in each subinterval. Since $f(x) = 3x$ is an increasing function, the absolute maximum occurs at the right endpoint $x_i = i\left(\frac{10}{n}\right) = \frac{10i}{n}$ of each subinterval. So, $C_i = x_i$, $f(C_i) = 3x_i$, and

$$S_n = \sum_{i=1}^{n} f(C_i)\Delta x = \sum_{i=1}^{n} 3x_i \cdot \frac{10}{n} = \sum_{i=1}^{n} \left(3 \cdot \frac{10i}{n} \cdot \frac{10}{n}\right) = \sum_{i=1}^{n} \frac{300}{n^2} i$$

$$\Delta x = \frac{10}{n} \qquad x_i = \frac{10i}{n}$$

**NEED TO REVIEW?** Summation properties are discussed in Section P.8, pp. 72–73.

**RECALL** $\displaystyle\sum_{i=1}^{n} i = \dfrac{n(n+1)}{2}$ (p. 72)

Using summation properties, we get

$$S_n = \sum_{i=1}^{n} \frac{300}{n^2} i = \frac{300}{n^2} \sum_{i=1}^{n} i = \frac{300}{n^2} \cdot \frac{n(n+1)}{2} = 150\left(\frac{n+1}{n}\right) = 150\left(1 + \frac{1}{n}\right)$$

Then

$$A = \lim_{n\to\infty} S_n = \lim_{n\to\infty}\left[150\left(1 + \frac{1}{n}\right)\right] = 150 \lim_{n\to\infty}\left(1 + \frac{1}{n}\right) = 150$$

The area $A$ under the graph of $f(x) = 3x$ from 0 to 10 is 150 square units. ■

**NOTE** The region under the graph of $f(x) = 3x$ from 0 to 10 is a triangle. So, we can verify that $A = 150$ by using the formula for the area $A$ of a triangle with base $b$ and height $h$:

$$A = \frac{1}{2}bh = \frac{1}{2}(10)(30) = 150$$

CALC
CLIP

**EXAMPLE 4  Finding Area Using Lower Sums**

Find the area $A$ under the graph of $f(x) = 16 - x^2$ from 0 to 4 by using lower sums $s_n$ (rectangles that lie below the graph of $f$). That is, find $A = \lim_{n\to\infty} s_n$.

**Solution** Figure 15 shows the area under the graph of $f$ and a typical rectangle that lies below the graph. Partition the closed interval $[0, 4]$ into $n$ subintervals

$$[x_0, x_1], [x_1, x_2], \ldots, [x_{i-1}, x_i], \ldots, [x_{n-1}, x_n]$$

where

$$0 = x_0 < x_1 < \cdots < x_i < \cdots < x_{n-1} < x_n = 4$$

and each interval is of length

$$\Delta x = \frac{4 - 0}{n} = \frac{4}{n}$$

As Figure 16 illustrates, the endpoints of each subinterval, written in terms of $n$, are

$$x_0 = 0, \ x_1 = 1\left(\frac{4}{n}\right), \ x_2 = 2\left(\frac{4}{n}\right), \ldots, x_{i-1} = (i-1)\left(\frac{4}{n}\right),$$

$$x_i = i\left(\frac{4}{n}\right), \ldots, x_n = n\left(\frac{4}{n}\right) = 4$$

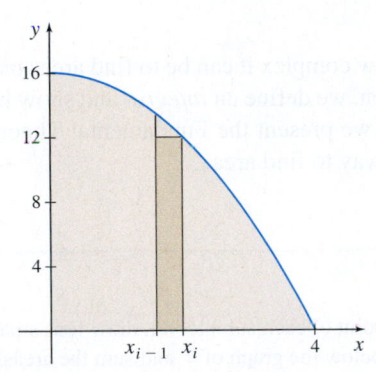

**Figure 15** $f(x) = 16 - x^2, 0 \le x \le 4$

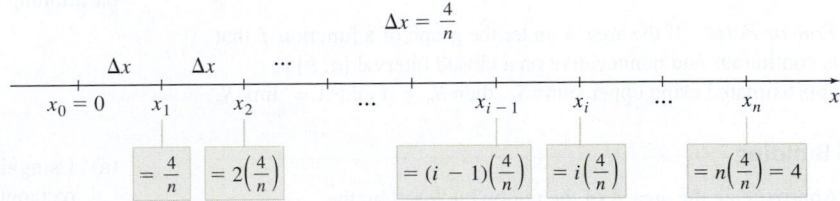

**Figure 16**

To find $A$ using lower sums $s_n$ (rectangles that lie below the graph of $f$), we must find the absolute minimum value of $f$ on each subinterval. Since the function $f$ is a decreasing function, the absolute minimum occurs at the right endpoint of each subinterval.

$$\text{Since } c_i = i\left(\frac{4}{n}\right) = \frac{4i}{n} \quad \text{and} \quad \Delta x = \frac{4}{n},$$

$$s_n = \sum_{i=1}^{n} f(c_i)\Delta x = \sum_{i=1}^{n} \left[16 - \left(\frac{4i}{n}\right)^2\right]\left(\frac{4}{n}\right) \qquad c_i = \frac{4i}{n}; \, f(c_i) = 16 - c_i^2; \, \Delta x = \frac{4}{n}$$

$$= \sum_{i=1}^{n} \left[\frac{64}{n} - \frac{64i^2}{n^3}\right]$$

$$= \frac{64}{n}\sum_{i=1}^{n} 1 - \frac{64}{n^3}\sum_{i=1}^{n} i^2$$

**RECALL** $\displaystyle\sum_{i=1}^{n} 1 = n$

$$= \frac{64}{n}(n) - \frac{64}{n^3}\left[\frac{n(n+1)(2n+1)}{6}\right] = 64 - \frac{32}{3n^2}\left[2n^2 + 3n + 1\right]$$

$\displaystyle\sum_{i=1}^{n} i^2 = \frac{n(n+1)(2n+1)}{6}$ (p. 72)

$$= 64 - \frac{64}{3} - \frac{32}{n} - \frac{32}{3n^2} = \frac{128}{3} - \frac{32}{n} - \frac{32}{3n^2}$$

Then

$$A = \lim_{n\to\infty} s_n = \lim_{n\to\infty}\left(\frac{128}{3} - \frac{32}{n} - \frac{32}{3n^2}\right) = \frac{128}{3}$$

The area $A$ under the graph of $f(x) = 16 - x^2$ from 0 to 4 is $\dfrac{128}{3}$ square units. ∎

**NOW WORK** Problem 29.

The previous two examples illustrate just how complex it can be to find areas using lower sums and/or upper sums. In the next section, we define an *integral* and show how it can be used to find area. Then in Section 5.3, we present the Fundamental Theorem of Calculus, which provides a relatively simple way to find area.

## 5.1 Assess Your Understanding

### Concepts and Vocabulary

1. Explain how rectangles can be used to approximate the area of the region enclosed by the graph of a function $y = f(x) \geq 0$, the $x$-axis, and the lines $x = a$ and $x = b$.

2. *True or False* When a closed interval $[a, b]$ is partitioned into $n$ subintervals each of the same length, the length of each subinterval is $\dfrac{a+b}{n}$.

3. If the closed interval $[-2, 4]$ is partitioned into 12 subintervals, each of the same length, then the length of each subinterval is _____.

4. *True or False* If the area $A$ under the graph of a function $f$ that is continuous and nonnegative on a closed interval $[a, b]$ is approximated using upper sums $S_n$, then $S_n \geq A$ and $A = \lim_{n\to\infty} S_n$.

### Skill Building

5. Approximate the area $A$ of the region enclosed by the graph of $f(x) = \dfrac{1}{2}x + 3$, the $x$-axis, and the lines $x = 2$ and $x = 4$ by partitioning the closed interval $[2, 4]$ into four subintervals:

$$\left[2, \frac{5}{2}\right], \left[\frac{5}{2}, 3\right], \left[3, \frac{7}{2}\right], \left[\frac{7}{2}, 4\right]$$

(a) Using the left endpoint of each subinterval, draw four small rectangles that lie below the graph of $f$ and sum the areas of the four rectangles.

(b) Using the right endpoint of each subinterval, draw four small rectangles that extend above the graph of $f$ and sum the areas of the four rectangles.

(c) Compare the answers from parts (a) and (b) to the exact area $A = 9$ and to the estimates obtained in Example 1.

6. Approximate the area $A$ of the region enclosed by the graph of $f(x) = 6 - 2x$, the $x$-axis, and the lines $x = 1$ and $x = 3$ by partitioning the closed interval $[1, 3]$ into four subintervals:

$$\left[1, \frac{3}{2}\right], \left[\frac{3}{2}, 2\right], \left[2, \frac{5}{2}\right], \left[\frac{5}{2}, 3\right]$$

(a) Using the right endpoint of each subinterval, draw four small rectangles that lie below the graph of $f$ and sum the areas of the four rectangles.

(b) Using the left endpoint of each subinterval, draw four small rectangles that extend above the graph of $f$ and sum the areas of the four rectangles.

(c) Compare the answers from parts (a) and (b) to the exact area $A = 4$.

---

**1.** = NOW WORK problem    ⧨ = Graphing technology recommended    CAS = Computer Algebra System recommended

*In Problems 7 and 8, refer to the graphs below. Approximate the shaded area under the graph of f:*

**(a)** *By constructing rectangles using the left endpoint of each subinterval.*

**(b)** *By constructing rectangles using the right endpoint of each subinterval.*

**7.**

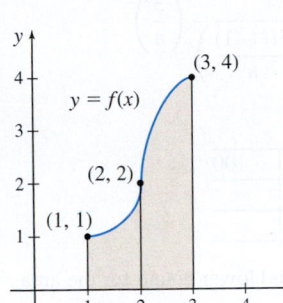

**8.**

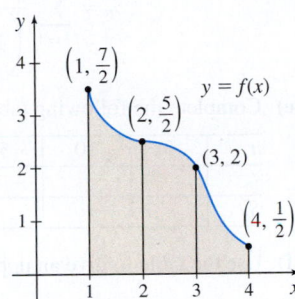

*In Problems 9–12, partition each interval into n subintervals each of the same length.*

**9.** $[1, 4]$ with $n = 3$

**10.** $[0, 9]$ with $n = 9$

**11.** $[-1, 4]$ with $n = 10$

**12.** $[-4, 4]$ with $n = 16$

*In Problems 13 and 14, refer to the graphs. Using the indicated subintervals, approximate the shaded area:*

**(a)** *By using lower sums $s_n$ (rectangles that lie below the graph of f).*

**(b)** *By using upper sums $S_n$ (rectangles that extend above the graph of f).*

**13.**

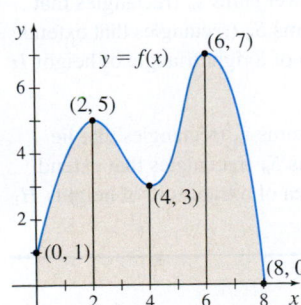

**14.**

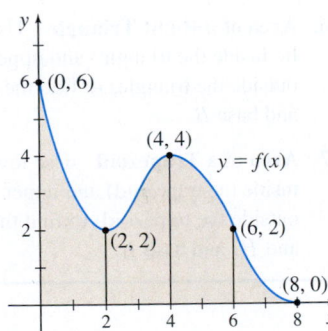

**15. Area Under a Graph**

**(a)** Graph $y = x$ and indicate the area under the graph from 0 to 3.

**(b)** Partition the interval $[0, 3]$ into $n$ subintervals each of equal length.

**(c)** Show that $s_n = \displaystyle\sum_{i=1}^{n} (i-1)\left(\frac{3}{n}\right)^2$.

**(d)** Show that $S_n = \displaystyle\sum_{i=1}^{n} i\left(\frac{3}{n}\right)^2$.

**(e)** Show that $\displaystyle\lim_{n\to\infty} s_n = \lim_{n\to\infty} S_n = \frac{9}{2}$.

**16. Area Under a Graph**

**(a)** Graph $y = 4x$ and indicate the area under the graph from 0 to 5.

**(b)** Partition the interval $[0, 5]$ into $n$ subintervals each of equal length.

**(c)** Show that $s_n = \displaystyle\sum_{i=1}^{n} (i-1)\frac{100}{n^2}$.

**(d)** Show that $S_n = \displaystyle\sum_{i=1}^{n} i\,\frac{100}{n^2}$.

**(e)** Show that $\displaystyle\lim_{n\to\infty} s_n = \lim_{n\to\infty} S_n = 50$.

*In Problems 17–22, approximate the area A under the graph of each function f from a to b for $n = 4$ and $n = 8$ subintervals:*

**(a)** *By using lower sums $s_n$ (rectangles that lie below the graph of f).*

**(b)** *By using upper sums $S_n$ (rectangles that extend above the graph of f).*

**17.** $f(x) = -x + 10$ on $[0, 8]$

**18.** $f(x) = 2x + 5$ on $[2, 6]$

**19.** $f(x) = 16 - x^2$ on $[0, 4]$

**20.** $f(x) = x^3$ on $[0, 8]$

**21.** $f(x) = \cos x$ on $\left[-\dfrac{\pi}{2}, \dfrac{\pi}{2}\right]$

**22.** $f(x) = \sin x$ on $[0, \pi]$

**23.** Rework Example 3 (p. 362) by using lower sums $s_n$ (rectangles that lie below the graph of $f$).

**24.** Rework Example 4 (p. 363) by using upper sums $S_n$ (rectangles that extend above the graph of $f$).

*In Problems 25–32, find the area A under the graph of f from a to b:*

**(a)** *By using lower sums $s_n$ (rectangles that lie below the graph of f).*

**(b)** *By using upper sums $S_n$ (rectangles that extend above the graph of f).*

**(c)** *Compare the work required in (a) and (b). Which is easier? Could you have predicted this?*

**25.** $f(x) = 2x + 1$ from $a = 0$ to $b = 4$

**26.** $f(x) = 3x + 1$ from $a = 0$ to $b = 4$

**27.** $f(x) = 12 - 3x$ from $a = 0$ to $b = 4$

**28.** $f(x) = 5 - x$ from $a = 0$ to $b = 4$

**29.** $f(x) = 4x^2$ from $a = 0$ to $b = 2$

**30.** $f(x) = \dfrac{1}{2}x^2$ from $a = 0$ to $b = 3$

**31.** $f(x) = 4 - x^2$ from $a = 0$ to $b = 2$

**32.** $f(x) = 12 - x^2$ from $a = 0$ to $b = 3$

## Applications and Extensions

*In Problems 33–38, find the area under the graph of f from a to b. Partition the closed interval $[a, b]$ into n subintervals*

$$[x_0, x_1], [x_1, x_2], \ldots, [x_{i-1}, x_i], \ldots, [x_{n-1}, x_n],$$

*where $a = x_0 < x_1 < \cdots < x_i < \cdots < x_{n-1} < x_n = b$,*

*and each subinterval is of length $\Delta x = \dfrac{b-a}{n}$. As the* *figure below illustrates, the endpoints of each subinterval, written in terms of n, are*

$$x_0 = a, \quad x_1 = a + \frac{b-a}{n}, \quad x_2 = a + 2\left(\frac{b-a}{n}\right), \ldots,$$

$$x_{i-1} = a + (i-1)\left(\frac{b-a}{n}\right), \quad x_i = a + i\left(\frac{b-a}{n}\right), \ldots,$$

$$x_n = a + n\left(\frac{b-a}{n}\right) = b$$

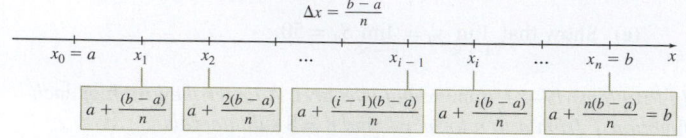

**33.** $f(x) = x + 3$ from $a = 1$ to $b = 3$

**34.** $f(x) = 3 - x$ from $a = 1$ to $b = 3$

**35.** $f(x) = 2x + 5$ from $a = -1$ to $b = 2$

**36.** $f(x) = 2 - 3x$ from $a = -2$ to $b = 0$

**37.** $f(x) = 2x^2 + 1$ from $a = 1$ to $b = 3$

**38.** $f(x) = 4 - x^2$ from $a = 1$ to $b = 2$

*In Problems 39–42, approximate the area A under the graph of each function f by partitioning [a, b] into 20 subintervals of equal length and using an upper sum.*

**39.** $f(x) = xe^x$ on $[0, 8]$

**40.** $f(x) = \ln x$ on $[1, 3]$

**41.** $f(x) = \dfrac{1}{x}$ on $[1, 5]$

**42.** $f(x) = \dfrac{1}{x^2}$ on $[2, 6]$

**43. (a)** Graph $y = \dfrac{4}{x}$ from $x = 1$ to $x = 4$ and shade the area under its graph.

**(b)** Partition the interval $[1, 4]$ into $n$ subintervals of equal length.

**(c)** Show that the lower sum $s_n$ is

$$s_n = \sum_{i=1}^{n} \frac{4}{\left(1 + \dfrac{3i}{n}\right)}\left(\frac{3}{n}\right)$$

**(d)** Show that the upper sum $S_n$ is

$$S_n = \sum_{i=1}^{n} \frac{4}{\left(1 + \dfrac{3(i-1)}{n}\right)}\left(\frac{3}{n}\right)$$

**(e)** Complete the following table:

| $n$ | 5 | 10 | 50 | 100 |
|-----|---|----|----|----|
| $s_n$ | | | | |
| $S_n$ | | | | |

**(f)** Use the table to give an upper and lower bound for the area.

## Challenge Problems

**44. Area Under a Graph** Approximate the area under the graph of $f(x) = x$ from $a \geq 0$ to $b$ by using lower sums $s_n$ and upper sums $S_n$ for a partition of $[a, b]$ into $n$ subintervals, each of length $\dfrac{b-a}{n}$. Show that

$$s_n < \frac{b^2 - a^2}{2} < S_n$$

**45. Area Under a Graph** Approximate the area under the graph of $f(x) = x^2$ from $a \geq 0$ to $b$ by using lower sums $s_n$ and upper sums $S_n$ for a partition of $[a, b]$ into $n$ subintervals, each of length $\dfrac{b-a}{n}$. Show that

$$s_n < \frac{b^3 - a^3}{3} < S_n$$

**46. Area of a Right Triangle** Use lower sums $s_n$ (rectangles that lie inside the triangle) and upper sums $S_n$ (rectangles that extend outside the triangle) to find the area of a right triangle of height $H$ and base $B$.

**47. Area of a Trapezoid** Use lower sums $s_n$ (rectangles that lie inside the trapezoid) and upper sums $S_n$ (rectangles that extend outside the trapezoid) to find the area of a trapezoid of heights $H_1$ and $H_2$ and base $B$.

# 5.2 The Definite Integral

**OBJECTIVES** *When you finish this section, you should be able to:*

**1** Form Riemann sums (p. 367)

**2** Define a definite integral as the limit of Riemann sums (p. 368)

**3** Approximate a definite integral using Riemann sums (p. 370)

**4** Know conditions that guarantee a definite integral exists (p. 371)

**5** Find a definite integral using the limit of Riemann sums (p. 372)

**6** Form Riemann sums from a table (p. 374)

The area $A$ under the graph of $y = f(x)$ from $a$ to $b$ is obtained by finding

$$A = \lim_{n \to \infty} s_n = \lim_{n \to \infty} \sum_{i=1}^{n} f(c_i)\Delta x = \lim_{n \to \infty} S_n = \lim_{n \to \infty} \sum_{i=1}^{n} f(C_i)\Delta x \qquad (1)$$

where the following assumptions are made:

- The function $f$ is continuous on the closed interval $[a, b]$.
- The function $f$ is nonnegative on the closed interval $[a, b]$.
- The closed interval $[a, b]$ is partitioned into $n$ subintervals, each of length

$$\Delta x = \frac{b - a}{n}$$

- $f(c_i)$ is the absolute minimum value of $f$ on the $i$th subinterval, $i = 1, 2, \ldots, n$.
- $f(C_i)$ is the absolute maximum value of $f$ on the $i$th subinterval, $i = 1, 2, \ldots, n$.

In Section 5.1, we found the area $A$ under the graph of $f$ from $a$ to $b$ by choosing either the number $c_i$, where $f$ has an absolute minimum on the $i$th subinterval, or the number $C_i$, where $f$ has an absolute maximum on the $i$th subinterval. Suppose we arbitrarily choose a number $u_i$ in each subinterval $[x_{i-1}, x_i]$, and draw rectangles of height $f(u_i)$ and length $\Delta x$. Then from the definitions of absolute minimum value and absolute maximum value

$$f(c_i) \le f(u_i) \le f(C_i)$$

and, since $\Delta x > 0$,

$$f(c_i)\Delta x \le f(u_i)\Delta x \le f(C_i)\Delta x$$

Then

$$\sum_{i=1}^{n} f(c_i)\Delta x \le \sum_{i=1}^{n} f(u_i)\Delta x \le \sum_{i=1}^{n} f(C_i)\Delta x$$

$$s_n \le \sum_{i=1}^{n} f(u_i)\Delta x \le S_n$$

Since $\lim\limits_{n \to \infty} s_n = \lim\limits_{n \to \infty} S_n = A$, by the Squeeze Theorem, we have

$$\lim_{n \to \infty} \sum_{i=1}^{n} f(u_i)\Delta x = A$$

In other words, we can use any number $u_i$ in the $i$th subinterval to find the area $A$.

## 1 Form Riemann Sums

We now investigate sums of the form

$$\boxed{\sum_{i=1}^{n} f(u_i)\Delta x_i}$$

using the following more general assumptions:

- The function $f$ is defined on a closed interval $[a, b]$.
- The function $f$ is not necessarily continuous on $[a, b]$.
- The function $f$ is not necessarily nonnegative on $[a, b]$.
- The lengths $\Delta x_i = x_i - x_{i-1}$ of the subintervals $[x_{i-1}, x_i]$, $i = 1, 2, \ldots, n$ of $[a, b]$ are not necessarily equal.
- The number $u_i$ may be any number in the subinterval $[x_{i-1}, x_i]$, $i = 1, 2, \ldots, n$.

The sums $\sum_{i=1}^{n} f(u_i)\Delta x_i$, called **Riemann sums** for $f$ on $[a, b]$, are the foundation of integral calculus.

**ORIGINS** Riemann sums are named after the German mathematician Georg Friedrich Bernhard Riemann (1826–1866). Early in his life, Riemann was home schooled. At age 14, he was sent to a lyceum (high school) and then the University of Göttingen to study theology. Once at Göttingen, he asked for and received permission from his father to study mathematics. He completed his Ph.D. under Karl Friedrich Gauss (1777–1855). In his thesis, Riemann used topology to analyze complex functions. Later he developed a theory of geometry to describe real space. His ideas were far ahead of their time and were not truly appreciated until they provided the mathematical framework for Einstein's Theory of Relativity.

**NEED TO REVIEW?** The Squeeze Theorem is discussed in Section 1.4, pp. 115–116.

## EXAMPLE 1  Finding Riemann Sums

For the function $f(x) = x^2 - 3$, $0 \leq x \leq 6$, partition the interval $[0, 6]$ into four subintervals $[0, 1]$, $[1, 2]$, $[2, 4]$, $[4, 6]$ and find the Riemann sum for which

**(a)** $u_i$ is the left endpoint of each subinterval.

**(b)** $u_i$ is the midpoint of each subinterval.

**Solution** In forming Riemann sums $\sum_{i=1}^{n} f(u_i) \Delta x_i$, $n$ is the number of subintervals in the partition, $f(u_i)$ is the value of $f$ at the number $u_i$ chosen in the $i$th subinterval, and $\Delta x_i$ is the length of the $i$th subinterval. The 4 subintervals have length

$$\Delta x_1 = 1 - 0 = 1 \qquad \Delta x_2 = 2 - 1 = 1 \qquad \Delta x_3 = 4 - 2 = 2 \qquad \Delta x_4 = 6 - 4 = 2$$

**(a)** Figure 17 shows the graph of $f$, the partition of the interval $[0, 6]$ into the 4 subintervals, and values of $f(u_i)$ at the left endpoint of each subinterval, namely,

$$f(u_1) = f(0) = -3 \qquad f(u_2) = f(1) = -2$$
$$f(u_3) = f(2) = 1 \qquad f(u_4) = f(4) = 13$$

The Riemann sum is found by adding the products $f(u_i) \Delta x_i$ for $i = 1, 2, 3, 4$.

$$\sum_{i=1}^{4} f(u_i) \Delta x_i = f(u_1) \Delta x_1 + f(u_2) \Delta x_2 + f(u_3) \Delta x_3 + f(u_4) \Delta x_4$$
$$= -3 \cdot 1 + (-2) \cdot 1 + 1 \cdot 2 + 13 \cdot 2 = 23$$

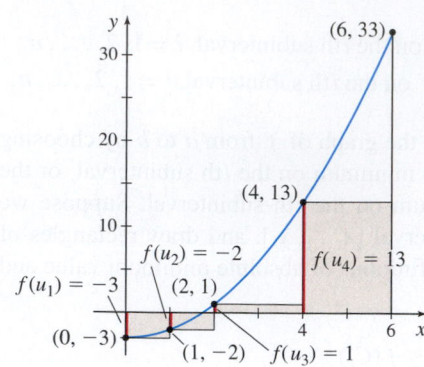

**Figure 17** $f(x) = x^2 - 3, 0 \leq x \leq 6$

**(b)** See Figure 18. If $u_i$ is chosen as the midpoint of each subinterval, then the value of $f(u_i)$ at the midpoint of each subinterval is

$$f(u_1) = f\left(\frac{1}{2}\right) = -\frac{11}{4} \qquad f(u_2) = f\left(\frac{3}{2}\right) = -\frac{3}{4}$$
$$f(u_3) = f(3) = 6 \qquad f(u_4) = f(5) = 22$$

The Riemann sum is found by adding the products $f(u_i) \Delta x_i$ for $i = 1, 2, 3, 4$.

$$\sum_{i=1}^{4} f(u_i) \Delta x_i = f(u_1) \Delta x_1 + f(u_2) \Delta x_2 + f(u_3) \Delta x_3 + f(u_4) \Delta x_4$$
$$= -\frac{11}{4} \cdot 1 + \left(-\frac{3}{4}\right) \cdot 1 + 6 \cdot 2 + 22 \cdot 2 = \frac{105}{2} = 52.5 \qquad \blacksquare$$

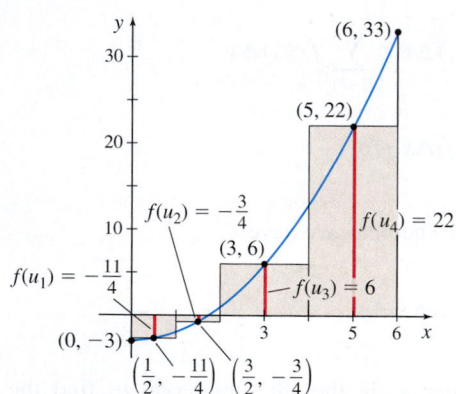

**DF** **Figure 18** $f(x) = x^2 - 3, 0 \leq x \leq 6$

**NOW WORK** Problem 9.

## 2 Define a Definite Integral as the Limit of Riemann Sums

Suppose a function $f$ is defined on a closed interval $[a, b]$, and we partition the interval $[a, b]$ into $n$ subintervals

$$[x_0, x_1], [x_1, x_2], [x_2, x_3], \ldots, [x_{i-1}, x_i], \ldots, [x_{n-1}, x_n]$$

where

$$a = x_0 < x_1 < x_2 < \cdots < x_{i-1} < x_i < \cdots < x_{n-1} < x_n = b$$

These subintervals are not necessarily of the same length. Denote the length of the first subinterval by $\Delta x_1 = x_1 - x_0$, the length of the second subinterval by $\Delta x_2 = x_2 - x_1$, and so on. In general, the length of the $i$th subinterval is

$$\boxed{\Delta x_i = x_i - x_{i-1}}$$

for $i = 1, 2, \ldots, n$. This set of subintervals of the interval $[a, b]$ is called a **partition** of $[a, b]$. The length of the largest subinterval in a partition is called the **norm** of the partition and is denoted by **max $\Delta x_i$**.

## DEFINITION Definite Integral

Let $f$ be a function defined on the closed interval $[a, b]$. Partition $[a, b]$ into $n$ subintervals of length $\Delta x_i = x_i - x_{i-1}$, $i = 1, 2, \ldots, n$. Choose a number $u_i$ in each subinterval, evaluate $f(u_i)$, and form the Riemann sums $\sum_{i=1}^{n} f(u_i) \Delta x_i$.

If $\lim\limits_{\max \Delta x_i \to 0} \sum_{i=1}^{n} f(u_i) \Delta x_i = I$ exists and does not depend on the choice of the partition or on the choice of $u_i$, then the number $I$ is called the **Riemann integral** or **definite integral** of $f$ from $a$ to $b$ and is denoted by the symbol $\int_a^b f(x)\,dx$. That is,

$$\int_a^b f(x)\,dx = \lim_{\max \Delta x_i \to 0} \sum_{i=1}^{n} f(u_i) \Delta x_i$$

When the above limit exists, then we say that $f$ is **integrable over** $[a, b]$ or that $\int_a^b f(x)\,dx$ **exists**.

If $f$ is integrable over $[a, b]$, then $\lim\limits_{\max \Delta x_i \to 0} \sum_{i=1}^{n} f(u_i) \Delta x_i$ exists for any choice of $u_i$ in the $i$th subinterval, so we are free to choose the $u_i$ any way we please. The choices could be the left endpoint of each subinterval, or the right endpoint, or the midpoint, or any other number in each subinterval. Also, $\lim\limits_{\max \Delta x_i \to 0} \sum_{i=1}^{n} f(u_i) \Delta x_i$ is independent of the partition of the closed interval $[a, b]$, provided $\max \Delta x_i$ can be made as close as we please to 0. It is this flexibility that makes the definite integral so important in engineering, physics, chemistry, geometry, and economics.

---

**EXAMPLE 2** **Expressing the Limit of Riemann Sums as a Definite Integral**

The Riemann sums for $f(x) = x^2 - 3$ on the closed interval $[0, 6]$ are

$$\sum_{i=1}^{n} f(u_i) \Delta x_i = \sum_{i=1}^{n} \left(u_i^2 - 3\right) \Delta x_i$$

where $[0, 6]$ is partitioned into $n$ subintervals $[x_{i-1}, x_i]$ of length $\Delta x_i$ and $u_i$ is some number in the subinterval $[x_{i-1}, x_i]$, $i = 1, 2, \ldots, n$. Assuming that the limit of the Riemann sums exists as $\max \Delta x_i \to 0$, express the limit as a definite integral.

**Solution** Since $\lim\limits_{\max \Delta x_i \to 0} \sum_{i=1}^{n} \left(u_i^2 - 3\right) \Delta x_i$ exists, then

$$\lim_{\max \Delta x_i \to 0} \sum_{i=1}^{n} \left(u_i^2 - 3\right) \Delta x_i = \int_0^6 (x^2 - 3)\,dx \qquad \blacksquare$$

**NOW WORK** Problem 19.

**NOTE** The integral sign $\int$ is an elongated "S" to remind you of summation.

For the definite integral $\int_a^b f(x)\,dx$, the number $a$ is called the **lower limit of integration**, the number $b$ is called the **upper limit of integration**, the symbol $\int$ is called the **integral sign**, $f(x)$ is called the **integrand**, and $dx$ is the differential of the independent variable $x$. The variable used in the definite integral is an *artificial* or a *dummy* variable because it may be replaced by any other symbol. For example,

$$\int_a^b f(x)\,dx \qquad \int_a^b f(t)\,dt \qquad \int_a^b f(s)\,ds \qquad \int_a^b f(\theta)\,d\theta$$

all denote the definite integral of $f$ from $a$ to $b$, and if any one of them exists, they all exist and equal the same number.

---

**EXAMPLE 3** Expressing a Definite Integral as the Limit of Riemann Sums

Express $\int_{-1}^{4}(x^3 - 2)\,dx$ as the limit of Riemann sums.

**Solution** The integrand is $f(x) = x^3 - 2$. The lower limit of integration is $-1$ and the upper limit of integration is 4, so the interval $[-1, 4]$ is partitioned. Partition the interval $[-1, 4]$ into $n$ subintervals of length $\Delta x_i$, $i = 1, 2, \ldots, n$. Let max $\Delta x_i$ denote the largest of these lengths. If $u_i$ is any number in the $i$th subinterval, the Riemann sums of $f$ on $[-1, 4]$ are

$$\sum_{i=1}^{n} f(u_i)\Delta x_i = \sum_{i=1}^{n}(u_i^3 - 2)\Delta x_i$$

Then

$$\int_{-1}^{4}(x^3 - 2)\,dx = \lim_{\max \Delta x_i \to 0} \sum_{i=1}^{n}(u_i^3 - 2)\Delta x_i \qquad \blacksquare$$

**NOW WORK** Problem 33.

In defining the definite integral $\int_a^b f(x)\,dx$, we assumed that $a < b$. To remove this restriction, we give the following definitions.

**DEFINITION**

- If $f(a)$ is defined, then

$$\int_a^a f(x)\,dx = 0 \qquad (2)$$

- If $a > b$ and if $\int_b^a f(x)\,dx$ exists, then

$$\int_a^b f(x)\,dx = -\int_b^a f(x)\,dx \qquad (3)$$

**IN WORDS** Interchanging the limits of integration reverses the sign of the definite integral.

---

**EXAMPLE 4** Using (2) and (3)

**(a)** $\int_1^1 x^2\,dx = 0$      **(b)** $\int_3^2 x^2\,dx = -\int_2^3 x^2\,dx$      $\blacksquare$

**NOW WORK** Problem 17.

## 3 Approximate a Definite Integral Using Riemann Sums

Suppose $f$ is integrable over the closed interval $[a, b]$. Then

$$\int_a^b f(x)\,dx = \lim_{\max \Delta x_i \to 0} \sum_{i=1}^{n} f(u_i)\Delta x_i$$

To approximate $\int_a^b f(x)\,dx$ using Riemann sums, we usually partition $[a, b]$ into $n$ subintervals, each of the same length $\Delta x = \dfrac{b-a}{n}$. Then $\Delta x_i = \Delta x$ for each $i = 1, 2, 3, \ldots, n$ and

$$\int_a^b f(x)\,dx \approx \sum_{i=1}^{n} f(u_i)\Delta x$$

where $u_i$ is usually chosen as the left endpoint, the right endpoint, or the midpoint of the $i$th subinterval $[x_{i-1}, x_i]$.

NEED TO REVIEW? Continuity is discussed in Section 1.3, pp. 102–109.

## NOTE

- Riemann sums formed using the left endpoint of each subinterval are called **Left Riemann sums.**
- Riemann sums formed using the midpoint of each subinterval are called **Midpoint Riemann sums.**
- Riemann sums formed using the right endpoint of each subinterval are called **Right Riemann sums.**

**EXAMPLE 5  Approximating a Definite Integral Using Riemann Sums**

Approximate $\int_1^9 (x-3)^2\, dx$ by partitioning $[1, 9]$ into four subintervals each of length 2

(a) using a Left Riemann sum. (Choose $u_i$ as the left endpoint of the $i$th subinterval.)

(b) using a Right Riemann sum. (Choose $u_i$ as the right endpoint of the $i$th subinterval.)

(c) using a Midpoint Riemann sum. (Choose $u_i$ as the midpoint of the $i$th subinterval.)

**Solution** For the given partition $\Delta x = \dfrac{9-1}{4} = 2$ and the four subintervals are

$$[1, 3], [3, 5], [5, 7], [7, 9]$$

(a) We choose $u_i$ to be the left endpoint of the $i$th subinterval. Then $u_1 = 1$, $u_2 = 3$, $u_3 = 5$, $u_4 = 7$, and

$$\int_1^9 (x-3)^2\, dx \approx \sum_{i=1}^4 f(u_i)\Delta x = [f(1) + f(3) + f(5) + f(7)]\Delta x$$
$$= [4 + 0 + 4 + 16] \cdot 2 = 48$$

(b) We choose $u_i$ to be the right endpoint of the $i$th subinterval. Then $u_1 = 3$, $u_2 = 5$, $u_3 = 7$, $u_4 = 9$, and

$$\int_1^9 (x-3)^2\, dx \approx \sum_{i=1}^4 f(u_i)\Delta x = [f(3) + f(5) + f(7) + f(9)]\Delta x$$
$$= [0 + 4 + 16 + 36] \cdot 2 = 112$$

(c) We choose $u_i$ to be the midpoint of the $i$th subinterval. Then $u_1 = 2$, $u_2 = 4$, $u_3 = 6$, $u_4 = 8$, and

$$\int_1^9 (x-3)^2\, dx \approx \sum_{i=1}^4 f(u_i)\Delta x = [f(2) + f(4) + f(6) + f(8)]\Delta x$$
$$= [1 + 1 + 9 + 25] \cdot 2 = 72 \qquad \blacksquare$$

**NOW WORK** Problem 39(a) and (b).

## 4 Know Conditions That Guarantee a Definite Integral Exists

Below we state two theorems that give conditions that guarantee a function $f$ is integrable. The proofs of these two results may be found in advanced calculus texts.

### THEOREM  Existence of the Definite Integral

If a function $f$ is continuous on a closed interval $[a, b]$, then the definite integral $\int_a^b f(x)\, dx$ exists.

**IN WORDS** If a function $f$ is continuous on $[a, b]$, then it is integrable over $[a, b]$. But if $f$ is not continuous on $[a, b]$, then $f$ may or may not be integrable over $[a, b]$.

The hypothesis of the theorem deserves special attention. First, $f$ is defined on a *closed* interval, and second, $f$ is *continuous* on that interval. There are some functions that are continuous on an open interval (or even a half-open interval) for which the integral does not exist. Also, there are many examples of discontinuous functions for which the integral exists.

The second theorem that guarantees a function $f$ is integrable requires that the function be *bounded*.

### DEFINITION  Bounded Function

A **function $f$ is bounded on an interval** if there are two numbers $m$ and $M$ for which $m \le f(x) \le M$ for all $x$ in the interval.

For example, the function $f(x) = x^2$ is bounded on the interval $[-1, 3]$ because there are two numbers, $m = 0$ and $M = 9$ for which $0 \le f(x) \le 9$ for all $x$ in the interval.

RECALL If $f$ is integrable on $[a, b]$ then $\int_a^b f(x)\,dx$ exists.

**THEOREM  Existence of the Definite Integral**

If a function $f$ is bounded on a closed interval $[a, b]$ and has at most a finite number of points of discontinuity on $[a, b]$, then $\int_a^b f(x)\,dx$ exists.

In most of our work with applications of definite integrals, the integrand will be continuous on a closed interval $[a, b]$, guaranteeing the integral exists.

For example, if $f$ is continuous and nonnegative on a closed interval $[a, b]$, the area under the graph of $f$ exists and equals $\int_a^b f(x)\,dx$.

## 5 Find a Definite Integral Using the Limit of Riemann Sums

Suppose $f$ is integrable over the closed interval $[a, b]$. To find

$$\int_a^b f(x)\,dx = \lim_{\max \Delta x_i \to 0} \sum_{i=1}^n f(u_i)\Delta x_i$$

using Riemann sums, we usually partition $[a, b]$ into $n$ subintervals, each of the same length $\Delta x = \dfrac{b-a}{n}$. Such a partition is called a **regular partition**. For a regular partition, the norm of the partition is

$$\max \Delta x_i = \frac{b-a}{n}$$

Since $\lim\limits_{n \to \infty} \dfrac{b-a}{n} = 0$, it follows that for a regular partition, the two statements

$$\max \Delta x_i \to 0 \qquad \text{and} \qquad n \to \infty$$

are interchangeable. As a result, for regular partitions $\Delta x = \dfrac{b-a}{n}$,

$$\int_a^b f(x)\,dx = \lim_{\max \Delta x_i \to 0} \sum_{i=1}^n f(u_i)\Delta x_i = \lim_{n \to \infty} \sum_{i=1}^n f(u_i)\Delta x$$

The next theorem uses Riemann sums to establish a formula to find the definite integral of a constant function.

**THEOREM**

If $f(x) = h$, where $h$ is some constant, then

$$\int_a^b h\,dx = h(b-a)$$

**Proof** The constant function $f(x) = h$ is continuous on the set of real numbers and so is integrable. Form the Riemann sums for $f$ on the closed interval $[a, b]$ using a regular partition. Then $\Delta x_i = \Delta x = \dfrac{b-a}{n}$, $i = 1, 2, \ldots, n$. The Riemann sums of $f$ on the interval $[a, b]$ are

$$\sum_{i=1}^n f(u_i)\Delta x_i = \sum_{i=1}^n f(u_i)\Delta x = \sum_{i=1}^n h\,\Delta x = \sum_{i=1}^n h\left(\frac{b-a}{n}\right)$$

$$= h\left(\frac{b-a}{n}\right) \sum_{i=1}^n 1 = h\left(\frac{b-a}{n}\right) \cdot n = h(b-a)$$

RECALL $\sum\limits_{i=1}^n 1 = n$ (p. 72)

Then

$$\lim_{n \to \infty} \sum_{i=1}^n f(u_i)\Delta x = \lim_{n \to \infty} [h(b-a)] = h(b-a)$$

So,

$$\int_a^b h\,dx = h(b-a)$$

∎

**EXAMPLE 6** Integrating a Constant Function

(a) $\displaystyle\int_1^2 3\,dx = 3(2-1) = 3$

(b) $\displaystyle\int_2^6 dx = 1(6-2) = 4$

(c) $\displaystyle\int_{-3}^4 (-2)\,dx = (-2)[4-(-3)] = -14$

$\blacksquare$

**NOW WORK** Problem 13.

CALC
$\circledcirc$ CLIP **EXAMPLE 7** Finding a Definite Integral Using the Limit of Riemann Sums

(a) Find $\displaystyle\int_0^3 (3x-8)\,dx$.

(b) Determine whether the integral represents an area.

**Solution (a)** The integrand $f(x) = 3x - 8$ is continuous on the closed interval $[0, 3]$, so the function $f$ is integrable over $[0, 3]$. Since we can use any partition of $[0, 3]$ whose norm can be made as close to $0$ as we please, and we can choose any $u_i$ in each subinterval, we use a regular partition and choose $u_i$ as the right endpoint of each subinterval. This will result in a simple expression for the Riemann sums.

Partition $[0, 3]$ into $n$ subintervals, each of length $\Delta x = \dfrac{3-0}{n} = \dfrac{3}{n}$. The endpoints of each subinterval of the partition, written in terms of $n$, are

$$x_0 = 0, \quad x_1 = \frac{3}{n}, \quad x_2 = 2\left(\frac{3}{n}\right), \ldots, x_{i-1} = (i-1)\left(\frac{3}{n}\right),$$

$$x_i = i\left(\frac{3}{n}\right), \ldots, x_n = n\left(\frac{3}{n}\right) = 3$$

The Riemann sums of $f(x) = 3x - 8$ from $0$ to $3$, using $u_i = x_i = \dfrac{3i}{n}$ (the right endpoint) and $\Delta x = \dfrac{3}{n}$, are

$$\sum_{i=1}^n f(u_i)\,\Delta x_i = \underset{\substack{\uparrow \\ u_i = x_i}}{\sum_{i=1}^n f(x_i)\,\Delta x} = \underset{\substack{\uparrow \\ \Delta x = \frac{3}{n}}}{\sum_{i=1}^n (3x_i - 8)\frac{3}{n}} = \underset{\substack{\uparrow \\ x_i = \frac{3i}{n}}}{\sum_{i=1}^n \left[3\left(\frac{3i}{n}\right) - 8\right]\frac{3}{n}}$$

**RECALL** $\displaystyle\sum_{i=1}^n i = \frac{n(n+1)}{2}$

$\displaystyle\sum_{i=1}^n 1 = n$

(p. 72)

$$= \sum_{i=1}^n \left(\frac{27i}{n^2} - \frac{24}{n}\right) = \frac{27}{n^2}\sum_{i=1}^n i - \frac{24}{n}\sum_{i=1}^n 1$$

$$= \frac{27}{n^2} \cdot \frac{n(n+1)}{2} - \frac{24}{n} \cdot n = \frac{27}{2} + \frac{27}{2n} - 24 = -\frac{21}{2} + \frac{27}{2n}$$

Now

$$\int_0^3 (3x-8)\,dx = \lim_{n\to\infty}\left(-\frac{21}{2} + \frac{27}{2n}\right) = -\frac{21}{2}$$

**(b)** Figure 19 shows the graph of $f(x) = 3x - 8$ on $[0, 3]$. Since $f$ is not nonnegative on $[0, 3]$, $\int_0^3 (3x-8)\,dx$ does not represent an area. The fact that the integral is negative is further evidence that this is not an area problem. $\blacksquare$

When finding a definite integral, do not presume it represents area. As you will see the definite integral has many interpretations. Interestingly enough, definite integrals are used to find the volume of a solid of revolution, the length of a graph, the work done by a variable force, and other quantities.

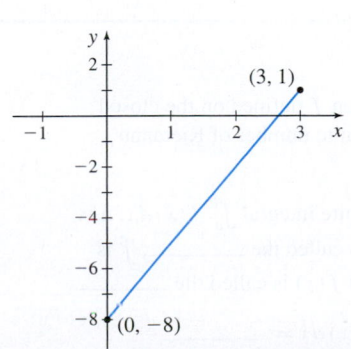

**DF** Figure 19 $f(x) = 3x - 8,\ 0 \le x \le 3$

**NOW WORK** Problem 51.

### EXAMPLE 8    Finding a Definite Integral Using Technology

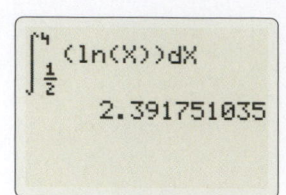

(a)  Use a graphing calculator to find $\int_{1/2}^{4} \ln x\, dx$.

(b)  Use a computer algebra system to find $\int_{1/2}^{4} \ln x\, dx$.

**Figure 20**

**Solution** (a)  A graphing calculator provides only an approximate numerical answer to the integral $\int_{1/2}^{4} \ln x\, dx$. As shown in Figure 20,

$$\int_{1/2}^{4} \ln x\, dx \approx 2.391751035$$

(b)  Because a computer algebra system manipulates symbolically, it can find an exact value of the definite integral.

$$\int_{1/2}^{4} \ln x\, dx = \frac{17}{2}\ln 2 - \frac{7}{2}$$

An approximate numerical value of the definite integral is 2.3918.  ■

**NOW WORK** Problem 39(c).

## 6 Form Riemann Sums from a Table

When a function is defined using a table, it is still possible to form a Riemann sum for the function.

### EXAMPLE 9    Finding a Riemann Sum from a Table

An object in rectilinear motion moves with velocity $v$. Table 1 displays $v$ (in meters per second) of the object for select times $t$, in seconds.

**TABLE 1**

| $t$ | 1 | 2.5 | 3 | 5 | 8 |
|-----|---|-----|---|---|---|
| $v(t)$ | 6 | 0 | −4 | 2 | 5 |

(a)  Find a Riemann sum for $v$ on the interval $[1, 8]$.

(b)  What are the units of the Riemann sum?

(c)  Interpret the Riemann sum in the context of the problem.

**Solution** (a)  Partition the interval $[1, 8]$ into four subintervals $[1, 2.5]$, $[2.5, 3]$, $[3, 5]$, and $[5, 8]$. On each subinterval choose $u_i$, $i = 1, 2, 3, 4$ as the left endpoint of the $i$th subinterval. (Alternatively the right endpoints or any combination of left and right endpoints could be chosen.) The Riemann sum of $v$ over the interval $[1, 8]$ is

$$\sum_{i=1}^{4} v(u_i)\Delta t_i = v(1)\Delta t_1 + v(2.5)\Delta t_2 + v(3)\Delta t_3 + v(5)\Delta t_4$$

$$= 6(2.5 - 1) + 0(3 - 2.5) + (-4)(5 - 3) + 2(8 - 5)$$

$$= (6)(1.5) + (0)(0.5) + (-4)(2) + (2)(3) = 7$$

(b)  Each term in the Riemann sum is the product of the velocity (in m/s) and the time (in s) so the units of the Riemann sum are meters.

(c)  The Riemann sum approximates the **net displacement** (the sum of the signed distances) of the object over the interval $[1, 8]$.  ■

**NOW WORK** Problem 63.

## 5.2 Assess Your Understanding

**Concepts and Vocabulary**

1.  *True or False*  In a Riemann sum $\sum_{i=1}^{n} f(u_i)\Delta x_i$, $u_i$ is always the left endpoint, the right endpoint, or the midpoint of the $i$th subinterval, $i = 1, 2, \ldots, n$.

2.  *Multiple Choice*  In a regular partition of $[0, 40]$ into 20 subintervals, $\Delta x =$ [(a) 20  (b) 0.5  (c) 2  (d) 4].

3.  *True or False*  A function $f$ defined on the closed interval $[a, b]$ has an infinite number of Riemann sums.

4.  In the notation for a definite integral $\int_{a}^{b} f(x)\, dx$, $a$ is called the _____; $b$ is called the _____; $\int$ is called the _____; and $f(x)$ is called the _____.

5.  If $f(a)$ is defined, $\int_{a}^{a} f(x)\, dx =$ _____.

**1.** = NOW WORK problem     = Graphing technology recommended     CAS = Computer Algebra System recommended

**6.** *True or False*  If a function $f$ is integrable over a closed interval $[a, b]$, then $\int_a^b f(x)\,dx = \int_b^a f(x)\,dx$.

**7.** *True or False*  If a function $f$ is continuous on a closed interval $[a, b]$, then the definite integral $\int_a^b f(x)\,dx$ exists.

**8.** *Multiple Choice*  Since $\int_0^2 (3x - 8)\,dx = -10$, then $\int_2^0 (3x - 8)\,dx =$  [**(a)** $-10$  **(b)** $10$  **(c)** $5$  **(d)** $0$].

## Skill Building

*In Problems 9 and 10, find the Riemann sum for $f(x) = x$, $0 \le x \le 2$, for the partition and the numbers $u_i$ given below.*

**9.** Partition: $\left[0, \frac{1}{4}\right]$, $\left[\frac{1}{4}, \frac{1}{2}\right]$, $\left[\frac{1}{2}, \frac{3}{4}\right]$, $\left[\frac{3}{4}, 1\right]$, $[1, 2]$

$u_1 = \frac{1}{8}, u_2 = \frac{3}{8}, u_3 = \frac{5}{8}, u_4 = \frac{7}{8}, u_5 = \frac{9}{8}$

**10.** Partition: $\left[0, \frac{1}{2}\right]$, $\left[\frac{1}{2}, 1\right]$, $\left[1, \frac{3}{2}\right]$, $\left[\frac{3}{2}, 2\right]$

$u_1 = \frac{1}{2}, u_2 = 1, u_3 = \frac{3}{2}, u_4 = 2$

*In Problems 11 and 12, find the Riemann sum for $f(x) = 3x + 1$, $0 \le x \le 8$, by partitioning the interval $[0, 8]$ into four subintervals, each of the same length.*

**11.** Choose $u_i$ as the left endpoint of each subinterval.

**12.** Choose $u_i$ as the right endpoint of each subinterval.

*In Problems 13–18, find each definite integral.*

**13.** $\displaystyle\int_{-3}^4 e\,dx$    **14.** $\displaystyle\int_0^3 (-\pi)\,dx$    **15.** $\displaystyle\int_3^0 (-\pi)\,dt$

**16.** $\displaystyle\int_7^2 2\,ds$    **17.** $\displaystyle\int_4^4 2\theta\,d\theta$    **18.** $\displaystyle\int_{-1}^{-1} 8\,dr$

*In Problems 19–26, write the limit of the Riemann sums as a definite integral. Here $u_i$ is in the subinterval $[x_{i-1}, x_i]$, $i = 1, 2, \ldots n$ and $\Delta x_i = x_i - x_{i-1}$, $i = 1, 2, \ldots n$. Assume each limit exists.*

**19.** $\displaystyle\lim_{\max \Delta x_i \to 0} \sum_{i=1}^n (e^{u_i} + 2)\Delta x_i$ on $[0, 2]$

**20.** $\displaystyle\lim_{\max \Delta x_i \to 0} \sum_{i=1}^n \ln u_i\,\Delta x_i$ on $[1, 8]$

**21.** $\displaystyle\lim_{\max \Delta x_i \to 0} \sum_{i=1}^n \cos u_i\,\Delta x_i$ on $[0, 2\pi]$

**22.** $\displaystyle\lim_{\max \Delta x_i \to 0} \sum_{i=1}^n (\cos u_i + \sin u_i)\,\Delta x_i$ on $[0, \pi]$

**23.** $\displaystyle\lim_{\max \Delta x_i \to 0} \sum_{i=1}^n \frac{2}{u_i^2}\,\Delta x_i$ on $[1, 4]$

**24.** $\displaystyle\lim_{\max \Delta x_i \to 0} \sum_{i=1}^n u_i^{1/3}\,\Delta x_i$ on $[0, 8]$

**25.** $\displaystyle\lim_{\max \Delta x_i \to 0} \sum_{i=1}^n u_i \ln u_i\,\Delta x_i$ on $[1, e]$

**26.** $\displaystyle\lim_{\max \Delta x_i \to 0} \sum_{i=1}^n \ln(u_i + 1)\Delta x_i$ on $[0, e]$

*In Problems 27–30, the graph of a function is shown. Express the shaded area as a definite integral.*

**27.**

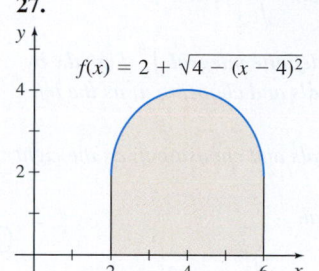

$f(x) = 2 + \sqrt{4 - (x - 4)^2}$

**28.**

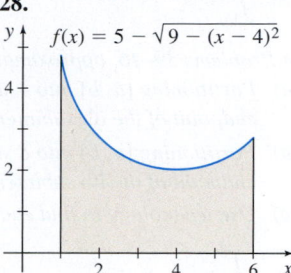

$f(x) = 5 - \sqrt{9 - (x - 4)^2}$

**29.**

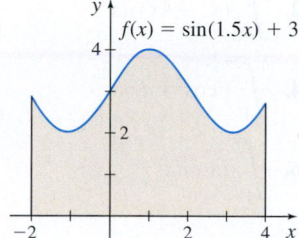

$f(x) = \sin(1.5x) + 3$

**30.**

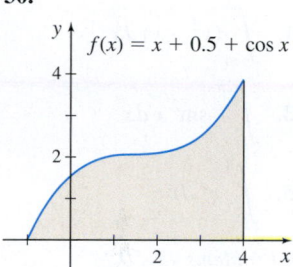

$f(x) = x + 0.5 + \cos x$

*In Problems 31 and 32, the graph of a function $f$ defined on an interval $[a, b]$ is given.*

**(a)** *Partition the interval $[a, b]$ into six subintervals (not necessarily of the same size) using the points shown on each graph.*

**(b)** *Using Riemann sums, approximate $\int_a^b f(x)\,dx$ by choosing $u_i$ as the left endpoint of each subinterval.*

**(c)** *Using Riemann sums, approximate $\int_a^b f(x)\,dx$ by choosing $u_i$ as the right endpoint of each subinterval.*

**31.**

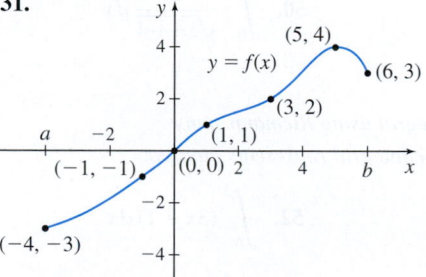

**32.**

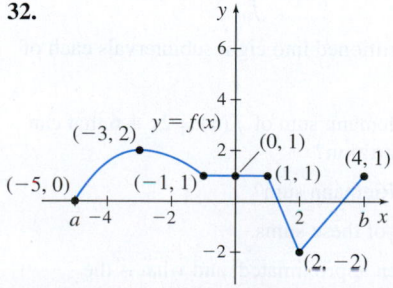

*In Problems 33–38, express each definite integral as the limit of Riemann sums.*

**33.** $\displaystyle\int_0^\pi \sin x \, dx$

**34.** $\displaystyle\int_{-\pi/4}^{\pi/4} \tan x \, dx$

**35.** $\displaystyle\int_1^4 (x-2)^{1/3} dx$

**36.** $\displaystyle\int_1^4 (x+2)^{1/3} dx$

**37.** $\displaystyle\int_1^4 (|x|-2) \, dx$

**38.** $\displaystyle\int_{-2}^4 |x| \, dx$

*In Problems 39–46, approximate each definite integral $\int_a^b f(x)\,dx$ by*

(a) *Partitioning $[a,b]$ into 4 subintervals and choosing $u_i$ as the left endpoint of the ith subinterval.*

(b) *Partitioning $[a,b]$ into 8 subintervals and choosing $u_i$ as the right endpoint of the ith subinterval.*

(c) *Use technology to find each integral.*

**39.** $\displaystyle\int_0^8 (x^2-4) \, dx$

**40.** $\displaystyle\int_1^9 (9-x^2) \, dx$

**41.** $\displaystyle\int_{-4}^4 (x^2-x) \, dx$

**42.** $\displaystyle\int_{-2}^6 (x^2-4x) \, dx$

**43.** $\displaystyle\int_0^{2\pi} \sin^2 x \, dx$

**44.** $\displaystyle\int_0^{2\pi} \cos^2 x \, dx$

**45.** $\displaystyle\int_0^8 e^x \, dx$

**46.** $\displaystyle\int_1^9 \ln x \, dx$

*In Problems 47–50,*

(a) *Approximate each definite integral by completing the table of Riemann sums using a regular partition of $[a,b]$.*

| $n$ | 10 | 50 | 100 |
|---|---|---|---|
| Using left endpoints | | | |
| Using right endpoints | | | |
| Using the midpoint | | | |

(b) *Use technology to find each definite integral.*

**47.** $\displaystyle\int_1^5 (2+\sqrt{x}) \, dx$

**48.** $\displaystyle\int_{-1}^3 (e^x + e^{-x}) \, dx$

**49.** $\displaystyle\int_{-1}^1 \frac{3}{1+x^2} \, dx$

**50.** $\displaystyle\int_0^2 \frac{1}{\sqrt{x^2+4}} \, dx$

*In Problems 51–54,*

(a) *Find each definite integral using Riemann sums.*

(b) *Determine whether the integral represents an area.*

**51.** $\displaystyle\int_0^1 (x-4) \, dx$

**52.** $\displaystyle\int_0^3 (3x-1) \, dx$

**53.** $\displaystyle\int_0^2 (2x+1) \, dx$

**54.** $\displaystyle\int_0^4 (1-x) \, dx$

**55.** The interval $[-3,5]$ is partitioned into eight subintervals each of the same length.

   (a) What is the largest Riemann sum of $f(x)=2x+6$ that can be found using this partition?

   (b) What is the smallest Riemann sum?

   (c) Compute the average of these sums.

   (d) What integral has been approximated, and what is the integral's exact value?

**56.** The interval $[2,7]$ is partitioned into five subintervals, each of the same length.

   (a) What is the largest Riemann sum of $g(x)=28-4x$ that can be formed using this partition?

   (b) What is the smallest Riemann sum?

   (c) Find the average of these sums.

   (d) What integral has been approximated and what is the integral's exact value?

## Applications and Extensions

*In Problems 57–60, for the given function f:*

(a) *Graph f.*

(b) *Express the area under the graph of f as an integral.*

(c) *Evaluate the integral.*

(d) *Confirm the answer to (c) using geometry.*

**57.** $f(x)=\sqrt{9-x^2}, 0 \le x \le 3$

**58.** $f(x)=\sqrt{25-x^2}, -5 \le x \le 5$

**59.** $f(x)=3-\sqrt{6x-x^2}, 0 \le x \le 6$

**60.** $f(x)=\sqrt{4x-x^2}+2, 0 \le x \le 4$

**61.** Find an approximate value of $\displaystyle\int_1^2 \frac{1}{x} \, dx$ by finding Riemann sums corresponding to a partition of $[1,2]$ into four subintervals, each of the same length, and evaluating the integrand at the midpoint of each subinterval. Compare your answer with the actual value, $\ln 2 = 0.6931\ldots$.

**62.** (a) Find the approximate value of $\int_0^2 \sqrt{4-x^2} \, dx$ by finding Riemann sums corresponding to a partition of $[0,2]$ into 16 subintervals, each of the same length, and evaluating the integrand at the left endpoint of each subinterval.

   (b) Can $\int_0^2 \sqrt{4-x^2} \, dx$ be interpreted as area? If it can, describe the area; if it cannot, explain why.

   (c) Find the actual value of $\int_0^2 \sqrt{4-x^2} \, dx$ by graphing $y=\sqrt{4-x^2}$ and using a familiar formula from geometry.

**63.** An object in rectilinear motion has acceleration $a$. The table below shows $a$ (in meters per second squared) of the object for select times $t$, in seconds.

| $t$ | 1 | 2 | 3.5 | 5 | 7 |
|---|---|---|---|---|---|
| $a(t)$ | 6 | 4 | 4 | 2 | 3 |

   (a) Find a Left Riemann sum for $a$ on the interval $[1,7]$.

   (b) What are the units of the Riemann sum?

   (c) Interpret the Riemann sum in the context of the problem.

**64.** An object in rectilinear motion moves with velocity $v$. The table below shows $v$ (in meters per second) for select times $t$, in seconds.

| $t$ | 0 | 1 | 3.5 | 6 | 9 |
|---|---|---|---|---|---|
| $v(t)$ | 4 | 5 | 9 | 6 | 5 |

   (a) Find a Right Riemann sum for $v$ on the interval $[0,9]$.

   (b) What are the units of the Riemann sum?

   (c) Interpret the Riemann sum in the context of the problem.

**65.** The table below gives the rate of change of revenue $R$ (in millions of dollars per year) of a manufacturing company for selected times $t$, in years.

| $t$ | 1 | 2 | 3 | 4 | 5 |
|---|---|---|---|---|---|
| $\dfrac{dR}{dt}$ | 6 | 10 | 12 | 10 | 14 |

   **(a)** Find a Right Riemann sum for $\dfrac{dR}{dt}$ on the interval $[1, 5]$.

   **(b)** What are the units of the Riemann sum?

   **(c)** Interpret the Riemann sum in the context of the problem.

**66.** The table below shows the rate of change of volume $V$ with respect to time $t$ (in liters per minute) of a balloon for selected times $t$, in minutes.

| $t$ | 1 | 2.5 | 3 | 4 | 5 |
|---|---|---|---|---|---|
| $\dfrac{dV}{dt}$ | 6 | 5 | 4 | 3 | 1 |

   **(a)** Find a Left Riemann sum for $\dfrac{dV}{dt}$ on the interval $[1, 5]$.

   **(b)** What are the units of the Riemann sum?

   **(c)** Interpret the Riemann sum in the context of the problem.

**67. Units of an Integral**   In the definite integral $\int_0^5 F(x)\,dx$, $F$ represents a force measured in newtons and $x$, $0 \le x \le 5$, is measured in meters. What are the units of $\int_0^5 F(x)\,dx$?

**68. Units of an Integral**   In the definite integral $\int_0^{50} C(x)\,dx$, $C$ represents the concentration of a drug in grams per liter and $x$, $0 \le x \le 50$, is measured in liters of alcohol. What are the units of $\int_0^{50} C(x)\,dx$?

**69. Units of an Integral**   In the definite integral $\int_a^b v(t)\,dt$, $v$ represents velocity measured in meters per second and time $t$ is measured in seconds. What are the units of $\int_a^b v(t)\,dt$?

**70. Units of an Integral**   In the definite integral $\int_a^b S(t)\,dt$, $S$ represents the rate of sales of a corporation measured in millions of dollars per year and time $t$ is measured in years. What are the units of $\int_a^b S(t)\,dt$?

*In Problems 71–74, write the limit of the Riemann sums as an integral. The partition is into $n$ subintervals each of the same length.*

**71.** $\displaystyle \lim_{n \to \infty} \frac{1}{n} \left[ \frac{1}{n} + \frac{2}{n} + \cdots + \frac{n}{n} \right]$

**72.** $\displaystyle \lim_{n \to \infty} \frac{1}{n} \left[ \left( \frac{1}{n} \right)^3 + \left( \frac{2}{n} \right)^3 + \cdots + \left( \frac{n}{n} \right)^3 \right]$

**73.** $\displaystyle \lim_{n \to \infty} \frac{3}{n} \left[ \sqrt{\frac{3}{n}} + \sqrt{\frac{6}{n}} + \cdots + \sqrt{\frac{3n}{n}} \right]$

**74.** $\displaystyle \lim_{n \to \infty} \frac{4}{n} [e^{4/n} + e^{8/n} + \cdots + e^{4n/n}]$

**Challenge Problems** ────────────────

**75.** Consider the **Dirichlet function** $f$, where

$$f(x) = \begin{cases} 1 & \text{if} \quad x \text{ is rational} \\ 0 & \text{if} \quad x \text{ is irrational} \end{cases}$$

Show that $\int_0^1 f(x)\,dx$ does not exist. (*Hint*: Evaluate the Riemann sums in two different ways: first by using rational numbers for $u_i$ and then by using irrational numbers for $u_i$.)

**76.** It can be shown (with a certain amount of work) that if $f(x)$ is integrable on the interval $[a, b]$, then so is $|f(x)|$. Is the converse true?

**77.** If only regular partitions are allowed, then we could not always partition an interval $[a, b]$ in a way that automatically partitions subintervals $[a, c]$ and $[c, b]$ for $a < c < b$. Why not?

# 5.3 The Fundamental Theorem of Calculus

**OBJECTIVES**   *When you finish this section, you should be able to:*

**1** Use Part 1 of the Fundamental Theorem of Calculus (p. 378)

**2** Use Part 2 of the Fundamental Theorem of Calculus (p. 379)

**3** Interpret the integral of a rate of change (p. 381)

**4** Interpret the integral as an accumulation function (p. 382)

In this section, we discuss the Fundamental Theorem of Calculus, a method for finding integrals more easily, avoiding the need to find the limit of Riemann sums. The Fundamental Theorem is aptly named because it links the two branches of calculus: differential calculus and integral calculus. As it turns out, the Fundamental Theorem of Calculus has two parts, each of which relates an integral to an antiderivative.

   We begin with a function $f$ that is continuous on a closed interval $[a, b]$. Then the definite integral $\int_a^b f(x)\,dx$ exists and is equal to a real number. Now if $x$ denotes any number in $[a, b]$, the definite integral $\int_a^x f(t)\,dt$ exists and depends on $x$. That is, $\int_a^x f(t)\,dt$ is a function of $x$, which we name $I$, for "integral."

$$I(x) = \int_a^x f(t)\,dt$$

NEED TO REVIEW? Antiderivatives are discussed in Section 4.8, pp. 342–345.

The domain of $I$ is the closed interval $[a, b]$. The integral that defines $I$ has a *variable upper limit of integration* $x$. The $t$ that appears in the integrand is a dummy variable. Part 1 of the Fundamental Theorem of Calculus states that if we differentiate $I$ with respect to $x$, we get back the original function $f$. That is, $\int_a^x f(t)\,dt$ *is an antiderivative of* $f$.

---

**THEOREM  Fundamental Theorem of Calculus, Part 1**

Let $f$ be a function that is continuous on a closed interval $[a, b]$. The function $I$ defined by

$$I(x) = \int_a^x f(t)\,dt$$

has the properties that it is continuous on $[a, b]$ and differentiable on $(a, b)$. Moreover,

$$I'(x) = \frac{d}{dx}\left[\int_a^x f(t)\,dt\right] = f(x)$$

for all $x$ in $(a, b)$.

---

The proof of Part 1 of the Fundamental Theorem of Calculus is given in Appendix B. However, if the integral $\int_a^x f(t)\,dt$ represents area, we can provide justification for the theorem using geometry.

Figure 21 shows the graph of a function $f$ that is nonnegative and continuous on a closed interval $[a, b]$. Then $I(x) = \int_a^x f(t)\,dt$ equals the area under the graph of $f$ from $a$ to $x$.

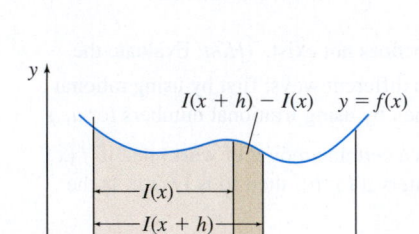

**DF** Figure 21

$$I(x) = \int_a^x f(t)\,dt = \text{the area under the graph of } f \text{ from } a \text{ to } x$$

$$I(x+h) = \int_a^{x+h} f(t)\,dt = \text{the area under the graph of } f \text{ from } a \text{ to } x+h$$

$$I(x+h) - I(x) = \text{the area under the graph of } f \text{ from } x \text{ to } x+h$$

$$\frac{I(x+h) - I(x)}{h} = \frac{\text{the area under the graph of } f \text{ from } x \text{ to } x+h}{h} \qquad (1)$$

Based on the definition of a derivative,

$$\lim_{h \to 0}\left[\frac{I(x+h) - I(x)}{h}\right] = I'(x) \qquad (2)$$

Since $f$ is continuous, $\lim\limits_{h \to 0} f(x+h) = f(x)$. As $h \to 0$, the area under the graph of $f$ from $x$ to $x+h$ gets closer to the area of a rectangle with width $h$ and height $f(x)$. That is,

$$\lim_{h \to 0}\frac{\text{the area under the graph of } f \text{ from } x \text{ to } x+h}{h} = \lim_{h \to 0}\frac{h[f(x)]}{h} = f(x) \qquad (3)$$

Combining (1), (2), and (3), it follows that $I'(x) = f(x)$.

## 1 Use Part 1 of the Fundamental Theorem of Calculus

**EXAMPLE 1   Using Part 1 of the Fundamental Theorem of Calculus**

(a) $\dfrac{d}{dx}\displaystyle\int_0^x \sqrt{t+1}\,dt = \sqrt{x+1}$    (b) $\dfrac{d}{dx}\displaystyle\int_2^x \dfrac{s^3 - 1}{2s^2 + s + 1}\,ds = \dfrac{x^3 - 1}{2x^2 + x + 1}$    ∎

**NOW WORK** Problem 5.

**CALC CLIP**

**EXAMPLE 2  Using Part 1 of the Fundamental Theorem of Calculus**

Find $\dfrac{d}{dx}\displaystyle\int_4^{3x^2+1}\sqrt{e^t+t}\,dt$.

**NEED TO REVIEW?** The Chain Rule is discussed in Section 3.1, pp. 208–211.

**Solution** The upper limit of integration is a function of $x$, so we use the Chain Rule along with Part 1 of the Fundamental Theorem of Calculus.

Let $y=\displaystyle\int_4^{3x^2+1}\sqrt{e^t+t}\,dt$ and $u(x)=3x^2+1$. Then $y=\displaystyle\int_4^{u}\sqrt{e^t+t}\,dt$ and

$$\frac{d}{dx}\int_4^{3x^2+1}\sqrt{e^t+t}\,dt=\underset{\underset{\text{Chain Rule}}{\uparrow}}{\frac{dy}{dx}}=\frac{dy}{du}\cdot\frac{du}{dx}=\left[\frac{d}{du}\int_4^{u}\sqrt{e^t+t}\,dt\right]\cdot\frac{du}{dx}$$

$$=\underset{\underset{\substack{\text{Use the Fundamental}\\\text{Theorem.}}}{\uparrow}}{\sqrt{e^u+u}}\cdot\underset{\underset{u=3x^2+1;\ \frac{du}{dx}=6x}{\uparrow}}{\frac{du}{dx}}=\sqrt{e^{(3x^2+1)}+3x^2+1}\cdot 6x\qquad\blacksquare$$

**NOW WORK** Problem 11.

**EXAMPLE 3  Using Part 1 of the Fundamental Theorem of Calculus**

Find $\dfrac{d}{dx}\displaystyle\int_{x^3}^{5}(t^4+1)^{1/3}\,dt$.

**Solution** To use Part 1 of the Fundamental Theorem of Calculus, the variable must be part of the upper limit of integration. So, we use the fact that $\int_a^b f(x)\,dx=-\int_b^a f(x)\,dx$ to interchange the limits of integration.

$$\frac{d}{dx}\int_{x^3}^{5}(t^4+1)^{1/3}\,dt=\frac{d}{dx}\left[-\int_5^{x^3}(t^4+1)^{1/3}\,dt\right]=-\frac{d}{dx}\int_5^{x^3}(t^4+1)^{1/3}\,dt$$

Now use the Chain Rule. Let $y=\displaystyle\int_5^{x^3}(t^4+1)^{1/3}\,dt$ and $u(x)=x^3$.

$$\frac{d}{dx}\int_{x^3}^{5}(t^4+1)^{1/3}\,dt=-\frac{d}{dx}\int_5^{x^3}(t^4+1)^{1/3}\,dt=-\underset{\underset{\text{Chain Rule}}{\uparrow}}{\frac{dy}{dx}}=-\frac{dy}{du}\cdot\frac{du}{dx}$$

$$=-\frac{d}{du}\int_5^{u}(t^4+1)^{1/3}\,dt\cdot\frac{du}{dx}$$

**NOTE** In these examples, the differentiation is with respect to the variable that appears in the upper or lower limit of integration and the answer is a function of that variable.

$$=-(u^4+1)^{1/3}\cdot\frac{du}{dx}\qquad\text{Use the Fundamental Theorem.}$$

$$=-(x^{12}+1)^{1/3}\cdot 3x^2\qquad u=x^3;\ \frac{du}{dx}=3x^2$$

$$=-3x^2(x^{12}+1)^{1/3}\qquad\blacksquare$$

**NOW WORK** Problem 15.

## 2  Use Part 2 of the Fundamental Theorem of Calculus

Part 1 of the Fundamental Theorem of Calculus establishes a relationship between the derivative and the definite integral. Part 2 of the Fundamental Theorem of Calculus provides a method for finding a definite integral without using Riemann sums.

**THEOREM  Fundamental Theorem of Calculus, Part 2**

Let $f$ be a function that is continuous on a closed interval $[a,b]$. If $F$ is any antiderivative of $f$ on $[a,b]$, then

$$\int_a^b f(x)\,dx=F(b)-F(a)$$

**IN WORDS** If $F'(x)=f(x)$, then $\int_a^b f(x)\,dx=F(b)-F(a)$. That is, if you can find an antiderivative of $f$, you can integrate $f$.

**Proof** Let $I(x) = \int_a^x f(t)\,dt$. Then from Part 1 of the Fundamental Theorem of Calculus, $I = I(x)$ is continuous for $a \le x \le b$ and differentiable for $a < x < b$. So,

$$\frac{d}{dx}\int_a^x f(t)\,dt = f(x) \qquad a < x < b$$

RECALL Any two antiderivatives of a function differ by a constant.

That is, $\int_a^x f(t)\,dt$ is an antiderivative of $f$. So, if $F$ is any antiderivative of $f$, then

$$F(x) = \int_a^x f(t)\,dt + C$$

where $C$ is some constant. Since $F$ is continuous on $[a, b]$, we have

$$F(a) = \int_a^a f(t)\,dt + C \qquad F(b) = \int_a^b f(t)\,dt + C$$

NOTE For any constant $C$, $[F(b) + C] - [F(a) + C] = F(b) - F(a)$. In other words, it does not matter which antiderivative of $f$ is chosen when using Part 2 of the Fundamental Theorem of Calculus, since the same answer is obtained for every antiderivative.

Since, $\int_a^a f(t)\,dt = 0$, subtracting $F(a)$ from $F(b)$ gives

$$F(b) - F(a) = \int_a^b f(t)\,dt$$

Since $t$ is a dummy variable, we can replace $t$ by $x$ and the result follows. ∎

As an aid in computation, we introduce the notation

$$\int_a^b f(x)\,dx = \Big[F(x)\Big]_a^b = F(b) - F(a)$$

The notation $\Big[F(x)\Big]_a^b$ also suggests that to find $\int_a^b f(x)\,dx$, we first find an antiderivative $F(x)$ of $f(x)$. Then we write $\Big[F(x)\Big]_a^b$ to represent $F(b) - F(a)$.

---

**EXAMPLE 4   Using Part 2 of the Fundamental Theorem of Calculus**

Use Part 2 of the Fundamental Theorem of Calculus to find:

NOTE Table 6 in Section 4.8, page 344, provides a list of common antiderivatives.

(a) $\displaystyle\int_{-2}^1 x^2\,dx$ 

(b) $\displaystyle\int_0^{\pi/6} \cos x\,dx$

(c) $\displaystyle\int_0^{\sqrt{3}/2} \frac{1}{\sqrt{1-x^2}}\,dx$ 

(d) $\displaystyle\int_1^2 \frac{1}{x}\,dx$

**Solution** (a) The function $f(x) = x^2$ is continuous on the closed interval $[-2, 1]$. An antiderivative of $f$ is $F(x) = \dfrac{x^3}{3}$. By Part 2 of the Fundamental Theorem of Calculus,

$$\int_{-2}^1 x^2\,dx = \left[\frac{x^3}{3}\right]_{-2}^1 = \frac{1^3}{3} - \frac{(-2)^3}{3} = \frac{1}{3} + \frac{8}{3} = \frac{9}{3} = 3$$

(b) The function $f(x) = \cos x$ is continuous on the closed interval $\left[0, \dfrac{\pi}{6}\right]$. An antiderivative of $f$ is $F(x) = \sin x$. By Part 2 of the Fundamental Theorem of Calculus,

$$\int_0^{\pi/6} \cos x\,dx = \Big[\sin x\Big]_0^{\pi/6} = \sin\frac{\pi}{6} - \sin 0 = \frac{1}{2}$$

(c) The function $f(x) = \dfrac{1}{\sqrt{1-x^2}}$ is continuous on the closed interval $\left[0, \dfrac{\sqrt{3}}{2}\right]$. An antiderivative of $f$ is $F(x) = \sin^{-1} x$, provided $|x| < 1$. By Part 2 of the Fundamental Theorem of Calculus,

$$\int_0^{\sqrt{3}/2} \frac{1}{\sqrt{1-x^2}}\,dx = \Big[\sin^{-1} x\Big]_0^{\sqrt{3}/2} = \sin^{-1}\frac{\sqrt{3}}{2} - \sin^{-1} 0 = \frac{\pi}{3} - 0 = \frac{\pi}{3}$$

**(d)** The function $f(x) = \dfrac{1}{x}$ is continuous on the closed interval $[1, 2]$. An antiderivative of $f$ is $F(x) = \ln |x|$. By Part 2 of the Fundamental Theorem of Calculus,

$$\int_1^2 \frac{1}{x}\, dx = \left[\ln |x|\right]_1^2 = \ln 2 - \ln 1 = \ln 2 - 0 = \ln 2 \qquad \blacksquare$$

**NOW WORK** Problems **25** and **31**.

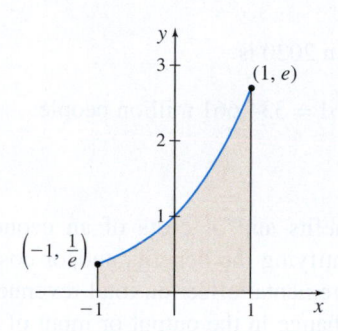

**Figure 22** $f(x) = e^x,\ -1 \le x \le 1$

**EXAMPLE 5** **Finding the Area Under a Graph**

Find the area under the graph of $f(x) = e^x$ from $-1$ to $1$.

**Solution** Figure 22 shows the graph of $f(x) = e^x$ on the closed interval $[-1, 1]$.

Since $f$ is continuous on the closed interval $[-1, 1]$, the area $A$ under the graph of $f$ from $-1$ to $1$ is given by

$$A = \int_{-1}^1 e^x\, dx = \left[e^x\right]_{-1}^1 = e^1 - e^{-1} = e - \frac{1}{e} \qquad \blacksquare$$

**NOW WORK** Problem **53**.

## 3 Interpret the Integral of a Rate of Change

Part 2 of the Fundamental Theorem of Calculus states that, under certain conditions, if $F$ is an antiderivative of $f$, then

$$\int_a^b f(x)\, dx = F(b) - F(a)$$

Since $F$ is an antiderivative of $f$, then $F' = f$ so

$$\int_a^b F'(x)\, dx = F(b) - F(a)$$

In other words,

> The integral from $a$ to $b$ of the rate of change of $F$ equals the change in $F$ from $a$ to $b$.

That is,

$$\boxed{\int_a^b F'(x)\, dx = F(b) - F(a)} \tag{4}$$

**EXAMPLE 6** **Interpreting an Integral Whose Integrand Is a Rate of Change**

The population of the United States is growing at the rate of $P'(t) = 2.867(1.009)^t$ million people per year, where $t$ is the number of years since 2015.

**(a)** Find $\displaystyle\int_0^5 P'(t)\, dt$.

**(b)** Interpret $\displaystyle\int_0^5 P'(t)\, dt$ in the context of the problem.

**(c)** If the population of the United States was 320 million people in 2015, what is the projected population in 2020?

*Source*: U.S. Census Bureau

**Solution (a)** We use the Fundamental Theorem of Calculus, Part 2.

$$\int_0^5 2.867(1.009)^t dt = 2.867 \left[\frac{1.009^t}{\ln 1.009}\right]_0^5 \qquad \text{An antiderivative of } a^t \text{ is } \frac{a^t}{\ln a}.$$

$$= 2.867 \left[\frac{1.009^5}{\ln 1.009} - \frac{1.009^0}{\ln 1.009}\right]$$

$$= 2.867 \left[\frac{1.009^5 - 1}{\ln 1.009}\right] \approx 14.661$$

**(b)** Since $P'(t)$ is the rate of change of the population $P$ (in millions) with respect to time $t$, in years,

$$\int_0^5 P'(t)\,dt = P(5) - P(0)$$

is the projected change in the population (in millions) of the United States from 2015 ($t = 0$) to 2020 ($t = 5$). The population of the United States is projected to increase by 14.661 million people from 2015 to 2020.

**(c)** The projected population of the United States in 2020 is

$$P(5) = P(0) + \int_0^5 P'(t)\,dt = 320 + 14.661 = 334.661 \text{ million people} \quad \blacksquare$$

**NOW WORK** Problem 63.

Economists use calculus to identify the benefits and/or costs of an economic decision. *Marginal analysis* is the process of identifying the benefits and/or costs of an economic decision based on examining the incremental effect on total revenue and total cost caused by a very small (just one unit) change in the output or input of each alternative.

**Marginal profit** $P'(x)$ is the rate of change in profit that results from producing and selling one more unit. Based on statement (4),

$$\int_{x_1}^{x_2} P'(t)\,dt = P(x_2) - P(x_1) = \text{change in profit from } x_1 \text{ to } x_2$$

---

**EXAMPLE 7**  **Interpreting an Integral Whose Integrand Is Marginal Profit**

The marginal profit $P'$ in dollars per unit for producing and selling $x$ units of a product is given by

$$P'(x) = \sqrt{x}$$

Suppose 100 units have been produced and sold. Find the additional profit $P$ from producing and selling 21 more units.

**Solution** After 100 units have been sold, the profit from producing and selling 21 additional units is given by

$$\int_{100}^{121} \sqrt{x}\,dx = \int_{100}^{121} x^{1/2}\,dx = \left[\frac{x^{3/2}}{\frac{3}{2}}\right]_{100}^{121} = \left[\frac{2x^{3/2}}{3}\right]_{100}^{121} = \frac{2}{3}(121^{3/2}) - \frac{2}{3}(100^{3/2})$$

$$= \frac{2}{3}(1331 - 1000) \approx 220.667$$

The company earns \$220.67 from selling an additional 21 units. $\blacksquare$

The profit earned from producing and selling the first hundred units is

$$\int_0^{100} P'(x)\,dx = \frac{2}{3} \cdot 100^{3/2} = 666.67, \text{ or } \$6.67 \text{ per unit. The profit earned from}$$

producing and selling the additional 21 units is \$220.67, or \$10.51 per additional unit. The profit margin is improving as sales increase.

**NOW WORK** Problem 65.

## 4 Interpret the Integral as an Accumulation Function

If $f$ is a function that is continuous and nonnegative on an interval $[a, b]$, then the definite integral $\int_a^b f(x)\,dx$ exists and equals the area $A$ under the graph of $f$ from $a$ to $b$. See Figure 23(a).

Suppose we define the function $F$ to be

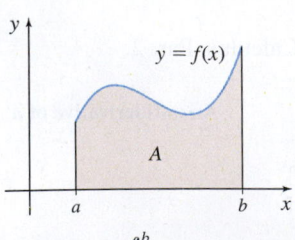

**Figure 23** (a) $A = \displaystyle\int_a^b f(x)\,dx$

$$F(x) = \int_a^x f(t)\,dt \quad a \le x \le b$$

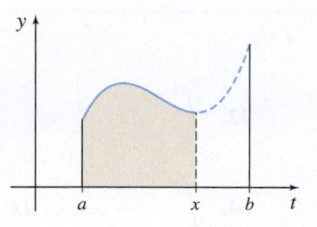

**Figure 23** (b) $F(x) = \displaystyle\int_a^x f(t)\,dt,\ a \le x \le b$

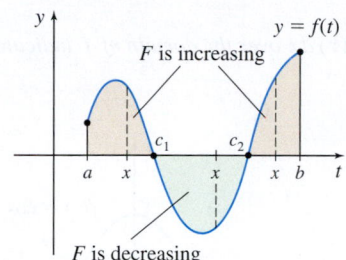

**Figure 24** $F(x) = \displaystyle\int_a^x f(t)\,dt$

Since the independent variable $x$ of $F$ is the upper limit of integration of the definite integral, we can interpret $F$ as the area under the graph of $f$ from $a$ to the number $x$, as shown in Figure 23(b). Notice that when $x = a$, then $F(a) = \int_a^a f(t)\,dt = 0$ and when $x = b$, then $F(b) = \int_a^b f(t)\,dt = A$. Notice also that as $x$ increases from $a$ to $b$, the function $F$ also increases, accumulating more area as $x$ goes from $a$ to $b$. It is for this reason we can describe $F(x) = \int_a^x f(t)\,dt$ as an *accumulation function*.

Now suppose $f$ is not required to be nonnegative on the interval $[a, b]$. Then $f$ may be nonnegative on some subinterval(s) of $[a, b]$ and negative on others. We examine how this affects $F(x) = \int_a^x f(t)\,dt$.

See Figure 24. On the interval $[a, c_1]$, the function $f$ is nonnegative. As $x$ increases from $a$ to $c_1$, $F(x) = \int_a^x f(t)\,dt$ accumulates value and is increasing. On the interval $[c_1, c_2]$, $f(t) \le 0$. As a result, the Riemann sums $\displaystyle\sum_{i=1}^{n} f(u_i)\Delta t_i$ for $f$ on the interval $[c_1, c_2]$ are negative.

So as $x$ increases from $a$ to $c_2$, the accumulation function $F$ increases from $a$ to $c_1$ and then starts to decrease as $x$ goes from $c_1$ to $c_2$, possibly becoming negative. Then from $c_2$ to $b$, the accumulation function $F$ starts to increase again. The net accumulation from $a$ to $b$, given by $\int_a^b f(t)\,dt$, can be a positive number, a negative number, or zero.

---

**DEFINITION  Accumulation Function**

Let $f$ be a function that is continuous on the closed interval $[a, b]$. The **accumulation function** $F$ **associated with** $f$ over the interval $[a, b]$ is defined as

$$F(x) = \int_a^x f(t)\,dt \qquad a \le x \le b$$

---

**EXAMPLE 8  Interpreting an Integral as an Accumulation Function**

The graph of $f(t) = \sin t,\ 0 \le t \le 2\pi$ is shown in Figure 25.

**(a)** Find the accumulation function $F$ associated with $f$ over the interval $[0, 2\pi]$.

**(b)** Find $F\left(\dfrac{\pi}{2}\right)$, $F(\pi)$, $F\left(\dfrac{3\pi}{2}\right)$, and $F(2\pi)$.

**(c)** Interpret the results found in (b).

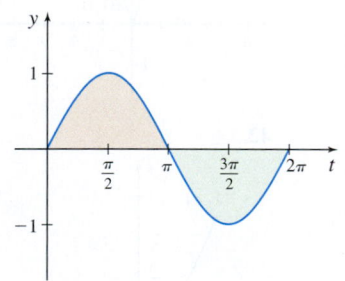

**Figure 25** $f(t) = \sin t,\ 0 \le t \le 2\pi$

**Solution  (a)** The accumulation function $F$ associated with $f(t) = \sin t$ over the interval $[0, 2\pi]$ is

$$F(x) = \int_0^x \sin t\,dt = \big[-\cos t\big]_0^x = -\cos x - (-\cos 0) = -\cos x + 1$$

**(b)** Now use $F(x) = -\cos x + 1$ to determine $F$ at the required numbers.

$$F\left(\frac{\pi}{2}\right) = -\cos\frac{\pi}{2} + 1 = 1$$

$$F(\pi) = -\cos \pi + 1 = 1 + 1 = 2$$

$$F\left(\frac{3\pi}{2}\right) = -\cos\frac{3\pi}{2} + 1 = 0 + 1 = 1$$

$$F(2\pi) = -\cos(2\pi) + 1 = -1 + 1 = 0$$

**(c)** The sine function is positive over the interval $(0, \pi)$ and is negative over the interval $(\pi, 2\pi)$. The accumulation function $F$ of $f(t) = \sin t$ increases from 0 to 1 to 2 as $x$ increases from 0 to $\pi$ and then will decrease from 2 to 1 to 0 as $x$ goes from $\pi$ to $2\pi$. The net accumulation over the interval $[0, 2\pi]$ equals 0. ∎

**NOW WORK** Problem **45.**

## 5.3 Assess Your Understanding

### Concepts and Vocabulary

1. According to Part 1 of the Fundamental Theorem of Calculus, if a function $f$ is continuous on a closed interval $[a, b]$, then

$$\frac{d}{dx}\left[\int_a^x f(t)\,dt\right] = \underline{\hspace{2cm}} \text{ for all numbers } x \text{ in } (a, b).$$

2. **True or False** By Part 2 of the Fundamental Theorem of Calculus, $\int_a^b x\,dx = b - a$.

3. **True or False** By Part 2 of the Fundamental Theorem of Calculus, $\int_a^b f(x)\,dx = f(b) - f(a)$.

4. **True or False** $\int_a^b F'(x)\,dx$ can be interpreted as the rate of change in $F$ from $a$ to $b$.

### Skill Building

*In Problems 5–18, find each derivative using Part 1 of the Fundamental Theorem of Calculus.*

5. $\dfrac{d}{dx}\left[\displaystyle\int_1^x \sqrt{t^2+1}\,dt\right]$

6. $\dfrac{d}{dx}\left[\displaystyle\int_3^x \dfrac{t+1}{t}\,dt\right]$

7. $\dfrac{d}{dt}\left[\displaystyle\int_0^t (3+x^2)^{3/2}dx\right]$

8. $\dfrac{d}{dx}\left[\displaystyle\int_{-4}^x (t^3+8)^{1/3}\,dt\right]$

9. $\dfrac{d}{dx}\left[\displaystyle\int_1^x \ln u\,du\right]$

10. $\dfrac{d}{dt}\left[\displaystyle\int_4^t e^x\,dx\right]$

11. $\dfrac{d}{dx}\left[\displaystyle\int_1^{2x^3} \sqrt{t^2+1}\,dt\right]$

12. $\dfrac{d}{dx}\left[\displaystyle\int_1^{\sqrt{x}} \sqrt{t^4+5}\,dt\right]$

13. $\dfrac{d}{dx}\left[\displaystyle\int_2^{x^5} \sec t\,dt\right]$

14. $\dfrac{d}{dx}\left[\displaystyle\int_3^{1/x} \sin^5 t\,dt\right]$

15. $\dfrac{d}{dx}\left[\displaystyle\int_x^5 \sin(t^2)\,dt\right]$

16. $\dfrac{d}{dx}\left[\displaystyle\int_x^3 (t^2-5)^{10}\,dt\right]$

17. $\dfrac{d}{dx}\left[\displaystyle\int_{5x^2}^5 (6t)^{2/3}\,dt\right]$

18. $\dfrac{d}{dx}\left[\displaystyle\int_{x^2}^0 e^{10t}\,dt\right]$

*In Problems 19–36, use Part 2 of the Fundamental Theorem of Calculus to find each definite integral.*

19. $\displaystyle\int_{-2}^3 dx$

20. $\displaystyle\int_{-2}^3 2\,dx$

21. $\displaystyle\int_{-1}^2 x^3\,dx$

22. $\displaystyle\int_1^3 \dfrac{1}{x^3}\,dx$

23. $\displaystyle\int_0^1 \sqrt{u}\,du$

24. $\displaystyle\int_1^8 \sqrt[3]{y}\,dy$

25. $\displaystyle\int_{\pi/6}^{\pi/2} \csc^2 x\,dx$

26. $\displaystyle\int_0^{\pi/2} \cos x\,dx$

27. $\displaystyle\int_0^{\pi/4} \sec x \tan x\,dx$

28. $\displaystyle\int_{\pi/6}^{\pi/2} \csc x \cot x\,dx$

29. $\displaystyle\int_{-1}^0 e^x\,dx$

30. $\displaystyle\int_{-1}^0 e^{-x}\,dx$

31. $\displaystyle\int_1^e \dfrac{1}{x}\,dx$

32. $\displaystyle\int_e^1 \dfrac{1}{x}\,dx$

33. $\displaystyle\int_0^1 \dfrac{1}{1+x^2}\,dx$

34. $\displaystyle\int_0^{\sqrt{2}/2} \dfrac{1}{\sqrt{1-x^2}}\,dx$

35. $\displaystyle\int_{-1}^8 x^{2/3}\,dx$

36. $\displaystyle\int_0^4 x^{3/2}\,dx$

*In Problems 37–42, find $\int_a^b f(x)\,dx$ over the domain of $f$ indicated in the graph.*

37.

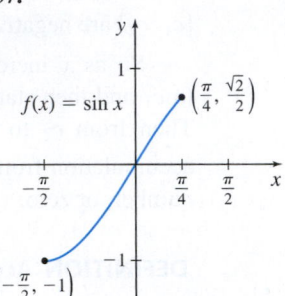

38.

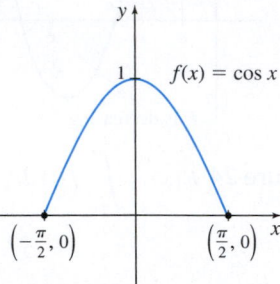

39.

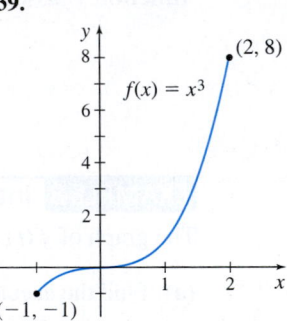

40.

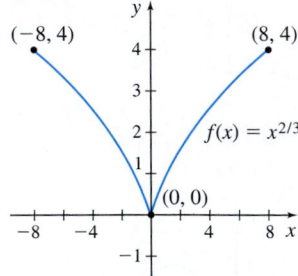

41.

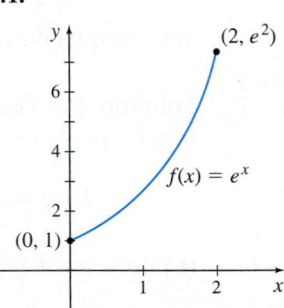

42.

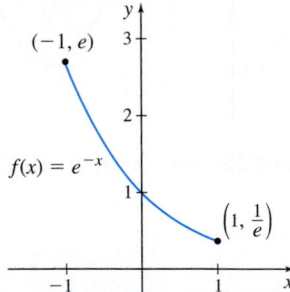

*In Problems 43–50,*

(a) Write the accumulation function $F$ associated with $f$ over the indicated interval as an integral.

(b) Find $F$ at the given numbers.

(c) Graph $F$.

43. $f(t) = \cos t,\ 0 \le t \le \dfrac{4\pi}{3};\ F\left(\dfrac{\pi}{6}\right),\ F\left(\dfrac{\pi}{2}\right),\ F\left(\dfrac{4\pi}{3}\right)$

44. $f(t) = \sin(2t),\ 0 \le t \le \dfrac{3\pi}{2};\ F\left(\dfrac{\pi}{6}\right),\ F\left(\dfrac{\pi}{2}\right),\ F\left(\dfrac{3\pi}{2}\right)$

---

**1.** = NOW WORK problem    = Graphing technology recommended    CAS = Computer Algebra System recommended

**45.** $f(t) = t^3, -2 \le t \le 3; F(-1), F(0), F(3)$

**46.** $f(t) = \sqrt[3]{t}, -8 \le t \le 8; F(-1), F(0), F(8)$

**47.** $f(t) = 4 - t^2, -1 \le t \le 3; F(0), F(2), F(3)$

**48.** $f(t) = 8 - t^3, 0 \le t \le 4; F(1), F(2), F(4)$

**49.** $f(t) = \sqrt{t} - 2, 0 \le t \le 9; F(1), F(4), F(9)$

**50.** $f(t) = t^2 - 2t - 3, 0 \le t \le 5; F(1), F(3), F(5)$

## Applications and Extensions

**51.** Given that $f(x) = (2x^3 - 3)^2$ and $f'(x) = 12x^2(2x^3 - 3)$, find $\int_0^2 [12x^2(2x^3 - 3)] \, dx$.

**52.** Given that $f(x) = (x^2 + 5)^3$ and $f'(x) = 6x(x^2 + 5)^2$, find $\int_{-1}^2 6x(x^2 + 5)^2 \, dx$.

**53. Area**   Find the area under the graph of $f(x) = \dfrac{1}{\sqrt{1 - x^2}}$ from 0 to $\dfrac{1}{2}$.

**54. Area**   Find the area under the graph of $f(x) = \cosh x$ from $-1$ to 1.

**55. Area**   Find the area under the graph of $f(x) = \dfrac{1}{x^2 + 1}$ from 0 to $\sqrt{3}$.

**56. Area**   Find the area under the graph of $f(x) = \dfrac{1}{1 + x^2}$ from 0 to $r$, where $r > 0$. What happens as $r \to \infty$?

**57. Area**   Find the area under the graph of $y = \dfrac{1}{\sqrt{x}}$ from $x = 1$ to $x = r$, where $r > 1$. Then examine the behavior of this area as $r \to \infty$.

**58. Area**   Find the area under the graph of $y = \dfrac{1}{x^2}$ from $x = 1$ to $x = r$, where $r > 1$. Then examine the behavior of this area as $r \to \infty$.

**59. Interpreting an Integral**   The function $R = R(t)$ models the rate of sales of a corporation measured in millions of dollars per year as a function of the time $t$ in years. Interpret the integral $\int_0^2 R(t) \, dt = 23$ in the context of the problem.

**60. Interpreting an Integral**   The function $v = v(t)$ models the speed $v$ in meters per second of an object at time $t$ in seconds. Interpret the integral $\int_0^{10} v(t) \, dt = 4.8$ in the context of the problem.

**61. Interpreting an Integral**   Helium is leaking from a large advertising balloon at a rate of $H(t)$ cubic centimeters per minute, where $t$ is measured in minutes.

(a) Write an integral that models the change in the amount of helium in the balloon over the interval $a \le t \le b$.

(b) What are the units of the integral from (a)?

(c) Interpret $\int_0^{300} H(t) \, dt = -100$ in the context of the problem.

**62. Interpreting an Integral**   Water is being added to a reservoir at a rate of $w(t)$ kiloliters per hour, where $t$ is measured in hours.

(a) Write an integral that models the change in the amount of water in the reservoir over the interval $a \le t \le b$.

(b) What are the units of the integral from (a)?

(c) Interpret $\int_0^{36} w(t) \, dt = 100$ in the context of the problem.

**63. Population Growth**   The growth rate of a colony of bacteria is $B'(t) = 3.455(1.259)^t$ grams per hour, where $t$ is the number of hours since time $t = 0$.

(a) Find $\int_0^6 B'(t) \, dt$.

(b) Interpret $\int_0^6 B'(t) \, dt$ in the context of the problem.

(c) If 15 grams of bacteria are present at time $t = 0$, how many grams of bacteria are present after 6 h?

**64. Return on Investment**   An investment in a hedge fund is growing at a continuous rate of $A'(t) = 1105.17(1.105)^t$ dollars per year.

(a) Find $\int_0^{10} A'(t) \, dt$.

(b) Interpret $\int_0^{10} A'(t) \, dt$ in the context of the problem.

(c) If initially \$1000 is invested, what is the value of the investment after 10 years.

**65. Increase in Revenue**   The marginal revenue function $R'(x)$ for selling $x$ units of a product is $R'(x) = \sqrt[3]{x}$ (in hundreds of dollars per unit).

(a) Interpret $\int_a^b R'(x) \, dx$ in the context of the problem.

(b) How much additional revenue is attained if sales increase from 40 to 50 units?

**66. Increase in Cost**   The marginal cost function $C'(x)$ of producing $x$ thousand units of a product is $C'(x) = 2x + 6$ thousand dollars.

(a) Interpret $\int_a^b C'(x) \, dx$ in the context of the problem.

(b) What does it cost the company to increase production from 10,000 units to 13,000 units?

**67. Free Fall**   The speed $v$ of an object dropped from rest is given by $v(t) = 9.8t$, where $v$ is in meters per second and time $t$ is in seconds.

(a) Express the distance traveled in the first 5.2 s as an integral.

(b) Find the distance traveled in 5.2 s.

**68. Area**   Find $h$ so that the area under the graph of $y^2 = x^3, 0 \le x \le 4, y \ge 0$, is equal to the area of a rectangle of base 4 and height $h$.

**69. Area**   If $P$ is a polynomial that is positive for $x > 0$, and for each $k > 0$ the area under the graph of $P$ from $x = 0$ to $x = k$ is $k^3 + 3k^2 + 6k$, find $P$.

**70. Put It Together**   If $f(x) = \displaystyle\int_0^x \dfrac{1}{\sqrt{t^3 + 2}} \, dt$, which of the following is *false*?

(a) $f$ is continuous at $x$ for all $x \ge 0$

(b) $f(1) > 0$

(c) $f(0) = \dfrac{1}{\sqrt{2}}$

(d) $f'(1) = \dfrac{1}{\sqrt{3}}$

*In Problems 71–74:*

(a) *Use Part of 2 the Fundamental Theorem of Calculus to find each definite integral.*

(b) *Determine whether the integrand is an even function, an odd function, or neither.*

(c) *Can you make a conjecture about the definite integrals in (a) based on the analysis from (b)?*

**71.** $\int_0^4 x^2 dx$ and $\int_{-4}^4 x^2 dx$    **72.** $\int_0^4 x^3 dx$ and $\int_{-4}^4 x^3 dx$

**73.** $\int_0^{\pi/4} \sec^2 x \, dx$ and $\int_{-\pi/4}^{\pi/4} \sec^2 x \, dx$

**74.** $\int_0^{\pi/4} \sin x \, dx$ and $\int_{-\pi/4}^{\pi/4} \sin x \, dx$

**75. Area** Find $c$, $0 < c < 1$, so that the area under the graph of $y = x^2$ from 0 to $c$ equals the area under the same graph from $c$ to 1.

**76. Area** Let $A$ be the area under the graph of $y = \dfrac{1}{x}$

from $x = m$ to $x = 2m$, $m > 0$. Which of the following is true about the area $A$?

(a) $A$ is independent of $m$.

(b) $A$ increases as $m$ increases.

(c) $A$ decreases as $m$ increases.

(d) $A$ decreases as $m$ increases when $m < \dfrac{1}{2}$ and increases as $m$ increases when $m > \dfrac{1}{2}$.

(e) $A$ increases as $m$ increases when $m < \dfrac{1}{2}$ and decreases as $m$ increases when $m > \dfrac{1}{2}$.

**77. Put It Together** If $F$ is a function whose derivative is continuous for all real $x$, find

$$\lim_{h \to 0} \frac{1}{h} \int_c^{c+h} F'(x) \, dx$$

**78.** Suppose the closed interval $\left[0, \dfrac{\pi}{2}\right]$ is partitioned into $n$ subintervals, each of length $\Delta x$, and $u_i$ is an arbitrary number in the subinterval $[x_{i-1}, x_i]$, $i = 1, 2, \ldots, n$. Explain why

$$\lim_{n \to \infty} \sum_{i=1}^n [(\cos u_i) \, \Delta x] = 1$$

**79.** The interval $[0, 4]$ is partitioned into $n$ subintervals, each of length $\Delta x$, and a number $u_i$ is chosen in the subinterval

$[x_{i-1}, x_i]$, $i = 1, 2, \ldots, n$. Find $\lim\limits_{n \to \infty} \sum\limits_{i=1}^n (e^{u_i} \Delta x)$

**80.** If $u$ and $v$ are differentiable functions and $f$ is a continuous function, find a formula for

$$\frac{d}{dx} \left[ \int_{u(x)}^{v(x)} f(t) \, dt \right]$$

**81.** Suppose that the graph of $y = f(x)$ contains the points $(0, 1)$ and $(2, 5)$. Find $\int_0^2 f'(x) \, dx$. (Assume that $f'$ is continuous.)

**82.** If $f'$ is continuous on the interval $[a, b]$, show that

$$\int_a^b f(x) f'(x) \, dx = \frac{1}{2} \left\{ [f(b)]^2 - [f(a)]^2 \right\}$$

$$\left[ \text{Hint: Look at the derivative of } F(x) = \frac{[f(x)]^2}{2}. \right]$$

**83.** If $f''$ is continuous on the interval $[a, b]$, show that

$$\int_a^b x f''(x) \, dx = b f'(b) - a f'(a) - f(b) + f(a)$$

$$\left[ \text{Hint: Look at the derivative of } F(x) = x f'(x) - f(x). \right]$$

## Challenge Problems

**84.** What conditions on $f$ and $f'$ guarantee that

$$f(x) = \int_0^x f'(t) \, dt?$$

**85.** Suppose that $F$ is an antiderivative of $f$ on the interval $[a, b]$. Partition $[a, b]$ into $n$ subintervals, each of length

$$\Delta x_i = x_i - x_{i-1}, \quad i = 1, 2, \ldots, n.$$

(a) Apply the Mean Value Theorem for derivatives to $F$ in each subinterval $[x_{i-1}, x_i]$ to show that there is a number $u_i$ in the subinterval for which $F(x_i) - F(x_{i-1}) = f(u_i) \Delta x_i$.

(b) Show that $\sum\limits_{i=1}^n [F(x_i) - F(x_{i-1})] = F(b) - F(a)$.

(c) Use parts (a) and (b) to explain why

$$\int_a^b f(x) \, dx = F(b) - F(a).$$

(In this alternate proof of Part 2 of the Fundamental Theorem of Calculus, the continuity of $f$ is not assumed.)

**86.** Given $y = \sqrt{x^2 - 1}\,(4 - x)$, $1 \le x \le a$, for what number $a$ will $\int_1^a y \, dx$ have a maximum value?

**87.** Find $a > 0$, so that the area under the graph of $y = x + \dfrac{1}{x}$ from $a$ to $(a + 1)$ is minimum.

**88.** If $n$ is a known positive integer, for what number $c$ is

$$\int_1^c x^{n-1} \, dx = \frac{1}{n}$$

**89.** Let $f(x) = \int_0^x \dfrac{dt}{\sqrt{1 - t^2}}$, $0 < x < 1$.

(a) Find $\dfrac{d}{dx} f(\sin x)$.

(b) Is $f$ one-to-one?

(c) Does $f$ have an inverse?

# 5.4 Properties of the Definite Integral

**OBJECTIVES** *When you finish this section, you should be able to:*

1 Use properties of the definite integral (p. 387)
2 Work with the Mean Value Theorem for Integrals (p. 390)
3 Find the average value of a function (p. 391)
4 Interpret integrals involving rectilinear motion (p. 393)

We have seen that there are properties of limits that make it easier to find limits, properties of continuity that make it easier to determine continuity, and properties of derivatives that make it easier to find derivatives. Here we investigate several properties of the definite integral that will make it easier to find integrals.

## 1 Use Properties of the Definite Integral

Properties (1) through (4) in this section require that a function be integrable.

Remember, in Section 5.2 we stated two theorems that guarantee a function $f$ is integrable.

- A function $f$ that is continuous on a closed interval $[a, b]$ is integrable over $[a, b]$.
- A function $f$ that is bounded on a closed interval $[a, b]$ and has at most a finite number of discontinuities on $[a, b]$ is integrable over $[a, b]$.

**NOTE** From now on, we will refer to Parts 1 and 2 of the Fundamental Theorem of Calculus simply as the **Fundamental Theorem of Calculus.**

Proofs of most of the properties in this section require the use of the definition of the definite integral. However, if the condition is added that the integrand has an antiderivative, then the Fundamental Theorem of Calculus can be used to establish the properties. We will use this added condition in the proofs included here.

---

**THEOREM** The Integral of the Sum of Two Functions

If two functions $f$ and $g$ are integrable on the closed interval $[a, b]$, then

$$\int_a^b [f(x) + g(x)] \, dx = \int_a^b f(x) \, dx + \int_a^b g(x) \, dx \tag{1}$$

---

**IN WORDS** The integral of a sum equals the sum of the integrals.

**Proof** For the proof, we assume that the functions $f$ and $g$ are continuous on $[a, b]$ and that $F$ and $G$ are antiderivatives of $f$ and $g$, respectively, on $(a, b)$. Then $F' = f$ and $G' = g$ on $(a, b)$.

Also, since $(F + G)' = F' + G' = f + g$, then $F + G$ is an antiderivative of $f + g$ on $(a, b)$. Now

$$\int_a^b [f(x) + g(x)] \, dx = \left[ F(x) + G(x) \right]_a^b = [F(b) + G(b)] - [F(a) + G(a)]$$

$$= [F(b) - F(a)] + [G(b) - G(a)]$$

$$= \int_a^b f(x) \, dx + \int_a^b g(x) \, dx \qquad \blacksquare$$

---

**EXAMPLE 1** Using Property (1) of the Definite Integral

$$\int_0^1 (x^2 + e^x) \, dx = \int_0^1 x^2 \, dx + \int_0^1 e^x \, dx = \left[ \frac{x^3}{3} \right]_0^1 + \left[ e^x \right]_0^1$$

$$= \left[ \frac{1^3}{3} - 0 \right] + \left[ e^1 - e^0 \right] = \frac{1}{3} + e - 1 = e - \frac{2}{3} \qquad \blacksquare$$

**NOW WORK** Problem **17**.

**THEOREM  The Integral of a Constant Times a Function**

If a function $f$ is integrable on the closed interval $[a, b]$ and $k$ is a constant, then

$$\int_a^b kf(x)\,dx = k\int_a^b f(x)\,dx \tag{2}$$

**IN WORDS** A constant factor can be factored out of an integral.

You are asked to prove this theorem in Problem 105 if $f$ is continuous on the interval $[a, b]$ and $f$ has an antiderivative on the interval $(a, b)$.

---

**EXAMPLE 2  Using Property (2) of the Definite Integral**

$$\int_1^e \frac{3}{x}\,dx = 3\underset{\substack{\uparrow \\ \text{Property (2)}}}{\int_1^e \frac{1}{x}}\,dx = 3\left[\ln|x|\right]_1^e = 3\,(\ln e - \ln 1) = 3\,(1 - 0) = 3 \qquad \blacksquare$$

**NOW WORK** Problem 19.

Properties (1) and (2) can be extended as follows:

**THEOREM**

Suppose each of the functions $f_1, f_2, \ldots, f_n$ is integrable on the closed interval $[a, b]$. If $k_1, k_2, \ldots, k_n$ are constants, then

$$\int_a^b \left[k_1 f_1(x) + k_2 f_2(x) + \cdots + k_n f_n(x)\right] dx$$
$$= k_1 \int_a^b f_1(x)\,dx + k_2 \int_a^b f_2(x)\,dx + \cdots + k_n \int_a^b f_n(x)\,dx \tag{3}$$

You are asked to prove this theorem in Problem 106 if each function is continuous on the interval $[a, b]$ and each function has an antiderivative on the interval $(a, b)$.

---

**EXAMPLE 3  Using Property (3) of the Definite Integral**

Find $\displaystyle\int_1^2 \frac{3x^3 - 6x^2 - 5x + 4}{2x}\,dx$.

**Solution** The function $f(x) = \dfrac{3x^3 - 6x^2 - 5x + 4}{2x}$ is continuous on the closed interval $[1, 2]$ so it is integrable on $[1, 2]$. Using algebra and properties of the definite integral,

$$\int_1^2 \frac{3x^3 - 6x^2 - 5x + 4}{2x}\,dx = \int_1^2 \left[\frac{3}{2}x^2 - 3x - \frac{5}{2} + \frac{2}{x}\right] dx$$

$$= \frac{3}{2}\int_1^2 x^2\,dx - 3\int_1^2 x\,dx - \frac{5}{2}\int_1^2 dx + 2\int_1^2 \frac{1}{x}\,dx \quad \text{Use property (3).}$$

$$= \frac{3}{2}\left[\frac{x^3}{3}\right]_1^2 - 3\left[\frac{x^2}{2}\right]_1^2 - \frac{5}{2}\left[x\right]_1^2 + 2\left[\ln|x|\right]_1^2 \quad \begin{array}{l}\text{Use the Fundamental} \\ \text{Theorem of Calculus.}\end{array}$$

$$= \frac{1}{2}(8 - 1) - \frac{3}{2}(4 - 1) - \frac{5}{2}(2 - 1) + 2(\ln 2 - \ln 1)$$

$$= -\frac{7}{2} + 2\ln 2 \qquad \blacksquare$$

**NOW WORK** Problem 29.

The next property states that a definite integral of a function $f$ from $a$ to $b$ can be evaluated in pieces.

## THEOREM

If a function $f$ is integrable on an interval containing the numbers $a$, $b$, and $c$, then

$$\int_a^b f(x)\,dx = \int_a^c f(x)\,dx + \int_c^b f(x)\,dx \tag{4}$$

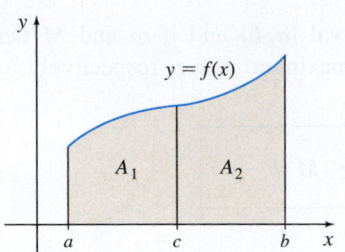

$A$ = area under the graph of $f$ from $a$ to $b$
= area $A_1$ + area $A_2$

$$\int_a^b f(x)\,dx = \int_a^c f(x)\,dx + \int_c^b f(x)\,dx$$

**Figure 26**

A proof of this theorem for a function $f$ that is continuous on $[a, b]$ is given in Appendix B. A proof for the more general case can be found in most advanced calculus books.

In particular, if $f$ is continuous and nonnegative on a closed interval $[a, b]$ and if $c$ is a number between $a$ and $b$, then property (4) has a simple geometric interpretation, as seen in Figure 26.

**EXAMPLE 4** **Using Property (4) of the Definite Integral**

**(a)** If $f$ is integrable on the closed interval $[2, 7]$, then

$$\int_2^7 f(x)\,dx = \int_2^4 f(x)\,dx + \int_4^7 f(x)\,dx$$

**(b)** If $g$ is integrable on the closed interval $[3, 25]$, then

$$\int_3^{10} g(x)\,dx = \int_3^{25} g(x)\,dx + \int_{25}^{10} g(x)\,dx$$    ∎

Example 4(b) illustrates that the number $c$ need not lie between $a$ and $b$.

**NOW WORK** Problem 47.

Property (4) is useful when integrating piecewise-defined functions.

**CALC**
**CLIP**

**EXAMPLE 5** **Using Property (4) to Find the Area Under the Graph of a Piecewise-defined Function**

Find the area $A$ under the graph of

$$f(x) = \begin{cases} x^2 & \text{if } 0 \le x < 10 \\ 100 & \text{if } 10 \le x \le 15 \end{cases}$$

from 0 to 15.

**Solution** See Figure 27. Since $f$ is continuous and nonnegative on the closed interval $[0, 15]$, then $\int_0^{15} f(x)\,dx$ exists and equals the area $A$ under the graph of $f$ from 0 to 15. We break the integral at $x = 10$, where the rule for the function changes.

$$\int_0^{15} f(x)\,dx = \int_0^{10} f(x)\,dx + \int_{10}^{15} f(x)\,dx = \int_0^{10} x^2\,dx + \int_{10}^{15} 100\,dx$$

$$= \left[\frac{x^3}{3}\right]_0^{10} + \left[100x\right]_{10}^{15} = \frac{1000}{3} + 500 = \frac{2500}{3}$$

The area under the graph of $f$ is approximately 833.333 square units.    ∎

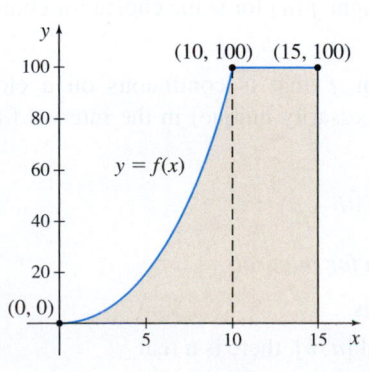

**Figure 27** $A = \displaystyle\int_0^{15} f(x)\,dx$

**NOW WORK** Problem 39.

The next property establishes bounds on a definite integral.

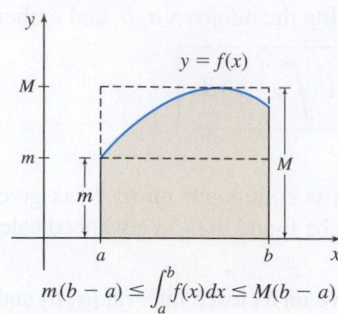

**Figure 28**

**THEOREM  Bounds on an Integral**

If a function $f$ is continuous on a closed interval $[a, b]$ and if $m$ and $M$ denote the absolute minimum value and the absolute maximum value, respectively, of $f$ on $[a, b]$, then

$$m(b - a) \leq \int_a^b f(x)\,dx \leq M(b - a) \tag{5}$$

A proof of this theorem is given in Appendix B.

If $f$ is nonnegative on $[a, b]$, then the inequalities in (5) have a geometric interpretation. In Figure 28, the area of the shaded region is $\int_a^b f(x)\,dx$. The rectangle of width $b - a$ and height $m$ has area $m(b - a)$. The rectangle of width $b - a$ and height $M$ has area $M(b - a)$. The areas of the three regions are numerically related by the inequalities in the theorem.

**EXAMPLE 6  Using Property (5) of Definite Integrals**

(a) Find an upper bound and a lower bound for the area $A$ under the graph of $f(x) = \sin x$ from 0 to $\pi$.

(b) Find the actual area under the graph.

**Solution** The graph of $f$ is shown in Figure 29. Since $f(x) \geq 0$ for all $x$ in the closed interval $[0, \pi]$, the area $A$ under its graph is given by the definite integral, $\int_0^\pi \sin x\,dx$.

(a) From the Extreme Value Theorem, $f$ has an absolute minimum value and an absolute maximum value on the interval $[0, \pi]$. The absolute maximum of $f$ occurs at $x = \dfrac{\pi}{2}$, and its value is $f\left(\dfrac{\pi}{2}\right) = \sin \dfrac{\pi}{2} = 1$. The absolute minimum occurs at $x = 0$ and at $x = \pi$; the absolute minimum value is $f(0) = \sin 0 = 0 = f(\pi)$. Using the inequalities in (5), the area under the graph of $f$ is bounded as follows:

$$0 \leq \int_0^\pi \sin x\,dx \leq \pi$$

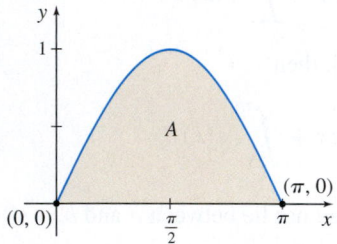

**Figure 29** $f(x) = \sin x, 0 \leq x \leq \pi$

(b) The actual area under the graph is

$$A = \int_0^\pi \sin x\,dx = \left[-\cos x\right]_0^\pi = -\cos \pi + \cos 0 = 1 + 1 = 2 \text{ square units} \quad \blacksquare$$

**NEED TO REVIEW?** The Extreme Value Theorem is discussed in Section 4.2, p. 274.

**NOW WORK** Problem 57.

## 2  Work with the Mean Value Theorem for Integrals

Suppose $f$ is a function that is continuous and nonnegative on a closed interval $[a, b]$. Figure 30 suggests that the area under the graph of $f$ from $a$ to $b$, $\int_a^b f(x)\,dx$, is equal to the area of some rectangle of width $b - a$ and height $f(u)$ for some choice (or choices) of $u$ in the interval $[a, b]$.

In more general terms, for every function $f$ that is continuous on a closed interval $[a, b]$, there is some number $u$ (not necessarily unique) in the interval $[a, b]$ for which

$$\int_a^b f(x)\,dx = f(u)(b - a)$$

This result is known as the *Mean Value Theorem for Integrals*.

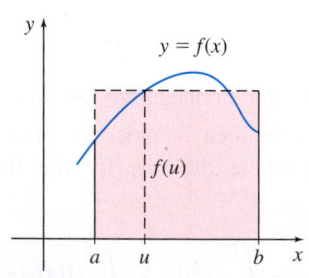

**Figure 30** $\int_a^b f(x)\,dx = f(u)(b - a)$

**THEOREM  Mean Value Theorem for Integrals**

If a function $f$ is continuous on a closed interval $[a, b]$, there is a real number $u$, $a \leq u \leq b$, for which

$$\int_a^b f(x)\,dx = f(u)(b - a) \tag{6}$$

**Proof** Let $f$ be a function that is continuous on a closed interval $[a, b]$.

If $f$ is a constant function, say, $f(x) = k$, on $[a, b]$, then

$$\int_a^b f(x)\,dx = \int_a^b k\,dx = k(b-a) = f(u)(b-a)$$

for any choice of $u$ in $[a, b]$.

If $f$ is not a constant function on $[a, b]$, then by the Extreme Value Theorem, $f$ has an absolute maximum and an absolute minimum on $[a, b]$. Suppose $f$ assumes its absolute minimum at the number $c$ so that $f(c) = m$; and suppose $f$ assumes its absolute maximum at the number $C$ so that $f(C) = M$. Then by the Bounds on an Integral Theorem (5), we have

$$m(b-a) \le \int_a^b f(x)\,dx \le M(b-a) \qquad \text{for all } x \text{ in } [a, b]$$

Divide each part by $(b-a)$ and replace $m$ by $f(c)$ and $M$ by $f(C)$. Then

$$f(c) \le \frac{1}{b-a}\int_a^b f(x)\,dx \le f(C)$$

**NEED TO REVIEW?** The Intermediate Value Theorem is discussed in Section 1.3, pp. 109–110.

Since $\dfrac{1}{b-a}\displaystyle\int_a^b f(x)\,dx$ is a real number between $f(c)$ and $f(C)$, it follows from the Intermediate Value Theorem that there is a real number $u$ between $c$ and $C$, for which

$$f(u) = \frac{1}{b-a}\int_a^b f(x)\,dx$$

That is, there is a real number $u$, $a \le u \le b$, for which

$$\int_a^b f(x)\,dx = f(u)(b-a) \qquad\blacksquare$$

---

**EXAMPLE 7** **Using the Mean Value Theorem for Integrals**

Find the number(s) $u$ guaranteed by the Mean Value Theorem for Integrals for $\int_2^6 x^2 dx$.

**Solution** Since the function $f(x) = x^2$ is continuous on the closed interval $[2, 6]$, the Mean Value Theorem for Integrals guarantees there is a number $u$, $2 \le u \le 6$, for which

$$\int_2^6 x^2 dx = f(u)(6-2) = 4u^2 \qquad f(u) = u^2$$

Integrating, we have

$$\int_2^6 x^2 dx = \left[\frac{x^3}{3}\right]_2^6 = \frac{1}{3}(216-8) = \frac{208}{3}$$

Then

$$\frac{208}{3} = 4u^2$$

$$u^2 = \frac{52}{3} \qquad\qquad 2 \le u \le 6$$

$$u = \sqrt{\frac{52}{3}} \approx 4.163 \qquad \text{Disregard the negative solution since } u > 0. \qquad\blacksquare$$

**NOW WORK** Problem 59.

## 3 Find the Average Value of a Function

We know from the Mean Value Theorem for Integrals that if a function $f$ is continuous on a closed interval $[a, b]$, there is a real number $u$, $a \le u \le b$, for which

$$\int_a^b f(x)\,dx = f(u)(b-a)$$

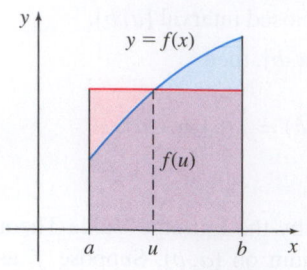

**Figure 31** $\displaystyle\int_a^b f(x)\,dx = f(u)(b-a)$

This means that if the function $f$ is also nonnegative on the closed interval $[a, b]$, the area enclosed by a rectangle of height $f(u)$ and width $b - a$ equals the area under the graph of $f$ from $a$ to $b$. See Figure 31. Consequently, $f(u)$ can be thought of as an *average value*, or *mean value*, of $f$ over $[a, b]$.

We can obtain the average value of $f$ over $[a, b]$ for any function $f$ that is continuous on the closed interval $[a, b]$ by partitioning $[a, b]$ into $n$ subintervals

$$[a, x_1], \quad [x_1, x_2], \quad \ldots, \quad [x_{i-1}, x_i], \quad \ldots, \quad [x_{n-1}, b]$$

each of length $\Delta x = \dfrac{b-a}{n}$, and choosing a number $u_i$ in each of the $n$ subintervals. Then an approximation of the average value of $f$ over the interval $[a, b]$ is

$$\frac{f(u_1) + f(u_2) + \cdots + f(u_n)}{n} \tag{7}$$

Now multiply (7) by $\dfrac{b-a}{b-a}$ to obtain

$$\frac{f(u_1) + f(u_2) + \cdots + f(u_n)}{n} = \frac{1}{b-a}\left[ f(u_1)\frac{b-a}{n} + f(u_2)\frac{b-a}{n} + \cdots + f(u_n)\frac{b-a}{n}\right]$$

$$= \frac{1}{b-a}[f(u_1)\,\Delta x + f(u_2)\,\Delta x + \cdots + f(u_n)\,\Delta x]$$

$$= \frac{1}{b-a}\sum_{i=1}^{n} f(u_i)\,\Delta x$$

This sum approximates the average value of $f$. As the length of each subinterval gets smaller, the sums become better approximations to the average value of $f$ on $[a, b]$. Furthermore, $\displaystyle\sum_{i=1}^{n} f(u_i)\,\Delta x$ are Riemann sums, so $\displaystyle\lim_{n\to\infty}\sum_{i=1}^{n} f(u_i)\,\Delta x$ is a definite integral. This suggests the following definition:

**DEFINITION** Average Value of a Function over an Interval

Let $f$ be a function that is continuous on the closed interval $[a, b]$. The **average value $\bar{y}$ of $f$ over $[a, b]$** is

$$\boxed{\bar{y} = \frac{1}{b-a}\int_a^b f(x)\,dx} \tag{8}$$

**IN WORDS** The average value $\bar{y}$ of a function $f$ equals the value $f(u)$ in the Mean Value Theorem for Integrals.

**EXAMPLE 8    Finding the Average Value of a Function**

Find the average value of $f(x) = x^3$ on the closed interval $[0, 2]$.

**Solution** The average value of $f(x) = x^3$ on the closed interval $[0, 2]$ is given by

$$\bar{y} = \frac{1}{b-a}\int_a^b f(x)\,dx = \frac{1}{2-0}\int_0^2 x^3\,dx = \frac{1}{2}\left[\frac{x^4}{4}\right]_0^2 = \frac{1}{2}\left(\frac{16}{4}\right) = 2$$

The average value of $f$ on $[0, 2]$ is $\bar{y} = 2$. ∎

The function $f$ and its average value are graphed in Figure 32.

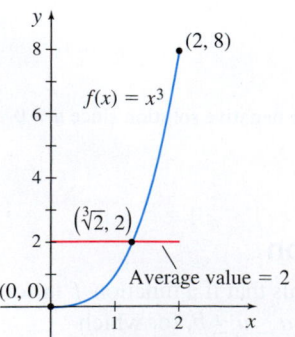

**Figure 32** $f(x) = x^3,\, 0 \le x \le 2$

**NOW WORK** Problem 65.

## 4 Interpret Integrals Involving Rectilinear Motion

For an object in rectilinear motion moving with velocity $v$,

- The **net displacement** of an object whose motion begins at $t = a$ and ends at $t = b$ is given by

$$\int_a^b v(t)\, dt$$

- The **total distance** traveled by the object from $t = a$ to $t = b$ is given by

$$\int_a^b |v(t)|\, dt$$

- If $v(t) \geq 0$, for $a \leq t \leq b$, then the net displacement and the total distance traveled are equal.

Remember speed and velocity are not the same thing. Speed measures how fast an object is moving regardless of direction. Velocity measures both the speed and the direction of an object. Because of this, velocity can be positive, negative, or zero. Speed is the absolute value of velocity. So $\int_a^b v(t)\, dt$ is a sum of signed distances and $\int_a^b |v(t)|dt$ is a sum of actual distances traveled.

**EXAMPLE 9  Interpreting an Integral Involving Rectilinear Motion**

An object in rectilinear motion is moving along a horizontal line with velocity $v(t) = 3t^2 - 18t + 15$ (in meters per minute) for time $0 \leq t \leq 6$ min.

**(a)** Find the net displacement of the object from $t = 0$ to $t = 6$. Interpret the result in the context of the problem.

**(b)** Find the total distance traveled by the object from $t = 0$ to $t = 6$ and interpret the result.

**Solution** **(a)** The net displacement of the object from $t = 0$ to $t = 6$ is given by

$$\int_0^6 v(t)\, dt = \int_0^6 (3t^2 - 18t + 15)\, dt = \left[t^3 - 9t^2 + 15t\right]_0^6 = 6^3 - 9 \cdot 6^2 + 15 \cdot 6 - 0 = -18$$

After moving from $t = 0$ to $t = 6$, the object is 18 m to the left of where it was at $t = 0$.

**(b)** The total distance traveled by the object from $t = 0$ to $t = 6$ min is given by $\int_0^6 |v(t)|dt$.

Since $\quad v(t) = 3t^2 - 18t + 15 = 3(t-1)(t-5)$

we find that $v(t) \geq 0$ if $0 \leq t \leq 1$ or if $5 \leq t \leq 6$ and $v(t) \leq 0$ if $1 \leq t \leq 5$. Now express $|v(t)|$ as the piecewise-defined function

$$|v(t)| = \begin{cases} v(t) = 3t^2 - 18t + 15 & \text{if } 0 \leq t \leq 1 \\ -v(t) = -(3t^2 - 18t + 15) & \text{if } 1 < t < 5 \\ v(t) = 3t^2 - 18t + 15 & \text{if } 5 \leq t \leq 6 \end{cases}$$

Then

$$\int_0^6 |v(t)|dt = \int_0^1 v(t)\, dt + \int_1^5 [-v(t)]\, dt + \int_5^6 v(t)\, dt$$

$$= [t^3 - 9t^2 + 15t]_0^1 - [t^3 - 9t^2 + 15t]_1^5 + [t^3 - 9t^2 + 15t]_5^6$$

$$= (7 - 0) - (-25 - 7) + [-18 - (-25)] = 7 + 32 + 7 = 46$$

The object traveled a total distance of 46 m from $t = 0$ to $t = 6$ min. ∎

Figure 33 shows the distinction between the net displacement and the total distance traveled by the object assuming that the object is at the origin when the motion starts.

**NOW WORK** Problem 79.

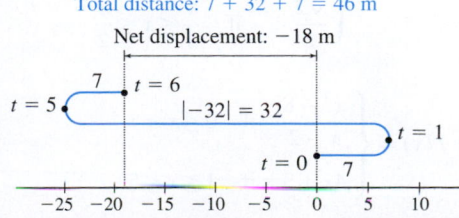

Total distance: $7 + 32 + 7 = 46$ m

Net displacement: $-18$ m

**Figure 33**

## 5.4 Assess Your Understanding

### Concepts and Vocabulary

**1.** *True or False*   $\displaystyle\int_2^3 (x^2 + x)\,dx = \int_2^3 x^2\,dx + \int_2^3 x\,dx$

**2.** *True or False*   $\displaystyle\int_0^3 5e^{x^2}\,dx = \int_0^3 5\,dx \cdot \int_0^3 e^{x^2}\,dx$

**3.** *True or False*
$$\int_0^5 (x^3 + 1)\,dx = \int_0^{-3} (x^3 + 1)\,dx + \int_{-3}^5 (x^3 + 1)\,dx$$

**4.** If $f$ is continuous on an interval containing the numbers $a$, $b$, and $c$, and if $\int_a^c f(x)\,dx = 3$ and $\int_c^b f(x)\,dx = -5$, then $\int_a^b f(x)\,dx =$ _____.

**5.** If a function $f$ is continuous on the closed interval $[a, b]$, then
$$\bar{y} = \frac{1}{b-a}\int_a^b f(x)\,dx \text{ is the } \text{_____ of } f \text{ over } [a, b].$$

**6.** *True or False*   If a function $f$ is continuous on a closed interval $[a, b]$ and if $m$ and $M$ denote the absolute minimum value and the absolute maximum value, respectively, of $f$ on $[a, b]$, then
$$m \le \int_a^b f(x)\,dx \le M$$

### Skill Building

*In Problems 7–16, find each definite integral given that*
$\int_1^3 f(x)\,dx = 5$, $\int_1^3 g(x)\,dx = -2$, $\int_3^5 f(x)\,dx = 2$, $\int_3^5 g(x)\,dx = 1$.

**7.** $\displaystyle\int_1^3 [f(x) - g(x)]\,dx$   **8.** $\displaystyle\int_1^3 [f(x) + g(x)]\,dx$

**9.** $\displaystyle\int_1^3 [5f(x) - 3g(x)]\,dx$   **10.** $\displaystyle\int_1^3 [3f(x) + 4g(x)]\,dx$

**11.** $\displaystyle\int_1^5 [2f(x) - 3g(x)]\,dx$   **12.** $\displaystyle\int_1^5 [f(x) - g(x)]\,dx$

**13.** $\displaystyle\int_1^3 [5f(x) - 3g(x) + 7]\,dx$   **14.** $\displaystyle\int_1^3 [f(x) + 4g(x) + 2x]\,dx$

**15.** $\displaystyle\int_1^5 [2f(x) - g(x) + 3x^2]\,dx$   **16.** $\displaystyle\int_1^5 [5f(x) - 3g(x) - 4]\,dx$

*In Problems 17–36, find each definite integral using the Fundamental Theorem of Calculus and properties of the definite integral.*

**17.** $\displaystyle\int_0^1 (t^2 - t^{3/2})\,dt$   **18.** $\displaystyle\int_{-2}^0 (x + x^2)\,dx$

**19.** $\displaystyle\int_{\pi/2}^{\pi} 4\sin x\,dx$   **20.** $\displaystyle\int_0^{\pi/2} 3\cos x\,dx$

**21.** $\displaystyle\int_{-\pi/4}^{\pi/4} (1 + 2\sec x\,\tan x)\,dx$   **22.** $\displaystyle\int_0^{\pi/4} (1 + \sec^2 x)\,dx$

**23.** $\displaystyle\int_1^4 (\sqrt{x} - 4x)\,dx$   **24.** $\displaystyle\int_0^1 (\sqrt[5]{t^2} + 1)\,dt$

**25.** $\displaystyle\int_{-2}^3 [(x-1)(x+3)]\,dx$   **26.** $\displaystyle\int_0^1 (z^2 + 1)^2\,dz$

**27.** $\displaystyle\int_1^2 \frac{x^2 - 12}{x^4}\,dx$   **28.** $\displaystyle\int_1^e \frac{5s^2 + s}{s^2}\,ds$

**29.** $\displaystyle\int_0^1 \frac{e^{2x} - 1}{e^x}\,dx$   **30.** $\displaystyle\int_0^1 \frac{e^{2x} + 4}{e^{-x}}\,dx$

**31.** $\displaystyle\int_{\pi/3}^{\pi/2} \frac{x\sin x + 2}{x}\,dx$   **32.** $\displaystyle\int_{\pi/6}^{\pi/2} \frac{x\cos x - 4}{x}\,dx$

**33.** $\displaystyle\int_1^2 \frac{x^2\sinh x + 2}{x^2}\,dx$   **34.** $\displaystyle\int_1^4 \frac{x\cosh x - \sqrt{x}}{x}\,dx$

**35.** $\displaystyle\int_0^{1/2} \left(5 + \frac{1}{\sqrt{1-x^2}}\right)\,dx$   **36.** $\displaystyle\int_0^1 \left(1 + \frac{5}{1+x^2}\right)\,dx$

*In Problems 37–42, use properties of integrals and the Fundamental Theorem of Calculus to find each integral.*

**37.** $\displaystyle\int_{-2}^1 f(x)\,dx$, where $f(x) = \begin{cases} 1 & \text{if } x < 0 \\ x^2 + 1 & \text{if } x \ge 0 \end{cases}$

**38.** $\displaystyle\int_{-1}^2 f(x)\,dx$, where $f(x) = \begin{cases} x + 1 & \text{if } x < 0 \\ x^2 + 1 & \text{if } x \ge 0 \end{cases}$

**39.** $\displaystyle\int_{-2}^2 f(x)\,dx$, where $f(x) = \begin{cases} 3x & \text{if } -2 \le x < 0 \\ 2x^2 & \text{if } 0 \le x \le 2 \end{cases}$

**40.** $\displaystyle\int_0^4 h(x)\,dx$, where $h(x) = \begin{cases} x - 2 & \text{if } 0 \le x \le 2 \\ 2 - x & \text{if } 2 < x \le 4 \end{cases}$

**41.** $\displaystyle\int_{-\pi/2}^{\pi/2} f(x)\,dx$, where $f(x) = \begin{cases} x^2 + x & \text{if } -\dfrac{\pi}{2} \le x \le 0 \\ \sin x & \text{if } 0 < x < \dfrac{\pi}{4} \\ \dfrac{\sqrt{2}}{2} & \text{if } \dfrac{\pi}{4} \le x \le \dfrac{\pi}{2} \end{cases}$

**42.** $\displaystyle\int_0^5 f(x)\,dx$, where $f(x) = \begin{cases} \dfrac{x^3 - 8}{x - 2} & \text{if } 0 \le x < 1 \\ 4x + 3 & \text{if } 1 \le x \le 5 \end{cases}$

---

**1.** = NOW WORK problem   ⊿ = Graphing technology recommended   [CAS] = Computer Algebra System recommended

*In Problems 43–46, the domain of f is a closed interval [a, b].*
Find $\int_a^b f(x)\,dx$.

**43.**

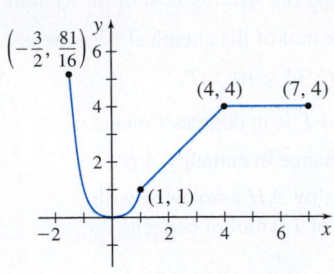

$$f(x) = \begin{cases} x^4 & \text{if } -\dfrac{3}{2} \le x < 1 \\ x & \text{if } 1 \le x < 4 \\ 4 & \text{if } 4 \le x \le 7 \end{cases}$$

**44.**

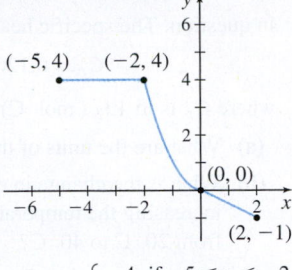

$$f(x) = \begin{cases} 4 & \text{if } -5 \le x < -2 \\ x^2 & \text{if } -2 \le x \le 0 \\ -\dfrac{x}{2} & \text{if } 0 < x \le 2 \end{cases}$$

**45.**

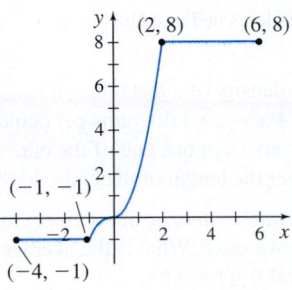

$$f(x) = \begin{cases} -1 & \text{if } -4 \le x \le -1 \\ x^3 & \text{if } -1 < x < 2 \\ 8 & \text{if } 2 \le x \le 6 \end{cases}$$

**46.**

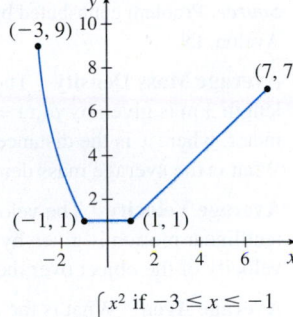

$$f(x) = \begin{cases} x^2 & \text{if } -3 \le x \le -1 \\ 1 & \text{if } -1 < x \le 1 \\ x & \text{if } 1 < x \le 7 \end{cases}$$

*In Problems 47–50, use properties of definite integrals to verify each statement. Assume that all integrals involved exist.*

**47.** $\displaystyle\int_3^{11} f(x)\,dx - \int_7^{11} f(x)\,dx = \int_3^7 f(x)\,dx$

**48.** $\displaystyle\int_{-2}^{6} f(x)\,dx - \int_3^6 f(x)\,dx = \int_{-2}^3 f(x)\,dx$

**49.** $\displaystyle\int_0^4 f(x)\,dx - \int_6^4 f(x)\,dx = \int_0^6 f(x)\,dx$

**50.** $\displaystyle\int_{-1}^3 f(x)\,dx - \int_5^3 f(x)\,dx = \int_{-1}^5 f(x)\,dx$

*In Problems 51–58, use the Bounds on an Integral Theorem to obtain a lower estimate and an upper estimate for each integral.*

**51.** $\displaystyle\int_1^3 (5x+1)\,dx$

**52.** $\displaystyle\int_0^1 (1-x)\,dx$

**53.** $\displaystyle\int_{\pi/4}^{\pi/2} \sin x\,dx$

**54.** $\displaystyle\int_{\pi/6}^{\pi/3} \cos x\,dx$

**55.** $\displaystyle\int_0^1 \sqrt{1+x^2}\,dx$

**56.** $\displaystyle\int_{-1}^1 \sqrt{1+x^4}\,dx$

**57.** $\displaystyle\int_0^1 e^x\,dx$

**58.** $\displaystyle\int_1^{10} \frac{1}{x}\,dx$

*In Problems 59–64, for each integral find the number(s) u guaranteed by the Mean Value Theorem for Integrals.*

**59.** $\displaystyle\int_0^3 (2x^2+1)\,dx$

**60.** $\displaystyle\int_0^2 (2-x^3)\,dx$

**61.** $\displaystyle\int_0^4 x^2\,dx$

**62.** $\displaystyle\int_0^4 (-x)\,dx$

**63.** $\displaystyle\int_0^{2\pi} \cos x\,dx$

**64.** $\displaystyle\int_{-\pi/4}^{\pi/4} \sec x \tan x\,dx$

*In Problems 65–74, find the average value of each function f over the given interval.*

**65.** $f(x) = e^x$ over $[0, 1]$

**66.** $f(x) = \dfrac{1}{x}$ over $[1, e]$

**67.** $f(x) = x^{2/3}$ over $[-1, 1]$

**68.** $f(x) = \sqrt{x}$ over $[0, 4]$

**69.** $f(x) = \sin x$ over $\left[0, \dfrac{\pi}{2}\right]$

**70.** $f(x) = \cos x$ over $\left[0, \dfrac{\pi}{2}\right]$

**71.** $f(x) = 1 - x^2$ over $[-1, 1]$

**72.** $f(x) = 16 - x^2$ over $[-4, 4]$

**73.** $f(x) = e^x - \sin x$ over $\left[0, \dfrac{\pi}{2}\right]$

**74.** $f(x) = x + \cos x$ over $\left[0, \dfrac{\pi}{2}\right]$

*In Problems 75–78, find:*

**(a)** The area under the graph of the function over the indicated interval.

**(b)** The average value of each function over the indicated interval.

**(c)** Interpret the results geometrically.

**75.** $[-1, 2]$

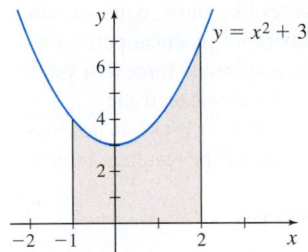

**76.** $[-2, 1]$

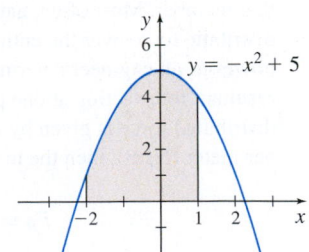

**77.** $[-1, 2]$

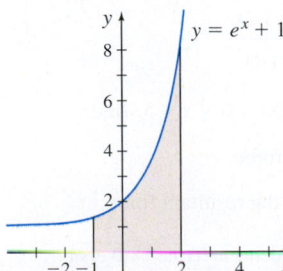

**78.** $\left[0, \dfrac{3\pi}{4}\right]$

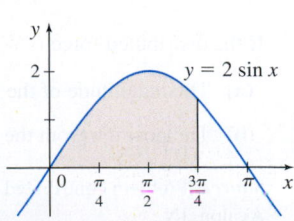

*In Problems 79–82, an object in rectilinear motion is moving along a horizontal line with velocity v over the indicated interval.*

**(a)** *Find the net displacement of the object. Interpret the result in the context of the problem.*

**(b)** *Find the total distance traveled by the object and interpret the result.*

**79.** $v(t) = 6t^2 + 12t - 18$ m/min; $0 \le t \le 3$

**80.** $v(t) = t^2 - 2t - 8$ m/min; $0 \le t \le 6$

**81.** $v(t) = t^3 + 2t^2 - 8t$ m/s; $-2 \le t \le 3$

**82.** $v(t) = t^3 + 5t^2 - 6t$ m/s; $-1 \le t \le 3$

## Applications and Extensions

*In Problems 83–86, find the indicated derivative.*

**83.** $\dfrac{d}{dx} \displaystyle\int_x^{4x} \sqrt{t^2 + 4}\, dt$

**84.** $\dfrac{d}{dx} \displaystyle\int_x^{x^2} (t^2 + 4)^{3/2}\, dt$

**85.** $\dfrac{d}{dx} \displaystyle\int_{x^2}^{x^3} \ln t\, dt$

**86.** $\dfrac{d}{dx} \displaystyle\int_{-x}^{x^2} e^{t^2}\, dt$

*In Problems 87–90, find each definite integral using the Fundamental Theorem of Calculus and properties of definite integrals.*

**87.** $\displaystyle\int_{-2}^{3} (x + |x|)\, dx$

**88.** $\displaystyle\int_0^3 |x - 1|\, dx$

**89.** $\displaystyle\int_0^2 |3x - 1|\, dx$

**90.** $\displaystyle\int_0^5 |2 - x|\, dx$

**91. Average Temperature** A rod 3 m long is heated to $25x\,°$C, where $x$ is the distance in meters from one end of the rod. Find the average temperature of the rod.

**92. Average Daily Rainfall** The rainfall per day, $x$ days after the beginning of the year, is modeled by the function $r(x) = 0.00002(6511 + 366x - x^2)$, measured in centimeters. Find the average daily rainfall for the first 180 days of the year.

**93. Structural Engineering** A structural engineer designing a member of a structure must consider the forces that will act on that member. Most often, natural forces like snow, wind, or rain distribute force over the entire member. For practical purposes, however, an engineer determines the distributed force as a single resultant force acting at one point on the member. If the distributed force is given by the function $W = W(x)$, in newtons per meter (N/m), then the magnitude $F_R$ of the resultant force is

$$F_R = \int_a^b W(x)\, dx$$

The position $\bar{x}$ of the resultant force measured in meters from the origin is given by

$$\bar{x} = \frac{\int_a^b x W(x)\, dx}{\int_a^b W(x)\, dx}$$

If the distributed force is $W(x) = 0.75x^3$, $0 \le x \le 5$, find:

**(a)** The magnitude of the resultant force.

**(b)** The position from the origin of the resultant force.

*Source*: Problem contributed by the students at Trine University, Avalon, IN.

**94. Chemistry: Enthalpy** In chemistry, **enthalpy** is a measure of the total energy of a system. For a nonreactive process with no phase change, the change in enthalpy $\Delta H$ is given by $\Delta H = \int_{T_1}^{T_2} C_p\, dT$, where $C_p$ is the specific heat of the system in question. The specific heat per mol of the chemical benzene is

$$C_p = 0.126 + (2.34 \times 10^{-6})T,$$

where $C_p$ is in kJ/ (mol °C), and $T$ is in degrees Celsius.

**(a)** What are the units of the change in enthalpy $\Delta H$?

**(b)** What is the change in enthalpy $\Delta H$ associated with increasing the temperature of 1.0 mol of benzene from 20 °C to 40 °C?

**(c)** What is the change in enthalpy $\Delta H$ associated with increasing the temperature of 1.0 mol of benzene from 20 °C to 60 °C?

**(d)** Does the enthalpy of benzene increase, decrease, or remain constant as the temperature increases?

*Source*: Problem contributed by the students at Trine University, Avalon, IN.

**95. Average Mass Density** The mass density of a metal bar of length 3 m is given by $\rho(x) = 1000 + x - \sqrt{x}$ kilograms per cubic meter, where $x$ is the distance in meters from one end of the bar. What is the average mass density over the length of the entire bar?

**96. Average Velocity** The velocity at time $t$, in m/s, of an object in rectilinear motion is given by $v(t) = 4\pi \cos t$. What is the average velocity of the object over the interval $0 \le t \le \pi$?

**97. Average Area** What is the average area of all circles whose radii are between 1 and 3 m?

**98. Area**

**(a)** Use properties of integrals and the Fundamental Theorem of Calculus to find the area under the graph of $y = 3 - |x|$ from $-3$ to 3.

**(b)** Check your answer by using elementary geometry.

**99. Area**

**(a)** Use properties of integrals and the Fundamental Theorem of Calculus to find the area under the graph of $y = 1 - \left|\dfrac{1}{2}x\right|$ from $-2$ to 2.

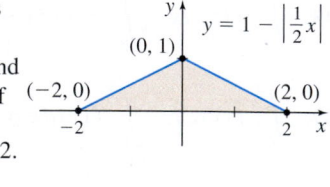

**(b)** Check your answer by using elementary geometry. See the figure.

**100. Area** Let $A$ be the area in the first quadrant that is enclosed by the graphs of

$y = 3x^2$, $y = \dfrac{3}{x}$, the x-axis, and the line $x = k$, where $k > 1$, as shown in the figure.

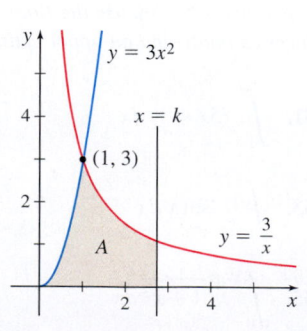

**(a)** Find the area $A$ as a function of $k$.

**(b)** When the area is 7, what is $k$?

**(c)** If the area $A$ is increasing at the constant rate of 5 square units per second, at what rate is $k$ increasing when $k = 15$?

**101. Rectilinear Motion**   A car starting from rest accelerates at the rate of 3 m/s². Find its average speed over the first 8 s.

**102. Rectilinear Motion**   A car moving at a constant speed of 80 miles per hour begins to decelerate at the rate of 10 mi/h². Find its average speed over the next 10 min.

**103. Average Slope**

    **(a)**  Use the definition of average value of a function to find the *average slope* of the graph of $y = f(x)$, where $a \leq x \leq b$. (Assume that $f'$ is continuous.)

    **(b)**  Give a geometric interpretation.

**104.** What theorem guarantees that the average slope found in Problem 103 is equal to $f'(u)$ for some $u$ in $[a, b]$? What *different* theorem guarantees the same thing? (Do you see the connection between these theorems?)

**105.** Prove that if a function $f$ is continuous on a closed interval $[a, b]$ and if $k$ is a constant, then $\int_a^b kf(x)\, dx = k \int_a^b f(x)\, dx$. Assume $f$ has an antiderivative.

**106.** Prove that if the functions $f_1, f_2, \ldots, f_n$ are continuous on a closed interval $[a, b]$ and if $k_1, k_2, \ldots, k_n$ are constants, then

$$\int_a^b [k_1 f_1(x) + k_2 f_2(x) + \cdots + k_n f_n(x)]\, dx$$

$$= k_1 \int_a^b f_1(x)\, dx + k_2 \int_a^b f_2(x)\, dx + \cdots + k_n \int_a^b f_n(x)\, dx$$

Assume each function $f$ has an antiderivative.

**107. Area**   The area under the graph of $y = \cos x$ from $-\dfrac{\pi}{2}$ to $\dfrac{\pi}{2}$ is separated into two parts by the line $x = k$, $-\dfrac{\pi}{2} < k < \dfrac{\pi}{2}$, as shown in the figure. If the area under the graph of $y$ from $-\dfrac{\pi}{2}$ to $k$ is three times the area under the graph of $y$ from $k$ to $\dfrac{\pi}{2}$, find $k$.

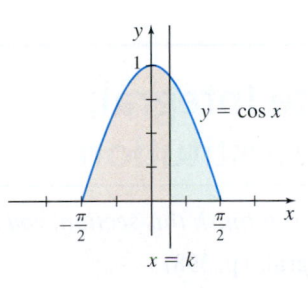

**108. Displacement of a Damped Spring**   The net displacement $x$ in meters of a damped spring from its equilibrium position at time $t$ seconds is given by

$$x(t) = \frac{\sqrt{15}}{10} e^{-t} \sin\left(\sqrt{15}t\right) + \frac{3}{2} e^{-t} \cos\left(\sqrt{15}t\right)$$

    **(a)**  What is the net displacement of the spring at $t = 0$?

     **(b)**  Graph the displacement for the first 2 s of the spring's motion.

    **(c)**  Find the average displacement of the spring for the first 2 s of its motion.

**109. Area**   Let

$$f(x) = |x^4 + 3.44x^3 - 0.5041x^2 - 5.0882x + 1.1523|$$

be defined on the interval $[-3, 1]$. Find the area under the graph of $f$.

**110.** If $f$ is continuous on $[a, b]$, show that the functions defined by

$$F(x) = \int_c^x f(t)\, dt \qquad G(x) = \int_d^x f(t)\, dt$$

for any choice of $c$ and $d$ in $(a, b)$ always differ by a constant. Also show that

$$F(x) - G(x) = \int_c^d f(t)\, dt$$

**111. Put It Together**   Suppose $a < c < b$ and the function $f$ is continuous on $[a, b]$ and differentiable on $(a, b)$. Which of the following is *not* necessarily true?

    **(I)** $\displaystyle\int_a^b f(x)\, dx = \int_a^c f(x)\, dx + \int_c^b f(x)\, dx.$

    **(II)**  There is a number $d$ in $(a, b)$ for which

$$f'(d) = \frac{f(b) - f(a)}{b - a}$$

    **(III)** $\displaystyle\int_a^b f(x)\, dx \geq 0.$

    **(IV)** $\displaystyle\lim_{x \to c} f(x) = f(c).$

    **(V)**  If $k$ is a real number, then $\int_a^b kf(x)\, dx = k \int_a^b f(x)\, dx$.

**112. Minimizing Area**   Find $b > 0$ so that the area enclosed in the first quadrant by the graph of $y = 1 + b - bx^2$ and the coordinate axes is a minimum.

**113. Area**   Find the area enclosed by the graph of $\sqrt{x} + \sqrt{y} = 1$ and the coordinate axes.

**Challenge Problems** ────────────

**114. Average Speed**   For a freely falling object starting from rest, $v_0 = 0$, find:

    **(a)**  The average speed $\bar{v}_t$ with respect to the time $t$ in seconds over the closed interval $[0, 5]$.

    **(b)**  The average speed $\bar{v}_s$ with respect to the distance $s$ of the object from its position at $t = 0$ over the closed interval $[0, s_1]$, where $s_1$ is the distance the object falls from $t = 0$ to $t = 5$ s.

    (*Hint:*  The derivation of the formulas for freely falling objects is given in Section 4.8, pp. 349–350.)

**115. Average Speed**   If an object falls from rest for 3 s, find:

    **(a)**  Its average speed with respect to time.

    **(b)**  Its average speed with respect to the distance it travels in 3 s.

    Express the answers in terms of $g$, the acceleration due to gravity.

**116. Free Fall**   For a freely falling object starting from rest, $v_0 = 0$, find:

    **(a)**  The average velocity $\bar{v}_t$ with respect to the time $t$ over the closed interval $[0, t_1]$.

    **(b)**  The average velocity $\bar{v}_s$ with respect to the distance $s$ of the object from its position at $t = 0$ over the closed interval $[0, s_1]$, where $s_1$ is the distance the object falls in time $t_1$. Assume $s(0) = 0$.

**117. Put It Together**

(a) What is the domain of $f(x) = 2|x - 1|x^2$?

(b) What is the range of $f$?

(c) For what values of $x$ is $f$ continuous?

(d) For what values of $x$ is the derivative of $f$ continuous?

(e) Find $\int_0^1 f(x)\,dx$.

**118. Probability**   A function $f$ that is continuous on the closed interval $[a, b]$, and for which (i) $f(x) \geq 0$ for numbers $x$ in $[a, b]$ and 0 elsewhere and (ii) $\int_a^b f(x)\,dx = 1$, is called a **probability density function**. If $a \leq c < d \leq b$, the probability of obtaining a value between $c$ and $d$ is defined as $\int_c^d f(x)\,dx$.

(a) Find a constant $k$ so that $f(x) = kx$ is a probability density function on $[0, 2]$.

(b) Find the probability of obtaining a value between 1 and 1.5.

**119. Cumulative Probability Distribution**   Refer to Problem 118. If $f$ is a probability density function, the **cumulative distribution function** $F$ for $f$ is defined as

$$F(x) = \int_a^x f(t)\,dt \qquad a \leq x \leq b$$

Find the cumulative distribution function $F$ for the probability density function $f(x) = kx$ of Problem 118(a).

**120.** For the cumulative distribution function $F(x) = x - 1$, on the interval $[1, 2]$:

(a) Find the probability density function $f$ corresponding to $F$.

(b) Find the probability of obtaining a value between 1.5 and 1.7.

**121.** Let $f(x) = x^3 - 6x^2 + 11x - 6$. Find $\int_1^3 |f(x)|\,dx$.

**122.** (a) Prove that if functions $f$ and $g$ are continuous on a closed interval $[a, b]$ and if $f(x) \geq g(x)$ on $[a, b]$, then

$$\int_a^b f(x)\,dx \geq \int_a^b g(x)\,dx.$$

(b) Show that for $x > 1$, $\ln x < 2(\sqrt{x} - 1)$.

**123.** Prove that the average value of a line segment $y = m(x - x_1) + y_1$ on the interval $[x_1, x_2]$ equals the $y$-coordinate of the midpoint of the line segment from $x_1$ to $x_2$.

**124.** Prove that if a function $f$ is continuous on a closed interval $[a, b]$ and if $f(x) \geq 0$ on $[a, b]$, then $\int_a^b f(x)\,dx \geq 0$.

**125.** (a) Prove that if $f$ is continuous on a closed interval $[a, b]$ and $\int_a^b f(x)\,dx = 0$, there is at least one number $c$ in $[a, b]$ for which $f(c) = 0$.

(b) Give a counter example to the statement above if $f$ is not required to be continuous.

**126.** Prove that if $f$ is continuous on $[a, b]$, then

$$\left| \int_a^b f(x)\,dx \right| \leq \int_a^b |f(x)|\,dx.$$

Give a geometric interpretation of the inequality.

# 5.5 The Indefinite Integral; Method of Substitution

**OBJECTIVES**   *When you finish this section, you should be able to:*

**1** Find indefinite integrals (p. 398)

**2** Use properties of indefinite integrals (p. 400)

**3** Find an indefinite integral using substitution (p. 401)

**4** Find a definite integral using substitution (p. 405)

**5** Integrate even and odd functions (p. 407)

The Fundamental Theorem of Calculus establishes an important relationship between definite integrals and antiderivatives: the definite integral $\int_a^b f(x)\,dx$ can be found easily if an antiderivative of $f$ can be found. Because of this, it is customary to use the integral symbol $\int$ as an instruction to find all the antiderivatives of a function.

## 1 Find Indefinite Integrals

**DEFINITION  Indefinite Integral**

The expression $\int f(x)\,dx$, called the **indefinite integral of $f$**, is defined as,

**IN WORDS**  The indefinite integral $\int f(x)\,dx$ is a symbol for all the antiderivatives of $f$.

$$\int f(x)\,dx = F(x) + C$$

where $F$ is any function for which $\dfrac{d}{dx}F(x) = f(x)$ and $C$ is a number, called the **constant of integration**.

**CAUTION** In writing an indefinite integral $\int f(x)\,dx$, remember to include the "$dx$."

For example,

$$\int (x^2+1)\,dx = \frac{x^3}{3}+x+C \qquad \frac{d}{dx}\left(\frac{x^3}{3}+x+C\right) = \frac{3x^2}{3}+1+0 = x^2+1$$

The process of finding either the indefinite integral $\int f(x)\,dx$ or the definite integral $\int_a^b f(x)\,dx$ is called **integration**, and in both cases the function $f$ is called the **integrand**.

It is important to distinguish between the definite integral $\int_a^b f(x)\,dx$ and the indefinite integral $\int f(x)\,dx$. The definite integral is a *number* that depends on the limits of integration $a$ and $b$. In contrast, the indefinite integral of $f$ is a *family* of functions $F(x)+C$, $C$ a constant, for which $F'(x)=f(x)$. For example,

**IN WORDS** The definite integral $\int_a^b f(x)\,dx$ is a number; the indefinite integral $\int f(x)\,dx$ is a family of functions.

$$\int_0^2 x^2\,dx = \left[\frac{x^3}{3}\right]_0^2 = \frac{8}{3} \qquad \int x^2\,dx = \frac{x^3}{3}+C$$

We list the indefinite integrals of some important functions in Table 2. Each entry results from a differentiation formula.

---

**TABLE 2 Basic Integral Formulas**

---

$$\int dx = x+C$$

$$\int x^a\,dx = \frac{x^{a+1}}{a+1}+C; \quad a \neq -1$$

$$\int x^{-1}\,dx = \int \frac{1}{x}\,dx = \ln|x|+C$$

$$\int e^x\,dx = e^x+C$$

$$\int a^x\,dx = \frac{a^x}{\ln a}+C; \quad a>0,\, a\neq 1$$

$$\int \sin x\,dx = -\cos x+C$$

$$\int \cos x\,dx = \sin x+C$$

$$\int \sec^2 x\,dx = \tan x+C$$

$$\int \sec x \tan x\,dx = \sec x+C$$

$$\int \csc x \cot x\,dx = -\csc x+C$$

$$\int \csc^2 x\,dx = -\cot x+C$$

$$\int \frac{1}{\sqrt{1-x^2}}\,dx = \sin^{-1}x+C, \quad |x|<1$$

$$\int \frac{1}{1+x^2}\,dx = \tan^{-1}x+C$$

$$\int \frac{1}{x\sqrt{x^2-1}}\,dx = \sec^{-1}x+C, \quad |x|>1$$

$$\int \sinh x\,dx = \cosh x+C$$

$$\int \cosh x\,dx = \sinh x+C$$

---

## EXAMPLE 1  Finding Indefinite Integrals

Find:

**(a)** $\displaystyle\int x^4\,dx$    **(b)** $\displaystyle\int \sqrt{x}\,dx$    **(c)** $\displaystyle\int \frac{\sin x}{\cos^2 x}\,dx$

**Solution (a)** The antiderivatives of $f(x)=x^4$ are $F(x)=\dfrac{x^5}{5}+C$, so

$$\int x^4\,dx = \frac{x^5}{5}+C$$

**(b)** The antiderivatives of $f(x)=\sqrt{x}=x^{1/2}$ are $F(x)=\dfrac{x^{3/2}}{\dfrac{3}{2}}+C = \dfrac{2x^{3/2}}{3}+C$, so

$$\int \sqrt{x}\,dx = \frac{2x^{3/2}}{3}+C$$

**NEED TO REVIEW?** Trigonometric identities are discussed in Appendix A.4, pp. A-32 to A-35.

**(c)** No integral in Table 2 corresponds to $\int \dfrac{\sin x}{\cos^2 x}\,dx$, so we begin by using trigonometric identities to rewrite $\dfrac{\sin x}{\cos^2 x}$ in a form whose antiderivative is recognizable.

$$\frac{\sin x}{\cos^2 x} = \frac{\sin x}{\cos x \cdot \cos x} = \frac{1}{\cos x} \cdot \frac{\sin x}{\cos x} = \sec x \tan x$$

Then

$$\int \frac{\sin x}{\cos^2 x}\,dx = \int \sec x \tan x \,dx = \sec x + C$$

The antiderivatives of<br>$\sec x \tan x$ are $\sec x + C$.

**NOW WORK** Problems 9 and 11.

## 2 Use Properties of Indefinite Integrals

Since the definite integral and the indefinite integral are closely related, properties of indefinite integrals are very similar to those of definite integrals:

- **Derivative of an Indefinite Integral:**

$$\frac{d}{dx}\left[ \int f(x)\,dx \right] = f(x) \tag{1}$$

Property (1) is a consequence of the definition of $\int f(x)\,dx$. For example,

$$\frac{d}{dx} \int \sqrt{x^2 + 1}\,dx = \sqrt{x^2 + 1} \qquad\qquad \frac{d}{dt} \int e^t \cos t \,dt = e^t \cos t$$

- **Indefinite Integral of the Sum of Two Functions:**

**IN WORDS** The indefinite integral of a sum of two functions equals the sum of the indefinite integrals.

$$\int [f(x) + g(x)]\,dx = \int f(x)\,dx + \int g(x)\,dx \tag{2}$$

The proof of property (2) follows directly from properties of derivatives, and is left as an exercise. See Problem 144.

- **Indefinite Integral of a Constant Times a Function:** If $k$ is a constant,

$$\int kf(x)\,dx = k \int f(x)\,dx \tag{3}$$

To prove property (3), differentiate the right side of (3).

**IN WORDS** To find the indefinite integral of a constant $k$ times a function $f$, find the indefinite integral of $f$ and then multiply by $k$.

$$\frac{d}{dx}\left[ k \int f(x)\,dx \right] = k \left[ \frac{d}{dx} \int f(x)\,dx \right] = kf(x)$$

**Constant Multiple Rule**       **Property (1)**

---

**EXAMPLE 2** **Using Properties of Indefinite Integrals**

**(a)**
$$\int (2x^{1/3} + 5x^{-1})\,dx = \int 2x^{1/3}\,dx + \int \frac{5}{x}\,dx = 2 \int x^{1/3}\,dx + 5 \int \frac{1}{x}\,dx$$

$$= 2 \cdot \frac{x^{4/3}}{\dfrac{4}{3}} + 5 \ln |x| + C = \frac{3x^{4/3}}{2} + 5 \ln |x| + C$$

**(b)** $\int \left( \dfrac{12}{x^5} + \dfrac{1}{\sqrt{x}} \right) dx = 12 \int \dfrac{1}{x^5} dx + \int \dfrac{1}{\sqrt{x}} dx = 12 \int x^{-5} dx + \int x^{-1/2} dx$

$$= 12 \left( \dfrac{x^{-4}}{-4} \right) + \dfrac{x^{1/2}}{\dfrac{1}{2}} + C = -\dfrac{3}{x^4} + 2\sqrt{x} + C \qquad \blacksquare$$

**NOW WORK** Problem 17.

Sometimes an appropriate algebraic manipulation is required before integrating.

---

**EXAMPLE 3    Using Properties of Indefinite Integrals**

$$\int \dfrac{x^2 + 6}{x^2 + 1} dx = \int \dfrac{(x^2 + 1) + 5}{x^2 + 1} dx = \int \left[ \dfrac{x^2 + 1}{x^2 + 1} + \dfrac{5}{x^2 + 1} \right] dx$$

$$= \int \left[ 1 + \dfrac{5}{x^2 + 1} \right] dx \underset{\underset{\text{Sum Property}}{\uparrow}}{=} \int dx + \int \dfrac{5}{x^2 + 1} dx$$

$$= \int dx + 5 \int \dfrac{1}{x^2 + 1} dx \underset{\uparrow}{=} x + 5 \tan^{-1} x + C \qquad \blacksquare$$

The antiderivatives of

$\dfrac{1}{x^2 + 1}$ are $\tan^{-1} x + C.$

**NOW WORK** Problem 13.

## 3  Find an Indefinite Integral Using Substitution

Indefinite integrals that cannot be found using Table 2 sometimes can be found using the *method of substitution*. In the method of substitution, we change the variable to transform the integrand into a form that is in the table.

For example, to find $\int (x^2 + 5)^3 2x\, dx$, we use the substitution $u = x^2 + 5$. The differential of $u = x^2 + 5$ is $du = 2x\, dx$. Now we write $(x^2 + 5)^3 2x\, dx$ in terms of $u$ and $du$, and integrate the simpler integral.

> **NEED TO REVIEW?** Differentials are discussed in Section 3.5, pp. 241–242.

> **NOTE** When using the method of substitution, once the substitution $u$ is chosen, write the original integral in terms of $u$ and $du$.

$$\int \underset{u}{\underbrace{(x^2 + 5)^3}}\, \underset{du}{\underbrace{2x\, dx}} = \int u^3 du = \dfrac{u^4}{4} + C \underset{\underset{u = x^2 + 5}{\uparrow}}{=} \dfrac{(x^2 + 5)^4}{4} + C$$

We can verify the result by differentiating it using the Power Rule for Functions.

$$\dfrac{d}{dx} \left[ \dfrac{(x^2 + 5)^4}{4} + C \right] = \dfrac{1}{4} \left[ 4(x^2 + 5)^3 (2x) + 0 \right] = (x^2 + 5)^3 2x$$

> **NEED TO REVIEW?** The Chain Rule is discussed in Section 3.1, pp. 208–211.

The method of substitution is based on the Chain Rule, which states that if $f$ and $g$ are differentiable functions, then for the composite function $f \circ g$,

$$\dfrac{d}{dx}[(f \circ g)(x)] = \dfrac{d}{dx} f(g(x)) = f'(g(x))\, g'(x)$$

The Chain Rule provides a template for finding integrals of the form

$$\int f'(g(x)) g'(x)\, dx$$

If, in the integral, we let $u = g(x)$, then the differential $du = g'(x)\, dx$, and we have

$$\int \underset{u}{\underbrace{f'(g(x))}}\, \underset{du}{\underbrace{g'(x)\, dx}} = \int f'(u)\, du = f(u) + C \underset{\underset{u = g(x)}{\uparrow}}{=} f(g(x)) + C$$

Replacing $g(x)$ by $u$ and $g'(x)dx$ by $du$ is called **substitution**. Substitution is a strategy for finding indefinite integrals when the integrand is a composite function.

EXAMPLE 4    Finding Indefinite Integrals Using Substitution

Find:

**(a)** $\displaystyle\int \sin(3x + 2)\, dx$    **(b)** $\displaystyle\int x\sqrt{x^2 + 1}\, dx$    **(c)** $\displaystyle\int \frac{e^{\sqrt{x}}}{\sqrt{x}}\, dx$

**Solution  (a)** Since we know $\int \sin x\, dx$, we let $u = 3x + 2$. Then $du = 3\, dx$ so $dx = \dfrac{du}{3}$.

$$\int \sin(\underbrace{3x + 2}_{u})\, \underbrace{dx}_{\frac{du}{3}} = \int \sin u\, \frac{du}{3} = \frac{1}{3}\int \sin u\, du$$

$$= \frac{1}{3}(-\cos u) + C = \underset{\substack{\uparrow \\ u = 3x + 2}}{-\frac{1}{3}\cos(3x + 2) + C}$$

**(b)** Since we know $\int \sqrt{x}\, dx$, let $u = x^2 + 1$. Then $du = 2x\, dx$, so $x\, dx = \dfrac{du}{2}$.

$$\int x\sqrt{x^2 + 1}\, dx = \int \sqrt{x^2 + 1}\, x\, dx = \int \sqrt{u}\, \frac{du}{2} = \frac{1}{2}\int u^{1/2}\, du = \frac{1}{2}\left(\frac{u^{3/2}}{\frac{3}{2}}\right) + C$$

$$= \frac{(x^2 + 1)^{3/2}}{3} + C$$

**(c)** Let $u = \sqrt{x} = x^{1/2}$. Then $du = \dfrac{1}{2}x^{-1/2}dx = \dfrac{dx}{2\sqrt{x}}$, so $\dfrac{dx}{\sqrt{x}} = 2\, du$.

$$\int \frac{e^{\sqrt{x}}}{\sqrt{x}}\, dx = \int e^{\sqrt{x}}\cdot \frac{dx}{\sqrt{x}} = \int e^u \cdot 2\, du = 2e^u + C = 2e^{\sqrt{x}} + C \qquad\blacksquare$$

NOW WORK  Problems 21 and 27.

When an integrand equals the product of an expression involving a function and its derivative (or a multiple of its derivative), then substitution is often a good strategy. In Example 4(c), $\displaystyle\int \frac{e^{\sqrt{x}}}{\sqrt{x}}\, dx$, we used the substitution $u = \sqrt{x}$, because

$$\frac{du}{dx} = \frac{d}{dx}\sqrt{x} = \frac{1}{2\sqrt{x}} \text{ is a multiple of } \frac{1}{\sqrt{x}}$$

Similarly, in Example 4(b), $\int x\sqrt{x^2 + 1}\, dx$, the factor $x$ in the integrand makes the substitution $u = x^2 + 1$ work. On the other hand, if we try to use this same substitution to integrate $\int \sqrt{x^2 + 1}\, dx$, then

$$\int \sqrt{x^2 + 1}\, dx = \int \sqrt{u}\, \frac{du}{2x} = \underset{\substack{\uparrow \\ x = \sqrt{u - 1}}}{\int \frac{\sqrt{u}}{2\sqrt{u - 1}}\, du}$$

and the resulting integral is *more* complicated than the original integral.

The idea behind substitution is to obtain an integral $\int h(u)\, du$ that is simpler than the original integral $\int f(x)\, dx$. When a substitution does not simplify the integral, try other substitutions. If none of these work, other integration methods should be tried. Some of these methods are explored in Chapter 7.

EXAMPLE 5    Finding Indefinite Integrals Using Substitution

Find:

**(a)** $\displaystyle\int \frac{e^x}{e^x + 4}\, dx$    **(b)** $\displaystyle\int \frac{5x^2\, dx}{4x^3 - 1}$

**Solution (a)** Here the numerator equals the derivative of the denominator. So, we use the substitution $u = e^x + 4$. Then $du = e^x dx$.

$$\int \frac{e^x}{e^x + 4} dx = \int \frac{1}{e^x + 4} \cdot e^x dx = \int \frac{1}{u} du = \ln |u| + C = \ln(e^x + 4) + C$$
$$\underset{\underset{u = e^x + 4 > 0}{\uparrow}}{}$$

**(b)** Notice that the numerator equals the derivative of the denominator, except for a constant factor. So, we try the substitution $u = 4x^3 - 1$. Then $du = 12x^2 dx$, so $5x^2 dx = \dfrac{5}{12} du$.

$$\int \frac{5x^2 dx}{4x^3 - 1} = \int \frac{\frac{5}{12} du}{u} = \frac{5}{12} \int \frac{du}{u} = \frac{5}{12} \ln |u| + C = \frac{5}{12} \ln \left|4x^3 - 1\right| + C \quad \blacksquare$$

In Example 5(a) and 5(b), the integrands had a numerator that was the derivative of the denominator. In general, we have the following formula:

$$\int \frac{g'(x)}{g(x)} dx = \ln |g(x)| + C \tag{4}$$

Notice that in formula (4) the integral equals the natural logarithm of the *absolute value of the function g*. The absolute value is necessary since the domain of the logarithm function is the set of positive real numbers. When $g$ is known to be positive, as in Example 5(a), the absolute value is not required.

**NOW WORK** Problem 33.

> **IN WORDS** If the numerator of the integrand equals the derivative of the denominator, then the integral equals a logarithmic function.

**EXAMPLE 6  Using Substitution to Establish an Integration Formula**

Show that:

**(a)**
$$\int \tan x \, dx = -\ln |\cos x| + C = \ln | \sec x| + C$$

**(b)**
$$\int \sec x \, dx = \ln | \sec x + \tan x| + C$$

**Solution (a)** Since $\tan x = \dfrac{\sin x}{\cos x}$, let $u = \cos x$. Then $du = -\sin x \, dx$ and

$$\int \tan x \, dx = \int \frac{\sin x}{\cos x} dx = \int -\frac{du}{u} = -\ln |u| + C = -\ln |\cos x| + C$$

$$= \underset{\underset{r \ln x = \ln x^r}{\uparrow}}{\ln |\cos x|^{-1}} + C = \ln \left| \frac{1}{\cos x} \right| + C = \ln |\sec x| + C$$

**(b)** To find $\int \sec x \, dx$, multiply the integrand by $\dfrac{\sec x + \tan x}{\sec x + \tan x}$.

$$\int \sec x \, dx = \int \sec x \cdot \frac{\sec x + \tan x}{\sec x + \tan x} dx = \int \frac{\sec^2 x + \sec x \tan x}{\sec x + \tan x} dx$$

Now the numerator equals the derivative of the denominator. So by (4)

$$\int \sec x \, dx = \ln |\sec x + \tan x| + C \quad \blacksquare$$

**NOW WORK** Problem 43.

As we saw in Example 6(b), sometimes we need to manipulate an integral so that a basic integration formula can be used. Unlike differentiation, integration has no prescribed method; some ingenuity and a lot of practice are required. To illustrate, two different substitutions are used to solve Example 7.

CALC
CLIP

▶ EXAMPLE 7  Finding an Indefinite Integral Using Substitution

Find $\int x\sqrt{4+x}\,dx$.

**Solution** *Substitution I*  Let $u = 4 + x$. Then $x = u - 4$ and $du = dx$. Substituting gives

$$\int x\sqrt{4+x}\,dx = \int \underbrace{(u-4)}_{x}\,\underbrace{\sqrt{u}}_{\uparrow}\,\underbrace{du}_{dx} = \int (u^{3/2} - 4u^{1/2})\,du$$
$$\qquad\qquad\qquad\qquad 4+x$$

$$= \frac{u^{5/2}}{\dfrac{5}{2}} - 4 \cdot \frac{u^{3/2}}{\dfrac{3}{2}} + C$$

$$= \frac{2(4+x)^{5/2}}{5} - \frac{8(4+x)^{3/2}}{3} + C$$

*Substitution II*  Let $u = \sqrt{4+x}$, so $u^2 = 4 + x$ and $x = u^2 - 4$. Then $dx = 2u\,du$ and

$$\int x\sqrt{4+x}\,dx = \int \underbrace{(u^2-4)}_{x}(u)\underbrace{(2u\,du)}_{dx} = 2\int (u^4 - 4u^2)\,du = 2\left[\frac{u^5}{5} - \frac{4u^3}{3}\right] + C$$

$$= \frac{2}{5}\left(\sqrt{4+x}\right)^5 - \frac{8}{3}\left(\sqrt{4+x}\right)^3 + C = \frac{2(4+x)^{5/2}}{5} - \frac{8(4+x)^{3/2}}{3} + C \qquad ■$$

**NOW WORK** Problem 51.

EXAMPLE 8  Finding Indefinite Integrals Using Substitution

Find:

**(a)** $\displaystyle\int \frac{dx}{\sqrt{4-x^2}}$    **(b)** $\displaystyle\int \frac{dx}{9+4x^2}$

**Solution** **(a)** $\displaystyle\int \frac{dx}{\sqrt{4-x^2}}$ resembles $\displaystyle\int \frac{1}{\sqrt{1-x^2}}\,dx = \sin^{-1}x + C$. We begin by rewriting the integrand as

$$\frac{1}{\sqrt{4-x^2}} = \frac{1}{\sqrt{4\left(1 - \dfrac{x^2}{4}\right)}} = \frac{1}{2\sqrt{1 - \left(\dfrac{x}{2}\right)^2}}$$

Now let $u = \dfrac{x}{2}$. Then $du = \dfrac{dx}{2}$.

$$\int \frac{dx}{\sqrt{4-x^2}} = \int \frac{dx}{2\sqrt{1 - \left(\dfrac{x}{2}\right)^2}} \underset{\substack{\uparrow \\ u = \frac{x}{2} \\ du = \frac{dx}{2}}}{=} \int \frac{du}{\sqrt{1-u^2}} = \sin^{-1}u + C$$

$$= \sin^{-1}\left(\frac{x}{2}\right) + C$$

**(b)** $\displaystyle\int \frac{dx}{9+4x^2}$ resembles $\displaystyle\int \frac{1}{1+x^2}dx = \tan^{-1}x + C$. Rewrite the integrand as

$$\frac{1}{9+4x^2} = \frac{1}{9\left(1+\dfrac{4x^2}{9}\right)} = \frac{1}{9\left[1+\left(\dfrac{2x}{3}\right)^2\right]}$$

Now let $u = \dfrac{2x}{3}$. Then $du = \dfrac{2}{3}dx$, so $dx = \dfrac{3}{2}\,du$.

$$\int \frac{dx}{9+4x^2} = \int \frac{dx}{9\left[1+\left(\dfrac{2x}{3}\right)^2\right]} = \int \frac{\dfrac{3}{2}du}{9(1+u^2)} = \frac{1}{6}\int \frac{du}{1+u^2}$$

$$= \frac{1}{6}\tan^{-1}u + C = \frac{1}{6}\tan^{-1}\left(\frac{2x}{3}\right) + C \qquad\blacksquare$$

NOW WORK **Problem 55.**

## 4 Find a Definite Integral Using Substitution

Two approaches can be used to find a definite integral using substitution:

- Method 1: Find the related indefinite integral using substitution, and then use the Fundamental Theorem of Calculus.
- Method 2: Find the definite integral directly by making a substitution in the integrand and using the substitution to *change the limits of integration.*

**EXAMPLE 9   Finding a Definite Integral Using Substitution**

Find $\displaystyle\int_0^2 x\sqrt{4-x^2}\,dx$.

**Solution**

**Method 1:** Use the related indefinite integral and then use the Fundamental Theorem of Calculus. The related indefinite integral $\displaystyle\int x\sqrt{4-x^2}\,dx$ can be found using the substitution $u = 4-x^2$. Then $du = -2x\,dx$, so $x\,dx = -\dfrac{du}{2}$.

$$\int x\sqrt{4-x^2}\,dx = \int \sqrt{u}\left(-\frac{du}{2}\right) = -\frac{1}{2}\int u^{1/2}du = -\frac{1}{2}\cdot\frac{u^{3/2}}{\dfrac{3}{2}} + C$$

$$= -\frac{1}{3}(4-x^2)^{3/2} + C$$

Then by the Fundamental Theorem of Calculus,

$$\int_0^2 x\sqrt{4-x^2}\,dx = -\frac{1}{3}\left[(4-x^2)^{3/2}\right]_0^2 = -\frac{1}{3}\left[0 - 4^{3/2}\right] = \frac{8}{3}$$

**Method 2:** Find the definite integral directly by making a substitution in the integrand and changing the limits of integration. Let $u = 4 - x^2$; then $du = -2x\,dx$. Now use the function $u = 4 - x^2$ to change the limits of integration.

- The lower limit of integration is $x = 0$ so, in terms of $u$, it changes to $u = 4 - 0^2 = 4$.
- The upper limit of integration is $x = 2$ so the upper limit changes to $u = 4 - 2^2 = 0$.

Then

$$\int_0^2 x\sqrt{4 - x^2}\,dx \underset{\substack{\uparrow \\ u = 4 - x^2 \\ x\,dx = -\frac{1}{2}du}}{=} \int_4^0 \sqrt{u}\left(-\frac{du}{2}\right) = -\frac{1}{2}\int_4^0 \sqrt{u}\,du = -\frac{1}{2}\cdot\left[\frac{u^{3/2}}{\frac{3}{2}}\right]_4^0$$

**CAUTION** When using substitution to find a definite integral directly, remember to change the limits of integration.

$$= -\frac{1}{3}(0 - 8) = \frac{8}{3} \qquad \blacksquare$$

**NOW WORK** Problem 61.

---

**EXAMPLE 10** Finding a Definite Integral Using Substitution

Find $\displaystyle\int_0^{\pi/2} \frac{1 - \cos(2\theta)}{2}\,d\theta$.

**Solution** Use properties of integrals to simplify before integrating.

$$\int_0^{\pi/2} \frac{1 - \cos(2\theta)}{2}\,d\theta = \frac{1}{2}\int_0^{\pi/2} [1 - \cos(2\theta)]\,d\theta$$

$$= \frac{1}{2}\left[\int_0^{\pi/2} d\theta - \int_0^{\pi/2} \cos(2\theta)\,d\theta\right]$$

$$= \frac{1}{2}\int_0^{\pi/2} d\theta - \frac{1}{2}\int_0^{\pi/2} \cos(2\theta)\,d\theta$$

$$= \frac{1}{2}\Big[\theta\Big]_0^{\pi/2} - \frac{1}{2}\int_0^{\pi/2} \cos(2\theta)\,d\theta$$

$$= \frac{\pi}{4} - \frac{1}{2}\int_0^{\pi/2} \cos(2\theta)\,d\theta$$

In the integral on the right, use the substitution $u = 2\theta$. Then $du = 2\,d\theta$ so $d\theta = \dfrac{du}{2}$. Now change the limits of integration:

- when $\theta = 0$      then $u = 2(0) = 0$
- when $\theta = \dfrac{\pi}{2}$      then $u = 2\left(\dfrac{\pi}{2}\right) = \pi$

Now

$$\int_0^{\pi/2} \cos(2\theta)\,d\theta = \int_0^{\pi} \cos u\,\frac{du}{2} = \frac{1}{2}\Big[\sin u\Big]_0^{\pi} = \frac{1}{2}(\sin\pi - \sin 0) = 0$$

Then,

$$\int_0^{\pi/2} \frac{1 - \cos(2\theta)}{2}\,d\theta = \frac{\pi}{4} - \frac{1}{2}\int_0^{\pi/2} \cos(2\theta)\,d\theta = \frac{\pi}{4} \qquad \blacksquare$$

**NOW WORK** Problem 69.

**RECALL** A function $f$ is even if $f(-x) = f(x)$ and is odd if $f(-x) = -f(x)$ for all $x$ in the domain of $f$.

## 5 Integrate Even and Odd Functions

Integrals of even and odd functions can be simplified due to symmetry. Figure 34 illustrates the conclusions of the theorem that follows.

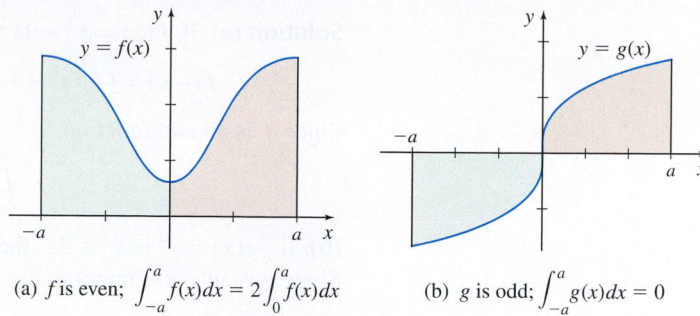

(a) $f$ is even; $\displaystyle\int_{-a}^{a} f(x)dx = 2\int_{0}^{a} f(x)dx$      (b) $g$ is odd; $\displaystyle\int_{-a}^{a} g(x)dx = 0$

**Figure 34**

---

**THEOREM  The Integrals of Even and Odd Functions**

Let a function $f$ be continuous on a closed interval $[-a, a]$, $a > 0$.

• If $f$ is an even function, then

$$\int_{-a}^{a} f(x)\,dx = 2\int_{0}^{a} f(x)\,dx$$

• If $f$ is an odd function, then

$$\int_{-a}^{a} f(x)\,dx = 0$$

---

The property for even functions is proved here; the proof for odd functions is left as an exercise. See Problem 135.

**Proof** $f$ is an even function:  Since $f$ is continuous on the closed interval $[-a, a]$, $a > 0$, and $0$ is in the interval $[-a, a]$, we have

$$\int_{-a}^{a} f(x)\,dx = \int_{-a}^{0} f(x)\,dx + \int_{0}^{a} f(x)\,dx = -\int_{0}^{-a} f(x)\,dx + \int_{0}^{a} f(x)\,dx \quad (5)$$

In $-\int_{0}^{-a} f(x)\,dx$, we use the substitution $u = -x$. Then $du = -dx$. Also, if $x = 0$, then $u = 0$, and if $x = -a$, then $u = a$. Therefore,

$$-\int_{0}^{-a} f(x)\,dx = \int_{0}^{a} f(-u)\,du = \underset{\substack{\uparrow \\ f \text{ is even} \\ f(-u) = f(u)}}{} \int_{0}^{a} f(u)\,du = \int_{0}^{a} f(x)\,dx \quad (6)$$

Combining (5) and (6), we obtain

$$\int_{-a}^{a} f(x)\,dx = \int_{0}^{a} f(x)\,dx + \int_{0}^{a} f(x)\,dx = 2\int_{0}^{a} f(x)\,dx \quad \blacksquare$$

To use the theorem involving even or odd functions, three conditions must be met:

• The function $f$ must be even or odd.
• The function $f$ must be continuous on the closed interval $[-a, a]$, $a > 0$.
• The limits of integration must be $a$ and $a$, $a > 0$.

EXAMPLE 11   Integrating an Even or Odd Function

Find:

(a) $\displaystyle\int_{-3}^{3} (x^7 - 4x^3 + x)\, dx$     (b) $\displaystyle\int_{-2}^{2} (x^4 - x^2 + 3)\, dx$

**Solution (a)** If $f(x) = x^7 - 4x^3 + x$, then

$$f(-x) = (-x)^7 - 4(-x)^3 + (-x) = -(x^7 - 4x^3 + x) = -f(x)$$

Since $f$ is an odd function,

$$\int_{-3}^{3} (x^7 - 4x^3 + x)\, dx = 0$$

**(b)** If $g(x) = x^4 - x^2 + 3$, then $g(-x) = (-x)^4 - (-x)^2 + 3 = x^4 - x^2 + 3 = g(x)$.
Since $g$ is an even function,

$$\int_{-2}^{2} (x^4 - x^2 + 3)\, dx = 2\int_{0}^{2} (x^4 - x^2 + 3)\, dx = 2\left[\frac{x^5}{5} - \frac{x^3}{3} + 3x\right]_{0}^{2}$$

$$= 2\left[\frac{32}{5} - \frac{8}{3} + 6\right] = \frac{292}{15} \qquad\blacksquare$$

NOW WORK   Problems 79 and 81.

EXAMPLE 12   Using Properties of Integrals

If $f$ is an even function and $\int_0^2 f(x)\, dx = -6$ and $\int_{-5}^{0} f(x)\, dx = 8$, find $\int_2^5 f(x)\, dx$.

**Solution** $\int_2^5 f(x)\, dx = \int_2^0 f(x)\, dx + \int_0^5 f(x)\, dx$

Now $\int_2^0 f(x)\, dx = -\int_0^2 f(x)\, dx = 6$.

Since $f$ is even, $\int_0^5 f(x)\, dx = \int_{-5}^{0} f(x)\, dx = 8$. Then

$$\int_2^5 f(x)\, dx = \int_2^0 f(x)\, dx + \int_0^5 f(x)\, dx = 6 + 8 = 14 \qquad\blacksquare$$

NOW WORK   Problem 87.

## 5.5 Assess Your Understanding

### Concepts and Vocabulary

1. $\dfrac{d}{dx}\left[\displaystyle\int f(x)\, dx\right] = $ _____

2. *True or False*   If $k$ is a constant, then
   $\int kf(x)\, dx = \int k\, dx \cdot \int f(x)\, dx$

3. If $a$ is a real number, $a \neq -1$, then $\int x^a\, dx = $ _____.

4. *True or False*   When finding the indefinite integral of a function $f$, a constant of integration $C$ is added to the result because $\int f(x)\, dx$ denotes all the antiderivatives of $f$.

5. If the substitution $u = 2x + 3$ is used with $\int \sin(2x + 3)\, dx$, the result is $\int$ _____ $du$.

6. *True or False*   If the substitution $u = x^2 + 3$ is used with $\int_0^1 x(x^2 + 3)^3\, dx$, then $\int_0^1 x(x^2 + 3)^3\, dx = \frac{1}{2}\int_0^1 u^3\, du$.

7. *Multiple Choice* $\int_{-4}^{4} x^3\, dx = $ [(a) 128 (b) 4 (c) 0 (d) 64]

8. *True or False* $\int_0^5 x^2\, dx = \frac{1}{2}\int_{-5}^{5} x^2\, dx$

### Skill Building

*In Problems 9–20, find each indefinite integral.*

9. $\displaystyle\int x^{2/3}\, dx$

10. $\displaystyle\int t^{-4}\, dt$

11. $\displaystyle\int \frac{1}{\sqrt{1 - x^2}}\, dx$

12. $\displaystyle\int \frac{1}{1 + x^2}\, dx$

13. $\displaystyle\int \frac{5x^2 + 2xe^x - 1}{x}\, dx$

14. $\displaystyle\int \frac{xe^x + 1}{x}\, dx$

15. $\displaystyle\int \frac{\tan x}{\cos x}\, dx$

16. $\displaystyle\int \frac{1}{\sin^2 x}\, dx$

17. $\displaystyle\int \frac{2}{5\sqrt{1 - x^2}}\, dx$

18. $\displaystyle\int -\frac{4}{x\sqrt{x^2 - 1}}\, dx$

19. $\displaystyle\int \sinh t\, dt$

20. $\displaystyle\int \cosh t\, dt$

*In Problems 21–26, find each indefinite integral using the given substitution.*

21. $\displaystyle\int e^{3x+1}\, dx$; let $u = 3x + 1$     22. $\displaystyle\int \frac{dx}{x \ln x}$; let $u = \ln x$

---

**1.** = NOW WORK problem        〰 = Graphing technology recommended        CAS = Computer Algebra System recommended

**23.** $\displaystyle\int (1-t^2)^6 t\, dt$; let $u=(1-t^2)$

**24.** $\displaystyle\int \sin^5 x \cos x\, dx$; let $u=\sin x$

**25.** $\displaystyle\int \frac{x^2\, dx}{\sqrt{1-x^6}}$; let $u=x^3$

**26.** $\displaystyle\int \frac{e^{-x}}{6+e^{-x}}\, dx$; let $u=6+e^{-x}$

*In Problems 27–60, find each indefinite integral.*

**27.** $\displaystyle\int \sin(3x)\, dx$

**28.** $\displaystyle\int x \sin x^2\, dx$

**29.** $\displaystyle\int \sin x \cos^2 x\, dx$

**30.** $\displaystyle\int \tan^2 x \sec^2 x\, dx$

**31.** $\displaystyle\int \frac{e^{1/x}}{x^2}\, dx$

**32.** $\displaystyle\int \frac{e^{\sqrt[3]{x}}}{\sqrt[3]{x^2}}\, dx$

**33.** $\displaystyle\int \frac{x\, dx}{x^2-1}$

**34.** $\displaystyle\int \frac{5x\, dx}{1-x^2}$

**35.** $\displaystyle\int \frac{e^x}{\sqrt{1+e^x}}\, dx$

**36.** $\displaystyle\int \frac{dx}{x(\ln x)^7}$

**37.** $\displaystyle\int \frac{1}{\sqrt{x}(1+\sqrt{x})^4}\, dx$

**38.** $\displaystyle\int \frac{dx}{\sqrt{x}(1+\sqrt{x})}$

**39.** $\displaystyle\int \frac{3e^x}{\sqrt[4]{e^x-1}}\, dx$

**40.** $\displaystyle\int \frac{[\ln(5x)]^3}{x}\, dx$

**41.** $\displaystyle\int \frac{\cos x\, dx}{2\sin x - 1}$

**42.** $\displaystyle\int \frac{\cos(2x)\, dx}{\sin(2x)}$

**43.** $\displaystyle\int \sec(5x)\, dx$

**44.** $\displaystyle\int \tan(2x)\, dx$

**45.** $\displaystyle\int \sqrt{\tan x}\, \sec^2 x\, dx$

**46.** $\displaystyle\int (2+3\cot x)^{3/2}\, \csc^2 x\, dx$

**47.** $\displaystyle\int \frac{\sin x}{\cos^2 x}\, dx$

**48.** $\displaystyle\int \frac{\cos x}{\sin^2 x}\, dx$

**49.** $\displaystyle\int \sin x \cdot e^{\cos x}\, dx$

**50.** $\displaystyle\int \sec^2 x \cdot e^{\tan x}\, dx$

**51.** $\displaystyle\int x\sqrt{x+3}\, dx$

**52.** $\displaystyle\int x\sqrt{4-x}\, dx$

**53.** $\displaystyle\int [\sin x + \cos(3x)]\, dx$

**54.** $\displaystyle\int [x^2 + \sqrt{3x+2}]\, dx$

**55.** $\displaystyle\int \frac{dx}{x^2+25}$

**56.** $\displaystyle\int \frac{\cos x}{1+\sin^2 x}\, dx$

**57.** $\displaystyle\int \frac{dx}{\sqrt{9-x^2}}$

**58.** $\displaystyle\int \frac{dx}{\sqrt{16-9x^2}}$

**59.** $\displaystyle\int \sinh x \cosh x\, dx$

**60.** $\displaystyle\int \mathrm{sech}^2 x \tanh x\, dx$

*In Problems 61–68, find each definite integral two ways:*

**(a)** *By finding the related indefinite integral and then using the Fundamental Theorem of Calculus;*

**(b)** *By making a substitution in the integrand and using the substitution to change the limits of integration.*

**(c)** *Which method did you prefer? Why?*

**61.** $\displaystyle\int_{-2}^{0} \frac{x}{(x^2+3)^2}\, dx$

**62.** $\displaystyle\int_{-1}^{1} (x^2-1)^5 x\, dx$

**63.** $\displaystyle\int_{0}^{1} x^2 e^{x^3+1}\, dx$

**64.** $\displaystyle\int_{0}^{1} x e^{x^2-2}\, dx$

**65.** $\displaystyle\int_{1}^{6} x\sqrt{x+3}\, dx$

**66.** $\displaystyle\int_{2}^{6} x^2\sqrt{x-2}\, dx$

**67.** $\displaystyle\int_{0}^{2} x \cdot 3^{2x^2}\, dx$

**68.** $\displaystyle\int_{0}^{1} x \cdot 10^{-x^2}\, dx$

*In Problems 69–78, find each definite integral.*

**69.** $\displaystyle\int_{1}^{3} \frac{1}{x^2}\sqrt{1-\frac{1}{x}}\, dx$

**70.** $\displaystyle\int_{0}^{\pi/4} \frac{\sin(2x)}{\sqrt{5-2\cos(2x)}}\, dx$

**71.** $\displaystyle\int_{0}^{2} \frac{e^{2x}}{e^{2x}+1}\, dx$

**72.** $\displaystyle\int_{1}^{3} \frac{e^{3x}}{e^{3x}-1}\, dx$

**73.** $\displaystyle\int_{2}^{3} \frac{dx}{x \ln x}$

**74.** $\displaystyle\int_{2}^{3} \frac{dx}{x(\ln x)^2}$

**75.** $\displaystyle\int_{0}^{\pi} e^x \cos(e^x)\, dx$

**76.** $\displaystyle\int_{0}^{\pi} e^{-x} \cos(e^{-x})\, dx$

**77.** $\displaystyle\int_{0}^{1} \frac{x\, dx}{1+x^4}$

**78.** $\displaystyle\int_{0}^{1} \frac{e^x}{1+e^{2x}}\, dx$

*In Problems 79–86, use properties of even and odd functions to find each integral.*

**79.** $\displaystyle\int_{-2}^{2} (x^2-4)\, dx$

**80.** $\displaystyle\int_{-1}^{1} (x^3-2x)\, dx$

**81.** $\displaystyle\int_{-\pi/2}^{\pi/2} \frac{1}{3} \sin\theta\, d\theta$

**82.** $\displaystyle\int_{-\pi/4}^{\pi/4} \sec^2 x\, dx$

**83.** $\displaystyle\int_{-1}^{1} \frac{3}{1+x^2}\, dx$

**84.** $\displaystyle\int_{-5}^{5} (x^{1/3}+x)\, dx$

**85.** $\displaystyle\int_{-5}^{5} |2x|\, dx$

**86.** $\displaystyle\int_{-1}^{1} [|x|-3]\, dx$

**87.** If $f$ is an odd function and $\int_0^3 f(x)\, dx = 6$ and $\int_3^5 f(x)\, dx = 2$, find $\int_{-3}^{5} f(x)\, dx$.

**88.** If $f$ is an odd function and $\int_{-5}^{10} f(x)\, dx = 8$, find $\int_{5}^{10} f(x)\, dx$.

**89.** If $f$ is an even function and $\int_0^4 f(x)\, dx = -2$ and $\int_0^6 f(x)\, dx = 6$, find $\int_{-4}^{6} f(x)\, dx$.

**90.** If $f$ is an even function and $\int_{-3}^{0} f(x)\, dx = 4$ and $\int_{-5}^{0} f(x)\, dx = 1$, find $\int_{3}^{5} f(x)\, dx$.

**Applications and Extensions** ———————————

*In Problems 91 and 92, find each indefinite integral by*

**(a)** *Using substitution.*

**(b)** *Expanding the integrand.*

**91.** $\displaystyle\int (x+1)^2\, dx$

**92.** $\displaystyle\int (x^2+1)^2 x\, dx$

In Problems 93 and 94, find each integral three ways:

(a) By using substitution.

(b) By using even-odd properties of the definite integral.

(c) By using trigonometry to simplify the integrand before integrating.

(d) Verify the results are equivalent.

**93.** $\displaystyle\int_{-\pi/2}^{\pi/2} \cos(2x + \pi)\,dx$    **94.** $\displaystyle\int_{-\pi/4}^{\pi/4} \sin(7\theta - \pi)\,d\theta$

**95. Area**   Find the area under the graph of $f(x) = \dfrac{x^2}{\sqrt{2x+1}}$ from 0 to 4.

**96. Area**   Find the area under the graph of $f(x) = \dfrac{x}{(x^2+1)^2}$ from 0 to 2.

**97. Area**   Find the area under the graph of $y = \dfrac{1}{3x^2+1}$ from $x = 0$ to $x = 1$.

**98. Area**   Find the area under the graph of $y = \dfrac{1}{x\sqrt{x^2-4}}$ from $x = 3$ to $x = 4$.

**99. Area**   Find the area under the graph of the catenary,
$$y = a\cosh\frac{x}{a} + b - a$$
from $x = 0$ to $x = a$.

**100. Area**   Find $b$ so that the area under the graph of
$$y = (x+1)\sqrt{x^2 + 2x + 4}$$
is $\dfrac{56}{3}$ for $0 \le x \le b$.

**101. Average Value**   Find the average value of $y = \tan x$ on the interval $\left[0, \dfrac{\pi}{4}\right]$.

**102. Average Value**   Find the average value of $y = \sec x$ on the interval $\left[0, \dfrac{\pi}{4}\right]$.

**103.** If $\displaystyle\int_0^2 f(x-3)\,dx = 8$, find $\displaystyle\int_{-3}^{-1} f(x)\,dx$.

**104.** If $\displaystyle\int_{-2}^{1} f(x+1)\,dx = \dfrac{5}{2}$, find $\displaystyle\int_{-1}^{2} f(x)\,dx$.

**105.** If $\displaystyle\int_0^4 g\left(\dfrac{x}{2}\right)dx = 8$, find $\displaystyle\int_0^2 g(x)\,dx$.

**106.** If $\displaystyle\int_0^1 g(3x)\,dx = 6$, find $\displaystyle\int_0^3 g(x)\,dx$.

In Problems 107–112, find each integral.
(Hint: Begin by using a Change of Base formula.)

**107.** $\displaystyle\int \dfrac{dx}{x\log_{10}x}$    **108.** $\displaystyle\int \dfrac{dx}{x\log_3\sqrt[5]{x}}$    **109.** $\displaystyle\int_{10}^{100} \dfrac{dx}{x\log x}$

**110.** $\displaystyle\int_8^{32} \dfrac{dx}{x\log_2 x}$    **111.** $\displaystyle\int_3^9 \dfrac{dx}{x\log_3 x}$    **112.** $\displaystyle\int_{10}^{100} \dfrac{dx}{x\log_5 x}$

**113.** If $\displaystyle\int_1^b t^2(5t^3 - 1)^{1/2}\,dt = \dfrac{38}{45}$, find $b$.

**114.** If $\displaystyle\int_a^3 t\sqrt{9 - t^2}\,dt = 6$, find $a$.

In Problems 115–128, find each integral.

**115.** $\displaystyle\int \dfrac{x+1}{x^2+1}\,dx$    **116.** $\displaystyle\int \dfrac{2x-3}{1+x^2}\,dx$

**117.** $\displaystyle\int \left(2\sqrt{x^2+3} - \dfrac{4}{x} + 9\right)^6 \left(\dfrac{x}{\sqrt{x^2+3}} + \dfrac{2}{x^2}\right)dx$

**118.** $\displaystyle\int \left[\sqrt{(z^2+1)^4 - 3}\right]\left[z(z^2+1)^3\right]dz$

**119.** $\displaystyle\int \dfrac{x + 4x^3}{\sqrt{x}}\,dx$    **120.** $\displaystyle\int \dfrac{z\,dz}{z + \sqrt{z^2+4}}$

**121.** $\displaystyle\int \sqrt{t}\sqrt{4 + t\sqrt{t}}\,dt$    **122.** $\displaystyle\int_0^1 \dfrac{x+1}{x^2+3}\,dx$

**123.** $\displaystyle\int 3^{2x+1}\,dx$    **124.** $\displaystyle\int 2^{3x+5}\,dx$

**125.** $\displaystyle\int \dfrac{\sin x}{\sqrt{4 - \cos^2 x}}\,dx$    **126.** $\displaystyle\int \dfrac{\sec^2 x\,dx}{\sqrt{1 - \tan^2 x}}$

**127.** $\displaystyle\int_0^1 \dfrac{(z^2+5)(z^3 + 15z - 3)}{196 - (z^3 + 15z - 3)^2}\,dz$

**128.** $\displaystyle\int_2^{17} \dfrac{dx}{\sqrt{\sqrt{x-1} + (x-1)^{5/4}}}$

**129.** Use an appropriate substitution to show that
$$\int_0^1 x^m(1-x)^n\,dx = \int_0^1 x^n(1-x)^m\,dx$$
where $m$, $n$ are positive integers.

**130.** Find $\displaystyle\int_{-1}^1 f(x)\,dx$, for the function given below
$$f(x) = \begin{cases} x+1 & \text{if } x < 0 \\ \cos(\pi x) & \text{if } x \ge 0 \end{cases}$$

**131.** If $f$ is continuous on $[a, b]$, show that
$$\int_a^b f(x)\,dx = \int_a^b f(a + b - x)\,dx$$

**132.** If $\displaystyle\int_0^1 f(x)\,dx = 2$, find

(a) $\displaystyle\int_0^{0.5} f(2x)\,dx$    (b) $\displaystyle\int_0^3 f\left(\dfrac{1}{3}x\right)dx$

(c) $\displaystyle\int_0^{1/5} f(5x)\,dx$

(d) Find the upper and lower limits of integration so that
$$\int_a^b f\left(\dfrac{x}{4}\right)dx = 8$$

(e) Generalize (d) so that $\displaystyle\int_a^b f(kx)\,dx = \dfrac{1}{k}\cdot 2$ for $k > 0$.

**133.** If $\int_0^2 f(s)\,ds = 5$, find

**(a)** $\displaystyle\int_{-1}^{1} f(s+1)\,ds$    **(b)** $\displaystyle\int_{-3}^{-1} f(s+3)\,ds$

**(c)** $\displaystyle\int_{4}^{6} f(s-4)\,ds$

**(d)** Find upper and lower limits of integration so that

$$\int_a^b f(s-2)\,ds = 5$$

**(e)** Generalize (d) so that $\int_a^b f(s-k)\,ds = 5$ for $k > 0$.

**134.** Find $\int_0^b |2x|\,dx$ for any real number $b$.

**135.** If $f$ is an odd function, show that $\int_{-a}^{a} f(x)\,dx = 0$.

**136.** Find the constant $k$, where $0 \le k \le 3$, for which

$$\int_0^3 \frac{x}{\sqrt{x^2+16}}\,dx = \frac{3k}{\sqrt{k^2+16}}$$

**137.** If $n$ is a positive integer, for what number $c > 0$ is

$$\int_0^c x^{n-1}\,dx = \frac{1}{n}?$$

**138.** If $f$ is a continuous function defined on the interval $[0, 1]$, show that

$$\int_0^\pi x\,f(\sin x)\,dx = \frac{\pi}{2}\int_0^\pi f(\sin x)\,dx$$

**139.** Prove that $\int \csc x\,dx = \ln|\csc x - \cot x| + C$.
[*Hint:* Multiply and divide by $(\csc x - \cot x)$.]

**140.** Describe a method for finding $\int_a^b |f(x)|\,dx$ in terms of $F(x) = \int f(x)\,dx$ when $f$ has finitely many zeros.

**141.** Find $\int \sqrt[n]{a+bx}\,dx$ where $a$ and $b$ are real numbers, $b \ne 0$, and $n \ge 2$ is an integer.

**142.** If $f$ is continuous for all $x$, which of the following integrals have the same value?

**(a)** $\displaystyle\int_a^b f(x)\,dx$    **(b)** $\displaystyle\int_0^{b-a} f(x+a)\,dx$

**(c)** $\displaystyle\int_{a+c}^{b+c} f(x+c)\,dx$

**143.** **Area**    Let $f(x) = k\sin(kx)$, where $k$ is a positive constant.

**(a)** Find the area of the region under one arch of the graph of $f$.

**(b)** Find the area of the triangle formed by the $x$-axis and the tangent lines to one arch of $f$ at the points where the graph of $f$ crosses the $x$-axis.

**144.** Prove that $\int [f(x) + g(x)]\,dx = \int f(x)\,dx + \int g(x)\,dx$.

**145.** Derive the integration formula $\displaystyle\int a^x\,dx = \frac{a^x}{\ln a} + C,\ a > 0$,
$a \ne 1$. (*Hint:* Begin with the derivative of $y = a^x$.)

**146.** Use the formula from Problem 145 to find

**(a)** $\displaystyle\int 2^x\,dx$    **(b)** $\displaystyle\int 3^x\,dx$

**147.** **Electric Potential**    The electric field strength a distance $z$ from the axis of a ring of radius $R$ carrying a charge $Q$ is given by the formula

$$E(z) = \frac{Qz}{(R^2+z^2)^{3/2}}$$

If the electric potential $V$ is related to $E$ by $E = -\dfrac{dV}{dz}$, what is $V(z)$?

**148.** **Impulse During a Rocket Launch**    The impulse $J$ due to a force $F$ is the product of the force times the amount of time $t$ for which the force acts. When the force varies over time,

$$J = \int_{t_1}^{t_2} F(t)\,dt.$$

We can model the force acting on a rocket during launch by an exponential function $F(t) = Ae^{bt}$, where $A$ and $b$ are constants that depend on the characteristics of the engine. At the instant lift-off occurs ($t = 0$), the force must equal the weight of the rocket.

NASA/Rick Wetherington and Tony Gray

**(a)** Suppose the rocket weighs 25,000 N (a mass of about 2500 kg or a weight of 5500 lb), and 30 s after lift-off the force acting on the rocket equals twice the weight of the rocket. Find $A$ and $b$.

**(b)** Find the impulse delivered to the rocket during the first 30 seconds after the launch.

## Challenge Problems

**149.** **Air Resistance on a Falling Object**    If an object of mass $m$ is dropped, the air resistance on it when it has speed $v$ can be modeled as $F_{air} = -kv$, where the constant $k$ depends on the shape of the object and the condition of the air. The minus sign is necessary because the direction of the force is opposite to the velocity. Using Newton's Second Law of Motion, this force leads to a downward acceleration $a(t) = ge^{-kt/m}$. (You are asked to prove this in Problem 150.) Using the equation for $a(t)$, find:

**(a)** $v(t)$, if the object starts from rest $v_0 = v(0) = 0$, with the positive direction downward.

**(b)** $s(t)$, if the object starts from the position $s_0 = s(0) = 0$, with the positive direction downward.

**(c)** What limits do $a(t)$, $v(t)$, and $s(t)$ approach if the object falls for a very long time ($t \to \infty$)? Interpret each result and explain if it is physically reasonable.

**(d)** Graph $a = a(t)$, $v = v(t)$, $s = s(t)$. Do the graphs support the conclusions obtained in part (c)? Use $g = 9.8$ m/s$^2$, $k = 5$, and $m = 10$ kg.

**150. Air Resistance on a Falling Object** (Refer to Problem 149.) If an object of mass $m$ is dropped, the air resistance on it when it has speed $v$ can be modeled as

$$F_{air} = -kv$$

where the constant $k$ depends on the shape of the object and the condition of the air. The minus sign is necessary because the direction of the force is opposite to the velocity. Using Newton's Second Law of Motion, show that the downward acceleration of the object is

$$a(t) = ge^{-kt/m}$$

where $g$ is the acceleration due to gravity.
(*Hint:* The velocity of the object obeys the differential equation

$$m\frac{dv}{dt} = mg - kv$$

Solve the differential equation for $v$ and use the fact that $ma = mg - kv$.)

**151.** Find $\displaystyle\int \frac{x^6 + 3x^4 + 3x^2 + x + 1}{(x^2 + 1)^2}\, dx$.

**152.** Find $\displaystyle\int \frac{\sqrt[4]{x}}{\sqrt{x} + \sqrt[3]{x}}\, dx$.

**153.** Find $\displaystyle\int \frac{3x + 2}{x\sqrt{x + 1}}\, dx$.

**154.** Find $\displaystyle\int \frac{dx}{(x \ln x)[\ln(\ln x)]}$.

**155. (a)** Find $y'$ if $y = \ln\left[\tan\left(\dfrac{x}{2} + \dfrac{\pi}{4}\right)\right]$.

**(b)** Use the result to show that

$$\int \sec x\, dx = \ln\left[\tan\left(\frac{x}{2} + \frac{\pi}{4}\right)\right] + C$$

**(c)** Show that $\ln\left[\tan\left(\dfrac{x}{2} + \dfrac{\pi}{4}\right)\right] = \ln|\sec x + \tan x|$.

**156. (a)** Find $y'$ if $y = x \sin^{-1} x + \sqrt{1 - x^2}$.

**(b)** Use the result to show that

$$\int \sin^{-1} x\, dx = x \sin^{-1} x + \sqrt{1 - x^2} + C$$

**157. (a)** Find $y'$ if $y = \dfrac{1}{2}x\sqrt{a^2 - x^2} + \dfrac{1}{2}a^2 \sin^{-1}\left(\dfrac{x}{a}\right)$, $a > 0$.

**(b)** Use the result to show that

$$\int \sqrt{a^2 - x^2}\, dx = \frac{1}{2}x\sqrt{a^2 - x^2} + \frac{1}{2}a^2 \sin^{-1}\left(\frac{x}{a}\right) + C$$

**158. (a)** Find $y'$ if $y = \ln|\csc x - \cot x|$.

**(b)** Use the result to show that

$$\int \csc x\, dx = \ln|\csc x - \cot x| + C$$

**159. Gudermannian Function**

**(a)** Graph $y = \text{gd}(x) = \tan^{-1}(\sinh x)$; this is called the **gudermannian** of $x$ (named after Christoph Gudermann).

**(b)** If $y = \text{gd}(x)$, show that $\cos y = \text{sech } x$ and $\sin y = \tanh x$.

**(c)** Show that if $y = \text{gd}(x)$, then $y$ satisfies the differential equation $y' = \cos y$.

**(d)** Use the differential equation of (c) to obtain the formula

$$\int \sec y\, dy = \text{gd}^{-1}(y) + C$$

Compare this to $\int \sec x\, dx = \ln|\sec x + \tan x| + C$.

**160.** The formula $\dfrac{d}{dx}\displaystyle\int f(x)\, dx = f(x)$ states that if a function is integrated and the result is differentiated, the original function is returned. What about the other way around? Is the formula $\int f'(x)\, dx = f(x)$ correct? Be sure to justify your answer.

# 5.6 Separable First-Order Differential Equations; Uninhibited and Inhibited Growth and Decay Models

**OBJECTIVES** *When you finish this section, you should be able to:*

1 Solve a separable first-order differential equation (p. 413)
2 Solve differential equations involving uninhibited growth and decay (p. 414)
3 Solve differential equations involving inhibited growth and decay (p. 417)
   Application: Newton's Law of Cooling (p. 418)

A **first-order differential equation** is of the form

$$\frac{dy}{dx} = f(x, y) \tag{1}$$

where $f$ is continuous on its domain. As the name suggests, first-order differential equations only contain the first derivative of $y$ with respect to $x$. In this section we discuss how to solve a certain class of first-order differential equations, called *separable first-order differential equations*, and discuss two applications involving unhibited growth and decay and inhibited growth and decay.

## 1 Solve a Separable First-Order Differential Equation

In equation (1), suppose $f(x,y)$ can be written in the form

$$f(x, y) = \frac{M(x)}{N(y)}$$

where $M$ is a function of $x$ alone, $N$ is a function of $y$ alone, and both $M$ and $N$ are continuous on their domains. Then the first-order differential equation (1) is called a separable *first-order differential equation.*

**DEFINITION  Separable First-Order Differentiable Equation**

A first-order differential equation is said to be **separable** if it can be written in the form

$$\frac{dy}{dx} = \frac{M(x)}{N(y)} \qquad (2)$$

where $M$ is a function of $x$ alone, $N$ is a function of $y$ alone, and both $M$ and $N$ are continuous on their domains. In terms of differentials, equation (2) takes the form

$$N(y)dy = M(x)dx \qquad (3)$$

For example, the differential equation

$$\frac{dy}{dx} = x \sec y$$

is separable since if we treat $dy$ and $dx$ as differentials, $\dfrac{dy}{dx} = x \sec y$ can be written as

$$\frac{dy}{\sec y} = x \, dx$$

The differential equation

$$\frac{dy}{dx} = e^x \sin(xy)$$

is not separable since it cannot be written in the form $N(y)dy = M(x)dx$.

To solve the differential equation (2), we "*separate*" the variables and express $\dfrac{dy}{dx} = \dfrac{M(x)}{N(y)}$ in terms of differentials as

$$N(y)dy = M(x)dx$$

Now integrate both sides of the separated equation

$$\int N(y) \, dy = \int M(x) \, dx$$

to obtain the general solution.

CALC
CLIP

**EXAMPLE 1  Solving a Separable First-Order Differential Equation**

**(a)** Find the general solution of the separable first-order differential equation

$$\frac{dy}{dx} = 3xy^2$$

**(b)** Find the particular solution that $y = 2$ when $x = 1$.

**NOTE** In the differential equation

$$\frac{dy}{dx} = 3xy^2$$

$y$ cannot be 0.

**Solution** (a) Begin with $\frac{dy}{dx} = 3xy^2$ and separate the variables to express $\frac{dy}{dx}$ using differentials.

$$\frac{dy}{y^2} = 3x\,dx$$

$$\int \frac{dy}{y^2} = \int 3x\,dx \qquad \text{Integrate both sides.}$$

$$-\frac{1}{y} = \frac{3x^2}{2} + C$$

This is the general solution to the differential equation, although it can be written in many equivalent forms. The domain of the general solution is restricted to points $(x, y)$, where $y \neq 0$.

(b) We use the general solution with $x = 1$ and $y = 2$ to find the particular solution of the differential equation.

$$-\frac{1}{y} = \frac{3x^2}{2} + C$$

$$-\frac{1}{2} = \frac{3 \cdot 1^2}{2} + C \qquad x = 1, \, y = 2$$

$$-\frac{1}{2} = \frac{3}{2} + C$$

$$C = -2$$

The particular solution that $y = 2$ when $x = 1$ is

$$-\frac{1}{y} = \frac{3x^2}{2} - 2$$

which we can write as

$$3x^2 y - 4y + 2 = 0 \qquad y \neq 0 \qquad \blacksquare$$

NOW WORK Problems 3 and 15.

## 2 Solve Differential Equations Involving Uninhibited Growth and Decay

There are situations in science, nature, and economics such as radioactive decay, population growth, and interest paid on an investment, in which a quantity $A$ varies with time $t$ in such a way that the rate of change of $A$ with respect to $t$ is proportional to $A$ itself. These situations can be modeled by the differential equation

$$\boxed{\frac{dA}{dt} = kA} \qquad (4)$$

where $k \neq 0$ is a real number.

- If $k > 0$ and $\frac{dA}{dt} = kA$, the rate of change of $A$ with respect to $t$ is positive, and the amount $A$ is increasing. (Then $\frac{dA}{dt}$ is a **growth model**.)

- If $k < 0$ and $\frac{dA}{dt} = kA$, the rate of change of $A$ with respect to $t$ is negative, and the amount $A$ is decreasing. (Then $\frac{dA}{dt}$ is a **decay model**.)

Suppose the rate of change of a quantity $A$ is proportional to $A$. If the initial amount $A_0$ of the quantity is known, then we have the boundary condition, or initial condition, $A = A(0) = A_0$ when $t = 0$.

We solve the differential equation $\dfrac{dA}{dt} = kA$ by separating the variables.

$$\frac{dA}{dt} = k\, dt \qquad \text{Separate the variables.}$$

$$\int \frac{1}{A}\, dA = \int k\, dt \qquad \text{Integrate both sides.}$$

$$\ln |A| = kt + C$$

$$\ln A = kt + C \qquad A > 0$$

The initial condition requires that $A = A_0$ when $t = 0$. Then $\ln A_0 = C$, so

$$\ln A = kt + \ln A_0$$

$$\ln A - \ln A_0 = kt$$

$$\ln \frac{A}{A_0} = kt$$

$$\frac{A}{A_0} = e^{kt}$$

$$A = A_0 e^{kt}$$

The solution to the differential equation $\dfrac{dA}{dt} = kA$ is

$$\boxed{A = A_0 e^{kt}} \tag{5}$$

where $A_0$ is the initial amount.

Functions $A = A(t)$ whose rates of change are $\dfrac{dA}{dt} = kA$ are said to follow the **exponential law,** or the **law of uninhibited growth or decay**—or in a business context, **the law of continuously compounded interest**. Figure 35 shows the graphs of the function $A(t) = A_0 e^{kt}$ for both $k > 0$ and $k < 0$.

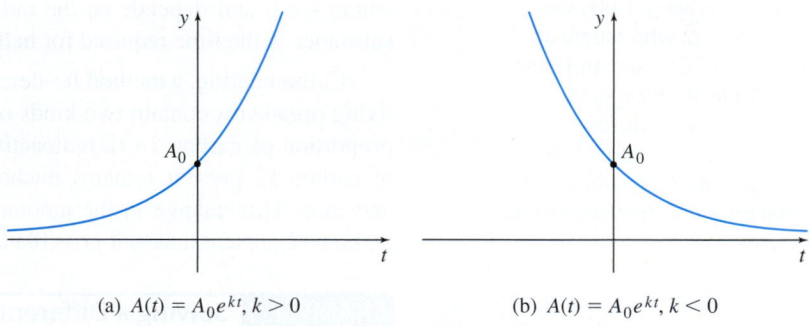

(a)  $A(t) = A_0 e^{kt}, k > 0$          (b)  $A(t) = A_0 e^{kt}, k < 0$

**Figure 35**

---

**EXAMPLE 2** **Solving a Differential Equation: Uninhibited Growth**

Assume that a colony of bacteria grows at a rate proportional to the number of bacteria present. If the number of bacteria doubles in 5 hours (h), how long will it take for the number of bacteria to triple?

**NOTE** Example 2 is a model of uninhibited growth; it accurately reflects growth in early stages. After a time, bacterial growth no longer continues at a rate proportional to the number present. Factors, such as disease, lack of space, and dwindling food supply, begin to affect the rate of growth.

**Solution** Let $N(t)$ be the number of bacteria present at time $t$. Then the assumption that this colony of bacteria grows at a rate proportional to the number present can be modeled by

$$\frac{dN}{dt} = kN$$

where $k$ is a positive constant of proportionality. This differential equation is of form (4), and its solution is given by (5). So, we have

$$N(t) = N_0 \, e^{kt}$$

where $N_0$ is the initial number of bacteria in the colony. To find $k$, we use the fact that the initial number of bacteria $N_0$ doubles to $2N_0$ in 5 h. That is, $N(5) = 2N_0$.

$$N(5) = N_0 \, e^{5k} = 2N_0$$
$$e^{5k} = 2$$
$$5k = \ln 2$$
$$k = \frac{1}{5} \ln 2$$

The time $t$ required for this colony to triple obeys the equation

$$N(t) = 3N_0$$
$$N_0 e^{kt} = 3N_0$$
$$e^{kt} = 3$$
$$t = \frac{1}{k} \ln 3 = 5 \frac{\ln 3}{\ln 2} \approx 8$$
$$\underset{\underset{k = \frac{1}{5} \ln 2}{\uparrow}}{}$$

The number of bacteria will triple in about 8 h. ∎

NOW WORK Problem **23.**

For a radioactive substance, the **rate of decay** is proportional to the amount of substance present at a given time $t$. That is, if $A = A(t)$ represents the amount of a radioactive substance at time $t$, then

$$\boxed{\frac{dA}{dt} = kA}$$

where $k < 0$ and depends on the radioactive substance. The **half-life** of a radioactive substance is the time required for half of the substance to decay.

Carbon dating, a method for determining the age of an artifact, uses the fact that all living organisms contain two kinds of carbon: carbon-12 (a stable carbon) and a small proportion of carbon-14 (a radioactive isotope). When an organism dies, the amount of carbon-12 present remains unchanged, while the amount of carbon-14 begins to decrease. This change in the amount of carbon-14 present relative to the amount of carbon-12 present makes it possible to calculate how long ago the organism died.

**ORIGINS** Willard F. Libby (1908–1980) grew up in California and went to college and graduate school at UC Berkeley. Libby was a physical chemist who taught at the University of Chicago and later at UCLA. While at Chicago, he developed the methods for using natural carbon-14 to date archaeological artifacts. Libby won the Nobel Prize in Chemistry in 1960 for this work.

### EXAMPLE 3 Solving a Differential Equation: Uninhibited Decay

The skull of an animal found in an archaeological dig contains about 20% of the original amount of carbon-14. If the half-life of carbon-14 is 5730 years, how long ago did the animal die?

**Solution** Let $A = A(t)$ be the amount of carbon-14 present in the skull at time $t$. Then $A$ satisfies the differential equation $\dfrac{dA}{dt} = kA$, whose solution is

$$A = A_0 e^{kt}$$

where $A_0$ is the amount of carbon-14 present at time $t = 0$. To determine the constant $k$, use the fact that when $t = 5730$, half of the original amount $A_0$ remains.

$$\frac{1}{2}A_0 = A_0 e^{5730k}$$

$$\frac{1}{2} = e^{5730k}$$

$$5730k = \ln\frac{1}{2} = -\ln 2$$

$$k = -\frac{\ln 2}{5730}$$

The relationship between the amount $A$ of carbon-14 present and the time $t$ is

$$A(t) = A_0 e^{(-\ln 2/5730)t}$$

In this skull, 20% of the original amount of carbon-14 remains, so $A(t) = 0.20A_0$.

$$0.20A_0 = A_0 e^{(-\ln 2/5730)t}$$

$$0.20 = e^{(-\ln 2/5730)t}$$

Now take the natural logarithm of both sides.

$$\ln 0.20 = -\frac{\ln 2}{5730} \cdot t$$

$$t = -5730 \cdot \frac{\ln 0.20}{\ln 2} \approx 13,300$$

The animal died approximately 13,300 years ago. ∎

**NOW WORK** Problem 31.

## 3 Solve Differential Equations Involving Inhibited Growth and Decay

In uninhibited growth or decay, the amount $y$ either grows without bound or decays to zero. In many situations there is an upper value $M$ that the amount $y$ cannot exceed (inhibited growth), or there is a lower value $M$ that $y$ cannot go below (inhibited decay). For inhibited growth or decay models the rate of change of $y$ with respect to time $t$ satisfies a different differential equation.

- Inhibited Growth Model:

$$\frac{dy}{dt} = k(M - y) \qquad y < M$$

where $k > 0$ and $M > 0$ are constants. Since $k > 0$ and $y < M$, the derivative $\frac{dy}{dt} > 0$ and $y = y(t)$ is increasing. Also, $y < M$, guarantees that the amount $y$ will never be greater than $M$. Since $y''(t) < 0$, the graph of $y = y(t)$ will be concave down.

- Inhibited Decay Model:

$$\frac{dy}{dt} = k(y - M) \qquad y > M$$

where $k < 0$ and $M > 0$ are constants. Since $k < 0$ and $y > M$, the derivative $\frac{dy}{dt} < 0$ and $y = y(t)$ is decreasing. Also, $y > M$ guarantees that the amount $y$ will never be less than $M$. Since $y''(t) > 0$, the graph of $y = y(t)$ will be concave up.

Figures 36 (a) and (b) show the graphs of the solutions of the differential equations for inhibited growth and decay. In each graph the line $y = M$ is a horizontal asymptote.

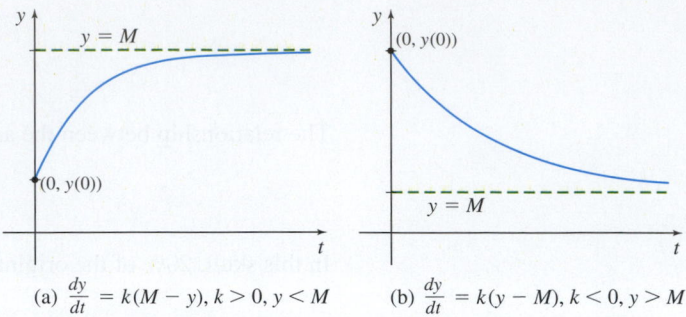

(a) $\dfrac{dy}{dt} = k(M - y), k > 0, y < M$     (b) $\dfrac{dy}{dt} = k(y - M), k < 0, y > M$

**Figure 36**

Newton's Law of Cooling is an example of an inhibited decay model.

## Application: Newton's Law of Cooling

Suppose an object is heated to a temperature $u_0$. Then at time $t = 0$, the object is put into a medium with a constant lower temperature causing the object to cool. Newton's Law of Cooling states that the rate of change of the temperature of the object with respect to time is continuous and proportional to the difference between the temperature of the object and the ambient temperature (the temperature of the surrounding medium). That is, if $u = u(t)$ is the temperature of the object at time $t$ and if $T$ is the (constant) ambient temperature, then Newton's Law of Cooling is modeled by the differential equation

$$\boxed{\frac{du}{dt} = k[u(t) - T]} \tag{6}$$

where $k$ is a constant that depends on the object. Since the ambient temperature $T$ is lower than $u(0) = u_0$, the object cools and its temperature decreases so that $\dfrac{du}{dt} < 0$. Then, since $T < u(t)$, $k$ is a negative constant.

We find $u$ as a function of $t$ by separating the variables and solving the differential equation $\dfrac{du}{dt} = k(u - T)$.

$$\frac{du}{u - T} = k\,dt \qquad \text{Separate the variables.}$$

$$\int \frac{du}{u - T} = \int k\,dt \qquad \text{Integrate both sides.}$$

$$\ln |u - T| = kt + C$$

To find $C$, use the boundary condition that at time $t = 0$, the initial temperature of the object is $u(0) = u_0$, Then

$$\ln |u_0 - T| = k \cdot 0 + C$$

$$C = \ln |u_0 - T|$$

Using this expression for $C$, we obtain

$$\ln|u - T| = kt + \ln|u_0 - T|$$

$$\ln|u - T| - \ln|u_0 - T| = kt$$

$$\ln\left|\frac{u - T}{u_0 - T}\right| = kt$$

$$\frac{u - T}{u_0 - T} = e^{kt} \qquad\qquad T < u_0; \ T < u$$

$$u - T = (u_0 - T)e^{kt}$$

$$\boxed{u(t) = (u_0 - T)\,e^{kt} + T} \tag{7}$$

### EXAMPLE 4    Using Newton's Law of Cooling

An object is heated to $90\,°\text{C}$ and allowed to cool in a room with a constant ambient temperature of $20\,°\text{C}$. If after 10 min the temperature of the object is $60\,°\text{C}$, what will its temperature be after 20 min?

**Solution**  When $t = 0$, $u(0) = u_0 = 90\,°\text{C}$, and when $t = 10$ min, $u(10) = 60\,°\text{C}$. Given that the ambient temperature $T$ is $20\,°\text{C}$, we substitute these values into equation (7).

$$u(t) = (u_0 - T)e^{kt} + T \qquad (7)$$

$$60 = (90 - 20)e^{10k} + 20 \qquad u = 60 \text{ when } t = 10; \ T = 20; \ u_0 = 90$$

$$\frac{40}{70} = e^{10k}$$

$$k = \frac{1}{10}\ln\frac{4}{7} = 0.10\ln\frac{4}{7}$$

The temperature $u = u(t)$ is

$$u(t) = 70e^{[0.1\ln(4/7)]t} + 20$$

Then when $t = 20$, the temperature $u = u(t)$ of the object is

$$u(20) = 70e^{[0.1\ln(4/7)](20)} + 20 = 70e^{2\ln(4/7)} + 20 \approx 42.86\,°\text{C}$$

See Figure 37 for the graph of $u = u(t)$. ∎

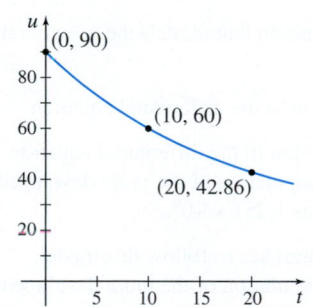

**Figure 37** $u(t) = 70e^{[0.1\ln(4/7)]t} + 20,$
$0 \le t \le 20$

NOW WORK Problem 37.

## 5.6 Assess Your Understanding

### Concepts and Vocabulary

1. *True or False* The differential equation $\dfrac{dy}{dx} = x$ can be solved by separating the variables.

2. *True or False* Every first-order differential equation can be solved by separating the variables.

### Skill Building

*In Problems 3–14, find the general solution of each differential equation.*

3. $\dfrac{dy}{dx} = xy$

4. $\dfrac{dy}{dx} = 5x^2 y^3$

5. $\dfrac{dy}{dx} = \sqrt{xy^2}$

6. $\dfrac{dy}{dx} = \sqrt{xy}$

7. $\dfrac{dy}{dx} = xe^y$

8. $\dfrac{dy}{dx} = 3ye^x$

9. $\dfrac{dy}{dx} = e^x \sin(2y)$

10. $\dfrac{dy}{dx} = e^{3y}\cos x$

11. $\dfrac{dy}{dx} = 5e^{2x-y}$

12. $\dfrac{dy}{dx} = -2e^{5x+y}$

13. $\dfrac{dy}{dx} = x^2(1+y^2)$

14. $\dfrac{dy}{dx} = x^3\sqrt{1-y^2}$

---

**1.** = NOW WORK problem          = Graphing technology recommended          CAS = Computer Algebra System recommended

In Problems 15–22, find the particular solution of each differential equation using the given boundary condition.

**15.** $\dfrac{dy}{dx} = xy^{1/2}$, $y = 1$ when $x = 2$

**16.** $\dfrac{dy}{dx} = x^{1/2}y$, $y = 1$ when $x = 0$

**17.** $\dfrac{dy}{dx} = \dfrac{y-1}{x-1}$, $y = 2$ when $x = 2$

**18.** $\dfrac{dy}{dx} = \dfrac{y}{x}$, $y = 2$ when $x = 1$

**19.** $\dfrac{dy}{dx} = \dfrac{x}{\cos y}$, $y = \pi$ when $x = 2$

**20.** $\dfrac{dy}{dx} = y\sin x$, $y = e$ when $x = 0$

**21.** $\dfrac{dy}{dx} = \dfrac{4e^x}{y}$, $y = 2$ when $x = 0$

**22.** $\dfrac{dy}{dx} = 5ye^x$, $y = 1$ when $x = 0$

## Applications and Extensions

**23. Uninhibited Growth**   The population of a colony of mosquitoes obeys the uninhibited growth equation $\dfrac{dN}{dt} = kN$. If there are 1500 mosquitoes initially, and there are 2500 mosquitoes after 24 h, what is the mosquito population after 3 days?

**24. Radioactive Decay**   A radioactive substance follows the decay equation $\dfrac{dA}{dt} = kA$. If 25% of the substance disappears in 10 years, what is its half-life?

**25. Population Growth**   The population of a suburb grows at a rate proportional to the population. Suppose the population doubles in size from 4000 to 8000 in an 18-month period and continues at the current rate of growth.

(a) Write a differential equation that models the population $P$ at time $t$ in months.

(b) Find the general solution to the differential equation.

(c) Find the particular solution to the differential equation with the initial condition $P(0) = 4000$.

(d) What will the population be in 4 years [$t = 48$]?

**26. Uninhibited Growth**   At any time $t$ in hours, the rate of increase in the area, in millimeters squared ($\text{mm}^2$), of a culture of bacteria is twice the current area $A$ of the culture.

(a) Write a differential equation that models the area of the culture at time $t$.

(b) Find the general solution to the differential equation.

(c) Find the particular solution to the differential equation if $A = 10\,\text{mm}^2$, when $t = 0$.

**27. Radioactive Decay**   The amount $A$ of the radioactive element radium in a sample decays at a rate proportional to the amount of radium present. The half-life of radium is 1690 years.

(a) Write a differential equation that models the amount $A$ of radium present at time $t$.

(b) Find the general solution to the differential equation.

(c) Find the particular solution to the differential equation with the initial condition $A(0) = 8$ g.

(d) How much radium will be present in the sample in 100 years?

**28. Radioactive Decay**   Carbon-14 is a radioactive element present in living organisms. After an organism dies, the amount $A$ of carbon-14 present begins to decline at a rate proportional to the amount present at the time of death. The half-life of carbon-14 is 5730 years.

(a) Write a differential equation that models the rate of decay of carbon-14.

(b) Find the general solution to the differential equation.

(c) A piece of fossilized charcoal is found that contains 30% of the carbon-14 that was present when the tree it came from died. How long ago did the tree die?

**29. World Population Growth**   Barring disasters (human-made or natural), the population $P$ of humans grows at a rate proportional to its current size. According to the U.N. World Population studies, from 2010 to 2015 the population of the more developed regions of the world (Europe, North America, Australia, New Zealand, and Japan) grew at an annual rate of 0.289% per year.

(a) Write a differential equation that models the growth rate of the population.

(b) Find the general solution to the differential equation.

(c) Find the particular solution to the differential equation if in 2015 ($t = 0$), the population of the more developed regions of the world was $1.251 \times 10^9$.

(d) If the rate of growth continues to follow this model, what is the projected population of the more developed regions in 2025?

*Source*: U.N. World Population Prospects: The 2015 Revision.

**30. Population Growth in Ecuador**   Barring disasters (human-made or natural), the population $P$ of humans grows at a rate proportional to its current size. According to the U.N. World Population studies, from 2010 to 2015 the population of Ecuador grew at an annual rate of 1.558% per year. Assuming this growth rate continues:

(a) Write a differential equation that models the growth rate of the population.

(b) Find the general solution to the differential equation.

(c) Find the particular solution to the differential equation if in 2015 ($t = 0$), the population of Ecuador was $1.614 \times 10^7$.

(d) If the rate of growth continues to follow this model, when will the projected population of Ecuador reach 20 million persons?

*Source*: U.N. World Population Prospects: The 2015 Revision.

**31. Oetzi the Iceman** was found in 1991 by a German couple who were hiking in the Alps near the border of Austria and Italy. Carbon-14 testing determined that Oetzi died 5300 years ago. Assuming the half-life of carbon-14 is 5730 years, what percent of carbon-14 was left in his body? (An interesting note: In September 2010 the complete genome mapping of Oetzi was completed.)

**32. Uninhibited Decay** Radioactive beryllium is sometimes used to date fossils found in deep-sea sediments. (Carbon-14 dating cannot be used for fossils that lived underwater.) The decay of beryllium satisfies the equation $\dfrac{dA}{dt} = -\alpha A$, where $\alpha = 1.5 \times 10^{-7}$ and $t$ is measured in years. What is the half-life of beryllium?

**33. Decomposition of Sucrose** Reacting with water in an acidic solution at $35\,°C$, sucrose ($C_{12}H_{22}O_{11}$) decomposes into glucose ($C_6H_{12}O_6$) and fructose ($C_6H_{12}O_6$) according to the law of uninhibited decay. An initial amount of 0.4 mol of sucrose decomposes to 0.36 mol in 30 min. How much sucrose will remain after 2 h? How long will it take until 0.10 mol of sucrose remains?

**34. Chemical Dissociation** Salt (NaCl) dissociates in water into sodium ($Na^+$) and chloride ($Cl^-$) ions at a rate proportional to its mass. The initial amount of salt is 25 kg, and after 10 h, 15 kg are left.

   (a) How much salt will be left after 1 day?

   (b) After how many hours will there be less than $\dfrac{1}{2}$ kg of salt left?

**35. Voltage Drop** The voltage of a certain condenser decreases at a rate proportional to the voltage. If the initial voltage is 20, and 2 s later it is 10, what is the voltage at time $t$? When will the voltage be 5?

**36. Uninhibited Growth** The rate of change in the number of bacteria in a culture is proportional to the number present. In a certain laboratory experiment, a culture has 10,000 bacteria initially, 20,000 bacteria at time $t_1$ minutes, and 100,000 bacteria at $(t_1 + 10)$ minutes.

   (a) In terms of $t$ only, find the number of bacteria in the culture at any time $t$ minutes ($t \geq 0$).

   (b) How many bacteria are there after 20 min?

   (c) At what time are 20,000 bacteria observed? That is, find the value of $t_1$.

**37. Newton's Law of Heating** Newton's Law of Heating states that the rate of change of temperature with respect to time is proportional to the difference between the temperature of the object and the ambient temperature. A thermometer that reads $4\,°C$ is brought into a room that is $30\,°C$.

   (a) Write the differential equation that models the temperature $u = u(t)$ of the thermometer at time $t$ in minutes (min).

   (b) Find the general solution of the differential equation.

   (c) If the thermometer reads $10\,°C$ after 2 min, determine the temperature reading 5 min after the thermometer is first brought into the room.

**38. Newton's Law of Cooling** A thermometer reading $70\,°F$ is taken outside where the ambient temperature is $22\,°F$. Four minutes later the reading is $32\,°F$.

   (a) Write the differential equation that models the temperature $u = u(t)$ of the thermometer at time $t$.

   (b) Find the general solution of the differential equation.

   (c) Find the particular solution to the differential equation, using the initial condition that when $t = 0$ min, then $u = 70\,°F$.

   (d) Find the thermometer reading 7 min after the thermometer was brought outside.

   (e) Find the time it takes for the reading to change from $70\,°F$ to within $\dfrac{1}{2}\,°F$ of the air temperature.

**39. Forensic Science** At 4 p.m., a body was found floating in $12\,°C$ water. When the woman was alive, her body temperature was $37\,°C$ and now it is $20\,°C$. Suppose the rate of change of the temperature $u = u(t)$ of the body with respect to the time $t$ in hours (h) is proportional to $u(t) - T$, where $T$ is the water temperature and the constant of proportionality is $-0.159$.

   (a) Write a differential equation that models the temperature $u = u(t)$ of the body at time $t$.

   (b) Find the general solution of the differential equation.

   (c) Find the particular solution of the differential equation, using the initial condition that at the time of death, when $t = 0$ h, her body temperature was $u = 37\,°C$.

   (d) At what time did the woman drown?

   (e) How long does it take for the woman's body to cool to $15°C$?

**40. Newton's Law of Cooling** A pie is removed from a $350\,°F$ oven to cool in a room whose temperature is $72\,°F$.

   (a) Write the differential equation that models the temperature $u = u(t)$ of the pie at time $t$.

   (b) Find the general solution of the differential equation.

   (c) Find the particular solution to the differential equation, using the initial condition that when $t = 0$ min, then $u = 350\,°F$.

   (d) If $u(5) = 200\,°F$, what is the temperature of the pie after 15 min?

   (e) How long will it take for the pie to be $100\,°F$ and ready to eat?

# Chapter Review

## THINGS TO KNOW

### 5.1 Area

**Definitions:**

- Partition of an interval $[a, b]$ (p. 359)
- Area $A$ under the graph of a function $f$ from $a$ to $b$ (p. 361)

### 5.2 The Definite Integral

**Definitions:**

- Riemann sums (p. 367)
- The definite integral (p. 369)
- $\displaystyle\int_a^a f(x)\,dx = 0$ (p. 370)
- $\displaystyle\int_a^b f(x)\,dx = -\int_b^a f(x)\,dx$ (p. 370)
- Bounded Function (p. 371)

**Theorems:**

- If a function $f$ is continuous on a closed interval $[a, b]$, then the definite integral $\int_a^b f(x)\,dx$ exists. (p. 371)
- If a function $f$ is bounded on a closed interval $[a, b]$ and has at most a finite number of points of discontinuity on $[a, b]$, then $\int_a^b f(x)\,dx$ exists. (p. 372)
- $\displaystyle\int_a^b h\,dx = h(b - a)$, $h$ a constant (p. 372)

### 5.3 The Fundamental Theorem of Calculus

**Fundamental Theorem of Calculus:**    Let $f$ be a function that is continuous on a closed interval $[a, b]$.

- Part 1: The function $I$ defined by $I(x) = \int_a^x f(t)\,dt$ has the properties that it is continuous on $[a, b]$ and differentiable on $(a, b)$. Moreover, $I'(x) = \dfrac{d}{dx}\left[\int_a^x f(t)\,dt\right] = f(x)$, for all $x$ in $(a, b)$. (p. 378)

- Part 2: If $F$ is any antiderivative of $f$ on $[a, b]$, then

$$\int_a^b f(x)\,dx = F(b) - F(a) \text{ (p. 379)}$$

- The integral from $a$ to $b$ of the rate of change of $F$ equals the change in $F$ from $a$ to $b$. That is, $\int_a^b F'(x)\,dx = F(b) - F(a)$. (p. 381)
- The accumulation function $F$ associated with a function $f$ that is continuous on a closed interval $[a, b]$ is

$$F(x) = \int_a^x f(t)\,dt \quad a \leq x \leq b \text{ (p. 383)}$$

### 5.4 Properties of the Definite Integral

**Properties of definite integrals:**

If the functions $f$ and $g$ are integrable on the closed interval $[a, b]$ and $k$ is a constant, then

- Integral of a sum:

$$\int_a^b [f(x) + g(x)]\,dx = \int_a^b f(x)\,dx + \int_a^b g(x)\,dx \text{ (p. 387)}$$

- Integral of a constant times a function:

$$\int_a^b kf(x)\,dx = k\int_a^b f(x)\,dx \text{ (p. 388)}$$

- $\displaystyle\int_a^b [k_1 f_1(x) + k_2 f_2(x) + \cdots + k_n f_n(x)]\,dx$

$$= k_1 \int_a^b f_1(x)\,dx + k_2 \int_a^b f_2(x)\,dx$$

$$+ \cdots + k_n \int_a^b f_n(x)\,dx \qquad \text{(p. 388)}$$

- If $f$ is integrable on an interval containing the numbers $a$, $b$, and $c$, then

$$\int_a^b f(x)\,dx = \int_a^c f(x)\,dx + \int_c^b f(x)\,dx \text{ (p. 389)}$$

- *Bounds on an Integral:* If a function $f$ is continuous on a closed interval $[a, b]$ and if $m$ and $M$ denote the absolute minimum and absolute maximum values, respectively, of $f$ on $[a, b]$, then

$$m(b - a) \leq \int_a^b f(x)\,dx \leq M(b - a) \quad \text{(p. 390)}$$

- *Mean Value Theorem for Integrals:* If a function $f$ is continuous on a closed interval $[a, b]$, then there is a real number $u$, $a \leq u \leq b$, for which

$$\int_a^b f(x)\,dx = f(u)(b - a) \quad \text{(p. 390)}$$

**Definitions:**

- Let $f$ be a function that is continuous on a closed interval $[a, b]$. The average value $\bar{y}$ of $f$ over $[a, b]$ is

$$\bar{y} = \frac{1}{b - a} \int_a^b f(x)\,dx \quad \text{(p. 392)}$$

- Net displacement and total distance (p. 393)

### 5.5 The Indefinite Integral; Method of Substitution

*The indefinite integral of $f$:* $\displaystyle\int f(x)\,dx = F(x) + C$

where $F$ is any function for which $\dfrac{d}{dx}[F(x) + C] = f(x)$, where $C$ is the constant of integration. (p. 398)

**Basic integration formulas:**    See Table 2.    (p. 399)

**Properties of indefinite integrals:**

- Derivative of an integral:

$$\frac{d}{dx}\left[\int f(x)\,dx\right] = f(x) \quad \text{(p. 400)}$$

- Integral of a sum:

$$\int [f(x) + g(x)]\,dx = \int f(x)\,dx + \int g(x)\,dx \quad \text{(p. 400)}$$

- Integral of a constant $k$ times a function:

$$\int kf(x)\,dx = k\int f(x)\,dx \quad \text{(p. 400)}$$

*Integration formulas:*

- $\displaystyle\int \frac{g'(x)}{g(x)}\,dx = \ln|g(x)| + C$   (p. 403)

- $\displaystyle\int \tan x\,dx = \ln|\sec x| + C$   (p. 403)

- $\displaystyle\int \sec x\,dx = \ln|\sec x + \tan x| + C$   (p. 403)

*Method of substitution (definite integrals):*

- Method 1: Find the related indefinite integral using substitution. Then use the Fundamental Theorem of Calculus.
- Method 2: Find the definite integral directly by making a substitution in the integrand and using the substitution to change the limits of integration. (p. 405)

*Theorems:*

- If $f$ is an even function, then
$$\int_{-a}^{a} f(x)\,dx = 2\int_{0}^{a} f(x)\,dx \quad \text{(p. 407)}$$

- If $f$ is an odd function, then $\displaystyle\int_{-a}^{a} f(x)\,dx = 0.$   (p. 407)

## 5.6 Separable First-Order Differential Equations; Uninhibited and Inhibited Growth and Decay Models

*Definitions:*

- First-order differential equation $\dfrac{dy}{dx} = f(x, y)$ (p. 412)

- Separable first-order differentiable equation. (p. 413)

- Uninhibited growth model: $\dfrac{dA}{dt} = kA,\, k > 0$   (p. 414)

- Uninhibited decay model: $\dfrac{dA}{dt} = kA,\, k < 0$   (p. 414)

- Inhibited growth model:
$$\frac{dy}{dt} = k(M - y),\, y < M,\, k > 0,\, M > 0 \quad \text{(p. 417)}$$

- Inhibited decay model:
$$\frac{dy}{dt} = k(y - M),\, y > M,\, k < 0,\, M > 0 \quad \text{(p. 417)}$$

## OBJECTIVES

| Section | You should be able to ... | Examples | Review Exercises |
|---|---|---|---|
| 5.1 | 1  Approximate the area under the graph of a function (p. 358) | 1, 2 | 1, 2 |
|  | 2  Find the area under the graph of a function (p. 362) | 3, 4 | 3, 4 |
| 5.2 | 1  Form Riemann sums (p. 367) | 1 | 5(a) |
|  | 2  Define a definite integral as the limit of Riemann sums (p. 368) | 2–4 | 5(b), 67 |
|  | 3  Approximate a definite integral using Riemann sums (p. 370) | 5 | 63, 64 |
|  | 4  Know conditions that guarantee a definite integral exists (p. 371) |  | 65 |
|  | 5  Find a definite integral using the limit of Riemann sums (p. 372) | 6–8 | 5(c), 66 |
|  | 6  Form Riemann sums from a table (p. 374) | 9 | 57 |
| 5.3 | 1  Use Part 1 of the Fundamental Theorem of Calculus (p. 378) | 1–3 | 7-10, 52, 53 |
|  | 2  Use Part 2 of the Fundamental Theorem of Calculus (p. 379) | 4, 5 | 5(d), 11–13, 17–18, 54 |
|  | 3  Interpret the integral of a rate of change (p. 381) | 6, 7 | 6, 21, 22, 57 |
|  | 4  Interpret the integral as an accumulation function (p. 382) | 8 | 68 |
| 5.4 | 1  Use properties of the definite integral (p. 387) | 1–6 | 15, 16, 23, 24, 27, 28, 45 |
|  | 2  Work with the Mean Value Theorem for Integrals (p. 390) | 7 | 29, 30 |
|  | 3  Find the average value of a function (p. 391) | 8 | 31–34 |
|  | 4  Interpret integrals involving rectilinear motion (p. 393) | 9 | 56 |
| 5.5 | 1  Find indefinite integrals (p. 398) | 1 | 14 |
|  | 2  Use properties of indefinite integrals (p. 400) | 2, 3 | 19, 20, 35, 36 |
|  | 3  Find an indefinite integral using substitution (p. 401) | 4–8 | 37–39, 42, 43, 46, 49 |
|  | 4  Find a definite integral using substitution (p. 405) | 9, 10 | 40, 41, 44, 45, 47, 48, 52, 53 |
|  | 5  Integrate even and odd functions (p. 407) | 11, 12 | 25, 26 |
| 5.6 | 1  Solve a separable first-order differential equation (p. 413) | 1 | 61, 62 |
|  | 2  Solve differential equations involving uninhibited growth and decay (p. 414) | 2, 3 | 59, 60 |
|  | 3  Solve differential equations involving inhibited growth and decay (p. 417) | 4 | 58 |

## REVIEW EXERCISES

1. **Area** Approximate the area under the graph of $f(x) = 2x + 1$ from 0 to 4 by finding lower sums $s_n$ and upper sums $S_n$ for $n = 4$ and $n = 8$ subintervals.

2. **Area** Approximate the area under the graph of $f(x) = x^2$ from 0 to 8 by finding lower sums $s_n$ and upper sums $S_n$ for $n = 4$ and $n = 8$ subintervals.

3. **Area** Find the area $A$ under the graph of $y = f(x) = 9 - x^2$ from 0 to 3 by using lower sums $s_n$ (rectangles that lie below the graph of $f$).

4. **Area** Find the area $A$ under the graph of $y = f(x) = 8 - 2x$ from 0 to 4 using upper sums $S_n$ (rectangles that extend above the graph of $f$).

5. **Riemann Sums**

    (a) Find the Riemann sum of $f(x) = x^2 - 3x + 3$ on the closed interval $[-1, 3]$ using a regular partition with four subintervals and the numbers $u_1 = -1$, $u_2 = 0$, $u_3 = 2$, and $u_4 = 3$.

    (b) Find the Riemann sums of $f$ by partitioning $[-1, 3]$ into $n$ subintervals of equal length and choosing $u_i$ as the right endpoint of the $i$th subinterval $[x_{i-1}, x_i]$. Write the limit of the Riemann sums as a definite integral. Do not evaluate.

    (c) Find the limit as $n$ approaches $\infty$ of the Riemann sums found in (b).

    (d) Find the definite integral from (b) using the Fundamental Theorem of Calculus. Compare the answer to the limit found in (c).

6. **Units of an Integral** In the definite integral $\int_a^b a(t)\,dt$, where $a$ represents acceleration measured in meters per second squared and $t$ is measured in seconds, what are the units of $\int_a^b a(t)\,dt$?

*In Problems 7–10, find each derivative using the Fundamental Theorem of Calculus.*

7. $\dfrac{d}{dx}\displaystyle\int_0^x t^{2/3}\sin t\,dt$

8. $\dfrac{d}{dx}\displaystyle\int_e^x \ln t\,dt$

9. $\dfrac{d}{dx}\displaystyle\int_{x^2}^1 \tan t\,dt$

10. $\dfrac{d}{dx}\displaystyle\int_0^{2\sqrt{x}} \dfrac{t}{t^2+1}\,dt$

*In Problems 11–20, find each integral.*

11. $\displaystyle\int_1^{\sqrt{2}} x^{-2}\,dx$

12. $\displaystyle\int_1^{e^2} \dfrac{1}{x}\,dx$

13. $\displaystyle\int_0^1 \dfrac{1}{1+x^2}\,dx$

14. $\displaystyle\int \dfrac{1}{x\sqrt{x^2-1}}\,dx$

15. $\displaystyle\int_0^{\ln 2} 4e^x\,dx$

16. $\displaystyle\int_0^2 (x^2 - 3x + 2)\,dx$

17. $\displaystyle\int_1^4 2^x\,dx$

18. $\displaystyle\int_0^{\pi/4} \sec x \tan x\,dx$

19. $\displaystyle\int \dfrac{1 + 2xe^x}{x}\,dx$

20. $\displaystyle\int \dfrac{1}{2}\sin x\,dx$

21. **Interpreting an Integral** The function $v = v(t)$ is the speed $v$, in kilometers per hour, of a train at a time $t$, in hours. Interpret the integral $\int_0^{16} v(t)\,dt = 460$.

22. **Interpreting an Integral** The function $f$ is the rate of change of the volume $V$ of oil, in liters per hour, draining from a storage tank at time $t$ (in hours). Interpret the integral $\int_0^2 f(t)\,dt = 100$.

*In Problems 23–26, find each integral.*

23. $\displaystyle\int_{-2}^2 f(x)\,dx$, where $f(x) = \begin{cases} 3x + 2 & \text{if } -2 \le x < 0 \\ 2x^2 + 2 & \text{if } 0 \le x \le 2 \end{cases}$

24. $\displaystyle\int_{-1}^4 |x|\,dx$

25. $\displaystyle\int_{-\pi/2}^{\pi/2} \sin x\,dx$

26. $\displaystyle\int_{-3}^3 \dfrac{x^2}{x^2+9}\,dx$

**Bounds on an Integral** *In Problems 27 and 28, find lower and upper bounds for each integral.*

27. $\displaystyle\int_0^2 e^{x^2}\,dx$

28. $\displaystyle\int_0^1 \dfrac{1}{1+x^2}\,dx$

*In Problems 29 and 30, for each integral find the number(s) u guaranteed by the Mean Value Theorem for Integrals.*

29. $\displaystyle\int_0^\pi \sin x\,dx$

30. $\displaystyle\int_{-3}^3 (x^3 + 2x)\,dx$

*In Problems 31–34, find the average value of each function over the given interval.*

31. $f(x) = \sin x$ over $\left[-\dfrac{\pi}{2}, \dfrac{\pi}{2}\right]$

32. $f(x) = x^3$ over $[1, 4]$

33. $f(x) = e^x$ over $[-1, 1]$

34. $f(x) = 6x^{2/3}$ over $[0, 8]$

35. Find $\dfrac{d}{dx}\displaystyle\int \sqrt{\dfrac{1}{1+4x^2}}\,dx$

36. Find $\dfrac{d}{dx}\displaystyle\int \ln x\,dx$.

*In Problems 37–49, find each integral.*

37. $\displaystyle\int \dfrac{y\,dy}{(y-2)^3}$

38. $\displaystyle\int \dfrac{x}{(2-3x)^3}\,dx$

39. $\displaystyle\int \sqrt{\dfrac{1+x}{x^5}}\,dx,\ x > 0$

40. $\displaystyle\int_{\pi^2/4}^{4\pi^2} \dfrac{1}{\sqrt{x}}\sin\sqrt{x}\,dx$

41. $\displaystyle\int_1^2 \dfrac{1}{t^4}\left(1 - \dfrac{1}{t^3}\right)^3 dt$

42. $\displaystyle\int \dfrac{e^x + 1}{e^x - 1}\,dx$

43. $\displaystyle\int \dfrac{dx}{\sqrt{x}\,(1 - 2\sqrt{x})}$

44. $\displaystyle\int_{1/5}^3 \dfrac{\ln(5x)}{x}\,dx$

45. $\displaystyle\int_{-1}^1 \dfrac{5^{-x}}{2^x}\,dx$

46. $\displaystyle\int e^{x + e^x}\,dx$

**47.** $\displaystyle\int_0^1 \frac{x\,dx}{\sqrt{2-x^4}}$    **48.** $\displaystyle\int_4^5 \frac{dx}{x\sqrt{x^2-9}}$

**49.** $\displaystyle\int \sqrt[3]{x^3+3\cos x}\,(x^2-\sin x)\,dx$

**50.** Find $f''(x)$ if $f(x)=\int_0^x \sqrt{1-t^2}\,dt$.

**51.** Suppose that $F(x)=\int_0^x \sqrt{t}\,dt$ and $G(x)=\int_1^x \sqrt{t}\,dt$. Explain why $F(x)-G(x)$ is constant. Find the constant.

**52.** If $\int_0^2 f(x+2)\,dx=3$, find $\int_2^4 f(x)\,dx$.

**53.** If $\int_1^2 f(x-c)\,dx=5$, where $c$ is a constant, find $\int_{1-c}^{2-c} f(x)\,dx$.

**54. Area** Find the area under the graph of $y=\cosh x$ from $x=0$ to $x=2$.

**55.** The table below shows the rate of change of area $A$ with respect to time $t$ (in square meters per hour) of an oil spill as a function of time $t$, in hours.

| $t$ | 1 | 2.5 | 3 | 4 | 5 |
|---|---|---|---|---|---|
| $\dfrac{dA}{dt}$ | 10 | 21 | 30 | 36 | 40 |

(a) Find a Riemann sum for $\dfrac{dA}{dt}$ on the interval $[1,5]$.

(b) What are the units of the Riemann sum?

(c) Interpret the Riemann sum in the context of the problem.

**56.** An object in rectilinear motion is moving along a horizontal line with velocity $v(t)=6t-t^2$ m/s over the interval $[0,9]$.

(a) Find the net displacement of the object over the interval $[0,9]$. Interpret the result in the context of the problem.

(b) Find the total distance traveled by the object over the interval $[0,9]$ and interpret the result.

**57. Water Supply** A sluice gate of a dam is opened and water is released from the reservoir at a rate of $r(t)=100+\sqrt{t}$ gallons per minute, where $t$ measures time in minutes since the gate has been opened. If the gate is opened at 7 a.m. and is left open until 9:24 a.m., how much water is released?

**58. Forensic Science** A body was found in a meat locker whose ambient temperature is $10\,°C$. When the person was alive, his body temperature was $37\,°C$ and now it is $25\,°C$. Suppose the rate of change of the temperature $u=u(t)$ of the body with respect time $t$ in hours (h) is proportional to $u(t)-T$, where $T$ is the ambient temperature and the constant of proportionality is $-0.294$.

(a) Write a differential equation that models the temperature $u=u(t)$ of the body at time $t$.

(b) Find the general solution of the differential equation.

(c) Find the particular solution of the differential equation, using the initial condition that at the time of death, $u(0)=37\,°C$.

(d) If the body was found at 1 a.m., when did the person die?

(e) How long will it take for the body to cool to $12\,°C$?

**59. Radioactive Decay** The amount $A$ of the radioactive element radium in a sample decays at a rate proportional to the amount of radium present. Given the half-life of radium is 1690 years:

(a) Write a differential equation that models the amount $A$ of radium present at time $t$.

(b) Find the general solution of the differential equation.

(c) Find the particular solution of the differential equation with the initial condition $A(0)=10$ g.

(d) How much radium will be present in the sample at $t=300$ years?

**60. Population Growth in China** Barring disasters (human-made or natural), the population $P$ of humans grows at a rate proportional to its current size. According to the U.N. World Population studies, from 2010 to 2015 the population of China grew at an annual rate of 0.516% per year.

(a) Write a differential equation that models the growth rate of the population.

(b) Find the general solution of the differential equation.

(c) Find the particular solution of the differential equation if in 2015 ($t=0$), the population of China was $1.376\times10^9$.

(d) If the rate of growth continues to follow this model, when will the projected population of China reach 2 billion persons?

*Source: U.N. World Population Prospects: The 2015 Revision.*

*In Problems 61 and 62, solve each differential equation using the given boundary condition.*

**61.** $\dfrac{dy}{dx}=3xy;\quad y=4$ when $x=0$

**62.** $\cos y\dfrac{dy}{dx}=\dfrac{\sin y}{x};\quad y=\dfrac{\pi}{3}$ when $x=-1$

**63.** Approximate $\int_0^4 (x^2+1)\,dx$ with a Left Riemann sum formed by partitioning $[0,4]$ into 4 subintervals each of length 1.

**64.** Approximate $\displaystyle\int_{-2}^6 \frac{2x}{x+3}\,dx$ with a Midpoint Riemann sum formed by partitioning $[-2,6]$ into 4 subintervals each of length 2.

**65.** State a condition that guarantees that a function $f$ is integrable over an interval $[a,b]$.

**66.** Find the definite integral of $f(x)=2x-1$ from 0 to 2 by finding

$$\lim_{\max \Delta x_i\to 0}\sum_{i=1}^n f(u_i)\Delta x_i$$

**67.** Write the limit of the Riemann sums

$$\lim_{n\to\infty}\frac{3}{n}\left[\left(\frac{3}{n}\right)^2+\left(\frac{6}{n}\right)^2+\cdots+\left(\frac{3n}{n}\right)^2\right]$$ as an integral.

The partition is into $n$ subintervals each of the same length.

**68.** (a) Write the accumulation function $F$ associated with $f(x)=x^3-2x^2-3x$ over the interval $[-2,4]$ as an integral.

(b) Find $F(-1)$, $F(0)$, $F(2)$, $F(3)$, and $F(4)$.

(c) Interpret the results found in (b).

## CHAPTER 5 PROJECT   Managing the Klamath River

There is a gauge on the Klamath River, just downstream from the dam at Keno, Oregon. The U.S. Geological Survey has posted flow rates for this gauge every month since 1930. The averages of these monthly measurements since 1930 are given in Table 1. Notice that the data in Table 1 measure the rate of change of the volume $V$ in cubic feet of water each second over one year; that is, the table gives $\dfrac{dV}{dt} = V'(t)$ in cubic feet per second, where $t$ is in months.

### TABLE 1

| Month ($t$) | Flow Rate (ft$^3$/s) |
|---|---|
| January (1) | 1911.79 |
| February (2) | 2045.40 |
| March (3) | 2431.73 |
| April (4) | 2154.14 |
| May (5) | 1592.73 |
| June (6) | 945.17 |
| July (7) | 669.46 |
| August (8) | 851.97 |
| September (9) | 1107.30 |
| October (10) | 1325.12 |
| November (11) | 1551.70 |
| December (12) | 1766.33 |

*Source*: USGS Surface-Water Monthly Statistics, available at http://waterdata.usgs.gov/or/nwis/monthly

1. Find the factor that will convert the data in Table 1 from ft$^3$/s to ft$^3$/day. *Hint:* 1 day $= 1$ d $= 24$ hours $= 24$ h
   1 h $= 60$ min; 1 min $= 60$ s

2. If we assume February has 28.25 days, to account for a leap year, then 1 year $= 365.25$ days. If $V'(t)$ is the rate of flow of water, in cubic feet per day, the total flow of water over 1 year is given by

$$V = V(t) = \int_0^{365.25} V'(t)\,dt$$

   Approximate the total annual flow using a Riemann sum. *Hint:* Use $\Delta t_1 = 31$, $\Delta t_2 = 28.25$, etc.

3. The solution to Problem 2 finds the sum of 12 rectangles whose widths are $\Delta t_i$, $1 \le i \le 12$, and whose heights are the flow rate for the $i$th month. Using the horizontal axis for time and the vertical axis for flow rate in ft$^3$/day, plot the points of Table 1 as follows: (January 1, flow rate for January), (February 1, flow rate for February), ..., (December 1, flow rate for December) and add the point (December 31, flow rate for January). Beginning with the point at January 1, connect each consecutive pair of points with a

line segment, creating 12 trapezoids whose bases are $\Delta t_i$, $1 \le i \le 12$. Approximate the total annual flow $V = V(t) = \int_0^{365.25} V'(t)\,dt$ by summing the areas of these trapezoids.

4. Using the horizontal axis for time and the vertical axis for flow rate in ft$^3$/day, plot the points of Table 1 as follows: (January 31, flow rate for January), (February 28, flow rate for February), ..., (December 31, flow rate for December). Then add the point (January 1, flow rate for December) to the left of (January 31, flow rate for January). Connect consecutive points with a line segment, creating 12 trapezoids whose bases are $\Delta t_i$, $1 \le i \le 12$.

   Approximate the total annual flow $V = V(t) = \int_0^{365.25} V'(t)\,dt$ by summing the areas of these trapezoids.

5. Why did we add the extra point in Problems 3 and 4? How do you justify the choice?

6. Compare the three approximations. Discuss which might be the most accurate.

7. Another way to approximate $V = V(t) = \int_0^{365.25} V'(t)\,dt$ is to fit a polynomial function to the data. We could find a polynomial of degree 11 that passes through every point of the data, but a polynomial of degree 6 is sufficient to capture the essence of the behavior. The polynomial function $f$ of degree 6 is

$$f(t) = 2.2434817 \times 10^{-10} t^6 - 2.5288956 \times 10^{-7} t^5$$
$$+ 0.00010598313 t^4 - 0.019872628 t^3 + 1.557403 t^2$$
$$- 39.387734 t + 2216.2455$$

   Find the total annual flow using $f(t) = V'(t)$.

8. Use technology to graph the polynomial function $f$ over the closed interval $[0, 12]$. How well does the graph fit the data?

9. A manager could approximate the rate of flow of the river for every minute of every day using the function

$$g(t) = 1529.403 + 510.330 \sin \frac{2\pi t}{365.25} + 489.377 \cos \frac{2\pi t}{365.25}$$
$$- 47.049 \sin \frac{4\pi t}{365.25} - 249.059 \cos \frac{4\pi t}{365.25}$$

   where $t$ represents the day of the year in the interval $[0, 365.25]$. The function $g$ represents the best fit to the data that has the form of a sum of trigonometric functions with the period 1 year. We could fit the data perfectly using more terms, but the improvement in results would not be worth the extra work in handling that approximation. Use $g$ to approximate the total annual flow.

10. Use technology to graph the function $g$ over the closed interval $[0, 12]$. How well does the graph fit the data?

11. Compare the five approximations to the annual flow of the river. Discuss the advantages and disadvantages of using one over another. What method would you recommend to measure the annual flow of the Klamath River?

# 6

# Applications of the Integral

**CHAPTER 6 PROJECT** The Chapter Project on page 497 investigates the design of cooling towers and determines the quantity and cost needed to build one.

Everett Collection Inc/Alamy Stock Photo

## The Cooling Tower at Callaway Nuclear Generating Station

The Callaway Nuclear Generating Station is located in Callaway County, Missouri. Power stations often use cooling towers like the one seen in the photograph above to remove waste and to heat and cool the water that cycles through the system. The Callaway station's energy-producing turbines are powered by steam. Once that steam travels through the turbine system, it passes over a series of tubes carrying cooled water from the tower. As heat from the steam is absorbed by the cooling water, the steam condenses back into water, which is then pumped back into steam generators, and the cycle repeats itself.

The cooling tower at Callaway is 553 ft tall and 430 ft wide at the base and has the capacity to cool 585,000 gallons of water per minute. Water enters the tower with a temperature of about 125 °F and the tower cools it down to 95 °F. Cooling towers like this one are examples of hyperboloid structures.

The applications of the definite integral in this chapter rely on two facts:

- If a function $f$ is continuous on a closed interval $[a, b]$, then $f$ is integrable over $[a, b]$. That is,

$$\int_a^b f(x)\, dx = \lim_{n \to \infty} \sum_{i=1}^{n} f(u_i) \Delta x = \text{a number where } \Delta x = \frac{b-a}{n}$$

- If $F$ is any antiderivative of $f$ on $(a, b)$, then $\int_a^b f(x)\, dx = F(b) - F(a)$

We use a definite integral to solve geometry problems, finding:

- the area of a region enclosed by the graphs of two or more functions;
- the volume of a solid;
- the length of the graph of a function;
- the surface area of a solid of revolution.

We also use a definite integral to find:

- the work done by a variable force;
- hydrostatic pressure and force;
- the centroid of a lamina.

In each application the words are different, but the melody is the same. We partition a quantity into small segments and use the segments to form Riemann sums that approximate the quantity. Then we allow the number of segments to grow without bound and express the quantity as a definite integral.

# 6.1 Area Between Graphs

**OBJECTIVES** *When you finish this section, you should be able to:*

**1** Find the area between the graphs of two functions by partitioning the $x$-axis (p. 428)

**2** Find the area between the graphs of two functions by partitioning the $y$-axis (p. 431)

When a function $f$ is continuous and nonnegative on a closed interval $[a, b]$, then the definite integral

$$\int_a^b f(x)\, dx$$

equals the area under the graph of $y = f(x)$ from $a$ to $b$. In this section, we relax the restriction that $f(x)$ is nonnegative and extend the interpretation of the definite integral to find the area enclosed by the graphs of two functions.

## 1 Find the Area Between the Graphs of Two Functions by Partitioning the $x$-Axis

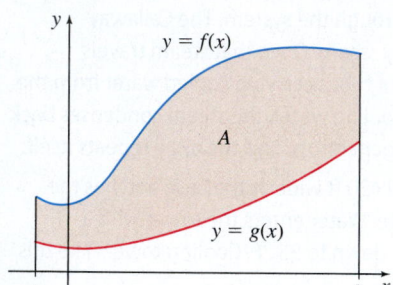

**Figure 1** $A$ is the area of the region enclosed by the graphs of $f$ and $g$ and the lines $x = a$ and $x = b$.

Suppose we want to find the area $A$ of the region enclosed by the graphs of $y = f(x)$ and $y = g(x)$ and the lines $x = a$ and $x = b$. Occasionally, the area can be found using geometry formulas, and calculus is not needed. But more often the region is irregular, and its area is found using definite integrals.

Assume that the functions $f$ and $g$ are continuous on the closed interval $[a, b]$ and that $f(x) \geq g(x)$ for all numbers $x$ in $[a, b]$, as illustrated in Figure 1. To find the area $A$ of the region enclosed by the two graphs, we partition the interval $[a, b]$ on the $x$-axis into $n$ subintervals:

$$[x_0, x_1], [x_1, x_2], \ldots, [x_{i-1}, x_i], \ldots, [x_{n-1}, x_n] \qquad x_0 = a \quad x_n = b$$

each of width $\Delta x = \dfrac{b - a}{n}$. For each $i$, $i = 1, 2, \ldots, n$, we select a number $u_i$ in the subinterval $[x_{i-1}, x_i]$. Then we construct $n$ rectangles, each of width $\Delta x$ and height $f(u_i) - g(u_i)$. The area of the $i$th rectangle is $[f(u_i) - g(u_i)]\Delta x$. See Figure 2.

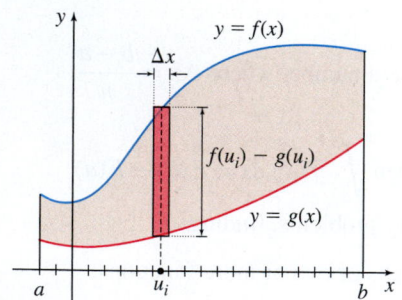

**Figure 2** The area of the $i$th rectangle is $[f(u_i) - g(u_i)]\Delta x$

The sum of the areas of the $n$ rectangles, $\sum\limits_{i=1}^{n} [f(u_i) - g(u_i)]\, \Delta x$, approximates the area $A$. As the number $n$ of subintervals increases, these sums become a better approximation to the area $A$, and

$$A = \lim_{n \to \infty} \sum_{i=1}^{n} [f(u_i) - g(u_i)]\, \Delta x$$

Since the approximating sums are Riemann sums, and $f$ and $g$ are continuous on the interval $[a, b]$, then the limit is a definite integral and

$$A = \int_a^b [f(x) - g(x)]\, dx$$

**AREA**

The area $A$ of the region enclosed by the graphs of $y = f(x)$ and $y = g(x)$, and the lines $x = a$ and $x = b$, where $f$ and $g$ are continuous on the interval $[a, b]$ and $f(x) \geq g(x)$ for all numbers $x$ in $[a, b]$, is

$$A = \int_a^b [f(x) - g(x)]\, dx \qquad (1)$$

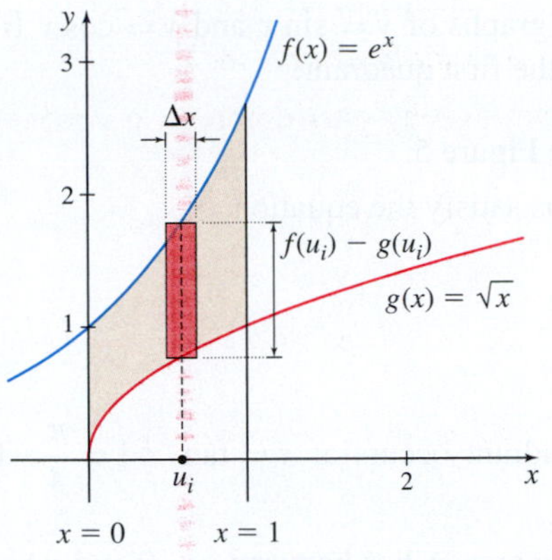

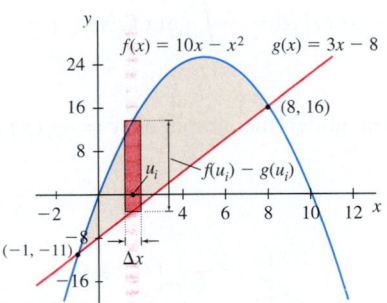

**Figure 3** The region enclosed by the graphs of $f(x) = e^x$, $g(x) = \sqrt{x}$, and the lines $x = 0$ and $x = 1$.

**EXAMPLE 1  Finding the Area Between the Graphs of Two Functions**

Find the area $A$ of the region enclosed by the graphs of $f(x) = e^x$ and $g(x) = \sqrt{x}$ and the lines $x = 0$ and $x = 1$.

**Solution** We begin by graphing the two functions and identifying the area $A$ to be found. See Figure 3.

From the graph, we see that $f(x) \geq g(x)$ on the interval $[0, 1]$. Then, using the definition of area, we have

$$A = \int_a^b [f(x) - g(x)]\,dx = \int_0^1 (e^x - \sqrt{x})\,dx$$

$$= \int_0^1 e^x\,dx - \int_0^1 x^{1/2}\,dx = \left[e^x\right]_0^1 - \left[\frac{x^{3/2}}{\frac{3}{2}}\right]_0^1$$

$$= (e^1 - e^0) - \frac{2}{3}(1 - 0) = e - 1 - \frac{2}{3} = e - \frac{5}{3} \text{ square units} \quad \blacksquare$$

**NOW WORK** Problems **3** and **7**.

Formula (1) for area holds whether the graphs of $f$ and $g$ lie above the $x$-axis, below the $x$-axis, or partially above and partially below the $x$-axis as long as $f(x) \geq g(x)$ on $[a, b]$. It is critical to graph $f$ and $g$ on $[a, b]$ to determine the relationship between the graphs of $f$ and $g$ before setting up the integral. The key is to always subtract the smaller value from the larger value. This ensures that the height of each rectangle is positive.

**EXAMPLE 2  Finding the Area Between the Graphs of Two Functions**

Find the area $A$ of the region enclosed by the graphs of

$$f(x) = 10x - x^2 \quad \text{and} \quad g(x) = 3x - 8$$

**Solution** First we graph the two functions. See Figure 4.

The region lies between the graphs of $f$ and $g$, and the $x$-values of the points of intersection identify the limits of integration. Solve the equation $f(x) = g(x)$ to find these values.

$$10x - x^2 = 3x - 8 \qquad\qquad f(x) = g(x)$$

$$x^2 - 7x - 8 = 0$$

$$(x + 1)(x - 8) = 0$$

$$x = -1 \quad \text{or} \quad x = 8$$

**Figure 4** The graphs intersect at $(-1, -11)$ and $(8, 16)$. So, the limits of integration are $-1$ and $8$.

The limits of integration are $a = -1$ and $b = 8$. Since $f(x) \geq g(x)$ on $[-1, 8]$, the area $A$ is given by

$$A = \int_a^b [f(x) - g(x)]\,dx = \int_{-1}^8 [(10x - x^2) - (3x - 8)]\,dx$$

$$= \int_{-1}^8 (-x^2 + 7x + 8)\,dx = \left[\frac{-x^3}{3} + \frac{7x^2}{2} + 8x\right]_{-1}^8$$

$$= \left(-\frac{512}{3} + 224 + 64\right) - \left(\frac{1}{3} + \frac{7}{2} - 8\right) = \frac{243}{2} = 121.5 \text{ square units} \quad \blacksquare$$

**NOW WORK** Problem **21**.

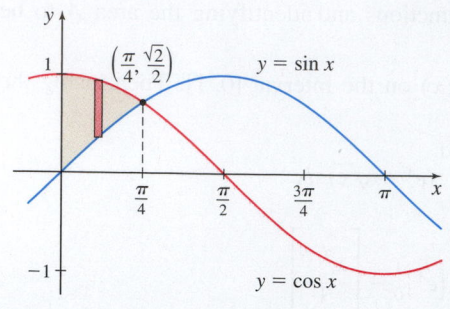

**Figure 5**

### EXAMPLE 3   Finding the Area Between the Graphs of Two Functions

Find the area $A$ of the region enclosed by the graphs of $y = \sin x$ and $y = \cos x$ from the $y$-axis to their first point of intersection in the first quadrant.

**Solution** First we graph the two functions. See Figure 5.

The points of intersection of the two graphs satisfy the equation

$$\sin x = \cos x$$
$$\tan x = 1$$

The first point of intersection in the first quadrant occurs at $x = \tan^{-1} 1 = \dfrac{\pi}{4}$. The graphs intersect at the point $\left(\dfrac{\pi}{4}, \dfrac{\sqrt{2}}{2}\right)$, so the area $A$ lies between $x = 0$ and $x = \dfrac{\pi}{4}$.

Since $\cos x \geq \sin x$ on $\left[0, \dfrac{\pi}{4}\right]$, the area $A$ is given by

$$A = \int_0^{\pi/4} (\cos x - \sin x)\, dx = \big[\sin x + \cos x\big]_0^{\pi/4} = \left(\frac{\sqrt{2}}{2} + \frac{\sqrt{2}}{2}\right) - (0 + 1)$$

$$= \sqrt{2} - 1 \text{ square units}$$
■

**NOW WORK** Problem 11.

Now suppose $y = g(x) \leq 0$ for all numbers $x$ in the interval $[a, b]$, as illustrated in Figure 6(a). Then by (1), the area $A$ of the region enclosed by the graph of $f(x) = 0$ (the $x$-axis), the graph of $g$, and the lines $x = a$ and $x = b$, is given by

$$A = \int_a^b [f(x) - g(x)]\, dx = \int_a^b [0 - g(x)]\, dx = -\int_a^b g(x)\, dx$$

By symmetry, this area $A$ is equal to the area under the graph of $y = -g(x) \geq 0$ from $a$ to $b$. See Figure 6(b).

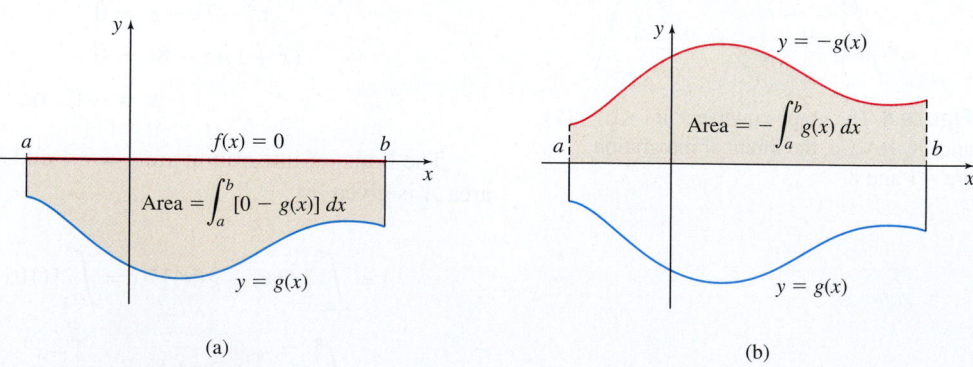

(a)                    (b)

**Figure 6**

### EXAMPLE 4   Finding the Area Between a Graph and the $x$-Axis

Find the area $A$ of the region enclosed by the graph of $f(x) = x^2 - 4$, the $x$-axis, and the lines $x = 0$ and $x = 5$.

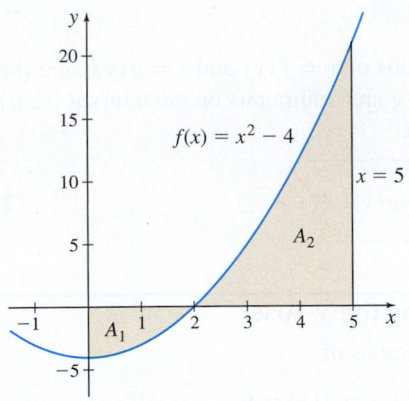

**Figure 7** The area $A$ enclosed by the graph of $f$, the $x$-axis, and the lines $x = 0$ and $x = 5$ is the sum of the areas $A_1$ and $A_2$.

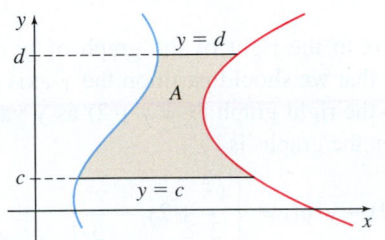

**Figure 8**

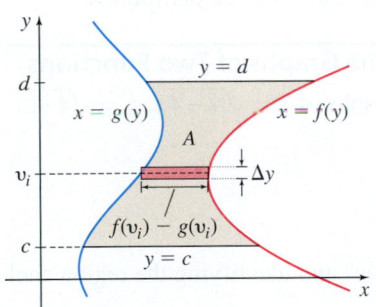

**Figure 9**

**Solution** First graph the function $f$. On the interval $[0, 5]$, the graph of $f$ intersects the $x$-axis at $x = 2$. As shown in Figure 7, $f(x) \leq 0$ on the interval $[0, 2]$, and $f(x) \geq 0$ on the interval $[2, 5]$. So, the area $A$ is the sum of the areas $A_1$ and $A_2$, where

$$A_1 = \int_0^2 [0 - f(x)] \, dx = \int_0^2 -(x^2 - 4) \, dx = \left[ -\frac{x^3}{3} + 4x \right]_0^2 = -\frac{8}{3} + 8 = \frac{16}{3}$$

$$A_2 = \int_2^5 f(x) \, dx = \int_2^5 (x^2 - 4) \, dx = \left[ \frac{x^3}{3} - 4x \right]_2^5 = \left( \frac{125}{3} - 20 \right) - \left( \frac{8}{3} - 8 \right)$$

$$= \frac{81}{3} = 27$$

The area is $A = A_1 + A_2 = \dfrac{16}{3} + \dfrac{81}{3} = \dfrac{97}{3}$ square units. ∎

## 2 Find the Area Between the Graphs of Two Functions by Partitioning the $y$-Axis

In the previous examples, we found the area by partitioning the $x$-axis. Sometimes it is necessary to partition the $y$-axis. Look at Figure 8. The region is enclosed by the graphs and the horizontal lines $y = c$ and $y = d$.

The graphs that form the left and right borders of the region are not the graphs of functions (they fail the Vertical-line Test). Finding the area $A$ of the region by partitioning the $x$-axis is not practical. However, both graphs in Figure 8 can be represented as functions of $y$ since each satisfies the Horizontal-line Test.

**THEOREM  Horizontal-line Test**

A set of points in the $xy$-plane is the graph of a function of the form $x = f(y)$ if and only if every horizontal line intersects the graph in at most one point.

The graphs in Figure 9 satisfy the Horizontal-line Test, so each is a function of $y$. In Figure 9, the graphs are labeled $x = g(y)$ and $x = f(y)$, where $x = g(y)$ is the horizontal distance from the $y$-axis to $g(y)$, and $x = f(y)$ is the horizontal distance from the $y$-axis to $f(y)$. Since the graph of $f$ lies to the right of the graph of $g$, we know $f(y) \geq g(y)$.

Now to find the area $A$ of the region enclosed by the two graphs and the horizontal lines $y = c$ and $y = d$, we partition the interval $[c, d]$ on the $y$-axis into $n$ subintervals:

$$[y_0, y_1], [y_1, y_2], \ldots, [y_{i-1}, y_i], \ldots, [y_{n-1}, y_n] \qquad y_0 = c \quad y_n = d$$

each of width $\Delta y = \dfrac{d - c}{n}$. For each $i$, $i = 1, 2, \ldots n$, we select a number $v_i$ in the subinterval $[y_{i-1}, y_i]$. Then we construct $n$ rectangles, each of height $\Delta y$ and width $f(v_i) - g(v_i)$. The area of the $i$th rectangle is $[f(v_i) - g(v_i)]\Delta y$.

The sum of the areas of the $n$ rectangles, $\displaystyle\sum_{i=1}^{n} [f(v_i) - g(v_i)]\Delta y$, approximates the area $A$. As the number of subintervals increases, the approximation to the area improves, and

$$A = \lim_{n \to \infty} \sum_{i=1}^{n} [f(v_i) - g(v_i)] \, \Delta y$$

The approximating sums are Riemann sums, so if $f$ and $g$ are continuous on the closed interval $[c, d]$, then the limit is a definite integral and

$$A = \int_c^d [f(y) - g(y)] \, dy$$

**AREA** (Partitioning the $y$-axis)

The area $A$ of the region enclosed by the graphs of $x = f(y)$ and $x = g(y)$, and the horizontal lines $y = c$ and $y = d$, where $f$ and $g$ are continuous on the interval $[c, d]$ and $f(y) \geq g(y)$ for all numbers $y$ in $[c, d]$, is

$$A = \int_c^d [f(y) - g(y)]\, dy \qquad (2)$$

**EXAMPLE 5  Finding Area by Partitioning the $y$-Axis**

Find the area $A$ of the region enclosed by the graphs of

$$x = f(y) = y + 2 \quad \text{and} \quad x = g(y) = y^2$$

**Solution** First graph the two functions and identify the region enclosed by the two graphs. See Figure 10.

The graphs intersect when $y + 2 = y^2$. Then $y^2 - y - 2 = (y - 2)(y + 1) = 0$. So, $y = 2$ or $y = -1$. When $y = 2$, $x = 4$; when $y = -1$, $x = 1$. The graphs intersect at the points $(4, 2)$ and $(1, -1)$.

Notice in Figure 10 that the graph of $f$ is to the right of the graph of $g$; that is, $f(y) \geq g(y)$ for $-1 \leq y \leq 2$. This indicates that we should partition the $y$-axis and form rectangles from the left graph ($x = y^2$) to the right graph ($x = y + 2$) as $y$ varies from $-1$ to $2$. The area $A$ of the region between the graphs is

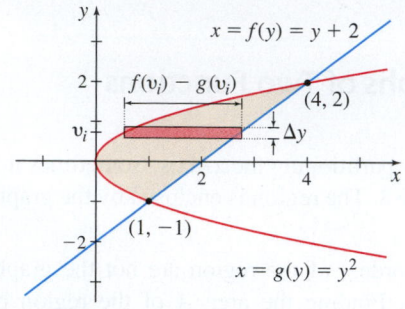

**Figure 10**

$$A \underset{\substack{\uparrow \\ \text{Use (2).}}}{=} \int_{-1}^2 [f(y) - g(y)]\, dy = \int_{-1}^2 [(y + 2) - y^2]\, dy = \left[ \frac{y^2}{2} + 2y - \frac{y^3}{3} \right]_{-1}^2$$

$$= \left( 2 + 4 - \frac{8}{3} \right) - \left( \frac{1}{2} - 2 + \frac{1}{3} \right) = 4.5 \text{ square units} \qquad \blacksquare$$

**NOW WORK** Problem 15.

There are times when either the $x$-axis or the $y$-axis can be partitioned.

**EXAMPLE 6  Finding the Area Between the Graphs of Two Functions**

Find the area $A$ of the region enclosed by the graphs of $y = \sqrt{4 - 4x}$, $y = \sqrt{4 - x}$, and the $x$-axis:

(a) by partitioning the $x$-axis.

(b) by partitioning the $y$-axis.

**Solution (a)** Begin by graphing the two equations and identifying the region enclosed by the graphs. See Figure 11.

If we partition the $x$-axis, the area $A$ must be expressed as the sum of the two areas $A_1$ and $A_2$ marked in the figure. [Do you see why? The bottom graph changes at $x = 1$ from $y = \sqrt{4 - 4x}$ to $y = 0$ (the $x$-axis).]

Area $A_1$ is the region enclosed by the graphs of $y = \sqrt{4 - x}$, and $y = \sqrt{4 - 4x}$, and the line $x = 1$. Area $A_2$ is the region enclosed by the graph $y = \sqrt{4 - x}$, the $x$-axis, and the line $x = 1$. Then

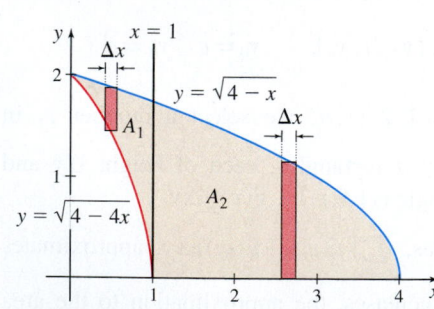

**Figure 11** The area $A$ is the sum of the areas $A_1$ and $A_2$.

$$A = A_1 + A_2 = \int_0^1 (\sqrt{4 - x} - \sqrt{4 - 4x})\, dx + \int_1^4 \sqrt{4 - x}\, dx$$

$$= \int_0^1 \sqrt{4 - x}\, dx - \int_0^1 \sqrt{4 - 4x}\, dx + \int_1^4 \sqrt{4 - x}\, dx$$

$$= \int_0^4 \sqrt{4 - x}\, dx - \int_0^1 \sqrt{4 - 4x}\, dx \qquad \int_0^1 \sqrt{4 - x}\, dx + \int_1^4 \sqrt{4 - x}\, dx = \int_0^4 \sqrt{4 - x}\, dx$$

To find the first integral, use the substitution $u = 4 - x$. Then $du = -dx$, and

$$\int_0^4 \sqrt{4 - x}\, dx = -\int_4^0 u^{1/2} du = \int_0^4 u^{1/2}\, du = \left[ \frac{2}{3} u^{3/2} \right]_0^4 = \frac{2}{3}(8 - 0) = \frac{16}{3}$$

For the other integral, use the substitution $u = 4 - 4x$. Then $du = -4\,dx$, or equivalently, $dx = -\dfrac{du}{4}$, and

$$\int_0^1 \sqrt{4-4x}\,dx = -\frac{1}{4}\int_4^0 u^{1/2}\,du = \frac{1}{4}\int_0^4 u^{1/2}\,du = \frac{1}{4}\left[\frac{2}{3}u^{3/2}\right]_0^4 = \frac{1}{6}(8-0) = \frac{4}{3}$$

The area $A = \dfrac{16}{3} - \dfrac{4}{3} = 4$ square units.

**(b)** Figure 12 shows the region enclosed by the graphs of $y = \sqrt{4-4x}$, $y = \sqrt{4-x}$, and the $x$-axis. Since the graphs of $y = \sqrt{4-4x}$ and $y = \sqrt{4-x}$ satisfy the Horizontal-line Test for $0 \le y \le 2$, we can express $y = \sqrt{4-4x}$ as a function $x = f(y)$ and $y = \sqrt{4-x}$ as a function $x = g(y)$. To find $x = f(y)$, we solve $y = \sqrt{4-4x}$ for $x$:

$$y = \sqrt{4-4x}$$
$$y^2 = 4 - 4x$$
$$x = \frac{4-y^2}{4}$$
$$x = f(y) = 1 - \frac{y^2}{4}$$

To express $y = \sqrt{4-x}$ as a function $x = g(y)$, we solve for $x$:

$$y = \sqrt{4-x}$$
$$y^2 = 4 - x$$
$$x = g(y) = 4 - y^2$$

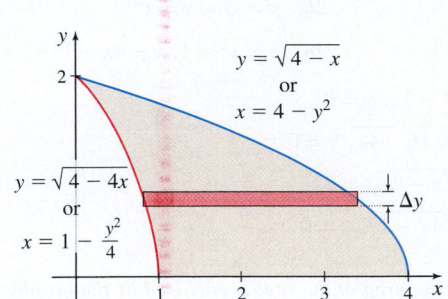

**Figure 12**

The graph of $x = g(y) = 4 - y^2$ is to the right of the graph of $x = f(y) = 1 - \dfrac{y^2}{4}$, $0 \le y \le 2$. So, $g(y) \ge f(y)$. Partitioning the $y$-axis, we have

$$A = \int_0^2 [g(y) - f(y)]\,dy = \int_0^2 \left[(4 - y^2) - \left(1 - \frac{y^2}{4}\right)\right] dy = \int_0^2 \left(3 - \frac{3y^2}{4}\right) dy$$

$$= \left[3y - \frac{y^3}{4}\right]_0^2 = 6 - \frac{8}{4} = 4 \text{ square units} \qquad \blacksquare$$

**NOW WORK** Problem 31.

When the graphs of functions of $x$ form the top and bottom borders of the area, partitioning the $x$-axis is usually easier, provided it is easy to find the integrals. If the graphs of functions of $y$ form the left and right borders of the area, partitioning the $y$-axis is usually easier, provided the integrals with respect to $y$ are easily found. Example 7 shows that sometimes the first choice of a partition does not work.

**EXAMPLE 7   Finding the Area Under a Graph**

Find the area $A$ of the region enclosed by the graph of $y = \ln x$, the $x$-axis, and the line $x = e$.

**Solution** Begin by graphing $y = \ln x$ and $x = e$ and identifying the region enclosed by the graphs. See Figure 13.

Since the graph of $y = \ln x$ forms the top border of the region, it appears that partitioning the $x$-axis from 1 to $e$ is easier. This leads to the integral

$$A = \int_1^e \ln x\,dx$$

But at this place in the text, we do not have the tools to find $\int \ln x\,dx$.

So instead, we partition the $y$-axis from 0 to 1. The left graph is $y = \ln x$ or, equivalently, $x = e^y$, the right graph is $x = e$, and $0 \le y \le 1$.

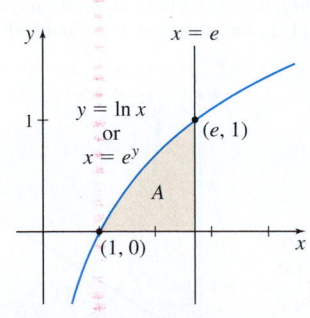

**Figure 13**

$$A = \int_0^1 (e - e^y)\,dy = [ey - e^y]_0^1 = (e - e) - (0 - 1) = 1 \text{ square unit} \qquad \blacksquare$$

**NOW WORK** Problem 37.

## 6.1 Assess Your Understanding

### Concepts and Vocabulary

1. Express the area between the graphs of $y = x^2$ and $y = \sqrt{x}$ as an integral using a partition of the $x$-axis. Do not find the integral.

2. Express the area between the graph of $x = y^2$ and the line $x = 1$ as an integral using a partition of the $y$-axis. Do not find the integral.

### Skill Building

*In Problems 3–12, find the area of the region enclosed by the graphs of the given equations by partitioning the x-axis.*

3. $y = x$, $y = 2x$, $x = 1$

4. $y = x$, $y = 3x$, $x = 3$

5. $y = x^2$, $y = x$

6. $y = x^2$, $y = 4x$

7. $y = e^x$, $y = e^{-x}$, $x = \ln 2$

8. $y = e^x$, $y = -x + 1$, $x = 2$

9. $y = x^2$, $y = x^4$

10. $y = x$, $y = x^3$

11. $y = \cos x$, $y = \dfrac{1}{2}$, $0 \le x \le \dfrac{\pi}{3}$

12. $y = \sin x$, $y = \dfrac{1}{2}$, $\dfrac{\pi}{6} \le x \le \dfrac{5\pi}{6}$

*In Problems 13–20, find the area of the region enclosed by the graphs of the given equations by partitioning the y-axis.*

13. $x = y^2$, $x = 2 - y$

14. $x = y^2$, $x = y + 2$

15. $x = 9 - y^2$, $x = 5$

16. $x = 16 - y^2$, $x = 7$

17. $x = y^2 + 4$, $y = x - 6$

18. $x = y^2 + 6$, $y = 8 - x$

19. $y = \ln x$, $x = 1$, $y = 2$

20. $y = \ln x$, $x = e$, $y = 0$

*In Problems 21–24, find the area of the shaded region in the graph.*

21.

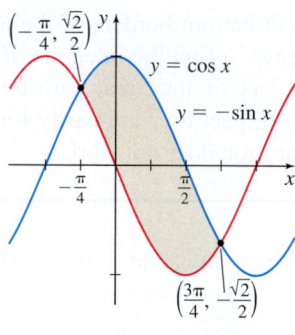

22.

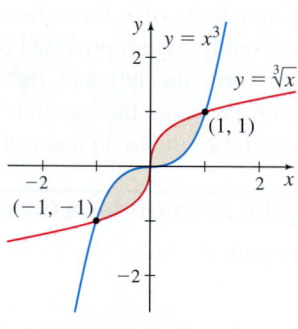

23.

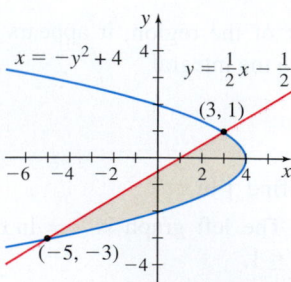

24.

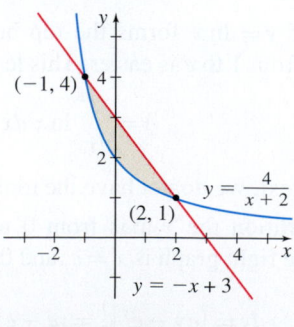

*In Problems 25–32, find the area A of the region enclosed by the graphs of the given equations:*
(a) *by partitioning the x-axis.*
(b) *by partitioning the y-axis.*

25. $y = \sqrt{x}$, $y = x^3$

26. $y = \sqrt{x}$, $y = x^2$

27. $y = x^2 + 1$, $y = x + 1$

28. $y = x^2 + 1$, $y = 4x + 1$

29. $y = \sqrt{9 - x}$, $y = \sqrt{9 - 3x}$, $y = 0$

30. $y = \sqrt{16 - 2x}$, $y = \sqrt{16 - 4x}$, $y = 0$

31. $y = \sqrt{2x - 6}$, $y = \sqrt{x - 2}$, $y = 0$

32. $y = \sqrt{2x - 5}$, $y = \sqrt{4x - 17}$, $y = 0$

*In Problems 33–46, find the area of the region enclosed by the graphs of the given equations.*

33. $y = 4 - x^2$, $y = x^2$

34. $y = 9 - x^2$, $y = x^2$

35. $x = y^2 - 4$, $x = 4 - y^2$

36. $x = y^2$, $x = 16 - y^2$

37. $y = \ln x^2$, $y = 0$, and the line $x = e$

38. $y = \ln x$, $y = 1 - x$, and the line $y = 1$

39. $y = \cos x$, $y = 1 - \dfrac{3}{\pi}x$, $x = \dfrac{\pi}{3}$

40. $y = \sin x$, $y = 1$, $0 \le x \le \dfrac{\pi}{2}$

41. $y = e^{2x}$ and the lines $x = 1$ and $y = 1$

42. $y = e^x$, $y = e^{3x}$, $x = 2$

43. $y^2 = 4x$, $4x - 3y - 4 = 0$

44. $y^2 = 4x + 1$, $x = y + 1$

45. $y = \sin x$, $y = \dfrac{2x}{\pi}$, $x \ge 0$

46. $y = \cos x$, $x \ge 0$, $y = \dfrac{3x}{\pi}$

### Applications and Extensions

47. **An Archimedean Result** Show that the area of the shaded region in the figure is two-thirds of the area of the parallelogram $ABCD$. (This illustrates a result due to Archimedes concerning sectors of parabolas.)

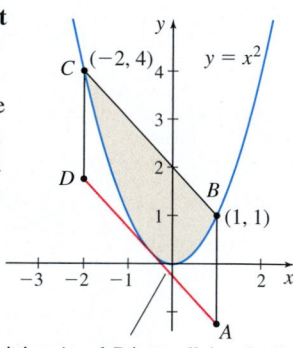

The line joining $A$ and $D$ is parallel to the line joining $(-2, 4)$ and $(1, 1)$ and is tangent to the graph of $y = x^2$.

48. **Equal Areas** Find $h$ so that the area of the region enclosed by the graphs of

$$y = x, \quad y = 8x, \quad \text{and} \quad y = \dfrac{1}{x^2}$$

is equal to that of an isosceles triangle of base 1 and height $h$. See the figure.

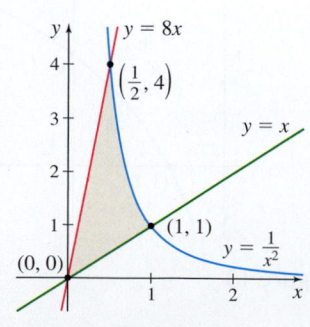

**49.** The graph of the region $R$ bounded by the graphs of $y = \cos^2 x$ and $y = \sin^2 x$ is shown below.

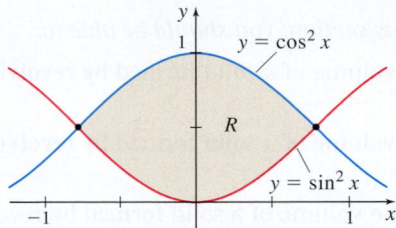

(a) Find the points of intersection of the two graphs.

(b) Find the area of the region in the first quadrant bounded by the graphs of $y = \cos^2 x$ and $y = \sin^2 x$ and the $y$-axis.

**50. Area** Find the area in the first quadrant enclosed by the graphs of $y = \sin(2x)$ and $y = \cos(2x)$, $0 \le x \le \dfrac{\pi}{8}$.

**51. Area** Find the area of the region enclosed by the graphs of

$$y = \sin^{-1} x, \ y = x, \ 0 \le x \le \frac{1}{2}$$

Which axis did you choose to partition? Explain your choice.

**52. Area**

(a) Graph $y = x^2$ and $y = \sin x$.

(b) Find the points of intersection of the graphs.

(c) Find the area enclosed by the graphs of $y = x^2$ and $y = \sin x$ using a partition of the $x$-axis.

(d) Find the area enclosed by the graphs of $y = x^2$ and $y = \sin x$ using a partition of the $y$-axis.

**53.** (a) Graph $y = \sin^{-1} x$, $x + y = 1$, and $y = 0$.

(b) Find the points of intersection of the graphs.

(c) Find the area enclosed by the graphs.

**54.** (a) Graph $y = \cos^{-1} x$, $y = x^3 + 1$, and $y = 0$.

(b) Find the points of intersection of the graphs.

(c) Find the area enclosed by the graphs.

**55.** (a) Graph $2x + y = 3$, $y = \dfrac{1}{1 + x^2}$, and $y = 1$.

(b) Find the points of intersection of the graphs.

(c) Find the area enclosed by the graphs.

**56.** (a) Graph $y = \dfrac{5}{1 + x^2}$, $x = \sqrt{y + 2}$, and $x = 0$.

(b) Find the points of intersection of the graphs.

(c) Find the area enclosed by the graphs.

**57.** (a) Graph $y = \sin^{-1} x$, $y = 1 - x^2$, and $y = 0$.

(b) Find the points of intersection of the graphs.

(c) Find the area enclosed by the graphs.

**58. Cost of Health Care** The cost of health care varies from one country to another. Between 2000 and 2010, the average cost of health insurance for a family of four in the United States was modeled by

$$A(x) = 8020.6596(1.0855^x) \qquad 0 \le x \le 10$$

where $x = 0$ corresponds to the year 2000 and $A(x)$ is measured in U.S. dollars. During the same years, the average cost of health care in Canada was given by

$$C(x) = 4944.6424(1.0711^x) \qquad 0 \le x \le 10$$

where $x = 0$ corresponds to the year 2000 and $C(x)$ is measured in U.S. dollars.

(a) Find the area between the graphs from $x = 0$ to $x = 10$. Round the answer to the nearest dollar.

(b) Interpret the answer to (a).

## Challenge Problems

**59.** (a) Express the area $A$ of the region in the first quadrant enclosed by the $y$-axis and the graphs of $y = \tan x$ and $y = k$ for $k > 0$ as a function of $k$.

(b) What is the value of $A$ when $k = 1$?

(c) If the line $y = k$ is moving upward at the rate of $\dfrac{1}{10}$ unit per second, at what rate is $A$ changing when $k = 1$?

**60.** The area $A$ of the shaded region in the figure is $A = \dfrac{t}{2}$. Prove this as follows:

(a) Let $A^*$ be twice the area outlined in the figure, and $(x, y)$ be the coordinates of $P$. Explain why

$$A^* = x\sqrt{x^2 - 1} - 2 \int_1^x \sqrt{u^2 - 1}\, du$$

(b) Differentiate $A^*$ to show that $\dfrac{dA^*}{dx} = \dfrac{1}{\sqrt{x^2 - 1}}$.

(c) Show that $A^* = \cosh^{-1} x + C$, and explain why $C = 0$.

(d) Why does it follow that $A^* = t$?

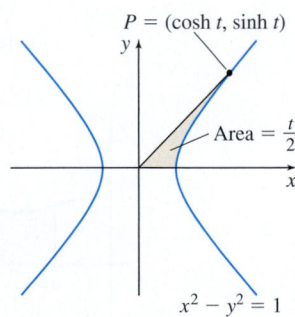

**61.** Find the area enclosed by the graph of $y^2 = x^2 - x^4$.

# 6.2 Volume of a Solid of Revolution: Disks and Washers

OBJECTIVES *When you finish this section, you should be able to:*

1 Use the disk method to find the volume of a solid formed by revolving a region about the $x$-axis (p. 437)

2 Use the disk method to find the volume of a solid formed by revolving a region about the $y$-axis (p. 439)

3 Use the washer method to find the volume of a solid formed by revolving a region about the $x$-axis (p. 440)

4 Use the washer method to find the volume of a solid formed by revolving a region about the $y$-axis (p. 442)

5 Find the volume of a solid formed by revolving a region about a line parallel to a coordinate axis (p. 443)

An example of a solid of revolution is a *right circular cylinder*, which is generated by revolving the region bounded by a horizontal line $y = A$, $A > 0$, the $x$-axis, and the lines $x = a$ and $x = b$ about the $x$-axis. See Figure 14(a). Another familiar example of a solid of revolution, pictured in Figure 14(b), is a *right circular cone* that is generated by revolving the region in the first quadrant bounded by a line $y = mx$, $m > 0$, the $x$-axis, and the line $x = h$ about the $x$-axis.

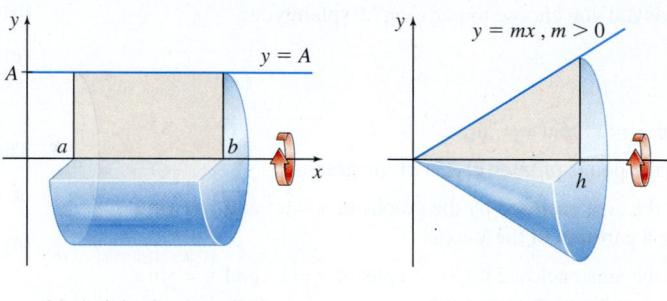

(a) A right circular cylinder       (b) A right circular cone

**Figure 14**

The volume of these simple solids of revolution can be found using geometry formulas, but finding the volume of most solids of revolution requires calculus.

Consider a region that is bounded by the graph of a function $y = f(x)$ that is continuous and nonnegative on the interval $[a, b]$, the $x$-axis, and the lines $x = a$ and $x = b$. See Figure 15(a). Suppose this region is revolved about the $x$-axis. The result is a **solid of revolution**. See Figure 15(b).

To find the volume of a solid of revolution, we begin by partitioning the interval $[a, b]$ into $n$ subintervals:

$$[a, x_1], [x_1, x_2], \ldots, [x_{i-1}, x_i], \ldots, [x_{n-1}, b]$$

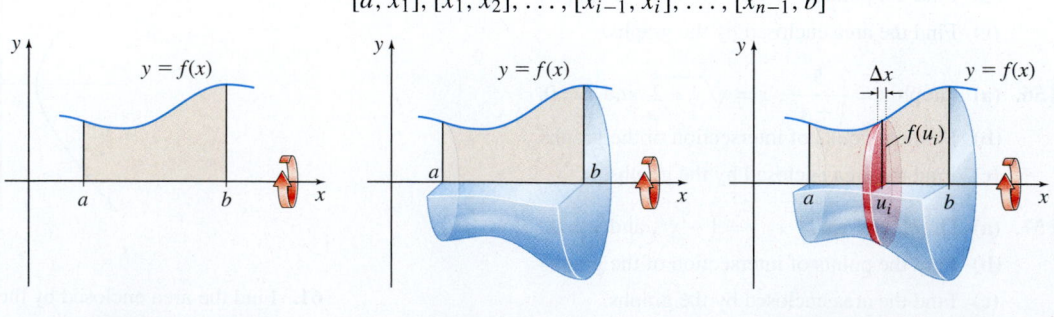

(a) Region to be revolved about the $x$-axis.

(b) Solid of revolution

(c) Radius = $f(u_i)$
Area of circle = $\pi[f(u_i)]^2$
Thickness of disk = $\Delta x$
Volume of disk = $\pi[f(u_i)]^2 \Delta x$

**Figure 15**

each of width $\Delta x = \dfrac{b-a}{n}$ and selecting a number $u_i$ in each subinterval. Refer to Figure 15(c). The circle obtained by slicing the solid at $u_i$ has radius $r_i = f(u_i)$ and area $A_i = \pi r_i^2 = \pi[f(u_i)]^2$. The volume $V_i$ of the disk obtained from the $i$th subinterval is $V_i = \pi[f(u_i)]^2 \Delta x$, and an approximation to the volume $V$ of the solid of revolution is

$$V \approx \pi \sum_{i=1}^{n} [f(u_i)]^2 \, \Delta x$$

**IN WORDS** The volume $V$ of a disk is
$V =$ (Area of the circular cross section)
$\quad \times$ (Thickness)
$\quad = \pi$ (Radius)$^2$ (Thickness)

**NEED TO REVIEW?** Riemann sums and the definite integral are discussed in Section 5.2, pp. 366–370.

As the number $n$ of subintervals increases, the Riemann sums $\pi \sum_{i=1}^{n} [f(u_i)]^2 \Delta x$ become better approximations to the volume $V$ of the solid, and

$$V = \pi \lim_{n \to \infty} \sum_{i=1}^{n} [f(u_i)]^2 \, \Delta x$$

Since the sums are Riemann sums, if $f$ is continuous on the interval $[a, b]$, then the limit is a definite integral and

$$V = \pi \int_a^b [f(x)]^2 \, dx$$

This approach to finding the volume of a solid of revolution is called the **disk method**.

> **Volume of a Solid of Revolution Using the Disk Method**
>
> If a function $f$ is continuous and nonnegative on a closed interval $[a, b]$, then the volume $V$ of the solid of revolution obtained by revolving the region bounded by the graph of $f$, the $x$-axis, and the lines $x = a$ and $x = b$ about the $x$-axis is
>
> $$V = \pi \int_a^b [f(x)]^2 \, dx$$

## 1 Use the Disk Method to Find the Volume of a Solid Formed by Revolving a Region About the $x$-Axis

**EXAMPLE 1**   **Using the Disk Method: Revolving About the $x$-Axis**

Find the volume of the solid of revolution generated by revolving the region bounded by the graph of $y = \sqrt{x}$, the $x$-axis, and the line $x = 5$ about the $x$-axis.

**Solution** We begin by graphing the region to be revolved. See Figure 16(a). Figure 16(b) shows a typical disk and Figure 16(c) shows the solid of revolution. Using the disk method, the volume $V$ of the solid of revolution is

$$V = \pi \int_0^5 (\sqrt{x})^2 \, dx = \pi \int_0^5 x \, dx = \pi \left[\frac{x^2}{2}\right]_0^5 = \frac{25}{2}\pi \text{ cubic units}$$

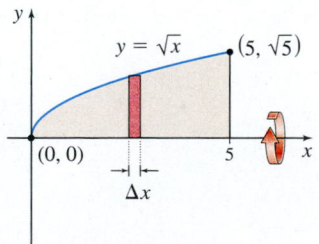

(a) The region to be revolved about the $x$-axis.

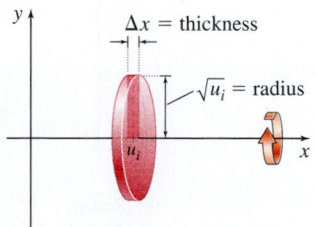

(b) Radius $= \sqrt{u_i}$
Area of circle $= \pi(\sqrt{u_i})^2$
Thickness of disk $= \Delta x$
Volume of disk $= \pi(\sqrt{u_i})^2 \Delta x$

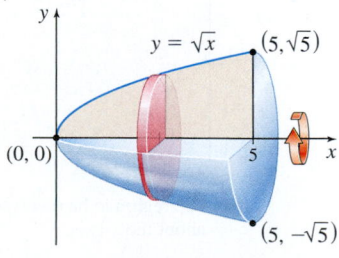

(c) $V = \pi \int_0^5 (\sqrt{x})^2 \, dx$

**DF Figure 16**

**NOW WORK** Problem 5.

The disk method does not require the function $f$ to be nonnegative. As Figure 17 illustrates, the volume $V$ of the solid of revolution obtained by revolving about the $x$-axis the region bounded by the graph of a function $f$ that is continuous on the interval $[a, b]$, the $x$-axis, and the lines $x = a$, and $x = b$ equals the volume of the solid of revolution obtained by revolving about the $x$-axis the region bounded by the graph of $y = |f(x)|$, the $x$-axis, and the lines $x = a$ and $x = b$. Since $|f(x)|^2 = [f(x)]^2$, the formula for $V$ does not change.

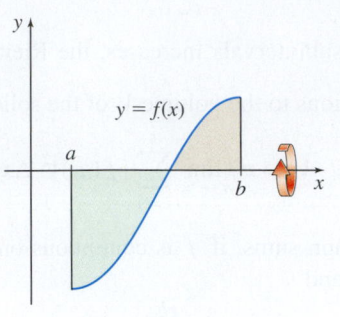

(a) The region to be revolved about the $x$-axis.

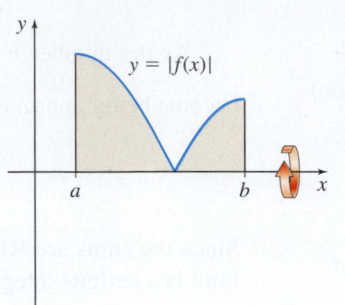

(b) The absolute value of the region to be revolved about the $x$-axis.

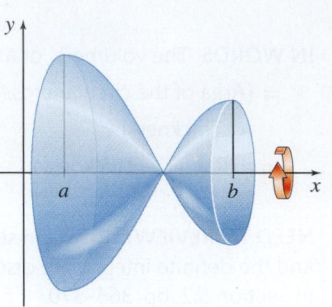

(c) The volume $V$ is the same in both cases:

$$V = \pi \int_a^b [f(x)]^2 \, dx$$

**Figure 17**

EXAMPLE 2  **Using the Disk Method: Revolving About the $x$-Axis**

Find the volume of the solid of revolution generated by revolving the region bounded by the graph of $y = x^3$, the $x$-axis, and the lines $x = -1$ and $x = 2$ about the $x$-axis.

**Solution** Figure 18(a) shows the graph of the region to be revolved about the $x$-axis. Figure 18(b) illustrates a typical disk and Figure 18(c) shows the solid of revolution.

Using the disk method, the volume $V$ of the solid of revolution is

$$V = \pi \int_{-1}^{2} x^6 \, dx = \pi \left[ \frac{x^7}{7} \right]_{-1}^{2} = \frac{\pi}{7}(128 + 1) = \frac{129}{7}\pi \text{ cubic units}$$

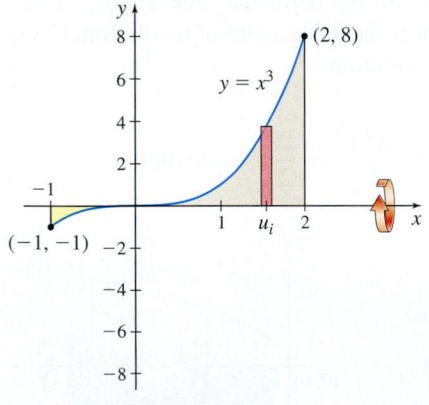

(a) The region to be revolved about the $x$-axis.

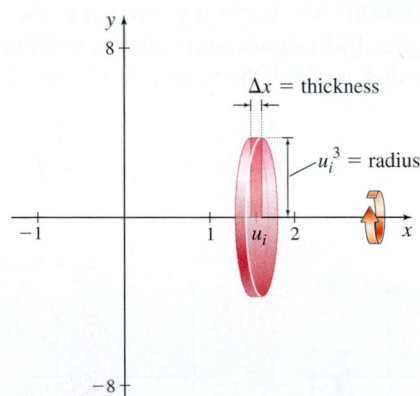

(b) Radius $= u_i^3$
Area of circle $= \pi(u_i^3)^2 = \pi u_i^6$
Thickness of disk $= \Delta x$
Volume of disk $= \pi u_i^6 \Delta x$

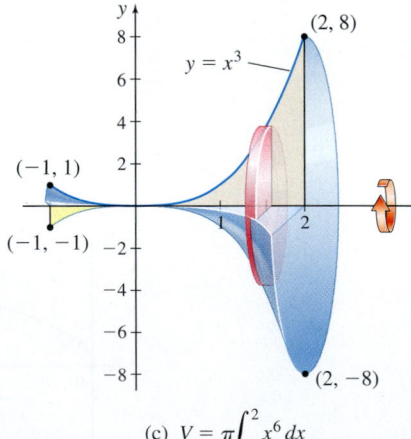

(c) $V = \pi \int_{-1}^{2} x^6 \, dx$

**Figure 18**

■

## 2 Use the Disk Method to Find the Volume of a Solid Formed by Revolving a Region About the $y$-Axis

A solid of revolution can be generated by revolving a region about any line. In particular, the volume $V$ of the solid of revolution generated by revolving the region bounded by the graph of $x = g(y)$, where $g$ is continuous and nonnegative on the closed interval $[c, d]$, the $y$-axis, and the horizontal lines $y = c$ and $y = d$ about the $y$-axis can be obtained by partitioning the $y$-axis and taking slices perpendicular to the $y$-axis of thickness $\Delta y$. See Figure 19(a).

Again, the cross sections are circles, as shown in Figure 19(b). The radius of a typical cross section is $r_i = g(v_i)$, and its area is $A_i = \pi r_i^2 = \pi [g(v_i)]^2$. The volume of the disk of radius $r_i$ and thickness $\Delta y = \dfrac{d - c}{n}$ is $V_i = \pi r_i^2 \Delta y = \pi [g(v_i)]^2 \, \Delta y$. By summing the volumes of all the disks and taking the limit, the volume $V$ of the solid of revolution is given by

$$V = \pi \int_c^d [g(y)]^2 \, dy$$

See Figure 19(c).

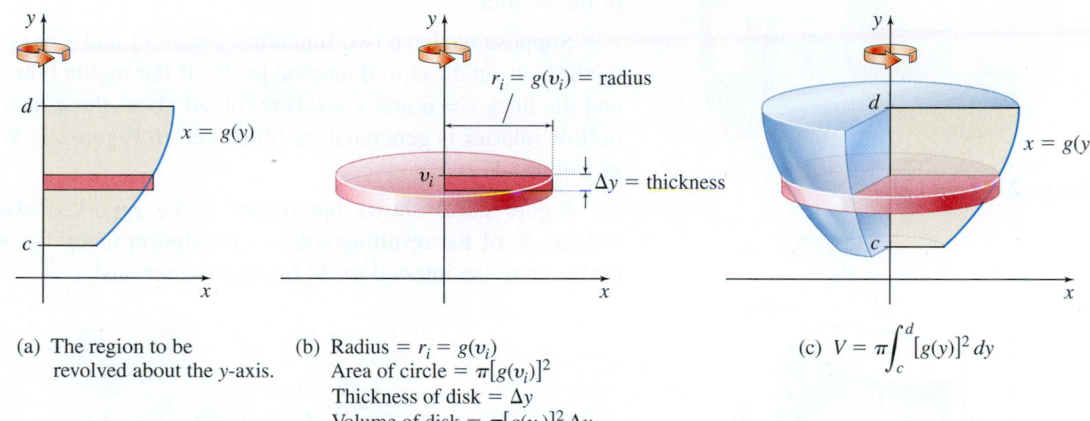

(a) The region to be revolved about the $y$-axis.

(b) Radius $= r_i = g(v_i)$
Area of circle $= \pi [g(v_i)]^2$
Thickness of disk $= \Delta y$
Volume of disk $= \pi [g(v_i)]^2 \, \Delta y$

(c) $V = \pi \displaystyle\int_c^d [g(y)]^2 \, dy$

**DF** **Figure 19**

**EXAMPLE 3   Using the Disk Method: Revolving About the $y$-Axis**

Use the disk method to find the volume of the solid of revolution generated by revolving the region bounded by the graph of $y = x^3$, the $y$-axis, and the lines $y = 1$ and $y = 8$ about the $y$-axis.

**Solution** Figure 20(a) shows the region to be revolved. Since the solid is formed by revolving the region about the $y$-axis, we write $y = x^3$ as $x = \sqrt[3]{y} = y^{1/3}$.

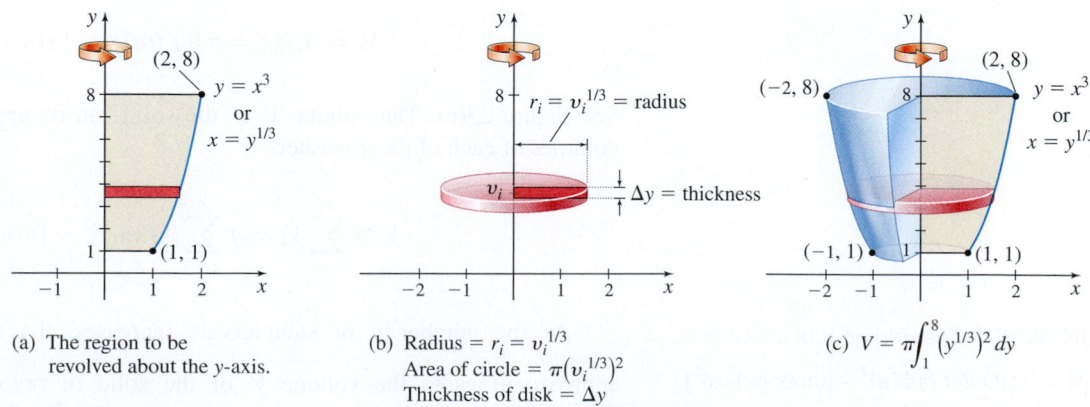

(a) The region to be revolved about the $y$-axis.

(b) Radius $= r_i = v_i^{1/3}$
Area of circle $= \pi (v_i^{1/3})^2$
Thickness of disk $= \Delta y$
Volume of disk $= \pi (v_i^{1/3})^2 \, \Delta y$

(c) $V = \pi \displaystyle\int_1^8 (y^{1/3})^2 \, dy$

**Figure 20**

Figure 20(b) illustrates a typical disk, and Figure 20(c) shows the solid of revolution. Using the disk method, the volume $V$ of the solid of revolution is

$$V = \pi \int_1^8 [y^{1/3}]^2 \, dy = \pi \int_1^8 y^{2/3} \, dy = \pi \left[ \frac{y^{5/3}}{\frac{5}{3}} \right]_1^8$$

$$= \frac{3\pi}{5}(32-1) = \frac{93}{5}\pi \text{ cubic units} \qquad \blacksquare$$

**NOW WORK** Problem 15.

## 3 Use the Washer Method to Find the Volume of a Solid Formed by Revolving a Region About the $x$-Axis

A **washer** is a thin, flat ring with a hole in the middle. It can be represented by two concentric circles, an outer circle of radius $R$ and an inner circle of radius $r$, as shown in Figure 21. The volume of a washer is found by finding the area of the outer circle, $\pi R^2$, subtracting the area of the inner circle, $\pi r^2$, then multiplying the result by the thickness of the washer.

Suppose we have two functions $y = f(x)$ and $y = g(x)$, $f(x) \geq g(x) \geq 0$, that are continuous on the closed interval $[a, b]$. If the region bounded by the graphs of $f$ and $g$ and the lines $x = a$ and $x = b$ is revolved about the $x$-axis, a solid of revolution with a hollow interior is generated, as illustrated in Figure 22. We seek a formula for finding its volume $V$.

Figure 23(a) shows the region to be revolved about the $x$-axis. To find the volume $V$ of the resulting solid of revolution using the washer method, we begin by partitioning the interval $[a, b]$ into $n$ subintervals:

$$[a, x_1], [x_1, x_2], \ldots, [x_{i-1}, x_i], \ldots, [x_{n-1}, b]$$

each of width $\Delta x = \dfrac{b-a}{n}$. In each subinterval $[x_{i-1}, x_i]$, $i = 1, 2, \ldots, n$, we select a number $u_i$. We then slice the solid at $x = u_i$. The slice is a washer that has outer radius $R_i = f(u_i)$, inner radius $r_i = g(u_i)$, and thickness $\Delta x$. A typical washer is shown in Figure 23(b). The area $A_i$ between the concentric circles that form the washer is the difference between the area of the outer circle and the area of the inner circle.

$$A_i = \pi R_i^2 - \pi r_i^2 = \pi[f(u_i)]^2 - \pi[g(u_i)]^2 = \pi\{[f(u_i)]^2 - [g(u_i)]^2\}$$

The volume $V_i$ of the washer is

$$V_i = A_i \Delta x = \pi\{[f(u_i)]^2 - [g(u_i)]^2\}\Delta x$$

See Figure 23(c). The volume $V$ of the solid can be approximated by summing the volumes of each of the $n$ washers.

$$V \approx \sum_{i=1}^n V_i = \pi \sum_{i=1}^n \{[f(u_i)]^2 - [g(u_i)]^2\}\Delta x$$

As the number $n$ of subintervals increases, the sums $\sum_{i=1}^n V_i$ become better approximations to the volume $V$ of the solid of revolution. Since the sums are Riemann sums and since $f$ and $g$ are continuous, the limit as $n \to \infty$ is a definite integral.

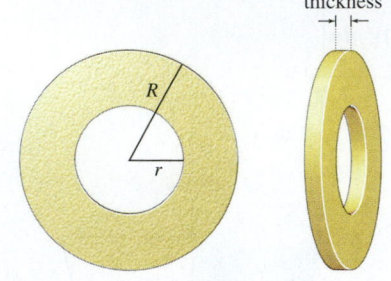

thickness

**Figure 21** A washer.

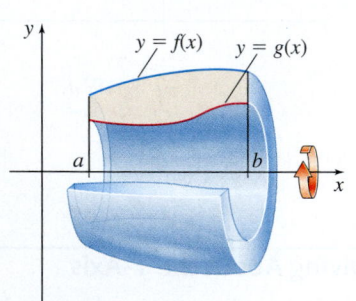

$y = f(x)$    $y = g(x)$

**Figure 22** Solid formed by revolving the region bounded by the graphs of $f$ and $g$ and the lines $x = a$ and $x = b$ about the $x$-axis. Notice that the interior of the solid is hollow.

**IN WORDS** The volume $V$ of a washer is

$V = \pi[(\text{Outer radius})^2 - (\text{Inner radius})^2]$
      $\times (\text{Thickness})$

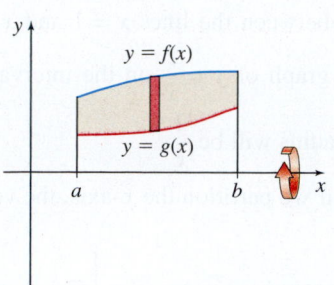 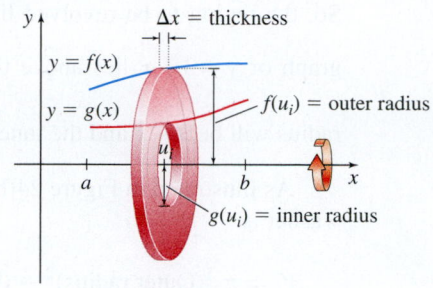 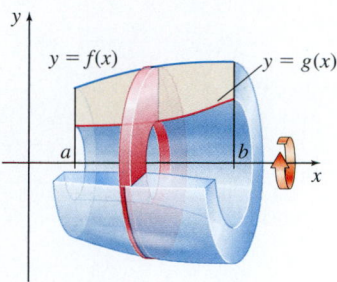

(a) The region to be revolved about the $x$-axis.

(b) Outer radius $= f(u_i)$
Inner radius $= g(u_i)$
Thickness of washer $= \Delta x$
Volume $= \pi\{[f(u_i)]^2 - [g(u_i)]^2\} \Delta x$

(c) $V = \pi \displaystyle\int_a^b \{[f(x)]^2 - [g(x)]^2\} \, dx$

**Figure 23**

## Volume of a Solid of Revolution Using the Washer Method

If the functions $y = f(x)$ and $y = g(x)$ are continuous on the closed interval $[a, b]$ and if $f(x) \geq g(x) \geq 0$ on $[a, b]$, then the volume $V$ of the solid of revolution obtained by revolving the region bounded by the graphs of $f$ and $g$ and the lines $x = a$ and $x = b$ about the $x$-axis is

$$V = \pi \int_a^b \{[f(x)]^2 - [g(x)]^2\} \, dx$$

CALC
CLIP
**EXAMPLE 4**   **Using the Washer Method: Revolving About the $x$-Axis**

Find the volume $V$ of the solid of revolution generated by revolving the region bounded by the graphs of $y = \dfrac{2}{x}$ and $y = 3 - x$ about the $x$-axis.

**Solution** Begin by graphing the two functions. See Figure 24(a). The $x$-coordinates of the points of intersection of the graphs satisfy the equation

$$\frac{2}{x} = 3 - x$$
$$x^2 - 3x + 2 = 0$$
$$(x - 1)(x - 2) = 0$$

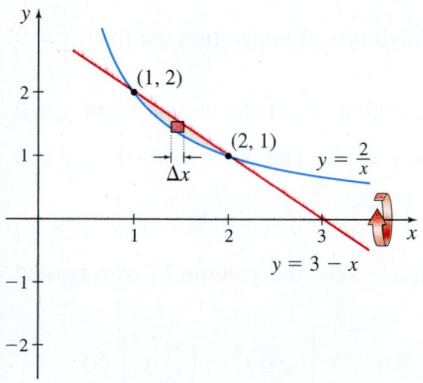 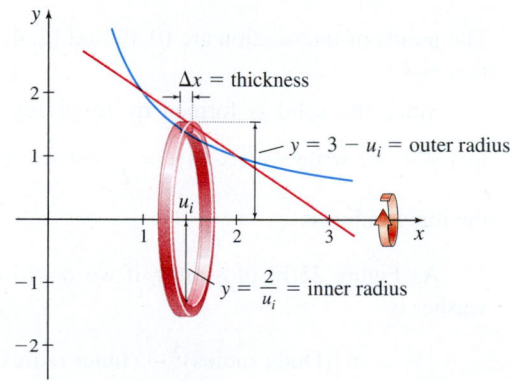 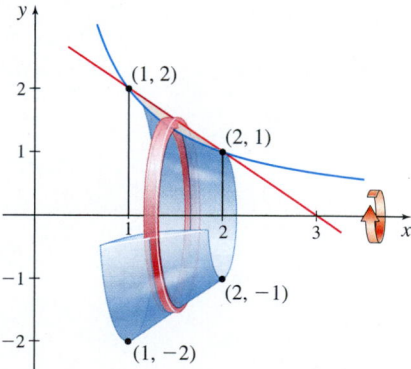

(a) The region to be revolved about the $x$-axis.

(b) Outer radius $= 3 - u_i$
Inner radius $= \dfrac{2}{u_i}$
Thickness of washer $= \Delta x$

Volume $= \pi\left[(3 - u_i)^2 - \left(\dfrac{2}{u_i}\right)^2\right] \Delta x$

(c) $V = \pi \displaystyle\int_1^2 \left[(3 - x)^2 - \left(\dfrac{2}{x}\right)^2\right] dx$

DF **Figure 24**

So, the region to be revolved lies between the lines $x = 1$ and $x = 2$. Notice that the graph of $y = 3 - x$ lies above the graph of $y = \dfrac{2}{x}$ on the interval $[1, 2]$ so the outer radius will be $3 - x$ and the inner radius will be $\dfrac{2}{x}$.

As illustrated in Figure 24(b), if we partition the $x$-axis, the volume $V_i$ of a typical washer is

$$V_i = \pi \left[ (\text{Outer radius})^2 - (\text{Inner radius})^2 \right] \Delta x = \pi \left[ (3 - u_i)^2 - \left( \frac{2}{u_i} \right)^2 \right] \Delta x$$

The volume $V$ of the solid of revolution is

$$V = \pi \int_1^2 \left[ (3 - x)^2 - \left( \frac{2}{x} \right)^2 \right] dx = \pi \int_1^2 \left( 9 - 6x + x^2 - \frac{4}{x^2} \right) dx$$

$$= \pi \left[ 9x - 3x^2 + \frac{x^3}{3} + \frac{4}{x} \right]_1^2 = \frac{\pi}{3} \text{ cubic units} \qquad\blacksquare$$

NOW WORK Problem 33.

## 4 Use the Washer Method to Find the Volume of a Solid Formed by Revolving a Region About the $y$-Axis

If we use the washer method and the region bounded by the graphs of two functions is revolved about the $y$-axis, we partition the $y$-axis, and the thickness of a typical washer will be $\Delta y$.

### EXAMPLE 5    Using the Washer Method: Revolving About the $y$-Axis

Find the volume $V$ of the solid of revolution generated by revolving the region enclosed by the graphs of $y = 2x$ and $y = x^2$ about the $y$-axis.

**Solution** Begin by graphing the two functions. See Figure 25(a). The $x$-coordinates of the points of intersection of the graphs satisfy the equation

$$2x = x^2$$
$$x^2 - 2x = 0$$
$$x(x - 2) = 0$$
$$x = 0 \quad \text{or} \quad x = 2$$

The points of intersection are $(0, 0)$ and $(2, 4)$. The limits of integration are from $y = 0$ to $y = 4$.

Since the solid is formed by revolving the region about the $y$-axis from $y = 0$ to $y = 4$, we write $y = 2x$ as $x = \dfrac{y}{2}$ and $y = x^2$ as $x = \sqrt{y}$. The outer radius is $\sqrt{y}$ and the inner radius is $\dfrac{y}{2}$.

As Figure 25(b) illustrates, if we partition the $y$-axis, the volume $V_i$ of a typical washer is

$$V_i = \pi \left[ (\text{Outer radius})^2 - (\text{Inner radius})^2 \right] \Delta y = \pi \left[ (\sqrt{v_i})^2 - \left( \frac{v_i}{2} \right)^2 \right] \Delta y$$

The volume $V$ of the solid of revolution shown in Figure 25(c) is

$$V = \pi \int_0^4 \left[ (\sqrt{y})^2 - \left( \frac{y}{2} \right)^2 \right] dy = \pi \int_0^4 \left( y - \frac{y^2}{4} \right) dy$$

$$= \pi \left[ \frac{y^2}{2} - \frac{y^3}{12} \right]_0^4 = \frac{8\pi}{3} \text{ cubic units}$$

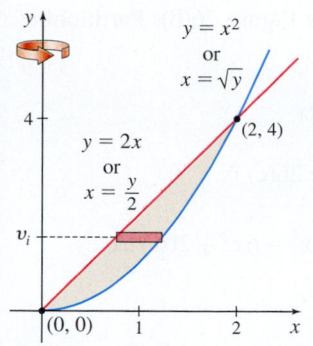

(a) The region to be revolved about the $y$-axis.

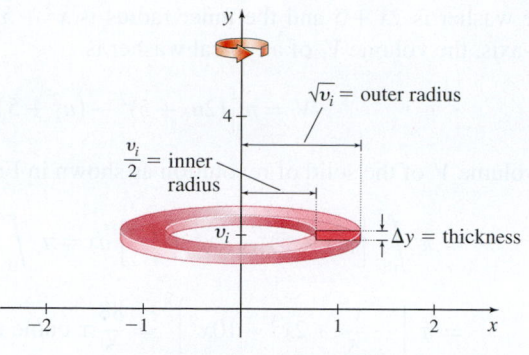

(b) Outer radius = $\sqrt{v_i}$
Inner radius = $\dfrac{v_i}{2}$
Thickness of washer = $\Delta y$
Volume = $\pi\left[\left(\sqrt{v_i}\right)^2 - \left(\dfrac{v_i}{2}\right)^2\right]\Delta y$

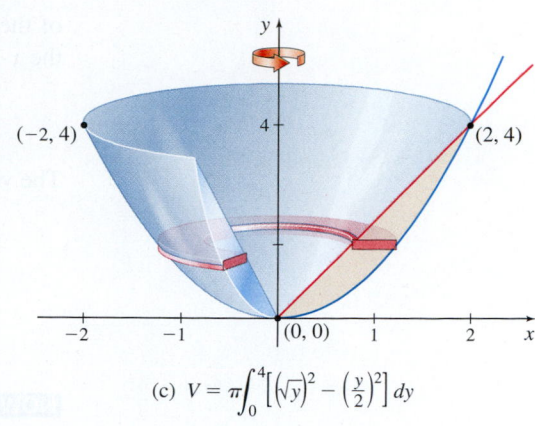

(c) $V = \pi\displaystyle\int_0^4\left[\left(\sqrt{y}\right)^2 - \left(\dfrac{y}{2}\right)^2\right]dy$

**DF** **Figure 25**

**NOW WORK** Problem 21.

## 5 Find the Volume of a Solid Formed by Revolving a Region About a Line Parallel to a Coordinate Axis

Earlier we stated that a solid of revolution can be generated by revolving a region about any line. Now we investigate how to find the volume of a solid formed by revolving a region about a line parallel to either the $x$-axis or the $y$-axis.

**EXAMPLE 6** **Using the Washer Method: Revolving About a Horizontal Line**

Find the volume $V$ of the solid of revolution generated by revolving the region bounded by the graphs of $y = 2x$ and $y = x^2$ about the line $y = -5$.

**Solution** The region to be revolved is the same as in Example 5, but it is now being revolved about the line $y = -5$. Figure 26 illustrates the region, a typical washer, and the solid of revolution.

In Figure 26(a) notice that the line $y = 2x$ lies $2x + 5$ units above the line $y = -5$ and the parabola $y = x^2$ lies $x^2 + 5$ units above the line $y = -5$. So the outer radius

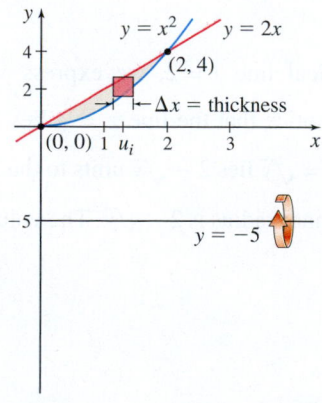

(a) The region to be revolved about the line $y = -5$.

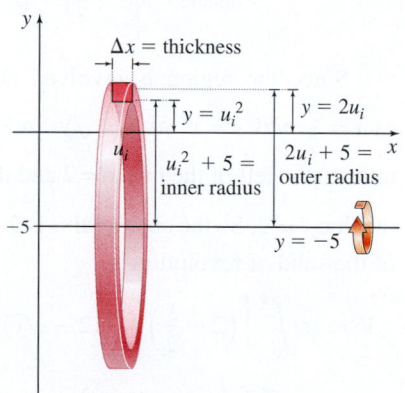

(b) Outer radius = $2u_i + 5$
Inner radius = $u_i^2 + 5$
Thickness of washer = $\Delta x$
Volume = $\pi[(2u_i + 5)^2 - (u_i^2 + 5)^2]\Delta x$

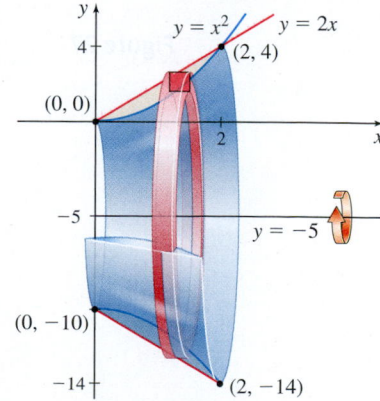

(c) $V = \pi\displaystyle\int_0^2[(2x + 5)^2 - (x^2 + 5)^2]\,dx$

**Figure 26**

of the washer is $2x + 5$ and the inner radius is $x^2 + 5$. See Figure 26(b). Partitioning the $x$-axis, the volume $V_i$ of a typical washer is

$$V_i = \pi \left[ (2u_i + 5)^2 - \left(u_i^2 + 5\right)^2 \right] \Delta x$$

The volume $V$ of the solid of revolution as shown in Figure 26(c) is

$$V = \pi \int_0^2 \left[ (2x + 5)^2 - (x^2 + 5)^2 \right] dx = \pi \int_0^2 (-x^4 - 6x^2 + 20x) \, dx$$

$$= \pi \left[ -\frac{x^5}{5} - 2x^3 + 10x^2 \right]_0^2 = \frac{88}{5} \pi \text{ cubic units} \quad \blacksquare$$

**NOW WORK** Problem 39.

---

**EXAMPLE 7**  Using the Washer Method: Revolving About a Vertical Line

Find the volume of the solid of revolution generated by revolving the region bounded by the graphs of $y = 2x$ and $y = x^2$ about the line $x = 2$.

**Solution**  This example is similar to Example 5 except that the region is revolved about the line $x = 2$. Figure 27 shows the graph of the region, a typical washer, and the solid of revolution.

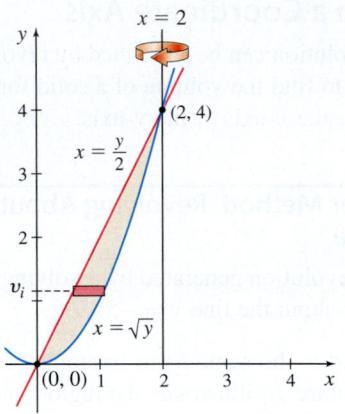

(a) The region to be revolved about the line $x = 2$.

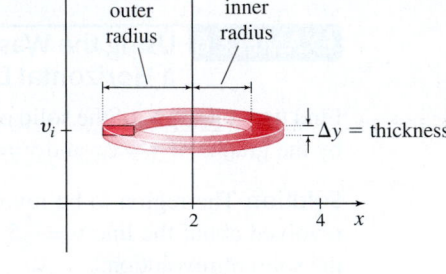

(b) Outer radius $= 2 - \dfrac{v_i}{2}$
Inner radius $= 2 - \sqrt{v_i}$
Thickness of washer $= \Delta y$
Volume $= \pi \left[ \left(2 - \dfrac{v_i}{2}\right)^2 - \left(2 - \sqrt{v_i}\right)^2 \right] \Delta y$

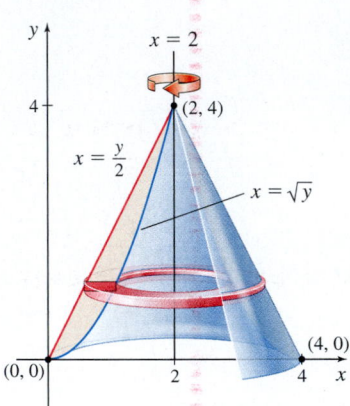

(c) $V = \pi \int_0^4 \left[ \left(2 - \dfrac{y}{2}\right)^2 - \left(2 - \sqrt{y}\right)^2 \right] dy$

**Figure 27**

Since the region is revolved about the vertical line $x = 2$, we express $y = 2x$ as $x = \dfrac{y}{2}$ and $y = x^2$ as $x = \sqrt{y}$. In Figure 27(a), notice that the line $x = \dfrac{y}{2}$ lies $2 - \dfrac{y}{2}$ units to the left of the line $x = 2$ and the graph of $x = \sqrt{y}$ lies $2 - \sqrt{y}$ units to the left of the line $x = 2$. So the outer radius is $2 - \dfrac{y}{2}$ and the inner radius is $2 - \sqrt{y}$. The volume $V$ of the solid of revolution is

$$V = \pi \int_0^4 \left[ \left(2 - \frac{y}{2}\right)^2 - (2 - \sqrt{y})^2 \right] dy$$

$$= \pi \int_0^4 \left[ \left(4 - 2y + \frac{y^2}{4}\right) - (4 - 4\sqrt{y} + y) \right] dy$$

$$= \pi \int_0^4 \left( \frac{y^2}{4} - 3y + 4\sqrt{y} \right) dy = \pi \left[ \frac{y^3}{12} - \frac{3y^2}{2} + \frac{8y^{3/2}}{3} \right]_0^4 = \frac{8}{3} \pi \text{ cubic units} \quad \blacksquare$$

**NOW WORK** Problem 41.

# 6.2 Assess Your Understanding

## Concepts and Vocabulary

1. If a function $f$ is continuous on a closed interval $[a, b]$, then the volume $V$ of the solid of revolution obtained by revolving the region bounded by the graph of $f$, the $x$-axis, and the lines $x = a$ and $x = b$ about the $x$-axis, is found using the formula $V = $ _____.

2. *True or False*  When the region bounded by the graphs of the functions $f$ and $g$ and the lines $x = a$ and $x = b$ is revolved about the $x$-axis, $f(x) \geq g(x) \geq 0$, the cross section exposed by making a slice at $u_i$ perpendicular to the $x$-axis is two concentric circles, and the area $A_i$ between the circles is

$$A_i = \pi [f(u_i) - g(u_i)]^2$$

3. *True or False*  If the functions $f$ and $g$ are continuous on the closed interval $[a, b]$ and if $f(x) \geq g(x) \geq 0$ on the interval, then the volume $V$ of the solid of revolution obtained by revolving the region bounded by the graphs of $f$ and $g$ and the lines $x = a$ and $x = b$ about the $x$-axis is

$$V = \pi \int_a^b [f(x) - g(x)]^2 \, dx$$

4. *True or False*  If the region bounded by the graphs of $y = x^2$ and $y = 2x$ is revolved about the line $y = 6$, the volume $V$ of the solid of revolution generated is found by finding the integral

$$V = \pi \int_6^7 (4x^2 - x^4) \, dx$$

## Skill Building

*In Problems 5–10, find the volume of the solid of revolution generated by revolving the region shown below about the indicated axis.*

5. $y = 2\sqrt{x}$ about the $x$-axis

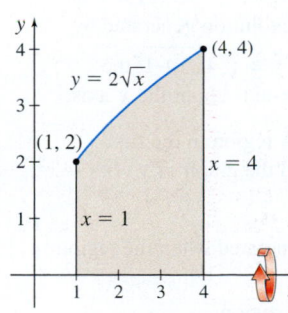

6. $y = x^4$ about the $y$-axis

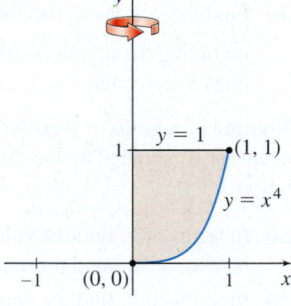

7. $y = \dfrac{1}{x}$ about the $y$-axis

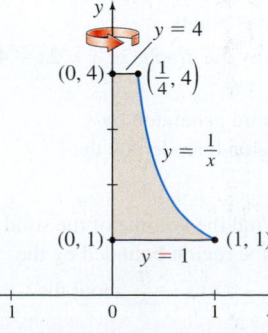

8. $y = x^{2/3}$ about the $y$-axis

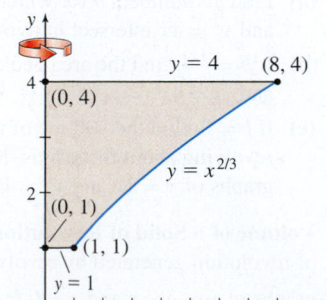

9. $y = \sec x$ about the $x$-axis

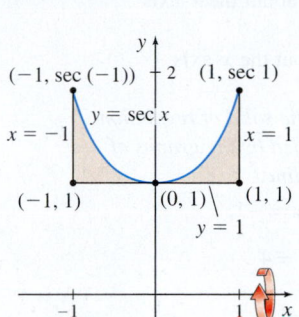

10. $y = x^2$ about the $y$-axis

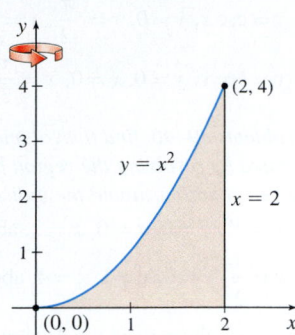

*In Problems 11–16, use the disk method to find the volume of the solid of revolution generated by revolving the region bounded by the graphs of the given equations about the indicated axis.*

11. $y = 2x^2$, the $x$-axis, $x = 1$; about the $x$-axis

12. $y = \sqrt{x}$, the $x$-axis, $x = 4$, $x = 9$; about the $x$-axis

13. $y = e^{-x}$, the $x$-axis, $x = 0$, $x = 2$; about the $x$-axis

14. $y = e^x$, the $x$-axis, $x = -1$, $x = 1$; about the $x$-axis

15. $y = x^2$, $x \geq 0$, $y = 1$, $y = 4$; about the $y$-axis

16. $y = 2\sqrt{x}$, the $y$-axis, $y = 4$; about the $y$-axis

*In Problems 17–22, use the washer method to find the volume of the solid of revolution generated by revolving the region bounded by the graphs of the given equations about the indicated axis.*

17. $y = x^2$, $x \geq 0$, the $y$-axis, $y = 4$; about the $x$-axis

18. $y = 2x^2$, $x \geq 0$, the $y$-axis, $y = 2$; about the $x$-axis

19. $y = 2\sqrt{x}$, the $y$-axis, $y = 4$; about the $x$-axis

20. $y = x^{2/3}$, the $x$-axis, $x = 8$; about the $y$-axis

21. $y = x^3$, the $x$-axis, $x = 1$; about the $y$-axis

22. $y = 2x^4$, the $x$-axis, $x = 1$; about the $y$-axis

*In Problems 23–38, find the volume of the solid of revolution generated by revolving the region bounded by the graphs of the given equations about the indicated axis.*

23. $y = \dfrac{1}{x}$, the $x$-axis, $x = 1$, $x = 2$; about the $x$-axis

24. $y = \dfrac{1}{x}$, the $x$-axis, $x = 1$, $x = 2$; about the $y$-axis

25. $y = \sqrt{x}$, the $y$-axis, $y = 9$; about the $y$-axis

26. $y = \sqrt{x}$, the $y$-axis, $y = 9$; about the $x$-axis

27. $y = (x - 2)^3$, the $x$-axis, $x = 0$, $x = 3$; about the $x$-axis

28. $y = (x - 2)^3$, the $x$-axis, $x = 0$, $x = 3$; about the $y$-axis

29. $y = (x + 1)^2$, $x \geq 0$, $y = 16$; about the $y$-axis

30. $y = (x + 1)^2$, $x \leq 0$, $y = 16$; about the $x$-axis

31. $x = y^4 - 1$, the $y$-axis; about the $y$-axis

32. $y = x^4 - 1$, the $x$-axis; about the $x$-axis

33. $y = 4x$, $y = x^3$, $x \geq 0$, about the $x$-axis

---

**1.** = NOW WORK problem      = Graphing technology recommended      **CAS** = Computer Algebra System recommended

**34.** $y = 2x + 1$, $y = x$, $x = 0$, $x = 3$; about the $x$-axis

**35.** $y = 1 - x$, $y = e^x$, $x = 1$; about the $x$-axis

**36.** $y = \cos x$, $y = \sin x$, $x = 0$, $x = \dfrac{\pi}{4}$; about the $x$-axis

**37.** $y = \csc x$, $y = 0$, $x = \dfrac{\pi}{2}$, $x = \dfrac{3\pi}{4}$; about the $x$-axis

**38.** $y = \sec x$, $y = 0$, $x = 0$, $x = \dfrac{\pi}{3}$; about the $x$-axis

*In Problems 39–46, find the volume of the solid of revolution generated by revolving the region bounded by the graphs of the given equations about the indicated line.*

**39.** $y = e^x$, $y = 0$, $x = 0$, $x = 2$; about $y = -1$

**40.** $y = \dfrac{1}{x}$, $y = 0$, $x = 1$, $x = 4$; about $y = 4$

**41.** $y = x^2$, the $x$-axis, $x = 1$; about $x = 1$

**42.** $y = x^3$, $x = 0$, $y = 1$; about $x = -1$

**43.** $y = \sqrt{x}$, the $x$-axis, $x = 4$; about $x = -4$

**44.** $y = \dfrac{1}{\sqrt{x}}$, the $x$-axis, $x = 1$, $x = 4$; about $x = 4$

**45.** $y = \dfrac{1}{x^2}$, $y = 0$, $x = 1$, $x = 4$; about $y = 4$

**46.** $y = \sqrt{x}$, $y = 0$, $0 \le x \le 4$; about $y = -4$

*In Problems 47–50, find the volume of the solid of revolution generated by revolving the indicated region about each line.*

**47.** The region bounded by $y = x^2$, the $x$-axis, and $x = 3$

   **(a)** About the $x$-axis

   **(b)** About the line $y = -1$

   **(c)** About the line $y = 10$

   **(d)** About the line $y = a$, $a \ge 9$

**48.** The region bounded by $y = x^2$, the $x$-axis, and $x = 3$

   **(a)** About the $y$-axis

   **(b)** About the line $x = -5$

   **(c)** About the line $x = 5$

   **(d)** About the line $x = b$, $b \ge 3$

**49.** The region bounded by $y = x^2$, the $y$-axis, and $y = 4$

   **(a)** About the $y$-axis

   **(b)** About the line $x = -5$

   **(c)** About the line $x = 5$

   **(d)** About the line $x = b$, $b \ge 2$

**50.** The region bounded by $y = x^2$, the $y$-axis, and $y = 4$

   **(a)** About the $x$-axis

   **(b)** About the line $y = -1$

   **(c)** About the line $y = 4$

   **(d)** About the line $y = a$, $a \ge 4$

## Applications and Extensions

**51. Volume of a Sphere**

   **(a)** Graph $y = \sqrt{a^2 - x^2}$.

   **(b)** Revolve the region bounded by $y = \sqrt{a^2 - x^2}$ and the $x$-axis about the $x$-axis to generate a sphere of radius $a$. Use the Disk Method to show that the volume $V$ of the sphere is $\dfrac{4}{3}\pi a^3$.

**52. Volume of a Cone**

   **(a)** Find an equation of the line segment from the point $(0, h)$, $h > 0$, to the point $(a, 0)$, $a > 0$.

   **(b)** Graph the line segment from (a).

   **(c)** Revolve the region in the first quadrant bounded by the line segment from (a), the $x$-axis, and the $y$-axis, about the $y$-axis to generate a cone of height $h$ and radius $a$. Use the Disk Method to show that the volume $V$ of the cone is $\dfrac{1}{3}\pi a^2 h$.

**53. Volume of a Solid of Revolution**

The figure shows the solid of revolution generated by revolving the region bounded by the graph of $y = \dfrac{1}{x^2 + 4}$ and the $x$-axis from $x = 0$ to $x = 1$ about the $x$-axis.

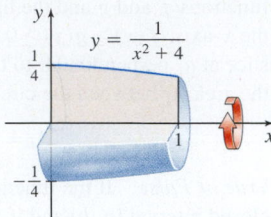

   **(a)** Express the volume of the solid of revolution as an integral.

   **(b)** Use technology to find the volume.

**54. Volume of a Solid of Revolution** A solid of revolution is generated by revolving the region bounded by the graph of $y = \ln x$, the line $x = e$, and the $x$-axis about the $x$-axis.

   **(a)** Express the volume of the solid of revolution as an integral.

   **(b)** Use technology to find the volume.

**55. Volume of an Ellipsoid**

   **(a)** Graph $y = \sqrt{9 - 4x^2}$, the upper half of an ellipse.

   **(b)** Find the volume of the solid of revolution generated by revolving the region bounded by $y = \sqrt{9 - 4x^2}$ and the $x$-axis about the $x$-axis.

**56. Volume of a Solid of Revolution**

   **(a)** Graph $y = \sqrt{4x^2 + 1}$, the upper portion of a hyperbola.

   **(b)** Find the volume of the solid of revolution generated by revolving the region bounded by $y = \sqrt{4x^2 + 1}$, the lines $x = -1$ and $x = 1$, and the $x$-axis about the $x$-axis.

**57. Volume of a Solid of Revolution** A region in the first quadrant is bounded by the $x$-axis and the graph of $y = kx - x^2$, where $k > 0$.

   **(a)** In terms of $k$, find the volume generated when the region is revolved around the $x$-axis.

   **(b)** In terms of $k$, find the area of the region.

**58. Volume of a Solid of Revolution**

   **(a)** Find all numbers $b$ for which the graphs of $y = 2x + b$ and $y^2 = 4x$ intersect in two distinct points.

   **(b)** If $b = -4$, find the area enclosed by the graphs of $y = 2x - 4$ and $y^2 = 4x$.

   **(c)** If $b = 0$, find the volume of the solid generated by revolving about the $x$-axis the region bounded by the graphs of $y = 2x$ and $y^2 = 4x$.

**59. Volume of a Solid of Revolution** Find the volume of the solid of revolution generated by revolving the region bounded by the graphs of $y = \cos x$ and $x = 0$ from $x = 0$ to $x = \dfrac{\pi}{2}$ about the line $y = 1$. *Hint:* $\cos^2 x = \dfrac{1 + \cos(2x)}{2}$.

**60. Volume of a Solid of Revolution** Find the volume of the solid of revolution generated by revolving the region bounded by the graphs of $y = \cos x$ and $x = 0$ from $x = 0$ to $x = \dfrac{\pi}{2}$ about the line $y = -1$. (See the hint in Problem 59.)

### Challenge Problems

**61. Volume of a Solid of Revolution** The graph of the function $P(x) = kx^2$ is symmetric with respect to the $y$-axis and contains the points $(0, 0)$ and $(b, e^{-b^2})$, where $b > 0$.

(a) Find $k$ and write an equation for $P = P(x)$.

(b) The region bounded by the graph of $P$, the $y$-axis, and the line $y = e^{-b^2}$ is revolved about the $y$-axis to form a solid. Find its volume.

(c) For what number $b$ is the volume of the solid in (b) a maximum? Justify your answer.

**62. Volume of a Solid of Revolution** Find the volume of the solid generated by revolving the region bounded by the catenary $y = a \cosh\left(\dfrac{x}{a}\right) + b - a$, the $x$-axis, $x = 0$, and $x = 1$ about the $x$-axis. *Hint:* $\cosh^2 x = \dfrac{\cosh(2x) + 1}{2}$.

# 6.3 Volume of a Solid of Revolution: Cylindrical Shells

**OBJECTIVES** *When you finish this section, you should be able to:*

1 Use the shell method to find the volume of a solid formed by revolving a region about the $y$-axis (p. 447)

2 Use the shell method to find the volume of a solid formed by revolving a region about the $x$-axis (p. 451)

3 Use the shell method to find the volume of a solid formed by revolving a region about a line parallel to a coordinate axis (p. 453)

**NOTE** Sewer lines and the water pipes in a house are cylindrical shells.

**Figure 28** $V = \pi R^2 h - \pi r^2 h$

**NEED TO REVIEW?** The volume of a right circular cylinder of height $h$ and radius $r$ is $V = \pi r^2 h$

There are solids of revolution for which the volume is difficult to find using the disk or washer method. In these situations, the volume can often be found using *cylindrical shells*.

A **cylindrical shell** is the solid between two concentric cylinders, as shown in Figure 28. If the inner radius of the cylinder is $r$ and the outer radius is $R$, the volume $V$ of a cylindrical shell of height $h$ is

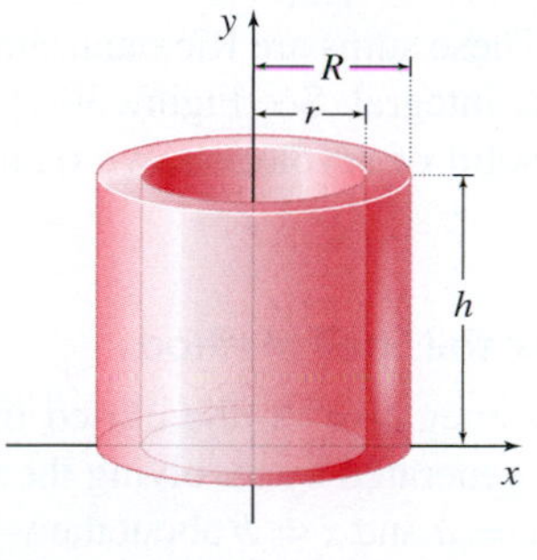

$$\boxed{V = \pi R^2 h - \pi r^2 h}$$

That is, the volume of a cylindrical shell equals the volume of the larger cylinder, which has radius $R$, minus the volume of the smaller cylinder, which has radius $r$. It is convenient to write this formula as

$$V = \pi(R^2 - r^2)h = \pi(R + r)(R - r)h = 2\pi\left(\frac{R + r}{2}\right)h(R - r)$$

$$\boxed{V = 2\pi\left(\frac{R + r}{2}\right)h(R - r)}$$

$$V = 2\pi \text{ (Average radius) (Height) (Thickness)}$$

## 1 Use the Shell Method to Find the Volume of a Solid Formed by Revolving a Region About the $y$-Axis

Suppose a function $y = f(x)$ is nonnegative and continuous on the closed interval $[a, b]$, where $a \geq 0$. We seek the volume $V$ of the solid generated by revolving the region bounded by the graph of $f$, the $x$-axis, and the lines $x = a$ and $x = b$ about the $y$-axis. See Figure 29.

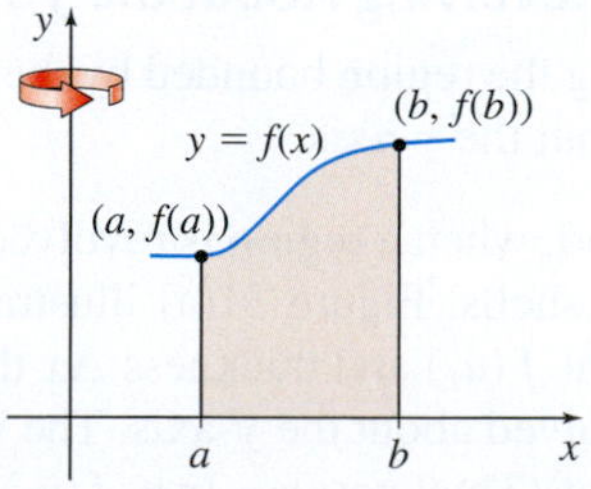

We begin by partitioning the interval $[a, b]$ into $n$ subintervals:

$$[a, x_1], [x_1, x_2], \ldots, [x_{i-1}, x_i], \ldots, [x_{n-1}, b]$$

**Figure 29**

each of width $\Delta x = \dfrac{b - a}{n}$. We concentrate on the rectangle whose base is the subinterval $[x_{i-1}, x_i]$ and whose height is $f(u_i)$, where $u_i = \dfrac{x_{i-1} + x_i}{2}$ is the midpoint of the subinterval. See Figure 30(a). When this rectangle is revolved about the $y$-axis, it generates a cylindrical shell of average radius $u_i$, height $f(u_i)$, and thickness $\Delta x$, as shown in Figure 30(b). The volume $V_i$ of this cylindrical shell is

$$V_i = 2\pi \, (\text{Average radius})(\text{Height})(\text{Thickness}) = 2\pi u_i f(u_i)\Delta x$$

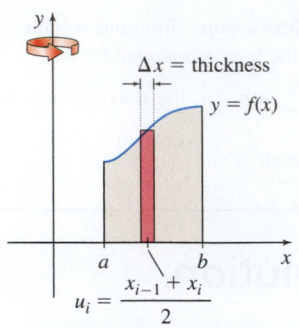

(a) The region to be revolved about the $y$-axis.

$$u_i = \frac{x_{i-1} + x_i}{2}$$

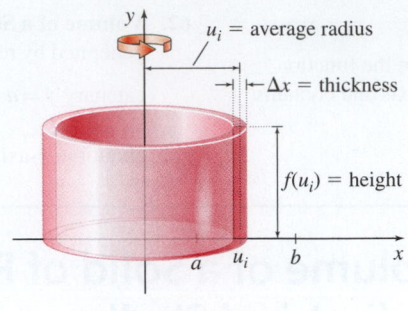

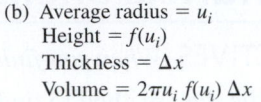

(b) Average radius $= u_i$
Height $= f(u_i)$
Thickness $= \Delta x$
Volume $= 2\pi u_i \, f(u_i) \, \Delta x$

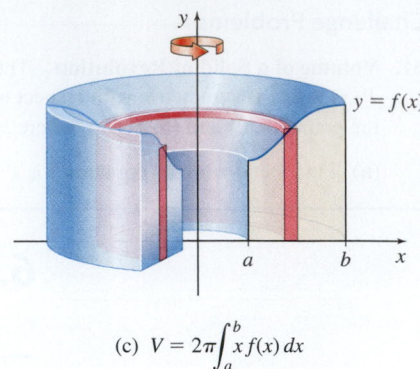

(c) $V = 2\pi \displaystyle\int_a^b xf(x)\,dx$

**Figure 30**

**IN WORDS** The volume $V$ of a cylindrical shell is

$V = (\text{circumference})(\text{height})(\text{thickness})$

$\quad = (2\pi \text{ radius})(\text{height})(\text{thickness})$

The sum of the volumes of the $n$ cylindrical shells approximates the volume $V$ of the solid generated by revolving the region bounded by the graph of $y = f(x)$, the $x$-axis, and the lines $x = a$ and $x = b$ about the $y$-axis. That is,

$$V \approx \sum_{i=1}^{n} [2\pi u_i f(u_i)\Delta x] = 2\pi \sum_{i=1}^{n} [u_i f(u_i)\Delta x]$$

As the number $n$ of subintervals increases, the sums $2\pi \sum_{i=1}^{n} [u_i f(u_i)\Delta x]$ become better approximations to the volume $V$ of the solid. These sums are Riemann sums, and since $f$ is continuous on $[a, b]$, the limit is a definite integral. See Figure 30(c). When cylindrical shells are used to find the volume of a solid of revolution, we refer to the process as the **shell method**.

### Volume of a Solid of Revolution About the $y$-Axis: the Shell Method

If $y = f(x)$ is a function that is continuous and nonnegative on the closed interval $[a, b]$, where $a \geq 0$, then the volume $V$ of the solid generated by revolving the region bounded by the graph of $f$, the $x$-axis, and the lines $x = a$ and $x = b$ about the $y$-axis is

$$V = 2\pi \int_a^b xf(x)\,dx$$

It can be shown that the shell method and the washer method of Section 6.2 are equivalent; that is, they both give the same answer.* The advantage of having two equivalent, yet different, formulas is flexibility. There are times when one of the two methods is easier to use, as Example 1 illustrates.

### EXAMPLE 1  Finding the Volume of a Solid: Revolving About the $y$-Axis

Find the volume $V$ of the solid generated by revolving the region bounded by the graphs of $f(x) = x^2 + 2x$, the $x$-axis, and the line $x = 1$ about the $y$-axis.

**Solution** *Using the shell method:* In the shell method, when a region is revolved about the $y$-axis, we partition the $x$-axis and use vertical shells. Figure 31(a) illustrates the region to be revolved and a typical rectangle of height $f(u_i)$ and thickness $\Delta x$ that will become a shell with average radius $u_i$ when it is revolved about the $y$-axis. The volume of a typical shell is $V_i = 2\pi \, (\text{Average radius})(\text{Height})(\text{Thickness}) = 2\pi u_i f(u_i)\Delta x$, as

*This topic is discussed in detail in an article by Charles A. Cable (February 1984), "The Disk and Shell Method," *American Mathematical Monthly*, 91(2), 139.

shown in Figure 31(b). Figure 31(c) illustrates the solid of revolution. The volume $V$ of the solid of revolution is

$$V = 2\pi \int_0^1 x f(x)\, dx = 2\pi \int_0^1 [x(x^2 + 2x)]\, dx = 2\pi \int_0^1 (x^3 + 2x^2)\, dx$$

$$= 2\pi \left[\frac{x^4}{4} + \frac{2x^3}{3}\right]_0^1 = \frac{11\pi}{6} \text{ cubic units}$$

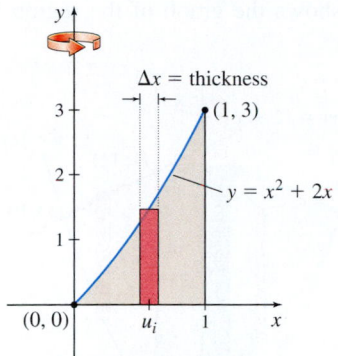

(a) The region to be revolved about the $y$-axis.

(b) Average radius $= u_i$
Height $= f(u_i) = u_i^2 + 2u_i$
Thickness $= \Delta x$
Volume $= 2\pi u_i (u_i^2 + 2u_i)\Delta x$

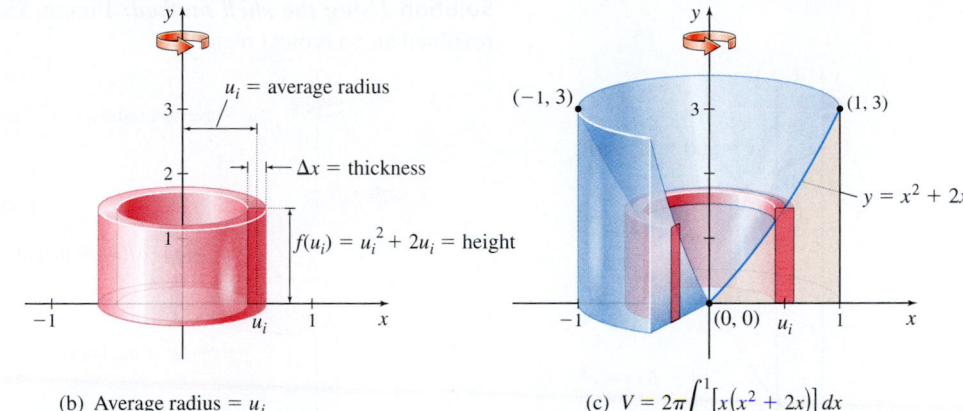

(c) $V = 2\pi \int_0^1 [x(x^2 + 2x)]\, dx$

**DF Figure 31** The shell method

**Using the washer method:** Using the washer method, a revolution about the $y$-axis requires integration with respect to $y$. This means we need to find the inverse function of $y = f(x)$. We treat $x^2 + 2x - y = 0$ as a quadratic equation in the variable $x$ and use the quadratic formula with $a = 1$, $b = 2$, and $c = -y$ to obtain $x = g(y) = -1 \pm \sqrt{1 + y}$. Since $x \geq 0$, we use the $+$ sign.

See Figure 32. The volume of a typical washer is

$$V_i = \pi[(\text{Outer radius})^2 - (\text{Inner radius})^2] \times (\text{Thickness}).$$

The volume $V$ of the solid of revolution is

$$V = \pi \int_0^3 \left[1^2 - \left(-1 + \sqrt{1 + y}\right)^2\right] dy = \pi \int_0^3 [1 - (1 - 2\sqrt{1 + y} + 1 + y)]\, dy$$

$$= \pi \int_0^3 [2\sqrt{1 + y} - 1 - y]\, dy = \pi \int_0^3 (2\sqrt{1 + y})\, dy - \pi \int_0^3 (1 + y)\, dy$$

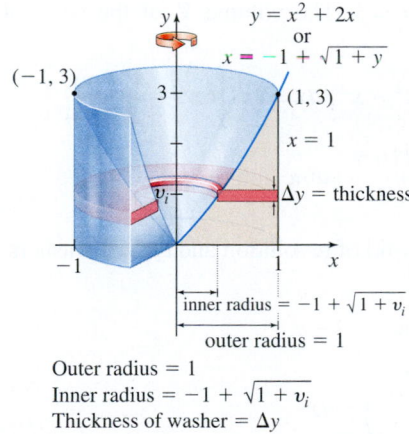

Outer radius $= 1$
Inner radius $= -1 + \sqrt{1 + v_i}$
Thickness of washer $= \Delta y$
Volume $= \pi\left[1^2 - \left(-1 + \sqrt{1 + v_i}\right)^2\right]\Delta y$

**Figure 32** The washer method

The two integrals are found as follows:

- $\pi \displaystyle\int_0^3 (2\sqrt{1 + y})\, dy = 2\pi \int_1^4 u^{1/2}\, du = 2\pi \left[\frac{u^{3/2}}{\frac{3}{2}}\right]_1^4 = \frac{28\pi}{3}$

  Let $u = 1 + y$;
  then $du = dy$

- $\pi \displaystyle\int_0^3 (1 + y)\, dy = \pi \left[y + \frac{y^2}{2}\right]_0^3 = \frac{15\pi}{2}$

The volume $V$ is

$$V = \frac{28\pi}{3} - \frac{15\pi}{2} = \frac{11\pi}{6} \text{ cubic units} \qquad \blacksquare$$

Example 1 gives a clue to when the shell method is preferable to the washer method. When it is difficult, or impossible, to solve $y = f(x)$ for $x$, we use the shell method. For example, if the function in Example 1 had been $y = f(x) = x^5 + x^2 + 1$, we would not have been able to solve for $x$, so the practical choice is the shell method.

**NOW WORK** Problem 5.

The next example illustrates the importance of sketching a graph before using a formula. Notice the limits of integration and how we determined them when we use the washer method.

**CALC CLIP**

## EXAMPLE 2    Finding the Volume of a Solid: Revolving About the *y*-Axis

Find the volume $V$ of the solid generated by revolving the region bounded by the graphs of $f(x) = x^2$ and $g(x) = 12 - x$ to the right of $x = 1$ about the *y*-axis.

**Solution  *Using the shell method:*** Figure 33(a) shows the graph of the region to be revolved and a typical rectangle.

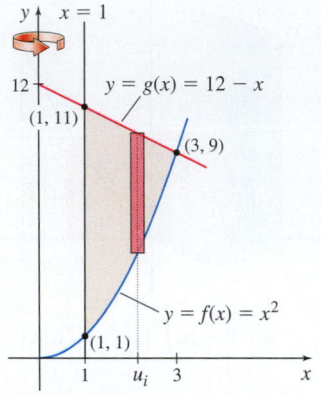

(a) Region to be revolved about the *y*-axis.

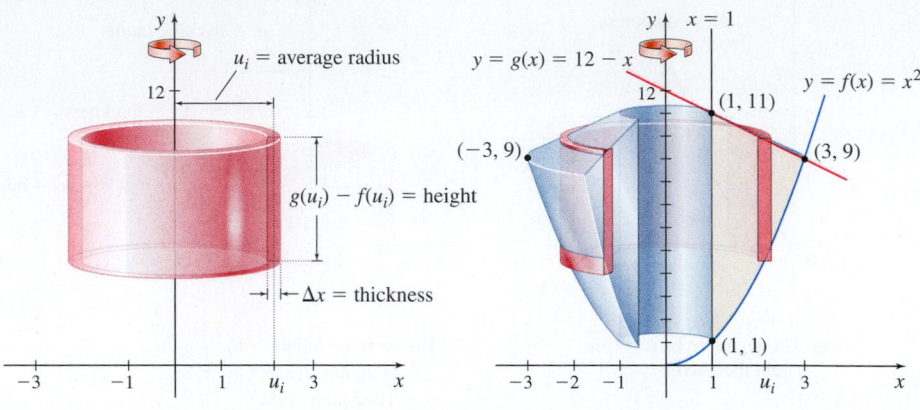

(b) Average radius = $u_i$
Height = $g(u_i) - f(u_i) = (12 - u_i) - u_i^2$
Thickness of shell = $\Delta x$
Volume = $2\pi u_i (12 - u_i - u_i^2) \Delta x$

(c) $V = 2\pi \displaystyle\int_1^3 x(12 - x - x^2)\, dx$

**Figure 33** The shell method

As shown in Figure 33(b), in the shell method we partition the *x*-axis and use vertical shells. A typical shell has height $= g(u_i) - f(u_i) = (12 - u_i) - u_i^2 = 12 - u_i - u_i^2$, and volume $V_i = 2\pi u_i (12 - u_i - u_i^2)\Delta x$. Figure 33(c) shows the solid of revolution. Notice that the integration takes place from $x = 1$ to $x = 3$. The volume $V$ of the solid of revolution is

$$V = 2\pi \int_1^3 x(12 - x - x^2)\, dx = 2\pi \int_1^3 (12x - x^2 - x^3)\, dx = 2\pi \left[ 6x^2 - \frac{x^3}{3} - \frac{x^4}{4} \right]_1^3$$

$$= 2\pi \left[ \left( 54 - 9 - \frac{81}{4} \right) - \left( 6 - \frac{1}{3} - \frac{1}{4} \right) \right] = \frac{116\,\pi}{3} \text{ cubic units}$$

***Using the washer method:*** Figure 34 shows the solid of revolution and typical washers.

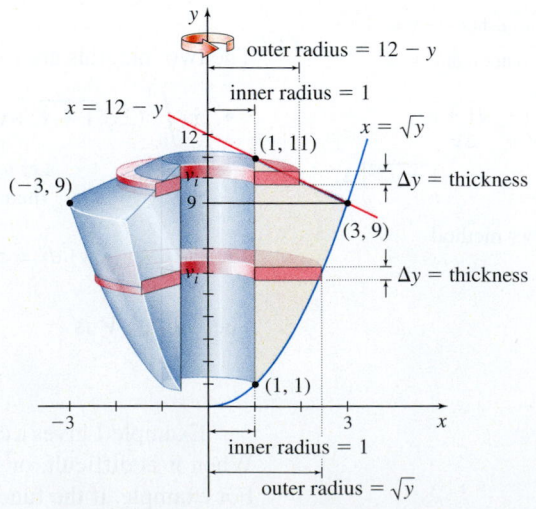

$$V_i = \pi[(\text{Outer radius})^2 - (\text{Inner radius})^2](\text{Thickness})$$

**Figure 34** The washer method

In the washer method, we partition the interval $[1, 11]$ on the $y$-axis and use horizontal washers. At $y = 9$, the function on the right changes. The volume of a typical washer in the interval $[1, 9]$ is

$$V_i = \pi\left[\left(\sqrt{v_i}\right)^2 - 1^2\right]\Delta y = \pi(v_i - 1)\Delta y$$

The volume of a typical washer in the interval $[9, 11]$ is

$$V_i = \pi\left[(12 - v_i)^2 - 1^2\right]\Delta y = \pi\left(143 - 24v_i + v_i^2\right)\Delta y$$

The volume $V$ of the solid of revolution is

$$V = \pi\int_1^9 (y - 1)\, dy + \pi\int_9^{11} (143 - 24y + y^2)\, dy$$

$$= \pi\left[\frac{y^2}{2} - y\right]_1^9 + \pi\left[143y - 12y^2 + \frac{y^3}{3}\right]_9^{11}$$

$$= \pi\left[\left(\frac{81}{2} - 9\right) - \left(\frac{1}{2} - 1\right)\right] + \pi\left(143\,(2) - (12)(121 - 81) + \frac{11^3}{3} - \frac{9^3}{3}\right)$$

$$= 32\pi + \frac{20}{3}\pi = \frac{116\,\pi}{3} \text{ cubic units}$$   ∎

**NOW WORK** Problem 15.

## 2 Use the Shell Method to Find the Volume of a Solid Formed by Revolving a Region About the $x$-Axis

Figure 35(a) shows a region that is to be revolved about the $x$-axis, Figure 35(b) shows a typical shell, and Figure 35(c) shows the solid of revolution.

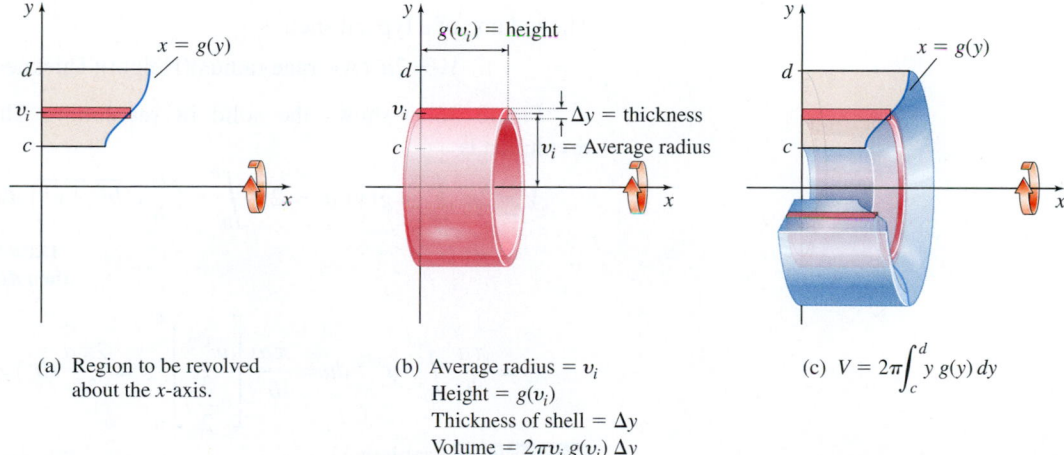

(a)  Region to be revolved about the $x$-axis.

(b)  Average radius $= v_i$
Height $= g(v_i)$
Thickness of shell $= \Delta y$
Volume $= 2\pi v_i\, g(v_i)\, \Delta y$

(c)  $V = 2\pi\displaystyle\int_c^d y\, g(y)\, dy$

**Figure 35**

### Volume of a Solid of Revolution About the $x$-Axis: the Shell Method

If $x = g(y)$ is a function that is continuous and nonnegative on the closed interval $[c, d]$, $c \geq 0$, the volume $V$ of the solid generated by revolving the region bounded by the graphs of $x = g(y)$, the $y$-axis, and the lines $y = c$ and $y = d$ about the $x$-axis is

$$\boxed{V = 2\pi\int_c^d y\, g(y)\, dy}$$

**EXAMPLE 3**  **Using the Shell Method: Revolving About the $x$-Axis**

Find the volume $V$ of the solid generated by revolving the region bounded by the graph of $\dfrac{x^2}{a^2} + \dfrac{y^2}{b^2} = 1$, $a > 0$, $b > 0$, in the first quadrant, about the $x$-axis.

**NEED TO REVIEW?** The equation
of an ellipse is discussed in
Appendix A.3, pp. A-23 to A-24.

**Solution** The equation $\dfrac{x^2}{a^2} + \dfrac{y^2}{b^2} = 1$ defines an ellipse. The intercepts of its graph
are $(a, 0)$, $(0, b)$, $(-a, 0)$, and $(0, -b)$. The region to be revolved is the shaded region
in the first quadrant shown in Figure 36(a).

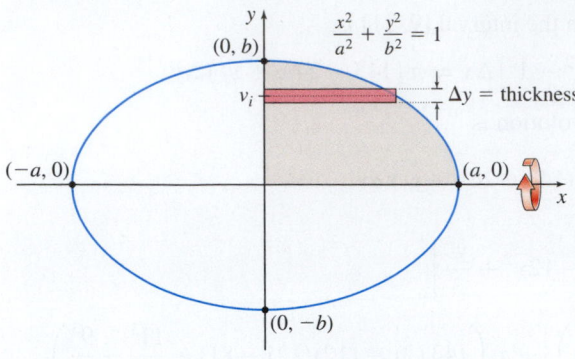

(a) Region enclosed by the ellipse in the first quadrant.

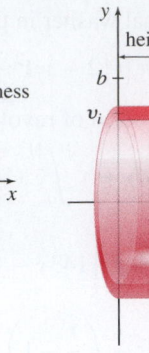

(b) Average radius $= v_i$
Height $= g(v_i) = \dfrac{a}{b}\sqrt{b^2 - v_i^2}$
Thickness of shell $= \Delta y$
Volume $= 2\pi\, v_i \cdot \dfrac{a}{b}\sqrt{b^2 - v_i^2}\ \Delta y$

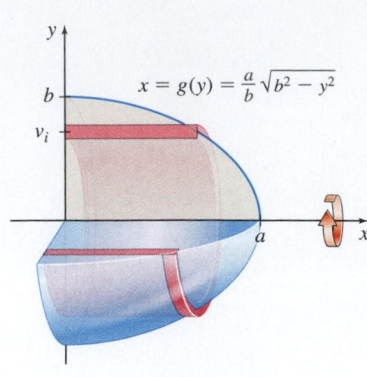

(c) $V = 2\pi \displaystyle\int_0^b y\left(\dfrac{a}{b}\sqrt{b^2 - y^2}\right) dy$

**Figure 36** The shell method

In the shell method, when the region is revolved about the $x$-axis, partition
the $y$-axis and use horizontal shells. See Figure 36(b). A revolution about the $x$-axis
requires integration with respect to $y$, so we express the equation of the ellipse as

$$x = g(y) = \frac{a}{b}\sqrt{b^2 - y^2}$$

The volume of a typical shell is

$$V_i = 2\pi \text{ (Average radius)(Height)(Thickness)} = 2\pi v_i g(v_i)\Delta y$$

Figure 36(c) shows the solid of revolution. The volume $V$ of the solid of
revolution is

$$V = 2\pi \int_0^b y\, g(y)\, dy = 2\pi \int_0^b y\left(\frac{a}{b}\sqrt{b^2 - y^2}\right) dy = 2\pi \frac{a}{b}\left(-\frac{1}{2}\right)\int_{b^2}^0 \sqrt{u}\, du$$

Let $u = b^2 - y^2$;
then $du = -2y\, dy$

$$= \frac{\pi a}{b}\int_0^{b^2} u^{1/2}\, du = \frac{\pi a}{b}\left[\frac{u^{3/2}}{\frac{3}{2}}\right]_0^{b^2} = \frac{2\pi a}{3b}(b^3) = \frac{2\pi ab^2}{3} \text{ cubic units} \qquad \blacksquare$$

**NOW WORK** Problem 11.

Using symmetry, the volume $V$ of the **ellipsoid** generated by revolving the region
bounded by the graph of $y = \dfrac{b}{a}\sqrt{a^2 - x^2}$, $-a \le x \le a$, about the $x$-axis is twice the
volume found in Example 3. That is,

$$\boxed{V = \frac{4}{3}\pi ab^2}$$

If in Example 3, $a = b = R$, then the solid generated is a **hemisphere** of radius $R$
whose volume $V$ is

$$\boxed{V = \frac{2}{3}\pi R^3}$$

By symmetry, the volume $V$ of a **sphere** of radius $R$, $(a = b = R)$, is twice the volume
of a hemisphere. That is,

$$\boxed{V = \frac{4}{3}\pi R^3}$$

### 3 Use the Shell Method to Find the Volume of a Solid Formed by Revolving a Region About a Line Parallel to a Coordinate Axis

**EXAMPLE 4** Using the Shell Method: Revolving About the Line $x = 2$

Find the volume $V$ of the solid of revolution generated by revolving the region bounded by the graph of $y = 2x - 2x^2$ and the $x$-axis about the line $x = 2$.

**Solution** The region bounded by the graph of $y = 2x - 2x^2$ and the $x$-axis is illustrated in Figure 37(a). A typical shell formed by revolving the region about the line $x = 2$, as shown in Figure 37(b), has an average radius of $2 - u_i$, height $f(u_i) = 2u_i - 2u_i^2$, and thickness $\Delta x$. The solid of revolution is depicted in Figure 37(c).

The volume $V$ of the solid is

$$V = 2\pi \int_0^1 (2 - x)(2x - 2x^2)\, dx = 4\pi \int_0^1 (x^3 - 3x^2 + 2x)\, dx$$

$$= 4\pi \left[ \frac{x^4}{4} - x^3 + x^2 \right]_0^1 = \pi \text{ cubic units}$$

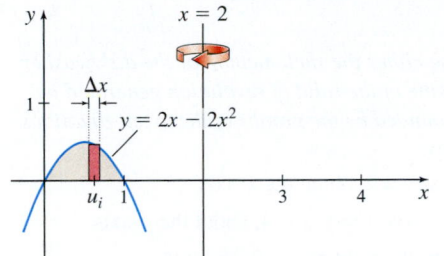

(a) Region to be revolved about the line $x = 2$.

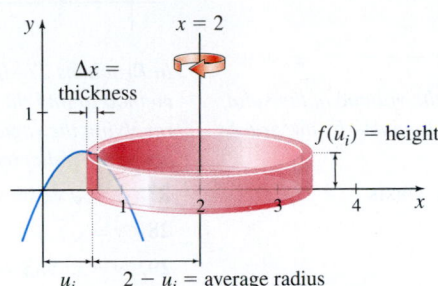
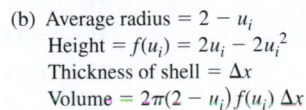

(b) Average radius $= 2 - u_i$
Height $= f(u_i) = 2u_i - 2u_i^2$
Thickness of shell $= \Delta x$
Volume $= 2\pi(2 - u_i) f(u_i) \Delta x$

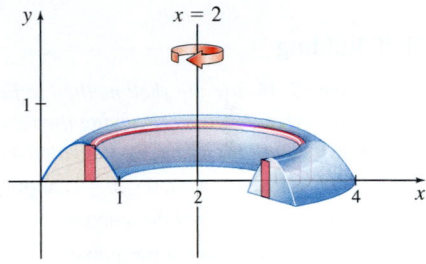

(c) $V = 2\pi \int_0^1 (2 - x)(2x - 2x^2)\, dx$

**DF** **Figure 37**

**NOW WORK** Problem 17.

## Summary

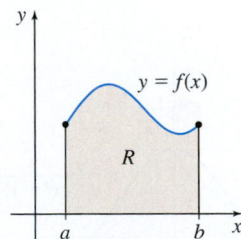

**Volume $V$ of the solid obtained by revolving the region $R$**

| about the $x$-axis | about the $y$-axis |
|---|---|
| **Disk Method** | **Shell Method** |
| $V = \pi \int_a^b [f(x)]^2\, dx$ | $V = 2\pi \int_a^b x f(x)\, dx$ |

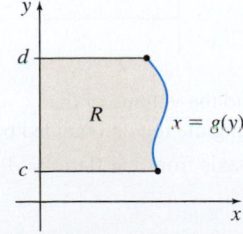

**Volume $V$ of the solid obtained by revolving the region $R$**

| about the $y$-axis | about the $x$-axis |
|---|---|
| **Disk Method** | **Shell Method** |
| $V = \pi \int_c^d [g(y)]^2\, dy$ | $V = 2\pi \int_c^d y g(y)\, dy$ |

## 6.3 Assess Your Understanding

### Concepts and Vocabulary

1. *True or False*  In using the shell method to find the volume of a solid revolved about the $x$-axis, the integration is with respect to $x$.

2. *True or False*  The volume of a cylindrical shell of outer radius $R$, inner radius $r$, and height $h$ is given by $V = \pi r^2 h$.

3. *True or False*  The volume of a solid of revolution can be found using either washers or cylindrical shells only if the region is revolved about the $y$-axis.

4. *True or False*  If $y = f(x)$ is a function that is continuous and nonnegative on the closed interval $[a, b]$, $a \geq 0$, then the volume $V$ of the solid generated by revolving the region bounded by the graph of $f$ and the $x$-axis from $x = a$ to $x = b$ about the $y$-axis is

$$V = 2\pi \int_a^b x f(x)\, dx$$

### Skill Building

*In Problems 5–16, use the shell method to find the volume of the solid of revolution generated by revolving the region bounded by the graphs of the given equations about the indicated axis.*

5.  $y = x^2 + 1$, the $x$-axis, $0 \leq x \leq 1$; about the $y$-axis

6.  $y = x^3$, $y = x^2$; about the $y$-axis

7.  $y = \sqrt{x}$, $y = x^2$; about the $y$-axis

8.  $y = \dfrac{1}{x}$, the $x$-axis, $x = 1$, $x = 4$; about the $y$-axis

9.  $y = x^3$, the $y$-axis, $y = 8$; about the $x$-axis

10. $y = \sqrt{x}$, the $y$-axis, $y = 2$; about the $x$-axis

11. $x = \sqrt{y}$, the $y$-axis, $y = 1$; about the $x$-axis

12. $x = 4\sqrt{y}$, the $y$-axis, $y = 4$; about the $x$-axis

13. $y = x$, $y = x^2$; about the $x$-axis

14. $y = x$, $y = x^3$ in the first quadrant; about the $x$-axis

15. $y = e^{-x^2}$ and the $x$-axis, from $x = 0$ to $x = 2$; about the $y$-axis

16. $y = x^3 + x$ and the $x$-axis, $0 \leq x \leq 1$; about the $y$-axis

*In Problems 17–22, use the shell method to find the volume of the solid of revolution generated by revolving the region bounded by the graphs of the given equations about the indicated line.*

17. $y = x^2$, $y = 4x - x^2$; about $x = 4$

18. $y = x^2$, $y = 4x - x^2$; about $x = -3$

19. $y = x^2$, $y = 0$, $x = 1$, $x = 2$; about $x = 1$

20. $y = x^2$, $y = 0$, $x = 1$, $x = 2$; about $x = -2$

21. $x = y - y^2$, the $y$-axis; about $y = -1$

22. $x = y - y^2$, the $y$-axis; about $y = 1$

*In Problems 23–26, use either the shell method or the disk/washer method to find the volume of the solid of revolution generated by revolving the shaded region in each graph:*

(a) about the $x$-axis.
(b) about the $y$-axis.
(c) Explain why you chose the method you used.

23.

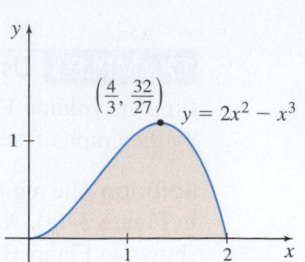

24.

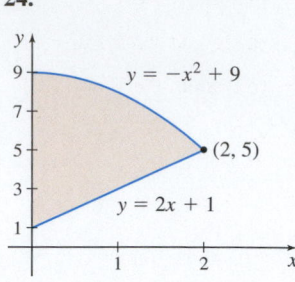

25.

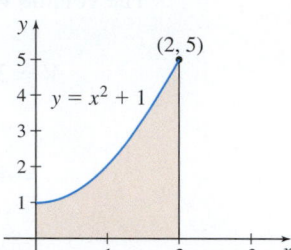

26.

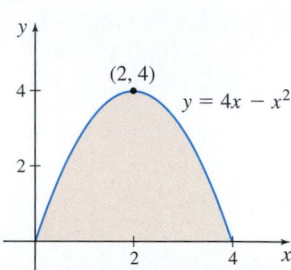

*In Problems 27–38, use either the shell method or the disk/washer method to find the volume of the solid of revolution generated by revolving the region bounded by the graphs of the given equations about the indicated axis.*

27. $y = \sqrt[3]{x}$, the $y$-axis, $y = 2$; about the $x$-axis

28. $y = \sqrt{x} + x$, the $x$-axis, $x = 1$, $x = 4$; about the $y$-axis

29. $y = x^3$ and $y = x$ to the right of $x = 0$; about the $y$-axis

30. $y = x^3$, $y = x^2$; about the $x$-axis

31. $y = 3x^2$ and $y = 30 - x$ to the right of $x = 1$; about the $y$-axis

32. $y = 3x^2$ and $y = 30 - x$ to the right of $x = 1$; about the $x$-axis

33. $y = x^2$ and $y = 8 - x^2$ to the right of $x = 1$; about the $x$-axis

34. $y = x^2$ and $y = 8 - x^2$ to the right of $x = 1$; about the $y$-axis

35. $y = x^{2/3}$, $y = x$; about the $y$-axis

36. $y = x^{2/3} + x^{1/3}$, the $x$-axis, $x = 1$, $x = 8$; about the $y$-axis

37. $y = \sqrt{x}$ and $y = 18 - x^2$ to the right of $x = 1$; about the $y$-axis

38. $y = \sqrt{x}$ and $y = 18 - x^2$ to the right of $x = 1$; about the $x$-axis

### Applications and Extensions

39. **Volume of a Solid of Revolution**
Find the volume of the solid of revolution generated by revolving the region bounded by the graph of $y = \dfrac{1}{(x^2 + 1)^2}$ and the $x$-axis from $x = 0$ to $x = 1$ about the $y$-axis as shown in the figure.

40. **Volume of a Solid of Revolution**  Find the volume of the solid of revolution generated by revolving the region bounded by the graph of $y = \sqrt{x + 1} + x$ and the $x$-axis from $x = 0$ to $x = 3$ about the $y$-axis.

---

**1.** = NOW WORK problem        ◣ = Graphing technology recommended        CAS = Computer Algebra System recommended

**41. Volume of a Solid of Revolution** Find the volume of the solid of revolution generated by revolving the region in the first quadrant bounded by the x-axis and the graph of $y = 2x - x^2$:

   (a) about the x-axis.
   (c) about the line $x = 3$.

   (b) about the y-axis.
   (d) about the line $y = 1$.

**42. Volume of a Solid of Revolution** Find the volume of the solid of revolution generated by revolving the region in the first quadrant bounded by the x-axis and the graph of $y = x\sqrt{9 - x}$.

   (a) about the x-axis.
   (c) about the line $y = -3$.

   (b) about the y-axis.
   (d) about the line $x = -2$.

**43. Volume of a Solid of Revolution** Suppose $f(x) \geq 0$ for $x \geq 0$, and the region bounded by the graph of $f$ and the x-axis from $x = 0$ to $x = k$, $k > 0$, is revolved about the x-axis. If the volume of the resulting solid is $\dfrac{1}{5}k^5 + k^4 + \dfrac{4}{3}k^3$, find $f$.

**44. Volume of a Solid of Revolution** Find the volume of the solid generated by revolving about the y-axis the region bounded by the graph of $y = e^{-x^2}$ and the y-axis from $x = 0$ to the positive x-coordinate of the point of inflection of $y = e^{-x^2}$.

**45. Volume of a Solid of Revolution** Show that the volume $V$ of the solid generated by revolving the region bounded by the graph of $y = \sin^3 x$ and the x-axis from $x = 0$ to $x = \pi$ about the line $x = -\pi$ is given by $V = 4\pi^2$.

**46. Volume of a Cone**

   (a) Find an equation of the line containing the points $(0, h)$, $h > 0$, and $(a, 0)$, $a > 0$.

   (b) Graph the line from (a).

   (c) Revolve the region in the first quadrant bounded by the line from (a), and the x- and y-axes about the y-axis to generate a cone of height $h$ and radius $a$.

   (d) Use the Shell Method to show that the volume $V$ of the cone is $\dfrac{1}{3}\pi a^2 h$.

**47. Volume of a Solid of Revolution** The figure below shows the solid of revolution generated by revolving the region bounded by the graph of $y = \cos x$ and the x-axis from $x = 0$ to $x = \dfrac{\pi}{2}$ about the y-axis.

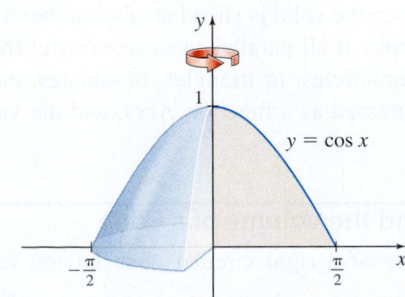

   (a) Express the volume of the solid of revolution as an integral.

   (b) Use technology to find the volume.

**48. Volume of a Solid of Revolution** The figure below shows the solid of revolution generated by revolving the region bounded by the graph of $y = \sin x$ and the x-axis from $x = 0$ to $x = \dfrac{\pi}{2}$ about the y-axis.

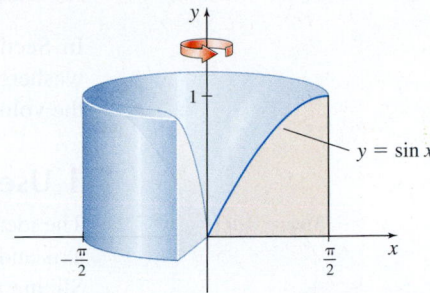

   (a) Express the volume of the solid of revolution as an integral.

   (b) Use technology to find the volume.

## Challenge Problems

**49.** Suppose a plane region of area $A$ that lies to the right of the y-axis is revolved about the y-axis, generating a solid of volume $V$. If this same region is revolved about the line $x = -k$, $k > 0$, show that the solid generated has volume $V + 2\pi k A$.

**50.** Find the volume generated by revolving the region bounded by the parabola $y^2 = 8x$ and its latus rectum about the latus rectum:

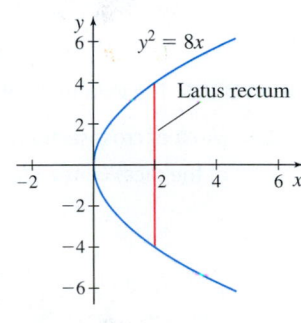

   (a) by using the disk method.

   (b) by using the shell method.

(The **latus rectum** is the chord through the focus perpendicular to the axis of the parabola.) See the figure.

**51. Comparing Volume Formulas** In the figure on the right, the region bounded by the graphs of $y = g(x)$, $y = f(x)$, and the y-axis is to be revolved about the y-axis. Show that the resulting volume $V$ is given by:

   (a) $V = \pi a^2[f(a) - g(a)]$

$$+ 2\pi \int_a^b x[f(x) - g(x)]\, dx$$

if the shell method is used.

   (b) $V = \pi \int_{g(a)}^{g(b)} [g^{-1}(y)]^2 dy + \pi \int_{g(b)}^{f(a)} [f^{-1}(y)]^2 dy$

if the disk method is used

# 6.4 Volume of a Solid: Slicing

**OBJECTIVE** *When you finish this section, you should be able to:*

1 Use slicing to find the volume of a solid (p. 456)

In Sections 6.2 and 6.3, we found the volumes of solids of revolution using disks, washers, and cylindrical shells. We now investigate a method, called *slicing*, to find the volume of a solid that is not necessarily a solid of revolution.

## 1 Use Slicing to Find the Volume of a Solid

The idea behind slicing is to cut a solid into thin slices using planes perpendicular to an axis and then to add the volumes of all the slices to obtain the total volume of the solid. Slicing relies on the fact that the area of each cross section obtained by slicing can be expressed as a function of the position of the slice. See Figure 38.

We begin by partitioning the interval $[a, b]$ on the $x$-axis into $n$ subintervals:

$$[a, x_1], \ [x_1, x_2], \ldots, [x_{i-1}, x_i], \ldots, [x_{n-1}, b]$$

each of width $\Delta x = \dfrac{b-a}{n}$. In the $i$th subinterval, we select a number $u_i$ and slice through the solid at $x = u_i$ using a plane that is perpendicular to the $x$-axis. Suppose the area of the cross section that results is $A(u_i)$.

The volume $V_i$ of the thin slice from $x_{i-1}$ to $x_i$ is approximately $V_i = A(u_i) \Delta x$, and the sum of the volumes from each subinterval is

$$\sum_{i=1}^{n} V_i = \sum_{i=1}^{n} [A(u_i) \Delta x]$$

These sums approximate the volume $V$ of the solid from $x = a$ to $x = b$. As the number $n$ of subintervals increases, the sums $\sum_{i=1}^{n} V_i = \sum_{i=1}^{n} [A(u_i) \Delta x]$ become better approximations to the volume $V$ of the solid. These sums are Riemann sums. So, if the area of the cross sections $A(u_i)$ vary continuously with $x$, then the limit is a definite integral.

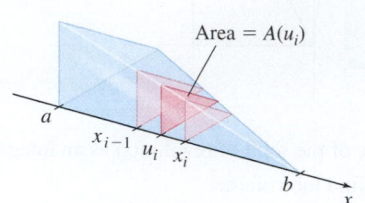

**Figure 38** A solid with a triangular cross section.

**IN WORDS** The volume $V_i$ of a slice is

$V_i =$ (Area of cross section) (Thickness of the slice) $= A(u_i) \Delta x$

### Volume of a Solid Using Slicing

If for each $x$ in $[a, b]$, the area $A(x)$ of the cross section of a solid is known and is continuous on $[a, b]$, then the volume $V$ of the solid is

$$V = \int_a^b A(x) \, dx \tag{1}$$

The key behind slicing is that when the solid is sliced at any number $x$, the area of the slice is a function of $x$. For example, if all parallel cross sections of the solid have the same geometric shape (all are semicircles, or triangles, or squares, etc.), then the area of each cross section can be expressed as a function $A(x)$, and the volume of the solid can be found using (1).*

**NEED TO REVIEW?** Geometry formulas are discussed in Appendix A.2, p. A-15.

---

**EXAMPLE 1** **Using Slicing to Find the Volume of a Cone**

Use slicing to verify that the volume of a right circular cone having radius $R$ and height $h$ is $V = \dfrac{1}{3} \pi R^2 h$.

---

*To find the volume of most solids requires the use of a double integral or a triple integral.

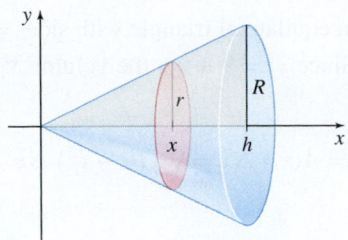

**Figure 39**

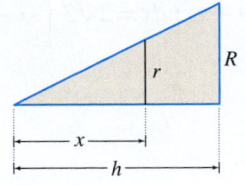

**Figure 40**

**NEED TO REVIEW?** Similar triangles are discussed in Appendix A.2, pp. A-13 to A-14.

**NOTE** A right circular cone is a solid of revolution, so the formula for its volume also can be verified using the disk method or the shell method.

**Solution** The mathematics is simplified if we position the cone with its vertex at the origin and its axis on the $x$-axis, as shown in Figure 39.

The cone extends from $x = 0$ to $x = h$. The cross section of the cone at any number $x$ is a circle. To obtain its area $A$, we need the radius $r$ of the circle. Embedded in Figure 39 are two similar triangles, one with sides $x$ and $r$; the other with sides $h$ and $R$, as shown in Figure 40. Because these triangles are similar (AAA), corresponding sides are in proportion. That is,

$$\frac{r}{x} = \frac{R}{h}$$

$$r = \frac{R}{h}x$$

So, $r$ is a function of $x$ and the area $A$ of the circular cross section is

$$A = A(x) = \pi[r(x)]^2 = \pi\left(\frac{R}{h}x\right)^2 = \frac{\pi R^2}{h^2}x^2$$

Since $A$ is a continuous function of $x$ (where the slice was made), we can use slicing. The volume $V$ of the right circular cone is

$$V = \int_a^b A(x)\,dx = \int_0^h \frac{\pi R^2}{h^2}x^2\,dx = \frac{\pi R^2}{h^2}\int_0^h x^2\,dx$$

$$= \frac{\pi R^2}{h^2}\left[\frac{x^3}{3}\right]_0^h = \frac{\pi R^2 h}{3} \text{ cubic units} \qquad\blacksquare$$

**NOW WORK** Problem 13.

In Example 1, the cone was positioned so that its axis coincided with the $x$-axis and its vertex was at the origin. The way in which a solid is positioned relative to the $x$-axis is important if the area $A$ of the slice is to be easily found, expressed as a function of $x$, and integrated.

**EXAMPLE 2** Using Slicing to Find the Volume of a Solid

A solid has a circular base of radius 3 units. Find the volume $V$ of the solid if every plane cross section that is perpendicular to a fixed diameter is an equilateral triangle.

**Solution** Position the circular base so that its center is at the origin, and the fixed diameter is along the $x$-axis. See Figure 41(a). Then the equation of the circular base

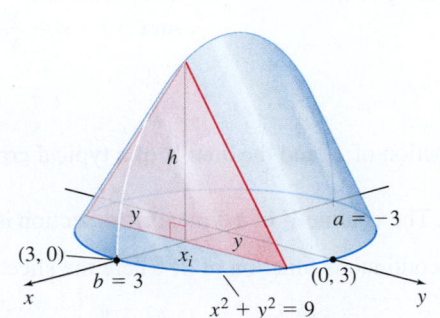

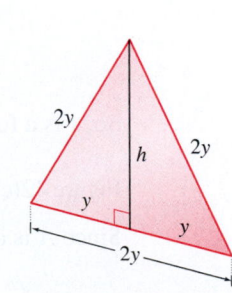

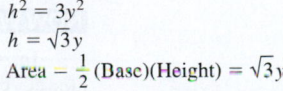

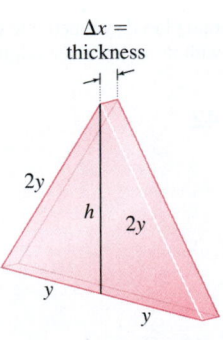

(a) Solid placed with the center of its circular base at the origin and its fixed diameter on the $x$-axis.

(b) Cross section:
$h^2 + y^2 = 4y^2$
$h^2 = 3y^2$
$h = \sqrt{3}\,y$
Area $= \frac{1}{2}$(Base)(Height) $= \sqrt{3}\,y^2$

(c) Slice:
$V = $ (Area)(Thickness)
$= \sqrt{3}\,y^2\,\Delta x = \sqrt{3}\left(9 - x^2\right)\Delta x$

**DF Figure 41**

is $x^2 + y^2 = 9$. Each cross section of the solid is an equilateral triangle with sides $= 2y$, height $h$, and area $A = \sqrt{3}y^2$. See Figure 41(b). Since $y^2 = 9 - x^2$, the volume $V_i$ of a typical slice is

$$V_i = (\text{Area of the cross section})(\text{Thickness}) = A(x_i)\,\Delta x = \sqrt{3}\left(9 - x_i^2\right)\Delta x$$

as shown in Figure 41(c).

The volume $V$ of the solid is

**NEED TO REVIEW?** Even and odd functions are discussed in Section P.1, pp. 10–11, and the integrals of even and odd functions are discussed in Section 5.5, pp. 407–408.

$$V = \int_a^b A(x)\,dx = \int_{-3}^{3} \sqrt{3}(9 - x^2)\,dx \underset{\underset{\substack{\text{The integrand is}\\\text{an even function.}}}{\uparrow}}{=} 2\sqrt{3}\int_0^3 (9 - x^2)\,dx = 2\sqrt{3}\left[9x - \frac{x^3}{3}\right]_0^3$$

$$= 36\sqrt{3} \text{ cubic units}$$    ∎

**NOW WORK** Problem 3.

---

**EXAMPLE 3**  **Using Slicing to Find the Volume of a Pyramid**

Find the volume $V$ of a pyramid of height $h$ with a square base, each side of length $b$.

**Solution** We position the pyramid with its vertex at the origin and its axis along the positive $x$-axis. Then the area $A$ of a typical cross section at $x$ is a square. Let $s$ denote the length of the side of the square at $x$. See Figure 42(a). We form two triangles: one with height $x$ and side $s$, the other with height $h$ and side $b$. These triangles are similar $(AAA)$, as shown in Figure 42(b). Then we have

$$\frac{x}{s} = \frac{h}{b} \qquad \text{or} \qquad s = \frac{b}{h}x$$

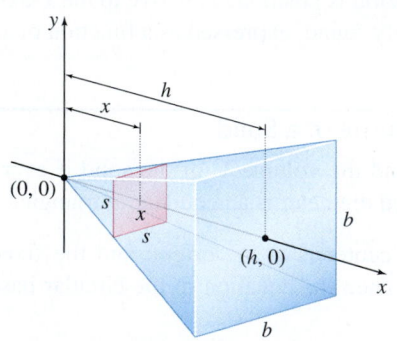

(a) Pyramid placed symmetric to the $x$-axis and with its vertex at the origin.

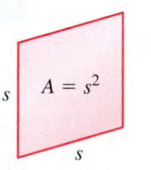

(c) Cross section at $x$:
Area $= A = s^2 = \dfrac{b^2}{h^2}x^2$

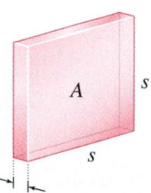

(d) Volume $= V = A\,\Delta x = \dfrac{b^2}{h^2}x^2\Delta x$

(b) Similar triangles

**Figure 42**

So, $s$ is a function of $x$, and the area $A$ of a typical cross section is $A = s^2 = \dfrac{b^2}{h^2}x^2$. See Figure 42(c). The volume $V$ of a typical cross section is $V = \dfrac{b^2}{h^2}x^2\Delta x$. See Figure 43(d). Since $A$ is a continuous function of $x$, where the slice occurred, we have

$$V = \int_0^h A(x)\,dx = \int_0^h \frac{b^2}{h^2}x^2\,dx = \frac{b^2}{h^2}\int_0^h x^2\,dx = \frac{b^2}{h^2}\left[\frac{x^3}{3}\right]_0^h = \frac{1}{3}b^2 h \text{ cubic units}$$    ∎

**NOW WORK** Problem 11.

In each example we have done so far, the base has been a recognizable geometric shape. Slicing can also be used to find the volume of a solid with an irregular base, provided information about the slices of the solid is known.

**CALC CLIP**

### ▶ EXAMPLE 4  Using Slicing to Find the Volume of a Solid

A region $R$ in the $xy$-plane is bounded by the graphs of $y = \sqrt{x}$ and $y = \dfrac{1}{8}x^2$.

(a) Find the volume $V$ of a solid whose base is $R$, if slices perpendicular to the $x$-axis have cross sections that are squares.

(b) Find the volume $V$ of a solid whose base is $R$, if slices perpendicular to the $y$-axis have cross sections that are squares.

**Solution**  Begin by graphing the region $R$ bounded by the graphs of $y = \sqrt{x}$ and $y = \dfrac{1}{8}x^2$ as shown in Figure 43. The points of intersection of the two graphs satisfy the equation

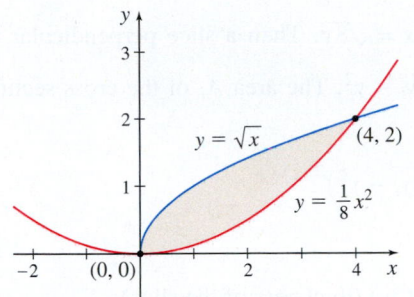

$$\sqrt{x} = \frac{1}{8}x^2$$
$$x = \frac{1}{64}x^4$$
$$x^4 - 64x = 0$$
$$x(x^3 - 64) = 0$$
$$x = 0 \text{ or } x = 4$$

**Figure 43**

The two graphs intersect at the points $(0, 0)$ and $(4, 2)$.

(a) The solid with slices perpendicular to the $x$-axis that are squares is shown in Figure 44(a). See Figure 44(b). A slice perpendicular to the $x$-axis at $x_i$ is a square with side $s_i = \sqrt{x_i} - \dfrac{1}{8}x_i^2$. The area $A_i$ of the cross section at $x_i$ is

$$A_i = s_i^2 = \left( \sqrt{x_i} - \frac{1}{8}x_i^2 \right)^2$$

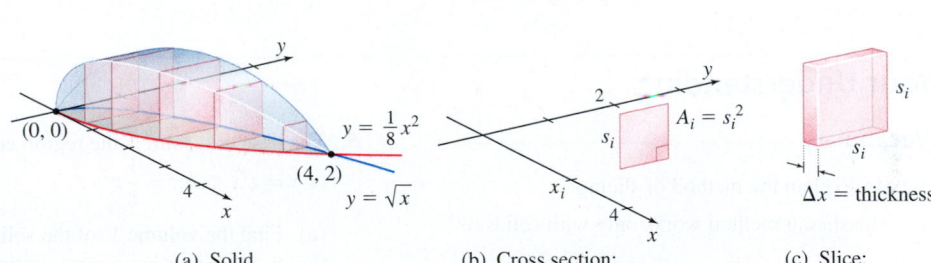

(a) Solid

(b) Cross section:
Area $A_i = s_i^2 = \left( \sqrt{x_i} - \frac{1}{8}x_i^2 \right)^2$

(c) Slice:
Volume $V_i = $ (Area)(Thickness)
$= \left( \sqrt{x_i} - \frac{1}{8}x_i^2 \right)^2 \Delta x$

**DF Figure 44**

See Figure 44(c). The volume $V_i$ of a typical slice is

$$V_i = (\text{Area of the cross section at } x_i)\,(\text{thickness of the slice})$$
$$= A(x_i)\Delta x = \left( \sqrt{x_i} - \frac{1}{8}x_i^2 \right)^2 \Delta x$$

The volume $V$ of the solid is

$$V = \int_0^4 A(x)\,dx = \int_0^4 \left( \sqrt{x} - \frac{1}{8}x^2 \right)^2 dx = \int_0^4 \left( x - \frac{1}{4}x^{5/2} + \frac{1}{64}x^4 \right) dx$$

$$= \left[ \frac{x^2}{2} - \frac{1}{14}x^{7/2} + \frac{1}{320}x^5 \right]_0^4 = 8 - \frac{128}{14} + \frac{1024}{320} = \frac{72}{35} \text{ cubic units}$$

(b) To find the volume of a solid when slices are made perpendicular to the $y$-axis, follow the same reasoning as in (a), but use a partition of the $y$-axis. For each $y$ in the closed interval $[c, d]$, the area $A(y)$ of the cross section of the solid is known and is continuous on $[c, d]$. Then the volume $V$ of the solid is

$$V = \int_d^c A(y)\, dy \qquad (2)$$

To use (2), we must express $y = \sqrt{x}$ and $y = \dfrac{1}{8}x^2$ as functions of $y$. That is, we write $y = \sqrt{x}$ as $x = y^2$ and $y = \dfrac{1}{8}x^2$ as $x = \sqrt{8y}$. Then a slice perpendicular to the $y$-axis at $y_i$ is a square with side $s_i = \sqrt{8y_i} - y_i^2$. The area $A_i$ of the cross section at $y_i$ is

$$A_i = s_i^2 = \left( \sqrt{8y_i} - y_i^2 \right)^2$$

and the volume $V_i$ of a typical slice is

$$V_i = (\text{Area of the cross section at } y_i)(\text{thickness of the slice})$$

$$= A(y_i)\Delta y = \left( \sqrt{8y_i} - y_i^2 \right)^2 \Delta y$$

The volume $V$ of the solid is

$$V = \int_0^2 A(y)\,dy = \int_0^2 \left( \sqrt{8y} - y^2 \right)^2 dy = \int_0^2 \left( 8y - 4\sqrt{2}\,y^{5/2} + y^4 \right) dy$$

$$= \left[ 4y^2 - \frac{8\sqrt{2}}{7}y^{7/2} + \frac{y^5}{5} \right]_0^2 = 16 - \frac{128}{7} + \frac{32}{5} = \frac{144}{35} \text{ cubic units} \quad \blacksquare$$

NOW WORK Problem 5.

---

# 6.4 Assess Your Understanding

## Concepts and Vocabulary

1. In your own words, explain the method of slicing.

2. *True or False* The slicing method works only with solids of revolution.

## Skill Building

*In Problems 3–10, find the volume of each solid by the method of slicing.*

3. Find the volume of the solid whose base is a circle of radius of 2, if slices made perpendicular to the base are squares. See the figure.

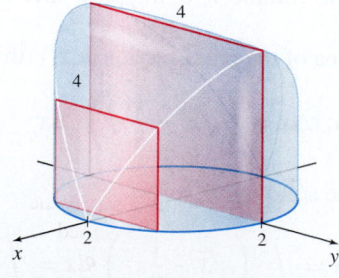

4. Find the volume of the solid whose base is a circle of radius 2, if slices made perpendicular to the base are isosceles right triangles with one leg on the base.

5. The base of a solid is the region enclosed by the graphs of $y = \sqrt{x}$ and $y = \dfrac{1}{8}x^2$.

   (a) Find the volume $V$ of the solid if slices made perpendicular to the $x$-axis have cross sections that are semicircles. See the figure.

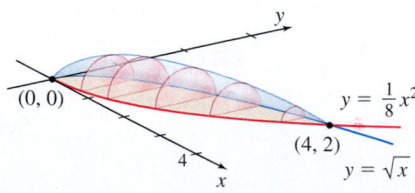

   (b) Find the volume $V$ of the solid if slices made perpendicular to the $x$-axis have cross sections that are equilateral triangles.

6. The base of a solid is the region enclosed by the graphs of $y = \sqrt{x}$ and $y = \dfrac{1}{8}x^2$.

   (a) Find the volume $V$ of the solid if slices made perpendicular to the $y$-axis have cross sections that are semicircles.

   (b) Find the volume $V$ of the solid if slices made perpendicular to the $y$-axis have cross sections that are equilateral triangles.

---

1. = NOW WORK problem      ⟥⟦ = Graphing technology recommended      CAS = Computer Algebra System recommended

**7.** The base of a solid is the region enclosed by the graphs of $y = x^2$, $x = 2$, and $y = 0$.

   **(a)** Find the volume $V$ of the solid if slices made perpendicular to the $x$-axis are squares.

   **(b)** Find the volume $V$ of the solid if slices made perpendicular to the $x$-axis are semicircles.

   **(c)** Find the volume $V$ of the solid if slices made perpendicular to the $x$-axis are equilateral triangles.

**8.** The base of a solid is the region enclosed by the graphs of $y = x^2$, $x = 2$, and $y = 0$.

   **(a)** Find the volume $V$ of the solid if slices made perpendicular to the $y$-axis are squares.

   **(b)** Find the volume $V$ of the solid if slices made perpendicular to the $y$-axis are semicircles.

   **(c)** Find the volume $V$ of the solid if slices made perpendicular to the $y$-axis are equilateral triangles.

**9.** The base of a solid is the region enclosed by the graphs of $y = 3\sqrt{3x}$ and $y = x^2$.

   **(a)** Find the volume $V$ of the solid if slices made perpendicular to the $y$-axis are squares.

   **(b)** Find the volume $V$ of the solid if slices made perpendicular to the $y$-axis are semicircles.

   **(c)** Find the volume $V$ of the solid if slices made perpendicular to the $y$-axis are equilateral triangles.

**10.** The base of a solid is the region enclosed by the graphs of $y = 3\sqrt{3x}$ and $y = x^2$.

   **(a)** Find the volume $V$ of the solid if slices made perpendicular to the $x$-axis are squares.

   **(b)** Find the volume $V$ of the solid if slices made perpendicular to the $x$-axis are semicircles.

   **(c)** Find the volume $V$ of the solid if slices made perpendicular to the $x$-axis are equilateral triangles.

### Applications and Extensions

*In Problems 11 and 12, find the volume of each solid.*

**11.** The solid is a pyramid 40 m high whose horizontal cross section $h$ meters from the top is a square with sides of length $2h$ meters.

**12.** The solid is a pyramid 20 m high whose horizontal cross section $h$ meters from the top is a rectangle with sides of length $2h$ and $h$ meters.

**13. Verifying a Geometry Formula** Use slicing to verify that the volume $V$ of a sphere of radius $R$ is $V = \dfrac{4}{3}\pi R^3$.

**14.** The base of a solid is the region enclosed by the graphs of $y = \sqrt{x}$ and $y = \dfrac{1}{8}x^2$.

   **(a)** Find the volume $V$ of the solid if slices made perpendicular to the $x$-axis have cross sections that are triangles whose base is the distance between the graphs and whose height is 3 times the base.

   **(b)** Find the volume $V$ of the solid if slices made perpendicular to the $y$-axis have cross sections that are triangles whose base is the distance between the graphs and whose height is 3 times the base.

*In Problems 15 and 16, find the volume of each solid.*

**15.** The solid is horn-shaped; slices taken perpendicular to the $x$ axis are circles whose diameters extend from the graph of $y = x^{1/2}$ to the graph of $y = \dfrac{4}{3}x^{1/3}$, $0 \le x \le 1$.

**16.** The solid is horn-shaped; slices taken perpendicular to the $x$-axis are circles whose diameters extend from the graph of $y = x^{1/3}$ to $y = \dfrac{3}{2}x^{1/3}$, $0 \le x \le 1$.

**17. Volume of a Solid** Find the volume of the cylindrical solid with a bulge in the middle, if slices taken perpendicular to the $x$-axis are circles whose diameters extend from the graph of $y = e^{-x^2}$ to $y = -e^{-x^2}$, $-1 \le x \le 1$.

### Challenge Problems

**18. Volume of Water** A hemispherical bowl of radius $R$ contains water to the depth $h$. Find the volume of the water in the bowl.

**19. Volume of Water Left in a Glass** Suppose a cylindrical glass full of water is tipped until the water level bisects the base and touches the rim. What is the volume of the water remaining?

**20. Volume of a Removed Sector** Suppose a wedge is cut from a solid right circular cylinder of diameter 10 m (like a wedge cut in a tree by an axe), where one side of the wedge is horizontal and the other is inclined at 30°. See the figure. If the horizontal part of the wedge penetrates 5 m into the cylinder and the two cuts

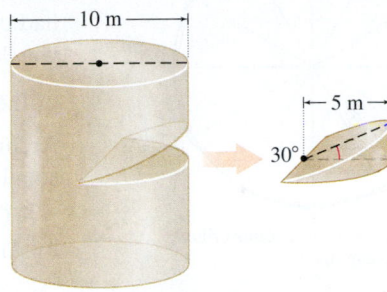

meet along a vertical line through the center of the cylinder, find the volume of the wedge removed.

[*Hint:* Vertical cross sections of the wedge are right triangles.]

**21. Volume of a Bore** A hole of radius 2 cm is bored completely through a solid metal sphere of radius 5 cm. If the axis of the hole passes through the center of the sphere, find the volume of the metal removed by the drilling. See the figure.

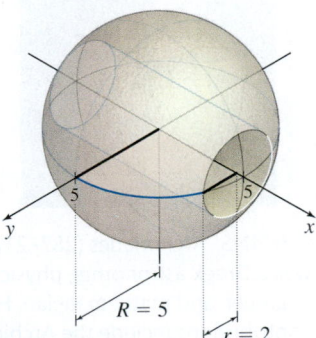

**22. Volume** The axes of two pipes of equal radii $r$ intersect at right angles. Find their common volume.

**23. Volume of a Cone** Find the volume of a cone with height $h$ and an elliptical base whose major axis has length $2a$ and minor axis has length $2b$. (*Hint:* The area of this ellipse is $\pi ab$.)

**24. Volume of a Solid** Find the volume of a parallelepiped with edge lengths $a$, $b$, and $c$, where the edges having lengths $a$ and $b$ make an acute angle $\theta$ with each other, and the edge of length $c$ makes an acute angle of $\phi$ with the diagonal of the parallelogram formed by $a$ and $b$.

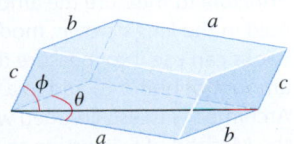

# 6.5 Arc Length; Surface Area of a Solid of Revolution

**OBJECTIVES** *When you finish this section, you should be able to:*

1. Find the arc length of the graph of a function $y = f(x)$ (p. 462)
2. Find the arc length of the graph of a function using a partition of the $y$-axis (p. 466)
3. Find the surface area of a solid of revolution (p. 467)

We have already seen that a definite integral can be used to find the area of a plane region and the volume of certain solids. In this section, we will see that a definite integral can be used to find the length of a graph and the surface area of a solid of revolution.

The idea of finding the length of a graph had its beginning with Archimedes. Ancient people knew that the ratio of the circumference $C$ of any circle to its diameter $d$ equaled the constant that we call $\pi$. That is, $\frac{C}{d} = \pi$. But Archimedes is credited as being the first person to analytically investigate the numerical value of $\pi$.* He drew a circle of diameter 1 unit, then inscribed (and circumscribed) the circle with regular polygons, and computed the perimeters of the polygons. He began with two hexagons (6 sides) and then two dodecagons (12 sides). See Figure 45. He continued until he had two polygons, each with 96 sides. He knew that the circumference $C$ of the circle was larger than the perimeter of the inscribed polygon and smaller than the perimeter of the circumscribed polygon. In this way, he proved that $3\frac{10}{71} < \pi < 3\frac{1}{7}$. In essence, Archimedes approximated the length of the circle by representing it by smaller and smaller line segments and summing the lengths of the line segments.

Our approach to finding the length of the graph of a function is similar to that used by Archimedes and follows the ideas we used to find area and volume.

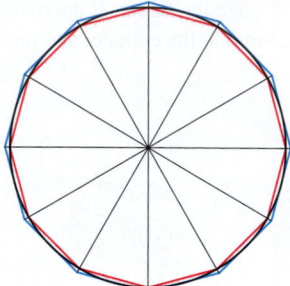

(a) Inscribed and circumscribed hexagons

(b) Inscribed and circumscribed dodecagons

**Figure 45**

**ORIGINS** Archimedes (287–212 BC) was a Greek astronomer, physicist, engineer, and mathematician. His contributions include the Archimedes screw, which is still used to lift water or grain, and the Archimedes principle, which states that the weight of an object can be determined by measuring the volume of water it displaces. Archimedes used the principle to measure the amount of gold in the king's crown; modern-day cooks can use it to measure the amount of butter to add to a recipe. Archimedes is also credited with using the Method of Exhaustion to find the area under a parabola and to approximate the value of $\pi$.

## 1 Find the Arc Length of the Graph of a Function $y = f(x)$

Suppose we want to find the length of the graph of a function $y = f(x)$ from $x = a$ to $x = b$. We assume $f$ is continuous on $[a, b]$ and the derivative $f'$ of $f$ is continuous on some interval containing $a$ and $b$. See Figure 46.

We begin by partitioning the closed interval $[a, b]$ into $n$ subintervals:

$$[a, x_1], [x_1, x_2], \ldots, [x_{i-1}, x_i], \ldots, [x_{n-1}, b] \qquad x_0 = a \quad x_n = b$$

each of width $\Delta x = \dfrac{b-a}{n}$. Corresponding to each number $a, x_1, x_2, \ldots, x_{i-1}, x_i, \ldots,$ $x_{n-1}, b$ in the partition, there is a point $P_0, P_1, P_2, \ldots, P_{i-1}, P_i, \ldots, P_{n-1}, P_n$ on the graph of $f$, as illustrated in Figure 47.

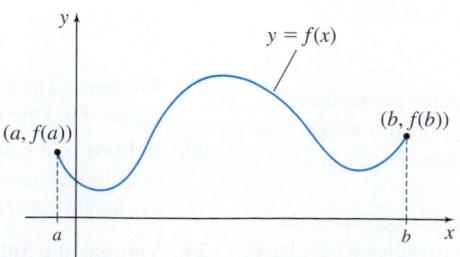

**Figure 46** $f$ is continuous on $[a, b]$; $f'$ is continuous on an interval containing $a$ and $b$.

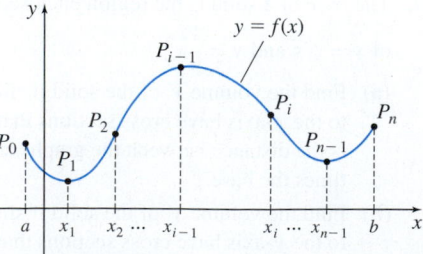

**Figure 47** Points $P_0, P_1, P_2, \ldots, P_{n-1}, P_n$ correspond to the numbers $a, x_1, x_2, \ldots, x_{n-1}, b$, respectively.

---

*Beckmann, Petr, *The History of $\pi$*. St Martin's Press, N.Y. 1971 p. 62.

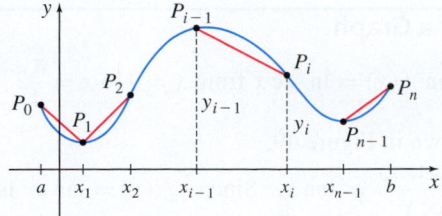

**Figure 48** After the interval $[a, b]$ is partitioned, connect consecutive points with line segments.

**NEED TO REVIEW?** The distance formula is discussed in Appendix A.3, p. A-16.

When each point $P_{i-1} = (x_{i-1}, y_{i-1})$ is connected to the next point $P_i = (x_i, y_i)$ by a line segment, the sum $L$ of the lengths of the line segments approximates the length of the graph of $y = f(x)$ from $x = a$ to $x = b$. The sum $L$ is given by

$$L = d(P_0, P_1) + d(P_1, P_2) + \cdots + d(P_{n-1}, P_n) = \sum_{i=1}^{n} d(P_{i-1}, P_i)$$

where $d(P_{i-1}, P_i)$ denotes the length of the line segment joining $P_{i-1} = (x_{i-1}, y_{i-1})$ to $P_i = (x_i, y_i)$. See Figure 48.

Using the distance formula, the length of the $i$th line segment is

$$d(P_{i-1}, P_i) = \sqrt{(x_i - x_{i-1})^2 + (y_i - y_{i-1})^2} = \sqrt{(\Delta x)^2 + (\Delta y_i)^2}$$

$$= \sqrt{\frac{(\Delta x)^2 + (\Delta y_i)^2}{(\Delta x)^2}} \cdot \underset{\underset{\Delta x > 0}{\uparrow}}{\sqrt{(\Delta x)^2}} = \sqrt{1 + \left(\frac{\Delta y_i}{\Delta x}\right)^2} \, \Delta x$$

where $\Delta x = \dfrac{b - a}{n}$ and $\Delta y_i = y_i - y_{i-1} = f(x_i) - f(x_{i-1})$. Then the sum $L$ of the lengths of the line segments is

$$L = \sum_{i=1}^{n} \left[ \sqrt{1 + \left(\frac{\Delta y_i}{\Delta x}\right)^2} \, \Delta x \right]$$

This sum is not a Riemann sum. To put this sum in the form of a Riemann sum, we need to express $\sqrt{1 + \left(\dfrac{\Delta y_i}{\Delta x}\right)^2}$ in terms of $u_i$, $x_{i-1} \le u_i \le x_i$. To do this, we first observe that the function $f$ has a derivative $f'$ that is continuous on an interval containing $a$ and $b$. It follows that $f$ has a derivative on each subinterval $(x_{i-1}, x_i)$ of $[a, b]$. Now use the Mean Value Theorem. Then in each open subinterval $(x_{i-1}, x_i)$, there is a number $u_i$ for which

**NEED TO REVIEW?** The Mean Value Theorem is discussed in Section 4.3, pp. 285–287.

$$f(x_i) - f(x_{i-1}) = f'(u_i)(x_i - x_{i-1}) \qquad \text{Use the Mean Value Theorem.}$$

$$\Delta y_i = f'(u_i) \, \Delta x \qquad \qquad \boldsymbol{\Delta y_i = f(x_i) - f(x_{i-1}); \quad \Delta x = x_i - x_{i-1}}$$

$$\frac{\Delta y_i}{\Delta x} = f'(u_i) \qquad \qquad \text{Divide both sides by } \boldsymbol{\Delta x.}$$

Now the sum $L$ of the lengths of the line segments can be written as

$$L = \sum_{i=1}^{n} \left[ \sqrt{1 + [f'(u_i)]^2} \, \Delta x \right]$$

where $u_i$ is some number in the open subinterval $(x_{i-1}, x_i)$.

This sum $L$ is a Riemann sum. As the number of subintervals increases, that is, as $n \to \infty$, the sum $L$ becomes a better approximation to the arc length of the graph of $y = f(x)$ from $x = a$ to $x = b$. Since $f'$ is continuous on $[a, b]$, the limit of the sum exists and is a definite integral.

### Arc Length Formula

Suppose a function $y = f(x)$ is continuous on $[a, b]$ and has a derivative that is continuous on an interval containing $a$ and $b$. The **arc length** $s$ of the graph of $f$ from $x = a$ to $x = b$ is given by

$$s = \int_a^b \sqrt{1 + [f'(x)]^2} \, dx \tag{1}$$

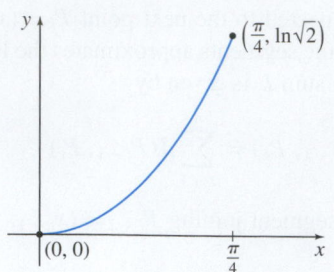

DF **Figure 49** $f(x) = \ln \sec x, 0 \le x \le \dfrac{\pi}{4}$

CALC
CLIP
▶ **EXAMPLE 1**  **Finding the Arc Length of a Graph**

Find the arc length of the graph of the function $f(x) = \ln \sec x$ from $x = 0$ to $x = \dfrac{\pi}{4}$.

**Solution** The graph of $f(x) = \ln \sec x$ is shown in Figure 49.

The derivative of $f$ is $f'(x) = \dfrac{\sec x \tan x}{\sec x} = \tan x$. Since $f'(x) = \tan x$ is

continuous on the open interval $\left(-\dfrac{\pi}{2}, \dfrac{\pi}{2}\right)$, which contains 0 and $\dfrac{\pi}{4}$, we can use the

arc length formula (1) to find the arc length $s$ from 0 to $\dfrac{\pi}{4}$.

$$s = \int_0^{\pi/4} \sqrt{1 + [f'(x)]^2} \, dx = \int_0^{\pi/4} \sqrt{1 + \tan^2 x} \, dx = \int_0^{\pi/4} \sqrt{\sec^2 x} \, dx$$

$$= \int_0^{\pi/4} \sec x \, dx = \Big[\ln|\sec x + \tan x|\Big]_0^{\pi/4}$$

$$= \ln\left|\sec \frac{\pi}{4} + \tan \frac{\pi}{4}\right| - \ln|\sec 0 + \tan 0| = \ln|\sqrt{2} + 1| \quad\blacksquare$$

**NOW WORK** Problem 9.

**EXAMPLE 2**  **Finding the Arc Length of a Graph**

Find the arc length of the graph of the function $f(x) = x^{2/3}$ from $x = 1$ to $x = 8$.

**Solution** We begin by graphing $f(x) = x^{2/3}$. See Figure 50.

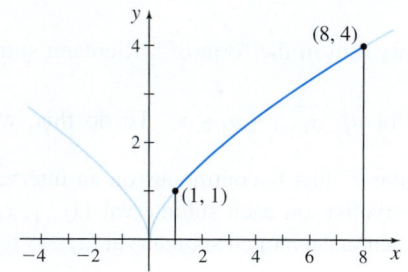

**Figure 50** $f(x) = x^{2/3}, 1 \le x \le 8$

The derivative of $f(x) = x^{2/3}$ is $f'(x) = \dfrac{2}{3}x^{-1/3} = \dfrac{2}{3x^{1/3}}$. Notice that $f'$ is not

continuous at 0. However, since $f'$ is continuous on an interval containing 1 and 8

(use $\left[\dfrac{1}{2}, 9\right]$, for example, which avoids 0), we can use the arc length formula (1) to find

the arc length $s$ from $x = 1$ to $x = 8$.

$$s = \int_1^8 \sqrt{1 + [f'(x)]^2} \, dx = \int_1^8 \sqrt{1 + \left(\frac{2}{3x^{1/3}}\right)^2} \, dx = \int_1^8 \sqrt{1 + \frac{4}{9x^{2/3}}} \, dx$$

$$= \int_1^8 \sqrt{\frac{9x^{2/3} + 4}{9x^{2/3}}} \, dx = \underset{\substack{\uparrow \\ x > 0 \text{ on } [1, 8], \\ \text{so } \sqrt{x^{2/3}} = x^{1/3}}}{\frac{1}{3} \int_1^8 \left(x^{-1/3}\sqrt{9x^{2/3} + 4}\right) dx}$$

Use the substitution $u = 9x^{2/3} + 4$. Then $du = 6x^{-1/3}dx$ and $x^{-1/3}dx = \dfrac{du}{6}$. The limits

of integration change to $u = 13$ when $x = 1$, and to $u = 40$ when $x = 8$. Then

$$s = \frac{1}{3}\int_1^8 \left[x^{-1/3}\sqrt{9x^{2/3} + 4}\right] dx = \frac{1}{3}\int_{13}^{40} \sqrt{u}\,\frac{du}{6} = \frac{1}{18}\left[\frac{u^{3/2}}{\frac{3}{2}}\right]_{13}^{40}$$

$$= \frac{1}{27}\left(80\sqrt{10} - 13\sqrt{13}\right) \quad\blacksquare$$

**NOW WORK** Problem 17.

**EXAMPLE 3**  **Finding the Arc Length of an Ellipse**

For the ellipse $x^2 + 4y^2 = 4$,

**(a)** Set up, but do not evaluate, the integral for the arc length $s$ of the ellipse in the first quadrant from the point $(0,1)$ to the point of intersection of the ellipse and the line $y = x$.

〽 **(b)** Use technology to evaluate the integral from (a).

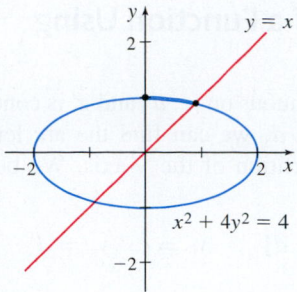

**Figure 51**

**Solution (a)** We begin by graphing the ellipse and the line $y = x$. See Figure 51.

The intersection of the ellipse $x^2 + 4y^2 = 4$ and the line $y = x$ satisfies the equation

$$x^2 + 4x^2 = 4 \qquad y = x$$
$$5x^2 = 4$$
$$x = \sqrt{\frac{4}{5}} \qquad x > 0 \text{ in the first quadrant.}$$
$$x = \frac{2\sqrt{5}}{5}$$

In the first quadrant, the ellipse and the line $y = x$ intersect at the point $\left(\dfrac{2\sqrt{5}}{5}, \dfrac{2\sqrt{5}}{5}\right)$.

To find the arc length $s$, we need to express the equation of the ellipse in the explicit form $y = f(x)$ and find the derivative $f'(x)$.

In the first quadrant both $x > 0$ and $y > 0$, so the ellipse $x^2 + 4y^2 = 4$ can be expressed explicitly as

$$x^2 + 4y^2 = 4$$
$$4y^2 = 4 - x^2$$
$$y^2 = \frac{4 - x^2}{4}$$
$$y = \sqrt{\frac{4 - x^2}{4}} \qquad y > 0 \text{ in the first quadrant.}$$
$$y = \frac{1}{2}\sqrt{4 - x^2}$$

Then,

$$y' = \frac{d}{dx}\left[\frac{1}{2}\sqrt{4 - x^2}\right] = \frac{d}{dx}\left[\frac{1}{2}(4 - x^2)^{1/2}\right]$$
$$= \frac{1}{2}\left[\frac{1}{2}(4 - x^2)^{-1/2}(-2x)\right] \qquad \text{\textcolor{blue}{Use the Power Rule for Functions.}}$$
$$= -\frac{x}{2(4 - x^2)^{1/2}} \qquad \text{\textcolor{blue}{Simplify.}}$$

and

$$1 + (y')^2 = 1 + \frac{x^2}{4(4 - x^2)}$$

The arc length $s$ of the ellipse from the point $(0, 1)$ to the point $\left(\dfrac{2\sqrt{5}}{5}, \dfrac{2\sqrt{5}}{5}\right)$ is given by

$$s = \int_0^{2\sqrt{5}/5} \sqrt{1 + (y')^2}\, dx = \int_0^{2\sqrt{5}/5} \sqrt{1 + \frac{x^2}{4(4 - x^2)}}\, dx$$

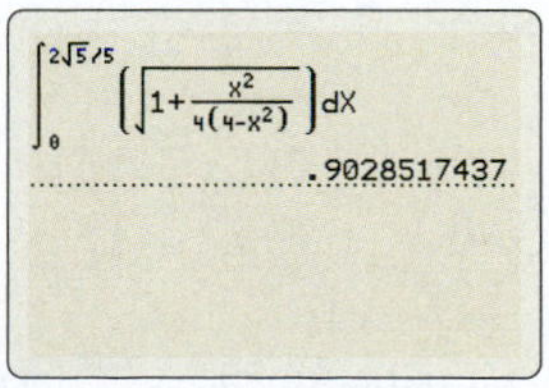

**(b)** As Figure 52 shows, $s \approx 0.903$. ∎

**Figure 52**

$$\int_0^{2\sqrt{5}/5}\left[\sqrt{1 + \tfrac{x^2}{4(4-x^2)}}\right]dX$$
.............................
                    .9028517437

**NOW WORK** Problem 29.

An observation about Example 3 is noteworthy. Notice that $y' = f'(x)$ is not defined at $x = 2$ so $f'$ is not continuous at 2. The arc length formula (1) cannot be used to compute any arc length that includes $x = 2$ (or $x = -2$). But we can find the arc length of the ellipse from the point $\left(\dfrac{2\sqrt{5}}{5}, \dfrac{2\sqrt{5}}{5}\right)$ to the point $(2, 0)$ by partitioning the $y$-axis.

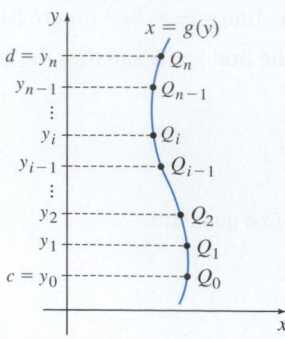

**Figure 53** $[c, d]$ is partitioned into $n$ subintervals of width $\dfrac{d-c}{n}$.

## 2 Find the Arc Length of the Graph of a Function Using a Partition of the $y$-Axis

For a function defined by $x = g(y)$, where $g$ is continuous on $[c, d]$ and $g'$ is continuous on an open interval containing the numbers $c$ and $d$, we can find the arc length of the graph of $g$ from $y = c$ to $y = d$ by using a partition of the $y$-axis. We begin by partitioning the interval $[c, d]$ into $n$ subintervals:

$$[c, y_1], [y_1, y_2], \ldots, [y_{i-1}, y_i], \ldots, [y_{n-1}, d] \qquad y_0 = c \quad y_n = d$$

each of width $\Delta y = \dfrac{d-c}{n}$. Corresponding to each number $c, y_1, y_2, \ldots, y_{i-1}, y_i, \ldots,$ $y_{n-1}, d$, there are points $Q_0, Q_1, Q_2, \ldots, Q_{i-1}, Q_i, \ldots, Q_{n-1}, Q_n$ on the graph of $g$, as shown in Figure 53. By forming line segments connecting each point $Q_{i-1}$ to $Q_i$, for $i = 1, 2, 3, \ldots, n$, and following the same process as described for a partition of the $x$-axis, we obtain the following result:

### Arc Length Formula (Partitioning the $y$-axis)

For a function defined by $x = g(y)$, where $g$ is continuous on $[c, d]$ and $g'$ is continuous on an interval containing $c$ and $d$, the arc length $s$ of the graph of $g$ from $y = c$ to $y = d$ is given by

$$s = \int_c^d \sqrt{1 + [g'(y)]^2}\, dy \tag{2}$$

The result in (2) can sometimes be used to find the length of the graph of a function $y = f(x)$ from $a$ to $b$ when its derivative $f'$ is not continuous at some number in $[a, b]$.

### EXAMPLE 4 Finding the Arc Length of a Function

Find the arc length of $f(x) = x^{2/3}$ from $x = -1$ to $x = 8$.

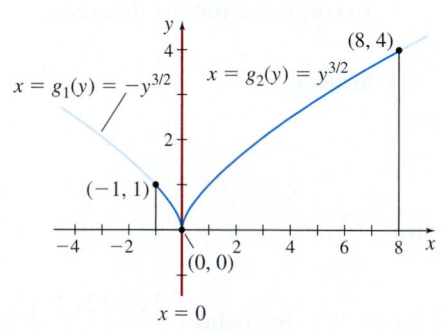

**Figure 54** $f(x) = x^{2/3}$, $-1 \le x \le 8$

**Solution** Since the derivative $f'(x) = \dfrac{2}{3x^{1/3}}$ does not exist at $x = 0$, we cannot use formula (1) to find the length of the graph from $x = -1$ to $x = 8$.

From Figure 54, we observe that the length of the graph of $f$ from $x = -1$ to $x = 8$ is the same as the length of the graph of $x = g_1(y) = -y^{3/2}$ from $y = 0$ to $y = 1$ plus the length of the graph of $x = g_2(y) = y^{3/2}$ from $y = 0$ to $y = 4$.

We investigate $x = g_1(y) = -y^{3/2}$ first. Since $g_1'(y) = -\dfrac{3}{2}y^{1/2}$ is continuous for all $y \ge 0$. we can use arc length formula (2) to find the arc length $s_1$ of $g_1$ from $y = 0$ to $y = 1$.

$$s_1 = \int_0^1 \sqrt{1 + [g_1'(y)]^2}\, dy = \int_0^1 \sqrt{1 + \left(-\frac{3}{2}y^{1/2}\right)^2}\, dy$$

$$\underset{\displaystyle g_1'(y) = -\frac{3}{2}y^{1/2}}{\uparrow}$$

$$= \int_0^1 \sqrt{1 + \frac{9}{4}y}\, dy = \frac{1}{2}\int_0^1 \sqrt{4 + 9y}\, dy$$

Use the substitution $u = 4 + 9y$. Then $du = 9\, dy$ or, equivalently, $dy = \dfrac{du}{9}$. The limits of integration are $u = 4$ when $y = 0$, and $u = 13$ when $y = 1$.

$$s_1 = \frac{1}{2}\int_4^{13} \sqrt{u}\, \frac{du}{9} = \frac{1}{18}\left[\frac{u^{3/2}}{\frac{3}{2}}\right]_4^{13} = \frac{1}{27}\left(13\sqrt{13} - 8\right)$$

Now we investigate $x = g_2(y) = y^{3/2}$. Since $g_2'(y) = \dfrac{3}{2}y^{1/2}$ is continuous for all $y \ge 0$, we can use arc length formula (2) to find the arc length $s_2$ of $g_2$ from $y = 0$ to $y = 4$.

$$s_2 = \int_0^4 \sqrt{1 + [g_2'(y)]^2} \, dy = \int_0^4 \sqrt{1 + \left(\frac{3}{2}y^{1/2}\right)^2} \, dy$$

$$= \int_0^4 \sqrt{1 + \frac{9}{4}y} \, dy = \frac{1}{2}\int_0^4 \sqrt{4 + 9y} \, dy \qquad \begin{array}{l} \text{Let } u = 4 + 9y \\ du = 9\,dy \end{array}$$

$$= \frac{1}{2}\int_4^{40} \sqrt{u} \, \frac{du}{9} = \frac{1}{18}\left[\frac{u^{3/2}}{\frac{3}{2}}\right]_4^{40} = \frac{1}{27}\left(80\sqrt{10} - 8\right)$$

The arc length $s$ of $y = f(x) = x^{2/3}$ from $x = -1$ to $x = 8$ is the sum

$$s = s_1 + s_2 = \frac{1}{27}\left(13\sqrt{13} - 8\right) + \frac{1}{27}\left(80\sqrt{10} - 8\right)$$

$$= \frac{1}{27}\left(80\sqrt{10} + 13\sqrt{13} - 16\right) \qquad \blacksquare$$

**NOW WORK** Problem 23.

## 3 Find the Surface Area of a Solid of Revolution

The **surface area** of a solid of revolution measures the outer surface of the solid. The formula for the surface area of a solid of revolution depends on the formula for the surface area of a *frustum* of a right circular cone.

**NOTE** A frustum is a portion of a solid of revolution that lies between two parallel planes.

Consider a line segment of length $L$ that lies above the $x$-axis. See Figure 55(a). If the region bounded by the line segment, the $x$-axis, and the lines $x = a$ and $x = b$, is revolved about an axis, the resulting solid of revolution is a **frustum**. A frustum of a right circular cone has slant height $L$ and base radii $r_1$ and $r_2$. See Figure 55(b).

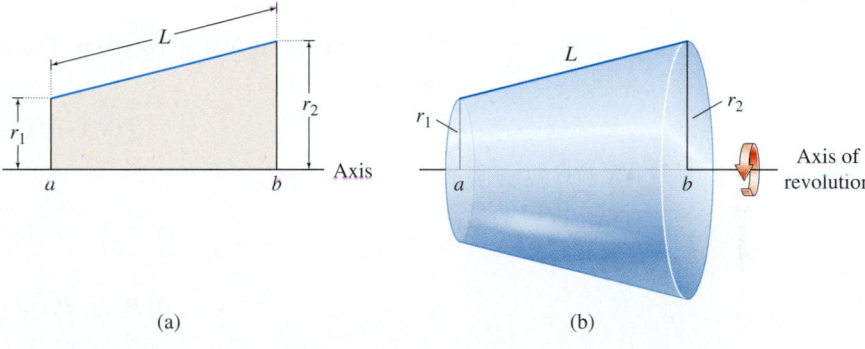

(a)                                    (b)

**Figure 55**

The **surface area** $S$ of a frustum of a right circular cone equals the product of $2\pi$, the average radius of the frustum, and the slant height of the frustum. That is,

$$\boxed{S = 2\pi\left(\frac{r_1 + r_2}{2}\right)L}$$

Suppose a solid of revolution is formed by revolving the region shown in Figure 56 bounded by the graph of a function $y = f(x)$, where $f(x) \geq 0$, the $x$-axis, and the lines $x = a$ and $x = b$, about the $x$-axis. We assume the derivative $f'$ of $f$ is continuous on some interval containing $a$ and $b$. To find the surface area of this solid of revolution, we partition the closed interval $[a, b]$ into $n$ subintervals

$$[a, x_1], [x_1, x_2], \ldots, [x_{i-1}, x_i], \ldots, [x_{n-1}, b]$$

each of width $\Delta x = \dfrac{b - a}{n}$. Corresponding to each number $a, x_1, x_2, \ldots, x_{i-1}, x_i, \ldots,$ $x_{n-1}, b$, there is a point $P_0, P_1, \ldots, P_{i-1}, P_i, \ldots, P_{n-1}, P_n$ on the graph of $f$. We join each point $P_{i-1}$ to the next point $P_i$ with a line segment and focus on the line

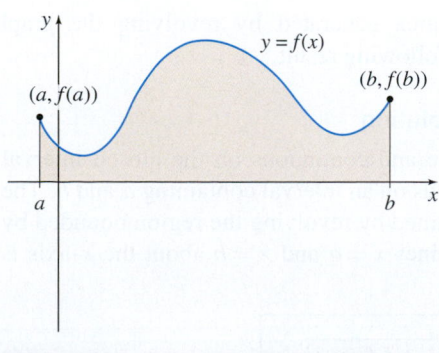

**Figure 56**

segment joining the points $P_{i-1}$ and $P_i$. See Figure 57(a). When the region bounded by the line segment of length $d(P_{i-1}, P_i)$ and the $x$-axis, is revolved about the $x$-axis, it generates a frustum of a right circular cone whose surface area $S_i$ is

$$S_i = 2\pi \left[\frac{y_{i-1} + y_i}{2}\right] [d(P_{i-1}, P_i)] \tag{3}$$

See Figure 57(b).

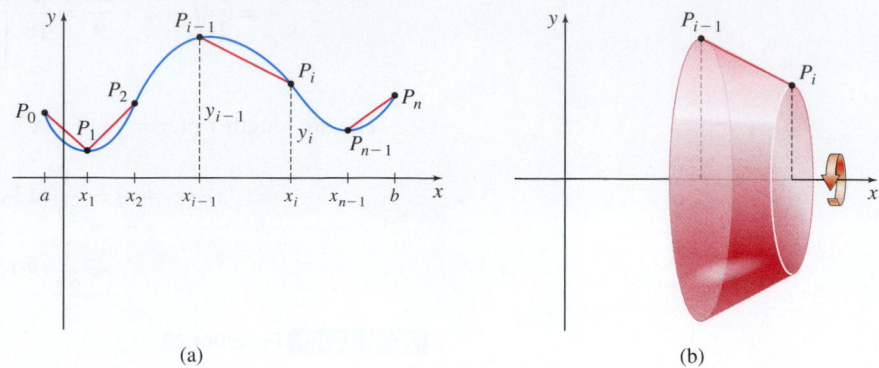

(a)                                     (b)

**Figure 57**

Now follow the same reasoning used for finding the arc length of the graph of a function.

From the distance formula, the length of the $i$th line segment is

$$d(P_{i-1}, P_i) = \sqrt{(x_i - x_{i-1})^2 + (y_i - y_{i-1})^2}$$

$$= \sqrt{(\Delta x)^2 + (\Delta y_i)^2} = \sqrt{1 + \left(\frac{\Delta y_i}{\Delta x}\right)^2} \, \Delta x$$

Now use the Mean Value Theorem. There is a number $u_i$ in the open interval $(x_{i-1}, x_i)$ for which

$$f(x_i) - f(x_{i-1}) = f'(u_i)\Delta x$$

$$\frac{\Delta y_i}{\Delta x} = f'(u_i)$$

So,

$$d(P_{i-1}, P_i) = \sqrt{1 + [f'(u_i)]^2} \, \Delta x$$

where $u_i$ is a number in the $i$th subinterval.

Now replace $d(P_{i-1}, P_i)$ in equation (3) with $\sqrt{1 + [f'(u_i)]^2} \, \Delta x$. Then the surface area generated by the sum of the line segments is

$$\sum_{i=1}^{n} S_i = \sum_{i=1}^{n} 2\pi \frac{y_{i-1} + y_i}{2} \sqrt{1 + [f'(u_i)]^2} \, \Delta x$$

These sums approximate the surface area generated by revolving the graph of $y = f(x)$ about the $x$-axis and lead to the following result.

**THEOREM   Surface Area of a Solid of Revolution**

Suppose a function $y = f(x)$ is nonnegative and continuous on the closed interval $[a, b]$ and has a derivative $f'$ that is continuous on an interval containing $a$ and $b$. The surface area $S$ of the solid of revolution obtained by revolving the region bounded by the graph of $y = f(x)$, the $x$-axis, and the lines $x = a$ and $x = b$ about the $x$-axis is given by

$$S = 2\pi \int_a^b f(x)\sqrt{1 + [f'(x)]^2} \, dx \tag{4}$$

EXAMPLE 5   **Finding the Surface Area of a Solid of Revolution**

Find the surface area $S$ of the solid generated by revolving the region bounded by the graph of $y = \sqrt{x}$, the $x$-axis, and the lines $x = 1$ and $x = 4$, about the $x$-axis.

**Solution** We begin with the graph of $y = \sqrt{x}$, $1 \le x \le 4$, and revolve it about the $x$-axis, as shown in Figure 58.

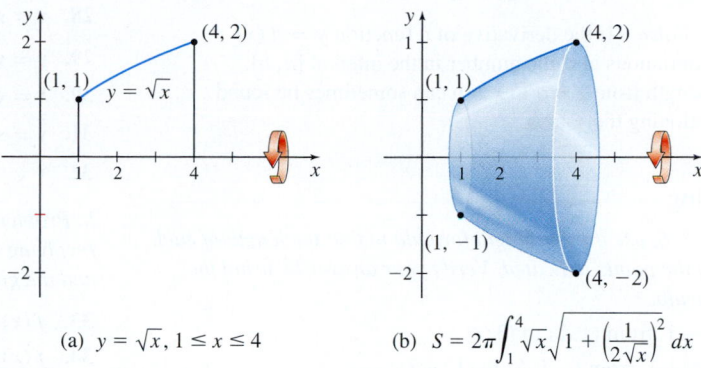

(a) $y = \sqrt{x},\ 1 \le x \le 4$

(b) $S = 2\pi \int_1^4 \sqrt{x} \sqrt{1 + \left(\dfrac{1}{2\sqrt{x}}\right)^2}\, dx$

**Figure 58**

Use formula (4) with $f(x) = \sqrt{x}$ and $f'(x) = \dfrac{1}{2\sqrt{x}}$. The surface area $S$ is

$$S = 2\pi \int_a^b f(x)\sqrt{1 + [f'(x)]^2}\, dx = 2\pi \int_1^4 \sqrt{x}\sqrt{1 + \left(\frac{1}{2\sqrt{x}}\right)^2}\, dx$$

$$= 2\pi \int_1^4 \sqrt{x\left(1 + \frac{1}{4x}\right)}\, dx = 2\pi \int_1^4 \frac{1}{2}\sqrt{4x + 1}\, dx$$

$$\underset{\substack{\uparrow \\ u = 4x + 1;\ \frac{du}{4} = dx \\ \text{when } x = 1,\, u = 5;\ \text{when } x = 4,\, u = 17}}{=} \pi \int_5^{17} \sqrt{u}\left(\frac{du}{4}\right) = \frac{\pi}{4}\left[\frac{u^{3/2}}{\frac{3}{2}}\right]_5^{17} = \frac{\pi}{6}(17^{3/2} - 5^{3/2})$$

The surface area of the solid of revolution is $\dfrac{\pi}{6}(17^{3/2} - 5^{3/2}) \approx 30.846$ square units.  ∎

NOW WORK  **Problem 35.**

## Summary

Suppose $y = f(x)$ is a nonnegative function that is continuous on a closed interval $[a, b]$, and suppose the derivative $f'$ of $f$ is continuous on some interval containing $a$ and $b$.

- **Arc Length:** The arc length $s$ of the graph of $f$ from $a$ to $b$ is

$$s = \int_a^b \sqrt{1 + [f'(x)]^2}\, dx$$

- **Surface Area:** The surface area $S$ of the solid obtained by revolving the region bounded by $y = f(x)$, the lines $x = a$ and $x = b$, and the $x$-axis about the $x$-axis is

$$S = 2\pi \int_a^b f(x)\sqrt{1 + [f'(x)]^2}\, dx$$

## 6.5 Assess Your Understanding

### Concepts and Vocabulary

1. *True or False*  If a function $f$ has a derivative that is continuous on an interval containing $a$ and $b$, the length $s$ of the graph of $y = f(x)$ from $x = a$ to $x = b$ is given by the formula $s = \int_a^b \sqrt{1 + [f'(x)]^2}\, dx$.

2. *True or False*  If the derivative of a function $y = f(x)$ is not continuous at some number in the interval $[a, b]$, its arc length from $x = a$ to $x = b$ can sometimes be found by partitioning the $y$-axis.

### Skill Building

*In Problems 3–6, use the arc length formula to find the length of each line between the points indicated. Verify your answer by using the distance formula.*

3. $y = 3x - 1$, from $(1, 2)$ to $(3, 8)$

4. $y = -4x + 1$, from $(-1, 5)$ to $(1, -3)$

5. $2x - 3y + 4 = 0$, from $(1, 2)$ to $(4, 4)$

6. $3x + 4y - 12 = 0$, from $(0, 3)$ to $(4, 0)$

*In Problems 7–22, find the arc length of each graph by partitioning the x-axis.*

7. $y = x^{2/3} + 1$, from $x = 1$ to $x = 8$

8. $y = x^{2/3} + 6$, from $x = 1$ to $x = 8$

9. $y = x^{3/2}$, from $x = 0$ to $x = 4$

10. $y = x^{3/2} + 4$, from $x = 1$ to $x = 4$

11. $9y^2 = 4x^3$, from $x = 0$ to $x = 1$; $y \geq 0$

12. $y = \dfrac{x^3}{6} + \dfrac{1}{2x}$, from $x = 1$ to $x = 3$

13. $y = \dfrac{2}{3}(x^2 + 1)^{3/2}$, from $x = 1$ to $x = 4$

14. $y = \dfrac{1}{3}(x^2 + 2)^{3/2}$, from $x = 2$ to $x = 4$

15. $y = \dfrac{2}{9}\sqrt{3}(3x^2 + 1)^{3/2}$, from $x = -1$ to $x = 2$

16. $y = (1 - x^{2/3})^{3/2}$, from $x = \dfrac{1}{8}$ to $x = 1$

17. $8y = x^4 + \dfrac{2}{x^2}$, from $x = 1$ to $x = 2$

18. $9y^2 = 4(1 + x^2)^3$, $y \geq 0$, from $x = 0$ to $x = 2\sqrt{2}$

19. $y = \ln(\sin x)$, from $x = \dfrac{\pi}{6}$ to $x = \dfrac{\pi}{3}$

20. $y = \ln(\cos x)$, from $x = \dfrac{\pi}{6}$ to $x = \dfrac{\pi}{3}$

21. $(x + 1)^3 = 4y^2$, $y \geq 0$, from $x = -1$ to $x = 16$

22. $y = x^{3/2} + 8$, from $x = 0$ to $x = 4$

*In Problems 23–26, find the arc length of each graph by partitioning the y-axis.*

23. $y = x^{2/3}$, from $x = 0$ to $x = 1$

24. $y = x^{2/3}$, from $x = -1$ to $x = 0$

25. $(x + 1)^2 = 4y^3$, $x \geq -1$, from $y = 0$ to $y = 1$

26. $x = \dfrac{2}{3}(y - 5)^{3/2}$, from $y = 5$ to $y = 6$

*In Problems 27–32, (a) use the arc length formula (1) to set up the integral for arc length.*

**(b)** *If you have access to a graphing calculator or a CAS, find the length. Do not attempt to integrate by hand.*

27. $y = x^2$, from $x = 0$ to $x = 2$

28. $x = y^2$, from $y = 1$ to $y = 3$

29. $y = \sqrt{25 - x^2}$, from $x = 0$ to $x = 4$

30. $x = \sqrt{4 - y^2}$, from $y = 0$ to $y = 1$

31. $y = \sin x$, from $x = 0$ to $x = \dfrac{\pi}{2}$

32. $x = y + \ln y$, from $y = 1$ to $y = 4$

*In Problems 33–42, find the surface area of the solid formed by revolving the region bounded by the graph of $y = f(x)$, the x-axis, and the given lines about the x-axis.*

33. $f(x) = 3x + 5$; lines: $x = 0$ and $x = 2$

34. $f(x) = -x + 5$; lines: $x = 1$ and $x = 5$

35. $f(x) = \sqrt{3x}$; lines: $x = 1$ and $x = 2$

36. $f(x) = \sqrt{2x}$; lines: $x = 2$ and $x = 8$

37. $f(x) = 2\sqrt{x}$; lines: $x = 1$ and $x = 4$

38. $f(x) = 4\sqrt{x}$; lines: $x = 4$ and $x = 9$

39. $f(x) = x^3$; lines: $x = 0$ and $x = 2$

40. $f(x) = \dfrac{1}{3}x^3$; lines: $x = 1$ and $x = 3$

41. $f(x) = \sqrt{4 - x^2}$; lines: $x = 0$ and $x = 1$

42. $f(x) = \sqrt{9 - 4x^2}$; lines: $x = 0$ and $x = 1$

*In Problems 43–50, (a) set up the integral that represents the surface area of the solid formed by revolving the region bounded by the graph of $y = f(x)$, the x-axis, and the given lines about the x-axis.*

**(b)** *Use technology to evaluate the integral found in (a).*

43. $f(x) = x^2$; lines: $x = 1$ and $x = 3$

44. $f(x) = 4 - x^2$; lines: $x = 0$ and $x = 2$

45. $f(x) = x^{2/3}$; lines: $x = 1$ and $x = 8$

46. $f(x) = x^{3/2}$; lines: $x = 0$ and $x = 4$

47. $f(x) = e^x$; lines: $x = 0$ and $x = 3$

48. $f(x) = \ln x$; lines: $x = 1$ and $x = e$

49. $f(x) = \sin x$; lines: $x = 0$ and $x = \dfrac{\pi}{2}$

50. $f(x) = \tan x$; lines: $x = 0$ and $x = \dfrac{\pi}{4}$

### Applications and Extensions

51. Find the length of the graph of $F(x) = \int_0^x \sqrt{16t^2 - 1}\, dt$ from $x = 0$ to $x = 2$.

52. Find the length of the graph of $F(x) = \int_0^x \sqrt{4t - 1}\, dt$ from $x = 0$ to $x = 2$.

53. Find the length of the graph of $F(x) = \int_0^{3x} \sqrt{e^t - \dfrac{1}{9}}\, dt$ from $x = 0$ to $x = 4$.

54. Find the length of the graph of $F(x) = \int_0^{4x} \sqrt{t^4 - \dfrac{1}{4}}\, dt$ from $x = 1$ to $x = 2$.

**1.** = NOW WORK problem    = Graphing technology recommended    CAS = Computer Algebra System recommended

**55. Length of a Hypocycloid**  Find the total length of the hypocycloid $x^{2/3} + y^{2/3} = a^{2/3}$, $a > 0$.

**56. Distance Along a Curved Path**  Find the distance between $(1, 1)$ and $(3, 3\sqrt{3})$ along the graph of $y^2 = x^3$.

**57. Perimeter**  Find the perimeter of the region bounded by the graphs of $y^3 = x^2$ and $y = x$.

**58. Perimeter**  Find the perimeter of the region bounded by the graphs of $y = 3(x-1)^{3/2}$ and $y = 3(x-1)$.

**59. Length of a Graph**  Find the length of the graph of $6xy = y^4 + 3$ from $y = 1$ to $y = 2$.

**60. Length of a Graph**  Find the length of the hyperbolic function $y = \cosh x$ from $(0, 1)$ to $(2, \cosh 2)$.

**61. Length of an Graph**  Find the length of $y = \ln(\csc x)$ from $x = \dfrac{\pi}{4}$ to $x = \dfrac{\pi}{2}$.

**62. Length of an Elliptical Arc**  Set up, but do not attempt to evaluate, the integral for the arc length of the ellipse $\dfrac{x^2}{a^2} + \dfrac{y^2}{b^2} = 1$ from $x = 0$ to $x = \dfrac{a}{2}$ in quadrant I. This integral, which is approximated by numerical techniques (see Chapter 7), is called an **elliptical integral of the second kind**.

**63. (a)** Set up, but do not attempt to evaluate, the integral for the arc length of the ellipse $x^2 + 4y^2 = 4$ in the first quadrant from the point of intersection of the ellipse and the line $y = x$ to the point $(2, 0)$.

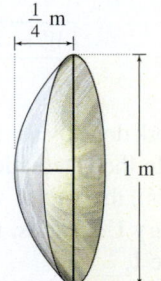 **(b)** Use technology to evaluate the integral found in (a).

**(c)** Use the result obtained in Example 3, p. 464, to find the perimeter of the ellipse $x^2 + 4y^2 = 4$.

**64. Surface Area of a Cone**

**(a)** Find an equation of the line from the origin $(0, 0)$ to the point $(h, a)$, $a > 0$, $h > 0$.

**(b)** Graph the line found in (a).

**(c)** Revolve the region bounded by the line from (a), the $x$-axis, and the line $x = h$ about the $x$-axis to generate a cone of height $h$ and radius $a$.

**(d)** Show that the surface area $S$ of the cone is $S = \pi a \sqrt{a^2 + h^2}$.

**65. Search Light**  The reflector of a search light is formed by revolving an arc of a parabola about its axis. Find the surface area of the reflector, if it measures 1 m across its widest point and is $\dfrac{1}{4}$ m deep.

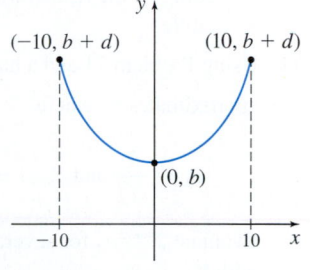

**66. Modeling a Ski Slope**  A ski slope is built on a mountainside and curves upward from ground level to a height $h$. The shape of the ski slope is modeled by the equation $y = Ax^{3/2}$, where $x$ is the horizontal distance from the bottom of the ski slope measured along the base of the mountain and $y$ is the vertical height of the ski slope at the distance $x$. See the figure (top, right column).

**(a)** Find an expression, in terms of $A$ and $h$, for the length of the ski slope.

**(b)** Find $A$ if the ski slope is 150 m high and has a horizontal distance of 250 m along the base.

**(c)** If a skier skis directly downhill from the top of the ski slope to the bottom, how far does she travel?

**(d)** Describe a simple way to check if the distance obtained in part (c) is reasonable.

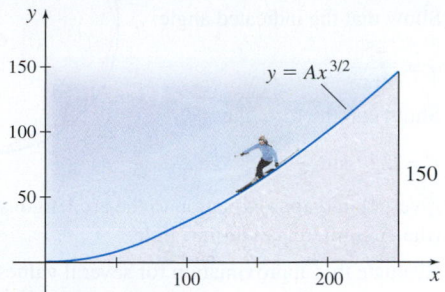

**67. Length of a Cable**  A cable hangs in the shape of a catenary between two supports at the same height 20 m apart. See the figure. The slope of the tangent line to the cable at the right-hand support is $\dfrac{3}{4}$. The equation of a catenary is $y = a \cosh \dfrac{x}{a} + b - a$, where $a$ is a constant that depends on the linear density of the cable and the tension on the cable and $b$ is the distance from the $x$-axis to the lowest point of the cable.

**(a)** What is the height of the supports? (*Hint:* Let $d$ represent the height of the supports above the lowest point $b$.)

**(b)** Find the length of the cable.

**68. Equation of a Catenary**  A rope of length $L$ is supported at two points, $(c, d)$ and $(-c, d)$. The middle of the rope is located at the point $(0, b)$. The shape of the rope is given by $y = a \cosh \left( \dfrac{x}{a} \right) + b - a$. Find $a$.

*Hint:* Show that $(d - b + a)^2 - a^2 = \dfrac{L^2}{4}$.

**69. Surface Area of a Catenoid**  When an arc of a catenary $y = \cosh x$, $a \le x \le b$, is revolved about the $x$-axis, it generates a surface called a **catenoid**, which has the least surface area of all surfaces generated by rotating curves having the same endpoints. Find its surface area. See the figure.

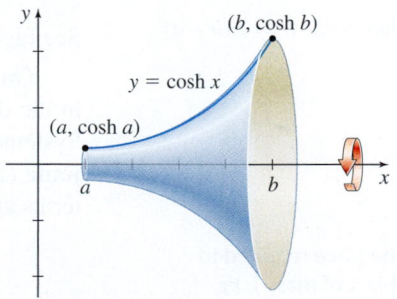

**70. Projectile Motion**  An object is launched from level ground, travels in a parabolic trajectory, and lands 150 m away. It reaches a maximum height of 46 m.

**(a)** Fit a parabola of the form $h = h(x)$ to the path of the object, where $h$ is the height and $x$ is the horizontal distance traveled.

CAS **(b)** Find the arc length of the trajectory.

## Challenge Problems

**71.** Inscribe a regular polygon of $2^n$ sides in a circle of radius 1. The figure illustrates the situation for $n = 3$.

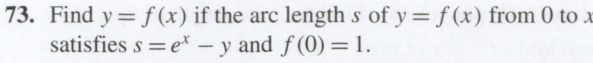

**(a)** Show that the indicated angle

$$\alpha = \frac{45°}{2^{n-2}}.$$

**(b)** Show that the formula

$$s = 2^{n+1} \sin \frac{45°}{2^{n-2}}$$

gives a good approximation to the arc length $s$ (the circumference) of the circle.

**(c)** Evaluate this approximation for several values of $n$, and compare the results with the actual circumference of the circle.

**72.** Using Problem 71 and a half-angle formula, show that $2^{n+2}a_n$ approximates $\pi$, where

$$a_0 = \frac{1}{\sqrt{2}} \quad \text{and} \quad a_{n+1} = \sqrt{\frac{1 - \sqrt{1 - a_n^2}}{2}}, \quad n = 0, 1, 2, \dots$$

Evaluate $2^{n+2}a_n$ for several values of $n$ and compare your answer with $\pi$.

**73.** Find $y = f(x)$ if the arc length $s$ of $y = f(x)$ from 0 to $x$ satisfies $s = e^x - y$ and $f(0) = 1$.

**74. Length of a Circular Arc** In each case below, $P_1 = (x_1, y_1)$ and $P_2 = (x_2, y_2)$ are points on the circle $x^2 + y^2 = 1$ that do not lie on a coordinate axis. Express the length of the counterclockwise arc $P_1 P_2$ in terms of integrals of the form

$$\int_u^v \frac{1}{\sqrt{1 - t^2}} \, dt, \qquad -1 < u < v < 1$$

**(a)** when $P_1$ is in quadrant I and $P_2$ is in quadrant II.

**(b)** when $P_1$ and $P_2$ are both in quadrant III and $y_1 < y_2$.

**(c)** when $P_1$ is in quadrant II and $P_2$ is in quadrant IV.

**75. Arc Length** The arc length formula cannot be used to find the arc length from $x = -\dfrac{\pi}{2}$ to $x = \dfrac{\pi}{2}$ of the function $f$ given below. Why not?

$$f(x) = \begin{cases} x \sin \dfrac{1}{x} & \text{if } x \neq 0 \\ 0 & \text{if } x = 0 \end{cases}$$

# 6.6 Work

**OBJECTIVES** *When you finish this section, you should be able to:*

**1** Find the work done by a variable force (p. 473)

**2** Find the work done by a spring force (p. 474)

**3** Find the work done to pump a liquid (p. 476)

Application: gravitational force (p. 478)

In physics, **work** is defined as the energy transferred to or from an object by a force acting on the object. The work $W$ done by a *constant* force $F$ in moving an object a distance $x$ along a straight line in the direction of $F$ is defined to be

$$\boxed{W = Fx}$$

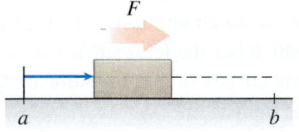

**Figure 59** Constant force $W = F(b - a)$

See Figure 59.

One unit of work is the work done by a unit force in moving an object a unit distance in the direction of the force. In the International System of Units (abbreviated SI, for Système International d'Unités), the unit of work is a newton-meter, which is called a **joule** (J). In terms of customary U.S. units, the unit of work is the foot-pound. These terms are summarized in Table 1.

**NOTE** 1 N is the force required to accelerate an object of mass 1 kg at 1 m/s². Also, 1 J ≈ 0.7376 ft lb and 1 ft lb ≈ 1.356 J.

**TABLE 1**

| | Work Units | Force Units | Distance Units |
|---|---|---|---|
| SI | joule (J) | newton (N) | meter (m) |
| U.S. | foot-pound (ft lb) | pound (lb) | foot (ft) |

For example, the work $W$ required to lift an object weighing 80 lb a distance of 5 ft would be $W = 80 \cdot 5 = 400$ ft lb.

**ORIGINS** Joule, the unit of energy, is named for James Prescott Joule, a British physicist and brewer who lived in the 19th century.

When a force $F$ acts in the same direction as the motion, the work done is positive; if a force $F$ acts in a direction opposite to the motion, the work done is negative.

In some cases, a force $F$ acts along the line of motion of an object, but the magnitude of the force *varies* depending on the position of the object. For example, the force required to raise a cable depends on the length of the cable, and the force required to stretch a spring depends on how far the spring has already been stretched from its normal length. These are examples of **variable forces**.

## 1 Find the Work Done by a Variable Force

Suppose a variable force $F = F(x)$ acts on an object, where $x$ is the distance of the object from the origin and $F$ is a function that is continuous on a closed interval $[a, b]$. We seek a formula for finding the work done by the force $F$ in moving the object from $x = a$ to $x = b$. We begin by partitioning the interval $[a, b]$ into $n$ subintervals:

$$[a, x_1], [x_1, x_2], \ldots, [x_{i-1}, x_i], \ldots, [x_{n-1}, b] \qquad a = x_0, b = x_n$$

each of width $\Delta x = \dfrac{b-a}{n}$.

Now consider the $i$th subinterval $[x_{i-1}, x_i]$, and choose a number $u_i$ in $[x_{i-1}, x_i]$. If the width of the subinterval is small, the force $F = F(x)$ acting on the object will not change much over the interval; that is, $F$ can be treated as a constant force. Then the work $W$ done by $F$ to move the object from $x_{i-1}$ to $x_i$, a distance $\Delta x = x_i - x_{i-1}$, can be approximated by

$$\boxed{\begin{array}{c} W_i \;=\; F(u_i)\,\Delta x \\ \uparrow \\ W = Fx \end{array}}$$

Summing the work done over all the subintervals approximates the total work $W$ done by the force $F$ in moving the object from $a$ to $b$. That is, the total work $W$ is approximated by the Riemann sums

$$W \approx F(u_1)\Delta x + F(u_2)\Delta x + \cdots + F(u_n)\Delta x = \sum_{i=1}^{n} F(u_i)\Delta x$$

As the number of subintervals increases, that is, as $n \to \infty$, the approximation improves, and, if $F = F(x)$ is continuous on $[a, b]$, then

$$W = \lim_{n \to \infty} \sum_{i=1}^{n} F(u_i)\Delta x = \int_{a}^{b} F(x)\,dx$$

### Work

The **work** $W$ done by a continuously varying force $F$ acting on an object, which moves the object along a straight line in the direction of $F$ from $x = a$ to $x = b$, is

$$\boxed{W = \int_{a}^{b} F(x)\,dx}$$

Consider a freely hanging cable or chain that is being lengthened or shortened. The weight of the cable is a function of its length. As the cable is let out, its weight increases proportionally to its length, and when it is pulled in, it decreases proportionally to its length.

### EXAMPLE 1  Finding the Work Done in Pulling in a Rope

A 60-ft rope weighing 8 lb per linear foot is used for mooring a cruise ship. As the ship prepares to leave port, the rope is released, and it hangs freely over the side of the ship. How much work is done by the deckhand who winds in the rope?

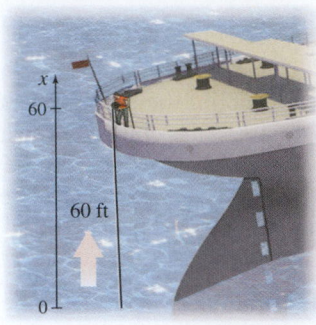

**Figure 60**

**Solution** We position an $x$-axis parallel to the side of the ship with the origin $O$ of the axis even with the bottom of the rope and $x = 60$ even with the ship's deck. See Figure 60. The work done by the deckhand depends on the weight of the rope and the length of rope hanging over the edge.

Partition the interval $[0, 60]$ into $n$ subintervals, each of length $\Delta x = \dfrac{60}{n}$, and choose a number $u_i$ in each subinterval. Now think of the rope as $n$ short segments, each of length $\Delta x$. Then

$$\text{Weight of the } i\text{th segment} = 8\Delta x \text{ lb} \qquad \boldsymbol{F}$$
$$\text{Distance the } i\text{th segment is lifted} = (60 - u_i) \text{ ft} \qquad \boldsymbol{x}$$
$$\text{Work done in lifting the } i\text{th segment} = 8(60 - u_i)\Delta x \text{ ft lb} \qquad \boldsymbol{W = Fx}$$

The work $W$ required to lift the 60 ft of rope is

$$W = \int_0^{60} 8(60 - x)\, dx = \left[480x - 4x^2\right]_0^{60} = 14{,}400 \text{ ft lb} \qquad \blacksquare$$

**NOW WORK** Problem 11.

## 2 Find the Work Done by a Spring Force

A common example of the work done by a variable force is found in stretching or compressing a spring that is fixed at one end. A spring is said to be in **equilibrium** when it is neither extended nor compressed. The **spring force** $F$ needed to extend or compress a spring depends on the *stiffness* of the spring, which is measured by its **spring constant** $k$, a positive real number. Since a spring always attempts to return to equilibrium, a spring force $F$ is often called a **restoring force**. A spring force $F$ varies with the distance $x$ that the free end of the spring is moved from its equilibrium length and obeys **Hooke's Law**:

$$\boxed{F(x) = -kx}$$

where $k$ is the spring constant measured in newtons per meter in SI units or pounds per foot in U.S. units. The minus sign in Hooke's Law indicates that the direction of a spring force is opposite from the direction of the displacement.

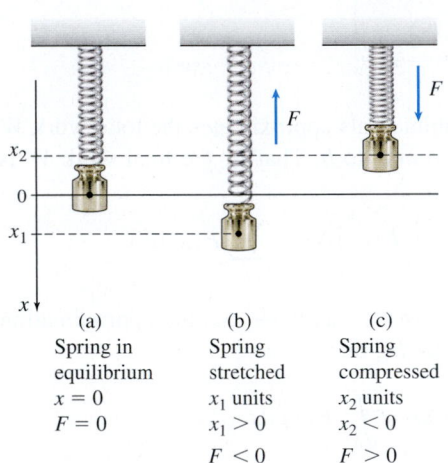

(a)
Spring in
equilibrium
$x = 0$
$F = 0$

(b)
Spring
stretched
$x_1$ units
$x_1 > 0$
$F < 0$

(c)
Spring
compressed
$x_2$ units
$x_2 < 0$
$F > 0$

**Figure 61**

As Figure 61 illustrates, the distance $x$ in Hooke's Law is measured from the equilibrium, or unstretched, position of the spring. This distance $x$ is positive if the spring is stretched from its equilibrium position and is negative if the spring is compressed from its equilibrium position. As a result, a spring force $F$ is negative if the spring is stretched and $F$ is positive if the spring is compressed.

In applied problems involving springs, the value of $k$, the spring constant, is often unknown. When we know information about how the spring behaves, the value of $k$ can be found. The next example illustrates this.

**EXAMPLE 2** Analyzing a Spring Force

Suppose a spring in equilibrium is 0.8 m long and a spring force of $-2$ N stretches the spring to a length of 1.2 m.

**(a)** Find the spring constant $k$ and the spring force $F$.

**(b)** What spring force is required to stretch the spring to a length of 3 m?

**(c)** How much work is done by the spring force in stretching it from equilibrium to 3 m?

**Solution** We position an axis parallel to the spring and place the origin at the free end of the spring in equilibrium, as in Figure 62.

**(a)** When the spring is stretched to a length of 1.2 m due to a spring force of $-2$ N, then $x = 0.4$ and $F = -2$. Using Hooke's Law, we get

$$-2 = -k(0.4) \qquad \text{Hooke's law: } F(x) = -kx; \ F = -2; \ x = 0.4$$
$$k = \frac{2}{0.4} = 5 \text{ N/m}$$

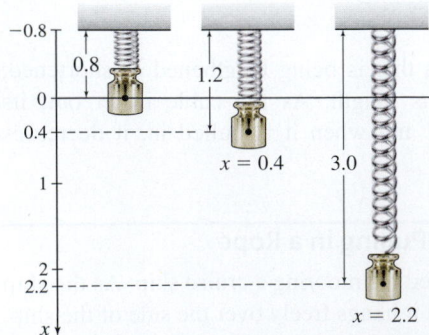

**Figure 62**

The spring constant is $k = 5$. The spring force $F$ is $F = -kx = -5x$.

**(b)** The spring force $F$ required to stretch the spring to a length of 3 m, that is, a distance $x = 3 - 0.8 = 2.2$ m from equilibrium, is

$$F = -5x = (-5)(2.2) = -11 \text{ N}$$

**(c)** The work $W$ done by the spring force $F$ when stretching the spring from equilibrium ($x = 0$) to 3 m ($x = 2.2$) is

$$W = \int_0^{2.2} F(x) \, dx = \underset{\underset{F(x) = -5x}{\uparrow}}{-5 \int_0^{2.2} x \, dx} = -5 \left[ \frac{x^2}{2} \right]_0^{2.2} = -\frac{5}{2}(4.84) = -12.1 \text{ J} \quad \blacksquare$$

---

**EXAMPLE 3**  **Finding the Work Done by a Spring Force**

Suppose an 0.8 m-long spring has a spring constant of $k = 5$ N/m.

**(a)** What spring force is required to compress the spring from its equilibrium position to a length of 0.5 m?

**(b)** How much work is done by the spring force when compressing the spring from equilibrium to a length of 0.5 m?

**(c)** How much work is done by the spring force when compressing the spring from 1.2 to 0.5 m?

**(d)** How much work is done by the spring force when compressing the spring from 1 to 0.6 m?

**Solution** Begin by positioning an axis parallel to the spring and placing the origin at the free end of the spring in equilibrium. See Figure 63.

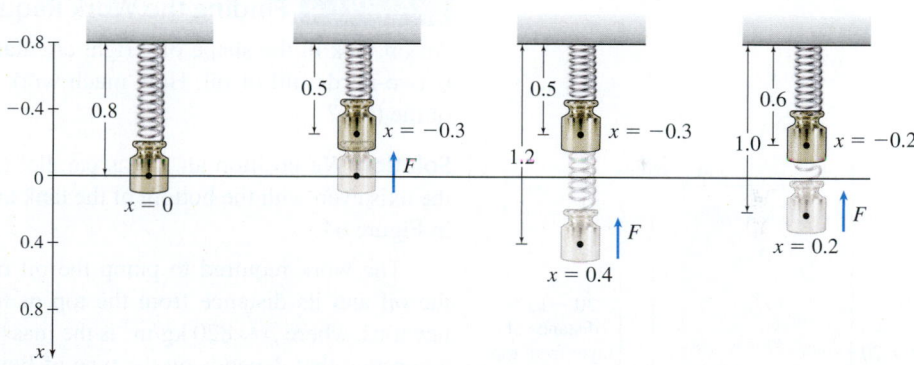

**Figure 63**

**(a)** By Hooke's Law, the spring force is $F = -5x$. When the spring is compressed to a length of 0.5 m, then $x = -0.3$. The spring force $F$ required to compress the spring to 0.5 m is

$$F = -kx = -5(-0.3) = 1.5 \text{ N}$$

**(b)** The work $W$ done by the spring force $F$ when compressing the spring from equilibrium ($x = 0$) to 0.5 m ($x = -0.3$) is

$$W = \int_0^{-0.3} F(x) \, dx = \underset{\underset{F(x) = -5x}{\uparrow}}{\int_0^{-0.3} (-5x) \, dx} = 5 \int_{-0.3}^0 x \, dx = 5 \left[ \frac{x^2}{2} \right]_{-0.3}^0$$

$$= 0 - \frac{5 \, (-0.3)^2}{2} = -0.225 \text{ J}$$

(c)   The work $W$ done by the spring force $F$ when compressing the spring from 1.2 m $(x = 0.4)$ to 0.5 m $(x = -0.3)$ is

$$W = \int_{0.4}^{-0.3} (-5x)\, dx = 5 \int_{-0.3}^{0.4} x\, dx = 5 \left[ \frac{x^2}{2} \right]_{-0.3}^{0.4} = \frac{5}{2}[0.4^2 - (-0.3)^2] = 0.175\, \text{J}$$

(d)   The work $W$ done when compressing the spring from 1 m $(x = 0.2)$ to 0.6 m $(x = -0.2)$ is

$$W = \int_{0.2}^{-0.2} (-5x)\, dx = 5 \int_{-0.2}^{0.2} x\, dx = 5 \left[ \frac{x^2}{2} \right]_{-0.2}^{0.2} = \frac{5}{2}[0.2^2 - (-0.2)^2] = 0\, \text{J}$$   ∎

Observe that the work $W$ done by a spring force can be positive, negative, or zero. If the force brings the spring closer to its equilibrium position $(x = 0)$, then $W > 0$; if the force brings the spring away from its equilibrium position, then $W < 0$. If the spring ends up the same distance from the equilibrium, then no work is done; that is, $W = 0$.

NOW WORK Problem 15.

### 3 Find the Work Done to Pump a Liquid

Another example of work done by a variable force is found in the work needed to pump a liquid out of a tank. The idea used is that the work needed to lift an object a given distance is the product of the weight (force) of the object and the distance it is lifted, that is,

$$\text{Work} = (\text{Weight of object})(\text{Distance lifted})$$

EXAMPLE 4   **Finding the Work Required to Pump Oil Out of a Tank**

An oil tank in the shape of a right circular cylinder, with height 30 m and radius 5 m, is two-thirds full of oil. How much work is required to pump all the oil over the top of the tank?

**Solution** We position an $x$-axis parallel to the side of the cylinder with the origin of the axis even with the bottom of the tank and $x = 30$ at the top of the tank, as illustrated in Figure 64.

The work required to pump the oil over the top is the product of the weight of the oil and its distance from the top of the tank. The weight of the oil equals $\rho g V$ newtons, where $\rho \approx 820$ kg/m$^3$ is the mass density of petroleum (mass per unit volume, a constant that depends on the type of liquid involved), $g \approx 9.8$ m/s$^2$ (the acceleration due to gravity), and $V$ is the volume of the liquid to be moved. The weight of the oil is

$$\text{Weight} = \rho g V \approx (820)(9.8)V = 8036V\ \text{N}$$

The oil fills the tank from $x = 0$ m to $x = 20$ m. We partition the interval $[0, 20]$ into $n$ subintervals, each of width $\Delta x = \dfrac{20}{n}$. Consider the oil in the $i$th subinterval as a thin layer of thickness $\Delta x$. Now choose a number $u_i$ in the $i$th subinterval. Then

$$\text{Volume } V_i \text{ of } i\text{th layer} = (\text{Area of layer})(\text{Thickness}) = \pi r^2 \Delta x = 25\pi\, \Delta x$$

$$\text{Weight of } i\text{th layer} = \rho g V_i \approx (8036)(25\pi\, \Delta x) = 200{,}900\pi\, \Delta x$$

$$\text{Distance } i\text{th layer is lifted} = 30 - u_i$$

$$\text{Work } W_i \text{ done in lifting } i\text{th layer} \approx (200{,}900\pi\, \Delta x)(30 - u_i)$$

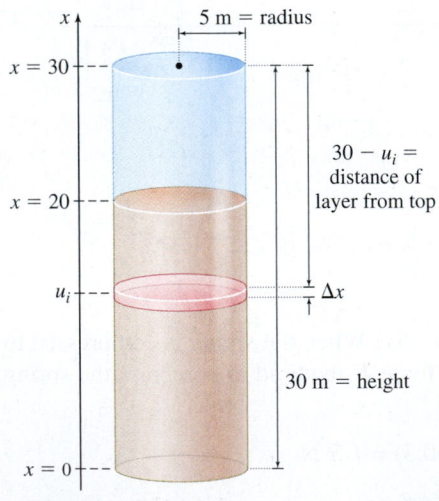

DF **Figure 64**

In the figure: $x$, $5\,\text{m} = \text{radius}$, $x = 30$, $x = 20$, $u_i$, $x = 0$, $30 - u_i = $ distance of layer from top, $\Delta x$, $30\,\text{m} = \text{height}$

The layers of oil are from 0 to 20 m. So, the work $W$ required to pump all the oil over the top is

$$W = \int_0^{20} 200{,}900\pi(30-x)\,dx = 200{,}900\pi \int_0^{20} (30-x)\,dx$$

$$= (200{,}900\,\pi)\left[30x - \frac{x^2}{2}\right]_0^{20}$$

$$= (200{,}900\,\pi)(600-200) = 80{,}360{,}000\pi \approx 2.525 \times 10^8 \text{ J} \qquad \blacksquare$$

The choice of $x=0$ for the position of the bottom of the tank is one of convenience. The work will be the same for other choices of $x$ for the bottom of the tank.

**NOW WORK** Problem 17(a).

**CALC** ▶ **CLIP**

**EXAMPLE 5** **Finding the Work Required to Pump Water Out of a Tank**

A tank in the shape of a hemisphere of radius 2 m is full of water. How much work is required to pump all the water to a level 3 m above the tank? (Density $\rho = 1000 \text{ kg/m}^3$)

**Solution** We position an $x$-axis so the bottom of the tank is at $x=0$ and the top of the tank is at $x=2$, as illustrated in Figure 65(a).

The work required to pump the water to a level 3 m above the top of the tank depends on the weight of the water and its distance from a level 3 m above the tank. The water fills the container from $x=0$ to $x=2$.

Partition the interval $[0, 2]$ into $n$ subintervals, each of width $\Delta x = \dfrac{2}{n}$, and choose a number $u_i$ in each subinterval. Now think of the water in the tank as $n$ circular layers, each of thickness $\Delta x$. As Figure 65(b) illustrates, the radius of the circular layer $u_i$ meters from the bottom of the tank is $\sqrt{4u_i - u_i^2}$. Then

$$\text{Volume } V_i \text{ of } i\text{th layer} = \pi(\text{Radius})^2(\text{Thickness})$$

$$= \pi\left(\sqrt{4u_i - u_i^2}\right)^2 \Delta x = \pi\left(4u_i - u_i^2\right)\Delta x$$

The density of water is $\rho = 1000 \text{ kg/m}^3$, so

$$\text{Weight of } i\text{th layer} = \rho g V_i = (1000)(9.8)\pi\left(4u_i - u_i^2\right)\Delta x$$

$$\text{Distance } i\text{th layer is lifted} = 5 - u_i$$

$$\text{Work done in lifting } i\text{th layer} = 9800\,\pi\left(4u_i - u_i^2\right)(5 - u_i)\,\Delta x$$

The work $W$ required to lift all the water from the tank to a level 3 m above the top of the tank is given by

$$W = \int_0^2 9800\,\pi(4x - x^2)(5 - x)\,dx = 9800\pi \int_0^2 (x^3 - 9x^2 + 20x)\,dx$$

$$= 9800\,\pi\left[\frac{x^4}{4} - 3x^3 + 10x^2\right]_0^2 = 196{,}000\pi \approx 615{,}752 \text{ J} \qquad \blacksquare$$

In general, to find the work required to pump liquid from a container, think of the liquid as thin layers of thickness $\Delta x$ and area $A(x)$. If the liquid is to be lifted a height $h$ above the bottom of the tank, the work required is

$$\boxed{W = \int_a^b \rho g\, A(x)(h - x)\,dx}$$

where $\rho$ is the mass density of the liquid, $g$ is the acceleration due to gravity, and the liquid to be lifted fills the container from $x=a$ to $x=b$. See Figure 66.

**NOW WORK** Problems 17(b) and 31.

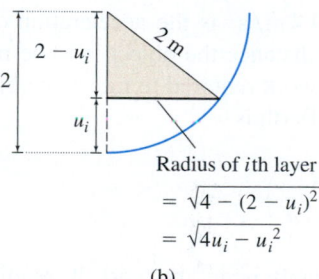

(a)

2 − $u_i$    2 m    2    $u_i$

Radius of $i$th layer
$= \sqrt{4 - (2 - u_i)^2}$
$= \sqrt{4u_i - u_i^2}$

(b)

**DF** **Figure 65**

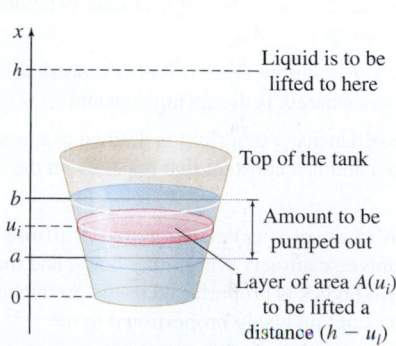

Liquid is to be lifted to here

Top of the tank

Amount to be pumped out

Layer of area $A(u_i)$ to be lifted a distance $(h - u_i)$

**Figure 66**

## Application: Gravitational Force

In the mid-1600s, Isaac Newton proposed his theory of gravity. **Newton's Law of Universal Gravitation** states that every body in the universe attracts every other body; he called this force **gravitation**. Newton theorized that the gravitational force $F$ attracting two bodies is proportional to the masses of both bodies and inversely proportional to the square of the distance $x$ between them. That is,

$$F = F(x) = G\frac{m_1 m_2}{x^2}$$

where $G$ is the **gravitational constant**. The widely accepted value of $G$,

$$G = 6.67 \times 10^{-11}\,\frac{\text{Nm}^2}{\text{kg}^2}$$

is a result of the work done by Henry Cavendish in 1798.

ORIGINS  Henry Cavendish (1731–1810) was an English chemist and physicist. Known for his precision and accuracy, Cavendish calculated the density of Earth by measuring the force of the attraction between pairs of lead balls. An immediate result of his experiments was the first calculation of the gravitational constant $G$.

In a paper appearing in *Science* (2007, January 5), 315(5808), 74–77, the measurement was reevaluated using atom interferometry to be $G = 6.693 \times 10^{-11}\,\frac{\text{N m}^2}{\text{kg}^2}$.

One of the conclusions of Newton's Law of Universal Gravitation is that the force required to move an object, say, a rocket of mass $m$ kilograms that is at a point $x$ meters above the center of Earth, is $F(x) = G\dfrac{Mm}{x^2}$, where $M$ kilograms is the mass of Earth. Then the work required to move an object of mass $m$ from the surface of Earth to a distance $r$ meters from the center of Earth (where $R$ meters is the radius of Earth) is

$$W = \int_R^r \left[ G\frac{Mm}{x^2} \right] dx = -G\left[\frac{Mm}{x}\right]_R^r = G\frac{Mm}{R} - G\frac{Mm}{r}$$
$$= GMm\left(\frac{1}{R} - \frac{1}{r}\right)\text{ J}$$

Physicists know that $GM = gR^2$, where $g \approx 9.8\,\text{m/s}^2$ is the acceleration due to gravity of Earth and $R \approx 6.37 \times 10^6$ m. If $d$ is the distance the object is to be moved above the surface of Earth, then $r = R + d$. So, the work required to move an object of mass $m$ to a distance $d$ meters above the surface of Earth is

$$W = gRm\left(1 - \frac{R}{R+d}\right) \tag{1}$$

Observe that although the distance $d$ may be extremely large, the work $W$ required to move an object $d$ meters will never be greater than $gRm \approx (6.24 \times 10^7)m$ J.

---

## 6.6 Assess Your Understanding

### Concepts and Vocabulary

1. *True or False*  Work is the energy transferred to or from an object by a force acting on the object.

2. The work $W$ done by a constant force $F$ in moving an object a distance $x$ along a straight line in the direction of $F$ is _____.

3. A unit of work is called a _____ in SI units and a _____ in the customary U.S. system of units.

4. The work $W$ done by a continuously varying force $F = F(x)$ acting on an object, which moves the object along a straight line in the direction of $F$ from $x = a$ to $x = b$, is given by the definite integral _____.

5. A spring is said to be in _____ when it is neither extended nor compressed.

6. *True or False*  The force $F$ required to extend or compress a spring $x$ units is $F = -kx$, where $k$ is the spring constant.

7. *True or False*  The mass density $\rho$ of a fluid is defined as mass per unit volume ($\text{kg/m}^3$) and is a constant that depends on the type of fluid.

8. *True or False*  Newton's Law of Universal Gravitation affirms that every body in the universe attracts every other body, and that the force $F$ attracting two bodies is proportional to the product of the masses of both bodies and inversely proportional to the square of the distance between them.

---

1. = NOW WORK problem        = Graphing technology recommended        CAS = Computer Algebra System recommended

**9.** How much work is done by a variable force $F(x) = (40 - x)$ N that moves an object along a straight line in the direction of $F$ from $x = 5$ m to $x = 20$ m?

**10.** How much work is done by a variable force $F(x) = \dfrac{1}{x}$ N that moves an object along a straight line in the direction of $F$ from $x = 1$ m to $x = 2$ m?

**11.** A 40-m chain weighing 3 kg/m hangs over the side of a bridge. How much work is done by a winch that winds the entire chain in?

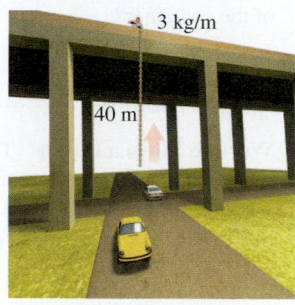

**12.** A 120-ft chain weighing 240 lb is dangling from the roof of an apartment building. How much work is done in pulling the entire chain up to the roof?

**13.** A force of 3 N is required to keep a spring extended $\dfrac{1}{4}$ m beyond its equilibrium length. What is its spring constant?

**14.** A force of 6 lb is required to keep a spring compressed to $\dfrac{1}{2}$ ft shorter than its equilibrium length. What is its spring constant?

**15.** A spring with spring constant $k = 5$ N/m has an equilibrium length of 0.8 m. How much work is required to stretch the spring to 1.4 m?

**16.** A spring with spring constant $k = 0.3$ N/m has an equilibrium length of 1.2 m. How much work is required to compress the spring to 1 m?

**17. Pumping Water out of a Pool**   A swimming pool in the shape of a right circular cylinder, with height 4 ft and radius 12 ft, is full of water. See the figure below.

   **(a)** How much work is required to pump all the water over the top of the pool?

   **(b)** How much work is required to pump all the water to a level 5 ft above the pool?

Note: The weight of water is 62.42 lb/ft$^3$.

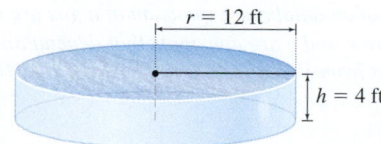

$r = 12$ ft

$h = 4$ ft

**18. Pumping Gasoline out of a Tank**   A storage tank in the shape of a right circular cylinder, with height 10 m and radius 8 m, is full of gasoline.

   **(a)** How much work is required to pump all the gasoline over the top of the tank?

   **(b)** How much is required to pump all the gasoline to a level 5 m above the tank?

Note: The density of gasoline is $\rho = 720$ kg/m$^3$.

**19. Pumping Corn Slurry**   A container in the shape of an inverted pyramid with a square base of 2 m by 2 m and height of 5 m is filled to a depth of 4 m with corn slurry.

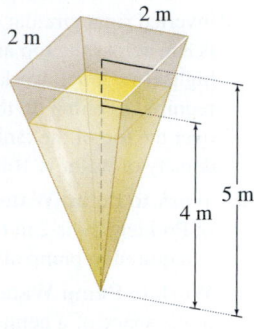

   **(a)** How much work is required to pump all the slurry over the top of the container?

   **(b)** How much work is required to pump all the slurry to a level 3 m above the container?

Note: The density of corn slurry is $\rho = 17.9$ kg/m$^3$.

**20. Pumping Olive Oil from a Vat**   A vat in the shape of an inverted pyramid with a rectangular base that measures 2 m by 0.5 m and height that measures 4 m is filled to a depth of 2 m with olive oil.

   **(a)** How much work is required to pump all the olive oil over the top of the vat?

   **(b)** How much is required to pump all the olive oil to a level 2 m above the vat?

Note: The density of olive oil is $\rho = 0.9$ g/cm$^3$.

**21. Work to Lift an Elevator**   How much work is required if six cables, each weighing 0.36 lb/in., lift a 10,000-lb elevator 400 ft? Assume the cables work together and equally share the weight of the elevator.

**22. Work by Gravity**   A cable with a uniform linear mass density of 9 kg/m is being unwound from a cylindrical drum. If 15 m are already unwound, what is the work done by gravity in unwinding another 60 m of the cable?

**23. Work to Lift a Bucket and Chain**   A uniform chain 10 m long and with mass 20 kg is hanging from the top of a building 10 m high. If a bucket filled with cement of mass 75 kg is attached to the end of the chain, how much work is required to pull the bucket and chain to the top of the building?

**24. Work to Lift a Bucket and Chain**   In Problem 23, if a uniform chain 10 m long and with mass 15 kg is used instead, how much work is required to pull the bucket and chain to the top of the building?

**25. Work of a Spring**   A spring, whose equilibrium length is 1 m, extends to a length of 3 m when a force of 3 N is applied. Find the work needed to extend the spring to a length of 2 m from its equilibrium length.

**26. Work of a Spring**   A spring, whose equilibrium length is 2 m, is compressed to a length of $\dfrac{1}{2}$ m when a force of 10 N is applied. Find the work required to compress the spring to a length of 1 m.

**27. Work of a Spring**   A spring, whose equilibrium length is 4 ft, extends to a length of 8 ft when a force of 2 lb is applied. If 9 ft lb of work is required to extend this spring from its equilibrium position, what is its total length?

**28. Work of a Spring**   If 8 ft lb of work is used on the spring in Problem 27, how far is it extended?

29. **Work to Pump Water**   A full
water tank in the shape of an
inverted right circular cone
is 8 m across the top and 4 m
high. How much work is
required to pump all the water
over the top of the tank? (The
density of water is 1000 kg/m³.)

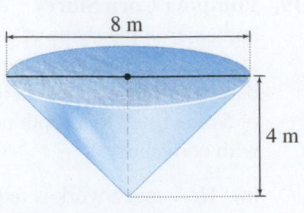

30. **Work to Pump Water**   If the surface of the water in the tank
of Problem 29 is 2 m below the top of the tank, how much work
is required to pump all the water over the top of the tank?

31. **Work to Pump Water**   A tank
in the shape of a hemispherical
bowl of radius 4 m is filled with
water to a depth of 2 m. How
much work is required to pump
all the water over the top of
the tank?

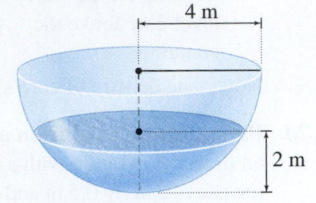

32. **Work to Pump Water**   If the tank in Problem 31 is completely
filled with water, how much work is required to
pump all the water to a height 2 m above the tank?

33. **Work to Pump Water**   A cylindrical tank, 4 m in diameter
and 6 m high, is full of water. (The density of water
is 1000 kg/m³.)

   **(a)** How much work is required to pump half the water over
   the top of the tank?

   **(b)** How much work is required to pump half the water to a
   level 3 m above the tank?

34. **Work to Pump Water**   A swimming pool in the shape of a
rectangular parallelepiped 6 ft deep, 30 ft long, and 20 ft
wide is filled with water to a depth of 5 ft. How much work is
required to pump all the water over the top? (The weight of water
is $\rho g = 62.42$ lb/ft³.)

35. **Work to Pump Water**   A 1 hp motor can do 550 ft lb of work
per second. The motor is used to pump the water out of a
swimming pool in the shape of a rectangular parallelepiped 5 ft
deep, 25 ft long, and 15 ft wide. How long does it take for the
pump to empty the pool if the pool is filled to a depth of 4 ft? (The
weight of water is 62.42 lb/ft³.)

**Newton's Law of Universal Gravitation**   *In Problems 36 and 37,
use formula (1) on page 478.*

36. The minimum energy required to move an object of mass 30 kg a
distance 30 km above the surface of Earth is equal to the work
required to accomplish this. Find the work required. (Earth's
radius $R$ is approximately 6370 km.)

37. The minimum energy required to move a rocket of mass 1000 kg
a distance of 800 km above Earth's surface is equal to the work
required to do this. Find the work required.

38. **Coulomb's Law**   By **Coulomb's Law**, a positive charge $m$ of
electricity repels a unit of positive charge at a distance $x$ with the
force $\dfrac{m}{x^2}$. What is the work done when the unit charge is carried
from $x = 2a$ to $x = a$, $a > 0$?

39. **Work to Lift a Leaky Load**   In raising a leaky bucket from the
bottom of a well 25 ft deep, one-fourth of the water is lost. If
the bucket weighs 1.5 lb, the water in the bucket at the start
weighs 20 lb, and the amount that has leaked out is assumed to be
proportional to the distance the bucket is lifted, find the work
done in raising the bucket. Ignore the weight of the rope used to
lift the bucket.

40. **Work of a Spring**   The spring constant on a bumping post in a
freight yard is 300,000 N/m. Find the work done in compressing
the spring 0.1 m.

41. **Work to Pump Water**   A container is formed by
revolving the region bounded by the graph of $y = x^2$, and
the $x$-axis, $0 \le x \le 2$, about the $y$-axis. How much work is
required to fill the container with water from a source 2 units
below the $x$-axis by pumping through a hole in the bottom
of the container?

## Challenge Problems

42. **Work to Move a Piston**   The force exerted by a gas in a
cylinder on a piston whose area is $A$, is given by $F = pA$,
where $p$ is the force per unit area, or **pressure**. The work $W$ in
displacing the piston from $x_1$ to $x_2$ is

$$W = \int_{x_1}^{x_2} F \, dx = \int_{x_1}^{x_2} pA \, dx = \int_{V_1}^{V_2} p \, dV$$

where $dV$ is the accompanying infinitesimal change of volume of
the gas.

   **(a)** During expansion of a gas at constant temperature
   (isothermal), the pressure $p$ depends on the volume $V$
   according to the relation

   $$p = \frac{nRT}{V}$$

   where $n$ and $R$ are constants and $T$ is the constant
   temperature. Calculate the work in expanding the gas
   isothermally from volume $V_1$ to volume $V_2$.

   **(b)** During expansion of a gas at constant entropy (adiabatic),
   the pressure depends on the volume according to the
   relation

   $$p = \frac{K}{V^\gamma}$$

   where $K$ and $\gamma \ne 1$ are constants. Calculate the work in
   expanding the gas adiabatically from $V_1$ to $V_2$.

**Expanding Gases**   *Problems 43 and 44 use the following discussion.
The pressure p (in pounds/square inch, lb/in²) and the volume V
(in cubic inches) of an adiabatic expansion of a gas are related
by $pV^k = c$, where k and c are constants that depend on the gas.
If the gas expands from $V = a$ to $V = b$, the work done (in
inch-pounds) is*

$$W = \int_a^b p \, dV.$$

43. **Work of Expanding Gas**   The pressure $p$ of 1 lb of a gas
is 100 lb/in² and the volume $V$ is 2 ft³. Find the work done by
the gas in expanding to double its volume according to the
law $pV^{1.4} = c$.

44. **Work of Expanding Gas**   The pressure $p$ and volume $V$ of a
certain gas obey the law $pV^{1.2} = 120$ (in inch-pounds).
Find the work done when the gas expands from $V = 2.4$
to $V = 4.6$ in³.

# 6.7 Hydrostatic Pressure and Force

**OBJECTIVE** *When you finish this section, you should be able to:*

1  Find hydrostatic pressure and force (p. 481)

**NOTE** Fluids are substances that can flow. Gases and liquids are fluids.

**IN WORDS** Pressure equals force per unit area.

When engineers design containers to hold fluids in place, it is important to know the *pressure P* caused by the force *F* of the fluid on the sides of the container. **Pressure P** is defined as the force *F* exerted per unit area *A*.

$$P = \frac{F}{A}$$

## 1 Find Hydrostatic Pressure and Force

The weight of a fluid is given by the formula $F = \rho g V$, where $\rho$ is the mass density of the fluid, $g \approx 9.8 \, \text{m/s}^2 \approx 32.2 \, \text{ft/s}^2$ is the acceleration due to gravity, and $V$ is the volume of the fluid. Then, at a depth $h$ below the surface of the fluid, the pressure $P$ is

$$P = \frac{\rho g V}{A} = \rho g h$$

where $\rho g$ is the weight density of the fluid.

As the formula indicates, the pressure $P$ is directly proportional to the depth $h$, so the pressure increases as the depth increases.

$P$ is often called **hydrostatic pressure**, and the force $F$ a **hydrostatic force**, because they are the result of fluids, such as water (*hydro*), that are at rest (*static*). Table 2 summarizes the units of measure used for these calculations.

**TABLE 2**

|      | Force, $F$ | Pressure, $P$ | Mass Density, $\rho$ | $g$ | Depth, $h$ |
|------|------------|---------------|----------------------|-----|------------|
| SI   | newton (N) | pascal (Pa) = newton/meter$^2$ (Pa = N/m$^2$) | kilogram/meter$^3$ (kg/m$^3$) | 9.8 m/s$^2$ | meter (m) |
| U.S. | pound (lb) | pound/foot$^2$ (lb/ft$^2$) | slug/foot$^3$ (slug/ft$^3$) | 32.2 ft/s$^2$ | foot (ft) |

**NOTE** When using U.S. units, the weight density $\rho g$ is often given in pounds per cubic foot, and the unit slug is not apparent. For example, the weight density of water is about 62.5 lb/ft$^3$ and the pressure $P$ is given in lb/ft$^2$.

Suppose a thin, flat plate of area $A$ is suspended horizontally in water at a depth $h$. The force $F$ of the water on the bottom face of the plate equals the weight of the water above the plate.

$$F = \rho g V = \rho g h A$$

The weight density $\rho g$ of water is about 62.5 lb/ft$^3$, so if the plate were suspended horizontally at a depth $h = 4$ ft, the weight of the water on the plate is

$$F = \rho g h A = (62.5)(4)A = 250A \, \text{lb}$$

and the pressure of the water on the plate is

$$P = \frac{F}{A} = 250 \, \text{lb/ft}^2$$

**IN WORDS** The pressure $P$ due to a static fluid on a horizontal plate depends on the depth of the plate in the fluid.

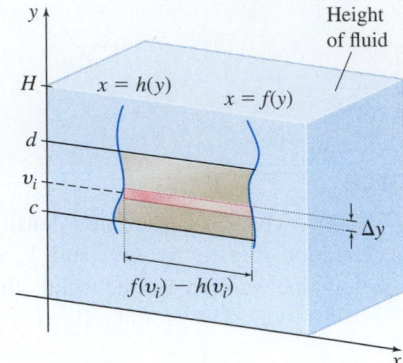

**Figure 67** A plate suspended vertically in fluid.

If the plate has an area of 2 ft$^2$, the force $F$ of the water exerted on one side of the plate is $F = AP = (2)(250) = 500$ lb. In fact, as the horizontal plate at a depth $h$ changes in size and or shape, the hydrostatic force $F$ varies, but the hydrostatic pressure $P$ remains constant. Nevertheless, the deeper the plate is in the fluid, the greater the pressure is on the plate.

If a plate is suspended vertically in a fluid, the pressure at the bottom of the plate is greater than the pressure at the top. To find the force of the fluid on one side of the plate, suppose the plate is suspended vertically in a fluid of mass density $\rho$. Suppose further that the plate is placed in a rectangular coordinate system and is enclosed by $y = c$, $y = d$, $x = h(y)$, and $x = f(y)$, where $f$ and $h$ are functions that are continuous on the closed interval $[c, d]$ and $f(y) \geq h(y)$ for all numbers $y$ in $[c, d]$. See Figure 67.

The surface of the fluid is at the line $y = H$, where $H \geq d$, so that the top of the plate is at a depth of $(H - d)$ and the bottom of the plate is at a depth of $(H - c)$.

Partition the interval $[c, d]$ into $n$ subintervals:

$$[c, y_1], [y_1, y_2], \ldots, [y_{i-1}, y_i], \ldots, [y_{n-1}, d] \qquad c = y_0 \quad d = y_n$$

each of length $\Delta y = \dfrac{d - c}{n}$, and let $v_i$ be a number in the $i$th subinterval $[y_{i-1}, y_i]$, $i = 1, 2, 3, \ldots, n$. If the length $\Delta y$ is small, then all points in the horizontal $i$th slice of the plate are roughly the same distance $(H - v_i)$ from the surface, and the pressure $P_i$ of the fluid on this portion of the plate is approximately $P_i = \rho g(H - v_i)$. The force $F_i$ due to hydrostatic pressure on the $i$th subinterval of the plate is approximately

$$F_i = \rho g h A_i \approx \rho g(H - v_i)[f(v_i) - h(v_i)]\Delta y \qquad A_i = [f(v_i) - h(v_i)]\Delta y$$

An approximation to the total force $F$ on the plate can be found by summing the forces from each subinterval. That is,

$$F \approx \sum_{i=1}^{n} \underbrace{\rho g}_{\text{Weight density of slice}} \underbrace{(H - v_i)}_{\text{Depth of slice}} \underbrace{[f(v_i) - h(v_i)]\,\Delta y}_{\text{Area of the slice}}$$

As the number of subintervals increases, that is, as $n \to \infty$, the sums $F$ become better approximations of the force due to fluid pressure on the plate. Since these sums are Riemann sums and the functions $f$ and $h$ are continuous on $[c, d]$, the limit is a definite integral. That is,

$$F = \lim_{n \to \infty} \sum_{i=1}^{n} \rho g(H - v_i)[f(v_i) - h(v_i)]\Delta y = \int_{c}^{d} \rho g(H - y)[f(y) - h(y)]\,dy$$

**Hydrostatic Force**

The hydrostatic force $F$ due to the pressure exerted by a fluid of weight density $\rho g$ on a plate of the type illustrated in Figure 67, where the functions $f$ and $h$ are continuous on the closed interval $[c, d]$, and $H$ is the depth of the fluid, is

$$F = \int_{c}^{d} \rho g(H - y)[f(y) - h(y)]\,dy$$

**CALC CLIP**

**EXAMPLE 1  Finding Hydrostatic Force**

A trough, whose cross section is a trapezoid that measures 2 m across at the bottom and 4 m across at the top, is 2 m deep. If the trough is filled with a liquid of weight density $\rho g$, what is the force due to hydrostatic pressure on one end of the trough?

**Solution** Position the trough in a rectangular coordinate system, as shown in Figure 68.

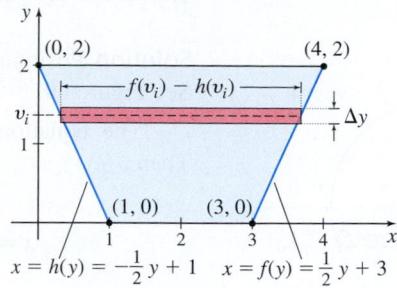

**Figure 68**

The sides of the end of the trough are lines that pass through the points $(0, 2)$, $(1, 0)$ and the points $(3, 0)$, $(4, 2)$, respectively. The equations of these lines are

$$y - 2 = \frac{0 - 2}{1 - 0}(x - 0) = -2x \qquad \text{or equivalently} \qquad x = h(y) = -\frac{1}{2}y + 1$$

and

$$y - 0 = \frac{2 - 0}{4 - 3}(x - 3) = 2x - 6 \qquad \text{or equivalently} \qquad x = f(y) = \frac{1}{2}y + 3$$

Then the hydrostatic force on the $i$th interval of the trough is

$$F_i = \underbrace{\rho \cdot g}_{\text{Weight}} \; \underbrace{(H - v_i)}_{\text{Depth}} \; \underbrace{[f(v_i) - h(v_i)]\Delta y}_{\text{Area}}$$

where $\rho$ is the mass density of the fluid and $g = 9.8$ m/s$^2$ is the acceleration due to gravity.

The liquid fills the trough from $y = 0$ to $y = 2$, so $H = 2$ and the hydrostatic force $F$ due to the pressure of the liquid on an end of the trough is

$$F = \int_c^d \rho g (H - y)[f(y) - h(y)] \, dy$$

$$= \int_0^2 \rho g (2 - y) \left[ \left( \frac{1}{2}y + 3 \right) - \left( -\frac{1}{2}y + 1 \right) \right] dy$$

$$= \rho g \int_0^2 (2 - y)(y + 2) \, dy = \rho g \int_0^2 (-y^2 + 4) \, dy = \rho g \left[ -\frac{y^3}{3} + 4y \right]_0^2$$

$$= \rho g \left( -\frac{8}{3} + 8 \right) = \frac{16}{3}\rho g \text{ N} \qquad \blacksquare$$

**NOW WORK** Problem 13.

So far we have positioned the coordinate system so that the submerged plate is located in the first quadrant. As the next example illustrates, the coordinates may be placed in any convenient position. Keep in mind that the essential idea behind the formula for force due to hydrostatic pressure is

$$\boxed{\text{Hydrostatic force} = \rho g \times \text{Depth} \times \text{Area}}$$

EXAMPLE 2 **Finding Hydrostatic Force**

A cylindrical sewer pipe of radius 2 m is half full of water. A gate used to seal off the sewer is placed perpendicular to the pipe opening. Find the hydrostatic force exerted on one side of the gate.

**Solution** Position a cross section of the pipe (a circle) so its center is at the origin. See Figure 69.

The equation of the circle with center at $(0, 0)$ and radius 2 is $x^2 + y^2 = 4$. Then

$$x = h(y) = -\sqrt{4 - y^2} \quad \text{and} \quad x = f(y) = \sqrt{4 - y^2}$$

The height of the water is at $H = 0$. Since the water fills the cylinder from $y = -2$ to $y = 0$, the hydrostatic force $F$ exerted on one side of the gate by the pressure of the water is

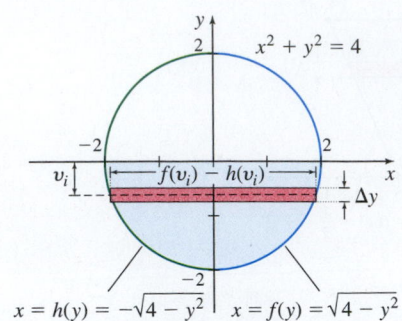

**Figure 69**

**NOTE** In SI units,

$\rho$ = mass density of water
$\quad = 1000 \text{ kg/m}^3$
$g = 9.8 \text{ m/s}^2$

$F$ is given in newtons (N).

$$F = \int_c^d \rho g (H - y)[f(y) - h(y)] \, dy$$

$$= \int_{-2}^0 \rho g (0 - y)\left[\sqrt{4 - y^2} - \left(-\sqrt{4 - y^2}\right)\right] dy$$

$$= 9800 \int_{-2}^0 -2y\sqrt{4 - y^2} \, dy \qquad \rho = 1000 \text{ kg/m}^3; \, g = 9.8 \text{ m/s}^2$$

$$= 9800 \int_0^4 \sqrt{u} \, du = 9800 \left[\frac{2u^{3/2}}{3}\right]_0^4 \qquad \begin{array}{l} \text{Let } u = 4 - y^2; \\ \text{then } du = -2y \, dy. \\ \text{When } y = -2, \text{ then } u = 0; \\ \text{when } y = 0, \text{ then } u = 4. \end{array}$$

$$= 9800 \left(\frac{16}{3}\right) \approx 52,267 \text{ N}$$

NOW WORK **Problem 15.**

## 6.7 Assess Your Understanding

### Concepts and Vocabulary

1. Pressure is defined as the ————— exerted per unit area.

2. *True or False* Hydrostatic pressure on a plate increases as the depth of the plate in a fluid increases.

3. *Multiple Choice* In SI units, the [(**a**) newton (**b**) pascal (**c**) joule (**d**) slug] is the unit measure of pressure.

4. *True or False* The hydrostatic force $F$ due to the pressure exerted by a fluid of mass density $\rho$ on a plate of area $A$ suspended horizontally in the fluid at a depth $h$ is $F = \rho g h A$, where $g$ is the acceleration due to gravity.

### Skill Building

*In Problems 5–8, the figure shows a vertical cross section of a container filled with water. Find the hydrostatic force of the water on the end of each container shown.*

5.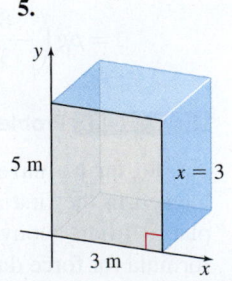

6.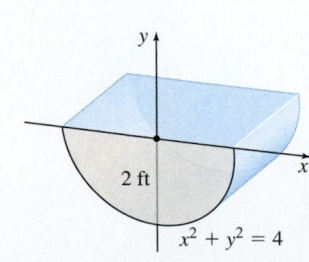

---

**1.** = NOW WORK problem ⟋⟍ = Graphing technology recommended CAS = Computer Algebra System recommended

**7.**

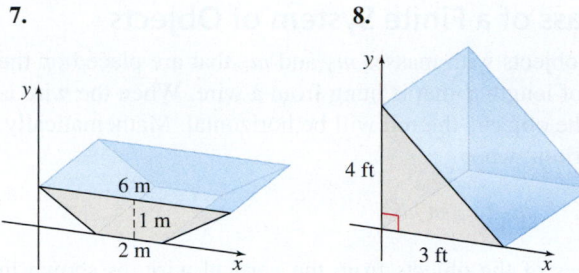

**8.**

**9.** A rectangular plate of width 2 m and height 6 m is suspended vertically in a pool of water so that the top of the plate is even with the surface of the water. What is the force due to water pressure on one side of the plate?

**10.** If the plate in Problem 9 is suspended in the water so that the top of the plate is 1 m below the water surface, what is the force due to water pressure on one side of the plate?

## Applications and Extensions

**11. Hydrostatic Force in a Swimming Pool** A swimming pool is in the shape of a rectangular parallelepiped 6 ft deep, 30 ft long, and 20 ft wide. If the pool is full of water, what is the force due to water pressure on one short side of the pool?
The weight density of water is $62.5 \text{ lb/ft}^2$.

**12. Hydrostatic Force in a Swimming Pool** For the pool in Problem 11, what is the force due to water pressure on one long side of the pool?

**13. Hydrostatic Force in a Trough** A trough whose cross section is a trapezoid is 1 m across at the bottom, 5 m across at the top, and 2 m deep. If the trough is filled to 1 m with water, what is the force due to water pressure on one end of the trough?

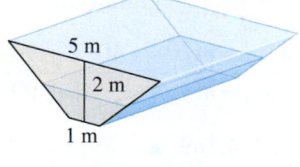

**14. Hydrostatic Force in a Trough** A trough whose cross section is an equilateral triangle with side 2 m long is filled with water. What is the force due to water pressure on one end of the trough?

**15. Hydrostatic Force in a Trough** A trough whose cross section is a semicircle of radius 2 m is filled with water. What is the force due to water pressure on one end of the trough?

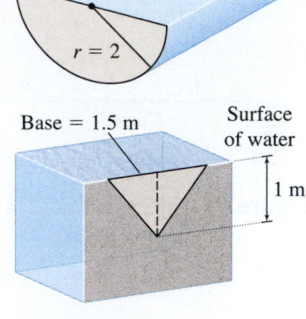

**16. Hydrostatic Force on a Floodgate** Find the hydrostatic force on the face of a vertical floodgate in the shape of an isosceles triangle whose base is 1.5 m and whose height is 1 m, if its base is on the surface of the water. See the figure.

**17. Hydrostatic Force on a Dam** A vertical masonry dam in the form of an isosceles trapezoid is 200 m long at the surface of the water, 150 m long at the bottom, and 60 m high. What force must it withstand?

**18. Hydrostatic Force in an Oil Tank** A tank filled with a half load of fuel oil is in the shape of a right circular cylinder lying on its side. If the radius of the cylinder is 4 ft and the weight density of the fuel oil is $60 \text{ lb/ft}^2$, find the hydrostatic force due to the pressure of the oil on one end of the cylinder.

**19. Hydrostatic Force on a Submarine** A vertical viewing plate in a submarine is a circle of radius 1 m. If the depth of the center of the viewing plate is 5 m below the surface of the water, what is the hydrostatic force due to water pressure on one side of the plate? The mass density of sea water is $1025 \text{ kg/m}^3$.

**20. Hydrostatic Force on a Submarine** A vertical viewing plate of a submarine has a perimeter given by the **fat circle** $x^4 + y^4 = 1$. If the depth of the center of the viewing plate is 5 m below the surface of the water, find the hydrostatic force due to water pressure on the plate. The mass density of sea water is $1025 \text{ kg/m}^3$.

**21. Hydrostatic Force in a Gas Tank** The gas tank in a sports car is a cylinder lying on its side. If the diameter of the tank is 0.5 m and if the tank is filled with gasoline to within 0.25 m of the top, find the force on one end of the tank. The density of gasoline is $690 \text{ kg/m}^3$.

**22. Hydrostatic Force in a Trough** A trough whose cross section is a semicircle of radius 4 m is filled with water to a depth of 2 m. What is the force due to water pressure on one end of the trough?

---

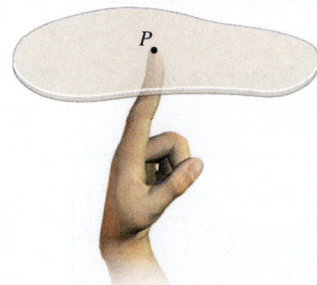

**Figure 70** The center of mass of an object is the point where the object is balanced.

# 6.8 Center of Mass; Centroid; The Pappus Theorem

**OBJECTIVES** *When you finish this section, you should be able to:*

1  Find the center of mass of a finite system of objects (p. 486)

2  Find the centroid of a homogeneous lamina (p. 487)

3  Find the volume of a solid of revolution using the Pappus Theorem (p. 491)

Anytime you have been able to balance an object on your fingertip, you have located its *center of mass*. See Figure 70. The **center of mass** of an object or a system of objects is the point that acts as if all the mass is concentrated at that point.

## 1 Find the Center of Mass of a Finite System of Objects

We begin with a system of two objects with masses $m_1$ and $m_2$ that are placed on the ends of a nearly weightless rod of length $d$ that is hung from a wire. When the wire is placed at the center of mass of the objects, the rod will be horizontal. Mathematically, the rod is balanced, or in equilibrium, when

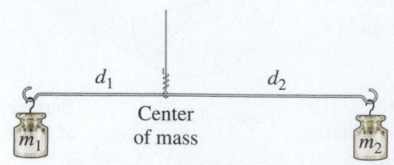

**Figure 71** The rod is balanced when $m_1 d_1 = m_2 d_2$

$$m_1 d_1 = m_2 d_2$$

where $d_1$ and $d_2$ are the distances of the objects from the vertical wire, as shown in Figure 71. The quantities $m_1 d_1$ and $m_2 d_2$, called **moments**, represent the tendency of the objects to rotate about the center of mass. When $m_1 d_1 = m_2 d_2$, the tendency of the objects to rotate is equal, so no rotation occurs and equilibrium is attained.

If the rod were placed on a positive $x$-axis, as in Figure 72, with mass $m_1$ at $x_1$, mass $m_2$ at $x_2$, and the center of mass at $\bar{x}$, then the rod is balanced when

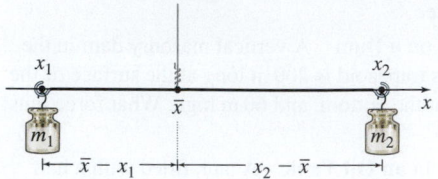

**Figure 72** The center of mass is located at $\bar{x}$.

$$m_1(\bar{x} - x_1) = m_2(x_2 - \bar{x})$$
$$m_1\bar{x} - m_1 x_1 = m_2 x_2 - m_2\bar{x}$$
$$(m_1 + m_2)\bar{x} = m_1 x_1 + m_2 x_2$$

The center of mass $\bar{x}$ of the two objects satisfies the equation

$$\boxed{\bar{x} = \frac{m_1 x_1 + m_2 x_2}{m_1 + m_2}}$$

---

**EXAMPLE 1    Finding the Center of Mass of a System of Objects on a Line**

Find the center of mass of the system when a mass of 90 kg is placed at 6 and a mass of 40 kg is placed at 2.

**DF Figure 73** At the center of mass $\bar{x}$ the system is in equilibrium.

**Solution** The system is shown in Figure 73, where the two masses are placed on a weightless seesaw. The center of mass $\bar{x}$ will be at some number where a fulcrum balances the two masses. Then for equilibrium,

$$\bar{x} = \frac{m_1 x_1 + m_2 x_2}{m_1 + m_2} = \frac{40 \cdot 2 + 90 \cdot 6}{40 + 90} = \frac{620}{130} \approx 4.769$$

The center of mass is at $\bar{x} \approx 4.769$. ∎

**NOW WORK** Problem 7.

The formula for the center of mass of two objects can be extended to any finite number of objects.

### Center of Mass of a System of $n$ Objects on a Line

If $n$ objects with masses $m_1, m_2, \ldots, m_n$ are placed on a line at $x_1, x_2, \ldots, x_n$, respectively, then for equilibrium

$$(m_1 + m_2 + \cdots + m_n)\bar{x} = m_1 x_1 + m_2 x_2 + \cdots + m_n x_n$$

and the **center of mass** is $\bar{x}$, where

$$\bar{x} = \frac{m_1 x_1 + m_2 x_2 + \cdots + m_n x_n}{m_1 + m_2 + \cdots + m_n} = \frac{\sum\limits_{i=1}^{n}(m_i x_i)}{\sum\limits_{i=1}^{n} m_i} = \frac{\sum\limits_{i=1}^{n}(m_i x_i)}{M}$$

where $M = \sum\limits_{i=1}^{n} m_i$ is the total mass of the system.

The numbers $m_1x_1, m_2x_2, \ldots, m_nx_n$ are called the **moments about the origin** of the masses $m_1, m_2, \ldots, m_n$. So, the center of mass $\bar{x}$ is found by adding the moments about the origin and dividing by the total mass $M$ of all the objects.

Now suppose the objects are not in a line but are scattered in a plane.

### Center of Mass of a System of $n$ Objects in a Plane

If $n$ objects with masses $m_1, m_2, \ldots, m_n$ are located at the points $(x_1, y_1)$, $(x_2, y_2), \ldots, (x_n, y_n)$ in a plane, then the **center of mass of the system** is located at the point $(\bar{x}, \bar{y})$, where

$$\bar{x} = \frac{\sum\limits_{i=1}^{n}(m_i x_i)}{M} \qquad \bar{y} = \frac{\sum\limits_{i=1}^{n}(m_i y_i)}{M}$$

and $M = \sum\limits_{i=1}^{n} m_i$ is the total mass of the system.

The sum $M_y = \sum\limits_{i=1}^{n}(m_i x_i)$ is called the **moment of the system about the y-axis**

and the sum $M_x = \sum\limits_{i=1}^{n}(m_i y_i)$ is called the **moment of the system about the x-axis**.

The center of mass formulas then can be written as

$$\bar{x} = \frac{M_y}{M} \qquad \bar{y} = \frac{M_x}{M} \tag{1}$$

Physically, $M_y$ measures the tendency of the system to rotate about the $y$-axis; $M_x$ measures the tendency of the system to rotate about the $x$-axis.

---

**EXAMPLE 2**  **Finding the Center of Mass of a System of Objects in a Plane**

Find the center of mass of the system of objects having masses 4, 6, and 9 kg, located at the points $(-2, 1)$, $(3, -2)$, and $(4, 3)$, respectively.

**Solution** See Figure 74. Where is a good estimate for the center of mass? Certainly, it will lie within the rectangle $-2 \leq x \leq 4; -2 \leq y \leq 3$.

To find the exact position of the center of mass, we first find the moment of the system about the $y$-axis, $M_y$, and the moment of the system about the $x$-axis, $M_x$.

$$M_y = \sum_{i=1}^{3} m_i x_i = 4(-2) + 6 \cdot 3 + 9 \cdot 4 = 46$$

$$M_x = \sum_{i=1}^{3} m_i y_i = 4 \cdot 1 + 6(-2) + 9 \cdot 3 = 19$$

Since the total mass $M$ of the system is $M = 4 + 6 + 9 = 19$, we have

$$\bar{x} = \frac{M_y}{M} = \frac{46}{19} \qquad \bar{y} = \frac{M_x}{M} = \frac{19}{19} = 1$$

The center of mass of these objects is at the point $\left(\dfrac{46}{19}, 1\right)$. Notice that the center of mass lies in the rectangle $-2 \leq x \leq 4; -2 \leq y \leq 3$. ∎

**NOW WORK** Problem 13.

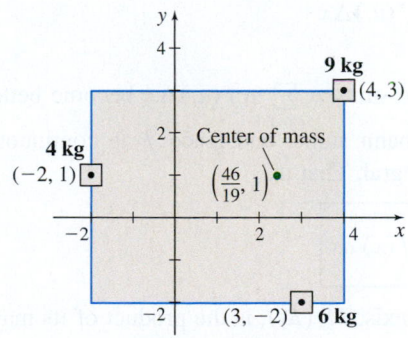

**Figure 74** The center of mass of the system is $\left(\dfrac{46}{19}, 1\right)$.

## 2 Find the Centroid of a Homogeneous Lamina

With the formulas (1), we can approximate the center of mass of a thin, flat sheet of material, called a **lamina**, and express its center of mass in terms of definite integrals. We will assume that matter is continuously distributed throughout the lamina; that is, the mass density function $\rho$ is continuous on the domain of the lamina. In the special case where the mass density function $\rho$ is a constant function, the lamina is called **homogeneous**.

The mass of a homogeneous lamina is $\rho A$, where $A$ is the area of the lamina and $\rho$ is its constant mass density. The center of mass of a homogeneous lamina is located at the **centroid**, the geometric center of the lamina.

Suppose a homogeneous lamina is determined by a region $R$ bounded by the graph of a function $f$, the $x$-axis, and the lines $x = a$ and $x = b$. Also suppose that $f$ is continuous and nonnegative on the closed interval $[a, b]$. See Figure 75.

As before, we partition the interval $[a, b]$ into $n$ subintervals:

$$[a, x_1], [x_1, x_2], \ldots, [x_{i-1}, x_i], \ldots, [x_{n-1}, b] \quad a = x_0 \quad b = x_n \quad i = 1, 2, \ldots, n$$

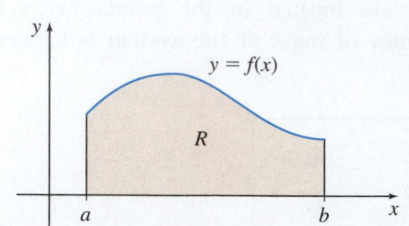

**Figure 75**

each of width $\Delta x = \dfrac{b - a}{n}$, and select a number $u_i$ in each subinterval. Here we choose $u_i$ to be the midpoint of the $i$th subinterval. That is,

$$u_i = \frac{x_{i-1} + x_i}{2}$$

The lamina is now partitioned into $n$ nonoverlapping rectangular regions, $R_i$, where $i = 1, 2, \ldots, n$, each of which is a homogeneous lamina. From the symmetry of a rectangle, the centroid of $R_i$ is $\left( u_i, \dfrac{1}{2} f(u_i) \right)$. See Figure 76.

The mass $m_i$ of $R_i$ is

$$m_i = \rho A_i = \rho f(u_i) \Delta x$$

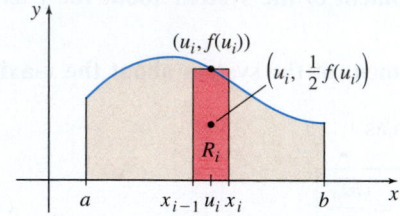

**Figure 76** By symmetry, the centroid of the $i$th rectangle is at $\left( u_i, \dfrac{1}{2} f(u_i) \right)$.

The moment of $R_i$ about the $y$-axis, $M_y(R_i)$, is the product of the mass of $R_i$ and the distance from $\left( u_i, \dfrac{1}{2} f(u_i) \right)$ to the $y$-axis, which is $u_i$. Then

$$M_y(R_i) = m_i u_i = [\rho f(u_i) \Delta x] u_i = \rho u_i f(u_i) \Delta x$$

Summing the $n$ moments gives an approximation of the moment $M_y$ of the lamina about the $y$-axis. That is,

$$M_y \approx \rho \sum_{i=1}^{n} u_i f(u_i) \Delta x$$

As the number $n$ of subintervals increases, the sums $\rho \sum_{i=1}^{n} u_i f(u_i) \Delta x$ become better approximations of $M_y$. These sums are Riemann sums, and since $f$ is continuous on $[a, b]$, the limit of the sums is a definite integral. That is,

$$\boxed{M_y = \rho \int_a^b x f(x)\, dx}$$

Similarly, the moment of $R_i$ about the $x$-axis, $M_x(R_i)$, is the product of its mass and the distance from the point $\left( u_i, \dfrac{1}{2} f(u_i) \right)$ to the $x$-axis, which is $\dfrac{1}{2} f(u_i)$.

$$M_x(R_i) = m_i \left[ \frac{1}{2} f(u_i) \right] = \rho f(u_i) \Delta x \left[ \frac{1}{2} f(u_i) \right] = \frac{1}{2} \rho [f(u_i)]^2 \Delta x$$

Again, adding these moments gives an approximation of the moment $M_x$ of the lamina about the $x$-axis, and as the number of subintervals increases, the sums $\dfrac{1}{2} \rho \sum_{i=1}^{n} [f(u_i)]^2 \Delta x$ become better approximations of $M_x$. Since the sums are Riemann sums and $f$ is continuous on $[a, b]$, we have

$$\boxed{M_x = \frac{1}{2} \rho \int_a^b [f(x)]^2\, dx}$$

For the region $R$, the mass of a homogeneous lamina is

$$M = \rho \int_a^b f(x)\, dx$$

Then the centroid $(\bar{x}, \bar{y})$ is given by

$$\bar{x} = \frac{M_y}{M} = \frac{\rho \displaystyle\int_a^b xf(x)\, dx}{\rho \displaystyle\int_a^b f(x)\, dx} = \frac{\displaystyle\int_a^b xf(x)\, dx}{\displaystyle\int_a^b f(x)\, dx} = \frac{\displaystyle\int_a^b x\, f(x)\, dx}{A}$$

$$\bar{y} = \frac{M_x}{M} = \frac{\dfrac{1}{2}\rho \displaystyle\int_a^b [f(x)]^2\, dx}{\rho \displaystyle\int_a^b f(x)\, dx} = \frac{\dfrac{1}{2}\displaystyle\int_a^b [f(x)]^2\, dx}{\displaystyle\int_a^b f(x)\, dx} = \frac{\displaystyle\int_a^b [f(x)]^2\, dx}{2A}$$

where $A = \int_a^b f(x)\, dx$ is the area under the graph of $f$ from $a$ to $b$.

### The Centroid of a Homogeneous Lamina

Let $R$ be a lamina with a constant mass density $\rho$. If $R$ is bounded by the graph of a function $f$, the $x$-axis, and the lines $x = a$ and $x = b$, where $f$ is continuous and nonnegative on the closed interval $[a, b]$, then the centroid $(\bar{x}, \bar{y})$ of $R$ is

$$\bar{x} = \frac{1}{A}\int_a^b xf(x)\, dx \qquad \bar{y} = \frac{1}{2A}\int_a^b [f(x)]^2\, dx \qquad (2)$$

where

$$A = \int_a^b f(x)\, dx$$

is the area under the graph of $f$ from $a$ to $b$.

**CALC CLIP**

**EXAMPLE 3**  **Finding the Centroid of a Homogeneous Lamina**

Find the centroid of the homogeneous lamina bounded by the graph of $f(x) = x^2$, the $x$-axis, and the line $x = 1$.

**Solution** The lamina is shown in Figure 77.

The area $A$ of the lamina is

$$A = \int_0^1 x^2 dx = \left[\frac{x^3}{3}\right]_0^1 = \frac{1}{3}$$

Using formulas (2), the centroid of the lamina is

$$\bar{x} = \frac{1}{A}\int_0^1 xf(x)\, dx = \frac{1}{\frac{1}{3}}\int_0^1 x \cdot x^2\, dx = 3\int_0^1 x^3\, dx = 3\left[\frac{x^4}{4}\right]_0^1 = \frac{3}{4}$$

$$\bar{y} = \frac{1}{2A}\int_0^1 [f(x)]^2\, dx = \frac{1}{2 \cdot \frac{1}{3}}\int_0^1 (x^2)^2\, dx = \frac{3}{2}\int_0^1 x^4\, dx = \frac{3}{2}\left[\frac{x^5}{5}\right]_0^1 = \frac{3}{10}$$

The centroid of the lamina is $\left(\dfrac{3}{4}, \dfrac{3}{10}\right)$. ∎

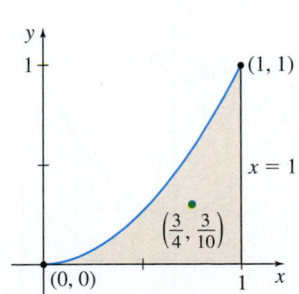

**Figure 77** $f(x) = x^2, 0 \le x \le 1$

**NOW WORK** Problem **15**.

Example 4 illustrates the **symmetry principle**: If a homogeneous lamina is symmetric about a line $L$ or a point $P$, then the centroid of the lamina lies on $L$ or at the point $P$.

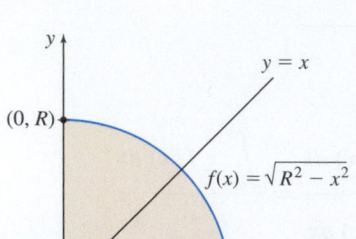

(a) The region is symmetric with respect to the line $y = x$. The centroid lies on the line $y = x$.

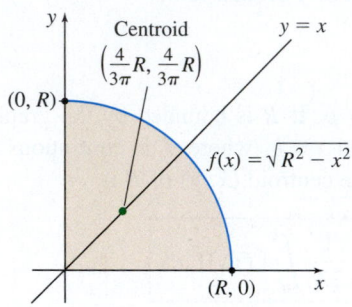

(b) The centroid is $\left(\frac{4}{3\pi}R, \frac{4}{3\pi}R\right)$.

**Figure 78**

### EXAMPLE 4  Finding the Centroid of a Homogeneous Lamina

Find the centroid of one-quarter of a circular plate of radius $R$.

**Solution** We place the quarter-circle in the first quadrant, as shown in Figure 78(a). The equation of the quarter circle can be expressed as $f(x) = \sqrt{R^2 - x^2}$, where $0 \le x \le R$.

Because the quarter of the circular plate is symmetric with respect to the line $y = x$, the centroid lies on this line. So, $\bar{x} = \bar{y}$. The area $A$ of the quarter circular region is $A = \dfrac{\pi R^2}{4}$.

$$\bar{x} = \frac{1}{A}\int_a^b [xf(x)]\,dx = \frac{1}{A}\int_0^R x\sqrt{R^2 - x^2}\,dx$$

$$\bar{y} = \frac{1}{2A}\int_a^b [f(x)]^2\,dx = \frac{1}{2A}\int_0^R (R^2 - x^2)\,dx$$

Since $\bar{x} = \bar{y}$, and $\bar{y}$ is easier to find, we evaluate $\bar{y}$.

$$\bar{x} = \bar{y} = \frac{1}{2A}\int_0^R (R^2 - x^2)\,dx = \frac{1}{2\left(\dfrac{\pi R^2}{4}\right)}\left[R^2 x - \frac{x^3}{3}\right]_0^R = \frac{2}{\pi R^2}\left(\frac{2}{3}R^3\right) = \frac{4}{3\pi}R$$

The centroid of the lamina, as shown in Figure 78(b), is $(\bar{x}, \bar{y}) = \left(\dfrac{4}{3\pi}R, \dfrac{4}{3\pi}R\right)$. ∎

### EXAMPLE 5  Finding the Centroid of a Homogeneous Lamina

Find the centroid of the lamina bounded by the graph of $y = f(x) = x^2 + 1$, the $x$-axis, and the lines $x = -2$ and $x = 2$.

**Solution** Figure 79(a) shows the graph of the lamina.

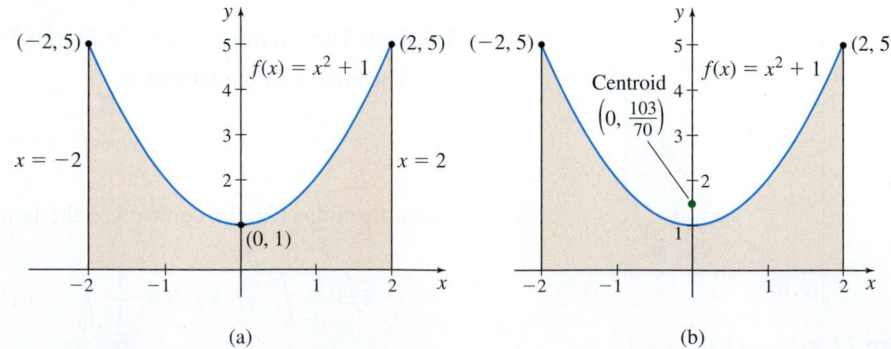

(a)             (b)

**Figure 79**

We notice two properties of $f$.

- The graph of $f$ is symmetric about the $y$-axis, so by the symmetry principle, $\bar{x} = 0$.

- $f$ is an even function, so $\displaystyle\int_{-2}^2 f(x)\,dx = 2\int_0^2 f(x)\,dx$.

The area $A$ of the region is

$$A = 2 \int_0^2 (x^2 + 1)\, dx = 2 \left[ \frac{x^3}{3} + x \right]_0^2 = 2 \left( \frac{8}{3} + 2 \right) = \frac{28}{3}$$

Using (2), we get

$$\bar{y} = \frac{1}{2A} \int_a^b [f(x)]^2\, dx = \frac{1}{2A} \int_{-2}^2 [f(x)]^2\, dx = \frac{1}{2A} \cdot 2 \int_0^2 [f(x)]^2\, dx$$

$$= \frac{3}{28} \int_0^2 (x^2 + 1)^2\, dx = \frac{3}{28} \int_0^2 (x^4 + 2x^2 + 1)\, dx$$

$$= \frac{3}{28} \left[ \frac{x^5}{5} + \frac{2x^3}{3} + x \right]_0^2 = \frac{3}{28} \left( \frac{32}{5} + \frac{16}{3} + 2 \right) = \frac{103}{70} \approx 1.471$$

The centroid of the lamina, as shown in Figure 79(b), is $(\bar{x}, \bar{y}) = \left( 0, \dfrac{103}{70} \right)$. ∎

Notice that the centroid in Example 5 does not lie within the lamina.

**NOW WORK** Problem 19.

In general, laminas are not homogeneous. The Challenge Problems at the end of the section investigate the center of mass of a lamina for which the density of the material varies with respect to $x$. The more general case, where the density of a lamina varies with both $x$ and $y$, requires double integration.

## 3 Find the Volume of a Solid of Revolution Using the Pappus Theorem

> **THEOREM  The Pappus Theorem for Volume**
>
> Let $R$ be a plane region of area $A$ and let $V$ be the volume of the solid of revolution obtained by revolving $R$ about a line that does not intersect $R$. Then the volume $V$ of the solid of revolution is
>
> $$\boxed{V = 2\pi A d}$$
>
> where $d$ is the distance from the centroid of $R$ to the line.

**IN WORDS** The volume of a solid formed by revolving a plane region around an axis equals the product of area of the region and the distance its centroid travels around the axis.

**Proof** We give a proof for the special case where the region $R$ is bounded by the graph of a function $f$ that is continuous and nonnegative on the interval $[a, b]$, the $x$-axis, and the lines $x = a$ and $x = b$, and where $R$ is revolved about the $y$-axis, as shown in Figure 80. Using the shell method to find the volume of the solid of revolution and the centroid formula (2) for $\bar{x}$, we find

$$V = 2\pi \underset{\substack{\uparrow \\ \text{Shell Method}}}{\int_a^b x f(x)\, dx} = \underset{\substack{\uparrow \\ \text{(2)}}}{2\pi (A\bar{x})} = 2\pi A d$$

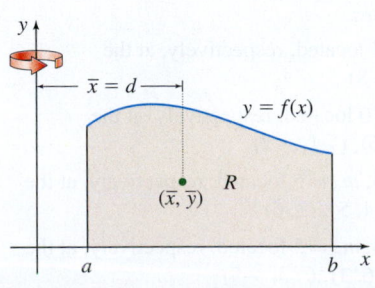

**Figure 80** The region $R$ to be revolved about the $y$-axis.

where $\bar{x} = d$ is the distance of the centroid $(\bar{x}, \bar{y})$ of $R$ from the $y$-axis. ∎

## EXAMPLE 6  Using the Pappus Theorem for Volume

Use the Pappus Theorem to find the volume of the solid formed by revolving the region enclosed by the circle $(x-3)^2 + y^2 = 1$ about the $y$-axis.

**ORIGINS** The Greek mathematician Pappus of Alexandria (c. 300 AD) produced a mathematical collection containing a record of much of classical Greek mathematics. In it he shows a relationship between volume and centroids.

**Solution** By symmetry, the centroid of a circular region is the center of the circle. Here the centroid is the point $(3, 0)$. See Figure 81(a).

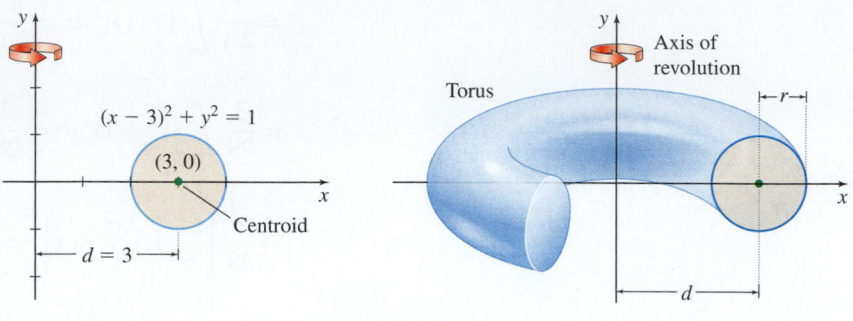

**NOTE** A **torus** is a doughnut-shaped surface.

(a) $(x-3)^2 + y^2 = 1$  (b) The solid of revolution is called a **solid torus**.

**DF** Figure 81

The distance $d$ from the centroid to the axis of revolution, which is the $y$-axis, is $d = 3$. The area of the circle is $A = \pi R^2 = \pi \cdot 1^2 = \pi$. It follows from the Pappus Theorem that the volume $V$ of the solid of revolution [Figure 81(b)] is

**IN WORDS** A disk moving in a circle generates a solid torus. The volume of the torus equals the product of the circumference of the circle traveled by the centroid of the generating disk and the area of the disk.

$$V = 2\pi \cdot (\text{Area of the circle}) \cdot d = 2\pi \cdot \pi \cdot 3 = 6\pi^2 \approx 59.218 \text{ cubic units.}$$ ∎

**NOW WORK** Problem 23.

---

## 6.8 Assess Your Understanding

### Concepts and Vocabulary

1. *Multiple Choice* The [(a) midpoint of mass (b) mass point (c) center of mass] of a system is the point that acts as if all the mass is concentrated at that point.

2. *Multiple Choice* If an object of mass $m$ is located on the $x$-axis a distance $d$ from the origin, then the moment of the mass about the origin is [(a) 0 (b) $d$ (c) $md$ (d) $m$].

3. *True or False* A homogeneous lamina is made of material that has a constant density.

4. *Multiple Choice* If a homogeneous lamina is symmetric about a line $L$, then its [(a) centroid (b) moment of mass (c) axis] lies on $L$.

5. *True or False* The centroid of a lamina $R$ always lies within $R$.

6. *Multiple Choice* The center of mass of a homogeneous lamina $R$ is located at the [(a) centroid (b) midpoint (c) equilibrium point] of $R$, the geometric center of the lamina.

### Skill Building

*In Problems 7–10, find the center of mass of each system of masses.*

7. $m_1 = 20$, $m_2 = 50$ located, respectively, at 4 and 10

8. $m_1 = 10$, $m_2 = 3$ located, respectively, at $-2$ and 3

9. $m_1 = 4$, $m_2 = 3$, $m_3 = 3$, $m_4 = 5$ located, respectively, at $-1, 2, 4, 3$

10. $m_1 = 7$, $m_2 = 3$, $m_3 = 2$, $m_4 = 4$ located, respectively, at $6, -2, -4, -1$

*In Problems 11–14, find the moments $M_x$ and $M_y$ and the center of mass of each system of masses.*

11. $m_1 = 4$, $m_2 = 8$, $m_3 = 1$ located, respectively, at the points $(0, 2)$, $(2, 1)$, $(4, 8)$

12. $m_1 = 6$, $m_2 = 2$, $m_3 = 10$ located, respectively, at the points $(-1, -1)$, $(12, 6)$, $(-1, -2)$

13. $m_1 = 4$, $m_2 = 3$, $m_3 = 3$, $m_4 = 5$ located, respectively, at the points $(-1, 2)$, $(2, 3)$, $(4, 5)$, $(3, 6)$

14. $m_1 = 8$, $m_2 = 6$, $m_3 = 3$, $m_4 = 5$ located, respectively, at the points $(-4, 4)$, $(0, 5)$, $(6, 4)$, $(-3, -5)$

---

**1.** = NOW WORK problem       **△** = Graphing technology recommended       **CAS** = Computer Algebra System recommended

In Problems 15–22, find the centroid of each homogeneous lamina bounded by the graphs of the given equations.

15. $y = 2x + 3$, $y = 0$, $x = -1$, $x = 2$

16. $y = \dfrac{3 - x}{2}$, $y = 0$, $x = -1$, $x = 3$

17. $y = x^2$, $y = 0$, $x = 3$

18. $y = x^3$, $x = 2$, $y = 0$

19. $y = 4x - x^2$ and the $x$-axis

20. $y = x^2 + x + 1$, $x = 0$, $x = 4$, and the $x$-axis

21. $y = \sqrt{x}$, $x = 4$, $y = 0$

22. $y = \sqrt[3]{x}$, $x = 8$, $y = 0$

In Problems 23 and 24, use the Pappus Theorem to find the volume of the solid of revolution.

23. The solid torus formed by revolving the circle $(x - 4)^2 + y^2 = 9$ about the $y$-axis

24. The solid torus formed by revolving the circle $x^2 + (y - 2)^2 = 1$ about the $x$-axis

## Applications and Extensions

**Centroid**  In Problems 25–30, find the centroid of each homogeneous lamina. (Hint: The moments of the union of two or more nonoverlapping regions equal the sum of the moments of the individual regions.)

25.

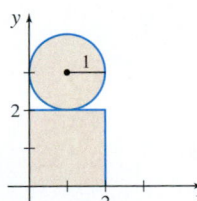

26.

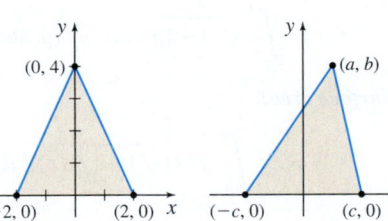

27.

28.

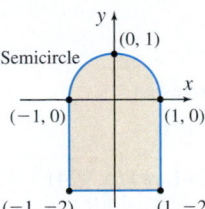

29.

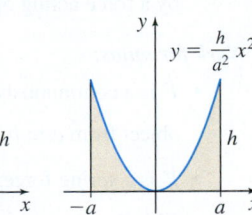

30.

31. **Center of Mass of a Baseball Bat**  Inspection of a baseball bat shows that it gets progressively thicker, that is, more massive, beginning at the handle and ending at the top. If the bat is aligned so that the $x$-axis runs through the center of the bat, the density $\lambda$, in kilograms per meter, can be modeled by $\lambda = kx$, where $k$ is the constant of proportionality. The mass $M$ of a bat of length $L$ meters is $\int_0^L \lambda \, dx$. The center of mass $\bar{x}$ of the bat is $\dfrac{\int_0^L x\lambda \, dx}{M}$.

   (a) Find the mass of the bat.

   (b) Use the result of (a) to find the constant of proportionality $k$.

(c) Find the center of mass of the bat.

(d) Where is the "sweet spot" of the bat (the best place to make contact with the ball)? Explain why.

(e) Give an explanation for the representation of the mass $M$ by $\int_0^L \lambda \, dx$.

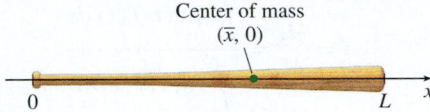

Center of mass
$(\bar{x}, 0)$

32. **The Pappus Theorem for Volume**  Find the volume of the solid formed by revolving the region bounded by the graphs of $y = \sqrt{x - 1}$, $y = 0$, and $x = 2$ about the $y$-axis.

33. **The Pappus Theorem for Volume**  Find the volume of the solid formed by revolving the triangular region bounded by the lines $x = 3$, $y = 2$, and $x + 2y = 9$ about the $y$-axis.

34. **The Pappus Theorem for Volume**  Find the volume of the solid formed by revolving the triangular region bounded by $x = 1$, $y = 5$, and $y = x - 1$ about the $y$-axis.

35. **The Pappus Theorem for Volume**  Find the volume of the solid formed by revolving the triangular region whose vertices are at the points $(1, 1)$, $(5, 3)$, and $(3, 3)$ about the $x$-axis.

36. **Centroid of a Right Triangle**  Find the centroid of a triangular region with vertices at $(0, 0)$, $(0, 4)$, and $(3, 0)$.

## Challenge Problems

37. **Centroid of a Triangle**

   (a) Show that the centroid of a triangular region is located at the intersection of the medians. [Hint: Place the vertices of the triangle at $(a, 0)$, $(b, 0)$, and $(0, c)$, $a < 0$, $b > 0$, $c > 0$.]

   (b) Show that the centroid of a triangular region divides each median into two segments whose lengths are in the ratio 2:1.

38. Derive formulas for finding the centroid of a lamina bounded by the graphs of $y = f(x)$ and $y = g(x)$ and the lines $x = a$ and $x = b$, as shown in the figure below.

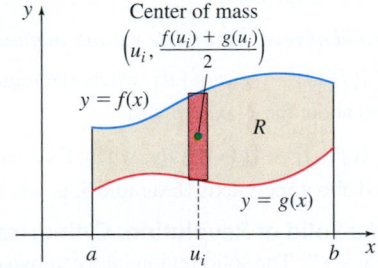

Center of mass
$\left(u_i, \dfrac{f(u_i) + g(u_i)}{2}\right)$

$y = f(x)$

$R$

$y = g(x)$

In Problems 39–44, use the result of Problem 38 to find the centroid of each homogeneous lamina enclosed by the graphs of the given equations.

39. $y = x^{2/3}$ and $y = x^2$ from $x = -1$ to $x = 1$

40. $y = \dfrac{1}{x}$, and the lines $y = \dfrac{1}{4}$, and $x = 1$

41. $y = -x^2 + 2$ and $y = |x|$

42. $y = \sqrt{x}$ and $y = x^2$

43. $y = 4 - x^2$ and $y = x + 2$

44. $y = 9 - x^2$ and $y = |x|$.

**45. Laminas with Variable Density**   Suppose a lamina is determined by a region $R$ enclosed by the graph of $y = f(x)$, the $x$-axis, and the lines $x = a$ and $x = b$, where $f$ is continuous and nonnegative on the closed interval $[a, b]$. Suppose further that the density of the material at $(x, y)$ is $\rho = \rho(x)$, where $\rho$ is continuous on $[a, b]$. Show that the center of mass $(\bar{x}, \bar{y})$ of $R$ is given by

$$\bar{x} = \frac{M_y}{M} = \frac{\displaystyle\int_a^b \rho(x) x f(x)\, dx}{\displaystyle\int_a^b \rho(x) f(x)\, dx}$$

$$\bar{y} = \frac{M_x}{M} = \frac{\dfrac{1}{2}\displaystyle\int_a^b \rho(x)[f(x)]^2\, dx}{\displaystyle\int_a^b \rho(x) f(x)\, dx}$$

(*Hint*: Partition the interval $[a, b]$ into $n$ subintervals, and use an argument similar to the one used for the centroid of a homogeneous lamina. Assume that if $\Delta x$ is small, then $R_i$ is homogeneous.)

**46. Lamina with Variable Density**   Use the results of Problem 45 to find the center of mass of a lamina enclosed by $f(x) = \dfrac{1}{2}x^2$, the $x$-axis, and the lines $x = 1$ and $x = e$. Suppose the density at any point $(x, y)$ of the lamina is $\rho(x) = \dfrac{1}{x^3}$.

**47. Lamina with Variable Density**   Use the results of Problem 45 to find the center of mass of the lamina enclosed by $y = \sqrt{x + 1}$, and the lines $x = 0$ and $y = 3$, where the density of the lamina is $\rho(x) = x$.

**Laminas with Variable Density**   *In Problems 48–51, use the results of Problem 45 to find the center of mass of each lamina enclosed by the graph of $f$ on the given interval and having density $\rho = \rho(x)$.*

**48.** $f(x) = x$; $[0, 3]$; $\rho(x) = x$

**49.** $f(x) = 3x - 1$; $[1, 4]$; $\rho(x) = x$

**50.** $f(x) = 2x$; $[0, 1]$; $\rho(x) = x + 1$

**51.** $f(x) = x$; $[1, 4]$; $\rho(x) = x + 1$

# Chapter Review

## THINGS TO KNOW

### 6.1 Area Between Graphs

*Area $A$ between two graphs:*

- $A = \int_a^b [f(x) - g(x)]\, dx$, where $f(x) \geq g(x)$ for all $x$ in the interval $[a, b]$  (p. 428)
- $A = \int_c^d [f(y) - g(y)]\, dy$, where $f(y) \geq g(y)$ for all $y$ in the interval $[c, d]$  (p. 432)

### 6.2 Volume of a Solid of Revolution: Disks and Washers

*Volume $V$ of a solid of revolution: The disk method:*

- $V = \pi \int_a^b [f(x)]^2\, dx$, where the region is revolved about the $x$-axis (p. 437)
- $V = \pi \int_c^d [g(y)]^2\, dy$, where the region is revolved about the $y$-axis (p. 439)

*Volume $V$ of a solid of revolution: The washer method:*

- $V = \pi \int_a^b \{[f(x)]^2 - [g(x)]^2\}\, dx$, where the region is revolved about the $x$-axis (p. 441)
- $V = \pi \int_c^d \{[f(y)]^2 - [g(y)]^2\}\, dy$, where the region is revolved about the $y$-axis (Example 5, p. 442)

### 6.3 Volume of a Solid of Revolution: Cylindrical Shells

- Cylindrical shell: The solid region between two concentric cylinders (p. 447)

*Volume $V$ of a solid of revolution: The shell method:*

- $V = 2\pi \int_a^b x f(x)\, dx$, where the region is revolved about the $y$-axis (p. 448)
- $V = 2\pi \int_c^d y g(y)\, dy$, where the region is revolved about the $x$-axis (p. 451)

### 6.4 Volume of a Solid: Slicing

*Volume $V$ of a solid:*

- $V = \int_a^b A(x)\, dx$, where the area $A = A(x)$ of the cross section of a solid is known and is continuous on $[a, b]$ (p. 456)

### 6.5 Arc Length; Surface Area of a Solid of Revolution

*Arc length formulas:*

- $s = \int_a^b \sqrt{1 + [f'(x)]^2}\, dx$ (p. 463)
- $s = \int_c^d \sqrt{1 + [g'(y)]^2}\, dy$ (p. 466)

*Surface area:*

- $S = 2\pi \int_a^b f(x) \sqrt{1 + [f'(x)]^2}\, dx$ (p. 468)

### 6.6 Work

- **Work** is the energy transferred to or from an object by a force acting on the object. (p. 472)

*Work formulas:*

- $F$ is a continuously varying force that moves an object from $a$ to $b$: $W = \int_a^b F(x)\, dx$  (p. 473)
- $F$ is a spring force: $F(x) = -kx$ (Hooke's Law) (p. 474)
- $W$ is the work required to pump a liquid out of a container: $W = \int_a^b \rho g A(x)(h - x)\, dx$  (p. 477)
- $F$ is the attraction between two bodies: $F(x) = G\dfrac{m_1 m_2}{x^2}$  (p. 478)

### 6.7 Hydrostatic Pressure and Force

- Pressure $P$ is the force $F$ exerted per unit area $A$: $P = \dfrac{F}{A}$ (p. 481)

*Hydrostatic force:*

- $F = \int_c^d \rho g (H - y)[f(y) - h(y)]\, dy$ (p. 482)

## 6.8 Center of Mass; Centroid; The Pappus Theorem

- The center of mass of an object or system of objects is the point that acts as if all the mass is concentrated at that point. (p. 485)
- The moments of a system represent the tendency of the system to rotate about the center of mass. (p. 486)
- A lamina is a thin, flat sheet of material. (p. 487)
- If a lamina has constant density, it is homogeneous and its center of mass is located at the centroid. (p. 488)

*The centroid of a homogeneous lamina:*

- A lamina of area $A$ enclosed by the graph of $f$, the $x$-axis, and the lines $x = a$ and $x = b$:

$$\bar{x} = \frac{1}{A} \int_a^b [x f(x)] \, dx \qquad \bar{y} = \frac{1}{2A} \int_a^b [f(x)]^2 \, dx,$$

where $A = \int_a^b f(x) \, dx$ (p. 489)

- Symmetry principle: If a homogeneous lamina is symmetric about a line $L$ or a point $P$, then the centroid of the lamina lies on $L$ or is at the point $P$. (p. 490)

*The Pappus Theorem for Volume*

Let $R$ be a plane region of area $A$ and let $V$ be the volume of the solid of revolution obtained by revolving $R$ about a line that does not intersect $R$. Then the volume $V$ of the solid of revolution is $V = 2\pi A d$, where $d$ is the distance from the centroid of $R$ to the line. (p. 491)

## OBJECTIVES

| Section | You should be able to ... | Examples | Review Exercises |
|---------|---------------------------|----------|------------------|
| 6.1 | 1 Find the area between the graphs of two functions by partitioning the $x$-axis (p. 428) | 1–4 | 1–5, 21 |
| | 2 Find the area between the graphs of two functions by partitioning the $y$-axis (p. 431) | 5–7 | 3, 4, 21 |
| 6.2 | 1 Use the disk method to find the volume of a solid formed by revolving a region about the $x$-axis (p. 437) | 1, 2 | 9, 11 |
| | 2 Use the disk method to find the volume of a solid formed by revolving a region about the $y$-axis (p. 439) | 3 | 8 |
| | 3 Use the washer method to find the volume of a solid formed by revolving a region about the $x$-axis (p. 440) | 4 | 6, 14, 38(a) |
| | 4 Use the washer method to find the volume of a solid formed by revolving a region about the $y$-axis (p. 442) | 5 | 13 |
| | 5 Find the volume of a solid formed by revolving a region about a line parallel to a coordinate axis (p. 443) | 6, 7 | 10, 12, 15, 38(c), (d) |
| 6.3 | 1 Use the shell method to find the volume of a solid formed by revolving a region about the $y$-axis (p. 447) | 1, 2 | 7, 8, 13, 38(b) |
| | 2 Use the shell method to find the volume of a solid formed by revolving a region about the $x$-axis (p. 451) | 3 | 9, 23 |
| | 3 Use the shell method to find the volume of a solid formed by revolving a region about a line parallel to a coordinate axis (p. 453) | 4 | 12, 15, 38(e), (f) |
| 6.4 | 1 Use slicing to find the volume of a solid (p. 456) | 1–4 | 22, 24, 25 |
| 6.5 | 1 Find the arc length of the graph of a function $y = f(x)$ (p. 462) | 1–3 | 16, 17, 26 |
| | 2 Find the arc length of the graph of a function using a partition of the $y$-axis (p. 466) | 4 | 18 |
| | 3 Find the surface area of a solid of revolution (p. 467) | 5 | 39, 40 |
| 6.6 | 1 Find the work done by a variable force (p. 473) | 1 | 27, 28 |
| | 2 Find the work done by a spring force (p. 474) | 2, 3 | 30, 31 |
| | 3 Find the work done to pump a liquid (p. 476) | 4, 5 | 29 |
| 6.7 | 1 Find hydrostatic pressure and force (p. 481) | 1, 2 | 32–34 |
| 6.8 | 1 Find the center of mass of a finite system of objects (p. 486) | 1, 2 | 19, 20 |
| | 2 Find the centroid of a homogeneous lamina (p. 487) | 3–5 | 35 |
| | 3 Find the volume of a solid of revolution using the Pappus Theorem (p. 491) | 6 | 36, 37 |

## REVIEW EXERCISES

**Area**    *In Problems 1–5, find the area A of the region bounded by the graphs of the given equations.*

1. $y = e^x$, $x = 0$, $y = 4$        2. $y = x^2$, $y = 18 - x^2$

3. $x = 2y^2$, $x = 2$        4. $y = \dfrac{1}{x}$, $x + y = 4$

5. $y = 4 - x^2$, $y = 3x$

**Volume of a Solid of Revolution**    *In Problems 6–15, find the volume of the solid of revolution generated by revolving the region bounded by the graphs of the given equations about the given line.*

6. $y = x^2$, $y = 4x - x^2$; about the x-axis

7. $y = x^2 - 5x + 6$, $y = 0$; about the y-axis

8. $x = y^2 - 4$, $x = 0$; about the y-axis

9. $xy = 1$, $x = 1$, $x = 2$, $y = 0$; about the x-axis

10. $y = x^2 - 4$, $y = 0$; about the line $y = -4$

11. $y = 4x - x^2$ and the x-axis; about the x-axis

12. $y^2 = 8x$, $y \geq 0$, and $x = 2$; about the line $x = 2$

13. $y = \dfrac{x^3}{2}$, $y = 0$, $x = 2$; about the y-axis

14. $y = e^x$, $y = 1$, $x = 1$; about the x-axis

15. $y^2 = x^3$, $y = 8$, $x = 0$; about the line $x = 4$

**Arc Length**    *In Problems 16–18, find the arc length of each graph.*

16. $y = x^{3/2} + 4$ from $x = 2$ to $x = 5$

17. $y = \dfrac{x^3}{6} + \dfrac{1}{2x}$ from $x = 2$ to $x = 6$

18. $2y^3 = x^2$ from $y = 0$ to $y = 2$

19. **Center of Mass**    Find the center of mass of the system of masses: $m_1 = 1$, $m_2 = 3$, $m_3 = 8$, $m_4 = 1$ located, respectively, at $-1, 2, 14,$ and $0$.

20. **Moments and Center of Mass**    Find the moments $M_x$ and $M_y$ and the center of mass of the system of masses: $m_1 = 2$, $m_2 = 2$, $m_3 = 3$, $m_4 = 2$ located, respectively, at the points $(-4, 4)$, $(2, 3)$, $(4, 4)$, $(-3, -5)$.

21. **Area of a Triangle**    Use integration to find the area of the triangle formed by the lines $x - y + 1 = 0$, $3x + y - 13 = 0$, and $x + 3y - 7 = 0$.

22. **Volume**    A solid has a circular base of radius 4 units. Slices taken perpendicular to a fixed diameter are equilateral triangles. Find its volume.

23. **Volume**    Find the volume of the solid generated by revolving the region bounded by the graph of $4x^2 + 9y^2 = 36$ in the first quadrant about the x-axis.

24. **Volume of a Cone**    Find the volume of an elliptical cone with base $\dfrac{x^2}{4} + y^2 = 1$ and height 5. (*Hint:* The area $A$ of an ellipse with semi-axes $a$ and $b$ is $A = \pi ab$.)

25. **Volume**    The base of a solid is enclosed by the graphs of $4x + 5y = 20$, $x = 0$, $y = 0$. Every cross section perpendicular to the base along the x-axis is a semicircle. Find the volume of the solid.

26. **Arc Length**    Find the point $P$ on the graph of $y = \dfrac{2}{3}x^{3/2}$ to the right of the y-axis so that the length of the graph from $(0, 0)$ to $P$ is $\dfrac{52}{3}$.

27. **Work**    Find the work done in raising an 800-lb anchor 150 ft with a chain weighing 20 lb/ft.

28. **Work**    Find the work done in raising a container of 1000 kg of silver ore from a mine 1200 m deep with a cable weighing 3 kg/m.

29. **Work Pumping Water**    A hemispherical water tank has a diameter of 12 m. It is filled to a depth of 4 m. How much work is done in pumping all the water over the edge? (Use $\rho = 1000 \, \text{kg/m}^3$.)

30. **Work of a Spring**    A spring with an unstretched length of 0.6 m requires a force of 4 N to stretch it to 0.8 m. How much work is done in stretching it to 1.4 m?

31. **Work of a Spring**    Find the unstretched length of a spring if the work required to stretch the spring from 1.0 to 1.4 m is half the work required to stretch it from 1.2 to 1.8 m.

32. **Hydrostatic Force**    A trough in the shape of a trapezoid, is 2 ft wide at the bottom, 4 ft wide at the top, and 3 ft deep. What is the force due to liquid pressure on the end, if it is full of water?

33. **Hydrostatic Force**    A cylindrical tank is on its side. It has a diameter of 10 m and is full to a depth of 5 m with gasoline that has a density of $737 \, \text{kg/m}^3$. What is the force due to liquid pressure on the end?

34. **Hydrostatic Force**    A dam is built in the shape of a trapezoid 1000 ft long at the top, 700 ft long at the bottom, and 80 ft deep. Determine the force of water on the dam if:

   (a) the reservoir behind the dam is full.

   (b) the reservoir behind the dam has a depth of 60 ft.

35. **Centroid**    Find the centroid of the homogeneous lamina bounded by the graphs of $y = \sqrt{x}$, $y = 0$, and $x = 9$.

36. **Pappus Theorem for Volume**    Find the volume of the torus with an outer diameter of 5 cm and an inner diameter of 2 cm.

37. **Volume of a Solid of Revolution**    Find the volume generated when the triangular region bounded by the lines $x = 3$, $y = 0$, and $2x + y - 12 = 0$ is revolved about the y-axis.

38. **Volume of a Solid of Revolution**    Find the volume generated when the region bounded by the graphs of $y = 3\sqrt{x}$ and $y = -x^2 + 6x - 2$ is revolved about each of the following lines:

   (a) $y = 0$    (b) $x = 0$    (c) $y = -2$

   (d) $y = 8$    (e) $x = -3$    (f) $x = 5$

**Surface Area of a Solid of Revolution**    *In Problems 39 and 40, find the surface area of the solid formed by revolving the region bounded by the graph of $y = f(x)$, the x-axis, and the given lines about the x-axis.*

39. $f(x) = 4\sqrt{2x}$; lines: $x = 1$ and $x = 4$

40. $f(x) = \dfrac{5}{3}x^3$; $x = 0$ and the line $x = 4$

| **CHAPTER 6 PROJECT** | **Determining the Amount of Concrete Needed for a Cooling Tower** |

A common design for cooling towers is modeled by a branch of a hyperbola rotated about an axis. The design is used because of its strength and efficiency. Not only is the shape stronger than either a cone or cylinder, it takes less material to build. This shape also maximizes the natural upward draft of hot air without the need for fans.*

How much concrete is needed to build a cooling tower that has a base 460 ft wide that is 442 ft below the vertex if the top of the tower is 310 ft wide and 123 ft above the vertex? Assume the walls are a constant 5 in = 0.42 ft thick. See Figure 82.

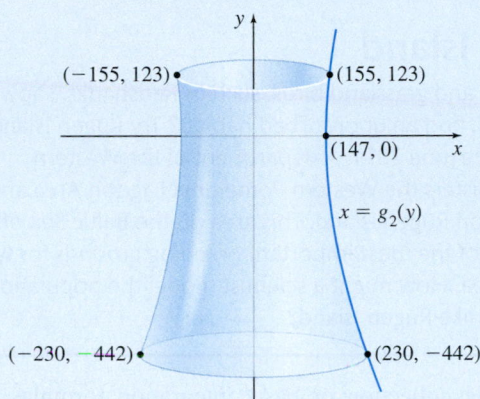

**Figure 82**

1. Show that the equation of the right branch of the hyperbola in Figure 82 is given by

$$x = g_2(y) = \sqrt{147^2 + 0.16y^2}$$

Verify that the point $(230, -442)$ satisfies $x = g_2(y)$.
*Hint:* Begin with the equation of a hyperbola

$$\frac{x^2}{a^2} - \frac{y^2}{b^2} = 1$$

Use the points $(155, 123)$ and $(147, 0)$ to find $a$ and $b$.

2. Find the volume of the solid of revolution obtained by revolving the area enclosed by $x = g_2(y)$ from $y = -442$ to $y = 123$ about the $y$-axis.

3. Using the fact that the walls are 0.42 ft thick, show that the equation of the interior hyperbola is given by

$$x = g_1(y) = \sqrt{146.58^2 + 0.16y^2}$$

Verify that the point $(220.58, -442)$ satisfies $g_1(y)$. See Figure 83.

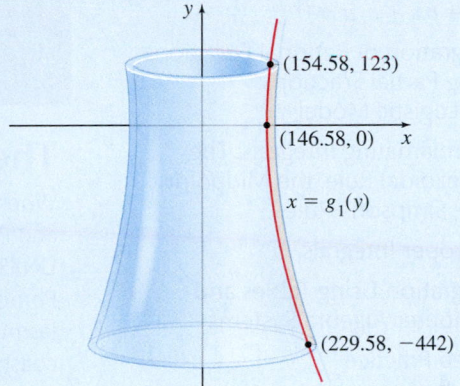

**Figure 83**

4. Find the volume of the solid of revolution obtained by revolving the area enclosed by $x = g_1(y)$ from $y = -442$ to $y = 123$ about the $y$-axis.

5. Now determine the volume (in cubic feet) of concrete required for the cooling tower.

6. Do an Internet search to find the cost of concrete. How much will the concrete used in the cooling tower cost?

7. Write a report for an audience that is not familiar with cooling towers summarizing the findings in Problems 1–6.

---

*The design of an unsupported, reinforced concrete hyperbolic cooling tower was patented in 1918 (UK patent 198,863) by Frederic von Herson and Gerard Kupeers of the Netherlands.

# 7

# Techniques of Integration

Arterra/UIG via Getty Images

## The Birds of Rügen Island

Where else can a variety of shore and grassland birds, such as Redshanks, Lapwings, and Pied Avocet (pictured above), find an undisturbed habitat? Try Rügen Island, a UNESCO World Heritage Site since June 2011. A department of the Western Pomeranian National Park administers the Western Pomerania Lagoon Area and Jasmund National Park, which is on Rügen Island. This area on the Baltic Sea off the coast of Germany includes some of the most important breeding grounds for water and mud flat birds of the Baltic Sea. How might a scientist model the population of birds that live in a protected area like Rügen Island?

**CHAPTER 7 PROJECT** See the Chapter Project on page 572 for a discussion of how we can use calculus to model bird populations.

We have developed a sizable collection of basic integration formulas that are listed below. Some of the integrals are the result of simple antidifferentiation. Others, such as $\int \tan x \, dx$, were found using substitution, a *technique of integration*. In this chapter, we expand the list of basic integrals by exploring more techniques of integration.

Integration, unlike differentiation, has no hard and fast rules. It is often difficult, sometimes even impossible, to integrate a function that appears to be simple.

As you study the techniques in this chapter, it is important to recognize the form of the integrand for each technique. Then, with practice, integration becomes easier.

$$\int x^a \, dx = \frac{x^{a+1}}{a+1} + C, a \neq -1$$

$$\int \frac{1}{x} \, dx = \ln |x| + C$$

$$\int e^x \, dx = e^x + C$$

$$\int a^x \, dx = \frac{a^x}{\ln a} + C, a>0, a\neq 1$$

$$\int \sin x \, dx = -\cos x + C$$

$$\int \cos x \, dx = \sin x + C$$

$$\int \sec^2 x \, dx = \tan x + C$$

$$\int \csc^2 x \, dx = -\cot x + C$$

$$\int \sec x \tan x \, dx = \sec x + C$$

$$\int \csc x \cot x \, dx = -\csc x + C$$

$$\int \tan x \, dx = \ln |\sec x| + C$$

$$\int \cot x \, dx = \ln |\sin x| + C$$

$$\int \sec x \, dx = \ln |\sec x + \tan x| + C$$

$$\int \csc x \, dx = \ln |\csc x - \cot x| + C$$

$$\int \frac{dx}{\sqrt{a^2-x^2}} = \sin^{-1} \frac{x}{a} + C, \ a>0$$

$$\int \frac{dx}{x\sqrt{x^2-a^2}} = \frac{1}{a} \sec^{-1} \frac{x}{a} + C, a>0$$

$$\int \frac{dx}{a^2+x^2} = \frac{1}{a} \tan^{-1} \frac{x}{a} + C, a>0$$

$$\int \sinh x \, dx = \cosh x + C$$

$$\int \cosh x \, dx = \sinh x + C$$

$$\int \text{sech}^2 x \, dx = \tanh x + C$$

$$\int \text{csch}^2 x \, dx = -\coth x + C$$

$$\int \text{sech} \, x \tanh x \, dx = -\text{sech} \, x + C$$

$$\int \text{csch} \, x \coth x \, dx = -\text{csch} \, x + C$$

# 7.1 Integration by Parts

**OBJECTIVES**  *When you finish this section, you should be able to:*

**1** Integrate by parts (p. 499)

**2** Find a definite integral using integration by parts (p. 503)

**3** Derive a general formula using integration by parts (p. 504)

*Integration by parts* is a technique of integration based on the Product Rule for derivatives: If $u = f(x)$ and $v = g(x)$ are functions that are differentiable on an open interval $(a, b)$, then

$$\frac{d}{dx}[f(x) \cdot g(x)] = f(x)\,g'(x) + f'(x)\,g(x)$$

Integrating both sides gives

$$\int \frac{d}{dx}[f(x) \cdot g(x)]\,dx = \int [f(x)\,g'(x) + f'(x)\,g(x)]\,dx$$

$$f(x) \cdot g(x) = \int f(x)\,g'(x)\,dx + \int f'(x)g(x)\,dx \qquad (1)$$

Solving equation (1) for $\int f(x)g'(x)\,dx$ yields

$$\boxed{\int f(x)\,g'(x)\,dx = f(x) \cdot g(x) - \int f'(x)\,g(x)\,dx}$$

which is known as the **integration by parts formula**.

Let $u = f(x)$, $v = g(x)$. Then we can use their differentials, $du = f'(x)\,dx$ and $dv = g'(x)\,dx$ to obtain the integration by parts formula in the form we usually use:

**IN WORDS**  The integral of $u\,dv$ equals $uv$ minus the integral of $v\,du$.

$$\boxed{\int u\,dv = uv - \int v\,du}$$

## 1 Integrate by Parts

The goal of integration by parts is to choose $u$ and $dv$ so that $\int v\,du$ is easier to integrate than $\int u\,dv$.

### EXAMPLE 1  Integrating by Parts

Find $\displaystyle\int x\,e^x\,dx$.

**Solution**  Choose $u$ and $dv$ so that

$$\int u\,dv = \int x\,e^x\,dx$$

Suppose we choose

$$u = x \qquad \text{and} \qquad dv = e^x\,dx$$

Then

$$du = dx \qquad \text{and} \qquad v = \int dv = \int e^x\,dx = e^x$$

Notice that we did not add a constant. Only a particular antiderivative of $dv$ is required at this stage; we add the constant of integration at the end. Using the integration by parts formula, we have

$$\int \underbrace{x}_{u}\,\underbrace{e^x\,dx}_{dv} = \underbrace{x\,e^x}_{uv} - \int \underbrace{e^x}_{v}\,\underbrace{dx}_{du} = x\,e^x - e^x + C \qquad \blacksquare$$

We intentionally chose $u = x$ and $dv = e^x dx$ so that $\int v\,du$ in the formula is easy to integrate. Suppose, instead, we chose

$$u = e^x \qquad \text{and} \qquad dv = x\,dx$$

Then

$$du = e^x \, dx \quad \text{and} \quad v = \int x \, dx = \frac{x^2}{2}$$

and the integration by parts formula yields

$$\int x \, e^x \, dx = \int \underbrace{e^x}_{u} \underbrace{x \, dx}_{dv} = \underbrace{e^x \frac{x^2}{2}}_{uv} - \int \underbrace{\frac{x^2}{2}}_{v} \underbrace{e^x \, dx}_{du}$$

**IN WORDS** We choose $dv$ so that it can be easily integrated and choose $u$ so that $\int v \, du$ is simpler than $\int u \, dv$.

For this choice of $u$ and $v$, the integral on the right is more complicated than the original integral, indicating an unwise choice of $u$ and $dv$.

NOW WORK Problem 3.

EXAMPLE 2   Integrating by Parts

Find $\displaystyle\int x \sin x \, dx$.

**Solution** Use the integration by parts formula with

$$u = x \quad \text{and} \quad dv = \sin x \, dx \qquad \int u \, dv = \int x \sin x \, dx$$

Then

$$du = dx \quad \text{and} \quad v = \int \sin x \, dx = -\cos x$$

Now

$$\int x \sin x \, dx = -x \cos x + \int \cos x \, dx = -x \cos x + \sin x + C$$
$$\underset{\uparrow}{\int u \, dv = uv - \int v \, du}$$

NOW WORK Problem 5.

Unfortunately, there are no exact rules for choosing $u$ and $dv$. But the following guidelines are helpful:

**Integration by Parts: General Guidelines for Choosing $u$ and $dv$**

- $dx$ is always part of $dv$.
- $dv$ should be easy to integrate.
- $u$ and $dv$ are chosen so that $\int v \, du$ is no more difficult to integrate than the original integral $\int u \, dv$.
- If the new integral is more complicated, try different choices for $u$ and $dv$.

Table 1 provides additional guidelines to help choose $u$ and $dv$ for several types of integrals that are found using integration by parts. In the table, $n$ is a positive integer.

**TABLE 1** Guidelines for Choosing $u$ and $dv$

| Integral; $n$ is a positive integer | $u$ | $dv$ |
|---|---|---|
| $\int x^n e^{ax} \, dx$ <br> $\int x^n \sin(ax) \, dx \quad \int x^n \cos(ax) \, dx$ <br> $\int x^n \sec^2(ax) \, dx \quad \int x^n \sec(ax) \tan(ax) \, dx$ | $u = x^n$ | $dv = \text{what remains}$ |
| $\int x^n \sin^{-1} x \, dx$ | $u = \sin^{-1} x$ | |
| $\int x^n \cos^{-1} x \, dx$ | $u = \cos^{-1} x$ | $dv = x^n \, dx$ |
| $\int x^n \tan^{-1} x \, dx$ | $u = \tan^{-1} x$ | |
| $\int x^m (\ln x)^n \, dx$; $m$ is a real number, $m \neq -1$ | $u = (\ln x)^n$ | $dv = x^m \, dx$ |

**EXAMPLE 3**  **Integrating by Parts to Find** $\int \ln x \, dx$

Derive the formula

$$\int \ln x \, dx = x \ln x - x + C$$

**Solution** Look at the last row of Table 1. With $n = 1$ and $m = 0$ the integral in the table reduces to $\int \ln x \, dx$. So we use the integration by parts formula with

$$u = \ln x \qquad \text{and} \qquad dv = dx$$

Then

$$du = \frac{1}{x} \, dx \qquad \text{and} \qquad v = \int dx = x$$

Now

**NOTE** The integral $\int \ln x \, dx$ can be found in the list of integrals at the back of the book and is considered a basic integral.

$$\underbrace{\int \ln x \, dx}_{} = x \cdot \ln x - \int x \cdot \frac{1}{x} \, dx = x \ln x - \int dx = x \ln x - x + C$$

$$\int u \, dv = uv - \int v \, du$$

∎

**EXAMPLE 4**  **Integrating by Parts to Find** $\int \tan^{-1} x \, dx$

Derive the formula

$$\int \tan^{-1} x \, dx = x \tan^{-1} x - \frac{1}{2} \ln(1 + x^2) + C$$

**Solution** Look again at Table 1. Then use the integration by parts formula with

$$u = \tan^{-1} x \qquad \text{and} \qquad dv = dx$$

Then

$$du = \frac{1}{1 + x^2} \, dx \qquad \text{and} \qquad v = \int dx = x$$

Now

$$\underbrace{\int \tan^{-1} x \, dx}_{} = x \cdot \tan^{-1} x - \int \frac{x}{1 + x^2} \, dx$$

$$\int u \, dv = uv - \int v \, du$$

**NEED TO REVIEW?** The method of substitution is discussed in Section 5.5, pp. 401–405.

To find the integral $\int \dfrac{x}{1 + x^2} \, dx$, use the substitution $t = 1 + x^2$. Then $dt = 2x \, dx$, or equivalently, $x \, dx = \dfrac{dt}{2}$.

$$\int \frac{x}{1 + x^2} \, dx = \frac{1}{2} \int \frac{dt}{t} = \frac{1}{2} \ln|t| = \frac{1}{2} \ln(1 + x^2)$$

**NOTE** Since $1 + x^2 > 0$, we can drop the absolute value bars and write $\ln(1 + x^2)$.

As a result, $\int \tan^{-1} x \, dx = x \tan^{-1} x - \dfrac{1}{2} \ln(1 + x^2) + C$.    ∎

**NOW WORK** Problem 9.

The next two examples show that sometimes it is necessary to integrate by parts more than once.

**EXAMPLE 5** **Integrating by Parts**

Find $\displaystyle\int x^2 e^x \, dx$.

**Solution** Use the integration by parts formula with

$$u = x^2 \qquad \text{and} \qquad dv = e^x \, dx$$

Then

$$du = 2x \, dx \qquad \text{and} \qquad v = \int e^x \, dx = e^x$$

Now

$$\int x^2 e^x \, dx = x^2 e^x - 2 \int x e^x \, dx$$

$$\underset{\uparrow}{\phantom{x}}$$

$$\int u \, dv = uv - \int v \, du$$

The integral on the right is simpler than the original integral. To find it, use integration by parts a second time. (Use the result from Example 1.)

$$\int x e^x \, dx = x e^x - e^x$$

Then

$$\int x^2 e^x \, dx = x^2 e^x - 2(x e^x - e^x) + C = e^x (x^2 - 2x + 2) + C \qquad \blacksquare$$

**NOW WORK** Problem 13.

**EXAMPLE 6** **Integrating by Parts**

Find $\displaystyle\int e^{2x} \cos(3x) \, dx$

**Solution** The choice for $u$ and $dv$ is not immediately evident. Both $e^{2x}$ and $\cos(3x)$ are easy to integrate and neither the derivative of $e^{2x}$ nor the derivative of $\cos(3x)$ is simpler. We try the following and see what happens.

$$u = e^{2x} \qquad \text{and} \qquad dv = \cos(3x) \, dx$$

Then

$$du = 2e^{2x} dx \qquad \text{and} \qquad v = \int \cos(3x) \, dx = \frac{1}{3} \sin(3x)$$

Now

$$\int e^{2x} \cos(3x) \, dx = e^{2x} \left[ \frac{1}{3} \sin(3x) \right] - \int \left[ \frac{1}{3} \sin(3x) \right] [2 e^{2x} dx] \qquad \int u \, dv = uv - \int v \, du$$

$$\int e^{2x} \cos(3x) \, dx = \frac{1}{3} e^{2x} \sin(3x) - \frac{2}{3} \int e^{2x} \sin(3x) \, dx \qquad (2)$$

The integral on the right of equation (2) is not simpler than the original integral, but it is no more difficult than the original. To find it, integrate by parts a second time using

$$u = e^{2x} \qquad \text{and} \qquad dv = \sin(3x) \, dx$$

Then

$$du = 2e^{2x} dx \qquad \text{and} \qquad v = \int \sin(3x) \, dx = -\frac{1}{3} \cos(3x)$$

Now

$$\int e^{2x} \sin(3x) \, dx = e^{2x} \left[ -\frac{1}{3} \cos(3x) \right] - \int -\frac{1}{3} \cos(3x) \cdot (2 e^{2x}) \, dx$$

$$= -\frac{1}{3} e^{2x} \cos(3x) + \frac{2}{3} \int e^{2x} \cos(3x) \, dx$$

Then from (2)

$$\int e^{2x} \cos(3x)\, dx = \frac{1}{3}e^{2x}\sin(3x) - \frac{2}{3}\left[-\frac{1}{3}e^{2x}\cos(3x) + \frac{2}{3}\int e^{2x}\cos(3x)\, dx\right]$$

$$\int e^{2x} \cos(3x)\, dx = \frac{1}{3}e^{2x}\sin(3x) + \frac{2}{9}e^{2x}\cos(3x) - \frac{4}{9}\int e^{2x}\cos(3x)\, dx \qquad (3)$$

The integral we are trying to find now appears on both sides of equation (3). By adding $\dfrac{4}{9}\displaystyle\int e^{2x}\cos(3x)\, dx$ to both sides of (3) and simplifying, we obtain

$$\int e^{2x} \cos(3x)\, dx + \frac{4}{9}\int e^{2x}\cos(3x)\, dx = \frac{1}{3}e^{2x}\sin(3x) + \frac{2}{9}e^{2x}\cos(3x)$$

$$\frac{13}{9}\int e^{2x}\cos(3x)\, dx = \frac{1}{3}e^{2x}\sin(3x) + \frac{2}{9}e^{2x}\cos(3x)$$

$$\int e^{2x}\cos(3x)\, dx = \frac{9}{13}\left[\frac{1}{3}e^{2x}\sin(3x) + \frac{2}{9}e^{2x}\cos(3x)\right]$$

Adding the constant of integration,

$$\int e^{2x}\cos(3x)\, dx = \frac{3}{13}e^{2x}\sin(3x) + \frac{2}{13}e^{2x}\cos(3x) + C \qquad \blacksquare$$

We started the solution by choosing $u = e^{2x}$ and $dv = \cos(3x)\, dx$, but we could have started with $u = \cos(3x)$ and $dv = e^{2x}dx$. The result would be the same. You should try this approach on your own.

**NOW WORK** Problem 31.

## 2 Find a Definite Integral Using Integration by Parts

There are two approaches for finding a definite integral using integration by parts. They are similar to the approaches used in Section 5.5 for finding a definite integral using substitution.

**NEED TO REVIEW?** Finding a definite integral using substitution is discussed in Section 5.5, pp. 405–406.

- Method 1: Find the related indefinite integral using integration by parts, and then use the Fundamental Theorem of Calculus.
- Method 2: Find the definite integral directly by using the limits of integration at each step of the process. That is,

$$\int_a^b u\, dv = [uv]_a^b - \int_a^b v\, du$$

**EXAMPLE 7**  Finding a Definite Integral Using Integration by Parts

Find $\displaystyle\int_0^{\pi/4} x\sec^2 x\, dx$

**Solution**  We use Method 2 to find the definite integral. Using the guidelines in Table 1, choose

$$u = x \quad\text{and}\quad dv = \sec^2 x\, dx$$

Then

$$du = dx \quad\text{and}\quad v = \int \sec^2 x\, dx = \tan x$$

Now

$$\int_0^{\pi/4} x\sec^2 x\, dx = [x\tan x]_0^{\pi/4} - \int_0^{\pi/4}\tan x\, dx \qquad \int_a^b u\, dv = [uv]_a^b - \int_a^b v\, du$$

$$= [x\tan x]_0^{\pi/4} - [\ln|\sec x|]_0^{\pi/4}$$

$$= \left(\frac{\pi}{4} - 0\right) - (\ln\sqrt{2} - 0) = \frac{\pi}{4} - \ln\sqrt{2} \qquad \blacksquare$$

**NOW WORK** Problem 43.

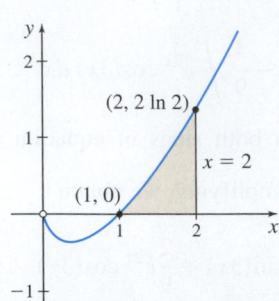

**Figure 1** $f(x) = x \ln x$

CALC
CLIP

**EXAMPLE 8  Finding the Area Under the Graph of $f(x) = x \ln x$**

Find the area under the graph of $f(x) = x \ln x$ from 1 to 2.

**Solution** See Figure 1 for the graph of $f(x) = x \ln x$. The area $A$ under the graph of $f$ from 1 to 2 is $A = \int_1^2 x \ln x \, dx$. We use Method 1 and the integration by parts formula to find $\int x \ln x \, dx$. Using the guidelines in Table 1, choose

$$u = \ln x \qquad \text{and} \qquad dv = x \, dx$$

Then

$$du = \frac{1}{x} \, dx \qquad \text{and} \qquad v = \int x \, dx = \frac{x^2}{2}$$

$$\underbrace{\int x \ln x \, dx}_{\uparrow} = (\ln x) \left( \frac{x^2}{2} \right) - \int \frac{x^2}{2} \cdot \frac{1}{x} \, dx = \frac{x^2}{2} \ln x - \frac{1}{2} \int x \, dx = \frac{x^2}{2} \ln x - \frac{x^2}{4}$$

$$\int u \, dv = uv - \int v \, du$$

Now use the Fundamental Theorem of Calculus.

$$A = \int_1^2 x \ln x \, dx = \left[ \frac{x^2}{2} \ln x - \frac{x^2}{4} \right]_1^2 = (2 \ln 2 - 1) - \left( 0 - \frac{1}{4} \right) = 2 \ln 2 - \frac{3}{4} \qquad \blacksquare$$

**NOW WORK** Problem 47.

## 3 Derive a General Formula Using Integration by Parts

Integration by parts is also used to derive general formulas involving integrals.

**EXAMPLE 9  Deriving a General Formula**

**(a)** Derive the formula

**NOTE** If $b = 0$, formula (4) still works provided $a \neq 0$.

$$\int e^{ax} \, dx = \frac{1}{a} e^{ax} + C$$

$$\boxed{\int e^{ax} \cos(bx) \, dx = \frac{e^{ax} [b \sin(bx) + a \cos(bx)]}{a^2 + b^2} + C \qquad b \neq 0} \qquad (4)$$

**(b)** Use formula (4) to find $\int e^{4x} \cos(5x) \, dx$.

**Solution (a)** Use the integration by parts formula with

$$u = e^{ax} \qquad \text{and} \qquad dv = \cos(bx) \, dx$$

Then

$$du = a e^{ax} \, dx \qquad \text{and} \qquad v = \int \cos(bx) \, dx = \frac{1}{b} \sin(bx)$$

Now

$$\int e^{ax} \cos(bx) \, dx = e^{ax} \frac{\sin(bx)}{b} - \frac{a}{b} \int e^{ax} \sin(bx) \, dx \qquad (5)$$

The new integral on the right, $\int e^{ax} \sin(bx) \, dx$, is different from the original integral, but it is essentially of the same form. Use integration by parts again with this integral by choosing

$$u = e^{ax} \qquad \text{and} \qquad dv = \sin(bx) \, dx$$

Then

$$du = a e^{ax} \, dx \qquad \text{and} \qquad v = \int \sin(bx) \, dx = -\frac{1}{b} \cos(bx)$$

Now

$$\int e^{ax} \sin(bx) \, dx = -\frac{1}{b} e^{ax} \cos(bx) + \frac{a}{b} \int e^{ax} \cos(bx) \, dx \qquad (6)$$

Substituting the result from (6) into (5), we obtain

$$\int e^{ax} \cos(bx)\,dx = e^{ax}\frac{\sin(bx)}{b} - \frac{a}{b}\left[-\frac{1}{b}e^{ax}\cos(bx) + \frac{a}{b}\int e^{ax}\cos(bx)\,dx\right]$$

$$\int e^{ax} \cos(bx)\,dx = \frac{1}{b}e^{ax}\sin(bx) + \frac{a}{b^2}e^{ax}\cos(bx) - \frac{a^2}{b^2}\int e^{ax}\cos(bx)\,dx$$

Now solve for $\int e^{ax}\cos(bx)\,dx$ and simplify.

$$\int e^{ax}\cos(bx)\,dx + \frac{a^2}{b^2}\int e^{ax}\cos(bx)\,dx = \frac{1}{b}e^{ax}\sin(bx) + \frac{a}{b^2}e^{ax}\cos(bx)$$

$$\left(1 + \frac{a^2}{b^2}\right)\int e^{ax}\cos(bx)\,dx = \frac{1}{b^2}e^{ax}[b\sin(bx) + a\cos(bx)]$$

$$\int e^{ax}\cos(bx)\,dx = \frac{e^{ax}[b\sin(bx) + a\cos(bx)]}{a^2 + b^2} + C$$

**(b)** To find $\int e^{4x}\cos(5x)\,dx$, use (4) with $a = 4$ and $b = 5$.

$$\int e^{4x}\cos(5x)\,dx = \frac{e^{4x}[5\sin(5x) + 4\cos(5x)]}{41} + C \qquad \blacksquare$$

**NOW WORK** Problem 83.

---

**EXAMPLE 10**  **Deriving a General Formula**

Derive the formula

$$\boxed{\int \sec^n x\,dx = \frac{\sec^{n-2} x\,\tan x}{n-1} + \frac{n-2}{n-1}\int \sec^{n-2} x\,dx \qquad n \ge 3} \qquad (7)$$

**Solution** We begin by writing $\sec^n x = \sec^{n-2} x \sec^2 x$, and choosing

$$u = \sec^{n-2} x \qquad \text{and} \qquad dv = \sec^2 x\,dx$$

This choice makes $\int dv$ easy to integrate. Then

$$du = [(n-2)\sec^{n-3} x \cdot \sec x \tan x]\,dx = [(n-2)\sec^{n-2} x\,\tan x]\,dx$$

$$v = \int \sec^2 x\,dx = \tan x$$

Using integration by parts, we get

$$\int \sec^n x\,dx = \sec^{n-2} x\,\tan x - (n-2)\int \sec^{n-2} x\,\tan^2 x\,dx$$

To express the integrand on the right in terms of $\sec x$, use the trigonometric identity, $\tan^2 x + 1 = \sec^2 x$, and replace $\tan^2 x$ by $\sec^2 x - 1$, obtaining

$$\int \sec^n x\,dx = \sec^{n-2} x\,\tan x - (n-2)\int \sec^{n-2} x(\sec^2 x - 1)\,dx$$

$$\int \sec^n x\,dx = \sec^{n-2} x\,\tan x - (n-2)\int \sec^n x\,dx + (n-2)\int \sec^{n-2} x\,dx$$

Adding $(n-2)\int \sec^n x\,dx$ to each side and simplifying, we obtain

$$(n-1)\int \sec^n x\,dx = \sec^{n-2} x\,\tan x + (n-2)\int \sec^{n-2} x\,dx$$

Finally, divide both sides by $n - 1$:

$$\int \sec^n x\,dx = \frac{\sec^{n-2} x\,\tan x}{n-1} + \frac{n-2}{n-1}\int \sec^{n-2} x\,dx \qquad \blacksquare$$

Formula (7) is called a **reduction formula** because repeated applications of the formula eventually lead to an elementary integral. For this reduction formula, when $n$ is even, repeated applications lead eventually to

$$\int \sec^2 x \, dx = \tan x + C$$

When $n$ is odd, repeated applications eventually lead to the integral

$$\int \sec x \, dx = \ln|\sec x + \tan x| + C$$

For example, if $n = 3$,

$$\int \sec^3 x \, dx = \frac{\sec x \tan x}{2} + \frac{1}{2}\int \sec x \, dx = \frac{\sec x \tan x}{2} + \frac{1}{2}\ln|\sec x + \tan x| + C$$

NOW WORK **Problem 73.**

## 7.1 Assess Your Understanding

### Concepts and Vocabulary

1. *True or False* Integration by parts is based on the Product Rule for derivatives.

2. The integration by parts formula states that $\int u \, dv = $ _____.

### Skill Building

*In Problems 3–34, use integration by parts to find each integral.*

3. $\displaystyle\int xe^{2x}\,dx$

4. $\displaystyle\int x\,e^{-3x}\,dx$

5. $\displaystyle\int x\cos x\,dx$

6. $\displaystyle\int x\sin(3x)\,dx$

7. $\displaystyle\int \sqrt{x}\ln x\,dx$

8. $\displaystyle\int x^{-2}\ln x\,dx$

9. $\displaystyle\int \cot^{-1}x\,dx$

10. $\displaystyle\int \sin^{-1}x\,dx$

11. $\displaystyle\int (\ln x)^2\,dx$

12. $\displaystyle\int x(\ln x)^2\,dx$

13. $\displaystyle\int x^2\sin x\,dx$

14. $\displaystyle\int x^2\cos x\,dx$

15. $\displaystyle\int x\cos^2 x\,dx$

16. $\displaystyle\int x\sin^2 x\,dx$

17. $\displaystyle\int x\sinh x\,dx$

18. $\displaystyle\int x\cosh x\,dx$

19. $\displaystyle\int \cosh^{-1}x\,dx$

20. $\displaystyle\int \sinh^{-1}x\,dx$

21. $\displaystyle\int \sin(\ln x)\,dx$

22. $\displaystyle\int \cos(\ln x)\,dx$

23. $\displaystyle\int (\ln x)^3\,dx$

24. $\displaystyle\int (\ln x)^4\,dx$

25. $\displaystyle\int x^2(\ln x)^2\,dx$

26. $\displaystyle\int x^3(\ln x)^2\,dx$

27. $\displaystyle\int x^2\tan^{-1}x\,dx$

28. $\displaystyle\int x\tan^{-1}x\,dx$

29. $\displaystyle\int 7^x x\,dx$

30. $\displaystyle\int 2^{-x}x\,dx$

31. $\displaystyle\int e^{-x}\cos(2x)\,dx$

32. $\displaystyle\int e^{-2x}\sin(3x)\,dx$

33. $\displaystyle\int e^{2x}\sin x\,dx$

34. $\displaystyle\int e^{3x}\cos(5x)\,dx$

*In Problems 35–44, use integration by parts to find each definite integral.*

35. $\displaystyle\int_0^\pi e^x\cos x\,dx$

36. $\displaystyle\int_0^{\pi/2} e^{-x}\sin x\,dx$

37. $\displaystyle\int_0^2 x^2\,e^{-3x}\,dx$

38. $\displaystyle\int_0^1 x^2\,e^{-x}\,dx$

39. $\displaystyle\int_0^{\pi/4} x\sec x\tan x\,dx$

40. $\displaystyle\int_0^{\pi/4} x\tan^2 x\,dx$

41. $\displaystyle\int_1^9 \ln\sqrt{x}\,dx$

42. $\displaystyle\int_{\pi/4}^{3\pi/4} x\csc^2 x\,dx$

43. $\displaystyle\int_1^e (\ln x)^2\,dx$

44. $\displaystyle\int_0^{\pi/4} x\sec^2 x\,dx$

### Applications and Extensions

**Area Between Two Graphs** *In Problems 45 and 46, find the area of the region enclosed by the graphs of $f$ and $g$.*

45. $f(x) = 3\ln x$ and $g(x) = x\ln x$, $x \geq 1$

46. $f(x) = 4x\ln x$ and $g(x) = x^2\ln x$, $x \geq 1$

47. **Area Under a Graph** Find the area under the graph of $y = e^x\sin x$ from 0 to $\pi$.

48. **Area Under a Graph** Find the area under the graph of $y = x\cos x$ from $x = 0$ to $x = \dfrac{\pi}{2}$.

49. **Area Under a Graph** Find the area under the graph of $y = xe^{-x}$ from $x = 0$ to $x = 1$.

50. **Area Under a Graph** Find the area under the graph of $y = xe^{3x}$ from $x = 0$ to $x = 2$.

---

**1.** = NOW WORK problem        ⟨N⟩ = Graphing technology recommended        CAS = Computer Algebra System recommended

**51. Rectilinear Motion** The acceleration $a$ of an object in rectilinear motion is given by

$$a(t) = e^{-2t} \sin t \text{ m/s}^2$$

(a) Find the velocity $v$ of the object if the initial velocity is $v(0) = 8$ m/s.

(b) Find the position $s$ of the object if the initial position is $s(0) = 0$ m.

**52. Rectilinear Motion** The acceleration $a$ of an object in rectilinear motion is given by

$$a(t) = t^2 e^{-t} \text{ ft/s}^2$$

(a) Find the velocity $v$ of the object if the initial velocity is $v(0) = 5$ ft/s.

(b) Find the position $s$ of the object if the initial position is $s(0) = 0$ ft.

**53. Volume of a Solid of Revolution** Find the volume of the solid of revolution generated by revolving the region bounded by the graph of $y = \sin x$ and the $x$-axis from $x = 0$ to $x = \dfrac{\pi}{2}$ about the $y$-axis.

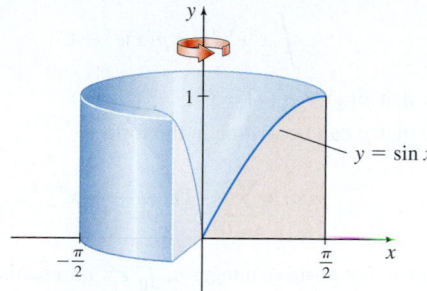

**54. Volume of a Solid of Revolution** Find the volume of the solid of revolution generated by revolving the region bounded by the graph of $y = \cos x$ and the $x$-axis from $x = 0$ to $x = \dfrac{\pi}{2}$ about the $y$-axis. See the figure below.

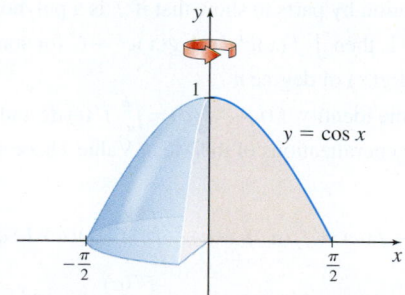

**55. Volume of a Solid of Revolution** Find the volume of the solid of revolution generated by revolving the region bounded by the graph of $y = \ln x$ and the $x$-axis from $x = 1$ to $x = e$ about the $x$-axis.

**56. Volume of a Solid of Revolution** Find the volume of the solid of revolution generated by revolving the region bounded by the graph of $y = x\sqrt{\sin x}$ and the $x$-axis from $x = 0$ to $x = \dfrac{\pi}{2}$ about the $x$-axis.

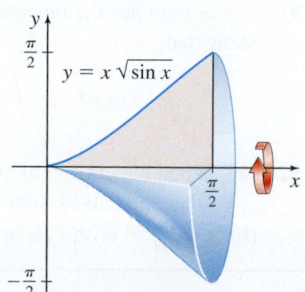

**57. Volume of a Solid of Revolution** Find the volume of the solid of revolution generated by revolving the region bounded by the graph of $y = x\sqrt{\ln x}$, the $x$-axis, and the lines $x = 1$ and $x = e^2$ about the $x$-axis.

**58. Volume of a Solid of Revolution** Find the volume of the solid of revolution generated by revolving the region bounded by the graph of $y = xe^{3x}$, the $x$-axis, and the line $x = 2$ about the $y$-axis.

**59.** A function $y = f(x)$ is continuous and differentiable on the interval $(0, 7)$. If $\int_1^5 f(x)\,dx = 10$, $f(1) = 2$, and $f(5) = -5$, find $\int_1^5 xf'(x)\,dx$.

**60.** A function $y = f(x)$ is continuous and differentiable on the interval $(2, 6)$. If $\int_3^5 f(x)\,dx = 18$ and $f(3) = 8$ and $f(5) = 11$, then find $\int_3^5 xf'(x)\,dx$.

**61.** The region $R$ is bounded by the graph of $y = \ln x$, the line $x = e$ and the $x$-axis.

(a) Find the area of the region $R$.

(b) Find the volume of the solid generated by revolving $R$ about the $y$-axis.

**62.** Use integration by parts to show that if $f'$ is continuous on a closed interval $[a, b]$ then

$$\int_a^b f(x)f'(x)\,dx = \frac{1}{2}\left\{[f(b)]^2 - [f(a)]^2\right\}$$

*In Problems 63–68, find each integral by first making a substitution and then integrating by parts.*

**63.** $\displaystyle \int \sin\sqrt{x}\,dx$

**64.** $\displaystyle \int e^{\sqrt{x}}\,dx$

**65.** $\displaystyle \int \cos x \ln(\sin x)\,dx$

**66.** $\displaystyle \int e^x \ln(2 + e^x)\,dx$

**67.** $\displaystyle \int e^{4x} \cos e^{2x}\,dx$

**68.** $\displaystyle \int \cos x \tan^{-1}(\sin x)\,dx$

**69.** Find $\displaystyle \int x^3 e^{x^2}\,dx$. (*Hint:* Let $u = x^2$, $dv = xe^{x^2}\,dx$.)

**70.** Find $\displaystyle \int x^n \ln x\,dx$; $n \neq -1$, $n$ real.

**71.** Find $\displaystyle \int xe^x \cos x\,dx$.

**72.** Find $\displaystyle \int xe^x \sin x\,dx$.

*In Problems 73–76, derive each reduction formula where $n > 1$ is an integer.*

**73.** $\displaystyle \int x^n \sin^{-1} x\,dx = \frac{x^{n+1}}{n+1}\sin^{-1}x - \frac{1}{n+1}\int \frac{x^{n+1}}{\sqrt{1-x^2}}\,dx$

**74.** $\displaystyle \int \frac{dx}{(x^2+1)^{n+1}} = \left(1 - \frac{1}{2n}\right)\int \frac{dx}{(x^2+1)^n} + \frac{x}{2n(x^2+1)^n}$

**75.** $\displaystyle \int \sin^n x\,dx = -\frac{\sin^{n-1}x \cos x}{n} + \frac{n-1}{n}\int \sin^{n-2}x\,dx$

**76.** $\displaystyle\int \sin^n x \cos^m x\, dx = -\frac{\sin^{n-1} x \cos^{m+1} x}{n+m}$

$$+ \frac{n-1}{n+m} \int \sin^{n-2} x \cos^m x\, dx$$

where $m \neq -n,\, m \neq -1$

**77. (a)** Find $\int x^2 e^{5x}\, dx$.

**(b)** Using integration by parts, derive a reduction formula for $\int x^n e^{kx}\, dx$, where $k \neq 0$ and $n \geq 2$ is an integer, in which the resulting integrand involves $x^{n-1}$.

**78. (a)** Assuming there is a function $p$ for which $\int x^3 e^x\, dx = p(x)e^x$, show that $p(x) + p'(x) = x^3$.

**(b)** Use integration by parts to find a polynomial $p$ of degree 3 for which $\int x^3 e^x\, dx = p(x)e^x + C$.

**79. (a)** Use integration by parts with $u = \sin x$ and $dv = \cos x\, dx$ to find a function $f$ for which $\int \sin x \cos x\, dx = f(x) + C_1$.

**(b)** Use integration by parts with $u = \cos x$ and $dv = \sin x\, dx$ to find a function $g$ for which $\int \sin x \cos x\, dx = g(x) + C_2$.

**(c)** Use the trigonometric identity $\sin(2x) = 2 \sin x \cos x$ and substitution to find a function $h$ for which

$$\int \sin x \cos x\, dx = h(x) + C_3.$$

**(d)** Compare the functions $f$ and $g$. Find a relationship between $C_1$ and $C_2$.

**(e)** Compare the functions $f$ and $h$. Find a relationship between $C_1$ and $C_3$.

**80. (a)** Graph the functions $f(x) = x^3 e^{-3x}$ and $g(x) = x^2 e^{-3x}$ on the same set of coordinate axes.

**(b)** Find the area enclosed by the graphs of $f$ and $g$.

**81. Damped Spring** The displacement $x$ of a damped spring at time $t$, $0 \leq t \leq 5$, is given by

$$x = x(t) = 3e^{-t} \cos(2t) + 2e^{-t} \sin(2t)$$

**(a)** Graph $x = x(t)$.

**(b)** Find the least positive number $t$ that satisfies $x(t) = 0$.

**(c)** Find the area under the graph of $x = x(t)$ from $t = 0$ to the value of $t$ found in (b).

**82.** Derive the formula

$$\int \ln\left(x + \sqrt{x^2 + a^2}\right) dx$$

$$= x \ln\left(x + \sqrt{x^2 + a^2}\right) - \sqrt{x^2 + a^2} + C$$

**83.** Derive the formula

$$\int e^{ax} \sin(bx)\, dx = \frac{e^{ax}[a \sin(bx) - b \cos(bx)]}{a^2 + b^2} + C,\, a > 0,\, b > 0$$

**84.** Suppose $F(x) = \int_0^x t\, g'(t)\, dt$ for all $x \geq 0$.

Show that $F(x) = x g(x) - \int_0^x g(t)\, dt$.

**85.** Use **Wallis' formulas**, given below, to find each definite integral.

$$\int_0^{\pi/2} \sin^n x\, dx = \int_0^{\pi/2} \cos^n x\, dx \quad n > 1 \text{ an integer}$$

$$= \begin{cases} \dfrac{(n-1)(n-3)\cdots(4)(2)}{n(n-2)\cdots(5)(3)(1)} & \text{if } n > 1 \text{ is odd} \\[2mm] \dfrac{(n-1)(n-3)\cdots(5)(3)(1)}{n(n-2)\cdots(4)(2)} \left(\dfrac{\pi}{2}\right) & \text{if } n > 1 \text{ is even} \end{cases}$$

**(a)** $\displaystyle\int_0^{\pi/2} \sin^6 x\, dx$     **(b)** $\displaystyle\int_0^{\pi/2} \sin^5 x\, dx$

**(c)** $\displaystyle\int_0^{\pi/2} \cos^8 x\, dx$     **(d)** $\displaystyle\int_0^{\pi/2} \cos^6 x\, dx$

## Challenge Problems

**86.** Derive Wallis' formulas given in Problem 85.
*Hint:* Use the result of Problem 75.

**87. (a)** If $n$ is a positive integer, use integration by parts to show that there is a polynomial $p$ of degree $n$ for which

$$\int x^n e^x\, dx = p(x)e^x + C$$

**(b)** Show that $p(x) + p'(x) = x^n$.

**(c)** Show that $p$ can be written in the form

$$p(x) = \sum_{k=0}^n (-1)^k \frac{n!}{(n-k)!} x^{n-k}$$

**88.** Show that for any positive integer $n$, $\int_0^1 e^{x^2}\, dx$ equals

$$e \cdot \left[1 - \frac{2}{3} + \frac{4}{15} - \frac{8}{105} + \cdots + \frac{(-1)^n 2^n}{(2n+1)(2n-1)\cdots 3 \cdot 1}\right]$$

$$+ (-1)^{n+1} \cdot \frac{2^{n+1}}{(2n+1)(2n-1)\cdots 3 \cdot 1} \int_0^1 x^{2n+2} e^{x^2}\, dx$$

**89.** Use integration by parts to show that if $f$ is a polynomial of degree $n \geq 1$, then $\int f(x)e^x\, dx = g(x)e^x + C$ for some polynomial $g(x)$ of degree $n$.

**90.** Start with the identity $f(b) - f(a) = \int_a^b f'(t)\, dt$ and derive the following generalizations of the Mean Value Theorem for Integrals:

**(a)** $\displaystyle f(b) - f(a) = f'(a)(b-a) - \int_a^b f''(t)(t-b)\, dt$

**(b)** $\displaystyle f(b) - f(a) = f'(a)(b-a) + \frac{f''(a)}{2}(b-a)^2$

$$+ \int_a^b \frac{f'''(t)}{2}(t-b)^2\, dt$$

**91.** If $y = f(x)$ has the inverse function given by $x = f^{-1}(y)$, show that

$$\int_a^b f(x)\, dx + \int_{f(a)}^{f(b)} f^{-1}(y)\, dy = bf(b) - af(a)$$

**92. (a)** When integration by parts is used to find $\int e^x \cosh x\, dx$, what happens? Explain.

**(b)** Find $\int e^x \cosh x\, dx$ without using integration by parts.

# 7.2 Integrals Containing Trigonometric Functions

**OBJECTIVES** *When you finish this section, you should be able to:*

1. Find integrals of the form $\int \sin^n x \, dx$ or $\int \cos^n x \, dx$, $n \geq 2$ an integer (p. 509)
2. Find integrals of the form $\int \sin^m x \cos^n x \, dx$ (p. 511)
3. Find integrals of the form $\int \tan^m x \sec^n x \, dx$ or $\int \cot^m x \csc^n x \, dx$ (p. 512)
4. Find integrals of the form $\int \sin(ax) \sin(bx) \, dx$, $\int \sin(ax) \cos(bx) \, dx$, or $\int \cos(ax) \cos(bx) \, dx$ (p. 515)

In this section, we develop techniques to find certain trigonometric integrals. When studying these techniques, concentrate on the strategies used in the examples rather than trying to memorize the results.

## 1 Find Integrals of the Form $\int \sin^n x \, dx$ or $\int \cos^n x \, dx$, $n \geq 2$ an Integer

Although we could use integration by parts to obtain reduction formulas for integrals of the form $\int \sin^n x \, dx$ or $\int \cos^n x \, dx$, $n \geq 2$ an integer, these integrals can be found more easily using trigonometric identities. We consider two cases:

- $n \geq 3$ an odd integer
- $n \geq 2$ an even integer.

Suppose we want to find $\int \sin^n x \, dx$ or $\int \cos^n x \, dx$ when $n \geq 3$ is an odd integer. We begin by factoring and writing each integral in the form

$$\int \sin^n x \, dx = \int \sin^{n-1} x \sin x \, dx \quad \text{or} \quad \int \cos^n x \, dx = \int \cos^{n-1} x \cos x \, dx$$

Since $n$ is odd, $(n-1)$ is even. Then use the Pythagorean identity $\sin^2 x + \cos^2 x = 1$.

- For $\int \sin^n x \, dx$, $n$ odd, the substitution $u = \cos x$, $du = -\sin x \, dx$, leads to an integral involving integer powers of $u$.
- For $\int \cos^n x \, dx$, $n$ odd, the substitution $u = \sin x$, $du = \cos x \, dx$ also leads to an integral involving integer powers of $u$.

CALC
CLIP
**EXAMPLE 1** Finding the Integral $\int \sin^5 x \, dx$

Find $\int \sin^5 x \, dx$.

**Solution** Since the exponent 5 is odd, write $\int \sin^5 x \, dx = \int \sin^4 x \sin x \, dx$, and use the identity $\sin^2 x = 1 - \cos^2 x$.

$$\int \sin^5 x \, dx = \int \sin^4 x \sin x \, dx = \int (\sin^2 x)^2 \sin x \, dx = \int (1 - \cos^2 x)^2 \sin x \, dx$$

$$= \int (1 - 2\cos^2 x + \cos^4 x) \sin x \, dx$$

**NEED TO REVIEW?** The method of substitution is discussed in Section 5.5, pp. 401–405.

Now use the substitution $u = \cos x$. Then $du = -\sin x \, dx$, and

$$\int \sin^5 x \, dx = -\int (1 - 2u^2 + u^4) \, du = -u + \frac{2}{3}u^3 - \frac{1}{5}u^5 + C$$

$$= -\cos x + \frac{2}{3}\cos^3 x - \frac{1}{5}\cos^5 x + C$$

■

EXAMPLE 2 Finding the Integral $\int \cos^3 x \, dx$

Find $\int \cos^3 x \, dx$.

Solution
$$\int \cos^3 x \, dx = \int \cos^2 x \cos x \, dx = \int (1 - \sin^2 x) \cos x \, dx$$

$$= \int (1 - u^2) \, du = u - \frac{u^3}{3} + C = \sin x - \frac{\sin^3 x}{3} + C$$

$$\underset{\substack{\uparrow \\ u = \sin x \\ du = \cos x \, dx}}{}$$

NOW WORK Problem 3.

To find $\int \sin^n x \, dx$ or $\int \cos^n x \, dx$ when $n \geq 2$ is an even integer, the preceding strategy does not work. (Try it for yourself.) Instead, we use one of the identities below:

NEED TO REVIEW? Trigonometric identities are discussed in Appendix A.4, pp. A-32 to A-35.

$$\sin^2 x = \frac{1 - \cos(2x)}{2} \qquad \cos^2 x = \frac{1 + \cos(2x)}{2}$$

to obtain a simpler integrand.

EXAMPLE 3 Finding the Integral $\int \sin^2 x \, dx$

Find $\int \sin^2 x \, dx$.

Solution Since the exponent of $\sin x$ is an even integer, we use the identity

$$\sin^2 x = \frac{1 - \cos(2x)}{2}$$

Then

$$\int \sin^2 x \, dx = \frac{1}{2} \int [1 - \cos(2x)] \, dx = \frac{1}{2} \int dx - \frac{1}{2} \int \cos(2x) \, dx$$

$$= \frac{1}{2} x + C_1 - \frac{1}{2} \int \cos u \frac{du}{2} \qquad u = 2x, \, du = 2 \, dx$$

$$= \frac{1}{2} x + C_1 - \frac{1}{4} \sin(2x) + C_2$$

NOTE Usually we will add the constant of integration at the end to avoid letting $C = C_1 + C_2$.

Since $C_1$ and $C_2$ are constants, we write the solution as

$$\int \sin^2 x \, dx = \frac{1}{2} x - \frac{1}{4} \sin(2x) + C$$

where $C = C_1 + C_2$. ∎

NOW WORK Problem 5.

EXAMPLE 4 Finding the Average Value of a Function

Find the average value $\bar{y}$ of the function $f(x) = \cos^4 x$ over the closed interval $[0, \pi]$.

NEED TO REVIEW? The average value of a function is discussed in Section 5.4, pp. 391–392.

Solution The average value $\bar{y}$ of a function $f$ over $[a, b]$ is $\bar{y} = \frac{1}{b - a} \int_a^b f(x) \, dx$.

For $f(x) = \cos^4 x$ on $[0, \pi]$, we have

$$\bar{y} = \frac{1}{\pi - 0} \int_0^\pi \cos^4 x \, dx = \frac{1}{\pi} \int_0^\pi (\cos^2 x)^2 \, dx = \frac{1}{4\pi} \int_0^\pi [1 + \cos(2x)]^2 \, dx$$

$$\underset{\substack{\uparrow \\ \cos^4 x = (\cos^2 x)^2}}{} \qquad \underset{\substack{\cos^2 x = \frac{1 + \cos(2x)}{2}}}{}$$

$$= \frac{1}{4\pi} \int_0^\pi [1 + 2\cos(2x) + \cos^2(2x)] \, dx$$

$$= \frac{1}{4\pi} \left[ \int_0^\pi dx + 2 \int_0^\pi \cos(2x) \, dx + \int_0^\pi \cos^2(2x) \, dx \right] \tag{1}$$

Now

$$\int_0^\pi dx = \pi - 0 = \pi \quad \text{and} \quad \underset{\underset{du=2\,dx}{u=2x}}{\int_0^\pi \cos(2x)\,dx = \int_0^{2\pi} \cos u\,\frac{du}{2}} = \left[\frac{1}{2}\sin u\right]_0^{2\pi} = 0$$

To find $\displaystyle\int_0^\pi \cos^2(2x)\,dx$, we use the identity $\cos^2\theta = \dfrac{1+\cos(2\theta)}{2}$ again to write $\cos^2(2x) = \dfrac{1+\cos(4x)}{2}$. Then

$$\int_0^\pi \cos^2(2x)\,dx = \int_0^\pi \frac{1+\cos(4x)}{2}\,dx = \frac{1}{2}\left[\int_0^\pi dx + \int_0^\pi \cos(4x)\,dx\right]$$

$$= \frac{1}{2}\left[\pi + \int_0^{4\pi} \cos u\,\frac{du}{4}\right] \qquad u = 4x;\ du = 4\,dx$$

$$= \frac{\pi}{2} + \frac{1}{8}[\sin u]_0^{4\pi} = \frac{\pi}{2}$$

So, from (1),

$$\bar{y} = \frac{1}{4\pi}\left[\pi + 0 + \frac{\pi}{2}\right] = \frac{3}{8} \qquad\blacksquare$$

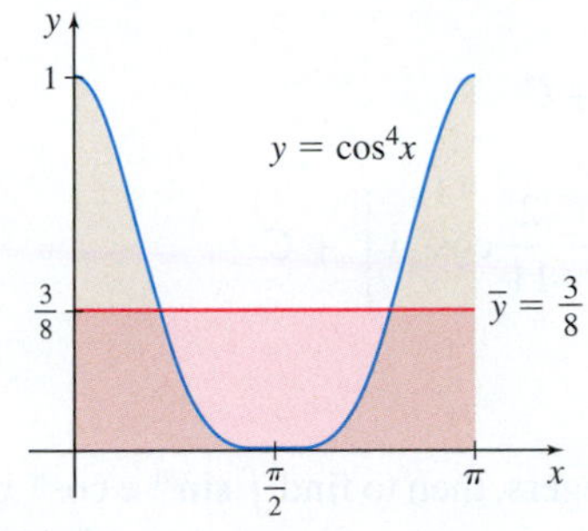

$y = \cos^4 x$

$\bar{y} = \dfrac{3}{8}$

**Figure 2** $\bar{y} \cdot \pi = \displaystyle\int_0^\pi \cos^4 x\,dx$

If $f$ is nonnegative on an interval $[a, b]$, the average value of $f$ over the interval $[a, b]$ represents the height of a rectangle with width $b - a$ whose area equals the area under the graph of $f$ from $a$ to $b$. Figure 2 shows the graph of $f$ from Example 4 and the rectangle of height $\bar{y} = \dfrac{3}{8}$ and base $\pi$ whose area is equal to the area under the graph of $f$.

**NOW WORK** Problem 61.

**Summary**    Integrals of the Form $\int \sin^n x\,dx$ or $\int \cos^n x\,dx$, $n \geq 2$ an Integer

|  | Integral | Use the Identity | u-substitution |
|---|---|---|---|
| $n$ odd | $\int \sin^n x\,dx = \int \sin^{n-1} x\,\sin x\,dx$ | $\sin^2 x = 1 - \cos^2 x$ | $u = \cos x;\quad du = -\sin x\,dx$ |
|  | $\int \cos^n x\,dx = \int \cos^{n-1} x\,\cos x\,dx$ | $\cos^2 x = 1 - \sin^2 x$ | $u = \sin x;\quad du = \cos x\,dx$ |
| $n$ even | $\int \sin^n x\,dx = \int \left[\dfrac{1-\cos(2x)}{2}\right]^{n/2} dx$ | $\sin^2 x = \dfrac{1-\cos(2x)}{2}$ | $u = 2x;\quad du = 2\,dx$ |
|  | $\int \cos^n x\,dx = \int \left[\dfrac{1+\cos(2x)}{2}\right]^{n/2} dx$ | $\cos^2 x = \dfrac{1+\cos(2x)}{2}$ | $u = 2x;\quad du = 2\,dx$ |

## 2 Find Integrals of the Form $\displaystyle\int \sin^m x \cos^n x\,dx$

Integrals of the form $\int \sin^m x \cos^n x\,dx$ are found using variations of previous techniques. We discuss two cases:

- At least one of the exponents $m$ or $n$ is a positive odd integer.
- Both exponents are positive even integers.

If at least one of the exponents is odd, the goal is to write the integral with only one sine factor and the other factors involving the cosine function, or with only one cosine factor and the other factors involving the sine function.

**EXAMPLE 5**    **Finding $\displaystyle\int \sin^m x \cos^n x\,dx$, One of the Exponents Is Odd**

Find $\displaystyle\int \sin^5 x\,\sqrt{\cos x}\,dx = \int \sin^5 x\,\cos^{1/2} x\,dx$.

**Solution** The exponent of $\sin x$ is 5, a positive, odd integer. We factor $\sin x$ from $\sin^5 x$ and write the rest of the integrand in terms of cosines.

$$\int \sin^5 x \cos^{1/2} x \, dx = \int \sin^4 x \, \cos^{1/2} x \sin x \, dx = \int (\sin^2 x)^2 \cos^{1/2} x \sin x \, dx$$

$$= \int (1 - \cos^2 x)^2 \cos^{1/2} x \sin x \, dx$$

Now use the substitution $u = \cos x$.

$$\int \sin^5 x \, \cos^{1/2} x \, dx = \underset{\substack{\uparrow \\ u = \cos x \\ du = -\sin x \, dx}}{\int (1 - \cos^2 x)^2 \cos^{1/2} x \, \sin x \, dx} = \int (1 - u^2)^2 u^{1/2} (-du)$$

$$= -\int (u^{1/2} - 2u^{5/2} + u^{9/2}) \, du = -\frac{2}{3} u^{3/2} + \frac{4}{7} u^{7/2} - \frac{2}{11} u^{11/2} + C$$

$$= u^{3/2} \left[ -\frac{2}{3} + \frac{4}{7} u^2 - \frac{2}{11} u^4 \right] + C$$

$$= \underset{\substack{\uparrow \\ u = \cos x}}{(\cos x)^{3/2}} \left[ -\frac{2}{3} + \frac{4}{7} \cos^2 x - \frac{2}{11} \cos^4 x \right] + C \qquad \blacksquare$$

**NOW WORK** Problem 11.

If both exponents $m$ and $n$ are positive even integers, then to find $\int \sin^m x \cos^n x \, dx$, use the Pythagorean identity $\sin^2 x + \cos^2 x = 1$ to obtain a sum of integrals, each integral involving only even powers of either $\sin x$ or $\cos x$. For example,

$$\int \sin^2 x \cos^4 x \, dx = \int (1 - \cos^2 x) \cos^4 x \, dx = \int \cos^4 x \, dx - \int \cos^6 x \, dx$$

The two integrals on the right are now of the form $\int \cos^n x \, dx$, $n$ a positive even integer, and we can integrate them using the techniques discussed in Examples 3 and 4.

---

**Summary**  Integrals of the form $\int \sin^m x \cos^n x \, dx$

| | Integral | Use the Identity | $u$-substitution |
|---|---|---|---|
| $m$ odd | $\int \sin^m x \cos^n x \, dx = \int \sin^{m-1} x \cos^n x \sin x \, dx$ | $\sin^2 x = 1 - \cos^2 x$ | $u = \cos x$<br>$du = -\sin x \, dx$ |
| $n$ odd | $\int \sin^m x \cos^n x \, dx = \int \sin^m x \cos^{n-1} x \cos x \, dx$ | $\cos^2 x = 1 - \sin^2 x$ | $u = \sin x$<br>$du = \cos x \, dx$ |
| $m$ and $n$ both even | $\int \sin^m x \cos^n x \, dx = \int (1 - \cos^2 x)^{m/2} \cos^n x \, dx$<br>or<br>$\int \sin^m x \cos^n x \, dx = \int \sin^m x (1 - \sin^2 x)^{n/2} \, dx$ | $\cos^2 x = \dfrac{1 + \cos(2x)}{2}$<br>$\sin^2 x = \dfrac{1 - \cos(2x)}{2}$ | $u = 2x; \quad du = 2\, dx$<br>$u = 2x; \quad du = 2\, dx$ |

---

## 3 Find Integrals of the Form $\displaystyle\int \tan^m x \, \sec^n x \, dx$

## or $\displaystyle\int \cot^m x \, \csc^n x \, dx$

We consider three cases involving integrals of the form $\int \tan^m x \, \sec^n x \, dx$:

- The tangent function is raised to a positive odd integer.
- The secant function is raised to a positive even integer.
- The tangent function is raised to a positive even integer $m$ and the secant function is raised to a positive odd integer $n$.

The idea is to express the integrand so that we can use either the substitution $u = \tan x$ and $du = \sec^2 x\, dx$ or the substitution $u = \sec x$ and $du = \sec x \tan x\, dx$, while leaving an even power of one of the functions. Then we use a Pythagorean identity to express the integrand in terms of only one trigonometric function.

**EXAMPLE 6**  Finding the Integral $\displaystyle\int \tan^3 x\ \sec^4 x\, dx$

Find $\displaystyle\int \tan^3 x\ \sec^4 x\, dx$.

**Solution** Here $\tan x$ is raised to the odd power 3. We factor $\tan x$ from $\tan^3 x$ and use the identity $\tan^2 x = \sec^2 x - 1$.

$$
\begin{aligned}
\int \tan^3 x\ \sec^4 x\, dx &= \int \tan^2 x\ \tan x\ \sec^4 x\, dx && \text{Factor } \tan x \text{ from } \tan^3 x. \\[2mm]
&= \int (\sec^2 x - 1)\tan x\ \sec^4 x\, dx && \tan^2 x = \sec^2 x - 1 \\[2mm]
&= \int (\sec^2 x - 1)\sec^3 x\ \sec x \tan x\, dx && \text{Factor } \sec x \text{ from } \sec^4 x. \\[2mm]
&= \int (u^2 - 1)u^3\, du && \begin{array}{l}\text{Substitute } u = \sec x; \\ du = \sec x \tan x\, dx.\end{array} \\[2mm]
&= \int (u^5 - u^3)\, du = \frac{u^6}{6} - \frac{u^4}{4} + C \\[2mm]
&= \frac{\sec^6 x}{6} - \frac{\sec^4 x}{4} + C && u = \sec x \qquad\blacksquare
\end{aligned}
$$

**NOW WORK** Problem **19**.

**EXAMPLE 7**  Finding the Integral $\displaystyle\int \tan^2 x\ \sec^6 x\, dx$

Find $\displaystyle\int \tan^2 x\ \sec^6 x\, dx$.

**Solution** Here $\sec x$ is raised to a positive even power. We factor $\sec^2 x$ from $\sec^6 x$ and use the identity $\sec^2 x = 1 + \tan^2 x$. Then

$$
\begin{aligned}
\int \tan^2 x\ \sec^6 x\, dx &= \int \tan^2 x\ \sec^4 x \cdot \sec^2 x\, dx && \text{Factor } \sec^2 x \text{ from } \sec^6 x. \\[2mm]
&= \int \tan^2 x(1 + \tan^2 x)^2 \sec^2 x\, dx && \sec^2 x = 1 + \tan^2 x \\[2mm]
&= \int u^2(1 + u^2)^2\, du && \begin{array}{l}\text{Substitute } u = \tan x; \\ du = \sec^2 x\, dx.\end{array} \\[2mm]
&= \int (u^2 + 2u^4 + u^6)\, du = \frac{u^3}{3} + \frac{2u^5}{5} + \frac{u^7}{7} + C \\[2mm]
&= \frac{\tan^3 x}{3} + 2\frac{\tan^5 x}{5} + \frac{\tan^7 x}{7} + C && u = \tan x \qquad\blacksquare
\end{aligned}
$$

**NOW WORK** Problem **21**.

When the tangent function is raised to a positive even integer $m$ and the secant function is raised to a positive odd integer $n$, the approach is slightly different. Rather than factoring, we begin by using the identity $\tan^2 x = \sec^2 x - 1$.

EXAMPLE 8 **Finding $\int \tan^m x \sec^n x\, dx$ Where $m$ Is Even and $n$ Is Odd**

Find $\int \tan^2 x \sec x\, dx$.

**Solution** Here $\tan x$ is raised to an even power and $\sec x$ to an odd power. We use the identity $\tan^2 x = \sec^2 x - 1$ to write

$$\int \tan^2 x \sec x\, dx = \int (\sec^2 x - 1) \sec x\, dx = \int (\sec^3 x - \sec x)dx$$

$$= \int \sec^3 x\, dx - \int \sec x\, dx \qquad (2)$$

Next we integrate $\int \sec^3 x\, dx$ by parts. Choose

$$u = \sec x \qquad\qquad \text{and} \qquad dv = \sec^2 x\, dx$$
$$du = \sec x \tan x\, dx \qquad\qquad v = \int \sec^2 x\, dx = \tan x$$

Then

$$\int \sec^3 x\, dx = \sec x \tan x - \int \tan^2 x \sec x\, dx \qquad \color{blue}{\int u\, dv = uv - \int v\, du}$$

$$= \sec x \tan x - \int (\sec^2 x - 1) \sec x\, dx \qquad \color{blue}{\tan^2 x = \sec^2 x - 1}$$

$$= \sec x \tan x - \int \sec^3 x\, dx + \int \sec x\, dx \qquad \color{blue}{\text{Write the integral as the sum of two integrals.}}$$

$$2\int \sec^3 x\, dx = \sec x \tan x + \int \sec x\, dx \qquad \color{blue}{\text{Add } \int \sec^3 x\, dx \text{ to both sides.}}$$

$$\int \sec^3 x\, dx = \frac{1}{2}[\sec x \tan x + \ln |\sec x + \tan x|] \qquad \color{blue}{\text{Solve for } \int \sec^3 x\, dx;}$$
$$\color{blue}{\int \sec x\, dx = \ln |\sec x + \tan x|.}$$

**NOTE** Substituting $n = 3$ into the reduction formula for $\int \sec^n x\, dx$ (derived in Example 10 of Section 7.1) could also have been used to find $\int \sec^3 x\, dx$.

Now substitute this result in (2).

$$\int \tan^2 x \sec x\, dx = \int \sec^3 x\, dx - \int \sec x\, dx$$

$$= \frac{1}{2}[\sec x \tan x + \ln |\sec x + \tan x|] - \ln |\sec x + \tan x| + C$$

$$= \frac{1}{2}[\sec x \tan x - \ln |\sec x + \tan x|] + C \qquad ■$$

NOW WORK Problem 51.

To find integrals of the form $\int \cot^m x \csc^n x\, dx$, use the same strategies, but with the identity $\csc^2 x = 1 + \cot^2 x$.

---

## Summary    Integrals of the form $\int \tan^m x \sec^n x\, dx$

| | The Integral | Use the identity | Method |
|---|---|---|---|
| $m$ odd | $\int \tan^m x \sec^n x\, dx = \int \tan^{m-1} x \sec^{n-1} x \sec x \tan x\, dx$ | $\tan^2 x = \sec^2 x - 1$ | $u = \sec x$ <br> $du = \sec x \tan x\, dx$ |
| $n$ even | $\int \tan^m x \sec^n x\, dx = \int \tan^m x \sec^{n-2} x \sec^2 x\, dx$ | $\sec^2 x = 1 + \tan^2 x$ | $u = \tan x$ <br> $du = \sec^2 x\, dx$ |
| $m$ even; $n$ odd | $\int \tan^m x \sec^n x\, dx = \int (\sec^2 x - 1)^{m/2} \sec^n x\, dx$ | $\tan^2 x = \sec^2 x - 1$ | Use integration by parts |

## 4 Find Integrals of the Form $\int \sin(ax)\sin(bx)\,dx$, $\int \sin(ax)\cos(bx)\,dx$, or $\int \cos(ax)\cos(bx)\,dx$

Trigonometric integrals of the form

$$\int \sin(ax)\sin(bx)\,dx \qquad \int \sin(ax)\cos(bx)\,dx \qquad \int \cos(ax)\cos(bx)\,dx$$

are integrated using the product-to-sum identities:

- $2\sin A \sin B = \cos(A-B) - \cos(A+B)$
- $2\sin A \cos B = \sin(A+B) + \sin(A-B)$
- $2\cos A \cos B = \cos(A-B) + \cos(A+B)$

These identities transform the integrand into a sum of sines and/or cosines.

**EXAMPLE 9**  Finding the Integral $\int \sin(3x)\sin(2x)\,dx$

Find $\int \sin(3x)\sin(2x)\,dx$.

**Solution**  Use the product-to-sum identity $2\sin A \sin B = \cos(A-B) - \cos(A+B)$.
Then

$$2\sin(3x)\sin(2x) = \cos(3x-2x) - \cos(3x+2x) \qquad A=3x,\ B=2x$$

$$\sin(3x)\sin(2x) = \frac{1}{2}[\cos x - \cos(5x)]$$

Then

$$\int \sin(3x)\sin(2x)\,dx = \frac{1}{2}\int [\cos x - \cos(5x)]\,dx = \frac{1}{2}\int \cos x\,dx - \frac{1}{2}\int \cos(5x)\,dx$$

$$\underset{\substack{\uparrow \\ u=5x \\ du=5\,dx}}{=} \frac{1}{2}\sin x - \frac{1}{2}\int \cos u\,\frac{du}{5} = \frac{1}{2}\sin x - \frac{1}{10}\sin(5x) + C \qquad\blacksquare$$

NOW WORK Problem 27.

## 7.2 Assess Your Understanding

### Concepts and Vocabulary

**1.** *True or False*  To find $\int \cos^5 x\,dx$, factor out $\cos x$ and use the identity $\cos^2 x = 1 - \sin^2 x$.

**2.** *True or False*  To find $\int \sin(2x)\cos(3x)\,dx$, use a product-to-sum identity.

### Skill Building

*In Problems 3–10, find each integral.*

**3.** $\displaystyle\int \cos^5 x\,dx$

**4.** $\displaystyle\int \sin^3 x\,dx$

**5.** $\displaystyle\int \sin^6 x\,dx$

**6.** $\displaystyle\int \cos^4 x\,dx$

**7.** $\displaystyle\int \sin^2(\pi x)\,dx$

**8.** $\displaystyle\int \cos^4(2x)\,dx$

**9.** $\displaystyle\int_0^\pi \cos^5 x\,dx$

**10.** $\displaystyle\int_{-\pi/3}^{\pi/3} \sin^3 x\,dx$

*In Problems 11–18, find each integral.*

**11.** $\displaystyle\int \sin^3 x \cos^2 x\,dx$

**12.** $\displaystyle\int \sin^4 x \cos^3 x\,dx$

**13.** $\displaystyle\int_0^{\pi/2} \sin^3 x(\cos x)^{3/2}\,dx$

**14.** $\displaystyle\int_0^{\pi/2} \cos^3 x\,\sqrt{\sin x}\,dx$

**15.** $\displaystyle\int \sin^3 x \cos^{1/3} x\,dx$

**16.** $\displaystyle\int \cos^3 x \sin^{1/2} x\,dx$

**17.** $\displaystyle\int \sin^2\left(\frac{x}{2}\right)\cos^3\left(\frac{x}{2}\right)\,dx$

**18.** $\displaystyle\int \sin^3(4x)\cos^3(4x)\,dx$

*In Problems 19–26, find each integral.*

**19.** $\displaystyle\int \tan^3 x \sec^3 x\,dx$

**20.** $\displaystyle\int \tan x \sec^5 x\,dx$

**21.** $\displaystyle\int \tan^{3/2} x \sec^4 x\,dx$

**22.** $\displaystyle\int \tan^5 x \sec x\,dx$

**23.** $\displaystyle\int \tan^3 x(\sec x)^{3/2}\,dx$

**24.** $\displaystyle\int \sec^4 x\,\sqrt{\tan x}\,dx$

**25.** $\displaystyle\int \cot^3 x \csc x\,dx$

**26.** $\displaystyle\int \cot^3 x \csc^2 x\,dx$

**1.** = NOW WORK problem          📈 = Graphing technology recommended          CAS = Computer Algebra System recommended

*In Problems 27–34, find each integral.*

**27.** $\displaystyle\int \sin(3x)\cos x\,dx$

**28.** $\displaystyle\int \sin x\cos(3x)\,dx$

**29.** $\displaystyle\int \cos x\cos(3x)\,dx$

**30.** $\displaystyle\int \cos(2x)\cos x\,dx$

**31.** $\displaystyle\int \sin(2x)\sin(4x)\,dx$

**32.** $\displaystyle\int \sin(3x)\sin x\,dx$

**33.** $\displaystyle\int_0^{\pi/2} \sin(2x)\sin x\,dx$

**34.** $\displaystyle\int_0^{\pi} \cos x\cos(4x)\,dx$

*In Problems 35–56, find each integral.*

**35.** $\displaystyle\int_0^{\pi/2} \sin^2 x\cos^5 x\,dx$

**36.** $\displaystyle\int_0^{\pi/2} \sin^3 x\cos^{1/2} x\,dx$

**37.** $\displaystyle\int \frac{\sin^3 x\,dx}{\cos^2 x}$

**38.** $\displaystyle\int \frac{\cos x\,dx}{\sin^4 x}$

**39.** $\displaystyle\int_0^{\pi/3} \cos^3(3x)\,dx$

**40.** $\displaystyle\int_0^{\pi/3} \sin^5(3x)\,dx$

**41.** $\displaystyle\int_0^{\pi} \sin^3 x\cos^5 x\,dx$

**42.** $\displaystyle\int_0^{\pi/2} \sin^3 x\cos^3 x\,dx$

**43.** $\displaystyle\int \tan^3 x\,dx$

**44.** $\displaystyle\int \cot^5 x\,dx$

**45.** $\displaystyle\int \frac{\sec^6 x}{\tan^3 x}\,dx$

**46.** $\displaystyle\int \tan^{1/2} x\sec^2 x\,dx$

**47.** $\displaystyle\int \csc^4 x\cot^3 x\,dx$

**48.** $\displaystyle\int \cot^3 x\csc^2 x\,dx$

**49.** $\displaystyle\int \cot(2x)\csc^4(2x)\,dx$

**50.** $\displaystyle\int \cot^2(2x)\csc^3(2x)\,dx$

**51.** $\displaystyle\int_0^{\pi/4} \tan^4 x\sec^3 x\,dx$

**52.** $\displaystyle\int_0^{\pi/4} \tan^2 x\sec x\,dx$

**53.** $\displaystyle\int_0^{\pi/2} \sin\frac{x}{2}\cos\frac{3x}{2}\,dx$

**54.** $\displaystyle\int_0^{\pi/4} \cos(-x)\sin(4x)\,dx$

**55.** $\displaystyle\int \sin\frac{x}{2}\sin\frac{3x}{2}\,dx$

**56.** $\displaystyle\int \cos(\pi x)\cos(3\pi x)\,dx$

## Applications and Extensions

**57. Volume of a Solid of Revolution** Find the volume of the solid of revolution generated by revolving the region bounded by the graph of $y = \sin x$ and the $x$-axis from $x = 0$ to $x = \pi$ about the $x$-axis. See the figure below.

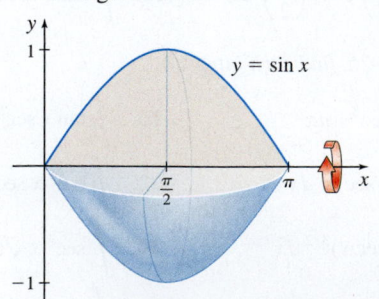

**58. Volume of a Solid of Revolution** Find the volume of the solid of revolution generated by revolving the region bounded by the graphs of $y = \cos x$, $y = \sin x$, and $x = 0$ from $x = 0$ to $x = \dfrac{\pi}{4}$ about the $x$-axis.

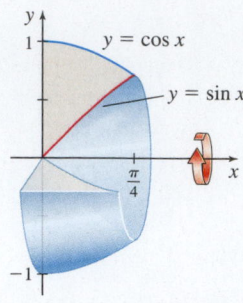

**59. Volume of a Solid of Revolution** Find the volume of the solid of revolution generated by revolving the region bounded by the graph of $y = \sin x(\cos x)^{3/2}$ and the $x$-axis from $x = 0$ to $x = \dfrac{\pi}{2}$ about the $x$-axis.

**60. Volume of a Solid of Revolution** Find the volume of the solid of revolution generated by revolving the region bounded by the graph of $y = (\sin x)^{3/2}\cos^2 x$ and the $x$-axis from $x = 0$ to $x = \dfrac{\pi}{2}$ about the $x$-axis.

**61. Average Value**

(a) Find the average value of $f(x) = \sin^3 x$ over the interval $\left[0, \dfrac{\pi}{2}\right]$.

(b) Give a geometric interpretation to the average value.

(c) Use technology to graph $f$ and the average value on the same screen.

**62. Average Value**

(a) Find the average value of $f(x) = \sin(4x)\cos(2x)$ over the interval $\left[0, \dfrac{\pi}{2}\right]$.

(b) Give a geometric interpretation to the average value.

(c) Use technology to graph $f$ and the average value on the same screen.

**63. Rectilinear Motion** An object in rectilinear motion is moving along a horizontal line with velocity $v(t) = \sin^2 t\cos^2 t$ (in meters per second) for time $0 \le t \le 2\pi$ seconds. Find the net displacement of the object from the origin from $t = 0$ to $t = 2\pi$.

**64. Rectilinear Motion** The acceleration $a$ of an object at time $t$ is given by $a(t) = \cos^3 t\sin^2 t$ m/s$^2$. At $t = 0$, its velocity is 5 m/s. Find its velocity at any time $t$.

**65. Area and Volume** Let $A$ be the area of the region in the first quadrant bounded by the graph of $y = \tan x$ and the lines $x = 0$ and $y = 1$.

(a) Find $A$.

(b) Find the volume of the solid of revolution generated by revolving the region about the $x$-axis.

**66. Area and Volume** Let $A$ be the area of the region in the first quadrant bounded by the graph of $y = \cot x$ from $x = \dfrac{\pi}{4}$ to $x = \dfrac{\pi}{2}$.

(a) Find $A$.

(b) Find the volume of the solid of revolution generated by revolving the region about the $x$-axis.

**67.** Find $\int \sin^4 x \, dx$.

    **(a)** Using the methods of this section.

    **(b)** Using the reduction formula given in Problem 75 in Section 7.1.

    **(c)** Verify that both results are equivalent.

    **CAS (d)** Use a CAS to find $\int \sin^4 x \, dx$.

**68.** **(a)** Use technology to graph the function $f(x) = \sin^n x$, $0 \le x \le \pi$, for $n = 5$, $n = 10$, $n = 20$, and $n = 50$.

    **(b)** Find $\int_0^\pi \sin^n x \, dx$ correct to three decimal places for $n = 5$, $n = 10$, $n = 20$, and $n = 50$.

    **(c)** What does (a) suggest about the shape of the graph of $f(x) = \sin^n x$, $0 \le x \le \pi$, as $n \to \infty$?

    **CAS (d)** Find $\lim_{n \to \infty} \int_0^\pi \sin^n x \, dx$.

**69.** Derive a formula for $\int \sin(mx) \sin(nx) \, dx$, $m \ne n$.

**70.** Derive a formula for $\int \sin(mx) \cos(nx) \, dx$, $m \ne n$.

**71.** Derive a formula for $\int \cos(mx) \cos(nx) \, dx$, $m \ne n$.

**Challenge Problems** ———

**72.** Use an appropriate substitution to show that

$$\int_0^{\pi/2} \sin^n \theta \, d\theta = \int_0^{\pi/2} \cos^n \theta \, d\theta$$

**73.** Use the substitution $\sqrt{x} = \sin y$ to find $\int_0^{1/2} \dfrac{\sqrt{x}}{\sqrt{1-x}} \, dx$.

**74.** **(a)** What is wrong with the following?

$$\int_0^\pi \cos^4 x \, dx = \int_0^\pi (\cos x)^3 \cos x \, dx = \int_0^\pi (\cos^2 x)^{3/2} \cos x \, dx$$

$$= \int_0^\pi (1 - \sin^2 x)^{3/2} \cos x \, dx = \int_0^0 (1 - u^2)^{3/2} du = 0$$

    **(b)** Find $\int_0^\pi \cos^4 x \, dx$.

# 7.3 Integration Using Trigonometric Substitution: Integrands Containing $\sqrt{a^2 - x^2}$, $\sqrt{x^2 + a^2}$, or $\sqrt{x^2 - a^2}$, $a > 0$

**OBJECTIVES** *When you finish this section, you should be able to:*

**1** Integrate a function containing $\sqrt{a^2 - x^2}$ (p. 518)

**2** Integrate a function containing $\sqrt{x^2 + a^2}$ (p. 519)

**3** Integrate a function containing $\sqrt{x^2 - a^2}$ (p. 520)

**4** Use trigonometric substitution to find definite integrals (p. 521)

When an integrand contains a square root of the form $\sqrt{a^2 - x^2}$, $\sqrt{x^2 + a^2}$, or $\sqrt{x^2 - a^2}$, $a > 0$, an appropriate trigonometric substitution will eliminate the square root and sometimes transform the integral into a trigonometric integral like those studied earlier.

    The substitutions to use for each of the three types of square roots are given in Table 2.

**TABLE 2** Trigonometric Substitution

| Integrand Contains | Substitution | Based on the Identity |
|---|---|---|
| $\sqrt{a^2 - x^2}$ | $x = a \sin\theta$, $-\dfrac{\pi}{2} \le \theta \le \dfrac{\pi}{2}$ | $1 - \sin^2\theta = \cos^2\theta$ |
| $\sqrt{x^2 + a^2}$ | $x = a \tan\theta$, $-\dfrac{\pi}{2} < \theta < \dfrac{\pi}{2}$ | $\tan^2\theta + 1 = \sec^2\theta$ |
| $\sqrt{x^2 - a^2}$ | $x = a \sec\theta$, $0 \le \theta < \dfrac{\pi}{2}$, $\pi \le \theta < \dfrac{3\pi}{2}$ | $\sec^2\theta - 1 = \tan^2\theta$ |

**CAUTION** Be sure to use the restrictions on each substitution. They guarantee the substitution is a one-to-one function, which is a requirement for using substitution.

**NEED TO REVIEW?** Right triangle trigonometry is discussed in Appendix A.4, pp. A-28 to A-31.

    Although the substitutions to use can be memorized, it is often easier to draw a right triangle and derive them as needed. Each substitution is based on the Pythagorean Theorem. By placing the sides $a$ and $x$ on a right triangle appropriately, the third side of the triangle will represent one of the three types of square roots, as shown in Figure 3.

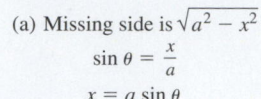
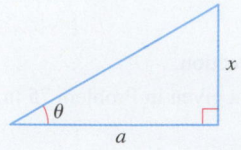
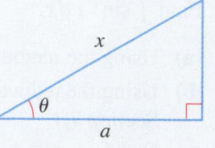

(a) Missing side is $\sqrt{a^2 - x^2}$
$$\sin\theta = \frac{x}{a}$$
$$x = a\sin\theta$$

(b) Missing side is $\sqrt{x^2 + a^2}$
$$\tan\theta = \frac{x}{a}$$
$$x = a\tan\theta$$

(c) Missing side is $\sqrt{x^2 - a^2}$
$$\sec\theta = \frac{x}{a}$$
$$x = a\sec\theta$$

**Figure 3**

## 1 Integrate a Function Containing $\sqrt{a^2 - x^2}$

When an integrand contains a square root of the form $\sqrt{a^2 - x^2}$, $a > 0$, we use the substitution $x = a\sin\theta$, $-\dfrac{\pi}{2} \le \theta \le \dfrac{\pi}{2}$. This substitution eliminates the square root as follows:

$$\sqrt{a^2 - x^2} = \sqrt{a^2 - a^2\sin^2\theta} \qquad \text{Let } x = a\sin\theta;\ -\frac{\pi}{2} \le \theta \le \frac{\pi}{2}.$$

$$= a\sqrt{1 - \sin^2\theta} \qquad \text{Factor out } a^2;\ a > 0.$$

$$= a\sqrt{\cos^2\theta} \qquad 1 - \sin^2\theta = \cos^2\theta$$

$$= a\cos\theta \qquad \cos\theta \ge 0 \text{ since } -\frac{\pi}{2} \le \theta \le \frac{\pi}{2}$$

**EXAMPLE 1** Integrating a Function That Contains $\sqrt{4 - x^2}$

Find $\displaystyle\int \frac{dx}{x^2\sqrt{4 - x^2}}$.

**Solution** The integrand contains the square root $\sqrt{4 - x^2}$ that is of the form $\sqrt{a^2 - x^2}$, where $a = 2$. Use the substitution $x = 2\sin\theta$, $-\dfrac{\pi}{2} < \theta < \dfrac{\pi}{2}$. Then $dx = 2\cos\theta\,d\theta$. Since

$$\underset{\underset{x=2\sin\theta}{\uparrow}}{\sqrt{4 - x^2}} = \sqrt{4 - 4\sin^2\theta} = 2\sqrt{1 - \sin^2\theta} = 2\underset{\underset{\cos\theta>0 \text{ since } -\frac{\pi}{2}<\theta<\frac{\pi}{2}}{\uparrow}}{\sqrt{\cos^2\theta}} = 2\cos\theta$$

**NOTE** We exclude $\theta = -\dfrac{\pi}{2}$ and $\theta = \dfrac{\pi}{2}$ because they lead to $x = \pm 2$, resulting in $\sqrt{4 - x^2} = 0$.

we have

$$\int \frac{dx}{x^2\sqrt{4 - x^2}} = \int \frac{2\cos\theta\,d\theta}{(4\sin^2\theta)(2\cos\theta)} = \int \frac{d\theta}{4\sin^2\theta} = \frac{1}{4}\int \csc^2\theta\,d\theta = -\frac{1}{4}\cot\theta + C$$

The original integral is a function of $x$, but the solution above is a function of $\theta$. To express $\cot\theta$ in terms of $x$, refer to the right triangles drawn in Figure 4.

Using the Pythagorean Theorem, the third side of each triangle is $\sqrt{4 - x^2}$. So,

$$\cot\theta = \frac{\sqrt{4 - x^2}}{x} \qquad -\frac{\pi}{2} < \theta < \frac{\pi}{2}$$

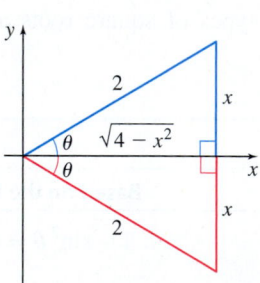

**Figure 4** $\sin\theta = \dfrac{x}{2}$, $-\dfrac{\pi}{2} < \theta < \dfrac{\pi}{2}$

Then

$$\int \frac{dx}{x^2\sqrt{4 - x^2}} = -\frac{1}{4}\cot\theta + C = -\frac{1}{4}\frac{\sqrt{4 - x^2}}{x} + C = -\frac{\sqrt{4 - x^2}}{4x} + C \qquad \blacksquare$$

Alternatively, trigonometric identities can be used to express $\cot\theta$ in terms of $x$. Using identities,

$$\cot\theta = \frac{\cos\theta}{\sin\theta} \underset{\substack{\uparrow \\ \cos^2\theta = 1 - \sin^2\theta \\ \cos\theta > 0}}{=} \frac{\sqrt{1 - \sin^2\theta}}{\sin\theta} \underset{\substack{\uparrow \\ x = 2\sin\theta \\ \sin\theta = \frac{x}{2}}}{=} \frac{\sqrt{1 - \left(\frac{x}{2}\right)^2}}{\frac{x}{2}} = \frac{2\sqrt{1 - \frac{x^2}{4}}}{x} = \frac{\sqrt{4 - x^2}}{x}$$

**NOW WORK** Problem 7.

## 2 Integrate a Function Containing $\sqrt{x^2 + a^2}$

When an integrand contains a square root of the form $\sqrt{x^2 + a^2}$, $a > 0$, use the substitution $x = a \tan\theta$, $-\dfrac{\pi}{2} < \theta < \dfrac{\pi}{2}$. Substituting $x = a \tan\theta$, $-\dfrac{\pi}{2} < \theta < \dfrac{\pi}{2}$, in the expression $\sqrt{x^2 + a^2}$ eliminates the square root as follows:

$$\sqrt{x^2 + a^2} = \underset{\underset{x = a \tan\theta}{\uparrow}}{\sqrt{a^2 \tan^2\theta + a^2}} = \underset{\underset{\substack{\text{Factor out } a,\\ a > 0}}{\uparrow}}{a\sqrt{\tan^2\theta + 1}} = a\sqrt{\sec^2\theta} = \underset{\underset{\substack{\sec\theta > 0\\ \text{since } -\frac{\pi}{2} < \theta < \frac{\pi}{2}}}{\uparrow}}{a \sec\theta}$$

---

**EXAMPLE 2**  Integrating a Function That Contains $\sqrt{x^2 + 9}$

Find $\displaystyle\int \frac{dx}{(x^2 + 9)^{3/2}}$.

**Solution**  The integral contains a square root $(x^2 + 9)^{3/2} = \left(\sqrt{x^2 + 9}\right)^3$ that is of the form $\sqrt{x^2 + a^2}$, where $a = 3$. Use the substitution $x = 3 \tan\theta$, $-\dfrac{\pi}{2} < \theta < \dfrac{\pi}{2}$.

Then $dx = 3 \sec^2\theta \, d\theta$. Since

$$(x^2 + 9)^{3/2} = \underset{\underset{x = 3 \tan\theta}{\uparrow}}{(9 \tan^2\theta + 9)^{3/2}} = 9^{3/2}(\tan^2\theta + 1)^{3/2} = 27\underset{\underset{\tan^2\theta + 1 = \sec^2\theta}{\uparrow}}{(\sec^2\theta)^{3/2}} = 27\underset{\underset{\sec\theta > 0}{\uparrow}}{\sec^3\theta}$$

we have

$$\int \frac{dx}{(x^2 + 9)^{3/2}} = \int \frac{3 \sec^2\theta \, d\theta}{27 \sec^3\theta} = \frac{1}{9}\int \frac{d\theta}{\sec\theta} = \frac{1}{9}\int \cos\theta \, d\theta = \frac{1}{9}\sin\theta + C$$

To express the solution in terms of $x$, use either the right triangles in Figure 5 or identities.

From the right triangles, the hypotenuse is $\sqrt{x^2 + 9}$. So, $\sin\theta = \dfrac{x}{\sqrt{x^2 + 9}}$. Then

$$\int \frac{dx}{(x^2 + 9)^{3/2}} = \frac{1}{9}\sin\theta + C = \frac{x}{9\sqrt{x^2 + 9}} + C$$  ■

**NOW WORK** Problem 15.

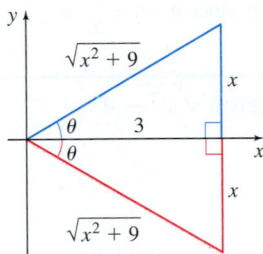

**Figure 5** $\tan\theta = \dfrac{x}{3}$, $-\dfrac{\pi}{2} < \theta < \dfrac{\pi}{2}$

---

**EXAMPLE 3**  Finding $\displaystyle\int (4x^2 + 9)^{1/2}\, dx$

Find $\displaystyle\int (4x^2 + 9)^{1/2} dx$.

**RECALL** $\displaystyle\int \sec^3\theta \, d\theta = \dfrac{1}{2}\left[\sec\theta \tan\theta + \ln|\sec\theta + \tan\theta|\right] + C$
Either integrate by parts, or use the reduction formula.

**Solution** $\displaystyle\int (4x^2 + 9)^{1/2} dx = \int \sqrt{(2x)^2 + 3^2}\, dx$

Use the substitution $2x = 3 \tan\theta$, $-\dfrac{\pi}{2} < \theta < \dfrac{\pi}{2}$. Then $dx = \dfrac{3}{2}\sec^2\theta \, d\theta$ and

$$\int (4x^2 + 9)^{1/2} dx = \frac{3}{2}\int \sqrt{9 \tan^2\theta + 9}\, \sec^2\theta \, d\theta$$

$$= \frac{9}{2}\int \sqrt{\tan^2\theta + 1}\, \sec^2\theta \, d\theta = \frac{9}{2}\int \sec^3\theta \, d\theta$$

$$= \frac{9}{2}\cdot\frac{1}{2}\left[\sec\theta \tan\theta + \ln|\sec\theta + \tan\theta|\right] + C$$

To express the solution in terms of $x$, refer to the right triangles drawn in Figure 6.

Using the Pythagorean Theorem, the hypotenuse of each triangle is $\sqrt{4x^2 + 9}$. So,

$$\sec\theta = \frac{\sqrt{4x^2 + 9}}{3} \quad\text{and}\quad \tan\theta = \frac{2x}{3} \qquad -\frac{\pi}{2} < \theta < \frac{\pi}{2}$$

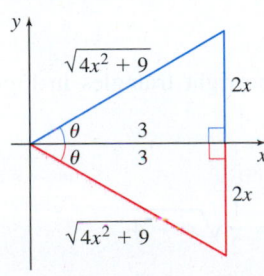

**Figure 6** $\tan\theta = \dfrac{2x}{3}$, $-\dfrac{\pi}{2} < \theta < \dfrac{\pi}{2}$

**NOTE** We can also find an integral containing $\sqrt{x^2 + a^2}$, $a > 0$, using the substitution $x = a \sinh \theta$, since

$$\sqrt{x^2 + a^2} = \sqrt{a^2 \sinh^2 \theta + a^2}$$

$$= a\sqrt{\sinh^2 \theta + 1}$$

$$= a\sqrt{\cosh^2 \theta} = a\cosh \theta$$

Try it!

Then

$$\int (4x^2 + 9)^{1/2} dx = \frac{9}{4}\left[\sec\theta\tan\theta + \ln|\sec\theta + \tan\theta|\right] + C$$

$$= \frac{9}{4}\left[\frac{\sqrt{4x^2 + 9}}{3} \cdot \frac{2x}{3} + \ln\left|\frac{\sqrt{4x^2 + 9}}{3} + \frac{2x}{3}\right|\right] + C$$

$$= \frac{9}{4}\left[\frac{2x\sqrt{4x^2 + 9}}{9} + \ln\frac{2x + \sqrt{4x^2 + 9}}{3}\right] + C \qquad ■$$

In general, if the integral contains $\sqrt{b^2x^2 + a^2}$, use the substitution $bx = a\tan\theta$ $\left(x = \dfrac{a}{b}\tan\theta\right)$, $-\dfrac{\pi}{2} < \theta < \dfrac{\pi}{2}$.

**NOW WORK** Problem **45**.

## 3 Integrate a Function Containing $\sqrt{x^2 - a^2}$

When an integrand contains $\sqrt{x^2 - a^2}$, $a > 0$, use the substitution $x = a\sec\theta$, $0 \le \theta < \dfrac{\pi}{2}$, $\pi \le \theta < \dfrac{3\pi}{2}$. Then

**NOTE** The substitution $x = a\cosh\theta$ can also be used for integrands containing $\sqrt{x^2 - a^2}$.

$$\underset{\underset{x = a\sec\theta}{\uparrow}}{\sqrt{x^2 - a^2}} = \underset{\underset{a > 0}{\uparrow}}{\sqrt{a^2\sec^2\theta - a^2}} = a\sqrt{\sec^2\theta - 1} = a\sqrt{\tan^2\theta} = \underset{\underset{\tan\theta \ge 0,\ \text{since}\ 0 \le \theta < \frac{\pi}{2},\ \pi \le \theta < \frac{3\pi}{2}}{\uparrow}}{a\tan\theta}$$

---

**EXAMPLE 4**    Integrating a Function That Contains $\sqrt{x^2 - 4}$

Find $\displaystyle\int \frac{\sqrt{x^2 - 4}}{x}\, dx$.

**Solution** The integrand contains the square root $\sqrt{x^2 - 4}$ which is of the form $\sqrt{x^2 - a^2}$, where $a = 2$. Use the substitution $x = 2\sec\theta$, $0 \le \theta < \dfrac{\pi}{2}$, $\pi \le \theta < \dfrac{3\pi}{2}$. Then $dx = 2\sec\theta\tan\theta\, d\theta$. Since

$$\underset{\underset{x = 2\sec\theta}{\uparrow}}{\sqrt{x^2 - 4}} = \sqrt{4\sec^2\theta - 4} = 2\sqrt{\sec^2\theta - 1} = 2\sqrt{\tan^2\theta} = \underset{\substack{\uparrow \\ \tan\theta \ge 0 \\ \text{since } 0 \le \theta < \frac{\pi}{2},\ \pi \le \theta < \frac{3\pi}{2}}}{2\tan\theta}$$

we have

$$\int \frac{\sqrt{x^2 - 4}}{x} dx = \int \frac{(2\tan\theta)(2\sec\theta\tan\theta\, d\theta)}{2\sec\theta} \qquad \begin{array}{l} x = 2\sec\theta \\ dx = 2\sec\theta\tan\theta\, d\theta \end{array}$$

$$= 2\int \tan^2\theta\, d\theta = 2\underset{\underset{\tan^2\theta = \sec^2\theta - 1}{\uparrow}}{\int (\sec^2\theta - 1)\, d\theta}$$

$$= 2\tan\theta - 2\theta + C$$

To express the solution in terms of $x$, use either the right triangles in Figure 7 or trigonometric identities.

Using identities, we find

$$\tan\theta = \underset{\substack{\uparrow \\ \tan^2\theta = \sec^2\theta - 1 \\ \tan\theta \ge 0}}{\sqrt{\sec^2\theta - 1}} = \underset{\substack{\uparrow \\ \sec\theta = \frac{x}{2}}}{\sqrt{\frac{x^2}{4} - 1}} = \frac{1}{2}\sqrt{x^2 - 4}$$

**Figure 7** $\sec\theta = \dfrac{x}{2}$,

$0 < \theta < \dfrac{\pi}{2}$, $\pi < \theta < \dfrac{3\pi}{2}$

Also since $\sec\theta = \dfrac{x}{2}$, $0 \leq \theta < \dfrac{\pi}{2}$, $\pi \leq \theta < \dfrac{3\pi}{2}$, the inverse function $\theta = \sec^{-1}\dfrac{x}{2}$ is defined.

Then

$$\int \frac{\sqrt{x^2-4}}{x}\,dx = 2\tan\theta - 2\theta + C = \sqrt{x^2-4} - 2\sec^{-1}\frac{x}{2} + C \qquad \blacksquare$$

**NOW WORK** Problem 29.

## 4 Use Trigonometric Substitution to Find Definite Integrals

Trigonometric substitution also can be used to find certain types of definite integrals.

**EXAMPLE 5   Finding the Area Enclosed by an Ellipse**

Find the area $A$ enclosed by the ellipse $\dfrac{x^2}{4} + \dfrac{y^2}{9} = 1$.

**Solution** Figure 8 shows the ellipse. Since the ellipse is symmetric with respect to both the $x$-axis and the $y$-axis, the total area $A$ of the ellipse is four times the shaded area in the first quadrant, where $0 \leq x \leq 2$ and $0 \leq y \leq 3$.

We begin by expressing $y$ as a function of $x$.

$$\frac{x^2}{4} + \frac{y^2}{9} = 1$$

$$\frac{y^2}{9} = 1 - \frac{x^2}{4} = \frac{4-x^2}{4}$$

$$y^2 = \frac{9}{4}(4 - x^2)$$

$$y = \frac{3}{2}\sqrt{4 - x^2} \qquad y \geq 0$$

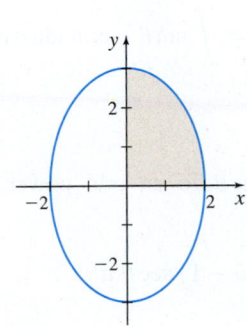

**Figure 8** $\dfrac{x^2}{4} + \dfrac{y^2}{9} = 1$

So, the area $A$ of the ellipse is four times the area under the graph of $y = \dfrac{3}{2}\sqrt{4-x^2}$, $0 \leq x \leq 2$. That is,

$$A = 4\int_0^2 \frac{3}{2}\sqrt{4-x^2}\,dx = 6\int_0^2 \sqrt{4-x^2}\,dx$$

Since the integrand contains a square root of the form $\sqrt{a^2 - x^2}$ with $a = 2$, we use the substitution $x = 2\sin\theta$, $-\dfrac{\pi}{2} \leq \theta \leq \dfrac{\pi}{2}$. Then $dx = 2\cos\theta\,d\theta$. The new limits of integration are:

- When $x = 0$,   $2\sin\theta = 0$, so $\theta = 0$.
- When $x = 2$,   $2\sin\theta = 2$, so $\sin\theta = 1$ and $\theta = \dfrac{\pi}{2}$.

Then

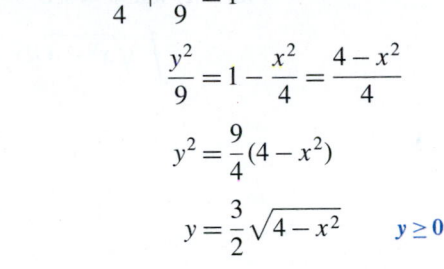

$$A = 6\int_0^2 \sqrt{4-x^2}\,dx = 6\int_0^{\pi/2} \sqrt{4 - 4\sin^2\theta}\cdot 2\cos\theta\,d\theta$$

$$= 6\int_0^{\pi/2} 2\sqrt{1 - \sin^2\theta}\cdot 2\cos\theta\,d\theta = 24\int_0^{\pi/2} \sqrt{\cos^2\theta}\cdot\cos\theta\,d\theta$$

$$\underset{\underset{\cos\theta\,\geq\,0}{\uparrow}}{= 24\int_0^{\pi/2} \cos^2\theta\,d\theta} = \underset{\underset{\cos^2\theta\,=\,\frac{1+\cos(2\theta)}{2}}{\uparrow}}{\frac{24}{2}\int_0^{\pi/2} [1 + \cos(2\theta)]\,d\theta}$$

$$= 12\left[\theta + \frac{1}{2}\sin(2\theta)\right]_0^{\pi/2} = 12\left(\frac{\pi}{2} + 0\right) = 6\pi$$

The area of the ellipse is $6\pi$ square units.   $\blacksquare$

**NOW WORK** Problem 63.

**NEED TO REVIEW?** The two approaches to finding a definite integral using substitution are discussed in Section 5.5, pp. 405–406.

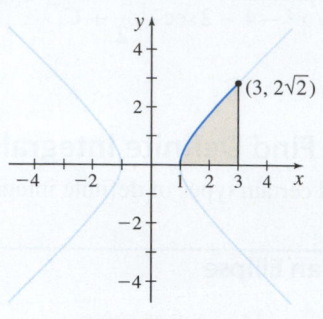

**Figure 9** $y = \sqrt{x^2 - 1}$, $1 \le x \le 3$

In Example 5, we changed the limits of integration to find the definite integral, so there was no need to change back to the variable $x$. But it is not always easy to obtain new limits of integration, as we see in the next example.

### EXAMPLE 6    Using Trigonometric Substitution to Find a Definite Integral

Find the area $A$ under the graph of $y = \sqrt{x^2 - 1}$ (the upper half of the right branch of the hyperbola $y^2 = x^2 - 1$) from 1 to 3. See Figure 9.

**Solution** The area $A = \int_1^3 \sqrt{x^2 - 1}\, dx$. The integral contains a square root of the form $\sqrt{x^2 - a^2}$, where $a = 1$, so we use the trigonometric substitution $x = \sec\theta$, $0 \le \theta < \dfrac{\pi}{2}$, $\pi \le \theta < \dfrac{3\pi}{2}$. Then $dx = \sec\theta \tan\theta\, d\theta$. Since the upper limit $x = 3$ does not result in a nice angle ($\theta = \sec^{-1} 3$), we find the indefinite integral first and then use the Fundamental Theorem of Calculus.

With $x = \sec\theta$ and $dx = \sec\theta \tan\theta\, d\theta$, we have

$$\int \sqrt{x^2 - 1}\, dx = \int \sqrt{\sec^2\theta - 1}\, \sec\theta \tan\theta\, d\theta = \int \tan\theta \cdot \sec\theta \tan\theta\, d\theta$$

$$= \int \tan^2\theta \sec\theta\, d\theta$$

Since $\tan\theta$ is raised to an even power and $\sec\theta$ to an odd power, use the identity $\tan^2\theta = \sec^2\theta - 1$. Then

$$\int \sqrt{x^2 - 1}\, dx = \int \tan^2\theta \sec\theta\, d\theta = \int (\sec^2\theta - 1)\sec\theta\, d\theta$$

$$= \int \sec^3\theta\, d\theta - \int \sec\theta\, d\theta$$

$$= \frac{1}{2}[\sec\theta \tan\theta + \ln|\sec\theta + \tan\theta|] - \ln|\sec\theta + \tan\theta|$$

$$= \frac{1}{2}\sec\theta \tan\theta - \frac{1}{2}\ln|\sec\theta + \tan\theta|$$

$$= \frac{1}{2}x\sqrt{x^2 - 1} - \frac{1}{2}\ln|x + \sqrt{x^2 - 1}|$$

$$\underset{\substack{\uparrow \\ \sec\theta = x \\ \tan\theta = \sqrt{x^2 - 1}}}{}$$

Then

$$A = \int_1^3 \sqrt{x^2 - 1}\, dx = \left[\frac{1}{2}x\sqrt{x^2 - 1} - \frac{1}{2}\ln\left|x + \sqrt{x^2 - 1}\right|\right]_1^3$$

$$= \frac{3}{2}\sqrt{8} - \frac{1}{2}\ln(3 + \sqrt{8}) = 3\sqrt{2} - \frac{1}{2}\ln(3 + 2\sqrt{2}) \qquad \blacksquare$$

**NOW WORK** Problems 53 and 65.

## 7.3 Assess Your Understanding

**Concepts and Vocabulary**

1. *True or False* To find $\int \sqrt{a^2 - x^2}\, dx$, $a > 0$, the substitution $x = a\sin\theta$, $-\dfrac{\pi}{2} \le \theta \le \dfrac{\pi}{2}$, can be used.

2. *Multiple Choice* To find $\int \sqrt{x^2 + 16}\, dx$, use the substitution $x = [$ **(a)** $4\sin\theta$ **(b)** $\tan\theta$ **(c)** $4\sec\theta$ **(d)** $4\tan\theta]$.

3. *Multiple Choice* To find $\int \sqrt{x^2 - 9}\, dx$, use the substitution $x = [$ **(a)** $\sec\theta$ **(b)** $3\sin\theta$ **(c)** $3\sec\theta$ **(d)** $3\tan\theta]$.

4. *Multiple Choice* To find $\int \sqrt{25 - 4x^2}\, dx$, use the substitution $x = \left[ \textbf{(a)}\ \dfrac{5}{2}\tan\theta\ \textbf{(b)}\ \dfrac{5}{2}\sin\theta\ \textbf{(c)}\ \dfrac{2}{5}\sin\theta\ \textbf{(d)}\ \dfrac{2}{5}\sec\theta \right]$.

---

**1.** = NOW WORK problem        〔⚠〕 = Graphing technology recommended        〔CAS〕 = Computer Algebra System recommended

**Skill Building**

*In Problems 5–14, find each integral. Each integral contains a term of the form $\sqrt{a^2 - x^2}$.*

**5.** $\displaystyle\int \sqrt{4 - x^2}\,dx$

**6.** $\displaystyle\int \sqrt{16 - x^2}\,dx$

**7.** $\displaystyle\int \frac{x^2}{\sqrt{16 - x^2}}\,dx$

**8.** $\displaystyle\int \frac{x^2}{\sqrt{36 - x^2}}\,dx$

**9.** $\displaystyle\int \frac{\sqrt{4 - x^2}}{x^2}\,dx$

**10.** $\displaystyle\int \frac{\sqrt{9 - x^2}}{x^2}\,dx$

**11.** $\displaystyle\int x^2\sqrt{4 - x^2}\,dx$

**12.** $\displaystyle\int x^2\sqrt{1 - 16x^2}\,dx$

**13.** $\displaystyle\int \frac{dx}{(4 - x^2)^{3/2}}$

**14.** $\displaystyle\int \frac{dx}{(1 - x^2)^{3/2}}$

*In Problems 15–26, find each integral. Each integral contains a term of the form $\sqrt{x^2 + a^2}$.*

**15.** $\displaystyle\int \sqrt{4 + x^2}\,dx$

**16.** $\displaystyle\int \sqrt{1 + x^2}\,dx$

**17.** $\displaystyle\int \frac{dx}{\sqrt{x^2 + 16}}$

**18.** $\displaystyle\int \frac{dx}{\sqrt{x^2 + 25}}$

**19.** $\displaystyle\int \sqrt{1 + 9x^2}\,dx$

**20.** $\displaystyle\int \sqrt{9 + 4x^2}\,dx$

**21.** $\displaystyle\int \frac{x^2}{\sqrt{4 + 9x^2}}\,dx$

**22.** $\displaystyle\int \frac{x^2}{\sqrt{x^2 + 16}}\,dx$

**23.** $\displaystyle\int \frac{dx}{x^2\sqrt{x^2 + 4}}$

**24.** $\displaystyle\int \frac{dx}{x^2\sqrt{4x^2 + 1}}$

**25.** $\displaystyle\int \frac{dx}{(x^2 + 4)^{3/2}}$

**26.** $\displaystyle\int \frac{dx}{(x^2 + 1)^{3/2}}$

*In Problems 27–36, find each integral. Each integral contains a term of the form $\sqrt{x^2 - a^2}$.*

**27.** $\displaystyle\int \frac{x^2}{\sqrt{x^2 - 25}}\,dx$

**28.** $\displaystyle\int \frac{x^2}{\sqrt{x^2 - 16}}\,dx$

**29.** $\displaystyle\int \frac{\sqrt{x^2 - 1}}{x}\,dx$

**30.** $\displaystyle\int \frac{\sqrt{x^2 - 1}}{x^2}\,dx$

**31.** $\displaystyle\int \frac{dx}{x^2\sqrt{x^2 - 36}}$

**32.** $\displaystyle\int \frac{dx}{x^2\sqrt{x^2 - 9}}$

**33.** $\displaystyle\int \frac{dx}{\sqrt{4x^2 - 9}}$

**34.** $\displaystyle\int \frac{dx}{\sqrt{9x^2 - 4}}$

**35.** $\displaystyle\int \frac{dx}{(x^2 - 9)^{3/2}}$

**36.** $\displaystyle\int \frac{dx}{(25x^2 - 1)^{3/2}}$

*In Problems 37–48, find each integral.*

**37.** $\displaystyle\int \frac{x^2\,dx}{(x^2 - 9)^{3/2}}$

**38.** $\displaystyle\int \frac{x^2\,dx}{(x^2 - 4)^{3/2}}$

**39.** $\displaystyle\int \frac{x^2\,dx}{16 + x^2}$

**40.** $\displaystyle\int \frac{x^2\,dx}{1 + 16x^2}$

**41.** $\displaystyle\int \sqrt{4 - 25x^2}\,dx$

**42.** $\displaystyle\int \sqrt{9 - 16x^2}\,dx$

**43.** $\displaystyle\int \frac{dx}{(4 - 25x^2)^{3/2}}$

**44.** $\displaystyle\int \frac{dx}{(1 - 9x^2)^{3/2}}$

**45.** $\displaystyle\int \sqrt{4 + 25x^2}\,dx$

**46.** $\displaystyle\int \sqrt{9 + 16x^2}\,dx$

**47.** $\displaystyle\int \frac{dx}{x^3\sqrt{x^2 - 16}}$

**48.** $\displaystyle\int \frac{dx}{x^3\sqrt{x^2 - 1}}$

*In Problems 49–58, find each definite integral.*

**49.** $\displaystyle\int_0^1 \sqrt{1 - x^2}\,dx$

**50.** $\displaystyle\int_0^{1/2} \sqrt{1 - 4x^2}\,dx$

**51.** $\displaystyle\int_0^1 \sqrt{1 + x^2}\,dx$

**52.** $\displaystyle\int_0^2 \frac{x^2}{\sqrt{9 + x^2}}\,dx$

**53.** $\displaystyle\int_4^5 \frac{x^2}{\sqrt{x^2 - 9}}\,dx$

**54.** $\displaystyle\int_1^2 \frac{x^2}{\sqrt{4x^2 - 1}}\,dx$

**55.** $\displaystyle\int_0^2 \frac{x^2\,dx}{(16 - x^2)^{3/2}}$

**56.** $\displaystyle\int_0^1 \frac{x^2\,dx}{(25 - x^2)^{3/2}}$

**57.** $\displaystyle\int_0^3 \frac{x^2\,dx}{9 + x^2}$

**58.** $\displaystyle\int_0^1 \frac{x^2}{25 + x^2}\,dx$

**Applications and Extensions**

**59. Volume of a Solid of Revolution**  Find the volume of the solid of revolution generated by revolving the region bounded by the graph of $y = \dfrac{1}{x^2 + 4}$ and the $x$-axis from $x = 0$ to $x = 1$ about the $x$-axis. See the figure.

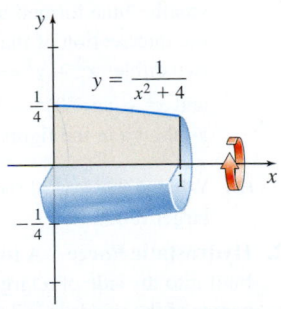

**60. Volume of a Solid of Revolution**  Find the volume of the solid of revolution generated by revolving the region bounded by the graphs of $y = \dfrac{1}{\sqrt{9 - x^2}}$, $y = 0$, $x = 0$, and $x = 2$ about the $x$-axis.

**61. Average Value**  Find the average value of the function $f(x) = \dfrac{1}{\sqrt{9 - 4x^2}}$ over the interval $\left[0, \dfrac{1}{2}\right]$.

**62. Average Value** Find the average value of the function $f(x) = \sqrt{x^2 - 4}$ over the interval $[2, 7]$.

**63. Area Under a Graph** Find the area under the graph of $y = \dfrac{1}{\sqrt{9 - x^2}}$ from $x = 0$ to $x = 2$.

**64. Area Under a Graph** Find the area under the graph of $y = x^2\sqrt{16 - x^2}$, $x \geq 0$.

**65. Area Under a Graph** Find the area under the graph of $y = \dfrac{x^2}{\sqrt{x^2 - 1}}$ from $x = 3$ to $x = 5$.

**66. Area Under a Graph** Find the area under the graph of $y = \sqrt{16x^2 + 9}$ from $x = 0$ to $x = 1$.

**67. Area** Find the area of the region enclosed by the hyperbola $\dfrac{y^2}{9} - x^2 = 1$ and

(a) the line $y = 4$.
(b) the line $y = -7$.

**68. Area** Find the area of the region enclosed by the hyperbola $\dfrac{x^2}{9} - \dfrac{y^2}{16} = 1$ and

(a) the line $x = 6$.
(b) the line $x = -8$.

**69. Arc Length** Find the length of the graph of the parabola $y = 5x - x^2$ that lies above the $x$-axis.

**70. Arc Length** Find the length of the graph of the part of the parabola $x = 6y - 3y^2$ that lies in the first quadrant.

**71. Area of a Lune** A **lune** is a crescent-shaped region formed when two circles intersect.

(a) Find the area of the smaller lune formed by the intersection of the two circles $x^2 + y^2 = 4$ and $x^2 + (y - 2)^2 = 1$, as shown in the figure.

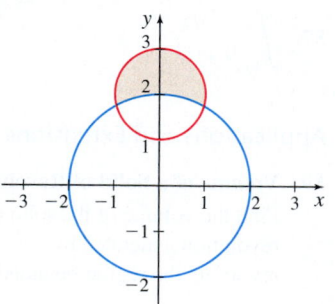

(b) What is the area of the larger lune?

**72. Hydrostatic Force** A round window of radius 2 meters (m) is built into the side of a large, fresh-water aquarium tank. If the center of the window is 3 m below the water line, find the force due to hydrostatic pressure on the window.
*Hint:* The mass density of fresh water is $\rho = 1000\,\text{kg/m}^3$.

*In Problems 73–78, find each integral.*
*Hint: Begin with a substitution.*

**73.** $\displaystyle\int \dfrac{dx}{\sqrt{1 - (x - 2)^2}}$

**74.** $\displaystyle\int \sqrt{4 - (x + 2)^2}\, dx$

**75.** $\displaystyle\int \dfrac{dx}{\sqrt{(2x - 1)^2 - 4}}$

**76.** $\displaystyle\int \dfrac{dx}{(3x - 2)\sqrt{(3x - 2)^2 + 9}}$

**77.** $\displaystyle\int e^x \sqrt{25 - e^{2x}}\, dx$

**78.** $\displaystyle\int e^x \sqrt{4 + e^{2x}}\, dx$

*In Problems 79 and 80, use integration by parts and then the methods of this section to find each integral.*

**79.** $\displaystyle\int x \sin^{-1} x\, dx$

**80.** $\displaystyle\int x \cos^{-1} x\, dx$

**81.** Find $\int \sqrt{x^2 + a^2}\, dx$

(a) By using a trigonometric substitution
(b) By using substitution with a hyperbolic function

**82. Area of an Ellipse** Find the area of the region enclosed by the ellipse $\dfrac{x^2}{a^2} + \dfrac{y^2}{b^2} = 1$, where $a > b > 0$.

*In Problems 83–87, use a trigonometric substitution to derive each formula. Assume $a > 0$.*

**83.** $\displaystyle\int \dfrac{dx}{\sqrt{a^2 - x^2}} = \sin^{-1} \dfrac{x}{a} + C$

**84.** $\displaystyle\int \dfrac{dx}{a^2 + x^2} = \dfrac{1}{a}\tan^{-1}\dfrac{x}{a} + C$

**85.** $\displaystyle\int \dfrac{dx}{x\sqrt{x^2 - a^2}} = \dfrac{1}{a}\sec^{-1}\dfrac{x}{a} + C$

**86.** $\displaystyle\int \dfrac{dx}{\sqrt{x^2 - a^2}} = \ln\left|\dfrac{x + \sqrt{x^2 - a^2}}{a}\right| + C$

**87.** $\displaystyle\int \dfrac{dx}{\sqrt{x^2 + a^2}} = \ln\left|x + \sqrt{x^2 + a^2}\right| + C$

**Challenge Problems**

**88.** Find $\displaystyle\int \dfrac{dx}{\sqrt{3x - x^2}}$.

**89.** Derive the formula
$$\int \sqrt{x^2 - a^2}\, dx = \dfrac{1}{2}x\sqrt{x^2 - a^2} - \dfrac{1}{2}a^2\ln\left|x + \sqrt{x^2 - a^2}\right| + C,$$
$a > 0$.

**90.** Find $\displaystyle\int \dfrac{dx}{\sqrt{x^2 + a^2}}$, $a > 0$, using the substitution $u = \sinh^{-1}\dfrac{x}{a}$.
Express your answer in logarithmic form.

**91.** Find $\displaystyle\int \dfrac{\sec^2 x}{\sqrt{\tan^2 x - 6\tan x + 8}}\, dx$.

# 7.4 Integrands Containing $ax^2 + bx + c, a \neq 0$

**OBJECTIVE** *When you finish this section, you should be able to:*

**1** Integrate a function that contains a quadratic expression (p. 525)

## 1 Integrate a Function That Contains a Quadratic Expression

**NEED TO REVIEW?** Completing the square is discussed in Appendix A.1, pp. A-2 to A-3.

Integrals containing $ax^2 + bx + c$, where $a \neq 0$, $b$ and $c$ are real numbers, can often be integrated by completing the square. We complete the square of $ax^2 + bx + c$ as follows:

$$ax^2 + bx + c = a\left(x^2 + \frac{b}{a}x\right) + c \qquad \text{Factor } a \text{ from the first two terms.}$$

$$= a\left(x^2 + \frac{b}{a}x + \frac{b^2}{4a^2}\right) + c - \frac{b^2}{4a} \qquad \text{Add and subtract } a\left(\frac{1}{2} \cdot \frac{b}{a}\right)^2 = \frac{b^2}{4a}.$$

$$= a\left(x + \frac{b}{2a}\right)^2 + \left(c - \frac{b^2}{4a}\right) \qquad \text{Factor } x^2 + \frac{b}{a}x + \frac{b^2}{4a^2} = \left(x + \frac{b}{2a}\right)^2.$$

After completing the square, use the substitution

$$u = x + \frac{b}{2a}$$

to express the original quadratic expression $ax^2 + bx + c$ in the simpler form $au^2 + r$, where $r = c - \dfrac{b^2}{4a}$.

---

**EXAMPLE 1** **Integrating a Function That Contains $x^2 + 6x + 10$**

Find $\displaystyle\int \frac{dx}{x^2 + 6x + 10}$.

**Solution** The integrand contains the quadratic expression $x^2 + 6x + 10$. So, complete the square.

$$x^2 + 6x + 10 = (x^2 + 6x + 9) + 1 = (x + 3)^2 + 1$$

Now write the integral as

$$\int \frac{dx}{x^2 + 6x + 10} = \int \frac{dx}{(x + 3)^2 + 1}$$

and use the substitution $u = x + 3$. Then $du = dx$, and

**RECALL** $\displaystyle\int \frac{du}{u^2 + a^2} = \frac{1}{a}\tan^{-1}\frac{u}{a} + C$

$$\int \frac{dx}{x^2 + 6x + 10} = \int \frac{dx}{(x + 3)^2 + 1} = \int \frac{du}{u^2 + 1} = \tan^{-1} u + C = \tan^{-1}(x + 3) + C$$

$\blacksquare$

**NOW WORK** Problem 1.

---

**EXAMPLE 2** **Integrating a Function That Contains $x^2 + x + 1$**

Find **(a)** $\displaystyle\int \frac{dx}{x^2 + x + 1}$ **(b)** $\displaystyle\int \frac{x\,dx}{x^2 + x + 1}$

**Solution** **(a)** The integrand contains the quadratic expression $x^2 + x + 1$. So, complete the square in the denominator.

$$\int \frac{dx}{x^2 + x + 1} = \int \frac{dx}{\left(x^2 + x + \frac{1}{4}\right) + \left(1 - \frac{1}{4}\right)} = \int \frac{dx}{\left(x + \frac{1}{2}\right)^2 + \frac{3}{4}}$$

Complete the square

Now use the substitution $u = x + \dfrac{1}{2}$. Then $du = dx$.

$$\int \frac{dx}{x^2 + x + 1} = \int \frac{dx}{\left(x + \dfrac{1}{2}\right)^2 + \dfrac{3}{4}} = \int \frac{du}{u^2 + \dfrac{3}{4}}$$

$$= \frac{1}{\dfrac{\sqrt{3}}{2}} \tan^{-1} \frac{u}{\dfrac{\sqrt{3}}{2}} + C = \frac{2}{\sqrt{3}} \tan^{-1} \frac{2u}{\sqrt{3}} + C$$

$$= \frac{2\sqrt{3}}{3} \tan^{-1} \frac{2x + 1}{\sqrt{3}} + C \qquad\qquad u = x + \dfrac{1}{2}$$

**NOTE** We could also complete the square and let $u = x + \dfrac{1}{2}$.

**(b)** Here we use a manipulation that causes the derivative of the denominator to appear in the numerator. This helps simplify the integration.

$$\int \frac{x\,dx}{x^2 + x + 1} = \frac{1}{2} \int \frac{2x\,dx}{x^2 + x + 1} \qquad\qquad \text{Multiply the integrand by } \dfrac{2}{2}.$$

$$= \frac{1}{2} \int \frac{[(2x + 1) - 1]dx}{x^2 + x + 1} \qquad\qquad \begin{array}{l}\text{Add and subtract 1 in} \\ \text{the numerator to get } 2x + 1.\end{array}$$

$$= \frac{1}{2} \int \frac{(2x + 1)dx}{x^2 + x + 1} - \frac{1}{2} \int \frac{dx}{x^2 + x + 1} \qquad\qquad \begin{array}{l}\text{Write the integral as the} \\ \text{sum of two integrals.}\end{array}$$

The integral on the right was found in (a). In the integral on the left, the numerator equals the derivative of the denominator.

**RECALL** $\displaystyle\int \frac{g'(x)}{g(x)}\, dx = \ln |g(x)| + C$

$$\frac{1}{2} \int \frac{(2x + 1)dx}{x^2 + x + 1} = \frac{1}{2} \ln \left| x^2 + x + 1 \right|$$

So,

$$\int \frac{x\,dx}{x^2 + x + 1} = \frac{1}{2} \int \frac{(2x + 1)\,dx}{x^2 + x + 1} - \frac{1}{2} \int \frac{dx}{x^2 + x + 1}$$

$$= \frac{1}{2} \ln \left| x^2 + x + 1 \right| - \frac{1}{2} \left[ \frac{2\sqrt{3}}{3} \tan^{-1} \frac{2x + 1}{\sqrt{3}} \right] + C$$

$$= \frac{1}{2} \ln \left| x^2 + x + 1 \right| - \frac{\sqrt{3}}{3} \tan^{-1} \frac{2x + 1}{\sqrt{3}} + C \qquad\blacksquare$$

**NOW WORK** Problem 7.

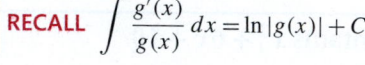

**EXAMPLE 3** Integrating a Function That Contains $2x - x^2$

Find $\displaystyle\int \frac{dx}{\sqrt{2x - x^2}}$.

**Solution** The integrand contains the quadratic expression $2x - x^2$, so complete the square.

$$2x - x^2 = -x^2 + 2x = -(x^2 - 2x) = -(x^2 - 2x + 1) + 1$$

$$= -(x - 1)^2 + 1 = 1 - (x - 1)^2$$

Then

**RECALL** $\displaystyle\int \frac{dx}{\sqrt{a^2 - x^2}} = \sin^{-1} \frac{x}{a} + C$

$$\int \frac{dx}{\sqrt{2x - x^2}} = \int \frac{dx}{\sqrt{1 - (x - 1)^2}} = \int \frac{du}{\sqrt{1 - u^2}} = \sin^{-1} u + C = \sin^{-1}(x - 1) + C$$
$$\underset{\substack{\uparrow \\ u = x - 1 \\ du = dx}}{}$$

$\blacksquare$

**NOW WORK** Problem 23.

## 7.4 Assess Your Understanding

### Skill Building

*In Problems 1–32, find each integral.*

1. $\int \dfrac{dx}{x^2+4x+5}$

2. $\int \dfrac{dx}{x^2+2x+5}$

3. $\int \dfrac{dx}{x^2+4x+8}$

4. $\int \dfrac{dx}{x^2-6x+10}$

5. $\int \dfrac{2\,dx}{3+2x+2x^2}$

6. $\int \dfrac{3\,dx}{x^2+6x+10}$

7. $\int \dfrac{x\,dx}{2x^2+2x+3}$

8. $\int \dfrac{3x\,dx}{x^2+6x+10}$

9. $\int \dfrac{dx}{\sqrt{8+2x-x^2}}$

10. $\int \dfrac{dx}{\sqrt{5-4x-2x^2}}$

11. $\int \dfrac{dx}{\sqrt{4x-x^2}}$

12. $\int \dfrac{dx}{\sqrt{x^2-6x-10}}$

13. $\int \dfrac{dx}{(x+1)\sqrt{x^2+2x+2}}$

14. $\int \dfrac{dx}{(x-4)x^2-8x+17}$

15. $\int \dfrac{dx}{\sqrt{24-2x-x^2}}$

16. $\int \dfrac{dx}{\sqrt{9x^2+6x+10}}$

17. $\int \dfrac{x-5}{\sqrt{x^2-2x+5}}\,dx$

18. $\int \dfrac{x+1}{x^2-4x+3}\,dx$

19. $\int_{1}^{3} \dfrac{dx}{\sqrt{x^2-2x+5}}$

20. $\int_{1/2}^{1} \dfrac{x^2\,dx}{\sqrt{2x-x^2}}$

21. $\int \dfrac{e^x\,dx}{\sqrt{e^{2x}+e^x+1}}$

22. $\int \dfrac{\cos x\,dx}{\sqrt{\sin^2 x+4\sin x+3}}$

23. $\int \dfrac{2x-3}{\sqrt{4x-x^2-3}}\,dx$

24. $\int \dfrac{x+3}{\sqrt{x^2+2x+2}}\,dx$

25. $\int \dfrac{dx}{(x^2-2x+10)^{3/2}}$

26. $\int \dfrac{dx}{\sqrt{x^2-2x+10}}$

27. $\int \dfrac{dx}{\sqrt{x^2+2x-3}}$

28. $\int x\sqrt{x^2-4x-1}\,dx$

29. $\int \dfrac{\sqrt{5+4x-x^2}}{x-2}\,dx$

30. $\int \sqrt{5+4x-x^2}\,dx$

31. $\int \dfrac{x\,dx}{\sqrt{x^2+2x-3}}$

32. $\int \dfrac{x\,dx}{\sqrt{x^2-4x+3}}$

### Applications and Extensions

33. Show that if $k>0$, then
$$\int \dfrac{dx}{\sqrt{(x+h)^2+k}} = \ln\left[\sqrt{(x+h)^2+k}+x+h\right]+C$$

34. Show that if $a>0$ and $b^2-4ac>0$, then
$$\int \dfrac{dx}{\sqrt{ax^2+bx+c}} = \dfrac{1}{\sqrt{a}}\ln\left|\sqrt{ax^2+bx+c}+\sqrt{a}x+\dfrac{b}{2\sqrt{a}}\right|+C$$

### Challenge Problem

35. Find $\displaystyle\int \sqrt{\dfrac{a+x}{a-x}}\,dx$, where $a>0$.

# 7.5 Integration of Rational Functions Using Partial Fractions; the Logistic Model

**OBJECTIVES** *When you finish this section, you should be able to:*

1 Integrate a proper rational function whose denominator contains only distinct linear factors (p. 529)

2 Integrate a proper rational function whose denominator contains a repeated linear factor (p. 531)

3 Integrate a proper rational function whose denominator contains a distinct irreducible quadratic factor (p. 532)

4 Integrate a proper rational function whose denominator contains a repeated irreducible quadratic factor (p. 533)

5 Work with the logistic model (p. 534)

In this section, we integrate rational functions using a technique called *partial fractions.* Although the integral of any rational function can be found using partial fractions, the process requires being able to factor the denominator of the rational function.

**1.** = NOW WORK problem     = Graphing technology recommended     CAS = Computer Algebra System recommended

**NEED TO REVIEW?** Rational functions are discussed in Section P.2, p. 22.

Recall that a rational function $R$ in lowest terms is the ratio of two polynomial functions $p$ and $q \neq 0$, where $p$ and $q$ have no common factors. The domain of $R$ is the set of all real numbers for which $q \neq 0$. The rational function $R$ is called **proper** when the degree of the polynomial $p$ is less than the degree of the polynomial $q$; otherwise, $R$ is an **improper** rational function.

Using polynomial division, every improper rational function $R$ can be written as the sum of a polynomial function and a proper rational function.

---

**EXAMPLE 1** **Integrating an Improper Rational Function**

Find $\displaystyle\int \frac{x^2}{x-1}\, dx$.

**Solution** The integrand is an improper rational function. Using polynomial division, the integrand can be expressed as the sum of a polynomial and a proper rational function.

$$
\begin{array}{r}
x + 1 + \dfrac{1}{x-1} \\[2pt]
\hline
x - 1 \,\big)\ x^2 \phantom{xxxxxxxx} \\
\underline{x^2 - x} \phantom{xxxxxxx} \\
x \phantom{xxxxx} \\
\underline{x - 1} \phantom{xx} \\
1 \phantom{xx}
\end{array}
$$

Then

$$
\int \frac{x^2}{x-1}\, dx = \int \left( x + 1 + \frac{1}{x-1} \right) dx = \int x\, dx + \int dx + \int \frac{1}{x-1}\, dx
$$

$$
= \frac{x^2}{2} + x + \ln |x - 1| + C \qquad\blacksquare
$$

**NOW WORK** Problem **17.**

Since we know how to integrate polynomial functions, once we know how to integrate proper rational functions, we will be able to integrate any rational function.

For this reason, the discussion that follows deals only with the integration of proper rational functions in lowest terms. Such rational functions can be written as the sum of simpler functions, called **partial fractions.**

For example, since $\dfrac{2}{x-1} + \dfrac{3}{x+4} = \dfrac{5x+5}{(x-1)(x+4)}$, the rational function

$$
R(x) = \frac{5x+5}{(x-1)(x+4)}
$$

can be expressed as the sum $R(x) = \dfrac{2}{x-1} + \dfrac{3}{x+4}$. Then

$$
\int \frac{5x+5}{(x-1)(x+4)}\, dx = \int \left( \frac{2}{x-1} + \frac{3}{x+4} \right) dx
$$

$$
= 2 \int \frac{1}{x-1}\, dx + 3 \int \frac{1}{x+4}\, dx
$$

$$
= 2 \ln |x-1| + 3 \ln |x+4| + C
$$

We use a technique called **partial fraction decomposition** to write a proper rational function $R$ as the sum of partial fractions. The partial fractions depend on the nature of the factors of the denominator $q$. It can be shown that any polynomial $q$ whose coefficients are real numbers can be factored (over the real numbers) into products of linear and/or irreducible quadratic factors. As we shall see, this means *the integral of every rational function can be expressed in terms of algebraic, logarithmic, and/or inverse trigonometric functions.*

## 1 Integrate a Proper Rational Function Whose Denominator Contains Only Distinct Linear Factors

*Case 1:* If the denominator $q$ contains only distinct linear factors, say, $x - a_1$, $x - a_2, \ldots, x - a_n$, then $\dfrac{p}{q}$ can be written as

$$\frac{p(x)}{q(x)} = \frac{A_1}{x - a_1} + \frac{A_2}{x - a_2} + \cdots + \frac{A_n}{x - a_n} \tag{1}$$

where $A_1, A_2, \ldots, A_n$ are real numbers.

To find $\displaystyle\int \frac{p(x)}{q(x)} dx$, integrate both sides of (1). Then

$$\int \frac{p(x)}{q(x)} dx = \int \frac{A_1}{x - a_1} dx + \int \frac{A_2}{x - a_2} dx + \cdots + \int \frac{A_n}{x - a_n} dx$$

$$= A_1 \ln |x - a_1| + A_2 \ln |x - a_2| + \cdots + A_n \ln |x - a_n| + C$$

All that remains is to find the numbers $A_1, \ldots, A_n$.

A procedure for finding the numbers $A_1, \ldots, A_n$ is illustrated in Example 2.

**NOTE** Case 1 type integrands lead to sums of natural logarithms.

---

**EXAMPLE 2** **Integrating a Proper Rational Function Whose Denominator Contains Only Distinct Linear Factors**

Find $\displaystyle\int \frac{x \, dx}{x^2 - 5x + 6}$.

**Solution** The integrand is a proper rational function in lowest terms. We begin by factoring the denominator: $x^2 - 5x + 6 = (x - 2)(x - 3)$. Since the factors are linear and distinct, this is a Case 1 type integrand and can be written using the terms $\dfrac{A}{x - 2}$ and $\dfrac{B}{x - 3}$. From (1),

$$\frac{x}{(x - 2)(x - 3)} = \frac{A}{x - 2} + \frac{B}{x - 3}$$

Now clear fractions by multiplying both sides of the equation by $(x - 2)(x - 3)$.

$$x = A(x - 3) + B(x - 2)$$
$$x = (A + B)x - (3A + 2B) \qquad \text{Group like terms.}$$

This is an identity in $x$, so the coefficients of like powers of $x$ must be equal.

$$1 = A + B \qquad \text{The coefficient of } x \text{ equals 1.}$$
$$0 = -3A - 2B \qquad \text{The constant term on the left is 0.}$$

This is a system of two equations containing two variables. Solving the second equation for $B$, we get $B = -\dfrac{3}{2}A$. Substituting for $B$ in the first equation produces the solution $A = -2$ from which $B = 3$. So,

$$\frac{x}{(x - 2)(x - 3)} \underset{\substack{\uparrow \\ A = -2,\ B = 3}}{=} \frac{-2}{x - 2} + \frac{3}{x - 3}$$

Then

$$\int \frac{x}{(x - 2)(x - 3)} dx = \int \frac{-2}{x - 2} dx + \int \frac{3}{x - 3} dx$$

$$= -2 \ln |x - 2| + 3 \ln |x - 3| + C = \ln \left| \frac{(x - 3)^3}{(x - 2)^2} \right| + C \qquad \blacksquare$$

**NOW WORK** Problem 23.

Alternatively, we can find the unknown numbers in the decomposition of $\dfrac{p}{q}$ by substituting convenient values of $x$ into the identity obtained after clearing fractions.* In Example 2, after clearing fractions, the identity is

$$x = A(x-3) + B(x-2)$$

When $x = 3$, the term involving $A$ drops out, leaving $3 = B \cdot 1$, so that $B = 3$. When $x = 2$, the term involving $B$ drops out, leaving $2 = A \cdot (-1)$, so that $A = -2$.

### EXAMPLE 3   Deriving Formulas Using Partial Fractions

Derive these formulas:

(a)
$$\int \frac{dx}{x^2 - a^2} = \frac{1}{2a} \ln \left| \frac{x-a}{x+a} \right| + C \quad a \neq 0$$

(b)
$$\int \frac{dx}{a^2 - x^2} = \frac{1}{2a} \ln \left| \frac{x+a}{x-a} \right| + C \quad a \neq 0$$

**Solution (a)** The factored denominator $x^2 - a^2 = (x-a)(x+a)$ contains only distinct linear factors. So, $\dfrac{1}{x^2 - a^2}$ can be decomposed into partial fractions of the form

$$\frac{1}{x^2 - a^2} = \frac{1}{(x-a)(x+a)} = \frac{A}{x-a} + \frac{B}{x+a}$$

$$1 = A(x+a) + B(x-a) \qquad \text{\color{blue}Multiply both sides by } (x-a)(x+a).$$

This is an identity in $x$.

- When $x = a$, the term involving $B$ drops out. Then $1 = A(2a)$, so $A = \dfrac{1}{2a}$.
- When $x = -a$, the term involving $A$ drops out. Then $1 = B(-2a)$,

  so $B = -\dfrac{1}{2a}$.

Then

$$\frac{1}{x^2 - a^2} = \frac{A}{x-a} + \frac{B}{x+a} = \frac{1}{2a(x-a)} - \frac{1}{2a(x+a)}$$

$$\int \frac{dx}{x^2 - a^2} = \frac{1}{2a} \int \frac{dx}{x-a} - \frac{1}{2a} \int \frac{dx}{x+a} = \frac{1}{2a} \left( \int \frac{dx}{x-a} - \int \frac{dx}{x+a} \right)$$

$$= \frac{1}{2a} \left( \ln |x-a| - \ln |x+a| \right) + C$$

$$= \frac{1}{2a} \ln \left| \frac{x-a}{x+a} \right| + C$$

**(b)** Using the result from (a), we get

$$\int \frac{dx}{a^2 - x^2} = -\int \frac{dx}{x^2 - a^2} = -\frac{1}{2a} \ln \left| \frac{x-a}{x+a} \right| + C = \frac{1}{2a} \ln \left| \frac{x-a}{x+a} \right|^{-1} + C$$

$$= \frac{1}{2a} \ln \left| \frac{x+a}{x-a} \right| + C \qquad\qquad \blacksquare$$

**NOW WORK** Problem 49.

---

*This method is discussed in detail in H. J. Straight & R. Dowds (1984, June–July), *American Mathematical Monthly*, 91(6), 365.

## 2 Integrate a Proper Rational Function Whose Denominator Contains a Repeated Linear Factor

**Case 2**: If the denominator $q$ has a repeated linear factor $(x - a)^n$, $n \geq 2$ an integer, then the decomposition of $\dfrac{p}{q}$ includes the terms

$$\frac{A_1}{x - a}, \frac{A_2}{(x - a)^2}, \ldots, \frac{A_n}{(x - a)^n}$$

where $A_1, A_2, \ldots, A_n$ are real numbers.

---

**EXAMPLE 4**  **Integrating a Proper Rational Function Whose Denominator Contains a Repeated Linear Factor**

Find $\displaystyle\int \frac{dx}{x(x - 1)^2}$.

**Solution** Since $x$ is a distinct linear factor of the denominator $q$, and $(x - 1)^2$ is a repeated linear factor of the denominator, the decomposition of $\dfrac{1}{x(x - 1)^2}$ into partial fractions has the three terms $\dfrac{A}{x}$, $\dfrac{B}{x - 1}$, and $\dfrac{C}{(x - 1)^2}$.

$$\frac{1}{x(x - 1)^2} = \frac{A}{x} + \frac{B}{x - 1} + \frac{C}{(x - 1)^2} \qquad \text{Write the identity.}$$

$$1 = A(x - 1)^2 + B \cdot x(x - 1) + C \cdot x \qquad \text{Multiply both sides by } x(x - 1)^2.$$

We find $A$, $B$, and $C$ by choosing values of $x$ that cause one or more terms to drop out. When $x = 1$, we have $1 = C \cdot 1$, so $C = 1$. When $x = 0$, we have $1 = A(0 - 1)^2$, so $A = 1$. Now using $A = 1$ and $C = 1$, we have

$$1 = (x - 1)^2 + B \cdot x(x - 1) + 1 \cdot x$$

Let $x = 2$. (Any choice other than 0 and 1 will also work.) Then

$$1 = 1 + 2B + 2$$

$$B = -1$$

Then

$$\frac{1}{x(x - 1)^2} = \frac{1}{x} + \frac{-1}{(x - 1)} + \frac{1}{(x - 1)^2} \qquad A = 1 \quad B = -1 \quad C = 1$$

**NOTE** Case 2 type integrands lead to sums of natural logarithms and rational functions.

So,

$$\int \frac{dx}{x(x - 1)^2} = \int \frac{dx}{x} - \int \frac{dx}{x - 1} + \int \frac{dx}{(x - 1)^2}$$

**NOTE** To avoid confusion, we use $K_1$ for the constant of integration whenever $C$ appears in the partial fraction decomposition.

$$= \ln|x| - \ln|x - 1| - \frac{1}{x - 1} + K_1 \qquad \blacksquare$$

**NOW WORK** Problem **27**.

## 3 Integrate a Proper Rational Function Whose Denominator Contains a Distinct Irreducible Quadratic Factor

**NEED TO REVIEW?** The discriminant of a quadratic equation is discussed in Appendix A.1, p. A-3.

A quadratic polynomial $ax^2 + bx + c$ is called **irreducible** if it cannot be factored into real linear factors. This happens if the discriminant $b^2 - 4ac < 0$. For example, $x^2 + x + 1$ and $x^2 + 4$ are irreducible.

> **Case 3:** If the denominator $q$ contains a nonrepeated irreducible quadratic factor $ax^2 + bx + c$, then the decomposition of $\dfrac{p}{q}$ includes the term
>
> $$\boxed{\dfrac{Ax + B}{ax^2 + bx + c}}$$
>
> where $A$ and $B$ are real numbers.

**CALC**
▶ **EXAMPLE 5** **Integrating a Proper Rational Function Whose Denominator Contains a Distinct Irreducible Quadratic Factor**
**CLIP**

Find $\displaystyle\int \frac{3x}{x^3 - 1}\, dx$.

**Solution** The denominator, $x^3 - 1 = (x - 1)(x^2 + x + 1)$, contains a nonrepeated linear factor, $x - 1$. Then by Case 1 the decomposition of $\dfrac{3x}{x^3 - 1} = \dfrac{3x}{(x - 1)(x^2 + x + 1)}$ has the term $\dfrac{A}{x - 1}$.

The discriminant of the quadratic expression $x^2 + x + 1$ is negative, so $x^2 + x + 1$ is an irreducible quadratic factor of the denominator $q$, and the decomposition of $\dfrac{p}{q}$ also contains the term $\dfrac{Bx + C}{x^2 + x + 1}$. Then

$$\frac{3x}{x^3 - 1} = \frac{A}{x - 1} + \frac{Bx + C}{x^2 + x + 1}$$

Clearing the denominators, we have

$$3x = A(x^2 + x + 1) + (Bx + C)(x - 1)$$

This is an identity in $x$. When $x = 1$, we have $3 = 3A$, so $A = 1$. With $A = 1$, the identity becomes

$$3x = (x^2 + x + 1) + (Bx + C)(x - 1)$$
$$-x^2 + 2x - 1 = (Bx + C)(x - 1)$$
$$-(x - 1)^2 = (Bx + C)(x - 1)$$
$$-(x - 1) = Bx + C$$
$$B = -1 \quad C = 1$$

So,

$$\frac{3x}{x^3 - 1} = \frac{1}{x - 1} + \frac{-x + 1}{x^2 + x + 1}$$

$$\int \frac{3x}{x^3 - 1}\, dx = \int \left(\frac{1}{x - 1} + \frac{-x + 1}{x^2 + x + 1}\right) dx = \int \frac{1}{x - 1}\, dx + \int \frac{-x + 1}{x^2 + x + 1}\, dx$$

$$= \ln|x - 1| - \int \frac{x - 1}{x^2 + x + 1}\, dx \tag{2}$$

To find the integral on the right, complete the square in the denominator and use substitution.

$$\int \frac{x-1}{x^2+x+1}\,dx = \int \frac{x-1}{\left(x+\frac{1}{2}\right)^2+\frac{3}{4}}\,dx \underset{\underset{u=x+\frac{1}{2}}{\uparrow}}{=} \int \frac{u-\frac{3}{2}}{u^2+\frac{3}{4}}\,du = \int \frac{u}{u^2+\frac{3}{4}}\,du - \frac{3}{2}\int \frac{du}{u^2+\frac{3}{4}}$$

$$= \frac{1}{2}\ln\left(u^2+\frac{3}{4}\right) - \frac{3}{2}\left[\frac{2}{\sqrt{3}}\tan^{-1}\left(\frac{2}{\sqrt{3}}u\right)\right] \qquad \int \frac{du}{u^2+a^2} = \frac{1}{a}\tan^{-1}\frac{u}{a}$$

$$= \frac{1}{2}\ln\left(u^2+\frac{3}{4}\right) - \sqrt{3}\tan^{-1}\left(\frac{2}{\sqrt{3}}u\right)$$

$$= \frac{1}{2}\ln(x^2+x+1) - \sqrt{3}\tan^{-1}\frac{2x+1}{\sqrt{3}} \qquad 2u = 2x+1$$

Then from (2),

$$\int \frac{3x}{x^3-1}\,dx = \ln|x-1| - \frac{1}{2}\ln(x^2+x+1) + \sqrt{3}\tan^{-1}\frac{2x+1}{\sqrt{3}} + K_1 \qquad \blacksquare$$

NOW WORK **Problem 33.**

**NOTE** Case 3 type integrands lead to sums of natural logarithms and/or inverse tangent functions.

## 4 Integrate a Proper Rational Function Whose Denominator Contains a Repeated Irreducible Quadratic Factor

*Case 4:* If the denominator $q$ contains a repeated irreducible quadratic polynomial $(x^2+bx+c)^n$, $n \geq 2$ an integer, then the decomposition of $\frac{p}{q}$ includes the terms

$$\frac{A_1 x + B_1}{x^2+bx+c}, \quad \frac{A_2 x + B_2}{(x^2+bx+c)^2}, \quad \dots, \quad \frac{A_n x + B_n}{(x^2+bx+c)^n}$$

where $A_1, B_1, A_2, B_2, \dots, A_n, B_n$ are real numbers.

---

EXAMPLE 6 **Integrating a Proper Rational Function Whose Denominator Contains a Repeated Irreducible Quadratic Factor**

Find $\displaystyle\int \frac{x^3+1}{(x^2+4)^2}\,dx$.

**Solution** The denominator is a repeated, irreducible quadratic, so the decomposition of $\dfrac{x^3+1}{(x^2+4)^2}$ is

$$\frac{x^3+1}{(x^2+4)^2} = \frac{Ax+B}{x^2+4} + \frac{Cx+D}{(x^2+4)^2}$$

Clearing fractions and combining terms give

$$x^3+1 = (Ax+B)(x^2+4) + Cx + D$$

$$x^3+1 = Ax^3 + Bx^2 + (4A+C)x + 4B+D$$

Equating coefficients,

$$A=1 \qquad B=0 \qquad 4A+C=0 \qquad 4B+D=1$$
$$C=-4 \qquad\qquad D=1$$

Then

$$\frac{x^3+1}{(x^2+4)^2} = \frac{x}{x^2+4} + \frac{-4x+1}{(x^2+4)^2}$$

and

$$\int \frac{x^3+1}{(x^2+4)^2}\,dx = \int \frac{x}{x^2+4}\,dx + \int \frac{-4x+1}{(x^2+4)^2}\,dx$$

$$= \int \frac{x}{x^2+4}\,dx - 4\int \frac{x}{(x^2+4)^2}\,dx + \int \frac{dx}{(x^2+4)^2} \quad (3)$$

In the first two integrals on the right in (3), use the substitution $u = x^2 + 4$, $x \geq 0$. Then $du = 2x\,dx$, and

$$\bullet \qquad \int \frac{x}{x^2+4}\,dx = \frac{1}{2}\int \frac{du}{u} = \frac{1}{2}\ln|u| = \frac{1}{2}\ln(x^2+4) \quad (4)$$

$$\bullet \quad -4\int \frac{x}{(x^2+4)^2}\,dx = -2\int \frac{du}{u^2} = \frac{2}{u} = \frac{2}{x^2+4} \quad (5)$$

In the third integral on the right in (3), use the trigonometric substitution $x = 2\tan\theta$, $-\frac{\pi}{2} < \theta < \frac{\pi}{2}$. Then $dx = 2\sec^2\theta\,d\theta$, and

$$\int \frac{dx}{(x^2+4)^2} = \int \frac{2\sec^2\theta\,d\theta}{(4\tan^2\theta+4)^2} = \frac{2}{16}\int \frac{\sec^2\theta\,d\theta}{(\sec^2\theta)^2}$$

$$= \frac{1}{8}\int \cos^2\theta\,d\theta = \frac{1}{8}\int \frac{1+\cos(2\theta)}{2}\,d\theta$$

$$= \frac{1}{16}\left[\theta + \frac{1}{2}\sin(2\theta)\right] = \frac{1}{16}(\theta + \sin\theta\cos\theta)$$

**Figure 10** $\tan\theta = \dfrac{x}{2}$, $-\dfrac{\pi}{2} < \theta < \dfrac{\pi}{2}$

To express the solution in terms of $x$, either use the triangles in Figure 10 or use trigonometric identities as follows:

$$\sin\theta\cos\theta = \frac{\sin\theta}{\cos\theta}\cdot\cos^2\theta = \frac{\tan\theta}{\sec^2\theta} = \frac{\tan\theta}{\tan^2\theta+1} = \frac{\dfrac{x}{2}}{\dfrac{x^2}{4}+1} = \frac{2x}{x^2+4}$$

Then

$$\int \frac{dx}{(x^2+4)^2} = \frac{1}{16}(\theta + \sin\theta\cos\theta) = \frac{1}{16}\left(\tan^{-1}\frac{x}{2} + \frac{2x}{x^2+4}\right) \quad (6)$$

$$\underset{x=2\tan\theta;\ \theta=\tan^{-1}\frac{x}{2}}{\uparrow}$$

Now combine the results of (3), (4), (5), and (6):

$$\int \frac{x^3+1}{(x^2+4)^2}\,dx = \frac{1}{2}\ln(x^2+4) + \frac{2}{x^2+4} + \frac{1}{16}\tan^{-1}\frac{x}{2} + \frac{x}{8(x^2+4)} + K_1 \quad \blacksquare$$

**NOTE** Case 4 type integrands lead to sums of natural logarithms, inverse tangents, and/or rational functions.

NOW WORK Problem **35**.

## 5 Work with the Logistic Model

**NEED TO REVIEW?** The uninhibited growth model is discussed in Section 5.6, pp. 414–417.

The uninhibited growth model $A(t) = A_0 e^{kt}$, $k > 0$, assumes that a population always grows at a rate proportional to its size. The assumption is valid for the early stages of growth. But as a population grows and matures, limited resources and other natural phenomena cause the growth rate to slow and taper off, making the uninhibited growth model inadequate.

In 1845, Pierre Verhulst proposed a differential equation known as the *logistic differential equation* to model world population. Verhulst's model assumes that a population initially exhibits uninhibited growth. The rate of uninhibited growth is called the **maximum population growth rate** because in Verhulst's model, the population continues to increase over time, but the rate of increase decreases. Further, the model assumes that population is bounded from above by some number, called the **carrying capacity**.

The **logistic differential equation** for a population $P$ that follows a logistic model over time $t$ is given by

$$\frac{dP}{dt} = kP\left(1 - \frac{P}{M}\right) \tag{7}$$

where $k$ is the maximum population growth rate and $M$ is the carrying capacity.

Notice that when $P$ is small relative to $M$, then $\frac{P}{M}$ is close to 0 and $\frac{dP}{dt} \approx kP$ so the logistic model approximates the uninhibited growth model. On the other hand, when $P$ is close to the carrying capacity $M$, then $\frac{P}{M}$ is close to 1 and $\frac{dP}{dt}$ approaches 0, indicating slow growth.

The logistic differential equation (7) can be rewritten in the form

$$\frac{dP}{dt} = kP\left(1 - \frac{P}{M}\right) = \frac{k}{M}P(M - P)$$

In a population in which some members have a disease and some do not, this differential equation states that the rate of change of the segment of the population with the disease is proportional to the product of the segment with the disease and the segment of the population without the disease.

The solution $P = P(t)$ of the logistic differential equation (7) is called a **logistic function** and its graph is called a **logistic curve**. The logistic function $P = P(t)$ can be found by separating the variables.

**NEED TO REVIEW?** Separation of variables is discussed in Section 5.6, pp. 413–414.

$$\frac{dP}{dt} = kP\left(1 - \frac{P}{M}\right) \qquad \text{Equation (7)}$$

$$\frac{dP}{P\left(1 - \frac{P}{M}\right)} = k\,dt \qquad \text{Separate the variables.}$$

$$\int \frac{dP}{P\left(1 - \frac{P}{M}\right)} = \int k\,dt \qquad \text{Integrate both sides.} \tag{8}$$

The integrand on the left is a rational function with distinct linear factors (Case 1). Express the integrand as

$$\frac{1}{P\left(1 - \frac{P}{M}\right)} = \frac{M}{P(M - P)} \qquad \text{Multiply by } \frac{M}{M}.$$

$$\frac{M}{P(M - P)} = \frac{A}{P} + \frac{B}{M - P} \qquad \text{Use Case 1.}$$

$$M = A(M - P) + BP$$

When $P = 0$, the term involving $B$ drops out, leaving $A = 1$. When $P = M$, the term involving $A$ drops out, leaving $B = 1$. Then

$$\frac{M}{P(M-P)} = \frac{1}{P} + \frac{1}{M-P}$$

$$\int \frac{dP}{P\left(1-\dfrac{P}{M}\right)} = \int \frac{M}{P(M-P)}\,dP = \int \left(\frac{1}{P} + \frac{1}{M-P}\right)dP$$

$$= \ln|P| - \ln|M-P| = \ln\left|\frac{P}{M-P}\right| \qquad (9)$$

Returning to (8), we have

$$\int \frac{dP}{P\left(1-\dfrac{P}{M}\right)} = \int k\,dt \qquad \text{(8)}$$

$$\ln\left|\frac{P}{M-P}\right| = kt + C \qquad \textcolor{blue}{\text{From (9)}}$$

$$\ln\left|\frac{M-P}{P}\right| = -kt - C$$

$$\frac{M-P}{P} = e^{-kt} \cdot e^{-C}$$

$$\frac{M-P}{P} = ae^{-kt} \qquad (10)$$

where $a = e^{-C}$. Now solve for $P$.

$$M - P = P(ae^{-kt})$$
$$M = P + P(ae^{-kt})$$
$$P = P(t) = \frac{M}{1+ae^{-kt}}$$

The value of $a$ is found using the initial condition $P(0) = P_0$ in equation (10). If $t = 0$, then $e^{-kt} = 1$ and

$$a = \frac{M-P_0}{P_0}$$

---

**THEOREM** Logistic Function

The solution to the logistic differential equation

$$\frac{dP}{dt} = kP\left(1-\frac{P}{M}\right)$$

is

$$\boxed{P(t) = \frac{M}{1+ae^{-kt}}}$$

where $P(t)$ is the population at time $t$, $M$ is the carrying capacity, $k$ is the maximum population growth rate, $P_0 = P(0)$, and $a = \dfrac{M-P_0}{P_0}$.

Figure 11 shows the graphs of $P = P(t)$ for a growth model $(k > 0)$ and a decay model $(k < 0)$.

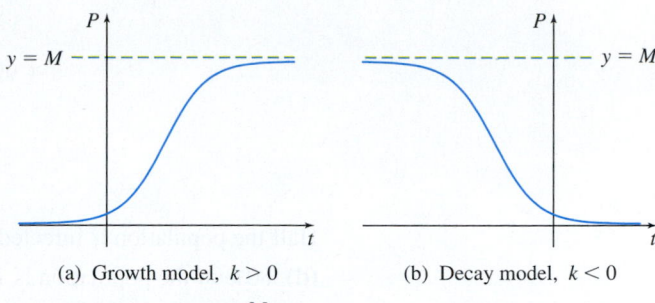

(a) Growth model, $k > 0$          (b) Decay model, $k < 0$

**Figure 11** Logistic curves $P(t) = \dfrac{M}{1 + ae^{-kt}}$.

An analysis of the logistic function $P(t) = \dfrac{M}{1 + ae^{-kt}}$ reveals several of its properties.

- The domain of the logistic function $P$ is the set of all real numbers.

- Since $\dfrac{dP}{dt} = kP\left(1 - \dfrac{P}{M}\right) > 0$ if $k > 0$ and $\dfrac{dP}{dt} < 0$ if $k < 0$, the logistic function
  $P = P(t)$ increases if $k > 0$ (growth model) and decreases if $k < 0$ (decay model).

- The second derivative of $P = P(t)$ is

$$\frac{d^2 P}{dt^2} = k\left(1 - \frac{2P}{M}\right)\frac{dP}{dt}$$

- $P = P(t)$ has an inflection point when the population equals $\dfrac{M}{2}$.

- The function $P = P(t)$ has two horizontal asymptotes $P = 0$ and $P = M$.

- From the information above, we also conclude that $0 < P(t) < M$ for all $t$.

---

**EXAMPLE 7**  **Analyzing a Logistic Model: Spread of Disease**

In an experiment involving the spread of disease, one infected mouse is placed in a group of 99 healthy mice. The rate of change of the number of infected mice with respect to time $t$ (in days) follows a logistic model.

**(a)** If $P$ is the number of infected mice and the maximum daily growth rate of the disease is 10%, write a differential equation that models the experiment.

**(b)** Write the logistic function that satisfies the differential equation.

**(c)** Find the time it takes for $\dfrac{1}{2}$ of the population to become infected.

**(d)** How long does it take for 80% of the population to become infected?

**Solution** **(a)** The total population of mice is 100, the carrying capacity. When $t = 0$, one mouse is infected. The maximum daily growth rate is $k = 0.10$. The differential equation for the experiment is

$$\frac{dP}{dt} = 0.10P\left(1 - \frac{P}{100}\right) \qquad P(0) = 1$$

**(b)** From (a), we know that the carrying capacity $M$ equals 100 mice and the maximum daily growth rate is $k = 0.10$. Since $P(0) = 1$, then

$$a = \frac{M - P_0}{P_0} = \frac{100 - 1}{1} = 99$$

So, the logistic function that satisfies the differential equation is

$$P(t) = \frac{M}{1 + ae^{-kt}} = \frac{100}{1 + 99e^{-0.1t}}$$

(c) Half the population means $P = 50$. Then

$$50 = \frac{100}{1 + 99e^{-0.1t}}$$

$$1 + 99e^{-0.1t} = 2$$

$$-0.1t = \ln\frac{1}{99}$$

$$t = -10(-\ln 99) \approx 45.951$$

Half the population is infected on the 46th day.

(d) 80% of the population is $P = 80$ mice.

$$80 = \frac{100}{1 + 99e^{-0.1t}}$$

$$1 + 99e^{-0.1t} = \frac{5}{4}$$

$$99e^{-0.1t} = \frac{1}{4}$$

$$e^{-0.1t} = \frac{1}{396}$$

$$t = -10\left(\ln\frac{1}{396}\right) \approx 59.814$$

By 60 days, 80% of the mouse population is infected. ∎

Figure 12 shows the graph of $P = P(t)$

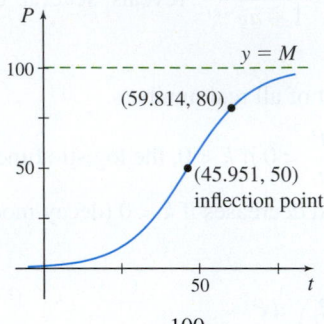

**Figure 12** $P(t) = \dfrac{100}{1 + 99e^{-0.1t}}$

NOW WORK **Problem 73.**

## 7.5 Assess Your Understanding

### Concepts and Vocabulary

**1.** *Multiple Choice*   A rational function $R(x) = \dfrac{p(x)}{q(x)}$ is proper when the degree of $p$ is [(a) less than (b) equal to (c) greater than] the degree of $q$.

**2.** *True or False*   Every improper rational function can be written as the sum of a polynomial and a proper rational function.

**3.** *True or False*   Sometimes the integration of a proper rational function leads to a logarithm.

**4.** *True or False*   The decomposition of $\dfrac{7x + 1}{(x + 1)^4}$ into partial fractions has three terms: $\dfrac{A}{x + 1} + \dfrac{B}{(x + 1)^2} + \dfrac{C}{(x + 1)^3}$, where $A$, $B$, and $C$ are real numbers.

### Skill Building

*In Problems 5–14, write each improper rational function as the sum of a polynomial function and a proper rational function.*

**5.** $R(x) = \dfrac{x^2 + 1}{x + 1}$

**6.** $R(x) = \dfrac{x^2 + 4}{x - 2}$

**7.** $R(x) = \dfrac{x^3 + 3x - 4}{x - 2}$

**8.** $R(x) = \dfrac{x^3 - 3x^2 + 4}{x + 3}$

**9.** $R(x) = \dfrac{2x^3 + 3x^2 - 17x - 27}{x^2 - 9}$

**10.** $R(x) = \dfrac{3x^3 - 2x^2 - 3x + 2}{x^2 - 1}$

**11.** $R(x) = \dfrac{x^4 - 1}{x(x + 4)}$

**12.** $R(x) = \dfrac{x^4 + x^2 + 1}{x(x - 2)}$

**13.** $R(x) = \dfrac{2x^4 + x^2 - 2}{x^2 + 4}$

**14.** $R(x) = \dfrac{3x^4 + x^2}{x^2 + 9}$

*In Problems 15–20, find each integral by first writing the integrand as the sum of a polynomial and a proper rational function.*

**15.** $\displaystyle\int \frac{x^2 + 1}{x + 1}\, dx$

**16.** $\displaystyle\int \frac{x^2 + 4}{x - 2}\, dx$

**17.** $\displaystyle\int \frac{x^3 + 3x - 4}{x - 2}\, dx$

**18.** $\displaystyle\int \frac{x^3 - 3x^2 + 4}{x + 3}\, dx$

**19.** $\displaystyle\int \frac{2x^3 + 3x^2 - 17x - 27}{x^2 - 9}\, dx$

**20.** $\displaystyle\int \frac{3x^3 - 2x^2 - 3x + 2}{x^2 - 1}\, dx$

*In Problems 21–26, find each integral. (Hint: Each of the denominators contains only distinct linear factors.)*

**21.** $\displaystyle\int \frac{dx}{(x - 2)(x + 1)}$

**22.** $\displaystyle\int \frac{dx}{(x + 4)(x - 1)}$

**23.** $\displaystyle\int \frac{x\, dx}{(x - 1)(x - 2)}$

**24.** $\displaystyle\int \frac{3x\, dx}{(x + 2)(x - 4)}$

**25.** $\displaystyle\int \frac{x\, dx}{(3x - 2)(2x + 1)}$

**26.** $\displaystyle\int \frac{dx}{(2x + 3)(4x - 1)}$

---

**1.** = NOW WORK problem        📈 = Graphing technology recommended        CAS = Computer Algebra System recommended

*In Problems 27–30, find each integral. (Hint: Each of the denominators contains a repeated linear factor.)*

**27.** $\displaystyle\int \frac{x-3}{(x+2)(x+1)^2}\,dx$

**28.** $\displaystyle\int \frac{x+1}{x^2(x-2)}\,dx$

**29.** $\displaystyle\int \frac{x^2\,dx}{(x-1)^2(x+1)}$

**30.** $\displaystyle\int \frac{x^2+x}{(x+2)(x-1)^2}\,dx$

*In Problems 31–34, find each integral. (Hint: Each of the denominators contains an irreducible quadratic factor.)*

**31.** $\displaystyle\int \frac{dx}{x(x^2+1)}$

**32.** $\displaystyle\int \frac{dx}{(x+1)(x^2+4)}$

**33.** $\displaystyle\int \frac{x^2+2x+3}{(x+1)(x^2+2x+4)}\,dx$

**34.** $\displaystyle\int \frac{x^2-11x-18}{x(x^2+3x+3)}\,dx$

*In Problems 35–38, find each integral. (Hint: Each of the denominators contains a repeated irreducible quadratic factor.)*

**35.** $\displaystyle\int \frac{2x+1}{(x^2+16)^2}\,dx$

**36.** $\displaystyle\int \frac{x^2+2x+3}{(x^2+4)^2}\,dx$

**37.** $\displaystyle\int \frac{x^3\,dx}{(x^2+16)^3}$

**38.** $\displaystyle\int \frac{x^2\,dx}{(x^2+4)^3}$

*In Problems 39–48, find each integral.*

**39.** $\displaystyle\int \frac{x\,dx}{x^2+2x-3}$

**40.** $\displaystyle\int \frac{x^2-x-8}{(x+1)(x^2+5x+6)}\,dx$

**41.** $\displaystyle\int \frac{10x^2+2x}{(x-1)^2(x^2+2)}\,dx$

**42.** $\displaystyle\int \frac{x+4}{x^2(x^2+4)}\,dx$

**43.** $\displaystyle\int \frac{7x+3}{x^3-2x^2-3x}\,dx$

**44.** $\displaystyle\int \frac{x^5+1}{x^6-x^4}\,dx$

**45.** $\displaystyle\int \frac{x^2}{(x-2)(x-1)^2}\,dx$

**46.** $\displaystyle\int \frac{x^2+1}{(x+3)(x-1)^2}\,dx$

**47.** $\displaystyle\int \frac{2x+1}{x^3-1}\,dx$

**48.** $\displaystyle\int \frac{dx}{x^3-8}$

*In Problems 49–52, find each definite integral.*

**49.** $\displaystyle\int_0^1 \frac{dx}{x^2-9}$

**50.** $\displaystyle\int_2^4 \frac{dx}{x^2-25}$

**51.** $\displaystyle\int_{-2}^3 \frac{dx}{16-x^2}$

**52.** $\displaystyle\int_1^2 \frac{dx}{9-x^2}$

## Applications and Extensions

*In Problems 53–66, find each integral. (Hint: Make a substitution before using partial fraction decomposition.)*

**53.** $\displaystyle\int \frac{\cos\theta}{\sin^2\theta+\sin\theta-6}\,d\theta$

**54.** $\displaystyle\int \frac{\sin x}{\cos^2 x-2\cos x-8}\,dx$

**55.** $\displaystyle\int \frac{\sin\theta}{\cos^3\theta+\cos\theta}\,d\theta$

**56.** $\displaystyle\int \frac{4\cos\theta}{\sin^3\theta+2\sin\theta}\,d\theta$

**57.** $\displaystyle\int \frac{e^t}{e^{2t}+e^t-2}\,dt$

**58.** $\displaystyle\int \frac{e^x}{e^{2x}+e^x-6}\,dx$

**59.** $\displaystyle\int \frac{e^x}{e^{2x}-1}\,dx$

**60.** $\displaystyle\int \frac{dx}{e^x-e^{-x}}$

**61.** $\displaystyle\int \frac{dt}{e^{2t}+1}$

**62.** $\displaystyle\int \frac{dt}{e^{3t}+e^t}$

**63.** $\displaystyle\int \frac{\sin x\cos x}{(\sin x-1)^2}\,dx$

**64.** $\displaystyle\int \frac{\cos x\sin x}{(\cos x-2)^2}\,dx$

**65.** $\displaystyle\int \frac{\cos x}{(\sin^2 x+9)^2}\,dx$

**66.** $\displaystyle\int \frac{\sin x}{(\cos^2 x+4)^2}\,dx$

**67. Area** Find the area under the graph of $y=\dfrac{4}{x^2-4}$ from $x=3$ to $x=5$, as shown in the figure below.

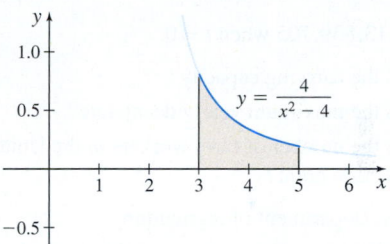

**68. Area** Find the area under the graph of $y=\dfrac{x-4}{(x+3)^2}$ from $x=4$ to $x=6$, as shown in the figure below.

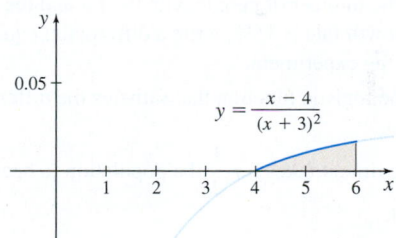

**69. Area** Find the area under the graph of $y=\dfrac{8}{x^3+1}$ from $x=0$ to $x=2$, as shown in the figure below.

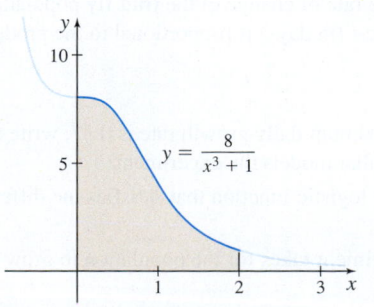

**70. Volume of a Solid of Revolution**  Find the volume of the solid of revolution generated by revolving the region bounded by the graph of $y = \dfrac{x}{x^2 - 4}$ and the $x$-axis from $x = 3$ to $x = 5$ about the $x$-axis, as shown in the figure below.

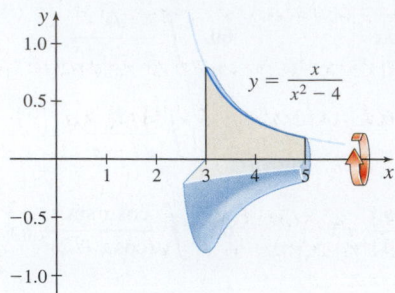

**71. Arc Length**  Find the arc length of the graph of $y = \ln x$ from $x = 1$ to $x = e$.

**72. Farmers**  The number $W$ of farm workers in the United States $t$ years after 1910 follows the logistic decay model

$$\frac{dW}{dt} = -0.057W \left( 1 - \frac{W}{14{,}656{,}248} \right)$$

where $W = 13{,}839{,}705$ when $t = 0$.

**(a)** What is the carrying capacity?

**(b)** What is the maximum yearly decay rate?

**(c)** What is the number of farm workers in the United States at the inflection point?

*Source:* U.S. Department of Agriculture

**73. Spread of Flu**  Suppose one person with the flu is placed in a group of 49 people without the flu. The rate of change of those with the flu with respect to time $t$ (in days) is proportional to the product of the number of those with the flu and the number without flu.

**(a)** If $P$ is the number of people with the flu and the maximum daily growth rate is 15%, write a differential equation that models the experiment.

**(b)** Write the logistic function that satisfies the differential equation.

**(c)** Find the time it takes for $\dfrac{1}{2}$ the population to become infected.

**(d)** How long does it take for 80% of the population to become infected?

**74. Fruit Fly Population**  Four fruit flies are placed in a half-pint bottle with a banana (for food) and yeast (for food and to provide a stimulus to lay eggs). The carrying capacity of the bottle is 230 fruit flies. The rate of change of the fruit fly population $P$ with respect to time $t$ (in days) is proportional to the product of $P$ and $1 - \dfrac{P}{230}$.

**(a)** If the maximum daily growth rate is 0.37, write a differential equation that models the experiment.

**(b)** Write the logistic function that satisfies the differential equation.

**(c)** Find the time it takes for the population to grow to 115 fruit flies.

**(d)** How long does it take for the population to grow to 180 fruit flies?

**75.** In a certain swamp in Florida, 100 insects were released into the swamp. The daily maximum population growth rate is 20%. Entomologists have determined that the swamp can support a colony of 600,000 insects and that the population will follow a logistic growth model.

**(a)** Write a logistic differential equation that models the insect population.

**(b)** Solve the differential equation.

**(c)** Find the time it takes for the population to exceed 100,000 insects.

**76.** A population $P$ grows according to the logistic differential equation

$$\frac{dP}{dt} = 0.0005 P (800 - P)$$

**(a)** What is the carrying capacity of the population?

**(b)** What is the maximum population growth rate of the population?

**(c)** How large is the population at the inflection point?

**77. (a)** Find the zeros of $q(x) = x^3 + 3x^2 - 10x - 24$.

**(b)** Factor $q$.

**(c)** Find the integral $\displaystyle \int \frac{3x - 7}{x^3 + 3x^2 - 10x - 24}\, dx$.

## Challenge Problems

*In Problems 78–89, simplify each integrand as follows: If the integrand involves fractional powers such as $x^{p/q}$ and $x^{r/s}$, make the substitution $x = u^n$, where $n$ is the least common denominator of $\dfrac{p}{q}$ and $\dfrac{r}{s}$. Then find each integral.*

**78.** $\displaystyle \int \frac{x\, dx}{3 + \sqrt{x}}$

**79.** $\displaystyle \int \frac{dx}{\sqrt{x} + 2}$

**80.** $\displaystyle \int \frac{dx}{x - \sqrt[3]{x}}$

**81.** $\displaystyle \int \frac{x\, dx}{\sqrt[3]{x} - 1}$

**82.** $\displaystyle \int \frac{dx}{\sqrt{x} + \sqrt[3]{x}}$

**83.** $\displaystyle \int \frac{dx}{3\sqrt{x} - \sqrt[3]{x}}$

**84.** $\displaystyle \int \frac{dx}{\sqrt[3]{2 + 3x}}$

**85.** $\displaystyle \int \frac{dx}{\sqrt[4]{1 + 2x}}$

**86.** $\displaystyle \int \frac{x\, dx}{(1 + x)^{3/4}}$

**87.** $\displaystyle \int \frac{dx}{(1 + x)^{2/3}}$

**88.** $\displaystyle \int \frac{\sqrt[3]{x} + 1}{\sqrt[3]{x} - 1}\, dx$

**89.** $\displaystyle \int \frac{dx}{\sqrt{x}(1 + \sqrt[3]{x})^2}$

**Weierstrass Substitution**  *In Problems 90–105, use the following substitution, called a **Weierstrass substitution**. If an integrand is a rational expression of $\sin x$ or $\cos x$ or both, the substitution*

$$z = \tan \frac{x}{2} \qquad -\frac{\pi}{2} < \frac{x}{2} < \frac{\pi}{2}$$

*or equivalently,*

$$\cos x = \frac{1 - z^2}{1 + z^2} \qquad dx = \frac{2\, dz}{1 + z^2}$$

*will transform the integrand into a rational function of $z$.*

**90.** $\displaystyle\int \frac{dx}{1-\sin x}$

**91.** $\displaystyle\int \frac{dx}{1+\sin x}$

**92.** $\displaystyle\int \frac{dx}{1-\cos x}$

**93.** $\displaystyle\int \frac{dx}{3+2\cos x}$

**94.** $\displaystyle\int \frac{2\,dx}{\sin x+\cos x}$

**95.** $\displaystyle\int \frac{dx}{1-\sin x+\cos x}$

**96.** $\displaystyle\int \frac{\sin x}{3+\cos x}dx$

**97.** $\displaystyle\int \frac{dx}{\tan x-1}$

**98.** $\displaystyle\int \frac{dx}{\tan x-\sin x}$

**99.** $\displaystyle\int \frac{\sec x}{\tan x-2}dx$

**100.** $\displaystyle\int \frac{\cot x}{1+\sin x}dx$

**101.** $\displaystyle\int \frac{\sec x}{1+\sin x}dx$

**102.** $\displaystyle\int_0^{\pi/2} \frac{dx}{\sin x+1}$

**103.** $\displaystyle\int_{\pi/4}^{\pi/3} \frac{\csc x}{3+4\tan x}dx$

**104.** $\displaystyle\int_0^{\pi/2} \frac{\cos x}{2-\cos x}dx$

**105.** $\displaystyle\int_0^{\pi/4} \frac{4\,dx}{\tan x+1}$

**106.** Use a Weierstrass substitution to derive the formula

$$\int \csc x\,dx = \ln\sqrt{\frac{1-\cos x}{1+\cos x}}+C$$

**107.** Show that the result obtained in Problem 106 is equivalent to

$$\int \csc x\,dx = \ln|\csc x-\cot x|+C$$

**108.** Since $\dfrac{d}{dx}\tanh^{-1}x = \dfrac{1}{1-x^2}$, we might expect that

$$\int_2^3 \frac{dx}{1-x^2} = \tanh^{-1}3-\tanh^{-1}2.$$

Why is this incorrect? What is the correct result?

**109.** Show that the two formulas below are equivalent.

$$\int \sec x\,dx = \ln|\sec x+\tan x|+C$$

$$\int \sec x\,dx = \ln\left|\frac{1+\tan\dfrac{x}{2}}{1-\tan\dfrac{x}{2}}\right|+C$$

*Hint:*  $\tan\dfrac{x}{2} = \dfrac{\sin\dfrac{x}{2}}{\cos\dfrac{x}{2}} = \dfrac{\sin^2\left(\dfrac{x}{2}\right)}{\sin\dfrac{x}{2}\cos\dfrac{x}{2}} = \dfrac{1-\cos x}{\sin x}.$

**110.** Use the methods of this section to find $\displaystyle\int \frac{dx}{1+x^4}$.

*Hint:* Factor $1+x^4$ into irreducible quadratics.

# 7.6 Approximating Integrals: The Trapezoidal Rule, the Midpoint Rule, Simpson's Rule

**OBJECTIVES** *When you finish this section, you should be able to:*

**1** Approximate an integral using the Trapezoidal Rule (p. 542)

**2** Approximate an integral using the Midpoint Rule (p. 547)

**3** Approximate an integral using Simpson's Rule (p. 549)

Most of the functions we have encountered so far belong to the class known as **elementary functions**. These functions include polynomial, exponential, logarithmic, trigonometric, inverse trigonometric, hyperbolic, and inverse hyperbolic functions, as well as functions formed by combining one or more of these functions using addition, subtraction, multiplication, division, or composition.

We have found that the derivatives $f'$, $f''$, and so on, of an elementary function $f$ are also elementary functions. The integral of an elementary function $\int f(x)\,dx$, however, is not always an elementary function. Some examples of integrals of elementary functions that cannot be expressed in terms of elementary functions are

$$\int e^{x^2}\,dx \qquad \int e^{-x^2}\,dx \qquad \int \frac{\sin x}{x}\,dx$$

$$\int \frac{\cos x}{x}\,dx \qquad \int \frac{e^x\,dx}{x} \qquad \int \frac{dx}{\sqrt{1-x^3}}$$

Recall that to find a definite integral using the Fundamental Theorem of Calculus requires an antiderivative of the integrand $f$. When it is not possible to express an antiderivative of $f$ in terms of elementary functions, or when the integrand $f$ is defined by a table of data or by a graph, the Fundamental Theorem is not useful. In such situations, there are numerical techniques we can use to approximate the definite integral.

NEED TO REVIEW? Approximating a definite integral using Riemann sums is discussed in Section 5.2, pp. 368–371.

In Chapter 5, Section 5.2, we used Riemann sums to approximate a definite integral. We began with a function $y = f(x)$ defined on a closed interval $[a, b]$ and partitioned $[a, b]$ into $n$ subintervals each of width $\Delta x = \dfrac{b - a}{n}$. In each subinterval we chose a number $u_i$ and evaluated $f(u_i)$. While the $u_i$ could be chosen in any way, we usually used the left endpoint, the midpoint, or the right endpoint of the interval. The Riemann sums $\sum\limits_{i=1}^{n} f(u_i)\Delta x$ gave an approximation to the definite integral $\int_a^b f(x)\, dx$.

In this section we discuss three other ways to approximate a definite integral: the *Trapezoidal Rule, the Midpoint Rule*, and *Simpson's Rule*. A fifth method, based on *series*, is discussed in Chapter 8.

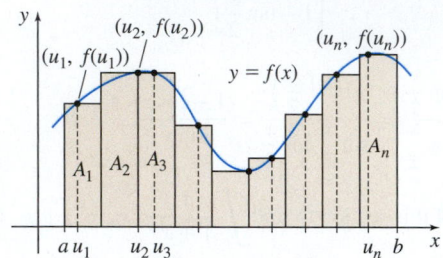

**Figure 13** $A = \int_a^b f(x)\, dx$

$$\approx \sum_{i=1}^{n} A_i = \sum_{i=1}^{n} f(u_i)\Delta x$$

## 1 Approximate an Integral Using the Trapezoidal Rule

As we learned in Section 5.2, if a function $f$ is nonnegative and continuous on an interval $[a, b]$, then the area $A$ under the graph of $f$ from $a$ to $b$ equals the definite integral $\int_a^b f(x)\, dx$. We can approximate $A$ by partitioning $[a, b]$ into $n$ subintervals, each of width $\Delta x = \dfrac{b - a}{n}$; choosing a number $u_i$ from each subinterval $[x_{i-1}, x_i]$, $i = 1, 2, \ldots, n$; and forming $n$ rectangles of height $f(u_i)$ and width $\Delta x$. The sum of the areas of these rectangles approximates the integral $\int_a^b f(x)\, dx$. See Figure 13.

The Trapezoidal Rule approximates the definite integral $\int_a^b f(x)\, dx$ by replacing the rectangles with trapezoids. Then the area $A$ under the graph of $f$ from $a$ to $b$, $\int_a^b f(x)\, dx$, is approximated by the sum of the areas of these trapezoids.

**The Trapezoidal Rule**

If a function $f$ is continuous on the closed interval $[a, b]$, then

$$\int_a^b f(x)\, dx \approx \frac{b - a}{2n} \left[ f(x_0) + 2f(x_1) + 2f(x_2) + \cdots + 2f(x_{n-1}) + f(x_n) \right]$$

where the closed interval $[a, b]$ is partitioned into $n$ subintervals $[x_0, x_1], [x_1, x_2], \ldots,$ $[x_{n-1}, x_n]$, each of width $\Delta x = \dfrac{b - a}{n}$.

**RECALL** The area $A$ of a trapezoid with width $\Delta x$ and parallel heights $l_1$ and $l_2$ is

$$A = \frac{1}{2}(l_1 + l_2)\Delta x.$$

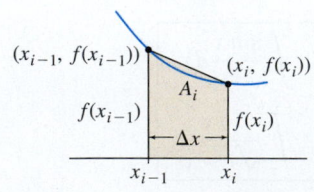

**Figure 14**

Although the Trapezoidal Rule does not require $f$ to be nonnegative on the interval $[a, b]$, the proof given here assumes that $f(x) \geq 0$. Proofs of the Trapezoidal Rule with the nonnegative restriction relaxed can be found in numerical analysis books.

**Proof** Assume $f$ is nonnegative and continuous on $[a, b]$. Partition the closed interval $[a, b]$ into $n$ subintervals, $[a, x_1], [x_1, x_2], \ldots, [x_{i-1}, x_i], \ldots, [x_{n-1}, b]$, each of width $\Delta x = \dfrac{b - a}{n}$. The $y$-coordinates corresponding to the numbers $x_0 = a, x_1, x_2, \ldots, x_{i-1},$ $x_i, \ldots, x_n = b$ are $f(x_0), f(x_1), f(x_2), \ldots, f(x_{i-1}), f(x_i), \ldots, f(x_{n-1}), f(x_n)$. Connect each pair of consecutive points $(x_{i-1}, f(x_{i-1}))$ and $(x_i, f(x_i))$, $i = 1, 2, \ldots, n$, with a line segment to form $n$ trapezoids.

The trapezoid whose base is $\Delta x = x_i - x_{i-1} = \dfrac{b - a}{n}$ has parallel sides of lengths $f(x_{i-1})$ and $f(x_i)$ as shown in Figure 14. Its area $A_i$ is

$$A_i = \frac{1}{2} \left[ f(x_{i-1}) + f(x_i) \right] \Delta x$$

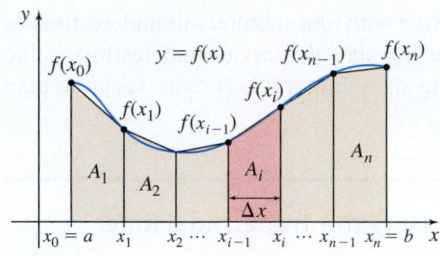

**Figure 15** $\displaystyle\sum_{i=1}^{n} A_i \approx \int_a^b f(x)\,dx$

The sum of the areas of the $n$ trapezoids approximates the area $A$ under the graph of $f$. See Figure 15.

Then

$$A \approx \frac{1}{2}\left[f(x_0)+f(x_1)\right]\Delta x + \frac{1}{2}\left[f(x_1)+f(x_2)\right]\Delta x + \cdots + \frac{1}{2}\left[f(x_{n-1})+f(x_n)\right]\Delta x$$

$$\approx \frac{1}{2}[f(x_0)+2f(x_1)+2f(x_2)+\cdots+2f(x_{n-1})+f(x_n)]\Delta x$$

Now substitute $\Delta x = \dfrac{b-a}{n}$ to obtain the Trapezoidal Rule. ∎

---

**EXAMPLE 1**

**Approximating** $\displaystyle\int_0^\pi \sin x\,dx$ **Using the Trapezoidal Rule**

**(a)** Use the Trapezoidal Rule with $n=4$ and $n=6$ to approximate $\int_0^\pi \sin x\,dx$, rounded to three decimal places.

**(b)** Compare each approximation to the exact value of $\int_0^\pi \sin x\,dx$.

**Solution (a)** Partition $[0, \pi]$ into four subintervals, each of width $\Delta x = \dfrac{\pi-0}{4} = \dfrac{\pi}{4}$.

$$\left[0, \frac{\pi}{4}\right] \qquad \left[\frac{\pi}{4}, \frac{\pi}{2}\right] \qquad \left[\frac{\pi}{2}, \frac{3\pi}{4}\right] \qquad \left[\frac{3\pi}{4}, \pi\right]$$

The value of $f(x) = \sin x$ corresponding to each endpoint is

$$f(0)=0 \qquad f\!\left(\frac{\pi}{4}\right)=\frac{\sqrt{2}}{2} \qquad f\!\left(\frac{\pi}{2}\right)=1 \qquad f\!\left(\frac{3\pi}{4}\right)=\frac{\sqrt{2}}{2} \qquad f(\pi)=0$$

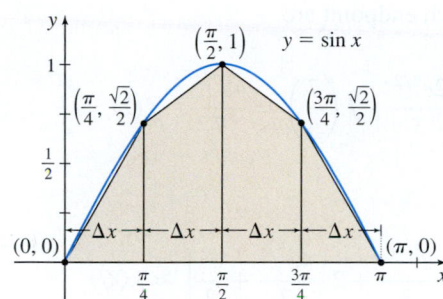

**DF Figure 16** $n=4$

See Figure 16. Now use the Trapezoidal Rule:

$$\int_0^\pi \sin x\,dx \approx \frac{\pi-0}{2\cdot 4}\left[\sin 0 + 2\sin\frac{\pi}{4} + 2\sin\frac{\pi}{2} + 2\sin\frac{3\pi}{4} + \sin\pi\right]$$

$$= \frac{\pi}{8}\left[0 + 2\cdot\frac{\sqrt{2}}{2} + 2\cdot 1 + 2\cdot\frac{\sqrt{2}}{2} + 0\right] = \frac{\pi}{8}\left(2+2\sqrt{2}\right)$$

$$= \frac{\pi}{4}\left(1+\sqrt{2}\right) \approx 1.896$$

To approximate $\int_0^\pi \sin x\,dx$ using six subintervals, partition $[0, \pi]$ into six subintervals, each of width $\Delta x = \dfrac{\pi-0}{6} = \dfrac{\pi}{6}$, namely,

$$\left[0, \frac{\pi}{6}\right] \quad \left[\frac{\pi}{6}, \frac{\pi}{3}\right] \quad \left[\frac{\pi}{3}, \frac{\pi}{2}\right] \quad \left[\frac{\pi}{2}, \frac{2\pi}{3}\right] \quad \left[\frac{2\pi}{3}, \frac{5\pi}{6}\right] \quad \left[\frac{5\pi}{6}, \pi\right]$$

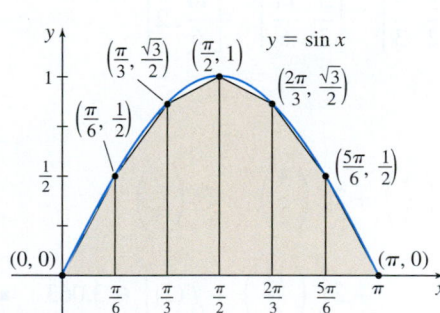

**DF Figure 17** $n=6$

See Figure 17. Then use the Trapezoidal Rule.

$$\int_0^\pi \sin x\,dx \approx \frac{\pi}{2\cdot 6}\left[0 + 2\cdot\frac{1}{2} + 2\cdot\frac{\sqrt{3}}{2} + 2\cdot 1 + 2\cdot\frac{\sqrt{3}}{2} + 2\cdot\frac{1}{2} + 0\right]$$

$$= \frac{\pi}{12}\left(4+2\sqrt{3}\right) = \frac{\pi}{6}\left(2+\sqrt{3}\right) \approx 1.954$$

**(b)** The exact value of the integral is

$$\int_0^\pi \sin x\,dx = \left[-\cos x\right]_0^\pi = -\cos\pi + \cos 0 = 1 + 1 = 2 \qquad ∎$$

The approximation using the Trapezoidal Rule with four subintervals underestimates the exact value by 0.104. The approximation using six subintervals underestimates the exact value by 0.046. The approximation using six subintervals is more accurate than the approximation using four subintervals.

---

**EXAMPLE 2** Approximating $\displaystyle\int_1^2 \frac{e^x}{x}\,dx$ Using the Trapezoidal Rule

Use the Trapezoidal Rule with $n=4$ and $n=6$ to approximate $\displaystyle\int_1^2 \frac{e^x}{x}\,dx$. Express the answer rounded to three decimal places.

**Solution** Partition $[1, 2]$ into four subintervals, each of width $\Delta x = \dfrac{2-1}{4} = \dfrac{1}{4}$:

$$\left[1, \frac{5}{4}\right] \qquad \left[\frac{5}{4}, \frac{3}{2}\right] \qquad \left[\frac{3}{2}, \frac{7}{4}\right] \qquad \left[\frac{7}{4}, 2\right]$$

The values of $f(x) = \dfrac{e^x}{x}$ corresponding to each endpoint are

$$f(1)=e \quad f\left(\frac{5}{4}\right)=\frac{4e^{5/4}}{5} \quad f\left(\frac{3}{2}\right)=\frac{2e^{3/2}}{3} \quad f\left(\frac{7}{4}\right)=\frac{4e^{7/4}}{7} \quad f(2)=\frac{e^2}{2}$$

Then, using the Trapezoidal Rule,

$$\int_1^2 \frac{e^x}{x}\,dx \approx \frac{1}{2\cdot 4}\left[e + 2\cdot\frac{4e^{5/4}}{5} + 2\cdot\frac{2e^{3/2}}{3} + 2\cdot\frac{4e^{7/4}}{7} + \frac{e^2}{2}\right] \approx 3.069$$

To approximate $\displaystyle\int_1^2 \frac{e^x}{x}\,dx$ using six subintervals, partition $[1, 2]$ into six subintervals, each of width $\Delta x = \dfrac{2-1}{6} = \dfrac{1}{6}$:

$$\left[1, \frac{7}{6}\right] \quad \left[\frac{7}{6}, \frac{4}{3}\right] \quad \left[\frac{4}{3}, \frac{3}{2}\right] \quad \left[\frac{3}{2}, \frac{5}{3}\right] \quad \left[\frac{5}{3}, \frac{11}{6}\right] \quad \left[\frac{11}{6}, 2\right]$$

Then, using the Trapezoidal Rule,

$$\int_1^2 \frac{e^x}{x}\,dx \approx \frac{1}{2\cdot 6}\left[f(1) + 2f\left(\frac{7}{6}\right) + 2f\left(\frac{4}{3}\right) + 2f\left(\frac{3}{2}\right) + 2f\left(\frac{5}{3}\right)\right.$$

$$\left. + 2f\left(\frac{11}{6}\right) + f(2)\right] \approx 3.063 \quad \blacksquare$$

Since $f(x) = \dfrac{e^x}{x} > 0$ on the interval $[1, 2]$, the integral $\displaystyle\int_1^2 \frac{e^x}{x}\,dx$ equals the area under the graph of $f$ from 1 to 2. Figure 18 shows the graph of $y = \dfrac{e^x}{x}$, the area $A = \displaystyle\int_1^2 \frac{e^x}{x}\,dx$, and the approximation to the area using the Trapezoidal Rule with four subintervals.

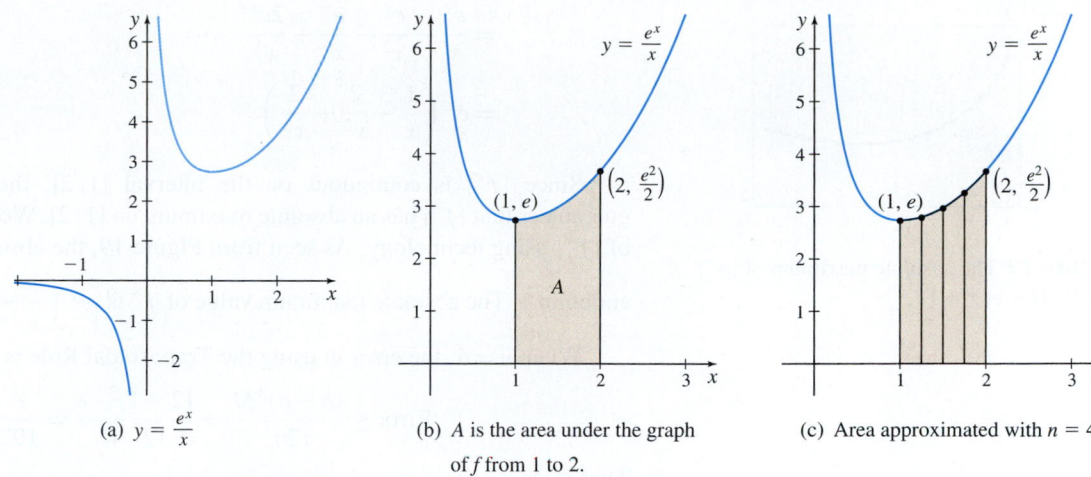

(a) $y = \dfrac{e^x}{x}$

(b) $A$ is the area under the graph of $f$ from 1 to 2.

(c) Area approximated with $n = 4$.

**Figure 18**

NOW WORK Problem 9(a).

We cannot find the exact value of the integral $\displaystyle\int_1^2 \dfrac{e^x}{x}\,dx$. The significance of the Trapezoidal Rule is that not only can it be used to approximate such integrals, but it also provides an estimate of the *error*, the difference between the exact value of the integral and the approximate value.

> **THEOREM  Error in the Trapezoidal Rule**
>
> Let $f$ be a function for which $f''$ is continuous on some open interval containing $a$ and $b$. The error in using the Trapezoidal Rule to approximate $\int_a^b f(x)\,dx$ can be estimated using the formula
>
> $$\boxed{\text{Error} \le \dfrac{(b-a)^3 M}{12n^2}}$$
>
> where $M$ is the absolute maximum value of $|f''|$ on the closed interval $[a, b]$.

**NOTE** The error formula gives an upper bound to the error. The usefulness of this result is that we can find an upper bound to the error without knowing the exact value of the integral.

A derivation of the error formula, which uses an extension of the Mean Value Theorem, can be found in numerical analysis books.

To find $M$ in the error estimate usually involves a lot of computation, so we will use technology to obtain $M$.

**EXAMPLE 3  Estimating the Error in Using the Trapezoidal Rule**

Estimate the error that results from using the Trapezoidal Rule with $n = 4$ and with $n = 6$ to approximate $\displaystyle\int_1^2 \dfrac{e^x}{x}\,dx$. See Example 2.

**NOTE** A CAS can be used to find $f''(x)$.

**Solution** To estimate the error resulting from approximating $\displaystyle\int_1^2 \dfrac{e^x}{x}\,dx$ using the Trapezoidal Rule, we need to find the absolute maximum value of $|f''|$ on the interval $[1, 2]$. We begin by finding $f''$:

$$f(x) = \dfrac{e^x}{x}$$

$$f'(x) = \dfrac{d}{dx}\dfrac{e^x}{x} = \dfrac{xe^x - e^x}{x^2} = \dfrac{e^x}{x} - \dfrac{e^x}{x^2}$$

$$f''(x) = \dfrac{d}{dx}\left(\dfrac{e^x}{x} - \dfrac{e^x}{x^2}\right) = \dfrac{d}{dx}\dfrac{e^x}{x} - \dfrac{d}{dx}\dfrac{e^x}{x^2} = \left(\dfrac{e^x}{x} - \dfrac{e^x}{x^2}\right) - \dfrac{e^x x^2 - 2xe^x}{x^4}$$

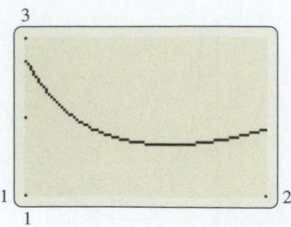

**Figure 19** The absolute maximum of $|f''|$ on $[1, 2]$ is at $x = 1$.

$$= \frac{e^x}{x} - \frac{e^x}{x^2} - \frac{e^x}{x^2} + \frac{2e^x}{x^3}$$

$$= e^x \left( \frac{1}{x} - \frac{2}{x^2} + \frac{2}{x^3} \right)$$

Since $|f''|$ is continuous on the interval $[1, 2]$, the Extreme Value Theorem guarantees that $|f''|$ has an absolute maximum on $[1, 2]$. We find the absolute maximum of $|f''|$ using technology. As seen from Figure 19, the absolute maximum is at the left endpoint 1. The absolute maximum value of $f''$ is $\left| e^1 \left( \frac{1}{1} - \frac{2}{1^2} + \frac{2}{1^3} \right) \right| = e$. So, $M = e$.

When $n = 4$, the error in using the Trapezoidal Rule is

$$\text{Error} \leq \frac{(b-a)^3 M}{12n^2} = \frac{(2-1)^3 \cdot e}{12 \cdot 4^2} = \frac{e}{192} \approx 0.014$$

That is,

$$3.069 - 0.014 \leq \int_1^2 \frac{e^x}{x} dx \leq 3.069 + 0.014$$

$$3.055 \leq \int_1^2 \frac{e^x}{x} dx \leq 3.083$$

When $n = 6$, the error in using the Trapezoidal Rule is

$$\text{Error} \leq \frac{(b-a)^3 M}{12n^2} = \frac{(2-1)^3 \cdot e}{12 \cdot 6^2} = \frac{e}{432} \approx 0.006$$

That is,

$$3.063 - 0.006 \leq \int_1^2 \frac{e^x}{x} dx \leq 3.063 + 0.006$$

$$3.057 \leq \int_1^2 \frac{e^x}{x} dx \leq 3.069$$

■

**NOW WORK** Problem 9(b).

---

**EXAMPLE 4**  **Obtaining a Desired Accuracy Using the Trapezoidal Rule**

Find the number of subintervals $n$ needed to guarantee that the Trapezoidal Rule approximates $\int_1^2 \frac{e^x}{x} dx$ with an error less than or equal to 0.0001.

**Solution** For the approximation of $\int_1^2 \frac{e^x}{x} dx$ to have an error less than or equal to 0.0001, requires that

$$\text{Error} \leq \frac{(b-a)^3 M}{12n^2} = \frac{(2-1)^3 M}{12n^2} \underset{\underset{M = e}{\uparrow}}{=} \frac{e}{12n^2} \leq 0.0001$$

$$n^2 \geq \frac{e}{(0.0001)(12)} = \frac{e}{0.0012}$$

$$n \geq \sqrt{\frac{e}{0.0012}} \approx 47.594$$

To ensure that the error is less than or equal to 0.0001, we round up to 48 subintervals. ■

**NOW WORK** Problem 9(c).

The Trapezoidal Rule is very useful in applications when only experimental (empirical) data are available.

**EXAMPLE 5** **Using the Trapezoidal Rule with Empirical Data**

A 140-foot (ft) tree trunk is cut into 20-ft logs. The diameter of each cross-sectional cut is measured and its area $A$ recorded in the table below. ($x$ is the distance in feet of the cut from the top of the trunk.)

| $x$ (ft) | 0 | 20 | 40 | 60 | 80 | 100 | 120 | 140 |
|---|---|---|---|---|---|---|---|---|
| $A$ (ft$^2$) | 120 | 124 | 128 | 130 | 132 | 136 | 144 | 158 |

Find the approximate volume of the tree trunk.

**Solution** The volume of the tree trunk is $V = \int_0^{140} A(x)\,dx$, where $A(x)$ is the area of a slice at $x$. Since only eight data points are given, the function $A(x)$ is not explicitly known. To approximate the volume, partition the interval $[0, 140]$ into seven subintervals, each of width $\Delta x = \dfrac{140}{7} = 20$. This partition corresponds to the given data. Using the Trapezoidal Rule, the approximate volume of the tree trunk is

$$
V = \int_0^{140} A(x)\,dx \approx \frac{1}{2}(20)\,[A(0) + 2\,A\,(20) + 2\,A\,(40) + 2\,A\,(60) + 2\,A\,(80)
$$
$$
+ 2\,A(100) + 2\,A(120) + A\,(140)]
$$
$$
V \approx 10\,[120 + 2 \cdot 124 + 2 \cdot 128 + 2 \cdot 130 + 2 \cdot 132 + 2 \cdot 136 + 2 \cdot 144 + 158]
$$
$$
= 18{,}660
$$

The volume of the tree trunk is approximately $18{,}660\,\text{ft}^3$. ∎

**NOW WORK** Problem **31**.

## 2 Approximate an Integral Using the Midpoint Rule

Recall that if a function $f$ is continuous on a closed interval $[a, b]$, then $\int_a^b f(x)\,dx$ exists. The integral can be approximated by partitioning $[a, b]$ into $n$ subintervals each of length $\Delta x = \dfrac{b - a}{n}$, choosing any number $u_i$ in the $i$th subinterval, $i = 1, 2, \ldots, n$, and forming Riemann sums $\displaystyle\sum_{i=1}^{n} f(u_i)\Delta x$. The *Midpoint Rule* approximates the definite integral $\int_a^b f(x)\,dx$ by using the Midpoint Riemann sum $\displaystyle\sum_{i=1}^{n} f(\bar{x}_i)\Delta x$, where $\bar{x}_i$ is the midpoint of the $i$th subinterval $[x_{i-1}, x_i]$.

**The Midpoint Rule**

If a function $f$ is continuous on the closed interval $[a, b]$, then

$$
\int_a^b f(x)\,dx \approx \frac{b - a}{n}[f(\bar{x}_1) + f(\bar{x}_2) + \cdots + f(\bar{x}_{n-1}) + f(\bar{x}_n)]
$$

where $\bar{x}_i = \dfrac{x_{i-1} + x_i}{2}$, $i = 1, 2, \ldots, n$, and the closed interval $[a, b]$ is partitioned into $n$ subintervals $[x_0, x_1]$, $[x_1, x_2]$, $\ldots$, $[x_{n-1}, x_n]$, each of width $\Delta x = \dfrac{b - a}{n}$.

**EXAMPLE 6**

**Approximating $\displaystyle\int_0^{\pi} \sin x\,dx$ Using the Midpoint Rule**

(a) Use the Midpoint Rule with $n = 4$ and $n = 6$ to approximate $\int_0^{\pi} \sin x\,dx$.

(b) Compare each approximation to the results of the Trapezoidal Rule (Example 1) and to the exact value of $\int_0^{\pi} \sin x\,dx$.

**Solution (a)** With $n = 4$, we have $\Delta x = \dfrac{\pi - 0}{4} = \dfrac{\pi}{4}$. Partition the interval $[0, \pi]$ into the four subintervals: $\left[0, \dfrac{\pi}{4}\right]$ $\left[\dfrac{\pi}{4}, \dfrac{\pi}{2}\right]$ $\left[\dfrac{\pi}{2}, \dfrac{3\pi}{4}\right]$ $\left[\dfrac{3\pi}{4}, \pi\right]$.

The values of $f(x) = \sin x$ corresponding to the midpoint of each subinterval are

$$f\left(\frac{\pi}{8}\right) = \sin\frac{\pi}{8} \quad f\left(\frac{3\pi}{8}\right) = \sin\frac{3\pi}{8} \quad f\left(\frac{5\pi}{8}\right) = \sin\frac{5\pi}{8} \quad f\left(\frac{7\pi}{8}\right) = \sin\frac{7\pi}{8}$$

Now use the Midpoint Rule:

$$\int_0^\pi \sin x \, dx \approx \frac{\pi}{4}\left(\sin\frac{\pi}{8} + \sin\frac{3\pi}{8} + \sin\frac{5\pi}{8} + \sin\frac{7\pi}{8}\right) \approx 2.052$$

rounded to three decimal places.

With $n = 6$, we have $\Delta x = \dfrac{\pi - 0}{6} = \dfrac{\pi}{6}$. Partition $[0, \pi]$ into the six subintervals:

$$\left[0, \frac{\pi}{6}\right] \quad \left[\frac{\pi}{6}, \frac{\pi}{3}\right] \quad \left[\frac{\pi}{3}, \frac{\pi}{2}\right] \quad \left[\frac{\pi}{2}, \frac{2\pi}{3}\right] \quad \left[\frac{2\pi}{3}, \frac{5\pi}{6}\right] \quad \left[\frac{5\pi}{6}, \pi\right]$$

Using the Midpoint Rule

$$\int_0^\pi \sin x \, dx \approx \frac{\pi}{6}\left(\sin\frac{\pi}{12} + \sin\frac{3\pi}{12} + \sin\frac{5\pi}{12} + \sin\frac{7\pi}{12} + \sin\frac{9\pi}{12} + \sin\frac{11\pi}{12}\right) \approx 2.023$$

**(b)** The exact value of the integral is $\int_0^\pi \sin x \, dx = [-\cos x]_0^\pi = -\cos\pi + \cos 0 = 2$.

Using four subintervals: the Midpoint Rule overestimates the exact value of the integral by $0.052$, while the Trapezoidal Rule underestimates the integral by $0.104$.

Using six subintervals: the Midpoint Rule overestimates the exact value of the integral by $0.023$, while the Trapezoidal Rule underestimates it by $0.046$.

Notice in both instances the magnitude of the error in using the Midpoint Rule is half that of the Trapezoidal Rule. ∎

NOW WORK Problem 13(a).

EXAMPLE 7
**Approximating $\displaystyle\int_1^2 \frac{e^x}{x}\,dx$ Using the Midpoint Rule**

Use the Midpoint Rule with $n = 4$ to approximate $\displaystyle\int_1^2 \frac{e^x}{x}\,dx$.

**Solution** The function $f$ is continuous on the interval $[1, 2]$, so the integral exists. With $n = 4$, partition the interval $[1, 2]$ into four subintervals:

$$\left[1, \frac{5}{4}\right] \quad \left[\frac{5}{4}, \frac{3}{2}\right] \quad \left[\frac{3}{2}, \frac{7}{4}\right] \quad \left[\frac{7}{4}, 2\right]$$

each of width $\Delta x = \dfrac{1}{4}$. The values of $f(x) = \dfrac{e^x}{x}$ corresponding to the midpoint of each subinterval are

$$f\left(\frac{9}{8}\right) = \frac{8e^{9/8}}{9} \quad f\left(\frac{11}{8}\right) = \frac{8e^{11/8}}{11} \quad f\left(\frac{13}{8}\right) = \frac{8e^{13/8}}{13} \quad f\left(\frac{15}{8}\right) = \frac{8e^{15/8}}{15}$$

Then using the Midpoint Rule,

$$\int_1^2 \frac{e^x}{x}\,dx \approx \frac{1}{4}\left[f\left(\frac{9}{8}\right) + f\left(\frac{11}{8}\right) + f\left(\frac{13}{8}\right) + f\left(\frac{15}{8}\right)\right]$$

$$= \frac{1}{4}\left(\frac{8e^{9/8}}{9} + \frac{8e^{11/8}}{11} + \frac{8e^{13/8}}{13} + \frac{8e^{15/8}}{15}\right) \approx 3.054 \qquad ∎$$

NOW WORK Problem 15(a).

The next theorem, whose proof can be found in numerical analysis books, states that the upper bound of the error using the Midpoint Rule is $\frac{1}{2}$ the upper bound of the error using the Trapezoidal Rule.

---

**THEOREM** **Error in the Midpoint Rule**

Let $f$ be a function for which $f''$ is continuous on some open interval containing $a$ and $b$. The error in using the Midpoint Rule to approximate $\int_a^b f(x)\,dx$ can be estimated using the formula

$$\boxed{\text{Error} \leq \frac{(b-a)^3 M}{24n^2}}$$

where $M$ is the absolute maximum of $|f''|$ on the closed interval $[a, b]$.

---

Because the upper bound of the error using the Midpoint Rule is one half that of the Trapezoidal Rule, the Midpoint Rule usually gives a better approximation to $\int_a^b f(x)\,dx$ than the Trapezoidal Rule.

## 3 Approximate an Integral Using Simpson's Rule

Simpson's Rule is another way to approximate the definite integral of a function $f$ that is continuous on a closed interval $[a, b]$. Simpson's Rule approximates $\int_a^b f(x)\,dx$ using parabolic arcs, instead of line segments as in the Trapezoidal Rule.

Before stating Simpson's Rule, we derive a formula for the area $A$ under a parabolic arc. Let the graph in Figure 20 represent the parabola $y = ax^2 + bx + c$. Draw the lines $x = -h$ and $x = h$, so that the points $(-h, y_0)$, $(0, y_1)$, and $(h, y_2)$ lie on the parabola.

The area $A$ enclosed by the parabola, the $x$-axis, and the lines $x = -h$ and $x = h$ is given by

$$A = \int_{-h}^{h} y\,dx = \int_{-h}^{h} (ax^2 + bx + c)\,dx = \left[ a\frac{x^3}{3} + b\frac{x^2}{2} + cx \right]_{-h}^{h}$$

$$= \frac{2}{3}ah^3 + 2ch = \frac{h}{3}(2ah^2 + 6c) \tag{1}$$

Since the parabola contains the points $(-h, y_0)$, $(0, y_1)$, and $(h, y_2)$, each point must satisfy the equation $y = ax^2 + bx + c$, and we have the system of equations:

$$\begin{aligned} y_0 &= ah^2 - bh + c & x = -h \\ y_1 &= c & x = 0 \\ y_2 &= ah^2 + bh + c & x = h \end{aligned}$$

Adding the first and third equations gives

$$y_0 + y_2 = 2ah^2 + 2c$$

Add $4y_1 = 4c$ to each side.

$$y_0 + y_2 + 4y_1 = 2ah^2 + 6c$$

Then substitute $y_0 + y_2 + 4y_1$ for $2ah^2 + 6c$ in formula (1) for the area $A$ to obtain

$$A = \frac{h}{3}(y_0 + 4y_1 + y_2) \tag{2}$$

This formula for $A$ depends on only $y_0$, $y_1$, $y_2$, and the distance $h$.

Now assume that a function $f$ is nonnegative* and continuous on $[a, b]$. Then the area $A$ under the graph of $f$, $A = \int_a^b f(x)\,dx$, can be approximated by partitioning the interval $[a, b]$ into an even number $n$ of subintervals, each of width $\Delta x = \frac{b-a}{n}$. The $y$-coordinates corresponding to the numbers $x_0 = a$, $x_1$, $x_2$, ..., $x_{i-1}$, $x_i$, ..., $x_n = b$ are $f(x_0)$, $f(x_1)$, $f(x_2)$, ..., $f(x_{i-1})$, $f(x_i)$, ..., $f(x_{n-1})$, $f(x_n)$. Connect each triple

**NOTE** The Trapezoidal Rule approximates the graph of a function $f$ with line segments (first degree polynomials), whereas Simpson's Rule approximates the graph of $f$ with parabolic arcs (second degree polynomials).

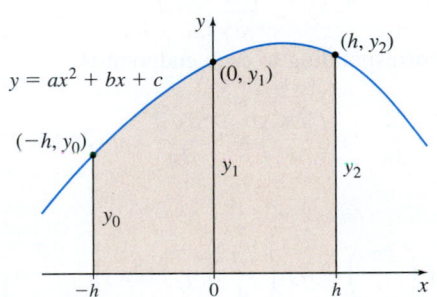

**Figure 20**

---

*A proof of Simpson's Rule that does not assume $f$ is nonnegative can be found in most numerical analysis books.

of points $(x_{i-1}, f(x_{i-1}))$, $(x_i, f(x_i))$, and $(x_{i+1}, f(x_{i+1}))$, $i = 1, 3, 5, \ldots, n-1$ with parabolic arcs, and use (2) to find the area $A_i$ enclosed by the arc centered at $(x_i, f(x_i))$ and the lines $x = x_{i-1}$ and $x = x_{i+1}$. With $h = \Delta x = \dfrac{b-a}{n}$, we have

$$A_i = \frac{b-a}{3n}[f(x_{i-1}) + 4f(x_i) + f(x_{i+1})]$$

The sum of these areas gives an approximation to $A = \int_a^b f(x)\, dx$.

### Simpson's Rule

**ORIGINS** *Simpson's Rule* is named after the British mathematician Thomas Simpson, who lived from 1710 to 1761.

If a function $f$ is continuous on the closed interval $[a, b]$, then

$$\int_a^b f(x)\, dx \approx \frac{b-a}{3n}[f(x_0) + 4f(x_1) + 2f(x_2) + 4f(x_3) + 2f(x_4)$$
$$+ \cdots + 2f(x_{n-2}) + 4f(x_{n-1}) + f(x_n)]$$

where the closed interval $[a, b]$ is partitioned into an even number $n$ of subintervals $[x_0, x_1], [x_1, x_2], \ldots, [x_{n-1}, x_n]$, each of width $\dfrac{b-a}{n}$.

---

**EXAMPLE 8**  **Approximating an Integral Using Simpson's Rule**

Use Simpson's Rule with $n = 4$ to approximate $\displaystyle\int_\pi^{2\pi} \frac{\sin x}{x}\, dx$. Express the answer rounded to three decimal places.

**Solution** Partition the interval $[\pi, 2\pi]$ into the four subintervals

$$\left[\pi, \frac{5\pi}{4}\right] \qquad \left[\frac{5\pi}{4}, \frac{3\pi}{2}\right] \qquad \left[\frac{3\pi}{2}, \frac{7\pi}{4}\right] \qquad \left[\frac{7\pi}{4}, 2\pi\right]$$

each of width $\dfrac{\pi}{4}$. The value of $f(x) = \dfrac{\sin x}{x}$ corresponding to each endpoint is

$$f(\pi) = 0 \quad f\left(\frac{5\pi}{4}\right) = -\frac{2\sqrt{2}}{5\pi} \quad f\left(\frac{3\pi}{2}\right) = -\frac{2}{3\pi} \quad f\left(\frac{7\pi}{4}\right) = -\frac{2\sqrt{2}}{7\pi} \quad f(2\pi) = 0$$

Then using Simpson's Rule,

$$\int_\pi^{2\pi} \frac{\sin x}{x}\, dx \approx \frac{2\pi - \pi}{3 \cdot 4}\left[f(\pi) + 4f\left(\frac{5\pi}{4}\right) + 2f\left(\frac{3\pi}{2}\right) + 4f\left(\frac{7\pi}{4}\right) + f(2\pi)\right]$$

$$= \frac{\pi}{12}\left[0 + 4\left(-\frac{2\sqrt{2}}{5\pi}\right) + 2\left(-\frac{2}{3\pi}\right) + 4\left(-\frac{2\sqrt{2}}{7\pi}\right) + 0\right] \approx -0.434 \qquad \blacksquare$$

The graph of $y = \dfrac{\sin x}{x}$ is shown in Figure 21.

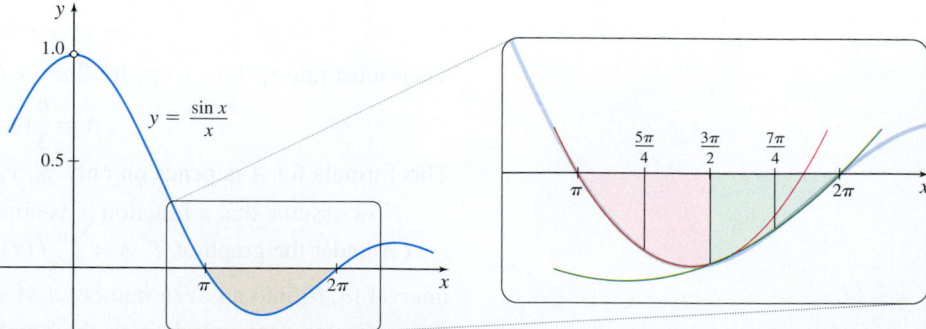

**DF Figure 21** The expanded graph shows the two parabolic arcs, one centered at $\dfrac{5\pi}{4}$ (in red), the other at $\dfrac{7\pi}{4}$ (in green).

**THEOREM  Error in Simpson's Rule**

Let $f$ be a function for which $f^{(4)}$ is continuous on an open interval containing $a$ and $b$. The error in using Simpson's Rule to approximate $\int_a^b f(x)\,dx$ can be estimated by the formula

$$\boxed{\text{Error} \leq \frac{(b-a)^5 M}{180n^4}}$$

where $M$ is the absolute maximum value of $|f^{(4)}(x)|$ on the closed interval $[a, b]$.

As with the Trapezoidal Rule and the Midpoint Rule, the error formula gives an upper bound to the error. A derivation of the error formula may be found in numerical analysis books.

To find $M$ in the error estimate usually involves a lot of computation. So, we use technology to find $M$.

**EXAMPLE 9  Estimating the Error in Using Simpson's Rule**

Estimate the error that results from using Simpson's Rule with $n = 4$ to approximate $\int_\pi^{2\pi} \dfrac{\sin x}{x}\,dx$. (See Example 8.)

**NOTE** A CAS can be used to find higher-order derivatives.

**Solution** To estimate the error, we need to find the absolute maximum value of $|f^{(4)}|$ on the interval $[\pi, 2\pi]$. We begin by finding $f^{(4)}$:

$$f(x) = \frac{\sin x}{x}$$

$$f'(x) = \frac{x\cos x - \sin x}{x^2} = \frac{\cos x}{x} - \frac{\sin x}{x^2}$$

$$f''(x) = \left[\frac{-x\sin x - \cos x}{x^2}\right] - \left[\frac{x^2\cos x - 2x\sin x}{x^4}\right] = \frac{2\sin x}{x^3} - \frac{\sin x}{x} - \frac{2\cos x}{x^2}$$

$$f'''(x) = \left[\frac{2x^3\cos x - 6x^2\sin x}{x^6}\right] - \left[\frac{x\cos x - \sin x}{x^2}\right] - \left[\frac{-2x^2\sin x - 4x\cos x}{x^4}\right]$$

$$= \frac{2\cos x}{x^3} - \frac{6\sin x}{x^4} - \frac{\cos x}{x} + \frac{\sin x}{x^2} + \frac{2\sin x}{x^2} + \frac{4\cos x}{x^3}$$

$$= \frac{6\cos x}{x^3} - \frac{\cos x}{x} - \frac{6\sin x}{x^4} + \frac{3\sin x}{x^2}$$

$$f^{(4)}(x) = \left[\frac{-6x^3\sin x - 18x^2\cos x}{x^6}\right] - \left[\frac{-x\sin x - \cos x}{x^2}\right]$$

$$+ \left[\frac{3x^2\cos x - 6x\sin x}{x^4}\right] - \left[\frac{6x^4\cos x - 24x^3\sin x}{x^8}\right]$$

$$= \frac{4\cos x}{x^2} - \frac{24\cos x}{x^4} + \frac{\sin x}{x} - \frac{12\sin x}{x^3} + \frac{24\sin x}{x^5}$$

We use technology to find the absolute maximum value of $|f^{(4)}|$. See Figure 22. On the interval $[\pi, 2\pi]$ the maximum value of $|f^{(4)}| < 0.176$, so we use $M = 0.176$. The error that results from using Simpson's Rule to approximate $\int_\pi^{2\pi} \dfrac{\sin x}{x}\,dx$ is

$$\text{Error} \leq \frac{(b-a)^5 M}{180n^4} = \frac{(2\pi - \pi)^5\,0.176}{180 \cdot 4^4} \approx 0.001$$

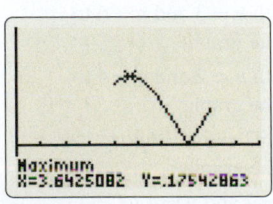

Maximum
X=3.6425082  Y=.17542863

**Figure 22** $y = |f^{(4)}|$

That is,

$$-0.434 - 0.001 \leq \int_\pi^{2\pi} \frac{\sin x}{x}\, dx \leq -0.434 + 0.001$$

$$-0.435 \leq \int_\pi^{2\pi} \frac{\sin x}{x}\, dx \leq -0.433 \qquad \blacksquare$$

**NOW WORK** Problem 19(a) and (b).

**EXAMPLE 10  Obtaining a Desired Accuracy Using Simpson's Rule**

Find the number of subintervals needed to guarantee that Simpson's Rule approximates $\int_\pi^{2\pi} \frac{\sin x}{x}\, dx$ with an error less than or equal to 0.0001.

**Solution** For the approximation of $\int_\pi^{2\pi} \frac{\sin x}{x}\, dx$ to have an error less than or equal to 0.0001 requires that

$$\text{Error} \leq \frac{(b-a)^5 M}{180 n^4} = \frac{(2\pi - \pi)^5 M}{180 n^4} = \underset{\underset{M\,=\,0.176}{\uparrow}}{\frac{(\pi^5)(0.176)}{180 n^4}} \leq 0.0001$$

$$n^4 \geq \frac{0.176\,\pi^5}{(0.0001)\,(180)}$$

$$n \geq \sqrt[4]{2992.192} \approx 7.396$$

Since Simpson's Rule requires $n$ to be even, eight subintervals are needed to guarantee that Simpson's Rule approximates $\int_\pi^{2\pi} \frac{\sin x}{x}\, dx$ with an error less than or equal to 0.0001. $\blacksquare$

**NOW WORK** Problem 19(c).

## 7.6 Assess Your Understanding

### Concepts and Vocabulary

1. *True or False*   The Trapezoidal Rule approximates an integral $\int_a^b f(x)\, dx$ by replacing the graph of $f$ with line segments.

2. *True or False*   Simpson's Rule approximates an integral by using parabolic arcs.

### Skill Building

*For Problems 3 and 5, use the graph below to approximate the area A. Round answers to three decimal places.*

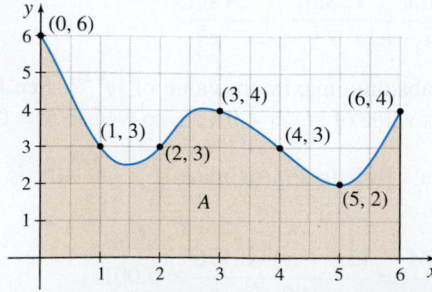

*For Problems 4 and 6, use the graph below to approximate the area A. Round answers to three decimal places.*

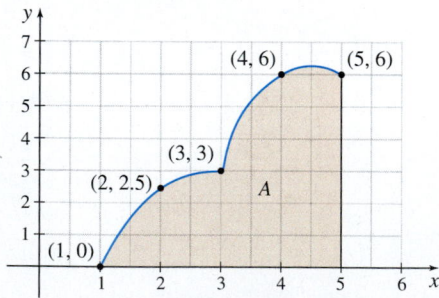

3. Use the Trapezoidal Rule with $n = 3$ and $n = 6$ to approximate the area under the graph.

4. Use the Trapezoidal Rule with $n = 2$ and $n = 4$ to approximate the area under the graph.

5. Use Simpson's Rule with $n = 2$ and $n = 6$ to approximate the area under the graph.

6. Use Simpson's Rule with $n = 2$ and $n = 4$ to approximate the area under the graph.

---

**1.** = NOW WORK problem          📉 = Graphing technology recommended          [CAS] = Computer Algebra System recommended

*In Problems 7–12:*

*(a)  Use the Trapezoidal Rule to approximate each integral.*

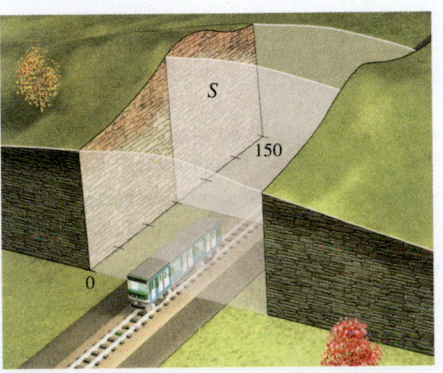 *(b)  Estimate the error in using the approximation. Express the answer rounded to three decimal places.*

*(c)  Determine the number of subintervals needed to guarantee that an approximation has an error less than or equal to 0.0001.*

**7.**  $\int_{\pi/2}^{\pi} \frac{\sin x}{x}\,dx; \quad n=3$    **8.**  $\int_{3\pi/2}^{2\pi} \frac{\cos x}{x}\,dx; \quad n=3$

**9.**  $\int_{0}^{1} e^{-x^2}\,dx; \quad n=4$    **10.**  $\int_{0}^{1} e^{x^2}\,dx; \quad n=4$

**11.**  $\int_{-1}^{0} \frac{dx}{\sqrt{1-x^3}}; \quad n=4$    **12.**  $\int_{0}^{1} \frac{dx}{\sqrt{1+x^3}}; \quad n=3$

*In Problems 13–18:*

*(a)  Use the Midpoint Rule to approximate each integral.*

*(b)  Estimate the error in using the approximation. Express the answer rounded to three decimal places.*

*(c)  Determine the number of subintervals needed to guarantee that an approximation has an error less than or equal to 0.0001.*

**13.**  $\int_{\pi/2}^{\pi} \frac{\sin x}{x}\,dx; \quad n=3$    **14.**  $\int_{3\pi/2}^{2\pi} \frac{\cos x}{x}\,dx; \quad n=3$

**15.**  $\int_{0}^{1} e^{-x^2}\,dx; \quad n=4$    **16.**  $\int_{0}^{1} e^{x^2}\,dx; \quad n=4$

**17.**  $\int_{-1}^{0} \frac{dx}{\sqrt{1-x^3}}; \quad n=4$    **18.**  $\int_{0}^{1} \frac{dx}{\sqrt{1+x^3}}; \quad n=3$

*In Problems 19–24:*

*(a)  Use Simpson's Rule to approximate each integral.*

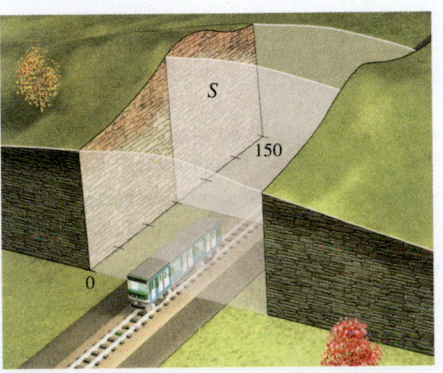 *(b)  Estimate the error in using the approximation. Express the answer rounded to three decimal places.*

*(c)  Determine the number of subintervals needed to guarantee that an approximation has an error less than or equal to 0.0001.*

**19.**  $\int_{1}^{2} \frac{e^x}{x}\,dx; \quad n=4$    **20.**  $\int_{3\pi/2}^{2\pi} \frac{\cos x}{x}\,dx; \quad n=4$

**21.**  $\int_{0}^{1} e^{-x^2}\,dx; \quad n=4$    **22.**  $\int_{0}^{1} e^{x^2}\,dx; \quad n=4$

**23.**  $\int_{-1}^{0} \frac{dx}{\sqrt{1-x^3}}; \quad n=4$    **24.**  $\int_{0}^{1} \frac{dx}{\sqrt{1+x^2}}; \quad n=4$

## Applications and Extensions

**25.  (a)**  Show that  $\int_{1}^{2} \frac{dx}{x} = \ln 2.$

**(b)**  Use the Trapezoidal Rule with $n=5$ to approximate  $\int_{1}^{2} \frac{dx}{x}.$

**(c)**  Use the Midpoint Rule with $n=5$ to approximate  $\int_{1}^{2} \frac{dx}{x}.$

**(d)**  Use Simpson's Rule with $n=6$ to approximate  $\int_{1}^{2} \frac{dx}{x}.$

**26.  Area**  Selected measurements of a continuous function $f$ are given in the table below. Use Simpson's Rule to approximate the area of the region enclosed by the graph of $f$, the $x$-axis, and the lines $x=2$ and $x=4.4$.

| $x$ | 2.0 | 2.4 | 2.8 | 3.2 | 3.6 | 4.0 | 4.4 |
|---|---|---|---|---|---|---|---|
| $y$ | 3.03 | 4.61 | 5.80 | 6.59 | 7.76 | 8.46 | 9.19 |

**27.  Arc Length**  Approximate the arc length of the graph of $y=\sin x$ from $x=0$ to $x=\dfrac{\pi}{2}$.

**(a)**  using Simpson's Rule with $n=4$.

**(b)**  using the Trapezoidal Rule with $n=3$.

**(c)**  using the Midpoint Rule with $n=3$.

**28.  Arc Length**  Approximate the arc length of the graph of $y=e^x$ from $x=0$ to $x=4$

**(a)**  using Simpson's Rule with $n=4$.

**(b)**  using the Trapezoidal Rule with $n=8$.

**(c)**  using the Midpoint Rule with $n=8$.

**29.  Work**  A gas expands from a volume of 1 cubic inch ($\text{in.}^3$) to $2.5\ \text{in.}^3$; values of the volume $V$ and pressure $p$ (in pounds per square inch) during the expansion are given in the table below. Use Simpson's rule to approximate the total work $W$ done in the expansion. (*Hint:* $W = \int_{a}^{b} p\,dV$.)

| $V$ | 1 | 1.25 | 1.5 | 1.75 | 2 | 2.25 | 2.5 |
|---|---|---|---|---|---|---|---|
| $p$ | 68.7 | 55.0 | 45.8 | 39.3 | 34.4 | 30.5 | 27.5 |

**30.  Work**  In the table below, $F$ is the force in pounds acting on an object in its direction of motion and $x$ is the displacement of the object in feet. Use the Trapezoidal Rule to approximate the work done by the force in moving the object from $x=0$ to $x=50$.

| $x$ | 0 | 5 | 10 | 15 | 20 | 25 | 30 | 35 | 40 | 45 | 50 |
|---|---|---|---|---|---|---|---|---|---|---|---|
| $F$ | 100 | 80 | 66 | 56 | 50 | 45 | 40 | 36 | 33 | 30 | 28 |

**31.  Volume**  In the table below, $S$ is the area in square meters of the cross section of a railroad track cutting through a mountain, and $x$ meters is the corresponding distance along the line. Use the Trapezoidal Rule to find the number of cubic meters of earth removed to make the cutting from $x=0$ to $x=150$. See the figure below.

| $x$ | 0 | 25 | 50 | 75 | 100 | 125 | 150 |
|---|---|---|---|---|---|---|---|
| $S$ | 105 | 118 | 142 | 120 | 110 | 90 | 78 |

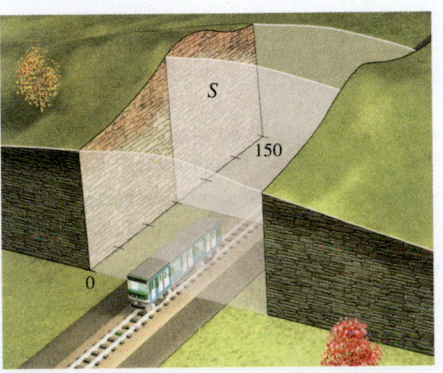

**32. Area** Use Simpson's Rule to approximate the surface area of the pond pictured in the figure.

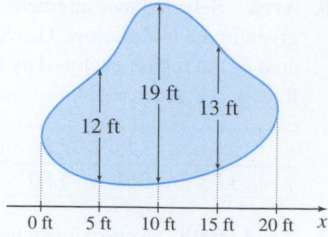

**33. Volume** The area of the horizontal section of a reservoir is $A$ square meters at a height $x$ meters from the bottom. Corresponding values of $A$ and $x$ are given in the table below. Approximate the volume of water in the reservoir using the Trapezoidal Rule and also using Simpson's Rule. See the figure.

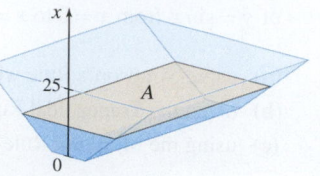

| $x$ | 0 | 2.5 | 5 | 7.5 | 10 | 12.5 | 15 | 17.5 | 20 | 22.5 | 25 |
|-----|---|-----|---|-----|----|------|----|------|----|------|----|
| $A$ | 0 | 2510 | 3860 | 4870 | 5160 | 5590 | 5810 | 6210 | 6890 | 7680 | 8270 |

**34. Area** A series of soundings taken across a river channel is given in the table below, where $x$ meters is the distance from one shore and $y$ meters is the corresponding depth of the water.

Use the Midpoint Rule to approximate $\int_0^{80} y\,dx$.
*Hint:* Use $x = 10, 30, 50,$ and $70$ as midpoints.

| $x$ | 0 | 10 | 20 | 30 | 40 | 50 | 60 | 70 | 80 |
|-----|---|----|----|----|----|----|----|----|----|
| $y$ | 5 | 10 | 13.2 | 15 | 15.6 | 12 | 6 | 4 | 0 |

**35. Volume of a Solid of Revolution** Use the Trapezoidal Rule with $n = 3$ to approximate the volume of the solid of revolution formed by revolving the region shown in the figure below about the $x$-axis.

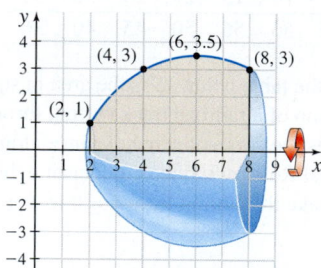

**36. Distance Traveled** The speed $v$, in meters per second, of an object in rectilinear motion at time $t$ is given in the table below.

| $t$ | 0 | 0.5 | 1 | 1.5 | 2 | 2.5 | 3 |
|-----|---|-----|---|-----|---|-----|---|
| $v$ | 5.1 | 5.3 | 5.6 | 6.1 | 6.8 | 6.7 | 6.5 |

**(a)** Use the Trapezoidal Rule to approximate the total distance $s$ traveled by the object from $t = 0$ to $t = 3$.

**(b)** Use Simpson's Rule to approximate the total distance $s$ traveled by the object from $t = 0$ to $t = 3$.

**Approximating Integrals with Nonregular Partitions** *In Problems 37 and 38, use the following discussion:*

*Often tabular data are not given at equal intervals. In these cases none of the numerical approximations (the Trapezoidal Rule, the Midpoint Rule, and Simpson's Rule) discussed in this section can be used to approximate $\int_a^b f(x)\,dx$ because each of these rules assumes that the values of $f$ are available at equal intervals. But as long as the underlying function $f$ is continuous on an interval containing the data, then $\int_a^b f(x)\,dx$ can be approximated using the trapezoids formed by the averaging the Left Riemann sum and the Right Riemann sum. Then*

$$\int_a^b f(x)\,dx \approx \frac{1}{2}\left[\sum_{i=1}^{n} f(x_{i-1})\Delta x_i + \sum_{i=1}^{n} f(x_i)\Delta x_i\right]$$

**37. Approximating Net Displacement** The velocity $v$, in meters per second, of an object in rectilinear motion is measured at the times $t$ shown in the table below.

| $t$ | 1 | 2 | 4 | 8 | 10 |
|-----|---|---|---|---|----|
| $v(t)$ | 4 | 7 | $-2$ | 0 | 3 |

Approximate the net displacement of the object from $t = 1$ to $t = 10$.

**38. Approximating Volume** The table below shows the rate of change of volume $V$ with respect to time $t$ (in liters per hour) of water in a tank for selected times $t$ in hours.

| $t$ | 0 | 2 | 5 | 10 | 12 |
|-----|---|---|---|----|----|
| $\dfrac{dV}{dt}$ | 16 | 12 | 10 | 12 | 8 |

Approximate the change in the volume $V$ of water in the tank from $t = 0$ to $t = 12$ hours.

**39. Volume of a Solid of Revolution** Approximate the volume of the solid of revolution in the figure below generated by revolving the region bounded by the graph of $y = \sin x$ and the $y$-axis from $x = 0$ to $x = \dfrac{\pi}{2}$ about the $y$-axis

**(a)** using Simpson's Rule with $n = 4$.

**(b)** using the Trapezoidal Rule with $n = 3$.

**(c)** using the Midpoint Rule with $n = 3$.

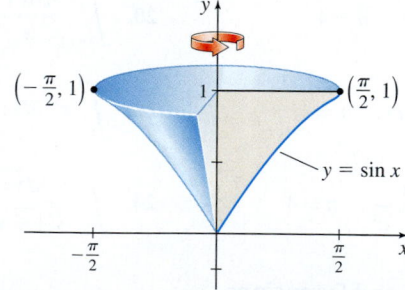

**40. Arc Length** Use the Trapezoidal Rule to approximate the arc length of the ellipse $9x^2 + 100y^2 = 900$ in the first quadrant from $x = 0$ to $x = 8$. Partition the interval into four equal subintervals, and round the answer to three decimal places.

**41. Approximate** $\displaystyle\int_0^{\pi} f(x)\,dx$ if $f(x) = \begin{cases} \dfrac{\sin x}{x} & \text{if } x \neq 0 \\ 1 & \text{if } x = 0 \end{cases}$

**(a)** Use the Trapezoidal Rule with $n = 6$.

**(b)** Use the Midpoint Rule with $n = 6$.

**(c)** Use Simpson's Rule with $n = 6$.

▧ **42.** Approximate $\int_{-1}^{1} 5e^{-x^2}\,dx$.

(a) Use the Trapezoidal Rule with $n = 20$ subintervals.

(b) Use the Midpoint Rule with $n = 20$ subintervals.

(c) Use Simpson's Rule with $n = 20$ subintervals.

(d) Use technology to find the integral.

**Challenge Problems**

**43.** Let $T_n$ be the approximation to $\int_a^b f(x)\,dx$ given by the Trapezoidal Rule with $n$ subintervals. Without using the error formula given in the text, show that $\lim_{n \to \infty} T_n = \int_a^b f(x)\,dx$.

**44.** Show that if $f(x) = Ax^3 + Bx^2 + Cx + D$, then Simpson's Rule gives the exact value of $\int_a^b f(x)\,dx$.

# 7.7 Improper Integrals

**OBJECTIVES** *When you finish this section, you should be able to:*

**1** Find integrals with an infinite limit of integration (p. 556)

**2** Interpret an improper integral geometrically (p. 557)

**3** Integrate functions over $[a, b]$ that are not defined at an endpoint (p. 558)

**4** Use the Comparison Test for improper integrals (p. 560)

In Chapter 5, the definition of the definite integral $\int_a^b f(x)\,dx$ required that both $a$ and $b$ be real numbers. We also required that the function $f$ be defined on the closed interval $[a, b]$. Here we take up instances for which:

- Integrals have infinite limits of integration:

$$\int_1^\infty \frac{1}{\sqrt{x}}\,dx \qquad \int_{-\infty}^0 \frac{x-3}{x^3-8}\,dx \qquad \int_{-\infty}^\infty \frac{x}{(x^2+1)^2}\,dx$$

- The integrand is not defined at a number in the interval of integration:

$$\int_0^1 \frac{1}{\sqrt{x}}\,dx \qquad \int_0^{\pi/2} \tan x\,dx \qquad \int_{-1}^1 \frac{1}{x^3}\,dx$$

not defined at 0     not defined at $\frac{\pi}{2}$     not defined at 0

Integrals like these are called *improper integrals.*

---

**DEFINITION** Improper Integral

If a function $f$ is continuous on the interval $[a, \infty)$, then $\int_a^\infty f(x)\,dx$, called an **improper integral**, is defined as

$$\int_a^\infty f(x)\,dx = \lim_{b \to \infty} \int_a^b f(x)\,dx$$

provided the limit exists and is a real number. If $\lim_{b \to \infty} \int_a^b f(x)\,dx$ exists and is a real number, the improper integral $\int_a^\infty f(x)\,dx$ is said to **converge**. If the limit does not exist or if the limit is infinite, the improper integral $\int_a^\infty f(x)\,dx$ is said to **diverge**.

If a function $f$ is continuous on the interval $(-\infty, b]$, then $\int_{-\infty}^b f(x)\,dx$, called an **improper integral**, is defined as

$$\int_{-\infty}^b f(x)\,dx = \lim_{a \to -\infty} \int_a^b f(x)\,dx$$

provided the limit exists and is a real number. If $\lim_{a \to -\infty} \int_a^b f(x)\,dx$ exists and is a real number, the improper integral $\int_{-\infty}^b f(x)\,dx$ **converges**. If the limit does not exist or if the limit is infinite, the improper integral $\int_{-\infty}^b f(x)\,dx$ **diverges**.

## 1 Find Integrals with an Infinite Limit of Integration

CALC
CLIP

**EXAMPLE 1** Integrating Functions over Infinite Intervals

Determine whether the following improper integrals converge or diverge:

(a) $\displaystyle\int_1^\infty \frac{1}{x}\,dx$    (b) $\displaystyle\int_{-\infty}^0 e^x\,dx$    (c) $\displaystyle\int_{\pi/2}^\infty \sin x\,dx$

**NEED TO REVIEW?** Limits at infinity are discussed in Section 1.5, pp. 128–135.

**Solution** (a) $\displaystyle\int_1^\infty \frac{1}{x}\,dx$: $\displaystyle\lim_{b\to\infty}\int_1^b \frac{1}{x}\,dx = \lim_{b\to\infty}\Big[\ln |x|\Big]_1^b = \lim_{b\to\infty}[\ln b - \ln 1]=\infty$.

The limit is infinite, so $\displaystyle\int_1^\infty \frac{1}{x}\,dx$ diverges.

(b) $\displaystyle\int_{-\infty}^0 e^x\,dx$: $\displaystyle\lim_{a\to-\infty}\int_a^0 e^x\,dx = \lim_{a\to-\infty}\Big[e^x\Big]_a^0 = \lim_{a\to-\infty}(1 - e^a)=1-0=1$.  Since

the limit exists, $\displaystyle\int_{-\infty}^0 e^x\,dx$ converges and equals 1.

(c) $\displaystyle\int_{\pi/2}^\infty \sin x\,dx$: $\displaystyle\lim_{b\to\infty}\int_{\pi/2}^b \sin x\,dx = \lim_{b\to\infty}\Big[-\cos x\Big]_{\pi/2}^b = \lim_{b\to\infty}[-\cos b + 0]$

$$= -\lim_{b\to\infty}\cos b$$

This limit does not exist, since as $b\to\infty$, the value of $\cos b$ oscillates between $-1$

and 1. So, $\displaystyle\int_{\pi/2}^\infty \sin x\,dx$ diverges.  ∎

**NOW WORK** Problem 17.

If both limits of integration are infinite, then the following definition is used.

**DEFINITION**

If a function $f$ is continuous for all $x$ and if, for any number $c$, *both* improper integrals $\int_{-\infty}^c f(x)\,dx$ and $\int_c^\infty f(x)\,dx$ converge, then the improper integral $\int_{-\infty}^\infty f(x)\,dx$ **converges,** and

$$\int_{-\infty}^\infty f(x)\,dx = \int_{-\infty}^c f(x)\,dx + \int_c^\infty f(x)\,dx$$

If *either* or *both* of the integrals on the right diverge, then the improper integral $\int_{-\infty}^\infty f(x)\,dx$ **diverges.**

**EXAMPLE 2** Integrating Functions over Infinite Intervals

Determine whether $\int_{-\infty}^\infty 4x^3\,dx$ converges or diverges.

**NOTE** The decision to use 0 to break up the integral is arbitrary. Usually, the choice made simplifies finding the integral.

**Solution** We begin by writing $\int_{-\infty}^\infty 4x^3\,dx = \int_{-\infty}^0 4x^3\,dx + \int_0^\infty 4x^3\,dx$ and evaluate each improper integral on the right.

$$\int_{-\infty}^0 4x^3\,dx: \quad \lim_{a\to-\infty}\int_a^0 4x^3\,dx = \lim_{a\to-\infty}\Big[x^4\Big]_a^0 = \lim_{a\to-\infty}(0-a^4)=-\infty$$

There is no need to continue. $\int_{-\infty}^\infty 4x^3\,dx$ diverges.  ∎

**CAUTION** The definition requires that two improper integrals converge in order for $\int_{-\infty}^\infty f(x)\,dx$ to converge. Do not set $\int_{-\infty}^\infty f(x)\,dx = \lim_{a\to\infty}\int_{-a}^a f(x)\,dx$. If this were done in Example 2, the result would have been $\int_{-a}^a 4x^3\,dx = \Big[x^4\Big]_{-a}^a = a^4 - a^4 = 0$, and we would have incorrectly concluded that $\int_{-\infty}^\infty f(x)\,dx$ converges and equals 0.

**NOW WORK** Problem 23.

## 2 Interpret an Improper Integral Geometrically

If a function $f$ is continuous and nonnegative on the interval $[a, \infty)$, then $\int_a^\infty f(x)\,dx$ can be interpreted geometrically. For each number $b > a$, the definite integral $\int_a^b f(x)\,dx$ represents the area under the graph of $y = f(x)$ from $a$ to $b$, as shown in Figure 23(a). As $b \to \infty$, this area approaches the area under the graph of $y = f(x)$ over the interval $[a, \infty)$, as shown in Figure 23(b). The area under the graph of $y = f(x)$ to the right of $a$ is defined by $\int_a^\infty f(x)\,dx$, provided the improper integral converges. If it diverges, there is no area defined.

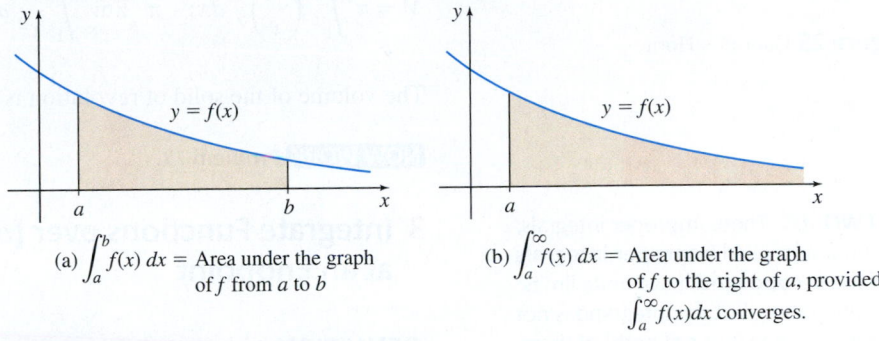

(a) $\displaystyle\int_a^b f(x)\,dx =$ Area under the graph of $f$ from $a$ to $b$

(b) $\displaystyle\int_a^\infty f(x)\,dx =$ Area under the graph of $f$ to the right of $a$, provided $\int_a^\infty f(x)\,dx$ converges.

**Figure 23**

---

EXAMPLE 3  **Determining Whether the Area Under a Graph Is Defined**

Determine whether the area under the graph of $y = \dfrac{1}{x^2}$ to the right of $x = 1$ is defined.

**Solution** See Figure 24. To determine whether the area is defined, we examine

$$\int_1^\infty \frac{1}{x^2}\,dx: \quad \lim_{b \to \infty} \int_1^b \frac{1}{x^2}\,dx = \lim_{b \to \infty}\left[-\frac{1}{x}\right]_1^b = \lim_{b \to \infty}\left(-\frac{1}{b} + 1\right) = 1$$

The area under the graph $f(x) = \dfrac{1}{x^2}$ to the right of 1 is defined and equals 1 sq. unit. ∎

NOW WORK  **Problem 71.**

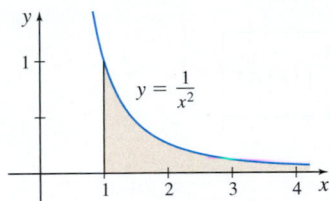

**Figure 24**

---

**THEOREM**

$\displaystyle\int_1^\infty \frac{dx}{x^p}$ converges if $p > 1$ and diverges if $p \le 1$.

**Proof** We consider two cases: $p = 1$ and $p \ne 1$.

$$p = 1; \quad \int_1^\infty \frac{dx}{x^p} = \int_1^\infty \frac{dx}{x}: \quad \lim_{b \to \infty}\int_1^b \frac{dx}{x} = \lim_{b \to \infty} \ln b = \infty \quad \text{From Example 1(a)}$$

$$p \ne 1; \quad \int_1^\infty \frac{dx}{x^p}: \quad \lim_{b \to \infty}\int_1^b \frac{dx}{x^p} = \lim_{b \to \infty}\left[\frac{x^{-p+1}}{-p+1}\right]_1^b = \lim_{b \to \infty}\left[\frac{1}{1-p}(b^{-p+1} - 1)\right]$$

$$= \frac{1}{1-p}\lim_{b \to \infty}\left(\frac{1}{b^{p-1}} - 1\right)$$

**NOTE** The area under the graph of $y = \dfrac{1}{x}$ to the right of 1 is not defined [see Example 1(a)], but the volume obtained by revolving the region about the $x$-axis is defined. (See Example 4.)

If $p > 1$, then $p - 1 > 0$, and $\displaystyle\lim_{b \to \infty}\frac{1}{b^{p-1}} = 0$. So,

$$\int_1^\infty \frac{dx}{x^p} = \frac{1}{1-p}(0 - 1) = \frac{1}{p-1}$$

and the improper integral $\displaystyle\int_1^\infty \frac{dx}{x^p}$, $p > 1$, converges.

If $p < 1$, then $p - 1 < 0$, and $\displaystyle\lim_{b \to \infty}\frac{1}{b^{p-1}} = \lim_{b \to \infty} b^{1-p} = \infty$. So, the improper integral $\displaystyle\int_1^\infty \frac{dx}{x^p}$, $p < 1$, diverges. ∎

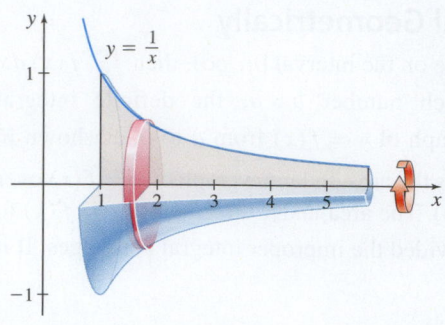

**Figure 25** Gabriel's Horn

NOW WORK Problem 73.

**EXAMPLE 4   Finding the Volume of Gabriel's Horn**

Find the volume of the solid of revolution, called **Gabriel's Horn,** that is generated by revolving the region bounded by the graph of $y = \dfrac{1}{x}$ and the $x$-axis to the right of 1 about the $x$-axis.

**Solution** Figure 25 illustrates the region being revolved and the solid of revolution that it generates. Using the disk method, the volume $V$ is

$$V = \pi \int_1^\infty \left(\frac{1}{x}\right)^2 dx: \quad \pi \lim_{b \to \infty} \int_1^b \frac{1}{x^2}\, dx = \pi \lim_{b \to \infty} \left[-\frac{1}{x}\right]_1^b = \pi \lim_{b \to \infty}\left(-\frac{1}{b}+1\right) = \pi$$

The volume of the solid of revolution is $\pi$ cubic units. ∎

## 3 Integrate Functions over $[a, b]$ That Are Not Defined at an Endpoint

**IN WORDS** These improper integrals, sometimes called **improper integrals of the second kind**, have finite limits of integration, but the integrand is not defined at one (but not both) of them.

**DEFINITION   Improper Integral**

If a function $f$ is continuous on $(a, b]$, but is not defined at $a$, then $\int_a^b f(x)\, dx$ is an **improper integral**, and

$$\int_a^b f(x)\, dx = \lim_{t \to a^+} \int_t^b f(x)\, dx$$

provided the limit exists and is a real number.

If a function $f$ is continuous on $[a, b)$, but is not defined at $b$, then $\int_a^b f(x)\, dx$ is an **improper integral**, and

$$\int_a^b f(x)\, dx = \lim_{t \to b^-} \int_a^t f(x)\, dx$$

provided the limit exists and is a real number.

When the limit exists and is a real number, the improper integral is said to **converge**; otherwise, it is said to **diverge**.

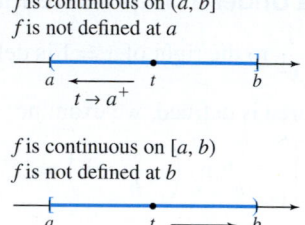

**Figure 26**

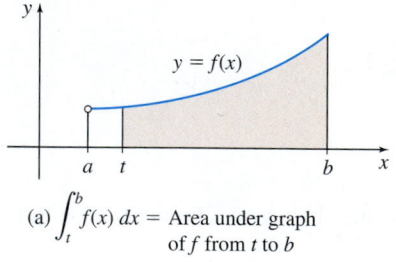

(a) $\int_t^b f(x)\, dx$ = Area under graph of $f$ from $t$ to $b$

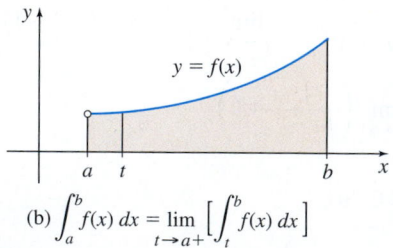

(b) $\int_a^b f(x)\, dx = \lim\limits_{t \to a^+}\left[\int_t^b f(x)\, dx\right]$

**Figure 27**

Be sure to use the correct one-sided limit. If $f$ is not defined at the left endpoint $a$, then for $t$ to be in the interval of integration, $t > a$ and $t$ must approach $a$ from the right. Similarly, if $f$ is not defined at the right endpoint $b$, then for $t$ to be in the interval of integration, $t < b$ and $t$ must approach $b$ from the left. Figure 26 may help you remember the correct one-sided limit to use.

If a function $f$ is continuous and nonnegative on the interval $(a, b]$, then the improper integral $\int_a^b f(x)\, dx$ may be interpreted geometrically. For each number $t$, $a < t \leq b$, the integral $\int_t^b f(x)\, dx$ represents the area under the graph of $y = f(x)$ from $t$ to $b$, as shown in Figure 27(a). As $t \to a^+$, this area approaches the area under the graph of $y = f(x)$ from $a$ to $b$, as shown in Figure 27(b). That is, if $\int_a^b f(x)\, dx$ converges, its value is defined to be the area $A$ under the graph of $f$ from $a$ to $b$.

**EXAMPLE 5   Determining If the Area Under a Graph Is Defined**

Determine if the area under the graph of $y = \dfrac{1}{\sqrt{x}}$ from 0 to 4 is defined.

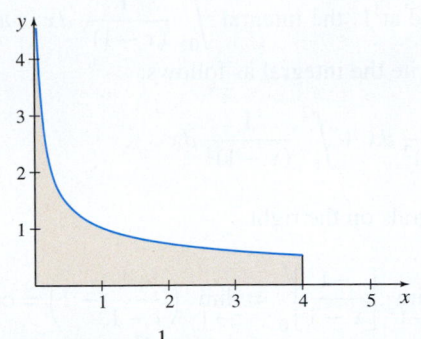

**Figure 28** $y = \dfrac{1}{\sqrt{x}}, 0 < x \le 4$

**Solution** Figure 28 shows the graph of $y = \dfrac{1}{\sqrt{x}}$. Since $y = \dfrac{1}{\sqrt{x}}$ is continuous on $(0, 4]$ but is not defined at 0, $\displaystyle\int_0^4 \dfrac{1}{\sqrt{x}}\, dx$ is an improper integral. We determine whether the integral converges or diverges.

$$\int_0^4 \frac{1}{\sqrt{x}}\, dx : \quad \lim_{t\to 0^+} \int_t^4 \frac{1}{\sqrt{x}}\, dx = \lim_{t\to 0^+} \int_t^4 x^{-1/2}\, dx = \lim_{t\to 0^+} \left[ \frac{x^{1/2}}{\frac{1}{2}} \right]_t^4$$

$$= \lim_{t\to 0^+} \left( 2\cdot 2 - 2\sqrt{t} \right) = 4 - 2\lim_{t\to 0^+}\sqrt{t} = 4$$

Since $\displaystyle\int_0^4 \dfrac{1}{\sqrt{x}}\, dx$ converges, the area under the graph of $y = \dfrac{1}{\sqrt{x}}$ from 0 to 4 is defined and equals 4. ∎

**NOW WORK** Problem **27.**

---

**EXAMPLE 6  Determining Whether an Improper Integral Converges or Diverges**

Determine whether $\int_0^{\pi/2} \tan x\, dx$ converges or diverges.

**Solution** The function $f(x) = \tan x$ is continuous on $\left[0, \dfrac{\pi}{2}\right)$ but is not defined at $\dfrac{\pi}{2}$, so $\int_0^{\pi/2} \tan x\, dx$ is an improper integral.

$$\int_0^{\pi/2} \tan x\, dx: \quad \lim_{t\to (\pi/2)^-} \int_0^t \tan x\, dx = \lim_{t\to (\pi/2)^-} \Big[ \ln|\sec x| \Big]_0^t$$

$$= \lim_{t\to (\pi/2)^-} \big[ \ln|\sec t| - \ln|\sec 0| \big]$$

$$= \lim_{t\to (\pi/2)^-} \ln(\sec t) = \infty$$

So, $\int_0^{\pi/2} \tan x\, dx$ diverges. ∎

**NOW WORK** Problem **43.**

Another type of improper integral occurs when the integrand is not defined at some number $c$, $a < c < b$, in the interval $[a, b]$.

**DEFINITION**

If a function $f$ is continuous on a closed interval $[a, b]$, except at the number $c$, $a < c < b$, where $f$ is not defined, the integral $\int_a^b f(x)\, dx$ is an **improper integral** and

$$\boxed{\int_a^b f(x)\, dx = \int_a^c f(x)\, dx + \int_c^b f(x)\, dx}$$

provided that both improper integrals on the right converge.

If either improper integral $\int_a^c f(x)\, dx$ or $\int_c^b f(x)\, dx$ diverges, then the improper integral $\int_a^b f(x)\, dx$ also **diverges**.

---

**EXAMPLE 7  Determining Whether an Improper Integral Converges or Diverges**

Determine whether $\displaystyle\int_0^2 \dfrac{1}{(x-1)^2}\, dx$ converges or diverges.

**Solution** Since $f(x) = \dfrac{1}{(x-1)^2}$ is not defined at 1, the integral $\displaystyle\int_0^2 \dfrac{1}{(x-1)^2}\,dx$ is an improper integral on the interval $[0, 2]$. We write the integral as follows:

$$\int_0^2 \frac{1}{(x-1)^2}\,dx = \int_0^1 \frac{1}{(x-1)^2}\,dx + \int_1^2 \frac{1}{(x-1)^2}\,dx$$

and investigate each of the two improper integrals on the right.

$$\int_0^1 \frac{1}{(x-1)^2}\,dx: \quad \lim_{t\to 1^-}\int_0^t \frac{1}{(x-1)^2}\,dx = \lim_{t\to 1^-}\left[\frac{-1}{x-1}\right]_0^t = \lim_{t\to 1^-}\left(\frac{-1}{t-1} - 1\right) = \infty$$

$\displaystyle\int_0^1 \dfrac{1}{(x-1)^2}\,dx$ diverges, so there is no need to investigate the second integral.

The improper integral $\displaystyle\int_0^2 \dfrac{1}{(x-1)^2}\,dx$ diverges. ∎

**CAUTION** It is a common mistake to look at an improper integral

like $\displaystyle\int_0^2 \dfrac{dx}{(x-1)^2}$ and not notice that

the integrand is undefined at 1. If that happens, and you use the Fundamental Theorem of Calculus, you obtain

$$\int_0^2 \frac{dx}{(x-1)^2} = \left[\frac{-1}{x-1}\right]_0^2 = -1 - 1 = -2$$

which is an incorrect answer. *Always check the domain of the integrand before attempting to integrate.*

**NOW WORK** Problem 31.

## 4 Use the Comparison Test for Improper Integrals

So far we have been able to determine whether an improper integral converges or diverges by finding an antiderivative of the integrand. If we cannot find an antiderivative, we may still be able to determine whether the improper integral converges or diverges. Then, if it converges, we can use numerical techniques such as the Trapezoidal Rule, the Midpoint Rule, or Simpson's Rule to approximate the integral. One way to determine if an improper integral converges or diverges is by using the *Comparison Test for Improper Integrals.*

**THEOREM Comparison Test for Improper Integrals**

Let $f$ and $g$ be two functions that are nonnegative and continuous on the interval $[a, \infty)$, and suppose

$$f(x) \ge g(x)$$

for all numbers $x > c$, where $c \ge a$.

- If $\displaystyle\int_a^\infty f(x)\,dx$ converges, then $\displaystyle\int_a^\infty g(x)\,dx$ also converges.
- If $\displaystyle\int_a^\infty g(x)\,dx$ diverges, then $\displaystyle\int_a^\infty f(x)\,dx$ also diverges.

You are asked to prove the theorem in Problem 100. The proof follows from a property of definite integrals: If the functions $f$ and $g$ are continuous on a closed interval $[a, b]$ and if $f(x) \ge g(x)$ on $[a, b]$, then $\displaystyle\int_a^b f(x)\,dx \ge \int_a^b g(x)\,dx$. (See Section 5.4, Problem 122, p. 398.)

**EXAMPLE 8 Using the Comparison Test for Improper Integrals**

Determine whether $\displaystyle\int_1^\infty e^{-x^2}\,dx$ converges or diverges.

**Solution** By definition, $\displaystyle\int_1^\infty e^{-x^2}\,dx = \lim_{b\to\infty}\int_1^b e^{-x^2}\,dx$ converges if the limit exists and equals a real number. Since $e^{-x^2}$ has no antiderivative, we use the Comparison Test for Improper Integrals. We proceed as follows: For $x \ge 1$,

$$x^2 \ge x$$
$$-x^2 \le -x$$
$$0 < e^{-x^2} \le e^{-x} \qquad \text{Since } e > 1, \text{ if } a \le b, \text{ then } e^a \le e^b.$$

Figure 29 illustrates this.

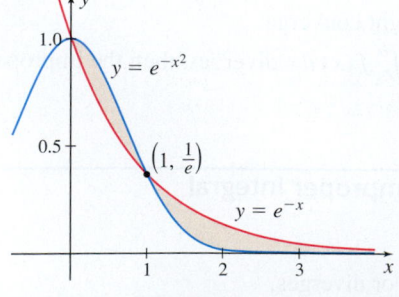

**Figure 29**

Based on the Comparison Test, if $\int_1^\infty e^{-x}dx$ converges, so does $\int_1^\infty e^{-x^2}dx$. We investigate $\int_1^\infty e^{-x}dx$.

$$\int_1^\infty e^{-x}dx : \lim_{b\to\infty}\int_1^b e^{-x}dx = \lim_{b\to\infty}\left[-e^{-x}\right]_1^b = \lim_{b\to\infty}[-e^{-b}+e^{-1}] = \lim_{b\to\infty}\left[\frac{1}{e}-\frac{1}{e^b}\right]$$

$$= \lim_{b\to\infty}\frac{1}{e} - \lim_{b\to\infty}\frac{1}{e^b} = \frac{1}{e}$$

Since $\int_1^\infty e^{-x}dx$ converges, we conclude that $\int_1^\infty e^{-x^2}dx$ converges. ∎

Notice that the Comparison Test for Improper Integrals does not give the value of $\int_1^\infty e^{-x^2}dx$. To find $\int_1^\infty e^{-x^2}dx$ requires numerical techniques. The Comparison Test does, however, tell us that $0\le\int_1^\infty e^{-x^2}dx\le\frac{1}{e}$.

NOW WORK **Problem 67.**

# 7.7 Assess Your Understanding

## Concepts and Vocabulary

1. *Multiple Choice* If a function $f$ is continuous on the interval $[a, \infty)$, then $\int_a^\infty f(x)\,dx$ is called [(a) a definite (b) an infinite (c) an improper (d) a proper] integral.

2. *Multiple Choice* If the $\lim_{b\to\infty}\int_a^b f(x)\,dx$ does not exist, the improper integral $\int_a^\infty f(x)\,dx$ [(a) converges (b) diverges (c) equals $ab$].

3. *True or False* If a function $f$ is continuous for all $x$, then the improper integral $\int_{-\infty}^\infty f(x)\,dx$ always converges.

4. *True or False* If a function $f$ is continuous and nonnegative on the interval $[a, \infty)$, and $\int_a^\infty f(x)\,dx$ converges, then $\int_a^\infty f(x)\,dx$ equals the area under the graph of $y = f(x)$ for $x \ge a$.

5. *True or False* If a function $f$ is continuous for all $x$, then the improper integral $\int_{-\infty}^\infty f(x)\,dx = \lim_{a\to\infty}\int_{-a}^a f(x)\,dx$.

6. To determine whether the improper integral $\int_a^b f(x)\,dx$ converges or diverges, where $f$ is continuous on $[a, b)$, but is not defined at $b$, requires finding what limit?

## Skill Building

*In Problems 7–14, determine whether each integral is improper. For those that are improper, state the reason.*

7. $\int_0^\infty x^2\,dx$

8. $\int_0^5 x^3\,dx$

9. $\int_2^3 \frac{dx}{x-1}$

10. $\int_1^2 \frac{dx}{x-1}$

11. $\int_0^1 \frac{1}{x}\,dx$

12. $\int_{-1}^1 \frac{x\,dx}{x^2+1}$

13. $\int_0^1 \frac{x}{x^2-1}\,dx$

14. $\int_0^\infty e^{-2x}\,dx$

*In Problems 15–24, determine whether each improper integral converges or diverges. If it converges, find its value.*

15. $\int_1^\infty \frac{dx}{x^3}$

16. $\int_{-\infty}^{-10} \frac{dx}{x^2}$

17. $\int_0^\infty e^{2x}\,dx$

18. $\int_0^\infty e^{-x}\,dx$

19. $\int_{-\infty}^{-1} \frac{4}{x}\,dx$

20. $\int_1^\infty \frac{4}{x}\,dx$

21. $\int_3^\infty \frac{dx}{(x-1)^4}$

22. $\int_{-\infty}^0 \frac{dx}{(x-1)^4}$

23. $\int_{-\infty}^\infty \frac{dx}{x^2+4}$

24. $\int_{-\infty}^\infty \frac{dx}{x^2+1}$

*In Problems 25–32, determine whether each improper integral converges or diverges. If it converges, find its value.*

25. $\int_0^1 \frac{dx}{x^2}$

26. $\int_0^1 \frac{dx}{x^3}$

27. $\int_0^1 \frac{dx}{x}$

28. $\int_4^6 \frac{dx}{x-4}$

29. $\int_0^4 \frac{dx}{\sqrt{4-x}}$

30. $\int_1^5 \frac{x\,dx}{\sqrt{5-x}}$

31. $\int_{-1}^1 \frac{dx}{\sqrt[3]{x}}$

32. $\int_0^3 \frac{dx}{(x-2)^2}$

*In Problems 33–62, determine whether each improper integral converges or diverges. If it converges, find its value.*

33. $\int_0^\infty \cos x\,dx$

34. $\int_0^\infty \sin(\pi x)\,dx$

35. $\int_{-\infty}^0 e^x\,dx$

36. $\int_{-\infty}^0 e^{-x}\,dx$

37. $\int_0^{\pi/2} \frac{x\,dx}{\sin x^2}$

38. $\int_0^1 \frac{\ln x\,dx}{x}$

---

**1.** = NOW WORK problem      ⚠ = Graphing technology recommended      CAS = Computer Algebra System recommended

**39.** $\displaystyle\int_0^1 \frac{dx}{1-x^2}$

**40.** $\displaystyle\int_1^2 \frac{dx}{\sqrt{x^2-1}}$

**41.** $\displaystyle\int_0^1 \frac{x}{\sqrt{1-x^2}}\,dx$

**42.** $\displaystyle\int_0^4 \frac{dx}{\sqrt{4-x}}$

**43.** $\displaystyle\int_0^{\pi/4} \tan(2x)\,dx$

**44.** $\displaystyle\int_0^{\pi/2} \csc x\,dx$

**45.** $\displaystyle\int_0^\infty \frac{x\,dx}{(x+1)^{5/2}}$

**46.** $\displaystyle\int_2^\infty \frac{dx}{x\sqrt{x^2-1}}$

**47.** $\displaystyle\int_{-\infty}^\infty \frac{dx}{x^2+4x+5}$

**48.** $\displaystyle\int_{-\infty}^\infty \frac{dx}{e^x+e^{-x}}$

**49.** $\displaystyle\int_{-\infty}^2 \frac{dx}{\sqrt{4-x}}$

**50.** $\displaystyle\int_{-\infty}^1 \frac{x\,dx}{\sqrt{2-x}}$

**51.** $\displaystyle\int_2^4 \frac{2x\,dx}{\sqrt[3]{x^2-4}}$

**52.** $\displaystyle\int_0^\pi \frac{1}{1-\cos x}\,dx$

**53.** $\displaystyle\int_{-1}^1 \frac{1}{x^3}\,dx$

**54.** $\displaystyle\int_0^2 \frac{dx}{x-1}$

**55.** $\displaystyle\int_0^2 \frac{dx}{(x-1)^{1/3}}$

**56.** $\displaystyle\int_{-1}^1 \frac{dx}{x^{5/3}}$

**57.** $\displaystyle\int_1^2 \frac{dx}{(2-x)^{3/4}}$

**58.** $\displaystyle\int_0^4 \frac{dx}{\sqrt{8x-x^2}}$

**59.** $\displaystyle\int_1^3 \frac{2x\,dx}{(x^2-1)^{3/2}}$

**60.** $\displaystyle\int_0^3 \frac{x\,dx}{(9-x^2)^{3/2}}$

**61.** $\displaystyle\int_0^\infty xe^{-x^2}\,dx$

**62.** $\displaystyle\int_0^\infty e^{-x}\sin x\,dx$

*In Problems 63–70:*

*(a)* *Use the Comparison Test for Improper Integrals to determine whether each improper integral converges or diverges. (Hint: Use the fact that* $\displaystyle\int_1^\infty \frac{dx}{x^p}$ *converges if* $p>1$ *and diverges if* $p\le 1$.)

**CAS** *(b) If the integral converges, use a CAS to find its value.*

**63.** $\displaystyle\int_1^\infty \frac{1}{\sqrt{x^2-1}}\,dx$

**64.** $\displaystyle\int_2^\infty \frac{2}{\sqrt{x^2-4}}\,dx$

**65.** $\displaystyle\int_1^\infty \frac{1+e^{-x}}{x}\,dx$

**66.** $\displaystyle\int_1^\infty \frac{3e^{-x}}{x}\,dx$

**67.** $\displaystyle\int_1^\infty \frac{\sin^2 x}{x^2}\,dx$

**68.** $\displaystyle\int_1^\infty \frac{\cos^2 x}{x^2}\,dx$

**69.** $\displaystyle\int_1^\infty \frac{dx}{(x+1)\sqrt{x}}$

**70.** $\displaystyle\int_1^\infty \frac{dx}{x\sqrt{1+x^2}}$

**Applications and Extensions**

**71. Area Between Graphs** Find the area, if it is defined, of the region enclosed by the graphs of $y=\dfrac{1}{x+1}$ and $y=\dfrac{1}{x+2}$ on the interval $[0,\infty)$. See the figure.

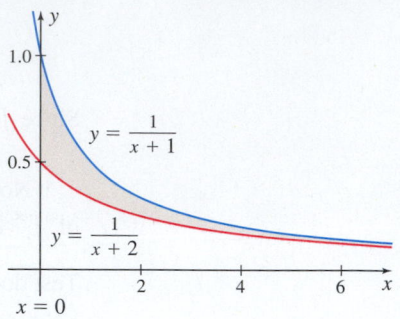

**72. Area Between Graphs** Find the area, if it is defined, under the graph of $y=\dfrac{1}{1+x^2}$ to the right of $x=0$. See the figure below.

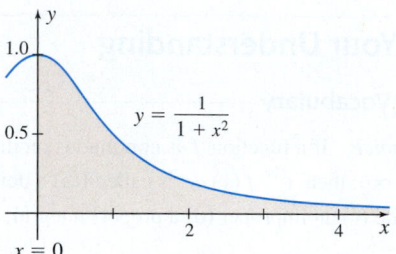

**73. Volume of a Solid of Revolution** Find the volume, if it is defined, of the solid of revolution generated by revolving the region bounded by the graph of $y=e^{-x}$ and the $x$-axis to the right of $x=0$ about the $x$-axis. See the figure below.

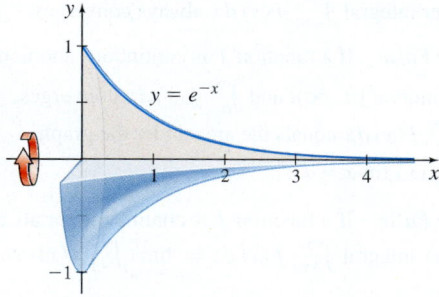

**74. Volume of a Solid of Revolution** Find the volume, if it is defined, of the solid of revolution generated by revolving the region bounded by the graph of $y=\dfrac{1}{\sqrt{x}}$ and the $x$-axis to the right of $x=1$ about the $x$-axis. See the figure below.

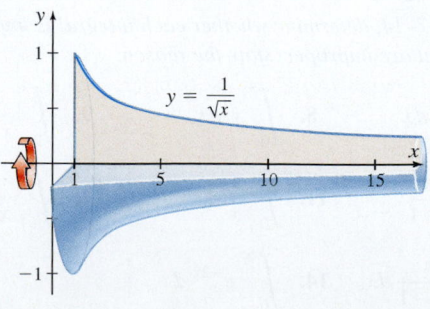

**75. Area Under a Graph**   Find the area, if it is defined, enclosed by the graph of $y = \dfrac{8a^3}{x^2 + 4a^2}$, $a > 0$, $x \geq 0$ and the $x$-axis.

**76. Drug Reaction**   The rate of reaction $r$ to a given dose of a drug at time $t$ hours after administration is given by $r(t) = t\,e^{-t^2}$ (measured in appropriate units).

   (a)  Why is it reasonable to define the total *reaction* as the area under the graph of $y = r(t)$ on $[0, \infty)$?

   (b)  Find the total reaction to the given dose of the drug.

**77. Present Value of Money**   The present value $PV$ of a capital asset that provides a perpetual stream of revenue that flows continuously at a rate of $R(t)$ dollars per year is given by

$$PV = \int_0^\infty R(t)e^{-rt}\,dt$$

where $r$, expressed as a decimal, is the annual rate of interest compounded continuously.

   (a)  Find the present value of an asset if it provides a constant return of \$100 per year and $r = 8\%$.

   (b)  Find the present value of an asset if it provides a return of $R(t) = 1000 + 80t$ dollars per year and $r = 7\%$.

**78. Electrical Engineering**   In a problem in electrical theory, the integral $\int_0^\infty Ri^2\,dt$ occurs, where the current $i = Ie^{-Rt/L}$, $t$ is time, and $R$, $I$, and $L$ are positive constants. Find the integral.

**79. Magnetic Potential**   The magnetic potential $u$ at a point on the axis of a circular coil is given by

$$u = \frac{2\pi N I r}{10} \int_x^\infty \frac{dy}{(r^2 + y^2)^{3/2}}$$

where $N$, $I$, $r$, and $x$ are constants. Find the integral.

**80. Electrical Engineering**   The field intensity $F$ around a long ("infinite") straight wire carrying electric current is given by the integral

$$F = \frac{rIm}{10} \int_{-\infty}^\infty \frac{dy}{(r^2 + y^2)^{3/2}}$$

where $r$, $I$, and $m$ are constants. Find the integral.

**81. Work**   The force $F$ of gravitational attraction between two point masses $m$ and $M$ that are $r$ units apart is $F = \dfrac{GmM}{r^2}$, where $G$ is the universal gravitational constant. Find the work done in moving the mass $m$ along a straight-line path from $r = 1$ unit to $r = \infty$.

**82.** For what numbers $a$ does $\int_0^1 x^a\,dx$ converge?

*In Problems 83–90, find each improper integral.*

**83.** $\displaystyle\int_0^\infty xe^{-x}\,dx$

**84.** $\displaystyle\int_0^1 x\ln x\,dx$

**85.** $\displaystyle\int_0^\infty e^{-x}\cos x\,dx$

**86.** $\displaystyle\int_0^\infty \tan^{-1}x\,dx$

**87.** $\displaystyle\int_0^\infty \frac{dx}{(x^2 + 4)^{3/2}}$

**88.** $\displaystyle\int_4^\infty \frac{dx}{(x^2 - 9)^{3/2}}$

**89.** $\displaystyle\int_1^\infty \frac{dx}{(x + 1)\sqrt{2x + x^2}}$

**90.** $\displaystyle\int_2^\infty \frac{dx}{(x + 2)^2\sqrt{4x + x^2}}$

**91.** Show that $\int_0^\infty \sin x\,dx$ and $\int_{-\infty}^0 \sin x\,dx$ each diverge, yet $\displaystyle\lim_{t \to \infty} \int_{-t}^t \sin x\,dx = 0$.

**92.** Find a function $f$ for which $\int_0^\infty f(x)\,dx$ and $\int_{-\infty}^0 f(x)\,dx$ each diverge, yet $\displaystyle\lim_{t \to \infty} \int_{-t}^t f(x)\,dx = 1$.

**93.** Use the Comparison Test for Improper Integrals to show that $\displaystyle\int_0^\infty \frac{1}{\sqrt{2 + \sin x}}\,dx$ diverges.

**94.** Use the Comparison Test for Improper Integrals to show that $\displaystyle\int_2^\infty \frac{\ln x}{\sqrt{x^2 - 1}}\,dx$ diverges.

**95.** If $n$ is a positive integer, show that:

   (a)  $\displaystyle\int_0^\infty x^n e^{-x}\,dx = n \int_0^\infty x^{n-1} e^{-x}\,dx$

   (b)  $\displaystyle\int_0^\infty x^n e^{-x}\,dx = n!$

**96.** Show that $\displaystyle\int_e^\infty \frac{dx}{x(\ln x)^p}$ converges if $p > 1$ and diverges if $p \leq 1$.

**97.** Show that $\displaystyle\int_a^b \frac{dx}{(x - a)^p}$ converges if $0 < p < 1$ and diverges if $p \geq 1$.

**98.** Show that $\displaystyle\int_a^b \frac{dx}{(b - x)^p}$ converges if $0 < p < 1$ and diverges if $p \geq 1$.

**99.** Refer to Problems 97 and 98. Discuss the convergence or divergence of the integrals if $p \leq 0$. Support your explanation with an example.

**100. Comparison Test for Improper Integrals**   Show that if two functions $f$ and $g$ are nonnegative and continuous on the interval $[a, \infty)$, and if $f(x) \geq g(x)$ for all numbers $x > c$, where $c \geq a$, then

- If $\int_a^\infty f(x)\,dx$ converges, then $\int_a^\infty g(x)\,dx$ also converges.
- If $\int_a^\infty g(x)\,dx$ diverges, then $\int_a^\infty f(x)\,dx$ also diverges.

*Laplace transforms are useful in solving a special class of differential equations. The **Laplace transform** $L[f(x)]$ **of a function** $f$ is defined as*

$$L[f(x)] = \int_0^\infty e^{-sx} f(x)\,dx, \quad x \geq 0,\ s \text{ a complex number}$$

*In Problems 101–106, find the Laplace transform of each function.*

**101.** $f(x) = x$

**102.** $f(x) = \cos x$

**103.** $f(x) = \sin x$

**104.** $f(x) = e^x$

**105.** $f(x) = e^{ax}$

**106.** $f(x) = 1$

## Challenge Problems

**107.** Find the arc length of $y = \sqrt{x - x^2} - \sin^{-1} \sqrt{x}$.

**108.** Find $\displaystyle\int_{-\infty}^{a} e^{(x - e^x)} \, dx$.

**109.** Find $\displaystyle\int_{-\infty}^{\infty} e^{(x - e^x)} \, dx$.

**110. (a)** Show that the area of the region in the first quadrant bounded by the graph of $y = e^{-x}$ and the $x$-axis is divided into two equal parts by the line $x = \ln 2$.

**(b)** If the two regions with equal areas described in (a) are rotated about the $x$-axis, are the resulting volumes equal? If they are unequal, which one is larger and by how much?

*In Problems 111 and 112, use the following definition: A **probability density function** is a function $f$, whose domain is the set of all real numbers, with the following properties:*

- $f(x) \geq 0$ for all $x$

- $\int_{-\infty}^{\infty} f(x) \, dx = 1$

**111. Uniform Density Function**   Show that the function $f$ below is a probability density function.

$$f(x) = \begin{cases} 0 & \text{if } x < a \\ \dfrac{1}{b - a} & \text{if } a \leq x \leq b, a < b \\ 0 & \text{if } x > b \end{cases}$$

**112. Exponential Density Function**   Show that the function $f$ is a probability density function for $a > 0$.

$$f(x) = \begin{cases} \dfrac{1}{a} e^{-x/a} & \text{if } x \geq 0 \\ 0 & \text{if } x < 0 \end{cases}$$

*In Problems 113 and 114, use the following definition: The **expected value** or **mean** $\mu$ associated with a probability density function $f$ is defined by*

$$\mu = \int_{-\infty}^{\infty} x f(x) \, dx$$

*The expected value can be thought of as a weighted average of its various probabilities.*

**113.** Find the expected value $\mu$ of the uniform density function $f$ given in Problem 111.

**114.** Find the expected value $\mu$ of the exponential probability density function $f$ defined in Problem 112.

*In Problems 115 and 116, use the following definitions: The **variance** $\sigma^2$ of a probability density function $f$ is defined as*

$$\sigma^2 = \int_{-\infty}^{\infty} (x - \mu)^2 f(x) \, dx$$

*The variance is the average of the squared deviation from the mean. The **standard deviation** $\sigma$ of a probability density function $f$ is the square root of its variance $\sigma^2$.*

**115.** Find the variance $\sigma^2$ and standard deviation $\sigma$ of the uniform density function defined in Problem 111.

**116.** Find the variance $\sigma^2$ and standard deviation $\sigma$ of the exponential density function defined in Problem 112.

# 7.8 Integration Using Tables and Computer Algebra Systems

**OBJECTIVES**  *When you finish this section, you should be able to:*

1  Use a Table of Integrals (p. 564)

2  Use a computer algebra system (p. 566)

While it is important to be able to use techniques of integration, to save time or to check one's work, a Table of Integrals or a computer algebra system (CAS) is often useful.

## 1 Use a Table of Integrals

The inserts in the back of the book contain a list of integration formulas called a **Table of Integrals.** Many of the integration formulas in the list were derived in this chapter. Although the table seems long, it is far from complete. A more comprehensive table of integrals may be found in Daniel Zwillinger (Ed.), *Standard Mathematical Tables and Formulae*, 32nd ed., Boca Raton, FL: CRC Press.

### EXAMPLE 1   Using a Table of Integrals

Use a Table of Integrals to find $\displaystyle\int \frac{dx}{\sqrt{\left(4x - x^2\right)^3}}$.

**Solution** Look through the headings in the Table of Integrals until you locate *Integrals Containing $\sqrt{2ax - x^2}$*. Continue in the subsection until you find a form that closely resembles the integrand given. You should find Integral 80:

$$\int \frac{dx}{\left(2ax - x^2\right)^{3/2}} = \frac{x - a}{a^2 \sqrt{2ax - x^2}} + C$$

This is the integral we seek with $a = 2$. So,

$$\int \frac{dx}{\sqrt{\left(4x - x^2\right)^3}} = \frac{x - 2}{4\sqrt{4x - x^2}} + C$$  ∎

**NOW WORK** Problem 3.

Some integrals in the table are reduction formulas.

### EXAMPLE 2   Using a Table of Integrals

Use a Table of Integrals to find $\displaystyle\int x^2 \tan^{-1} x \, dx$.

**Solution** Find the subsection of the table titled *Integrals Containing Inverse Trigonometric Functions*. Then look for an integral whose form closely resembles the problem. You should find Integral 112:

$$\int x^n \tan^{-1} x \, dx = \frac{1}{n+1}\left(x^{n+1} \tan^{-1} x - \int \frac{x^{n+1} \, dx}{1 + x^2}\right) \qquad n \neq -1$$

This is the integral we seek with $n = 2$.

$$\int x^2 \tan^{-1} x \, dx = \frac{1}{3}\left(x^3 \tan^{-1} x - \int \frac{x^3 \, dx}{1 + x^2}\right)$$

We find the integral on the right by using the substitution $u = 1 + x^2$. Then $du = 2x \, dx$ and

$$\int \frac{x^3}{1 + x^2} dx = \int \frac{x^2 x \, dx}{1 + x^2} = \int \frac{u - 1}{u} \frac{du}{2} = \frac{1}{2} \int \left(1 - \frac{1}{u}\right) du = \frac{1}{2} u - \frac{1}{2} \ln |u|$$

$$= \frac{1 + x^2}{2} - \frac{\ln(1 + x^2)}{2}$$

So,

$$\int x^2 \tan^{-1} x \, dx = \frac{1}{3}\left(x^3 \tan^{-1} x - \int \frac{x^3}{1 + x^2} dx\right)$$

$$= \frac{1}{3}\left[x^3 \tan^{-1} x - \frac{1 + x^2}{2} + \frac{1}{2} \ln(1 + x^2)\right] + C$$  ∎

**NOW WORK** Problem 11.

Sometimes the given integral is found in the tables after a substitution is made.

CALC
CLIP

**EXAMPLE 3** **Using a Table of Integrals**

Use a Table of Integrals to find $\displaystyle\int \frac{3x+5}{\sqrt{3x+6}}\,dx$.

**Solution** Find the subsection of the table titled *Integrals Containing* $\sqrt{ax+b}$ (the square root of a linear expression). Then look for an integral whose form closely resembles the problem. The closest one is an integral with $x$ in the numerator and $\sqrt{ax+b}$ in the denominator,

Integral 40: $\displaystyle\int \frac{x\,dx}{\sqrt{a+bx}} = \frac{2}{3b^2}(bx-2a)\sqrt{a+bx} + C$    (1)

To express the given integral as one with a single variable in the numerator, use the substitution $u = 3x + 5$. Then $du = 3\,dx$. Since

$$\sqrt{3x+6} = \sqrt{(3x+5)+1} = \sqrt{u+1}$$

we find

$$\int \frac{3x+5}{\sqrt{3x+6}}\,dx \underset{\substack{\uparrow \\ u=3x+5,\ \frac{1}{3}du=dx}}{=} \frac{1}{3}\int \frac{u\,du}{\sqrt{u+1}}$$

which is in the form of (1) with $a = 1$ and $b = 1$ So,

$$\int \frac{3x+5}{\sqrt{3x+6}}\,dx = \frac{1}{3}\int \frac{u\,du}{\sqrt{u+1}} \underset{\uparrow \atop (1)}{=} \frac{1}{3}\cdot\frac{2}{3}(u-2)\sqrt{1+u} + C$$

$$\underset{\substack{\uparrow \\ u=3x+5}}{=} \frac{2}{9}[(3x+5)-2]\sqrt{1+(3x+5)} + C = \frac{2}{3}(x+1)\sqrt{3x+6} + C \qquad\blacksquare$$

NOW WORK Problem 33.

## 2 Use a Computer Algebra System

A computer algebra system (or CAS) is computer software that can perform symbolic manipulation of mathematical expressions. Some graphing calculators such as the TI Nspire, also have a CAS capability. CAS software packages such as Mathematica, Maple, and Matlab offer more extensive symbolic manipulation capabilities, as well as tools for visualization and numerical approximation. These packages offer intuitive interfaces so that the user can obtain results without needing to write a program. An online CAS, based on Mathematica, can be found at WolframAlpha.com.

A CAS is often used instead of a Table of Integrals. When using a CAS to find an integral, keep in mind:

- A CAS can find indefinite integrals or definite integrals.
- For indefinite integrals, the constant of integration is often omitted.
- For definite integrals, there is often an option to specify whether the solution should be expressed in exact form or as an approximate solution, and to what precision.
- Absolute value symbols are often omitted from logarithmic answers.
- The symbol "log" is often used to mean ln.
- An indefinite integral is sometimes expressed in a different, but equivalent, algebraic form from that found in a table or by hand. With simplification, integrals found by hand, with a Table of Integrals, or with a CAS will be the same.
- When the CAS fails to find a result, either because the integral is infeasible or beyond the capability of the CAS, most systems show this by repeating the integral.
- Many CAS products have the capability to handle improper integrals.

EXAMPLE 4   **Finding an Integral Using a CAS**

Find $\int x^2(2x^3 - 4)^5 dx$.

**Solution** Using WolframAlpha to find the integral, input

$$\text{integrate } x\wedge2((2x\wedge3) - 4)\wedge5$$

The output is

$$\int x^2(2x^3 - 4)^5 dx = 32\left(\frac{x^{18}}{18} - \frac{2x^{15}}{3} + \frac{10x^{12}}{3} - \frac{80x^9}{9} + \frac{40x^6}{3} - \frac{32x^3}{3}\right) + C$$

Using Mathematica returns the same result without the constant $C$. ∎

We can find $\int x^2\left(2x^3 - 4\right)^5 dx$ using the substitution $u = 2x^3 - 4$.
Then $du = 6x^2\,dx$, and

$$\int x^2(2x^3 - 4)^5 dx = \frac{1}{6}\int u^5\,du = \frac{1}{6}\cdot\frac{u^6}{6} = \frac{(2x^3 - 4)^6}{36} + C$$

If this solution is expanded using the Binomial Theorem, it will differ from the CAS answer by a constant.

NOW WORK   Problems **39, 47,** and **69** using a CAS and compare your answers to Problems **3, 11,** and **33.**

# 7.8 Assess Your Understanding

**Skill Building**

*In Problems 1–36, find each integral using the Table of Integrals found at the back of the book.*

**1.** $\int e^{2x}\cos x\,dx$

**2.** $\int e^{5x}\sin(2x + 3)\,dx$

**3.** $\int x(2 + 3x)^4\,dx$

**4.** $\int \frac{x}{(5 + 2x)^2}\,dx$

**5.** $\int \frac{x^2\,dx}{\sqrt{6 + 3x}}$

**6.** $\int \frac{\sqrt{1 + x}}{x}\,dx$

**7.** $\int \frac{\sqrt{x^2 + 4}}{x}\,dx$

**8.** $\int \frac{x^2}{\sqrt{x^2 - 9}}\,dx$

**9.** $\int \frac{dx}{(x^2 - 4)^{3/2}}$

**10.** $\int (x^2 + 9)^{3/2}dx$

**11.** $\int x^3(\ln x)^2 dx$

**12.** $\int \frac{dx}{x\ln x}$

**13.** $\int \sqrt{x^2 - 16}\,dx$

**14.** $\int \frac{\sqrt{x^2 + 3}}{x}\,dx$

**15.** $\int (6 - x^2)^{3/2}\,dx$

**16.** $\int \frac{dx}{x^2\sqrt{10 - x^2}}$

**17.** $\int \sqrt{10x - x^2}\,dx$

**18.** $\int \frac{dx}{x\sqrt{6x - x^2}}$

**19.** $\int \cos(3x)\cos(8x)\,dx$

**20.** $\int \sin(2x)\sin(5x)\,dx$

**21.** $\int x\tan^{-1} x\,dx$

**22.** $\int x\sin^{-1} x\,dx$

**23.** $\int x^4\ln x\,dx$

**24.** $\int (\ln x)^3 dx$

**25.** $\int \sinh^2 x\,dx$

**26.** $\int \tanh^2 x\,dx$

**27.** $\int \frac{dx}{x^2\sqrt{8x - x^2}}$

**28.** $\int \frac{\sqrt{12x - x^2}}{x^2}\,dx$

**29.** $\int \frac{dx}{(4 + x^2)^2}$

**30.** $\int \frac{dx}{(x^2 - 25)^3}$

**31.** $\int x\sinh x\,dx$

**32.** $\int x^2\sinh x\,dx$

**33.** $\int (x + 1)\sqrt{4x + 5}\,dx$

**34.** $\int \frac{dx}{[(2x + 3)^2 - 1]^{3/2}}$

**35.** $\int_1^2 \frac{x^3}{\sqrt{3x - x^2}}\,dx$

**36.** $\int_1^e \frac{1}{x^2\sqrt{x^2 + 2}}\,dx$

---

**1.** = NOW WORK problem          📈 = Graphing technology recommended          CAS = Computer Algebra System recommended

CAS In Problems 37–72, redo Problems 1–36 using a CAS. (Answers may vary.)

CAS In Problems 73–78, use a CAS to find each integral. (Answers may vary.)

**73.** $\displaystyle\int \sqrt{1+x^3}\,dx$

**74.** $\displaystyle\int \sqrt{1+\sin x}\,dx$

**75.** $\displaystyle\int e^{-x^2}\,dx$

**76.** $\displaystyle\int \frac{\cos x}{x}\,dx$

**77.** $\displaystyle\int x\tan x\,dx$

**78.** $\displaystyle\int \sqrt{1+e^x}\,dx$

# 7.9 Mixed Practice

**OBJECTIVE** *When you finish this section, you should be able to:*

1  Recognize the form of an integrand and find its integral (p. 568)

## 1 Recognize the Form of an Integrand and Find Its Integral

Throughout this chapter we have discussed many integration techniques. As you now realize, integration, unlike differentiation, is an art that needs practice and often some ingenuity. The techniques studied in each section were based on the form of the integrand. In the *Skill Building* problems at the end of each section, all the integrands followed the forms discussed in the section. Here we have a short section with exercises that "mix up" the integrals.

Always begin by looking at the form of the integrand. Then choose an integration technique that fits that form. Only then attempt to integrate. If the chosen technique does not work, reconsider the form and try again.

## 7.9 Assess Your Understanding

### Skill Building

*Find each integral. If the integral is improper, determine whether it converges or diverges. If it converges, find its value.*

**1.** $\displaystyle\int x^2 \sin^2(x^3)\,dx$

**2.** $\displaystyle\int \cot^3 x \csc x\,dx$

**3.** $\displaystyle\int (5x-1)\cos x\,dx$

**4.** $\displaystyle\int \frac{x^2}{(4+x^2)^{3/2}}\,dx$

**5.** $\displaystyle\int \frac{dx}{\sqrt{4x^2+25}}$

**6.** $\displaystyle\int_0^1 \frac{dx}{(x+1)(x^2+9)}$

**7.** $\displaystyle\int_0^1 \frac{4x\,dx}{x^2+5x+6}$

**8.** $\displaystyle\int x^3 \ln x\,dx$

**9.** $\displaystyle\int \tan x \sec^3 x\,dx$

**10.** $\displaystyle\int \frac{dx}{x^2+2x+10}$

**11.** $\displaystyle\int \tan^2 x \sec^4 x\,dx$

**12.** $\displaystyle\int_1^\infty \frac{\ln x}{x}\,dx$

**13.** $\displaystyle\int \frac{dx}{x\sqrt{x^2-16}}$

**14.** $\displaystyle\int \frac{x+3}{x(x^2+1)^2}\,dx$

**15.** $\displaystyle\int \tan^{1/3} x \sec^4 x\,dx$

**16.** $\displaystyle\int \sin(5x)\cos(2x)\,dx$

**17.** $\displaystyle\int (2x^3+1)\sin(x^4+2x)\,dx$

**18.** $\displaystyle\int_1^2 \frac{1}{(x-1)^2}\,dx$

**19.** $\displaystyle\int x^2 \sin^{-1} x\,dx$

**20.** $\displaystyle\int \sqrt{16x^2-1}\,dx$

**21.** $\displaystyle\int_0^{\pi/2} \sin(2x)\sin\left(\frac{1}{2}x\right)\,dx$

**22.** $\displaystyle\int \sin^5(\pi x)\cos^3(\pi x)\,dx$

**23.** $\displaystyle\int \sqrt{36-x^2}\,dx$

**24.** $\displaystyle\int \cos^5(2x)\,dx$

**25.** $\displaystyle\int \frac{x}{x^2+2x+5}\,dx$

**26.** $\displaystyle\int \frac{x^2-4}{x(x-1)^2}\,dx$

**27.** $\displaystyle\int_0^2 \frac{dx}{\sqrt{4-x^2}}$

**28.** $\displaystyle\int x(x+4)^5\,dx$

**29.** $\displaystyle\int \frac{2x\,dx}{(x-1)(x^2+1)}$

**30.** $\displaystyle\int_0^1 xe^{-3x}\,dx$

**31.** $\displaystyle\int \sin^3 x \sqrt[3]{\cos x}\,dx$

**32.** $\displaystyle\int \frac{\sqrt{4-x^2}}{x}\,dx$

---

**1.** = NOW WORK problem      📈 = Graphing technology recommended      CAS = Computer Algebra System recommended

# Chapter Review

## THINGS TO KNOW

### 7.1 Integration by Parts

- Integration by parts formula $\int u\, dv = uv - \int v\, du$ (p. 499)
- Guidelines for choosing $u$ and $dv$ (p. 500 and Table 1, p. 500)

**Basic Integral** (p. 501):
$$\int \ln x\, dx = x \ln x - x + C$$

### 7.2 Integrals Containing Trigonometric Functions

*Integrals of the Form* $\int \sin^n x\, dx$ *or* $\int \cos^n x\, dx$, $n \geq 2$ *an Integer* (p. 511)

| | Integral | Use the Identity | $u$-substitution |
|---|---|---|---|
| $n$ odd | $\int \sin^n x\, dx = \int \sin^{n-1} x \sin x\, dx$ | $\sin^2 x = 1 - \cos^2 x$ | $u = \cos x;\quad du = -\sin x\, dx$ |
| | $\int \cos^n x\, dx = \int \cos^{n-1} x \cos x\, dx$ | $\cos^2 x = 1 - \sin^2 x$ | $u = \sin x;\quad du = \cos x\, dx$ |
| $n$ even | $\int \sin^n x\, dx = \int \left[\dfrac{1 - \cos(2x)}{2}\right]^{n/2} dx$ | $\sin^2 x = \dfrac{1 - \cos(2x)}{2}$ | $u = 2x;\quad du = 2\, dx$ |
| | $\int \cos^n x\, dx = \int \left[\dfrac{1 + \cos(2x)}{2}\right]^{n/2} dx$ | $\cos^2 x = \dfrac{1 + \cos(2x)}{2}$ | $u = 2x;\quad du = 2\, dx$ |

*Integrals of the Form* $\int \sin^m x \cos^n x\, dx$ (p. 512)

| | Integral | Use the Identity | $u$-substitution |
|---|---|---|---|
| $m$ odd | $\int \sin^m x \cos^n x\, dx = \int \sin^{m-1} x \cos^n x \sin x\, dx$ | $\sin^2 x = 1 - \cos^2 x$ | $u = \cos x$<br>$du = -\sin x\, dx$ |
| $n$ odd | $\int \sin^m x \cos^n x\, dx = \int \sin^m x \cos^{n-1} x \cos x\, dx$ | $\cos^2 x = 1 - \sin^2 x$ | $u = \sin x$<br>$du = \cos x\, dx$ |
| $m$ and $n$ both even | $\int \sin^m x \cos^n x\, dx = \int (1 - \cos^2 x)^{m/2} \cos^n x\, dx$<br>or<br>$\int \sin^m x \cos^n x\, dx = \int \sin^m x (1 - \sin^2 x)^{n/2}\, dx$ | $\cos^2 x = \dfrac{1 + \cos(2x)}{2}$<br><br>$\sin^2 x = \dfrac{1 - \cos(2x)}{2}$ | $u = 2x;\quad du = 2\, dx$<br><br>$u = 2x;\quad du = 2\, dx$ |

*Integrals of the Form* $\int \tan^m x \sec^n x\, dx$ (p. 514)

| | The Integral | Use the Identity | Method |
|---|---|---|---|
| $m$ odd | $\int \tan^m x \sec^n x\, dx = \int \tan^{m-1} x \sec^{n-1} x \sec x \tan x\, dx$ | $\tan^2 x = \sec^2 x - 1$ | $u = \sec x$<br>$du = \sec x \tan x\, dx$ |
| $n$ even | $\int \tan^m x \sec^n x\, dx = \int \tan^m x \sec^{n-2} x \sec^2 x\, dx$ | $\sec^2 x = 1 + \tan^2 x$ | $u = \tan x$<br>$du = \sec^2 x\, dx$ |
| $m$ even; $n$ odd | $\int \tan^m x \sec^n x\, dx = \int (\sec^2 x - 1)^{m/2} \sec^n x\, dx$ | $\tan^2 x = \sec^2 x - 1$ | Use integration by parts |

*Other Integrals Containing Trigonometric Functions* (p. 515)

- $\displaystyle \int \sin(ax)\sin(bx)\, dx = \frac{1}{2} \int [\cos(ax - bx) - \cos(ax + bx)]\, dx$

- $\displaystyle \int \cos(ax)\cos(bx)\, dx = \frac{1}{2} \int [\cos(ax - bx) + \cos(ax + bx)]\, dx$

- $\displaystyle \int \sin(ax)\cos(bx)\, dx = \frac{1}{2} \int [\sin(ax + bx) + \sin(ax - bx)]\, dx$

## 7.3 Integration Using Trigonometric Substitution: Integrands Containing $\sqrt{a^2 - x^2}$, $\sqrt{x^2 + a^2}$, or $\sqrt{x^2 - a^2}$

See Table 2 (p. 517):

- Integrands containing $\sqrt{a^2 - x^2}$: Use the substitution $x = a \sin \theta$, $-\dfrac{\pi}{2} \leq \theta \leq \dfrac{\pi}{2}$. (p. 518)

- Integrands containing $\sqrt{a^2 + x^2}$: Use the substitution $x = a \tan \theta$, $-\dfrac{\pi}{2} < \theta < \dfrac{\pi}{2}$. (p. 519)

- Integrands containing $\sqrt{x^2 - a^2}$: Use the substitution $x = a \sec \theta$, $0 \leq \theta < \dfrac{\pi}{2}$ or $\pi \leq \theta < \dfrac{3\pi}{2}$. (p. 520)

## 7.4 Integrands Containing $ax^2 + bx + c$, $a \neq 0$ (p. 525)

## 7.5 Integration of Rational Functions Using Partial Fractions; the Logistic Model

**Definitions:**

- Proper and improper rational functions (p. 528)
- Partial fractions (p. 528)
- Irreducible quadratic factor (p. 532)
- Logistic differential equation (p. 535)
- Logistic function and curve (pp. 536–537)

**Partial Fraction Decomposition**  $R(x) = \dfrac{p(x)}{q(x)}$, $q(x) \neq 0$

- Case 1: a proper rational function whose denominator contains only distinct linear factors (p. 529)
- Case 2: a proper rational function whose denominator contains a repeated linear factor $(x - a)^n$, $n \geq 2$ (p. 531)
- Case 3: a proper rational function whose denominator contains a distinct irreducible quadratic factor $x^2 + bx + c$ (p. 532)
- Case 4: a proper rational function whose denominator contains a repeated irreducible quadratic factor $(x^2 + bx + c)^n$, $n \geq 2$ an integer (p. 533)

## 7.6 Approximating Integrals: The Trapezoidal Rule, The Midpoint Rule, Simpson's Rule

- Trapezoidal Rule: (p. 542)
- Error in the Trapezoidal Rule $\leq \dfrac{(b-a)^3 M}{12n^2}$, where $M$ is the absolute maximum value of $|f''|$ on the closed interval $[a, b]$. (p. 545)
- The Midpoint Rule (p. 547)
- Error in the Midpoint Rule $\leq \dfrac{(b-a)^3 M}{24n^2}$, where $M$ is the absolute maximum value of $|f''|$ on the closed interval $[a, b]$. (p. 549)
- Simpson's Rule: (p. 550)
- Error in Simpson's Rule $\leq \dfrac{(b-a)^5 M}{180n^4}$, where $M$ is the absolute maximum value of $|f^{(4)}(x)|$ on the closed interval $[a, b]$. (p. 551)

## 7.7 Improper Integrals

- Improper integrals of the form $\int_a^{\infty} f(x)\, dx$; $\int_{-\infty}^b f(x)\, dx$; and $\int_{-\infty}^{\infty} f(x)\, dx$ (pp. 555, 556)
- Converge; diverge (pp. 555, 556, 558, 559)
- Improper integrals $\int_a^b f(x)\, dx$, where $f(a)$, $f(b)$, or $f(c)$, $a < c < b$, is not defined. (pp. 558, 559)

**Theorems**

- $\displaystyle\int_1^{\infty} \dfrac{dx}{x^p}$ converges if $p > 1$ and diverges if $p \leq 1$. (p. 557)
- Comparison Test for Improper Integrals (p. 560)

## 7.8 Integration Using Tables and Computer Algebra Systems (p. 564)

## 7.9 Mixed Practice (p. 568)

## OBJECTIVES

| Section | You should be able to ... | Examples | Review Exercises |
|---|---|---|---|
| 7.1 | 1  Integrate by parts (p. 499) | 1–6 | 9, 16, 21, 25, 47 |
|  | 2  Find a definite integral using integration by parts (p. 503) | 7, 8 | 8, 19, 48 |
|  | 3  Derive a general formula using integration by parts (p. 504) | 9, 10 | 37, 38 |
| 7.2 | 1  Find integrals of the form $\int \sin^n x\, dx$ or $\int \cos^n x\, dx$, $n \geq 2$ an integer (p. 509) | 1–4 | 5, 31 |
|  | 2  Find integrals of the form $\int \sin^m x \cos^n x\, dx$ (p. 511) | 5 | 10, 28 |
|  | 3  Find integrals of the form $\int \tan^m x \sec^n x\, dx$ or $\int \cot^m x \csc^n x\, dx$ (p. 512) | 6–8 | 3, 4 |
|  | 4  Find integrals of the form $\int \sin(ax) \sin(bx)\, dx$, $\int \sin(ax) \cos(bx)\, dx$, or $\int \cos(ax) \cos(bx)\, dx$ (p. 515) | 9 | 32, 33 |
| 7.3 | 1  Integrate a function containing $\sqrt{a^2 - x^2}$ (p. 518) | 1 | 6, 11 |
|  | 2  Integrate a function containing $\sqrt{x^2 + a^2}$ (p. 519) | 2, 3 | 18, 24, 26 |
|  | 3  Integrate a function containing $\sqrt{x^2 - a^2}$ (p. 520) | 4 | 7, 30 |
|  | 4  Use trigonometric substitution to find definite integrals (p. 521) | 5, 6 | 34, 35 |
| 7.4 | 1  Integrate a function that contains a quadratic expression (p. 525) | 1–3 | 1, 14, 20 |

| Section | You should be able to ... | Examples | Review Exercises |
|---|---|---|---|
| 7.5 | 1 Integrate a proper rational function whose denominator contains only distinct linear factors (p. 529) | 2, 3 | 12, 27, 29 |
| | 2 Integrate a proper rational function whose denominator contains a repeated linear factor (p. 531) | 4 | 17, 23 |
| | 3 Integrate a proper rational function whose denominator contains a distinct irreducible quadratic factor (p. 532) | 5 | 2, 15 |
| | 4 Integrate a proper rational function whose denominator contains a repeated irreducible quadratic factor (p. 533) | 6 | 22 |
| | 5 Work with the logistic model (p. 534) | 7 | 53 |
| 7.6 | 1 Approximate an integral using the Trapezoidal Rule (p. 542) | 1–5 | 49(a), 50 |
| | 2 Approximate an integral using the Midpoint Rule (p. 547) | 6, 7 | 49(b), 54 |
| | 3 Approximate an integral using Simpson's Rule (p. 549) | 8–10 | 49(c) |
| 7.7 | 1 Find integrals with an infinite limit of integration (p. 556) | 1, 2 | 39, 42, 44 |
| | 2 Interpret an improper integral geometrically (p. 557) | 3, 4 | 51, 52 |
| | 3 Integrate functions over $[a, b]$ that are not defined at an endpoint (p. 558) | 5–7 | 40, 41, 43 |
| | 4 Use the Comparison Test for improper integrals (p. 560) | 8 | 45, 46 |
| 7.8 | 1 Use a Table of Integrals (p. 564) | 1–3 | 36(a) |
| | 2 Use a computer algebra system (p. 566) | 4 | 36(b) |

## REVIEW EXERCISES

*In Problems 1–35, find each integral.*

**1.** $\displaystyle\int \frac{dx}{x^2 + 4x + 20}$

**2.** $\displaystyle\int \frac{y+1}{y^2 + y + 1}\,dy$

**3.** $\displaystyle\int \sec^3 \phi \tan \phi \, d\phi$

**4.** $\displaystyle\int \cot^2 \theta \, \csc \theta \, d\theta$

**5.** $\displaystyle\int \sin^3 \phi \, d\phi$

**6.** $\displaystyle\int \frac{x^2}{\sqrt{4-x^2}}\,dx$

**7.** $\displaystyle\int \frac{dx}{\sqrt{(x+2)^2 - 1}}$

**8.** $\displaystyle\int_0^{\pi/4} x \sin(2x)\,dx$

**9.** $\displaystyle\int v \csc^2 v \, dv$

**10.** $\displaystyle\int \sin^2 x \cos^3 x \, dx$

**11.** $\displaystyle\int (4 - x^2)^{3/2}\,dx$

**12.** $\displaystyle\int \frac{3x^2 + 1}{x^3 + 2x^2 - 3x}\,dx$

**13.** $\displaystyle\int \frac{e^{2t}\,dt}{e^t - 2}$

**14.** $\displaystyle\int \frac{dy}{5 + 4y + 4y^2}$

**15.** $\displaystyle\int \frac{x\,dx}{x^4 - 16}$

**16.** $\displaystyle\int x^3 e^{x^2}\,dx$

**17.** $\displaystyle\int \frac{y^2\,dy}{(y+1)^3}$

**18.** $\displaystyle\int \frac{dx}{x^2 \sqrt{x^2 + 25}}$

**19.** $\displaystyle\int x \sec^2 x \, dx$

**20.** $\displaystyle\int \frac{dx}{\sqrt{16 + 4x - 2x^2}}$

**21.** $\displaystyle\int \ln(1-y)\,dy$

**22.** $\displaystyle\int \frac{x^3 - 2x - 1}{(x^2 + 1)^2}\,dx$

**23.** $\displaystyle\int \frac{3x^2 + 2}{x^3 - x^2}\,dx$

**24.** $\displaystyle\int \frac{dy}{\sqrt{2 + 3y^2}}$

**25.** $\displaystyle\int x^2 \sin^{-1} x \, dx$

**26.** $\displaystyle\int \sqrt{16 + 9x^2}\,dx$

**27.** $\displaystyle\int \frac{dx}{x^2 + 2x}$

**28.** $\displaystyle\int \sin^4 y \cos^4 y \, dy$

**29.** $\displaystyle\int \frac{w-2}{1-w^2}\,dw$

**30.** $\displaystyle\int \frac{x}{\sqrt{x^2 - 4}}\,dx$

**31.** $\displaystyle\int \frac{1}{\sqrt{x}} \cos^2 \sqrt{x}\,dx$

**32.** $\displaystyle\int \sin\left(\frac{\pi}{2}x\right) \sin(\pi x)\,dx$

**33.** $\displaystyle\int \sin x \cos(2x)\,dx$

**34.** $\displaystyle\int_0^1 \frac{x^2}{\sqrt{4-x^2}}\,dx$

**35.** $\displaystyle\int_0^{\sqrt{3}} \frac{x\,dx}{\sqrt{1+x^2}}$

**36.** (a) Find $\displaystyle\int \frac{\cos^2(2x)\,dx}{\sin^3(2x)}$ using a Table of Integrals.

(b) Find $\displaystyle\int \frac{\cos^2(2x)\,dx}{\sin^3(2x)}$ using a computer algebra system (CAS).

(c) Verify the results from (a) and (b) are equivalent.

*In Problems 37 and 38, derive each formula where $n > 1$ is an integer.*

**37.** $\displaystyle\int x^n \tan^{-1} x \, dx = \frac{x^{n+1}}{n+1} \tan^{-1} x - \frac{1}{n+1} \int \frac{x^{n+1}}{1+x^2}\,dx$

**38.** $\displaystyle\int x^n (ax+b)^{1/2}\,dx = \frac{2x^n(ax+b)^{3/2}}{(2n+3)a}$
$\qquad\qquad - \frac{2bn}{(2n+3)a} \int x^{n-1}(ax+b)^{1/2}\,dx$

*In Problems 39–42, determine whether each improper integral converges or diverges. If it converges, find its value.*

**39.** $\displaystyle\int_1^\infty \frac{e^{-\sqrt{x}}}{\sqrt{x}}\,dx$

**40.** $\displaystyle\int_0^1 \frac{\sin \sqrt{x}}{\sqrt{x}}\,dx$

**41.** $\displaystyle\int_0^1 \frac{x\,dx}{\sqrt{1-x^2}}$

**42.** $\displaystyle\int_{-\infty}^0 x e^x \, dx$

**43.** Show that $\displaystyle\int_0^{\pi/2} \frac{\sin x}{\cos x}\,dx$ diverges.

**44.** Show that $\displaystyle\int_1^\infty \frac{\sqrt{1+x^{1/8}}}{x^{3/4}}\,dx$ diverges.

*In Problems 45 and 46, use the Comparison Test for Improper Integrals to determine whether each improper integral converges or diverges.*

**45.** $\displaystyle\int_1^\infty \frac{1+e^{-x}}{x}\,dx$

**46.** $\displaystyle\int_0^\infty \frac{x}{(1+x)^3}\,dx$

**47.** If $\int x^2 \cos x\,dx = f(x) - \int 2x \sin x\,dx$, find $f$.

**48. Area and Volume**

(a) Find the area $A$ of the region $R$ bounded by the graphs of $y = \ln x$, the $x$-axis, and the line $x = e$.

(b) Find the volume of the solid of revolution generated by revolving $R$ about the $x$-axis.

(c) Find the volume of the solid of revolution generated by revolving $R$ about the $y$-axis.

**49. Arc Length** Approximate the arc length of $y = \cos x$ from $x = 0$ to $x = \dfrac{\pi}{2}$.

(a) using the Trapezoidal Rule with $n = 3$.

(b) using the Midpoint Rule with $n = 3$.

(c) using Simpson's Rule with $n = 4$.

**50. Distance** The velocity $v$ (in meters per second) of an object in rectilinear motion at time $t$ is given in the table. Use the Trapezoidal Rule to approximate the distance traveled from $t = 1$ to $t = 4$.

| $t$ (s)   | 1 | 1.5 | 2   | 2.5 | 3   | 3.5 | 4   |
|-----------|---|-----|-----|-----|-----|-----|-----|
| $v$ (m/s) | 3 | 4.3 | 4.6 | 5.1 | 5.8 | 6.2 | 6.6 |

**51. Area** Find the area, if it exists, of the region bounded by the graphs of $y = x^{-2/3}$, $y = 0$, $x = 0$, and $x = 1$.

**52. Volume** Find the volume, if it exists, of the solid of revolution generated when the region bounded by the graphs of $y = x^{-2/3}$, $y = 0$, $x = 0$, and $x = 1$ is revolved about the $x$-axis.

**53. Logistic Model** A population $P$ grows according to the logistic differential equation

$$\frac{dP}{dt} = 0.0024 P(100 - P)$$

(a) What is the carrying capacity of the population?

(b) What is the maximum population growth rate of the population?

(c) How large is the population at the inflection point?

**54.** Use the Midpoint Rule with $n = 4$ to approximate the arc length of the graph of $y = e^x$ from $x = 2$ to $x = 10$.

---

**CHAPTER 7 PROJECT**    **The Birds of Rügen Island**

Let $P = P(t)$ denote the population of a species of rare birds on Rügen Island, where $t$ is the time, in years. Suppose $M$ equals the maximum sustainable number of birds and $m$ equals the minimum population, below which the species becomes extinct. The population $P$ can be modeled by the differential equation

$$\frac{dP}{dt} = k(M - P)(P - m)$$

where $k$ is a positive constant.

**1.** Suppose the maximum population $M$ is 1200 birds and the minimum population $m$ is 100 birds. If $k = 0.001$, write the differential equation that models the population $P = P(t)$.

**2.** Solve the differential equation from Problem 1.

**3.** If the population at time $t = 0$ is 300 birds, find the particular solution of the differential equation.

**4.** How many birds will exist in 5 years?

**5.** Using technology, graph the solution found in Problem 3.

**6.** The graph from Problem 5 seems to have an inflection point. Verify this and find it.

**7.** What conclusions can you draw about the rate of change of population based on your answer to Problem 6?

**8.** Write an essay about using the given differential equation to model the bird population. What assumptions are being made? What situations are being ignored?

# Infinite Series

**CHAPTER 8 PROJECT** In the
Chapter 8 Project on page 678 we
explore how computers obtain
lightning-fast approximations.

Shane Stillings/Shutterstock

## How Calculators Calculate

When you want to find the sine of a number, you probably type it into your calculator or computer and use the number that comes out. Have you ever thought about how the calculator/computer obtains the result? Since we are only good at computing integer powers of numbers, we have no direct way to evaluate transcendental functions, such as the exponential function or the sine function. We know that $\sin \pi = 0$, but what is sin 3? Since sin 3 is an irrational number, it is represented by a nonrepeating, nonterminating decimal. So, computers and calculators work with approximations for transcendental functions that involve integer powers.

Most transcendental functions can be expressed as *power series*. As we shall see in Section 8.8, a power series is an infinite sum of monomials with integer exponents. Since we are good at raising numbers to integer powers, we can truncate these series and find good polynomial approximations to transcendental functions on restricted domains. This is how computers evaluate many transcendental functions.

T he idea of adding an infinite collection of numbers has long intrigued mathematicians. In this chapter, we explore the consequences of a definition for an infinite sum. Beginning with sequences, we are led to an *infinite series*, an infinite sum of numbers. We examine various tests to determine whether an infinite series has a sum, *converges*, or whether it does not, *diverges*. Finally, we investigate *power series*, representations of functions by infinite series.

# 8.1 Sequences

**OBJECTIVES** *When you finish this section, you should be able to:*

1 Write several terms of a sequence (p. 575)
2 Find the *n*th term of a sequence (p. 575)
3 Use properties of convergent sequences (p. 578)
4 Use a related function or the Squeeze Theorem to show a sequence converges (p. 579)
5 Determine whether a sequence converges or diverges (p. 581)

The study of infinite sums of numbers has important applications in physics and engineering since it provides an alternate way of representing functions. In particular, *infinite series* may be used to approximate irrational numbers, such as $e$, $\pi$, and $\ln 2$. The theory of infinite series is developed through the use of a special kind of function called a *sequence*.

**NEED TO REVIEW?** Sequences are discussed in Section P.8, pp. 68–71.

### DEFINITION

A **sequence** is a function whose domain is the set of positive integers and whose range is a subset of the real numbers.

To get an idea of what this means, consider the graph of the function $f(x) = \dfrac{1}{x}$ for $x > 0$, as shown in Figure 1(a). If all the points on the graph are removed *except* the points $(1, 1)$, $\left(2, \dfrac{1}{2}\right)$, $\left(3, \dfrac{1}{3}\right)$, and so on, as shown in Figure 1(b), then these points are the *graph of a sequence*. Notice the graph consists of points, one point for each positive integer.

A sequence is often represented by listing its values in order. For example, the sequence in Figure 1(b) can be written as

$$f(1), f(2), f(3), f(4), f(5), \ldots$$

or as the list

$$1, \frac{1}{2}, \frac{1}{3}, \frac{1}{4}, \frac{1}{5}, \ldots$$

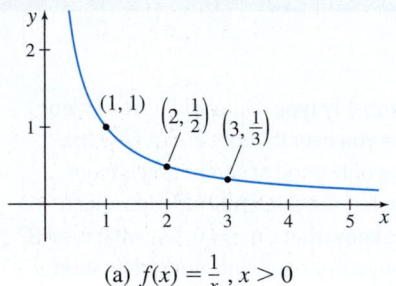

(a) $f(x) = \dfrac{1}{x}$, $x > 0$

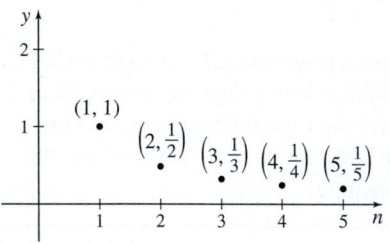

(b) $f(n) = \dfrac{1}{n}$, $n$ is a positive integer

**Figure 1**

The list never ends, as the three dots at the end (called **ellipsis**) indicate. The numbers in the list are called the **terms** of the sequence. Using subscripted letters to represent the terms of a sequence, this sequence can be written as

$$s_1 = f(1) = 1 \qquad s_2 = f(2) = \frac{1}{2} \qquad s_3 = f(3) = \frac{1}{3} \cdots \qquad s_n = f(n) = \frac{1}{n} \cdots$$

It is easy to obtain any term of this sequence because $s_n = f(n) = \dfrac{1}{n}$. In general, whenever a rule for the **nth term** of a sequence is known, then any term of the sequence can be found. We also use the *n*th term to identify the sequence. When the *n*th term is enclosed in braces, it represents the sequence. The notation $\{s_n\} = \left\{\dfrac{1}{n}\right\}$

**IN WORDS** $\{s_n\}_{n=1}^{\infty} = \left\{\dfrac{1}{n}\right\}_{n=1}^{\infty}$

or $\{s_n\} = \left\{\dfrac{1}{n}\right\}$ is the name of the sequence; $s_n = \dfrac{1}{n}$ is the *n*th term of the sequence.

or $\{s_n\}_{n=1}^{\infty} = \left\{\dfrac{1}{n}\right\}_{n=1}^{\infty}$ both represent the sequence $1, \dfrac{1}{2}, \dfrac{1}{3}, \dfrac{1}{4}, \dfrac{1}{5}, \ldots$.

## 1 Write Several Terms of a Sequence

EXAMPLE 1  **Writing Terms of a Sequence**

Write the first three terms of each sequence:

**(a)** $\{b_n\}_{n=1}^{\infty} = \left\{ \dfrac{1}{3n-2} \right\}_{n=1}^{\infty}$    **(b)** $\{c_n\} = \left\{ \dfrac{2n-1}{n^3} \right\}$

**Solution (a)** The $n$th term of this sequence is $b_n = \dfrac{1}{3n-2}$. The first three terms are

$$b_1 = \frac{1}{3 \cdot 1 - 2} = 1 \qquad b_2 = \frac{1}{3 \cdot 2 - 2} = \frac{1}{4} \qquad b_3 = \frac{1}{3 \cdot 3 - 2} = \frac{1}{7}$$

**(b)** The $n$th term of this sequence is $c_n = \dfrac{2n-1}{n^3}$. Then

$$c_1 = \frac{2 \cdot 1 - 1}{1^3} = 1 \qquad c_2 = \frac{2 \cdot 2 - 1}{2^3} = \frac{3}{8} \qquad c_3 = \frac{2 \cdot 3 - 1}{3^3} = \frac{5}{27} \qquad \blacksquare$$

NOW WORK  Problem 15.

For simplicity, from now on we use the notation $\{s_n\}$, rather than $\{s_n\}_{n=1}^{\infty}$, to represent a sequence.

EXAMPLE 2  **Writing Terms of a Sequence**

Write the first five terms of the sequence

$$\{a_n\} = \left\{ (-1)^n \left( \frac{1}{2} \right)^n \right\}$$

**Solution** The $n$th term of this sequence is $a_n = (-1)^n \left( \dfrac{1}{2} \right)^n$. So,

$$a_1 = (-1)^1 \left( \frac{1}{2} \right)^1 = -\frac{1}{2} \qquad a_2 = (-1)^2 \left( \frac{1}{2} \right)^2 = \frac{1}{4}$$

$$a_3 = (-1)^3 \left( \frac{1}{2} \right)^3 = -\frac{1}{8} \qquad a_4 = \frac{1}{16} \qquad a_5 = -\frac{1}{32}$$

The first five terms of the sequence $\{a_n\}$ are $-\dfrac{1}{2}, \dfrac{1}{4}, -\dfrac{1}{8}, \dfrac{1}{16}, -\dfrac{1}{32}$.  $\blacksquare$

Notice the terms of the sequence $\{a_n\}$ given in Example 2 *alternate* between positive and negative due to the factor $(-1)^n$, which equals $-1$ when $n$ is odd and equals $1$ when $n$ is even. Sequences of the form $\{(-1)^n a_n\}$, $\{(-1)^{n-1} a_n\}$, or $\{(-1)^{n+1} a_n\}$, where $a_n > 0$ for all $n$, are called **alternating sequences**.

NOW WORK  Problem 17.

## 2 Find the $n$th Term of a Sequence

The rule defining a sequence $\{s_n\}$ is often expressed by an explicit formula for its $n$th term $s_n$. There are times, however, when a sequence is indicated using the first few terms, suggesting a natural choice for the $n$th term.

EXAMPLE 3 **Finding the $n$th Term of a Sequence**

Find the $n$th term of each of the following sequences. Assume that the indicated pattern continues.

| Sequence | Solution | $n$th term |
|---|---|---|
| **(a)** $\quad e, \dfrac{e^2}{2}, \dfrac{e^3}{3}, \dots$ | | $a_n = \dfrac{e^n}{n}$ |
| **(b)** $\quad 1, \dfrac{1}{3}, \dfrac{1}{9}, \dfrac{1}{27}, \dots$ | | $b_n = \left(\dfrac{1}{3}\right)^{n-1}$ |
| **(c)** $\quad 1, 4, 9, 16, 25, \dots$ | | $c_n = n^2$ |
| **(d)** $\quad \dfrac{2}{2}, \dfrac{4}{3}, \dfrac{6}{4}, \dfrac{8}{5}, \dots$ | | $d_n = \dfrac{2n}{n+1}$ |
| **(e)** $\quad 1, -\dfrac{1}{2}, \dfrac{1}{3}, -\dfrac{1}{4}, \dfrac{1}{5}, \dots$ | | $e_n = \dfrac{(-1)^{n+1}}{n}$ |
| **(f)** $\quad 1, \dfrac{1}{2}, 1, \dfrac{1}{4}, 1, \dfrac{1}{6}, \dots$ | | $f_n = \begin{cases} 1 & \text{if } n \text{ is odd} \\ \dfrac{1}{n} & \text{if } n \text{ is even} \end{cases}$ ∎ |

The graphs of the sequences (d)–(f) are given in Figure 2.

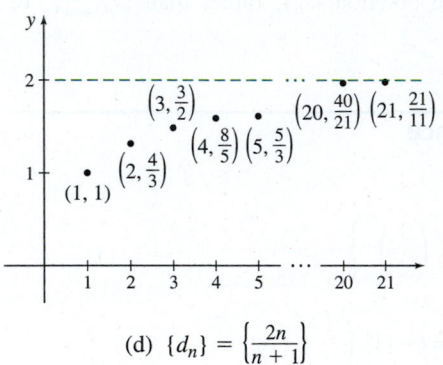

(d) $\{d_n\} = \left\{\dfrac{2n}{n+1}\right\}$

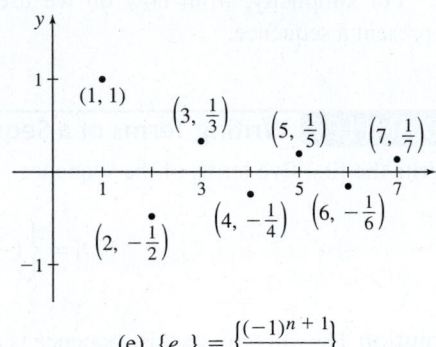

(e) $\{e_n\} = \left\{\dfrac{(-1)^{n+1}}{n}\right\}$

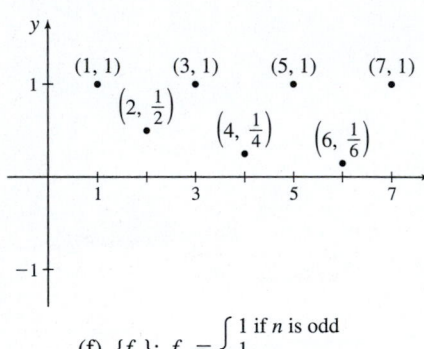

(f) $\{f_n\}; \ f_n = \begin{cases} 1 & \text{if } n \text{ is odd} \\ \dfrac{1}{n} & \text{if } n \text{ is even} \end{cases}$

**DF Figure 2**

NOW WORK **Problem 25.**

### Convergent Sequences

As we look from left to right, the points of the graph of the sequence $\{d_n\} = \left\{\dfrac{2n}{n+1}\right\}$

in Figure 2(d) get closer to the line $y = 2$. If we could look far enough to the right, the points would *appear* to be *on* the line $y = 2$ (although, in reality, the points are always some very small distance below the line). In fact, as Table 1 suggests, $\dfrac{2n}{n+1}$ can be made as close as we please to 2 by taking $n$ sufficiently large.

**TABLE 1**

| $n$ | 1 | 9 | 99 | 999 | 9999 | 999,999 |
|---|---|---|---|---|---|---|
| $d_n = \dfrac{2n}{n+1}$ | 1 | 1.8 | 1.98 | 1.998 | 1.9998 | 1.999998 |

We describe this behavior by saying that the sequence $\{d_n\} = \left\{\dfrac{2n}{n+1}\right\}$ *converges to* 2,

and we write $\displaystyle\lim_{n \to \infty} \dfrac{2n}{n+1} = 2$.

**DEFINITION** Convergent Sequence

Let $L$ be a real number and let $\{s_n\}$ be a sequence. The sequence $\{s_n\}$ **converges** to the real number $L$ if, for any number $\varepsilon > 0$, there is a positive integer $N$ so that

$$|s_n - L| < \varepsilon \qquad \text{for all integers } n > N$$

If $\{s_n\}$ converges to $L$, then $L$ is called the **limit of the sequence** and we write $\lim\limits_{n\to\infty} s_n = L$. If a sequence converges, it is said to be **convergent**; otherwise, it is said to be **divergent**.

**IN WORDS** A sequence $\{s_n\}$ converges if $\lim\limits_{n\to\infty} s_n = L$, a number.

Figure 3 provides a geometric interpretation of the statement $\lim\limits_{n\to\infty} s_n = L$. The figure also illustrates that the beginning terms of a sequence do not affect the convergence (or divergence) of the sequence. That is, **the beginning terms of a sequence can be ignored when determining convergence or divergence.**

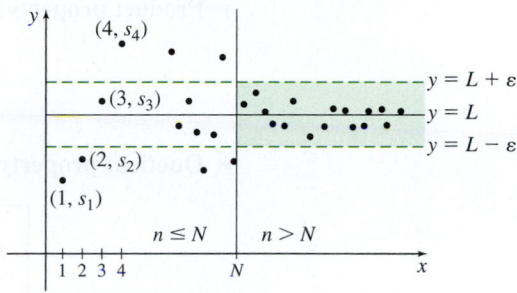

DF **Figure 3**

When $n \le N$, the distance between the point $(n, s_n)$ and the line $y = L$ may or may not be less than $\varepsilon$; these terms have no effect on the convergence or divergence of the sequence. But if a sequence $\{s_n\}$ converges, then for any $\varepsilon > 0$, there is an integer $N$ so that for all $n > N$, the distance between the point $(n, s_n)$ and the line $y = L$ remains less than $\varepsilon$, that is, $|s_n - L| < \varepsilon$ for all $n > N$.

**EXAMPLE 4** Showing a Sequence Converges

Show that:

**(a)** $\lim\limits_{n\to\infty} c = c$ **(b)** $\lim\limits_{n\to\infty} \dfrac{1}{n} = 0$

**Solution (a)** The graph of the sequence $\{s_n\} = \{c\}$ suggests that $\{s_n\}$ converges to $c$. See Figure 4.

To show the sequence converges to $c$, we look at $|s_n - c|$. Then for any $\varepsilon > 0$,

$$|s_n - c| = |c - c| = 0 < \varepsilon \qquad \text{for all } n$$

The sequence $\{s_n\} = \{c\}$ converges to $c$.

**(b)** The graph of the sequence $\{s_n\} = \left\{\dfrac{1}{n}\right\}$ shown in Figure 5 suggests that $\{s_n\}$ converges to 0.

To show the sequence $\{s_n\} = \left\{\dfrac{1}{n}\right\}$ converges to 0, we look at

$$|s_n - 0| = \left|\frac{1}{n} - 0\right| = \frac{1}{n}$$

For any $\varepsilon > 0$, choose any integer $N > \dfrac{1}{\varepsilon}$. Then for all $n > N > \dfrac{1}{\varepsilon}$, we have $|s_n - 0| = \dfrac{1}{n} < \dfrac{1}{N} < \varepsilon$, so the sequence $\{s_n\} = \left\{\dfrac{1}{n}\right\}$ converges to 0. ∎

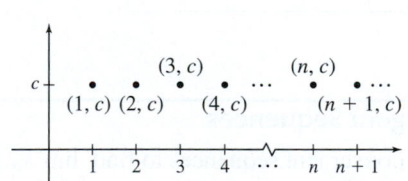

**Figure 4** $\{s_n\} = \{c\}$

**Figure 5** $\{s_n\} = \left\{\dfrac{1}{n}\right\}$

## 3 Use Properties of Convergent Sequences

We state, without proof, some properties of convergent sequences.

> **THEOREM  Properties of Convergent Sequences**
>
> If $\{s_n\}$ and $\{t_n\}$ are convergent sequences and if $c$ is a number, then
>
> • **Constant multiple property:**
>
> $$\lim_{n \to \infty} (cs_n) = c \lim_{n \to \infty} s_n \qquad (1)$$
>
> • **Sum and difference properties:**
>
> $$\lim_{n \to \infty} (s_n \pm t_n) = \lim_{n \to \infty} s_n \pm \lim_{n \to \infty} t_n \qquad (2)$$
>
> • **Product property:**
>
> $$\lim_{n \to \infty} (s_n \cdot t_n) = \left( \lim_{n \to \infty} s_n \right) \left( \lim_{n \to \infty} t_n \right) \qquad (3)$$
>
> • **Quotient property:**
>
> $$\lim_{n \to \infty} \frac{s_n}{t_n} = \frac{\lim\limits_{n \to \infty} s_n}{\lim\limits_{n \to \infty} t_n} \qquad \text{provided} \qquad \lim_{n \to \infty} t_n \neq 0 \qquad (4)$$
>
> • **Power property:**
>
> $$\lim_{n \to \infty} s_n^p = \left[ \lim_{n \to \infty} s_n \right]^p \qquad p \geq 2 \text{ is an integer} \qquad (5)$$
>
> • **Root property:**
>
> $$\lim_{n \to \infty} \sqrt[p]{s_n} = \sqrt[p]{\lim_{n \to \infty} s_n} \qquad p \geq 2 \text{ and } s_n > 0 \text{ if } p \text{ is even} \qquad (6)$$

---

**EXAMPLE 5  Using Properties of Convergent Sequences**

Use the results of Example 4, and properties of convergent sequences to find $\lim\limits_{n \to \infty} s_n$.

**(a)** $\{s_n\} = \left\{ \dfrac{2}{n} + 3 \right\}$ **(b)** $\{s_n\} = \left\{ \dfrac{4}{n^2} \right\}$ **(c)** $\{s_n\} = \left\{ \sqrt[3]{\dfrac{16n^2 + 3n}{2n^2}} \right\}$

**Solution (a)** $\lim\limits_{n \to \infty} s_n = \lim\limits_{n \to \infty} \left( \dfrac{2}{n} + 3 \right) = \lim\limits_{n \to \infty} \dfrac{2}{n} + \lim\limits_{n \to \infty} 3$

$$= 2 \lim_{n \to \infty} \frac{1}{n} + \lim_{n \to \infty} 3$$

$$= 2 \cdot 0 + 3 = 3 \qquad \lim_{n \to \infty} \frac{1}{n} = 0; \ \lim_{n \to \infty} 3 = 3$$

**(b)** $\lim\limits_{n \to \infty} s_n = \lim\limits_{n \to \infty} \dfrac{4}{n^2} = 4 \lim\limits_{n \to \infty} \dfrac{1}{n^2} = 4 \lim\limits_{n \to \infty} \left( \dfrac{1}{n} \right)^2 = 4 \left( \lim\limits_{n \to \infty} \dfrac{1}{n} \right)^2 = 4 \cdot 0^2 = 0$

$\uparrow$ Power property $\qquad\qquad \uparrow$ $\lim\limits_{n \to \infty} \dfrac{1}{n} = 0$

(c) $\displaystyle\lim_{n\to\infty} s_n = \lim_{n\to\infty} \sqrt[3]{\frac{16n^2+3n}{2n^2}} = \sqrt[3]{\lim_{n\to\infty}\left(\frac{16n^2+3n}{2n^2}\right)} = \sqrt[3]{\lim_{n\to\infty}\left(8+\frac{3}{2n}\right)}$

$\displaystyle = \sqrt[3]{\lim_{n\to\infty}8 + \lim_{n\to\infty}\frac{3}{2n}} = \sqrt[3]{8+\frac{3}{2}\lim_{n\to\infty}\frac{1}{n}} = \sqrt[3]{8+\frac{3}{2}\cdot 0} = \sqrt[3]{8} = 2$  ■

$$\underset{\displaystyle \lim_{n\to\infty}8=8}{\uparrow} \qquad \underset{\displaystyle \lim_{n\to\infty}\frac{1}{n}=0}{\uparrow}$$

**NOW WORK** Problem 37.

The next result is also useful for showing a sequence converges. You are asked to prove it in Problem 142.

**THEOREM**

Let $\{s_n\}$ be a sequence of real numbers. If $\displaystyle\lim_{n\to\infty} s_n = L$ and if $f$ is a function that is continuous at $L$ and is defined for all numbers $s_n$, then $\displaystyle\lim_{n\to\infty} f(s_n) = f(L)$.

**EXAMPLE 6    Showing a Sequence Converges**

Show $\left\{\ln\left(\dfrac{2}{n}+3\right)\right\}$ converges and find its limit.

**Solution** Since $\displaystyle\lim_{n\to\infty}\left(\frac{2}{n}+3\right)=3$ [from Example 5(a)], the sequence $\{s_n\}=\left\{\dfrac{2}{n}+3\right\}$

converges to 3. The function $f(x)=\ln x$ is continuous on its domain, $\{x\,|\,x>0\}$, so it is continuous at 3. Then

$$\lim_{n\to\infty}f(s_n)=\lim_{n\to\infty}f\left(\frac{2}{n}+3\right)=\lim_{n\to\infty}\ln\left(\frac{2}{n}+3\right)=\ln\left[\lim_{n\to\infty}\left(\frac{2}{n}+3\right)\right]=\ln 3$$

So, the sequence $\left\{\ln\left(\dfrac{2}{n}+3\right)\right\}$ converges to $\ln 3$.  ■

**NOW WORK** Problem 45.

## 4 Use a Related Function or the Squeeze Theorem to Show a Sequence Converges

Sometimes a sequence $\{s_n\}$ can be associated with a *related function* $f$, which can be helpful in determining whether the sequence converges.

**DEFINITION** Related Function of a Sequence

A **related function** $f$ of the sequence $\{s_n\}$ has the following two properties:

- $f$ is defined on the open interval $(0,\infty)$; that is, the domain of $f$ is the set of positive real numbers.
- $f(n)=s_n$ for all integers $n\geq 1$.

**EXAMPLE 7    Identifying a Related Function of a Sequence**

If $\{s_n\}=\left\{\dfrac{n}{e^n}\right\}$, then a related function is given by $f(x)=\dfrac{x}{e^x}$, where $x>0$, as shown in Figure 6.  ■

There is a connection between the convergence of certain sequences $\{s_n\}$ and the behavior at infinity of a related function $f$ of the sequence $\{s_n\}$. The following result, which we state without proof, explains this connection.

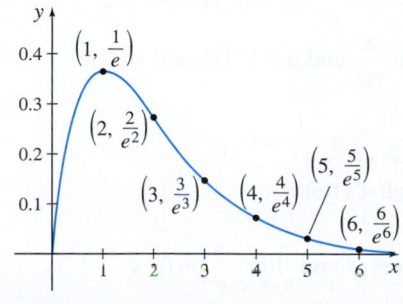

**DF** Figure 6 $f(x)=\dfrac{x}{e^x}$; $\{s_n\}=\left\{\dfrac{n}{e^n}\right\}$

**THEOREM**

Let $\{s_n\}$ be a sequence of real numbers and let $f$ be a related function of $\{s_n\}$. Suppose $L$ is a real number.

$$\text{If } \lim_{x \to \infty} f(x) = L \qquad \text{then} \qquad \lim_{n \to \infty} s_n = L \qquad (7)$$

**NEED TO REVIEW?** Limits at infinity are discussed in Section 1.5, pp. 128–135.

**EXAMPLE 8** **Using a Related Function to Show a Sequence Converges**

Show that $\left\{ \dfrac{3n^2 + 5n - 2}{6n^2 - 6n + 5} \right\}$ converges and find its limit.

**Solution** The function

$$f(x) = \frac{3x^2 + 5x - 2}{6x^2 - 6x + 5} \qquad x > 0$$

is a related function of the sequence $\left\{ \dfrac{3n^2 + 5n - 2}{6n^2 - 6n + 5} \right\}$. Since

$$\lim_{x \to \infty} f(x) = \lim_{x \to \infty} \frac{3x^2 + 5x - 2}{6x^2 - 6x + 5} = \lim_{x \to \infty} \frac{3x^2}{6x^2} = \frac{1}{2}$$

the sequence $\left\{ \dfrac{3n^2 + 5n - 2}{6n^2 - 6n + 5} \right\}$ converges and $\lim_{n \to \infty} \dfrac{3n^2 + 5n - 2}{6n^2 - 6n + 5} = \dfrac{1}{2}$. ∎

Be careful! A related function $f$ can be used to show a sequence $\{s_n\}$ converges only if $\lim_{x \to \infty} f(x) = L$, where $L$ is a *real number*.

- If $\lim_{x \to \infty} f(x)$ is infinite, then $\{s_n\}$ diverges.
- If $\lim_{x \to \infty} f(x)$ does not exist, then the theorem relating $\lim_{n \to \infty} s_n$ and $\lim_{x \to \infty} f(x)$ provides no information about $\lim_{n \to \infty} s_n$.

**NOW WORK** Problem 51.

**NEED TO REVIEW?** L'Hôpital's Rule is discussed in Section 4.5, pp. 312–315.

We can sometimes find the limit of a sequence $\{s_n\}$ by applying L'Hôpital's Rule to its related function $f$, provided $f$ meets the necessary requirements.

**CALC CLIP**

**EXAMPLE 9** **Using L'Hôpital's Rule to Show a Sequence Converges**

Show that $\left\{ \dfrac{n}{e^n} \right\}$ converges and find its limit.

**Solution** We begin with the related function $f(x) = \dfrac{x}{e^x}$, $x > 0$. To find $\lim_{x \to \infty} f(x)$, we note that $\dfrac{x}{e^x}$ is an indeterminate form of the type $\dfrac{\infty}{\infty}$ and use L'Hôpital's Rule.

$$\lim_{x \to \infty} f(x) = \lim_{x \to \infty} \frac{x}{e^x} \underset{\underset{\text{Use L'Hôpital's Rule}}{\uparrow}}{=} \lim_{x \to \infty} \frac{1}{e^x} = 0$$

Since $\lim_{x \to \infty} f(x) = 0$, the sequence $\left\{ \dfrac{n}{e^n} \right\}$ converges and $\lim_{n \to \infty} \dfrac{n}{e^n} = 0$. ∎

**NOW WORK** Problem 57.

**NEED TO REVIEW?** The Squeeze Theorem for functions is discussed in Section 1.4, pp. 115–116.

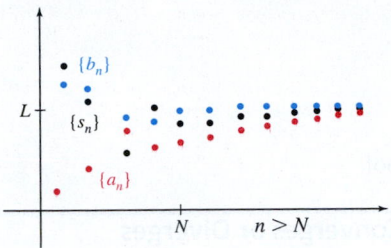

**Figure 7** $a_n \le s_n \le b_n$ for $n > N$

**THEOREM The Squeeze Theorem for Sequences**

Suppose $\{a_n\}$, $\{b_n\}$, and $\{s_n\}$ are sequences and $N$ is a positive integer. If $a_n \le s_n \le b_n$ for every integer $n > N$, and if $\lim\limits_{n\to\infty} a_n = \lim\limits_{n\to\infty} b_n = L$, then $\lim\limits_{n\to\infty} s_n = L$.

Figure 7 illustrates the Squeeze Theorem for sequences.

**EXAMPLE 10  Using the Squeeze Theorem for Sequences**

Show that $\{s_n\} = \left\{ (-1)^n \dfrac{1}{n} \right\}$ converges and find its limit.

**Solution** We seek two sequences that "squeeze" $\{s_n\} = \left\{ (-1)^n \dfrac{1}{n} \right\}$ as $n$ becomes large. We begin with $|s_n|$:

$$|s_n| = \left| (-1)^n \frac{1}{n} \right| \le \frac{1}{n}$$

$$-\frac{1}{n} \le (-1)^n \frac{1}{n} \le \frac{1}{n}$$

Notice that $s_n$ is bounded by $\{a_n\} = \left\{ -\dfrac{1}{n} \right\}$ and $\{b_n\} = \left\{ \dfrac{1}{n} \right\}$. Since $a_n \le s_n \le b_n$ for all $n$ and $\lim\limits_{n\to\infty} a_n = \lim\limits_{n\to\infty} \left( -\dfrac{1}{n} \right) = 0$ and $\lim\limits_{n\to\infty} b_n = \lim\limits_{n\to\infty} \dfrac{1}{n} = 0$, then by the Squeeze Theorem for sequences, the sequence $\{s_n\} = \left\{ (-1)^n \dfrac{1}{n} \right\}$ converges and $\lim\limits_{n\to\infty} s_n = 0$. ∎

**NOW WORK** Problem 59.

## 5 Determine Whether a Sequence Converges or Diverges

A sequence $\{s_n\}$ diverges if $\lim\limits_{n\to\infty} s_n$ does not exist. This can happen if

- there is no single number $L$ that the terms of the sequence approach as $n \to \infty$

- $\lim\limits_{n\to\infty} s_n = \infty$

**DEFINITION Divergence of a Sequence to Infinity**

The sequence $\{s_n\}$ **diverges to infinity**, that is,

$$\lim_{n\to\infty} s_n = \infty$$

if, given any positive number $M$, there is a positive integer $N$ so that whenever $n > N$, then $s_n > M$.

**EXAMPLE 11  Showing a Sequence Diverges**

Show that the following sequences diverge:

**(a)** $\{1 + (-1)^n\}$ **(b)** $\{n\}$

**Solution (a)** The terms of the sequence are $0,\ 2,\ 0,\ 2,\ 0,\ 2,\dots$. See Figure 8. Since the terms alternate between 0 and 2, the terms of the sequence $\{1 + (-1)^n\}$ do not approach a single number $L$ as $n \to \infty$. So, the sequence $\{1 + (-1)^n\}$ is divergent.

**(b)** The terms of the sequence $\{s_n\} = \{n\}$ are $1,\ 2,\ 3,\ 4,\dots$. Given any positive number $M$, we choose a positive integer $N > M$. Then whenever $n > N$, we have $s_n - n > N > M$. That is, the sequence $\{n\}$ diverges to infinity. See Figure 9. ∎

**NOW WORK** Problem 63.

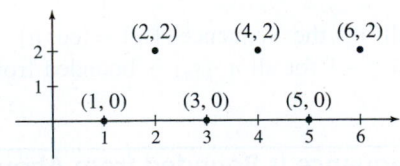

**Figure 8** $\{1 + (-1)^n\}$

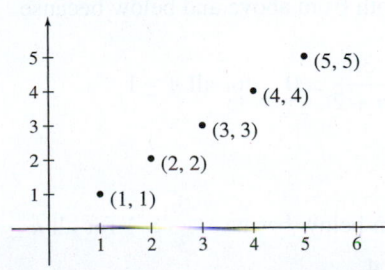

**Figure 9** $\{s_n\} = \{n\}$

The next results are useful throughout the chapter.

> **THEOREM** Convergence/Divergence of a Geometric Sequence
>
> The geometric sequence $\{s_n\} = \{r^n\}$, where $r$ is a real number,
>
> - converges to 0, if $-1 < r < 1$.
> - converges to 1, if $r = 1$.
> - diverges for all other numbers.

Problems 138–141 provide an outline of the proof.

**EXAMPLE 12** Determine Whether $\{r^n\}$ Converges or Diverges

Determine whether the following geometric sequences converge or diverge:

**(a)** $\{s_n\} = \left\{ \left( \dfrac{3}{4} \right)^n \right\}$    **(b)** $\{t_n\} = \left\{ \left( \dfrac{4}{3} \right)^n \right\}$

**Solution (a)** The sequence $\{s_n\} = \left\{ \left( \dfrac{3}{4} \right)^n \right\}$ converges to 0 because $-1 < \dfrac{3}{4} < 1$.

**(b)** The sequence $\{t_n\} = \left\{ \left( \dfrac{4}{3} \right)^n \right\}$ diverges because $\dfrac{4}{3} > 1$. ∎

**NOW WORK** Problem 67.

There are other ways to show that a sequence converges or diverges. To explore these, we need to define a *bounded sequence* and a *monotonic sequence*.

## Bounded Sequences

A sequence $\{s_n\}$ is **bounded from above** if every term of the sequence is less than or equal to some number $M$. That is,

$$\boxed{s_n \leq M \qquad \text{for all } n}$$

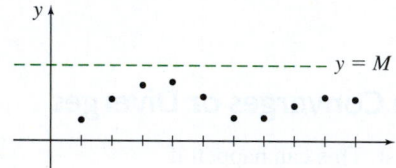

**Figure 10** The sequence is bounded from above.

See Figure 10.

Similarly, a sequence $\{s_n\}$ is **bounded from below** if every term of the sequence is greater than or equal to some number $m$. That is,

$$\boxed{s_n \geq m \qquad \text{for all } n}$$

See Figure 11.

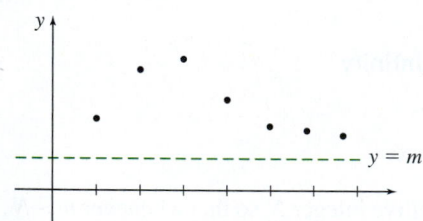

**Figure 11** The sequence is bounded from below.

For example, since $s_n = \cos n \leq 1$ for all $n$, the sequence $\{s_n\} = \{\cos n\}$ is bounded from above by 1, and since $s_n = \cos n \geq -1$ for all $n$, $\{s_n\}$ is bounded from below by $-1$.

**EXAMPLE 13** Determining Whether a Sequence Is Bounded from Above or Bounded from Below

**(a)** The sequence $\{s_n\} = \left\{ \dfrac{3n}{n+2} \right\}$ is bounded both from above and below because

$$\frac{3n}{n+2} = \frac{3}{1 + \dfrac{2}{n}} < 3 \quad \text{and} \quad \frac{3n}{n+2} > 0 \quad \text{for all } n \geq 1$$

See Figure 12(a).

**(b)** The sequence $\{a_n\} = \left\{ \dfrac{4n}{3} \right\}$ is bounded from below because $\dfrac{4n}{3} > 1$ for all $n \geq 1$.

It is not bounded from above because $\displaystyle\lim_{n \to \infty} \frac{4n}{3} = \frac{4}{3} \lim_{n \to \infty} n = \infty$. See Figure 12(b).

**(c)** The sequence $\{b_n\} = \{(-1)^{n+1}n\}$ is neither bounded from above nor bounded from below. If $n$ is odd, $\lim\limits_{n\to\infty} b_n = \lim\limits_{n\to\infty} n = \infty$, and if $n$ is even, $\lim\limits_{n\to\infty} b_n = \lim\limits_{n\to\infty} (-n) = -\infty$. See Figure 12(c).

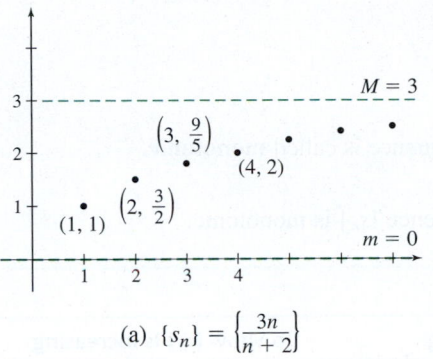 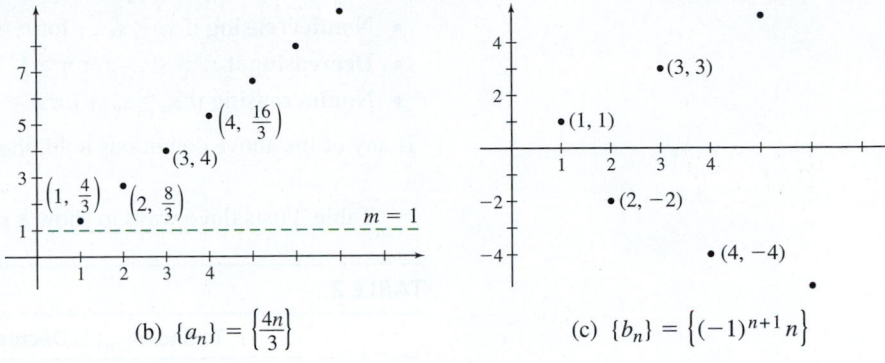

(a) $\{s_n\} = \left\{\dfrac{3n}{n+2}\right\}$    (b) $\{a_n\} = \left\{\dfrac{4n}{3}\right\}$    (c) $\{b_n\} = \{(-1)^{n+1}n\}$

**DF** Figure 12    ∎

**NOW WORK** Problem 73.

A sequence $\{s_n\}$ is **bounded** if it is bounded both from above and from below. For a bounded sequence $\{s_n\}$, there is a positive number $K$ for which

$$\boxed{|s_n| \le K \qquad \text{for all } n \ge 1}$$

For example, the sequence $\{s_n\} = \left\{\dfrac{3n}{n+2}\right\}$ [from Example 13(a)] is bounded, since $\left|\dfrac{3n}{n+2}\right| \le 3$ for all integers $n \ge 1$.

### THEOREM  Boundedness Theorem

A convergent sequence is bounded.

**CAUTION** The converse of the Boundedness Theorem is not true. Bounded sequences may converge or they may diverge. For example, the sequence $\{1 + (-1)^n\}$ in Example 11(a) is bounded, but it diverges.

**NEED TO REVIEW?** The Triangle Inequality is discussed in Appendix A.1, p. A-7

**Proof**  If $\{s_n\}$ is a convergent sequence, there is a number $L$ for which $\lim\limits_{n\to\infty} s_n = L$. We use the definition of the limit of a sequence with $\varepsilon = 1$. Then there is a positive integer $N$ so that

$$|s_n - L| < 1 \qquad \text{for all } n > N$$

Then for all $n > N$,

$$|s_n| = |s_n - L + L| \le \underset{\underset{\text{Triangle Inequality}}{\uparrow}}{|s_n - L| + |L|} < 1 + |L|$$

If we choose $K$ to be the largest number in the finite collection

$$|s_1|, |s_2|, |s_3|, \ldots, |s_N|, 1 + |L|$$

it follows that $|s_n| \le K$ for *all* integers $n \ge 1$. That is, the sequence $\{s_n\}$ is bounded. ∎

A restatement of the boundedness theorem provides a test for divergent sequences.

### THEOREM  Test for Divergence of a Sequence

If a sequence is not bounded from above or if it is not bounded from below, then it diverges.

For example, the sequence $\{-n\}$ is not bounded from below, and the sequences $\left\{\dfrac{4n}{3}\right\}$, $\{2^n\}$, and $\{\ln n\}$ are not bounded from above, so each sequence is divergent.

## Monotonic Sequences

**DEFINITION**

A sequence $\{s_n\}$ is said to be:

- **Increasing** if $s_n < s_{n+1}$ for $n \geq 1$.
- **Nondecreasing** if $s_n \leq s_{n+1}$ for $n \geq 1$.
- **Decreasing** if $s_n > s_{n+1}$ for $n \geq 1$.
- **Nonincreasing** if $s_n \geq s_{n+1}$ for $n \geq 1$.

If any of the above conditions hold, the sequence is called **monotonic.**

Table 2 lists three ways to show a sequence $\{s_n\}$ is monotonic.

**TABLE 2**

| | To Show $\{s_n\}$ Is Decreasing | To Show $\{s_n\}$ Is Increasing |
|---|---|---|
| **Algebraic Difference** | Show $s_{n+1} - s_n < 0$ for all $n \geq 1$. | Show $s_{n+1} - s_n > 0$ for all $n \geq 1$. |
| **Algebraic Ratio** | If $s_n > 0$ for all $n \geq 1$, show $\dfrac{s_{n+1}}{s_n} < 1$ for all $n \geq 1$. | If $s_n > 0$ for all $n \geq 1$, show $\dfrac{s_{n+1}}{s_n} > 1$ for all $n \geq 1$. |
| **Derivative** | Show the derivative of a related function $f$ of $\{s_n\}$ is negative for all $x \geq 1$. | Show the derivative of a related function $f$ of $\{s_n\}$ is positive for all $x \geq 1$. |

These tests can be extended to show a sequence is nonincreasing or nondecreasing.

**EXAMPLE 14** **Showing a Sequence Is Monotonic**

Show that each of the following sequences is monotonic by determining whether it is increasing, nondecreasing, decreasing, or nonincreasing:

**(a)** $\{s_n\} = \left\{ \dfrac{n}{n+1} \right\}$  **(b)** $\{s_n\} = \left\{ \dfrac{e^n}{n!} \right\}$  **(c)** $\{s_n\} = \{\ln n\}$

**Solution (a)** We use the algebraic difference test.

$$s_{n+1} - s_n = \frac{n+1}{n+2} - \frac{n}{n+1} = \frac{n^2 + 2n + 1 - n^2 - 2n}{(n+2)(n+1)}$$

$$= \frac{1}{(n+2)(n+1)} > 0 \qquad \text{for all } n \geq 1$$

So, $\{s_n\}$ is an increasing sequence.

**(b)** When the sequence contains a factorial, the algebraic ratio test is usually easiest to use.

**RECALL** $n! = n(n-1)(n-2) \cdots 2 \cdot 1$

So $\dfrac{n!}{(n+1)!} = \dfrac{1}{n+1}$.

$$\frac{s_{n+1}}{s_n} = \frac{\dfrac{e^{n+1}}{(n+1)!}}{\dfrac{e^n}{n!}} = \left( \frac{e^{n+1}}{e^n} \right) \frac{n!}{(n+1)!} = \frac{e}{n+1} < 1 \qquad \text{for all } n \geq 2$$

After the first term, $\{s_n\} = \left\{ \dfrac{e^n}{n!} \right\}$ is a decreasing sequence.

**(c)** Here we use the derivative of the related function $f(x) = \ln x$ of the sequence $\{s_n\} = \{\ln n\}$. Since $\dfrac{d}{dx} \ln x = \dfrac{1}{x} > 0$ for all $x > 0$, it follows that $f$ is an increasing function and so the sequence $\{\ln n\}$ is an increasing sequence. ∎

**NOW WORK** Problem **81.**

Not every sequence is monotonic. For example, the sequences $\{s_n\} = \left\{\sin\left(\dfrac{\pi}{2}n\right)\right\}$ and $\{t_n\} = \left\{1 + \dfrac{(-1)^n}{n^2}\right\}$, shown in Figures 13 and 14, are not monotonic. Notice that although both $\{s_n\}$ and $\{t_n\}$ are bounded sequences, $\{s_n\}$ diverges and $\{t_n\}$ converges.

The sequence $\{u_n\} = \{n\}$ shown in Figure 15 is monotonic, but it is not bounded from above. So, $\{u_n\}$ is divergent.

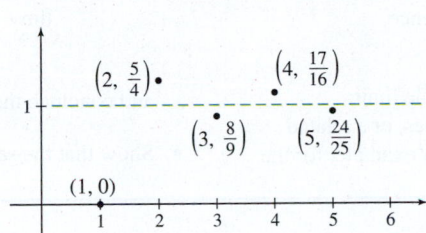

Bounded, not monotonic; diverges

**Figure 13** $\{s_n\} = \left\{\sin\left(\dfrac{\pi}{2}n\right)\right\}$

**Figure 14** $\{t_n\} = \left\{1 + \dfrac{(-1)^n}{n^2}\right\}$

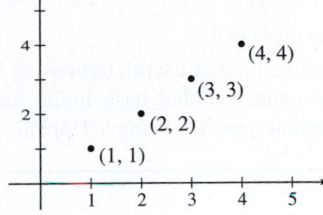

Not bounded, monotonic; diverges

**Figure 15** $\{u_n\} = \{n\}$

So, there are examples of monotonic sequences that diverge and examples of bounded sequences that diverge. However, a sequence that is both monotonic and bounded always converges.

**THEOREM**

An increasing (or nondecreasing) sequence $\{s_n\}$ that is bounded from above converges. A decreasing (or nonincreasing) sequence $\{s_n\}$ that is bounded from below converges.

**IN WORDS** A bounded, monotonic sequence converges.

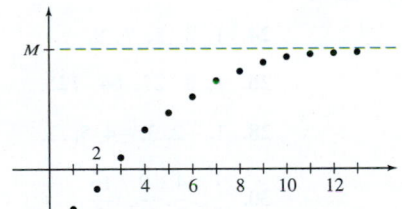

**Figure 16**

Since the convergence of a sequence $\{s_n\}$ is about the behavior of $\{s_n\}$ for large values of $n$, if a sequence *eventually increases* and is bounded from above, it is convergent. Similar remarks hold for sequences that *eventually decrease* and are bounded from below.

The proof of this theorem can be found in Appendix B. Figure 16 illustrates the theorem for a sequence that is increasing and bounded from above.

EXAMPLE 15  **Determining Whether a Sequence Converges or Diverges**

Determine whether the sequence $\{s_n\} = \left\{\dfrac{2^n}{n!}\right\}$ converges or diverges.

**Solution** To see if $\left\{\dfrac{2^n}{n!}\right\}$ is monotonic, find the algebraic ratio $\dfrac{s_{n+1}}{s_n}$:

$$\frac{s_{n+1}}{s_n} = \frac{\dfrac{2^{n+1}}{(n+1)!}}{\dfrac{2^n}{n!}} = \frac{2^{n+1}\, n!}{(n+1)!\, 2^n} = \frac{2}{n+1} \le 1 \qquad \text{for all } n \ge 1$$

Since $s_{n+1} \le s_n$ for $n \ge 1$, the sequence $\{s_n\}$ is nonincreasing.

Next, since each term of the sequence is positive, $s_n > 0$ for $n \ge 1$, the sequence $\{s_n\}$ is bounded from below.

Since $\{s_n\}$ is nonincreasing and bounded from below, it converges. ∎

**NOTE** Although the sequence $\{s_n\}$ converges, the theorem does not tell us what $\lim\limits_{n \to \infty} s_n$ equals.

NOW WORK **Problem 89.**

## Summary

To determine whether a sequence converges:

- Look at a few terms of the sequence to see if a trend is developing. For example, the first five terms of the sequence

$$\left\{1 + \frac{(-1)^n}{n^2}\right\} \text{ are } 1 - 1, \ 1 + \frac{1}{4}, \ 1 - \frac{1}{9}, \ 1 + \frac{1}{16},$$

and $1 - \dfrac{1}{25}$. The pattern suggests that the sequence converges to 1.

- Find the limit of the $n$th term using any available limit technique, including basic limits, limit properties, or a related function (possibly using L'Hôpital's Rule). For example, for the

sequence $\left\{\dfrac{\ln n}{n}\right\}$, we examine the limit of the related

function $f(x) = \dfrac{\ln x}{x}$.

$$\lim_{x \to \infty} f(x) = \lim_{x \to \infty}\left(\frac{\ln x}{x}\right) = \lim_{x \to \infty} \frac{\frac{1}{x}}{1} = 0$$

and conclude the sequence $\left\{\dfrac{\ln n}{n}\right\}$ converges to 0.

- Show that the sequence is bounded and monotonic.

---

## 8.1 Assess Your Understanding

### Concepts and Vocabulary

1. *True or False*  A sequence is a function whose domain is the set of positive real numbers.

2. *True or False*  If the sequence $\{s_n\}$ is convergent, then $\lim\limits_{n \to \infty} s_n = 0$.

3. *True or False*  If $f(x)$ is a related function of the sequence $\{s_n\}$ and there is a real number $L$ for which $\lim\limits_{x \to \infty} f(x) = L$, then $\{s_n\}$ converges.

4. *Multiple Choice*  If there is a positive number $K$ for which $|s_n| \le K$ for all integers $n \ge 1$, then $\{s_n\}$ is [(a) increasing (b) bounded (c) decreasing (d) convergent].

5. *True or False*  A bounded sequence is convergent.

6. *True or False*  An unbounded sequence is divergent.

7. *True or False*  A sequence $\{s_n\}$ is decreasing if and only if $s_n \le s_{n+1}$ for all integers $n \ge 1$.

8. *True or False*  A sequence must be monotonic to be convergent.

9. *True or False*  To use an algebraic ratio to show that the sequence $\{s_n\}$ is increasing, show that $\dfrac{s_{n+1}}{s_n} \ge 0$ for all $n \ge 1$.

10. *Multiple Choice*  If the derivative of a related function $f$ of a sequence $\{s_n\}$ is negative, then the sequence $\{s_n\}$ is [(a) bounded (b) decreasing (c) increasing (d) convergent].

11. *True or False*  When determining whether a sequence $\{s_n\}$ converges or diverges, the beginning terms of the sequence can be ignored.

12. *True or False*  Sequences that are both bounded and monotonic diverge.

### Skill Building

*In Problems 13–22, the $n$th term of a sequence $\{s_n\}$ is given. Write the first four terms of each sequence.*

13. $s_n = \dfrac{n+1}{n}$

14. $s_n = \dfrac{2}{n^2}$

15. $s_n = \ln n$

16. $s_n = \dfrac{n}{\ln(n+1)}$

17. $s_n = \dfrac{(-1)^{n+1}}{2n+1}$

18. $s_n = \dfrac{1-(-1)^n}{2}$

19. $s_n = \begin{cases} (-1)^{n+1} & \text{if } n \text{ is even} \\ 1 & \text{if } n \text{ is odd} \end{cases}$

20. $s_n = \begin{cases} n^2 + n & \text{if } n \text{ is even} \\ 4n+1 & \text{if } n \text{ is odd} \end{cases}$

21. $s_n = \dfrac{n!}{2^n}$

22. $s_n = \dfrac{n!}{n^2}$

*In Problems 23–32, the first few terms of a sequence are given. Find an expression for the $n$th term of each sequence, assuming the indicated pattern continues for all $n$.*

23. $2, 4, 6, 8, 10, \ldots$

24. $1, 3, 5, 7, 9, \ldots$

25. $2, 4, 8, 16, 32, \ldots$

26. $1, 8, 27, 64, 125, \ldots$

27. $\dfrac{1}{2}, -\dfrac{1}{3}, \dfrac{1}{4}, -\dfrac{1}{5}, \dfrac{1}{6}, \ldots$

28. $1, -2, 3, -4, 5, \ldots$

29. $\dfrac{1}{2}, \dfrac{2}{3}, \dfrac{3}{4}, \dfrac{4}{5}, \ldots$

30. $\dfrac{1}{2}, \dfrac{4}{3}, \dfrac{9}{4}, \dfrac{16}{5}, \ldots$

31. $1, 1, 2, 6, 24, 120, 720, \ldots$

32. $1, 1, \dfrac{1}{2}, \dfrac{1}{6}, \dfrac{1}{24}, \dfrac{1}{120}, \ldots$

*In Problems 33–44, use properties of convergent sequences to find the limit of each sequence.*

33. $\left\{\dfrac{3}{n}\right\}$

34. $\left\{\dfrac{-2}{n}\right\}$

35. $\left\{1 - \dfrac{1}{n}\right\}$

36. $\left\{\dfrac{1}{n} + 4\right\}$

37. $\left\{\dfrac{4n+2}{n}\right\}$

38. $\left\{\dfrac{2n+1}{n}\right\}$

39. $\left\{\left(\dfrac{2-n}{n^2}\right)^4\right\}$

40. $\left\{\left(\dfrac{n^3 - 2n}{n^3}\right)^2\right\}$

41. $\left\{\sqrt{\dfrac{n+1}{n^2}}\right\}$

42. $\left\{\sqrt[3]{8 - \dfrac{1}{n}}\right\}$

43. $\left\{\left(1 - \dfrac{1}{n}\right)\left(1 - \dfrac{1}{n^2}\right)\right\}$

44. $\left\{\left(1 - \dfrac{1}{n}\right)\left(1 - \dfrac{1}{n^2}\right)\left(1 - \dfrac{1}{n^3}\right)\right\}$

---

**1.** = NOW WORK problem      $\boxed{\wedge}$ = Graphing technology recommended      $\boxed{\text{CAS}}$ = Computer Algebra System recommended

*In Problems 45–50, show that each sequence converges. Find its limit.*

**45.** $\left\{\ln\dfrac{n+1}{3n}\right\}$    **46.** $\left\{\ln\dfrac{n^2+2}{2n^2+3}\right\}$    **47.** $\left\{e^{(4/n)-2}\right\}$

**48.** $\left\{e^{3+(6/n)}\right\}$    **49.** $\left\{\sin\dfrac{1}{n}\right\}$    **50.** $\left\{\cos\dfrac{1}{n}\right\}$

*In Problems 51–62, use a related function or the Squeeze Theorem for sequences to show each sequence converges. Find its limit.*

**51.** $\left\{\dfrac{n^2-4}{n^2+n-2}\right\}$    **52.** $\left\{\dfrac{n+2}{n^2+6n+8}\right\}$

**53.** $\left\{\dfrac{n^2}{2n+1}-\dfrac{n^2}{2n-1}\right\}$    **54.** $\left\{\dfrac{6n^4-5}{7n^4+3}\right\}$

**55.** $\left\{\dfrac{\sqrt{n}+2}{\sqrt{n}+5}\right\}$    **56.** $\left\{\dfrac{\sqrt{n}}{e^n}\right\}$    **57.** $\left\{\dfrac{n^2}{3^n}\right\}$

**58.** $\left\{\dfrac{(n-1)^2}{e^n}\right\}$    **59.** $\left\{\dfrac{(-1)^n}{3n^2}\right\}$    **60.** $\left\{\dfrac{(-1)^n}{\sqrt{n}}\right\}$

**61.** $\left\{\dfrac{\sin n}{n}\right\}$    **62.** $\left\{\dfrac{\cos n}{n}\right\}$

*In Problems 63–72, determine whether each sequence converges or diverges.*

**63.** $\{\cos(\pi n)\}$    **64.** $\left\{\cos\left(\dfrac{\pi}{2}n\right)\right\}$    **65.** $\{\sqrt{n}\}$

**66.** $\{n^2\}$    **67.** $\left\{\left(-\dfrac{1}{3}\right)^n\right\}$    **68.** $\left\{\left(\dfrac{1}{3}\right)^n\right\}$

**69.** $\left\{\left(\dfrac{5}{4}\right)^n\right\}$    **70.** $\left\{\left(\dfrac{\pi}{2}\right)^n\right\}$    **71.** $\left\{\dfrac{n+(-1)^n}{n}\right\}$

**72.** $\left\{\dfrac{1}{n}+(-1)^n\right\}$

*In Problems 73–80, determine whether each sequence is bounded from above, bounded from below, both, or neither.*

**73.** $\left\{\dfrac{\ln n}{n}\right\}$    **74.** $\left\{\dfrac{\sin n}{n}\right\}$    **75.** $\left\{n+\dfrac{1}{n}\right\}$

**76.** $\left\{\dfrac{3}{n+1}\right\}$    **77.** $\left\{\dfrac{n^2}{n+1}\right\}$    **78.** $\left\{\dfrac{2^n}{n^2}\right\}$

**79.** $\left\{\left(-\dfrac{1}{2}\right)^n\right\}$    **80.** $\{n^{1/2}\}$

*In Problems 81–88, determine whether each sequence is monotonic. If the sequence is monotonic, is it increasing, nondecreasing, decreasing, or nonincreasing?*

**81.** $\left\{\dfrac{3^n}{(n+1)^3}\right\}$    **82.** $\left\{\dfrac{2n+1}{n}\right\}$    **83.** $\left\{\dfrac{\ln n}{\sqrt{n}}\right\}$

**84.** $\left\{\dfrac{\sqrt{n+1}}{n}\right\}$    **85.** $\left\{\left(\dfrac{1}{3}\right)^n\right\}$    **86.** $\left\{\dfrac{n^2}{5^n}\right\}$

**87.** $\left\{\dfrac{n!}{3^n}\right\}$    **88.** $\left\{\dfrac{n!}{n^2}\right\}$

*In Problems 89–94, show that each sequence converges by showing it is either increasing (nondecreasing) and bounded from above or decreasing (nonincreasing) and bounded from below.*

**89.** $\{ne^{-n}\}$    **90.** $\{\tan^{-1}n\}$    **91.** $\left\{\dfrac{n}{n+1}\right\}$

**92.** $\left\{\dfrac{n}{n^2+1}\right\}$    **93.** $\left\{2-\dfrac{1}{n}\right\}$    **94.** $\left\{\dfrac{n}{2^n}\right\}$

*In Problems 95–114, determine whether each sequence converges or diverges. If it converges, find its limit.*

**95.** $\left\{\dfrac{3}{n}+6\right\}$    **96.** $\left\{2-\dfrac{4}{n}\right\}$

**97.** $\left\{\ln\dfrac{n^2+4}{3n^2}\right\}$    **98.** $\left\{\cos\left(n\pi+\dfrac{\pi}{2}\right)\right\}$

**99.** $\{(-1)^n\sqrt{n}\}$    **100.** $\left\{\dfrac{(-1)^n}{2n}\right\}$

**101.** $\left\{\dfrac{3^n+1}{4^n}\right\}$    **102.** $\left\{n+\sin\dfrac{1}{n}\right\}$

**103.** $\left\{\dfrac{\ln(n+1)}{n+1}\right\}$    **104.** $\left\{\dfrac{\ln(n+1)}{\sqrt{n}}\right\}$

**105.** $\{0.5^n\}$    **106.** $\{(-2)^n\}$

**107.** $\left\{\cos\dfrac{\pi}{n}\right\}$    **108.** $\left\{\sin\dfrac{\pi}{n}\right\}$

**109.** $\left\{\cos\left(\dfrac{n}{e^n}\right)\right\}$    **110.** $\left\{\sin\left(\dfrac{(n+1)^3}{e^n}\right)\right\}$

**111.** $\{e^{1/n}\}$    **112.** $\left\{\dfrac{1}{ne^{-n}}\right\}$

**113.** $\left\{1+\left(\dfrac{1}{2}\right)^n\right\}$    **114.** $\left\{1-\left(\dfrac{1}{2}\right)^n\right\}$

*In Problems 115 and 116, use a related function to find the limit of each sequence.*

**115.** $\left\{\dfrac{(\ln n)^2}{n}\right\}$    **116.** $\left\{\sqrt{n}\ln\dfrac{n+1}{n}\right\}$

## Applications and Extensions

*In Problems 117–126, determine whether each sequence converges or diverges.*

**117.** $\left\{\dfrac{n^2\tan^{-1}n}{n^2+1}\right\}$    **118.** $\left\{n\sin\dfrac{1}{n}\right\}$

**119.** $\left\{\dfrac{n+\sin n}{n+\cos(4n)}\right\}$    **120.** $\left\{\dfrac{n^2}{2n+1}\sin\dfrac{1}{n}\right\}$

**121.** $\{\ln n-\ln(n+1)\}$    **122.** $\left\{\ln n^2+\ln\dfrac{1}{n^2+1}\right\}$

**123.** $\left\{\dfrac{n^2}{\sqrt{n^2+1}}\right\}$    **124.** $\left\{\dfrac{5^n}{(n+1)^2}\right\}$

**125.** $\left\{\dfrac{2^n}{(2)(4)(6)\cdots(2n)}\right\}$    **126.** $\left\{\dfrac{3^{n+1}}{(3)(6)(9)\cdots(3n)}\right\}$

**127.** The $n$th term of a sequence is $s_n=\dfrac{1}{n^2+n\cos n+1}$. Does the sequence $\{s_n\}$ converge or diverge? (*Hint:* Show that the derivative of $\dfrac{1}{x^2+x\cos x+1}$ is negative for $x>1$.)

**128. Fibonacci Sequence** The famous **Fibonacci sequence** $\{u_n\}$ is defined recursively as

$$u_1 = 1 \qquad u_2 = 1 \qquad u_{n+2} = u_n + u_{n+1} \qquad n \geq 1$$

(a) Write the first eight terms of the Fibonacci sequence.

(b) Verify that the $n$th term is given by

$$u_n = \frac{(1+\sqrt{5})^n - (1-\sqrt{5})^n}{2^n \sqrt{5}}$$

*Hint:* Show that $u_1 = 1$, $u_2 = 1$, and $u_{n+2} = u_n + u_{n+1}$.

**129. Stocking a Lake** Mirror Lake is stocked with rainbow trout. Considering fish reproduction and natural death, along with vigorous efforts by fishermen to decimate the population, managers find that some ratio $r$, $0 < r < 1$, of the population persists from one stocking period to the next. If the lake is stocked with $h$ fish each year, the fish population $p_n$, in year $n$ of the stocking program, is approximately $p_n = r p_{n-1} + h$. If $p_0$ is 3000, write a general expression for the $n$th term of the sequence in terms of $r$ and $h$ only. Does this sequence converge?

**130. Electronics: A Discharging Capacitor** A capacitor is an electronic device that stores an electrical charge. When connected across a resistor, it loses the charge (discharges) in such a way that during a fixed time interval, called the **time constant**, the charge stored in the capacitor is $\dfrac{1}{e}$ of the charge at the beginning of that interval.

(a) Develop a sequence for the charge remaining after $n$ time constants if the initial charge is $Q_0$.

(b) Does this sequence converge? If yes, to what?

**131. Reflections in a Mirror** A highly reflective mirror reflects 95% of the light that falls on it. In a light box having walls made of this mirror, the light will reflect back-and-forth between the mirrors.

(a) If the original intensity of the light is $I_0$ before it falls on a mirror, develop a sequence to describe the intensity of the light after $n$ reflections.

(b) How many reflections are needed to reduce the light intensity by at least 98%?

**132. A Fission Chain Reaction** A chain reaction is any sequence of events for which each event causes one or more additional events to occur. For example, in chain-reaction auto accidents, one car rear-ends another car, that car rear-ends another, and so on. In one type of nuclear fission chain reaction, a uranium-235 nucleus is struck by a neutron, causing it to break apart and release several more neutrons. Each of these neutrons strikes another nucleus, causing it to break apart and release additional neutrons, resulting in a chain reaction. In the fission of uranium-235 in nuclear reactors, each fission event releases an average of $2\dfrac{1}{2}$ neutrons, and each of these neutrons causes another fission event. The first fission is triggered by a single free neutron.

(a) Develop a sequence for the average number of neutrons, that is, fission events, that occur at the $n$th event if we start with one such event.

(b) Does the sequence converge or diverge?

(c) Interpret the answer found in (b).

*In Problems 133–136, find the limit of each sequence.*

**133.** $\left\{ \left(1 + \dfrac{2}{n}\right)^n \right\}$

**134.** $\left\{ \left(1 - \dfrac{4}{n}\right)^n \right\}$

**135.** $\left\{ \left(1 + \dfrac{1}{n}\right)^{3n} \right\}$

**136.** $\left\{ \left(1 + \dfrac{1}{n}\right)^{-2n} \right\}$

**137.** Use the Squeeze Theorem to show that the sequence

$$\left\{ (-1)^n \frac{1}{n!} \right\}$$ converges to 0.

## Challenge Problems

**138.** Show that if $0 < r < 1$, then $\lim\limits_{n \to \infty} r^n = 0$. *Hint:* Let $r = \dfrac{1}{1+p}$, where $p > 0$. Then by the Binomial Theorem,

$$r^n = \frac{1}{(1+p)^n} = \frac{1}{1 + np + n(n-1)\dfrac{p^2}{2} + \cdots + p^n} < \frac{1}{np}.$$

**139.** Use the result of Problem 138 to show that if $-1 < r < 0$, then $\lim\limits_{n \to \infty} r^n = 0$.

**140.** Show that if $r > 1$, then $\lim\limits_{n \to \infty} r^n = \infty$. *Hint:* Let $r = 1 + p$, where $p > 0$. Then by the Binomial Theorem,

$$r^n = (1+p)^n = 1 + np + n(n-1)\frac{p^2}{2} + \cdots + p^n > np.$$

**141.** Use the result of Problem 140 to show that if $r < -1$, then $\lim\limits_{n \to \infty} r^n$ does not exist.

*Hint:* $r^n$ oscillates between positive and negative values.

**142.** Suppose $\{s_n\}$ is a sequence of real numbers. Show that if $\lim\limits_{n \to \infty} s_n = L$ and if $f$ is a function that is continuous at $L$ and is defined for all numbers $s_n$, then $\lim\limits_{n \to \infty} f(s_n) = f(L)$.

**143.** Show that if $\lim\limits_{n \to \infty} s_n = L$, then $\lim\limits_{n \to \infty} |s_n|$ exists and $\lim\limits_{n \to \infty} |s_n| = |L|$. Is the converse true?

**144. The Limit of a Sequence Is Unique** Show that a convergent sequence $\{s_n\}$ cannot have two distinct limits.

**145.** Review the definition of the limit at infinity of a function from Section 1.6. Write a paragraph that compares and contrasts the limit at infinity of a function $f$ and the limit of a sequence $\{s_n\}$.

**146. (a)** Show that the sequence $\{\ln n\}$ is increasing.

(b) Show that the sequence $\{\ln n\}$ is unbounded from above.

(c) Conclude $\{\ln n\}$ diverges.

(d) Find the smallest number $N$ so that $\ln N > 20$.

(e) Graph $y = \ln x$ and zoom in for $x$ large.

(f) Does the graph confirm the result in (c)?

**147.** Let $a_1 > 0$ and $b_1 > 0$ be two real numbers for which $a_1 > b_1$. Define sequences $\{a_n\}$ and $\{b_n\}$ as

$$a_{n+1} = \frac{a_n + b_n}{2} \qquad b_{n+1} = \sqrt{a_n b_n}$$

(a) Show that $b_n < b_{n+1} < a_1$ for all $n$.

(b) Show that $b_1 < a_{n+1} < a_n$ for all $n$.

(c) Show that $0 < a_{n+1} - b_{n+1} < \dfrac{a_1 - b_1}{2^n}$.

(d) Show that $\lim\limits_{n \to \infty} a_n$ and $\lim\limits_{n \to \infty} b_n$ each exist and are equal.

*In Problems 148–150, determine whether each sequence converges or diverges.*

**148.** $s_n = \dfrac{2^{n-1} \cdot 4^n}{n!}$    **149.** $s_n = \dfrac{n!}{3^n \cdot 4^n}$    **150.** $s_n = \dfrac{n!}{3^n + 8n}$

**151.** Show that $\{(3^n + 5^n)^{1/n}\}$ converges.

**152.** Let $N$ be a fixed positive number and define a sequence by $\{a_{n+1}\} = \left\{ \dfrac{1}{2} \left[ a_n + \dfrac{N}{a_n} \right] \right\}$, where $a_1$ is a positive number.

  **(a)** Show that the sequence $\{a_n\}$ converges to $\sqrt{N}$.

  **(b)** Use this sequence to approximate $\sqrt{28}$ rounded to three decimal places. How accurate is $a_3$? $a_6$?

**153.** Show that $\{s_n\} = \left\{ \dfrac{1 \cdot 3 \cdot 5 \cdot \cdots \cdot (2n-1)}{2 \cdot 4 \cdot 6 \cdot \cdots \cdot 2n} \right\}$ is bounded and monotonic.

**154.** Show that $\{s_n\} = \left\{ \left( 1 + \dfrac{1}{n} \right)^n \right\}$ is increasing and bounded from above.

  *Hint:* Use the Binomial Theorem to expand $\left( 1 + \dfrac{1}{n} \right)^n$.

**155.** Let $\{s_n\}$ be a convergent sequence, and suppose the $n$th term of the sequence $\{a_n\}$ is the arithmetic mean (average) of the first $n$ terms of $\{s_n\}$. That is, $a_n = \dfrac{1}{n}(s_1 + s_2 + \cdots + s_n)$.

  Show that $\{a_n\}$ converges and has the same limit as $\{s_n\}$.

**156. Area** Let $A_n$ be the area enclosed by a regular $n$-sided polygon inscribed in a circle of radius $R$. Show that:

  **(a)** $A_n = \dfrac{n}{2} R^2 \sin \dfrac{2\pi}{n}$.

  **(b)** $\displaystyle \lim_{n \to \infty} A_n = \lim_{n \to \infty} \left( \dfrac{n}{2} R^2 \sin \dfrac{2\pi}{n} \right) = \pi R^2$

  (the area of a circle of radius $R$).

**157. Area** Let $A_n$ be the area enclosed by a regular $n$-sided polygon circumscribed around a circle of radius $r$. Show that:

  **(a)** $A_n = nr^2 \tan \dfrac{\pi}{n}$.

  **(b)** $\displaystyle \lim_{n \to \infty} A_n = \pi r^2$   (the area of a circle of radius $r$).

  *Hint:* $r$, called the **apothem** of the polygon, is the perpendicular distance from the center of the polygon to the midpoint of a side.

**158. Perimeter** Suppose $P_n$ is the perimeter of a regular $n$-sided polygon inscribed in a circle of radius $R$. Show that:

  **(a)** $P_n = 2nR \sin \dfrac{\pi}{n}$.

  **(b)** $\displaystyle \lim_{n \to \infty} P_n = 2\pi R$ (the circumference of a circle of radius $R$).

**159. Cauchy Sequence** A sequence $\{s_n\}$ is said to be a **Cauchy sequence** if and only if for each $\varepsilon > 0$, there exists a positive integer $N$ for which

$$|s_n - s_m| < \varepsilon \qquad \text{for all } n, m > N$$

  Show that every convergent sequence is a Cauchy sequence.

**160. (a)** Show that the sequence $\{e^{n/(n+2)}\}$ converges.

  **(b)** Find $\displaystyle \lim_{n \to \infty} e^{n/(n+2)}$.

  **(c)** Graph $y = e^{n/(n+2)}$. Does the graph confirm the results of (a) and (b)?

---

# 8.2 Infinite Series

**OBJECTIVES** *When you finish this section, you should be able to:*

**1** Determine whether a series has a sum (p. 590)

**2** Analyze a geometric series (p. 593)

**3** Analyze the harmonic series (p. 597)

  Application: Using a geometric series in biology (p. 598)

Is it possible for the sum of an infinite collection of nonzero numbers to be finite? Look at Figure 17. The square in the figure has sides of length 1 unit, making its area 1 square unit. If we divide the square into two rectangles of equal area, each rectangle has an area of $\dfrac{1}{2}$ square unit. If one of these rectangles is divided in half, the result is two squares, the area of each equaling $\dfrac{1}{4}$ square unit. If we were to continue the process of dividing one of the smallest regions in half, we would obtain a decomposition of the original area of 1 square unit into regions of area $\dfrac{1}{2}, \dfrac{1}{4}, \dfrac{1}{8}, \dfrac{1}{16}$, and so forth. Therefore,

$$1 = \frac{1}{2} + \frac{1}{4} + \frac{1}{8} + \frac{1}{16} + \cdots$$

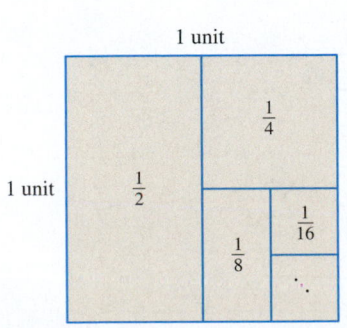

1 unit

1 unit    $\frac{1}{2}$    $\frac{1}{4}$    $\frac{1}{16}$    $\frac{1}{8}$

**Figure 17**

Surprised?

Now look at this result from a different point of view by starting with the infinite sum

$$\frac{1}{2} + \frac{1}{4} + \frac{1}{8} + \frac{1}{16} + \cdots \tag{1}$$

One way we might add the fractions is by using *partial sums* to see whether a trend develops. The first five partial sums are

$$\frac{1}{2} = 0.5$$

$$\frac{1}{2} + \frac{1}{4} = \frac{3}{4} = 0.75$$

$$\frac{1}{2} + \frac{1}{4} + \frac{1}{8} = \frac{3}{4} + \frac{1}{8} = \frac{7}{8} = 0.875$$

$$\frac{1}{2} + \frac{1}{4} + \frac{1}{8} + \frac{1}{16} = \frac{7}{8} + \frac{1}{16} = \frac{15}{16} = 0.9375$$

$$\frac{1}{2} + \frac{1}{4} + \frac{1}{8} + \frac{1}{16} + \frac{1}{32} = \frac{15}{16} + \frac{1}{32} = \frac{31}{32} = 0.96875$$

Each of these sums uses more terms from (1), and each sum seems to be getting closer to 1. The infinite sum in (1) is an example of an *infinite series*.

> **DEFINITION** Infinite Series
>
> If $a_1, a_2, \ldots, a_n, \ldots$ is an infinite collection of numbers, the expression
>
> $$\sum_{k=1}^{\infty} a_k = a_1 + a_2 + \cdots + a_n + \cdots$$
>
> is called an **infinite series** or, simply, a **series**.

**NEED TO REVIEW?** Sums and summation notation are discussed in Section P.8, pp. 71–73.

The numbers $a_1, a_2, \ldots, a_n, \ldots$ are called the **terms** of the series, and the number $a_n$ is called the **nth term** or **general term** of the series. The symbol $\sum$ stands for summation; $k$ is the **index of summation**. Although the index of summation can begin at any integer, in most of our work with series, it will begin at 1.

## 1 Determine Whether a Series Has a Sum

To define a sum of an infinite series $\sum_{k=1}^{\infty} a_k$, we make use of the sequence $\{S_n\}$ defined by

$$S_1 = a_1$$

$$S_2 = a_1 + a_2 = \sum_{k=1}^{2} a_k$$

$$S_3 = a_1 + a_2 + a_3 = \sum_{k=1}^{3} a_k$$

$$\vdots$$

$$S_n = a_1 + a_2 + \cdots + a_n = \sum_{k=1}^{n} a_k$$

$$\vdots$$

This sequence $\{S_n\}$ is called the **sequence of partial sums** of the series $\sum_{k=1}^{\infty} a_k$.

EXAMPLE 1    **Finding the Sequence of Partial Sums**

Find the sequence of partials sums of the series

$$\sum_{k=1}^{\infty} \frac{1}{2^k} = \frac{1}{2} + \frac{1}{2^2} + \frac{1}{2^3} + \frac{1}{2^4} + \cdots = \frac{1}{2} + \frac{1}{4} + \frac{1}{8} + \frac{1}{16} + \cdots$$

**Solution** As it turns out, the partial sums $S_n$ can each be written as 1 minus a power of $\frac{1}{2}$, as follows:

$$S_1 = a_1 = \frac{1}{2} = 1 - \frac{1}{2}$$

$$S_2 = S_1 + a_2 = \left(1 - \frac{1}{2}\right) + \frac{1}{4} = 1 - \frac{1}{4} = 1 - \frac{1}{2^2}$$

$$S_3 = S_2 + a_3 = \left(1 - \frac{1}{4}\right) + \frac{1}{8} = 1 - \frac{1}{8} = 1 - \frac{1}{2^3}$$

$$S_4 = S_3 + a_4 = \left(1 - \frac{1}{8}\right) + \frac{1}{16} = 1 - \frac{1}{16} = 1 - \frac{1}{2^4}$$

$$\vdots$$

$$S_n = 1 - \frac{1}{2^n}$$

$$\vdots$$

The $n$th partial sum is $S_n = 1 - \frac{1}{2^n}$, and as $n$ increases, the sequence $\{S_n\}$ of partial sums approaches a limit. That is,

$$\lim_{n\to\infty} S_n = \lim_{n\to\infty}\left(1 - \frac{1}{2^n}\right) = \lim_{n\to\infty} 1 - \lim_{n\to\infty} \frac{1}{2^n} = 1 - 0 = 1$$

We agree to call this limit the *sum of the series*, and we write

$$\boxed{\sum_{k=1}^{\infty} \frac{1}{2^k} = \frac{1}{2} + \frac{1}{4} + \frac{1}{8} + \frac{1}{16} + \cdots = 1}$$

**DEFINITION Convergence/Divergence of an Infinite Series**

If the sequence $\{S_n\}$ of partial sums of an infinite series $\sum_{k=1}^{\infty} a_k$ has a limit $S$, then the series **converges** and is said to have the **sum** $S$. That is, if $\lim_{n\to\infty} S_n = S$, then

$$\boxed{\sum_{k=1}^{\infty} a_k = a_1 + a_2 + \cdots + a_n + \cdots = S}$$

- If $\lim_{n\to\infty} S_n = \infty$, then the infinite series **diverges**.
- If $\lim_{n\to\infty} S_n$ does not exist, then the infinite series **diverges**.

EXAMPLE 2    **Finding the Sum of a Series**

Show that

$$\sum_{k=1}^{\infty} \frac{1}{k(k+1)} = \frac{1}{1\cdot 2} + \frac{1}{2\cdot 3} + \frac{1}{3\cdot 4} + \cdots = \frac{1}{2} + \frac{1}{6} + \frac{1}{12} + \cdots = 1$$

**Solution** We begin with the sequence $\{S_n\}$ of partial sums,

$$S_1 = \frac{1}{1 \cdot 2}$$

$$S_2 = \frac{1}{1 \cdot 2} + \frac{1}{2 \cdot 3}$$

$$S_3 = \frac{1}{1 \cdot 2} + \frac{1}{2 \cdot 3} + \frac{1}{3 \cdot 4}$$

$$\vdots$$

$$S_n = \frac{1}{1 \cdot 2} + \frac{1}{2 \cdot 3} + \frac{1}{3 \cdot 4} + \cdots + \frac{1}{n(n+1)}$$

$$\vdots$$

Using partial fractions, we can write $\dfrac{1}{n(n+1)}$ as

$$\frac{1}{n(n+1)} = \frac{1}{n} - \frac{1}{n+1}$$

Then

$$S_n = \frac{1}{1 \cdot 2} + \frac{1}{2 \cdot 3} + \frac{1}{3 \cdot 4} + \cdots + \frac{1}{n(n+1)}$$

can be written as

$$S_n = \left(\frac{1}{1} - \frac{1}{2}\right) + \left(\frac{1}{2} - \frac{1}{3}\right) + \left(\frac{1}{3} - \frac{1}{4}\right) + \cdots + \left(\frac{1}{n-1} - \frac{1}{n}\right) + \left(\frac{1}{n} - \frac{1}{n+1}\right)$$

$$\frac{1}{1 \cdot 2} = \frac{1}{1} - \frac{1}{2} \qquad \frac{1}{2 \cdot 3} = \frac{1}{2} - \frac{1}{3} \qquad \frac{1}{3 \cdot 4} = \frac{1}{3} - \frac{1}{4}$$

Notice that all the terms except the first and last cancel, so that

$$S_n = 1 - \frac{1}{n+1}$$

Then

$$\lim_{n \to \infty} S_n = \lim_{n \to \infty}\left(1 - \frac{1}{n+1}\right) = 1$$

**NOTE** Sums for which the middle terms cancel, as in Example 1, are called **telescoping sums**.

The series $\displaystyle\sum_{k=1}^{\infty} \frac{1}{k(k+1)}$ converges, and its sum is 1. ∎

**NOW WORK** Problem 11.

**EXAMPLE 3** **Showing a Series Diverges**

Show that the series $\displaystyle\sum_{k=1}^{\infty} (-1)^k = -1 + 1 - 1 + \cdots$ diverges.

**Solution** The sequence $\{S_n\}$ of partial sums is

$$S_1 = -1$$
$$S_2 = -1 + 1 = 0$$
$$S_3 = -1 + 1 - 1 = -1$$
$$S_4 = -1 + 1 - 1 + 1 = 0$$

$$\vdots$$

$$S_n = \begin{cases} -1 & \text{if } n \text{ is odd} \\ 0 & \text{if } n \text{ is even} \end{cases}$$

Since $\displaystyle\lim_{n \to \infty} S_n$ does not exist, the sequence $\{S_n\}$ of partial sums diverges. Therefore, the series diverges. ∎

**NOW WORK** Problem 49.

**EXAMPLE 4** Determining Whether a Series Converges or Diverges

Determine whether the series $\displaystyle\sum_{k=1}^{\infty} k = 1 + 2 + 3 + \cdots$ converges or diverges.

**Solution** The sequence $\{S_n\}$ of partial sums is

$$S_1 = 1$$
$$S_2 = 1 + 2$$
$$S_3 = 1 + 2 + 3$$
$$\vdots$$
$$S_n = 1 + 2 + 3 + \cdots + n$$

To express $S_n$ in a way that will make it easy to find $\displaystyle\lim_{n \to \infty} S_n$, we use the formula for the sum of the first $n$ integers:

$$S_n = \sum_{k=1}^{n} k = 1 + 2 + 3 + \cdots + n = \frac{n(n+1)}{2}$$

> **RECALL**
> $$\sum_{k=1}^{n} k = 1 + 2 + \cdots + n = \frac{n(n+1)}{2}.$$
> (Section P.8, p. 72.)

Since $\displaystyle\lim_{n \to \infty} S_n = \lim_{n \to \infty} \frac{n(n+1)}{2} = \infty$, the sequence $\{S_n\}$ of partial sums diverges.

So, the series $\displaystyle\sum_{k=1}^{\infty} k$ diverges. ∎

**NOW WORK** Problem 43.

## 2 Analyze a Geometric Series

Geometric series occur in a large variety of applications including biology, finance, and probability. They are also useful in analyzing other infinite series.

In a geometric series, the ratio $r$ of any two consecutive terms is a fixed real number.

> **DEFINITION** Geometric Series
>
> A series of the form
>
> $$\sum_{k=0}^{\infty} ar^k = \sum_{k=1}^{\infty} ar^{k-1} = a + ar + ar^2 + \cdots + ar^{n-1} + \cdots$$
>
> where $a \neq 0$, is called a **geometric series** with ratio $r$.

To investigate the conditions for convergence of a geometric series, we examine the $n$th partial sum:

$$S_n = a + ar + ar^2 + \cdots + ar^{n-1} \qquad (2)$$

- If $r = 0$, the $n$th partial sum is $S_n = a$ and $\displaystyle\lim_{n \to \infty} S_n = a$. The sequence of partial sums converges when $r = 0$.

- If $r = 1$, the series becomes $\displaystyle\sum_{k=1}^{\infty} a = a + a + \cdots + a + \cdots$, and the $n$th partial sum is

$$S_n = a + a + \cdots + a = na$$

Since $a \neq 0$, $\displaystyle\lim_{n \to \infty} S_n = \infty$ or $-\infty$, so the sequence $\{S_n\}$ of partial sums diverges when $r = 1$.

- If $r = -1$, the series is $\sum_{k=1}^{\infty} a(-1)^{k-1} = a - a + a - a + \cdots$ and the $n$th partial sum is

$$S_n = \begin{cases} 0 & \text{if } n \text{ is even} \\ a & \text{if } n \text{ is odd} \end{cases}$$

Since $a \neq 0$, $\lim_{n \to \infty} S_n$ does not exist. The sequence $\{S_n\}$ of partial sums diverges when $r = -1$.

- Suppose $r \neq 0$, $r \neq 1$, and $r \neq -1$. Since $r \neq 0$, we multiply both sides of (2) by $r$ to obtain

$$r S_n = ar + ar^2 + \cdots + ar^n$$

Now subtract $r S_n$ from $S_n$.

$$S_n - r S_n = (a + ar + ar^2 + \cdots + ar^{n-1}) - (ar + ar^2 + \cdots + ar^{n-1} + ar^n)$$

$$= a - ar^n$$

$$S_n(1 - r) = a(1 - r^n)$$

Since $r \neq 1$, the $n$th partial sum of the geometric series can be expressed as

$$S_n = \frac{a(1 - r^n)}{1 - r} = \frac{a - ar^n}{1 - r} = \frac{a}{1 - r} - \frac{ar^n}{1 - r}$$

Now

$$\lim_{n \to \infty} S_n = \lim_{n \to \infty} \left[ \frac{a}{1 - r} - \frac{ar^n}{1 - r} \right] = \lim_{n \to \infty} \frac{a}{1 - r} - \lim_{n \to \infty} \frac{ar^n}{1 - r} = \frac{a}{1 - r} - \frac{a}{1 - r} \lim_{n \to \infty} r^n$$

Now use the fact that if $|r| < 1$, then $\lim_{n \to \infty} r^n = 0$ (refer to page 582 in Section 8.1). We conclude that if $|r| < 1$, then $\lim_{n \to \infty} S_n = \frac{a}{1 - r}$. So, a geometric series converges to $S = \frac{a}{1 - r}$ if $-1 < r < 1$.

If $|r| > 1$, use the fact that $\lim_{n \to \infty} r^n$ does not exist to conclude that a geometric series diverges if $r < -1$ or $r > 1$.

This proves the following theorem:

---

**THEOREM** Convergence of a Geometric Series

- If $|r| < 1$, the geometric series $\sum_{k=1}^{\infty} ar^{k-1}$ converges, and its sum is

$$\boxed{\sum_{k=1}^{\infty} ar^{k-1} = \frac{a}{1 - r}}$$

- If $|r| \geq 1$, the geometric series $\sum_{k=1}^{\infty} ar^{k-1}$ diverges.

---

CALC CLIP

**EXAMPLE 5** Determining Whether a Geometric Series Converges

Determine whether each geometric series converges or diverges. If it converges, find its sum.

(a) $\sum_{k=1}^{\infty} 8 \left( \frac{2}{5} \right)^{k-1}$     (b) $\sum_{k=1}^{\infty} \left( -\frac{5}{9} \right)^{k-1}$     (c) $\sum_{k=1}^{\infty} 3 \left( \frac{3}{2} \right)^{k-1}$

(d) $\sum_{k=1}^{\infty} \frac{1}{2^k}$     (e) $\sum_{k=0}^{\infty} \left( \frac{1}{3} \right)^{k-1}$     (f) $\sum_{k=1}^{\infty} \frac{10^{k+1}}{9^{k-2}}$

**Solution** Compare each series to $\displaystyle\sum_{k=1}^{\infty} ar^{k-1}$.

**(a)** $\displaystyle\sum_{k=1}^{\infty} 8\left(\frac{2}{5}\right)^{k-1}$    Here $a = 8$ and $r = \dfrac{2}{5}$. Since $|r| = \dfrac{2}{5} < 1$, the series converges and

$$\sum_{k=1}^{\infty} 8\left(\frac{2}{5}\right)^{k-1} = \frac{8}{1 - \dfrac{2}{5}} = 8 \cdot \frac{5}{3} = \frac{40}{3}$$

**(b)** $\displaystyle\sum_{k=1}^{\infty}\left(-\frac{5}{9}\right)^{k-1}$    Here $a = 1$ and $r = -\dfrac{5}{9}$. Since $|r| = \dfrac{5}{9} < 1$, the series converges and

$$\sum_{k=1}^{\infty}\left(-\frac{5}{9}\right)^{k-1} = \frac{1}{1 - \left(-\dfrac{5}{9}\right)} = \frac{9}{14}$$

**(c)** $\displaystyle\sum_{k=1}^{\infty} 3\left(\frac{3}{2}\right)^{k-1}$    Here $a = 3$ and $r = \dfrac{3}{2}$. Since $|r| = \dfrac{3}{2} > 1$, $\displaystyle\sum_{k=1}^{\infty} 3\left(\frac{3}{2}\right)^{k-1}$ diverges.

**(d)** $\displaystyle\sum_{k=1}^{\infty}\frac{1}{2^k}$ is not in the form $\displaystyle\sum_{k=1}^{\infty} ar^{k-1}$. To place it in this form, we proceed as follows:

$$\sum_{k=1}^{\infty}\frac{1}{2^k} = \sum_{k=1}^{\infty}\left(\frac{1}{2}\right)^k = \underset{\uparrow k=1}{\sum_{k=1}^{\infty}}\left[\frac{1}{2}\cdot\left(\frac{1}{2}\right)^{k-1}\right]$$

Write in the form $\displaystyle\sum_{k=1}^{\infty} ar^{k-1}$

So, $\displaystyle\sum_{k=1}^{\infty}\frac{1}{2^k}$ is a geometric series with $a = \dfrac{1}{2}$ and $r = \dfrac{1}{2}$. Since $|r| < 1$, the series converges, and its sum is

$$\sum_{k=1}^{\infty}\frac{1}{2^k} = \frac{\dfrac{1}{2}}{1 - \dfrac{1}{2}} = 1$$

which agrees with the sum we found earlier (p. 591).

**(e)** $\displaystyle\sum_{k=0}^{\infty}\left(\frac{1}{3}\right)^{k-1}$ starts at $0$, so it is not in the form, $\displaystyle\sum_{k=1}^{\infty} ar^{k-1}$. To start the series at $1$, write

$$\sum_{k=0}^{\infty}\left(\frac{1}{3}\right)^{k-1} = \left(\frac{1}{3}\right)^{-1} + \sum_{k=1}^{\infty}\left(\frac{1}{3}\right)^{k-1} = 3 + \sum_{k=1}^{\infty}\left(\frac{1}{3}\right)^{k-1}$$

Now, $\displaystyle\sum_{k=1}^{\infty}\left(\frac{1}{3}\right)^{k-1}$ is a geometric series with $a = 1$ and $r = \dfrac{1}{3}$. Since $|r| < 1$, the series converges, and the sum

$$\sum_{k=0}^{\infty}\left(\frac{1}{3}\right)^{k-1} = 3 + \frac{1}{1 - \dfrac{1}{3}} = \frac{9}{2}$$

**(f)** We put $\displaystyle\sum_{k=1}^{\infty}\frac{10^{k+1}}{9^{k-2}}$ in the form $\displaystyle\sum_{k=1}^{\infty} ar^{k-1}$ by using the facts that

$$10^{k+1} = 10^2 \cdot 10^{k-1} = 100 \cdot 10^{k-1} \quad\text{and}\quad 9^{k-2} = \frac{1}{9}\cdot 9^{k-1}$$

Then

$$\sum_{k=1}^{\infty}\frac{10^{k+1}}{9^{k-2}} = \sum_{k=1}^{\infty}\frac{100 \cdot 10^{k-1}}{\dfrac{1}{9}\cdot 9^{k-1}} = \sum_{k=1}^{\infty} 900 \cdot\left(\frac{10}{9}\right)^{k-1}$$

Since the ratio $r = \dfrac{10}{9} > 1$, the series diverges. ∎

**NOW WORK** Problems **21** and **29**.

EXAMPLE 6  **Writing a Repeating Decimal as a Fraction**

Express the repeating decimal $0.090909\ldots$ as a quotient of two integers.

**Solution** We write the infinite decimal $0.090909\ldots$ as an infinite series:

$$0.090909\ldots = 0.09 + 0.0009 + 0.000009 + 0.00000009 + \cdots$$

$$= \frac{9}{100} + \frac{9}{10000} + \frac{9}{1000000} + \cdots$$

$$= \frac{9}{100}\left(1 + \frac{1}{100} + \frac{1}{10000} + \frac{1}{1000000} + \cdots\right)$$

$$= \sum_{k=1}^{\infty} \frac{9}{100}\left(\frac{1}{100}\right)^{k-1}$$

This is a geometric series with $a = \dfrac{9}{100}$ and $r = \dfrac{1}{100}$. Since $|r| < 1$, the series converges and its sum is

$$\sum_{k=1}^{\infty} \frac{9}{100}\left(\frac{1}{100}\right)^{k-1} = \frac{\dfrac{9}{100}}{1 - \dfrac{1}{100}} = \frac{9}{99} = \frac{1}{11}$$

So, $0.090909\ldots = \dfrac{1}{11}$.  ∎

NOW WORK  **Problem 59.**

EXAMPLE 7  **Using a Geometric Series with a Bouncing Ball**

A ball is dropped from a height of 12 m. Each time it strikes the ground, it bounces back to a height three-fourths the distance from which it fell. Find the total distance traveled by the ball. See Figure 18.

**Solution** Let $h_n$ denote the height of the ball on the $n$th bounce. Then

$$h_0 = 12$$

$$h_1 = \frac{3}{4}\cdot 12$$

$$h_2 = \frac{3}{4}\left[\frac{3}{4}\cdot 12\right] = \left(\frac{3}{4}\right)^2 \cdot 12$$

$$\vdots$$

$$h_n = \left(\frac{3}{4}\right)^n \cdot 12$$

After the first bounce, the ball travels up a distance $h_1 = \dfrac{3}{4}\cdot 12$ and then the same distance back down. Between the first and the second bounce, the total distance traveled is therefore $h_1 + h_1 = 2h_1$. The *total* distance $H$ traveled by the ball is

$$H = h_0 + 2h_1 + 2h_2 + 2h_3 + \cdots = h_0 + \sum_{k=1}^{\infty}(2h_k) = 12 + \sum_{k=1}^{\infty} 2\left[12\left(\frac{3}{4}\right)^k\right]$$

$$= 12 + \sum_{k=1}^{\infty} 24\left[\frac{3}{4}\left(\frac{3}{4}\right)^{k-1}\right]$$

$$= 12 + \sum_{k=1}^{\infty} 18\left(\frac{3}{4}\right)^{k-1}$$

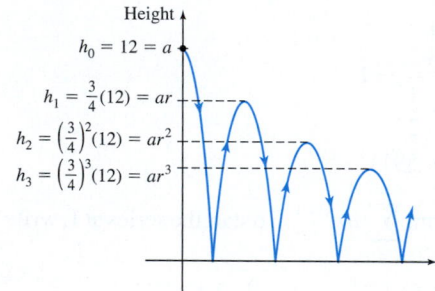

Height

$h_0 = 12 = a$

$h_1 = \frac{3}{4}(12) = ar$

$h_2 = \left(\frac{3}{4}\right)^2(12) = ar^2$

$h_3 = \left(\frac{3}{4}\right)^3(12) = ar^3$

DF **Figure 18**

The sum is a geometric series with $a = 18$ and $r = \dfrac{3}{4}$. The series converges and

$$H = 12 + \sum_{k=1}^{\infty} 18 \left(\frac{3}{4}\right)^{k-1} = 12 + \frac{18}{1 - \dfrac{3}{4}} = 84$$

The ball travels a total distance of 84 m. ∎

**NOW WORK** Problem 63.

### 3 Analyze the Harmonic Series

Another useful series, even though it diverges, is the *harmonic series*.

**DEFINITION** Harmonic Series

The infinite series

$$\boxed{\sum_{k=1}^{\infty} \frac{1}{k} = 1 + \frac{1}{2} + \frac{1}{3} + \cdots}$$

is called the **harmonic series**.

**THEOREM**

The harmonic series $\displaystyle\sum_{k=1}^{\infty} \frac{1}{k}$ diverges.

**Proof** To show that the harmonic series diverges, we look at the partial sums whose indexes of summation are powers of 2. That is, we investigate the sequence $S_1, S_2, S_4, S_8$, and so on. We can show that this sequence is not bounded as follows:

$$S_1 = 1 > \frac{1}{2} = 1\left(\frac{1}{2}\right)$$

$$S_2 = 1 + \frac{1}{2} > \frac{1}{2} + \frac{1}{2} = 2\left(\frac{1}{2}\right)$$

$$S_4 = 1 + \frac{1}{2} + \left(\frac{1}{3} + \frac{1}{4}\right) > 2\left(\frac{1}{2}\right) + \left(\frac{1}{4} + \frac{1}{4}\right) = 3\left(\frac{1}{2}\right)$$

$$S_8 = 1 + \frac{1}{2} + \left(\frac{1}{3} + \frac{1}{4}\right) + \left(\frac{1}{5} + \frac{1}{6} + \frac{1}{7} + \frac{1}{8}\right)$$

$$> 3\left(\frac{1}{2}\right) + \left(\frac{1}{8} + \frac{1}{8} + \frac{1}{8} + \frac{1}{8}\right) = 4\left(\frac{1}{2}\right)$$

$$\vdots$$

$$S_{2^{n-1}} > n\left(\frac{1}{2}\right)$$

**RECALL** An unbounded sequence diverges (p. 583).

We conclude that the sequence $\{S_{2^{n-1}}\}$ is not bounded, so the sequence $\{S_n\}$ of partial sums is not bounded. It follows that the sequence $\{S_n\}$ of partial sums diverges. Therefore, the harmonic series $\displaystyle\sum_{k=1}^{\infty} \frac{1}{k}$ diverges. ∎

## Summary

- An infinite series is an expression of the form

$$\sum_{k=1}^{\infty} a_k = a_1 + a_2 + a_3 + \cdots + a_n + \cdots$$

where $a_k$ are numbers. Here $a_n$ is the **nth term** or the **general term** of the series.

- The $n$th term of the sequence $\{S_n\}$ of partial sums of the infinite series $\displaystyle\sum_{k=1}^{\infty} a_k$ is

$$S_n = \sum_{k=1}^{n} a_k = a_1 + a_2 + a_3 + \cdots + u_n$$

- If $\lim_{n \to \infty} S_n = S$, then the series $\sum_{k=1}^{\infty} a_k$ converges and $\sum_{k=1}^{\infty} a_k = S$.

- If $\lim_{n \to \infty} S_n$ does not exist or if $\lim_{n \to \infty} S_n = \infty$, then the series $\sum_{k=1}^{\infty} a_k$ diverges.

- The geometric series $\sum_{k=1}^{\infty} ar^{k-1} = a + ar + ar^2 + ar^3 + \cdots$
  - converges to $\dfrac{a}{1-r}$ if $-1 < r < 1$.
  - diverges if $|r| \geq 1$.

- The harmonic series $\sum_{k=1}^{n} \dfrac{1}{k} = 1 + \dfrac{1}{2} + \dfrac{1}{3} + \cdots$ diverges.

## Application: Using a Geometric Series in Biology*

This application deals with the rate of occurrence of *retinoblastoma*, a rare type of eye cancer in children. An *allele (allelomorph)* is a gene that gives rise to one of a pair of contrasting characteristics, such as smooth or rough, tall or short. Each person normally has two such genes for each characteristic. An individual may have two "tall" genes, two "short" genes, or one of each. In reproduction, each parent gives one of the two types to the child.

The tendency to develop retinoblastoma apparently depends on the mutation of both copies of a gene, called RB1. The mutation rate from a normal RB1 allele to the mutant RB1 in each generation is approximately $m = 0.00002 = 2 \times 10^{-5}$. In this example, we ignore the very unlikely possibility of mutation from an abnormal RB1 to a normal RB1 gene. At the beginning of the twentieth century, retinoblastoma was nearly always fatal, but by the early 1950s, approximately 70% of children affected with the disease survived, although they usually became blind in one or both eyes. The current (2017) survival rate is about 95%, and the goals of treatment are to prevent the tumor cells from growing and spreading and to preserve vision.

Assume that survivors reproduce at about half the normal rate. (The assumption is based on scientific guesswork.) Then the productive proportion of persons affected with the retinoblastoma in 2017 was $r = (0.5)(0.95) = 0.475$. This rate is remarkable, considering that in 1900, $r \approx 0$, and in 1950 $r \approx 0.35$.

Starting with zero inherited cases in an early generation, for the $n$th consecutive generation, we obtain a rate of

| | |
|---|---|
| $m$ | due to mutation in the $n$th generation |
| $mr$ | due to mutation in the $(n-1)$st generation |
| $mr^2$ | due to mutation in the $(n-2)$nd generation |
| $\vdots$ | |
| $mr^n$ | due to mutation in the zero (original) generation |

Then the total rate of occurrence of the disease in the $n$th generation is

$$p_n = m + mr + \cdots + mr^n = \frac{m(1 - r^{n+1})}{1 - r}$$

from which

$$p = \lim_{n \to \infty} p_n = \frac{m}{1 - r} = \frac{2 \times 10^{-5}}{1 - 0.475} = 3.810 \times 10^{-5}$$

indicating that the total rate of persons affected with the disease will be almost twice the mutation rate.

*Adapted from

J. L. Young & M. A. Smith (1999), Retinoblastoma. In L. A. G. Ries, M. A. Smith, J. G. Gurney, M. Linet, T. Tamra, J. L. Young, & G. R. Bunin (Eds.), *Cancer incidence and survival among children and adolescents: United States SEER Program, 1975–1995*, NIH Pub. No. 99-4649, Bethesda, MD: National Cancer Institute.

Children's Hospital of Philadelphia (2017), *Retinoblastoma*, http://www.CHOP.edu.

J. V. Neel & W. J. Schull (1958), *Human heredity*, 3rd ed. (pp. 333–334), Chicago: University of Chicago Press.

Notice that if $r = 0$, as in 1900, then $p = m = 2 \times 10^{-5}$, and that if $r = 0.35$, as in 1950, then $p = 3.08 \times 10^{-5}$. We see that with better medical care, retinoblastoma has become more frequent. As medical care improves, the rate of occurrence of the disease can be expected to become even greater.

## 8.2 Assess Your Understanding

### Concepts and Vocabulary

**1. Multiple Choice** If $a_1, a_2, \ldots, a_n, \ldots$ is an infinite collection of numbers, the expression $\displaystyle\sum_{k=1}^{\infty} a_k = a_1 + a_2 + \cdots + a_n + \cdots$ is called [(**a**) an infinite sequence (**b**) an infinite series (**c**) a partial sum].

**2. Multiple Choice** If $\displaystyle\sum_{k=1}^{\infty} a_k$ is an infinite series, then the sequence $\{S_n\}$ where $S_n = \displaystyle\sum_{k=1}^{n} a_k$, is called the sequence of [(**a**) fractional parts (**b**) early terms (**c**) completeness (**d**) partial sums] of the infinite series.

**3. True or False** A series converges if and only if its sequence of partial sums converges.

**4. True or False** A geometric series $\displaystyle\sum_{k=1}^{\infty} ar^{k-1}$, $a \neq 0$, converges if $|r| \leq 1$.

**5.** The sum of a convergent geometric series $\displaystyle\sum_{k=1}^{\infty} ar^{k-1}$, $a \neq 0$, is $S =$ _____.

**6. True or False** The harmonic series $\displaystyle\sum_{k=1}^{\infty} \dfrac{1}{k}$ converges because $\displaystyle\lim_{n \to \infty} \dfrac{1}{n} = 0$.

### Skill Building

*In Problems 7–10, find the fourth partial sum of each series.*

**7.** $\displaystyle\sum_{k=1}^{\infty} \left(\dfrac{3}{4}\right)^{k-1}$

**8.** $\displaystyle\sum_{k=1}^{\infty} \dfrac{(-1)^{k+1}}{3^{k-1}}$

**9.** $\displaystyle\sum_{k=1}^{\infty} k$

**10.** $\displaystyle\sum_{k=1}^{\infty} \ln k$

*In Problems 11–16, find the sum of each telescoping series.*

**11.** $\displaystyle\sum_{k=1}^{\infty} \left(\dfrac{1}{k+2} - \dfrac{1}{k+3}\right)$

**12.** $\displaystyle\sum_{k=1}^{\infty} \left[\dfrac{1}{k^2} - \dfrac{1}{(k+1)^2}\right]$

**13.** $\displaystyle\sum_{k=1}^{\infty} \left(\dfrac{1}{3^{k+1}} - \dfrac{1}{3^k}\right)$

**14.** $\displaystyle\sum_{k=1}^{\infty} \left(\dfrac{1}{4^{k+1}} - \dfrac{1}{4^k}\right)$

**15.** $\displaystyle\sum_{k=1}^{\infty} \dfrac{1}{4k^2 - 1}$  *Hint:* $\dfrac{1}{4k^2 - 1} = \dfrac{1}{2}\left(\dfrac{1}{2k-1} - \dfrac{1}{2k+1}\right)$

**16.** $\displaystyle\sum_{k=2}^{\infty} \dfrac{2}{k^2 - 1}$

*In Problems 17–38, determine whether each geometric series converges or diverges. If it converges, find its sum.*

**17.** $\displaystyle\sum_{k=1}^{\infty} (\sqrt{2})^{k-1}$

**18.** $\displaystyle\sum_{k=1}^{\infty} (0.33)^{k-1}$

**19.** $\displaystyle\sum_{k=1}^{\infty} 5\left(\dfrac{1}{6}\right)^{k-1}$

**20.** $\displaystyle\sum_{k=1}^{\infty} 4\,(1.1)^{k-1}$

**21.** $\displaystyle\sum_{k=0}^{\infty} 7\left(\dfrac{1}{3}\right)^{k}$

**22.** $\displaystyle\sum_{k=0}^{\infty} \left(\dfrac{7}{4}\right)^{k}$

**23.** $\displaystyle\sum_{k=1}^{\infty} (-0.38)^{k-1}$

**24.** $\displaystyle\sum_{k=1}^{\infty} (-0.38)^{k}$

**25.** $\displaystyle\sum_{k=0}^{\infty} \dfrac{2^{k+1}}{3^k}$

**26.** $\displaystyle\sum_{k=0}^{\infty} \dfrac{5^k}{6^{k+1}}$

**27.** $\displaystyle\sum_{k=0}^{\infty} \dfrac{1}{4^{k+1}}$

**28.** $\displaystyle\sum_{k=0}^{\infty} \dfrac{4^{k+1}}{3^k}$

**29.** $\displaystyle\sum_{k=1}^{\infty} \left(\sin\dfrac{\pi}{2}\right)^{k-1}$

**30.** $\displaystyle\sum_{k=1}^{\infty} \left(\tan\dfrac{\pi}{4}\right)^{k-1}$

**31.** $\displaystyle\sum_{k=1}^{\infty} \left(-\dfrac{3}{2}\right)^{k-1}$

**32.** $\displaystyle\sum_{k=1}^{\infty} \left(-\dfrac{2}{3}\right)^{k-1}$

**33.** $1 + \dfrac{1}{3} + \dfrac{1}{9} + \cdots + \left(\dfrac{1}{3}\right)^{n} + \cdots$

**34.** $1 + \dfrac{1}{4} + \dfrac{1}{16} + \cdots + \left(\dfrac{1}{4}\right)^{n} + \cdots$

**35.** $1 + 2 + 4 + \cdots + 2^n + \cdots$

**36.** $1 - \dfrac{1}{2} + \dfrac{1}{4} - \dfrac{1}{8} + \cdots + \dfrac{(-1)^{n-1}}{2^{n-1}} + \cdots$

**37.** $\left(\dfrac{1}{7}\right)^{2} + \left(\dfrac{1}{7}\right)^{3} + \cdots + \left(\dfrac{1}{7}\right)^{n} + \cdots$

**38.** $\left(\dfrac{3}{4}\right)^{5} + \left(\dfrac{3}{4}\right)^{6} + \cdots + \left(\dfrac{3}{4}\right)^{n} + \cdots$

*In Problems 39–58, determine whether each series converges or diverges. If it converges, find its sum.*

**39.** $\displaystyle\sum_{k=0}^{\infty} \dfrac{1}{k+1}$

**40.** $\displaystyle\sum_{k=4}^{\infty} k^{-1}$

**41.** $\displaystyle\sum_{k=1}^{\infty} \dfrac{1}{100^k}$

**42.** $\displaystyle\sum_{k=1}^{\infty} e^{-k}$

**1.** = NOW WORK problem      ◨ = Graphing technology recommended      CAS = Computer Algebra System recommended

**43.** $\displaystyle\sum_{k=1}^{\infty}(-10k)$

**44.** $\displaystyle\sum_{k=1}^{\infty}\frac{3k}{5}$

**45.** $\displaystyle\sum_{k=1}^{\infty}\left(\cos\frac{2\pi}{3}\right)^{k-1}$

**46.** $\displaystyle\sum_{k=1}^{\infty}\left(\sin\frac{\pi}{6}\right)^{k-1}$

**47.** $\displaystyle\sum_{k=1}^{\infty}\frac{\left(\tan\frac{\pi}{4}\right)^{k}}{k}$

**48.** $\displaystyle\sum_{k=1}^{\infty}\frac{\left(\sin\frac{\pi}{2}\right)^{k}}{k}$

**49.** $\displaystyle\sum_{k=1}^{\infty}\cos(\pi k)$

**50.** $\displaystyle\sum_{k=1}^{\infty}\sin\frac{\pi k}{2}$

**51.** $\displaystyle\sum_{k=1}^{\infty}2^{-k}3^{k+1}$

**52.** $\displaystyle\sum_{k=1}^{\infty}3^{1-k}2^{1+k}$

**53.** $\displaystyle\sum_{k=1}^{\infty}\left(-\frac{1}{3}\right)^{k}$

**54.** $\displaystyle\sum_{k=1}^{\infty}\frac{\pi}{3^{k}}$

**55.** $\displaystyle\sum_{k=1}^{\infty}\ln\frac{k}{k+1}$

**56.** $\displaystyle\sum_{k=1}^{\infty}\left[e^{2k-1}-e^{2(k+1)^{-1}}\right]$

**57.** $\displaystyle\sum_{k=1}^{\infty}\left(\sin\frac{1}{k}-\sin\frac{1}{k+1}\right)$

**58.** $\displaystyle\sum_{k=1}^{\infty}\left(\tan\frac{1}{k}-\tan\frac{1}{k+1}\right)$

*In Problems 59–62, express each repeating decimal as a rational number by using a geometric series.*

**59.** $0.5555\ldots$

**60.** $0.727272\ldots$

**61.** $4.28555\ldots$ *Hint:* $4.28555\ldots=4.28+0.00555\ldots$

**62.** $7.162162\ldots$

## Applications and Extensions

**63. Distance a Ball Travels** A ball is dropped from a height of 18 ft. Each time it strikes the ground, it bounces back to two-thirds of the previous height. Find the total distance traveled by the ball.

**64. Diminishing Returns** A rich man promises to give you $1000 on January 1, 2020. Each day thereafter he will give you $\frac{9}{10}$ of what he gave you the previous day.

(a) What is the total amount you will receive?

(b) What is the first date on which the amount you receive is less than 1 cent?

**65. Stocking a Lake** Mirror Lake is stocked periodically with rainbow trout. In year $n$ of the stocking program, the population is given by $p_n = 3000r^n + h\displaystyle\sum_{k=1}^{n}r^{k-1}$, where $h$ is the number of fish added by the program per year and $r$, $0 < r < 1$, is the percent of fish removed each year.

(a) What does a manager expect the steady rainbow trout population to be as $n \to \infty$?

(b) If $r = 0.5$, how many fish $h$ should be added annually to obtain a steady population of 4000 rainbow trout?

**66. Marginal Propensity to Consume** Suppose that individuals in the United States spend 90% of every additional dollar that they earn. Then according to economists, an individual's **marginal propensity to consume** is 0.90. For example, if Jane earns an additional dollar, she will spend $0.9(1) = \$0.90$ of it. The individual who earns Jane's $0.90 will spend 90% of it or $0.81. The process of spending continues and results in the series

$$\sum_{k=1}^{\infty}0.90^{k-1}=1+0.90+0.90^{2}+0.90^{3}+\cdots$$

The sum of this series is called the **multiplier**. What is the multiplier if the marginal propensity to consume is 90%?

**67. Stock Pricing** One method of pricing a stock is to discount the stream of future dividends of the stock. Suppose a stock currently pays $\$P$ annually in dividends, and historically, the dividend has increased by $i\%$ annually. If an investor wants an annual rate of return of $r\%$, $r > i$, this method of pricing a stock states that the stock should be priced at the present value of an infinite stream of payments:

$$\text{Price}=P+P\frac{1+i}{1+r}+P\left(\frac{1+i}{1+r}\right)^{2}+P\left(\frac{1+i}{1+r}\right)^{3}+\cdots$$

(a) Find the price of a stock priced using this method.

(b) Suppose an investor desires a 9% return on a stock that currently pays an annual dividend of $4.00, and, historically, the dividend has been increased by 3% annually. What is the highest price the investor should pay for the stock?

**68. The Koch Snowflake** The area inside the fractal known as the **Koch snowflake** can be described as the sum of the areas of infinitely many equilateral triangles. See the figure. For all but the center (largest) triangle, a triangle in the Koch snowflake is $\frac{1}{9}$ the area of the next largest triangle in the fractal. Suppose the largest (center) triangle has an area of 1 square unit. Then the area of the snowflake is given by the series

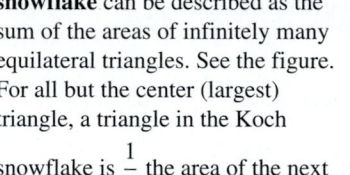

$$1+3\left(\frac{1}{9}\right)+12\left(\frac{1}{9}\right)^{2}+48\left(\frac{1}{9}\right)^{3}+192\left(\frac{1}{9}\right)^{4}+\cdots$$

Find the area of the Koch snowflake by finding the sum of the series.

**69. Zeno's Paradox** is about a race between Achilles and a tortoise. The tortoise is allowed a certain lead at the start of the race. Zeno claimed the tortoise must win such a race. He reasoned that for Achilles to overtake the tortoise, at some time he must cover $\frac{1}{2}$ of the distance that originally separated them. Then when he covers another $\frac{1}{4}$ of the original distance separating them, he will still have $\frac{1}{4}$ of that distance remaining, and so on. Therefore by Zeno's reasoning, Achilles never catches the tortoise. Use a series argument to explain this paradox. Assume that the difference in speed between Achilles and the tortoise is a constant $v$ meters per second.

**70. Probability** A coin-flipping game involves two people who successively flip a coin. The first person to obtain a head is the winner. In probability, it turns out that the person who flips first has the probability of winning given by the series below. Find this probability.

$$\frac{1}{2} + \frac{1}{8} + \frac{1}{32} + \cdots + \frac{1}{2^{2n-1}} + \cdots$$

**71. Controlling *Salmonella*** *Salmonella* is a common enteric bacterium infecting both humans and farm animals with salmonellosis. Barn surfaces contaminated with *Salmonella* can be the major source of salmonellosis spread in a farm. While cleaning barn surfaces is used as a control measure on pig and cattle farms, the efficiency of cleaning has been a concern. Suppose, on average, there are $p$ kilograms (kg) of feces produced each day and cleaning is performed with the constant efficiency $e$, $0 < e < 1$. At the end of each day, $(1 - e)$ kg of feces from the previous day is added to the amount of the present day. Let $T(n)$ be the total accumulated fecal material on day $n$. Farmers are concerned when $T(n)$ exceeds the threshold level $L$.

(a) Express $T(n)$ as a geometric series.

(b) Find $\lim\limits_{n \to \infty} T(n)$.

(c) Determine the minimum cleaning efficiency $e_{\min}$ required to guarantee $T(n) \le L$ for all $n$.

(d) Suppose that $p = 120$ kg and $L = 180$ kg. Using (b) and (c), find $e_{\min}$ and $T(365)$ for $e = \frac{4}{5}$.

*Source*: R. Gautam, G. Lahodny, M. Bani-Yaghoub, & R. Ivanek Based on their paper "Understanding the role of cleaning in the control of *Salmonella Typhimurium* in a grower finisher pig herd: a modeling approach."

**72.** Show that $0.9999\ldots = 1$.

*In Problems 73 and 74, use a geometric series to prove the given statement.*

**73.** $\dfrac{x}{x-1} = \displaystyle\sum_{k=1}^{\infty} \dfrac{1}{x^{k-1}}$ for $|x| > 1$

**74.** $\dfrac{1}{1+x} = \displaystyle\sum_{k=0}^{\infty} (-1)^k x^k$ for $|x| < 1$

**75.** Find the smallest number $n$ for which $\displaystyle\sum_{k=1}^{n} \frac{1}{k} \ge 3$.

**76.** Find the smallest number $n$ for which $\displaystyle\sum_{k=1}^{n} \frac{1}{k} \ge 4$.

**77.** Show that the series $\displaystyle\sum_{k=1}^{\infty} \frac{\sqrt{k+1} - \sqrt{k}}{\sqrt{k(k+1)}}$ converges and has the sum 1.

**78.** Show that $\displaystyle\sum_{k=1}^{\infty} \frac{1}{k(k+2)} = \frac{3}{4}$.

**79.** Show that $\displaystyle\sum_{k=1}^{\infty} \frac{1}{k(k+1)(k+2)} = \frac{1}{4}$.

**80.** Show that $\displaystyle\sum_{k=1}^{\infty} \frac{1}{k(k+1)(k+2)(k+3)} = \frac{1}{18}$.

**81.** Show that $\displaystyle\sum_{k=1}^{\infty} \frac{1}{k(k+1)(k+2) \cdots (k+a)} = \frac{1}{a} \cdot \frac{1}{a!}$; $a \ge 1$ is an integer.

**82.** Solve for $x$: $\dfrac{x}{2+2x} = x + x^2 + x^3 + \cdots$, $|x| < 1$.

**83.** Show that the sum of any convergent geometric series whose first term and common ratio are rational is rational.

**84.** The sum $S_n$ of the first $n$ terms of a geometric series is given by the formula

$$S_n = a + ar + ar^2 + \cdots + ar^{n-1} = \frac{a(r^n - 1)}{r - 1} \qquad a > 0, \quad r \ne 1$$

Find $\lim\limits_{r \to 1} \dfrac{a(r^n - 1)}{r - 1}$ and compare the result with a geometric series in which $r = 1$.

## Challenge Problems

*The following discussion relates to Problems 85 and 86.*

An interesting relationship between the $n$th partial sum of the harmonic series and $\ln n$ was discovered by Euler. In particular, he showed that

$$\gamma = \lim_{n \to \infty} \left( 1 + \frac{1}{2} + \frac{1}{3} + \cdots + \frac{1}{n} - \ln n \right)$$

exists and is approximately equal to 0.5772. The **Euler–Mascheroni constant**, as $\gamma$ is called, appears in many interesting areas of mathematics. For example, it is involved in the evaluation of the exponential integral, $\displaystyle\int_x^{\infty} \frac{e^{-t}}{t} \, dt$, which is important in applied mathematics. It is also related to two special functions—the gamma function and Riemann's zeta function (see Challenge Problem 85, Section 8.3). Surprisingly, it is still unknown whether $\gamma$ is rational or irrational.

**85.** The harmonic series diverges quite slowly. For example, the partial sums $S_{10}$, $S_{20}$, $S_{50}$, and $S_{100}$ have approximate values 2.92897, 3.59774, 4.49921, and 5.18738, respectively. In fact, the sum of the first million terms of the harmonic series is about 14.4. With this in mind, what would you conjecture about the rate of convergence of the limit defining $\gamma$? Test your conjecture by calculating approximate values for $\gamma$ by using the partial sums given above.

**86.** Use $\gamma \approx 0.5772$ to approximate

$$1 + \frac{1}{2} + \frac{1}{3} + \cdots + \frac{1}{1,000,000,000}$$

**87.** Show that a real number has a repeating decimal if and only if it is rational.

**88.** (a) Suppose $\displaystyle\sum_{k=1}^{\infty} s_k$ is a series with the property that $s_n \ge 0$ for all integers $n \ge 1$. Show that $\displaystyle\sum_{k=1}^{\infty} s_k$ converges if and only if the sequence $\{S_n\}$ of partial sums is bounded.

(b) Use the result of (a) to show the harmonic series diverges.

## 8.3 Properties of Series; Series with Positive Terms; the Integral Test

**OBJECTIVES** *When you finish this section, you should be able to:*

1 Use the Test for Divergence (p. 603)
2 Work with properties of series (p. 603)
3 Use the Integral Test (p. 605)
4 Analyze a *p*-series (p. 607)
5 Approximate the sum of a convergent series (p. 608)

We have been determining whether a series $\sum_{k=1}^{\infty} a_k$ converges or diverges by finding a single compact expression for the sequence $\{S_n\}$ of partial sums as a function of $n$, and then examining $\lim_{n \to \infty} S_n$. For most series $\sum_{k=1}^{\infty} a_k$, however, this is not possible. As a result, we develop alternate methods for determining whether $\sum_{k=1}^{\infty} a_k$ is convergent or divergent. Most of the alternate methods only tell us whether a series converges or diverges but provide no information about the sum of a convergent series. Fortunately, in many applications involving series, it is more important to know whether or not a series converges. Knowing the sum of a convergent series, although desirable, is not always necessary.

The next result gives a property of convergent series that is used often.

---

**THEOREM**

If the infinite series $\sum_{k=1}^{\infty} a_k$ converges, then $\lim_{n \to \infty} a_n = 0$.

---

**Proof** The *n*th partial sum of $\sum_{k=1}^{\infty} a_k$ is $S_n = \sum_{k=1}^{n} a_k$. Since $S_{n-1} = \sum_{k=1}^{n-1} a_k$, it follows that

$$a_n = S_n - S_{n-1}$$

Since the series $\sum_{k=1}^{\infty} a_k$ converges, the sequence $\{S_n\}$ of partial sums has a limit $S$. Then $\lim_{n \to \infty} S_n = S$ and $\lim_{n \to \infty} S_{n-1} = S$, so

$$\lim_{n \to \infty} a_n = \lim_{n \to \infty} (S_n - S_{n-1}) = \lim_{n \to \infty} S_n - \lim_{n \to \infty} S_{n-1} = S - S = 0 \qquad \blacksquare$$

**IN WORDS**

- If $\sum_{k=1}^{\infty} a_k$ converges, then $\lim_{n \to \infty} a_n$ equals 0.
- If $\lim_{n \to \infty} a_n$ equals 0, then the series $\sum_{k=1}^{\infty} a_k$ may converge or diverge.

So if a series $\sum_{k=1}^{\infty} a_k$ converges, then $\lim_{n \to \infty} a_n = 0$. But there are many divergent series $\sum_{k=1}^{\infty} a_k$ for which $\lim_{n \to \infty} a_n = 0$. For example, the limit of the *n*th term of the harmonic series $\sum_{k=1}^{\infty} \frac{1}{k}$ is $\lim_{n \to \infty} \frac{1}{n} = 0$, and yet it diverges.

By restating the theorem, we obtain a useful test for divergence.

---

**THEOREM Test for Divergence**

The infinite series $\sum_{k=1}^{\infty} a_k$ diverges if $\lim_{n \to \infty} a_n \neq 0$.

## 1 Use the Test for Divergence

**EXAMPLE 1   Using the Test for Divergence**

**(a)** $\sum\limits_{k=1}^{\infty} 87$ diverges, since $\lim\limits_{n \to \infty} 87 = 87 \neq 0$.

**(b)** $\sum\limits_{k=1}^{\infty} k$ diverges, since $\lim\limits_{n \to \infty} n = \infty \neq 0$.

**(c)** $\sum\limits_{k=1}^{\infty} (-1)^k$ diverges, since $\lim\limits_{n \to \infty} (-1)^n$ does not exist.

**(d)** $\sum\limits_{k=1}^{\infty} 2^k$ diverges, since $\lim\limits_{n \to \infty} 2^n = \infty \neq 0$.    ∎

Be careful! In testing a series $\sum\limits_{k=1}^{\infty} a_k$ for convergence/divergence

- if $\lim\limits_{n \to \infty} a_n \neq 0$, the series diverges.

- if $\lim\limits_{n \to \infty} a_n = 0$, no information about the convergence or divergence of the series is obtained.

**NOW WORK** Problem 17.

## 2 Work with Properties of Series

Next we investigate some properties of convergent and divergent series. Knowing these properties can help to determine whether a series converges or diverges.

**THEOREM**

If two infinite series are identical after a certain term, then either both series converge or both series diverge. If both series converge, they do not necessarily have the same sum.

**Proof**   Consider the two series

$$\sum_{k=1}^{\infty} a_k = a_1 + a_2 + \cdots + a_p + a_{p+1} + \cdots + a_n + \cdots$$

$$\sum_{k=1}^{\infty} b_k = b_1 + b_2 + \cdots + b_p + a_{p+1} + \cdots + a_n + \cdots$$

Notice that after the first $p$ terms, the terms of both series are identical. The $n$th partial sum $S_n$ of $\sum\limits_{k=1}^{\infty} a_k$ and the $n$th partial sum $T_n$ of $\sum\limits_{k=1}^{\infty} b_k$ are given by

$$S_n = a_1 + a_2 + \cdots + a_p + a_{p+1} + \cdots + a_n \qquad \text{\color{blue}After the } p\text{th term, the}$$
$$T_n = b_1 + b_2 + \cdots + b_p + a_{p+1} + \cdots + a_n \qquad \text{\color{blue}terms are the same.}$$
$$S_n - T_n = (a_1 + \cdots + a_p) - (b_1 + \cdots + b_p)$$
$$S_n = T_n + (a_1 + \cdots + a_p) - (b_1 + \cdots + b_p) \qquad \text{\color{blue}For } n > p$$
$$\lim_{n \to \infty} S_n = \lim_{n \to \infty} \left[ T_n + (a_1 + \cdots + a_p) - (b_1 + \cdots + b_p) \right]$$
$$\lim_{n \to \infty} S_n = \lim_{n \to \infty} T_n + \lim_{n \to \infty} \left[ (a_1 + \cdots + a_p) - (b_1 + \cdots + b_p) \right]$$
$$\lim_{n \to \infty} S_n = \lim_{n \to \infty} T_n + k$$

where $k = (a_1 + \cdots + a_p) - (b_1 + \cdots + b_p)$ is some number. Consequently, either both limits exist (both series converge) or neither limit exists (both series diverge). No other possibility can occur.   ∎

**THEOREM Sum and Difference of Convergent Series**

If $\sum_{k=1}^{\infty} a_k = S$ and $\sum_{k=1}^{\infty} b_k = T$ are two convergent series, then the series $\sum_{k=1}^{\infty} (a_k + b_k)$

and the series $\sum_{k=1}^{\infty} (a_k - b_k)$ also converge. Moreover,

$$\sum_{k=1}^{\infty}(a_k + b_k) = \sum_{k=1}^{\infty} a_k + \sum_{k=1}^{\infty} b_k = S + T$$

$$\sum_{k=1}^{\infty}(a_k - b_k) = \sum_{k=1}^{\infty} a_k - \sum_{k=1}^{\infty} b_k = S - T$$

The proof of the sum part of this theorem is left as an exercise. See Problem 75.

**THEOREM Constant Multiple of a Series**

Let $c$ be a nonzero real number. If $\sum_{k=1}^{\infty} a_k = S$ is a convergent series, then the

series $\sum_{k=1}^{\infty} (ca_k)$ also converges. Moreover,

$$\sum_{k=1}^{\infty}(ca_k) = c\sum_{k=1}^{\infty} a_k = cS$$

If the series $\sum_{k=1}^{\infty} a_k$ diverges, then the series $\sum_{k=1}^{\infty} (ca_k)$ also diverges.

**IN WORDS** Multiplying each term of a series by a nonzero constant does not affect the convergence (or divergence) of the series.

The proof of the convergent part of this theorem is left as an exercise. See Problem 76.

**EXAMPLE 2 Using Properties of Series**

Determine whether each series converges or diverges. If it converges, find its sum.

**(a)** $\sum_{k=4}^{\infty} \frac{1}{k}$  **(b)** $\sum_{k=1}^{\infty} \frac{2}{k}$  **(c)** $\sum_{k=1}^{\infty} \left( \frac{1}{2^{k-1}} + \frac{1}{3^{k-1}} \right)$

**Solution** **(a)** Except for the first three terms, the series $\sum_{k=4}^{\infty} \frac{1}{k} = \frac{1}{4} + \frac{1}{5} + \frac{1}{6} + \cdots$

is identical to the harmonic series, which diverges. So, it follows that $\sum_{k=4}^{\infty} \frac{1}{k}$ also diverges.

**(b)** Since the harmonic series $\sum_{k=1}^{\infty} \frac{1}{k}$ diverges, the series $\sum_{k=1}^{\infty} \left( 2 \cdot \frac{1}{k} \right) = \sum_{k=1}^{\infty} \frac{2}{k}$ diverges.

**(c)** Since the series $\sum_{k=1}^{\infty} \frac{1}{2^{k-1}}$ and the series $\sum_{k=1}^{\infty} \frac{1}{3^{k-1}}$ are both convergent geometric

series, the series defined by $\sum_{k=1}^{\infty} \left( \frac{1}{2^{k-1}} + \frac{1}{3^{k-1}} \right)$ is also convergent. The sum is

$$\sum_{k=1}^{\infty} \left( \frac{1}{2^{k-1}} + \frac{1}{3^{k-1}} \right) = \sum_{k=1}^{\infty} \frac{1}{2^{k-1}} + \sum_{k=1}^{\infty} \frac{1}{3^{k-1}} = \frac{1}{1-\frac{1}{2}} + \frac{1}{1-\frac{1}{3}} = 2 + \frac{3}{2} = \frac{7}{2}$$

**NOW WORK** Problem 39.

## Series with Positive Terms

For series that have only positive terms, it is possible to construct tests for convergence that use only the $n$th term of the series and do not require knowledge of the form of the sequence $\{S_n\}$ of partial sums.

For example, consider an infinite series

$$\sum_{k=1}^{\infty} a_k = a_1 + a_2 + \cdots + a_n + \cdots$$

where each term $a_k > 0$. The general term $S_n$ of the sequence of partial sums $\{S_n\}$ is

$$S_n = (a_1 + a_2 + a_3 + \cdots + a_{n-1}) + a_n = S_{n-1} + a_n > S_{n-1}$$
$$\uparrow$$
$$a_n > 0$$

That is, the sequence of partial sums is increasing.

There are two possibilities.

* If $\{S_n\}$ is unbounded, then $\{S_n\}$ diverges and the series diverges. This is how we showed that the harmonic series $\sum_{k=1}^{\infty} \dfrac{1}{k}$ diverges. (p. 597)

**RECALL** A bounded, monotonic sequence converges.

* If $\{S_n\}$ is bounded, then $\{S_n\}$ converges and the series converges.

We formalize this conclusion as the *General Convergence Test*.

> **THEOREM  General Convergence Test**
>
> An infinite series of positive terms converges if and only if its sequence of partial sums is bounded. The sum of such an infinite series will not exceed an upper bound.

We use this theorem to develop other tests for convergence of series with only positive terms; the first one we discuss is the *Integral Test*.

## 3 Use the Integral Test

> **THEOREM  Integral Test**
>
> Let $f$ be a function that is continuous, positive, and nonincreasing on the interval $[1, \infty)$. Let $a_k = f(k)$ for all positive integers $k$. Then
>
> * if the improper integral $\int_1^{\infty} f(x)\, dx$ converges, the series $\sum_{k=1}^{\infty} a_k$ converges.
>
> * if the improper integral $\int_1^{\infty} f(x)\, dx$ diverges, the series $\sum_{k=1}^{\infty} a_k$ diverges.

**NEED TO REVIEW?** Improper Integrals are discussed in Section 7.7, pp. 555–558.

A proof of the Integral Test is given at the end of the section.

The Integral Test is used when the $n$th term of the series is related to a function $f$ that is not only continuous, positive, and nonincreasing on the interval $[1, \infty)$ but also has an antiderivative that can be readily found.

CALC
CLIP

**EXAMPLE 3  Using the Integral Test**

Determine whether the series $\sum_{k=1}^{\infty} a_k = \sum_{k=1}^{\infty} \dfrac{4}{k^2 + 1}$ converges or diverges.

**Solution** On the interval $[1, \infty)$, the related function $f(x) = \dfrac{4}{x^2 + 1}$ is continuous,

positive, and decreasing. ($f$ is decreasing because $f'(x) = -\dfrac{8x}{(x^2 + 1)^2} < 0$ for $x \geq 1$.)

Also, $a_k = f(k)$ for all positive integers. To use the Integral Test, we need to determine

whether the improper integral $\displaystyle\int_1^\infty \dfrac{4}{x^2 + 1}\, dx$ converges or diverges.

**CAUTION** In Example 3 the fact
that $\displaystyle\int_1^\infty \dfrac{4}{x^2 + 1}\, dx = \pi$ does not
mean that the sum $S$ of the series is $\pi$.
Do not confuse the value of the
improper integral $\int_1^\infty f(x)\, dx$ with
the sum $S$ of the series. In general,
they are *not* equal.

$$\int_1^\infty \frac{4}{x^2 + 1}\, dx: \quad \lim_{b \to \infty} \int_1^b \frac{4}{x^2 + 1}\, dx = \lim_{b \to \infty} \left[ 4 \int_1^b \frac{1}{x^2 + 1}\, dx \right] = 4 \lim_{b \to \infty} \left[ \tan^{-1} x \right]_1^b$$

$$= 4 \lim_{b \to \infty} \left[ \tan^{-1} b - \frac{\pi}{4} \right] = 4 \left[ \frac{\pi}{2} - \frac{\pi}{4} \right] = \pi$$

Since the improper integral converges, the series $\displaystyle\sum_{k=1}^\infty \frac{4}{k^2 + 1}$ converges. ∎

---

**EXAMPLE 4  Using the Integral Test**

Determine whether the series $\displaystyle\sum_{k=1}^\infty a_k = \sum_{k=1}^\infty \frac{2k}{k^2 + 1}$ converges or diverges.

**Solution** On the interval $[1, \infty)$, the related function $f(x) = \dfrac{2x}{x^2 + 1}$ is continuous,

positive, and decreasing. ($f$ is decreasing because $f'(x) = \dfrac{2(1 - x^2)}{(x^2 + 1)^2} < 0$ for $x > 1$.)

Also, $a_k = f(k)$ for all positive integers $k$. Now investigate the improper

integral $\displaystyle\int_1^\infty \frac{2x}{x^2 + 1}\, dx$.

$$\int_1^\infty \frac{2x}{x^2 + 1}\, dx: \quad \lim_{b \to \infty} \int_1^b \frac{2x}{x^2 + 1}\, dx = \lim_{b \to \infty} \left[ \ln(x^2 + 1) \right]_1^b$$

$$= \lim_{b \to \infty} \left[ \ln(b^2 + 1) - \ln 2 \right] = \infty$$

Since the improper integral diverges, by the Integral Test the series $\displaystyle\sum_{k=1}^\infty \frac{2k}{k^2 + 1}$ also

diverges. ∎

**NOW WORK** Problem 23.

To use the Integral Test, the lower limit of integration does not need to be 1, as we see in the next example.

---

**EXAMPLE 5  Using the Integral Test**

Determine whether the series $\displaystyle\sum_{k=2}^\infty \frac{1}{k(\ln k)^2}$ converges or diverges.

**Solution** On the interval $[2, \infty)$, the function $f(x) = \dfrac{1}{x(\ln x)^2}$ is continuous,

positive, and decreasing. (Check this for yourself.) To investigate the improper

integral $\displaystyle\int_2^\infty \frac{1}{x(\ln x)^2}\, dx$, we first find an antiderivative of $\dfrac{1}{x(\ln x)^2}$. Using the

substitution $u = \ln x$, $du = \dfrac{1}{x}\, dx$, we find

$$\int \frac{dx}{x(\ln x)^2} = \int \frac{du}{u^2} = -\frac{1}{u} + C = -\frac{1}{\ln x} + C$$

Then

$$\int_2^\infty \frac{dx}{x(\ln x)^2}: \quad \lim_{b \to \infty} \left[ -\frac{1}{\ln x} \right]_2^b = \lim_{b \to \infty} \left( -\frac{1}{\ln b} + \frac{1}{\ln 2} \right) = \frac{1}{\ln 2}$$

Since the improper integral $\int_2^\infty \dfrac{dx}{x(\ln x)^2}$ converges, by the Integral Test the series $\sum\limits_{k=2}^\infty \dfrac{1}{k(\ln k)^2}$ converges. ∎

NOW WORK **Problem 27.**

## 4 Analyze a p-Series

Another important series of positive terms is the *p-series*.

> **DEFINITION** p-Series
>
> A **p-series** is an infinite series of the form
>
> $$\sum_{k=1}^\infty \frac{1}{k^p} = 1 + \frac{1}{2^p} + \frac{1}{3^p} + \cdots + \frac{1}{n^p} + \cdots$$
>
> where $p$ is a positive real number.

A *p*-series is sometimes referred to as a **hyperharmonic series**, since the harmonic series is a special case of a *p*-series when $p = 1$. Some examples of *p*-series are

$$p = 1: \quad \sum_{k=1}^\infty \frac{1}{k} = 1 + \frac{1}{2} + \frac{1}{3} + \cdots + \frac{1}{n} + \cdots \qquad \text{The harmonic series}$$

$$p = \frac{1}{2}: \quad \sum_{k=1}^\infty \frac{1}{k^{1/2}} = 1 + \frac{1}{2^{1/2}} + \frac{1}{3^{1/2}} + \cdots + \frac{1}{n^{1/2}} + \cdots$$

$$p = 3: \quad \sum_{k=1}^\infty \frac{1}{k^3} = 1 + \frac{1}{2^3} + \frac{1}{3^3} + \cdots + \frac{1}{n^3} + \cdots$$

The following theorem establishes the values of $p$ for which a *p*-series converges and for which it diverges.

> **THEOREM** Convergence/Divergence of a p-Series
>
> The *p*-series
>
> $$\sum_{k=1}^\infty \frac{1}{k^p} = 1 + \frac{1}{2^p} + \frac{1}{3^p} + \cdots + \frac{1}{n^p} + \cdots$$
>
> converges if $p > 1$ and diverges if $0 < p \le 1$.

**Proof** First, notice that if $p = 1$, the series $\sum\limits_{k=1}^\infty \dfrac{1}{k^p} = \sum\limits_{k=1}^\infty \dfrac{1}{k}$ is the harmonic series, which diverges.

Next, on the interval $[1, \infty)$, the function $f(x) = \dfrac{1}{x^p}$, $p > 0$, is continuous, positive, and decreasing, and $f(k) = \dfrac{1}{k^p}$ for all positive integers $k$. By the Integral Test, the series $\sum\limits_{k=1}^\infty \dfrac{1}{k^p}$ converges if and only if the improper integral $\int_1^\infty \dfrac{1}{x^p}\,dx$ converges.

**NOTE** The improper integral $\int_1^\infty \dfrac{1}{x^p}\,dx$ is discussed in Section 7.7, pp. 557–558.

So we investigate $\lim\limits_{b \to \infty} \int_1^b \dfrac{1}{x^p}\,dx,\ p > 0,\ p \ne 1$.

$$\lim_{b \to \infty} \int_1^b \frac{1}{x^p}\,dx = \lim_{b \to \infty} \left[ \frac{x^{-p+1}}{-p+1} \right]_1^b = \frac{1}{1-p} \lim_{b \to \infty} \left[ x^{1-p} \right]_1^b = \frac{1}{1-p} \left[ \lim_{b \to \infty} b^{1-p} - 1 \right]$$

- If $0 < p < 1$, then $1 - p > 0$ and $\lim\limits_{b \to \infty} b^{1-p} = \infty$. So $\int_1^\infty \dfrac{1}{x^p}\,dx$ diverges.

- If $p > 1$, then $1 - p < 0$ and $\lim\limits_{b \to \infty} b^{1-p} = 0$. So $\int_1^\infty \dfrac{1}{x^p}\,dx$ converges.

So, the *p*-series $\sum\limits_{k=1}^\infty \dfrac{1}{k^p}$ converges if $p > 1$ and diverges if $0 < p \le 1$. ∎

**EXAMPLE 6**  **Analyzing a *p*-Series**

(a) The series

$$\sum_{k=1}^{\infty} \frac{1}{k^3} = 1 + \frac{1}{2^3} + \frac{1}{3^3} + \cdots + \frac{1}{n^3} + \cdots$$

converges, since it is a *p*-series where $p = 3$ ($p > 1$).

(b) The series

$$\sum_{k=1}^{\infty} \frac{1}{\sqrt{k}} = 1 + \frac{1}{\sqrt{2}} + \frac{1}{\sqrt{3}} + \cdots + \frac{1}{\sqrt{n}} + \cdots$$

diverges, since it is a *p*-series where $p = \dfrac{1}{2}$ ($0 < p \le 1$). ∎

**NOW WORK** Problem 35.

## 5 Approximate the Sum of a Convergent Series

For a convergent *p*-series, we can find a lower bound and an upper bound for the sum.

**THEOREM**  **Bounds for the Sum of a Convergent *p*-Series**

If $p > 1$, then

$$\frac{1}{p-1} < \sum_{k=1}^{\infty} \frac{1}{k^p} < 1 + \frac{1}{p-1} \tag{1}$$

A proof of this result is given after the solution to Example 8.

**IN WORDS** The theorem gives an interval of width 1 that contains the sum of a convergent *p*-series. But it gives no information about the actual sum.

**EXAMPLE 7**  **Finding Bounds for the Sum of a Convergent *p*-Series**

Find bounds for the sum of each *p*-series.

(a) $\displaystyle\sum_{k=1}^{\infty} \frac{1}{k^3}$   (b) $1 + \dfrac{1}{2^\pi} + \dfrac{1}{3^\pi} + \dfrac{1}{4^\pi} + \cdots$

**Solution** (a) $\displaystyle\sum_{k=1}^{\infty} \frac{1}{k^3}$ is a convergent *p*-series with $p = 3$. Based on (1),

$$\frac{1}{2} < \sum_{k=1}^{\infty} \frac{1}{k^3} < \frac{3}{2} \qquad \frac{1}{p-1} = \frac{1}{3-1} = \frac{1}{2}; \quad 1 + \frac{1}{p-1} = 1 + \frac{1}{2} = \frac{3}{2}$$

(b) $\displaystyle\sum_{k=1}^{\infty} \frac{1}{k^\pi}$, is a convergent *p*-series with $p = \pi$. Based on (1),

$$\frac{1}{\pi-1} < \sum_{k=1}^{\infty} \frac{1}{k^\pi} < 1 + \frac{1}{\pi-1} \quad \text{or} \quad 0.466 < \sum_{k=1}^{\infty} \frac{1}{k^\pi} < 1.467 \qquad ∎$$

Since $\displaystyle\sum_{k=1}^{\infty} \frac{1}{k^p}$ converges for $p > 1$, the series has a sum $S$. However, the exact sum has been found only for positive even integers $p$. In 1752, Euler was the first to show that $\displaystyle\sum_{k=1}^{\infty} \frac{1}{k^2} = \frac{\pi^2}{6}$. But the sum of other convergent *p*-series, such as $\displaystyle\sum_{k=1}^{\infty} \frac{1}{k^3}$, is not known.

**NOW WORK** Problem 55.

The Bounds for a Convergent $p$-series theorem can be extended to any convergent series of positive terms.

> **THEOREM Bounds for the Sum of a Convergent Series**
>
> Consider the series $\sum_{k=1}^{\infty} a_k$, where $a_k > 0$ for all positive integers $k$. Suppose $f$ is a continuous, positive, decreasing function on the interval $[1, \infty)$ and $a_k = f(k)$ for all positive integers $k$. If the improper integral $\int_1^{\infty} f(x)dx$ converges, then the sum of the series is bounded by
>
> $$\int_1^{\infty} f(x)dx < \sum_{k=1}^{\infty} a_k < a_1 + \int_1^{\infty} f(x)dx \qquad (2)$$

**EXAMPLE 8  Finding Bounds for the Sum of a Convergent Series**

Find bounds for the sum of the series $\sum_{k=1}^{\infty} \dfrac{4}{k^2+1}$.

**Solution** In Example 3 we showed that the series $\sum_{k=1}^{\infty} \dfrac{4}{k^2+1}$ converges by using the Integral Test and showing $\int_1^{\infty} \dfrac{4}{x^2+1}dx$ converges to $\pi$. Based on (2), the sum of the series $\sum_{k=1}^{\infty} \dfrac{4}{k^2+1}$ has the bounds

$$\pi < \sum_{k=1}^{\infty} \dfrac{4}{k^2+1} < 2+\pi \quad a_1 = \dfrac{4}{2} = 2 \qquad \blacksquare$$

**NOW WORK** Problem 63.

**Proof of the Bounds for the Sum of a Convergent $p$-Series Theorem**

From Figure 19, we see that the area under the graph of $f(x) = \dfrac{1}{x^p}$ is less than the area of the sums of the rectangles. That is,

$$\left(\text{Area under the graph of } y = \dfrac{1}{x^p}\right) = \int_1^{\infty} \dfrac{dx}{x^p} < 1 + \dfrac{1}{2^p} + \dfrac{1}{3^p} + \cdots = \sum_{k=1}^{\infty} \dfrac{1}{k^p} \qquad (3)$$

From Figure 20 we have

$$1 + \left(\text{Area under the graph of } y = \dfrac{1}{x^p}\right) = 1 + \int_1^{\infty} \dfrac{dx}{x^p} > 1 + \dfrac{1}{2^p} + \dfrac{1}{3^p} + \cdots = \sum_{k=1}^{\infty} \dfrac{1}{k^p} \qquad (4)$$

Combining (3) and (4), we have

$$\int_1^{\infty} \dfrac{dx}{x^p} < \sum_{k=1}^{\infty} \dfrac{1}{k^p} < 1 + \int_1^{\infty} \dfrac{dx}{x^p} \qquad (5)$$

Since $p > 1$,

$$\int_1^{\infty} \dfrac{dx}{x^p} = \lim_{b \to \infty} \int_1^b \dfrac{dx}{x^p} = \lim_{b \to \infty} \left[\dfrac{x^{-p+1}}{-p+1}\right]_1^b = \lim_{b \to \infty} \dfrac{b^{1-p}-1}{1-p} \underset{\underset{p>1}{\uparrow}}{=} \dfrac{0-1}{1-p} = \dfrac{1}{p-1}$$

Using this result in (5), we get

$$\dfrac{1}{p-1} < \sum_{k=1}^{\infty} \dfrac{1}{k^p} < 1 + \dfrac{1}{p-1} \qquad \blacksquare$$

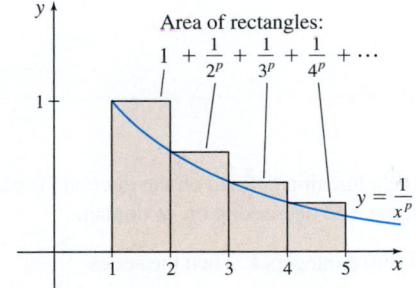

Area of rectangles:
$1 + \dfrac{1}{2^p} + \dfrac{1}{3^p} + \dfrac{1}{4^p} + \cdots$

$y = \dfrac{1}{x^p}$

**Figure 19**

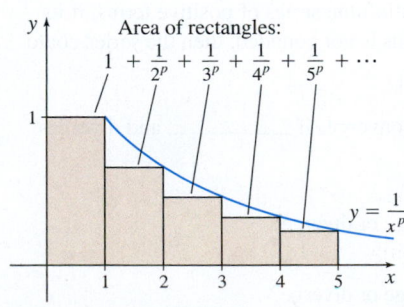

Area of rectangles:
$1 + \dfrac{1}{2^p} + \dfrac{1}{3^p} + \dfrac{1}{4^p} + \dfrac{1}{5^p} + \cdots$

$y = \dfrac{1}{x^p}$

**Figure 20**

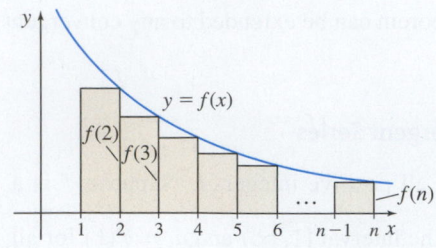

**Figure 21**

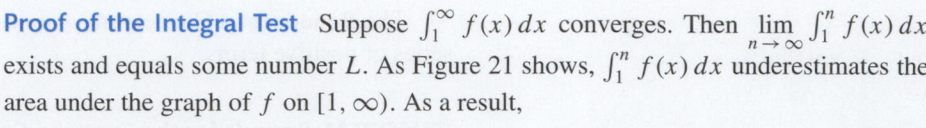

**Proof of the Integral Test** Suppose $\int_1^\infty f(x)\,dx$ converges. Then $\lim\limits_{n \to \infty} \int_1^n f(x)\,dx$ exists and equals some number $L$. As Figure 21 shows, $\int_1^n f(x)\,dx$ underestimates the area under the graph of $f$ on $[1, \infty)$. As a result,

$$\int_1^n f(x)\,dx < L \qquad \text{for any number } n$$

But $f$ is decreasing on the interval $[1, n]$. So, the sum of the areas of the $n-1$ rectangles drawn in Figure 21 underestimates the area under the graph of $y = f(x)$ from 1 to $n$. That is,

$$f(2) + f(3) + \cdots + f(n) < \int_1^n f(x)\,dx < L$$

Now add $f(1) = a_1$ to each part, and use the fact that $a_1 = f(1)$, $a_2 = f(2)$, …, $a_n = f(n)$. Then

$$a_1 + a_2 + a_3 + \cdots + a_n = f(1) + f(2) + \cdots + f(n) < f(1) + \int_1^n f(x)\,dx < f(1) + L$$

This means that the partial sums of $\sum\limits_{k=1}^\infty a_k$ are bounded from above. Since the partial sums are also increasing, $\sum\limits_{k=1}^\infty a_k$ converges.

Now suppose $\int_1^\infty f(x)\,dx$ diverges. Since $f$ is decreasing on $[1, n]$, the sum of the areas of the $n-1$ rectangles shown in Figure 22 overestimates the area under the graph of $y = f(x)$ from 1 to $n$. That is,

$$f(1) + f(2) + \cdots + f(n-1) > \int_1^n f(x)\,dx$$

Since $\int_1^n f(x)\,dx \to \infty$ as $n \to \infty$, the $n$th partial sum $S_n$ must also approach infinity, and it follows that the series $\sum\limits_{k=1}^\infty a_k$ diverges. ∎

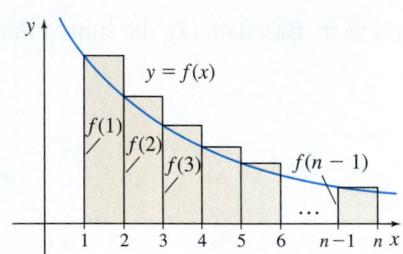

**Figure 22**

---

## 8.3 Assess Your Understanding

**Concepts and Vocabulary**

1. *Multiple Choice* If the series $\sum\limits_{k=1}^\infty a_k$, $a_k > 0$, converges then $\lim\limits_{n \to \infty} a_n = $ [(**a**) 0 (**b**) $a_1$ (**c**) $a_n$ (**d**) $\infty$].

2. *True or False* If $\lim\limits_{n \to \infty} a_n = 0$, then the series $\sum\limits_{k=1}^\infty a_k$ converges.

3. *True or False* The series $\sum\limits_{k=1}^\infty k^3$ diverges.

4. *True or False* If the first 100 terms of two infinite series are different, but from the 101st term on they are identical, then either both series converge or both series diverge.

5. *True or False* If $\sum\limits_{k=1}^\infty (a_k + b_k)$ converges, then $\sum\limits_{k=1}^\infty a_k$ converges and $\sum\limits_{k=1}^\infty b_k$ converges.

6. *True or False* If $\sum\limits_{k=1}^\infty a_k = S$ is a convergent series and $c$ is a nonzero real number, then $\sum\limits_{k=1}^\infty (ca_k) = cS$.

7. *True or False* Let $f$ be a function defined on the interval $[1, \infty)$ that is continuous, positive, and decreasing on its domain. Let $a_k = f(k)$ for all positive integers $k$. Then the series $\sum\limits_{k=1}^\infty a_k$ converges if and only if the improper integral $\int_1^\infty f(x)\,dx$ converges.

8. *True or False* For an infinite series of positive terms, if its sequence of partial sums is not bounded, then the series could converge or diverge.

9. The $p$-series $\sum\limits_{k=1}^\infty \dfrac{1}{k^p}$ converges if _____ and diverges if _____.

10. Does $\sum\limits_{k=1}^\infty \dfrac{1}{k^{3/2}}$ converge or diverge?

11. Does $\sum\limits_{k=1}^\infty \dfrac{1}{k^{1/2}}$ converge or diverge?

12. *True or False* If $p > 1$, then the convergent $p$-series $\sum\limits_{k=1}^\infty \dfrac{1}{k^p}$ is bounded by $\dfrac{1}{p-1} < \sum\limits_{k=1}^\infty \dfrac{1}{k^p} < 1$.

---

**1.** = NOW WORK problem        〔╱〕 = Graphing technology recommended        〔CAS〕 = Computer Algebra System recommended

## Skill Building

*In Problems 13–18, use the Test for Divergence to show each series diverges.*

**13.** $\displaystyle\sum_{k=1}^{\infty} 16$

**14.** $\displaystyle\sum_{k=1}^{\infty} \frac{k+9}{k}$

**15.** $\displaystyle\sum_{k=1}^{\infty} \ln k$

**16.** $\displaystyle\sum_{k=1}^{\infty} e^k$

**17.** $\displaystyle\sum_{k=1}^{\infty} \frac{k^2}{k^2+4}$

**18.** $\displaystyle\sum_{k=1}^{\infty} \frac{k^2+3}{\sqrt{k}}$

*In Problems 19–28, use the Integral Test to determine whether each series converges or diverges.*

**19.** $\displaystyle\sum_{k=1}^{\infty} \frac{1}{k^{1.01}}$

**20.** $\displaystyle\sum_{k=1}^{\infty} \frac{1}{k^{0.9}}$

**21.** $\displaystyle\sum_{k=1}^{\infty} \frac{\sqrt{\ln k}}{k}$

**22.** $\displaystyle\sum_{k=2}^{\infty} \frac{1}{k\sqrt{\ln k}}$

**23.** $\displaystyle\sum_{k=1}^{\infty} ke^{-k^2}$

**24.** $\displaystyle\sum_{k=1}^{\infty} ke^{-k}$

**25.** $\displaystyle\sum_{k=1}^{\infty} \frac{1}{k^2+1}$

**26.** $\displaystyle\sum_{k=2}^{\infty} \frac{1}{k\sqrt{k^2-1}}$

**27.** $\displaystyle\sum_{k=2}^{\infty} \frac{1}{k\ln k}$

**28.** $\displaystyle\sum_{k=2}^{\infty} \frac{1}{k(\ln k)^3}$

*In Problems 29–38, determine whether each p-series converges or diverges.*

**29.** $\displaystyle\sum_{k=1}^{\infty} \frac{1}{k^2}$

**30.** $\displaystyle\sum_{k=1}^{\infty} \frac{1}{k^4}$

**31.** $\displaystyle\sum_{k=1}^{\infty} \frac{1}{k^{1/3}}$

**32.** $\displaystyle\sum_{k=1}^{\infty} \frac{1}{k^{2/3}}$

**33.** $\displaystyle\sum_{k=1}^{\infty} \frac{1}{k^e}$

**34.** $\displaystyle\sum_{k=1}^{\infty} \frac{1}{k^{\sqrt{2}}}$

**35.** $1 + \dfrac{1}{2\sqrt{2}} + \dfrac{1}{3\sqrt{3}} + \dfrac{1}{4\sqrt{4}} + \cdots$

**36.** $1 + \dfrac{1}{\sqrt[3]{2}} + \dfrac{1}{\sqrt[3]{3}} + \dfrac{1}{\sqrt[3]{4}} + \cdots$

**37.** $1 + \dfrac{1}{4\sqrt{2}} + \dfrac{1}{9\sqrt{3}} + \dfrac{1}{16\sqrt{4}} + \cdots$

**38.** $1 + \dfrac{1}{8} + \dfrac{1}{27} + \dfrac{1}{64} + \cdots$

*In Problems 39–54, determine whether each series converges or diverges.*

**39.** $\displaystyle\sum_{k=1}^{\infty} \frac{10}{k}$

**40.** $\displaystyle\sum_{k=1}^{\infty} \frac{2}{1+k}$

**41.** $\displaystyle\sum_{k=1}^{\infty} \frac{k^2+1}{4k+1}$

**42.** $\displaystyle\sum_{k=1}^{\infty} \frac{k^3}{k^3+3}$

**43.** $\displaystyle\sum_{k=1}^{\infty} \left(k+\frac{1}{k}\right)$

**44.** $\displaystyle\sum_{k=1}^{\infty} \left(k-\frac{10}{k}\right)$

**45.** $\displaystyle\sum_{k=1}^{\infty} \left(\frac{1}{3k}-\frac{1}{4k}\right)$

**46.** $\displaystyle\sum_{k=1}^{\infty} \left(\frac{1}{3^k}-\frac{1}{4^k}\right)$

**47.** $\displaystyle\sum_{k=1}^{\infty} \sin\left(\frac{\pi}{2}k\right)$

**48.** $\displaystyle\sum_{k=1}^{\infty} \sec(\pi k)$

**49.** $\displaystyle\sum_{k=3}^{\infty} \frac{k+1}{k-2}$

**50.** $\displaystyle\sum_{k=5}^{\infty} \frac{2k^5+3}{k^5-4k^4}$

**51.** $\displaystyle\sum_{k=2}^{\infty} \frac{1}{k(\ln k)^{1/2}}$

**52.** $\displaystyle\sum_{k=2}^{\infty} \frac{1}{k(\ln k)^2}$

**53.** $\displaystyle\sum_{k=3}^{\infty} \frac{2k}{k^2-4}$

**54.** $\displaystyle\sum_{k=1}^{\infty} \frac{1}{(2k-1)(2k)}$

*In Problems 55–64, find bounds for the sum of each convergent series.*

**55.** $\displaystyle\sum_{k=1}^{\infty} \frac{1}{k^2}$

**56.** $\displaystyle\sum_{k=1}^{\infty} \frac{1}{k^4}$

**57.** $\displaystyle\sum_{k=1}^{\infty} \frac{1}{k^e}$

**58.** $\displaystyle\sum_{k=1}^{\infty} \frac{1}{k^{\sqrt{2}}}$

**59.** $1 + \dfrac{1}{2\sqrt{2}} + \dfrac{1}{3\sqrt{3}} + \dfrac{1}{4\sqrt{4}} + \cdots$

**60.** $1 + \dfrac{1}{4\sqrt{2}} + \dfrac{1}{9\sqrt{3}} + \dfrac{1}{16\sqrt{4}} + \cdots$

**61.** $\displaystyle\sum_{k=1}^{\infty} ke^{-k^2}$

**62.** $\displaystyle\sum_{k=1}^{\infty} k^2 e^{-3k^3}$

**63.** $\displaystyle\sum_{k=1}^{\infty} \frac{1}{k^2+9}$

**64.** $\displaystyle\sum_{k=1}^{\infty} \frac{1}{4k^2+1}$

## Applications and Extensions

**65. Integral Test** Use the Integral Test to show that the series $\displaystyle\sum_{k=2}^{\infty} \frac{1}{k(\ln k)^p}$ converges if and only if $p > 1$.

**66. Integral Test** Use the Integral Test to show that the series $\displaystyle\sum_{k=3}^{\infty} \frac{1}{k(\ln k)[\ln(\ln k)^p]}$ converges if and only if $p > 1$.

**67. Faulty Logic** Let $S = 1 + 2 + 4 + 8 + \cdots$. Then

$$2S = 2 + 4 + 8 + 16 + \cdots = -1 + (1 + 2 + 4 + \cdots) = -1 + S$$

Therefore,

$$S = 1 + 2 + 4 + 8 + \cdots = -1$$

What went wrong here?

**68.** Find examples to show that the series $\displaystyle\sum_{k=1}^{\infty} (a_k + b_k)$ and $\displaystyle\sum_{k=1}^{\infty} (a_k - b_k)$ may converge or diverge if $\displaystyle\sum_{k=1}^{\infty} a_k$ and $\displaystyle\sum_{k=1}^{\infty} b_k$ each diverge.

**69. Approximating $\pi^2$** The *p*-series $\displaystyle\sum_{k=1}^{\infty} \frac{1}{k^2}$ converges to $\dfrac{\pi^2}{6}$. Use the first hundred terms of the series to approximate $\pi^2$.

**70.** The $p$-series $\displaystyle\sum_{k=1}^{\infty}\frac{1}{k^3}$ converges.

    **(a)** Find an interval of width 1 that contains the sum $\displaystyle\sum_{k=1}^{\infty}\frac{1}{k^3}$.

    **(b)** Use the first hundred terms of the series to approximate the sum.

**CAS** *In Problems 71–73, use the Integral Test to determine whether each series converges or diverges.*

**71.** $\displaystyle\sum_{k=1}^{\infty}\left(k^6 e^{-k}\right)$    **72.** $\displaystyle\sum_{k=1}^{\infty}\frac{k+3}{k^2+6k+7}$    **73.** $\displaystyle\sum_{k=2}^{\infty}\frac{5k+6}{k^3-1}$

**CAS** **74.** The $p$-series $\displaystyle\sum_{k=1}^{\infty}\frac{1}{k^p}$ converges for $p > 1$ and diverges

    for $0 < p \le 1$. The series $\displaystyle\sum_{k=1}^{\infty}a_k = \sum_{k=1}^{\infty}\frac{1}{k^{0.99}}$ diverges and the

    series $\displaystyle\sum_{k=1}^{\infty}b_k = \sum_{k=1}^{\infty}\frac{1}{k^{1.01}}$ converges.

    **(a)** Find the partial sums $S_{10}$, $S_{1000}$, and $S_{100,000}$ for each series.

    **(b)** Explain the results found in (a).

**75.** Show that the sum of two convergent series is a convergent series.

**76.** Show that for a nonzero real number $c$, if $\displaystyle\sum_{k=1}^{\infty}a_k = S$ is a

    convergent series, then the series $\displaystyle\sum_{k=1}^{\infty}(ca_k) = cS$.

**77.** Suppose $\displaystyle\sum_{k=N+1}^{\infty}a_k = S$ and $a_1 + a_2 + \cdots + a_N = K$.

    Prove that $\displaystyle\sum_{k=1}^{\infty}a_k$ converges and its sum is $S + K$.

**78.** If $\displaystyle\sum_{k=1}^{\infty}a_k$ converges and $\displaystyle\sum_{k=1}^{\infty}b_k$ diverges, then prove

    that $\displaystyle\sum_{k=1}^{\infty}(a_k + b_k)$ diverges.

**79.** Suppose $\displaystyle\sum_{k=1}^{\infty}a_k$ converges, and $a_n > 0$ for all $n$.

    Show that $\displaystyle\sum_{k=1}^{\infty}\frac{a_k}{1+a_k}$ converges.

### Challenge Problems

**80.** Determine whether the series $\displaystyle\sum_{k=1}^{\infty}\frac{1}{k\,\ln\left(1+\dfrac{1}{k}\right)}$ converges.

**81.** **Integral Test** Use the Integral Test to show that the

    series $\displaystyle\sum_{k=2}^{\infty}\frac{1}{(\ln k)^p}$ diverges for all numbers $p$.

**82.** For what positive integers $p$ and $q$ does the series $\displaystyle\sum_{k=2}^{\infty}\frac{(\ln k)^q}{k^p}$

    converge?

**83.** Find all real numbers $x$ for which $\displaystyle\sum_{k=1}^{\infty}k^x$ converges. Express your

    answer using interval notation.

**84.** Consider the finite sum $\displaystyle S_n = \sum_{k=1}^{n}\frac{1}{1+k^2}$.

    **(a)** By comparing $S_n$ with an appropriate integral, show that $S_n \le \tan^{-1}n$ for $n \ge 1$.

    **(b)** Use (a) to deduce that $\displaystyle\sum_{k=1}^{\infty}\frac{1}{1+k^2}$ converges.

    **(c)** Prove that $\dfrac{\pi}{4} \le \displaystyle\sum_{k=1}^{\infty}\frac{1}{1+k^2} \le \dfrac{\pi}{2}$.

**85.** **Riemann's zeta function** is defined as

$$\zeta(s) = 1 + \frac{1}{2^s} + \frac{1}{3^s} + \cdots \qquad \text{for } s > 1$$

    As mentioned on page 608, Euler showed that $\zeta(2) = \dfrac{\pi^2}{6}$. He also found the value of the zeta function for many other even values of $s$. As of now, no one knows the value of the zeta function for odd values of $s$. However, it is not too difficult to approximate these values, as this problem demonstrates.

    **(a)** Find $\displaystyle\sum_{k=1}^{10}\frac{1}{k^3}$.

    **(b)** Using integrals in a way analogous to their use in the proof of the Integral Test, find upper and lower bounds for $\displaystyle\sum_{k=1}^{\infty}\frac{1}{k^3}$.

    **(c)** What can you conclude about $\zeta(3)$?

**86.** **(a)** By considering graphs like those shown below, show that if $f$ is decreasing, positive, and continuous, then

$$f(n+1)+\cdots+f(m) \le \int_n^m f(x)\,dx \le f(n)+\cdots+f(m-1)$$

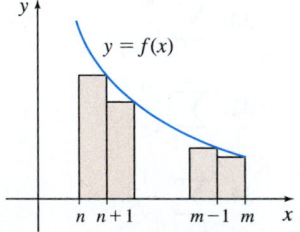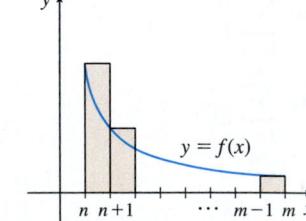

    **(b)** Under the assumption of (a), prove that if $\displaystyle\sum_{k=1}^{\infty}f(k)$ converges, then

$$\sum_{k=n+1}^{\infty}f(k) \le \int_n^{\infty}f(x)\,dx \le \sum_{k=n}^{\infty}f(k)$$

    **(c)** Let $f(x) = \dfrac{1}{x^2}$. Use the inequality in (b) to determine *exactly* how many terms of the series $\displaystyle\sum_{k=1}^{\infty}\frac{1}{k^2}$ one must use to have $|\,\text{Error}\,| < \dfrac{1}{2}\cdot 10^{-2}$. How many terms must one take to have $|\,\text{Error}\,| < \dfrac{1}{2}\cdot 10^{-10}$?

    *Source*: Based on an article by R. P. Boas, Jr., *American Mathematical Monthly*, "Partial Sums of Infinite Series, and How They Grow," Vol. 84, No. 4 (April 1977), pp. 237–258.

# 8.4 Comparison Tests

**OBJECTIVES** *When you finish this section, you should be able to:*

1 Use Comparison Tests for Convergence and Divergence (p. 613)
2 Use the Limit Comparison Test (p. 614)

## 1 Use Comparison Tests for Convergence and Divergence

We can determine whether a series converges or diverges by comparing it to a series whose behavior we already know. Suppose $\sum_{k=1}^{\infty} b_k$ is a series of positive terms that is known to converge to $B$, and we want to determine if the series of positive terms $\sum_{k=1}^{\infty} a_k$ converges. If term-by-term $a_k \leq b_k$, that is, if

$$a_1 \leq b_1, \quad a_2 \leq b_2, \ldots, \quad a_n \leq b_n, \ldots$$

then it follows that

$$\left( n\text{th partial sum of } \sum_{k=1}^{\infty} a_k \right) \leq \left( n\text{th partial sum of } \sum_{k=1}^{\infty} b_k \right) < B$$

**RECALL** The General Convergence Test: An infinite series of positive terms converges if and only if its sequence of partial sums is bounded.

This shows that the sequence of partial sums $\{S_n\}$ of $\sum_{k=1}^{\infty} a_k$ is bounded, so by the General Convergence Test, $\sum_{k=1}^{\infty} a_k$ must also converge. This proves the *Comparison Test for Convergence.*

> **THEOREM Comparison Test for Convergence**
>
> If $0 < a_k \leq b_k$ for all $k$ and $\sum_{k=1}^{\infty} b_k$ converges, then $\sum_{k=1}^{\infty} a_k$ converges.

It is important to remember that the early terms in a series have no effect on the convergence or divergence of a series. In fact, the Comparison Test for Convergence is true if $0 < a_n \leq b_n$ for all $n \geq N$, where $N$ is some suitably selected integer. We use this in the next example by ignoring the first term.

 CALC CLIP

**EXAMPLE 1 Using the Comparison Test for Convergence**

Show that the series below converges:

$$\sum_{k=1}^{\infty} \frac{1}{k^k} = 1 + \frac{1}{2^2} + \frac{1}{3^3} + \cdots + \frac{1}{n^n} + \cdots$$

**Solution** We know that the geometric series $\sum_{k=1}^{\infty} \frac{1}{2^k}$ converges since $|r| = \frac{1}{2} < 1$.

Now $\frac{1}{n^n} \leq \frac{1}{2^n}$ for all $n \geq 2$. So, except for the first term, each term of the series $\sum_{k=1}^{\infty} \frac{1}{k^k}$ is less than or equal to the corresponding term of a convergent geometric series. By the Comparison Test for Convergence, the series $\sum_{k=1}^{\infty} \frac{1}{k^k}$ converges. ∎

**NOTE** Do you see why $\frac{1}{n^n} \leq \frac{1}{2^n}$?

$$n \geq 2 \Rightarrow n^n \geq 2^n \Rightarrow \frac{1}{n^n} \leq \frac{1}{2^n}.$$

**NOW WORK** Problem 5.

In Example 1, you may have asked, "How did you know that the given series should be compared to $\sum_{k=1}^{\infty} \frac{1}{2^k}$?" The easy answer is to try a series that you know converges, such as a geometric series with $|r| < 1$ or a $p$-series with $p > 1$. The honest answer is that only practice and experience will guide you.

There is also a comparison test for divergence. Suppose $\sum\limits_{k=1}^{\infty} c_k$ is a series of positive terms we know diverges, and $\sum\limits_{k=1}^{\infty} a_k$ is the series of positive terms to be tested. If term-by-term $a_k \geq c_k$, that is, if

$$a_1 \geq c_1, \ a_2 \geq c_2, \ldots, \ a_n \geq c_n, \ldots$$

then it follows that

$$\left( n\text{th partial sum of } \sum_{k=1}^{\infty} a_k \right) \geq \left( n\text{th partial sum of } \sum_{k=1}^{\infty} c_k \right)$$

But we know that the sequence of partial sums of $\sum\limits_{k=1}^{\infty} c_k$ is unbounded. So, the sequence of partial sums of $\sum\limits_{k=1}^{\infty} a_k$ is also unbounded, and $\sum\limits_{k=1}^{\infty} a_k$ diverges.

**THEOREM Comparison Test for Divergence**

If $0 < c_k \leq a_k$ for all $k$ and $\sum\limits_{k=1}^{\infty} c_k$ diverges, then $\sum\limits_{k=1}^{\infty} a_k$ diverges.

**EXAMPLE 2  Using the Comparison Test for Divergence**

Show that the series $\sum\limits_{k=1}^{\infty} \dfrac{k+3}{k(k+2)}$ diverges.

**Solution** Since $\dfrac{n+3}{n+2} > 1$, it follows that

$$\frac{n+3}{n(n+2)} = \frac{n+3}{n+2} \cdot \frac{1}{n} > \frac{1}{n}$$

Term by term, the terms of the series $\sum\limits_{k=1}^{\infty} \dfrac{k+3}{k(k+2)}$ are greater than the terms of the harmonic series $\sum\limits_{k=1}^{\infty} \dfrac{1}{k}$, which diverges. It follows from the Comparison Test for Divergence that $\sum\limits_{k=1}^{\infty} \dfrac{k+3}{k(k+2)}$ diverges. ∎

**NOW WORK** Problem 9.

## 2 Use the Limit Comparison Test

The Comparison Tests for Convergence and Divergence are algebraic tests that require certain inequalities hold. Obtaining such inequalities can be challenging. The *Limit Comparison Test* requires certain *conditions* on the limit of a ratio. You may find this comparison test easier to use.

**THEOREM Limit Comparison Test**

Suppose $\sum\limits_{k=1}^{\infty} a_k$ and $\sum\limits_{k=1}^{\infty} b_k$ are both series of positive terms.

- If $\lim\limits_{n \to \infty} \dfrac{a_n}{b_n} = L, 0 < L < \infty$, then both series converge or both series diverge.

- If $\lim\limits_{n \to \infty} \dfrac{a_n}{b_n} = 0$ and if $\sum\limits_{k=1}^{\infty} b_k$ converges, then $\sum\limits_{k=1}^{\infty} a_k$ converges.

- If $\lim\limits_{n \to \infty} \dfrac{a_n}{b_n} = \infty$ and if $\sum\limits_{k=1}^{\infty} b_k$ diverges, then $\sum\limits_{k=1}^{\infty} a_k$ diverges.

**EXAMPLE 3**   **Using the Limit Comparison Test**

Determine whether the series $\sum\limits_{k=1}^{\infty} \dfrac{k+1}{k(k+2)}$ converges or diverges.

**Solution** It is tempting to try the same technique that we used in Example 2 and compare $\sum\limits_{k=1}^{\infty} \dfrac{k+1}{k(k+2)}$ to the harmonic series $\sum\limits_{k=1}^{\infty} \dfrac{1}{k}$. Let's see what happens.

$$\frac{n+1}{n(n+2)} = \frac{n+1}{n+2} \cdot \frac{1}{n} < \frac{1}{n} \quad \text{since} \quad \frac{n+1}{n+2} < 1$$

The conditions required for the Comparison Test for Divergence are not met, so no information about $\sum\limits_{k=1}^{\infty} \dfrac{k+1}{k(k+2)}$ is obtained.

Now use the Limit Comparison Test and compare the series $\sum\limits_{k=1}^{\infty} \dfrac{k+1}{k(k+2)}$ to the harmonic series $\sum\limits_{k=1}^{\infty} \dfrac{1}{k}$. Then with $a_n = \dfrac{n+1}{n(n+2)}$ and $b_n = \dfrac{1}{n}$, we have

$$\lim_{n\to\infty} \frac{a_n}{b_n} = \lim_{n\to\infty} \frac{\dfrac{n+1}{n(n+2)}}{\dfrac{1}{n}} = \lim_{n\to\infty} \frac{n+1}{n+2} = 1$$

Since the limit is a positive number, and $\sum\limits_{k=1}^{\infty} \dfrac{1}{k}$ diverges, by the Limit Comparison Test, $\sum\limits_{k=1}^{\infty} \dfrac{k+1}{k(k+2)}$ also diverges. ∎

**EXAMPLE 4**   **Using the Limit Comparison Test with a $p$-Series**

Determine whether the series $\sum\limits_{k=1}^{\infty} \dfrac{1}{2k^{3/2}+5}$ converges or diverges.

**Solution** We choose an appropriate $p$-series to use for comparison by examining the behavior of the series for large values of $n$:

$$\frac{1}{2n^{3/2}+5} = \frac{1}{n^{3/2}\left(2 + \dfrac{5}{n^{3/2}}\right)} = \frac{1}{n^{3/2}}\left(\frac{1}{2 + \dfrac{5}{n^{3/2}}}\right) \approx \underset{\substack{\uparrow \\ \text{for large } n}}{\frac{1}{n^{3/2}}}\left(\frac{1}{2}\right)$$

This leads us to choose the $p$-series $\sum\limits_{k=1}^{\infty} \dfrac{1}{k^{3/2}}$, which converges, and use the Limit Comparison Test with

$$a_n = \frac{1}{2n^{3/2}+5} \qquad \text{and} \qquad b_n = \frac{1}{n^{3/2}}$$

$$\lim_{n\to\infty} \frac{a_n}{b_n} = \lim_{n\to\infty} \frac{\dfrac{1}{2n^{3/2}+5}}{\dfrac{1}{n^{3/2}}} = \lim_{n\to\infty} \frac{n^{3/2}}{2n^{3/2}+5} = \lim_{n\to\infty} \frac{1}{2 + \dfrac{5}{n^{3/2}}} = \frac{1}{2}$$

Since the limit is a positive number and the $p$-series $\sum\limits_{k=1}^{\infty} \dfrac{1}{k^{3/2}}$ converges, then by the Limit Comparison Test, $\sum\limits_{k=1}^{\infty} \dfrac{1}{2k^{3/2}+5}$ also converges. ∎

**NOW WORK** Problem 15.

EXAMPLE 5   Using the Limit Comparison Test with a $p$-Series

Determine whether the series $\displaystyle\sum_{k=1}^{\infty} \frac{3\sqrt{k}+2}{\sqrt{k^3+3k^2+1}}$ converges or diverges.

**Solution** We choose a $p$-series for comparison by examining how the terms of the series behave for large values of $n$:

$$\frac{3\sqrt{n}+2}{\sqrt{n^3+3n^2+1}} = \frac{\sqrt{n}\left(3+\dfrac{2}{\sqrt{n}}\right)}{\sqrt{n^3}\left(\sqrt{1+\dfrac{3}{n}+\dfrac{1}{n^3}}\right)} = \frac{n^{1/2}}{n^{3/2}}\frac{3+\dfrac{2}{\sqrt{n}}}{\sqrt{1+\dfrac{3}{n}+\dfrac{1}{n^3}}} \underset{\substack{\uparrow \\ \text{for large } n}}{\approx} \frac{1}{n}(3)$$

So, we compare the series $\displaystyle\sum_{k=1}^{\infty} \frac{3\sqrt{k}+2}{\sqrt{k^3+3k^2+1}}$ to the harmonic series $\displaystyle\sum_{k=1}^{\infty}\frac{1}{k}$, which diverges, and use the Limit Comparison Test with

$$a_n = \frac{3\sqrt{n}+2}{\sqrt{n^3+3n^2+1}} \qquad \text{and} \qquad b_n = \frac{1}{n}$$

$$\lim_{n\to\infty} \frac{a_n}{b_n} = \lim_{n\to\infty} \frac{\dfrac{3\sqrt{n}+2}{\sqrt{n^3+3n^2+1}}}{\dfrac{1}{n}} = \lim_{n\to\infty}\frac{n\left(3\sqrt{n}+2\right)}{\sqrt{n^3+3n^2+1}} = \lim_{n\to\infty}\frac{3n^{3/2}+2n}{\sqrt{n^3+3n^2+1}}$$

$$= \lim_{n\to\infty}\frac{3n^{3/2}}{n^{3/2}} = 3$$

Since the limit is a positive real number and the $p$-series $\displaystyle\sum_{k=1}^{\infty}\frac{1}{k}$ diverges, then by the

Limit Comparison Test, $\displaystyle\sum_{k=1}^{\infty}\frac{3\sqrt{k}+2}{\sqrt{k^3+3k^2+1}}$ also diverges. ∎

The Limit Comparison Test is quite versatile for comparing algebraically complex series to a $p$-series. To find the correct choice of the $p$-series to use in the comparison, examine the behavior of the terms of the series for large values of $n$. For example,

• To test $\displaystyle\sum_{k=1}^{\infty}\frac{1}{3k^2+5k+2}$, use the $p$-series $\displaystyle\sum_{k=1}^{\infty}\frac{1}{k^2}$, because for large $n$,

$$\frac{1}{3n^2+5n+2} = \frac{1}{n^2\left(3+\dfrac{5}{n}+\dfrac{2}{n^2}\right)} \underset{\substack{\uparrow \\ \text{for large } n}}{\approx} \frac{1}{n^2}\left(\frac{1}{3}\right)$$

• To test $\displaystyle\sum_{k=1}^{\infty}\frac{2k^2+5}{3k^3-5k^2+2}$, use the $p$-series $\displaystyle\sum_{k=1}^{\infty}\frac{1}{k}$, because for large $n$,

$$\frac{2n^2+5}{3n^3-5n^2+2} = \frac{n^2\left(2+\dfrac{5}{n^2}\right)}{n^3\left(3-\dfrac{5}{n}+\dfrac{2}{n^3}\right)} = \frac{1}{n}\left(\frac{2+\dfrac{5}{n^2}}{3-\dfrac{5}{n}+\dfrac{2}{n^3}}\right) \underset{\substack{\uparrow \\ \text{for large } n}}{\approx} \frac{1}{n}\left(\frac{2}{3}\right)$$

• To test $\displaystyle\sum_{k=1}^{\infty}\frac{\sqrt{3k+1}}{\sqrt{4k^2-2k+1}}$, use the $p$-series $\displaystyle\sum_{k=1}^{\infty}\frac{1}{k^{1/2}}$, because for large $n$,

$$\frac{\sqrt{3n+1}}{\sqrt{4n^2-2n+1}} = \frac{\sqrt{n}\sqrt{3+\dfrac{1}{n}}}{\sqrt{n^2}\sqrt{4-\dfrac{2}{n}+\dfrac{1}{n^2}}} = \frac{1}{\sqrt{n}}\left(\frac{\sqrt{3+\dfrac{1}{n}}}{\sqrt{4-\dfrac{2}{n}+\dfrac{1}{n^2}}}\right) \underset{\substack{\uparrow \\ \text{for large } n}}{\approx} \frac{1}{n^{1/2}}\left(\frac{\sqrt{3}}{2}\right)$$

NOW WORK Problem 17.

The geometric series $\sum_{k=1}^{\infty} ar^{k-1}$ is also a valuable series to use when the series to be tested has the index $k$ as an exponent.

---

**EXAMPLE 6** **Using the Limit Comparison Test with a Geometric Series**

Determine whether the series $\sum_{k=1}^{\infty} \dfrac{2^k}{3^k + 5}$ converges or diverges.

**Solution** The index $k$ of the series $\sum_{k=1}^{\infty} \dfrac{2^k}{3^k + 5}$ is an exponent. Since for large $n$,

$$\frac{2^n}{3^n + 5} \approx \frac{2^n}{3^n} = \left(\frac{2}{3}\right)^n$$

compare the series $\sum_{k=1}^{\infty} \dfrac{2^k}{3^k + 5}$ to the geometric series $\sum_{k=1}^{\infty} \left(\dfrac{2}{3}\right)^k$, which converges. With

$$a_n = \frac{2^n}{3^n + 5} \quad \text{and} \quad b_n = \left(\frac{2}{3}\right)^n$$

$$\lim_{n \to \infty} \frac{a_n}{b_n} = \lim_{n \to \infty} \frac{\dfrac{2^n}{3^n + 5}}{\left(\dfrac{2}{3}\right)^n} = \lim_{n \to \infty} \left(\frac{2^n}{3^n + 5} \cdot \frac{3^n}{2^n}\right) = \lim_{n \to \infty} \frac{3^n}{3^n + 5} = \lim_{n \to \infty} \frac{1}{1 + \dfrac{5}{3^n}} = 1$$

Since the limit is a positive number and the geometric series $\sum_{k=1}^{\infty} \left(\dfrac{2}{3}\right)^k$ converges, by the Limit Comparison Test the series $\sum_{k=1}^{\infty} \dfrac{2^k}{3^k + 5}$ also converges. ∎

**NOW WORK** Problem 25.

Using comparison tests to determine whether a series converges or diverges requires that you know convergent and divergent series to use in the comparison. Table 3 lists some series we have already encountered that are useful for this purpose.

**TABLE 3 Summary**

| Series | Convergent | Divergent |
|---|---|---|
| The geometric series $\sum_{k=1}^{\infty} ar^{k-1}$ | $\lvert r \rvert < 1$ | $\lvert r \rvert \geq 1$ |
| The harmonic series $\sum_{k=1}^{\infty} \dfrac{1}{k}$ | | Divergent |
| The $p$-series $\sum_{k=1}^{\infty} \dfrac{1}{k^p}$ | $p > 1$ | $0 < p \leq 1$ |
| The series $\sum_{k=1}^{\infty} \dfrac{1}{k^k}$ | Convergent | |

**Proof of the Limit Comparison Test**  If $\lim\limits_{n \to \infty} \dfrac{a_n}{b_n} = L > 0$, then for any $\varepsilon > 0$, there is

a positive real number $N$ so that $\left| \dfrac{a_n}{b_n} - L \right| < \varepsilon$, for all $n > N$. If we choose $\varepsilon = \dfrac{L}{2}$, then

$$\left| \frac{a_n}{b_n} - L \right| < \frac{L}{2} \qquad \text{for all } n > N$$

$$-\frac{L}{2} < \frac{a_n}{b_n} - L < \frac{L}{2} \qquad \text{for all } n > N$$

$$\frac{L}{2} < \frac{a_n}{b_n} < \frac{3L}{2} \qquad \text{for all } n > N$$

Since $b_n > 0$,

$$\frac{L}{2} b_n < a_n < \frac{3L}{2} b_n \qquad \text{for all } n > N$$

**RECALL** If $c$ is a nonzero real number and if $\sum\limits_{k=1}^{\infty} a_k = S$ is a convergent series, then the series $\sum\limits_{k=1}^{\infty} (ca_k)$ also converges. Furthermore,

$$\sum_{k=1}^{\infty} (ca_k) = c \sum_{k=1}^{\infty} a_k = cS$$

If $\sum\limits_{k=1}^{\infty} a_k$ converges, by the Comparison Test for Convergence, the series $\sum\limits_{k=1}^{\infty} \left( \dfrac{L}{2} b_k \right) = \dfrac{L}{2} \sum\limits_{k=1}^{\infty} b_k$ also converges. It follows that the series $\sum\limits_{k=1}^{\infty} b_k$ is also convergent.

If $\sum\limits_{k=1}^{\infty} b_k$ converges, then so does $\sum\limits_{k=1}^{\infty} \left( \dfrac{3L}{2} b_k \right)$. Since $a_n < \dfrac{3L}{2} b_n$, for all $n > N$, by the Comparison Test for Convergence, $\sum\limits_{k=1}^{\infty} a_k$ also converges.

Therefore, $\sum\limits_{k=1}^{\infty} a_k$ and $\sum\limits_{k=1}^{\infty} b_k$ converge together. Since one cannot converge and the other diverge, they must diverge together. ∎

The rest of the proof is left as an exercise (Problems 59 and 60).

# 8.4 Assess Your Understanding

## Concepts and Vocabulary

1. *Multiple Choice*  If each term of a series $\sum\limits_{k=1}^{\infty} a_k$ of positive terms is greater than or equal to the corresponding term of a known divergent series $\sum\limits_{k=1}^{\infty} c_k$ of positive terms, then the series $\sum\limits_{k=1}^{\infty} a_k$ is [**(a)** convergent **(b)** divergent].

2. *True or False*  Suppose $\sum\limits_{k=1}^{\infty} a_k$ and $\sum\limits_{k=1}^{\infty} b_k$ are both series of positive terms. The series $\sum\limits_{k=1}^{\infty} a_k$ and the series $\sum\limits_{k=1}^{\infty} b_k$ both converge or both diverge if $\lim\limits_{n \to \infty} \dfrac{a_n}{b_n} = 0$.

3. *True or False*  If $\sum\limits_{k=1}^{\infty} a_k$ and $\sum\limits_{k=1}^{\infty} b_k$ are both series of positive terms and if $\lim\limits_{n \to \infty} \dfrac{a_n}{b_n} = L$, where $L$ is a positive real number, then the series to be tested converges.

4. *True or False*  Since the $p$-series $\sum\limits_{k=1}^{\infty} \dfrac{1}{k^{3/2}}$ converges and $\lim\limits_{n \to \infty} \dfrac{\dfrac{1}{2n^{3/2} + 5}}{\dfrac{1}{n^{3/2}}} = \dfrac{1}{2}$, then by the Limit Comparison Test, the series $\sum\limits_{k=1}^{\infty} \dfrac{1}{2k^{3/2} + 5}$ converges to $\dfrac{1}{2}$.

## Skill Building

*In Problems 5–14, use the Comparison Tests for Convergence or Divergence to determine whether each series converges or diverges.*

5. $\sum\limits_{k=1}^{\infty} \dfrac{1}{k(k+1)}$ : by comparing it with $\sum\limits_{k=1}^{\infty} \dfrac{1}{k^2}$

6. $\sum\limits_{k=1}^{\infty} \dfrac{1}{(k+2)^2}$ : by comparing it with $\sum\limits_{k=1}^{\infty} \dfrac{1}{k^2}$

7. $\sum\limits_{k=2}^{\infty} \dfrac{4^k}{7^k + 1}$ : by comparing it with $\sum\limits_{k=2}^{\infty} \left( \dfrac{4}{7} \right)^k$

---

**1.** = NOW WORK problem          = Graphing technology recommended          CAS = Computer Algebra System recommended

8. $\displaystyle\sum_{k=1}^{\infty} \frac{1}{(2k-1)(2^k)}$ : by comparing it with $\displaystyle\sum_{k=1}^{\infty} \frac{1}{2^k}$

9. $\displaystyle\sum_{k=2}^{\infty} \frac{1}{\sqrt{k(k-1)}}$ : by comparing it with $\displaystyle\sum_{k=2}^{\infty} \frac{1}{k}$

10. $\displaystyle\sum_{k=2}^{\infty} \frac{\sqrt{k}}{k-1}$ : by comparing it with $\displaystyle\sum_{k=2}^{\infty} \frac{1}{\sqrt{k}}$

11. $\displaystyle\sum_{k=1}^{\infty} \frac{1}{k(k+1)(k+2)}$

12. $\displaystyle\sum_{k=1}^{\infty} \frac{6}{5k-2}$

13. $\displaystyle\sum_{k=1}^{\infty} \frac{\sin^2 k}{k^\pi}$

14. $\displaystyle\sum_{k=1}^{\infty} \frac{\cos^2 k}{k^2+1}$

*In Problems 15–36, use the Limit Comparison Test to determine whether each series converges or diverges.*

15. $\displaystyle\sum_{k=1}^{\infty} \frac{1}{(k+1)(k+2)}$

16. $\displaystyle\sum_{k=1}^{\infty} \frac{1}{k^2+1}$

17. $\displaystyle\sum_{k=1}^{\infty} \frac{1}{\sqrt{k^2+1}}$

18. $\displaystyle\sum_{k=1}^{\infty} \frac{\sqrt{k}}{k+4}$

19. $\displaystyle\sum_{k=1}^{\infty} \frac{3\sqrt{k}+2}{2k^2+5}$

20. $\displaystyle\sum_{k=2}^{\infty} \frac{3\sqrt{k}+2}{2k-3}$

21. $\displaystyle\sum_{k=2}^{\infty} \frac{1}{k\sqrt{k^2-1}}$

22. $\displaystyle\sum_{k=1}^{\infty} \frac{k}{(2k-1)^2}$

23. $\displaystyle\sum_{k=1}^{\infty} \frac{3^k}{2^k+5}$

24. $\displaystyle\sum_{k=1}^{\infty} \frac{8^k}{5+3^k}$

25. $\displaystyle\sum_{k=1}^{\infty} \frac{3^k+4}{5^k+3}$

26. $\displaystyle\sum_{k=1}^{\infty} \frac{2^k+1}{7^k+4}$

27. $\displaystyle\sum_{k=1}^{\infty} \frac{2\cdot 3^k+5}{4\cdot 8^k}$

28. $\displaystyle\sum_{k=1}^{\infty} \frac{-4\cdot 5^k}{2\cdot 6^k}$

29. $\displaystyle\sum_{k=1}^{\infty} \frac{8\cdot 3^k+4\cdot 5^k}{8^{k-1}}$

30. $\displaystyle\sum_{k=1}^{\infty} \frac{2\cdot 5^k+3^k}{2^{k-1}}$

31. $\displaystyle\sum_{k=1}^{\infty} \frac{3k+4}{k2^k}$

32. $\displaystyle\sum_{k=2}^{\infty} \frac{k-1}{k2^k}$

33. $\displaystyle\sum_{k=1}^{\infty} \frac{1}{2^k+1}$

34. $\displaystyle\sum_{k=1}^{\infty} \frac{5}{3^k+2}$

35. $\displaystyle\sum_{k=1}^{\infty} \frac{k+5}{k^k+1}$

36. $\displaystyle\sum_{k=1}^{\infty} \frac{5}{k^k+1}$

*In Problems 37–48, use any of the comparison tests to determine whether each series converges or diverges.*

37. $\displaystyle\sum_{k=1}^{\infty} \frac{6k}{5k^2+2}$

38. $\displaystyle\sum_{k=2}^{\infty} \frac{6k+3}{2k^3-2}$

39. $\displaystyle\sum_{k=1}^{\infty} \frac{7+k}{(1+k^2)^4}$

40. $\displaystyle\sum_{k=1}^{\infty} \left(\frac{7+k}{1+k^2}\right)^4$

41. $\displaystyle\sum_{k=1}^{\infty} \frac{e^{1/k}}{k}$

42. $\displaystyle\sum_{k=1}^{\infty} \frac{1}{1+e^k}$

43. $\displaystyle\sum_{k=1}^{\infty} \frac{\left(1+\dfrac{1}{k}\right)^2}{e^k}$

44. $\displaystyle\sum_{k=1}^{\infty} \frac{1}{k\,2^k}$

45. $\displaystyle\sum_{k=1}^{\infty} \frac{1+\sqrt{k}}{k}$

46. $\displaystyle\sum_{k=1}^{\infty} \frac{1+3\sqrt{k}}{k^2}$

47. $\displaystyle\sum_{k=1}^{\infty} \left(\frac{1}{2}\right)^k \sin^2 k$

48. $\displaystyle\sum_{k=1}^{\infty} \frac{\tan^{-1} k}{k^3}$

## Applications and Extensions

*In Problems 49–56, determine whether each series converges or diverges.*

49. $\displaystyle\sum_{k=2}^{\infty} \frac{2}{k^3 \ln k}$

50. $\displaystyle\sum_{k=2}^{\infty} \frac{1}{\sqrt{k}(\ln k)^4}$

51. $\displaystyle\sum_{k=2}^{\infty} \frac{\ln k}{k+3}$

52. $\displaystyle\sum_{k=2}^{\infty} \frac{(\ln k)^2}{k^{5/2}}$

53. $\displaystyle\sum_{k=1}^{\infty} \sin \frac{1}{k}$

54. $\displaystyle\sum_{k=1}^{\infty} \tan \frac{1}{k}$

55. $\displaystyle\sum_{k=1}^{\infty} \frac{1}{k!}$

56. $\displaystyle\sum_{k=1}^{\infty} \frac{k!}{k^k}$

57. It is known that $\displaystyle\sum_{k=1}^{\infty} \frac{1}{k^2}$ is a convergent *p*-series.

 (a) Use the Comparison Test for Convergence with $\displaystyle\sum_{k=1}^{\infty} \frac{1}{k^2}$ to show that $\displaystyle\sum_{k=1}^{\infty} \frac{1}{k^2+1}$ converges.

 (b) Explain why the Comparison Test for Convergence and $\displaystyle\sum_{k=2}^{\infty} \frac{1}{k^2}$ cannot be used to show $\displaystyle\sum_{k=2}^{\infty} \frac{1}{k^2-1}$ converges.

 (c) Can the Limit Comparison Test be used to show $\displaystyle\sum_{k=2}^{\infty} \frac{1}{k^2-1}$ converges? If it cannot, explain why.

58. Show that any series of the form $\displaystyle\sum_{k=1}^{\infty} \frac{d_k}{10^k}$, where the $d_k$ are digits $(0, 1, 2, \ldots, 9)$, converges.

59. Suppose the series $\displaystyle\sum_{k=1}^{\infty} a_k$ of positive terms is to be tested for convergence or divergence, and the series $\displaystyle\sum_{k=1}^{\infty} b_k$ of positive terms diverges. Show that if $\displaystyle\lim_{n\to\infty} \frac{a_n}{b_n} = \infty$, then $\displaystyle\sum_{k=1}^{\infty} a_k$ diverges.

60. Suppose the series $\displaystyle\sum_{k=1}^{\infty} a_k$ of positive terms is to be tested for convergence or divergence, and the series $\displaystyle\sum_{k=1}^{\infty} b_k$ of positive terms converges. Show that if $\displaystyle\lim_{n\to\infty} \frac{a_n}{b_n} = 0$, then $\displaystyle\sum_{k=1}^{\infty} a_k$ converges.

*In Problems 61–64, use the results of Problems 59 and 60 to determine whether each of the following series converges or diverges.*

61. $\displaystyle\sum_{k=2}^{\infty} \frac{1}{\ln k}$

62. $\displaystyle\sum_{k=2}^{\infty} \left(\frac{1}{\ln k}\right)^2$

63. $\displaystyle\sum_{k=1}^{\infty} \frac{\ln k}{k^2}$

64. $\displaystyle\sum_{k=2}^{\infty} \frac{1}{(k \ln k)^2}$

**65. (a)** Show that $\displaystyle\sum_{k=1}^{\infty} \frac{\ln k}{k^p}$ converges for $p > 1$.

**(b)** Show that $\displaystyle\sum_{k=1}^{\infty} \frac{(\ln k)^r}{k^p}$, where $r$ is a positive number, converges for $p > 1$.

**66. (a)** Show that $\displaystyle\sum_{k=1}^{\infty} \frac{\ln k}{k^p}$ diverges for $0 < p \le 1$.

**(b)** Show that $\displaystyle\sum_{k=1}^{\infty} \frac{(\ln k)^r}{k^p}$, where $r$ is a positive real number and $0 < p \le 1$, diverges.

*In Problems 67–70, use the results of Problems 65 and 66, to determine whether each of the following series converges or diverges.*

**67.** $\displaystyle\sum_{k=1}^{\infty} \frac{\sqrt{\ln k}}{\sqrt{k}}$

**68.** $\displaystyle\sum_{k=1}^{\infty} \frac{\ln k}{k}$

**69.** $\displaystyle\sum_{k=1}^{\infty} \frac{(\ln k)^2}{\sqrt{k^3}}$

**70.** $\displaystyle\sum_{k=2}^{\infty} \frac{\ln k}{k^3}$

**71.** Use the Comparison Tests for Convergence or Divergence to show that the $p$-series $\displaystyle\sum_{k=1}^{\infty} \frac{1}{k^p}$:

**(a)** converges if $p > 1$.

**(b)** diverges if $0 < p \le 1$.

**72.** If the series $\displaystyle\sum_{k=1}^{\infty} a_k$ of positive terms converges, show that the series $\displaystyle\sum_{k=1}^{\infty} \frac{a_k}{k}$ also converges.

**73.** Show that $\displaystyle\sum_{k=1}^{\infty} \frac{1}{1 + 2^k}$ converges.

**Challenge Problems**

*In Problems 74 and 75, determine whether each series converges or diverges.*

**74.** $\displaystyle\sum_{k=2}^{\infty} \frac{\ln(2k+1)}{\sqrt{k^2 - 2}\sqrt{k^3 - 2k - 3}}$

**75.** $\displaystyle\sum_{k=1}^{\infty} \frac{\sqrt{k}}{\sqrt{(k^3 - k + 1)\ln(2k + 1)}}$

**76.** Show that the series $\displaystyle\sum_{k=1}^{\infty} \frac{1 + \sin k}{4^k}$ converges.

**77.** Show that the series $\displaystyle\sum_{k=1}^{\infty} \frac{1}{k^{1 + 1/k}}$ diverges.

**78. (a)** Determine whether the series $\displaystyle\sum_{k=1}^{\infty} \frac{k^2 - 3k - 2}{k^2(k+1)^2}$ converges or diverges.

**(b)** If it converges, find its sum.

# 8.5 Alternating Series; Absolute Convergence

**OBJECTIVES** *When you finish this section, you should be able to:*

**1** Determine whether an alternating series converges or diverges (p. 621)

**2** Approximate the sum of a convergent alternating series (p. 623)

**3** Use the Absolute Convergence Test (p. 625)

Sections 8.3 and 8.4, deal with series whose terms are all positive. We now investigate series that have some positive and some negative terms. The properties of series with positive terms often help to explain the properties of series that include terms with mixed signs.

The most common series with both positive and negative terms are *alternating series,* in which the terms alternate between positive and negative.

**DEFINITION** Alternating Series

An **alternating series** is a series either of the form

$$\sum_{k=1}^{\infty} (-1)^{k+1} a_k = a_1 - a_2 + a_3 - a_4 + \cdots$$

or of the form

$$\sum_{k=1}^{\infty} (-1)^k a_k = -a_1 + a_2 - a_3 + a_4 - \cdots$$

where $a_k > 0$ for all integers $k \ge 1$.

Two examples of alternating series are

$$1 - \frac{1}{2} + \frac{1}{3} - \frac{1}{4} + \cdots = \sum_{k=1}^{\infty} \frac{(-1)^{k+1}}{k} \quad \text{and} \quad -1 + \frac{1}{3!} - \frac{1}{5!} + \frac{1}{7!} - \cdots = \sum_{k=1}^{\infty} \frac{(-1)^k}{(2k-1)!}$$

## 1 Determine Whether an Alternating Series Converges or Diverges

To determine whether an alternating series converges, we use the *Alternating Series Test*.

---

**THEOREM** Alternating Series Test

If the numbers $a_k$, where $a_k > 0$, of an alternating series

$$\sum_{k=1}^{\infty} (-1)^{k+1} a_k \quad \text{or} \quad \sum_{k=1}^{\infty} (-1)^k a_k$$

satisfy the two conditions:

- $\displaystyle\lim_{n \to \infty} a_n = 0$
- the $a_k$ are nonincreasing; that is, $a_1 \geq a_2 \geq a_3 \geq \cdots \geq a_n \geq a_{n+1} \geq \cdots$

then the alternating series converges.

---

**NOTE** The Alternating Series Test is credited to Leibniz and is often called the **Leibniz Test** or **Leibniz Criterion.**

The Alternating Series Test also works if the $a_k$ are *eventually* nonincreasing.

**Proof** First consider partial sums with an even number of terms, and group them into pairs:

$$S_{2n} = a_1 - a_2 + a_3 - a_4 + \cdots + a_{2n-1} - a_{2n}$$
$$= (a_1 - a_2) + (a_3 - a_4) + \cdots + (a_{2n-1} - a_{2n})$$

Since $a_1 \geq a_2 \geq a_3 \geq a_4 \geq \cdots \geq a_{2n-1} \geq a_{2n} \geq a_{2n+1} \geq \cdots$, the difference within each pair of parentheses is either 0 or positive. So, the sequence $\{S_{2n}\}$ of partial sums is nondecreasing. That is,

$$S_2 \leq S_4 \leq S_6 \leq \cdots \leq S_{2n} \leq S_{2n+2} \leq \cdots \tag{1}$$

For partial sums with an odd number of terms, group the terms a little differently, and write

$$S_{2n+1} = a_1 - a_2 + a_3 - a_4 + a_5 - \cdots - a_{2n} + a_{2n+1}$$
$$= a_1 - (a_2 - a_3) - (a_4 - a_5) - \cdots - (a_{2n} - a_{2n+1})$$

Again, the difference within each pair of parentheses is either 0 or positive. But this time the difference is subtracted from the previous term, so the sequence $\{S_{2n+1}\}$ of partial sums is nonincreasing. That is,

$$S_1 \geq S_3 \geq S_5 \geq \cdots \geq S_{2n-1} \geq S_{2n+1} \geq \cdots \tag{2}$$

Now since $S_{2n+1} - S_{2n} = a_{2n+1} > 0$, we have

$$S_{2n} < S_{2n+1} \tag{3}$$

Combine (1), (2), and (3) as follows:

$$\underbrace{S_2 \leq S_4 \leq S_6 \leq \cdots \leq S_{2n}}_{(1)} \underbrace{<}_{(3)} \underbrace{S_{2n+1} \leq \cdots \leq S_5 \leq S_3 \leq S_1}_{(2)}$$

Figure 23 shows the sequences $\{S_{2n}\}$ and $\{S_{2n+1}\}$.

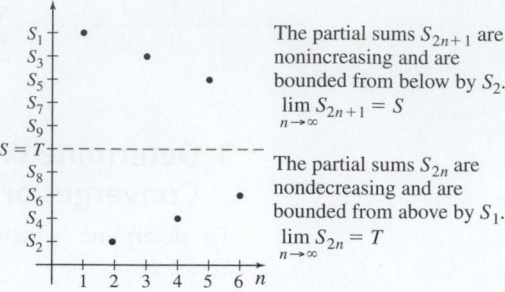

The partial sums $S_{2n+1}$ are nonincreasing and are bounded from below by $S_2$.
$\lim\limits_{n \to \infty} S_{2n+1} = S$

The partial sums $S_{2n}$ are nondecreasing and are bounded from above by $S_1$.
$\lim\limits_{n \to \infty} S_{2n} = T$

**Figure 23**

**RECALL** A bounded, monotonic sequence converges.

The sequence $S_2, S_4, S_6, \ldots$ is nondecreasing and bounded above by $S_1$. So, this sequence has a limit; call it $T$. Similarly, the sequence $S_1, S_3, S_5, \ldots$ is nonincreasing and is bounded from below by $S_2$, so it, too, has a limit; call it $S$. Since $S_{2n+1} - S_{2n} = a_{2n+1}$, we have

$$S - T = \lim_{n \to \infty} S_{2n+1} - \lim_{n \to \infty} S_{2n} = \lim_{n \to \infty} (S_{2n+1} - S_{2n}) = \lim_{n \to \infty} a_{2n+1}$$

Since $\lim\limits_{n \to \infty} a_{2n+1} = 0$, we find $S - T = 0$. This means that $S = T$, and the sequences $\{S_{2n}\}$ and $\{S_{2n+1}\}$ both converge to $S$. It follows that

$$\lim_{n \to \infty} S_n = S$$

and so the alternating series converges. ∎

**EXAMPLE 1    Showing the Alternating Harmonic Series Converges**

Show that the **alternating harmonic series** $\sum\limits_{k=1}^{\infty} \dfrac{(-1)^{k+1}}{k} = 1 - \dfrac{1}{2} + \dfrac{1}{3} - \dfrac{1}{4} + \cdots$ converges.

**Solution** We check the two conditions of the Alternating Series Test. Since $\lim\limits_{n \to \infty} a_n = \lim\limits_{n \to \infty} \dfrac{1}{n} = 0$, the first condition is met. Since $a_{n+1} = \dfrac{1}{n+1} < \dfrac{1}{n} = a_n$, the $a_n$ are nonincreasing, so the second condition is met. By the Alternating Series Test, the series converges. ∎

**NOTE** In Section 8.8 we show that the sum of the alternating harmonic series is ln 2.

**NOW WORK** Problem 7.

When using the Alternating Series Test, check the limit of the $n$th term first. If $\lim\limits_{n \to \infty} a_n \neq 0$, the Test for Divergence tells us immediately that the series diverges. For example, in testing the alternating series

$$\sum_{k=2}^{\infty} (-1)^k \frac{k}{k-1} = 2 - \frac{3}{2} + \frac{4}{3} - \frac{5}{4} + \cdots$$

we find $\lim\limits_{n \to \infty} a_n = \lim\limits_{n \to \infty} \dfrac{n}{n-1} = 1$. There is no need to look further. By the Test for Divergence, the series diverges.

**EXAMPLE 2    Using the Alternating Series Test**

Determine whether the series $\sum\limits_{k=1}^{\infty} (-1)^k \dfrac{2k^2 + 9}{k^2 + 1}$ converges or diverges.

**Solution** The series $\sum\limits_{k=1}^{\infty} (-1)^k \dfrac{2k^2 + 9}{k^2 + 1}$ is an alternating series, where

$$a_n = \frac{2n^2 + 9}{n^2 + 1}$$

We check $\lim\limits_{n \to \infty} a_n$.

$$\lim_{n \to \infty} a_n = \lim_{n \to \infty} \frac{2n^2 + 9}{n^2 + 1} = \lim_{n \to \infty} \frac{2 + \dfrac{9}{n^2}}{1 + \dfrac{1}{n^2}} = \frac{2}{1} = 2$$

Since $\lim\limits_{n \to \infty} a_n \neq 0$, by the Test for Divergence, the series diverges. ∎

**NOW WORK** Problem 9.

If $\lim\limits_{n \to \infty} a_n = 0$, then check whether the terms $a_k$ are nonincreasing. Recall there are three ways to verify whether the $a_k$ are nonincreasing:

- Use the Algebraic Difference test and show that $a_{n+1} - a_n \leq 0$ for all $n \geq 1$.
- Use the Algebraic Ratio test and show that $\dfrac{a_{n+1}}{a_n} \leq 1$ for all $n \geq 1$.
- Use the derivative of the related function $f$, defined for $x > 0$ and for which $a_n = f(n)$ for all $n$, and show that $f'(x) \leq 0$ for all $x > 0$.

**NOTE** If $\lim\limits_{n \to \infty} a_n = 0$, for an alternating series and if the terms are not nonincreasing, the Alternating Series Test gives no information. In these cases the alternating series may converge (Problem 61) or diverge (Problem 62).

---

**EXAMPLE 3** **Showing an Alternating Series Converges**

Show that the alternating series $\sum\limits_{k=0}^{\infty} \dfrac{(-1)^k}{(2k)!} = 1 - \dfrac{1}{2} + \dfrac{1}{24} - \dfrac{1}{720} + \cdots$ converges.

**Solution** First, $\lim\limits_{n \to \infty} a_n = \lim\limits_{n \to \infty} \dfrac{1}{(2n)!} = 0$, so the first condition is met. Next using the Algebraic Ratio test, we verify that the terms $a_k = \dfrac{1}{(2k)!}$ are nonincreasing. Since

$$\frac{a_{n+1}}{a_n} = \frac{\dfrac{1}{[2(n+1)]!}}{\dfrac{1}{(2n)!}} = \frac{(2n)!}{(2n+2)!} = \frac{(2n)!}{(2n+2)(2n+1)(2n)!}$$

$$= \frac{1}{(2n+2)(2n+1)} < 1 \qquad \text{for all } n \geq 1$$

the terms $a_k$ are nonincreasing. By the Alternating Series Test, the series converges. ∎

**NOW WORK** Problem 13.

## 2 Approximate the Sum of a Convergent Alternating Series

An important property of a convergent alternating series that satisfies the two conditions of the Alternating Series Test is that we can estimate the error made by using a partial sum $S_n$ to approximate the sum $S$ of the series.

**THEOREM** **Error Estimate for a Convergent Alternating Series**

For a convergent alternating series $\sum\limits_{k=1}^{\infty} (-1)^{k+1} a_k$ that satisfies the two conditions of the Alternating Series Test, the error $E_n$ in using the sum $S_n$ of the first $n$ terms as an approximation to the sum $S$ of the series is numerically less than or equal to the $(n+1)$st term of the series. That is,

$$\boxed{|E_n| \leq a_{n+1}}$$

**NOTE** The error estimate given here is only applicable to alternating series that satisfy the two conditions required in the Alternating Series Test.

**Proof** We follow the ideas used in the proof of the Alternating Series Test.

If $n$ is even, $S_n < S$, and

$$0 < S - S_n = a_{n+1} - (a_{n+2} - a_{n+3}) - \cdots \le a_{n+1}$$

If $n$ is odd, $S_n > S$, and

$$0 < S_n - S = a_{n+1} - (a_{n+2} - a_{n+3}) - \cdots \le a_{n+1}$$

In either case,

$$|E_n| = |S - S_n| \le a_{n+1} \qquad \blacksquare$$

In Example 1, we found that the alternating harmonic series $\sum\limits_{k=1}^{\infty} \dfrac{(-1)^{k+1}}{k}$ converges. Table 4 shows how slowly the series converges. For example, using an additional 900 terms from 99 to 999 reduces the estimated error by only $|E_{999} - E_{99}| = 0.009$.

**TABLE 4**

| Series | Number of Terms | Estimate of the Sum | Maximum Error |
|---|---|---|---|
| $\sum\limits_{k=1}^{\infty} \dfrac{(-1)^{k+1}}{k}$ | $n = 3$ | $S_3 = 1 - \dfrac{1}{2} + \dfrac{1}{3} \approx 0.833$ | $\|E_3\| \le \dfrac{1}{4} = 0.25$ |
| | $n = 9$ | $S_9 = \sum\limits_{k=1}^{9} \dfrac{(-1)^{k+1}}{k} \approx 0.746$ | $\|E_9\| \le \dfrac{1}{10} = 0.1$ |
| | $n = 99$ | $S_{99} = \sum\limits_{k=1}^{99} \dfrac{(-1)^{k+1}}{k} \approx 0.698$ | $\|E_{99}\| \le \dfrac{1}{100} = 0.01$ |
| | $n = 999$ | $S_{999} = \sum\limits_{k=1}^{999} \dfrac{(-1)^{k+1}}{k} \approx 0.694$ | $\|E_{999}\| \le \dfrac{1}{1000} = 0.001$ |

**EXAMPLE 4    Approximating the Sum of a Convergent Alternating Series**

Approximate the sum $S$ of the alternating series so that the error is less than or equal to 0.0001.

$$\sum_{k=0}^{\infty} \frac{(-1)^k}{(2k)!} = 1 - \frac{1}{2!} + \frac{1}{4!} - \frac{1}{6!} + \frac{1}{8!} - \cdots$$

**Solution** We proved in Example 3 that the series $\sum\limits_{k=0}^{\infty} \dfrac{(-1)^k}{(2k)!}$ converges by showing that it satisfies the conditions of the Alternating Series Test. The fifth term of the series, $\dfrac{1}{8!} = \dfrac{1}{40,320} \approx 0.000025$, is the first term less than or equal to 0.0001. This term represents an upper estimate to the error when the sum $S$ of the series is approximated by adding the first four terms. So,

$$S \approx \sum_{k=0}^{3} \frac{(-1)^k}{(2k)!} = 1 - \frac{1}{2!} + \frac{1}{4!} - \frac{1}{6!} = 1 - \frac{1}{2} + \frac{1}{24} - \frac{1}{720} = \frac{389}{720} \approx 0.540$$

approximates the sum with an error less than or equal to 0.0001. $\blacksquare$

**NOW WORK** Problems **23** and **29**.

## Absolute and Conditional Convergence

The ideas of *absolute convergence* and *conditional convergence* are used to describe the convergence or divergence of a series $\sum_{k=1}^{\infty} a_k$ in which the terms $a_k$ are sometimes positive and sometimes negative (not necessarily alternating).

> **DEFINITION  Absolute Convergence**
>
> A series $\sum_{k=1}^{\infty} a_k$ is **absolutely convergent** if the series
>
> $$\sum_{k=1}^{\infty} |a_k| = |a_1| + |a_2| + \cdots + |a_n| + \cdots$$
>
> is convergent.

## 3 Use the Absolute Convergence Test

> **THEOREM  Absolute Convergence Test**
>
> If a series $\sum_{k=1}^{\infty} a_k$ is absolutely convergent, then it is convergent.

**Proof** For any $n$,

$$-|a_n| \leq a_n \leq |a_n|$$

Adding $|a_n|$ yields

$$0 \leq a_n + |a_n| \leq 2 |a_n|$$

Since $\sum_{k=1}^{\infty} |a_k|$ converges, then $\sum_{k=1}^{\infty} (2|a_k|) = 2 \sum_{k=1}^{\infty} |a_k|$ also converges. By the Comparison Test for Convergence, $\sum_{k=1}^{\infty} (a_k + |a_k|)$ converges. But $a_n = (a_n + |a_n|) - |a_n|$.

Since $\sum_{k=1}^{\infty} a_k$ is the difference of two convergent series, it also converges. ∎

---

**EXAMPLE 5  Using the Absolute Convergence Test**

Determine whether the series $1 - \dfrac{1}{2} - \dfrac{1}{4} + \dfrac{1}{8} - \dfrac{1}{16} - \dfrac{1}{32} + \dfrac{1}{64} - \cdots$ converges.

**Solution** First note that the given series is not an alternating series, so the Alternating Series Test cannot be used. Next, the series of the absolute values of each term, $1 + \dfrac{1}{2} + \dfrac{1}{4} + \cdots + \dfrac{1}{2^{n-1}} + \cdots = \sum_{k=1}^{\infty} \left(\dfrac{1}{2}\right)^{k-1}$, is a geometric series with $r = \dfrac{1}{2}$, which converges. So by the Absolute Convergence Test, the series

$$1 - \dfrac{1}{2} - \dfrac{1}{4} + \dfrac{1}{8} - \dfrac{1}{16} - \dfrac{1}{32} + \dfrac{1}{64} - \cdots \text{ converges. } \blacksquare$$

**NOW WORK** Problem 37.

---

**EXAMPLE 6  Using the Absolute Convergence Test**

Determine whether the series $\sum_{k=1}^{\infty} \dfrac{\sin k}{k^2} = \dfrac{\sin 1}{1^2} + \dfrac{\sin 2}{2^2} + \dfrac{\sin 3}{3^2} + \cdots$ converges.

**Solution** This series has both positive and negative terms, but it is not an alternating series. We investigate the series of the absolute values of each term $\sum\limits_{k=1}^{\infty} \left|\dfrac{\sin k}{k^2}\right|$. Since $|\sin n| \le 1$,

$$\left|\frac{\sin n}{n^2}\right| \le \frac{1}{n^2}$$

for all $n$, and since $\sum\limits_{k=1}^{\infty} \dfrac{1}{k^2}$ is a convergent $p$-series, then by the Comparison Test for Convergence, the series $\sum\limits_{k=1}^{\infty} \left|\dfrac{\sin k}{k^2}\right|$ converges. Since $\sum\limits_{k=1}^{\infty} \dfrac{\sin k}{k^2}$ is absolutely convergent, it follows that $\sum\limits_{k=1}^{\infty} \dfrac{\sin k}{k^2}$ is convergent. ■

**NOW WORK** Problem 39.

The converse of the Absolute Convergence Test is not true. That is, if a series is convergent, the series may or may not be absolutely convergent. There are convergent series that are not absolutely convergent. For instance, we have shown that the alternating harmonic series

$$\sum_{k=1}^{\infty} \frac{(-1)^{k+1}}{k} = 1 - \frac{1}{2} + \frac{1}{3} - \frac{1}{4} + \frac{1}{5} - \cdots$$

**IN WORDS** If a series is absolutely convergent, then it is convergent. But just because a series converges does not mean it is absolutely convergent.

converges. The series of absolute values is the harmonic series $\sum\limits_{k=1}^{\infty} \dfrac{1}{k}$, which is divergent.

> **DEFINITION** Conditional Convergence
>
> A series that is convergent without being absolutely convergent is called **conditionally convergent**.

**CALC**
**CLIP**

**EXAMPLE 7** Determining Whether a Series Is Absolutely Convergent, Conditionally Convergent, or Divergent

Determine the numbers $p$ for which the series $\sum\limits_{k=1}^{\infty} \dfrac{(-1)^{k+1}}{k^p}$ is absolutely convergent, conditionally convergent, or divergent.

**Solution** We begin by testing the series for absolute convergence. The series of absolute values is $\sum\limits_{k=1}^{\infty} \left|\dfrac{(-1)^{k+1}}{k^p}\right| = \sum\limits_{k=1}^{\infty} \dfrac{1}{k^p}$. This is a $p$-series, which converges if $p > 1$ and diverges if $p \le 1$. So, $\sum\limits_{k=1}^{\infty} \dfrac{(-1)^{k+1}}{k^p}$ is absolutely convergent if $p > 1$.

It remains to determine what happens when $p \le 1$. We use the Alternating Series Test, and begin by investigating $\lim\limits_{n \to \infty} a_n$:

- $\lim\limits_{n \to \infty} \dfrac{1}{n^p} = 0$   $0 < p \le 1$
- $\lim\limits_{n \to \infty} \dfrac{1}{n^p} = 1$   $p = 0$
- $\lim\limits_{n \to \infty} \dfrac{1}{n^p} = \infty$   $p < 0$

Consequently, $\sum\limits_{k=1}^{\infty} \dfrac{(-1)^{k+1}}{k^p}$ diverges if $p \le 0$.

Continuing, we check the second condition of the Alternating Series Test when $0 < p \le 1$. Using the related function $f(x) = \dfrac{1}{x^p}$, for $x > 0$, we have $f'(x) = -\dfrac{p}{x^{p+1}}$. Since $f'(x) < 0$ for $0 < p \le 1$, the second condition of the Alternating Series Test is satisfied. We conclude that $\sum\limits_{k=1}^{\infty} \dfrac{(-1)^{k+1}}{k^p}$ is conditionally convergent if $0 < p \le 1$. ■

To summarize, the alternating $p$-series, $\displaystyle\sum_{k=1}^{\infty} \frac{(-1)^{k+1}}{k^p}$ is absolutely convergent if $p > 1$, conditionally convergent if $0 < p \le 1$, and divergent if $p \le 0$.

**NOW WORK** Problem **43**.

As illustrated in Figure 24, a series $\displaystyle\sum_{k=1}^{\infty} a_k$ is either convergent or divergent. If it is convergent, it is either absolutely convergent or conditionally convergent. The flowchart in Figure 25 can be used to determine whether a series is absolutely convergent, conditionally convergent, or divergent.

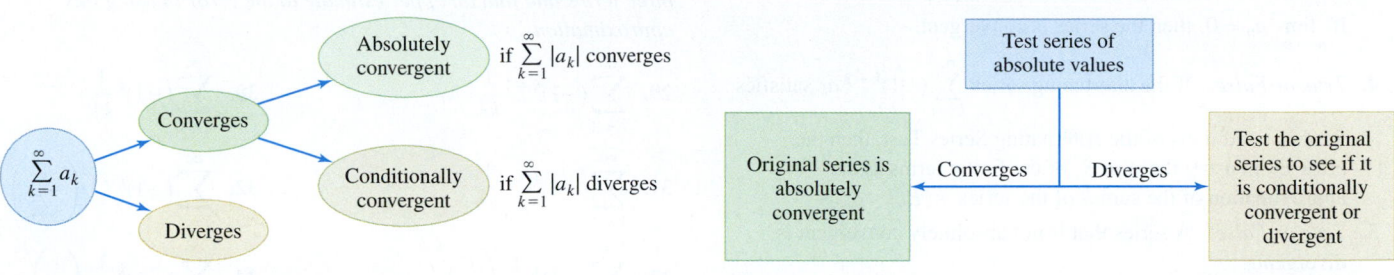

**Figure 24**                 **Figure 25**

Absolute convergence of a series is stronger than conditional convergence, and absolutely convergent series and conditionally convergent series have different properties. We state, but do not prove, some of these properties now.

---

### Properties of Absolutely Convergent and Conditionally Convergent Series

- If a series $\displaystyle\sum_{k=1}^{\infty} a_k$ is absolutely convergent, any rearrangement of its terms results in a series that is also absolutely convergent. Both series will have the same sum. See Problem 78.

- If two series $\displaystyle\sum_{k=1}^{\infty} a_k$ and $\displaystyle\sum_{k=1}^{\infty} b_k$ are absolutely convergent, the series $\displaystyle\sum_{k=1}^{\infty} (a_k + b_k)$ is also absolutely convergent. Moreover,

$$\sum_{k=1}^{\infty} (a_k + b_k) = \sum_{k=1}^{\infty} a_k + \sum_{k=1}^{\infty} b_k$$

  See Problem 74.

- If two series $\displaystyle\sum_{k=1}^{\infty} a_k$ and $\displaystyle\sum_{k=1}^{\infty} b_k$ are absolutely convergent, then the series

$$\sum_{k=1}^{\infty} c_k = \sum_{k=1}^{\infty} (a_1 b_k + a_2 b_{k-1} + \cdots + a_k b_1)$$

  is also absolutely convergent.

  The sum of the series $\displaystyle\sum_{k=1}^{\infty} c_k$ equals the product of the sums of the two original series.

- If a series is absolutely convergent, then the series consisting of just the positive terms converges, as does the series consisting of just the negative terms. See Problem 75.

- If a series is conditionally convergent, then the series consisting of just the positive terms diverges, as does the series consisting of just the negative terms. See Problems 63, 64, and 76.

- If a series $\displaystyle\sum_{k=1}^{\infty} a_k$ is conditionally convergent, its terms may be rearranged to form a new series that converges to *any* sum we like. In fact, they may be rearranged to form a *divergent series*. See Problems 65–67.

# 8.5 Assess Your Understanding

## Concepts and Vocabulary

1. *True or False* The series $\sum_{k=1}^{\infty} (-1)^k \cos(k\pi)$ is an alternating series.

2. *True or False* $\sum_{k=1}^{\infty} [1 + (-1)^k]$ is an alternating series.

3. *True or False* In an alternating series, $\sum_{k=1}^{\infty} (-1)^k a_k$, $a_k > 0$, if $\lim_{n \to \infty} a_n = 0$, then the series is convergent.

4. *True or False* If the alternating series $\sum_{k=1}^{\infty} (-1)^{k+1} a_k$ satisfies the two conditions of the Alternating Series Test, then the error $E_n$ in using the sum $S_n$ of the first $n$ terms as an approximation of the sum $S$ of the series is $|E_n| \le a_n$.

5. *True or False* A series that is not absolutely convergent is divergent.

6. *True or False* If a series is absolutely convergent, then the series converges.

## Skill Building

*In Problems 7–22, use the Alternating Series Test to determine whether each alternating series converges or diverges.*

7. $\sum_{k=1}^{\infty} (-1)^{k+1} \dfrac{1}{k^2}$

8. $\sum_{k=1}^{\infty} (-1)^{k+1} \dfrac{1}{2\sqrt{k}}$

9. $\sum_{k=1}^{\infty} (-1)^{k+1} \dfrac{k}{2k+1}$

10. $\sum_{k=1}^{\infty} (-1)^{k+1} \dfrac{k+1}{k}$

11. $\sum_{k=1}^{\infty} (-1)^{k+1} \dfrac{k^2}{5k^2+2}$

12. $\sum_{k=1}^{\infty} (-1)^{k+1} \dfrac{k+1}{k^2}$

13. $\sum_{k=1}^{\infty} \dfrac{(-1)^{k+1}}{(k+1)2^k}$

14. $\sum_{k=2}^{\infty} (-1)^k \dfrac{1}{k \ln k}$

15. $\sum_{k=2}^{\infty} (-1)^k \dfrac{1}{1+2^{-k}}$

16. $\sum_{k=0}^{\infty} (-1)^k \dfrac{1}{k!}$

17. $\sum_{k=1}^{\infty} (-1)^{k+1} \left( \dfrac{k}{k+1} \right)^k$

18. $\sum_{k=1}^{\infty} (-1)^{k+1} \dfrac{k^2}{(k+1)^3}$

19. $\sum_{k=1}^{\infty} (-1)^k e^{-k}$

20. $\sum_{k=1}^{\infty} (-1)^k k e^{-k}$

21. $\sum_{k=1}^{\infty} (-1)^k \tan^{-1} k$

22. $\sum_{k=1}^{\infty} (-1)^k e^{2/k}$

*In Problems 23–28:*

(a) *Show that each alternating series satisfies the conditions of the Alternating Series Test.*

(b) *How many terms must be added so the error in using the sum $S_n$ of the first $n$ terms as an approximation to the sum $S$ of the series is less than or equal to 0.001?*

(c) *Approximate the sum $S$ of the series so that the error is less than or equal to 0.001.*

23. $\sum_{k=1}^{\infty} \dfrac{(-1)^{k+1}}{k^5}$

24. $\sum_{k=1}^{\infty} \dfrac{(-1)^{k+1}}{k^4}$

25. $\sum_{k=1}^{\infty} \dfrac{(-1)^{k+1}}{k!}$

26. $\sum_{k=1}^{\infty} \dfrac{(-1)^{k+1}}{k^k}$

27. $\sum_{k=1}^{\infty} \dfrac{(-1)^{k+1}}{2^k}$

28. $\sum_{k=1}^{\infty} \dfrac{(-1)^{k+1}}{3^k}$

*In Problems 29–36, approximate the sum of each series using the first three terms and find an upper estimate to the error in using this approximation.*

29. $\sum_{k=1}^{\infty} (-1)^{k+1} \dfrac{1}{k^2}$

30. $\sum_{k=0}^{\infty} (-1)^k \dfrac{1}{k!}$

31. $\sum_{k=1}^{\infty} (-1)^{k+1} \dfrac{1}{k^4}$

32. $\sum_{k=1}^{\infty} (-1)^{k+1} \left( \dfrac{1}{\sqrt{k}} \right)^k$

33. $\sum_{k=0}^{\infty} (-1)^k \dfrac{1}{k!} \left( \dfrac{1}{3} \right)^k$

34. $\sum_{k=0}^{\infty} (-1)^k \dfrac{1}{k!} \left( \dfrac{1}{2} \right)^k$

35. $\sum_{k=0}^{\infty} (-1)^k \dfrac{1}{2k+1} \left( \dfrac{1}{3} \right)^{2k+1}$

36. $\sum_{k=1}^{\infty} (-1)^{k+1} \dfrac{1}{k^k}$

*In Problems 37–40, use the Absolute Convergence Test to show that each series converges.*

37. $\sum_{k=1}^{\infty} \dfrac{(-1)^{k+1}}{k^2}$

38. $\sum_{k=1}^{\infty} \dfrac{(-1)^{k+1}}{k^3}$

39. $\sum_{k=1}^{\infty} (-1)^{k+1} \dfrac{\sin k}{k^2+1}$

40. $\sum_{k=1}^{\infty} (-1)^{k+1} \dfrac{\cos k}{k^2}$

*In Problems 41–52, determine whether each series is absolutely convergent, conditionally convergent, or divergent.*

41. $\sum_{k=1}^{\infty} (-1)^{k+1} \left( \dfrac{1}{5} \right)^k$

42. $\sum_{k=1}^{\infty} (-1)^{k+1} \dfrac{5^k}{6^{k+1}}$

43. $\sum_{k=1}^{\infty} (-1)^{k+1} \dfrac{e^k}{k}$

44. $\sum_{k=1}^{\infty} \dfrac{(-1)^{k+1} 2^k}{k^2}$

45. $\sum_{k=1}^{\infty} \dfrac{(-1)^{k+1}}{k(k+1)}$

46. $\sum_{k=1}^{\infty} \dfrac{(-1)^{k+1}}{k \sqrt{k+3}}$

47. $\sum_{k=1}^{\infty} \dfrac{(-1)^{k+1} \sqrt{k}}{k^2+1}$

48. $\sum_{k=1}^{\infty} \dfrac{(-1)^{k+1} \sqrt{k}}{k+1}$

49. $\sum_{k=1}^{\infty} (-1)^{k+1} \dfrac{\ln k}{k}$

50. $\sum_{k=1}^{\infty} \dfrac{(-1)^{k+1} \ln k}{k^3}$

51. $\sum_{k=1}^{\infty} (-1)^{k+1} \dfrac{1}{k e^k}$

52. $\sum_{k=1}^{\infty} \dfrac{(-1)^{k+1}}{e^k}$

## Applications and Extensions

*In Problems 53–60, determine whether each series is absolutely convergent, conditionally convergent, or divergent.*

53. $\sum_{k=2}^{\infty} \dfrac{(-1)^k}{k \ln k}$

54. $\sum_{k=2}^{\infty} \dfrac{(-1)^k}{k (\ln k)^2}$

---

**1.** = NOW WORK problem      〰️ = Graphing technology recommended      CAS = Computer Algebra System recommended

**55.** $\displaystyle\sum_{k=1}^{\infty} \frac{\cos k}{3^{k-1}}$

**56.** $\displaystyle\sum_{k=1}^{\infty} \frac{(-1)^{k+1} \tan^{-1} k}{k}$

**57.** $\displaystyle\sum_{k=1}^{\infty} \frac{(-1)^{k+1}}{k^{1/k}}$

**58.** $\displaystyle\sum_{k=1}^{\infty}(-1)^{k+1}\left(\frac{k}{k+1}\right)^{k}$

**59.** $1 - \dfrac{1}{2!} + \dfrac{1}{3!} - \dfrac{1}{4!} + \dfrac{1}{5!} - \cdots$

**60.** $1 - \dfrac{1}{3^2} + \dfrac{1}{5^2} - \dfrac{1}{7^2} + \cdots$

**61. (a)** Show that the Alternating Series Test is not applicable to the series
$$\frac{1}{2} - \frac{1}{3} + \frac{1}{2^2} - \frac{1}{3^2} + \frac{1}{2^3} - \frac{1}{3^3} + \cdots$$

**(b)** Show the series converges.

**(c)** Find the sum of the series.

**62. (a)** Show that the Alternating Series Test is not applicable to the series
$$\sum_{k=1}^{\infty}(-1)^k a_k \text{ where } a_k = \begin{cases} \dfrac{1}{k} & \text{if } k \text{ is odd} \\[2mm] \dfrac{1}{k^2} & \text{if } k \text{ is even} \end{cases}$$

**(b)** Show the series diverges.

**63.** Show that the positive terms of $\displaystyle\sum_{k=1}^{\infty} \frac{(-1)^{k+1}}{k}$ diverge.

**64.** Show that the negative terms of $\displaystyle\sum_{k=1}^{\infty} \frac{(-1)^{k+1}}{k}$ diverge.

**65.** Show that the terms of the series $\displaystyle\sum_{k=1}^{\infty} \frac{(-1)^{k+1}}{k}$ can be rearranged so the resulting series converges to 0.

**66.** Show that the terms of the series $\displaystyle\sum_{k=1}^{\infty} \frac{(-1)^{k+1}}{k}$ can be rearranged so the resulting series converges to 2.

**67.** Show that the terms of the series $\displaystyle\sum_{k=1}^{\infty} \frac{(-1)^{k+1}}{k}$ can be rearranged so the resulting series diverges.

**68.** Show that the series
$$e^{-x}\cos x + e^{-2x}\cos(2x) + e^{-3x}\cos(3x) + \cdots$$
is absolutely convergent for all positive values of $x$.
*Hint:* Use the fact that $|\cos\theta| \le 1$.

**69.** Determine whether the series below converges (absolutely or conditionally) or diverges.
$$1 + r\cos\theta + r^2\cos(2\theta) + r^3\cos(3\theta) + \cdots$$

**70.** What is wrong with the following argument?
$$A = 1 - \frac{1}{2} + \frac{1}{3} - \frac{1}{4} + \frac{1}{5} - \frac{1}{6} + \frac{1}{7} - \frac{1}{8} + \cdots$$
$$\frac{1}{2}A = \frac{1}{2} - \frac{1}{4} + \frac{1}{6} - \frac{1}{8} + \cdots$$
So,
$$A + \frac{1}{2}A = 1 + \frac{1}{3} - \frac{1}{2} + \frac{1}{5} + \frac{1}{7} - \frac{1}{4} + \cdots$$
The series on the right is a rearrangement of the terms of the series $A$. So its sum is $A$, meaning
$$A + \frac{1}{2}A = A$$
$$A = 0$$

But,
$$A = \left(1 - \frac{1}{2}\right) + \left(\frac{1}{3} - \frac{1}{4}\right) + \left(\frac{1}{5} - \frac{1}{6}\right) + \cdots > 0$$

**Bernoulli's Error**   *In Problems 71–73, consider an incorrect argument given by Jakob Bernoulli to prove that*
$$\sum_{k=1}^{\infty} \frac{1}{k(k+1)} = \frac{1}{2} + \frac{1}{6} + \frac{1}{12} + \cdots = 1$$

*Bernoulli's argument went as follows:*

*Let* $N = 1 + \dfrac{1}{2} + \dfrac{1}{3} + \dfrac{1}{4} + \cdots$. *Then*
$$N - 1 = \frac{1}{2} + \frac{1}{3} + \frac{1}{4} + \frac{1}{5} + \cdots.$$

*Now subtract term-by-term to get*
$$N - (N - 1) = \left(1 - \frac{1}{2}\right) + \left(\frac{1}{2} - \frac{1}{3}\right) + \left(\frac{1}{3} - \frac{1}{4}\right) + \cdots \quad (4)$$
$$1 = \frac{1}{2} + \frac{1}{6} + \frac{1}{12} + \cdots$$

**71.** What is wrong with Bernoulli's argument?

**72.** In general, what can be said about the convergence or divergence of a series formed by taking the term-by-term difference (or sum) of two divergent series? Support your answer with examples.

**73.** Although the method is wrong. Bernoulli's conclusion is correct; that is, it is true that $\displaystyle\sum_{k=1}^{\infty} \frac{1}{k(k+1)} = 1$. Prove it!

*Hint:* Look at the partial sums using the form of the series in (4).

**74.** Show that if two series $\displaystyle\sum_{k=1}^{\infty} a_k$ and $\displaystyle\sum_{k=1}^{\infty} b_k$ are absolutely convergent, the series $\displaystyle\sum_{k=1}^{\infty}(a_k + b_k)$ is also absolutely convergent.

Moreover, $\displaystyle\sum_{k=1}^{\infty}(a_k + b_k) = \sum_{k=1}^{\infty} a_k + \sum_{k=1}^{\infty} b_k$.

**75.** Show that if a series is absolutely convergent, then the series consisting of just the positive terms converges, as does the series consisting of just the negative terms.

**76.** Show that if a series is conditionally convergent, then the series consisting of just the positive terms diverges, as does the series consisting of just the negative terms.

**77.** Determine whether the series $\displaystyle\sum_{k=1}^{\infty} c_k$, where
$$c_k = \begin{cases} \dfrac{1}{a^k} & \text{if } k \text{ is even} \\[3mm] -\dfrac{1}{b^k} & \text{if } k \text{ is odd} \end{cases} \quad a > 1, \ b > 1,$$
converges absolutely, converges conditionally, or diverges.

**78.** Prove that if a series $\displaystyle\sum_{k=1}^{\infty} a_k$ is absolutely convergent, any rearrangement of its terms results in a series that is also absolutely convergent.

*Hint:* Use the triangle inequality: $|a + b| \le |a| + |b|$.

## Challenge Problems

*In Problems 79–84, determine whether each series is absolutely convergent, conditionally convergent, or divergent.*

**79.** $\displaystyle\sum_{k=2}^{\infty} \frac{(-1)^k}{\sqrt[p]{k^3+1}}, \quad p > 2$

**80.** $\displaystyle\sum_{k=2}^{\infty} (-1)^k \ln \frac{k+1}{k}$

**81.** $\displaystyle\sum_{k=2}^{\infty} \frac{(-1)^k}{(\ln k)^{\ln k}}$

**82.** $\displaystyle\sum_{k=2}^{\infty} (-1)^k k^{(1-k)/k}$

**83.** $\displaystyle\sum_{k=1}^{\infty} (-1)^{k+1} \frac{k}{(k+1)^2}$

**84.** $\displaystyle\sum_{k=1}^{\infty} (-1)^{k+1} \frac{\sqrt{k}}{(k+1)}$

**85.** Let $\{a_n\}$ be a sequence that is decreasing and is bounded from below by 0. Define

$$R_n = \sum_{k=n+1}^{\infty} (-1)^{k+1} a_k \quad \text{and} \quad \Delta a_k = a_k - a_{k+1}$$

Suppose that the sequence $\{\Delta a_k\}$ decreases.

**(a)** Show that the series $\displaystyle\sum_{k=1}^{\infty} (-1)^{k+1} \Delta a_k$ is a convergent alternating series.

**(b)** Show that

$$|R_n| = \frac{a_n}{2} + \frac{1}{2} \sum_{k=1}^{\infty} (-1)^k \Delta a_{k+n-1}$$

and $\displaystyle\sum_{k=1}^{\infty} (-1)^k \Delta a_{k+n-1} < 0$. Deduce $|R_n| < \dfrac{1}{2} a_n$.

**(c)** Show that

$$|R_n| = \frac{a_{n+1}}{2} + \frac{1}{2} \sum_{k=1}^{\infty} (-1)^{k+1} \Delta a_{k+n}$$

**(d)** Conclude that $\dfrac{a_{n+1}}{2} < |R_n|$.

*Source*: Based on R. Johnsonbaugh, Summing an alternating series, *American Mathematical Monthly*, Vol. 86, No. 8 (Oct. 1979), pp. 637–648.

# 8.6 Ratio Test; Root Test

**OBJECTIVES** *When you finish this section, you should be able to:*

1 Use the Ratio Test (p. 630)
2 Use the Root Test (p. 633)

One of the most practical tests for convergence of a series of nonzero terms makes use of the ratio of two consecutive terms.

## 1 Use the Ratio Test

> **THEOREM** Ratio Test
>
> Let $\displaystyle\sum_{k=1}^{\infty} a_k$ be a series of nonzero terms.
>
> **1.** Let $\displaystyle\lim_{n \to \infty} \left| \frac{a_{n+1}}{a_n} \right| = L$, a number.
>
> - If $L < 1$, then the series $\displaystyle\sum_{k=1}^{\infty} a_k$ converges absolutely and so $\displaystyle\sum_{k=1}^{\infty} a_k$ is convergent.
> - If $L = 1$, the test provides no information about whether the series converges or diverges.
> - If $L > 1$, then the series $\displaystyle\sum_{k=1}^{\infty} a_k$ diverges.
>
> **2.** If $\displaystyle\lim_{n \to \infty} \left| \frac{a_{n+1}}{a_n} \right| = \infty$, then the series $\displaystyle\sum_{k=1}^{\infty} a_k$ diverges.

A proof of the Ratio Test is given at the end of the section.

CALC
CLIP

**EXAMPLE 1  Using the Ratio Test**

Use the Ratio Test to determine whether each series converges or diverges.

(a) $\displaystyle\sum_{k=1}^{\infty} \frac{k}{4^k}$   (b) $\displaystyle\sum_{k=1}^{\infty} \frac{2^k}{k}$   (c) $\displaystyle\sum_{k=1}^{\infty} \frac{3k+1}{k^2}$   (d) $\displaystyle\sum_{k=1}^{\infty} \frac{1}{k!}$

**Solution** (a) $\displaystyle\sum_{k=1}^{\infty} \frac{k}{4^k}$ is a series of nonzero terms; $a_{n+1} = \dfrac{n+1}{4^{n+1}}$ and $a_n = \dfrac{n}{4^n}$. The absolute value of their ratio is

$$\left| \frac{a_{n+1}}{a_n} \right| = \frac{\dfrac{n+1}{4^{n+1}}}{\dfrac{n}{4^n}} = \frac{n+1}{4^{n+1}} \cdot \frac{4^n}{n} = \frac{n+1}{4n}$$

Then

$$\lim_{n \to \infty} \left| \frac{a_{n+1}}{a_n} \right| = \lim_{n \to \infty} \frac{n+1}{4n} = \frac{1}{4} < 1$$

Since the limit is less than 1, the series converges.

(b) $\displaystyle\sum_{k=1}^{\infty} \frac{2^k}{k}$ is a series of nonzero terms; $a_{n+1} = \dfrac{2^{n+1}}{n+1}$ and $a_n = \dfrac{2^n}{n}$. The absolute value of their ratio is

$$\left| \frac{a_{n+1}}{a_n} \right| = \frac{2^{n+1}}{n+1} \cdot \frac{n}{2^n} = \frac{2n}{n+1}$$

Then

$$\lim_{n \to \infty} \left| \frac{a_{n+1}}{a_n} \right| = \lim_{n \to \infty} \frac{2n}{n+1} = 2 > 1$$

Since the limit is greater than 1, the series diverges.

(c) $\displaystyle\sum_{k=1}^{\infty} \frac{3k+1}{k^2}$ is a series of nonzero terms; $a_{n+1} = \dfrac{3(n+1)+1}{(n+1)^2} = \dfrac{3n+4}{(n+1)^2}$ and $a_n = \dfrac{3n+1}{n^2}$. The absolute value of their ratio is

$$\left| \frac{a_{n+1}}{a_n} \right| = \left[ \frac{3n+4}{(n+1)^2} \right] \left( \frac{n^2}{3n+1} \right) = \left( \frac{3n+4}{n^2+2n+1} \right) \left( \frac{n^2}{3n+1} \right) = \frac{3n^3+4n^2}{3n^3+7n^2+5n+1}$$

Then

$$\lim_{n \to \infty} \left| \frac{a_{n+1}}{a_n} \right| = \lim_{n \to \infty} \frac{3n^3+4n^2}{3n^3+7n^2+5n+1} = 1$$

The Ratio Test provides no information about this series. Another test must be used.

$\left( \text{You can show that the series diverges by comparing it to the harmonic series } \displaystyle\sum_{k=1}^{\infty} \frac{1}{k}. \right)$

(d) $\displaystyle\sum_{k=1}^{\infty} \frac{1}{k!}$ is a series of nonzero terms; $a_{n+1} = \dfrac{1}{(n+1)!}$ and $a_n = \dfrac{1}{n!}$. The absolute value of their ratio is

$$\left| \frac{a_{n+1}}{a_n} \right| = \frac{n!}{(n+1)!} = \frac{1}{n+1}$$

**NOTE** In Section 8.9, we show that $\sum\limits_{k=1}^{\infty} \dfrac{1}{k!} = e$.

Then

$$\lim_{n\to\infty}\left|\frac{a_{n+1}}{a_n}\right| = \lim_{n\to\infty}\frac{1}{n+1} = 0$$

Since the limit is less than 1, the series $\sum\limits_{k=1}^{\infty} \dfrac{1}{k!}$ converges. ∎

NOW WORK **Problem 7.**

As Example 1 illustrates, the Ratio Test is useful in determining whether a series containing factorials and/or powers converges or diverges.

EXAMPLE 2 **Using the Ratio Test**

Use the Ratio Test to determine whether the series $\sum\limits_{k=1}^{\infty} \dfrac{k!}{k^k}$ converges or diverges.

**Solution** $\sum\limits_{k=1}^{\infty} \dfrac{k!}{k^k}$ is a series of nonzero terms. Since $a_{n+1} = \dfrac{(n+1)!}{(n+1)^{n+1}}$ and $a_n = \dfrac{n!}{n^n}$, the absolute value of their ratio is

**NEED TO REVIEW?** The number $e$ expressed as a limit is discussed in Section 3.4, pp. 237–239.

$$\left|\frac{a_{n+1}}{a_n}\right| = \frac{\dfrac{(n+1)!}{(n+1)^{n+1}}}{\dfrac{n!}{n^n}} = \frac{(n+1)!}{(n+1)^{n+1}} \cdot \frac{n^n}{n!} = \frac{n^n}{(n+1)^n}$$

$$= \left(\frac{n}{n+1}\right)^n = \left(\frac{1}{1+\dfrac{1}{n}}\right)^n = \frac{1}{\left(1+\dfrac{1}{n}\right)^n}$$

**CAUTION** In Example 2, the ratio $\left|\dfrac{a_{n+1}}{a_n}\right|$ converges to $\dfrac{1}{e}$. This does not mean that $\sum\limits_{k=1}^{\infty} \dfrac{k!}{k^k}$ converges to $\dfrac{1}{e}$. In fact, the sum of the series $\sum\limits_{k=1}^{\infty} \dfrac{k!}{k^k}$ is *not* known; all that is known is that the series converges.

So,

$$\lim_{n\to\infty}\left|\frac{a_{n+1}}{a_n}\right| = \lim_{n\to\infty}\frac{1}{\left(1+\dfrac{1}{n}\right)^n} = \frac{\lim\limits_{n\to\infty} 1}{\lim\limits_{n\to\infty}\left(1+\dfrac{1}{n}\right)^n} = \frac{1}{e}$$

Since $\dfrac{1}{e} < 1$, the series converges. ∎

NOW WORK **Problem 15.**

EXAMPLE 3 **Using the Ratio Test**

Use the Ratio Test to determine whether the series $\sum\limits_{k=1}^{\infty} (-1)^k \dfrac{k!}{10^k}$ converges or diverges.

**Solution** $\sum\limits_{k=1}^{\infty} (-1)^k \dfrac{k!}{10^k}$ is a series of nonzero terms.

$$a_{n+1} = (-1)^{n+1}\frac{(n+1)!}{10^{n+1}} \qquad a_n = (-1)^n\frac{n!}{10^n}$$

$$\lim_{n\to\infty}\left|\frac{a_{n+1}}{a_n}\right| = \lim_{n\to\infty}\frac{\dfrac{(n+1)!}{10^{n+1}}}{\dfrac{n!}{10^n}} = \lim_{n\to\infty}\frac{(n+1)!\,10^n}{n!\,10^{n+1}} = \lim_{n\to\infty}\frac{n+1}{10} = \infty$$

By the Ratio Test, the series $\sum\limits_{k=1}^{\infty} (-1)^k \dfrac{k!}{10^k}$ diverges. ∎

NOW WORK **Problem 19.**

We conclude the discussion of the Ratio Test with these observations:

- To test $\sum\limits_{k=1}^{\infty} a_k$ for convergence, it is important to check whether the *limit of the ratio* $\left| \dfrac{a_{n+1}}{a_n} \right|$, not the ratio itself, is less than 1.

   For example, for the harmonic series $\sum\limits_{k=1}^{\infty} \dfrac{1}{k}$, which diverges,

   the ratio $\left| \dfrac{a_{n+1}}{a_n} \right| = \left| \dfrac{n}{n+1} \right| < 1$, but $\lim\limits_{n \to \infty} \left| \dfrac{a_{n+1}}{a_n} \right| = \lim\limits_{n \to \infty} \dfrac{n}{n+1} = 1$.

- For divergence, it is sufficient to show that the ratio $\left| \dfrac{a_{n+1}}{a_n} \right| > 1$ for all $n$.

- The Ratio Test provides no information if $\lim\limits_{n \to \infty} \left| \dfrac{a_{n+1}}{a_n} \right| = 1$. It also provides no information if $\lim\limits_{n \to \infty} \left| \dfrac{a_{n+1}}{a_n} \right| \neq \infty$ does not exist.

- If the general term $a_n$ of an infinite series involves $n$, either exponentially or factorially, the Ratio Test often answers the question of convergence or divergence.

## 2 Use the Root Test

The *Root Test* works well for series of nonzero terms whose $n$th term involves an $n$th power.

---

**THEOREM  Root Test**

Let $\sum\limits_{k=1}^{\infty} a_k$ be a series of nonzero terms. Suppose $\lim\limits_{n \to \infty} \sqrt[n]{|a_n|} = L$, a number.

- If $L < 1$, then $\sum\limits_{k=1}^{\infty} a_k$ is absolutely convergent, so the series $\sum\limits_{k=1}^{\infty} a_k$ converges.

- If $L > 1$, then $\sum\limits_{k=1}^{\infty} a_k$ diverges.

- If $L = 1$, the test provides no information.

---

The proof of the Root Test is similar to the proof of the Ratio Test. It is left as an exercise (Problem 67).

---

**EXAMPLE 4  Using the Root Test**

Use the Root Test to determine whether the series $\sum\limits_{k=1}^{\infty} \dfrac{e^k}{k^k}$ converges or diverges.

**Solution** $\sum\limits_{k=1}^{\infty} \dfrac{e^k}{k^k}$ is a series of nonzero terms. The $n$th term is $a_n = \dfrac{e^n}{n^n} = \left( \dfrac{e}{n} \right)^n$.

Since $a_n$ involves an $n$th power, we use the Root Test.

$$\lim\limits_{n \to \infty} \sqrt[n]{|a_n|} = \lim\limits_{n \to \infty} \sqrt[n]{\left( \dfrac{e}{n} \right)^n} = \lim\limits_{n \to \infty} \dfrac{e}{n} = 0 < 1$$

The series $\sum\limits_{k=1}^{\infty} \dfrac{e^k}{k^k}$ converges. ∎

**NOW WORK** Problem 33.

EXAMPLE 5  **Using the Root Test**

Use the Root Test to determine whether the series $\displaystyle\sum_{k=1}^{\infty}\left(\frac{8k+3}{5k-2}\right)^{k}$ converges or diverges.

**Solution** $\displaystyle\sum_{k=1}^{\infty}\left(\frac{8k+3}{5k-2}\right)^{k}$ is a series of nonzero terms. The $n$th term is $a_{n}=\left(\dfrac{8n+3}{5n-2}\right)^{n}$. Since $a_{n}$ involves an $n$th power, we use the Root Test.

$$\lim_{n\to\infty}\sqrt[n]{|a_{n}|}=\lim_{n\to\infty}\sqrt[n]{\left(\frac{8n+3}{5n-2}\right)^{n}}=\lim_{n\to\infty}\frac{8n+3}{5n-2}=\frac{8}{5}>1$$

The series diverges. ∎

NOW WORK  Problem 29.

**Proof of the Ratio Test**  Let $\displaystyle\sum_{k=1}^{\infty}a_{k}$ be a series of nonzero terms.

*Case 1:* $\displaystyle\lim_{n\to\infty}\left|\frac{a_{n+1}}{a_{n}}\right|=L$, a number.

- $0\leq L<1$  Let $r$ be any number for which $L<r<1$. Since $\displaystyle\lim_{n\to\infty}\left|\frac{a_{n+1}}{a_{n}}\right|=L$ and $L<r$, then by the definition of the limit of a sequence, we can find a number $N$ so that for any number $n>N$, the ratio $\left|\dfrac{a_{n+1}}{a_{n}}\right|$ can be made as close as we please to $L$ and be less than $r$. Then

$$\left|\frac{a_{N+1}}{a_{N}}\right|<r \qquad \text{or} \qquad |a_{N+1}|<r\cdot|a_{N}|$$

$$\left|\frac{a_{N+2}}{a_{N+1}}\right|<r \qquad \text{or} \qquad |a_{N+2}|<r\cdot|a_{N+1}|<r^{2}\cdot|a_{N}|$$

$$\left|\frac{a_{N+3}}{a_{N+2}}\right|<r \qquad \text{or} \qquad |a_{N+3}|<r\cdot|a_{N+2}|<r^{3}\cdot|a_{N}|$$

Each term of the series $|a_{N+1}|+|a_{N+2}|+\cdots$ is less than the corresponding term of the geometric series $|a_{N}|\,r+|a_{N}|\,r^{2}+|a_{N}|\,r^{3}+\cdots$. Since $|r|<1$, the geometric series converges. By the Comparison Test for Convergence, the series $|a_{N+1}|+|a_{N+2}|+\cdots$ also converges. So, the series $\displaystyle\sum_{k=1}^{\infty}|a_{k}|$ converges. By the Absolute Convergence Test, the series $\displaystyle\sum_{k=1}^{\infty}a_{k}$ converges.

- $L=1$  To show that the test provides no information for $L=1$, we exhibit two series, one that diverges and another that converges, to show that no conclusion can be drawn. Consider $\displaystyle\sum_{k=1}^{\infty}\frac{1}{k}$ and $\displaystyle\sum_{k=1}^{\infty}\frac{1}{k^{2}}$. The first is the harmonic series, which diverges. The second is a $p$-series with $p>1$, which converges. It is left to you to show that $\displaystyle\lim_{n\to\infty}\left|\frac{a_{n+1}}{a_{n}}\right|=1$ in each case. (See Problems 57 and 58.)

- $L > 1$   Let $r$ be any number for which $1 < r < L$. Since $\lim\limits_{n \to \infty} \left| \dfrac{a_{n+1}}{a_n} \right| = L$, there

  is a number $N$ so that for any number $n > N$, the ratio $\left| \dfrac{a_{n+1}}{a_n} \right|$ can be made as

  close as we please to $L$ and will be greater than $r$. That is, for all numbers $n > N$,

  the ratio $\left| \dfrac{a_{n+1}}{a_n} \right| > r > 1$ so that $|a_{n+1}| > |a_n|$. After the $N$th term, the absolute

  value of the terms are positive and increasing. So, $\lim\limits_{n \to \infty} |a_n| \neq 0$ and, therefore,

  $\lim\limits_{n \to \infty} a_n \neq 0$. By the Test for Divergence, the series diverges.

*Case 2:* $\lim\limits_{n \to \infty} \left| \dfrac{a_{n+1}}{a_n} \right| = \infty$   The proof that this series diverges is left as an exercise

(Problem 66). ■

# 8.6 Assess Your Understanding

## Concepts and Vocabulary

**1.** *True or False*   The Ratio Test can be used to show that the

series $\sum\limits_{k=1}^{\infty} \cos(k\pi)$ diverges.

**2.** *True or False*   In using the Ratio Test, if $\lim\limits_{n \to \infty} \left| \dfrac{a_{n+1}}{a_n} \right| = L < 1$,

then the sum of the series $\sum\limits_{k=1}^{\infty} a_k$ equals $L$.

**3.** *True or False*   In using the Ratio Test, if $\lim\limits_{n \to \infty} \left| \dfrac{a_{n+1}}{a_n} \right| = 1$,

then the Ratio Test indicates that the series $\sum\limits_{k=1}^{\infty} a_k$ converges.

**4.** *True or False*   The Root Test works well if the $n$th term of a series of nonzero terms involves an $n$th root.

## Skill Building

*In Problems 5–28, use the Ratio Test to determine whether each series converges or diverges or state that it provides no information.*

**5.** $\sum\limits_{k=1}^{\infty} \dfrac{4k^2 - 1}{2^k}$

**6.** $\sum\limits_{k=1}^{\infty} \dfrac{1}{(2k+1)2^k}$

**7.** $\sum\limits_{k=1}^{\infty} k \left( \dfrac{2}{3} \right)^k$

**8.** $\sum\limits_{k=1}^{\infty} \dfrac{5^k}{k^2}$

**9.** $\sum\limits_{k=1}^{\infty} \dfrac{10^k}{(2k)!}$

**10.** $\sum\limits_{k=1}^{\infty} \dfrac{(2k)!}{5^k 3^{k-1}}$

**11.** $\sum\limits_{k=1}^{\infty} \dfrac{k}{(2k-2)!}$

**12.** $\sum\limits_{k=1}^{\infty} \dfrac{(k+1)!}{3^k}$

**13.** $\sum\limits_{k=1}^{\infty} \dfrac{2^k}{k(k+1)}$

**14.** $\sum\limits_{k=1}^{\infty} \dfrac{k!}{k^2(k+1)^2}$

**15.** $\sum\limits_{k=1}^{\infty} \dfrac{k^3}{k!}$

**16.** $\sum\limits_{k=1}^{\infty} \dfrac{k!}{k^{k+1}}$

**17.** $\sum\limits_{k=1}^{\infty} (-1)^k \dfrac{k^2}{2^k}$

**18.** $\sum\limits_{k=1}^{\infty} (-1)^k \dfrac{k}{e^k}$

**19.** $\sum\limits_{k=2}^{\infty} (-1)^k \dfrac{\ln k}{3^k}$

**20.** $\sum\limits_{k=2}^{\infty} (-1)^k \dfrac{\ln k}{k!}$

**21.** $\sum\limits_{k=1}^{\infty} (-1)^k \dfrac{5^k}{3^{k-1}}$

**22.** $\sum\limits_{k=1}^{\infty} (-1)^k \dfrac{3^k + 4}{2^{k-1}}$

**23.** $\sum\limits_{k=1}^{\infty} \dfrac{3^{k-1}}{k \cdot 2^k}$

**24.** $\sum\limits_{k=1}^{\infty} \dfrac{k(k+2)}{3^k}$

**25.** $\sum\limits_{k=1}^{\infty} \dfrac{k}{e^k}$

**26.** $\sum\limits_{k=1}^{\infty} \dfrac{e^k}{k^3}$

**27.** $\sum\limits_{k=1}^{\infty} k \cdot 2^k$

**28.** $\sum\limits_{k=1}^{\infty} \dfrac{4^k}{k}$

*In Problems 29–46, use the Root Test to determine whether each series converges or diverges or state that it provides no information.*

**29.** $\sum\limits_{k=1}^{\infty} \left( \dfrac{2k+1}{5k+1} \right)^k$

**30.** $\sum\limits_{k=1}^{\infty} \left( \dfrac{3k-1}{2k+1} \right)^k$

**31.** $\sum\limits_{k=1}^{\infty} \left( \dfrac{k}{5} \right)^k$

**32.** $\sum\limits_{k=1}^{\infty} \dfrac{\pi^{2k}}{k^k}$

**33.** $\sum\limits_{k=2}^{\infty} \left( \dfrac{\ln k}{k} \right)^k$

**34.** $\sum\limits_{k=2}^{\infty} \left( \dfrac{1}{\ln k} \right)^k$

**35.** $\sum\limits_{k=2}^{\infty} (-1)^k \dfrac{(\ln k)^k}{2^k}$

**36.** $\sum\limits_{k=1}^{\infty} (-1)^k \dfrac{2^{k+1}}{e^k}$

**37.** $\sum\limits_{k=1}^{\infty} (-1)^k \left( \dfrac{k+1}{k} \right)^k$

**38.** $\sum\limits_{k=1}^{\infty} (-1)^k \left( \dfrac{2k+3}{k+1} \right)^k$

**39.** $\sum\limits_{k=1}^{\infty} (-1)^k \dfrac{k^k}{e^{2k}}$

**40.** $\sum\limits_{k=2}^{\infty} (-1)^k \dfrac{k \ln k}{e^k}$

---

**1.** = NOW WORK problem        〔∿〕 = Graphing technology recommended        〔CAS〕 = Computer Algebra System recommended

**41.** $\displaystyle\sum_{k=1}^{\infty} \left(\frac{\sqrt{k^2+1}}{3k}\right)^k$

**42.** $\displaystyle\sum_{k=1}^{\infty} \left(\frac{\sqrt{4k^2+1}}{k}\right)^k$

**43.** $\displaystyle\sum_{k=1}^{\infty} \frac{k^2}{2^k}$

**44.** $\displaystyle\sum_{k=1}^{\infty} \frac{k^3}{3^k}$

**45.** $\displaystyle\sum_{k=1}^{\infty} \frac{k^4}{5^k}$

**46.** $\displaystyle\sum_{k=1}^{\infty} \frac{k}{3^k}$

*In Problems 47–56, use the Ratio Test or Root Test to determine whether each series converges or diverges or state that it provides no information.*

**47.** $\displaystyle\sum_{k=1}^{\infty} \frac{10}{(3k+1)^k}$

**48.** $\displaystyle\sum_{k=1}^{\infty} \left(1+\frac{1}{k}\right)^{k^2}$

**49.** $\displaystyle\sum_{k=1}^{\infty} \frac{(k+1)(k+2)}{k!}$

**50.** $\displaystyle\sum_{k=1}^{\infty} \frac{k!}{(3k+1)!}$

**51.** $\displaystyle\sum_{k=1}^{\infty} \frac{k \ln k}{2^k}$

**52.** $\displaystyle\sum_{k=1}^{\infty} \left[\ln\left(e^3+\frac{1}{k}\right)\right]^k$

**53.** $\displaystyle\sum_{k=1}^{\infty} \sin^k\left(\frac{1}{k}\right)$

**54.** $\displaystyle\sum_{k=1}^{\infty} \frac{k^k}{2^{k^2}}$

**55.** $\displaystyle\sum_{k=1}^{\infty} \frac{\left(1+\frac{1}{k}\right)^{2k}}{e^k}$

**56.** $\displaystyle\sum_{k=2}^{\infty} \frac{2^k(k+1)}{k^2(k+2)}$

## Applications and Extensions

**57.** For the divergent series $\displaystyle\sum_{k=1}^{\infty} \frac{1}{k}$, show that $\displaystyle\lim_{n\to\infty} \left|\frac{a_{n+1}}{a_n}\right| = 1$.

**58.** For the convergent series $\displaystyle\sum_{k=1}^{\infty} \frac{1}{k^2}$, show that $\displaystyle\lim_{n\to\infty} \left|\frac{a_{n+1}}{a_n}\right| = 1$.

**59.** Give an example of a convergent series $\displaystyle\sum_{k=1}^{\infty} a_k$ for which $\displaystyle\lim_{n\to\infty} \left|\frac{a_{n+1}}{a_n}\right| \neq \infty$ does not exist.

**60.** Give an example of a divergent series $\displaystyle\sum_{k=1}^{\infty} a_k$ for which $\displaystyle\lim_{n\to\infty} \left|\frac{a_{n+1}}{a_n}\right| \neq \infty$ does not exist.

**61.** **(a)** Show that the series $\displaystyle\sum_{k=1}^{\infty} \frac{(-1)^k 3^k}{k!}$ converges.

      CAS **(b)** Use technology to find the sum of the series.

**62.** Determine whether the following series is convergent or divergent:

$$\frac{1}{3} - \frac{2^3}{3^2} + \frac{3^3}{3^3} - \frac{4^3}{3^4} + \cdots + \frac{(-1)^{n-1}n^3}{3^n} + \cdots$$

**63.** Show that $\displaystyle\lim_{n\to\infty} \frac{n!}{n^n} = 0$, where $n$ denotes a positive integer.

**64.** Show that the Root Test provides no information for $\displaystyle\sum_{k=1}^{\infty} \frac{1}{k}$ and $\displaystyle\sum_{k=1}^{\infty} \frac{1}{k^2}$.

**65.** Use the Ratio Test to find the real numbers $x$ for which the series $\displaystyle\sum_{k=1}^{\infty} \frac{x^k}{k^2}$ converges or diverges.

**66.** Prove that $\displaystyle\sum_{k=1}^{\infty} a_k$ diverges if $\displaystyle\lim_{n\to\infty} \left|\frac{a_{n+1}}{a_n}\right| = \infty$.

**67.** Prove the Root Test.

**68.** The terms of the series

$$\frac{1}{4} + \frac{1}{2} + \frac{1}{8} + \frac{1}{4} + \frac{1}{16} + \frac{1}{8} + \frac{1}{32} + \cdots$$

are $a_{2k} = \dfrac{1}{2^k}$ and $a_{2k-1} = \dfrac{1}{2^{k+1}}$.

  **(a)** Show that using the Ratio Test to determine whether the series converges provides no information.

  **(b)** Show that using the Root Test to determine whether the series converges is conclusive.

  **(c)** Does the series converge?

## Challenge Problems

**69.** Show that the following series converges:

$$1 + \frac{2}{2^2} + \frac{3}{3^3} + \frac{1}{4^4} + \frac{2}{5^5} + \frac{3}{6^6} + \cdots$$

**70.** Show that $\displaystyle\sum_{k=1}^{\infty} \frac{(k+1)^2}{(k+2)!}$ converges. *Hint:* Use the Limit Comparison Test and the convergent series $\displaystyle\sum_{k=1}^{\infty} \frac{1}{k!}$.

**71.** Suppose $0 < a < b < 1$. Use the Root Test to show that the series

$$a + b + a^2 + b^2 + a^3 + b^3 + \cdots$$

converges.

**72.** Show that if the Ratio Test indicates a series converges, then so will the Root Test. The converse is not true. Refer to Problem 68.

# 8.7 Summary of Tests

**OBJECTIVE** *When you finish this section, you should be able to:*

**1** Choose an appropriate test to determine whether a series converges (p. 637)

In the previous sections, we discussed a variety of tests that can be used to determine whether a series converges or diverges. In the exercises following each section, the series under consideration used the tests for convergence/divergence discussed in that section. In practice, such information is not provided. This section summarizes the tests that we have discussed and gives some clues as to what test has the best chance of answering the fundamental question, "Does the series converge or diverge?"

## 1 Choose an Appropriate Test to Determine Whether a Series Converges

The following outline is a guide to help you choose a test to use when determining the convergence or divergence of a series. Table 5 (on the next page) lists the tests we have discussed. Table 6 (on page 639) describes important series we have analyzed.

### Guide to Choosing a Test to Determine Whether a Series Is Convergent

**1.** Check to see if the series is a geometric series or a $p$-series. If yes, then use the conclusion given for these series in Table 6.

**2.** Find $\lim\limits_{n \to \infty} a_n$ of the series $\sum\limits_{k=1}^{\infty} a_k$. If $\lim\limits_{n \to \infty} a_n \neq 0$, then by the Test for Divergence, the series diverges.

**3.** If the series $\sum\limits_{k=1}^{\infty} a_k$ has only positive terms and meets the conditions of the Integral Test, find the related function $f$. Use the Integral Test if $\int_1^{\infty} f(x)\,dx$ is easy to find.

**4.** If the series $\sum\limits_{k=1}^{\infty} a_k$ has only positive terms and the $n$th term is a quotient of sums or differences of powers of $n$, the Limit Comparison Test with an appropriate $p$-series will usually work.

**5.** If the series $\sum\limits_{k=1}^{\infty} a_k$ has only positive terms and the preceding attempts fail, then try the Comparison Test for Convergence or the Comparison Test for Divergence.

**6.** *Series with some negative terms.*

- For an alternating series, use the Alternating Series Test. It is sometimes better to use the Absolute Convergence Test first.
- For other series containing negative terms, always use the Absolute Convergence Test first.

**7.** If the series $\sum\limits_{k=1}^{\infty} a_k$ has nonzero terms that involve products, factorials, or powers, the Ratio Test is a good choice.

**8.** If the series $\sum\limits_{k=1}^{\infty} a_k$ has nonzero terms and the $n$th term involves an $n$th power, try the Root Test.

**TABLE 5** Tests for Convergence and Divergence of Series

| Test Name | Description | Comment |
|---|---|---|
| **Test for Divergence** for all series (p. 602) | $\sum_{k=1}^{\infty} a_k$ diverges if $\lim_{n \to \infty} a_n \neq 0$. | No information is obtained about convergence if $\lim_{n \to \infty} a_n = 0$. |
| **Integral Test** for series of positive terms (p. 605) | $\sum_{k=1}^{\infty} a_k$ converges (diverges) if $\int_1^{\infty} f(x)\,dx$ converges (diverges), where $f$ is continuous, positive, and nonincreasing for $x \geq 1$; and $f(k) = a_k$ for all $k$. | Good to use if $f$ is easy to integrate. |
| **Comparison Test for Convergence** for series of positive terms (p. 613) | $\sum_{k=1}^{\infty} a_k$ converges if $0 < a_k \leq b_k$ and the series $\sum_{k=1}^{\infty} b_k$ converges. | $\sum_{k=1}^{\infty} b_k$ must have positive terms and be convergent. |
| **Comparison Test for Divergence** for series of positive terms (p. 614) | $\sum_{k=1}^{\infty} a_k$ diverges if $a_k \geq c_k > 0$ and the series $\sum_{k=1}^{\infty} c_k$ diverges. | $\sum_{k=1}^{\infty} c_k$ must have positive terms and be divergent. |
| **Limit Comparison Test** for series of positive terms (p. 614) | $\sum_{k=1}^{\infty} a_k$ converges (diverges) if $\sum_{k=1}^{\infty} b_k$ converges (diverges), and $\lim_{n \to \infty} \dfrac{a_n}{b_n} = L$, a positive real number. | $\sum_{k=1}^{\infty} b_k$ must have positive terms, whose convergence (divergence) can be determined. |
| **Alternating Series Test** (p. 621) | $\sum_{k=1}^{\infty} (-1)^{k+1} a_k, \ a_k > 0$, converges if <br> • $\lim_{n \to \infty} a_n = 0$ and <br> • the $a_k$ are nonincreasing. | The error made by using the $n$th partial sum to approximate the sum $S$ of the series is less than or equal to $|a_{n+1}|$. |
| **Absolute Convergence Test** (p. 625) | If $\sum_{k=1}^{\infty} |a_k|$ converges, then $\sum_{k=1}^{\infty} a_k$ converges. | The converse is not true. That is, if $\sum_{k=1}^{\infty} |a_k|$ diverges, $\sum_{k=1}^{\infty} a_k$ may or may not converge. |
| **Ratio Test** for series with nonzero terms (p. 630) | $\sum_{k=1}^{\infty} a_k$ converges if $\lim_{n \to \infty} \left| \dfrac{a_{n+1}}{a_n} \right| < 1$. <br><br> $\sum_{k=1}^{\infty} a_k$ diverges if $\lim_{n \to \infty} \left| \dfrac{a_{n+1}}{a_n} \right| > 1$ or if $\lim_{n \to \infty} \left| \dfrac{a_{n+1}}{a_n} \right| = \infty$. | Good to use if $a_n$ includes factorials or powers. It provides no information if $\lim_{n \to \infty} \left| \dfrac{a_{n+1}}{a_n} \right| = 1$ or if $\lim_{n \to \infty} \left| \dfrac{a_{n+1}}{a_n} \right| \neq \infty$ does not exist. |
| **Root Test** for series with nonzero terms (p. 633) | $\sum_{k=1}^{\infty} a_k$ converges if $\lim_{n \to \infty} \sqrt[n]{|a_n|} < 1$. <br><br> $\sum_{k=1}^{\infty} a_k$ diverges if $\lim_{n \to \infty} \sqrt[n]{|a_n|} > 1$ or if $\lim_{n \to \infty} \sqrt[n]{|a_n|} = \infty$. | Good to use if $a_n$ involves $n$th powers. It provides no information if $\lim_{n \to \infty} \sqrt[n]{|a_n|} = 1$. |

**TABLE 6** Important Series

| Series Name | Series Description | Comments |
|---|---|---|
| *Geometric series* (pp. 593–594) | $\sum_{k=1}^{\infty} ar^{k-1} = a + ar + ar^2 + \cdots, \ a \neq 0$ | Converges to $\dfrac{a}{1-r}$ if $|r| < 1$; diverges if $|r| \geq 1$. |
| *Harmonic series* (p. 597) | $\sum_{k=1}^{\infty} \dfrac{1}{k} = 1 + \dfrac{1}{2} + \dfrac{1}{3} + \cdots$ | Diverges. |
| *p-series* (p. 607) | $\sum_{k=1}^{\infty} \dfrac{1}{k^p} = 1 + \dfrac{1}{2^p} + \dfrac{1}{3^p} + \cdots$ | Converges if $p > 1$; diverges if $0 < p \leq 1$. |
| *k-to-the-k series* (p. 613) | $\sum_{k=1}^{\infty} \dfrac{1}{k^k} = 1 + \dfrac{1}{2^2} + \dfrac{1}{3^3} + \dfrac{1}{4^4} + \cdots$ | Converges. |
| *Alternating harmonic series* (p. 622) | $\sum_{k=1}^{\infty} \dfrac{(-1)^{k+1}}{k} = 1 - \dfrac{1}{2} + \dfrac{1}{3} - \dfrac{1}{4} + \cdots$ | Converges. |

# 8.7 Assess Your Understanding

## Concepts and Vocabulary

**1.** *True or False* The series $\sum_{k=1}^{\infty} \dfrac{1}{k^p}$ converges if $p \geq 1$.

**2.** *True or False* According to the Test for Divergence, an infinite series $\sum_{k=1}^{\infty} a_k$ converges if $\lim_{n \to \infty} a_n = 0$.

**3.** *True or False* If a series is absolutely convergent, then it is convergent.

**4.** *True or False* If a series is not absolutely convergent, then it is divergent.

**5.** *True or False* According to the Ratio Test, a series $\sum_{k=1}^{\infty} a_k$ of nonzero terms converges if $\left| \dfrac{a_{n+1}}{a_n} \right| < 1$.

**6.** To use the Comparison Test for Convergence to show that a series $\sum_{k=1}^{\infty} a_k$ of positive terms converges, find a series $\sum_{k=1}^{\infty} b_k$ that is known to converge and show that $0 < \underline{\quad} \leq \underline{\quad}$.

## Skill Building

*In Problems 7–39, determine whether each series converges (absolutely or conditionally) or diverges. Use any applicable test.*

**7.** $\sum_{k=1}^{\infty} \dfrac{9k^3 + 5k^2}{k^{5/2} + 4}$

**8.** $\sum_{k=1}^{\infty} \dfrac{(-1)^{k+1}}{\sqrt{2k+1}}$

**9.** $6 + 2 + \dfrac{2}{3} + \dfrac{2}{9} + \dfrac{2}{27} + \cdots$

**10.** $\sum_{k=1}^{\infty} \dfrac{1}{k^2} \sin \dfrac{\pi}{k}$

**11.** $\sum_{k=1}^{\infty} \dfrac{3k+2}{k^3+1}$

**12.** $1 + \dfrac{2^2+1}{2^3+1} + \dfrac{3^2+1}{3^3+1} + \dfrac{4^2+1}{4^3+1} + \cdots$

**13.** $\sum_{k=1}^{\infty} \dfrac{k+4}{k\sqrt{3k-2}}$

**14.** $\sum_{k=1}^{\infty} \dfrac{\sin k}{k^3}$

**15.** $\sum_{k=1}^{\infty} \dfrac{3^{2k-1}}{k^2 + 2k}$

**16.** $\sum_{k=1}^{\infty} \dfrac{5^k}{k!}$

**17.** $\sum_{k=1}^{\infty} \left(1 + \dfrac{2}{k}\right)^k$

**18.** $\sum_{k=1}^{\infty} \dfrac{k^2 + 4}{e^k}$

**19.** $\dfrac{2}{3} - \dfrac{3}{4} \cdot \dfrac{1}{2} + \dfrac{4}{5} \cdot \dfrac{1}{3} - \dfrac{5}{6} \cdot \dfrac{1}{4} + \cdots$

**20.** $2 + \dfrac{3}{2} \cdot \dfrac{1}{4} + \dfrac{4}{3} \cdot \dfrac{1}{4^2} + \dfrac{5}{4} \cdot \dfrac{1}{4^3} + \cdots$

**21.** $1 + \dfrac{1 \cdot 3}{2!} + \dfrac{1 \cdot 3 \cdot 5}{3!} + \dfrac{1 \cdot 3 \cdot 5 \cdot 7}{4!} + \cdots$

**22.** $\dfrac{1}{\sqrt{1 \cdot 2 \cdot 3}} + \dfrac{1}{\sqrt{2 \cdot 3 \cdot 4}} + \dfrac{1}{\sqrt{3 \cdot 4 \cdot 5}} + \cdots$

**23.** $\sum_{k=1}^{\infty} \dfrac{k!}{(2k)!}$

**24.** $\sum_{k=1}^{\infty} k^3 e^{-k^4}$

**25.** $\sum_{k=1}^{\infty} \dfrac{1}{\sqrt{k} + 100}$

**26.** $\sum_{k=1}^{\infty} \dfrac{k^2 + 5k}{3 + 5k^2}$

**27.** $\sum_{k=1}^{\infty} \dfrac{1}{\sqrt[3]{k^4 + 4}}$

**28.** $\sum_{k=1}^{\infty} \dfrac{1}{11} \left(\dfrac{-3}{2}\right)^k$

**29.** $\dfrac{1}{3} - \dfrac{2}{4} + \dfrac{3}{5} - \dfrac{4}{6} + \cdots$

**30.** $\sum_{k=1}^{\infty} \dfrac{k(-4)^{3k}}{5^k}$

**31.** $\sum_{k=1}^{\infty} \left(-\dfrac{1}{k}\right)^k$

**32.** $\sum_{k=1}^{\infty} \dfrac{5}{2^k + 1}$

**33.** $\sum_{k=1}^{\infty} e^{-k^2}$

**34.** $\dfrac{\sin\sqrt{1}}{1^{3/2}} + \dfrac{\sin\sqrt{2}}{2^{3/2}} + \dfrac{\sin\sqrt{3}}{3^{3/2}} + \cdots$

**35.** $\displaystyle\sum_{k=2}^{\infty} \dfrac{(-1)^{k-1}}{k(\ln k)^3}$

**36.** $\displaystyle\sum_{k=1}^{\infty} \dfrac{1}{(2k)^k}$

**37.** $\displaystyle\sum_{k=2}^{\infty} \left(\dfrac{\ln k}{1000}\right)^k$

**38.** $\displaystyle\sum_{k=1}^{\infty} \dfrac{1}{\cosh^2 k}$

**39.** $\displaystyle\sum_{k=1}^{\infty} \dfrac{\tan^{-1} k}{k^2}$

*In Problems 40–42, determine whether each series converges or diverges. If it converges, find its sum.*

**40.** $\displaystyle\sum_{k=1}^{\infty} \left(\sqrt{k+1} - \sqrt{k}\right)$

**41.** $\displaystyle\sum_{k=4}^{\infty} \left(\dfrac{1}{k-3} - \dfrac{1}{k}\right)$

**42.** $\displaystyle\sum_{k=2}^{\infty} \ln \dfrac{k}{k+1}$

**43.** Determine whether $1 + \dfrac{1\cdot 2}{1\cdot 3} + \dfrac{1\cdot 2\cdot 3}{1\cdot 3\cdot 5} + \dfrac{1\cdot 2\cdot 3\cdot 4}{1\cdot 3\cdot 5\cdot 7} + \cdots$ converges or diverges.

**44.** **(a)** Show that the series $\displaystyle\sum_{k=1}^{\infty} \left[\left(\dfrac{2}{3}\right)^k - \dfrac{2}{k^2 + 2k}\right]$ converges.

**(b)** Find the sum of the series.

**45.** **(a)** Show that the series $\displaystyle\sum_{k=1}^{\infty} \left[\left(-\dfrac{1}{4}\right)^k + \dfrac{3}{k(k+1)}\right]$ converges.

**(b)** Find the sum of the series.

**Challenge Problems**

**46.** **(a)** Determine whether the series
$$1 - 1 - \dfrac{1}{2} + \dfrac{1}{3} + \dfrac{1}{3} - \dfrac{1}{9} - \dfrac{1}{4} + \dfrac{1}{27} + \dfrac{1}{5} - \dfrac{1}{81} - \cdots$$
converges or diverges.

**(b)** Find the sum of the series if it converges.

*In Problems 47 and 48, determine whether each series converges or diverges.*

**47.** $\displaystyle\sum_{k=1}^{\infty} \dfrac{\ln k}{2k^3 - 1}$

**48.** $\displaystyle\sum_{k=1}^{\infty} \sin^3\left(\dfrac{1}{k}\right)$

# 8.8 Power Series

**OBJECTIVES** *When you finish this section, you should be able to:*

1. Determine whether a power series converges (p. 641)
2. Find the interval of convergence of a power series (p. 643)
3. Define a function using a power series (p. 645)
4. Use properties of power series (p. 647)

In this section, we study series with variable terms, called *power series*. Just as a polynomial is the sum of a *finite* number of monomials, a power series is the sum of an *infinite* number of monomials.

---

**DEFINITION** Power Series

If $x$ is a variable, then a series of the form

$$\sum_{k=0}^{\infty} a_k x^k = a_0 + \sum_{k=1}^{\infty} a_k x^k = a_0 + a_1 x + a_2 x^2 + \cdots$$

where the coefficients $a_0, a_1, a_2, \ldots$ are constants, is called a **power series in $x$** or a **power series centered at 0**.

A series of the form

$$\sum_{k=0}^{\infty} a_k (x-c)^k = a_0 + \sum_{k=1}^{\infty} a_k (x-c)^k = a_0 + a_1(x-c) + a_2(x-c)^2 + \cdots$$

where $c$ is a constant, is called a **power series in $(x-c)$** or a **power series centered at $c$**.

---

**IN WORDS** A power series $\displaystyle\sum_{k=0}^{\infty} a_k x^k$ is a sum of an infinite number of monomials.

If $x = 0$ in a power series $\displaystyle\sum_{k=0}^{\infty} a_k x^k$, or if $x = c$ in a power series $\displaystyle\sum_{k=0}^{\infty} a_k (x-c)^k$, then the power series equals $a_0$.

## 1 Determine Whether a Power Series Converges

For a particular value of $x$, a power series in $x$ reduces to a series of real numbers like the series studied so far. For example, $\sum\limits_{k=1}^{\infty} \dfrac{x^k}{k}$ is a power series in $x$. The series converges (to 0) if $x = 0$. If $x = 1$, it becomes the harmonic series $\sum\limits_{k=1}^{\infty} \dfrac{1}{k}$, which is divergent. If $x = -1$, it becomes the alternating harmonic series $\sum\limits_{k=1}^{\infty} (-1)^k \dfrac{1}{k}$, which is convergent. To find all numbers $x$ for which a power series in $x$ is convergent, the Ratio Test (p. 630) or the Root Test (p. 633) is usually used since, in a power series, $x$ is raised to a power.

---

**EXAMPLE 1** **Determining Whether a Power Series Converges**

Find all numbers $x$ for which each power series in $x$ converges.

**(a)** $\sum\limits_{k=0}^{\infty} \dfrac{x^k}{k!} = 1 + x + \dfrac{x^2}{2!} + \dfrac{x^3}{3!} + \cdots$

**(b)** $\sum\limits_{k=0}^{\infty} \dfrac{kx^k}{4^k} = \dfrac{x}{4} + \dfrac{2x^2}{4^2} + \dfrac{3x^3}{4^3} + \cdots$

**(c)** $\sum\limits_{k=0}^{\infty} k!\,x^k = 1 + x + 2!\,x^2 + 3!\,x^3 + \cdots$

**Solution (a)** For the series $\sum\limits_{k=0}^{\infty} \dfrac{x^k}{k!}$, we use the Ratio Test with

$$a_n = \frac{x^n}{n!} \qquad \text{and} \qquad a_{n+1} = \frac{x^{n+1}}{(n+1)!}$$

Then

$$\lim_{n\to\infty} \left| \frac{a_{n+1}}{a_n} \right| = \lim_{n\to\infty} \left| \frac{\frac{x^{n+1}}{(n+1)!}}{\frac{x^n}{n!}} \right| = \lim_{n\to\infty} \frac{|x|^{n+1}\, n!}{(n+1)!\, |x|^n} = |x| \lim_{n\to\infty} \frac{1}{n+1} = 0$$

**NOTE** Since $\sum\limits_{k=0}^{\infty} \dfrac{x^k}{k!}$ converges absolutely for every number $x$, the limit of the $n$th term equals 0. That is, $\lim\limits_{n\to\infty} \dfrac{x^n}{n!} = 0$ for every number $x$.

Since the limit is less than 1 for every number $x$, it follows from the Ratio Test that the power series $\sum\limits_{k=0}^{\infty} \dfrac{x^k}{k!}$ is absolutely convergent for all real numbers.

**(b)** For $\sum\limits_{k=0}^{\infty} \dfrac{kx^k}{4^k}$, we use the Ratio Test with $a_n = \dfrac{nx^n}{4^n}$ and $a_{n+1} = \dfrac{(n+1)x^{n+1}}{4^{n+1}}$. Then

$$\lim_{n\to\infty} \left| \frac{a_{n+1}}{a_n} \right| = \lim_{n\to\infty} \frac{\frac{(n+1)\,|x|^{n+1}}{4^{n+1}}}{\frac{n\,|x|^n}{4^n}} = \lim_{n\to\infty} \frac{(n+1)\,|x|^{n+1}\cdot 4^n}{4^{n+1}\cdot n\,|x|^n}$$

$$= |x| \lim_{n\to\infty} \frac{n+1}{4n} = \frac{|x|}{4}$$

By the Ratio Test, the series converges absolutely if $\dfrac{|x|}{4} < 1$, or equivalently if $|x| < 4$.

It diverges if $\dfrac{|x|}{4} > 1$ or equivalently if $|x| > 4$. The Ratio Test gives no information when $\dfrac{|x|}{4} = 1$, that is, when $x = -4$ or $x = 4$. However, we can check these values directly by replacing $x$ by 4 and $-4$.

For $x = 4$, the series becomes

$$\sum_{k=0}^{\infty} \frac{k4^k}{4^k} = \sum_{k=1}^{\infty} k = 1 + 2 + \cdots$$

which diverges. For $x = -4$, the series becomes

$$\sum_{k=0}^{\infty} \frac{k(-4)^k}{4^k} = \sum_{k=0}^{\infty} \frac{(-1)^k k(4^k)}{4^k} = \sum_{k=0}^{\infty} (-1)^k k = -1 + 2 - 3 + \cdots$$

which also diverges. (Look at the sequence of partial sums.)

The series $\displaystyle\sum_{k=0}^{\infty} \frac{kx^k}{4^k}$ converges absolutely for $-4 < x < 4$ and diverges for $|x| \geq 4$.

(c) For $\displaystyle\sum_{k=0}^{\infty} k!x^k$, we use the Ratio Test with $a_n = n!x^n$ and $a_{n+1} = (n+1)!x^{n+1}$. Then

$$\lim_{n \to \infty} \left| \frac{a_{n+1}}{a_n} \right| = \lim_{n \to \infty} \frac{(n+1)!|x|^{n+1}}{n!\,|x|^n} = |x| \lim_{n \to \infty} (n+1) = \begin{cases} 0 & \text{if } x = 0 \\ \infty & \text{if } x \neq 0 \end{cases}$$

We conclude that the power series $\displaystyle\sum_{k=0}^{\infty} k!\,x^k$ converges only when $x = 0$. For any other number $x$, the power series diverges. ∎

NOW WORK Problem 15.

The next theorem gives more information about the numbers for which a power series converges or diverges.

> **THEOREM  Convergence/Divergence of a Power Series**
>
> (a) If the power series $\displaystyle\sum_{k=0}^{\infty} a_k x^k$ converges for a number $x_0 \neq 0$, then it converges absolutely for all numbers $x$ for which $|x| < |x_0|$.
>
> (b) If the power series $\displaystyle\sum_{k=0}^{\infty} a_k x^k$ diverges for a number $x_1$, then it diverges for all numbers $x$ for which $|x| > |x_1|$.

**Proof  Part (a)**  Assume that $\displaystyle\sum_{k=0}^{\infty} a_k x_0^k$ converges. Then

$$\lim_{n \to \infty} \left( a_n x_0^n \right) = 0$$

**RECALL** If a series $\displaystyle\sum_{k=0}^{\infty} a_k$ converges, then $\displaystyle\lim_{n \to \infty} a_n = 0$.

Using the definition of the limit of a sequence and choosing $\varepsilon = 1$, there is a positive integer $N$ for which $|a_n x_0^n| < 1$ for all $n > N$. Now for any number $x$ for which $|x| < |x_0|$, we have

$$|a_n x^n| = \left| \frac{a_n x^n x_0^n}{x_0^n} \right| = |a_n x_0^n| \left| \frac{x}{x_0} \right|^n < \left| \frac{x}{x_0} \right|^n \qquad \text{for } n > N$$

$$\underset{\substack{\uparrow \\ |a_n x_0^n| < 1}}{}$$

Since $|x| < |x_0|$, then $\left| \dfrac{x}{x_0} \right| < 1$, so the series $\displaystyle\sum_{k=0}^{\infty} \left| \frac{x}{x_0} \right|^k$ is a convergent geometric series. Therefore, by the Comparison Test for Convergence, the series $\displaystyle\sum_{k=0}^{\infty} |a_k x^k|$ converges, and so the power series $\displaystyle\sum_{k=0}^{\infty} a_k x^k$ converges absolutely for all numbers $x$ for which $|x| < |x_0|$.

**Part (b)**  Suppose the series converges for some number $x$, $|x| > |x_1|$. Then it must converge for $x_1$ [by Part (a)], which contradicts the hypothesis of the theorem. Therefore, the series diverges for all $x$ for which $|x| > |x_1|$. ∎

**EXAMPLE 2** Using the Convergence/Divergence of a Power Series Theorem

(a) If the power series $\sum_{k=0}^{\infty} a_k x^k$ converges for $x = 6$, then it converges absolutely for $x = 3$.

(b) If the power series $\sum_{k=0}^{\infty} a_k x^k$ diverges for $x = 4$, then it also diverges for $x = -5$. ∎

**NOW WORK** Problems 9 and 51.

The next result is a consequence of the previous theorem. It states that every power series belongs to one of three categories.

**THEOREM**

For a power series $\sum_{k=0}^{\infty} a_k (x - c)^k$, exactly one of the following is true:

- The series converges only if $x = c$.
- The series converges absolutely for all $x$.
- There is a positive number $R$ for which the series converges absolutely for all $x$, $|x - c| < R$, and diverges for all $x$, $|x - c| > R$. The behavior of the series at $|x - c| = R$ must be determined separately.

In the theorem, the number $R$ is called the **radius of convergence.** If the series converges only for $x = c$, then $R = 0$; if the series converges absolutely for all $x$, then $R = \infty$. If the series converges absolutely for $|x - c| < R$, $0 < R < \infty$, we call the set of all numbers $x$ for which the power series converges the **interval of convergence** of the power series. Once the radius $R$ of convergence is determined, we test the endpoints $x = c - R$ and $x = c + R$ to find the interval of convergence.

As Example 1 illustrates, the Ratio Test is a useful method for determining the radius of convergence of a power series. However, the test gives no information about convergence or divergence at the endpoints of the interval of convergence. At an endpoint, a power series may be convergent or divergent.

For example, the series $\sum_{k=0}^{\infty} \dfrac{kx^k}{4^k}$ in Example 1(b) converges absolutely for $|x| < 4$ and diverges for $|x| \geq 4$. So, the radius of convergence is $R = 4$, and the interval of convergence is the open interval $(-4, 4)$, as shown in Figure 26.

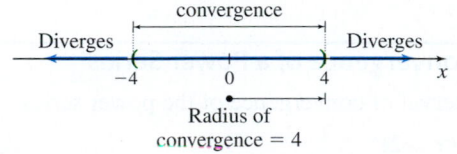

**Figure 26** Convergence/divergence of $\sum_{k=0}^{\infty} \dfrac{k\,x^k}{4^k}$

**NOW WORK** Problems 4 and 11.

## 2 Find the Interval of Convergence of a Power Series

**CALC CLIP**

**EXAMPLE 3** Finding the Interval of Convergence of a Power Series

Find the radius of convergence and the interval of convergence of the power series

$$\sum_{k=1}^{\infty} \frac{x^{2k}}{k}$$

**Solution** We use the Ratio Test with $a_n = \dfrac{x^{2n}}{n}$ and $a_{n+1} = \dfrac{x^{2(n+1)}}{n+1}$. Then

$$\lim \left| \frac{a_{n+1}}{a_n} \right| = \lim_{n \to \infty} \left| \frac{\frac{x^{2(n+1)}}{n+1}}{\frac{x^{2n}}{n}} \right| = \lim_{n \to \infty} \left| \frac{x^{2n+2}}{n+1} \cdot \frac{n}{x^{2n}} \right| = x^2 \lim_{n \to \infty} \frac{n}{n+1} = x^2$$

The series converges absolutely if $x^2 < 1$, or equivalently, if $-1 < x < 1$. The radius of convergence is $R = 1$.

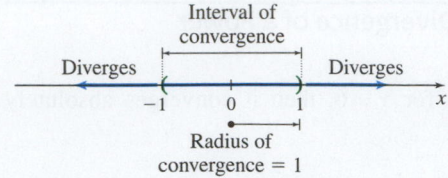

**Figure 27** Convergence/divergence

of $\sum\limits_{k=0}^{\infty} \dfrac{x^{2k}}{k}$

To find the interval of convergence, we test the endpoints: $x = -1$ and $x = 1$.

When $x = 1$ or $x = -1$, $\dfrac{x^{2k}}{k} = \dfrac{(x^2)^k}{k} = \dfrac{1^k}{k} = \dfrac{1}{k}$, so the series reduces to the harmonic

series $\sum\limits_{k=1}^{\infty} \dfrac{1}{k}$, which diverges. Consequently, the interval of convergence is $-1 < x < 1$,

as shown in Figure 27. ∎

**NOW WORK** Problem 19(a).

The power series discussed next converges for all real numbers.

**EXAMPLE 4** **Finding the Interval of Convergence of a Power Series**

Find the radius of convergence $R$ and the interval of convergence of the power series

$$\sum_{k=0}^{\infty} \frac{x^k}{(k+2)^{2k}}$$

**Solution** We use the Root Test. Then

$$\lim_{n\to\infty} \sqrt[n]{\left| \frac{x^n}{(n+2)^{2n}} \right|} = \lim_{n\to\infty} \frac{|x|}{(n+2)^2} = |x| \lim_{n\to\infty} \frac{1}{(n+2)^2} = 0$$

The series converges absolutely for all $x$. The radius of convergence is $R = \infty$, and
the interval of convergence is $(-\infty, \infty)$. ∎

**NOW WORK** Problems 19(b) and 21.

The methods used in Examples 3 and 4 can also be used to find the radius of
convergence of a power series in $x - c$.

**EXAMPLE 5** **Finding the Interval of Convergence of a Power Series**

Find the radius of convergence $R$ and the interval of convergence of the power series

$$\sum_{k=0}^{\infty} (-1)^k \frac{(x-2)^k}{k+1}$$

**Solution** $\sum\limits_{k=0}^{\infty} (-1)^k \dfrac{(x-2)^k}{k+1}$ is a power series centered at 2. Use the Ratio Test

with $a_n = (-1)^n \dfrac{(x-2)^n}{n+1}$ and $a_{n+1} = (-1)^{n+1} \dfrac{(x-2)^{n+1}}{n+2}$. Then

$$\lim_{n\to\infty} \left| \frac{a_{n+1}}{a_n} \right| = \lim_{n\to\infty} \left| \frac{\dfrac{(-1)^{n+1}(x-2)^{n+1}}{n+2}}{\dfrac{(-1)^n (x-2)^n}{n+1}} \right| = \lim_{n\to\infty} \left| \frac{(n+1)(x-2)}{n+2} \right|$$

$$= |x-2| \lim_{n\to\infty} \frac{n+1}{n+2} = |x-2|$$

The series converges absolutely if $|x-2| < 1$, or equivalently if $1 < x < 3$. The
radius of convergence is $R = 1$. To find the interval of convergence, test the endpoints.
If $x = 1$,

$$\sum_{k=0}^{\infty} (-1)^k \frac{(x-2)^k}{k+1} = \sum_{k=0}^{\infty} (-1)^k \frac{(-1)^k}{k+1} = \sum_{k=0}^{\infty} \frac{(-1)^{2k}}{k+1} = \sum_{k=0}^{\infty} \frac{1}{k+1}$$

$$= 1 + \frac{1}{2} + \frac{1}{3} + \cdots + \frac{1}{n+1} + \cdots$$

which is the divergent harmonic series.

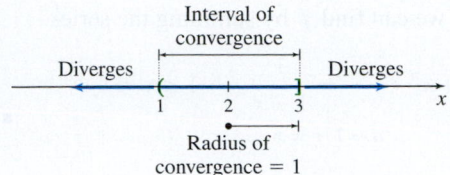

**Figure 28** Convergence/divergence

of $\displaystyle\sum_{k=0}^{\infty}(-1)\frac{(x-2)^k}{k+1}$

If $x=3$,

$$\sum_{k=0}^{\infty}(-1)^k\frac{(x-2)^k}{k+1}=\sum_{k=0}^{\infty}(-1)^k\frac{1}{k+1}=1-\frac{1}{2}+\frac{1}{3}-\frac{1}{4}+\cdots+\frac{(-1)^n}{n+1}+\cdots$$

which is the convergent alternating harmonic series.

The series $\displaystyle\sum_{k=0}^{\infty}(-1)^k\frac{(x-2)^k}{k+1}$ converges for $1<x\le 3$, as shown in Figure 28. ∎

NOW WORK Problem 29.

## 3 Define a Function Using a Power Series

A power series $\displaystyle\sum_{k=0}^{\infty}a_kx^k$ defines a function

$$f(x)=a_0+a_1x+a_2x^2+\cdots+a_nx^n+\cdots$$

The domain of $f$ is the interval of convergence of the power series.

If $f$ is defined by the power series $\displaystyle\sum_{k=0}^{\infty}a_kx^k$, whose interval of convergence is $I$,

and if $x_0$ is a number in $I$, then $f$ can be evaluated at $x_0$ by finding the sum of the series

$$f(x_0)=\sum_{k=0}^{\infty}a_kx_0^k=a_0+a_1x_0+a_2x_0^2+\cdots+a_nx_0^n+\cdots$$

---

EXAMPLE 6 **Analyzing a Function Defined by a Power Series**

A function $f$ is defined by the power series $\displaystyle f(x)=\sum_{k=0}^{\infty}x^k$.

**(a)** Find the domain of $f$.

**(b)** Evaluate $f\left(\dfrac{1}{2}\right)$ and $f\left(-\dfrac{1}{3}\right)$.

**(c)** Find $f$ by summing the series.

Solution **(a)** $\displaystyle\sum_{k=0}^{\infty}x^k$ is a power series centered at 0 with $a_k=1$. Then

$$f(x)=1+x+x^2+x^3+x^4+\cdots$$

The domain of $f$ equals the interval of convergence of the power series. The series $\displaystyle\sum_{k=0}^{\infty}x^k$ is a geometric series, so it converges for $|x|<1$. The radius of convergence is 1, and the interval of convergence is $(-1,1)$. The domain of $f$ is the open interval $(-1,1)$.

**(b)** The numbers $\dfrac{1}{2}$ and $-\dfrac{1}{3}$ are in the interval $(-1,1)$, so they are in the domain of $f$.

Then $f\left(\dfrac{1}{2}\right)$ is a geometric series with $r=\dfrac{1}{2}$, $a=1$, and

$$f\left(\frac{1}{2}\right)=1+\frac{1}{2}+\left(\frac{1}{2}\right)^2+\left(\frac{1}{2}\right)^3+\cdots=\frac{a}{1-r}=\frac{1}{1-\dfrac{1}{2}}=2$$

Similarly,

$$f\left(-\frac{1}{3}\right)=1-\frac{1}{3}+\left(-\frac{1}{3}\right)^2+\left(-\frac{1}{3}\right)^3+\cdots=\frac{1}{1+\dfrac{1}{3}}=\frac{3}{4}$$

**(c)** Since $f$ is defined by a geometric series, we can find $f$ by summing the series.

$$f(x) = \sum_{k=0}^{\infty} x^k = 1 + x + x^2 + \cdots + x^n + \cdots = \underset{\underset{a=1;\, r=x}{\uparrow}}{\frac{1}{1-x}} \qquad -1 < x < 1$$

■

In Example 6, the function $f$ defined by the power series $\sum\limits_{k=0}^{\infty} x^k$ was found to be

$$f(x) = \frac{1}{1-x} = 1 + x + x^2 + x^3 + \cdots = \sum_{k=0}^{\infty} x^k \qquad -1 < x < 1 \qquad (1)$$

We say that the function $f$ is **represented** by the power series.

NOW WORK Problem 45.

Many other functions can be represented by power series formed from variations of the function $f(x) = \dfrac{1}{1-x} = \sum\limits_{k=1}^{\infty} x^k, -1 < x < 1.$

EXAMPLE 7 **Representing a Function by a Power Series Centered at 0**

Each of the following functions is a variation of the function $f(x) = \dfrac{1}{1-x}$. Represent each function by a power series centered at 0:

**(a)** $h(x) = \dfrac{1}{1-2x^2}$  **(b)** $g(x) = \dfrac{1}{3+x}$  **(c)** $F(x) = \dfrac{x^2}{1-x}$

**Solution**

**(a)** In the power series representation for $f(x) = \dfrac{1}{1-x}$, statement (1), replace $x$ by $2x^2$. The resulting series

$$\sum_{k=0}^{\infty} (2x^2)^k = 1 + 2x^2 + (2x^2)^2 + \cdots + (2x^2)^n + \cdots$$

converges if $|2x^2| < 1$, or equivalently if $-\dfrac{\sqrt{2}}{2} < x < \dfrac{\sqrt{2}}{2}$. Then on the open interval $\left(-\dfrac{\sqrt{2}}{2}, \dfrac{\sqrt{2}}{2}\right)$, the function $h(x) = \dfrac{1}{1-2x^2}$ is represented by the power series

$$h(x) = \frac{1}{1-2x^2} = 1 + (2x^2) + (2x^2)^2 + (2x^2)^3 + \cdots$$

$$= 1 + 2x^2 + 4x^4 + 8x^6 + \cdots + 2^n x^{2n} + \cdots = \sum_{k=0}^{\infty} (2x^2)^k = \sum_{k=0}^{\infty} 2^k x^{2k}$$

**(b)** We begin by writing

$$g(x) = \frac{1}{3+x} = \frac{1}{3}\left(\frac{1}{1+\dfrac{x}{3}}\right) = \frac{1}{3}\left[\frac{1}{1-\left(-\dfrac{x}{3}\right)}\right]$$

Now in the power series representation for $f(x) = \dfrac{1}{1-x}$, replace $x$ by $-\dfrac{x}{3}$. The resulting series

$$\sum_{k=0}^{\infty}\left(-\frac{x}{3}\right)^k = 1 + \left(-\frac{x}{3}\right) + \left(-\frac{x}{3}\right)^2 + \cdots + \left(-\frac{x}{3}\right)^n + \cdots$$

converges if $\left| -\dfrac{x}{3} \right| < 1$, or equivalently if $-3 < x < 3$. Then in the open interval $(-3, 3)$,

$g(x) = \dfrac{1}{3+x}$ is represented by the power series

$$g(x) = \frac{1}{3+x} = \frac{1}{3} \left[ 1 + \left( -\frac{x}{3} \right) + \left( -\frac{x}{3} \right)^2 + \left( -\frac{x}{3} \right)^3 + \cdots \right] = \frac{1}{3} \sum_{k=0}^{\infty} (-1)^k \left( \frac{x}{3} \right)^k$$

$$= \sum_{k=0}^{\infty} \frac{(-1)^k x^k}{3^{k+1}}$$

**(c)** $F(x) = \dfrac{x^2}{1-x} = x^2 \left( \dfrac{1}{1-x} \right)$. Now for all numbers in the interval $(-1, 1)$,

$$\frac{1}{1-x} = 1 + x + x^2 + \cdots + x^n + \cdots$$

So for any number $x$ in the interval of convergence, $-1 < x < 1$, we have

**NOTE** Notice that the power series representation for $F$ begins at $k = 2$.

$$F(x) = x^2 \left( 1 + x + x^2 + \cdots + x^n + \cdots \right) = x^2 + x^3 + x^4 + \cdots + x^{n+2} + \cdots = \sum_{k=2}^{\infty} x^k$$ ∎

**NOW WORK** Problem 55.

## 4 Use Properties of Power Series

The function $f$ represented by a power series has properties similar to those of a polynomial. We state three of these properties without proof.

**THEOREM Properties of Power Series**

Let $\displaystyle\sum_{k=0}^{\infty} a_k x^k$ be a power series in $x$ having a nonzero radius of convergence $R$. Define the function $f$ as

$$f(x) = \sum_{k=0}^{\infty} a_k x^k = a_0 + a_1 x + a_2 x^2 + \cdots + a_n x^n + \cdots \qquad -R < x < R$$

- *Continuity property:*

$$\lim_{x \to x_0} \left( \sum_{k=0}^{\infty} a_k x^k \right) = \sum_{k=0}^{\infty} \left( \lim_{x \to x_0} a_k x^k \right) = \sum_{k=0}^{\infty} a_k x_0^k \qquad -R < x_0 < R$$

- *Differentiation property:*

$$\frac{d}{dx} \left( \sum_{k=0}^{\infty} a_k x^k \right) = \sum_{k=0}^{\infty} \left( \frac{d}{dx} a_k x^k \right) = \sum_{k=1}^{\infty} k a_k x^{k-1}$$

- *Integration property:*

$$\int_0^x \left( \sum_{k=0}^{\infty} a_k t^k \right) dt = \sum_{k=0}^{\infty} \left( \int_0^x a_k t^k \, dt \right) = \sum_{k=0}^{\infty} \frac{a_k x^{k+1}}{k+1}$$

**CAUTION** The theorem states that the *radii of convergence* of the three series $\displaystyle\sum_{k=0}^{\infty} a_k x^k, \sum_{k=1}^{\infty} k a_k x^{k-1}$, and $\displaystyle\sum_{k=0}^{\infty} \frac{a_k x^{k+1}}{k+1}$ are the same. This does not imply that the *intervals of convergence* are the same; the endpoints of the interval must be checked separately for each series. For example, the interval of convergence of $\displaystyle\sum_{k=1}^{\infty} \frac{x^k}{k}$ is $[-1, 1)$, but the interval of convergence of its derivative $\displaystyle\sum_{k=2}^{\infty} x^{k-1}$ is $(-1, 1)$.

The differentiation and integration properties of power series state that a power series can be differentiated and integrated term-by-term and that the resulting series represent the derivative and integral, respectively, of the function represented by the original power series. Moreover, it can be shown that the power series obtained by differentiating (or integrating) a power series whose radius of convergence is $R$, converges and has the same radius of convergence $R$ as the original power series. (See Problem 86.)

The differentiation and integration properties can be used to obtain new functions defined by a power series.

### EXAMPLE 8   Using the Differentiation Property of Power Series

Use the differentiation property on the power series $f(x) = \dfrac{1}{1-x} = \sum\limits_{k=0}^{\infty} x^k$ to find a power series representation for $f'(x) = \dfrac{1}{(1-x)^2}$.

**Solution** The function $f(x) = \dfrac{1}{1-x}$, defined on the open interval $(-1, 1)$, is represented by the power series

$$f(x) = \frac{1}{1-x} = 1 + x + x^2 + \cdots + x^n + \cdots = \sum_{k=0}^{\infty} x^k$$

Using the differentiation property,

$$f'(x) = \frac{1}{(1-x)^2} = 1 + 2x + 3x^2 + \cdots + nx^{n-1} + \cdots = \sum_{k=1}^{\infty} kx^{k-1}$$

whose radius of convergence is 1. ∎

**NOW WORK** Problem 59(a).

### EXAMPLE 9   Finding the Power Series Representation for $\ln \dfrac{1}{1-x}$

(a) Find the power series representation for $\ln \dfrac{1}{1-x}$.

(b) Find its interval of convergence.

(c) Find $\ln 2$.

**Solution** (a) If $y = \ln \dfrac{1}{1-x}$, then $y' = \dfrac{1}{1-x}$. That is, $y'$ is represented by the geometric series $\sum\limits_{k=0}^{\infty} x^k$, which converges on the interval $(-1, 1)$. So, if we use the integration property of power series for $y' = \dfrac{1}{1-x}$, we obtain a series for $y = \ln \dfrac{1}{1-x}$.

$$y' = \frac{1}{1-x} = 1 + x + x^2 + \cdots + x^n + \cdots = \sum_{k=0}^{\infty} x^k$$

$$\int_0^x \frac{1}{1-t}\,dt = \int_0^x (1 + t + t^2 + \cdots + t^n + \cdots)\,dt$$

$$\ln \frac{1}{1-x} = x + \frac{x^2}{2} + \frac{x^3}{3} + \cdots + \frac{x^{n+1}}{n+1} + \cdots = \sum_{k=0}^{\infty} \frac{x^{k+1}}{k+1}$$

The radius of convergence of this series is 1.

(b) To find the interval of convergence, we investigate the endpoints. When $x = 1$,

$$x + \frac{x^2}{2} + \frac{x^3}{3} + \cdots + \frac{x^{n+1}}{n+1} + \cdots = 1 + \frac{1}{2} + \frac{1}{3} + \cdots$$

the harmonic series, which diverges. When $x = -1$,

$$x + \frac{x^2}{2} + \frac{x^3}{3} + \cdots + \frac{x^{n+1}}{n+1} + \cdots = -1 + \frac{1}{2} - \frac{1}{3} + \cdots + (-1)^{n+1}\frac{1}{n+1} + \cdots$$

an alternating harmonic series, which converges. The interval of convergence of the power series $\sum_{k=0}^{\infty} \dfrac{x^{k+1}}{k+1}$ is $[-1, 1)$. So,

$$\ln \frac{1}{1-x} = x + \frac{x^2}{2} + \frac{x^3}{3} + \cdots = \sum_{k=0}^{\infty} \frac{x^{k+1}}{k+1} \qquad -1 \le x < 1 \tag{2}$$

**(c)** To find $\ln 2$, notice that when $x = -1$, we have $\ln \dfrac{1}{1-x} = \ln \dfrac{1}{2} = -\ln 2$. In (2) let $x = -1$. Then

$$-\ln 2 = -1 + \frac{1}{2} - \frac{1}{3} + \cdots$$

$$\ln 2 = 1 - \frac{1}{2} + \frac{1}{3} \cdots = \sum_{k=0}^{\infty} \frac{(-1)^k}{k+1}$$

The sum of the alternating harmonic series is $\ln 2$. ∎

**NOW WORK** Problems **59(b)** and **65**.

## EXAMPLE 10 Finding the Power Series Representation for $\tan^{-1} x$

Show that the power series representation for $\tan^{-1} x$ is

$$\tan^{-1} x = x - \frac{x^3}{3} + \frac{x^5}{5} - \frac{x^7}{7} + \cdots + (-1)^n \frac{x^{2n+1}}{2n+1} + \cdots = \sum_{k=0}^{\infty} \frac{(-1)^k x^{2k+1}}{2k+1}$$

Find the radius of convergence and the interval of convergence.

**Solution** If $y = \tan^{-1} x$, then $y' = \dfrac{1}{1+x^2}$, which is the sum of the geometric series $\sum_{k=0}^{\infty} (-1)^k x^{2k}$. That is,

$$\frac{1}{1+x^2} = \sum_{k=0}^{\infty} (-1)^k x^{2k} = 1 - x^2 + x^4 - x^6 + \cdots$$

This series converges when $|x^2| < 1$, or equivalently for $-1 < x < 1$. Now use the integration property of power series to integrate $y' = \dfrac{1}{1+x^2}$ and obtain a series for $y = \tan^{-1} x$.

$$\int_0^x \frac{dt}{1+t^2} = \int_0^x (1 - t^2 + t^4 - \cdots)\, dt$$

$$\tan^{-1} x = x - \frac{x^3}{3} + \frac{x^5}{5} - \frac{x^7}{7} + \cdots + (-1)^n \frac{x^{2n+1}}{2n+1} + \cdots = \sum_{k=0}^{\infty} (-1)^k \frac{x^{2k+1}}{2k+1}$$

The radius of convergence is 1. To find the interval of convergence, we check $x = -1$ and $x = 1$. For $x = -1$,

$$-1 + \frac{1}{3} - \frac{1}{5} + \frac{1}{7} - \cdots$$

For $x = 1$, we get

$$1 - \frac{1}{3} + \frac{1}{5} - \frac{1}{7} + \cdots$$

Both of these series satisfy the two conditions of the Alternating Series Test, and so each one converges. The interval of convergence is the closed interval $[-1, 1]$. ∎

**ORIGINS** The power series representation for $\tan^{-1} x$ is called **Gregory's series** (or Gregory–Leibniz series or Madhava–Gregory series). James Gregory (1638–1675) was a Scottish mathematician. His mother, Janet Anderson, was his teacher and taught him geometry. Gregory, like many other mathematicians of his time, was searching for a good way to approximate $\pi$, a result of which was Gregory's series. Gregory was also a major contributor to the theory of optics, and he is credited with inventing the reflective telescope.

Since $\tan^{-1} 1 = \dfrac{\pi}{4}$, we can use Gregory's series to approximate $\pi$. Then

$$\frac{\pi}{4} = \sum_{k=0}^{\infty} (-1)^k \frac{1^{2k+1}}{2k+1} = \sum_{k=0}^{\infty} \frac{(-1)^k}{2k+1} = 1 - \frac{1}{3} + \frac{1}{5} - \cdots$$

While we now have an approximation for $\pi$, unfortunately the series converges very slowly, requiring many terms to get close to $\pi$. See Problem 82.

NOW WORK Problem 69.

## 8.8 Assess Your Understanding

### Concepts and Vocabulary

1. *True or False*   Every power series $\sum_{k=0}^{\infty} a_k(x - c)^k$ converges for at least one number.

2. *True or False*   Let $b_n$ denote the $n$th term of the power series $\sum_{k=0}^{\infty} a_k x^k$. If $\lim\limits_{n \to \infty} \left| \dfrac{b_{n+1}}{b_n} \right| < 1$ for every number $x$, then $\sum_{k=0}^{\infty} a_k x^k$ is absolutely convergent on the interval $(-\infty, \infty)$.

3. *True or False*   If the radius of convergence of a power series $\sum_{k=0}^{\infty} a_k x^k$ is 0, then the power series converges only for $x = 0$.

4. *True or False*   If a power series converges at one endpoint of its interval of convergence, then it must converge at its other endpoint.

5. *True or False*   The power series $\sum_{k=0}^{\infty} a_k x^k$ and $\sum_{k=0}^{\infty} a_k(x - 3)^k$ have the same radius of convergence.

6. *True or False*   The power series $\sum_{k=0}^{\infty} a_k x^k$ and $\sum_{k=0}^{\infty} a_k(x - 3)^k$ have the same interval of convergence.

7. *True or False*   If the power series $\sum_{k=0}^{\infty} a_k x^k$ converges for $x = 8$, then it converges for $x = -8$.

8. *True or False*   If the power series $\sum_{k=0}^{\infty} a_k x^k$ converges for $x = 3$, then it converges for $x = 1$.

9. *True or False*   If the power series $\sum_{k=0}^{\infty} a_k x^k$ converges for $x = -4$, then it converges for $x = 3$.

10. *True or False*   If the power series $\sum_{k=0}^{\infty} a_k x^k$ converges for $x = 3$, then it diverges for $x = 5$.

11. *True or False*   A possible interval of convergence for the power series $\sum_{k=0}^{\infty} a_k x^k$ is $[-2, 4]$.

12. *True or False*   If the power series $\sum_{k=0}^{\infty} a_k x^k$ diverges for a number $x_1$, then it converges for all numbers $x$ for which $|x| < |x_1|$.

### Skill Building

*In Problems 13–16, find all numbers x for which each power series converges.*

13. $\sum_{k=0}^{\infty} k x^k$

14. $\sum_{k=0}^{\infty} \dfrac{k x^k}{3^k}$

15. $\sum_{k=0}^{\infty} \dfrac{(x+1)^k}{3^k}$

16. $\sum_{k=1}^{\infty} \dfrac{(x-2)^k}{k^2}$

*In Problems 17–26:*

**(a)**  Use the Ratio Test to find the radius of convergence and the interval of convergence of each power series.

**(b)**  Use the Root Test to find the radius of convergence and the interval of convergence of each power series.

**(c)**  Which test, the Ratio Test or the Root Test, did you find easier to use? Give the reasons why.

17. $\sum_{k=0}^{\infty} \dfrac{x^k}{2^k(k+1)}$

18. $\sum_{k=0}^{\infty} (-1)^k \dfrac{x^k}{2^k(k+1)}$

19. $\sum_{k=0}^{\infty} \dfrac{x^k}{k+5}$

20. $\sum_{k=0}^{\infty} \dfrac{x^k}{1+k^2}$

21. $\sum_{k=0}^{\infty} \dfrac{k^2 x^k}{3^k}$

22. $\sum_{k=0}^{\infty} \dfrac{2^k x^k}{3^k}$

23. $\sum_{k=0}^{\infty} \dfrac{k x^k}{2k+1}$

24. $\sum_{k=0}^{\infty} (6x)^k$

25. $\sum_{k=0}^{\infty} (x-3)^k$

26. $\sum_{k=0}^{\infty} \dfrac{k(2x)^k}{3^k}$

*In Problems 27–44, find the radius of convergence and the interval of convergence of each power series.*

27. $\sum_{k=1}^{\infty} \dfrac{x^k}{k^3}$

28. $\sum_{k=2}^{\infty} \dfrac{x^k}{\ln k}$

29. $\sum_{k=1}^{\infty} \dfrac{(x-2)^k}{k^3}$

30. $\sum_{k=0}^{\infty} \dfrac{k(x-2)^k}{3^k}$

31. $\sum_{k=0}^{\infty} \dfrac{(-1)^k}{(2k+1)!} x^{2k+1}$

32. $\sum_{k=1}^{\infty} (kx)^k$

---

1. = NOW WORK problem        [N] = Graphing technology recommended        [CAS] = Computer Algebra System recommended

**33.** $\displaystyle\sum_{k=1}^{\infty} \frac{kx^k}{\ln(k+1)}$

**34.** $\displaystyle\sum_{k=1}^{\infty} \frac{x^k}{\ln(k+1)}$

**35.** $\displaystyle\sum_{k=0}^{\infty} \frac{k(k+1)x^k}{4^k}$

**36.** $\displaystyle\sum_{k=1}^{\infty} \frac{(-1)^k(x-5)^k}{k(k+1)}$

**37.** $\displaystyle\sum_{k=0}^{\infty} (-1)^k \frac{(x-3)^{2k}}{9^k}$

**38.** $\displaystyle\sum_{k=0}^{\infty} \frac{x^k}{e^k}$

**39.** $\displaystyle\sum_{k=0}^{\infty} (-1)^k \frac{(2x)^k}{k!}$

**40.** $\displaystyle\sum_{k=0}^{\infty} \frac{(x+1)^k}{k!}$

**41.** $\displaystyle\sum_{k=0}^{\infty} (-1)^k \frac{(x-1)^{4k}}{k!}$

**42.** $\displaystyle\sum_{k=1}^{\infty} \frac{(x+1)^k}{k(k+1)(k+2)}$

**43.** $\displaystyle\sum_{k=1}^{\infty} \frac{k^k x^k}{k!}$

**44.** $\displaystyle\sum_{k=0}^{\infty} \frac{3^k(x-2)^k}{k!}$

**45.** A function $f$ is defined by the power series $f(x) = \displaystyle\sum_{k=0}^{\infty} \frac{x^k}{3^k}$.

    **(a)** Find the domain of $f$.

    **(b)** Evaluate $f(2)$ and $f(-1)$.

    **(c)** Find $f$ by summing the series.

**46.** A function $f$ is defined by the power

    series $f(x) = \displaystyle\sum_{k=0}^{\infty} (-1)^k \left(\frac{x}{2}\right)^k$.

    **(a)** Find the domain of $f$.

    **(b)** Evaluate $f(0)$ and $f(1)$.

    **(c)** Find $f$ by summing the series.

**47.** A function $f$ is defined by the power

    series $f(x) = \displaystyle\sum_{k=0}^{\infty} \frac{(x-2)^k}{2^k}$.

    **(a)** Find the domain of $f$.

    **(b)** Evaluate $f(1)$ and $f(2)$.

    **(c)** Find $f$ by summing the series.

**48.** A function $f$ is defined by the power

    series $f(x) = \displaystyle\sum_{k=0}^{\infty} (-1)^k(x+3)^k$.

    **(a)** Find the domain of $f$.

    **(b)** Evaluate $f(-3)$ and $f(-2.5)$.

    **(c)** Find $f$ by summing the series.

**49.** If $\displaystyle\sum_{k=0}^{\infty} a_k x^k$ converges for $x=3$, what, if anything, can be said about the convergence at $x=2$? Can anything be said about the convergence at $x=5$?

**50.** If $\displaystyle\sum_{k=0}^{\infty} a_k(x-2)^k$ converges for $x=6$, at what other numbers $x$ must the series necessarily converge?

**51.** If the series $\displaystyle\sum_{k=0}^{\infty} a_k x^k$ converges for $x=6$ and diverges for $x=-8$, what, if anything, can be said about the truth of the following statements?

    **(a)** The series converges for $x=2$.

    **(b)** The series diverges for $x=7$.

    **(c)** The series is absolutely convergent for $x=6$.

    **(d)** The series converges for $x=-6$.

    **(e)** The series diverges for $x=10$.

    **(f)** The series is absolutely convergent for $x=4$.

**52.** If the radius of convergence of the power series $\displaystyle\sum_{k=0}^{\infty} a_k(x-3)^k$

    is $R=5$, what, if anything, can be said about the truth of the following statements?

    **(a)** The series converges for $x=2$.

    **(b)** The series diverges for $x=7$.

    **(c)** The series diverges for $x=8$.

    **(d)** The series converges for $x=-6$.

    **(e)** The series converges for $x=-2$.

*In Problems 53–58:*

*(a) Use a geometric series to represent each function as a power series centered at 0.*

*(b) Determine the radius of convergence and the interval of convergence of each series.*

**53.** $f(x) = \dfrac{1}{1+x^3}$

**54.** $f(x) = \dfrac{1}{1-x^2}$

**55.** $f(x) = \dfrac{1}{6-2x}$

**56.** $f(x) = \dfrac{4}{x+2}$

**57.** $f(x) = \dfrac{x}{1+x^3}$

**58.** $f(x) = \dfrac{4x^2}{x+2}$

*In Problems 59–62:*

*(a) Use the differentiation property of power series to find a power series representation for $f'(x)$.*

*(b) Use the integration property of power series to find a power series representation for the indefinite integral of $f$.*

**59.** $f(x) = \displaystyle\sum_{k=0}^{\infty} \frac{(-1)^k x^{2k+1}}{(2k+1)!}$

**60.** $f(x) = \displaystyle\sum_{k=0}^{\infty} \frac{(-1)^k x^{2k}}{(2k)!}$

**61.** $f(x) = \displaystyle\sum_{k=0}^{\infty} \frac{x^k}{k!}$

**62.** $f(x) = \displaystyle\sum_{k=0}^{\infty} \frac{(-1)^k x^k}{k!}$

*In Problems 63–70, find a power series representation of $f$. Use a geometric series and properties of a power series.*

**63.** $f(x) = \dfrac{1}{(1+x)^2}$

**64.** $f(x) = \dfrac{1}{(1-x)^3}$

**65.** $f(x) = \dfrac{2}{3(1-x)^2}$

**66.** $f(x) = \dfrac{1}{(1-x)^4}$

**67.** $f(x) = \ln\dfrac{1}{1+x}$

**68.** $f(x) = \ln(1-2x)$

**69.** $f(x) = \ln(1-x^2)$

**70.** $f(x) = \ln(1+x^2)$

## Applications and Extensions

*In Problems 71–78, find all $x$ for which each power series converges.*

**71.** $\displaystyle\sum_{k=1}^{\infty} \frac{x^k}{k}$

**72.** $\displaystyle\sum_{k=1}^{\infty} \frac{(x-4)^k}{k}$

**73.** $\displaystyle\sum_{k=1}^{\infty} \frac{x^k}{2k+1}$

**74.** $\displaystyle\sum_{k=1}^{\infty} \frac{x^k}{k^2}$

**75.** $\displaystyle\sum_{k=0}^{\infty} x^{k^2}$

**76.** $\displaystyle\sum_{k=1}^{\infty} \frac{k^a}{a^k}(x-a)^k, \quad a \neq 0$

**77.** $\displaystyle\sum_{k=0}^{\infty} \frac{(k!)^2}{(2k)!}(x-1)^k$

**78.** $\displaystyle\sum_{k=0}^{\infty} \frac{\sqrt{k!}}{(2k)!} x^k$

**79. (a)** In the geometric series $\dfrac{1}{1-x} = \displaystyle\sum_{k=0}^{\infty} x^k,\ -1 < x < 1,$

replace $x$ by $x^2$ to obtain the power series representation for $\dfrac{1}{1-x^2}$.

**(b)** What is its interval of convergence?

**80. (a)** Integrate the power series found in Problem 79 for $\dfrac{1}{1-x^2}$

to obtain the power series representation for $\dfrac{1}{2}\ln\dfrac{1+x}{1-x}$.

**(b)** What is its interval of convergence?

**81.** Use the power series found in Problem 80 to get an approximation for $\ln 2$ correct to three decimal places.

**82.** Use the first 1000 terms of Gregory's series to approximate $\dfrac{\pi}{4}$. What is the approximation for $\pi$?

**83.** If $R > 0$ is the radius of convergence of $\displaystyle\sum_{k=1}^{\infty} a_k x^k$, show

that $\displaystyle\lim_{n \to \infty} \left|\dfrac{a_{n+1}}{a_n}\right| = \dfrac{1}{R}$, provided this limit exists.

**84.** If $R$ is the radius of convergence of $\displaystyle\sum_{k=1}^{\infty} a_k x^k$, show that the

radius of convergence of $\displaystyle\sum_{k=1}^{\infty} a_k x^{2k}$ is $\sqrt{R}$.

**85.** Prove that if a power series is absolutely convergent at one endpoint of its interval of convergence, then the power series is absolutely convergent at the other endpoint.

**86.** Suppose $\displaystyle\sum_{k=0}^{\infty} a_k x^k$ converges for $|x| < R$ and that $\displaystyle\lim_{n \to \infty} \left|\dfrac{a_{n+1}}{a_n}\right|$

exists. Show that $\displaystyle\sum_{k=1}^{\infty} k a_k x^{k-1}$ and $\displaystyle\sum_{k=0}^{\infty} \dfrac{a_k}{k+1} x^{k+1}$ also converge

for $|x| < R$.

**Challenge Problems**

**87.** Consider the differential equation

$$(1+x^2)\,y'' - 4xy' + 6y = 0$$

Assuming there is a solution $y(x) = \displaystyle\sum_{k=0}^{\infty} a_k x^k$, substitute and

obtain a formula for $a_k$. Your answer should have the form

$$y(x) = a_0(1 - 3x^2) + a_1\left(x - \frac{1}{3}x^3\right) \qquad a_0,\ a_1 \text{ real numbers}$$

**88.** If the series $\displaystyle\sum_{k=0}^{\infty} a_k 3^k$ converges, show that the series $\displaystyle\sum_{k=1}^{\infty} k a_k 2^k$ also converges.

**89.** Find the interval of convergence of the series $\displaystyle\sum_{k=1}^{\infty} \dfrac{(x-2)^k}{k(3^k)}$.

**90.** Let a power series $S(x)$ be convergent for $|x| < R$. Assume

that $S(x) = \displaystyle\sum_{k=0}^{\infty} a_k x^k$ with partial sums $S_n(x) = \displaystyle\sum_{k=0}^{n} a_k x^k$.

Suppose for any number $\varepsilon > 0$, there is a number $N$ so that

when $n > N$, $|S(x) - S_n(x)| < \dfrac{\varepsilon}{3}$ for all $|x| < R$.

Show that $S(x)$ is continuous for all $|x| < R$.

**91.** Find the power series in $x$, denoted by $f(x)$, for which $f''(x) + f(x) = 0$ and $f(0) = 0$, $f'(0) = 1$. What is the radius of convergence of the series?

**92.** The **Bessel function of order** $m$ of the first kind, where $m$ is a nonnegative integer, is defined as

$$J_m(x) = \sum_{k=0}^{\infty} (-1)^k \frac{1}{(k+m)!\,k!} \left(\frac{x}{2}\right)^{2k+m}$$

Show that:

**(a)** $J_0(x) = x^{-1} \dfrac{d}{dx}(x J_1(x))$

**(b)** $J_1(x) = x^{-2} \dfrac{d}{dx}(x^2 J_2(x))$

# 8.9 Taylor Series; Maclaurin Series

**OBJECTIVES** *When you finish this section, you should be able to:*

**1** Express a function as a Taylor series or a Maclaurin series (p. 654)

**2** Determine the convergence of a Taylor/Maclaurin series (p. 655)

**3** Find Taylor/Maclaurin expansions (p. 658)

**4** Work with a binomial series (p. 660)

We saw in Section 8.8 that it is often possible to obtain a power series representation for a function by starting with a known series and differentiating, integrating, or substituting. But what if you have no initial series? In other words, so far we have seen that functions can be represented by power series, and if we know the sum of the power series, then we know the function. In this section, we investigate what the power series representation of a function must look like *if it has a power series representation*.

Consider the power series in $(x - c)$:

$$\sum_{k=0}^{\infty} a_k(x - c)^k = a_0 + a_1(x - c) + a_2(x - c)^2 + \cdots + a_n(x - c)^n + \cdots$$

and suppose its interval of convergence is the open interval $(c - R, \; c + R)$, $R > 0$. We define the function $f$ as the series

$$f(x) = \sum_{k=0}^{\infty} a_k(x - c)^k = a_0 + a_1(x - c) + a_2(x - c)^2 + \cdots + a_n(x - c)^n + \cdots \quad (1)$$

The coefficients $a_0$, $a_1$, ... can be expressed in terms of $f$ and its derivatives in the following way. Repeatedly differentiate the function using the differentiation property of a power series,

$$f'(x) = \sum_{k=1}^{\infty} k\, a_k(x - c)^{k-1} = a_1 + 2a_2(x - c) + 3a_3(x - c)^2 + 4a_4(x - c)^3 + \cdots$$

$$f''(x) = \sum_{k=2}^{\infty} [k(k - 1)]\, a_k(x - c)^{k-2}$$

$$= (2 \cdot 1)\, a_2 + (3 \cdot 2)\, a_3(x - c) + (4 \cdot 3)\, a_4(x - c)^2 + \cdots$$

$$f'''(x) = \sum_{k=3}^{\infty} [k(k - 1)(k - 2)] a_k(x - c)^{k-3}$$

$$= (3 \cdot 2 \cdot 1)\, a_3 + (4 \cdot 3 \cdot 2)\, a_4(x - c) + \cdots$$

and for any positive integer $n$,

$$f^{(n)}(x) = \sum_{k=n}^{\infty} k(k - 1)(k - 2) \cdots (k - n + 1)a_k(x - c)^{k-n}$$

$$= [n \cdot (n - 1) \cdot \; \ldots \; \cdot 1]\, a_n + [(n + 1) \cdot n \cdot (n - 1) \cdot \; \ldots \; \cdot 2]\, a_{n+1}(x - c) + \cdots$$

$$= n!\, a_n + (n + 1)!\, a_{n+1}(x - c) + \cdots$$

Now let $x = c$ in each derivative and solve for $a_k$.

$$f(c) = a_0 \qquad\qquad a_0 = f(c)$$

$$f'(c) = a_1 \qquad\qquad a_1 = f'(c)$$

$$f''(c) = 2a_2 \qquad\qquad a_2 = \frac{f''(c)}{2!}$$

$$f'''(c) = 3!\, a_3 \qquad\qquad a_3 = \frac{f'''(c)}{3!}$$

$$\vdots \qquad\qquad\qquad \vdots$$

$$f^{(n)}(c) = n!\, a_n \qquad\qquad a_n = \frac{f^{(n)}(c)}{n!}$$

If we substitute for $a_k$ in (1), we obtain

$$f(x) = f(c) + f'(c)(x - c) + \frac{f''(c)}{2!}(x - c)^2 + \cdots + \frac{f^{(n)}(c)}{n!}(x - c)^n + \cdots$$

and have proved the following result.

**THEOREM** Taylor Series; Maclaurin Series

Suppose $f$ is a function that has derivatives of all orders on the open interval $(c - R, c + R)$, $R > 0$. If $f$ can be represented by the power series $\sum_{k=0}^{\infty} a_k (x - c)^k$, whose radius of convergence is $R$, then

$$f(x) = f(c) + f'(c)(x - c) + \frac{f''(c)}{2!}(x - c)^2 + \cdots + \frac{f^{(n)}(c)}{n!}(x - c)^n + \cdots$$

$$= \sum_{k=0}^{\infty} \frac{f^{(k)}(c)}{k!}(x - c)^k \qquad (2)$$

for all numbers $x$ in the open interval $(c - R, c + R)$. A power series that has the form of equation (2) is called a **Taylor series** of the function $f$.

When $c = 0$, the Taylor series

$$f(x) = f(0) + f'(0)x + \frac{f''(0)}{2!}x^2 + \cdots + \frac{f^{(n)}(0)}{n!}x^n + \cdots = \sum_{k=0}^{\infty} \frac{f^{(k)}(0)}{k!}x^k$$

is called a **Maclaurin series**.

The Taylor series in $(x - c)$ of a function $f$ is referred to as the **Taylor expansion of $f$ about $c$**; the Maclaurin series of a function $f$ is called the **Maclaurin expansion of $f$ about 0**.

In a Taylor series, all the derivatives are evaluated at $c$, and the interval of convergence has its center at $c$. In a Maclaurin series, all the derivatives are evaluated at 0, and the interval of convergence has its center at 0.

## 1 Express a Function as a Taylor Series or a Maclaurin Series

The next example shows what a Maclaurin expansion of $f(x) = e^x$ must look like (if there is one).

CALC
CLIP
**EXAMPLE 1  Expressing a Function as a Maclaurin Series**

Assuming that $f(x) = e^x$ can be represented by a power series in $x$, find its Maclaurin series.

**Solution** To express a function $f$ as a Maclaurin series, we begin by evaluating $f$ and its derivatives at 0.

$$f(x) = e^x \qquad f(0) = 1$$
$$f'(x) = e^x \qquad f'(0) = 1$$
$$f''(x) = e^x \qquad f''(0) = 1$$
$$\vdots \qquad\qquad \vdots$$

Then use the definition of a Maclaurin series.

$$f(x) = \sum_{k=0}^{\infty} \frac{f^{(k)}(0)}{k!}x^k = 1 + x + \frac{x^2}{2!} + \frac{x^3}{3!} + \cdots + \frac{x^n}{n!} + \cdots = \sum_{k=0}^{\infty} \frac{x^k}{k!} \qquad (3)$$

$$\uparrow$$
$$f^{(k)}(0) = 1$$

NOW WORK Problem 3. ■

But how can we be sure $f(x) = e^x$ can be represented by a power series? We know [Example 1(a), p. 641] that the power series $\sum_{k=0}^{\infty} \frac{x^k}{k!}$ converges absolutely for

ORIGINS  Colin Maclaurin (1698–1746) was a Scottish mathematician. An orphan, he and his brother were raised by an uncle. Maclaurin was 11 years old when he entered the University of Glasgow, and within a year he had taught himself from Euclid's Elements (geometry). Maclaurin was 19 when he was appointed professor of Mathematics at the University of Aberdeen, Scotland. Maclaurin was an avid supporter of Newton's mathematics. In response to a claim that Newton's calculus lacked rigor, Maclaurin wrote the first systematic treatise on Newton's methods. In it he used Taylor series centered about 0, which are now known as Maclaurin series.

MACLAURIN.

all numbers $x$. But does the series $\sum\limits_{k=0}^{\infty} \dfrac{x^k}{k!}$ converge to $e^x$? To answer these questions, we need to investigate the convergence of a Taylor series.

## 2 Determine the Convergence of a Taylor/Maclaurin Series

Suppose a function $f$ has derivatives of all orders on the open interval $(c - R, c + R)$, $R > 0$. If $f$ can be represented by a power series $\sum\limits_{k=0}^{\infty} a_k (x - c)^k$ whose radius of convergence is $R$, then

$$f(x) = f(c) + f'(c)(x - c) + \frac{f''(c)}{2!}(x - c)^2 + \cdots + \frac{f^{(n)}(c)}{n!}(x - c)^n + \cdots = \sum_{k=0}^{\infty} \frac{f^{(k)}(c)}{k!}(x - c)^k \quad (4)$$

for all $x$ in the interval $(c - R, c + R)$. We seek conditions on $f$ that guarantee the power series (4) converges to $f$.

We rewrite (4) as follows

$$f(x) = P_n(x) + R_n$$

where $P_n(x) = f(c) + f'(c)(x - c) + \dfrac{f''(c)}{2!}(x - c)^2 + \cdots + \dfrac{f^{(n)}(c)}{n!}(x - c)^n$, called the **Taylor polynomial** $P_n(x)$ of $f$ of degree $n$, and $R_n$ is the **remainder**.

> **THEOREM** Taylor's Formula with Remainder
>
> Let $f$ be a function whose first $n + 1$ derivatives are continuous on an open interval $I$ containing the number $c$. Then for every $x$ in $I$, there is a number $u$ between $x$ and $c$ for which
>
> $$f(x) = f(c) + f'(c)(x - c) + \frac{f''(c)}{2!}(x - c)^2 + \cdots + \frac{f^{(n)}(c)}{n!}(x - c)^n + R_n(x)$$
>
> where
>
> $$R_n(x) = \frac{f^{(n+1)}(u)}{(n+1)!}(x - c)^{n+1}$$
>
> is the remainder after $n + 1$ terms.

**NOTE** The remainder $R_n(x)$ contains the $(n + 1)$st derivative of $f$ evaluated at $u$.

The proof of this theorem is in Appendix B.

The Taylor series in $x - c$ of a function $f$ is $\sum\limits_{k=0}^{\infty} \dfrac{f^{(k)}(c)}{k!}(x - c)^k$. Notice that the $n$th partial sum of the Taylor series in $x - c$ of $f$ is precisely the Taylor polynomial $P_n(x)$ of $f$ at $c$. If the Taylor series $\sum\limits_{k=0}^{\infty} \dfrac{f^{(k)}(c)}{k!}(x - c)^k$ converges to $f(x)$, it follows that

$$f(x) = \lim_{n \to \infty} P_n(x)$$

But Taylor's Formula with Remainder states that

$$f(x) = P_n(x) + R_n(x)$$

So, if the Taylor series converges, we must have

$$f(x) = \lim_{n \to \infty} [P_n(x) + R_n(x)] = \lim_{n \to \infty} P_n(x) + \lim_{n \to \infty} R_n(x) = f(x) + \lim_{n \to \infty} R_n(x)$$

That is, $\lim\limits_{n \to \infty} R_n(x) = 0$.

Archive Photos/Getty Images

**ORIGINS** The Taylor polynomial is named after the English mathematician Brook Taylor (1685–1731). Taylor grew up in an affluent but strict home. He was home-schooled in the arts and the classics until he attended Cambridge University in 1703, where he studied mathematics. He was an accomplished musician and painter as well as a mathematician.

**THEOREM** Convergence of a Taylor Series

If a function $f$ has derivatives of all orders in an open interval $I = (c - R, c + R)$, $R > 0$, centered at $c$, and if

$$\lim_{n \to \infty} R_n(x) = \lim_{n \to \infty} \frac{f^{(n+1)}(u)}{(n+1)!}(x - c)^{n+1} = 0 \tag{5}$$

for all numbers $x$ in $I$, where $u$ is a number between $c$ and $x$, then

$$f(x) = \sum_{k=0}^{\infty} \frac{f^{(k)}(c)}{k!}(x - c)^k = f(c) + f'(c)(x - c) + \frac{f''(c)}{2!}(x - c)^2$$
$$+ \cdots + \frac{f^{(n)}(c)}{n!}(x - c)^n + \cdots \tag{6}$$

for all numbers $x$ in $I$.

At first glance, the convergence theorem appears simple, but in practice it is not always easy to show that $\lim_{n \to \infty} R_n(x) = 0$. One reason is that the term $f^{(n+1)}(u)$, which appears in $R_n(x)$, depends on $u$, making the limit difficult to find.

---

**EXAMPLE 2** Determining the Convergence of a Maclaurin Series

Show that $1 + x + \dfrac{x^2}{2!} + \dfrac{x^3}{3!} + \cdots + \dfrac{x^n}{n!} + \cdots$ converges to $e^x$ for every number $x$. That is, prove that

$$e^x = 1 + \frac{x}{1!} + \frac{x^2}{2!} + \frac{x^3}{3!} + \cdots + \frac{x^n}{n!} + \cdots = \sum_{k=0}^{\infty} \frac{x^k}{k!} \tag{7}$$

for all real numbers.

**Solution** To prove that $1 + \dfrac{x}{1!} + \dfrac{x^2}{2!} + \dfrac{x^3}{3!} + \cdots = e^x$ for every number $x$, we need to show that $\lim_{n \to \infty} R_n(x) = 0$. Since $f^{(n+1)}(x) = e^x$, we have

$$R_n(x) = \frac{f^{(n+1)}(u)x^{n+1}}{(n+1)!} = \frac{e^u x^{n+1}}{(n+1)!}$$

where $u$ is between $0$ and $x$. To show that $\lim_{n \to \infty} R_n(x) = 0$, we consider two cases: $x > 0$ and $x < 0$.

*Case 1:* When $x > 0$, then $0 < u < x$, so that $1 < e^u < e^x$ and, for every positive integer $n$,

$$0 < R_n(x) = \frac{e^u x^{n+1}}{(n+1)!} < \frac{e^x x^{n+1}}{(n+1)!}$$

By the Ratio Test, the series $\displaystyle\sum_{k=0}^{\infty} \frac{x^{k+1}}{(k+1)!}$ converges for all $x$. It follows that $\lim_{n \to \infty} \dfrac{x^{n+1}}{(n+1)!} = 0$, and, therefore,

$$\lim_{n \to \infty} \frac{e^x x^{n+1}}{(n+1)!} = e^x \lim_{n \to \infty} \frac{x^{n+1}}{(n+1)!} = 0$$

By the Squeeze Theorem, $\lim_{n \to \infty} R_n(x) = 0$.

**NEED TO REVIEW?** The Squeeze Theorem is discussed in Section 1.4, pp. 115–116.

*Case 2:* When $x < 0$, then $x < u < 0$ and $e^x < e^u < 1$, so that

$$0 \leq |R_n(x)| = \frac{e^u \cdot |x|^{n+1}}{(n+1)!} < \frac{|x|^{n+1}}{(n+1)!}$$

Since $\lim\limits_{n \to \infty} \dfrac{|x|^{n+1}}{(n+1)!} = 0$, by the Squeeze Theorem, $\lim\limits_{n \to \infty} R_n(x) = 0$.

So for all $x$, $\lim\limits_{n \to \infty} R_n(x) = 0$. As a result,

$$e^x = 1 + x + \frac{x^2}{2!} + \frac{x^3}{3!} + \cdots + \frac{x^n}{n!} + \cdots = \sum_{k=0}^{\infty} \frac{x^k}{k!}$$

for all numbers $x$.                                                          ■

---

**EXAMPLE 3** **Finding the Maclaurin Expansion for $f(x) = \sin x$**

(a) Find the Maclaurin expansion for $f(x) = \sin x$.

(b) Show that it converges to $\sin x$ for all numbers $x$.

**Solution (a)** The value of $f$ and its derivatives at 0 are

$$
\begin{array}{ll}
f(x) = \sin x & f(0) = 0 \\
f'(x) = \cos x & f'(0) = 1 \\
f''(x) = -\sin x & f''(0) = 0 \\
f'''(x) = -\cos x & f'''(0) = -1
\end{array}
$$

Higher-order derivatives follow this same pattern, so if $f(x) = \sin x$ can be represented by a power series in $x$, then

$$\sin x = f(0) + \frac{f'(0)}{1!} x + \frac{f''(0)}{2!} x^2 + \frac{f'''(0)}{3!} x^3 + \cdots = x - \frac{x^3}{3!} + \frac{x^5}{5!} - \frac{x^7}{7!} + \cdots$$

$$= \sum_{k=0}^{\infty} (-1)^k \frac{x^{2k+1}}{(2k+1)!}$$

**(b)** To prove that the series actually converges to $\sin x$ for all $x$, we need to show that the remainders $R_{2n+1}(x)$ and $R_{2n}(x)$ approach zero. For $R_{2n+1}(x)$ we have

$$R_{2n+1}(x) = (-1)^{n+1} \frac{f^{(2n+2)}(u) \, x^{2n+2}}{(2n+2)!}$$

where $u$ is between 0 and $x$. Since $|f^{(2n+2)}(u)| = |\sin u| \leq 1$ for every number $u$, then

$$0 \leq |R_{2n+1}(x)| = \frac{\left| f^{(2n+2)}(u) \right|}{(2n+2)!} |x|^{2n+2} \leq \frac{|x|^{2n+2}}{(2n+2)!}$$

By the Ratio Test, the series $\sum\limits_{k=0}^{\infty} \dfrac{|x|^{2k+2}}{(2k+2)!}$ converges for all $x$, so

$$\lim_{n \to \infty} \frac{|x|^{2n+2}}{(2n+2)!} = 0$$

By the Squeeze Theorem, $\lim\limits_{n \to \infty} |R_{2n+1}(x)| = 0$ and therefore $\lim\limits_{n \to \infty} R_{2n+1}(x) = 0$ for all $x$.

A similar argument holds for the remainder $|R_{2n}(x)| = \dfrac{|\cos u| \, x^{2n+1}}{(2n+1)!}$.

We conclude that the series $\sum\limits_{k=0}^{\infty} (-1)^k \dfrac{x^{2k+1}}{(2k+1)!}$ converges to $\sin x$ for all $x$. That is,

$$\boxed{\sin x = x - \frac{x^3}{3!} + \frac{x^5}{5!} - \cdots + (-1)^n \frac{x^{2n+1}}{(2n+1)!} + \cdots = \sum_{k=0}^{\infty} (-1)^k \frac{x^{2k+1}}{(2k+1)!}} \qquad (8)$$

■

## 3 Find Taylor/Maclaurin Expansions

The task of finding the Taylor expansion for a function $f$ by taking successive derivatives and then showing that $\lim_{n \to \infty} R_n(x) = 0$ can be challenging. Consequently, it is usually easier to use a known series and properties of power series, such as the Differentiation Property and the Integration Property (see page 647), to find the Taylor expansion of a function $f$.

For example, we can find the Maclaurin expansion for $f(x) = \cos x$ by differentiating the Maclaurin series for $f(x) = \sin x$.

**EXAMPLE 4    Finding the Maclaurin Expansion for $f(x) = \cos x$**

Find the Maclaurin expansion for $f(x) = \cos x$.

**NOTE** The Maclaurin expansion for $f(x) = \cos x$ could also have been found by integrating the Maclaurin expansion for $f(x) = \sin x$.

**Solution** Use the differentiation property of power series and the Maclaurin expansion for $\sin x$ [equation (8)].

$$\frac{d}{dx} \sin x = \frac{d}{dx} \left( x - \frac{x^3}{3!} + \frac{x^5}{5!} - \cdots + (-1)^n \frac{x^{2n+1}}{(2n+1)!} + \cdots \right)$$

$$= \frac{d}{dx} \sum_{k=0}^{\infty} (-1)^k \frac{x^{2k+1}}{(2k+1)!}$$

Then

$$\boxed{\cos x = 1 - \frac{x^2}{2!} + \frac{x^4}{4!} - \cdots + (-1)^n \frac{x^{2n}}{(2n)!} + \cdots = \sum_{k=0}^{\infty} (-1)^k \frac{x^{2k}}{(2k)!}} \tag{9}$$

for all numbers $x$. ∎

**EXAMPLE 5    Finding the Maclaurin Expansion for $f(x) = \sin(2x)$**

Find the Maclaurin expansion for $f(x) = \sin(2x)$.

**Solution** Use the Maclaurin expansion for $\sin x$, equation (8), and substitute $2x$ for $x$.

$$\sin(2x) = 2x - \frac{(2x)^3}{3!} + \frac{(2x)^5}{5!} - \cdots + (-1)^n \frac{(2x)^{2n+1}}{(2n+1)!} + \cdots$$

$$= \sum_{k=0}^{\infty} (-1)^k \frac{(2x)^{2k+1}}{(2k+1)!}$$

∎

**NOW WORK** Problem 9.

We can use known Taylor expansions to obtain the power series representations of other functions. For example, if in the Maclaurin expansion for $e^x$, the variable $x$ is replaced by $-x$, then

$$f(x) = e^{-x} = 1 - x + \frac{x^2}{2!} - \frac{x^3}{3!} + \cdots + (-1)^n \frac{x^n}{n!} + \cdots$$

for all numbers $x$. We use this in the next example to find the Maclaurin expansion for the hyperbolic cosine function.

**EXAMPLE 6    Finding the Maclaurin Expansion for $f(x) = \cosh x$**

Find the Maclaurin expansion for $f(x) = \cosh x$.

**Solution** Since

$$\cosh x = \frac{e^x + e^{-x}}{2}$$

its Maclaurin expansion can be found by adding corresponding terms of the Maclaurin expansions for $e^x$ and $e^{-x}$ and then dividing by 2. The result is

$$\cosh x = \frac{e^x + e^{-x}}{2} = \frac{\left(1 + x + \dfrac{x^2}{2!} + \dfrac{x^3}{3!} + \cdots + \dfrac{x^n}{n!} + \cdots\right) + \left(1 - x + \dfrac{x^2}{2!} - \dfrac{x^3}{3!} + \cdots + (-1)^n \dfrac{x^n}{n!} + \cdots\right)}{2}$$

$$= \frac{1+1}{2} + \frac{x-x}{2} + \frac{x^2+x^2}{2\cdot 2!} + \frac{x^3-x^3}{2\cdot 3!} + \cdots + \frac{x^n + (-1)^n x^n}{2\cdot n!} + \cdots$$

$$\boxed{\cosh x = 1 + \frac{x^2}{2!} + \frac{x^4}{4!} + \frac{x^6}{6!} + \cdots + \frac{x^{2n}}{(2n)!} + \cdots = \sum_{k=0}^{\infty} \frac{x^{2k}}{(2k)!}} \qquad (10)$$

for all $x$. ■

**NOW WORK** Problem 13.

**EXAMPLE 7** Finding the Maclaurin Expansion for $f(x) = e^x \cos x$

Find the first five terms of the Maclaurin expansion for $f(x) = e^x \cos x$.

**Solution** The Maclaurin expansion for $f(x) = e^x \cos x$ is obtained by multiplying the Maclaurin expansion for $e^x$ by the Maclaurin expansion for $\cos x$. That is,

$$e^x \cos x = \left(1 + x + \frac{x^2}{2!} + \frac{x^3}{3!} + \frac{x^4}{4!} + \frac{x^5}{5!} + \cdots\right)\left(1 - \frac{x^2}{2!} + \frac{x^4}{4!} - \cdots\right)$$

**NOTE** In multiplying the two expressions in Example 7, we only included powers of $x$ less than six.

Then for all $x$,

$$e^x \cos x = 1\left(1 - \frac{x^2}{2!} + \frac{x^4}{4!} - \cdots\right) + x\left(1 - \frac{x^2}{2!} + \frac{x^4}{4!} - \cdots\right)$$

$$+ \frac{x^2}{2!}\left(1 - \frac{x^2}{2!} + \frac{x^4}{4!} - \cdots\right) + \frac{x^3}{3!}\left(1 - \frac{x^2}{2!} + \frac{x^4}{4!} - \cdots\right)$$

$$+ \frac{x^4}{4!}\left(1 - \frac{x^2}{2!} + \frac{x^4}{4!} - \cdots\right) + \frac{x^5}{5!}\left(1 - \frac{x^2}{2!} + \frac{x^4}{4!} - \cdots\right) + \cdots$$

$$= \left(1 - \frac{x^2}{2} + \frac{x^4}{24}\right) + \left(x - \frac{x^3}{2} + \frac{x^5}{24}\right) + \left(\frac{x^2}{2} - \frac{x^4}{4}\right) + \left(\frac{x^3}{6} - \frac{x^5}{12}\right)$$

$$+ \frac{x^4}{24} + \frac{x^5}{120} + \cdots$$

$$= 1 + x + \left(-\frac{1}{2} + \frac{1}{2}\right)x^2 + \left(-\frac{1}{2} + \frac{1}{6}\right)x^3 + \left(\frac{1}{24} - \frac{1}{4} + \frac{1}{24}\right)x^4$$

$$+ \left(\frac{1}{24} - \frac{1}{12} + \frac{1}{120}\right)x^5 + \cdots$$

$$= 1 + x - \frac{1}{3}x^3 - \frac{1}{6}x^4 - \frac{1}{30}x^5 + \cdots$$

■

**NOW WORK** Problem 31.

EXAMPLE 8 **Finding the Taylor Expansion for $f(x) = \cos x$ about $\dfrac{\pi}{2}$**

Find the Taylor expansion for $f(x) = \cos x$ about $\dfrac{\pi}{2}$.

**Solution** To express $f(x) = \cos x$ as a Taylor expansion about $\dfrac{\pi}{2}$, we evaluate $f$ and its derivatives at $\dfrac{\pi}{2}$.

$$f(x) = \cos x \qquad f\left(\frac{\pi}{2}\right) = 0$$

$$f'(x) = -\sin x \qquad f'\left(\frac{\pi}{2}\right) = -1$$

$$f''(x) = -\cos x \qquad f''\left(\frac{\pi}{2}\right) = 0$$

$$f'''(x) = \sin x \qquad f'''\left(\frac{\pi}{2}\right) = 1$$

For derivatives of odd order, $f^{(2n+1)}\left(\dfrac{\pi}{2}\right) = (-1)^{n+1}$. For derivatives of even order, $f^{(2n)}\left(\dfrac{\pi}{2}\right) = 0$. The Taylor expansion for $f(x) = \cos x$ about $\dfrac{\pi}{2}$ is

$$f(x) = \cos x = f\left(\frac{\pi}{2}\right) + f'\left(\frac{\pi}{2}\right)\left(x - \frac{\pi}{2}\right) + \frac{f''\left(\frac{\pi}{2}\right)}{2!}\left(x - \frac{\pi}{2}\right)^2$$

$$+ \frac{f'''\left(\frac{\pi}{2}\right)}{3!}\left(x - \frac{\pi}{2}\right)^3 + \cdots$$

$$= -\left(x - \frac{\pi}{2}\right) + \frac{1}{3!}\left(x - \frac{\pi}{2}\right)^3 - \frac{1}{5!}\left(x - \frac{\pi}{2}\right)^5 + \cdots$$

$$= \sum_{k=0}^{\infty} \frac{(-1)^{k+1}}{(2k+1)!}\left(x - \frac{\pi}{2}\right)^{2k+1}$$

The radius of convergence is $\infty$; the interval of convergence is $(-\infty, \infty)$. ∎

NOW WORK **Problem 39.**

## 4 Work with a Binomial Series

In algebra, the Binomial Theorem states that if $m$ is a positive integer and $a$ and $b$ are real numbers, then

$$(a+b)^m = a^m + \binom{m}{1}a^{m-1}b + \binom{m}{2}a^{m-2}b^2 + \cdots + \binom{m}{m-2}a^2 b^{m-2}$$

$$+ \binom{m}{m-1}ab^{m-1} + b^m = \sum_{k=0}^{m}\binom{m}{k}a^{m-k}b^k$$

where

**NEED TO REVIEW?** The Binomial Theorem is discussed in Section P.8, pp. 73–74.

$$\binom{m}{k} = \frac{m!}{k!\,(m-k)!} = \frac{m(m-1)\cdots(m-k+1)}{k!}$$

In the Binomial Theorem, $m$ is a positive integer. To generalize the result, we find a Maclaurin series for the function $f(x) = (1+x)^m$, where $m$ is *any real number*.

EXAMPLE 9  **Finding the Maclaurin Expansion for $f(x) = (1+x)^m$**

Find the Maclaurin series for $f(x) = (1+x)^m$, where $m$ is any real number.

**Solution** We begin by finding the derivatives of $f$ at 0:

$$f(x) = (1+x)^m \qquad\qquad\qquad f(0) = 1$$

$$f'(x) = m(1+x)^{m-1} \qquad\qquad f'(0) = m$$

$$f''(x) = m(m-1)(1+x)^{m-2} \qquad f''(0) = m(m-1)$$

$$\vdots \qquad\qquad\qquad\qquad\qquad \vdots$$

$$f^{(n)}(x) = m(m-1)(m-2)\cdots(m-n+1)(1+x)^{m-n} \qquad f^{(n)}(0) = m(m-1)(m-2)\cdots(m-n+1)$$

The Maclaurin expansion for $f$ is

$$(1+x)^m = 1 + mx + \frac{m(m-1)}{2!}x^2 + \cdots$$

$$+ \frac{m(m-1)(m-2)\cdots(m-n+1)}{n!}x^n + \cdots$$

$$= \binom{m}{0} + \binom{m}{1}x + \binom{m}{2}x^2 + \cdots + \binom{m}{n}x^n + \cdots = \sum_{k=0}^{\infty} \binom{m}{k}x^k$$

where

$$\binom{m}{0} = 1 \quad \text{and} \quad \binom{m}{k} = \frac{m(m-1)(m-2)\cdots(m-k+1)}{k!}$$

The series

$$(1+x)^m = \sum_{k=0}^{\infty} \binom{m}{k}x^k$$

$$= 1 + mx + \frac{m(m-1)}{2!}x^2 + \frac{m(m-1)(m-2)}{3!}x^3 + \cdots + \binom{m}{n}x^n + \cdots$$

is called a **binomial series** because of its similarity in form to the Binomial Theorem, and $\binom{m}{k}$ is called the **binomial coefficient of $x^k$**.

The following theorem, which we state without proof, gives the conditions under which the binomial series $(1+x)^m$ converges.

**THEOREM  Convergence of a Binomial Series**

The binomial series

$$(1+x)^m = \sum_{k=0}^{\infty} \binom{m}{k}x^k$$

converges

- for all $x$ if $m$ is a nonnegative integer. (In this case, there are only $m+1$ nonzero terms.)
- on the open interval $(-1, 1)$ if $m \leq -1$.
- on the half-open interval $(-1, 1]$ if $-1 < m < 0$.
- on the closed interval $[-1, 1]$ if $m > 0$, but $m$ is not an integer.

**EXAMPLE 10** Using a Binomial Series

Represent the function $f(x) = \sqrt{x+1}$ as a Maclaurin series, and find its interval of convergence.

**Solution** Write $\sqrt{x+1} = (1+x)^{1/2}$ and use the binomial series with $m = \dfrac{1}{2}$. The result is

$$(1+x)^{1/2} = \sum_{k=0}^{\infty} \binom{1/2}{k} x^k = 1 + \frac{1}{2}x + \frac{\left(\dfrac{1}{2}\right)\left(-\dfrac{1}{2}\right)}{2!} x^2 + \frac{\dfrac{1}{2}\left(-\dfrac{1}{2}\right)\left(-\dfrac{3}{2}\right)}{3!} x^3 + \cdots$$

$$= 1 + \frac{1}{2}x - \frac{1}{8}x^2 + \frac{1}{16}x^3 - \cdots$$

Since $m = \dfrac{1}{2} > 0$ and $m$ is not an integer, the series converges on the closed interval $[-1, 1]$. ∎

**NOW WORK** Problem 25.

**EXAMPLE 11** Using a Binomial Series

Represent the function $f(x) = \sin^{-1} x$ by a Maclaurin series.

**Solution** Recall that

$$\sin^{-1} x = \int_0^x \frac{dt}{\sqrt{1-t^2}} = \int_0^x (1-t^2)^{-1/2} dt$$

Write the integrand as a binomial series, with $x = -t^2$ and $m = -\dfrac{1}{2}$.

$$(1-t^2)^{-1/2} = \sum_{k=0}^{\infty} \binom{-1/2}{k} (-t^2)^k = 1 + \left(-\frac{1}{2}\right)(-t^2) + \frac{\left(-\dfrac{1}{2}\right)\left(-\dfrac{3}{2}\right)}{2!}(-t^2)^2$$

$$+ \frac{\left(-\dfrac{1}{2}\right)\left(-\dfrac{3}{2}\right)\left(-\dfrac{5}{2}\right)}{3!}(-t^2)^3 + \cdots$$

$$= 1 + \frac{1}{2}t^2 + \frac{3}{8}t^4 + \frac{5}{16}t^6 + \cdots$$

Now use the integration property of a power series to obtain

$$\int_0^x \frac{dt}{\sqrt{1-t^2}} = \int_0^x \left(1 + \frac{1}{2}t^2 + \frac{3}{8}t^4 + \frac{5}{16}t^6 + \cdots\right) dt$$

$$\sin^{-1} x = x + \frac{1}{2} \cdot \frac{x^3}{3} + \frac{3}{8} \cdot \frac{x^5}{5} + \frac{5}{16} \cdot \frac{x^7}{7} + \cdots$$

$$= x + \frac{x^3}{6} + \frac{3}{40}x^5 + \frac{5}{112}x^7 + \cdots \quad ∎$$

In Problem 59, you are asked to show that the interval of convergence is $[-1, 1]$.

**TABLE 7** Summary

| Series | Comment |
|---|---|
| $\dfrac{1}{1-x} = \displaystyle\sum_{k=0}^{\infty} x^k = 1 + x + x^2 + x^3 + \cdots$ | Converges for $|x| < 1$. |
| $\ln \dfrac{1}{1-x} = \displaystyle\sum_{k=1}^{\infty} \dfrac{x^k}{k} = x + \dfrac{x^2}{2} + \dfrac{x^3}{3} + \cdots + \dfrac{x^{n+1}}{n+1} + \cdots$ | Converges on the interval $[-1, 1)$. |
| $\ln(1+x) = \displaystyle\sum_{k=0}^{\infty} \dfrac{(-1)^k x^{k+1}}{k+1}$ | Converges on the interval $(-1, 1]$. |
| $\tan^{-1} x = \displaystyle\sum_{k=0}^{\infty} (-1)^k \dfrac{x^{2k+1}}{2k+1} = x - \dfrac{x^3}{3} + \dfrac{x^5}{5} - \cdots$ | Converges on the interval $[-1, 1]$. |
| $e^x = \displaystyle\sum_{k=0}^{\infty} \dfrac{x^k}{k!} = 1 + \dfrac{x}{1!} + \dfrac{x^2}{2!} + \dfrac{x^3}{3!} + \cdots$ | Converges for all real numbers $x$. |
| $\sin x = \displaystyle\sum_{k=0}^{\infty} \dfrac{(-1)^k x^{2k+1}}{(2k+1)!} = x - \dfrac{x^3}{3!} + \dfrac{x^5}{5!} - \dfrac{x^7}{7!} + \cdots$ | Converges for all real numbers $x$. |
| $\cos x = \displaystyle\sum_{k=0}^{\infty} \dfrac{(-1)^k x^{2k}}{(2k)!} = 1 - \dfrac{x^2}{2!} + \dfrac{x^4}{4!} - \dfrac{x^6}{6!} + \cdots$ | Converges for all real numbers $x$. |
| $\cosh x = \displaystyle\sum_{k=0}^{\infty} \dfrac{x^{2k}}{(2k)!} = 1 + \dfrac{x^2}{2!} + \dfrac{x^4}{4!} + \dfrac{x^6}{6!} + \cdots$ | Converges for all real numbers $x$. |
| $(1+x)^m = \displaystyle\sum_{k=0}^{\infty} \binom{m}{k} x^k$ | For convergence, see the theorem on page 661. |

## 8.9 Assess Your Understanding

### Concepts and Vocabulary

1. The series representation of a function $f$ given by the power series $f(x) = f(c) + f'(c)(x-c) + \dfrac{f''(c)(x-c)^2}{2!} + \cdots + \dfrac{f^{(n)}(c)(x-c)^n}{n!} + \cdots$ is called a(n) _____ about $c$.

2. If $c = 0$ in the Taylor expansion of a function $f$, then the expansion is called a(n) _____ expansion.

3. Write the first four terms of the Maclaurin expansion of $f$ if
$$f(0) = 2,\ f'(0) = -4,\ f''(0) = 3,\ f'''(0) = -2.$$

4. Write the first four terms of the Taylor expansion of $f$ centered at 5, if
$$f(5) = 3,\ f'(5) = 2,\ f''(5) = -1,\ f'''(5) = -4.$$

### Skill Building

In Problems 5–8, find the Taylor expansion of each function about the given number $c$. Comment on the result.

5. $f(x) = 3x^3 + 2x^2 + 5x - 6;\quad c = 0$

6. $f(x) = 4x^4 - 2x^3 - x;\quad c = 0$

7. $f(x) = 3x^3 + 2x^2 + 5x - 6;\quad c = 1$

8. $f(x) = 4x^4 - 2x^3 + x;\quad c = 1$

In Problems 9–20, find the Maclaurin expansion for each function and determine the interval of convergence.

9. $f(x) = e^{x^2}$

10. $f(x) = e^{-x^2}$

11. $f(x) = xe^x$

12. $f(x) = xe^{-x}$

13. $f(x) = x^2 \sin x$

14. $f(x) = x^2 \cos x$

15. $f(x) = \dfrac{1}{1-x^2}$

16. $f(x) = \dfrac{1}{1+x^2}$

17. $f(x) = \ln \dfrac{1}{1-3x}$

18. $f(x) = \ln \dfrac{1}{1+x}$

19. $f(x) = \ln(1+x^2)$

20. $f(x) = \ln(1-2x)$

In Problems 21–30, use a binomial series to represent each function as a Maclaurin series, and find the interval of convergence.

21. $f(x) = (1+x)^{-3}$

22. $f(x) = (1+x)^{-4}$

23. $f(x) = \sqrt{1+x^2}$

24. $f(x) = \dfrac{1}{\sqrt{1-x}}$

25. $f(x) = (1+x)^{1/5}$

26. $f(x) = (1-x)^{5/3}$

27. $f(x) = \dfrac{1}{(1+x^2)^{1/2}}$

28. $f(x) = \dfrac{1}{(1+x)^{3/4}}$

29. $f(x) = \dfrac{2x}{\sqrt{1-x}}$

30. $f(x) = \dfrac{x}{1+x^3}$

**1.** = NOW WORK problem      📈 = Graphing technology recommended      CAS = Computer Algebra System recommended

*In Problems 31–38, find the first five nonzero terms of the Maclaurin expansion.*

**31.** $f(x) = e^{-x} + \cos x$

**32.** $f(x) = e^x + \sin x$

**33.** $f(x) = \dfrac{e^x}{1-x}$

**34.** $f(x) = \dfrac{e^x}{1+x}$

**35.** $f(x) = \dfrac{\cos x}{1-x}$

**36.** $f(x) = \dfrac{\sin x}{1+x}$

**37.** $f(x) = e^{-x} \sin x$

**38.** $f(x) = e^{-x} \cos x$

*In Problems 39–48, find the Taylor expansion of each function about the given number c.*

**39.** $f(x) = \sqrt{x} \quad c = 1$

**40.** $f(x) = \sqrt[3]{x} \quad c = 1$

**41.** $f(x) = \ln x; \quad c = 1$

**42.** $f(x) = \ln(x-1); \quad c = 2$

**43.** $f(x) = \dfrac{1}{x}; \quad c = 1$

**44.** $f(x) = \dfrac{1}{\sqrt{x}}; \quad c = 4$

**45.** $f(x) = \sin x; \quad c = \dfrac{\pi}{6}$

**46.** $f(x) = \cos x; \quad c = -\dfrac{\pi}{2}$

**47.** $f(x) = e^{2x}; \quad c = 1$

**48.** $f(x) = e^{-3x}; \quad c = 2$

## Applications and Extensions

*In Problems 49–52, use a Maclaurin expansion to find each integral.*

**49.** $\displaystyle \int e^{x^2}\, dx$

**50.** $\displaystyle \int e^{\sqrt{x}}\, dx$

**51.** $\displaystyle \int \sin x^2\, dx$

**52.** $\displaystyle \int \cos x^2\, dx$

**53.** Find the first five nonzero terms of the Maclaurin expansion for $f(x) = \sec x$.

**54. Probability** The **standard normal distribution**

$$p(x) = \frac{1}{\sqrt{2\pi}} e^{-x^2/2}$$ is important in probability and statistics. If a random variable $Z$ has a standard normal distribution, then the probability that an observation of $Z$ is between $Z = a$ and $Z = b$ is given by

$$P(a \le Z \le b) = \frac{1}{\sqrt{2\pi}} \int_a^b e^{-x^2/2}\, dx$$

(a) Find the Maclaurin expansion for $p(x) = \dfrac{1}{\sqrt{2\pi}} e^{-x^2/2}$.

(b) Use properties of power series to find a power series representation for $P$.

(c) Use the first four terms of the series representation for $P$ to approximate $P(-0.5 \le Z \le 0.3)$.

**CAS** (d) Use technology to approximate $P(-0.5 \le Z \le 0.3)$.

**55. Even Functions** Show that if $f$ is an even function, then the Maclaurin expansion for $f$ has only even powers of $x$.

**56. Odd Functions** Show that if $f$ is an odd function, then the Maclaurin expansion for $f$ has only odd powers of $x$.

**57.** Show that $(1+x)^m = \displaystyle\sum_{k=0}^{\infty} \binom{m}{k} x^k$, when $m$ is a nonnegative integer, by showing that $R_n(x) \to 0$ as $n \to \infty$.

**58.** Show that the series $\displaystyle\sum_{k=0}^{\infty} \binom{m}{k} x^k$ converges absolutely for $|x| < 1$ and diverges for $|x| > 1$ if $m < 0$. (*Hint:* Use the Ratio Test.)

**59.** Show that the interval of convergence of the Maclaurin expansion for $f(x) = \sin^{-1} x$ is $[-1, 1]$.

**60. Euler's Error** Euler believed $\dfrac{1}{2} = 1 - 1 + 1 - 1 + 1 - 1 + \cdots$.

He based his argument to support this equation on his belief in the identification of a series and the values of the function from which it was derived.

(a) Write the Maclaurin expansion for $\dfrac{1}{1+x}$. Do this without calculating any derivatives.

(b) Evaluate both sides of the equation you derived in (a) at $x = 1$ to arrive at the formula above.

(c) Criticize the procedure used in (b).

## Challenge Problems

**61.** Find the exact sum of the infinite series:

$$\frac{x^3}{1\cdot 3} - \frac{x^5}{3\cdot 5} + \frac{x^7}{5\cdot 7} - \frac{x^9}{7\cdot 9} + \cdots \quad \text{for } x = 1$$

**62.** Find an elementary expression for $\displaystyle\sum_{k=1}^{\infty} \frac{x^{k+1}}{k(k+1)}$.

*Hint:* Integrate the series for $\ln \dfrac{1}{1-x}$.

**63.** Show that $\displaystyle\sum_{k=1}^{\infty} \frac{k}{(k+1)!} = 1$

**64.** Let $s_n = \dfrac{1}{1!} + \dfrac{1}{2!} + \cdots + \dfrac{1}{n!}$, $n = 1, 2, 3, \dots$.

(a) Show that $n! \ge 2^{n-1}$.

(b) Show that $0 < s_n \le 1 + \dfrac{1}{2} + \left(\dfrac{1}{2}\right)^2 + \cdots + \left(\dfrac{1}{2}\right)^{n-1}$.

(c) Show that $0 < s_n < s_{n+1} < 2$. Then conclude that $S = \displaystyle\lim_{n\to\infty} s_n$ and $S \le 2$.

(d) Let $t_n = \left[1 + \dfrac{1}{n}\right]^n$. Show that

$$t_n = 1 + 1 + \frac{1}{2!}\left[1 - \frac{1}{n}\right] + \frac{1}{3!}\left[1 - \frac{1}{n}\right]\left[1 - \frac{2}{n}\right] + \cdots$$

$$+ \frac{1}{n!}\left[1 - \frac{1}{n}\right]\left[1 - \frac{2}{n}\right]\cdots\left[1 - \frac{n-1}{n}\right] < s_n + 1$$

(e) Show that $0 < t_n < t_{n+1} < 3$. Then conclude that $e = \displaystyle\lim_{n\to\infty} t_n \le 3$.

**65.** Show that $\left[1 + \dfrac{1}{n}\right]^n < e$ for all $n > 0$.

**66.** From the fact that $\sin t \le t$ for all $t \ge 0$, use integration repeatedly to prove

$$1 - \frac{x^2}{2!} \le \cos x \le 1 - \frac{x^2}{2!} + \frac{x^4}{4!} \quad \text{for all } x \ge 0$$

**67.** Find the first four nonzero terms of the Maclaurin expansion for $f(x) = (1+x)^x$.

**68.** Show that $f(x) = \begin{cases} e^{-1/x^2} & x \ne 0 \\ 0 & x = 0 \end{cases}$ has a Maclaurin expansion at $x = 0$. Then show that the Maclaurin series does not converge to $f$.

# 8.10 Approximations Using Taylor/Maclaurin Expansions

**OBJECTIVES** *When you finish this section, you should be able to:*

**1** Approximate a function and its graph using a Taylor polynomial (p. 665)

**2** Approximate the number *e*; approximate logarithms (p. 668)

**3** Approximate definite integrals (p. 670)

We begin by restating Taylor's Formula with Remainder.

Suppose $f$ is a function whose first $n + 1$ derivatives are continuous on an open interval $I$ containing the number $c$. Then for every $x$ in $I$, there is a number $u$ between $x$ and $c$ for which

$$f(x) = f(c) + f'(c)(x - c) + \frac{f''(c)}{2!}(x - c)^2 + \cdots + \frac{f^{(n)}(c)}{n!}(x - c)^n + R_n(x) \quad (1)$$

where

$$R_n(x) = \frac{f^{(n+1)}(u)}{(n+1)!}(x - c)^{n+1} \quad (2)$$

is the remainder after $n + 1$ terms.

In (1), the sum of the first $n + 1$ terms on the right is a polynomial of degree $n$ called the **Taylor Polynomial** $P_n(x)$ **of degree** $n$ **of** $f$ **at** $c$. The remainder function $R_n$ in (2) is the **Lagrange form** of the remainder.

If $f$ has derivatives of all orders in the open interval $I$ and if $\lim_{n \to \infty} R_n(x) = 0$, then the infinite series $\sum_{k=0}^{\infty} \frac{f^{(k)}(c)}{k!}(x - c)^k$ converges to $f(x)$.

If we know a Taylor series converges to a function $f$, then we can use a Taylor polynomial to approximate both the function and its graph.

## 1 Approximate a Function and Its Graph Using a Taylor Polynomial

In Section 3.5, we found that if a function $f$ is differentiable on an interval $(a, b)$, then we can approximate $f$ at a number $c$ in the interval using the linear approximation

$$L(x) = f(c) + f'(c)(x - c)$$

where $y = L(x)$ is the tangent line to the graph of $f$ at the point $(c, f(c))$. This linear approximation for $f$ is good if $x$ is close enough to $c$, but as the interval widens, the accuracy of the estimate decreases.

Notice that $y = L(x)$ is a first degree polynomial whose terms are identical to the first two terms of the Taylor polynomial evaluated at $c$. If $f$ is twice differentiable on the interval $(a, b)$, we can improve the approximation by using a second degree polynomial $P_2$ whose graph has the same tangent line as $f$ at $c$ and whose second derivative $P''(c) = f''(c)$. This polynomial at $c$ is

$$P_2(x) = f(c) + f'(c)(x - c) + \frac{f''(c)}{2}(x - c)^2$$

Continuing in this fashion, if the function $f$ has derivatives of all orders at $c$, and if the Taylor series converges to $f$, then the Taylor polynomial

$$P_n(x) = f(c) + f'(c)(x - c) + \frac{f''(c)}{2!}(x - c)^2 + \cdots + \frac{f^{(n)}(c)}{n!}(x - c)^n$$

can be used to approximate $f$ near $c$.

**EXAMPLE 1** **Finding a Taylor Polynomial for** $f(x) = e^x$

Find the Taylor polynomial $P_n(x)$ for $f(x) = e^x$ at 0.

**Solution** The Taylor polynomial $P_n(x)$ at 0 equals the first $n+1$ terms of the Maclaurin expansion of $f(x) = e^x$. Since the Maclaurin expansion of $f(x) = e^x$ is

$$f(x) = \sum_{k=0}^{\infty} \frac{f^{(k)}(0)}{k!} x^k = 1 + x + \frac{x^2}{2!} + \frac{x^3}{3!} + \cdots + \frac{x^n}{n!} + \cdots = \sum_{k=0}^{\infty} \frac{x^k}{k!}$$

the Taylor polynomial $P_n(x)$ at 0 for $f(x) = e^x$ is

$$P_n(x) = 1 + x + \frac{x^2}{2!} + \frac{x^3}{3!} + \cdots + \frac{x^n}{n!}$$ ∎

Figure 29 illustrates how the graphs of the Taylor polynomials $P_1(x)$, $P_2(x)$, and $P_3(x)$ approximate the graph of $f(x) = e^x$ near 0.

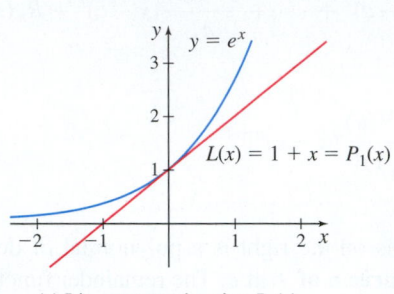

(a) Linear approximation $P_1(x)$

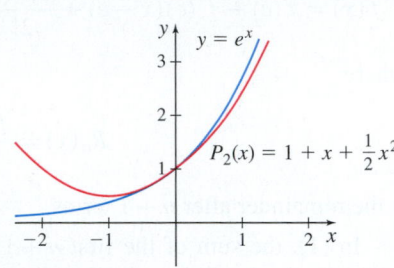

(b) Taylor polynomial $P_2(x)$

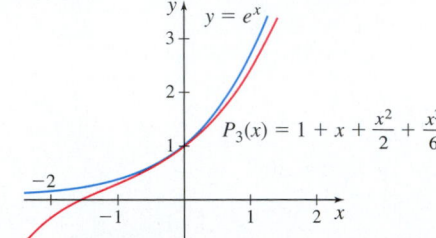

(c) Taylor polynomial $P_3(x)$

**DF Figure 29**

---

CALC
▶ CLIP
**EXAMPLE 2** **Approximating** $y = \sin x$ **Using a Taylor Polynomial**

**(a)** Approximate $y = \sin x$ using the Taylor polynomial $P_7(x)$ of $y = \sin x$ at 0.

**(b)** Graph $y = \sin x$ and the Taylor polynomial $P_7(x)$.

**(c)** Use (a) to approximate $\sin 0.1$.

**(d)** What is the error in using this approximation?

**Solution** **(a)** The Maclaurin expansion for $y = \sin x$ found in Example 3 (p. 657) of Section 8.9 is

$$y = \sin x = x - \frac{x^3}{3!} + \frac{x^5}{5!} - \frac{x^7}{7!} + \cdots = \sum_{k=0}^{\infty} (-1)^k \frac{x^{2k+1}}{(2k+1)!}$$

The Taylor polynomial $P_7(x)$ of degree 7 for $y = \sin x$ at 0 is

$$\sin x \approx P_7(x) = x - \frac{x^3}{3!} + \frac{x^5}{5!} - \frac{x^7}{7!} \tag{3}$$

**(b)** The graphs of $y = \sin x$ and $y = P_7(x)$ are given in Figure 30.

**(c)** Using (3), we get

$$\sin 0.1 \approx 0.1 - \frac{0.1^3}{3!} + \frac{0.1^5}{5!} - \frac{0.1^7}{7!} \approx 0.0998$$

**(d)** Since the Maclaurin expansion for $y = \sin x$ at $x = 0$ is an alternating series that satisfies the conditions of the Alternating Series Test, the error $E$ in using the first four nonzero terms as an approximation is less than or equal to the absolute value of the 5th nonzero term at $x = 0.1$. That is,

$$E \leq \left| \frac{0.1^9}{9!} \right| = 2.756 \times 10^{-15}$$ ∎

**DF Figure 30**

$P_7(x) = x - \frac{x^3}{3!} + \frac{x^5}{5!} - \frac{x^7}{7!}$

**NOW WORK** Problem **13.**

As the next example shows, we can find a Taylor polynomial for a function without using its Taylor expansion.

**EXAMPLE 3**  **Finding a Taylor Polynomial Without Using a Taylor Expansion**

**(a)**  Find the Taylor polynomial $P_3(x)$ for $f(x) = \sqrt{x}$ at 1.
**(b)**  Graph $y = \sqrt{x}$ and $y = P_3(x)$ on the same pair of axes.

**Solution (a)**  The first three derivatives of $f(x) = \sqrt{x}$ are:

$$f'(x) = \frac{1}{2\sqrt{x}} \quad f''(x) = \frac{d}{dx}\frac{x^{-1/2}}{2} = -\frac{1}{4x^{3/2}} \quad f'''(x) = \frac{d}{dx}\left(-\frac{x^{-3/2}}{4}\right) = \frac{3}{8x^{5/2}}$$

Then

$$f(1) = 1 \quad f'(1) = \frac{1}{2} \quad f''(1) = -\frac{1}{4} \quad f'''(1) = \frac{3}{8}$$

The Taylor polynomial $P_3(x)$ for $f(x) = \sqrt{x}$ at 1 is

$$P_3(x) = f(1) + f'(1)(x-1) + \frac{f''(1)}{2!}(x-1)^2 + \frac{f'''(1)}{3!}(x-1)^3$$

$$= 1 + \frac{x-1}{2} - \frac{(x-1)^2}{8} + \frac{(x-1)^3}{16}$$

**(b)**  The graphs of $y = \sqrt{x}$ and $y = P_3(x)$ are shown in Figure 31. ∎

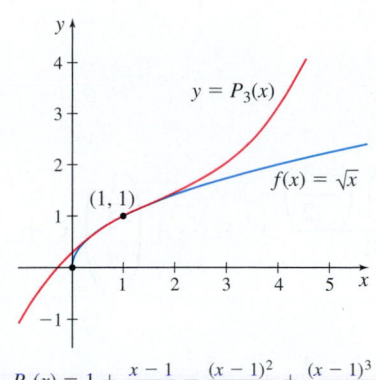

$$P_3(x) = 1 + \frac{x-1}{2} - \frac{(x-1)^2}{8} + \frac{(x-1)^3}{16}$$

**Figure 31**

Notice how the graphs of $y = P_3(x)$ and $y = f(x)$ are virtually identical for values of $x$ near 1.

**NOW WORK** Problem 3.

**EXAMPLE 4**  **Finding a Taylor Polynomial for $f(x) = \ln x$**

**(a)**  Find the Taylor polynomial $P_4(x)$ for $f(x) = \ln x$ at 1.
**(b)**  Graph $y = \ln x$ and $y = P_4(x)$ on the same pair of axes.

**Solution (a)**  $f$ and the first four derivatives of $f(x) = \ln x$ at 1 are:

$$f(x) = \ln x \qquad\qquad f(1) = \ln 1 = 0$$

$$f'(x) = \frac{1}{x} \qquad\qquad f'(1) = \frac{1}{1} = 1$$

$$f''(x) = -\frac{1}{x^2} \qquad\qquad f''(1) = -\frac{1}{1^2} = -1$$

$$f'''(x) = \frac{2}{x^3} \qquad\qquad f'''(1) = \frac{2}{1^3} = 2$$

$$f^{(4)}(x) = -\frac{6}{x^4} \qquad\qquad f^{(4)}(1) = -\frac{6}{1^4} = -6$$

The Taylor polynomial $P_4(x)$ for $f(x) = \ln x$ at 1 is

$$P_4(x) = f(1) + f'(1)(x-1) + \frac{f''(1)}{2!}(x-1)^2 + \frac{f'''(1)}{3!}(x-1)^3 + \frac{f^{(4)}(1)}{4!}(x-1)^4$$

$$= 0 + 1(x-1) + \frac{(-1)}{2!}(x-1)^2 + \frac{2}{3!}(x-1)^3 + \frac{(-6)}{4!}(x-1)^4$$

$$= (x-1) - \frac{(x-1)^2}{2} + \frac{(x-1)^3}{3} - \frac{(x-1)^4}{4}$$

**(b)**  The graphs of $y = \ln x$ and $y = P_4(x)$ are shown in Figure 32. ∎

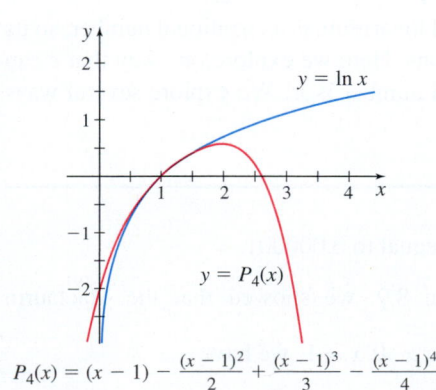

$$P_4(x) = (x-1) - \frac{(x-1)^2}{2} + \frac{(x-1)^3}{3} - \frac{(x-1)^4}{4}$$

**Figure 32**

**NOW WORK** Problem 5.

EXAMPLE 5   **Approximating $y = \sqrt{1+x}$ Using a Taylor Polynomial**

(a) Write the Maclaurin expansion for $y = \sqrt{1+x}$ using a binomial series.

(b) Approximate $\sqrt{1+x}$ using the Taylor polynomial $P_4(x)$ at 0.

(c) Graph $y = \sqrt{1+x}$ and the Taylor polynomial $P_4(x)$.

(d) Use $P_4(x)$ to approximate $\sqrt{1.2}$.

(e) What is the error in using this approximation?

**Solution** (a) For $-1 \le x \le 1$, we have

$$\sqrt{1+x} = (1+x)^{1/2} = \sum_{k=0}^{\infty} \binom{\frac{1}{2}}{k} x^k$$

$$= 1 + \frac{1}{2}x + \frac{\frac{1}{2}\left(-\frac{1}{2}\right)}{2!}x^2 + \frac{\frac{1}{2}\left(-\frac{1}{2}\right)\left(-\frac{3}{2}\right)}{3!}x^3 + \cdots + \binom{\frac{1}{2}}{n}x^n + \cdots$$

(b) From part (a), we obtain

$$\sqrt{1+x} \approx P_4(x) = 1 + \frac{1}{2}x - \frac{1}{8}x^2 + \frac{1}{16}x^3 - \frac{5}{128}x^4$$

(c) The graphs of $y = \sqrt{1+x}$ and $y = P_4(x)$ are given in Figure 33. On the interval of convergence $[-1, 1]$, the graphs are almost identical. Outside this interval, the graphs diverge.

(d) Using $P_4(x)$ and $x = 0.2$, we have

$$\sqrt{1.2} \approx P_4(x) = 1 + \frac{1}{2}(0.2) - \frac{1}{8}(0.2)^2 + \frac{1}{16}(0.2)^3 - \frac{5}{128}(0.2)^4 \approx 1.0954375$$

(e) Since the binomial series for $y = \sqrt{1+x}$ at $x = 0.2$ is an alternating series (after the first term) that satisfies the conditions of the Alternating Series Test, the error $E$ in using $P_4(x)$ to approximate $\sqrt{1.2}$ is

$$E \le \left| \frac{\frac{1}{2}\left(-\frac{1}{2}\right)\left(-\frac{3}{2}\right)\left(-\frac{5}{2}\right)\left(-\frac{7}{2}\right)}{5!} x^5 \right|_{\substack{\uparrow \\ x = 0.2}} = \frac{7}{256}(0.2)^5 \approx 8.75 \times 10^{-6}$$   ∎

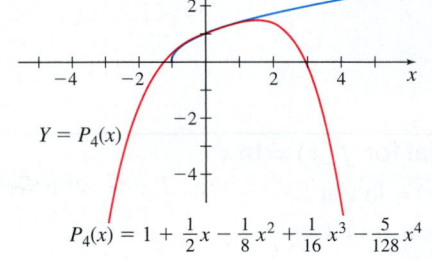

$y$

$y = \sqrt{1+x}$

$Y = P_4(x)$

$P_4(x) = 1 + \frac{1}{2}x - \frac{1}{8}x^2 + \frac{1}{16}x^3 - \frac{5}{128}x^4$

**Figure 33**

NOW WORK **Problem 15.**

## 2 Approximate the Number $e$; Approximate Logarithms

The important number $e$, the base of the natural logarithm, is an irrational number, so its decimal expansion neither terminates nor repeats. Here we explore one way that $e$ can be approximated. Another important irrational number is $\pi$. We explore several ways to approximate $\pi$ in Problems 34, 35, and 41.

EXAMPLE 6   **Approximating $e$**

Approximate $e$ so that the error is less than or equal to 0.000001.

**Solution** In Example 2 (p. 656) of Section 8.9, we showed that the Maclaurin series $\sum_{k=0}^{\infty} \frac{x^k}{k!}$ converges to $e^x$ for every number $x$. If $x = 1$, we have

$$e = e^1 = 1 + 1 + \frac{1}{2!} + \frac{1}{3!} + \cdots + \frac{1}{n!} + \cdots = \sum_{k=0}^{\infty} \frac{1}{k!}$$

But this series converges very slowly.

For faster convergence, we can use $x = -1$. Then

$$e^{-1} = 1 - 1 + \frac{1}{2!} - \frac{1}{3!} + \cdots + \frac{(-1)^n}{n!} + \cdots = \sum_{k=0}^{\infty} \frac{(-1)^k}{k!}$$

Since this series is an alternating series that satisfies the conditions of the Alternating Series Test, we can use the error estimate for an alternating series. Since we want the error $\leq 0.000001$, we need $\frac{1}{n!} \leq 0.000001$, or equivalently, $n! \geq 10^6$. Since $10! = 3.6288 \times 10^6 > 10^6$, but $9! = 362{,}880 < 10^6$, using 10 terms of the Maclaurin series will approximate $e^{-1}$ so that the error is less than or equal to 0.000001.

$$\sum_{k=0}^{9} \frac{(-1)^k}{k!} = 1 - 1 + \frac{1}{2} - \frac{1}{3!} + \frac{1}{4!} - \frac{1}{5!} + \frac{1}{6!} - \frac{1}{7!} + \frac{1}{8!} - \frac{1}{9!} = \frac{16{,}687}{45{,}360}$$

Then

$$e = \frac{1}{e^{-1}} = \left(\frac{16{,}687}{45{,}360}\right)^{-1} \approx 2.71828 \qquad \blacksquare$$

**NOW WORK** Problem **35**.

## Approximate Logarithms

In Example 9 (p. 648) of Section 8.8, we applied the integration property to the geometric series $\frac{1}{1-x} = \sum_{k=0}^{\infty} x^k$, and found that

$$\ln \frac{1}{1-x} = x + \frac{x^2}{2} + \frac{x^3}{3} + \cdots + \frac{x^n}{n} + \cdots \tag{4}$$

for $-1 \leq x < 1$. It might appear that this expansion can be used to compute the logarithms of numbers, but the series converges only for $-1 \leq x < 1$, and unless $x$ is close to 0, the series converges so slowly that too many terms would be required for practical use.

A more useful formula for computing logarithms is obtained by multiplying (4) by $(-1)$ and using the fact that $-\ln \frac{1}{A} = \ln \left(\frac{1}{A}\right)^{-1} = \ln A$. Then

$$\ln(1-x) = -x - \frac{x^2}{2} - \frac{x^3}{3} - \cdots \tag{5}$$

Now replace $x$ by $-x$. The result is

$$\ln(1+x) = x - \frac{x^2}{2} + \frac{x^3}{3} - \cdots \tag{6}$$

Subtracting (5) from (6),

$$\ln(1+x) - \ln(1-x) = \ln \frac{1+x}{1-x} = 2\left(x + \frac{x^3}{3} + \frac{x^5}{5} + \cdots\right) \tag{7}$$

This series converges for $-1 < x < 1$.

If $N$ is a positive integer, let $x = \frac{1}{2N+1}$. Then $0 < x < 1$, and

$$\frac{1+x}{1-x} = \frac{1 + \dfrac{1}{2N+1}}{1 - \dfrac{1}{2N+1}} = \frac{2N+1+1}{2N+1-1} = \frac{N+1}{N}$$

Substituting into (7), we obtain

$$\ln \frac{N+1}{N} = 2\left(x + \frac{x^3}{3} + \frac{x^5}{5} + \cdots\right)$$

Since $x = \dfrac{1}{2N+1}$,

$$\ln(N+1) = \ln N + 2\left[\frac{1}{2N+1} + \frac{1}{3}\left(\frac{1}{2N+1}\right)^3 + \frac{1}{5}\left(\frac{1}{2N+1}\right)^5 + \cdots\right] \qquad (8)$$

This series converges for all positive integers $N$.

**EXAMPLE 7** **Approximating a Logarithm**

(a) Approximate $\ln 2$ using the first three terms of (8).

(b) Approximate $\ln 3$ using the first three terms of (8).

**Solution** (a) In (8), we let $N = 1$ and use the first three terms of the series; then

$$\ln 2 \approx 2\left[\frac{1}{3} + \frac{1}{3}\left(\frac{1}{3}\right)^3 + \frac{1}{5}\left(\frac{1}{3}\right)^5\right] \approx 0.693004$$

(The first three terms of this series approximates $\ln 2$ correct to within 0.001.)

(b) To find $\ln 3$, let $N = 2$ and use $\ln 2 = 0.693004$ in (8).

$$\ln 3 \approx \ln 2 + 2\left[\frac{1}{5} + \frac{1}{3}\left(\frac{1}{5}\right)^3 + \frac{1}{5}\left(\frac{1}{5}\right)^5\right]$$

$$\approx 0.693004 + 2\left[\frac{1}{5} + \frac{1}{3}\left(\frac{1}{5}\right)^3 + \frac{1}{5}\left(\frac{1}{5}\right)^5\right] \approx 1.098465 \qquad \blacksquare$$

Continuing in this way, a table of natural logarithms can be obtained.

**NOW WORK** Problem 27.

## 3 Approximate Definite Integrals

We have already discussed the approximation of definite integrals using Riemann sums, the trapezoidal rule, the Midpoint Rule, and Simpson's rule. Taylor/Maclaurin series can also be used to approximate definite integrals.

**EXAMPLE 8** **Using a Maclaurin Expansion to Approximate an Integral**

Use a Maclaurin expansion to approximate $\int_0^{1/2} e^{-x^2}\, dx$ with an error less than or equal to 0.001.

**Solution** Replace $x$ by $-x^2$ in the Maclaurin expansion for $e^x$ to obtain the Maclaurin expansion for $e^{-x^2}$.

$$e^x = 1 + x + \frac{x^2}{2!} + \frac{x^3}{3!} + \cdots$$

$$e^{-x^2} = 1 - x^2 + \frac{x^4}{2!} - \frac{x^6}{3!} + \frac{x^8}{4!} - \cdots$$

Now use the integration property of power series.

$$\int_0^{1/2} e^{-x^2}\, dx = \int_0^{1/2}\left(1 - x^2 + \frac{x^4}{2!} - \frac{x^6}{3!} + \cdots\right) dx = \left[x - \frac{x^3}{3} + \frac{x^5}{2!\,5} - \frac{x^7}{3!\,7} + \cdots\right]_0^{1/2}$$

$$= \frac{1}{2} - \frac{1}{3(2)^3} + \frac{1}{2!\,5(2)^5} - \frac{1}{3!\,7(2)^7} + \cdots \approx 0.5 - 0.041666 + 0.003125 - 0.000186 + \cdots$$

Since this series satisfies the two conditions of the Alternating Series Test, the error due to using the first three terms as an approximation is less than or equal to the absolute value of the 4th term. Since $\dfrac{1}{3!\,7(2)^7} = 1.\,860\,119\,048 \times 10^{-4} < 0.001$,

$$\int_0^{1/2} e^{-x^2}\,dx \approx \frac{1}{2} - \frac{1}{3(2)^3} + \frac{1}{2!\,5(2)^5} \approx 0.461458$$

with an error less than or equal to 0.001. ∎

Notice that only three terms were needed to obtain the desired accuracy. To obtain this same accuracy using Simpson's rule, the Midpoint Rule, or the trapezoidal rule would have required a very fine partition of $\left[0, \dfrac{1}{2}\right]$. In practice, efficiency of technique is an important consideration for solving numerical problems.

**NOW WORK** Problem 21.

## 8.10 Assess Your Understanding

### Skill Building

*In Problems 1–12, for each function $f$ find the Taylor Polynomial $P_5(x)$ for $f$ at $c$.*

**1.** $f(x) = \ln x^2$ at $c = 1$

**2.** $f(x) = \ln(1+x)$ at $c = 0$

**3.** $f(x) = \dfrac{1}{x}$ at $c = 1$

**4.** $f(x) = \dfrac{1}{x^2}$ at $c = 1$

**5.** $f(x) = \cos x$ at $c = \dfrac{\pi}{2}$

**6.** $f(x) = \sin x$ at $c = \dfrac{\pi}{4}$

**7.** $f(x) = e^{2x}$ at $c = 0$

**8.** $f(x) = e^{-x}$ at $c = 0$

**9.** $f(x) = \dfrac{1}{1-2x}$ at $c = 0$

**10.** $f(x) = \dfrac{1}{1+x}$ at $c = 0$

**11.** $f(x) = x \ln x$ at $c = 1$

**12.** $f(x) = xe^x$ at $c = 1$

**13.** (a) Approximate $y = \cos x$ using the Taylor polynomial $P_6(x)$ of $y = \cos x$ at 0.

(b) Graph $y = \cos x$ and $y = P_6(x)$.

(c) Use $P_6(x)$ to approximate $\cos \dfrac{\pi}{90}$, (2°).

(d) What is the error in using this approximation?

(e) If $P_6(x)$ is used to approximate $y = \cos \dfrac{\pi}{90}$, for what values of $x$ is the error less than or equal to 0.0001?

**14.** (a) Approximate $y = e^x$ using the Taylor polynomial $P_3(x)$ of $y = e^x$ at 0.

(b) Graph $y = e^x$ and $y = P_3(x)$.

(c) Use $P_3(x)$ to approximate $e^{1/2}$.

(d) What is the error in using this approximation?

(e) If $P_3(x)$ is used to approximate $y = e^{1/2}$, for what values of $x$ is the error less than or equal to 0.0001?

**15.** (a) Represent the function $f(x) = \sqrt[3]{1+x}$ as a Maclaurin series.

(b) What is the interval of convergence?

(c) Approximate $y = \sqrt[3]{1+x}$ using the Taylor polynomial $P_4(x)$ of $y = \sqrt[3]{1+x}$ at 0.

(d) Graph $y = \sqrt[3]{1+x}$ and $y = P_4(x)$.

(e) Comment on the graphs in (d) and the result of the approximation.

(f) Use $P_4(x)$ to approximate $\sqrt[3]{0.9}$. What is the error in using this approximation?

**16.** (a) Represent $y = \dfrac{1}{\sqrt{4+x}}$ as a Maclaurin series.

(b) What is the interval of convergence?

(c) Approximate $y = \dfrac{1}{\sqrt{4+x}}$ using the Taylor polynomial $P_4(x)$ of $y = \dfrac{1}{\sqrt{4+x}}$ at 0.

(d) Graph $y = \dfrac{1}{\sqrt{4+x}}$ and $y = P_4(x)$.

(e) Comment on the graphs in (d) and the result of the approximation.

(f) Use $P_4(x)$ to approximate $\dfrac{1}{\sqrt{4.2}}$. What is the error in using this approximation?

**17.** (a) Represent $y = \tan^{-1} x$ as a Maclaurin series.

(b) What is the interval of convergence?

(c) Approximate $y = \tan^{-1} x$ using the Taylor polynomial $P_9(x)$ of $y = \tan^{-1} x$ at 0.

(d) Graph $y = \tan^{-1} x$ and $y = P_9(x)$.

(e) Comment on the graphs in (d) and the result of the approximation.

**18. (a)** Represent $y = \dfrac{1}{1-x}$ as a Maclaurin series.

**(b)** What is the interval of convergence?

**(c)** Approximate $y = \dfrac{1}{1-x}$ using the Taylor polynomial $P_3(x)$

of $y = \dfrac{1}{1-x}$ at 0.

**(d)** Graph $y = \dfrac{1}{1-x}$ and $y = P_3(x)$.

**(e)** Comment on the graphs in (d) and the result of the approximation.

*In Problems 19–26, approximate each integral using the first four terms of a Maclaurin series.*

**19.** $\displaystyle\int_0^1 \sin x^2\, dx$

**20.** $\displaystyle\int_0^1 \cos x^2\, dx$

**21.** $\displaystyle\int_0^{0.1} \dfrac{e^x}{1-x}\, dx$

**22.** $\displaystyle\int_0^{0.1} \dfrac{\cos x}{1-x}\, dx$

**23.** $\displaystyle\int_0^{0.2} \sqrt[3]{1+x^4}\, dx$

**24.** $\displaystyle\int_0^{1/2} \sqrt[3]{1+x}\, dx$

**25.** $\displaystyle\int_0^{1/2} \dfrac{1}{\sqrt[3]{1+x^2}}\, dx$

**26.** $\displaystyle\int_0^{0.2} \dfrac{1}{\sqrt{1+x^3}}\, dx$

## Applications and Extensions

**27.** Use the recursive formula (8) for $\ln(N+1)$ to show that

$$\ln 4 \approx 1.38629$$

**28. (a)** Write the first three nonzero terms and the general term of the Maclaurin series for $f(x) = 3\sin\left(\dfrac{x}{2}\right)$.

**(b)** What is the interval of convergence for the series found in (a)?

**(c)** What is the minimum number of terms of the series in (a) that are necessary to approximate $f$ on the interval $(-2,\ 2)$ with an error less than or equal to 0.1?

CAS *In Problems 29–32,*

**(a)** *find the indicated Taylor polynomial for each function $f$.*

**(b)** *Graph $f$ and the Taylor polynomial found in (a).*

**29. Uninhibited Decay** $f(x) = 0.34e^{[(-\ln 2)/5600]x}$ at $x = 0$, $P_4(x)$

**30. Uninhibited Growth** $f(x) = 5000e^{0.04x}$ at $x = 0$, $P_4(x)$

**31. Logistic Population Growth Model** $f(x) = \dfrac{100}{1+30.2e^{-0.2x}}$ at $x = 0$, $P_3(x)$

**32. Gompertz Population Growth Model** $f(x) = 100e^{-3e^{-0.2x}}$ at $x = 0$, $P_3(x)$

**33. Calculators Are Perfect, Right?** They always get the right answer with what looks like little or no effort at all. This is very misleading because calculators use a lot of the ideas in this chapter to obtain these answers. One advantage we have over a calculator is that we can approximate answers up to any accuracy we choose. So let's prove you are smarter than any calculator by approximating each of the following so that the error is less than or equal to 0.001 using a Maclaurin series.

**(a)** $F(x) = \sin x$, $x = 4°$

**(b)** $F(x) = \cos x$, $x = 15°$

**(c)** $F(x) = \tan^{-1} x$, $x = 0.05$

*Source*: Contributed by the students at Lander University, Greenwood, SC.

**34. Leibniz Formula for $\pi$** Leibniz derived the following formula for $\dfrac{\pi}{4}$: $\dfrac{\pi}{4} = 1 - \dfrac{1}{3} + \dfrac{1}{5} - \dfrac{1}{7} + \dfrac{1}{9} - \cdots$.

**(a)** Find $\displaystyle\int_0^1 \dfrac{1}{1+x^2}\, dx$.

**(b)** Expand the integrand in (a) into a power series and integrate it term-by-term to get Leibniz's formula.

**(c)** Find the sum of the first 10 terms in the above series. Does it appear that Leibniz's formula is useful for approximating $\pi$?

**(d)** How many terms are required to approximate $\pi$ correct to 10 decimal places?

**35. Approximating $\pi$** The series approximation of $\pi$ using Gregory's series converges very slowly. (See Problem 82, Section 8.8. p. 652.) A more rapidly convergent series is obtained by using the identity

$$\tan^{-1} 1 = \tan^{-1}\left(\dfrac{1}{2}\right) + \tan^{-1}\left(\dfrac{1}{3}\right)$$

Use $x = \dfrac{1}{2}$ and $x = \dfrac{1}{3}$ in **Gregory's series**, together with this identity, to approximate $\pi$ using the first four terms.

**36. Faster than light?** At low speeds, the kinetic energy $K$, that is, the energy due to the motion of an object of mass $m$ and speed $v$, is given by the formula $K = K(v) = \dfrac{1}{2}mv^2$. But this formula is only an approximation to the general formula, and works only for speeds much less than the speed of light, $c$. The general formula, which holds for all speeds, is

$$K_{\text{gen}}(v) = mc^2\left(\dfrac{1}{\sqrt{1 - \dfrac{v^2}{c^2}}} - 1\right)$$

The formula for $K$ was used very successfully for many years before Einstein arrived at the general formula, so $K$ must be essentially correct for low speeds. Use a binomial expansion to show that $\dfrac{1}{2}mv^2$ is a first approximation to $K_{\text{gen}}$ for $v$ close to 0.

## Challenge Problems

**37.** The graphs of $y = \sin x$ and $y = \lambda x$ intersect near $x = \pi$ if $\lambda$ is small. Let $f(x) = \sin x - \lambda x$. Find the Taylor polynomial $P_2(x)$ for $f$ at $\pi$, and use it to show that an approximate solution of the equation $\sin x = \lambda x$ is $x = \dfrac{\pi}{1+\lambda}$.

**38.** The graphs of $y = \cot x$ and $y = \lambda x$ intersect near $x = \dfrac{\pi}{2}$ if $\lambda$ is small. Let $f(x) = \cot x - \lambda x$. Find the Taylor polynomial $P_2(x)$ for $f$ at $\dfrac{\pi}{2}$, and use it to find an approximate solution of the equation $\cot x = \lambda x$.

**39.** Let $a_k = (-1)^{k+1} \int_0^{\pi/k} \sin(kx)\,dx$.

   **(a)** Find $a_k$.

   **(b)** Show that the infinite series $\sum\limits_{k=1}^{\infty} a_k$ converges.

   **(c)** Show that $1 \le \sum\limits_{k=1}^{\infty} a_k \le \dfrac{3}{2}$.

**40.** Find the Maclaurin expansion for $f(x) = xe^{x^3}$.

**41.** **(a)** Use the Maclaurin expansion for $f(x) = \sin^{-1} x$ to find numerical series for $\dfrac{\pi}{2}$ and for $\dfrac{\pi}{6}$.

   **(b)** Use the result of (a) to approximate $\dfrac{\pi}{2}$ so that the error is less than or equal to 0.001.

   **(c)** Use the result of (a) to approximate $\dfrac{\pi}{6}$ so that the error is less than or equal to 0.001.

# Chapter Review

## THINGS TO KNOW

### 8.1 Sequences

***Definitions:***

- A sequence is a function whose domain is the set of positive integers and whose range is a subset of real numbers. (p. 574)
- $n$th term of a sequence (p. 574)
- Limit of a sequence (p. 577)
- Convergence; divergence of a sequence (p. 577)
- Related function of a sequence (p. 579)
- Divergence of a sequence to infinity (p. 581)
- Bounded sequence (p. 583)
- Monotonic sequence (p. 584)

***Properties of a Convergent Sequence:***    (p. 578)

If $\{s_n\}$ and $\{t_n\}$ are convergent sequences and if $c$ is a number, then

- Constant multiple property: $\lim\limits_{n \to \infty} (cs_n) = c \lim\limits_{n \to \infty} s_n$
- Sum and difference properties:

$$\lim_{n \to \infty} (s_n \pm t_n) = \lim_{n \to \infty} s_n \pm \lim_{n \to \infty} t_n$$

- Product property: $\lim\limits_{n \to \infty} (s_n \cdot t_n) = \left( \lim\limits_{n \to \infty} s_n \right)\left( \lim\limits_{n \to \infty} t_n \right)$
- Quotient property: $\lim\limits_{n \to \infty} \dfrac{s_n}{t_n} = \dfrac{\lim\limits_{n \to \infty} s_n}{\lim\limits_{n \to \infty} t_n}$

   provided $\lim\limits_{n \to \infty} t_n \ne 0$

- Power property: $\lim\limits_{n \to \infty} s_n^p = \left[ \lim\limits_{n \to \infty} s_n \right]^p$ where $p \ge 2$

   is an integer

- Root property: $\lim\limits_{n \to \infty} \sqrt[p]{s_n} = \sqrt[p]{\lim\limits_{n \to \infty} s_n}$, where $p \ge 2$

   and $s_n > 0$ if $p$ is even

***Theorems:***

- Let $\{s_n\}$ be a sequence of real numbers. If $\lim\limits_{n \to \infty} s_n = L$ and if $f$ is a function that is continuous at $L$ and is defined for all numbers $s_n$, then $\lim\limits_{n \to \infty} f(s_n) = f(L)$. (p. 579)
- Let $\{s_n\}$ be a sequence and let $f$ be a related function of $\{s_n\}$. Suppose $L$ is a real number. If $\lim\limits_{x \to \infty} f(x) = L$, then $\lim\limits_{n \to \infty} s_n = L$. (p. 580)
- The Squeeze Theorem for sequences (p. 581)
- The geometric sequence $\{r^n\}$, where $r$ is a real number,
  - converges to 0, for $-1 < r < 1$.
  - converges to 1, for $r = 1$.
  - diverges for all other numbers. (p. 582)

- A convergent sequence is bounded (p. 583)
- If a sequence is not bounded from above or if it is not bounded from below, then it diverges. (p. 583)
- An increasing (or nondecreasing) sequence $\{s_n\}$ that is bounded from above converges. (p. 585)
- A decreasing (or nonincreasing) sequence $\{s_n\}$ that is bounded from below converges. (p. 585)

***Procedure:***   Ways to show a sequence is monotonic: Table 2 (p. 584)

***Summary:***   How to determine if a sequence converges (p. 586)

### 8.2 Infinite Series

- If $a_1, a_2, \ldots, a_n, \ldots$ is an infinite collection of numbers, the expression $\sum\limits_{k=1}^{\infty} a_k = a_1 + a_2 + \cdots + a_n + \cdots$ is called an infinite series or, simply, a series. (p. 590)
- $n$th term or general term of a series (p. 590)
- Partial sum $S_n = \sum\limits_{k=1}^{n} a_k$, where $S_n$ is the sum of the first $n$ terms of the series $\sum\limits_{k=1}^{\infty} a_k$ (p. 590)
- Convergence, divergence of a series (p. 591)
- Geometric series $\sum\limits_{k=1}^{\infty} ar^{k-1} = a + ar + ar^2 + \cdots, a \ne 0$ (p. 593)
  - $\sum\limits_{k=1}^{\infty} ar^{k-1}$ converges if $|r| < 1$, and its sum is $\dfrac{a}{1-r}$
  - $\sum\limits_{k=1}^{\infty} ar^{k-1}$ diverges if $|r| \ge 1$. (p. 594)
- Harmonic series $\sum\limits_{k=1}^{\infty} \dfrac{1}{k} = 1 + \dfrac{1}{2} + \dfrac{1}{3} + \cdots$ (p. 597)

The harmonic series diverges. (p. 597)

***Summary:***   Series and convergence of series (pp. 597–598)

### 8.3 Properties of Series; Series with Positive Terms; the Integral Test

- If the series $\sum\limits_{k=1}^{\infty} a_k$ converges, then $\lim\limits_{n \to \infty} a_n = 0$. (p. 602)
- The Test for Divergence:

The infinite series $\sum\limits_{k=1}^{\infty} a_k$ diverges if $\lim\limits_{n \to \infty} a_n \ne 0$. (p. 602)

- If two infinite series are identical after a certain term, then either both series converge or both series diverge. If both series converge, they do not necessarily have the same sum. (p. 603)

- The General Convergence Test (p. 605)
- The Integral Test (p. 605)

    Let $f$ be a function that is continuous, positive, and nonincreasing on the interval $[1, \infty)$ and $a_k = f(k)$ for all positive integers $k$.

    - If the improper integral $\int_1^\infty f(x)dx$ converges, the series $\sum_{k=1}^\infty a_k$ converges.

    - If the improper integral $\int_1^\infty f(x)dx$ diverges, the series $\sum_{k=1}^\infty a_k$ diverges.

- A $p$-series $\sum_{k=1}^\infty \dfrac{1}{k^p} = 1 + \dfrac{1}{2^p} + \dfrac{1}{3^p} + \cdots + \dfrac{1}{n^p} + \cdots,$

    where $p$ is a positive real number, converges if $p > 1$ and diverges if $0 < p \le 1$. (p. 607)

- Bounds for the sum of a convergent $p$-series: (p. 608)

    If $p > 1$, then $\dfrac{1}{p-1} < \sum_{k=1}^\infty \dfrac{1}{k^p} < 1 + \dfrac{1}{p-1}.$

- Bounds for the sum of a convergent series: (p. 609)

    For the series $\sum_{k=1}^\infty a_k$, $a_k > 0$, suppose $f$ is a function that is continuous, positive, and decreasing on $[1, \infty)$, and $a_k = f(k)$ for all positive integers $k$. If the improper integral $\int_1^\infty f(x)dx$ converges, the sum of the series $\sum_{k=1}^\infty a_k$ is bounded by

    $$\int_1^\infty f(x)dx < \sum_{k=1}^\infty a_k < a_1 + \int_1^\infty f(x)dx$$

**Properties of Convergent Series:** If $\sum_{k=1}^\infty a_k$ and $\sum_{k=1}^\infty b_k$ are two convergent series and if $c \ne 0$ is a number, then

- Sum and difference properties:

    $$\sum_{k=1}^\infty (a_k \pm b_k) = \sum_{k=1}^\infty a_k \pm \sum_{k=1}^\infty b_k \quad \text{(p. 604)}$$

- Constant multiple property:

    $$\sum_{k=1}^\infty (ca_k) = c\sum_{k=1}^\infty a_k \quad c \ne 0 \quad \text{(p. 604)}$$

- If $\sum_{k=1}^\infty a_k$ diverges, then $\sum_{k=1}^\infty (ca_k)$ also diverges. (p. 604)

## 8.4 Comparison Tests
**Theorems:**

- Comparison Test for Convergence: If $0 < a_k \le b_k$ for all $k$ and $\sum_{k=1}^\infty b_k$ converges, then $\sum_{k=1}^\infty a_k$ converges. (p. 613)
- Comparison Test for Divergence: If $0 < c_k \le a_k$ for all $k$ and $\sum_{k=1}^\infty c_k$ diverges, then $\sum_{k=1}^\infty a_k$ diverges. (p. 614)
- Limit Comparison Test: Suppose $\sum_{k=1}^\infty a_k$ and $\sum_{k=1}^\infty b_k$ are both series of positive terms.
    - If $\lim_{n \to \infty} \dfrac{a_n}{b_n} = L$, $0 < L < \infty$, then both series converge or both diverge. (p. 614)

- If $\lim_{n \to \infty} \dfrac{a_n}{b_n} = 0$ and if $\sum_{k=1}^\infty b_k$ converges, then $\sum_{k=1}^\infty a_k$ converges.
- If $\lim_{n \to \infty} \dfrac{a_n}{b_n} = \infty$ and if $\sum_{k=1}^\infty b_k$ diverges, then $\sum_{k=1}^\infty a_k$ diverges.

**Summary:**   Table 3: Series often used for comparisons (p. 617)

## 8.5 Alternating Series; Absolute Convergence
**Definitions:**

- Alternating series (p. 620)
- A series $\sum_{k=1}^\infty a_k$ is absolutely convergent if the series $\sum_{k=1}^\infty |a_k|$ is convergent. (p. 625)
- A series that is convergent without being absolutely convergent is conditionally convergent. (p. 626)

**Theorems:**

- Alternating Series Test: (p. 621)
- The alternating harmonic series $\sum_{k=1}^\infty \dfrac{(-1)^{k+1}}{k}$ converges. (p. 622).
- Error estimate for a convergent alternating series (p. 623)
- Absolute Convergence Test: If a series $\sum_{k=1}^\infty a_k$ is absolutely convergent, then it is convergent. (p. 625)

**Properties of Absolutely Convergent and Conditionally Convergent Series:**   (p. 627)

## 8.6 Ratio Test, Root Test
- Ratio Test (p. 630)
- Root Test (p. 633)

## 8.7 Summary of Tests
- Guide to choosing a test (p. 637)
- Tests for convergence and divergence (Table 5; p. 638)
- Important Series (Table 6; p. 639)

## 8.8 Power Series
**Definitions:**

- Power series: $\sum_{k=0}^\infty a_k x^k$ or $\sum_{k=0}^\infty a_k(x-c)^k$, where $c$ is a constant. (p. 640)
- Radius of convergence (p. 643)
- Interval of convergence (p. 643)

**Theorems:**

- If a power series centered at 0 converges for a number $x_0 \ne 0$, then it converges absolutely for all numbers $x$ for which $|x| < |x_0|$. (p. 642)
- If a power series centered at 0 diverges for a number $x_1$, then it diverges for all numbers $x$ for which $|x| > |x_1|$. (p. 642)
- For a power series centered at $c$, exactly one of the following is true (p. 643):
    - The series converges for only $x = c$.
    - The series converges absolutely for all $x$.
    - There is a positive number $R$ for which the series converges absolutely for all $x$, $|x - c| < R$, and diverges for all $x$, $|x - c| > R$.

**Properties of Power Series:**   (p. 647)

Let $f(x) = \sum\limits_{k=0}^{\infty} a_k x^k$ be a power series in $x$ having a nonzero radius of convergence $R$.

- *Continuity property*: If $-R < x_0 < R$,

$$\lim_{x \to x_0} \left( \sum_{k=0}^{\infty} a_k x^k \right) = \sum_{k=0}^{\infty} \left( \lim_{x \to x_0} a_k x^k \right) = \sum_{k=0}^{\infty} a_k x_0^k$$

- *Differentiation property*:

$$\frac{d}{dx} \left( \sum_{k=0}^{\infty} a_k x^k \right) = \sum_{k=0}^{\infty} \left( \frac{d}{dx} a_k x^k \right) = \sum_{k=1}^{\infty} k a_k x^{k-1}$$

- *Integration property*:

$$\int_0^x \left( \sum_{k=0}^{\infty} a_k t^k \right) dt = \sum_{k=0}^{\infty} \left( \int_0^x a_k t^k \, dt \right) = \sum_{k=0}^{\infty} \frac{a_k x^{k+1}}{k+1}$$

- Taylor series:   (p. 654)

$$f(x) = f(c) + f'(c)(x-c) + \frac{f''(c)}{2!}(x-c)^2$$
$$+ \cdots + \frac{f^{(n)}(c)}{n!}(x-c)^n + \cdots = \sum_{k=0}^{\infty} \frac{f^{(k)}(c)}{k!}(x-c)^k$$

- Maclaurin series (p. 654)

$$f(x) = f(0) + f'(0)x + \frac{f''(0)x^2}{2!} + \cdots + \frac{f^{(n)}(0)x^n}{n!} + \cdots$$
$$= \sum_{k=0}^{\infty} \frac{f^{(k)}(0)}{k!}x^k$$

- Taylor's formula with remainder (p. 655)
- Convergence of a Taylor series (p. 656)
- Binomial series (p. 661)
- Convergence of a binomial series (p. 661)

***Table 7:***   A list of important Maclaurin expansions (p. 663)

## 8.9 Taylor Series; Maclaurin Series

***Theorems:***   (top, right column)

## 8.10 Approximations Using Taylor/Maclaurin Expansions
(pp. 665–671)

Taylor polynomial $P_n(x)$ of degree $n$ of $f$ at $c$. (p. 665)

## OBJECTIVES

| Section | You should be able to ... | Examples | Review Exercises |
|---|---|---|---|
| 8.1 | 1 Write several terms of a sequence (p. 575) | 1, 2 | 1, 2 |
| | 2 Find the $n$th term of a sequence (p. 575) | 3, 4 | 3 |
| | 3 Use properties of convergent sequences (p. 578) | 5, 6 | 4, 5 |
| | 4 Use a related function or the Squeeze Theorem to show a sequence converges (p. 579) | 7–10 | 6, 7 |
| | 5 Determine whether a sequence converges or diverges (p. 581) | 11–15 | 8–13 |
| 8.2 | 1 Determine whether a series has a sum (p. 590) | 1–4 | 14, 15 |
| | 2 Analyze a geometric series (p. 593) | 5–7 | 17–20 |
| | 3 Analyze the harmonic series (p. 597) | | 16 |
| 8.3 | 1 Use the Test for Divergence (p. 603) | 1 | 21 |
| | 2 Work with properties of series (p. 603) | 2 | 25–27 |
| | 3 Use the Integral Test (p. 605) | 3–5 | 22, 23 |
| | 4 Analyze a $p$-series (p. 607) | 6 | 24 |
| | 5 Approximate the Sum of a Convergent Series (p. 608) | 7, 8 | 24, 53, 54 |
| 8.4 | 1 Use Comparison Tests for Convergence and Divergence (p. 613) | 1, 2 | 28 |
| | 2 Use the Limit Comparison Test (p. 614) | 3–6 | 28–30 |
| 8.5 | 1 Determine whether an alternating series converges (p. 621) | 1–3 | 31–33 |
| | 2 Approximate the sum of a convergent alternating series (p. 623) | 4 | 31–33 |
| | 3 Use the Absolute Convergence test (p. 625) | 5–7 | 34–37 |
| 8.6 | 1 Use the Ratio Test (p. 630) | 1–3 | 38, 39 |
| | 2 Use the Root Test (p. 633) | 4, 5 | 40, 41 |
| 8.7 | 1 Choose an appropriate test to determine whether a series converges (p. 637) | | 42–52 |
| 8.8 | 1 Determine whether a power series converges (p. 641) | 1, 2 | 55(a)–60(a) |
| | 2 Find the interval of convergence of a power series (p. 643) | 3–5 | 55(b)–60(b) |
| | 3 Define a function using a power series (p. 645) | 6, 7 | 61, 62 |
| | 4 Use properties of power series (p. 647) | 8–10 | 63 |
| 8.9 | 1 Express a function as a Taylor series or a Maclaurin series (p. 654) | 1 | 66 |
| | 2 Determine the convergence of a Taylor/Maclaurin series (p. 655) | 2, 3 | |
| | 3 Find Taylor/Maclaurin expansions (p. 658) | 4–8 | 64, 65, 66, 67, 68 |
| | 4 Work with a binomial series (p. 660) | 9–11 | 69–71 |
| 8.10 | 1 Approximate a function and its graph using a Taylor polynomial (p. 665) | 1–5 | 72 |
| | 2 Approximate the number $e$; approximate logarithms (p. 668) | 6, 7 | 73 |
| | 3 Approximate definite integrals (p. 670) | 8 | 74, 75 |

## REVIEW EXERCISES

*In Problems 1 and 2, the nth term of a sequence $\{s_n\}$ is given. Write the first five terms of each sequence.*

**1.** $s_n = \dfrac{(-1)^{n+1}}{n^4}$

**2.** $s_n = \dfrac{2^n}{3^n}$

**3.** Find an expression for the $n$th term of the sequence,

$$2, -\frac{3}{2}, \frac{9}{8}, -\frac{27}{32}, \frac{81}{128}, \ldots$$

assuming the indicated pattern continues for all $n$.

*In Problems 4 and 5, use properties of convergent sequences to find the limit of each sequence.*

**4.** $\left\{ 1 + \dfrac{n}{n^2 + 1} \right\}$

**5.** $\left\{ \ln \dfrac{n+2}{n} \right\}$

*In Problems 6 and 7, use a related function or the Squeeze Theorem for sequences to show each sequence converges. Find its limit.*

**6.** $\left\{ \tan^{-1} n \right\}$

**7.** $\left\{ \dfrac{(-1)^n}{(n+1)^2} \right\}$

**8.** Determine if the sequence $\left\{ \dfrac{e^n}{(n+2)^2} \right\}$ is monotonic. If it is monotonic, is it increasing, nondecreasing, decreasing, or nonincreasing? Is it bounded from above and/or from below? Does it converge?

*In Problems 9–12, determine whether each sequence converges or diverges. If it converges, find its limit.*

**9.** $\{ n! \}$

**10.** $\left\{ \left( \dfrac{5}{8} \right)^n \right\}$

**11.** $\left\{ \left( -\dfrac{1}{2} \right)^n \right\}$

**12.** $\left\{ (-1)^n + e^{-n} \right\}$

**13.** Show that sequence $\left\{ 1 + \dfrac{2}{n} \right\}$ converges by showing it is either bounded from above and increasing or is bounded from below and decreasing.

**14.** Find the fifth partial sum of $\displaystyle\sum_{k=1}^{\infty} \dfrac{(-1)^k}{4^{k-1}}$.

**15.** Find the sum of the telescoping series $\displaystyle\sum_{k=1}^{\infty} \left( \dfrac{4}{k+4} - \dfrac{4}{k+5} \right)$.

*In Problems 16–19, determine whether each series converges or diverges. If it converges, find its sum.*

**16.** $\displaystyle\sum_{k=1}^{\infty} \dfrac{\cos^2(k\pi)}{k}$

**17.** $\displaystyle\sum_{k=1}^{\infty} -(\ln 2)^k$

**18.** $\displaystyle\sum_{k=0}^{\infty} \dfrac{e}{3^k}$

**19.** $\displaystyle\sum_{k=1}^{\infty} (4^{1/3})^k$

**20.** Express $0.123123123\ldots$ as a rational number using a geometric series.

**21.** Show that the series $\displaystyle\sum_{k=1}^{\infty} \dfrac{3k - 2}{k}$ diverges.

*In Problems 22 and 23, use the Integral Test to determine whether each series converges or diverges.*

**22.** $\displaystyle\sum_{k=1}^{\infty} \dfrac{\ln k}{k^2}$

**23.** $\displaystyle\sum_{k=1}^{\infty} \dfrac{1}{4k^2 + 9}$

**24.** Determine whether the $p$-series $\displaystyle\sum_{k=1}^{\infty} \dfrac{1}{k^{5/2}}$ converges or diverges. If it converges, find bounds for the sum.

*In Problems 25–27, determine whether each series converges or diverges.*

**25.** $\displaystyle\sum_{k=5}^{\infty} \left[ \dfrac{1}{k^5} \cdot \dfrac{1}{2^k} \right]$

**26.** $\displaystyle\sum_{k=1}^{\infty} \left[ \dfrac{3}{5^k} - \left( \dfrac{2}{3} \right)^{k-1} \right]$

**27.** $\displaystyle\sum_{k=1}^{\infty} \dfrac{3}{k^5}$

*In Problems 28–30, use a Comparison Test to determine whether each series converges or diverges.*

**28.** $\displaystyle\sum_{k=1}^{\infty} \dfrac{1}{\sqrt{k+1}}$

**29.** $\displaystyle\sum_{k=1}^{\infty} \dfrac{k+1}{k^k + 1}$

**30.** $\displaystyle\sum_{k=1}^{\infty} \dfrac{4}{k \, 3^k}$

*In Problems 31–33, determine whether each alternating series converges or diverges. If it converges and satisfies the conditions of the Alternating Series Test, approximate the sum of the series so that the error is less than or equal to 0.001.*

**31.** $\displaystyle\sum_{k=1}^{\infty} (-1)^{k+1} \dfrac{k+2}{k(k+1)}$

**32.** $\displaystyle\sum_{k=1}^{\infty} (-1)^{k+1} \dfrac{k^2}{e^k}$

**33.** $\displaystyle\sum_{k=1}^{\infty} (-1)^k \dfrac{3}{\sqrt[3]{k}}$

*In Problems 34–37, determine whether each series converges (absolutely or conditionally) or diverges.*

**34.** $\displaystyle\sum_{k=1}^{\infty} \sin \left( \dfrac{\pi}{2} k \right)$

**35.** $\displaystyle\sum_{k=1}^{\infty} \dfrac{(-1)^{k+1}}{\sqrt{k}}$

**36.** $\displaystyle\sum_{k=1}^{\infty} \dfrac{\cos k}{k^3}$

**37.** $\dfrac{1}{2} - \dfrac{4}{2^3 + 1} + \dfrac{9}{3^3 + 1} - \dfrac{16}{4^3 + 1} + \cdots$

*In Problems 38 and 39, use the Ratio Test to determine whether each series converges or diverges.*

**38.** $\displaystyle\sum_{k=1}^{\infty} \dfrac{2^k}{k!}$

**39.** $\displaystyle\sum_{k=1}^{\infty} \dfrac{k!}{e^{k^2}}$

*In Problems 40 and 41, use the Root Test to determine whether each series converges or diverges.*

**40.** $\displaystyle\sum_{k=1}^{\infty} \frac{2^k}{(k+3)^{k+1}}$

**41.** $\displaystyle\sum_{k=1}^{\infty} \left[ \ln\left(e^4 + \frac{1}{k^2}\right) \right]^k$

*In Problems 42–52, determine whether each series converges or diverges.*

**42.** $\displaystyle\sum_{k=1}^{\infty} (-1)^{k+1} \frac{2^{k+1}}{3^k}$

**43.** $\displaystyle\sum_{k=1}^{\infty} \ln\left(1 + \frac{1}{k}\right)$

**44.** $\displaystyle\sum_{k=5}^{\infty} \frac{3}{k\sqrt{k-4}}$

**45.** $\displaystyle\sum_{k=1}^{\infty} \frac{1}{\left(1 + \dfrac{k^2+1}{k^2}\right)^k}$

**46.** $\displaystyle\sum_{k=1}^{\infty} \frac{2 \cdot 4 \cdot 6 \cdots (2k)}{1 \cdot 3 \cdot 5 \cdots (2k-1)}$

**47.** $\displaystyle\sum_{k=1}^{\infty} \frac{k^2}{(1+k^3) \ln \sqrt[3]{1+k^3}}$

**48.** $\displaystyle\sum_{k=1}^{\infty} \frac{k^{10}}{2^k}$

**49.** $\displaystyle\sum_{k=1}^{\infty} \frac{\left(1 + \dfrac{1}{k^2}\right)^{k^2}}{2^k}$

**50.** $\displaystyle\sum_{k=1}^{\infty} \left(\frac{k^2+1}{k}\right)^k$

**51.** $\displaystyle\sum_{k=1}^{\infty} \frac{k!}{3^k k^k}$

**52.** $\displaystyle\sum_{k=1}^{\infty} (-1)^{k+1} \frac{k+2}{3k-2}$

**53.** Find bounds for the sum of the convergent series $\displaystyle\sum_{k=2}^{\infty} \frac{6}{k(\ln k)^3}$.

**54.** Find bounds for the sum of the convergent series $\displaystyle\sum_{k=1}^{\infty} \frac{2}{k^{4/3}}$.

*In Problems 55–60,*

*(a) Find the radius of convergence of each power series.*

*(b) Find the interval of convergence of each power series.*

**55.** $\displaystyle\sum_{k=1}^{\infty} \frac{(x-3)^{3k-1}}{k^2}$

**56.** $\displaystyle\sum_{k=1}^{\infty} \frac{x^k}{\sqrt[3]{k}}$

**57.** $\displaystyle\sum_{k=0}^{\infty} (-1)^k \frac{1}{k!(k+1)} \left(\frac{x}{2}\right)^{2k+1}$

**58.** $\displaystyle\sum_{k=1}^{\infty} \frac{k^k}{(k!)^2} x^k$

**59.** $\displaystyle\sum_{k=1}^{\infty} \frac{(x-1)^k}{k}$

**60.** $\displaystyle\sum_{k=0}^{\infty} \frac{3^k x^k}{5^k}$

*In Problems 61 and 62, express each function as a power series centered at 0.*

**61.** $f(x) = \dfrac{2}{x+3}$

**62.** $f(x) = \dfrac{1}{1-3x}$

**63. (a)** Use properties of a power series to find the power series representation for $\displaystyle\int \frac{1}{1-3x^2}\, dx$.

   **(b)** Use (a) to approximate $\displaystyle\int_0^{1/2} \frac{1}{1-3x^2}\, dx$ correct to within 0.001.

**64.** Find the Taylor expansion of $f(x) = \dfrac{1}{1-2x}$ about $c = 1$.

**65.** Find the Taylor expansion of $f(x) = e^{x/2}$ about $c = 1$.

**66.** Find the Maclaurin expansion of $f(x) = 2x^3 - 3x^2 + x + 5$. Comment on the result.

**67.** Find the Taylor expansion of $f(x) = \tan x$ about $c = \dfrac{\pi}{4}$.

**68.** Find the first five terms of the Maclaurin expansion for

$$f(x) = e^{-x} \sin x$$

*In Problems 69–71, use a binomial series to represent each function. Then determine its interval of convergence.*

**69.** $f(x) = \dfrac{1}{(x+1)^4}$

**70.** $f(x) = \sqrt[3]{x^2 - 1}$

**71.** $f(x) = \dfrac{1}{\sqrt{1-x}}$

**72. (a)** Approximate $y = \cos x$ using the Taylor polynomial $P_8(x)$ of $y = \cos x$ at $\dfrac{\pi}{2}$.

   **(b)** Graph $y = \cos x$ and $y = P_8(x)$.

   **(c)** Use $P_8(x)$ to approximate $\cos 88°$.

   **(d)** What is the error in using this approximation?

   **(e)** If the approximation in (a) is used, for what values of $x$ is the error less than or equal to 0.0001?

**73.** Use the Maclaurin expansion for $f(x) = e^x$ to approximate $e^{0.3}$ correct to three decimal points

*In Problems 74 and 75, use properties of power series to approximate each integral using the first four terms of a Maclaurin series.*

**74.** $\displaystyle\int_0^{1/2} \frac{dx}{\sqrt{1-x^3}}$

**75.** $\displaystyle\int_0^{1/2} e^{x^2} dx$

## CHAPTER 8 PROJECT    How Calculators Calculate

The sine function is used in many scientific applications, so a calculator/computer must be able to evaluate it with lightning-fast speed.

While we know how to find the exact value of the sine function for many numbers, such as 0, $\frac{\pi}{6}$, $\frac{\pi}{2}$, and so on, we have no method for finding the exact value of sin 3 (which should be close to sin $\pi$) or sin 1.5 (which should be close to sin $\frac{\pi}{2}$). Since the sine function can be evaluated at any real number, we first use some of its properties to restrict its domain to something more manageable.

1. Explain why we can evaluate $\sin x$ for any $x$ using only the interval $\left[-\frac{\pi}{2}, \frac{\pi}{2}\right]$. (We could restrict that domain further, but this will work for now. See Problem 6 below.)

2. Use the Maclaurin expansion for $\sin x$ to find an approximation for $\sin \frac{1}{2}$ correct to within $10^{-5}$. Compare your approximation to the one your calculator/computer provides. How many terms of the series do you need to obtain this accuracy?

3. Use the Maclaurin expansion for $\sin x$ to find an approximation for $\sin \frac{3}{2}$ correct to within $10^{-5}$. How many terms of the series do you need to obtain this accuracy?

4. Explain why the approximation in Problem 3 requires more terms than that of Problem 2.

5. Represent $\sin x$ as a Taylor expansion about $\frac{\pi}{4}$.

6. Explain why we can evaluate $\sin x$ for any $x$ using only the interval $\left[0, \frac{\pi}{2}\right]$.

7. Use the result of Problem 5 to find an approximation for $\sin \frac{1}{2}$ correct to within $10^{-5}$. Compare the result with the values your calculator/computer supplies for $\sin \frac{1}{2}$, as well as with the result from the Maclaurin approximation obtained in Problem 2. How many terms of the series do you need for the approximation?

8. Use the Taylor expansion for $\sin x$ at $\frac{\pi}{2}$ to find an approximation for $\sin \frac{3}{2}$ correct to within $10^{-5}$. Compare the result with the value your calculator/computer supplies for $\sin \frac{3}{2}$, as well as with the result from the Maclaurin approximation obtained in Problem 3. How many terms of the series do you need for the approximation?

The answers to Problems 3 and 8 reveal why Maclaurin series or Taylor series are not used to approximate the value of most functions. But often the methods used are similar. For example, a Chebyshev polynomial approximation to the sine function on the interval $\left[-\frac{\pi}{2}, \frac{\pi}{2}\right]$ still has the form of a Maclaurin series, but it was designed to converge more uniformly than the Maclaurin series, so that it can be expected to give answers near $\frac{\pi}{2}$ that are roughly as accurate as those near zero.

Chebyshev polynomials are commonly found in mathematical libraries for calculators/computers. For example, the widely used GNU Compiler Collection* uses Chebyshev polynomials to evaluate trigonometric functions. The Chebyshev polynomial approximation of degree 7 for the sine function is

$$S_7(x) = 0.9999966013x - 0.1666482357x^3$$
$$+ 0.008306286146x^5 - 0.1836274858 \times 10^{-3}x^7 \quad (1)$$

The Chebyshev polynomials are designed to remain close to a function across an entire closed interval. They seek to keep the approximation within a specified distance of the function being approximated at every point of that interval. If $S_n$ is a Chebyshev approximation of degree $n$ to the sine function on $\left[-\frac{\pi}{2}, \frac{\pi}{2}\right]$, then the error estimate in using $S_n(x)$ is given by

$$\max_{-\pi/2 \le x \le \pi/2} |\sin x - S_n(x)| \le \frac{\left(\frac{\pi}{2}\right)^{n+1}}{2^n (n+1)!} \quad (2)$$

Like most error estimates of this type, it gives an upper bound to the error.

9. Use the Chebyshev polynomial approximation in (1) for $x = \frac{1}{2}$ and $x = \frac{3}{2}$. Compare the results with the values your calculator/computer supplies for $\sin \frac{1}{2}$ and $\sin \frac{3}{2}$, as well as with the results from the Maclaurin approximations and the Taylor series approximations obtained in Problems 2, 3, 7, and 8.

10. Define $E_7(x) = |\sin x - S_7(x)|$. Use technology to graph $E_7$ on $\left[-\frac{\pi}{2}, \frac{\pi}{2}\right]$.

11. Find the local maximum and local minimum values of $E_7$ on $\left[-\frac{\pi}{2}, \frac{\pi}{2}\right]$. Compare these numbers with the error estimate in equation (2), and discuss the characteristics of the error.

12. Which approximation for the sine function would be preferable: the Maclaurin approximation, the Taylor approximation, or the Chebyshev approximation? Why?

---

*For more information on the GNU Compiler Collection (GCC), go to https://www.gcc.gnu.org.

# 9

# Parametric Equations; Polar Equations

**CHAPTER 9 PROJECT** The Chapter Project on page 732 examines the design of several different microphones and the directions from where the sounds they amplify originate.

Snob Mine Olivari/EyeEm/Getty Images

## Polar Graphs and Microphones

If you have ever been early to a concert or performance and seen the staff setting up the sound system, you understand a bit about the work that goes into obtaining optimal sound quality. Different microphones are designed to pick up sound from different directions, and sound technicians know that the selection and placement of microphones and monitors (speakers that are aimed back at the musicians so that they can hear the other instruments) vary depending on the specific purpose and venue. The same choices factor into the construction of cell phones and Bluetooth devices, which need to amplify the sound coming out of your mouth and minimize the barking dogs, loud trucks, and random conversations that may be happening around you.

The choice of the right microphone and speaker and their placement to achieve a specific goal requires an understanding of the direction of the sound and the area where it is being picked up.

Until now, we have worked primarily with explicitly defined functions $y = f(x)$, where $x$ and $y$ represent rectangular coordinates. Because the graphs of such functions satisfy the Vertical-line Test, many graphs, such as circles, cannot be represented using a single, explicitly defined function. To address this issue, we introduce *parametric equations*, a way to represent graphs that are not necessarily those of functions. Additionally, parametric equations are particularly useful because they allow us to model motion along a curve and to determine not only the location of an object (a point) but also the time it is there.

The chapter continues with a discussion of an alternate coordinate system, *polar coordinates*. Using polar coordinates, we can represent graphs that would be extremely complicated, or even impossible, to represent in rectangular coordinates.

# 9.1 Parametric Equations

**OBJECTIVES** *When you finish this section, you should be able to:*

1 Graph parametric equations (p. 680)
2 Find a rectangular equation for a curve represented parametrically (p. 681)
3 Use time as the parameter in parametric equations (p. 683)
4 Convert a rectangular equation to parametric equations (p. 684)

The graph of an equation of the form $y = f(x)$, where $f$ is a function, is intersected no more than once by any vertical line. There are many graphs, however, such as circles and some parabolas, that do not pass the Vertical-line Test. To study these graphs requires a different model. One model uses a pair of equations and a third variable, called a *parameter*.

> **DEFINITION**
>
> Suppose $x = x(t)$ and $y = y(t)$ are two functions of a third variable $t$, called the **parameter**, that are defined on the same interval $I$. Then the equations
>
> $$x = x(t) \qquad y = y(t)$$
>
> where $t$ is in $I$, are called **parametric equations**, and the graph of the points defined by
>
> $$(x, y) = (x(t), y(t))$$
>
> is called a **plane curve**.

## 1 Graph Parametric Equations

Parametric equations are particularly useful for describing motion along a curve. Suppose an object moves along a curve represented by the parametric equations

$$x = x(t) \qquad y = y(t)$$

where each function is defined over the interval $a \le t \le b$. For a given value of $t$, the values of $x = x(t)$ and $y = y(t)$ determine a point $(x, y)$ on the curve. In fact, as $t$ varies over the interval from $t = a$ to $t = b$, successive values of $t$ determine the direction of the motion of the object moving along the curve. See Figure 1. The arrows show the direction, or the **orientation**, of the object as it moves from $A$ to $B$.

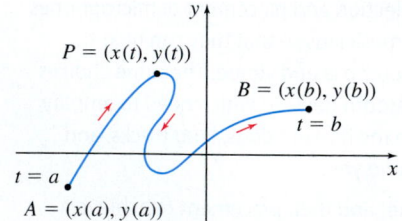

**DF** Figure 1 $x = x(t),\ y = y(t),\ a \le t \le b$

**EXAMPLE 1** Graphing a Plane Curve

Graph the plane curve represented by the parametric equations

$$x(t) = 3t^2 \qquad y(t) = 2t \qquad -2 \le t \le 2$$

Indicate the orientation of the curve.

**Solution** Corresponding to each number $t$, $-2 \le t \le 2$, there is a number $x$ and a number $y$ that are the coordinates of a point $(x, y)$ on the curve. We form Table 1 by listing various choices of the parameter $t$ and the corresponding values for $x$ and $y$.

The motion begins when $t = -2$ at the point $(12, -4)$ and ends when $t = 2$ at the point $(12, 4)$. Figure 2 illustrates the plane curve whose parametric equations are $x(t) = 3t^2$ and $y(t) = 2t$. The arrows indicate the orientation of the plane curve for increasing values of the parameter $t$. ∎

**TABLE 1**

| $t$ | $x(t) = 3t^2$ | $y(t) = 2t$ | $(x, y)$ |
|---|---|---|---|
| $-2$ | 12 | $-4$ | $(12, -4)$ |
| $-1$ | 3 | $-2$ | $(3, -2)$ |
| 0 | 0 | 0 | $(0, 0)$ |
| 1 | 3 | 2 | $(3, 2)$ |
| 2 | 12 | 4 | $(12, 4)$ |

## 2 Find a Rectangular Equation for a Curve Represented Parametrically

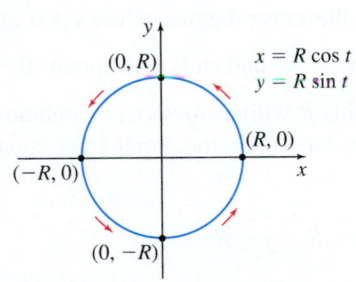

**Figure 2** $x(t) = 3t^2$, $y(t) = 2t$, $-2 \le t \le 2$

The plane curve in Figure 2 should look familiar. To identify it, we find the corresponding rectangular equation by eliminating the parameter $t$ from the parametric equations

$$x(t) = 3t^2 \qquad y(t) = 2t \qquad -2 \le t \le 2$$

We begin by solving for $t$ in $y = 2t$, obtaining $t = \dfrac{y}{2}$. Then we substitute $t = \dfrac{y}{2}$ in the equation, $x = 3t^2$.

$$x = 3t^2 = 3\left(\frac{y}{2}\right)^2 = \frac{3y^2}{4}$$

The equation $x = \dfrac{3y^2}{4}$ is a parabola with its vertex at the origin and its axis of symmetry along the $x$-axis. We refer to this equation as the **rectangular equation** of the curve to distinguish it from the parametric equations.

Notice that the plane curve represented by the parametric equations $x(t) = 3t^2$, $y(t) = 2t$, $-2 \le t \le 2$, is only part of the parabola $x = \dfrac{3y^2}{4}$. In general, the graph of the rectangular equation obtained by eliminating the parameter will contain more points than the plane curve defined using parametric equations. Therefore, when graphing a plane curve represented by a rectangular equation, we may have to restrict the graph to match the parametric equations.

### EXAMPLE 2 Finding a Rectangular Equation for a Plane Curve Represented Parametrically

Find a rectangular equation of the curve whose parametric equations are

$$x(t) = R \cos t \qquad y(t) = R \sin t$$

where $R > 0$ is a constant. Graph the plane curve and indicate its orientation.

**Solution** The presence of the sine and cosine functions in the parametric equations suggests using the Pythagorean Identity $\cos^2 t + \sin^2 t = 1$. Then

$$\left(\frac{x}{R}\right)^2 + \left(\frac{y}{R}\right)^2 = 1 \qquad \cos t = \frac{x}{R}, \quad \sin t = \frac{y}{R}$$

$$x^2 + y^2 = R^2$$

**Figure 3** The orientation is counterclockwise.

The graph of the rectangular equation is a circle with center at the origin and radius $R$. In the parametric equations, as the parameter $t$ increases, the points $(x, y)$ on the circle are traced out in the counterclockwise direction, as shown in Figure 3. ∎

**NOW WORK** Problems **9** and **15**.

When analyzing the parametric equations in Example 2, notice there are no restrictions on $t$, so $t$ varies from $-\infty$ to $\infty$. As a result, the circle is repeated each time $t$ increases by $2\pi$.

If we want to describe a curve consisting of exactly one revolution in the counterclockwise direction, the domain must be restricted to an interval of length $2\pi$. For example, we could use

$$x(t) = R \cos t \qquad y(t) = R \sin t \qquad 0 \le t \le 2\pi$$

Now the plane curve starts when $t = 0$ at $(R, 0)$ and ends when $t = 2\pi$ at $(R, 0)$.

If we wanted the curve to consist of exactly three revolutions in the counterclockwise direction, the domain must be restricted to an interval of length $6\pi$. For example, we could use $x(t) = R \cos t$, $y(t) = R \sin t$, with $-2\pi \le t \le 4\pi$, or $0 \le t \le 6\pi$, or $2\pi \le t \le 8\pi$, or any arbitrary interval of length $6\pi$.

CALC
▶ **EXAMPLE 3** **Finding a Rectangular Equation for a Plane Curve**
CLIP **Represented Parametrically**

Find rectangular equations for the plane curves represented by each of the parametric equations. Graph each curve and indicate its orientation.

**(a)** $x(t) = R \cos t \quad y(t) = R \sin t \quad 0 \le t \le \pi$ and $R > 0$

**(b)** $x(t) = R \sin t \quad y(t) = R \cos t \quad 0 \le t \le \pi$ and $R > 0$

**Solution** **(a)** We eliminate the parameter $t$ using a Pythagorean Identity.

$$\cos^2 t + \sin^2 t = 1$$

$$\left(\frac{x}{R}\right)^2 + \left(\frac{y}{R}\right)^2 = 1 \qquad \cos t = \frac{x}{R}, \quad \sin t = \frac{y}{R}$$

$$x^2 + y^2 = R^2 \tag{1}$$

The rectangular equation represents a circle with radius $R$ and center at the origin. In the parametric equations, $0 \le t \le \pi$, so the curve begins when $t = 0$ at the point $(R, 0)$, passes through the point $(0, R)$ when $t = \dfrac{\pi}{2}$, and ends when $t = \pi$ at the point $(-R, 0)$.

The curve is an upper semicircle of radius $R$ with counterclockwise orientation, as shown in Figure 4. If we solve equation (1) for $y$, we obtain the rectangular equation of the semicircle

$$y = \sqrt{R^2 - x^2} \qquad \text{where } -R \le x \le R$$

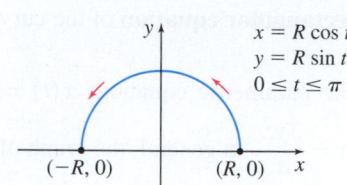

$x = R \cos t$
$y = R \sin t$
$0 \le t \le \pi$

$(-R, 0) \qquad (R, 0) \quad x$

**Figure 4** $y = \sqrt{R^2 - x^2},\ -R \le x \le R$

**(b)** Eliminate the parameter $t$ as in (a), to obtain

$$x^2 + y^2 = R^2 \tag{2}$$

The rectangular equation represents a circle with radius $R$ and center at $(0, 0)$. But in the parametric equations, $0 \le t \le \pi$, so now the curve begins when $t = 0$ at the point $(0, R)$, passes through the point $(R, 0)$ when $t = \dfrac{\pi}{2}$, and ends at the point $(0, -R)$ when $t = \pi$. The curve is a right semicircle of radius $R$ with a *clockwise* orientation, as shown in Figure 5. If we solve equation (2) for $x$, we obtain the rectangular equation of the semicircle

$$x = \sqrt{R^2 - y^2} \qquad \text{where } -R \le y \le R \qquad \blacksquare$$

$(0, R)$
$x = R \sin t$
$y = R \cos t$
$0 \le t \le \pi$

$(0, -R)$

**Figure 5** $x = \sqrt{R^2 - y^2},\ -R \le y \le R$

**NOTE** In Example 3(a) we expressed the rectangular equation as a function $y = f(x)$, and in (b) we expressed the rectangular equation as a function $x = g(y)$.

Parametric equations are not unique; that is, different parametric equations can represent the same graph. Two examples of other parametric equations that represent the graph in Figure 5 are

- $x(t) = R \sin t \qquad y(t) = -R \cos(\pi - t) \quad 0 \le t \le \pi$
- $x(t) = R \sin(3t) \qquad y(t) = R \cos(3t) \qquad 0 \le t \le \dfrac{\pi}{3}$

There are other examples.

NOW WORK **Problem 17.**

**EXAMPLE 4** **Finding a Rectangular Equation for a Plane Curve**
**Represented Parametrically**

**(a)** Find a rectangular equation of the plane curve whose parametric equations are

$$x(t) = \cos(2t) \qquad y(t) = \sin t \qquad -\frac{\pi}{2} \le t \le \frac{\pi}{2}$$

**(b)** Graph the rectangular equation.

(c) Determine the restrictions on $x$ and $y$ so the graph corresponding to the rectangular equation is identical to the plane curve described by

$$x = x(t) \qquad y = y(t) \qquad -\frac{\pi}{2} \le t \le \frac{\pi}{2}$$

(d) Graph the plane curve whose parametric equations are

$$x = \cos(2t) \qquad y = \sin t \qquad -\frac{\pi}{2} \le t \le \frac{\pi}{2}$$

**Solution (a)** Here we cannot use a Pythagorean identity to eliminate the parameter $t$ because $\cos(2t)$ and $\sin t$ have different arguments. Instead use a trigonometric identity that involves $\cos(2t)$ and $\sin t$, namely, $\sin^2 t = \dfrac{1 - \cos(2t)}{2}$. Then

$$y^2 = \underset{\substack{\uparrow \\ y(t)=\sin t}}{\sin^2 t} = \frac{1 - \cos(2t)}{2} = \underset{\substack{\uparrow \\ x(t)=\cos(2t)}}{\frac{1 - x}{2}}$$

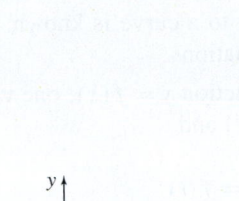

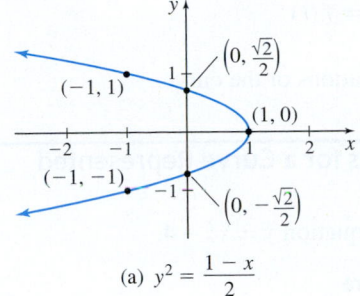

(a) $y^2 = \dfrac{1-x}{2}$

(b) The curve represented by the rectangular equation $y^2 = \dfrac{1-x}{2}$ is the parabola shown in Figure 6(a).

(c) The plane curve represented by the parametric equations does not include all the points on the parabola. Since $x(t) = \cos(2t)$ and $-1 \le \cos(2t) \le 1$, then $-1 \le x \le 1$. Also, since $y(t) = \sin t$, then $-1 \le y \le 1$. Finally, the curve is traced out exactly once in the counterclockwise direction from the point $(-1, -1)$ (when $t = -\dfrac{\pi}{2}$) to the point $(-1, 1)$ (when $t = \dfrac{\pi}{2}$).

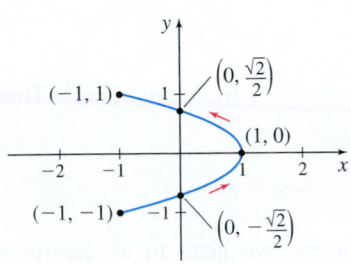

(b) $x = \cos(2t),\ y = \sin t,$
$-\dfrac{\pi}{2} \le t \le \dfrac{\pi}{2}$

**Figure 6**

(d) The plane curve represented by the given parametric equations is the part of the parabola shown in Figure 6(b). ∎

NOW WORK **Problem 21.**

### 3 Use Time as the Parameter in Parametric Equations

If the parameter $t$ represents time, then the parametric equations $x = x(t)$ and $y = y(t)$ specify how the $x$- and $y$-coordinates of a moving object vary with time. This motion is sometimes referred to as **curvilinear motion**. Describing curvilinear motion using parametric equations specifies not only *where* the object is, that is, its location $(x, y)$, but also *when* it is there, that is, the time $t$. A rectangular equation provides only the location of the object.

**EXAMPLE 5** Using Time as the Parameter in Parametric Equations

Describe the motion of an object that moves along a curve so that at time $t$ it has coordinates

$$x(t) = 3\cos t \qquad y(t) = 4\sin t \qquad 0 \le t \le 2\pi$$

**Solution** Eliminate the parameter $t$ using the Pythagorean Identity $\cos^2 t + \sin^2 t = 1$.

$$\frac{x^2}{9} + \frac{y^2}{16} = 1 \qquad \cos t = \frac{x}{3}, \qquad \sin t = \frac{y}{4}$$

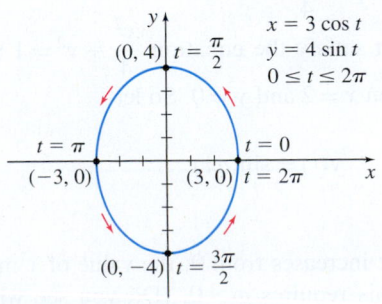

The plane curve is the ellipse shown in Figure 7. When $t = 0$, the object is at the point $(3, 0)$. As $t$ increases, the object moves around the ellipse in a counterclockwise direction, reaching the point $(0, 4)$ when $t = \dfrac{\pi}{2}$, the point $(-3, 0)$ when $t = \pi$, the point $(0, -4)$ when $t = \dfrac{3\pi}{2}$, and returning to its starting point $(3, 0)$ when $t = 2\pi$. ∎

**Figure 7** $\dfrac{x^2}{9} + \dfrac{y^2}{16} = 1$

NOW WORK **Problem 41.**

## 4 Convert a Rectangular Equation to Parametric Equations

Suppose the rectangular equation corresponding to a curve is known. Then the curve can be represented by a variety of parametric equations.

For example, if a curve is defined by the function $y = f(x)$, one way of obtaining parametric equations is to let $x = t$. Then $y = f(t)$ and

$$x(t) = t \qquad y(t) = f(t)$$

where $t$ is in the domain of $f$, are parametric equations of the curve.

> **EXAMPLE 6** **Finding Parametric Equations for a Curve Represented by a Rectangular Equation**

Find parametric equations corresponding to the equation $y = x^2 - 4$.

**Solution** Let $x = t$. Then parametric equations are

$$x(t) = t \qquad y(t) = t^2 - 4 \qquad -\infty < t < \infty$$

We can also find parametric equations for $y = x^2 - 4$ by letting $x = t^3$. Then the parametric equations are

$$x(t) = t^3 \qquad y(t) = t^6 - 4 \qquad -\infty < t < \infty$$

Although the choice of $x$ may appear arbitrary, we need to be careful when choosing the substitution for $x$. The substitution must be a function that allows $x$ to take on all the values in the domain of $f$. For example, if we let $x(t) = t^2$, then $y(t) = t^4 - 4$. But $x(t) = t^2$, $y(t) = t^4 - 4$ are not parametric equations for all of $y = x^2 - 4$, since only points for which $x \geq 0$ are obtained.

**NOW WORK** Problem 47.

> **EXAMPLE 7** **Finding Parametric Equations for an Object in Motion**

Find parametric equations that trace out the ellipse

$$\frac{x^2}{4} + y^2 = 1$$

where the parameter $t$ is time (in seconds) and:

**(a)** The motion around the ellipse is counterclockwise, begins at the point $(2, 0)$, and requires 1 s for a complete revolution.

**(b)** The motion around the ellipse is clockwise, begins at the point $(0, 1)$, and requires 2 s for a complete revolution.

**Solution** **(a)** Figure 8 shows the graph of the ellipse $\dfrac{x^2}{4} + y^2 = 1$.

The choice for $x = x(t)$ and $y = y(t)$ must satisfy the equation $\dfrac{x^2}{4} + y^2 = 1$ and also satisfy the requirement that when $t = 0$, then $x = 2$ and $y = 0$. So let

$$x(t) = 2\cos(\omega t) \qquad \text{and} \qquad y(t) = \sin(\omega t)$$

for some constant $\omega$.

For the motion to be counterclockwise, as $t$ increases from 0, the value of $x$ must decrease and the value of $y$ must increase. This requires $\omega > 0$. [Do you see why? If $\omega > 0$, then $x = 2\cos(\omega t)$ is decreasing when $t > 0$ is near zero, and $y = \sin(\omega t)$ is increasing when $t > 0$ is near zero.]

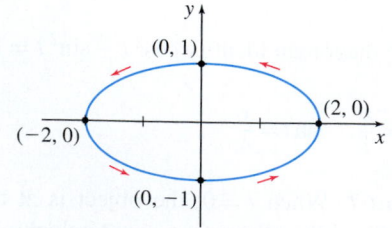

**Figure 8** $\dfrac{x^2}{4} + y^2 = 1$

**NEED TO REVIEW?** The period of a sinusoidal graph is discussed in Section P. 6, p. 57.

Finally, since 1 revolution takes 1 s, the period is $\dfrac{2\pi}{\omega} = 1$, so $\omega = 2\pi$. Parametric equations that satisfy the conditions given in (a) are

$$x(t) = 2\cos(2\pi t) \qquad y(t) = \sin(2\pi t) \qquad 0 \le t \le 1$$

**(b)** Figure 9 shows the graph of the ellipse $\dfrac{x^2}{4} + y^2 = 1$.

The choice for $x = x(t)$ and $y = y(t)$ must satisfy the equation $\dfrac{x^2}{4} + y^2 = 1$ and also satisfy the requirement that when $t = 0$, then $x = 0$ and $y = 1$. So let

$$x(t) = 2\sin(\omega t) \qquad \text{and} \qquad y(t) = \cos(\omega t)$$

for some constant $\omega$.

For the motion to be clockwise, as $t$ increases from 0, the value of $x$ must increase and the value of $y$ must decrease. This requires that $\omega > 0$.

Finally, since 1 revolution takes 2 s, the period is $\dfrac{2\pi}{\omega} = 2$, or $\omega = \pi$. Parametric equations that satisfy the conditions given in (b) are

$$x(t) = 2\sin(\pi t) \qquad y(t) = \cos(\pi t) \qquad 0 \le t \le 2$$
∎

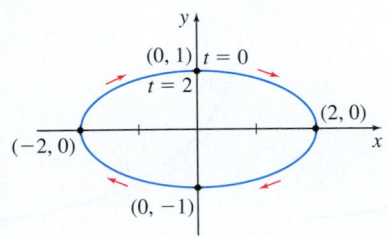

**Figure 9** $\dfrac{x^2}{4} + y^2 = 1$

**NOW WORK** Problem 55.

## The Cycloid

Suppose a circle rolls along a horizontal line without slipping. As the circle rolls along the line, a point $P$ on the circle traces out a curve called a **cycloid**, as shown in Figure 10. Deriving the equation of a cycloid in rectangular coordinates is complicated, but the derivation in terms of parametric equations is relatively easy.

We begin with a circle of radius $a$. Suppose the fixed line on which the circle rolls is the $x$-axis. Let the origin be one of the points at which the point $P$ comes into contact with the $x$-axis. Figure 10 shows the position of this point $P$ after the circle has rolled a bit. The angle $t$ (in radians) measures the angle through which the circle has rolled. Since the circle does not slip, it follows that

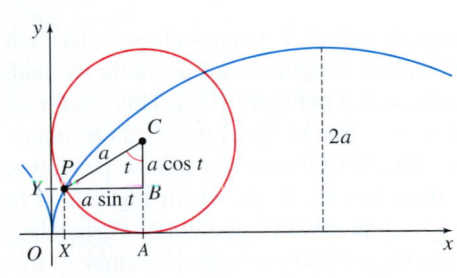

**DF Figure 10**

$$\text{Arc } AP = d(O, A)$$

**RECALL** For a circle of radius $r$, a central angle of $\theta$ radians subtends an arc whose length $s$ is $s = r\theta$.

The length of the arc $AP$ is

$$at = d(O, A) \qquad s = r\theta, \quad r = a, \quad \theta = t$$

The $x$-coordinate of the point $P = (x, y)$ is

$$x = d(O, X) = d(O, A) - d(X, A) = at - a\sin t = a(t - \sin t)$$

The $y$-coordinate of the point $P$ is

$$y = d(O, Y) = d(A, C) - d(B, C) = a - a\cos t = a(1 - \cos t)$$

**THEOREM Parametric Equations of a Cycloid**

Parametric equations of a cycloid are

$$\boxed{x(t) = a(t - \sin t) \qquad y(t) = a(1 - \cos t) \qquad -\infty < t < \infty}$$

**NOTE** In Greek, *brachistochrone* means "the shortest time," and *tautochrone* "equal time."

## Applications to Mechanics

If $a < 0$ in the parametric equations of the cycloid, the result is an inverted cycloid, as shown in Figure 11(a). The inverted cycloid arises as a result of some remarkable applications in the field of mechanics. We discuss two of them: the *brachistochrone* and *tautochrone*.

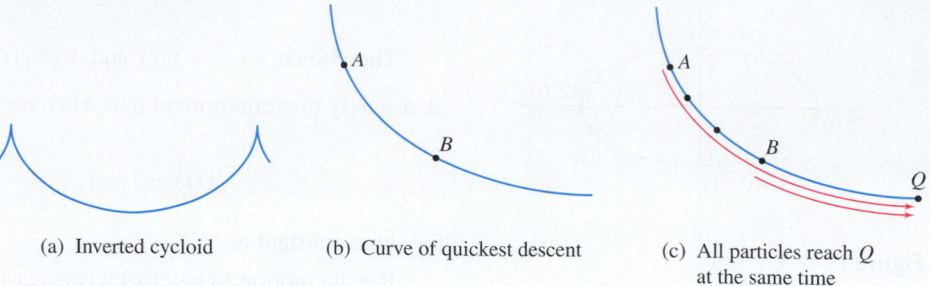

(a) Inverted cycloid  (b) Curve of quickest descent  (c) All particles reach $Q$ at the same time

**DF** **Figure 11**

The **brachistochrone** is the curve of quickest descent. If an object is constrained to follow some path from a point $A$ to a lower point $B$ (not on the same vertical line) and is acted on only by gravity, the time needed to make the descent is minimized if the path is an inverted cycloid. See Figure 11(b). For example, in sliding packages from a loading dock onto a ship, a ramp in the shape of an inverted cycloid might be used so the packages get to the ship in the least amount of time. This discovery, which is attributed to many famous mathematicians (including Johann Bernoulli and Blaise Pascal), was a significant step in creating the branch of mathematics known as the *calculus of variations*.

Suppose $Q$ is the lowest point on an inverted cycloid. If several objects placed at various positions on an inverted cycloid simultaneously begin to slide down the cycloid, they will reach the point $Q$ at the same time, as indicated in Figure 11(c). This is referred to as the **tautochrone property** of the cycloid. It was used by the Dutch mathematician, physicist, and astronomer Christian Huygens (1629–1695) to construct a pendulum clock with a bob that swings along an inverted cycloid, as shown in Figure 12. In Huygens' clock, the bob was made to swing along an inverted cycloid by suspending the bob on a thin wire constrained by two plates shaped like cycloids. In a clock of this design, the period of the pendulum is independent of its amplitude.

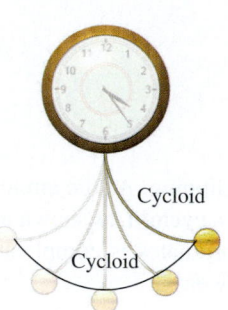

Cycloid

Cycloid

**Figure 12**

## 9.1 Assess Your Understanding

### Concepts and Vocabulary

1. Let $x = x(t)$ and $y = y(t)$ be two functions whose common domain is the interval $I$. The collection of points defined by $(x, y) = (x(t), y(t))$ is called a _____. The variable $t$ is called a(n) _____.

2. *Multiple Choice* The parametric equations $x(t) = 2 \sin t$, $y(t) = 3 \cos t$ define a(n)
   [(a) line (b) hyperbola (c) ellipse (d) parabola].

3. *Multiple Choice* The parametric equations $x(t) = a \sin t$, $y(t) = a \cos t$, $a > 0$, define a
   [(a) line (b) hyperbola (c) parabola (d) circle].

4. If a circle rolls along a horizontal line without slipping, a point $P$ on the circle traces out a curve called a(n) _____.

5. *True or False* The parametric equations defining a curve are unique.

6. *True or False* Plane curves represented using parametric equations have an orientation.

### Skill Building

*In Problems 7–20:*

(a) *Find the rectangular equation of each plane curve with the given parametric equations.*

(b) *Graph the plane curve represented by the parametric equations and indicate its orientation.*

7. $x(t) = 2t + 1$, $y(t) = t + 2$;  $-\infty < t < \infty$

8. $x(t) = t - 2$, $y(t) = 3t + 1$;  $-\infty < t < \infty$

9. $x(t) = 2t + 1$, $y(t) = t + 2$;  $0 \le t \le 2$

10. $x(t) = t - 2$, $y(t) = 3t + 1$;  $0 \le t \le 2$

11. $x(t) = e^t$, $y(t) = t$;  $-\infty < t < \infty$

12. $x(t) = t$, $y(t) = \dfrac{1}{t}$;  $-\infty < t < \infty, t \ne 0$

13. $x(t) = \sin t$, $y(t) = \cos t$;  $0 \le t \le 2\pi$

14. $x(t) = \cos t$, $y(t) = \sin t$;  $0 \le t \le \pi$

---

**1.** = NOW WORK problem      |/\\| = Graphing technology recommended      **CAS** = Computer Algebra System recommended

**15.** $x(t) = 2\sin t,\ y(t) = 3\cos t;\quad 0 \le t \le 2\pi$

**16.** $x(t) = 4\cos t,\ y(t) = 3\sin t;\quad 0 \le t \le 2\pi$

**17.** $x(t) = 2\sin t - 3,\ y(t) = 2\cos t + 1;\quad 0 \le t \le \pi$

**18.** $x(t) = 4\cos t + 1,\ y(t) = 4\sin t - 3;\quad 0 \le t \le \pi$

**19.** $x(t) = t^2,\ y(t) = 2t;\quad -\infty < t < \infty$

**20.** $x(t) = 4t + 1,\ y(t) = 2t;\quad -\infty < t < \infty$

*In Problems 21–36:*

(a) *Find a rectangular equation of each plane curve with the given parametric equations.*

(b) *Graph the rectangular equation.*

(c) *Determine the restrictions on x and y so that the graph corresponding to the rectangular equation is identical to the plane curve.*

(d) *Graph the plane curve represented by the parametric equations.*

**21.** $x(t) = 2t,\ y(t) = t^2 + 4;\quad t > 0$

**22.** $x(t) = t + 3,\ y(t) = t^3;\quad -4 \le t \le 4$

**23.** $x(t) = t + 5,\ y(t) = \sqrt{t};\quad t \ge 0$

**24.** $x(t) = 2t^2,\ y(t) = 2t^3;\quad 0 \le t \le 3$

**25.** $x(t) = t^{1/2} + 1,\ y(t) = t^{3/2};\quad t \ge 1$

**26.** $x(t) = 2e^t,\ y(t) = 1 - e^t;\quad t \ge 0$

**27.** $x(t) = \sec t,\ y(t) = \tan t;\quad -\dfrac{\pi}{2} < t < \dfrac{\pi}{2}$

**28.** $x(t) = 3\sinh t,\ y(t) = 2\cosh t;\quad -\infty < t < \infty$

**29.** $x(t) = t^4,\quad y(t) = t^2$

**30.** $x(t) = t^2,\quad y(t) = t^4$

**31.** $x(t) = t^2,\quad y(t) = 2t - 1$

**32.** $x(t) = e^{2t},\quad y(t) = 2e^t - 1$

**33.** $x(t) = \left(\dfrac{1}{t}\right)^2,\ y(t) = \dfrac{2}{t} - 1$

**34.** $x(t) = t^4,\ y(t) = 2t^2 - 1$

**35.** $x(t) = 3\sin^2 t - 2,\ y(t) = 2\cos t;\quad 0 \le t \le \pi$

**36.** $x(t) = 1 + 2\sin^2 t,\ y(t) = 2 - \cos t;\quad 0 \le t \le 2\pi$

**Using Time as the Parameter** *In Problems 37–42, describe the motion of an object that moves along a curve so that at time t it has coordinates $(x(t), y(t))$.*

**37.** $x(t) = \dfrac{1}{t^2},\quad y(t) = \dfrac{2}{t^2 + 1};\quad t > 0$

**38.** $x(t) = \dfrac{3t}{\sqrt{t^2 + 1}},\quad y(t) = \dfrac{3}{\sqrt{t^2 + 1}}$

**39.** $x(t) = \dfrac{4}{\sqrt{4 - t^2}},\quad y(t) = \dfrac{4t}{\sqrt{4 - t^2}};\quad 0 \le t < 2$

**40.** $x(t) = \sqrt{t - 3},\quad y(t) = \sqrt{t + 1};\quad t \ge 3$

**41.** $x(t) = \sin t - 2,\quad y(t) = 4 - 2\cos t;\quad 0 \le t \le 2\pi$

**42.** $x(t) = 2 + \tan t,\quad y(t) = 3 - 2\sec t;\quad -\dfrac{\pi}{2} < t < \dfrac{\pi}{2}$

*In Problems 43–50, find two different pairs of parametric equations corresponding to each rectangular equation.*

**43.** $y = 4x - 2$

**44.** $y = -8x + 3$

**45.** $y = -2x^2 + 1$

**46.** $y = x^2 + 1$

**47.** $y = 4x^3$

**48.** $y = 2x^2$

**49.** $x = \dfrac{1}{3}\sqrt{y} - 3$

**50.** $x = y^{3/2}$

*In Problems 51–54, find parametric equations that represent the plane curve shown.*

**51.**
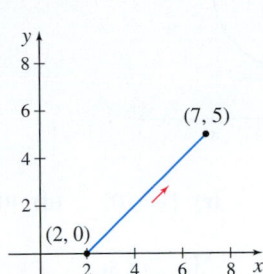

**52.**
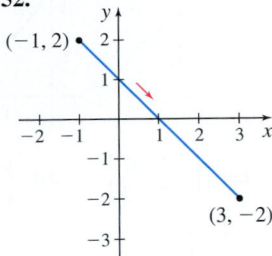

**53.**
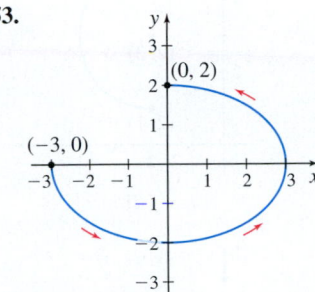

**54.**
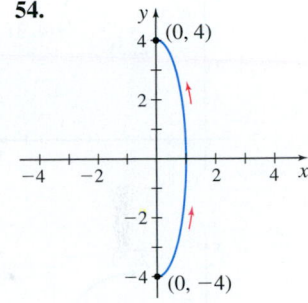

**Motion of an Object** *In Problems 55–58, find parametric equations for an object that moves along the ellipse $\dfrac{x^2}{9} + \dfrac{y^2}{4} = 1$, where the parameter is time (in seconds) if:*

**55.** the motion begins at $(3, 0)$, is counterclockwise, and requires 3 s for 1 revolution.

**56.** the motion begins at $(3, 0)$, is clockwise, and requires 3 s for 1 revolution.

**57.** The motion begins at $(0, 2)$, is clockwise, and requires 2 s for 1 revolution.

**58.** the motion begins at $(0, 2)$, is counterclockwise, and requires 1 s for 1 revolution.

*In Problems 59 and 60, the parametric equations of four plane curves are given. Graph each curve, indicate its orientation, and compare the graphs.*

**59.** (a) $x(t) = t,\ y(t) = t^2;\quad -4 \le t \le 4$

(b) $x(t) = \sqrt{t},\ y(t) = t;\quad 0 \le t \le 16$

(c) $x(t) = e^t,\ y(t) = e^{2t};\quad 0 \le t \le \ln 4$

(d) $x(t) = \cos t,\ y(t) = 1 - \sin^2 t;\quad 0 \le t \le \pi$

**60.** (a) $x(t) = t,\ y(t) = \sqrt{1 - t^2};\quad -1 \le t \le 1$

(b) $x(t) = \sin t,\ y(t) = \cos t;\quad 0 \le t \le 2\pi$

(c) $x(t) = \cos t,\ y(t) = \sin t;\quad 0 \le t \le 2\pi$

(d) $x(t) = \sqrt{1 - t^2},\ y(t) = t;\quad -1 \le t \le 1$

## Applications and Extensions

*In Problems 61–64, the parametric equations of a plane curve and its graph are given. Match each graph in I–IV to the restricted domain given in (a)–(d). Also indicate the orientation of each graph.*

**61.**

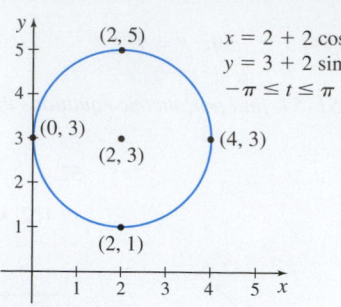

$$x = 2 + 2 \cos t$$
$$y = 3 + 2 \sin t$$
$$-\pi \le t \le \pi$$

**(a)** $\left[-\dfrac{\pi}{2}, \dfrac{\pi}{2}\right]$ **(b)** $\left[-\pi, \dfrac{\pi}{2}\right]$ **(c)** $[-\pi, 0]$ **(d)** $[0, \pi]$

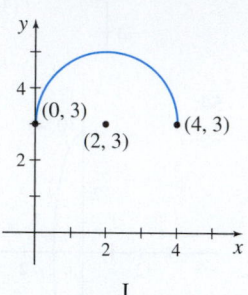

I

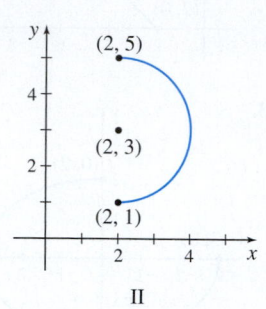

II

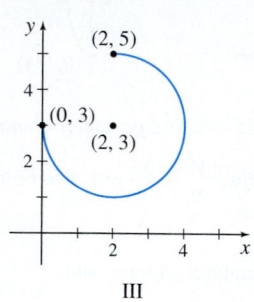

III

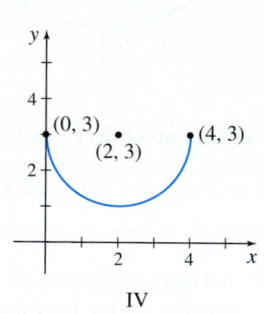

IV

**62.**

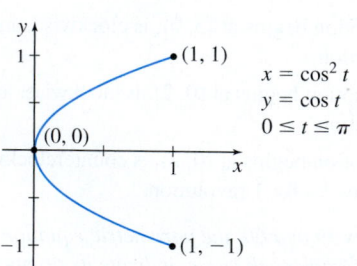

$$x = \cos^2 t$$
$$y = \cos t$$
$$0 \le t \le \pi$$

**(a)** $\left[\dfrac{\pi}{2}, \pi\right]$ **(b)** $\left[0, \dfrac{\pi}{3}\right]$ **(c)** $\left[0, \dfrac{\pi}{2}\right]$ **(d)** $\left[\dfrac{\pi}{3}, \pi\right]$

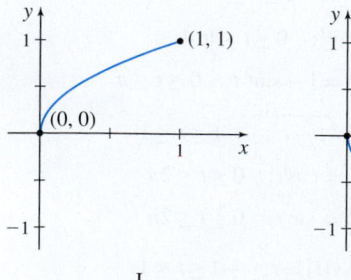

I

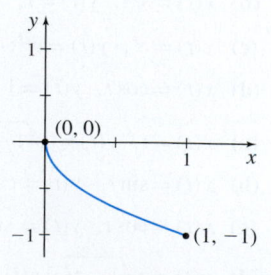

II

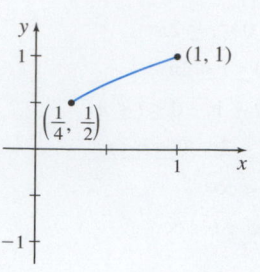

III

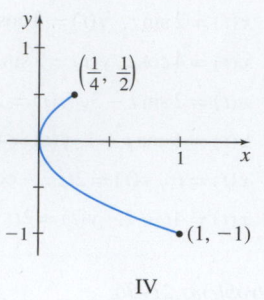

IV

**63.**

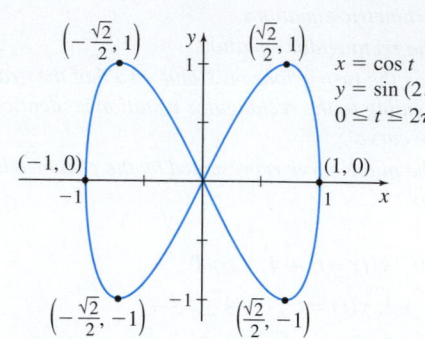

$$x = \cos t$$
$$y = \sin(2t)$$
$$0 \le t \le 2\pi$$

**(a)** $\left[\dfrac{\pi}{2}, \dfrac{3\pi}{2}\right]$ **(b)** $[\pi, 2\pi]$ **(c)** $[0, \pi]$ **(d)** $\left[\dfrac{5\pi}{4}, 2\pi\right]$

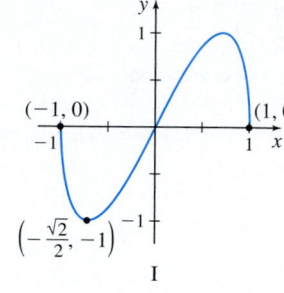

I

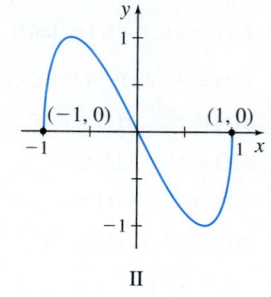

II

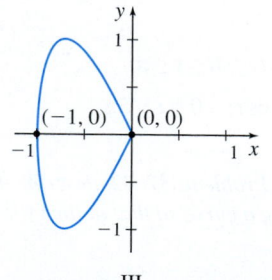

III

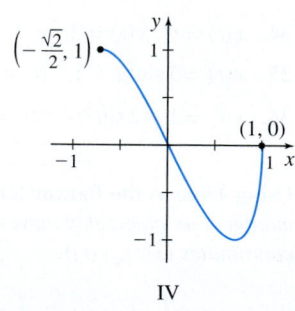

IV

**64.**

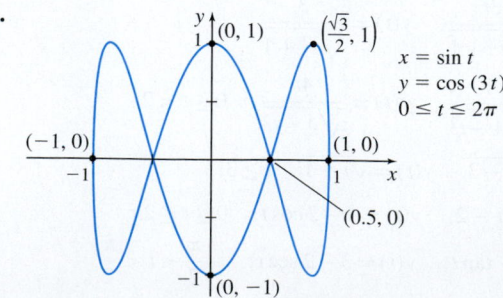

$$x = \sin t$$
$$y = \cos(3t)$$
$$0 \le t \le 2\pi$$

(a) $\left[0, \dfrac{5\pi}{6}\right]$    (b) $[\pi, 2\pi]$    (c) $\left[0, \dfrac{\pi}{2}\right]$    (d) $\left[\dfrac{\pi}{6}, \dfrac{5\pi}{6}\right]$

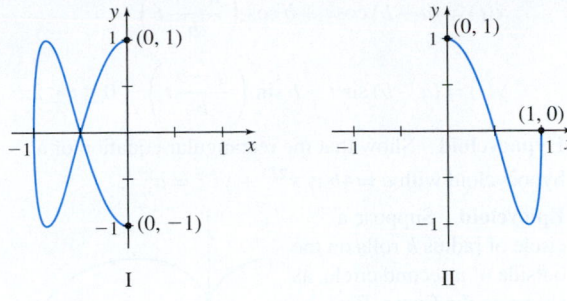

I    II

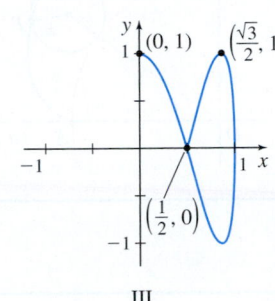    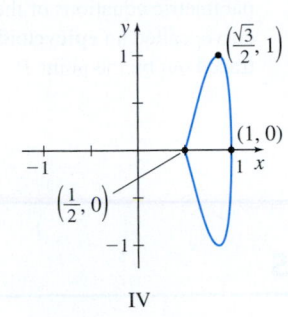

III    IV

**65. (a)** Graph the plane curve represented by the parametric equations $x(t) = 7(t - \sin t)$, $y(t) = 7(1 - \cos t)$, $0 \le t \le 2\pi$.

**(b)** Find the coordinates of the point $(x, y)$ on the curve when $t = 2.1$.

**66. (a)** Graph the plane curve represented by the parametric equations $x(t) = t - e^t$, $y(t) = 2e^{t/2}$, $-8 \le t \le 2$.

**(b)** Find the coordinates of the point $(x, y)$ on the curve when $t = 1.5$.

**67.** Find parametric equations for the ellipse $\dfrac{x^2}{a^2} + \dfrac{y^2}{b^2} = 1$.

**68.** Find parametric equations for the hyperbola $\dfrac{x^2}{a^2} - \dfrac{y^2}{b^2} = 1$.

*Problems 69–71 involve projectile motion. When an object is propelled upward at an inclination $\theta$ to the horizontal with initial speed $v_0$, the resulting motion is called **projectile motion**. See the figure. Parametric equations that model the path of the projectile, ignoring air resistance, are given by*

$$x(t) = (v_0 \cos \theta)t \qquad y(t) = -\frac{1}{2}gt^2 + (v_0 \sin \theta)t + h$$

*where t is time, g is the constant acceleration due to gravity (approximately 32 ft/s$^2$ or 9.8 m/s$^2$), and h is the height from which the projectile is released.*

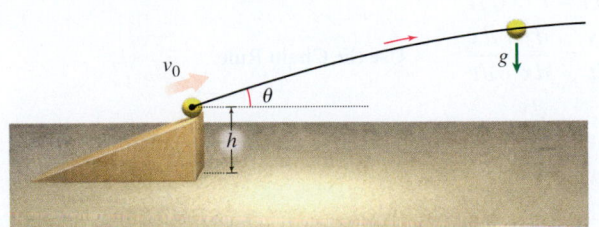

**69. Trajectory of a Baseball**    A baseball is hit with an initial speed of 125 ft/s at an angle of 40° to the horizontal. The ball is hit at a height of 3 ft above the ground.

**(a)** Find parametric equations that model the position of the ball as a function of the time $t$ in seconds.

**(b)** What is the height of the baseball after 2 s?

**(c)** What horizontal distance $x$ has the ball traveled after 2 s?

**(d)** How long does it take the baseball to travel $x = 300$ ft?

**(e)** What is the height of the baseball at the time found in (d)?

**(f)** How long is the ball in the air before it hits the ground?

**(g)** How far has the baseball traveled horizontally when it hits the ground?

**70. Trajectory of a Baseball**    A pitcher throws a baseball with an initial speed of 145 ft/s at an angle of 20° to the horizontal. The ball leaves his hand at a height of 5 ft.

**(a)** Find parametric equations that model the position of the ball as a function of the time $t$ in seconds.

**(b)** What is the height of the baseball after $\dfrac{1}{2}$ s?

**(c)** What horizontal distance $x$ has the ball traveled after $\dfrac{1}{2}$ s?

**(d)** How long does it take the baseball to travel $x = 60$ ft?

**(e)** What is the height of the baseball at the time found in (d)?

**71. Trajectory of a Football**    A quarterback throws a football with an initial speed of 80 ft/s at an angle of 35° to the horizontal. The ball leaves the quarterback's hand at a height of 6 ft.

**(a)** Find parametric equations that model the position of the ball as a function of time $t$ in seconds.

**(b)** What is the height of the football after 1 s?

**(c)** What horizontal distance $x$ has the ball traveled after 1 s?

**(d)** How long does it take the football to travel $x = 120$ ft?

**(e)** What is the height of the football at the time found in (d)?

**72.** The plane curve represented by the parametric equations $x(t) = t$, $y(t) = t^2$, and the plane curve represented by the parametric equations $x(t) = t^2$, $y(t) = t^4$ (where, in each case, time $t$ is the parameter) appear to be identical, but they differ in an important aspect. Identify the difference, and explain its meaning.

**73.** The circle $x^2 + y^2 = 4$ can be represented by the parametric equations $x(\theta) = 2\cos\theta$, $y(\theta) = 2\sin\theta$, $0 \le \theta \le 2\pi$, or by the parametric equations $x(\theta) = 2\sin\theta$, $y(\theta) = 2\cos\theta$, $0 \le \theta \le 2\pi$. But the plane curves represented by each pair of parametric equations are different. Identify and explain the difference.

**74. Uniform Motion**    A train leaves a station at 7:15 a.m. and accelerates at the rate of 3 mi/h$^2$. Mary, who can run 6 mi/h, arrives at the station 2 s after the train has left. Find parametric equations that model the motion of the train and of Mary as a function of time. *Hint:* The position $s$ at time $t$ of an object having acceleration $a$ is $s = \dfrac{1}{2}at^2$.

## Challenge Problems

**75.** Find parametric equations for the circle $x^2 + y^2 = R^2$, using as the parameter the slope $m$ of the line through the point $(-R, 0)$ and a general point $P = (x, y)$ on the circle.

**76.** Find parametric equations for the parabola $y = x^2$, using as the parameter the slope $m$ of the line joining the point $(1, 1)$ to a general point $P = (x, y)$ on the parabola.

**77. Hypocycloid**   Let a circle of radius $b$ roll, without slipping, inside a fixed circle with radius $a$, where $a > b$. A fixed point $P$ on the circle of radius $b$ traces out a curve, called a **hypocycloid**, as shown in the figure. If $A = (a, 0)$ is the initial position of the point $P$ and if $t$ denotes the angle from

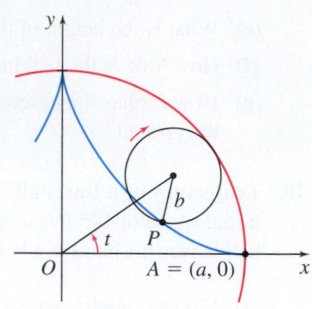

the positive $x$-axis to the line segment from the origin to the center of the circle, show that the parametric equations of the hypocycloid are

$$x(t) = (a - b)\cos t + b\cos\left(\frac{a - b}{b}t\right)$$

$$y(t) = (a - b)\sin t - b\sin\left(\frac{a - b}{b}t\right) \qquad 0 \le t \le 2\pi$$

**78. Hypocycloid**   Show that the rectangular equation of a hypocycloid with $a = 4b$ is $x^{2/3} + y^{2/3} = a^{2/3}$.

**79. Epicycloid**   Suppose a circle of radius $b$ rolls on the outside of a second circle, as shown in the figure. Find the parametric equations of the curve, called an **epicycloid**, traced out by the point $P$.

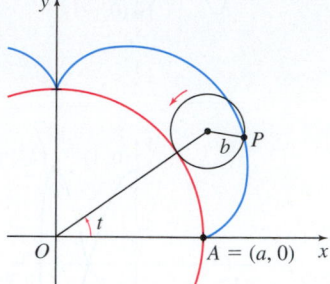

# 9.2 Tangent Lines

**OBJECTIVES**   *When you finish this section, you should be able to:*

**1  Find an equation of the tangent line at a point on a plane curve (p. 691)**

Recall that if a function $f$ is differentiable, the derivative $f'(x) = \dfrac{dy}{dx}$ is the slope of the tangent line to the graph of $f$ at a point $(x, y)$. To obtain a formula for the slope of the tangent line to a curve when it is represented by parametric equations, $x = x(t)$, $y = y(t)$, $a \le t \le b$, we require the curve to be *smooth*.

> **DEFINITION** Smooth Curve
>
> Let $C$ denote a plane curve represented by the parametric equations
>
> $$x = x(t) \qquad y = y(t) \qquad a \le t \le b$$
>
> Suppose each function $x(t)$ and $y(t)$ is continuous on the closed interval $[a, b]$ and differentiable on the open interval $(a, b)$. If both $\dfrac{dx}{dt}$ and $\dfrac{dy}{dt}$ are continuous and are never simultaneously 0 on $(a, b)$, then $C$ is called a **smooth curve** on $[a, b]$.

A smooth curve $x = x(t)$, $y = y(t)$, for which $\dfrac{dx}{dt}$ is never 0, can be represented by the rectangular equation $y = F(x)$, where $F$ is differentiable. (You are asked to prove this in Problem 33.)

Then

$$y = F(x)$$

$$y(t) = F(x(t))$$

$$\frac{dy}{dt} = \frac{dy}{dx} \cdot \frac{dx}{dt} \qquad \text{Use the Chain Rule.}$$

$$\frac{dy}{dx} = \frac{\dfrac{dy}{dt}}{\dfrac{dx}{dt}} \qquad \text{Divide by } \frac{dx}{dt} \ne 0.$$

Since $\dfrac{dy}{dx}$ is the slope of the tangent line to the graph of $y = F(x)$, we have proved the following theorem.

**THEOREM  Slope of the Tangent Line to a Smooth Curve**

For a smooth curve $C$ represented by the parametric equations $x = x(t)$, $y = y(t)$, $a \leq t \leq b$, the slope of the tangent line to $C$ at the point $(x, y)$ is given by

$$\frac{dy}{dx} = \frac{\dfrac{dy}{dt}}{\dfrac{dx}{dt}} \tag{1}$$

provided $\dfrac{dx}{dt} \neq 0$.

- At a number $t$ where $\dfrac{dx}{dt} = 0$, but $\dfrac{dy}{dt} \neq 0$, a smooth curve $C$ has a **vertical tangent line**.

- At a number $t$ where $\dfrac{dy}{dt} = 0$, but $\dfrac{dx}{dt} \neq 0$, a smooth curve $C$ has a **horizontal tangent line**.

## 1 Find an Equation of the Tangent Line at a Point on a Plane Curve

**NEED TO REVIEW?** Equations of tangent lines are discussed in Section 2.1, pp. 162–163.

Now that we have a way to find the slope of the tangent line to a smooth curve $C$ at a point, we can use the point-slope form of a line to obtain an equation of the tangent line.

**EXAMPLE 1  Finding an Equation of the Tangent Line to a Smooth Curve**

(a)  Find an equation of the tangent line to the plane curve with parametric equations $x(t) = 3t^2$, $y(t) = 2t$, when $t = 1$.

(b)  Find all the points on the plane curve at which the tangent line is vertical.

**Solution** (a)  The curve is smooth $\left( \dfrac{dy}{dt} = 2 \text{ is never zero} \right)$. Since $\dfrac{dx}{dt} = 6t$ is not 0

at $t = 1$, the slope of the tangent line to the curve is

$$\frac{dy}{dx} = \frac{\dfrac{dy}{dt}}{\dfrac{dx}{dt}} = \frac{\dfrac{d}{dt}(2t)}{\dfrac{d}{dt}(3t^2)} = \frac{2}{6t} = \frac{1}{3t}$$

When $t = 1$, the slope of the tangent line is $\dfrac{1}{3}$. Since $x(1) = 3$ and $y(1) = 2$, an equation of the tangent line is

$$y - 2 = \frac{1}{3}(x - 3) \qquad y - y_1 = m(x - x_1); \quad x_1 = 3, \, y_1 = 2, \, m = \frac{1}{3}$$

$$y = \frac{1}{3}x + 1$$

(b)  A vertical tangent line occurs when $\dfrac{dx}{dt} = 0$ and $\dfrac{dy}{dt} \neq 0$. Since $\dfrac{dx}{dt} = 6t = 0$

and $\dfrac{dy}{dt} = 2$ when $t = 0$, there is a vertical tangent line to the curve at the point $(0, 0)$. ∎

Figure 13 illustrates the results of Example 1.

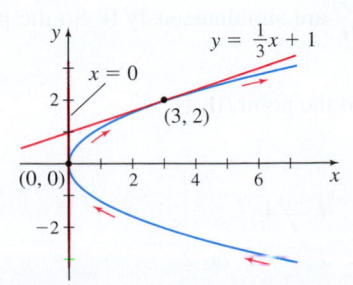

**Figure 13** $x(t) = 3t^2$, $y(t) = 2t$

**NOW WORK** Problem **15**.

**EXAMPLE 2**  **Finding the Slope of the Tangent Line to a Cycloid**

Consider the cycloid

$$x(t) = a(t - \sin t) \qquad y(t) = a(1 - \cos t) \qquad 0 < t < 2\pi \qquad a > 0$$

(a) Show that the slope of any tangent line to the cycloid is given by $\dfrac{\sin t}{1 - \cos t}$.

(b) Find any points where the tangent line to the cycloid is horizontal.

(c) Find an equation of the tangent line to the cycloid at the point(s) found in (b).

**Solution**  **(a)**  $\quad x(t) = at - a \sin t \qquad y(t) = a - a \cos t$

$$\frac{dx}{dt} = a - a \cos t \qquad \frac{dy}{dt} = a \sin t$$

For $0 < t < 2\pi$, $\dfrac{dx}{dt} = a(1 - \cos t) \neq 0$, so the cycloid is a smooth curve. Then the slope of any tangent line is

$$\frac{dy}{dx} = \frac{\dfrac{dy}{dt}}{\dfrac{dx}{dt}} = \frac{a \sin t}{a(1 - \cos t)} = \frac{\sin t}{1 - \cos t}$$

**(b)**  The cycloid has a horizontal tangent line when $\dfrac{dy}{dt} = a \sin t = 0$, but $\dfrac{dx}{dt} \neq 0$.

For $0 < t < 2\pi$, we have $a \sin t = 0$ when $t = \pi$. Since $\dfrac{dx}{dt} \neq 0$ for $0 < t < 2\pi$, the cycloid has a horizontal tangent line when $t = \pi$, at the point $(\pi a, 2a)$.

**(c)**  An equation of the horizontal tangent line when $t = \pi$ is $y = 2a$. See Figure 14. ■

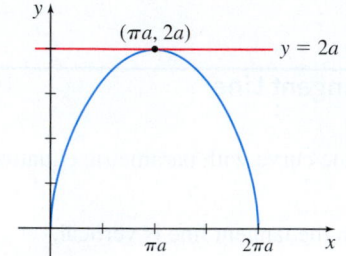

**Figure 14**  $x(t) = a(t - \sin t)$,
$y(t) = a(1 - \cos t), 0 < t < 2\pi$

**NOW WORK** Problem 27.

**EXAMPLE 3**  **Finding an Equation of a Tangent Line to a Smooth Curve**

The plane curve represented by the parametric equations $x(t) = t^3 - 4t$, $y(t) = t^2$ crosses itself at the point $(0, 4)$, and so has two tangent lines there, as shown in Figure 15. Find equations for these tangent lines.

**Solution**  Since $\dfrac{dx}{dt} = 3t^2 - 4 = 0$ only for $t = \dfrac{2\sqrt{3}}{3}$ and for $t = -\dfrac{2\sqrt{3}}{3}$, and $\dfrac{dy}{dt} = 2t = 0$ only for $t = 0$, there is no $t$ for which $\dfrac{dx}{dt}$ and $\dfrac{dy}{dt}$ are simultaneously 0. So the plane curve is a smooth curve.

Next we find the numbers $t$ that correspond to the point $(0, 4)$.

$$x = 0 \qquad\qquad y = 4$$
$$t^3 - 4t = 0 \qquad\qquad t^2 = 4$$
$$t(t^2 - 4) = 0 \qquad\qquad t = -2 \text{ or } 2$$
$$t = 0, -2, \text{ or } 2$$

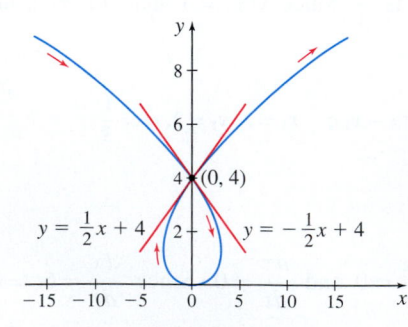

**DF** **Figure 15** $x(t) = t^3 - 4t, y(t) = t^2$

We exclude $t = 0$, since if $t = 0$, then $y \neq 4$. So we investigate only $t = -2$ and $t = 2$. Since $\dfrac{dx}{dt} = 3t^2 - 4 \neq 0$ for $t = -2$ and $t = 2$, the slopes of the tangent lines are given by

$$\frac{dy}{dx} = \frac{\dfrac{dy}{dt}}{\dfrac{dx}{dt}} = \frac{\dfrac{d}{dt}(t^2)}{\dfrac{d}{dt}(t^3 - 4t)} = \frac{2t}{3t^2 - 4}$$

When $t = -2$, the slope of the tangent line is

$$\frac{dy}{dx} = \frac{2(-2)}{3(-2)^2 - 4} = \frac{-4}{8} = -\frac{1}{2}$$

and an equation of the tangent line at the point $(0, 4)$ is

$$y - 4 = -\frac{1}{2}(x - 0)$$

$$y = -\frac{1}{2}x + 4$$

When $t = 2$, the slope of the tangent line at the point $(0, 4)$ is

$$\frac{dy}{dx} = \frac{2 \cdot 2}{3 \cdot 2^2 - 4} = \frac{4}{8} = \frac{1}{2}$$

and an equation of the tangent line at $(0, 4)$ is

$$y - 4 = \frac{1}{2}(x - 0)$$

$$y = \frac{1}{2}x + 4$$

NOW WORK Problem 31.

EXAMPLE 4  **Projectile Motion**

A projectile is fired at an angle $\theta$, $0 < \theta < \dfrac{\pi}{2}$, to the horizontal with an initial speed of $v_0$ m/s. Assuming no air resistance, the position of the projectile after $t$ seconds is given by the parametric equations $x(t) = (v_0 \cos \theta)t$, $y(t) = (v_0 \sin \theta)t - \dfrac{1}{2}gt^2$, $t \geq 0$, where $g$ is the acceleration due to gravity.

(a)  Express the slope of the tangent line to the motion of the projectile as a function of $t$.

(b)  At what time $t$ is the projectile at its maximum height?

**Solution**  (a)  The slope of the tangent line is given by $\dfrac{dy}{dx}$.

$$\frac{dy}{dx} = \frac{\dfrac{dy}{dt}}{\dfrac{dx}{dt}} = \frac{\dfrac{d}{dt}\left[(v_0 \sin \theta)\, t - \dfrac{1}{2}gt^2\right]}{\dfrac{d}{dt}\left[(v_0 \cos \theta)\, t\right]} = \frac{v_0 \sin \theta - gt}{v_0 \cos \theta}$$

(b)  The projectile is at its maximum height when the slope of the tangent line equals 0. That is, when

$$\frac{dy}{dx} = \frac{v_0 \sin \theta - gt}{v_0 \cos \theta} = 0$$

$$v_0 \sin \theta - gt = 0$$

$$t = \frac{v_0 \sin \theta}{g}$$

## 9.2 Assess Your Understanding

### Concepts and Vocabulary

1. *Multiple Choice* Let $C$ denote a plane curve represented by the parametric equations $x = x(t)$, $y = y(t)$, $a \le t \le b$, where each function $x(t)$ and $y(t)$ is continuous on the closed interval $[a, b]$ and differentiable on the open interval $(a, b)$. If both $\dfrac{dx}{dt}$ and $\dfrac{dy}{dt}$ are continuous and never simultaneously 0 on $(a, b)$, then $C$ is called a [(a) smooth (b) differentiable (c) parametric] curve.

2. *True or False* If $C$ is a smooth curve, represented by the parametric equations $x = x(t)$ and $y = y(t)$, $a \le t \le b$, then the slope of the tangent line to $C$ at the point $(x, y)$ is given by the formula $\dfrac{dy}{dx} = \dfrac{\frac{dy}{dt}}{\frac{dx}{dt}}$, provided $\dfrac{dx}{dt} \ne 0$.

3. If in the formula for the slope of a tangent line, $\dfrac{dy}{dt} = 0$, but $\dfrac{dx}{dt} \ne 0$, then the curve has a(n) _____ tangent line at the point $(x(t), y(t))$. If $\dfrac{dx}{dt} = 0$, but $\dfrac{dy}{dt} \ne 0$, then the curve has a(n) _____ tangent line at the point $(x(t), y(t))$.

4. *True or False* A smooth curve $x = x(t)$, $y = y(t)$, $a \le t \le b$, can be represented by a rectangular equation $y = F(x)$ if $\dfrac{dy}{dt}$ is never 0 on $(a, b)$.

### Skill Building

*In Problems 5–12, find* $\dfrac{dy}{dx}$. *Assume* $\dfrac{dx}{dt} \ne 0$.

5. $x(t) = e^t \cos t$, $\quad y(t) = e^t \sin t$
6. $x(t) = 1 + e^{-t}$, $\quad y(t) = e^{3t}$
7. $x(t) = t + \dfrac{1}{t}$, $\quad y(t) = 4 + t$
8. $x(t) = t + \dfrac{1}{t}$, $\quad y(t) = t - \dfrac{1}{t}$
9. $x(t) = \cos t + t \sin t$, $\quad y(t) = \sin t - t \cos t$
10. $x(t) = \cos^3 t$, $\quad y(t) = \sin^3 t$
11. $x(t) = \cot^2 t$, $\quad y(t) = \cot t$
12. $x(t) = \sin t$, $\quad y(t) = \sec^2 t$

*In Problems 13–26, for each pair of parametric equations:*
(a) *Find an equation of the tangent line to the curve at the given number.*
(b) *Graph the curve and the tangent line.*

13. $x(t) = 2t^2$, $\quad y(t) = t$ at $t = 2$
14. $x(t) = t$, $\quad y(t) = 3t^2$ at $t = -2$
15. $x(t) = 3t$, $\quad y(t) = 2t^2 - 1$ at $t = 1$
16. $x(t) = 2t$, $\quad y(t) = t^2 - 2$ at $t = 2$
17. $x(t) = \sqrt{t}$, $\quad y(t) = \dfrac{1}{t}$ at $t = 4$
18. $x(t) = \dfrac{2}{t^2}$, $\quad y(t) = \dfrac{1}{t}$ at $t = 1$
19. $x(t) = \dfrac{t}{t+2}$, $\quad y(t) = \dfrac{4}{t+2}$ at $t = 0$
20. $x(t) = \dfrac{t^2}{1+t}$, $\quad y(t) = \dfrac{1}{1+t}$ at $t = 0$

21. $x(t) = e^t$, $\quad y(t) = e^{-t}$ at $t = 0$
22. $x(t) = e^{2t}$, $\quad y(t) = e^t$ at $t = 0$
23. $x(t) = \sin t$, $\quad y(t) = \cos t$ at $t = \dfrac{\pi}{4}$
24. $x(t) = \sin^2 t$, $\quad y(t) = \cos t$ at $t = \dfrac{\pi}{4}$
25. $x(t) = 4 \sin t$, $\quad y(t) = 3 \cos t$ at $t = \dfrac{\pi}{3}$
26. $x(t) = 2 \sin t - 1$, $\quad y(t) = \cos t + 2$ at $t = \dfrac{\pi}{6}$

*In Problems 27–30, for each smooth curve, find any points where the tangent line is either horizontal or vertical.*

27. $x(t) = t^2$, $\quad y(t) = t^3 - 4t$
28. $x(t) = t^3 - 9t$, $\quad y(t) = t^2$
29. $x(t) = 1 - \cos t$, $\quad y(t) = 1 - \sin t$, $\quad 0 \le t \le 2\pi$
30. $x(t) = -3 \cos t + \cos(3t)$, $\quad y(t) = \sin t$, $\quad 0 \le t \le 2\pi$

### Applications and Extensions

31. **Tangent Lines**
    (a) Find all the points on the plane curve $C$ represented by
    $$x(t) = t^2 + 2 \quad y(t) = t^3 - 4t$$
    where the tangent line is horizontal or where it is vertical.
    (b) Show that $C$ has two tangent lines at the point $(6, 0)$.
    (c) Find equations of these tangent lines.
    📈 (d) Graph $C$.

32. **Tangent Lines**
    (a) Find all the points on the plane curve $C$ represented by
    $$x(t) = 2t^2 \quad y(t) = 8t - t^3$$
    where the tangent line is horizontal or where it is vertical.
    (b) Show that $C$ has two tangent lines at the point $(16, 0)$.
    (c) Find equations of these tangent lines.
    📈 (d) Graph $C$.

### Challenge Problems

33. Show that a smooth curve $x = f(t)$, $y = g(t)$, for which $\dfrac{dx}{dt}$ is never 0, can be represented by a rectangular equation $y = F(x)$, where $F$ is differentiable. *Hint:* Use the fact that $t = f^{-1}(x)$ exists and is differentiable.

34. **Higher-Order Derivatives** Find an expression for $\dfrac{d^2 y}{dx^2}$ if $x = f(t)$, $y = g(t)$, where $f$ and $g$ have second-order derivatives.

35. Find $\dfrac{d^2 y}{dx^2}$ if $x(\theta) = a \cos^3 \theta$, $y(\theta) = a \sin^3 \theta$.

36. Consider the cycloid
    $$x(t) = a(t - \sin t) \quad y(t) = a(1 - \cos t) \quad a > 0$$
    Discuss the behavior of the tangent line to the cycloid at $t = 0$.

---

**1.** = NOW WORK problem    📈 = Graphing technology recommended    [CAS] = Computer Algebra System recommended

# 9.3 Arc Length; Surface Area of a Solid of Revolution

**OBJECTIVES** *When you finish this section, you should be able to:*

1 Find the arc length of a plane curve (p. 695)
2 Find the surface area of a solid of revolution obtained from parametric equations (p. 697)

## 1 Find the Arc Length of a Plane Curve

For a function $y = f(x)$ that has a derivative that is continuous on some interval containing $a$ and $b$, the arc length $s$ of the graph of $f$ from $a$ to $b$ is given by $s = \int_a^b \sqrt{1 + [f'(x)]^2}\, dx$. For a smooth curve represented by the parametric equations $x = x(t)$, $y = y(t)$, we have the following formula for arc length.

> **NEED TO REVIEW?** Arc length is discussed in Section 6.5, pp. 506–511.

**THEOREM Arc Length Formula for Parametric Equations**

For a smooth curve $C$ represented by the parametric equations

$$x = x(t) \qquad y = y(t) \qquad a \le t \le b$$

the arc length $s$ of $C$ from $t = a$ to $t = b$ is given by the formula

$$s = \int_a^b \sqrt{\left(\frac{dx}{dt}\right)^2 + \left(\frac{dy}{dt}\right)^2}\, dt$$

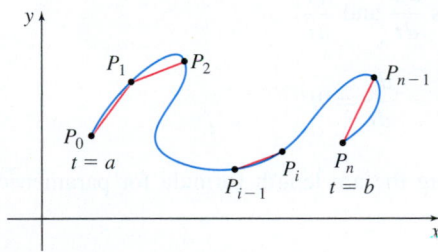

**DF Figure 16** $x = x(t),\ y = y(t),\ a \le t \le b$

**Partial Proof of the Arc Length Formula** The proof for parametric equations is similar to the one we used in Chapter 6 to prove the arc length formula for a function $y = f(x)$, $a \le x \le b$.

First, we partition the closed interval $[a, b]$ into $n$ subintervals:

$$[a, t_1], [t_1, t_2], \ldots, [t_{i-1}, t_i], \ldots, [t_{n-1}, b] \qquad a = t_0, b = t_n$$

each of length $\Delta t = \dfrac{b - a}{n}$. Corresponding to each number $a, t_1, t_2, \ldots, t_{n-1}, b$, there is a point $P_0, P_1, P_2, \ldots, P_n$, on the plane curve, as shown in Figure 16.

We join each point $P_{i-1}$ to the next point $P_i$ with a line segment. The sum of the lengths of the line segments approximates the length of the curve from $t = a$ to $t = b$. This sum can be written as

$$d(P_0, P_1) + d(P_1, P_2) + \cdots + d(P_{n-1}, P_n) = \sum_{i=1}^{n} d(P_{i-1}, P_i)$$

where $d(P_{i-1}, P_i)$ is the length of the line segment joining the points $P_{i-1}$ and $P_i$. Using the distance formula,

$$d(P_{i-1}, P_i) = \sqrt{[x(t_i) - x(t_{i-1})]^2 + [y(t_i) - y(t_{i-1})]^2}$$

the sum of the lengths of the line segments is

$$\sum_{i=1}^{n} d(P_{i-1}, P_i) = \sum_{i=1}^{n} \sqrt{[x(t_i) - x(t_{i-1})]^2 + [y(t_i) - y(t_{i-1})]^2}$$

Since the curve is smooth, the functions $x = x(t)$ and $y = y(t)$ satisfy the conditions of the Mean Value Theorem on $[a, b]$, and so satisfy the same conditions on each subinterval $[t_{i-1}, t_i]$. This means that there are numbers $u_i$ and $v_i$ in each open interval $(t_{i-1}, t_i)$ for which

> **NEED TO REVIEW?** The Mean Value Theorem is discussed in Section 4.3, pp. 302–305.

$$x(t_i) - x(t_{i-1}) = \left[\frac{dx}{dt}(u_i)\right]\Delta t \qquad y(t_i) - y(t_{i-1}) = \left[\frac{dy}{dt}(v_i)\right]\Delta t$$

The sum of the lengths of the segments can be written as

$$\sum_{i=1}^{n} d(P_{i-1}, P_i) = \sum_{i=1}^{n} \sqrt{\left[\frac{dx}{dt}(u_i)\right]^2 + \left[\frac{dy}{dt}(v_i)\right]^2} \, \Delta t$$

These sums are not Riemann sums, since the numbers $u_i$ and $v_i$ are not necessarily equal. However, there is a result (usually given in advanced calculus) that states the limit

of the sums $\sum_{i=1}^{n} \sqrt{\left[\frac{dx}{dt}(u_i)\right]^2 + \left[\frac{dy}{dt}(v_i)\right]^2} \, \Delta t$, as $\Delta t \to 0$, is a definite integral.* That

is, the length $s$ of the curve from $t = a$ to $t = b$ is

$$s = \int_a^b \sqrt{\left(\frac{dx}{dt}\right)^2 + \left(\frac{dy}{dt}\right)^2} \, dt$$

■

 **CALC CLIP**  **EXAMPLE 1   Finding Arc Length for Parametric Equations**

Find the length $s$ of the smooth curve represented by the parametric equations

$$x(t) = t^3 + 2 \qquad y(t) = 2t^{9/2}$$

from the point where $t = 1$ to the point where $t = 3$. Figure 17 shows the graph of the curve.

**Solution**   We begin by finding the derivatives $\dfrac{dx}{dt}$ and $\dfrac{dy}{dt}$.

$$\frac{dx}{dt} = 3t^2 \qquad \text{and} \qquad \frac{dy}{dt} = 9t^{7/2}$$

The curve is smooth for $1 \le t \le 3$. Now using the arc length formula for parametric equations, we have

$$s = \int_a^b \sqrt{\left(\frac{dx}{dt}\right)^2 + \left(\frac{dy}{dt}\right)^2} \, dt = \int_1^3 \sqrt{(3t^2)^2 + (9t^{7/2})^2} \, dt$$

$$= \int_1^3 \sqrt{9t^4 + 81t^7} \, dt = \int_1^3 3t^2 \sqrt{1 + 9t^3} \, dt$$

Use the substitution $u = 1 + 9t^3$. Then $du = 27t^2 dt$, or equivalently, $3t^2 dt = \dfrac{du}{9}$. Changing the limits of integration, we find that when $t = 1$, then $u = 10$; and when $t = 3$, then $u = 1 + 9 \cdot 3^3 = 244$. The arc length $s$ is

$$s = \int_1^3 3t^2 \sqrt{1 + 9t^3} \, dt = \int_{10}^{244} \sqrt{u} \left(\frac{du}{9}\right) = \frac{1}{9} \int_{10}^{244} u^{1/2} du = \frac{1}{9} \left[\frac{u^{3/2}}{\frac{3}{2}}\right]_{10}^{244}$$

$$= \frac{2}{27} \left[244^{3/2} - 10^{3/2}\right] = \frac{4}{27} \left[244\sqrt{61} - 5\sqrt{10}\right]$$

■

**NEED TO REVIEW?**  The substitution method for definite integrals is discussed in Section 5.5, pp. 437–439.

**NOW WORK**  Problem **5.**

y axis with gridlines at 100, 200, 300; point (29, 162√3); point (3, 2); x axis marked at 5, 10, 15, 20, 25, 30.

**Figure 17**  $x(t) = t^3 + 2$, $y(t) = 2t^{9/2}$, $1 \le t \le 3$

---

*This is where the continuity of the derivatives is used, to guarantee the existence of the definite integral.

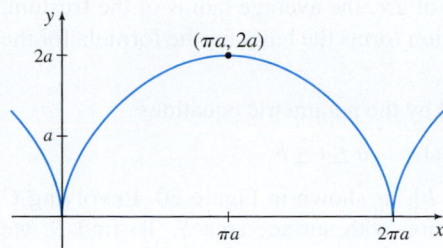

**Figure 18** $x(t) = a(t - \sin t)$,
$y(t) = a(1 - \cos t)$, $a > 0$

**EXAMPLE 2  Finding the Arc Length of a Cycloid**

Find the length $s$ of one arch of the cycloid:

$$x(t) = a(t - \sin t) \qquad y(t) = a(1 - \cos t) \qquad a > 0$$

Figure 18 shows the graph of the cycloid.

**Solution**  One arch of the cycloid is obtained when $t$ varies from 0 to $2\pi$. Since

$$\frac{dx}{dt} = a(1 - \cos t) \qquad \text{and} \qquad \frac{dy}{dt} = a \sin t$$

are both continuous and are never simultaneously 0 on $(0, 2\pi)$, the cycloid is smooth on $[0, 2\pi]$. The arc length $s$ is

$$s = \int_a^b \sqrt{\left(\frac{dx}{dt}\right)^2 + \left(\frac{dy}{dt}\right)^2}\, dt = \int_0^{2\pi} \sqrt{a^2(1 - \cos t)^2 + a^2 \sin^2 t}$$

$$= a \int_0^{2\pi} \sqrt{(1 - 2\cos t + \cos^2 t) + \sin^2 t}\, dt = a \int_0^{2\pi} \sqrt{1 - 2\cos t + 1}\, dt = \sqrt{2}a \int_0^{2\pi} \sqrt{1 - \cos t}\, dt$$

To integrate $\sqrt{1 - \cos t}$ we use a half-angle identity. Since $\sin \dfrac{t}{2} \geq 0$ if $0 \leq t \leq 2\pi$, we have $\sin \dfrac{t}{2} = \sqrt{\dfrac{1 - \cos t}{2}}$. Then

$$\sqrt{1 - \cos t} = \sqrt{2} \sin \frac{t}{2}$$

Now the arc length $s$ from $t = 0$ to $t = 2\pi$ is

$$s = \sqrt{2}a \int_0^{2\pi} \sqrt{1 - \cos t}\, dt = \sqrt{2}a \int_0^{2\pi} \sqrt{2} \sin \frac{t}{2}\, dt = 2a \left[ -2 \cos \frac{t}{2} \right]_0^{2\pi} = 8a \qquad \blacksquare$$

**NOW WORK** Problem 11.

## 2  Find the Surface Area of a Solid of Revolution Obtained from Parametric Equations

**NEED TO REVIEW?**  The derivation of the surface area formula using a rectangular equation is in Section 6.5, pp. 511–513.

In Chapter 6, we used a definite integral to find the *surface area* of a solid of revolution. The derivation of the surface area formula using parametric equations follows the same reasoning used for rectangular equations. Consider a line segment of length $L$ that lies above the $x$-axis. See Figure 19(a). If the region bounded by the line segment and an axis from $a$ to $b$ is revolved about the axis, the resulting solid of revolution is a **frustum**, with slant height $L$ and base radii $r_1$ and $r_2$. See Figure 19(b).

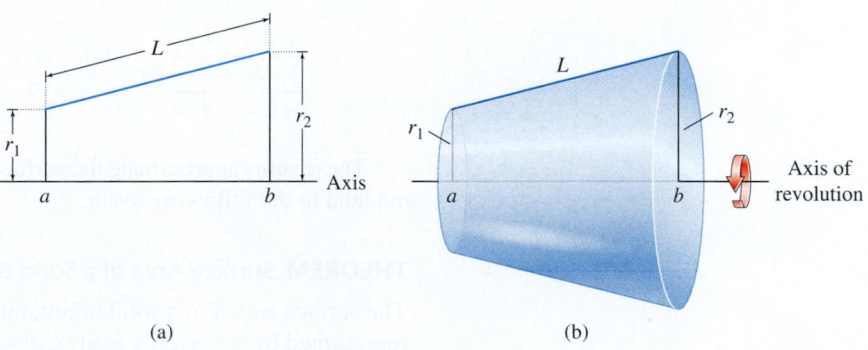

(a)                                        (b)

**Figure 19**

The surface area $S$ of the frustum is

$$\boxed{S = 2\pi \left( \frac{r_1 + r_2}{2} \right) L}$$

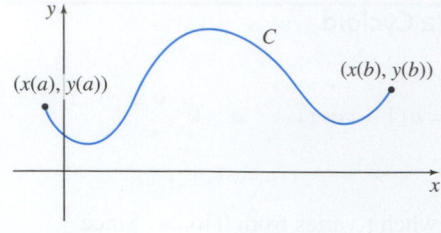

**Figure 20**

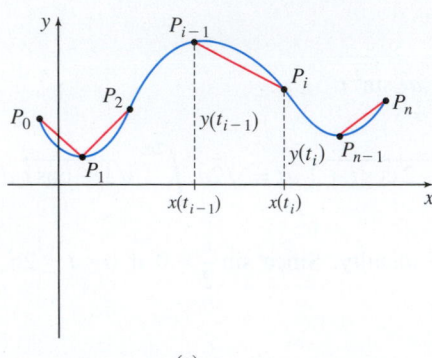

(a)

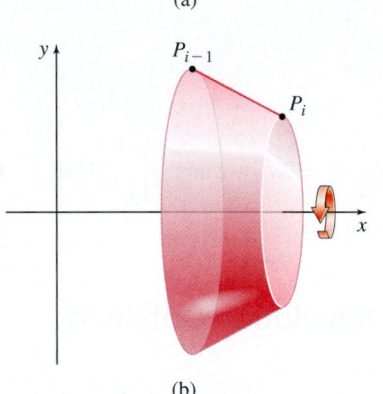

(b)

**Figure 21**

That is, the surface area $S$ equals the product of $2\pi$, the average radius of the frustum, and the slant height of the frustum. This equation forms the basis for the formula for the surface area of a solid of revolution.

Suppose $C$ is a smooth curve represented by the parametric equations

$$x = x(t) \qquad y = y(t) \qquad a \le t \le b$$

where $y = y(t) \ge 0$ on the closed interval $[a, b]$, as shown in Figure 20. Revolving $C$ about the $x$-axis generates a solid of revolution with surface area $S$. To find $S$, we partition the interval $[a, b]$ into $n$ subintervals:

$$[a, t_1], [t_1, t_2], \dots, [t_{i-1}, t_i], \dots, [t_{n-1}, b] \qquad a = t_0, b = t_n$$

each of length $\Delta t = \dfrac{b-a}{n}$. Corresponding to each number $a, t_1, t_2, \dots, t_{i-1}, t_i, \dots,$ $t_{n-1}, b$, there is a point $P_0, P_1, \dots, P_{i-1}, P_i, \dots, P_{n-1}, P_n$ on the curve $C$. We join each point $P_{i-1}$ to the next point $P_i$ with a line segment and focus on the line segment joining the points $P_{i-1}$ and $P_i$. See Figure 21(a). When this line segment of length $d(P_{i-1}, P_i)$ is revolved about the $x$-axis, it generates a frustum of a right circular cone whose surface area $S_i$ is

$$S_i = 2\pi \left[ \frac{y(t_{i-1}) + y(t_i)}{2} \right] [d(P_{i-1}, P_i)] \tag{1}$$

See Figure 21(b).

We follow the same reasoning used for finding the arc length of a smooth curve. Using the distance formula, the length of the $i$th line segment is

$$d(P_{i-1}, P_i) = \sqrt{[x(t_i) - x(t_{i-1})]^2 + [y(t_i) - y(t_{i-1})]^2}$$

Now apply the Mean Value Theorem to $x(t)$ and $y(t)$. There are numbers $u_i$ and $v_i$ in each open interval $(t_{i-1}, t_i)$ for which

$$x(t_i) - x(t_{i-1}) = \left[ \frac{dx}{dt}(u_i) \right] \Delta t \qquad \text{and} \qquad y(t_i) - y(t_{i-1}) = \left[ \frac{dy}{dt}(v_i) \right] \Delta t$$

So,

$$d(P_{i-1}, P_i) = \sqrt{ \left\{ \left[ \frac{dx}{dt}(u_i) \right] \Delta t \right\}^2 + \left\{ \left[ \frac{dy}{dt}(v_i) \right] \Delta t \right\}^2 } = \sqrt{ \left[ \frac{dx}{dt}(u_i) \right]^2 + \left[ \frac{dy}{dt}(v_i) \right]^2 } \, \Delta t$$

where $u_i$ and $v_i$ are numbers in the $i$th subinterval.

Now replace $d(P_{i-1}, P_i)$ in equation (1) with $\sqrt{ \left[ \dfrac{dx}{dt}(u_i) \right]^2 + \left[ \dfrac{dy}{dt}(v_i) \right]^2 } \, \Delta t$.

Then the surface area generated by the sum of the line segments is

$$\sum_{i=1}^{n} S_i = \sum_{i=1}^{n} 2\pi \left[ \frac{y(t_{i-1}) + y(t_i)}{2} \right] \sqrt{ \left[ \frac{dx}{dt}(u_i) \right]^2 + \left[ \frac{dy}{dt}(v_i) \right]^2 } \, \Delta t$$

These sums approximate the surface area generated by revolving $C$ about the $x$-axis and lead to the following result.

**THEOREM  Surface Area of a Solid of Revolution**

The surface area $S$ of a solid of revolution generated by revolving the smooth curve $C$ represented by $x = x(t)$, $y = y(t)$, $a \le t \le b$, where $y(t) \ge 0$, about the $x$-axis is

$$S = 2\pi \int_a^b y(t) \sqrt{ \left( \frac{dx}{dt} \right)^2 + \left( \frac{dy}{dt} \right)^2 } \, dt \tag{2}$$

**CALC**
**CLIP**
**EXAMPLE 3** **Finding the Surface Area of a Solid of Revolution Obtained from Parametric Equations**

Find the surface area of the solid generated by revolving the smooth curve $C$ represented by the parametric equations $x(t) = 2t^3$, $y(t) = 3t^2$, $0 \leq t \leq 1$, about the $x$-axis.

**Solution** We begin by graphing the smooth curve $C$ and revolving it about the $x$-axis. See Figure 22.

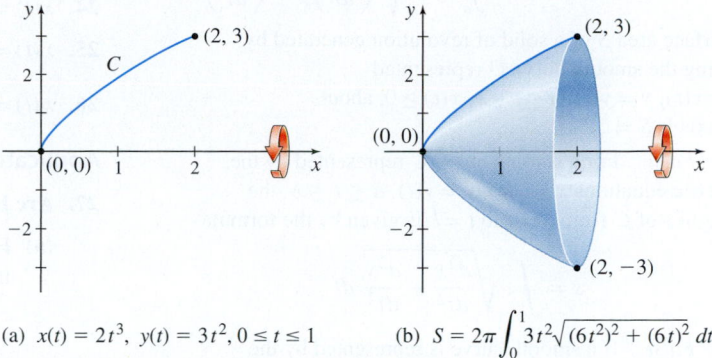

(a) $x(t) = 2t^3$, $y(t) = 3t^2$, $0 \leq t \leq 1$     (b) $S = 2\pi \int_0^1 3t^2 \sqrt{(6t^2)^2 + (6t)^2} \, dt$

**Figure 22**

Now use formula (2) with $\dfrac{dx}{dt} = \dfrac{d}{dt}(2t^3) = 6t^2$ and $\dfrac{dy}{dt} = \dfrac{d}{dt}(3t^2) = 6t$. Then

$$S = 2\pi \int_a^b y(t) \sqrt{\left(\frac{dx}{dt}\right)^2 + \left(\frac{dy}{dt}\right)^2} \, dt = 2\pi \int_0^1 3t^2 \sqrt{(6t^2)^2 + (6t)^2} \, dt$$

$$= 2\pi \int_0^1 3t^2 \sqrt{36t^4 + 36t^2} \, dt = 36\pi \int_0^1 t^3 \sqrt{t^2 + 1} \, dt = \underset{\underset{\substack{u = t^2 + 1; \; du = 2t \, dt \\ \text{when } t = 0, \, u = 1; \; \text{when } t = 1, \, u = 2}}{\uparrow}}{\frac{36\pi}{2} \int_1^2 (u - 1)\sqrt{u} \, du}$$

$$= 18\pi \left[\frac{2}{5} u^{5/2} - \frac{2}{3} u^{3/2}\right]_1^2 = \frac{24\pi}{5}(\sqrt{2} + 1)$$

The surface area of the solid of revolution is $\dfrac{24\pi}{5}(\sqrt{2} + 1) \approx 36.405$ square units. ∎

**NOW WORK** Problem 21.

To help remember the formula for the surface area of a solid of revolution, think of the integrand as the product of the slant height $\sqrt{\left(\dfrac{dx}{dt}\right)^2 + \left(\dfrac{dy}{dt}\right)^2}$ and the circumference $2\pi y$ of the circle traced by a point $(x, y)$ on the corresponding subarc. Also keep in mind that the limits of integration are parameter values, not $x$ values.

If the curve is revolved about the $y$-axis, we have a similar formula for the surface area of the solid of revolution.

**THEOREM Surface Area of a Solid of Revolution**

The surface area $S$ of a solid of revolution generated by revolving the smooth curve $C$ represented by $x = x(t)$, $y = y(t)$, $a \leq t \leq b$, where $x = x(t) \geq 0$, about the $y$-axis is

$$S = 2\pi \int_a^b x(t) \sqrt{\left(\frac{dx}{dt}\right)^2 + \left(\frac{dy}{dt}\right)^2} \, dt$$

## 9.3 Assess Your Understanding

### Concepts and Vocabulary

1. *True or False* When a smooth curve $C$ represented by the parametric equations $x = x(t)$, $y = y(t)$, $y \geq 0$, $a \leq t \leq b$, is revolved about the $x$-axis, the surface area $S$ of the solid of revolution is given by $S = 2\pi \int_a^b x(t) \sqrt{\left(\dfrac{dx}{dt}\right)^2 + \left(\dfrac{dy}{dt}\right)^2}\, dt$.

2. The surface area $S$ of a solid of revolution generated by revolving the smooth curve $C$ represented by $x = x(t)$, $y = y(t)$, $a \leq t \leq b$, $x(t) \geq 0$, about the $y$-axis is $S = $ _____.

3. *True or False* For a smooth curve $C$ represented by the parametric equations $x = x(t)$, $y = y(t)$, $a \leq t \leq b$, the arc length $s$ of $C$ from $t = a$ to $t = b$ is given by the formula
$$s = \int_a^b \sqrt{\dfrac{d^2 x}{dt^2} + \dfrac{d^2 y}{dt^2}}\, dt$$

4. *True or False* If a smooth curve is represented by the parametric equations $x = x(t)$, $y = y(t)$, $a \leq t \leq b$, and if $\dfrac{dx}{dt} = 0$ for any number $t$, $a \leq t \leq b$, then the arc length formula cannot be used.

### Skill Building

*In Problems 5–12, find the arc length of each curve on the given interval.*

5. $x(t) = t^3$, $y(t) = t^2$; $0 \leq t \leq 2$

6. $x(t) = 3t^2 + 1$, $y(t) = t^3 - 1$; $0 \leq t \leq 2$

7. $x(t) = t - 1$, $y(t) = \dfrac{1}{2}t^2$; $0 \leq t \leq 2$

8. $x(t) = t^2$, $y(t) = 2t$; $1 \leq t \leq 3$

9. $x(t) = 4 \sin t$, $y(t) = 4 \cos t$; $-\dfrac{\pi}{2} \leq t \leq \dfrac{\pi}{2}$

10. $x(t) = 6 \sin t$, $y(t) = 6 \cos t$; $-\dfrac{\pi}{2} \leq t \leq \dfrac{\pi}{2}$

11. $x(t) = 2 \sin t - 1$, $y(t) = 2 \cos t + 1$; $0 \leq t \leq 2\pi$

12. $x(t) = e^t \sin t$, $y(t) = e^t \cos t$, $0 \leq t \leq \pi$

*In Problems 13–18,*

(a) Use the arc length formula for parametric equations to set up the integral for finding the length of each curve on the given interval.

(b) Find the length $s$ of each curve on the given interval.

(c) Graph each curve over the given interval.

13. $x(t) = 2 \cos(2t)$, $y(t) = t^2$, $0 \leq t \leq 2\pi$

14. $x(t) = t^2$, $y(t) = \sin t$, $0 \leq t \leq 2\pi$

15. $x(t) = t^2$, $y(t) = \sqrt{t + 2}$, $-2 \leq t \leq 2$

16. $x(t) = t - \cos t$, $y(t) = 1 - \sin t$, $0 \leq t \leq \pi$

17. $x(t) = 3 \cos t + \cos(3t)$, $y(t) = 3 \sin t - \sin(3t)$, $0 \leq t \leq 2\pi$
(A hypocycloid)

18. $x(t) = 5 \cos t - \cos(5t)$, $y(t) = 5 \sin t - \sin(5t)$, $0 \leq t \leq 2\pi$
(An epicycloid)

*In Problems 19–22, find the surface area of the solid generated by revolving each curve about the $x$-axis.*

19. $x(t) = 3t^2$, $y(t) = 6t$; $0 \leq t \leq 1$

20. $x(t) = t^2$, $y(t) = 2t$; $0 \leq t \leq 3$

21. $x(\theta) = \cos^3 \theta$, $y(\theta) = \sin^3 \theta$; $0 \leq \theta \leq \dfrac{\pi}{2}$

22. $x(t) = t - \sin t$, $y(t) = 1 - \cos t$; $0 \leq t \leq \pi$

*In Problems 23–26, find the surface area of the solid generated by revolving each curve about the $y$-axis.*

23. $x(t) = 3t^2$, $y(t) = 2t^3$; $0 \leq t \leq 1$

24. $x(t) = 2t + 1$, $y(t) = t^2 + 3$; $0 \leq t \leq 3$

25. $x(t) = 2 \sin t$, $y(t) = 2 \cos t$; $0 \leq t \leq \dfrac{\pi}{2}$

26. $x(t) = 3 \cos t$, $y(t) = 2 \sin t$; $0 \leq t \leq \dfrac{\pi}{2}$

### Applications and Extensions

27. **Arc Length**

(a) Find the arc length of one arch of the four-cusped hypocycloid
$$x(t) = b \sin^3 t \quad y(t) = b \cos^3 t \quad 0 \leq t \leq \dfrac{\pi}{2}$$

(b) Graph the portion of the curve for $0 \leq t \leq \dfrac{\pi}{2}$.

28. **Arc Length**

(a) Find the arc length of the spiral
$$x(t) = t \cos t \quad y(t) = t \sin t \quad 0 \leq t \leq \pi$$

(b) Graph the portion of the curve for $0 \leq t \leq \pi$.

**Distance Traveled** *In Problems 29–34, find the distance a particle travels along the given path over the indicated time interval.*

29. $x(t) = 3t$, $y(t) = t^2 - 3$; $0 \leq t \leq 2$

30. $x(t) = t^2$, $y(t) = 3t$; $0 \leq t \leq 2$

31. $x(t) = \dfrac{t^2}{2} + 1$, $y(t) = \dfrac{1}{3}(2t + 3)^{3/2}$; $0 \leq t \leq 2$

32. $x(t) = a \cos t$, $y(t) = a \sin t$, $a > 0$; $0 \leq t \leq \pi$

33. $x(t) = \cos(2t)$, $y(t) = \sin^2 t$; $0 \leq t \leq \dfrac{\pi}{2}$

34. $x(t) = \dfrac{1}{t}$, $y(t) = \ln t$; $1 \leq t \leq 2$

35. **Surface Area of a Cycloid** Find the surface area of the solid of revolution obtained by revolving one arch of the cycloid $x(t) = 6(t - \sin t)$, $y(t) = 6(1 - \cos t)$ about the $x$-axis.

36. **Surface Area of a Sphere** Develop a formula for the surface area of a sphere of radius $R$.

**Using Differentials to Approximate Arc Length** Problems 37–40 use the following discussion. For a smooth curve C represented by the parametric equations $x = x(t)$, $y = y(t)$, $a \leq t \leq b$, the arc length $s$ satisfies the equation $\dfrac{ds}{dt} = \sqrt{\left(\dfrac{dx}{dt}\right)^2 + \left(\dfrac{dy}{dt}\right)^2}$ so that
$$\left(\dfrac{ds}{dt}\right)^2 = \left(\dfrac{dx}{dt}\right)^2 + \left(\dfrac{dy}{dt}\right)^2$$

In terms of differentials, this can be written as
$$(ds)^2 = (dx)^2 + (dy)^2$$
$$ds = \sqrt{(dx)^2 + (dy)^2}$$

---

**1.** = NOW WORK problem      = Graphing technology recommended      CAS = Computer Algebra System recommended

Geometrically, the differential $ds = \sqrt{(dx)^2 + (dy)^2}$ is the length of the hypotenuse of a right triangle with sides of lengths $dx$ and $dy$ as shown in the figure.

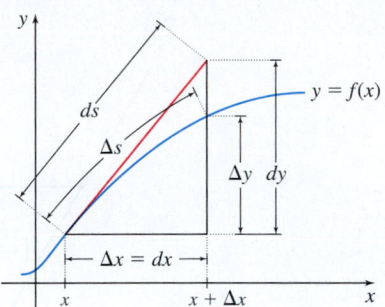

Because $\dfrac{dy}{dx}$ is the slope of the tangent line, the hypotenuse lies on the tangent line to the curve at $x$. Then the differential $ds$ can be used to approximate the arc length $s$ between two nearby points. Using a

differential to approximate arc length is particularly useful when it is difficult, or impossible, to find the arc length using

$$s = \int_a^b \sqrt{\left(\frac{dx}{dt}\right)^2 + \left(\frac{dy}{dt}\right)^2}\, dt$$

*Use the differential ds to approximate each arc length.*

**37.** $x(t) = \sqrt{t}$, $y(t) = t^3$ from $t = 1$ to $t = 1.2$

**38.** $x(t) = t^{1/3}$, $y(t) = t^2$ from $t = 1$ to $t = 1.1$

**39.** $x(t) = e^{at}$, $y(t) = e^{bt}$ from $t = 0$ to $t = 0.2$

**40.** $x(t) = a \sin t$, $y(t) = b \cos t$ from $t = 0$ to $t = 0.1$

**Challenge Problem**

**41.** Find the point on the curve $x = \dfrac{4}{3}t^3 + 3t^2$, $y = t^3 - 4t^2$, for which the length of the curve from $(0,0)$ to $(x, y)$ is $\dfrac{80\sqrt{2} - 40}{3}$.

# 9.4 Polar Coordinates

**OBJECTIVES** *When you finish this section, you should be able to:*

**1** Plot points using polar coordinates (p. 701)

**2** Convert between rectangular coordinates and polar coordinates (p. 703)

**3** Identify and graph polar equations (p. 706)

In a rectangular coordinate system, there are two perpendicular axes, one horizontal (the $x$-axis) and one vertical (the $y$-axis). The point of intersection of the axes is the origin and is labeled $O$. A point is represented by a pair of numbers $(x, y)$, where $x$ and $y$ equal the signed distance of the point from the $y$-axis and the $x$-axis, respectively.

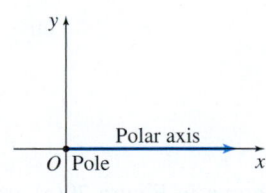

**Figure 23**

A **polar coordinate system** is constructed by selecting a point $O$, called the **pole,** and a ray with its vertex at the pole, called the **polar axis**. It is customary to have the pole coincide with the origin and the polar axis coincide with the positive $x$-axis of the rectangular coordinate system, as shown in Figure 23.

## 1 Plot Points Using Polar Coordinates

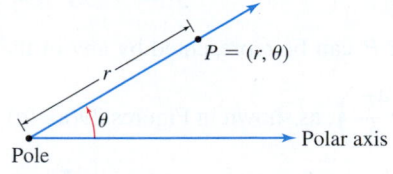

**Figure 24** Polar Coordinates.

A point $P$ in the polar coordinate system is represented by an ordered pair of numbers $(r, \theta)$, called the **polar coordinates** of $P$. If $r > 0$, then $r$ is the distance of the point from the pole (the origin), and $\theta$ is an angle (measured in radians or degrees) whose initial side is the polar axis (the positive $x$-axis) and whose terminal side is a ray from the pole through the point $P$. See Figure 24.

As an example, suppose the polar coordinates of a point are $\left(3, \dfrac{2\pi}{3}\right)$. We locate the point by drawing an angle of $\dfrac{2\pi}{3}$ radians with its vertex at the pole and its initial side along the polar axis. Then $\left(3, \dfrac{2\pi}{3}\right)$ is the point on the terminal side of the angle that is 3 units from the pole. See Figure 25.

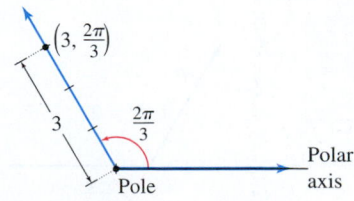

**Figure 25**

If $r = 0$, then the point $P = (0, \theta)$ is at the pole for any $\theta$.

If $r < 0$, the point $P = (r, \theta)$ is *not* on the terminal side of $\theta$. Instead, we extend the terminal side of $\theta$ in the opposite direction. The point $P$ is on this extended ray a distance $|r|$ from the pole. See Figure 26 on page 702.

**NOTE** The points $(r, \theta)$ and $(-r, \theta)$ are reflections about the pole.

So, the point $\left(-3, \dfrac{2\pi}{3}\right)$ is located 3 units from the pole on the extension in the opposite direction of the ray that forms the angle $\dfrac{2\pi}{3}$ with the polar axis. See Figure 27.

**Figure 26**    **Figure 27**

---

### EXAMPLE 1   Plotting Points Using Polar Coordinates

Plot the points with the following polar coordinates:

**(a)** $\left(3, \dfrac{5\pi}{3}\right)$    **(b)** $\left(2, -\dfrac{\pi}{4}\right)$    **(c)** $(3, 0)$    **(d)** $\left(-2, \dfrac{\pi}{4}\right)$

**Solution**  Figure 28 shows the points.

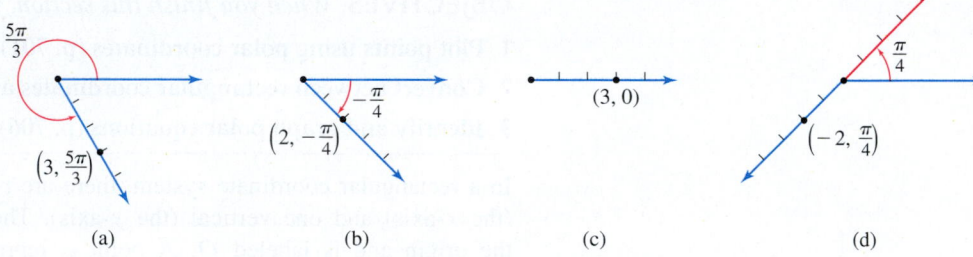

(a)     (b)     (c)     (d)

**Figure 28**

**NOW WORK** Problems **17** and **23**.

The polar coordinates $\left(3, \dfrac{2\pi}{3}\right)$ of the point $P$ shown in Figure 29(a) are one of many possible polar coordinates of $P$. For example, since the angles $\dfrac{8\pi}{3}$ and $-\dfrac{4\pi}{3}$ have the same terminal side as the angle $\dfrac{2\pi}{3}$, the point $P$ can be represented by any of the polar coordinates $\left(3, \dfrac{2\pi}{3}\right)$, $\left(3, \dfrac{8\pi}{3}\right)$, and $\left(3, -\dfrac{4\pi}{3}\right)$, as shown in Figures 29(a)–(c). The point $\left(3, \dfrac{2\pi}{3}\right)$ can also be represented by the polar coordinates $\left(-3, -\dfrac{\pi}{3}\right)$, as illustrated in Figure 29(d).

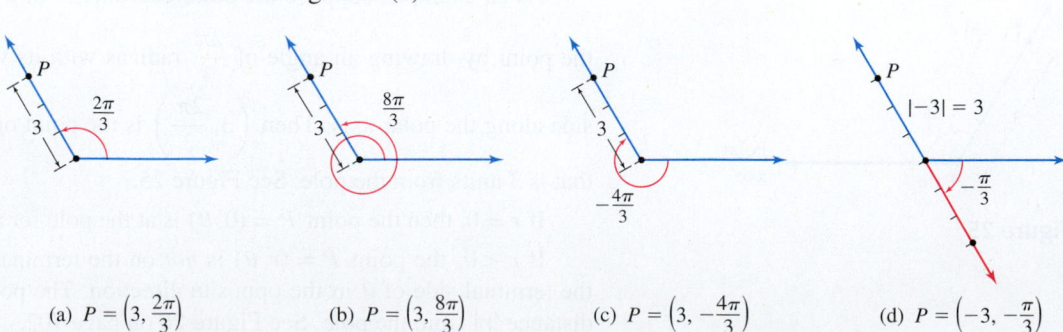

(a) $P = \left(3, \dfrac{2\pi}{3}\right)$    (b) $P = \left(3, \dfrac{8\pi}{3}\right)$    (c) $P = \left(3, -\dfrac{4\pi}{3}\right)$    (d) $P = \left(-3, -\dfrac{\pi}{3}\right)$

**Figure 29**

So, there is a major difference between rectangular and polar coordinate systems. In a rectangular system, each point in the plane corresponds to exactly one pair of rectangular coordinates; in a polar coordinate system, every point in the plane can be represented by infinitely many polar coordinates.

### EXAMPLE 2  Plotting a Point Using Polar Coordinates

Plot the point $P$ whose polar coordinates are $\left(-2, -\dfrac{3\pi}{4}\right)$. Then find three other polar coordinates of the same point with the properties:

(a) $r > 0$,  $0 < \theta < 2\pi$   (b) $r > 0$,  $-2\pi < \theta < 0$   (c) $r < 0$,  $0 < \theta < 2\pi$

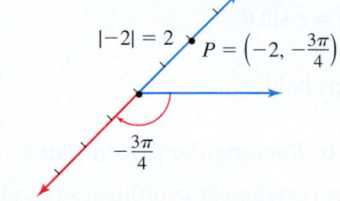

$|-2| = 2$

$P = \left(-2, -\dfrac{3\pi}{4}\right)$

$-\dfrac{3\pi}{4}$

**Figure 30**

**Solution**  The point $\left(-2, -\dfrac{3\pi}{4}\right)$ is located by first drawing the angle $-\dfrac{3\pi}{4}$. Then $P$ is on the extension of the terminal side of $\theta$ in the opposite direction a distance 2 units from the pole, as shown in Figure 30.

(a)  The point $P = (r, \theta)$, $r > 0$, $0 < \theta < 2\pi$ is $\left(2, \dfrac{\pi}{4}\right)$, as shown in Figure 31(a).

(b)  The point $P = (r, \theta)$, $r > 0$, $-2\pi < \theta < 0$ is $\left(2, -\dfrac{7\pi}{4}\right)$, as shown in Figure 31(b).

(c)  The point $P = (r, \theta)$, $r < 0$, $0 < \theta < 2\pi$ is $\left(-2, \dfrac{5\pi}{4}\right)$, as shown in Figure 31(c).

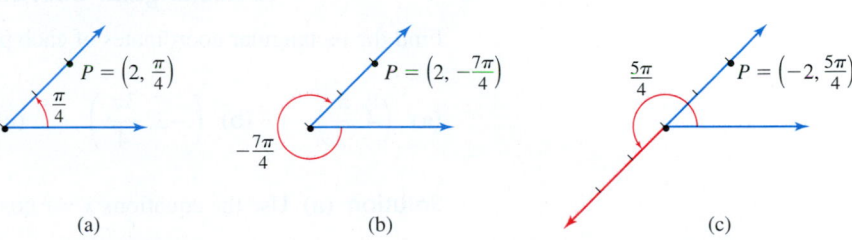

$P = \left(2, \dfrac{\pi}{4}\right)$

$\dfrac{\pi}{4}$

(a)

$P = \left(2, -\dfrac{7\pi}{4}\right)$

$-\dfrac{7\pi}{4}$

(b)

$\dfrac{5\pi}{4}$

$P = \left(-2, \dfrac{5\pi}{4}\right)$

(c)

**Figure 31**

NOW WORK Problem **25**.

## Summary

- A point $P$ with polar coordinates $(r, \theta)$ also can be represented by either

$$(r, \theta + 2n\pi) \qquad \text{or} \qquad (-r, \theta + (2n+1)\pi) \quad n \text{ an integer}$$

- The polar coordinates of the pole are $(0, \theta)$, where $\theta$ is any angle.

## 2  Convert Between Rectangular Coordinates and Polar Coordinates

It is sometimes useful to transform coordinates or equations in rectangular form into polar form, or vice versa. To do this, recall that the origin in the rectangular coordinate system coincides with the pole in the polar coordinate system and the positive $x$-axis in the rectangular system coincides with the polar axis in the polar system. The positive $y$-axis in the rectangular system is the ray $\theta = \dfrac{\pi}{2}$ in the polar system.

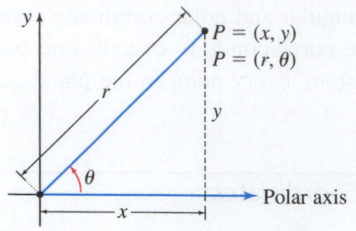

**Figure 32**

**NEED TO REVIEW?** The definitions of the trigonometric functions and their properties are discussed in Appendix A.4, pp. A-27–A-32.

Suppose a point $P$ has the polar coordinates $(r, \theta)$ and the rectangular coordinates $(x, y)$, as shown in Figure 32. If $r > 0$, then $P$ is on the terminal side of $\theta$, and

$$\cos \theta = \frac{x}{r} \quad \text{and} \quad \sin \theta = \frac{y}{r}$$

If $r < 0$, then $P = (r, \theta)$ can be represented as $(-r, \pi + \theta)$, where $-r > 0$. Since

$$\cos(\theta + \pi) = -\cos \theta = \frac{x}{-r} \quad \text{and} \quad \sin(\theta + \pi) = -\sin \theta = \frac{y}{-r}$$

then whether $r > 0$ or $r < 0$,

$$x = r \cos \theta \quad \text{and} \quad y = r \sin \theta$$

If $r = 0$, then $P$ is the pole and the same relationships hold.

---

**THEOREM  Conversion from Polar Coordinates to Rectangular Coordinates**

If $P$ is a point with polar coordinates $(r, \theta)$, then the rectangular coordinates $(x, y)$ of $P$ are given by

$$\boxed{x = r \cos \theta \quad \text{and} \quad y = r \sin \theta}$$

---

**EXAMPLE 3  Converting from Polar Coordinates to Rectangular Coordinates**

Find the rectangular coordinates of each point whose polar coordinates are:

**(a)** $\left(4, \dfrac{\pi}{3}\right)$     **(b)** $\left(-2, \dfrac{3\pi}{4}\right)$     **(c)** $\left(-3, -\dfrac{5\pi}{6}\right)$

**Solution  (a)** Use the equations $x = r \cos \theta$ and $y = r \sin \theta$ with $r = 4$ and $\theta = \dfrac{\pi}{3}$.

$$x = 4 \cos \frac{\pi}{3} = 4\left(\frac{1}{2}\right) = 2 \quad \text{and} \quad y = 4 \sin \frac{\pi}{3} = 4\left(\frac{\sqrt{3}}{2}\right) = 2\sqrt{3}$$

The rectangular coordinates are $(2, 2\sqrt{3})$.

**(b)** Use the equations $x = r \cos \theta$ and $y = r \sin \theta$ with $r = -2$ and $\theta = \dfrac{3\pi}{4}$.

$$x = -2 \cos \frac{3\pi}{4} = -2\left(-\frac{\sqrt{2}}{2}\right) = \sqrt{2} \quad \text{and} \quad y = -2 \sin \frac{3\pi}{4} = -2\left(\frac{\sqrt{2}}{2}\right) = -\sqrt{2}$$

The rectangular coordinates are $(\sqrt{2}, -\sqrt{2})$.

**(c)** Use the equations $x = r \cos \theta$ and $y = r \sin \theta$ with $r = -3$ and $\theta = -\dfrac{5\pi}{6}$.

$$x = -3 \cos \left(-\frac{5\pi}{6}\right) = -3\left(-\frac{\sqrt{3}}{2}\right) = \frac{3\sqrt{3}}{2}$$

$$y = -3 \sin \left(-\frac{5\pi}{6}\right) = -3\left(-\frac{1}{2}\right) = \frac{3}{2}$$

The rectangular coordinates are $\left(\dfrac{3\sqrt{3}}{2}, \dfrac{3}{2}\right)$.  ∎

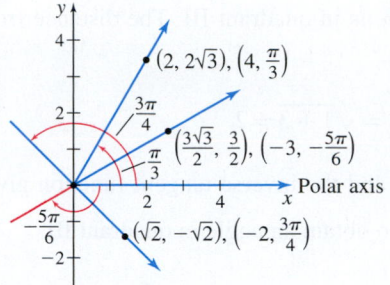

**Figure 33**

The points $\left(4, \dfrac{\pi}{3}\right)$, $\left(-2, \dfrac{3\pi}{4}\right)$, and $\left(-3, -\dfrac{5\pi}{6}\right)$ and the points $(2, 2\sqrt{3})$,

$(\sqrt{2}, -\sqrt{2})$, and $\left(\dfrac{3\sqrt{3}}{2}, \dfrac{3}{2}\right)$ are graphed in their respective coordinate systems in

Figure 33.

**NOW WORK** Problems 33 and 39.

Now suppose $P$ has the rectangular coordinates $(x, y)$. To represent $P$ in polar coordinates, use the fact that $x = r \cos \theta$, $y = r \sin \theta$. Then

$$x^2 + y^2 = r^2 \cos^2 \theta + r^2 \sin^2 \theta = r^2$$

If $x \neq 0$, then $\dfrac{y}{x} = \dfrac{r \sin \theta}{r \cos \theta} = \tan \theta$.

If $x = 0$, the point $P = (x, y)$ is on the $y$-axis. So, $r = y$ and $\theta = \dfrac{\pi}{2}$.

> **THEOREM** Conversion from Rectangular Coordinates to Polar Coordinates
>
> If $P$ is any point in the plane with rectangular coordinates $(x, y)$, the polar coordinates $(r, \theta)$ of $P$ are given by
>
> $$r = \sqrt{x^2 + y^2} \qquad \text{and} \qquad \tan \theta = \frac{y}{x} \quad \text{if } x \neq 0$$
>
> $$r = y \qquad \text{and} \qquad \theta = \frac{\pi}{2} \quad \text{if } x = 0$$

**NEED TO REVIEW?** The inverse tangent function is discussed in Section P.7, p. 63.

Be careful when applying this theorem. If $x \neq 0$, $\tan \theta = \dfrac{y}{x}$ and $\theta = \tan^{-1} \dfrac{y}{x}$, $-\dfrac{\pi}{2} < \theta < \dfrac{\pi}{2}$, placing $\theta$ in quadrants I or IV. If the point lies in quadrant II or III, we must find $\tan^{-1}\left(\dfrac{y}{x}\right)$ in quadrant I or IV, respectively, and then add $\pi$ to the result. It is advisable to plot the point $(x, y)$ at the start to identify the quadrant that contains the point.

**EXAMPLE 4** Converting from Rectangular Coordinates to Polar Coordinates

Find polar coordinates of each point whose rectangular coordinates are:

**(a)** $(4, -4)$ **(b)** $(-1, -\sqrt{3})$ **(c)** $(4, -1)$

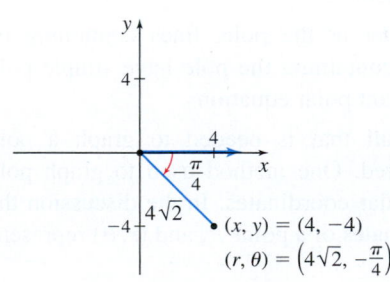

**Figure 34**

**Solution (a)** The point $(4, -4)$, plotted in Figure 34, is in quadrant IV. The distance from the pole to the point $(4, -4)$ is

$$r = \sqrt{x^2 + y^2} = \sqrt{4^2 + (-4)^2} = \sqrt{32} = 4\sqrt{2}$$

Since the point $(4, -4)$ is in quadrant IV, $-\dfrac{\pi}{2} < \theta < 0$. So,

$$\theta = \tan^{-1}\left(\frac{y}{x}\right) = \tan^{-1}\left(\frac{-4}{4}\right) = \tan^{-1}(-1) = -\frac{\pi}{4}$$

A pair of polar coordinates for this point is $\left(4\sqrt{2}, -\dfrac{\pi}{4}\right)$. Other possible representations include $\left(-4\sqrt{2}, \dfrac{3\pi}{4}\right)$ and $\left(4\sqrt{2}, \dfrac{7\pi}{4}\right)$.

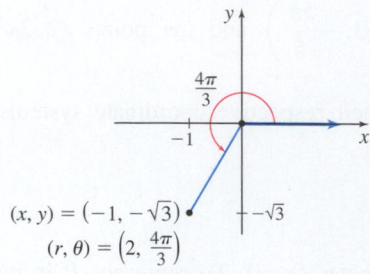

$(x, y) = (-1, -\sqrt{3})$
$(r, \theta) = \left(2, \frac{4\pi}{3}\right)$

**Figure 35**

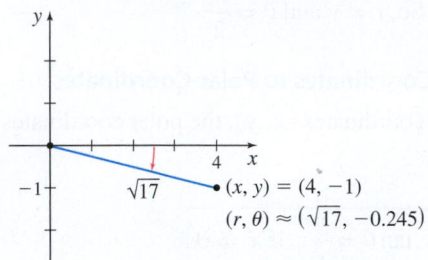

$(x, y) = (4, -1)$
$(r, \theta) \approx (\sqrt{17}, -0.245)$

**Figure 36**

**(b)** The point $(-1, -\sqrt{3})$, plotted in Figure 35, is in quadrant III. The distance from the pole to the point $(-1, -\sqrt{3})$ is

$$r = \sqrt{(-1)^2 + (-\sqrt{3})^2} = \sqrt{1 + 3} = 2$$

Since the point $(-1, -\sqrt{3})$ lies in quadrant III and the inverse tangent function gives an angle in quadrant I, we add $\pi$ to $\tan^{-1}\left(\frac{y}{x}\right)$ to obtain an angle in quadrant III.

$$\theta = \tan^{-1}\left(\frac{-\sqrt{3}}{-1}\right) + \pi = \tan^{-1}(\sqrt{3}) + \pi = \frac{\pi}{3} + \pi = \frac{4\pi}{3}$$

A pair of polar coordinates for the point is $\left(2, \frac{4\pi}{3}\right)$. Other possible representations include $\left(-2, \frac{\pi}{3}\right)$ and $\left(2, -\frac{2\pi}{3}\right)$.

**(c)** The point $(4, -1)$, plotted in Figure 36, lies in quadrant IV. The distance from the pole to the point $(4, -1)$ is

$$r = \sqrt{x^2 + y^2} = \sqrt{4^2 + (-1)^2} = \sqrt{17}$$

Since the point $(4, -1)$ is in quadrant IV, $-\frac{\pi}{2} < \theta < 0$. So,

$$\theta = \tan^{-1}\left(\frac{y}{x}\right) = \tan^{-1}\left(\frac{-1}{4}\right) \approx -0.245 \text{ radians}$$

A pair of polar coordinates for this point is $(\sqrt{17}, -0.245)$. Other possible representations for the point include $\left(\sqrt{17}, \tan^{-1}\left(-\frac{1}{4}\right) + 2\pi\right) \approx (\sqrt{17}, 6.038)$ and $\left(-\sqrt{17}, \tan^{-1}\left(-\frac{1}{4}\right) + \pi\right) \approx (-\sqrt{17}, 2.897)$. ∎

**NOW WORK** Problems **41** and **47**.

## 3 Identify and Graph Polar Equations

Just as a rectangular grid is used to plot points given by rectangular coordinates, a *polar grid* is used to plot points given by polar coordinates. A **polar grid** consists of concentric circles centered at the pole and of rays with vertices at the pole, as shown in Figure 37. An equation whose variables are polar coordinates is called a **polar equation**, and the **graph of a polar equation** is the set of all points for which at least one of the polar coordinate representations satisfies the equation.

Polar equations of circles with their center at the pole, lines containing the pole, horizontal and vertical lines, and circles containing the pole have simple polar equations. In Section 9.5, we graph other important polar equations.

There are occasions when geometry is all that is needed to graph a polar equation. But usually other methods are required. One method used to graph polar equations is to convert the equation to rectangular coordinates. In the discussion that follows, $(x, y)$ represents the rectangular coordinates of a point $P$, and $(r, \theta)$ represents polar coordinates of the point $P$.

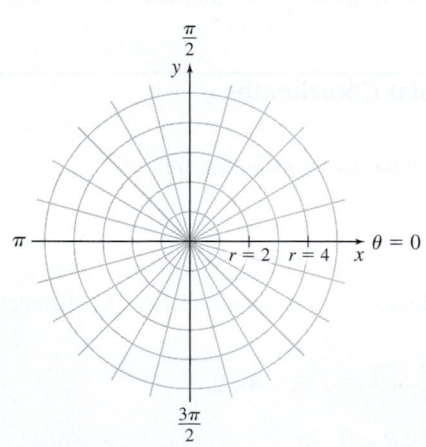

**Figure 37** Polar grid

### EXAMPLE 5  Identifying and Graphing a Polar Equation

Identify and graph each equation:

**(a)** $r = 3$      **(b)** $\theta = \frac{\pi}{4}$

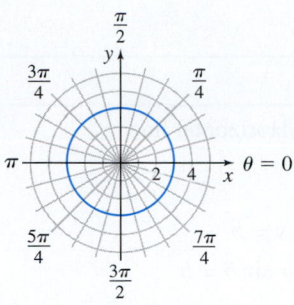

**Figure 38** $r = 3$ or $x^2 + y^2 = 9$

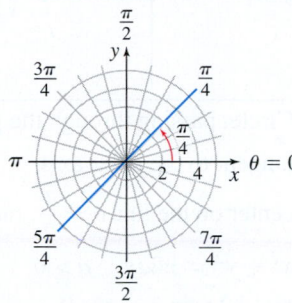

**Figure 39** $\theta = \dfrac{\pi}{4}$ or $y = x$

**Solution** **(a)** If $r$ is fixed at 3 and $\theta$ is allowed to vary, the graph is a circle with its center at the pole and radius 3, as shown in Figure 38. To confirm this, we convert the polar equation $r = 3$ to a rectangular equation.

$$r = 3$$
$$r^2 = 9 \qquad \text{Square both sides.}$$
$$x^2 + y^2 = 9 \qquad r^2 = x^2 + y^2$$

**(b)** If $\theta$ is fixed at $\dfrac{\pi}{4}$ and $r$ is allowed to vary, the result is a line containing the pole, making an angle of $\dfrac{\pi}{4}$ with the polar axis. That is, the graph of $\theta = \dfrac{\pi}{4}$ is a line containing the pole with slope $\tan\theta = \tan\dfrac{\pi}{4} = 1$, as shown in Figure 39. To confirm this, we convert the polar equation to a rectangular equation.

$$\theta = \frac{\pi}{4}$$
$$\tan\theta = \tan\frac{\pi}{4}$$
$$\frac{y}{x} = 1$$
$$y = x \qquad \blacksquare$$

NOW WORK Problems 51 and 61.

---

CALC
▶ **EXAMPLE 6** **Identifying and Graphing Polar Equations**
CLIP
Identify and graph the equations:

**(a)** $r\sin\theta = 2$        **(b)** $r = 4\sin\theta$

**Solution** **(a)** Here both $r$ and $\theta$ are allowed to vary, so the graph of the equation is not as obvious. If we use the fact that $y = r\sin\theta$, the equation $r\sin\theta = 2$ becomes $y = 2$. So, the graph of $r\sin\theta = 2$ is the horizontal line $y = 2$ that lies 2 units above the pole, as shown in Figure 40.

**(b)** To convert the equation $r = 4\sin\theta$ to rectangular coordinates, multiply the equation by $r$ to obtain

$$r^2 = 4r\sin\theta$$

Since $r^2 = x^2 + y^2$ and $y = r\sin\theta$, we have

$$r^2 = 4r\sin\theta$$
$$x^2 + y^2 = 4y \qquad r^2 = x^2 + y^2, \quad r\sin\theta = y$$
$$x^2 + (y^2 - 4y) = 0$$
$$x^2 + (y^2 - 4y + 4) = 4 \qquad \text{Complete the square in } y.$$
$$x^2 + (y - 2)^2 = 4 \qquad \text{Factor.}$$

This is the standard form of the equation of a circle with its center at the point $(0, 2)$ and radius 2 in rectangular coordinates. See Figure 41. Notice that the circle passes through the pole. $\blacksquare$

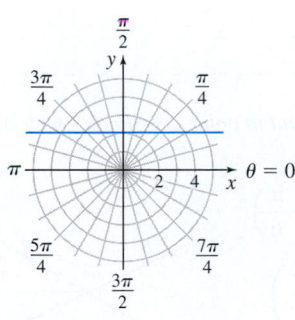

**Figure 40** $r\sin\theta = 2$ or $y = 2$

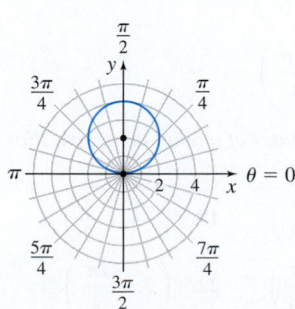

**Figure 41** $r = 4\sin\theta$ or $x^2 + (y - 2)^2 = 4$

NOW WORK Problems 65 and 67.

Table 2 summarizes and extends the results of Examples 5 and 6.

**TABLE 2**

| Description | Line passing through the pole making an angle $\alpha$ with the polar axis | Vertical line | Horizontal line |
|---|---|---|---|
| Rectangular equation | $y = (\tan \alpha)x$ | $x = a$ | $y = b$ |
| Polar equation | $\theta = \alpha$ | $r \cos \theta = a$ | $r \sin \theta = b$ |
| Typical graph | 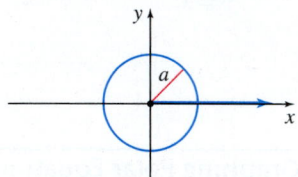 | 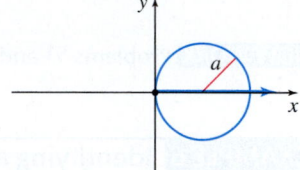 | 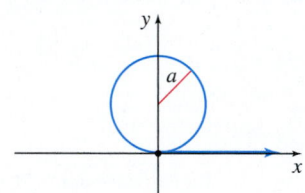 |

| Description | Circle, center at the pole, radius $a$ | Circle, passing through the pole, tangent to the line $\theta = \dfrac{\pi}{2}$, center on the polar axis, radius $a$ | Circle, passing through the pole, tangent to the polar axis, center on the line $\theta = \dfrac{\pi}{2}$, radius $a$ |
|---|---|---|---|
| Rectangular equation | $x^2 + y^2 = a^2, \quad a > 0$ | $x^2 + y^2 = \pm 2ax, \quad a > 0$ | $x^2 + y^2 = \pm 2ay, \quad a > 0$ |
| Polar equation | $r = a, \quad a > 0$ | $r = \pm 2a \cos \theta, \quad a > 0$ | $r = \pm 2a \sin \theta, \quad a > 0$ |
| Typical graph | | | |

## 9.4 Assess Your Understanding

### Concepts and Vocabulary

**1.** In a polar coordinate system, the origin is called the _____, and the _____ coincides with the positive $x$-axis of the rectangular coordinate system.

**2.** *True or False*  Another representation in polar coordinates for the point $\left(2, \dfrac{\pi}{3}\right)$ is $\left(2, \dfrac{4\pi}{3}\right)$.

**3.** *True or False*  In a polar coordinate system, each point in the plane has exactly one pair of polar coordinates.

**4.** *True or False*  In a rectangular coordinate system, each point in the plane has exactly one pair of rectangular coordinates.

**5.** *True or False*  In polar coordinates $(r, \theta)$, the number $r$ can be negative.

**6.** *True or False*  If $(r, \theta)$ are the polar coordinates of the point $P$, then $|r|$ is the distance of the point $P$ from the pole.

**7.** To convert the point $(r, \theta)$ in polar coordinates to a point $(x, y)$ in rectangular coordinates, use the formulas $x = $ _____ and $y = $ _____.

**8.** An equation whose variables are polar coordinates is called a(n) _____.

### Skill Building

*In Problems 9–16, match each point in polar coordinates with A, B, C, or D on the graph.*

**9.** $\left(2, -\dfrac{11\pi}{6}\right)$    **10.** $\left(-2, -\dfrac{\pi}{6}\right)$

**11.** $\left(-2, \dfrac{\pi}{6}\right)$    **12.** $\left(2, \dfrac{7\pi}{6}\right)$

**13.** $\left(2, \dfrac{5\pi}{6}\right)$    **14.** $\left(-2, \dfrac{5\pi}{6}\right)$

**15.** $\left(-2, \dfrac{7\pi}{6}\right)$    **16.** $\left(2, \dfrac{11\pi}{6}\right)$

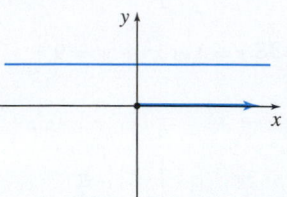

*In Problems 17–24, polar coordinates of a point are given. Plot each point in a polar coordinate system.*

**17.** $\left(4, \dfrac{\pi}{3}\right)$    **18.** $\left(-4, \dfrac{\pi}{3}\right)$    **19.** $\left(-4, -\dfrac{\pi}{3}\right)$

**20.** $\left(4, -\dfrac{\pi}{3}\right)$    **21.** $\left(\sqrt{2}, \dfrac{\pi}{4}\right)$    **22.** $\left(7, -\dfrac{7\pi}{4}\right)$

**23.** $\left(-6, \dfrac{4\pi}{3}\right)$    **24.** $\left(5, \dfrac{\pi}{2}\right)$

---

**1.** = NOW WORK problem         = Graphing technology recommended        CAS = Computer Algebra System recommended

*In Problems 25–32, polar coordinates of a point are given. Find other polar coordinates $(r, \theta)$ of the point for which:*

(a) $r > 0, -2\pi \leq \theta < 0$

(b) $r < 0, 0 \leq \theta < 2\pi$

(c) $r > 0, 2\pi \leq \theta < 4\pi$

**25.** $\left(5, \dfrac{2\pi}{3}\right)$    **26.** $\left(4, \dfrac{3\pi}{4}\right)$    **27.** $(-2, 3\pi)$

**28.** $(-3, 4\pi)$    **29.** $\left(1, \dfrac{\pi}{2}\right)$    **30.** $(2, \pi)$

**31.** $\left(-3, -\dfrac{\pi}{4}\right)$    **32.** $\left(-2, -\dfrac{2\pi}{3}\right)$

*In Problems 33–40, polar coordinates of a point are given. Find the rectangular coordinates of each point.*

**33.** $\left(6, \dfrac{\pi}{6}\right)$    **34.** $\left(-6, \dfrac{\pi}{6}\right)$    **35.** $\left(-6, -\dfrac{\pi}{6}\right)$

**36.** $\left(6, -\dfrac{\pi}{6}\right)$    **37.** $\left(5, \dfrac{\pi}{2}\right)$    **38.** $\left(8, \dfrac{\pi}{4}\right)$

**39.** $\left(2\sqrt{2}, -\dfrac{\pi}{4}\right)$    **40.** $\left(-5, -\dfrac{\pi}{3}\right)$

*In Problems 41–50, rectangular coordinates of a point are given. Plot the point. Find polar coordinates $(r, \theta)$ of each point for which $r > 0$ and $0 \leq \theta < 2\pi$.*

**41.** $(5, 0)$    **42.** $(2, -2)$    **43.** $(-2, 2)$    **44.** $(-2, -2\sqrt{3})$

**45.** $(\sqrt{3}, 1)$    **46.** $(0, -3)$    **47.** $(-\sqrt{3}, 1)$    **48.** $(3\sqrt{2}, -3\sqrt{2})$

**49.** $(3, 2)$    **50.** $(-6.5, 1.2)$

*In Problems 51–58, match each of the graphs (A) through (H) to one of the following polar equations.*

**51.** $r = 2$    **52.** $\theta = \dfrac{\pi}{4}$    **53.** $r = 2\cos\theta$

**54.** $r\cos\theta = 2$    **55.** $r = -2\cos\theta$    **56.** $r = 2\sin\theta$

**57.** $\theta = \dfrac{3\pi}{4}$    **58.** $r\sin\theta = 2$

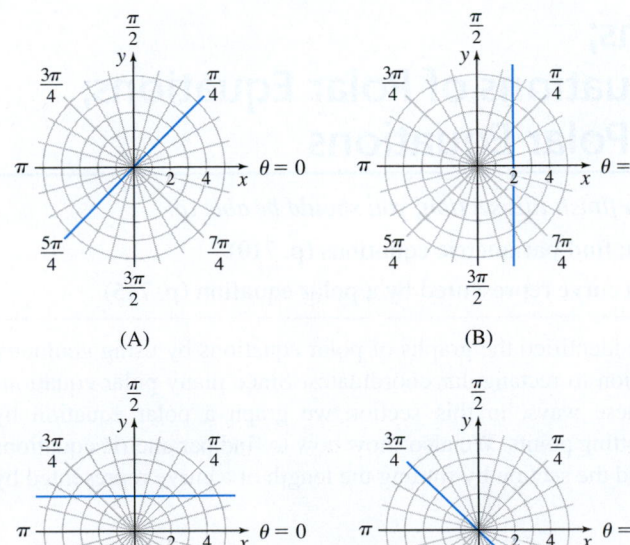

(A)    (B)

(C)    (D)

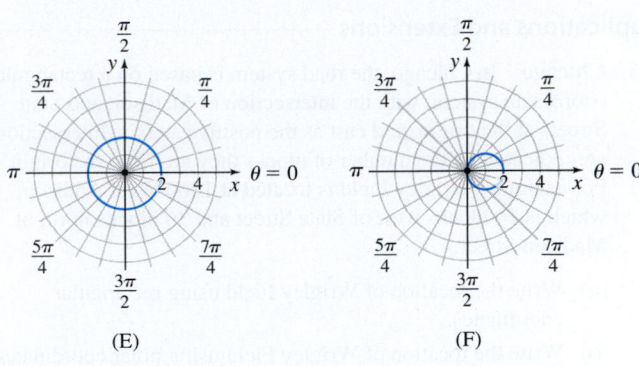

(E)    (F)

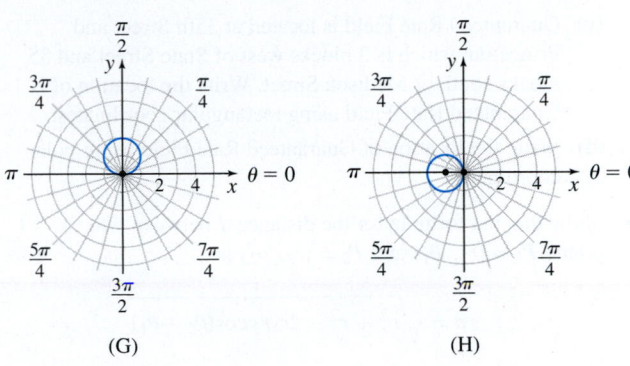

(G)    (H)

*In Problems 59–74, identify and graph each polar equation. Convert to a rectangular equation if necessary.*

**59.** $r = 4$    **60.** $r = 2$    **61.** $\theta = \dfrac{\pi}{3}$

**62.** $\theta = -\dfrac{\pi}{4}$    **63.** $r\sin\theta = 4$    **64.** $r\cos\theta = 4$

**65.** $r\cos\theta = -2$    **66.** $r\sin\theta = -2$    **67.** $r = 2\cos\theta$

**68.** $r = 2\sin\theta$    **69.** $r = -4\sin\theta$    **70.** $r = -4\cos\theta$

**71.** $r\sec\theta = 4$    **72.** $r\csc\theta = 8$    **73.** $r\csc\theta = -2$

**74.** $r\sec\theta = -4$

*In Problems 75–82, the letters $x$ and $y$ represent rectangular coordinates. Write each equation in polar coordinates $r$ and $\theta$.*

**75.** $\dfrac{x^2}{4} + \dfrac{y^2}{9} = 1$    **76.** $x - 4y + 4 = 0$

**77.** $x^2 + y^2 - 4x = 0$    **78.** $y = -6$

**79.** $x^2 = 1 - 4y$    **80.** $y^2 = 1 - 4x$

**81.** $xy = 1$    **82.** $x^2 + y^2 - 2x + 4y = 0$

*In Problems 83–94, the letters $r$ and $\theta$ represent polar coordinates. Write each equation in rectangular coordinates $x$ and $y$.*

**83.** $r = \cos\theta$    **84.** $r = 2 + \cos\theta$    **85.** $r^2 = \sin\theta$

**86.** $r^2 = 1 - \sin\theta$    **87.** $r = \dfrac{4}{1 - \cos\theta}$    **88.** $r = \dfrac{3}{3 - \cos\theta}$

**89.** $r^2 = \theta$    **90.** $\theta = -\dfrac{\pi}{4}$    **91.** $r = 2$

**92.** $r = -5$    **93.** $\tan\theta = 4$    **94.** $\cot\theta = 3$

## Applications and Extensions

**95. Chicago** In Chicago, the road system is based on a rectangular coordinate system, with the intersection of Madison and State Streets at the origin, and east as the positive $x$-axis. Intersections are indicated by the number of blocks they are from the origin. For example, Wrigley Field is located at 1060 West Addison, which is 10 blocks west of State Street and 36 blocks north of Madison Street.

  **(a)** Write the location of Wrigley Field using rectangular coordinates.

  **(b)** Write the location of Wrigley Field using polar coordinates. Use east as the polar axis.

  **(c)** Guaranteed Rate Field is located at 35th Street and Princeton, which is 3 blocks west of State Street and 35 blocks south of Madison Street. Write the location of Guaranteed Rate Field using rectangular coordinates.

  **(d)** Write the location of Guaranteed Rate Field using polar coordinates.

**96.** Show that the formula for the distance $d$ between two points $P_1 = (r_1, \theta_1)$ and $P_2 = (r_2, \theta_2)$ is

$$d = \sqrt{r_1^2 + r_2^2 - 2r_1 r_2 \cos(\theta_2 - \theta_1)}$$

**97. Horizontal Line** Show that the graph of the equation $r \sin \theta = a$ is a horizontal line $a$ units above the pole if $a > 0$, and $|a|$ units below the pole if $a < 0$.

**98. Vertical Line** Show that the graph of the equation $r \cos \theta = a$ is a vertical line $a$ units to the right of the pole if $a > 0$, and $|a|$ units to the left of the pole if $a < 0$.

**99. Circle** Show that the graph of the equation $r = 2a \sin \theta$, $a > 0$, is a circle of radius $a$ with its center at the rectangular coordinates $(0, a)$.

**100. Circle** Show that the graph of the equation $r = -2a \sin \theta$, $a > 0$, is a circle of radius $a$ with its center at the rectangular coordinates $(0, -a)$.

**101. Circle** Show that the graph of the equation $r = 2a \cos \theta$, $a > 0$, is a circle of radius $a$ with its center at the rectangular coordinates $(a, 0)$.

**102. Circle** Show that the graph of the equation $r = -2a \cos \theta$, $a > 0$, is a circle of radius $a$ with its center at the rectangular coordinates $(-a, 0)$.

**103. Exploring Using Technology**

  **(a)** Use a square screen to graph $r_1 = \sin \theta$, $r_2 = 2 \sin \theta$, and $r_3 = 3 \sin \theta$.

  **(b)** Describe how varying the constant $a$, $a > 0$, alters the graph of $r = a \sin \theta$.

  **(c)** Graph $r_1 = -\sin \theta$, $r_2 = -2 \sin \theta$, and $r_3 = -3 \sin \theta$.

  **(d)** Describe how varying the constant $a$, $a < 0$, alters the graph of $r = a \sin \theta$.

**104. Exploring Using Technology**

  **(a)** Use a square screen to graph $r_1 = \cos \theta$, $r_2 = 2 \cos \theta$, and $r_3 = 3 \cos \theta$.

  **(b)** Describe how varying the constant $a$, $a > 0$, alters the graph of $r = a \cos \theta$.

  **(c)** Graph $r_1 = -\cos \theta$, $r_2 = -2 \cos \theta$, and $r_3 = -3 \cos \theta$.

  **(d)** Describe how varying the constant $a$, $a < 0$, alters the graph of $r = a \cos \theta$.

## Challenge Problems

**105.** Show that $r = a \sin \theta + b \cos \theta$, $a$, $b$ not both zero, is the equation of a circle. Find the center and radius of the circle.

**106.** Express $r^2 = \cos(2\theta)$ in rectangular coordinates free of radicals.

**107.** Prove that the area of the triangle with vertices $(0, 0)$, $(r_1, \theta_1)$, $(r_2, \theta_2)$ is

$$A = \frac{1}{2} r_1 r_2 \, \sin(\theta_2 - \theta_1) \qquad 0 \le \theta_1 < \theta_2 \le \pi$$

# 9.5 Polar Equations; Parametric Equations of Polar Equations; Arc Length of Polar Equations

**OBJECTIVES** *When you finish this section, you should be able to:*

**1** Graph a polar equation; find parametric equations (p. 710)

**2** Find the arc length of a curve represented by a polar equation (p. 715)

In the previous section, we identified the graphs of polar equations by using geometry or by converting the equation to rectangular coordinates. Since many polar equations cannot be identified in these ways, in this section we graph a polar equation by constructing a table and plotting points. We also show how to find parametric equations for a polar equation. We end the section by finding the length of a curve represented by polar coordinates.

## 1 Graph a Polar Equation; Find Parametric Equations

**EXAMPLE 1** **Graphing a Polar Equation (Cardioid); Finding Parametric Equations**

  **(a)** Graph the polar equation $r = 1 - \sin \theta$, $0 \le \theta \le 2\pi$.

  **(b)** Find parametric equations for $r = 1 - \sin \theta$.

**TABLE 3**

| $\theta$ | $r = 1 - \sin\theta$ | $(r, \theta)$ |
|---|---|---|
| $0$ | $1 - 0 = 1$ | $(1, 0)$ |
| $\dfrac{\pi}{6}$ | $1 - \dfrac{1}{2} = \dfrac{1}{2}$ | $\left(\dfrac{1}{2}, \dfrac{\pi}{6}\right)$ |
| $\dfrac{\pi}{2}$ | $1 - 1 = 0$ | $\left(0, \dfrac{\pi}{2}\right)$ |
| $\dfrac{5\pi}{6}$ | $1 - \dfrac{1}{2} = \dfrac{1}{2}$ | $\left(\dfrac{1}{2}, \dfrac{5\pi}{6}\right)$ |
| $\pi$ | $1 - 0 = 1$ | $(1, \pi)$ |
| $\dfrac{7\pi}{6}$ | $1 - \left(-\dfrac{1}{2}\right) = \dfrac{3}{2}$ | $\left(\dfrac{3}{2}, \dfrac{7\pi}{6}\right)$ |
| $\dfrac{3\pi}{2}$ | $1 - (-1) = 2$ | $\left(2, \dfrac{3\pi}{2}\right)$ |
| $\dfrac{11\pi}{6}$ | $1 - \left(-\dfrac{1}{2}\right) = \dfrac{3}{2}$ | $\left(\dfrac{3}{2}, \dfrac{11\pi}{6}\right)$ |
| $2\pi$ | $1 - 0 = 1$ | $(1, 2\pi)$ |

**NOTE** Graphs of polar equations of the form

| $r = a(1 + \cos\theta)$ | $r = a(1 + \sin\theta)$ |
|---|---|
| $r = a(1 - \cos\theta)$ | $r = a(1 - \sin\theta)$ |

where $a > 0$, are called **cardioids**. A cardioid contains the pole and is heart-shaped (giving the curve its name).

**NOTE** Limaçon (pronounced "leema sown") is a French word for "snail."

**TABLE 4**

| $\theta$ | $r = 3 + 2\cos\theta$ | $(r, \theta)$ |
|---|---|---|
| $0$ | $3 + 2(1) = 5$ | $(5, 0)$ |
| $\dfrac{\pi}{3}$ | $3 + 2\left(\dfrac{1}{2}\right) = 4$ | $\left(4, \dfrac{\pi}{3}\right)$ |
| $\dfrac{\pi}{2}$ | $3 + 2(0) = 3$ | $\left(3, \dfrac{\pi}{2}\right)$ |
| $\dfrac{2\pi}{3}$ | $3 + 2\left(-\dfrac{1}{2}\right) = 2$ | $\left(2, \dfrac{2\pi}{3}\right)$ |
| $\pi$ | $3 + 2(-1) = 1$ | $(1, \pi)$ |
| $\dfrac{4\pi}{3}$ | $3 + 2\left(-\dfrac{1}{2}\right) = 2$ | $\left(2, \dfrac{4\pi}{3}\right)$ |
| $\dfrac{3\pi}{2}$ | $3 + 2(0) = 3$ | $\left(3, \dfrac{3\pi}{2}\right)$ |
| $\dfrac{5\pi}{3}$ | $3 + 2\left(\dfrac{1}{2}\right) = 4$ | $\left(4, \dfrac{5\pi}{3}\right)$ |
| $2\pi$ | $3 + 2(1) = 5$ | $(5, 2\pi)$ |

**Solution (a)** The polar equation $r = 1 - \sin\theta$ contains $\sin\theta$, which has period $2\pi$. We construct Table 3 using common values of $\theta$ that range from 0 to $2\pi$, plot the points $(r, \theta)$, and trace out the graph, beginning at the point $(1, 0)$ and ending at the point $(1, 2\pi)$, as shown in Figure 42(a). Figure 42(b) shows the graph using technology.

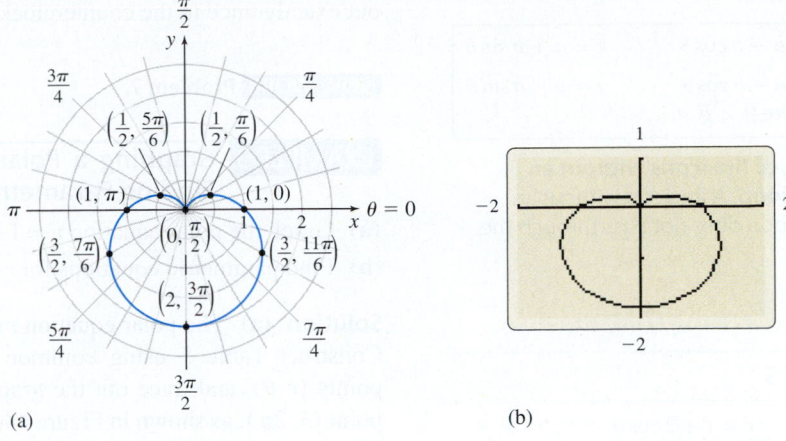

**DF Figure 42** The cardioid $r = 1 - \sin\theta$.

**(b)** We obtain parametric equations for $r = 1 - \sin\theta$ by using the conversion formulas $x = r\cos\theta$ and $y = r\sin\theta$:

$$x = r\cos\theta = (1 - \sin\theta)\cos\theta \qquad y = r\sin\theta = (1 - \sin\theta)\sin\theta$$

Here $\theta$ is the parameter, and if $0 \le \theta \le 2\pi$, then the graph starts at $(1, 0)$ and is traced out exactly once in the counterclockwise direction. ∎

**NOW WORK** Problem **5**.

**EXAMPLE 2**  **Graphing a Polar Equation (Limaçon Without an Inner Loop); Finding Parametric Equations**

**(a)** Graph the polar equation $r = 3 + 2\cos\theta$, $0 \le \theta \le 2\pi$.

**(b)** Find parametric equations for $r = 3 + 2\cos\theta$.

**Solution (a)** The polar equation $r = 3 + 2\cos\theta$ contains $\cos\theta$, which has period $2\pi$. We construct Table 4 using common values of $\theta$ that range from 0 to $2\pi$, plot the points $(r, \theta)$, and trace out the graph, beginning at the point $(5, 0)$ and ending at the point $(5, 2\pi)$, as shown in Figure 43(a). Figure 43(b) shows the graph using technology.

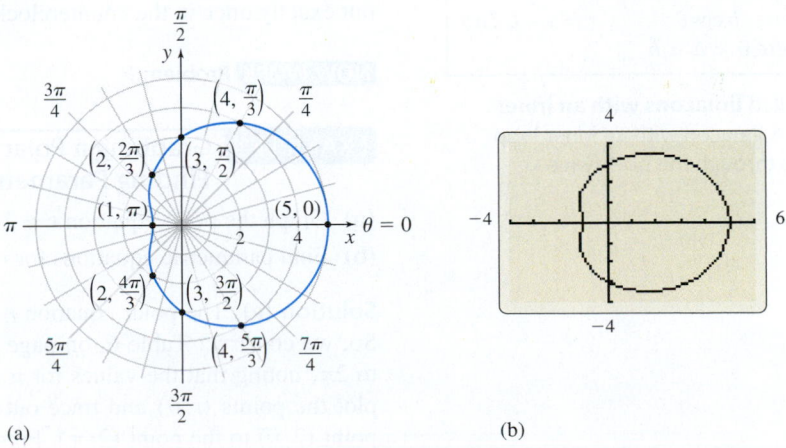

**Figure 43** The limaçon $r = 3 + 2\cos\theta$.

**(b)** We obtain parametric equations for $r = 3 + 2\cos\theta$ by using the conversion formulas $x = r\cos\theta$ and $y = r\sin\theta$:

$$x = r\cos\theta = (3 + 2\cos\theta)\cos\theta \qquad y = r\sin\theta = (3 + 2\cos\theta)\sin\theta$$

Here $\theta$ is the parameter, and if $0 \le \theta \le 2\pi$, then the graph starts at $(5, 0)$ and is traced out exactly once in the counterclockwise direction. ∎

**NOTE** Graphs of polar equations of the form

| $r = a + b\cos\theta$ | $r = a + b\sin\theta$ |
|---|---|
| $r = a - b\cos\theta$ | $r = a - b\sin\theta$ |
| where $0 < b < a$ | |

are called **limaçons without an inner loop**. A limaçon without an inner loop does not pass through the pole.

**NOW WORK** Problem 7.

**EXAMPLE 3** **Graphing a Polar Equation (Limaçon with an Inner Loop); Finding Parametric Equations**

**(a)** Graph the polar equation $r = 1 + 2\cos\theta$, $0 \le \theta \le 2\pi$.

**(b)** Find parametric equations for $r = 1 + 2\cos\theta$.

**Solution** **(a)** The polar equation $r = 1 + 2\cos\theta$ contains $\cos\theta$, which has period $2\pi$. Construct Table 5 using common values of $\theta$ that range from 0 to $2\pi$, plot the points $(r, \theta)$, and trace out the graph, beginning at the point $(3, 0)$ and ending at the point $(3, 2\pi)$, as shown in Figure 44(a). Figure 44(b) shows the graph using technology.

**TABLE 5**

| $\theta$ | $r = 1 + 2\cos\theta$ | $(r, \theta)$ |
|---|---|---|
| $0$ | $1 + 2(1) = 3$ | $(3, 0)$ |
| $\dfrac{\pi}{3}$ | $1 + 2\left(\dfrac{1}{2}\right) = 2$ | $\left(2, \dfrac{\pi}{3}\right)$ |
| $\dfrac{\pi}{2}$ | $1 + 2(0) = 1$ | $\left(1, \dfrac{\pi}{2}\right)$ |
| $\dfrac{2\pi}{3}$ | $1 + 2\left(-\dfrac{1}{2}\right) = 0$ | $\left(0, \dfrac{2\pi}{3}\right)$ |
| $\pi$ | $1 + 2(-1) = -1$ | $(-1, \pi)$ |
| $\dfrac{4\pi}{3}$ | $1 + 2\left(-\dfrac{1}{2}\right) = 0$ | $\left(0, \dfrac{4\pi}{3}\right)$ |
| $\dfrac{3\pi}{2}$ | $1 + 2(0) = 1$ | $\left(1, \dfrac{3\pi}{2}\right)$ |
| $\dfrac{5\pi}{3}$ | $1 + 2\left(\dfrac{1}{2}\right) = 2$ | $\left(2, \dfrac{5\pi}{3}\right)$ |
| $2\pi$ | $1 + 2(1) = 3$ | $(3, 2\pi)$ |

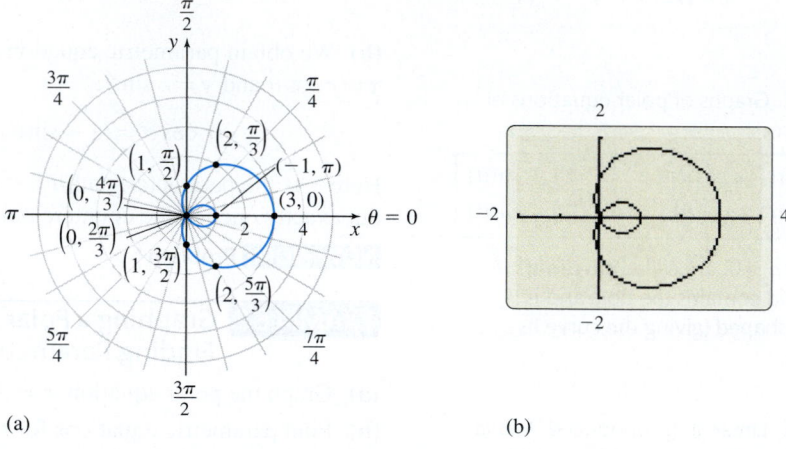

**Figure 44** The limaçon $r = 1 + 2\cos\theta$.

**(b)** We obtain parametric equations for $r = 1 + 2\cos\theta$ by using the conversion formulas $x = r\cos\theta$ and $y = r\sin\theta$:

$$x = r\cos\theta = (1 + 2\cos\theta)\cos\theta \qquad y = r\sin\theta = (1 + 2\cos\theta)\sin\theta$$

Here $\theta$ is the parameter, and if $0 \le \theta \le 2\pi$, then the graph starts at $(3, 0)$ and is traced out exactly once in the counterclockwise direction. ∎

**NOTE** Graphs of polar equations of the form

| $r = a + b\cos\theta$ | $r = a + b\sin\theta$ |
|---|---|
| $r = a - b\cos\theta$ | $r = a - b\sin\theta$ |
| where $0 < a < b$ | |

are called **limaçons with an inner loop**. A limaçon with an inner loop passes through the pole twice.

**NOW WORK** Problem 9.

**EXAMPLE 4** **Graphing a Polar Equation (Rose); Finding Parametric Equations**

**(a)** Graph the polar equation $r = 2\cos(2\theta)$, $0 \le \theta \le 2\pi$.

**(b)** Find parametric equations for $r = 2\cos(2\theta)$.

**Solution** **(a)** The polar equation $r = 2\cos(2\theta)$ contains $\cos(2\theta)$, which has period $\pi$. So, we construct Table 6, on page 713, using common values of $\theta$ that range from 0 to $2\pi$, noting that the values for $\pi \le \theta \le 2\pi$ repeat the values for $0 \le \theta \le \pi$. Then we plot the points $(r, \theta)$ and trace out the graph. Figure 45(a) shows the graph from the point $(2, 0)$ to the point $(2, \pi)$. Figure 45(b) completes the graph from the point $(2, \pi)$ to the point $(2, 2\pi)$.

**TABLE 6**

| $\theta$ | $r = 2\cos(2\theta)$ | $(r, \theta)$ | $\theta$ | $r = 2\cos(2\theta)$ | $(r, \theta)$ |
|---|---|---|---|---|---|
| $0$ | $2(1) = 2$ | $(2, 0)$ | | | |
| $\dfrac{\pi}{6}$ | $2\left(\dfrac{1}{2}\right) = 1$ | $\left(1, \dfrac{\pi}{6}\right)$ | $\dfrac{7\pi}{6}$ | $2\left(\dfrac{1}{2}\right) = 1$ | $\left(1, \dfrac{7\pi}{6}\right)$ |
| $\dfrac{\pi}{4}$ | $2(0) = 0$ | $\left(0, \dfrac{\pi}{4}\right)$ | $\dfrac{5\pi}{4}$ | $2(0) = 0$ | $\left(0, \dfrac{5\pi}{4}\right)$ |
| $\dfrac{\pi}{3}$ | $2\left(-\dfrac{1}{2}\right) = -1$ | $\left(-1, \dfrac{\pi}{3}\right)$ | $\dfrac{4\pi}{3}$ | $2\left(-\dfrac{1}{2}\right) = -1$ | $\left(-1, \dfrac{4\pi}{3}\right)$ |
| $\dfrac{\pi}{2}$ | $2(-1) = -2$ | $\left(-2, \dfrac{\pi}{2}\right)$ | $\dfrac{3\pi}{2}$ | $2(-1) = -2$ | $\left(-2, \dfrac{3\pi}{2}\right)$ |
| $\dfrac{2\pi}{3}$ | $2\left(-\dfrac{1}{2}\right) = -1$ | $\left(-1, \dfrac{2\pi}{3}\right)$ | $\dfrac{5\pi}{3}$ | $2\left(-\dfrac{1}{2}\right) = -1$ | $\left(-1, \dfrac{5\pi}{3}\right)$ |
| $\dfrac{3\pi}{4}$ | $2(0) = 0$ | $\left(0, \dfrac{3\pi}{4}\right)$ | $\dfrac{7\pi}{4}$ | $2(0) = 0$ | $\left(0, \dfrac{7\pi}{4}\right)$ |
| $\dfrac{5\pi}{6}$ | $2\left(\dfrac{1}{2}\right) = 1$ | $\left(1, \dfrac{5\pi}{6}\right)$ | $\dfrac{11\pi}{6}$ | $2\left(\dfrac{1}{2}\right) = 1$ | $\left(1, \dfrac{11\pi}{6}\right)$ |
| $\pi$ | $2(1) = 2$ | $(2, \pi)$ | $2\pi$ | $2(1) = 2$ | $(2, 2\pi)$ |

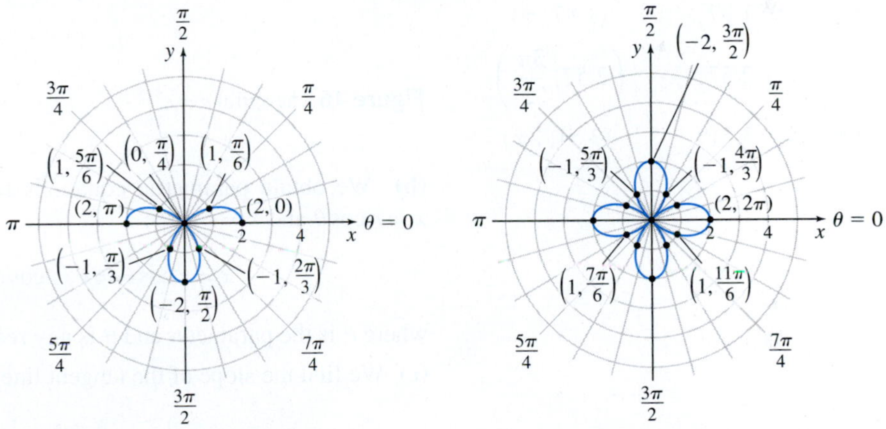

(a) $r = 2\cos(2\theta),\ 0 \le \theta \le \pi$

(b) $r = 2\cos(2\theta),\ 0 \le \theta \le 2\pi$

**Figure 45** A rose with four petals.

**NOTE** Graphs of polar equations of the form $r = a\cos(n\theta)$ or $r = a\sin(n\theta)$, $a > 0$, $n$ an integer, are called **roses**. If $n$ is an even integer, the rose has $2n$ petals. If $n$ is an odd integer, the rose has $n$ petals.

**(b)** Parametric equations for $r = 2\cos(2\theta)$ are

$$x = r\cos\theta = 2\cos(2\theta)\cos\theta \qquad y = r\sin\theta = 2\cos(2\theta)\sin\theta$$

where $\theta$ is the parameter. If $0 \le \theta \le 2\pi$, then the graph starts at $(2, 0)$ and is traced out exactly once in the counterclockwise direction. ∎

**NOW WORK** Problem 11.

**NOTE** Graphs of polar equations of the form $r = e^{\theta/a}$, $a > 0$, are called **logarithmic spirals**, since the equation can be written as $\theta = a \ln r$. A logarithmic spiral spirals infinitely both toward the pole and away from it.

**CALC**
**CLIP**

**EXAMPLE 5** **Graphing a Polar Equation (Spiral); Finding Parametric Equations and Tangent Lines**

**(a)** Graph the equation $r = e^{\theta/5}$.

**(b)** Find parametric equations for $r = e^{\theta/5}$.

**(c)** Find an equation of the tangent line to the spiral at $\theta = 0$.

**Solution (a)** The polar equation $r = e^{\theta/5}$ lacks the symmetry you may have observed in the previous examples. Since there is no number $\theta$ for which $r = 0$, the graph does not contain the pole. Also observe that:

- $r$ is positive for all $\theta$.
- $r$ increases as $\theta$ increases.
- $r \to 0$ as $\theta \to -\infty$.
- $r \to \infty$ as $\theta \to \infty$.

Use a calculator to construct Table 7. Figure 46(a) shows part of the graph $r = e^{\theta/5}$. Figure 46(b) shows the graph for $\theta = -\dfrac{3\pi}{2}$ to $\theta = 2\pi$ using technology.

**TABLE 7**

| $\theta$ | $r = e^{\theta/5}$ | $(r, \theta)$ |
|---|---|---|
| $-\dfrac{3\pi}{2}$ | 0.39 | $\left(0.39, -\dfrac{3\pi}{2}\right)$ |
| $-\pi$ | 0.53 | $(0.53, -\pi)$ |
| $-\dfrac{\pi}{2}$ | 0.73 | $\left(0.73, -\dfrac{\pi}{2}\right)$ |
| $-\dfrac{\pi}{4}$ | 0.85 | $\left(0.85, -\dfrac{\pi}{4}\right)$ |
| $0$ | 1 | $(1, 0)$ |
| $\dfrac{\pi}{4}$ | 1.17 | $\left(1.17, \dfrac{\pi}{4}\right)$ |
| $\dfrac{\pi}{2}$ | 1.37 | $\left(1.37, \dfrac{\pi}{2}\right)$ |
| $\pi$ | 1.87 | $(1.87, \pi)$ |
| $\dfrac{3\pi}{2}$ | 2.57 | $\left(2.57, \dfrac{3\pi}{2}\right)$ |
| $2\pi$ | 3.51 | $(3.51, 2\pi)$ |

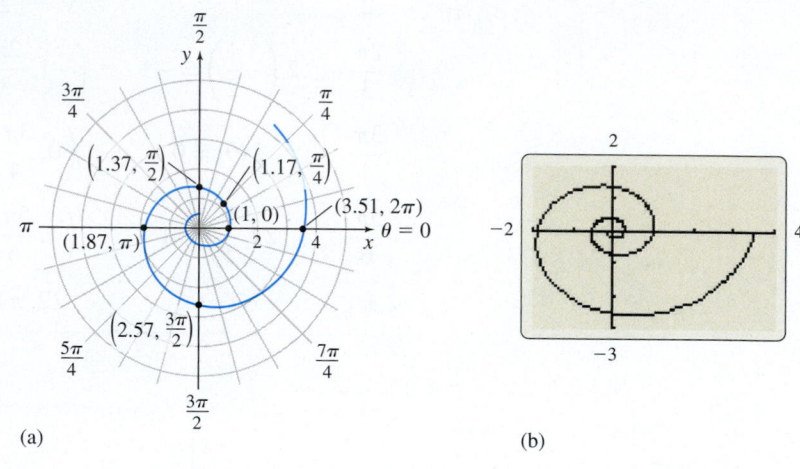

(a)    (b)

**Figure 46** The spiral $r = e^{\theta/5}$.

**(b)** We obtain parametric equations for $r = e^{\theta/5}$ by using the conversion formulas $x = r \cos \theta$ and $y = r \sin \theta$:

$$x = r \cos \theta = e^{\theta/5} \cos \theta \qquad y = r \sin \theta = e^{\theta/5} \sin \theta$$

where $\theta$ is the parameter and $\theta$ is any real number.

**(c)** We find the slope of the tangent line using the parametric equations from (b).

$$x = e^{\theta/5} \cos \theta \qquad y = e^{\theta/5} \sin \theta$$

Then

$$\frac{dx}{d\theta} = \frac{d}{d\theta}(e^{\theta/5} \cos \theta) = \frac{1}{5} e^{\theta/5} \cos \theta - e^{\theta/5} \sin \theta$$

$$\frac{dy}{d\theta} = \frac{d}{d\theta}(e^{\theta/5} \sin \theta) = \frac{1}{5} e^{\theta/5} \sin \theta + e^{\theta/5} \cos \theta$$

and

$$\frac{dy}{dx} = \frac{\dfrac{1}{5} e^{\theta/5} \sin \theta + e^{\theta/5} \cos \theta}{\dfrac{1}{5} e^{\theta/5} \cos \theta - e^{\theta/5} \sin \theta} = \frac{\dfrac{1}{5} \sin \theta + \cos \theta}{\dfrac{1}{5} \cos \theta - \sin \theta} = \frac{\sin \theta + 5 \cos \theta}{\cos \theta - 5 \sin \theta}$$

At $\theta = 0$, the slope of the tangent line is 5. Since $x = 1$ and $y = 0$ at $\theta = 0$, an equation of the tangent line to the spiral is $y = 5(x - 1) = 5x - 5$. ∎

**NOW WORK** Problem **27.**

**TABLE 8** Library of Polar Equations

| Name | Cardioid | Limaçon without inner loop | Limaçon with inner loop |
|---|---|---|---|
| Polar equations | $r = a \pm a \cos\theta, \quad a > 0$ <br> $r = a \pm a \sin\theta, \quad a > 0$ | $r = a \pm b \cos\theta, \quad 0 < b < a$ <br> $r = a \pm b \sin\theta, \quad 0 < b < a$ | $r = a \pm b \cos\theta, \quad 0 < a < b$ <br> $r = a \pm b \sin\theta, \quad 0 < a < b$ |
| Typical graph | | | |

| Name | Lemniscate (See p. 717) | Rose with three petals | Rose with four petals |
|---|---|---|---|
| Polar equations | $r^2 = a^2 \cos(2\theta), \quad a > 0$ <br> $r^2 = a^2 \sin(2\theta), \quad a > 0$ | $r = a \sin(3\theta), \quad a > 0$ <br> $r = a \cos(3\theta), \quad a > 0$ | $r = a \sin(2\theta), \quad a > 0$ <br> $r = a \cos(2\theta), \quad a > 0$ |
| Typical graph | | | |

## 2 Find the Arc Length of a Curve Represented by a Polar Equation

Suppose a curve $C$ is represented by the polar equation $r = f(\theta)$, $\alpha \le \theta \le \beta$, where both $f$ and its derivative $f'(\theta) = \dfrac{dr}{d\theta}$ are continuous on an interval containing $\alpha$ and $\beta$. Using $\theta$ as the parameter, parametric equations for the curve $C$ are

$$x(\theta) = r\cos\theta = f(\theta)\cos\theta \qquad y(\theta) = r\sin\theta = f(\theta)\sin\theta$$

Then

$$\frac{dx}{d\theta} = -f(\theta)\sin\theta + f'(\theta)\cos\theta \qquad \frac{dy}{d\theta} = f(\theta)\cos\theta + f'(\theta)\sin\theta$$

After simplification,

$$\left(\frac{dx}{d\theta}\right)^2 + \left(\frac{dy}{d\theta}\right)^2 = [f(\theta)]^2 + [f'(\theta)]^2 = r^2 + \left(\frac{dr}{d\theta}\right)^2 \qquad r = f(\theta)$$

Since we are using parametric equations, the length $s$ of $C$ from $\theta = \alpha$ to $\theta = \beta$ is

$$s = \int_{\alpha}^{\beta} \sqrt{\left(\frac{dx}{d\theta}\right)^2 + \left(\frac{dy}{d\theta}\right)^2}\, d\theta = \int_{\alpha}^{\beta} \sqrt{r^2 + \left(\frac{dr}{d\theta}\right)^2}\, d\theta$$

**THEOREM  Arc Length of the Graph of a Polar Equation**

If a curve $C$ is represented by the polar equation $r = f(\theta)$, $\alpha \le \theta \le \beta$, and if $f'(\theta) = \dfrac{dr}{d\theta}$ is continuous on an interval containing $\alpha$ and $\beta$, then the arc length $s$ of $C$ from $\theta = \alpha$ to $\theta = \beta$ is

$$s = \int_{\alpha}^{\beta} \sqrt{r^2 + \left(\frac{dr}{d\theta}\right)^2}\, d\theta$$

EXAMPLE 6   **Finding the Arc Length of a Logarithmic Spiral**

Find the arc length $s$ of the logarithmic spiral represented by $r = f(\theta) = e^{3\theta}$ from $\theta = 0$ to $\theta = 2$.

**Solution**   We use the arc length formula $s = \displaystyle\int_{\alpha}^{\beta} \sqrt{r^2 + \left(\dfrac{dr}{d\theta}\right)^2}\, d\theta$ with $r = e^{3\theta}$.

Then $\dfrac{dr}{d\theta} = 3e^{3\theta}$ and

$$s = \int_{0}^{2} \sqrt{(e^{3\theta})^2 + (3e^{3\theta})^2}\, d\theta = \int_{0}^{2} \sqrt{10e^{6\theta}}\, d\theta = \sqrt{10} \int_{0}^{2} e^{3\theta}\, d\theta$$

$$= \sqrt{10}\left[\frac{e^{3\theta}}{3}\right]_{0}^{2} = \frac{\sqrt{10}}{3}(e^6 - 1)$$

NOW WORK   Problem 19.

# 9.5 Assess Your Understanding

## Concepts and Vocabulary

**1.** *True or False*   A cardioid passes through the pole.

**2.** *Multiple Choice*   The equations for cardioids and limaçons are very similar. They all have the form

$$r = a \pm b\cos\theta \quad \text{or} \quad r = a \pm b\sin\theta, \quad a > 0, b > 0$$

The equations represent a limaçon with an inner loop if [(a) $a < b$ (b) $a > b$ (c) $a = b$];
a cardioid if [(a) $a < b$ (b) $a > b$ (c) $a = b$];
a limaçon without an inner loop if [(a) $a < b$ (b) $a > b$ (c) $a = b$].

**3.** *True or False*   The graph of $r = \sin(4\theta)$ is a rose.

**4.** The rose $r = \cos(3\theta)$ has _____ petals.

## Skill Building

*In Problems 5–12, for each polar equation:*
*(a) Graph the equation.*
*(b) Find parametric equations that represent the equation.*

**5.** $r = 2 + 2\cos\theta$

**6.** $r = 3 - 3\sin\theta$

**7.** $r = 4 - 2\cos\theta$

**8.** $r = 2 + \sin\theta$

**9.** $r = 1 + 2\sin\theta$

**10.** $r = 2 - 3\cos\theta$

**11.** $r = \sin(3\theta)$

**12.** $r = 4\cos(4\theta)$

*In Problems 13–18, graph each pair of polar equations on the same polar grid. Find polar coordinates of the point(s) of intersection and label the point(s) on the graph.*

**13.** $r = 8\cos\theta, \quad r = 2\sec\theta$

**14.** $r = 8\sin\theta, \quad r = 4\csc\theta$

**15.** $r = \sin\theta, \quad r = 1 + \cos\theta$

**16.** $r = 3, \quad r = 2 + 2\cos\theta$

**17.** $r = 1 + \sin\theta, \quad r = 1 + \cos\theta$

**18.** $r = 1 + \cos\theta, \quad r = 3\cos\theta$

*In Problems 19–22, find the arc length of each curve.*

**19.** $r = f(\theta) = e^{\theta/2}$ from $\theta = 0$ to $\theta = 2$

**20.** $r = f(\theta) = e^{2\theta}$ from $\theta = 0$ to $\theta = 2$

**21.** $r = f(\theta) = \cos^2\dfrac{\theta}{2}$ from $\theta = 0$ to $\theta = \pi$

**22.** $r = f(\theta) = \sin^2\dfrac{\theta}{2}$ from $\theta = 0$ to $\theta = \pi$

## Applications and Extensions

*In Problems 23–26, the polar equation for each graph is either $r = a \pm b\cos\theta$ or $r = a \pm b\sin\theta$, $a > 0, b > 0$. Select the correct equation and find the values of a and b.*

**23.**

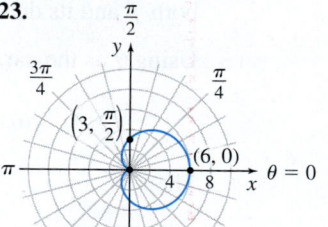

**24.**

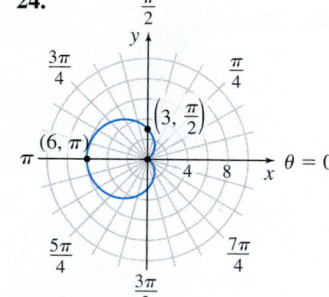

**25.**

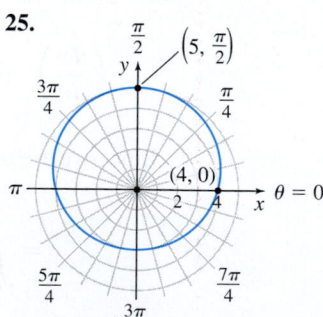

**26.**

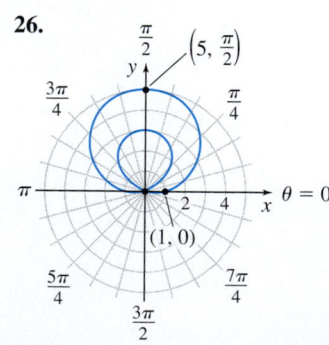

*In Problems 27–32, find an equation of the tangent line to each curve at the given value of $\theta$. Hint: Find parametric equations that represent each polar equation.*

**27.** $r = 2\cos(3\theta)$ at $\theta = \dfrac{\pi}{6}$

**28.** $r = 3\sin(3\theta)$ at $\theta = \dfrac{\pi}{3}$

**29.** $r = 2 + \cos\theta$ at $\theta = \dfrac{\pi}{4}$

**30.** $r = 3 - \sin\theta$ at $\theta = \dfrac{\pi}{6}$

**31.** $r = 4 + 5\sin\theta$ at $\theta = \dfrac{\pi}{4}$

**32.** $r = 1 - 2\cos\theta$ at $\theta = \dfrac{\pi}{4}$

---

**1.** = NOW WORK problem          ⟪∿⟫ = Graphing technology recommended          CAS = Computer Algebra System recommended

**Lemniscates**   Graphs of polar equations of the form $r^2 = a^2 \cos(2\theta)$ or $r^2 = a^2 \sin(2\theta)$, where $a \neq 0$, are called **lemniscates**. A lemniscate passes through the pole twice and is shaped like the infinity symbol ∞.

*In Problems 33–36, for each equation:*
*(a) Graph the lemniscate.*
*(b) Find parametric equations that represent the equation.*

**33.** $r^2 = 4 \sin(2\theta)$          **34.** $r^2 = 9 \cos(2\theta)$

**35.** $r^2 = \cos(2\theta)$          **36.** $r^2 = 16 \sin(2\theta)$

*In Problems 37–48:*
*(a) Graph each polar equation.*
*(b) Find parametric equations that represent each equation.*

**37.** $r = \dfrac{2}{1 - \cos\theta}$ (parabola)     **38.** $r = \dfrac{1}{1 - \cos\theta}$ (parabola)

**39.** $r = \dfrac{1}{3 - 2\cos\theta}$ (ellipse)     **40.** $r = \dfrac{2}{1 - 2\cos\theta}$ (hyperbola)

**41.** $r = \theta;\ \theta \geq 0$ (spiral of Archimedes)

**42.** $r = \dfrac{3}{\theta};\ \theta > 0$ (reciprocal spiral)

**43.** $r = \csc\theta - 2;\ \ 0 < \theta < \pi$ (conchoid)

**44.** $r = 3 - \dfrac{1}{2}\csc\theta$ (conchoid)

**45.** $r = \sin\theta\tan\theta$ (cissoid)     **46.** $r = \cos\dfrac{\theta}{2}$

**47.** $r = \tan\theta$ (kappa curve)     **48.** $r = \cot\theta$ (kappa curve)

**49.** Show that $r = 4(\cos\theta + 1)$ and $r = 4(\cos\theta - 1)$ have the same graph.

**50.** Show that $r = 5(\sin\theta + 1)$ and $r = 5(\sin\theta - 1)$ have the same graph.

**51. Arc Length**   Find the arc length of the spiral $r = \theta$ from $\theta = 0$ to $\theta = 2\pi$.

**52. Arc Length**   Find the arc length of the spiral $r = 3\theta$ from $\theta = 0$ to $\theta = 2\pi$.

**53. Perimeter**   Find the perimeter of the cardioid
$$r = f(\theta) = 1 - \cos\theta \quad -\pi \leq \theta \leq \pi$$

**54. Exploring Using Technology**

(a) Graph $r_1 = 2\cos(4\theta)$. Clear the screen and graph $r_2 = 2\cos(6\theta)$. How many petals does each of the graphs have?

(b) Clear the screen and graph, in order, each on a clear screen, $r_1 = 2\cos(3\theta)$, $r_2 = 2\cos(5\theta)$, and $r_3 = 2\cos(7\theta)$. What do you notice about the number of petals? Do the results support the definition of a rose?

**55. Exploring Using Technology**   Graph $r_1 = 3 - 2\cos\theta$. Clear the screen and graph $r_2 = 3 + 2\cos\theta$. Clear the screen and graph $r_3 = 3 + 2\sin\theta$. Clear the screen and graph $r_4 = 3 - 2\sin\theta$. Describe the pattern.

**56. Horizontal and Vertical Tangent Lines**   Find the horizontal and vertical tangent lines of the cardioid $r = 1 - \sin\theta$ discussed in Example 1.

**57. Horizontal and Vertical Tangent Lines**   Find the horizontal and vertical tangent lines of the cardioid $r = 3 + 3\cos\theta$.

**CAS  58. Horizontal and Vertical Tangent Lines**   Find the horizontal and vertical tangent lines of the limaçon with an inner loop $r = 1 + 2\cos\theta$ discussed in Example 3.

**CAS  59. Horizontal and Vertical Tangent Lines**   Find the horizontal and vertical tangent lines of the rose with four petals $r = 2\cos(2\theta)$, $0 \leq \theta \leq 2\pi$, discussed in Example 4.

**CAS  60. Horizontal and Vertical Tangent Lines**   Find the horizontal and vertical tangent lines of the spiral $r = e^{\theta/5}$ discussed in Example 5.

**CAS  61. Horizontal and Vertical Tangent Lines**   Find the horizontal and vertical tangent lines of the lemniscate $r^2 = 4\sin(2\theta)$.

**62. Test for Symmetry**   Symmetry with respect to the polar axis can be tested by replacing $\theta$ with $-\theta$. If an equivalent equation results, the graph is symmetric with respect to the polar axis.

(a) Explain why this test is valid.

(b) Use the test to show that $r = 3 + 2\cos\theta$ is symmetric with respect to the polar axis.

**63. Test for Symmetry**   Symmetry with respect to the pole can be tested by replacing $r$ by $-r$ or by replacing $\theta$ by $\theta + \pi$. If either substitution produces an equivalent equation, the graph is symmetric with respect to the pole.

(a) Explain why these tests are valid.

(b) Show that $r^2 = 4\sin(2\theta)$ is symmetric with respect to the pole.

**64. Test for Symmetry**   Symmetry with respect to the line $\theta = \dfrac{\pi}{2}$ can be tested by replacing $\theta$ by $\pi - \theta$. If an equivalent equation results, the graph is symmetric with respect to the line $\theta = \dfrac{\pi}{2}$.

(a) Explain why this test is valid.

(b) Use the test to show that $r = 2\cos(2\theta)$ is symmetric with respect to the line $\theta = \dfrac{\pi}{2}$.

## Challenge Problems

**Tests for Symmetry**   The three tests for symmetry described in Problems 62–64 are *sufficient* conditions for symmetry, but they are not *necessary* conditions. That is, an equation may fail these tests and still have a graph that is symmetric with respect to the polar axis, the line $\theta = \dfrac{\pi}{2}$, or the pole. See Problem 65.

**65. Testing for Symmetry**   The graph of $r = \sin(2\theta)$ (a rose with four petals) is symmetric with respect to the polar axis, the line $\theta = \dfrac{\pi}{2}$, and the pole. Show that the test for symmetry with respect to the pole (see Problem 63) works, but the test for symmetry with respect to the polar axis fails (see Problem 62).

**66. Arc Length**   Find the entire arc length of the curve $r = a\sin^3\dfrac{\theta}{3}$, $a > 0$. *Hint:* Use parametric equations.

**67. Arc Length of a Rose Petal**

(a) Graph the rose $r = 4\sin(5\theta)$.

(b) The petal in the first quadrant begins when $\theta = 0$ and ends when $\theta = \alpha$. Find $\alpha$.

(c) Find the length $s$ of the petal described in (b).

# 9.6 Area in Polar Coordinates

**OBJECTIVES** *When you finish this section, you should be able to:*

1 Find the area of a region enclosed by the graph of a polar equation (p. 718)
2 Find the area of a region enclosed by the graphs of two polar equations (p. 720)
3 Find the surface area of a solid of revolution obtained from the graph of a polar equation (p. 721)

**RECALL** The area $A$ of the sector of a circle of radius $r$ formed by a central angle of $\theta$ radians is $A = \dfrac{1}{2}r^2\theta$.

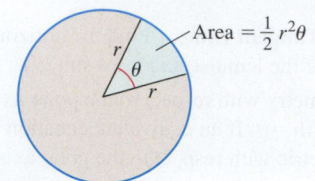

**Figure 47**

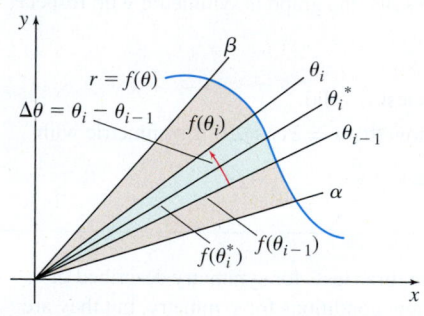

**Figure 48**

**DF Figure 49**

**NEED TO REVIEW?** The definite integral is discussed in Section 5.2, pp. 394–400.

In this section, we find the area of a region enclosed by the graph of a polar equation and two rays that have the pole as a common vertex. The technique used is similar to that used in Chapter 5, except, instead of approximating the area using rectangles, we approximate the area using sectors of a circle. Figure 47 illustrates the area of a sector of a circle.

## 1 Find the Area of a Region Enclosed by the Graph of a Polar Equation

In Figure 48, $r = f(\theta)$ is a function that is nonnegative and continuous on the interval $\alpha \le \theta \le \beta$. Let $A$ denote the area of the region enclosed by the graph of $r = f(\theta)$ and the rays $\theta = \alpha$ and $\theta = \beta$, where $0 \le \alpha < \beta \le 2\pi$. It is helpful to think of the region as being "swept out" by rays, beginning with the ray $\theta = \alpha$ and continuing to the ray $\theta = \beta$.

Partition the closed interval $[\alpha, \beta]$ into $n$ subintervals:

$$[\alpha, \theta_1], [\theta_1, \theta_2], \dots, [\theta_{i-1}, \theta_i], \dots, [\theta_{n-1}, \beta]$$

each of length $\Delta\theta = \dfrac{\beta - \alpha}{n}$. As shown in Figure 49, we select an angle $\theta_i{}^*$ in each subinterval $[\theta_{i-1}, \theta_i]$. The quantity $\dfrac{1}{2}[f(\theta_i{}^*)]^2\Delta\theta$ is the area of the circular sector with radius $r = f(\theta_i{}^*)$ and central angle $\Delta\theta$. The sums of the areas of these sectors $\displaystyle\sum_{i=1}^{n} \dfrac{1}{2}[f(\theta_i{}^*)]^2\Delta\theta$ are an approximation of the area $A$. As the number $n$ of subintervals increases, the sums $\displaystyle\sum_{i=1}^{n} \dfrac{1}{2}[f(\theta_i{}^*)]^2\,\Delta\theta$ become better approximations to $A$. Since the sums $\displaystyle\sum_{i=1}^{n}\dfrac{1}{2}[f(\theta_i{}^*)]^2\Delta\theta$ are Riemann sums, and since $r = f(\theta)$ is continuous, the limit of the sums exists and equals a definite integral.

$$\lim_{n \to \infty} \sum_{i=1}^{n} \frac{1}{2}[f(\theta_i{}^*)]^2\Delta\theta = \int_{\alpha}^{\beta} \frac{1}{2}[f(\theta)]^2 d\theta = \int_{\alpha}^{\beta} \frac{1}{2}r^2 d\theta$$

**THEOREM Area in Polar Coordinates**

If $r = f(\theta)$ is nonnegative and continuous on the closed interval $[\alpha, \beta]$, where $\alpha < \beta$ and $\beta - \alpha \le 2\pi$, then the area $A$ of the region enclosed by the graph of $r = f(\theta)$ and the rays $\theta = \alpha$ and $\theta = \beta$ is given by

$$A = \int_{\alpha}^{\beta} \frac{1}{2}r^2\,d\theta$$

**NOTE** Be sure to graph the equation $r = f(\theta)$, $\alpha \le \theta \le \beta$, before using the formula. In drawing the graph, include the rays $\theta = \alpha$, indicating the start, and $\theta = \beta$, indicating the end, of the region whose area is to be found. These rays determine the limits of integration in the area formula. Be sure that $r = f(\theta)$ is nonnegative on the interval $[\alpha, \beta]$. Otherwise an incorrect result may be obtained.

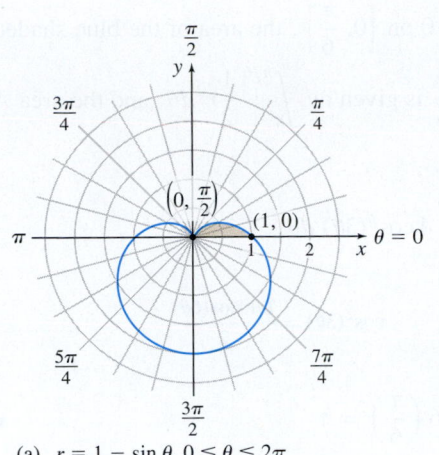

(a)  $r = 1 - \sin\theta, 0 \le \theta \le 2\pi$

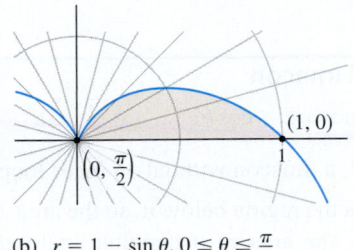

(b)  $r = 1 - \sin\theta, 0 \le \theta \le \dfrac{\pi}{2}$

**Figure 50**

EXAMPLE 1 **Finding the Area Enclosed by a Part of a Cardioid**

Find the area of the region enclosed by the cardioid $r = 1 - \sin\theta, 0 \le \theta \le \dfrac{\pi}{2}$.

**Solution** The cardioid represented by $r = 1 - \sin\theta$, shown in Figure 50(a), is nonnegative on $\left[0, \dfrac{\pi}{2}\right]$. The region enclosed by the cardioid from $\theta = 0$ to $\theta = \dfrac{\pi}{2}$ is shaded.

The region is swept out beginning with the ray $\theta = 0$ and ending with the ray $\theta = \dfrac{\pi}{2}$, as shown in Figure 50(b). The limits of integration are 0 and $\dfrac{\pi}{2}$ and the area $A$ is

$$A = \int_{\alpha}^{\beta} \frac{1}{2} r^2 \, d\theta = \int_0^{\pi/2} \frac{1}{2} (1 - \sin\theta)^2 \, d\theta = \frac{1}{2} \int_0^{\pi/2} (1 - 2\sin\theta + \sin^2\theta) \, d\theta$$

$$= \frac{1}{2} \int_0^{\pi/2} \left\{ 1 - 2\sin\theta + \frac{1}{2}[1 - \cos(2\theta)] \right\} d\theta \quad \sin^2\theta = \frac{1 - \cos(2\theta)}{2}$$

$$= \frac{1}{2} \int_0^{\pi/2} \left[ \frac{3}{2} - 2\sin\theta - \frac{1}{2}\cos(2\theta) \right] d\theta$$

$$= \frac{1}{2} \left[ \frac{3}{2}\theta + 2\cos\theta - \frac{1}{4}\sin(2\theta) \right]_0^{\pi/2} = \frac{3\pi - 8}{8} \qquad \blacksquare$$

NOW WORK **Problem 9.**

EXAMPLE 2 **Finding the Area Enclosed by a Cardioid**

Find the area $A$ enclosed by the cardioid $r = 1 - \sin\theta$.

**Solution** Look again at the cardioid in Figure 50(a). The region enclosed by the cardioid is swept out beginning with the ray $\theta = 0$ and ending with the ray $\theta = 2\pi$. So, the limits of integration are 0 and $2\pi$, and the area $A$ is

$$A = \int_0^{2\pi} \frac{1}{2} (1 - \sin\theta)^2 \, d\theta = \frac{1}{2} \left[ \frac{3}{2}\theta + 2\cos\theta - \frac{1}{4}\sin(2\theta) \right]_0^{2\pi} = \frac{3\pi}{2} \qquad \blacksquare$$

NOW WORK **Problem 13.**

The area formula requires that $r \ge 0$. When there are intervals on which $r < 0$, symmetry can sometimes be used.

CALC
CLIP
EXAMPLE 3 **Finding the Area Enclosed by a Rose**

Find the area $A$ enclosed by the graph of $r = 2\cos(3\theta)$, a rose with three petals.

**Solution** Figure 51 shows the rose. Using symmetry, the area of the blue shaded region in quadrant I equals one-sixth of the area $A$ enclosed by the rose. The blue shaded region is swept out beginning with the ray $\theta = 0$ [the point $(2, 0)$]. It ends at the point $(0, \theta)$, $0 < \theta < \dfrac{\pi}{2}$, where $\theta$ is the solution of the equation $2\cos(3\theta) = 0$. Then

$$\cos(3\theta) = 0 \qquad 0 < \theta < \frac{\pi}{2}$$

$$3\theta = \frac{\pi}{2} + k\pi$$

$$\theta = \frac{\pi}{6} + \frac{k\pi}{3}$$

**Figure 51** $r = 2\cos(3\theta)$

**NEED TO REVIEW?** Solving trigonometric equations is discussed in Section P.7, pp. 65–66.

Since $0 < \theta < \dfrac{\pi}{2}$, we have $\theta = \dfrac{\pi}{6}$. Since $r \geq 0$ on $\left[0, \dfrac{\pi}{6}\right]$, the area of the blue shaded region swept out by the rays $\theta = 0$ and $\theta = \dfrac{\pi}{6}$ is given by $\displaystyle\int_0^{\pi/6} \dfrac{1}{2} r^2 d\theta$, and the area $A$ enclosed by the rose is 6 times this area.

$$A = 6 \int_0^{\pi/6} \frac{1}{2} r^2 \, d\theta = 3 \int_0^{\pi/6} 4\cos^2(3\theta)\, d\theta$$

$$= 12 \int_0^{\pi/6} \frac{1 + \cos(6\theta)}{2} \, d\theta \qquad \cos^2(3\theta) = \frac{1+\cos(6\theta)}{2}$$

$$= 6\left[\theta + \frac{1}{6}\sin(6\theta)\right]_0^{\pi/6} = 6\left(\frac{\pi}{6}\right) = \pi \qquad \blacksquare$$

**NOW WORK** Problem 17.

---

**EXAMPLE 4    Finding the Area Enclosed by a Limaçon**

Find the area $A$ of the region enclosed by the limaçon $r = 2 + \cos\theta$.

**Solution** Figure 52 shows the graph of $r = 2 + \cos\theta$, a limaçon without an inner loop.

We see that the region above the polar axis equals the region below it, so the area $A$ of the region enclosed by the limaçon equals twice the area of the region enclosed by $r = 2 + \cos\theta$ and swept out by the rays $\theta = 0$ and $\theta = \pi$ [from the point $(3, 0)$ to the point $(1, \pi)$].

$$A = 2 \int_0^\pi \frac{1}{2} r^2 d\theta = \int_0^\pi (2 + \cos\theta)^2 \, d\theta = \int_0^\pi (4 + 4\cos\theta + \cos^2\theta)\, d\theta$$

$$= \int_0^\pi \left[4 + 4\cos\theta + \frac{1 + \cos(2\theta)}{2}\right] d\theta \qquad \cos^2\theta = \frac{1 + \cos(2\theta)}{2}$$

$$= \left[4\theta + 4\sin\theta + \frac{\theta}{2} + \frac{1}{4}\sin(2\theta)\right]_0^\pi = \frac{9\pi}{2} \qquad \blacksquare$$

**NOW WORK** Problem 21.

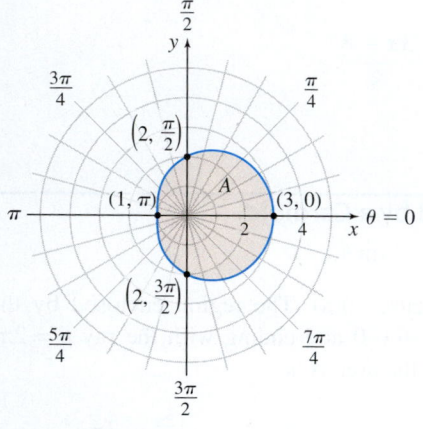

**Figure 52** $r = 2 + \cos\theta$

## 2 Find the Area of a Region Enclosed by the Graphs of Two Polar Equations

To find the area $A$ of the region enclosed by the graphs of two polar equations, we begin by graphing the equations and finding their points of intersection, if any.

---

**EXAMPLE 5    Finding the Area of the Region Enclosed by the Graphs of Two Polar Equations**

Find the area $A$ of the region that lies outside the cardioid $r = 1 + \cos\theta$ and inside the circle $r = 3\cos\theta$.

**Solution** We begin by graphing each equation. See Figure 53(a) on page 721. Then we find the points of intersection of the two graphs by solving the equation,

$$3\cos\theta = 1 + \cos\theta$$

$$2\cos\theta = 1$$

$$\cos\theta = \frac{1}{2}$$

$$\theta = -\frac{\pi}{3} \quad \text{or} \quad \theta = \frac{\pi}{3}$$

The graphs intersect at the points $\left(\dfrac{3}{2}, -\dfrac{\pi}{3}\right)$ and $\left(\dfrac{3}{2}, \dfrac{\pi}{3}\right)$.

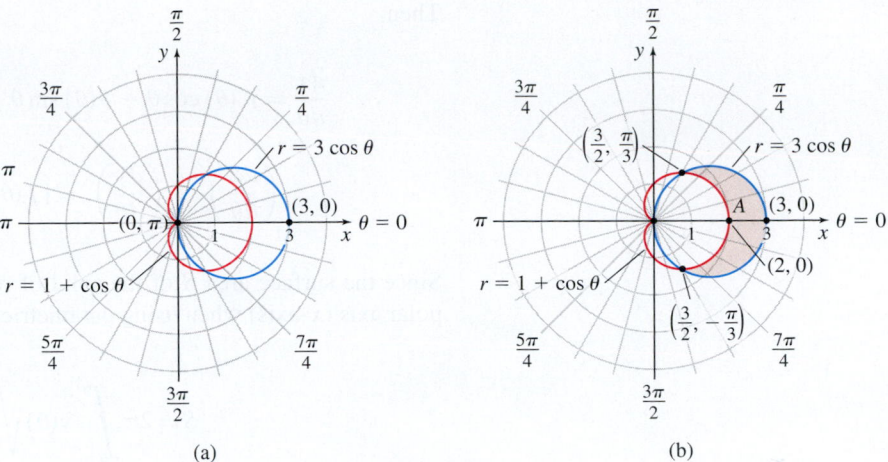

**Figure 53**

The area $A$ of the region that lies outside the cardioid and inside the circle is shown as the shaded portion in Figure 53(b). Notice that the area $A$ is the difference between the area of the region enclosed by the circle $r = 3 \cos \theta$ swept out by the rays $\theta = -\dfrac{\pi}{3}$ and $\theta = \dfrac{\pi}{3}$, and the area of the region enclosed by the cardioid $r = 1 + \cos \theta$ swept out by the same rays. So,

$$A = \int_{-\pi/3}^{\pi/3} \frac{1}{2}(3 \cos \theta)^2 \, d\theta - \int_{-\pi/3}^{\pi/3} \frac{1}{2}(1 + \cos \theta)^2 \, d\theta = \frac{1}{2} \int_{-\pi/3}^{\pi/3} [9 \cos^2 \theta - (1 + 2 \cos \theta + \cos^2 \theta)] \, d\theta$$

$$= \frac{1}{2} \int_{-\pi/3}^{\pi/3} (8 \cos^2 \theta - 1 - 2 \cos \theta) \, d\theta = \frac{1}{2} \int_{-\pi/3}^{\pi/3} \left[ 8 \left( \frac{1 + \cos(2\theta)}{2} \right) - 1 - 2 \cos \theta \right] \, d\theta$$

$$= \frac{1}{2} \int_{-\pi/3}^{\pi/3} [3 + 4 \cos(2\theta) - 2 \cos \theta] \, d\theta = \frac{1}{2} \Big[ 3\theta + 2 \sin(2\theta) - 2 \sin \theta \Big]_{-\pi/3}^{\pi/3} = \pi \qquad \blacksquare$$

**CAUTION**   The circle and the cardioid shown in Figure 53(a) actually intersect in a third point, the pole. The pole is not identified when we solve the equation to find the points of intersection because the pole has coordinates $(0, \pi)$ on $r = 1 + \cos \theta$, but it has coordinates $\left( 0, \dfrac{\pi}{2} \right)$ and $\left( 0, \dfrac{3\pi}{2} \right)$ on $r = 3 \cos \theta$. This demonstrates the importance of graphing polar equations when looking for their points of intersection. Since the pole presents particular difficulties, let $r = 0$ in each equation to determine whether the graph passes through the pole.

**NOW WORK** Problem 23.

## 3 Find the Surface Area of a Solid of Revolution Obtained from the Graph of a Polar Equation

Suppose a smooth curve $C$ is given by the polar equation $r = f(\theta)$, $\alpha \le \theta \le \beta$. Parametric equations for this curve are

$$x(\theta) = r \cos \theta = f(\theta) \cos \theta \qquad y(\theta) = r \sin \theta = f(\theta) \sin \theta$$

Then

$$\frac{dx}{d\theta} = f'(\theta)\cos\theta - f(\theta)\sin\theta \qquad \frac{dy}{d\theta} = f'(\theta)\sin\theta + f(\theta)\cos\theta$$

$$\left(\frac{dx}{d\theta}\right)^2 + \left(\frac{dy}{d\theta}\right)^2 = [f(\theta)]^2 + [f'(\theta)]^2 = r^2 + \left(\frac{dr}{d\theta}\right)^2$$

Since the surface area $S$ of the solid of revolution obtained by revolving $C$ about the polar axis ($x$-axis) when using parametric equations is

$$S = 2\pi \int_\alpha^\beta y(\theta)\sqrt{\left(\frac{dx}{d\theta}\right)^2 + \left(\frac{dy}{d\theta}\right)^2}\, d\theta$$

the surface area $S$ using a polar equation is

$$S = 2\pi \int_\alpha^\beta r\sin\theta\sqrt{r^2 + \left(\frac{dr}{d\theta}\right)^2}\, d\theta = 2\pi \int_\alpha^\beta f(\theta)\sin\theta\sqrt{[f(\theta)]^2 + [f'(\theta)]^2}\, d\theta \quad (1)$$

### EXAMPLE 6   Finding the Surface Area of a Solid of Revolution

Find the surface area of the solid of revolution generated by revolving the arc of the circle $r = a$, $a > 0$, $0 \le \theta \le \dfrac{\pi}{4}$, about the polar axis.

**Solution**  See Figure 54. We find the surface area $S$ using formula (1). Since $r = f(\theta) = a$, $f'(\theta) = 0$. Then

$$S = 2\pi \int_\alpha^\beta f(\theta)\sin\theta\sqrt{[f(\theta)]^2 + [f'(\theta)]^2}\, d\theta$$

$$= 2\pi \int_0^{\pi/4} a\sin\theta\sqrt{a^2}\, d\theta = 2\pi a^2 \int_0^{\pi/4} \sin\theta\, d\theta$$

$$= 2\pi a^2 \big[-\cos\theta\big]_0^{\pi/4} = 2\pi a^2\left(-\frac{\sqrt{2}}{2} + 1\right) = \pi a^2(2 - \sqrt{2})$$

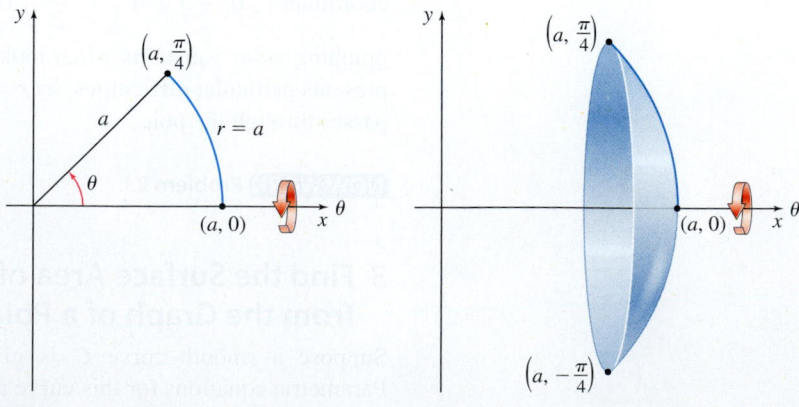

**Figure 54**

NOW WORK Problem 27.

# 9.6 Assess Your Understanding

## Concepts and Vocabulary

1. The area $A$ of the sector of a circle of radius $r$ and central angle $\theta$ radians is $A =$ _____.

2. *True or False* The area enclosed by the graph of a polar equation and two rays that have the pole as a common vertex is found by approximating the area using sectors of a circle.

3. *True or False* The area $A$ enclosed by the graph of the equation $r = f(\theta)$, $r \geq 0$, and the rays $\theta = \alpha$ and $\theta = \beta$, is given by

$$A = \int_{\alpha}^{\beta} f(\theta)\, d\theta$$

4. *True or False* If $x(\theta) = r \cos \theta$, $y(\theta) = r \sin \theta$ are parametric equations of the polar equation $r = f(\theta)$, then

$$\left(\frac{dx}{d\theta}\right)^2 + \left(\frac{dy}{d\theta}\right)^2 = r^2 + \left(\frac{dr}{d\theta}\right)^2$$

## Skill Building

*In Problems 5–8, find the area of the shaded region.*

5. $r = \cos(2\theta)$

6. $r = 2\sin(3\theta)$

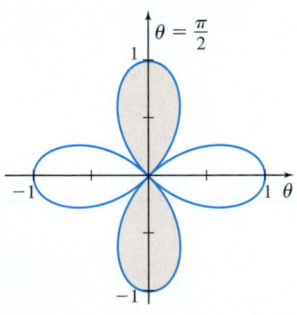

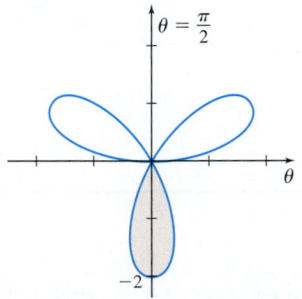

7. $r = 2 + 2\sin\theta$

8. $r = 3 - 3\cos\theta$

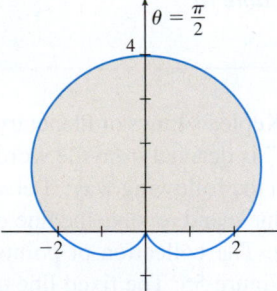

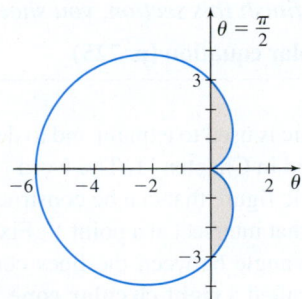

*In Problems 9–12, find the area of the region enclosed by the graph of each polar equation swept out by the given rays.*

9. $r = 3\cos\theta$;  $\theta = 0$ to $\theta = \dfrac{\pi}{3}$

10. $r = 3\sin\theta$;  $\theta = 0$ to $\theta = \dfrac{\pi}{4}$

11. $r = a\theta$;  $\theta = 0$ to $\theta = 2\pi$

12. $r = e^{a\theta}$;  $\theta = 0$ to $\theta = \dfrac{\pi}{2}$

*In Problems 13–18, find the area of the region enclosed by the graph of each polar equation.*

13. $r = 1 + \cos\theta$

14. $r = 2 - 2\sin\theta$

15. $r = 3 + \sin\theta$

16. $r = 3(2 - \sin\theta)$

17. $r = 8\sin(3\theta)$

18. $r = \cos(4\theta)$

*In Problems 19–22, find the area of the region enclosed by one loop of the graph of each polar equation.*

19. $r = 4\sin(2\theta)$

20. $r = 5\cos(3\theta)$

21. $r^2 = 4\cos(2\theta)$

22. $r = a^2\cos(2\theta)$

*In Problems 23–26, find the area of each region described.*

23. Inside $r = 2\sin\theta$; outside $r = 1$

24. Inside $r = 4\cos\theta$; outside $r = 2$

25. Inside $r = \sin\theta$; outside $r = 1 - \cos\theta$

26. Inside $r^2 = 4\cos(2\theta)$; outside $r = \sqrt{2}$

*In Problems 27–30, find the surface area of the solid of revolution generated by revolving each curve about the polar axis.*

27. $r = \sin\theta$,  $0 \leq \theta \leq \dfrac{\pi}{2}$

28. $r = 1 + \cos\theta$,  $0 \leq \theta \leq \pi$

29. $r = e^\theta$,  $0 \leq \theta \leq \pi$

30. $r = 2a\cos\theta$,  $0 \leq \theta \leq \dfrac{\pi}{2}$

## Applications and Extensions

*In Problems 31–48, find the area of the region:*

31. enclosed by the small loop of the limaçon $r = 1 + 2\cos\theta$.

32. enclosed by the small loop of the limaçon $r = 1 + 2\sin\theta$.

33. enclosed by the loop of the graph of $r = 2 - \sec\theta$.

34. enclosed by the loop of the graph of $r = 5 + \sec\theta$.

35. enclosed by $r = 2\sin^2\dfrac{\theta}{2}$.

36. enclosed by $r = 6\cos^2\theta$.

37. inside the circle $r = 8\cos\theta$ and to the right of the line $r = 2\sec\theta$.

38. inside the circle $r = 10\sin\theta$ and above the line $r = 2\csc\theta$.

39. outside the circle $r = 3$ and inside the cardioid $r = 2 + 2\cos\theta$.

40. inside the circle $r = \sin\theta$ and outside the cardioid

$$r = 1 + \cos\theta$$

41. common to the circle $r = \cos\theta$ and the cardioid $r = 1 - \cos\theta$.

42. common to the circles $r = \cos\theta$ and $r = \sin\theta$.

43. common to the inside of the cardioid $r = 1 + \sin\theta$ and the outside of the cardioid $r = 1 + \cos\theta$.

44. common to the inside of the lemniscate $r^2 = 8\cos(2\theta)$ and the outside of the circle $r = 2$.

45. enclosed by the rays $\theta = 0$ and $\theta = 1$ and $r = e^{-\theta}$, $0 \leq \theta \leq 1$.

46. enclosed by the rays $\theta = 0$ and $\theta = 1$ and $r = e^\theta$, $0 \leq \theta \leq 1$.

47. enclosed by the rays $\theta = 1$ and $\theta = \pi$ and $r = \dfrac{1}{\theta}$, $1 \leq \theta \leq \pi$.

---

**1.** = NOW WORK problem    = Graphing technology recommended    CAS = Computer Algebra System recommended

**48.** inside the outer loop but outside the inner loop of

$$r = 1 + 2\sin\theta$$

**49. Area**   Find the area of the loop of the graph of $r = \sec\theta + 2$.

**50. Surface Area of a Sphere**   Develop a formula for the surface area of a sphere of radius $R$.

**51. Surface Area of a Bead**   A sphere of radius $R$ has a hole of radius $a < R$ drilled through it. See the figure. The axis of the hole coincides with a diameter of the sphere. Find the surface area of that part of the sphere that remains.

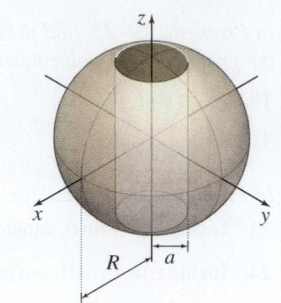

**52. Surface Area of a Plug**   A plug is made to repair the hole in the sphere in Problem 51. What is the surface area of the plug?

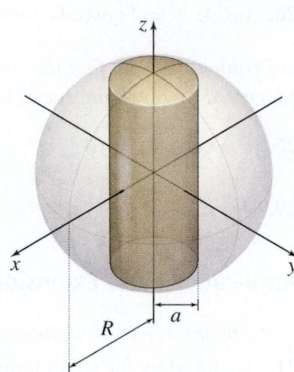

**53. Area**   Find the area enclosed by the loop of the **strophoid**

$$r = \sec\theta - 2\cos\theta$$
$$-\frac{\pi}{2} < \theta < \frac{\pi}{2}$$

as shown in the figure.

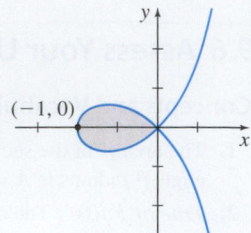

**54. Area**

**(a)** Graph the limaçon $r = 2 - 3\cos\theta$.

**(b)** Find the area enclosed by the inner loop. Round the answer to three decimal places.

**Challenge Problems**

**55.** Show that the area enclosed by the graph of $r\theta = a$ and the rays $\theta = \theta_1$ and $\theta = \theta_2$ is proportional to the difference of the radii, $r_1 - r_2$, where $r_1 = \dfrac{a}{\theta_1}$ and $r_2 = \dfrac{a}{\theta_2}$.

**56.** Find the area of the region that lies outside the circle $r = 1$ and inside the rose $r = 3\sin(3\theta)$.

**57.** Find the area of the region that lies inside the circle $r = 2$ and outside the rose $r = 3\sin(2\theta)$.

---

# 9.7 The Polar Equation of a Conic

**OBJECTIVE**   *When you finish this section, you should be able to:*

**1** Express a conic as a polar equation (p. 725)

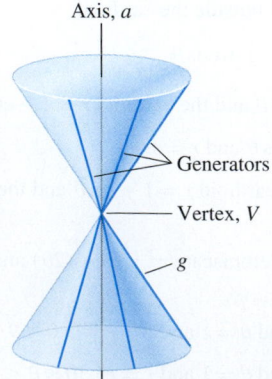

**Figure 55**

The polar equation of a conic is used to explain and to derive Kepler's Laws of Planetary Motion, which are discussed in Chapter 11. The word "conic" is derived from the word "cone," which is a geometric figure that can be constructed in the following way: Let $a$ and $g$ be two distinct lines that intersect at a point $V$. Fix the line $a$ and revolve the line $g$ about $a$ while keeping the angle between the lines constant. The collection of points swept out by the line $g$ is called a **right circular cone**. See Figure 55. The fixed line $a$ is called the **axis** of the cone; the point $V$ is its **vertex**. Any line passing through $V$ that makes the same angle with $a$ as the original line $g$ is called a **generator** of the cone. Each generator lies entirely on the cone. The cone consists of two parts, called **nappes**, that intersect at the vertex.

   **Conics**, an abbreviation for **conic sections**, are curves that result when a right circular cone and a plane intersect. As shown in Figure 56, on page 725, the conics discussed here arise when the plane does not contain the vertex of the cone. If the plane contains the vertex, the intersection of the plane and the cone is a point, a line, or a pair of lines. These are called **degenerate cases**.

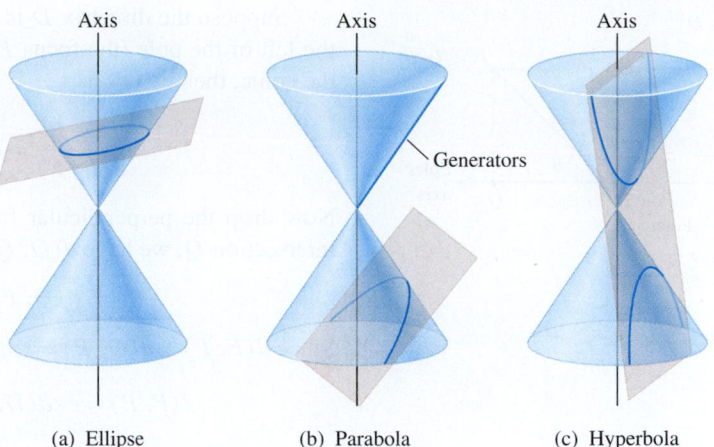

(a) Ellipse      (b) Parabola      (c) Hyperbola

**Figure 56**

A conic is:

- an **ellipse**, when the plane intersects the axis $a$ at an angle greater than the angle between $a$ and $g$. An ellipse lies on only one nappe of the cone. See Figure 56(a).
- a **parabola**, when the plane intersects the axis $a$ at the same angle as the line $g$; that is, the plane is parallel to exactly one generator. A parabola lies on only one nappe of the cone. See Figure 56(b).
- a **hyperbola**, when the plane intersects the axis $a$ at an angle smaller than the angle between $a$ and $g$. A hyperbola lies on both nappes of the cone. See Figure 56(c).

In Appendix A.3, pp. A-22 to A-25, we discuss the rectangular equations of a parabola, an ellipse, and a hyperbola. To obtain the polar equations, we use a unified definition that simultaneously defines all three conics.

---

**DEFINITION** Conic

Let $D$ denote a fixed line called the **directrix**; let $F$ denote a fixed point called the **focus**, which is not on $D$; and let $e$ be a fixed positive number called the **eccentricity**. A **conic** is the set of points $P$ in the plane for which the ratio of the distance from $F$ to $P$ to the distance from $D$ to $P$ equals $e$. That is, a **conic** is the collection of points $P$ for which

$$\boxed{\frac{d(F, P)}{d(D, P)} = e} \tag{1}$$

- If $e = 1$, the conic is a parabola.
- If $e < 1$, the conic is an ellipse.
- If $e > 1$, the conic is a hyperbola.

---

Figure 57 illustrates the definition.

- In a parabola, the **axis** is the line through the focus perpendicular to the directrix.
- In an ellipse, the **major axis** is the line through the focus perpendicular to the directrix.
- In a hyperbola, the **transverse axis** is the line through the focus perpendicular to the directrix.

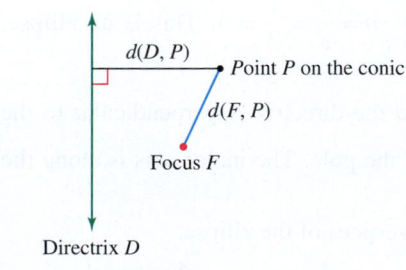

**Figure 57** $\dfrac{d(F, P)}{d(D, P)} = e$

## 1 Express a Conic as a Polar Equation

The equations for the conics in polar coordinates are derived by positioning the focus $F$ at the pole (the origin) and the directrix $D$ either parallel to or perpendicular to the polar axis.

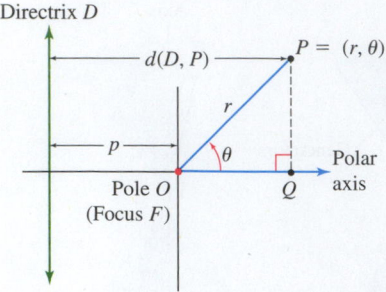

**Figure 58**

Suppose the directrix $D$ is perpendicular to the polar axis at a distance $p$ units to the left of the pole (the focus $F$), as shown in Figure 58. If $P = (r, \theta)$ is any point on the conic, then by (1),

$$\frac{d(F, P)}{d(D, P)} = e \qquad \text{or} \qquad d(F, P) = e \cdot d(D, P)$$

Now drop the perpendicular from the point $P$ to the polar axis. Using the point of intersection $Q$, we have $d(O, Q) = r \cos \theta$. Then

$$d(D, P) = p + d(O, Q) = p + r \cos \theta$$

Since $d(F, P) = d(O, P) = r$,

$$d(F, P) = e \cdot d(D, P)$$

$$r = e(p + r \cos \theta) \qquad \boldsymbol{d(F, P) = r; \, d(D, P) = p + r \cos \theta}$$

$$r - er \cos \theta = ep$$

$$r = \frac{ep}{1 - e \cos \theta}$$

---

**THEOREM  The Polar Equation of a Conic**

The polar equation of a conic with focus at the pole and directrix perpendicular to the polar axis at a distance $p$ to the left of the pole is

$$\boxed{r = \frac{ep}{1 - e \cos \theta}} \tag{2}$$

where $e$ is the eccentricity of the conic.

---

**CALC CLIP**

**EXAMPLE 1   Identifying and Graphing the Polar Equation of a Conic**

(a) Identify and graph the equation $r = \dfrac{4}{2 - \cos \theta}$.

(b) Convert the polar equation to a rectangular equation.

(c) Find parametric equations for the polar equation.

**Solution**  (a) Divide the numerator and the denominator by 2 to express the equation in the form $r = \dfrac{ep}{1 - e \cos \theta}$.

$$r = \frac{4}{2 - \cos \theta} = \frac{2}{1 - \dfrac{1}{2} \cos \theta} \qquad r = \frac{ep}{1 - e \cos \theta}$$

Now we see that $e = \dfrac{1}{2}$. Since $ep = 2$, we find $p = \dfrac{2}{e} = \dfrac{2}{\dfrac{1}{2}} = 4$. This is an ellipse, because $e = \dfrac{1}{2} < 1$. One focus is at the pole, and the directrix is perpendicular to the polar axis a distance of $p = 4$ units to the left of the pole. The major axis is along the polar axis.

By letting $\theta = 0$ and $\theta = \pi$, we can find the vertices of the ellipse.

$$r = \frac{4}{2 - \cos 0} = \frac{4}{2 - 1} = 4 \qquad \text{and} \qquad r = \frac{4}{2 - \cos \pi} = \frac{4}{2 - (-1)} = \frac{4}{3}$$

So, the vertices of the ellipse are the points whose polar coordinates are $(4, 0)$ and $\left(\dfrac{4}{3}, \pi\right)$. The $y$-intercepts of the ellipse are at $\theta = \dfrac{\pi}{2}$ and $\theta = \dfrac{3\pi}{2}$, which give rise to the points $\left(2, \dfrac{\pi}{2}\right)$ and $\left(2, \dfrac{3\pi}{2}\right)$. The graph of the ellipse is shown in Figure 59.

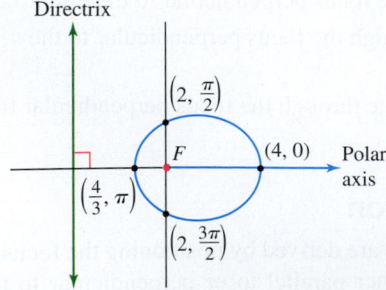

**Figure 59**

**(b)** To obtain a rectangular equation of the ellipse, we proceed as follows.

$$r = \frac{4}{2 - \cos\theta}$$

$r(2 - \cos\theta) = 4$        **Eliminate the fraction.**

$2r - r\cos\theta = 4$

$2r = 4 + r\cos\theta$

$4r^2 = (4 + r\cos\theta)^2$        **Square both sides.**

$4(x^2 + y^2) = (4 + x)^2$        $r^2 = x^2 + y^2, \quad x = r\cos\theta$

$4x^2 + 4y^2 = 16 + 8x + x^2$

$3\left(x^2 - \dfrac{8}{3}x\right) + 4y^2 = 16$

$3\left(x - \dfrac{4}{3}\right)^2 + 4y^2 = 16 + 3\left(\dfrac{16}{9}\right) = \dfrac{64}{3}$        **Complete the square in $x$.**

This is the equation of an ellipse in rectangular coordinates, with its center at the point $\left(\dfrac{4}{3}, 0\right)$ in rectangular coordinates.

**(c)** Parametric equations of the ellipse $r = \dfrac{4}{2 - \cos\theta}$ are

$$x(\theta) = r\cos\theta = \frac{4\cos\theta}{2 - \cos\theta} \qquad y(\theta) = r\sin\theta = \frac{4\sin\theta}{2 - \cos\theta} \qquad 0 \le \theta \le 2\pi$$

where $\theta$ is the parameter. ∎

There are four possibilities for the position of the directrix relative to the polar axis. These are summarized in Table 9.

**TABLE 9** Polar Equations of Conics (Focus at the Pole, Eccentricity $e$)

| Equation | Description |
|---|---|
| $r = \dfrac{ep}{1 - e\cos\theta}$ | Directrix is perpendicular to the polar axis a distance $p$ units to the left of the pole. |
| $r = \dfrac{ep}{1 + e\cos\theta}$ | Directrix is perpendicular to the polar axis a distance $p$ units to the right of the pole. |
| $r = \dfrac{ep}{1 + e\sin\theta}$ | Directrix is parallel to the polar axis a distance $p$ units above the pole. |
| $r = \dfrac{ep}{1 - e\sin\theta}$ | Directrix is parallel to the polar axis a distance $p$ units below the pole. |

**Eccentricity**

If $e = 1$, the conic is a parabola; the axis of symmetry is perpendicular to the directrix.

If $e < 1$, the conic is an ellipse; the major axis is perpendicular to the directrix.

If $e > 1$, the conic is a hyperbola; the transverse axis is perpendicular to the directrix.

**NOW WORK** Problems 5 and 13.

**EXAMPLE 2** **Analyzing the Orbit of an Exoplanet**

Many hundreds of planets beyond our solar system have been discovered. They are known as **exoplanets**. One of these exoplanets, HD 190360b, orbits the star HD 190360 (most stars do not have names) in an elliptical orbit given by the polar equation

$$r = \frac{3.575}{1 - 0.316\cos\theta}$$

where the star HD 190360 is at the pole, the major axis is along the polar axis, and $r$ is measured in astronomical units (AU). (One AU $= 1.5 \times 10^{11}$ m, which is the average distance from Earth to the Sun.)

(a) What is the eccentricity of the exoplanet HD 190360b's orbit?

(b) Find the distance from the exoplanet HD 190360b to the star HD 190360 at periapsis (shortest distance).

(c) Find the distance from the exoplanet HD 190360b to the star HD 190360 at apoapsis (greatest distance).

**Solution** (a) The equation of HD 190360b's orbit is in the form $r = \dfrac{ep}{1 - e\cos\theta}$. So, the eccentricity of the orbit is $e = 0.316$. See Figure 60.

(b) Periapsis occurs when $r$ is a minimum. Since $r$ is minimum when $\cos\theta = -1$, periapsis occurs for

$$r = \frac{3.575}{1 - 0.316\cos\theta} = \frac{3.575}{1 - 0.316(-1)} = \frac{3.575}{1.316} \approx 2.717$$

The periapsis is at the point $(2.717, \pi)$. The exoplanet HD 190360b is approximately 2.717 AU from the star HD 190360 at periapsis.

(c) The apoapsis occurs when $r$ is a maximum. Since $r$ is maximum when $\cos\theta = 1$, the apoapsis occurs for

$$r = \frac{3.575}{1 - 0.316\cos\theta} = \frac{3.575}{1 - 0.316(1)} = \frac{3.575}{0.684} \approx 5.227$$

The apoapsis is at the point $(5.227, 0)$. The exoplanet HD 190360b is approximately 5.227 AU from the star HD 190360 at apoapsis. ∎

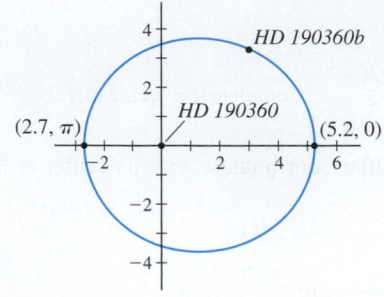

**Figure 60** $r = \dfrac{3.575}{1 - 0.316\cos\theta}$

The orbit of exoplanet HD 190360b about star HD 190360.

**NOW WORK** Problem 33.

## 9.7 Assess Your Understanding

### Concepts and Vocabulary

1. *Multiple Choice* A(n) [(a) parabola (b) ellipse (c) hyperbola] is the set of points $P$ in the plane for which the distance from a fixed point called the focus to $P$ equals the distance from a fixed line called the directrix to $P$.

2. In your own words, explain what is meant by the eccentricity $e$ of a conic.

3. Identify the graphs of each of these polar equations:

$$r = \frac{2}{1 + \sin\theta} \quad \text{and} \quad r = \frac{2}{1 + \cos\theta}$$

How are they the same? How are they different?

4. *True or False* The polar equation of a conic with focus at the pole and directrix perpendicular to the polar axis at a distance $p$ to the left of the pole is $r = \dfrac{ep}{1 - p\cos\theta}$, where $e$ is the eccentricity of the conic.

### Skill Building

*In Problems 5–12, identify each conic. Find its eccentricity $e$ and the position of its directrix.*

5. $r = \dfrac{1}{1 + \cos\theta}$

6. $r = \dfrac{3}{1 - \sin\theta}$

7. $r = \dfrac{4}{2 - 3\sin\theta}$

8. $r = \dfrac{2}{1 + 2\cos\theta}$

9. $r = \dfrac{3}{4 - 2\cos\theta}$

10. $r = \dfrac{6}{8 + 2\cos\theta}$

11. $r = \dfrac{4}{3 + 3\sin\theta}$

12. $r = \dfrac{1}{6 + 2\sin\theta}$

*In Problems 13–20, for each polar equation:*

(a) *Identify and graph the equation.*

(b) *Convert the polar equation to a rectangular equation.*

(c) *Find parametric equations for the polar equation.*

13. $r = \dfrac{8}{4 + 3\sin\theta}$

14. $r = \dfrac{10}{5 + 4\cos\theta}$

15. $r = \dfrac{9}{3 - 6\cos\theta}$

16. $r = \dfrac{12}{4 + 8\sin\theta}$

17. $r(3 - 2\sin\theta) = 6$

18. $r(2 - \cos\theta) = 2$

19. $r = \dfrac{6\sec\theta}{2\sec\theta - 1}$

20. $r = \dfrac{3\csc\theta}{\csc\theta - 1}$

### Applications and Extensions

*In Problems 21–26, find the slope of the tangent line to the graph of each conic at $\theta$.*

21. $r = \dfrac{9}{4 - \cos\theta}$, $\theta = 0$

22. $r = \dfrac{3}{1 - \sin\theta}$, $\theta = 0$

23. $r = \dfrac{8}{4 + \sin\theta}$, $\theta = \dfrac{\pi}{2}$

24. $r = \dfrac{10}{5 + 4\sin\theta}$, $\theta = \pi$

25. $r(2 + \cos\theta) = 4$, $\theta = \pi$

26. $r(3 - 2\sin\theta) = 6$, $\theta = \dfrac{\pi}{2}$

*In Problems 27–32, find a polar equation for each conic. For each equation, a focus is at the pole.*

27. $e = \dfrac{4}{5}$; directrix is perpendicular to the polar axis 3 units to the left of the pole.

---

**1.** = NOW WORK problem        〰 = Graphing technology recommended        CAS = Computer Algebra System recommended

**28.** $e = \dfrac{2}{3}$; directrix is parallel to the polar axis 3 units above the pole.

**29.** $e = 1$; directrix is parallel to the polar axis 1 unit above the pole.

**30.** $e = 1$; directrix is parallel to the polar axis 2 units below the pole.

**31.** $e = 6$; directrix is parallel to the polar axis 2 units below the pole.

**32.** $e = 5$; directrix is perpendicular to the polar axis 3 units to the right of the pole.

**33. Halley's Comet**   As with most comets, Halley's comet has a highly elliptical orbit about the Sun, given by the polar equation

$$r = \frac{1.155}{1 - 0.967 \cos \theta}$$

where the Sun is at the pole, the semimajor axis is along the polar axis, and $r$ is measured in AU (astronomical unit). One $AU = 1.5 \times 10^{11}$ m, which is the average distance from Earth to the Sun.

Stocktrek/Corbis

**(a)** What is the eccentricity of the comet's orbit?

**(b)** Find the distance from Halley's comet to the Sun at perihelion (shortest distance from the Sun).

**(c)** Find the distance from Halley's comet to the Sun at aphelion (greatest distance from the Sun).

**(d)** Graph the orbit of Halley's comet.

**34. Orbit of Mercury**   The planet Mercury travels around the Sun in an elliptical orbit given approximately by

$$r = \frac{(3.442)\,10^7}{1 - 0.206 \cos \theta}$$

where $r$ is measured in miles and the Sun is at the pole.

**(a)** What is the eccentricity of Mercury's orbit?

**(b)** Find the distance from Mercury to the Sun at perihelion (shortest distance from the Sun).

**(c)** Find the distance from Mercury to the Sun at aphelion (greatest distance from the Sun).

**(d)** Graph the orbit of Mercury.

**35. The Effect of Eccentricity**

**(a)** Graph the conic $r = \dfrac{2e}{1 - e \cos \theta}$ for the following values of $e$: (i) $e = 0.2$, (ii) $e = 0.6$, (iii) $e = 0.9$, (iv) $e = 1$, (v) $e = 2$, (vi) $e = 4$.

**(b)** Describe how the shape of the conic changes as $e > 1$ gets larger.

**(c)** Describe how the shape of the conic changes as $e < 1$ gets closer to 0.

**36.** Show that the polar equation for a conic with its focus at the pole and whose directrix is perpendicular to the polar axis at a distance $p$ units to the right of the pole is given by

$$r = \frac{ep}{1 + e \cos \theta}$$

**37.** Show that the polar equation for a conic with its focus at the pole and whose directrix is parallel to the polar axis at a distance $p$ units above the pole is given by $r = \dfrac{ep}{1 + e \sin \theta}$.

**38.** Show that the polar equation for a conic with its focus at the pole and whose directrix is parallel to the polar axis at a distance $p$ units below the pole is given by $r = \dfrac{ep}{1 - e \sin \theta}$.

## Challenge Problems

**39.** Show that the surface area of the solid generated by revolving the first-quadrant arc of the ellipse $\dfrac{x^2}{a^2} + \dfrac{y^2}{b^2} = 1$, $x \geq 0$, $y \geq 0$, about the $x$-axis is

$$S = \pi b^2 + \frac{\pi ab}{e}\, \sin^{-1} e$$

where $e$ is the eccentricity of the ellipse.

**40.** In this section, one focus of each conic has been at the pole. Write the general equation for a conic in polar coordinates if no focus is at the pole. That is, suppose the focus $F$ has polar coordinates $(r_1, \theta_1)$, and the directrix $D$ is given by $r \cos(\theta + \theta_0) = -d$, where $d > 0$. Let the eccentricity be $e$.

# Chapter Review

## THINGS TO KNOW

### 9.1 Parametric Equations

- Parametric equations $x = x(t)$, $y = y(t)$, where $t$ is the parameter (p. 680)

- Plane curve: the graph of the points $(x, y) = (x(t), y(t))$ (p. 680)

- Orientation: the direction of the motion along the curve defined for $a \leq t \leq b$ moves as $t$ varies from $a$ to $b$ (p. 680)

- Convert between parametric equations and rectangular equations (pp. 681–685)

- Cycloid: the curve represented by $x(t) = a(t - \sin t)$, $y(t) = a(1 - \cos t)$ (p. 685)

### 9.2 Tangent Lines

- Smooth curve (p. 690)

- Slope of a tangent line to a smooth curve

$$\frac{dy}{dx} = \frac{\dfrac{dy}{dt}}{\dfrac{dx}{dt}} \quad \frac{dx}{dt} \neq 0 \text{ (p. 691)}$$

- Vertical tangent line to a smooth curve

$$\frac{dx}{dt} = 0, \text{ but } \frac{dy}{dt} \neq 0 \text{ (p. 691)}$$

- Horizontal tangent line to a smooth curve

$$\frac{dy}{dt} = 0, \text{ but } \frac{dx}{dt} \neq 0 \text{ (p. 691)}$$

## 9.3 Arc Length; Surface Area of a Solid of Revolution

• Arc length formula for parametric equations:

$$s = \int_a^b \sqrt{\left(\frac{dx}{dt}\right)^2 + \left(\frac{dy}{dt}\right)^2}\, dt \text{ (p. 695)}$$

The surface area $S$ of the solid of revolution generated by revolving a smooth curve $C$ represented by the parametric equations $x = x(t)$, $y = y(t)$, $a \le t \le b$

• about the $x$-axis: $S = 2\pi \displaystyle\int_a^b y(t)\sqrt{\left(\frac{dx}{dt}\right)^2 + \left(\frac{dy}{dt}\right)^2}\, dt$
(p. 698)

• about the $y$-axis: $S = 2\pi \displaystyle\int_a^b x(t)\sqrt{\left(\frac{dx}{dt}\right)^2 + \left(\frac{dy}{dt}\right)^2}\, dt$
(p. 699)

## 9.4 Polar Coordinates

• Polar coordinates $(r, \theta)$; pole $O$; polar axis (p. 701)
• A point $P$ with polar coordinates $(r, \theta)$ also can be represented by $(r, \theta + 2n\pi)$ or $(-r, \theta + (2n+1)\pi)$, $n$ an integer. (p. 703)
• The polar coordinates of the pole are $(0, \theta)$, where $\theta$ is any angle. (p. 703)
• For polar coordinates $(r, \theta)$ and rectangular coordinates $(x, y)$:

 • $x = r\cos\theta$, $y = r\sin\theta$ (p. 704)
 • $r = \sqrt{x^2 + y^2}$ and $\tan\theta = \dfrac{y}{x}$, $x \ne 0$ (p. 705)
 • $r = y$ and $\theta = \dfrac{\pi}{2}$ if $x = 0$ (p. 705)

• Table 2 gives polar equations for some lines and circles. (p. 708)

## 9.5 Polar Equations; Parametric Equations of Polar Equations; Arc Length of Polar Equations

• Library of Polar Equations (Table 8) (p. 715)
• Arc length $s$ of a curve represented by a polar equation from

$$\theta = \alpha \text{ to } \theta = \beta, \ \alpha \le \theta \le \beta: \ s = \int_\alpha^\beta \sqrt{r^2 + \left(\frac{dr}{d\theta}\right)^2}\, d\theta$$
(p. 715)

## 9.6 Area in Polar Coordinates

• The area $A$ enclosed by the graph of $r = f(\theta)$, $r \ge 0$, and the rays $\theta = \alpha$ and $\theta = \beta$: $A = \displaystyle\int_\alpha^\beta \frac{1}{2}r^2\, d\theta$ (p. 718)

• The surface area $S$ of the solid of revolution obtained by revolving $r = f(\theta)$, $\alpha \le \theta \le \beta$, about the polar axis:

$$S = 2\pi \int_\alpha^\beta r\sin\theta \sqrt{r^2 + \left(\frac{dr}{d\theta}\right)^2}\, d\theta$$

$$= 2\pi \int_\alpha^\beta f(\theta)\sin\theta \sqrt{[f(\theta)]^2 + [f'(\theta)]^2}\, d\theta$$
(p. 722)

## 9.7 The Polar Equation of a Conic

• Definition of a conic (p. 725)
• Eccentricity: Parabola: $e = 1$; ellipse: $e < 1$; hyperbola: $e > 1$ (p. 725)
• The polar equation of a conic: Table 9 (p. 727)

## OBJECTIVES

| Section | You should be able to … | Examples | Review Exercises |
|---|---|---|---|
| 9.1 | 1 Graph parametric equations (p. 680) | 1 | 1(b)–6(b) |
| | 2 Find a rectangular equation for a curve represented parametrically (p. 681) | 2–4 | 1(a)–6(a), 1(c)–6(c) |
| | 3 Use time as the parameter in parametric equations (p. 683) | 5 | 13, 48, 49 |
| | 4 Convert a rectangular equation to parametric equations (p. 684) | 6, 7 | 11, 12 |
| 9.2 | 1 Find an equation of the tangent line at a point on a plane curve (p. 691) | 1–4 | 7–10, 50, 51 |
| 9.3 | 1 Find the arc length of a plane curve (p. 695) | 1, 2 | 52–55 |
| | 2 Find the surface area of a solid of revolution obtained from parametric equations (p. 697) | 3 | 63, 64 |
| 9.4 | 1 Plot points using polar coordinates (p. 701) | 1, 2 | 14–17 |
| | 2 Convert between rectangular coordinates and polar coordinates (p. 703) | 3, 4 | 14–31 |
| | 3 Identify and graph polar equations (p. 706) | 5, 6 | 32–35 |
| 9.5 | 1 Graph a polar equation; find parametric equations (p. 710) | 1–5 | 36–38, 40–45 |
| | 2 Find the arc length of a curve represented by a polar equation (p. 715) | 6 | 56–59 |
| 9.6 | 1 Find the area of a region enclosed by the graph of a polar equation (p. 718) | 1–4 | 60 |
| | 2 Find the area of a region enclosed by the graphs of two polar equations (p. 720) | 5 | 61, 62 |
| | 3 Find the surface area of a solid of revolution obtained from the graph of a polar equation (p. 721) | 6 | 65 |
| 9.7 | 1 Express a conic as a polar equation (p. 725) | 1, 2 | 39, 46, 47 |

## REVIEW EXERCISES

*In Problems 1–6:*

*(a) Find the rectangular equation of each curve.*

*(b) Graph each plane curve whose parametric equations are given and show its orientation.*

*(c) Determine the restrictions on $x$ and $y$ that make the rectangular equation identical to the plane curve.*

1. $x(t) = 4t - 2$, $y(t) = 1 - t$; $\quad -\infty < t < \infty$
2. $x(t) = 2t^2 + 6$, $y(t) = 5 - t$; $\quad -\infty < t < \infty$
3. $x(t) = e^t$, $y(t) = e^{-t}$; $\quad -\infty < t < \infty$
4. $x(t) = \ln t$, $y(t) = t^3$; $\quad t > 0$
5. $x(t) = \sec^2 t$, $y(t) = \tan^2 t$; $\quad 0 \le t \le \dfrac{\pi}{4}$
6. $x(t) = t^{3/2}$, $y(t) = 2t + 4$; $\quad t \ge 0$

*In Problems 7–10, for the parametric equations below:*

**(a)** *Find an equation of the tangent line to the curve at t.*
**(b)** *Graph the curve and the tangent line.*

**7.** $x(t) = t^2 - 4$, $y(t) = t$ at $t = 1$

**8.** $x(t) = 3 \sin t$, $y(t) = 4 \cos t + 2$ at $t = \dfrac{\pi}{4}$

**9.** $x(t) = \dfrac{1}{t^2}$, $y(t) = \sqrt{t^2 + 1}$ at $t = 3$

**10.** $x(t) = \dfrac{t^2}{1+t}$, $y(t) = \dfrac{t}{1+t}$ at $t = 0$

*In Problems 11 and 12, find two different pairs of parametric equations for each rectangular equation.*

**11.** $y = -2x + 4$          **12.** $y = 2x$

**13.** Describe the motion of an object that moves so that at time $t$ (in seconds) it has coordinates

$$x(t) = 2 \cos t \quad y(t) = \sin t \quad 0 \le t \le 2\pi$$

*In Problems 14–17, the polar coordinates of a point are given. Plot each point in a polar coordinate system, and find its rectangular coordinates.*

**14.** $\left(3, \dfrac{\pi}{6}\right)$          **15.** $\left(-2, \dfrac{4\pi}{3}\right)$

**16.** $\left(3, -\dfrac{\pi}{2}\right)$          **17.** $\left(-4, -\dfrac{\pi}{4}\right)$

*In Problems 18–21, the rectangular coordinates of a point are given. Find two pairs of polar coordinates $(r, \theta)$ for each point, one with $r > 0$ and the other with $r < 0, 0 \le \theta < 2\pi$.*

**18.** $(2, 0)$     **19.** $(3, 4)$     **20.** $(-5, 12)$     **21.** $(-3, 3)$

*In Problems 22–27, the letters $r, \theta$ represent polar coordinates. Write each equation in terms of the rectangular coordinates $x, y$.*

**22.** $r = 4 \sin(2\theta)$     **23.** $r = e^{\theta/2}$     **24.** $r = \dfrac{1}{1 + 2 \cos \theta}$

**25.** $r = a - \sin \theta$     **26.** $r^2 = 4 \cos(2\theta)$     **27.** $r = \theta$

*In Problems 28–31, the letters $x, y$ represent rectangular coordinates. Write each equation in terms of the polar coordinates $r, \theta$.*

**28.** $x^2 + y^2 = x$          **29.** $(x^2 + y^2)^2 = x^2 - y^2$

**30.** $y^2 = (x^2 + y^2) \cos^2[(x^2 + y^2)^{1/2}]$

**31.** $\dfrac{x^2}{2^2} + \dfrac{y^2}{3^2} = 1$

*In Problems 32–35, identify and graph each polar equation. Convert it to a rectangular equation if necessary.*

**32.** $r \sin \theta = 1$          **33.** $r \sec \theta = 2$

**34.** $r = \sin \theta$          **35.** $r = -5 \cos \theta$

*In Problems 36–45, for each equation:*

**(a)** *Graph the equation.*
**(b)** *Find parametric equations that represent the equation.*

**36.** $r = 1 - \sin \theta$          **37.** $r = 4 \cos(2\theta)$

**38.** $r = \dfrac{1}{2} - \sin \theta$          **39.** $r = \dfrac{4}{1 - 2 \cos \theta}$

**40.** $r = 4 \sin(3\theta)$          **41.** $r = 2 - 2 \cos \theta$

**42.** $r = 2 - \sin \theta$          **43.** $r = e^{0.5\theta}$

**44.** $r^2 = 1 - \sin^2 \theta$          **45.** $r^2 = 1 + \sin^2 \theta$

*In Problems 46 and 47, for each polar equation:*

**(a)** *Identify and graph the equation.*
**(b)** *Convert the polar equation to a rectangular equation.*
**(c)** *Find parametric equations for the polar equation.*

**46.** $r = \dfrac{2}{1 - \cos \theta}$          **47.** $r = \dfrac{1}{1 - \dfrac{1}{6} \cos \theta}$

*In Problems 48 and 49, find parametric equations for an object that moves along the ellipse $\dfrac{x^2}{16} + \dfrac{y^2}{9} = 1$ with the motion described.*

**48.** The motion begins at $(4, 0)$, is counterclockwise, and requires 4 s for a complete revolution.

**49.** The motion begins at $(0, 3)$, is clockwise, and requires 5 s for a complete revolution.

*In Problems 50 and 51, find the points (if any) on the curve at which the tangent line is vertical or horizontal.*

**50.** $x(t) = t^3 - 1$, $y(t) = 2t^2 + 1$

**51.** $x(t) = 1 - \sin t$, $y(t) = 2 + 3 \cos t$, $0 \le t \le 2\pi$

*In Problems 52–55, find the arc length of each plane curve.*

**52.** $x(t) = \sinh^{-1} t$, $y(t) = \sqrt{t^2 + 1}$ from $t = 0$ to $t = 1$

**53.** $x(t) = \tan t$, $y(t) = \dfrac{1}{3}(\sec^2 t + 1)$ from $t = 0$ to $t = \dfrac{\pi}{4}$

**54.** $x(t) = e^t$, $y(t) = \dfrac{1}{2}e^{2t} - \dfrac{1}{4}t$ from $t = 0$ to $t = 2$

**55.** $x = \dfrac{1}{2}y^2 - \dfrac{1}{4} \ln y$ from $y = 1$ to $y = 2$

*In Problems 56–59, find the arc length of each curve represented by a polar equation.*

**56.** $r = 2 \sin \theta$ from $\theta = 0$ to $\theta = \pi$

**57.** $r = e^{-\theta}$ from $\theta = 0$ to $\theta = 2\pi$

**58.** $r = 3\theta$ from $\theta = 0$ to $\theta = 2\pi$

**59.** $r = 2 \sin^2 \dfrac{\theta}{2}$ from $\theta = -\dfrac{\pi}{2}$ to $\theta = \dfrac{\pi}{2}$

**60. Area**   Find the area of the region inside the circle $r = 4 \sin \theta$ and above the line $r = 3 \csc \theta$.

**61. Area**   Find the area of the region that lies inside the graph of $r^2 = 4 \cos(2\theta)$ and outside the circle $r = \sqrt{2}$.

**62. Area**   Find the area of the region common to the graphs of $r = \cos \theta$ and $r = 1 - \cos \theta$.

**63. Surface Area**   Find the surface area of the solid generated by revolving the smooth curve represented by the parametric equations $x(t) = \sinh^{-1} t$, $y(t) = \sqrt{t^2 + 1}$, $0 \le t \le 1$, about the $x$-axis.

**64. Surface Area**   Find the surface area of the solid generated by revolving the smooth curve represented by $x(t) = e^t - t$, $y(t) = 4e^{t/2}$ from $t = 0$ to $t = 1$ about the $x$-axis.

**65. Surface Area**   Find the surface area of the solid of revolution generated by revolving the arc of the circle $r = 4$, $0 \le \theta \le \dfrac{\pi}{3}$, about the $x$-axis.

## CHAPTER 9 PROJECT    Polar Graphs and Microphones

Microphones can be configured to have different sensitivities, which are best modeled using various polar patterns. The oldest example is an **omnidirectional microphone**, which records sound equally from all directions. Omnidirectional microphones can be a good option when recording jungle sounds or general ambient noise, or when recording for your cell phone. The recording pattern for this microphone is virtually circular, and a good polar model is

$$r = k$$

where $k > 0$ is the sensitivity parameter. Notice that by adjusting $k$ (say using 1, 4, or 8) we obtain concentric circles. Rotating any one of these circles around a diameter results in a sphere. Notice that the microphone receiver is at the center of the sphere and that sound from all directions is received equally.

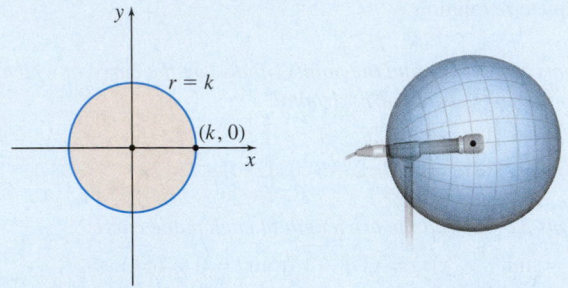

For this model, say when $k = 1$, we can convert the polar equation to parametric equations obtaining

$$x = \cos t \quad y = \sin t \quad 0 \le t \le 2\pi$$

Sound technicians have several things to consider when using microphones, including finding the best location to place the microphone and the best location to place any monitors. Since an omnidirectional microphone collects sound equally from all directions, the microphone should be placed directly in the center of the sounds being recorded with the most important sounds positioned closest to the microphone. Unfortunately, monitors cannot be used with an omnidirectional microphone since there is no place to put them where the microphone does not pick up their sound. To help remedy this and other possible issues, different types of microphone models are used.

A **cardioid microphone** is used when we want to pick up sounds mostly from the front of the microphone. Cardioid microphones are considered "unidirectional." A polar model for a cardioid microphone is

$$r = k(1 + \cos\theta) \quad 0 \le \theta \le 2\pi \quad k > 0$$

Notice that the microphone is placed with the transducer at the pole and pointing toward the polar axis. A cardioid microphone is formed by rotating the cardioid about the polar axis. See the figure.

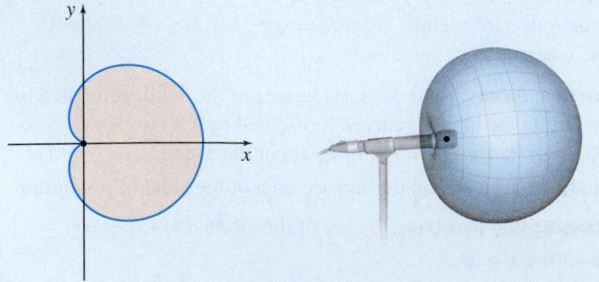

1. Sketch several cardioids using $k = 1, 4, 8$.

2. Explain why any noise created by the person's hand or the microphone stand is unlikely to be picked up by a cardioid microphone.

3. Using $k = 1$, convert the cardioid polar equation to a pair of parametric equations.

4. To determine the percentage of the sound that comes from the "front" of the microphone, we consider the sound coming at the microphone with angles in the polar wedge $-\dfrac{\pi}{4} \le \theta \le \dfrac{\pi}{4}$. With $k = 1$, determine the proportion of the surface area coming in from the front relative to that coming in from all directions.

5. Determine the proportion of the sound that is picked up from behind the microphone in the polar wedge $\dfrac{3\pi}{4} \le \theta \le \dfrac{5\pi}{4}$.

Since less than 1% of all the sound received comes from the backside of a cardioid microphone, monitors can be placed directly behind the microphone.

A **hypercardioid microphone** is used to pick up sounds mostly from in front of the microphone, but it also receives some sound from the rear. A polar model for such a hypercardioid microphone is

$$r = \frac{k\,|1 + 2\cos\theta|}{3} \quad 0 \le \theta \le 2\pi \quad k > 0$$

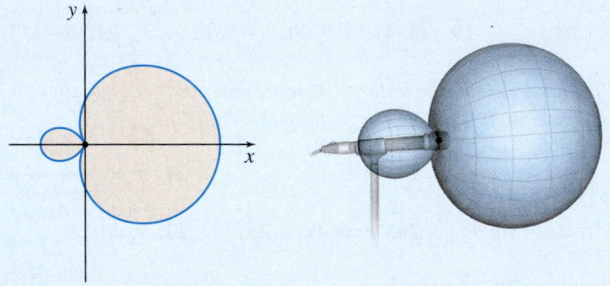

6. Sketch several hypercardioid graphs using $k = 1, 4,$ and $8$.

7. Express the hypercardioid polar equation with $k = 1$, as a pair of parametric equations.

8. Determine the proportion of the sound that is picked up from the front of a hypercardioid microphone $r = \dfrac{|1 + 2\cos\theta|}{3}$ by finding the percentage of sound coming at the microphone from the polar wedge $-\dfrac{\pi}{4} \le \theta \le \dfrac{\pi}{4}$.

9. Determine the proportion of the sound that is picked up from behind the hypercardioid microphone in the polar wedge $\dfrac{3\pi}{4} \le \theta \le \dfrac{5\pi}{4}$.

10. Compare the results of Problems 4, 5, 8, and 9, and describe how the microphones differ.

Friso Gentsch/picture-alliance/dpa/AP Images

**CHAPTER 10 PROJECT** The Chapter Project on page 796 uses vector algebra and properties of vectors to model the Hall effect.

## The Hall Effect

In our ever-growing electronic environment, sensors of various types, shapes, and sizes have been developed to advance the detection and analysis of information. Sensors often collect data that are then fed to other electronic components that can be programmed to react based on the results. Detecting the presence and strength of a magnetic field can be accomplished by using a sensor, called a Hall probe, a device that operates based on an electromagnetic property of a material— the *Hall effect*. The Hall effect was discovered by physicist Edwin H. Hall in the late nineteenth century, decades before semiconductors and other technologies allowed for the full breadth of applications of this theory to be realized. Today we see the Hall effect put into use in disk drives, laptops and mobile phones, vending machines, automobile sensors, camera equipment, electronic locks, magnetic card sensors, spacecraft thrusters, prosthetic limbs (such as the i-limb ultra, pictured above), and countless other areas.

The Hall effect is observed when an electric current is run through a material that is in the presence of an external magnetic field. The voltage related to the strength of the magnetic field is measured on either side of the material. Alternatively, a materials scientist could make a similar measurement of this voltage while controlling an external magnetic field in order to benchmark the ability of a particular material to carry a current, and identify the number density and type of charge carriers.

In this chapter, we extend the rectangular coordinate system from two dimensions, the Cartesian plane, to three dimensions, *space*. Just as lines, circles, and conics are important in two dimensions, lines, planes, spheres, and *quadric surfaces* play an important role in space.

We also introduce *vectors*, both in the plane and in space. Vectors, which are entities that have both magnitude and direction, are particularly useful in describing the way forces behave, such as the force of the wind on an airplane. Vectors are also used to find the work done by a constant force in moving an object when the direction of the force is not along the line of motion.

# 10.1 Rectangular Coordinates in Space

**OBJECTIVES** *When you finish this section, you should be able to:*

1 Locate points in space (p. 734)
2 Find the distance between two points in space (p. 735)
3 Find the equation of a sphere (p. 736)

A two-dimensional rectangular coordinate system is represented by a plane in which every point $P$ corresponds to exactly one ordered pair of real numbers $(x, y)$, the coordinates of $P$. In space every point $P$ corresponds to exactly one *ordered triple* of real numbers $(x, y, z)$.

## 1 Locate Points in Space

We begin by selecting a fixed point called the **origin** $O$. Through the origin, draw three mutually perpendicular number lines, called the **coordinate axes.** The coordinate axes are usually labeled the **x-axis**, the **y-axis**, and the **z-axis.** On each axis choose one direction as positive and select an appropriate scale, as shown in Figure 1.

Notice in Figure 1 that the positive $z$-axis points upward, the positive $y$-axis points toward the right, and the positive $x$-axis points forward. This is known as a **right-handed** system because it conforms to the *right-hand rule*. The **right-hand rule** states: if the index finger of the right hand points in the direction of the positive $x$-axis and the other fingers point in the direction of the positive $y$-axis, then the thumb will point in the direction of the positive $z$-axis, as shown in Figure 2.

**NOTE** There is also a left-handed system and a left-handed rule; we use only a right-handed system.

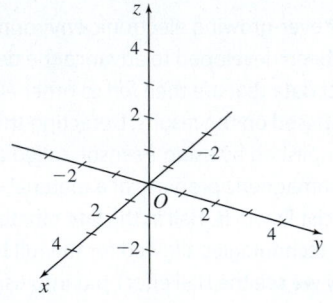

**Figure 1**

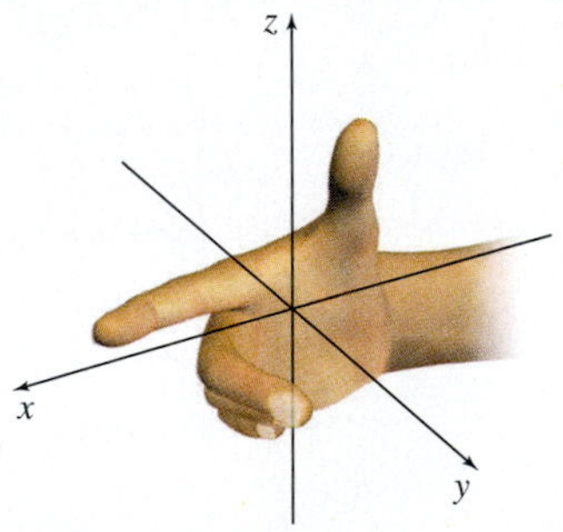

**Figure 2** Right-hand rule

As in the plane, we assign coordinates to each point $P$ in space, but in space we use an ordered triple of real numbers $(x, y, z)$. For example, the point $(3, 5, 7)$ is the point for which 3 is the $x$-coordinate, 5 is the $y$-coordinate, and 7 is the $z$-coordinate. The point $(3, 5, 7)$ is plotted by starting at the origin, moving 3 units along the positive $x$-axis, 5 units in the direction of the positive $y$-axis, and 7 units in the direction of the positive $z$-axis. See Figure 3.

Figure 3 also shows the points $(3, 0, 0)$, $(0, 5, 0)$, $(0, 0, 7)$, $(3, 5, 0)$, $(3, 0, 7)$, $(0, 5, 7)$, and $(0, 0, 0)$, which form a box with the points $(0, 0, 0)$ and $(3, 5, 7)$ as opposite vertices.

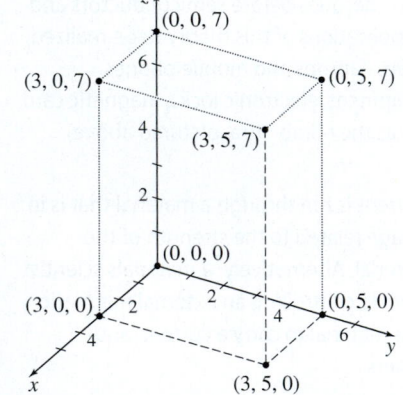

**Figure 3**

**NOW WORK** Problems **11** and **15.**

Points of the form $(x, 0, 0)$ lie on the $x$-axis, while points of the form $(0, y, 0)$ and $(0, 0, z)$ lie on the $y$-axis and $z$-axis, respectively.

Points of the form $(x, y, 0)$ lie on a plane called the **xy-plane.** The $xy$-plane is perpendicular to the $z$-axis and its equation is $z = 0$. This is the plane used in the familiar two-dimensional rectangular coordinate system.

Points of the form $(x, 0, z)$ lie on the **xz-plane.** The $xz$-plane is perpendicular to the $y$-axis, and its equation is $y = 0$. Finally, points of the form $(0, y, z)$ lie on the **yz-plane**, which is perpendicular to the $x$-axis, and its equation is $x = 0$. Collectively, these three planes are referred to as **coordinate planes.** See Figure 4.

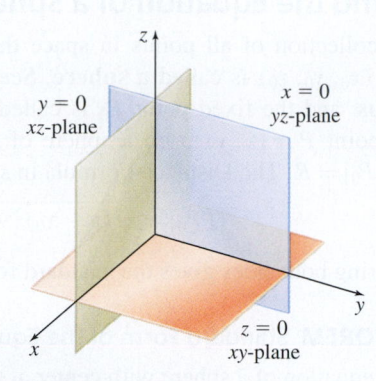

**DF** Figure 4

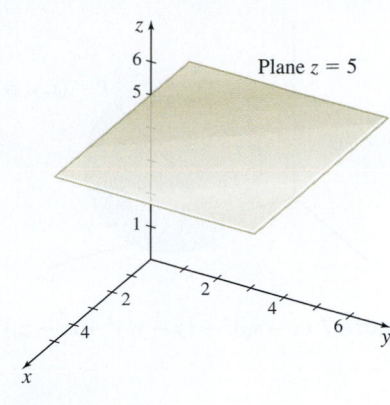

Figure 5 $z = 5$

In general, for any real number $k$, the graph of $x = k$ is a plane parallel to the $yz$-plane; the graph of $y = k$ is a plane parallel to the $xz$-plane; and the graph of $z = k$ is a plane parallel to the $xy$-plane. As shown in Figure 5, the graph of the equation $z = 5$ is a plane parallel to and 5 units above the $xy$-plane.

**NOW WORK** Problem 19.

## 2 Find the Distance Between Two Points in Space

**NEED TO REVIEW?** The distance formula in the plane is given in Appendix A.3, p. A-16.

Figure 6

The formula for the distance between two points in space is an extension of the distance formula in the plane.

To find the distance $|P_1 P_2|$ between two points $P_1 = (x_1, y_1, z_1)$ and $P_2 = (x_2, y_2, z_2)$, we use the Pythagorean Theorem twice. Figure 6 shows the points $P_1$ and $P_2$ and a third point $A = (x_2, y_2, z_1)$. The first application of the Pythagorean Theorem involves observing that the triangle $P_1 A P_2$ is a right triangle where the side of length $|P_1 P_2|$ is the hypotenuse. As a result,

$$|P_1 P_2| = \sqrt{|P_1 A|^2 + |A P_2|^2} \qquad \text{\color{blue}Pythagorean Theorem}$$

The points $P_1$ and $A$ lie in a plane parallel to the $xy$-plane. Do you see why? To find $|P_1 A|$, we use the distance formula in the plane:

$$|P_1 A| = \sqrt{(x_2 - x_1)^2 + (y_2 - y_1)^2}$$

The points $P_2$ and $A$ lie on a line parallel to the $z$-axis, so that $|A P_2| = |z_2 - z_1|$ and $|A P_2|^2 = (z_2 - z_1)^2$. Combining these results, we obtain a formula for the distance between two points in space.

**THEOREM** Distance Formula in Space

The distance between two points $P_1 = (x_1, y_1, z_1)$ and $P_2 = (x_2, y_2, z_2)$ in space, denoted by $|P_1 P_2|$, is

$$|P_1 P_2| = \sqrt{(x_2 - x_1)^2 + (y_2 - y_1)^2 + (z_2 - z_1)^2}$$

**EXAMPLE 1** Finding the Distance Between Two Points in Space

Find the distance between $P_1 = (1, 3, -2)$ and $P_2 = (2, -1, -3)$.

**Solution**

$$|P_1 P_2| = \sqrt{(2-1)^2 + (-1-3)^2 + (-3+2)^2} = \sqrt{1 + 16 + 1} = \sqrt{18} = 3\sqrt{2} \qquad \blacksquare$$

**NOW WORK** Problem 35.

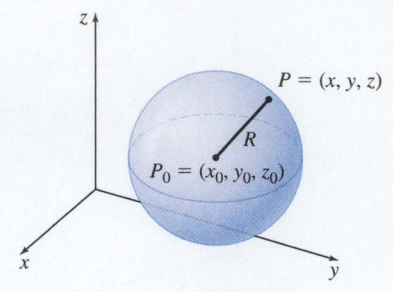

**Figure 7** $(x - x_0)^2 + (y - y_0)^2 + (z - z_0)^2 = R^2$

## 3 Find the Equation of a Sphere

The collection of all points in space that are a fixed distance $R$ from a fixed point $P_0 = (x_0, y_0, z_0)$ is called a **sphere**. See Figure 7. The fixed distance $R$ is called the **radius**, and the fixed point $P_0$ is called the **center** of the sphere. The distance from any point $P = (x, y, z)$ on a sphere of radius $R$ to the center point $P_0 = (x_0, y_0, z_0)$ is $|PP_0| = R$. The Distance Formula in space shows that

$$|PP_0| = \sqrt{(x - x_0)^2 + (y - y_0)^2 + (z - z_0)^2} = R$$

Squaring both sides gives the standard form of the equation of a sphere.

> **THEOREM  Standard Form of the Equation of a Sphere**
>
> The equation of a sphere with center at the point $P_0 = (x_0, y_0, z_0)$ and radius $R$ is
>
> $$\boxed{(x - x_0)^2 + (y - y_0)^2 + (z - z_0)^2 = R^2}$$

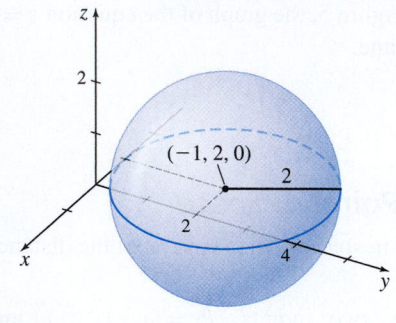

**Figure 8** $(x + 1)^2 + (y - 2)^2 + z^2 = 4$

**CALC CLIP** ▶ **EXAMPLE 2  Finding the Equation of a Sphere**

The standard form of the equation of the sphere shown in Figure 8, with radius 2 and center at $(-1, 2, 0)$, is

$$(x + 1)^2 + (y - 2)^2 + z^2 = 4$$    ∎

**NOW WORK** Problem 41.

**EXAMPLE 3  Finding the Center and Radius of a Sphere**

Show that

$$x^2 + y^2 + z^2 + 2x + 4y - 2z = 10$$

is the equation of a sphere. Find its center and radius.

**Solution** We begin by writing the equation as

$$(x^2 + 2x) + (y^2 + 4y) + (z^2 - 2z) = 10$$

and then we complete the square three times. The result is

$$(x^2 + 2x + 1) + (y^2 + 4y + 4) + (z^2 - 2z + 1) = 10 + 1 + 4 + 1$$

$$(x + 1)^2 + (y + 2)^2 + (z - 1)^2 = 16$$

This is the equation of a sphere with radius 4 and center at $(-1, -2, 1)$. ∎

**NEED TO REVIEW?** Completing the square is discussed in Appendix A.1, pp. A-2 to A-3.

**NOW WORK** Problem 43.

---

## 10.1 Assess Your Understanding

### Concepts and Vocabulary

1. *True or False*  Each point in space is denoted by an ordered triple of real numbers.

2. *True or False*  The point $(0, 4, 0)$ lies on the $z$-axis.

3. *True or False*  The right-hand rule states that if your index finger points toward the positive $x$-axis, and your other fingers point toward the positive $y$-axis, then your thumb points toward the positive $z$-axis.

4. *True or False*  In space, the equation $y = 3$ describes a plane.

5. In space, the set of all points a fixed distance $R$ from a fixed point $(x_0, y_0, z_0)$ is called a(n) _____.

6. The equation $x^2 + y^2 + z^2 = 8$ describes a sphere whose center is at the point _____.

### Skill Building

*In Problems 7–12, plot each point in space.*

7. $(1, 1, 1)$      8. $(0, 0, 1)$      9. $(0, 2, 5)$

10. $(-1, 0, 5)$    11. $(-3, 1, 0)$    12. $(4, -1, -3)$

---

**1.** = NOW WORK problem      📈 = Graphing technology recommended      **CAS** = Computer Algebra System recommended

*In Problems 13–18, opposite vertices of a rectangular box whose edges are parallel to the coordinate axes are given. List the coordinates of the other six vertices of the box.*

**13.** $(0, 0, 0)$; $(2, 1, 3)$  **14.** $(0, 0, 0)$; $(4, 2, 2)$

**15.** $(1, 2, 3)$; $(3, 4, 5)$  **16.** $(5, 6, 1)$; $(3, 8, 2)$

**17.** $(-1, 0, 2)$; $(4, 2, 5)$  **18.** $(-2, -3, 0)$; $(-6, 7, 1)$

*In Problems 19–32, describe in words the set of all points $(x, y, z)$ that satisfy the given statements.*

**19.** $y = -2$  **20.** $z = -3$

**21.** $x = 0$  **22.** $z = 5$

**23.** $x = 1$ and $y = 0$  **24.** $x = y$ and $z = 0$

**25.** $x > -2$  **26.** $0 \leq y \leq 4$

**27.** $x^2 + y^2 + z^2 \leq 1$  **28.** $x^2 + y^2 + z^2 \geq 25$

**29.** $xy = 0$  **30.** $xyz = 0$

**31.** $(x - 1)(z + 3) = 0$  **32.** $(y + 2)(z + 1) = 0$

*In Problems 33–38, find the distance between each pair of points.*

**33.** $(3, 2, 5)$ and $(-1, 2, 2)$  **34.** $(12, -5, 16)$ and $(21, 15, 4)$

**35.** $(-1, 2, -3)$ and $(4, -2, 1)$

**36.** $(1.0, 3.2, 4.5)$ and $(1.4, 1.0, 6.5)$

**37.** $(4, -2, -2)$ and $(3, 2, 1)$  **38.** $(2, -3, -3)$ and $(4, 1, -1)$

*In Problems 39–42, find the standard equation of a sphere with radius $R$ and center at $P_0$.*

**39.** $R = 1$; $P_0 = (3, 1, 1)$  **40.** $R = 2$; $P_0 = (1, 2, 2)$

**41.** $R = 3$; $P_0 = (-1, 1, 2)$  **42.** $R = 1$; $P_0 = (-3, 1, -1)$

*In Problems 43–48, find the radius and center of each sphere.*

**43.** $x^2 + y^2 + z^2 + 2x - 2y = 2$

**44.** $x^2 + y^2 + z^2 + 2x - 2z = -1$

**45.** $x^2 + y^2 + z^2 + 4x - 4y + 2z = 0$

**46.** $x^2 + y^2 + z^2 - 4x = 0$

**47.** $2x^2 + 2y^2 + 2z^2 - 8x + 5z + 1 = 0$

**48.** $3x^2 + 3y^2 + 3z^2 + 6x - y - 3 = 0$

## Applications and Extensions

*In Problems 49–54, write the equation of each sphere.*

**49.** The endpoints of a diameter are $(-2, 0, 4)$ and $(2, 6, 8)$.

**50.** The endpoints of a diameter are $(1, 3, 6)$ and $(-3, 1, 4)$.

**51.** The center is at $(-3, 2, 1)$, and the sphere passes through the point $(4, -1, 3)$.

**52.** The center is at $(0, -3, 4)$, and the sphere passes through the point $(2, 1, 1)$.

**53.** The center is at $(2, 1, -2)$ and the sphere is tangent to the $xy$-plane.

**54.** The center is at $(1, 5, 4)$ and the sphere is tangent to the $yz$-plane.

**55.** **Chemistry: Iron Crystals**   In its solid form, iron forms a crystal lattice known as a body-centered cubic (bcc). This crystal consists of eight iron atoms at the corners of a cube plus one more atom at the center of the cube. The sides of the cube are 0.287 nm long ($1$ nm $= 10^{-9}$ m). Put one corner of the cube at the origin, and place three of the faces along the positive coordinate planes.

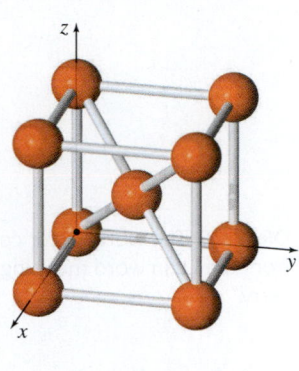

(a) Assign coordinates to each atom in the crystal.

(b) What are the coordinates of the atom that is farthest from the origin?

(c) What are the coordinates of the atom at the center of the cube?

(d) Find the distance between the iron atom at the center of the cube and the atom in the $xz$-plane that is farthest from the origin.

**56.** **Equilateral Triangle**   Show that the points $(2, 4, 2)$, $(2, 1, 5)$, and $(5, 1, 2)$ are the vertices of an equilateral triangle.

**57.** **Sphere**

(a) Find an equation of the sphere with center at $(4, 0, -2)$ and radius 5.

(b) Describe the intersection of the sphere and the $xy$-plane.

**58.** **Right Triangle Inscribed in a Sphere**

(a) Show that the points $(-2, 6, 0)$, $(4, 9, 1)$, and $(-3, 2, 18)$ are the vertices of a right triangle.

(b) Find an equation of the sphere that passes through the vertices of the right triangle and that has a diameter along its hypotenuse.

**59.** **Distance**   Find the shortest distance from the point $(4, 2, -1)$ to the $xy$-plane.

**60.** **Midpoint Formula**   Derive a formula to find the midpoint of the line segment joining two points in space.

## Challenge Problems

**61.** **Sphere**   Find an equation of the sphere passing through $(3, 0, 0)$, $(0, 0, 1)$, $(-1, -3, 1)$, and $(2, 0, 1)$.

**62.** **Sphere**

(a) Show that the graph of

$$Ax^2 + Ay^2 + Az^2 + Dx + Ey + Fz = G \qquad A > 0$$

is a sphere, a point, or the empty set.

(b) When is it a sphere? What are its center and radius?

# 10.2 Introduction to Vectors

**OBJECTIVES** *When you finish this section, you should be able to:*

1 Represent vectors geometrically (p. 738)
2 Use properties of vectors (p. 741)

**ORIGINS** The word *vector* comes from the Latin word meaning "to carry."

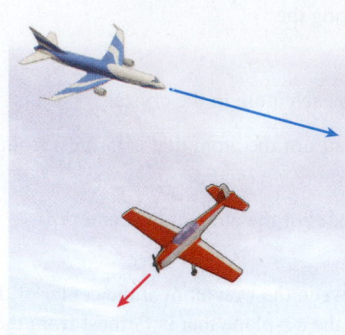

**Figure 9**

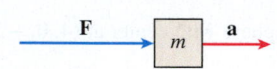

**Figure 10** $\mathbf{F} = m\mathbf{a}$

Geometrically, a **vector** is an entity that has both magnitude and direction, and is often represented by an arrow pointing in the *direction* of the vector. The length of the arrow represents the *magnitude* of the vector.

Velocity is a vector. The velocity of an airplane can be represented by an arrow that points in the direction of its flight; the length of the arrow represents its speed (the magnitude of the velocity). See Figure 9. If the airplane speeds up, the length of the arrow increases.

When working with vectors, we refer to real numbers as *scalars*. **Scalars** are quantities that have magnitude but no direction. Physical examples of scalar quantities are time, mass, temperature, and speed. In print, boldface letters are used to denote vectors, to distinguish them from scalars. For handwritten work, an arrow is placed over a letter to denote a vector.

*Force* and *acceleration* are also vectors. If a force $\mathbf{F}$ is exerted on an object of mass $m$, the force causes the object to accelerate in the direction of the force. See Figure 10. If $\mathbf{a}$ is the acceleration, then $\mathbf{F}$ and $\mathbf{a}$ are related by the equation $\mathbf{F} = m\mathbf{a}$, Newton's Second Law of Motion.

## 1 Represent Vectors Geometrically

Vectors are closely related to the geometric idea of a *directed line segment*. If $P$ and $Q$ are distinct points, there is exactly one line containing both $P$ and $Q$. See Figure 11(a). The points $P$ and $Q$, and those on the line between $P$ and $Q$, form the **line segment** $\overline{PQ}$. See Figure 11(b). If the points are ordered so they begin with $P$ and end with $Q$, then we have a **directed line segment from** $P$ **to** $Q$, denoted by $\overrightarrow{PQ}$. For the directed line segment $\overrightarrow{PQ}$, the point $P$ is the **initial point** and the point $Q$ is the **terminal point**. See Figure 11(c).

The **magnitude** of the directed line segment $\overrightarrow{PQ}$ is the distance from the point $P$ to the point $Q$; that is, it is the length of the line segment $\overline{PQ}$. The **direction** of $\overrightarrow{PQ}$ is from $P$ to $Q$. If a vector $\mathbf{v}$ has the same magnitude and the same direction as the directed line segment $\overrightarrow{PQ}$, then

$$\mathbf{v} = \overrightarrow{PQ}$$

The vector whose magnitude is 0 is called the **zero vector 0**. The zero vector is assigned no direction.

Two vectors $\mathbf{v}$ and $\mathbf{w}$ are **equal**, written $\mathbf{v} = \mathbf{w}$, if they have the same magnitude and the same direction. For example, the vectors in Figure 12 are equal even though they have different initial points and terminal points.

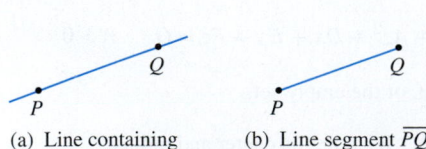

(a) Line containing $P$ and $Q$
(b) Line segment $\overline{PQ}$

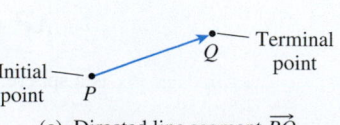

(c) Directed line segment $\overrightarrow{PQ}$

**Figure 11**

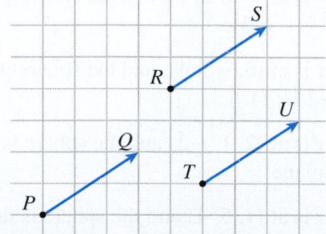

**Figure 12** The vectors $\overrightarrow{PQ}$, $\overrightarrow{RS}$, and $\overrightarrow{TU}$ are equal.

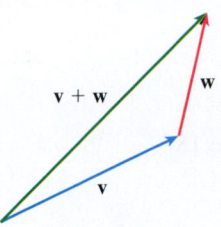

**Figure 13** The sum of vectors **v** and **w** is found by positioning the initial point of **w** at the terminal point of **v**.

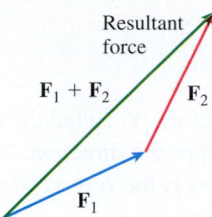

**Figure 14** $\mathbf{F}_1 + \mathbf{F}_2$ is the resultant force.

As a result, it is useful to think of a vector as an arrow, keeping in mind that two arrows (vectors) are equal if they have the same direction and the same length (magnitude).

Since location does not affect a vector, we **add** two vectors **v** and **w** by positioning vector **w** so that its initial point coincides with the terminal point of **v**. The **vector sum v + w** is the unique vector represented by the arrow from the initial point of **v** to the terminal point of **w**, as shown in Figure 13.

We said that forces can be represented by vectors. But how do we know that forces "add" the same way vectors do? Well, physicists say they do, and laboratory experiments bear it out. So if $\mathbf{F}_1$ and $\mathbf{F}_2$ are two forces simultaneously acting on an object, the vector sum $\mathbf{F}_1 + \mathbf{F}_2$ is equal to the force that produces the same effect as that obtained when the forces $\mathbf{F}_1$ and $\mathbf{F}_2$ are applied at the same time to the same object. The force $\mathbf{F}_1 + \mathbf{F}_2$ is called the **resultant force**. See Figure 14.

**NOW WORK** Problem **23.**

If **w** is a vector describing the velocity of the wind relative to Earth and if **v** is a vector describing the velocity of an airplane in the air, then **w + v** is a vector describing the velocity of the airplane relative to Earth. See Figure 15.

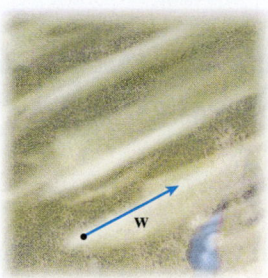

(a) Velocity of the wind relative to Earth

(b) Velocity of an airplane relative to the air

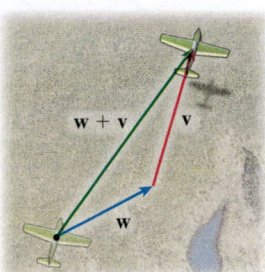

(c) Resultant equals the velocity of an airplane relative to Earth

**Figure 15**

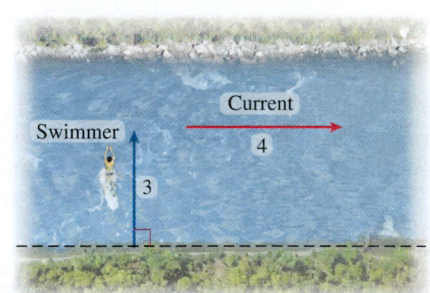

**Figure 16**

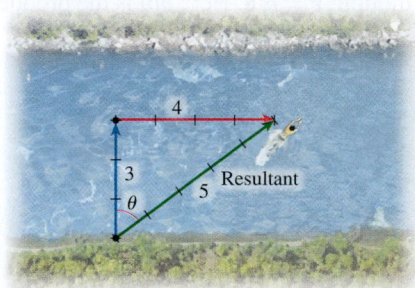

**Figure 17**

**EXAMPLE 1** **Graphing a Resultant Vector**

A swimmer swims at a constant speed of 3 miles per hour (mi/h) and heads directly across a river whose current is 4 mi/h.

**(a)** Draw vectors representing the swimmer and the current.

**(b)** Draw the resultant vector.

**(c)** Interpret the result.

**Solution (a)** We represent the swimmer by a vector pointing directly across the river and the current by a vector pointing parallel to a straight shore line. Since the swimmer's speed is 3 mi/h, and the current's speed is 4 mi/h, the length of the vector representing the swimmer is 3 units, and the length of the vector representing the current is 4 units. See Figure 16.

**(b)** The resultant vector is the sum of the vectors in (a). See Figure 17.

**(c)** The resultant vector shows the true speed of the swimmer and true direction of the swim. In fact, the true speed is 5 mi/h, and the true direction can be described in terms of the angle $\theta$ as $\theta = \tan^{-1} \dfrac{4}{3} \approx 53.1°$. ∎

**NOW WORK** Problem **25.**

Next, we examine properties of vectors. Many of these properties are similar to properties of real numbers.

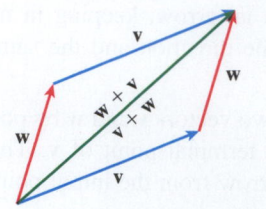

**DF Figure 18** $\mathbf{w} + \mathbf{v} = \mathbf{v} + \mathbf{w}$

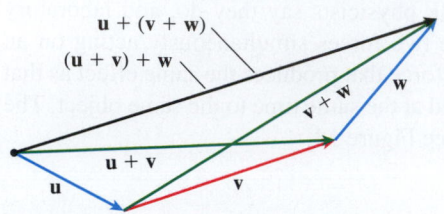

**Figure 19** $\mathbf{u} + (\mathbf{v} + \mathbf{w}) = (\mathbf{u} + \mathbf{v}) + \mathbf{w}$

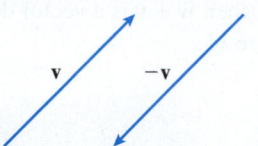

**Figure 20** $\mathbf{v} + (-\mathbf{v}) = \mathbf{0}$

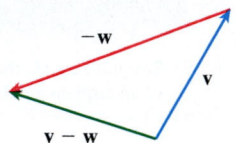

**Figure 21** $\mathbf{v} - \mathbf{w} = \mathbf{v} + (-\mathbf{w})$

## Properties of Vector Addition

Let $\mathbf{u}$, $\mathbf{v}$, and $\mathbf{w}$ represent any three vectors.

- **Commutative property of vector addition** (Figure 18):

$$\mathbf{w} + \mathbf{v} = \mathbf{v} + \mathbf{w}$$

- **Associative property of vector addition** (Figure 19):

$$\mathbf{u} + (\mathbf{v} + \mathbf{w}) = (\mathbf{u} + \mathbf{v}) + \mathbf{w}$$

- **Additive identity property**  The zero vector $\mathbf{0}$ is called the **additive identity**. The sum of any vector $\mathbf{v}$ and the zero vector $\mathbf{0}$ equals $\mathbf{v}$.

$$\mathbf{v} + \mathbf{0} = \mathbf{0} + \mathbf{v} = \mathbf{v}$$

- **Additive inverse property**  For any vector $\mathbf{v}$, the vector $-\mathbf{v}$, called the **additive inverse** of $\mathbf{v}$, has the same magnitude as $\mathbf{v}$, but the opposite direction, as shown in Figure 20. The sum of $\mathbf{v}$ and its additive inverse $-\mathbf{v}$ is the zero vector $\mathbf{0}$.

$$\mathbf{v} + (-\mathbf{v}) = (-\mathbf{v}) + \mathbf{v} = \mathbf{0}$$

**DEFINITION  Difference of Two Vectors**

The **difference** of two vectors $\mathbf{v} - \mathbf{w}$ is defined as

$$\mathbf{v} - \mathbf{w} = \mathbf{v} + (-\mathbf{w})$$

The difference $\mathbf{v} - \mathbf{w}$ between two vectors can be obtained by positioning the initial point of $-\mathbf{w}$ with the terminal point of $\mathbf{v}$. Then the vector from the initial point of $\mathbf{v}$ to the terminal point of $-\mathbf{w}$ equals the vector $\mathbf{v} - \mathbf{w}$, as shown in Figure 21.

The product of a vector $\mathbf{v}$ and a scalar $a$ produces a vector called the *scalar multiple* of $\mathbf{v}$.

**DEFINITION  Scalar Multiple**

If $a$ is a scalar and $\mathbf{v}$ is a vector, the product $a\mathbf{v}$, called a **scalar multiple** of $\mathbf{v}$, is the vector with magnitude $|a|$ times the magnitude of $\mathbf{v}$, in the same direction as $\mathbf{v}$ when $a > 0$ and in the opposite direction from $\mathbf{v}$ when $a < 0$.

If either $a = 0$ or $\mathbf{v} = \mathbf{0}$, then $a\mathbf{v} = \mathbf{0}$.

See Figure 22 for illustrations of scalar multiples of $\mathbf{v}$. Notice that the additive inverse of $\mathbf{v}$ can be represented as the scalar multiple $(-1)\mathbf{v}$.

We saw earlier that if $\mathbf{a}$ is the acceleration of an object of mass $m$ due to a force $\mathbf{F}$ being exerted, then by Newton's second law of motion, $\mathbf{F} = m\mathbf{a}$. Here $m\mathbf{a}$ is the product of the scalar $m$ and the vector $\mathbf{a}$.

Scalar multiples have the following properties:

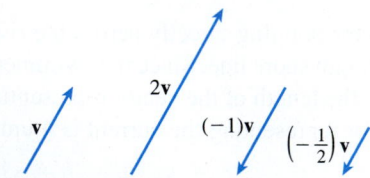

**Figure 22** Scalar multiples of $\mathbf{v}$

## Properties of Scalar Multiplication

If $\mathbf{v}$ and $\mathbf{w}$ are any two vectors and $\mathbf{0}$ is the zero vector, then:

- $0\mathbf{v} = \mathbf{0}$
- $1\mathbf{v} = \mathbf{v}$
- $(-1)\mathbf{v} = -\mathbf{v}$
- $(a + b)\mathbf{v} = a\mathbf{v} + b\mathbf{v}$
- $a(\mathbf{v} + \mathbf{w}) = a\mathbf{v} + a\mathbf{w}$
- $a(b\mathbf{v}) = (ab)\mathbf{v}$

## 2  Use Properties of Vectors

CALC
▶ **EXAMPLE 2**  **Using Properties of Vectors**
CLIP

Use the vectors illustrated in Figure 23 to graph each of the following vectors:

(a)  $\mathbf{v} - \mathbf{w}$    (b)  $2\mathbf{v} - \mathbf{w} + \mathbf{u}$    (c)  $\dfrac{2}{3}\mathbf{u} + \dfrac{1}{2}(\mathbf{v} - \mathbf{w})$

**Solution**  We reposition the vectors in Figure 23 as shown in Figure 24 to graph the desired vectors.

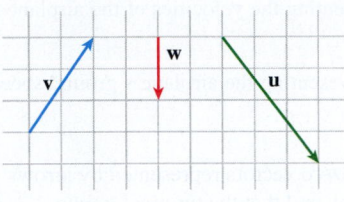

**Figure 23**

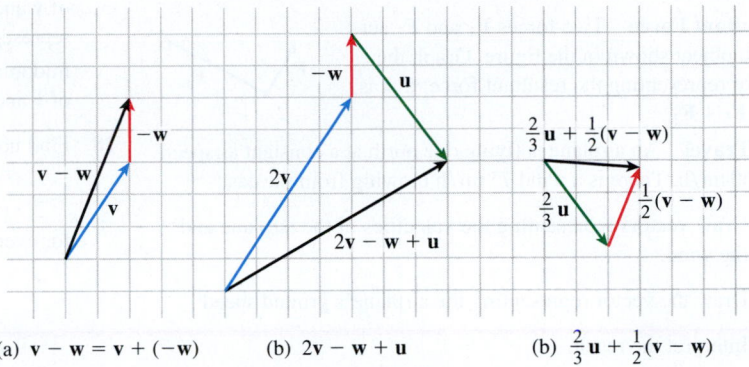

(a)  $\mathbf{v} - \mathbf{w} = \mathbf{v} + (-\mathbf{w})$    (b)  $2\mathbf{v} - \mathbf{w} + \mathbf{u}$    (b)  $\dfrac{2}{3}\mathbf{u} + \dfrac{1}{2}(\mathbf{v} - \mathbf{w})$

**Figure 24**

∎

NOW WORK  Problems **7** and **11**.

# 10.2  Assess Your Understanding

## Concepts and Vocabulary

**1.** *True or False*  A vector is an entity that has magnitude and direction.

**2.** Scalars are quantities that have only _____.

**3.** The product of a scalar *a* and a vector **v** is called a(n) _____ of **v**.

**4.** *Multiple Choice*  The vectors $-\mathbf{v}$ and $\mathbf{v}$ have
[**(a)** the same **(b)** different] magnitude and the
[**(c)** same **(d)** opposite] direction.

## Skill Building

**5.** State which of the following are scalars and which are vectors:

  (a) Volume  (b) Speed  (c) Force  (d) Work  (e) Mass
  (f) Distance  (g) Age  (h) Velocity  (i) Acceleration

*In Problems 6–14, use the vectors in the figure (top, right) to graph each of the following vectors:*

**6.** $2\mathbf{v}$        **7.** $-2\mathbf{v}$        **8.** $\mathbf{v} + \mathbf{w}$

**9.** $\mathbf{v} - \mathbf{w}$        **10.** $\mathbf{w} - \mathbf{v}$        **11.** $\mathbf{v} - 2\mathbf{w}$

**12.** $(\mathbf{v} + \mathbf{w}) + 3\mathbf{u}$    **13.** $\mathbf{v} + (\mathbf{w} + 3\mathbf{u})$    **14.** $2\mathbf{u} - \dfrac{1}{3}(\mathbf{v} - \mathbf{w})$

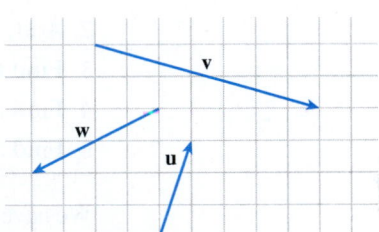

*In Problems 15–22, use the vectors in the figure below.*

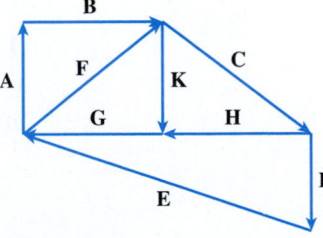

**15.** Find the vector **x** if $\mathbf{x} + \mathbf{B} = \mathbf{F}$.

**16.** Find the vector **x** if $\mathbf{x} + \mathbf{K} = \mathbf{C}$.

**17.** Write **C** in terms of **E**, **D**, and **F**.

**18.** Write **G** in terms of **C**, **D**, **E**, and **K**.

---

**1.** = NOW WORK problem     = Graphing technology recommended    CAS = Computer Algebra System recommended

**19.** Write **E** in terms of **G**, **H**, and **D**.

**20.** Write **E** in terms of **A**, **B**, **C**, and **D**.

**21.** What is **A** + **B** + **K** + **G**?

**22.** What is **A** + **B** + **C** + **H** + **G**?

### Applications and Extensions

**23. Resultant Force**  Two forces **F₁** and **F₂** act on an object shown in the figure. Graph the vector representing the resultant force; that is, find **F₁** + **F₂**.

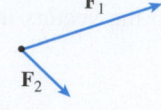

**24. Resultant Force**  Two forces **F₁** and **F₂** act on an object shown in the figure. Graph the vector representing the resultant force; that is, find **F₁** + **F₂**.

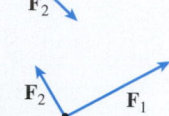

**25. Air Travel**  An airplane is flying due north at a constant airspeed of 560 mi/h. There is a wind 75 mi/h blowing from the east.

(a) Draw vectors representing the velocities of the airplane and the wind.

(b) Draw the vector representing the airplane's ground speed.

(c) Interpret the result.

**26. Air Travel**  An airplane maintains a constant airspeed of 500 km/h headed due west. There is a tail wind blowing at 130 km/h.

(a) Draw vectors representing the velocities of the airplane and the tail wind.

(b) Draw the vector representing the airplane's ground speed.

(c) Interpret the result.

**27.** Suppose **v** and **w** are nonzero vectors represented by arrows with the same initial point, and that the terminal points of **v** and **w** are $P$ and $Q$, respectively. Suppose the vector **u** is represented by an arrow from the initial point of **v** to the midpoint of the directed line segment $\overrightarrow{PQ}$. Write **u** in terms of **v** and **w**.

**28.** Find nonzero scalars $a$ and $b$ so that

$$4\mathbf{v} + a(\mathbf{v} - \mathbf{w}) + b(\mathbf{v} + \mathbf{w}) = \mathbf{0}$$

for every pair of vectors **v** and **w**.

# 10.3 Vectors in the Plane and in Space

**OBJECTIVES**  *When you finish this section, you should be able to:*

**1** Represent a vector algebraically (p. 742)

**2** Add, subtract, and find scalar multiples of vectors (p. 744)

**3** Find the magnitude of a vector (p. 746)

**4** Find a unit vector (p. 747)

**5** Find a vector in the plane from its direction and magnitude (p. 748)

We have been treating vectors geometrically, but to use vectors in a meaningful way, we need to look at vectors algebraically as well. To do this, we use a rectangular coordinate system: with two dimensions for vectors in the plane and with three dimensions for vectors in space.

## 1 Represent a Vector Algebraically

If **v** is a vector in a plane, and if its initial point is at the origin and its terminal point is at the point $(v_1, v_2)$, then we may represent **v** by the ordered pair.

$$\mathbf{v} = \langle v_1, v_2 \rangle$$

The numbers $v_1$ and $v_2$ are called the **components** of **v**.

Similarly, if **v** is a vector in space with its initial point at the origin and its terminal point at the point $(v_1, v_2, v_3)$, then **v** is represented by the ordered triple.

$$\mathbf{v} = \langle v_1, v_2, v_3 \rangle$$

The numbers $v_1$, $v_2$, and $v_3$ are called the **components** of **v**.

The vectors defined here, as well as any other vector whose initial point is at the origin, are called **position vectors**. See Figure 25.

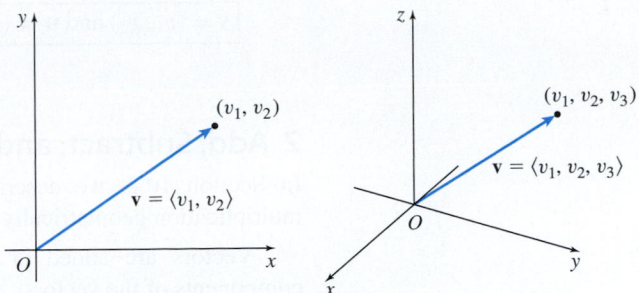

**Figure 25**

Any vector whose initial point is not at the origin is equal to a unique position vector. If $\mathbf{v}$ is a vector in the plane with initial point $P_1 = (x_1, y_1)$ and terminal point $P_2 = (x_2, y_2)$, then $\mathbf{v}$ equals the position vector

$$\mathbf{v} = \langle x_2 - x_1, \ y_2 - y_1 \rangle$$

Similarly, if $\mathbf{v}$ is a vector in space with initial point $P_1 = (x_1, y_1, z_1)$ and terminal point $P_2 = (x_2, y_2, z_2)$, then $\mathbf{v}$ equals the position vector

$$\mathbf{v} = \langle x_2 - x_1, \ y_2 - y_1, \ z_2 - z_1 \rangle$$

**IN WORDS** Every vector can be represented by a unique position vector.

As Figure 26 illustrates, this feature allows us to move vectors around in applications.

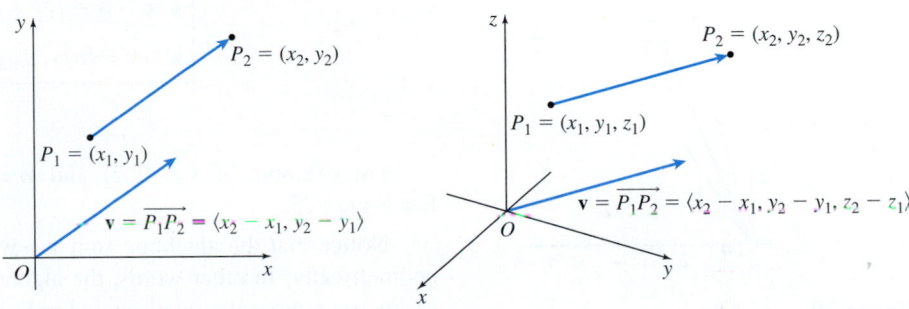

**Figure 26**

If both the initial point and the terminal point of vector $\mathbf{v}$ are at the origin, then $\mathbf{v}$ is the **zero vector**. As a result,

$$\mathbf{0} = \langle 0, 0 \rangle \text{ in the plane} \qquad \text{and} \qquad \mathbf{0} = \langle 0, 0, 0 \rangle \text{ in space}$$

**EXAMPLE 1**  **Finding a Position Vector**

Find the unique position vector $\mathbf{v}$ of a vector whose initial point is $P_1$ and whose terminal point is $P_2$.

**(a)** $P_1 = (-1, 2), \quad P_2 = (4, 6)$  **(b)** $P_1 = (3, 4, 2), \quad P_2 = (7, -8, 5)$

**Solution** To find $\mathbf{v}$, subtract corresponding components.

**(a)** $\mathbf{v} = \langle 4 - (-1), 6 - 2 \rangle = \langle 5, 4 \rangle$, as shown in Figure 27.

**(b)** $\mathbf{v} = \langle 7 - 3, -8 - 4, 5 - 2 \rangle = \langle 4, -12, 3 \rangle$  ∎

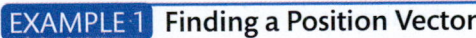

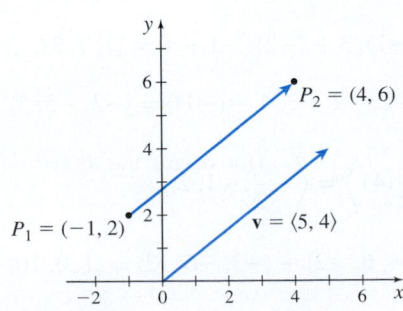

**Figure 27**

**NOW WORK** Problems 9(a) and 13(a).

Two position vectors **v** and **w** are equal if and only if the terminal point of **v** is the same as the terminal point of **w**.

$\mathbf{v} = \langle v_1, v_2 \rangle$ and $\mathbf{w} = \langle w_1, w_2 \rangle$ are equal if and only if $v_1 = w_1$ and $v_2 = w_2$.

## 2 Add, Subtract, and Find Scalar Multiples of Vectors

In Section 10.2, we described the operations of addition, subtraction, and scalar multiplication geometrically. Here we present them algebraically.

Vectors are added or subtracted and scalar multiples are formed using the components of the vectors.

**DEFINITION** Operations with Vectors

If $\mathbf{v} = \langle v_1, v_2 \rangle$ and $\mathbf{w} = \langle w_1, w_2 \rangle$ are two vectors in the plane and if $a$ is a scalar, then

- $\mathbf{v} + \mathbf{w} = \langle v_1 + w_1, v_2 + w_2 \rangle$
- $\mathbf{v} - \mathbf{w} = \langle v_1 - w_1, v_2 - w_2 \rangle$
- $a\mathbf{v} = \langle av_1, av_2 \rangle$

If $\mathbf{v} = \langle v_1, v_2, v_3 \rangle$ and $\mathbf{w} = \langle w_1, w_2, w_3 \rangle$ are two vectors in space and if $a$ is a scalar, then

- $\mathbf{v} + \mathbf{w} = \langle v_1 + w_1, v_2 + w_2, v_3 + w_3 \rangle$
- $\mathbf{v} - \mathbf{w} = \langle v_1 - w_1, v_2 - w_2, v_3 - w_3 \rangle$
- $a\mathbf{v} = \langle av_1, av_2, av_3 \rangle$

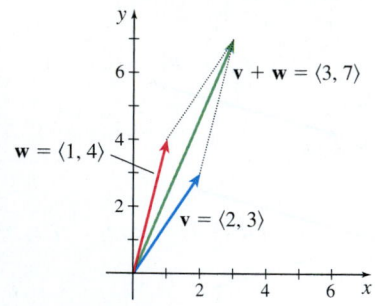

For example, if $\mathbf{v} = \langle 2, 3 \rangle$ and $\mathbf{w} = \langle 1, 4 \rangle$, then $\mathbf{v} + \mathbf{w} = \langle 2 + 1, 3 + 4 \rangle = \langle 3, 7 \rangle$. See Figure 28.

**Figure 28**

Notice that the algebraic sum $\mathbf{v} + \mathbf{w}$ is the same as the resultant vector obtained geometrically. In other words, the algebraic representation of vectors agrees with the geometric representation discussed in Section 10.2.

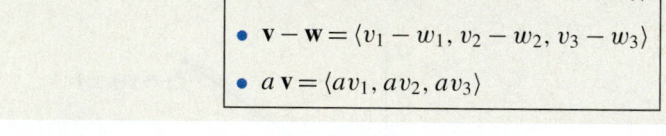

**EXAMPLE 2** Adding and Subtracting Vectors Algebraically

If $\mathbf{v} = \langle 2, 3, -1 \rangle$ and $\mathbf{w} = \langle -1, -2, 4 \rangle$, find:

**(a)** $\mathbf{v} + \mathbf{w}$      **(b)** $\mathbf{w} - \mathbf{v}$      **(c)** $\dfrac{1}{2}\mathbf{w}$      **(d)** $2\mathbf{v} + 3\mathbf{w}$

**Solution**

**(a)** $\mathbf{v} + \mathbf{w} = \langle 2, 3, -1 \rangle + \langle -1, -2, 4 \rangle = \langle 2 + (-1), 3 + (-2), -1 + 4 \rangle = \langle 1, 1, 3 \rangle$

**(b)** $\mathbf{w} - \mathbf{v} = \langle -1, -2, 4 \rangle - \langle 2, 3, -1 \rangle = \langle -1 - 2, -2 - 3, 4 - (-1) \rangle = \langle -3, -5, 5 \rangle$

**(c)** $\dfrac{1}{2}\mathbf{w} = \dfrac{1}{2} \langle -1, -2, 4 \rangle = \left\langle \dfrac{1}{2}(-1), \dfrac{1}{2}(-2), \dfrac{1}{2}(4) \right\rangle = \left\langle -\dfrac{1}{2}, -1, 2 \right\rangle$

**(d)** $2\mathbf{v} + 3\mathbf{w} = 2 \langle 2, 3, -1 \rangle + 3 \langle -1, -2, 4 \rangle = \langle 4, 6, -2 \rangle + \langle -3, -6, 12 \rangle = \langle 1, 0, 10 \rangle$

■

NOW WORK **Problem 25.**

The properties listed below for vectors are similar to the properties of real numbers.

---

**THEOREM  Properties of Vector Addition and Scalar Multiplication**

If **u**, **v**, and **w** are vectors and if $a$ and $b$ are scalars, then the following properties hold:

- **Commutative property of vector addition:**

$$\boxed{\mathbf{u} + \mathbf{v} = \mathbf{v} + \mathbf{u}}$$

- **Associative property of vector addition:**

$$\boxed{\mathbf{u} + (\mathbf{v} + \mathbf{w}) = (\mathbf{u} + \mathbf{v}) + \mathbf{w}}$$

- **Additive identity 0:**

$$\boxed{\mathbf{v} + \mathbf{0} = \mathbf{0} + \mathbf{v} = \mathbf{v}}$$

- **Additive inverse property:**

$$\boxed{\mathbf{v} + (-\mathbf{v}) = \mathbf{0}}$$

- **Distributive property of scalar multiplication over vector addition:**

$$\boxed{a(\mathbf{u} + \mathbf{v}) = a\,\mathbf{u} + a\,\mathbf{v}}$$

- **Distributive property of scalar multiplication over scalar addition:**

$$\boxed{(a + b)\mathbf{v} = a\,\mathbf{v} + b\,\mathbf{v}}$$

- **Linearity property of scalar multiplication:**

$$\boxed{a(b\,\mathbf{v}) = (ab)\mathbf{v}}$$

- **Scalar multiplication identity:**

$$\boxed{1\mathbf{v} = \mathbf{v}}$$

---

We proved some of these properties geometrically. For example, in Figure 18, we gave a geometric proof of the commutative property of addition. Here we algebraically prove the commutative property for vectors in the plane and the scalar multiplication identity property for vectors in space. The remaining proofs are left as exercises. See Problems 106 and 107.

**Proof** Commutative property of vector addition (in the plane): If $\mathbf{u} = \langle u_1, u_2 \rangle$ and if $\mathbf{v} = \langle v_1, v_2 \rangle$, then

$$\mathbf{u} + \mathbf{v} = \langle u_1, u_2 \rangle + \langle v_1, v_2 \rangle = \langle u_1 + v_1, u_2 + v_2 \rangle = \langle v_1 + u_1, v_2 + u_2 \rangle$$

$$= \langle v_1, v_2 \rangle + \langle u_1, u_2 \rangle = \mathbf{v} + \mathbf{u}$$

Scalar multiplication identity (in space):  If $\mathbf{v} = \langle v_1, v_2, v_3 \rangle$, then

$$1\mathbf{v} = \langle 1v_1, 1v_2, 1v_3 \rangle = \langle v_1, v_2, v_3 \rangle = \mathbf{v} \quad \blacksquare$$

---

**DEFINITION  Parallel Vectors**

Let **v** and **w** be two distinct nonzero vectors. If there is a nonzero scalar $a$ for which $\mathbf{v} = a\,\mathbf{w}$, then **v** and **w** are called **parallel vectors**.

**EXAMPLE 3** Showing Two Vectors Are Parallel

**(a)** The vectors

$$\mathbf{v} = \langle 2, 3, -1 \rangle \qquad \text{and} \qquad \mathbf{w} = \langle 4, 6, -2 \rangle$$

are parallel, since $\mathbf{v} = \dfrac{1}{2}\mathbf{w}$. In this case, $\mathbf{v}$ and $\mathbf{w}$ have the same direction.

**(b)** The vectors

$$\mathbf{v} = \langle 1, 2, 3 \rangle \qquad \text{and} \qquad \mathbf{w} = \langle -3, -6, -9 \rangle$$

are parallel, since $\mathbf{w} = -3\,\mathbf{v}$. In this case, $\mathbf{v}$ and $\mathbf{w}$ have opposite directions. ∎

NOW WORK Problem 75.

## 3 Find the Magnitude of a Vector

The representation of $\mathbf{v}$ as the vector $\langle v_1, v_2 \rangle$ gives the direction of $\mathbf{v}$. The definition below gives the magnitude of $\mathbf{v}$.

**DEFINITION**

The **magnitude** or **length** of a vector $\mathbf{v}$, denoted by the symbol $\|\mathbf{v}\|$, is the distance from the initial point to the terminal point of $\mathbf{v}$. $\|\mathbf{v}\|$ is also called the **norm** of the vector $\mathbf{v}$.

**THEOREM** Magnitude of a Vector

If $\mathbf{v} = \langle v_1, v_2 \rangle$ is a vector in the plane, the magnitude of $\mathbf{v}$ is

$$\|\mathbf{v}\| = \sqrt{v_1^2 + v_2^2}$$

If $\mathbf{v} = \langle v_1, v_2, v_3 \rangle$ is a vector in space, the magnitude of $\mathbf{v}$ is

$$\|\mathbf{v}\| = \sqrt{v_1^2 + v_2^2 + v_3^2}$$

The proof follows from the Distance Formula.

**EXAMPLE 4** Finding the Magnitude of a Vector

Find the magnitude of $\mathbf{v}$ if:

**(a)** $\mathbf{v} = \langle 3, -4 \rangle$    **(b)** $\mathbf{v} = \langle 2, 3, -1 \rangle$

**Solution (a)** $\|\mathbf{v}\| = \sqrt{v_1^2 + v_2^2} = \sqrt{3^2 + (-4)^2} = \sqrt{9 + 16} = 5$

**(b)** $\|\mathbf{v}\| = \sqrt{v_1^2 + v_2^2 + v_3^2} = \sqrt{2^2 + 3^2 + (-1)^2} = \sqrt{4 + 9 + 1} = \sqrt{14}$ ∎

NOW WORK Problems 21 and 29.

**THEOREM** Magnitude of the Scalar Multiple of a Vector

If $a$ is scalar and $\mathbf{v}$ is a vector, then

$$\|a\mathbf{v}\| = |a|\,\|\mathbf{v}\|$$

We prove the theorem for a vector $\mathbf{v}$ in the plane. The proof for a vector $\mathbf{v}$ in space is left as an exercise (Problem 108).

**Proof**  Suppose $\mathbf{v} = \langle v_1, v_2 \rangle$. Then the scalar multiple of $a$ and $\mathbf{v}$ is $a\mathbf{v} = \langle av_1, av_2 \rangle$ and

$$\|a\mathbf{v}\| = \sqrt{(av_1)^2 + (av_2)^2} = \sqrt{a^2 v_1^2 + a^2 v_2^2} = \sqrt{a^2 \left(v_1^2 + v_2^2\right)}$$

$$= \sqrt{a^2} \sqrt{v_1^2 + v_2^2} = |a| \|\mathbf{v}\| \qquad \blacksquare$$

## 4 Find a Unit Vector

**DEFINITION**

A vector with magnitude 1 is called a **unit vector**.

In many applications, it is necessary to find the unit vector with the same direction as a given vector $\mathbf{v}$.

**THEOREM**  Unit Vector in the Direction of v

For any nonzero vector $\mathbf{v}$, the vector

$$\boxed{\mathbf{u} = \frac{\mathbf{v}}{\|\mathbf{v}\|}}$$

is the unit vector that has the same direction as $\mathbf{v}$.

**Proof**  Since $\dfrac{1}{\|\mathbf{v}\|}$ is a positive scalar, the vector $\mathbf{u} = \dfrac{\mathbf{v}}{\|\mathbf{v}\|} = \dfrac{1}{\|\mathbf{v}\|}\mathbf{v}$ has the same direction as $\mathbf{v}$. Next we show that the magnitude of $\mathbf{u}$ is 1.

$$\|\mathbf{u}\| = \left\| \frac{\mathbf{v}}{\|\mathbf{v}\|} \right\| = \left\| \frac{1}{\|\mathbf{v}\|}\mathbf{v} \right\| = \frac{1}{\|\mathbf{v}\|}\|\mathbf{v}\| = 1$$

Since $\|\mathbf{u}\| = 1$, $\mathbf{u}$ is a unit vector in the direction of $\mathbf{v}$.  $\blacksquare$

Multiplying a nonzero vector $\mathbf{v}$ by $\dfrac{1}{\|\mathbf{v}\|}$ to obtain a unit vector that has the same direction as $\mathbf{v}$ is called **normalizing v**.

CALC
CLIP

**EXAMPLE 5**  **Normalizing a Vector**

Normalize each vector. That is, find a unit vector $\mathbf{u}$ that has the same direction as:

**(a)** $\mathbf{v} = \langle 3, -4 \rangle$     **(b)** $\mathbf{v} = \langle -1, 2, -2 \rangle$     **(c)** $\mathbf{v} = \langle 3, 0, 2 \rangle$

**Solution (a)**  Since $\mathbf{v} = \langle 3, -4 \rangle$, then $\|\mathbf{v}\| = \sqrt{9 + 16} = 5$. The unit vector $\mathbf{u}$ in the same direction as $\mathbf{v}$ is

$$\mathbf{u} = \frac{\mathbf{v}}{\|\mathbf{v}\|} = \frac{\mathbf{v}}{5} = \left\langle \frac{3}{5}, -\frac{4}{5} \right\rangle$$

**(b)**  Since $\mathbf{v} = \langle -1, 2, -2 \rangle$, then $\|\mathbf{v}\| = \sqrt{1 + 4 + 4} = 3$. The unit vector $\mathbf{u}$ in the same direction as $\mathbf{v}$ is

$$\mathbf{u} = \frac{\mathbf{v}}{\|\mathbf{v}\|} = \frac{\mathbf{v}}{3} = \left\langle -\frac{1}{3}, \frac{2}{3}, -\frac{2}{3} \right\rangle$$

**(c)**  Since $\mathbf{v} = \langle 3, 0, 2 \rangle$, then $\|\mathbf{v}\| = \sqrt{3^2 + 0^2 + 2^2} = \sqrt{13}$. The unit vector $\mathbf{u}$ in the same direction as $\mathbf{v}$ is

$$\mathbf{u} = \frac{\mathbf{v}}{\|\mathbf{v}\|} = \frac{\mathbf{v}}{\sqrt{13}} = \left\langle \frac{3}{\sqrt{13}}, \frac{0}{\sqrt{13}}, \frac{2}{\sqrt{13}} \right\rangle = \left\langle \frac{3\sqrt{13}}{13}, 0, \frac{2\sqrt{13}}{13} \right\rangle \qquad \blacksquare$$

NOW WORK  Problem 43.

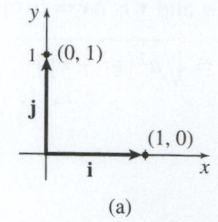

(a)

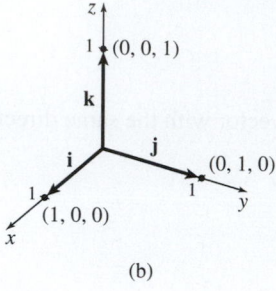

(b)

**Figure 29**

Unit vectors directed along the positive coordinate axes are called **standard basis vectors**. In the plane, the standard basis vectors are

$$\mathbf{i} = \langle 1, 0 \rangle \qquad \mathbf{j} = \langle 0, 1 \rangle$$

In space, the standard basis vectors are

$$\mathbf{i} = \langle 1, 0, 0 \rangle \qquad \mathbf{j} = \langle 0, 1, 0 \rangle \qquad \mathbf{k} = \langle 0, 0, 1 \rangle$$

See Figure 29(a) and (b).

**THEOREM  Writing a Vector in Terms of Standard Basis Vectors**

Every vector $\mathbf{v} = \langle v_1, v_2 \rangle$ in the plane can be written in terms of the standard basis vectors $\mathbf{i}$ and $\mathbf{j}$ as

$$\mathbf{v} = v_1 \mathbf{i} + v_2 \mathbf{j}$$

Every vector $\mathbf{v} = \langle v_1, v_2, v_3 \rangle$ in space can be written in terms of the standard basis vectors $\mathbf{i}$, $\mathbf{j}$, and $\mathbf{k}$ as

$$\mathbf{v} = v_1 \mathbf{i} + v_2 \mathbf{j} + v_3 \mathbf{k}$$

The proof for vectors in the plane follows. The proof for vectors in space is left as an exercise (Problem 109).

**Proof**  $\mathbf{v} = \langle v_1, v_2 \rangle = \langle v_1, 0 \rangle + \langle 0, v_2 \rangle = v_1 \langle 1, 0 \rangle + v_2 \langle 0, 1 \rangle = v_1 \mathbf{i} + v_2 \mathbf{j}$   ∎

**EXAMPLE 6  Writing a Vector Using Standard Basis Vectors**

(a)  $\langle 4, 3 \rangle = 4\mathbf{i} + 3\mathbf{j}$   (b)  $\langle 2, -3 \rangle = 2\mathbf{i} + (-3)\mathbf{j} = 2\mathbf{i} - 3\mathbf{j}$

(c)  $\langle 1, -3, 2 \rangle = \mathbf{i} - 3\mathbf{j} + 2\mathbf{k}$   (d)  $\langle 0, 3, 4 \rangle = 0\mathbf{i} + 3\mathbf{j} + 4\mathbf{k} = 3\mathbf{j} + 4\mathbf{k}$   ∎

**NOW WORK** Problems 9(b) and 13(b).

**EXAMPLE 7  Working with Standard Basis Vectors**

(a)  If $\mathbf{v} = 2\mathbf{i} - 3\mathbf{j}$ and $\mathbf{w} = -\mathbf{i} + 2\mathbf{j}$, find $4\mathbf{v} - \mathbf{w}$.

(b)  If $\mathbf{v} = 2\mathbf{i} + 3\mathbf{j} - \mathbf{k}$ and $\mathbf{w} = -\mathbf{i} - 2\mathbf{j} + 4\mathbf{k}$, find $2\mathbf{v} - 3\mathbf{w}$.

**Solution** (a)  $4\mathbf{v} - \mathbf{w} = 4(2\mathbf{i} - 3\mathbf{j}) - (-\mathbf{i} + 2\mathbf{j}) = (8\mathbf{i} - 12\mathbf{j}) + (\mathbf{i} - 2\mathbf{j}) = 9\mathbf{i} - 14\mathbf{j}$

(b)  $2\mathbf{v} - 3\mathbf{w} = 2(2\mathbf{i} + 3\mathbf{j} - \mathbf{k}) - 3(-\mathbf{i} - 2\mathbf{j} + 4\mathbf{k})$

$$= (4\mathbf{i} + 6\mathbf{j} - 2\mathbf{k}) + (3\mathbf{i} + 6\mathbf{j} - 12\mathbf{k}) = 7\mathbf{i} + 12\mathbf{j} - 14\mathbf{k}$$   ∎

**NOW WORK** Problems 55 and 65.

## 5 Find a Vector in the Plane from Its Direction and Magnitude

Often a vector is described by its magnitude and direction, rather than by its components. For example, a weather report warning of 40 mi/h southwesterly winds (from the southwest) describes a velocity vector, but to work with it requires finding its components.

A nonzero vector $\mathbf{v}$ in the plane can be described by specifying the angle $\theta$, $0 \le \theta < 2\pi$, between $\mathbf{v}$ and the positive $x$-axis, which gives its direction, and by specifying its magnitude $\|\mathbf{v}\|$. See Figure 30. If $\mathbf{v} = x\mathbf{i} + y\mathbf{j}$, then $x = \|\mathbf{v}\| \cos \theta$ and $y = \|\mathbf{v}\| \sin \theta$. So, $\mathbf{v} = \|\mathbf{v}\| \cos \theta \, \mathbf{i} + \|\mathbf{v}\| \sin \theta \, \mathbf{j} = \|\mathbf{v}\|(\cos \theta \, \mathbf{i} + \sin \theta \, \mathbf{j})$.

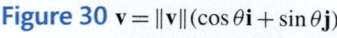

**Figure 30** $\mathbf{v} = \|\mathbf{v}\|(\cos \theta \mathbf{i} + \sin \theta \mathbf{j})$

**THEOREM**

If $\mathbf{v}$ is a nonzero vector, and if $\theta$ is the angle between $\mathbf{v}$ and the positive $x$-axis, then

$$\mathbf{v} = \|\mathbf{v}\| \, (\cos \theta \, \mathbf{i} + \sin \theta \, \mathbf{j})$$

**EXAMPLE 8** **Using Standard Basis Vectors to Model a Problem**

An airplane has an air speed of 400 km/h and is headed east. There is a northwesterly wind of 80 km/h. (Northwesterly winds blow toward the southeast.)

**(a)** Find a vector representing the velocity of the airplane in the air.

**(b)** Find a vector representing the wind velocity.

**(c)** Find a vector representing the velocity of the airplane relative to the ground.

**(d)** Find the true speed of the airplane.

**NOTE** Wind direction is measured from where it blows.

**Solution** We use a two-dimensional coordinate system with the direction north along the positive $y$-axis. Then the direction east is along the positive $x$-axis.

Using a scale of 1 unit $= 1$ km/h, define

$$\mathbf{v}_a = \text{Velocity of the airplane in the air}$$

$$\mathbf{v}_w = \text{Velocity of the wind}$$

$$\mathbf{v}_g = \text{Velocity of the airplane relative to the ground}$$

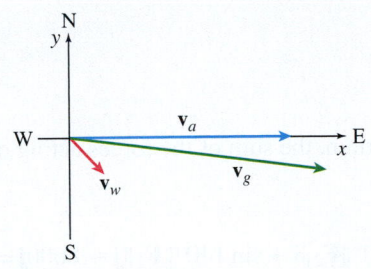

**Figure 31**

Figure 31 shows the vectors $\mathbf{v}_a$, $\mathbf{v}_w$, $\mathbf{v}_g$.

**(a)** The vector $\mathbf{v}_a$ has magnitude 400 and direction $\mathbf{i}$, so $\mathbf{v}_a = 400\mathbf{i}$.

**(b)** The vector $\mathbf{v}_w$ has magnitude 80 and makes an angle of $\dfrac{7\pi}{4}$ with the positive $x$-axis, as shown in Figure 32. Since the magnitude of $\mathbf{v}_w$ is 80,

$$\mathbf{v}_w = 80 \left( \cos \frac{7\pi}{4}\mathbf{i} + \sin \frac{7\pi}{4}\mathbf{j} \right) = 80 \left( \frac{\sqrt{2}}{2}\mathbf{i} - \frac{\sqrt{2}}{2}\mathbf{j} \right) = 40\sqrt{2}\mathbf{i} - 40\sqrt{2}\mathbf{j}$$

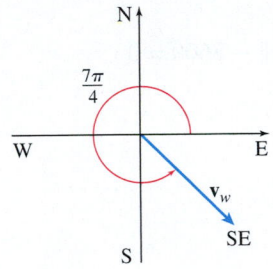

**Figure 32**

**(c)** The velocity $\mathbf{v}_g$ of the airplane relative to the ground is the resultant of $\mathbf{v}_a$ and $\mathbf{v}_w$.

$$\mathbf{v}_g = \mathbf{v}_a + \mathbf{v}_w = 400\,\mathbf{i} + \left[ 40\sqrt{2}\mathbf{i} - 40\sqrt{2}\mathbf{j} \right] = (400 + 40\sqrt{2})\mathbf{i} - 40\sqrt{2}\mathbf{j}$$

**(d)** The true speed of the airplane is the magnitude of the vector $\mathbf{v}_g$.

$$\|\mathbf{v}_g\| = \sqrt{(400 + 40\sqrt{2})^2 + (40\sqrt{2})^2} \approx 460.060$$

The true speed of the airplane is approximately 460.060 km/h. ∎

The direction of the airplane relative to the ground can be found using trigonometry, but we provide an easier method in the next section.

**NOW WORK** Problem 93.

When multiple forces $\mathbf{F}_1, \mathbf{F}_2, \ldots, \mathbf{F}_n$ simultaneously act on an object, the vector sum $\mathbf{F}_1 + \mathbf{F}_2 + \cdots + \mathbf{F}_n$ is the **resultant force**. The resultant force produces the same effect on the object as that obtained when each of the forces $\mathbf{F}_1, \mathbf{F}_2, \ldots, \mathbf{F}_n$ acts on the same object at the same time. An object is in **static equilibrium** if the object is at rest and the resultant force is **0**.

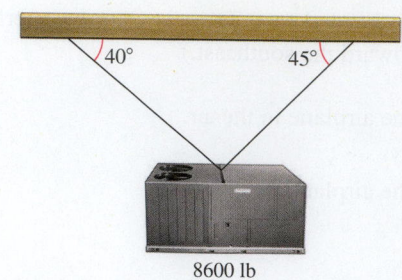

**Figure 33**

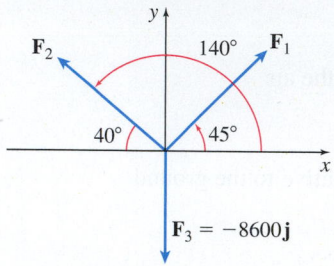

**Figure 34**

## EXAMPLE 9 Finding the Tension in a Cable

A commercial air conditioner weighing 8600 lb is suspended from two cables, as shown in Figure 33. What is the tension in each cable?

**Solution** We represent the cables and the air conditioner with a force diagram, as shown in Figure 34.

The tension in each cable is the magnitude of $\|\mathbf{F}_1\|$ and $\|\mathbf{F}_2\|$, respectively. The weight of the air conditioner is $\|\mathbf{F}_3\| = 8600$ lb.

$$\mathbf{F}_1 = \|\mathbf{F}_1\| (\cos 45°\mathbf{i} + \sin 45°\mathbf{j}) = \|\mathbf{F}_1\| \left( \frac{\sqrt{2}}{2}\mathbf{i} + \frac{\sqrt{2}}{2}\mathbf{j} \right)$$

$$\mathbf{F}_2 = \|\mathbf{F}_2\| (\cos 140°\mathbf{i} + \sin 140°\mathbf{j})$$

$$\mathbf{F}_3 = -8600\mathbf{j}$$

For the air conditioner to be in static equilibrium, the sum of the forces acting on it must equal $\mathbf{0}$. That is,

$$\mathbf{F}_1 + \mathbf{F}_2 + \mathbf{F}_3 = \frac{\sqrt{2}}{2}\|\mathbf{F}_1\|\mathbf{i} + \frac{\sqrt{2}}{2}\|\mathbf{F}_1\|\mathbf{j} + \cos 140°\|\mathbf{F}_2\|\mathbf{i} + \sin 140°\|\mathbf{F}_2\|\mathbf{j} - 8600\mathbf{j} = \mathbf{0}$$

Since both the $\mathbf{i}$ and $\mathbf{j}$ components equal 0, we have a system of two equations:

$$\begin{cases} \dfrac{\sqrt{2}}{2}\|\mathbf{F}_1\| + \cos 140°\|\mathbf{F}_2\| = 0 & (1) \\[2mm] \dfrac{\sqrt{2}}{2}\|\mathbf{F}_1\| + \sin 140°\|\mathbf{F}_2\| - 8600 = 0 & (2) \end{cases}$$

Solve equation (1) for $\|\mathbf{F}_1\|$:

$$\|\mathbf{F}_1\| = -\frac{2\cos 140°\|\mathbf{F}_2\|}{\sqrt{2}} \qquad (3)$$

and substitute the result into equation (2).

$$\frac{\sqrt{2}}{2} \left( -\frac{2\cos 140°\|\mathbf{F}_2\|}{\sqrt{2}} \right) + \sin 140°\|\mathbf{F}_2\| - 8600 = 0$$

$$\left[ -\cos 140° + \sin 140° \right]\|\mathbf{F}_2\| = 8600$$

$$\|\mathbf{F}_2\| = \frac{8600}{\sin 140° - \cos 140°} \approx 6104$$

Then from (3)

$$\|\mathbf{F}_1\| = -\frac{2\cos 140°\|\mathbf{F}_2\|}{\sqrt{2}} \approx -\frac{2\cos 140°}{\sqrt{2}}(6104) \approx 6613$$

The left cable has a tension of approximately 6104 lb and the right cable has a tension of approximately 6613 lb. ∎

**NOW WORK** Problem 97.

## Vectors in $n$ Dimensions

In a space of $n$ dimensions, the rectangular coordinates of a point are given by an ordered $n$-tuple $(v_1, v_2, \ldots, v_n)$ of real numbers. A vector $\mathbf{v}$ in $n$-space whose initial point is at the origin and whose terminal point is at $(v_1, v_2, \ldots, v_n)$ is defined as

$$\mathbf{v} = \langle v_1, v_2, \ldots, v_n \rangle$$

where the numbers $v_1, v_2, \ldots, v_n$ are the **components** of $\mathbf{v}$. In $n$-space there are $n$ coordinate axes and, along each positive axis, a standard basis vector is defined as follows:

$$\mathbf{e}_1 = \langle 1, 0, \ldots, 0 \rangle, \quad \mathbf{e}_2 = \langle 0, 1, \ldots, 0 \rangle, \quad \ldots, \quad \mathbf{e}_n = \langle 0, 0, \ldots, 1 \rangle$$

Any vector $\mathbf{v} = \langle v_1, v_2, \ldots, v_n \rangle$ in $n$-space can be written in terms of the standard basis vectors $\mathbf{e}_1, \mathbf{e}_2, \ldots, \mathbf{e}_n$ as

$$\mathbf{v} = v_1 \mathbf{e}_1 + v_2 \mathbf{e}_2 + \cdots + v_n \mathbf{e}_n$$

We add two vectors in $n$-space by adding respective components. The scalar multiple $a\mathbf{v}$ of a vector is found by multiplying each component by the scalar $a$. The magnitude of the vector $\mathbf{v} = v_1 \mathbf{e}_1 + v_2 \mathbf{e}_2 + \cdots + v_n \mathbf{e}_n$ is $\|\mathbf{v}\| = \sqrt{v_1^2 + v_2^2 + \cdots + v_n^2}$.

# 10.3 Assess Your Understanding

## Concepts and Vocabulary

**1.** In space, the standard basis vectors are $\mathbf{i} =$ _____, $\mathbf{j} =$ _____, and $\mathbf{k} =$ _____.

**2.** A vector whose magnitude is 1 is called a(n) _____ vector.

**3.** $v_1$, $v_2$, and $v_3$ are called the _____ of $\mathbf{v} = \langle v_1, v_2, v_3 \rangle$.

**4.** *True or False*  If $\mathbf{w} = 4\mathbf{v}$, then $\mathbf{w}$ has the same direction as $\mathbf{v}$.

**5.** *True or False*  The magnitude of a vector is found by adding its components.

**6.** *True or False*  In the plane, if the angle between a nonzero vector $\mathbf{v}$ and the unit vector $\mathbf{i}$ is $\theta$, then $\mathbf{v} = \|\mathbf{v}\|(\cos\theta\,\mathbf{i} + \sin\theta\,\mathbf{j})$.

**7.** If $\mathbf{v} = \mathbf{i} + \mathbf{j} + \mathbf{k}$, the unit vector $\mathbf{u}$ in the direction of $\mathbf{v}$ is $\mathbf{u} =$ _____.

**8.** *True or False*  For a vector $\mathbf{v}$ in the plane,

$$\|-2\mathbf{v}\| = -2\|\mathbf{v}\|$$

## Skill Building

*In Problems 9–16, the vector $\mathbf{v}$ has initial point $P_1$ and terminal point $P_2$.*

**(a)**  *Write each vector $\mathbf{v}$ as a position vector.*

**(b)**  *Write each vector $\mathbf{v}$ in terms of the standard basis vectors.*

**(c)**  *In Problems 9–12, graph the vector $\mathbf{v}$ and the corresponding position vector.*

**9.** $P_1 = (2, 3)$ and $P_2 = (6, -2)$

**10.** $P_1 = (4, -1)$ and $P_2 = (-3, 2)$

**11.** $P_1 = (0, 5)$ and $P_2 = (-1, 6)$

**12.** $P_1 = (-2, 0)$ and $P_2 = (3, 2)$

**13.** $P_1 = (6, 2, 1)$ and $P_2 = (3, 0, 2)$

**14.** $P_1 = (4, 7, 0)$ and $P_2 = (0, 5, 6)$

**15.** $P_1 = (-1, 0, 1)$ and $P_2 = (2, 0, 0)$

**16.** $P_1 = (6, 2, 2)$ and $P_2 = (2, 6, 2)$

*In Problems 17–24, use $\mathbf{v} = \langle 3, -1 \rangle$ and $\mathbf{w} = \langle -3, 2 \rangle$ to find each quantity.*

**17.** $2\mathbf{v} - \mathbf{w}$

**18.** $\mathbf{v} + 5\mathbf{w}$

**19.** $\dfrac{1}{3}\mathbf{v} + \dfrac{1}{2}\mathbf{w}$

**20.** $\dfrac{2}{3}\mathbf{v} - \dfrac{1}{2}\mathbf{w}$

**21.** $\|\mathbf{v}\|$

**22.** $\|\mathbf{w}\|$

**23.** $\|2\mathbf{v} - \mathbf{w}\|$

**24.** $\|\mathbf{v} + \mathbf{w}\|$

*In Problems 25–32, use $\mathbf{u} = \langle 1, 1, -6 \rangle$, $\mathbf{v} = \langle 3, 0, 8 \rangle$, and $\mathbf{w} = \langle -3, -1, 2 \rangle$ to find each quantity.*

**25.** $3\mathbf{v} - 2\mathbf{w}$

**26.** $-2\mathbf{v} + \mathbf{w}$

**27.** $2\mathbf{u} - 3\mathbf{v} + 4\mathbf{w}$

**28.** $\mathbf{u} + 3(\mathbf{v} - 4\mathbf{w})$

**29.** $\|\mathbf{u}\|$

**30.** $\|\mathbf{v}\|$

**31.** $\|5\mathbf{u} - \mathbf{v} + \mathbf{w}\|$

**32.** $\|5\mathbf{u}\| - \|\mathbf{v}\| + \|\mathbf{w}\|$

*In Problems 33–42, find the magnitude of each vector $\mathbf{v}$.*

**33.** $\mathbf{v} = 4\mathbf{i} - 3\mathbf{j}$

**34.** $\mathbf{v} = \mathbf{i} - 12\mathbf{j}$

**35.** $\mathbf{v} = \mathbf{i} + \mathbf{j}$

**36.** $\mathbf{v} = \mathbf{i} - \mathbf{j}$

**37.** $\mathbf{v} = 4\mathbf{i} + 2\mathbf{j} - \mathbf{k}$

**38.** $\mathbf{v} = \mathbf{i} - \mathbf{j} + \mathbf{k}$

**39.** $\mathbf{v} = \mathbf{j} + \mathbf{k}$

**40.** $\mathbf{v} = 2\mathbf{i} - \mathbf{k}$

**41.** $\mathbf{v} = \cos\theta\,\mathbf{i} + \sin\theta\,\mathbf{j}$

**42.** $\mathbf{v} = \cos\theta\,\mathbf{i} + \sin\theta\,\mathbf{j} + \mathbf{k}$

*In Problems 43–54, normalize each vector $\mathbf{v}$. What is the unit vector in the opposite direction of $\mathbf{v}$?*

**43.** $\mathbf{v} = \langle 5, 12 \rangle$

**44.** $\mathbf{v} = \langle 3, 4 \rangle$

**45.** $\mathbf{v} = 2\mathbf{i} + \mathbf{j}$

**46.** $\mathbf{v} = \mathbf{i} - \mathbf{j}$

**47.** $\mathbf{v} = \dfrac{1}{2}\mathbf{i} + \dfrac{\sqrt{3}}{2}\mathbf{j}$

**48.** $\mathbf{v} = \dfrac{\sqrt{2}}{2}(\mathbf{i} - \mathbf{j})$

**49.** $\mathbf{v} = \langle 1, 1, 1 \rangle$

**50.** $\mathbf{v} = \langle 1, -1, -1 \rangle$

**51.** $\mathbf{v} = \cos\theta\,\mathbf{i} + \sin\theta\,\mathbf{j} + 2\sqrt{2}\mathbf{k}$

**52.** $\mathbf{v} = \sin\theta\,\mathbf{i} + \cos\theta\,\mathbf{j} + \mathbf{k}$

**53.** $\mathbf{v} = 3\mathbf{i} - 4\mathbf{j} + 12\mathbf{k}$

**54.** $\mathbf{v} = \dfrac{1}{2}\mathbf{i} - \dfrac{1}{2}\mathbf{j} + \dfrac{\sqrt{2}}{2}\mathbf{k}$

**1.** = NOW WORK problem        = Graphing technology recommended        CAS = Computer Algebra System recommended

*In Problems 55–60, use $v = 2i - 3j$ and $w = i + 2j$ to find the indicated quantity.*

**55.** $2v - w$

**56.** $v + 5w$

**57.** $\dfrac{1}{3}v + \dfrac{1}{2}w$

**58.** $\dfrac{2}{3}v - \dfrac{1}{2}w$

**59.** $\|v - w\|$

**60.** $\|v + w\|$

*In Problems 61–70, use $u = i - 2j + 3k$, $v = 3i + j - k$, and $w = 6i + j + k$ to find the indicated quantity.*

**61.** $3v - 2w$

**62.** $-2v + w$

**63.** $2u - 3v + 4w$

**64.** $5u - v + w$

**65.** $\|v - w\|$

**66.** $\|2v + w\|$

**67.** $\|u\| + \|v\| + \|w\|$

**68.** $\|u + v + w\|$

**69.** $\dfrac{u + v}{\|w\|}$

**70.** $\dfrac{u}{\|v + w\|}$

*In Problems 71–76, determine whether the vectors $v$ and $w$ are parallel. If the vectors are parallel, are their directions the same or opposite?*

**71.** $v = 2i + 3j$, $\quad w = 3i + \dfrac{9}{2}j$

**72.** $v = 3i + 2j$, $\quad w = 6i - 4j$

**73.** $v = \langle 2, 6, -12 \rangle$, $\quad w = \langle -1, -3, 4 \rangle$

**74.** $v = \langle \sqrt{2}, 1, 3 \rangle$, $\quad w = \langle 2, 1, 9 \rangle$

**75.** $v = \langle 12, 3, -9 \rangle$, $\quad w = \langle -4, -1, 3 \rangle$

**76.** $v = \langle -6, 8, 2 \rangle$, $\quad w = \langle -3, 4, 0 \rangle$

## Applications and Extensions

**77.** Find a vector whose magnitude is 4 that is parallel to $2i - 3j + 4k$.

**78.** Find a vector whose magnitude is 2 that is parallel to $-i + j + 2k$.

**79.** Find $a$ so that the vectors $2i + j - k$ and $2ai + j + k$ have the same magnitude.

**80.** Find $a$ so that $\|ai + (a + 1)j + 2k\| = 3$.

**81.** If $v = 2i + j$ and $w = xi + 3j$, find all numbers $x$ so that $\|v + w\| = 5$.

**82.** A vector $v$ has initial point $P_1 = (-3, 1)$ and terminal point $P_2 = (x, 4)$. Find all numbers $x$ so that $v$ has length 5.

*In Problems 83–88, find all vectors $v = ai + bj$ that have the indicated properties.*

**83.** Magnitude of $v$ is 4; $v$ makes an angle of $30°$ with the positive $x$-axis.

**84.** Magnitude of $v$ is 2; $v$ makes an angle of $45°$ with the positive $x$-axis.

**85.** Magnitude of $v$ is 5; the $i$ component is twice the $j$ component.

**86.** Magnitude of $v$ is 5; the $i$ component equals the $j$ component.

**87.** The $j$ component of $v$ is 1; $v$ makes an angle of $135°$ with the positive $x$-axis.

**88.** The $i$ component of $v$ is $-3$; $v$ makes an angle of $210°$ with the positive $x$-axis.

**89. Acceleration** A force $F = 10i - 15j + 25k$, measured in newtons (N), is applied to an object of mass 10 kg, causing it to accelerate in the direction of $F$. Find the magnitude of the acceleration of the object (in newtons per kilogram).

**90. Computer Graphics** In computer graphics, vectors are used to translate points. For example, if the point $(-3, 2)$ is translated by the vector $v = \langle 5, 2 \rangle$, then the new location will be $u' = u + v = \langle -3, 2 \rangle + \langle 5, 2 \rangle = \langle 2, 4 \rangle$, as shown in the figure.

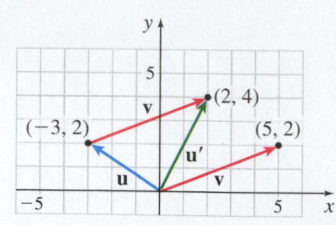

*Source*: Phil Dadd. *Vectors and matrices: A primer.* http://www.gamedev.net/reference/articles/article1832.asp.

**(a)** Determine the new coordinates of $(3, -1)$ if it is translated by $v = \langle -4, 5 \rangle$.

**(b)** Illustrate the translation graphically.

**91. Computer Graphics** The points $(-3, 0)$, $(-1, -2)$, $(3, 1)$, and $(1, 3)$ are the vertices of a parallelogram $ABCD$.

**(a)** If $ABCD$ is translated by $v = \langle 3, -2 \rangle$, find the vertices of the new parallelogram $A'B'C'D'$.

**(b)** If $ABCD$ is translated by $\dfrac{1}{2}v$, find the vertices of the new parallelogram $A'B'C'D'$.

**92. Ground Speed of an Airplane** An airplane has an air speed of 500 km/h in an easterly direction. If there is a southeast wind of 60 km/h, find the speed of the airplane relative to the ground.

**93. Position of an Airplane** An airplane travels 200 mi due west and then travels 150 mi in a direction $60°$ north of west. Find the vector that gives the position of the airplane.

**94. Average Air Speed of an Airplane** An airplane, after 1 h of flying, arrives at a point 200 mi due south of the departure point. If, during the flight, there was a steady northwest wind of 20 mi/h, what was the airplane's average air speed?

**95. Static Equilibrium** Forces $F_1, F_2, F_3, F_4$ act on an object $Q$ as shown in the figure. Draw the force $F$ needed to prevent $Q$ from moving.

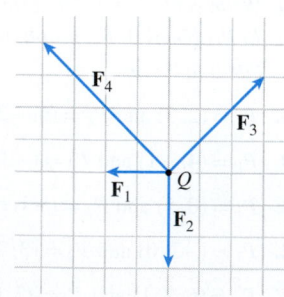

**96. Overcoming a Force** Two friends are trying to pull a car out of a ditch. They apply forces, measured in pounds, of $F_1 = 200i + 350j + 100k$ and $F_2 = 250i + 300j + 150k$, respectively, to the car, but it does not move. Find the third force $F_3$, due to the mud and ground, acting on the car and preventing it from moving.

**97. Static Equilibrium**
A weight of 1000 lb is suspended from two cables, as shown in the figure. What is the tension in each cable?

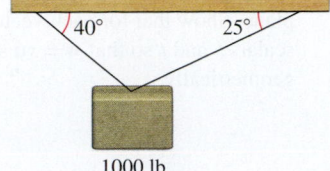

1000 lb

**98. Static Equilibrium**
A weight of 800 lb is suspended from two cables attached to a horizontal steel beam, as shown in the figure. What is the tension in each cable?

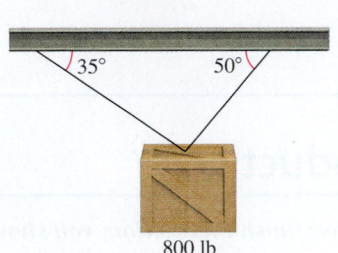

800 lb

**99. Tug of War** Three pieces of rope are knotted together for use in a three-team tug of war. Each rope is pulled parallel to the ground. As viewed from above, Team A pulls with a force of 1100 lb at an angle of 0°, Team B pulls with a force of 1050 lb at an angle of 120°, and Team C pulls with a force of 1000 lb at an angle of 240°.

  **(a)** Draw a force diagram representing the forces acting on the knot.

  **(b)** Find the net force acting on the knot.

**100. Pushing a Wheelbarrow** A man pushes a wheelbarrow up an incline of 20° to the ground with a force of 100 lb. Represent graphically this force.

**101. Pulling a Wagon** A child pulls a wagon with a force of 40 lb. The handle of the wagon makes an angle of 30° with the ground. Represent graphically this force.

**102. Orthopedics** A patient with a broken tibia has the lower half of his leg suspended horizontally by a cable attached to the center of his foot. See the figure. The tension in the cable needs to exert a force of 25 N horizontally for traction and 40 N vertically upward to support the leg.
Find the tension in the cable.
*Hint:* Put the center of the patient's foot at the origin of an *xy*-plane.

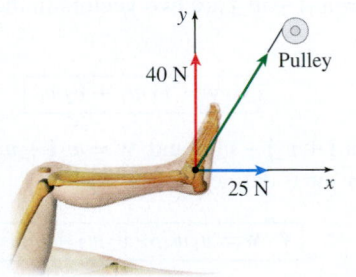

40 N
Pulley
25 N    *x*

**103. Chemistry: Gold Crystals**
In its solid form, gold forms a crystal lattice known as a face-centered cubic (fcc). The crystal consists of eight gold atoms at the corners of a cube plus six more atoms, one at the center of each face of the cube, as shown in the figure. The sides of the cube are 0.408 nm long (1 nm = $10^{-9}$ m). Put one corner of the cube at the origin, and place three of the faces along the positive coordinate planes.

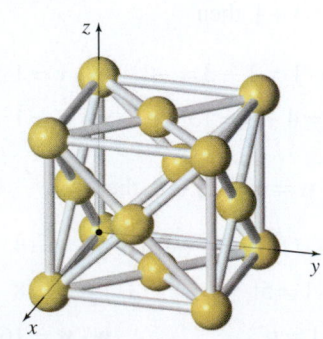

  **(a)** Assign coordinates to each atom in the crystal.

  **(b)** Find a vector that represents each of the following atoms:

  **(i)** The atom farthest from the origin

  **(ii)** The atom in the center of the cube face that is parallel to the *xz*-plane, but not in that plane

  **(iii)** The atom in the center of the cube face that is parallel to the *xy*-plane, but not in that plane

  **(c)** What is the magnitude of each of the vectors found in (b)?

**104. Theoretical Speed** An airplane travels in a northeasterly direction at 250 km/h relative to the ground, due to the fact that there is an easterly wind of 50 km/h relative to the ground. How fast would the plane be going if there were no wind?

**105. Wind Velocity** A woman on a bike traveling east at 8 mi/h finds that the wind appears to be coming from the north. Upon doubling her speed, she finds that the wind appears to be coming from the northeast. Find the velocity of the wind.

**106.** Show that the properties of vector addition and scalar multiplication (p. 745) hold for vectors in the plane.

**107.** Show that the properties of vector addition and scalar multiplication (p. 745) hold for vectors in space.

**108.** Show that if *a* is scalar and **v** is a vector in space, then

$$\|a\mathbf{v}\| = |a|\|\mathbf{v}\|$$

**109.** Show that every vector $\mathbf{v} = \langle v_1, v_2, v_3 \rangle$ in space can be written in terms of the vectors **i**, **j**, and **k** as $\mathbf{v} = v_1\mathbf{i} + v_2\mathbf{j} + v_3\mathbf{k}$.

**Challenge Problems** ─────────────

**110. Geometry** Let *A*, *B*, *C*, and *D* be the vertices of a tetrahedron (triangular pyramid) in space. Let $\mathbf{b} = \overrightarrow{AB}$, $\mathbf{c} = \overrightarrow{AC}$, and $\mathbf{d} = \overrightarrow{AD}$. Express the directed edges $\overrightarrow{BC}$, $\overrightarrow{BD}$, and $\overrightarrow{CD}$ of the tetrahedron in terms of the vectors **b**, **c**, and **d**.

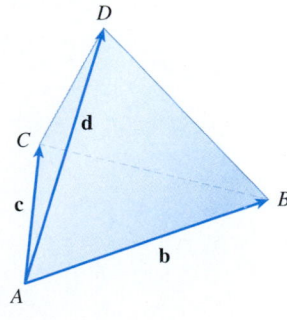

**111. Geometry** Let *A*, *B*, *C*, *D*, *E*, *F*, *G*, and *H* be the vertices of a parallelepiped in space, whose faces are the parallelograms *ABCD*, *ABFE*, *AEHD*, and so forth. Let $\mathbf{b} = \overrightarrow{AB}$, $\mathbf{e} = \overrightarrow{AE}$, and $\mathbf{d} = \overrightarrow{AD}$. Express the vectors $\overrightarrow{AC}$, $\overrightarrow{AF}$, $\overrightarrow{AG}$, $\overrightarrow{FG}$, and $\overrightarrow{EG}$ in terms of **b**, **e**, and **d**.

**112.** Let points $P_1 = (x_1, y_1, z_1)$ and $P_2 = (x_2, y_2, z_2)$ have respective position vectors $\mathbf{r}_1$ and $\mathbf{r}_2$. Show that the point $P = (x, y, z)$ that divides the segment $P_1 P_2$ in the ratio $t = \dfrac{\overline{P_1 P}}{\overline{P_1 P_2}}$ has position vector $\mathbf{r} = \mathbf{r}_1 + t(\mathbf{r}_2 - \mathbf{r}_1)$. From this, conclude that

$$x = x_1 + t(x_2 - x_1)$$
$$y = y_1 + t(y_2 - y_1)$$
$$z = z_1 + t(z_2 - z_1)$$

**113.** Let $\mathbf{u}$ and $\mathbf{v}$ be two nonparallel, nonzero vectors in the plane. Show that for each vector $\mathbf{w}$ there exist unique scalars $s$ and $t$ so that $\mathbf{w} = s\mathbf{u} + t\mathbf{v}$. Interpret the result geometrically.

# 10.4 The Dot Product

**OBJECTIVES** *When you finish this section, you should be able to:*

**1** Find the dot product of two vectors (p. 754)

**2** Find the angle between two vectors (p. 755)

**3** Determine whether two vectors are orthogonal (p. 757)

**4** Express a vector in space using its magnitude and direction (p. 757)

**5** Find the projection of a vector (p. 758)

**6** Compute work (p. 759)

When working with vectors there are several types of multiplication. In Sections 10.2 and 10.3, we defined the scalar multiple of a vector $\mathbf{v}$ as the *product of a scalar and a vector*. The resulting product is a vector. Now we discuss a second type of multiplication, the *dot product*. The dot product is the *product of two vectors*, but the result is a *scalar*. In Section 10.5, we investigate a third type of multiplication, called the *cross product*.

---

**DEFINITION** Dot Product

If $\mathbf{v} = v_1 \mathbf{i} + v_2 \mathbf{j}$ and $\mathbf{w} = w_1 \mathbf{i} + w_2 \mathbf{j}$ are two vectors in the plane, the **dot product $\mathbf{v} \cdot \mathbf{w}$** is defined as

$$\boxed{\mathbf{v} \cdot \mathbf{w} = v_1 w_1 + v_2 w_2}$$

Similarly, if $\mathbf{v} = v_1 \mathbf{i} + v_2 \mathbf{j} + v_3 \mathbf{k}$ and $\mathbf{w} = w_1 \mathbf{i} + w_2 \mathbf{j} + w_3 \mathbf{k}$ are two vectors in space, the dot product $\mathbf{v} \cdot \mathbf{w}$ is defined as

$$\boxed{\mathbf{v} \cdot \mathbf{w} = v_1 w_1 + v_2 w_2 + v_3 w_3}$$

---

**NOTE** The dot product of two vectors is not a vector. It is a scalar. The dot product is sometimes called the **scalar product**.

## 1 Find the Dot Product of Two Vectors

**EXAMPLE 1** Finding the Dot Product of Two Vectors

**(a)** If $\mathbf{v} = 2\mathbf{i} - 3\mathbf{j}$ and $\mathbf{w} = \mathbf{i} + \mathbf{j}$, then

$$\mathbf{v} \cdot \mathbf{w} = 2 \cdot 1 + (-3) \cdot 1 = 2 - 3 = -1 \qquad \mathbf{w} \cdot \mathbf{v} = 1 \cdot 2 + 1 \cdot (-3) = 2 - 3 = -1$$
$$\mathbf{v} \cdot \mathbf{v} = 2^2 + (-3)^2 = 4 + 9 = 13 \qquad \mathbf{w} \cdot \mathbf{w} = 1^2 + 1^2 = 2$$

**(b)** If $\mathbf{v} = 2\mathbf{i} - \mathbf{j} + \mathbf{k}$ and $\mathbf{w} = 4\mathbf{i} + 2\mathbf{j} - \mathbf{k}$, then

$$\mathbf{v} \cdot \mathbf{w} = 2 \cdot 4 + (-1) \cdot 2 + 1 \cdot (-1) \qquad \mathbf{w} \cdot \mathbf{v} = 4 \cdot 2 + 2 \cdot (-1) + (-1) \cdot 1$$
$$= 8 - 2 - 1 = 5 \qquad\qquad\qquad = 8 - 2 - 1 = 5$$
$$\mathbf{v} \cdot \mathbf{v} = 4 + 1 + 1 = 6 \qquad\qquad\quad \mathbf{w} \cdot \mathbf{w} = 16 + 4 + 1 = 21$$

**NOW WORK** Problem 7(a).

The results from Example 1 suggest some of the properties of the dot product.

> **THEOREM** Properties of the Dot Product
>
> If **u**, **v**, and **w** are vectors and $a$ is any scalar, then:
>
> - **Magnitude property:**
> $$\mathbf{v} \cdot \mathbf{v} = \|\mathbf{v}\|^2$$
>
> - **Commutative property:**
> $$\mathbf{u} \cdot \mathbf{v} = \mathbf{v} \cdot \mathbf{u}$$
>
> - **Distributive property:**
> $$\mathbf{u} \cdot (\mathbf{v} + \mathbf{w}) = \mathbf{u} \cdot \mathbf{v} + \mathbf{u} \cdot \mathbf{w}$$
>
> - **Linearity property:**
> $$a(\mathbf{u} \cdot \mathbf{v}) = (a\mathbf{u}) \cdot \mathbf{v}$$
>
> - **Zero-vector property:**
> $$\mathbf{0} \cdot \mathbf{v} = 0$$

We prove the magnitude property $\mathbf{v} \cdot \mathbf{v} = \|\mathbf{v}\|^2$ and the commutative property for vectors in space. The remaining proofs are left as exercises (see Problems 84–88).

**Proof**  To show $\mathbf{v} \cdot \mathbf{v} = \|\mathbf{v}\|^2$, we let $\mathbf{v} = v_1\mathbf{i} + v_2\mathbf{j} + v_3\mathbf{k}$. Then

$$\mathbf{v} \cdot \mathbf{v} = v_1 v_1 + v_2 v_2 + v_3 v_3 = v_1^2 + v_2^2 + v_3^2 = \|\mathbf{v}\|^2$$

Commutative Property: Let $\mathbf{u} = u_1\mathbf{i} + u_2\mathbf{j} + u_3\mathbf{k}$ and $\mathbf{v} = v_1\mathbf{i} + v_2\mathbf{j} + v_3\mathbf{k}$. Then

$$\mathbf{u} \cdot \mathbf{v} = u_1 v_1 + u_2 v_2 + u_3 v_3 = v_1 u_1 + v_2 u_2 + v_3 u_3 = \mathbf{v} \cdot \mathbf{u} \qquad \blacksquare$$

A consequence of the property $\mathbf{v} \cdot \mathbf{v} = \|\mathbf{v}\|^2$ is that for any vector **v**,

$$\mathbf{v} \cdot \mathbf{v} \geq 0 \quad \text{and} \quad \mathbf{v} \cdot \mathbf{v} = 0 \quad \text{if and only if} \quad \mathbf{v} = \mathbf{0}$$

One important use of the dot product is to find the angle between two vectors.

## 2 Find the Angle Between Two Vectors

If **v** and **w** are two nonzero vectors, we can position **v** and **w** so they have the same initial point. Then the vectors **v**, **w**, and **w** − **v** form a triangle, as shown in Figure 35. We want to find an expression for the angle $\theta$ between the vectors **v** and **w**.

Since the sides of the triangle have lengths $\|\mathbf{v}\|$, $\|\mathbf{w}\|$, and $\|\mathbf{w} - \mathbf{v}\|$, we can use the Law of Cosines to find the cosine of the angle $\theta$.

$$\|\mathbf{w} - \mathbf{v}\|^2 = \|\mathbf{v}\|^2 + \|\mathbf{w}\|^2 - 2\|\mathbf{v}\|\|\mathbf{w}\|\cos\theta \qquad (1)$$

Now use properties of the dot product to write this equation in terms of dot products. Since $\mathbf{v} \cdot \mathbf{v} = \|\mathbf{v}\|^2$, we can write (1) in the form

$$(\mathbf{w} - \mathbf{v}) \cdot (\mathbf{w} - \mathbf{v}) = \mathbf{v} \cdot \mathbf{v} + \mathbf{w} \cdot \mathbf{w} - 2\|\mathbf{v}\|\|\mathbf{w}\|\cos\theta \qquad (2)$$

Using the distributive property twice on the term $(\mathbf{w} - \mathbf{v}) \cdot (\mathbf{w} - \mathbf{v})$, we obtain

$$(\mathbf{w} - \mathbf{v}) \cdot (\mathbf{w} - \mathbf{v}) = \mathbf{w} \cdot (\mathbf{w} - \mathbf{v}) - \mathbf{v} \cdot (\mathbf{w} - \mathbf{v})$$
$$= \mathbf{w} \cdot \mathbf{w} - \mathbf{w} \cdot \mathbf{v} - \mathbf{v} \cdot \mathbf{w} + \mathbf{v} \cdot \mathbf{v}$$
$$= \mathbf{w} \cdot \mathbf{w} + \mathbf{v} \cdot \mathbf{v} - 2(\mathbf{v} \cdot \mathbf{w})$$

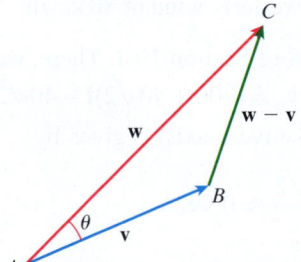

**Figure 35**

**NEED TO REVIEW?**  The Law of Cosines states that $c^2 = a^2 + b^2 - 2ab\cos\theta$, where $a$, $b$, and $c$ are the lengths of the sides of a triangle and $\theta$ is the angle opposite the side of length $c$. The Law of Cosines is discussed in Appendix A.4, p. A-37.

Substitute this result back into equation (2) and simplify. This yields

$$\mathbf{w} \cdot \mathbf{w} + \mathbf{v} \cdot \mathbf{v} - 2(\mathbf{v} \cdot \mathbf{w}) = \mathbf{v} \cdot \mathbf{v} + \mathbf{w} \cdot \mathbf{w} - 2\|\mathbf{v}\|\|\mathbf{w}\| \cos \theta$$

$$\mathbf{v} \cdot \mathbf{w} = \|\mathbf{v}\|\|\mathbf{w}\| \cos \theta$$

We have proved the following theorem.

---

**THEOREM  Angle Between Two Nonzero Vectors**

The angle $\theta$, where $0 \leq \theta \leq \pi$, between two nonzero vectors $\mathbf{v}$ and $\mathbf{w}$ is given by the formula

$$\boxed{\cos \theta = \frac{\mathbf{v} \cdot \mathbf{w}}{\|\mathbf{v}\|\|\mathbf{w}\|}}$$

**NOTE** If $\mathbf{v} \cdot \mathbf{w} > 0$, the angle between $\mathbf{v}$ and $\mathbf{w}$ is acute. If $\mathbf{v} \cdot \mathbf{w} < 0$, the angle between $\mathbf{v}$ and $\mathbf{w}$ is obtuse.

---

**EXAMPLE 2  Finding the Angle Between Two Vectors**

Find the angle between the vectors $\mathbf{v} = 2\mathbf{i} - \mathbf{j} + \mathbf{k}$ and $\mathbf{w} = -\mathbf{i} + \mathbf{j}$.

**Solution** We find $\mathbf{v} \cdot \mathbf{w}$, $\|\mathbf{v}\|$, and $\|\mathbf{w}\|$.

$$\mathbf{v} \cdot \mathbf{w} = 2 \cdot (-1) + (-1) \cdot 1 + 1 \cdot 0 = -2 - 1 + 0 = -3$$

$$\|\mathbf{v}\| = \sqrt{2^2 + (-1)^2 + 1^2} = \sqrt{6}$$

$$\|\mathbf{w}\| = \sqrt{(-1)^2 + 1^2} = \sqrt{2}$$

Then if $\theta$ is the angle between $\mathbf{v}$ and $\mathbf{w}$,

$$\cos \theta = \frac{\mathbf{v} \cdot \mathbf{w}}{\|\mathbf{v}\|\|\mathbf{w}\|} = \frac{-3}{(\sqrt{6})(\sqrt{2})} = \frac{-3}{\sqrt{12}} = -\frac{\sqrt{3}}{2}$$

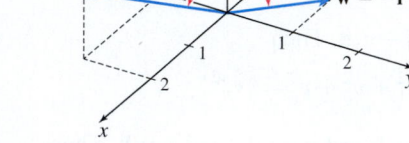

Since $0 \leq \theta \leq \pi$, the angle $\theta$ between $\mathbf{v}$ and $\mathbf{w}$ is $\dfrac{5\pi}{6}$ radians. See Figure 36. ■

**Figure 36**

**NOW WORK** Problem 7(b).

---

**EXAMPLE 3  Finding the True Direction of an Airplane**

An airplane has an air speed of 400 km/h and is headed east. Find the true direction of the airplane relative to the ground if there is a northwesterly wind of 80 km/h.

**Solution** This is the same situation from Example 8 of Section 10.3. There, we found the velocity of the airplane relative to the ground is $\mathbf{v}_g = (400 + 40\sqrt{2})\mathbf{i} - 40\sqrt{2}\mathbf{j}$.

The angle $\theta$ between $\mathbf{v}_g$ and the vector $\mathbf{i}$ (the positive $x$-axis) is given by

$$\cos \theta = \frac{\mathbf{v}_g \cdot \mathbf{i}}{\|\mathbf{v}_g\|\|\mathbf{i}\|} = \frac{400 + 40\sqrt{2}}{460.06} \approx 0.9924$$

$$\theta \approx \cos^{-1}(0.9924) \approx 7.07°$$

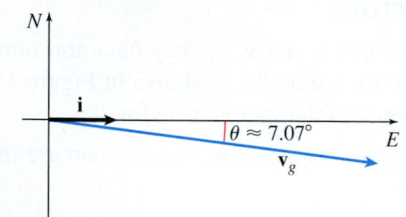

The true direction of the plane is approximately $7.07°$ south of east. See Figure 37. ■

**Figure 37**

**NOW WORK** Problem 41.

Two nonzero vectors are **orthogonal** if the angle between them is a right angle ($90°$). Whether two nonzero vectors are parallel or orthogonal is determined by the angle $\theta$ between the two vectors.

---

If the angle $\theta$ between two nonzero vectors $\mathbf{v}$ and $\mathbf{w}$ is $0$ or $\pi$, the vectors $\mathbf{v}$ and $\mathbf{w}$ are parallel.

If the angle $\theta$ between two nonzero vectors $\mathbf{v}$ and $\mathbf{w}$ is $\dfrac{\pi}{2}$, the vectors $\mathbf{v}$ and $\mathbf{w}$ are orthogonal.

**THEOREM  Orthogonal Vectors**

Two nonzero vectors **v** and **w** are orthogonal if and only if

$$\boxed{\mathbf{v} \cdot \mathbf{w} = 0}$$

**Proof**  We use the formula for the angle between two nonzero vectors, namely

$$\cos \theta = \frac{\mathbf{v} \cdot \mathbf{w}}{\|\mathbf{v}\| \, \|\mathbf{w}\|}$$

If the vectors **v** and **w** are orthogonal, then the angle $\theta$ between **v** and **w** is $\dfrac{\pi}{2}$.

Since $\cos \dfrac{\pi}{2} = 0$, it follows that $\mathbf{v} \cdot \mathbf{w} = 0$.

Conversely, if $\mathbf{v} \cdot \mathbf{w} = 0$, then $\mathbf{v} = \mathbf{0}$, $\mathbf{w} = \mathbf{0}$, or $\cos \theta = 0$. Since **v** and **w** are nonzero vectors, $\cos \theta = 0$, so $\theta = \dfrac{\pi}{2}$ and **v** and **w** are orthogonal. ∎

> **NOTE** *Orthogonal, perpendicular,* and *normal* are all terms that mean "meet at a right angle." It is customary to refer to two vectors as *orthogonal,* two lines as *perpendicular,* and a line and a plane or a vector and a plane as *normal.*

## 3 Determine Whether Two Vectors Are Orthogonal

**EXAMPLE 4**  **Showing Two Vectors Are Orthogonal**

The vectors $\mathbf{v} = 2\mathbf{i} - \mathbf{j} + 5\mathbf{k}$ and $\mathbf{w} = 3\mathbf{i} + \mathbf{j} - \mathbf{k}$ are orthogonal, since

$$\mathbf{v} \cdot \mathbf{w} = 6 - 1 - 5 = 0$$ ∎

**NOW WORK**  Problem 17.

Since $\mathbf{i} \cdot \mathbf{j} = 1 \cdot 0 + 0 \cdot 1 = 0$, the standard basis vectors **i** and **j** in the plane are orthogonal. The standard basis vectors **i**, **j**, and **k** in space are mutually orthogonal, since $\mathbf{i} \cdot \mathbf{j} = 0$, $\mathbf{j} \cdot \mathbf{k} = 0$, and $\mathbf{k} \cdot \mathbf{i} = 0$.

**EXAMPLE 5**  **Making Two Vectors Orthogonal**

Find a scalar $a$ so that the vectors $\mathbf{v} = 2a\mathbf{i} + \mathbf{j} - \mathbf{k}$ and $\mathbf{w} = \mathbf{i} - a\mathbf{j} + \mathbf{k}$ are orthogonal.

**Solution**  The vectors **v** and **w** are orthogonal if $\mathbf{v} \cdot \mathbf{w} = 0$. So,

$$\mathbf{v} \cdot \mathbf{w} = 2a - a - 1 = 0$$
$$a = 1$$

The vectors $\mathbf{v} = 2\mathbf{i} + \mathbf{j} - \mathbf{k}$ and $\mathbf{w} = \mathbf{i} - \mathbf{j} + \mathbf{k}$ are orthogonal. ∎

**NOW WORK**  Problem 23.

## 4 Express a Vector in Space Using Its Magnitude and Direction

A nonzero vector **v** in space can be described by specifying its direction, given by three *direction angles* $\alpha$, $\beta$, $\gamma$, and its magnitude $\|\mathbf{v}\|$. The **direction angles** as shown in Figure 38 are defined as

$$\alpha = \text{Angle between } \mathbf{v} \text{ and } \mathbf{i}, \text{ the positive } x\text{-axis, } 0 \le \alpha \le \pi$$
$$\beta = \text{Angle between } \mathbf{v} \text{ and } \mathbf{j}, \text{ the positive } y\text{-axis, } 0 \le \beta \le \pi$$
$$\gamma = \text{Angle between } \mathbf{v} \text{ and } \mathbf{k}, \text{ the positive } z\text{-axis, } 0 \le \gamma \le \pi$$

We want expressions for $\alpha$, $\beta$, and $\gamma$ in terms of the components of a nonzero vector **v**. Let $\mathbf{v} = v_1 \mathbf{i} + v_2 \mathbf{j} + v_3 \mathbf{k}$ denote a nonzero vector. The angle $\alpha$ between **v** and **i**, which lies on the positive $x$-axis, is given by

$$\cos \alpha = \frac{\mathbf{v} \cdot \mathbf{i}}{\|\mathbf{v}\| \, \|\mathbf{i}\|} = \frac{v_1}{\|\mathbf{v}\|}$$

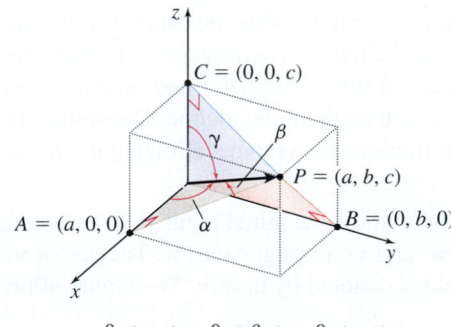

$0 \le \alpha \le \pi, \ 0 \le \beta \le \pi, \ 0 \le \gamma \le \pi$

**Figure 38**

Similarly,

$$\cos \beta = \frac{\mathbf{v} \cdot \mathbf{j}}{\|\mathbf{v}\| \, \|\mathbf{j}\|} = \frac{v_2}{\|\mathbf{v}\|} \qquad \text{and} \qquad \cos \gamma = \frac{\mathbf{v} \cdot \mathbf{k}}{\|\mathbf{v}\| \, \|\mathbf{k}\|} = \frac{v_3}{\|\mathbf{v}\|}$$

The components of **v** are

$$v_1 = \|\mathbf{v}\| \cos \alpha \qquad v_2 = \|\mathbf{v}\| \cos \beta \qquad v_3 = \|\mathbf{v}\| \cos \gamma$$

**THEOREM**

If $\mathbf{v} = v_1 \mathbf{i} + v_2 \mathbf{j} + v_3 \mathbf{k}$ is a nonzero vector in space, then

$$\cos \alpha = \frac{v_1}{\sqrt{v_1^2 + v_2^2 + v_3^2}} = \frac{v_1}{\|\mathbf{v}\|} \qquad \cos \beta = \frac{v_2}{\sqrt{v_1^2 + v_2^2 + v_3^2}} = \frac{v_2}{\|\mathbf{v}\|}$$

$$\cos \gamma = \frac{v_3}{\sqrt{v_1^2 + v_2^2 + v_3^2}} = \frac{v_3}{\|\mathbf{v}\|}$$

$$\mathbf{v} = \|\mathbf{v}\| \, [\cos \alpha \mathbf{i} + \cos \beta \mathbf{j} + \cos \gamma \mathbf{k}]$$

This result gives the vector **v** when its magnitude and direction are known. The numbers $\cos \alpha$, $\cos \beta$, and $\cos \gamma$ are called the **direction cosines** of the vector **v**. In Problem 68, you are asked to show that $\cos^2 \alpha + \cos^2 \beta + \cos^2 \gamma = 1$. In other words, the vector $\cos \alpha \mathbf{i} + \cos \beta \mathbf{j} + \cos \gamma \mathbf{k}$ is a unit vector.

**EXAMPLE 6**  **Finding the Direction Cosines of a Vector**

(a)  Find the magnitude and the direction cosines of $\mathbf{v} = -3\mathbf{i} + 2\mathbf{j} - 6\mathbf{k}$.

(b)  Write **v** in terms of its magnitude and its direction cosines.

**Solution** (a)  The magnitude of **v** is

$$\|\mathbf{v}\| = \sqrt{(-3)^2 + 2^2 + (-6)^2} = \sqrt{49} = 7$$

The direction cosines of the vector **v** are

$$\cos \alpha = \frac{v_1}{\|\mathbf{v}\|} = \frac{-3}{7} \qquad \cos \beta = \frac{v_2}{\|\mathbf{v}\|} = \frac{2}{7} \qquad \cos \gamma = \frac{v_3}{\|\mathbf{v}\|} = \frac{-6}{7}$$

(b)  Now $\mathbf{v} = \|\mathbf{v}\| \, [\cos \alpha \mathbf{i} + \cos \beta \mathbf{j} + \cos \gamma \mathbf{k}] = 7 \left( -\dfrac{3}{7}\mathbf{i} + \dfrac{2}{7}\mathbf{j} - \dfrac{6}{7}\mathbf{k} \right)$. ∎

**NOW WORK** Problem 25.

## 5 Find the Projection of a Vector

In many applications, it is important to find "how much" of a vector is applied along a given direction. In Figure 39, the force **F** due to gravity pulls the block toward the center of Earth. To study the effect of gravity on the block, it is necessary to determine how much of **F** is actually pulling the block down the incline ($\mathbf{F}_1$) and how much is pressing the block against the incline ($\mathbf{F}_2$) at a right angle to the incline. Decomposing **F** into $\mathbf{F}_1$ and $\mathbf{F}_2$ allows us to determine when friction is overcome, allowing the block to slide down the incline.

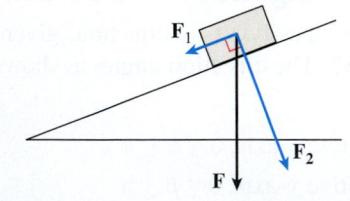

**Figure 39** $\mathbf{F} = \mathbf{F}_1 + \mathbf{F}_2$

Suppose **v** and **w** are two nonzero vectors with the same initial point $P$. We want to decompose **v** into two vectors: $\mathbf{v}_1$, parallel to **w**, and $\mathbf{v}_2$, orthogonal to **w**. The vector $\mathbf{v}_1$ is called the **vector projection of v onto w** and is denoted by $\text{proj}_\mathbf{w} \mathbf{v}$. See Figure 40(a) and (b).

We obtain the vector $\mathbf{v}_1$ as follows: We drop a perpendicular from the terminal point of **v** to the line containing **w**. The vector $\mathbf{v}_1$ is the vector from $P$ to the intersection of the line containing **w** and the perpendicular. Since $\mathbf{v} = \mathbf{v}_1 + \mathbf{v}_2$, the dot product $\mathbf{v} \cdot \mathbf{w}$ is

$$\mathbf{v} \cdot \mathbf{w} = (\underbrace{\mathbf{v}_1 + \mathbf{v}_2}_{\mathbf{v} = \mathbf{v}_1 + \mathbf{v}_2}) \cdot \mathbf{w} = \underbrace{\mathbf{v}_1 \cdot \mathbf{w} + \mathbf{v}_2 \cdot \mathbf{w}}_{\text{Distribute}} \qquad (1)$$

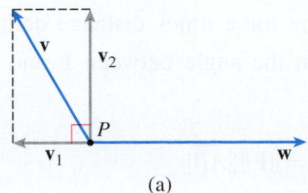

(a)

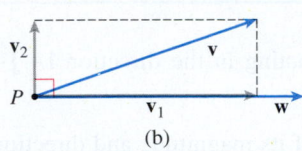

(b)

**Figure 40** $\mathbf{v} = \mathbf{v_1} + \mathbf{v_2}$

Since $\mathbf{v_2}$ is orthogonal to $\mathbf{w}$, the dot product $\mathbf{v_2} \cdot \mathbf{w} = 0$. Since $\mathbf{v_1}$ is parallel to $\mathbf{w}$, there is a scalar $a$ for which $\mathbf{v_1} = a\mathbf{w}$. We make these substitutions in equation (1).

$$\mathbf{v} \cdot \mathbf{w} = a\,\mathbf{w} \cdot \mathbf{w} + 0 = a\|\mathbf{w}\|^2$$

$$a = \frac{\mathbf{v} \cdot \mathbf{w}}{\|\mathbf{w}\|^2}$$

Then

$$\mathbf{v_1} = a\mathbf{w} = \frac{\mathbf{v} \cdot \mathbf{w}}{\|\mathbf{w}\|^2}\,\mathbf{w}$$

**THEOREM  Vector Projection**

If $\mathbf{v}$ and $\mathbf{w}$ are two nonzero vectors, the vector projection of $\mathbf{v}$ onto $\mathbf{w}$ is

$$\boxed{\operatorname{proj}_\mathbf{w}\mathbf{v} = \frac{\mathbf{v} \cdot \mathbf{w}}{\|\mathbf{w}\|^2}\,\mathbf{w}}$$

The decomposition of $\mathbf{v}$ into $\mathbf{v_1}$ and $\mathbf{v_2}$, where $\mathbf{v_1}$ is parallel to $\mathbf{w}$ and $\mathbf{v_2}$ is orthogonal to $\mathbf{w}$, is

$$\boxed{\mathbf{v_1} = \frac{\mathbf{v} \cdot \mathbf{w}}{\|\mathbf{w}\|^2}\,\mathbf{w} \qquad \mathbf{v_2} = \mathbf{v} - \mathbf{v_1}}$$

**CALC CLIP**

**▶ EXAMPLE 7  Decomposing a Vector into Two Orthogonal Vectors**

Find the vector projection of $\mathbf{v} = 2\mathbf{i} - \mathbf{j} + \mathbf{k}$ onto $\mathbf{w} = \mathbf{i} + \mathbf{j} + \mathbf{k}$. Decompose $\mathbf{v}$ into two vectors $\mathbf{v_1}$ and $\mathbf{v_2}$, where $\mathbf{v_1}$ is parallel to $\mathbf{w}$ and $\mathbf{v_2}$ is orthogonal to $\mathbf{w}$.

**Solution** We use the formula for the projection of $\mathbf{v}$ onto $\mathbf{w}$.

$$\mathbf{v_1} = \operatorname{proj}_\mathbf{w}\mathbf{v} = \frac{\mathbf{v} \cdot \mathbf{w}}{\|\mathbf{w}\|^2}\mathbf{w} = \frac{2}{(\sqrt{3})^2}\mathbf{w} = \frac{2}{3}(\mathbf{i} + \mathbf{j} + \mathbf{k}) = \frac{2}{3}\mathbf{i} + \frac{2}{3}\mathbf{j} + \frac{2}{3}\mathbf{k}$$
$$\uparrow$$
$$\mathbf{v} \cdot \mathbf{w} = 2 - 1 + 1 = 2$$
$$\|\mathbf{w}\| = \sqrt{1^2 + 1^2 + 1^2} = \sqrt{3}$$

$$\mathbf{v_2} = \mathbf{v} - \mathbf{v_1} = (2\mathbf{i} - \mathbf{j} + \mathbf{k}) - \left(\frac{2}{3}\mathbf{i} + \frac{2}{3}\mathbf{j} + \frac{2}{3}\mathbf{k}\right) = \frac{4}{3}\mathbf{i} - \frac{5}{3}\mathbf{j} + \frac{1}{3}\mathbf{k} \qquad ■$$

**NOW WORK** Problem 33.

## 6 Compute Work

**Work** is defined as the energy transferred to or from an object by a **force** acting on the object.

The work $W$ done by a *constant* force $\mathbf{F}$ in moving an object from $A$ to $B$ along a straight line in the direction of $\mathbf{F}$ is defined to be

$$W = \|\mathbf{F}\|\,\|\overrightarrow{AB}\| \qquad \text{Work} = \text{force} \times \text{distance}$$

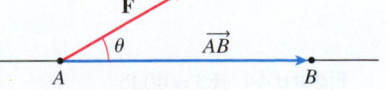

**Figure 41** $W = \|\mathbf{F}\|\,\|\overrightarrow{AB}\|$

In this definition, it is assumed that the constant force $\mathbf{F}$ is applied along the line of motion $\overrightarrow{AB}$, as shown in Figure 41.

If the constant force $\mathbf{F}$ is not along the line of motion, but instead is at an angle $\theta$ to the direction of motion, as in Figure 42, then the **work** $W$ **done by** $\mathbf{F}$ in moving an object from $A$ to $B$ is defined as the dot product.

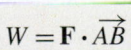

**Figure 42** $W = \mathbf{F} \cdot \overrightarrow{AB}$

$$\boxed{W = \mathbf{F} \cdot \overrightarrow{AB}}$$

This definition of work is compatible with the force times distance definition, since, if the force **F** is in the direction of $\overrightarrow{AB}$, then the angle between **F** and $\overrightarrow{AB}$ is zero, and

$$W = \mathbf{F} \cdot \overrightarrow{AB} = \|\mathbf{F}\| \|\overrightarrow{AB}\| \cos 0 = \|\mathbf{F}\| \|\overrightarrow{AB}\|$$

### EXAMPLE 8   Computing Work

Find the work done by a force of 2 newtons (N) acting in the direction $\mathbf{i} + \mathbf{j} + \mathbf{k}$ in moving an object 1 m from $(0, 0, 0)$ to $(1, 0, 0)$.

**Solution** We need to express the force **F** in terms of its magnitude and direction. The unit vector **u** in the direction $\mathbf{v} = \mathbf{i} + \mathbf{j} + \mathbf{k}$ is

$$\mathbf{u} = \frac{\mathbf{v}}{\|\mathbf{v}\|} = \frac{\mathbf{i} + \mathbf{j} + \mathbf{k}}{\sqrt{3}} = \frac{1}{\sqrt{3}}\mathbf{i} + \frac{1}{\sqrt{3}}\mathbf{j} + \frac{1}{\sqrt{3}}\mathbf{k}$$

Since the force vector **F** has magnitude 2, we have

$$\mathbf{F} = 2\left(\frac{1}{\sqrt{3}}\mathbf{i} + \frac{1}{\sqrt{3}}\mathbf{j} + \frac{1}{\sqrt{3}}\mathbf{k}\right) = \frac{2}{\sqrt{3}}(\mathbf{i} + \mathbf{j} + \mathbf{k})$$

The line of motion of the object from $(0, 0, 0)$ to $(1, 0, 0)$ is $\overrightarrow{AB} = \mathbf{i}$. The work $W$ is therefore

$$W = \mathbf{F} \cdot \overrightarrow{AB} = \frac{2}{\sqrt{3}}(\mathbf{i} + \mathbf{j} + \mathbf{k}) \cdot \mathbf{i} = \frac{2}{\sqrt{3}} \text{ joules} \qquad \blacksquare$$

> **RECALL** 1 joule = 1 newton·meter;
> 1 J = 1 N · m.

**NOW WORK** Problem 49.

### EXAMPLE 9   Computing Work

Figure 43 shows a man pushing on a lawn mower handle with a force of 40 lb. How much work is done in moving the lawn mower a distance of 75 ft if the handle makes an angle of $60°$ with the ground?

**Solution** Set up the coordinate system so that the lawn mower is moved from $(0, 0)$ to $(75, 0)$. Then the motion occurs along $\overrightarrow{AB} = 75\mathbf{i}$. The force vector **F**, as shown in Figure 44, makes an angle of $300°$ to the positive $x$-axis. Since $\|\mathbf{F}\| = 40$, the work $W$ done in moving the lawn mower 75 ft is

$$W = \mathbf{F} \cdot \overrightarrow{AB} = \|\mathbf{F}\| \|\overrightarrow{AB}\| \cos 300° = 40 \cdot 75 \cdot \frac{1}{2} = 1500 \text{ ft-lb}$$

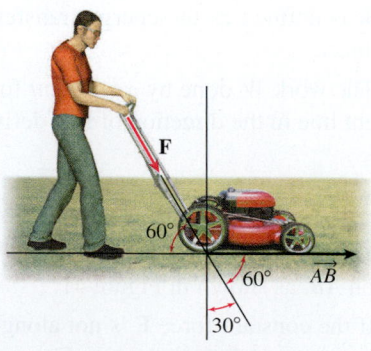

**Figure 43** $\overrightarrow{AB} = 75\mathbf{i}$

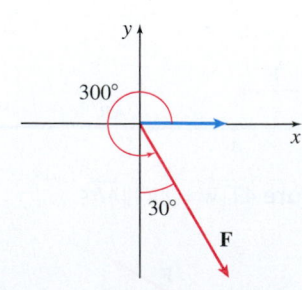

**Figure 44** $\|\mathbf{F}\| = 40$ lb

## 10.4 Assess Your Understanding

### Concepts and Vocabulary

1. *True or False*   If **v** is a nonzero vector, then $\mathbf{v} \cdot \mathbf{v} = \|\mathbf{v}\|$.

2. *True or False*   If two nonzero vectors **v** and **w** are parallel, then $\mathbf{v} \cdot \mathbf{w} = 0$.

3. If $\theta$ is the angle between two nonzero vectors **v** and **w**, then $\cos \theta =$ _____.

4. *True or False*   If $\mathbf{v} = v_1 \mathbf{i} + v_2 \mathbf{j} + v_3 \mathbf{k}$ is a nonzero vector in space, then $\mathbf{v} = \|\mathbf{v}\| [\cos \alpha \mathbf{i} + \cos \beta \mathbf{j} + \cos \gamma \mathbf{k}]$, where $\cos \alpha = \dfrac{v_1}{\|\mathbf{v}\|}$, $\cos \beta = \dfrac{v_2}{\|\mathbf{v}\|}$, and $\cos \gamma = \dfrac{v_3}{\|\mathbf{v}\|}$.

5. *True or False*   The dot product of two vectors is sometimes called the vector product.

6. *True or False*   For any two nonzero vectors **v** and **w**, the vector **v** can be decomposed into two vectors, one parallel to **w** and the other orthogonal to **w**.

### Skill Building

*In Problems 7–14, for each pair of vectors* **v** *and* **w**:
(a) *Find the dot product* $\mathbf{v} \cdot \mathbf{w}$.
(b) *Find the angle* $\theta$ *between* **v** *and* **w**.

7. $\mathbf{v} = 2\mathbf{i} - 3\mathbf{j} + \mathbf{k}$,   $\mathbf{w} = \mathbf{i} - \mathbf{j} + \mathbf{k}$

8. $\mathbf{v} = -3\mathbf{i} + 2\mathbf{j} - \mathbf{k}$,   $\mathbf{w} = 2\mathbf{i} + \mathbf{j} - \mathbf{k}$

9. $\mathbf{v} = \mathbf{i} - \mathbf{j}$,   $\mathbf{w} = \mathbf{j} + \mathbf{k}$

10. $\mathbf{v} = \mathbf{j} - \mathbf{k}$,   $\mathbf{w} = \mathbf{i} + \mathbf{k}$

11. $\mathbf{v} = 3\mathbf{i} + \mathbf{j} - \mathbf{k}$,   $\mathbf{w} = -2\mathbf{i} - \mathbf{j} + \mathbf{k}$

12. $\mathbf{v} = \mathbf{i} - 3\mathbf{j} + 4\mathbf{k}$,   $\mathbf{w} = 4\mathbf{i} - \mathbf{j} + 3\mathbf{k}$

13. $\mathbf{v} = \mathbf{i} - \mathbf{j}$,   $\mathbf{w} = \mathbf{i} + \mathbf{j}$

14. $\mathbf{v} = 3\mathbf{i} + 4\mathbf{j}$,   $\mathbf{w} = -6\mathbf{i} - 8\mathbf{j}$

*In Problems 15–20, determine whether the vectors* **v** *and* **w** *are orthogonal. If they are not orthogonal, find the angle between them.*

15. $\mathbf{v} = 2\mathbf{i} - 8\mathbf{j}$,   $\mathbf{w} = 4\mathbf{i} + \mathbf{j}$

16. $\mathbf{v} = 3\mathbf{i} + \mathbf{j}$,   $\mathbf{w} = 2\mathbf{i} + 2\mathbf{j}$

17. $\mathbf{v} = -4\mathbf{i} - \mathbf{j} - 2\mathbf{k}$,   $\mathbf{w} = -2\mathbf{i} - 2\mathbf{j} + 5\mathbf{k}$

18. $\mathbf{v} = -3\mathbf{i} + 6\mathbf{j} + 4\mathbf{k}$,   $\mathbf{w} = 2\mathbf{i} + \dfrac{1}{3}\mathbf{j} + \mathbf{k}$

19. $\mathbf{v} = \mathbf{i} + 2\mathbf{j} + \mathbf{k}$,   $\mathbf{w} = 4\mathbf{i} + 3\mathbf{j}$

20. $\mathbf{v} = -2\mathbf{j} - 3\mathbf{k}$,   $\mathbf{w} = 2\mathbf{i} + 6\mathbf{j} - 4\mathbf{k}$

*In Problems 21–24, find a scalar a so that the vectors* **v** *and* **w** *are orthogonal.*

21. $\mathbf{v} = 2a\mathbf{i} + \mathbf{j} - \mathbf{k}$,   $\mathbf{w} = \mathbf{i} - \mathbf{j} + \mathbf{k}$

22. $\mathbf{v} = \mathbf{i} + 2a\mathbf{j} - \mathbf{k}$,   $\mathbf{w} = \mathbf{i} - \mathbf{j} + \mathbf{k}$

23. $\mathbf{v} = a\mathbf{i} + \mathbf{j} + \mathbf{k}$,   $\mathbf{w} = \mathbf{i} + a\mathbf{j} + 4\mathbf{k}$

24. $\mathbf{v} = \mathbf{i} - a\mathbf{j} + 2\mathbf{k}$,   $\mathbf{w} = 2a\mathbf{i} + \mathbf{j} + \mathbf{k}$

*In Problems 25–32:*
(a) *Find the magnitude and the direction cosines of each vector* **v**.
(b) *Write* **v** *in terms of its magnitude and its direction cosines.*

25. $\mathbf{v} = 3\mathbf{i} - 6\mathbf{j} - 2\mathbf{k}$     26. $\mathbf{v} = -6\mathbf{i} + 12\mathbf{j} + 4\mathbf{k}$

27. $\mathbf{v} = \mathbf{i} + \mathbf{j} + \mathbf{k}$     28. $\mathbf{v} = \mathbf{i} - \mathbf{j} - \mathbf{k}$

29. $\mathbf{v} = \mathbf{i} - \mathbf{k}$     30. $\mathbf{v} = \mathbf{j} + \mathbf{k}$

31. $\mathbf{v} = 3\mathbf{i} - 5\mathbf{j} + 2\mathbf{k}$     32. $\mathbf{v} = 2\mathbf{i} + 3\mathbf{j} - 4\mathbf{k}$

*In Problems 33–38:*
(a) *Find the vector projection of* **v** *onto* **w**.
(b) *Decompose* **v** *into two vectors* $\mathbf{v}_1$ *and* $\mathbf{v}_2$*, where* $\mathbf{v}_1$ *is parallel to* **w** *and* $\mathbf{v}_2$ *is orthogonal to* **w**.

33. $\mathbf{v} = 2\mathbf{i} - 3\mathbf{j} + \mathbf{k}$,   $\mathbf{w} = \mathbf{i} - \mathbf{j} + \mathbf{k}$

34. $\mathbf{v} = -3\mathbf{i} + 2\mathbf{j} - \mathbf{k}$,   $\mathbf{w} = 2\mathbf{i} + \mathbf{j} - \mathbf{k}$

35. $\mathbf{v} = \mathbf{i} - \mathbf{j}$,   $\mathbf{w} = \mathbf{j} + \mathbf{k}$     36. $\mathbf{v} = \mathbf{j} - \mathbf{k}$,   $\mathbf{w} = \mathbf{i} + \mathbf{k}$

37. $\mathbf{v} = 3\mathbf{i} + \mathbf{j} - \mathbf{k}$,   $\mathbf{w} = -2\mathbf{i} - \mathbf{j} + \mathbf{k}$

38. $\mathbf{v} = \mathbf{i} - 3\mathbf{j} + 4\mathbf{k}$,   $\mathbf{w} = 4\mathbf{i} - \mathbf{j} + 3\mathbf{k}$

39. Find a scalar $a$ so that the angle between the vectors $\mathbf{v} = a\mathbf{i} + \mathbf{j} + \mathbf{k}$ and $\mathbf{w} = \mathbf{i} + a\mathbf{j} + \mathbf{k}$ is $\dfrac{\pi}{3}$.

40. Find a scalar $a$ so that the angle between the vectors $\mathbf{v} = a\mathbf{i} - \mathbf{j} + \mathbf{k}$ and $\mathbf{w} = \mathbf{i} - \mathbf{j} + a\mathbf{k}$ is $\dfrac{2\pi}{3}$.

### Applications and Extensions

41. **Ground Speed and Direction**   An airplane flying due north has an air speed of 500 km/h. There is a northwesterly wind of 60 km/h. Find the true speed and direction of the plane relative to the ground.

42. **Ground Speed and Direction**   A bird flying west has an air speed of 12 km/h. If there is a southeasterly wind of 5 km/h, find the true speed and direction of the bird relative to the ground.

43. **Steering a Boat**   A stream 1 km wide has a constant current of 5 km/h. At what angle to the shore should a person navigate a boat, which is maintaining a constant speed of 15 km/h, in order to reach a point directly opposite?

44. **Swimming to a Destination**   A river is 500 m wide and has a current of 1 km/h. If a swimmer can swim at a constant speed of 2 km/h, at what angle to the shore should she swim if she wishes to reach the shore at a point directly opposite? How long will it take her to cross the river?

45. **Braking Load**   A car with a gross weight of 5300 lb is parked on a street with an 8° grade, as shown in the figure below. Find the magnitude of the force required to keep the car from rolling down the street. What is the magnitude of the force perpendicular to the street?

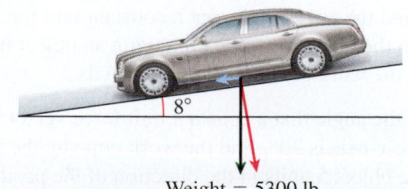

Weight = 5300 lb

46. **Ramp Angle** Billy stops while rolling a 250-lb piano up a ramp. If the angle of inclination of the ramp is $20°$, how many pounds of force must Billy exert to hold the piano in position?

47. **Work** A wagon is pulled horizontally by exerting a force of 20 lb on the handle at an angle of $30°$ with the horizontal. How much work is done in moving the wagon 100 ft?

48. **Work** Find the work done by a force of 1 N acting in the direction $2\mathbf{i} + 2\mathbf{j} + \mathbf{k}$ in moving an object 3 m from $(0, 0, 0)$ to $(1, 2, 2)$.

49. **Work** Find the work done by a force of 3 N acting in the direction $2\mathbf{i} + \mathbf{j} + 2\mathbf{k}$ in moving an object 2 m from $(0, 0, 0)$ to $(0, 2, 0)$.

50. **Work** Find the work done by gravity when an object of mass $m$ moves once around the triangle shown in the figure. (The force of gravity is $mg$, where $g = 9.80 \text{ m/s}^2$.)

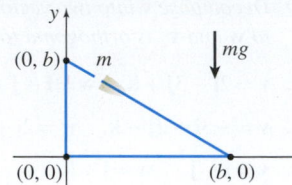

51. **Work** Two workers, A and B, are each pulling identical crates of tools a distance of 10 m along a level floor. Worker A pulls the crate with a force of 850 N at an angle of $30°$ with respect to the floor, and Worker B also pulls the crate with a force of 850 N, but at an angle of $45°$ with respect to the floor. Assume that in both cases, the crates are moved from $(0, 0, 0)$ to $(10, 0, 0)$, and the z-axis points upward, perpendicular to the floor.

   (a) Find the amount of work done by each person in moving the crate.

   (b) Which person do you think took less time moving the crate? Explain your reasoning.

52. A horse pulls a plow a distance of 15 m along the ground from $(0, 0)$ to $(15, 0)$, during which a force of 725 N is applied at an angle of $15°$ with respect to the ground.

   (a) What is the work done by the horse?

   (b) During this motion, the force of friction is also acting on the plow. Friction is directed parallel to the surface and opposes the motion. If the friction force has a magnitude of 625 N, what is the work done by friction on the plow?

53. **Parallelogram** Show that the points $(2, 2, 2)$, $(0, 1, 2)$, $(-1, 3, 3)$, and $(3, 0, 1)$ are the vertices of a parallelogram.

54. **Rectangle** Show that the points $(2, 2, 2)$, $(2, 0, 1)$, $(4, 1, -1)$, and $(4, 3, 0)$ are the vertices of a rectangle.

55. **Right Triangle** Show that the points $(-2, 6, 0)$, $(4, 9, 1)$, and $(-3, 2, 18)$ are the vertices of a right triangle.

56. **Right Triangle** Find all scalars $c$ so that the triangle with vertices $A = (1, -1, 0)$, $B = (-2, 2, 1)$, and $C = (1, 2, c)$ is a right triangle with its right angle at $C$.

57. **Work** Find the acute angle that a constant unit force vector makes with the positive x-axis in moving an object from $(0, 0)$ to $(4, 0)$ if the work done by the force equals 2.

58. **Work** If the angle that a constant unit force vector makes with the positive x-axis is $30°$, find the work done by the force in moving the object 5 units in the direction of the positive x-axis.

59. Let $\mathbf{u}$ and $\mathbf{w}$ be two unit vectors and let $\theta$ be the angle between them. Find $\left\| \dfrac{1}{2}\mathbf{u} - \mathbf{w} \right\|$ in terms of $\theta$.

60. Show that the vector projection of $\mathbf{v}$ on $\mathbf{i}$ is $(\mathbf{v} \cdot \mathbf{i})\mathbf{i}$. Then show that a vector $\mathbf{v}$ can always be written as
$$\mathbf{v} = (\mathbf{v} \cdot \mathbf{i})\mathbf{i} + (\mathbf{v} \cdot \mathbf{j})\mathbf{j} + (\mathbf{v} \cdot \mathbf{k})\mathbf{k}$$

61. If $\|\mathbf{v}\| = 2$, $\|\mathbf{w}\| = 6$, and the angle between $\mathbf{v}$ and $\mathbf{w}$ is $\dfrac{\pi}{3}$, find $\|\mathbf{v} + \mathbf{w}\|$ and $\|\mathbf{v} - \mathbf{w}\|$.

62. Find all numbers $a$ and $b$ for which the vectors $\mathbf{v} = 2a\mathbf{i} - 2\mathbf{j} + \mathbf{k}$ and $\mathbf{w} = b\mathbf{i} + 2\mathbf{j} + 2\mathbf{k}$ are orthogonal and have the same magnitude.

63. Suppose $\mathbf{v}$ and $\mathbf{w}$ are two nonzero vectors. Show that the vector $\mathbf{v} - a\mathbf{w}$ is orthogonal to $\mathbf{w}$ if $a = \dfrac{\mathbf{v} \cdot \mathbf{w}}{\|\mathbf{w}\|^2}$.

64. Let $\mathbf{v}$ and $\mathbf{w}$ be nonzero vectors. Show that the vectors $\|\mathbf{w}\|\mathbf{v} + \|\mathbf{v}\|\mathbf{w}$ and $\|\mathbf{w}\|\mathbf{v} - \|\mathbf{v}\|\mathbf{w}$ are orthogonal.

65. Let $\mathbf{w}$ be a nonzero vector and let $\mathbf{u}$ be a unit vector. Show that the unit vector $\mathbf{u}$ that makes $\mathbf{w} \cdot \mathbf{u}$ a maximum is the unit vector pointing in the same direction as $\mathbf{w}$.

66. **Cauchy–Schwarz Inequality** Show that if $\mathbf{v}$ and $\mathbf{w}$ are two vectors, then $|\mathbf{v} \cdot \mathbf{w}| \leq \|\mathbf{v}\| \|\mathbf{w}\|$. Under what conditions is the Cauchy–Schwarz inequality an equality?

67. **Triangle Inequality** Show that if $\mathbf{v}$ and $\mathbf{w}$ are vectors, then
$$\|\mathbf{v} + \mathbf{w}\| \leq \|\mathbf{v}\| + \|\mathbf{w}\|$$
   *Hint:* Use the Cauchy–Schwarz inequality, Problem 66.

68. Show that if $\alpha$, $\beta$, and $\gamma$ are the direction angles of a nonzero vector $\mathbf{v}$ in space, then $\cos^2 \alpha + \cos^2 \beta + \cos^2 \gamma = 1$.

69. The vector $\mathbf{v}$ makes an angle of $\alpha = \dfrac{\pi}{3}$ with the positive x-axis, an angle of $\beta = \dfrac{\pi}{3}$ with the positive y-axis, and an acute angle $\gamma$ with the positive z-axis. Find $\gamma$.
   *Hint:* Use the result from Problem 68.

*In Problems 70–73, use the result from Problem 68 to find a vector $\mathbf{v}$ in space that has the given magnitude and direction angles.*

70. $\|\mathbf{v}\| = 3$, $\alpha = \dfrac{\pi}{3}$, $\beta = \dfrac{\pi}{4}$, $0 < \gamma < \dfrac{\pi}{2}$

71. $\|\mathbf{v}\| = 2$, $\alpha = \dfrac{\pi}{4}$, $\dfrac{\pi}{2} < \beta < \pi$, $\gamma = \dfrac{\pi}{3}$

72. $\|\mathbf{v}\| = 3$, the direction angles are equal, and $\mathbf{v}$ has positive components.

73. $\|\mathbf{v}\| = \dfrac{1}{2}$, $\cos \alpha > 0$, $\cos \beta = \dfrac{1}{4}$, $\cos \gamma = \sqrt{\dfrac{7}{8}}$

74. **Orthogonal Vectors** Let $\mathbf{v} = 2\mathbf{i} - \mathbf{j} + \mathbf{k}$ and $\mathbf{w} = 4\mathbf{i} + 2\mathbf{j} - \mathbf{k}$ be two nonzero vectors that are not parallel. Find a vector $\mathbf{u} \neq \mathbf{0}$ that is orthogonal to both $\mathbf{v}$ and $\mathbf{w}$.

75. If $\mathbf{v}$ is a vector for which $\mathbf{v} \cdot \mathbf{i} = 0$, $\mathbf{v} \cdot \mathbf{j} = 0$, and $\mathbf{v} \cdot \mathbf{k} = 0$, find $\mathbf{v}$.

76. Solve for $\mathbf{x}$ in terms of $a$, $\mathbf{a}$, $\mathbf{b}$, and $\mathbf{c}$, if $a\mathbf{x} + (\mathbf{x} \cdot \mathbf{b})\mathbf{a} = \mathbf{c}$, $a \neq 0$, $a + \mathbf{a} \cdot \mathbf{b} \neq 0$. *Hint:* First find $\mathbf{x} \cdot \mathbf{b}$, then $\mathbf{x}$.

**77.** Show that:

    **(a)** There is no vector with direction angles $\alpha = \dfrac{\pi}{6}$ and $\beta = \dfrac{\pi}{4}$.

    **(b)** If $\alpha$ and $\beta$ are positive acute direction angles of a vector, then $\alpha + \beta \geq \dfrac{\pi}{2}$.

**78. Geometry**

    **(a)** Show that the vectors $\mathbf{u} = 3\mathbf{i} + \mathbf{j} - 2\mathbf{k}$, $\mathbf{v} = -\mathbf{i} + 3\mathbf{j} + 4\mathbf{k}$, and $\mathbf{w} = 4\mathbf{i} - 2\mathbf{j} - 6\mathbf{k}$ form the sides of a triangle.

    **(b)** Find the lengths of the medians of the triangle.

**79.** Suppose $(r, \theta)$ are polar coordinates of a point, where $r \geq 0$. Let

$$\mathbf{u}_r = (\cos\theta)\,\mathbf{i} + (\sin\theta)\,\mathbf{j} \qquad \mathbf{u}_\theta = (-\sin\theta)\,\mathbf{i} + (\cos\theta)\,\mathbf{j}$$

    **(a)** Show that $\mathbf{u}_r$ and $\mathbf{u}_\theta$ are unit vectors.

    **(b)** Show that $\mathbf{u}_r$ and $\mathbf{u}_\theta$ are orthogonal.

    **(c)** Show that $\mathbf{u}_r$ has the same direction as the ray from the origin to $(r, \theta)$ and that $\mathbf{u}_\theta$ is $90°$ counterclockwise from $\mathbf{u}_r$.

**80.** Find an example of three nonzero vectors **a**, **b**, and **c** for which $\mathbf{b} \neq \mathbf{c}$, neither **b** nor **c** is orthogonal to **a**, and $\mathbf{a} \cdot \mathbf{b} = \mathbf{a} \cdot \mathbf{c}$. Discuss whether there is a Cancellation Property for dot products.

**81.** If $\theta$ is the angle between the nonzero vectors **v** and **w**, where $0 \leq \theta \leq \pi$, show that

$$\sin^2\theta = \frac{\|\mathbf{v}\|^2\|\mathbf{w}\|^2 - (\mathbf{v}\cdot\mathbf{w})^2}{\|\mathbf{v}\|^2\|\mathbf{w}\|^2}$$

**82.** Let $\theta$ be the angle between the nonzero vectors $\mathbf{u} = u_1\mathbf{i} + u_2\mathbf{j} + u_3\mathbf{k}$ and $\mathbf{v} = v_1\mathbf{i} + v_2\mathbf{j} + v_3\mathbf{k}$. Show that

$$\sin^2\theta = \frac{(u_2v_3 - u_3v_2)^2 + (u_1v_3 - u_3v_1)^2 + (u_1v_2 - u_2v_1)^2}{\|\mathbf{u}\|^2\|\mathbf{v}\|^2}$$

    *Hint:* Begin with $\sin^2\theta = 1 - \cos^2\theta$.

**83.** Suppose that $\mathbf{v} = c_1\mathbf{u}_1 + c_2\mathbf{u}_2$, where $\mathbf{u}_1$, $\mathbf{u}_2$ are nonzero, nonparallel vectors, is a vector in the plane. Find $c_1$ and $c_2$ in terms of $\mathbf{u}_1$ and $\mathbf{u}_2$, and **v**. What are $c_1$ and $c_2$ if $\mathbf{u}_1$ and $\mathbf{u}_2$ are orthogonal unit vectors?

**84.** Prove the distributive property of the dot product: $\mathbf{u} \cdot (\mathbf{v} + \mathbf{w}) = \mathbf{u} \cdot \mathbf{v} + \mathbf{u} \cdot \mathbf{w}$ for vectors in the plane and in space.

**85.** Prove the linearity property: $a\,(\mathbf{u} \cdot \mathbf{v}) = (a\mathbf{u}) \cdot \mathbf{v}$ for vectors in the plane and in space.

**86.** Prove the zero-vector property: $\mathbf{0} \cdot \mathbf{v} = 0$ for vectors in the plane and in space.

**87.** Prove the commutative property of the dot product: $\mathbf{u} \cdot \mathbf{v} = \mathbf{v} \cdot \mathbf{u}$ for vectors in the plane.

**88.** Prove the magnitude property: $\mathbf{v} \cdot \mathbf{v} = \|\mathbf{v}\|^2$ for vectors in the plane.

**89.** The dot product of two vectors $\mathbf{v} = v_1\mathbf{e}_1 + v_2\mathbf{e}_2 + \cdots + v_n\mathbf{e}_n$ and $\mathbf{w} = w_1\mathbf{e}_1 + w_2\mathbf{e}_2 + \cdots + w_n\mathbf{e}_n$ in $n$-space is defined as

$$\mathbf{v} \cdot \mathbf{w} = v_1w_1 + v_2w_2 + \cdots + v_nw_n$$

Show that the five properties of the dot product on p. 755 hold for vectors in $n$-space.

## Challenge Problems

**90.** Show that $\mathbf{w} = \|\mathbf{v}\|\mathbf{u} + \|\mathbf{u}\|\mathbf{v}$ bisects the angle between **u** and **v**.

**91.** Prove the **polarization identity**: $\|\mathbf{u} + \mathbf{v}\|^2 - \|\mathbf{u} - \mathbf{v}\|^2 = 4(\mathbf{u} \cdot \mathbf{v})$.

**92. (a)** If **u** and **v** are two vectors having the same magnitude, show that $\mathbf{u} + \mathbf{v}$ and $\mathbf{u} - \mathbf{v}$ are orthogonal.

    **(b)** Use this to prove that an angle inscribed in a semicircle is a right angle (see the figure).

**93.** Suppose **v** and **w** are unit vectors in the plane, making angles $\alpha$ and $\beta$, respectively, with the positive $x$-axis.

    **(a)** Show that $\mathbf{v} = (\cos\alpha)\mathbf{i} + (\sin\alpha)\mathbf{j}$ and $\mathbf{w} = (\cos\beta)\mathbf{i} + (\sin\beta)\mathbf{j}$.

    **(b)** Use the dot product $\mathbf{v} \cdot \mathbf{w}$ to prove the sum and difference formulas

$$\cos(\alpha - \beta) = \cos\alpha\cos\beta + \sin\alpha\sin\beta$$
$$\cos(\alpha + \beta) = \cos\alpha\cos\beta - \sin\alpha\sin\beta$$

**94. Gram–Schmidt Orthogonalization Process**    Let $\mathbf{u}_1$, $\mathbf{u}_2$, and $\mathbf{u}_3$ be three noncoplanar vectors. Let $\mathbf{w}_1 = \dfrac{\mathbf{u}_1}{\|\mathbf{u}_1\|}$,

$$\mathbf{v}_2 = \mathbf{u}_2 - (\mathbf{u}_2 \cdot \mathbf{w}_1)\mathbf{w}_1, \quad \mathbf{w}_2 = \frac{\mathbf{v}_2}{\|\mathbf{v}_2\|},$$

$\mathbf{v}_3 = \mathbf{u}_3 - (\mathbf{u}_3 \cdot \mathbf{w}_1)\mathbf{w}_1 - (\mathbf{u}_3 \cdot \mathbf{w}_2)\mathbf{w}_2$, and $\mathbf{w}_3 = \dfrac{\mathbf{v}_3}{\|\mathbf{v}_3\|}$.

Show that $\mathbf{w}_1$, $\mathbf{w}_2$, and $\mathbf{w}_3$ are mutually orthogonal unit vectors.

**95. Orthogonal Vectors**   Use the Gram–Schmidt orthogonalization process (Problem 94) to transform $-\mathbf{i} + \mathbf{j}$, $2\mathbf{i} + \mathbf{k}$, and $3\mathbf{i} - \mathbf{j} + 2\mathbf{k}$ into a set of mutually orthogonal unit vectors.

**96.** Generalize the Gram–Schmidt orthogonalization process (Problem 94) to vectors in $n$ dimensions.

**97.** Let **u** and **v** be fixed vectors and define $g(t) = \|\mathbf{u} - t\mathbf{v}\|^2$. Find the minimum value of $g$ and deduce the Cauchy–Schwarz inequality $|\mathbf{u} \cdot \mathbf{v}| \leq \|\mathbf{u}\|\|\mathbf{v}\|$.

**98.** Show that two vectors in the plane are parallel if and only if their projections on two fixed mutually orthogonal vectors are proportional.

**99.** Sketch the set of points in the plane with position vector **r** satisfying $\mathbf{r} \cdot (\mathbf{i} + \mathbf{j}) \geq 0$.

**100.** Show that $(\mathbf{r} - \mathbf{b}) \cdot (\mathbf{r} + \mathbf{b}) = 0$ is a vector equation of the sphere, with **b** and $-\mathbf{b}$ as endpoints of a diameter. Show that $(\mathbf{r} - \mathbf{r}_0 - \mathbf{b}) \cdot (\mathbf{r} - \mathbf{r}_0 + \mathbf{b}) = 0$ is a vector equation of the sphere with center $\mathbf{r}_0$ and radius $\|\mathbf{b}\|$.

# 10.5 The Cross Product

**OBJECTIVES** *When you finish this section, you should be able to:*

**1** Find the cross product of two vectors (p. 765)

**2** Prove algebraic properties of the cross product (p. 766)

**3** Use geometric properties of the cross product (p. 767)

**NOTE** The cross product of two vectors is defined only for vectors in space.

The *cross product* of two vectors, defined only for vectors in space, is a third type of vector multiplication. It is of special interest for those studying physics, particularly mechanics, electricity, and magnetism, since it is used to describe angular velocity, torque, angular momentum, and magnetic forces. (See the Chapter Project on the Hall effect.)

Since each of these physical quantities is a vector that is orthogonal to two other vectors, it would be helpful to have a formula for finding a vector that is orthogonal to two given vectors.

Suppose we are given two nonzero, nonparallel vectors $\mathbf{v} = v_1\mathbf{i} + v_2\mathbf{j} + v_3\mathbf{k}$ and $\mathbf{w} = w_1\mathbf{i} + w_2\mathbf{j} + w_3\mathbf{k}$.* To determine what a vector $\mathbf{N} = n_1\mathbf{i} + n_2\mathbf{j} = n_3\mathbf{k}$ that is orthogonal to both $\mathbf{v}$ and $\mathbf{w}$ might look like, we use the fact that two vectors are orthogonal if and only if their dot product equals zero. This requires

$$\mathbf{v} \cdot \mathbf{N} = 0 \quad \text{and} \quad \mathbf{w} \cdot \mathbf{N} = 0$$
$$v_1n_1 + v_2n_2 + v_3n_3 = 0 \qquad w_1n_1 + w_2n_2 + w_3n_3 = 0$$

This is a system of two equations containing three variables $n_1$, $n_2$, and $n_3$. We begin by multiplying $\mathbf{v} \cdot \mathbf{N}$ by $w_3$ and $\mathbf{w} \cdot \mathbf{N}$ by $v_3$.

$$v_1w_3n_1 + v_2w_3n_2 + v_3w_3n_3 = 0 \tag{1}$$
$$v_3w_1n_1 + v_3w_2n_2 + v_3w_3n_3 = 0 \tag{2}$$

Now subtract (2) from (1).

$$(v_1w_3n_1 + v_2w_3n_2 + v_3w_3n_3) - (v_3w_1n_1 + v_3w_2n_2 + v_3w_3n_3) = 0$$
$$(v_1w_3 - v_3w_1)n_1 + (v_2w_3 - v_3w_2)n_2 = 0$$

One solution is

$$n_1 = v_2w_3 - v_3w_2 \quad \text{and} \quad n_2 = -(v_1w_3 - v_3w_1) = v_3w_1 - v_1w_3$$

Back-substitute these results into $\mathbf{v} \cdot \mathbf{N} = 0$.

$$v_1n_1 + v_2n_2 + v_3n_3 = 0 \qquad \mathbf{v} \cdot \mathbf{N} = 0$$
$$v_1(v_2w_3 - v_3w_2) + v_2(v_3w_1 - v_1w_3) + v_3n_3 = 0$$
$$-v_1v_3w_2 + v_2v_3w_1 + v_3n_3 = 0$$
$$n_3 = v_1w_2 - v_2w_1 \qquad v_3 \neq 0$$

We conclude the vector

$$\mathbf{N} = n_1\mathbf{i} + n_2\mathbf{j} + n_3\mathbf{k} = (v_2w_3 - v_3w_2)\mathbf{i} + (v_3w_1 - v_1w_3)\mathbf{j} + (v_1w_2 - v_2w_1)\mathbf{k}$$

is orthogonal to both vectors $\mathbf{v}$ and $\mathbf{w}$.

The form of the vector $\mathbf{N}$ is the basis for defining an operation, called the *cross product*, performed on two nonzero vectors in space.

### DEFINITION Cross Product

If $\mathbf{v} = v_1\mathbf{i} + v_2\mathbf{j} + v_3\mathbf{k}$ and $\mathbf{w} = w_1\mathbf{i} + w_2\mathbf{j} + w_3\mathbf{k}$ are two vectors in space, the **cross product** $\mathbf{v} \times \mathbf{w}$ is defined as the vector

$$\mathbf{v} \times \mathbf{w} = (v_2w_3 - v_3w_2)\mathbf{i} - (v_1w_3 - v_3w_1)\mathbf{j} + (v_1w_2 - v_2w_1)\mathbf{k}$$

**NOTE** Because the cross product $\mathbf{v} \times \mathbf{w}$ of two vectors is a vector, it is sometimes referred to as the **vector product**.

To be sure the cross product produces the desired result, we show that $\mathbf{v} \times \mathbf{w}$ is orthogonal to both $\mathbf{v}$ and $\mathbf{w}$.

*Here we assume that $v_3 \neq 0$, $w_3 \neq 0$, and $v_3 \neq w_3$.

Let $\mathbf{v} = v_1\mathbf{i} + v_2\mathbf{j} + v_3\mathbf{k}$ and $\mathbf{w} = w_1\mathbf{i} + w_2\mathbf{j} + w_3\mathbf{k}$. Then

$$\mathbf{v} \times \mathbf{w} = (v_2 w_3 - w_2 v_3)\mathbf{i} - (v_1 w_3 - w_1 v_3)\mathbf{j} + (v_1 w_2 - w_1 v_2)\mathbf{k}$$

We show that $\mathbf{v} \times \mathbf{w}$ is orthogonal to $\mathbf{v}$ by computing the dot product $\mathbf{v} \cdot (\mathbf{v} \times \mathbf{w})$.

$$\mathbf{v} \cdot (\mathbf{v} \times \mathbf{w}) = v_1(v_2 w_3 - w_2 v_3) - v_2(v_1 w_3 - w_1 v_3) + v_3(v_1 w_2 - w_1 v_2)$$
$$= v_1 v_2 w_3 - v_1 w_2 v_3 - v_2 v_1 w_3 + v_2 w_1 v_3 + v_3 v_1 w_2 - v_3 w_1 v_2 = 0$$

So, $\mathbf{v} \times \mathbf{w}$ is orthogonal to $\mathbf{v}$.

Now we show $\mathbf{v} \times \mathbf{w}$ is orthogonal to $\mathbf{w}$ by computing the dot product $\mathbf{w} \cdot (\mathbf{v} \times \mathbf{w})$.

$$\mathbf{w} \cdot (\mathbf{v} \times \mathbf{w}) = w_1(v_2 w_3 - w_2 v_3) - w_2(v_1 w_3 - w_1 v_3) + w_3(v_1 w_2 - w_1 v_2)$$
$$= w_1 v_2 w_3 - w_1 w_2 v_3 - w_2 v_1 w_3 + w_2 w_1 v_3 + w_3 v_1 w_2 - w_3 w_1 v_2 = 0$$

So, $\mathbf{v} \times \mathbf{w}$ is orthogonal to $\mathbf{w}$. ∎

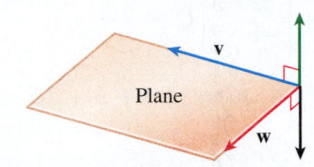

**Figure 45**

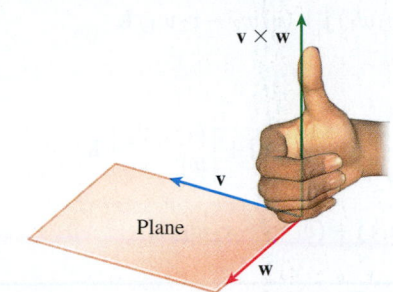

**Figure 46** Right-hand rule

Any two nonparallel vectors $\mathbf{v}$ and $\mathbf{w}$ lie in a single plane. Since the vector $\mathbf{v} \times \mathbf{w}$ is orthogonal to both vectors $\mathbf{v}$ and $\mathbf{w}$, it follows that the vector $\mathbf{v} \times \mathbf{w}$ is normal to the plane containing $\mathbf{v}$ and $\mathbf{w}$. As Figure 45 illustrates, there actually are two directions that are normal to the plane containing $\mathbf{v}$ and $\mathbf{w}$; one directed up and the other directed down. Which is the direction of $\mathbf{v} \times \mathbf{w}$? It can be shown that the direction of $\mathbf{v} \times \mathbf{w}$ is determined by the thumb of the right hand when the other fingers of the right hand are cupped so that they point in the direction from $\mathbf{v}$ to $\mathbf{w}$, as shown in Figure 46. As stated in Section 10.1, this is usually referred to as the **right-hand rule**.

## 1 Find the Cross Product of Two Vectors

**EXAMPLE 1**  **Finding the Cross Product of Two Vectors**

**(a)**  Find the cross product of the vectors $\mathbf{v} = 2\mathbf{i} - \mathbf{j} + \mathbf{k}$ and $\mathbf{w} = 4\mathbf{i} + 2\mathbf{j} - \mathbf{k}$.

**(b)**  Verify that the cross product $\mathbf{v} \times \mathbf{w}$ is orthogonal to both vectors $\mathbf{v}$ and $\mathbf{w}$.

**Solution (a)** Using the definition of cross product, we find

$$\mathbf{v} \times \mathbf{w} = [(-1)(-1) - 1 \cdot 2]\mathbf{i} - [2 \cdot (-1) - 1 \cdot 4]\mathbf{j} + [2 \cdot 2 - (-1) \cdot 4]\mathbf{k}$$
$$= -\mathbf{i} + 6\mathbf{j} + 8\mathbf{k}$$

**(b)** To verify that $\mathbf{v} \times \mathbf{w}$ is orthogonal to $\mathbf{v}$ and $\mathbf{w}$, we find the dot products.

$$(\mathbf{v} \times \mathbf{w}) \cdot \mathbf{v} = (-\mathbf{i} + 6\mathbf{j} + 8\mathbf{k}) \cdot (2\mathbf{i} - \mathbf{j} + \mathbf{k}) = -2 - 6 + 8 = 0$$
$$(\mathbf{v} \times \mathbf{w}) \cdot \mathbf{v} = (-\mathbf{i} + 6\mathbf{j} + 8\mathbf{k}) \cdot (4\mathbf{i} + 2\mathbf{j} - \mathbf{k}) = -4 + 12 - 8 = 0$$

Since both $(\mathbf{v} \times \mathbf{w}) \cdot \mathbf{v} = 0$ and $(\mathbf{v} \times \mathbf{w}) \cdot \mathbf{w} = 0$, the vector $\mathbf{v} \times \mathbf{w}$ is orthogonal to both $\mathbf{v}$ and $\mathbf{w}$. ∎

NOW WORK **Problems 13 and 29.**

*Determinants* are helpful for finding a cross product. A **2 by 2 determinant** is a number or expression associated with an array containing two rows and two columns and is denoted by

$$\begin{vmatrix} a & b \\ c & d \end{vmatrix} = ad - bc$$

A **3 by 3 determinant** is a number or expression associated with an array containing three rows and three columns. Its value is given by

$$\begin{vmatrix} A & B & C \\ a_1 & b_1 & c_1 \\ a_2 & b_2 & c_2 \end{vmatrix} = \begin{vmatrix} b_1 & c_1 \\ b_2 & c_2 \end{vmatrix} A - \begin{vmatrix} a_1 & c_1 \\ a_2 & c_2 \end{vmatrix} B + \begin{vmatrix} a_1 & b_1 \\ a_2 & b_2 \end{vmatrix} C$$

$$= (b_1 c_2 - b_2 c_1)A - (a_1 c_2 - a_2 c_1)B + (a_1 b_2 - a_2 b_1)C$$

**NOTE**  Notice the minus sign in the middle term.

**EXAMPLE 2** Evaluating Determinants

(a) $\begin{vmatrix} 2 & 3 \\ -1 & 2 \end{vmatrix} = 2 \cdot 2 - (-1) \cdot 3 = 4 + 3 = 7$

(b) $\begin{vmatrix} A & B & C \\ 2 & 3 & 1 \\ -1 & 2 & -1 \end{vmatrix} = \begin{vmatrix} 3 & 1 \\ 2 & -1 \end{vmatrix} A - \begin{vmatrix} 2 & 1 \\ -1 & -1 \end{vmatrix} B + \begin{vmatrix} 2 & 3 \\ -1 & 2 \end{vmatrix} C$

$$= (-3 - 2)A - (-2 + 1)B + (4 + 3)C = -5A + B + 7C \qquad \blacksquare$$

**NOW WORK** Problem 11.

The cross product of the vectors $\mathbf{v} = v_1 \mathbf{i} + v_2 \mathbf{j} + v_3 \mathbf{k}$ and $\mathbf{w} = w_1 \mathbf{i} + w_2 \mathbf{j} + w_3 \mathbf{k}$:

$$\mathbf{v} \times \mathbf{w} = (v_2 w_3 - v_3 w_2)\mathbf{i} - (v_1 w_3 - v_3 w_1)\mathbf{j} + (v_1 w_2 - v_2 w_1)\mathbf{k}$$

can be written symbolically as the determinant:

$$\mathbf{v} \times \mathbf{w} = \begin{vmatrix} \mathbf{i} & \mathbf{j} & \mathbf{k} \\ v_1 & v_2 & v_3 \\ w_1 & w_2 & w_3 \end{vmatrix} = \begin{vmatrix} v_2 & v_3 \\ w_2 & w_3 \end{vmatrix} \mathbf{i} - \begin{vmatrix} v_1 & v_3 \\ w_1 & w_3 \end{vmatrix} \mathbf{j} + \begin{vmatrix} v_1 & v_2 \\ w_1 & w_2 \end{vmatrix} \mathbf{k}$$

$$= (v_2 w_3 - v_3 w_2)\mathbf{i} - (v_1 w_3 - v_3 w_1)\mathbf{j} + (v_1 w_2 - v_2 w_1)\mathbf{k}$$

**EXAMPLE 3** Finding a Cross Product Using Determinants

If $\mathbf{v} = 2\mathbf{i} - \mathbf{j} + \mathbf{k}$ and $\mathbf{w} = 4\mathbf{i} + 2\mathbf{j} - \mathbf{k}$, find

**(a)** $\mathbf{v} \times \mathbf{w}$     **(b)** $\mathbf{w} \times \mathbf{v}$     **(c)** $\mathbf{v} \times \mathbf{v}$

**Solution (a)** $\mathbf{v} \times \mathbf{w} = \begin{vmatrix} \mathbf{i} & \mathbf{j} & \mathbf{k} \\ 2 & -1 & 1 \\ 4 & 2 & -1 \end{vmatrix} = (1-2)\mathbf{i} - (-2-4)\mathbf{j} + (4+4)\mathbf{k} = -\mathbf{i} + 6\mathbf{j} + 8\mathbf{k}$

**(b)** $\mathbf{w} \times \mathbf{v} = \begin{vmatrix} \mathbf{i} & \mathbf{j} & \mathbf{k} \\ 4 & 2 & -1 \\ 2 & -1 & 1 \end{vmatrix} = (2-1)\mathbf{i} - (4+2)\mathbf{j} + (-4-4)\mathbf{k} = \mathbf{i} - 6\mathbf{j} - 8\mathbf{k} = -(\mathbf{v} \times \mathbf{w})$

**NOTE** The cross product is not a commutative operation. That is, $\mathbf{v} \times \mathbf{w} \neq \mathbf{w} \times \mathbf{v}$.

**(c)** $\mathbf{v} \times \mathbf{v} = \begin{vmatrix} \mathbf{i} & \mathbf{j} & \mathbf{k} \\ 2 & -1 & 1 \\ 2 & -1 & 1 \end{vmatrix} = 0\mathbf{i} - 0\mathbf{j} + 0\mathbf{k} = \mathbf{0}$    $\blacksquare$

**NOW WORK** Problem 15.

## 2 Prove Algebraic Properties of the Cross Product

Example 3 illustrates two properties of the cross product. Parts (a) and (b) suggest that $\mathbf{v} \times \mathbf{w} = -(\mathbf{w} \times \mathbf{v})$ and (c) suggests that the cross product of a vector with itself equals the zero vector $\mathbf{0}$. These and other properties of the cross product are listed in the theorem below.

---

**THEOREM Algebraic Properties of the Cross Product**

If $\mathbf{u}$, $\mathbf{v}$, and $\mathbf{w}$ are vectors in space and if $a$ is a scalar, then

- $\mathbf{v} \times \mathbf{v} = \mathbf{0}$           (3)
- $\mathbf{v} \times \mathbf{w} = -(\mathbf{w} \times \mathbf{v})$        (4)
- $a(\mathbf{v} \times \mathbf{w}) = (a\mathbf{v}) \times \mathbf{w} = \mathbf{v} \times (a\mathbf{w})$        (5)
- $\mathbf{v} \times (\mathbf{w} + \mathbf{u}) = (\mathbf{v} \times \mathbf{w}) + (\mathbf{v} \times \mathbf{u})$      (6)
- $\|\mathbf{v} \times \mathbf{w}\|^2 = \|\mathbf{v}\|^2 \|\mathbf{w}\|^2 - (\mathbf{v} \cdot \mathbf{w})^2$      (7)
- $\mathbf{u} \cdot (\mathbf{v} \times \mathbf{w}) = (\mathbf{u} \times \mathbf{v}) \cdot \mathbf{w}$         (8)

---

We prove (3), (4), and (7), and leave the proofs of (5), (6), and (8) as exercises (Problems 60, 61, and 62).

**Proof** Let $\mathbf{v} = v_1\mathbf{i} + v_2\mathbf{j} + v_3\mathbf{k}$ and $\mathbf{w} = w_1\mathbf{i} + w_2\mathbf{j} + w_3\mathbf{k}$.

(3) $\mathbf{v} \times \mathbf{v} = (v_2v_3 - v_2v_3)\mathbf{i} - (v_1v_3 - v_1v_3)\mathbf{j} + (v_1v_2 - v_1v_2)\mathbf{k} = \mathbf{0}$

(4) $\mathbf{v} \times \mathbf{w} = (v_2w_3 - w_2v_3)\mathbf{i} - (v_1w_3 - w_1v_3)\mathbf{j} + (v_1w_2 - w_1v_2)\mathbf{k}$

$\mathbf{w} \times \mathbf{v} = (w_2v_3 - v_2w_3)\mathbf{i} - (w_1v_3 - v_1w_3)\mathbf{j} + (w_1v_2 - v_1w_2)\mathbf{k}$

Consequently, $\mathbf{v} \times \mathbf{w} = -(\mathbf{w} \times \mathbf{v})$.

(7) First we find $\|\mathbf{v} \times \mathbf{w}\|^2$.

$$\mathbf{v} \times \mathbf{w} = (v_2w_3 - w_2v_3)\mathbf{i} - (v_1w_3 - w_1v_3)\mathbf{j} + (v_1w_2 - w_1v_2)\mathbf{k}$$

$$\|\mathbf{v} \times \mathbf{w}\|^2 = (v_2w_3 - w_2v_3)^2 + (v_1w_3 - w_1v_3)^2 + (v_1w_2 - w_1v_2)^2$$

$$= [(v_2\,w_3)^2 - 2v_2\,w_2\,v_3\,w_3 + (w_2\,v_3)^2] + [(v_1w_3)^2 - 2v_1w_1v_3w_3 + (w_1v_3)^2]$$

$$+ [(v_1\,w_2)^2 - 2\,v_1\,w_1\,v_2\,w_2 + (w_1v_2)^2]$$

Then we find $\|\mathbf{v}\|^2\|\mathbf{w}\|^2 - (\mathbf{v} \cdot \mathbf{w})^2$.

$$\|\mathbf{v}\|^2\|\mathbf{w}\|^2 - (\mathbf{v} \cdot \mathbf{w})^2 = \left[(v_1^2 + v_2^2 + v_3^2)(w_1^2 + w_2^2 + w_3^2)\right] - (v_1w_1 + v_2w_2 + v_3w_3)^2$$

$$= (v_1w_2)^2 + (v_1w_3)^2 + (v_2w_1)^2 + (v_2w_3)^2 + (v_3w_1)^2 + (v_3w_2)^2$$

$$- 2v_1v_2w_1w_2 - 2v_1v_3w_1w_3 - 2v_2v_3w_2w_3$$

Reordering the terms in each equation proves the property. ∎

The cross products of the unit vectors $\mathbf{i}$, $\mathbf{j}$, and $\mathbf{k}$ are particularly useful:

$$\mathbf{i} \times \mathbf{j} = \begin{vmatrix} \mathbf{i} & \mathbf{j} & \mathbf{k} \\ 1 & 0 & 0 \\ 0 & 1 & 0 \end{vmatrix} = 0\mathbf{i} + 0\mathbf{j} + \mathbf{k} = \mathbf{k}$$

$$\mathbf{j} \times \mathbf{k} = \begin{vmatrix} \mathbf{i} & \mathbf{j} & \mathbf{k} \\ 0 & 1 & 0 \\ 0 & 0 & 1 \end{vmatrix} = \mathbf{i}$$

$$\mathbf{k} \times \mathbf{i} = \begin{vmatrix} \mathbf{i} & \mathbf{j} & \mathbf{k} \\ 0 & 0 & 1 \\ 1 & 0 & 0 \end{vmatrix} = \mathbf{j}$$

Notice the cyclic pattern for the cross products of $\mathbf{i}$, $\mathbf{j}$, and $\mathbf{k}$:

$$\boxed{\mathbf{i} \times \mathbf{j} = \mathbf{k} \qquad \mathbf{j} \times \mathbf{k} = \mathbf{i} \qquad \mathbf{k} \times \mathbf{i} = \mathbf{j}}$$

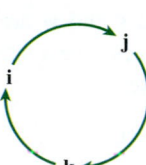

**Figure 47**

Figure 47 may be helpful in remembering this pattern.

The remaining cross products involving the unit vectors $\mathbf{i}$, $\mathbf{j}$, and $\mathbf{k}$ are listed below. They are consequences of algebraic properties (3) and (4) of the cross product.

$$\boxed{\begin{array}{ccc} \mathbf{j} \times \mathbf{i} = -\mathbf{k} & \mathbf{k} \times \mathbf{j} = -\mathbf{i} & \mathbf{i} \times \mathbf{k} = -\mathbf{j} \\ \mathbf{i} \times \mathbf{i} = \mathbf{0} & \mathbf{j} \times \mathbf{j} = \mathbf{0} & \mathbf{k} \times \mathbf{k} = \mathbf{0} \end{array}}$$

## 3 Use Geometric Properties of the Cross Product

Besides useful algebraic properties, the cross product has important geometric properties.

> **THEOREM Geometric Properties of the Cross Product**
>
> If $\mathbf{v}$ and $\mathbf{w}$ are two nonzero vectors in space, then
>
> - $\mathbf{v} \times \mathbf{w}$ is orthogonal to both $\mathbf{v}$ and $\mathbf{w}$     (9)
> - $\|\mathbf{v} \times \mathbf{w}\| = \|\mathbf{v}\|\,\|\mathbf{w}\|\sin\theta$, where $\theta$ is the angle between $\mathbf{v}$ and $\mathbf{w}$   (10)
> - $\|\mathbf{v} \times \mathbf{w}\|$ is the area of the parallelogram having $\mathbf{v}$ and $\mathbf{w}$ as adjacent sides   (11)
> - $\mathbf{v} \times \mathbf{w} = \mathbf{0}$ if and only if $\mathbf{v}$ and $\mathbf{w}$ are parallel vectors     (12)
> - $|\mathbf{u} \cdot (\mathbf{v} \times \mathbf{w})|$ is the volume of a parallelepiped having $\mathbf{u}$, $\mathbf{v}$, and $\mathbf{w}$ as adjacent sides   (13)

**Proof of (10)** Let $\theta$ be the angle between the nonzero vectors $\mathbf{v}$ and $\mathbf{w}$. Then from algebraic property (7), we have

$$\|\mathbf{v} \times \mathbf{w}\|^2 = \|\mathbf{v}\|^2 \|\mathbf{w}\|^2 - (\mathbf{v} \cdot \mathbf{w})^2$$

Since $\cos \theta = \dfrac{\mathbf{v} \cdot \mathbf{w}}{\|\mathbf{v}\|\,\|\mathbf{w}\|}$, we have $\mathbf{v} \cdot \mathbf{w} = \|\mathbf{v}\|\,\|\mathbf{w}\| \cos \theta$ and

$$
\begin{aligned}
\|\mathbf{v} \times \mathbf{w}\|^2 &= \|\mathbf{v}\|^2 \|\mathbf{w}\|^2 - \|\mathbf{v}\|^2 \|\mathbf{w}\|^2 \cos^2 \theta && (\mathbf{v} \cdot \mathbf{w})^2 = \|\mathbf{v}\|^2 \|\mathbf{w}\|^2 \cos^2 \theta \\
&= \|\mathbf{v}\|^2 \|\mathbf{w}\|^2 (1 - \cos^2 \theta) && \text{Factor out } \|\mathbf{v}\|^2 \|\mathbf{w}\|^2 \\
&= \|\mathbf{v}\|^2 \|\mathbf{w}\|^2 \sin^2 \theta && 1 - \cos^2 \theta = \sin^2 \theta
\end{aligned}
$$

Since $0 \le \theta \le \pi$, we know $\sin \theta \ge 0$. Now take the square root of both sides to obtain $\|\mathbf{v} \times \mathbf{w}\| = \|\mathbf{v}\|\|\mathbf{w}\| \sin \theta$. ∎

Property (10) is used to prove Property (11): $\|\mathbf{v} \times \mathbf{w}\|$ equals the area of the parallelogram with adjacent sides $\mathbf{v}$ and $\mathbf{w}$.

**Proof of (11)** See Figure 48. If $\mathbf{v}$ and $\mathbf{w}$ are nonzero vectors and $\theta$ is the angle between $\mathbf{v}$ and $\mathbf{w}$, then $\|\mathbf{v}\|$ and $\|\mathbf{w}\|$ represent the lengths of the sides of a parallelogram. The base of the parallelogram is $\|\mathbf{v}\|$ and its altitude is $\|\mathbf{w}\| \sin \theta$, so its area $A$ is

$$A = (\text{base})(\text{altitude}) = \|\mathbf{v}\|\,\|\mathbf{w}\| \sin \theta$$

Since $\|\mathbf{v} \times \mathbf{w}\| = \|\mathbf{v}\|\|\mathbf{w}\| \sin \theta$, the magnitude of $\mathbf{v} \times \mathbf{w}$ is the area of the parallelogram whose sides are the vectors $\mathbf{v}$ and $\mathbf{w}$. ∎

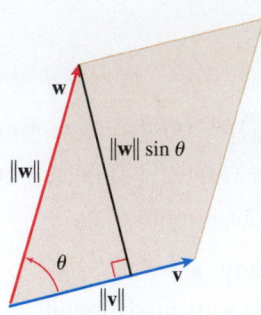

**Figure 48**    Area = base × altitude
$= \|\mathbf{v}\|\|\mathbf{w}\| \sin \theta$

---

**CALC CLIP**

**EXAMPLE 4   Finding the Area of a Parallelogram**

Find the area $A$ of the parallelogram shown in Figure 49, which has vertices at the points $P_1 = (0, 0, 0)$, $P_2 = (-1, 2, 4)$, $P_3 = (1, 1, 8)$, and $P_4 = (2, -1, 4)$.

**Solution** The area $A$ of a parallelogram is $\|\mathbf{v} \times \mathbf{w}\|$, where $\mathbf{v}$ and $\mathbf{w}$ are two adjacent sides of the parallelogram. Be careful! Not all pairs of vertices give a side. It is helpful to sketch the parallelogram before choosing $\mathbf{v}$ and $\mathbf{w}$. Choose

$$\mathbf{v} = \overrightarrow{P_1 P_2} = \langle -1,\ 2, 4 \rangle = -\mathbf{i} + 2\mathbf{j} + 4\mathbf{k}$$

and

$$\mathbf{w} = \overrightarrow{P_1 P_4} = \langle 2, -1, 4 \rangle = 2\mathbf{i} - \mathbf{j} + 4\mathbf{k}$$

Then

$$\mathbf{v} \times \mathbf{w} = \begin{vmatrix} \mathbf{i} & \mathbf{j} & \mathbf{k} \\ -1 & 2 & 4 \\ 2 & -1 & 4 \end{vmatrix} = 12\mathbf{i} + 12\mathbf{j} - 3\mathbf{k}$$

The area of the parallelogram is

$$\|\mathbf{v} \times \mathbf{w}\| = \sqrt{144 + 144 + 9} = \sqrt{297} \approx 17.234 \text{ square units}$$ ∎

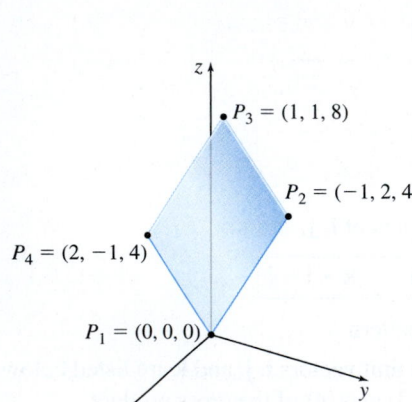

**Figure 49**

**NOW WORK** Problems 41 and 49.

Property (12): $\mathbf{v} \times \mathbf{w} = \mathbf{0}$ if and only if $\mathbf{v}$ and $\mathbf{w}$ are parallel nonzero vectors.

**Proof of (12)** If $\mathbf{v}$ and $\mathbf{w}$ are parallel nonzero vectors, then the angle $\theta$ between the vectors $\mathbf{v}$ and $\mathbf{w}$ is either $\theta = 0$ or $\theta = \pi$. For either angle, $\sin \theta = 0$, and

$$\|\mathbf{v} \times \mathbf{w}\| = \|\mathbf{v}\|\|\mathbf{w}\| \sin \theta = 0 \qquad \text{so} \quad \mathbf{v} \times \mathbf{w} = \mathbf{0}$$

Now suppose $\mathbf{v} \times \mathbf{w} = \mathbf{0}$. Since $\|\mathbf{v} \times \mathbf{w}\| = \|\mathbf{v}\|\|\mathbf{w}\| \sin \theta$, we have

$$\|\mathbf{v}\|\,\|\mathbf{w}\| \sin \theta = 0$$

$$\sin \theta = 0 \qquad\qquad\qquad \mathbf{v} \ne \mathbf{0}, \quad \mathbf{w} \ne \mathbf{0}$$

$$\theta = 0 \quad \text{or} \quad \theta = \pi$$

That is, the vectors $\mathbf{v}$ and $\mathbf{w}$ are parallel. ∎

**EXAMPLE 5**  **Verifying Two Vectors Are Parallel**

Show that the vectors $\mathbf{v} = 2\mathbf{i} + \mathbf{j} - \mathbf{k}$ and $\mathbf{w} = -4\mathbf{i} - 2\mathbf{j} + 2\mathbf{k}$ are parallel.

**Solution** We show that $\mathbf{v} \times \mathbf{w} = \mathbf{0}$.

$$\mathbf{v} \times \mathbf{w} = \begin{vmatrix} \mathbf{i} & \mathbf{j} & \mathbf{k} \\ 2 & 1 & -1 \\ -4 & -2 & 2 \end{vmatrix} = \begin{vmatrix} 1 & -1 \\ -2 & 2 \end{vmatrix} \mathbf{i} - \begin{vmatrix} 2 & -1 \\ -4 & 2 \end{vmatrix} \mathbf{j} + \begin{vmatrix} 2 & 1 \\ -4 & -2 \end{vmatrix} \mathbf{k}$$

$$= 0\mathbf{i} + 0\mathbf{j} + 0\mathbf{k} = \mathbf{0}$$  ∎

For the vectors $\mathbf{v}$ and $\mathbf{w}$ in Example 5, notice that $\mathbf{w} = -2\mathbf{v}$. So, the magnitude of $\mathbf{w}$ is twice that of $\mathbf{v}$, and $\mathbf{v}$ and $\mathbf{w}$ are opposite in direction.

Problem 78 asks you to prove Property (13).

## Application: Angular Velocity

Figure 50 illustrates a rigid body that is rotating about a fixed axis $l$ with a constant angular speed $\omega$. The **angular velocity** $\omega$ is defined as the vector of magnitude $\omega$ whose direction is parallel to the axis $l$, so that if the fingers of the right hand are wrapped about $l$ in the direction of the rotation, the thumb will point in the direction of $\boldsymbol{\omega}$.

We seek a formula for the velocity $\mathbf{v}$ of an object on the rigid body. Let the origin $O$ be on the axis $l$ of rotation, let $\mathbf{r}$ denote the position of the object, and let $\theta$ be the angle between $\mathbf{r}$ and the axis $l$. The motion of the object is circular. Since $\|\mathbf{r}\| \sin\theta$ equals the distance of the object from the axis $l$, $\|\mathbf{r}\| \sin\theta$ is the radius of the circle. As a result, we have

$$\|\mathbf{v}\| = \|\boldsymbol{\omega}\|\,\|\mathbf{r}\|\sin\theta \qquad v = r\omega \qquad \text{(linear speed} = \text{radius times angular speed)}$$

Since the velocity is tangent to the circle of motion, the velocity $\mathbf{v}$ is normal to the plane formed by $\mathbf{r}$ and $\boldsymbol{\omega}$. That is, $\mathbf{v}$ is parallel to $\boldsymbol{\omega} \times \mathbf{r}$. Using the right-hand rule to define $\boldsymbol{\omega}$, the velocity $\mathbf{v}$ is in the direction of $\boldsymbol{\omega} \times \mathbf{r}$. Since the magnitude of $\boldsymbol{\omega} \times \mathbf{r}$ is

$$\|\boldsymbol{\omega} \times \mathbf{r}\| = \|\boldsymbol{\omega}\|\,\|\mathbf{r}\|\sin\theta = \|\mathbf{v}\|$$

it follows that

$$\boxed{\mathbf{v} = \boldsymbol{\omega} \times \mathbf{r}}$$

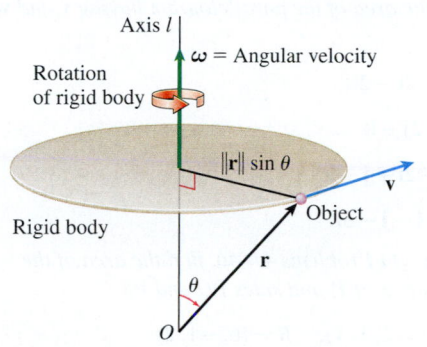

Axis $l$

$\boldsymbol{\omega}$ = Angular velocity

Rotation of rigid body

$\|\mathbf{r}\|\sin\theta$

$\mathbf{v}$

Object

Rigid body

$\mathbf{r}$

$\theta$

$O$

**DF** **Figure 50**

**NEED TO REVIEW?**  Circular motion is discussed in Appendix A.4, p. A-27.

**EXAMPLE 6**  **Finding the Speed of an Object in Circular Motion**

A rigid body rotates with a constant angular speed of $\omega$ radians per second about a line through the origin in the direction of the vector $2\mathbf{i} + \mathbf{j} + 2\mathbf{k}$. Find the speed of an object on this body at the instant the object passes through the point $(1, 3, 5)$. Assume the distance is in meters.

**Solution** The vector $\boldsymbol{\omega}$ is parallel to the axis of rotation of the rigid body, and so it is in the direction of the vector $2\mathbf{i} + \mathbf{j} + 2\mathbf{k}$. Since $\dfrac{2\mathbf{i} + \mathbf{j} + 2\mathbf{k}}{\sqrt{2^2 + 1^2 + 2^2}} = \dfrac{2}{3}\mathbf{i} + \dfrac{1}{3}\mathbf{j} + \dfrac{2}{3}\mathbf{k}$ is a unit vector in the direction of $2\mathbf{i} + \mathbf{j} + 2\mathbf{k}$, the angular velocity $\boldsymbol{\omega}$ is

$$\boldsymbol{\omega} = \omega\left(\frac{2}{3}\mathbf{i} + \frac{1}{3}\mathbf{j} + \frac{2}{3}\mathbf{k}\right)$$

The position of the object is $\mathbf{r} = \mathbf{i} + 3\mathbf{j} + 5\mathbf{k}$. So, the velocity $\mathbf{v}$ of the object is

$$\mathbf{v} = \boldsymbol{\omega} \times \mathbf{r} = \left[\omega\left(\frac{2}{3}\mathbf{i} + \frac{1}{3}\mathbf{j} + \frac{2}{3}\mathbf{k}\right)\right] \times [\mathbf{i} + 3\mathbf{j} + 5\mathbf{k}] = \omega\begin{vmatrix} \mathbf{i} & \mathbf{j} & \mathbf{k} \\ \dfrac{2}{3} & \dfrac{1}{3} & \dfrac{2}{3} \\ 1 & 3 & 5 \end{vmatrix}$$

$$= \omega\left(-\frac{1}{3}\mathbf{i} - \frac{8}{3}\mathbf{j} + \frac{5}{3}\mathbf{k}\right)$$

The speed $v$ of the object is $v = \|\mathbf{v}\| = \omega\sqrt{\dfrac{1}{9} + \dfrac{64}{9} + \dfrac{25}{9}} = \sqrt{10}\,\omega$ m/s.  ∎

**NOW WORK** Problem **51**.

## 10.5 Assess Your Understanding

### Concepts and Vocabulary

1. *True or False* The direction of $\mathbf{v} \times \mathbf{w}$ is determined by the right-hand rule.

2. *True or False* Since $\mathbf{i}$ and $\mathbf{j}$ are orthogonal, $\mathbf{i} \times \mathbf{j} = \mathbf{0}$.

3. *Multiple Choice* The cross product $\mathbf{v} \times \mathbf{w}$ is [(**a**) parallel (**b**) orthogonal] to the vector $\mathbf{v}$ and to the vector $\mathbf{w}$.

4. *True or False* The cross product is a commutative operation.

5. *Multiple Choice* If $\mathbf{u}$ and $\mathbf{v}$ are parallel, then $\mathbf{u} \times \mathbf{v} = [(\mathbf{a})\ \mathbf{i}\ (\mathbf{b})\ \mathbf{j}\ (\mathbf{c})\ \mathbf{k}\ (\mathbf{d})\ \mathbf{0}]$

6. *True or False* If $\mathbf{v}$ and $\mathbf{w}$ are two nonzero vectors, then
$$(\mathbf{v} \times \mathbf{w}) + (\mathbf{w} \times \mathbf{v}) = \mathbf{0}$$

7. *Multiple Choice* The area of a parallelogram having vectors $\mathbf{v}$ and $\mathbf{w}$ as adjacent sides equals [(**a**) $\|\mathbf{v} + \mathbf{w}\|$ (**b**) $\|\mathbf{v} \cdot \mathbf{w}\|$ (**c**) $\|\mathbf{v} \times \mathbf{w}\|$]

8. *True or False* If $\theta$ is the angle between two nonzero vectors $\mathbf{v}$ and $\mathbf{w}$, then $\|\mathbf{v} \times \mathbf{w}\| = \|\mathbf{v}\| \|\mathbf{w}\| \cos \theta$.

### Skill Building

*In Problems 9–12, find the value of each determinant.*

9. $\begin{vmatrix} 3 & 4 \\ 1 & 2 \end{vmatrix}$

10. $\begin{vmatrix} -2 & 4 \\ 2 & -1 \end{vmatrix}$

11. $\begin{vmatrix} 1 & -2 & 2 \\ 1 & 2 & 1 \\ 2 & 1 & 4 \end{vmatrix}$

12. $\begin{vmatrix} 7 & 0 & 1 \\ 0 & 2 & 3 \\ 0 & 1 & 3 \end{vmatrix}$

*In Problems 13–22:*
(a) Find the cross product $\mathbf{v} \times \mathbf{w}$.
(b) Check your answer by showing that $\mathbf{v}$ and $\mathbf{w}$ are each orthogonal to $\mathbf{v} \times \mathbf{w}$.

13. $\mathbf{v} = 2\mathbf{i} + \mathbf{j} - \mathbf{k}, \quad \mathbf{w} = \mathbf{i} - \mathbf{j} + \mathbf{k}$

14. $\mathbf{v} = 4\mathbf{i} - \mathbf{j} + 2\mathbf{k}, \quad \mathbf{w} = 2\mathbf{i} + \mathbf{j} + \mathbf{k}$

15. $\mathbf{v} = \mathbf{i} + \mathbf{j}, \quad \mathbf{w} = \mathbf{i} - \mathbf{k}$

16. $\mathbf{v} = \mathbf{j} - \mathbf{k}, \quad \mathbf{w} = \mathbf{i} - \mathbf{j}$

17. $\mathbf{v} = 3\mathbf{i} - 2\mathbf{j} + \mathbf{k}, \quad \mathbf{w} = \mathbf{i} + \mathbf{j}$

18. $\mathbf{v} = 2\mathbf{i} - \mathbf{j}, \quad \mathbf{w} = \mathbf{i} + \mathbf{j} - 3\mathbf{k}$

19. $\mathbf{v} = -\mathbf{i} + 8\mathbf{j} + 3\mathbf{k}, \quad \mathbf{w} = 7\mathbf{i} + 2\mathbf{j}$

20. $\mathbf{v} = 2\mathbf{j} - \mathbf{k}, \quad \mathbf{w} = -3\mathbf{i} + \mathbf{j} + \mathbf{k}$

21. $\mathbf{v} = 2\mathbf{i} + 3\mathbf{j} - 4\mathbf{k}, \quad \mathbf{w} = -\mathbf{i} + \mathbf{j} - 4\mathbf{k}$

22. $\mathbf{v} = (\cos \theta)\,\mathbf{i} - (\sin \theta)\,\mathbf{j}, \quad \mathbf{w} = (\sin \theta)\,\mathbf{i} + (\cos \theta)\,\mathbf{j}$

*In Problems 23–34, $\mathbf{u} = 2\mathbf{i} + 3\mathbf{j} - 4\mathbf{k}$, $\mathbf{v} = -3\mathbf{i} + \mathbf{j} - 4\mathbf{k}$, $\mathbf{w} = \mathbf{i} + \mathbf{j} - 3\mathbf{k}$. Find the following.*

23. $\mathbf{u} \times \mathbf{u}$
24. $\mathbf{w} \times \mathbf{w}$
25. $\mathbf{v} \times \mathbf{w}$

26. $\mathbf{u} \times \mathbf{v}$
27. $(3\mathbf{v}) \times \mathbf{w}$
28. $\mathbf{u} \times (-\mathbf{v})$

29. a vector orthogonal to both $\mathbf{v}$ and $\mathbf{w}$

30. a vector orthogonal to both $\mathbf{w}$ and $\mathbf{u}$

31. a vector orthogonal to both $\mathbf{w}$ and $\mathbf{i} + \mathbf{j} - \mathbf{k}$

32. a vector orthogonal to both $\mathbf{u}$ and $\mathbf{k}$

33. a unit vector normal to the plane containing $\mathbf{u}$ and $\mathbf{v}$

34. a unit vector normal to the plane containing $\mathbf{u}$ and $\mathbf{w}$

35. Find a unit vector normal to the plane containing $\mathbf{v} = 2\mathbf{i} - 6\mathbf{j} - 3\mathbf{k}$ and $\mathbf{w} = 4\mathbf{i} + 3\mathbf{j} - \mathbf{k}$.

36. Find a unit vector normal to the plane containing $\mathbf{v} = \mathbf{i} + \mathbf{j} - 2\mathbf{k}$ and $\mathbf{w} = 3\mathbf{i} + 2\mathbf{j} - \mathbf{k}$.

*In Problems 37–40, find the area of the parallelogram having $\mathbf{v}$ and $\mathbf{w}$ as adjacent sides.*

37. $\mathbf{v} = \mathbf{i} + \mathbf{j} + 2\mathbf{k}, \mathbf{w} = -2\mathbf{i} - 2\mathbf{k}$

38. $\mathbf{v} = 2\mathbf{i} + \mathbf{j}, \mathbf{w} = -\mathbf{i} + 2\mathbf{j} + \mathbf{k}$

39. $\mathbf{v} = \mathbf{i} - \mathbf{j} + 5\mathbf{k}, \mathbf{w} = -2\mathbf{i} + \mathbf{j} + \mathbf{k}$

40. $\mathbf{v} = 2\mathbf{i} - 3\mathbf{j} - \mathbf{k}, \mathbf{w} = \mathbf{i} - \mathbf{j} - 3\mathbf{k}$

**Area of a Parallelogram** *In Problems 41–46, find the area of the parallelogram with one vertex at $P$ and sides $PQ$ and $PR$.*

41. $P = (1, -3, 7); \quad Q = (2, 1, 1); \quad R = (6, -1, 2)$

42. $P = (0, 1, 1); \quad Q = (2, 0, -4); \quad R = (-3, -2, 1)$

43. $P = (-2, 1, 6); \quad Q = (2, 1, -7); \quad R = (4, 1, 1)$

44. $P = (0, 0, 3); \quad Q = (2, -5, 3); \quad R = (1, 1, -2)$

45. $P = (1, 1, -6); \quad Q = (5, -3, 0); \quad R = (-2, 4, 1)$

46. $P = (-4, 6, 3); \quad Q = (1, 1, -5); \quad R = (2, 2, 2)$

**Area of a Parallelogram** *In Problems 47–50, find the area of the parallelogram whose vertices are $P_1$, $P_2$, $P_3$, and $P_4$.*

47. $P_1 = (0, 0, 0); \quad P_2 = (1, 2, 3); \quad P_3 = (3, 1, 4); \quad P_4 = (2, -1, 1)$

48. $P_1 = (0, 0, 0); \quad P_2 = (-1, 2, 0); \quad P_3 = (1, 5, -4); \quad P_4 = (2, 3, -4)$

49. $P_1 = (-2, 1, 6); \quad P_2 = (2, 1, -7); \quad P_3 = (4, 1, 1); \quad P_4 = (8, 1, -12)$

50. $P_1 = (-1, 1, 1); \quad P_2 = (-1, 2, 2); \quad P_3 = (-3, 5, -4); \quad P_4 = (-3, 4, -5)$

### Applications and Extensions

51. **Angular Velocity** A rigid body rotates about an axis through the origin with a constant angular speed of 30 radians per second. The angular velocity $\boldsymbol{\omega}$ points in the direction of $\mathbf{i} + \mathbf{j} + \mathbf{k}$. Find the speed of an object at the instant it passes through the point $(-1, 2, 3)$. Assume the distance scale is in meters.

52. **Angular Velocity** A rigid body rotates with constant angular speed $\omega$ about a line through the origin parallel to $3\mathbf{i} + \mathbf{j} - 2\mathbf{k}$.

    (a) Find the speed of an object at the instant that it passes through the point $(4, 4, 0)$. Assume that distance is measured in meters.

    (b) Find $\omega$ if the speed of the object at the point $(4, 4, 0)$ is $8\sqrt{14}$ m/s.

**53. Area of a Triangle** Show that the area of the triangle whose vertices are the endpoints of the vectors **u**, **v**, and **w** is

$$A = \frac{1}{2} \|(\mathbf{v} - \mathbf{u}) \times (\mathbf{w} - \mathbf{u})\|$$

**54. Area of a Triangle** Use the result of Problem 53 to find the area of the triangle with vertices $(0, 0, 0)$, $(2, 3, -2)$, and $(-1, 1, 4)$.

*Problems 55 and 56 use the following discussion. When a force **F** acts on an object at some distance r from a point P, called the pivot point, it can cause the object to rotate rather than translate. The ability of a force to cause rotation is called the **torque** about P due to the force **F** and is symbolized by the Greek lowercase letter tau **τ**. Torque equals the cross product between the applied force **F** and the distance vector* $\mathbf{r} = \overrightarrow{PQ}$ *that points from P to the point Q where **F** acts on the object. That is,*

$$\boldsymbol{\tau} = \mathbf{r} \times \mathbf{F}$$

**55. Using a Wrench** The figure below shows a wrench that is being used to turn a nut and three separate forces that could be applied to the end of the wrench. The point $P$ is at the origin of the coordinate system and the point $Q$ is at the end of the wrench at $(0.25, 0, 0)$, where the forces are applied.

(a) Describe the direction of the positive $z$-axis, given the axes shown in the figure.

(b) Find the torque about $P$ due to each of the three forces shown in the figure. Assume that the magnitude of each force is 250 N and the wrench is 0.25 m long.

(c) Which force would be the most effective at causing the wrench to rotate as much as possible?

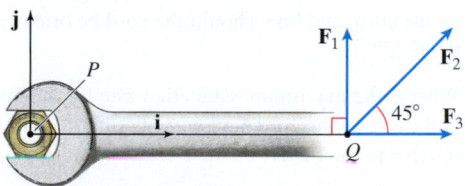

**56. Exercise Science** Suppose a person lifting a weight extends his arm upward at 40° above the horizontal while holding a 75-N (about 17-lb) weight in his hand. His arm is 65 cm long from the shoulder socket to the weight. What is the magnitude of the torque?

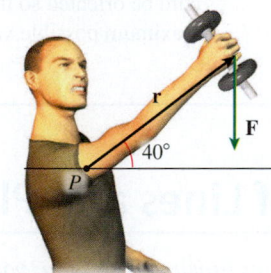

**57. Lorentz Force** The **Lorentz force** on an electric charge moving through electric and magnetic fields is given by the formula $\mathbf{F} = q\,[\mathbf{E} + (\mathbf{v} \times \mathbf{B})]$. When the charge is expressed in coulombs (C), the electric field in newtons per coulomb (N/C), the magnetic field in teslas (T), and the velocity in meters per second (m/s), then the unit of force is newtons (N). A positive charge $(q = 2.5 \text{ C})$ is moving with velocity $\mathbf{v} = (0.0050\,\mathbf{i} + 0.0035\,\mathbf{j})$ m/s through a region of space where there is an electric field $\mathbf{E} = (0.0064\,\mathbf{i} - 0.0075\,\mathbf{j} - 0.0023\,\mathbf{k})$ N/C and a magnetic field $\mathbf{B} = (0.47\,\mathbf{i} + 0.50\,\mathbf{j} - 0.25\,\mathbf{k})$ T. Find the Lorentz force acting on this charge.

**58.** Show that if **u** and **v** are orthogonal vectors, then $\|\mathbf{u} \times \mathbf{v}\| = \|\mathbf{u}\|\,\|\mathbf{v}\|$.

**59.** Show that if **u** and **v** are orthogonal unit vectors, then $\mathbf{u} \times \mathbf{v}$ is also a unit vector.

**60. Algebraic Properties of the Cross Product** Show that if **v** and **w** are vectors and $a$ is a scalar, then $a(\mathbf{v} \times \mathbf{w}) = (a\mathbf{v}) \times \mathbf{w} = \mathbf{v} \times (a\mathbf{w})$ [property (5)].

**61. Algebraic Properties of the Cross Product** Show that if **u**, **v**, and **w** are vectors, then $\mathbf{v} \times (\mathbf{w} + \mathbf{u}) = (\mathbf{v} \times \mathbf{w}) + (\mathbf{v} \times \mathbf{u})$ [property (6)].

**62. Algebraic Properties of the Cross Product** Show that if **u**, **v**, and **w** are vectors, then $\mathbf{u} \cdot (\mathbf{v} \times \mathbf{w}) = (\mathbf{u} \times \mathbf{v}) \cdot \mathbf{w}$. [Property (8)]

**63. Algebraic Properties of the Cross Product** If $\mathbf{v} \times \mathbf{w} = \mathbf{0}$ and $\mathbf{v} \cdot \mathbf{w} = 0$, can you draw any conclusions about **v** and/or **w**? Explain.

**64. Algebraic Properties of the Cross Product** Give an example to show that the cross product is not associative. That is, find vectors **u**, **v**, and **w** so that $\mathbf{u} \times (\mathbf{v} \times \mathbf{w}) \neq (\mathbf{u} \times \mathbf{v}) \times \mathbf{w}$.

**Volume of a Parallelepiped** *In Problems 65 and 66, use Property (13) to find the volume of the parallelepiped whose adjacent sides are the vectors **u**, **v**, and **w**. See the figure.*

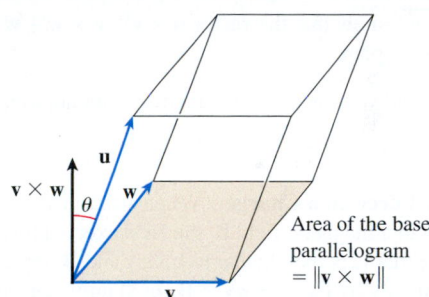

**65.** $\mathbf{u} = 8\mathbf{i} - 6\mathbf{j} + 5\mathbf{k}$, $\mathbf{v} = 2\mathbf{i} + 3\mathbf{j} - 8\mathbf{k}$, and $\mathbf{w} = \mathbf{i} + 6\mathbf{k}$

**66.** $\mathbf{u} = 2\mathbf{i} + \mathbf{j} - 2\mathbf{k}$, $\mathbf{v} = 3\mathbf{i} - 2\mathbf{j} + 4\mathbf{k}$, and $\mathbf{w} = 3\mathbf{i} + 6\mathbf{j} - 2\mathbf{k}$

**67.** Show that the vector $2\mathbf{v} \times 3\mathbf{w}$ is orthogonal to both the vector **v** and the vector **w**.

**68.** Show that $\mathbf{u} \cdot (\mathbf{v} \times \mathbf{w}) = \mathbf{v} \cdot (\mathbf{w} \times \mathbf{u}) = \mathbf{w} \cdot (\mathbf{u} \times \mathbf{v})$.

**Triple Vector Product** *Problems 69 and 70 use the following definition. If **u**, **v**, and **w** are three vectors, the expression* $\mathbf{u} \times (\mathbf{v} \times \mathbf{w})$ *is called the **triple vector product**.*

**69.** Show that $\mathbf{u} \times (\mathbf{v} \times \mathbf{w}) + \mathbf{v} \times (\mathbf{w} \times \mathbf{u}) + \mathbf{w} \times (\mathbf{u} \times \mathbf{v}) = \mathbf{0}$.

**70.** Show that $\mathbf{u} \times (\mathbf{v} \times \mathbf{w}) = (\mathbf{u} \cdot \mathbf{w})\mathbf{v} - (\mathbf{u} \cdot \mathbf{v})\mathbf{w}$.

**71.** Prove that $(\mathbf{a} \times \mathbf{b}) \times (\mathbf{c} \times \mathbf{d}) = [\mathbf{a} \cdot (\mathbf{b} \times \mathbf{d})]\mathbf{c} - [\mathbf{a} \cdot (\mathbf{b} \times \mathbf{c})]\mathbf{d}$.

**72.** Prove **Lagrange's identity**: $(\mathbf{a} \times \mathbf{b}) \cdot (\mathbf{c} \times \mathbf{d}) = (\mathbf{a} \cdot \mathbf{c})(\mathbf{b} \cdot \mathbf{d}) - (\mathbf{a} \cdot \mathbf{d})(\mathbf{b} \cdot \mathbf{c})$

**Geometry Proofs** *In Problems 73 and 74, use vector methods to prove each statement.*

**73.** The altitudes of a triangle meet at one point; the medians of a triangle meet at one point.

**74.** The diagonals of a parallelogram are perpendicular if and only if the parallelogram is a rhombus.

## Challenge Problems

**75.** If $\mathbf{v}, \mathbf{w}, \mathbf{u}$ and $\mathbf{v}', \mathbf{w}', \mathbf{u}'$ are vectors for which the following identities hold,

$$\mathbf{v}' \cdot \mathbf{v} = \mathbf{w}' \cdot \mathbf{w} = \mathbf{u}' \cdot \mathbf{u} = 1$$

$$\mathbf{v}' \cdot \mathbf{w} = \mathbf{v}' \cdot \mathbf{u} = \mathbf{w}' \cdot \mathbf{v} = \mathbf{w}' \cdot \mathbf{u} = \mathbf{u}' \cdot \mathbf{v} = \mathbf{u}' \cdot \mathbf{w} = 0$$

show that

$$\mathbf{v}' = \frac{\mathbf{w} \times \mathbf{u}}{\mathbf{v} \cdot (\mathbf{w} \times \mathbf{u})} \qquad \mathbf{w}' = \frac{\mathbf{u} \times \mathbf{v}}{\mathbf{v} \cdot (\mathbf{w} \times \mathbf{u})} \qquad \mathbf{u}' = \frac{\mathbf{v} \times \mathbf{w}}{\mathbf{v} \cdot (\mathbf{w} \times \mathbf{u})}$$

**76.** Solve for $\mathbf{x}$ in terms of $a, \mathbf{a},$ and $\mathbf{b}$ if $a\mathbf{x} + (\mathbf{x} \times \mathbf{a}) = \mathbf{b}, \ a \neq 0$. *Hint:* First find $\mathbf{x} \cdot \mathbf{a}$, then $\mathbf{x} \times \mathbf{a}$.

**77.** Prove that if $\mathbf{u} = u_1\mathbf{i} + u_2\mathbf{j} + u_3\mathbf{k}, \ \mathbf{v} = v_1\mathbf{i} + v_2\mathbf{j} + v_3\mathbf{k}$, and $\mathbf{w} = w_1\mathbf{i} + w_2\mathbf{j} + w_3\mathbf{k}$ are three vectors, the triple scalar product $\mathbf{u} \cdot (\mathbf{v} \times \mathbf{w})$ is given by

$$\mathbf{u} \cdot (\mathbf{v} \times \mathbf{w}) = \begin{vmatrix} u_1 & u_2 & u_3 \\ v_1 & v_2 & v_3 \\ w_1 & w_2 & w_3 \end{vmatrix}$$

**78.** Show that the volume $V$ of a parallelepiped whose adjacent sides are the vectors $\mathbf{u}, \mathbf{v},$ and $\mathbf{w}$ is $V = |\mathbf{u} \cdot (\mathbf{v} \times \mathbf{w})|$.

**79.** The points $A, B,$ and $C$ determine a plane. If $\mathbf{u} = \overrightarrow{OA}, \mathbf{v} = \overrightarrow{OB}$, and $\mathbf{w} = \overrightarrow{OC}$, show that the vector $\mathbf{u} \times \mathbf{v} + \mathbf{v} \times \mathbf{w} + \mathbf{w} \times \mathbf{u}$ is normal to the plane.

**80.** Show that the volume of the tetrahedron with adjacent edges $\mathbf{u}, \mathbf{v}$, and $\mathbf{w}$ is $\frac{1}{6} |\mathbf{u} \cdot (\mathbf{v} \times \mathbf{w})|$.

**81.** **Magnetic Force on a Charge**   When a charge $q$ moves with velocity $\mathbf{v}$ in a magnetic field $\mathbf{B}$, the field exerts a force $\mathbf{F}$ on the charge that is perpendicular to both $\mathbf{v}$ and $\mathbf{B}$, represented mathematically by $\mathbf{F}_{mag} = q\mathbf{v} \times \mathbf{B}$. In SI units, the charge $q$ is measured in coulombs (C), the velocity $\mathbf{v}$ is in meters per second (m/s), the force $\mathbf{F}$ is in newtons (N), and the magnetic field $\mathbf{B}$ is in teslas (T). Suppose a proton with a charge of $q = 1.60 \times 10^{-19}$ C enters a magnetic field of 1.20 T that is pointing vertically downward and, as a result, experiences a horizontal force of $1.63 \times 10^{-13}$ N to the left.

**(a)** What are the magnitude and direction of the minimum velocity $\mathbf{v}$ the charge could have had as it entered the field?

**(b)** Why is the velocity in (a) the minimum velocity?

**82.** **Magnetic Effect on a TV**   In older analog televisions, the image on the screen was formed by a beam of electrons with charge $q$ hitting the screen and causing it to glow. In some cases, the electrons reach speeds of $v = 2.0 \times 10^6$ m/s. Earth produces a magnetic field $\mathbf{B}$ that exerts a magnetic force $\mathbf{F}_{mag}$ on the moving electrons. The force $\mathbf{F}_{mag}$ is given by $\mathbf{F}_{mag} = q\mathbf{v} \times \mathbf{B}$. Earth's magnetic field is about $5.0 \times 10^{-3}$ T and generally points horizontally from south to north. If the charge on an electron is $q = 1.60 \times 10^{-19}$ C, find the magnitude and direction of the magnetic force $\mathbf{F}_{mag}$ on the electrons in a TV if the set is oriented so that the electrons move

**(a)** from north to south.

**(b)** from west to east.

**83.** **Magnetic Force on an Extension Cord**   If a straight wire of length $L$ carrying an electric current $I$ is in a magnetic field $\mathbf{B}$, there will be a magnetic force $\mathbf{F}_{mag}$ on the wire. The magnetic force $\mathbf{F}_{mag}$ is orthogonal to both the wire and the field $\mathbf{B}$ and is given by $\mathbf{F}_{mag} = I\mathbf{L} \times \mathbf{B}$, where the direction of the vector $\mathbf{L}$ is the direction in which the current is flowing and its magnitude is $L$. An extension cord 2 m long carries a typical current $I = 5.0$ amperes (A) and lies in a horizontal plane. Earth's magnetic field $\mathbf{B}$ has a magnitude of $5.0 \times 10^{-3}$ T and points horizontally from south to north. The magnetic force $\mathbf{F}_{mag}$ is in newtons.

**(a)** What is the minimum force Earth's magnetic field can exert on the cord, and how should the cord be oriented relative to the field?

**(b)** What is the maximum magnetic force Earth's magnetic field will exert on the cord, and how should the cord be oriented relative to the field?

**(c)** At what angle $\theta$ relative to the magnetic field $\mathbf{B}$ should the cord be oriented so that the magnetic force $\mathbf{F}_{mag}$ is 50% of its maximum possible value?

# 10.6 Equations of Lines and Planes in Space

**OBJECTIVES**   *When you finish this section, you should be able to:*

**1** Find a vector equation of a line in space (p. 772)

**2** Find parametric equations of a line in space (p. 773)

**3** Find symmetric equations of a line in space (p. 774)

**4** Determine whether two distinct lines are skew, parallel, or intersecting (p. 775)

**5** Find an equation of a plane (p. 776)

**6** Determine whether two distinct planes are parallel or intersecting (p. 778)

**7** Find the distance from a point to a plane and from a point to a line (p. 779)

## 1 Find a Vector Equation of a Line in Space

In space, a line is determined when we know a point on the line and its direction. If $P_0 = (x_0, y_0, z_0)$ and $P_1 = (x_1, y_1, z_1)$ are two distinct points on a line, then the

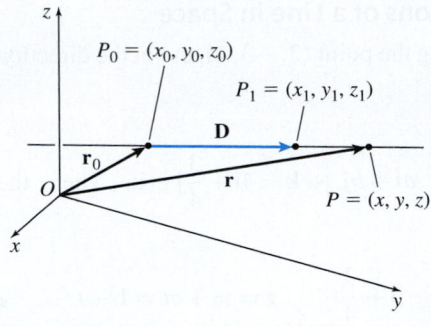

**Figure 51**

vector $\mathbf{D}$ represented by the directed line segment $\overrightarrow{P_0P_1}$ is a nonzero vector that gives the direction of the line. Note that $\mathbf{D}$ is not unique. Any two points on the line can be used to find the direction of the line. See Figure 51.

If $\mathbf{r}_0$ denotes the vector $\overrightarrow{OP_0}$ and $\mathbf{r}$ denotes the vector $\overrightarrow{OP}$ of any point $P = (x, y, z)$ on the line, then the vector $\mathbf{r} - \mathbf{r}_0$ is parallel to the vector $\mathbf{D}$. That is,

$$\mathbf{r} - \mathbf{r}_0 = t\mathbf{D} \qquad \text{for some scalar } t$$

$$\mathbf{r} = \mathbf{r}_0 + t\mathbf{D}$$

**THEOREM  Vector Equation of a Line**

A vector equation of a line in the direction $\mathbf{D} \neq \mathbf{0}$ and containing the point $P_0 = (x_0, y_0, z_0)$ is

$$\boxed{\mathbf{r} = \mathbf{r}_0 + t\mathbf{D}}$$

where $\mathbf{r}_0$ is the vector $\overrightarrow{OP_0}$, $\mathbf{r}$ is the vector $\overrightarrow{OP}$, and $P = (x, y, z)$ is any point on the line.

In the vector equation $\mathbf{r} = \mathbf{r}_0 + t\mathbf{D}$ of a line, both the vector $\mathbf{r}_0$ and the vector $\mathbf{D}$ are known. To locate other points $P$ on the line, we assign values to the scalar $t$. For each real number $t$, we obtain a vector whose terminal point is on the line. Vector equations of a line are not unique since the choice of a point on the line and the choice of the vector representing the direction of the line can vary.

### EXAMPLE 1  Finding a Vector Equation of a Line in Space

Find a vector equation of the line containing the points $P_0 = (1, 2, -1)$ and $P_1 = (4, 3, -2)$.

**Solution** The vector $\mathbf{D}$ in the direction from $P_0 = (1, 2, -1)$ to $P_1 = (4, 3, -2)$ is

$$\mathbf{D} = \overrightarrow{P_0P_1} = 3\mathbf{i} + \mathbf{j} - \mathbf{k}$$

If we let $\mathbf{r}_0 = \mathbf{i} + 2\mathbf{j} - \mathbf{k}$, then a vector equation of the line is

$$\mathbf{r} = \mathbf{r}_0 + t\mathbf{D} = \mathbf{i} + 2\mathbf{j} - \mathbf{k} + t(3\mathbf{i} + \mathbf{j} - \mathbf{k}) = (1 + 3t)\mathbf{i} + (2 + t)\mathbf{j} + (-1 - t)\mathbf{k} \qquad \blacksquare$$

**NOW WORK** Problem 11(a).

## 2  Find Parametric Equations of a Line in Space

**NEED TO REVIEW?** Parametric equations are discussed in Section 9.1, pp. 680–685.

The vector equation of a line in the direction $\mathbf{D} = a\mathbf{i} + b\mathbf{j} + c\mathbf{k}$ and containing the point $P_0 = (x_0, y_0, z_0)$ can be written as $\mathbf{r} = \mathbf{r}_0 + t\mathbf{D}$, where $\mathbf{r} = x\mathbf{i} + y\mathbf{j} + z\mathbf{k}$ and $\mathbf{r}_0 = x_0\mathbf{i} + y_0\mathbf{j} + z_0\mathbf{k}$. Then

$$x\mathbf{i} + y\mathbf{j} + z\mathbf{k} = (x_0\mathbf{i} + y_0\mathbf{j} + z_0\mathbf{k}) + t(a\mathbf{i} + b\mathbf{j} + c\mathbf{k}) \qquad \mathbf{r} = \mathbf{r}_0 + t\mathbf{D}$$

$$= (x_0 + at)\mathbf{i} + (y_0 + bt)\mathbf{j} + (z_0 + ct)\mathbf{k}$$

Since two vectors are equal if and only if their corresponding components are equal, we have

$$\boxed{x = x_0 + at \qquad y = y_0 + bt \qquad z = z_0 + ct}$$

These equations are **parametric equations** of the line, and the variable $t$ is the parameter. Points $(x, y, z)$ on the line are obtained by assigning values to the parameter $t$. For example, when $t = 0$, we obtain the point $(x_0, y_0, z_0)$ on the line.

---

**EXAMPLE 2    Finding Parametric Equations of a Line in Space**

Find parametric equations of the line containing the point $(2, -3, 1)$ and in the direction of the vector $4\mathbf{i} + \dfrac{3}{4}\mathbf{j} - \mathbf{k}$.

**Solution**    Let $(x_0, y_0, z_0) = (2, -3, 1)$ and $a\mathbf{i} + b\mathbf{j} + c\mathbf{k} = 4\mathbf{i} + \dfrac{3}{4}\mathbf{j} - \mathbf{k}$. Then the parametric equations of the line are

$$x = x_0 + at = 2 + 4t \qquad y = y_0 + bt = -3 + \frac{3}{4}t \qquad z = z_0 + ct = 1 - t \qquad \blacksquare$$

**NOW WORK** Problem 9(b).

## 3  Find Symmetric Equations of a Line in Space

In the parametric equations $x = x_0 + at$, $y = y_0 + bt$, and $z = z_0 + ct$, if the numbers $a$, $b$, and $c$ (the components of the vector $\mathbf{D}$) are all nonzero, we can solve for $t$, obtaining

$$t = \frac{x - x_0}{a} \qquad t = \frac{y - y_0}{b} \qquad t = \frac{z - z_0}{c}$$

from which

> **NOTE** The symmetric equations actually represent a pair of equations
> $\dfrac{x - x_0}{a} = \dfrac{y - y_0}{b}$ and $\dfrac{x - x_0}{a} = \dfrac{z - z_0}{c}$.

$$\boxed{\dfrac{x - x_0}{a} = \dfrac{y - y_0}{b} = \dfrac{z - z_0}{c}}$$

These equations are referred to as **symmetric equations of a line**.

---

**EXAMPLE 3    Finding Symmetric Equations of a Line in Space**

Find symmetric equations of the line containing the point $(1, -1, 2)$ and in the direction of the vector $5\mathbf{i} - 2\mathbf{j} + 3\mathbf{k}$.

**Solution**    The components of the vector $5\mathbf{i} - 2\mathbf{j} + 3\mathbf{k}$ are all nonzero. So, we use $a = 5$, $b = -2$, and $c = 3$ and the coordinates of the point $(1, -1, 2)$ to obtain the symmetric equations

$$\frac{x - 1}{5} = \frac{y + 1}{-2} = \frac{z - 2}{3} \qquad \blacksquare$$

Parametric and symmetric equations of a line are not unique. For example, since the vector $-10\mathbf{i} + 4\mathbf{j} - 6\mathbf{k}$ is parallel to $5\mathbf{i} - 2\mathbf{j} + 3\mathbf{k}$, the symmetric equations of the line described in Example 3 can be written as

$$\frac{x - 1}{-10} = \frac{y + 1}{4} = \frac{z - 2}{-6}$$

**NOW WORK** Problem 9(c).

If one of the components of the direction $\mathbf{D}$ of a line equals 0, the symmetric equations have a different form. For example, if $\mathbf{D} = a\mathbf{i} + b\mathbf{j} + c\mathbf{k}$ and $a = 0$, but $b \neq 0$ and $c \neq 0$, the symmetric equations of the line are written as

$$x = x_0 \qquad \frac{y - y_0}{b} = \frac{z - z_0}{c}$$

where $P_0 = (x_0, y_0, z_0)$ is a point on the line. This particular line lies in the plane $x = x_0$.

**EXAMPLE 4    Finding Symmetric Equations of a Line**

Find symmetric equations of the line that contains the point $(5, -2, 3)$ and is in the direction of the vector $\mathbf{D} = 3\mathbf{i} - 2\mathbf{k}$.

**Solution** For $\mathbf{D} = 3\mathbf{i} - 2\mathbf{k}$, $a = 3$, $b = 0$, and $c = -2$. So, symmetric equations in the direction $\mathbf{D}$ are

$$\frac{x - 5}{3} = \frac{z - 3}{-2} \qquad y = -2$$                                                     ∎

NOW WORK **Problem 17.**

CALC
▶ CLIP

**EXAMPLE 5    Analyzing Symmetric Equations of a Line**

A line is defined by the symmetric equations $\dfrac{x - 6}{3} = \dfrac{y + 2}{1} = \dfrac{z + 3}{-2}$.

**(a)** Find a vector in the direction of the line.

**(b)** Find two points on the line.

**Solution (a)** From the denominators of the symmetric equations, we find $a = 3$, $b = 1$, and $c = -2$. So, $\mathbf{D} = 3\mathbf{i} + \mathbf{j} - 2\mathbf{k}$ is a vector in the direction of the line.

**(b)** Since the line is defined by the symmetric equations $\dfrac{x - 6}{3} = \dfrac{y + 2}{1} = \dfrac{z + 3}{-2}$, one

point on the line is $(6, -2, -3)$. To find a second point, we assign a value to $x$, say $x = 0$. Then

$$\frac{0 - 6}{3} = \frac{y + 2}{1} = \frac{z + 3}{-2}$$

$$-2 = \frac{y + 2}{1} = \frac{z + 3}{-2}$$

Now we solve for $y$ and $z$, and find $y = -4$ and $z = 1$. So, another point on the line is $(0, -4, 1)$. ∎

NOW WORK **Problem 15.**

## 4 Determine Whether Two Distinct Lines Are Skew, Parallel, or Intersecting

In the plane, two distinct lines either intersect or are parallel. In space, two distinct lines *intersect*, are *parallel*, or are *skew*. The lines **intersect** when they have exactly one point in common. The lines are **parallel** when they lie in the same plane but do not intersect. The lines are **skew** if they do not intersect and are not parallel. Figure 52 illustrates these possibilities.

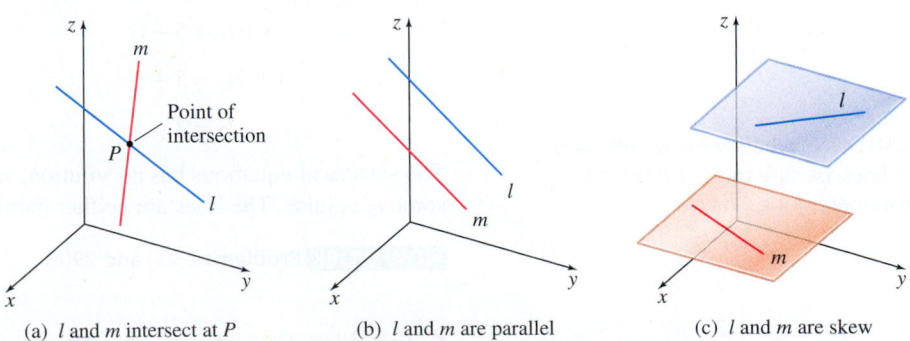

(a) $l$ and $m$ intersect at $P$      (b) $l$ and $m$ are parallel      (c) $l$ and $m$ are skew

**Figure 52**

As an example of skew lines, think of two airplanes, one flying at an altitude of 1000 m traveling north and the other flying at an altitude of 3000 m traveling east. Since they are flying at different altitudes, they will never collide (intersect). Since they are flying in different directions, they are not parallel.

**EXAMPLE 6  Determining Whether Two Lines Are Skew, Parallel, or Intersecting**

Determine whether the lines given below intersect, are parallel, or are skew.

(a)  $l_1$:    $\mathbf{r}_1 = (3\mathbf{i} + 3\mathbf{j} + \mathbf{k}) + t_1(\mathbf{i} - 2\mathbf{j} + \mathbf{k})$

     $l_2$:    $\mathbf{r}_2 = (5\mathbf{i} + \mathbf{j} + \mathbf{k}) + t_2(\mathbf{i} - 4\mathbf{j} + 3\mathbf{k})$

(b)  $l_1$:    $\mathbf{r}_1 = (3\mathbf{i} + 2\mathbf{j} + \mathbf{k}) + t_1(\mathbf{i} - \mathbf{j} + \mathbf{k})$

     $l_2$:    $\mathbf{r}_2 = (5\mathbf{i} + 6\mathbf{j} + \mathbf{k}) + t_2(\mathbf{i} - \mathbf{j} + 2\mathbf{k})$

**Solution (a)**  The line $l_1$ is in the direction of the vector $\mathbf{i} - 2\mathbf{j} + \mathbf{k}$ and the line $l_2$ is in the direction of the vector $\mathbf{i} - 4\mathbf{j} + 3\mathbf{k}$. Since these vectors are not parallel (do you know why?), $l_1$ and $l_2$ either intersect or are skew.

Suppose the lines intersect. Then there is some value of the parameter $t_1$ and some value of the parameter $t_2$ for which $\mathbf{r}_1 = \mathbf{r}_2$. Since $\mathbf{r}_1 = (3 + t_1)\mathbf{i} + (3 - 2t_1)\mathbf{j} + (1 + t_1)\mathbf{k}$ and $\mathbf{r}_2 = (5 + t_2)\mathbf{i} + (1 - 4t_2)\mathbf{j} + (1 + 3t_2)\mathbf{k}$, for $\mathbf{r}_1$ to equal $\mathbf{r}_2$, we have

$$(3 + t_1)\mathbf{i} + (3 - 2t_1)\mathbf{j} + (1 + t_1)\mathbf{k} = (5 + t_2)\mathbf{i} + (1 - 4t_2)\mathbf{j} + (1 + 3t_2)\mathbf{k} \qquad \mathbf{r}_1 = \mathbf{r}_2$$

Equate the components of each vector to obtain

$$3 + t_1 = 5 + t_2 \qquad 3 - 2t_1 = 1 - 4t_2 \qquad 1 + t_1 = 1 + 3t_2$$

The result is a system of three equations containing two variables. From the third equation, $t_1 = 3t_2$. Now substitute $t_1 = 3t_2$ into each of the first two equations.

$$3 + t_1 = 5 + t_2 \qquad\quad 3 - 2t_1 = 1 - 4t_2$$

$$3 + 3t_2 = 5 + t_2 \qquad 3 - 6t_2 = 1 - 4t_2 \qquad t_1 = 3t_2$$

$$t_2 = 1 \qquad\qquad\qquad t_2 = 1$$

Back-substituting, we conclude $t_1 = 3$ and $t_2 = 1$. The point of intersection is found by substituting $t_1 = 3$ in $l_1$ or $t_2 = 1$ in $l_2$. The result is $\mathbf{r}_1 = \mathbf{r}_2 = 6\mathbf{i} - 3\mathbf{j} + 4\mathbf{k}$. The point of intersection is $(6, -3, 4)$.

**(b)**  The line $l_1$ is in the direction of the vector $\mathbf{i} - \mathbf{j} + \mathbf{k}$, and the line $l_2$ is in the direction of the vector $\mathbf{i} - \mathbf{j} + 2\mathbf{k}$. These vectors are not parallel, so the lines $l_1$ and $l_2$ either intersect or are skew. As before, suppose they intersect. Then $\mathbf{r}_1 = \mathbf{r}_2$ so that

$$(3 + t_1)\mathbf{i} + (2 - t_1)\mathbf{j} + (1 + t_1)\mathbf{k} = (5 + t_2)\mathbf{i} + (6 - t_2)\mathbf{j} + (1 + 2t_2)\mathbf{k}$$

$$3 + t_1 = 5 + t_2 \qquad 2 - t_1 = 6 - t_2 \qquad 1 + t_1 = 1 + 2t_2 \qquad \text{Equate components.}$$

Now substitute $t_1 = 2t_2$ (from the third equation) into each of the first two equations.

$$3 + t_1 = 5 + t_2 \qquad\qquad\quad 2 - t_1 = 6 - t_2$$

$$3 + 2t_2 = 5 + t_2 \qquad\qquad 2 - 2t_2 = 6 - t_2 \qquad t_1 = 2t_2$$

$$t_2 = 2 \qquad\qquad\qquad\quad t_2 = -4$$

**CAUTION**  When working with pairs of lines, be sure to use a different parameter for each line.

The system of equations has no solution, so the assumption that $\mathbf{r}_1 = \mathbf{r}_2$, for some $t_1$ and some $t_2$ is false. The lines are neither parallel nor intersecting, so they are skew.  ∎

**NOW WORK**  Problems 23(a) and 29(a).

## 5 Find an Equation of a Plane

The vector equation of a line is determined once a point on the line and a vector in the direction of the line are known. To determine the vector equation of a plane, we use a point on the plane and a vector perpendicular to the plane, a **normal** to the plane.

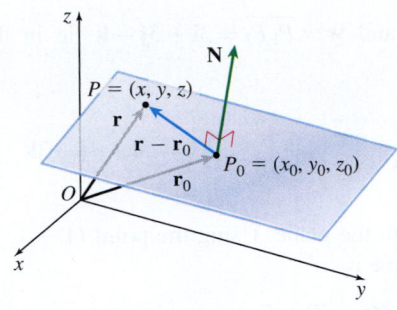

**Figure 53** $(\mathbf{r} - \mathbf{r}_0) \cdot \mathbf{N} = 0$

**NOTE** The normal vector $\mathbf{N}$ is orthogonal to every vector in the plane.

Suppose the point $P_0 = (x_0, y_0, z_0)$ is on a plane and suppose the nonzero vector $\mathbf{N}$ is a normal to the plane. See Figure 53. Let $P = (x, y, z)$ be any other point on the plane. Denote the vectors $\overrightarrow{OP_0}$ and $\overrightarrow{OP}$ by $\mathbf{r}_0$ and $\mathbf{r}$, respectively. Since $P$ is a point on the plane, the vector $\mathbf{r} - \mathbf{r}_0$ is in the plane and is orthogonal to the vector $\mathbf{N}$. Then $(\mathbf{r} - \mathbf{r}_0) \cdot \mathbf{N} = 0$.

### THEOREM  Vector Equation of a Plane

The vector equation of the plane containing the point $P_0$ is

$$(\mathbf{r} - \mathbf{r}_0) \cdot \mathbf{N} = 0$$

where $\mathbf{r}_0$ is the vector $\overrightarrow{OP_0}$, $\mathbf{r}$ is the vector $\overrightarrow{OP}$ of any other point $P$ on the plane, and $\mathbf{N}$ is normal to the plane.

If the coordinates of $P_0$ are $(x_0, y_0, z_0)$ and the vector $\mathbf{N} = A\mathbf{i} + B\mathbf{j} + C\mathbf{k}$, then

$$[(x - x_0)\mathbf{i} + (y - y_0)\mathbf{j} + (z - z_0)\mathbf{k}] \cdot (A\mathbf{i} + B\mathbf{j} + C\mathbf{k}) = 0$$

$$A(x - x_0) + B(y - y_0) + C(z - z_0) = 0$$

If $D = Ax_0 + By_0 + Cz_0$, we can write the equation as

$$Ax + By + Cz = D$$

Notice that the coefficients of $x$, $y$, and $z$ in the equation are the components of the normal vector $\mathbf{N}$.

### THEOREM  General Equation of a Plane

An equation of the form

$$\boxed{Ax + By + Cz = D}$$

where $A$, $B$, $C$, and $D$ are real numbers, and at least one of the numbers $A$, $B$, $C$ is not 0, is called the **general equation of a plane**, and the vector $A\mathbf{i} + B\mathbf{j} + C\mathbf{k}$ is normal to this plane.

---

**EXAMPLE 7**  **Finding the General Equation of a Plane**

Find the general equation of the plane containing the point $(1, 2, -1)$ if the vector $\mathbf{N} = 2\mathbf{i} + 3\mathbf{j} - 4\mathbf{k}$ is normal to the plane. Then use the intercepts to graph the plane.

**Solution** The general equation of a plane is $Ax + By + Cz = D$, where $A$, $B$, and $C$ are the components of a normal vector to the plane. Since the normal is $\mathbf{N} = 2\mathbf{i} + 3\mathbf{j} - 4\mathbf{k}$, we have

$$2x + 3y - 4z = D$$

To find the number $D$, we use the point $(1, 2, -1)$. Then $D = 2 \cdot 1 + 3 \cdot 2 - 4 \cdot (-1) = 12$. The general equation of the plane is

$$2x + 3y - 4z = 12$$

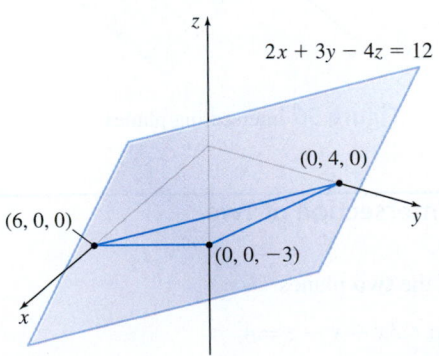

**Figure 54** $2x + 3y - 4z = 12$

We find the intercepts by letting two variables equal zero and then solving for the third variable. For example, to find the $x$-intercept, we let $y = 0$ and $z = 0$. Then $2x = 12$ or $x = 6$. Similarly, the $y$-intercept is 4 and the $z$-intercept is $-3$. Figure 54 illustrates how the intercepts are used to graph the plane. ∎

**NOW WORK** Problem 31.

---

**EXAMPLE 8**  **Finding the General Equation of a Plane**

Find the general equation of the plane containing the points

$$P_1 = (1, -1, 2) \quad P_2 = (3, 0, 0) \quad P_3 = (4, 2, 1)$$

**Solution** The vectors $\mathbf{v} = \overrightarrow{P_1P_2} = 2\mathbf{i} + \mathbf{j} - 2\mathbf{k}$ and $\mathbf{w} = \overrightarrow{P_1P_3} = 3\mathbf{i} + 3\mathbf{j} - \mathbf{k}$ lie in the plane. The vector

$$\mathbf{N} = \mathbf{v} \times \mathbf{w} = \begin{vmatrix} \mathbf{i} & \mathbf{j} & \mathbf{k} \\ 2 & 1 & -2 \\ 3 & 3 & -1 \end{vmatrix} = \begin{vmatrix} 1 & -2 \\ 3 & -1 \end{vmatrix} \mathbf{i} - \begin{vmatrix} 2 & -2 \\ 3 & -1 \end{vmatrix} \mathbf{j} + \begin{vmatrix} 2 & 1 \\ 3 & 3 \end{vmatrix} \mathbf{k} = 5\mathbf{i} - 4\mathbf{j} + 3\mathbf{k}$$

is orthogonal to both $\mathbf{v}$ and $\mathbf{w}$ and so is normal to the plane. Using the point $(1, -1, 2)$ and the vector $\mathbf{N}$, the general equation of the plane is

$$5(x - 1) - 4(y + 1) + 3(z - 2) = 0$$
$$5x - 5 - 4y - 4 + 3z - 6 = 0$$
$$5x - 4y + 3z = 15$$

■

**NOW WORK** Problem 43(a).

## 6 Determine Whether Two Distinct Planes Are Parallel or Intersecting

**NOTE** The planes $p_1: 2x + y - z = 3$ and $p_2: 4x + 2y - 2z = 6$ are identical. Do you see why?

Two distinct planes are **parallel** if they have parallel normal vectors, that is, if $\mathbf{N}_1 = k\mathbf{N}_2$ for some scalar $k \neq 0$. For example, the distinct planes

$$p_1: \quad 2x + y - z = 3 \quad \text{and} \quad p_2: 2x + y - z = 4$$

are parallel because $\mathbf{N}_1 = 2\mathbf{i} + \mathbf{j} - \mathbf{k} = \mathbf{N}_2$ and $D_1 = 3$ is not equal to $D_2 = 4$. See Figure 55.

If the normal vectors of two planes are not parallel, then the planes intersect in a line, as shown in Figure 56. The line of intersection lies in both planes and so is perpendicular to both $\mathbf{N}_1$ and $\mathbf{N}_2$.

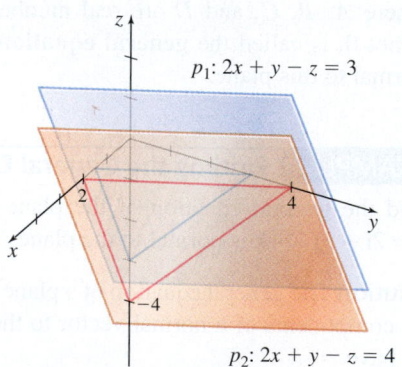

**Figure 55** Parallel planes

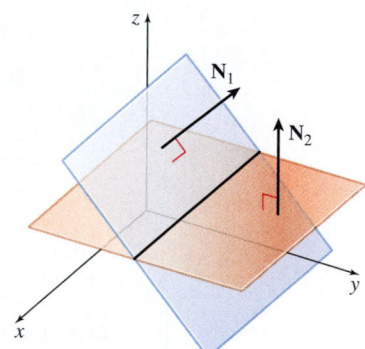

**Figure 56** Intersecting planes

**EXAMPLE 9** Determining the Line of Intersection of Two Nonparallel Planes

Find an equation of the line of intersection of the two planes

$$x - 2y + z = -1 \quad \text{and} \quad 3x + y - z = 4$$

**Solution** First notice that the normals $\mathbf{N}_1 = \mathbf{i} - 2\mathbf{j} + \mathbf{k}$ and $\mathbf{N}_2 = 3\mathbf{i} + \mathbf{j} - \mathbf{k}$ of the two planes are not parallel. To find an equation of the line of intersection, we need a point on the line and the direction of the line. Since the line of intersection is perpendicular to both $\mathbf{N}_1$ and $\mathbf{N}_2$, it is parallel to the vector $\mathbf{D} = \mathbf{N}_1 \times \mathbf{N}_2$.

$$\mathbf{D} = \mathbf{N}_1 \times \mathbf{N}_2 = \begin{vmatrix} \mathbf{i} & \mathbf{j} & \mathbf{k} \\ 1 & -2 & 1 \\ 3 & 1 & -1 \end{vmatrix} = \mathbf{i} + 4\mathbf{j} + 7\mathbf{k}$$

The vector $\mathbf{D}$ gives the direction of the line. We can find a point on the line by locating any point common to both planes. For example, if $z = 0$, then $x - 2y = -1$ and $3x + y = 4$. Solving these equations simultaneously, we find $x = 1$ and $y = 1$. So, the point $(1, 1, 0)$ is on the line, and symmetric equations of the line of intersection are

$$\frac{x-1}{1} = \frac{y-1}{4} = \frac{z}{7}$$

**NOW WORK** Problem 47.

### 7 Find the Distance from a Point to a Plane and from a Point to a Line

To find the distance from a point $P_0$ to a plane, we use projections. As Figure 57 illustrates, if $P_1$ is a point on a plane with a normal vector $\mathbf{N}$, then the distance from $P_0$ to the plane is the magnitude of the projection of $\overrightarrow{P_1 P_0}$ onto $\mathbf{N}$.

Suppose a plane is given by the general equation

$$Ax + By + Cz = D$$

A normal vector $\mathbf{N}$ to the plane is

$$\mathbf{N} = A\mathbf{i} + B\mathbf{j} + C\mathbf{k}$$

The distance from the point $P_0 = (x_0, y_0, z_0)$ to the plane is

$$\begin{bmatrix} \text{Distance from} \\ P_0 \text{ to the plane} \end{bmatrix} = \left\| \text{proj}_{\mathbf{N}} \overrightarrow{P_1 P_0} \right\| = \frac{\left| \overrightarrow{P_1 P_0} \cdot \mathbf{N} \right|}{\| \mathbf{N} \|}$$

where $P_1 = (x_1, y_1, z_1)$ is a point on the plane.

Since $\overrightarrow{P_1 P_0} = (x_0 - x_1)\mathbf{i} + (y_0 - y_1)\mathbf{j} + (z_0 - z_1)\mathbf{k}$, then

$$\begin{bmatrix} \text{Distance from} \\ P_0 \text{ to the plane} \end{bmatrix} = \frac{|A(x_0 - x_1) + B(y_0 - y_1) + C(z_0 - z_1)|}{\sqrt{A^2 + B^2 + C^2}}$$

$$= \frac{|(Ax_0 + By_0 + Cz_0) - (Ax_1 + By_1 + Cz_1)|}{\sqrt{A^2 + B^2 + C^2}}$$

$$\underset{\substack{\uparrow \\ P_1 \text{ is on the plane, so} \\ Ax_1 + By_1 + Cz_1 = D}}{=} \frac{|Ax_0 + By_0 + Cz_0 - D|}{\sqrt{A^2 + B^2 + C^2}}$$

**THEOREM  Distance from a Point to a Plane**

The distance $d$ from the point $P_0 = (x_0, y_0, z_0)$ to the plane $Ax + By + Cz = D$ is

$$\boxed{d = \frac{|Ax_0 + By_0 + Cz_0 - D|}{\sqrt{A^2 + B^2 + C^2}}} \tag{1}$$

---

**EXAMPLE 10  Finding the Distance from a Point to a Plane**

Find the distance $d$ from the point $(2, 3, -1)$ to the plane $x + 4y + z = 5$.

**Solution** We use (1) with $A = 1$, $B = 4$, $C = 1$, $D = 5$, and $P_0 = (2, 3, -1)$. Then

$$d = \frac{|1 \cdot 2 + 4 \cdot 3 + 1 \cdot (-1) - 5|}{\sqrt{1 + 16 + 1}} = \frac{8}{3\sqrt{2}} = \frac{4\sqrt{2}}{3}$$

**NOW WORK** Problem 51.

**NOTE** We did not have to choose $z = 0$; a value could have been assigned to any of the variables and the resulting two equations solved. Different choices would give equivalent equations of the line.

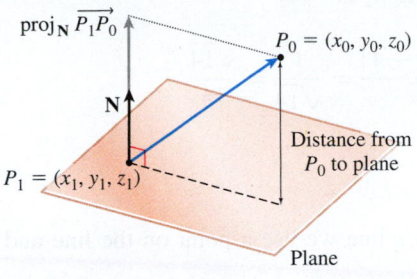

$\text{proj}_{\mathbf{N}} \overrightarrow{P_1 P_0}$

$P_0 = (x_0, y_0, z_0)$

$\mathbf{N}$

Distance from $P_0$ to plane

$P_1 = (x_1, y_1, z_1)$

Plane

**DF** Figure 57

EXAMPLE 11 **Finding the Distance Between Two Parallel Planes**

Find the distance $d$ between the parallel planes

$$p_1 : \ 2x + 3y - z = 3 \quad \text{and} \quad p_2 : \ 2x + 3y - z = 4$$

**Solution** The planes $p_1$ and $p_2$ are parallel because $\mathbf{N}_1 = 2\mathbf{i} + 3\mathbf{j} - \mathbf{k} = \mathbf{N}_2$ and $D_1 = 3$ is not equal to $D_2 = 4$.

We find the distance between the planes by choosing a point on one plane and using the formula for the distance from that point to the other plane. A point on $p_1$ can be found by letting $x = 0$ and $y = 0$. Then $z = -3$, and $(0, 0, -3)$ is a point on $p_1$. The distance $d$ from the point $(0, 0, -3)$ to the plane $p_2$ is

$$d = \frac{|2 \cdot 0 + 3 \cdot 0 - 1 \cdot (-3) - 4|}{\sqrt{2^2 + 3^2 + (-1)^2}} = \frac{1}{\sqrt{14}} = \frac{\sqrt{14}}{14}$$

∎

**NOW WORK** Problem **55**.

To find the distance $d$ from a point $P$ to a line we use a point on the line and a direction vector $\mathbf{D}$ of the line.

EXAMPLE 12 **Finding the Distance from a Point to a Line**

Find the distance $d$ from the point $P = (4, 1, 5)$ to the line

$$\mathbf{r}(t) = (1 + 2t)\mathbf{i} + (2 - t)\mathbf{j} + (3 + t)\mathbf{k}$$

**Solution** The line $\mathbf{r}$ has direction $\mathbf{D} = 2\mathbf{i} - \mathbf{j} + \mathbf{k}$ and contains the point $P_0 = (1, 2, 3)$. Using the point $P = (4, 1, 5)$, we have $\overrightarrow{P_0P} = 3\mathbf{i} - \mathbf{j} + 2\mathbf{k}$. Figure 58 shows the vector $\overrightarrow{P_0P}$, the direction $\mathbf{D}$ of the line, and the distance $d = \left\| \overrightarrow{P_0P} \right\| \sin\theta$ from $P$ to the line.

Using Property (10) on page 767, if $\theta$ is the angle between the vectors $\overrightarrow{P_0P}$ and $\mathbf{D}$, then $\left\| \overrightarrow{P_0P} \times \mathbf{D} \right\| = \left\| \overrightarrow{P_0P} \right\| \|\mathbf{D}\| \sin\theta$. As a result,

$$\frac{\left\| \overrightarrow{P_0P} \times \mathbf{D} \right\|}{\|\mathbf{D}\|} = \left\| \overrightarrow{P_0P} \right\| \sin\theta$$

and

$$\boxed{d = \left\| \overrightarrow{P_0P} \right\| \sin\theta = \frac{\left\| \overrightarrow{P_0P} \times \mathbf{D} \right\|}{\|\mathbf{D}\|}}$$

Now

$$\overrightarrow{P_0P} \times \mathbf{D} = \begin{vmatrix} \mathbf{i} & \mathbf{j} & \mathbf{k} \\ 3 & -1 & 2 \\ 2 & -1 & 1 \end{vmatrix} = \mathbf{i} + \mathbf{j} - \mathbf{k}$$

$$\left\| \overrightarrow{P_0P} \times \mathbf{D} \right\| = \sqrt{3} \quad \text{and} \quad \|\mathbf{D}\| = \sqrt{4 + 1 + 1} = \sqrt{6}$$

Then $d = \dfrac{\left\| \overrightarrow{P_0P} \times \mathbf{D} \right\|}{\|\mathbf{D}\|} = \dfrac{\sqrt{3}}{\sqrt{6}} = \sqrt{\dfrac{1}{2}} = \dfrac{\sqrt{2}}{2}$. The point $(4, 1, 5)$ is $\dfrac{\sqrt{2}}{2}$ units from the line $\mathbf{r}(t) = (1 + 2t)\mathbf{i} + (2 - t)\mathbf{j} + (3 + t)\mathbf{k}$. ∎

**NOW WORK** Problem **63**.

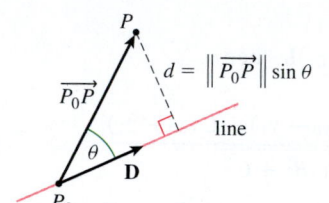

**Figure 58**

# 10.6 Assess Your Understanding

## Concepts and Vocabulary

1. *True or False*   The symmetric equations $\dfrac{x-3}{2} = \dfrac{y-4}{1} = \dfrac{z+2}{-1}$ represent a line parallel to the vector $3\mathbf{i} + 4\mathbf{j} - 2\mathbf{k}$.

2. *True or False*   Skew lines can sometimes lie in the same plane.

3. *True or False*   If two distinct planes are not parallel, they intersect in a line.

4. *True or False*   The two planes $p_1$: $2x + 3y - z = 3$ and $p_2$: $6x + 3y - 3z = 9$ are parallel.

5. *True or False*   The planes $x + 2y + z = 3$ and $x + 2y + z = 4$ are parallel and are 1 unit apart.

6. Which of the three representations of a line in space do you like best? Give reasons for your choice.

7. Which of the two representations of a plane in space do you like better? Give reasons for your choice.

8. Two distinct lines in space that do not intersect and are not parallel are called _____.

## Skill Building

*In Problems 9–12, find:*
*(a) A vector equation*
*(b) Parametric equations*
*(c) Symmetric equations of the line*

9. containing the point $(1, 2, 3)$ and in the direction of $2\mathbf{i} - \mathbf{j} + \mathbf{k}$.

10. containing the point $(4, -1, 6)$ and in the direction of $\mathbf{i} + \mathbf{j}$.

11. containing the points $P_0 = (1, -1, 3)$ and $P_1 = (4, 2, 1)$.

12. containing the points $P_0 = (-2, 3, 0)$ and $P_1 = (1, -1, 2)$.

13. Find parametric equations of a line containing the point $(-1, 5, 6)$ and in the direction of the line $\dfrac{x+1}{5} = \dfrac{y-2}{4} = \dfrac{z-3}{-3}$.

14. Find parametric equations of a line containing the point $(1, -2, -3)$ and in the direction of the line $\dfrac{x+1}{6} = \dfrac{y+2}{2} = \dfrac{z}{-1}$.

15. Let $\dfrac{x-4}{2} = \dfrac{y+1}{-1} = \dfrac{z-2}{2}$ be symmetric equations of a line. Find a vector in the direction of the line, and find two points on this line.

16. Let $x + 1 = y + 3 = \dfrac{z+4}{2}$ be symmetric equations of a line. Find a vector in the direction of the line, and find two points on this line.

*In Problems 17–22, find symmetric equations of the line containing the point $P_0$.*

17. $P_0 = (4, 2, 1)$ and the line is in the direction of the vector $\mathbf{D} = 2\mathbf{i} + \mathbf{k}$.

18. $P_0 = (-1, 2, 0)$ and the line is in the direction of the vector $\mathbf{D} = 2\mathbf{k} - \mathbf{j}$.

19. $P_0 = (0, 0, 0)$ and the line is perpendicular to each of the lines $\dfrac{x+1}{2} = \dfrac{y-1}{3} = \dfrac{z}{-2}$ and $\dfrac{x-3}{3} = \dfrac{y}{1} = \dfrac{z+1}{2}$.

20. $P_0 = (0, 0, 0)$ and the line is perpendicular to each of the lines $\dfrac{x+2}{4} = \dfrac{y-1}{2} = z + 1$ and $\dfrac{x+5}{5} = \dfrac{y+1}{3} = z$.

21. $P_0 = (1, 2, -1)$ and the line is perpendicular to the vectors $\mathbf{u} = 2\mathbf{i} + 4\mathbf{j} - 2\mathbf{k}$ and $\mathbf{v} = -3\mathbf{i} - 2\mathbf{j} + \mathbf{k}$.

22. $P_0 = (-1, 3, 2)$ and the line is perpendicular to the vectors $\mathbf{u} = -2\mathbf{i} + 2\mathbf{j} - 3\mathbf{k}$ and $\mathbf{v} = 4\mathbf{i} + 2\mathbf{j} + \mathbf{k}$.

*In Problems 23–30:*
*(a) Determine whether the lines $l_1$ and $l_2$ intersect, are parallel, or are skew.*
**CAS** *(b) Graph the lines. Does the graph confirm the answer to (a)?*

23. $l_1$:  $\mathbf{r}_1 = 2\mathbf{i} - 2\mathbf{k} + t_1(-3\mathbf{i} + 6\mathbf{j} + 6\mathbf{k})$
    $l_2$:  $\mathbf{r}_2 = 6\mathbf{i} + 2\mathbf{j} + 5\mathbf{k} + t_2(-\mathbf{i} + 2\mathbf{j} + 2\mathbf{k})$

24. $l_1$:  $\mathbf{r}_1 = 3\mathbf{i} + 3\mathbf{k} + t_1(3\mathbf{i} + 6\mathbf{j} - 2\mathbf{k})$
    $l_2$:  $\mathbf{r}_2 = 3\mathbf{i} + 3\mathbf{k} + t_2(-2\mathbf{i} + 4\mathbf{j} + 7\mathbf{k})$

25. $l_1$:  $\mathbf{r}_1 = 4\mathbf{i} + 3\mathbf{j} + t_1(\mathbf{i} - \mathbf{j} + 6\mathbf{k})$
    $l_2$:  $\mathbf{r}_2 = 4\mathbf{i} + 3\mathbf{j} + 2\mathbf{k} + t_2(\mathbf{i} - \mathbf{j} - 2\mathbf{k})$

26. $l_1$:  $\mathbf{r}_1 = 2\mathbf{i} + 7\mathbf{j} + t_1(-2\mathbf{i} + 8\mathbf{j} - 6\mathbf{k})$
    $l_2$:  $\mathbf{r}_2 = 6\mathbf{i} - 5\mathbf{j} + t_2(\mathbf{i} - 4\mathbf{j} + 3\mathbf{k})$

27. $l_1$:  $\dfrac{x-3}{2} = \dfrac{y+2}{3} = \dfrac{z-1}{4}$
    $l_2$:  $\dfrac{x+4}{-4} = \dfrac{y-3}{-6} = \dfrac{z+4}{-8}$

28. $l_1$:  $\dfrac{x}{3} = \dfrac{y-2}{4} = \dfrac{z+4}{1}$
    $l_2$:  $\dfrac{x+6}{3} = \dfrac{y+2}{4} = \dfrac{z-3}{2}$

29. $l_1$:  $\dfrac{x+1}{5} = \dfrac{y-2}{4} = \dfrac{z-3}{-3}$
    $l_2$:  $\dfrac{x+1}{6} = \dfrac{y-2}{3} = \dfrac{z+3}{2}$

30. $l_1$:  $\dfrac{x+5}{6} = \dfrac{y-2}{3} = \dfrac{z-4}{-1}$
    $l_2$:  $x = \dfrac{y-2}{3} = \dfrac{z-8}{2}$

*In Problems 31–40, find the general equation of each plane.*

31. Containing the point $(1, -1, 2)$ and normal to the vector $2\mathbf{i} - \mathbf{j} + \mathbf{k}$

32. Containing the point $(-3, 2, 1)$ and normal to the vector $\mathbf{i} + \mathbf{j} - 2\mathbf{k}$

33. Containing the point $(0, 5, -2)$ and parallel to the plane $x + 2y - z = 6$

34. Containing the point $(1, -2, 0)$ and parallel to the plane $2x - y + 3z = 10$

35. Containing the point $(2, 3, -1)$ and normal to the line
$$\dfrac{x-1}{2} = \dfrac{y-3}{5} = \dfrac{z+1}{-2}$$

---

**1.** = NOW WORK problem      **⩗** = Graphing technology recommended      **CAS** = Computer Algebra System recommended

**36.** Containing the point $(-1, 2, 3)$ and normal to the line

$$\frac{x+5}{3} = \frac{y+2}{4} = \frac{z-4}{4}$$

**37.** Parallel to the $xy$-plane and containing the point $(0, 0, 4)$

**38.** Parallel to the $yz$-plane and containing the point $(2, 0, 0)$

**39.** Parallel to the $xz$-plane and containing the point $(1, -2, 3)$

**40.** Parallel to the $xy$-plane and containing the point $(1, -3, 4)$

*In Problems 41–46:*
**(a)** *Find the general equation of the plane containing the points $P_1$, $P_2$, and $P_3$.*
CAS **(b)** *Graph the plane.*

**41.** $P_1 = (0, 0, 0)$; $P_2 = (1, 2, -1)$; $P_3 = (-1, 1, 0)$

**42.** $P_1 = (0, 0, 0)$; $P_2 = (3, -1, 2)$; $P_3 = (-3, 1, 0)$

**43.** $P_1 = (1, 2, 1)$; $P_2 = (3, 2, 2)$; $P_3 = (4, -1, -1)$

**44.** $P_1 = (-1, 2, 0)$; $P_2 = (3, 4, -1)$; $P_3 = (-2, -1, 0)$

**45.** $P_1 = (6, 8, -2)$; $P_2 = (4, -1, 0)$; $P_3 = (1, 0, 0)$

**46.** $P_1 = (-3, -4, 0)$; $P_2 = (6, -7, 2)$; $P_3 = (0, 0, 1)$

*In Problems 47–50, find symmetric equations of the line of intersection of the two planes.*

**47.** $p_1: x + y - z = -5$ and $p_2: 2x + 3y - 4z = -1$

**48.** $p_1: 2x - y + z = 2$ and $p_2: x + y + z = 3$

**49.** $p_1: x - y = 2$ and $p_2: y - z = 2$

**50.** $p_1: 2x - 3y + z = 1$ and $p_2: 2x - 3y + 4z = 2$

*In Problems 51–54, find the distance from the point to the plane.*

**51.** from $(1, 2, -1)$ to $2x - y + z = 1$

**52.** from $(-1, 3, -2)$ to $x + 2y - 3z = 4$

**53.** from $(2, -1, 1)$ to $-x + y - 3z = 6$

**54.** from $(-2, 1, 1)$ to $-3x + 2y + z = 1$

*In Problems 55–58, find the distance between the two parallel planes.*

**55.** $p_1: 2x + y - 2z = -1$ and $p_2: 2x + y - 2z = 3$

**56.** $p_1: x + 2y - 2z = -3$ and $p_2: x + 2y - 2z = 3$

**57.** $p_1: x - 2z = -1$ and $p_2: x - 2z = 3$

**58.** $p_1: -x + y + 2z = -4$ and $p_2: -x + y + 2z = -1$

*In Problems 59–62, find the point of intersection of the plane and the line.*

**59.** Plane $2x + y - z = 5$, line $\dfrac{x-1}{2} = \dfrac{y+3}{4} = \dfrac{z-1}{1}$

**60.** Plane $x + y - 2z = 8$, line $\dfrac{x+1}{2} = \dfrac{y-3}{1} = \dfrac{z-4}{-2}$

**61.** Plane $2x + 3y + z = 5$, line $\dfrac{x-3}{1} = \dfrac{y+4}{2} = \dfrac{z-1}{2}$

**62.** Plane $x + y - z = 3$; line $\dfrac{x+2}{2} = \dfrac{y-3}{1} = \dfrac{z}{2}$

*In Problems 63–66, find the distance from the point $P$ to the line $\mathbf{r} = \mathbf{r}(t)$.*

**63.** $P = (1, 1, 4)$; $\mathbf{r}(t) = (3 + 2t)\mathbf{i} + (3 + t)\mathbf{j} + (-1 - 2t)\mathbf{k}$

**64.** $P = (2, 0, 1)$; $\mathbf{r}(t) = (4 + t)\mathbf{i} + (1 + 2t)\mathbf{j} + (1 + 2t)\mathbf{k}$

**65.** $P = (-2, 1, 1)$; $\mathbf{r}(t) = (-3 + 4t)\mathbf{i} + (2 + t)\mathbf{j} + (1 + t)\mathbf{k}$

**66.** $P = (3, -1, 0)$; $\mathbf{r}(t) = (2 + 2t)\mathbf{i} + (1 - 2t)\mathbf{j} + (2 + 3t)\mathbf{k}$

## Applications and Extensions

*In Problems 67–70, find the point of intersection of $l_1$ and $l_2$ and the angle between $l_1$ and $l_2$.*

**67.** $l_1: \mathbf{r}_1 = (2 - t_1)\mathbf{i} + (4 + 2t_1)\mathbf{j} + (-5 + t_1)\mathbf{k}$

$l_2: \mathbf{r}_2 = (4 - t_2)\mathbf{i} + (3 + t_2)\mathbf{j} + (-13 + 3t_2)\mathbf{k}$

**68.** $l_1: \mathbf{r}_1 = (5 + 2t_1)\mathbf{i} + (6 - t_1)\mathbf{j} + 2t_1\mathbf{k}$

$l_2: \mathbf{r}_2 = (7 + 3t_2)\mathbf{i} + (5 - 2t_2)\mathbf{j} + (2 - t_2)\mathbf{k}$

**69.** $l_1: \mathbf{r}_1 = t_1\mathbf{i} + (1 + 2t_1)\mathbf{j} + (-3 + t_1)\mathbf{k}$

$l_2: \mathbf{r}_2 = (-3 + t_2)\mathbf{i} + (1 - 4t_2)\mathbf{j} + (2 - 7t_2)\mathbf{k}$

**70.** $l_1: \mathbf{r}_1 = (2 - 3t_1)\mathbf{i} + 6t_1\mathbf{j} + (-2 + 5t_1)\mathbf{k}$

$l_2: \mathbf{r}_2 = -2t_2\mathbf{i} + (1 + t_2)\mathbf{j} + 2t_2\mathbf{k}$

**71. (a)** Find parametric equations of the line containing the point $(1, 2, -1)$ and normal to the plane $2x - y + z - 6 = 0$.

CAS **(b)** Graph the line and the plane.

**72. (a)** Find parametric equations of the line containing the point $(2, 3, -1)$ and normal to the plane $x + y - z - 3 = 0$.

CAS **(b)** Graph the line and the plane.

**73.** What property can you assign to the lines $l_1$ and $l_2$, given below, if $a_1b_1 + a_2b_2 + a_3b_3 = 0$?

$$l_1: \quad \frac{x - x_1}{a_1} = \frac{y - y_1}{a_2} = \frac{z - z_1}{a_3}$$

$$l_2: \quad \frac{x - x_1}{b_1} = \frac{y - y_1}{b_2} = \frac{z - z_1}{b_3}$$

**74.** What property can you assign to the distinct lines $l_1$ and $l_2$, given below, if $\dfrac{a_1}{b_1} = \dfrac{a_2}{b_2} = \dfrac{a_3}{b_3}$?

$$l_1: \quad \frac{x - x_1}{a_1} = \frac{y - y_1}{a_2} = \frac{z - z_1}{a_3}$$

$$l_2: \quad \frac{x - x_2}{b_1} = \frac{y - y_2}{b_2} = \frac{z - z_2}{b_3}$$

*Problems 75–80, use the following discussion.* When two planes intersect, the **angle $\theta$ between the planes** is defined as the non-obtuse angle between their normals. See the figure. If $\mathbf{N}_1$ and $\mathbf{N}_2$ are the normals of two intersecting planes, the angle $\theta$ between these planes is given by

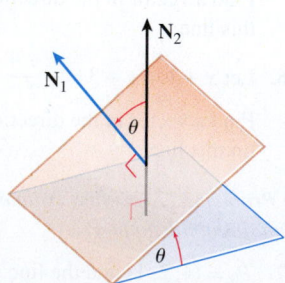

$$\cos\theta = \frac{|\mathbf{N}_1 \cdot \mathbf{N}_2|}{\|\mathbf{N}_1\| \|\mathbf{N}_2\|} \qquad 0 \le \theta \le \frac{\pi}{2}$$

*In Problems 75–80, the two planes intersect. Find the angle between them.*

**75.** $p_1$: $2x - y + z = 2$ and $p_2$: $x + y + z = 3$

**76.** $p_1$: $x + y - z = 5$; and $p_2$: $2x + 3y - 4z = 1$

**77.** $p_1$: $2x - 3y + z = 1$ and $p_2$: $2x - 3y + 4z = 2$

**78.** $p_1$: $x - y = 2$ and $p_2$: $y - z = 2$

**79.** $p_1$: $2x - y + z = 3$ and $p_2$: $4x - y + 6z = 7$

**80.** $p_1$: $x + y - z = 1$ and $p_2$: $2x - 2y + 2z = -2$

**81. Paths of Spacecrafts** Two unidentified flying objects are at the points $(t, -t, 1 - t)$ and $(t - 3, 2t, 4t - 1)$ at time $t$, $t \geq 0$.

(a) Describe the paths of the objects.

(b) Find the acute angle between the paths.

(c) Find where the paths intersect (or determine that they do not).

(d) Will the objects collide?

**82.** Find symmetric equations of the line passing through the centers of the spheres

$$x^2 + y^2 + z^2 - 2x - 4y + 4z = 8$$

and

$$x^2 + y^2 + z^2 + 2x + 6y + 4z = 20$$

**83.** Find parametric equations of the line perpendicular to the lines

$$l_1: \quad x = 1 - t, \quad y = t, \quad z = 2t - 1$$

and

$$l_2: \quad x = t + 1, \quad y = -t, \quad z = t - 1$$

at their point of intersection. Why is this line parallel to the $xy$-plane?

**84.** Explain why the set of points $(x, y, z)$ equidistant from the points $(1, 3, 0)$ and $(-1, 1, 2)$ is a plane. Then find its equation in two ways, as follows:

(a) Use the distance formula to equate the distances between $(x, y, z)$ and the given points, simplifying the result to obtain an equation of the plane.

(b) Find a point on the plane and a vector normal to the plane, and use the answer to find an equation of the plane.

**85.** Find symmetric equations of the line of intersection of the planes $2x + y - z = 6$ and $x - y + 3z = 4$.

**86.** Find symmetric equations of the line that contains $(2, 0, -3)$, is perpendicular to $\mathbf{i} + 2\mathbf{j} - \mathbf{k}$, and is parallel to the plane $2x + 3y - z = 1$.

**87.** Find the point of intersection of the line through the points $(0, 2, -2)$, and $(2, 1, -3)$, and the plane through the points $(0, 4, -2)$, $(1, 3, -2)$, and $(2, 2, -3)$.

**88.** Find an equation of the plane parallel to the line $\mathbf{r} = 2\mathbf{i} + t(-\mathbf{i} + \mathbf{j} + 2\mathbf{k})$ and containing the points $(2, 2, -1)$ and $(1, 0, 1)$.

**89.** Find an equation of the plane tangent to the sphere $(x - 1)^2 + (y + 2)^2 + (z - 2)^2 = 6$ at the point $(2, -1, 0)$.

**90. Sphere** Find an equation of the sphere with its center at $(-2, 1, 5)$ and tangent to $x + 2y - 2z = 8$.

**91.** Find symmetric equations of a line normal to the plane containing the lines

$$x - 2 = \frac{y + 1}{2} = \frac{z - 1}{2} \qquad \text{and} \qquad x + 1 = \frac{y + 8}{3} = \frac{z}{-3}$$

at their point of intersection.

## Challenge Problems

**92.** Find an equation for the plane containing the origin that is perpendicular to the plane $x - 2y - z = 0$ and makes an angle of $60°$ with the positive $y$-axis.

**93.** The distance between the point $P_0 = (x_0, y_0, z_0)$ and the plane $Ax + By + Cz = D$ is

$$d = \frac{|Ax_0 + By_0 + Cz_0 - D|}{\sqrt{A^2 + B^2 + C^2}}$$

Derive this formula differently, as follows:

(a) Show that the line through $P_0$ normal to the plane has the parametric equations $x = x_0 + At$, $y = y_0 + Bt$, $z = z_0 + Ct$.

(b) If $P = (x, y, z)$ is the point of intersection of the line in (a) with the plane, show that $x - x_0 = At$, $y - y_0 = Bt$, $z - z_0 = Ct$, where

$$t = \frac{-(Ax_0 + By_0 + Cz_0) + D}{A^2 + B^2 + C^2}$$

(c) Explain why $d$ is the distance between $P_0$ and $P$, and use the distance formula to finish the proof.

**94.** Prove that the distance between the parallel planes $Ax + By + Cz = D_1$ and $Ax + By + Cz = D_2$ is

$$d = \frac{|D_2 - D_1|}{\sqrt{A^2 + B^2 + C^2}}$$

**95.** Let two skew lines have respective direction vectors $\mathbf{D}_1$ and $\mathbf{D}_2$. Let $A$ and $B$ be points on the respective lines, and let $\mathbf{w} = \vec{AB}$. Show that the distance between the two lines is the magnitude of the vector projection of $\mathbf{w}$ onto $\mathbf{D}_1 \times \mathbf{D}_2$:

$$|\text{proj}_{\mathbf{D}_1 \times \mathbf{D}_2} \mathbf{w}| = \frac{|\mathbf{w} \cdot (\mathbf{D}_1 \times \mathbf{D}_2)|}{\|\mathbf{D}_1 \times \mathbf{D}_2\|}$$

(The shortest distance is measured along the common perpendicular to the two lines.)

**96.** Find the minimum distance between the lines

$$\frac{x - 3}{2} = \frac{y}{3} = z \qquad \text{and} \qquad x = \frac{y + 1}{-2} = \frac{z - 2}{-1}$$

**97.** What is the minimum distance between the skew lines

$$x - 1 = y - 2 = z + 6 \text{ and } \frac{x - 1}{2} = \frac{y + 2}{-3} = z - 10.$$

Locate the points on each line at which the distance is minimum.

# 10.7 Quadric Surfaces

**OBJECTIVES** *When you finish this section, you should be able to:*

**1** Identify quadric surfaces based on an ellipse (p. 784)

**2** Identify quadric surfaces based on a hyperbola (p. 786)

**3** Identify cylinders (p. 788)

**4** Graph quadric surfaces (p. 790)

The graph of a second degree equation in two variables

$$Ax^2 + Bxy + Cy^2 + Dx + Ey + F = 0$$

> **NEED TO REVIEW?** Conics are discussed in Appendix A.3, pp. A-22 to A-25.

is a conic (a parabola, an ellipse or a hyperbola), except for degenerate cases.

The graph of a second-degree equation in three variables

$$Ax^2 + By^2 + Cz^2 + Dxy + Exz + Fyz + Gx + Hy + Iz + J = 0 \qquad (1)$$

is called a **quadric surface**. There are nine different types of quadric surfaces, excluding the *degenerate cases*. Examples of degenerate quadric surfaces include $x^2 + y^2 + z^2 = 0$, whose graph is the point $(0, 0, 0)$, and $2x^2 + 3y^2 + 4z^2 = -2$, whose graph has no points.

As with conics, when the equation of a quadric surface is written in the form of equation (1), it is in **general form**. On the other hand, if we complete the squares, it is in **standard form**. The equations we consider here are in standard form and the corresponding quadric surfaces are centered at the origin of an $xyz$-coordinate system. Variations of these surfaces can be found using transformations (discussed in Chapter P, Section 3).

A quadric surface is characterized by listing its intercepts, and by finding its *traces*. A **trace** is the intersection of the surface with a coordinate plane or a plane parallel to a coordinate plane. The term **section** (or *cross section*) is sometimes used when the intersection is with a plane other than a coordinate plane.

## 1 Identify Quadric Surfaces Based on an Ellipse

The first three quadric surfaces we investigate are based on an ellipse. Each of these surfaces has at least one trace that is an ellipse.

> **DEFINITION** Ellipsoid
>
> The graph of an **ellipsoid** has an equation of the form
>
> $$\frac{x^2}{a^2} + \frac{y^2}{b^2} + \frac{z^2}{c^2} = 1 \qquad a > 0, \ b > 0, \ c > 0$$

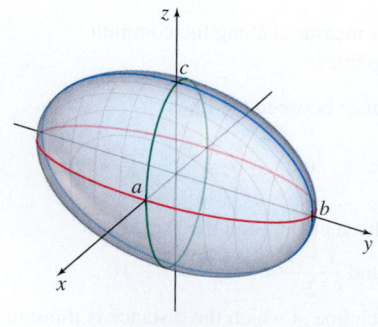

**DF** **Figure 59** Ellipsoid:
$$\frac{x^2}{a^2} + \frac{y^2}{b^2} + \frac{z^2}{c^2} = 1$$

See Figure 59. The intercepts of an ellipsoid are the points $(\pm a, 0, 0)$, $(0, \pm b, 0)$, and $(0, 0, \pm c)$, and every trace is an ellipse. For example, if $z = 0$, the trace in the $xy$-plane is the ellipse $\dfrac{x^2}{a^2} + \dfrac{y^2}{b^2} = 1$. The traces parallel to the $xy$-plane are the ellipses $\dfrac{x^2}{a^2} + \dfrac{y^2}{b^2} = 1 - \dfrac{k^2}{c^2}$, where $|z| = |k| < c$. Traces parallel to the $xz$- and $yz$-planes are the ellipses $\dfrac{x^2}{a^2} + \dfrac{z^2}{c^2} = 1 - \dfrac{k^2}{b^2}$, where $|y| = |k| < b$, and $\dfrac{y^2}{b^2} + \dfrac{z^2}{c^2} = 1 - \dfrac{k^2}{a^2}$, where $|x| = |k| < a$, respectively.

An ellipsoid centered at the origin is symmetric with respect to each coordinate plane. The symmetry holds because all three variables are squared. If two of the three numbers $a$, $b$, or $c$ are equal, the ellipsoid is called an **ellipsoid of revolution**. If all three numbers $a$, $b$, and $c$ are equal, the ellipsoid is a sphere. See Figure 60.

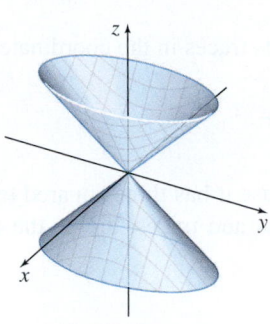

(a) Elliptic cone

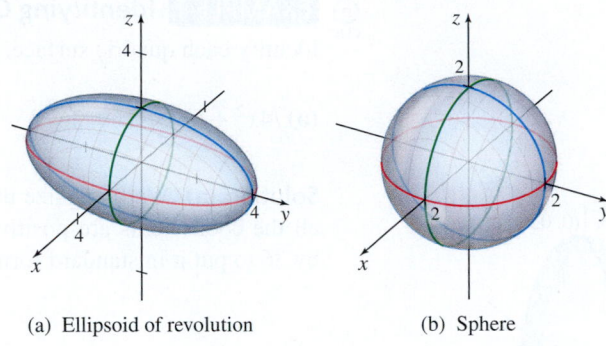

(a) Ellipsoid of revolution     (b) Sphere

**Figure 60**

The second quadric surface based on an ellipse is called an *elliptic cone*.

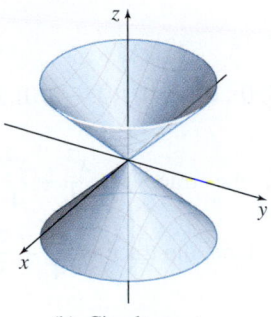

(b) Circular cone

**Figure 61**

**DEFINITION Elliptic Cone**

The graph of an **elliptic cone** has an equation of the form

$$z^2 = \frac{x^2}{a^2} + \frac{y^2}{b^2} \qquad a > 0, b > 0$$

See Figure 61(a). The only intercept of an elliptic cone is the origin $(0, 0, 0)$, called the **vertex** of the cone. The trace of an elliptic cone in the $xy$-plane is the origin. Traces parallel to the $xy$-plane are ellipses centered on the $z$-axis. The trace in the $yz$-plane consists of the pair of intersecting lines $z = \pm \frac{y}{b}$, and the trace in the $xz$-plane consists of the pair of intersecting lines $z = \pm \frac{x}{a}$. When these lines are rotated about the $z$-axis they **generate** the elliptic cone.

If $a = b$, the elliptic cone becomes a *circular cone* and its traces are circles. See Figure 61(b).

Other forms of an elliptic cone are

$$x^2 = \frac{y^2}{b^2} + \frac{z^2}{c^2} \qquad y^2 = \frac{x^2}{a^2} + \frac{z^2}{c^2} \qquad a > 0, b > 0, c > 0$$

The third quadric surface based on the ellipse is the *elliptic paraboloid*.

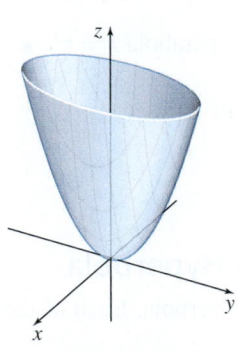

(a) Elliptic paraboloid

**DEFINITION Elliptic Paraboloid**

The graph of an **elliptic paraboloid** has an equation of the form

$$z = \frac{x^2}{a^2} + \frac{y^2}{b^2} \qquad a > 0, b > 0$$

See Figure 62(a). An elliptic paraboloid has only one intercept, the origin $(0, 0, 0)$, called the **vertex**. The trace of an elliptic paraboloid in the $xy$-plane is the vertex. The traces parallel to the $xy$-plane are ellipses. The traces in the other two coordinate planes are parabolas: $z = \frac{y^2}{b^2}$ in the $yz$-plane, and $z = \frac{x^2}{a^2}$ in the $xz$-plane. Traces parallel to the $xz$- and $yz$-coordinate planes are parabolas. Since $z \geq 0$, the surface (except for the origin) lies above the $xy$-plane.

If $a = b$, the surface is a **paraboloid of revolution**, and the traces parallel to the $xy$-plane are circles. See Figure 62(b).

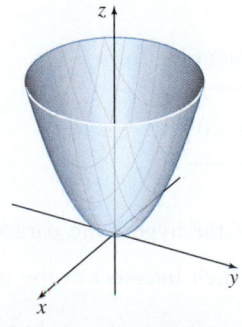

(b) Paraboloid of revolution

DF **Figure 62**

Other forms of an elliptic paraboloid are

$$x = \frac{y^2}{b^2} + \frac{z^2}{c^2} \qquad y = \frac{x^2}{a^2} + \frac{z^2}{c^2} \qquad a > 0, b > 0, c > 0$$

CALC
CLIP

## EXAMPLE 1  Identifying Quadric Surfaces

Identify each quadric surface. List its intercepts and its traces in the coordinate planes.

(a) $4x^2 + 9y^2 + z^2 = 36$  (b) $x = \dfrac{y^2}{4} + z^2$

**Solution** (a) We recognize this as an ellipsoid because it has three squared terms and all the coefficients are positive. To find its intercepts and traces, divide the equation by 36 to put it in standard form.

$$\frac{x^2}{9} + \frac{y^2}{4} + \frac{z^2}{36} = 1$$

$$\frac{x^2}{3^2} + \frac{y^2}{2^2} + \frac{z^2}{6^2} = 1$$

The intercepts of the ellipsoid are $(3, 0, 0)$, $(-3, 0, 0)$, $(0, 2, 0)$, $(0, -2, 0)$, $(0, 0, 6)$, and $(0, 0, -6)$.

The traces are all ellipses. In the $xy$-plane, the trace is $\dfrac{x^2}{9} + \dfrac{y^2}{4} = 1$; in the $xz$-plane, the trace is $\dfrac{x^2}{9} + \dfrac{z^2}{36} = 1$; and in the $yz$-plane, the trace is $\dfrac{y^2}{4} + \dfrac{z^2}{36} = 1$.

Figure 63 shows the graph of the ellipsoid.

(b) This equation defines an elliptic paraboloid. Its only intercept (vertex) is $(0, 0, 0)$.

The trace in the $yz$-plane is the vertex $(0, 0, 0)$ and traces parallel to the $yz$-plane are ellipses, provided $x > 0$.

To find the trace in the $xy$-plane, let $z = 0$. The trace is the parabola $x = \dfrac{y^2}{4}$. To find the trace in the $xz$-plane, let $y = 0$. The trace is the parabola $x = z^2$. ∎

Figure 64 shows the graph of the elliptic paraboloid.

**NOW WORK** Problem 21.

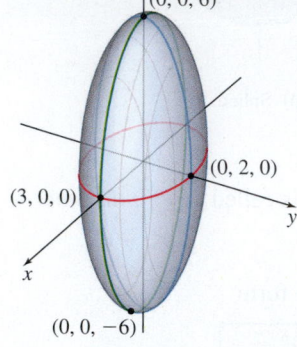

**Figure 63** $4x^2 + 9y^2 + z^2 = 36$

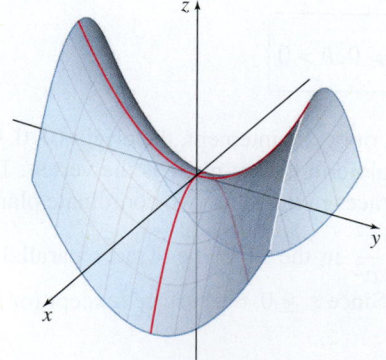

**Figure 64** $x = \dfrac{y^2}{4} + z^2$

## 2 Identify Quadric Surfaces Based on a Hyperbola

There are three quadric surfaces that are based on a hyperbola. Each of these surfaces has at least one trace that is a hyperbola.

### DEFINITION  Hyperbolic Paraboloid

The graph of a **hyperbolic paraboloid** has an equation of the form

$$z = \frac{y^2}{b^2} - \frac{x^2}{a^2} \qquad a > 0, b > 0$$

See Figure 65. The origin is the only intercept of the hyperbolic paraboloid. The trace in the $xy$-plane is the pair of lines $\dfrac{y}{b} = \pm\dfrac{x}{a}$, which intersect at the origin. The traces parallel to the $xy$-plane are hyperbolas given by the equations $\dfrac{y^2}{b^2} - \dfrac{x^2}{a^2} = k$, where $z = k$.

DF **Figure 65** Hyperbolic paraboloid

The trace in the $xz$-plane is the parabola $z = -\dfrac{x^2}{a^2}$, which opens down. Traces parallel to the $xz$-plane are also parabolas that open down. The trace in the $yz$-plane is the parabola $z = \dfrac{y^2}{b^2}$, which opens up. Traces parallel to the $yz$-plane are also parabolas that open up.

Notice that the origin is a minimum point for the trace in the $yz$-plane and is a maximum point for the trace in the $xz$-plane. Because of the graph's appearance, the origin is called a **saddle point** of the surface.

Other forms of a hyperbolic paraboloid are

$$x = \frac{y^2}{b^2} - \frac{z^2}{c^2} \qquad y = \frac{x^2}{a^2} - \frac{z^2}{c^2} \qquad a > 0, b > 0, c > 0$$

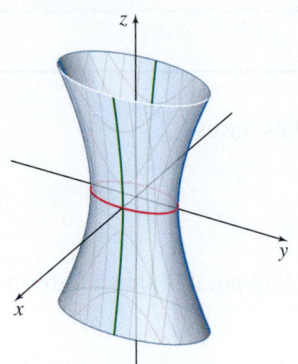

(a) Hyperboloid of one sheet

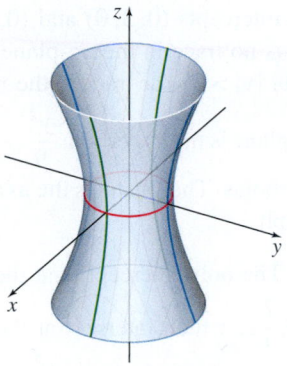

(b) Hyperboloid of revolution

**Figure 66**

**NOTE** Remember the cooling towers from Chapter 6? They are physical examples of a hyperboloid of revolution.

### DEFINITION  Hyperboloid of One Sheet

The graph of a **hyperboloid of one sheet** has an equation of the form

$$\frac{x^2}{a^2} + \frac{y^2}{b^2} - \frac{z^2}{c^2} = 1 \qquad a > 0, b > 0, c > 0$$

See Figure 66(a). The intercepts a hyperboloid of one sheet are $(\pm a, 0, 0)$ and $(0, \pm b, 0)$. The trace in the $xy$-plane is the ellipse $\dfrac{x^2}{a^2} + \dfrac{y^2}{b^2} = 1$, and all traces parallel to the $xy$-plane are also ellipses. The trace in the $yz$-plane is the hyperbola $\dfrac{y^2}{b^2} - \dfrac{z^2}{c^2} = 1$; the trace in the $xz$-plane is the hyperbola $\dfrac{x^2}{a^2} - \dfrac{z^2}{c^2} = 1$. Traces parallel to the $yz$- and $xz$-planes are also hyperbolas. The $z$-axis is called the **axis** of the hyperboloid of one sheet.

If $a = b$, the surface is a **hyperboloid of revolution**. See Figure 66(b).

Other forms of a hyperboloid of one sheet are

$$-\frac{x^2}{a^2} + \frac{y^2}{b^2} + \frac{z^2}{c^2} = 1 \qquad \frac{x^2}{a^2} - \frac{y^2}{b^2} + \frac{z^2}{c^2} = 1 \qquad a > 0, b > 0, c > 0$$

### DEFINITION  Hyperboloid of Two Sheets

The graph of a **hyperboloid of two sheets** has an equation of the form

$$\frac{x^2}{a^2} + \frac{y^2}{b^2} - \frac{z^2}{c^2} = -1 \qquad a > 0, b > 0, c > 0$$

**NOTE** The equation defining a hyperboloid of one sheet has a 1 on the right side; the equation defining a *hyperboloid of two sheets* has a $-1$ on the right.

A hyperboloid of two sheets is shown in Figure 67. It has two intercepts $(0, 0, c)$ and $(0, 0, -c)$. Also, the surface consists of two parts, one for which $z \geq c$, and the other for which $z \leq -c$.

If $z = 0$, the equation has no real solution, so there is no trace in the $xy$-plane. Traces parallel to the $xy$-plane are ellipses and are defined for $|z| > c$. The traces in the $yz$-plane and the $xz$-plane are the hyperbolas $\dfrac{z^2}{c^2} - \dfrac{y^2}{b^2} = 1$ and $\dfrac{z^2}{c^2} - \dfrac{x^2}{a^2} = 1$, respectively. Traces parallel to these coordinate planes are also hyperbolas. The $z$-axis is called the **axis** of the hyperboloid of two sheets.

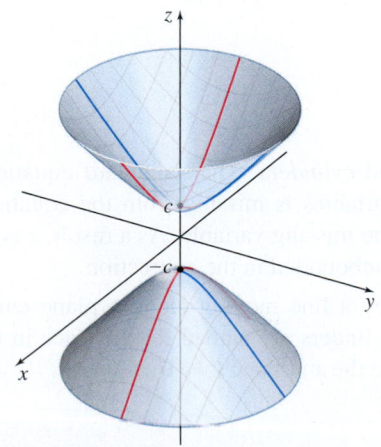

**Figure 67** A hyperboloid of two sheets

Other forms of a hyperboloid of two sheets are

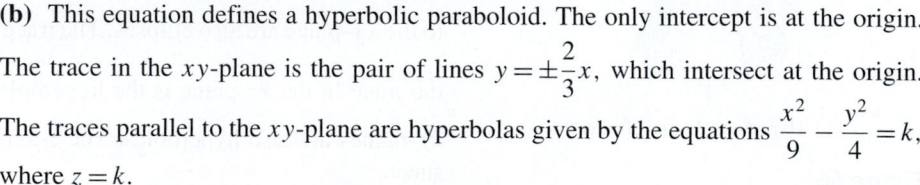

$$\frac{x^2}{a^2} - \frac{y^2}{b^2} + \frac{z^2}{c^2} = -1 \qquad -\frac{x^2}{a^2} + \frac{y^2}{b^2} + \frac{z^2}{c^2} = -1 \qquad a > 0, b > 0, c > 0$$

### EXAMPLE 2    Identifying Quadric Surfaces

Identify each quadric surface. List its intercepts and its traces.

**(a)** $x^2 - y^2 + z^2 = -9$     **(b)** $z = \dfrac{x^2}{9} - \dfrac{y^2}{4}$

**Solution (a)** We begin by dividing the equation by 9 to put it in standard form.

$$\frac{x^2}{9} - \frac{y^2}{9} + \frac{z^2}{9} = -1$$

This is a hyperboloid of two sheets. It has two intercepts $(0, 3, 0)$ and $(0, -3, 0)$. If $y = 0$, the equation has no real solution, so there is no trace in the $xz$-plane. Traces parallel to the $xz$-plane are ellipses and are defined for $|y| > 3$. The trace in the $xy$-plane is the hyperbola $\dfrac{y^2}{9} - \dfrac{x^2}{9} = 1$ and the trace in the $yz$-plane is the hyperbola $\dfrac{y^2}{9} - \dfrac{z^2}{9} = 1$. Traces parallel to the $xy$- and $yz$-plane are also hyperbolas. The $y$-axis is the axis of this hyperboloid of two sheets. See Figure 68 for the graph.

**(b)** This equation defines a hyperbolic paraboloid. The only intercept is at the origin. The trace in the $xy$-plane is the pair of lines $y = \pm\dfrac{2}{3}x$, which intersect at the origin. The traces parallel to the $xy$-plane are hyperbolas given by the equations $\dfrac{x^2}{9} - \dfrac{y^2}{4} = k$, where $z = k$.

The trace in the $xz$-plane is the parabola $z = \dfrac{x^2}{9}$, which opens up. Traces parallel to the $xz$-plane are also parabolas that open up. The trace in the $yz$-plane is the parabola $z = -\dfrac{y^2}{4}$, which opens down. Traces parallel to the $yz$-plane are also parabolas that open down.

The origin is the saddle point of the hyperbolic paraboloid. See Figure 69 for the graph. ∎

NOW WORK Problem 7.

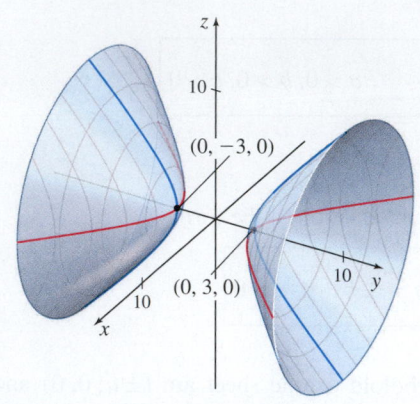

**Figure 68** $x^2 - y^2 + z^2 = -9$

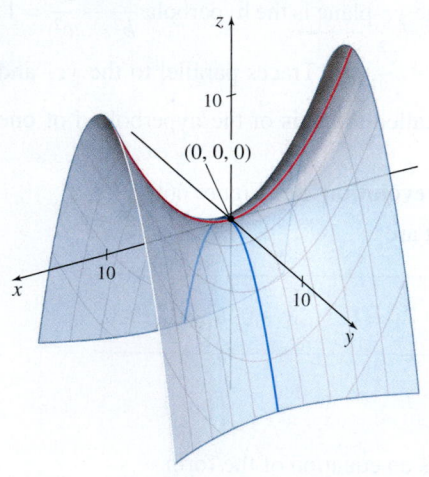

**Figure 69** $z = \dfrac{x^2}{9} - \dfrac{y^2}{4}$

## 3 Identify Cylinders

The three remaining quadric surfaces are called *cylinders*. Their standard equations are characterized by the fact that one of the variables is missing from the equation. In naming these surfaces, we have chosen $z$ as the missing variable. As a result, $z$ is an unrestricted variable, so these cylinders will be unbounded in the $z$ direction.

Cylinders are surfaces that are generated by a line moving along a plane curve while remaining perpendicular* to the plane. Cylinders are named for the trace in the plane in which the unrestricted variable is 0; here the $xy$-plane $(z = 0)$.

**NOTE** Any of the three variables $x$, $y$, or $z$ could be missing. The corresponding graphs are unbounded in the direction of the missing variable.

---

*In general, the line moving along the plane curve can meet the curve at any constant angle. We only discuss cylinders for which this angle is 90°.

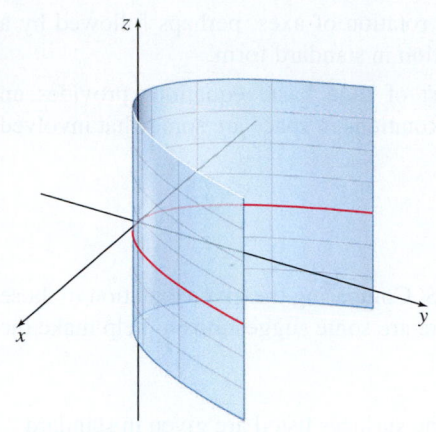

**Figure 70** Parabolic cylinder: $x^2 = 4ay$

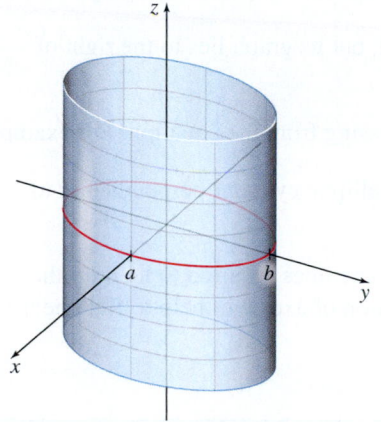

**Figure 71** Elliptic cylinder:
$$\frac{x^2}{a^2} + \frac{y^2}{b^2} = 1$$

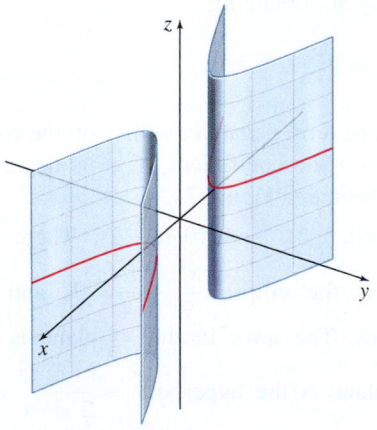

**Figure 72** Hyperbolic cylinder:
$$\frac{x^2}{a^2} - \frac{y^2}{b^2} = 1$$

**DEFINITION Parabolic Cylinder**

The graph of a **parabolic cylinder** has an equation of the form

$$x^2 = 4ay \qquad a \neq 0$$

See Figure 70. This parabolic cylinder does not include the variable $z$, so the trace in the $xy$-plane is the parabola $x^2 = 4ay$. The surface is generated by moving a line along the parabola while remaining perpendicular to the plane containing the parabola. It can be visualized by taking a piece of paper and bending it so that an edge of the paper traces out a portion of a parabola.

**DEFINITION Elliptic Cylinder**

The graph of an **elliptic cylinder** has an equation of the form

$$\frac{x^2}{a^2} + \frac{y^2}{b^2} = 1 \qquad a > 0, b > 0$$

See Figure 71. The trace of an elliptic cylinder in the $xy$-plane is an ellipse with its center at the origin. The surface is generated by moving a line along the ellipse $\frac{x^2}{a^2} + \frac{y^2}{b^2} = 1$ while remaining perpendicular to the plane containing the ellipse. An elliptic cylinder can be visualized by taking a piece of paper and attaching the two opposite edges in the shape of an ellipse.

If $a = b$, then the trace in the $xy$-plane is a circle, and the cylinder is a right circular cylinder with radius $a$.

**DEFINITION Hyperbolic Cylinder**

The graph of a **hyperbolic cylinder** has an equation of the form

$$\frac{x^2}{a^2} - \frac{y^2}{b^2} = 1 \qquad a > 0, b > 0$$

See Figure 72. The surface is generated by moving a line along one branch of the hyperbola $\frac{x^2}{a^2} - \frac{y^2}{b^2} = 1$ while remaining perpendicular to the plane containing the hyperbola and then repeating the procedure on the second branch. The intercepts are $(\pm a, 0, 0)$. The hyperbolic cylinder consists of two parts, neither of which touches the $z$-axis. A hyperbolic cylinder can be visualized by bending two pieces of paper so that the curved edges trace out a portion of a hyperbola.

**EXAMPLE 3  Identifying a Cylinder**

Identify the cylinder defined by the equation $y^2 = 8z$. Describe how it can be generated.

**Solution** $y^2 = 8z$ is an equation of a parabolic cylinder. Since $x$ is the missing variable, the cylinder is generated by moving a line that is perpendicular to the $yz$-plane along the parabola $y^2 = 8z$. See Figure 73 on page 790. ∎

NOW WORK Problem 11.

The quadric surfaces classified here are in standard position, with centers (or vertices) at the origin and symmetries with respect to the coordinate axes. When the center (or vertex) of a quadric surface is not located at the origin, but there is symmetry with respect to the lines parallel to the coordinate axes, a translation of axes may be applied to put the equation in standard form. The correct translation is obtained by completing squares. If the symmetry of the quadric surface is with respect to the lines

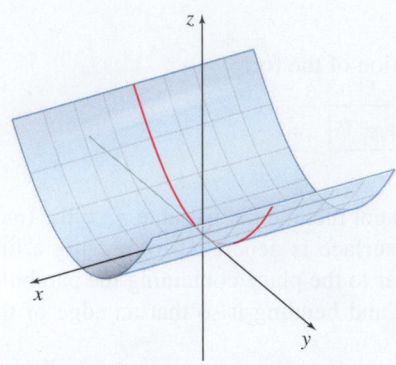

**Figure 73** Parabolic cylinder: $y^2 = 8z$

that are not parallel to the coordinate axes, a rotation of axes, perhaps followed by a translation of axes, is required to put the equation in standard form.

Using rotations and translations, the list of nine basic equations provides an exhaustive classification of quadric surfaces. Rotations in space are somewhat involved and are not discussed here.

## 4 Graph Quadric Surfaces

Quadric surfaces are best graphed using a CAS. Comparing the given equation to those discussed here helps to identify the graph. Here are some suggestions to help make the comparison:

- Remember that the equations of the quadric surfaces listed are given in standard form. Interchanging the variables does not affect the classification; it only affects the orientation of the quadric surface. For example, $z = \dfrac{x^2}{4} + \dfrac{y^2}{9}$ is the equation of an elliptic paraboloid whose graph lies above the $xy$-plane. The graph of $y = \dfrac{x^2}{9} + \dfrac{z^2}{4}$ is also an elliptic paraboloid, but its graph lies to the right of the $xz$-plane.
- Cylinders are characterized by a variable missing from the equation. For example, the graph of the equation $\dfrac{y^2}{4} + \dfrac{z^2}{9} = 1$ is an elliptic cylinder perpendicular to the $yz$-plane.
- The technique of completing the square is sometimes required to identify the given quadric surface. In this case, a translation of axes will help to find the correct location of the graph.

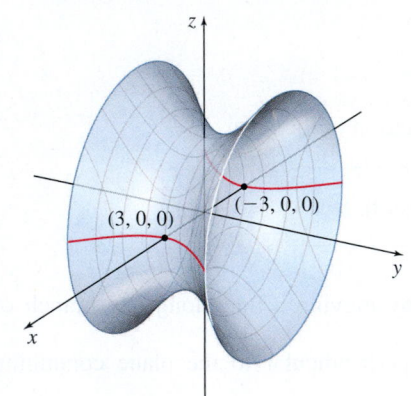

**Figure 74** $\dfrac{x^2}{9} - \dfrac{y^2}{2} + \dfrac{z^2}{4} = 1$

**CAS** **EXAMPLE 4**  **Graphing Quadric Surfaces Using a CAS**

Identify each equation and use a CAS to obtain its graph.

**(a)** $4x^2 - 18y^2 + 9z^2 = 36$    **(b)** $4y^2 + 9z^2 - 36x^2 = 0$    **(c)** $x^2 + z^2 - 4x = 0$

**Solution (a)** Divide both sides of the equation by 36, obtaining

$$\frac{x^2}{9} - \frac{y^2}{2} + \frac{z^2}{4} = 1$$

This equation has three squared variables, one term negative, with 1 on the right, so it represents a hyperboloid of one sheet. Since the coefficient of $y^2$ is negative, the $y$-axis is the axis of the hyperboloid of one sheet. See Figure 74.

The intercepts are $(3, 0, 0)$, $(-3, 0, 0)$, $(0, 0, 2)$, and $(0, 0, -2)$. There are no $y$-intercepts. The trace in the $xz$-plane is the ellipse $\dfrac{x^2}{9} + \dfrac{z^2}{4} = 1$, and all traces parallel to the $xz$-plane are also ellipses. The trace in the $xy$-plane is the hyperbola $\dfrac{x^2}{9} - \dfrac{y^2}{2} = 1$. The trace in the $yz$-plane is the hyperbola $-\dfrac{y^2}{2} + \dfrac{z^2}{4} = 1$. Traces parallel to the $xy$- and $yz$-planes are also hyperbolas. The $y$-axis is the axis of the hyperboloid of one sheet.

**(b)** We begin by writing the equation as

$$x^2 = \frac{y^2}{9} + \frac{z^2}{4}$$

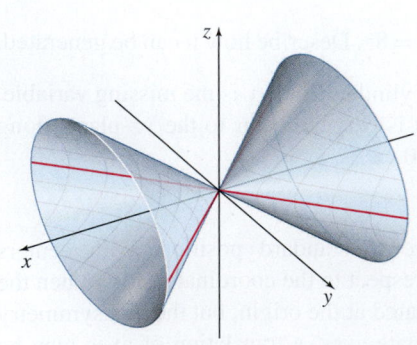

**Figure 75** $x^2 = \dfrac{y^2}{9} + \dfrac{z^2}{4}$

Observe that the equation has three variables, all of which are squared, and no constant. This is the equation of an elliptic cone. See Figure 75.

The vertex of the cone is at the origin $(0, 0, 0)$.

The trace of the elliptic cone in the $yz$-plane is the origin. Traces parallel to the $yz$-plane are ellipses centered on the $x$-axis. The trace in the $xy$-plane consists of the pair of intersecting lines $x = \pm\dfrac{y}{3}$, and the trace in the $xz$-plane consists of the pair of intersecting lines $x = \pm\dfrac{z}{2}$.

**(c)** Since the $y$ variable is missing, the surface is a cylinder. To identify the cylinder, we complete the square in $x$:

$$x^2 + z^2 - 4x = 0$$
$$(x - 2)^2 + z^2 = 4$$

This is the equation of a right circular cylinder with a radius 2. The trace in the $xz$-plane is a circle of radius 2 and with its center at the point $(2, 0, 0)$. See Figure 76. ■

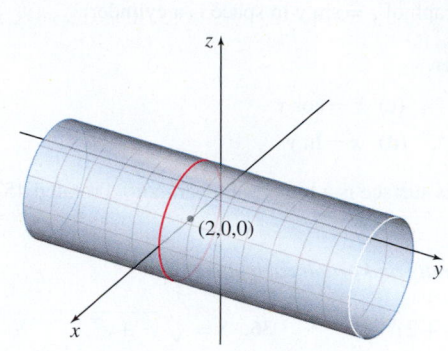

**Figure 76** $(x - 2)^2 + z^2 = 4$

**NOW WORK** Problems **37** and **45.**

## 10.7 Assess Your Understanding

### Concepts and Vocabulary

**1.** *Multiple Choice*   The trace in the $xy$-plane of the graph of the equation $\dfrac{x^2}{2} - \dfrac{y^2}{3} + \dfrac{z^2}{4} = 1$ is

$\left[ \textbf{(a)}\ \dfrac{x^2}{2} + \dfrac{z^2}{4} = 1\ \textbf{(b)}\ \dfrac{x^2}{2} - \dfrac{y^2}{3} = 0\ \textbf{(c)}\ \dfrac{x^2}{2} - \dfrac{y^2}{3} = 1 \right]$

**2.** The intercept(s) of the graph of $\dfrac{x^2}{4} - \dfrac{y^2}{9} - z = 4$ is (are) _____.

**3.** *True or False*   A cylinder is formed when a line moves along a plane curve while remaining perpendicular to the plane containing the curve.

**4.** *Multiple Choice*   The quadric surface $z^2 = x^2 + \dfrac{y^2}{4}$ is called a(n)

[**(a)** elliptic cylinder **(b)** elliptic cone **(c)** elliptic paraboloid **(d)** hyperboloid].

**5.** The quadric surface $y^2 - x^2 = 4$ is called a(n) _____.

**6.** The point $(0, 0, 0)$ on the hyperbolic paraboloid $z = \dfrac{y^2}{2^2} - \dfrac{x^2}{5^2}$ is called a(n) _____.

### Skill Building

*In Problems 7–12, use the Figures A–F to match each graph to an equation.*

**7.** $z = 4y^2 - x^2$

**8.** $2z = x^2 + 4y^2$

**9.** $2x^2 + y^2 - z^2 = 1$

**10.** $2x^2 + y^2 + 3z^2 = 1$

**11.** $y^2 = 4x$

**12.** $x^2 - z^2 = y$

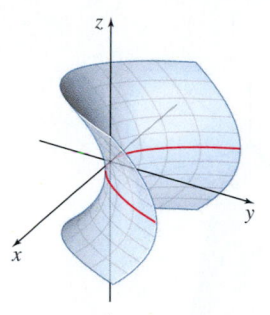

**A**

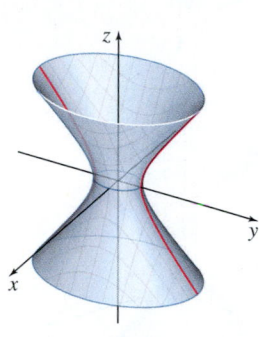

**B**

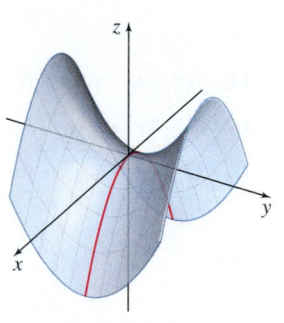

**C**

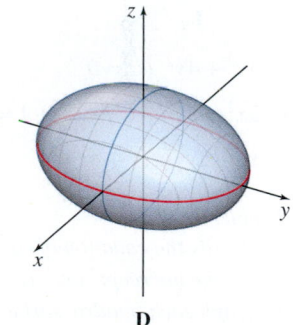

**D**

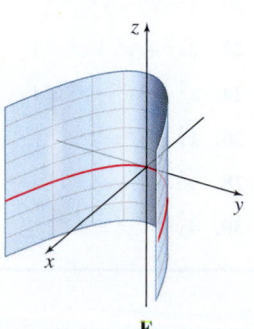

**E**

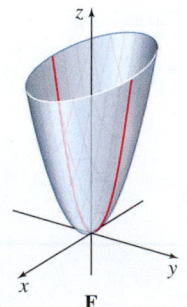

**F**

**1.** = NOW WORK problem         = Graphing technology recommended        CAS = Computer Algebra System recommended

*Figures G–L are graphs of quadric surfaces. In Problems 13–18, match each equation with its graph.*

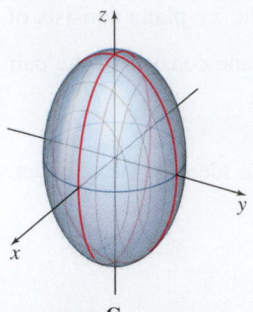

**G**

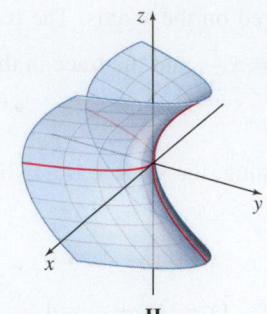

**H**

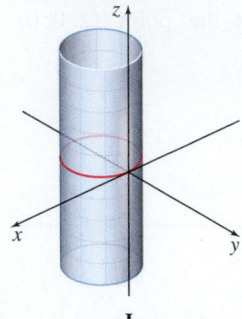

**I**

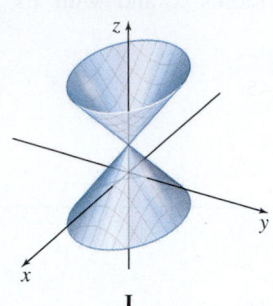

**J**

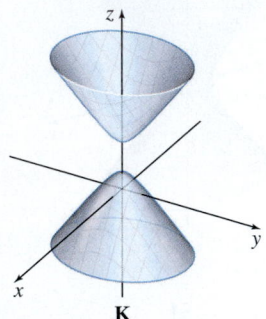

**K**

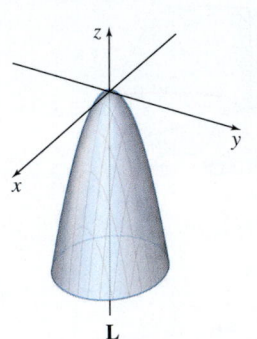

**L**

**13.** $3x^2 + 4y^2 + z = 0$

**14.** $3x^2 + 4y^2 + 4y = 0$

**15.** $3x^2 + 2y^2 - (z - 2)^2 + 1 = 0$

**16.** $z^2 - 4x^2 = 3y$

**17.** $x^2 + 2y^2 - z^2 + 4z = 4$

**18.** $3x^2 + 3y^2 + z^2 = 1$

*In Problems 19–30:*
*(a) Identify the equation of each quadric surface.*
*(b) List the intercepts and traces.*
*(c) Graph each quadric surface.*

**19.** $z = x^2 + y^2$

**20.** $z = x^2 - y^2$

**21.** $4x^2 + y^2 + 4z^2 = 4$

**22.** $2x^2 + y^2 + z^2 = 1$

**23.** $z^2 = x^2 + 2y^2$

**24.** $x^2 + 2y^2 - z^2 = 1$

**25.** $x = 4z^2$

**26.** $x^2 + y^2 = 1$

**27.** $x^2 + 2y^2 - z^2 = -4$

**28.** $y^2 - x^2 = 4$

**29.** $2x = y^2$

**30.** $4y^2 - x^2 = 1$

## Applications and Extensions

**31.** Explain why the graph of $xy = 1$ in space is a cylinder.

**32.** Explain why the graph of $z = \sin y$ in space is a cylinder.

**33.** Graph each cylinder.

    (a) $xy = 1$         (c) $z = \cos y$

    (b) $z = \sin y$     (d) $x = \ln y$

**CAS** **34.** Graph $z = xy$. (This surface is a hyperbolic paraboloid rotated $45°$ about the $z$-axis.)

*In Problems 35–42, identify and graph each equation.*

**35.** $(x - 4)^2 = \dfrac{y^2}{4} + (z + 2)^2$

**36.** $x = \sqrt{y^2 + z^2}$

**37.** $x^2 + (y + 1)^2 = 1$

**38.** $x^2 = 12z$

**39.** $y = (x - 2)^2 + (z + 1)^2$

**40.** $2z = \sqrt{4 - x^2 - y^2}$

**41.** $y = \sqrt{x^2 + z^2 + 1}$

**42.** $y = \sqrt{x^2 + z^2 - 1}$

*In Problems 43–48, identify each quadric surface by completing the squares and putting the equation in standard form.*

**43.** $x^2 + y^2 + z^2 - 2x + 8y - 4 = 0$

**44.** $4x^2 = y^2 + 4z^2 - 24z + 36$

**45.** $4x = y^2 - 2y + z^2 + 1$

**46.** $16z = x^2 + 4x + 4y^2 + 4$

**47.** $4x^2 - 8x + y^2 + 4y - z^2 + 2 = 0$

**48.** $x^2 + 4x - 4y^2 + 4z^2 - 8z = -12$

**49.** Find an equation of the surface consisting of all points that are equidistant from the point $(0, 3, 0)$ and containing the origin.

**50.** Find an equation of the surface consisting of all points that are equidistant from the point $(2, 0, 0)$ and the plane $x = -2$.

## Challenge Problem

**51.** Graph the region bounded by the surface $z = x^2 + y^2$ and the cylinder $x^2 + y^2 = 4$, $0 \le z \le 4$.

**52.** Show that through each point on the hyperboloid of one sheet

$$\frac{x^2}{a^2} + \frac{y^2}{b^2} - \frac{z^2}{c^2} = 1$$

there are two lines lying entirely on the surface.

*Hint:* Write the equation as $\dfrac{x^2}{a^2} - \dfrac{z^2}{c^2} = 1 - \dfrac{y^2}{b^2}$ and factor.

# Chapter Review

## THINGS TO KNOW

### 10.1 Rectangular Coordinates in Space
- The right-hand rule (p. 734)
- The distance formula in space: (p. 735)

$$|P_1 P_2| = \sqrt{(x_2 - x_1)^2 + (y_2 - y_1)^2 + (z_2 - z_1)^2}$$

- The equation of a sphere: (p. 736)

$$(x - x_0)^2 + (y - y_0)^2 + (z - z_0)^2 = R^2$$

Center: $(x_0, y_0, z_0)$; Radius: $R$

### 10.2 Introduction to Vectors
- Vector (p. 738)   • Scalar (p. 738)

#### Properties of vector addition: (p. 740)
- Commutative property:  $\mathbf{w} + \mathbf{v} = \mathbf{v} + \mathbf{w}$
- Associative property:  $\mathbf{u} + (\mathbf{v} + \mathbf{w}) = (\mathbf{u} + \mathbf{v}) + \mathbf{w}$
- Additive identity property:  $\mathbf{v} + \mathbf{0} = \mathbf{0} + \mathbf{v} = \mathbf{v}$
- Additive inverse property:  $\mathbf{v} + (-\mathbf{v}) = (-\mathbf{v}) + \mathbf{v} = \mathbf{0}$

#### Properties of scalar multiplication: (p. 740)
If $\mathbf{v}$ and $\mathbf{w}$ are any two vectors, and $a$ and $b$ are scalars, then:

- $0\mathbf{v} = \mathbf{0}$
- $1\mathbf{v} = \mathbf{v}$
- $(-1)\mathbf{v} = -\mathbf{v}$
- $(a + b)\mathbf{v} = a\mathbf{v} + b\mathbf{v}$
- $a(\mathbf{v} + \mathbf{w}) = a\mathbf{v} + a\mathbf{w}$
- $a(b\mathbf{v}) = (ab)\mathbf{v}$

### 10.3 Vectors in the Plane and in Space
- Position vector (p. 743)
- Two vectors are equal if and only if their components are equal. (p. 744)
- Two distinct nonzero vectors $\mathbf{v}$ and $\mathbf{w}$ are parallel if there is a nonzero scalar $a$ for which $\mathbf{v} = a\,\mathbf{w}$. (p. 745)
- Magnitude of a vector $\mathbf{v}$ (p. 746)
  in the plane:

$$\|\mathbf{v}\| = \sqrt{v_1^2 + v_2^2} \quad \text{where } \mathbf{v} = v_1\mathbf{i} + v_2\mathbf{j}$$

  in space:

$$\|\mathbf{v}\| = \sqrt{v_1^2 + v_2^2 + v_3^2} \quad \text{where } \mathbf{v} = v_1\mathbf{i} + v_2\mathbf{j} + v_3\mathbf{k}$$

- A unit vector $\mathbf{u}$ has magnitude 1; that is, $\|\mathbf{u}\| = 1$. (p. 747)
- For any nonzero vector $\mathbf{v}$, $\mathbf{u} = \dfrac{\mathbf{v}}{\|\mathbf{v}\|}$ is a unit vector that has the same direction as $\mathbf{v}$. (p. 747)
- Standard basis vectors (p. 748)
  in the plane:

$$\mathbf{i} = \langle 1, 0 \rangle \quad \mathbf{j} = \langle 0, 1 \rangle$$

  in space:

$$\mathbf{i} = \langle 1, 0, 0 \rangle \quad \mathbf{j} = \langle 0, 1, 0 \rangle \quad \mathbf{k} = \langle 0, 0, 1 \rangle$$

- If $\mathbf{v}$ is a nonzero vector in the plane, then $\mathbf{v} = \|\mathbf{v}\| \, (\cos\theta\,\mathbf{i} + \sin\theta\,\mathbf{j})$, where $\theta$ is the angle between $\mathbf{v}$ and the positive $x$-axis. (p. 748)

### 10.4 The Dot Product
#### Dot product: (pp. 754)
- If $\mathbf{v} = v_1\mathbf{i} + v_2\mathbf{j}$ and $\mathbf{w} = w_1\mathbf{i} + w_2\mathbf{j}$, then

$$\mathbf{v} \cdot \mathbf{w} = v_1 w_1 + v_2 w_2$$

- If $\mathbf{v} = v_1\mathbf{i} + v_2\mathbf{j} + v_3\mathbf{k}$ and $\mathbf{w} = w_1\mathbf{i} + w_2\mathbf{j} + w_3\mathbf{k}$, then

$$\mathbf{v} \cdot \mathbf{w} = v_1 w_1 + v_2 w_2 + v_3 w_3$$

#### Properties of the dot product: (p. 755)
- Magnitude property:  $\mathbf{v} \cdot \mathbf{v} = \|\mathbf{v}\|^2$
- Commutative property:  $\mathbf{u} \cdot \mathbf{v} = \mathbf{v} \cdot \mathbf{u}$
- Distributive property:  $\mathbf{u} \cdot (\mathbf{v} + \mathbf{w}) = \mathbf{u} \cdot \mathbf{v} + \mathbf{u} \cdot \mathbf{w}$
- Linearity property:  $a(\mathbf{u} \cdot \mathbf{v}) = (a\mathbf{u}) \cdot \mathbf{v}$
- Zero-vector property:  $\mathbf{0} \cdot \mathbf{v} = 0$
- The angle $\theta$ between two nonzero vectors $\mathbf{v}$ and $\mathbf{w}$:

$$\cos\theta = \frac{\mathbf{v} \cdot \mathbf{w}}{\|\mathbf{v}\| \, \|\mathbf{w}\|} \quad 0 \le \theta \le \pi \text{ (p. 756)}$$

- Two nonzero vectors $\mathbf{v}$ and $\mathbf{w}$ are orthogonal if and only if

$$\mathbf{v} \cdot \mathbf{w} = 0 \text{ (p. 757)}$$

- Direction angles $\alpha$, $\beta$, $\gamma$ of a nonzero vector

$$\mathbf{v} = v_1\mathbf{i} + v_2\mathbf{j} + v_3\mathbf{k} \text{ (p. 757)}$$

$$\cos\alpha = \frac{v_1}{\sqrt{v_1^2 + v_2^2 + v_3^2}} = \frac{v_1}{\|\mathbf{v}\|}$$

$$\cos\beta = \frac{v_2}{\sqrt{v_1^2 + v_2^2 + v_3^2}} = \frac{v_2}{\|\mathbf{v}\|}$$

$$\cos\gamma = \frac{v_3}{\sqrt{v_1^2 + v_2^2 + v_3^2}} = \frac{v_3}{\|\mathbf{v}\|}$$

- For a vector $\mathbf{v}$ in space: $\mathbf{v} = \|\mathbf{v}\| \, [\cos\alpha\,\mathbf{i} + \cos\beta\,\mathbf{j} + \cos\gamma\,\mathbf{k}]$ (p. 758)
- Decomposition of a nonzero vector $\mathbf{v}$: (p. 759)

$$\text{Projection of } \mathbf{v} \text{ onto } \mathbf{w} \text{ is } \mathbf{v}_1 = \text{proj}_{\mathbf{w}}\mathbf{v} = \frac{\mathbf{v} \cdot \mathbf{w}}{\|\mathbf{w}\|^2}\mathbf{w}$$

$$\text{The vector } \mathbf{v}_2 \text{ orthogonal to } \mathbf{w} \text{ is } \mathbf{v}_2 = \mathbf{v} - \mathbf{v}_1$$

- Work $W = \mathbf{F} \cdot \overrightarrow{AB}$ (p. 759)

### 10.5 The Cross Product
#### Cross product: (p. 764)
If $\mathbf{v} = v_1\mathbf{i} + v_2\mathbf{j} + v_3\mathbf{k}$ and $\mathbf{w} = w_1\mathbf{i} + w_2\mathbf{j} + w_3\mathbf{k}$, then
$\mathbf{v} \times \mathbf{w} = (v_2 w_3 - v_3 w_2)\mathbf{i} - (v_1 w_3 - v_3 w_1)\mathbf{j} + (v_1 w_2 - v_2 w_1)\mathbf{k}$.

#### Algebraic properties of the cross product: (p. 766)
If $\mathbf{u}$, $\mathbf{v}$, and $\mathbf{w}$ are vectors in space and if $a$ is a scalar, then
- $\mathbf{v} \times \mathbf{v} = \mathbf{0}$
- $\mathbf{v} \times \mathbf{w} = -(\mathbf{w} \times \mathbf{v})$
- $a(\mathbf{v} \times \mathbf{w}) = (a\mathbf{v}) \times \mathbf{w} = \mathbf{v} \times (a\mathbf{w})$
- $\mathbf{v} \times (\mathbf{w} + \mathbf{u}) = (\mathbf{v} \times \mathbf{w}) + (\mathbf{v} \times \mathbf{u})$
- $\|\mathbf{v} \times \mathbf{w}\|^2 = \|\mathbf{v}\|^2 \, \|\mathbf{w}\|^2 - (\mathbf{v} \cdot \mathbf{w})^2$
- $\mathbf{u} \cdot (\mathbf{v} \times \mathbf{w}) = (\mathbf{u} \times \mathbf{v}) \cdot \mathbf{w}$

#### Geometric properties of the cross product: (p. 767)
If $\mathbf{v}$ and $\mathbf{w}$ are nonzero vectors in space, then
- $\mathbf{v} \times \mathbf{w}$ is orthogonal to both $\mathbf{v}$ and $\mathbf{w}$
- $\|\mathbf{v} \times \mathbf{w}\| = \|\mathbf{v}\|\|\mathbf{w}\| \sin\theta$, where $\theta$ is the angle between $\mathbf{v} \ne \mathbf{0}$ and $\mathbf{w} \ne \mathbf{0}$.
- $\|\mathbf{v} \times \mathbf{w}\|$ is the area of the parallelogram having $\mathbf{v} \ne \mathbf{0}$ and $\mathbf{w} \ne \mathbf{0}$ as adjacent sides.

- $\mathbf{v} \times \mathbf{w} = \mathbf{0}$ if and only if $\mathbf{v}$ and $\mathbf{w}$ are parallel vectors.
- $|\mathbf{u} \cdot (\mathbf{v} \times \mathbf{w})|$ is the volume of a parallelepiped having $\mathbf{u}$, $\mathbf{v}$, and $\mathbf{w}$ as adjacent sides.

**Angular Velocity** $\mathbf{v} = \boldsymbol{\omega} \times \mathbf{r}$   (p. 769)

- Distance $d$ from the point $(x_0, y_0, z_0)$ to the plane $Ax + By + Cz = D$:

$$d = \frac{|Ax_0 + By_0 + Cz_0 - D|}{\sqrt{A^2 + B^2 + C^2}} \quad \text{(p. 779)}$$

## 10.6 Equations of Lines and Planes in Space

***Lines:*** The direction of a line is $\mathbf{D} = a\mathbf{i} + b\mathbf{j} + c\mathbf{k} \neq \mathbf{0}$ and $\mathbf{r}_0 = x_0\mathbf{i} + y_0\mathbf{j}$ is the position vector of a point on the line.

- Vector equation of the line:  $\mathbf{r} = \mathbf{r}_0 + t\mathbf{D}$ (p. 773)
- Parametric equations of the line  (p. 773)
  $x = x_0 + at, \; y = y_0 + bt, \; z = z_0 + ct$
- Symmetric equations of the line (p. 774)

$$\frac{x - x_0}{a} = \frac{y - y_0}{b} = \frac{z - z_0}{c} \qquad abc \neq 0$$

- Distance $d$ from a point $P$ to a line $\mathbf{r} = \mathbf{r}(t) = \mathbf{r}_0 + t\mathbf{D}, \mathbf{D} \neq \mathbf{0}$, where $P_0$ is a point on $\mathbf{r}$, is given by

$$d = \frac{\|\overrightarrow{P_0 P} \times \mathbf{D}\|}{\|\mathbf{D}\|} \quad \text{(p. 780)}$$

***Planes:*** $\mathbf{r}_0$ is the position vector of a point on the plane and $\mathbf{N} = a\mathbf{i} + b\mathbf{j} + c\mathbf{k} \neq \mathbf{0}$ is a normal vector to the plane.

- Vector equation of the plane: $(\mathbf{r} - \mathbf{r}_0) \cdot \mathbf{N} = 0$  (p. 777)
- General equation of the plane: $Ax + By + Cz = D$, where at least one number $A, B, C$ is not 0. (p. 777)

## 10.7 Quadric Surfaces

- Ellipsoid:  $\dfrac{x^2}{a^2} + \dfrac{y^2}{b^2} + \dfrac{z^2}{c^2} = 1$   (p. 784)

- Elliptic cone:   $z^2 = \dfrac{x^2}{a^2} + \dfrac{y^2}{b^2}$   (p. 785)

- Elliptic paraboloid:   $z = \dfrac{x^2}{a^2} + \dfrac{y^2}{b^2}$   (p.785)

- Hyperbolic paraboloid:   $z = \dfrac{y^2}{b^2} - \dfrac{x^2}{a^2}$   (p. 786)

- Hyperboloid of one sheet:   $\dfrac{x^2}{a^2} + \dfrac{y^2}{b^2} - \dfrac{z^2}{c^2} = 1$   (p. 787)

- Hyperboloid of two sheets:   $\dfrac{x^2}{a^2} + \dfrac{y^2}{b^2} - \dfrac{z^2}{c^2} = -1$   (p. 787)

- Parabolic cylinder:   $x^2 = 4ay$   (p. 789)

- Elliptic cylinder:   $\dfrac{x^2}{a^2} + \dfrac{y^2}{b^2} = 1$   (p. 789)

- Hyperbolic cylinder:   $\dfrac{x^2}{a^2} - \dfrac{y^2}{b^2} = 1$   (p. 789)

## OBJECTIVES

| Section | You should be able to ... | Examples | Review Exercises |
|---|---|---|---|
| 10.1 | 1 Locate points in space (p. 734) | | 1, 2 |
| | 2 Find the distance between two points in space (p. 735) | 1 | 5, 24 |
| | 3 Find the equation of a sphere (p. 736) | 2, 3 | 6, 7, 24 |
| 10.2 | 1 Represent vectors geometrically (p. 738) | 1 | 27 |
| | 2 Use properties of vectors (p. 741) | 2 | 27 |
| 10.3 | 1 Represent a vector algebraically (p. 742) | 1 | 3(a) |
| | 2 Add, subtract, and find scalar multiples of vectors (p. 744) | 2, 3 | 8, 12, 16 |
| | 3 Find the magnitude of a vector (p. 746) | 4 | 4, 9 |
| | 4 Find a unit vector (p. 747) | 5–7 | 3(b), 10 |
| | 5 Find a vector in the plane from its direction and magnitude (p. 748) | 8, 9 | 11 |
| 10.4 | 1 Find the dot product of two vectors (p. 754) | 1 | 14 |
| | 2 Find the angle between two vectors (p. 755) | 2, 3 | 17, 23 |
| | 3 Determine whether two vectors are orthogonal (p. 757) | 4, 5 | 23 |
| | 4 Express a vector in space using its magnitude and direction (p. 757) | 6 | 13 |
| | 5 Find the projection of a vector (p. 758) | 7 | 14 |
| | 6 Compute work (p. 759) | 8, 9 | 15 |
| 10.5 | 1 Find the cross product of two vectors (p. 765) | 1–3 | 21, 31 |
| | 2 Prove algebraic properties of the cross product (p. 766) | | |
| | 3 Use geometric properties of the cross product (p. 767) | 4–6 | 20 |
| 10.6 | 1 Find a vector equation of a line in space (p. 772) | 1 | 21, 22, 25, 32 |
| | 2 Find parametric equations of a line in space (p. 773) | 2 | 21, 22, 25, 32 |
| | 3 Find symmetric equations of a line in space (p. 774) | 3–5 | 21, 22, 25, 32 |
| | 4 Determine whether two distinct lines are skew, parallel, or intersecting (p. 775) | 6 | 28, 30 |
| | 5 Find an equation of a plane (p. 776) | 7, 8 | 26, 29 |
| | 6 Determine whether two distinct planes are parallel or intersecting (p. 778) | 9 | 25 |
| | 7 Find the distance from a point to a plane and from a point to a line (p. 779) | 10–12 | 18, 19 |
| 10.7 | 1 Identify quadric surfaces based on an ellipse  (p. 784) | 1 | 36, 37 |
| | 2 Identify quadric surfaces based on a hyperbola  (p. 786) | 2 | 34, 35 |
| | 3 Identify cylinders (p. 788) | 3 | 33 |
| | 4 Graph quadric surfaces (p. 790) | 4 | 33(c)–37(c) |

## REVIEW EXERCISES

**1.** Opposite vertices of a rectangular box whose edges are parallel to the coordinate axes are $(1, 0, 2)$ and $(2, 3, 4)$. List the coordinates of the other six vertices of the box.

**2.** Describe in words the set of all points $(x, y, z)$ that satisfy $y = 3$.

**3.** The vector $\mathbf{v}$ has initial point $P_1 = (5, 0, -2)$ and terminal point $P_2 = (7, 1, 3)$.

   **(a)** Write the vector $\mathbf{v}$ in terms of its components.

   **(b)** Write the vector $\mathbf{v}$ in terms of standard basis vectors.

**4.** Find the magnitude of the vector $\mathbf{v} = \mathbf{i} + 2\mathbf{j} + 3\mathbf{k}$.

**5.** Find the distance between the points $(1, 2, 3)$ and $(7, 5, 1)$.

**6.** Write the equation of the sphere with center at the point $(1, 2, 3)$ and radius 4.

**7.** Find the radius and center of the sphere

$$x^2 + y^2 + z^2 - 4x + 8y = 5$$

**8.** If $\mathbf{v} = \langle -2, -1, 3\rangle$ and $\mathbf{w} = \langle 5, 4, -2\rangle$, find $3\mathbf{v} - 2\mathbf{w}$.

**9.** If $\mathbf{v} = 2\mathbf{i} + 3\mathbf{j} - \mathbf{k}$ and $\mathbf{w} = -\mathbf{i} - 2\mathbf{j} + 3\mathbf{k}$, find $\|\mathbf{v} + \mathbf{w}\|$.

**10.** Find a unit vector in the direction of $\mathbf{v} = 4\mathbf{i} + 12\mathbf{j} - 3\mathbf{k}$.

**11.** Find a vector $\mathbf{v}$ in the plane with magnitude 25 if the angle between $\mathbf{v}$ and $\mathbf{i}$ is $120°$.

**12.** Find $\mathbf{v}$ and $\mathbf{w}$ if

$$\mathbf{w} - 3\mathbf{v} = 2\mathbf{i} + \mathbf{j} - \mathbf{k} \quad \text{and} \quad 2\mathbf{w} + \mathbf{v} = \mathbf{i} + \mathbf{j}$$

**13.** **(a)** Find the magnitude and direction cosines of the vector $\mathbf{v} = 5\mathbf{i} + 8\mathbf{j} - 2\mathbf{k}$.

   **(b)** Write $\mathbf{v}$ in terms of its magnitude and direction cosines.

**14.** Find the vector projection of $\mathbf{v} = 3\mathbf{i} + \mathbf{j} - 2\mathbf{k}$ onto $\mathbf{w} = 5\mathbf{i} - \mathbf{j} + \mathbf{k}$. Decompose $\mathbf{v}$ into two vectors $\mathbf{v}_1$ and $\mathbf{v}_2$, where $\mathbf{v}_1$ is parallel to $\mathbf{w}$ and $\mathbf{v}_2$ is orthogonal to $\mathbf{w}$.

**15.** Find the work done in moving an object along a vector $\mathbf{u} = 3\mathbf{i} + 2\mathbf{j} - 5\mathbf{k}$ if the applied force is $\mathbf{F} = 2\mathbf{i} - \mathbf{j} - \mathbf{k}$. Use meters for distance and newtons for force.

**16.** Three forces are applied to an object in mutually orthogonal directions: 12 units along the $x$-axis, 16 units along the $y$-axis, and 15 units along the $z$-axis. What is the resultant force?

**17.** Find the angle between the vectors $\mathbf{u} = \mathbf{i} + 2\mathbf{j} + 3\mathbf{k}$ and $\mathbf{v} = 2\mathbf{i} - 3\mathbf{j} - \mathbf{k}$.

**18.** **(a)** Find the distance from the point $(1, 2, 1)$ to the plane $3x + 6y + 2z = 2$.

   **(b)** Find the distance from the point $(3, 3, -2)$ to the line $\mathbf{r}(t) = (2 + t)\mathbf{i} + (3 - 2t)\mathbf{j} - (4 - 2t)\mathbf{k}$.

**19.** Find the distance between the parallel planes

$$p_1: \quad x + y - z = 1$$
$$p_2: \quad -2x - 2y + 2z = 2$$

**20.** A rigid body rotates with constant angular speed of $\omega$ radians per second about a line through the origin and parallel to $2\mathbf{i} + 3\mathbf{j} + 6\mathbf{k}$. Find the speed of an object on this body at the instant it passes through the point $(1, 2, 3)$. Assume the distance is in meters.

**21.** Find an equation of the line perpendicular to the lines

$$l: \quad x = 1 - t, \qquad y = t, \qquad z = 2t - 1$$
$$m: \quad x = t + 1, \qquad y = -t, \qquad z = t - 1$$

**22.** Find an equation of the line containing the point $(1, 2, 3)$ in the direction of $\mathbf{v} = 2\mathbf{i} - \mathbf{j} - 4\mathbf{k}$.

**23.** Find the scalar $a$ so that $\mathbf{v} = 2\mathbf{i} - \mathbf{j} - 4\mathbf{k}$ and $\mathbf{w} = a\mathbf{i} + a\mathbf{j} + 3\mathbf{k}$ are orthogonal.

**24.** Find an equation of the sphere with $(2, 2, -3)$ and $(-2, 6, 5)$ as endpoints of a diameter. Graph the sphere.

**25.** Find an equation of the line of intersection of the two planes

$$p_1: x - y + z = 3 \qquad p_2: 2x + 2y - z = 1$$

**26.** Find an equation of the plane with nonzero intercepts $(a, 0, 0)$, $(0, b, 0)$, $(0, 0, c)$, where $abc \neq 0$.

**27.** Use the vectors in the figure to graph each of the following vectors:

   **(a)** $(2 + 3)\mathbf{v}$     **(b)** $2(\mathbf{u} - \mathbf{w})$     **(c)** $(\mathbf{u} + \mathbf{v}) + \mathbf{w}$

   **(d)** $\mathbf{v} + (-\mathbf{v})$     **(e)** $\mathbf{u} - 3\mathbf{v}$

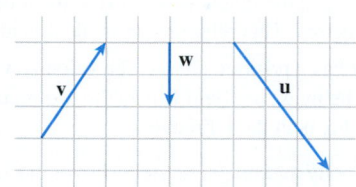

**28.** Determine whether the lines $l_1$ and $l_2$ intersect, are parallel, or are skew.

$$l_1: \frac{x + 5}{6} = \frac{y - 2}{3} = \frac{z + 4}{7}$$
$$l_2: \frac{x + 5}{6} = \frac{y + 1}{3} = \frac{z - 2}{7}$$

**29.** Find the general equation of the plane containing the point $(2, -1, 4)$ and normal to the line $\dfrac{x + 3}{2} = \dfrac{y - 3}{3} = z$.

**30.** Determine whether the lines $l_1$ and $l_2$ intersect, are parallel, or are skew.

$$l_1: x = t + 1, \qquad y = 3t, \qquad z = t - 1$$
$$l_2: \frac{x + 3}{2} = \frac{y - 3}{3} = z$$

**31.** Find an equation of the line normal to the plane containing the lines

$$x - 2 = \frac{y + 1}{2} = \frac{z - 1}{2} \quad \text{and} \quad x + 1 = \frac{y + 8}{3} = \frac{z}{-3}$$

at their point of intersection.

32. Find an equation of the line containing the points $P_1 = (4, 0, 7)$ and $P_2 = (5, 3, 6)$.

*In Problems 33–37:*

(a) *Identify each quadric surface.*

(b) *List its intercepts and traces.*

CAS (c) *Graph each quadric surface.*

33. $\dfrac{x^2}{4} - \dfrac{y^2}{9} = 1$

34. $x^2 + y^2 - \dfrac{z^2}{4} = -1$

35. $z = \dfrac{x^2}{4} - \dfrac{y^2}{9}$

36. $z^2 = \dfrac{x^2}{4} + \dfrac{y^2}{9}$

37. $\dfrac{x^2}{4} + \dfrac{y^2}{9} + z^2 = 1$

---

## CHAPTER 10 PROJECT    The Hall Effect

Friso Gentsch/picture-alliance/dpa/AP Images

While investigating the effects of magnets on electric current in 1879, E. H. Hall observed a phenomenon that is now used to measure the density of electric charge carriers in semiconductors. The **Hall effect** describes the creation of a voltage, or potential difference, on opposite sides of a conductor in the presence of an external magnetic field, created when an electric current flows through it. The magnetic field exerts a force on the moving carriers that tends to push them to one side of the conductor. The buildup of charges at the sides of the conductor balances the magnetic influence and produces a voltage between the two sides of the conductor.

The geometry of this situation is based on the relationships among the initial velocity $\mathbf{v}_0$ of the charges, the magnetic field $\mathbf{B}$, and the resulting magnetic force $\mathbf{F}_m$ on the charges.

When an electric charge $q$ moves in the presence of a magnetic field, it experiences a magnetic force given by $\mathbf{F}_m = q\mathbf{v} \times \mathbf{B}$. Note that because the magnetic force is proportional to the cross product of the velocity $\mathbf{v}$ and the magnetic field $\mathbf{B}$, it must be orthogonal to each of these vectors. This relationship is illustrated in Figure 77.

Assuming $q$ has a positive charge, the velocity $\mathbf{v}$ is in the $\mathbf{j}$ direction, and the magnetic field $\mathbf{B}$ is in the $\mathbf{k}$ direction, then the resulting force $\mathbf{F}_m$ is in the $\mathbf{i}$ direction.

1. Given the velocity and magnetic field shown in Figure 77, in what direction would the magnetic force point if $q$ has a negative charge?

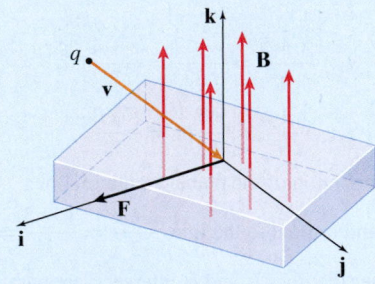

**Figure 77**

As charges continue to flow into the region of the magnetic field, they are deflected toward the side of the conductor in the direction of the magnetic force and collect there. At the same time, the opposite side of the material can be thought of as being deficient in the amount of charge carriers. This creates a voltage, or potential difference, between the two sides. A consequence of this is that a static electric field is created within the conductor. In the example depicted in Figure 77, the electric field $\mathbf{E}$ would point in the $-\mathbf{i}$ direction, as positive charges collect on the side of the material in the direction of $\mathbf{F}_m$. See Figure 78. Note that the magnetic field is still present as charges continue to build up and make the electric field stronger and stronger.

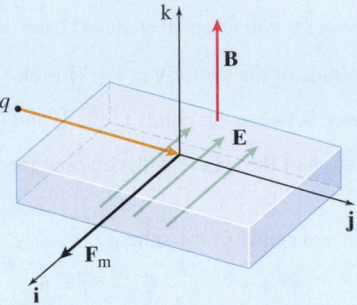

**Figure 78** The polarity resulting from positive charge carriers, as happens in many semiconductors.

You might ask, "Won't this electric field just keep increasing as charges continue to flow through the conductor and are deflected by the magnetic field?" This would be true if not for a competing effect—the electric field itself. The charges also experience an electric force, $\mathbf{F}_e$, as the electric field $\mathbf{E}$ is created and becomes stronger. The electric force on a charge in an electric field is given by the relationship $\mathbf{F}_e = q\mathbf{E}$. Note that the electric force on a positive charge would be in the same direction as the electric field and the directions of the magnetic and electric forces are opposed. Then as the electric field increases, subsequent charges moving into the material experience less and less of a net force, until the electric field becomes strong enough that subsequent charges experience no net force at all. When this happens, the potential difference across the conductor is established (by the electric field) and is measured. This potential difference is often referred to as the **Hall voltage** for the conductor.

2. A stream of electrons, each with a charge $q = -1.6 \times 10^{-19}$ C, has an initial velocity $\mathbf{v}_0$ in meters per second (m/s) of $\mathbf{v}_0 = (1.0 \times 10^{-4})\mathbf{i} + (2.0 \times 10^{-4})\mathbf{j} - (4.0 \times 10^{-4})\mathbf{k}$. The charges move along the length of a semiconducting material and enter a region of magnetic field measured in teslas (T) with $\mathbf{B} = 0.5\mathbf{i} + 1.0\mathbf{j} + 0.625\mathbf{k}$. Find the magnitude of the initial velocity and the magnetic field.

3. Find unit vectors in the direction of the velocity and the magnetic field.

4. Verify that the velocity $\mathbf{v}$ and the magnetic field $\mathbf{B}$ are orthogonal, as required for the Hall effect.

5. Find the magnetic force $\mathbf{F}_m$ on the electrons as they enter the region of the magnetic field. The units of the magnetic force are newtons (N).

6. The electric force $\mathbf{F}_e$ will grow until it is finally equal in magnitude to the magnetic force $\mathbf{F}_m$ but in the opposite direction. Write an expression for the final electric force. What are the units of the electric force?

7. Find the electric field $\mathbf{E}$ responsible for the electric force $\mathbf{F}_e$ found in Problem 6. Verify that $\mathbf{E}$ points in one of the two possible directions found in Problem 6. What are the units of the electric field?

# 11

# Vector Functions

supergenijalac/Getty Images

**CHAPTER 11 PROJECT** In the Chapter project on page 850, we investigate four different scenarios in which forces act on a vehicle to keep it on a road.

## How to Design a Safe Road

The Innerbelt Curve is a stretch of road in downtown Cleveland that has been given the ominous nickname of "Deadman's Curve." For decades Ohio transportation department officials have struggled to find ways to help drivers safely negotiate the sharp 90-degree curve, which has a posted speed limit of 35 mph, as they connect from one 50 mph roadway to another. Improvements have included changing the speed limit, posting warning signs, introducing rumble strips to create more friction, and modifying the bank of the curve, but accidents are still common.

Though perhaps among the more famous, Cleveland's Deadman's Curve is hardly the only dangerous curve among the country's interstates and highways. And while transportation officials and engineers can look at statistical evidence to identify places where accidents are abnormally frequent, engineers who are designing new exit ramps or roadways that curve around geographical features need to understand the forces that act on an automobile of a given size and moving at a given speed so that it can safely travel the road.

Until now the domain and range of a function have both been subsets of real numbers. Such functions are referred to as *real-valued functions of a real variable* or, simply, as *real functions*. In this chapter, we discuss functions whose domains remain a subset of real numbers, but whose range consists of vectors in the plane or vectors in space. Such functions are called *vector-valued functions of a real variable* or, simply, *vector functions*.

Vector functions are particularly useful for describing the motion of a particle as it moves through space. As a result, vector functions have diverse applications from describing Kepler's laws of planetary motion to modeling DNA to producing lifelike, animated films with computer graphics.

# 11.1 Vector Functions and Their Derivatives

**OBJECTIVES**  *When you finish this section, you should be able to:*

1  Find the domain of a vector function (p. 798)
2  Graph a vector function (p. 798)
3  Find the limit and determine the continuity of a vector function (p. 801)
4  Find the derivative of a vector function (p. 802)
5  Find the derivative of a vector function using derivative rules (p. 803)

A *vector function* is denoted by $\mathbf{r} = \mathbf{r}(t)$, where $t$ is a real number and $\mathbf{r}$ is a vector. It is customary to use $t$ as the independent variable, because in applications it often represents time.

> **DEFINITION  Vector Function**
>
> A **vector-valued function of a real variable**, or simply, a **vector function**, $\mathbf{r} = \mathbf{r}(t)$, is a function whose domain is a subset of the set of real numbers and whose range is a set of vectors.

For a vector function $\mathbf{r} = \mathbf{r}(t)$ in the plane whose domain is the closed interval $a \le t \le b$, the components of $\mathbf{r}$ are two real functions $x = x(t)$, $y = y(t)$, each defined on the interval $a \le t \le b$. So,

$$\boxed{\mathbf{r} = \mathbf{r}(t) = \langle x(t), y(t) \rangle = x(t)\mathbf{i} + y(t)\mathbf{j}}$$

For a vector function $\mathbf{r} = \mathbf{r}(t)$ in space whose domain is the closed interval $a \le t \le b$, the components of $\mathbf{r}$ are three real functions $x = x(t)$, $y = y(t)$, and $z = z(t)$, each defined on the interval $a \le t \le b$. So,

$$\boxed{\mathbf{r} = \mathbf{r}(t) = \langle x(t), y(t), z(t) \rangle = x(t)\mathbf{i} + y(t)\mathbf{j} + z(t)\mathbf{k}}$$

**NOTE**  As with vectors and scalars, the symbol for a vector function is written using boldface roman type and the symbol for a real function is written using lightface italic type.

If no domain is specified for a vector function, it is assumed that the domain consists of all real numbers for which each of the component functions is defined.

## 1  Find the Domain of a Vector Function

**EXAMPLE 1    Finding the Domain of a Vector Function**

If $\mathbf{r}(t) = (2t + 2)\,\mathbf{i} + \sqrt{t}\,\mathbf{j} - \ln(t + 3)\,\mathbf{k}$, then the components of $\mathbf{r} = \mathbf{r}(t)$ are

$$x(t) = 2t + 2 \qquad y(t) = \sqrt{t} \qquad z(t) = -\ln(t + 3)$$

Since no domain is specified, it is assumed that the domain consists of all real numbers $t$ for which each of the component functions is defined. Since $x = x(t)$ is defined for all real numbers, $y = y(t)$ is defined for all real numbers $t \ge 0$, and $z = z(t)$ is defined for all real numbers $t > -3$, the domain of the vector function $\mathbf{r} = \mathbf{r}(t)$ is the intersection of these three sets, namely, the set of nonnegative real numbers, $\{t \mid t \ge 0\}$.  ∎

**NOW WORK**  Problem 17.

## 2  Graph a Vector Function

**NEED TO REVIEW?**  Parametric equations are discussed in Section 9.1, pp. 680–685.

In general, the graph of a vector function $\mathbf{r} = \mathbf{r}(t)$ with domain $a \le t \le b$ is a curve $C$ in the plane or in space traced out by the tip of the position vector $\mathbf{r} = \mathbf{r}(t)$ as $t$ varies over the interval $a \le t \le b$. The components of the vector function are the parametric equations of the curve. We refer to the curve $C$, shown in Figure 1, traced out by the tip of $\mathbf{r}(t) = x(t)\mathbf{i} + y(t)\mathbf{j}$, $a \le t \le b$, as a **plane curve**. The curve $C$, shown in Figure 2, traced out by the tip of $\mathbf{r}(t) = x(t)\mathbf{i} + y(t)\mathbf{j} + z(t)\mathbf{k}$, $a \le t \le b$, is referred to as a **space curve**.

**NOTE** When graphing a vector function $\mathbf{r} = \mathbf{r}(t)$, we treat $\mathbf{r}$ as a position vector. Also, from now on we just write, "the curve $C$ is traced out by the vector function $\mathbf{r} = \mathbf{r}(t)$" rather than "traced out by the tip of the position vector of $\mathbf{r}$."

The curve $C$ traced out by a vector function $\mathbf{r} = \mathbf{r}(t)$, $a \le t \le b$, has a **direction**, or an **orientation**. We show the orientation of $C$ by placing arrows on its graph in the direction the vector $\mathbf{r} = \mathbf{r}(t)$ moves from $t = a$ to $t = b$. See Figures 1, 2, and 3.

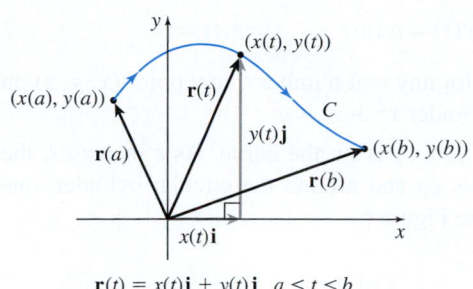

**DF** **Figure 1** A plane curve $C$.

$$\mathbf{r}(t) = x(t)\mathbf{i} + y(t)\mathbf{j} \quad a \le t \le b$$

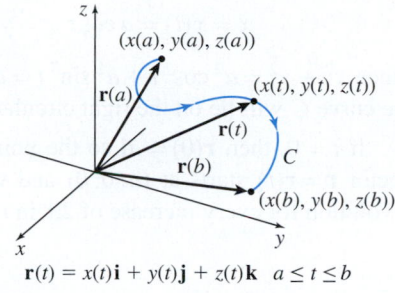

**Figure 2** A space curve $C$.

$$\mathbf{r}(t) = x(t)\mathbf{i} + y(t)\mathbf{j} + z(t)\mathbf{k} \quad a \le t \le b$$

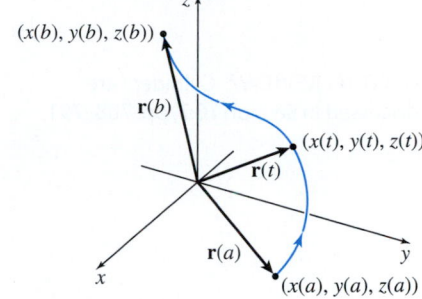

**DF** **Figure 3** $\mathbf{r} = \mathbf{r}(t) \quad a \le t \le b$

---

**EXAMPLE 2** **Graphing a Vector Function**

Discuss the graph of the vector function

$$\mathbf{r}(t) = \cos t\,\mathbf{i} + \sin t\,\mathbf{j} \quad 0 \le t \le 2\pi$$

**Solution** The components of $\mathbf{r} = \mathbf{r}(t)$ are the parametric equations

$$x = x(t) = \cos t \qquad y = y(t) = \sin t \quad 0 \le t \le 2\pi$$

Since $x^2 + y^2 = \cos^2 t + \sin^2 t = 1$, the vector function $\mathbf{r} = \mathbf{r}(t)$ traces out a circle with its center at the origin and radius 1. See Figure 4. The curve begins at $\mathbf{r}(0) = \cos 0\,\mathbf{i} + \sin 0\,\mathbf{j} = \mathbf{i}$, contains the vectors $\mathbf{r}\left(\dfrac{\pi}{2}\right) = \cos\dfrac{\pi}{2}\,\mathbf{i} + \sin\dfrac{\pi}{2}\,\mathbf{j} = \mathbf{j}$, $\mathbf{r}(\pi) = \cos\pi\,\mathbf{i} + \sin\pi\,\mathbf{j} = -\mathbf{i}$, and $\mathbf{r}\left(\dfrac{3\pi}{2}\right) = -\mathbf{j}$. It ends at $\mathbf{r}(2\pi) = \mathbf{i}$, so the positive direction of the circle is counterclockwise. ∎

**Figure 4** $\mathbf{r}(t) = \cos t\,\mathbf{i} + \sin t\,\mathbf{j} \quad 0 \le t \le 2\pi$

The domain of the vector function in Example 2 is restricted to $0 \le t \le 2\pi$. If $t$ were unrestricted, the domain would be all real numbers. The graph would look the same, as would its orientation. The graph would just keep repeating over intervals of length $2\pi$.

**NOW WORK** Problems **11(a)** and **31**.

**NEED TO REVIEW?** Parametric equations and symmetric equations of a line are discussed in Section 10.6, pp. 773–775.

**EXAMPLE 3** **Graphing a Vector Function**

Graph the curve $C$ traced out by the vector function

$$\mathbf{r}(t) = (2 + 3t)\mathbf{i} + (3 - t)\mathbf{j} + 2t\mathbf{k}$$

**Solution** The components of $\mathbf{r} = \mathbf{r}(t)$ are the parametric equations

$$x(t) = 2 + 3t \qquad y(t) = 3 - t \qquad z(t) = 2t$$

These are the parametric equations of a line in space containing the point $(2, 3, 0)$ (corresponding to $t = 0$) and in the direction of the vector $3\mathbf{i} - \mathbf{j} + 2\mathbf{k}$. Solving for $t$, we obtain the symmetric equations of this line

$$\frac{x - 2}{3} = \frac{y - 3}{-1} = \frac{z}{2}$$

Since $\mathbf{r}(0) = 2\mathbf{i} + 3\mathbf{j}$ and $\mathbf{r}(1) = 5\mathbf{i} + 2\mathbf{j} + 2\mathbf{k}$, the positive direction of $C$ is given by $\mathbf{r}(1) - \mathbf{r}(0) = 3\mathbf{i} - \mathbf{j} + 2\mathbf{k}$. See Figure 5. ∎

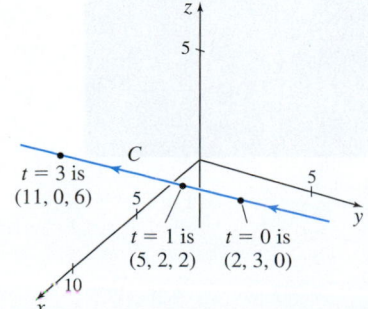

**Figure 5** $\mathbf{r}(t) = (2 + 3t)\mathbf{i} + (3 - t)\mathbf{j} + 2t\mathbf{k}$

EXAMPLE 4    **Graphing a Vector Function**

Graph the curve $C$ traced out by the vector function

$$\mathbf{r}(t) = a \cos t \mathbf{i} + a \sin t \mathbf{j} + t \mathbf{k} \quad t \geq 0$$

where $a$ is a positive constant.

**Solution** The parametric equations of the curve $C$ are

$$x = x(t) = a \cos t \qquad y = y(t) = a \sin t \qquad z = z(t) = t$$

**NEED TO REVIEW?** Cylinders are discussed in Section 10.7, pp. 788–791.

Since $x^2 + y^2 = a^2 \cos^2 t + a^2 \sin^2 t = a^2$, for any real number $t$, any point $(x, y, z)$ on the curve $C$ will lie on the right circular cylinder $x^2 + y^2 = a^2$.

If $t = 0$, then $\mathbf{r}(0) = a\mathbf{i}$ so the point $(a, 0, 0)$ is on the curve. As $t$ increases, the vector $\mathbf{r} = \mathbf{r}(t)$ starts at $(a, 0, 0)$ and winds up and around the circular cylinder, one revolution for every increase of $2\pi$ in $t$. See Figure 6.

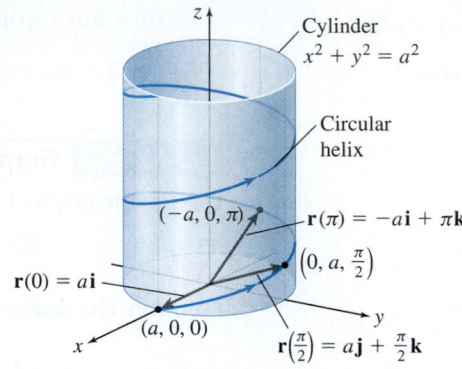

**Figure 6** $\mathbf{r}(t) = a \cos t \mathbf{i} + a \sin t \mathbf{j} + t \mathbf{k} \quad t \geq 0$

The space curve shown in Figure 6 is a **circular helix** and may be viewed as a coiled spring, much like a Slinky®. One of the many occurrences of a helix in nature is in the model of deoxyribonucleic acid (DNA), which consists of a *double helix*, that is, one helix intertwined with another helix, as shown in Figure 7.

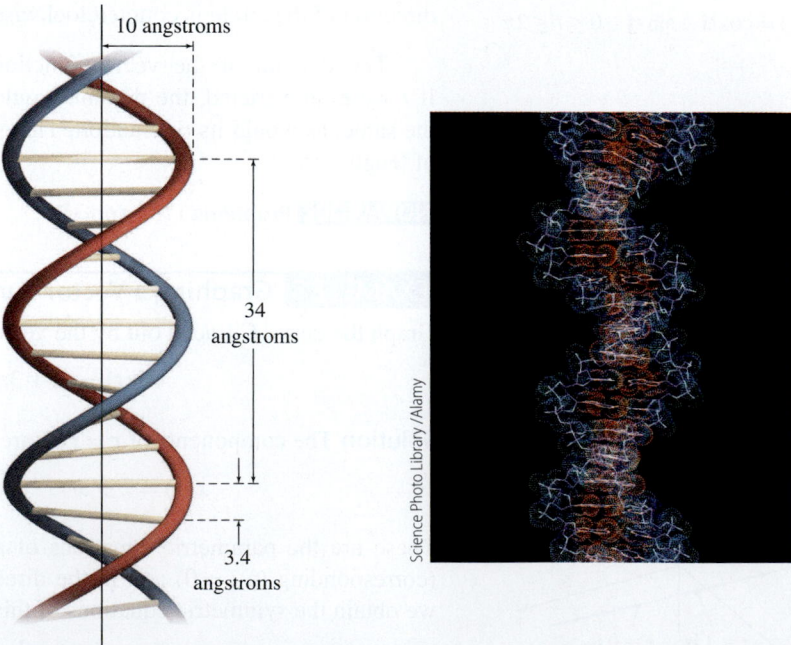

Science Photo Library / Alamy

**Figure 7** Models of DNA's double helix.

NOW WORK    **Problems 37 and 43.**

## 3 Find the Limit and Determine the Continuity of a Vector Function

We use the components of a vector function to define limits, continuity, and the derivative of a vector function.

---

**DEFINITION  Limit of a Vector Function**

For a vector function $\mathbf{r}(t) = x(t)\mathbf{i} + y(t)\mathbf{j}$ in the plane

$$\lim_{t \to t_0} \mathbf{r}(t) = \left[ \lim_{t \to t_0} x(t) \right] \mathbf{i} + \left[ \lim_{t \to t_0} y(t) \right] \mathbf{j}$$

provided $\lim_{t \to t_0} x(t)$ and $\lim_{t \to t_0} y(t)$ both exist.

For a vector function $\mathbf{r}(t) = x(t)\mathbf{i} + y(t)\mathbf{j} + z(t)\mathbf{k}$ in space

$$\lim_{t \to t_0} \mathbf{r}(t) = \left[ \lim_{t \to t_0} x(t) \right] \mathbf{i} + \left[ \lim_{t \to t_0} y(t) \right] \mathbf{j} + \left[ \lim_{t \to t_0} z(t) \right] \mathbf{k}$$

provided $\lim_{t \to t_0} x(t)$, $\lim_{t \to t_0} y(t)$, and $\lim_{t \to t_0} z(t)$ all exist.

---

**EXAMPLE 5  Finding the Limit of a Vector Function**

$$\lim_{t \to \pi/2} [2 \cos t\mathbf{i} + 2 \sin(2t)\mathbf{j} + 5\mathbf{k}] = \left[ \lim_{t \to \pi/2} (2 \cos t) \right] \mathbf{i} + \left[ \lim_{t \to \pi/2} [2 \sin(2t)] \right] \mathbf{j} + \left[ \lim_{t \to \pi/2} 5 \right] \mathbf{k}$$

$$= \left[ 2 \lim_{t \to \pi/2} \cos t \right] \mathbf{i} + \left[ 2 \lim_{t \to \pi/2} \sin(2t) \right] \mathbf{j} + 5\mathbf{k}$$

$$= 2 \cdot 0\mathbf{i} + 2 \cdot 0\mathbf{j} + 5\mathbf{k} = 5\mathbf{k}$$  ∎

NOW WORK Problem 11(b).

**NEED TO REVIEW?** Continuity of a real-valued function is discussed in Section 1.3, pp. 102–108.

The definition of continuity of a vector function at a number is almost identical to the definition for real functions. To determine if a vector function is continuous at a real number, we consider the continuity of its components.

---

**Continuity of a Vector Function at a Number**

A vector function $\mathbf{r} = \mathbf{r}(t)$ is **continuous at a real number** $t_0$ if:

- $\mathbf{r}(t_0)$ is defined.
- $\lim_{t \to t_0} \mathbf{r}(t)$ exists.
- $\lim_{t \to t_0} \mathbf{r}(t) = \mathbf{r}(t_0)$.

---

**EXAMPLE 6  Determining Whether a Vector Function Is Continuous**

Determine whether the vector function

$$\mathbf{r}(t) = t^2\mathbf{i} + (1+t)\mathbf{j} + \sin t\mathbf{k}$$

**IN WORDS** A vector function is continuous at a real number $t_0$ if, and only if, each of its component functions is continuous at $t_0$.

is continuous at 0.

**Solution** We check the three conditions for a vector function to be continuous at a number.

- $\mathbf{r}(0) = 0^2\mathbf{i} + (1+0)\mathbf{j} + \sin 0\mathbf{k} = \mathbf{j}$
- $\lim_{t \to 0} \mathbf{r}(t) = \lim_{t \to 0}[t^2\mathbf{i} + (1+t)\mathbf{j} + \sin t\mathbf{k}] = \left[ \lim_{t \to 0} t^2 \right]\mathbf{i} + \left[ \lim_{t \to 0}(1+t) \right]\mathbf{j} + \left[ \lim_{t \to 0} \sin t \right]\mathbf{k}$

$$= 0\mathbf{i} + 1\mathbf{j} + 0\mathbf{k} = \mathbf{j}$$

- $\lim_{t \to 0} \mathbf{r}(t) = \mathbf{r}(0) = \mathbf{j}$

Since all three conditions are met, $\mathbf{r} = \mathbf{r}(t)$ is continuous at $t_0$. ∎

NOW WORK Problem 11(c).

The definition of continuity on an interval for vector functions follows the pattern developed in Chapter 1 for real functions. That is, $\mathbf{r} = \mathbf{r}(t)$ is continuous on an interval $I$, provided each of its components is continuous on $I$. For example, the vector function $\mathbf{r}(t) = t^2\mathbf{i} + (1 + t)\mathbf{j} + \sin t\,\mathbf{k}$ is continuous for all real numbers since the real functions $x(t) = t^2$, $y(t) = 1 + t$, and $z(t) = \sin t$ are continuous for all real numbers.

NOW WORK Problem 49.

## 4 Find the Derivative of a Vector Function

The definition of a derivative for a vector function is very similar to the definition for a real function.

NEED TO REVIEW? The definition of the derivative of a function is discussed in Section 2.2, pp. 163–164.

DEFINITION Derivative of a Vector Function

The **derivative** of the vector function $\mathbf{r} = \mathbf{r}(t)$, denoted by $\mathbf{r}'(t) = \dfrac{d\mathbf{r}}{dt}$, is

$$\mathbf{r}'(t) = \frac{d\mathbf{r}}{dt} = \lim_{h \to 0} \frac{\mathbf{r}(t + h) - \mathbf{r}(t)}{h}$$

provided the limit exists. If the limit exists, then we say $\mathbf{r}$ is **differentiable**.

The next theorem provides a simple method for differentiating a vector function.

THEOREM Differentiating a Vector Function

If $\mathbf{r}(t) = x(t)\mathbf{i} + y(t)\mathbf{j}$ and if $x = x(t)$ and $y = y(t)$ are both differentiable real functions, then

$$\mathbf{r}'(t) = \frac{d\mathbf{r}}{dt} = x'(t)\mathbf{i} + y'(t)\mathbf{j}$$

If $\mathbf{r}(t) = x(t)\mathbf{i} + y(t)\mathbf{j} + z(t)\mathbf{k}$ and if $x = x(t)$, $y = y(t)$, and $z = z(t)$ are each differentiable real functions, then

$$\mathbf{r}'(t) = \frac{d\mathbf{r}}{dt} = x'(t)\mathbf{i} + y'(t)\mathbf{j} + z'(t)\mathbf{k}$$

IN WORDS To find the derivative of a vector function, differentiate each of its components.

We prove this theorem for a vector function in space.

Proof Use the definition of the derivative of a vector function.

$$\mathbf{r}'(t) = \lim_{h \to 0} \frac{\mathbf{r}(t + h) - \mathbf{r}(t)}{h}$$

$$= \lim_{h \to 0} \frac{[x(t + h) - x(t)]\mathbf{i} + [y(t + h) - y(t)]\mathbf{j} + [z(t + h) - z(t)]\mathbf{k}}{h}$$

$$= \lim_{h \to 0} \left[\frac{x(t + h) - x(t)}{h}\right]\mathbf{i} + \lim_{h \to 0} \left[\frac{y(t + h) - y(t)}{h}\right]\mathbf{j} + \lim_{h \to 0} \left[\frac{z(t + h) - z(t)}{h}\right]\mathbf{k}$$

$$= x'(t)\mathbf{i} + y'(t)\mathbf{j} + z'(t)\mathbf{k} \qquad \blacksquare$$

In Leibniz notation, the derivative of a vector function **r** has the form

$$\frac{d\mathbf{r}}{dt} = \frac{dx}{dt}\mathbf{i} + \frac{dy}{dt}\mathbf{j} \quad \text{or} \quad \frac{d\mathbf{r}}{dt} = \frac{dx}{dt}\mathbf{i} + \frac{dy}{dt}\mathbf{j} + \frac{dz}{dt}\mathbf{k}$$

---

**EXAMPLE 7** **Finding the Derivative of a Vector Function**

Find the derivative of each vector function.

**(a)** $\mathbf{r}(t) = 2\sin t\,\mathbf{i} + 3\cos t\,\mathbf{j}$     **(b)** $\mathbf{r}(t) = e^t\mathbf{i} + (1+t)\mathbf{j} + \sin t\,\mathbf{k}$

**Solution** **(a)** Find the derivative of each component. Since $\dfrac{d}{dt}(2\sin t) = 2\cos t$

and $\dfrac{d}{dt}(3\cos t) = -3\sin t$,

$$\mathbf{r}'(t) = \frac{d\mathbf{r}}{dt} = 2\cos t\,\mathbf{i} - 3\sin t\,\mathbf{j}$$

**(b)** $\mathbf{r}'(t) = \dfrac{d\mathbf{r}}{dt} = \dfrac{d}{dt}e^t\mathbf{i} + \dfrac{d}{dt}(1+t)\mathbf{j} + \dfrac{d}{dt}\sin t\,\mathbf{k} = e^t\mathbf{i} + \mathbf{j} + \cos t\,\mathbf{k}$   ∎

**NOW WORK** Problem **59** (find $\mathbf{r}'(t)$).

### Higher-Order Derivatives

Since the derivative $\mathbf{r}'$ is a vector function derived from the vector function $\mathbf{r} = \mathbf{r}(t)$, the derivative of $\mathbf{r}'$ (if it exists) is also a function, called the **second derivative** of **r**, and denoted by $\mathbf{r}''$. If the second derivative exists, then **r** is called **twice differentiable**. Continuing in this fashion, we can find the **third derivative** $\mathbf{r}'''$, the **fourth derivative** $\mathbf{r}^{(4)}$, and so on, provided these derivatives exist. Collectively, these derivatives are called **higher-order derivatives** of the vector function $\mathbf{r} = \mathbf{r}(t)$.

---

**EXAMPLE 8** **Finding $\mathbf{r}''$ for a Vector Function**

Find $\mathbf{r}''(t)$ for the vector function $\mathbf{r}(t) = e^t\mathbf{i} + \ln t\,\mathbf{j} - \cos t\,\mathbf{k}$.

**Solution** First find $\mathbf{r}'(t)$.

$$\mathbf{r}'(t) = \frac{d}{dt}e^t\mathbf{i} + \frac{d}{dt}\ln t\,\mathbf{j} - \frac{d}{dt}\cos t\,\mathbf{k} = e^t\mathbf{i} + \frac{1}{t}\mathbf{j} + \sin t\,\mathbf{k}$$

Then

$$\mathbf{r}''(t) = \frac{d}{dt}e^t\mathbf{i} + \frac{d}{dt}\frac{1}{t}\mathbf{j} + \frac{d}{dt}\sin t\,\mathbf{k} = e^t\mathbf{i} - \frac{1}{t^2}\mathbf{j} + \cos t\,\mathbf{k}$$   ∎

**NOW WORK** Problem **59** (find $\mathbf{r}''(t)$).

## 5 Find the Derivative of a Vector Function Using Derivative Rules

The derivative of a vector function $\mathbf{r} = \mathbf{r}(t)$ is defined in much the same way as the derivative of a real function $y = f(x)$. For this reason, the sum and scalar multiplication rules for derivatives of vector functions are similar to the rules for real functions. However, for vector functions, there are rules for the derivatives of the dot product and the cross product (in space). For vector functions, there is no rule analogous to the Quotient Rule.

**THEOREM** Derivative Rules

If the vector functions $\mathbf{u} = \mathbf{u}(t)$ and $\mathbf{v} = \mathbf{v}(t)$ are differentiable and the real function $y = f(t)$ is differentiable, then

- **Sum Rule:**

$$\frac{d}{dt}[\mathbf{u}(t) + \mathbf{v}(t)] = \frac{d}{dt}\mathbf{u}(t) + \frac{d}{dt}\mathbf{v}(t) = \mathbf{u}'(t) + \mathbf{v}'(t)$$

- **Scalar Multiplication Rule:**

$$\frac{d}{dt}[f(t)\mathbf{u}(t)] = \left[\frac{d}{dt}f(t)\right]\mathbf{u}(t) + f(t)\left[\frac{d}{dt}\mathbf{u}(t)\right] = f'(t)\mathbf{u}(t) + f(t)\mathbf{u}'(t)$$

- **Dot Product Rule:**

$$\frac{d}{dt}[\mathbf{u}(t) \cdot \mathbf{v}(t)] = [\mathbf{u}(t) \cdot \mathbf{v}(t)]' = \mathbf{u}'(t) \cdot \mathbf{v}(t) + \mathbf{u}(t) \cdot \mathbf{v}'(t)$$

- **Cross Product Rule:**

$$\frac{d}{dt}[\mathbf{u}(t) \times \mathbf{v}(t)] = [\mathbf{u}(t) \times \mathbf{v}(t)]' = \mathbf{u}'(t) \times \mathbf{v}(t) + \mathbf{u}(t) \times \mathbf{v}'(t)$$

where $\mathbf{u}$ and $\mathbf{v}$ are vector functions in space.

- **Chain Rule:**

$$\frac{d}{dt}\mathbf{v}(f(t)) = \mathbf{v}'(f(t))f'(t)$$

Notice that if $f(t) = c$, where $c$ is a constant, and $\mathbf{u} = \mathbf{u}(t)$ is a differentiable vector function, then

$$\frac{d}{dt}[c\mathbf{u}(t)] = c\left[\frac{d}{dt}\mathbf{u}(t)\right] = c\mathbf{u}'(t)$$

The Sum, Scalar Multiplication, and Dot Product Rules, and the Chain Rule hold for vector functions in the plane and in space. The Cross Product Rule holds only for vectors in space. The proof for the Scalar Multiplication Rule for a vector function in space is provided after Example 9. The proofs of the other four rules are left as exercises. See Problems 99–102.

**CALC** ▶ **CLIP**

**EXAMPLE 9** Finding the Derivative of a Dot Product and a Cross Product

For:

$$\mathbf{u}(t) = \cos t\,\mathbf{i} + \sin t\,\mathbf{j} + t\mathbf{k} \qquad \text{and} \qquad \mathbf{v}(t) = t\mathbf{i} + \ln t\,\mathbf{j} + \mathbf{k}$$

**(a)** Find the derivative of $\mathbf{u}(t) \cdot \mathbf{v}(t)$.
**(b)** Find the derivative of $\mathbf{u}(t) \times \mathbf{v}(t)$.

**Solution** The derivatives of $\mathbf{u}$ and $\mathbf{v}$ are

$$\mathbf{u}'(t) = -\sin t\,\mathbf{i} + \cos t\,\mathbf{j} + \mathbf{k} \qquad \text{and} \qquad \mathbf{v}'(t) = \mathbf{i} + \frac{1}{t}\mathbf{j}$$

**(a)** To find the derivative of $\mathbf{u}(t) \cdot \mathbf{v}(t)$, use the Dot Product Rule.

$$\frac{d}{dt}[\mathbf{u}(t) \cdot \mathbf{v}(t)] = [\mathbf{u}(t) \cdot \mathbf{v}(t)]' = \mathbf{u}'(t) \cdot \mathbf{v}(t) + \mathbf{u}(t) \cdot \mathbf{v}'(t)$$

$$= (-\sin t\,\mathbf{i} + \cos t\,\mathbf{j} + \mathbf{k}) \cdot (t\mathbf{i} + \ln t\,\mathbf{j} + \mathbf{k})$$

$$+ (\cos t\,\mathbf{i} + \sin t\,\mathbf{j} + t\mathbf{k}) \cdot \left(\mathbf{i} + \frac{1}{t}\mathbf{j}\right)$$

$$= -t \sin t + (\cos t)(\ln t) + 1 + \cos t + \frac{\sin t}{t}$$

**(b)** Use the Cross Product Rule.

$$\frac{d}{dt}[\mathbf{u}(t) \times \mathbf{v}(t)] = [\mathbf{u}(t) \times \mathbf{v}(t)]' = \mathbf{u}'(t) \times \mathbf{v}(t) + \mathbf{u}(t) \times \mathbf{v}'(t)$$

$$= \begin{vmatrix} \mathbf{i} & \mathbf{j} & \mathbf{k} \\ -\sin t & \cos t & 1 \\ t & \ln t & 1 \end{vmatrix} + \begin{vmatrix} \mathbf{i} & \mathbf{j} & \mathbf{k} \\ \cos t & \sin t & t \\ 1 & \dfrac{1}{t} & 0 \end{vmatrix}$$

$$= \left[(\cos t - \ln t)\mathbf{i} + (\sin t + t)\mathbf{j} - [(\sin t)(\ln t) + t\cos t]\mathbf{k}\right]$$

$$+ \left[-\mathbf{i} + t\mathbf{j} + \left(\frac{1}{t}\cos t - \sin t\right)\mathbf{k}\right]$$

$$= (\cos t - \ln t - 1)\mathbf{i} + (\sin t + 2t)\mathbf{j}$$

$$+ \left[\frac{1}{t}\cos t - \sin t - (\sin t)(\ln t) - t\cos t\right]\mathbf{k} \qquad \blacksquare$$

We can also work Example 7 by first finding the dot product or the cross product of the vector functions **u** and **v** and then differentiating. Try it, and compare the two solutions.

NOW WORK Problems **73** and **79**.

**Proof (Scalar Multiplication Rule)** Let $\mathbf{u}(t) = x(t)\mathbf{i} + y(t)\mathbf{j} + z(t)\mathbf{k}$. Then

$$f(t)\mathbf{u}(t) = [f(t)x(t)]\mathbf{i} + [f(t)y(t)]\mathbf{j} + [f(t)z(t)]\mathbf{k}$$

Use the Product Rule and differentiate each component.

$$[f(t)\mathbf{u}(t)]' = [f(t)x(t)]'\mathbf{i} + [f(t)y(t)]'\mathbf{j} + [f(t)z(t)]'\mathbf{k}$$

$$= [f(t)x'(t) + f'(t)x(t)]\mathbf{i} + [f(t)y'(t) + f'(t)y(t)]\mathbf{j}$$
$$+ [f(t)z'(t) + f'(t)z(t)]\mathbf{k}$$

$$= f'(t)[x(t)\mathbf{i} + y(t)\mathbf{j} + z(t)\mathbf{k}] + f(t)[x'(t)\mathbf{i} + y'(t)\mathbf{j} + z'(t)\mathbf{k}]$$

$$= f'(t)\mathbf{u}(t) + f(t)\mathbf{u}'(t) \qquad \blacksquare$$

# 11.1 Assess Your Understanding

## Concepts and Vocabulary

1. The domain of $\mathbf{r}(t) = (4t + 1)\mathbf{i} + \sqrt{4 - t}\mathbf{j} + \mathbf{k}$ is _____.

2. *True or False* A vector function is continuous at a real number $t_0$ if at least one of its component functions is continuous at $t_0$.

3. *True or False* To find the derivative of a vector function $\mathbf{r} = \mathbf{r}(t)$, find the derivative of each of its component functions.

4. *True or False* The derivative of a dot product is a scalar function.

5. *True or False* If $y = f(t)$ is a differentiable real (scalar) function and $\mathbf{r} = \mathbf{r}(t)$ is a differentiable vector function, then

$$\frac{d}{dt}[f(t)\mathbf{r}(t)] = f'(t)\frac{d}{dt}\mathbf{r}(t)$$

6. If $\mathbf{u} = \mathbf{u}(t)$ and $\mathbf{v} = \mathbf{v}(t)$ are vector functions in space that are differentiable, then $\dfrac{d}{dt}(\mathbf{u} \times \mathbf{v}) = $ _____.

## Skill Building

*In Problems 7–14,*
(a) Find the value of each vector function at $t_0$.
(b) Find the limit of each vector function as $t \to t_0$.
(c) Determine whether each vector function is continuous at $t_0$.

7. $\mathbf{r}(t) = t^2\mathbf{i} - 2t\mathbf{j}$ at $t_0 = 1$       8. $\mathbf{r}(t) = t^3\mathbf{i} + 2t\mathbf{j}$ at $t_0 = 2$

9. $\mathbf{v}(t) = \sin t\mathbf{i} - \cos t\mathbf{j}$ at $t_0 = \dfrac{\pi}{4}$

10. $\mathbf{v}(t) = \tan t\mathbf{i} + \cos(2t)\mathbf{j}$ at $t_0 = 0$

11. $\mathbf{u}(t) = e^{2t}\mathbf{i} + t^2\mathbf{j} - 2\mathbf{k}$ at $t_0 = 0$

12. $\mathbf{u}(t) = \ln t\mathbf{i} - \mathbf{j} + t^3\mathbf{k}$ at $t_0 = 1$

13. $\mathbf{g}(t) = t\mathbf{i} - \cos\dfrac{\pi t}{4}\mathbf{j} + 2t\mathbf{k}$ at $t_0 = 4$

14. $\mathbf{f}(t) = \sin\dfrac{3\pi t}{4}\mathbf{i} + 3\mathbf{j} - 3t^2\mathbf{k}$ at $t_0 = 1$

---

**1.** = NOW WORK problem          = Graphing technology recommended          CAS = Computer Algebra System recommended

*In Problems 15–24, find the domain of each vector function.*

**15.** $\mathbf{r}(t) = \cos t\mathbf{i} + \sin t\mathbf{j} + 2\mathbf{k}$

**16.** $\mathbf{r}(t) = \cos(2t)\mathbf{i} + \sin t\mathbf{j} + 2t\mathbf{k}$

**17.** $\mathbf{f}(t) = \sqrt{t}\mathbf{i} + t\mathbf{j} - \mathbf{k}$

**18.** $\mathbf{r}(t) = t^2\mathbf{i} + \dfrac{1}{\sqrt{t}}\mathbf{j} - t\mathbf{k}$

**19.** $\mathbf{u}(t) = \dfrac{t^2 + 1}{t}\mathbf{i} + 3t\mathbf{j} - \dfrac{2}{t}\mathbf{k}$

**20.** $\mathbf{v}(t) = e^t\mathbf{i} + (5 + 3t)\mathbf{j} - \dfrac{2}{t}\mathbf{k}$

**21.** $\mathbf{v}(t) = \ln t\mathbf{i} + (t + 1)\mathbf{j} + 4t\mathbf{k}$

**22.** $\mathbf{v}(t) = \sqrt{t}\mathbf{i} + \ln t\mathbf{j} + t\mathbf{k}$

**23.** $\mathbf{r}(t) = \ln(t - 1)\mathbf{i} + 3t\mathbf{j} + \sqrt{4 - t}\mathbf{k}$  **24.** $\mathbf{r}(t) = \sqrt{t^2 - 4}\mathbf{i} + t\mathbf{j} + \dfrac{1}{t}\mathbf{k}$

*In Problems 25–36, graph the curve C traced out by each vector function and show its orientation.*

**25.** $\mathbf{r}(t) = 2t\mathbf{i} + t^2\mathbf{j}$   $t \geq 0$

**26.** $\mathbf{r}(t) = t^2\mathbf{i} - 4t\mathbf{j}$   $t \leq 0$

**27.** $\mathbf{r}(t) = t\mathbf{i}$   $-1 \leq t \leq 1$

**28.** $\mathbf{r}(t) = t\mathbf{j}$   $-1 \leq t \leq 1$

**29.** $\mathbf{r}(t) = 3t\mathbf{i} - 2t\mathbf{j}$

**30.** $\mathbf{r}(t) = t\mathbf{i} + t^2\mathbf{j}$

**31.** $\mathbf{r}(t) = \cos t\mathbf{i} - \sin t\mathbf{j}$   $0 \leq t \leq \dfrac{\pi}{2}$

**32.** $\mathbf{r}(t) = \cos t\mathbf{i} + \sin t\mathbf{j}$   $0 \leq t \leq \dfrac{\pi}{2}$

**33.** $\mathbf{r}(t) = \sin^2 t\mathbf{i} + \cos^2 t\mathbf{j}$   $0 \leq t \leq \dfrac{\pi}{2}$

**34.** $\mathbf{r}(t) = \sin^2 t\mathbf{i} - \cos^2 t\mathbf{j}$   $0 \leq t \leq \dfrac{\pi}{2}$

**35.** $\mathbf{r}(t) = \sin t\mathbf{i} + t^2\mathbf{j}$   $0 \leq t \leq 4\pi$

**36.** $\mathbf{r}(t) = \sin^3 t\mathbf{i} + \cos^3 t\mathbf{j}$   $0 \leq t \leq 2\pi$

*In Problems 37–40, match each vector function to its plane curve.*

**37.** $\mathbf{r}(t) = t^2\mathbf{i} + \cos t\mathbf{j}$   $0 \leq t \leq 2\pi$

**38.** $\mathbf{r}(t) = \cos t\mathbf{i} + t^2\mathbf{j}$   $0 \leq t \leq 2\pi$

**39.** $\mathbf{r}(t) = \sqrt{t}\mathbf{i} + \cos t\mathbf{j}$   $0 \leq t \leq 2\pi$

**40.** $\mathbf{r}(t) = \cos t\mathbf{i} - t\mathbf{j}$   $0 \leq t \leq 2\pi$

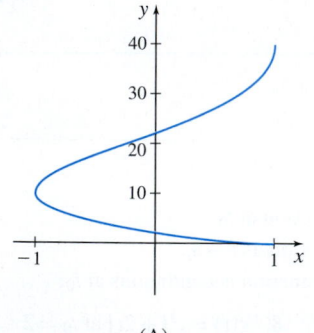

(A)

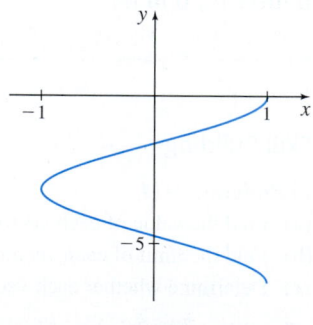

(B)

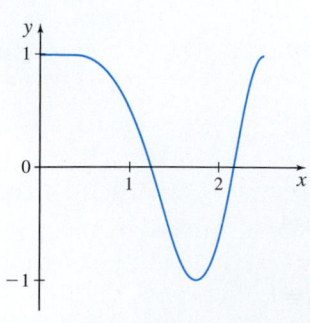

(C)

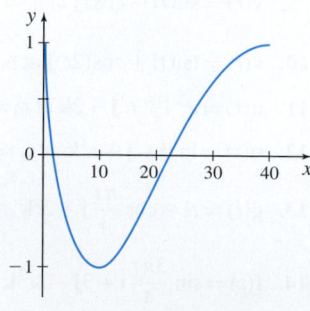

(D)

**CAS**  *In Problems 41–48, graph the curve C traced out by each vector function.*

**41.** $\mathbf{r}(t) = (2t + 1)\mathbf{i} + \dfrac{t}{3}\mathbf{j} - (t + 1)\mathbf{k}$

**42.** $\mathbf{r}(t) = t\mathbf{i} + t\mathbf{j} - (t + 1)\mathbf{k}$

**43.** $\mathbf{r}(t) = 2\cos t\mathbf{i} + t\mathbf{j} + 2\sin t\mathbf{k}$   $-2\pi \leq t \leq 2\pi$

**44.** $\mathbf{r}(t) = \dfrac{t}{2}\mathbf{i} + 3\sin t\mathbf{j} + 2\cos t\mathbf{k}$   $-2\pi \leq t \leq 2\pi$

**45.** $\mathbf{r}(t) = 4\cos t\mathbf{i} + \sin t\mathbf{j} + e^t\mathbf{k}$   $-\pi \leq t \leq 2\pi$

**46.** $\mathbf{r}(t) = t\mathbf{i} + \sin t\mathbf{j} + e^t\mathbf{k}$   $-5 \leq t \leq 5$

**47.** $\mathbf{r}(t) = 3\cos t\mathbf{i} + 3\sin t\mathbf{j} + \sin(2t)\mathbf{k}$   $0 \leq t \leq 2\pi$

**48.** $\mathbf{r}(t) = t\mathbf{i} + t\cos(3t)\mathbf{j} + t\sin(3t)\mathbf{k}$   $0 \leq t \leq 2\pi$

*In Problems 49–56, determine where each vector function is continuous.*

**49.** $\mathbf{r}(t) = t^2\mathbf{i} + \dfrac{2}{1 - t^2}\mathbf{j} + \sin t\mathbf{k}$   **50.** $\mathbf{r}(t) = \ln t\mathbf{i} + e^{-t}\mathbf{j}$

**51.** $\mathbf{r}(t) = \sec t\mathbf{i} - \cos t\mathbf{j}$

**52.** $\mathbf{r}(t) = \sqrt{t + 1}\mathbf{i} + \tan t\mathbf{j} - \dfrac{\sin t - 1}{t}\mathbf{k}$

**53.** $\mathbf{r}(t) = \dfrac{t}{t + 1}\mathbf{i} - t\mathbf{j}$   **54.** $\mathbf{r}(t) = \dfrac{t}{t^2 + 1}\mathbf{i} + t\mathbf{j}$

**55.** $\mathbf{r}(t) = \sin t\mathbf{i} - \tan t\mathbf{j} + \mathbf{k}$   **56.** $\mathbf{r}(t) = \sec t\mathbf{i} + \csc t\mathbf{j} + \mathbf{k}$

*In Problems 57–72, find $\mathbf{r}'(t)$ and $\mathbf{r}''(t)$.*

**57.** $\mathbf{r}(t) = 4t^2\mathbf{i} - 2t^3\mathbf{j}$   **58.** $\mathbf{r}(t) = 8t\mathbf{i} + 4t^3\mathbf{j}$

**59.** $\mathbf{r}(t) = 4\sqrt{t}\mathbf{i} + 2e^t\mathbf{j}$   **60.** $\mathbf{r}(t) = e^{3t}\mathbf{i} + \sqrt[3]{t}\mathbf{j}$

**61.** $\mathbf{r}(t) = \mathbf{i} + t\mathbf{j} + t^2\mathbf{k}$   **62.** $\mathbf{r}(t) = \mathbf{i} - \mathbf{j} + t\mathbf{k}$

**63.** $\mathbf{r}(t) = t^2\mathbf{i} + t^3\mathbf{j} - t\mathbf{k}$   **64.** $\mathbf{r}(t) = (1 + t)\mathbf{i} - 3t^2\mathbf{j} + t\mathbf{k}$

**65.** $\mathbf{r}(t) = \sin^2 t\mathbf{i} - \cos^2 t\mathbf{j}$   **66.** $\mathbf{r}(t) = \sin(2t)\mathbf{i} - \cos(2t)\mathbf{j}$

**67.** $\mathbf{r}(t) = 5e^{t^2 + t}\mathbf{i} - \ln e^{t^2 + t}\mathbf{j}$

**68.** $\mathbf{r}(t) = \sin(3t^3 - t)\mathbf{i} - \cos^2(3t^3 - t)\mathbf{j}$

**69.** $\mathbf{r}(t) = e^t\cos t\mathbf{i} + e^t\sin t\mathbf{j} + t\mathbf{k}$

**70.** $\mathbf{r}(t) = e^{-t}\cos t\mathbf{i} + e^{-t}\sin t\mathbf{j} - t\mathbf{k}$

**71.** $\mathbf{r}(t) = (t - t^3)\mathbf{i} + (t + t^3)\mathbf{j} - t\mathbf{k}$

**72.** $\mathbf{r}(t) = (t^2 - t)\mathbf{i} + (t^2 + t)\mathbf{j} + t\mathbf{k}$

*In Problems 73–78, find $\dfrac{d}{dt}[\mathbf{u}(t) \cdot \mathbf{v}(t)]$.*

**73.** $\mathbf{u}(t) = t^2\mathbf{i} - t\mathbf{j}$   and   $\mathbf{v}(t) = t\mathbf{i} + t^2\mathbf{j}$

**74.** $\mathbf{u}(t) = t^3\mathbf{i} + t\mathbf{j}$   and   $\mathbf{v}(t) = t\mathbf{i} - 2t\mathbf{j}$

**75.** $\mathbf{u}(t) = e^t\mathbf{i} + e^{-t}\mathbf{j}$   and   $\mathbf{v}(t) = t\mathbf{i} - t^2\mathbf{j}$

**76.** $\mathbf{u}(t) = t\mathbf{i} + 4\sqrt{t}\mathbf{j}$   and   $\mathbf{v}(t) = 2t\mathbf{i} - t^2\mathbf{j}$

**77.** $\mathbf{u}(t) = \sin(\omega t)\mathbf{i} + \cos(\omega t)\mathbf{j}$   and   $\mathbf{v}(t) = \mathbf{i} + \mathbf{j}$

**78.** $\mathbf{u}(t) = \sin^2 t\mathbf{i} - \cos^2 t\mathbf{j}$   and   $\mathbf{v}(t) = \mathbf{i} - \mathbf{j}$

*In Problems 79–82, find $\dfrac{d}{dt}[\mathbf{u}(t) \cdot \mathbf{v}(t)]$ and $\dfrac{d}{dt}[\mathbf{u}(t) \times \mathbf{v}(t)]$.*

**79.** $\mathbf{u}(t) = 2t\mathbf{i} + t^2\mathbf{j} - 5\mathbf{k}$   and   $\mathbf{v}(t) = t^2\mathbf{i} + 2t\mathbf{j} + \mathbf{k}$

**80.** $\mathbf{u}(t) = t^3\mathbf{i} - t^2\mathbf{j} + t\mathbf{k}$   and   $\mathbf{v}(t) = t\mathbf{i} - t^2\mathbf{j} + t^3\mathbf{k}$

**81.** $\mathbf{u}(t) = \cos(2t)\mathbf{i} + \sin(2t)\mathbf{j} + \mathbf{k}$   and   $\mathbf{v}(t) = \cos t\mathbf{i} + \sin t\mathbf{j} + \mathbf{k}$

**82.** $\mathbf{u}(t) = e^{2t}\mathbf{i} + e^{-2t}\mathbf{j} + t\mathbf{k}$   and   $\mathbf{v}(t) = e^{-t}\mathbf{i} + e^{-2t}\mathbf{j} - t\mathbf{k}$

## Applications and Extensions

**83.** Given the vector function $\mathbf{u}(t) = \cos(\omega t)\mathbf{i} + \sin(\omega t)\mathbf{j}$,

find $\dfrac{d\mathbf{u}}{dt}$ and $\left\|\dfrac{d\mathbf{u}}{dt}\right\|$.

**84.** Given the vector function $\mathbf{v}(t) = t\mathbf{i} + t^2\mathbf{j} + t^3\mathbf{k}$,

find $\dfrac{d^2\mathbf{v}}{dt^2}$ and $\left\|\dfrac{d^2\mathbf{v}}{dt^2}\right\|$.

**85.** For $\mathbf{f}(t) = \sin t\,\mathbf{i} - \cos t\,\mathbf{j}$, show that $\mathbf{f}(t)$ and $\mathbf{f}''(t)$ are parallel.

**86.** For $\mathbf{f}(t) = e^{3t}\mathbf{i} + e^{-3t}\mathbf{j}$, show that $\mathbf{f}(t)$ and $\mathbf{f}''(t)$ are parallel.

*In Problems 87–94, a vector function $\mathbf{r} = \mathbf{r}(t)$ defining a curve in the plane is given. Graph each curve, indicating its orientation. Include in the graph the vectors $\mathbf{r}(0)$ and $\mathbf{r}'(0)$. Draw $\mathbf{r}'(0)$ so its initial point is at the terminal point of $\mathbf{r}(0)$.*

**87.** $\mathbf{r}(t) = t\mathbf{i} + t^2\mathbf{j}$

**88.** $\mathbf{r}(t) = 2t^2\mathbf{i} - t\mathbf{j}$

**89.** $\mathbf{r}(t) = t\mathbf{i} + e^t\mathbf{j}$

**90.** $\mathbf{r}(t) = t\mathbf{i} + \ln(1+t)\mathbf{j}$

**91.** $\mathbf{r}(t) = 3\sin t\,\mathbf{i} - 3\cos t\,\mathbf{j}$

**92.** $\mathbf{r}(t) = 4\sin t\,\mathbf{i} + 4\cos t\,\mathbf{j}$

**93.** $\mathbf{r}(t) = 2\cos t\,\mathbf{i} - 3\sin t\,\mathbf{j}$

**94.** $\mathbf{r}(t) = -\cos t\,\mathbf{i} + 2\sin t\,\mathbf{j}$

*In Problems 95–98, the position of an object at time $t$ is given.*
**(a)** *Eliminate the parameter $t$ to find $y$ as a function of $x$.*
**(b)** *Graph $\mathbf{r} = \mathbf{r}(t)$ and indicate the orientation.*

**95.** $\mathbf{r}(t) = (1 - t^2)\mathbf{i} + t^2\mathbf{j}$

**96.** $\mathbf{r}(t) = t^2\mathbf{i} + (4t^2 - t^4)\mathbf{j}$

**97.** $\mathbf{r}(t) = \sin^2 t\,\mathbf{i} + \tan t\,\mathbf{j}$  $0 < t < \dfrac{\pi}{2}$

**98.** $\mathbf{r}(t) = e^t\mathbf{i} + e^{-t}\mathbf{j}$

**99. Proof of the Sum Rule**  If the vector functions $\mathbf{u} = \mathbf{u}(t)$ and $\mathbf{v} = \mathbf{v}(t)$ are differentiable, show that

$$[\mathbf{u}(t) + \mathbf{v}(t)]' = \mathbf{u}'(t) + \mathbf{v}'(t)$$

**100. Proof of the Dot Product Rule**  If the vector functions $\mathbf{u} = \mathbf{u}(t)$ and $\mathbf{v} = \mathbf{v}(t)$ are differentiable, show that

$$[\mathbf{u}(t) \cdot \mathbf{v}(t)]' = \mathbf{u}'(t) \cdot \mathbf{v}(t) + \mathbf{u}(t) \cdot \mathbf{v}'(t)$$

**101. Proof of the Cross Product Rule**  If the vector functions $\mathbf{u} = \mathbf{u}(t)$ and $\mathbf{v} = \mathbf{v}(t)$ in space are differentiable, show that

$$[\mathbf{u}(t) \times \mathbf{v}(t)]' = \mathbf{u}'(t) \times \mathbf{v}(t) + \mathbf{u}(t) \times \mathbf{v}'(t)$$

**102. Proof of the Chain Rule**  If the vector function $\mathbf{v} = \mathbf{v}(t)$ is differentiable and the real function $y = f(t)$ is differentiable, show that

$$\frac{d}{dt}\mathbf{v}(f(t)) = \mathbf{v}'(f(t))f'(t)$$

**103.** Suppose the vector function $\mathbf{r} = \mathbf{r}(t)$ is twice differentiable; show that $[\mathbf{r}(t) \times \mathbf{r}'(t)]' = \mathbf{r}(t) \times \mathbf{r}''(t)$.

**104.** Show that $\lim\limits_{t \to t_0} \mathbf{r}(t) = \mathbf{L}$ if and only if $\lim\limits_{t \to t_0} \|\mathbf{r}(t) - \mathbf{L}\| = 0$.

**105.** If the vector function $\mathbf{r} = \mathbf{r}(t)$ is differentiable, show that

$$\frac{d}{dt}\|\mathbf{r}(t)\| = \frac{\mathbf{r}(t) \cdot \mathbf{r}'(t)}{\|\mathbf{r}(t)\|}.$$ Use this result to show that $\|\mathbf{r}(t)\|$ is constant if and only if $\mathbf{r}(t) \cdot \mathbf{r}'(t) = 0$ for all $t$.

## Challenge Problems

**106. (a)** Discuss the curve $C$ traced out by the vector function
$$\mathbf{r}(t) = e^t \sin t\,\mathbf{i} + e^t \cos t\,\mathbf{j} + e^t\mathbf{k}.$$

**(b)** Name the surface on which the curve $C$ lies.

**107.** If the vector function $\mathbf{r}$ is differentiable, show that

$$\frac{d}{dt}\frac{\mathbf{r}(t)}{\|\mathbf{r}(t)\|} = \frac{\mathbf{r}'(t)}{\|\mathbf{r}(t)\|} - \frac{\mathbf{r}(t) \cdot \mathbf{r}'(t)}{\|\mathbf{r}(t)\|^3}\mathbf{r}(t)$$

# 11.2 Unit Tangent and Principal Unit Normal Vectors; Arc Length

**OBJECTIVES** *When you finish this section, you should be able to:*

**1** Interpret the derivative of a vector function geometrically (p. 807)
**2** Find the unit tangent vector and the principal unit normal vector of a smooth curve (p. 809)
**3** Find the arc length of a curve traced out by a vector function (p. 811)

**RECALL** The terms "normal," "perpendicular," and "orthogonal" all mean "meet at right angles."

A vector function $\mathbf{r} = \mathbf{r}(t)$ that is defined on a closed interval $a \le t \le b$ traces out a curve as $t$ varies over the interval. In this section, we analyze the curve geometrically by defining a *tangent vector* and a *normal vector* to the curve. We also express the arc length of a curve in terms of the unit tangent vector $\mathbf{T}$.

## 1 Interpret the Derivative of a Vector Function Geometrically

Suppose a vector function $\mathbf{r} = \mathbf{r}(t)$ is defined on an interval $[a, b]$ and is differentiable on $(a, b)$. Then $\mathbf{r} = \mathbf{r}(t)$ traces out a curve $C$ as $t$ varies over the interval. For a number $t_0$, $a < t_0 < b$, there is a point $P_0$ on $C$ given by the vector $\mathbf{r} = \mathbf{r}(t_0)$. If $h \ne 0$, there is a point $Q$ on $C$ different from $P_0$ given by the vector $\mathbf{r}(t_0 + h)$.

The vector $\mathbf{r}(t_0 + h) - \mathbf{r}(t_0)$ can be thought of as the secant vector from $P_0$ to $Q$, as shown in Figure 8. Then the scalar multiple $\dfrac{1}{h}$ of the secant vector from $P_0$ to $Q$, $\dfrac{\mathbf{r}(t_0 + h) - \mathbf{r}(t_0)}{h}$, is also a vector in the direction from $P_0$ to $Q$. Notice that whether $h < 0$ or $h > 0$, the direction of this vector follows the orientation of $C$.

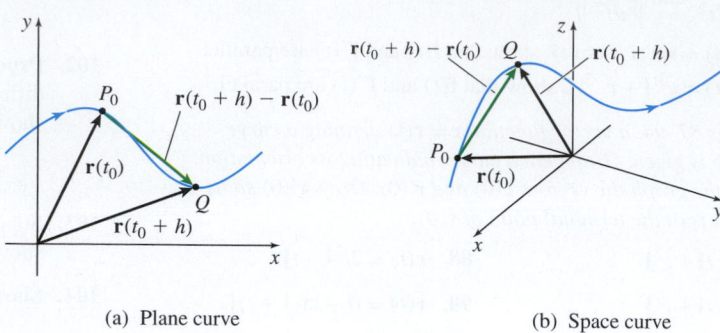

(a) Plane curve  (b) Space curve

**Figure 8** Secant vectors

As $h \to 0$, the vector $\dfrac{\mathbf{r}(t_0 + h) - \mathbf{r}(t_0)}{h}$ approaches the derivative $\mathbf{r}'(t_0)$. Also, as $h \to 0$, the vectors in the direction from $P_0$ to $Q$ move along the curve $C$ toward $P_0$, getting closer and closer to the vector that is tangent to $C$ at $P_0$. The direction of this tangent vector follows the orientation of $C$. See Figure 9. This leads to the following definition.

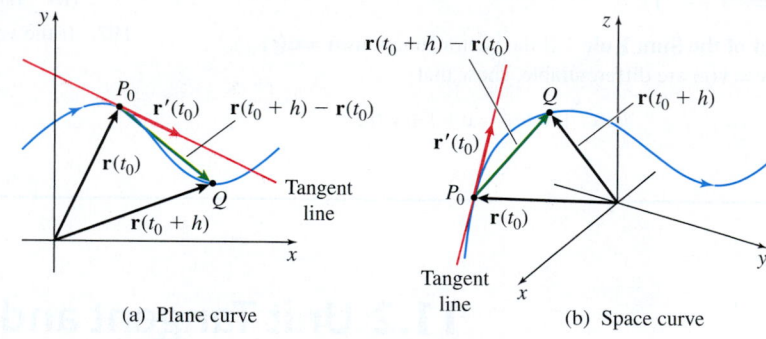

(a) Plane curve  (b) Space curve

**DF** **Figure 9** Tangent vectors; $\mathbf{r}'(t_0) = \lim\limits_{h \to 0} \dfrac{\mathbf{r}(t_0 + h) - \mathbf{r}(t_0)}{h}$

**DEFINITION** Tangent Vector to a Curve $C$

Suppose $\mathbf{r} = \mathbf{r}(t)$ is a vector function defined on the interval $[a, b]$ and differentiable on the interval $(a, b)$. Let $P_0$ be the point on the curve $C$ traced out by $\mathbf{r} = \mathbf{r}(t)$ corresponding to $t = t_0$, $a < t_0 < b$. If $\mathbf{r}'(t_0) \neq \mathbf{0}$, then the vector $\mathbf{r}'(t_0)$ is a **tangent vector** to the curve at $t_0$. The line containing $P_0$ in the direction of $\mathbf{r}'(t_0)$ is the **tangent line** to the curve traced out by $\mathbf{r} = \mathbf{r}(t)$ at $t_0$.

**NOTE** There is no tangent vector defined if $\mathbf{r}'(t_0) = \mathbf{0}$.

**EXAMPLE 1**  **Finding the Angle Between a Tangent Vector to a Helix and the Direction k**

**RECALL** The angle between two nonzero vectors $\mathbf{v}$ and $\mathbf{w}$ is determined by $\cos\theta = \dfrac{\mathbf{v} \cdot \mathbf{w}}{\|\mathbf{v}\| \|\mathbf{w}\|}$, where $0 \leq \theta \leq \pi$.

Show that the acute angle between the tangent vector to the helix

$$\mathbf{r}(t) = \cos t\,\mathbf{i} + \sin t\,\mathbf{j} + t\,\mathbf{k} \quad 0 \leq t \leq 2\pi$$

and the direction $\mathbf{k}$ is $\dfrac{\pi}{4}$ radian.

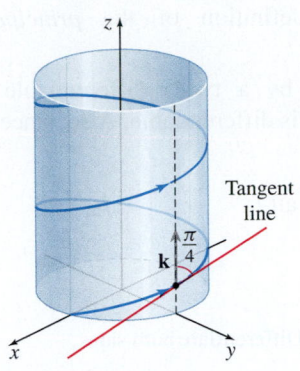

**Figure 10** $\mathbf{r}(t) = \cos t\mathbf{i} + \sin t\mathbf{j} + t\mathbf{k}$
$0 \le t \le 2\pi$

**Solution** A tangent vector at any point on the helix is given by

$$\mathbf{r}'(t) = -\sin t\mathbf{i} + \cos t\mathbf{j} + \mathbf{k}$$

Then $\|\mathbf{r}'(t)\| = \sqrt{(-\sin t)^2 + \cos^2 t + 1} = \sqrt{1 + 1} = \sqrt{2}$.

The cosine of the acute angle $\theta$ between $\mathbf{r}'(t)$ and $\mathbf{k}$ is

$$\cos\theta = \frac{\mathbf{r}'(t) \cdot \mathbf{k}}{\|\mathbf{r}'(t)\|\|\mathbf{k}\|} = \frac{1}{\sqrt{2}} = \frac{\sqrt{2}}{2} \qquad \mathbf{r}'(t) \cdot \mathbf{k} = 1$$

So, $\theta = \dfrac{\pi}{4}$ radian. See Figure 10. ∎

NOW WORK Problems 9 and 15.

## 2 Find the Unit Tangent Vector and the Principal Unit Normal Vector of a Smooth Curve

We now state the definition of a smooth curve in terms of vector functions. You may wish to compare this definition with the one given in Section 9.2, p. 690.

> **DEFINITION Smooth Curve**
>
> Let $C$ denote a curve traced out by a vector function $\mathbf{r} = \mathbf{r}(t)$ that is continuous on $a \le t \le b$. If the derivative $\mathbf{r}'(t)$ exists and is continuous on the interval $(a, b)$ and if $\mathbf{r}'(t)$ is never $\mathbf{0}$ on $(a, b)$, then $C$ is called a **smooth curve**.

Since $\mathbf{r}'(t)$ is never $\mathbf{0}$ for a smooth curve, such curves have tangent vectors at every point.

The *unit tangent vector* and the *unit normal vector* are important for analyzing the geometry of a curve (Section 11.3) and for describing motion along a curve (Section 11.4).

For a smooth curve $C$ traced out by the vector function $\mathbf{r} = \mathbf{r}(t)$, $a \le t \le b$, the **unit tangent vector** $\mathbf{T}(t)$ to $C$ at $t$ is

$$\boxed{\mathbf{T}(t) = \frac{\mathbf{r}'(t)}{\|\mathbf{r}'(t)\|}}$$

CALC
CLIP **EXAMPLE 2   Finding a Unit Tangent Vector to a Curve**

Show that the unit tangent vector $\mathbf{T}(t)$ to the circle of radius $R$

$$\mathbf{r}(t) = R\cos t\mathbf{i} + R\sin t\mathbf{j} \quad 0 \le t \le 2\pi$$

is everywhere orthogonal to $\mathbf{r}(t)$. Graph $\mathbf{r} = \mathbf{r}(t)$ and $\mathbf{T} = \mathbf{T}(t)$.

**Solution** We begin by finding $\mathbf{r}'(t)$ and $\|\mathbf{r}'(t)\|$:

$$\mathbf{r}'(t) = \frac{d}{dt}(R\cos t)\mathbf{i} + \frac{d}{dt}(R\sin t)\mathbf{j} = -R\sin t\mathbf{i} + R\cos t\mathbf{j}$$

$$\|\mathbf{r}'(t)\| = \sqrt{(-R\sin t)^2 + (R\cos t)^2} = R$$

Then the unit tangent vector is

$$\mathbf{T}(t) = \frac{\mathbf{r}'(t)}{\|\mathbf{r}'(t)\|} = \frac{-R\sin t\mathbf{i} + R\cos t\mathbf{j}}{R} = -\sin t\mathbf{i} + \cos t\mathbf{j}$$

To determine whether $\mathbf{r}(t)$ is orthogonal to $\mathbf{T}(t)$, we find their dot product.

$$\mathbf{T}(t) \cdot \mathbf{r}(t) = (-\sin t\mathbf{i} + \cos t\mathbf{j}) \cdot (R\cos t\mathbf{i} + R\sin t\mathbf{j}) = -R\sin t\cos t + R\sin t\cos t = 0$$

for all $t$. That is, $\mathbf{T}(t)$ is everywhere orthogonal to $\mathbf{r}(t)$, as shown in Figure 11. ∎

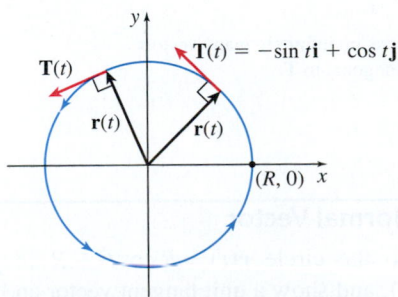

**Figure 11** $\mathbf{r}(t) = R\cos t\mathbf{i} + R\sin t\mathbf{j}$

The following discussion leads to the definition of the *principal unit normal vector.*

Suppose a smooth curve $C$ is traced out by a twice differentiable vector function $\mathbf{r} = \mathbf{r}(t)$. Then the unit tangent vector $\mathbf{T}(t)$ is differentiable. Also, since $\mathbf{T}(t)$ is a unit vector,

$$\mathbf{T}(t) \cdot \mathbf{T}(t) = 1 \quad \text{for all } t$$

If we differentiate this expression, we find

$$\frac{d}{dt}[\mathbf{T}(t) \cdot \mathbf{T}(t)] = \frac{d}{dt}(1) \qquad \text{Differentiate both sides.}$$

$$\mathbf{T}'(t) \cdot \mathbf{T}(t) + \mathbf{T}(t) \cdot \mathbf{T}'(t) = 0 \qquad \text{Use the Dot Product Rule.}$$

$$\mathbf{T}(t) \cdot \mathbf{T}'(t) = 0 \qquad \text{Simplify.}$$

We conclude that $\mathbf{T}'(t)$ is a vector that is orthogonal to $\mathbf{T}(t)$ at every point on the curve $C$. If $\mathbf{T}'(t) \neq \mathbf{0}$, the vector $\dfrac{\mathbf{T}'(t)}{\|\mathbf{T}'(t)\|}$ is a unit vector that is orthogonal to $\mathbf{T}(t)$ at every point on the curve $C$.

---

**DEFINITION**  Principal Unit Normal Vector

For a smooth curve $C$ traced out by a vector function $\mathbf{r} = \mathbf{r}(t)$, $a \leq t \leq b$, that is twice differentiable for $a < t < b$, the **principal unit normal vector** $\mathbf{N}(t)$ to $C$ at $t$ is

$$\boxed{\mathbf{N}(t) = \frac{\mathbf{T}'(t)}{\|\mathbf{T}'(t)\|}}$$

provided $\mathbf{T}'(t) \neq \mathbf{0}$.

---

Notice that the unit normal vector $\mathbf{N}$ is not defined if $\mathbf{T}' = \mathbf{0}$. So, for example, lines do not have unit normal vectors.

If $C$ is a plane curve, there are two unit vectors that are orthogonal to the unit tangent vector $\mathbf{T}$, as shown in Figure 12(a). If $C$ is a curve in space, there are infinitely many unit vectors that are orthogonal to the unit tangent vector $\mathbf{T}$, as shown in Figure 12(b). The definition of the principal unit normal vector identifies exactly one of the orthogonal vectors and gives it a name.

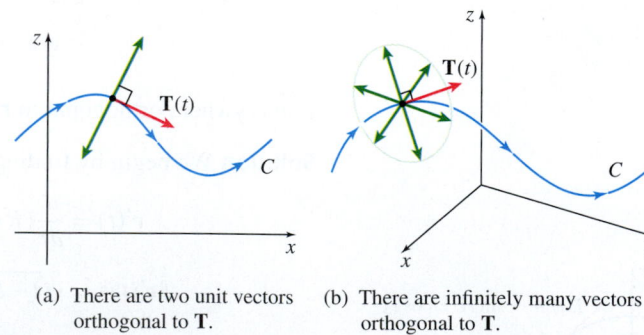

(a) There are two unit vectors orthogonal to $\mathbf{T}$.

(b) There are infinitely many vectors orthogonal to $\mathbf{T}$.

**Figure 12**

---

**EXAMPLE 3**  Finding the Principal Unit Normal Vector

Find the principal unit normal vector $\mathbf{N}(t)$ to the circle $\mathbf{r}(t) = R \cos t\,\mathbf{i} + R \sin t\,\mathbf{j}$, $0 \leq t \leq 2\pi$. (Refer to Example 2.) Graph $\mathbf{r} = \mathbf{r}(t)$, and show a unit tangent vector and a unit normal vector.

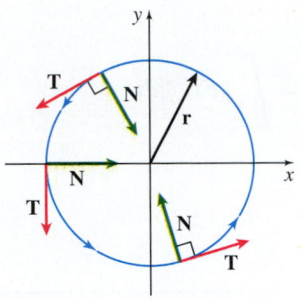

**DF** **Figure 13** $\mathbf{r}(t) = R\cos t\,\mathbf{i} + R\sin t\,\mathbf{j}$
$0 \le t \le 2\pi$

**Solution** From Example 2, the unit tangent vector $\mathbf{T}(t)$ is

$$\mathbf{T}(t) = -\sin t\,\mathbf{i} + \cos t\,\mathbf{j}$$

Since $\mathbf{T}'(t) = -\cos t\,\mathbf{i} - \sin t\,\mathbf{j}$, the principal unit normal vector $\mathbf{N}(t)$ is

$$\mathbf{N}(t) = \frac{\mathbf{T}'(t)}{\|\mathbf{T}'(t)\|} = \frac{-\cos t\,\mathbf{i} - \sin t\,\mathbf{j}}{\sqrt{(-\cos t)^2 + (-\sin t)^2}} = -\cos t\,\mathbf{i} - \sin t\,\mathbf{j}$$

Since $\mathbf{r}(t) = R(\cos t\,\mathbf{i} + \sin t\,\mathbf{j})$, the vector $\mathbf{N}(t)$ is a unit vector opposite in direction to the vector $\mathbf{r}(t)$, so $\mathbf{N}$ is directed toward the center of the circle, as shown in Figure 13. ∎

**NOW WORK** Problem 29.

**EXAMPLE 4**  **Analyzing the Principal Unit Normal Vector of a Helix**

Show that the principal unit normal vector $\mathbf{N}(t)$ of the helix

$$\mathbf{r}(t) = \cos t\,\mathbf{i} + \sin t\,\mathbf{j} + t\,\mathbf{k}$$

is orthogonal to the $z$-axis.

**Solution** We begin by finding $\mathbf{r}'(t)$ and $\|\mathbf{r}'(t)\|$.

$$\mathbf{r}'(t) = -\sin t\,\mathbf{i} + \cos t\,\mathbf{j} + \mathbf{k} \qquad \|\mathbf{r}'(t)\| = \sqrt{(-\sin t)^2 + (\cos t)^2 + 1} = \sqrt{2}$$

Then the unit tangent vector is $\mathbf{T}(t) = \dfrac{\mathbf{r}'(t)}{\|\mathbf{r}'(t)\|} = \dfrac{-\sin t\,\mathbf{i} + \cos t\,\mathbf{j} + \mathbf{k}}{\sqrt{2}}$.

Since

$$\mathbf{T}'(t) = \frac{-\cos t\,\mathbf{i} - \sin t\,\mathbf{j}}{\sqrt{2}} \qquad \text{and} \qquad \|\mathbf{T}'(t)\| = \sqrt{\left(-\frac{\cos t}{\sqrt{2}}\right)^2 + \left(-\frac{\sin t}{\sqrt{2}}\right)^2} = \frac{1}{\sqrt{2}}$$

we have

$$\mathbf{N}(t) = \frac{\mathbf{T}'(t)}{\|\mathbf{T}'(t)\|} = -\cos t\,\mathbf{i} - \sin t\,\mathbf{j}$$

The direction of the $z$-axis is $\mathbf{k}$, so it follows that $\mathbf{N}(t) \cdot \mathbf{k} = 0$ for all $t$. The principal unit normal vector $\mathbf{N}(t)$ is always orthogonal to the $z$-axis, as shown in Figure 14. ∎

**NOW WORK** Problem 37.

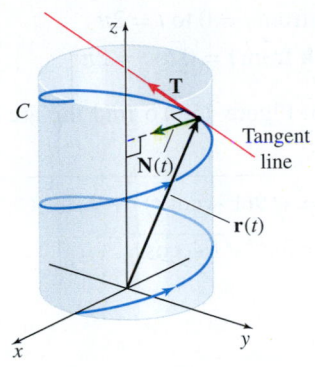

**Figure 14** $\mathbf{r}(t) = \cos t\,\mathbf{i} + \sin t\,\mathbf{j} + t\,\mathbf{k}$

**NEED TO REVIEW?** The arc length of a smooth curve is discussed in Section 9.3, pp. 695–697.

### 3 Find the Arc Length of a Curve Traced Out by a Vector Function

A formula for the arc length of a smooth plane curve was derived in Chapter 9. The formula for the arc length of a smooth curve traced out by a vector function is given next.

**THEOREM  Arc Length of a Vector Function**

If a smooth curve $C$ is traced out by the vector function $\mathbf{r} = \mathbf{r}(t)$, $a \le t \le b$, the **arc length** $s$ along $C$ from $t = a$ to $t = b$ is

$$s = \int_a^b \|\mathbf{r}'(t)\|\,dt$$

**Proof** For a smooth plane curve $C$, traced out by

$$\mathbf{r}(t) = x(t)\mathbf{i} + y(t)\mathbf{j} \qquad a \le t \le b$$

we have

$$\mathbf{r}'(t) = \frac{dx}{dt}\mathbf{i} + \frac{dy}{dt}\mathbf{j} \qquad \text{and} \qquad \|\mathbf{r}'(t)\| = \sqrt{\left(\frac{dx}{dt}\right)^2 + \left(\frac{dy}{dt}\right)^2}$$

The arc length of $C$ from $a$ to $b$ is given by

$$s = \int_a^b \sqrt{\left(\frac{dx}{dt}\right)^2 + \left(\frac{dy}{dt}\right)^2}\, dt = \int_a^b \|\mathbf{r}'(t)\|\, dt$$

For a smooth space curve $C$, traced out by

$$\mathbf{r}(t) = x(t)\mathbf{i} + y(t)\mathbf{j} + z(t)\mathbf{k} \qquad a \le t \le b$$

we have

$$\mathbf{r}'(t) = \frac{dx}{dt}\mathbf{i} + \frac{dy}{dt}\mathbf{j} + \frac{dz}{dt}\mathbf{k} \qquad \text{and} \qquad \|\mathbf{r}'(t)\| = \sqrt{\left(\frac{dx}{dt}\right)^2 + \left(\frac{dy}{dt}\right)^2 + \left(\frac{dz}{dt}\right)^2}$$

The arc length of $C$ from $a$ to $b$ is given by

$$s = \int_a^b \sqrt{\left(\frac{dx}{dt}\right)^2 + \left(\frac{dy}{dt}\right)^2 + \left(\frac{dz}{dt}\right)^2}\, dt = \int_a^b \|\mathbf{r}'(t)\|\, dt$$    ∎

### EXAMPLE 5    Finding the Arc Length of a Cycloid and a Helix

Find the arc length of:

(a)  The cycloid $\mathbf{r}(t) = 3(t - \sin t)\mathbf{i} + 3(1 - \cos t)\mathbf{j}$ from $t = 0$ to $t = 2\pi$

(b)  The circular helix $\mathbf{r}(t) = R\cos t\,\mathbf{i} + R\sin t\,\mathbf{j} + t\mathbf{k}$ from $t = 0$ to $t = 2\pi$

**Solution** (a) The graph of the cycloid is shown in Figure 15. To find the arc length from $t = 0$ to $t = 2\pi$, we first find $\mathbf{r}'(t)$ and $\|\mathbf{r}'(t)\|$:

$$\mathbf{r}'(t) = 3(1 - \cos t)\mathbf{i} + 3\sin t\,\mathbf{j} \qquad \|\mathbf{r}'(t)\| = \sqrt{9(1 - \cos t)^2 + 9\sin^2 t}$$
$$= 3\sqrt{2}\sqrt{1 - \cos t}$$

Now use the formula for arc length.

$$s = \int_a^b \|\mathbf{r}'(t)\|\, dt = 3\sqrt{2}\int_0^{2\pi} \sqrt{1 - \cos t}\, dt$$

Since $\sin\dfrac{t}{2} \ge 0$, for $0 \le t \le 2\pi$, use the half-angle identity $\sin\dfrac{t}{2} = \sqrt{\dfrac{1 - \cos t}{2}}$ to integrate $\sqrt{1 - \cos t}$.

$$s = 3\sqrt{2}\int_0^{2\pi} \sqrt{1 - \cos t}\, dt = 3\sqrt{2}\int_0^{2\pi} \sqrt{2}\sin\frac{t}{2}\, dt$$
$$= 6\int_0^{2\pi} \sin\frac{t}{2}\, dt = 12\left[-\cos\frac{t}{2}\right]_0^{2\pi} = 12(1 + 1) = 24$$

(b) We begin by finding $\mathbf{r}'(t)$ and $\|\mathbf{r}'(t)\|$.

$$\mathbf{r}'(t) = -R\sin t\,\mathbf{i} + R\cos t\,\mathbf{j} + \mathbf{k},$$
$$\|\mathbf{r}'(t)\| = \sqrt{(-R\sin t)^2 + (R\cos t)^2 + 1^2} = \sqrt{R^2 + 1}$$

Then

$$s = \int_a^b \|\mathbf{r}'(t)\|\, dt = \int_0^{2\pi} \sqrt{R^2 + 1}\, dt = \left[\sqrt{R^2 + 1}\, t\right]_0^{2\pi} = 2\pi\sqrt{R^2 + 1}$$

See Figure 16. ∎

NOW WORK Problem 51.

For most vector functions, finding arc length requires technology.

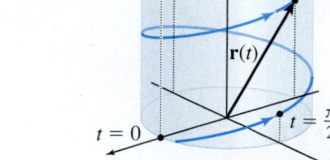

**Figure 15** $\mathbf{r}(t) = 3(t - \sin t)\mathbf{i} + 3(1 - \cos t)\mathbf{j}$

**Figure 16** $\mathbf{r}(t) = R\cos t\,\mathbf{i} + R\sin t\,\mathbf{j} + t\mathbf{k}$
$0 \le t \le 2\pi$

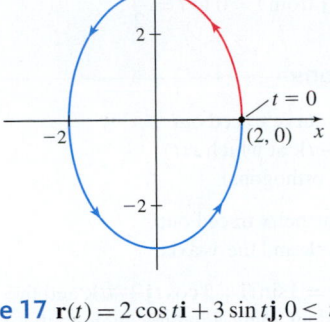

**Figure 17** $\mathbf{r}(t) = 2\cos t\mathbf{i} + 3\sin t\mathbf{j}, 0 \le t \le \dfrac{\pi}{2}$

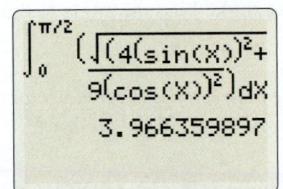

**Figure 18**

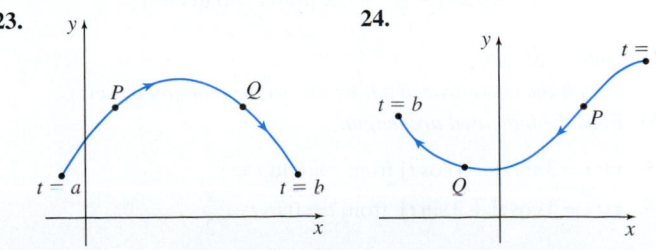

**EXAMPLE 6  Using Technology to Find the Arc Length of an Ellipse**

Use technology to find the arc length $s$ of the ellipse shown in Figure 17 traced out by the vector function $\mathbf{r}(t) = 2\cos t\mathbf{i} + 3\sin t\mathbf{j}$ from $t = 0$ to $t = \dfrac{\pi}{2}$.

**Solution** We begin by finding $\mathbf{r}'(t)$ and $\|\mathbf{r}'(t)\|$.

$$\mathbf{r}'(t) = -2\sin t\mathbf{i} + 3\cos t\mathbf{j} \qquad \|\mathbf{r}'(t)\| = \sqrt{4\sin^2 t + 9\cos^2 t}$$

Now use the formula for arc length.

$$s = \int_a^b \|\mathbf{r}'(t)\|dt = \int_0^{\pi/2} \sqrt{4\sin^2 t + 9\cos^2 t}\,dt$$

This is an integral that has no antiderivative in terms of elementary functions. To obtain a numerical approximation to the arc length $s$, we use a graphing calculator. Then

$$s = \int_0^{\pi/2} \sqrt{4\sin^2 t + 9\cos^2 t}\,dt \approx 3.96636$$

The screen shot is given in Figure 18.  ■

**NOW WORK** Problem 55.

# 11.2 Assess Your Understanding

## Concepts and Vocabulary

1. *True or False*  If a vector function $\mathbf{r} = \mathbf{r}(t)$ is differentiable at $t_0$, $a < t_0 < b$, and $P_0$ is the point on the curve traced out by $\mathbf{r} = \mathbf{r}(t)$ corresponding to $t = t_0$, then the tangent line to the curve at $t_0$ contains the point $P_0$ and has the direction $\mathbf{r}(t_0)$.

2. *True or False*  A curve traced out by a vector function $\mathbf{r} = \mathbf{r}(t)$ is called a smooth curve on the interval $[a, b]$ if $\mathbf{r} = \mathbf{r}(t)$ is continuous on the interval $[a, b]$.

3. *True or False*  If $C$ is a smooth curve traced out by a vector function $\mathbf{r} = \mathbf{r}(t)$, $a \le t \le b$, then $\mathbf{T}(t) = \mathbf{r}'(t)$ is the unit tangent vector to $C$ at $t$.

4. *True or False*  If $C$ is a smooth curve traced out by a twice differentiable vector function $\mathbf{r} = \mathbf{r}(t)$, $a \le t \le b$, and if $\mathbf{T}$ is the unit tangent vector to $C$, then $\mathbf{N}(t) = \dfrac{\mathbf{T}(t)}{\|\mathbf{T}(t)\|}$ is the principal unit normal vector to $C$ at $t$.

5. *True or False*  For a smooth space curve $C$ traced out by $\mathbf{r} = \mathbf{r}(t)$, $a \le t \le b$, the integral $\displaystyle\int_a^b \|\mathbf{r}'(t)\|dt$ equals the arc length along $C$ from $t = a$ to $t = b$.

6. *True or False*  For a smooth curve traced out by a twice differentiable vector function, the unit tangent vector is orthogonal to the principal unit normal vector.

## Skill Building

*In Problems 7–14:*

*(a)  Find the tangent vector to each curve at $t$.*

*(b)  Graph the curve.*

*(c)  Add the graph of the tangent vector to (b).*

7. $\mathbf{r}(t) = t\mathbf{i} - t^2\mathbf{j}$  $t = 1$

8. $\mathbf{r}(t) = t\mathbf{i} - t^3\mathbf{j}$  $t = -1$

9. $\mathbf{r}(t) = (t^2 + 1)\mathbf{i} + (1 - t)\mathbf{j}$  $t = 1$

10. $\mathbf{r}(t) = t^3\mathbf{i} + 3t\mathbf{j}$  $t = 1$

11. $\mathbf{r}(t) = 4t\mathbf{i} - \sqrt{t}\mathbf{j}$  $t = 1$

12. $\mathbf{r}(t) = \sqrt{t}\mathbf{i} + \dfrac{1}{2}t\mathbf{j}$  $t = 4$

13. $\mathbf{r}(t) = e^t\mathbf{i} + e^{-t}\mathbf{j}$  $t = 0$

14. $\mathbf{r}(t) = e^{2t}\mathbf{i} + e^{-t}\mathbf{j}$  $t = 0$

*In Problems 15–22, find the tangent vector to each curve at $t$.*

15. $\mathbf{r}(t) = (1 - 3t)\mathbf{i} + 2t\mathbf{j} - (5 + t)\mathbf{k}$  $t = 0$

16. $\mathbf{r}(t) = (2 + t)\mathbf{i} + (2 - t)\mathbf{j} + 3t\mathbf{k}$  $t = 0$

17. $\mathbf{r}(t) = \cos(2t)\mathbf{i} + \sin(2t)\mathbf{j} - 5\mathbf{k}$  $t = \dfrac{\pi}{4}$

18. $\mathbf{r}(t) = 3\mathbf{i} + \cos t\mathbf{j} + \sin t\mathbf{k}$  $t = \dfrac{\pi}{6}$

19. $\mathbf{r}(t) = 2\cos t\mathbf{i} + \mathbf{j} + 2\sin t\mathbf{k}$  $t = \dfrac{\pi}{6}$

20. $\mathbf{r}(t) = 2\cos(2t)\mathbf{i} + 2\sin(2t)\mathbf{j} + 5\mathbf{k}$  $t = \dfrac{\pi}{2}$

21. $\mathbf{r}(t) = e^t\cos t\mathbf{i} + e^t\sin t\mathbf{j} + e^t\mathbf{k}$  $t = 0$

22. $\mathbf{r}(t) = e^{-t}\cos t\mathbf{i} + e^{-t}\sin t\mathbf{j} - e^{-t}\mathbf{k}$  $t = 0$

*In Problems 23–26, draw the unit tangent vector $\mathbf{T}$ and principal unit normal vector $\mathbf{N}$ of each curve at the points $P$ and $Q$. Be sure to use the orientation of the curve. Assume the length of a unit vector is one inch.*

23.

24.

---

1. = NOW WORK problem      = Graphing technology recommended      **CAS** = Computer Algebra System recommended

**25.**

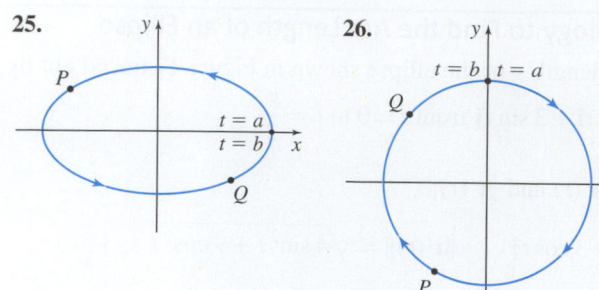

**26.**

**57.** $\mathbf{r}(t) = 2\sin t\mathbf{i} + 4\cos(2t)\mathbf{j}$ from $t = 0$ to $t = 2\pi$

**58.** $\mathbf{r}(t) = 4\sin(2t)\mathbf{i} - 6\cos t\mathbf{j}$ from $t = 0$ to $t = \dfrac{\pi}{2}$

## Applications and Extensions

**59.** Find all the points on the curve traced out by $\mathbf{r}(t) = t^2\mathbf{i} + (t^2 - 1)\mathbf{j} - t\mathbf{k}$ at which $\mathbf{r}(t)$ and its tangent vector are orthogonal.

**60.** Find the angle between the helix traced out by $\mathbf{r}(t) = \cos t\mathbf{i} + t\mathbf{j} + \sin t\mathbf{k}$ and the $y$-axis.

**61.** The helix defined by $\mathbf{r}(t) = 3\sin t\mathbf{i} + 3\cos t\mathbf{j} + 4t\mathbf{k}$ and the direction $\mathbf{k}$ meet at a constant angle. Find the angle to the nearest degree.

**62.** The helix defined by $\mathbf{r}(t) = 2\cos t\mathbf{i} + 2\sin t\mathbf{j} + t\mathbf{k}$ and the direction $\mathbf{k}$ meet at a constant angle. Find the angle to the nearest degree.

*In Problems 27–42, for each vector function $\mathbf{r} = \mathbf{r}(t)$, find the unit tangent vector $\mathbf{T}$ and the principal unit normal vector $\mathbf{N}$ at $t$.*

**27.** $\mathbf{r}(t) = t\mathbf{i} + t^2\mathbf{j}$   $t = 1$

**28.** $\mathbf{r}(t) = t\mathbf{i} - t^3\mathbf{j}$   $t = -1$

**29.** $\mathbf{r}(t) = (t^2 + 1)\mathbf{i} + (1 - t)\mathbf{j}$   $t = 1$

**30.** $\mathbf{r}(t) = t^3\mathbf{i} + 3t\mathbf{j}$   $t = 1$

**31.** $\mathbf{r}(t) = \dfrac{2t}{t + 1}\mathbf{i} - \dfrac{t^2}{t + 1}\mathbf{j}$   $t = 1$

**32.** $\mathbf{r}(t) = \sqrt{t}\mathbf{i} + \dfrac{1}{2}t\mathbf{j}$   $t = 4$

**33.** $\mathbf{r}(t) = e^t\mathbf{i} + e^{-t}\mathbf{j}$   $t = 0$

**34.** $\mathbf{r}(t) = e^{2t}\mathbf{i} + e^{-t}\mathbf{j}$   $t = 0$

**35.** $\mathbf{r}(t) = (1 - 3t)\mathbf{i} + 2t\mathbf{j} - (5 + t)\mathbf{k}$   $t = 0$

**36.** $\mathbf{r}(t) = (2 + t)\mathbf{i} + (2 - t)\mathbf{j} + 3t\mathbf{k}$   $t = 0$

**37.** $\mathbf{r}(t) = 2\cos t\mathbf{i} + \mathbf{j} + 2\sin t\mathbf{k}$   $t = \dfrac{\pi}{6}$

**38.** $\mathbf{r}(t) = 3\mathbf{i} + \cos t\mathbf{j} + \sin t\mathbf{k}$   $t = \dfrac{\pi}{6}$

**39.** $\mathbf{r}(t) = \cos(2t)\mathbf{i} + \sin(2t)\mathbf{j} - 5\mathbf{k}$   $t = \dfrac{\pi}{4}$

**40.** $\mathbf{r}(t) = 2\cos(2t)\mathbf{i} + 2\sin(2t)\mathbf{j} + 5\mathbf{k}$   $t = \dfrac{\pi}{2}$

**41.** $\mathbf{r}(t) = e^t\cos t\mathbf{i} + e^t\sin t\mathbf{j} + e^t\mathbf{k}$   $t = 0$

**42.** $\mathbf{r}(t) = e^{-t}\cos t\mathbf{i} + e^{-t}\sin t\mathbf{j} - e^{-t}\mathbf{k}$   $t = 0$

*In Problems 43–54, find the arc length $s$ of each vector function.*

**43.** $\mathbf{r}(t) = t\mathbf{i} + (t^{3/2} + 1)\mathbf{j}$ from $t = 1$ to $t = 8$

**44.** $\mathbf{r}(t) = t\mathbf{i} + t^{3/2}\mathbf{j}$ from $t = 0$ to $t = 4$

**45.** $\mathbf{r}(t) = 8t\mathbf{i} + \dfrac{t^6 + 2}{t^2}\mathbf{j}$ from $t = 1$ to $t = 2$

**46.** $\mathbf{r}(t) = t\mathbf{i} + (1 - t^{2/3})^{3/2}\mathbf{j}$ from $t = \dfrac{1}{8}$ to $t = 1$

**47.** $\mathbf{r}(t) = t\mathbf{i} + 2t\mathbf{j} + t\mathbf{k}$ from $t = 0$ to $t = 1$

**48.** $\mathbf{r}(t) = 2t\mathbf{i} + t\mathbf{j} + 3t\mathbf{k}$ from $t = 1$ to $t = 3$

**49.** $\mathbf{r}(t) = \sin(2t)\mathbf{i} + \cos(2t)\mathbf{j} + t\mathbf{k}$ from $t = 0$ to $t = \pi$

**50.** $\mathbf{r}(t) = \sin t\mathbf{i} + \cos t\mathbf{j} + bt\mathbf{k}$ from $t = 0$ to $t = 2\pi$

**51.** $\mathbf{r}(t) = e^t\mathbf{i} + e^{-t}\mathbf{j} + \sqrt{2}t\mathbf{k}$ from $t = 0$ to $t = 1$

**52.** $\mathbf{r}(t) = \cos^3 t\mathbf{i} + \sin^3 t\mathbf{j} + \mathbf{k}$ from $t = 0$ to $t = \dfrac{\pi}{2}$

**53.** $\mathbf{r}(t) = e^t\cos t\mathbf{i} + e^t\sin t\mathbf{j} + e^t\mathbf{k}$ from $t = 0$ to $t = 2\pi$

**54.** $\mathbf{r}(t) = t^2\mathbf{i} - 2\sqrt{2}\,t\mathbf{j} + (t^2 - 1)\mathbf{k}$ from $t = 0$ to $t = 1$

*In Problems 63–66, find the angle between the tangent vectors to the curves traced out by $\mathbf{r}_1(t_1)$ and $\mathbf{r}_2(t_2)$ at the point of intersection given.*

**63.** $\mathbf{r}_1(t_1) = t_1\mathbf{i} + \sin(\pi t_1)\mathbf{j} + \mathbf{k}$ and $\mathbf{r}_2(t_2) = \mathbf{i} + t_2\mathbf{j} + (1 + t_2)\mathbf{k}$ at $(1, 0, 1)$

**64.** $\mathbf{r}_1(t_1) = \sin(2t_1)\mathbf{i} + \sin t_1\mathbf{j} + t_1\mathbf{k}$ and $\mathbf{r}_2(t_2) = t_2^2\mathbf{i} + t_2\mathbf{j} + t_2^3\mathbf{k}$ at $(0, 0, 0)$

**65.** $\mathbf{r}_1(t_1) = (e^{t_1} - 1)\mathbf{i} - \cos(\pi t_1)\mathbf{j} + t_1\mathbf{k}$ and $\mathbf{r}_2(t_2) = (1 - t_2)\mathbf{i} - \mathbf{j} + (t_2 - 1)\mathbf{k}$ at $(0, -1, 0)$

**66.** $\mathbf{r}_1(t_1) = t_1^2\mathbf{i} - (t_1 - 3)\mathbf{j} - (t_1 - 3)\mathbf{k}$ and $\mathbf{r}_2(t_2) = (t_2^2 + 3)\mathbf{i} - (t_2 - 2)\mathbf{j} + t_2\mathbf{k}$ at $(4, 1, 1)$

*If $\mathbf{r} = \mathbf{r}(t)$ is a vector function in space that is twice differentiable, then the cross product of the unit tangent vector and the principal unit normal vector, $\mathbf{T} \times \mathbf{N}$, defines a third unit vector $\mathbf{B}(t)$. This vector $\mathbf{B}$, called the **binormal vector**, is orthogonal to both the unit tangent vector $\mathbf{T}$ and the principal unit normal vector $\mathbf{N}$, as shown in the figure.*

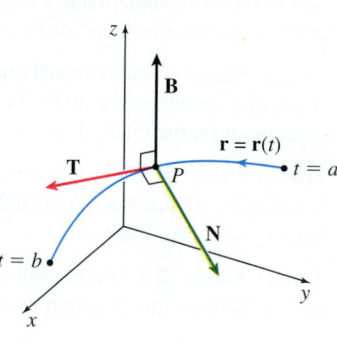

*In Problems 67–70, find the unit tangent vector $\mathbf{T}$, the principal unit normal vector $\mathbf{N}$, and the binormal vector $\mathbf{B}$ to each vector function.*

**67.** $\mathbf{r}(t) = t^2\mathbf{i} - (t - 3)\mathbf{j} - (t - 3)\mathbf{k}$

**68.** $\mathbf{r}(t) = (t + 5)\mathbf{i} - (t + 5)\mathbf{j} + t^2\mathbf{k}$

**69.** $\mathbf{r}(t) = \cos t\mathbf{i} + \sin t\mathbf{j} + t\mathbf{k}$

**70.** $\mathbf{r}(t) = 2\sin t\mathbf{i} + 2\cos t\mathbf{j} + 3t\mathbf{k}$

**71.** Suppose the vector function $\mathbf{r} = \mathbf{r}(t)$ in space is differentiable and has a nonzero tangent vector at $t_0$. Find the direction cosines for the tangent vector at $t_0$.

*In Problems 55–58,*

**(a)** *Graph the curve traced out by the vector function $\mathbf{r} = \mathbf{r}(t)$.*

**(b)** *Find the indicated arc length.*

**55.** $\mathbf{r}(t) = 3\sin t\mathbf{i} - 4\cos t\mathbf{j}$ from $t = 0$ to $t = \pi$

**56.** $\mathbf{r}(t) = 5\cos t\mathbf{i} + 3\sin t\mathbf{j}$ from $t = 0$ to $t = \dfrac{\pi}{2}$

**72.** Show that at any point, the tangent vectors to a helix $\mathbf{r}(t) = a\cos t\mathbf{i} + a\sin t\mathbf{j} + bt\mathbf{k}$, where $a$ and $b$ are real numbers, intersect the direction of the $z$-axis at a constant angle.

**Challenge Problems**

73. Use Simpson's Rule with $n = 4$ to approximate the length of the curve $\mathbf{r}(t) = t^2\mathbf{i} + t^3\mathbf{j} + (2t + 3)\mathbf{k}$, $0 \le t \le 2$.

74. (a) Approximate the arc length of $\mathbf{r}(t) = 3\sin t\mathbf{i} - 4\cos t\mathbf{j}$ from $t = 0$ to $t = \pi$ using Simpson's Rule with $n = 4$.

    (b) Approximate the arc length of $\mathbf{r}(t) = 3\sin t\mathbf{i} - 4\cos t\mathbf{j}$ from $t = 0$ to $t = \pi$ using the Trapezoidal Rule with $n = 6$.

    (c) Compare the answers to (a) and (b) to the result given by technology (see Problem 55).

75. (a) Approximate the arc length of $\mathbf{r}(t) = 2\sin t\mathbf{i} + 4\cos(2t)\mathbf{j}$ from $t = 0$ to $t = 2\pi$ using Simpson's Rule with $n = 4$.

    (b) Approximate the arc length of $\mathbf{r}(t) = 2\sin t\mathbf{i} + 4\cos(2t)\mathbf{j}$ from $t = 0$ to $t = 2\pi$ using the Trapezoidal Rule with $n = 6$.

    (c) Compare the answers to (a) and (b) to the result given by technology (see Problem 57).

76. Approximate the length of the curve $\mathbf{r}(t) = \dfrac{1}{3}t^3\mathbf{i} + (t - 1)\mathbf{j} + 2\mathbf{k}$, $0 \le t \le \dfrac{1}{2}$, correct to three decimal places, by expanding the integrand in a power series.

# 11.3 Arc Length as Parameter; Curvature

**OBJECTIVES** *When you finish this section, you should be able to:*

1 Determine whether the parameter used in a vector function is arc length (p. 815)
2 Find the curvature of a curve (p. 817)
3 Find the curvature of a space curve (p. 819)
4 Find the curvature of a plane curve given by $y = f(x)$ (p. 820)
5 Find an osculating circle (p. 821)

In this section, we discuss geometric properties of curves.

We have seen that the same curve can be defined using different parametric equations. For example, the upper half of a unit circle can be given by

$$\mathbf{r}(t) = t\mathbf{i} + \sqrt{1 - t^2}\,\mathbf{j} \quad -1 \le t \le 1$$

where $x(t) = t$ and $y(t) = \sqrt{1 - t^2}$. Here the parameter $t$ is the position on the $x$-axis.

This same curve can be given by

$$\mathbf{r}(t) = \cos t\mathbf{i} + \sin t\mathbf{j} \quad 0 \le t \le \pi$$

where $x(t) = \cos t$ and $y(t) = \sin t$. Here the parameter $t$ is the angle between the positive $x$-axis and $\mathbf{r} = \mathbf{r}(t)$. Since the curve is the upper half of the unit circle, the arc length $s$ subtended by the angle $t$ is $s = r\theta = t$. That is, the parameter $t$ equals the arc length along the circle.

Just as time is the preferred parameter for motion along a curve, arc length along a curve is the preferred parameter for discussing a curve's geometric properties—for several reasons.

First, the arc length of a curve is a natural consequence of the curve's shape. Further, arc length is independent of the representation of the curve. For example, the arc length of the semi-circle, $x = 2t$, $y = 2\sqrt{1 - t^2}$, $-1 \le t \le 1$, is $2\pi$. The arc length of the same semi-circle, $x = 2\sin\theta$, $y = 2\cos\theta$, $-\dfrac{\pi}{2} \le \theta \le \dfrac{\pi}{2}$, is also $2\pi$. No matter how the curve is defined, its arc length is the same.

## 1 Determine Whether the Parameter Used in a Vector Function Is Arc Length

So how do we determine whether the parameter of a curve represents arc length? And, if the parameter does not represent arc length, how can we change it so it does represent arc length? The next result provides a means for determining whether the parameter of a vector function is arc length.

Let's examine the relationship between arc length $s$ along a curve $C$ and the parameter $t$ used to define $C$. The arc length $s$ of a smooth curve $C$ traced out by a vector function $\mathbf{r} = \mathbf{r}(t)$, $a \leq t \leq b$, from $a$ to an arbitrary parameter value $t$ in $[a, b]$ is

$$s = s(t) = \int_a^t \|\mathbf{r}'(u)\| \, du \tag{1}$$

Then by the Fundamental Theorem of Calculus,

$$\boxed{\frac{ds}{dt} = \|\mathbf{r}'(t)\|} \tag{2}$$

**NEED TO REVIEW?** The Fundamental Theorem of Calculus, Part 1 is discussed in Section 5.3, pp. 378–379.

---

**THEOREM** Criterion for Arc Length as Parameter

Suppose a smooth curve $C$ is traced out by the vector function

$$\mathbf{r} = \mathbf{r}(t) \quad a \leq t \leq b$$

The parameter $t$ is the arc length along $C$ if and only if

$$\boxed{\|\mathbf{r}'(t)\| = 1 \quad \text{for all } t}$$

---

**Proof** If $\|\mathbf{r}'(t)\| = 1$ for all $t$, then from equation (1) the arc length $s$ is

$$s = s(t) = \int_a^t du = t - a$$

That is, the parameter $t$ is the arc length $s$ measured along $C$ from $t = a$ to $t = b$. Notice that when $t = a$, the length $s$ is zero, as expected.

Conversely, if the parameter $t$ is arc length $s$, then $s(t) = t$ and

$$\frac{ds}{dt} = 1 = \|\mathbf{r}'(t)\| \qquad \text{for all } t \qquad \blacksquare$$

---

**EXAMPLE 1** Determining Whether the Parameter Used in a Vector Function Is Arc Length

Determine whether the parameter used in each vector function is arc length.

**(a)** $C_1$: $\mathbf{r}(t) = 2 \sin \dfrac{t}{2} \mathbf{i} + 2 \cos \dfrac{t}{2} \mathbf{j}$ $\quad 0 \leq t \leq 2\pi$

**(b)** $C_2$: $\mathbf{r}(t) = \cos t \, \mathbf{i} + \sin t \, \mathbf{j} + t \mathbf{k}$ $\quad t \geq 0$

**Solution (a)** We begin by finding $\mathbf{r}'(t)$ and $\|\mathbf{r}'(t)\|$.

$$\mathbf{r}'(t) = \cos \frac{t}{2} \mathbf{i} - \sin \frac{t}{2} \mathbf{j} \qquad \|\mathbf{r}'(t)\| = \sqrt{\cos^2 \frac{t}{2} + \sin^2 \frac{t}{2}} = 1 \quad \text{for all } t$$

Since $\|\mathbf{r}'(t)\| = 1$ for all $t$, the parameter $t$ is arc length as measured along $C_1$.

**(b)** We begin by finding $\mathbf{r}'(t)$ and $\|\mathbf{r}'(t)\|$.

$$\mathbf{r}'(t) = -\sin t \, \mathbf{i} + \cos t \, \mathbf{j} + \mathbf{k} \qquad \|\mathbf{r}'(t)\| = \sqrt{\sin^2 t + \cos^2 t + 1} = \sqrt{2}$$

Since $\|\mathbf{r}'(t)\| \neq 1$ for all $t$, the parameter $t$ does not measure arc length along $C_2$. $\blacksquare$

**NOW WORK** Problem 13.

---

**EXAMPLE 2** Changing the Parameter to Arc Length

Represent the helix in Example 1(b)

$$\mathbf{r}(t) = \cos t \, \mathbf{i} + \sin t \, \mathbf{j} + t \mathbf{k} \quad t \geq 0$$

so the parameter is arc length.

**Solution** The arc length function $s(t)$ along the curve $\mathbf{r} = \mathbf{r}(t)$, $a \leq t \leq b$, from $t = a$ to an arbitrary $t$ is given by the integral

$$s(t) = \int_a^t \|\mathbf{r}'(u)\| \, du$$

Since the graph of the helix starts at $t = 0$, and $\|\mathbf{r}'(t)\| = \sqrt{2}$ for all $t$ (Example 1(b)), we have

$$s(t) = \int_0^t \sqrt{2} \, du = \left[\sqrt{2}\, u\right]_0^t = \sqrt{2}\, t$$

Then $t = \dfrac{s}{\sqrt{2}}$. The helix using arc length $s$ as the parameter is expressed as

$$\mathbf{r}(s) = \cos \frac{s}{\sqrt{2}} \mathbf{i} + \sin \frac{s}{\sqrt{2}} \mathbf{j} + \frac{s}{\sqrt{2}} \mathbf{k} \quad s \geq 0 \qquad \blacksquare$$

As Example 2 shows, to change the parameter $t$ of a curve $C$ traced out by the vector function $\mathbf{r} = \mathbf{r}(t)$, $a \leq t \leq b$, to arc length $s$ requires finding the integral

$$s(t) = \int_a^t \|\mathbf{r}'(u)\| \, du$$

Since $\|\mathbf{r}'(u)\|$ is continuous on $(a, b)$ and $\mathbf{r}'(t)$ is never zero on $(a, b)$, then $s = s(t)$ exists for all $t \geq a$. Moreover, $s'(t) = \|\mathbf{r}'(t)\| > 0$, so $s$ is increasing and has an inverse function. To find it, we need to solve $s = s(t)$ for $t$. For most curves, finding the integral and solving for $t$ is not possible. Regardless, for smooth curves, we know an arc length parametrization exists. In the discussion that follows, we make use of this fact and assume the curve is expressed using arc length as the parameter.

**IN WORDS** If a vector function $\mathbf{r} = \mathbf{r}(t)$ traces out a smooth curve $C$, then an arc length parametrization of $C$ exists.

## 2 Find the Curvature of a Curve

A curve may change direction quickly at some points and slowly at others. For example, the curve in Figure 19 bends sharply at $P$, but at $Q$ the curve bends only a little. *Curvature* is a measure of how sharply a curve bends, that is, curvature measures how much the direction of the curve is changing. For the curve in Figure 19, the curvature at $P$ is greater than the curvature at $Q$.

The amount a curve changes direction at a point can be quantified using the unit tangent vector $\mathbf{T}$ at the point. If we move a little away from $Q$ and there is little or no change in the direction of $\mathbf{T}$, the curve has not bent much. If we move a little away from $P$, the curve bends a lot, and the direction of $\mathbf{T}$ changes dramatically. The curvature at $Q$ is less than the curvature at $P$ because the tangent vectors near $P$ have turned more than the tangent vectors near $Q$. We use the rate of change of the tangent vector to measure curvature.

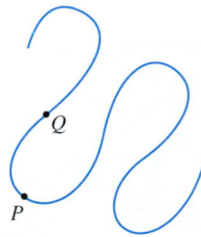

**Figure 19**

**NOTE** $\kappa$ is the Greek letter kappa.

**DEFINITION Curvature**

Suppose a smooth curve $C$ is traced out by a twice differentiable vector function $\mathbf{r} = \mathbf{r}(s)$, where the parameter is the arc length $s$ along $C$. The **curvature** $\kappa = \kappa(s)$ at a point $P$ on $C$ is

$$\kappa = \left\| \frac{d\mathbf{T}}{ds} \right\|$$

**IN WORDS** Curvature is measured by the magnitude of the rate of change of the unit tangent vector $\mathbf{T}$ with respect to arc length.

In the next example, we find the curvature of a line. Since a line has constant direction, the tangent vectors never turn so the curvature of a line is 0.

**EXAMPLE 3** Finding the Curvature of a Line

Show that the curvature of a line is 0.

**Solution** For a line, the unit tangent vector $\mathbf{T}$ is along the constant direction of the line, so $\mathbf{T}$ is a constant and $\dfrac{d\mathbf{T}}{ds} = \mathbf{0}$. The curvature $\kappa = \left\| \dfrac{d\mathbf{T}}{ds} \right\|$ of a line is 0. ∎

Earlier we stated that it is not easy to express a curve using arc length as the parameter. So, we seek expressions for the curvature $\kappa$ that do not require the parameter to be arc length. Suppose $\mathbf{T} = \mathbf{T}(t)$ is the unit tangent vector of a curve $C$ traced out by the vector function $\mathbf{r} = \mathbf{r}(t)$, $a \leq t \leq b$. If $s$ is the arc length along $C$, then we can use the Chain Rule to obtain

$$\frac{d\mathbf{T}}{dt} = \frac{d\mathbf{T}}{ds}\frac{ds}{dt}$$

Since $\dfrac{ds}{dt} = \|\mathbf{r}'(t)\| \neq 0$, then

$$\kappa = \left\| \frac{d\mathbf{T}}{ds} \right\| = \left\| \frac{\mathbf{T}'(t)}{\dfrac{ds}{dt}} \right\|$$

Since $\dfrac{ds}{dt} = \|\mathbf{r}'(t)\|$, we have the following expression for the curvature $\kappa$ of a curve $C$ in terms of any parametrization of $C$:

$$\boxed{\kappa = \frac{\|\mathbf{T}'(t)\|}{\|\mathbf{r}'(t)\|}} \qquad (3)$$

The next example finds the curvature of a circle. Since a circle has a tangent line that turns uniformly, its curvature is a constant. What that constant is may surprise you.

**EXAMPLE 4** Finding the Curvature of a Circle

Find the curvature of a circle of radius $R$.

**Solution** A circle of radius $R$ can be expressed by

$$\mathbf{r} = \mathbf{r}(t) = R \cos t\, \mathbf{i} + R \sin t\, \mathbf{j} \qquad 0 \leq t \leq 2\pi$$

Now we find $\mathbf{r}'(t)$ and $\|\mathbf{r}'(t)\|$.

$$\mathbf{r}'(t) = -R \sin t\, \mathbf{i} + R \cos t\, \mathbf{j} \qquad \|\mathbf{r}'(t)\| = \sqrt{R^2 \sin^2 t + R^2 \cos^2 t} = R$$

The unit tangent vector $\mathbf{T}$ and its derivative $\mathbf{T}'$ are

$$\mathbf{T}(t) = \frac{\mathbf{r}'(t)}{\|\mathbf{r}'(t)\|} = \frac{-R \sin t\, \mathbf{i} + R \cos t\, \mathbf{j}}{R} = -\sin t\, \mathbf{i} + \cos t\, \mathbf{j}$$

$$\mathbf{T}'(t) = -\cos t\, \mathbf{i} - \sin t\, \mathbf{j} \qquad \|\mathbf{T}'(t)\| = 1$$

Now use formula (3) to find the curvature $\kappa$.

$$\kappa = \frac{\|\mathbf{T}'(t)\|}{\|\mathbf{r}'(t)\|} = \frac{1}{R}$$

**NOTE** The larger a circle, the smaller its curvature. As $R \to \infty$, then $\kappa \to 0$ and the circle will look more and more like a line.

The curvature of a circle of radius $R$ is the reciprocal of its radius. ∎

**NOW WORK** Problem **21.**

## 3 Find the Curvature of a Space Curve

It is easier to find the curvature of a space curve using the following formula rather than $\kappa = \dfrac{\|\mathbf{T}'(t)\|}{\|\mathbf{r}'(t)\|}$.

> **THEOREM  Curvature of a Space Curve**
>
> If a smooth curve $C$ in space is traced out by a twice differentiable vector function $\mathbf{r} = \mathbf{r}(t)$, then its curvature $\kappa$ is given by the formula
>
> $$\kappa = \frac{\|\mathbf{r}'(t) \times \mathbf{r}''(t)\|}{\|\mathbf{r}'(t)\|^3} \tag{4}$$

**NEED TO REVIEW?** The cross product and its properties are discussed in Section 10.5, pp. 764–769.

**Proof**  Since $\mathbf{T}(t) = \dfrac{\mathbf{r}'(t)}{\|\mathbf{r}'(t)\|}$ and $\|\mathbf{r}'(t)\| = \dfrac{ds}{dt}$ [from (2) on p. 816], then

$$\mathbf{r}'(t) = \|\mathbf{r}'(t)\|\,\mathbf{T}(t) = \frac{ds}{dt}\,\mathbf{T}(t) \tag{5}$$

Now use the scalar multiplication rule to find $\mathbf{r}''(t)$.

$$\mathbf{r}''(t) = \frac{d}{dt}\mathbf{r}'(t) = \frac{d}{dt}\left(\frac{ds}{dt}\mathbf{T}(t)\right) = \frac{d^2s}{dt^2}\mathbf{T}(t) + \frac{ds}{dt}\mathbf{T}'(t) \tag{6}$$

Using (5) and (6),

$$\mathbf{r}'(t) \times \mathbf{r}''(t) = \frac{ds}{dt}\mathbf{T} \times \left(\frac{d^2s}{dt^2}\mathbf{T} + \frac{ds}{dt}\mathbf{T}'\right) = \frac{ds}{dt}\frac{d^2s}{dt^2}(\mathbf{T}\times\mathbf{T}) + \left(\frac{ds}{dt}\right)^2(\mathbf{T}\times\mathbf{T}')$$

Since $\mathbf{T} \times \mathbf{T} = \mathbf{0}$, this reduces to

$$\mathbf{r}'(t) \times \mathbf{r}''(t) = \left(\frac{ds}{dt}\right)^2(\mathbf{T}\times\mathbf{T}') \tag{7}$$

**RECALL** $\|\mathbf{u} \times \mathbf{v}\| = \|\mathbf{u}\|\,\|\mathbf{v}\|\sin\theta$, where $\theta$ is the angle between $\mathbf{u}$ and $\mathbf{v}$.

Since $\mathbf{T}'$ is parallel to $\mathbf{N}$ and $\mathbf{T}$ is orthogonal to $\mathbf{N}$, it follows that $\mathbf{T}$ and $\mathbf{T}'$ are orthogonal. This means $\|\mathbf{T} \times \mathbf{T}'\| = \|\mathbf{T}\|\,\|\mathbf{T}'\|\sin\dfrac{\pi}{2} = \|\mathbf{T}'\|$, so

$$\|\mathbf{r}'(t) \times \mathbf{r}''(t)\| = \left(\frac{ds}{dt}\right)^2\|\mathbf{T}'\|$$

Now solve the above equation for $\|\mathbf{T}'\|$ and use the fact that $\dfrac{ds}{dt} = \|\mathbf{r}'(t)\|$. Then

$$\|\mathbf{T}'\| = \frac{\|\mathbf{r}'(t) \times \mathbf{r}''(t)\|}{\|\mathbf{r}'(t)\|^2} \tag{8}$$

From equation (3), the curvature $\kappa$ is

$$\kappa = \underset{\underset{(3)}{\uparrow}}{\frac{\|\mathbf{T}'\|}{\|\mathbf{r}'(t)\|}} = \underset{\underset{(8)}{\uparrow}}{\frac{\|\mathbf{r}'(t) \times \mathbf{r}''(t)\|}{\|\mathbf{r}'(t)\|^3}} \qquad\blacksquare$$

CALC
CLIP
▶ **EXAMPLE 5** Finding the Curvature of a Space Curve

Find the curvature $\kappa$ of the space curve $\mathbf{r}(t) = t\mathbf{i} + t^2\mathbf{j} + t^3\mathbf{k}$.

**Solution** To use formula (4), we need to find $\mathbf{r}'(t)$, $\|\mathbf{r}'(t)\|$, $\mathbf{r}''(t)$, $\mathbf{r}'(t) \times \mathbf{r}''(t)$, and $\|\mathbf{r}'(t) \times \mathbf{r}''(t)\|$.

$$\mathbf{r}'(t) = \mathbf{i} + 2t\mathbf{j} + 3t^2\mathbf{k} \qquad \|\mathbf{r}'(t)\| = \sqrt{1 + 4t^2 + 9t^4} \qquad \mathbf{r}''(t) = 2\mathbf{j} + 6t\mathbf{k}$$

$$\mathbf{r}'(t) \times \mathbf{r}''(t) = \begin{vmatrix} \mathbf{i} & \mathbf{j} & \mathbf{k} \\ 1 & 2t & 3t^2 \\ 0 & 2 & 6t \end{vmatrix} = 6t^2\mathbf{i} - 6t\mathbf{j} + 2\mathbf{k}$$

$$\|\mathbf{r}'(t) \times \mathbf{r}''(t)\| = \sqrt{36t^4 + 36t^2 + 4} = 2\sqrt{9t^4 + 9t^2 + 1}$$

Now use formula (4). The curvature $\kappa$ is

$$\kappa = \frac{\|\mathbf{r}'(t) \times \mathbf{r}''(t)\|}{\|\mathbf{r}'(t)\|^3} = \frac{2\sqrt{9t^4 + 9t^2 + 1}}{(1 + 4t^2 + 9t^4)^{3/2}}$$

∎

**NOW WORK** Problem 31.

## 4 Find the Curvature of a Plane Curve Given by $y = f(x)$

The curvature formula $\kappa = \dfrac{\|\mathbf{r}'(t) \times \mathbf{r}''(t)\|}{\|\mathbf{r}'(t)\|^3}$ for a space curve can be used to find the curvature of a plane curve given by a twice differentiable function $y = f(x)$. We use $x$ as the parameter and define the plane curve $y = f(x)$ using the vector equation

$$\mathbf{r} = \mathbf{r}(x) = x\mathbf{i} + f(x)\mathbf{j} + 0\mathbf{k}$$

Notice that a $\mathbf{k}$-component of 0 is used. Then

$$\mathbf{r}'(x) = \mathbf{i} + f'(x)\mathbf{j} \qquad \|\mathbf{r}'(x)\| = \sqrt{1 + [f'(x)]^2} \qquad \mathbf{r}''(x) = f''(x)\mathbf{j}$$

$$\mathbf{r}'(x) \times \mathbf{r}''(x) = \begin{vmatrix} \mathbf{i} & \mathbf{j} & \mathbf{k} \\ 1 & f'(x) & 0 \\ 0 & f''(x) & 0 \end{vmatrix} = f''(x)\mathbf{k}$$

Now use the curvature formula (4).

$$\boxed{\kappa = \frac{\|\mathbf{r}'(x) \times \mathbf{r}''(x)\|}{\|\mathbf{r}'(x)\|^3} = \frac{|f''(x)|}{(1 + [f'(x)]^2)^{3/2}}} \qquad (9)$$

**EXAMPLE 6** Finding the Curvature of Plane Curves

(a) Find an equation of the tangent line to the graph of the parabola $f(x) = \dfrac{1}{4}x^2$ at the point $(2, 1)$.

(b) What is the curvature of the graph of $f$ at the point $(2, 1)$?

(c) Find an equation of the tangent line to the graph of $g(x) = \dfrac{x^3 + 4}{12}$ at the point $(2, 1)$.

(d) What is the curvature of the graph of $g$ at the point $(2, 1)$?

(e) Graph both curves and their tangent lines.

**Solution (a)** $f(x) = \dfrac{1}{4}x^2$;  $f'(x) = \dfrac{1}{2}x$.

An equation of the tangent line to the graph of $f$ at the point $(2, 1)$ is

$$y - 1 = 1 \cdot (x - 2) \qquad f'(2) = 1$$

$$y = x - 1$$

**(b)** $f''(x) = \dfrac{1}{2}$;  $f'(2) = 1$;  $f''(2) = \dfrac{1}{2}$. Now use formula (9) to find $\kappa$ at $x = 2$.

$$\kappa = \frac{|f''(2)|}{\left(1 + [f'(2)]^2\right)^{3/2}} = \frac{\dfrac{1}{2}}{(1 + 1)^{3/2}} = \frac{1}{2 \cdot 2\sqrt{2}} = \frac{1}{4\sqrt{2}} \approx 0.177$$

**(c)** $g(x) = \dfrac{x^3 + 4}{12}$;  $g'(x) = \dfrac{3x^2}{12} = \dfrac{x^2}{4}$.

An equation of the tangent line to the graph of $g$ at the point $(2, 1)$ is

$$y - 1 = 1 \cdot (x - 2) \qquad g'(2) = 1$$

$$y = x - 1$$

**(d)** $g''(x) = \dfrac{x}{2}$;  $g'(2) = 1$;  $g''(2) = 1$. The curvature $\kappa$ at $x = 2$ is

$$\kappa = \frac{|g''(2)|}{(1 + [g'(2)]^2)^{3/2}} = \frac{1}{(1 + 1)^{3/2}} = \frac{1}{2\sqrt{2}} \approx 0.354$$

**(e)** The functions $f$ and $g$ are graphed in Figure 20. Both graphs have the same tangent line at $(2, 1)$, but the tangent line to the graph of $y = \dfrac{1}{4}x^2$ at $(2, 1)$ (in green) turns more slowly ($\kappa \approx 0.177$) than the tangent line to the graph of $y = \dfrac{x^3 + 4}{12}$ at $(2, 1)$ (in blue) ($\kappa \approx 0.354$). ∎

 **Problem 39.**

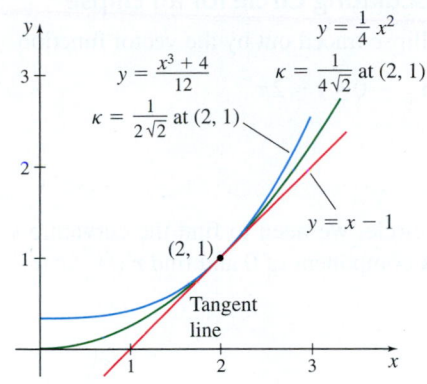

**Figure 20**

### 5 Find an Osculating Circle

Provided the curvature $\kappa \neq 0$, each point $P$ on a smooth curve $C$ traced out by a twice differentiable vector function $\mathbf{r} = \mathbf{r}(t)$ can be associated with a circle, called the **osculating circle** of $C$ at $P$. The osculating circle of $C$ at $P$ has the following properties:

**NOTE** The term *osculating* means "kissing."

- The osculating circle contains the point $P$.
- The tangent line to $C$ at $P$ and the tangent line to the osculating circle at $P$ are the same.
- The radius $\rho$ of the osculating circle is $\rho = \dfrac{1}{\kappa}$, where $\kappa$ is the curvature of $C$ at $P$. That is, the osculating circle and the curve $C$ have the same curvature at $P$.

  **NOTE** $\rho$ is the Greek letter rho, pronounced "roe."

- The center of the osculating circle lies on a line in the direction of the principal unit normal $\mathbf{N}$ to $C$ at the point $P$.

Figure 21 illustrates these properties.

The osculating circle of a curve $C$ is sometimes called the **circle of curvature**. Its radius $\rho = \dfrac{1}{\kappa}$ is called the **radius of curvature**, and its center is called the **center of curvature**.

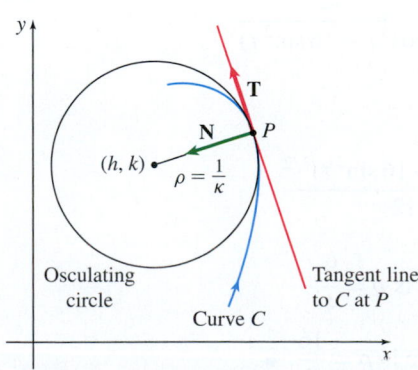

**Figure 21** Osculating circle

Since the osculating circle and the curve share a common tangent at each point and have the same curvature at each point, sometimes the osculating circle is used to approximate the curve $C$. See Figure 22.

**NOTE** The osculating circle at a point $P$ is the limiting circle that results from the circle containing three points $P$, $Q$, and $R$ on the curve $C$ as the points $Q$ and $R$ are made to approach point $P$.

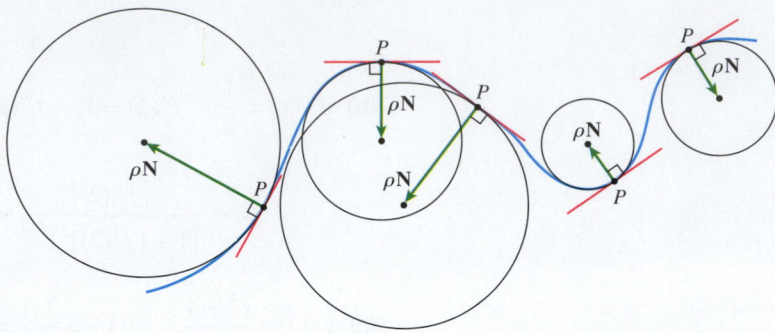

**DF** **Figure 22**

**EXAMPLE 7**  **Finding the Radius of an Osculating Circle for an Ellipse**

Find the radius of the osculating circle of the ellipse traced out by the vector function

$$\mathbf{r}(t) = 3 \sin t\,\mathbf{i} + 4 \cos t\,\mathbf{j} \qquad 0 \le t \le 2\pi$$

**(a)** At $t = 0$    **(b)** At $t = \dfrac{\pi}{2}$

**Solution** To find the radius of the osculating circle, we need to find the curvature $\kappa$. We treat the ellipse as a curve in space with a $\mathbf{k}$ component of 0 and find $\mathbf{r}'(t)$, $\|\mathbf{r}'(t)\|$, $\mathbf{r}''(t)$, $\mathbf{r}'(t) \times \mathbf{r}''(t)$, and $\|\mathbf{r}'(t) \times \mathbf{r}''(t)\|$.

$$\mathbf{r}(t) = 3 \sin t\,\mathbf{i} - 4 \cos t\,\mathbf{j} + 0\,\mathbf{k}$$

$$\mathbf{r}'(t) = 3 \cos t\,\mathbf{i} - 4 \sin t\,\mathbf{j} + 0\,\mathbf{k} \qquad \|\mathbf{r}'(t)\| = \sqrt{9 \cos^2 t + 16 \sin^2 t}$$

$$\mathbf{r}''(t) = -3 \sin t\,\mathbf{i} - 4 \cos t\,\mathbf{j} + 0\,\mathbf{k}$$

$$\mathbf{r}'(t) \times \mathbf{r}''(t) = \begin{vmatrix} \mathbf{i} & \mathbf{j} & \mathbf{k} \\ 3 \cos t & -4 \sin t & 0 \\ -3 \sin t & -4 \cos t & 0 \end{vmatrix} = (-12 \cos^2 t - 12 \sin^2 t)\mathbf{k} = -12\mathbf{k}$$

$$\|\mathbf{r}'(t) \times \mathbf{r}''(t)\| = 12$$

Using formula (4), the curvature $\kappa$ of $C$ is

$$\kappa = \frac{\|\mathbf{r}'(t) \times \mathbf{r}''(t)\|}{\|\mathbf{r}'(t)\|^3} = \frac{12}{(9 \cos^2 t + 16 \sin^2 t)^{3/2}}$$

The radius $\rho$ of the osculating circle is

$$\rho = \frac{1}{\kappa} = \frac{(9 \cos^2 t + 16 \sin^2 t)^{3/2}}{12}$$

**(a)** At $t = 0$, the radius of the osculating circle is $\rho = \dfrac{9}{4}$.

**(b)** At $t = \dfrac{\pi}{2}$, the radius of the osculating circle is $\rho = \dfrac{16}{3}$.  ■

Figures 23(a) and 23(b) illustrate the graph of the ellipse and the osculating circles at the points $P = (0, 4)$ and $P = (3, 0)$ on the ellipse.

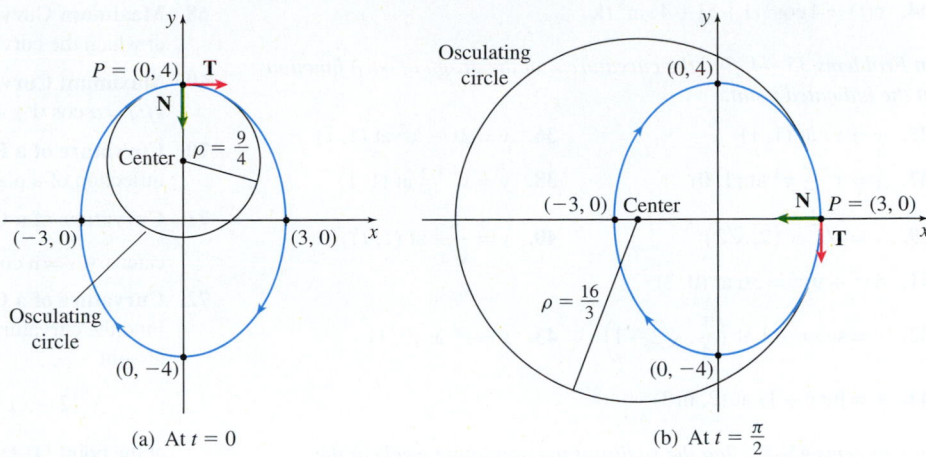

(a) At $t = 0$    (b) At $t = \frac{\pi}{2}$

**Figure 23**  $\mathbf{r}(t) = 3 \sin t \mathbf{i} + 4 \cos t \mathbf{j}, 0 \le t \le 2\pi$

NOW WORK **Problem 57.**

## 11.3 Assess Your Understanding

### Concepts and Vocabulary

1. *Multiple Choice* If a smooth curve $C$ is traced out by the vector function $\mathbf{r} = \mathbf{r}(t)$, $a \le t \le b$, then the parameter $t$ is the arc length if and only if [(**a**) $\|\mathbf{r}(t)\| = 1$ (**b**) $\|\mathbf{r}'(t)\| = 0$

   (**c**) $\|\mathbf{r}''(t)\| = 1$ (**d**) $\|\mathbf{r}'(t)\| = 1$] for all $t$.

2. *True or False* The curvature $\kappa$ of a straight line equals the slope of the line.

3. *True or False* The curvature $\kappa$ of a circle equals the radius of the circle.

4. *True or False* The curvature $\kappa$ of a smooth curve traced out by a twice differentiable vector function $\mathbf{r} = \mathbf{r}(t)$ equals the magnitude of the rate of change of the unit tangent vector $\mathbf{T}$ with respect to arc length.

5. *True or False* The curvature $\kappa$ of a smooth curve traced out by a twice differentiable vector function $\mathbf{r} = \mathbf{r}(t)$

   is $\kappa = \dfrac{\|\mathbf{r}'(t) \times \mathbf{r}''(t)\|}{\|\mathbf{r}(t)\|^3}$.

6. The curvature of a twice differentiable function $y = f(x)$ is $\kappa = $ _____.

7. The curvature of the circle $x^2 + (y - 2)^2 = 9$ is $\kappa = $ _____.

8. *True or False* The radius $\rho$ of the osculating circle at a point $P$ of a smooth curve $C$ equals $\kappa$, provided $\kappa \ne 0$.

### Skill Building

*In Problems 9–18, determine whether the parameter used for each vector function is arc length.*

9. $\mathbf{r}(t) = 4 \cos t \mathbf{i} - 4 \sin t \mathbf{j}$   $0 \le t \le 2\pi$

10. $\mathbf{r}(t) = \sin(3t)\mathbf{i} + \cos(3t)\mathbf{j}$   $0 \le t \le 2\pi$

11. $\mathbf{r}(t) = t^2 \mathbf{i} + t \mathbf{j}$   $0 \le t \le 4$

12. $\mathbf{r}(t) = t \mathbf{i} + t^3 \mathbf{j}$   $0 \le t \le 2$

13. $\mathbf{r}(t) = (2t + 1)\mathbf{i} + (3t - 2)\mathbf{j}$   $0 \le t \le 5$

14. $\mathbf{r}(t) = \mathbf{i} + t \mathbf{j}$   $0 \le t \le 1$

15. $\mathbf{r}(t) = \left( \dfrac{2}{\sqrt{13}} t + 1 \right) \mathbf{i} + \left( \dfrac{3}{\sqrt{13}} t - 2 \right) \mathbf{j}$   $0 \le t \le 5\sqrt{13}$

16. $\mathbf{r}(t) = \mathbf{i} + t^2 \mathbf{j}$   $0 \le t \le 1$      17. $\mathbf{r}(t) = \sin t \mathbf{i} + \cos t \mathbf{j} + t \mathbf{k}$

18. $\mathbf{r}(t) = a \sin t \mathbf{i} + a \cos t \mathbf{j} + \sqrt{1 - a^2}\, t \mathbf{k}$

*In Problems 19 and 20, rank the curvature of each curve at points $P$, $Q$, and $R$ from greatest to least.*

19. 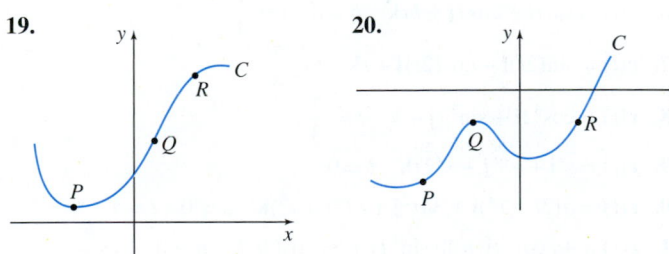   20.

*In Problems 21–26, find the curvature $\kappa$ of each plane curve traced out by the vector function $\mathbf{r} = \mathbf{r}(t)$.*

21. $\mathbf{r}(t) = t^2 \mathbf{i} + \dfrac{2}{t} \mathbf{j}$      22. $\mathbf{r}(t) = 2t \mathbf{i} + t^3 \mathbf{j}$

23. $\mathbf{r}(t) = 2 \sin t \mathbf{i} + 2 \cos t \mathbf{j}$      24. $\mathbf{r}(t) = \cos t \mathbf{i} + 2 \sin t \mathbf{j}$

25. $\mathbf{r}(t) = (3t - t^3)\mathbf{i} + 3t^2 \mathbf{j}$      26. $\mathbf{r}(t) = (3t - t^3)\mathbf{i} + (3t + t^3)\mathbf{j}$

*In Problems 27–34, find the curvature $\kappa$ of each space curve traced out by the vector function $\mathbf{r} = \mathbf{r}(t)$.*

27. $\mathbf{r}(t) = t \mathbf{i} + 2t \mathbf{j} + t \mathbf{k}$      28. $\mathbf{r}(t) = 2t \mathbf{i} + t \mathbf{j} + 3t \mathbf{k}$

29. $\mathbf{r}(t) = \sin(2t)\mathbf{i} + \cos(2t)\mathbf{j} + t \mathbf{k}$

30. $\mathbf{r}(t) = \sin t \mathbf{i} + \cos t \mathbf{j} + bt \mathbf{k}$

**1.** = NOW WORK problem       ⟨Ⓐ⟩ = Graphing technology recommended       ⟨CAS⟩ = Computer Algebra System recommended

**31.** $\mathbf{r}(t) = e^t\mathbf{i} + e^{-t}\mathbf{j} + \sqrt{2}t\mathbf{k}$

**32.** $\mathbf{r}(t) = e^t\mathbf{i} + e^{2t}\mathbf{j} + e^{-t}\mathbf{k}$

**33.** $\mathbf{r}(t) = \cos^3 t\mathbf{i} + \sin^3 t\mathbf{j} + \mathbf{k}$

**34.** $\mathbf{r}(t) = 4\cos^3 t\mathbf{i} + 3\mathbf{j} + 4\sin^3 t\mathbf{k}$

*In Problems 35–44, find the curvature $\kappa$ of the graph of each function at the indicated point.*

**35.** $y = x^2$ at $(1, 1)$

**36.** $y = 2x - x^2$ at $(1, 1)$

**37.** $y = x^2 - x^3$ at $(1, 0)$

**38.** $y = x^{-3/2}$ at $(1, 1)$

**39.** $y = \sqrt{x}$ at $\left(2, \sqrt{2}\right)$

**40.** $y = \dfrac{1}{\sqrt{x}}$ at $(1, 1)$

**41.** $4x^2 + 9y^2 = 36$ at $(0, 2)$

**42.** $y = \sec x - 1$ at $\left(\dfrac{\pi}{4}, \sqrt{2} - 1\right)$

**43.** $y = e^x$ at $(0, 1)$

**44.** $y = \ln(x + 1)$ at $(2, \ln 3)$

*In Problems 45–52, find the radius of the osculating circle at the indicated point.*

**45.** $y = x^3 - 6x$ at $(1, -5)$

**46.** $y = \dfrac{1}{x^2}$ at $(-1, 1)$

**47.** $y = \sin x$ at $\left(\dfrac{\pi}{2}, 1\right)$

**48.** $y = e^{-x}$ at $(0, 1)$

**49.** $x^2 + xy + y^2 = 3$ at $(1, 1)$

**50.** $y^2 - y + x = 0$ at $(0, 0)$

**51.** $y = \ln(\sec x)$ at $\left(\dfrac{\pi}{4}, \ln\sqrt{2}\right)$

**52.** $y = \cosh x$ at $(0, 1)$

*In Problems 53–62, find the radius of the osculating circle at the point corresponding to $t$ on the curve $C$ traced out by the vector function $\mathbf{r} = \mathbf{r}(t)$.*

**53.** $\mathbf{r}(t) = 3t^2\mathbf{i} + (3t - t^3)\mathbf{j} \quad t = 1$

**54.** $\mathbf{r}(t) = t\mathbf{i} + t^2\mathbf{j} + t^3\mathbf{k} \quad t = 1$

**55.** $\mathbf{r}(t) = \sin t\mathbf{i} + \cos(2t)\mathbf{j} \quad t = \dfrac{\pi}{4}$

**56.** $\mathbf{r}(t) = \sin t\mathbf{i} + \cos t\mathbf{j} + bt\mathbf{k} \quad b > 0 \quad t = \dfrac{\pi}{4}$

**57.** $\mathbf{r}(t) = \sin(2t)\mathbf{i} + \cos(2t)\mathbf{j} + t\mathbf{k} \quad t = \dfrac{\pi}{4}$

**58.** $\mathbf{r}(t) = \cos^3 t\mathbf{i} + \sin^3 t\mathbf{j} + \mathbf{k} \quad t = \dfrac{\pi}{3}$

**59.** $\mathbf{r}(t) = e^t\mathbf{i} + e^{-t}\mathbf{j} + \sqrt{2}t\mathbf{k} \quad t = 0$

**60.** $\mathbf{r}(t) = a(3t - t^3)\mathbf{i} + 3at^2\mathbf{j} + a(3t + t^3)\mathbf{k} \quad a > 0 \quad t = 1$

**61.** $\mathbf{r}(t) = 4a\cos^3 t\mathbf{i} + 4a\sin^3 t\mathbf{j} + 3a\cos(2t)\mathbf{k} \quad a > 0 \quad t = \dfrac{\pi}{4}$

**62.** $\mathbf{r}(t) = t\mathbf{i} + 2t\mathbf{j} + \sqrt{1 - 5t^2}\mathbf{k} \quad -\dfrac{\sqrt{5}}{5} < t < \dfrac{\sqrt{5}}{5} \quad t = 0$

## Applications and Extensions

**63. Radius of Curvature** Show that the radius of curvature of the osculating circle of the parabola $y = ax^2 + bx + c$ is a minimum at its vertex.

**64. Radius of Curvature** Show that the radii of curvature of the osculating circles at the ends of the axes of the ellipse

$$b^2x^2 + a^2y^2 = a^2b^2 \text{ are } \dfrac{b^2}{a} \text{ and } \dfrac{a^2}{b}, \quad a > 0, b > 0.$$

**65. Maximum Curvature** Find the point on the curve $y = \ln x$ at which the curvature is maximum.

**66. Maximum Curvature** Find the point on the curve $y = e^x$ at which the curvature is maximum.

**67. Maximum Curvature** Find the point(s) on the curve $y = \dfrac{1}{3}x^3$ at which the curvature is maximum.

**68. Maximum Curvature** Find the point(s) on the curve $y = \sin x$ at which the curvature is maximum.

**69. Maximum Curvature** Find $\alpha > 0$ so that $\mathbf{r}(t) = \alpha\cos t\mathbf{i} + \alpha\sin t\mathbf{j} + t\mathbf{k}$ has maximum curvature.

**70. Curvature of a Plane Curve** What is the curvature at a point of inflection of a plane curve?

**71. Curvature of a Catenary** Show that the curvature of the catenary $y = a\cosh\dfrac{x}{a}, a > 0$, at any point $(x, y)$ is $\dfrac{a}{y^2}$.

**72. Curvature of a Cissoid** Find the curvature of the cissoid

$$y^2(2 - x) = x^3$$

at the point $(1, 1)$. See the figure.

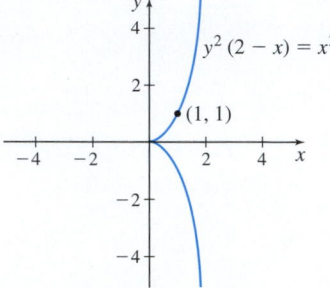

**73. Curvature of a Cycloid** Find the curvature of the cycloid

$$x(\theta) = \theta - \sin\theta$$
$$y(\theta) = 1 - \cos\theta$$

at the highest point of an arch. See the figure.

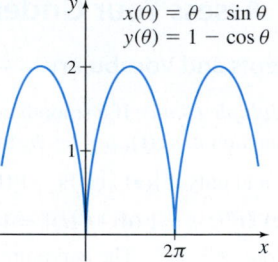

**74. (a)** Find the curvature of the curve traced out by $\mathbf{r}(t) = (1 - t^3)\mathbf{i} + t^2\mathbf{j}$.

**(b)** Graph $\mathbf{r} = \mathbf{r}(t)$. Where is the curvature undefined? Do you see any geometric reason for this?

**75. Curvature of a Spiral** Find the curvature of the spiral $\mathbf{r}(t) = e^{-t}\cos t\mathbf{i} + e^{-t}\sin t\mathbf{j}$ shown in the figure. How does the curvature behave when $t \to \infty$? Do you see any geometric reason for this?

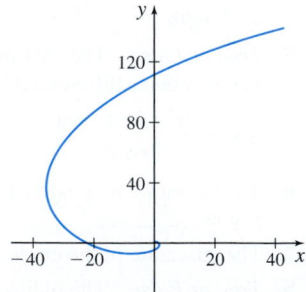

**76. Curvature** Find the curvature $\kappa$ of the curve traced out by $\mathbf{r}(t) = 2a\cos t\mathbf{i} + 2a\sin t\mathbf{j} + bt^2\mathbf{k}, \quad a > 0, b > 0$.

**77. Curvature** Show that the curvature of an ellipse $\dfrac{x^2}{a^2} + \dfrac{y^2}{b^2} = 1$, $a > b$, is a maximum at the points $(\pm a, 0)$ and is a minimum at the points $(0, \pm b)$.

*Source*: Contributed by students at the University of Missouri.

**78. Curvature** Show that the circular helix $\mathbf{r}(t) = a\cos t\mathbf{i} + a\sin t\mathbf{j} + t\mathbf{k}$ has constant curvature.

*Source*: Contributed by students at the University of Missouri.

**79.** Write a vector equation for the curve $C$ traced out by

$$\mathbf{r}(t) = 2t\mathbf{i} + (2t - 1)\mathbf{j} + t\mathbf{k}, \ 0 \le t \le 2$$

using arc length $s$ as the parameter. *Hint:* For each $t$, $0 \le t \le 2$, calculate the length $s(t)$ of the curve from 0 to $t$.

**80.** Compare the solutions of Problems 13 and 15. Then show how to change the parameter $t$ of the line $\mathbf{r}(t) = (at + b)\mathbf{i} + (ct + d)\mathbf{j}$, where either $a \neq 0$ or $c \neq 0$, to one that is arc length as measured along the line.

**81.** Suppose a smooth curve $C$ is traced out by a twice differentiable vector function $\mathbf{r} = \mathbf{r}(t)$, $a \le t \le b$. If the curvature $\kappa \neq 0$ at a point $P$ on $C$, show that the position vector $\mathbf{C}$ of the center of the osculating circle at $P$ is given by

$$\mathbf{C}(t) = \mathbf{r}(t) + \rho\mathbf{N}(t)$$

where $\mathbf{N}$ is the principal unit normal vector to $C$ at $P$, and $\rho = \dfrac{1}{\kappa}$.

**82.** Use the result of Problem 81 to find the center and radius of the osculating circle for the helix

$$\mathbf{r}(t) = a \sin t\mathbf{i} + a \cos t\mathbf{j} + a^2 t\mathbf{k} \quad a > 0$$

**(a)** at $t = \dfrac{\pi}{2}$    **(b)** at $t = \pi$

*Use the following discussion for Problems 83–87. Suppose $C$ is a smooth curve traced out by the twice differentiable vector function $\mathbf{r} = \mathbf{r}(s)$, $a \le s \le b$, where $s$ is arc length as measured along $C$. Define the **binormal vector** $\mathbf{B}$ of $C$ as $\mathbf{B}(s) = \mathbf{T}(s) \times \mathbf{N}(s)$.*

**83.** Show that the three vectors $\mathbf{T}$, $\mathbf{N}$, and $\mathbf{B}$ form a collection of mutually orthogonal unit vectors at each point on $C$.

**84.** Show that $\dfrac{d\mathbf{T}}{ds} = \kappa(s)\mathbf{N}(s)$.

**85.** Show that $\dfrac{d\mathbf{B}}{ds}$ is orthogonal to both $\mathbf{B}(s)$ and $\mathbf{T}(s)$.

**86.** If the **torsion** $\tau(s)$ of $C$ is defined by the equation $\dfrac{d\mathbf{B}}{ds} = -\tau\mathbf{N}$, show that $\dfrac{d\mathbf{N}}{ds} = \tau\mathbf{B} - \kappa\mathbf{T}$.

**87.** Find $\kappa$, $\mathbf{T}$, $\mathbf{N}$, and $\mathbf{B}$ for $\mathbf{r}(s) = \dfrac{1}{\sqrt{2}}[\sin s\mathbf{i} + \cos s\mathbf{j} + s\mathbf{k}]$.

## Challenge Problems

**88. Curvature of a Polar Curve**   Show that the formula for the curvature of a polar curve $r = f(\theta)$ is

$$\kappa = \frac{\left| r^2 + 2\left(\dfrac{dr}{d\theta}\right)^2 - r\left(\dfrac{d^2 r}{d\theta^2}\right) \right|}{\left[ r^2 + \left(\dfrac{dr}{d\theta}\right)^2 \right]^{3/2}}$$

*In Problems 89–94, use the result of Problem 88 to find the curvature of each polar curve.*

**89.** $r = 2\cos(2\theta)$ at $\theta = \dfrac{\pi}{12}$     **90.** $r = e^{a\theta}$ at $\theta = \dfrac{\pi}{2}$, $a > 0$

**91.** $r = a\theta$ at $\theta = 1$ and $a > 0$     **92.** $r = 1 - \cos\theta$ at $\theta = 0$

**93.** $r = 3 - 2\sin\theta$ at $\theta = \dfrac{\pi}{6}$     **94.** $r = 2 + 3\cos\theta$ at $\theta = \dfrac{\pi}{3}$

**95.** Use the figure below to show that the coordinates $(h, k)$ of the center of curvature of $y = f(x)$ are

$$h = x - \rho\sin\phi \qquad k = y + \rho\cos\phi$$

where $\rho$ is the radius of curvature. Show that

$$\sin\phi = \frac{y'}{\sqrt{1 + (y')^2}} \qquad \text{and} \qquad \cos\phi = \frac{1}{\sqrt{1 + (y')^2}}$$

so

$$h = x - \frac{y'\left[1 + (y')^2\right]}{y''} \qquad \text{and} \qquad k = y + \frac{1 + (y')^2}{y''}$$

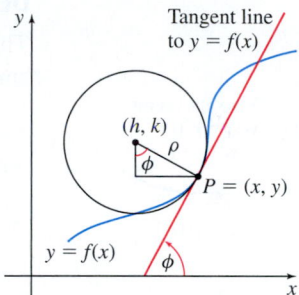

*In Problems 96–100, use the result obtained in Problem 95 to find the center of curvature of each function.*

**96.** $y = x^2$ at $x = 1$     **97.** $y = \sin x$ at $x = \dfrac{\pi}{2}$

**98.** $y = \dfrac{x}{x + 1}$ at $(0, 0)$     **99.** $x^3 + y^3 = 4xy$ at $(2, 2)$

**100.** $xy = 4$ at $x = 2$

**101.** As a point $P$ moves along a curve $C$, the center of curvature corresponding to $P$ traces out a curve $C_1$ called the **evolute** of $C$; conversely, $C$ is the **involute** of $C_1$. Show that parametric equations of the evolute of $y = \dfrac{1}{2}x^2$ are $h = -x^3, k = \dfrac{3}{2}x^2 + 1$. Then eliminate the parameter $x$ to obtain

$$h^2 = \frac{8}{27}(k - 1)^3$$

**102.** Refer to Problem 101. Find parametric equations and a rectangular equation for the evolute of

$$\mathbf{r}(t) = a \cos t\mathbf{i} + b \sin t\mathbf{j} \quad a > 0 \quad b > 0 \quad 0 \le t \le 2\pi$$

# 11.4 Motion Along a Curve

**OBJECTIVES** *When you finish this section, you should be able to:*

**1** Find the velocity, acceleration, and speed of a moving particle (p. 826)

**2** Express the acceleration vector using tangential and normal components (p. 829)

An important physical application of vector functions and their derivatives is the study of moving objects. If we think of the mass of an object as being concentrated at the object's center of gravity, then the object can be thought of as a particle, and it can be represented as a point on a graph. As the particle moves through space, its coordinates $x$, $y$, and $z$ are each functions of time $t$, and its position at time $t$ is given by the vector function:

$$\mathbf{r}(t) = x(t)\mathbf{i} + y(t)\mathbf{j} + z(t)\mathbf{k}$$

## 1 Find the Velocity, Acceleration, and Speed of a Moving Particle

If the vector function $\mathbf{r} = \mathbf{r}(t)$ represents the position vector of a particle at time $t$, we can find the velocity, acceleration, and speed of the particle at any time $t$.

**DEFINITION Position; Velocity; Acceleration; Speed**

The **position**, **velocity**, **acceleration**, and **speed** of a particle whose motion is along a smooth curve traced out by the twice differentiable vector function

$$\mathbf{r}(t) = x(t)\mathbf{i} + y(t)\mathbf{j} + z(t)\mathbf{k} \quad a \le t \le b$$

are defined as

| Position | $\mathbf{r}(t) = x(t)\mathbf{i} + y(t)\mathbf{j} + z(t)\mathbf{k}$ |
|---|---|
| Velocity | $\mathbf{v}(t) = \mathbf{r}'(t) = \dfrac{dx}{dt}\mathbf{i} + \dfrac{dy}{dt}\mathbf{j} + \dfrac{dz}{dt}\mathbf{k}$ |
| Acceleration | $\mathbf{a}(t) = \mathbf{r}''(t) = \dfrac{d^2x}{dt^2}\mathbf{i} + \dfrac{d^2y}{dt^2}\mathbf{j} + \dfrac{d^2z}{dt^2}\mathbf{k}$ |
| Speed | $v(t) = \|\mathbf{v}(t)\| = \|\mathbf{r}'(t)\| = \sqrt{\left(\dfrac{dx}{dt}\right)^2 + \left(\dfrac{dy}{dt}\right)^2 + \left(\dfrac{dz}{dt}\right)^2}$ |

For plane curves, the definitions are similar. See Figure 24.

**Figure 24** $\mathbf{v} = \mathbf{r}'(t) = \dfrac{dx}{dt}\mathbf{i} + \dfrac{dy}{dt}\mathbf{j}$

$$v = \|\mathbf{v}\| = \sqrt{\left(\dfrac{dx}{dt}\right)^2 + \left(\dfrac{dy}{dt}\right)^2}$$

**NOTE** Since $\mathbf{v}(t) = \mathbf{r}'(t)$, the velocity vector is directed along the tangent vector.

**EXAMPLE 1 Finding Velocity, Acceleration, and Speed**

Find the velocity $\mathbf{v}$, acceleration $\mathbf{a}$, and speed $v$ of a particle that is moving along:

**(a)** the plane curve $\mathbf{r}(t) = \left(\dfrac{1}{2}t^2 + t\right)\mathbf{i} + t^3\mathbf{j}$ from $t = 0$ to $t = 2$.

**(b)** the space curve $\mathbf{r}(t) = t\mathbf{i} + t^2\mathbf{j} + t^3\mathbf{k}$ from $t = 0$ to $t = 2$.

For each curve, graph the motion of the particle and the vectors $\mathbf{v}(1)$ and $\mathbf{a}(1)$.

**Solution (a)** The velocity, acceleration, and speed are

$$\mathbf{v}(t) = \mathbf{r}'(t) = \frac{d}{dt}\left(\frac{1}{2}t^2 + t\right)\mathbf{i} + \frac{d}{dt}t^3\mathbf{j} = (t+1)\mathbf{i} + 3t^2\mathbf{j}$$

$$\mathbf{a}(t) = \mathbf{r}''(t) = \frac{d}{dt}\mathbf{r}'(t) = \frac{d}{dt}(t+1)\mathbf{i} + \frac{d}{dt}(3t^2)\mathbf{j} = \mathbf{i} + 6t\mathbf{j}$$

$$v(t) = \|\mathbf{v}(t)\| = \sqrt{(t+1)^2 + (3t^2)^2} = \sqrt{9t^4 + t^2 + 2t + 1}$$

At $t = 1$, the velocity is $\mathbf{v}(1) = 2\mathbf{i} + 3\mathbf{j}$ and the acceleration is $\mathbf{a}(1) = \mathbf{i} + 6\mathbf{j}$. Figure 25 illustrates the graph of $\mathbf{r} = \mathbf{r}(t)$ and the vectors $\mathbf{v}(1)$ and $\mathbf{a}(1)$.

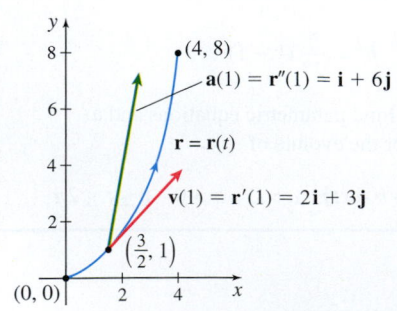

**Figure 25** $\mathbf{r}(t) = \left(\dfrac{1}{2}t^2 + t\right)\mathbf{i} + t^3\mathbf{j}$

$0 \le t \le 2$

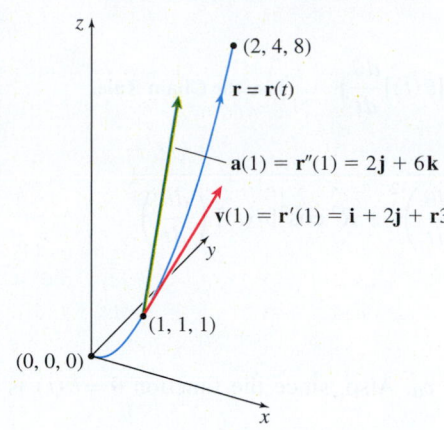

**Figure 26** $\mathbf{r}(t) = t\mathbf{i} + t^2\mathbf{j} + t^3\mathbf{k}$   $0 \le t \le 2$

**(b)** The velocity, acceleration, and speed of the particle are

$$\mathbf{v}(t) = \mathbf{r}'(t) = \frac{d}{dt}t\mathbf{i} + \frac{d}{dt}t^2\mathbf{j} + \frac{d}{dt}t^3\mathbf{k} = \mathbf{i} + 2t\mathbf{j} + 3t^2\mathbf{k}$$

$$\mathbf{a}(t) = \mathbf{r}''(t) = \frac{d}{dt}\mathbf{r}'(t) = \frac{d}{dt}\mathbf{i} + \frac{d}{dt}(2t)\mathbf{j} + \frac{d}{dt}(3t^2)\mathbf{k} = 2\mathbf{j} + 6t\mathbf{k}$$

$$v(t) = \|\mathbf{v}(t)\| = \sqrt{1^2 + (2t)^2 + (3t^2)^2} = \sqrt{1 + 4t^2 + 9t^4}$$

At $t = 1$, the velocity is $\mathbf{v}(1) = \mathbf{i} + 2\mathbf{j} + 3\mathbf{k}$ and the acceleration is $\mathbf{a}(1) = 2\mathbf{j} + 6\mathbf{k}$. See Figure 26.  ∎

**NOW WORK** Problems 7 and 11.

## EXAMPLE 2  Finding the Force Acting on a Particle Moving Along an Ellipse

The position of a particle of mass $m$ that is moving along an ellipse with a constant angular speed $\omega$ is given by the vector function

$$\mathbf{r}(t) = A\cos(\omega t)\mathbf{i} + B\sin(\omega t)\mathbf{j}    0 \le t \le 2\pi$$

Find the force $\mathbf{F}$ acting on the particle at any time $t$. Graph $\mathbf{r} = \mathbf{r}(t)$ and $\mathbf{F} = \mathbf{F}(t)$.

**Solution** To find the force $\mathbf{F}$, we use Newton's Second Law of Motion, $\mathbf{F} = m\mathbf{a}$. We begin by finding the acceleration $\mathbf{a}$ of the particle.

$$\mathbf{r}(t) = A\cos(\omega t)\mathbf{i} + B\sin(\omega t)\mathbf{j}$$

$$\mathbf{v}(t) = \mathbf{r}'(t) = -A\omega\sin(\omega t)\mathbf{i} + B\omega\cos(\omega t)\mathbf{j}$$

$$\mathbf{a}(t) = \mathbf{r}''(t) = -A\omega^2\cos(\omega t)\mathbf{i} - B\omega^2\sin(\omega t)\mathbf{j} = -\omega^2\mathbf{r}(t)$$

Then by Newton's Law,

$$\mathbf{F}(t) = m\mathbf{a}(t) = -m\omega^2\mathbf{r}(t)$$

The direction of the force vector $\mathbf{F}$ is opposite to that of the vector $\mathbf{r}$ at any time $t$, as shown in Figure 27.  ∎

**NOW WORK** Problem 23.

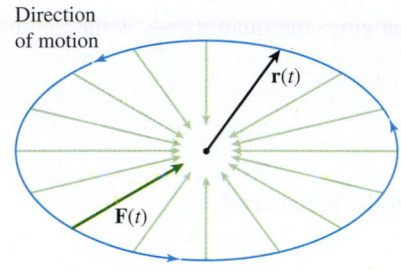

**Figure 27** $\mathbf{F}(t) = -m\omega^2\mathbf{r}(t)$

**NOTE** The force $\mathbf{F}$ acting on the particle in Example 2 is directed toward the center of the ellipse at a time $t$. Because of this, it is called a **centripetal** (center-seeking) **force**.

## EXAMPLE 3  Finding the Position, Velocity, and Acceleration of a Particle Moving Along a Circle

**(a)** Find the position $\mathbf{r} = \mathbf{r}(t)$ of a particle that moves counterclockwise along a circle of radius $R$ with a constant speed $v_0$.

**(b)** Express the velocity and acceleration of the particle as vector functions of $t$.

**(c)** Express the magnitude of the acceleration in terms of $v_0$ and $R$.

**Solution** For convenience, we place the circle of radius $R$ in the $xy$-plane, with its center at the origin, and assume that at time $t = 0$ the particle is on the positive $x$-axis.

**(a)** The particle is moving counterclockwise along the circle, as shown in Figure 28. If $\theta(t)$ is the angle between the positive $x$-axis and the position vector of the particle is $\mathbf{r} = \mathbf{r}(t)$, then the vector $\mathbf{r} = \mathbf{r}(t)$ is

$$\mathbf{r}(t) = R\cos[\theta(t)]\mathbf{i} + R\sin[\theta(t)]\mathbf{j} \qquad (1)$$

Notice that $\mathbf{r}(0) = R\mathbf{i}$, as required. Also for the motion to be counterclockwise, the function $\theta = \theta(t)$ must be increasing.

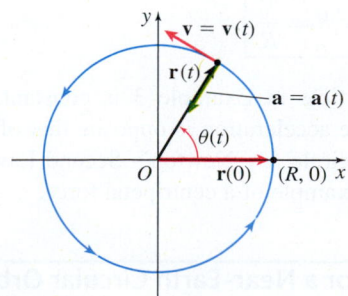

**Figure 28**

**(b)** The velocity **v** of the particle is

$$\mathbf{v}(t) = \frac{d\mathbf{r}}{dt} = -R\sin[\theta(t)]\frac{d\theta}{dt}\mathbf{i} + R\cos[\theta(t)]\frac{d\theta}{dt}\mathbf{j} \qquad \text{Use the Chain Rule.}$$

and the speed $v$ of the particle is

$$v(t) = \|\mathbf{v}(t)\| = \sqrt{R^2\sin^2[\theta(t)]\left(\frac{d\theta}{dt}\right)^2 + R^2\cos^2[\theta(t)]\left(\frac{d\theta}{dt}\right)^2}$$

$$= \sqrt{R^2\left(\frac{d\theta}{dt}\right)^2} = R\left|\frac{d\theta}{dt}\right|$$

Since we know the speed is constant, $v(t) = v_0$. Also, since the function $\theta = \theta(t)$ is increasing, $\dfrac{d\theta}{dt} > 0$. As a result,

$$v_0 = R\frac{d\theta}{dt}$$

$$\frac{d\theta}{dt} = \frac{v_0}{R}$$

Since $\dfrac{d\theta}{dt}$ is the rate at which the angle $\theta$ is changing, the quantity $\dfrac{v_0}{R}$ is the angular speed $\omega$ of the particle. That is,

$$\frac{d\theta}{dt} = \omega$$

Solve this differential equation, using $\theta(0) = 0$ as the initial condition.

$$d\theta = \omega dt$$

$$\theta(t) = \omega t + k$$

$$\theta(0) = k = 0$$

$$\theta(t) = \omega t$$

Substitute $\theta(t) = \omega t$ into the vector function $\mathbf{r} = \mathbf{r}(t)$ [statement (1)] to obtain

$$\mathbf{r}(t) = R\cos(\omega t)\mathbf{i} + R\sin(\omega t)\mathbf{j}$$

Then the velocity **v** and acceleration **a** of the particle are

$$\mathbf{v} = \mathbf{v}(t) = -R\omega\sin(\omega t)\mathbf{i} + R\omega\cos(\omega t)\mathbf{j}$$

$$\mathbf{a} = \mathbf{a}(t) = -R\omega^2\cos(\omega t)\mathbf{i} - R\omega^2\sin(\omega t)\mathbf{j} = -\omega^2[R\cos(\omega t)\mathbf{i} + R\sin(\omega t)\mathbf{j}]$$

$$= -\omega^2\mathbf{r}(t)$$

**(c)** Since $\omega = \dfrac{v_0}{R}$, the magnitude of the acceleration is

$$\boxed{\|\mathbf{a}(t)\| = \omega^2\|\mathbf{r}(t)\| = \omega^2 R = \frac{v_0^2}{R}} \qquad (2)$$

∎

Observe that although the speed of the particle in Example 3 is constant, its velocity **v** is not. Furthermore, the direction of the acceleration is opposite that of the vector **r** and is directed toward the center of the circle. By Newton's Second Law of Motion, $\mathbf{F} = m\mathbf{a}$, so the force vector **F** is another example of a centripetal force.

**NOTE** Near-Earth orbits are above 100 miles (out of Earth's atmosphere) up to an altitude of approximately 15,000 miles.

**EXAMPLE 4  Finding the Speed Required for a Near-Earth Circular Orbit**

Find the speed required to maintain a satellite in a near-Earth circular orbit. (The gravitational attraction of other bodies is ignored.)

**Solution** Let $R$ be the distance of a satellite from the center of Earth. Then from (2), the magnitude of the acceleration of the satellite whose motion is circular is

$$\|\mathbf{a}(t)\| = \frac{v_0^2}{R}$$

**NOTE** The acceleration due to gravity at near-Earth orbits is somewhat less than $g \approx 32.2 \, \text{ft/s}^2 \approx 79{,}036 \, \text{mi/h}^2$, the acceleration due to gravity at Earth's surface.

For the satellite to remain in orbit, the magnitude of the acceleration $\|\mathbf{a}(t)\|$ of the satellite at any time $t$ must equal $g$, the acceleration due to gravity for Earth. As a result,

$$\frac{v_0^2}{R} = g$$

$$v_0 = \sqrt{gR}$$

The speed $v_0$ required to maintain a near-Earth circular orbit is

$$\boxed{v_0 = \sqrt{gR}}$$

where $R$ is the distance of the satellite from the center of Earth and $g$ is the acceleration due to gravity. ∎

For example, the speed required of a communications satellite whose circular orbit must be 4500 mi from the center of Earth is

$$v_0 = \sqrt{79{,}036 \cdot 4500} \approx 18{,}859 \, \text{mi/h}$$

**NOW WORK** Problem 63.

---

**EXAMPLE 5** **Finding the Frictional Force Necessary to Prevent Skidding**

A motorcycle with a mass of 150 kg is driven at a constant speed of 120 km/h on a circular track whose radius is 100 m. To keep the motorcycle from skidding, what frictional force must be exerted by the track on the tires?

**Solution** By Newton's Second Law of Motion, the force $\mathbf{F}$ required to keep an object of mass $m$ traveling along a curve traced out by $\mathbf{r} = \mathbf{r}(t)$ is $\mathbf{F} = m\mathbf{a}$. The magnitude of the frictional force exerted by the tires must therefore equal

$$\|\mathbf{F}\| = m\|\mathbf{a}\|$$

In this example, the motion is circular. So, using (2), we have

**NOTE** The SI unit of force is the newton (N): $\text{N} = \dfrac{\text{kg} \cdot \text{m}}{\text{s}^2}$.

$$\|\mathbf{F}\| = m\left(\frac{v_0^2}{R}\right) = 150 \, \text{kg} \left[\frac{(120 \, \text{km/h})^2}{100 \, \text{m}}\right] = 150 \cdot 144 \cdot \underset{\underset{1 \, \text{h} = 3600 \, \text{s}}{\uparrow}}{\frac{1000^2}{3600^2}} \frac{\text{kg} \cdot \text{m}}{\text{s}^2} \approx 1667 \, \text{N}$$

The track must exert a force of magnitude 1667 N or greater to prevent skidding. ∎

**NOW WORK** Problem 53.

---

## 2 Express the Acceleration Vector Using Tangential and Normal Components

To continue our investigation of the motion of a particle, we show that the acceleration vector $\mathbf{a}$ lies in the plane determined by the unit tangent vector $\mathbf{T}$ and the principal unit normal vector $\mathbf{N}$.

Consider a particle whose motion is along a smooth curve $C$ traced out by a twice differentiable vector function $\mathbf{r} = \mathbf{r}(t)$. The velocity vector $\mathbf{v}(t) = \mathbf{r}'(t)$ can be expressed in terms of the unit tangent vector $\mathbf{T}(t)$ of $C$ as follows:

**RECALL** The unit tangent vector is

$$\mathbf{T}(t) = \frac{\mathbf{r}'(t)}{\|\mathbf{r}'(t)\|}$$

$$\boxed{\mathbf{v}(t) = \mathbf{r}'(t) = \|\mathbf{r}'(t)\|\mathbf{T}(t) = \|\mathbf{v}(t)\|\mathbf{T}(t) = v(t)\mathbf{T}(t)} \tag{3}$$

where $v(t) = \|\mathbf{v}(t)\|$ is the speed of the particle.

By differentiating $\mathbf{v} = \mathbf{v}(t)$, we obtain the acceleration vector $\mathbf{a} = \mathbf{a}(t)$.

$$\mathbf{a} = \mathbf{a}(t) = \frac{d}{dt}\mathbf{v}(t) = \frac{d}{dt}[v(t)\mathbf{T}(t)] = v'(t)\mathbf{T}(t) + v(t)\mathbf{T}'(t)$$

**RECALL** The principal unit normal vector is $\mathbf{N}(t) = \dfrac{\mathbf{T}'(t)}{\|\mathbf{T}'(t)\|}$.

Since $\mathbf{T}'(t) = \|\mathbf{T}'(t)\|\mathbf{N}(t)$, $\mathbf{a}(t)$ can be written as

$$\mathbf{a}(t) = v'(t)\mathbf{T}(t) + v(t)\|\mathbf{T}'(t)\|\mathbf{N}(t) \tag{4}$$

Recall that the curvature $\kappa$ of a curve $C$ traced out by $\mathbf{r} = \mathbf{r}(t)$ can be expressed as $\kappa = \dfrac{\|\mathbf{T}'(t)\|}{\|\mathbf{r}'(t)\|}$, so

$$\|\mathbf{T}'(t)\| = \|\mathbf{r}'(t)\|\kappa = v(t)\kappa \tag{5}$$

Combining (4) and (5), we have

$$\mathbf{a}(t) = v'(t)\mathbf{T}(t) + [v(t)]^2\kappa\mathbf{N}(t)$$

> **THEOREM The Acceleration Vector Expressed Using T and N**
>
> The acceleration vector $\mathbf{a} = \mathbf{a}(t)$ of a particle moving along a smooth curve $C$ traced out by a twice differentiable vector function $\mathbf{r} = \mathbf{r}(t)$ lies in the plane formed by the unit tangent vector $\mathbf{T}(t)$ and the principal unit normal vector $\mathbf{N}(t)$ to $C$. Moreover,
>
> $$\boxed{\mathbf{a}(t) = v'(t)\mathbf{T}(t) + [v(t)]^2\kappa\mathbf{N}(t)} \tag{6}$$
>
> where $v(t)$ is the speed of the particle and $\kappa$ is the curvature of the curve $C$ traced out by $\mathbf{r} = \mathbf{r}(t)$.

See Figure 29.

The acceleration vector

$$\mathbf{a}(t) = \underbrace{v'(t)}_{\substack{\text{tangential} \\ \text{component}}}\mathbf{T}(t) + \underbrace{[v(t)]^2\kappa}_{\substack{\text{normal} \\ \text{component}}}\mathbf{N}(t)$$

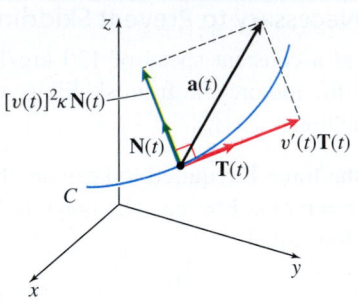

**Figure 29** $\mathbf{a}(t) = v'(t)\mathbf{T}(t) + [v(t)]^2\kappa\mathbf{N}(t)$

has a **tangential component**, denoted by $a_{\mathbf{T}}$, and a **normal component**, denoted by $a_{\mathbf{N}}$, namely

$$\boxed{a_{\mathbf{T}} = v'(t) = \frac{dv}{dt} \qquad a_{\mathbf{N}} = [v(t)]^2\kappa}$$

Then the acceleration vector can be written as

$$\boxed{\mathbf{a}(t) = a_{\mathbf{T}}\mathbf{T}(t) + a_{\mathbf{N}}\mathbf{N}(t)}$$

Let's look more closely at this equation by examining what happens for motion that is linear and for motion that is circular.

Suppose the motion is along a line. Then the curvature $\kappa = 0$, and the normal component of the acceleration $a_N = 0$. Therefore, the acceleration is along the tangent vector. That is, the acceleration is along the line of motion, and the force due to the acceleration is also along the line of motion.

For example, if a car accelerates on a straight stretch of road, the force due to the acceleration moves the car forward. If a car maintains a constant speed on a straight stretch of road, then $\dfrac{dv}{dt} = 0$ and $\mathbf{a} = \mathbf{0}$. There is no force acting on the car in this case.

Now suppose the motion is along a circle. Then the curvature $\kappa$ is a non-zero constant. If the speed is constant, then $\dfrac{dv}{dt} = 0$, and the acceleration is along the normal to the circle. So the force due to the acceleration is directed toward the center of the circle.

For example, if a car maintains a constant speed on a circular ramp, the acceleration is directed toward the center of the circle. The force due to this acceleration is directed toward the center of the circle. If a car speeds up on a circular ramp, then $\dfrac{dv}{dt} > 0$, and the force on the car is in a direction somewhere between the tangent line to the circle and the direction toward the center of the circle.

For motion along a curve, the tangential component of the acceleration equals the rate of change of speed. For a car, this means the car is speeding up (accelerating), slowing down (decelerating), or neither (a constant speed). The normal component of the acceleration depends on the curvature of the curve. For a car, this means the greater the curvature of a turn, the greater the force will be in the normal direction.

**CALC CLIP**

**EXAMPLE 6  Finding the Tangential and Normal Components of Acceleration for an Elliptical Path**

Find the tangential component $a_{\mathbf{T}}$ and normal component $a_{\mathbf{N}}$ of the acceleration of a particle moving along the ellipse

$$\mathbf{r}(t) = 3\sin t\,\mathbf{i} + 4\cos t\,\mathbf{j} \quad 0 \le t \le 2\pi$$

Graph the ellipse, showing $\mathbf{a} = \mathbf{a}(t)$, $a_{\mathbf{T}}$, and $a_{\mathbf{N}}$ when $t = 0$, $\dfrac{\pi}{4}$, and $\dfrac{\pi}{2}$.

**Solution** We begin by finding the velocity, speed, and acceleration of the particle.

$$\mathbf{v}(t) = \mathbf{r}'(t) = 3\cos t\,\mathbf{i} - 4\sin t\,\mathbf{j}$$

$$v(t) = \|\mathbf{v}(t)\| = \sqrt{(3\cos t)^2 + (-4\sin t)^2} = \sqrt{9\cos^2 t + 16\sin^2 t}$$

$$\mathbf{a}(t) = \mathbf{r}''(t) = -3\sin t\,\mathbf{i} - 4\cos t\,\mathbf{j}$$

The tangential component of the acceleration is

$$a_{\mathbf{T}} = v'(t) = \frac{d}{dt}\sqrt{9\cos^2 t + 16\sin^2 t}$$

$$= \frac{1}{2}(9\cos^2 t + 16\sin^2 t)^{-1/2}[18\cos t(-\sin t) + 32\sin t(\cos t)]$$

$$= \frac{7\sin t\cos t}{\sqrt{9\cos^2 t + 16\sin^2 t}}$$

To find the normal component of the acceleration, we need to find the curvature $\kappa$ of $\mathbf{r} = \mathbf{r}(t)$.

First we find the cross product $\mathbf{r}'(t) \times \mathbf{r}''(t)$.

$$\mathbf{r}'(t) \times \mathbf{r}''(t) = \begin{vmatrix} \mathbf{i} & \mathbf{j} & \mathbf{k} \\ 3\cos t & -4\sin t & 0 \\ -3\sin t & -4\cos t & 0 \end{vmatrix} = 0\mathbf{i} + 0\mathbf{j} + (-12\cos^2 t - 12\sin^2 t)\mathbf{k} = -12\mathbf{k}$$

Now the curvature $\kappa$ is

$$\kappa = \frac{\|\mathbf{r}'(t) \times \mathbf{r}''(t)\|}{\|\mathbf{r}'(t)\|^3} = \frac{\|-12\mathbf{k}\|}{(9\cos^2 t + 16\sin^2 t)^{3/2}} = \frac{12}{(9\cos^2 t + 16\sin^2 t)^{3/2}}$$

Then

$$a_{\mathbf{N}} = v^2(t)\kappa = (9\cos^2 t + 16\sin^2 t)\frac{12}{(9\cos^2 t + 16\sin^2 t)^{3/2}} = \frac{12}{\sqrt{9\cos^2 t + 16\sin^2 t}}$$

The motion of the particle on the ellipse starts at $\mathbf{r}(0) = 4\mathbf{j}$. As $t$ increases, $x(t) = 3\sin t$ increases and $y(t) = 4\cos t$ decreases, so the motion of the particle is clockwise.

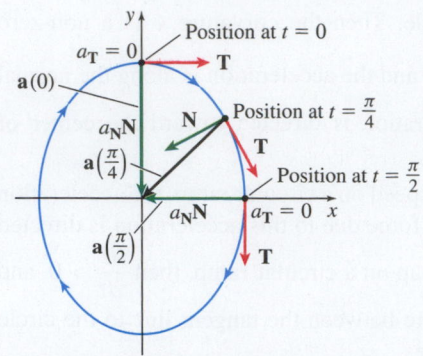

y

Position at $t = 0$

$a_T = 0$  T

$\mathbf{a}(0)$

$a_N\mathbf{N}$  N

$\mathbf{a}\left(\frac{\pi}{4}\right)$  Position at $t = \frac{\pi}{4}$

T

Position at $t = \frac{\pi}{2}$

$a_N\mathbf{N}$  $a_T = 0$  x

$\mathbf{a}\left(\frac{\pi}{2}\right)$

T

**DF Figure 30** $\mathbf{r}(t) = 3 \sin t\mathbf{i} + 4 \cos t\mathbf{j}$

At $t = 0$,

$$\mathbf{a}(0) = -3 \cdot 0\mathbf{i} - 4 \cdot 1\mathbf{j} = -4\mathbf{j} \qquad a_T = \frac{7 \cdot 0 \cdot 1}{3} = 0 \qquad a_N = \frac{12}{3} = 4$$

At $t = \frac{\pi}{4}$, $\mathbf{r}\left(\frac{\pi}{4}\right) = \frac{3\sqrt{2}}{2}\mathbf{i} + 2\sqrt{2}\,\mathbf{j}$, and

$$\mathbf{a}\left(\frac{\pi}{4}\right) = -\frac{3\sqrt{2}}{2}\mathbf{i} - 2\sqrt{2}\,\mathbf{j} \qquad a_T = \frac{7\sqrt{2}}{10} \qquad a_N = \frac{12\sqrt{2}}{5}$$

At $t = \frac{\pi}{2}$, $\mathbf{r}\left(\frac{\pi}{2}\right) = 3\mathbf{i} + 4 \cdot 0\mathbf{j} = 3\mathbf{i}$, and

$$\mathbf{a}\left(\frac{\pi}{2}\right) = -3 \cdot 1\mathbf{i} - 4 \cdot 0\mathbf{j} = -3\mathbf{i} \qquad a_T = 0 \qquad a_N = \frac{12}{4} = 3$$

See Figure 30. ∎

**NOW WORK** Problem 27.

Shutterstock

**EXAMPLE 7** Analyzing the Acceleration of a Car

A car on the ramp of a multistory parking garage travels along a curve traced out by $\mathbf{r}(t) = 10 \cos t\mathbf{i} + 10 \sin t\mathbf{j} + 3t\mathbf{k}$, where $t$ is the time in hours and distance is in miles. Find the tangential component $a_T$ and normal component $a_N$ of the acceleration of the car. What are the magnitude and direction of the force on the driver?

**Solution** We begin by finding the velocity $\mathbf{v}$, speed $v$, and acceleration $\mathbf{a}$ of the car.

$$\mathbf{v}(t) = \mathbf{r}'(t) = \frac{d}{dt}(10 \cos t)\mathbf{i} + \frac{d}{dt}(10 \sin t)\mathbf{j} + \frac{d}{dt}(3t)\mathbf{k} = -10 \sin t\mathbf{i} + 10 \cos t\mathbf{j} + 3\mathbf{k}$$

$$v(t) = \|\mathbf{v}(t)\| = \|\mathbf{r}'(t)\| = \sqrt{(-10 \sin t)^2 + (10 \cos t)^2 + 9} = \sqrt{109} \approx 10.440 \, \text{mi/h}$$

$$\mathbf{a}(t) = \mathbf{r}''(t) = \frac{d}{dt}\mathbf{v}(t) = -10 \cos t\mathbf{i} - 10 \sin t\mathbf{j}$$

Since the speed is constant, the tangential component of acceleration $a_T$ is

$$a_T = \frac{dv}{dt} = 0 \, \text{mi/h}^2$$

The normal component of acceleration $a_N$ is

$$a_N = v^2 \kappa = v^2 \frac{\|\mathbf{r}'(t) \times \mathbf{r}''(t)\|}{\|\mathbf{r}'(t)\|^3} \underset{\underset{v = \|\mathbf{r}'(t)\| = \sqrt{109}}{\uparrow}}{=} \frac{\|\mathbf{r}'(t) \times \mathbf{r}''(t)\|}{\sqrt{109}}$$

Since

$$\mathbf{r}'(t) \times \mathbf{r}''(t) = \begin{vmatrix} \mathbf{i} & \mathbf{j} & \mathbf{k} \\ -10 \sin t & 10 \cos t & 3 \\ -10 \cos t & -10 \sin t & 0 \end{vmatrix} = 30 \sin t\mathbf{i} - 30 \cos t\mathbf{j} + 100\mathbf{k}$$

the normal component is

$$a_N = \frac{\|30 \sin t\mathbf{i} - 30 \cos t\mathbf{j} + 100\mathbf{k}\|}{\sqrt{109}} = \frac{\sqrt{900 \sin^2 t + 900 \cos^2 t + 10000}}{\sqrt{109}}$$

$$= \frac{\sqrt{10900}}{\sqrt{109}} = 10 \, \text{mi/h}^2$$

As the car travels on the ramp at a constant speed, the force

$$\mathbf{F} = m\mathbf{a} = ma_N\mathbf{N} = 10m\mathbf{N}$$

pulls the car and driver toward the center of the ramp, with a magnitude about 10 times the mass of the car. ∎

**NOW WORK** Problem 69.

## Application: The UVW-Axes in Orbital Mechanics

A spacecraft in orbit about Earth follows a curve in space, which is modeled as the graph of a vector function of time $t$. The spacecraft carries a gyroscope whose three axes remain fixed in direction throughout time and are aligned with the **i**, **j**, **k** unit vectors of the $xyz$-coordinate system. The $xyz$-coordinate system has its origin at the center of Earth, providing the same basic frame of reference for both Earth-bound observers and the spacecraft. See Figure 31.

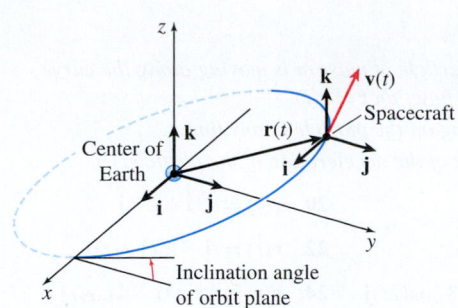

**Figure 31**

Suppose the position of the spacecraft is given by the vector function

$$\mathbf{r}(t) = x(t)\mathbf{i} + y(t)\mathbf{j} + z(t)\mathbf{k}$$

Then the velocity vector $\mathbf{v} = \mathbf{v}(t)$ is tangent to the orbit: $\mathbf{v}(t) = \mathbf{r}'(t)$.

If Earth is taken to have all its mass concentrated at the center, then the laws of Kepler and Newton tell us that any orbit remains in a *fixed plane* containing the center. Furthermore, a closed orbit describes an ellipse (or circle) in that plane with Earth's center at *one of the foci*.

The UVW system of axes is an orthogonal set of unit vectors $\mathbf{i_U}(t)$, $\mathbf{i_V}(t)$, $\mathbf{i_W}(t)$ that are "attached" to the spacecraft's orbit at each point. This set of vectors is particularly useful in manned spacecraft, since it corresponds to the axes of *yaw, roll,* and *pitch* customarily used in airplanes flying over the surface of Earth. These time-dependant unit vectors are defined as

$$\mathbf{i_U} = \frac{\mathbf{r}}{\|\mathbf{r}\|}, \qquad \mathbf{i_W} = \frac{\mathbf{r} \times \mathbf{v}}{\|\mathbf{r} \times \mathbf{v}\|}, \qquad \mathbf{i_V} = \mathbf{i_W} \times \mathbf{i_U} \qquad (7)$$

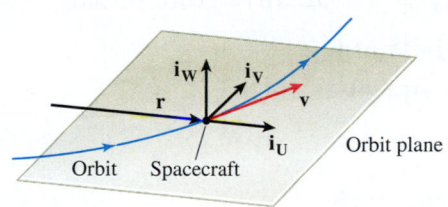

**Figure 32**

Notice that $\mathbf{i_U}(t)$ lies in the orbit plane and points radially *outward*. Since both **r** and **v** are vectors in the orbit plane and $\mathbf{r} \times \mathbf{v}$ is orthogonal to both **r** and **v**, the vector $\mathbf{i_W}(t)$ is *normal* to the orbit plane. The vector $\mathbf{i_V}(t)$ completes the set of axes and lies in the orbit plane. See Figure 32.

---

**EXAMPLE 8** **Finding the UVW-Axes of an Orbit**

A satellite in a circular orbit intersects the $x$-axis and is inclined to the $xy$-plane at an angle of $30°$. Suppose the orbit has a radius $a$, and the motion has angular speed $\omega$. Then the position of the satellite at time $t$ is

$$\mathbf{r}(t) = a\left(\cos(\omega t)\mathbf{i} + \frac{\sqrt{3}}{2}\sin(\omega t)\mathbf{j} + \frac{1}{2}\sin(\omega t)\mathbf{k}\right)$$

The velocity vector **v** is

$$\mathbf{v}(t) = \mathbf{r}'(t) = a\omega\left(-\sin(\omega t)\mathbf{i} + \frac{\sqrt{3}}{2}\cos(\omega t)\mathbf{j} + \frac{1}{2}\cos(\omega t)\mathbf{k}\right)$$

In this case, using (7), the UVW system of axes is

$$\mathbf{i_U} = \cos(\omega t)\mathbf{i} + \frac{\sqrt{3}}{2}\sin(\omega t)\mathbf{j} + \frac{1}{2}\sin(\omega t)\mathbf{k} \qquad \mathbf{i_U} = \frac{\mathbf{r}}{\|\mathbf{r}\|}$$

$$\mathbf{i_W} = -\frac{1}{2}\mathbf{j} + \frac{\sqrt{3}}{2}\mathbf{k} \qquad \mathbf{i_W} = \frac{\mathbf{r} \times \mathbf{v}}{\|\mathbf{r} \times \mathbf{v}\|}$$

$$\mathbf{i_V} = -\sin(\omega t)\mathbf{i} + \frac{\sqrt{3}}{2}\cos(\omega t)\mathbf{j} + \frac{1}{2}\cos(\omega t)\mathbf{k} \qquad \mathbf{i_V} = \mathbf{i_W} \times \mathbf{i_U}$$

The directions of the onboard gyroscope axes UVW are now expressed in terms of **i**, **j**, and **k**. ∎

# 11.4 Assess Your Understanding

## Concepts and Vocabulary

1. *True or False*   If a twice differentiable vector function $r(t) = x(t)i + y(t)j + z(t)k$ represents the position of a particle moving along a smooth curve, then the velocity vector is $v(t) = r'(t)$ and the acceleration vector is $a(t) = r''(t)$.

2. *True or False*   If a particle moving along a smooth curve traced out by a twice differentiable vector function $r = r(t)$ travels at a constant speed, then the acceleration of the particle is directed along the unit tangent vector $T$.

3. *True or False*   The acceleration vector $a = a(t)$ of a particle moving along a smooth curve traced out by a twice differentiable vector function $r = r(t)$ can be written as the sum of two orthogonal vectors: one vector in the direction of the velocity and the other vector perpendicular to the direction of the velocity.

4. When the acceleration vector $a(t)$ of a particle moving along a smooth curve is written as $a(t) = v'(t)T(t) + [v(t)]^2 \kappa N(t)$, then $a_T = v'(t)$ is called the _____ component and $a_N = [v(t)]^2 \kappa$ is called the _____ component of the acceleration vector $a$.

## Skill Building

*In Problems 5–10:*

(a) Find the velocity, acceleration, and speed of a particle whose motion is along the plane curve traced out by the vector function $r = r(t)$.

(b) For each curve, graph the motion of the particle and the vectors $v(0)$ and $a(0)$.

5.  $r(t) = ti + t^2 j$

6.  $r(t) = ti - t^3 j$

7.  $r(t) = (t^2 + 1)i + (1 - t)j$

8.  $r(t) = t^3 i + 3t j$

9.  $r(t) = 4ti - t^3 j$

10. $r(t) = t^3 i + \dfrac{1}{2} t j$

*In Problems 11–16, find the velocity, acceleration, and speed of a particle whose motion is along the space curve traced out by the vector function $r = r(t)$.*

11. $r(t) = t^2 i + t j - 3t^3 k$

12. $r(t) = (t + 1)i + 6t j + t^2 k$

13. $r(t) = \sqrt{16 - t^2} i + t^2 j + t k$

14. $r(t) = \dfrac{1}{t^2} i + \dfrac{t + 1}{t^2} j + t k$

15. $r(t) = 2 \cos t i + \sin t j + t k$

16. $r(t) = t i + \sin(4t) j + \cos(4t) k$

*In Problems 17 and 18:*

(a) Find the velocity, acceleration, and speed of a particle whose motion is along the space curve traced out by the vector function $r = r(t)$.

(b) Evaluate $v(t)$, $a(t)$, and $v(t)$ at the given $t$.

CAS (c) Graph the space curve traced out by $r = r(t)$. On the same axes graph the velocity vector and acceleration vector at the given value of $t$.

17. $r(t) = \sin t i + \cos t j + \sin(2t) k$   $t = \dfrac{\pi}{2}$

18. $r(t) = \sin t i + \cos t j + \sqrt{t} k$   $t = \pi$

*In Problems 19–26, a particle of mass m is moving along the curve traced out by the vector function $r = r(t)$.*

(a) Find the force acting on the particle at any time $t$.

(b) Find the magnitude of the acceleration of the particle.

19. $r(t) = e^t i + e^{-t} j$

20. $r(t) = e^{2t} i + e^{-t} j$

21. $r(t) = t i + e^t j$

22. $r(t) = t i + \ln(1 + t) j$

23. $r(t) = 3 \sin(2t) i + 3 \cos(2t) j$

24. $r(t) = 4 \sin t i - 4 \cos t j$

25. $r(t) = 2 \cos t i - 3 \sin t j$

26. $r(t) = -\cos(3t) i + 2 \sin(3t) j$

*In Problems 27–44:*

(a) Find the velocity, acceleration, and speed of a particle traveling along the curve traced out by the vector function $r = r(t)$.

(b) Find the tangential and normal components of the acceleration.

27. $r(t) = 2t i + (t + 1) j$

28. $r(t) = (1 - 3t)i + 2t j$

29. $r(t) = e^t i + e^{2t} j$

30. $r(t) = e^{-t} i + e^{-2t} j$

31. $r(t) = 2 \sin t i + \cos t j$

32. $r(t) = 2 \cos t i + 3 \sin t j$

33. $r(t) = (1 - 3t)i + 2t j - (5 + t)k$

34. $r(t) = (2 + t)i + (2 - t)j + 3t k$

35. $r(t) = t i + t^2 j + t^3 k$

36. $r(t) = \dfrac{t^2}{t + 1} i + \dfrac{1}{t + 1} j + \dfrac{t}{t + 1} k$

37. $r(t) = 3i + \cos t j + \sin t k$          38. $r(t) = t i + \sin t j + \cos t k$

39. $r(t) = \ln t i + \sqrt{t} j + t^{3/2} k$   $t > 0$

40. $r(t) = \cos(2t) i + \sin(2t) j - 5k$

41. $r(t) = e^t \cos t i + e^t \sin t j + e^t k$

42. $r(t) = e^{-t} \cos t i + e^{-t} \sin t j - e^{-t} k$

43. $r(t) = a \cos t i + b \sin t j + ct k$   $a > 0$   $b > 0$   $c > 0$

44. $r(t) = \cosh t i + \sinh t j + t k$

## Applications and Extensions

45. **Velocity and Acceleration**   Find the velocity and acceleration of a particle moving on the cycloid
$$r(t) = [\pi t - \sin(\pi t)]i + [1 - \cos(\pi t)]j$$

46. **Velocity and Acceleration**

(a) Find the velocity and acceleration of a particle moving on the parabola $r(t) = (t^2 - 2t)i + t j$.

(b) At what time $t$ is the speed of the particle 0?

47. **Acceleration**   A particle moves along the path $y = 3x^2 - x^3$ with the horizontal component of the velocity equal to $\dfrac{1}{3}$. Find the acceleration at the points where the velocity $v$ is horizontal. Graph the motion and indicate $v$ and $a$ at these points.

48. Show that if the speed of a particle along a curve is constant, then the velocity and acceleration vectors are orthogonal.

49. If a particle moves along the graph of $y = f(x)$, show that $a_N = 0$ at a point of inflection of the graph.

---

**1.** = NOW WORK problem          〰 = Graphing technology recommended          CAS = Computer Algebra System recommended

**50.** Suppose that the vector function $\mathbf{r}(t) = e^t\mathbf{i} + e^{-t}\mathbf{j}$ gives the position of a particle at time $t$.

  **(a)** Show that the force on the particle is directed away from the origin.

  **(b)** What is the minimum speed of the particle and where does it occur?

  **(c)** Find the tangential and normal components of the acceleration at the point found in (b).

  **(d)** The answers to (c) are $a_T = 0$ and $a_N = ||\mathbf{a}(t)|| = \sqrt{2}$. How could these have been predicted?

**51.** A particle moves along the graph $y = \dfrac{1}{2}x^2$ with constant speed $v_0 = 2$. Find $a_T$ and $a_N$ in terms of $x$ alone at a general point $\left(x, \dfrac{1}{2}x^2\right)$ on the graph.

**52.** If the motion of a particle is along the curve traced out by $\mathbf{r}(t) = \alpha \cosh t\mathbf{i} + \beta \sinh t\mathbf{j}$, where $\alpha > 0$ and $\beta > 0$, show that the force acting on the particle is in the direction of $\mathbf{r} = \mathbf{r}(t)$.

**53. Centripetal Force**  Refer to Example 5.

  **(a)** If the speed of the motorcycle is increased by 10%, by how much does the frictional force on the tires need to increase to keep the motorcycle from skidding on the track?

  **(b)** If the radius of the circular track is halved, how much slower must the motorcycle be driven so the frictional force doesn't change?

**54. Frictional Force**  A race car with mass 1000 kg is driven at a constant speed of 200 km/h around a circular track whose radius is 75 m. What frictional force must be exerted by the track on the tires to keep the car from skidding?

**55. Particle Collider**  The Large Hadron Collider (LHC), buried 100 m beneath Earth's surface, is located near Geneva, Switzerland, and spans the border between France and Switzerland. The LHC is 26,659 m in circumference and, at full power, protons travel at a speed of 299,792,455.3 m/s (99.9999991% the speed of light) around the circle. Find the magnitude of force necessary to keep a proton of mass $m$ moving at this speed.

  *Source*: http://public.web.cern.ch.

**56. Particle Collider**  At the Fermi National Accelerator Laboratory in Batavia, Illinois, protons were accelerated along a circular route with radius 1 km. Find the magnitude of the force necessary to give a proton of mass $m$ a constant speed of 280,000 km/s.

**57. Magnitude of a Force**  Find the maximum magnitude of the force acting on a particle of mass $m$ whose motion is along the curve $\mathbf{r}(t) = 4\cos t\mathbf{i} - 2\cos(2t)\mathbf{j}$.

**58. Curvature of the Path of a Golf Ball**  A golf ball is hit with an initial speed $v_0$ at an angle $\theta$ to the horizontal and travels on a smooth curve traced out by the vector function

$$\mathbf{r} = \mathbf{r}(t) = (v_0\cos\theta)t\mathbf{i} + \left[(v_0\sin\theta)t - \frac{1}{2}gt^2\right]\mathbf{j},$$

where $g = 9.8\,\text{m/s}^2$ and $0 \le \theta \le \dfrac{\pi}{2}$.

  **(a)** At the highest point in the ball's trajectory, find the curvature of its path.

  **(b)** At the highest point in its trajectory, express the ball's acceleration vector using tangential and normal components. Are these components physically reasonable? Explain why.

**59. Velocity**  A particle moves on the circle $x^2 + y^2 = 1$ so that at time $t \ge 0$ the position is given by the vector

$$\mathbf{r}(t) = \frac{1 - t^2}{1 + t^2}\mathbf{i} + \frac{2t}{1 + t^2}\mathbf{j}$$

  **(a)** Find the velocity vector.

  **(b)** Is the particle ever at rest? Justify your answer.

  **(c)** Find the coordinates of the point that the particle approaches as $t$ increases without bound.

**60. Velocity**  The position of a particle at time $t \ge 0$ is given by

$$\mathbf{r}(t) = e^{-t}\cos t\mathbf{i} + e^{-t}\sin t\mathbf{j}$$

  **(a)** Show that the path of the particle is part of the spiral $r = e^{-\theta}$ (in polar coordinates).

  **(b)** Graph the path of the particle and indicate its direction of travel.

  **(c)** Show that the angle between the vector $\mathbf{r}$ and velocity vector $\mathbf{v}$ is always $135°$.

**61. Oscillating Motion**  Suppose a particle moves along the curve traced out by $\mathbf{r}(t) = 4\cos t\mathbf{i} - 2\cos(2t)\mathbf{j}$.

  **(a)** Show that the particle oscillates on an arc of a parabola.

  **(b)** Graph the path.

  **(c)** Find the acceleration $\mathbf{a}$ at points of zero velocity.

**62.** If $\mathbf{r}'(t) = \mathbf{b} \times \mathbf{r}(t)$ for all $t$, where $\mathbf{b}$ is a constant vector, show that the acceleration $\mathbf{a}(t)$ is perpendicular to $\mathbf{b}$ and that the speed is constant.

**63. Mars Orbit**  NASA's *Odyssey* orbiter's mission is to relay directly back to Earth the UHF telemetry taken by the rovers on Mars. *Odyssey* is in a "near-Mars" orbit at a height of 400 km above its surface. Find the speed of the *Odyssey* if the acceleration due to gravity on Mars is 3.71 m/s² and Mars' radius is 3390 km.

NASA/JPL

**64. Physics**  The **power** expended by a force $\mathbf{F}(t)$ acting on an object of mass $m$ with velocity $\mathbf{v}(t)$ is the dot product

$$\text{Power} = \mathbf{F}(t) \cdot \mathbf{v}(t)$$

The **kinetic energy** of an object of mass $m$ and velocity $\mathbf{v}(t)$ equals one-half the product of its mass times the square of its speed:

$$\text{Kinetic energy} = \frac{1}{2}m\|\mathbf{v}(t)\|^2$$

Show that the power of an object equals the rate of change of its kinetic energy with respect to time.

**65. Momentum**  In classical mechanics, the **momentum** $\mathbf{p}(t)$ of an object of mass $m$ at time $t$ is defined as $\mathbf{p}(t) = m\mathbf{v}(t)$, where $\mathbf{v}(t)$ is the velocity of the object at time $t$. Show that force equals the rate of change of momentum with respect to time.

**66. Angular Momentum** The **torque** $\tau$ (Greek letter *tau*) produced by a force $\mathbf{F}(t)$ acting on an object whose position at time $t$ is $\mathbf{r}(t)$ is defined as $\tau(t) = \mathbf{r}(t) \times \mathbf{F}(t)$. Torque measures the twist imparted on the object by the force. See the figure below. The **angular momentum L** of an object of mass $m$ and velocity $\mathbf{v}(t)$ whose position at time $t$ is $\mathbf{r}(t)$ is $\mathbf{L}(t) = \mathbf{r}(t) \times m\mathbf{v}(t)$. Show that the rate of change of angular momentum with respect to time equals the torque.

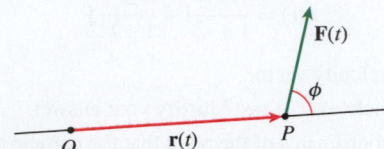

**67. Central Fields of Force** A **central field of force** $\mathbf{F}(t)$ is one whose direction is parallel to that of the position $\mathbf{r} = \mathbf{r}(t)$ of the object it acts upon; that is, $\mathbf{F}(t) = u(t)\mathbf{r}(t)$, where $u(t)$ is a scalar function. Show that a central field of force produces zero torque.

**68. Hydrogen Atom** The **Bohr model of an atom** views it as a miniature solar system, with negative electrons in circular orbits around the positive nucleus. A hydrogen atom consists of a single electron in orbit around a single proton. The electric force $F_{\text{elec}}$ that the proton of charge $e = 1.60 \times 10^{-19}$ C exerts on the electron, also of charge $e = 1.60 \times 10^{-19}$ C , is $F_{\text{elec}} = \dfrac{ke^2}{r^2}$, where $r$ is the distance in meters between the nucleus and the electron and $k = 9.0 \times 10^9$ N $\cdot$ m²/C is a constant. Apply Newton's Second Law of Motion to the electron in a hydrogen atom and show that its speed is $v = \sqrt{\dfrac{ke^2}{mr}}$, where $m = 9.11 \times 10^{-31}$ kg is the mass of the electron.

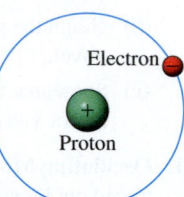

**69. Analyzing Forces** Use the formula

$$\mathbf{F} = m\mathbf{a} = m\frac{dv}{dt}\mathbf{T} + mv^2\kappa\mathbf{N}$$

to discuss the forces on a passenger in a car.

**(a)** What force corresponds to the push against the seat experienced by the passenger when the accelerator is depressed?

**(b)** What force corresponds to the push against the door experienced by the passenger when the car is going around a curve?

**(c)** How does this second force vary with the curvature of the road?

**(d)** How does it vary with the speed of the car?

**70. Analyzing Motion** Suppose $\mathbf{r} = \mathbf{r}(t)$ gives the position of a particle at time $t$. If the normal component of acceleration equals 0 at any time $t$, explain why the motion of the particle must be in a straight line.

**71.** Suppose the position of a particle in space is given by the vector function $\mathbf{r} = \mathbf{r}(t)$. Show that if $\mathbf{r} = \mathbf{r}(t)$ lies on a sphere, then the tangent vector is always orthogonal to $\mathbf{r} = \mathbf{r}(t)$.

## Challenge Problems

**72. Coriolis Acceleration** A particle moves on a disk from the center directly toward the edge. See the figure below. The disk has radius 1 and is revolving in the counterclockwise direction at a constant angular speed $\omega$, so the position of a point on the edge of the disk is given by $\mathbf{R}(t) = \cos(\omega t)\mathbf{i} + \sin(\omega t)\mathbf{j}$. Suppose the position of the particle at time $t$ is $\mathbf{r}(t) = t\mathbf{R}(t)$.

**(a)** Show that the velocity $\mathbf{v}$ of the particle is

$$\mathbf{v} = \cos(\omega t)\mathbf{i} + \sin(\omega t)\mathbf{j} + t\mathbf{v_d}$$

where $\mathbf{v_d} = \mathbf{R}'(t)$ is the velocity of the rotating disk.

**(b)** Also show that the acceleration $\mathbf{a}$ of the particle is

$$\mathbf{a} = 2\mathbf{v_d} + t\mathbf{a_d}$$

where $\mathbf{a_d} = \mathbf{R}''(t)$ is the acceleration of the rotating disk. The extra term $2\mathbf{v_d}$ is called the **Coriolis acceleration**, which results from the interaction of the rotation of the disk and the motion of the particle on the disk.

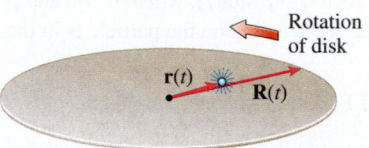

**73. Coriolis Acceleration** Refer to Problem 72.

**(a)** Find the velocity and acceleration of a particle revolving on a rotating disk according to

$$\mathbf{r}(t) = t^2\cos(\omega t)\mathbf{i} + t^2\sin(\omega t)\mathbf{j}$$

**(b)** What is the Coriolis acceleration?

**74.** Suppose a curve $C$ is traced out by the vector function $\mathbf{r} = \mathbf{r}(t)$. Find two nonparallel vectors that are orthogonal to $\mathbf{T}(t)$ other than the principal unit normal vector. Assume $\mathbf{T}(t) \neq \mathbf{0}$ and $\mathbf{N}(t) \neq \mathbf{0}$.

**75.** Let $\mathbf{u}_r = \cos\theta\,\mathbf{i} + \sin\theta\,\mathbf{j}$ and $\mathbf{u}_\theta = -\sin\theta\,\mathbf{i} + \cos\theta\,\mathbf{j}$.

**(a)** Show that $\mathbf{u}_r$ and $\mathbf{u}_\theta$ are unit orthogonal vectors.

**(b)** Show that $\dfrac{d\mathbf{u}_r}{d\theta} = \mathbf{u}_\theta$ and $\dfrac{d\mathbf{u}_\theta}{d\theta} = -\mathbf{u}_r$.

**(c)** Suppose that $\mathbf{r}(t)$ is the position vector of a plane curve and that the tip of $\mathbf{r}(t)$ has polar coordinates $(r(t),\ \theta(t))$. Show that $\mathbf{r}(t) = \|\mathbf{r}(t)\|\mathbf{u}_r = r(t)\mathbf{u}_r(t)$.

**(d)** Use the Chain Rule to show that

$$\mathbf{v}(t) = \frac{dr}{dt}\mathbf{u}_r + r\frac{d\theta}{dt}\mathbf{u}_\theta = r'\mathbf{u}_r + r\theta'\mathbf{u}_\theta$$

**(e)** Show that $\mathbf{a}(t) = [r'' - r(\theta')^2]\mathbf{u}_r + [r\theta'' + 2r'\theta']\mathbf{u}_\theta$.

**76.** Express the velocity $\mathbf{v}$ and acceleration $\mathbf{a}$ in terms of $\mathbf{u}_r$ and $\mathbf{u}_\theta$ for a motion along the polar curve $r = 2 + \cos t$, where $\theta = 2t$.

**77.** Find polar coordinate equations for $\mathbf{r}(t) = e^{2t}\cos t\,\mathbf{i} + e^{2t}\sin t\,\mathbf{j}$ and express the velocity $\mathbf{v}$ and acceleration $\mathbf{a}$ in terms of $\mathbf{u}_r$ and $\mathbf{u}_\theta$.

**78. (a)** Show that if $\mathbf{r}'(t) \cdot \mathbf{r}(t) = 0$ for all $t$, then the motion $\mathbf{r}(t)$ is on the surface of a sphere centered at the origin.

**(b)** What can be said about a motion for which $\mathbf{r}''(t) \cdot \mathbf{r}'(t) = 0$ for all $t$?

**(c)** What can be said about a motion for which $\mathbf{r}'(t) \times \mathbf{r}(t) = 0$ for all $t$?

# 11.5 Integrals of Vector Functions; Projectile Motion

**OBJECTIVES**  *When you finish this section, you should be able to:*

1  **Integrate vector functions (p. 837)**
2  **Solve projectile motion problems (p. 839)**

Up to now we have been given the position function $\mathbf{r} = \mathbf{r}(t)$ of a moving particle, and we have used it to find the particle's velocity, speed, and acceleration. But in motion problems involving a force applied to an object, we often only know the mass $m$ of the object and the force vector $\mathbf{F}$ that is applied to it. By using Newton's Second Law of Motion, $\mathbf{F} = m\mathbf{a}$, we can find the acceleration $\mathbf{a}$ of the particle. But to find the motion of the particle, we also need its velocity $\mathbf{v}$ and its position $\mathbf{r}$. Since $\mathbf{a} = \dfrac{d\mathbf{v}}{dt}$ and $\mathbf{v} = \dfrac{d\mathbf{r}}{dt}$, we can find $\mathbf{v}$ and then $\mathbf{r}$ by integrating.

## 1  Integrate Vector Functions

Let $\mathbf{r} = \mathbf{r}(t)$ be a vector function whose domain is the closed interval $a \le t \le b$. The **indefinite integral of a vector function** $\mathbf{r} = \mathbf{r}(t)$ is found by integrating each component of the vector function separately. That is, if

$$\mathbf{r}(t) = x(t)\mathbf{i} + y(t)\mathbf{j}$$

then

$$\int \mathbf{r}(t)dt = \left( \int x(t)dt \right)\mathbf{i} + \left( \int y(t)dt \right)\mathbf{j}$$

and if

$$\mathbf{r}(t) = x(t)\mathbf{i} + y(t)\mathbf{j} + z(t)\mathbf{k}$$

then

$$\int \mathbf{r}(t)dt = \left( \int x(t)dt \right)\mathbf{i} + \left( \int y(t)dt \right)\mathbf{j} + \left( \int z(t)dt \right)\mathbf{k}$$

When finding the indefinite integral of a vector function, we insert a constant vector of integration after integrating each component.

---

**EXAMPLE 1**  **Finding the Indefinite Integral of a Vector Function**

Find $\int (e^t\mathbf{i} + \ln t\,\mathbf{j})dt$.

**Solution** Integrate each component. Remember to insert a constant vector of integration for each component.

$$\int (e^t\mathbf{i} + \ln t\,\mathbf{j})dt = \left( \int e^t dt \right)\mathbf{i} + \left( \int \ln t\, dt \right)\mathbf{j}$$

$$= (e^t + c_1)\mathbf{i} + (t\ln t - t + c_2)\mathbf{j} \quad c_1 \text{ and } c_2 \text{ are constants of integration.}$$

$$= e^t\mathbf{i} + (t\ln t - t)\mathbf{j} + \mathbf{c}$$

where $\mathbf{c} = c_1\mathbf{i} + c_2\mathbf{j}$ is a constant vector.  ∎

**NOW WORK** Problem 3.

The **definite integral of a vector function** $\mathbf{r} = \mathbf{r}(t)$, $a \leq t \leq b$, is defined similarly. That is, if

$$\mathbf{r}(t) = x(t)\mathbf{i} + y(t)\mathbf{j}$$

then

$$\int_a^b \mathbf{r}(t)dt = \left( \int_a^b x(t)dt \right)\mathbf{i} + \left( \int_a^b y(t)dt \right)\mathbf{j}$$

and if

$$\mathbf{r}(t) = x(t)\mathbf{i} + y(t)\mathbf{j} + z(t)\mathbf{k}$$

then

$$\int_a^b \mathbf{r}(t)dt = \left( \int_a^b x(t)dt \right)\mathbf{i} + \left( \int_a^b y(t)dt \right)\mathbf{j} + \left( \int_a^b z(t)dt \right)\mathbf{k}$$

---

### EXAMPLE 2    Integrating a Vector Function

Find each vector integral.

(a) $\displaystyle \int_1^2 \left[ (6t^2 - t)\mathbf{i} + \left( 4t + \frac{1}{t} \right)\mathbf{j} \right] dt$     (b) $\displaystyle \int (e^t\mathbf{i} + e^{3t}\mathbf{j})dt$

(c) $\displaystyle \int_0^1 \left( 4t^{1/3}\mathbf{i} + (6t + 1)\mathbf{j} + \frac{2}{t+1}\mathbf{k} \right) dt$     (d) $\displaystyle \int (\sin t\mathbf{i} + \cos t\mathbf{j} + 2\mathbf{k})dt$

**Solution** (a) Integrate each component.

$$\int_1^2 \left[ (6t^2 - t)\mathbf{i} + \left( 4t + \frac{1}{t} \right)\mathbf{j} \right] dt = \left[ \int_1^2 (6t^2 - t)dt \right]\mathbf{i} + \left[ \int_1^2 \left( 4t + \frac{1}{t} \right) dt \right]\mathbf{j}$$

$$= \left[ 2t^3 - \frac{t^2}{2} \right]_1^2 \mathbf{i} + \left[ 2t^2 + \ln|t| \right]_1^2 \mathbf{j}$$

$$= \left[ (2 \cdot 8 - 2) - \left( 2 - \frac{1}{2} \right) \right]\mathbf{i} + [(8 + \ln 2) - (2 + 0)]\mathbf{j}$$

$$= \frac{25}{2}\mathbf{i} + (6 + \ln 2)\mathbf{j}$$

(b) Integrate each component, remembering to add a constant vector of integration to each component.

$$\int (e^t\mathbf{i} + e^{3t}\mathbf{j})dt = \left( \int e^t dt \right)\mathbf{i} + \left( \int e^{3t} dt \right)\mathbf{j} = (e^t + c_1)\mathbf{i} + \left( \frac{e^{3t}}{3} + c_2 \right)\mathbf{j} = e^t\mathbf{i} + \frac{e^{3t}}{3}\mathbf{j} + \mathbf{c}$$

where $\mathbf{c} = c_1\mathbf{i} + c_2\mathbf{j}$ is a constant vector.

(c) Integrate each component.

$$\int_0^1 \left( 4t^{1/3}\mathbf{i} + (6t + 1)\mathbf{j} + \frac{2}{t+1}\mathbf{k} \right) dt = \left( \int_0^1 4t^{1/3} dt \right)\mathbf{i} + \left( \int_0^1 (6t + 1)dt \right)\mathbf{j} + \left( \int_0^1 \frac{2}{t+1}dt \right)\mathbf{k}$$

$$= \left[ 3t^{4/3} \right]_0^1 \mathbf{i} + \left[ 3t^2 + t \right]_0^1 \mathbf{j} + \left[ 2\ln|t + 1| \right]_0^1 \mathbf{k} = 3\mathbf{i} + 4\mathbf{j} + 2\ln 2\mathbf{k}$$

(d) Integrate each component, remembering to add a constant vector of integration to each component.

$$\int (\sin t\mathbf{i} + \cos t\mathbf{j} + 2\mathbf{k})dt = \left( \int \sin t dt \right)\mathbf{i} + \left( \int \cos t dt \right)\mathbf{j} + \left( \int 2dt \right)\mathbf{k}$$

$$= (-\cos t + c_1)\mathbf{i} + (\sin t + c_2)\mathbf{j} + (2t + c_3)\mathbf{k}$$

$$= -\cos t\mathbf{i} + \sin t\mathbf{j} + 2t\mathbf{k} + \mathbf{c}$$

where $\mathbf{c} = c_1\mathbf{i} + c_2\mathbf{j} + c_3\mathbf{k}$ is a constant vector. ∎

**NOW WORK** Problem 15.

**EXAMPLE 3**  **Solving a Vector Differential Equation**

Solve the vector differential equation $\mathbf{r}'(t) = 2t\mathbf{i} + e^t\mathbf{j} + e^{-t}\mathbf{k}$ with the initial condition $\mathbf{r}(0) = \mathbf{i} - \mathbf{j} + \mathbf{k}$.

**Solution** The general solution to the differential equation is

$$\mathbf{r}(t) = \int \mathbf{r}'(t)\,dt = \int (2t\mathbf{i} + e^t\mathbf{j} + e^{-t}\mathbf{k})\,dt$$

$$= \left(\int 2t\,dt\right)\mathbf{i} + \left(\int e^t\,dt\right)\mathbf{j} + \left(\int e^{-t}\,dt\right)\mathbf{k}$$

$$= (t^2 + c_1)\mathbf{i} + (e^t + c_2)\mathbf{j} + (-e^{-t} + c_3)\mathbf{k}$$

Now use the initial condition $\mathbf{r}(0) = \mathbf{i} - \mathbf{j} + \mathbf{k}$.

$$\mathbf{r}(0) = c_1\mathbf{i} + (1 + c_2)\mathbf{j} + (-1 + c_3)\mathbf{k} = \mathbf{i} - \mathbf{j} + \mathbf{k}$$

from which we find

$$c_1 = 1 \qquad 1 + c_2 = -1 \qquad -1 + c_3 = 1$$
$$c_2 = -2 \qquad c_3 = 2$$

The solution of the differential equation with the initial condition $\mathbf{r}(0) = \mathbf{i} - \mathbf{j} + \mathbf{k}$ is

$$\mathbf{r}(t) = (t^2 + 1)\mathbf{i} + (e^t - 2)\mathbf{j} + (2 - e^{-t})\mathbf{k} \qquad\blacksquare$$

**NOW WORK** Problem 21.

## 2 Solve Projectile Motion Problems

Newton's Second Law of Motion states that $\mathbf{F}(t) = m\mathbf{a}(t)$. Once we know the force $\mathbf{F}(t)$ acting on a particle of mass $m$, the acceleration vector $\mathbf{a}(t)$ is determined. The velocity vector $\mathbf{v}(t)$ is the integral of $\mathbf{a}(t)$. That is,

$$\mathbf{v}(t) = \int \mathbf{a}(t)\,dt$$

From this we can find the velocity and speed of the particle, provided an initial condition on $\mathbf{v}(t)$ is given. Furthermore, since

$$\mathbf{r}(t) = \int \mathbf{v}(t)\,dt$$

we can find the path of the particle if we know an initial condition on $\mathbf{r}(t)$.

CALC
CLIP

**EXAMPLE 4**  **Finding the Velocity and Position of a Particle**

Find the velocity vector and position vector of a particle if its acceleration is given by $\mathbf{a}(t) = -9.8\mathbf{j}\,\text{m/s}^2$, $\mathbf{v}(0) = 9.8\mathbf{j}\,\text{m/s}$, and $\mathbf{r}(0) = \mathbf{0}\,\text{m}$ for $0 \le t \le 2$.

**Solution** The velocity vector $\mathbf{v}(t)$ is

$$\mathbf{v}(t) = \int \mathbf{a}(t)\,dt = \left(\int -9.8\,dt\right)\mathbf{j} = c_1\mathbf{i} + (-9.8t + c_2)\mathbf{j}$$

Since $\mathbf{v}(0) = 9.8\mathbf{j}$, then $c_1 = 0$ and $c_2 = 9.8$, and the velocity vector is

$$\mathbf{v}(t) = (-9.8t + 9.8)\mathbf{j}$$

The position vector of the particle is given by

$$\mathbf{r}(t) = \int \mathbf{v}(t)\,dt = \left[\int (-9.8t + 9.8)\,dt\right]\mathbf{j} = d_1\mathbf{i} + (-4.9t^2 + 9.8t + d_2)\mathbf{j}$$

Since $\mathbf{r}(0) = \mathbf{0}$, then $d_1 = 0$ and $d_2 = 0$, and the position of the particle is given by

$$\mathbf{r}(t) = (-4.9t^2 + 9.8t)\mathbf{j} \qquad\blacksquare$$

**NOW WORK** Problem 27.

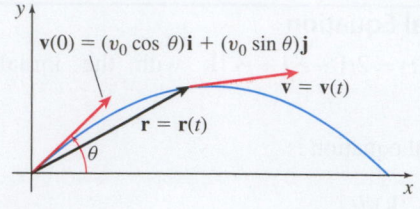

**DF** Figure 33

**NOTE** The assumption $\mathbf{F} = -mg\mathbf{j}$ is called the **flat Earth approximation**. It is a good approximation for short-range missiles. We also ignore air resistance here.

**RECALL** In SI units, $g \approx 9.8 \text{ m/s}^2$; in U.S. customary units $g \approx 32 \text{ ft/s}^2$.

Example 4 is a special case of the projectile problem: Find the path of a projectile fired at an angle $\theta$ to the horizontal with initial speed $v_0$, assuming that the only force acting on the projectile is gravity. We are interested in knowing the maximum height of the projectile, how far it travels, and how long it is in the air. To simplify the analysis, we choose to start the motion at the origin, so $\mathbf{r}(0) = \mathbf{0}$. The initial velocity vector $\mathbf{v}(0)$ has magnitude $v_0$ and is directed at an inclination $\theta$ to the horizontal axis, so $\mathbf{v}(0) = v_0(\cos\theta\mathbf{i} + \sin\theta\mathbf{j}) = (v_0\cos\theta)\mathbf{i} + (v_0\sin\theta)\mathbf{j}$.

In Figure 33, $\mathbf{r} = \mathbf{r}(t)$ represents the position of the projectile after time $t$ and $\mathbf{v} = \mathbf{v}(t)$ is the velocity vector. We assume the only force acting on the projectile is $\mathbf{F} = -mg\mathbf{j}$, where $g$ is the acceleration due to gravity and $m$ is the mass of the projectile. From Newton's Second Law of Motion, the acceleration $\mathbf{a}(t)$ of the projectile is given by

$$m\mathbf{a}(t) = -mg\mathbf{j} \qquad \mathbf{F} = m\mathbf{a}$$

$$\mathbf{a}(t) = -g\mathbf{j}$$

Integrating both sides with respect to $t$ gives

$$\mathbf{v}(t) = \int \mathbf{a}(t)\,dt = \left(\int -g\,dt\right)\mathbf{j} = -gt\mathbf{j} + \mathbf{c}$$

where $\mathbf{c}$ is a constant vector. When $t = 0$, $\mathbf{v}(0) = v_0\cos\theta\mathbf{i} + v_0\sin\theta\mathbf{j} = \mathbf{c}$. So, the velocity vector of the projectile is

$$\boxed{\mathbf{v}(t) = v_0\cos\theta\mathbf{i} + (v_0\sin\theta - gt)\mathbf{j}}$$

The position of the projectile is given by

$$\mathbf{r}(t) = \int \mathbf{v}(t)\,dt = (v_0\cos\theta)t\mathbf{i} + \left[(v_0\sin\theta)t - \frac{1}{2}gt^2\right]\mathbf{j} + \mathbf{d}$$

where $\mathbf{d}$ is a constant vector. When $t = 0$, the position of the projectile is at the origin, $\mathbf{r}(0) = \mathbf{0}$, so $\mathbf{d} = \mathbf{0}$. The position of the projectile at time $t$ is

$$\boxed{\mathbf{r}(t) = (v_0\cos\theta)t\mathbf{i} + \left[(v_0\sin\theta)t - \frac{1}{2}gt^2\right]\mathbf{j}}$$

The parametric equations of the motion of the projectile are

$$\boxed{x = (v_0\cos\theta)t \qquad \text{and} \qquad y = -\frac{1}{2}gt^2 + (v_0\sin\theta)t}$$

We use $t = \dfrac{x}{v_0\cos\theta}$ to eliminate $t$ from these equations. Then the rectangular equation of the motion is

$$\boxed{y = -\frac{g}{2v_0^2\cos^2\theta}x^2 + x\tan\theta}$$

which is the equation of a parabola. See Figure 34.

We are interested in knowing how far the projectile travels horizontally, how long it is in the air, and its maximum height.

The $x$-intercepts of the parabola are found by letting $y = 0$. As a result, the $x$-intercepts are

$$x = 0 \qquad \text{and} \qquad x = \frac{2v_0^2\cos^2\theta\tan\theta}{g} = \frac{v_0^2\sin(2\theta)}{g}$$

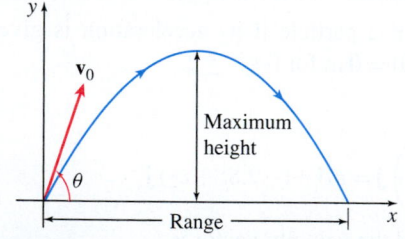

Figure 34

The number $\dfrac{v_0^2 \sin(2\theta)}{g}$, called the **range** of the projectile, tells us the horizontal distance the projectile travels. In Problem 47, you are asked to show that an inclination of $\theta = \dfrac{\pi}{4}$ gives the maximum range.

The projectile hits the ground when $y = 0$ and $t > 0$. In the parametric equation of the motion for $y$, let $y = 0$, and solve for $t$.

$$y = -\frac{1}{2}gt^2 + (v_0 \sin\theta)t$$

**NOTE** The other solution to the quadratic equation, $t = 0$, is a result of the initial condition.

$$0 = -\frac{1}{2}gt^2 + (v_0 \sin\theta)t$$

$$\frac{1}{2}gt = v_0 \sin\theta$$

$$t = \frac{2v_0 \sin\theta}{g}$$

The projectile is in the air for $\dfrac{2v_0 \sin\theta}{g}$ seconds.

The projectile reaches its maximum height when $\dfrac{dy}{dt} = -gt + v_0 \sin\theta = 0$. The maximum height is reached at $t = \dfrac{v_0 \sin\theta}{g}$ seconds. The maximum height is

$$y = -\frac{1}{2}g\left(\frac{v_0^2 \sin^2\theta}{g^2}\right) + (v_0 \sin\theta)\left(\frac{v_0 \sin\theta}{g}\right) = \frac{v_0^2 \sin^2\theta}{2g}$$

The formulas for the range of a projectile and its maximum height derived here are valid only for the initial conditions specified by $\mathbf{r}(0) = \mathbf{0}$ and $\mathbf{v}(0) = v_0 \cos\theta\,\mathbf{i} + v_0 \sin\theta\,\mathbf{j}$. If different initial conditions are given, the range and maximum height of the projectile will be different. In such instances, merely follow the same pattern of solution, adjusting where needed for different initial conditions.

NOW WORK **Problem 33.**

## 11.5 Assess Your Understanding

### Concepts and Vocabulary

1. *True or False* To integrate a vector function $\mathbf{r} = \mathbf{r}(t)$, integrate each component individually.

2. *True or False* When integrating a vector function $\mathbf{r} = \mathbf{r}(t)$, there is no need to add a constant of integration.

### Skill Building

*In Problems 3–18, find each integral.*

3. $\displaystyle\int (\sin t\mathbf{i} - \cos t\mathbf{j} + t\mathbf{k})\,dt$

4. $\displaystyle\int (\cos t\mathbf{i} + \sin t\mathbf{j} - \mathbf{k})\,dt$

5. $\displaystyle\int (t^2\mathbf{i} - t\mathbf{j} + e^t\mathbf{k})\,dt$

6. $\displaystyle\int (e^t\mathbf{i} - \sqrt{t}\mathbf{j} + t^2\mathbf{k})\,dt$

7. $\displaystyle\int (\ln t\mathbf{i} - t\ln t\mathbf{j} - 2\mathbf{k})\,dt$

8. $\displaystyle\int \left(\mathbf{i} + \ln t\mathbf{j} + \frac{1}{t}\mathbf{k}\right)dt$

9. $\displaystyle\int [(t-2)\mathbf{i} - (t-2)^2\mathbf{j} + \mathbf{k}]\,dt$

10. $\displaystyle\int [(3t+1)\mathbf{i} + (3t+1)^2\mathbf{j} + (3t+1)^{-1}\mathbf{k}]\,dt$

11. $\displaystyle\int_3^6 \left[(t-2)^{-3/2}\mathbf{i} + 2t\mathbf{j}\right]dt$

12. $\displaystyle\int_1^4 \left[t^{-1/2}\mathbf{i} + t^{1/2}\mathbf{j}\right]dt$

13. $\displaystyle\int_0^{\pi/2} [\cos(3t)\mathbf{i} + \sin(2t)\mathbf{j}]\,dt$

14. $\displaystyle\int_0^{\pi/4} [\sin t\cos t\mathbf{i} + \sec t\tan t\mathbf{j}]\,dt$

15. $\displaystyle\int_0^4 \left[(1 + 3\sqrt{t})\mathbf{i} + (3t^2 + 1)\mathbf{j} + (2t - 1)\mathbf{k}\right]dt$

16. $\displaystyle\int_1^8 \left[2t\mathbf{i} + \sqrt[3]{t}\mathbf{j} + \mathbf{k}\right]dt$

17. $\displaystyle\int_0^1 \left[te^{t^2}\mathbf{i} + e^t\mathbf{j} + 2e^{2t}\mathbf{k}\right]dt$

18. $\displaystyle\int_0^{\pi/4} [\sin(2t)\mathbf{i} + \cos(2t)\mathbf{j} + \mathbf{k}]\,dt$

**1.** = NOW WORK problem    [N] = Graphing technology recommended    CAS = Computer Algebra System recommended

*In Problems 19–24, solve each vector differential equation with the given condition.*

**19.** $\mathbf{r}'(t) = e^t \mathbf{i} - \ln t \mathbf{j} + 2t \mathbf{k}$  $\mathbf{r}(1) = \mathbf{j} + \mathbf{k}$

**20.** $\mathbf{r}'(t) = t \mathbf{i} + e^{-t} \mathbf{j} - \dfrac{1}{t} \mathbf{k}$  $\mathbf{r}(1) = \mathbf{i} - \mathbf{j} + 2\mathbf{k}$

**21.** $\mathbf{r}'(t) = 2 \sin t \mathbf{i} + \cos t \mathbf{j} + \mathbf{k}$  $\mathbf{r}(0) = \mathbf{i} - \mathbf{j}$

**22.** $\mathbf{r}'(t) = \cos(2t) \mathbf{i} + \sin(2t) \mathbf{j} + 2\mathbf{k}$  $\mathbf{r}(0) = \mathbf{i} + \mathbf{k}$

**23.** $\mathbf{r}'(t) = t^{-1} \mathbf{i} + t \mathbf{j} + t^2 \mathbf{k}$  $\mathbf{r}(1) = \mathbf{i} + \mathbf{j} + \mathbf{k}$

**24.** $\mathbf{r}'(t) = t^3 \mathbf{i} + \dfrac{1}{t+1} \mathbf{j} + \mathbf{k}$  $\mathbf{r}(0) = \mathbf{i} + \mathbf{j} + \mathbf{k}$

*In Problems 25–32, find the velocity, speed, and position of a particle having the given acceleration, initial velocity, and initial position.*

**25.** $\mathbf{a}(t) = -32\mathbf{k}$  $\mathbf{v}(0) = \mathbf{0}$  $\mathbf{r}(0) = \mathbf{0}$

**26.** $\mathbf{a}(t) = -32\mathbf{k}$  $\mathbf{v}(0) = \mathbf{i} + \mathbf{j}$  $\mathbf{r}(0) = \mathbf{0}$

**27.** $\mathbf{a}(t) = \cos t \mathbf{i} + \sin t \mathbf{j}$  $\mathbf{v}(0) = \mathbf{i}$  $\mathbf{r}(0) = \mathbf{j}$

**28.** $\mathbf{a}(t) = \cos t \mathbf{i} + \sin t \mathbf{j}$  $\mathbf{v}(0) = \mathbf{j}$  $\mathbf{r}(0) = \mathbf{i}$

**29.** $\mathbf{a}(t) = -9.8\mathbf{k}$  $\mathbf{v}(0) = \mathbf{i}$  $\mathbf{r}(0) = 5\mathbf{k}$

**30.** $\mathbf{a}(t) = -9.8\mathbf{k}$  $\mathbf{v}(0) = \mathbf{i} + \mathbf{j}$  $\mathbf{r}(0) = 2\mathbf{i}$

**31.** $\mathbf{a}(t) = e^{-t} \mathbf{i} + \mathbf{j}$  $\mathbf{v}(0) = \mathbf{i} + \mathbf{j}$  $\mathbf{r}(0) = \mathbf{i} - \mathbf{j}$

**32.** $\mathbf{a}(t) = t^2 \mathbf{i} - e^{-t} \mathbf{k}$  $\mathbf{v}(0) = \mathbf{i} - \mathbf{j}$  $\mathbf{r}(0) = \mathbf{k}$

## Applications and Extensions

**33. Projectile Motion**  A projectile is fired at an angle of 30° to the horizontal with an initial speed of 520 m/s. What are its range, the time of flight, and the greatest height reached?

**34. Projectile Motion**  A projectile is fired with an initial speed of 200 m/s at an inclination of 60° to the horizontal. What are its range, the time of flight, and the greatest height reached?

**35. Projectile Motion**  A projectile is fired with an initial speed of 100 m/s at an inclination of $\tan^{-1} \dfrac{5}{12}$ to the horizontal.

   **(a)** Find parametric equations of the path of the projectile.

   **(b)** Find the range.

   **(c)** How long is the projectile in the air?

   **(d)** Graph the trajectory of the projectile.

**36. Projectile Motion**  A projectile is fired with an initial speed of 120 m/s at an inclination of $\tan^{-1} \dfrac{3}{4}$ to the horizontal.

   **(a)** Find parametric equations of the path of the projectile.

   **(b)** Find the range.

   **(c)** How long is the projectile in the air?

   **(d)** Graph the trajectory of the projectile.

**37. Projectile Motion**  A projectile is fired up a hill that makes a 30° angle to the horizontal. Suppose the projectile is fired at an angle of inclination of 45° to the horizontal with an initial speed of 100 ft/s. See the figure.

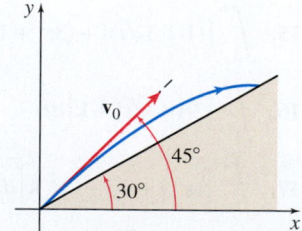

   **(a)** How far up the hill does the projectile land?

   **(b)** How long is the projectile in the air?

**38. Projectile Motion**  A projectile is propelled horizontally at a height of 3 m above the ground in order to hit a target 1 m high that is 30 m away. See the figure below. What should the initial velocity $\mathbf{v}_0$ of the projectile be?

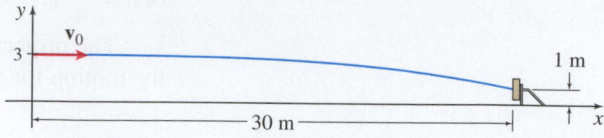

**39.** A force whose magnitude is 5 N and whose direction is along the positive $x$-axis is continuously applied to a projectile of mass $m = 1$ kg. At $t = 0$ s, the position of the object is the origin and its velocity is 3 m/s in the direction of the positive $y$-axis.

   **(a)** Find the velocity and speed of the projectile after $t$ seconds.

   **(b)** Find the position after the force has been applied for $t$ seconds.

   **(c)** How long is the projectile in the air?

   **(d)** Graph the path of the projectile.

**40.** An object of mass $m$ is propelled from the point $(1, 2)$ with initial velocity $\mathbf{v}_0 = 3\mathbf{i} + 4\mathbf{j}$. Thereafter, it is subjected only to the force

$$\mathbf{F} = \frac{m}{\sqrt{2}} (-\mathbf{i} - \mathbf{j}).$$

Find the vector equation for the position of the object at any time $t > 0$.

**41. Projectile Motion: Basketball**  In a Metro Conference men's basketball game on January 21, 1980, between Florida State University and Virginia Tech, a record was set. Les Henson, who is 6 ft 6 in. tall, made a basket from $89\dfrac{1}{4}$ ft down court to win the game for Virginia Tech by a score of 79 to 77. Assuming he released the ball at a height of 6 ft 6 in. and threw it at an angle of 45° (to maximize distance), with what initial velocity was the ball tossed? See the figure.

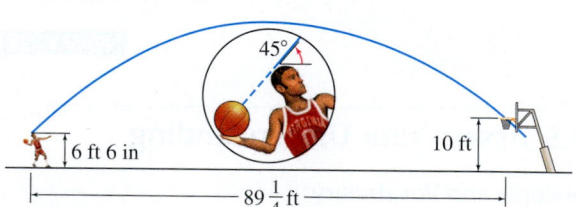

**42. Projectile Motion: Airplanes**  A plane is flying at an elevation of 4.0 km with a constant horizontal speed of 400 km/h toward a point directly above its target $T$. See the figure. At what angle of sight $\alpha$ should a package be released in order to strike the target? *Hint:* $g \approx 127{,}008$ km/h$^2$.

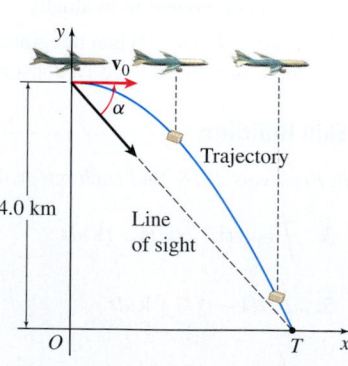

**43. Projectile Motion: Baseball**  A baseball is hit at an angle of 45° to the horizontal from an initial height of 3 ft. If the ball just clears the vines in front of the bleachers in Wrigley Field, which are 10 ft high and a distance of 400 ft from home plate, what was the initial speed of the ball? How long did it take the ball to reach the vines?

**44. Projectile Motion: Baseball**   An outfielder throws a baseball at an angle of 45° to the horizontal from an initial height of 6 ft. Suppose he can throw the ball with an initial velocity of 100 ft/s.

(a) What is the farthest the outfielder can be from home plate to ensure that the ball reaches home plate on the fly?

(b) How long is the ball in flight?

(c) Use (a) and (b) to determine how fast a player must run to get from third base to home plate on a fly ball and beat the throw to score a run.

*Hint:* It is 90 ft from third base to home plate, and the runner cannot leave the base until the ball is caught.

**45. Projectile Motion: Football**   In a field goal attempt on a flat field, a football is kicked at an angle of 30° to the horizontal with an initial speed of 65 ft/s.

(a) What horizontal distance does the football travel while it is in the air?

(b) To score a field goal, the ball must clear the cross bar of the goal post, which is 10 ft above the ground. What is the farthest from the goal post the kick can originate and score a field goal?

**46. Projectile Motion: Skeet Shooting**   A gun, lifted at an angle $\theta_0$ to the horizontal, is aimed at an elevated target T, which is released the moment the gun is fired. See the figure. No matter what the initial speed $v_0$ of the bullet is, show that it will always hit the falling target.

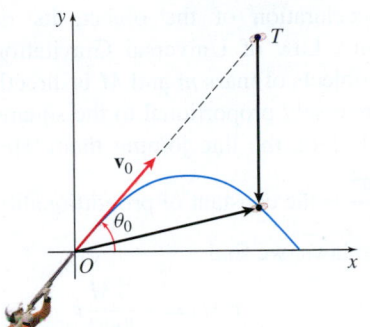

**47. (a)** Show that the maximum range of the projectile with position vector $\mathbf{r}(t) = (v_0 \cos\theta)t\mathbf{i} + \left[ (v_0 \sin\theta)t - \dfrac{1}{2}gt^2 \right]\mathbf{j}$ occurs when $\theta = \dfrac{\pi}{4}$.

**(b)** Show that the maximum range is $\dfrac{v_0^2}{g}$.

**48.** Show that the speed of the projectile whose position vector is given by $\mathbf{r}(t) = (v_0 \cos\theta)t\mathbf{i} + \left[ -\dfrac{1}{2}gt^2 + (v_0 \sin\theta)t \right]\mathbf{j}$ is least when the projectile is at its highest point.

**49.** Show that a particle subject to no outside forces is either stationary or moves with constant speed along a straight line.

**50.** Show that if $\mathbf{r}'(t) = \mathbf{0}$ for all $t$ on some interval $I$, then $\mathbf{r}(t) = \mathbf{c}$, a constant vector, for all $t$ in $I$.

**51.** Show that if $\mathbf{f}'(t) = \mathbf{g}'(t)$ for all $t$ in some interval $I$, then $\mathbf{f}(t) = \mathbf{g}(t) + \mathbf{c}$ for all $t$ in $I$.

**Challenge Problems**

**52.** If $\mathbf{c}$ is a constant vector and $\mathbf{r} = \mathbf{r}(t)$ is continuous on a closed interval $[a, b]$, show that
$$\int_a^b [\mathbf{c} \cdot \mathbf{r}(t)]dt = \mathbf{c} \cdot \int_a^b \mathbf{r}(t)\, dt$$

**53.** Use the result of Problem 52 to show that
$$\left\| \int_a^b \mathbf{r}(t)dt \right\| \leq \int_a^b \|\mathbf{r}(t)\|dt$$

*Hint:* Set $\mathbf{c} = \int_a^b \mathbf{r}(t)dt$ and find $\|\mathbf{c}\|^2$.

**54.** If $\mathbf{c}$ is a constant vector and $\mathbf{r} = \mathbf{r}(t)$ is continuous on a closed interval $[a, b]$, prove that
$$\int_a^b [\mathbf{c} \times \mathbf{r}(t)]dt = \mathbf{c} \times \int_a^b \mathbf{r}(t)dt$$

# 11.6 Application: Kepler's Laws of Planetary Motion

**OBJECTIVE**   *When you finish this section, you should be able to:*

**1** Discuss Kepler's Laws of Planetary Motion (p. 843)

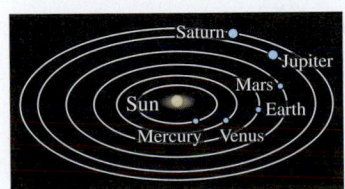

## 1 Discuss Kepler's Laws of Planetary Motion

In the sixteenth century, Nicolaus Copernicus (1473–1543) proposed the **Heliocentric Theory of Planetary Motion.** He postulated that the planets travel in circular orbits with the Sun at the center. This theory agreed better with observation than the earlier **Geocentric Theory,** which claimed the Sun, Moon, and planets travel in circular paths around Earth.

However, Copernicus' theory did not explain some observable facts. Early in the seventeenth century, Johannes Kepler (1571–1630) used the data of the Danish astronomer Tycho Brahe (1546–1601) to state three Laws of Planetary Motion.

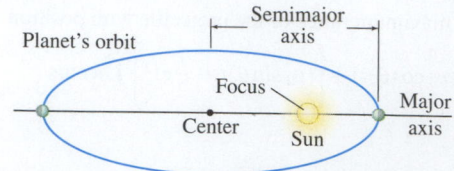

**DF Figure 35**

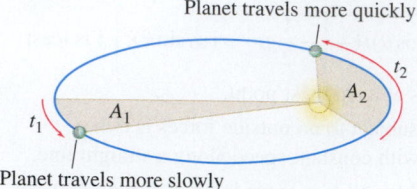

**Figure 36** $A_1 = A_2$; $t_1 = t_2$

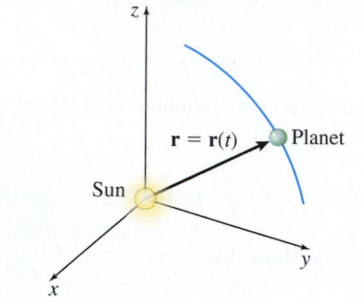

**Figure 37**

- The orbit of each planet is an ellipse with the Sun at a focus [postulated in (1609)]. See Figure 35.
- A line joining the Sun to a planet sweeps out equal areas in equal times (1609). See Figure 36.
- The square of the period of revolution of a planet is proportional to the cube of the length of the semimajor axis of the planet's elliptical orbit (1619).

Although Kepler's Laws of Planetary Motion fit Brahe's observations, the theory could not be proved until Newton used calculus to explain the motion of planetary bodies. Here we derive the first two of Kepler's laws using Newton's ideas and vector calculus. The proof of Kepler's third law for circular orbits is left as an exercise. See Problem 5.

Position a coordinate system with the Sun at the origin and let $\mathbf{r} = \mathbf{r}(t)$ denote the position of a planet in orbit about the Sun, as shown in Figure 37. Then $\mathbf{r}$ is subject to two laws:

Newton's Second Law of Motion:     $\mathbf{F} = m\mathbf{a}(t) = m\mathbf{r}''(t)$

Newton's Law of Universal Gravitation:     $\mathbf{F} = -\dfrac{GmM}{\|\mathbf{r}\|^2} \dfrac{\mathbf{r}}{\|\mathbf{r}\|}$

where $m$ is the mass of the planet, $M$ is the mass of the Sun, $G$ is the gravitational constant, and $\|\mathbf{r}\|$ is the distance between the Sun and the planet.

Newton's Second Law of Motion states that the force acting on an object is proportional to the acceleration of the object, its mass being the constant of proportionality. Newton's Law of Universal Gravitation states that the force $\mathbf{F}$ of attraction between two objects of mass $m$ and $M$ is directly proportional to the product $mM$ of the masses, is inversely proportional to the square of the distance between the objects, and is directed along the line joining them. Here the gravitational constant $G \approx 6.67 \times 10^{-11} \dfrac{\text{N} \cdot \text{m}^2}{\text{kg}^2}$ is the constant of proportionality.

From Newton's two laws, we find

$$\mathbf{r}''(t) = -\frac{GM}{\|\mathbf{r}\|^3}\mathbf{r}$$

If $\mathbf{v} = \mathbf{v}(t)$ is the velocity of a planet, then $\mathbf{a}(t) = \dfrac{d\mathbf{v}}{dt} = \mathbf{r}''(t)$, so

$$\boxed{\frac{d\mathbf{v}}{dt} = -\frac{GM}{\|\mathbf{r}\|^3}\mathbf{r}} \tag{1}$$

This shows that $\dfrac{d\mathbf{v}}{dt}$ and $\mathbf{r}$ are parallel, so

$$\mathbf{r} \times \frac{d\mathbf{v}}{dt} = \mathbf{0}$$

Now differentiate $\mathbf{r} \times \mathbf{v}$.

$$\frac{d}{dt}(\mathbf{r} \times \mathbf{v}) = \left(\frac{d\mathbf{r}}{dt} \times \mathbf{v}\right) + \left(\mathbf{r} \times \frac{d\mathbf{v}}{dt}\right) = (\mathbf{v} \times \mathbf{v}) + \mathbf{0} = \mathbf{0}$$
$$\underset{\dfrac{d\mathbf{r}}{dt} = \mathbf{v}}{\uparrow}$$

**NEED TO REVIEW?** Properties of the cross product are discussed in Section 10.5, pp. 766–769.

This means $\mathbf{r} \times \mathbf{v}$ is a constant vector, which we denote by $\mathbf{D}$. That is,

$$\boxed{\mathbf{r} \times \mathbf{v} = \mathbf{D}} \tag{2}$$

This fact implies that the motion of a planet lies in a plane.

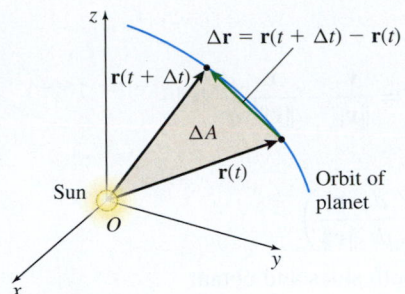

**Figure 38** $\Delta A \approx \frac{1}{2} \|\mathbf{r} \times \Delta \mathbf{r}\|$

We now prove Kepler's second law. A little later we prove Kepler's first law. As Figure 38 illustrates, if $A$ is the area swept out by the vector $\mathbf{r}$ in time $t$, then the area $\Delta A$ that is swept out in time $\Delta t$ can be approximated by a triangle. Since $\|\mathbf{r} \times \Delta \mathbf{r}\|$ is the area of a parallelogram with sides $\mathbf{r}$ and $\Delta \mathbf{r}$, the area of the triangle is $\frac{1}{2}$ the area of the parallelogram. That is,

$$\Delta A \approx \frac{1}{2} \|\mathbf{r} \times \Delta \mathbf{r}\|$$

and

$$\frac{\Delta A}{\Delta t} \approx \frac{1}{2} \left\| \mathbf{r} \times \frac{\Delta \mathbf{r}}{\Delta t} \right\|$$

As $\Delta t$ approaches 0, then $\dfrac{\Delta A}{\Delta t} \to \dfrac{dA}{dt}$ and $\dfrac{\Delta \mathbf{r}}{\Delta t} \to \dfrac{d\mathbf{r}}{dt}$. So,

$$\frac{dA}{dt} = \frac{1}{2} \left\| \mathbf{r} \times \frac{d\mathbf{r}}{dt} \right\| \underset{\underset{\dfrac{d\mathbf{r}}{dt} = \mathbf{v}}{\uparrow}}{=} \frac{1}{2} \|\mathbf{r} \times \mathbf{v}\| \underset{\underset{\text{By (2)}}{\uparrow}}{=} \frac{1}{2} \|\mathbf{D}\|$$

That is, $\dfrac{dA}{dt}$ is constant. This proves Kepler's second law.

**THEOREM Kepler's Second Law of Planetary Motion**

The speed of a planet varies so that the line joining the Sun to the planet sweeps out equal areas in equal times.

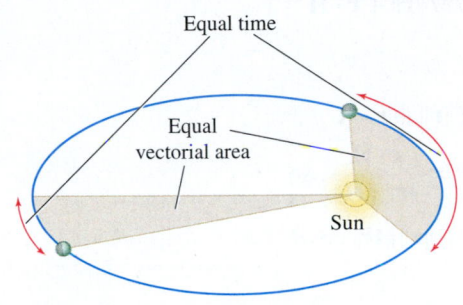

**Figure 39**

**NOTE** The Triple Vector Product is discussed in Section 10.5, Problems 69 and 70, p. 771.

See Figure 39.

To obtain Kepler's first law, use equation (1), $\dfrac{d\mathbf{v}}{dt} = -\dfrac{GM}{\|\mathbf{r}\|^3} \mathbf{r}$, and equation (2), $\mathbf{r} \times \mathbf{v} = \mathbf{D}$. We seek an expression for $\dfrac{d\mathbf{v}}{dt} \times \mathbf{D}$.

$$\frac{d\mathbf{v}}{dt} \times \mathbf{D} = -\frac{GM}{\|\mathbf{r}\|^3} \mathbf{r} \times (\mathbf{r} \times \mathbf{v}) \underset{\underset{\text{Triple Vector Product}}{\uparrow}}{=} -\frac{GM}{\|\mathbf{r}\|^3} [(\mathbf{r} \cdot \mathbf{v})\mathbf{r} - (\mathbf{r} \cdot \mathbf{r})\mathbf{v}] \tag{3}$$

Now differentiate the dot product $\mathbf{r} \cdot \mathbf{r}$.

$$\frac{d}{dt}(\mathbf{r} \cdot \mathbf{r}) = \frac{d\mathbf{r}}{dt} \cdot \mathbf{r} + \mathbf{r} \cdot \frac{d\mathbf{r}}{dt} = 2\mathbf{r} \cdot \underset{\underset{\dfrac{d\mathbf{r}}{dt} = \mathbf{v}}{\uparrow}}{\frac{d\mathbf{r}}{dt}} = 2\mathbf{r} \cdot \mathbf{v} \tag{4}$$

Since $\mathbf{r} \cdot \mathbf{r} = \|\mathbf{r}\|^2$, the following is also true:

$$\frac{d}{dt}(\mathbf{r} \cdot \mathbf{r}) = \frac{d}{dt}\|\mathbf{r}\|^2 = 2\|\mathbf{r}\| \frac{d}{dt}\|\mathbf{r}\| \tag{5}$$

Based on (4) and (5), we have

$$2\mathbf{r} \cdot \mathbf{v} = 2\|\mathbf{r}\| \frac{d}{dt}\|\mathbf{r}\|$$

$$\mathbf{r} \cdot \mathbf{v} = \|\mathbf{r}\| \frac{d}{dt}\|\mathbf{r}\|$$

Substitute this result in (3). Then

$$\frac{d\mathbf{v}}{dt} \times \mathbf{D} = -\frac{GM}{\|\mathbf{r}\|^3} [(\mathbf{r} \cdot \mathbf{v})\mathbf{r} - (\mathbf{r} \cdot \mathbf{r})\mathbf{v}] = -\frac{GM}{\|\mathbf{r}\|^3} \left[ \|\mathbf{r}\| \left( \frac{d}{dt}\|\mathbf{r}\| \right) \mathbf{r} - \|\mathbf{r}\|^2 \mathbf{v} \right]$$

$$= GM \left( \frac{\mathbf{v}}{\|\mathbf{r}\|} - \frac{\mathbf{r}}{\|\mathbf{r}\|^2} \frac{d}{dt}\|\mathbf{r}\| \right)$$

But,

$$\frac{d}{dt}\frac{\mathbf{r}}{\|\mathbf{r}\|} = \frac{\frac{d\mathbf{r}}{dt}}{\|\mathbf{r}\|} - \frac{\mathbf{r}}{\|\mathbf{r}\|^2}\frac{d}{dt}\|\mathbf{r}\| = \frac{\mathbf{v}}{\|\mathbf{r}\|} - \frac{\mathbf{r}}{\|\mathbf{r}\|^2}\frac{d}{dt}\|\mathbf{r}\|$$

As a result, we can write

$$\frac{d\mathbf{v}}{dt} \times \mathbf{D} = GM\left(\frac{d}{dt}\frac{\mathbf{r}}{\|\mathbf{r}\|}\right)$$

Since **D** is a constant vector, we can integrate both sides and obtain

$$\mathbf{v} \times \mathbf{D} = GM\frac{\mathbf{r}}{\|\mathbf{r}\|} + \mathbf{H}$$

where **H** is a constant vector.

To find $\|\mathbf{r}\|$, we first find the dot product $\mathbf{r} \cdot (\mathbf{v} \times \mathbf{D})$:

$$\mathbf{r} \cdot (\mathbf{v} \times \mathbf{D}) = \mathbf{r} \cdot \left(GM\frac{\mathbf{r}}{\|\mathbf{r}\|} + \mathbf{H}\right) = GM\|\mathbf{r}\| + \mathbf{r} \cdot \mathbf{H}$$

But $\mathbf{A} \cdot (\mathbf{B} \times \mathbf{C}) = (\mathbf{A} \times \mathbf{B}) \cdot \mathbf{C}$, so

$$(\mathbf{r} \times \mathbf{v}) \cdot \mathbf{D} = GM\|\mathbf{r}\| + \mathbf{r} \cdot \mathbf{H}$$

Since $\mathbf{r} \times \mathbf{v} = \mathbf{D}$, we have

$$\|\mathbf{D}\|^2 = GM\|\mathbf{r}\| + \mathbf{r} \cdot \mathbf{H}$$

If $\theta$ is the angle between **r** and **H**, then $\cos\theta = \dfrac{\mathbf{r} \cdot \mathbf{H}}{\|\mathbf{r}\|\,\|\mathbf{H}\|}$, so

$$\|\mathbf{D}\|^2 = GM\|\mathbf{r}\| + \|\mathbf{r}\|\,\|\mathbf{H}\|\cos\theta$$

Now solve for $\|\mathbf{r}\|$:

$$\|\mathbf{r}\| = \frac{\|\mathbf{D}\|^2}{GM + \|\mathbf{H}\|\cos\theta} = \frac{\|\mathbf{D}\|^2}{GM}\left(\frac{1}{1 + e\cos\theta}\right)$$

**NEED TO REVIEW?** The polar equations of conics are discussed in Section 9.7, pp. 724–727.

where $e = \dfrac{\|\mathbf{H}\|}{GM}$. This is the polar equation of a conic with a focus at the origin, the Sun. It is a hyperbola if $e > 1$, a parabola if $e = 1$, or an ellipse if $e < 1$. Since the planets have closed orbits, the only possibility is an ellipse, resulting in Kepler's First Law of Planetary Motion:

> **THEOREM Kepler's First Law of Planetary Motion**
>
> The orbit of each planet is an ellipse with the Sun at a focus.

## 11.6 Assess Your Understanding

### Applications and Extensions

1. **Exoplanets** As a result of NASA's Kepler mission, astronomers have discovered hundreds of planetary systems beyond our own. In one such planetary system, the planet GJ 667Cc has a circular orbit of radius $1.8 \times 10^{10}$ m that takes 28.1 days to revolve about its central star GJ 667C.

   Suppose another planet in circular orbit about central star GJ 667C is discovered. If this second planet is twice as far from the central star as planet GJ 667Cc, how many days would it take for the second planet to orbit the central star?

   *Source*: http://www.kepler.nasa.gov.

2. **Satellites of Jupiter** The giant planet Jupiter has many satellites. The four largest satellites were discovered by Galileo in 1609. One of them, Io, is the most volcanically active body in the solar system. It  is 422,000 km from Jupiter and takes 1.77 days for each orbit. Another satellite, Europa, probably has an ocean (and life?) beneath its icy surface. It is 671,000 km from Jupiter. A third satellite, Ganymede, is larger than the planet Mercury and takes 7.16 days for each orbit. Assuming the orbits of each of these satellites is circular,

   (a) How long does it take Europa to orbit Jupiter?

   (b) How far is Ganymede from Jupiter?

---

**1.** = NOW WORK problem    ◫ = Graphing technology recommended    CAS = Computer Algebra System recommended

3. Equation (2) on p. 844 states that $\mathbf{r} \times \mathbf{v}$ is a constant vector. Explain why this means the motion of the planet lies in a plane.

4. Explain why $\dfrac{dA}{dt}$ equal to a constant results in the statement given for Kepler's second law.

## Challenge Problems

5. **Kepler's Third Law**

   (a) Prove Kepler's Third Law of Planetary Motion for orbits that are circular.

   (b) Use the result from (a) to calculate the mass of the Sun.

   *Hint:* Earth is $1.5 \times 10^{11}$ m from the Sun, and the gravitational constant is $G = 6.67 \times 10^{-11}$ N $\cdot$ m$^2$/kg$^2$.

6. From the equation
$$\|\mathbf{r}\| = \frac{\|\mathbf{D}\|^2}{GM + \|\mathbf{H}\| \cos\theta} = \frac{\|\mathbf{D}\|^2}{GM}\left(\frac{1}{1 + e\cos\theta}\right),$$

deduce the following:

$$\text{Length of the semimajor axis} = \frac{\|\mathbf{D}\|^2}{GM}\left(\frac{1}{1 - e^2}\right)$$

$$\text{Length of the semiminor axis} = \frac{\|\mathbf{D}\|^2}{GM}\left(\frac{1}{\sqrt{1 - e^2}}\right)$$

$$\text{Area of the ellipse} = \frac{\pi\|\mathbf{D}\|^4}{G^2 M^2}\left[\frac{1}{(1 - e^2)^{3/2}}\right]$$

7. (a) For a particle moving under the influence of a central field of force directed toward the origin, show that $r\theta'' + 2r'\theta' = 0$.

   (b) Multiply the equation in part (a) by $r$ and integrate to obtain $r^2\theta' = c$, where $c$ is a constant.

   (c) Using the expression for area in polar coordinates, deduce Kepler's second law for any central field of force.

# Chapter Review

## THINGS TO KNOW

### 11.1 Vector Functions and Their Derivatives
- Vector function $\mathbf{r} = \mathbf{r}(t)$ (p. 798)
- The vector function $\mathbf{r} = \mathbf{r}(t) = \langle x(t), y(t)\rangle = x(t)\mathbf{i} + y(t)\mathbf{j}$, $a \le t \le b$, traces out a **plane curve**. (p. 798)
- The vector function $\mathbf{r} = \mathbf{r}(t) = \langle x(t), y(t), z(t)\rangle = x(t)\mathbf{i} + y(t)\mathbf{j} + z(t)\mathbf{k}$, $a \le t \le b$, traces out a **space curve**. (p. 798)
- Direction of a vector function (p. 799)
- Limit of a vector function (p. 801)
- A vector function $\mathbf{r} = \mathbf{r}(t)$ is **continuous at a real number** $t_0$ if:
- $\mathbf{r}(t_0)$ is defined
- $\lim\limits_{t \to t_0} \mathbf{r}(t)$ exists
- $\lim\limits_{t \to t_0} \mathbf{r}(t) = \mathbf{r}(t_0)$ (p. 801)
- The **derivative** of the vector function $\mathbf{r} = \mathbf{r}(t)$ is
$$\mathbf{r}'(t) = \frac{d\mathbf{r}}{dt} = \lim_{h \to 0}\frac{\mathbf{r}(t+h) - \mathbf{r}(t)}{h}, \text{ provided the limit}$$
exists. (p. 802)

*Derivative Rules:* (p. 804)
- **Sum Rule:** $\dfrac{d}{dt}[\mathbf{u}(t) + \mathbf{v}(t)] = \dfrac{d}{dt}\mathbf{u}(t) + \dfrac{d}{dt}\mathbf{v}(t) = \mathbf{u}'(t) + \mathbf{v}'(t)$
- **Scalar Multiplication Rule:**
$$\frac{d}{dt}[f(t)\mathbf{u}(t)] = \left[\frac{d}{dt}f(t)\right]\mathbf{u}(t) + f(t)\left[\frac{d}{dt}\mathbf{u}(t)\right]$$
$$= f'(t)\mathbf{u}(t) + f(t)\mathbf{u}'(t)$$
- **Dot Product Rule:**
$$\frac{d}{dt}[\mathbf{u}(t)\cdot\mathbf{v}(t)] = [\mathbf{u}(t)\cdot\mathbf{v}(t)]' = \mathbf{u}'(t)\cdot\mathbf{v}(t) + \mathbf{u}(t)\cdot\mathbf{v}'(t)$$
- **Cross Product Rule:**
$$\frac{d}{dt}[\mathbf{u}(t)\times\mathbf{v}(t)] = [\mathbf{u}(t)\times\mathbf{v}(t)]' = \mathbf{u}'(t)\times\mathbf{v}(t) + \mathbf{u}(t)\times\mathbf{v}'(t)$$
- **Chain Rule:** $\dfrac{d}{dt}\mathbf{v}(f(t)) = \mathbf{v}'(f(t))f'(t)$

### 11.2 Unit Tangent and Principal Unit Normal Vectors; Arc Length
- Tangent vector to a curve (p. 808)
- Smooth curve (p. 809)
- Unit tangent vector: $\mathbf{T}(t) = \dfrac{\mathbf{r}'(t)}{\|\mathbf{r}'(t)\|}$ (p. 809)
- Principal unit normal vector: $\mathbf{N}(t) = \dfrac{\mathbf{T}'(t)}{\|\mathbf{T}'(t)\|}$ (p. 810)
- Arc length $s$ of a vector function: $s = \int_a^b \|\mathbf{r}'(t)\|dt$ (p. 811)

### 11.3 Arc Length as Parameter; Curvature
- The parameter $t$ is arc length if and only if $\|\mathbf{r}'(t)\| = 1$. (p. 816)
- Curvature of a curve: $\kappa = \left\|\dfrac{d\mathbf{T}}{ds}\right\| = \dfrac{\|\mathbf{T}'(t)\|}{\|\mathbf{r}'(t)\|}$, where the parameter $s$ is arc length. (p. 817–818)
- Curvature of a space curve: $\kappa = \dfrac{\|\mathbf{r}'(t)\times\mathbf{r}''(t)\|}{\|\mathbf{r}'(t)\|^3}$ (p. 819)
- Curvature of a plane curve given by $y = f(x)$:
$$\kappa = \frac{|f''(x)|}{(1 + [f'(x)]^2)^{3/2}} \text{ (p. 820)}$$

*Properties of an osculating circle of a smooth curve C* (p. 821)
- The osculating circle contains a point $P$ on a curve $C$.
- The tangent line to $C$ at $P$ and the tangent line to the osculating circle at $P$ are the same.
- The radius $\rho$ of the osculating circle is $\rho = \dfrac{1}{\kappa}$, where $\kappa \ne 0$ is the curvature of the curve $C$.
- The center of the osculating circle lies on the line in the direction of the principal unit normal vector $\mathbf{N}$ to $C$ at $P$.

## 11.4 Motion Along a Curve

- Position: $\mathbf{r}(t) = x(t)\mathbf{i} + y(t)\mathbf{j} + z(t)\mathbf{k}$ (p. 826)

- Velocity: $\mathbf{v}(t) = \mathbf{r}'(t) = \dfrac{dx}{dt}\mathbf{i} + \dfrac{dy}{dt}\mathbf{j} + \dfrac{dz}{dt}\mathbf{k}$ (p. 826)

- Acceleration: $\mathbf{a}(t) = \mathbf{r}''(t) = \dfrac{d^2x}{dt^2}\mathbf{i} + \dfrac{d^2y}{dt^2}\mathbf{j} + \dfrac{d^2z}{dt^2}\mathbf{k}$ (p. 826)

- Speed: $v(t) = \|\mathbf{v}(t)\| = \|\mathbf{r}'(t)\|$ (p. 826)

- Tangential and normal components of an acceleration vector:

$$\mathbf{a}(t) = a_\mathbf{T}\mathbf{T}(t) + a_\mathbf{N}\mathbf{N}(t)$$

where

$$a_\mathbf{T} = v'(t) = \frac{dv}{dt} \quad \text{and} \quad a_\mathbf{N} = [v(t)]^2 \kappa \text{ (p. 830)}$$

## 11.5 Integrals of Vector Functions; Projectile Motion

- If $\mathbf{r}(t) = x(t)\mathbf{i} + y(t)\mathbf{j} + z(t)\mathbf{k}$, then

$$\int \mathbf{r}(t)dt = \left[\int x(t)dt\right]\mathbf{i} + \left[\int y(t)dt\right]\mathbf{j} + \left[\int z(t)dt\right]\mathbf{k}$$

$$\int_a^b \mathbf{r}(t)dt = \left[\int_a^b x(t)dt\right]\mathbf{i} + \left[\int_a^b y(t)dt\right]\mathbf{j} + \left[\int_a^b z(t)dt\right]\mathbf{k}$$

(pp. 837–838)

- Projectile motion (pp. 839–841)

## 11.6 Kepler's Laws of Planetary Motion

- The orbit of each planet is an ellipse with the Sun at a focus. (p. 846)
- A line joining the Sun to a planet sweeps out equal areas in equal times. (p. 845)
- The square of the period of revolution of a planet is proportional to the cube of the length of the semimajor axis of the planet's elliptical orbit. (p. 844)

## OBJECTIVES

| Section | You should be able to ... | Examples | Review Exercises |
|---|---|---|---|
| 11.1 | 1 Find the domain of a vector function (p. 798) | 1 | 1–4(a) |
| | 2 Graph a vector function (p. 798) | 2–4 | 1–4(b) |
| | 3 Find the limit and determine the continuity of a vector function (p. 801) | 5, 6 | 5–7 |
| | 4 Find the derivative of a vector function (p. 802) | 7, 8 | 1–4(c), 8, 9 |
| | 5 Find the derivative of a vector function using derivative rules (p. 803) | 9 | 10, 11 |
| 11.2 | 1 Interpret the derivative of a vector function geometrically (p. 807) | 1 | 12–14(a), 15 |
| | 2 Find the unit tangent vector and the principal unit normal vector of a smooth curve (p. 809) | 2–4 | 12–14(b), (c) |
| | 3 Find the arc length of a curve traced out by a vector function (p. 811) | 5, 6 | 16–18 |
| 11.3 | 1 Determine whether the parameter used in a vector function is arc length (p. 815) | 1, 2 | 19, 20 |
| | 2 Find the curvature of a curve (p. 817) | 3, 4 | 21, 30 |
| | 3 Find the curvature of a space curve (p. 819) | 5 | 22, 23 |
| | 4 Find the curvature of a plane curve given by $y = f(x)$ (p. 820) | 6 | 24, 25, 29 |
| | 5 Find an osculating circle (p. 821) | 7 | 26–28 |
| 11.4 | 1 Find the velocity, acceleration, and speed of a moving particle (p. 826) | 1–5 | 31–34(a), 35, 36 |
| | 2 Express the acceleration vector using tangential and normal components (p. 829) | 6–8 | 31–34(b) |
| 11.5 | 1 Integrate vector functions (p. 837) | 1–3 | 37–42 |
| | 2 Solve projectile motion problems (p. 839) | 4 | 43 |
| 11.6 | 1 Discuss Kepler's Laws of Planetary Motion (p. 843) | | 44 |

## REVIEW EXERCISES

*In Problems 1–4, for each vector function* $\mathbf{r} = \mathbf{r}(t)$

**(a)** *Find the domain.*

**(b)** *Graph the curve C traced out by* $\mathbf{r} = \mathbf{r}(t)$, *indicating its orientation.*

**(c)** *Find the value of* $\mathbf{r}'(t)$ *at the given number t.*

**1.** $\mathbf{r}(t) = t^2\mathbf{i} + 3t\mathbf{j}$    $t = 2$

**2.** $\mathbf{r}(t) = 3\cos t\mathbf{i} + \sin t\mathbf{j}$    $0 \leq t \leq \pi$    $t = \dfrac{\pi}{3}$

**3.** $\mathbf{r}(t) = t\mathbf{i} + 2\cos t\mathbf{j} + 2\sin t\mathbf{k}$    $t = 0$

**4.** $\mathbf{r}(t) = t\mathbf{i} + t\mathbf{j} - 2\mathbf{k}$    $t = 1$

**5.** Find $\displaystyle\lim_{t \to 3}\left[(t - 3)\mathbf{i} + \dfrac{t^2 - 3t}{t^2 - 9}\mathbf{j} - \dfrac{4}{t}\mathbf{k}\right]$.

*In Problems 6 and 7, determine if the vector function is continuous at each number in the given interval. Identify any numbers where* $\mathbf{r} = \mathbf{r}(t)$ *is discontinuous.*

**6.** $\mathbf{r}(t) = \dfrac{t}{t - 2}\mathbf{i} + \sin t\mathbf{j}$    interval $(0, 2\pi)$

**7.** $\mathbf{r}(t) = \dfrac{t}{t + 1}\mathbf{i} + \sqrt{t - 1}\mathbf{j} + 3t\mathbf{k}$    interval $[1, 3]$

*In Problems 8 and 9, find* $\mathbf{r}'(t)$ *and* $\mathbf{r}''(t)$.

**8.** $\mathbf{r}(t) = 3t\mathbf{i} - \ln t\mathbf{j} + 2e^t\mathbf{k}$

**9.** $\mathbf{r}(t) = 2\cos t\mathbf{i} + 3\cos t\mathbf{j} + t\mathbf{k}$

*In Problems 10 and 11, find* $[\mathbf{f}(t) \cdot \mathbf{g}(t)]'$ *and* $[\mathbf{f}(t) \times \mathbf{g}(t)]'$.

**10.** $\mathbf{f}(t) = t\mathbf{i} - \dfrac{t}{2}\mathbf{j} + \mathbf{k}$    and    $\mathbf{g}(t) = \sqrt{1 - t}\mathbf{i} + t^3\mathbf{j} + 2(t + 1)\mathbf{k}$

**11.** $\mathbf{f}(t) = t\mathbf{i} + \cos(2t)\mathbf{j} - 5\mathbf{k}$    and    $\mathbf{g}(t) = 2t\mathbf{i} + \cos t\mathbf{j} + \sin t\mathbf{k}$

*In Problems 12–14, for each curve C traced out by* $\mathbf{r} = \mathbf{r}(t)$:

*(a) Find a tangent vector to the curve C at* $t = 0$.

*(b) Find the unit tangent vector* $\mathbf{T}$ *at* $t = 0$.

*(c) Find the principal unit normal vector* $\mathbf{N}$ *at* $t = 0$.

**12.** $\mathbf{r}(t) = e^t \mathbf{i} + e^{-t} \mathbf{j} + \mathbf{k}$

**13.** $\mathbf{r}(t) = t\mathbf{i} + 2t\mathbf{j} + \sqrt{1 - 5t^2}\, \mathbf{k}$    $-\dfrac{\sqrt{5}}{5} < t < \dfrac{\sqrt{5}}{5}$

**14.** $\mathbf{r}(t) = e^t \sin t\mathbf{i} + e^t \cos t\mathbf{j} + e^t \mathbf{k}$

**15.** Find the acute angle between the tangent vector to the helix traced out by $\mathbf{r}(t) = \sin t\mathbf{i} + 3t\mathbf{j} + \cos t\mathbf{k}$ and the direction $\mathbf{j}$.

*In Problems 16–18, find the arc length of each vector function.*

**16.** $\mathbf{r}(t) = \cos^3 t\mathbf{i} + \sin^3 t\mathbf{j}$ from $t = 0$ to $t = 2\pi$

**17.** $\mathbf{r}(t) = t\mathbf{i} + 2t\mathbf{j} + \sqrt{1 - 5t^2}\, \mathbf{k}$ from $t = 0$ to $t = \dfrac{1}{4}$

**18.** $\mathbf{r}(t) = \cos t\mathbf{i} + \sin t\mathbf{j} + \dfrac{t}{5}\mathbf{k}$ from $t = 0$ to $t = 2\pi$

*In Problems 19 and 20, determine whether the parameter is arc length.*

**19.** $\mathbf{r}(t) = 3t\mathbf{i} + 4t\mathbf{j} + (5t + 1)\,\mathbf{k}$

**20.** $\mathbf{r}(t) = 3\cos t\mathbf{i} + 3\sin t\mathbf{j} + 8t\mathbf{k}$

*In Problems 21–23, find the curvature* $\kappa = \kappa(t)$ *of the curve C traced out by each vector function* $\mathbf{r} = \mathbf{r}(t)$.

**21.** $\mathbf{r}(t) = 4\cos t\mathbf{i} - 4\sin t\mathbf{j}$    **22.** $\mathbf{r}(t) = t\mathbf{i} + \cos t\mathbf{j} + \sin t\mathbf{k}$

**23.** $\mathbf{r}(t) = e^t \mathbf{i} + 4t\mathbf{j} + 5e^{-t}\mathbf{k}$

**24.** Find the curvature of the graph of $y = x^3 - 4$ at the point $(1, -3)$.

**25.** Find the curvature of the graph of $y = 4e^{-2x}$ at the point $(0, 4)$.

*In Problems 26–28, find the radius of the osculating circle at the point corresponding to t on the curve C traced out by each vector function.*

**26.** $\mathbf{r}(t) = (1 - t^2)\mathbf{i} + e^{-t}\mathbf{j}$    $t = 0$

**27.** $\mathbf{r}(t) = \cos^2 t\mathbf{i} + \sin^2 t\mathbf{j}$    $t = \dfrac{\pi}{3}$

**28.** $\mathbf{r}(t) = t\mathbf{i} + 2t\mathbf{j} + \sqrt{1 - t^2}\, \mathbf{k}$    $-1 < t < 1$    $t = 0$

**29.** Find the curvature of the curve $y = \sqrt[3]{x}$, $x > 0$.

**30.** Find the minimum curvature of the curve C traced out by the vector function $\mathbf{r}(t) = \cos^3 t\mathbf{i} + \sin^3 t\mathbf{j}$.

*In Problems 31–34:*

*(a) Find the velocity* $\mathbf{v}$, *acceleration* $\mathbf{a}$, *and speed* $v$ *of a particle traveling along the curve traced out by the vector function* $\mathbf{r} = \mathbf{r}(t)$.

*(b) Find the tangential and normal components of the acceleration.*

**31.** $\mathbf{r}(t) = 2\cos t\mathbf{i} + \sin t\mathbf{j}$    **32.** $\mathbf{r}(t) = e^t \sin t\mathbf{i} + e^{-t}\mathbf{j}$

**33.** $\mathbf{r}(t) = e^t \mathbf{i} + e^{-t}\mathbf{j} + \mathbf{k}$    **34.** $\mathbf{r}(t) = e^t \sin t\mathbf{i} + e^t \cos t\mathbf{j} + e^t \mathbf{k}$

**35.** Find the velocity and acceleration of a particle moving on the cycloid $\mathbf{r}(t) = [\pi t - \sin(\pi t)]\mathbf{i} + [1 - \cos(\pi t)]\mathbf{j}$.

**36. (a)** Find the velocity and acceleration of a particle moving on the parabola $\mathbf{r}(t) = (t^2 - 2t)\mathbf{i} + t\,\mathbf{j}$.

   **(b)** At what time $t$ is the speed of the particle 0?

*In Problems 37 and 38, find each integral.*

**37.** $\displaystyle\int [(t^2 - 2)\mathbf{i} - (t - 2)^2\mathbf{j} + e^{2t}\mathbf{k}]dt$

**38.** $\displaystyle\int_0^\pi \left( \cos\dfrac{t}{2}\mathbf{i} + \sin^2\dfrac{t}{2}\mathbf{j} + 3t\mathbf{k} \right) dt$

*In Problems 39–41, solve each vector differential equation with the given boundary condition.*

**39.** $\mathbf{r}'(t) = e^{2t}\mathbf{i} + \ln t\mathbf{j} + 2e^t\mathbf{k}$,    $\mathbf{r}(1) = \mathbf{j} + \mathbf{k}$

**40.** $\mathbf{r}'(t) = \sin t\cos t\mathbf{i} + \tan t\sec t\mathbf{j} + t\mathbf{k}$,    $\mathbf{r}(0) = \mathbf{i} + \mathbf{k}$

**41.** $\dfrac{d\mathbf{r}}{dt} = \cos t\mathbf{i} - \sin t\mathbf{j} + \dfrac{t}{2}\mathbf{k}$,    $\mathbf{r}(0) = \mathbf{i}$

**42.** Find the speed, velocity, and position vectors at time $t$ for a particle whose acceleration at time $t$ is given by $\mathbf{a}(t) = \tan t\sec t\mathbf{i} + \cos t\mathbf{j}$. Assume that the initial velocity and initial position vectors are $\mathbf{0}$.

**43. Projectile Motion**  A projectile is fired with a speed of 500 m/s at an inclination of 45° to the horizontal from a point 30 m above level ground. Find the point where the projectile strikes the ground.

**44.** State Kepler's three laws of Planetary Motion. Explain each law in your own words.

## CHAPTER 11 PROJECT    How to Design a Safe Road

When a car moves around a corner, we often feel ourselves being pulled outward. This "pull" is due to inertia, the tendency to move along straight-line paths. Without some sort of centripetal (center-seeking) force, a passenger would move in a tangent line to the motion while the car turned underneath. It is the frictional force exerted by the road on the tires that keeps the car from sliding off the road, and it is a combination of forces between a passenger and the seat/seatbelt which pushes the passenger into a circular path.

In this project we examine the forces acting on a moving vehicle under three different scenarios involving exiting from one road to another, and a fourth involving the forces affecting a vehicle during a high-speed turn. They are:

- Exiting from one road to another road perpendicular to the first road along a flat circular ramp.
- Exiting from one road to another road perpendicular to the first road along a flat circular ramp that is angled (banked) at an angle to the horizontal.
- Exiting from one road to another road that is approximately 28 ft above the first along an inclined, banked ramp.
- Traveling along a rural highway as the road curves.

1. Suppose a car weighing 3800 lb is traveling at a constant speed of 30 mi/h.

   (a) What is the speed of the car in ft/s?

   (b) What is the mass $m$ of the car?
   *Hint:* weight $w = mg$, where $g = 32\,\text{ft/s}^2$.

   Suppose the car exits from one road to another road perpendicular to it on a flat circular ramp of radius 100 ft given by the vector equation

   $$\mathbf{r} = \mathbf{r}(t) = 100\sin(\omega t)\mathbf{i} + 100\cos(\omega t)\mathbf{j} \qquad 0 \le t \le \frac{\pi}{2\omega} \qquad (1)$$

   where $\omega$ is the angular speed of the car.

2. Graph $\mathbf{r} = \mathbf{r}(t)$. Explain how the vector function $\mathbf{r} = \mathbf{r}(t)$ models a circular ramp connecting two roads that intersect at a right angle.

3. (a) Find $\omega$.

   (b) Find the velocity vector $\mathbf{v} = \mathbf{v}(t)$.

   (c) Find the acceleration vector $\mathbf{a} = \mathbf{a}(t)$.

4. What is the magnitude of the frictional force required to keep the car from sliding on the ramp?

   Now suppose the circular exit ramp leading from one road to the other is banked. That is, the ramp is pitched at an angle $\theta$ to the horizontal. This banking helps push the car into a circular path, much as friction did in the scenario where the road was flat. We wish to find the angle $\theta$ so that the car will travel in a circular path even on a very slick (frictionless) road. Since the road will not be frictionless, the angle of banking is often less than $\theta$.

   We wish to find $\theta$ so that there is no lateral centripetal force. In doing so, we ignore the frictional force exerted on the tires so that, in practice, the angle used could be smaller than $\theta$.

5. Suppose the same 3800-lb car, traveling at a constant speed of 30 mi/h, exits from one road to another road perpendicular to it on a banked, circular ramp of radius 100 ft (Equation (1)), that is pitched at an angle $\theta$, $0 < \theta < \dfrac{\pi}{2}$, to the horizontal.

   (a) If $\mathbf{n}$ is the normal force vector exerted by the car as it moves along the ramp, show that the weight of the car is $\mathbf{n}\cos\theta$ and the centripetal force is $\mathbf{F} = \mathbf{n}\sin\theta$.
   *Hint:* Draw a force diagram.

   (b) Find the magnitude of $\mathbf{F}$. *Hint:* Use $\mathbf{F} = m\mathbf{a}$ and the results found in Problems 3 and 4.

   (c) Use the results of (a) and (b) to find $\sin\theta$ and $\cos\theta$.

   (d) Find $\tan\theta$, and solve for $\theta$.

   Suppose a road with an E-W direction passes under a second road with a N-S direction, and the N-S road is 28.27 ft higher than the E-W road. You are driving East and want to head North. An inclined, banked exit ramp from the lower road to the higher road can be modeled by the vector equation

   $$\mathbf{r} = \mathbf{r}(t) = 100\sin(\omega t)\mathbf{i} + 100\cos(\omega t)\mathbf{j} + 6\omega t\mathbf{k} \qquad (2)$$
   $$0 \le t \le \frac{3\pi}{2\omega}$$

6. Graph $\mathbf{r} = \mathbf{r}(t)$ and explain how the graph satisfies the conditions required of the ramp.

7. Find the angle $\theta$ to bank the exit ramp so that a 3800-lb car travelling at a constant speed of 30 mi/h will travel in a circular path of constant radius. Explain your answer mathematically.

   The prior discussion can also be used to answer the following questions.
   A rural highway has a posted speed of 65 mi/h. At a certain point the highway curves and follows a flat circular path of radius 1000 ft given by

   $$\mathbf{r} = \mathbf{r}(t) = 1000\sin(\omega t)\mathbf{i} + 1000\cos(\omega t)\mathbf{j}$$

   Suppose the road is banked at an angle of 10°.

8. What speed limit should be set so that a car weighing 3800 lb would travel in a circular path of constant radius even if the road became slick? Express your answer in mi/h.

9. Repeat Problem 8 for a vehicle weighing 10,000 lb. Explain why the answer makes sense.

10. Narrowing the width of a lane from 12 ft to 11 ft increases the likelihood of an accident at approximately the same rate as decreasing the banking angle by 1°. Repeat Problem 8 if the banking angle is 9°. Explain how this might influence the choice of a banking angle versus an increase in lane width.

11. Describe the role of the curvature of the road used for an exit ramp. Assume the speed of the vehicle is constant.

*Source*: "Prediction of the Expected Safety Performance of Rural Two-Lane Highways"
http://www.fhwa.dot.gov/publications/research/safety/99207/99207.pdf

# 12

# Functions of Several Variables

Detlev van Ravenswaay/Science Source

**CHAPTER 12 PROJECT** In the Chapter 12 Project on page 906, we examine the transit method for discovering exoplanets and investigate the main source of error when it is used with telescopes on the ground.

## Searching for Exoplanets

In 1992, astronomers discovered the first planet outside our solar system. Since that time, astronomers have identified thousands of **extrasolar planets**, or **exoplanets**, which are of great interest in the search of potential extraterrestrial life forms. In February 2017, NASA announced the discovery of seven planets orbiting the TRAPPIST-1 star, all of which may have conditions suitable for life.

The first exoplanets discovered were large planets similar in size to Jupiter and Neptune. Larger planets are easier to detect but less likely to have an atmosphere hospitable for life. Since the light reflected by a planet is much fainter than the light emitted from the star it revolves around, most exoplanets cannot be observed directly. Instead, exoplanets are discovered through the use of indirect methods such as radial velocity or the Doppler method. A star with a planet will move in a small orbit in response to the planet's gravity. This creates a slight fluctuation in the radial velocity, the speed at which the star moves toward or away from earth. Due to the Doppler effect, these fluctuations can be measured, indicating the existence of an exoplanet.

In certain cases the exoplanet passes between Earth and the star once every orbit. This is called a **transit**. Since planets are cooler than stars, the transit causes the star's brightness to decrease slightly while the exoplanet is in front of the star. With very careful observations, this decrease in brightness can be recorded, even by amateur astronomers. The transit method has also been used from space by NASA's Kepler satellite to identify more than 3000 candidate exoplanets.

In Chapter 1, we discussed finding the limit of a function $y = f(x)$, a function of one variable, and investigated where such a function is continuous. In Chapters 2 and 3, we defined the derivative of a function $y = f(x)$, investigated properties of derivatives, and developed methods for finding derivatives that eliminated the need to actively use the definition. In this chapter, we extend these concepts to functions of more than one variable. Instead of a derivative, we define *partial derivatives* of such functions. And instead of a single chain rule, we discover functions of more than one variable have a variety of chain rules.

# 12.1 Functions of Two or More Variables and Their Graphs

**OBJECTIVES** *When you finish this section, you should be able to:*

1 Work with functions of two or three variables (p. 852)
2 Graph functions of two variables (p. 854)
3 Graph level curves (p. 854)
4 Describe level surfaces (p. 857)

## 1 Work with Functions of Two or Three Variables

Many applications of the physical sciences, biology, the social sciences, business and economics, and even sports require functions of two or more variables. For example, the monthly cost of using a cell phone depends on the megabytes of data used domestically, the megabytes of data used on a ship at sea, and the megabytes of data used on an airplane. The cost is a *function of three variables*. For another example, the volume $V = \pi R^2 h$ of water stored in a cylindrical tank depends on both the height $h$ of the water and the radius $R$ of the tank. The volume $V$ is a *function $f$ of two variables*, $R$ and $h$. In function notation,

$$V = f(R, h) = \pi R^2 h$$

So, if $R = 10$ centimeters (cm) and $h = 3$ cm, then the volume of water in the tank is $V = f(10, 3) = \pi \cdot 10^2 \cdot 3 = 300\pi$ cm$^3$.

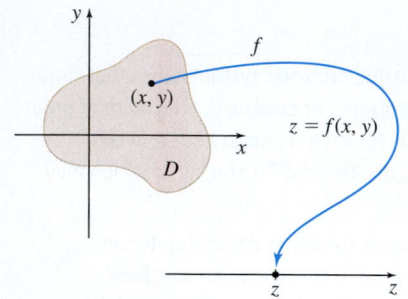

**Figure 1** $D$ is the domain of $f$; $z = f(x, y)$ is in the range of $f$.

> **DEFINITION  Function of Two Variables**
>
> Let $D$ be a nonempty subset of the $xy$-plane. A **function** $f$ **of two variables** $x$ and $y$ is a relation that associates with each point $(x, y)$ of $D$ a unique real number $z = f(x, y)$.

The function $z = f(x, y)$ has two **independent variables** $x$ and $y$, and one **dependent variable** $z$. The **domain** $D$ of the function $f$ is the set of points in the $xy$-plane for which the function is defined. The **range** of $f$ is the set of real numbers $z = f(x, y)$, where $(x, y)$ is in $D$. Figure 1 illustrates how the point $(x, y)$ is mapped to the number $z = f(x, y)$.

**EXAMPLE 1  Evaluating a Function of Two Variables**

Let $f(x, y) = \sqrt{x} + x\sqrt{y}$. Find:

**(a)** $f(1, 4)$   **(b)** $f(a^2, 9b^2), a > 0, b > 0$   **(c)** $f(x + \Delta x, y)$   **(d)** $f(x, y + \Delta y)$

**Solution (a)** $f(1, 4) = \sqrt{1} + 1\sqrt{4} = 1 + 2 = 3$        $x = 1; y = 4$

**(b)** $f(a^2, 9b^2) = \sqrt{a^2} + a^2\sqrt{9b^2} = a + 3a^2 b$        $x = a^2; y = 9b^2; a > 0; b > 0$

**(c)** $f(x + \Delta x, y) = \sqrt{x + \Delta x} + (x + \Delta x)\sqrt{y}$

**(d)** $f(x, y + \Delta y) = \sqrt{x} + x\sqrt{y + \Delta y}$ ∎

**NOW WORK** Problem 7.

> **DEFINITION  Function of Three Variables**
>
> Let $D$ be a nonempty set of points in space. A **function** $f$ **of three variables** $x, y,$ and $z$ is a relation that associates with each point $(x, y, z)$ of $D$ a unique real number $w = f(x, y, z)$.

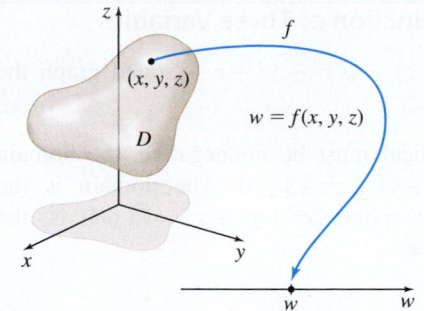

**Figure 2** $D$ is the domain of $f$; $w$ is the value of $f(x, y, z)$.

The function $w = f(x, y, z)$ has *three* independent variables $x$, $y$, and $z$ and one dependent variable $w$. The domain $D$ of the function $f$ is the set of points in space for which the function is defined, and the range of $f$ is the set of real numbers $w = f(x, y, z)$ for points $(x, y, z)$ in $D$. See Figure 2.

Functions of $n$ variables, where $n$ is a positive integer, are defined similarly. A **function** $f$ of $n$ variables $x_1, x_2, \ldots, x_n$ is a relation that associates a unique real number $z = f(x_1, x_2, \ldots, x_n)$ with each point $(x_1, x_2, \ldots, x_n)$ in a nonempty subset $D$ of $n$-dimensional space. Here $z$ is the dependent variable and $x_1, x_2, \ldots, x_n$ are the $n$ independent variables.

Collectively, functions of two or more variables are referred to as **functions of several variables**. As with functions of a single variable, a function of several variables is usually given by a formula, and unless the domain is specified, the domain is the largest set of points for which the dependent variable is a real number.

Functions of several variables are expressed **explicitly** when they are in the form

$$z = f(x, y) \qquad w = g(x, y, z) \qquad z = h(x_1, x_2, \ldots, x_n)$$

or **implicitly** when they are in the form

$$F(x, y, z) = 0 \qquad G(x, y, z, w) = 0 \qquad H(x_1, x_2, \ldots, x_n, z) = 0$$

---

**EXAMPLE 2** **Finding the Domain of a Function of Two Variables**

Find the domain of each of the following functions. Then graph the domain.

**(a)** $z = f(x, y) = \sqrt{16 - x^2 - y^2}$    **(b)** $z = f(x, y) = \dfrac{1}{(y - 1) \ln(x - 2)}$

**Solution (a)** Since the expression under the radical must be nonnegative, the domain of $f$ consists of all points in the plane for which

$$16 - x^2 - y^2 \geq 0$$
$$x^2 + y^2 \leq 16$$

The domain is all the points inside and on the circle $x^2 + y^2 = 16$. See Figure 3.

**(b)** Since the logarithmic function is defined only for positive numbers, $x - 2 > 0$, so $x > 2$. Also, for the function $f$ to be defined, the denominator must be nonzero. So, $x \neq 3$, (since $\ln(3 - 2) = \ln 1 = 0$), and $y \neq 1$.

The domain of $f$ consists of the half-plane $x > 2$, excluding the lines $x = 3$ and $y = 1$. The domain of $f$ is graphed in Figure 4. ∎

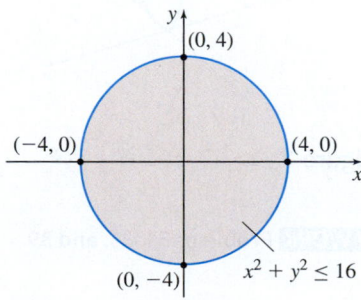

**Figure 3** The domain of
$$f(x, y) = \sqrt{16 - x^2 - y^2}$$

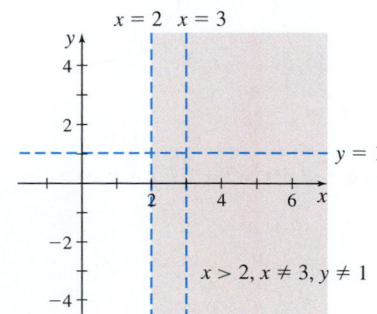

**Figure 4** The domain of
$$f(x, y) = \dfrac{1}{(y - 1) \ln(x - 2)}$$

**NOW WORK** **Problems 13 and 17.**

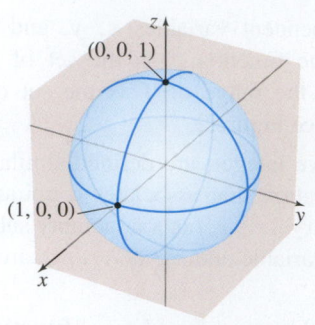

**Figure 5** $f(x, y, z) = \sqrt{x^2 + y^2 + z^2 - 1}$

**NEED TO REVIEW?** Planes in space are discussed in Section 10.6, pp. 774–775. Quadric surfaces are discussed in Section 10.7, pp. 784–791.

---

**EXAMPLE 3** **Finding the Domain of a Function of Three Variables**

Find the domain of the function $w = f(x, y, z) = \sqrt{x^2 + y^2 + z^2 - 1}$ and graph the domain.

**Solution** Since the expression under the radical must be nonnegative, the domain of $f$ consists of all points for which $x^2 + y^2 + z^2 - 1 \geq 0$. The domain is the set of all points on and outside of the unit sphere $x^2 + y^2 + z^2 = 1$; that is, the set $\{(x, y, z) \mid x^2 + y^2 + z^2 \geq 1\}$. See Figure 5. ∎

**NOW WORK** Problem 27.

## 2 Graph Functions of Two Variables

The graph of a function $z = f(x, y)$ of two variables is called a **surface** and consists of all points $(x, y, z)$ for which $z = f(x, y)$, and $(x, y)$ is in the domain of $f$. We use the graphing techniques introduced in Chapter 10 to graph functions of two variables.

---

**EXAMPLE 4** **Graphing a Function of Two Variables**

Graph each function:

**(a)** $z = f(x, y) = 1 - x - y$      **(b)** $z = f(x, y) = x^2 + 4y^2$

**(c)** $z = f(x, y) = \sqrt{x^2 + y^2}$

**Solution** **(a)** The graph of the equation $z = 1 - x - y$, or $x + y + z = 1$, is a plane. The intercepts are the points $(1, 0, 0)$, $(0, 1, 0)$, and $(0, 0, 1)$. See Figure 6.

**(b)** The graph of the equation $z = x^2 + 4y^2$ is an elliptic paraboloid whose vertex is at the origin. See Figure 7.

**(c)** The equation $z = f(x, y) = \sqrt{x^2 + y^2}$ is equivalent to $z^2 = x^2 + y^2$, where $z \geq 0$. The graph of the equation is part of a circular cone whose vertex is at the origin. Since $z \geq 0$, the graph of $f$ is the upper nappe of the cone. See Figure 8.

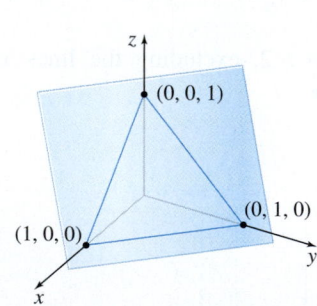

**Figure 6** $z = f(x, y) = 1 - x - y$

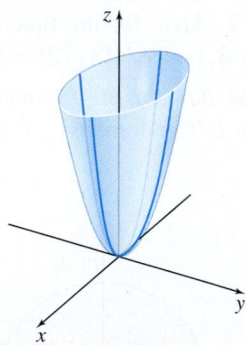

**Figure 7** $z = f(x, y) = x^2 + 4y^2$

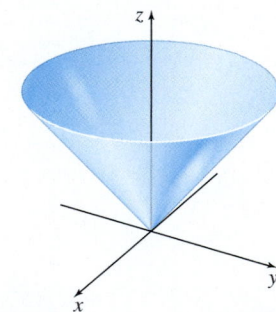

**Figure 8** $z = f(x, y) = \sqrt{x^2 + y^2}$

∎

**NOW WORK** Problems 31, 35, and 39.

## 3 Graph Level Curves

The graph of a function $f$ of two variables is usually difficult to draw by hand. In practice, such as in *topography*, the surface $z = f(x, y)$ is conveyed by drawing properly labeled curves for fixed values of $z$. These curves are called **contour lines**, and each contour line corresponds to the intersection of the surface $z = f(x, y)$ and a plane $z = c$, where $c$ is a constant.

**NOTE** A contour line is equivalent to a trace used to describe a quadric surface in Chapter 10. Recall that a trace is the intersection of a surface with a coordinate plane or a plane parallel to a coordinate plane.

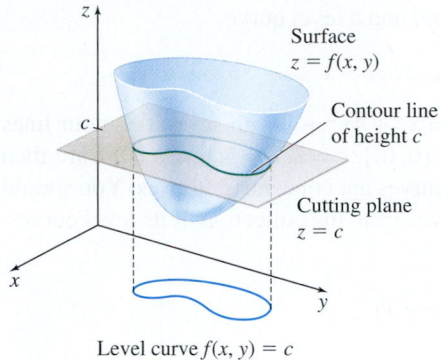

**Figure 9** Level curve of $z = f(x, y)$ at $z = c$.

For example, suppose the surface is a mountain whose height is measured in meters, and $z = 0$ represents sea level. Then by walking along the contour line $z = 500$, we would be walking on a level path 500 m above sea level.

When a contour line is projected onto the $xy$-plane, the resulting graph is called a **level curve**. See Figure 9.

In a topographical map, a series of level curves are drawn, each representing a contour line of height (or depth) $c$. For example, to represent a hilly terrain, a topographer draws level curves corresponding to contour lines for various heights measured at equal intervals, say, every 100 m. When the level curves are close to each other, the terrain is steep; when they are farther apart, the terrain is flatter. See Figure 10.

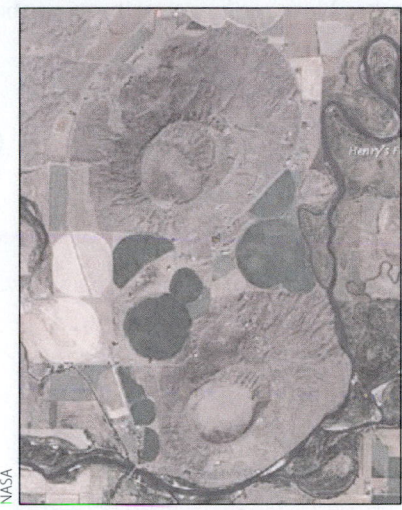
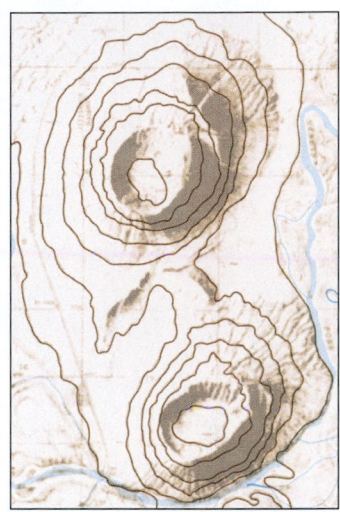

**Figure 10** Menan Buttes, Idaho. Contour intervals are shown for each 100-ft change in altitude. The north crater is about 100 ft deep, and the south crater is about 150 ft deep.

So by reversing the process, we can visualize a surface by mentally raising each level curve to its corresponding height $z$ and viewing the resulting contour lines. Figure 11 shows this process. Figure 11(a) is a U.S. geological survey topographic map of Devil's Tower, a monolith in Wyoming. Observe how the level curves are close together, indicating that the surface is steep on all sides. Figure 11(b) shows the contour lines drawn on the monolith. Figure 11(c) is a photograph of Devil's Tower.

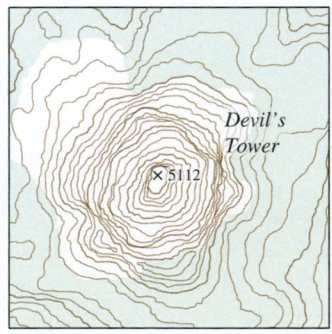
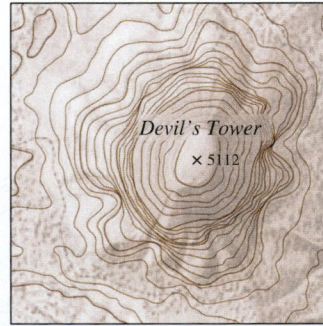

(a) Topographic map of Devil's Tower　(b) Contour lines drawn on Devil's Tower　(c) Photo of Devil's Tower

**Figure 11**

**IN WORDS** The prefix *iso-* is Greek meaning "same."

Level curves are used by meteorologists to indicate points at which barometric pressure is fixed (*isobars*), to illustrate places at which wind speed remains constant (*isolines*), and to show temperature bands (*isotherms*).

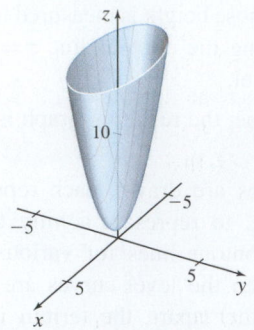

**Figure 12**

  EXAMPLE 5 **Graphing Level Curves**

Graph the level curves of the function $z = f(x, y) = x^2 + 4y^2 + 1$ for $c = 1, 2, 5$, and 17.

**Solution** Here we recognize the graph of $f$ to be an elliptic paraboloid, as shown in Figure 12. Since $z \geq 1$, the level curves of $f$ consist of the graphs of $x^2 + 4y^2 = c - 1$, $c \geq 1$. Figure 13(a) shows the elliptic paraboloid and a level curve

$$x^2 + 4y^2 + 1 = c \qquad c > 1$$

Figure 13(b) shows the elliptic paraboloid $z = x^2 + 4y^2 + 1$ with several contour lines marked. The level curves for $c = 1$ [the point $(0, 0)$], $c = 2$, $c = 5$, and $c = 17$ are then graphed in Figure 13(c). Notice that the level curves are concentric ellipses. You should be able to see how the elliptic paraboloid evolves from the collection of its level curves.

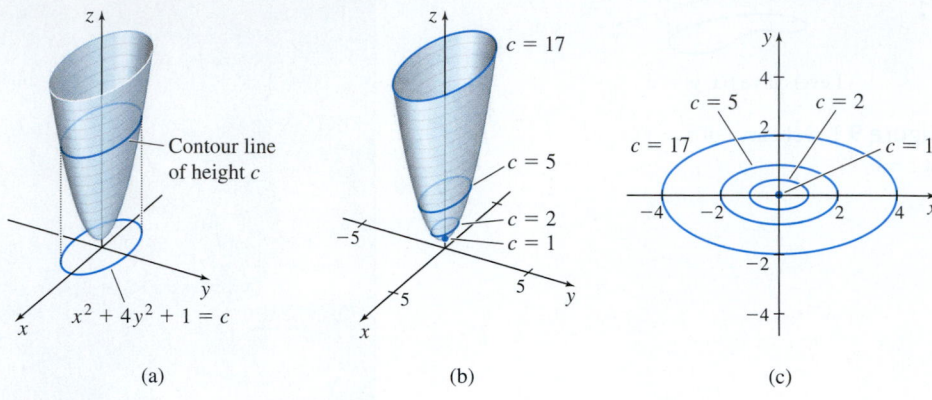

(a)  (b)  (c)

**DF Figure 13**

NOW WORK **Problem 41.**

EXAMPLE 6 **Graphing Level Curves**

Graph the level curves of the function $z = f(x, y) = e^{x^2 + y^2}$ for $c = 1, e, e^4$, and $e^{16}$.

**Solution** Because $x^2 + y^2 \geq 0$, it follows that $z \geq e^0 = 1$. The level curves satisfy the equation $e^{x^2 + y^2} = c$ or $x^2 + y^2 = \ln c$, where $c \geq 1$. For $c = 1$, the level curve is the point $(0, 0)$. If $c > 1$, the level curves are concentric circles. Figure 14 illustrates several level curves of $f$. A graph of the surface $z = e^{x^2 + y^2}$ is given in Figure 15. Do you see how the graph evolved from the collection of its level curves?

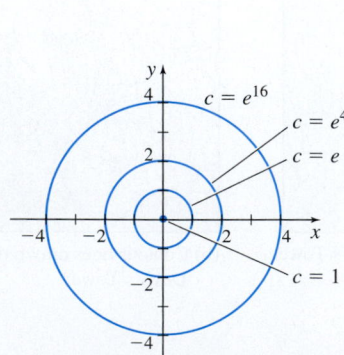

**Figure 14** Level curves of $z = e^{x^2 + y^2}$

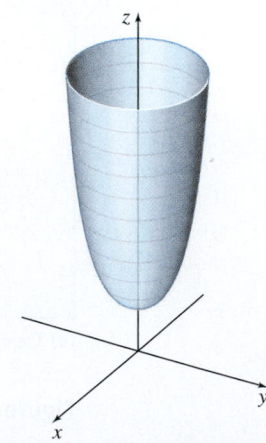

**Figure 15** The surface $z = e^{x^2 + y^2}$

NOW WORK **Problem 45.**

## 4 Describe Level Surfaces

The graph of a function $w = f(x, y, z)$ of three variables consists of all points $(x, y, z, w)$ for which $w = f(x, y, z)$ and $(x, y, z)$ is in the domain of $f$. We cannot draw the graph of a function of three variables because it requires four dimensions. But we can visualize the graph by examining its **level surfaces**, that is, the surfaces obtained by letting $w$ equal a constant.

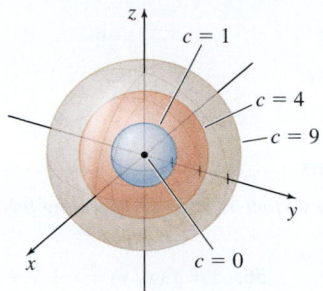

**Figure 16** Level surfaces $x^2 + y^2 + z^2 = c$

> **EXAMPLE 7    Describing Level Surfaces**
>
> Describe the level surfaces of the function $w = f(x, y, z) = x^2 + y^2 + z^2$.
>
> **Solution** Since $w \ge 0$, the level surfaces are the graphs of
>
> $$x^2 + y^2 + z^2 = c \qquad c \ge 0$$
>
> These are concentric spheres if $c > 0$ and the origin if $c = 0$. See Figure 16. ∎

**NOW WORK** Problem **49**.

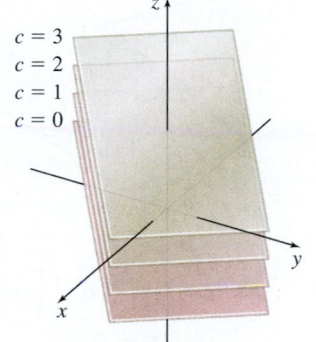

**Figure 17** Level surfaces $2x + 3y + z = c$

> **EXAMPLE 8    Describing Level Surfaces**
>
> Describe the level surfaces of the function $w = f(x, y, z) = 2x + 3y + z$.
>
> **Solution** The level surfaces are the graphs of
>
> $$2x + 3y + z = c$$
>
> This is a collection of parallel planes, each plane having the vector $\mathbf{N} = 2\mathbf{i} + 3\mathbf{j} + \mathbf{k}$ as normal. See Figure 17. ∎

**NOW WORK** Problem **51**.

## 12.1 Assess Your Understanding

### Concepts and Vocabulary

**1.** For the function $z = f(x, y) = \sqrt{16 - x^2 - y^2}$, $f(2, 3) = $ _____.

**2.** *True or False*  The domain of a function of two variables is a nonempty set of points $(x, y)$ in the $xy$-plane.

**3.** For a function $w = f(x, y, z)$, the independent variable(s) is(are) _____ and the dependent variable(s) is(are) _____.

**4.** *True or False*  A level curve of the graph of $z = f(x, y)$ is the set of points in the $xy$-plane for which $f(x, y) = c$.

### Skill Building

*In Problems 5–10, evaluate each function.*

**5.** $f(x, y) = 3x + 2y + xy$

    (a) $f(0, 1)$

    (b) $f(2, 1)$

    (c) $f(x + \Delta x, y)$

    (d) $f(x, y + \Delta y)$

**6.** $f(x, y) = x^2 y + x + 1$

    (a) $f(0, 1)$

    (b) $f(2, 1)$

    (c) $f(x + \Delta x, y)$

    (d) $f(x, y + \Delta y)$

**7.** $f(x, y) = \sqrt{xy} + x$

    (a) $f(0, 1)$

    (b) $f(a^2, t^2)$;  $a > 0, t > 0$

    (c) $f(x + \Delta x, y)$

    (d) $f(x, y + \Delta y)$

**8.** $f(x, y) = e^{x+y}$

    (a) $f(1, -1)$

    (b) $f(x + \Delta x, y)$

    (c) $f(x, y + \Delta y)$

**9.** $F(x, y, z) = \dfrac{3xy + z}{x^2 + y^2 + z^2}$

    (a) $F(0, 0, 1)$

    (b) $F(0, 1, 0)$

    (c) $F(\sin t, \cos t, 0)$

**10.** $F(x, y, z) = \dfrac{xyz}{x^2 + y^2 + z^2}$

    (a) $F(1, -1, 1)$

    (b) $F(a, a, a), a \ne 0$

    (c) $F(\sin t, \cos t, a)$

**11.** $f(x, y) = 3xy - x^2$, where $x = x(t) = \sqrt{t}$ and $y = y(t) = t^2$. Find:

    (a) $f(x(0), y(0))$     (b) $f(x(1), y(1))$     (c) $f(x(4), y(4))$

**12.** $f(x, y) = x^2 + xy + y^2$, where $x = x(t) = t$ and $y = y(t) = t^2$. Find:

    (a) $f(x(0), y(0))$     (b) $f(x(1), y(1))$     (c) $f(x(2), y(2))$

---

**1.** = NOW WORK problem          📈 = Graphing technology recommended          CAS = Computer Algebra System recommended

In Problems 13–26, find the domain of each function and graph the domain. Use a solid curve to indicate that the domain includes the curve and a dashed curve to indicate that the domain excludes the curve.

**13.** $f(x, y) = \dfrac{\sqrt{x}}{\sqrt{y}}$

**14.** $f(x, y) = \sqrt{\dfrac{x}{y}}$

**15.** $f(x, y) = \sqrt{xy}$

**16.** $f(x, y) = \sqrt{x}\sqrt{y}$

**17.** $f(x, y) = \dfrac{e^{1/x}}{\sin y}$

**18.** $f(x, y) = e^{x/y} \sin \dfrac{y}{x}$

**19.** $f(x, y) = \sqrt{\dfrac{x^2 + y^2}{x^2 - y^2}}$

**20.** $f(x, y) = \sqrt{\dfrac{x^2 + y^2}{xy}}$

**21.** $f(x, y) = \dfrac{\ln x}{\ln y}$

**22.** $f(x, y) = \ln(x^2 - y^2)$

**23.** $f(x, y) = \dfrac{y}{\sqrt{9 - x^2 - y^2}}$

**24.** $f(x, y) = \dfrac{xy}{x^2 + y^2 - 4}$

**25.** $f(x, y) = \sin^{-1} \dfrac{x^2}{y^2}$

**26.** $f(x, y) = \tan^{-1}(x^2 + y^2)$

In Problems 27–30, find the domain of each function.

**27.** $f(x, y, z) = \dfrac{x^2 + y^2}{z^2}$

**28.** $f(x, y, z) = e^z \ln(x^2 + y^2)$

**29.** $f(x, y, z) = \dfrac{z \sin x}{\cos y}$

**30.** $f(x, y, z) = \dfrac{xyz}{\sqrt{x^2 + y^2 + z^2}}$

In Problems 31–40, graph each surface.

**31.** $z = f(x, y) = 3 - x - y$

**32.** $z = f(x, y) = 2 + x - y$

**33.** $z = f(x, y) = x^2 + y^2$

**34.** $z = f(x, y) = x^2 - y^2$

**35.** $z = f(x, y) = \sqrt{4 - x^2 - y^2}$

**36.** $z = f(x, y) = \sqrt{x^2 + y^2 - 4}$

**37.** $z = f(x, y) = \sin y$

**38.** $z = f(x, y) = \cos x$

**39.** $z = f(x, y) = 4 - x^2 - y^2$

**40.** $z = f(x, y) = x^2 + y^2 - 4$

In Problems 41–48, for each function
*(a)* Graph the level curves corresponding to the given values of c.
**CAS** *(b)* Use a CAS to graph the surface.

**41.** $z = f(x, y) = x^2 - y^2$ at $c = 0, 1, 4, 9$

**42.** $z = f(x, y) = 2x^2 + y^2$ at $c = 0, 1, 4, 9$

**43.** $z = f(x, y) = \sqrt{9 - x^2 - y^2}$ at $c = 0, 1, 3$

**44.** $z = f(x, y) = \sqrt{x^2 + y^2 - 4}$ at $c = 0, 1, 4, 9$

**45.** $z = f(x, y) = x^2 - 2y$ at $c = -4, -1, 0, 1, 4$

**46.** $z = f(x, y) = y^2 - x$ at $c = -4, -1, 0, 1, 4$

**47.** $z = f(x, y) = x + \sin y$ at $c = 0, 2, 4, 8$

**48.** $z = f(x, y) = y - \ln x$ at $c = 1, 2, 4$

In Problems 49–54:
*(a)* Describe in words the level surfaces of each function $f$.
**CAS** *(b)* Graph at least two of the level surfaces.
*(c)* Does the graph match your verbal description?

**49.** $w = f(x, y, z) = x^2 + z^2$

**50.** $w = f(x, y, z) = x^2 + y^2$

**51.** $w = f(x, y, z) = z - 2x - 2y$

**52.** $w = f(x, y, z) = x + y - z$

**53.** $w = f(x, y, z) = 4 - x^2 - y^2$

**54.** $w = f(x, y, z) = z$

## Applications and Extensions

In Problems 55–60, match each surface to its corresponding level curves shown in (A)–(F).

**55.** $z = f(x, y) = 2x - y - 2$

**56.** $z = f(x, y) = x + y - 3$

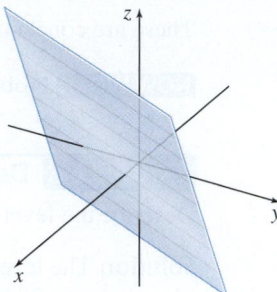

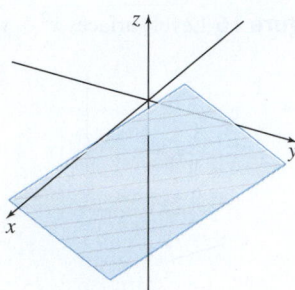

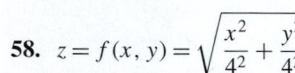

**57.** $z = f(x, y) = 4x^2 + y^2$

**58.** $z = f(x, y) = \sqrt{\dfrac{x^2}{4^2} + \dfrac{y^2}{4^2}}$

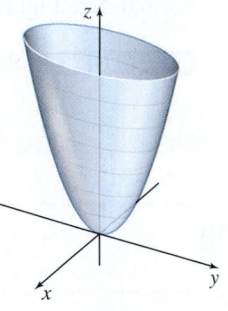

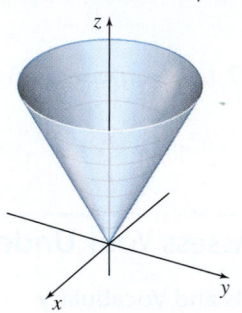

**59.** $z = f(x, y) = \dfrac{x^2}{2^2} - \dfrac{y^2}{1^2}$

**60.** $z = f(x, y) = \dfrac{xy(x^2 - y^2)}{x^2 + y^2}$

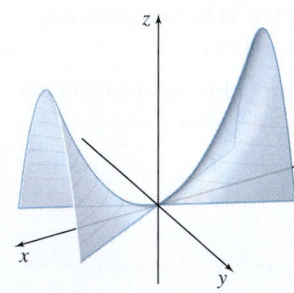

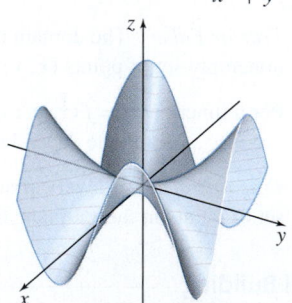

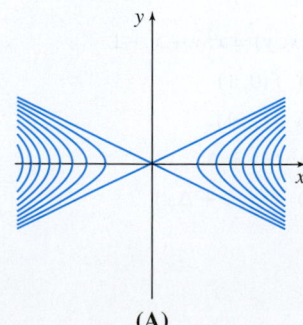

(A)

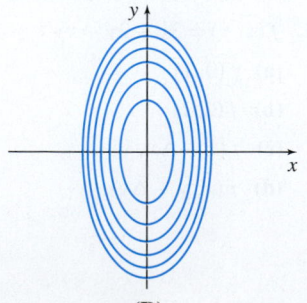

(B)

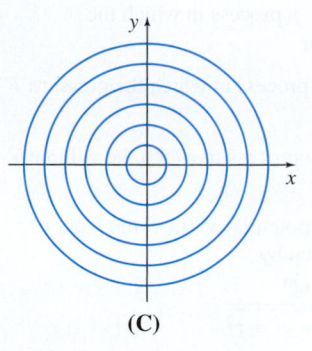

(C)

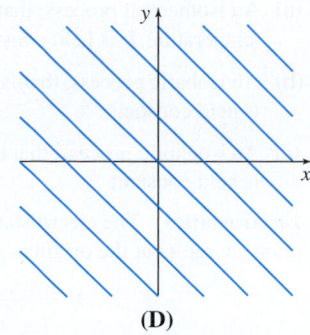

(D)

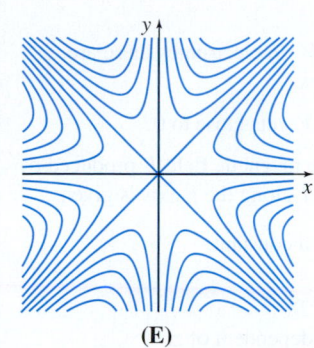

(E)

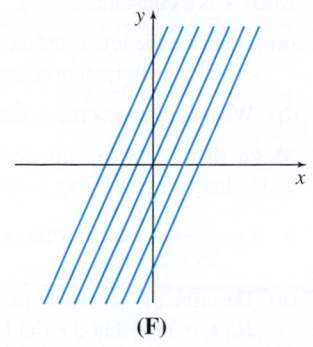

(F)

**61. Weather Maps**   A contour map of the temperatures in °F of the western United States made on a given day in December 2017 is shown. The level curves are called **isotherms**.

(a) Estimate the temperature at the Four Corners, the point where Utah, Colorado, Arizona, and New Mexico meet.

(b) Describe the change in temperature as you move south from the Four Corners.

(c) Describe the direction you would travel if you were at the Four Corners and wanted to move as quickly as possible to the next warmer temperature zone. Comment on how your path intersects the isotherm.

(d) A weather front is moving in if the isotherms are close together. Do you think a front is approaching the Four Corners? From what direction? Explain.

**62. Climbing a Mountain**
A topographical map of Mount Washington in New Hampshire is shown. Suppose a climber is at the base of the mountain and wants to climb to the summit. If she can begin at point A, B, C, or D, and climb straight up the mountain,

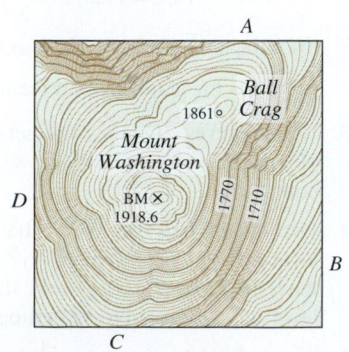

(a) At which point should she start to climb so the distance climbed is a minimum?

(b) At which point is the climb initially the steepest?

(c) Suppose someone is on any level curve of the mountain. Describe how he would walk so that no change in altitude occurs.

(d) Comment on how the paths in (a)–(c) intersect the contour lines.

**63. Social Science**   In psychology, the intelligence quotient (IQ) is measured by $IQ = f(M, C) = 100\dfrac{M}{C}$, where $M$ is a person's mental age and $C$ is the person's chronological or actual age, $0 < C \le 16$.

(a) Find the IQ of a 12-year-old child whose mental age is 10.

(b) Find the IQ of a 10-year-old child whose mental age is 12.

(c) If a 10-year-old girl has an IQ of 139, what is her mental age? Round answers to the nearest integer.

**64. International Mobile Data Cost**   The monthly cost $C$ (in dollars) of data for an AT&T subscriber traveling outside the United States without an international package is $C = C(x, y, z) = 2.05x + 8.19y + 10.24z$, where $x$ is the number of megabytes (mb) used on land, $y$ is the number of mb used on a cruise ship, and $z$ is the number of mb used on an airplane. Find the monthly bill (in dollars) of a subscriber using 210 mb of data on land, 15 mb of data on a cruise ship, and 250 mb of data in an airplane.
*Source*: AT&T Wireless.

**65. Baseball**   A pitcher's earned run average is calculated using the function $A(N, I) = 9\left(\dfrac{N}{I}\right)$, where $N$ is the total number of earned runs given up in $I$ innings pitched. Find:

(a) $A(3, 4)$     (b) $A(6, 3)$     (c) $A(2, 9)$     (d) $A(3, 18)$

**66. Field Goal Percentages in the NBA**   In the National Basketball Association (NBA), the adjusted field goal percentage is modeled by the function $f(x, y, s) = \dfrac{x + 1.5y}{s}$, where $x$ is the number of two-point field goals made, $y$ is the number of three-point field goals made, and $s$ is the total of all field goals attempted.

(a) During the 2016–2017 NBA season, Russell Westbrook of the Oklahoma City Thunder led the league in field goals. He made 624 out of 1358 two-point attempts and 200 out of 583 three-point attempts. Find his adjusted field goal percentage.

(b) During the 2016–2017 NBA season Stephen Curry of the Golden State Warriors led the league in three-point shooting. He made 351 out of 654 two-point attempts and 324 out of 789 three-point attempts. Find his adjusted field goal percentage.

**67. Meteorology**   The apparent temperature (in degrees Fahrenheit) is measured by the heat index $H$ according to the formula

$$H = H(t, r) = -42.379 + 2.04901523\,t$$
$$+ 10.1333127\,r - 0.22475541\,tr - 0.00683783\,t^2$$
$$- 0.05481717\,r^2 + 0.00122874\,t^2r + 0.00085282\,tr^2$$
$$- 0.00000199\,t^2r^2$$

where $H$ = the heat index, $t$ = the air temperature, and $r$ = the percent relative humidity (for example, $r = 75$ when the relative humidity is 75%).

(a) What is the heat index when the air temperature is 95° and the relative humidity is 50%?

(b) If the air temperature is 97°, what is the lowest relative humidity that will result in a heat index of 105°?

*Source*: Weather Information Center; 4WX.com.

68. **Economics**   The production function for a toy manufacturer is given by the equation $Q(L, M) = 400L^{0.3}M^{0.7}$, where $Q$ is the output in units, $L$ is the labor in hours, and $M$ is the number of machine hours. Find:

(a) $Q(19, 21)$          (b) $Q(21, 20)$

69. **Rectangular Box**   Write the equation for the surface area $S$ of an open box as a function of its length $x$, width $y$, and depth $z$.

70. **Cost Function**   The cost C of the bottom and top of a cylindrical tank is \$300 per square meter and the cost of the sides is \$500 per square meter. Find a function that models the total cost of constructing such a tank as a function of the radius $R$ and height $h$, both in meters.

71. **Cost Function**   Find a function that models the total cost C of constructing an open rectangular box if the cost per square centimeter of the material to be used for the bottom is \$4, for two of the opposite sides is \$3, and for the remaining pair of opposite sides is \$2.

72. **Cost Function**   Repeat Problem 71 for a closed rectangular box that has a top made of material costing \$5 per square centimeter.

73. **Electrical Potential**   The formula

$$V(x, y) = \frac{9}{\sqrt{4 - (x^2 + y^2)}}$$

gives the electrical potential V (in volts) at a point $(x, y)$ in the $xy$-plane. Draw the equipotential curves (level curves) for $V = 18, 9$, and $6$ volts. Describe the surface $z = V(x, y)$.

74. **Temperature**   The temperature $T$ in degrees Celsius at any point $(x, y)$ of a flat plate in the $xy$-plane is $T = 60 - 2x^2 - 3y^2$. Draw the isothermal curves (level curves) for $T = 60\,°C, 54\,°C, 48\,°C, 6\,°C, 0\,°C$. Describe the surface $z = T(x, y)$.

75. **Electric Field**   The strength $E$ of an electric field at a point $(x, y, z)$ resulting from an infinitely long charged wire lying along the $x$-axis is

$$E(x, y, z) = \frac{3}{\sqrt{y^2 + z^2}}$$

Describe the level surfaces of $E$.

76. **Gravitation**   The magnitude $F$ of the force of attraction between two objects, one located at the origin and the other at the point $(x, y, z) \neq (0, 0, 0)$, of masses $m$ and $M$ is given by

$$F = \frac{GmM}{x^2 + y^2 + z^2}$$

where $G = 6.67 \times 10^{-11}\,N\,m^2/kg^2$ is the gravitational constant. Describe the level surfaces of $F$.

77. **Thermodynamics**   The Ideal Gas Law, $PV = nRT$, is used to describe the relationship among the pressure $P$, volume $V$, and temperature $T$ of an ideal gas, where $n$ is the number of moles of gas and $R$ is the universal gas constant. Describe the level curves for each of the following thermodynamic processes on an ideal gas:

(a) An isothermal process, that is, a process in which the temperature $T$ is held constant

(b) An isobaric process, that is, a process in which the pressure $P$ is held constant

(c) An isochoric process, that is, a process in which the volume $V$ is held constant

78. **Electrostatics**   The electrostatic potential $V$ (in volts) from a point charge $Q$ at the origin is given by

$$V = \frac{kQ}{\sqrt{x^2 + y^2 + z^2}}$$

where $k$ is a constant.

(a) Describe the level surfaces of $V$. In electronics these are called the equipotential surfaces.

(b) What happens as the potential $V$ gets close to 0?

79. **Magnetism**   The magnitude of the magnetic field $B$ produced by a very long wire carrying a current $I$ along the $z$-axis is given by $B = \dfrac{\mu_0 I}{2\pi \sqrt{x^2 + y^2}}$, where $\mu_0$ is a constant.

(a) Describe the level surfaces for $B$.
    *Hint*: Notice that the field is independent of $z$.

(b) What happens as the magnitude of $B$ increases?

80. **Orbit of a Satellite**   The gravitational potential energy $U$ of a satellite in orbit around Earth is $U = -\dfrac{GM_{Earth}m_{satellite}}{\sqrt{x^2 + y^2 + z^2}}$, where the origin is placed at the center of Earth and $G$ is the universal gravitation constant.

(a) Describe the level surfaces for $U$.

(b) What happens as the gravitational potential energy $U$ gets close to 0?

[CAS] 81. (a) Graph the surface $z = f(x, y) = 5e^{-(x^2 + y^2)}$.

(b) Experiment by changing the perspectives and plotting options until a clear picture of the graph is obtained.

(c) Change the style of the graph to show contour lines.

(d) Plot the level curves $f(x, y) = c$   $c = 1, 2, 3, 4$.

(e) Explain how the curves obtained in (d) relate to the graph from (a).

82. (a) Find the curve of intersection of the surfaces

$$x^2 + y^2 + z^2 = 4 \text{ and } z = \frac{1}{3}(x^2 + y^2) \text{ above the } xy\text{-plane.}$$

[CAS] (b) Graph the surfaces from (a).

(c) Does the graph support the result from (a)?

83. **Level Curves**   On the same set of axes, graph the level curves of $f(x, y) = x^2 - y^2$ and $g(x, y) = xy$. Use $c = \pm 1, \pm 2, \pm 3$ for both $f$ and $g$.

84. **Orthogonal Curves**   Show that at each point $P_0 \neq (0, 0)$, the level curve of $f(x, y) = x^2 - y^2$ through $P_0$ is perpendicular to the level curve of $g(x, y) = xy$ through $P_0$. The two families of level curves are said to be **orthogonal**.

## Challenge Problem

85. Describe the set of points $(x, y, z)$ satisfying the conditions $x^2 + y^2 + z^2 < 1$ and $x^2 + y^2 < z^2$, where $z > 0$.

# 12.2 Limits and Continuity

**OBJECTIVES** *When you finish this section, you should be able to:*

**1** Define the limit of a function of several variables (p. 861)

**2** Find a limit using properties of limits (p. 863)

**3** Examine when limits exist (p. 865)

**4** Determine where a function is continuous (p. 867)

In Chapters 1 and 2, we saw the importance of the limit of a function of one variable in determining continuity and defining and finding the derivative. Limits of functions of several variables play a similar role.

## 1 Define the Limit of a Function of Several Variables

We begin by reviewing the definition of a limit (Chapter 1, pp. 86 and 141).

> Let $f$ be a function of one variable defined everywhere in an open interval containing $c$, except possibly at $c$. Then the **limit of $f(x)$ as $x$ approaches $c$ is $L$**, written
>
> $$\lim_{x \to c} f(x) = L$$
>
> if for any number $\varepsilon > 0$, there is a number $\delta > 0$, so that
>
> whenever $\quad 0 < |x - c| < \delta \qquad$ then $\qquad |f(x) - L| < \varepsilon$

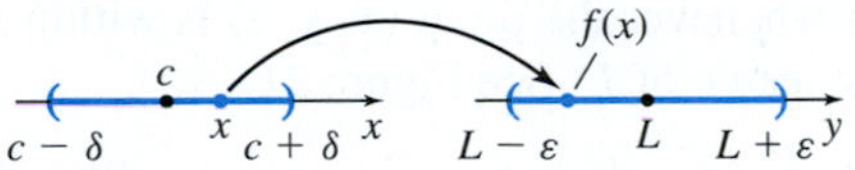

In this definition, the phrase "whenever $0 < |x - c| < \delta$" means "for all $x$, $x \neq c$, within a distance $\delta$ of $c$," and "$|f(x) - L| < \varepsilon$" means "the corresponding value of $f$ is within a distance $\varepsilon$ of $L$." See Figure 18.

The definition of a limit for functions of two or more variables is similar, but requires some preliminary definitions.

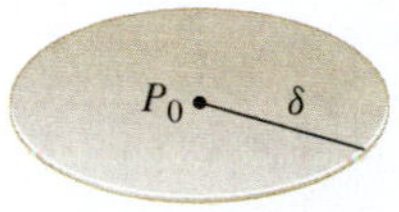

**Figure 18**

> **DEFINITION** $\delta$-Neighborhood; Disk; Ball
>
> Let $P_0$ be a point in the plane or in space. A **$\delta$-neighborhood** of $P_0$ consists of all the points $P$ that lie within a distance $\delta$ of $P_0$. That is, a $\delta$-neighborhood consists of all points $P$ for which $d(P, P_0) < \delta$, where $d(P, P_0)$ is the distance from $P$ to $P_0$.
>
> - In the plane, a $\delta$-neighborhood of $P_0$ is called a **disk of radius $\delta$ centered at $P_0$**.
> - In space, a $\delta$-neighborhood of $P_0$ is called a **ball of radius $\delta$ centered at $P_0$**.

(a) Disk of radius $\delta$ centered at $P_0$

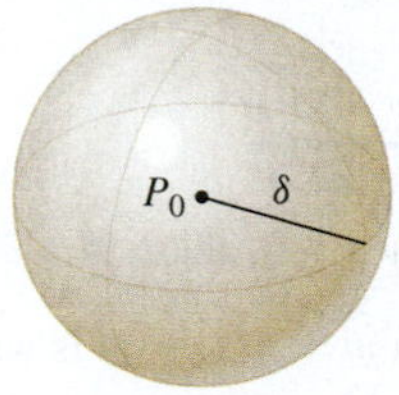

(b) Ball of radius $\delta$ centered at $P_0$

**Figure 19**

See Figure 19.

With these definitions, we are ready to define the limit of a function of several variables.

**NOTE** On a real line, a $\delta$-neighborhood of a number $P_0$ is an open interval with $P_0$ as midpoint.

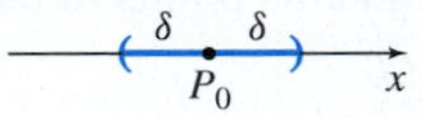

> **DEFINITION** Limit of a Function of Two Variables
>
> Let $f$ be a function of two variables. Suppose every disk centered at $(x_0, y_0)$, which may or may not be in the domain of $f$, contains points in the domain of $f$. Then the **limit of $f(x, y)$ as $(x, y)$ approaches $(x_0, y_0)$ is $L$** if, for any number $\varepsilon > 0$, there is a number $\delta > 0$, so that for all points $(x, y)$ in the domain of $f$
>
> whenever $\quad 0 < \sqrt{(x - x_0)^2 + (y - y_0)^2} < \delta \qquad$ then $\qquad |f(x, y) - L| < \varepsilon$
>
> Then we say, "The limit exists," and write
>
> $$\lim_{(x,y) \to (x_0, y_0)} f(x, y) = L$$

Figure 20 illustrates the definition.

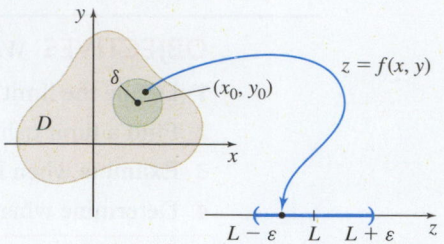

**DF** **Figure 20** $\displaystyle\lim_{(x,y)\to(x_0,y_0)} f(x,y) = L$ means that for any $\varepsilon > 0$, there is a $\delta > 0$, so that whenever $(x, y) \neq (x_0, y_0)$ is within $\delta$ of $(x_0, y_0)$, then $z = f(x, y)$ is within $\varepsilon$ of $L$.

### DEFINITION  Limit of a Function of Three Variables

Let $f$ be a function of three variables. Suppose every ball centered at $(x_0, y_0, z_0)$, which may or may not be in the domain of $f$, contains points in the domain of $f$. Then the **limit of** $f(x, y, z)$ **as** $(x, y, z)$ **approaches** $(x_0, y_0, z_0)$ **is** $L$ if, for any number $\varepsilon > 0$, there is a number $\delta > 0$, so that for all points $(x, y, z)$ in the domain of $f$

$$\text{whenever } 0 < \sqrt{(x - x_0)^2 + (y - y_0)^2 + (z - z_0)^2} < \delta \quad \text{then } |f(x, y, z) - L| < \varepsilon$$

Then we say, "The limit exists," and write

$$\lim_{(x,y,z)\to(x_0,y_0,z_0)} f(x, y, z) = L$$

The definition states that if you decide how close you want $f$ to be to $L$, that is, if you choose $\varepsilon$, you can find a number $\delta$ so that whenever the point $(x, y, z)$ is within a distance $\delta$ of $(x_0, y_0, z_0)$, then $f$ is within a distance $\varepsilon$ of $L$. See Figure 21.

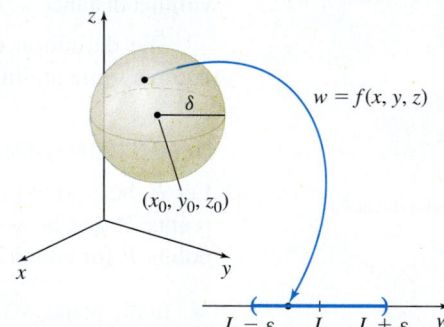

**DF** **Figure 21** $\displaystyle\lim_{(x,y,z)\to(x_0,y_0,z_0)} f(x, y, z) = L$ means that for any $\varepsilon > 0$, there is a $\delta > 0$, so that whenever $(x, y, z) \neq (x_0, y_0, z_0)$ is within $\delta$ of $(x_0, y_0, z_0)$, then $w = f(x, y, z)$ is within $\varepsilon$ of $L$.

Notice the similarity of the definitions for the limit of a function of one variable, two variables, and three variables. In fact, limits of functions of more than three variables are defined similarly, as follows.

### DEFINITION  Limit of a Function of Several Variables

Suppose $f$ is a function of one or more variables. Suppose every neighborhood of a point $P_0$, which may or may not be in the domain of $f$, contains points in the domain of $f$. Then

$$\lim_{P \to P_0} f(P) = L$$

if, for any number $\varepsilon > 0$, there is a number $\delta > 0$, so that for all points $P$ that are in the domain of $f$

$$\text{whenever}\quad 0 < d(P, P_0) < \delta \quad\text{then}\quad |f(P) - L| < \varepsilon$$

## 2 Find a Limit Using Properties of Limits

The algebra of limits for functions of one variable extends to that for functions of several variables. We state the algebraic properties of limits of functions of two variables without proof. These properties are also true for functions of three or more variables.

---

**THEOREM** **Algebraic Properties of Limits (Two Variables)**

Let $f$ and $g$ be functions of two variables for which

$$\lim_{(x,y) \to (x_0,y_0)} f(x, y) = L \qquad \text{and} \qquad \lim_{(x,y) \to (x_0,y_0)} g(x, y) = M$$

where $L$ and $M$ are two real numbers.

- **Limit of a sum:**

$$\lim_{(x,y) \to (x_0,y_0)} [f(x, y) + g(x, y)] = \lim_{(x,y) \to (x_0,y_0)} f(x, y) + \lim_{(x,y) \to (x_0,y_0)} g(x, y)$$
$$= L + M \tag{1}$$

- **Limit of a difference:**

$$\lim_{(x,y) \to (x_0,y_0)} [f(x, y) - g(x, y)] = \lim_{(x,y) \to (x_0,y_0)} f(x, y) - \lim_{(x,y) \to (x_0,y_0)} g(x, y)$$
$$= L - M \tag{2}$$

- **Limit of a constant times a function:** If $k$ is any real number, then

$$\lim_{(x,y) \to (x_0,y_0)} [k\, f(x, y)] = k \left[ \lim_{(x,y) \to (x_0,y_0)} f(x, y) \right] = kL \tag{3}$$

- **Limit of a product:**

$$\lim_{(x,y) \to (x_0,y_0)} [f(x, y)g(x, y)] = \left[ \lim_{(x,y) \to (x_0,y_0)} f(x, y) \right] \left[ \lim_{(x,y) \to (x_0,y_0)} g(x, y) \right]$$
$$= LM \tag{4}$$

- **Limit of a quotient:** If $\lim_{(x,y) \to (x_0,y_0)} g(x, y) = M \neq 0$,

$$\lim_{(x,y) \to (x_0,y_0)} \frac{f(x, y)}{g(x, y)} = \frac{\displaystyle\lim_{(x,y) \to (x_0,y_0)} f(x, y)}{\displaystyle\lim_{(x,y) \to (x_0,y_0)} g(x, y)} = \frac{L}{M} \tag{5}$$

- **Limit of a constant:** If $c$ is a constant,

$$\lim_{(x,y) \to (x_0,y_0)} c = c \tag{6}$$

- **Limit of a function of a single variable:**
  If $f(x, y) = h(x)$, then

$$\lim_{(x,y) \to (x_0,y_0)} f(x, y) = \lim_{x \to x_0} h(x) \tag{7}$$

  If $f(x, y) = g(y)$, then

$$\lim_{(x,y) \to (x_0,y_0)} f(x, y) = \lim_{y \to y_0} g(y) \tag{8}$$

EXAMPLE 1   **Using Properties of Limits to Find a Limit**

Find each limit:

**(a)** $\lim\limits_{(x,y)\to(1,2)}(3x^2+2xy+y^2)$

**(b)** $\lim\limits_{(x,y)\to(1,2)}\dfrac{xy}{x^2+y^2}$

**(c)** $\lim\limits_{(x,y)\to(3,-1)}\dfrac{x^2+2xy-3y^2}{x^2+9y^2}$

**(d)** $\lim\limits_{(x,y)\to(3,2)}\dfrac{y^2-4}{xy-2x}$

**Solution**

**(a)** $\lim\limits_{(x,y)\to(1,2)}(3x^2+2xy+y^2)\underset{\underset{\text{Property (1)}}{\uparrow}}{=}\lim\limits_{(x,y)\to(1,2)}(3x^2)+\lim\limits_{(x,y)\to(1,2)}(2xy)+\lim\limits_{(x,y)\to(1,2)}y^2$

$$\underset{\underset{\text{Properties (3), (4), (7), and (8)}}{\uparrow}}{=}3\lim\limits_{x\to1}x^2+\left[\lim\limits_{(x,y)\to(1,2)}(2x)\right]\left(\lim\limits_{(x,y)\to(1,2)}y\right)+\lim\limits_{y\to2}y^2$$

$$=3+2\left(\lim\limits_{x\to1}x\right)\left(\lim\limits_{y\to2}y\right)+4=3+2\cdot1\cdot2+4=11$$

**(b)** Since $\lim\limits_{(x,y)\to(1,2)}(x^2+y^2)=\lim\limits_{x\to1}x^2+\lim\limits_{y\to2}y^2=5\neq0$, we use the limit of a quotient, property (5):

$$\lim\limits_{(x,y)\to(1,2)}\dfrac{xy}{x^2+y^2}\underset{\underset{\text{Property (5)}}{\uparrow}}{=}\dfrac{\lim\limits_{(x,y)\to(1,2)}(xy)}{\lim\limits_{(x,y)\to(1,2)}(x^2+y^2)}\underset{\underset{\text{Properties (1), (4), (7), and (8)}}{\uparrow}}{=}\dfrac{\left(\lim\limits_{x\to1}x\right)\left(\lim\limits_{y\to2}y\right)}{\lim\limits_{x\to1}x^2+\lim\limits_{y\to2}y^2}=\dfrac{1\cdot2}{1+4}=\dfrac{2}{5}$$

**(c)** Since $\lim\limits_{(x,y)\to(3,-1)}(x^2+9y^2)=18$, we can use the limit of a quotient, property (5). Then

$$\lim\limits_{(x,y)\to(3,-1)}\dfrac{x^2+2xy-3y^2}{x^2+9y^2}\underset{\underset{\text{Property (5)}}{\uparrow}}{=}\dfrac{\lim\limits_{(x,y)\to(3,-1)}(x^2+2xy-3y^2)}{\lim\limits_{(x,y)\to(3,-1)}(x^2+9y^2)}$$

$$\underset{\underset{\text{Properties (1)–(4), (7), and (8)}}{\uparrow}}{=}\dfrac{\lim\limits_{x\to3}x^2+2\left(\lim\limits_{x\to3}x\right)\left(\lim\limits_{y\to-1}y\right)-3\lim\limits_{y\to-1}y^2}{18}$$

$$=\dfrac{9+2\cdot3\cdot(-1)-3\cdot1}{18}=0$$

**(d)** For the denominator of this function, $\lim\limits_{(x,y)\to(3,2)}(xy-2x)=0$, so property (5) cannot be used. As with functions of a single variable, we try factoring the numerator and the denominator.

$$\dfrac{y^2-4}{xy-2x}=\dfrac{(y-2)(y+2)}{x(y-2)}$$

Since $(3,2)$ is not in the domain of the function, $y\neq2$, and we can divide out the factor $y-2$. Then

$$\lim\limits_{(x,y)\to(3,2)}\dfrac{y^2-4}{xy-2x}=\lim\limits_{(x,y)\to(3,2)}\dfrac{(y-2)(y+2)}{x(y-2)}=\lim\limits_{(x,y)\to(3,2)}\dfrac{y+2}{x}$$

Now use property (5).

$$\lim\limits_{(x,y)\to(3,2)}\dfrac{y^2-4}{xy-2x}=\lim\limits_{(x,y)\to(3,2)}\dfrac{y+2}{x}\underset{\underset{\text{Property (5)}}{\uparrow}}{=}\dfrac{\lim\limits_{(x,y)\to(3,2)}(y+2)}{\lim\limits_{(x,y)\to(3,2)}x}\underset{\underset{\text{Properties (7) and (8)}}{\uparrow}}{=}\dfrac{\lim\limits_{y\to2}(y+2)}{\lim\limits_{x\to3}x}=\dfrac{4}{3}$$ ∎

NOW WORK **Problems 9, 21, and 49.**

For certain functions, converting to polar coordinates can make finding a limit easier. See Problems 77–80.

NOW WORK Problem 77.

## 3 Examine When Limits Exist

The limit of a function $f$ of single variable exists only if the left-hand limit equals the right-hand limit. That is, $\lim\limits_{x \to x_0} f(x)$ exists if the limit is the same no matter how the number $x_0$ is approached. For a single variable, there are only two ways to approach $x_0$: from the left or from the right. See Figure 22(a). When $z = f(x, y)$ is a function of two variables, however, there are infinitely many ways to approach a point $P_0 = (x_0, y_0)$, namely along any curve that passes through $P_0$. See Figure 22(b). So, for $\lim\limits_{(x,y) \to (x_0, y_0)} f(x, y)$ to exist, it is necessary that the same limit be obtained no matter what curve $C$ containing $P_0$ is used. Put another way, if two different curves that contain $(x_0, y_0)$ result in different limits, then $\lim\limits_{(x,y) \to (x_0, y_0)} f(x, y)$ does not exist.

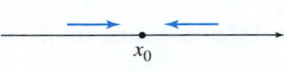

(a) Two ways to approch $x_0$

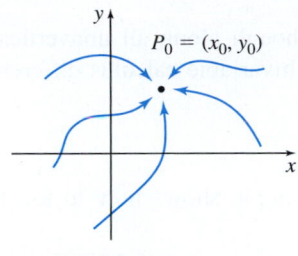

(b) Infinitely many ways to approch $P_0$

**Figure 22**

### THEOREM

If $\lim\limits_{P \to P_0} f(P)$ is computed along two different curves containing $P_0$ and if two different answers are obtained, then $\lim\limits_{P \to P_0} f(P)$ does not exist.

**CALC CLIP**

## EXAMPLE 2  Showing That a Limit Does Not Exist

Show that $\lim\limits_{(x,y) \to (0,0)} \dfrac{xy}{x^2 + y^2}$ does not exist.

**Solution** Look at the denominator. Since $\lim\limits_{(x,y) \to (0,0)} (x^2 + y^2) = 0$, we cannot use the limit of a quotient. We begin with two lines, say, $y = 2x$ and $y = -x$. These curves are easy to work with and both contain $(0, 0)$. We investigate the limit using these two curves that contain $(0, 0)$.

Using $y = 2x$: $\lim\limits_{(x,y) \to (0,0)} \dfrac{x\,(2x)}{x^2 + (2x)^2} \underset{\text{Property (7)}}{=} \lim\limits_{x \to 0} \dfrac{2x^2}{x^2 + 4x^2} = \lim\limits_{x \to 0} \dfrac{2x^2}{5x^2} = \dfrac{2}{5}$

Using $y = -x$: $\lim\limits_{(x,y) \to (0,0)} \dfrac{x\,(-x)}{x^2 + (-x)^2} \underset{\text{Property (7)}}{=} \lim\limits_{x \to 0} \dfrac{-x^2}{x^2 + x^2} = \lim\limits_{x \to 0} \dfrac{-x^2}{2x^2} = -\dfrac{1}{2}$

Since two different answers are obtained, the limit does not exist. ∎

Figure 23 shows the graph of $f(x, y) = \dfrac{xy}{x^2 + y^2}$.

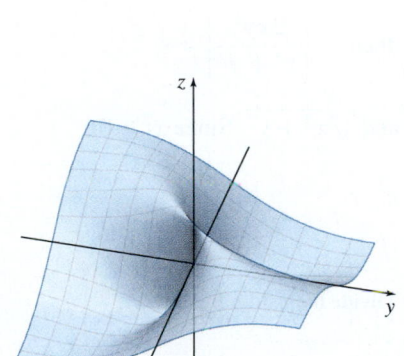

**DF Figure 23** $f(x, y) = \dfrac{xy}{x^2 + y^2}$

NOW WORK Problem 25.

**CAUTION** Even if the same limit is obtained along two curves that contain $P_0$ (or three, or even several hundred curves), we cannot conclude that the limit of the function exists. For instance, in Example 2, using the curves $y = 2x$ and $y = \dfrac{1}{2}x$ would both result in the same value for the limit, but we have shown the limit does not exist.

## EXAMPLE 3  Showing That a Limit Does Not Exist

Show that $\lim\limits_{(x,y) \to (0,0)} \dfrac{x^2 y}{x^4 + y^2}$ does not exist.

**Solution** We investigate the limit using curves that contain $(0, 0)$. By using $y = mx$, $m \neq 0$, we can test all nonvertical and nonhorizontal lines that contain $(0, 0)$ simultaneously. Then

$$\lim\limits_{(x,y) \to (0,0)} \dfrac{x^2\,(mx)}{x^4 + (mx)^2} \underset{\text{Property (7)}}{=} \lim\limits_{x \to 0} \dfrac{mx^3}{x^4 + m^2 x^2} = \lim\limits_{x \to 0} \dfrac{mx}{x^2 + m^2} = 0$$

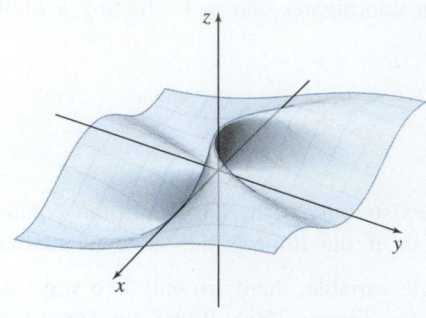

**Figure 24** $f(x, y) = \dfrac{x^2 y}{x^4 + y^2}$

Along every nonvertical and nonhorizontal line containing the point $(0, 0)$, the limit is 0. But this does not mean that $\lim\limits_{(x,y) \to (0,0)} \dfrac{x^2 y}{x^4 + y^2}$ is 0. Suppose we use the curve $y = x^2$. Then

Using $y = x^2$: $\quad \lim\limits_{(x,y) \to (0,0)} \dfrac{x^4}{x^4 + x^4} = \lim\limits_{x \to 0} \dfrac{x^4}{2x^4} = \dfrac{1}{2}$

Since two different curves that contain $(0, 0)$ result in different values for the limit, the limit does not exist. ∎

The graph of $f(x, y) = \dfrac{x^2 y}{x^4 + y^2}$ is shown in Figure 24.

In Example 3, the limit failed to exist even though along all nonvertical lines the limit equals 0. This kind of behavior makes multivariable calculus different from single-variable calculus.

**NOW WORK** Problem 33.

How do we know a limit exists? The next example shows how to use the $\varepsilon$-$\delta$ definition of a limit to prove that a limit exists.

---

**EXAMPLE 4** **Showing That a Limit Exists**

Use the $\varepsilon$-$\delta$ definition of a limit to show that $\lim\limits_{(x,y) \to (0,0)} \dfrac{2xy^2}{x^2 + y^2} = 0$.

**Solution** Given a number $\varepsilon > 0$, we seek a number $\delta > 0$, so that

$$\text{whenever} \quad 0 < \sqrt{(x - 0)^2 + (y - 0)^2} < \delta \quad \text{then} \quad \left| \dfrac{2xy^2}{x^2 + y^2} - 0 \right| < \varepsilon$$

That is,

$$\text{whenever} \quad 0 < \sqrt{x^2 + y^2} < \delta \quad \text{then} \quad \left| \dfrac{2xy^2}{x^2 + y^2} \right| < \varepsilon$$

We need to find a connection between $\dfrac{2xy^2}{x^2 + y^2}$ and $\sqrt{x^2 + y^2}$. Since $x^2 \geq 0$,

$$y^2 \leq x^2 + y^2$$

So, for all points $(x, y)$ not equal to $(0, 0)$,

$$\dfrac{y^2}{x^2 + y^2} \leq 1 \qquad \text{\color{blue}Divide both sides by } x^2 + y^2.$$

Now

$$\left| \dfrac{2xy^2}{x^2 + y^2} \right| = \dfrac{2|x|y^2}{x^2 + y^2} = 2|x| \cdot \dfrac{y^2}{x^2 + y^2} \leq 2|x| \cdot 1 = 2\sqrt{x^2} \leq 2\sqrt{x^2 + y^2}$$

Given $\varepsilon > 0$, choose $\delta \leq \dfrac{\varepsilon}{2}$. Then whenever $0 < \sqrt{x^2 + y^2} < \delta \leq \dfrac{\varepsilon}{2}$.

$$\left| \dfrac{2xy^2}{x^2 + y^2} \right| \leq 2\sqrt{x^2 + y^2} < 2\delta \leq \varepsilon$$

That is,

$$\lim\limits_{(x,y) \to (0,0)} \dfrac{2xy^2}{x^2 + y^2} = 0 \qquad ∎$$

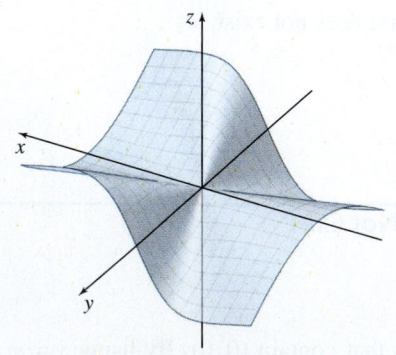

**Figure 25** $f(x, y) = \dfrac{2xy^2}{x^2 + y^2}$

Figure 25 shows the graph of $f(x, y) = \dfrac{2xy^2}{x^2 + y^2}$.

**NOW WORK** Problem 67.

As Example 4 illustrates, showing a limit exists using the $\varepsilon$-$\delta$ definition can be challenging. For most of our work, we will not be required to show that limits exist in this way. We will instead find limits using various properties of functions, such as continuity.

# 4 Determine Where a Function Is Continuous

When investigating continuity for a function of one variable in Chapter 1, we began by defining what it means for a function $f$ to be continuous at a number $c$, where $f$ is defined on an open interval containing $c$. Then we extended the definition to include continuity on intervals: open, closed, or neither.

**NEED TO REVIEW?** Continuity for functions of one variable is discussed in Section 1.3, pp. 102–108.

For functions of two or three variables, we begin by defining continuity at a point in the plane or in space. But first, we need some preliminary definitions.

---

**DEFINITION** Interior Point; Boundary Point; Open Set; Closed Set

Let $S$ be a set of points in the plane or in space.

- A point $P_0$ is called an **interior point of** $S$ if there is a $\delta$-neighborhood of $P_0$ that lies entirely in $S$.
- A point $P_0$ is called a **boundary point of** $S$ if every $\delta$-neighborhood of $P_0$ contains points both in $S$ and not in $S$. A boundary point of $S$ may be in $S$ or not in $S$.
- A set $S$ is called an **open set** if every point of $S$ is an interior point.
- A set $S$ is called a **closed set** if it contains all its boundary points.
- A set $S$ that contains some, but not all, of its boundary points is neither open nor closed.

---

Figure 26(a) shows an interior point $P_0$ in the set $S$. In Figure 26(b), the point $P_0$ is a boundary point of $S$. Figure 27 shows an open set (a), a closed set (b), and a set that is neither open nor closed (c).

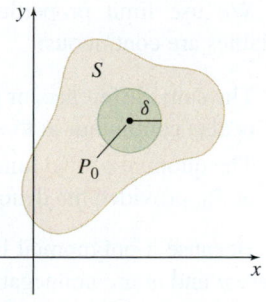
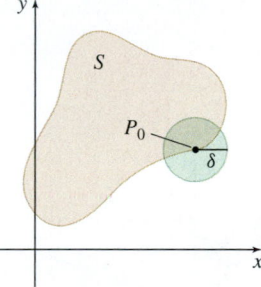

(a) There is a $\delta$-neighborhood of $P_0$ that lies entirely in $S$, so $P_0$ is an interior point of $S$.

(b) Every $\delta$-neighborhood of $P_0$ contains points both in $S$ and not in $S$, so $P_0$ is a boundary point of $S$.

**Figure 26**

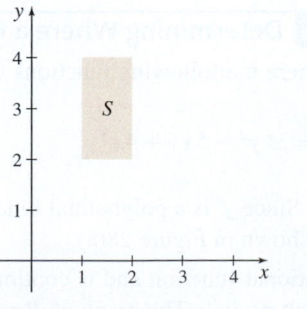
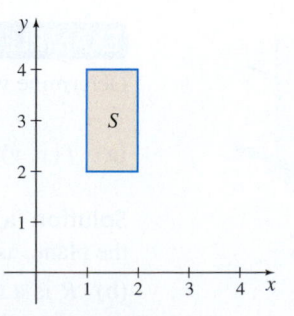
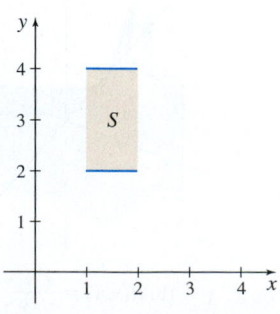

(a) $1 < x < 2$, $2 < y < 4$.
Every point in $S$ is an interior point, so $S$ is open.

(b) $1 \le x \le 2$, $2 \le y \le 4$.
$S$ contains all of its boundary points, so $S$ is closed.

(c) $1 < x < 2$, $2 \le y \le 4$.
$S$ contains some, but not all, of its boundary points, so $S$ is neither open nor closed.

**Figure 27**

**NOTE** You may find it helpful to compare the definition given here with the definition of continuity at a number given in Section 1.3, p. 102.

**DEFINITION** Continuity at a Point $P_0$

Let $f$ be a function whose domain $D$ is some subset of points in the plane, and let $P_0$ be an interior point of $D$. Then $f$ is **continuous** at $P_0$ if the following two conditions are met:

- $\lim\limits_{P \to P_0} f(P)$ exists
- $\lim\limits_{P \to P_0} f(P) = f(P_0)$

If either one of the two conditions is not met, then we say that $f$ is **discontinuous** at the point $P_0$.

If $P_0$ is a boundary point of the domain $D$ of $f$, then $f$ is continuous at $P_0$, provided $\lim\limits_{P \to P_0} f(P) = f(P_0)$ is taken to mean that given any $\varepsilon > 0$, there is a $\delta > 0$, so that whenever $P$ is in the domain of $f$ and $d(P, P_0) < \delta$, then $|f(P) - f(P_0)| < \varepsilon$.

---

**EXAMPLE 5** Showing a Function Is Continuous

Show that the function $f(x, y) = ax^n y^m$, where $n$ and $m$ are nonnegative integers and $a$ is a constant, is continuous at every point $(x_0, y_0)$ in the plane.

**Solution** For any point $(x_0, y_0)$ in the plane, we have

$$\lim_{(x,y) \to (x_0,y_0)} f(x,y) = \lim_{(x,y) \to (x_0,y_0)} (ax^n y^m) \underset{\underset{\text{Property (4)}}{\uparrow}}{=} \left[\lim_{(x,y) \to (x_0,y_0)} (ax^n)\right]\left[\lim_{(x,y) \to (x_0,y_0)} y^m\right]$$

$$\underset{\underset{\text{Properties (3), (7), and (8)}}{\uparrow}}{=} a\left[\lim_{x \to x_0} x^n\right]\left[\lim_{y \to y_0} y^m\right] = ax_0^n y_0^m$$

The conditions of the definition of continuity are met, so $f$ is continuous at $(x_0, y_0)$. ∎

We use limit properties to obtain information about where functions of two variables are continuous:

- The sum, difference, or product of two functions that are continuous at a point $P_0$ is also continuous at the point $P_0$.
- The quotient of two functions that are continuous at a point $P_0$ is also continuous at $P_0$, provided the denominator function is not 0 at $P_0$.

Because a polynomial in two variables is a sum of functions of the form $ax^n y^m$, where $n$ and $m$ are nonnegative integers and $a$ is a constant, it follows from Example 5 that

- Polynomial functions in two variables are continuous at every point $(x, y)$ in the plane.
- Rational functions, quotients of polynomial functions, are continuous on their domains, all points in the plane except those at which the denominator function is 0.

---

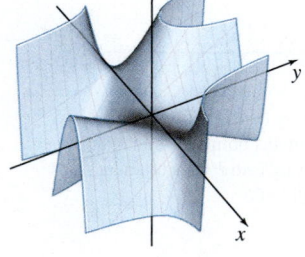

(a) $f(x, y) = 3xy^2 - 5xy + 4x^3y$

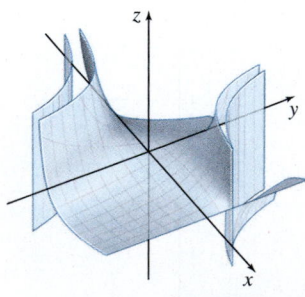

(b) $R(x, y) = \dfrac{x^2y - 2xy}{xy - 1}$

**Figure 28**

**EXAMPLE 6** Determining Where a Function Is Continuous

Determine where the following functions are continuous:

**(a)** $f(x, y) = 3xy^2 - 5xy + 4x^3y$      **(b)** $R(x, y) = \dfrac{x^2y - 2xy}{xy - 1}$

**Solution** **(a)** Since $f$ is a polynomial function, it is continuous at every point $(x, y)$ in the plane, as shown in Figure 28(a).

**(b)** $R$ is a rational function and is continuous at all points $(x, y)$ in the plane, except those for which $xy = 1$. The graph of $R$ will have no points on the cylinder $xy = 1$, as shown in Figure 28(b). ∎

**NOW WORK** Problems **37** and **39**.

**EXAMPLE 7** **Determining Where a Function Is Continuous**

(a) The function $z = e^t$ is continuous for all real numbers $t$, and the function $t = x^2 + y^2$ is continuous at every point in the plane, so the composite function $z = e^{x^2 + y^2}$ is continuous at every point in the plane.

(b) The sine function $z = \sin t$ is continuous for all real numbers $t$, and the function $t = e^{x^2 + y^2}$ is continuous for all points $(x, y)$ in the plane, so the composite function $z = \sin e^{x^2 + y^2}$ is continuous for all points $(x, y)$ in the plane.

(c) The function $z = \ln t$ is continuous for all real numbers $t > 0$, and the function $t = \dfrac{y}{x}$ is continuous for all $x \neq 0$. The composite function $z = \ln \dfrac{y}{x}$ is continuous for all $(x, y)$ for which $x > 0$, $y > 0$ or $x < 0$, $y < 0$. ∎

Figure 29 shows the graphs of the functions discussed in Example 7. Notice that the graph of the function $z = \ln \dfrac{x}{y}$ in Figure 29(c) has no points for which $x \geq 0$, $y \leq 0$ or $x \leq 0$, $y \geq 0$.

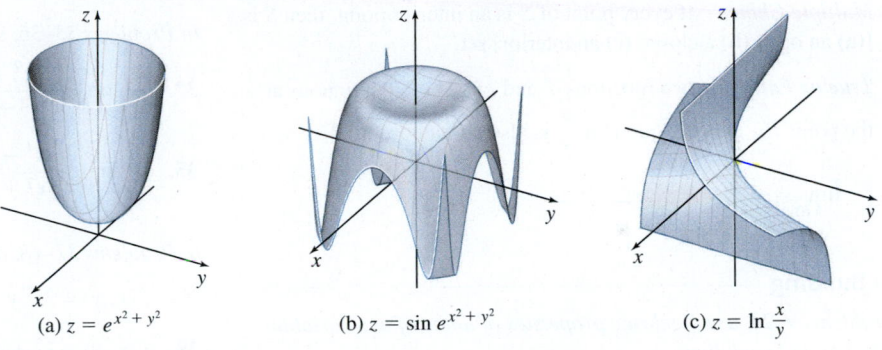

(a) $z = e^{x^2 + y^2}$          (b) $z = \sin e^{x^2 + y^2}$          (c) $z = \ln \dfrac{x}{y}$

**Figure 29**

**NOW WORK** Problem 43.

If a function is known to be continuous at a point $P_0$, its limit at $P_0$ can be found by substitution:

- If $z = f(x, y)$ is continuous at the point $(x_0, y_0)$, then

$$\lim_{(x,y) \to (x_0, y_0)} f(x, y) = f(x_0, y_0)$$

**EXAMPLE 8** **Finding Limits**

(a) $\displaystyle \lim_{(x,y) \to (0,0)} e^{x^2 + y^2} = e^{0+0} = 1$
 ↑
 Example 6(a)

(b) $\displaystyle \lim_{(x,y) \to (\sqrt{\pi},0)} \sin e^{x^2 + y^2} = \sin e^{\pi}$
 ↑
 Example 6(b)

(c) $\displaystyle \lim_{(x,y) \to (e,1)} \ln \frac{y}{x} = \ln \frac{1}{e} = \ln e^{-1} = -1$
 ↑
 Example 6(c)
 ∎

**NOW WORK** Problem 53.

## 12.2 Assess Your Understanding

### Concepts and Vocabulary

**1.** *Multiple Choice* The set of all points $P$ for which the distance $d(P, P_0) < \delta$ is called
[**(a)** a $\delta$ position **(b)** an $\varepsilon$-value **(c)** a $\delta$-neighborhood] of $P_0$.

**2.** $\lim\limits_{(x,y) \to (1,8)} (2y) = $ _____.

**3.** *True or False* One way to show that $\lim\limits_{(x,y) \to (0,0)} f(x) = L$ is to find two lines $y = m_1 x$ and $y = m_2 x$ for which $\lim\limits_{(x,y) \to (0,0)} f(x) = L$.

**4.** *True or False* In order for the limit of a function $f$ to exist at a point $P_0$, the limit of $f$ must be the same along every curve in the domain of $f$ that contains $P_0$.

**5.** *Multiple Choice* A point $P_0$ is
[**(a)** an interior **(b)** an exterior **(c)** a boundary **(d)** an isolated] point of $S$ if every $\delta$-neighborhood of $P_0$ contains points both in $S$ and not in $S$.

**6.** *Multiple Choice* If every point of $S$ is an interior point, then $S$ is
[**(a)** an open **(b)** a closed **(c)** an interior] set.

**7.** *True or False* If two functions $f$ and $g$ are both continuous at the point $P_0$, then the quotient $\dfrac{g}{f}$ is also continuous at $P_0$.

**8.** $\lim\limits_{(x,y) \to (0,0)} e^{x^2+y^2} = $ _____.

### Skill Building

*In Problems 9–24, use algebraic properties of limits of two variables (p. 863) to find each limit.*

**9.** $\lim\limits_{(x,y) \to (1,2)} (x^2 + xy - y^2 + 8)$

**10.** $\lim\limits_{(x,y) \to (-1,3)} (x^2 y + y^2 - 3xy - 2)$

**11.** $\lim\limits_{(x,y,z) \to (1,-1,2)} (3x^2 y + y^2 z)$    **12.** $\lim\limits_{(x,y,z) \to (0,1,-1)} (x^2 - y^2 z^2)$

**13.** $\lim\limits_{(x,y) \to (\pi/2,\pi)} (\sin x \cos y)$    **14.** $\lim\limits_{(x,y) \to (2,e)} (x^2 y \ln y)$

**15.** $\lim\limits_{(x,y) \to (1,5)} \dfrac{4x - xy + 4}{4y - y^2}$

**16.** $\lim\limits_{(x,y) \to (2,2)} \dfrac{x^2 + 2xy + y^2 - 9}{x + y - 3}$

**17.** $\lim\limits_{(x,y) \to (\pi,0)} \dfrac{\cos x \cos y}{x}$    **18.** $\lim\limits_{(x,y) \to (\pi,\pi)} \dfrac{\cos y(1 - \cos x)}{xy}$

**19.** $\lim\limits_{(x,y) \to (0,0)} \dfrac{e^x - 4}{e^y}$    **20.** $\lim\limits_{(x,y) \to (0,0)} \dfrac{e^x \cos y - \cos y}{e^y}$

**21.** $\lim\limits_{(x,y) \to (2,1)} \dfrac{x^2 + xy - 6y^2}{x^2 + 4y^2}$

**22.** $\lim\limits_{(x,y) \to (0,-2)} \dfrac{y^2 + xy + 4y + e^x}{xy - y + 2x - e^x}$

**23.** $\lim\limits_{(x,y) \to (2,0)} \dfrac{x^2 y + x}{x^3 y + 3xy^2 - 8}$

**24.** $\lim\limits_{(x,y) \to (0,1)} \dfrac{x^3 - 4x^2 y + 2}{xy + 4}$

*In Problems 25–32, find each limit by approaching $(0, 0)$ along:*
**(a)** *The $x$-axis.*
**(b)** *The $y$-axis.*
**(c)** *The line $y = x$.*
**(d)** *The line $y = 3x$.*
**(e)** *The parabola $y = x^2$.*

*What, if anything, can you conclude?*

**25.** $\lim\limits_{(x,y) \to (0,0)} \dfrac{3xy}{2x^2 + y^2}$    **26.** $\lim\limits_{(x,y) \to (0,0)} \dfrac{2xy}{x^2 + 3y^2}$

**27.** $\lim\limits_{(x,y) \to (0,0)} \dfrac{xy^2}{x^2 + y^3}$    **28.** $\lim\limits_{(x,y) \to (0,0)} \dfrac{2x^2 y}{3x^3 + y^2}$

**29.** $\lim\limits_{(x,y) \to (0,0)} \dfrac{3x^2 y^2}{x^4 + y^4}$    **30.** $\lim\limits_{(x,y) \to (0,0)} \dfrac{x^2}{x^2 + y^2}$

**31.** $\lim\limits_{(x,y) \to (0,0)} \dfrac{x^2 + xy}{x^2 + y^2}$    **32.** $\lim\limits_{(x,y) \to (0,0)} \dfrac{(x - y)^2}{x^2 + y^2}$

*In Problems 33–36, show that the limit does not exist.*

**33.** $\lim\limits_{(x,y) \to (0,0)} \dfrac{2x^2 + y^2}{x^2 + y^2}$    **34.** $\lim\limits_{(x,y) \to (0,0)} \dfrac{2xy}{x^2 + y^2}$

**35.** $\lim\limits_{(x,y) \to (0,0)} \dfrac{x^4 - y^2}{x^2 + y^2}$    **36.** $\lim\limits_{(x,y) \to (0,0)} \dfrac{x^2 + y^4}{x^2 + y^2}$

*In Problems 37–48, determine where each function is continuous.*

**37.** $f(x, y) = 3x^2 y - 4x^2 y^2 + 10xy^2 - 9$

**38.** $f(x, y) = x^3 + 2x^2 y + xy^2 - 4y^3$

**39.** $f(x, y) = \dfrac{x^2 - y^2}{x - y}$    **40.** $f(x, y) = \dfrac{2x^2 y + xy^2}{1 - xy}$

**41.** $f(x, y) = e^{x^2 - y^2}$    **42.** $f(x, y) = \ln(x^2 + y^2)$

**43.** $f(x, y) = \sin(x^2 - y)$    **44.** $f(x, y) = \cos \sqrt{x^2 - y}$

**45.** $f(x, y) = \sin(x + y) \cos(x - y)$

**46.** $f(x, y) = e^x \sin(xy)$    **47.** $f(x, y) = \dfrac{x + 3xy^2}{e^{x^2 - y^2}}$

**48.** $f(x, y) = \dfrac{x + 3xy^2}{\ln(x^2 + y^2)}$

*In Problems 49–62, find each limit.*

**49.** $\lim\limits_{(x,y) \to (1,0)} \dfrac{x^2 - y^2}{x - y}$    **50.** $\lim\limits_{(x,y) \to (0,0)} \dfrac{2x^2 y + xy^2}{1 - xy}$

**51.** $\lim\limits_{(x,y) \to (1,1)} e^{x^2 - y^2}$    **52.** $\lim\limits_{(x,y) \to (0,e)} \ln(x^2 + y^2)$

**53.** $\lim\limits_{(x,y) \to (\pi/2, \pi)} [\sin(x + y) \cos(x - y)]$

**54.** $\lim\limits_{(x,y) \to (0, \pi/2)} e^x \sin(xy)$    **55.** $\lim\limits_{(x,y) \to (\pi,0)} \dfrac{e^{x^2+y^2} \cos x^2}{\cos y^2}$

**56.** $\lim\limits_{(x,y) \to (\pi/2,4)} \dfrac{e^{x^2 y}}{\cos(2x)}$    **57.** $\lim\limits_{(x,y) \to (0,0)} \tan^{-1} \dfrac{e^{x+y}}{y^2 + 1}$

---

**1.** = NOW WORK problem    〔∿〕 = Graphing technology recommended    〔CAS〕 = Computer Algebra System recommended

**58.** $\lim\limits_{(x,y)\to(\pi,0)} \tan^{-1}[\cos(x+y)]$

**59.** $\lim\limits_{(x,y)\to(8,8)} \dfrac{x^{2/3} - y^{2/3}}{x^{1/3} - y^{1/3}}$

**60.** $\lim\limits_{(x,y)\to(1,1)} \dfrac{x^4 - y^4}{x^3 - y^3}$

**61.** $\lim\limits_{(x,y)\to(1,1)} \dfrac{\sqrt{xy} - 1}{xy - 1}$

**62.** $\lim\limits_{(x,y)\to(1,2)} \dfrac{\sqrt{y} - \sqrt{x+1}}{y - x - 1}$

### Applications and Extensions

*In Problems 63–66, find* $\lim\limits_{(x,y,z)\to(0,0,0)} \dfrac{2yz}{x^4 + y^2 + z^2}$ *along the indicated curves.*

**63.** the line $x = t$, $y = t$, $z = t$

**64.** the line $x = 2t$, $y = 3t$, $z = 4t$

**65.** the curve $x = t$, $y = t^2$, $z = t^2$

**66.** the line $x = at$, $y = bt$, $z = ct$, $\quad a^2 + b^2 + c^2 > 0$

*In Problems 67 and 68, use the $\varepsilon$-$\delta$ definition of a limit to prove each limit statement.*

**67.** $\lim\limits_{(x,y)\to(0,0)} \dfrac{x^2 y}{x^2 + y^2} = 0$

**68.** $\lim\limits_{(x,y)\to(0,0)} \dfrac{\sin(x^2 + y^2)}{x^2 + y^2} = 1$

*In Problems 69–76,*
*(a) Determine whether each function $f$ is continuous at $(0, 0)$.*
*(b) If $f$ is discontinuous at $(0, 0)$, is it possible to define $f(0, 0)$ so that $f$ would be continuous at $(0, 0)$?*
*(c) If the answer to (b) is yes, how should $f(0, 0)$ be defined?*

**69.** $f(x, y) = \dfrac{xy^2}{x^2 + y^2}$

**70.** $f(x, y) = \dfrac{x^2 y}{x^2 + y^2}$

**71.** $f(x, y) = \dfrac{2x^2 + y^2}{x^2 + y^2}$

**72.** $f(x, y) = \dfrac{x^4 - y^2}{x^2 + y^2}$

**73.** $f(x, y) = \begin{cases} \dfrac{3xy}{x^2 + y^2} & \text{if} \quad (x, y) \neq (0, 0) \\ 0 & \text{if} \quad (x, y) = (0, 0) \end{cases}$

**74.** $f(x, y) = \begin{cases} \dfrac{\sin(xy)}{x^2 + y^2} & \text{if} \quad (x, y) \neq (0, 0) \\ 1 & \text{if} \quad (x, y) = (0, 0) \end{cases}$

**75.** $f(x, y) = \begin{cases} \dfrac{\sin(x^2 + y^2)}{x^2 + y^2} & \text{if} \quad (x, y) \neq (0, 0) \\ 1 & \text{if} \quad (x, y) = (0, 0) \end{cases}$

**76.** $f(x, y) = \begin{cases} \dfrac{\sin(x^2 - y^2)}{x^2 + y^2} & \text{if} \quad (x, y) \neq (0, 0) \\ 1 & \text{if} \quad (x, y) = (0, 0) \end{cases}$

*In Problems 77–80, find each limit by converting to polar coordinates.*
*Hint: $r^2 = x^2 + y^2$ and $(x, y) \to (0, 0)$ is equivalent to $r \to 0$.*

**77.** $\lim\limits_{(x,y)\to(0,0)} \dfrac{x^2 y}{x^2 + y^2}$

**78.** $\lim\limits_{(x,y)\to(0,0)} \dfrac{xy^2}{x^2 + y^2}$

**79.** $\lim\limits_{(x,y)\to(0,0)} \dfrac{\cos(x^2 + y^2)}{x^2 + y^2}$

**80.** $\lim\limits_{(x,y)\to(0,0)} \dfrac{\sin(x^2 + y^2)}{x^2 + y^2}$

### Challenge Problems

*In Problems 81 and 82, show the limits do not exist.*

**81.** $\lim\limits_{(x,y,z)\to(0,0,0)} \dfrac{4xy}{x^2 + y^2 + z^2}$

**82.** $\lim\limits_{(x,y,z)\to(0,0,0)} \dfrac{xyz}{x^3 + y^3 + z^3}$

**83.** Find $\lim\limits_{(x,y)\to(0,0)} f(x, y)$, if

$$f(x, y) = \dfrac{x^3 - 4x^2 y + 4xy^2 + 5x - 10y}{x - 2y}$$

---

# 12.3 Partial Derivatives

**OBJECTIVES**  *When you finish this section, you should be able to:*

**1** Find the partial derivatives of a function of two variables (p. 871)
**2** Interpret partial derivatives as the slope of a tangent line (p. 874)
**3** Interpret partial derivatives as a rate of change (p. 875)
**4** Find second-order partial derivatives (p. 878)
**5** Find the partial derivatives of a function of $n$ variables (p. 879)

## 1 Find the Partial Derivatives of a Function of Two Variables

Suppose $z = f(x, y)$ is a function of two variables $x$ and $y$. If $y_0$ is a constant, then the function $z = f(x, y) = f(x, y_0)$ is a function of the single variable $x$. If $z = f(x, y_0)$ is differentiable at $x = x_0$, then the derivative of the function $f(x, y_0)$ at $x_0$ is called the *partial derivative of $f$ with respect to $x$ at $(x_0, y_0)$* and is denoted by $f_x(x_0, y_0)$.

Similarly, if $x_0$ is a constant, then $z = f(x_0, y)$ is a function of the single variable $y$. If $z = f(x_0, y)$ is differentiable at $y = y_0$, then its derivative is called the *partial derivative of $f$ with respect to $y$ at $(x_0, y_0)$* and is denoted by $f_y(x_0, y_0)$.

**DEFINITION Partial Derivatives**

Let $z = f(x, y)$ be a function of two variables, and let $(x, y)$ be any interior point* of the domain of $f$. The **first-order partial derivative of $f$ with respect to $x$** and the **first-order partial derivative of $f$ with respect to $y$** are the functions $f_x$ and $f_y$ defined by

$$f_x(x, y) = \lim_{\Delta x \to 0} \frac{f(x + \Delta x, y) - f(x, y)}{\Delta x}$$

$$f_y(x, y) = \lim_{\Delta y \to 0} \frac{f(x, y + \Delta y) - f(x, y)}{\Delta y}$$

provided these limits exist.

---

**EXAMPLE 1** Using the Definition to Find Partial Derivatives of a Function of Two Variables

Use the definition to find the partial derivatives of the function $f(x, y) = 2x^2 + 3xy$.

**Solution** Since the function $f$ is a polynomial, its domain is all the points in the $xy$-plane, so every point $(x, y)$ is an interior point of the domain of $f$. The partial derivatives of $f$ at any point $(x, y)$ are

$$f_x(x, y) = \lim_{\Delta x \to 0} \frac{f(x + \Delta x, y) - f(x, y)}{\Delta x} = \lim_{\Delta x \to 0} \frac{\left[2(x + \Delta x)^2 + 3(x + \Delta x)y\right] - (2x^2 + 3xy)}{\Delta x}$$

$$= \lim_{\Delta x \to 0} \frac{2x^2 + 4x\Delta x + 2\Delta x^2 + 3xy + 3\Delta xy - 2x^2 - 3xy}{\Delta x}$$

$$= \lim_{\Delta x \to 0} \frac{4x\Delta x + 2\Delta x^2 + 3\Delta xy}{\Delta x} = \lim_{\Delta x \to 0} \frac{(4x + 2\Delta x + 3y)\Delta x}{\Delta x}$$

$$= \lim_{\Delta x \to 0} (4x + 2\Delta x + 3y) = 4x + 3y$$

$$f_y(x, y) = \lim_{\Delta y \to 0} \frac{f(x, y + \Delta y) - f(x, y)}{\Delta y} = \lim_{\Delta y \to 0} \frac{\left[2x^2 + 3x(y + \Delta y)\right] - (2x^2 + 3xy)}{\Delta y}$$

$$= \lim_{\Delta y \to 0} \frac{2x^2 + 3xy + 3x\Delta y - 2x^2 - 3xy}{\Delta y} = \lim_{\Delta y \to 0} \frac{3x\Delta y}{\Delta y} = \lim_{\Delta y \to 0} 3x = 3x \qquad \blacksquare$$

**NOW WORK** Problem 7.

**NEED TO REVIEW?** The definition of the derivative is discussed in Section 2.2, p. 163.

It is not usually required to use the definition to find partial derivatives. Finding most partial derivatives is relatively easy once we notice the similarity between the definitions of partial derivatives above and the definition of a derivative as the limit of a difference quotient. In Chapter 2, we defined the derivative of a function $y = f(x)$ as $f'(x) = \lim_{h \to 0} \dfrac{f(x + h) - f(x)}{h}$. In $f_x(x, y)$, the increment $\Delta x$ is added to $x$, while $y$ is treated as a constant, and in $f_y(x, y)$, the increment $\Delta y$ is added to $y$, while $x$ is treated as a constant.

**IN WORDS** We find the partial derivative of $f$ with respect to $x$ by differentiating $f$ with respect to $x$ while treating $y$ as if it were a constant.

We find the partial derivative of $f$ with respect to $y$ by differentiating $f$ with respect to $y$ while treating $x$ as if it were a constant.

Example 2 below shows how this observation allows us to find partial derivatives. Just as the derivative of a function $y = f(x)$ is itself a function, the partial derivatives $f_x$ and $f_y$ of a function $z = f(x, y)$ are functions, generally of two variables. The Sum, Difference, Product, Quotient, and Chain Rules used for finding the derivative of a function $f$ of one variable also are used to find partial derivatives.

**EXAMPLE 2** Finding the Partial Derivatives of a Function of Two Variables

For each function $z = f(x, y)$, find $f_x(x, y)$ and $f_y(x, y)$.

**(a)** $f(x, y) = 3x^2y + 2x - 3y$      **(b)** $f(x, y) = x \sin y + y \sin x$

---

*Although partial derivatives may be defined at boundary points of the domain of $f$, they are more difficult to handle and are not discussed in this book.

**Solution (a)** To find $f_x(x, y)$, treat $y$ as a constant in $f(x, y) = 3x^2y + 2x - 3y$ and differentiate with respect to $x$. The result is

$$f_x(x, y) = 6xy + 2$$

To find $f_y(x, y)$, treat $x$ as a constant and differentiate with respect to $y$. The result is

$$f_y(x, y) = 3x^2 - 3$$

**(b)** For $f(x, y) = x \sin y + y \sin x$, we have

$$f_x(x, y) = \underset{\substack{\uparrow \\ \text{Treat } y \text{ as a constant.}}}{\sin y + y \cos x} \qquad\qquad f_y(x, y) = \underset{\substack{\uparrow \\ \text{Treat } x \text{ as a constant.}}}{x \cos y + \sin x} \qquad ■$$

NOW WORK Problem 19.

---

EXAMPLE 3  **Finding the Partial Derivatives of a Function at a Point $(x_0, y_0)$**

For $f(x, y) = x \sin y + y \sin x$, find $f_x\left(\dfrac{\pi}{3}, \dfrac{\pi}{6}\right)$ and $f_y\left(\dfrac{\pi}{3}, \dfrac{\pi}{6}\right)$.

**Solution** Use the results from Example 2(b):

$$f_x(x, y) = \sin y + y \cos x$$

$$f_x\left(\frac{\pi}{3}, \frac{\pi}{6}\right) = \sin \frac{\pi}{6} + \frac{\pi}{6} \cos \frac{\pi}{3} = \frac{1}{2} + \frac{\pi}{6} \cdot \frac{1}{2} = \frac{6 + \pi}{12}$$

$$f_y(x, y) = x \cos y + \sin x$$

$$f_y\left(\frac{\pi}{3}, \frac{\pi}{6}\right) = \frac{\pi}{3} \cos \frac{\pi}{6} + \sin \frac{\pi}{3} = \frac{\pi}{3} \cdot \frac{\sqrt{3}}{2} + \frac{\sqrt{3}}{2} = \frac{(\pi + 3)\sqrt{3}}{6} \qquad ■$$

**Another Notation**

Another notation used for the partial derivatives of a function $z = f(x, y)$ is

$$f_x(x, y) = \frac{\partial f}{\partial x} = \frac{\partial z}{\partial x} \qquad \text{and} \qquad f_y(x, y) = \frac{\partial f}{\partial y} = \frac{\partial z}{\partial y}$$

The symbols $\dfrac{\partial}{\partial x}$ and $\dfrac{\partial}{\partial y}$ are read as "the partial with respect to $x$" and "the partial with respect to $y$," respectively. They denote operations on a function to obtain the partial derivatives with respect to $x$ in the case of $\dfrac{\partial}{\partial x}$ and with respect to $y$ in the case of $\dfrac{\partial}{\partial y}$.

For example, $\dfrac{\partial}{\partial x}(e^x \cos y) = e^x \cos y$ and $\dfrac{\partial}{\partial y}(e^x \cos y) = -e^x \sin y$.

---

EXAMPLE 4  **Using the Chain Rule to Find Partial Derivatives**

Find the partial derivatives $f_x(x, y)$ and $f_y(x, y)$ of each function.

**(a)** $f(x, y) = e^{x^2 + y^2}$        **(b)** $f(x, y) = \tan^{-1}\left(1 + \dfrac{y}{x}\right)$

**Solution (a)** To find $f_x(x, y)$, treat $y$ as a constant and differentiate with respect to $x$ using the Chain Rule.

$$f_x(x, y) = e^{x^2 + y^2} \frac{\partial}{\partial x}(x^2 + y^2) = 2xe^{x^2 + y^2} \qquad \frac{d}{dx}e^{u(x)} = e^{u(x)}\frac{du}{dx}$$

To find $f_y(x, y)$, treat $x$ as a constant and differentiate with respect to $y$ using the Chain Rule.

$$f_y(x, y) = e^{x^2 + y^2} \frac{\partial}{\partial y}(x^2 + y^2) = 2ye^{x^2 + y^2}$$

**(b)** For $f(x, y) = \tan^{-1}\left(1 + \dfrac{y}{x}\right)$,

$$f_x(x, y) = \frac{1}{1 + \left(1 + \dfrac{y}{x}\right)^2} \cdot \frac{\partial}{\partial x}\left(1 + \frac{y}{x}\right) = \frac{1}{1 + \left(1 + \dfrac{y}{x}\right)^2} \cdot \left(-\frac{y}{x^2}\right)$$

$$= -\frac{1}{1 + 1 + 2\dfrac{y}{x} + \dfrac{y^2}{x^2}} \cdot \frac{y}{x^2} = -\frac{y}{2x^2 + 2xy + y^2}$$

and

$$f_y(x, y) = \frac{1}{1 + \left(1 + \dfrac{y}{x}\right)^2} \cdot \frac{\partial}{\partial y}\left(1 + \frac{y}{x}\right) = \frac{1}{1 + \left(1 + \dfrac{y}{x}\right)^2} \cdot \frac{1}{x}$$

$$= \frac{x}{2x^2 + 2xy + y^2}$$    ∎

**NOW WORK** Problem 21.

## 2 Interpret Partial Derivatives as the Slope of a Tangent Line

Suppose $z = f(x, y)$ is a function of two variables for which both $f_x(x, y)$ and $f_y(x, y)$ exist. When we find $f_x(x, y)$, the variable $y$ is held fixed, say, at $y = y_0$, and $f$ is differentiated with respect to $x$. Holding $y$ fixed at $y_0$ is equivalent to intersecting the surface $z = f(x, y)$ with the plane $y = y_0$. This intersection is the curve $z = f(x, y_0)$ shown in Figure 30. Since $f_x(x, y_0)$ is the derivative of the function $z = f(x, y_0)$ of one variable, then $f_x(x, y_0)$ is the slope of the tangent line to the curve $z = f(x, y_0)$.

Similarly, Figure 31 shows that $f_y(x_0, y)$ is the slope of the tangent line to the curve $z = f(x_0, y)$, the curve of intersection of the plane $x = x_0$ and the surface $z = f(x, y)$.

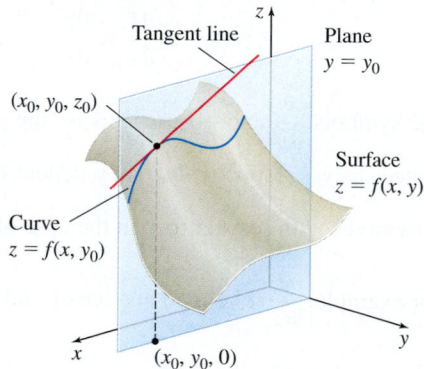

**DF** **Figure 30** $f_x(x_0, y_0)$ is the slope of the tangent line to the curve $z = f(x, y_0)$ at $(x_0, y_0, z_0)$.

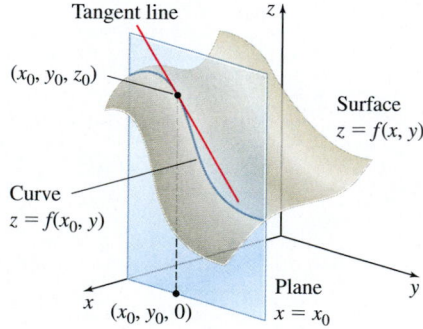

**DF** **Figure 31** $f_y(x_0, y_0)$ is the slope of the tangent line to the curve $z = f(x_0, y)$ at $(x_0, y_0, z_0)$.

**THEOREM** Slope of a Tangent Line

The partial derivative $f_x(x_0, y_0)$ equals the **slope of the tangent line** to the curve of intersection of the surface $z = f(x, y)$ and the plane $y = y_0$ at the point $(x_0, y_0, z_0)$ on the surface $z = f(x, y)$.

The partial derivative $f_y(x_0, y_0)$ equals the **slope of the tangent line** to the curve of intersection of the surface $z = f(x, y)$ and the plane $x = x_0$ at the point $(x_0, y_0, z_0)$ on the surface $z = f(x, y)$.

### EXAMPLE 5   Finding an Equation of a Tangent Line

**CALC CLIP**

Find an equation of the tangent line to the curve of intersection of the surface $z = f(x, y) = 16 - x^2 - y^2$:

(a) With the plane $y = 2$ at the point $(1, 2, 11)$.
(b) With the plane $x = 1$ at the point $(1, 2, 11)$.

**Solution** The surface $z = f(x, y) = 16 - x^2 - y^2$ is a paraboloid.

(a) The intersection of the plane $y = 2$ and the paraboloid $z = 16 - x^2 - y^2$ is the parabola $z = 12 - x^2$, $y = 2$. With $y$ held fixed at $y = 2$, the slope of the tangent line to the parabola at any point is $f_x(x, 2) = -2x$. At the point $(1, 2, 11)$, the slope of the tangent line is $f_x(1, 2) = -2 \cdot 1 = -2$. An equation of the tangent line to the parabola $z = 12 - x^2$, $y = 2$ is

$$z - 11 = -2(x - 1) \qquad y = 2$$

**NEED TO REVIEW?** Symmetric equations of a line in space are discussed in Section 10.6, pp. 774–775.

This representation of the equation of the tangent line is essentially the symmetric equations of a line in space. See Figure 32(a).

(b) The intersection of the plane $x = 1$ and the paraboloid $z = 16 - x^2 - y^2$ is the parabola $z = 15 - y^2$, $x = 1$. With $x$ held fixed at $x = 1$, the slope of the tangent line to the parabola at any point is $f_y(1, y) = -2y$. At the point $(1, 2, 11)$, the slope of the tangent line is $f_y(1, 2) = -2 \cdot 2 = -4$. An equation of the tangent line to the parabola $z = 15 - y^2$, $x = 1$ is

$$z - 11 = -4(y - 2) \qquad x = 1$$

See Figure 32(b). ∎

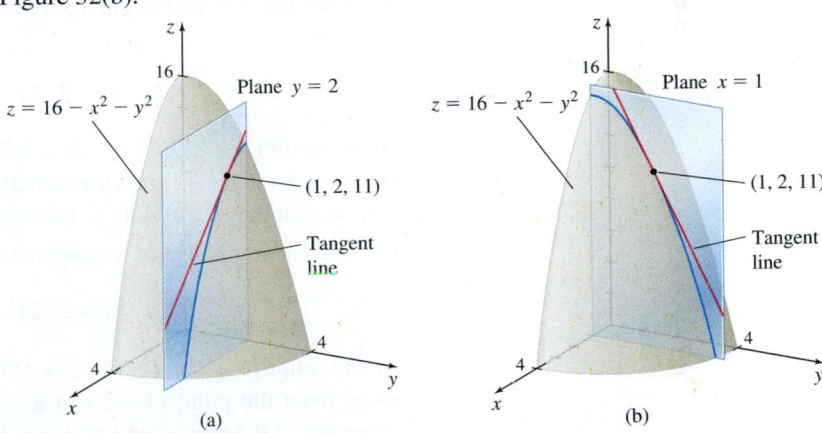

**Figure 32**

**NOW WORK** Problem 57.

## 3 Interpret Partial Derivatives as a Rate of Change

The definition of the partial derivatives of $z = f(x, y)$ is a generalization of the definition of the derivative of a function of one variable. It is not surprising, therefore, to see similarities in the interpretations. For example, $f_x(x, y)$ equals the rate of change of $f$ in the direction of the positive $x$-axis (while $y$ is held fixed), as shown in Figure 33.

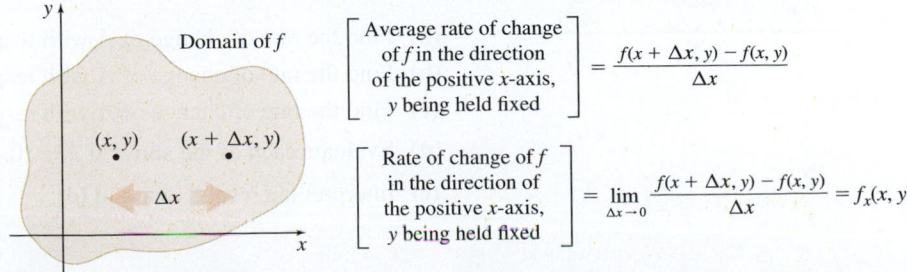

**Figure 33**

Similarly, $f_y(x, y)$ equals the rate of change of $f$ in the direction of the positive $y$-axis ($x$ is held fixed), as shown in Figure 34.

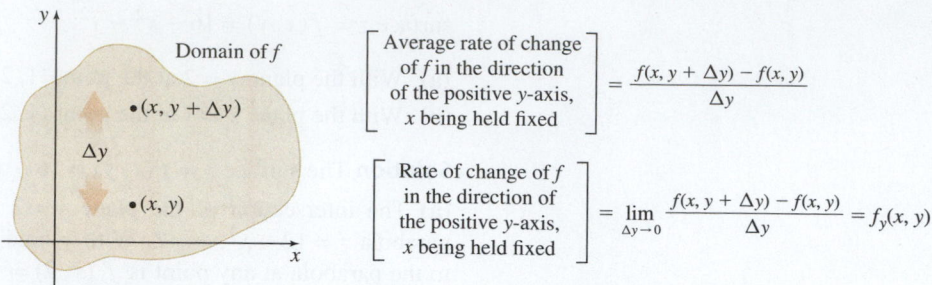

$$\begin{bmatrix} \text{Average rate of change} \\ \text{of } f \text{ in the direction} \\ \text{of the positive } y\text{-axis,} \\ x \text{ being held fixed} \end{bmatrix} = \frac{f(x, y + \Delta y) - f(x, y)}{\Delta y}$$

$$\begin{bmatrix} \text{Rate of change of } f \\ \text{in the direction of} \\ \text{the positive } y\text{-axis,} \\ x \text{ being held fixed} \end{bmatrix} = \lim_{\Delta y \to 0} \frac{f(x, y + \Delta y) - f(x, y)}{\Delta y} = f_y(x, y)$$

**Figure 34**

**NOTE** In Chapter 13, we discuss rates of change in any direction.

**EXAMPLE 6   Finding the Rate of Change of Temperature**

The temperature $T$ (in degrees Celsius) of a metal plate, located in the $xy$-plane, at any point $(x, y)$ is given by $T = T(x, y) = 24(x^2 + y^2)^2$.

**(a)** Find the rate of change of $T$ in the direction of the positive $x$-axis at the point $(1, -2)$. Interpret the result in the context of the problem.

**(b)** Find the rate of change of $T$ in the direction of the positive $y$-axis at the point $(1, -2)$. Interpret the result in the context of the problem.

**Solution (a)** The rate of change of temperature in the direction of the positive $x$-axis is given by

$$T_x(x, y) = 2 \cdot 24 \cdot (x^2 + y^2) \cdot 2x = 96x(x^2 + y^2)$$

At the point $(1, -2)$, $T_x(1, -2) = 96 \cdot 1 \cdot 5 = 480$. This means that as one moves away from the point $(1, -2)$ in a horizontal direction to the right, the temperature of the plate increases at the rate of $480\,°C$ per unit of distance.

**(b)** The rate of change of temperature in the direction of the positive $y$-axis is given by

$$T_y(x, y) = 2 \cdot 24 \cdot (x^2 + y^2) \cdot 2y = 96y(x^2 + y^2)$$

At the point $(1, -2)$, $T_y(1, -2) = 96(-2)(5) = -960$. This means that as one moves away from the point $(1, -2)$ in a vertical direction up, the temperature of the plate decreases at the rate of $960\,°C$ per unit of distance. ∎

**NOW WORK** Problem 63.

**EXAMPLE 7   Finding Rates of Change**

See Figure 35. The area $A$ of a triangle with sides $a$ and $b$ and included angle $\theta$ is given by

$$A = \frac{1}{2}ab \sin \theta$$

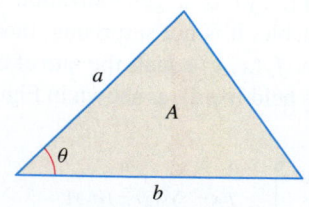

**Figure 35**

**(a)** Find the rate of change of $A$ with respect to $a$ when $b$ and $\theta$ are held fixed.

**(b)** Find the rate of change of $A$ with respect to $b$ when $a$ and $\theta$ are held fixed.

**(c)** Find the rate of change of $A$ with respect to $\theta$ when $a$ and $b$ are held fixed.

**(d)** Evaluate each of the above if $a = 20$, $b = 30$, and $\theta = \dfrac{\pi}{6}$.

**(e)** Interpret the results found in (d).

**Solution (a)** $\dfrac{\partial A}{\partial a} = \dfrac{1}{2} b \sin \theta$  **(b)** $\dfrac{\partial A}{\partial b} = \dfrac{1}{2} a \sin \theta$  **(c)** $\dfrac{\partial A}{\partial \theta} = \dfrac{1}{2} ab \cos \theta$

**(d)** When $a = 20$, $b = 30$, and $\theta = \dfrac{\pi}{6}$, then

$$\frac{\partial A}{\partial a} = \frac{1}{2} \cdot 30 \cdot \sin \frac{\pi}{6} = \frac{15}{2}$$

$$\frac{\partial A}{\partial b} = \frac{1}{2} \cdot 20 \cdot \sin \frac{\pi}{6} = 5$$

$$\frac{\partial A}{\partial \theta} = \frac{1}{2} \cdot 20 \cdot 30 \cdot \cos \frac{\pi}{6} = 150\sqrt{3}$$

**(e)** Each partial derivative equals the rate of change of area with respect to $a$, $b$, or $\theta$. The relative size of each rate of change tells us that the area increases most rapidly when $a$ and $b$ are fixed and $\theta$ varies. The area increases the least when $a$ and $\theta$ are fixed and $b$ varies. ∎

In business and economics, analysts are often interested in the rate of change of a process, called the **marginal** process. For example, if the amount $z$ of a good produced depends on the quantities of the $x$ and $y$ components used, then the **production function** $z = f(x, y)$ gives the amount of **output** for the amounts $x$ and $y$ of **input** used. The **marginal productivity** of $z$ with respect to an input $x$ is $f_x$ and the marginal productivity of $z$ with respect to an input $y$ is $f_y$.

### EXAMPLE 8  Finding Marginal Productivity: The Cobb–Douglas Model

**ORIGINS** Charles Wiggins Cobb (1875–1949) and Paul H. Douglas (1892–1976) were American scholars who studied the growth of the U.S. economy from 1899 to 1922. They produced a simple model, known as the Cobb–Douglas model, in which production output is a function of the amount of labor used and the amount of capital invested. Later, Cobb continued to publish in mathematics (he obtained a PhD from the University of Michigan) and in economics. Douglas went into politics; he represented Illinois in the U.S. Senate from 1949 to 1966.

In 1928, the mathematician Charles Cobb and the economist Paul Douglas empirically (from data) derived a production model for the manufacturing sector of the U.S. economy for the period 1899–1922. Using the model $P = aK^b L^{1-b}$, where $P$ is manufacturing productivity, $K$ is capital input, and $L$ is labor input, and multiple regression techniques, Cobb and Douglas determined that manufacturing productivity was represented by the function

$$P = 1.014651 K^{0.254124} L^{0.745876} \approx 1.01 K^{0.25} L^{0.75}$$

**(a)** Find the marginal productivity of manufacturing output with respect to capital input. Interpret the result.

**(b)** Find the marginal productivity of manufacturing output with respect to labor input. Interpret the result.

**Solution (a)** The marginal productivity of manufacturing output with respect to capital input $K$ is

$$\frac{\partial P}{\partial K} \approx 1.01(0.25 K^{-0.75}) L^{0.75} = 0.2525 \left( \frac{L}{K} \right)^{0.75}$$

For every unit increase in capital input, there is an increase of $0.2525 \left( \dfrac{L}{K} \right)^{0.75}$ units in manufacturing productivity.

**(b)** The marginal productivity of manufacturing output with respect to labor input $L$ is

$$\frac{\partial P}{\partial L} \approx 1.01 K^{0.25} \left( 0.75 L^{-0.25} \right) = 0.7575 \left( \frac{K}{L} \right)^{0.25}$$

For every unit increase in labor input, there is an increase of $0.7575 \left( \dfrac{K}{L} \right)^{0.25}$ units in manufacturing productivity. ∎

Marginal productivity is usually positive. That is, as the amount of one input increases (with the other input held constant), the output also increases. However, as the input of one component increases, the output usually increases at a decreasing rate until the point is reached where there is no further increase in output. This is called the **law of eventually diminishing marginal productivity**.

NOW WORK Problem 67.

## 4 Find Second-Order Partial Derivatives

If the first-order partial derivatives of a function $z = f(x, y)$ of two variables exist, and if it is possible to differentiate each of these with respect to $x$ or $y$, the result is four **second-order partial derivatives.** They are

$$f_{xx}(x, y) = \frac{\partial}{\partial x} f_x(x, y) = \frac{\partial}{\partial x} \frac{\partial z}{\partial x} = \frac{\partial^2 z}{\partial x^2} \qquad f_{xy}(x, y) = \frac{\partial}{\partial y} f_x(x, y) = \frac{\partial}{\partial y} \frac{\partial z}{\partial x} = \frac{\partial^2 z}{\partial y \, \partial x}$$

$$f_{yx}(x, y) = \frac{\partial}{\partial x} f_y(x, y) = \frac{\partial}{\partial x} \frac{\partial z}{\partial y} = \frac{\partial^2 z}{\partial x \, \partial y} \qquad f_{yy}(x, y) = \frac{\partial}{\partial y} f_y(x, y) = \frac{\partial}{\partial y} \frac{\partial z}{\partial y} = \frac{\partial^2 z}{\partial y^2}$$

The two second-order partial derivatives

$$\frac{\partial^2 z}{\partial x \, \partial y} = f_{yx}(x, y) \qquad \text{and} \qquad \frac{\partial^2 z}{\partial y \, \partial x} = f_{xy}(x, y)$$

are called **mixed partials.**

CAUTION    Notice the differences in the two mixed partial derivatives. To find $\dfrac{\partial^2 z}{\partial x \partial y} = f_{yx}$, first differentiate $f$ with respect to $y$ and then differentiate $f_y$ with respect to $x$, in that order. On the other hand, to find $\dfrac{\partial^2 z}{\partial y \, \partial x} = f_{xy}$, first differentiate $f$ with respect to $x$ and then differentiate $f_x$ with respect to $y$.

### EXAMPLE 9    Finding Second-Order Partial Derivatives

Find the four second-order partial derivatives of $z = f(x, y) = x \ln y + ye^x$.

**Solution** We begin by finding the first-order partial derivatives $f_x(x, y)$ and $f_y(x, y)$.

$$f_x(x, y) = \ln y + ye^x \qquad f_y(x, y) = \frac{x}{y} + e^x$$

Then the second-order partial derivatives are

$$f_{xx}(x, y) = \frac{\partial}{\partial x} f_x(x, y) = \frac{\partial}{\partial x} (\ln y + ye^x) = ye^x \qquad f_{xy}(x, y) = \frac{\partial}{\partial y} f_x(x, y) = \frac{\partial}{\partial y} (\ln y + ye^x) = \frac{1}{y} + e^x$$

$$f_{yx}(x, y) = \frac{\partial}{\partial x} f_y(x, y) = \frac{\partial}{\partial x} \left( \frac{x}{y} + e^x \right) = \frac{1}{y} + e^x \qquad f_{yy}(x, y) = \frac{\partial}{\partial y} f_y(x, y) = \frac{\partial}{\partial y} \left( \frac{x}{y} + e^x \right) = -\frac{x}{y^2} \qquad ■$$

Notice in Example 9 that the mixed partials $f_{xy}$ and $f_{yx}$ are equal. As it turns out, the mixed partials are equal for most functions we encounter. The next result gives conditions for the mixed partials to be equal.

NOW WORK Problem 39.

### THEOREM    Equality of Mixed Partials (Clairaut's Theorem)

Let $z = f(x, y)$ be a function of two variables whose domain is the set $D$. Let $(x_0, y_0)$ be an interior point of $D$. If the partial derivatives $f_x$, $f_y$, $f_{xy}$, and $f_{yx}$ exist at each point of some disk centered at $(x_0, y_0)$, and if $f_{xy}$ and $f_{yx}$ are continuous at $(x_0, y_0)$, then $f_{xy}(x_0, y_0) = f_{yx}(x_0, y_0)$.

A proof of Clairaut's Theorem may be found in most advanced calculus texts. For examples of functions for which the mixed partials are not equal, see Problems 93 and 94.

## Continuity and Partial Derivatives

In Chapter 2, we proved the important result that if a function $f$ of one variable has a derivative at $c$, then $f$ is continuous at $c$. This property is not necessarily true for a function $f$ of several variables. That is, if a function $f$ of several variables has partial derivatives at a point $P_0$, the function $f$ may fail to be continuous at $P_0$.

Smithsonian Institution Libraries/Science Source

CLAIRAUT.

**ORIGINS** Alexis Claude Clairaut (1713–1765) was home-schooled by his father, a respected mathematician. As a result, Clairaut completed algebra when he was 9, produced his first scholarly paper at age 13, and was the youngest person admitted to the Paris Academy of Sciences. When he was about 30, Clairaut began to study the three-body problem, seeking a solution to the motion of three masses that attract each other according to Newton's Law. He applied this theory to the orbit of Halley's comet, and was able to predict the return of the comet to within weeks.

**EXAMPLE 10** **Showing That the Existence of Partial Derivatives Does Not Imply Continuity**

Show that the function

$$f(x, y) = \begin{cases} \dfrac{xy}{x^2 + y^2} & \text{if } (x, y) \neq (0, 0) \\ 0 & \text{if } (x, y) = (0, 0) \end{cases}$$

has partial derivatives at $(0, 0)$, but is not continuous at $(0, 0)$.

**Solution** Use the definition to find the partial derivatives of $f$ at $(0, 0)$.

$$f_x(0, 0) = \lim_{\Delta x \to 0} \frac{f(0 + \Delta x, 0) - f(0, 0)}{\Delta x} = \lim_{\Delta x \to 0} \frac{f(\Delta x, 0)}{\Delta x} = \lim_{\Delta x \to 0} \frac{0}{\Delta x} = \lim_{\Delta x \to 0} 0 = 0$$

$$\uparrow f(0,0) = 0 \qquad f(\Delta x, 0) = \frac{\Delta x \cdot 0}{(\Delta x)^2} = 0$$

$$f_y(0, 0) = \lim_{\Delta y \to 0} \frac{f(0, 0 + \Delta y) - f(0, 0)}{\Delta y} = \lim_{\Delta y \to 0} \frac{f(0, \Delta y)}{\Delta y} = \lim_{\Delta y \to 0} \frac{0}{\Delta y} = \lim_{\Delta y \to 0} 0 = 0$$

$$\uparrow f(0,0) = 0 \qquad f(0, \Delta y) = \frac{0 \cdot \Delta y}{(\Delta y)^2} = 0$$

The partial derivatives $f_x(0, 0)$ and $f_y(0, 0)$ both exist.

But $\displaystyle\lim_{(x,y) \to (0,0)} f(x, y) = \lim_{(x,y) \to (0,0)} \frac{xy}{x^2 + y^2}$ does not exist. (See Example 2 from Section 12.2.) As a result, $f$ is not continuous at $(0, 0)$. ∎

**NOW WORK** Problem 51.

## 5 Find the Partial Derivatives of a Function of $n$ Variables

The idea of partial differentiation can be extended to functions of three or more variables. If $w = f(x, y, z)$ is a function of three variables, there will be three first-order partial derivatives:

- The partial derivative of $w$ with respect to $x$ is $\dfrac{\partial w}{\partial x} = f_x(x, y, z)$.

- The partial derivative of $w$ with respect to $y$ is $\dfrac{\partial w}{\partial y} = f_y(x, y, z)$.

- The partial derivative of $w$ with respect to $z$ is $\dfrac{\partial w}{\partial z} = f_z(x, y, z)$.

Each partial derivative is found by differentiating with respect to the indicated variable, while treating the other two variables as constants.

The function $f_x$ equals the rate of change of $w$ with respect to $x$; the function $f_y$ equals the rate of change of $w$ with respect to $y$; and the function $f_z$ equals the rate of change of $w$ with respect to $z$.

### EXAMPLE 11 Finding the Partial Derivatives of a Function of Three Variables

If $w = f(x, y, z) = 5x^2yz^3$, then

$$f_x(x, y, z) = 10xyz^3 \qquad f_y(x, y, z) = 5x^2z^3 \qquad f_z(x, y, z) = 15x^2yz^2 \qquad \blacksquare$$

**NOW WORK** Problem 43.

Higher-order partial derivatives of a function $w = f(x, y, z)$ of three variables are defined in a similar way to those of a function of two variables. For example,

$$f_{zy} = \frac{\partial}{\partial y}\left(\frac{\partial w}{\partial z}\right) = \frac{\partial^2 w}{\partial y\,\partial z} \qquad \text{and} \qquad f_{xz} = \frac{\partial}{\partial z}\left(\frac{\partial w}{\partial x}\right) = \frac{\partial^2 w}{\partial z\,\partial x}$$

As in the case of functions of two variables, for most functions $w = f(x, y, z)$, the mixed partials have the property that $f_{xy} = f_{yx}$ and $f_{xz} = f_{zx}$ and $f_{yz} = f_{zy}$.

If $z = f(x_1, x_2, \ldots, x_n)$ is a function of $n$ variables, there are $n$ partial derivatives of $f$. The partial derivative of $f$ with respect to $x_1$, $\dfrac{\partial f}{\partial x_1}$, is found by differentiating $f$ with respect to $x_1$, while treating the remaining variables as constants. The other partial derivatives $\dfrac{\partial f}{\partial x_2}, \dfrac{\partial f}{\partial x_3}, \ldots, \dfrac{\partial f}{\partial x_n}$ are defined similarly.

### EXAMPLE 12 Finding the Partial Derivatives of a Function of n Variables

If $z = f(x_1, x_2, \ldots, x_n) = x_1^2 + x_2^2 + \cdots + x_n^2$, then

$$\frac{\partial f}{\partial x_1} = 2x_1 \qquad \frac{\partial f}{\partial x_2} = 2x_2 \qquad \cdots \qquad \frac{\partial f}{\partial x_n} = 2x_n$$

Collectively, these partial derivatives can be written as

$$\frac{\partial f}{\partial x_i} = 2x_i \qquad \text{where } i = 1, 2, \ldots, n \qquad \blacksquare$$

## 12.3 Assess Your Understanding

### Concepts and Vocabulary

1. *True or False* To find the partial derivative $f_y(x, y)$, differentiate $f$ with respect to $x$ while treating $y$ as if it were a constant.

2. *Multiple Choice* The partial derivative $f_x(x_0, y_0)$ equals the slope of the tangent line to the curve of intersection of the surface $z = f(x, y)$ and the plane
   [(a) $x = x_0$ (b) $y = y_0$ (c) $z = z_0$ (d) $x + y + z = 0$]
   at the point $(x_0, y_0, f(x_0, y_0))$ on the surface.

3. *True or False* $f_x(x, y)$ equals the rate of change of $f$ in the direction of the positive $x$-axis.

4. The two second-order partial derivatives $\dfrac{\partial^2 z}{\partial x\,\partial y}$ and $\dfrac{\partial^2 z}{\partial y\,\partial x}$ are called _____ partials.

5. *True or False* If $f(x, y) = x \cos y$, then $f_x(x, y) = f_y(x, y)$.

6. For a function $w = f(x, y, z)$ of three variables, to find the partial derivative $f_y(x, y, z)$, treat the variables _____ and _____ as constants, and differentiate $f$ with respect to _____.

### Skill Building

*In Problems 7–12, find $f_x(x, y)$ and $f_y(x, y)$ using the definition of a partial derivative.*

7. $f(x, y) = x^2 + 3xy - y$

8. $f(x, y) = 5xy - 4y^2$

9. $f(x, y) = (6 - x^2)y + 3x$

10. $f(x, y) = (x - 4)y^2 + 2y$

11. $f(x, y) = xy^2 + \sin x$

12. $f(x, y) = x \cos y - xy$

*In Problems 13–24, find $f_x(x, y)$ and $f_y(x, y)$.*

13. $f(x, y) = x^2y + 6y^2$

14. $f(x, y) = 3x^2 + 6xy^3$

15. $f(x, y) = \dfrac{x - y}{x + y}$

16. $f(x, y) = \dfrac{x + y}{y^2}$

17. $f(x, y) = e^y \cos x + e^x \sin y$

18. $f(x, y) = x^2 \cos y + y^2 \sin x$

19. $f(x, y) = x^2 e^{xy}$

20. $f(x, y) = \cos(x^2y^3)$

21. $f(x, y) = \tan^{-1}\dfrac{x}{y}$

22. $f(x, y) = \sin^{-1}\dfrac{y}{x}$

23. $f(x, y) = x \ln(x^2 + y^2)$

24. $f(x, y) = ye^{x^2 + y^2}$

---

**1.** = NOW WORK problem     ⟅∿⟆ = Graphing technology recommended    CAS = Computer Algebra System recommended

In Problems 25–34, find $\dfrac{\partial z}{\partial x}$ and $\dfrac{\partial z}{\partial y}$.

**25.** $z = x^2 y + \cos y$

**26.** $z = y \cos x - x \sin y$

**27.** $z = \cos(xy + y^2)$

**28.** $z = \sin(2xy)$

**29.** $z = \tan^{-1}(xy^2)$

**30.** $z = x \tan^{-1} y^2$

**31.** $z = e^{\sqrt{x^2 + y^2}}$

**32.** $z = \ln \sqrt{x^2 + y^2}$

**33.** $z = \sin(e^{x^2 y})$

**34.** $z = \sin[\ln(x^2 + y^2)]$

In Problems 35–42, find the second-order partial derivatives $f_{xx}$, $f_{xy}$, $f_{yx}$, and $f_{yy}$. Verify that $f_{xy} = f_{yx}$.

**35.** $f(x, y) = 6x^2 - 8xy + 9y^2$

**36.** $f(x, y) = x^3 + 2x^2 y - 3xy^3$

**37.** $f(x, y) = (2x + 3y)(3x - 2y)$

**38.** $f(x, y) = (xy - 3x)(x^2 + y^2)$

**39.** $f(x, y) = \ln(x^3 + y^2)$

**40.** $f(x, y) = e^{2x + 3y}$

**41.** $f(x, y) = \cos(x^2 y^3)$

**42.** $f(x, y) = \sin^2(xy)$

In Problems 43–50, find $f_x(x, y, z)$, $f_y(x, y, z)$, and $f_z(x, y, z)$.

**43.** $f(x, y, z) = xy + yz + xz$

**44.** $f(x, y, z) = xe^y + ye^z + ze^x$

**45.** $f(x, y, z) = xy \sin z - yz \sin x$

**46.** $f(x, y, z) = \dfrac{1}{\sqrt{x^2 + y^2 + z^2}}$

**47.** $f(x, y, z) = z \tan^{-1} \dfrac{y}{x}$

**48.** $f(x, y, z) = \tan^{-1} \dfrac{xy}{z}$

**49.** $f(x, y, z) = \sin[\ln(x^2 + y^2 + z^2)]$

**50.** $f(x, y, z) = e^{x^2 + y^2} \ln z$

In Problems 51 and 52, use the definition of a partial derivative to find $f_x(0, 0)$ and $f_y(0, 0)$.

**51.** $f(x, y) = \begin{cases} \dfrac{x^3 + y^3}{x^2 + y^2} & \text{if } (x, y) \neq (0, 0) \\ 0 & \text{if } (x, y) = (0, 0) \end{cases}$

**52.** $f(x, y) = \begin{cases} \dfrac{x^2 y^2}{x^2 + 4y^3} & \text{if } (x, y) \neq (0, 0) \\ 0 & \text{if } (x, y) = (0, 0) \end{cases}$

In Problems 53–60, find an equation of the tangent line to the curve of intersection of each surface with the given plane at the indicated point.

**53.** $z = x^2 + y^2$ and $y = 2$ at $(1, 2, 5)$

**54.** $z = x^2 - y^2$ and $x = 3$ at $(3, 1, 8)$

**55.** $z = \sqrt{1 - x^2 - y^2}$ and $x = 0$ at $\left(0, \dfrac{1}{2}, \dfrac{\sqrt{3}}{2}\right)$

**56.** $z = \sqrt{16 - x^2 - y^2}$ and $y = 2$ at $(\sqrt{3}, 2, 3)$

**57.** $z = \sqrt{x^2 + y^2}$ and $x = 4$ at $(4, 2, 2\sqrt{5})$

**58.** $z = x^2 + \dfrac{y^2}{4}$ and $x = 2$ at $(2, 4, 8)$

**59.** $z = e^x \sin y$ and $x = 0$ at $\left(0, \dfrac{\pi}{3}, \dfrac{\sqrt{3}}{2}\right)$

**60.** $z = e^x \ln y$ and $y = e$ at $(0, e, 1)$

**61.** Find the rate of change of $z = \ln \sqrt{x^2 + y^2}$ at $(3, 4, \ln 5)$,

   **(a)** In the direction of the positive $x$-axis.

   **(b)** In the direction of the positive $y$-axis.

**62.** Find the rate of change of $z = e^y \sin x$ at $\left(\dfrac{\pi}{3}, 0, \dfrac{\sqrt{3}}{2}\right)$,

   **(a)** In the direction of the positive $x$-axis.

   **(b)** In the direction of the positive $y$-axis.

## Applications and Extensions

**63.** **Temperature Distribution**  The temperature distribution $T$ (in degrees Celsius) of a heated plate at a point $(x, y)$ in the $xy$-plane is modeled by

$$T = T(x, y) = \dfrac{100}{\ln 2} \cdot \ln(x^2 + y^2) \qquad 1 \leq x^2 + y^2 \leq 9$$

   **(a)** Show that $T = 0\,°\mathrm{C}$ on the circle $x^2 + y^2 = 1$, and $T = 200\,°\mathrm{C}$ on the circle $x^2 + y^2 = 4$.

   **(b)** Find the rate of change of $T$ in the direction of the positive $x$-axis at the point $(1, 2)$ and at the point $(2, 1)$. Interpret the rate of change in the context of the problem.

   **(c)** Find the rate of change of $T$ in the direction of the positive $y$-axis at the point $(2, 0)$ and at the point $(0, 2)$. Interpret the rate of change in the context of the problem.

**64.** **Temperature Distribution**  The temperature distribution (in degrees Celsius) of a heated plate at a point $(x, y)$ in the $xy$-plane is modeled by $T = T(x, y) = \dfrac{100}{\sqrt{x^2 + y^2}}$, $1 \leq x^2 + y^2 \leq 9$.

   **(a)** Show that $T = 100\,°\mathrm{C}$ on the circle $x^2 + y^2 = 1$, and $T = 50\,°\mathrm{C}$ on the circle $x^2 + y^2 = 4$.

   **(b)** Find the rate of change of $T$ in the direction of the positive $x$-axis at the point $(1, 0)$ and at the point $(0, 1)$. Interpret the rate of change in the context of the problem.

   **(c)** Find the rate of change of $T$ in the direction of the positive $y$-axis at the point $(2, 0)$ and at the point $(0, 2)$. Interpret the rate of change in the context of the problem.

**65.** **Thermodynamics**  The Ideal Gas Law $PV = nrT$ is used to describe the relationship between pressure $P$, volume $V$, and temperature $T$ of a confined gas, where $n$ is the number of moles of the gas and $r$ is the universal gas constant. Show that

$$\dfrac{\partial V}{\partial T} \cdot \dfrac{\partial T}{\partial P} \cdot \dfrac{\partial P}{\partial V} = -1$$

66. **Thermodynamics** The volume $V$ of a fixed amount of gas varies directly with the temperature $T$ and inversely with the pressure $P$. That is, $V = k\dfrac{T}{P}$, where $k > 0$ is a constant.

   (a) Find $\dfrac{\partial V}{\partial T}$ and $\dfrac{\partial V}{\partial P}$.

   (b) Show that $T\dfrac{\partial V}{\partial T} + P\dfrac{\partial V}{\partial P} = 0$.

67. **Economics** The data used to develop the **Cobb–Douglas productivity model** included capital input $K$ and labor input $L$ for each year during the period 1899–1922. Using the model $P = aK^b L^{1-b}$ and multiple regression techniques, Cobb and Douglas determined that manufacturing productivity was represented by the function

   $$P = 1.014651 K^{0.254124} L^{0.745876} \approx 1.01 K^{0.25} L^{0.75}$$

   (a) Find the marginal productivity with respect to capital input and the marginal productivity with respect to labor input in 1900 when $K = 107$ and $L = 105$.

   (b) Find the marginal productivity with respect to capital input and the marginal productivity with respect to labor input in 1920 when $K = 407$ and $L = 193$.

   (c) Compare the answers. What do you conclude about the change in manufacturing productivity in the United States during the 20-year period?

68. **Economics** The function

   $$z = f(x, y, r) = \dfrac{1 + (1 - x)y}{1 + r} - 1$$

   describes the net gain or loss of money invested, where $x =$ annual marginal tax rate, $y =$ annual effective yield on an investment, and $r =$ annual inflation rate.

   (a) Find the annual net gain or loss if money is invested at an effective yield of 4% when the marginal tax rate is 25% and the inflation rate is 5%; that is, find $f(0.25, 0.04, 0.05)$.

   (b) Find the rate of change of gain (or loss) of money with respect to the marginal tax rate when the effective yield is 4% and the inflation rate is 5%.

   (c) Find the rate of change of gain (or loss) of money with respect to the effective yield when the marginal tax rate is 25% and the inflation rate is 5%.

   (d) Find the rate of change of gain (or loss) of money with respect to the inflation rate when the marginal tax rate is 25% and the effective yield is 4%. Use $r = 5\%$.

69. **Economics** Let $w = 2x^{1/2} y^{1/3} z^{1/6}$ be a production function that depends on three inputs: $x$, $y$, and $z$. Find the marginal productivity with respect to $x$, the marginal productivity with respect to $y$, and the marginal productivity with respect to $z$.

70. **Speed of Sound** The speed $v$ of sound in a gas depends on the pressure $p$ and density $d$ of the gas and is modeled by the formula $v(p, d) = k\sqrt{\dfrac{p}{d}}$, where $k > 0$ is some constant. Find the rate of change of speed with respect to $p$ and with respect to $d$.

71. **Vibrating String** Suppose a vibrating string satisfies the equation $f(x, t) = 2\cos(5t)\sin x$, where $x$ is the horizontal distance of a point from one end of the string, $t$ is time, and $f(x, t)$ is the displacement. Show that $\dfrac{\partial^2 f}{\partial t^2} = 25\dfrac{\partial^2 f}{\partial x^2}$ at all points $(x, t)$.

72. **Temperature Distribution** Suppose a thin metal rod extends along the $x$-axis from $x = 0$ to $x = 20$, and for each $x$, where $0 \le x \le 20$, the temperature $T$ of the rod at time $t \ge 0$ and position $x$ is $T(t, x) = 40e^{-\lambda t}\sin\dfrac{\pi x}{20}$, where $\lambda > 0$ is a constant.

   (a) Show that $T_t = -\lambda T$, $T_{xx} = -\dfrac{\pi^2}{400}T$, and $T_t = \dfrac{1}{k^2}T_{xx}$ for some $k$.

   (b) Graph the initial temperature distribution, $y = T(0, x)$, where $0 \le x \le 20$.

73. Find $\dfrac{\partial x}{\partial r}$, $\dfrac{\partial x}{\partial \theta}$, $\dfrac{\partial y}{\partial r}$, and $\dfrac{\partial y}{\partial \theta}$ if $x = r\cos\theta$ and $y = r\sin\theta$.

74. Find $\dfrac{\partial r}{\partial x}$, $\dfrac{\partial \theta}{\partial x}$, $\dfrac{\partial r}{\partial y}$, and $\dfrac{\partial \theta}{\partial y}$ if $r = \sqrt{x^2 + y^2}$ and $\theta = \tan^{-1}\dfrac{y}{x}$, $x \ne 0$.

[CAS] 75. (a) Graph $f(x, y) = \dfrac{1}{2}x^2 + \dfrac{1}{3}y^2$ and the plane $y = 1$.

   (b) Find an equation of the tangent line to the curve of intersection of the surface and the plane at the point $\left(2, 1, \dfrac{7}{3}\right)$.

[CAS] 76. (a) Graph $f(x, y) = \dfrac{5}{\sqrt{x^2 + y^2 + 1}}$ and the plane $x = 1$.

   (b) Find an equation of the tangent line to the curve of intersection of the surface and the plane at the point $\left(1, -2, \dfrac{5\sqrt{6}}{6}\right)$.

77. Show that $\dfrac{\partial u}{\partial x} = \dfrac{\partial v}{\partial y}$ and $\dfrac{\partial u}{\partial y} = -\dfrac{\partial v}{\partial x}$ for $u = e^x\cos y$ and $v = e^x\sin y$.

78. Show that $\dfrac{\partial u}{\partial x} = \dfrac{\partial v}{\partial y}$ and $\dfrac{\partial u}{\partial y} = -\dfrac{\partial v}{\partial x}$ for $u = \ln\sqrt{x^2 + y^2}$ and $v = \tan^{-1}\dfrac{y}{x}$.

79. If $u = x^2 + 4y^2$, show that $x\dfrac{\partial u}{\partial x} + y\dfrac{\partial u}{\partial y} = 2u$.

80. If $u = xy^2$, show that $x\dfrac{\partial u}{\partial x} + y\dfrac{\partial u}{\partial y} = 3u$.

81. If $w = x^2 + y^2 - 3yz$, show that $x\dfrac{\partial w}{\partial x} + y\dfrac{\partial w}{\partial y} + z\dfrac{\partial w}{\partial z} = 2w$.

82. If $w = \dfrac{xz + y^2}{yz}$, show that $x\dfrac{\partial w}{\partial x} + y\dfrac{\partial w}{\partial y} + z\dfrac{\partial w}{\partial z} = 0$.

83. If $z = \cos(x + y) + \cos(x - y)$, show that $\dfrac{\partial^2 z}{\partial x^2} - \dfrac{\partial^2 z}{\partial y^2} = 0$.

84. If $z = \sin(x - y) + \ln(x + y)$, show that $\dfrac{\partial^2 z}{\partial x^2} = \dfrac{\partial^2 z}{\partial y^2}$.

**85.** Show that $u = e^{-\alpha^2 t} \sin(\alpha x)$ satisfies the equation $\dfrac{\partial u}{\partial t} = \dfrac{\partial^2 u}{\partial x^2}$ for all values of the constant $\alpha$.

**Laplace's Equation**  *The partial differential*

*equation* $\dfrac{\partial^2 z}{\partial x^2} + \dfrac{\partial^2 z}{\partial y^2} = 0$ *is called **Laplace's Equation**, and the*

*function $z = f(x, y)$ that satisfies Laplace's equation is called a*
*harmonic function.\* In Problems 86–89, show that each function is*
*a harmonic function.*

**86.** $z = \ln\sqrt{x^2 + y^2}$ $\qquad$ **87.** $z = e^{ax}\sin(ay)$

**88.** $z = \tan^{-1}\dfrac{y}{x}$ $\qquad$ **89.** $z = e^{ax}\cos(ay)$

**90. Harmonic Functions**  Suppose $u(x, y)$ and $v(x, y)$ have continuous second-order partial derivatives, $u_x = v_y$ and $u_y = -v_x$. Show that $u$ and $v$ are harmonic functions.

**91. Harmonic Functions**  If $u = z\tan^{-1}\dfrac{x}{y}$, show that

$$\frac{\partial^2 u}{\partial x^2} + \frac{\partial^2 u}{\partial y^2} + \frac{\partial^2 u}{\partial z^2} = 0$$

**92. Harmonic Functions** Show that $f(x, y, z) = (x^2 + y^2 + z^2)^{-1/2}$ satisfies the three-dimensional Laplace equation

$$f_{xx} + f_{yy} + f_{zz} = 0$$

**93.** Let $f(x, y) = \begin{cases} \dfrac{xy^3}{x^2 + y^2} & \text{if} \quad (x, y) \neq (0, 0) \\ 0 & \text{if} \quad (x, y) = (0, 0) \end{cases}$.

$\quad$ **(a)** Find $f_x$ and $f_y$. *Hint:* $f_x$ and $f_y$, where $(x, y) \neq (0, 0)$, can be found using derivative formulas. To find $f_x(0, 0)$ and $f_y(0, 0)$, use the definition of a partial derivative.

$\quad$ **(b)** Show that $f_{xy}(0, 0) \neq f_{yx}(0, 0)$.

**94.** Let $f(x, y) = \begin{cases} \dfrac{xy(x^2 - y^2)}{x^2 + y^2} & \text{if} \quad (x, y) \neq (0, 0) \\ 0 & \text{if} \quad (x, y) = (0, 0) \end{cases}$.

$\quad$ Show that:

$\quad$ **(a)** $f_x(0, y) = -y$ $\qquad$ **(b)** $f_y(x, 0) = x$

$\quad$ **(c)** $f_{xy}(0, 0) = -1$ $\qquad$ **(d)** $f_{yx}(0, 0) = 1$

**95.** If you are told that $f$ is a function of two variables whose partial derivatives are $f_x(x, y) = 3x - y$ and $f_y(x, y) = x - 3y$, should you believe it? Explain.

**96.** Show that there is no function $z = f(x, y)$ for which $f_x(x, y) = 2x - y$ and $f_y(x, y) = x - 2y$.

**97.** Use the definition of a partial derivative to show that the function $z = \sqrt{x^2 + y^2}$ does not have partial derivatives at $(0, 0)$. By discussing the graph of the function, give a geometric reason why this should be so.

**98.** If $z = f(x, y) = 4x^2 + 9y^2 - 12$, interpret $f_x\left(1, -\dfrac{1}{3}\right)$

and $f_y\left(1, -\dfrac{1}{3}\right)$ geometrically.

**99.** Show that $xf_x + yf_y + zf_z = 0$ for $f(x, y, z) = e^{x/y} + e^{y/z} + e^{z/x}$.

**100.** Find $a$ in terms of $b$ and $c$ so that $f(t, x, y) = e^{at}\sin(bx)\cos(cy)$ satisfies $f_t = f_{xx} + f_{yy}$.

**101. Wave Equation**  Show that $f(x, t) = \cos(x + ct)$ satisfies the one-dimensional wave equation $f_{tt} = c^2 f_{xx}$, where $c$ is a constant.

**102.** Find $f_x$ and $f_y$ if $f(x, y) = \int_x^y \ln(\cos\sqrt{t})\, dt$.

**103.** Find $\dfrac{\partial a}{\partial b}$, $\dfrac{\partial a}{\partial c}$, and $\dfrac{\partial a}{\partial A}$ for the Law of Cosines:

$$a^2 = b^2 + c^2 - 2bc\cos A$$

**104.** If $x = r\cos\theta$ and $y = r\sin\theta$, show that

$$\begin{vmatrix} \dfrac{\partial x}{\partial r} & \dfrac{\partial x}{\partial \theta} \\[2mm] \dfrac{\partial y}{\partial r} & \dfrac{\partial y}{\partial \theta} \end{vmatrix} = r \quad \text{and} \quad \begin{vmatrix} \dfrac{\partial r}{\partial x} & \dfrac{\partial r}{\partial y} \\[2mm] \dfrac{\partial \theta}{\partial x} & \dfrac{\partial \theta}{\partial y} \end{vmatrix} = \dfrac{1}{r}$$

**Challenge Problems**

*In Problems 105 and 106, find $f_x$ and $f_y$.*

**105.** $f(x, y) = x^y$ $\qquad\qquad$ **106.** $f(x, y) = x^{2x + 3y}$

*In Problems 107–110, find $f_x$, $f_y$, and $f_z$.*

**107.** $f(x, y, z) = x^{y + z}$ $\qquad$ **108.** $f(x, y, z) = x^{yz}$

**109.** $f(x, y, z) = (x + y)^z$ $\qquad$ **110.** $f(x, y, z) = (xy)^z$

**111.** Find $f_x$ and $f_y$ at $(0, 0)$ if

$$f(x, y) = \begin{cases} e^{-1/(x^2 + y^2)} & \text{if} \quad (x, y) \neq (0, 0) \\ 0 & \text{if} \quad (x, y) = (0, 0) \end{cases}$$

**112.** Find $f_x$, $f_y$, $f_{xx}$, $f_{yy}$, and $f_{xy}$ for $f(x, y) = (xy)^{xy}$. What is the domain of $f$?

**113.** Show that the following function has first partial derivatives at all points in the plane:

$$f(x, y) = \begin{cases} \dfrac{x^3 - y^3}{x^2 + y^2} & \text{if} \quad (x, y) \neq (0, 0) \\ 0 & \text{if} \quad (x, y) = (0, 0) \end{cases}$$

**114. Laplace's Equation in Polar Coordinates**  Show that the function $f(r, \theta) = r^n\sin(n\theta)$ satisfies the Laplace equation

$$f_{rr} + \frac{1}{r}f_r + \frac{1}{r^2}f_{\theta\theta} = 0$$

**115.** Let $u = r^m\cos(m\theta)$. Show that

$$\frac{\partial^2 u}{\partial r^2} + \frac{1}{r^2}\left(\frac{\partial^2 u}{\partial \theta^2}\right) + \frac{1}{r}\left(\frac{\partial u}{\partial r}\right) = 0 \text{ for all } m.$$

---

\*Laplace's equation is important in many applications, including fluid dynamics, heat, elasticity, and electricity.

**116. (a)** Find an equation of the tangent lines at $(x_0, y_0, f(x_0, y_0))$ to the curve of intersection of $z = f(x, y)$ and $y = y_0$, and the curve of intersection of $z = f(x, y)$ and $x = x_0$.

**(b)** Write an equation of the plane determined by these two lines.

**(c)** What is the geometric relationship of this plane to the surface $z = f(x, y)$?

**117.** Consider two coordinate systems as given in the figure.

**(a)** Let $f(x, y) = 3x^2 + 4y^3$.
Find $f_x(1, 6)$ and $f_y(1, 6)$.

**(b)** Let $(a, b)$ be the $x'y'$-coordinates of $(1, 6)$.
Let $\bar{f}(x', y') = f(x, y)$.
Find $\bar{f}_{x'}(a, b)$ and $\bar{f}_{y'}(a, b)$.

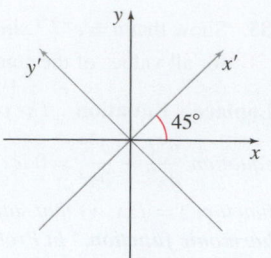

# 12.4 Differentiability and the Differential

**OBJECTIVES** *When you finish this section, you should be able to:*

**1** Find the change in $z = f(x, y)$ (p. 884)

**2** Show that a function of two variables is differentiable (p. 885)

**3** Use the differential as an approximating tool (p. 887)

**4** Find the differential of a function of three or more variables (p. 890)

For a function $y = f(x)$ of a single variable, we said that when $f$ has a derivative, then $f$ is differentiable. We also defined the differential of $y$ in terms of its derivative as $dy = f'(x)dx$, where $dx = \Delta x$. For functions of more than one variable, the situation is more involved. We begin by defining what it means for a function $z = f(x, y)$ to be *differentiable*, which involves the change $\Delta z$ in $z$. Then under suitable conditions on $z$, we will define $dz$, the differential of $z$.

## 1 Find the Change in $z = f(x, y)$

If $y = f(x)$ is a function of one variable, then the change $\Delta y$ in $y$ from $x_0$ to $x_0 + \Delta x$ is

$$\Delta y = f(x_0 + \Delta x) - f(x_0)$$

There is a similar definition for a function $z = f(x, y)$ of two variables.

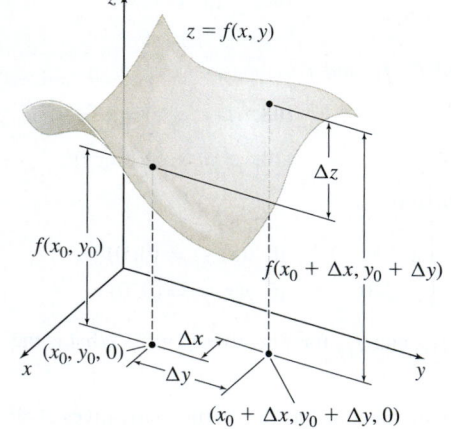

**Figure 36**

**DEFINITION** Change in $z = f(x, y)$

Let $z = f(x, y)$ be a function of two variables with domain $D$. Let $(x_0, y_0)$ be an interior point of $D$, and let $\Delta x$ and $\Delta y$ represent changes in $x$ and in $y$, respectively. If $\Delta x$ and $\Delta y$ are chosen so that the point $(x_0 + \Delta x, y_0 + \Delta y)$ is also in $D$, then the **change in $z$**, denoted by $\Delta z$, is defined as

$$\boxed{\Delta z = f(x_0 + \Delta x, y_0 + \Delta y) - f(x_0, y_0)}$$

See Figure 36.

**EXAMPLE 1** **Finding the Change in $z$**

**(a)** Find the change $\Delta z$ in $z = f(x, y) = x^2 y - 1$ from $(x_0, y_0)$ to $(x_0 + \Delta x, y_0 + \Delta y)$.

**(b)** Use the answer to calculate the change in $z$ from $(1, 2)$ to $(1.1, 1.9)$.

**Solution (a)** $\Delta z = f(x_0 + \Delta x, y_0 + \Delta y) - f(x_0, y_0)$
$$= [(x_0 + \Delta x)^2 (y_0 + \Delta y) - 1] - [x_0^2 y_0 - 1]$$

**(b)** Let $(x_0, y_0) = (1, 2)$ and $(x_0 + \Delta x, y_0 + \Delta y) = (1.1, 1.9)$. Then

$$\Delta z = [(1.1)^2 (1.9) - 1] - [(1)^2 (2) - 1] = 0.299$$

Since $z = f(1, 2) = 1$, the result found in (b) states that $z$ increases from 1 to 1.299, when $x$ increases from 1 to 1.1 and $y$ decreases from 2 to 1.9. ∎

**NOW WORK** Problem 7.

## 2 Show That a Function of Two Variables Is Differentiable

**NEED TO REVIEW?** The differentials $dx$ and $dy$ of a function $y = f(x)$ of one variable are discussed in Section 3.5, pp. 241–242.

Before defining the differentiability of a function of two variables, let's reexamine the differential $dy$ of a function $y = f(x)$ of a single variable. For a differentiable function $y = f(x)$ of one variable, the differentials $dx = \Delta x$ and $dy = f'(x)\,dx = f'(x)\Delta x$ can be used to approximate $\Delta y = f(x_0 + \Delta x) - f(x_0)$. That is,

$$dy \approx \Delta y = f(x_0 + \Delta x) - f(x_0)$$

provided $\Delta x$ is close to 0.

**NOTE** $\eta$ is the Greek letter eta.

If the error in using $dy$ to approximate $\Delta y$ is $\eta \Delta x$, then

$$\Delta y - dy = \eta \Delta x$$

$$\Delta y = f'(x_0)\Delta x + \eta \Delta x \qquad \boldsymbol{dy = f'(x_0)\Delta x} \qquad (1)$$

where $\eta$ is a function of $\Delta x$ for which $\lim\limits_{\Delta x \to 0} \eta = 0$.

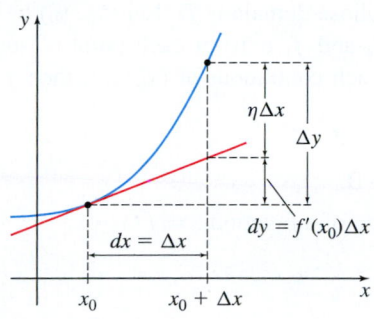

**DF** Figure 37

Figure 37 gives a geometric interpretation of equation (1). From Figure 37, we can see that $\eta \Delta x \to 0$ as $\Delta x \to 0$. But Figure 37 does not show something else suggested by equation (1), namely that $\eta \to 0$ as $\Delta x \to 0$. Do you see why? From (1), $\dfrac{\Delta y}{\Delta x} = f'(x_0) + \eta$. Since $f$ is differentiable, $\lim\limits_{\Delta x \to 0} \dfrac{\Delta y}{\Delta x} = f'(x_0)$. This means $\lim\limits_{\Delta x \to 0} \eta$ must be 0.

The discussion above provides background for the definition of differentiability of a function of two variables.

**DEFINITION  Differentiability of $z = f(x, y)$ at $(x_0, y_0)$**

Let $z = f(x, y)$ be a function of two variables whose domain is $D$. Let $(x_0, y_0)$ be an interior point of $D$ and let $\Delta x$ and $\Delta y$ be chosen so that the point $(x_0 + \Delta x, y_0 + \Delta y)$ is also in $D$. If the change $\Delta z$ from $(x_0, y_0)$ to $(x_0 + \Delta x, y_0 + \Delta y)$ can be expressed in the form

$$\boxed{\Delta z = f_x(x_0, y_0)\Delta x + f_y(x_0, y_0)\Delta y + \eta_1 \Delta x + \eta_2 \Delta y} \qquad (2)$$

where both $\eta_1$ and $\eta_2$ are functions of $\Delta x$ and $\Delta y$ for which

$$\lim_{(\Delta x, \Delta y) \to (0, 0)} \eta_1 = 0 \qquad \text{and} \qquad \lim_{(\Delta x, \Delta y) \to (0, 0)} \eta_2 = 0 \qquad (3)$$

then $z = f(x, y)$ is said to be **differentiable at the point** $(x_0, y_0)$.

If the domain $D$ of a function $z = f(x, y)$ is an open set and if $z = f(x, y)$ is differentiable at every point of $D$, then we say $f$ is **differentiable on $D$**.

**EXAMPLE 2  Showing That a Function of Two Variables Is Differentiable**

Show that $z = f(x, y) = x^2 y - 1$ is differentiable.

**Solution** The domain of $f$ is the $xy$-plane and the partial derivatives of $f$ are

$$f_x(x, y) = 2xy \qquad \text{and} \qquad f_y(x, y) = x^2$$

We find the change $\Delta z$ in $z$ and express it in the form of (2).

$$\Delta z = [(x + \Delta x)^2 (y + \Delta y) - 1] - [x^2 y - 1] \qquad \boldsymbol{\Delta z = f(x + \Delta x, y + \Delta y) - f(x, y)}$$

$$= [x^2 + 2x\Delta x + (\Delta x)^2](y + \Delta y) - x^2 y$$

$$= 2xy\Delta x + y(\Delta x)^2 + x^2 \Delta y + 2x\Delta x\Delta y + (\Delta x)^2 \Delta y$$

**NOTE** There are other ways to choose $\eta_1$ and $\eta_2$ so that equations (2) and (3) are satisfied.

$$= \underbrace{(2xy)}_{f_x(x,\,y)}\ \Delta x + \underbrace{(x^2)}_{f_y(x,\,y)}\ \Delta y + \underbrace{[y\Delta x]}_{\eta_1}\Delta x + \underbrace{[2x\Delta x + (\Delta x)^2]}_{\eta_2}\Delta y$$

$$= f_x(x, y)\Delta x + f_y(x, y)\Delta y + \eta_1 \Delta x + \eta_2 \Delta y$$

Equation (2) is satisfied. It remains to show that equation (3) is satisfied.

$$\lim_{(\Delta x, \Delta y) \to (0,0)} \eta_1 = \lim_{\Delta x \to 0} (y\Delta x) = 0 \quad \text{and} \quad \lim_{(\Delta x, \Delta y) \to (0,0)} \eta_2 = \lim_{\Delta x \to 0} [2x\Delta x + (\Delta x)^2] = 0$$

So, $z = f(x, y)$ is differentiable on its domain. ∎

**NOW WORK** Problem 29.

Using the method of Example 2 to determine whether a function $z = f(x, y)$ is differentiable can be challenging. The theorem below provides a relatively straightforward way to show a function is differentiable.

**THEOREM  Continuous Partial Derivatives Are Sufficient for Differentiability**

Let $z = f(x, y)$ be a function of two variables whose domain is $D$. Let $(x_0, y_0)$ be an interior point of $D$. If the partial derivatives $f_x$ and $f_y$ exist at each point of some disk centered at $(x_0, y_0)$, and if $f_x$ and $f_y$ are each continuous at $(x_0, y_0)$, then $f$ is differentiable at $(x_0, y_0)$.

The proof of this result is given in Appendix B.

Now we are ready to define the differential $dz$ of a function $z = f(x, y)$.

**DEFINITION  Differential $dz$ of $z = f(x, y)$**

Let $z = f(x, y)$ be a function of two variables whose domain is $D$. Let $(x_0, y_0)$ be an interior point of $D$ and let $\Delta x$ and $\Delta y$ be chosen so that the point $(x_0 + \Delta x, y_0 + \Delta y)$ is also in $D$. If $f$ is differentiable at the point $(x_0, y_0)$, then the **differentials** $dx$ and $dy$ are defined as

$$dx = \Delta x \quad \text{and} \quad dy = \Delta y$$

The **differential** $dz$, also called the **total differential** of $z = f(x, y)$ at $(x_0, y_0)$, is defined as

$$dz = f_x(x_0, y_0)\, dx + f_y(x_0, y_0)\, dy$$

**EXAMPLE 3  Finding the Differential $dz$ of $z = f(x, y)$**

Find the differential $dz$ of each function:

**(a)** $f(x, y) = e^x \cos y$ 　　　　**(b)** $f(x, y) = \dfrac{\ln x}{y}$

**Solution (a)** $f$ is defined everywhere in the $xy$-plane. The partial derivatives of $f$ are

$$f_x(x, y) = e^x \cos y \qquad f_y(x, y) = -e^x \sin y$$

Since $f_x$ and $f_y$ are continuous everywhere in the $xy$-plane, the function $z = f(x, y)$ is differentiable. The differential $dz$ is

$$dz = e^x \cos y\, dx - e^x \sin y\, dy$$

**(b)** The domain of $f$ is $\{(x, y) \mid x > 0, \ y \neq 0\}$. The partial derivatives of $f$ are

$$f_x(x, y) = \frac{1}{xy} \qquad f_y(x, y) = -\frac{\ln x}{y^2}$$

Since both partial derivatives exist and are continuous at every point $(x_0, y_0)$ in the domain of $f$, the function $z = f(x, y)$ is differentiable at every point $(x_0, y_0)$ in its domain. The differential $dz$ is

$$dz = \frac{1}{xy}\, dx - \frac{\ln x}{y^2}\, dy$$

∎

**NOW WORK** Problem 11.

If $z = f(x, y)$ is a differentiable function, then $\Delta z$ can be expressed in terms of the differential $dz$.

$$\Delta z = f_x(x_0, y_0)\Delta x + f_y(x_0, y_0)\Delta y + \eta_1\Delta x + \eta_2\Delta y$$

$$= \underbrace{f_x(x_0, y_0)\, dx + f_y(x_0, y_0)\, dy}_{dz} + \eta_1\Delta x + \eta_2\Delta y$$

$$\begin{array}{c} \uparrow \\ dx = \Delta x \\ dy = \Delta y \end{array}$$

$$\Delta z = dz + \eta_1\Delta x + \eta_2\Delta y$$

where $\displaystyle\lim_{(\Delta x, \Delta y) \to (0,0)} \eta_1 = 0$ and $\displaystyle\lim_{(\Delta x, \Delta y) \to (0,0)} \eta_2 = 0$.

When the differentials $dx = \Delta x$ and $dy = \Delta y$ are close to 0, then $\eta_1$ and $\eta_2$ are also close to 0, and the differential $dz$ is approximately equal to $\Delta z$. That is,

$$\boxed{\Delta z \approx dz = f_x(x_0, y_0)\, dx + f_y(x_0, y_0)\, dy} \tag{4}$$

The differential $dz$ is usually easier to calculate than $\Delta z$, making $dz$ a useful approximation to $\Delta z$. The error introduced in using $dz$ to approximate $\Delta z$ equals the expression $\eta_1\Delta x + \eta_2\Delta y$.

## 3 Use the Differential as an Approximating Tool

**EXAMPLE 4** **Using the Differential in Astronomy**

The luminosity $L$ (total power output in watts, W) of a star is given by the formula

$$L = L(R, T) = 4\pi R^2 \sigma T^4$$

where $R$ is the radius of the star (in meters), $T$ is its effective surface temperature (in kelvin, K), and $\sigma$ is the **Stefan–Boltzmann constant**. For the sun, $L_s = (3.90 \times 10^{26})$ W, $R_s = (6.94 \times 10^8)$ m, and $T_s = 4800$ K. Suppose in a billion years, the changes in the Sun will be $\Delta R_s = (0.08 \times 10^8)$ m and $\Delta T_s = 100$ K. What will be the resulting percent increase in luminosity?

**Solution** We begin with $L = 4\pi R^2 \sigma T^4$. Then

$$dL = 4\pi\sigma(2R)T^4 dR + 4\pi\sigma R^2(4T^3)dT = 8\pi\sigma RT^3(T\,dR + 2R\,dT)$$

The relative error in luminosity is

$$\frac{\Delta L}{L} \approx \frac{dL}{L} = \frac{8\pi\sigma RT^3}{4\pi R^2\sigma T^4}(T\,dR + 2R\,dT) = 2\left(\frac{dR}{R} + 2\frac{dT}{T}\right) = 2\frac{\Delta R}{R} + 4\frac{\Delta T}{T}$$

$$= \frac{2(0.08 \times 10^8)}{6.94 \times 10^8} + \frac{4 \cdot 100}{4800} \approx 0.106$$

The percent increase in luminosity will be approximately 10.6%. ∎

Incidentally, a reasonable, although rough, estimate of how this change would affect Earth's temperature is

$$\Delta T_e \approx \frac{1}{4} \cdot 0.106\, T_e = (0.0265)(290\text{ K}) \approx 7.69\text{ K}$$

That is, the average temperature of Earth's surface will change from 16.9 °C (62.4 °F) to 24.6 °C (76.3 °F). Such a change in temperature would be enough to modify Earth's climate.

3 cm

10 cm

CALC
CLIP

**EXAMPLE 5** Using the Differential in Error Analysis

A cola company requires a can in the shape of a right circular cylinder of height 10 cm and radius 3 cm. If the manufacturer of the cans claims a percentage error of no more than 0.2% in the height and no more than 0.1% in the radius, what is the approximate maximum variation in the volume of the can?

**Solution** The volume $V$ of a right circular cylinder of height $h$ cm and radius $R$ cm is $V = \pi R^2 h$ cm$^3$. We find the differential $dV$.

$$dV = \frac{\partial V}{\partial R} dR + \frac{\partial V}{\partial h} dh = 2\pi R h \, dR + \pi R^2 dh$$

The relative error in the radius $R$ is $\dfrac{|\Delta R|}{R} = \dfrac{|dR|}{R} = 0.001$, and the relative error in the height $h$ is $\dfrac{|\Delta h|}{h} = \dfrac{|dh|}{h} = 0.002$. The relative error in the volume $V$ is

$$\frac{|\Delta V|}{V} \approx \frac{|dV|}{V} = \frac{\left|2\pi R h \, dR + \pi R^2 dh\right|}{\pi R^2 h} = \left|2\frac{dR}{R} + \frac{dh}{h}\right|$$

$$= \left|2\frac{\Delta R}{R} + \frac{\Delta h}{h}\right| \le 2\frac{|\Delta R|}{R} + \frac{|\Delta h|}{h}$$

$$= 2(0.001) + 0.002 = 0.004$$

The maximum variation in the volume is approximately 0.4%, so the actual volume of the container varies as follows:

$$V = \pi R^2 h \pm 0.004(\pi R^2 h) = \pi R^2 h(1 \pm 0.004) = 90\pi(1 \pm 0.004) \text{ cm}^3$$

The volume $V$ is between $89.64\pi \approx 281.612$ cm$^3$ and $90.36\pi \approx 283.874$ cm$^3$. ∎

NOW WORK Problem 39.

**EXAMPLE 6** Using the Differential $dz$ to Approximate the Value of a Function $z$

For the function $f(x, y) = x^2 y - 1$, use the differential $dz$ to approximate $f(1.1, 1.9)$.

**Solution** Example 2 shows $f$ is differentiable and $f_x(x, y) = 2xy$ and $f_y(x, y) = x^2$.

Let $(x_0, y_0) = (1, 2)$ and $(x_0 + \Delta x, y_0 + \Delta y) = (1.1, 1.9)$. Then

$$f(x_0, y_0) = f(1, 2) = 1^2 \cdot 2 - 1 = 1 \qquad f(x_0 + \Delta x, y_0 + \Delta y) = f(1.1, 1.9)$$

$$f_x(x_0, y_0) = 2x_0 y_0 = 2 \cdot 1 \cdot 2 = 4 \qquad f_y(x_0, y_0) = (x_0)^2 = 1$$

$$dx = \Delta x = 1.1 - 1 = 0.1 \qquad dy = \Delta y = 1.9 - 2 = -0.1$$

Now $\Delta z = f(x_0 + \Delta x, y_0 + \Delta y) - f(x_0, y_0)$ and from statement (4), we have

$$\Delta z \approx dz = f_x(x_0, y_0)dx + f_y(x_0, y_0)dy$$

$$f(x_0 + \Delta x, y_0 + \Delta y) - f(x_0, y_0) \approx f_x(x_0, y_0)dx + f_y(x_0, y_0)dy$$

$$f(x_0 + \Delta x, y_0 + \Delta y) \approx f(x_0, y_0) + f_x(x_0, y_0)dx + f_y(x_0, y_0)dy$$

Then

$$f(1.1, 1.9) \approx 1 + 4 \cdot 0.1 + 1 \cdot (-0.1) = 1.3 \qquad ∎$$

The actual value of $f(1.1, 1.9) = (1.1)^2(1.9) - 1 = 1.299$, so the error in using the differential $dz$ to approximate $z$ is 0.001.

NOW WORK Problem 33.

## Continuity and Differentiability for Functions of Two Variables

Differentiable functions of a single variable are necessarily continuous. This result is also true for functions of two variables.

**THEOREM Differentiability Is Sufficient for Continuity**

Let $z = f(x, y)$ be a function of two variables whose domain is $D$. Let $(x_0, y_0)$ be an interior point of $D$. If $f$ is differentiable at $(x_0, y_0)$, then $f$ is continuous at $(x_0, y_0)$.

**Proof** The function $z = f(x, y)$ is continuous at $(x_0, y_0)$ if

$$\lim_{(x,y) \to (x_0, y_0)} f(x, y) = f(x_0, y_0)$$

This is equivalent to the statement $\lim\limits_{(x,y) \to (x_0, y_0)} \Delta z = 0$. Since $z = f(x, y)$ is differentiable at $(x_0, y_0)$, then $\Delta z$ can be expressed as

$$\Delta z = f_x(x_0, y_0)\Delta x + f_y(x_0, y_0)\Delta y + \eta_1 \Delta x + \eta_2 \Delta y$$

where $\lim\limits_{(\Delta x, \Delta y) \to (0,0)} \eta_1 = 0$ and $\lim\limits_{(\Delta x, \Delta y) \to (0,0)} \eta_2 = 0$. Then

$$\Delta z = [f_x(x_0, y_0) + \eta_1]\,\Delta x + \left[f_y(x_0, y_0) + \eta_2\right]\Delta y$$

Now let $\Delta x = x - x_0$ and $\Delta y = y - y_0$. Then $(\Delta x, \Delta y) \to (0, 0)$ is equivalent to $(x, y) \to (x_0, y_0)$ and

$$\lim_{(x,y) \to (x_0, y_0)} \Delta z = \lim_{(x,y) \to (x_0, y_0)} \{[f_x(x_0, y_0) + \eta_1](x - x_0)$$
$$+ [f_y(x_0, y_0) + \eta_2](y - y_0)\} = 0$$

That is, $f$ is continuous at $(x_0, y_0)$. ∎

For functions of two variables, differentiability implies continuity. However, the existence of partial derivatives at a point does not necessarily result in continuity at that point because $f_x$ and/or $f_y$ might not be continuous at the point. For example, the function

$$f(x, y) = \begin{cases} \dfrac{xy}{x^2 + y^2} & \text{if} \quad (x, y) \neq (0, 0) \\[2mm] 0 & \text{if} \quad (x, y) = (0, 0) \end{cases}$$

has partial derivatives at $(0, 0)$, but as shown in Example 10 on page 879, $f$ is not continuous at $(0, 0)$.

A function that is differentiable at $(x_0, y_0)$ is continuous at $(x_0, y_0)$. So, a function that is not continuous at $(x_0, y_0)$ is not differentiable at $(x_0, y_0)$.

The following corollary provides a condition for $z = f(x, y)$ to be continuous.

**COROLLARY Continuity of a Function of Two Variables**

Let $z = f(x, y)$ be a function of two variables whose domain is $D$. Let $(x_0, y_0)$ be an interior point of $D$. If the partial derivatives $f_x$ and $f_y$ exist at each point of some disk centered at $(x_0, y_0)$, and if $f_x$ and $f_y$ are each continuous at $(x_0, y_0)$, then $f$ is continuous at $(x_0, y_0)$.

Although the precise formulations are given as theorems, the following summary might be helpful:

If $z = f(x, y)$,

- The continuity of the partial derivatives $f_x$ and $f_y$ implies the differentiability of $f$.
- The differentiability of $f$ implies the continuity of $f$.
- The continuity of the partial derivatives $f_x$ and $f_y$ implies the continuity of $f$.
- The existence of the partial derivatives $f_x$ and $f_y$ does not necessarily mean $f$ is differentiable.
- The existence of the partial derivatives $f_x$ and $f_y$ does not necessarily mean $f$ is continuous.

## 4 Find the Differential of a Function of Three or More Variables

Under suitable conditions, the definitions and theorems involving the differentiability of a function of two variables extend to functions of three or more variables.

If $w = f(x, y, z)$ is a function of three variables, the function $f$ is **differentiable** at a point $(x_0, y_0, z_0)$ if the change $\Delta w$ in $w$ can be expressed in the form

$$\Delta w = f_x(x, y, z)\Delta x + f_y(x, y, z)\Delta y + f_z(x, y, z)\Delta z + \eta_1 \Delta x + \eta_2 \Delta y + \eta_3 \Delta z$$

where $\eta_1$, $\eta_2$, and $\eta_3$ are each functions of $\Delta x$, $\Delta y$, and $\Delta z$ for which

$$\lim_{(\Delta x, \Delta y, \Delta z) \to (0,0,0)} \eta_1 = 0 \quad \text{and} \quad \lim_{(\Delta x, \Delta y, \Delta z) \to (0,0,0)} \eta_2 = 0 \quad \text{and} \quad \lim_{(\Delta x, \Delta y, \Delta z) \to (0,0,0)} \eta_3 = 0$$

If $w = f(x, y, z)$ is differentiable at a point $(x_0, y_0, z_0)$, the differentials $dx$, $dy$, and $dz$ are defined as

$$dx = \Delta x \qquad dy = \Delta y \qquad dz = \Delta z$$

The **differential** $dw$ is defined as

$$dw = f_x(x_0, y_0, z_0)\, dx + f_y(x_0, y_0, z_0)\, dy + f_z(x_0, y_0, z_0)\, dz$$

It can be shown that if $w = f(x, y, z)$ is defined within a ball centered at $(x_0, y_0, z_0)$, and if the partial derivatives $f_x$, $f_y$, and $f_z$ exist in this ball and are continuous at $(x_0, y_0, z_0)$, then $f$ is differentiable at $(x_0, y_0, z_0)$.

---

**EXAMPLE 7** **Finding the Differential of a Function of Three Variables**

Find the differential $dw$ of the function $w = f(x, y, z) = 3x^2 \sin^2 y \cos z$.

**Solution** The function $f$ is defined everywhere in space. We begin by finding the partial derivatives of $f$.

$$f_x(x, y, z) = 6x \sin^2 y \cos z \qquad f_y(x, y, z) = 6x^2 \sin y \cos y \cos z$$
$$f_z(x, y, z) = -3x^2 \sin^2 y \sin z$$

Since the partial derivatives are continuous everywhere, we have

$$dw = f_x(x, y, z)\, dx + f_y(x, y, z)\, dy + f_z(x, y, z)\, dz$$
$$= 6x \sin^2 y \cos z\, dx + 6x^2 \sin y \cos y \cos z\, dy - 3x^2 \sin^2 y \sin z\, dz \qquad \blacksquare$$

**NOW WORK** Problem 23.

The discussion above extends to functions of more than three variables in an analogous way.

# 12.4 Assess Your Understanding

## Concepts and Vocabulary

**1.** *True or False*  If $z = x \sin y$, then the change in $z$ from $(0.9, 0)$ to $\left(1, \dfrac{\pi}{4}\right)$ is $\Delta z = \sin \dfrac{\pi}{4} - 0.9 \sin 0 = \dfrac{\sqrt{2}}{2}$.

**2.** *True or False*  If $f_x(x, y) = -\dfrac{y^2}{1-x}$ and $f_y(x, y) = 2y \ln(1-x)$, then $f$ is differentiable at $(0, 2)$.

**3.** *True or False*  Let $z = f(x, y) = e^x \cos y$. Then the differential $dz$ of $z$ at $\left(0, \dfrac{\pi}{2}\right)$ is $dz = -dy$.

**4.** *True or False*  If the partial derivatives of a function $z = f(x, y)$ are $f_x(x, y) = 2x + y$ and $f_y(x, y) = x$, then $f$ is continuous.

**5.** *True or False*  If the partial derivatives of a function $z = f(x, y)$ are $f_x(x, y) = 2x + y$ and $f_y(x, y) = x$, then $f$ is differentiable.

**6.** *True or False*  If $w = f(x, y, z) = e^{xyz}$, then $dw = yze^x dx + xze^y dy + xye^z dz$.

## Skill Building

*In Problems 7–10, for each function $z = f(x, y)$, find the change in $z$.*

**7.** $z = x^2 + y^2$ from $(1, 3)$ to $(1.1, 3.2)$

**8.** $z = 2x^2 + xy - y^2$ from $(2, -1)$ to $(2.1, -1.1)$

**9.** $z = e^x \ln(xy)$ from $(1, 2)$ to $(0.9, 2.1)$

**10.** $z = \dfrac{xy}{x + y}$ from $(-1, 2)$ to $(-0.9, 1.9)$

*In Problems 11–22, find the differential $dz$ of each function.*

**11.** $z = x^2 + y^2$

**12.** $z = 2x^2 + xy - y^2$

**13.** $z = x \sin y + y \sin x$

**14.** $z = e^x \cos y + e^{-x} \sin y$

**15.** $z = \tan^{-1} \dfrac{y}{x}$

**16.** $z = x \tan^{-1} y$

**17.** $z = \ln \dfrac{y}{x}$

**18.** $z = \ln(x^2 + y^2)$

**19.** $z = e^{xy}$

**20.** $z = e^{x^2 + y}$

**21.** $z = x^2 y + e^{y^2}$

**22.** $z = xy^3 - \ln x^2$

*In Problems 23–28, find the differential $dw$ of each function.*

**23.** $w = xe^{yz} + ye^{xz} + ze^{xy}$

**24.** $w = x^2 y + y^2 z + z^2 x$

**25.** $w = \ln(x^2 + y^2 + z^2)$

**26.** $w = \ln(xy) + \ln(xz) + \ln(yz)$

**27.** $w = xyz$

**28.** $w = \dfrac{xyz}{x + y + z}$

*In Problems 29–32, show that the function $z = f(x, y)$ is differentiable at any point $(x_0, y_0)$ in its domain by:*

(a) *Finding $\Delta z$.*

(b) *Finding $\eta_1$ and $\eta_2$ so that*
$\Delta z = f_x(x_0, y_0)\Delta x + f_y(x_0, y_0)\Delta y + \eta_1 \Delta x + \eta_2 \Delta y$ *holds.*

(c) *Showing that* $\displaystyle\lim_{(\Delta x, \Delta y) \to (0, 0)} \eta_1 = 0$ *and* $\displaystyle\lim_{(\Delta x, \Delta y) \to (0, 0)} \eta_2 = 0$.

**29.** $z = f(x, y) = xy^2 - 2xy$

**30.** $z = f(x, y) = 3x^2 + y^2$

**31.** $z = f(x, y) = \dfrac{y}{x}$

**32.** $z = f(x, y) = \dfrac{2x}{y}$

*In Problems 33–36, for each function $z = f(x, y)$ use $f(x_0, y_0)$ and the differential $dz$ to approximate the given value of $f$.*

**33.** Approximate $f(2.3, -2.1)$ if $z = \dfrac{16}{x^2 + y^2}$ and $f(x_0, y_0) = f(2, -2)$.

**34.** Approximate $f(-0.9, 1.9)$ if $z = \dfrac{xy}{x + y}$ and $f(x_0, y_0) = f(-1, 2)$.

**35.** Approximate $f(2.03, 0.1)$ if $z = x^2 y + e^{y^2}$ and $f(x_0, y_0) = f(2, 0)$.

**36.** Approximate $f(0.9, 2.1)$ if $z = e^x \ln(xy)$ and $f(x_0, y_0) = f(1, 2)$.

## Applications and Extensions

**37. Area of a Triangle**  Approximate the increase in the area of a triangle if its base is increased from 2 to 2.05 cm and its height is increased from 5 to 5.1 cm.

**38. Area of a Triangle**  Approximate the change in the area of a triangle if the base is increased from 5 to 5.1 cm and the height is decreased from 10 to 9.8 cm.

**39. Volume and Surface Area of a Cylinder**

(a) Use a differential to approximate the change in the volume of a right circular cylinder if the height changes from 2 to 2.1 cm and the radius changes from 0.5 to 0.49 cm.

(b) Approximate the change in the surface area of the cylinder. Assume that the cylinder is closed on the top and on the bottom.

**40. Electricity**  In a parallel circuit, the total resistance $R$ due to two sources of resistance $R_1$ and $R_2$ is given by $\dfrac{1}{R} = \dfrac{1}{R_1} + \dfrac{1}{R_2}$. If $R_1 = 50\,\Omega$ with a possible error of 1.2% and $R_2 = 75\,\Omega$ with a possible error of 1%, what is the approximate maximum variation in the total resistance?

**41. Specific Gravity**  The specific gravity of an object is defined as $s = \dfrac{a}{a - w}$, where $a$ is the weight of the object in the air and $w$ is its weight in water. If $a$ is found to be 6 lb with a possible error of 1% and $w$ is 5 lb with a possible error of 2%, what is the approximate maximum variation in the specific gravity?

**42. Volume**  A grain silo consists of a hemisphere mounted on a cylinder of the same radius (see the figure). The height and radius of the cylinder are measured as 14 m and 5 m, respectively. However, the device used to make this measurement was found to be in error by 1%. What is the approximate maximum variation in the volume of the silo?

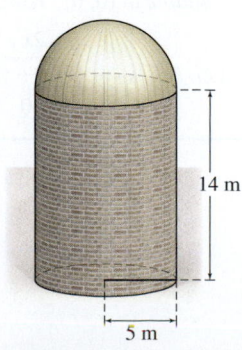

14 m

5 m

**43. Snell's Law**   The index of refraction is defined as $\mu = \dfrac{\sin i}{\sin r}$,
where $i$ is the angle of incidence and $r$ is the angle of refraction.
If $i = 30°$ and $r = 60°$, and each measurement is subject to a
possible error of 2%, what is the approximate maximum relative
error for $\mu$?

**44. Ideal Gas Law**   The equation $PV = kT$, where $k$ is a constant,
relates the pressure $P$, volume $V$, and temperature $T$ of one mole
of an ideal gas. If $P = 0.1\,\text{g/mm}^2$, $V = 12\,\text{mm}^3$, and $T = 32\,°\text{C}$,
approximate the change in $P$ if $V$ and $T$ change to $15\,\text{mm}^3$
and $29\,°\text{C}$, respectively.

**45. Determining Density**   In a laboratory experiment, the average
density $\rho$ of a spherical object is computed by measuring the
mass $m$ and diameter $d$ of the sphere. In all empirical
measurements, there is error caused by the inaccuracy of the tool,
the scientist, or both. Suppose the experimental measurements
are $m = 24.0\,\text{g} \pm 1\%$ and $d = 1.5\,\text{cm} \pm 5.0\%$.

(a) Find the approximate maximum change in the computed
density.

(b) Find the approximate maximum percentage error in the
computed density.

**46. Geometry**   Two sides, $b$ and $c$, and the included angle $\alpha$ of a
triangle are measured by a ruler (for sides) and protractor (for
angles), which are subject to errors of 2% and 3%, respectively.
The area of the triangle is then computed from the formula

$$A = \frac{1}{2}bc \sin \alpha$$

(a) Show that $\dfrac{dA}{A} = \dfrac{db}{b} + \dfrac{dc}{c} + \cot \alpha \, d\alpha$.

(b) If $\alpha = \dfrac{\pi}{4}$, what is the approximate maximum relative
variation in the computation of $A$?

**47. (a)** Find the partial derivatives $f_x$ and $f_y$ of the function
$f(x, y) = \sqrt{x^2 + y^2 - 2x - 6y + 10}$.

(b) Determine where the partial derivatives are continuous.

[CAS] (c) Graph the surface $f(x, y) = \sqrt{x^2 + y^2 - 2x - 6y + 10}$.

(d) Use the graph to give a geometric interpretation to the
discontinuity found in (b).

**48.** If $x = r \cos \theta$ and $y = r \sin \theta$, show that $x\,dy - y\,dx = r^2 d\theta$.

*In Problems 49 and 50, show that each function $f$ has partial
derivatives at $(0, 0)$. Also show that $f$ is not continuous at $(0, 0)$. Is $f$
differentiable at $(0, 0)$? Explain.*

**49.** Let $f(x, y) = \begin{cases} \dfrac{2xy}{x^2 + y^2} & \text{if} \quad (x, y) \neq (0, 0) \\ 0 & \text{if} \quad (x, y) = (0, 0) \end{cases}$

**50.** Let $f(x, y) = \begin{cases} \dfrac{xy(1 + y^2)}{x^2 + y^2} & \text{if} \quad (x, y) \neq (0, 0) \\ 0 & \text{if} \quad (x, y) = (0, 0) \end{cases}$

**51.** Let $f(x, y) = \begin{cases} \dfrac{xy - 1}{x^2 + y^2 - 2} & \text{if} \quad x^2 + y^2 \neq 2 \\ \dfrac{1}{2} & \text{if} \quad x^2 + y^2 = 2 \end{cases}$

Show that $f_x(1, 1)$ and $f_y(1, 1)$ each exist, but $f$ is not
differentiable at $(1, 1)$.

**52.** Let $f(x, y) = \begin{cases} \dfrac{x^2 y^2}{x^4 + y^4} & \text{if} \quad (x, y) \neq (0, 0) \\ 0 & \text{if} \quad (x, y) = (0, 0) \end{cases}$

Show that $f_x(0, 0)$ and $f_y(0, 0)$ each exist, but $f$ is not
differentiable at $(0, 0)$.

## Challenge Problems

**53. Electrical Resistance**   The electrical resistance $R$ of a wire is
given by $R = \dfrac{\rho L}{A}$, where $L$ is the length of the wire, $A$ is its
cross-sectional area, and $\rho$ is the resistivity of the wire. If the
temperature of the wire increases by $\Delta T$, the wire expands so $L$
and $A$ change. The resistivity $\rho$ also changes. The change in the
length $L$ is given by $\Delta L = L_0 \alpha \Delta T$, the change in the radius $r$
is $\Delta r = r_0 \alpha \Delta T$, and the change in the resistivity $\rho$
is $\Delta \rho = \rho_0 K \Delta T$, where $\alpha$ and $K$ are constants that depend on the
material being heated, $L_0$ is the initial length, $r_0$ is the initial
radius, and $\rho_0$ is the initial resistivity.

A copper wire with a circular cross-sectional area $A$ has a
resistance $R$ of $0.50\,\Omega$ at $0\,°\text{C}$. But suppose the temperature of the
wire increases to $40\,°\text{C}$ from the heat generated by the current
passing through the wire. For copper, $\alpha = 5.1 \times 10^{-5}\,(°\text{C})^{-1}$
and $K = 3.93 \times 10^{-3}\,(°\text{C})^{-1}$.

(a) Find the resistance $R$ of the wire at $40\,°\text{C}$.

(b) Find the approximate maximum percentage change in
the resistance $R$ of the wire caused by the change in
temperature.

(c) Which effect contributes more to the change in resistance:
the expansion of the wire or the increase in the
resistivity?

**54. (a)** Find the differential of $f(x, y, z) = x^{y^z}$.

(b) Find the differential of $g(x, y, z) = (x^y)^z$.

# 12.5 Chain Rules

**OBJECTIVES**  *When you finish this section, you should be able to:*

**1** Differentiate functions of several variables where each variable is a function of a single variable (p. 893)

**2** Differentiate functions of several variables where each variable is a function of two or more variables (p. 895)

**3** Differentiate an implicitly defined function of several variables (p. 898)

**4** Use a Chain Rule in a proof (p. 900)

## 1 Differentiate Functions of Several Variables Where Each Variable Is a Function of a Single Variable

**NEED TO REVIEW?**  The Chain Rule is discussed in Section 3.1, pp. 208–215.

Recall that the Chain Rule is used to find the derivative of a composite function. For functions of several variables, there is more than one version of the Chain Rule. The first Chain Rule is used when the independent variables of a composite function are each a function of a single variable $t$. For example, for the function $z = f(x, y)$ of two variables, if $x = x(t)$ and $y = y(t)$, then the composite function $z = f(x(t), y(t))$ is a function of a single independent variable $t$.

> **THEOREM  Chain Rule I: One Independent Variable**
>
> If $x = x(t)$ and $y = y(t)$ are differentiable functions of $t$, and if $z = f(x, y)$ is a differentiable function of $x$ and $y$, then $z = f(x(t), y(t))$ is a differentiable function of $t$. Moreover,
>
> $$\boxed{\dfrac{dz}{dt} = \dfrac{\partial z}{\partial x}\dfrac{dx}{dt} + \dfrac{\partial z}{\partial y}\dfrac{dy}{dt}}$$

**Proof**  Since $\dfrac{dz}{dt} = \lim\limits_{\Delta t \to 0} \dfrac{\Delta z}{\Delta t}$, we seek an expression for $\Delta z$. Since $z = f(x, y)$ is differentiable, the change $\Delta z$ is

$$\Delta z = \frac{\partial z}{\partial x}\Delta x + \frac{\partial z}{\partial y}\Delta y + \eta_1 \Delta x + \eta_2 \Delta y$$

where $\eta_1$ and $\eta_2$ are functions of $\Delta x$ and $\Delta y$ and $\lim\limits_{(\Delta x, \Delta y) \to (0,0)} \eta_1 = 0$ and $\lim\limits_{(\Delta x, \Delta y) \to (0,0)} \eta_2 = 0$. Next we divide both sides by $\Delta t$.

$$\frac{\Delta z}{\Delta t} = \frac{\partial z}{\partial x}\frac{\Delta x}{\Delta t} + \frac{\partial z}{\partial y}\frac{\Delta y}{\Delta t} + \eta_1 \frac{\Delta x}{\Delta t} + \eta_2 \frac{\Delta y}{\Delta t}$$

Then

$$\frac{dz}{dt} = \lim_{\Delta t \to 0} \frac{\Delta z}{\Delta t} = \lim_{\Delta t \to 0} \left[ \frac{\partial z}{\partial x}\frac{\Delta x}{\Delta t} + \frac{\partial z}{\partial y}\frac{\Delta y}{\Delta t} + \eta_1 \frac{\Delta x}{\Delta t} + \eta_2 \frac{\Delta y}{\Delta t} \right]$$

In the right-hand expression, $\dfrac{\partial z}{\partial x}$ and $\dfrac{\partial z}{\partial y}$ are evaluated at $(x(t), y(t))$ and do not depend on $\Delta t$. Also, since $x = x(t)$ and $y = y(t)$ are differentiable,

$$\lim_{\Delta t \to 0} \frac{\Delta x}{\Delta t} = \frac{dx}{dt} \qquad \text{and} \qquad \lim_{\Delta t \to 0} \frac{\Delta y}{\Delta t} = \frac{dy}{dt}$$

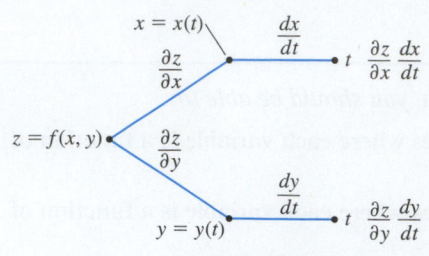

**Figure 38** $\dfrac{dz}{dt} = \dfrac{\partial z}{\partial x}\dfrac{dx}{dt} + \dfrac{\partial z}{\partial y}\dfrac{dy}{dt}$

Furthermore, as $\Delta t \to 0$, then $(\Delta x, \Delta y) \to (0, 0)$, so that $\eta_1 \to 0$ and $\eta_2 \to 0$. Putting this all together, we get

$$\frac{dz}{dt} = \frac{\partial z}{\partial x} \lim_{\Delta t \to 0} \frac{\Delta x}{\Delta t} + \frac{\partial z}{\partial y} \lim_{\Delta t \to 0} \frac{\Delta y}{\Delta t} + \lim_{\Delta t \to 0} \eta_1 \cdot \lim_{\Delta t \to 0} \frac{\Delta x}{\Delta t} + \lim_{\Delta t \to 0} \eta_2 \cdot \lim_{\Delta t \to 0} \frac{\Delta y}{\Delta t}$$

$$= \frac{\partial z}{\partial x}\frac{dx}{dt} + \frac{\partial z}{\partial y}\frac{dy}{dt} + 0 \cdot \frac{dx}{dt} + 0 \cdot \frac{dy}{dt} = \frac{\partial z}{\partial x}\frac{dx}{dt} + \frac{\partial z}{\partial y}\frac{dy}{dt} \qquad \blacksquare$$

The tree diagram in Figure 38 may help you to remember the form of Chain Rule I.

---

**CALC CLIP** ▶ **EXAMPLE 1** Differentiating a Function of Two Variables Where Each Variable Is a Function of $t$

Let $z = x^2 y - y^2 x$, where $x = \sin t$ and $y = e^t$. Find $\dfrac{dz}{dt}$.

**Solution** We begin by finding the partial derivatives $\dfrac{\partial z}{\partial x}$ and $\dfrac{\partial z}{\partial y}$, and the derivatives $\dfrac{dx}{dt}$ and $\dfrac{dy}{dt}$.

$$\frac{\partial z}{\partial x} = 2xy - y^2 \qquad \frac{\partial z}{\partial y} = x^2 - 2xy \qquad \frac{dx}{dt} = \cos t \qquad \frac{dy}{dt} = e^t$$

Then use Chain Rule I to find $\dfrac{dz}{dt}$.

$$\underbrace{\frac{dz}{dt}}_{\text{Chain Rule I}} = \frac{\partial z}{\partial x}\frac{dx}{dt} + \frac{\partial z}{\partial y}\frac{dy}{dt} = (2xy - y^2)(\cos t) + (x^2 - 2xy)(e^t) \tag{1}$$

Since $z$ is a function of $t$, we express $\dfrac{dz}{dt}$ in terms of $t$.

$$\frac{dz}{dt} = (2e^t \sin t - e^{2t})(\cos t) + (\sin^2 t - 2e^t \sin t)(e^t) \qquad x = \sin t,\ y = e^t$$

$$= e^t[\sin(2t) - e^t \cos t + \sin^2 t - 2e^t \sin t] \qquad 2\sin t \cos t = \sin(2t) \tag{2}$$

$\blacksquare$

When $\dfrac{dz}{dt}$ is expressed in terms of $x$, $y$, and $t$, as in (1), we say it is expressed in **mixed form**. When $\dfrac{dz}{dt}$ is expressed in terms of $t$ alone, as in (2), we say it is in **final form**. When computations become involved, we will leave our answers in mixed form.

**NOW WORK** Problem 3.

---

**EXAMPLE 2** Differentiating a Function of Two Variables Where Each Variable Is a Function of $t$

Let $z = e^x \sin y$, where $x = e^t$ and $y = \dfrac{\pi}{3}e^{-t}$. Find $\dfrac{dz}{dt}$ at $t = 0$.

**Solution** Begin by finding the partial derivatives $\dfrac{\partial z}{\partial x}$ and $\dfrac{\partial z}{\partial y}$, and the derivatives $\dfrac{dx}{dt}$ and $\dfrac{dy}{dt}$.

$$\frac{\partial z}{\partial x} = e^x \sin y \qquad \frac{\partial z}{\partial y} = e^x \cos y \qquad \frac{dx}{dt} = e^t \qquad \frac{dy}{dt} = -\frac{\pi}{3}e^{-t}$$

Then use Chain Rule I to find $\dfrac{dz}{dt}$.

$$\underset{\text{Chain Rule I}}{\dfrac{dz}{dt} = \dfrac{\partial z}{\partial x}\dfrac{dx}{dt} + \dfrac{\partial z}{\partial y}\dfrac{dy}{dt}} = (e^x \sin y)e^t + (e^x \cos y)\left(-\dfrac{\pi}{3}e^{-t}\right)$$

The above expression for $\dfrac{dz}{dt}$ is in mixed form. We can use it to evaluate $\dfrac{dz}{dt}$ at $t=0$. When $t=0$, then $x = e^0 = 1$ and $y = \dfrac{\pi}{3}e^0 = \dfrac{\pi}{3}$. So, when $t=0$,

$$\dfrac{dz}{dt}\bigg|_{t=0} = \left(e \sin \dfrac{\pi}{3}\right)(1) + \left(e \cos \dfrac{\pi}{3}\right)\left(-\dfrac{\pi}{3}\right) = \dfrac{e\sqrt{3}}{2} - \dfrac{\pi e}{6} = \dfrac{e}{6}(3\sqrt{3} - \pi) \qquad \blacksquare$$

**NOW WORK** Problem 41.

Chain Rule I extends to functions of three or more variables, where each of these variables is a function of a single variable $t$. If $z = f(x_1, x_2, \ldots, x_n)$ is differentiable and each variable $x_i = x_i(t)$, $i = 1, 2, \ldots, n$, is a differentiable function of $t$, then

$$\dfrac{dz}{dt} = \dfrac{\partial z}{\partial x_1}\dfrac{dx_1}{dt} + \dfrac{\partial z}{\partial x_2}\dfrac{dx_2}{dt} + \cdots + \dfrac{\partial z}{\partial x_n}\dfrac{dx_n}{dt}$$

where each of the partial derivatives $\dfrac{\partial z}{\partial x_1}, \ldots, \dfrac{\partial z}{\partial x_n}$ is expressed in terms of $t$.

**EXAMPLE 3** Differentiating a Function of Three Variables Where Each Variable Is a Function of $t$

Find $\dfrac{dw}{dt}$ if $w = x^2 y + y^2 z$, where $x = \sin t$, $y = \cos t$, and $z = 5t$.

**Solution** Use Chain Rule I since each variable is a function of a single variable $t$.

$$\dfrac{dw}{dt} = \dfrac{\partial w}{\partial x}\dfrac{dx}{dt} + \dfrac{\partial w}{\partial y}\dfrac{dy}{dt} + \dfrac{\partial w}{\partial z}\dfrac{dz}{dt}$$

$$= (2xy)(\cos t) + (x^2 + 2yz)(-\sin t) + y^2(5)$$

Now since $w$ is a function of $t$, write the derivative in terms of $t$ alone.

**NOTE** For functions of a single variable $t$, it is sometimes easier to express the function in terms of $t$, and then differentiate it.

$$\dfrac{dw}{dt} = 2 \sin t \cos^2 t - \sin^3 t - 10t \sin t \cos t + 5 \cos^2 t \qquad \blacksquare$$

**NOW WORK** Example 3 by expressing $w$ as a function of $t$.

**NOW WORK** Problem 17.

## 2 Differentiate Functions of Several Variables Where Each Variable Is a Function of Two or More Variables

A second version of the Chain Rule is used for differentiating $z = f(x, y)$, where $x$ and $y$ are each functions of two independent variables $u$ and $v$. For example, if $x = g(u, v)$ and $y = h(u, v)$, then the composite function $z = f(x, y) = f(g(u, v), h(u, v))$ is a function of the two variables $u$ and $v$. We seek the partial derivatives $\dfrac{\partial z}{\partial u}$ and $\dfrac{\partial z}{\partial v}$.

**THEOREM  Chain Rule II: Two Independent Variables**

Let $z = f(g(u, v), h(u, v))$ be the composite of $z = f(x, y)$, where $x = g(u, v)$ and $y = h(u, v)$. If $g$ and $h$ are each continuous and have continuous first-order partial derivatives at a point $(u, v)$ in the interior of the domains of both $g$ and $h$, and if $f$ is differentiable in some disk centered at the point $(x, y) = (g(u, v), h(u, v))$, then

$$\frac{\partial z}{\partial u} = \frac{\partial z}{\partial x}\frac{\partial x}{\partial u} + \frac{\partial z}{\partial y}\frac{\partial y}{\partial u} \qquad \text{and} \qquad \frac{\partial z}{\partial v} = \frac{\partial z}{\partial x}\frac{\partial x}{\partial v} + \frac{\partial z}{\partial y}\frac{\partial y}{\partial v}$$

**Proof** To find $\dfrac{\partial z}{\partial u}$, we hold $v$ fixed. Then $x = g(u, v)$ and $y = h(u, v)$ are functions of $u$ alone, and we can use Chain Rule I. Then $\dfrac{\partial z}{\partial u} = \dfrac{\partial z}{\partial x}\dfrac{dx}{du} + \dfrac{\partial z}{\partial y}\dfrac{dy}{du}$. Now replace $\dfrac{dx}{du}$ with $\dfrac{\partial x}{\partial u}$, and replace $\dfrac{dy}{du}$ with $\dfrac{\partial y}{\partial u}$.

A similar argument is used for finding $\dfrac{\partial z}{\partial v}$. ∎

The tree diagrams in Figures 39 and 40 may help you remember the form of Chain Rule II.

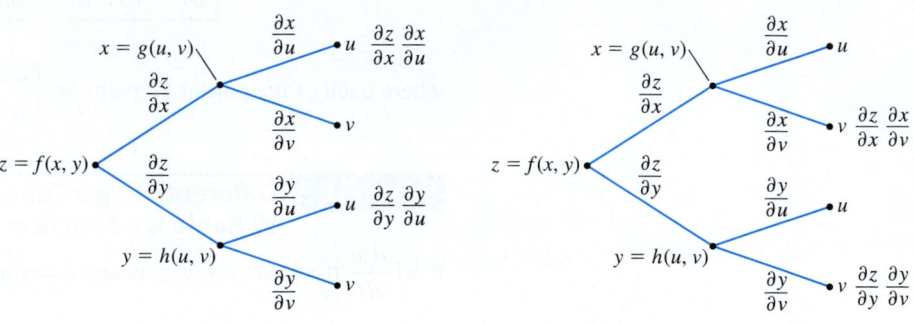

**Figure 39** $\dfrac{\partial z}{\partial u} = \dfrac{\partial z}{\partial x}\dfrac{\partial x}{\partial u} + \dfrac{\partial z}{\partial y}\dfrac{\partial y}{\partial u}$     **Figure 40** $\dfrac{\partial z}{\partial v} = \dfrac{\partial z}{\partial x}\dfrac{\partial x}{\partial v} + \dfrac{\partial z}{\partial y}\dfrac{\partial y}{\partial v}$

---

**EXAMPLE 4  Differentiating a Function of Two Variables Where Each Variable Is a Function of Two Variables**

Find $\dfrac{\partial z}{\partial u}$ and $\dfrac{\partial z}{\partial v}$ if $z = f(x, y) = x^2 + xy - y^2$ and $x = e^{2u+v}$ and $y = \ln\dfrac{v}{u}$.

**Solution** The function $z = f(x, y)$ is a composite function of two independent variables $u$ and $v$. So, we use Chain Rule II.

$$\frac{\partial z}{\partial u} = \frac{\partial z}{\partial x}\frac{\partial x}{\partial u} + \frac{\partial z}{\partial y}\frac{\partial y}{\partial u} \qquad \text{and} \qquad \frac{\partial z}{\partial v} = \frac{\partial z}{\partial x}\frac{\partial x}{\partial v} + \frac{\partial z}{\partial y}\frac{\partial y}{\partial v}$$

Since

$$\frac{\partial z}{\partial x} = 2x + y \qquad \frac{\partial z}{\partial y} = x - 2y \qquad \frac{\partial x}{\partial u} = 2e^{2u+v} \qquad \frac{\partial y}{\partial u} = \frac{\partial}{\partial u}(\ln v - \ln u) = -\frac{1}{u}$$

we have

$$\frac{\partial z}{\partial u} = \underbrace{(2x + y)}_{\frac{\partial z}{\partial x}}\underbrace{(2e^{2u+v})}_{\frac{\partial x}{\partial u}} + \underbrace{(x - 2y)}_{\frac{\partial z}{\partial y}}\underbrace{\left(-\frac{1}{u}\right)}_{\frac{\partial y}{\partial u}}$$

$$= \left(2e^{2u+v} + \ln\frac{v}{u}\right)(2e^{2u+v}) - \left(e^{2u+v} - 2\ln\frac{v}{u}\right)\left(\frac{1}{u}\right) \qquad x = e^{2u+v}; \, y = \ln\frac{v}{u}$$

Similarly, since $\dfrac{\partial x}{\partial v} = e^{2u+v}$ and $\dfrac{\partial y}{\partial v} = \dfrac{1}{v}$, we have

$$\dfrac{\partial z}{\partial v} = \underbrace{(2x+y)}_{\dfrac{\partial z}{\partial x}}\underbrace{(e^{2u+v})}_{\dfrac{\partial x}{\partial v}} + \underbrace{(x-2y)}_{\dfrac{\partial z}{\partial y}}\underbrace{\left(\dfrac{1}{v}\right)}_{\dfrac{\partial y}{\partial v}}$$

$$= \left(2e^{2u+v} + \ln\dfrac{v}{u}\right)(e^{2u+v}) + \left(e^{2u+v} - 2\ln\dfrac{v}{u}\right)\left(\dfrac{1}{v}\right) \qquad x = e^{2u+v};\ y = \ln\dfrac{v}{u} \qquad ■$$

Notice that since $z$ is a function of $u$ and $v$, the final form of $\dfrac{\partial z}{\partial u}$ and $\dfrac{\partial z}{\partial v}$ is expressed in terms of only $u$ and $v$.

NOW WORK Problem 31.

The form of Chain Rule II stays the same if $z$ is a function of $m \geq 3$ variables. That is, if $z = f(x_1, x_2, \ldots, x_m)$ is a differentiable function, and if each of the variables $x_1 = g_1(u_1, u_2, \ldots, u_n), x_2 = g_2(u_1, u_2, \ldots, u_n), \ldots, x_m = g_m(u_1, u_2, \ldots, u_n)$ has continuous first-order partial derivatives, then the composite function $z = f(g_1(u_1, \ldots, u_n), g_2(u_1, \ldots, u_n), \ldots, g_m(u_1, \ldots, u_n))$ is a function of $u_1, u_2, \ldots, u_n$, and the partial derivatives are found using an extension of Chain Rule II.

$$\dfrac{\partial z}{\partial u_1} = \dfrac{\partial z}{\partial x_1}\dfrac{\partial x_1}{\partial u_1} + \dfrac{\partial z}{\partial x_2}\dfrac{\partial x_2}{\partial u_1} + \cdots + \dfrac{\partial z}{\partial x_m}\dfrac{\partial x_m}{\partial u_1}$$

$$\dfrac{\partial z}{\partial u_2} = \dfrac{\partial z}{\partial x_1}\dfrac{\partial x_1}{\partial u_2} + \dfrac{\partial z}{\partial x_2}\dfrac{\partial x_2}{\partial u_2} + \cdots + \dfrac{\partial z}{\partial x_m}\dfrac{\partial x_m}{\partial u_2}$$

$$\vdots$$

$$\dfrac{\partial z}{\partial u_n} = \dfrac{\partial z}{\partial x_1}\dfrac{\partial x_1}{\partial u_n} + \dfrac{\partial z}{\partial x_2}\dfrac{\partial x_2}{\partial u_n} + \cdots + \dfrac{\partial z}{\partial x_m}\dfrac{\partial x_m}{\partial u_n}$$

These partial derivatives can be written more compactly as

$$\dfrac{\partial z}{\partial u_i} = \sum_{j=1}^{m} \dfrac{\partial z}{\partial x_j}\dfrac{\partial x_j}{\partial u_i} \qquad i = 1, 2, \ldots, n$$

### EXAMPLE 5   Differentiating a Function of Three Variables Where Each Variable Is a Function of Four Variables

Find $\dfrac{\partial w}{\partial u}, \dfrac{\partial w}{\partial v}, \dfrac{\partial w}{\partial s}, \dfrac{\partial w}{\partial t}$ for the function $w = f(x, y, z) = x^2 + y^2 + z^2$, where $x = uvst$, $y = e^{u+v+s+t}$, and $z = u + 2v + 3s + 4t$.

**Solution** We use an extension of Chain Rule II, where each variable $x$, $y$, and $z$ is a function of four variables: $u, v, s, t$.

$$\dfrac{\partial w}{\partial u} = \dfrac{\partial w}{\partial x}\dfrac{\partial x}{\partial u} + \dfrac{\partial w}{\partial y}\dfrac{\partial y}{\partial u} + \dfrac{\partial w}{\partial z}\dfrac{\partial z}{\partial u}$$

$$= (2x)(vst) + (2y)(e^{u+v+s+t}) + (2z)(1)$$

$$= 2uv^2s^2t^2 + 2e^{2(u+v+s+t)} + 2(u + 2v + 3s + 4t)$$

$$\dfrac{\partial w}{\partial v} = \dfrac{\partial w}{\partial x}\dfrac{\partial x}{\partial v} + \dfrac{\partial w}{\partial y}\dfrac{\partial y}{\partial v} + \dfrac{\partial w}{\partial z}\dfrac{\partial z}{\partial v}$$

$$= (2x)(ust) + (2y)(e^{u+v+s+t}) + (2z)(2)$$

$$= 2u^2vs^2t^2 + 2e^{2(u+v+s+t)} + 4(u + 2v + 3s + 4t)$$

$$\dfrac{\partial w}{\partial s} = \dfrac{\partial w}{\partial x}\dfrac{\partial x}{\partial s} + \dfrac{\partial w}{\partial y}\dfrac{\partial y}{\partial s} + \dfrac{\partial w}{\partial z}\dfrac{\partial z}{\partial s}$$

$$= (2x)(uvt) + (2y)(e^{u+v+s+t}) + (2z)(3)$$

$$= 2u^2v^2st^2 + 2e^{2(u+v+s+t)} + 6(u + 2v + 3s + 4t)$$

$$\dfrac{\partial w}{\partial t} = \dfrac{\partial w}{\partial x}\dfrac{\partial x}{\partial t} + \dfrac{\partial w}{\partial y}\dfrac{\partial y}{\partial t} + \dfrac{\partial w}{\partial z}\dfrac{\partial z}{\partial t}$$

$$= (2x)(uvs) + (2y)(e^{u+v+s+t}) + (2z)(4)$$

$$= 2u^2v^2s^2t + 2e^{2(u+v+s+t)} + 8(u + 2v + 3s + 4t)$$

Again, notice that the partial derivatives of $f$ in Example 5 are expressed in terms of $u$, $v$, $s$, and $t$ alone. ■

NOW WORK Problem 37.

## 3 Differentiate an Implicitly Defined Function of Several Variables

NEED TO REVIEW? Implicit differentiation is discussed in Section 3.2, pp. 219–224.

If a differentiable function $y = f(x)$ of one variable is defined implicitly by the equation $F(x, y) = 0$, then $F(x, f(x)) = 0$ for all $x$ in the domain of $f$. That is, $F(x, f(x)) \equiv 0$.

We can find the derivative $\dfrac{dy}{dx}$ as follows. Let

$$z = F(u, y) \qquad \text{where} \quad u = x \quad \text{and} \quad y = f(x)$$

Since $u$ and $y$ are functions of one independent variable $x$, we use Chain Rule I to find $\dfrac{dz}{dx}$.

$$\frac{dz}{dx} = \frac{\partial F}{\partial u} \cdot \frac{du}{dx} + \frac{\partial F}{\partial y} \cdot \frac{dy}{dx}$$

Since the composite function $z = F(u, y) = F(x, f(x)) \equiv 0$, the derivative $\dfrac{dz}{dx} = 0$. Also because $u = x$, $\dfrac{\partial F}{\partial u} = \dfrac{\partial F}{\partial x}$ and $\dfrac{du}{dx} = 1$. So,

$$0 = \frac{\partial F}{\partial x} \cdot 1 + \frac{\partial F}{\partial y} \cdot \frac{dy}{dx}$$

Now if $\dfrac{\partial F}{\partial y} \neq 0$, we can solve for $\dfrac{dy}{dx}$.

$$\frac{dy}{dx} = -\frac{\dfrac{\partial F}{\partial x}}{\dfrac{\partial F}{\partial y}} = -\frac{F_x}{F_y} \qquad \frac{\partial F}{\partial y} \neq 0$$

### Implicit Differentiation Formula I

Suppose $F$ is a differentiable function and $y = f(x)$ is a function defined implicitly by the equation $F(x, y) = 0$. Then

$$\boxed{\frac{dy}{dx} = -\frac{F_x(x, y)}{F_y(x, y)}} \tag{3}$$

provided $\dfrac{\partial F}{\partial y} = F_y(x, y) \neq 0$.

### EXAMPLE 6   Differentiating an Implicitly Defined Function

Find $\dfrac{dy}{dx}$ if $y = f(x)$ is defined implicitly by $F(x, y) = e^y \cos x - x - 1 = 0$.

**Solution** First find the partial derivatives of $F$.

$$F_x = \frac{\partial F}{\partial x} = -e^y \sin x - 1 \qquad \text{and} \qquad F_y = \frac{\partial F}{\partial y} = e^y \cos x$$

Then use (3). If $e^y \cos x \neq 0$,

$$\frac{dy}{dx} = -\frac{F_x}{F_y} = -\frac{-e^y \sin x - 1}{e^y \cos x} = \frac{e^y \sin x + 1}{e^y \cos x} \qquad \blacksquare$$

**NOW WORK** Problem 47.

If a differentiable function $z = f(x, y)$ of two variables is defined implicitly by the equation $F(x, y, z) = 0$, we can find the partial derivatives $\dfrac{\partial z}{\partial x}$ and $\dfrac{\partial z}{\partial y}$ by using Chain Rule II.

We begin by letting $w = F(u, v, z)$, where $u = x$, $v = y$, and $z = f(x, y)$. Since the composite function $w = F(x, y, f(x, y)) \equiv 0$, it follows that $\dfrac{\partial w}{\partial x} = 0$ and $\dfrac{\partial w}{\partial y} = 0$. To find an expression for $\dfrac{\partial w}{\partial x}$, use Chain Rule II.

$$\frac{\partial w}{\partial x} = \frac{\partial F}{\partial u} \cdot \frac{\partial u}{\partial x} + \frac{\partial F}{\partial v} \cdot \frac{\partial v}{\partial x} + \frac{\partial F}{\partial z} \cdot \frac{\partial z}{\partial x} = 0$$

Since $u = x$, then $\dfrac{\partial F}{\partial u} = \dfrac{\partial F}{\partial x}$ and $\dfrac{\partial u}{\partial x} = 1$, and since $v = y$, then $\dfrac{\partial F}{\partial v} = \dfrac{\partial F}{\partial y}$ and $\dfrac{\partial v}{\partial x} = 0$. So,

$$\frac{\partial F}{\partial x} \cdot 1 + \frac{\partial F}{\partial y} \cdot 0 + \frac{\partial F}{\partial z} \cdot \frac{\partial z}{\partial x} = 0$$

$$\frac{\partial F}{\partial x} + \frac{\partial F}{\partial z} \cdot \frac{\partial z}{\partial x} = 0$$

If $\dfrac{\partial F}{\partial z} \neq 0$, it follows that

$$\frac{\partial z}{\partial x} = -\frac{\dfrac{\partial F}{\partial x}}{\dfrac{\partial F}{\partial z}} = -\frac{F_x(x, y, z)}{F_z(x, y, z)}$$

In a similar way, it can be shown that

$$\frac{\partial z}{\partial y} = -\frac{F_y(x, y, z)}{F_z(x, y, z)}$$

**Implicit Differentiation Formulas II**

If a differentiable function $z = f(x, y)$ is defined implicitly by the equation $F(x, y, z) = 0$, then

$$\frac{\partial z}{\partial x} = -\frac{F_x(x, y, z)}{F_z(x, y, z)} \qquad \text{and} \qquad \frac{\partial z}{\partial y} = -\frac{F_y(x, y, z)}{F_z(x, y, z)} \tag{4}$$

provided $\dfrac{\partial F}{\partial z} = F_z(x, y, z) \neq 0$.

**EXAMPLE 7  Differentiating an Implicitly Defined Function**

Find $\dfrac{\partial z}{\partial x}$ and $\dfrac{\partial z}{\partial y}$ if $z = f(x, y)$ is defined implicitly by the function

$$F(x, y, z) = x^2 z^2 + y^2 - z^2 + 6yz - 10 = 0$$

**Solution** First find the partial derivatives of $F$.

$$F_x = \frac{\partial F}{\partial x} = 2xz^2 \qquad F_y = \frac{\partial F}{\partial y} = 2y + 6z \qquad F_z = \frac{\partial F}{\partial z} = 2x^2 z - 2z + 6y$$

Then use (4). If $F_z = 2x^2 z - 2z + 6y \neq 0$,

$$\frac{\partial z}{\partial x} = -\frac{2xz^2}{2x^2 z - 2z + 6y} = -\frac{xz^2}{x^2 z - z + 3y}$$

and

$$\frac{\partial z}{\partial y} = -\frac{2y + 6z}{2x^2 z - 2z + 6y} = -\frac{y + 3z}{x^2 z - z + 3y}$$

**NOW WORK** Problem 55.

## 4 Use a Chain Rule in a Proof

Chain Rules are often used in proofs involving functions of two or more variables.

**EXAMPLE 8    Using a Chain Rule in a Proof**

Let $p = f(v - w, v - u, u - w)$ be a differentiable function. Show that

$$\frac{\partial p}{\partial u} + \frac{\partial p}{\partial v} + \frac{\partial p}{\partial w} = 0$$

**Solution** Let $x = v - w$, $y = v - u$, and $z = u - w$. Then $p = f(x, y, z)$. We use an extension of Chain Rule II. Since $\dfrac{\partial x}{\partial u} = 0$, $\dfrac{\partial y}{\partial u} = -1$, and $\dfrac{\partial z}{\partial u} = 1$, we have

$$\frac{\partial p}{\partial u} = \frac{\partial p}{\partial x} \frac{\partial x}{\partial u} + \frac{\partial p}{\partial y} \frac{\partial y}{\partial u} + \frac{\partial p}{\partial z} \frac{\partial z}{\partial u} = \frac{\partial p}{\partial x}(0) + \frac{\partial p}{\partial y}(-1) + \frac{\partial p}{\partial z}(1) = -\frac{\partial p}{\partial y} + \frac{\partial p}{\partial z}$$

Since $\dfrac{\partial x}{\partial v} = 1$, $\dfrac{\partial y}{\partial v} = 1$, and $\dfrac{\partial z}{\partial v} = 0$, we have

$$\frac{\partial p}{\partial v} = \frac{\partial p}{\partial x} \frac{\partial x}{\partial v} + \frac{\partial p}{\partial y} \frac{\partial y}{\partial v} + \frac{\partial p}{\partial z} \frac{\partial z}{\partial v} = \frac{\partial p}{\partial x}(1) + \frac{\partial p}{\partial y}(1) + \frac{\partial p}{\partial z}(0) = \frac{\partial p}{\partial x} + \frac{\partial p}{\partial y}$$

Since $\dfrac{\partial x}{\partial w} = -1$, $\dfrac{\partial y}{\partial w} = 0$, and $\dfrac{\partial z}{\partial w} = -1$, we have

$$\frac{\partial p}{\partial w} = \frac{\partial p}{\partial x} \frac{\partial x}{\partial w} + \frac{\partial p}{\partial y} \frac{\partial y}{\partial w} + \frac{\partial p}{\partial z} \frac{\partial z}{\partial w} = \frac{\partial p}{\partial x}(-1) + \frac{\partial p}{\partial y}(0) + \frac{\partial p}{\partial z}(-1) = -\frac{\partial p}{\partial x} - \frac{\partial p}{\partial z}$$

Adding these, we get

$$\frac{\partial p}{\partial u} + \frac{\partial p}{\partial v} + \frac{\partial p}{\partial w} = \underbrace{\left(-\frac{\partial p}{\partial y} + \frac{\partial p}{\partial z}\right)}_{\frac{\partial p}{\partial u}} + \underbrace{\left(\frac{\partial p}{\partial x} + \frac{\partial p}{\partial y}\right)}_{\frac{\partial p}{\partial v}} + \underbrace{\left(-\frac{\partial p}{\partial x} - \frac{\partial p}{\partial z}\right)}_{\frac{\partial p}{\partial w}} = 0$$

∎

**NOW WORK** Problem 71.

---

# 12.5 Assess Your Understanding

## Concepts and Vocabulary

1. *True or False*   If a differentiable function $z$ is defined implicitly by the equation $F(x, y, z) = 0$, then $\dfrac{\partial z}{\partial x} = \dfrac{F_x(x, y, z)}{F_z(x, y, z)}$, provided $F_z(x, y, z) \neq 0$.

2. *True or False*   If $x = x(t)$ and $y = y(t)$ are differentiable functions of $t$ and if $z = f(x, y)$ is a differentiable function of $x$ and $y$, then $\dfrac{dz}{dt} = \dfrac{\partial z}{\partial x} + \dfrac{\partial z}{\partial y}$.

## Skill Building

Answers to Problems 3–39 are given in final form and in mixed form.

*In Problems 3–14, find $\dfrac{dz}{dt}$ using Chain Rule I.*

3. $z = x^2 + y^2$,    $x = \sin t$,    $y = \cos(2t)$

4. $z = x^2 - y^2$,    $x = \sin(2t)$,    $y = \cos t$

5. $z = x^2 + y^2$,    $x = te^t$,    $y = te^{-t}$

6. $z = x^2 - y^2$,    $x = te^{-t}$,    $y = t^2 e^{-t}$

7. $z = e^u \sin v$,    $u = \sqrt{t}$,    $v = \pi t$

8. $z = e^{u/v}$,    $u = \sqrt{t}$,    $v = t^3 + 1$

9. $z = e^{u/v}$,    $u = te^t$,    $v = e^{t^2}$

10. $z = \ln(uv)$,    $u = t^5$,    $v = \sqrt{t + 1}$

11. $z = e^{x^2 + y^2}$,    $x = \sin(2t)$,    $y = \cos t$

12. $z = e^{x^2 - y^2}$,    $x = \sin(2t)$,    $y = \cos(2t)$

13. $z = \dfrac{xy}{x^2 + y^2}$,    $x = \sin t$,    $y = \cos t$

14. $z = y \ln x + xy + \tan y$,    $x = \dfrac{t}{t + 1}$,    $y = t^3 - t$

*In Problems 15–22, find $\dfrac{dp}{dt}$.*

15. $p = x^2 + y^2 - z^2$,    $x = te^t$,    $y = te^{-t}$,    $z = e^{2t}$

16. $p = x^2 - y^2 - z^2$,    $x = te^{-t}$,    $y = t^2 e^{-t}$,    $z = e^{-t}$

---

**1.** = NOW WORK problem          ⃝ = Graphing technology recommended          **CAS** = Computer Algebra System recommended

**17.** $p = e^x \sin y \cos z$,   $x = \sqrt{t}$,   $y = \pi t$,   $z = \dfrac{t}{2}$

**18.** $p = \ln(xyz)$,   $x = t^5$,   $y = \sqrt{t+1}$,   $z = t^2$

**19.** $p = w \ln\dfrac{u}{v}$,   $u = te^t$,   $v = e^{t^2}$,   $w = e^{2t}$

**20.** $p = we^{u/v}$,   $u = \sqrt{t}$,   $v = t^3 + 1$,   $w = e^t$

**21.** $p = u^2 vw$,   $u = \sin t$,   $v = \cos t$,   $w = e^t$

**22.** $p = \sqrt{uvw}$,   $u = e^t$,   $v = te^t$,   $w = t^2 e^{2t}$

*In Problems 23–34, find $\dfrac{\partial z}{\partial u}$ and $\dfrac{\partial z}{\partial v}$ using Chain Rule II.*

**23.** $z = x^2 + y^2$,   $x = ue^v$,   $y = ve^u$

**24.** $z = x^2 - y^2$,   $x = u \ln v$,   $y = v \ln u$

**25.** $z = e^x \sin y$,   $x = u^2 v$,   $y = \ln(uv)$

**26.** $z = \dfrac{1}{y} \ln x$,   $x = \sqrt{uv}$,   $y = \dfrac{v}{u}$

**27.** $z = \ln(x^2 + y^2)$,   $x = \dfrac{v^2}{u}$,   $y = \dfrac{u}{v^2}$

**28.** $z = x \sin y - y \sin x$,   $x = u^2 v$,   $y = uv^2$

**29.** $z = x^2 + y^2$,   $x = \sin(u - v)$,   $y = \cos(u + v)$

**30.** $z = e^x + y$,   $x = \tan^{-1}\left(\dfrac{u}{v}\right)$,   $y = \ln(u + v)$

**31.** $z = se^r$,   $r = u^2 + v^2$,   $s = \dfrac{v}{u}$

**32.** $z = \sqrt{s^2 + r^2}$,   $s = \ln(uv)$,   $r = \sqrt{uv}$

**33.** $z = xy^2 w^3$,   $x = 2u + v$,   $y = 5u - 3v$,   $w = 2u + 3v$

**34.** $z = x^2 - y^2 + w$,   $x = e^{u+v}$,   $y = uv$,   $w = \dfrac{v}{u}$

*In Problems 35–40, find each partial derivative.*

**35.** Find $\dfrac{\partial f}{\partial u}, \dfrac{\partial f}{\partial v}, \dfrac{\partial f}{\partial w}$ if $f(x, y, z) = x^2 + y^2 + z^2$, $x = uv$,

$y = e^{u + 2v + 3w}$, $z = 2v + 3w$.

**36.** Find $\dfrac{\partial f}{\partial u}, \dfrac{\partial f}{\partial v}, \dfrac{\partial f}{\partial w}$ if $f(x, y, z) = x - y^2 + z^2$, $x = \sqrt{u + v}$,

$y = (u + w) \ln v$, $z = 2 - v + 3w$.

**37.** Find $\dfrac{\partial f}{\partial u}, \dfrac{\partial f}{\partial v}, \dfrac{\partial f}{\partial w}$ if $f(x, y, z) = x \cos y - z \cos x + x^2 yz$,

$x = uvw$, $y = u^2 + v^2 + w^2$, $z = w$.

**38.** Find $\dfrac{\partial f}{\partial u}, \dfrac{\partial f}{\partial v}, \dfrac{\partial f}{\partial w}$ if $f(x, y, z) = x^2 + y^2$, $x = \sin(u - v)$,

$y = \cos(u + v)$, $z = uw^2$.

**39.** Find $\dfrac{\partial f}{\partial u}, \dfrac{\partial f}{\partial v}, \dfrac{\partial f}{\partial w}, \dfrac{\partial f}{\partial t}$ if $f(x, y, z) = x + 2y^2 - z^2$, $x = ut$,

$y = e^{u + 2v + 3w + 4t}$, $z = u + \dfrac{1}{2} v + 4t$.

**40.** Find $\dfrac{\partial f}{\partial u}, \dfrac{\partial f}{\partial v}, \dfrac{\partial f}{\partial w}, \dfrac{\partial f}{\partial t}$ if $f(x, y, z) = x^2 + y^2 + z$,

$x = \sin(u + t)$, $y = \cos(v - t)$, $z = uw^2$.

*In Problems 41–44, for each function $z = f(x, y)$, find $\dfrac{dz}{dt}$ at $t = t_0$.*

**41.** $z = \sin(\pi x + y)$, where $x = e^t$ and $y = t^2$; $t_0 = 0$

**42.** $z = \cos(x + \pi y^2)$, where $x = t^3$ and $y = e^{t/2}$; $t_0 = 0$

**43.** $z = \sqrt{1 + 3x^2 + y^2}$, where $x = \cos t$ and $y = \sin t$; $t_0 = \dfrac{\pi}{6}$

**44.** $z = \sqrt{x^2 + y^2}$, where $x = \ln t$ and $y = \sqrt{t}$; $t_0 = 1$

*In Problems 45 and 46, $z = f(x, y)$, $x = u(t)$, and $y = v(t)$ are differentiable. Use the given information to find $\dfrac{dz}{dt}$ at $t = t_0$.*

**45.** $u(2) = 5$,   $u'(2) = \dfrac{1}{2}$,   $f_x(5, 1) = 4$

$v(2) = 1$,   $v'(2) = 3$,   $f_y(5, 1) = 1$

$t_0 = 2$

**46.** $u(1) = -3$,   $u'(1) = 2$,   $f_x(-3, 0) = 4$

$v(1) = 0$,   $v'(1) = -2$,   $f_y(-3, 0) = 5$

$t_0 = 1$

*In Problems 47–52, $y$ is a function of $x$. Find $\dfrac{dy}{dx}$.*

**47.** $F(x, y) = x^2 y - y^2 x + xy - 5 = 0$

**48.** $F(x, y) = x^3 y^2 - xy + x^2 y - 10 = 0$

**49.** $F(x, y) = x \sin y + y \sin x - 2 = 0$

**50.** $F(x, y) = xe^y + ye^x - xy = 0$

**51.** $F(x, y) = x^{1/3} + y^{1/3} - 1 = 0$

**52.** $F(x, y) = x^{2/3} + y^{2/3} - 1 = 0$

*In Problems 53–58, $z$ is a function of $x$ and $y$. Find $\dfrac{\partial z}{\partial x}$ and $\dfrac{\partial z}{\partial y}$.*

**53.** $F(x, y, z) = xz + 3yz^2 + x^2 y^3 - 5z = 0$

**54.** $F(x, y, z) = x^2 z + y^2 z + x^3 y - 10z = 0$

**55.** $F(x, y, z) = \sin z + y \cos z + xyz - 10 = 0$

**56.** $F(x, y, z) = x \sin y - \cos z + x^2 z = 0$

**57.** $F(x, y, z) = xe^{yz} + ye^{xz} + xyz = 0$

**58.** $F(x, y, z) = e^{yz} \ln x + ye^{xz} - yz = 0$

*In Problems 59 and 60, find $\dfrac{\partial w}{\partial x}, \dfrac{\partial w}{\partial y}$, and $\dfrac{\partial w}{\partial z}$.*

**59.** $w = (2x + 3y)^{4z}$        **60.** $w = (2x)^{3y + 4z}$

## Applications and Extensions

**61. Ideal Gas Law** One mole of a gas obeys the Ideal Gas Law $PV = 20T$, where $P$ is pressure, $V$ is volume, and $T$ is temperature in kelvin (K). If the temperature $T$ of the gas is increasing at the rate of 5 K/s and if, when the temperature is 353 K, the pressure $P$ is $10\,\text{N/m}^2$ and is decreasing at the rate of $2\,\dfrac{\text{N}}{\text{m}^2 \cdot \text{s}}$, find the rate of change of the volume $V$ with respect to time.

**62. Melting Ice** A block of ice of dimensions $l$, $w$, and $h$ is melting. When $l = 3$ m, $w = 2$ m, $h = 1$ m, these variables are changing so that $\dfrac{dl}{dt} = -1$ m/h, $\dfrac{dw}{dt} = -1$ m/h, and $\dfrac{dh}{dt} = -0.5$ m/h.

(a) What is the rate of change in the surface area of the block of ice?

(b) What is the rate of change in the volume of the block of ice?

**63. Wave Equation** The one-dimensional wave equation

$$\frac{\partial^2 f}{\partial x^2} = \frac{1}{v^2} \frac{\partial^2 f}{\partial t^2}$$ describes a wave traveling with speed $v$ along the $x$-axis. The function $f$ represents the displacement $x$ from the equilibrium of the wave at time $t$.

(a) Show that $z = f(x, t) = \sin(x + vt)$ satisfies the wave equation.

(b) Show that $z = f(x, t) = e^{x - vt}$ satisfies the wave equation.

(c) Show that $z = f(x, t) = \sin x + \sin(vt)$ does not satisfy the wave equation.

(d) Show that any twice-differentiable function of the form $f(x + vt)$ is a solution of the wave equation.

**64. Economics** A toy manufacturer's production function satisfies a **Cobb–Douglas model**, $Q(L, M) = 400L^{0.3} M^{0.7}$, where $Q$ is the output in thousands of units, $L$ is the labor in thousands of hours, and $M$ is the machine hours (in thousands). Suppose the labor hours are decreasing at a rate of 4000 h/yr and the machine hours are increasing at a rate of 2000 h/yr. Find the rate of change of production when

(a) $L = 19$ and $M = 21$      (b) $L = 21$ and $M = 20$

**65. Related Rates** Let $y = h(a, b)$, where $a = h(s, t)$ and $b = k(s, t)$. Suppose when $s = 1$ and $t = 3$, we know that

$$\frac{\partial h}{\partial s} = 4 \qquad \frac{\partial k}{\partial s} = -3 \qquad \frac{\partial h}{\partial t} = 1 \qquad \frac{\partial k}{\partial t} = -5$$

Also, suppose

$$h(1, 3) = 6 \qquad k(1, 3) = 2 \qquad f_a(6, 2) = 7 \qquad f_b(6, 2) = 2$$

What are $\dfrac{\partial y}{\partial s}$ and $\dfrac{\partial y}{\partial t}$ at the point $(1, 3)$?

**66. Related Rates** Let $z = f(x, y)$, where $x = g(s, t)$ and $y = h(s, t)$. Suppose when $s = 2$ and $t = -1$, we know that

$$\frac{\partial x}{\partial s} = 5 \qquad \frac{\partial y}{\partial s} = 2 \qquad \frac{\partial x}{\partial t} = -3 \qquad \frac{\partial y}{\partial t} = -2$$

Also, suppose

$$g(2, -1) = 3 \qquad h(2, -1) = 4 \qquad f_x(3, 4) = 12 \qquad f_y(3, 4) = 7$$

What are $\dfrac{\partial z}{\partial s}$ and $\dfrac{\partial z}{\partial t}$ at the point $(2, -1)$?

**67.** If $z = f(x, y)$, where $x = g(u, v)$ and $y = h(u, v)$, find expressions for $\dfrac{\partial^2 z}{\partial u^2}$, $\dfrac{\partial^2 z}{\partial u \, \partial v}$, and $\dfrac{\partial^2 z}{\partial v^2}$.

**68.** If $z = f(x, y)$ and $x = r \cos \theta$, $y = r \sin \theta$, show that

$$\left( \frac{\partial z}{\partial r} \right)^2 + \frac{1}{r^2} \left( \frac{\partial z}{\partial \theta} \right)^2 = \left( \frac{\partial z}{\partial x} \right)^2 + \left( \frac{\partial z}{\partial y} \right)^2$$

**69.** If $z = f(x, y)$, where $x = u \cos \theta - v \sin \theta$ and $y = u \sin \theta + v \cos \theta$, with $\theta$ a constant, show that

$$\left( \frac{\partial f}{\partial u} \right)^2 + \left( \frac{\partial f}{\partial v} \right)^2 = \left( \frac{\partial f}{\partial x} \right)^2 + \left( \frac{\partial f}{\partial y} \right)^2$$

**70.** If $z = f(u - v, v - u)$, show that $\dfrac{\partial z}{\partial u} + \dfrac{\partial z}{\partial v} = 0$.

*Hint:* Let $x = u - v$ and $y = v - u$.

**71.** If $z = vf(u^2 - v^2)$, show that $v \dfrac{\partial z}{\partial u} + u \dfrac{\partial z}{\partial v} = \dfrac{uz}{v}$.

**72.** If $w = f(u)$ and $u = \sqrt{x^2 + y^2 + z^2}$, show that

$$\left( \frac{\partial w}{\partial x} \right)^2 + \left( \frac{\partial w}{\partial y} \right)^2 + \left( \frac{\partial w}{\partial z} \right)^2 = \left( \frac{dw}{du} \right)^2$$

**73.** If $z = f\left( \dfrac{x}{y} \right)$, show that $x \dfrac{\partial z}{\partial x} + y \dfrac{\partial z}{\partial y} = 0$.

**74.** Show that if $z = f\left( \dfrac{u}{v}, \dfrac{v}{w} \right)$, then $u \dfrac{\partial z}{\partial u} + v \dfrac{\partial z}{\partial v} + w \dfrac{\partial z}{\partial w} = 0$.

**75.** Suppose we denote the expression $\dfrac{\partial^2}{\partial x^2} + \dfrac{\partial^2}{\partial y^2}$ by $\Delta$.

If $z = f(x, y)$, where $x = r \cos \theta$ and $y = r \sin \theta$, show that $\Delta f = \dfrac{\partial^2 f}{\partial r^2} + \dfrac{1}{r} \cdot \dfrac{\partial f}{\partial r} + \dfrac{1}{r^2} \cdot \dfrac{\partial^2 f}{\partial \theta^2}$

**76.** Prove that if $F(x, y, z) = 0$ is differentiable, then

$$\frac{\partial z}{\partial x} \cdot \frac{\partial x}{\partial y} \cdot \frac{\partial y}{\partial z} = -1$$

## Challenge Problems

**77. (a)** Suppose that $F(x, y)$ has continuous second-order partial derivatives and $F(x, y) = 0$ defines $y$ as a function of $x$ implicitly. Show that

$$\frac{d^2 y}{dx^2} = -\frac{F_y^2 F_{xx} - 2F_x F_y F_{xy} + F_x^2 F_{yy}}{F_y^3} \qquad F_y \neq 0$$

**(b)** Use the result from (a) to find $\dfrac{d^2 y}{dx^2}$ for $x^3 + 3xy - y^3 = 6$.

**78.** $f(t, x) = \int_0^{x/(2\sqrt{\lambda t})} e^{-u^2} \, du$. Show that $f_t = \lambda f_{xx}$.

# Chapter Review

## THINGS TO KNOW

### 12.1 Functions of Two or More Variables and Their Graphs

*Definitions:*

- Function $z = f(x, y)$ of two variables (p. 852)
- Function $w = f(x, y, z)$ of three variables (p. 852)
- Contour line: The curve formed by the intersection of a surface $z = f(x, y)$ and the plane $z = c$, where $c$ is a constant (p. 854)
- Level curve: the projection of a contour line onto the $xy$-plane (p. 855)

### 12.2 Limits and Continuity

*Definitions:*

- A $\delta$-neighborhood of $P_0$: the set of all points $P$ that lie within a distance $\delta$ of $P_0$ (p. 861)
  - In the plane, a $\delta$-neighborhood of $P_0$ is called a disk of radius $\delta$ centered at $P_0$.
  - In space, a $\delta$-neighborhood of $P_0$ is called a ball of radius $\delta$ centered at $P_0$.
- Limit of a function $f$ of two variables (p. 861)
- Limit of a function $f$ of three variables (p. 862)
- Interior point, boundary point, open set, closed set (p. 867)
- Continuity at a point (p. 868)

*Theorems:*

- Properties of limits (p. 863)
- Properties of continuous functions (p. 869)

### 12.3 Partial Derivatives

- First-order partial derivatives of a function $z = f(x, y)$:

$$f_x(x, y) = \frac{\partial f}{\partial x} = \frac{\partial z}{\partial x} = \lim_{\Delta x \to 0} \frac{f(x + \Delta x, y) - f(x, y)}{\Delta x}$$

$$f_y(x, y) = \frac{\partial f}{\partial y} = \frac{\partial z}{\partial y} = \lim_{\Delta y \to 0} \frac{f(x, y + \Delta y) - f(x, y)}{\Delta y}$$

provided the limits exist. (p. 872)

- Mixed partials: $\dfrac{\partial^2 z}{\partial x\, \partial y} = f_{yx}(x, y)$ and $\dfrac{\partial^2 z}{\partial y\, \partial x} = f_{xy}(x, y)$ (p. 878)
- Equality of mixed partials (Clairaut's Theorem) (p. 878)

### 12.4 Differentiability and the Differential

- Change in $z$: $\Delta z = f(x_0 + \Delta x, y_0 + \Delta y) - f(x_0, y_0)$ (p. 884)
- Differentiability of $z = f(x, y)$ at a point $(x_0, y_0)$ (p. 885)
- Differentials: $dx = \Delta x, \ dy = \Delta y$, $dz = f_x(x_0, y_0)\, dx + f_y(x_0, y_0)\, dy$ (p. 886)

*Summary:*  (p. 890) If $z = f(x, y)$,

- The continuity of the partial derivatives $f_x$ and $f_y$ implies the differentiability of $f$.
- The differentiability of $f$ implies the continuity of $f$.
- The continuity of the partial derivatives $f_x$ and $f_y$ implies the continuity of $f$.
- The existence of the partial derivatives $f_x$ and $f_y$ does not necessarily mean $f$ is differentiable.
- The existence of the partial derivatives $f_x$ and $f_y$ does not necessarily mean $f$ is continuous.

### 12.5 Chain Rules

- Chain Rule I: $\dfrac{dz}{dt} = \dfrac{\partial z}{\partial x} \dfrac{dx}{dt} + \dfrac{\partial z}{\partial y} \dfrac{dy}{dt}$ (p. 893)
- Chain Rule II: $\dfrac{\partial z}{\partial u} = \dfrac{\partial z}{\partial x} \dfrac{\partial x}{\partial u} + \dfrac{\partial z}{\partial y} \dfrac{\partial y}{\partial u}$

  $\dfrac{\partial z}{\partial v} = \dfrac{\partial z}{\partial x} \dfrac{\partial x}{\partial v} + \dfrac{\partial z}{\partial y} \dfrac{\partial y}{\partial v}$ (p. 896)

- Implicit differentiation formula I:

  If $F(x, y) = 0$, $\dfrac{dy}{dx} = -\dfrac{F_x(x, y)}{F_y(x, y)}$

  provided $F_y \neq 0$. (p. 898)

- Implicit differentiation formula II: If $F(x, y, z) = 0$,

  $\dfrac{\partial z}{\partial x} = -\dfrac{F_x(x, y, z)}{F_z(x, y, z)}$ and $\dfrac{\partial z}{\partial y} = -\dfrac{F_y(x, y, z)}{F_z(x, y, z)}$

  provided $F_z \neq 0$. (p. 899)

## OBJECTIVES

| Section | You should be able to ... | Example | Review Exercises |
|---|---|---|---|
| **12.1** | **1** Work with functions of two or three variables (p. 852) | 1–3 | 1–9 |
| | **2** Graph functions of two variables (p. 854) | 4 | 10–13 |
| | **3** Graph level curves (p. 854) | 5, 6 | 14–16 |
| | **4** Describe level surfaces (p. 857) | 7, 8 | 17, 18 |
| **12.2** | **1** Define the limit of a function of several variables (p. 861) | | |
| | **2** Find a limit using properties of limits (p. 863) | 1 | 19, 20 |
| | **3** Examine when limits exist (p. 865) | 2–4 | 21, 22 |
| | **4** Determine where a function is continuous (p. 867) | 5–8 | 23–25 |

| Section | You should be able to ... | Example | Review Exercises |
|---|---|---|---|
| **12.3** | **1** Find the partial derivatives of a function of two variables (p. 871) | 1–4 | 26–31, 41 |
|  | **2** Interpret partial derivatives as the slope of a tangent line (p. 874) | 5 | 32, 33 |
|  | **3** Interpret partial derivatives as a rate of change (p. 875) | 6–8 | 34(a), (b) |
|  | **4** Find second-order partial derivatives (p. 878) | 9, 10 | 35, 36 |
|  | **5** Find the partial derivatives of a function of $n$ variables (p. 879) | 11, 12 | 37–40 |
| **12.4** | **1** Find the change in $z = f(x, y)$ (p. 884) | 1 | 42 |
|  | **2** Show that a function of two variables is differentiable (p. 885) | 2, 3 | 43 |
|  | **3** Use the differential as an approximating tool (p. 887) | 4–6 | 44, 45, 48–51 |
|  | **4** Find the differential of a function of three or more variables (p. 890) | 7 | 46, 47 |
| **12.5** | **1** Differentiate functions of several variables where each variable is a function of a single variable (p. 893) | 1–3 | 34(c), 52, 53 |
|  | **2** Differentiate functions of several variables where each variable is a function of two or more variables (p. 895) | 4, 5 | 54, 55 |
|  | **3** Differentiate an implicitly defined function of several variables (p. 898) | 6, 7 | 56, 57 |
|  | **4** Use a Chain Rule in a proof (p. 900) | 8 | 58 |

## REVIEW EXERCISES

*In Problems 1–3, evaluate each function.*

**1.** $f(x, y) = e^x \ln y$

   **(a)** $f(1, 1)$

   **(b)** $f(x + \Delta x, y)$

   **(c)** $f(x, y + \Delta y)$

**2.** $f(x, y) = 2x^2 + 6xy - y^3$

   **(a)** $f(1, 1)$

   **(b)** $f(x + \Delta x, y)$

   **(c)** $f(x, y + \Delta y)$

**3.** $f(x, y, z) = e^x \sin^{-1}(y + 2z)$

   **(a)** $f\left(\ln 3, \dfrac{1}{2}, \dfrac{1}{4}\right)$   **(b)** $f\left(1, 0, \dfrac{1}{4}\right)$   **(c)** $f(0, 0, 0)$

*In Problems 4–7, find the domain of each function and graph the domain. Use a solid curve to indicate that the domain includes the curve, and a dashed curve to indicate that the domain excludes the curve.*

**4.** $z = f(x, y) = \ln(x^2 - 3y)$

**5.** $z = f(x, y) = \sqrt{9 - x^2 - 4y^2}$

**6.** $z = f(x, y) = \dfrac{25xy}{\sqrt{5 - y^2}}$

**7.** $z = f(x, y) = \dfrac{(y - 3x)^2}{x + 2y}$

**8.** Find the domain of $w = f(x, y, z) = \dfrac{y^2 + z^2}{x^2}$.

**9.** Find the domain of $w = f(x, y, z) = e^{x+y} \ln z$.

*In Problems 10–13, graph each surface.*

**10.** $z = f(x, y) = x - y + 5$

**11.** $z = f(x, y) = \sin x$

**12.** $z = f(x, y) = \ln x$

**13.** $z = f(x, y) = e^y$

**14.** For $z = f(x, y) = x^2 - 2y$, graph the level curves corresponding to $c = -4, -1, 0, 1, 4$.

**15.** For $z = f(x, y) = \sqrt{x^2 + y^2}$, graph the level curves corresponding to $c = 0, 1, 4, 9$.

**16.** For $z = f(x, y) = e^{4x^2 + y^2}$, graph the level curves corresponding to $c = 1, e, e^4, e^{16}$.

**17.** Describe the level surfaces associated with the function $w = f(x, y, z) = 4x^2 + y^2 + z^2$.

**18.** Describe the level surfaces associated with the function $w = f(x, y, z) = x + y + 2z$.

**19.** Find $\displaystyle\lim_{(x,y) \to (\pi/2, 0)} \dfrac{\sin x \cos y}{x}$.

**20.** Find $\displaystyle\lim_{(x,y) \to (1,2)} \dfrac{4x^2 + y^2}{2x + y}$.

**21.** Let $f(x, y) = \dfrac{3xy^2}{x^2 + y^4}$.

   **(a)** Show that $\displaystyle\lim_{(x,y) \to (0,0)} f(x, y) = 0$ along the lines $y = mx$.

   **(b)** Find $\displaystyle\lim_{(x,y) \to (0,0)} f(x, y)$ along the parabola $x = y^2$.

   **(c)** What can you conclude?

**22.** Show that $\displaystyle\lim_{(x,y)\to(0,0)} \dfrac{2y^2 - x^2}{x^2 + y^2}$ does not exist.

**23.** Determine where the function $f(x, y) = 2x^2y - y^2 + 3$ is continuous.

**24.** Determine where the function $R(x, y) = \dfrac{xy}{x^2 - y^2}$ is continuous.

**25.** (a) Determine where the function $f(x, y) = \tan^{-1}\dfrac{1}{x^2 + y^2}$ is continuous.

(b) Find $\displaystyle\lim_{(x, y) \to (0, 1)} \tan^{-1}\dfrac{1}{x^2 + y^2}$.

**26.** Find $f_x$ and $f_y$ for $z = f(x, y) = e^{x^2 + y^2}\sin(xy)$.

**27.** Find $f_x$ and $f_y$ for $z = f(x, y) = \dfrac{x + y}{y}$.

**28.** Find $\dfrac{\partial z}{\partial x}$ and $\dfrac{\partial z}{\partial y}$ for $z = f(x, y) = \sqrt{x - 2y^2}$.

**29.** Find $\dfrac{\partial z}{\partial x}$ and $\dfrac{\partial z}{\partial y}$ for $z = f(x, y) = e^x \ln(5x + 2y)$.

**30.** For $f(x, y) = \sqrt{x^2 - y^2}$, find $f_x(2, 1)$ and $f_y(2, -1)$.

**31.** For $F(x, y) = e^x \sin y$, find $F_x\left(0, \dfrac{\pi}{6}\right)$ and $F_y\left(0, \dfrac{\pi}{6}\right)$.

**32.** Find an equation for the tangent line to the curve of intersection of the ellipsoid $\dfrac{x^2}{24} + \dfrac{y^2}{12} + \dfrac{z^2}{6} = 1$, and the plane $y = 1$, where $x = 4$ and $z$ is positive.

**33.** Find an equation for the tangent line to the curve of intersection of the surface $z = 4x^2 - y^2 + 7$:

(a) With the plane $y = -2$ at the point $(1, -2, 7)$.

(b) With the plane $x = 1$ at the point $(1, -2, 7)$.

**34. Boyle's Law**   The volume $V$ of a gas varies directly with the temperature $T$ and inversely with the pressure $P$.

(a) Find the rate of change of the volume $V$ with respect to the temperature $T$.

(b) Find the rate of change of the volume $V$ with respect to the pressure $P$.

(c) If $T$ and $P$ are functions of $t$, what is $\dfrac{dV}{dt}$?

*In Problems 35 and 36, find the second-order partial derivatives $f_{xx}$, $f_{xy}$, $f_{yx}$, and $f_{yy}$.*

**35.** $z = f(x, y) = (x + y^2)e^{3x}$   **36.** $z = f(x, y) = \sec(xy)$

*In Problems 37–40, find $f_x$, $f_y$, and $f_z$.*

**37.** $w = f(x, y, z) = e^{xyz}$   **38.** $w = f(x, y, z) = ze^{xy}$

**39.** $f(x, y, z) = e^x \sin y + e^y \sin z$

**40.** $w = f(x, y, z) = z\tan^{-1}\dfrac{y}{x}$

**41.** For the function $z = x^3 y^2 - 2xy^4 + 3x^2 y^3$, show that $x\dfrac{\partial z}{\partial x} + y\dfrac{\partial z}{\partial y} = 5z$.

**42.** Find the change $\Delta z$ in $z = f(x, y) = xy^2 + 2$ from $(x_0, y_0)$ to $(x_0 + \Delta x, y_0 + \Delta y)$. Use the answer to calculate the change in $z$ from $(1, 0)$ to $(0.9, 0.2)$.

**43.** Show that the function $z = f(x, y) = xy - 5y^2$ is differentiable at any point $(x, y)$ in its domain by:

(a) Finding $\Delta z$.

(b) Finding $\eta_1$ and $\eta_2$ so that
$$\Delta z = f_x(x_0, y_0)\Delta x + f_y(x_0, y_0)\Delta y + \eta_1 \Delta x + \eta_2 \Delta y.$$

(c) Show that $\displaystyle\lim_{(\Delta x, \Delta y) \to (0, 0)} \eta_1 = 0$ and $\displaystyle\lim_{(\Delta x, \Delta y) \to (0, 0)} \eta_2 = 0$.

*In Problems 44 and 45, find the differential $dz$ of each function.*

**44.** $z = x\sqrt{1 + y^2}$   **45.** $z = \sin^{-1}\dfrac{x}{y}, \ y > 0$

*In Problems 46 and 47, find the differential $dw$ of each function.*

**46.** $w = ze^{xy}$   **47.** $w = \ln(xyz)$

**48.** Use differentials to estimate the change in $z = x\sqrt{1 + y^2}$ from $(4, 0)$ to $(4.1, 0.1)$.

**49.** Use differentials to estimate the change in $z = \sin^{-1}\dfrac{x}{y}$ from $(0, 1)$ to $(0.1, 1.1)$.

**50.** Use the differential of $f(x, y) = y^2 \cos x$ to approximate the value of $f(0.05, 1.98)$. (Compare your answer with a calculator result.)

**51. Electricity**   The electrical resistance $R$ of a wire is $R = k\dfrac{L}{D^2}$ where $L$ is the length of the wire, $D$ is the diameter of the wire, and $k$ is a constant. If $L$ has a 1% error and $D$ has a 2% error, what is the approximate maximum percentage error in the computation of $R$?

**52.** Find $\dfrac{dz}{dt}$ if $z = \sin(xy) - x\sin y$, where $x = e^t$ and $y = te^t$.

**53.** Find $\dfrac{dw}{dt}$ if $w = \dfrac{x}{y} + \dfrac{y}{z} + \dfrac{z}{x}$, where $x = \dfrac{1}{t}$, $y = \dfrac{1}{t^2}$, and $z = \dfrac{1}{t^3}$.

**54.** Find $\dfrac{\partial w}{\partial u}$ and $\dfrac{\partial w}{\partial v}$ if $w = xy + yz - xz$, where $x = u + v$, $y = uv$, and $z = v$. Express each answer in terms of $u$ and $v$.

**55.** Find $\dfrac{\partial u}{\partial r}$ and $\dfrac{\partial u}{\partial s}$ if $u = \sqrt{x^2 + y^2 + z^2}$, where $x = r\cos s$, $y = r\sin s$, and $z = \sqrt{r^2 + s^2}$. Express each answer in terms of $r$ and $s$.

**56.** Find $\dfrac{\partial z}{\partial x}$ and $\dfrac{\partial z}{\partial y}$ if $z = f(x, y)$ is defined implicitly by $F(x, y, z) = x^2 + y^2 - 2xyz = 0$.

**57.** Find $\dfrac{\partial z}{\partial x}$ and $\dfrac{\partial z}{\partial y}$ if $z = f(x, y)$ is defined implicitly by $F(x, y, z) = 2x\sin y + 2y\sin x + 2xyz = 0$.

**58.** If $z = uf(u^2 + v^2)$, show that $2v\dfrac{\partial z}{\partial u} - 2u\dfrac{\partial z}{\partial v} = \dfrac{2vz}{u}$.

## CHAPTER 12 PROJECT    Searching for Exoplanets

To find an exoplanet generally requires powerful telescopes, but some exoplanets can be observed by amateur astronomers with smaller telescopes and careful observation using the transit method. When using less powerful telescopes, very small errors can greatly affect the results of the observation since the change in apparent brightness of the star because of the transit is at most a few percent. Atmospheric scintillation, the main source of error, refers to variations, due to differences in the index of refraction of Earth's atmosphere, in the apparent brightness (magnitude) of an extrasolar object. When a star is viewed through a lens, the error $E$ in magnitude due to atmospheric scintillation is estimated by

$$E(D, Z, h, T) = 0.09 D^{-2/3} (2T)^{-1/2} \left(\frac{1}{\cos Z}\right)^{1.75} e^{(-h/8000)} \quad (1)$$

where

- $D$ is the diameter, in centimeters, of the lens of the telescope
- $Z$ is the zenithal angle, in degrees, of the planet
- $h$ is the altitude, in meters above sea level, of the telescope
- $T$ is the exposure time in seconds.

The zenithal angle is defined to be the angle from the highest point in the sky ($Z = 0°$) to an object in the heavens, and it is important because there is less atmosphere to get in the way of viewing a star that is higher in the sky. To obtain observations accurate enough to detect the small fluctuation that would indicate the transit of an exoplanet, the fluctuation in magnitude due to scintillation needs to be smaller than 0.001.

1. Suppose the diameter of the telescope lens is 7.1 cm, the zenithal angle is 20°, and the altitude of the telescope is 3000 m. What is the minimum exposure time needed to keep the error $E$ in magnitude below 0.001? Round the answer to the nearest second.

2. Repeat Problem 1 for $h = 4000$ m and $h = 5000$ m. Comment on the effect of altitude on the error. Explain why your result makes sense in terms of the effects of atmospheric scintillation.

3. Find $\lim\limits_{Z \to 0} E(7.1, Z, 3000, 180)$ and $\lim\limits_{Z \to 90} E(7.1, Z, 3000, 180)$. Comment on the effect of the zenithal angle on the error.

4. Find $\dfrac{\partial E}{\partial D}$. Interpret $\dfrac{\partial E}{\partial D}$.

5. Find $\dfrac{\partial E}{\partial Z}$. Interpret $\dfrac{\partial E}{\partial Z}$.

6. Find $\dfrac{\partial E}{\partial h}$. Interpret $\dfrac{\partial E}{\partial h}$.

7. Find $\dfrac{\partial E}{\partial T}$. Interpret $\dfrac{\partial E}{\partial T}$.

8. Find the differential $dE$. Interpret $dE$.

9. Suppose $D = 7.1$, $Z = 20$, $h = 3000$, and $T = 180$. Use the differential to determine which of the following reduces the error by a greater amount:

   (a) using a telescope with a lens of diameter of 7.3 cm, or

   (b) using a platform to increase the altitude of the telescope to 3200 m.

10. Suppose you are interested in astronomy and searching for exoplanets. You research the cost of telescopes with different-size lenses and the resulting effects that the different lenses will have on your observations. Do a cost benefit analysis, and then write a position paper supporting your decision to purchase the telescope you chose.

# 13 Directional Derivatives, Gradients, and Extrema

**CHAPTER 13 PROJECT** In the Chapter Project on page 949, we use two hypothetical models to measure the ice depth in a frozen lake.

Tom Martin/Getty Images

## Measuring Ice Thickness on Crystal Lake

Minnesota's nickname is the Land of 10,000 Lakes, and in the winter many of those lakes freeze over. While this might put an end to some water sports like swimming, it opens up a host of other activities for Minnesotans to enjoy—ice skating, ice fishing, ice hockey, ice golf, ice bowling, and ice driving, to name just a few. The annual Art Shanty Projects festival on Medicine Lake even features shacks (complete with radio and heaters), sculptures, printing presses, and concerts, all comfortably resting atop the lake's frozen surface.

Of course, a frozen lake can only host those people and events that its ice will support. Specific estimates vary, but the recommended ice depth that will safely support one ice skater is about 4 inches. A dozen or more ice skaters might require closer to 7 inches of ice. A small truck? That would need about 10.5 inches. But how do you determine the depth of the ice on a given lake? Trial and error is prohibitively risky. The Minnesota Department of Natural Resources suggests that you ask at local bait shops or resorts. In the chapter project, we examine another method for measuring ice depth.

I n Chapter 12 we defined the partial derivatives $\dfrac{\partial z}{\partial x} = f_x(x, y)$ and $\dfrac{\partial z}{\partial y} = f_y(x, y)$ of a function $z = f(x, y)$ of two variables. One interpretation of $f_x(x, y)$ and $f_y(x, y)$ is the rate of change of $z$ in the direction of the positive $x$-axis and in the direction of the positive $y$-axis, respectively. Here we generalize these ideas to obtain the rate of change of $z$ in any direction, a *directional derivative*.

Directional derivatives lead us to the *gradient*, a vector that is used in finding tangent planes to a surface. The gradient also provides information about paths of quickest ascent (or descent) on a surface.

Finally, we revisit optimization. Just as derivatives are used to find extreme values of functions of a single variable, partial derivatives are used to find extreme values of functions of several variables. We identify techniques for locating extreme values on a surface and investigate applications involving optimization of functions of two variables.

# 13.1 Directional Derivatives; Gradients

**OBJECTIVES** *When you finish this section, you should be able to:*

**1** Find the directional derivative of a function of two variables (p. 908)

**2** Find the gradient of a function of two variables (p. 911)

**3** Work with properties of the gradient (p. 912)

**4** Find the directional derivative and gradient of a function of three variables (p. 917)

## 1 Find the Directional Derivative of a Function of Two Variables

In Chapter 2, we saw that an important interpretation of the derivative of a function $y = f(x)$ is the rate of change of $f$. In Chapter 12, the partial derivatives of a function $z = f(x, y)$ at the point $(x_0, y_0)$ are defined as

$$f_x(x_0, y_0) = \lim_{\Delta x \to 0} \frac{f(x_0 + \Delta x, y_0) - f(x_0, y_0)}{\Delta x}$$

$$f_y(x_0, y_0) = \lim_{\Delta y \to 0} \frac{f(x_0, y_0 + \Delta y) - f(x_0, y_0)}{\Delta y}$$

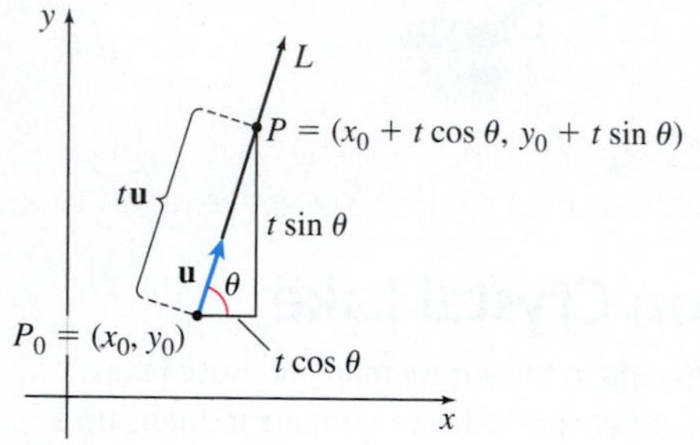

where $(x_0, y_0)$ is an interior point of the domain of $f$. The partial derivative $f_x(x_0, y_0)$ equals the rate of change of $f$ at $(x_0, y_0)$ in the *direction of the positive x-axis,* and the partial derivative $f_y(x_0, y_0)$ equals the rate of change of $f$ at $(x_0, y_0)$ in the *direction of the positive y-axis.* But what if we want the rate of change of $f$ in a direction other than the positive x-axis or the positive y-axis?

To lay the groundwork consider a function $z = f(x, y)$ whose graph is a surface in space. In Figure 1, $P_0 = (x_0, y_0)$ is an interior point in the domain of $f$, and $\mathbf{u} = \cos\theta\,\mathbf{i} + \sin\theta\,\mathbf{j}$ is a unit vector in the $xy$-plane with its initial point at $P_0$ making an angle $\theta$, $0 \le \theta < 2\pi$, with the positive x-axis. The vector $\mathbf{u}$ allows us to move away from $P_0$ in any direction.

Now draw a directed line segment $L$ with initial point $P_0$ in the direction of $\mathbf{u}$, and choose a point $P = (x, y)$ on $L$ different from $P_0$, so that the directed line segment $\overrightarrow{P_0P}$ is in the domain of $f$. Then $\overrightarrow{P_0P} = t\mathbf{u}$ for some $t$ and the coordinates of $P$ are $(x_0 + t\cos\theta, y_0 + t\sin\theta)$.

In Figure 2, we have drawn a surface $z = f(x, y)$ and have labeled the points $Q_0 = (x_0, y_0, f(x_0, y_0))$ and $Q = (x_0 + t\cos\theta, y_0 + t\sin\theta, f(x_0 + t\cos\theta, y_0 + t\sin\theta))$ on the surface. These points correspond to the points $P_0 = (x_0, y_0, 0)$ and $P = (x_0 + t\cos\theta, y_0 + t\sin\theta, 0)$ in the $xy$-plane. The average rate of change of $z$ with respect to $t$ in the direction of $\mathbf{u} = \cos\theta\,\mathbf{i} + \sin\theta\,\mathbf{j}$ is

$$\frac{\Delta z}{t} = \frac{f(x_0 + t\cos\theta, y_0 + t\sin\theta) - f(x_0, y_0)}{t} \qquad t \ne 0$$

By taking the limit as $t \to 0$, we obtain the rate of change of $f$ at $(x_0, y_0)$ in the direction of $\mathbf{u}$, provided the limit exists. This limit, denoted by $D_{\mathbf{u}}f(x_0, y_0)$, is called the *directional derivative of $f$ at $(x_0, y_0)$ in the direction of $\mathbf{u}$.*

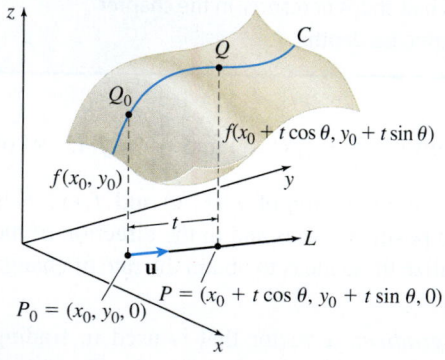

**Figure 1**

**NEED TO REVIEW?** Unit vectors are discussed in Section 10.3, pp. 747–748.

**Figure 2**

---

**DEFINITION** Directional Derivative of a Function of Two Variables

Suppose $z = f(x, y)$ is a function of two variables whose domain is $D$ and $(x_0, y_0)$ is an interior point of $D$. If $\mathbf{u} = \cos\theta\,\mathbf{i} + \sin\theta\,\mathbf{j}$, $0 \le \theta < 2\pi$, is a unit vector, the **directional derivative of $f$ at $(x_0, y_0)$ in the direction of u** is the number

$$\boxed{D_{\mathbf{u}}f(x_0, y_0) = \lim_{t \to 0} \frac{f(x_0 + t\cos\theta, y_0 + t\sin\theta) - f(x_0, y_0)}{t}}$$

provided the limit exists.

There are an infinite number of directions from $P_0$ and so there are an infinite number of directional derivatives of $f$ at the point $(x_0, y_0)$. The directional derivative $D_{\mathbf{u}} f(x_0, y_0)$ depends not only on the point $(x_0, y_0)$ but also on the angle $\theta$, the direction taken from $(x_0, y_0)$. The partial derivatives $f_x$ and $f_y$ of $f$ at $(x_0, y_0)$ are special cases of the directional derivative $D_{\mathbf{u}} f(x_0, y_0)$, where the direction taken is $\mathbf{i}$ for $f_x(x_0, y_0)$ and the direction taken is $\mathbf{j}$ for $f_y(x_0, y_0)$. For example, if $\theta = 0$, then

$$\mathbf{u} = \cos 0 \mathbf{i} + \sin 0 \mathbf{j} = \mathbf{i}$$

Then using the definition of the directional derivative of $f$ at $(x_0, y_0)$ in the direction $\mathbf{u} = \mathbf{i}$, we find

$$D_{\mathbf{u}} f(x_0, y_0) = D_{\mathbf{i}} f(x_0, y_0) = \lim_{\substack{t \to 0 \\ \theta = 0}} \frac{f(x_0 + t, y_0) - f(x_0, y_0)}{t} = f_x(x_0, y_0)$$

Similarly, if $\theta = \dfrac{\pi}{2}$, then

$$\mathbf{u} = \cos \frac{\pi}{2} \mathbf{i} + \sin \frac{\pi}{2} \mathbf{j} = \mathbf{j}$$

and

$$D_{\mathbf{j}} f(x_0, y_0) = f_y(x_0, y_0)$$

**NEED TO REVIEW?** Conditions under which a function $z = f(x, y)$ is differentiable are discussed in Section 12.4, pp. 884–888.

Finding directional derivatives using the definition can be challenging. However, for functions $z = f(x, y)$ that are differentiable, we can use the formula below to find directional derivatives.

**THEOREM  Directional Derivative of a Differentiable Function $f$**

If $z = f(x, y)$ is a differentiable function, then the directional derivative of $f$ at $(x_0, y_0)$ in the direction of the unit vector $\mathbf{u} = \cos \theta \mathbf{i} + \sin \theta \mathbf{j}$ is given by

$$D_{\mathbf{u}} f(x_0, y_0) = f_x(x_0, y_0) \cos \theta + f_y(x_0, y_0) \sin \theta \qquad (1)$$

**Proof** If $x = x_0 + t \cos \theta$, $y = y_0 + t \sin \theta$, and $\theta$ is fixed, we can express $z = f(x, y)$ as the function $z = f(x_0 + t \cos \theta, y_0 + t \sin \theta) = g(t)$. Then the derivative of $g$ at $t = 0$ is

$$g'(0) = \lim_{t \to 0} \frac{g(t) - g(0)}{t - 0} = \lim_{t \to 0} \frac{f(x_0 + t \cos \theta, y_0 + t \sin \theta) - f(x_0, y_0)}{t}$$

$$= D_{\mathbf{u}} f(x_0, y_0) \qquad (2)$$

Because $f$ is differentiable, we can use Chain Rule I to find $g'(t)$.

$$g'(t) = \frac{\partial f}{\partial x} \frac{dx}{dt} + \frac{\partial f}{\partial y} \frac{dy}{dt} \qquad \text{Chain Rule I}$$

$$= \frac{\partial f}{\partial x} \cos \theta + \frac{\partial f}{\partial y} \sin \theta \qquad \frac{dx}{dt} = \frac{d}{dt}(x_0 + t \cos \theta) = \cos \theta; \ \frac{dy}{dt} = \frac{d}{dt}(y_0 + t \sin \theta) = \sin \theta$$

If $t = 0$, then $x = x_0$ and $y = y_0$, and

$$g'(0) = f_x(x_0, y_0) \cos \theta + f_y(x_0, y_0) \sin \theta = D_{\mathbf{u}} f(x_0, y_0)$$
$$\uparrow$$
$$(2) \qquad \blacksquare$$

**EXAMPLE 1  Finding the Directional Derivative of a Function**

(a) Find the directional derivative $D_{\mathbf{u}} f(x, y)$ of $f(x, y) = x^2 y + y^2$ in the direction of $\mathbf{u} = \cos \dfrac{\pi}{4} \mathbf{i} + \sin \dfrac{\pi}{4} \mathbf{j}$.

(b) What is $D_{\mathbf{u}} f(1, 2)$?

(c) Interpret $D_{\mathbf{u}} f(1, 2)$ as a rate of change.

**Solution (a)** Since the partial derivatives of $f$,

$$f_x(x, y) = 2xy \qquad \text{and} \qquad f_y(x, y) = x^2 + 2y$$

are continuous, the function $f$ is differentiable. So, we can use formula (1) with the unit vector $\mathbf{u} = \cos\dfrac{\pi}{4}\mathbf{i} + \sin\dfrac{\pi}{4}\mathbf{j} = \dfrac{\sqrt{2}}{2}\mathbf{i} + \dfrac{\sqrt{2}}{2}\mathbf{j}$. Then

$$D_{\mathbf{u}}f(x, y) = f_x(x, y)\cos\theta + f_y(x, y)\sin\theta$$

$$= 2xy\cos\frac{\pi}{4} + (x^2 + 2y)\sin\frac{\pi}{4}$$

$$= \sqrt{2}xy + \frac{\sqrt{2}}{2}(x^2 + 2y)$$

**(b)** $D_{\mathbf{u}}f(1, 2) = 2\sqrt{2} + \dfrac{\sqrt{2}}{2}(1 + 4) = \dfrac{9\sqrt{2}}{2} \approx 6.364$

**(c)** When we are at the point $(1, 2)$ and moving in the direction $\mathbf{u} = \dfrac{\sqrt{2}}{2}\mathbf{i} + \dfrac{\sqrt{2}}{2}\mathbf{j}$, the function $f$ is changing at a rate of approximately $6.364$ units per unit length. ∎

**NOW WORK** Problem 11(a) and (b).

There is another (geometric) interpretation for a directional derivative. Look at Figure 3. If $P_0 = (x_0, y_0, 0)$ is an interior point in the domain of $z = f(x, y)$, then the point $Q_0 = (x_0, y_0, z_0) = (x_0, y_0, f(x_0, y_0))$ is a point on the surface of $z$. As we move in the domain of $f$ along the line $L$ in the direction of $\mathbf{u}$, points are traced out on a curve $C$ that lie on the surface $z = f(x, y)$. This curve $C$ is the intersection of the surface and the plane perpendicular to the $xy$-plane that contains the line $L$. It follows that for any point $P = (x_0 + t\cos\theta, y_0 + t\sin\theta, 0)$ on $L$ that is in the domain of $f$, there is a corresponding point $Q$ on $C$. Furthermore, the expression

$$\frac{\Delta z}{t} = \frac{f(x_0 + t\cos\theta, y_0 + t\sin\theta) - f(x_0, y_0)}{t} \qquad t \neq 0$$

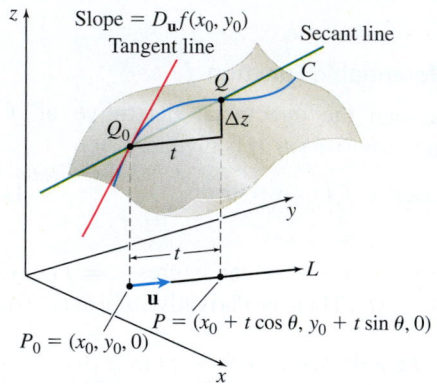

**DF Figure 3**

equals the slope of the secant line containing the two points $Q_0$ and $Q$ on $C$, and the limit as $t \to 0$ is the slope of the tangent line to $C$ at $Q_0$. This proves the following result.

**THEOREM The Directional Derivative as a Slope**

If $z = f(x, y)$ is a differentiable function, then the directional derivative $D_{\mathbf{u}}f(x_0, y_0)$ of $f$ at $(x_0, y_0)$ in the direction of $\mathbf{u}$ equals the slope of the tangent line to the curve $C$ at the point $(x_0, y_0, f(x_0, y_0))$ on the surface $z = f(x, y)$, where $C$ is the intersection of the surface with the plane perpendicular to the $xy$-plane and containing the line through $(x_0, y_0, 0)$ in the direction $\mathbf{u}$.

**EXAMPLE 2** Interpreting the Directional Derivative as a Slope

Find the directional derivative of $f(x, y) = x\sin y$ at the point $\left(2, \dfrac{\pi}{3}\right)$ in the direction of $\mathbf{a} = 3\mathbf{i} + 4\mathbf{j}$. Interpret the result as a slope.

**NOTE** The unit vector $\mathbf{u}$ in the direction of a given vector is defined in Section 10.3, p. 747.

**Solution** To find a directional derivative of a function $f$ in the direction of $\mathbf{a}$, where $\mathbf{a}$ is a nonzero vector, we must first find the unit vector in the direction of $\mathbf{a}$. The unit vector $\mathbf{u}$ in the direction of $\mathbf{a}$ is

$$\mathbf{u} = \frac{\mathbf{a}}{\|\mathbf{a}\|} = \frac{3\mathbf{i} + 4\mathbf{j}}{\sqrt{9 + 16}} = \frac{3}{5}\mathbf{i} + \frac{4}{5}\mathbf{j}$$

The partial derivatives of $f$

$$f_x(x, y) = \sin y \quad \text{and} \quad f_y(x, y) = x \cos y$$

are continuous throughout the $xy$-plane, so $f$ is differentiable. Using the unit vector $\mathbf{u} = \dfrac{3}{5}\mathbf{i} + \dfrac{4}{5}\mathbf{j} = \cos\theta\,\mathbf{i} + \sin\theta\,\mathbf{j}$, we find

$$D_{\mathbf{u}}f(x, y) = \frac{3}{5}\sin y + \frac{4}{5}x\cos y \qquad \color{blue}{D_{\mathbf{u}}f(x, y) = f_x(x, y)\cos\theta + f_y(x, y)\sin\theta}$$

At the point $\left(2, \dfrac{\pi}{3}\right)$,

$$D_{\mathbf{u}}f\left(2, \frac{\pi}{3}\right) = \frac{3}{5}\sin\frac{\pi}{3} + \frac{4}{5}\cdot 2\cos\frac{\pi}{3} = \frac{3}{5}\cdot\frac{\sqrt{3}}{2} + \frac{8}{5}\cdot\frac{1}{2} = \frac{3\sqrt{3}+8}{10} \approx 1.320$$

The slope of the tangent line to the curve $C$ formed by the intersection of the surface $f(x, y) = x\sin y$ and the plane perpendicular to the $xy$-plane that contains the line through the point $\left(2, \dfrac{\pi}{3}, 0\right)$ in the direction $\mathbf{a}$ is approximately 1.320. See Figure 4. ∎

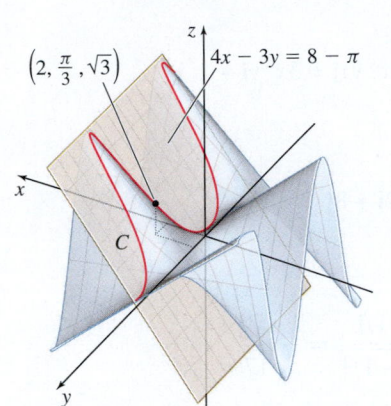

**Figure 4** $f(x, y) = x\sin y$

$\left(2, \frac{\pi}{3}, \sqrt{3}\right)$    $4x - 3y = 8 - \pi$    $C$

**NOW WORK** Problems 11(c) and 15.

## 2 Find the Gradient of a Function of Two Variables

We can express the directional derivative

$$D_{\mathbf{u}}f(x_0, y_0) = f_x(x_0, y_0)\cos\theta + f_y(x_0, y_0)\sin\theta$$

of a differentiable function $z = f(x, y)$ at a point $(x_0, y_0)$ in the direction of the unit vector $\mathbf{u} = \cos\theta\,\mathbf{i} + \sin\theta\,\mathbf{j}$ as the dot product of two vectors as follows:

$$
\begin{aligned}
D_{\mathbf{u}}f(x_0, y_0) &= f_x(x_0, y_0)\cos\theta + f_y(x_0, y_0)\sin\theta \\
&= [f_x(x_0, y_0)\mathbf{i} + f_y(x_0, y_0)\mathbf{j}] \cdot [\cos\theta\,\mathbf{i} + \sin\theta\,\mathbf{j}] \\
&= [f_x(x_0, y_0)\mathbf{i} + f_y(x_0, y_0)\mathbf{j}] \cdot \mathbf{u}
\end{aligned}
$$

> **NEED TO REVIEW?** The dot product is discussed in Section 10.4, pp. 754–755.

The vector $f_x(x_0, y_0)\mathbf{i} + f_y(x_0, y_0)\mathbf{j}$ is called the *gradient of $f$* at $(x_0, y_0)$.

---

**DEFINITION** Gradient of a Function of Two Variables

Let $z = f(x, y)$ be a differentiable function. The vector $\nabla f(x, y)$, read "del $f$," is called the **gradient** of $f$ and is defined as

$$\boxed{\nabla f(x, y) = f_x(x, y)\mathbf{i} + f_y(x, y)\mathbf{j}}$$

---

> **NOTE** "del $f$" is short for "delta $f$" and the symbol $\nabla$ is an upside down Greek letter delta. In some books, $\nabla f$ is written as "grad $f$." The symbol $\nabla$ has no meaning except when it operates on a function $z = f(x, y)$. So, $\nabla$ is referred to as an **operator**, just as $\dfrac{d}{dx}$ is an operator indicating "take the derivative of a function of $x$."

The directional derivative $D_{\mathbf{u}}f(x, y)$ of $f$ can be expressed in terms of the gradient of $f$.

---

**THEOREM** The Directional Derivative as a Dot Product

The dot product of the vector $\nabla f(x, y)$ and the unit vector $\mathbf{u} = \cos\theta\,\mathbf{i} + \sin\theta\,\mathbf{j}$ is a scalar that is equal to the directional derivative $D_{\mathbf{u}}f(x, y)$. That is,

$$\boxed{\nabla f(x, y) \cdot \mathbf{u} = D_{\mathbf{u}}f(x, y)} \tag{3}$$

---

**Proof** $\nabla f(x, y) \cdot \mathbf{u} = [f_x(x, y)\mathbf{i} + f_y(x, y)\mathbf{j}] \cdot [\cos\theta\,\mathbf{i} + \sin\theta\,\mathbf{j}]$

$$= f_x(x, y)\cos\theta + f_y(x, y)\sin\theta = D_{\mathbf{u}}f(x, y). \ ∎$$

CALC
CLIP
**EXAMPLE 3**  **Finding the Gradient of a Function of Two Variables**

(a) Find the gradient of $f(x, y) = x^3 y$ at the point $(2, 1)$.

(b) Use the gradient to find the directional derivative of $f$ at $(2, 1)$ in the direction from $(2, 1)$ to $(3, 5)$.

**Solution** (a) The gradient of $f$ at $(x, y)$ is

$$\nabla f(x, y) = f_x(x, y)\mathbf{i} + f_y(x, y)\mathbf{j} = 3x^2 y\mathbf{i} + x^3 \mathbf{j}$$

The gradient of $f$ at $(2, 1)$ is

$$\nabla f(2, 1) = 12\mathbf{i} + 8\mathbf{j}$$

(b) The unit vector $\mathbf{u}$ from $(2, 1)$ to $(3, 5)$ is

$$\mathbf{u} = \frac{(3-2)\mathbf{i} + (5-1)\mathbf{j}}{\sqrt{(3-2)^2 + (5-1)^2}} = \frac{\mathbf{i} + 4\mathbf{j}}{\sqrt{17}}$$

Now use formula (3) for the directional derivative to find $D_{\mathbf{u}}(2, 1)$.

$$D_{\mathbf{u}}(2, 1) = \nabla f(2, 1) \cdot \mathbf{u} = (12\mathbf{i} + 8\mathbf{j}) \cdot \frac{\mathbf{i} + 4\mathbf{j}}{\sqrt{17}} = \frac{44\sqrt{17}}{17}$$

The directional derivative of $f$ at $(2, 1)$ in the direction of $\mathbf{u}$ is $\dfrac{44\sqrt{17}}{17}$. ∎

NOW WORK Problem 25.

## 3 Work with Properties of the Gradient

The first property of a gradient involves the relationship between the gradient $\nabla f$ of a function $z = f(x, y)$ and the level curves of $f$.

**NEED TO REVIEW?** Level curves are discussed in Section 12.1, pp. 854–856.

> **THEOREM  The Gradient Is Normal to the Level Curve**
>
> Suppose the function $z = f(x, y)$ is differentiable at a point $P_0 = (x_0, y_0)$. If $\nabla f(x_0, y_0) \neq \mathbf{0}$, then $\nabla f(x_0, y_0)$ is normal to the level curve of $f$ at $P_0$.

**RECALL** A normal vector to a curve at a point $P$ is defined as a vector with initial point at $P$ and orthogonal (perpendicular) to the tangent line to the curve at $P$.

**Proof** Let $f(x, y) = k$ be the level curve through $P_0$. Suppose this level curve is represented parametrically by $x = x(t)$ and $y = y(t)$, with $x_0 = x(t_0)$ and $y_0 = y(t_0)$. Then

$$f(x(t), y(t)) = k$$

Differentiating $f(x(t), y(t))$ with respect to $t$, we get

$$f_x(x(t), y(t))\, x'(t) + f_y(x(t), y(t))\, y'(t) = 0 \qquad \text{Use Chain Rule I.}$$

or, equivalently,

$$\nabla f(x, y) \cdot [x'(t)\mathbf{i} + y'(t)\mathbf{j}] = 0$$

Since the dot product equals 0 and $x'(t)\mathbf{i} + y'(t)\mathbf{j}$ is tangent to the level curve, the vector $\nabla f(x, y)$ is normal to the level curve. In particular, when $t = t_0$, $\nabla f(x_0, y_0)$ is normal to the level curve of $f$ at $(x_0, y_0)$. ∎

**EXAMPLE 4**  **Using Properties of the Gradient**

For the function $f(x, y) = \sqrt{x^2 + y^2}$, graph the level curve containing the point $(3, 4)$ and graph the gradient $\nabla f(x, y)$ at this point.

**Solution** See Figure 5(a). The graph of the equation $z = \sqrt{x^2 + y^2}$ is the upper half of a circular cone whose traces are circles. So, the level curves are concentric circles centered at $(0, 0)$. Because $f(3, 4) = \sqrt{9 + 16} = 5$, the level curve through $(3, 4)$ is the circle $x^2 + y^2 = 25$. Since

$$\nabla f(x, y) = \frac{x}{\sqrt{x^2 + y^2}}\mathbf{i} + \frac{y}{\sqrt{x^2 + y^2}}\mathbf{j} \qquad \nabla f(x, y) = f_x(x, y)\mathbf{i} + f_y(x, y)\mathbf{j}$$

the gradient at $(3, 4)$ is

$$\nabla f(3, 4) = \frac{3}{5}\mathbf{i} + \frac{4}{5}\mathbf{j}$$

This vector is orthogonal to the level curve $x^2 + y^2 = 25$, as shown in Figure 5(b).

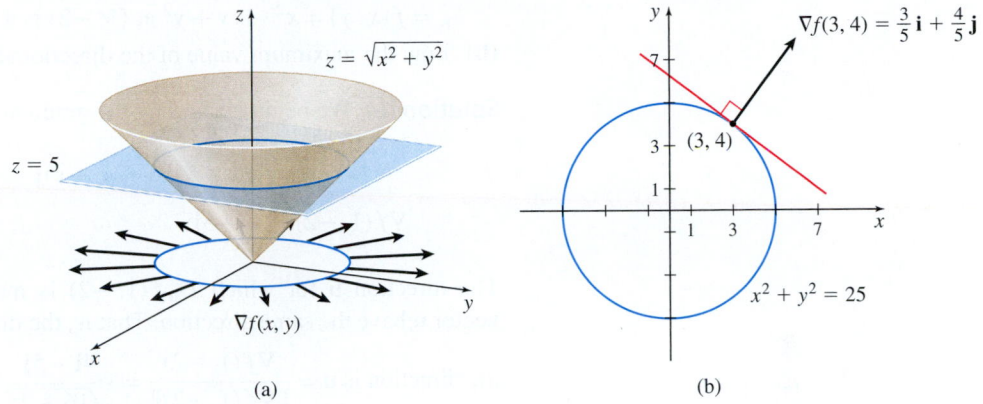

(a)                                (b)

**Figure 5**                                                          ■

**NOW WORK** Problem 45.

Let's look more closely at the directional derivative $D_{\mathbf{u}} f(x, y)$, when it is expressed as a dot product:

$$D_{\mathbf{u}} f(x, y) = \nabla f(x, y) \cdot \mathbf{u}$$

- If $\nabla f(x_0, y_0) = \mathbf{0}$, then for any direction $\mathbf{u}$,

$$D_{\mathbf{u}} f(x_0, y_0) = \nabla f(x_0, y_0) \cdot \mathbf{u} = \mathbf{0} \cdot \mathbf{u} = 0$$

- If $\nabla f(x_0, y_0) \neq \mathbf{0}$ and $\phi$, $0 \leq \phi \leq \pi$, is the angle between the unit vector $\mathbf{u}$ and the gradient $\nabla f(x_0, y_0)$, then

**NEED TO REVIEW?** The angle between two vectors is discussed in Section 10.4, pp. 755–757.

$$\boxed{D_{\mathbf{u}} f(x_0, y_0) = \nabla f(x_0, y_0) \cdot \mathbf{u} = \|\nabla f(x_0, y_0)\| \|\mathbf{u}\| \cos \phi = \|\nabla f(x_0, y_0)\| \cos \phi} \quad (4)$$

From (4) we conclude that the directional derivative $D_{\mathbf{u}} f(x_0, y_0)$ is largest when $\cos \phi = 1$. This happens if the angle $\phi$ between $\mathbf{u}$ and $\nabla f$ is 0, or equivalently, if $\mathbf{u}$ and $\nabla f(x_0, y_0)$ have the same direction. In this case, the *maximum value of the directional derivative* is

$$\boxed{\text{Maximum value of } D_{\mathbf{u}} f(x_0, y_0) = \|\nabla f(x_0, y_0)\|}$$

Similarly, the minimum value of $D_{\mathbf{u}} f(x_0, y_0)$ occurs when $\cos \phi = -1$, that is, when $\phi = \pi$. This happens when $\mathbf{u}$ and $\nabla f$ are opposite in direction. The *minimum value of the directional derivative* is

$$\boxed{\text{Minimum value of } D_{\mathbf{u}} f(x_0, y_0) = -\|\nabla f(x_0, y_0)\|}$$

**SUMMARY  Properties of the Gradient**

Suppose the function $z = f(x, y)$ is differentiable at $(x_0, y_0)$.

- If $\nabla f(x_0, y_0) = \mathbf{0}$, then $D_{\mathbf{u}} f(x_0, y_0) = 0$ for any direction $\mathbf{u}$.
- If $\nabla f(x_0, y_0) \neq \mathbf{0}$, then the directional derivative $D_{\mathbf{u}} f(x_0, y_0)$ of $f$ at $(x_0, y_0)$ is a maximum when $\mathbf{u}$ is in the direction of $\nabla f(x_0, y_0)$. The maximum value of $D_{\mathbf{u}} f(x_0, y_0)$ is $\|\nabla f(x_0, y_0)\|$.
- If $\nabla f(x_0, y_0) \neq \mathbf{0}$, then the directional derivative of $f$ at $(x_0, y_0)$ is a minimum when $\mathbf{u}$ is in the direction of $-\nabla f(x_0, y_0)$. The minimum value of $D_{\mathbf{u}} f(x_0, y_0)$ is $-\|\nabla f(x_0, y_0)\|$.

**EXAMPLE 5   Using Properties of the Gradient**

(a) Find the direction for which the directional derivative of
$z = f(x, y) = x^2 - xy + y^2$ at $(1, -2)$ is a maximum.
(b) Find the maximum value of the directional derivative.

**Solution** (a) We begin by finding the gradient of $f$ at $(1, -2)$.

$$\nabla f(x, y) = (2x - y)\mathbf{i} + (2y - x)\mathbf{j} \qquad \nabla f(x, y) = f_x(x, y)\mathbf{i} + f_y(x, y)\mathbf{j}$$
$$\nabla f(1, -2) = 4\mathbf{i} - 5\mathbf{j} \qquad\qquad x = 1; \; y = -2$$

The direction $\mathbf{u}$ for which $D_{\mathbf{u}} f(1, -2)$ is maximum occurs when $\nabla f$ and the unit vector $\mathbf{u}$ have the same direction. That is, the directional derivative is a maximum when

$$\text{the direction is } \mathbf{u} = \frac{\nabla f(1, -2)}{\|\nabla f(1, -2)\|} = \frac{4\mathbf{i} - 5\mathbf{j}}{\sqrt{16 + 25}} = \frac{4\sqrt{41}}{41}\mathbf{i} - \frac{5\sqrt{41}}{41}\mathbf{j}.$$

(b) The maximum value of the directional derivative at $(1, -2)$ equals the magnitude of the gradient,

$$\|\nabla f(1, -2)\| = \sqrt{16 + 25} = \sqrt{41} \qquad\qquad\blacksquare$$

**NOW WORK** Problem 35.

**NOTE** We discuss the consequences of $\nabla f(x_0, y_0) = \mathbf{0}$ in the next section.

The directional derivative $D_{\mathbf{u}} f(x_0, y_0)$ equals the rate of change of $z = f(x, y)$ at $(x_0, y_0)$ in the direction of $\mathbf{u}$. When $\mathbf{u}$ has the same direction as $\nabla f(x_0, y_0) \neq \mathbf{0}$, then the rate of change of $f$ at $(x_0, y_0)$ is a maximum. It follows that $f(x, y)$ will increase most rapidly in this direction.

**THEOREM**

Suppose $z = f(x, y)$ is differentiable at $(x_0, y_0)$ and $\nabla f(x_0, y_0) \neq \mathbf{0}$.

- The value of $z = f(x, y)$ at $(x_0, y_0)$ increases most rapidly in the direction of $\nabla f(x_0, y_0)$ and decreases most rapidly in the direction of $-\nabla f(x_0, y_0)$, that is, in directions normal to the level curves of $f$.
- The value of $z = f(x, y)$ remains the same for directions orthogonal to $\nabla f(x_0, y_0)$, that is, in directions tangent to the level curves of $f$.

The second bullet is a consequence of statement (4). If $\phi$ is the angle between $\mathbf{u}$ and $\nabla f$, then the rate of change of $z$ in the direction of $\mathbf{u}$ is 0 when

$$D_{\mathbf{u}} f(x_0, y_0) = \nabla f \cdot \mathbf{u} = \|\nabla f(x_0, y_0)\| \cos\phi = 0 \qquad 0 \leq \phi \leq \pi$$

If $\nabla f(x_0, y_0) \neq \mathbf{0}$, then $\cos\phi = 0$, or equivalently, $\phi = \dfrac{\pi}{2}$. In other words, the rate of change of $z$ is 0 when the direction of $\mathbf{u}$ is orthogonal to $\nabla f(x_0, y_0)$.

Look again at Example 5. Figure 6(a) shows a contour graph of the surface $z = f(x, y) = x^2 - xy + y^2$, and Figure 6(b) shows several of the level curves of $f$. Figure 6(c) shows the level curve containing the point $(1, -2)$. Notice also the directions at $(1, -2)$ for which the values of $z$ increase and decrease most rapidly are $\nabla f(1, -2) = 4\mathbf{i} - 5\mathbf{j}$ and $-\nabla f(1, -2) = -4\mathbf{i} + 5\mathbf{j}$, respectively. The directions at which $z = f(x, y)$ remains constant are tangent to the level curve.

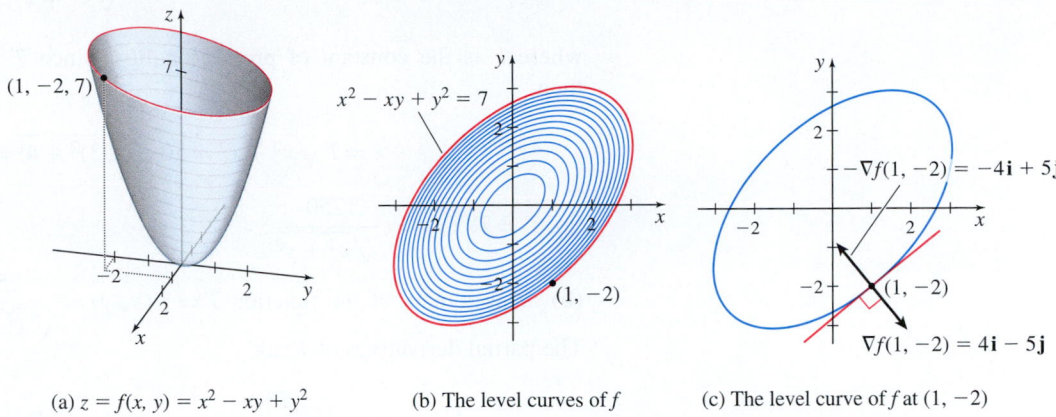

(a) $z = f(x, y) = x^2 - xy + y^2$ · (b) The level curves of $f$ · (c) The level curve of $f$ at $(1, -2)$

**Figure 6**

Suppose the surface $z = f(x, y)$ represents a mountain whose elevation above sea level is $z$. See Figure 7(a). The level curves of $f$ shown in Figure 7(b) correspond to contour lines along which the elevation remains fixed. So along the level curves, the value of $z$ does not change. The elevation increases (or decreases) most rapidly in directions perpendicular to the level curves. In other words, the most direct (but steepest) route to the top of the mountain is in the direction of $\nabla f$, which is orthogonal to the level curves. The shortest route down the mountain is in the direction of $-\nabla f$, which is also perpendicular to the level curves. A stream of water flows this way down the mountain. As water moves from one level curve to another, the gradient changes, giving the water a new direction. See Figure 8.

**NOTE** The grade of a mountain is a measure of its steepness. The name gradient given to $\nabla f$ means the direction of steepest ascent.

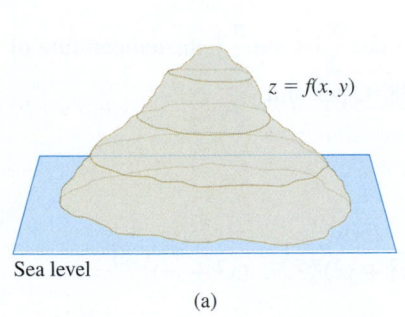

(a)

**DF Figure 7**

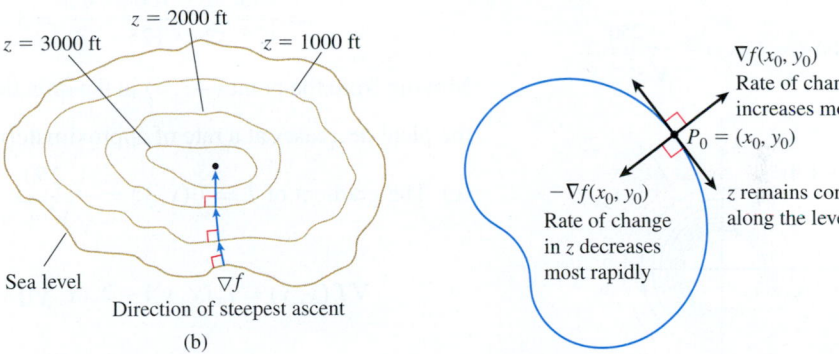

(b)

**Figure 8** Level curve of the graph of $z = f(x, y)$.

---

**EXAMPLE 6** Modeling Temperature Change

A metal plate is placed on the $xy$-plane in such a way that the temperature $T$ in degrees Celsius at any point $P = (x, y)$ is inversely proportional to the distance of $P$ from $(0, 0)$. Suppose the temperature of the plate at the point $(-3, 4)$ equals $50\,°C$.

**(a)** Find $T = T(x, y)$.

**(b)** Find the directional derivative of $T$ at the point $(-3, 4)$ in the
direction $\mathbf{u} = \cos\dfrac{\pi}{3}\mathbf{i} + \sin\dfrac{\pi}{3}\mathbf{j}$. Interpret the result in the context of the problem.

**(c)** Find the gradient of $T$ at the point $(-3, 4)$.

**(d)** In what direction does the temperature increase most rapidly?

**(e)** In what direction does the temperature decrease most rapidly?

**(f)** In what direction is the rate of change of $T$ at $(-3, 4)$ equal to 0?

**Solution (a)** The temperature $T$ at any point $(x, y)$ is inversely proportional to the distance of $(x, y)$ from the origin $(0, 0)$. We can model $T = T(x, y)$ as

$$T(x, y) = \frac{k}{\sqrt{x^2 + y^2}}$$

where $k$ is the constant of proportionality. Since $T = 50\,°C$ when $(x, y) = (-3, 4)$, then

$$k = T\sqrt{x^2 + y^2} = 50\sqrt{(-3)^2 + 4^2} = 50 \cdot 5 = 250$$

So, $T = T(x, y) = \dfrac{250}{\sqrt{x^2 + y^2}}$.

**(b)** The domain of the function $T = T(x, y) = \dfrac{250}{\sqrt{x^2 + y^2}}$ is $\{(x, y) \mid (x, y) \neq (0, 0)\}$. The partial derivatives of $T$ are

$$T_x(x, y) = -\frac{250x}{(x^2 + y^2)^{3/2}} \quad \text{and} \quad T_y(x, y) = -\frac{250y}{(x^2 + y^2)^{3/2}}.$$

Since the partial derivatives $T_x$ and $T_y$ are continuous at every point in the domain of $T$, the function $T$ is differentiable at all points $(x, y) \neq (0, 0)$. The directional derivative of $T$ at the point $(-3, 4)$ in the direction of $\mathbf{u} = \cos\dfrac{\pi}{3}\mathbf{i} + \sin\dfrac{\pi}{3}\mathbf{j}$ is

$$D_{\mathbf{u}}T(-3, 4) = T_x(-3, 4)\cos\frac{\pi}{3} + T_y(-3, 4)\sin\frac{\pi}{3} \qquad \text{Use (1).}$$

$$= -\frac{250 \cdot (-3)}{[(-3)^2 + 4^2]^{3/2}} \cdot \frac{1}{2} + \left\{-\frac{250 \cdot 4}{[(-3)^2 + 4^2]^{3/2}}\right\} \cdot \frac{\sqrt{3}}{2}$$

$$= \frac{750}{125} \cdot \frac{1}{2} - \frac{1000}{125} \cdot \frac{\sqrt{3}}{2} = 3 - 4\sqrt{3} \approx -3.928$$

Moving from the point $(-3, 4)$ in the direction $\mathbf{u} = \cos\dfrac{\pi}{3}\mathbf{i} + \sin\dfrac{\pi}{3}\mathbf{j}$, the temperature of the plate decreases at a rate of approximately $3.928\,°C$ per unit.

**(c)** The gradient of $T = T(x, y) = \dfrac{250}{\sqrt{x^2 + y^2}}$ is

$$\nabla T(x, y) = T_x(x, y)\mathbf{i} + T_y(x, y)\mathbf{j} = -\frac{250x}{(x^2 + y^2)^{3/2}}\mathbf{i} - \frac{250y}{(x^2 + y^2)^{3/2}}\mathbf{j}$$

At $(-3, 4)$,

$$\nabla T(-3, 4) = -\frac{250(-3)}{[(-3)^2 + 4^2]^{3/2}}\mathbf{i} - \frac{250 \cdot 4}{[(-3)^2 + 4^2]^{3/2}}\mathbf{j} = \frac{750}{125}\mathbf{i} - \frac{1000}{125}\mathbf{j} = 6\mathbf{i} - 8\mathbf{j}$$

**(d)** The temperature increases most rapidly in the direction $\nabla T(-3, 4) = 6\mathbf{i} - 8\mathbf{j}$.

**(e)** The temperature decreases most rapidly in the direction $-\nabla T(-3, 4) = -6\mathbf{i} + 8\mathbf{j}$.

**(f)** The rate of change in $T$ at $(-3, 4)$ equals $0$ for directions orthogonal to $\nabla T(-3, 4) = 6\mathbf{i} - 8\mathbf{j}$. That is, the temperature of the plate remains constant in either of the two directions orthogonal to $6\mathbf{i} - 8\mathbf{j}$, namely $\pm(8\mathbf{i} + 6\mathbf{j})$. ∎

Figure 9(a) shows the graph of the surface $T = T(x, y)$ with several contour lines. Figure 9(b) shows the level curves of $T$. Figure 9(c) shows the level curve containing the point $(-3, 4)$.

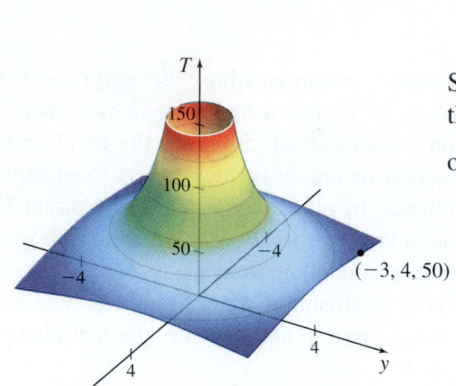

(a) $T = T(x, y) = \dfrac{250}{\sqrt{x^2 + y^2}}$

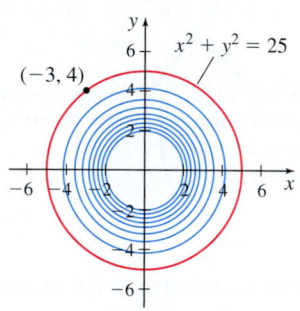

(b) The level curves of $T$

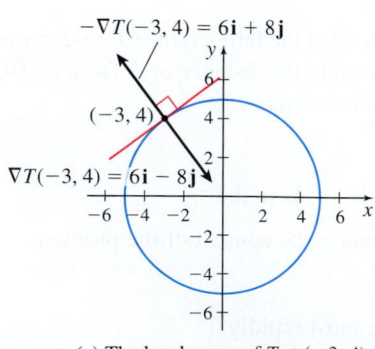

(c) The level curve of $T$ at $(-3, 4)$

**Figure 9**

**NOW WORK** Problem 47.

## 4 Find the Directional Derivative and Gradient of a Function of Three Variables

The concepts of directional derivative and gradient can be extended to functions $w = f(x, y, z)$ of three variables.

---

**DEFINITION** Directional Derivative of a Function of Three Variables

The **directional derivative** $D_\mathbf{u} f(x_0, y_0, z_0)$ of a function $w = f(x, y, z)$ at $(x_0, y_0, z_0)$ in the direction of a unit vector $\mathbf{u} = \cos \alpha \, \mathbf{i} + \cos \beta \, \mathbf{j} + \cos \gamma \, \mathbf{k}$ is given by

$$D_\mathbf{u} f(x_0, y_0, z_0) = \lim_{t \to 0} \frac{f(x_0 + t \cos \alpha, \, y_0 + t \cos \beta, \, z_0 + t \cos \gamma) - f(x_0, y_0, z_0)}{t}$$

provided the limit exists.

---

If the function $w = f(x, y, z)$ is differentiable, then the directional derivative of $f$ in the direction of $\mathbf{u} = \cos \alpha \, \mathbf{i} + \cos \beta \, \mathbf{j} + \cos \gamma \, \mathbf{k}$ equals

$$D_\mathbf{u} f(x_0, y_0, z_0) = f_x(x_0, y_0, z_0) \cos \alpha + f_y(x_0, y_0, z_0) \cos \beta + f_z(x_0, y_0, z_0) \cos \gamma$$

---

**DEFINITION** Gradient of a Function of Three Variables

Let $w = f(x, y, z)$ be a differentiable function. Then the **gradient** of $f$ at $(x_0, y_0, z_0)$ is the vector

$$\nabla f(x_0, y_0, z_0) = f_x(x_0, y_0, z_0) \mathbf{i} + f_y(x_0, y_0, z_0) \mathbf{j} + f_z(x_0, y_0, z_0) \mathbf{k}$$

---

Then, as for functions of two variables, the directional derivative of $f$ at $(x_0, y_0, z_0)$ in the direction of $\mathbf{u}$ can be found using the gradient as follows:

$$D_\mathbf{u} f(x_0, y_0, z_0) = \nabla f(x_0, y_0, z_0) \cdot \mathbf{u} = \| \nabla f(x_0, y_0, z_0) \| \cos \phi$$

where $\phi$ is the angle between $\nabla f(x_0, y_0, z_0)$ and $\mathbf{u}$.

As with functions of two variables, the directional derivative $D_\mathbf{u} f(x_0, y_0, z_0)$ equals the rate of change of $f$ at $(x_0, y_0, z_0)$ in the direction of $\mathbf{u}$. Also, if $\nabla f(x_0, y_0, z_0) \neq \mathbf{0}$, the value of $f$ increases most rapidly in the direction of the gradient $\nabla f(x_0, y_0, z_0)$, and decreases most rapidly in the direction of $-\nabla f(x_0, y_0, z_0)$.

For example, suppose $T = T(x, y, z)$ is the temperature of a homogeneous object at the point $(x, y, z)$. At the point $(x_0, y_0, z_0)$ on the object, heat will flow in the direction of greatest decrease in temperature, namely in the direction of $-\nabla T(x_0, y_0, z_0)$. This direction of greatest heat transfer is normal to the level surface containing the point $(x_0, y_0, z_0)$.

---

**THEOREM** The Gradient Is Normal to the Level Surface

**NOTE** This theorem is used in the next section where the tangent plane to a surface is discussed.

Suppose the function $w = f(x, y, z)$ is differentiable at a point $P_0 = (x_0, y_0, z_0)$. If $\nabla f(x_0, y_0, z_0) \neq \mathbf{0}$, the gradient $\nabla f(x_0, y_0, z_0)$ is normal to the level surface of $f$ through $P_0$.

---

You are asked to prove this in Problem 76.

## 13.1 Assess Your Understanding

### Concepts and Vocabulary

1. *True or False*    The directional derivative $D_\mathbf{u} f(x_0, y_0)$ equals the rate of change of $z = f(x, y)$ at $(x_0, y_0)$ in the direction of the unit vector $\mathbf{u}$.

2. *True or False*    For a function $z = f(x, y)$, the partial derivative $f_x(x_0, y_0)$ is the directional derivative at $(x_0, y_0)$ in the direction of $\mathbf{i}$.

3. *True or False*    Both the directional derivative $D_\mathbf{u} f(x, y)$ and the gradient $\nabla f(x, y)$ of a function $z = f(x, y)$ are vectors.

---

**1.** = NOW WORK problem    ⬜ = Graphing technology recommended    CAS = Computer Algebra System recommended

**4.** *True or False* If $z = f(x, y)$ is a differentiable function, then the directional derivative of $f$ at $(x_0, y_0)$ in the direction of $\mathbf{u} = \cos\theta\mathbf{i} + \sin\theta\mathbf{j}$ is given by

$$D_{\mathbf{u}} f(x_0, y_0) = f_x(x_0, y_0)\cos\theta + f_y(x_0, y_0)\sin\theta$$

**5.** *True or False* The vector $x\mathbf{i} + \sin y\mathbf{j}$ is the gradient of $f(x, y) = x\sin y$.

**6.** The directional derivative $D_{\mathbf{u}} f(x_0, y_0)$ of $f$ at $(x_0, y_0)$ is a maximum when $\mathbf{u} = $ _____.

**7.** The value of $z = f(x, y)$ at $(x_0, y_0)$ decreases most rapidly in the direction of _____.

**8.** For a differentiable function $z = f(x, y)$, the maximum value of $D_{\mathbf{u}} f(x_0, y_0)$ is _____.

## Skill Building

In Problems 9–16,

**(a)** Find the directional derivative of each function at the indicated point and in the indicated direction.

**(b)** Interpret the result as a rate of change.

**(c)** Interpret the result as a slope.

**9.** $f(x, y) = xy^2 + x^2$ at $(-1, 2)$ in the direction $\mathbf{u} = \dfrac{1}{2}\mathbf{i} + \dfrac{\sqrt{3}}{2}\mathbf{j}$

**10.** $f(x, y) = 3xy + y^2$ at $(2, 1)$ in the direction

$$\mathbf{u} = \frac{\sqrt{2}}{2}\mathbf{i} + \frac{\sqrt{2}}{2}\mathbf{j}$$

**11.** $f(x, y) = 2xy - y^2$ at $(-1, 3)$ in the direction $\theta = \dfrac{2\pi}{3}$

**12.** $f(x, y) = 2xy + x^2$ at $(0, 3)$ in the direction $\theta = \dfrac{4\pi}{3}$

**13.** $f(x, y) = xe^y + ye^x$ at $(0, 0)$ in the direction $\theta = \dfrac{\pi}{6}$

**14.** $f(x, y) = x\ln y$ at $(5, 1)$ in the direction $\theta = \dfrac{\pi}{4}$

**15.** $f(x, y) = \tan^{-1}\dfrac{y}{x}$ at $(1, 1)$ in the direction $\mathbf{a} = 3\mathbf{i} - 4\mathbf{j}$

**16.** $f(x, y) = \ln\sqrt{x^2 + y^2}$ at $(3, 4)$ in the direction $\mathbf{a} = 5\mathbf{i} + 12\mathbf{j}$

In Problems 17–32:

**(a)** Find the gradient of $f$ at the given point $P$.

**(b)** Use the gradient to find the directional derivative of $f$ at $P$ in the direction from $P$ to $Q$.

**17.** $f(x, y) = xy^2 + x^2$;   $P = (1, 2)$, $Q = (2, 4)$

**18.** $f(x, y) = 2xy + x^2$;   $P = (-1, 1)$, $Q = (1, 2)$

**19.** $f(x, y) = 2xy + x^2$;   $P = (0, 3)$, $Q = (4, 1)$

**20.** $f(x, y) = 3xy + y^2$;   $P = (2, 1)$, $Q = (4, 1)$

**21.** $f(x, y) = xy + \sin x$;   $P = (0, 1)$, $Q = (\pi, 2)$

**22.** $f(x, y) = e^{xy} + \sin y$;   $P = \left(0, \dfrac{\pi}{2}\right)$, $Q = (1, 0)$

**23.** $f(x, y) = \tan^{-1}\dfrac{y}{x}$;   $P = (1, 0)$, $Q = (4, \pi)$

**24.** $f(x, y) = \ln\sqrt{x^2 + y^2}$;   $P = (3, 4)$, $Q = (0, 5)$

**25.** $f(x, y) = x^2 e^y$;   $P = (2, 0)$, $Q = (3, 0)$

**26.** $f(x, y) = e^{x^2 + y^2}$;   $P = (1, 2)$, $Q = (2, 3)$

**27.** $f(x, y, z) = x^2 y - xyz^2$;   $P = (0, 1, 2)$, $Q = (1, 4, 3)$

**28.** $f(x, y, z) = x^2 y + y^2 z + z^2 x$;   $P = (1, 2, -1)$, $Q = (2, 0, 1)$

**29.** $f(x, y, z) = z\tan^{-1}\dfrac{y}{x}$;   $P = (1, 1, 3)$, $Q = (2, 0, -2)$

**30.** $f(x, y, z) = \sin x\cos(y + z)$;   $P = (1, 1, 1)$, $Q = (2, -1, 0)$

**31.** $f(x, y, z) = \sqrt{x^2 + y^2 + z^2}$;   $P = (3, 4, 0)$, $Q = (1, -1, 1)$

**32.** $f(x, y, z) = \dfrac{x}{\sqrt{x^2 + 2y^2 + 3z^2}}$;   $P = (1, 2, 1)$, $Q = (-1, 1, 1)$

In Problems 33–40:

**(a)** Find the direction for which the directional derivative of the function $f$ is a maximum.

**(b)** Find the maximum value of the directional derivative.

**33.** $f(x.y) = xy^2 + x^2$ at $P = (-1, 2)$

**34.** $f(x.y) = 3xy + y^2$ at $P = (2, 1)$

**35.** $f(x.y) = xe^y + ye^x$ at $P = (0, 0)$

**36.** $f(x.y) = x\ln y$ at $P = (5, 1)$

**37.** $f(x, y) = \dfrac{x}{x^2 + y^2}$ at $P = (1, 2)$

**38.** $f(x, y) = \sqrt{x^2 + y^2}$ at $P = (3, 4)$

**39.** $f(x.y, z) = z\tan^{-1}\dfrac{y}{x}$ at $P = (1, 1, 3)$

**40.** $f(x.y, z) = \sqrt{x^2 + y^2 + z^2}$ at $P = (3, 4, 0)$

In Problems 41–46, graph the level curve of $f$ containing the point $P$ and the gradient $\nabla f$ at that point.

**41.** $f(x, y) = x^2 + y^2$ containing $P = (3, 4)$

**42.** $f(x, y) = x^2 - y^2$ containing $P = (2, -1)$

**43.** $f(x, y) = x^2 - 4y^2$ containing $P = \left(3, \dfrac{\sqrt{5}}{2}\right)$

**44.** $f(x, y) = x^2 + 4y^2$ containing $P = (-2, 0)$

**45.** $f(x, y) = x^2 y$ containing $P = \left(3, \dfrac{1}{9}\right)$

**46.** $f(x, y) = xy$ containing $P = (1, 1)$

## Applications and Extensions

**47. Heat Transfer** A metal plate is placed on the $xy$-plane in such a way that the temperature $T$ at any point $(x, y)$ is given by $T = e^x(\sin x + \sin y)\,°\text{C}$.

**(a)** What is the directional derivative of $T$ at $(0, 0)$ in the direction of $3\mathbf{i} - 4\mathbf{j}$? Interpret the result in the context of the problem.

**(b)** At $(0, 0)$, in what direction is the temperature increasing most rapidly?

**(c)** In what direction is the temperature decreasing most rapidly?

**(d)** In what direction does the temperature remain the same?

**48. Electrical Potential** The electrical potential $V$ at any point $(x, y)$ is given by $V = \ln \sqrt{x^2 + y^2}$. Find the rate of change of potential $V$ at any point $(x, y) \neq (0, 0)$:

(a) In a direction toward $(0, 0)$.

(b) In the two directions orthogonal to a direction toward $(0, 0)$.

(c) In what direction is the potential $V$ increasing most rapidly?

(d) In what direction is the potential $V$ decreasing most rapidly?

**49. Geography** The surface of a hill is modeled by the equation $z = (8 - 2x^2 - y^2)$ m. See the figure below. If a freshwater spring is located at the point $(1, 2, 2)$, in what direction will the water flow?

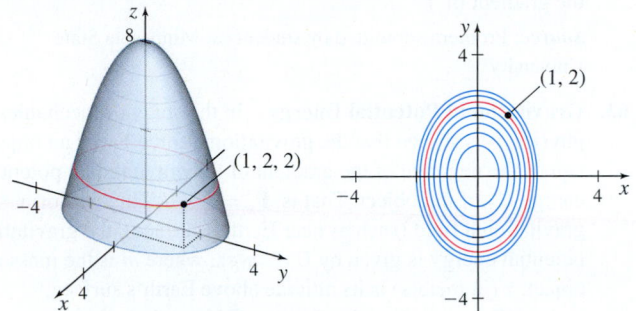

The graph of $z$ and its level curves.

**50. Temperature Change** Suppose that the temperature in degrees Celsius at each point of the coordinate plane is $T = (3x^2 + 4y^2 + 5)\,°\text{C}$.

(a) If we leave the point $(3, 4)$ heading for the point $(4, 3)$, how fast (in degrees per unit of distance) is the temperature changing at the point $(3, 4)$?

(b) After reaching $(4, 3)$, in what direction should we go if we want to cool off as fast as possible? How fast is the temperature changing at the point $(4, 3)$?

(c) Suppose that we want to move away from $(4, 3)$ along a path of constant temperature. What is an equation of our path and what is the temperature?

(d) A person at the origin may go in *any* direction and will experience the same rate of change of temperature. Why? What is the rate?

**51. Temperature Change** The temperature at any point $(x, y)$ of a rectangular plate lying in the $xy$-plane is given by $T = x \sin(2y)\,°\text{C}$. Find the rate of change of temperature at the point $\left(1, \dfrac{\pi}{4}\right)$ in the direction making an angle of $\dfrac{\pi}{6}$ with the positive $x$-axis.

**52. Rate of Change** Suppose that $z = xy^2$. In what direction(s) can we go from the point $(-1, 1)$ if we want the rate of change of $z$ to be 2?

**53. Rate of Change**

(a) Find the direction through the point $(2, 1)$ in which the function $z = 4x^2 + 9y^2$ has a maximum rate of change.

(b) Verify that this direction is that of the normal to the curve $4x^2 + 9y^2 = 25$ at the point $(2, 1)$. See the figure in the next column.

(c) Find the value of this maximum rate of change.

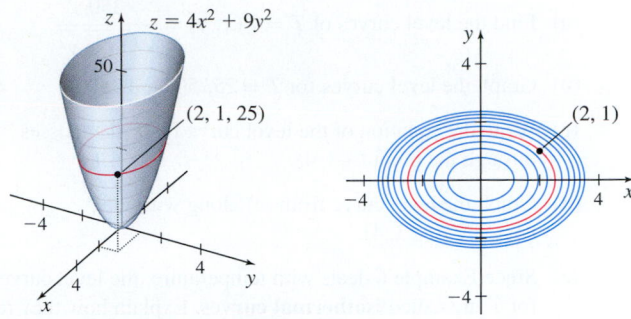

The curve $4x^2 + 9y^2 = 25$ appears in red.

**54.** Show that the level curves of $z = f(x, y) = x^2 - y^2$ are orthogonal to the level curves of $h(x, y) = xy$ for all $(x_0, y_0) \neq (0, 0)$. See the figure below.

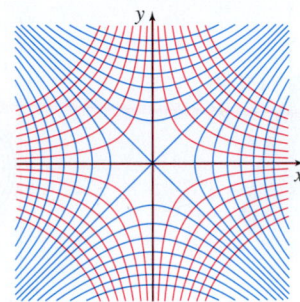

The level curves of $z$ are in blue. The level curves of $h$ are in red.

**55.** Find a unit vector $\mathbf{u}$ that is normal to the level curve of $f(x, y) = 4x^2 y$ through $P = (1, -2)$ at $P$. See the figure below.

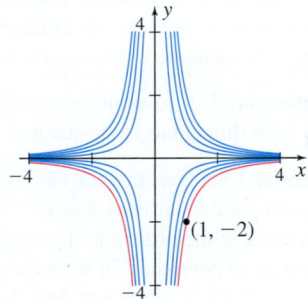

The level curves of $f(x, y) = 4x^2 y$.

**56.** Find a unit vector $\mathbf{u}$ that is normal to the level curve of $f(x, y) = 2x^2 + y^2 + 1$ through $P = (1, 1)$ at $P$. See the figure below.

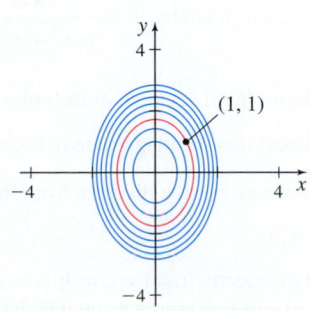

The level curves of $f(x, y) = 2x^2 + y^2 + 1$.

**57. Isothermal Curves** Using the information from Example 6:

(a) Find the level curves of $T = T(x, y) = \dfrac{250}{\sqrt{x^2 + y^2}}$.

(b) Graph the level curves for $T = 25$, $50$, and $250$.

(c) Write the equation of the level curve for $T$ that passes through the point $(-3, 4)$.

(d) Graph the level curve from (c) along with the vector $\nabla T(-3, 4)$.

(e) Since Example 6 deals with temperature, the level curves for $T$ are called **isothermal curves.** Explain how they relate to the temperature $T$.

**58. Mountaineering** Suppose that you are climbing a mountain whose surface is modeled by the equation $x^2 + y^2 - 5x + z = 0$. (The $x$-axis points east, the $y$-axis north, and the $z$-axis up; units are in thousands of feet.)

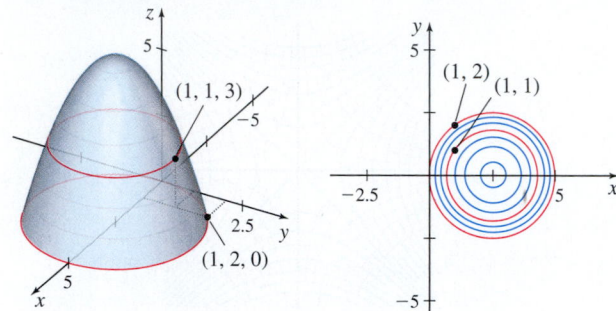

The graph of $z$ and its level curves.

(a) Your route is such that as you pass through the point $(1, 1, 3)$, you are heading northeast. At what rate (with respect to distance) is your altitude changing at that point?

(b) Another climber at $(1, 2, 0)$ wants to reach the summit in as short a distance as possible. In what direction should he start? At what rate is his altitude changing when he starts?

(c) A third climber at $(2, 1, 5)$ wants to remain at the same altitude. In what direction(s) may she go?

**59. Electrical Potential** Suppose that the electrical potential (voltage $V$) at each point in space is $V = e^{xyz}$ volts and that electric charges move in the direction of greatest *potential drop* (most rapid decrease of potential). In what direction does a charge at the point $(1, -1, 2)$ move? How fast does the potential change as the charge leaves this point?

**60. Electric Field** The electric field vector **E** is the negative of the gradient of the electrostatic potential $V$. That is, $\mathbf{E} = -\nabla V$. If $Q$ is a point charge at the origin, then the electrostatic potential $V$ at the point $P(x, y, z)$ is given by $V = \dfrac{kQ}{\sqrt{x^2 + y^2 + z^2}}$, where $k$ is a constant.

(a) Find the electric field vector **E** at the point $P(x, y, z)$.

(b) Use (a) to show that the magnitude of the electric field is $\|\mathbf{E}\| = \dfrac{kQ}{r^2}$, where $r$ is the distance from the origin to the point $P(x, y, z)$.

(c) Verify that the electric field vector **E** is normal to the equipotential surfaces (surfaces on which the electrostatic potential $V$ is constant).

**61. Modeling Temperature Change** The temperature at each point of the region $x^2 + y^2 + z^2 \leq 9$ is $T = \sqrt{9 - x^2 - y^2 - z^2}\,°$C. If we start at the point $(0, 1, 2)$ and move across the region in a straight path ending at $(2, 1, 2)$, find the rate of change of $T$ at an arbitrary point on the path.

**62. Chemotaxis** Chemotaxis is the phenomenon in which bodily cells, bacteria, and other organisms direct their movement according to a chemical stimulus. Suppose the concentration of a certain biochemical distribution at a wounded place is given by $w = f(x, y, z) = \dfrac{1}{x^2 + y^2 + z^2}$. If a cell that reacts by chemotaxis is located at the point $(1, 2, 3)$, find the direction it will move by chemotaxis if the direction of movement is along the gradient of $f$.

*Source:* Problem submitted by students at Minnesota State University

**63. Gravitational Potential Energy** In the study of mechanics in physics, it is shown that the gravitational force $\mathbf{F}_g$ on an object equals the negative of the gradient of the gravitational potential energy $U_g$ of the object. That is, $\mathbf{F}_g = -\nabla U_g$. In a uniform gravitational field (such as near Earth's surface), the gravitational potential energy is given by $U_g = mgz$, where $m$ is the mass of an object, $z$ (in meters) is its altitude above Earth's surface, and $g = 9.8 \text{ m/s}^2$. Use the gradient $\nabla U_g$ to show that the gravitational force is downward and has magnitude $mg$.

**64. Gravitational Field** Two objects of masses $M$ and $m$ are a distance $r$, in meters, apart. The gravitational potential energy between these objects is given by $U_g = -G\dfrac{mM}{r}$, where $G = 6.67 \times 10^{-11} \text{ Nm}^2/\text{kg}^2$. If one object is at the origin and the other at the point $(x, y, z)$, then the gravitational field $\mathbf{F}_g$ equals $-\nabla U_g$. Show that the magnitude of the force $\mathbf{F}_g$ equals $G\dfrac{mM}{r^2}$.

*Source:* Problem contributed by students at Minnesota State University

**65.** If the function $z = f(x, y)$ is differentiable at a point $P_0 = (x_0, y_0)$ and $\nabla f(x_0, y_0) \neq \mathbf{0}$, show that $D_{\mathbf{u}} f(x_0, y_0) = 0$ in the direction orthogonal to that of $\nabla f(x_0, y_0)$.

**66. Algebraic Properties of the Gradient** If $u = f(x, y)$ and $v = g(x, y)$ are differentiable, show that:

(a) $\nabla(ku) = k\nabla u$, where $k$ is a constant.

(b) $\nabla(u + v) = \nabla u + \nabla v$.

(c) $\nabla(uv) = u\nabla v + v\nabla u$.

(d) $\nabla\dfrac{u}{v} = \dfrac{v\nabla u - u\nabla v}{v^2}$.

(e) $\nabla u^a = au^{a-1}\nabla u$, where $a$ a real number.

**67.** Show that for a nonzero vector $\mathbf{a} = a_1\mathbf{i} + a_2\mathbf{j}$ and a differentiable function $z = f(x, y)$,

$$D_{\mathbf{u}} f(x, y) = \frac{a_1\dfrac{\partial f}{\partial x} + a_2\dfrac{\partial f}{\partial y}}{\sqrt{a_1^2 + a_2^2}}$$

where $\mathbf{u} = \dfrac{\mathbf{a}}{\|\mathbf{a}\|}$.

**68.** Let $F(x, y) = 0$ be the equation of a curve in the $xy$-plane. If $F$ is differentiable and if $(x_0, y_0)$ is a point on the curve, show that $\nabla F(x_0, y_0)$ is normal to the curve at $(x_0, y_0)$.

**69.** Use Problem 68 to show that the tangent line to the curve $F(x, y) = 0$ at $(x_0, y_0)$ is given by $a(x - x_0) + b(y - y_0) = 0$, where $a = F_x(x_0, y_0)$ and $b = F_y(x_0, y_0)$. (Assume that $a$ and $b$ are not both 0.)

**70.** Use Problem 69 to find the following tangent lines to the hyperbola $x^2 - y^2 = 16$:

    **(a)** The tangent line at $(5, 3)$. Check the result by finding $\dfrac{dy}{dx}$ at $(5, 3)$ using implicit differentiation.

    **(b)** The tangent line at $(4, 0)$. Note that this tangent line is vertical and requires special treatment in single variable calculus.

**71.** Suppose $z = f(x, y)$ has directional derivatives in all directions. Must $f$ be differentiable? Explain.

**72.** If $f(x, y, z) = z^3 + 3xz - y^2$, find the directional derivative of $f$ at $(1, 2, 1)$ in the direction of the line $x - 1 = y - 2 = z - 1$. Interpret the result as a rate of change.

### Challenge Problems

**73.** Let $\mathbf{r} = x\mathbf{i} + y\mathbf{j}$ and $r = \|\mathbf{r}\| = \sqrt{x^2 + y^2}$.

    **(a)** Show that $\nabla r^n = nr^{n-2}\mathbf{r}$.

    **(b)** Show that $\nabla g(r) = g'(r)\dfrac{\mathbf{r}}{r}$.

    **(c)** Show that $D_\mathbf{u}[g(r)] = \dfrac{g'(r)}{r}(\mathbf{u} \cdot \mathbf{r})$.

    **(d)** Show that $\nabla\dfrac{x}{r} = \dfrac{\mathbf{i}}{r} - \dfrac{x}{r} \cdot \dfrac{\mathbf{r}}{r^2}$.

**74.** Find the directional derivative of

$$f(x, y) = \begin{cases} \dfrac{\sin(x^2 + y^2)}{x^2 + y^2} & \text{if } (x, y) \neq (0, 0) \\ 1 & \text{if } (x, y) = (0, 0) \end{cases}$$

at $(0, 0)$ in the direction of $\mathbf{a} = \mathbf{i} + \mathbf{j}$.

**75.** Show that if $\nabla f(x, y) = c(x\mathbf{i} + y\mathbf{j})$, where $c$ is a constant, then $f(x, y)$ is constant on any circle of radius $k$, centered at $(0, 0)$.

**76. The Gradient Is Normal to the Level Surface**  Show that if the function $w = f(x, y, z)$ is differentiable at a point $P_0 = (x_0, y_0, z_0)$, and if $\nabla f(x_0, y_0, z_0) \neq \mathbf{0}$, then the gradient $\nabla f(x_0, y_0, z_0)$ is normal to the level surface of $f$ through $P_0$.

**77.** Let $z = f(x, y)$ have continuous second-order partial derivatives. If $\mathbf{u} = u_1\mathbf{i} + u_2\mathbf{j}$ is a unit vector, we have a directional derivative $D_\mathbf{u} f(x, y) = g(x, y)$. If $g(x, y)$ is differentiable and $\mathbf{v} = v_1\mathbf{i} + v_2\mathbf{j}$ is a unit vector, we have a directional derivative $D_\mathbf{v} g(x, y)$. We can view this second quantity as a second-order directional derivative for $z = f(x, y)$. Express it in terms of $f_{xx}$, $f_{xy}$, and $f_{yy}$ by showing that it has the value

$$u_1 v_1 f_{xx} + (u_1 v_2 + u_2 v_1) f_{xy} + u_2 v_2 f_{yy}$$

**78.** Assuming that $b \neq 0$ in Problem 69, show that the slope of the tangent line to the curve $F(x, y) = 0$ at $(x_0, y_0)$ is $m = -\dfrac{F_x(x_0, y_0)}{F_y(x_0, y_0)}$. (This is a proof of the Implicit Differentiation Formula I from Chapter 12, Section 12.5.)

---

# 13.2 Tangent Planes

**OBJECTIVES**  *When you finish this section, you should be able to:*

**1** Find an equation of a tangent plane to a surface (p. 922)

**2** Find an equation of a normal line to a tangent plane (p. 922)

**3** Find an equation of a tangent plane to a surface defined explicitly (p. 923)

**NEED TO REVIEW?** A smooth curve is discussed in Section 9.2, pp. 690–691.

In Chapter 3 we used the tangent line to the graph of a function $y = f(x)$ at $(x_0, y_0)$ to approximate the function at nearby points. For surfaces we use the tangent plane at a point $(x_0, y_0, z_0)$ on the surface of $z = f(x, y)$ to approximate the function at nearby points. For example, we live on a sphere, but for nearby calculations we use the "flat earth approximation" and treat Earth as if it were flat, that is, as if it were a plane.

**DEFINITION  Tangent Plane to a Surface**

Let $P_0 = (x_0, y_0, z_0)$ be a point on a surface $S$, and let $C$ be any smooth curve containing the point $P_0$ and lying entirely on $S$. If the tangent lines to all such curves $C$ at $P_0$ lie in the same plane, then this plane is called the **tangent plane to $S$ at $P_0$**. See Figure 10.

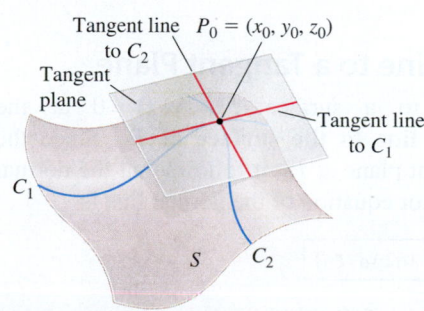

Tangent line to $C_2$   $P_0 = (x_0, y_0, z_0)$
Tangent plane
Tangent line to $C_1$
$C_1$
$S$   $C_2$

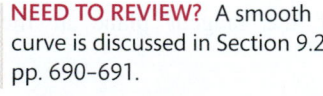
**Figure 10**

Now suppose the surface $S$ is defined implicitly by the equation $F(x, y, z) = 0$. Then $S$ is equivalent to a level surface of the function $w = F(x, y, z)$. Let $P_0 = (x_0, y_0, z_0)$ be a point on the surface $S$. Suppose $F$ is differentiable at $P_0$ and $\nabla F(x_0, y_0, z_0) \neq \mathbf{0}$. Then

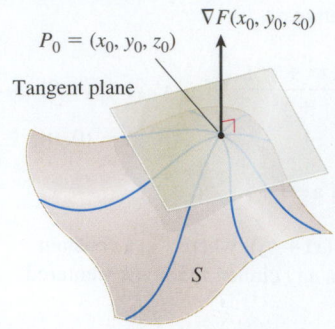

**Figure 11** $F(x, y, z) = 0$

**NEED TO REVIEW?** The equation of a plane is discussed in Section 10.6, pp. 776–778.

the gradient $\nabla F(x_0, y_0, z_0)$ is normal to the level surface that contains the point $P_0$. That is, for any smooth curve $C$ on the surface $S$ that contains the point $P_0$, the tangent lines to $C$ at $P_0$ are orthogonal to the vector $\nabla F(x_0, y_0, z_0)$. As a result, the tangent lines to $C$ at the point $P_0$ all lie in the same plane, and the surface $S$ has a tangent plane at $P_0$.

See Figure 11. Suppose $(x, y, z)$ is any point on the tangent plane to $S$ at the point $P_0 = (x_0, y_0, z_0)$. Since $\nabla F(x_0, y_0, z_0)$ is normal to the tangent plane at $P_0$, an equation of the tangent plane is

$$\nabla F(x_0, y_0, z_0) \cdot [(x - x_0)\mathbf{i} + (y - y_0)\mathbf{j} + (z - z_0)\mathbf{k}] = 0$$

where $(x - x_0)\mathbf{i} + (y - y_0)\mathbf{j} + (z - z_0)\mathbf{k}$ is a vector in the direction of the tangent line to a smooth curve $C$ at the point $P_0$. Since

$$\nabla F(x_0, y_0, z_0) = F_x(x_0, y_0, z_0)\mathbf{i} + F_y(x_0, y_0, z_0)\mathbf{j} + F_z(x_0, y_0, z_0)\mathbf{k}$$

an equation of the tangent plane is

$$F_x(x_0, y_0, z_0)(x - x_0) + F_y(x_0, y_0, z_0)(y - y_0) + F_z(x_0, y_0, z_0)(z - z_0) = 0$$

> **THEOREM  Equation of a Tangent Plane**
>
> If $F$ is differentiable at a point $P_0 = (x_0, y_0, z_0)$ on the surface $F(x, y, z) = 0$, and if $\nabla F(x_0, y_0, z_0) \neq \mathbf{0}$, then the surface has a tangent plane at $P_0$. An equation of the tangent plane at $P_0$ is
>
> $$\boxed{F_x(x_0, y_0, z_0)(x - x_0) + F_y(x_0, y_0, z_0)(y - y_0) + F_z(x_0, y_0, z_0)(z - z_0) = 0}$$

## 1 Find an Equation of a Tangent Plane to a Surface

 **EXAMPLE 1  Finding an Equation of a Tangent Plane to a Surface**

Find an equation of the tangent plane to the hyperboloid of one sheet $x^2 + y^2 - z^2 = 24$ at the point $(3, -4, 1)$.

**Solution** The surface is given by the function $F(x, y, z) = x^2 + y^2 - z^2 - 24 = 0$. The partial derivatives of $F$ are

$$F_x(x, y, z) = 2x \qquad F_y(x, y, z) = 2y \qquad F_z(x, y, z) = -2z$$

At the point $(3, -4, 1)$, the partial derivatives are

$$F_x(3, -4, 1) = 6 \qquad F_y(3, -4, 1) = -8 \qquad F_z(3, -4, 1) = -2$$

An equation of the tangent plane at $(3, -4, 1)$ is

$$6(x - 3) - 8(y + 4) - 2(z - 1) = 0$$
$$3x - 4y - z = 24 \qquad \blacksquare$$

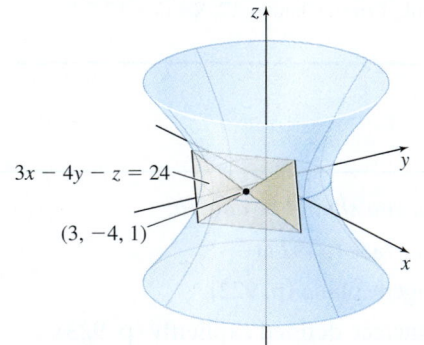

**Figure 12** $x^2 + y^2 - z^2 = 24$

Figure 12 shows the hyperboloid and part of the tangent plane at the point $(3, -4, 1)$.

**NOW WORK** Problems 3 and 9(a).

## 2 Find an Equation of a Normal Line to a Tangent Plane

**NEED TO REVIEW?** Lines in space are discussed in Section 10.6, pp. 772–775.

The line normal to the tangent plane to a surface $F(x, y, z) = 0$ at the point $P_0 = (x_0, y_0, z_0)$ is called the **normal line to the surface at** $P_0$. Since the gradient $\nabla F(x_0, y_0, z_0)$ is normal to the tangent plane at $P_0$, it follows that the normal line is in the direction of the gradient. The vector equation of the normal line is

$$\boxed{\mathbf{r}(t) = \mathbf{r}_0 + t\nabla F(x_0, y_0, z_0)}$$

where $\mathbf{r}_0 = x_0\mathbf{i} + y_0\mathbf{j} + z_0\mathbf{k}$ is the position vector of $P_0$ and $\mathbf{r}$ is the position vector of any point $P$ on the normal line.

The corresponding parametric equations of the normal line are

$$x = x_0 + at \qquad y = y_0 + bt \qquad z = z_0 + ct$$

where $a = F_x(x_0, y_0, z_0)$, $b = F_y(x_0, y_0, z_0)$, and $c = F_z(x_0, y_0, z_0)$.

If $abc \neq 0$, symmetric equations of the normal line are

$$\frac{x - x_0}{a} = \frac{y - y_0}{b} = \frac{z - z_0}{c}$$

---

**EXAMPLE 2**  **Finding an Equation of a Normal Line to a Tangent Plane**

Find symmetric equations of the normal line to the hyperboloid of one sheet defined by the equation $x^2 + y^2 - z^2 = 24$ at the point $(3, -4, 1)$.

**Solution**  The surface is given by $F(x, y, z) = x^2 + y^2 - z^2 - 24 = 0$. From Example 1, $\nabla F(3, -4, 1) = 6\mathbf{i} - 8\mathbf{j} - 2\mathbf{k}$. Since the normal line is in the direction of the gradient, symmetric equations of the normal line to the hyperboloid at $(3, -4, 1)$ are

$$\frac{x - 3}{6} = \frac{y + 4}{-8} = \frac{z - 1}{-2} \qquad \blacksquare$$

**NOW WORK** Problems 9(b) and 11.

## 3  Find an Equation of a Tangent Plane to a Surface Defined Explicitly

To find an equation of the tangent plane to the surface $z = f(x, y)$ at $(x_0, y_0, z_0)$, $z_0 = f(x_0, y_0)$, we write the equation of the surface implicitly as

$$F(x, y, z) = z - f(x, y) = 0$$

Then

$$F_x = -f_x \qquad F_y = -f_y \qquad F_z = 1$$

An equation of the tangent plane to $z = f(x, y)$ at $(x_0, y_0, z_0)$ is

$$F_x(x_0, y_0, z_0)(x - x_0) + F_y(x_0, y_0, z_0)(y - y_0) + F_z(x_0, y_0, z_0)(z - z_0) = 0$$

$$-f_x(x_0, y_0)(x - x_0) - f_y(x_0, y_0)(y - y_0) + (z - z_0) = 0$$

$$\boxed{z - z_0 = f_x(x_0, y_0)(x - x_0) + f_y(x_0, y_0)(y - y_0)}$$

where $z_0 = f(x_0, y_0)$.

The parametric equations of the normal line to $z = f(x, y)$ at the point $(x_0, y_0, z_0)$ are

$$x = x(t) = x_0 + t f_x(x_0, y_0) \qquad y = y(t) = y_0 + t f_y(x_0, y_0) \qquad z = z(t) = z_0 - t$$

The next result tells us about the tangent plane to a surface $z = f(x, y)$ at a point $(x_0, y_0)$ where $\nabla f(x_0, y_0) = \mathbf{0}$.

### THEOREM

If $\nabla f(x_0, y_0) = \mathbf{0}$, then the tangent plane to the surface $z = (x, y)$ at $(x_0, y_0)$ is $z = z_0 = f(x_0, y_0)$. That is, the tangent plane is parallel to the $xy$-plane.

We shall see a consequence of this theorem in the next section.

---

**EXAMPLE 3**  **Finding an Equation of a Tangent Plane for Surfaces Defined Explicitly**

For the surface $z = f(x, y) = x^2 + y^3 + 2xy - 3x - 4y + 3$

**(a)**  Find the tangent plane to the graph of $f$ at the point $(0, 0, 3)$.

**(b)**  Locate the points, if any, at which the tangent plane is parallel to the $xy$-plane.

**Solution** (a) Begin by finding $f_x(0, 0)$ and $f_y(0, 0)$.

$$f_x(x, y) = 2x + 2y - 3 \qquad f_x(0, 0) = -3$$
$$f_y(x, y) = 3y^2 + 2x - 4 \qquad f_y(0, 0) = -4$$

An equation of the tangent plane to the graph of $f$ at the point $(0, 0, 3)$ is

$$z - 3 = -3x - 4y \qquad z - z_0 = f_x(x_0, y_0)(x - x_0) + f_y(x_0, y_0)(y - y_0)$$
$$3x + 4y + z = 3$$

(b) The tangent plane is parallel to the $xy$-plane at any point for which $\nabla f(x, y) = \mathbf{0}$. This results in the system of equations

$$f_x(x, y) = 2x + 2y - 3 = 0$$
$$f_y(x, y) = 3y^2 + 2x - 4 = 0$$

We solve for $2x$ in $f_y(x, y) = 0$, and substitute into $f_x(x, y) = 0$.

$$f_y(x, y) = 3y^2 + 2x - 4 = 0$$
$$2x = 4 - 3y^2$$

Then

$$f_x(x, y) = 2x + 2y - 3 = 0$$
$$4 - 3y^2 + 2y - 3 = 0$$
$$3y^2 - 2y - 1 = 0$$
$$(3y + 1)(y - 1) = 0$$
$$y = -\frac{1}{3} \qquad y = 1$$

Using $2x = 4 - 3y^2$, or equivalently, $x = \frac{1}{2}(4 - 3y^2)$,

$$\text{if } y = -\frac{1}{3}, \text{ then } x = \frac{1}{2}\left(4 - 3 \cdot \frac{1}{9}\right) = \frac{11}{6}$$

$$\text{if } y = 1, \text{ then } x = \frac{1}{2}(4 - 3) = \frac{1}{2}$$

Finally, $z = f\left(\dfrac{11}{6}, -\dfrac{1}{3}\right) = \dfrac{101}{108}$ and $z = f\left(\dfrac{1}{2}, 1\right) = -\dfrac{1}{4}$

At the points $\left(\dfrac{11}{6}, -\dfrac{1}{3}, \dfrac{101}{108}\right)$ and $\left(\dfrac{1}{2}, 1, -\dfrac{1}{4}\right)$, the tangent plane to the surface is parallel to the $xy$-plane. ∎

**NOW WORK** Problem 25.

---

## 13.2 Assess Your Understanding

**Concepts and Vocabulary**

**1.** *True or False* If $F$ is differentiable at a point $P_0 = (x_0, y_0, z_0)$ on the surface $F(x, y, z) = 0$, and if $\nabla F(x_0, y_0, z_0) \neq \mathbf{0}$, then the surface has a tangent plane at $P_0$.

**2.** *True or False* The normal line to the surface $F(x, y, z) = 0$ at a point $P_0 = (x_0, y_0, z_0)$ is in the direction of the gradient $\nabla F(x_0, y_0, z_0)$, provided $\nabla F(x_0, y_0, z_0) \neq \mathbf{0}$.

**Skill Building**

*In Problems 3–6, find an equation of the tangent plane to each surface at the given point.*

**3.** $z = 10 - 2x^2 - y^2$ at $(0, 2, 6)$   **4.** $z = 4 + x^2 + y^2$ at $(1, -2, 9)$

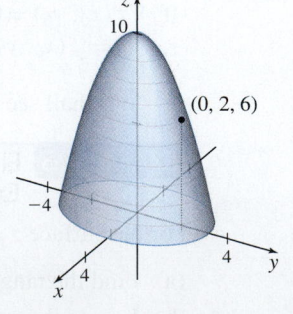

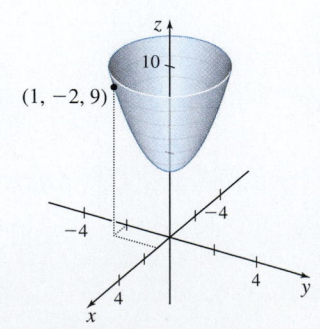

---

**1.** = NOW WORK problem    = Graphing technology recommended   **CAS** = Computer Algebra System recommended

**5.** $z = 4x^2 + 9y^2$ at $(1, 1, 13)$    **6.** $z = \sqrt{x^2 + 3y^2}$ at $(1, 1, 2)$

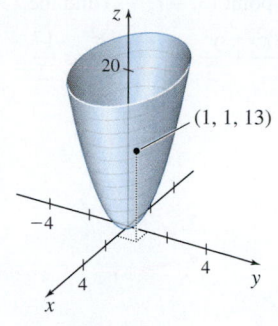

$(1, 1, 13)$

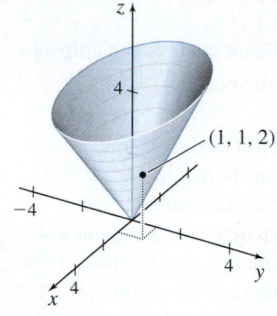

$(1, 1, 2)$

*In Problems 7–18:*

**(a)** *Find an equation of the tangent plane to each surface at the given point.*

**(b)** *Find symmetric equations of the normal line to each surface at the given point.*

**7.** $x^2 + y^2 + z^2 = 14$ at $(1, -2, 3)$

**8.** $2x^2 + 3y^2 + z^2 = 12$ at $(2, 1, -1)$

**9.** $4x^2 + y^2 - 2z^2 = 3$ at $(1, 1, -1)$

**10.** $z^2 = x^2 + 3y^2$ at $(1, -1, -2)$

**11.** $z = x^2 + y^2$ at $(-2, 1, 5)$

**12.** $z = 2x^2 - 3y^2$ at $(2, 1, 5)$

**13.** $4x^2 - y^2 = z^2$ at $(1, 0, 2)$

**14.** $x^2 + y^2 - 2z^2 = -13$ at $(2, 1, 3)$

**15.** $x^2 - 3y^2 - 4z^2 = 2$ at $(3, 1, 1)$

**16.** $x^2 + 2y^2 - 5z^2 = 4$ at $(1, 2, 1)$

**17.** $f(x, y) = \ln(\sec(x^2 + y^2))$ at $\left(\sqrt{\dfrac{\pi}{8}}, \sqrt{\dfrac{\pi}{8}}, \ln \sqrt{2}\right)$

**18.** $f(x, y) = \dfrac{x^2 - y^2}{x^2 + y^2}$ at $\left(1, 2, -\dfrac{3}{5}\right)$

*In Problems 19–22:*

**(a)** *Find an equation of the tangent plane to each surface at the given point.*

**(b)** *Find symmetric equations of the normal line to each surface at the given point.*

**CAS (c)** *Use technology to graph each surface and the tangent plane to the surface at the given point.*

**19.** $z = e^x \cos y$ at $\left(0, \dfrac{\pi}{2}, 0\right)$

**20.** $z = \ln(x^2 + y^2)$ at $(1, -1, \ln 2)$

**21.** $x^{2/3} + y^{2/3} + z^{2/3} = 9$ at $(1, 8, -8)$

**22.** $x^{1/2} + y^{1/2} + z^{1/2} = 6$ at $(1, 4, 9)$

## Applications and Extensions

**23.** Show that the same results would have been obtained in Example 1 by solving the equation $x^2 + y^2 - z^2 - 24 = 0$ for $z$ and using the function $z = f(x, y) = \sqrt{x^2 + y^2 - 24}$ whose graph is the part of the hyperboloid lying above the $xy$-plane.

**24.** Show that the same results would have been obtained in Example 2 by solving the equation $x^2 + y^2 - z^2 - 24 = 0$ for $z$ and using the function $z = f(x, y) = \sqrt{x^2 + y^2 - 24}$ whose graph is the part of the hyperboloid lying above the $xy$-plane.

*In Problems 25–32:*

**(a)** *determine the point(s) on the surface $z = f(x, y)$ at which the tangent plane is parallel to the $xy$-plane.*

**(b)** *Find an equation of each tangent plane parallel to the xy-plane.*

**25.** $z = 6x - 4y - x^2 - 2y^2$    **26.** $z = 4x - 2y + x^2 + y^2$

**27.** $z = e^{-y}(x^2 + y^2)$    **28.** $z = xy + \dfrac{4}{x} + \dfrac{2}{y}$

**29.** $z = x^2 + 2x + y^2$    **30.** $z = x^2 - 3y + y^2$

**31.** $z = 2x^4 - y^2 - x^2 - 2y$    **32.** $z = x^2 + y^4 - 4y^2 - 2x$

**33. Solar Panel** An engineer is building experimental, dome-shaped living quarters 4 m high and modeled by the function $z = 4 - x^2 - y^2$. She wants to bolt a flat solar panel to the dome at the point $(1, 1, 2)$. In what direction should the engineer drill?

**34. (a)** Two surfaces are said to be **tangent at a common point** $P_0$ if each has the same tangent plane at $P_0$. Show that the surfaces $x^2 + z^2 + 4y = 0$ and $x^2 + y^2 + z^2 - 6z + 7 = 0$ are tangent at the point $(0, -1, 2)$.

**CAS (b)** Graph each surface.

**35. (a)** Show that the surfaces $x^2 + y^2 + z^2 = 1$ and $x^2 + y^2 - z^2 = 1$ are tangent at every point of intersection.

**CAS (b)** Graph each surface.

**36. (a)** Two surfaces are **orthogonal at a common point** $P_0$ if their normal lines at $P_0$ are orthogonal. Show that the surfaces $x^2 + y^2 + z^2 = 4$ and $x^2 + y^2 - z^2 = 0$ are orthogonal at $(0, \sqrt{2}, \sqrt{2})$. In fact, show that the surfaces are orthogonal at every point of intersection.

**CAS (b)** Graph each surface.

**37. Normal Lines of a Sphere** Show that the normal lines of a sphere $x^2 + y^2 + z^2 = R^2$ pass through the center of the sphere.

**38. Normal Line of a Sphere** Let $P_0 = (x_0, y_0, z_0)$ be a point on the sphere $x^2 + y^2 + z^2 = R^2$. Show that the normal line to the sphere at $P_0$ passes through the point $(-x_0, -y_0, -z_0)$.

**39. Tangent Plane** Find equations of the tangent plane and normal line to $\sin(x + y) + \sin(y + z) = 1$ at the point $\left(0, \dfrac{\pi}{2}, \dfrac{\pi}{2}\right)$.

**40. Tangent Plane to an Ellipsoid** Find the points on the surface $\dfrac{x^2}{9} + \dfrac{y^2}{4} + \dfrac{z^2}{36} = 1$, where the tangent plane is parallel to the plane $2x - 3y + z = 4$.

**41. Tangent Plane to a Cylinder** Show that the tangent plane to the cylinder $x^2 + y^2 = a^2$ at the point $(x_0, y_0, z_0)$ is given by the equation $x_0 x + y_0 y = a^2$.

**42. Tangent Plane to a Cone**   Show that the tangent plane to the cone $\dfrac{z^2}{c^2} = \dfrac{x^2}{a^2} + \dfrac{y^2}{b^2}$ at the point $(x_0, y_0, z_0) \neq (0, 0, 0)$ is given by the equation $\dfrac{z_0}{c^2}z = \dfrac{x_0}{a^2}x + \dfrac{y_0}{b^2}y$. Why must the point of tangency be different from the origin?

**43.** Show that at any point, the sum of the intercepts on the coordinate axes of the tangent plane to the surface $x^{1/2} + y^{1/2} + z^{1/2} = a^{1/2}$, $a > 0$, is a constant.

## Challenge Problems

**44. Tangent Line**   Find an equation of the tangent line to the curve of intersection of the surfaces $x\sin(yz) = 1$ and $ze^{y^2 - x^2} = \dfrac{\pi}{2}$ at the point $\left(1, 1, \dfrac{\pi}{2}\right)$.

**45.** Find equations of the tangent plane and normal line to $(yz)^{xz} = 16$ at the point $(2, 1, 2)$.

**46.** Find the intersection of the normal line to the hyperbolic paraboloid $x = \dfrac{y^2}{2} - z^2 - \dfrac{35}{2}$ at the point $(3, -7, -2)$ and the tangent plane to the ellipsoid $\dfrac{(x-1)^2}{2} + \dfrac{y^2}{3} + \dfrac{(z+1)^2}{4} = \dfrac{17}{2}$ at the point $(4, 3, 1)$.

**47.** Consider the tangent plane to the graph of a differentiable function $z = f(x, y)$ at a point $(x_0, y_0)$. Suppose $\mathbf{v}_1$ is a tangent vector on the plane whose $\mathbf{j}$ component is 0 and $\mathbf{v}_2$ is a tangent vector on the plane whose $\mathbf{i}$ component is 0. Must $\mathbf{v}_1$ and $\mathbf{v}_2$ be at right angles? Justify the answer. See the figure.

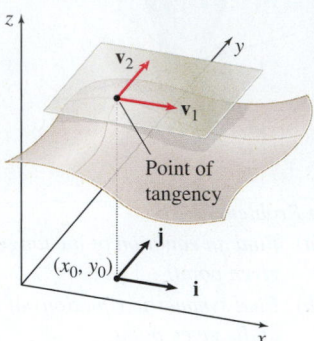

# 13.3 Extrema of Functions of Two Variables

**OBJECTIVES**  *When you finish this section, you should be able to:*

**1** Find critical points (p. 927)
**2** Use the Second Partial Derivative Test (p. 928)
**3** Find the absolute extrema of a function of two variables (p. 930)
**4** Solve optimization problems (p. 933)

An important application of the derivative of a function of a single variable is to identify local extrema and absolute extrema. In a similar way, partial derivatives are used to identify the extrema of a function of several variables. We begin with the definitions of the local maximum and the local minimum of a function $z = f(x, y)$ of two variables. Compare these definitions to those given for a function $f$ of a single variable.

**NEED TO REVIEW?**  Local extrema and absolute extrema of a function of a single variable are discussed in Section 4.2, pp. 271–280.

> **DEFINITION**  Local Maximum, Local Minimum
>
> Let $z = f(x, y)$ be a function with domain $D$ and let $(x_0, y_0)$ be an interior point of $D$. Then $f$ has a **local maximum** at $(x_0, y_0)$ if
>
> $$f(x_0, y_0) \geq f(x, y)$$
>
> for all points $(x, y)$ within some disk centered at $(x_0, y_0)$. The number $f(x_0, y_0)$ is called a **local maximum value.**
>
> The function $f$ has a **local minimum** at $(x_0, y_0)$ if
>
> $$f(x_0, y_0) \leq f(x, y)$$
>
> for all points $(x, y)$ within some disk centered at $(x_0, y_0)$. The number $f(x_0, y_0)$ is called a **local minimum value.**
>
> Collectively, the local maxima and local minima of $f$ are called **local extrema.**

> **DEFINITION**  Absolute Maximum, Absolute Minimum
>
> Let $z = f(x, y)$ be a function with domain $D$. If there is a point $(x_0, y_0)$ in $D$ for which $f(x_0, y_0) \geq f(x, y)$ for all $(x, y)$ in $D$, then $f$ has an **absolute maximum** at $(x_0, y_0)$ and $f(x_0, y_0)$ is called the **absolute maximum value** of $f$ on $D$.
>
> If there is a point $(x_0, y_0)$ in $D$ for which $f(x_0, y_0) \leq f(x, y)$ for all $(x, y)$ in $D$, then $f$ has an **absolute minimum** at $(x_0, y_0)$ and $f(x_0, y_0)$ is the **absolute minimum value** of $f$ on $D$.
>
> The absolute maximum and absolute minimum of $f$ are called the **absolute extrema** of the function.

Figure 13 illustrates these definitions.

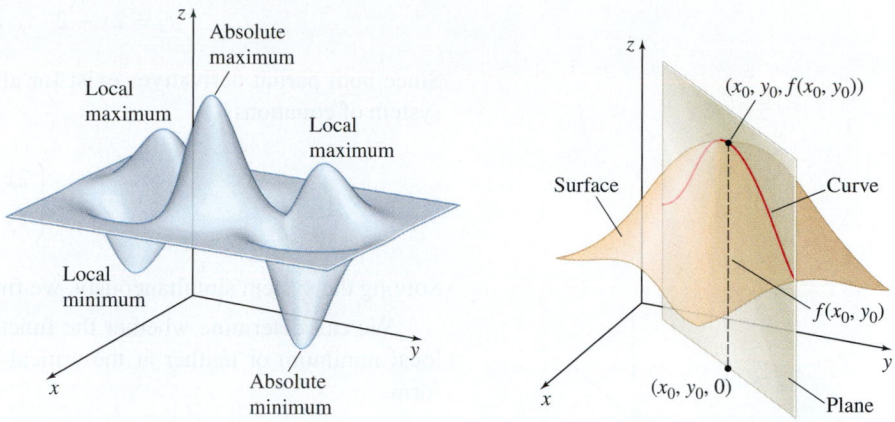

**Figure 13** $z = f(x, y)$          **Figure 14** $z = f(x, y)$

Let $z = f(x, y)$ be a function of two variables that is defined and continuous on an open set $D$. Suppose $f$ has a local maximum at the point $(x_0, y_0)$ in $D$. See Figure 14. Then the curve that results from the intersection of the surface $z = f(x, y)$ and any plane through $(x_0, y_0, 0)$ perpendicular to the $xy$-plane will have a local maximum at $(x_0, y_0)$.

In particular, the curve that results from intersecting $z = f(x, y)$ with the plane $x = x_0$ has this property. This means that if $f_y(x_0, y_0)$ exists, then $f_y(x_0, y_0) = 0$. By a similar argument, if the surface $z = f(x, y)$ is intersected by the plane $y = y_0$, then the curve that results has a local maximum at $(x_0, y_0)$, and if $f_x(x_0, y_0)$ exists, then $f_x(x_0, y_0) = 0$.

> **THEOREM  A Necessary Condition for Local Extrema**
>
> Let $z = f(x, y)$ be a function of two variables that is defined and continuous on an open set containing the point $(x_0, y_0)$. Suppose $f_x$ and $f_y$ each exist at $(x_0, y_0)$. If $f$ has a local extremum at $(x_0, y_0)$, then
>
> $$\boxed{f_x(x_0, y_0) = 0 \qquad \text{and} \qquad f_y(x_0, y_0) = 0}$$

**IN WORDS** If $f_x$ and $f_y$ each exist at $(x_0, y_0)$ and if $f$ has a local extremum at $(x_0, y_0)$, then $\nabla f(x_0, y_0) = \mathbf{0}$.

It can also happen that the local extrema of a function occur at points at which one or both of the partial derivatives fail to exist. For example, the half-cone $z = \sqrt{x^2 + y^2}$ has a local minimum at $(0, 0)$ where both $f_x$ and $f_y$ do not exist.

> **DEFINITION  Critical Point**
>
> Let $z = f(x, y)$ be a function of two variables whose domain is an open set $D$ containing the point $(x_0, y_0)$. The point $(x_0, y_0)$ is a **critical point** of $f$ if either of the conditions given below are met.
>
> - $f_x(x_0, y_0) = f_y(x_0, y_0) = 0$.
> - $f_x(x_0, y_0)$ does not exist, or $f_y(x_0, y_0)$ does not exist, or both $f_x(x_0, y_0)$ and $f_y(x_0, y_0)$ do not exist.

Using critical points, the necessary condition for local extrema is restated.

> **THEOREM  A Necessary Condition for Local Extrema**
>
> Let $z = f(x, y)$ be a function of two variables whose domain is an open set $D$. If $f$ has a local extremum at $(x_0, y_0)$, then $(x_0, y_0)$ is a critical point of $f$.

## 1 Find Critical Points

**EXAMPLE 1  Finding Critical Points**

Find the critical points of the function

$$z = f(x, y) = x^2 + y^2 - 2x + 4y$$

**Solution** The partial derivatives of $f$ are

$$f_x = 2x - 2 \qquad \text{and} \qquad f_y = 2y + 4$$

Since both partial derivatives exist for all $x$ and $y$, the critical points of $f$ satisfy the system of equations

$$\begin{cases} 2x - 2 = 0 \\ 2y + 4 = 0 \end{cases}$$

Solving the system simultaneously, we find that the only critical point is $(1, -2)$. ∎

We can determine whether the function in Example 1 has a local maximum or a local minimum or neither at the critical point by rewriting the function in a simpler form.

$$z = x^2 + y^2 - 2x + 4y$$
$$z = (x^2 - 2x + 1) + (y^2 + 4y + 4) - 5 \qquad \text{Complete the squares.}$$
$$z = (x - 1)^2 + (y + 2)^2 - 5 \qquad \text{Factor.}$$

Since $z \geq -5$, there is a local minimum at $(1, -2)$; the local minimum value is $-5$. In fact, the absolute minimum occurs at the point $(1, -2)$ and the absolute minimum value is also $-5$. See Figure 15.

NOW WORK **Problem 7.**

As with functions of one variable, it is possible for a function $z = f(x, y)$ to have a critical point at $(x_0, y_0)$, but no local extremum at $(x_0, y_0)$.

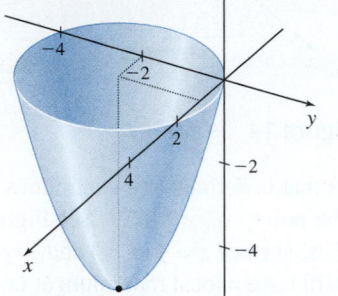

**Figure 15** $z = f(x, y) = x^2 + y^2 - 2x + 4y$ has a local minimum at $(1, -2)$.

---

EXAMPLE 2 **Finding Critical Points**

Consider the hyperbolic paraboloid defined by the function $z = f(x, y) = y^2 - x^2$. Show that $(0, 0)$ is the only critical point of $f$, but that $f$ has neither a local maximum nor a local minimum at $(0, 0)$.

**Solution** The partial derivatives of $f$ are

$$f_x(x, y) = -2x \qquad \text{and} \qquad f_y(x, y) = 2y$$

At $(0, 0)$, both partial derivatives equal 0, so $(0, 0)$ is the only critical point.

See Figure 16. If we consider the values of $f(x, 0) = -x^2$, the function has a maximum value of 0 at the origin. However, if we consider the values of $f(0, y) = y^2$, the function has a minimum value of 0 at the origin. In other words, at the critical point $(0, 0)$, the function appears to have a maximum when viewed in one direction and to have a minimum when viewed in another direction. As a result, $z = f(x, y) = y^2 - x^2$ has neither a maximum nor a minimum at $(0, 0)$. ∎

Example 2 and Figure 16 help explain the next definition.

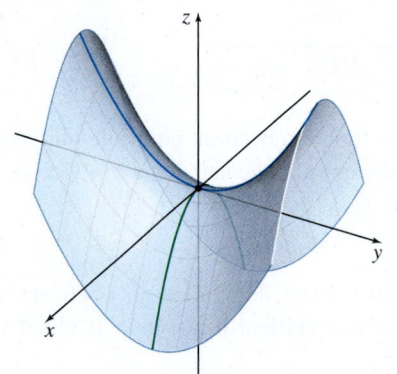

**Figure 16** $z = f(x, y) = y^2 - x^2$ has a saddle point at $(0, 0, 0)$.

**DEFINITION Saddle Point**

A point $(x_0, y_0, f(x_0, y_0))$ is called a **saddle point** of $z = f(x, y)$ if both $f_x(x_0, y_0) = 0$ and $f_y(x_0, y_0) = 0$, but $f$ does not have a local extremum at $(x_0, y_0)$.

## 2 Use the Second Partial Derivative Test

In Examples 1 and 2, we used algebraic and/or geometric arguments to determine whether a critical point resulted in a local maximum, a local minimum, or a saddle point. For most functions, such arguments cannot be used. The following test, which is stated without a proof, provides an analytic method to determine the nature of critical points. The test is analogous to the Second Derivative Test for functions of one variable. A proof of the theorem can be found in texts on advanced calculus.

> **THEOREM  Second Partial Derivative Test for Local Extrema**
>
> Let $z = f(x, y)$ be a function of two variables for which the first- and second-order partial derivatives are continuous in some disk containing the point $(x_0, y_0)$. Suppose $f_x(x_0, y_0) = 0$ and $f_y(x_0, y_0) = 0$. Let
>
> $$A = f_{xx}(x_0, y_0) \qquad B = f_{xy}(x_0, y_0) \qquad C = f_{yy}(x_0, y_0)$$
>
> - If $AC - B^2 > 0$ and $A > 0$, then $f$ has a local minimum at $(x_0, y_0)$.
> - If $AC - B^2 > 0$ and $A < 0$, then $f$ has a local maximum at $(x_0, y_0)$.
> - If $AC - B^2 < 0$, then $f$ has a saddle point at $(x_0, y_0)$.
> - If $AC - B^2 = 0$, then the test provides no guidance.*

**NOTE** The expression
$$AC - B^2 = f_{xx}f_{yy} - f_{xy}^2$$
is also called the **discriminant** or **Hessian** of $f$. It is often written as the determinant
$$AC - B^2 = f_{xx}f_{yy} - f_{xy}^2 = \begin{vmatrix} f_{xx} & f_{xy} \\ f_{xy} & f_{yy} \end{vmatrix}$$

**EXAMPLE 3  Using the Second Partial Derivative Test**

Find any local maxima, local minima, and saddle points of the elliptic paraboloid

$$z = f(x, y) = x^2 + xy + y^2 - 6x + 6$$

**Solution** We begin with the system of equations

$$\begin{cases} f_x(x, y) = 2x + y - 6 = 0 \\ f_y(x, y) = x + 2y = 0 \end{cases}$$

The only solution is $x = 4$ and $y = -2$, so $(4, -2)$ is the only critical point.

The second-order partial derivatives of $f$ are

$$f_{xx}(x, y) = 2 \qquad f_{xy}(x, y) = f_{yx}(x, y) = 1 \qquad f_{yy}(x, y) = 2$$

Since the first and second order partial derivatives are continuous for all points $(x, y)$, the function $f$ satisfies the requirements of the Second Partial Derivative Test at the critical point $(4, -2)$. Now

$$A = f_{xx}(4, -2) = 2 \qquad B = f_{xy}(4, -2) = 1 \qquad C = f_{yy}(4, -2) = 2$$
$$AC - B^2 = 2 \cdot 2 - 1^2 = 3 > 0$$

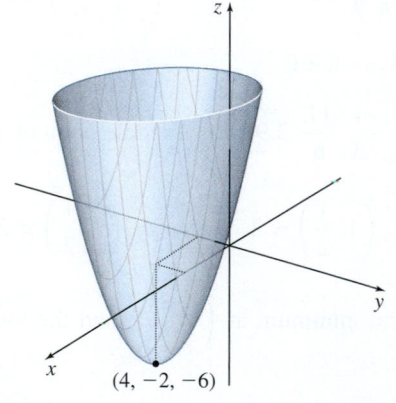

**Figure 17** $z = f(x, y)$
$= x^2 + xy + y^2 - 6x + 6$

Since $AC - B^2 > 0$ and $A = f_{xx}(4, -2) = 2 > 0$, the Second Partial Derivative Test states that $f$ has a local minimum at $(4, -2)$, and $f(4, -2) = -6$ is the local minimum value. See Figure 17. ∎

**NOW WORK** Problems **13** and **17**.

**EXAMPLE 4  Using the Second Partial Derivative Test**

Find any local maxima, local minima, and saddle points for

$$z = f(x, y) = x^3 + y^2 + 2xy - 4x - 3y + 5$$

**Solution** We begin with the system of equations

$$\begin{cases} f_x(x, y) = 3x^2 + 2y - 4 = 0 \\ f_y(x, y) = 2y + 2x - 3 = 0 \end{cases}$$

**NOTE** To use the Second Partial Derivative Test for Local Extrema we must solve the system of equations $f_x = 0$ and $f_y = 0$. If the system is nonlinear, no general method of solution is available and a CAS is generally used. If solving by hand, care must be taken because extraneous roots are sometimes introduced.

*If $AC - B^2 = 0$ at a critical point, the Second Partial Derivative Test for local extrema provides no guidance about the behavior of $z = f(x, y)$ at the critical point. Although there are tests that can be used when $AC - B^2 = 0$, we do not deal with them in this book.

We solve the system using substitution. Solving for $y$ in $f_y(x, y) = 0$, we obtain $y = \dfrac{1}{2}(3 - 2x)$. Then if we substitute for $y$ in $f_x(x, y) = 0$, the result is

$$3x^2 + 2\left[\frac{1}{2}(3 - 2x)\right] - 4 = 0$$

$$3x^2 - 2x - 1 = 0$$

$$(3x + 1)(x - 1) = 0$$

$$x = -\frac{1}{3} \qquad \text{or} \qquad x = 1$$

When $x = -\dfrac{1}{3}$, then $y = \dfrac{1}{2}(3 - 2x) = \dfrac{1}{2}\left(3 + \dfrac{2}{3}\right) = \dfrac{11}{6}$.

When $x = 1$, then $y = \dfrac{1}{2}(3 - 2x) = \dfrac{1}{2}$.

The critical points are $\left(-\dfrac{1}{3}, \dfrac{11}{6}\right)$ and $\left(1, \dfrac{1}{2}\right)$.

The second-order partial derivatives of $f$ are

$$f_{xx}(x, y) = 6x \qquad f_{xy}(x, y) = 2 \qquad f_{yy}(x, y) = 2$$

The requirements of the Second Partial Derivative Test are met at each critical point. At $\left(-\dfrac{1}{3}, \dfrac{11}{6}\right)$,

$$A = f_{xx}\left(-\frac{1}{3}, \frac{11}{6}\right) = -2 \qquad B = f_{xy}\left(-\frac{1}{3}, \frac{11}{6}\right) = 2 \qquad C = f_{yy}\left(-\frac{1}{3}, \frac{11}{6}\right) = 2$$

$$AC - B^2 = -4 - 4 = -8 < 0$$

Since $AC - B^2 < 0$ and $f\left(-\dfrac{1}{3}, \dfrac{11}{6}\right) \approx 2.94$, $\left(-\dfrac{1}{3}, \dfrac{11}{6}, 2.94\right)$ is a saddle point of $f$.

At $\left(1, \dfrac{1}{2}\right)$, $\quad A = f_{xx}\left(1, \dfrac{1}{2}\right) = 6 \quad B = f_{xy}\left(1, \dfrac{1}{2}\right) = 2 \quad C = f_{yy}\left(1, \dfrac{1}{2}\right) = 2$

Since $AC - B^2 = 8 > 0$ and $A > 0$, $f$ has a local minimum at $\left(1, \dfrac{1}{2}\right)$, and the local minimum value is $f\left(1, \dfrac{1}{2}\right) = \dfrac{7}{4}$. ∎

Figure 18 $z = f(x, y)$
$= x^3 + y^2 + 2xy - 4x - 3y + 5$

A graph of $z = f(x, y) = x^3 + y^2 + 2xy - 4x - 3y + 5$ is shown in Figure 18.

NOW WORK Problems 19 and 23.

## 3 Find the Absolute Extrema of a Function of Two Variables

For a function $f$ of one variable, the Extreme Value Theorem states that if $f$ is continuous on a closed interval $[a, b]$, then $f$ has an absolute maximum value and an absolute minimum value on that interval. For functions of two variables, there is a similar theorem. But first we need to define the set in the plane that is analogous to a closed interval on the real line.

Recall that a set $D$ is closed if it contains all its boundary points. We say that $D$ is **bounded** if $D$ can be enclosed by some disk (if $D$ is in the plane) or by some ball (if $D$ is in space). With this definition, we can state the following two-variable version of the Extreme Value Theorem:

**NEED TO REVIEW?** Open and closed sets are discussed in Section 12.2, p. 867.

**THEOREM** Extreme Value Theorem for Functions of Two Variables

Let $z = f(x, y)$ be a function of two variables. If $f$ is continuous on a closed, bounded set $D$, then $f$ has an absolute maximum and an absolute minimum on $D$.

To find the absolute extrema of a function satisfying the criteria of the Extreme Value Theorem for Functions of Two Variables, we note that if $f$ has an extreme value at $(x_0, y_0)$, then $(x_0, y_0)$ either is a critical point of $f$ or is a boundary point of $D$. As a result, we have the following test:

> **THEOREM  Test for Absolute Maximum and Absolute Minimum**
>
> Let $z = f(x, y)$ be a function of two variables defined on a closed, bounded set $D$. If $f$ is continuous on $D$, then the absolute maximum value and the absolute minimum value of $f$ are, respectively, the largest and smallest values found among the following:
>
> • The values of $f$ at the critical points of $f$ in $D$
> • The values of $f$ on the boundary of $D$

### CALC ▶ EXAMPLE 5  Finding Absolute Extrema
CLIP

Find the absolute maximum and the absolute minimum of

$$z = f(x, y) = 2x - 2xy + y^2$$

whose domain is the region defined by $0 \le x \le 4$ and $0 \le y \le 3$.

**Solution** The function $f$ is continuous on its domain, which is a closed, bounded set. Then the Extreme Value Theorem for Functions of Two Variables guarantees that $f$ has an absolute maximum and an absolute minimum on its domain. To find them, we need the critical points of $f$. Since $f_x$ and $f_y$ exist at every point in the domain of $f$, the critical points of $f$ are the solutions of the system of equations

$$\begin{cases} f_x(x, y) = 2 - 2y = 0 \\ f_y(x, y) = -2x + 2y = 0 \end{cases}$$

Solving, we find that the only critical point is $(1, 1)$. The value of $f$ at $(1, 1)$ is

$$f(1, 1) = 2 \cdot 1 - 2 \cdot 1 \cdot 1 + 1^2 = 1$$

The domain of $f$ is the set $0 \le x \le 4$, $0 \le y \le 3$. The boundary of the domain of $f$ consists of the four line segments $L_1$, $L_2$, $L_3$, and $L_4$ shown in Figure 19. We evaluate $f$ on each line segment.

• **On $L_1$:** $x$ is in the interval $[0, 4]$ and $y = 0$.
  The function $f(x, y) = f(x, 0) = 2x$ is increasing on $0 \le x \le 4$, so its extreme values occur at the endpoints 0 and 4.

$$\text{At } x = 0, \; f(0, 0) = 0 \qquad \text{and} \qquad \text{at } x = 4, \; f(4, 0) = 8$$

• **On $L_2$:** $x = 4$ and $y$ is in the interval $[0, 3]$.
  The function $f(x, y) = f(4, y) = 8 - 8y + y^2$. To find the extreme values of $f$ on $[0, 3]$, begin by finding the critical number(s) of the function $g(y) = 8 - 8y + y^2$. That is, we find where $g'(y) = 0$.

$$g'(y) = -8 + 2y = 0$$
$$y = 4$$

Since 4 is not in the interval $[0, 3]$, the extreme values occur at the endpoints 0 and 3.

$$\text{At } y = 0, \; f(4, 0) = 8 \qquad \text{and} \qquad \text{at } y = 3, \; f(4, 3) = -7$$

• **On $L_3$:** $x$ is in the interval $[0, 4]$ and $y = 3$.
  The function $f(x, y) = f(x, 3) = 2x - 6x + 9 = -4x + 9$ is decreasing on $0 \le x \le 4$, so its extreme values occur at the endpoints 0 and 4.

$$\text{At } x = 0, \; f(0, 3) = 9 \qquad \text{and} \qquad \text{at } x = 4, \; f(4, 3) = -7$$

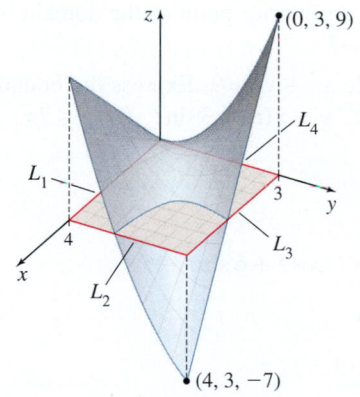

**Figure 19** $z = f(x, y) = 2x - 2xy + y^2$, $0 \le x \le 4$, $0 \le y \le 3$

- **On $L_4$:** $x = 0$ and $y$ is in the interval $[0, 3]$.
  The function $f(x, y) = f(0, y) = y^2$ is increasing on $0 \le y \le 3$, so its extreme values occur at the endpoints 0 and 3.

$$\text{At } y = 0, \ f(0, 0) = 0 \qquad \text{and} \qquad \text{at } y = 3, \ f(0, 3) = 9$$

The values of $f$ at the critical point $(1, 1)$ and at the extreme values on the boundary are

| Point | $(1, 1)$ | $(0, 0)$ | $(4, 0)$ | $(4, 3)$ | $(0, 3)$ |
|-------|----------|----------|----------|----------|----------|
| Value | 1 | 0 | 8 | $-7$ | 9 |

The absolute maximum value of $f$ is $f(0, 3) = 9$; the absolute minimum value is $f(4, 3) = -7$. ∎

**NOW WORK** Problem 37.

**EXAMPLE 6    Finding Absolute Extrema**

Find the absolute maximum and the absolute minimum of

$$z = f(x, y) = x^2 + y^2 - 2x + 2y - 5$$

whose domain is the disk $x^2 + y^2 \le 9$.

**Solution** The function $f$ is continuous on its domain, a closed, bounded set. The Extreme Value Theorem for Functions of Two Variables guarantees that $f$ has an absolute maximum and an absolute minimum on its domain. To find them, find the critical points of $f$. Since $f_x$ and $f_y$ exist at every point in the domain of $f$, the critical points of $f$ are the solutions of the system of equations

$$\begin{cases} f_x(x, y) = 2x - 2 = 0 \\ f_y(x, y) = 2y + 2 = 0 \end{cases}$$

The only solution is $x = 1$, $y = -1$, which is an interior point of the domain of $f$. The value of $f$ at the critical point is $f(1, -1) = -7$.

The boundary of the domain of $f$ is the circle $x^2 + y^2 = 9$. Express the boundary using the parametric equations $x = x(t) = 3 \cos t$, $y = y(t) = 3 \sin t$, $0 \le t \le 2\pi$, and evaluate $f$ on its boundary.

$$z = f(x, y) = f(3 \cos t, 3 \sin t)$$
$$= 9 \cos^2 t + 9 \sin^2 t - 6 \cos t + 6 \sin t - 5$$
$$= 4 - 6 \cos t + 6 \sin t$$

Now determine where $\dfrac{dz}{dt} = 0$.

$$\frac{dz}{dt} = 6 \sin t + 6 \cos t = 0$$
$$\tan t = -1$$
$$t = \frac{3\pi}{4} \qquad \text{or} \qquad t = \frac{7\pi}{4}$$

At $t = \dfrac{3\pi}{4}$, the value of $f$ is

$$f\left(3 \cos \frac{3\pi}{4}, 3 \sin \frac{3\pi}{4}\right) = 4 - 6 \cos \frac{3\pi}{4} + 6 \sin \frac{3\pi}{4} = 4 - 6 \left(-\frac{\sqrt{2}}{2}\right) + 6 \cdot \frac{\sqrt{2}}{2}$$
$$= 4 + 6\sqrt{2}$$

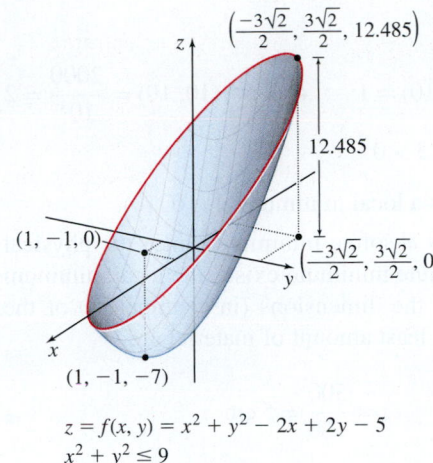

$\left(\dfrac{-3\sqrt{2}}{2}, \dfrac{3\sqrt{2}}{2}, 12.485\right)$

12.485

$(1, -1, 0)$

$\left(\dfrac{-3\sqrt{2}}{2}, \dfrac{3\sqrt{2}}{2}, 0\right)$

$(1, -1, -7)$

$z = f(x, y) = x^2 + y^2 - 2x + 2y - 5$
$x^2 + y^2 \le 9$

**Figure 20**

At $t = \dfrac{7\pi}{4}$, the value of $f$ is

$$f\left(3\cos\frac{7\pi}{4}, 3\sin\frac{7\pi}{4}\right) = 4 - 6\cos\frac{7\pi}{4} + 6\sin\frac{7\pi}{4} = 4 - 6 \cdot \frac{\sqrt{2}}{2} + 6\left(-\frac{\sqrt{2}}{2}\right)$$

$$= 4 - 6\sqrt{2}$$

The absolute maximum value of $f$, which occurs on the boundary of $f$, is $4 + 6\sqrt{2} \approx 12.485$. The absolute minimum value of $f$, which occurs at an interior point, is $-7$. ∎

Figure 20 shows the graph of $z = f(x, y) = x^2 + y^2 - 2x + 2y - 5$ with the absolute extrema marked.

**NOW WORK** Problem **39.**

## 4 Solve Optimization Problems

Now that we have the tools for finding the extrema of functions of two variables, we can investigate applied problems that involve optimization.

---

**EXAMPLE 7**  **Solving an Optimization Problem:**
**Minimizing Resources Used**

A manufacturer wants to make an open rectangular box of volume $V = 500$ cm³ using the least possible amount of material. Find the dimensions of the box.

**Solution** Let $x$ and $y$ be the dimensions of the base of the box, and let $z$ be the height of the box. Then

$$V = 500 = xyz \text{ cm}^3 \qquad x > 0 \quad y > 0 \quad z > 0$$

The manufacturer wants to minimize the amount of material used, which equals the surface area $S$ of the box. Since the box is open,

$$S = xy + 2xz + 2yz$$

If we solve the equation $500 = xyz$ for $z$ and substitute $z = \dfrac{500}{xy}$ into the formula for the surface area $S$ of the box, we can express $S$ as a function of two variables.

$$S = S(x, y) = xy + 2x \cdot \frac{500}{xy} + 2y \cdot \frac{500}{xy} = xy + \frac{1000}{y} + \frac{1000}{x}$$

This is the function to be minimized. The partial derivatives of $S$ are

$$S_x = y - \frac{1000}{x^2} \qquad \text{and} \qquad S_y = x - \frac{1000}{y^2}$$

Since $x > 0$ and $y > 0$, the critical points satisfy the system of equations

$$\begin{cases} y - \dfrac{1000}{x^2} = 0 \\[2mm] x - \dfrac{1000}{y^2} = 0 \end{cases}$$

Using substitution, we find $x = y = 1000^{1/3} = 10$. So the point $(10, 10)$ is the only critical point of $S$.

The second-order partial derivatives of $S$ are

$$S_{xx} = \frac{\partial}{\partial x}\left(y - \frac{1000}{x^2}\right) = \frac{2000}{x^3} \qquad S_{xy} = \frac{\partial}{\partial y}\left(y - \frac{1000}{x^2}\right) = 1$$

$$S_{yy} = \frac{\partial}{\partial y}\left(x - \frac{1000}{y^2}\right) = \frac{2000}{y^3}$$

At the critical point $(10, 10)$,

$$A = S_{xx}(10, 10) = \frac{2000}{10^3} = 2 \qquad B = S_{xy}(10, 10) = 1 \qquad C = S_{yy}(10, 10) = \frac{2000}{10^3} = 2$$

$$AC - B^2 = 3 > 0$$

From the Second Partial Derivative Test, $S$ has a local minimum at $(10, 10)$.

But, we want to know where $S$ attains its absolute minimum. Since the physical properties of the problem require that the absolute minimum exists, the local minimum is also the absolute minimum. Consequently, the dimensions (in centimeters) of the open box of volume $V = 500 \text{ cm}^3$ that uses the least amount of material are

$$x = 10 \text{ cm} \qquad y = 10 \text{ cm} \qquad z = \frac{500}{100} = 5 \text{ cm} \qquad\blacksquare$$

In Example 7, the conditions of the Extreme Value Theorem for Functions of Two Variables were not met, since the domain of $S$ is not bounded. Fortunately, in applied problems (such as Example 7), it is often possible to argue from physical or geometric properties that an absolute extremum must exist and that it must occur at one of the local extrema.

NOW WORK Problem 49.

---

EXAMPLE 8 **Solving an Optimization Problem: Maximizing Profit**

The demand functions for two products are

$$p = 12 - 2x \qquad \text{and} \qquad q = 20 - y$$

where $p$ and $q$ are the respective prices (in thousands of dollars) of each product, and $x$ and $y$ are the respective amounts (in thousands of units) of each sold. Suppose the joint cost function is

$$C(x, y) = x^2 + 2xy + 2y^2$$

(a) Find the revenue function $R = R(x, y)$ and the profit function $P = P(x, y)$.
(b) Determine the prices and amounts that will maximize profit.
(c) What is the maximum profit?

**Solution** (a) Revenue is the amount of money brought in. That is, revenue is the product of price and quantity sold. The revenue function $R = R(x, y)$ is the sum of the revenues from each product.

$$R(x, y) = xp + yq = x(12 - 2x) + y(20 - y)$$

Profit is the difference between revenue and cost. The profit function $P = P(x, y)$ is

$$P(x, y) = R(x, y) - C(x, y) = [x(12 - 2x) + y(20 - y)] - [x^2 + 2xy + 2y^2]$$

$$= 12x - 2x^2 + 20y - y^2 - x^2 - 2xy - 2y^2$$

$$= -3x^2 - 3y^2 - 2xy + 12x + 20y$$

(b) The partial derivatives of $P$ are

$$P_x(x, y) = \frac{\partial}{\partial x}(-3x^2 - 3y^2 - 2xy + 12x + 20y) = -6x - 2y + 12$$

$$P_y(x, y) = \frac{\partial}{\partial y}(-3x^2 - 3y^2 - 2xy + 12x + 20y) = -6y - 2x + 20$$

Find the critical points by solving the system of equations:

$$\begin{cases} -6x - 2y + 12 = 0 \\ -2x - 6y + 20 = 0 \end{cases}$$

We solve the first equation for $y$ and substitute the result into the second equation. Since $y = 6 - 3x$, then

$$-2x - 6(6 - 3x) + 20 = 0$$

$$x = 1$$

and $y = 6 - 3 \cdot 1 = 3$, so $(1, 3)$ is the only critical point.

The second-order partial derivatives are

$$P_{xx}(x, y) = -6 \qquad P_{xy}(x, y) = -2 \qquad P_{yy}(x, y) = -6$$

At the critical point $(1, 3)$,

$$P_{xx}(1, 3) = -6 < 0 \qquad P_{xy}(1, 3) = -2 \qquad P_{yy}(1, 3) = -6 \qquad P_{xx}(1, 3)P_{yy}(1, 3) - [P_{xy}(1, 3)]^2 = 36 - 4 = 32 > 0$$

From the Second Partial Derivative Test, $P$ has a local maximum at $(1, 3)$.

The domains of the demand functions $p = 12 - 2x$ and $q = 20 - y$ are $0 \le x \le 6$ and $0 \le y \le 20$, respectively. Since the domain of the profit function $P$ is $0 \le x \le 6$, $0 \le y \le 20$, the domain is closed and bounded, so a maximum profit exists. From the test for absolute maximum and absolute minimum, the maximum profit is found at a critical point or on the boundary of the domain.

Figure 21 shows the boundary of the profit function $P$, and Table 1 gives the maximum value of $P$ on its boundary and at the critical point $(1, 3)$.

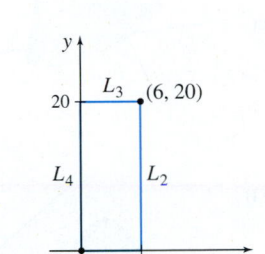

**Figure 21** The boundary of $P$

**TABLE 1**

|  |  |  | Maximum $P$ |
|---|---|---|---|
| $L_1$: | $0 \le x \le 6, y = 0$ | $P(x, 0) = -3x^2 + 12x$ | $P(2, 0) = 12$ |
| $L_2$: | $x = 6, x \le y \le 20$ | $P(6, y) = -3y^2 + 8y - 36$ | $P(6, y) < 0$ |
| $L_3$: | $0 \le x \le 6, y = 20$ | $P(x, 20) = -3x^2 - 28x - 800$ | $P(x, 20) < 0$ |
| $L_4$: | $x = 0, 0 \le y \le 20$ | $P(0, y) = -3y^2 + 20y$ | $P\left(0, \dfrac{10}{3}\right) = 33.33$ |
| Critical point $(1, 3)$: |  |  | $P(1, 3) = 36$ |

We conclude that by producing and selling 1000 units of product $x$ for the price $p = 12 - 2(1) = 10$ thousand dollars and 3000 units of product $y$ for the price $q = 20 - 3 = 17$ thousand dollars, the profit is maximized.

**(c)** The maximum profit is $P(1, 3) = \$36{,}000$. ∎

NOW WORK Problem 61.

# 13.3 Assess Your Understanding

## Concepts and Vocabulary

**1.** *Multiple Choice*  Suppose $z = f(x, y)$ is a function with domain $D$, and $(x_0, y_0)$ is an interior point of $D$. If $f(x_0, y_0) \ge f(x, y)$ for all points $(x, y)$ within some disk centered at $(x_0, y_0)$, then $f$ has [**(a)** a local maximum **(b)** a local minimum **(c)** a saddle point **(d)** none of these] at $(x_0, y_0)$.

**2.** *True or False*  If $z = f(x, y)$ is a function of two variables whose domain is an open set $D$ containing the point $(x_0, y_0)$, and if $(x_0, y_0)$ is a critical point of $f$, then $f_x(x_0, y_0) = f_y(x_0, y_0) = 0$.

**3.** *True or False*  It is possible for a function $z = f(x, y)$ defined on a open set to have a local extremum at a point $(x_0, y_0)$ that is not a critical point.

**4.** *Multiple Choice*  If $f_x(x_0, y_0) = 0$ and $f_y(x_0, y_0) = 0$, but $f$ does not have a local extremum at $(x_0, y_0)$, then the point $(x_0, y_0, f(x_0, y_0))$ is called [**(a)** a local maximum value **(b)** a local minimum value **(c)** a saddle point **(d)** none of these].

**5.** *True or False*  The Extreme Value Theorem for Functions of Two Variables guarantees that a function $z = f(x, y)$ of two variables, that is continuous on a closed, bounded set $D$, has an absolute maximum and an absolute minimum on $D$.

**6.** If $z - f(x, y)$ is a function of two variables that is continuous on a closed, bounded set $D$, then the absolute extrema are either at the critical points of $f$ in $D$ or on the _____ of $D$.

---

**1.** = NOW WORK problem       ◢◣ = Graphing technology recommended       CAS = Computer Algebra System recommended

## Skill Building

*In Problems 7–12, find all the critical points for each function.*

**7.** $f(x, y) = x^4 - 2x^2 + y^2 + 5$

**8.** $f(x, y) = x^2 - y^2 + 6x - 2y + 4$

**9.** $f(x, y) = 4xy - x^4 - y^4 + 2$

**10.** $f(x, y) = x^3 + 6xy + 3y^2 + 3$

**11.** $f(x, y) = e^{-x}(x^2 + y)$     **12.** $f(x, y) = xy + \dfrac{2}{x} + \dfrac{4}{y}$

*In Problems 13–40, find the local maxima, local minima, and saddle points, if any, for each function.*

**13.** $z = x^2 + y^2 - 2x + 4y + 2$     **14.** $z = x^2 + y^2 - 4x + 2y - 4$

**15.** $z = x^2 + 4y^2 - 4x + 8y - 1$

**16.** $z = 2x^2 - y^2 + 4x - 4y + 8$

**17.** $z = x^3 - 6xy + y^3$     **18.** $z = x^2 - 3xy - y^2$

**19.** $z = x^2 + 3xy - y^2 + 4y - 6x$

**20.** $z = 2x^2 + xy + y^2 - 2x + 3y + 6$

**21.** $z = x^3 + 3xy + y^3$     **22.** $z = x^3 - 3xy - y^3$

**23.** $z = \dfrac{4}{3}x^3 - xy^2 + y$     **24.** $z = 3y^3 - x^2y + x$

**25.** $z = \dfrac{y}{x + y}$     **26.** $z = \dfrac{x}{x + y}$

**27.** $z = x + y + \dfrac{1}{xy}$     **28.** $z = \dfrac{1}{x} + \dfrac{1}{y} + xy$

**29.** $z = \cos x + \cos y$     **30.** $z = x^2 + 4 - 4x \cos y$

**31.** $z = y^2 - 6y \cos x + 6$     **32.** $z = x^2 - 6x \sin y + 2$

**33.** $z = e^{xy}$     **34.** $z = xye^{-(x+y)}$

**35.** $z = e^{-y^2} \sin x \quad 0 \le x \le 2\pi$     **36.** $z = e^{x^2} \cos y \quad 0 \le y \le 2\pi$

*In Problems 37–42, find the absolute maximum and absolute minimum of f on the domain D.*

**37.** $z = f(x, y) = x^2 - 4xy + 4y; \quad D: 0 \le x \le 2, 0 \le y \le 1$

**38.** $z = f(x, y) = x^2 - 4xy + 4y; \quad D: 0 \le x \le 3, 0 \le y \le 2$

**39.** $z = f(x, y) = x^2 + 2x + y^2 - 2y; \quad D: x^2 + y^2 \le 4$

**40.** $z = f(x, y) = x^2 - 4x + y^2 - 4y; \quad D: x^2 + y^2 \le 9$

**41.** $z = f(x, y) = xy + 4x - 3y + 5; \quad D:$ the closed triangular region whose vertices are $(0, 0)$, $(0, 4)$, and $(5, 4)$

**42.** $z = f(x, y) = 3xy - 3x - 6y + 1; \quad D:$ the closed triangular region whose vertices are $(0, 0)$, $(5, 0)$, and $(0, 3)$

## Applications and Extensions

**43. (a)** Which point(s) on the plane $2x + y + z = 6$ is (are) closest to the origin?

**(b)** Find the shortest distance from the plane to the origin.

**44. (a)** Find the point on the plane $3x + 2y + z = 14$ that is closest the origin.

**(b)** What is the distance of the point found in (a) from the origin?

**45. (a)** Which point(s) on the surface $z^2 - xy = 4$ is (are) closest to the origin?

**(b)** Find the shortest distance from the surface to the origin.

**46. (a)** Which point(s) on the surface $z^2 - 3x^2y = 9$ is (are) closest to the origin?

**(b)** Find the shortest distance from the surface to the origin.

**47. Geography**   The surface of a mountain is modeled by the graph of the function $z = 2xy - 2x^2 - y^2 - 8x + 6y + 4$ where $z$ is the height in kilometers. See the figure. If sea level is the $xy$-plane, how high is the mountain above sea level?

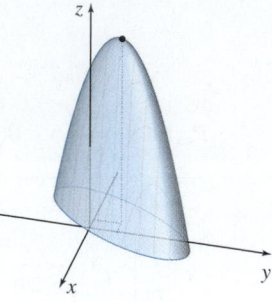

**48.** Find the highest point of the paraboloid $x^2 + y^2 - 6x + 4y + z + 12 = 0$ as follows:

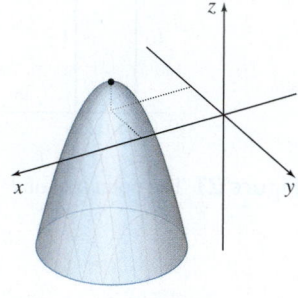

**(a)** Find the vertex of the paraboloid without using calculus.

**(b)** Express $z$ as a function of $x$ and $y$ and use calculus to find its maximum value. See the figure.

**49. Constructing a Box**   A rectangular box open at the top has a fixed surface area $S$. Find the dimensions that make its volume a maximum. Express the length, width, and depth of the box in terms of $S$.

**50. Constructing a Box**   Rework Problem 49 if the box is closed.

**51. Constructing a Box**   Rework Example 7 if the box is closed.

**52. Minimizing Cost**   Find the dimensions of an open rectangular box having a volume of 80 m³ if the cost per square meter of the material to be used is $4 for the bottom, $3 for one of the sides, and $2 for the three remaining sides, and the total cost is to be minimized.

**53. USPS Retail Ground Regulations**   The U.S. Postal Service regulations state that large packages can be up to 108 in. in combined length and girth (the perimeter of a cross section at the widest spot) before additional charges apply.

**(a)** Find the dimensions of the rectangular box of maximum volume that can be sent without additional charges.

**(b)** Find the dimensions with maximum volume, if the package is cylindrical.

*Source*: U.S. Postal Service, October, 2017, https://www.usps.com

**54. Pharmacology**   A patient's reaction to an injection of $x$ units of a certain drug, $t$ hours after the injection, is given by

$$y = x^2(a - x)te^{-t} \qquad 0 \le x \le a \quad 0 \le t$$

where $a > 0$ is a constant. Find the values of $x$ and $t$, if any, that maximize $y$.

**55. Pharmacology**   Two drugs A and B are used simultaneously to treat a disease. A patient's reaction $R$ to $x$ units of Drug A and $y$ units of Drug B is

$$R(x, y) = x^2 y^2 (a - x)(b - y) \qquad 0 \le x \le a \quad 0 \le y \le b$$

**(a)** For a fixed amount $x$ of Drug A, what amount $y$ of Drug B produces the maximum reaction?

**(b)** For a fixed amount $y$ of Drug B, what amount $x$ of Drug A produces the maximum reaction?

**(c)** If amounts $x$ and $y$ are both variable, what amount of each maximizes the reaction?

**56. Agriculture**   An open irrigation channel is to be made into a symmetric form with three straight sides, as indicated in the figure.

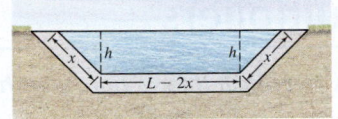

**(a)** If the total length of the three sides is $L$, find a channel design that will allow the maximum possible flow.

**(b)** Is this design preferable to an open semicircular channel? Give the reasons why.

**57. Fuel Cost**   See the figure. A fuel reservoir at $D$ is planned to service plants located at $A$, $B$, and $C$. The cost $F$, in thousands of dollars, of connecting the plants to $D$ is determined by the formula

$$F = 6x^2 + 6y^2 - 4x - 6y + 5$$

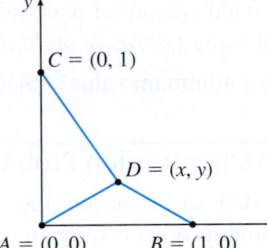

Find the location of $D$ that will minimize this cost.

**58. Minimizing Cost**   The cost of material for the sides of a rectangular shipping container is $a$ cents per square foot; the cost of the top and bottom material is $\frac{3}{2} a$ cents per square foot. If the volume is to be $\frac{3}{2}$ ft$^3$, what dimensions of the container will minimize its cost?

**59. Metal Detector**   A metal detector is used to locate an underground pipe. When several meter readings of the detector are compared, the reading at an arbitrary point $(x, y)$, where $x \ge 0$ and $y \ge 0$, is modeled by

$$M = y(x - x^2) - y^2$$

Find the point $(x, y)$ with the highest reading.

**60. Maximizing Profit**   A steel manufacturer produces two grades of steel, $x$ kilograms of Grade A and $y$ kilograms of Grade B. The cost $C$ and revenue $R$ in dollars are given by

$$C = \frac{1}{20} x^2 + 700x + y^2 - 150y - \frac{1}{2} xy$$

$$R = 2700x - \frac{3}{20} x^2 + 1000y - y^2 + \frac{1}{2} xy + 10,000$$

If $P = \text{Profit} = R - C$, find the production (in kilograms) of Grades A and B that maximizes the manufacturer's profit.

**61. Economics**   The demand functions for two products are $p = 12 - x$ and $q = 8 - y$, where $p$ and $q$ are the respective prices (in thousands of dollars) of each product, and $x$ and $y$ are the respective amounts (in thousands of units) of each product sold. If the joint cost function is

$$C(x, y) = x^2 + 2xy + 3y^2$$

find the quantities $x$ and $y$ and the prices $p$ and $q$ that maximize profit. What is the maximum profit?

**62. Economics**   The labor cost of a firm is modeled by the function

$$Q(x, y) = x^2 + y^2 - xy - 3x - 6y - 5$$

where $x$ is the number of workdays required by a skilled worker and $y$ is the number of workdays required by a semiskilled worker. Find the values of $x$ and $y$ for which the labor cost is a minimum.

**63. Second Partial Derivative Test**   Suppose

$$g(x, y) = Ax^2 + 2Bxy + Cy^2$$

where $A$, $B$, and $C$ are constants. Show that:

**(a)** If $AC - B^2 > 0$, $A \ne 0$, then $g(x, y)$ has a maximum at $(0, 0)$ if $A < 0$ and a minimum at $(0, 0)$ if $A > 0$.

**(b)** If $AC - B^2 < 0$, then $g(x, y)$ has both positive and negative values at points $(x, y)$ near $(0, 0)$, and $(0, 0, 0)$ is a saddle point for $z = g(x, y)$.

## Challenge Problems

**64.** Find the distance between the skew lines

$$L_1: x = t - 6, \ y = t, \ z = 2t \qquad L_2: x = t, \ y = t, \ z = -t$$

**65. Method of Least Squares**   A scientist plots the points $(x_1, y_1), \ldots, (x_n, y_n)$ from her experimental data. Her theory tells her that the points should lie on a straight line, but they do not; perhaps as a result of experimental error. She is looking for the line $y = mx + b$ that "best" fits the data. The most often used criterion is the *least-squares fit*, in which $m$ and $b$ are chosen to minimize

$$\sum_{k=1}^{n} (mx_k + b - y_k)^2$$

**(a)** Show that the minimizing values of $m$ and $b$ are

$$m = \frac{n \sum\limits_{k=1}^{n} x_k y_k - \sum\limits_{k=1}^{n} x_k \sum\limits_{k=1}^{n} y_k}{n \sum\limits_{k=1}^{n} x_k^2 - \left( \sum\limits_{k=1}^{n} x_k \right)^2}$$

$$b = \frac{1}{n} \left( \sum_{k=1}^{n} y_k - m \sum_{k=1}^{n} x_k \right)$$

**(b)** Find the least-squares estimate $y = mx + b$ for the points $(0, 2)$, $(1, 1)$, $(2, 2)$, $(3, 4)$, and $(4, 4)$. Plot the points and draw the line.

**66.** Let $z = f(x, y)$ have continuous second-order partial derivatives. If $\mathbf{u} = u_1\mathbf{i} + u_2\mathbf{j}$ is a unit vector, there is a directional derivative $D_{\mathbf{u}}f(x, y) = g(x, y)$. If $g(x, y)$ is differentiable and if $\mathbf{v} = v_1\mathbf{i} + v_2\mathbf{j}$ is a unit vector, then $D_{\mathbf{v}}g(x, y) = u_1v_1f_{xx} + (u_1v_2 + u_2v_1)f_{xy} + u_2v_2f_{yy}$. If we let $A = f_{xx}(x_0, y_0)$, $B = f_{xy}(x_0, y_0)$, and $C = f_{yy}(x_0, y_0)$, then we have an expression in the components $u_1$ and $u_2$ of $\mathbf{u}$, and $v_1$ and $v_2$ of $\mathbf{v}$. (See Section 13.1, Problem 77.)

**(a)** In particular, if $\mathbf{u} = \mathbf{v}$, show that the second-order directional derivative at $(x_0, y_0)$ is given by $Au_1^2 + 2Bu_1u_2 + Cu_2^2$.

**(b)** Let $f_x(x_0, y_0) = 0 = f_y(x_0, y_0)$. Show that if $A > 0$ and $AC - B^2 > 0$, then $z = f(x, y)$ has a local minimum at $(x_0, y_0)$; if $A < 0$ and $AC - B^2 > 0$, then $z = f(x, y)$ has a local maximum at $(x_0, y_0)$; and if $AC - B^2 < 0$, then $z = f(x, y)$ has a saddle point at $(x_0, y_0)$.

**(c)** Show by examples that if $AC - B^2 = 0$, we may have any of the above three possibilities.

# 13.4 Lagrange Multipliers

**OBJECTIVES** *When you finish this section, you should be able to:*

**1** Use Lagrange multipliers for an optimization problem with one constraint (p. 940)

**2** Use Lagrange multipliers for an optimization problem with two constraints (p. 943)

Many applications require finding maximum (or minimum) values of a function when a certain condition or constraint is present. For example, we might want to maximize the volume of a box using a fixed amount of material (its surface area). A solution to such problems can be obtained using *Lagrange multipliers.*

We begin by finding a minimum value subject to one constraint.

**EXAMPLE 1  Solve an Optimization Problem**

Find the point in the first quadrant on the hyperbola $xy = 4$ where the value of $z = 12x + 3y$ is a minimum. What is the minimum value?

**Solution** We want the minimum value of $z = 12x + 3y$ subject to the condition, or constraint, that $x$ and $y$ satisfy the equation $xy = 4$. Since we are looking for a point in the first quadrant, $x > 0$ and $y > 0$. We express this as a problem in one variable by solving $xy = 4$ for $y$ and substituting $y = \dfrac{4}{x}$ in the expression $z = 12x + 3y$.

$$z = 12x + 3y = 12x + \frac{12}{x} \qquad x > 0$$
$$\uparrow$$
$$y = \frac{4}{x}$$

Then

$$\frac{dz}{dx} = 12 - \frac{12}{x^2} = \frac{12x^2 - 12}{x^2} \qquad x > 0$$

The critical numbers of $z$ are $-1$ and $1$. We exclude $x = -1$, since $x > 0$. To examine $x = 1$, use the Second Derivative Test. Since $\dfrac{d^2z}{dx^2} = \dfrac{24}{x^3} > 0$ for $x = 1$, $z$ has a minimum when $x = 1$ and $y = \dfrac{4}{x} = 4$. The minimum value of $z$ is 24 at the point $(1, 4)$ on the hyperbola. ∎

**NEED TO REVIEW?** The Second Derivative Test is discussed in Section 4.4, pp. 302–304.

In Example 1, we solved the problem by eliminating the variable $y$ and treating the problem as a minimum problem in one variable. Sometimes, however, it is not easy, or perhaps it is impossible, to eliminate a variable. Then the Method of Lagrange Multipliers is useful. Before we use the method, let's look at a geometric rationale.

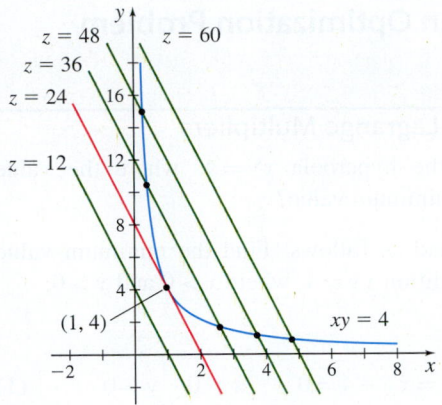

**Figure 22** Minimum value of $z = 12x + 3y$ subject to $xy = 4$ is $z = 24$.

## A Geometric Rationale for the Method of Lagrange Multipliers

Consider Example 1 again. We want to minimize the function $z = f(x, y) = 12x + 3y$, subject to the constraint $g(x, y) = xy - 4 = 0$. The graph of $g(x, y) = xy - 4 = 0$ is a curve (a hyperbola) in the $xy$-plane. We are interested in finding the smallest value of $z = f(x, y)$, where $(x, y)$ is a point in the first quadrant on the graph of $g(x, y) = 0$.

Figure 22 shows several level curves of the plane $z = 12x + 3y$. On each level curve, the value of $z$ is fixed. We want the smallest $z$ that coincides with the hyperbola $xy = 4$. Figure 22 shows that the minimum value of $z$ occurs precisely at the point where the line $12x + 3y = 24$ is tangent to the graph of the hyperbola $xy = 4$. At this point, the two curves have a common tangent line, so they must have the same normal line. Since the gradient gives the direction of the normal line to a curve, it follows that $\nabla f$ is parallel to $\nabla g$. That is, there is a scalar $\lambda \neq 0$, so that

$$\nabla f = \lambda \nabla g \qquad \text{and} \qquad g(x, y) = 0$$

This discussion leads to the following result.

### THEOREM  Lagrange's Theorem

Let $f$ and $g$ be functions of two variables with continuous partial derivatives at every point of some open set containing the smooth curve $g(x, y) = 0$. Suppose that $f$, when restricted to points on the curve $g(x, y) = 0$, has a local extremum at the point $(x_0, y_0)$ and that $\nabla g(x_0, y_0) \neq \mathbf{0}$. Then there is a number $\lambda$ for which

$$\boxed{\nabla f(x_0, y_0) = \lambda \nabla g(x_0, y_0)}$$

**Proof** Express the smooth curve $C$ defined by $g(x, y) = 0$ as a parametric equation in vector form $\mathbf{r}(t) = x(t)\mathbf{i} + y(t)\mathbf{j}$. Define the function $s(t) = f(x(t), y(t))$ and choose $t_0$ so that $x_0 = x(t_0)$, $y_0 = y(t_0)$. Now since $f$ has a local extremum at the point $(x_0, y_0)$, the extreme value is given by

$$s(t_0) = f(x(t_0), y(t_0)) = f(x_0, y_0)$$

Then $s'(t_0) = 0$. By Chain Rule I,

$$s'(t_0) = f_x(x_0, y_0)x'(t_0) + f_y(x_0, y_0)y'(t_0) \qquad \frac{ds}{dt} = \frac{\partial s}{\partial x}\frac{dx}{dt} + \frac{\partial s}{\partial y}\frac{dy}{dt}$$

$$= \nabla f(x_0, y_0) \cdot \mathbf{r}'(t_0) \qquad \qquad \text{Definition of the gradient}$$

$$= 0$$

If $\nabla f(x_0, y_0) \neq \mathbf{0}$, it is orthogonal to $\mathbf{r}'(t_0)$, which is a tangent vector to $C$ at $(x_0, y_0)$. Since the gradient is normal to the level curve at $(x_0, y_0)$, the vector $\nabla g(x_0, y_0)$ is also orthogonal to $\mathbf{r}'(t_0)$. If $\nabla f(x_0, y_0) = \mathbf{0}$, then let $\lambda = 0$. Then, in every case, there is a number $\lambda$ for which $\nabla f(x_0, y_0) = \lambda \nabla g(x_0, y_0)$. ∎

The number $\lambda$ in the equations $\nabla f(x, y) = \lambda \nabla g(x, y)$ is called a **Lagrange multiplier**. Since $\nabla f = \lambda \nabla g$, we have $f_x\mathbf{i} + f_y\mathbf{j} = \lambda(g_x\mathbf{i} + g_y\mathbf{j})$, so that $f_x(x, y) = \lambda g_x(x, y)$ and $f_y(x, y) = \lambda g_y(x, y)$. These equations and the equation $g(x, y) = 0$ form a system of three equations in the three unknowns $x$, $y$, and $\lambda$.

$$\boxed{f_x(x, y) = \lambda g_x(x, y) \qquad f_y(x, y) = \lambda g_y(x, y) \qquad g(x, y) = 0}$$

A solution $(x_0, y_0)$ of the system of equations is called a **test point**, since it is a candidate for the desired extrema. The maximum and minimum values, if they exist, are found by choosing the largest and smallest values of $z = f(x, y)$ at the test points.

Photo 12/Getty Images

**ORIGINS** Joseph-Louis Lagrange (1736–1813), a.k.a. Giuseppe Lodovico Lagrangia, was a self-taught mathematician and astronomer. He was born and began his career teaching in Turin, Italy, but later moved first to Berlin and then to Paris, where he held prestigious positions at the Berlin Academy and Académie des Sciences. After the French Revolution (1789), Lagrange was named the first professor of analysis at the École Polytechnique, in Paris. Lagrange is one of 72 names of mathematicians, engineers, and scientists inscribed on the Eiffel Tower.

## 1 Use Lagrange Multipliers for an Optimization Problem with One Constraint

**EXAMPLE 2** **Solving Example 1 Using Lagrange Multipliers**

Find the point in the first quadrant on the hyperbola $xy = 4$, where the value of $z = 12x + 3y$ is a minimum. What is the minimum value?

**Solution** We can restate the problem to read as follows: Find the minimum value of $z = f(x, y) = 12x + 3y$ subject to the condition $xy = 4$, where $x > 0$ and $y > 0$.

The test points satisfy the equations

$$\nabla f(x, y) = \lambda \nabla g(x, y) \qquad g(x, y) = xy - 4 = 0 \qquad x > 0 \quad y > 0 \qquad (1)$$

where $\lambda$ is a number.

Since $\nabla f(x, y) = 12\mathbf{i} + 3\mathbf{j}$ and $\nabla g(x, y) = y\mathbf{i} + x\mathbf{j}$, the equations in (1) lead to the following system of three equations containing three variables:

$$\begin{cases} 12 = \lambda y & f_x(x, y) = \lambda g_x(x, y) \\ 3 = \lambda x & f_y(x, y) = \lambda g_y(x, y) \\ xy - 4 = 0 & g(x, y) = 0 \end{cases}$$

**NOTE** In eliminating $\lambda$, we use the fact that $\lambda \neq 0$ since $\nabla f \neq \mathbf{0}$.

We solve this system by eliminating $\lambda$ from the first two equations. This results in $y = 4x$, so the third equation becomes

$$4x^2 - 4 = 0$$

$$x = 1 \qquad \text{or} \qquad x = -1$$

Since $x > 0$, we ignore $x = -1$. When $x = 1$, then $y = 4$, so the only test point is $(1, 4)$. The corresponding minimum value of $z = 12x + 3y$ is $z = 24$. ∎

**NOW WORK** Problem 5.

---

**Steps for Using Lagrange Multipliers**

**Step 1** Express the problem in the following form: Find the maximum (or minimum) value of $z = f(x, y)$ subject to the constraint $g(x, y) = 0$.

**Step 2** Check that the functions $f$ and $g$ satisfy Lagrange's Theorem.

**Step 3** Solve the equations $\nabla f(x, y) = \lambda \nabla g(x, y)$ and $g(x, y) = 0$ for the test points by solving the system of equations

$$\begin{cases} f_x(x, y) = \lambda g_x(x, y) \\ f_y(x, y) = \lambda g_y(x, y) \\ g(x, y) = 0 \end{cases}$$

**Step 4** Evaluate $z = f(x, y)$ at each test point found in Step 3. Choose the maximum (or minimum) value of $z$.

---

**EXAMPLE 3** **Using Lagrange Multipliers**

Find the maximum and minimum values of $z = f(x, y) = 3x - y + 1$ subject to the constraint $3x^2 + y^2 - 9 = 0$.

**Solution** Follow the steps for using Lagrange multipliers:

**Step 1** The problem is expressed in the required form.

**Step 2** The functions $f(x, y) = 3x - y + 1$ and $g(x, y) = 3x^2 + y^2 - 9 = 0$ each have continuous partial derivatives.

**Step 3** $\nabla f(x, y) = 3\mathbf{i} - \mathbf{j}$ and $\nabla g(x, y) = 6x\mathbf{i} + 2y\mathbf{j}$

The equations $\nabla f(x, y) = \lambda \nabla g(x, y)$ and $g(x, y) = 0$ lead to the system of equations

$$\begin{cases} 3 = 6\lambda x & f_x(x, y) = \lambda g_x(x, y) \\ -1 = 2\lambda y & f_y(x, y) = \lambda g_y(x, y) \\ 3x^2 + y^2 - 9 = 0 & g(x, y) = 0 \end{cases}$$

From the first two equations, we find that $x = \dfrac{1}{2\lambda}$ and $y = -\dfrac{1}{2\lambda}$ from

which $x = -y$. Substituting into the third equation, we have

$$4x^2 = 9$$

$$x = \pm \frac{3}{2}$$

Then $x = \pm\dfrac{3}{2}$, $y = \mp\dfrac{3}{2}$, and the test points are $\left(\dfrac{3}{2}, -\dfrac{3}{2}\right)$ and $\left(-\dfrac{3}{2}, \dfrac{3}{2}\right)$.

**Step 4** The values of $z$ corresponding to the test points are $z = \dfrac{9}{2} + \dfrac{3}{2} + 1 = 7$

and $z = -\dfrac{9}{2} - \dfrac{3}{2} + 1 = -5$. The maximum value of $z$ is 7 and the minimum

value is $-5$.  ∎

**NOW WORK** Problem 9.

CALC
▶ **EXAMPLE 4** **Using Lagrange Multipliers to Find Absolute Extrema**
CLIP

Find the absolute maximum and absolute minimum of $f(x, y) = x^2 + y^2 + 4x - 4y + 3$ subject to $x^2 + y^2 \leq 2$.

**Solution** From the Extreme Value Theorem for Functions of Two Variables, $f$ has both an absolute maximum and an absolute minimum on the closed, bounded set $x^2 + y^2 \leq 2$. Begin by finding any critical points of $f$.

$$f_x = 2x + 4 = 0 \qquad f_y = 2y - 4 = 0$$

$$x = -2 \qquad\qquad y = 2$$

The point $(-2, 2)$ lies outside the disk $x^2 + y^2 \leq 2$, so $f$ has no relevant critical points. The extrema must occur on the boundary of the domain, that is, on the curve $x^2 + y^2 = 2$. To find the extrema, we use Lagrange multipliers.

**Step 1** Using $g(x, y) = x^2 + y^2 - 2 = 0$ as the constraint, the problem is expressed in the required form.

**Step 2** The functions $f(x, y) = x^2 + y^2 + 4x - 4y + 3$ and $g(x, y) = x^2 + y^2 - 2 = 0$ each have continuous partial derivatives.

**Step 3** The equations $\nabla f(x, y) = \lambda \nabla g(x, y)$ and $g(x, y) = 0$ lead to the system of equations

$$\begin{cases} 2x + 4 = 2\lambda x & f_x(x, y) = \lambda g_x(x, y) \\ 2y - 4 = 2\lambda y & f_y(x, y) = \lambda g_y(x, y) \\ x^2 + y^2 = 2 & g(x, y) = 0 \end{cases}$$

Eliminate $\lambda$ from the first two equations.

$$2x + 4 = 2x\lambda \qquad\qquad 2y - 4 = 2y\lambda$$

$$\lambda = \frac{2x + 4}{2x} \qquad\qquad \lambda = \frac{2y - 4}{2y}$$

$$\lambda = 1 + \frac{2}{x} \qquad\qquad \lambda = 1 - \frac{2}{y}$$

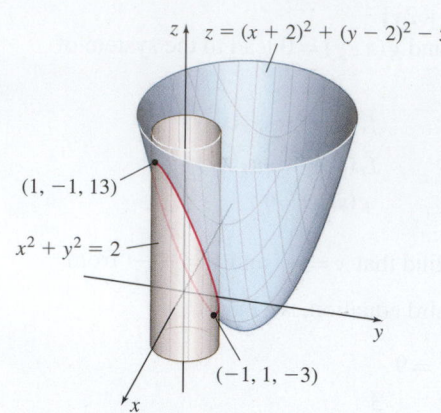

$$z = (x + 2)^2 + (y - 2)^2 - 5$$

(1, −1, 13)

$x^2 + y^2 = 2$

(−1, 1, −3)

**Figure 23**

Then $1 + \dfrac{2}{x} = 1 - \dfrac{2}{y}$, so $y = -x$.

If we substitute $y = -x$ into the third equation $x^2 + y^2 = 2$, we find $2x^2 = 2$. Then $x = \pm 1$ and $y = \mp 1$. There are two test points: $(1, -1)$ and $(-1, 1)$.

**Step 4** Evaluate $z = f(x, y)$ at each test point.

$$f(1, -1) = 13 \qquad \text{Maximum value}$$
$$f(-1, 1) = -3 \qquad \text{Minimum value}$$

The cylinder $x^2 + y^2 = 2$ and the elliptic paraboloid $z = (x + 2)^2 + (y - 2)^2 - 5$ are shown in Figure 23.

**NOW WORK** Problem 17.

Lagrange multipliers can be used for functions of three variables. The extreme values of $w = f(x, y, z)$ subject to the constraint $g(x, y, z) = 0$, if they exist, occur at the solutions $(x, y, z)$ of the system of equations.

$$\boxed{\nabla f(x, y, z) = \lambda \nabla g(x, y, z) \qquad \text{and} \qquad g(x, y, z) = 0}$$

These equations form a system of four equations in the four unknowns $x$, $y$, $z$, and $\lambda$.

**EXAMPLE 5** **Using Lagrange Multipliers with a Function of Three Variables**

A container in the shape of a rectangular box is open on top and has a fixed volume of 12 m³. The material used to make the bottom of the container costs \$3 per square meter, while the material used for the sides costs \$1 per square meter. What dimensions will minimize the cost of material?

**Solution** Let $x$ be the width of the container, $y$ be its depth, and $z$ be its height. Then the volume of the box is $xyz = 12$. The cost function $C = C(x, y, z)$, which is to be minimized, is

$$C(x, y, z) = 3xy + 2xz + 2yz$$

We use Lagrange multipliers.

**Step 1** We want to minimize $C$ subject to the constraint $g(x, y, z) = xyz - 12 = 0$.

**Step 2** The functions $C$ and $g$ each have continuous partial derivatives.

**Step 3** The equations $\nabla C(x, y, z) = \lambda \nabla g(x, y, z)$ and $g(x, y, z) = 0$ lead to the system of equations

$$\begin{cases} 3y + 2z = \lambda yz & C_x(x, y, z) = \lambda g_x(x, y, z) \quad (1) \\ 3x + 2z = \lambda xz & C_y(x, y, z) = \lambda g_y(x, y, z) \quad (2) \\ 2x + 2y = \lambda xy & C_z(x, y, z) = \lambda g_z(x, y, z) \quad (3) \\ xyz - 12 = 0 & g(x, y, z) = 0 \qquad\qquad\quad (4) \end{cases}$$

Now use the facts that $x > 0$, $y > 0$, and $z > 0$ to solve the first three equations for $\lambda$, obtaining

$$\lambda = \frac{3}{z} + \frac{2}{y} \quad (5) \qquad \lambda = \frac{3}{z} + \frac{2}{x} \quad (6) \qquad \lambda = \frac{2}{y} + \frac{2}{x} \quad (7)$$

From these, we find that

$$y = x \qquad \text{From equations (5) and (6)}$$

$$z = \frac{3}{2}x \qquad \text{From equations (5) and (7)}$$

Substituting into $xyz - 12 = 0$, we find

$$xyz - 12 = 0$$

$$x \cdot x \cdot \frac{3}{2}x - 12 = 0$$

$$\frac{3}{2}x^3 = 12$$

$$x = 2$$

Then $y = 2$, $z = 3$, and the test point is $(2, 2, 3)$.

**Step 4** The dimensions of the container that minimize the cost are 2 m by 2 m by 3 m. The minimum cost is $C(2, 2, 3) = 3 \cdot 2 \cdot 2 + 2 \cdot 2 \cdot 3 + 2 \cdot 2 \cdot 3 = \$36$. ∎

**NOW WORK** Problem 33.

## 2 Use Lagrange Multipliers for an Optimization Problem with Two Constraints

Lagrange multipliers can be used to solve problems involving more than one constraint. In the two-constraint situation, we seek the extreme values of a function $w = f(x, y, z)$ subject to two constraints $g(x, y, z) = 0$ and $h(x, y, z) = 0$. The extreme values of $f$, if they exist, occur at solutions $(x, y, z)$ of the system of five equations containing the five variables $x$, $y$, $z$, $\lambda_1$, and $\lambda_2$.

$$\begin{cases} \nabla f(x, y, z) = \lambda_1 \nabla g(x, y, z) + \lambda_2 \nabla h(x, y, z) \\ g(x, y, z) = 0 \\ h(x, y, z) = 0 \end{cases}$$

where $\lambda_1$ and $\lambda_2$ are Lagrange multipliers.

**EXAMPLE 6** Using Lagrange Multipliers with Two Constraints

Find the points of intersection of the ellipsoid $x^2 + y^2 + 9z^2 = 25$ and the plane $x + 3y - 2z = 0$ that are farthest from the origin. Also find the points that are closest to the origin.

**Solution** See Figure 24.

Let $C$ be the curve formed by intersecting the ellipsoid and the plane, and let $P = (x, y, z)$ be a point on $C$. We use the steps for Lagrange multipliers.

**Step 1** The function to be optimized is the distance $d$ from $P$ to the origin:

$$d = f(x, y, z) = \sqrt{x^2 + y^2 + z^2}$$

subject to the constraints

$$g(x, y, z) = x^2 + y^2 + 9z^2 - 25 = 0 \qquad \text{The point must be on the ellipsoid.}$$

$$h(x, y, z) = x + 3y - 2z = 0 \qquad \text{The point must be on the plane.}$$

**Step 2** The functions $f$, $g$, and $h$ are each functions of three variables, and each has continuous partial derivatives at every point of some open set containing the smooth curve defined by $g(x, y, z) = 0$ and $h(x, y, z) = 0$.

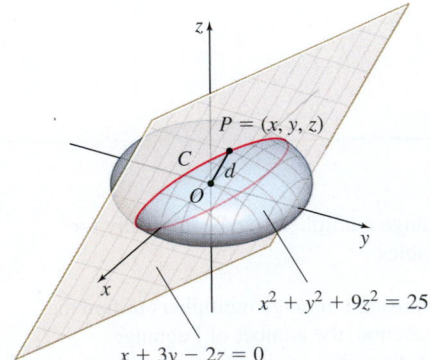

**Figure 24**

**Step 3**  We introduce two Lagrange multipliers and solve the system of equations $\nabla f(x, y, z) = \lambda_1 \nabla g(x, y, z) + \lambda_2 \nabla h(x, y, z)$, $g(x, y, z) = 0$, and $h(x, y, z) = 0$.

$$
\begin{cases}
\dfrac{x}{\sqrt{x^2 + y^2 + z^2}} = 2\lambda_1 x + \lambda_2 & f_x(x, y, z) = \lambda_1 g_x(x, y, z) + \lambda_2 h_x(x, y, z) \\[2ex]
\dfrac{y}{\sqrt{x^2 + y^2 + z^2}} = 2\lambda_1 y + 3\lambda_2 & f_y(x, y, z) = \lambda_1 g_y(x, y, z) + \lambda_2 h_y(x, y, z) \\[2ex]
\dfrac{z}{\sqrt{x^2 + y^2 + z^2}} = 18\lambda_1 z - 2\lambda_2 & f_z(x, y, z) = \lambda_1 g_z(x, y, z) + \lambda_2 h_z(x, y, z) \\[2ex]
x^2 + y^2 + 9z^2 - 25 = 0 & g(x, y, z) = 0 \\[1ex]
x + 3y - 2z = 0 & h(x, y, z) = 0
\end{cases}
$$

Eliminating $\lambda_1$ and $\lambda_2$ from the first three equations, we have two possibilities: $z = 0$ or $y = 3x$. Using each of these equations with the constraints $x^2 + y^2 + 9z^2 - 25 = 0$ and $x + 3y - 2z = 0$ results in the following:

$z = 0$:  $\left(-\dfrac{15}{\sqrt{10}}, \dfrac{5}{\sqrt{10}}, 0\right)$      **5 units from the origin**

$\left(\dfrac{15}{\sqrt{10}}, -\dfrac{5}{\sqrt{10}}, 0\right)$      **5 units from the origin**

$y = 3x$:  $\left(\dfrac{5}{\sqrt{235}}, \dfrac{15}{\sqrt{235}}, \dfrac{25}{\sqrt{235}}\right)$      $\sqrt{\dfrac{875}{235}} \approx$ **1.930 units from the origin**

$\left(-\dfrac{5}{\sqrt{235}}, -\dfrac{15}{\sqrt{235}}, -\dfrac{25}{\sqrt{235}}\right)$      $\sqrt{\dfrac{875}{235}} \approx$ **1.930 units from the origin**

The points $\left(-\dfrac{15}{\sqrt{10}}, \dfrac{5}{\sqrt{10}}, 0\right)$ and $\left(\dfrac{15}{\sqrt{10}}, -\dfrac{5}{\sqrt{10}}, 0\right)$ are farthest

from the origin, and the points $\left(\dfrac{5}{\sqrt{235}}, \dfrac{15}{\sqrt{235}}, \dfrac{25}{\sqrt{235}}\right)$ and

$\left(-\dfrac{5}{\sqrt{235}}, -\dfrac{15}{\sqrt{235}}, -\dfrac{25}{\sqrt{235}}\right)$ are closest to the origin. Since the intersection

of the ellipsoid and the plane is an ellipse, the points we have found are the endpoints of the major and minor axes of this ellipse.  ∎

**NOW WORK** Problem 19.

## 13.4 Assess Your Understanding

### Concepts and Vocabulary

1. *Multiple Choice*  The number $\lambda$ in the equations $\nabla f(x, y) = \lambda \nabla g(x, y)$ is called a
[(a) factor (b) Lagrangian (c) Lagrange multiplier].

2. *True or False*  Extreme values of the function $z = f(x, y)$ subject to the constraint $g(x, y) = 0$ are found as solutions of the system of equations $\nabla f(x, y) = \lambda \nabla g(x, y)$ and $g(x, y) = 0$.

3. *True or False*  Lagrange multipliers can be used only for functions of two variables.

4. *True or False*  When using Lagrange multipliers to find the extreme values of a function, the number of Lagrange multipliers introduced depends on the number of variables in the function.

---

**1.** = NOW WORK problem      〰 = Graphing technology recommended      CAS = Computer Algebra System recommended

## Skill Building

*In Problems 5–16, use Lagrange multipliers to find the maximum and minimum values of f subject to the constraint $g(x, y) = 0$.*

5. $f(x, y) = 3x + y$, $g(x, y) = xy - 8 = 0$

6. $f(x, y) = 3x + y + 4$, $g(x, y) = xy - 1 = 0$

7. $f(x, y) = 3x + y + 4$, $g(x, y) = x^2 + 4y^2 - 1 = 0$

8. $f(x, y) = 3x + y$, $g(x, y) = x^2 + y^2 - 4 = 0$

9. $f(x, y) = x - 2y^2$, $g(x, y) = x^2 + y^2 - 1 = 0$

10. $f(x, y) = x^2 + 4y^3$, $g(x, y) = x^2 + 2y^2 - 2 = 0$

11. $f(x, y) = 2xy$, $g(x, y) = x^2 + y^2 - 2 = 0$

12. $f(x, y) = xy$, $g(x, y) = 9x^2 + 4y^2 - 36 = 0$

13. $f(x, y) = x^2 - 4xy + 4y^2$, $g(x, y) = x^2 + y^2 - 4 = 0$

14. $f(x, y) = 9x^2 - 6xy + y^2$, $g(x, y) = x^2 + y^2 - 25 = 0$

15. $f(x, y, z) = 4x - 3y + 2z$, $g(x, y, z) = x^2 + y^2 - 6z = 0$

16. $f(x, y, z) = x^2 + 2y^2 + z^2$, $g(x, y, z) = 2x - 3y + z - 6 = 0$

17. Find the absolute maximum and the absolute minimum of the function $f(x, y) = x^2 + y^2 + 4xy$ subject to the constraint $x^2 + y^2 \leq 2$.

18. Find the absolute maximum and the absolute minimum of the function $f(x, y) = 2x^2 + y^2$ subject to the constraint $x^2 + y^2 \leq 4$.

19. Find the absolute maximum and the absolute minimum values of the function $f(x, y, z) = x^2 + y^2 + z^2$ subject to the constraints $z^2 = x^2 + y^2$ and $x + y - z + 1 = 0$.

20. Find the absolute minimum value of $w = x^2 + y^2 + z^2$ subject to the constraints $2x + y + 2z = 9$ and $5x + 5y + 7z = 29$.

## Applications and Extensions

21. **Minimizing Distance**   Find the point on the line $x - 3y = 6$ that is closest to the origin.

22. **Minimizing Distance**   Find the point on the plane $2x + y - 3z = 6$ that is closest to the origin.

23. Find the maximum product of two numbers $x$ and $y$ subject to $x + 2y = 21$.

24. Find the minimum quotient of two positive numbers $x$ and $y$ subject to $4x + 2y = 100$.

25. (a) Find the points on the intersection of $4x^2 + y^2 + z^2 = 16$ and the plane $3 - y = 0$ that are farthest from the origin.

   (b) Find the points that are closest to the origin.

26. Find the point on the intersection of the sphere $x^2 + 2x + y^2 + z^2 = 16$ and the plane $3x + y - z = 0$ that is farthest from the origin. Also find the point that is closest to the origin.

27. At which points on the ellipse $x^2 + 2y^2 = 2$ is the product $xy$ a maximum?

28. **Maximizing Volume**   Find the dimensions of an open-topped box that maximizes volume when the surface area is fixed at 48 square centimeters.

29. **Manufacturing and Design**   Find the optimal dimensions for a can in the shape of a right circular cylinder of fixed volume $V$. That is, find the height $h$ and the radius $r$ of the can in terms of $V$ so that the surface area is minimized. Assume the can is closed at the top and at the bottom.

   *Source*: Problem submitted by the students at Minnesota State University

30. **Minimizing Materials**   A manufacturer receives an order to build a closed rectangular container with a volume of 216 m³. What dimensions will minimize the amount of material needed to produce the container?

31. **Cost of a Box**   An open-topped box has a volume of 12 m³ and is to be made from material costing $1 per square meter. What dimensions minimize the cost?

32. **Cost of a Box**   A rectangular box is to have a bottom made from material costing $2 per square meter, while the top and sides are made from material costing $1 per square meter. If the volume of the box is to be 18 m³, what dimensions will minimize the cost of production?

33. **Carry-On Luggage Requirements**   The linear measurements (length + width + height) for luggage carried onto a Delta Airlines plane must not exceed 45 inches. Find the dimensions of the rectangular suitcase of greatest volume that meets this requirement.

   *Source*: Luggage measurements from Delta Airlines, 2017.

34. **Extreme Temperature**   Suppose that $T = T(x, y, z) = 100x^2yz$ is the temperature (in degrees Celsius) at any point $(x, y, z)$ on the sphere given by $x^2 + y^2 + z^2 = 1$. Find the points on the sphere where the temperature is greatest and least. What is the temperature at these points?

35. **Fencing**   A farmer has 340 m of fencing for enclosing two separate fields, one of which is to be a rectangle twice as long as it is wide and the other is a square. The square field must enclose at least 100 m², and the rectangular one must enclose at least 800 m².

   (a) If $x$ is the width of the rectangular field, what are the maximum and minimum values of $x$?

   (b) What is the greatest number of square meters that can be enclosed in the two fields?

36. **Fencing in an Area**   A Vinyl Fence Co. prices its Cape Cod Concave fence, which is 3 ft tall, at $26.62 per linear foot. A home builder has $5000 available to spend on enclosing a rectangular garden. What is the largest area that can be enclosed?

   *Source*: A Vinyl Fence Co. San Jose, California, 2017.

37. **Joint Cost Function**   Let $x$ and $y$ be the number of units (in thousands) of two products manufactured at a factory, and let $C = 18x^2 + 9y^2$ in thousands of dollars be the joint cost of production of the products. If $x + y = 5400$, find $x$ and $y$ that minimize production cost.

**38. Production Function**   The production function of a company is $P(x, y) = x^2 + 3xy - 6x$, where $x$ and $y$ represent two different types of input. Find the amounts of $x$ and $y$ that maximize production if $x + y = 40$.

**39. Economics: The Cobb–Douglas Model**   Use the Cobb–Douglas production model $P = 1.01K^{0.25}L^{0.75}$ as follows: Suppose that each unit of capital ($K$) has a value of $175 and each unit of labor ($L$) has a value of $125.

   **(a)** If there is a total of $175,000 to invest in the economy, use Lagrange multipliers to find the units of capital and the units of labor that maximize the total production in the manufacturing sector of the economy.

   **(b)** What are the maximum units of production that the manufacturing sector of the economy could generate under these conditions?

**40.** The surface $xyz = -1$ is cut by the plane $x + y + z = 1$, resulting in a curve $C$. Find the points on $C$ that are nearest to the origin and farthest from the origin.

**41.** Find the points on the curve of intersection of the cylinder $x^2 + y^2 = 4$ and the plane $2x + y + z = 2$ that are closest to and furthest from the origin.

**42. Maximizing Volume**   A closed rectangular box of fixed surface area and maximum volume is a cube. Use Lagrange multipliers to confirm this fact.

**43. Maximizing Volume**   A closed cylindrical can of fixed surface area and maximum volume has a height equal to the diameter of its base. Use Lagrange multipliers to confirm this fact.

**44.** Find the points of intersection on the plane $x + y + z = 1$ and the hyperboloid $x^2 + y^2 - z^2 = 1$ nearest the origin.

**45.** At what points on the union of the two curves $x^2 + y^2 = 1$ and $x^3 + y^3 = 1$ is the function $f(x, y) = x^4 + y^4 + 4$ a maximum? At what points is it a minimum?

**46.** Find the extreme values of $f(x, y, z) = xyz$ on the sphere $x^2 + y^2 + z^2 = 1$.

## Challenge Problems

**47.** Minimize $x^4 + y^4 + z^4$ subject to the constraint $Ax + By + Cz = D$.

**48.** Use Lagrange multipliers to show that the triangle of largest perimeter that can be inscribed in a circle of radius $R$ is an equilateral triangle.

**49.** Find the point on the paraboloid $z = 2 - x^2 - y^2$ that is closest to the point $(1, 1, 2)$.

**50.** What points on the surface $xy - z^2 - 6y + 36 = 0$ are closest to the origin?

# Chapter Review

## THINGS TO KNOW

### 13.1 Directional Derivative; Gradient

- The directional derivative of a differentiable function $z = f(x, y)$ at the point $(x_0, y_0)$ in the direction of the unit vector $\mathbf{u} = \cos\theta\mathbf{i} + \sin\theta\mathbf{j}$ is
$D_{\mathbf{u}}f(x_0, y_0) = f_x(x_0, y_0)\cos\theta + f_y(x_0, y_0)\sin\theta$. (p. 909)
- The directional derivative $D_{\mathbf{u}}f(x_0, y_0)$ of a differentiable function $f$ equals the slope of the tangent line to the curve $C$ at the point $(x_0, y_0, f(x_0, y_0))$ on the surface $z = f(x, y)$, where $C$ is the intersection of the surface with the plane perpendicular to the $xy$-plane and containing the line through $(x_0, y_0)$ in the direction $\mathbf{u}$. (p. 910)
- The gradient of a differentiable function $f$
   - of two variables: $\nabla f(x, y) = f_x(x, y)\mathbf{i} + f_y(x, y)\mathbf{j}$ (p. 911)
   - of three variables:
   $\nabla f(x, y, z) = f_x(x, y, z)\mathbf{i} + f_y(x, y, z)\mathbf{j} + f_z(x, y, z)\mathbf{k}$
   (p. 917)
   - The directional derivative $D_{\mathbf{u}}f(x, y) = \nabla f(x, y) \cdot \mathbf{u}$, where $\mathbf{u} = \cos\theta\mathbf{i} + \sin\theta\mathbf{j}$. (p. 911)
   - $D_{\mathbf{u}}f(x, y, z) = \nabla f(x, y, z) \cdot \mathbf{u}$, where $\mathbf{u} = \cos\alpha\mathbf{i} + \cos\beta\mathbf{j} + \cos\gamma\mathbf{k}$. (p. 917)

*Properties of the Gradient:*

- The gradient $\nabla f(x_0, y_0) \neq \mathbf{0}$ is normal to the level curve of $f$ at $P_0 = (x_0, y_0)$. (p. 912)
- If $\nabla f(x_0, y_0) = \mathbf{0}$, then $D_{\mathbf{u}}f(x_0, y_0) = 0$ for any direction $\mathbf{u}$. (p. 913)

- If $\nabla f(x_0, y_0) \neq \mathbf{0}$, then the directional derivative of $f$ at $(x_0, y_0)$ is a maximum when $\mathbf{u}$ is in the direction of $\nabla f(x_0, y_0)$. The maximum value of $D_{\mathbf{u}}f(x_0, y_0)$ is $\|\nabla f(x_0, y_0)\|$. (p. 913)

- If $\nabla f(x_0, y_0) \neq \mathbf{0}$, then the directional derivative of $f$ at $(x_0, y_0)$ is a minimum when $\mathbf{u}$ is in the direction of $-\nabla f(x_0, y_0)$. The minimum value of $D_{\mathbf{u}}f(x_0, y_0)$ is $-\|\nabla f(x_0, y_0)\|$. (p. 913)

- $z = f(x, y)$ increases most rapidly in the direction of $\nabla f(x_0, y_0) \neq \mathbf{0}$. (p. 914)

- $z = f(x, y)$ decreases most rapidly in the direction of $-\nabla f(x_0, y_0) \neq \mathbf{0}$. (p. 914)

- The value of $z = f(x, y)$ remains the same for directions orthogonal to $\nabla f(x_0, y_0) \neq \mathbf{0}$. (p. 914)

### 13.2 Tangent Planes

- Definition of a tangent plane to a surface (p. 921)

- Equation of a tangent plane to a surface (p. 922)

- Equations of the normal line to the surface $F(x, y, z) = 0$ at the point $P_0 = (x_0, y_0, z_0)$:

   - vector equation: $\mathbf{r}(t) = \mathbf{r}_0 + t\nabla F(x_0, y_0, z_0)$, where $\mathbf{r}_0 = x_0\mathbf{i} + y_0\mathbf{j} + z_0\mathbf{k}$ is the position vector of $P_0$ and $\mathbf{r}$ is the position vector of any point $P$ on the normal line. (p. 922)

- parametric equations: $x = x_0 + at$, $y = y_0 + bt$, and $z = z_0 + ct$, where $a = F_x(x_0, y_0, z_0)$, $b = F_y(x_0, y_0, z_0)$, and $c = F_z(x_0, y_0, z_0)$. (p. 923)

- symmetric equations: $\dfrac{x - x_0}{a} = \dfrac{y - y_0}{b} = \dfrac{z - z_0}{c}$ if $abc \neq 0$. (p. 923)

- If $\nabla f(x_0, y_0) = \mathbf{0}$, then the tangent plane to the surface $z = f(x, y)$ at $(x_0, y_0)$ is $z = z_0$. That is, the tangent plane is parallel to the $xy$-plane. (p. 923)

## 13.3 Extrema of Functions of Two Variables

*Definitions:*

- Local maximum; local minimum (p. 926)
- Absolute maximum; absolute minimum (p. 926)
- Critical point (p. 927)
- Saddle point (p. 928)

- *A Necessary Condition for Local Extrema*
  Let $z = f(x, y)$ be a function of two variables whose domain is an open set $D$. If $f$ has a local extremum at $(x_0, y_0)$, then $(x_0, y_0)$ is a critical point of $f$. (p. 927)

- *Second Partial Derivative Test:* (p. 929)
  Let $z = f(x, y)$ be a function of two variables for which the first- and second-order partial derivatives are continuous in some disk containing the point $(x_0, y_0)$. Suppose that $f_x(x_0, y_0) = 0$ and $f_y(x_0, y_0) = 0$. Let

  $A = f_{xx}(x_0, y_0)$    $B = f_{xy}(x_0, y_0)$    $C = f_{yy}(x_0, y_0)$

- If $AC - B^2 > 0$ and $f_{xx}(x_0, y_0) > 0$, then $f$ has a local minimum at $(x_0, y_0)$.
- If $AC - B^2 > 0$ and $f_{xx}(x_0, y_0) < 0$, then $f$ has a local maximum at $(x_0, y_0)$.
- If $AC - B^2 < 0$, then $(x_0, y_0, f(x_0, y_0))$ is a saddle point of $f$.
- If $AC - B^2 = 0$, then the test gives no guidance.

*Extreme Value Theorem for Functions of Two Variables:* (p. 930)

Let $z = f(x, y)$ be a function of two variables. If $f$ is continuous on a closed, bounded set $D$, then $f$ has an absolute maximum and an absolute minimum on $D$.

*Test for Absolute Maximum and Absolute Minimum:* (p. 931)

Let $z = f(x, y)$ be a function of two variables defined on a closed, bounded set $D$. If $f$ is continuous on $D$, then the absolute maximum and the absolute minimum of $f$ are, respectively, the largest and smallest values found among the following:

- The values of $f$ at the critical points of $f$ in $D$
- The values of $f$ on the boundary of $D$

## 13.4 Lagrange Multipliers

*Lagrange's Theorem* (p. 939)

*The Method of Lagrange Multipliers:* (p. 939)

The extreme values of $z = f(x, y)$ subject to the condition $g(x, y) = 0$, if they exist, occur at the solutions $(x, y)$ of the system equations.

$$\begin{cases} \nabla f(x, y) = \lambda \nabla g(x, y) \\ g(x, y) = 0 \end{cases}$$

*Steps for Using Lagrange Multipliers:* (p. 940)

## OBJECTIVES

| Section | You should be able to ... | Examples | Review Exercises |
|---|---|---|---|
| 13.1 | 1 Find the directional derivative of a function of two variables (p. 908) | 1, 2 | 1–3 |
| | 2 Find the gradient of a function of two variables (p. 911) | 3 | 1–3 |
| | 3 Work with properties of the gradient (p. 912) | 4–6 | 7–9 |
| | 4 Find the directional derivative and gradient of a function of three variables (p. 917) | | 4–6 |
| 13.2 | 1 Find an equation of a tangent plane to a surface (p. 922) | 1 | 10 |
| | 2 Find an equation of a normal line to a tangent plane (p. 922) | 2 | 10 |
| | 3 Find an equation of a tangent plane to a surface defined explicitly (p. 923) | 3 | 11, 12 |
| 13.3 | 1 Find critical points (p. 927) | 1, 2 | 13–15, 25 |
| | 2 Use the Second Partial Derivative Test (p. 928) | 3, 4 | 16–18 |
| | 3 Find the absolute extrema of a function of two variables (p. 930) | 5, 6 | 19, 20 |
| | 4 Solve optimization problems (p. 933) | 7, 8 | 26–28 |
| 13.4 | 1 Use Lagrange multipliers for an optimization problem with one constraint (p. 940) | 2–5 | 21–24, 29–31 |
| | 2 Use Lagrange multipliers for an optimization problem with two constraints (p. 943) | 6 | 32 |

## REVIEW EXERCISES

*In Problems 1–6, find the gradient and directional derivative of each function at the indicated point in the direction of* **a**.

1.  $f(x, y) = \ln(\sec(x^2 + y^2))$ at $\left(\sqrt{\dfrac{\pi}{8}}, \sqrt{\dfrac{\pi}{8}}\right)$;   $\mathbf{a} = \mathbf{i} + \mathbf{j}$

2.  $f(x, y) = \dfrac{x^2 - y^2}{x^2 + y^2}$ at $(1, 1)$;   $\mathbf{a} = \mathbf{i} - \mathbf{j}$

3.  $f(x, y) = y \sin^{-1} x$ at $(0, 1)$;   $\mathbf{a} = 3\mathbf{i} + \mathbf{j}$

4.  $f(x, y, z) = \ln(xyz)$ at $(1, 1, 3)$;   $\mathbf{a} = \mathbf{i} + \mathbf{j} + 3\mathbf{k}$

5.  $f(x, y, z) = ye^{-x}(x^2 + y^2 + z^2 + 1)$ at $(0, 0, 0)$;
    $\mathbf{a} = 2\mathbf{i} + \mathbf{j} + 2\mathbf{k}$

6.  $f(x, y, z) = (2x + y + z)^2 + xyz$ at $(1, 1, 1)$;
    $\mathbf{a} = -\mathbf{i} - \mathbf{j} - \mathbf{k}$

*In Problems 7–9:*

**(a)** *Find the direction for which the directional derivative of the function f is maximum.*

**(b)** *Find the maximum value and the minimum value of the directional derivative.*

7.  $f(x, y) = \sec(x^2 + y^2)$ at $\left(\sqrt{\dfrac{\pi}{8}}, \sqrt{\dfrac{\pi}{8}}\right)$

8.  $f(x, y) = \dfrac{x^2 - y^2}{x^2 + y^2}$ at $(1, 1)$

9.  $f(x, y) = y \sin^{-1} x$ at $(0, 1)$

*In Problems 10–12:*

**(a)** *Find an equation of the tangent plane to each surface at the given point.*

**(b)** *Find an equation of the normal line to the tangent plane of each surface at the given point.*

**(c)** *Find the points, if any, at which the tangent plane is parallel to the xy-plane.*

10.  $x^2 - y^2 + z^2 = 4$ at $(-1, 1, 2)$    11.  $2x^2 + y^2 = z$ at $(1, 0, 2)$

12.  $f(x, y) = y \sin^{-1} x$ at $(0, 1)$

*In Problems 13–15, find the critical points, if any, of each function.*

13.  $z = f(x, y) = x^2 + xy + y^2 + 6x$

14.  $z = f(x, y) = x^3 - y^3 + 3xy$    15.  $z = f(x, y) = xy$

*In Problems 16–18, use the Second Partial Derivative Test to find any local maxima, local minima, and saddle points for each function.*

16.  $z = f(x, y) = xe^{xy}$

17.  $z = f(x, y) = \sin x + \sin y$

18.  $z = f(x, y) = x^2 - 9y + y^2$

*In Problems 19 and 20, find the absolute maximum and absolute minimum of f on the domian D.*

19.  $z = f(x, y) = x^2 - 2xy + 2y$    $D: 0 \le x \le 3, 0 \le y \le 4$.

20.  $z = f(x, y) = 3xy^2$    $D: x^2 + y^2 \le 9$.

*In Problems 21–24, use Lagrange multipliers to find the maximum and minimum values of f subject to the constraint $g(x, y) = 0$.*

21.  $z = f(x, y) = 5x^2 + 3y^2 + xy$    $g(x, y) = 2x - y - 20 = 0$

22.  $z = f(x, y) = x\sqrt{y}$    $g(x, y) = 2x + y - 3000 = 0$

23.  $z = f(x, y) = x^2 + y^2$    $g(x, y) = 2x + y - 4 = 0$

24.  $z = f(x, y) = xy^2$    $g(x, y) = x^2 + y^2 - 1 = 0$

25.  **Heat Transfer**   A metal plate is placed on the $xy$-plane in such a way that the temperature $T$ at any point $(x, y)$ is given by $T = e^y(\sin x + \sin y)°$C.

   **(a)** Find the directional derivative of $T$ at $(0, 0)$ in the direction of $4\mathbf{i} + 3\mathbf{j}$. Interpret the result in the context of the problem.

   **(b)** At $(0, 0)$, in what direction is the temperature increasing most rapidly?

   **(c)** In what direction is the temperature decreasing most rapidly?

   **(d)** In what direction does the temperature remain the same?

26.  **Maximizing Profit**   A company produces two products at a total cost $C(x, y) = x^2 + 200x + y^2 + 100y - xy$, where $x$ and $y$ represent the units of each product produced and sold. The revenue function is $R(x, y) = 2000x - 2x^2 + 100y - y^2 + xy$.

   **(a)** Find the number of units of each product that will maximize profit.

   **(b)** What is the maximum profit?

27.  **Maximizing Profit**   A manufacturer introduces a new product with a Cobb–Douglas production function of $P(K, L) = 10K^{0.3}L^{0.7}$, where $K$ represents the units of capital and $L$ the units of labor needed to produce $P$ units of the product. A total of \$51,000 has been budgeted for production. Each unit of labor costs the manufacturer \$100 and each unit of capital costs \$50.

   **(a)** How should the \$51,000 be allocated between labor and capital to maximize production?

   **(b)** What is the maximum number of units that can be produced?

28.  **Volume**   Find the volume of the largest rectangular solid that can be inscribed in the interior of the surface

$$\frac{x^2}{a^2} + \frac{y^2}{b^2} + \frac{z^2}{c^2} = 1$$

   if the sides of the solid are parallel to the axes.

29.  Use Lagrange multipliers to find the points on the surface $xyz = 1$ closest to the origin.

30.  **Volume**   Use Lagrange multipliers to maximize the volume of a rectangular solid that has three faces on the coordinate planes and one vertex on the plane $\dfrac{x}{a} + \dfrac{y}{b} + \dfrac{z}{c} = 1$, $a > 0$, $b > 0$, $c > 0$.

31.  **Minimizing Cost**   The base of a rectangular box costs five times as much as the other five sides. Use Lagrange multipliers to find the proportions of the dimensions for the cheapest possible box of volume $V$.

32.  Find the extreme values of $f(x, y, z) = xyz$ subject to the constraints $x^2 + y^2 = 1$ and $y = 3z$.

## CHAPTER 13 PROJECT    **Measuring Ice Thickness on Crystal Lake**

Tom Martin/Getty Images

In this project we examine safety issues involving frozen fresh-water lakes. When a lake freezes over, the depth of the ice varies from place to place on the lake. If we can construct a model that gives the depth of ice at any position on a lake, then we can determine whether a given position is safe or not for a particular load/activity. Based on studies of ice depth, Table 1 lists minimum safe ice depths for various loads and activities on clear lake ice.

### TABLE 1

| Load/Activity | Minimum Safe Ice Depth (in inches) |
|---|---|
| Cross-country skiing (one person) | 3 |
| Ice fishing or skating (one person) | 4 |
| One snowmobile | 5 |
| Ice boating | 6 |
| Group activities | 7 |
| One car | 8 |
| Several snowmobiles | 9 |
| Light truck (2.5 tons) | 10.5 |
| Medium truck (3.5 tons) | 12 |
| 10-ton load | 15 |
| 25-ton load | 20 |

*Data from:* U.S. Army Corps of Engineers, Minnesota
Department of Natural Resources,
Pennsylvania Fish and Boat Commission, and
The Old Farmer's Almanac

Our study focuses on Crystal Lake in Otter Tail, Minnesota. Crystal Lake is roughly circular in shape, with a radius of 0.838 mi.

1. Place the center of the lake at the origin $(0, 0)$ of a rectangular coordinate system. Using rectangular coordinates $(x, y)$ measured in miles, write an equation for the shoreline of Crystal Lake. What are the minimum and maximum values of $x$ and $y$? Draw a graph of the lake with the direction north pointing up.

   A simple model for determining the depth $d$ of ice is

   $$d = d(x, y) = \alpha + \alpha \sin(3\pi x y)$$

   where $d$ is measured in inches, $(x, y)$ are the coordinates of a point on the lake, as given in Problem 1, and $\alpha$, measured in inches, is a nonnegative constant related to various conditions such as temperature, lake depth, currents, and so on. Suppose on a certain day in December, the value of $\alpha$ for Crystal Lake is $\alpha = 9$ in.

2. What is the domain of $d = d(x, y)$? What is the range? What are the minimum and maximum values of $d$?

3. What is the depth of the ice at the center of the lake?

4. Graph the level curves of $d(x, y) = c$, for $c = 0, 3, 6, 12, 15$, and 18.

5. (a) Using the model, determine whether a snowmobile can safely cross the lake along the path $y = x$.

   (b) Using the model, determine whether a car can safely cross the lake along the path $y = -x$.

6. (a) Find the rate of change of $d$ at the point $(0.4, -0.1)$ in the direction east (the direction of **i**).

   (b) Find the rate of change of $d$ at the point $(0.4, -0.1)$ in the direction north (the direction of **j**).

   (c) Interpret the results obtained in (a) and (b).

7. (a) Find the direction from the point $(0.4, -0.1)$ for which the ice depth increases most rapidly.

   (b) Travel a short distance, say 1 ft, in the direction found in (a). At the new point, recalculate the direction for which the ice depth increases most rapidly.

   (c) Explain the results obtained in (a) and (b).

8. (a) Find the direction from the point $(0.4, -0.1)$ that results in no change in ice depth $d$.

   (b) Travel a short distance, say 1 ft, in the direction found in (a), and at the new point, recalculate the direction that again results in no change in ice depth.

   (c) Explain the results obtained in (a) and (b).

   Another model for the depth of ice on Crystal Lake is given by

   $$h = h(x, y) = \alpha|x + y| + \beta|x - y|$$

   where $h$ is measured in inches and $\alpha$, $\beta$, measured in inches per mile, are nonnegative constants determined by local conditions. Suppose that on a certain day in January, the value of $\alpha = \beta = 9$ in/mi for Crystal Lake.

9. Sketch the level curves of $h(x, y) = c$, for $c = 0, 3, 6, 12$, and 15.

10. (a) Using this model, find the direction from the point $(0.4, -0.1)$ for which the ice depth increases most rapidly.

    (b) Travel a short distance, say 1 ft, in the direction found in (a). At the new point, recalculate the direction for which the ice depth increases most rapidly.

    (c) Explain the results obtained in (a) and (b).

11. (a) Using this model, find the direction from the point $(0.4, -0.1)$ that results in no change in ice depth $h$.

    (b) Travel a short distance, say 1 ft, in the direction found in (a), and at the new point, recalculate the direction for no increase in ice depth.

A year earlier, the values of $\alpha$ and $\beta$ were different. For that day in January, $\alpha = 5$ and $\beta = 10$.

12. Sketch the level curves of $h(x, y) = 5\,|x + y| + 10\,|x - y|$, for $c = 0, 3, 6, 12,$ and 15.

13. (a) Find the direction from the point $(0.4, -0.1)$ for which the ice depth increases most rapidly.

    (b) Travel a short distance, say 1 ft, in the direction found in (a). At the new point, recalculate the direction for which the ice depth increases most rapidly.

    (c) Explain the results obtained in (a) and (b).

14. (a) Find the direction from the point $(0.4, -0.1)$ that results in no change in ice depth.

    (b) Travel a short distance, say 1 ft, in the direction found in (a), and at the new point, recalculate the direction for no increase in ice depth.

    (c) Explain the results obtained in (a) and (b).

15. What conditions might have influenced the difference in $\alpha$ and $\beta$ from one year to the next?

16. Compare the results obtained from the two models. Which would you choose to model the ice depth of Crystal Lake? Write a brief report that supports your decision.

# 14 Multiple Integrals

**CHAPTER 14 PROJECT** The Chapter Project on page 1023 investigates models for determining the mass of a star.

Robert Gendler/Stocktrek Images/Science Source

## The Mass of Stars

Some of the most massive objects in the universe are not solid, but are large balls of gas—stars. The nearest star, the Sun, has a mass of $1.99 \times 10^{30}$ kg and is about 333,000 times more massive than Earth.

Stars are often depicted on a Hertzsprung–Russell diagram, or HR diagram, which compares the luminosity (brightness) and color (temperature) of stars. As centuries pass, the composition of a star changes because of the nuclear reactions at the center that generate the energy we see as light. Over billions of years changes in composition alter the star's structure and the star evolves, tracing a unique path in the HR diagram determined primarily by its mass.

In Chapter 5, we defined the definite integral of a function (of a single variable). When extended to functions of two or more variables, definite integrals are called *multiple integrals*. The integral of a function of two variables is called a *double integral*, and the integral of a function of three variables is a *triple integral*. For consistency, we refer to the integral of a function of one variable as a *single integral*.

In Chapter 6, we used a single integral to find the area of a region, the volume and surface area of a solid, arc length, and the work done by a variable force. Two of the many uses of multiple integrals are to find areas and volumes of more general regions than those considered in Chapter 6.

# 14.1 The Double Integral over a Rectangular Region

**OBJECTIVES** *When you finish this section, you should be able to:*

1 Find double Riemann sums of $z = f(x, y)$ over a closed rectangular region (p. 952)

2 Find the value of a double integral defined on a closed rectangular region (p. 954)

3 Find the volume under a surface and over a rectangular region (p. 956)

Historically, the investigation of the area problem led to the definite integral. Similarly, the search for methods to find the volume of an irregularly shaped solid led to the development of multiple integrals.

## 1 Find Double Riemann Sums of $z = f(x, y)$ over a Closed Rectangular Region

Since the development of multiple integrals parallels that of a single integral, we begin by reviewing the process that led to the definition of a single integral.

To define a single integral, we begin with a function $y = f(x)$ defined on a closed interval $[a, b]$. Then we:

• Partition $[a, b]$ into $n$ subintervals of lengths $\Delta x_1, \Delta x_2, \ldots, \Delta x_i, \ldots, \Delta x_n$. (Recall that the lengths of the subintervals need not be equal and the length of the largest subinterval is the norm, max $\Delta x_i$, of the partition.)

• Choose a number $u_i$ in each subinterval.

• Evaluate $f$ at $u_i$, multiply $f(u_i)$ by $\Delta x_i$, and form the Riemann sums $\sum_{i=1}^{n} f(u_i)\Delta x_i$.

If $\lim\limits_{\max \Delta x_i \to 0} \sum_{i=1}^{n} [f(u_i)\Delta x_i]$ exists, it is called the definite integral of $f$ from $a$ to $b$ and is denoted by

$$\int_a^b f(x)\, dx = \lim_{\max \Delta x_i \to 0} \sum_{i=1}^{n} f(u_i)\Delta x_i$$

For a function $f$ of two variables, the double integral of $f$ is defined similarly. We begin with a function $z = f(x, y)$ of two variables defined on a closed rectangular region $R$ defined by $a \le x \le b$ and $c \le y \le d$.

Partition the interval $[a, b]$ into $n$ subintervals of length $\Delta x_i$, $i = 1, 2, \ldots, n$, and the interval $[c, d]$ into $m$ subintervals of length $\Delta y_j$, $j = 1, 2, \ldots, m$. By drawing lines parallel to the coordinate axes through the endpoints of the subintervals, we form a **rectangular partition** $P$ of the region $R$ into $nm$ subrectangles $R_{ij}$.

Now we define the **norm**, $\|\Delta\|$, of the rectangular partition as the largest of the lengths of the subintervals $\Delta x_i$ or $\Delta y_j$. That is, $\|\Delta\| = \max(\Delta x_i, \Delta y_j)$. See Figure 1.

In each subrectangle $R_{ij}$, choose a point $(u_{ij}, v_{ij})$, evaluate the function $z = f(x, y)$ at $(u_{ij}, v_{ij})$, multiply $f(u_{ij}, v_{ij})$ by the area $\Delta A_{ij} = \Delta x_i \Delta y_j$, and form the sums

$$\sum_{j=1}^{m}\sum_{i=1}^{n} f(u_{ij}, v_{ij})\Delta A_{ij} = \sum_{j=1}^{m}\sum_{i=1}^{n} f(u_{ij}, v_{ij})\Delta x_i \Delta y_j$$

These sums are called the **double Riemann sums** of $f$ for the partition $P$.

Although the subintervals $\Delta x_i$, $i = 1, 2, \ldots, n$ need not be equal, we usually partition $[a, b]$ into $n$ subintervals each of the same length $\Delta x = \dfrac{b-a}{n}$. Similarly, we usually partition $[c, d]$ into $m$ subintervals each of the same length $\Delta y = \dfrac{d-c}{m}$.

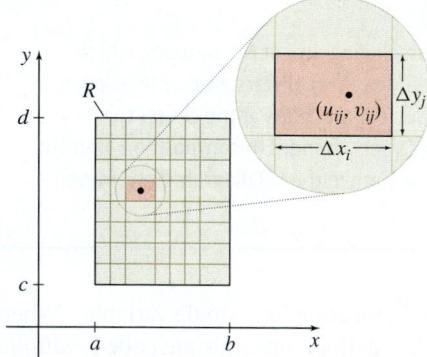

**Figure 1** A rectangular partition of the region $R$. The norm of the partition, $\|\Delta\|$, is the largest of the lengths of the subintervals $\Delta x_i$ or $\Delta y_j$.

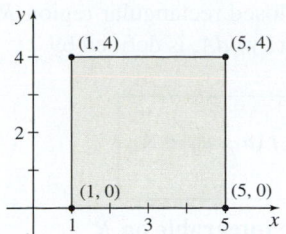

**DF** **Figure 2** $f(x, y) = x^2 y$ is defined on the square $1 \le x \le 5, 0 \le y \le 4$.

## EXAMPLE 1  Finding Double Riemann Sums

Let $f(x, y) = x^2 y$ be a function defined on the square having its lower left corner at $(1, 0)$ and its upper right corner at $(5, 4)$, as shown in Figure 2.

**(a)** Find a double Riemann sum of $f$ over this region by partitioning the square into four congruent subsquares. Choose the lower left corner of each subsquare as $(u_{ij}, v_{ij})$.

**(b)** Find a double Riemann sum of $f$ over this region by partitioning the square into eight congruent rectangles with sides of length $\Delta x_i = 2$, $i = 1, 2$, and $\Delta y_j = 1$, $j = 1, 2, 3, 4$. Choose the lower right corner of each rectangle as $(u_{ij}, v_{ij})$.

**Solution** **(a)** Begin by partitioning the square in Figure 2 into four congruent subsquares. We form four subsquares by partitioning the interval $[1, 5]$ into two subintervals each of length $\Delta x_i = \dfrac{5 - 1}{2} = 2$, $i = 1, 2$, and by partitioning the interval $[0, 4]$ into two subintervals each of length $\Delta y_j = \dfrac{4 - 0}{2} = 2$, $j = 1, 2$. So, $\Delta x_i \Delta y_j = 2 \cdot 2 = 4$ for $i = 1, 2$ and $j = 1, 2$. Now draw lines parallel to the coordinate axes through the endpoints of each subinterval. Figure 3 shows the lower left corner of each subsquare.

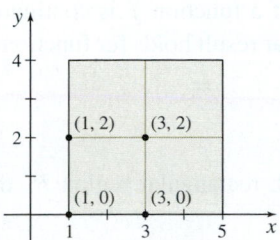

**Figure 3** The square is partitioned into four congruent subsquares.

The double Riemann sum for which $(u_{ij}, v_{ij})$ is the lower left corner of each subsquare is

$$\sum_{j=1}^{2} \sum_{i=1}^{2} f(u_{ij}, v_{ij}) \Delta x_i \Delta y_j = f(1, 0) \cdot 4 + f(3, 0) \cdot 4 + f(1, 2) \cdot 4 + f(3, 2) \cdot 4$$

$$= [f(1, 0) + f(3, 0) + f(1, 2) + f(3, 2)] \cdot 4$$

$$= [0 + 0 + 2 + 18] \cdot 4 = 80 \qquad f(x, y) = x^2 y$$

**(b)** Now partition the square in Figure 2 into eight congruent rectangles with sides $\Delta x_i = 2$, $i = 1, 2$, and $\Delta y_j = 1$, $j = 1, 2, 3, 4$, as shown in Figure 4. The double Riemann sum, for which $(u_{ij}, v_{ij})$ is the lower right corner of each subrectangle, is

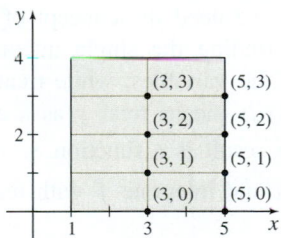

**Figure 4** The square is partitioned into eight congruent subrectangles.

$$\sum_{j=1}^{4} \sum_{i=1}^{2} f(u_{ij}, v_{ij}) \Delta x_i \Delta y_j = [f(3, 0) + f(5, 0) + f(3, 1) + f(5, 1)$$

$$+ f(3, 2) + f(5, 2) + f(3, 3) + f(5, 3)] \cdot 2 \cdot 1$$

$$= [0 + 0 + 9 + 25 + 18 + 50 + 27 + 75] \cdot 2 = 408 \qquad \blacksquare$$

**NOW WORK** Problem 5.

As Example 1 demonstrates, the double Riemann sums of a function $z = f(x, y)$ over a region $R$ depend both on the choice of the partition, which depends on $\Delta x_i$ and $\Delta y_j$, and on the choice of the points $(u_{ij}, v_{ij})$, $i = 1, 2, \ldots, n$, $j = 1, 2, \ldots, m$, in $R$. If

$$\lim_{\|\Delta\| \to 0} \sum_{j=1}^{m} \sum_{i=1}^{n} f(u_{ij}, v_{ij}) \Delta x_i \Delta y_j = \lim_{\|\Delta\| \to 0} \sum_{j=1}^{m} \sum_{i=1}^{n} f(u_{ij}, v_{ij}) \Delta A_{ij} = I$$

exists and does not depend on the choice of the partition or on the choice of $u_{ij}$ or $v_{ij}$, then the number $I$ is called the *double integral* of $z = f(x, y)$ over the region $R$ and is denoted by $\iint\limits_{R} f(x, y) dA$.

**DEFINITION  Double Integral**

Let $f$ be a function of two variables defined on a closed rectangular region $R$. Then the **double integral of $f$ over** $R$, denoted by $\iint\limits_R f(x, y)\, dA$, is defined by

$$\iint\limits_R f(x, y)\, dA = \lim_{\|\Delta\| \to 0} \sum_{j=1}^{m} \sum_{i=1}^{n} f(u_{ij}, v_{ij}) \Delta A_{ij}$$

provided this limit exists. In this case, $f$ is said to be **integrable on** $R$.

Since $\Delta x_i \Delta y_j = \Delta y_j \Delta x_i = \Delta A_{ij}$, other symbols for the double integral of $f$ over $R$ are

$$\iint\limits_R f(x, y)\, dx\, dy \qquad \text{and} \qquad \iint\limits_R f(x, y)\, dy\, dx$$

In Chapter 5, an important theorem states that if a function $f$ is continuous on a closed interval $[a, b]$, then $\int_a^b f(x)\, dx$ exists. A similar result holds for functions of two variables.

**THEOREM  Existence of the Double Integral**

If a function $z = f(x, y)$ is continuous on a closed, rectangular region $R$, then the double integral $\iint\limits_R f(x, y)\, dA$ exists.

A proof can be found in most advanced calculus books.

## 2  Find the Value of a Double Integral Defined on a Closed Rectangular Region

To find the value of a double integral of a function $f$ of two variables whose domain is a closed rectangle without using Riemann sums, we need the concept of *partial integration*. **Partial integration** is the process of finding the single integral of a function $f$ of two variables with respect to one of the variables, while treating the other variable as a constant. The symbol $\int_a^b f(x, y)\, dx$ means treat $y$ as a constant and integrate $f$ with respect to the variable $x$. The result is a function of $y$ alone. Similarly, $\int_c^d f(x, y)\, dy$ means treat $x$ as a constant and integrate $f$ with respect to the variable $y$. This result is a function of $x$ alone.

**IN WORDS**  Partial Integration is the reverse of partial differentiation.

---

EXAMPLE 2   **Using Partial Integration**

Use partial integration to find:

**(a)** $\displaystyle\int_{x=1}^{x=2} x^3 y^2\, dx$  **(b)** $\displaystyle\int_{y=0}^{y=4} x^3 y^2\, dy$

**Solution (a)** For $\int_{x=1}^{x=2} x^3 y^2\, dx$, treat $y$ as a constant and integrate with respect to $x$:

$$\int_{x=1}^{x=2} x^3 y^2\, dx = y^2 \int_{x=1}^{x=2} x^3\, dx = y^2 \left[\frac{x^4}{4}\right]_{x=1}^{x=2} = y^2 \left(4 - \frac{1}{4}\right) = \frac{15}{4} y^2$$

**(b)** For $\int_{y=0}^{y=4} x^3 y^2\, dy$, treat $x$ as a constant and integrate with respect to $y$:

$$\int_{y=0}^{y=4} x^3 y^2\, dy = x^3 \int_{y=0}^{y=4} y^2\, dy = x^3 \left[\frac{y^3}{3}\right]_{y=0}^{y=4} = \frac{64x^3}{3}$$

■

NOW WORK  Problem 11.

## EXAMPLE 3  Using Partial Integration

Use partial integration to find:

(a) $\displaystyle\int_{y=0}^{y=4}\left[\int_{x=1}^{x=2}x^{3}y^{2}\,dx\right]dy$ 

(b) $\displaystyle\int_{x=1}^{x=2}\left[\int_{y=0}^{y=4}x^{3}y^{2}\,dy\right]dx$

**Solution (a)**  Begin by using partial integration to find the integral within the bracket. Then integrate the result with respect to $y$.

$$\int_{y=0}^{y=4}\left[\int_{x=1}^{x=2}x^{3}y^{2}dx\right]dy=\int_{y=0}^{y=4}y^{2}\left[\int_{x=1}^{x=2}x^{3}dx\right]dy=\int_{y=0}^{y=4}\frac{15}{4}y^{2}dy=\frac{15}{4}\left[\frac{y^{3}}{3}\right]_{y=0}^{y=4}=80$$

↑ From Example 2(a)    ↑ Find the single integral.

**(b)**

$$\int_{x=1}^{x=2}\left[\int_{y=0}^{y=4}x^{3}y^{2}\,dy\right]dx=\int_{x=1}^{x=2}x^{3}\left[\int_{y=0}^{y=4}y^{2}dy\right]dx=\int_{x=1}^{x=2}\frac{64}{3}x^{3}dx=\frac{64}{3}\left[\frac{x^{4}}{4}\right]_{x=1}^{x=2}=80$$

↑ From Example 2(b)    ↑ Find the single integral.  ∎

**NOW WORK** Problem 29.

Integrals of the form

$$\int_{x=a}^{x=b}\left[\int_{y=c}^{y=d}f(x,y)\,dy\right]dx\qquad\text{and}\qquad\int_{y=c}^{y=d}\left[\int_{x=a}^{x=b}f(x,y)\,dx\right]dy$$

are called **iterated integrals**. In the iterated integral on the left, we first integrate the function $f$ partially with respect to $y$ from $c$ to $d$. The result is a function of $x$ that we then integrate with respect to $x$ from $a$ to $b$. In the iterated integral on the right, we first integrate $f$ partially with respect to $x$ from $a$ to $b$. The result is a function of $y$ that we then integrate with respect to $y$ from $c$ to $d$.

In writing iterated integrals, we will omit the variables $x$ and $y$ from the upper and lower limits of integration. Then the iterated integrals in Example 3 are written as $\int_{0}^{4}\left[\int_{1}^{2}x^{3}y^{2}\,dx\right]dy$ and $\int_{1}^{2}\left[\int_{0}^{4}x^{3}y^{2}\,dy\right]dx$.

Notice that the two iterated integrals in Example 3 are equal. This is a consequence of *Fubini's Theorem*, which provides a practical way to find certain double integrals.

**ORIGINS**  Guido Fubini (1879–1943) was an Italian mathematician born in Venice. His father was a mathematics teacher who encouraged Guido to study mathematics. A short man, Fubini made many contributions in diverse areas of math, earning him the nickname "the little giant." In 1939 the anti-Semitic policies of Mussolini caused Fubini and his family to immigrate to the United States. He accepted a position at the Institute for Advanced Study in Princeton, NJ, and also taught at Princeton University for several years. He died in New York in 1943.

**THEOREM  Fubini's Theorem**

If a function $z=f(x,y)$ is continuous on a closed rectangular region $R$ defined by $a\le x\le b$ and $c\le y\le d$, then

$$\iint\limits_{R}f(x,y)\,dA=\int_{a}^{b}\left[\int_{c}^{d}f(x,y)\,dy\right]dx=\int_{c}^{d}\left[\int_{a}^{b}f(x,y)\,dx\right]dy$$

Fubini's Theorem states that an iterated integral can be integrated in either order provided the function $z=f(x,y)$ is continuous on the closed rectangular region $R$.

## EXAMPLE 4  Finding the Value of a Double Integral Using Fubini's Theorem

Find the value of $\iint\limits_{R}(2x+y)\,dA$ if $R$ is the closed rectangular region defined by $1\le x\le 5$ and $0\le y\le 4$.

**Solution**  The function $f(x,y)=2x+y$ is continuous at every point $(x,y)$ in the $xy$-plane. Then by Fubini's Theorem, the double integral $\iint\limits_{R}(2x+y)\,dA$ equals either of the iterated integrals

$$\int_{1}^{5}\left[\int_{0}^{4}(2x+y)\,dy\right]dx\qquad\text{or}\qquad\int_{0}^{4}\left[\int_{1}^{5}(2x+y)\,dx\right]dy\qquad(1)$$

We use the integral on the left in (1) to find the double integral.

$$\iint\limits_{R} (2x + y)\, dA = \int_{1}^{5} \left[ \int_{0}^{4} (2x + y)\, dy \right] dx = \int_{1}^{5} \left[ 2xy + \frac{y^2}{2} \right]_{0}^{4} dx$$

$$\underset{\substack{\uparrow \\ \text{Integrate partially} \\ \text{with respect to } y.}}{}$$

$$= \int_{1}^{5} (8x + 8)\, dx = \left[ 4x^2 + 8x \right]_{1}^{5} = 140 - 12 = 128 \qquad \blacksquare$$

**NOW WORK** Example 4 using the integral on the right in (1).

**NOW WORK** Problem 41.

## 3 Find the Volume Under a Surface and over a Rectangular Region

The double integral $\iint\limits_{R} f(x, y)\, dA$ over a rectangular region $R$ has a geometric interpretation. If the function $z = f(x, y)$ is nonnegative on $R$, then $f(u_{ij}, v_{ij})$ represents the height of $z$ at the point $(u_{ij}, v_{ij})$, and the product $f(u_{ij}, v_{ij}) \Delta x_i \Delta y_j$ equals the volume of a small, rectangular cylinder of height $f(u_{ij}, v_{ij})$ and a base of area $\Delta x_i \Delta y_j$ as shown in Figure 5. The sum of the volumes of these small cylinders, $\sum\limits_{j=1}^{m} \sum\limits_{i=1}^{n} f(u_{ij}, v_{ij}) \Delta x_i \Delta y_j$, is an approximation for the volume of the solid under the graph of $z = f(x, y)$ and over $R$, as illustrated in Figure 6. If $f$ is continuous on $R$, then $\lim\limits_{\|\Delta\| \to 0} \sum\limits_{j=1}^{m} \sum\limits_{i=1}^{n} f(u_{ij}, v_{ij}) \Delta x_i \Delta y_j = \lim\limits_{\|\Delta\| \to 0} \sum\limits_{j=1}^{m} \sum\limits_{i=1}^{n} f(u_{ij}, v_{ij}) \Delta A_{ij}$ exists, and $\iint\limits_{R} f(x, y)\, dA$ can be interpreted as the volume of the solid under the graph of $z = f(x, y)$ over the region $R$.

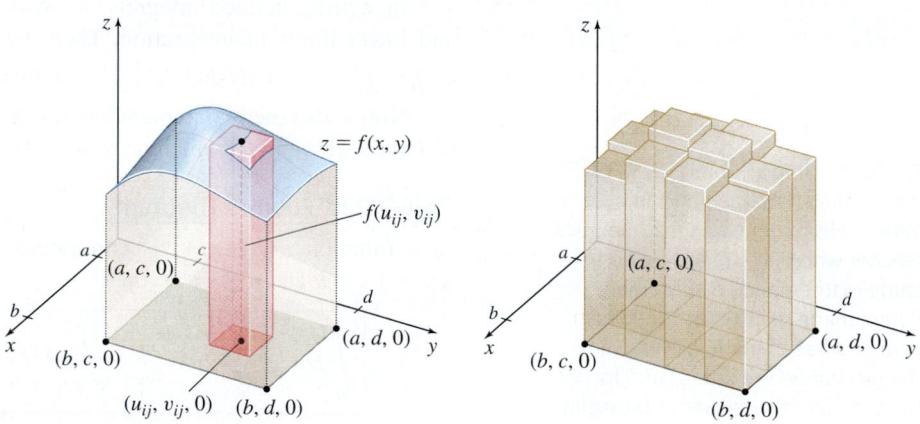

**DF Figure 5**                     **Figure 6**

**NOTE** It can be proved that this formula for volume is consistent with the formulas for volume given in Chapter 6.

### Volume Under a Surface

Let $z = f(x, y)$ be a function of two variables that is continuous and nonnegative on a closed, rectangular region $R$. The volume $V$ under the surface $z = f(x, y)$ and over the region $R$ is given by

$$V = \iint\limits_{R} f(x, y)\, dA$$

CALC
CLIP

### EXAMPLE 5   Finding the Volume of a Solid

Find the volume $V$ under the paraboloid $z = f(x, y) = 4 - x^2 - y^2$ and over the rectangular region $R$ defined by $-1 \le x \le 1$ and $0 \le y \le 1$.

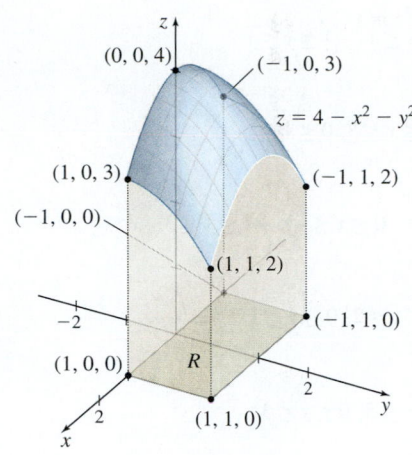

**Figure 7** $z = f(x, y) = 4 - x^2 - y^2$
$-1 \le x \le 1, 0 \le y \le 1$

**Solution** Begin by graphing $z = f(x, y)$ and the rectangular region $R$, as shown in Figure 7.

Since $z = f(x, y)$ is continuous and $z \ge 0$ on the rectangular region $-1 \le x \le 1$, $0 \le y \le 1$, the volume $V$ under the surface $z = f(x, y)$ and over $R$ is given by

$$V = \iint\limits_R f(x, y) \, dA = \iint\limits_R (4 - x^2 - y^2) \, dA$$

Using Fubini's Theorem, we have

$$V = \int_{-1}^{1} \left[ \int_0^1 (4 - x^2 - y^2) \, dy \right] dx = \int_{-1}^{1} \left[ 4y - x^2 y - \frac{y^3}{3} \right]_0^1 dx$$

$$= \int_{-1}^{1} \left( 4 - x^2 - \frac{1}{3} \right) dx = \int_{-1}^{1} \left( \frac{11}{3} - x^2 \right) dx = \left[ \frac{11}{3} x - \frac{x^3}{3} \right]_{-1}^{1}$$

$$= \left( \frac{11}{3} - \frac{1}{3} \right) - \left( -\frac{11}{3} + \frac{1}{3} \right) = \frac{20}{3} \text{ cubic units} \quad \blacksquare$$

**NOW WORK** Example 5 by finding $\int_0^1 \left[ \int_{-1}^1 (4 - x^2 - y^2) \, dx \right] dy$.

**NOW WORK** Problem 61.

## 14.1 Assess Your Understanding

### Concepts and Vocabulary

1. *True or False* $\int_0^2 \left[ \int_0^1 xy^2 dy \right] dx = \left[ \int_0^2 x \, dx \right] \cdot \left[ \int_0^1 y^2 dy \right]$.

2. *True or False* Fubini's Theorem states that if a function $z = f(x, y)$ is continuous on a closed rectangular region $R$ defined by $a \le x \le b$ and $c \le y \le d$, then $\int_c^d \left[ \int_a^b f(x, y) \, dx \right] dy = \int_a^b \left[ \int_c^d f(x, y) \, dx \right] dy$.

3. *Multiple Choice* The result of integrating $\int_1^4 x^2 \sqrt{y} \, dy$ is [(**a**) a number (**b**) a function of $y$ (**c**) a function of $x$ and $y$ (**d**) a function of $x$].

4. *True or False* If a function $z = f(x, y)$ is continuous on a closed, rectangular region $R$, then the double integral $\iint\limits_R f(x, y) \, dA$ exists.

### Skill Building

5. Let $f(x, y) = x(3 - y)$ be defined on the region $R$ shown in the figure.

    (a) Find the double Riemann sum of $f$ over $R$ by partitioning the region into nine congruent subsquares with sides $\Delta x_i = 1, i = 1, 2, 3$, and $\Delta y_j = 1, j = 1, 2, 3$. Choose the lower right corner of each subsquare as $(u_{ij}, v_{ij}), i = 1, 2, 3$ and $j = 1, 2, 3$.

    (b) Find the double Riemann sum of $f$ over the partition used in (a) but choose the upper left corner of each subsquare as $(u_{ij}, v_{ij}), i = 1, 2, 3$ and $j = 1, 2, 3$.

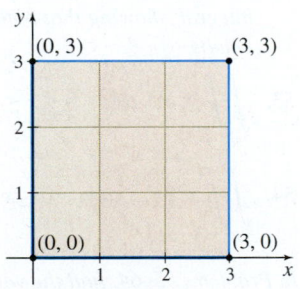

6. Let $f(x, y) = 3xy^2$ be defined on the rectangular region $R$ defined by $0 \le x \le 6, 0 \le y \le 4$.

(a) Find the double Riemann sum of $f$ over $R$ by partitioning the region into six congruent subsquares, with sides $\Delta x_i = 2$, $i = 1, 2, 3$, and $\Delta y_j = 2, j = 1, 2$. Choose the lower left corner of each subsquare as $(u_{ij}, v_{ij})$.

(b) Find the double Riemann sum of $f$ over the partition used in (a) but choose the lower right corner of each subsquare as $(u_{ij}, v_{ij})$.

7. Let $f(x, y) = x^2 + y$ be defined on the rectangular region $R$ defined by $1 \le x \le 5, 2 \le y \le 4$.

(a) Find the double Riemann sum of $f$ over $R$ by partitioning the region into four congruent subrectangles with sides $\Delta x_i = 2$, $i = 1, 2$, and $\Delta y_j = 1, j = 1, 2$. Choose the lower left corner of each subrectangle as $(u_{ij}, v_{ij})$.

(b) Find the double Riemann sum of $f$ over the partition used in (a) but choose the upper right corner of each subrectangle as $(u_{ij}, v_{ij})$.

8. Let $f(x, y) = x(1 - y)$ be defined on the rectangular region $R$ defined by $0 \le x \le 3, 0 \le y \le 4$.

(a) Find the double Riemann sum of $f$ over $R$ by partitioning the region into six congruent subrectangles with sides $\Delta x_i = 1$, $i = 1, 2, 3$, and $\Delta y_j = 2, j = 1, 2$. Choose the lower left corner of each subrectangle as $(u_{ij}, v_{ij})$.

(b) Find the double Riemann sum of $f$ over the partition used in (a) but choose the upper left corner of each subrectangle as $(u_{ij}, v_{ij})$.

9. Find the double Riemann sum for Example 1 if $R$ is divided into four congruent subsquares. Choose $(u_{ij}, v_{ij})$ as the center of the $ij$th subsquare.

10. Find the double Riemann sum for Example 1 if $R$ is divided into eight congruent subrectangles with $\Delta x_i = 1, i = 1, 2, 3, 4$, and $\Delta y_j = 2, j = 1, 2$. Choose $(u_{ij}, v_{ij})$ as the center of the $ij$th subrectangle.

---

**1.** = NOW WORK problem      📐 = Graphing technology recommended      CAS = Computer Algebra System recommended

*In Problems 11–16, find the indicated partial integral.*

**11.** $\int_1^e \dfrac{x}{y}\, dy$  **12.** $\int_0^2 \dfrac{x}{y}\, dx$  **13.** $\int_0^{\pi/2} x \sin y\, dx$

**14.** $\int_0^{\pi/2} x \sin y\, dy$  **15.** $\int_0^1 e^x\, dx$  **16.** $\int_0^1 e^x\, dy$

*In Problems 17–36, find each iterated integral.*

**17.** $\int_0^1 \left[ \int_0^2 x^2 y\, dy \right] dx$  **18.** $\int_0^2 \left[ \int_0^4 x^2 y\, dx \right] dy$

**19.** $\int_0^3 \left[ \int_0^2 3xy\, dy \right] dx$  **20.** $\int_0^2 \left[ \int_0^1 3xy\, dx \right] dy$

**21.** $\int_{-1}^1 \left[ \int_0^1 3x^2 y^2\, dx \right] dy$  **22.** $\int_0^1 \left[ \int_0^2 3x^2 y^2\, dy \right] dx$

**23.** $\int_0^2 \left[ \int_{-1}^1 2xy^2\, dx \right] dy$  **24.** $\int_{-1}^1 \left[ \int_1^2 2xy^2\, dy \right] dx$

**25.** $\int_0^{\pi/4} \left[ \int_0^2 x \cos y\, dx \right] dy$  **26.** $\int_0^2 \left[ \int_0^{\pi/3} x \cos y\, dy \right] dx$

**27.** $\int_0^3 \left[ \int_0^{\pi/3} (4x-3)^2 \sin y\, dy \right] dx$

**28.** $\int_0^{\pi/3} \left[ \int_0^3 (4x-3)^2 \sin y\, dx \right] dy$

**29.** $\int_0^1 \left[ \int_0^{\pi/2} e^x \cos y\, dy \right] dx$  **30.** $\int_0^{\pi/4} \left[ \int_0^1 e^x \cos y\, dx \right] dy$

**31.** $\int_0^2 \left[ \int_{-\pi/2}^{\pi/2} \dfrac{\sin y}{2x+1}\, dy \right] dx$  **32.** $\int_0^{\pi/2} \left[ \int_0^2 \dfrac{\sin y}{2x+1}\, dx \right] dy$

**33.** $\int_0^{\pi/2} \left[ \int_0^2 xe^x \sin y\, dx \right] dy$  **34.** $\int_0^1 \left[ \int_0^{\pi/2} xe^x \sin y\, dy \right] dx$

**35.** $\int_1^2 \left[ \int_0^{\pi/2} \dfrac{x \cos x}{y}\, dx \right] dy$  **36.** $\int_0^{\pi/3} \left[ \int_1^2 \dfrac{x \cos x}{y}\, dy \right] dx$

*In Problems 37–46, use Fubini's Theorem to find each double integral over the rectangular region R.*

**37.** $\displaystyle\iint_R (2x + y^2)\, dA, \quad 0 \le x \le 2, \, 0 \le y \le 3$

**38.** $\displaystyle\iint_R (x^2 - 3y)\, dA, \quad -1 \le x \le 2, \, 0 \le y \le 1$

**39.** $\displaystyle\iint_R x(x^2 + 5)\, dA, \quad 0 \le x \le 2, \, -1 \le y \le 1$

**40.** $\displaystyle\iint_R \sqrt{y}(3x^2 + x)\, dA, \quad 0 \le x \le 2, \, 0 \le y \le 4$

**41.** $\displaystyle\iint_R 2xe^y\, dA, \quad -3 \le x \le 2, \, 0 \le y \le 1$

**42.** $\displaystyle\iint_R e^{2x+y}\, dA, \quad 0 \le x \le 1, \, -1 \le y \le 1$

**43.** $\displaystyle\iint_R x \sec^2 y\, dA, \quad 0 \le x \le 3, \, 0 \le y \le \dfrac{\pi}{4}$

**44.** $\displaystyle\iint_R y^2 \sec x \tan x\, dA, \quad 0 \le x \le \dfrac{\pi}{3}, \, -1 \le y \le 3$

**45.** $\displaystyle\iint_R \dfrac{x}{2y+3}\, dA, \quad 0 \le x \le 2, \, 0 \le y \le 1$

**46.** $\displaystyle\iint_R \dfrac{y^2}{x-3}\, dA, \quad 4 \le x \le 5, \, 0 \le y \le 3$

## Applications and Extensions

*In Problems 47–52, use Fubini's Theorem to find each double integral over the given rectangular region.*

**47.** $\displaystyle\iint_R x \sin(xy)\, dA, \quad 0 \le x \le \dfrac{\pi}{2}, \, 0 \le y \le 1$

**48.** $\displaystyle\iint_R y \cos(xy)\, dA, \quad 0 \le x \le 1, \, \dfrac{\pi}{2} \le y \le 2\pi$

**49.** $\displaystyle\iint_R x^3 \cos(x^2 y)\, dA, \quad 0 \le x \le \dfrac{\pi}{2}, \, 0 \le y \le 1$

**50.** $\displaystyle\iint_R x^3 \sin(x^2 y)\, dA, \quad 0 \le x \le \dfrac{\pi}{2}, \, 0 \le y \le 1$

**51.** $\displaystyle\iint_R \dfrac{y^2}{(1 + xy^2)^3}\, dA, \quad 0 \le x \le 2, \, 0 \le y \le 3$

**52.** $\displaystyle\iint_R (x^3 + x)(x^2 y + y)\, dA, \quad 0 \le x \le 1, \, 0 \le y \le 1$

*In Problems 53 and 54:*

**(a)** Use Fubini's Theorem to find each double integral over the given rectangular region.

**(CAS) (b)** Change the order of integration and use a CAS to find the double integral, showing that both orders of integration yield the same results.

**53.** $\displaystyle\iint_R y^5 e^{xy^3}\, dA, \quad 0 \le x \le 1, \, 0 \le y \le 2$

**54.** $\displaystyle\iint_R x^5 y e^{x^3 y^2}\, dA, \quad 0 \le x \le 2, \, 0 \le y \le 1$

*In Problems 55–64, find the volume under the surface $z = f(x, y)$ and over the given rectangular region.*

**55.** $f(x, y) = x + 2y, \quad 0 \le x \le 1, \, 0 \le y \le 2$

**56.** $f(x, y) = 2x + 3y, \quad 0 \le x \le 2, \, 0 \le y \le 3$

**57.** $f(x, y) = x^2 + y^2, \quad 0 \le x \le 2, \, 0 \le y \le 1$

**58.** $f(x, y) = 4x^2 + 3y^2, \quad 0 \le x \le 3, -2 \le y \le 1$

**59.** $f(x, y) = \sin x, \quad 0 \le x \le \dfrac{\pi}{2}, 0 \le y \le 1$

**60.** $f(x, y) = \cos y, \quad 0 \le x \le 1, 0 \le y \le \dfrac{\pi}{2}$

**61.** $f(x, y) = \dfrac{x^2 + y^2}{xy} \quad 1 \le x \le 2, 1 \le y \le 2$

**62.** $f(x, y) = \dfrac{y^2}{x^2}, \quad 1 \le x \le 2, 1 \le y \le 2$

**63.** $f(x, y) = e^{x+y}, \quad 0 \le x \le 2, 0 \le y \le 1$

**64.** $f(x, y) = e^{x-y}, \quad 0 \le x \le 2, 1 \le y \le 2$

**65. Volume**   Find the volume of the solid below the paraboloid $z = x^2 + y^2$ and over the square in the $xy$-plane enclosed by the lines $x = \pm 1$ and $y = \pm 1$.

**66. Volume**   Find the volume of the solid under the elliptic paraboloid $z = 9 - x^2 - 3y^2$ and over the rectangle in the $xy$-plane enclosed by the coordinate axes and the lines $x = 2$ and $y = 1$.

*In Problems 67–70, use the following discussion: Suppose a function P is defined as*

$$P(x, y) = \begin{cases} f(x, y) & \text{for } x \text{ and } y \text{ in a closed rectangular region } R \\ 0 & \text{elsewhere} \end{cases}$$

*Then P is a **joint probability density function** of the continuous random variables X and Y if $f(x, y) \ge 0$, for all x, y in R and $\iint_R f(x, y)dA = 1$.*

*The probability $Pr((X, Y) \in D) = \iint_D f(x, y)\, dA$, where D is a subset of R.*

**67. Probability**   The joint probability density function $P$ for the random variables $X$ and $Y$ is given by

$$P(x, y) = \begin{cases} \dfrac{2}{3}(2x + y) & \text{on } R = \{(x, y)|0 \le x \le 1, 0 \le y \le 1\} \\ 0 & \text{elsewhere} \end{cases}$$

(a) Graph the region $R$.

(b) Verify that $P$ is a joint probability density function by showing that $P(x, y) \ge 0$ on $R$, and $\iint_R P(x, y)\, dA = 1$.

(c) Find the probability $Pr((X, Y) \in D)$ where $D$ is the region $\dfrac{1}{4} \le X \le 1, 0 \le Y \le \dfrac{1}{2}$.

**68. Probability**   The joint probability density function $P$ for the random variables $X$ and $Y$ is given by

$$P(x, y) = \begin{cases} xy & \text{on } R = \{(x, y)|0 \le x \le 2, 0 \le y \le 1\} \\ 0 & \text{elsewhere} \end{cases}$$

(a) Graph the region $R$.

(b) Verify that $P$ is a joint probability density function by showing that $P(x, y) \ge 0$ on $R$ and $\iint_R P(x, y)\, dA = 1$.

(c) Find the probability that $Pr((X, Y) \in D)$ where $D$ is the region $0 \le X \le 1, \dfrac{1}{2} \le Y \le 1$.

**69. Probability (a)** Find the number $c$ that makes the function $P$ a joint probability density function for the random variables $X$ and $Y$.

$$P(x, y) = \begin{cases} cx^2 y & 0 \le x \le 1, 0 \le y \le 1 \\ 0 & \text{elsewhere} \end{cases}$$

(b) Use the probability density function from (a) to find

$$Pr\left(0 \le X \le \dfrac{1}{2}, \dfrac{1}{4} \le Y \le 1\right)$$

**70. Probability (a)** Find the number $c$ that makes the function $P$ a joint probability density function for the random variables $X$ and $Y$.

$$P(x, y) = \begin{cases} cx(1 + y^2) & 0 \le x \le 1, 0 \le y \le 3 \\ 0 & \text{elsewhere} \end{cases}$$

(b) Use the probability density function from (a) to find $Pr(0 \le X \le 1, 1 \le Y \le 2)$.

**71.** Show that $\iint_R dA = \int_c^d \left[\int_a^b dx\right] dy = (b - a)(d - c)$. That is, show that the volume of a solid with height 1, $\iint_R dA$, defined over a rectangular region $a \le x \le b, c \le y \le d$ is numerically equal to the area of the rectangle.

## Challenge Problems

**72. Average Value of a Function**   Suppose that $z = f(x, y)$ is integrable over a closed, rectangular region $R$ in the $xy$-plane. Let $P$ be a partition of $R$ into $nm$ subrectangles of equal area $\Delta A$. Evaluate $f$ at the center $(u_{ij}, v_{ij})$ of the $ij$th subrectangle $(i = 1, 2, \ldots, n$ and $j = 1, 2, \ldots, m)$, and let $AVG$ be the average of these $nm$ values.

(a) Show that $AVG = \dfrac{1}{A} \sum_{j=1}^m \sum_{i=1}^n f(u_{ij}, v_{ij})\Delta x_i \Delta y_j$, where $A$ is the area of $R$.

(b) Explain why

$$\lim_{\|\Delta\| \to 0} AVG = \dfrac{1}{A} \iint_R f(x, y)\, dA$$

(This is called the **average value of $f$ over $R$**.)

*In Problems 73 and 74, find the average value of each function over the given rectangular region.*

**73.** $f(x, y) = \dfrac{xy}{x^2 + 1}, \quad 0 \le x \le 1, 0 \le y \le 2$

**74.** $f(x, y) = y \cos x, \quad 0 \le x \le \pi, 1 \le y \le 5$

# 14.2 The Double Integral over Nonrectangular Regions

**OBJECTIVES** *When you finish this section, you should be able to:*

1 Use Fubini's Theorem for an *x*-simple region (p. 962)
2 Use Fubini's Theorem for a *y*-simple region (p. 963)
3 Use properties of double integrals (p. 966)
4 Use double integrals to find area (p. 967)

In Section 14.1, we defined the double integral of a function $f$ over a closed rectangular region. Although single integrals are always integrated over an interval, for double integrals the region of integration is not always rectangular. In this section, we extend the definition of a double integral to functions defined over closed, bounded regions.

**RECALL** A region $R$ in the plane is closed if it contains all its boundary points. A region $R$ is bounded if it can be enclosed by some circle, or equivalently, some rectangle in the plane.

We begin with a function $z = f(x, y)$ defined on a closed, bounded region $R$. See Figure 8(a). Then $R$ can be enclosed by a rectangle $D$ defined by $a \le x \le b$ and $c \le y \le d$, as shown in Figure 8(b). Now partition the interval $[a, b]$ into $n$ subintervals of length $\Delta x_i$, $i = 1, 2, \ldots, n$, and the interval $[c, d]$ into $m$ subintervals of length $\Delta y_j$, $j = 1, 2, \ldots, m$, as shown in Figure 8(c). This partitions the rectangle $D$ into $mn$ subrectangles $D_{ij}$. The norm $\|\Delta\|$ of this partition is defined as the largest of the lengths of the subintervals $\Delta x_i$ or $\Delta y_j$.

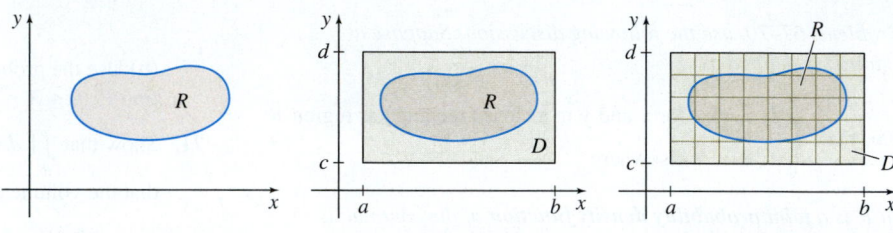

(a) $R$ is the domain of the function $z = f(x, y)$.

(b) $D$ is a rectangle that encloses $R$.

(c) Partition $D$ into subrectangles.

**Figure 8**

Now we introduce a function $P = P(x, y)$ whose domain is the rectangular region $D$. We define $P$ as follows:

$$P(x, y) = \begin{cases} f(x, y) & \text{if } (x, y) \text{ is in } R \\ 0 & \text{if } (x, y) \text{ is in } D, \text{ but not in } R \end{cases}$$

In other words, $P(x, y) = f(x, y)$ for points in $R$, and $P(x, y) = 0$ for points in $D$, but not in $R$.

Choose a point $(u_{ij}, v_{ij})$ in each subrectangle $D_{ij}$ and evaluate $P(u_{ij}, v_{ij})$. If $\Delta x_i \Delta y_j$ is the area $A_{ij}$ of the subrectangle $D_{ij}$, form the double Riemann sums $\sum_{j=1}^{m} \sum_{i=1}^{n} P(u_{ij}, v_{ij}) \Delta x_i \Delta y_j = \sum_{j=1}^{m} \sum_{i=1}^{n} P(u_{ij}, v_{ij}) \Delta A_{ij}$. Then the double integral of the function $z = P(x, y)$ over the rectangle $D$ is

$$\iint\limits_{D} P(x, y)\,dA = \lim_{\|\Delta\| \to 0} \sum_{j=1}^{m} \sum_{i=1}^{n} P(u_{ij}, v_{ij}) \Delta A_{ij} \qquad (1)$$

provided the limit exists and does not depend on the partition or on the choice of $(u_{ij}, v_{ij})$.

If $\iint\limits_{D} P(x, y)\,dA$ exists, we define the **double integral of $f$ over $R$** as

$$\iint\limits_{R} f(x, y)\,dA = \iint\limits_{D} P(x, y)\,dA$$

Notice that this definition makes sense because for points $(u_{ij}, v_{ij})$ in $D$, but not in $R$, $P(u_{ij}, v_{ij}) = 0$. So such points do not contribute to $\sum_{j=1}^{m} \sum_{i=1}^{n} P(u_{ij}, v_{ij}) \Delta A_{ij}$. Figure 9(a) and (b) illustrate this for $f(x, y) \ge 0$ on $R$.

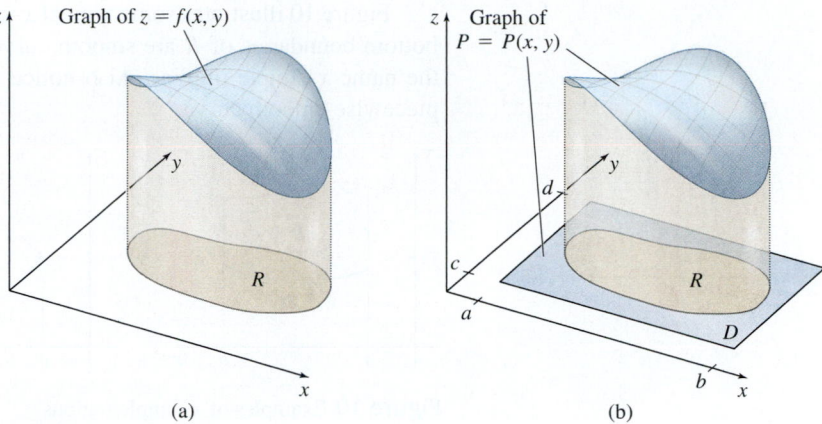

**Figure 9**

The limit of the double sums in (1) exists under certain conditions on the function $z = f(x, y)$ and the boundary of the region $R$. Before giving a theorem which guarantees the existence of a double integral, we need the following definition.

**NEED TO REVIEW?** Smooth curves are discussed in Section 11.2, p. 809.

Recall that a curve $C$ is smooth if the parametric equations $x = x(t)$, $y = y(t)$ defining $C$ have derivatives $\dfrac{dx}{dt}$ and $\dfrac{dy}{dt}$ that are continuous and are never simultaneously zero. If a curve $C$ is not smooth, but consists of a finite number of smooth curves that are joined end to end, then $C$ is a **piecewise-smooth curve**.

The next theorem, proved in most advanced calculus texts, is very similar to the theorem stated in Section 14.1 guaranteeing the existence of a double integral over a closed, rectangular region.

**THEOREM  Existence of a Double Integral**

If a function $z = f(x, y)$ is continuous on a closed, bounded region $R$ whose boundary is a piecewise-smooth curve, then $\iint\limits_R f(x, y)\,dA$ exists.

When $\iint\limits_R f(x, y)\,dA$ exists, we say $f$ is **integrable** on $R$.

As before, other representations for $\iint\limits_R f(x, y)\,dA$ are

$$\iint\limits_R f(x, y)\,dx\,dy \qquad \text{and} \qquad \iint\limits_R f(x, y)\,dy\,dx$$

If $z = f(x, y) \geq 0$ on a closed, bounded region $R$ whose boundary is a piecewise-smooth curve, then $\iint\limits_R f(x, y)\,dA$ equals the volume of the solid under the surface $z = f(x, y)$ and over the region $R$. Refer back to Figure 9.

**Volume Under a Surface and over a Region $R$**

Let $z = f(x, y)$ be a function of two variables that is continuous and nonnegative on a closed, bounded region $R$ whose boundary is a piecewise-smooth curve. The volume $V$ under the surface $z = f(x, y)$ and over the region $R$ is given by the double integral

$$\boxed{V = \iint\limits_R f(x, y)\,dA}$$

Certain types of regions have boundaries that make it possible to express a double integral as an iterated integral. One such type of region is called $x$-*simple*.

**DEFINITION  $x$-Simple Region**

Let $R$ be a closed, bounded region. $R$ is an **$x$-simple region** if the boundary of $R$ consists of two smooth curves, $y = g_1(x)$ and $y = g_2(x)$, each defined on a closed interval $[a, b]$, where $g_1(x) \leq g_2(x)$ on $[a, b]$. The boundary may also consist of a portion of the vertical lines $x = a$ and $x = b$.

Figure 10 illustrates some typical $x$-simple regions. Notice in each case the top and bottom boundaries of $R$ are smooth curves that can be expressed as functions of $x$, as the name $x$-simple implies. Also notice that the boundary of an $x$-simple region is a piecewise-smooth curve.

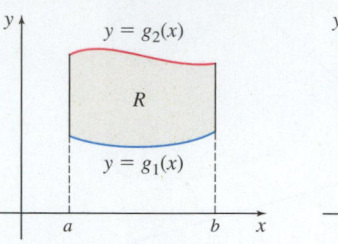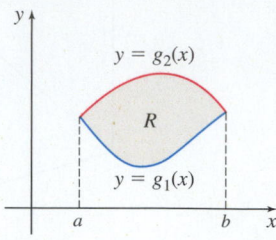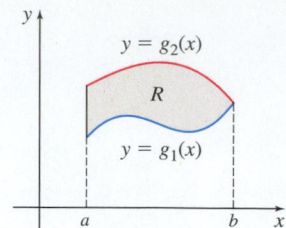

**Figure 10** Examples of $x$-simple regions

Fubini's Theorem can be extended to regions that are $x$-simple.

## 1 Use Fubini's Theorem for an $x$-Simple Region

**THEOREM** Fubini's Theorem for an $x$-Simple Region

If a function $z = f(x, y)$ is continuous on a closed, bounded region $R$ that is $x$-simple, then the double integral of $f$ over $R$ is given by

$$\iint\limits_{R} f(x, y)\, dA = \int_a^b \left[ \int_{g_1(x)}^{g_2(x)} f(x, y)\, dy \right] dx$$

where $y = g_1(x)$ and $y = g_2(x)$, $g_1(x) \le g_2(x)$, $a \le x \le b$, are two smooth curves that form the boundary of $R$.

The proof of Fubini's Theorem can be found in advanced calculus books. Here we give a geometric argument for Fubini's Theorem for $x$-simple regions when $f(x, y) \ge 0$ is continuous on $R$.

Suppose we choose a number $x_i$ in the closed interval $[a, b]$ and let $A(x_i)$ equal the area of the intersection of the plane $x = x_i$ and the solid under the surface $z = f(x, y)$ and over $R$. See Figure 11. Using slicing, the volume $V$ of the solid under $z = f(x, y)$ and over $R$ from $x = a$ to $x = b$ is

$$V = \int_a^b A(x)\, dx$$

Since the volume $V$ is also given by $V = \iint\limits_{R} f(x, y)\, dA$, we have

$$V = \iint\limits_{R} f(x, y)\, dA = \int_a^b A(x)\, dx \qquad (2)$$

But $A(x_i)$ is the area of the plane region under the surface $z = f(x_i, y)$ from $y = g_1(x_i)$ to $y = g_2(x_i)$. That is,

$$A(x) = \int_{g_1(x)}^{g_2(x)} f(x, y)\, dy$$

where $x$ is held constant. By substituting $\int_{g_1(x)}^{g_2(x)} f(x, y)\, dy$ for $A(x)$ in the volume formula (2), we find

$$V = \iint\limits_{R} f(x, y)\, dA = \int_a^b \underbrace{\left[ \int_{g_1(x)}^{g_2(x)} f(x, y)\, dy \right]}_{A(x)} dx$$

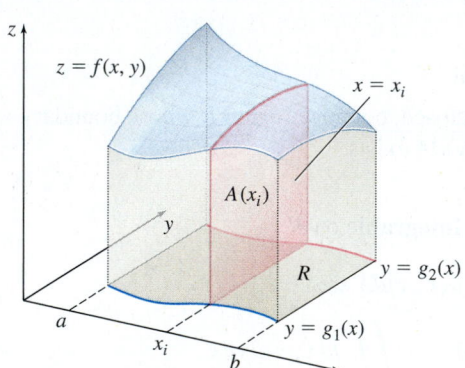

**DF Figure 11** $V = \displaystyle\int_a^b A(x)\, dx$

$$A(x) = \int_{g_1(x)}^{g_2(x)} f(x, y)\, dy$$

**NEED TO REVIEW?** Finding the volume of a solid using slicing is discussed in Section 6.4, pp. 456–460.

**EXAMPLE 1** Using Fubini's Theorem for an $x$-Simple Region

**(a)** Find $\iint\limits_{R} xy\, dA$ if $R$ is the region enclosed by $y = x^2$ and $y = \sqrt{x}$.

**(b)** Verify that $\iint\limits_{R} xy\, dA$ represents the volume of a solid. Describe the solid.

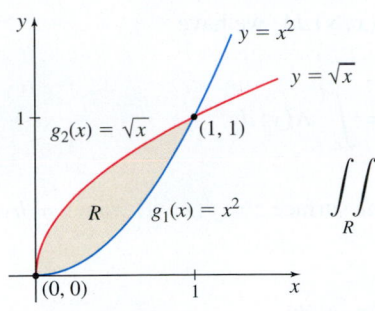

**Figure 12** The region $R$ is $x$-simple.

**Solution (a)** We begin by graphing the region $R$. See Figure 12. Observe that $R$ is a closed, bounded region. Also notice that the bottom and top boundaries of $R$ are smooth curves expressed as functions of $x$: $g_1(x) = x^2$, $g_2(x) = \sqrt{x}$, $0 \le x \le 1$. That is, $R$ is $x$-simple. Since $f(x, y) = xy$ is continuous on $R$, we can use Fubini's Theorem.

$$\iint\limits_R xy \, dA = \int_0^1 \left[ \int_{x^2}^{\sqrt{x}} xy \, dy \right] dx = \int_0^1 x \left[ \int_{x^2}^{\sqrt{x}} y \, dy \right] dx = \int_0^1 x \left[ \frac{y^2}{2} \right]_{x^2}^{\sqrt{x}} dx \quad \text{Integrate partially with respect to } y.$$

$$= \int_0^1 x \left( \frac{x - x^4}{2} \right) dx = \frac{1}{2} \int_0^1 (x^2 - x^5) \, dx$$

$$= \frac{1}{2} \left[ \frac{x^3}{3} - \frac{x^6}{6} \right]_0^1 = \frac{1}{12}$$

**(b)** Since $f(x, y) = xy$ is continuous and nonnegative on the region $R$, the double integral $\iint\limits_R xy \, dA$ can be interpreted as the volume of the solid under the surface $f(x, y) = xy$ and over the region $R$. The volume equals $\dfrac{1}{12}$ cubic unit. ∎

**NOW WORK** Problems **19** and **25**.

## 2 Use Fubini's Theorem for a $y$-Simple Region

Regions whose left and right boundaries consist of smooth curves expressed as functions of $y$ are referred to as *y-simple regions*.

### DEFINITION $y$-Simple Region

Let $R$ be a closed, bounded region. Then $R$ is a **$y$-simple region** if the boundary of $R$ consists of two smooth curves, $x = h_1(y)$ and $x = h_2(y)$, each defined on the closed interval $c \le y \le d$, where $h_1(y) \le h_2(y)$ on $[c, d]$. The boundary may also consist of a portion of the horizontal lines $y = c$ and $y = d$.

Figure 13 illustrates some typical $y$-simple regions.

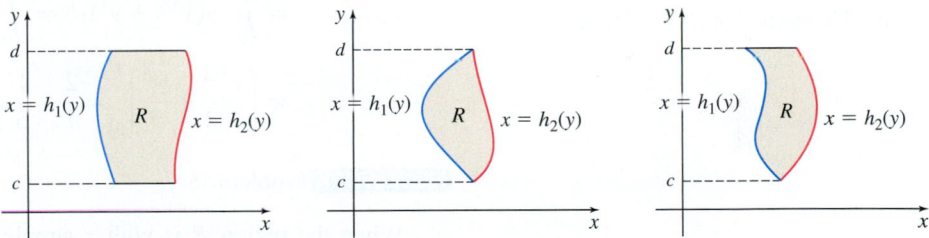

**Figure 13** Examples of $y$-simple regions

Fubini's Theorem also applies to closed, bounded regions that are $y$-simple.

### THEOREM Fubini's Theorem for a $y$-Simple Region

If a function $z = f(x, y)$ is continuous on a closed, bounded region $R$ that is $y$-simple, the double integral of $f$ over $R$ is given by

$$\boxed{\iint\limits_R f(x, y) \, dA = \int_c^d \left[ \int_{h_1(y)}^{h_2(y)} f(x, y) \, dx \right] dy}$$

where $x = h_1(y)$ and $x = h_2(y)$, $h_1(y) \le h_2(y)$, $c \le y \le d$, are two smooth curves that form the boundary of $R$.

Figure 14 provides justification for Fubini's Theorem for a $y$-simple region $R$ when $z = f(x, y) \ge 0$ is continuous on $R$. The volume $V$ of the solid under $z = f(x, y)$ from $y = c$ to $y = d$ is given by

$$V = \int_c^d A(y) \, dy$$

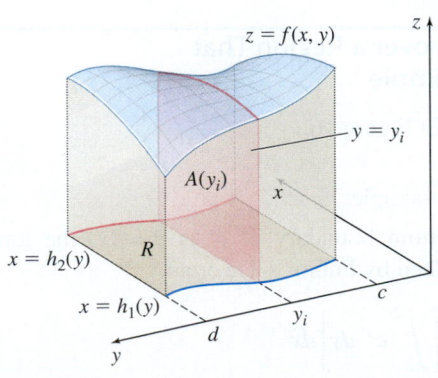

**Figure 14** $V = \displaystyle\int_c^d A(y) \, dy$,

$$A(y) = \int_{h_1(y)}^{h_2(y)} f(x, y) \, dx$$

Since this volume $V$ is also given by $V = \iint\limits_{R} f(x, y)\, dA$, we have

$$V = \iint\limits_{R} f(x, y)\, dA = \int_{c}^{d} A(y)\, dy \qquad (3)$$

But $A(y_i)$ is the area of the plane region under the surface $z = f(x, y_i)$ from $x = h_1(y_i)$ to $x = h_2(y_i)$. That is,

$$A(y) = \int_{h_1(y)}^{h_2(y)} f(x, y)\, dx$$

where $y$ is held constant. By substituting $\int_{h_1(y)}^{h_2(y)} f(x, y)\, dx$ for $A(y)$ in the volume formula (3), we find

$$V = \iint\limits_{R} f(x, y)\, dA = \int_{c}^{d} \underbrace{\left[ \int_{h_1(y)}^{h_2(y)} f(x, y)\, dx \right]}_{A(y)} dy$$

CALC
CLIP

## EXAMPLE 2  Using Fubini's Theorem for a $y$-Simple Region

Find $\iint\limits_{R} 3x^2 y\, dA$ if $R$ is the region bounded by the smooth curves $x = \sqrt{y}$ and $x = -y$, and the line $y = 1$.

**Solution** Always begin by graphing the region $R$. See Figure 15. Observe that $R$ is a closed, bounded region that is $y$-simple, where $h_1(y) = -y$ and $h_2(y) = \sqrt{y}$, $0 \le y \le 1$. Using Fubini's Theorem for a $y$-simple region, we have

$$\iint\limits_{R} 3x^2 y\, dA = \int_{0}^{1} \left[ \int_{-y}^{\sqrt{y}} 3x^2 y\, dx \right] dy = \int_{0}^{1} y \left[ \int_{-y}^{\sqrt{y}} 3x^2\, dx \right] dy = \int_{0}^{1} y \left[ x^3 \right]_{-y}^{\sqrt{y}} dy$$

$$= \int_{0}^{1} y(y^{3/2} + y^3)\, dy = \int_{0}^{1} (y^{5/2} + y^4)\, dy$$

$$= \left[ \frac{y^{7/2}}{7/2} + \frac{y^5}{5} \right]_{0}^{1} = \frac{2}{7} + \frac{1}{5} = \frac{17}{35} \qquad \blacksquare$$

**NOW WORK** Problem 15.

When the region $R$ is both $x$-simple and $y$-simple, we can *choose* the order of integration. In some cases, integration in one order uses simpler techniques than the techniques needed to integrate in the opposite order.

## EXAMPLE 3  Finding a Double Integral over a Region That Is Both $x$-Simple and $y$-Simple

Find $\iint\limits_{R} e^{y^2}\, dA$, where $R$ is the region shown in Figure 16.

**Solution** The region $R$ is both $x$-simple and $y$-simple.

If we treat $R$ as $x$-simple, then the bottom boundary of $R$ is $y = 2x$, the top boundary is $y = 2$, and $x$ varies from 0 to 1. Then by Fubini's Theorem,

$$\iint\limits_{R} e^{y^2}\, dA = \int_{0}^{1} \left[ \int_{2x}^{2} e^{y^2}\, dy \right] dx$$

Since the integral $\int e^{y^2}\, dx$ cannot be expressed in terms of elementary functions, we cannot find $\int_{2x}^{2} e^{y^2}\, dy$ using the Fundamental Theorem of Calculus.

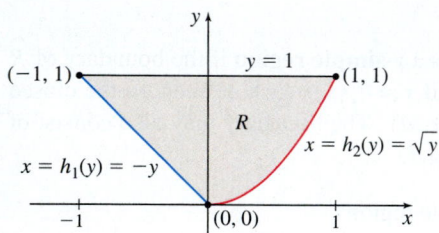

**Figure 15** The region $R$ is $y$-simple.

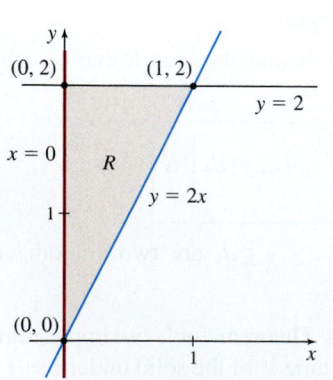

**Figure 16** The region $R$ is both $x$-simple and $y$-simple.

**NEED TO REVIEW?** Elementary functions are discussed in Section 7.6, pp. 541–542.

So, we treat the region $R$ as $y$-simple. Then the left boundary of $R$ is $x = 0$, the right boundary is $x = \dfrac{y}{2}$, and $y$ varies from 0 to 2. Then by Fubini's Theorem,

$$\iint\limits_{R} e^{y^2}\, dA = \int_0^2 \left[\int_0^{y/2} e^{y^2}\, dx\right] dy = \int_0^2 e^{y^2}\left[\int_0^{y/2} dx\right] dy = \int_0^2 e^{y^2}\left[x\right]_0^{y/2} dy = \int_0^2 \frac{y}{2}e^{y^2}\, dy$$

$$= \frac{1}{2}\int_0^2 ye^{y^2}\, dy = \frac{1}{2}\left[\frac{e^{y^2}}{2}\right]_0^2 = \frac{1}{4}(e^4 - 1)$$    ∎

**NOW WORK** Problem 39.

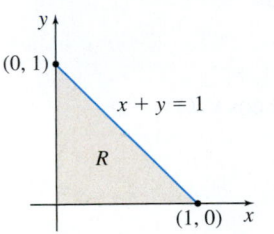

(a) The solid enclosed by $x + y + z = 1$

**EXAMPLE 4  Using a Double Integral to Find the Volume of a Solid**

Find the volume $V$ of the solid in the first octant enclosed by the plane $x + y + z = 1$.

**Solution** Figure 17(a) shows the solid in the first octant enclosed by the plane $x + y + z = 1$. The volume $V$ of the solid lies under the plane $z = f(x, y) = 1 - x - y$ and over the region $R$ in the $xy$-plane bounded by the lines $x = 0$, $y = 0$, and $x + y = 1$. See Figure 17(b). The region $R$ is $x$-simple and $y$-simple. If we treat $R$ as $x$-simple, then $y$ varies from $y = 0$ to $y = 1 - x$, $0 \le x \le 1$. The volume $V$ of the solid is

$$V = \iint\limits_{R} f(x, y)\, dA = \int_0^1\left[\int_0^{1-x}(1 - x - y)\, dy\right] dx = \int_0^1\left[(1-x)y - \frac{y^2}{2}\right]_0^{1-x} dx$$

$$= \int_0^1\left[(1-x)^2 - \frac{(1-x)^2}{2}\right] dx$$

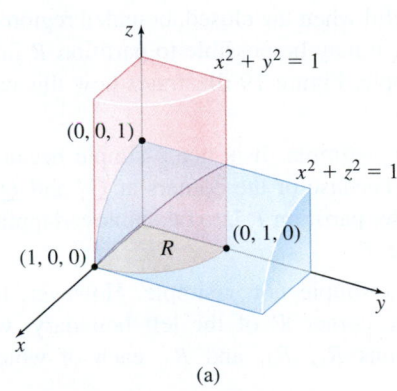

(b) The region $R$ in the $xy$-plane

**Figure 17**

$$= \frac{1}{2}\int_0^1 (1-x)^2\, dx = \frac{1}{2}\left[x - x^2 + \frac{x^3}{3}\right]_0^1 = \frac{1}{6} \text{ cubic unit}$$    ∎

**NOW WORK** Problem 55.

**EXAMPLE 5  Using a Double Integral to Find the Volume of a Solid**

Find the volume $V$ common to the two cylinders $x^2 + y^2 = 1$ and $x^2 + z^2 = 1$.

**Solution** Each cylinder has radius 1; their axes are perpendicular to each other and lie on the $z$-axis and $y$-axis, respectively. Figure 18(a) shows the portion of the solid lying in the first octant. If we find the volume of this portion of the solid, then since the volume in each octant is the same, the volume $V$ is eight times the volume in the first octant. That is, if $z = f(x, y) = \sqrt{1 - x^2}$, then

$$V = 8\iint\limits_{R} f(x, y)\, dA = 8\iint\limits_{R}(1 - x^2)^{1/2}\, dA$$

where $R$ is the region in the first quadrant inside the circle $x^2 + y^2 = 1$. See Figure 18(b).

Consider the integrand. It is simpler to integrate with respect to $y$ first. (Do you see why?) So, we treat $R$ as an $x$-simple region. (If this approach fails, we would try treating $R$ as a $y$-simple region.) If $R$ is $x$-simple, then $y$ varies from 0 to $(1 - x^2)^{1/2}$, $0 \le x \le 1$. The volume $V$ is

$$V = 8\iint\limits_{R}(1 - x^2)^{1/2}\, dy\, dx = 8\int_0^1\left[\int_0^{\sqrt{1-x^2}}(1 - x^2)^{1/2}\, dy\right] dx$$

$$= 8\int_0^1 (1 - x^2)^{1/2}\left[y\right]_0^{\sqrt{1-x^2}} dx$$

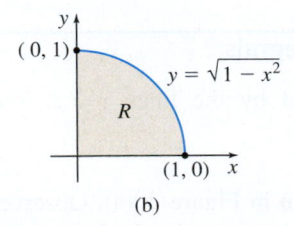

(b)

**Figure 18**

$$= 8\int_0^1 (1 - x^2)\, dx = 8\left[x - \frac{x^3}{3}\right]_0^1 = \frac{16}{3} \text{ cubic units}$$    ∎

**NOW WORK** Problems 53 and 59.

NEED TO REVIEW? Properties of single integrals are discussed in Section 5.4, pp. 387–390.

# 3 Use Properties of Double Integrals

Double integrals have properties very similar to properties of single integrals.

> **THEOREM  Properties of Double Integrals**
>
> Suppose $f$ and $g$ are functions of two variables that are continuous on a closed, bounded region $R$, whose boundary is a piecewise-smooth curve, so that both $\iint_R f(x, y)\, dA$ and $\iint_R g(x, y)\, dA$ exist. Then
>
> - $$\iint_R [f(x, y) \pm g(x, y)]\, dA = \iint_R f(x, y)\, dA \pm \iint_R g(x, y)\, dA.$$
>
> - $$\iint_R cf(x, y)\, dA = c \iint_R f(x, y)\, dA, \text{ where } c \text{ is a constant.}$$
>
> - If $R$ consists of two subregions, $R_1$ and $R_2$, that have no points in common except for points lying on portions of their common boundary, then
>
> $$\iint_R f(x, y)\, dA = \iint_{R_1} f(x, y)\, dA + \iint_{R_2} f(x, y)\, dA$$

---

**EXAMPLE 6  Using Properties of Double Integrals**

If $R$ is a closed, bounded region, then

**(a)** $$\iint_R (x^2 y + \sin x \cos y)\, dA = \iint_R x^2 y\, dA + \iint_R \sin x \cos y\, dA$$

**(b)** $$\iint_R 8(x^2 + y^2)\, dA = 8 \iint_R (x^2 + y^2)\, dA$$ ∎

**NOW WORK** Problem 47.

The third property of double integrals is useful when the closed, bounded region $R$ is neither $x$-simple nor $y$-simple. In such cases, it may be possible to partition $R$ into subregions, each of which is $x$-simple or $y$-simple. Figure 19 illustrates how this can be done.

In Figure 19(a), $R$ is neither $x$-simple nor $y$-simple. It is not $x$-simple because of the corner at the point $P$. It is not $y$-simple because of the corners at $Q_1$ and $Q_2$. However, by drawing a vertical line through $P$, we partition $R$ into two nonoverlapping subregions $R_1$ and $R_2$, each of which is $x$-simple.

In Figure 19(b), the region $R$ is neither $x$-simple nor $y$-simple. However, by drawing a vertical line that passes through the corner $P$ of the left boundary, we partition $R$ into three nonoverlapping subregions $R_1$, $R_2$, and $R_3$, each of which is $x$-simple.

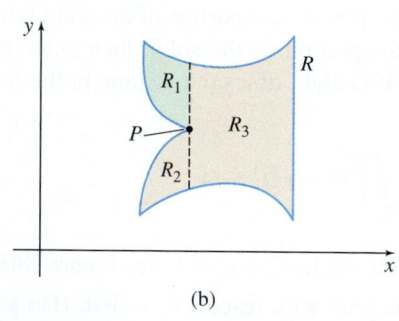

(a)

(b)

**Figure 19**

---

**EXAMPLE 7  Using Properties of Double Integrals**

Find $\iint_R x^2 y\, dA$, where $R$ is the region enclosed by the lines $y = x$, $y = x + 2$, $y = -x + 2$, and $y = -x + 4$.

**Solution** Begin by graphing the region $R$ as shown in Figure 20(a). Observe that $R$ is a closed, bounded region, but it is neither $x$-simple nor $y$-simple. It is not $x$-simple because of the corners at $(1, 3)$ and $(1, 1)$. It is not $y$-simple because of the corners at $(0, 2)$ and $(2, 2)$. But we can partition $R$ into two subregions $R_1$ and $R_2$, both of which are $x$-simple, by drawing a vertical line from $(1, 3)$ to $(1, 1)$. See Figure 20(b).

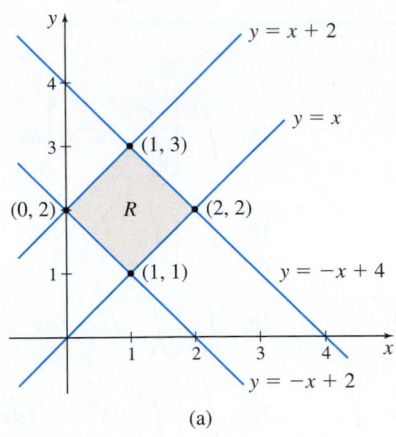

(a)

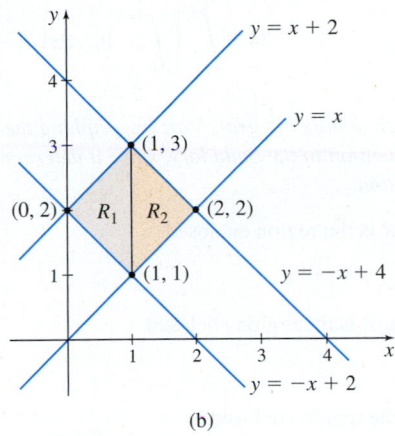

(b)

**Figure 20**

The subregion $R_1$ is $x$-simple with $g_1(x) = -x + 2$; $g_2(x) = x + 2$, $0 \le x \le 1$. The subregion $R_2$ is $x$-simple with $h_1(x) = x$; $h_2(x) = -x + 4$, $1 \le x \le 2$. Then

$$\iint\limits_R x^2 y \, dA = \iint\limits_{R_1} x^2 y \, dA + \iint\limits_{R_2} x^2 y \, dA$$

$$= \int_0^1 \left[ \int_{-x+2}^{x+2} x^2 y \, dy \right] dx + \int_1^2 \left[ \int_x^{-x+4} x^2 y \, dy \right] dx$$

$$= \int_0^1 x^2 \left[ \frac{y^2}{2} \right]_{-x+2}^{x+2} dx + \int_1^2 x^2 \left[ \frac{y^2}{2} \right]_x^{-x+4} dx$$

$$= \frac{1}{2} \int_0^1 x^2 \left[ (x+2)^2 - (-x+2)^2 \right] dx + \frac{1}{2} \int_1^2 x^2 \left[ (-x+4)^2 - x^2 \right] dx$$

$$= \frac{1}{2} \int_0^1 8x^3 \, dx + \frac{1}{2} \int_1^2 (-8x^3 + 16x^2) \, dx$$

$$= 4 \left[ \frac{x^4}{4} \right]_0^1 - 4 \left[ \frac{x^4}{4} \right]_1^2 + 8 \left[ \frac{x^3}{3} \right]_1^2 = 1 - 15 + \frac{56}{3} = \frac{14}{3} \qquad \blacksquare$$

In Example 7, we could have also partitioned $R$ into two subregions, both of which are $y$-simple.

**NOW WORK** Example 7 using $y$-simple subregions.

**NOW WORK** Problem **29**.

## 4 Use Double Integrals to Find Area

Double integrals can be used to find the area of a closed, bounded region $R$. To find area using a double integral, let $f(x, y) = 1$ and use the theorem for finding the volume under a surface. The volume under the surface $f(x, y) = 1$ and over the region $R$ is numerically equal to the area of the region $R$. That is,

$$\boxed{\text{Area of the region } R = \iint\limits_R dA}$$

**EXAMPLE 8** **Using a Double Integral to Find the Area of a Region**

Use a double integral to find the area $A$ of the region $R$ in the first quadrant enclosed by the parabola $y = 6x - x^2$ and the line $y = 4x - 8$.

**Solution** The graph of the region $R$ is shaded tan in Figure 21. Notice that $R$ is neither $x$-simple nor $y$-simple. It is not $x$-simple because of the corner at $(2, 0)$; it is not $y$-simple because of the corner at $(4, 8)$. If we partition $R$ by drawing a vertical line through the point $(2, 0)$, we obtain two $x$-simple subregions $R_1$ and $R_2$. Then the area $A$ of the region $R$ is

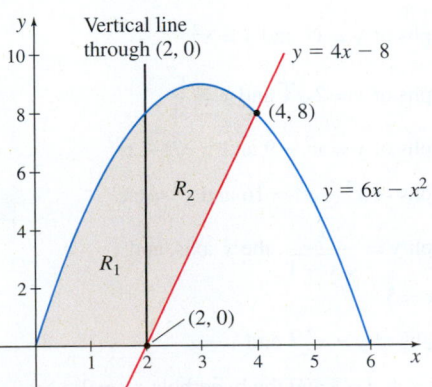

**Figure 21** Regions $R_1$ and $R_2$ are each $x$-simple.

$$A = \iint\limits_R dA = \iint\limits_{R_1} dA + \iint\limits_{R_2} dA = \int_0^2 \left[ \int_0^{6x - x^2} dy \right] dx + \int_2^4 \left[ \int_{4x-8}^{6x - x^2} dy \right] dx$$

$$= \int_0^2 (6x - x^2) \, dx + \int_2^4 (-x^2 + 2x + 8) \, dx$$

$$= \left[ 3x^2 - \frac{x^3}{3} \right]_0^2 + \left[ -\frac{x^3}{3} + x^2 + 8x \right]_2^4 = \frac{56}{3} \text{ square units} \qquad \blacksquare$$

**NOW WORK** Problems **31** and **51**.

## 14.2 Assess Your Understanding

### Concepts and Vocabulary

1. A closed, bounded region $R$ that has a boundary consisting of two smooth curves, $y = g_1(x)$ and $y = g_2(x)$, defined on $a \le x \le b$, where $g_1(x) \le g_2(x)$ on $[a, b]$ and possibly a portion of the vertical lines $x = a$ and $x = b$, is called a(n) _____ region.

2. *True or False* The double integral of a function that is continuous on a closed, bounded, $y$-simple region $R$ can be expressed as an iterated integral.

3. *True or False* The region $R$ enclosed by the parabola $y = x^2 + 1$ and the line $y = 3x + 1$ is both $x$-simple and $y$-simple.

4. Explain why $\int_0^1 \left[ \int_0^x f(x, y)\, dy \right] dx \ne \int_0^1 \left[ \int_0^y f(x, y)\, dx \right] dy$.

### Skill Building

In Problems 5–8, express $\iint\limits_R xy^2\, dA$ as an iterated integral
in two ways:
(a) one with the integration with respect to $x$ first;
(b) the other with the integration with respect to $y$ first;
(c) find each double integral over the region $R$.

**5.**

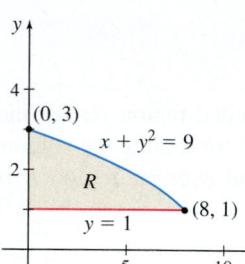

**6.**

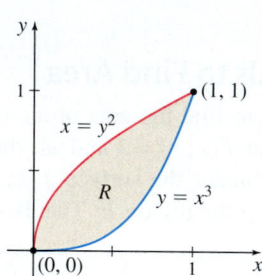

**7.**

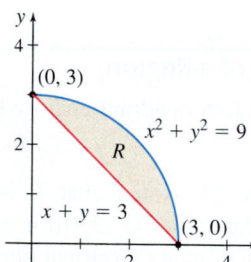

**8.**

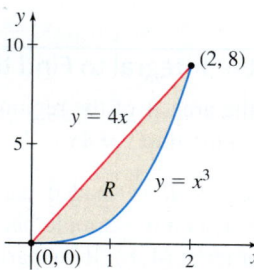

In Problems 9–24, find each iterated integral. Identify and graph the region $R$ associated with each integral.

**9.** $\displaystyle \int_0^1 \left[ \int_{x^2}^{\sqrt{x}} dy \right] dx$

**10.** $\displaystyle \int_0^1 \left[ \int_{y^2}^{\sqrt{y}} x\, dx \right] dy$

**11.** $\displaystyle \int_{-1}^2 \left[ \int_{y^2}^{y+2} dx \right] dy$

**12.** $\displaystyle \int_0^1 \left[ \int_x^{2x} y\, dy \right] dx$

**13.** $\displaystyle \int_0^1 x \left[ \int_x^{3x} y\, dy \right] dx$

**14.** $\displaystyle \int_0^1 y \left[ \int_y^{\sqrt{y}} x^2\, dx \right] dy$

**15.** $\displaystyle \int_0^1 \left[ \int_1^{e^y} \frac{y}{x}\, dx \right] dy$

**16.** $\displaystyle \int_0^1 \left[ \int_0^{x^2} xe^y\, dy \right] dx$

**17.** $\displaystyle \int_0^2 \left[ \int_y^{2y} xy\, dx \right] dy$

**18.** $\displaystyle \int_2^4 \left[ \int_1^{y^2} \frac{y}{x^2}\, dx \right] dy$

**19.** $\displaystyle \int_0^1 \left[ \int_{x^2}^x \sqrt{x}\sqrt{y}\, dy \right] dx$

**20.** $\displaystyle \int_0^1 \left[ \int_{x^2}^x \sqrt{x}\, dy \right] dx$

**21.** $\displaystyle \int_0^1 \left[ \int_x^1 \frac{1}{x^2+1}\, dy \right] dx$

**22.** $\displaystyle \int_0^1 \left[ \int_y^{\sqrt{y}} (x^2+y^2)\, dx \right] dy$

**23.** $\displaystyle \int_1^2 \left[ \int_0^{\ln x} xe^y\, dy \right] dx$

**24.** $\displaystyle \int_2^3 \left[ \int_0^{1/y} \ln y\, dx \right] dy$

In Problems 25–30, find each double integral. Start by graphing the region R. Pay particular attention to the boundary, since it determines the correct limits of integration.

**25.** $\displaystyle \iint\limits_R (x+y)\, dA$, where $R$ is the region enclosed by $y = x^2$ and $y^2 = 8x$

**26.** $\displaystyle \iint\limits_R (x^2 - y^2)\, dA$, where $R$ is the region enclosed by $y = x$ and $y = x^2$

**27.** $\displaystyle \iint\limits_R y^2\, dA$, where $R$ is the region enclosed by $y = 2 - x$ and $y = x^2$

**28.** $\displaystyle \iint\limits_R xy\, dA$, where $R$ is the region enclosed by $y^2 = x + 1$ and $y = 1 - x$

**29.** $\displaystyle \iint\limits_R x^2 y\, dA$ where $R$ is the triangular region enclosed by $y = 2x$, $y = -x$, and $y = 2$

**30.** $\displaystyle \iint\limits_R x^2 y\, dA$ where $R$ is the triangular region enclosed by $y = 2x$, $y = -x$, and $x = 1$

In Problems 31–38, use double integration to find the area of each region.

**31.** Enclosed by the graphs of $y = x^3$ and $y = x^2$

**32.** Enclosed by the graphs of $y = 2\sqrt{x}$ and $y = \dfrac{x^2}{4}$

**33.** Enclosed by the graphs of $y = x^2 - 9$ and $y = 9 - x^2$

**34.** Enclosed by the graphs of $x^2 + y^2 = 16$ and $y^2 = 6x$

**35.** Enclosed by the graph $y = \dfrac{1}{\sqrt{x-1}}$, the $x$-axis, and the lines $x = 2$ and $x = 5$

**36.** Enclosed by the graphs of $y = x^{3/2}$ and $y = x$

**37.** Enclosed by the line $x + y = 3$ and the hyperbola $xy = 2$

**38.** Enclosed by the hyperbola $xy = \sqrt{3}$ and the circle $x^2 + y^2 = 4$, in the first quadrant only

---

**1.** = NOW WORK problem    ◢◣ = Graphing technology recommended    [CAS] = Computer Algebra System recommended

*In Problems 39–46, change the order of integration of each iterated integral to obtain an equivalent expression.*

**39.** $\displaystyle\int_0^1\left[\int_0^x f(x,y)\,dy\right]dx$    **40.** $\displaystyle\int_0^2\left[\int_{y^2}^{2y} f(x,y)\,dx\right]dy$

**41.** $\displaystyle\int_0^a\left[\int_0^{\sqrt{a^2-y^2}} f(x,y)\,dx\right]dy$

**42.** $\displaystyle\int_0^{2\sqrt[3]{2}}\left[\int_{x^2/4}^{\sqrt{x}} f(x,y)\,dy\right]dx$

**43.** $\displaystyle\int_0^{16}\left[\int_{y/8}^{y^{1/4}} f(x,y)\,dx\right]dy$

**44.** $\displaystyle\int_2^5\left[\int_{x^2-6x+9}^{x-1} f(x,y)\,dy\right]dx$

**45.** $\displaystyle\int_1^2\left[\int_0^{\ln y} f(x,y)\,dx\right]dy$    **46.** $\displaystyle\int_1^e\left[\int_{\ln x}^1 f(x,y)\,dy\right]dx$

*In Problems 47–50, the functions f and g are continuous on a closed, bounded region R, $\iint_R f(x,y)\,dA=8$ and $\iint_R g(x,y)\,dA=-6$. Use properties of double integrals to find each double integral.*

**47.** $\displaystyle\iint_R [f(x,y)-g(x,y)]\,dA$    **48.** $\displaystyle\iint_R 4f(x,y)\,dA$

**49.** $\displaystyle\iint_R [3f(x,y)+4g(x,y)]\,dA$

**50.** $\displaystyle\iint_R [2f(x,y)+5g(x,y)]\,dA$

## Applications and Extensions

**51. Area**   Find the area of the region $R$ in the first quadrant enclosed by the graphs of $y=6x-x^2$ and $y=4x-8$ by partitioning $R$ with a horizontal line through the point $(4,8)$. Refer to Figure 21.

**52. Area**   Use double integration to find the area in the first quadrant enclosed by the parabola $x^2=9y$, the $y$-axis, and the circle $x^2+y^2=10$.

*In Problems 53 and 54, set up, but do not evaluate, an iterated double integral to find the volume under each surface over the given region.*

**53. Volume**   Find the volume of the solid under the surface $z=e^{x^2+y^2}$ and over the region enclosed by the graphs of $y=x+4$ and $y=x^2-2x$.

**54. Volume**   Find the volume of the solid under the surface $z=e^{x^2}+e^{y^2}$ enclosed by the graphs of $x=y^2-6y+5$ and $y=x+1$.

**Volume**   *In Problems 55–64, find the volume of each solid.*

**55.** The tetrahedron enclosed by the plane $x+2y+z=2$ and the coordinate planes

**56.** Enclosed by the elliptic paraboloid $4x^2+9y^2=36$ and the planes $z=0$ and $z=1$

**57.** Below the paraboloid $z=x^2+y^2$ and above the triangular region $R$ formed by the $x$- and $y$-axes and the line $x+y=1$, as shown in the figure below.

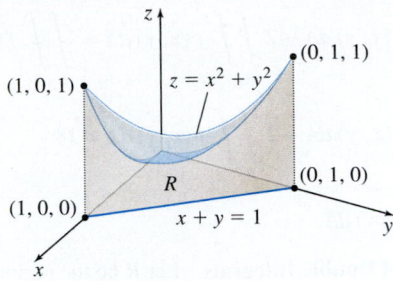

**58.** Enclosed by the cylinder $z=9-y^2$ and the planes $x=0$, $x=4$, and $z=0$

**59.** Enclosed by the surface $z=4-x^2-y^2$ and $z=0$

**60.** Enclosed by the surfaces $x^2+y^2=1$, $z=|x|$, and $z=1$

**61.** The solid bounded from above by the graph of $z=x^2+y^2$, from below by the $xy$-plane, and on the sides by the cylinder $x^2+y^2=1$, as shown in the figure.

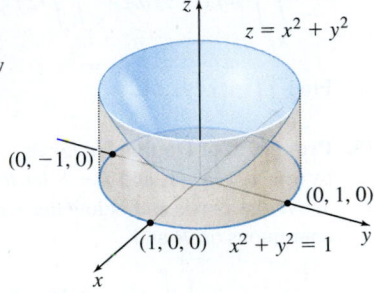

**62.** Enclosed by a portion of the parabolic cylinder $z=\dfrac{x^2}{2}$ and the four planes $y=0$, $y=x$, $x=2$, and $z=0$

**63.** Enclosed by the coordinate planes, the plane $y=3$, and the surface $z=y+1-x^2$

**64.** Enclosed by the surfaces $z=xe^y$, $z=0$, $y=0$, $x=1$, and $y=x^2$

*In Problems 65–72, (a) graph the region associated with each iterated integral, (b) reverse the order of integration, and (c) find the new iterated integral.*

**65.** $\displaystyle\int_0^{\sqrt{\pi/2}}\left[\int_y^{\sqrt{\pi/2}} \sin x^2\,dx\right]dy$    **66.** $\displaystyle\int_0^{1/2}\left[\int_{2x}^1 e^{y^2}\,dy\right]dx$

**67.** $\displaystyle\int_0^1\left[\int_0^{\sqrt{1-x^2}} \frac{1}{\sqrt{1-y^2}}\,dy\right]dx$

**68.** $\displaystyle\int_0^1\left[\int_y^1 \frac{\sin x}{x}\,dx\right]dy$

**69.** $\displaystyle\int_0^1\left[\int_y^1 \sqrt{2+x^2}\,dx\right]dy$    **70.** $\displaystyle\int_0^1\left[\int_{\sqrt[3]{x}}^1 \sqrt{1+y^4}\,dy\right]dx$

**71.** $\displaystyle\int_0^1\left[\int_{\sqrt{y}}^1 e^{y/x}\,dx\right]dy$    **72.** $\displaystyle\int_0^1\left[\int_{\sin^{-1}y}^{\pi/2} e^{\cos x}\,dx\right]dy$

**73. Properties of Double Integrals** Let $R$ be the region enclosed by $y = 1$, $y = -1$, $x = 0$, and $x = 1$; let $R_1$ and $R_2$ be the subregions of $R$ in the first and fourth quadrants, respectively. Suppose $f$ is continuous on $R$ and

$$\iint_R 3f(x, y)\, dA - 2\iint_{R_1} f(x, y)\, dA = \iint_{R_2} f(x, y)\, dA,$$

$$\iint_{R_2} 5f(x, y)\, dA - 2\iint_{R_1} f(x, y)\, dA = 18$$

Find $\iint_R f(x, y)\, dA$.

**74. Properties of Double Integrals** Let $R$ be the region enclosed by $y = 0$, $y = 2$, $x = -2$, and $x = 1$; let $R_1$ and $R_2$ be the subregions of $R$ in the first and second quadrants, respectively. Suppose $f$ is continuous on $R$ and

$$\iint_{R_2} 3f(x, y)\, dA - \iint_{R_1} f(x, y)\, dA = 2\iint_R f(x, y)\, dA,$$

$$\iint_{R_1} 4f(x, y)\, dA + \iint_{R_2} 2f(x, y)\, dA = 7$$

Find $\iint_R f(x, y)\, dA$.

**75. Properties of Double Integrals** Let $R$ be the region enclosed by $y = x$, $y = -1$, and $x = 2$; let $R_1$ and $R_2$ be the subregions of $R$ above the $x$-axis and below the $x$-axis, respectively. Suppose $f$ is continuous on $R$ and

$$2\iint_{R_2} f(x, y)\, dA - 7\iint_{R_1} f(x, y)\, dA = 17$$

$$\iint_{R_2} f(x, y)\, dA - 2\iint_{R_1} f(x, y)\, dA = 7$$

Find $\iint_R f(x, y)\, dA$.

**76. Properties of Double Integrals** Let $R$ be the region enclosed by $y = x^2 + 1$, $y = 0$, $x = -1$, and $x = 2$; let $R_1$ and $R_2$ be the subregions of $R$ in the first and second quadrants, respectively. Suppose $f$ is continuous on $R$ and

$$\iint_{R_2} 6f(x, y)\, dA - 3\iint_{R_1} f(x, y)\, dA = -12$$

$$\iint_{R_2} 4f(x, y)\, dA + 2 = \iint_{R_1} f(x, y)\, dA$$

Find $\iint_R f(x, y)\, dA$.

*In Problems 77–80, each of the iterated integrals represents the volume of a solid. Describe the solid.*

**77.** $\displaystyle \int_0^1 \left[ \int_0^{\sqrt{1-x^2}} \sqrt{1 - x^2 - y^2}\, dy \right] dx$

**78.** $\displaystyle \int_0^2 \left[ \int_0^{\sqrt{4-y^2}} (4 - x^2 - y^2)\, dx \right] dy$

**79.** $\displaystyle \int_0^4 \left[ \int_0^1 3\, dy \right] dx$

**80.** $\displaystyle \int_0^1 \left[ \int_0^2 2\, dy \right] dx$

**81.** Find:

  (a) $\iint_R x\, dA$, where $R$ is the region enclosed by the circle of radius 1 centered at the origin.

  (b) $4 \iint_{R_1} x\, dA$, where $R_1$ is the region in the first quadrant enclosed by the circle of radius 1 centered at the origin. (This shows that although the region over which we are integrating is symmetric about the $x$-axis and the $y$-axis, we must also consider properties of the integrand on this region before we use symmetry.)

**82.** (a) Find

$$\int_1^e \frac{1}{x} \left[ \int_0^{\ln x} dy \right] dx$$

  (b) Reverse the order of integration and find the resulting iterated integral.

**83.** (a) Find

$$\int_0^{\pi/2} \sin x \left[ \int_0^{\cos x} dy \right] dx$$

  (b) Reverse the order of integration and find the resulting iterated integral.

**84. Volume of a Liquid** A tank in the shape of a half-cylinder is lying on its flat side. The radius of the tank is 1 m and its length is 4 m. If the tank is filled with liquid to a depth of $a$ meters, what is the volume of the liquid? *Hint:* Position the half cylinder as shown in the figure below.

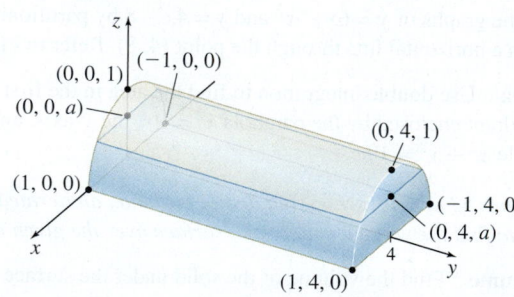

**85. Volume of a Liquid** Repeat Problem 84 if the tank is buried so its rounded base is 1 m under the ground and its flat side is at ground level. See the figure below.

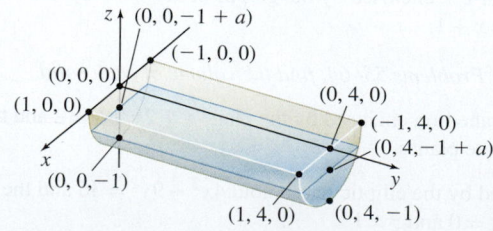

**86. Damming a Valley**   A dam is built in a V-shaped valley. The top of the dam is 200 m wide and the lowest point in the V is 150 m below the top, as shown in the figure. The pressure $P$ due to the water at a depth $d$ meters below the surface is $P = \rho g d$, where $\rho = 1000 \, \text{kg/m}^3$ and $g = 9.8 \, \text{m/s}^2$. What is the total outward force $F$ from the water that this dam must be able to withstand? Express your answer in both newtons and pounds using the fact that $1 \, \text{N} \approx 0.2248 \, \text{lb}$. (Recall that pressure is $P = \dfrac{F}{A}$, so the force $F = P A$.)

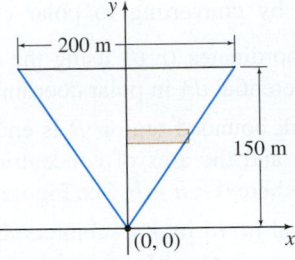

*In Problems 87–90, refer to the discussion of joint probability density functions on page 959.*

**87. Probability**   The joint probability density function $P$ for the random variables $X$ and $Y$ is given by

$$P(x, y) = \begin{cases} \dfrac{2}{3}(2x + y) & 0 \le x \le 1, 0 \le y \le 1 \\ 0 & \text{elsewhere} \end{cases}$$

**(a)** Find $Pr(X + Y \le 1)$.      **(b)** Find $Pr(Y \le X)$.

**88. Probability**   The joint probability density function $P$ for the random variables $X$ and $Y$ is given by

$$P(x, y) = \begin{cases} xy & 0 \le x \le 2, 0 \le y \le 1 \\ 0 & \text{elsewhere} \end{cases}$$

**(a)** Find $Pr(X \le Y)$.      **(b)** Find $Pr\left(\dfrac{1}{2} \le X \le Y\right)$.

**89. Probability (a)** Find the number $c$ that makes the function $P$ a joint probability density function for the random variables $X$ and $Y$.

$$P(x, y) = \begin{cases} cxy & 0 \le x \le 1, 0 \le y \le x \\ 0 & \text{elsewhere} \end{cases}$$

**(b)** Use the probability density function from (a) to find $Pr\left(\dfrac{1}{2} \le Y \le X\right)$.

**90. Probability (a)** Find the number $c$ that makes the function $P$ a joint probability density function for the random variables $X$ and $Y$.

$$P(x, y) = \begin{cases} cx(1 + y) & 0 \le x \le y, 0 \le y \le 1 \\ 0 & \text{elsewhere} \end{cases}$$

**(b)** Use the probability density function from (a) to find $Pr\left(\dfrac{1}{9} \le X \le Y\right)$.

**91.** Explain if the following equality is true or false without finding the exact value of the two integrals:

$$\int_{-3}^{3} \int_{-\sqrt{9 - y^2}}^{\sqrt{9 - y^2}} (7 - x) \, dx \, dy = 4 \int_{0}^{3} \int_{0}^{\sqrt{9 - y^2}} (7 - x) \, dx \, dy$$

**92. Volume**   Find the volume of the tetrahedron enclosed by the coordinate planes and the plane $\dfrac{x}{a} + \dfrac{y}{b} + \dfrac{z}{c} = 1$. (Assume that $a$, $b$, and $c$ are positive.)

**93.** Find $\iint\limits_{R} xy \, dA$, where $R$ is the region enclosed by $y = 4 - x^2$ and $y = x^2$ to the right of the $y$-axis.

**94.** Find $\iint\limits_{R} xy \, dA$, where $R$ is the region enclosed by $x = 4 - y^2$ and $x = y^2$ above the $x$-axis.

**95.** Find $\iint\limits_{R} e^{\sqrt{x}} \, dA$, where $R$ is the region enclosed by $y = x$, $x = 1$, $y = 0$.

**96. (a)** Graph the region $R$ defined by $0 \le a \le y \le b$, $0 \le a \le x \le y$.

**(b)** If $z = f(x)$ is continuous on $R$, show that

$$\iint\limits_{R} f(x) \, dA = \int_{a}^{b} (b - x) f(x) \, dx$$

**97. Properties of Double Integrals**   Using properties of double integrals, show that if the functions $f$ and $g$ are integrable on $R$ and $f(x, y) \le g(x, y)$ for all $(x, y)$ in $R$, then

$$\iint\limits_{R} f(x, y) \, dA \le \iint\limits_{R} g(x, y) \, dA$$

**Challenge Problem**

**98. Volume**   Find the volume of the solid under the graph of the elliptic paraboloid $z = 8 - 2x^2 - y^2$ and above the first quadrant of the $xy$-plane.
*Hint:* Use **Wallis's formula** from Section 7.1, Problem 85.

# 14.3 Double Integrals Using Polar Coordinates

**OBJECTIVES** *When you finish this section, you should be able to:*

**1** Find a double integral using polar coordinates (p. 972)

**2** Find area and volume using polar coordinates (p. 974)

Suppose the function $z = f(x, y)$ is continuous on a closed, bounded region $R$ whose boundary is a piecewise-smooth curve. Then the double integral $\iint\limits_{R} f(x, y) \, dA$ exists, and $dA = dx \, dy = dy \, dx$ is the differential of the area $A$ of $R$. If the boundary of $R$ consists of rays (lines through the origin) and parts of circles, it is often easier

**NEED TO REVIEW?** Polar coordinates are discussed in Section 9.4, pp. 701–708.

to find $\iint\limits_{R} f(x, y)\, dA$ by converting to polar coordinates. We convert a function $z = f(x, y)$ to polar coordinates $(r, \theta)$ using the equations $x = r\cos\theta$, $y = r\sin\theta$. It remains to find the differential $dA$ in polar coordinates.

Suppose the closed, bounded region $R$ is enclosed by the rays $\theta = \alpha$ and $\theta = \beta$, where $0 \le \alpha < \beta \le 2\pi$, and the arcs of concentric circles with radii $r = a$ and $r = b$ centered at the origin, where $0 < a < b$. See Figure 22.

Partition the interval $[a, b]$ into $n$ subintervals of radius $\Delta r_i$, $i = 1, 2, \ldots, n$, and the interval $[\alpha, \beta]$ into $m$ subintervals of measure $\Delta\theta_j$, $j = 1, 2, \ldots, m$. The result is a collection of $mn$ polar subregions $R_{ij}$ as shown in Figure 23.

Consider a typical polar subregion $R_{ij}$ as shown in Figure 24. The area $\Delta A_{ij}$ of this subregion is the difference of the areas of two circular sectors.

$$\Delta A_{ij} = \frac{1}{2}r_i^2(\theta_j - \theta_{j-1}) - \frac{1}{2}r_{i-1}^2(\theta_j - \theta_{j-1}) = \frac{1}{2}r_i^2\Delta\theta_j - \frac{1}{2}r_{i-1}^2\Delta\theta_j \qquad \text{Area } A \text{ of a sector} = \frac{1}{2}r^2\theta$$

$$= \frac{1}{2}(r_i^2 - r_{i-1}^2)\Delta\theta_j = \frac{1}{2}(r_i + r_{i-1})(r_i - r_{i-1})\Delta\theta_j$$

$$= \frac{1}{2}(r_i + r_{i-1})\Delta r_i \Delta\theta_j \qquad \Delta r_i = r_i - r_{i-1}$$

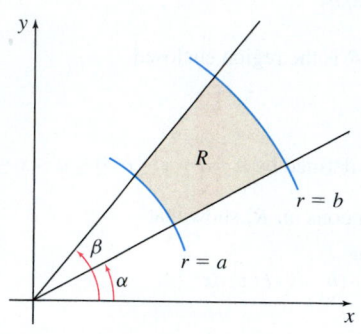

**Figure 22**

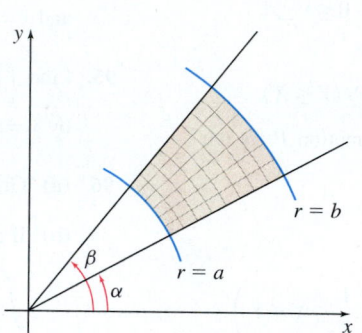

**Figure 23**

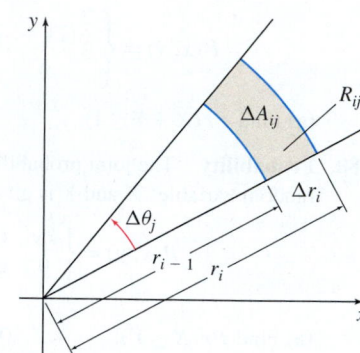

**Figure 24** $\Delta A_{ij} = \dfrac{1}{2}(r_i + r_{i-1})\Delta r_i \Delta\theta_j$

Next choose a point $(\bar{r}_{ij}, \bar{\theta}_{ij})$ in the polar subregion $R_{ij}$ so that $\bar{r}_{ij} = \dfrac{1}{2}(r_i + r_{i-1})$. Then the area $\Delta A_{ij}$ of $R_{ij}$ is

$$\boxed{\Delta A_{ij} = \bar{r}_{ij}\Delta r_i \Delta\theta_j}$$

and the differential $dA$ is

$$\boxed{dA = r\, dr\, d\theta}$$

It can be shown that if $f$ is continuous on the region $R$, then

$$\iint\limits_{R} f(x, y)\, dA = \iint\limits_{R} f(r\cos\theta, r\sin\theta)\, r\, dr\, d\theta$$

## 1 Find a Double Integral Using Polar Coordinates

As with double integrals in rectangular coordinates, certain regions in polar coordinates will lead to iterated integrals.

**THEOREM** Iterated Double Integral in Polar Coordinates

Let the function $z = f(x, y)$ be continuous on a closed, bounded region $R$. Suppose $R$ is enclosed by the rays $\theta = \alpha$ and $\theta = \beta$, where $0 \le \alpha < \beta \le 2\pi$, and the curves $r = r_1(\theta)$ and $r = r_2(\theta)$, where $0 \le r_1 \le r_2$. If $r = r_1(\theta)$ and $r = r_2(\theta)$ are continuous for $\alpha \le \theta \le \beta$, then

$$\iint\limits_{R} f(x, y)\, dA = \iint\limits_{R} f(r\cos\theta, r\sin\theta)\, dA = \int_{\alpha}^{\beta}\int_{r_1(\theta)}^{r_2(\theta)} f(r\cos\theta, r\sin\theta)\, r\, dr\, d\theta$$

Notice that we wrote the iterated double integral without using brackets to indicate the order of integration. In the iterated integral, $dr$ appears before $d\theta$; this indicates that we integrate with respect to $r$ before we integrate with respect to $\theta$. In finding the inside integral $\int_{r_1(\theta)}^{r_2(\theta)} f(r\cos\theta, r\sin\theta)\, r\, dr$, we treat $\theta$ as a constant and integrate partially with respect to $r$.

**EXAMPLE 1  Finding a Double Integral Using Polar Coordinates**

Find $\iint\limits_{R} \cos\theta\, dA$, where $R$ is the region enclosed by the rays $\theta = 0$ and $\theta = \dfrac{\pi}{4}$ and the circle $r = 4\cos\theta$.

**Solution** We begin with a graph of the region $R$. See Figure 25. Notice that $\theta$ varies from $0$ to $\dfrac{\pi}{4}$ and $r$ varies from $r_1 = 0$ to the circle $r_2 = 4\cos\theta$. Then

$$\iint\limits_{R} \cos\theta\, dA = \int_0^{\pi/4}\int_0^{4\cos\theta} r\cos\theta\, dr\, d\theta \qquad dA = r\, dr\, d\theta$$

$$= \int_0^{\pi/4} \cos\theta \left[\frac{r^2}{2}\right]_0^{4\cos\theta} d\theta = \int_0^{\pi/4} 8\cos^3\theta\, d\theta$$

$$= 8\int_0^{\pi/4}(1 - \sin^2\theta)\cos\theta\, d\theta = 8\int_0^{\sqrt{2}/2}(1 - u^2)\, du$$
$$\uparrow$$
$$u = \sin\theta;\ \ du = \cos\theta\, d\theta$$

$$= 8\left[u - \frac{u^3}{3}\right]_0^{\sqrt{2}/2} = 8\left[\frac{\sqrt{2}}{2} - \frac{\sqrt{2}}{12}\right] = \frac{10\sqrt{2}}{3} \qquad \blacksquare$$

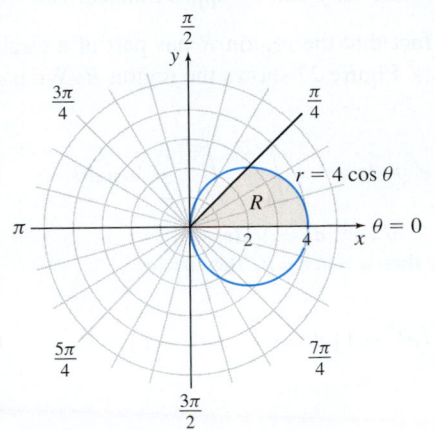

**Figure 25**

**NEED TO REVIEW?** Integrals containing trigonometric functions are discussed in Section 7.2, pp. 509–515.

**NOW WORK** Problems **15** and **25**.

When the integrand of a double integral $\iint\limits_{R} f(x, y)\, dA$ is a function of $x^2 + y^2$, it is sometimes easier to integrate using polar coordinates than using rectangular coordinates.

**EXAMPLE 2  Finding an Iterated Integral Using Polar Coordinates**

CALC
CLIP

Find the iterated integral $\int_0^2 \int_0^{\sqrt{4-x^2}} (x^2 + y^2)\, dy\, dx$.

**NOTE** The order $dy\, dx$ indicates the integration is done with respect to $y$ first, then with respect to $x$.

**Solution** The region $R$ is enclosed by the graph of $y = \sqrt{4 - x^2}$ (a portion of a circle of radius 2), $0 \le x \le 2$, and the positive $x$- and $y$-axes. See Figure 26. Since the boundary of $R$ consists of rays and arcs of circles, it may be easier to find the double integral using polar coordinates. In polar coordinates, this region $R$ is given by $0 \le r \le 2$ and $0 \le \theta \le \dfrac{\pi}{2}$. Then

$$\int_0^2 \int_0^{\sqrt{4-x^2}} (x^2 + y^2)\, dy\, dx = \iint\limits_{R} (x^2 + y^2)\, dA$$

$$= \iint\limits_{R} r^2 r\, dr\, d\theta \qquad x^2 + y^2 = r^2;\ \ dA = r\, dr\, d\theta$$

$$= \int_0^{\pi/2}\int_0^2 r^3\, dr\, d\theta = \int_0^{\pi/2}\left[\frac{r^4}{4}\right]_0^2 d\theta$$

$$= \int_0^{\pi/2} 4\, d\theta = \left[4\theta\right]_0^{\pi/2} = 2\pi \qquad \blacksquare$$

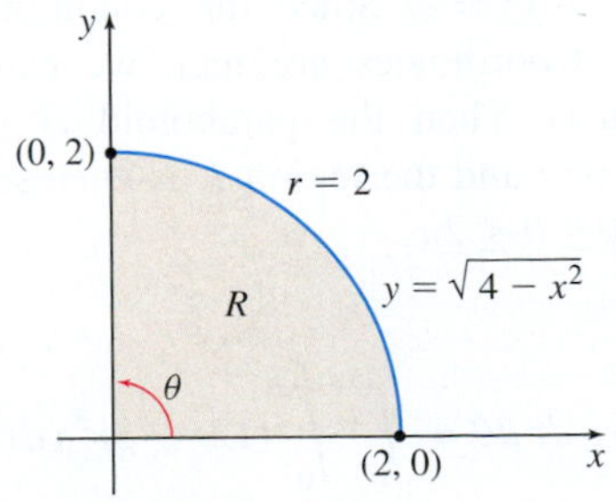

**Figure 26** $0 \le y \le \sqrt{4 - x^2},\ 0 \le x \le 2$
$0 \le r \le 2,\ 0 \le \theta \le \dfrac{\pi}{2}$

**NOW WORK** Problems **9** and **29**.

**NEED TO REVIEW?** Elementary functions are discussed in Section 7.6, p. 541–542.

There are occasions when using polar coordinates converts an integral that is not expressible in terms of elementary functions to an integral that can be written in terms of elementary functions.

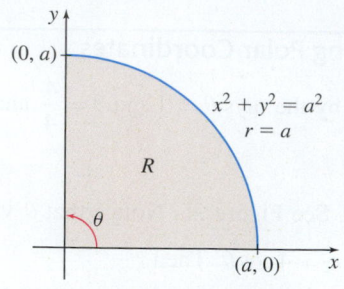

**Figure 27** $0 \leq r \leq a,\ 0 \leq \theta \leq \dfrac{\pi}{2}$

EXAMPLE 3 **Finding a Double Integral Using Polar Coordinates**

Find $\iint\limits_{R} e^{x^2+y^2}\,dA$, where $R$ is the region in the first quadrant inside the circle $x^2 + y^2 = a^2$.

**Solution** In rectangular coordinates, $\iint\limits_{R} e^{x^2+y^2}\,dA$ only can be approximated. But the presence of $x^2 + y^2$ in the integrand and the fact that the region $R$ has part of a circle as its boundary suggest using polar coordinates. Figure 27 shows the region $R$. We use polar coordinates to obtain

$$\iint\limits_{R} e^{x^2+y^2}\,dA = \underset{\substack{\uparrow \\ dA = r\,dr\,d\theta}}{\iint\limits_{R} e^{r^2} r\,dr\,d\theta} = \int_0^{\pi/2}\int_0^a e^{r^2} r\,dr\,d\theta = \underset{\substack{\uparrow \\ u=r^2,\ du=2r\,dr \\ r=0,\ \text{then}\ u=0;\ r=a,\ \text{then}\ u=a^2}}{\int_0^{\pi/2}\int_0^{a^2} \frac{1}{2} e^u\,du\,d\theta}$$

$$= \frac{1}{2}\int_0^{\pi/2} \left[e^u\right]_0^{a^2}\,d\theta = \frac{1}{2}\int_0^{\pi/2}\left(e^{a^2}-1\right)\,d\theta = \frac{\pi}{4}\cdot\left(e^{a^2}-1\right) \qquad \blacksquare$$

NOW WORK Problem 35.

## 2 Find Area and Volume Using Polar Coordinates

Suppose $z = f(x,y) \geq 0$, where $f$ is continuous over a closed, bounded region $R$ that satisfies the conditions of the theorem: Iterated Double Integral in Polar Coordinates. Then the double integral $\iint\limits_{R} f(x,y)\,dA = \iint\limits_{R} f(r\cos\theta,\ r\sin\theta)\,r\,dr\,d\theta$ represents the volume $V$ of the solid under the surface $z = f(x,y)$ and over the region $R$.

EXAMPLE 4 **Finding the Volume of a Solid Using Polar Coordinates**

Find the volume $V$ of the solid under the circular paraboloid $z = 12 - 3x^2 - 3y^2$ and over the $xy$-plane.

**Solution** Begin by graphing the paraboloid for $z \geq 0$, as shown in Figure 28.

The region $R$ is determined by the intersection of the surface $z$ and the $xy$-plane. So, we let $z = 0$ and simplify the equation.

$$12 - 3x^2 - 3y^2 = 0$$
$$3x^2 + 3y^2 = 12$$
$$x^2 + y^2 = 4$$

So $R$ is the region enclosed by the circle $x^2 + y^2 = 4$. Since the conditions of the theorem: Iterated Double Integral in Polar Coordinates are met, we can use polar coordinates. Let $x = r\cos\theta$ and $y = r\sin\theta$. Then the paraboloid is given by $z = 12 - 3x^2 - 3y^2 = 12 - 3(x^2+y^2) = 12 - 3r^2$, and the region $R$ is enclosed by the circle $r^2 = 4$. So, $R$ is given by $0 \leq r \leq 2$ and $0 \leq \theta \leq 2\pi$.

Then

$$V = \iint\limits_{R}(12 - 3x^2 - 3y^2)\,dA = \underset{\substack{\uparrow \\ 3x^2+3y^2=3r^2 \\ dA=r\,dr\,d\theta}}{\int_0^{2\pi}\int_0^2 (12-3r^2)\,r\,dr\,d\theta} = \int_0^{2\pi}\int_0^2 (12r - 3r^3)\,dr\,d\theta$$

$$= \int_0^{2\pi}\left[6r^2 - 3\frac{r^4}{4}\right]_0^2\,d\theta = \int_0^{2\pi} 12\,d\theta = \left[12\theta\right]_0^{2\pi} = 24\pi \text{ cubic units} \qquad \blacksquare$$

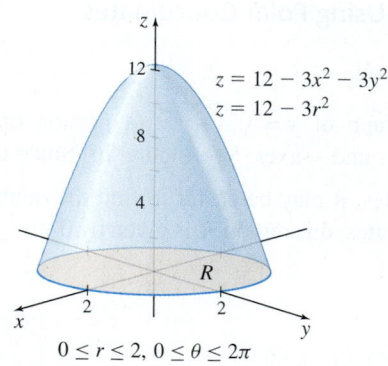

$0 \leq r \leq 2,\ 0 \leq \theta \leq 2\pi$

**Figure 28**

NOW WORK Problem 43.

**EXAMPLE 5**  **Finding the Volume of a Solid**

Find the volume $V$ of the solid under the surface $z = x^2 + y^2$, above the $xy$-plane, and inside the cylinder $x^2 + y^2 = 2y$.

**Solution** See Figure 29. We want the volume of the solid that lies above the region $R$ in the $xy$-plane enclosed by the circle $x^2 + y^2 = 2y$, of radius 1 and center at $(0, 1, 0)$. Since the conditions of the theorem: Iterated Double Integral in Polar Coordinates are met, we can use polar coordinates. To convert to polar coordinates, let $x = r \cos \theta$ and $y = r \sin \theta$. Then the region $R$ is enclosed by

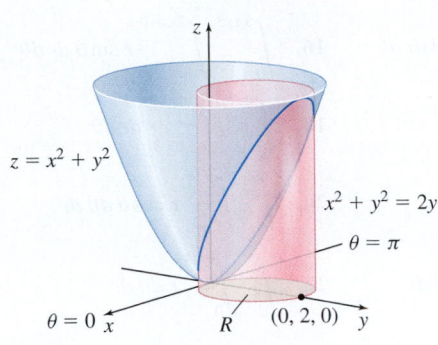

$$x^2 + y^2 = 2y$$

$$r^2 \cos^2 \theta + r^2 \sin^2 \theta = 2r \sin \theta$$

$$r^2 = 2r \sin \theta$$

$$r = 2 \sin \theta$$

**Figure 29**

The region $R$ is bounded by $0 \le r \le 2 \sin \theta$ and $0 \le \theta \le \pi$.

The volume $V$ under $z = x^2 + y^2 = r^2$, above the $xy$-plane, and over $R$ is

$$V = \iint\limits_R (x^2 + y^2)\, dA = \iint\limits_R r^2 r\, dr\, d\theta = \int_0^\pi \int_0^{2 \sin \theta} r^3\, dr\, d\theta = \int_0^\pi \left[ \frac{r^4}{4} \right]_0^{2 \sin \theta} d\theta$$

$$= 4 \int_0^\pi \sin^4 \theta\, d\theta = 4 \int_0^\pi \left( \frac{1 - \cos(2\theta)}{2} \right)^2 d\theta$$

$$= \int_0^\pi [1 - 2 \cos(2\theta) + \cos^2(2\theta)]\, d\theta = \int_0^\pi \left[ 1 - 2 \cos(2\theta) + \frac{1 + \cos(4\theta)}{2} \right] d\theta$$

$$= \left[ \theta - \sin(2\theta) + \frac{1}{2}\theta + \frac{1}{2} \cdot \frac{\sin(4\theta)}{4} \right]_0^\pi = \frac{3\pi}{2} \text{ cubic units} \qquad \blacksquare$$

**NOW WORK** Problem 45.

The double integral $\iint\limits_R dA = \iint\limits_R r\, dr\, d\theta$ represents the volume of a solid whose height is 1. As a result, the area $A$ of the region $R$ is numerically equal to

$$\boxed{A = \iint\limits_R r\, dr\, d\theta}$$

**EXAMPLE 6**  **Finding the Area Enclosed by a Cardioid**

Use double integration to find the area enclosed by the cardioid $r = a(1 - \cos \theta)$.

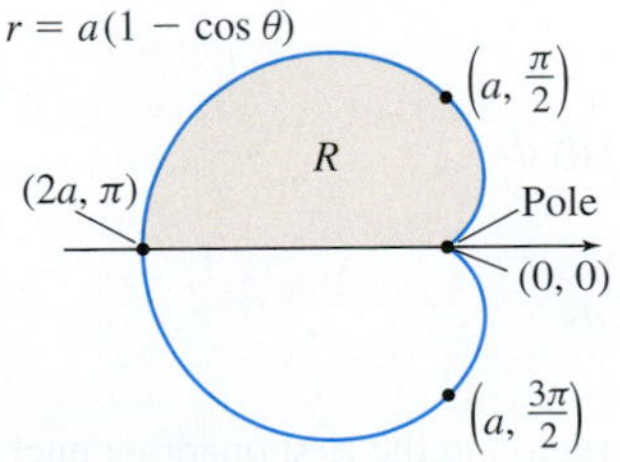

**Solution** The cardioid $r = a(1 - \cos \theta)$ is shown in Figure 30. Using symmetry, the area enclosed by the cardioid is twice the area of the shaded region $R$ in Figure 30. So,

$$A = 2 \iint\limits_R r\, dr\, d\theta$$

**Figure 30** The upper half of the cardioid is swept out by the rays going from $\theta = 0$ to $\theta = \pi$.

where the region $R$ is given by $0 \le r \le a(1 - \cos \theta)$ and $0 \le \theta \le \pi$. Then

$$A = 2 \int_0^\pi \int_0^{a(1 - \cos \theta)} r\, dr\, d\theta = 2 \int_0^\pi \left[ \frac{r^2}{2} \right]_0^{a(1 - \cos \theta)} d\theta$$

$$= \int_0^\pi a^2 (1 - \cos \theta)^2\, d\theta = a^2 \int_0^\pi (1 - 2 \cos \theta + \cos^2 \theta)\, d\theta$$

$$= a^2 \left[ \theta - 2 \sin \theta + \left( \frac{\theta}{2} + \frac{\sin(2\theta)}{4} \right) \right]_0^\pi = \frac{3}{2}\pi a^2 \text{ square units} \qquad \blacksquare$$

**NOW WORK** Problem 49.

## 14.3 Assess Your Understanding

### Concepts and Vocabulary

1. Convert the function $f(x, y) = \sqrt{4x^2 + 4y^2}$ to polar coordinates.

2. In polar coordinates, the differential $dA$ of area is $dA = $ _____.

3. *True or False*   Geometrically, if $f(x, y) \geq 0$ and if $f$ is continuous over a closed, bounded region $R$ that satisfies the conditions of the theorem: Iterated Double Integral in Polar Coordinates, then the double integral $\iint\limits_R f(r \cos \theta, r \sin \theta)\, dr\, d\theta$ represents the volume of the solid under the surface $z = f(x, y)$ and over the region $R$ in the $xy$-plane.

4. *True or False*   $\iint\limits_R r\, dr\, d\theta$ is numerically equal to the area $A$ of the closed, bounded region $R$.

### Skill Building

*In Problems 5–8, assume that the function $z = f(x, y)$ is continuous on the region $R$. Express the double integral $\iint\limits_R f(x, y)\, dA$ over $R$ as an iterated integral using polar coordinates.*

5.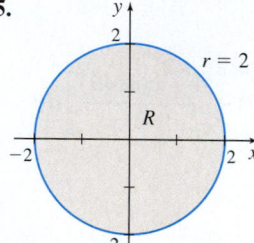

6.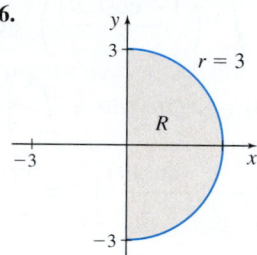

7.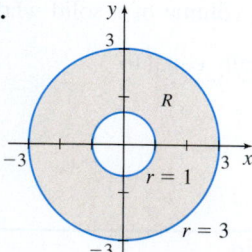

8.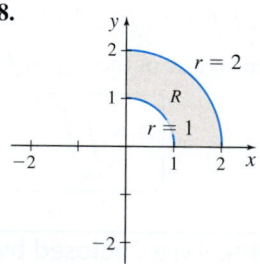

*In Problems 9–14, express each double integral over the region $R$ as an iterated integral using polar coordinates. Do not integrate.*

9. $\iint\limits_R (x + y)\, dA$, $R$ is enclosed by $x^2 + y^2 = 9$, $x \geq 0$, and $y \geq 0$.

10. $\iint\limits_R (x^2 + y^2)\, dA$, $R$ is enclosed by $y = \sqrt{1 - x^2}$ and the lines $y = x$ and $y = -x$.

11. $\iint\limits_R x\, dA$, $R$ is enclosed by $x = \sqrt{4 - y^2}$, the positive $x$-axis, and the line $y = x$.

12. $\iint\limits_R y\, dA$, $R$ is enclosed by $y = \sqrt{16 - x^2}$, in the first quadrant.

13. $\iint\limits_R (x + y^2)\, dA$, $R$ is enclosed by $x^2 + y^2 = 1$ and $x^2 + y^2 = 4$.

14. $\iint\limits_R \sqrt{x^2 + y^2}\, dA$, $R$ is enclosed by $x^2 + y^2 = \dfrac{1}{4}$ and $x^2 + y^2 = 5$.

*In Problems 15–22, find each iterated integral.*

15. $\displaystyle\int_0^{\pi/4} \int_0^{4\sin\theta} r \cos\theta\, dr\, d\theta$

16. $\displaystyle\int_0^{5\pi/3} \int_0^{2\cos\theta} r \sin\theta\, dr\, d\theta$

17. $\displaystyle\int_0^{\pi/2} \int_0^{\pi/3} r\, d\theta\, dr$

18. $\displaystyle\int_0^{\pi/3} \int_0^{\sin\theta} r\, dr\, d\theta$

19. $\displaystyle\int_0^{\pi/4} \int_0^{\cos\theta} r\, dr\, d\theta$

20. $\displaystyle\int_0^{4} \int_{-\pi/4}^{\pi/4} r \cos\theta\, d\theta\, dr$

21. $\displaystyle\int_0^{2} \int_{-\pi/3}^{5\pi/3} r \sin\theta\, d\theta\, dr$

22. $\displaystyle\int_0^{\pi} \int_0^{\pi/4} r\, d\theta\, dr$

*In Problems 23–28, graph the region $R$, then find each double integral.*

23. $\displaystyle\iint\limits_R 3 \sin\theta\, dA$,   $0 \leq r \leq 2$, $0 \leq \theta \leq \dfrac{\pi}{2}$

24. $\displaystyle\iint\limits_R 4 \cos\theta\, dA$,   $0 \leq r \leq 3$, $0 \leq \theta \leq \dfrac{\pi}{2}$

25. $\displaystyle\iint\limits_R 2r \sin\theta\, dA$,   $0 \leq r \leq 1$, $0 \leq \theta \leq \dfrac{\pi}{2}$

26. $\displaystyle\iint\limits_R 3r \cos\theta\, dA$,   $0 \leq r \leq 2$, $0 \leq \theta \leq \dfrac{\pi}{4}$

27. $\displaystyle\iint\limits_R 3r^2 \sin\theta\, dA$,   $0 \leq r \leq 2$, $-\dfrac{\pi}{2} \leq \theta \leq \dfrac{\pi}{2}$

28. $\displaystyle\iint\limits_R 2r^2 \cos\theta\, dA$,   $0 \leq r \leq 4$, $-\dfrac{\pi}{2} \leq \theta \leq \dfrac{\pi}{2}$

*In Problems 29–42, find each double integral by changing to polar coordinates.*

29. $\displaystyle\int_0^{1} \int_0^{\sqrt{1 - x^2}} dy\, dx$

30. $\displaystyle\int_0^{3} \int_0^{y} \sqrt{x^2 + y^2}\, dx\, dy$

31. $\displaystyle\int_{-2}^{2} \int_0^{\sqrt{4 - y^2}} (x^2 + y^2)\, dx\, dy$

32. $\displaystyle\int_0^{2} \int_0^{\sqrt{4 - x^2}} \sqrt{4 - x^2 - y^2}\, dy\, dx$

33. $\displaystyle\int_0^{1} \int_0^{\sqrt{1 - x^2}} \cos(x^2 + y^2)\, dy\, dx$

34. $\displaystyle\int_0^{1} \int_0^{\sqrt{1 - y^2}} e^{\sqrt{x^2 + y^2}}\, dx\, dy$

35. $\iint\limits_R e^{-(x^2 + y^2)}\, dA$, $R$ is the region in the first quadrant enclosed by the circles $x^2 + y^2 = 1$ and $x^2 + y^2 = 4$.

36. $\displaystyle\iint\limits_R \frac{y}{\sqrt{x^2 + y^2}}\, dA$, $R$ is the region in the first quadrant inside the circle $x^2 + y^2 = a^2$.

37. $\iint\limits_R x\, dA$, $R$ is the region enclosed by the circle $x^2 + y^2 = x$.

**38.** $\iint\limits_R y^2 \, dA$, $R$ is the region enclosed by the circle $x^2 + y^2 = 2y$.

**39.** $\iint\limits_R \sqrt{x^2 + y^2} \, dA$, $R$ is the region enclosed by $r = 3 + \cos\theta$.

**40.** $\iint\limits_R (x^2 + y^2) \, dA$, $R$ is the region enclosed by $r = 2(1 + \sin\theta)$.

**41.** $\displaystyle\int_{-2}^{2} \int_{-\sqrt{4-y^2}}^{\sqrt{4-y^2}} (x^2 + y^2)^2 \, dx \, dy$

**42.** $\displaystyle\int_{0}^{2} \int_{\sqrt{2x-x^2}}^{\sqrt{4x-x^2}} dy \, dx + \int_{2}^{4} \int_{0}^{\sqrt{4x-x^2}} dy \, dx$

## Applications and Extensions

*In Problems 43–52, use double integrals in polar coordinates.*

**43. Volume**  Find the volume of the solid enclosed by the ellipsoid $x^2 + y^2 + 4z^2 = 4$.

**44. Volume**  Find the volume of the solid enclosed by the paraboloid $x^2 + y^2 = az$, $a > 0$ the $xy$-plane, and the cylinder $x^2 + y^2 = a^2$.

**45. Volume**  Find the volume of the portion of the **ball** (a solid sphere) of radius 4 and center at the origin that lies above the region enclosed by the circle $r = 4\sin\theta$, as shown in the figure below.

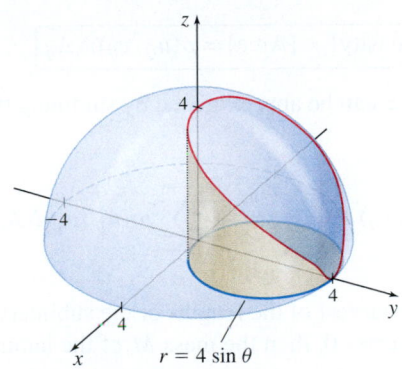

**46. Volume**  Find the volume of the solid cut from the ellipsoid $x^2 + y^2 + 4z^2 = 4$ by the cylinder $x^2 + y^2 = 1$.

**47. Area**  Find the area enclosed by one loop of $r^2 = 9\sin(2\theta)$.

**48. Area**  Find the area of the region that lies inside the circle $r = 4\cos\theta$ but outside the circle $r = \cos\theta$.

**49. Area**  Find the area of the region in the first quadrant that lies inside the limaçon $r = 3 + 2\sin\theta$ but outside the circle $r = 4\sin\theta$. See the figure.

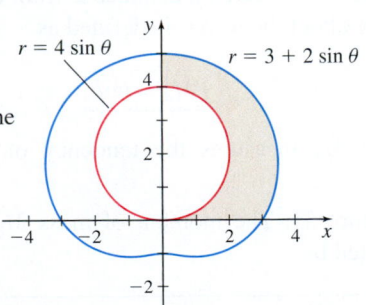

**50. Area**  Find the area of the region that lies inside the circle $r = 1$ but outside the cardioid $r = 1 + \cos\theta$.

**51. Area**  Find the area of the region that lies inside the cardioid $r = 1 + \cos\theta$ but outside the circle $r = \dfrac{1}{2}$.

**52. Area**  Find the area of the region that lies inside the limaçon $r = 3 - \cos\theta$ but outside the circle $r = 5\cos\theta$.

*In Problems 53 and 54, replace each iterated integral in polar coordinates with an iterated integral in rectangular coordinates. Do not find the integrals.*

**53.** $\displaystyle\int_{0}^{1/\sqrt{2}} \int_{0}^{\sin^{-1} r} r \, d\theta \, dr$

**54.** $\displaystyle\int_{0}^{1} \int_{\cos^{-1} r}^{\pi/2} r^2 \sin\theta \, d\theta \, dr$

**55.** A tank in the shape of a hemisphere of radius 1 m is lying on its flat side. If the tank is filled with liquid to a depth of $a$ meters, what is the volume of the liquid? *Hint:* Position the hemisphere as shown in the figure.

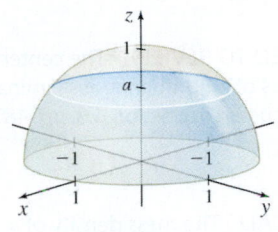

**56.** Repeat Problem 55 if the tank is configured as in the figure, with its flat side at ground level and its rounded base under the ground.

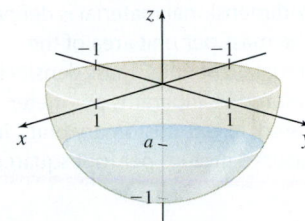

**57. Area**

(a) Graph $r = \theta$, $0 \le \theta \le 2\pi$.

(b) Find the area of the region enclosed by the graph of $r = \theta$ and the positive $x$-axis.

**58. Volume**  Find the volume of the solid cut from the sphere $x^2 + y^2 + z^2 = a^2$ by the cylinder $x^2 + y^2 = ay$, $a > 0$.

**59. Area**  Find the area enclosed by one leaf of the rose $r = \sin(3\theta)$.

## Challenge Problems

**60.** (a) Express $f(x, y) = \dfrac{1}{x^2\sqrt{x^2 + y^2}}$, using polar coordinates.

(b) Find $\displaystyle\int_{1}^{2} \int_{0}^{1} \dfrac{1}{x^2\sqrt{x^2 + y^2}} \, dy \, dx$.

**61.** To find $\int_{-\infty}^{\infty} e^{-x^2} \, dx$, let $I_a = \int_{0}^{a} e^{-x^2} \, dx$.

(a) Show that:

$$I_a^2 = \int_{0}^{a} e^{-x^2} \, dx \int_{0}^{a} e^{-y^2} \, dy = \int_{0}^{a} \int_{0}^{a} e^{-(x^2 + y^2)} \, dx \, dy$$

(b) Let $J_a = \iint\limits_R e^{-(x^2 + y^2)} \, dA$, where $R$ is the quarter circle

$$0 \le \theta \le \frac{\pi}{2}, \, 0 \le r \le a.$$ Show that

$$\left| I_a^2 - J_a \right| \le \frac{(4 - \pi)a^2}{4} e^{-a^2}$$

(c) Find $J_a$.

(d) Show that

$$\lim_{a \to \infty} J_a = \frac{\pi}{4} \quad \text{and} \quad \lim_{a \to \infty} \left| I_a^2 - J_a \right| = 0$$

(e) Show that $\int_{-\infty}^{\infty} e^{-x^2} \, dx = \sqrt{\pi}$. (This integral is of special importance in statistics.)

# 14.4 Center of Mass; Moment of Inertia

## 1 Find the Mass and the Center of Mass of a Lamina

**NEED TO REVIEW?** The center of mass of a homogeneous lamina is discussed in Section 6.8, pp. 487–491.

In many applications, thin sheets of material, such as copper stripping, are treated as if they were two-dimensional. A **lamina** is a plane area that represents a thin, flat sheet of material. If the mass density of the material is constant, the lamina is called **homogeneous**. The mass $M$ of a homogeneous lamina is $\rho A$, where $A$ is the area of the lamina and $\rho$ is its constant mass density.

**RECALL** The mass density of a two-dimensional material is defined as the mass per unit area of the material. In SI units, mass density is measured in kilograms per meter squared; in U.S. customary units, it is measured in slugs per foot squared.

However, materials usually are not homogeneous and so the mass density is variable. Suppose a lamina is represented by a rectangular region $R$ defined by $a \leq x \leq b$ and $c \leq y \leq d$, and its mass density $\rho = \rho(x, y)$ varies continuously over $R$. To find the mass $M$ of this lamina, we use double integration.

We begin by partitioning the interval $[a, b]$ into $n$ subintervals of length $\Delta x_i, i = 1, 2, \ldots, n$, and the interval $[c, d]$ into $m$ subintervals of length $\Delta y_j$, $j = 1, 2, \ldots, m$, as shown in Figure 31. Then in each rectangle $R_{ij}$ of area $\Delta A_{ij} = \Delta x_i \Delta y_j$, choose a point $(u_{ij}, v_{ij})$. An approximation to the mass $M_{ij}$ of the $ij$th rectangle is

$$M_{ij} = [\text{Mass density}] \times [\text{Area}] = \rho(u_{ij}, v_{ij})\Delta A_{ij}$$

The total mass $M$ of the lamina can be approximated by summing the masses of all the rectangles. That is,

$$M \approx \sum_{j=1}^{m} \sum_{i=1}^{n} \rho(u_{ij}, v_{ij})\Delta x_i \Delta y_j = \sum_{j=1}^{m} \sum_{i=1}^{n} \rho(u_{ij}, v_{ij})\Delta A_{ij}$$

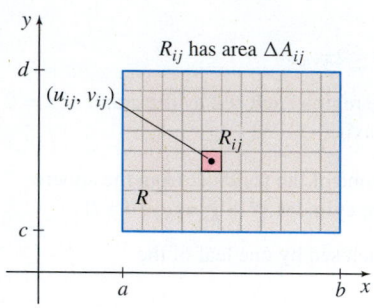

**Figure 31** Mass $M_{ij} = \rho(u_{ij}, v_{ij})\Delta A_{ij}$

If the norm, $\|\Delta\|$, defined to be the largest of the lengths of the subintervals $\Delta x_i$ or $\Delta y_j$ of the rectangular partition, approaches 0, then the mass $M$ of the lamina is given by a double integral. That is,

$$M = \lim_{\|\Delta\| \to 0} \sum_{j=1}^{m} \sum_{i=1}^{n} \rho(u_{ij}, v_{ij})\Delta A_{ij} = \iint\limits_{R} \rho(x, y)\, dA$$

**NOTE** Observe that if the mass density $\rho$ is constant, then the formula for the mass of the lamina reduces to
$M = \iint\limits_{R} \rho\, dA = \rho \iint\limits_{R} dA = \rho A$
where $A$ is the area of the region $R$.

If a particle of mass $m$ is located a distance $d$ from a fixed axis $a$, the **moment of mass** $M_a$ of the particle about the $a$-axis is defined as

$$M_a = md$$

The moment of mass $M_a$ measures the tendency of the particle to rotate about the $a$-axis.

**NEED TO REVIEW?** Moments of mass are discussed in Section 6.8, p. 459.

Look again at Figure 31. The moment of mass $M_{ij}$ about the $x$-axis of the $ij$th rectangle is approximated by

$$M_{ij} = [\text{Mass}] \times [\text{Distance from the } x\text{-axis}] = [\rho(u_{ij}, v_{ij})\Delta A_{ij}](v_{ij})$$

**NOTE** Sometimes $M_a$ is referred to as the first moment of the particle about the axis $a$.

By summing all the moments $M_{ij}$ about the $x$-axis and taking the limit as $\|\Delta\|$ approaches 0, we obtain the moment of mass $M_x$ about the $x$-axis of the lamina.

$$M_x = \lim_{\|\Delta\| \to 0} \sum_{j=1}^{m} \sum_{i=1}^{n} v_{ij}\, \rho(u_{ij}, v_{ij})\, \Delta A_{ij} = \iint\limits_{R} y\, \rho(x, y)\, dA$$

Similarly, the moment of mass $M_y$ about the $y$-axis of the lamina is given by

$$M_y = \lim_{\|\Delta\| \to 0} \sum_{j=1}^{m} \sum_{i=1}^{n} u_{ij}\, \rho(u_{ij}, v_{ij})\, \Delta A_{ij} = \iint_R x\, \rho(x, y)\, dA$$

It can be shown that the formulas for the mass and moments of a lamina, given above, also hold for any lamina represented by a closed, bounded region $R$ of the $xy$-plane whose boundary is a piecewise-smooth curve.

### THEOREM  Mass of a Lamina

Suppose a lamina of variable mass density $\rho = \rho(x, y)$ is represented by a closed, bounded region $R$ of the $xy$-plane whose boundary is a piecewise-smooth curve. If $\rho$ is continuous on $R$, then the **mass of the lamina** is

$$M = \iint_R \rho(x, y)\, dA$$

The **moment of mass about the the $x$-axis** is

$$M_x = \iint_R y\rho(x, y)\, dA$$

The **moment of mass about the $y$-axis** is

$$M_y = \iint_R x\rho(x, y)\, dA$$

The **center of mass of a lamina** is the point $(\bar{x}, \bar{y})$ whose coordinates satisfy the equations

$$\bar{x} = \frac{M_y}{M} = \frac{\displaystyle\iint_R x\rho(x, y)\, dA}{\displaystyle\iint_R \rho(x, y)\, dA} \qquad \bar{y} = \frac{M_x}{M} = \frac{\displaystyle\iint_R y\rho(x, y)\, dA}{\displaystyle\iint_R \rho(x, y)\, dA}$$

We distinguish between the *center of mass* (or *center of gravity*) of a lamina, which is defined by $(\bar{x}, \bar{y}) = \left(\dfrac{M_y}{M}, \dfrac{M_x}{M}\right)$, and the *centroid* of a lamina. The centroid is a purely geometric property of the lamina; it coincides with the center of mass in the case of a homogeneous lamina. To see the distinction, every square lamina has its centroid at the (geometric) center, but a square lamina with variable density will almost always have its center of mass off-center.

For a physical interpretation, suppose a piece of string is attached to the lamina at the center of mass $(\bar{x}, \bar{y})$. Then the lamina, when suspended from the string, will hang in a horizontal position.

**IN WORDS** The centroid of a lamina is its geometric center; the center of mass of a lamina is the point where the lamina is balanced.

### EXAMPLE 1  Finding the Mass and the Center of Mass of a Lamina

A lamina in the shape of an isosceles right triangle has variable density. Its mass density function $\rho = \rho(x, y)$ is directly proportional to the square of the distance from the vertex at the right angle. Find the mass and center of mass of the lamina.

**Solution** Position the lamina in the $xy$-plane so that the vertex at the right angle is at the origin and the two equal sides lie along the positive coordinate axes. See Figure 32. Suppose the equal sides of the triangle each measure $a$.

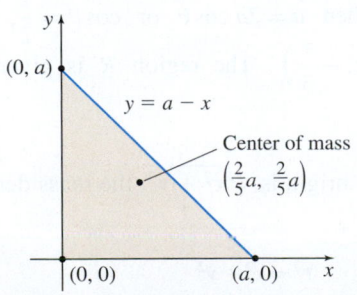

**DF** **Figure 32** $\rho(x, y) = k(x^2 + y^2)$

Since the distance of a point $(x, y)$ from the origin is $\sqrt{x^2 + y^2}$, the mass density of the lamina is

$$\rho(x, y) = k(x^2 + y^2)$$

where $k$ is the constant of proportionality. Let $R$ denote the region in the first quadrant enclosed by the triangle formed by the lamina. Since $\rho = \rho(x, y)$ is continuous on $R$, the mass $M$ of this lamina is given by a double integral.

$$M = \iint_R \rho(x, y)\, dA = \iint_R k(x^2 + y^2)\, dA = k \int_0^a \int_0^{a-x} (x^2 + y^2)\, dy\, dx$$

$$= k \int_0^a \left[ x^2 y + \frac{y^3}{3} \right]_0^{a-x} dx = k \int_0^a \left[ x^2 a - x^3 + \frac{(a-x)^3}{3} \right] dx$$

$$= k \int_0^a \frac{1}{3}(a^3 - 3a^2 x + 6ax^2 - 4x^3)\, dx = \frac{k}{3}\left[ a^3 x - 3a^2 \frac{x^2}{2} + 2ax^3 - x^4 \right]_0^a$$

$$= \frac{k}{3}\left( a^4 - \frac{3a^4}{2} + 2a^4 - a^4 \right) = \frac{ka^4}{6}$$

Due to the symmetry of both the region $R$ and the mass density, the center of mass $(\bar{x}, \bar{y})$ lies on the line $y = x$. So, once we find $\bar{x}$, we also know $\bar{y}$. The moment of mass with respect to the $y$-axis, $M_y$, is

$$M_y = \iint_R x\rho(x, y)\, dA = \iint_R xk(x^2 + y^2)\, dA = k \int_0^a \int_0^{a-x} (x^3 + xy^2)\, dy\, dx$$

$$= \frac{k}{3} \int_0^a (a^3 x - 3a^2 x^2 + 6ax^3 - 4x^4)\, dx = \frac{ka^5}{15}$$

The center of mass $(\bar{x}, \bar{y})$ is:

$$\bar{x} = \frac{M_y}{M} = \frac{\dfrac{ka^5}{15}}{\dfrac{ka^4}{6}} = \frac{2}{5}a = \bar{y}$$

The center of mass of the lamina is the point $\left( \dfrac{2}{5}a, \dfrac{2}{5}a \right)$. ∎

NOW WORK Problem 11.

CALC
CLIP
▶ EXAMPLE 2   **Finding the Mass and the Center of Mass of a Lamina**

Find the mass $M$ and the center of mass $(\bar{x}, \bar{y})$ of a lamina in the shape of a region $R$ in the $xy$-plane that lies outside the circle $x^2 + y^2 = a^2$ and inside the circle $x^2 + y^2 = 2ax$. The mass density $\rho$ is inversely proportional to the distance from the origin.

**Solution** Figure 33 illustrates the region $R$. Since the boundary of $R$ involves two circles, we use polar coordinates. Then the two circles are given by $r = a$ and by $r = 2a \cos \theta$. The circles intersect when $a = 2a \cos \theta$ or $\cos \theta = \dfrac{1}{2}$, and the points of intersection are $\left( a, \dfrac{\pi}{3} \right)$ and $\left( a, -\dfrac{\pi}{3} \right)$. The region $R$ is given by

$$a \le r \le 2a \cos \theta, \quad -\frac{\pi}{3} \le \theta \le \frac{\pi}{3}.$$

Since the distance of a point $(x, y)$ from the origin is $\sqrt{x^2 + y^2}$, the mass density of the lamina at any point $(x, y)$ in $R$ is

$$\rho(x, y) = \frac{k}{\sqrt{x^2 + y^2}} = \frac{k}{r} \qquad r^2 = x^2 + y^2$$

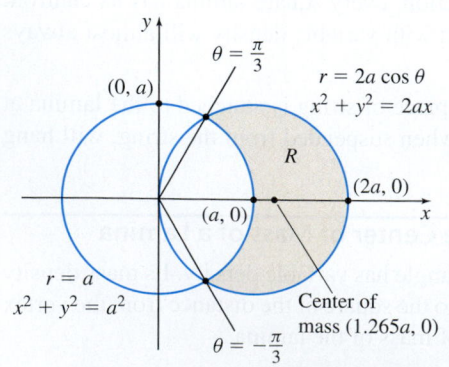

**Figure 33** $a \le r \le 2a \cos \theta, \ -\dfrac{\pi}{3} \le \theta \le \dfrac{\pi}{3}$

where $k$ is the constant of proportionality. Since $R$ does not contain the origin, the mass density $\rho = \rho(x, y)$ is continuous on $R$. The mass $M$ of the lamina is

$$M = \iint_R \rho(x, y)\, dA = \iint_R \frac{k}{r} \cdot r\, dr\, d\theta = k \int_{-\pi/3}^{\pi/3} \int_a^{2a\cos\theta} dr\, d\theta$$

$$= k \int_{-\pi/3}^{\pi/3} (2a\cos\theta - a)\, d\theta = k\left[2a\sin\theta - a\theta\right]_{-\pi/3}^{\pi/3} = k\left(2a\sqrt{3} - a\frac{2\pi}{3}\right)$$

$$= \frac{2ka}{3}(3\sqrt{3} - \pi)$$

Both the mass density function and the region $R$ are symmetric about the $x$-axis. So, the center of mass lies on the $x$-axis; that is, $\bar{y} = 0$. To find $\bar{x}$, we first find $M_y$.

$$M_y = \iint_R x\rho(x, y)\, dA = \iint_R r\cos\theta \cdot \frac{k}{r} \cdot r\, dr\, d\theta = k \int_{-\pi/3}^{\pi/3} \int_a^{2a\cos\theta} r\cos\theta\, dr\, d\theta$$

$$= \frac{k}{2} \int_{-\pi/3}^{\pi/3} (4a^2\cos^2\theta - a^2)\cos\theta\, d\theta = a^2\frac{k}{2} \int_{-\pi/3}^{\pi/3} (3 - 4\sin^2\theta)\cos\theta\, d\theta$$

$$= \frac{a^2 k}{2}\left[3\sin\theta - \frac{4\sin^3\theta}{3}\right]_{-\pi/3}^{\pi/3} = ka^2\sqrt{3}$$

Then

$$\bar{x} = \frac{M_y}{M} = \frac{ka^2\sqrt{3}}{\dfrac{2ka}{3}(3\sqrt{3} - \pi)} = \frac{3a\sqrt{3}}{2(3\sqrt{3} - \pi)} \approx 1.265a$$

The center of mass is approximately $(1.265a, 0)$. ∎

**NOW WORK** Problem **17.**

## 2 Find Moments of Inertia

In rectilinear motion, the mass $m$ of an object can be thought of as a measure of the natural resistance of the object to being accelerated. In rotational motion, the *inertia* $I$ of an object can be thought of as a measure of the resistance of the object to being rotated about an axis.

Suppose an object of mass $m$ is located a distance $d$ from a fixed axis $a$. Then its tendency to rotate about the $a$-axis, called the **moment of inertia** $I_a$, is defined as

$$\boxed{I_a = md^2}$$

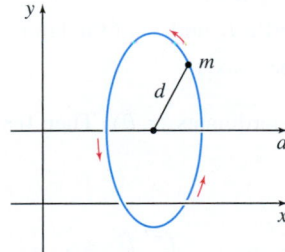

**Figure 34**

See Figure 34.

Proceeding as we did for moments of mass, the moment of inertia $I_x$ about the $x$-axis of a lamina of variable mass density $\rho(x, y)$ represented by a closed, bounded region $R$ of the $xy$-plane, whose boundary is a piecewise-smooth curve, is given by

$$I_x = \lim_{\|\Delta\| \to 0} \sum_{j=1}^m \sum_{i=1}^n [\underbrace{\rho(u_{ij}, v_{ij})\Delta A_{ij}}_{\text{Mass}} \cdot \underbrace{v_{ij}^2}_{\substack{\text{Distance from} \\ \text{$x$-axis squared}}}] = \iint_R y^2\rho(x, y)\, dA$$

**NOTE** The moment of inertia $I_a$ is often referred to as the **second moment** of the particle **about the axis** $a$. It is called the second moment because the distance is squared.

Similarly, the moment of inertia $I_y$ about the $y$-axis is given by

$$I_y = \lim_{\|\Delta\| \to 0} \sum_{j=1}^m \sum_{i=1}^n [\underbrace{\rho(u_{ij}, v_{ij})\Delta A_{ij}}_{\text{Mass}} \cdot \underbrace{u_{ij}^2}_{\substack{\text{Distance from} \\ \text{$y$-axis squared}}}] = \iint_R x^2\rho(x, y)\, dA$$

and the moment of inertia $I_O$ about the origin (the $z$-axis) is given by

$$I_O = \lim_{\|\Delta\| \to 0} \sum_{j=1}^{m} \sum_{i=1}^{n} \underbrace{\rho(u_{ij}, v_{ij}) \Delta A_{ij}}_{\text{Mass}} \cdot \underbrace{(u_{ij}^2 + v_{ij}^2)}_{\substack{\text{Distance from} \\ \text{origin squared}}} = \iint_R (x^2 + y^2)\rho(x, y)\, dA$$

The moment of inertia $I_O$ about the origin is referred to as the **polar moment of inertia** or the **polar second moment**.

Using the sum property of double integrals, the sum of the moments of inertia about the $x$-axis and the $y$-axis equals the moment of inertia about the origin. That is,

$$\boxed{I_x + I_y = I_O}$$

The polar moment of inertia $I_O$ is often used to find the moments $I_x$ and $I_y$ when there is symmetry, that is, when $I_x = I_y$.

The distance $k$ from a fixed axis $I$ (line or point) to a point where all the mass $M$ of the lamina could be concentrated without altering the moment of inertia of the lamina about the axis (line or point), is called the **radius of gyration** $k$, and is given by $k = \sqrt{\dfrac{I}{M}}$.

In particular, if $k_x = \sqrt{\dfrac{I_y}{M}}$ is the radius of gyration of a mass $M$ about the $y$-axis, and $k_y = \sqrt{\dfrac{I_x}{M}}$ is the radius of gyration of a mass $M$ about the $x$-axis, then the point $(k_x, k_y)$ is the point at which the mass of a lamina can be concentrated without affecting the moments of inertia about either of the coordinate axes.

---

**EXAMPLE 3   Finding Moments of Inertia**

(a) Find the polar moment of inertia $I_O$ of a homogeneous lamina of mass density $\rho$ in the shape of a region $R$ in the $xy$-plane enclosed by the circle $x^2 + y^2 = c^2$, $c > 0$.

(b) Use the polar moment to find the moments of inertia $I_x$ and $I_y$ of the lamina.

(c) Find the radius of gyration of the lamina about the $y$-axis.

**Solution** The region $R$ is circular so we use polar coordinates $(r, \theta)$. Then $0 \le r \le c$, $0 \le \theta \le 2\pi$, and $dA = r\, dr\, d\theta$.

**(a)** The polar moment of inertia is

$$I_O = \iint_R (x^2 + y^2)\rho(x, y)\, dA = \iint_R r^2 \rho r\, dr\, d\theta = \rho \int_0^{2\pi} \int_0^c r^3\, dr\, d\theta$$

$$\underset{\substack{\uparrow \\ x^2 + y^2 = r^2}}{} \qquad \underset{\substack{\uparrow \\ \rho \text{ is constant.}}}{}$$

$$= \rho \int_0^{2\pi} \frac{c^4}{4}\, d\theta = \frac{\pi c^4 \rho}{2}$$

**(b)** By symmetry, the moment of inertia about the $x$-axis, $I_x$, equals the moment of inertia about the $y$-axis, $I_y$. Then

$$I_x = I_y = \frac{1}{2} I_O = \frac{1}{2}\left(\frac{\pi c^4 \rho}{2}\right) = \frac{\pi c^4 \rho}{4}$$

**(c)** We begin by finding the mass $M$ of the lamina. Since the lamina is homogeneous, the mass density $\rho$ is constant. So $M = \pi \rho c^2$. Then the radius of gyration of the lamina about the $y$-axis is $k_x = \sqrt{\dfrac{I_y}{M}} = \sqrt{\dfrac{\dfrac{\pi c^4 \rho}{4}}{\pi \rho c^2}} = \dfrac{c}{2}$. The radius of gyration of the lamina about the $y$-axis is one half the radius of the lamina.  ∎

**NOW WORK** Problem 25.

In dynamics, the moment of inertia of a lamina occurs in connection with the study of rotational motion. If a rigid object (a lamina) is rotated about an axis $a$ with angular speed $\omega$, then its **rotational kinetic energy** $K$ is given by

$$K = \frac{1}{2} I_a \, \omega^2$$

For example, suppose the homogeneous lamina in Example 3 has mass density $\rho = 20 \, \text{kg/m}^2$ and radius $c = 1$ m. If it is rotated about the origin at the constant angular speed $\omega = 2\pi$ rad/s, then its kinetic energy $K$ is

$$K = \frac{1}{2} I_0 \omega^2 = \frac{1}{2} \left( \frac{\pi c^4 \rho}{2} \right) (2\pi)^2 \underset{\underset{c=1; \, \rho=20}{\uparrow}}{=} \frac{1}{2} (10\pi)(4\pi^2) = 20\pi^3 \, \frac{\text{kg/m}^2}{\text{s}^2}$$

Observe the similarity between the formula for rotational kinetic energy $K$ presented here and the formula for the kinetic energy $K$ of an object of mass $m$ moving in a straight line with speed $v$, namely, $K = \frac{1}{2} m v^2$.

---

## 14.4 Assess Your Understanding

### Concepts and Vocabulary

1. If a lamina represented by a closed, bounded region $R$ has a continuous mass density function $\rho = \rho(x, y)$, then the double integral $\iint\limits_R \rho(x, y) \, dA$ gives the _____ of the lamina.

2. *True or False*   If a lamina represented by a closed, bounded region $R$ has a continuous mass density function $\rho = \rho(x, y)$, then $\iint\limits_R y \, \rho(x, y) \, dA$ is called the moment of mass about the $y$-axis.

3. If a lamina represented by a closed, bounded region $R$ has a continuous mass density function $\rho = \rho(x, y)$, the moment of mass about the $y$-axis $M_y$ is given by the double integral _____.

4. *True or False*   Center of mass and centroid are synonyms.

5. *True or False*   The moment of mass $M_a$ of a particle about the $a$-axis measures the tendency of the particle to rotate about the $a$-axis.

6. *True or False*   The moment of inertia about the origin equals the difference between the moment of inertia about the $x$-axis and the moment of inertia about the $y$-axis.

### Skill Building

7. Find the mass $M$ and center of mass $(\bar{x}, \bar{y})$ of the triangular lamina shown in the figure. The mass density function of the lamina is $\rho = \rho(x, y) = x^2 y$.

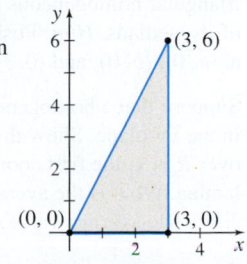

8. Find the mass $M$ and the center of mass $(\bar{x}, \bar{y})$ of the square lamina shown in the figure. The mass density function of the lamina is $\rho = \rho(x, y) = xy^2$.

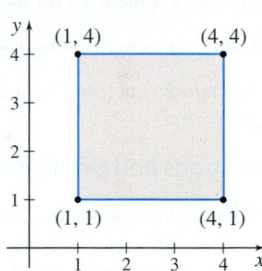

*In Problems 9–20, find the mass and center of mass of each lamina.*

9. The lamina enclosed by the lines $x = 2$ and $y = 4$, and the coordinate axes, and whose mass density is $\rho = \rho(x, y) = 3x^2 y$

10. The lamina enclosed by the lines $x = 1$ and $y = 2$, and the coordinate axes, and whose mass density is $\rho = \rho(x, y) = 2x^2 y^2$

11. The lamina in the first quadrant enclosed by $y^2 = x$, $x = 1$, and the $x$-axis, and whose mass density is $\rho = \rho(x, y) = 2x + 3y$

12. The lamina in the first quadrant enclosed by $y^2 = 4x$, $x = 1$, and the $x$-axis, and whose mass density is $\rho = \rho(x, y) = x + 1$

13. The lamina enclosed by $y^2 = x$ and $y = x$, and whose mass density $\rho = \rho(x, y)$ is proportional to the distance from the $y$-axis

14. The lamina enclosed by $y = \sin x$ and the $x$-axis, $0 \le x \le \pi$, and whose mass density $\rho = \rho(x, y)$ is proportional to the distance from the $x$-axis

15. The lamina enclosed by the lines $2x + 3y = 6$, $x = 0$, and $y = 0$, and whose mass density $\rho = \rho(x, y)$ is proportional to the sum of the distances from the coordinate axes

16. The lamina enclosed by the lines $3x + 4y = 12$, $x = 0$, and $y = 0$, and whose mass density $\rho = \rho(x, y)$ is proportional to the product of the distances from the coordinate axes

---

**1.** = NOW WORK problem        〔◿〕 = Graphing technology recommended        〔CAS〕 = Computer Algebra System recommended

**17.** The lamina enclosed by the cardioid $r = 1 + \sin\theta$, and whose mass density $\rho = \rho(r, \theta)$ is proportional to the distance from the pole

**18.** The lamina enclosed by $r = \cos\theta$, $-\dfrac{\pi}{2} \le \theta \le \dfrac{\pi}{2}$, whose mass density $\rho = \rho(r, \theta)$ is proportional to the distance from the pole

**19.** A lamina inside the graph of $r = 2a\sin\theta$ and outside the graph of $r = a$, whose mass density $\rho = \rho(r, \theta)$ is inversely proportional to the distance from the pole

**20.** A lamina outside the limaçon $r = 2 - \cos\theta$ and inside the circle $r = 4\cos\theta$, whose mass density $\rho = \rho(r, \theta)$ is inversely proportional to the distance from the pole

*In Problems 21–26, find the mass, the moment of inertia about the indicated axis, and the radius of gyration for each homogeneous lamina of mass density $\rho$.*

**21.** The lamina enclosed by the lines $2x + 3y = 6$, $x = 0$, and $y = 0$ about the $x$-axis

**22.** Rework Problem 21 for the moment of inertia about the $y$-axis.

**23.** The lamina in the first quadrant enclosed by the lines $x = a$ and $y = b$ about the $y$-axis

**24.** Rework Problem 23 for the moment of inertia about the $x$-axis.

**25.** The lamina enclosed by $y = x^2$ and $y = 2 - x^2$; about the $y$-axis

**26.** The lamina enclosed by the loop of $y^2 = x^2(4 - x)$; about the $y$-axis

## Applications and Extensions

**27. Mass** Find the mass of a flat circular washer with inner radius $a$ and outer radius $2a$ if its mass density $\rho = \rho(x, y)$ is inversely proportional to the square of the distance from the center. See the figure.

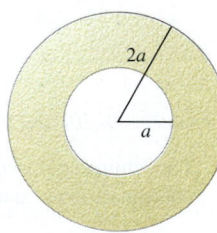

**28. Mass** Rework Problem 27 if the mass density $\rho = \rho(x, y)$ is inversely proportional to the distance from the center.

**29. Mass and Center of Mass** Find the mass and center of mass of a lamina in the shape of the region enclosed on the left by the line $x = a$, $a > 0$, and on the right by the circle $r = 2a\cos\theta$ if its mass density $\rho$ is inversely proportional to the distance from the $y$-axis.

**30. Mass and Center of Mass** Find the mass and center of mass of a lamina in the shape of the smaller region cut from the circle $r = 6$ by the line $r\cos\theta = 3$ if its mass density is $\rho = \rho(r, \theta) = \cos^2\theta$.

**31. Mass** Find the mass of the lamina enclosed by $y = x^2$ and $y = x^3$; the mass density is $\rho = \rho(x, y) = \sqrt{xy}$.

**32. Center of Mass** Find the center of mass of the lamina inside $r = 4\cos\theta$ and outside $r = 2\sqrt{3}$ if the mass density $\rho = \rho(r, \theta)$ is inversely proportional to the distance from the origin.

**33. Center of Mass** Find the mass and the center of mass of the lamina enclosed by $x = y - 2$ and $x = -y^2$, if the mass density is $\rho = \rho(x, y) = x^2$.

**34. Center of Mass** Find the center of mass of the lamina enclosed by $y = x^2$ and $x - 2y + 1 = 0$, if the mass density is $\rho = \rho(x, y) = 2x + 8y + 2$.

**35. Center of Mass** Find the center of mass of the homogeneous lamina enclosed by $bx^2 = a^2 y$ and $ay = bx$, $a > 0$, $b > 0$.

**CAS 36. Center of Mass** Find the center of mass of the homogeneous lamina enclosed by $y = \ln x$, the $x$-axis, and the line $x = e^2$.

**37. Moment of Inertia** A homogeneous lamina is in the shape of a square with sides $s$. Find the moment of inertia about
(a) a side    (b) a diagonal    (c) the centroid.

**38. Moment of Inertia** A homogeneous lamina is in the shape of an equilateral triangle with each side measuring $s$. Find the moment of inertia about
(a) a side    (b) an altitude.

**39. Moment of Inertia** The mass density at each point of a circular washer of inner radius $a$ and outer radius $b$ is inversely proportional to the square of the distance from the center.

(a) Find the mass of the washer. Then discuss its behavior as $a \to 0^+$. (Note that the mass density becomes unbounded as we approach the center of the washer.)

(b) Find the moment of inertia of the washer about its center, and show that (unlike the mass) it remains finite as $a \to 0^+$.

**40. Moment of Inertia**

(a) Graph the limaçon $r = 3 + 2\cos\theta$ and the circle $r = 2$.

(b) Set up the integral for the moment of inertia about the $x$-axis of the lamina inside the limaçon and outside the circle from (a) if the mass density $\rho$ of the lamina is inversely proportional to the square of the distance from the origin.

**CAS (c)** Find the integral in (b).

**41.** Show that the center of mass of a rectangular homogeneous lamina lies at the intersection of its diagonals.

**42.** A homogeneous lamina is in the shape of the region enclosed by a right triangle of base $b$ and height $h$. Show that its moment of inertia about the base is $\dfrac{1}{6}mh^2$, where $m$ is the mass of the lamina.
*Hint:* Position the triangle so that the base lies on the positive $x$-axis from $(0, 0)$ to $(b, 0)$.

## Challenge Problems

**43.** Show that the center of mass of the region enclosed by a triangular homogeneous lamina lies at the point of intersection of its medians. *Hint:* Position the triangle so that its vertices are at $(a, 0)$, $(b, 0)$, and $(0, c)$, where $a < 0$, $b > 0$, and $c > 0$

**44.** Suppose that a homogeneous lamina occupies a region $R$ in the $xy$-plane. Show that the average value of $f(x, y) = x$ over $R$ is $\bar{x}$, the first coordinate of the center of mass of the lamina. What is the average value of $g(x, y) = y$ over $R$?
*Hint:* The average value of $f$ over a region $R$ is defined to be the number $\dfrac{1}{A} \iint\limits_R f(x, y)\, dA$.

**45. (a)** Show that the moment of inertia about the $z$-axis of the thin flat plate in the figure equals the sum of its moments of inertia about the $x$- and $y$-axes.

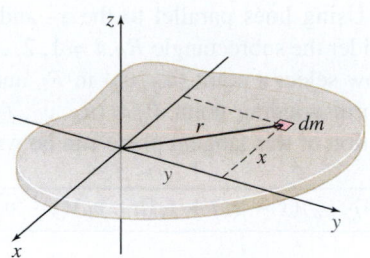

**(b)** Given that the moment of inertia of a homogeneous disk about an axis through its center and perpendicular to its plane is $\dfrac{mR^2}{2}$ (where $m$ is mass and $R$ is the radius), use (a) to find its moment of inertia about a diameter.

**(c)** What is the moment of inertia of a disk about an axis tangent to its edge?

# 14.5 Surface Area

**OBJECTIVE** *When you finish this section, you should be able to:*

**1** Find the area of a surface that lies above a region $R$ (p. 985)

So far we have used a double integral to find the volume under a surface $z = f(x, y) \geq 0$ and above a closed, bounded region $R$ of the $xy$-plane. In this section, we use a double integral to find the surface area of $z = f(x, y) \geq 0$ above a closed, bounded region $R$ whose boundary is a piecewise-smooth curve. In developing the formula for surface area, we use tangent planes to approximate surface area.

## 1 Find the Area of a Surface That Lies Above a Region $R$

We begin with the simple case where the surface is a plane $z = ax + by + c$ and the region $R$ is a rectangle with sides of lengths $\Delta x$ and $\Delta y$. The surface area $S_p$ is the area of the part of the plane that lies above the rectangle, in this case, a parallelogram, as shown in Figure 35. Notice that this parallelogram is the intersection of the plane and the cylinder formed by projecting the perimeter of the rectangle with sides $\Delta x$ and $\Delta y$ vertically upward.

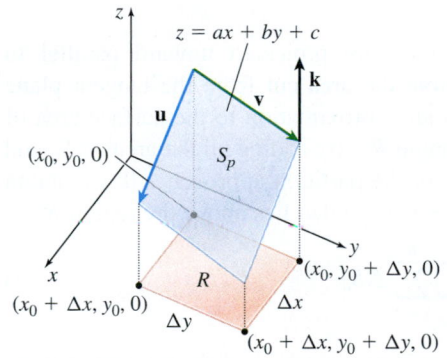

**Figure 35**

**NEED TO REVIEW?** Vectors are discussed in Section 10.2, pp. 738–741, and properties of the cross product are discussed in Section 10.5, pp. 764–769.

The adjacent sides of the parallelogram are given by the vectors $\mathbf{u}$ and $\mathbf{v}$. Both $\mathbf{u}$ and $\mathbf{v}$ have the same initial point $(x_0, y_0, ax_0 + by_0 + c)$ on the parallelogram. The terminal point of $\mathbf{u}$ is $(x_0 + \Delta x, y_0, a(x_0 + \Delta x) + by_0 + c)$ and the terminal point of $\mathbf{v}$ is $(x_0, y_0 + \Delta y, ax_0 + b(y_0 + \Delta y) + c)$. So, $\mathbf{u}$ and $\mathbf{v}$ are given by

$$\mathbf{u} = \Delta x \mathbf{i} + a \Delta x \mathbf{k}, \qquad \mathbf{v} = \Delta y \mathbf{j} + b \Delta y \mathbf{k}$$

The area of the parallelogram is

$$\|\mathbf{u} \times \mathbf{v}\| = \|\Delta x (\mathbf{i} + a\mathbf{k}) \times \Delta y (\mathbf{j} + b\mathbf{k})\| = \|(\mathbf{i} + a\mathbf{k}) \times (\mathbf{j} + b\mathbf{k})\| \, \Delta x \Delta y$$

$$= \|-a\mathbf{i} - b\mathbf{j} + \mathbf{k}\| \, \Delta x \Delta y = \sqrt{a^2 + b^2 + 1} \, \Delta x \Delta y$$

Let $R$ be a closed rectangular region in the $xy$-plane with sides of length $\Delta x$ and $\Delta y$. If the perimeter of $R$ is projected upward parallel to the $z$-axis, the result is a cylinder. Let $S_p$ denote the area cut from the plane $z = ax + by + c$ by this cylinder. Then the area $S_p$ is given by

$$\boxed{S_p = \sqrt{a^2 + b^2 + 1} \, \Delta x \Delta y} \tag{1}$$

We now develop a formula for the surface area of the graph of a nonnegative function $z = f(x, y)$ that is defined over a closed, rectangular region $R$. We assume that $f$ has first-order partial derivatives that are continuous at every point of the region $R$.

See Figure 36(a). Using lines parallel to the $x$- and $y$-axes, we partition $R$ into subrectangles and consider the subrectangle $R_{ij}$, $i = 1, 2, \ldots, n$, $j = 1, 2, \ldots, m$ having area $A_{ij} = \Delta x_i \Delta y_j$. Now select a point $(u_{ij}, v_{ij})$ in $R_{ij}$ and construct the tangent plane to the surface at the corresponding point $P_{ij} = (u_{ij}, v_{ij}, f(u_{ij}, v_{ij}))$ on the surface. See Figure 36(b). The equation of this tangent plane can be written as

$$z = f_x(u_{ij}, v_{ij})(x - u_{ij}) + f_y(u_{ij}, v_{ij})(y - v_{ij}) + f(u_{ij}, v_{ij}) \qquad (2)$$

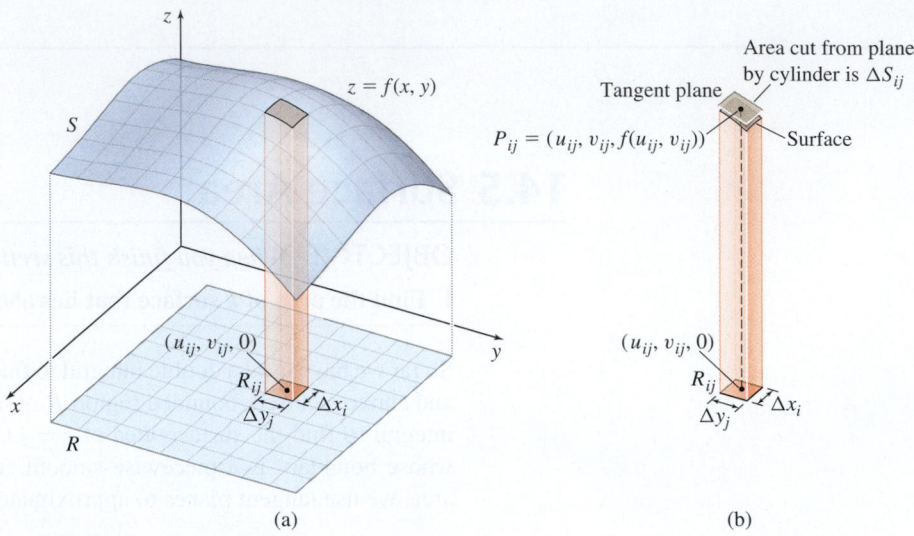

(a)    (b)

**Figure 36**

The sides of the rectangle $R_{ij}$, when they are projected upward parallel to the $z$-axis, result in a cylinder. Let $\Delta S_{ij}$ denote the area cut from the tangent plane at the point $P_{ij}$ by the cylinder. Then $\Delta S_{ij}$ is an approximation to the surface area of the part of the surface that lies above the rectangle $R_{ij}$. By adding all the areas $\Delta S_{ij}$ and taking the limit of the sums as the norm $\|\Delta\|$ of the partition approaches 0, we obtain the surface area $S$ of the part of the surface $z = f(x, y)$ that lies above the region $R$.

$$S = \lim_{\|\Delta\| \to 0} \sum_{j=1}^{m} \sum_{i=1}^{n} \Delta S_{ij} \qquad (3)$$

To obtain a formula for $\Delta S_{ij}$, we combine (2), the equation of the tangent plane to the surface at the point $P_{ij}$,

$$z = f_x(u_{ij}, v_{ij})(x - u_{ij}) + f_y(u_{ij}, v_{ij})(y - v_{ij}) + f(u_{ij}, v_{ij}) \qquad \textbf{Equation (2)}$$

and formula (1) for the surface area $S_p$ of a plane $z = ax + by + c$

$$S_p = \sqrt{a^2 + b^2 + 1}\, \Delta x \Delta y \qquad \textbf{Formula (1)}$$

The result is

$$\Delta S_{ij} = \sqrt{[f_x(u_{ij}, v_{ij})]^2 + [f_y(u_{ij}, v_{ij})]^2 + 1}\, \Delta A_{ij}$$

Now substitute this expression for $\Delta S_{ij}$ into (3) to obtain the surface area $S$ of the part of the surface that lies above the region $R$. Then

$$S = \lim_{\|\Delta\| \to 0} \sum_{j=1}^{m} \sum_{i=1}^{n} \sqrt{[f_x(u_{ij}, v_{ij})]^2 + [f_y(u_{ij}, v_{ij})]^2 + 1}\, \Delta x_i \Delta y_j$$

Since the first-order partial derivatives of $f$ are continuous on $R$, the limit exists and $S$ is a double integral.

$$S = \iint\limits_{R} \sqrt{[f_x(x, y)]^2 + [f_y(x, y)]^2 + 1}\, dA$$

It can be shown that the surface area $S$ of the graph of a nonnegative function $z = f(x, y)$ over a closed, bounded region $R$ whose boundary is a piecewise-smooth curve can be found using a double integral provided the partial derivatives $f_x$ and $f_y$ are continuous on $R$.

### Surface Area Above a Region $R$

Let $z = f(x, y)$ be a function of two variables defined on a closed, bounded region $R$ whose boundary is a piecewise-smooth curve. If $f_x$ and $f_y$ are continuous on $R$, then the surface area $S$ of the part of the surface that lies above $R$ is given by

$$S = \iint\limits_{R} \sqrt{[f_x(x, y)]^2 + [f_y(x, y)]^2 + 1}\, dA$$

**CALC CLIP** ▶ **EXAMPLE 1  Finding Surface Area**

Find the surface area of the part of the surface $z = f(x, y) = \dfrac{2}{3}(x^{3/2} + y^{3/2})$ that lies above the region $R$, a rectangle enclosed by the lines $x = 0$, $x = 1$, $y = 0$, and $y = 2$.

**Solution** Figure 37 shows the surface $z = f(x, y) = \dfrac{2}{3}(x^{3/2} + y^{3/2})$ and the surface area $S$ above the rectangle $R$. We begin by finding the partial derivatives of $z = f(x, y)$:

$$f_x(x, y) = x^{1/2} \qquad f_y(x, y) = y^{1/2}$$

Since $f_x$ and $f_y$ are continuous on $R$, the surface area $S$ above $R$ is

$$S = \iint\limits_{R} \sqrt{[f_x(x, y)]^2 + [f_y(x, y)]^2 + 1}\, dA = \iint\limits_{R} \sqrt{(x^{1/2})^2 + (y^{1/2})^2 + 1}\, dA$$

$$= \int_0^1 \int_0^2 \sqrt{x + y + 1}\, dy\, dx = \int_0^1 \left[ \frac{2}{3}(x + y + 1)^{3/2} \right]_0^2 dx$$

$$= \frac{2}{3} \int_0^1 \left[ (x + 3)^{3/2} - (x + 1)^{3/2} \right] dx = \frac{2}{3} \cdot \frac{2}{5} \cdot \left[ (x + 3)^{5/2} - (x + 1)^{5/2} \right]_0^1$$

$$= \frac{4}{15}(32 - 9\sqrt{3} - 4\sqrt{2} + 1) = \frac{4}{15}(33 - 9\sqrt{3} - 4\sqrt{2}) \qquad ∎$$

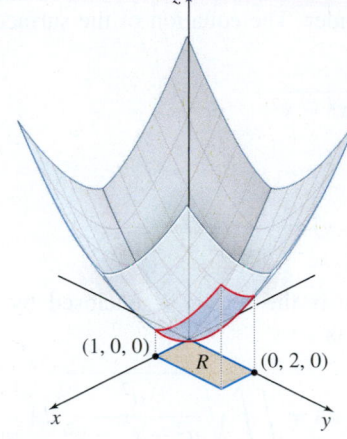

**Figure 37** $z = \dfrac{2}{3}(x^{3/2} + y^{3/2})$

(1, 0, 0)    $R$    (0, 2, 0)

**NOW WORK** Problem 5.

### EXAMPLE 2  Finding Surface Area

Find the surface area of the part of the paraboloid $z = f(x, y) = 1 - x^2 - y^2$ that lies above the $xy$-plane.

**Solution** Figure 38 shows the part of the surface $z = 1 - x^2 - y^2$ that lies above the $xy$-plane and its projection onto the $xy$-plane, the region $R$ enclosed by the circle $x^2 + y^2 = 1$. We begin by finding the partial derivatives of $z = f(x, y)$.

$$f_x(x, y) = -2x \qquad f_y(x, y) = -2y$$

Since $f_x$ and $f_y$ are continuous on $R$, the surface area $S$ of the part of the paraboloid $z = f(x, y) = 1 - x^2 - y^2$ that lies above $R$ is

$$S = \iint\limits_{R} \sqrt{[f_x(x, y)]^2 + [f_y(x, y)]^2 + 1}\, dA = \iint\limits_{R} \sqrt{(-2x)^2 + (-2y)^2 + 1}\, dA$$

$$= \iint\limits_{R} \sqrt{4(x^2 + y^2) + 1}\, dA$$

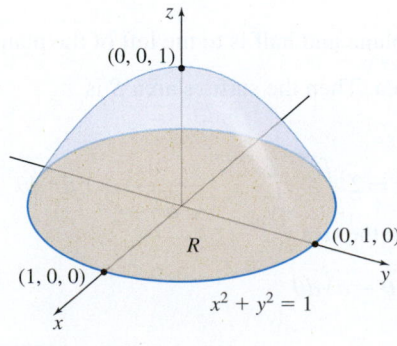

**Figure 38** $z = 1 - x^2 - y^2$, $z \geq 0$

(0, 0, 1)

$R$    (0, 1, 0)

(1, 0, 0)    $x^2 + y^2 = 1$

Since both the region $R$ (a circle) and the integrand involve $x^2 + y^2$, we use polar coordinates $(r, \theta)$. For the region $R$, we have $0 \le r \le 1$ and $0 \le \theta \le 2\pi$. Then

$$S = \iint_R \sqrt{4(x^2 + y^2) + 1} \, dA = \int_0^{2\pi} \int_0^1 \sqrt{4r^2 + 1} \, r \, dr \, d\theta \qquad \substack{x^2 + y^2 = r^2 \quad dA = r \, dr \, d\theta}$$

$$= \int_0^{2\pi} \frac{1}{12} \left[ (4r^2 + 1)^{3/2} \right]_0^1 d\theta \qquad \text{Let } u = 4r^2 + 1, \text{ then } du = 8r \, dr.$$

$$= \frac{1}{12}(5\sqrt{5} - 1) \int_0^{2\pi} d\theta = \frac{1}{12}(5\sqrt{5} - 1)2\pi$$

$$= \frac{\pi}{6}(5\sqrt{5} - 1) \qquad\qquad\qquad ∎$$

**NOW WORK** Problem 9.

---

**EXAMPLE 3    Finding Surface Area**

Find the surface area of the part of the sphere $x^2 + y^2 + z^2 = a^2$ that lies above the $xy$-plane and is contained within the cylinder $x^2 + y^2 = ax$, $a > 0$.

**Solution** Figure 39 shows the sphere and the cylinder. The equation of the surface in explicit form is

$$z = f(x, y) = \sqrt{a^2 - x^2 - y^2}$$

The partial derivatives of $z = f(x, y)$ are

$$f_x(x, y) = -\frac{x}{\sqrt{a^2 - x^2 - y^2}} \qquad f_y(x, y) = -\frac{y}{\sqrt{a^2 - x^2 - y^2}}$$

The projection of the cylinder onto the $xy$-plane is the region $R$ enclosed by the circle $x^2 + y^2 = ax$, $a > 0$. Then the surface area $S$ is

$$S = \iint_R \sqrt{\frac{x^2}{a^2 - x^2 - y^2} + \frac{y^2}{a^2 - x^2 - y^2} + 1} \, dA = \iint_R \sqrt{\frac{a^2}{a^2 - x^2 - y^2}} \, dA$$

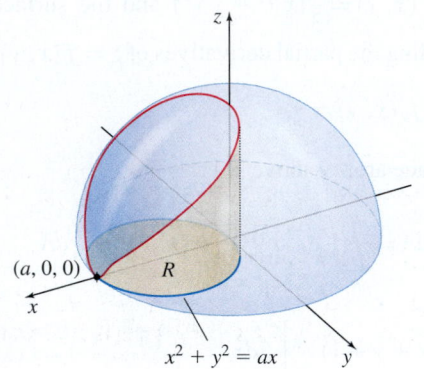

**Figure 39** $x^2 + y^2 + z^2 = a^2$, $z \ge 0$

$(a, 0, 0)$    $R$    $x^2 + y^2 = ax$

Since the integrand involves the expression $x^2 + y^2$ and the boundary of $R$ is the circle $x^2 + y^2 = ax$, we use polar coordinates $(r, \theta)$. We begin by finding the limits of integration. Since

$$x^2 + y^2 = ax$$

$$r^2 = ar \cos \theta$$

$$r = a \cos \theta$$

we find that $r$ varies from $0$ to $a \cos \theta$.

Since half the area is to the right of the $xz$-plane and half is to the left of the plane, we can let $\theta$ vary from $0$ to $\dfrac{\pi}{2}$ and double the area. Then the surface area $S$ is

$$S = \iint_R \sqrt{\frac{a^2}{a^2 - x^2 - y^2}} \, dA = \underset{a > 0}{\iint_R} \frac{a}{\sqrt{a^2 - r^2}} \, r \, dr \, d\theta = \underset{\text{Double the area}}{2 \int_0^{\pi/2} \int_0^{a \cos \theta} \frac{a}{\sqrt{a^2 - r^2}} \, r \, dr \, d\theta}$$

$$= -2a \int_0^{\pi/2} \left[ \sqrt{a^2 - r^2} \right]_0^{a \cos \theta} d\theta = -2a \int_0^{\pi/2} (a \sin \theta - a) \, d\theta$$

$$= -2a^2 \left[ -\cos \theta - \theta \right]_0^{\pi/2} = a^2(\pi - 2) \qquad\qquad ∎$$

**NOW WORK** Problem 13.

## 14.5 Assess Your Understanding

### Concepts and Vocabulary

1. *True or False* Suppose $R$ is a closed, rectangular region in the $xy$-plane with sides of length $\Delta x$ and $\Delta y$. If the perimeter of $R$ is projected upward parallel to the $z$-axis, the result is a cylinder. If $S_p$ is the area cut from the plane $z = ax + by + c$ by this cylinder, then $S_p = \sqrt{a^2 + b^2 + 1}\,\Delta x \Delta y$.

2. *True or False* Suppose $z = f(x, y)$ is a function of two variables defined on a closed, bounded region $R$. If $f_x$ and $f_y$ are continuous on $R$, then the area $S$ of the part of the surface that lies over $R$ is given by $S = \iint\limits_R \sqrt{f_x(x, y) + f_y(x, y) + 1}\, dA$.

### Skill Building

3. Find the surface area of the part of the plane $2x + 2y + z = 6$ that lies above the region in the $xy$-plane bounded by $x^2 + y^2 = 4$, the $x$-axis, and the $y$-axis, as shown in the figure.

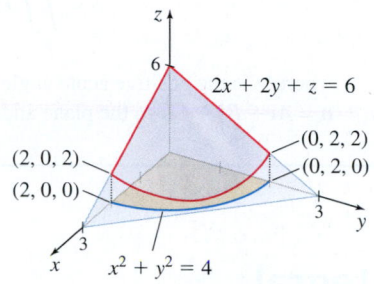

4. Find the surface area of the part of the plane $4z - 2x - y = 8$ that lies over the region enclosed by the triangle with vertices $(0, 0, 0)$, $(1, 0, 0)$, and $(1, 1, 0)$, as shown in the figure.

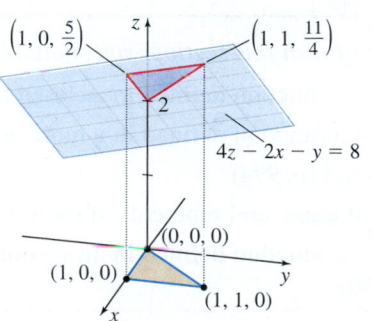

*In Problems 5–16, find the surface area described.*

5. The part of the surface $z = f(x, y) = \sqrt{9 - x^2}$ that lies above the rectangle enclosed by the lines $x = 0$, $x = 2$, $y = -1$, $y = 4$

6. The part of the surface $z = f(x, y) = \sqrt{8 - y^2}$ that lies above the rectangle enclosed by the lines $x = 0$, $x = 5$, $y = 0$, $y = 2$

7. The part of the surface $z = \dfrac{2}{3}(x^{3/2} + y^{3/2})$ that lies above the triangle enclosed by the lines $x = 0$, $y = 0$, and $2x + 3y = 6$

8. The part of the surface $z = \dfrac{2}{3}(x^{3/2} + y^{3/2})$ that lies above the triangle enclosed by $x = 0$, $y = 0$, and $3x + y = 3$

9. The part of the paraboloid $z = 4 - x^2 - y^2$ that lies above the $xy$-plane

10. The part of the cylinder $z = \sqrt{a^2 - x^2}$ that lies above the square defined by $-\dfrac{1}{2}a \le x \le \dfrac{1}{2}a$ and $-\dfrac{1}{2}a \le y \le \dfrac{1}{2}a$, $a > 0$

11. The part of the cone $z = \sqrt{x^2 + y^2}$ that lies inside the cylinder $x^2 + y^2 = 2x$

12. The part of the sphere $x^2 + y^2 + z^2 = 4z$ that lies within the paraboloid $x^2 + y^2 = 2z$

13. The part of the surface $z = xy$ in the first octant that lies within the cylinder $x^2 + y^2 = a^2$, $a > 0$

14. The part of the surface $z = x^2 - y^2$ in the first octant that lies within the cylinder $x^2 + y^2 = 4$

15. The part of the sphere $x^2 + y^2 + z^2 = 4z$ that lies between the planes $z = 1$ and $z = 3$

16. The part of the sphere $x^2 + y^2 + z^2 = 4a^2$ that lies inside the cylinder $x^2 + y^2 = 2ax$, $a > 0$

### Applications and Extensions

17. **Surface Area** Find the surface area cut from the hyperbolic paraboloid $y^2 - x^2 = 6z$ by the cylinder $x^2 + y^2 = 36$.

18. **Surface Area** Find the surface area of the part of $z = 4 - y^2$ that lies in the first octant and is enclosed by $z = 0$, $x = 0$, $x = y$, and $y = 2$.

19. **Surface Area** Find the surface area of the part of $x^2 = y$ that lies in the first octant under the plane $x + z = 3$.
    *Hint:* Project the surface onto the $xz$-plane.

20. **Surface Area** Find the surface area of the part of the paraboloid $z = 9 - x^2 - y^2$ that lies between the planes $z = 0$ and $z = 8$.

21. **Surface Area** Find the surface area of the part of the hemisphere $x^2 + y^2 + z^2 = 16$, $z \ge 0$, that lies inside the cylinder bounded by $x^2 + y^2 = 4$. See the figure.

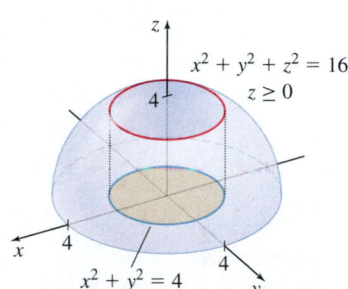

22. **Surface Area of the Pantheon** The Pantheon in Rome is the largest unreinforced concrete dome in the world. Its inner surface is a hemisphere of diameter 43.3 m with an open circular ocular 9.1 m in diameter cut from its apex. Find the inner surface area of the dome.

**CAS** 23. **Surface Area**

   (a) Graph the surface $z = 10e^{-(x^2 + y^2)}$ that lies inside the cylinder $x^2 + y^2 = 4$.

   (b) Find the surface area of the surface graphed in (a).

24. **Surface Area**

   (a) Set up, but do not evaluate, the integral to find the surface area of the part of the surface $z = 1 - x^4 - y^2$ that lies above the $xy$-plane.

   **CAS** (b) Graph the surface.

---

**1.** = NOW WORK problem    � = Graphing technology recommended    CAS = Computer Algebra System recommended

**25.** Derive the formula $S = \pi a \sqrt{a^2 + h^2}$ for the lateral surface area of a right circular cone of base radius $a$ and altitude $h$.

**26.** Show that the area of the first-octant portion of the plane $\dfrac{x}{a} + \dfrac{y}{b} + \dfrac{z}{c} = 1$ (where $a$, $b$, and $c$ are positive)

is $S = \dfrac{1}{2}\sqrt{a^2 b^2 + b^2 c^2 + c^2 a^2}$.

**27.** Use a double integral to derive the formula for the surface area of a sphere of radius $R$.

### Challenge Problems

**28.** Let $F$ denote a function of three variables that possesses continuous first-order partial derivatives at each point of its domain. Suppose also that $F_z(x, y, z)$ is never 0. If $S$ is the area of the part of the surface $F(x, y, z) = 0$ that lies over the closed, bounded region $R$, show that

$$S = \iint\limits_{R} \frac{\sqrt{[F_x(x, y, z)]^2 + [F_y(x, y, z)]^2 + [F_z(x, y, z)]^2}}{|F_z(x, y, z)|} \, dA$$

**29. Surface Area** The center of a sphere of radius $R$ is on the surface of a right circular cylinder of base radius $\dfrac{R}{2}$. Find the surface area of the sphere inside the cylinder. See the figure.

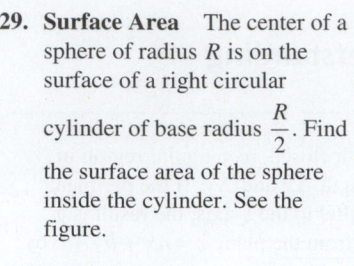
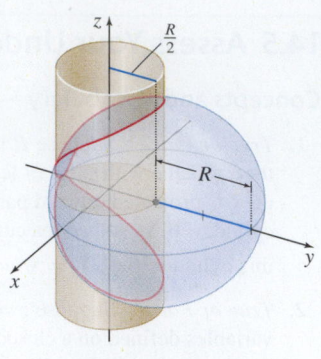

**30.** For the plane surface $F(x, y, z) = Ax + By + Cz - D = 0$, $C \neq 0$, if $z = f(x, y)$, show that the surface area $S = \iint\limits_{R} \sqrt{[f_x(x, y)]^2 + [f_y(x, y)]^2 + 1} \, dA$ can be written as

$$S = \iint\limits_{R} \sec \gamma \, dA$$

where $\gamma$ is the positive acute angle between the normal $\mathbf{n} = A\mathbf{i} + B\mathbf{j} + C\mathbf{k}$ to the plane and $\mathbf{k}$.

# 14.6 The Triple Integral

**OBJECTIVES** *When you finish this section, you should be able to:*

1 Find a triple integral of a function defined in a closed box (p. 991)
2 Find a triple integral of a function defined in a more general solid (p. 992)
3 Find the volume of a solid (p. 994)
4 Find the mass, center of mass, and moments of inertia of a solid (p. 994)
5 Find a triple integral of a function defined in an *xz*-simple or a *yz*-simple solid (p. 996)

Just as single integrals are used to integrate over closed intervals $[a, b]$, and double integrals are used to integrate over two-dimensional, closed, bounded regions $R$, triple integrals are used to integrate over three-dimensional, closed, bounded solids $E$. For example, we have seen that if $\rho = \rho(x)$ is the mass density of a long, thin wire, then the single integral $\int_a^b \rho(x)\, dx$ models the mass of the wire from $a$ to $b$. Also if $\rho = \rho(x, y)$ is the mass density of a lamina, then the double integral $\iint\limits_{R} \rho(x, y)\, dA$ models the mass $M$ of the lamina over $R$. Similarly, a triple integral can be used to find the mass of a solid with mass density $\rho = \rho(x, y, z)$.

We begin with a function $f$ of three variables defined in a box-shaped region $E$ of three-dimensional space. Partition $E$ into smaller rectangular boxes $E_{ijk}$, $i = 1, 2, \ldots, n$, $j = 1, 2, \ldots, m$, and $k = 1, 2, \ldots, l$, by drawing planes parallel to the three coordinate planes. See Figure 40.

The **norm** of the partition, denoted by $\|\Delta\|$, is defined as the largest of the lengths of the sides of the boxes $E_{ijk}$. If $\Delta x_i$, $\Delta y_j$, and $\Delta z_k$ denote the length, width, and height, respectively, of the $ijk$th box, then its volume is $\Delta V_{ijk} = \Delta x_i \Delta y_j \Delta z_k$. In each box $E_{ijk}$ we arbitrarily select a point $(u_{ijk}, v_{ijk}, w_{ijk})$, evaluate $f(u_{ijk}, v_{ijk}, w_{ijk})$, multiply $f(u_{ijk}, v_{ijk}, w_{ijk})$ by $\Delta x_i \Delta y_j \Delta z_k$, and form the sums

$$\sum_{k=1}^{l} \sum_{j=1}^{m} \sum_{i=1}^{n} f(u_{ijk}, v_{ijk}, w_{ijk})\, \Delta x_i \Delta y_j \Delta z_k$$

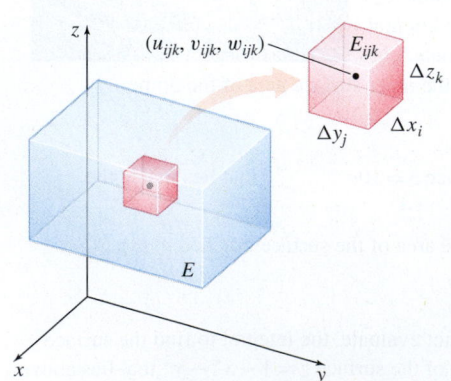

**Figure 40**

Sums of this form are referred to as **triple Riemann sums** of $f$ for the partition. If

$$\lim_{\|\Delta\| \to 0} \sum_{k=1}^{l} \sum_{j=1}^{m} \sum_{i=1}^{n} f(u_{ijk}, v_{ijk}, w_{ijk}) \Delta x_i \Delta y_j \Delta z_k = \lim_{\|\Delta\| \to 0} \sum_{k=1}^{l} \sum_{j=1}^{m} \sum_{i=1}^{n} f(u_{ijk}, v_{ijk}, w_{ijk}) \Delta V_{ijk}$$

exists and does not depend on the choice of the partition or on the choice of $(u_{ijk}, v_{ijk}, w_{ijk})$, then this limit is called the *triple integral of $f$ over $E$.*

---

**DEFINITION** Triple Integral

Let $f$ be a function of three variables defined in a closed, bounded, box-shaped solid $E$. The **triple integral of $f$ over $E$** is defined as

$$\iiint_E f(x, y, z)\, dV = \lim_{\|\Delta\| \to 0} \sum_{k=1}^{l} \sum_{j=1}^{m} \sum_{i=1}^{n} f(u_{ijk}, v_{ijk}, w_{ijk}) \Delta V_{ijk}$$

provided the limit exists.

---

As with single integrals and double integrals, it can be shown that if $f$ is continuous in $E$, then $\iiint_E f(x, y, z)\, dV$ exists.

There are six different orders of integration for the triple integral of $f$ in $E$. They are

$$\iiint_E f(x, y, z)\, dx\, dy\, dz \qquad \iiint_E f(x, y, z)\, dx\, dz\, dy \qquad \iiint_E f(x, y, z)\, dy\, dx\, dz$$

$$\iiint_E f(x, y, z)\, dy\, dz\, dx \qquad \iiint_E f(x, y, z)\, dz\, dx\, dy \qquad \iiint_E f(x, y, z)\, dz\, dy\, dx$$

Fubini's Theorem for triple integrals defined in a closed box-shaped solid $E$, uses iterated integrals to find $\iiint_E f(x, y, z)\, dV$.

---

**THEOREM** Fubini's Theorem for Triple Integrals

Let $f$ be a function of three variables defined in the closed box $E$, where $x_1 \le x \le x_2$, $y_1 \le y \le y_2$, and $z_1 \le z \le z_2$. If $f$ is continuous in $E$, then

$$\iiint_E f(x, y, z)\, dV = \int_{x_1}^{x_2} \left\{ \int_{y_1}^{y_2} \left[ \int_{z_1}^{z_2} f(x, y, z)\, dz \right] dy \right\} dx$$

---

The iterated integral on the right is shown first with respect to $z$, then $y$, and finally $x$. To find $\int_{z_1}^{z_2} f(x, y, z)\, dz$, we treat $x$ and $y$ as constants and integrate partially with respect to $z$. As is the case with iterated double integrals, the braces and brackets are usually omitted.

There are actually six different forms of Fubini's Theorem for Triple Integrals. Here we state the theorem in the order $dz\, dy\, dx$. But the same result is obtained no matter what order is used.

## 1 Find a Triple Integral of a Function Defined in a Closed Box

**EXAMPLE 1** Finding a Triple Integral of a Function Defined in a Closed Box

**(a)** If $E$ is the closed box $0 \le x \le 1$, $0 \le y \le 2$, and $0 \le z \le 3$ shown in Figure 41, express $\iiint_E 4xyz\, dV$ using six different orders of integration.

**(b)** Find $\iiint_E 4xyz\, dV$.

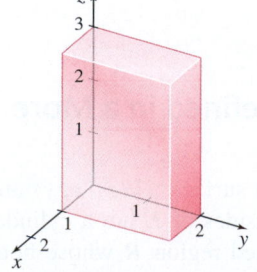

**Figure 41** The closed box $E$

**Solution** **(a)** Since $f(x, y, z) = 4xyz$ is continuous in $E$, we can find the triple integral using Fubini's Theorem for Triple Integrals.

If we use Fubini's Theorem in the order $dx\,dy\,dz$, then the limits of integration from left to right are $0 \le z \le 3, 0 \le y \le 2, 0 \le x \le 1$, and the iterated integral is

$$\iiint_E 4xyz\,dV = \int_0^3 \int_0^2 \int_0^1 4xyz\,dx\,dy\,dz = \int_0^3 \left\{ \int_0^2 \left[ \int_0^1 4x\,dx \right] y\,dy \right\} z\,dz$$

Notice how the integrals are nested: $\int_0^1 4x\,dx$ is on the inside, $\int_0^2 \ldots y\,dy$ is in the middle, and $\int_0^3 \ldots z\,dz$ is on the outside.

Using Fubini's Theorem in the order $dy\,dx\,dz$, the limits of integration from left to right are $0 \le z \le 3, 0 \le x \le 1, 0 \le y \le 2$ and the iterated integral is

$$\iiint_E 4xyz\,dV = \int_0^3 \int_0^1 \int_0^2 4xyz\,dy\,dx\,dz = \int_0^3 \left\{ \int_0^1 \left[ \int_0^2 4y\,dy \right] x\,dx \right\} z\,dz$$

Following the pattern, in the order $dz\,dy\,dx$, we obtain

$$\iiint_E 4xyz\,dV = \int_0^1 \int_0^2 \int_0^3 4xyz\,dz\,dy\,dx = \int_0^1 \left\{ \int_0^2 \left[ \int_0^3 4z\,dz \right] y\,dy \right\} x\,dx$$

In the order $dz\,dx\,dy$

$$\iiint_E 4xyz\,dV = \int_0^2 \int_0^1 \int_0^3 4xyz\,dz\,dx\,dy = \int_0^2 \left\{ \int_0^1 \left[ \int_0^3 4z\,dz \right] x\,dx \right\} y\,dy$$

In the order $dx\,dz\,dy$

$$\iiint_E 4xyz\,dV = \int_0^2 \int_0^3 \int_0^1 4xyz\,dx\,dz\,dy = \int_0^2 \left\{ \int_0^3 \left[ \int_0^1 4x\,dx \right] z\,dz \right\} y\,dy$$

In the order $dy\,dz\,dx$

$$\iiint_E 4xyz\,dV = \int_0^1 \int_0^3 \int_0^2 4xyz\,dy\,dz\,dx = \int_0^1 \left\{ \int_0^3 \left[ \int_0^2 4y\,dy \right] z\,dz \right\} x\,dx$$

**(b)** Using Fubini's Theorem in the order $dx\,dy\,dz$,

$$\iiint_E 4xyz\,dV = \int_0^3 \int_0^2 \int_0^1 4xyz\,dx\,dy\,dz = \int_0^3 \left\{ \int_0^2 \left[ \int_0^1 4x\,dx \right] y\,dy \right\} z\,dz$$

$$= \int_0^3 \int_0^2 \left[ 2x^2 \right]_0^1 yz\,dy\,dz = \int_0^3 \int_0^2 2yz\,dy\,dz$$

$$= \int_0^3 \left[ y^2 \right]_0^2 z\,dz = \int_0^3 4z\,dz = \left[ 2z^2 \right]_0^3 = 18 \qquad \blacksquare$$

**NOW WORK** Problem 9.

## 2 Find a Triple Integral of a Function Defined in a More General Solid

Consider a solid $E$ in space enclosed on the top by a surface $z = z_2(x, y)$ and on the bottom by a surface $z = z_1(x, y)$, where $z_1 \le z_2$. The sides of $E$ are a cylinder whose intersection with the $xy$-plane forms a closed, bounded region $R$ whose boundary is a piecewise-smooth curve. Also the functions $z_1$ and $z_2$ are continuous on $R$. See Figure 42. Since the projection of $E$ onto the $xy$-plane results in a closed, bounded region $R$, and $E$ lies between the two surfaces $z_1$ and $z_2$, the solid $E$ is called **xy-simple**. The next theorem, which we do not prove, provides a way to find triple integrals defined over solids that are $xy$-simple.

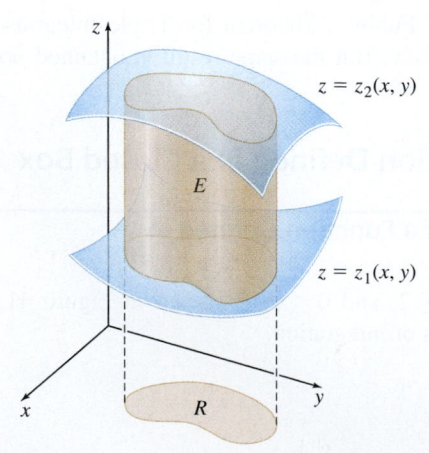

**Figure 42** $E$ is an $xy$-simple solid.

**THEOREM Triple Integral over an $xy$-Simple Solid**

Let $E$ be an $xy$-simple solid that lies between the two surfaces $z = z_1(x, y)$ and $z = z_2(x, y)$, where $z_1 \leq z_2$, and whose projection onto the $xy$-plane is a closed, bounded region $R$ whose boundary is a piecewise-smooth curve. If $f$ is a function of three variables that is continuous in $E$, then

$$\iiint\limits_E f(x, y, z)\, dV = \iint\limits_R \left[ \int_{z_1(x,y)}^{z_2(x,y)} f(x, y, z)\, dz \right] dA$$

If the region $R$ is $x$-simple, and $R$ is enclosed by $y = y_1(x)$ and $y = y_2(x)$, where $y_1 \leq y_2$ and $a \leq x \leq b$, then

$$\iiint\limits_E f(x, y, z)\, dV = \iint\limits_R \left[ \int_{z_1(x,y)}^{z_2(x,y)} f(x, y, z)\, dz \right] dA$$

$$= \int_a^b \int_{y_1(x)}^{y_2(x)} \int_{z_1(x,y)}^{z_2(x,y)} f(x, y, z)\, dz\, dy\, dx$$

If the region $R$ is $y$-simple, and $R$ is enclosed by $x = x_1(y)$ and $x = x_2(y)$, where $x_1 \leq x_2$ and $c \leq y \leq d$, then

$$\iiint\limits_E f(x, y, z)\, dV = \iint\limits_R \left[ \int_{z_1(x,y)}^{z_2(x,y)} f(x, y, z)\, dz \right] dA$$

$$= \int_c^d \int_{x_1(y)}^{x_2(y)} \int_{z_1(x,y)}^{z_2(x,y)} f(x, y, z)\, dz\, dx\, dy$$

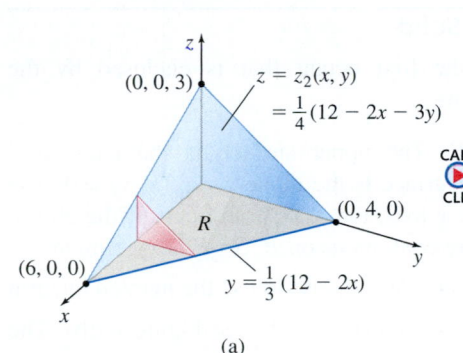

(a)

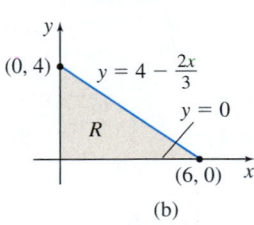

(b)

**Figure 43**

**CALC CLIP**

**EXAMPLE 2 Finding a Triple Integral of a Function Defined in an $xy$-Simple Solid**

Find the triple integral $\iiint\limits_E 4x\, dV$ over the tetrahedron formed by the coordinate planes and the plane $2x + 3y + 4z = 12$.

**Solution** The solid $E$ is shown in Figure 43(a). A typical plane section of $E$ using a plane perpendicular to the $x$-axis reveals that the upper surface is $z = z_2(x, y) = \frac{1}{4}(12 - 2x - 3y)$ and the lower surface is $z = z_1(x, y) = 0$. The solid $E$ lies between the two surfaces $z_1$ and $z_2$ and the projection of $E$ onto the $xy$-plane is a closed, bounded region $R$ on which $z_1$ and $z_2$ are continuous. So $E$ is $xy$-simple.

The region $R$ in the $xy$-plane is the triangle enclosed by the $x$-axis, the $y$-axis, and the line $y = \frac{1}{3}(12 - 2x) = 4 - \frac{2x}{3}$, as shown in Figure 43(b). The triangle $R$ is both $x$-simple and $y$-simple. If we treat $R$ as an $x$-simple region, so that $R$ is bounded by $y_1(x) = 0$, $y_2(x) = 4 - \frac{2x}{3}$, and $0 \leq x \leq 6$, we have

$$\iiint\limits_E 4x\, dV = \int_a^b \int_{y_1(x)}^{y_2(x)} \int_{z_1(x,y)}^{z_2(x,y)} 4x\, dz\, dy\, dx = \int_0^6 \int_0^{4-2x/3} \int_0^{(1/4)(12-2x-3y)} 4x\, dz\, dy\, dx$$

$$= \int_0^6 \int_0^{4-2x/3} 4x\, [z]_0^{(1/4)(12-2x-3y)}\, dy\, dx = \int_0^6 \int_0^{4-2x/3} 4x\left(\frac{12 - 2x - 3y}{4}\right) dy\, dx$$

$$= \int_0^6 \int_0^{4-2x/3} (12x - 2x^2 - 3xy)\, dy\, dx = \int_0^6 \left[ (12x - 2x^2)\, y - 3x\frac{y^2}{2} \right]_0^{4-2x/3} dx$$

$$= \int_0^6 \left( 24x - 8x^2 + \frac{2}{3}x^3 \right) dx = \left[ 12x^2 - \frac{8}{3}x^3 + \frac{1}{6}x^4 \right]_0^6 = 72 \qquad \blacksquare$$

In Example 2 we used Fubini's Theorem in the order $dz\,dy\,dx$. If we treat $R$ as a $y$-simple region, then $R$ is bounded by $x_1(y) = 0$, $x_2(y) = 6 - \dfrac{3}{2}y$, and $0 \le y \le 4$. Then Fubini's Theorem in the order $dz\,dx\,dy$ results in

$$\iiint_E 4x\,dV = \int_c^d \int_{x_1(y)}^{x_2(y)} \int_{z_1(x,y)}^{z_2(x,y)} 4x\,dz\,dx\,dy = \int_0^4 \int_0^{6-3y/2} \int_0^{(1/4)(12-2x-3y)} 4x\,dz\,dx\,dy$$

The result is the same, as you should verify.

NOW WORK Problems 11 and 21.

## 3 Find the Volume of a Solid

If $z = f(x, y) = 1$ on $R$, the double integral $\iint_R f(x, y)\,dA = \iint_R dA$ is numerically equal to the area of the region $R$. If $f(x, y, z) = 1$ in $E$, then the triple integral $\iiint_E f(x, y, z)\,dV = \iiint_E dV$ is numerically equal to the volume of the solid $E$. That is,

$$\boxed{\text{Volume } V \text{ of a solid } E: \quad V = \iiint_E dV}$$

### EXAMPLE 3 Finding the Volume of a Solid

Find the volume $V$ of the solid $E$ in the first octant that is enclosed by the paraboloid $z = 16 - 4x^2 - y^2$ and the $xy$-plane.

**Solution** Figure 44(a) shows the solid $E$. The upper surface is the paraboloid $z = z_2(x, y) = 16 - 4x^2 - y^2$ and the lower surface is the plane $z = z_1(x, y) = 0$. The region $R$ in the $xy$-plane is enclosed by the $x$-axis, the $y$-axis, and part of the ellipse $4x^2 + y^2 = 16$, $x \ge 0$, $y \ge 0$, and $z_1$ and $z_2$ are continuous on $R$. So $E$ is $xy$-simple.

The region $R$ is both $x$-simple and $y$-simple. We choose to use the iterated integral for an $x$-simple region, so $0 \le y \le \sqrt{16 - 4x^2}$ and $0 \le x \le 2$. See Figure 44(b). The volume $V$ of $E$ is given by

$$V = \iiint_E dV = \int_a^b \int_{y_1(x)}^{y_2(x)} \int_{z_1(x,y)}^{z_2(x,y)} dz\,dy\,dx = \int_0^2 \int_0^{\sqrt{16-4x^2}} \int_0^{16-4x^2-y^2} dz\,dy\,dx$$

$$= \int_0^2 \int_0^{\sqrt{16-4x^2}} (16 - 4x^2 - y^2)\,dy\,dx = \int_0^2 \left[(16 - 4x^2)y - \frac{y^3}{3}\right]_0^{\sqrt{16-4x^2}} dx$$

$$= \int_0^2 \left((16 - 4x^2)^{3/2} - \frac{(16 - 4x^2)^{3/2}}{3}\right) dx = \frac{16}{3} \int_0^2 (4 - x^2)^{3/2}\,dx$$

We use the Table of Integrals, Integral 67 with $a = 2$. Then

$$V = \frac{16}{3} \int_0^2 (4 - x^2)^{3/2}\,dx = \frac{16}{3} \left[\frac{x}{4}(4 - x^2)^{3/2} + \frac{3x}{2}\sqrt{4 - x^2} + 6\sin^{-1}\frac{x}{2}\right]_0^2$$

$$= \frac{16}{3} \cdot 6 \cdot \frac{\pi}{2} = 16\pi \qquad \blacksquare$$

NOW WORK Problem 37.

## 4 Find the Mass, Center of Mass, and Moments of Inertia of a Solid

Triple integrals can be used to find the mass, center of mass, and moments of inertia of solids. The formulas are obtained in a manner similar to those using double integrals, so we only state them here.

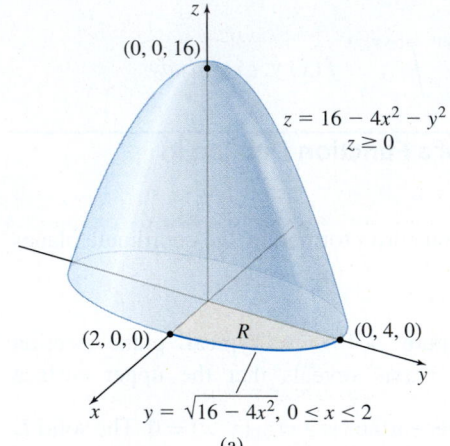

(0, 0, 16)

$z = 16 - 4x^2 - y^2$
$z \ge 0$

(2, 0, 0)   $R$   (0, 4, 0)

$y = \sqrt{16 - 4x^2}$, $0 \le x \le 2$

(a)

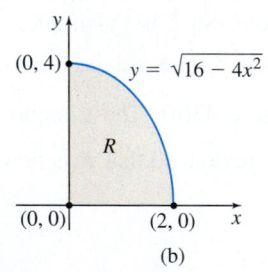

(0, 4)   $y = \sqrt{16 - 4x^2}$

$R$

(0, 0)   (2, 0)

(b)

**Figure 44**

NOTE Alternatively, we could use a CAS to find the integral or we could use the trigonometric substitution $x = 2\sin\theta$, $-\dfrac{\pi}{2} \le \theta \le \dfrac{\pi}{2}$.

The **mass** $M$ of a solid $E$ of volume $V$ with mass density $\rho = \rho(x, y, z)$ that is continuous in $E$ is given by

$$M = \iiint_E \rho(x, y, z)\, dV$$

The **moments of the solid about the coordinate planes** are

$$M_{xy} = \iiint_E z\rho(x, y, z)\, dV \qquad M_{xz} = \iiint_E y\rho(x, y, z)\, dV \qquad M_{yz} = \iiint_E x\rho(x, y, z)\, dV$$

The **center of mass of the solid** is the point $(\bar{x}, \bar{y}, \bar{z})$ whose coordinates are given by

$$\bar{x} = \frac{M_{yz}}{M} = \frac{\displaystyle\iiint_E x\rho(x, y, z)\, dV}{\displaystyle\iiint_E \rho(x, y, z)\, dV} \qquad \bar{y} = \frac{M_{xz}}{M} = \frac{\displaystyle\iiint_E y\rho(x, y, z)\, dV}{\displaystyle\iiint_E \rho(x, y, z)\, dV} \qquad \bar{z} = \frac{M_{xy}}{M} = \frac{\displaystyle\iiint_E z\rho(x, y, z)\, dV}{\displaystyle\iiint_E \rho(x, y, z)\, dV}$$

The **moment of inertia about an axis** $a$ is

$$I_a = \iiint_E r^2 \rho(x, y, z)\, dV$$

where $r$ is the distance from the point $(x, y, z)$ of the solid to the axis $a$ about which the moment is to be found.

CAS **EXAMPLE 4  Finding the Mass and Center of Mass of a Solid**

**(a)** Find the mass $M$ of a solid in the shape of a tetrahedron cut from the first octant by the plane $x + y + z = 1$ if the mass density $\rho = \rho(x, y, z)$ is proportional to the distance from the $yz$-plane.

**(b)** Find the center of mass $(\bar{x}, \bar{y}, \bar{z})$ of the tetrahedron.

**Solution** Figure 45(a) shows the solid $E$. It is enclosed by the surfaces $z = z_1(x, y) = 0$ and $z = z_2(x, y) = 1 - x - y$, and its projection onto the $xy$-plane is the closed, bounded $x$-simple region $R$ defined by the lines $y = 0$ and $y = 1 - x$, where $0 \leq x \leq 1$. See Figure 45(b). Since $z_1$ and $z_2$ are continuous on $R$, the solid $E$ is $xy$-simple.

**(a)**   Since the mass density $\rho = \rho(x, y, z)$ is proportional to the distance from the $yz$-plane, we have $\rho(x, y, z) = kx$, where $k$ is the constant of proportionality. Then the mass $M$ is

$$M = \iiint_E \rho(x, y, z)\, dV = \iiint_E kx\, dV = k \int_0^1 \int_0^{1-x} \int_0^{1-x-y} x\, dz\, dy\, dx$$

Using a CAS, $M = \dfrac{k}{24}$.

**(b)**  The center of mass $(\bar{x}, \bar{y}, \bar{z})$ is

$$\bar{x} = \frac{\displaystyle\iiint_E x\rho(x, y, z)\, dV}{M} = \frac{\displaystyle\iiint_E xkx\, dV}{\dfrac{k}{24}} = 24 \iiint_E x^2\, dV$$

$$= 24 \int_0^1 \int_0^{1-x} \int_0^{1-x-y} x^2\, dz\, dy\, dx = \frac{2}{5} \qquad \text{Use a CAS.}$$

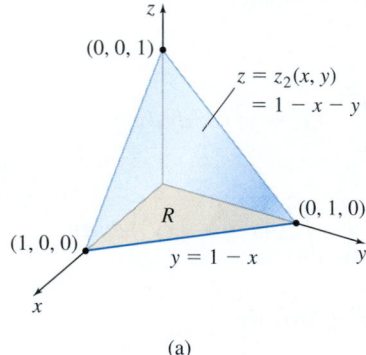

(a)

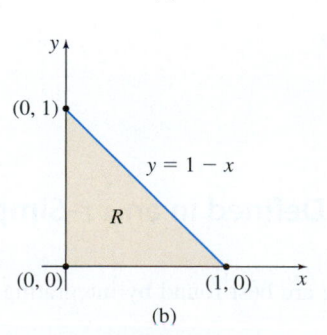

(b)

**Figure 45**

Similarly,

$$\bar{y} = \dfrac{\displaystyle\iiint_E y\rho(x,y,z)\,dV}{M} = \dfrac{\displaystyle\iiint_E ykx\,dV}{\dfrac{k}{24}} = 24\int_0^1\int_0^{1-x}\int_0^{1-x-y} xy\,dz\,dy\,dx = \dfrac{1}{5}$$

$$\bar{z} = \dfrac{\displaystyle\iiint_E z\rho(x,y,z)\,dV}{M} = \dfrac{\displaystyle\iiint_E zkx\,dV}{\dfrac{k}{24}} = 24\int_0^1\int_0^{1-x}\int_0^{1-x-y} xz\,dz\,dy\,dx = \dfrac{1}{5}$$

**NOTE** The integrals in (a) and (b) can be found by hand, but the integration is somewhat tedious.

The center of mass of the tetrahedron is located at the point $\left(\dfrac{2}{5}, \dfrac{1}{5}, \dfrac{1}{5}\right)$. ∎

**NOW WORK** Problem 49.

---

**EXAMPLE 5** Finding the Moment of Inertia About the $z$-axis of a Solid

Find the moment of inertia about the $z$-axis of the homogeneous solid of mass density $\rho$ in the first octant enclosed by the surface $z = 4xy$ and the planes $z = 0$, $x = 3$, and $y = 2$.

**Solution** The moment of inertia about the $z$-axis is

$$I_z = \iiint_E r^2\rho\,dV$$

where $r = \sqrt{x^2 + y^2}$ is the distance of the point $(x, y, z)$ from the $z$-axis.

Figure 46(a) shows the solid $E$. The upper surface is $z = z_2(x, y) = 4xy$ and the lower surface is the plane $z = z_1(x, y) = 0$. The region $R$ in the $xy$-plane is enclosed by the $x$-axis, the $y$-axis, the line $x = 3$, and the line $y = 2$, as shown in Figure 46(b). Since $z_1$ and $z_2$ are continuous on $R$, the solid $E$ is $xy$-simple.

The region $R$ is both $x$-simple and $y$-simple.

$$I_z = \iiint_E (x^2 + y^2)\rho\,dV = \rho\int_0^3\int_0^2\int_0^{4xy}(x^2 + y^2)\,dz\,dy\,dx$$

$$= \rho\int_0^3\int_0^2 (x^2 + y^2)\big[z\big]_0^{4xy}\,dy\,dx = \rho\int_0^3\int_0^2 (4x^3y + 4xy^3)\,dy\,dx$$

$$= \rho\int_0^3\big[2x^3y^2 + xy^4\big]_0^2\,dx = \rho\int_0^3 (8x^3 + 16x)\,dx$$

$$= \rho\big[2x^4 + 8x^2\big]_0^3 = \rho(162 + 72) = 234\rho \qquad ∎$$

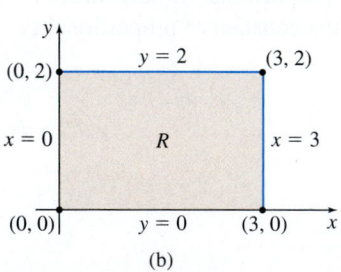

**Figure 46**

**NOW WORK** Problem 51.

---

## 5 Find a Triple Integral of a Function Defined in an $xz$-Simple or a $yz$-Simple Solid

Triple integrals on certain types of regions in space are best found by integrating first with respect to $x$ or $y$ rather than $z$.

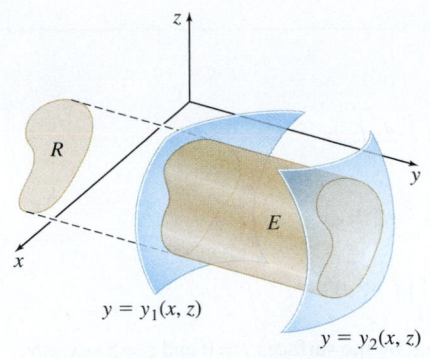

**Figure 47** An $xz$-simple solid; $R$ lies in the $xz$-plane.

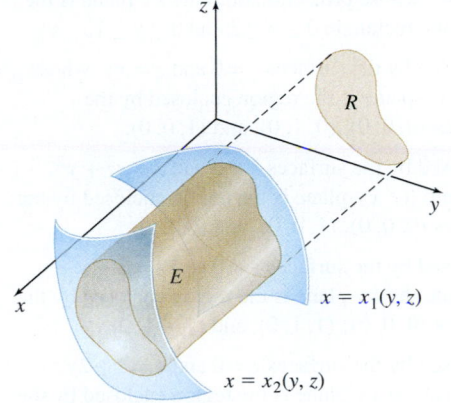

**Figure 48** A $yz$-simple solid; $R$ lies in the $yz$-plane.

Suppose a solid $E$ in space is enclosed by the two surfaces $y = y_1(x, z)$ and $y = y_2(x, z)$, where $y_1 \leq y_2$. If the projection of $E$ onto the $xz$-plane is a closed, bounded region $R$, and $y_1$ and $y_2$ are continuous on $R$, then $E$ is called an *xz*-**simple solid**. See Figure 47.

**THEOREM  Triple Integral over an $xz$-Simple Solid**

If $f$ is a function of three variables that is continuous in an $xz$-simple solid $E$, then

$$\iiint_E f(x, y, z)\, dV = \iint_R \left[ \int_{y_1(x,z)}^{y_2(x,z)} f(x, y, z)\, dy \right] dA$$

Similarly, suppose a solid $E$ in space is enclosed by the two surfaces $x = x_1(y, z)$ and $x = x_2(y, z)$, where $x_1 \leq x_2$. If the projection of $E$ onto the $yz$-plane is a closed, bounded region $R$, and $x_1$ and $x_2$ are continuous on $R$, then $E$ is called a *yz*-**simple solid**. See Figure 48.

**THEOREM  Triple Integral over an $yz$-Simple Solid**

If $f$ is a function of three variables that is continuous in a $yz$-simple solid $E$, then

$$\iiint_E f(x, y, z)\, dV = \iint_R \left[ \int_{x_1(y,z)}^{x_2(y,z)} f(x, y, z)\, dx \right] dA$$

**EXAMPLE 6  Finding the Volume of a Solid That Is $yz$-simple**

Find the volume $V$ of the solid $E$ that is enclosed by the cylinder $y^2 + z^2 = 4$, and the planes $x = 0$, $z = 0$, and $x + z = 5$.

**Solution** Figure 49(a) shows that the solid $E$ is $yz$-simple. The front surface is the plane $x_2(y, z) = 5 - z$ and the back surface is the plane $x_1(y, z) = 0$. The region $R$ in the $yz$-plane is enclosed by the $y$-axis and the semi-circle $z = \sqrt{4 - y^2}$; it is $y$-simple. So, $0 \leq z \leq \sqrt{4 - y^2}$ and $-2 \leq y \leq 2$. See Figure 49(b). The volume $V$ of $E$ is given by

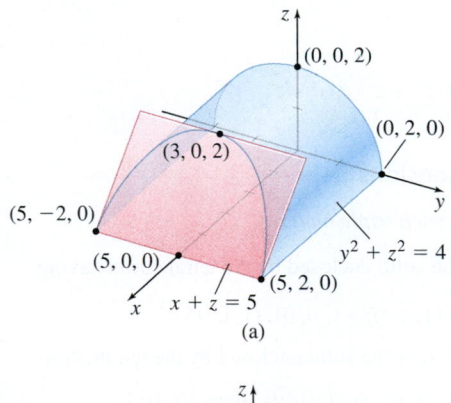

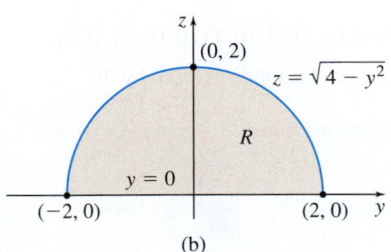

**Figure 49**

$$V = \iiint_E dV = \iint_R \left[ \int_{x_1(y,z)}^{x_2(y,z)} dx \right] dz\, dy = \int_{-2}^{2} \int_0^{\sqrt{4-y^2}} \int_0^{5-z} dx\, dz\, dy$$

$$= \int_{-2}^{2} \int_0^{\sqrt{4-y^2}} (5 - z)\, dz\, dy = \int_{-2}^{2} \left[ 5z - \frac{z^2}{2} \right]_0^{\sqrt{4-y^2}} dy$$

$$= \int_{-2}^{2} \left( 5\sqrt{4 - y^2} - \frac{4 - y^2}{2} \right) dy$$

$$= 5 \int_{-2}^{2} \sqrt{4 - y^2}\, dy - \int_{-2}^{2} \left( 2 - \frac{y^2}{2} \right) dy$$

$$= 5 \cdot 2\pi - \left[ 2y - \frac{y^3}{6} \right]_{-2}^{2} = 10\pi - \frac{16}{3} \qquad \blacksquare$$

**NOTE** $\int_{-2}^{2} \sqrt{4 - y^2}\, dy$ equals the area of a semi-circle of radius 2, namely $2\pi$.

NOW WORK Example 2 (p. 993) treating $E$ as an $xz$-simple region.

## 14.6 Assess Your Understanding

### Concepts and Vocabulary

1. *True or False* A solid $E$ in space is called $xy$-simple if it is enclosed on the top by a surface $z = z_2(x, y)$ and on the bottom by a surface $z = z_1(x, y)$, where $z_1 \leq z_2$ and the sides of $E$ are a cylinder whose intersection with the $xy$-plane forms a closed, bounded region $R$, whose boundary is a piecewise-smooth curve.

2. *True or False* The region $R$ resulting from the projection onto the $xy$-plane of an $xy$-simple solid $E$ is both $x$-simple and $y$-simple.

3. *True or False* The mass $M$ of a solid $E$ of volume $V$ with mass density $\rho = \rho(x, y, z)$ that is continuous in $E$ is given by $M = \iiint_E \rho(x, y, z)\, dV$.

4. *True or False* The triple integral $\iiint_E dV$ can be interpreted geometrically as the area of the region $R$ that is the projection of $E$ onto the $xy$-plane.

### Skill Building

*In Problems 5–10, (a) express each triple integral in six different ways. (b) Find the integral.*

5. $\iiint_E xy^2\, dV$,

   $E$ is the closed box $0 \leq x \leq 1, 0 \leq y \leq 2, 0 \leq z \leq 4$.

6. $\iiint_E x^2 yz\, dV$,

   $E$ is the closed box $0 \leq x \leq 2, 0 \leq y \leq 1, 0 \leq z \leq 3$.

7. $\iiint_E (x^2 + y^2 + z^2)\, dV$,

   $E$ is the closed box $0 \leq x \leq 1, 0 \leq y \leq 2, 0 \leq z \leq 3$.

8. $\iiint_E (x^2 - y^2 + z^2)\, dV$,

   $E$ is the closed box $0 \leq x \leq 3, 0 \leq y \leq 2, 0 \leq z \leq 1$.

9. $\iiint_E e^z \sin x \cos y\, dV$,

   $E$ is the closed box $0 \leq x \leq \dfrac{\pi}{4}, 0 \leq y \leq \dfrac{\pi}{2}, 0 \leq z \leq 1$.

10. $\iiint_E e^{-z} \cos x \cos y\, dV$,

    $E$ is the closed box $0 \leq x \leq \dfrac{\pi}{2}, 0 \leq y \leq \dfrac{\pi}{3}, 0 \leq z \leq 1$.

*In Problems 11–18, find each iterated triple integral.*

11. $\displaystyle\int_0^2 \int_0^{2-3x} \int_0^{x+y} x\, dz\, dy\, dx$

12. $\displaystyle\int_0^1 \int_0^{-x} \int_0^{2x+y} z\, dz\, dy\, dx$

13. $\displaystyle\int_0^3 \int_z^{z+2} \int_y^{y+z} (2x+1)\, dx\, dy\, dz$

14. $\displaystyle\int_0^2 \int_y^{y+2} \int_y^{x+y} (4z-1)\, dz\, dx\, dy$

15. $\displaystyle\int_0^2 \int_0^1 \int_0^y e^x\, dx\, dy\, dz$    16. $\displaystyle\int_1^2 \int_0^z \int_0^{\sqrt{y}} xe^{x^2}\, dx\, dy\, dz$

17. $\displaystyle\int_0^{\pi/2} \int_y^{\pi/2} \int_0^{xy} \sin \frac{z}{y}\, dz\, dx\, dy$

18. $\displaystyle\int_0^{\pi/2} \int_x^{\pi/2} \int_0^{xy} \cos \frac{z}{x}\, dz\, dy\, dx$

*In Problems 19–24, find $\iiint_E xy\, dV$.*

19. $E$ is the solid enclosed by the surfaces $z = 0$ and $z = 5 - x - y$, whose projection onto the $xy$-plane is the region enclosed by the rectangle $0 \leq x \leq 1$ and $0 \leq y \leq 3$.

20. $E$ is the solid enclosed by the surfaces $z = 0$ and $z = 16 - x^2 - y^2$, whose projection onto the $xy$-plane is the region enclosed by the rectangle $0 \leq x \leq 2$ and $0 \leq y \leq 1$.

21. $E$ is the solid enclosed by the surfaces $z = 0$ and $z = xy$, whose projection onto the $xy$-plane is the region enclosed by the triangle with vertices $(0, 0, 0)$, $(0, 1, 0)$, and $(1, 0, 0)$.

22. $E$ is the solid enclosed by the surfaces $z = 0$ and $z = x^2 + y^2$, whose projection onto the $xy$-plane is the region enclosed by the triangle with vertices $(0, 0, 0)$, $(1, 0, 0)$, and $(0, 2, 0)$.

23. $E$ is the solid enclosed by the surfaces $z = 0$ and $z = 3 - x - y$, whose projection onto the $xy$-plane is the region enclosed by the triangle with vertices $(0, 0, 0)$, $(1, 1, 0)$, and $(1, -1, 0)$.

24. $E$ is the solid enclosed by the surfaces $z = 0$ and $z = 3 + 2y$, whose projection onto the $xy$-plane is the region enclosed by the triangle with vertices $(-1, 0, 0)$, $(0, 1, 0)$, and $(2, 0, 0)$.

*In Problems 25 and 26, use a CAS to approximate each integral.*

CAS 25. $\displaystyle\int_0^1 \int_1^{x^2} \int_0^{\sqrt{x}} ye^x\, dz\, dy\, dx$

CAS 26. $\displaystyle\int_3^4 \int_2^3 \int_0^1 \ln(x^2 + y^2 + z^2)\, dx\, dy\, dz$

### Applications and Extensions

*In Problems 27–30, find each triple integral.*

27. $\iiint_E x\, dV$, if $E$ is the solid enclosed by the tetrahedron having vertices at $(0, 0, 0)$, $(1, 1, 0)$, $(1, 0, 0)$, $(1, 0, 1)$.

28. $\iiint_E (x^2 + z^2)\, dV$, if $E$ is the solid enclosed by the tetrahedron having vertices at $(0, 0, 0)$, $(1, 1, 0)$, $(1, 0, 0)$, $(1, 0, 1)$.

29. $\iiint_E (xy + 3y)\, dV$, if $E$ is the solid enclosed by the cylinder $x^2 + y^2 = 9$ and the planes $x + z = 3$, $y = 0$, and $z = 0$.

30. $\iiint_E xyz\, dV$, if $E$ is the solid enclosed by the cylinders $x^2 + y^2 = 1$ and $x^2 + z^2 = 1$. See the figure.

*x² + y² = 1*, *x² + z² = 1*

*In Problems 31–34, (a) describe the solid whose volume is given by each integral.*

**CAS** (b) Graph the solid.

**31.** $\int_{-2}^{2} \int_{0}^{\sqrt{4-y^2}} \int_{0}^{1} dz\, dx\, dy$

**32.** $\int_{0}^{2} \int_{0}^{\sqrt{4-y^2}} \int_{0}^{4-x^2-y^2} dz\, dx\, dy$

**33.** $\int_{0}^{1} \int_{0}^{x^2} \int_{0}^{y} dz\, dy\, dx$    **34.** $\int_{0}^{1} \int_{y^2}^{1} \int_{0}^{1-x} dz\, dx\, dy$

**35.** (a) Set up the triple integral $\iiint_{E} dV$ over the tetrahedron formed by the coordinate planes and the plane $x+2y+3z=6$ using six different orders of integration.

(b) Find the volume of the tetrahedron.

**36.** (a) Set up the triple integral $\iiint_{E} dV$ over the tetrahedron formed by the coordinate planes and the plane $x+y+z=3$ using six different orders of integration.

(b) Find the volume of the tetrahedron.

**37. Volume of a Solid** Find the volume $V$ of the solid enclosed by $y^2=z$, $x=0$, and $x=y-z$.

**38. Volume of a Solid** Find the volume $V$ of the solid enclosed by $z=4-y^2$, $z=9-x$, $x=0$, and $z=0$.

**39.** Set up, but do not find, $\iiint_{E} xy\, dV$, where $E$ is the solid enclosed by the surfaces $z=1-x-y$ and $z=3-x-y$, whose projection onto the $xy$-plane is the region enclosed by the circle $x^2+y^2=1$.

**40.** Set up, but do not find, $\iiint_{E} xy\, dV$, where $E$ is the solid enclosed by the surfaces $z=0$ and $z=x^2+y$, whose projection onto the $xy$-plane is the region enclosed by the circle $x^2+y^2=4$.

*In Problems 41–46, set up, but do not find, an iterated triple integral that equals the volume of the solid,*

**41.** Enclosed by $z=x^2+y^2$ and $z=16-x^2-y^2$

**42.** Enclosed by $z=x^2+y^2$ and $z=4-x^2-y^2$

**43.** Enclosed by $z=x^2+y^2$ and $z=2-x$

**44.** Enclosed by $z=x^2+y^2$ and $z=3-y$

**45.** Enclosed by $z^2=4x$ and $x^2+y^2=2x$

**46.** Enclosed by $z^2=4y$ and $x^2+y^2=2y$

**47. Mass** Set up, but do not find, the integral that equals the mass $M$ of an object in the shape of a tetrahedron cut from the first octant by the plane $x+y+z=1$, as shown in the figure, if its mass density is proportional to the product of the distances from the three coordinate planes.

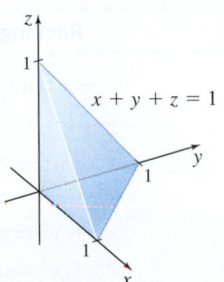

**48. Mass** Set up, but do not find, the integral that equals the mass $M$ of an object in the shape of a right circular cylinder of height $h$ and radius $a$, if its mass density is proportional to the square of the distance from the axis of the cylinder.

**49. Mass** Find the mass $M$ of an object in the shape of a cube of edge $a$ if its mass density is proportional to the square of the distance from one corner.

**CAS 50. Mass** A cylindrical bar of radius $R$ and length $2L$ is positioned with its axis along the $x$-axis and its center of mass at the origin. The mass density of the bar is given by $\rho(x,y,z)=kz^2$. Show that the mass $M$ of the bar is $M=\dfrac{1}{2}\pi k R^4 L$.

**51. Moments of Inertia** Set up, but do not find, the integrals that equal the moments of inertia $I_x$ and $I_y$ for the solid region enclosed by the hemisphere $z=\sqrt{9-x^2-y^2}$ and the $xy$-plane, if the mass density is proportional to the distance from the $xy$-plane.

**52.** Set up, but do not find, the integral of the function $f(x,y,z)=x^2yz$ over the solid enclosed by the cone $3x^2+3y^2=z^2$, $z\geq 0$, and the plane $z=3$.

**53. Volume of an Ellipsoid** Show that the volume of the ellipsoid $\dfrac{x^2}{a^2}+\dfrac{y^2}{b^2}+\dfrac{z^2}{c^2}=1$ is $\dfrac{4}{3}\pi abc$. (Assume that $a$, $b$, and $c$ are positive.) What does this formula reduce to if $a=b=c$?

## Challenge Problems

**54. Volume of a Solid** Find the volume of the solid in the first octant enclosed by the coordinate planes, $a^2y=b(a^2-x^2)$, and $a^2z=c(a^2-x^2)$, $a>0$, $b>0$, $c>0$.

**55. Volume of a Solid** Find the volume $V$ of the region enclosed by $z=0$, $z=1-x^2$, and $z=1-y^2$.

**56. Average Value of a Function** The average value of $z=f(x,y)$ over a region $R$ that is not necessarily rectangular is defined to be the number $\dfrac{1}{A}\iint_{R} f(x,y)\, dA$, where $A$ is the area of $R$.

(a) In single variable calculus the average value of a function $y=f(x)$ over the closed interval $[a,b]$ is defined to be the number $\dfrac{1}{b-a}\int_{a}^{b} f(x)\, dx$. In what sense is this a special case of the above definition of the average value of $z=f(x,y)$ over $R$?

(b) Let $w=f(x,y,z)$ be integrable over a solid $E$ in space. What definition would you give for the average value of $f$ in $E$?

**57.** Show that the following integrals represent the same volume. Do not find the integrals.

(a) $4\int_{0}^{4} \int_{0}^{\sqrt{16-x^2}} \int_{(x^2+y^2)/4}^{4} dz\, dy\, dx$

(b) $4\int_{0}^{4} \int_{0}^{2\sqrt{z}} \int_{0}^{\sqrt{4z-x^2}} dy\, dx\, dz$

(c) $4\int_{0}^{4} \int_{y^2/4}^{4} \int_{0}^{\sqrt{4z-y^2}} dx\, dz\, dy$

**58.** Show that:
$$\int_{a}^{b} \int_{a}^{z} \int_{a}^{y} f(x)\, dx\, dy\, dz = \int_{a}^{b} \frac{(b-x)^2}{2} f(x)\, dx$$

# 14.7 Triple Integrals Using Cylindrical Coordinates

**OBJECTIVES** *When you finish this section, you should be able to:*

1 Convert rectangular coordinates to cylindrical coordinates (p. 1000)
2 Find a triple integral using cylindrical coordinates (p. 1001)

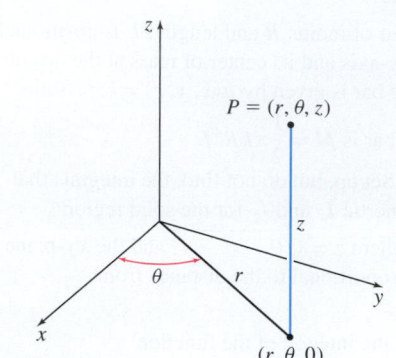

**Figure 50**

We have seen that there are instances where it is easier to find a double integral using polar coordinates than to find it using rectangular coordinates. For triple integrals, we give two alternatives to integration in rectangular coordinates: One uses *cylindrical coordinates* and the other uses *spherical coordinates* (Section 14.8).

## 1 Convert Rectangular Coordinates to Cylindrical Coordinates

If the rectangular coordinates of a point $P$ in three-dimensional space are $(x, y, z)$ and if $(r, \theta, 0)$ are the polar coordinates for the projection of $P$ onto the $xy$-plane, then the point $P$ can be located by the ordered triple $(r, \theta, z)$, called the **cylindrical coordinates of the point** $P$.

Figure 50 shows how to graph a point $P = (r, \theta, z)$. The algebraic relationship between the cylindrical coordinates $(r, \theta, z)$ and the rectangular coordinates $(x, y, z)$ of a point $P$ is given by the formulas

$$x = r \cos \theta \qquad y = r \sin \theta \qquad z = z$$

**EXAMPLE 1**   **Converting Between Cylindrical Coordinates and Rectangular Coordinates**

(a) Find the rectangular coordinates of a point $P$ whose cylindrical coordinates are $\left(6, \dfrac{\pi}{3}, -2\right)$.

(b) Find the cylindrical coordinates of a point $P$ whose rectangular coordinates are $(\sqrt{3}, 1, 5)$.

**Solution** (a) We use the equations $x = r \cos \theta$ and $y = r \sin \theta$ with $r = 6$ and $\theta = \dfrac{\pi}{3}$. Then

$$x = 6 \cos \frac{\pi}{3} = 3 \qquad y = 6 \sin \frac{\pi}{3} = 3\sqrt{3} \qquad z = -2$$

In rectangular coordinates, $P = (3, 3\sqrt{3}, -2)$.

(b) In cylindrical coordinates $r = \sqrt{x^2 + y^2} = \sqrt{3 + 1} = 2$. Then $\cos \theta = \dfrac{x}{r} = \dfrac{\sqrt{3}}{2}$ and $\sin \theta = \dfrac{y}{r} = \dfrac{1}{2}$, so $\theta = \dfrac{\pi}{6}$. Since $z$ remains the same, the point $P$ in cylindrical coordinates is $(r, \theta, z) = \left(2, \dfrac{\pi}{6}, 5\right)$.   ∎

**NOW WORK** Problems 5 and 13.

> **NOTE** The numbers $r$, $\theta$, and $z$ are called cylindrical coordinates because if $r$ is a constant, then the surface traced out by the set of points $(r, \theta, z)$ is a cylinder.

Table 1 lists the equations of several surfaces expressed in rectangular coordinates and in cylindrical coordinates. Figure 51 illustrates the surfaces.

**TABLE 1**

| Surface | Rectangular | Cylindrical |
|---|---|---|
| (a) Half-Plane | $y = x \tan k$ | $\theta = k, \quad -\dfrac{\pi}{2} < k < \dfrac{\pi}{2}$ |
| (b) Plane | $z = k$ | $z = k$ |
| (c) Cylinder | $x^2 + y^2 = a^2$ | $r = a, \quad a > 0$ |
| (d) Sphere | $x^2 + y^2 + z^2 = R^2$ | $r^2 + z^2 = R^2$ |
| (e) Circular cone | $x^2 + y^2 = a^2 z^2$ | $r = az, \quad a > 0$ |
| (f) Circular paraboloid | $x^2 + y^2 = az$ | $r^2 = az, \quad a > 0$ |

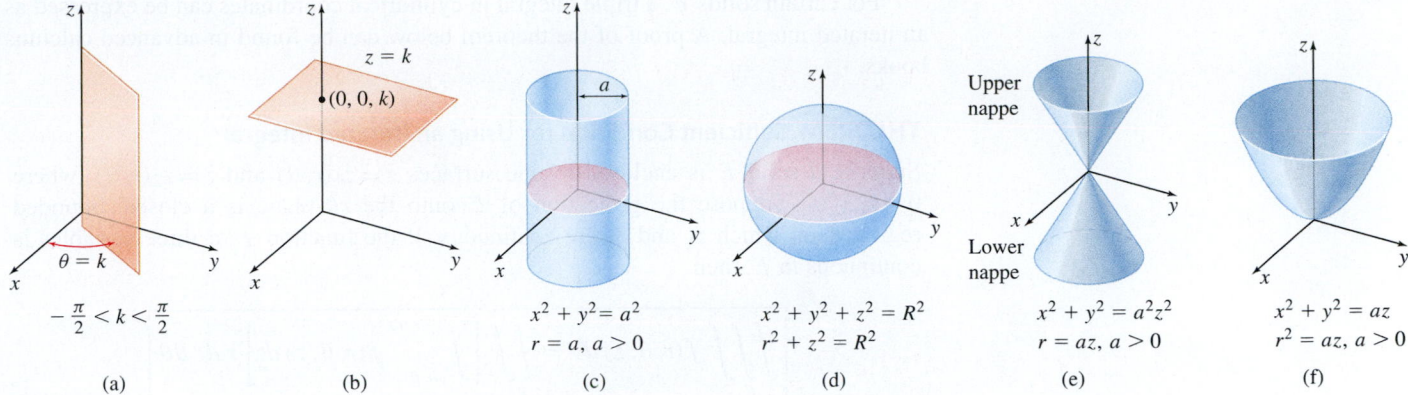

$\theta = k$

$-\dfrac{\pi}{2} < k < \dfrac{\pi}{2}$

(a)

$z = k$

$(0, 0, k)$

(b)

$a$

$x^2 + y^2 = a^2$
$r = a, \ a > 0$

(c)

$x^2 + y^2 + z^2 = R^2$
$r^2 + z^2 = R^2$

(d)

Upper nappe

Lower nappe

$x^2 + y^2 = a^2 z^2$
$r = az, \ a > 0$

(e)

$x^2 + y^2 = az$
$r^2 = az, \ a > 0$

(f)

**Figure 51**

Triple integrals that have symmetry about an axis, particularly integrals involving cylinders or cones whose axes are aligned with the $z$-axis, are often easier to find using cylindrical coordinates.

## 2 Find a Triple Integral Using Cylindrical Coordinates

When a solid $E$ is enclosed by parts of a circular paraboloid, a cylinder, or a cone which have symmetry with the $z$-axis, it is sometimes easier to find $\iiint\limits_{E} f(x, y, z)\,dV$ by converting to cylindrical coordinates. We convert a function $w = f(x, y, z)$ to cylindrical coordinates $(r, \theta, z)$ using the equations $x = r\cos\theta$, $y = r\sin\theta$, $z = z$. It remains to find the differential $dV$ in cylindrical coordinates.

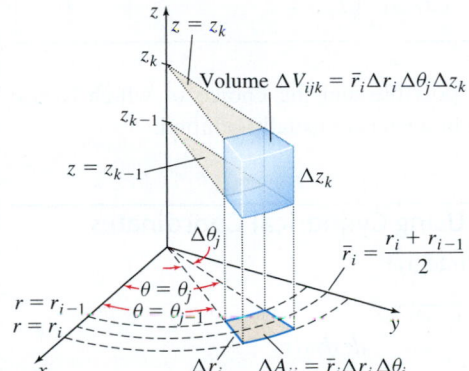

$z = z_k$

Volume $\Delta V_{ijk} = \bar{r}_i \Delta r_i \Delta\theta_j \Delta z_k$

$z = z_{k-1}$

$\Delta z_k$

$\Delta\theta_j$

$\bar{r}_i = \dfrac{r_i + r_{i-1}}{2}$

$r = r_{i-1}$   $\theta = \theta_j$
$r = r_i$   $\theta = \theta_{j-1}$

$\Delta r_i$   $\Delta A_{ij} = \bar{r}_i \Delta r_i \Delta\theta_j$

**Figure 52**

**NEED TO REVIEW?** The area of a sector of a circle is discussed in Appendix A.4, p. A-27.

We proceed in a way similar to that used for double integrals in polar coordinates. Suppose a solid $E$ is formed by part of a cylinder centered about the $z$-axis and two planes parallel to the $xy$-plane. We partition this solid using concentric, circular cylinders centered about the $z$-axis, vertical planes through the $z$-axis, and horizontal planes parallel to the $xy$-plane. The result is a collection of subsolids.

Suppose there are $n$ concentric circles of radii $r_i$, $i = 1, 2, \ldots, n$; $m$ vertical planes $\theta = \theta_j$, $j = 1, 2, \ldots, m$, through the $z$-axis; and $l$ horizontal planes $z = z_k$, $k = 1, 2, \ldots, l$ parallel to the $xy$-plane that form $lmn$ subsolids. Figure 52 illustrates a typical subsolid, $E_{ijk}$, whose dimensions are $\Delta r_i = r_i - r_{i-1}$, $\Delta\theta_j = \theta_j - \theta_{j-1}$, and $\Delta z_k = z_k - z_{k-1}$.

The projection of $E_{ijk}$ onto the $xy$-plane is a polar subregion whose area $\Delta A_{ij}$ is

$$\Delta A_{ij} = \text{outer sector - inner sector} = \frac{1}{2}r_i^2(\theta_j - \theta_{j-1}) - \frac{1}{2}r_{i-1}^2(\theta_j - \theta_{j-1})$$

$$= \frac{1}{2}(r_i^2 - r_{i-1}^2)(\theta_j - \theta_{j-1}) = \frac{1}{2}(r_i^2 - r_{i-1}^2)\Delta\theta_j$$

$$= \frac{1}{2}(r_i + r_{i-1})(r_i - r_{i-1})\Delta\theta_j = \bar{r}_i \Delta r_i \Delta\theta_j$$

where $\bar{r}_i = \dfrac{r_i + r_{i-1}}{2}$. The volume $\Delta V_{ijk}$ of the subsolid $E_{ijk}$ is

$$\Delta V_{ijk} = \Delta A_{ij} \cdot (z_k - z_{k-1}) = \Delta A_{ij}\Delta z_k = \bar{r}_i \Delta r_i \Delta\theta_j \Delta z_k$$

Then the differential $dV$ is

$$\boxed{dV = r\,dr\,d\theta\,dz}$$

It can be shown that if $z = f(x, y, z)$ is continuous in $E$, then

$$\iiint\limits_{E} f(x, y, z)\,dV = \iiint\limits_{E} f(r\cos\theta, r\sin\theta, z)\,dV = \iiint\limits_{E} f(r\cos\theta, r\sin\theta, z)\,r\,dr\,d\theta\,dz$$

For certain solids $E$, a triple integral in cylindrical coordinates can be expressed as an iterated integral. A proof of the theorem below can be found in advanced calculus books.

---

**THEOREM Sufficient Condition for Using an Iterated Integral**

Suppose a solid $E$ is enclosed by the surfaces $z = z_1(r, \theta)$ and $z = z_2(r, \theta)$, where $0 \le z_1 \le z_2$. Suppose the projection of $E$ onto the $r\theta$-plane is a closed, bounded region $R$ on which $z_1$ and $z_2$ are continuous. If the function $f$ of three variables is continuous in $E$, then

$$\iiint_E f(r, \theta, z)\, dV = \iint_R \left[ \int_{z_1(r,\theta)}^{z_2(r,\theta)} f(r, \theta, z)\, dz \right] r\, dr\, d\theta$$

---

**NOTE** The integrand contains a factor of $r$ because in cylindrical coordinates the differential $dV$ of volume is $dV = r\, dr\, d\theta\, dz$.

To find the value of the inside integral $\int_{z_1(r,\theta)}^{z_2(r,\theta)} f(r, \theta, z)\, dz$, we treat $r$ and $\theta$ as constants and integrate partially with respect to $z$.

Suppose the region $R$ is enclosed by the rays $\theta = \theta_1$ and $\theta = \theta_2$, and the curves $r = r_1(\theta)$ and $r = r_2(\theta)$, where $0 \le r_1 \le r_2$. If the functions $r_1$ and $r_2$ are continuous for $\theta_1 \le \theta \le \theta_2$, then

$$\iiint_E f(r, \theta, z)\, dV = \iiint_E f(r, \theta, z)\, r\, dr\, d\theta\, dz = \int_{\theta_1}^{\theta_2} \left\{ \int_{r_1(\theta)}^{r_2(\theta)} \left[ \int_{z_1(r,\theta)}^{z_2(r,\theta)} f(r, \theta, z)\, dz \right] r\, dr \right\} d\theta$$

Five other orders for the iteration are possible and the choice of which to use depends on $E$. As before, the braces and the brackets are usually omitted.

---

**EXAMPLE 2 Finding a Triple Integral Using Cylindrical Coordinates**

Give a geometric interpretation of the triple integral

$$\int_{-1}^{1} \int_{-\sqrt{1-x^2}}^{\sqrt{1-x^2}} \int_{0}^{2\sqrt{1-x^2-y^2}} dz\, dy\, dx$$

Then use cylindrical coordinates to find the triple integral.

**Solution** The solid $E$ and its projection onto the $xy$-plane can be described by the inequalities

$$0 \le z \le 2\sqrt{1-x^2-y^2} \qquad -\sqrt{1-x^2} \le y \le \sqrt{1-x^2} \qquad -1 \le x \le 1$$

The limits of integration on $z$ are $z = 0$ and $z = 2\sqrt{1-x^2-y^2} = \sqrt{4-4x^2-4y^2}$ or, equivalently, $z^2 = 4 - 4x^2 - 4y^2$, $z \ge 0$. We can interpret the integral as the volume of a solid $E$ that is the upper half of the ellipsoid $4x^2 + 4y^2 + z^2 = 4$, $z \ge 0$, as shown in Figure 53. From the $x$ and $y$ limits of integration, the projection onto the $xy$-plane is the region $R$ enclosed by the circle $x^2 + y^2 = 1$. Since the region $R$ is a circle, we use cylindrical coordinates to find the triple integral.

First, we convert the rectangular coordinates to cylindrical coordinates. Then

$$z = 2\sqrt{1-x^2-y^2} = 2\sqrt{1-(x^2+y^2)} = 2\sqrt{1-r^2}$$

The projection onto the $xy$-plane is the region $R$ enclosed by the circle $x^2 + y^2 = 1$ on which $z_1 = 0$ and $z_2 = \sqrt{1-x^2-y^2}$ are continuous. So in cylindrical coordinates, we have

$$0 \le z \le 2\sqrt{1-r^2} \qquad 0 \le r \le 1 \qquad 0 \le \theta \le 2\pi$$

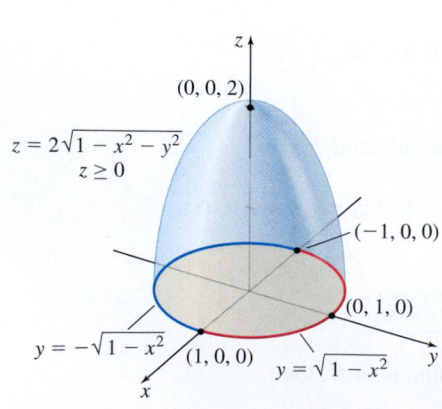

**Figure 53**

Then

$$\int_{-1}^{1}\int_{-\sqrt{1-x^2}}^{\sqrt{1-x^2}}\int_{0}^{2\sqrt{1-x^2-y^2}} dz\,dy\,dx = \int_{0}^{2\pi}\int_{0}^{1}\int_{0}^{2\sqrt{1-r^2}} dz\,r\,dr\,d\theta \qquad dx\,dy\,dz = r\,dr\,d\theta\,dz$$

$$= \int_{0}^{2\pi}\int_{0}^{1} [z]_{0}^{2\sqrt{1-r^2}} r\,dr\,d\theta = \int_{0}^{2\pi}\int_{0}^{1} 2r\sqrt{1-r^2}\,dr\,d\theta$$

$$= \int_{0}^{2\pi} \left[-\frac{2}{3}(1-r^2)^{3/2}\right]_{0}^{1} d\theta = \frac{2}{3}\int_{0}^{2\pi} d\theta = \frac{4\pi}{3}$$

The value of the triple integral $\dfrac{4\pi}{3}$ equals the volume $V$ of the solid. That is,

$$V = \frac{4\pi}{3} \text{ cubic units} \qquad\blacksquare$$

**NOW WORK** Problem 19.

---

CALC ⊙ CLIP **EXAMPLE 3**  **Finding the Volume of a Solid Using Cylindrical Coordinates**

Find the volume of the solid $E$ enclosed by the hemisphere $x^2 + y^2 + z^2 = 4$, $z \geq 0$, and the cylinder $(x-1)^2 + y^2 = 1$.

**Solution** Figure 54(a) shows the hemisphere and the part of the cylinder which encloses the solid $E$.

Because $x^2 + y^2$ appears in the equation of both the hemisphere and cylinder, we use cylindrical coordinates. The equation of the hemisphere is

Rectangular Coordinates:   $x^2 + y^2 + z^2 = 4,\ z \geq 0$

Cylindrical Coordinates:     $r^2 + z^2 = 4,\ z \geq 0$, or equivalently, $z = \sqrt{4-r^2}$

The equation of the region in the $xy$-plane below the surface is given by:

Rectangular Coordinates:   $(x-1)^2 + y^2 = 1$

$$x^2 + y^2 = 2x$$

Cylindrical Coordinates:      $r^2 = 2(r\cos\theta) \qquad -\dfrac{\pi}{2} \leq \theta \leq \dfrac{\pi}{2}$

$$r = 2\cos\theta$$

See Figure 54(b).

The solid $E$ is given by $0 \leq z \leq \sqrt{4-r^2}$, $0 \leq r \leq 2\cos\theta$, and $-\dfrac{\pi}{2} \leq \theta \leq \dfrac{\pi}{2}$.

The volume $V$ of the solid $E$ is

$$V = \iiint\limits_{E} dV = \iiint\limits_{E} r\,dr\,d\theta\,dz = 2\int_{0}^{\pi/2}\int_{0}^{2\cos\theta}\int_{0}^{\sqrt{4-r^2}} r\,dz\,dr\,d\theta$$

$$\underset{\underset{\text{Use symmetry}}{\uparrow}}{}$$

$$= 2\int_{0}^{\pi/2}\int_{0}^{2\cos\theta} r[z]_{0}^{\sqrt{4-r^2}} dr\,d\theta = 2\int_{0}^{\pi/2}\int_{0}^{2\cos\theta} r(4-r^2)^{1/2}\,dr\,d\theta$$

$$= -\int_{0}^{\pi/2}\left[\frac{2}{3}(4-r^2)^{3/2}\right]_{0}^{2\cos\theta} d\theta$$

$$= \frac{2}{3}\int_{0}^{\pi/2}[8 - (4 - 4\cos^2\theta)^{3/2}]\,d\theta = \frac{16}{3}\int_{0}^{\pi/2}(1 - \sin^3\theta)\,d\theta$$

$$= \frac{8\pi}{3} - \frac{32}{9} = \frac{24\pi - 32}{9} \text{ cubic units} \qquad\blacksquare$$

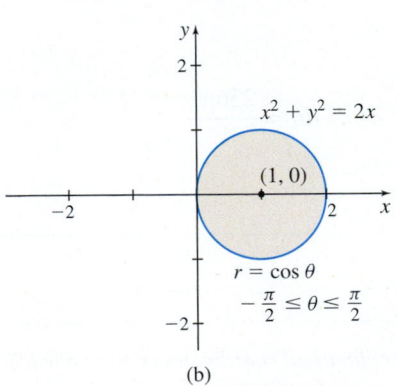

(a)

(b)

**Figure 54**

**NOW WORK** Problems 27 and 33.

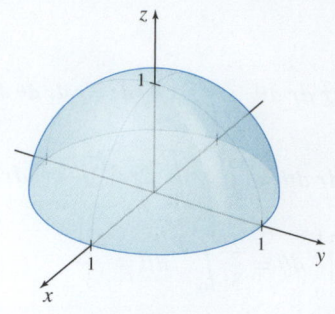

**Figure 55** $z = 1 - x^2 - y^2$

| **NOTE** Remember in cylindrical coordinates $dV = r\, dr\, d\theta\, dz$.

**EXAMPLE 4** Finding the Moment of Inertia of a Solid

Find the moment of inertia about the $z$-axis of a homogeneous solid $E$ of mass density $\rho$ enclosed by the paraboloid $z = 1 - x^2 - y^2$ and the $xy$-plane.

**Solution** In cylindrical coordinates, the solid $E$ is enclosed on the top by the paraboloid $z = 1 - x^2 - y^2 = 1 - r^2$ and on the bottom by $z = 0$. See Figure 55. In the $xy$-plane, the region is bounded by the circle $x^2 + y^2 = r^2 = 1$. So, we have

$$0 \le \theta \le 2\pi \qquad 0 \le r \le 1 \qquad 0 \le z \le 1 - r^2$$

The moment of inertia $I_z$ about the $z$-axis is

$$I_z = \iiint_E r^2 \rho\, dV = \rho \iiint_E r^2 r\, dr\, d\theta\, dz = \rho \int_0^{2\pi} \int_0^1 \int_0^{1-r^2} r^3\, dz\, dr\, d\theta$$

$$= \rho \int_0^{2\pi} \int_0^1 r^3(1 - r^2)\, dr\, d\theta = \rho \int_0^{2\pi} \left[ \frac{r^4}{4} - \frac{r^6}{6} \right]_0^1 d\theta = \rho \int_0^{2\pi} \frac{1}{12}\, d\theta = \frac{\pi}{6}\rho \quad \blacksquare$$

**NOW WORK** Problem 43.

**[CAS]** **EXAMPLE 5** Finding the Moment of Inertia of a Ball

(a) Set up an integral to find the moment of inertia of a homogeneous ball of radius 2 about a diameter.

(b) Use a CAS to find the moment of inertia from (a).

**Solution (a)** Let $\rho$ denote the constant mass density of the ball of radius 2. We position the ball so that its center is at the origin. The equation of the ball is then $x^2 + y^2 + z^2 = 4$, or, in cylindrical coordinates, $r^2 + z^2 = 4$. If we use the $z$-axis as the diameter, the moment of inertia about a diameter is given by the moment of inertia $I_z$ about the $z$-axis. Then

$$I_z = \iiint_E r^2 \rho\, dV = \rho \int_0^{2\pi} \int_0^2 \int_{-\sqrt{4-r^2}}^{\sqrt{4-r^2}} r^3\, dz\, dr\, d\theta$$

**(b)** We use a CAS to find

$$I_z = \rho \int_0^{2\pi} \int_0^2 \int_{-\sqrt{4-r^2}}^{\sqrt{4-r^2}} r^3\, dz\, dr\, d\theta = \frac{256\pi\rho}{15} \quad \blacksquare$$

## 14.7 Assess Your Understanding

### Concepts and Vocabulary

1. To convert a point $P = (x, y, z)$ from rectangular coordinates to cylindrical coordinates $(r, \theta, z)$, use the equations:
   $x = \underline{\hspace{1cm}}$, $y = \underline{\hspace{1cm}}$, and $z = \underline{\hspace{1cm}}$.

2. Expressed in cylindrical coordinates, the circular cone $x^2 + y^2 = 4z^2$ has the form $\underline{\hspace{1cm}}$.

3. In cylindrical coordinates $(r, \theta, z)$, the differential $dV$ of volume is $\underline{\hspace{1cm}}$.

4. *True or False* In cylindrical coordinates, $\iiint_E r\, dr\, d\theta\, dz$ is numerically equal to the volume of the solid $E$.

### Skill Building

*In Problems 5–12, find the cylindrical coordinates $(r, \theta, z)$ of each point with the given rectangular coordinates.*

5. $(-\sqrt{3}, -1, -5)$

6. $(-1, \sqrt{3}, 4)$

7. $(1, 1, \sqrt{2})$

8. $(2, -2, 4)$

9. $(2, 0, 4)$

10. $\left(-1, 0, \dfrac{1}{2}\right)$

11. $(0, 3, 4)$

12. $(0, 1, -3)$

---

**1. = NOW WORK problem**    **[N]** = Graphing technology recommended    **[CAS]** = Computer Algebra System recommended

*In Problems 13–18, find the rectangular coordinates $(x, y, z)$ of each point with the given cylindrical coordinates.*

**13.** $\left(2, \dfrac{\pi}{6}, -5\right)$    **14.** $\left(4, \dfrac{\pi}{3}, 3\right)$    **15.** $(1, 0, 8)$

**16.** $\left(4, \dfrac{\pi}{6}, 2\right)$    **17.** $\left(2, \dfrac{\pi}{2}, 0\right)$    **18.** $\left(-3, \dfrac{\pi}{2}, 1\right)$

*In Problems 19–22, give a geometric interpretation of each triple integral.*

**19.** $\displaystyle\int_{-1}^{1}\int_{-\sqrt{1-x^2}}^{\sqrt{1-x^2}}\int_{0}^{\sqrt{1-x^2-z^2}} dy\,dz\,dx$

**20.** $\displaystyle\int_{0}^{1}\int_{0}^{\sqrt{1-z^2}}\int_{0}^{\sqrt{1-y^2-z^2}} dx\,dy\,dz$

**21.** $\displaystyle\int_{0}^{2}\int_{-\sqrt{4-x^2}}^{\sqrt{4-x^2}}\int_{0}^{\sqrt{x^2+y^2}} dz\,dy\,dx$

**22.** $\displaystyle\int_{0}^{3}\int_{-\sqrt{9-x^2}}^{\sqrt{9-x^2}}\int_{0}^{\sqrt{x^2+z^2}} dy\,dz\,dx$

*In Problems 23–26, find each iterated integral.*

**23.** $\displaystyle\int_{\pi/6}^{\pi/2}\int_{0}^{3}\int_{0}^{r\sin\theta} r\csc^3\theta\,dz\,dr\,d\theta$

**24.** $\displaystyle\int_{\pi/6}^{\pi/2}\int_{0}^{1}\int_{0}^{\sin\theta} r\cos\theta\sin\theta\,dz\,dr\,d\theta$

**25.** $\displaystyle\int_{0}^{\pi/3}\int_{0}^{1}\int_{0}^{e-1} r\,dz\,dr\,d\theta$

**26.** $\displaystyle\int_{0}^{\pi/3}\int_{0}^{\sin\theta}\int_{0}^{r\sin\theta} r\,dz\,dr\,d\theta$

*In Problems 27–32, find each triple integral by converting to cylindrical coordinates.*

**27.** $\iiint_E dV$, where $E$ is the solid enclosed by the planes $z=1$ and $z=4$, and the cylinders $x^2+y^2=1$ and $x^2+y^2=9$.

**28.** $\iiint_E dV$, where $E$ is the solid enclosed by the $xy$-plane, $z=3$, and the cylinder $x^2+y^2=4$.

**29.** $\iiint_E y\,dV$, where $E$ is the solid enclosed by the planes $z=1$ and $z=x+3$, and the cylinders $x^2+y^2=1$ and $x^2+y^2=4$.

**30.** $\iiint_E x\,dV$, where $E$ is the solid enclosed by the planes $z=0$ and $z=x$, and the cylinder $x^2+y^2=9$.

**31.** $\iiint_E xy\,dV$, where $E$ is the solid enclosed by the surfaces $z=1-x-y$ and $z=3-x-y$, whose projection onto the $xy$-plane is the circle $x^2+y^2=1$.

**32.** $\iiint_E xy\,dV$, where $E$ is the solid enclosed by the surfaces $z=0$ and $z=x^2+y^2$, whose projection onto the $xy$-plane is the circle $x^2+y^2=4$.

## Applications and Extensions

**33. Volume**  Find the volume of the solid enclosed by the intersection of the sphere $x^2+y^2+z^2=9$ and the cylinder $x^2+y^2=2$.

**34. Volume**  Find the volume of the solid enclosed by the intersection of the sphere $x^2+y^2+z^2=4$ and the cylinder $x^2+y^2=2x$.

**35. Volume**  Find the volume $V$ of the solid enclosed by $z=x^2+y^2$ and $z=16-x^2-y^2$.

**36. Volume**  Find the volume $V$ of the solid enclosed by $z=x^2+y^2$ and $z=2-x$.

**37. Volume**  Find the volume $V$ of the solid enclosed by $z^2=4x$ and $x^2+y^2=2x$.

**38. Mass**  Find the mass of a homogeneous solid of mass density $\rho$ in the shape of a sphere of radius $a$.

**39. Mass**  Find the mass of a solid in the shape of a sphere of radius $a$, if the mass density $\rho$ is proportional to the square of the distance from the center.

**40. Mass**  Find the mass $M$ of an object in the shape of a right circular cylinder of height $h$ and radius $a$, if its mass density is proportional to the square of the distance from the axis of the cylinder.

**41. Moments of Inertia**  Find the moments of inertia $I_x$ and $I_y$ for the solid region enclosed by the hemisphere $z=\sqrt{9-x^2-y^2}$ and the $xy$-plane, if the mass density is proportional to the distance from the $xy$-plane.

**42. Center of Mass**  Find the center of mass of a homogeneous solid in the first octant enclosed by the surface $z=xy$ and the cylinder $x^2+y^2=4$.

**43. Center of Mass**  Find the center of mass of a homogeneous solid enclosed by the surface $x^2+y^2=4z$ and the plane $z=2$.

**44. Center of Mass**  Find the center of mass of a homogeneous solid enclosed by the inside of the sphere $x^2+y^2+z^2=12$ and above the paraboloid $z=x^2+y^2$.

**45. Center of Mass**  Find the center of mass of a homogeneous solid enclosed by the paraboloid $z=x^2+y^2$ and the plane $z=4$.

**46. Mass**  Use cylindrical coordinates to find the mass of the homogeneous solid bounded on the sides by $x^2+y^2=1$, on the bottom by the $xy$-plane, and on the top by $x^2+y^2+z^2=2$.

**47. Joint Between Two Rods**  Find the mass of the intersection of two rods with constant mass density $\rho$ that is formed by the cylinders $x^2+y^2=1$ and $x^2+z^2=1$.

**48. Volume of a Mountain**  The height of a mountain (in km) can be approximated by $z=5.3e^{-(x^2+y^2)}$.

(a) Sketch the mountain over the region $x^2+y^2\le 4$.

(CAS) (b) Find the volume of the mountain over the region $x^2+y^2\le 4$.

## Challenge Problems

**49. Volume** A circular hole of radius $r$ is drilled through a sphere of radius $R > r$, as shown in the figure.

(a) Find $r$ in terms of $R$ so that the hole removes exactly half of the volume of the sphere. *Hint:* set up an integral in cylindrical coordinates for the volume of the hole.

(b) Find $r$ rounded to three decimal places if $R = 10\,\text{cm}$.

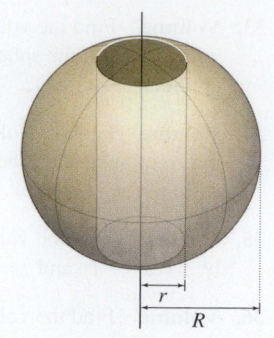

**50. Volume** Find the volume enclosed on the top by the sphere $x^2 + y^2 + z^2 = 5$ and on the bottom by the paraboloid $x^2 + y^2 = 4z$.

**51. Volume of a Joint** Two pipes intersect at right angles as shown in the figure. Find the inner radius $r$ of the pipes to ensure the volume of the intersecting joint is $10\,\text{m}^3$.

# 14.8 Triple Integrals Using Spherical Coordinates

**OBJECTIVES** *When you finish this section, you should be able to:*

**1** Convert rectangular coordinates to spherical coordinates (p. 1006)

**2** Find a triple integral using spherical coordinates (p. 1008)

Application: Spherical coordinates in navigation (p. 1011)

As we have seen in Section 14.7, there are times when it is easier to find a triple integral using an alternate coordinate system. When a triple integral involves all or a portion of a sphere centered at the origin or all or a portion of a cone, converting the problem to *spherical coordinates* often reduces the effort needed to find the integral.

## 1 Convert Rectangular Coordinates to Spherical Coordinates

Suppose $P$ is a point in space. If $P$ is not at the origin, we define the three numbers $\rho, \theta$, and $\phi$ as:

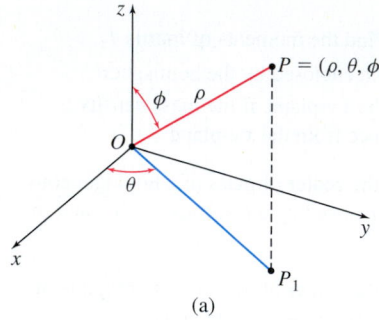

(a)

- $\rho = |OP|$ is the distance from the origin to the point $P$.
- $\theta, 0 \leq \theta < 2\pi$, is the angle between the positive $x$-axis and $OP_1$, where $P_1$ is the projection of $P$ onto the $xy$-plane.
- $\phi, 0 \leq \phi \leq \pi$, is the angle between the positive $z$-axis and the line segment $OP$.

The ordered triple $(\rho, \theta, \phi)$ is called the **spherical coordinates** of the point $P$. See Figure 56(a).

The algebraic relationship between the spherical coordinates $(\rho, \theta, \phi)$ of a point $P$ and the rectangular coordinates $(x, y, z)$ of the point $P$ is found using the congruent triangles $OQP$ and $OP_1P$ as shown in Figure 56(b).

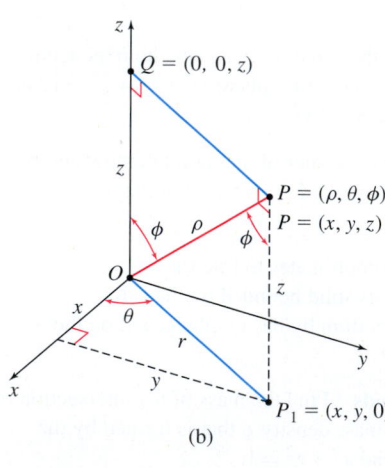

(b)

**Figure 56**

Since

$$x = r\cos\theta \quad y = r\sin\theta \quad r = \rho\sin\phi \quad z = \rho\cos\phi$$

it follows that

$$\boxed{x = \rho\sin\phi\cos\theta \qquad y = \rho\sin\phi\sin\theta \qquad z = \rho\cos\phi} \quad (1)$$

These equations are used to convert spherical coordinates $(\rho, \theta, \phi)$ to rectangular coordinates $(x, y, z)$.

Using the distance formula and the equations (1), we obtain

$$\boxed{\rho = \sqrt{x^2 + y^2 + z^2} \quad \tan\theta = \frac{y}{x}, \; x \neq 0, \; 0 \leq \theta < 2\pi \quad \cos\phi = \frac{z}{\rho}, \; 0 \leq \phi \leq \pi} \quad (2)$$

These equations are used to convert rectangular coordinates to spherical coordinates.

**EXAMPLE 1** **Converting Spherical Coordinates**

If the spherical coordinates of a point $P$ are $\left(4, \dfrac{\pi}{6}, \dfrac{2\pi}{3}\right)$, find the rectangular coordinates of $P$.

**Solution** We use the equations (1) to convert spherical coordinates to rectangular coordinates. Then

$$x = \rho \sin\phi \cos\theta = 4 \sin\frac{2\pi}{3} \cos\frac{\pi}{6} = 4 \cdot \frac{\sqrt{3}}{2} \cdot \frac{\sqrt{3}}{2} = 3 \qquad \rho = 4;\ \theta = \frac{\pi}{6},\ \phi = \frac{2\pi}{3}$$

$$y = \rho \sin\phi \sin\theta = 4 \sin\frac{2\pi}{3} \sin\frac{\pi}{6} = 4 \cdot \frac{\sqrt{3}}{2} \cdot \frac{1}{2} = \sqrt{3}$$

$$z = \rho \cos\phi = 4\cos\frac{2\pi}{3} = 4\left(-\frac{1}{2}\right) = -2$$

The rectangular coordinates of $P$ are $(3, \sqrt{3}, -2)$. ∎

**NOW WORK** Problem 7.

**EXAMPLE 2** **Converting Rectangular Coordinates to Spherical Coordinates**

If the rectangular coordinates of a point $P$ are $(1, \sqrt{3}, -2)$, find the spherical coordinates of $P$.

**Solution** We use the equations (2) to convert rectangular coordinates to spherical coordinates. Then

$$\rho = \sqrt{x^2 + y^2 + z^2} = \sqrt{1 + 3 + 4} = \sqrt{8} = 2\sqrt{2}$$

$\tan\theta = \dfrac{y}{x} = \dfrac{\sqrt{3}}{1}$, and $(1, \sqrt{3}, 0)$ is in the first quadrant of the $xy$-plane.
So $\theta = \tan^{-1}\sqrt{3} = \dfrac{\pi}{3}$

$$\cos\phi = \frac{z}{\rho} = \frac{-2}{2\sqrt{2}} = -\frac{\sqrt{2}}{2}, \text{ so } \phi = \cos^{-1}\left(-\frac{\sqrt{2}}{2}\right) = \frac{3\pi}{4} \qquad 0 \le \phi \le \pi$$

The spherical coordinates of $P$ are $\left(2\sqrt{2}, \dfrac{\pi}{3}, \dfrac{3\pi}{4}\right)$. ∎

**NOW WORK** Problem 9.

Spherical coordinates are useful in problems involving spheres, vertical planes, and cones because their spherical coordinate equations are relatively simple. See Table 2.

**NOTE** Points $(\rho, \theta, \phi)$ are called spherical coordinates because the surface formed when $\rho = a$, where $a > 0$ is a constant, is a sphere.

**TABLE 2**

| Surface | Equation in Spherical Coordinates |
|---|---|
| Sphere | $\rho = a \quad a > 0$ |
| Half-cone | $\phi = a \quad 0 < a < \pi, a \ne \dfrac{\pi}{2}$ |
| Vertical Half-plane | $\theta = a \quad -\dfrac{\pi}{2} < a < \dfrac{\pi}{2}$ |

The surfaces in Table 2 are illustrated in Figure 57. The half-cone defined by $\phi = a$, where $0 < a < \pi$ and $a \neq \dfrac{\pi}{2}$, is generated by revolving a ray with its initial point at the origin and forming an angle $a$ about the $z$-axis.

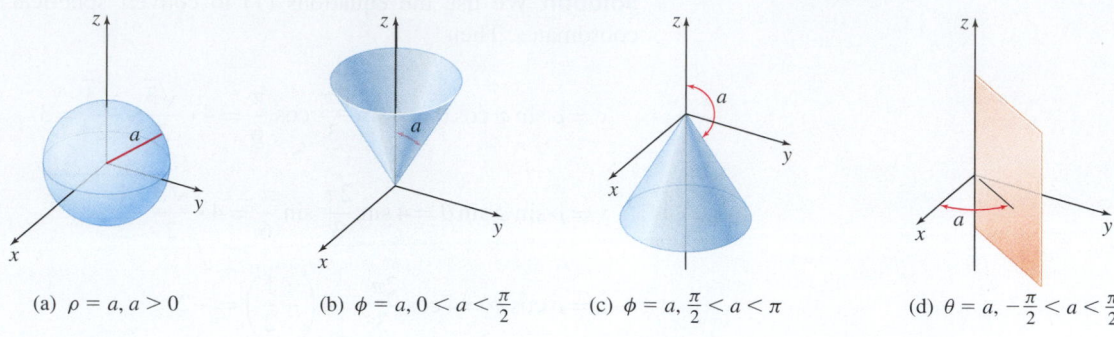

(a) $\rho = a, a > 0$     (b) $\phi = a, 0 < a < \dfrac{\pi}{2}$     (c) $\phi = a, \dfrac{\pi}{2} < a < \pi$     (d) $\theta = a, -\dfrac{\pi}{2} < a < \dfrac{\pi}{2}$

**Figure 57**

## 2 Find a Triple Integral Using Spherical Coordinates

When a solid $E$ is comprised of parts of spheres, cones, or vertical planes, it is sometimes easier to find $\iiint\limits_{E} f(x, y, z)dV$ by converting the rectangular coordinates to spherical coordinates. We convert a function $w = f(x, y, z)$ to spherical coordinates $(\rho, \theta, \phi)$ using the equations $x = \rho \sin\phi \cos\theta$, $y = \rho \sin\phi \sin\theta$, $z = \rho \cos\phi$. It remains to find the differential $dV$ in spherical coordinates.

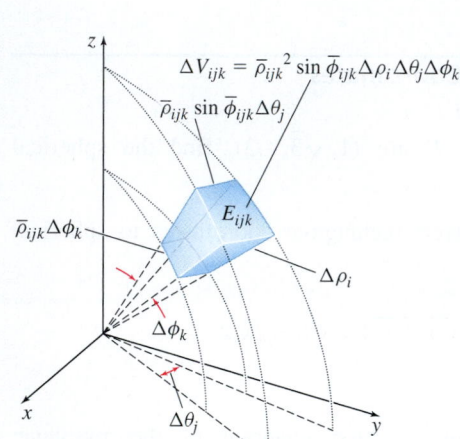

**Figure 58**

We proceed in a way similar to that used for triple integrals in cylindrical coordinates. Let $(\rho, \theta, \phi)$ be the spherical coordinates of a point in space, and let $E$ be a ball in space. We partition the ball using concentric spheres centered about the $z$-axis, vertical planes passing through the $z$-axis, and circular cones with vertex at the origin and axis along the $z$-axis. The result is a collection of subsolids.

Suppose there are $n$ concentric spheres of radii $\rho_i$, $i = 1, 2, \ldots, n$; $m$ vertical planes $\theta = \theta_j$, $j = 1, 2, \ldots, m$; and $l$ circular cones $\phi = \phi_k$, $k = 1, 2, \ldots, l$, that form $lmn$ subsolids $E_{ijk}$. The volume $\Delta V_{ijk}$ of a typical subsolid $E_{ijk}$ is found by treating it as if it were a rectangular parallelepiped as shown in Figure 58. Then there is a point $(\bar{\rho}_{ijk}, \bar{\theta}_{ijk}, \bar{\phi}_{ijk})$ in $E_{ijk}$ for which

$$\Delta V_{ijk} = (\bar{\rho}_{ijk} \sin \bar{\phi}_{ijk} \Delta \theta_j)(\Delta \rho_i)(\bar{\rho}_{ijk} \Delta \phi_k) = \bar{\rho}_{ijk}^2 \sin \bar{\phi}_{ijk} \Delta \rho_i \, \Delta \theta_j \, \Delta \phi_k$$

Then the differential $dV$ is

$$\boxed{dV = \rho^2 \sin\phi \, d\rho \, d\theta \, d\phi}$$

It can be shown that if $f$ is continuous in $E$, then

$$\iiint\limits_{E} f(x, y, z)dV = \iiint\limits_{E} f(\rho \sin\phi \cos\theta, \rho \sin\phi \sin\theta, \rho \cos\phi)dV$$

$$= \iiint\limits_{E} f(\rho \sin\phi \cos\theta, \rho \sin\phi \sin\theta, \rho \cos\phi)\rho^2 \sin\phi \, d\rho \, d\theta \, d\phi$$

---

**EXAMPLE 3**  **Finding an Iterated Integral Using Spherical Coordinates**

Use spherical coordinates to find $\int_{-2}^{2} \int_{0}^{\sqrt{4-x^2}} \int_{0}^{\sqrt{4-x^2-y^2}} y^2 \, dz \, dy \, dx$.

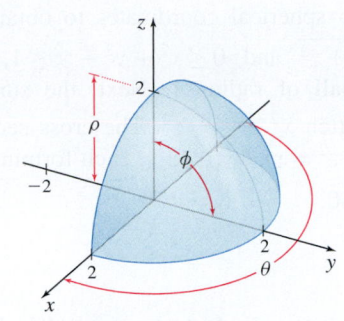

**Figure 59** $z = \sqrt{4 - x^2 - y^2},\, y \geq 0$

**Solution** From the limits of integration of the iterated integral, we see that $z$ varies from $z = 0$ to $z = \sqrt{4 - x^2 - y^2}$, which is a hemisphere of radius 2 centered at the origin. Also $y$ varies from 0 to $\sqrt{4 - x^2}$, and $-2 \leq x \leq 2$. The solid described by the limits of integration is the part of the hemisphere that lies above the region $0 \leq y \leq \sqrt{4 - x^2},\, -2 \leq x \leq 2$, in the $xy$-plane. See Figure 59.

To convert the limits of integration from rectangular coordinates to spherical coordinates, we begin with $z = \sqrt{4 - x^2 - y^2}$. Then $x^2 + y^2 + z^2 = 4$ so $\sqrt{x^2 + y^2 + z^2} = 2 = \rho$. The lower limit of integration is 0, so $0 \leq \rho \leq 2$.

Look again at Figure 59. The angle $\theta$ varies from 0 to $\pi$, and the angle $\phi$ between the positive $z$-axis and the line segment $\overline{OP}$ for any point $P$ on the surface varies from 0 to $\dfrac{\pi}{2}$.

Next, use $y = \rho \sin \phi \sin \theta$, and the fact that $dz\, dy\, dx = \rho^2 \sin \phi\, d\rho\, d\theta\, d\phi$, to express the iterated integral in spherical coordinates.

$$\int_{-2}^{2} \int_{0}^{\sqrt{4-x^2}} \int_{0}^{\sqrt{4-x^2-y^2}} y^2 \, dz\, dy\, dx = \int_{0}^{\pi/2} \int_{0}^{\pi} \int_{0}^{2} (\rho \sin \phi \sin \theta)^2 \rho^2 \sin \phi\, d\rho\, d\theta\, d\phi$$

$$= \int_{0}^{\pi/2} \int_{0}^{\pi} \int_{0}^{2} \rho^4 \sin^2 \theta \sin^3 \phi\, d\rho\, d\theta\, d\phi$$

$$= \int_{0}^{\pi/2} \left\{ \int_{0}^{\pi} \left[ \int_{0}^{2} \rho^4 d\rho \right] \sin^2 \theta\, d\theta \right\} \sin^3 \phi\, d\phi$$

$$= \int_{0}^{\pi/2} \left\{ \int_{0}^{\pi} \left[ \frac{\rho^5}{5} \right]_{0}^{2} \sin^2 \theta\, d\theta \right\} \sin^3 \phi\, d\phi$$

$$= \frac{32}{5} \int_{0}^{\pi/2} \left\{ \int_{0}^{\pi} \left[ \frac{1}{2} - \frac{\cos(2\theta)}{2} \right] d\theta \right\} \sin^3 \phi\, d\phi \qquad \sin^2 \theta = \frac{1}{2} - \frac{\cos(2\theta)}{2}$$

$$= \frac{32}{5} \int_{0}^{\pi/2} \left[ \frac{1}{2}\theta - \frac{1}{4} \sin(2\theta) \right]_{0}^{\pi} \sin^3 \phi\, d\phi$$

$$= \frac{32}{5} \cdot \frac{\pi}{2} \int_{0}^{\pi/2} \sin^3 \phi\, d\phi = \frac{16\pi}{5} \int_{0}^{\pi/2} (1 - \cos^2 \phi) \sin \phi\, d\phi$$

$$= \frac{16}{5}\pi \int_{1}^{0} (1 - u^2)(-du) \qquad u = \cos \phi,\, du = -\sin \phi\, d\phi$$

$$= \frac{16}{5}\pi \left[ \frac{u^3}{3} - u \right]_{1}^{0} = \frac{16}{5}\pi \left[ 0 - \left( \frac{1}{3} - 1 \right) \right] = \frac{32}{15}\pi \qquad \blacksquare$$

NOW WORK Problem **21**.

---

**EXAMPLE 4** **Finding a Triple Integral Using Spherical Coordinates**

Use spherical coordinates to find the integral

$$\iiint\limits_{E} z\, dx\, dy\, dz$$

where $E$ is the solid defined by the inequalities

$$0 \leq \sqrt{x^2 + y^2} \leq z \qquad 0 \leq x^2 + y^2 + z^2 \leq 1$$

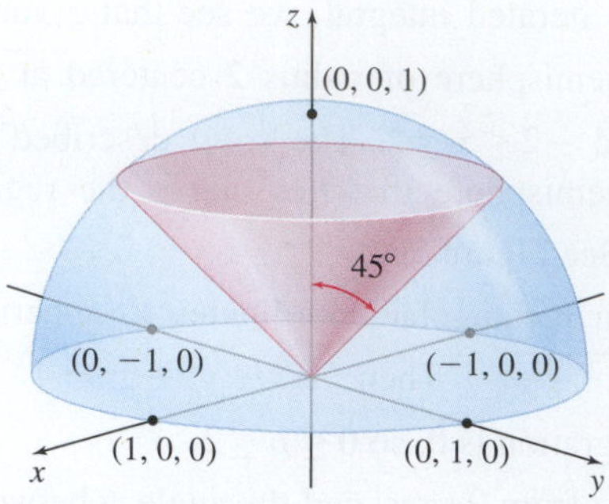

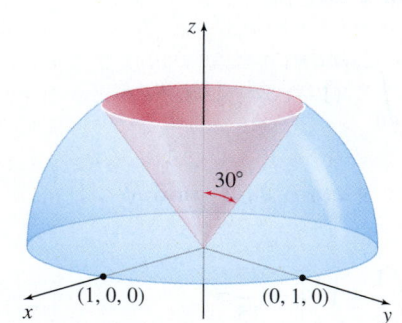

**Figure 60**

**Solution** Convert the rectangular coordinates to spherical coordinates to obtain $E$ in spherical coordinates. Since $\rho = \sqrt{x^2 + y^2 + z^2}$ and $0 \le x^2 + y^2 + z^2 \le 1$, we have $0 \le \rho \le 1$. This is the upper half of a ball of radius 1. Next, the surface $z = \sqrt{x^2 + y^2}$ (which is a half-cone) can be written $x^2 + y^2 = z^2$. The cross section of the cone with the $xz$-plane is the pair of lines $z^2 = x^2$ or $z = \pm x$, each forming an angle $\phi = \dfrac{\pi}{4}$ with the positive $z$-axis. See Figure 60.

Since $z = \rho \cos \phi$, the integral becomes

$$\iiint\limits_E z\, dx\, dy\, dz = \iiint\limits_E \rho \cos\phi \cdot \rho^2 \sin\phi\, d\rho\, d\phi\, d\theta \qquad \boxed{dx\,dy\,dz = \rho^2 \cos\phi\, d\rho\, d\phi\, d\theta}$$

$$= \int_0^{2\pi} \int_0^{\pi/4} \int_0^1 \rho^3 \sin\phi \cos\phi\, d\rho\, d\phi\, d\theta$$

$$= \int_0^{2\pi} \int_0^{\pi/4} \left[ \int_0^1 \rho^3 d\rho \right] \sin\phi \cos\phi\, d\phi\, d\theta = \int_0^{2\pi} \int_0^{\pi/4} \frac{1}{4} \sin\phi \cos\phi\, d\phi\, d\theta$$

$$= \frac{1}{4} \int_0^{2\pi} \left[ \int_0^{\pi/4} \sin\phi \cos\phi\, d\phi \right] d\theta = \frac{1}{4} \int_0^{2\pi} \left[ \int_1^{\sqrt{2}/2} -u\, du \right] d\theta$$

$$\uparrow$$
$$u = \cos\phi,\ du = -\sin\phi\, d\phi$$
$$\text{when } \phi = 0,\ u = 1,\ \text{when } \phi = \frac{\pi}{4},\ u = \frac{\sqrt{2}}{2}$$

$$= \frac{1}{4} \int_0^{2\pi} \left[ -\frac{u^2}{2} \right]_1^{\sqrt{2}/2} d\theta = \frac{1}{4} \int_0^{2\pi} \left( -\frac{1}{4} + \frac{1}{2} \right) d\theta = \frac{1}{16} \int_0^{2\pi} d\theta = \frac{\pi}{8} \qquad ■$$

**NOW WORK** Problem 25.

---

**EXAMPLE 5** **Finding the Volume of a Solid Using Spherical Coordinates**

Find the volume of the solid that is removed from a hemisphere of radius 1 when it is cut by a cone that makes an angle of 30° with the positive $z$-axis.

**Solution** Figure 61 shows the part of the hemisphere cut by the cone, namely, the volume under the sphere and inside the cone. The upper surface is the hemisphere of radius $r = 1$; the lower surface is the $xy$-plane ($z = 0$). The angle $\theta$ ranges from 0 to $2\pi$ and $\phi$ ranges from 0 to $\dfrac{\pi}{6}$.

$$0 \le \rho \le 1 \qquad 0 \le \theta \le 2\pi \qquad 0 \le \phi \le \frac{\pi}{6}$$

**Figure 61**

Then the volume $V$ of the solid is

$$V = \iiint\limits_E \rho^2 \sin\phi\, d\rho\, d\theta\, d\phi = \int_0^{\pi/6} \int_0^{2\pi} \int_0^1 \rho^2 \sin\phi\, d\rho\, d\theta\, d\phi$$

$$= \int_0^{\pi/6} \int_0^{2\pi} \left[ \int_0^1 \rho^2 d\rho \right] \sin\phi\, d\theta\, d\phi$$

$$= \int_0^{\pi/6} \int_0^{2\pi} \left[ \frac{\rho^3}{3} \right]_0^1 d\theta \sin\phi\, d\phi = \int_0^{\pi/6} \left[ \int_0^{2\pi} \frac{1}{3} d\theta \right] \sin\phi\, d\phi$$

$$= \frac{2\pi}{3} \int_0^{\pi/6} \sin\phi\, d\phi = \frac{2\pi}{3} [(-\cos\phi)]_0^{\pi/6} = \frac{2\pi}{3} \left( 1 - \frac{\sqrt{3}}{2} \right) \qquad ■$$

**NOW WORK** Problem 33.

CALC
CLIP

### EXAMPLE 6 Finding the Mass of a Ball

Find the mass $M$ of a ball if its mass density $\delta$ is proportional to the distance from the center of the ball.

**NOTE** Since $\rho$ is used as a spherical coordinate, we use $\delta$ to represent the mass density.

**Solution** Let $a$ be the radius of the ball, and position the ball so that its center is at the origin. In spherical coordinates, the equation of the ball is $\rho = a$. The mass density $\delta$ of the ball is $\delta = k\rho$, where $k$ is the constant of proportionality. The mass $M$ is

$$M = \iiint_E \delta \, dV = \iiint_E k\rho \rho^2 \sin\phi \, d\rho \, d\phi \, d\theta = k \iiint_E \rho^3 \sin\phi \, d\rho \, d\phi \, d\theta$$

$$= k \int_0^{2\pi} \int_0^{\pi} \left[ \int_0^a \rho^3 \, d\rho \right] \sin\phi \, d\phi \, d\theta$$

$$= k \int_0^{2\pi} \int_0^{\pi} \left[ \frac{\rho^4}{4} \right]_0^a \sin\phi \, d\phi \, d\theta = \frac{ka^4}{4} \int_0^{2\pi} \left[ \int_0^{\pi} \sin\phi \, d\phi \right] d\theta$$

$$= \frac{ka^4}{4} \int_0^{2\pi} \left[ -\cos\phi \right]_0^{\pi} d\theta$$

$$= \frac{ka^4}{2} \int_0^{2\pi} d\theta = k\pi a^4 \qquad \blacksquare$$

**NOW WORK** Problem 37.

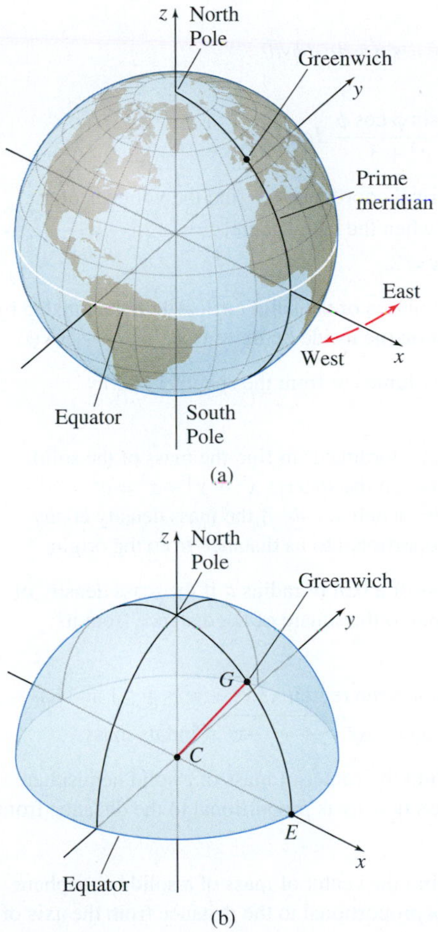

### Application: Spherical Coordinates in Navigation

If we assume that the Earth is a ball, there is a simple relationship between the spherical coordinates we have defined and the system of latitude and longitude measurements used in geography.

The origin is placed at the center of Earth and the $z$-axis is chosen to be the diameter through the North and South Poles. See Figure 62(a). The equator is then the great circle in the $xy$-plane. The $x$-axis is chosen so that the $xz$-plane passes through the Greenwich Observatory near London.

The longitude for a point $P$ on the surface of Earth is then the angle we have called $\theta$, except that degree measure is used and east and west are measured from the great circle through the poles and Greenwich. Greenwich has longitude $0°$.

The latitude for a point on the surface of Earth north of the equator is $90° - \phi$; latitude is measured from the equator rather than from the North Pole. For a point on Earth south of the equator, the latitude is $\phi - 90°$. Points on the equator have $\phi = 90°$ and latitude $0°$. For the North Pole, we write $\phi = 0°$ and north latitude $90°$; for the South Pole, we write $\phi = 180°$ and south latitude $90°$.

(a)

(b)

**Figure 62**

### EXAMPLE 7 Finding Longitude and Latitude in Spherical Coordinates

Greenwich, England, has a longitude of $0°$ and an approximate latitude of $51.5°$ N. If $R_E = 3960$ miles is used for the radius of the Earth and degrees are used to measure angles instead of radians, then the spherical coordinates for Greenwich, England, are $(3960, 0°, 38.5°)$. See Figure 62(b), where the angle $ECG = 51.5°$. $\blacksquare$

## 14.8 Assess Your Understanding

### Concepts and Vocabulary

1. In spherical coordinates, the number $\rho$ equals the distance from the _____ to the point $P$.

2. In spherical coordinates, $\phi$ is the angle between the positive _____ and the line segment $OP$.

3. *True or False*  To change spherical coordinates $(\rho, \theta, \phi)$ to rectangular coordinates, use the equations $x = \rho \sin\phi \cos\phi$, $y = \rho \sin\phi \sin\theta$, and $z = \rho \cos\theta$.

4. *True or False*  In spherical coordinates $(\rho, \theta, \phi)$, $\rho = \sqrt{x^2 + y^2 + z^2}$, where $(x, y, z)$ are the rectangular coordinates of the point $P$.

1. = NOW WORK problem     📐 = Graphing technology recommended     [CAS] = Computer Algebra System recommended

**5.** What is the differential $dV$ of the volume $V$ in spherical coordinates?

**6.** In spherical coordinates, the surface described by the equation $\rho = a$, where $a > 0$ is a constant, is a(n) _____.

## Skill Building

*In Problems 7 and 8, convert each point given in spherical coordinates to rectangular coordinates.*

**7.** (a) $\left(4, \dfrac{\pi}{6}, \pi\right)$    (b) $\left(2, \dfrac{2\pi}{3}, \dfrac{\pi}{2}\right)$

**8.** (a) $\left(2, \dfrac{\pi}{4}, \dfrac{2\pi}{3}\right)$    (b) $\left(4, \dfrac{5\pi}{6}, \dfrac{3\pi}{4}\right)$

*In Problems 9–16, convert each point given in rectangular coordinates to spherical coordinates.*

**9.** $(-\sqrt{2}, -\sqrt{2}, 2\sqrt{3})$    **10.** $(-1, \sqrt{3}, 2)$    **11.** $(1, 1, \sqrt{2})$

**12.** $(1, -\sqrt{3}, -2)$    **13.** $(1, 2, 3)$    **14.** $(1, -1, \sqrt{2})$

**15.** $(0, 3\sqrt{3}, 3)$    **16.** $(-5\sqrt{3}, 5, 0)$

*In Problems 17–20, find each iterated integral.*

**17.** $\displaystyle\int_0^{\pi/2} \int_0^{\sin\phi} \int_0^{\pi/4} \rho^2 \sin\phi\, d\theta\, d\rho\, d\phi$

**18.** $\displaystyle\int_0^{\pi/2} \int_0^{\pi/2} \int_0^{\sin\phi} \rho^2 \sin\phi \cos\phi\, d\rho\, d\theta\, d\phi$

**19.** $\displaystyle\int_0^{2\pi} \int_0^{\pi/4} \int_0^{\sec\phi} \rho^2 \sin^2\phi\, d\rho\, d\phi\, d\theta$

**20.** $\displaystyle\int_0^{\pi/4} \int_0^{\cos\phi} \int_0^{2\pi} \rho^2 \sin\phi\, d\theta\, d\rho\, d\phi$

*In Problems 21–24, use spherical coordinates to find each triple integral.*

**21.** $\displaystyle\int_{-2}^{2} \int_{-\sqrt{4-y^2}}^{\sqrt{4-y^2}} \int_{2}^{2+\sqrt{4-x^2-y^2}} y\, dz\, dx\, dy$

**22.** $\displaystyle\int_{-3}^{3} \int_{-\sqrt{9-x^2}}^{\sqrt{9-x^2}} \int_{-\sqrt{9-x^2-y^2}}^{\sqrt{9-x^2-y^2}} (x^2 z + y^2 z + z^3)\, dz\, dy\, dx$

**23.** $\displaystyle\int_0^1 \int_0^{\sqrt{1-y^2}} \int_{\sqrt{x^2+y^2}}^{\sqrt{1-x^2-y^2}} (x^2 + y^2 + z^2)^{3/2}\, dz\, dx\, dy$

**24.** $\displaystyle\int_0^1 \int_0^{\sqrt{1-y^2}} \int_0^{\sqrt{1-x^2-y^2}} e^{(x^2+y^2+z^2)^{3/2}}\, dz\, dx\, dy$

## Applications and Extensions

**25.** Use spherical coordinates to integrate $f(x, y, z) = \sqrt{x^2 + y^2 + z^2}$ over the solid outside the half-cone $z = -\sqrt{3x^2 + 3y^2}$ and inside the sphere $x^2 + y^2 + z^2 = 4$.

**26.** Use spherical coordinates to integrate $f(x, y, z) = \sqrt{x^2 + y^2 + z^2}$ over the solid between the spheres $x^2 + y^2 + z^2 = 1$ and $x^2 + y^2 + z^2 = 4$.

*In Problems 27 and 28, use either cylindrical or spherical coordinates to find each triple integral.*

**27.** $\displaystyle\int_0^2 \int_0^2 \int_0^{\sqrt{4-x^2}} \sqrt{x^2 + y^2}\, dy\, dx\, dz$

**28.** $\displaystyle\int_0^2 \int_0^{\sqrt{4-y^2}} \int_0^{\sqrt{4-x^2-y^2}} \dfrac{2z}{\sqrt{x^2 + y^2}}\, dz\, dx\, dy$

*In Problems 29 and 30, set up the triple integral $\iiint\limits_E f(x, y, z)\, dV$ for $E$ in rectangular, cylindrical, and spherical coordinates.*

**29.** $E$ is a sphere of radius $a$ with its center at the origin.

**30.** $E$ is the solid inside the cylinder $x^2 + y^2 = 4$ and inside the sphere $x^2 + y^2 + z^2 = 9$.

*The integrals in Problems 31 and 32 are given in spherical coordinates. Express each integral using rectangular coordinates. Do not find the integral.*

**31.** $\displaystyle\int_0^\pi \int_{3\pi/4}^\pi \int_0^4 \rho^5 \cos\phi \sin\phi\, d\rho\, d\phi\, d\theta$

**32.** $\displaystyle\int_{\pi/2}^{3\pi/2} \int_{\pi/2}^\pi \int_0^2 \dfrac{\rho^4 \sin\phi \cos\phi}{\rho^2 + 3}\, d\rho\, d\phi\, d\theta$

**33. Volume** Use spherical coordinates to find the volumes of the two solids obtained when the hemisphere, $z = \sqrt{25 - x^2 - y^2}$, is sliced by the plane $x = 2$.

**34. Volume** Find the volume of the solid enclosed on the outside by the sphere $\rho = 2$ and on the inside by the surface $\rho = 1 + \cos\phi$.

**35. Volume** Find the volume cut from the sphere $\rho = a$ by the cone $\phi = \alpha$.

**36. Mass** Use spherical coordinates to find the mass of the solid in the first octant between the spheres $x^2 + y^2 + z^2 = a^2$ and $x^2 + y^2 + z^2 = b^2$, where $a > b$, if the mass density at any point is inversely proportional to its distance from the origin.

**37. Mass** Find the mass of a ball of radius $a$ if the mass density of the ball is proportional to the square of the distance from its center.

**38. Mass** A solid occupies the region $\sqrt{x^2 + y^2} \le z \le 1$ and has mass density $\delta(x, y, z) = z\sqrt{x^2 + y^2 + z^2}$. Find its mass.

**39. Center of Mass** Find the center of mass of a solid hemisphere of radius $a$ if the mass density is proportional to the distance from the center.

**40. Center of Mass** Find the center of mass of a solid hemisphere if the mass density is proportional to the distance from the axis of symmetry.

**41. Center of Mass** Find the center of mass of a homogeneous solid enclosed from above by the sphere $x^2 + y^2 + z^2 = 9$ and from below by the half-cone $z = \sqrt{x^2 + y^2}$.

**42. Center of Mass** Find the center of mass of a homogeneous solid in the shape of the wedge bounded by $x^2 + y^2 = 16$, $z = 2y$, $y \ge 0$, $z \ge 0$.

**43. Moment of Inertia** Find the moment of inertia of a homogeneous solid in the shape of a ball of radius 2 about a diameter.

**44. Moment of Inertia**   Show that the moment of inertia of a homogeneous solid in the shape of a ball of radius $a$ about a diameter is $I = \dfrac{8}{15}a^5\pi$.

**45. Centroid**   Use spherical coordinates to find the centroid of a hemisphere of radius $a$, whose base is on the $xy$-plane.

**46. Gravitational Force**   The magnitude of the resultant gravitational force $F$ of a solid hemisphere $E$ of radius $a$ and constant mass density $\delta$ on a unit mass particle situated at the center of the base of the hemisphere is given by the triple integral

$$F = k\delta \iiint_E \frac{\cos\phi}{\rho^2}\,dV$$

where the center of the sphere is at the origin. Find the force $F$.

**47. Modeling a Moon of Saturn**
On December 12, 2011, the Cassini mission spacecraft made its closest approach to the **moon Dione**. As with the planets, Dione is denser inside than at the surface because gravity compresses the matter. Suppose on Dione the density at the center is $2.50\,\text{g/cm}^3 = 2500\,\text{kg/m}^3$,

NASA/JPL-Caltech/Space Science Institute

the density at the surface is $1.25\,\text{g/cm}^3 = 1250\,\text{kg/m}^3$, and the radius is 560 km. Model the density $D$ as $D = D_0 e^{-k\rho}$, where $\rho$ is the distance from the center and $k$ is a constant to be determined.

**(a)** Using this model, find $k$ and $D_0$.

**(b)** Use this model to find the mass of Dione. Test the reasonableness of the model by comparing the calculated mass with the measured mass, which is $1.095 \times 10^{21}$ kg.

**48. Modeling a Moon of Saturn**   Refer to Problem 47. Instead of an exponential function, model the density of Dione as a linear decrease in density from $2.50\,\text{g/cm}^3$ at the center of Dione to $1.25\,\text{g/cm}^3$ at its surface. Find the mass of Dione using this model and compare it with the measured value ($1.05 \times 10^{21}$ kg).

**Challenge Problem**

**49. Volume**   A **spherical cap** is a portion of height $h$ of a sphere of radius $R$. See the figure. Using a triple integral and an appropriate coordinate system, show that the volume of the cap is $V = \dfrac{1}{3}\pi h^2(3R - h)$.

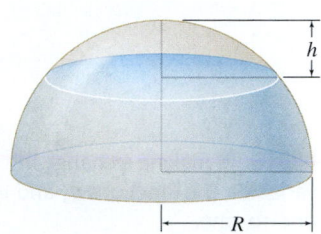

# 14.9 Change of Variables Using Jacobians

**OBJECTIVES**   *When you finish this section, you should be able to:*

**1** Find a Jacobian in two variables (p. 1014)

**2** Change the variables of a double integral using a Jacobian (p. 1015)

**3** Change the variables of a triple integral using a Jacobian (p. 1016)

We often find a single integral, $\int_a^b f(x)\,dx$, using the method of substitution. To use substitution, we let $x = g(u)$. Then $dx = g'(u)\,du$, and

$$\int_a^b f(x)\,dx = \int_c^d f(g(u))g'(u)\,du$$

where $a = g(c)$ and $b = g(d)$. In making this substitution, the variable $x$ is replaced by $g(u)$, and the factor $g'(u)$ is introduced into the integrand.

Sometimes it is easier to find a double integral $\iint_R f(x, y)\,dA$ by changing the variables $x$ and $y$ to the polar coordinates $r$ and $\theta$, by letting $x = r\cos\theta$, $y = r\sin\theta$, and $dA = r\,dr\,d\theta$. Then

$$\iint_R f(x, y)\,dA = \iint_{R^\#} f(r\cos\theta, r\sin\theta)\,r\,dr\,d\theta$$

where $R^\#$ is the region $R$ expressed in polar coordinates. In changing the variables from rectangular coordinates to polar coordinates, the factor $r$ is introduced into the integrand on the right.

When finding a triple integral $\iiint_E f(x, y, z)\,dV$, we discussed two types of change of variables: Sometimes the variables $x$, $y$, and $z$ were changed to cylindrical coordinates; other times the variables were changed to spherical coordinates.

**ORIGINS**   Jacobians are named after the German mathematician Carl Gustav Jacobi (1804–1851). Jacobi was a child prodigy. By age 12, he had met the entrance requirements for university admission. He studied mathematics, Latin, and Greek and became a professor at the University of Königsberg. Although he worked with functional determinants, he is most noted for his work with elliptic functions. Jacobi died of smallpox at age 46.

INTERFOTO/Alamy

Using cylindrical coordinates $(r, \theta, z)$, we get

$$\iiint_E f(x, y, z)\, dV = \iiint_{E^\#} f(r \cos \theta, r \sin \theta, z)\, r\, dr\, d\theta\, dz$$

where $E^\#$ is the solid $E$ expressed in cylindrical coordinates. In converting from rectangular coordinates to cylindrical coordinates, the factor $r$ is introduced into the integrand on the right.

Using spherical coordinates $(\rho, \theta, \phi)$, we have

$$\iiint_E f(x, y, z)\, dV = \iiint_{E^\#} f(\rho \sin \phi \cos \theta, \rho \sin \phi \sin \theta, \rho \cos \phi) \rho^2 \sin \phi\, d\rho\, d\theta\, d\phi$$

where $E^\#$ is the solid $E$ expressed in spherical coordinates. In this conversion, a factor of $\rho^2 \sin \phi$ is introduced into the integrand on the right.

In general, when the variables of an integral are changed, a factor, called the *Jacobian*, is introduced into the resulting integrand.

**NOTE** In making a change in variables, the functions $g_1$ and $g_2$ are one-to-one.

---

**DEFINITION** Jacobian: Two Variables

For a change of variables from $x$, $y$ to $u$, $v$ given by

$$\boxed{x = g_1(u, v) \qquad y = g_2(u, v)}$$

the **Jacobian of $x$ and $y$ with respect to $u$ and $v$** is

$$\frac{\partial(x, y)}{\partial(u, v)} = \begin{vmatrix} \dfrac{\partial x}{\partial u} & \dfrac{\partial x}{\partial v} \\[2ex] \dfrac{\partial y}{\partial u} & \dfrac{\partial y}{\partial v} \end{vmatrix} = \frac{\partial x}{\partial u} \frac{\partial y}{\partial v} - \frac{\partial x}{\partial v} \frac{\partial y}{\partial u}$$

---

**NEED TO REVIEW?** Determinants are discussed in Section 10.5, p. 765.

## 1 Find a Jacobian in Two Variables

**EXAMPLE 1** Finding a Jacobian

Show that in changing from rectangular coordinates $(x, y)$ to polar coordinates $(r, \theta)$, the Jacobian of $x$ and $y$ with respect to $r$ and $\theta$ is $r$.

**Solution** To change the variables from rectangular to polar coordinates, we use the equations

$$x = r \cos \theta \qquad y = r \sin \theta$$

The partial derivatives are

$$\frac{\partial x}{\partial r} = \cos \theta \qquad \frac{\partial x}{\partial \theta} = -r \sin \theta \qquad \frac{\partial y}{\partial r} = \sin \theta \qquad \frac{\partial y}{\partial \theta} = r \cos \theta$$

The Jacobian of $x$, $y$ with respect to $r$, $\theta$ is

$$\frac{\partial(x, y)}{\partial(r, \theta)} = \begin{vmatrix} \dfrac{\partial x}{\partial r} & \dfrac{\partial x}{\partial \theta} \\[2ex] \dfrac{\partial y}{\partial r} & \dfrac{\partial y}{\partial \theta} \end{vmatrix} = \begin{vmatrix} \cos \theta & -r \sin \theta \\ \sin \theta & r \cos \theta \end{vmatrix} = \cos \theta \cdot r \cos \theta - \sin \theta (-r \sin \theta)$$

$$= r \cos^2 \theta + r \sin^2 \theta = r$$

∎

NOW WORK Problem 9.

## 2 Change the Variables of a Double Integral Using a Jacobian

In changing variables to find a double integral, we make use of the following result, which we state without proof:

---

**THEOREM   Change in the Variables in a Double Integral**

Suppose $R$ is a closed, bounded region of the $xy$-plane and $R^{\#}$ is a closed, bounded region of the $uv$-plane, where each point in $R$ corresponds to one and only one point in $R^{\#}$ when the change of variables $x = g_1(u, v)$, $y = g_2(u, v)$ is made. If:

- $f$ is continuous on $R$,
- $g_1$ and $g_2$ have continuous partial derivatives on $R^{\#}$, and
- the Jacobian $\dfrac{\partial(x, y)}{\partial(u, v)} \neq 0$ on $R^{\#}$,

then

$$\iint\limits_{R} f(x, y)\, dA = \iint\limits_{R^{\#}} f(g_1(u, v), g_2(u, v)) \left| \frac{\partial(x, y)}{\partial(u, v)} \right| du\, dv$$

---

Notice that in the integral on the right, the absolute value of the Jacobian is introduced as a factor.

Changing variables can simplify the integration of $\iint\limits_{R} f(x, y)\, dA$ in two ways: either by simplifying the region $R$ or by simplifying the integrand $f(x, y)$.

**EXAMPLE 2   Using a Jacobian in a Double Integral**

Find $\iint\limits_{R} y\, dA$, where $R$ is the region enclosed by the lines $2x + y = 0$, $2x + y = 3$, $x - y = 0$, and $x - y = 2$.

**Solution** See Figure 63(a). The region $R$ in the $xy$-plane is neither $x$-simple nor $y$-simple. To change $R$ we use the change of variables $u = 2x + y$ and $v = x - y$. Then the lines $2x + y = 0$ and $2x + y = 3$ in the $xy$-plane correspond to the lines $u = 0$ and $u = 3$ in the $uv$-plane. The lines $x - y = 0$ and $x - y = 2$ in the $xy$-plane correspond to the lines $v = 0$ and $v = 2$ in the $uv$-plane. As Figure 63(b) shows, the region $R^{\#}$ in the $uv$-plane is a rectangle.

To find the Jacobian, we need to express $x$ and $y$ as functions of $u$ and $v$. We can achieve this by solving the system of equations $\begin{cases} u = 2x + y \\ v = x - y \end{cases}$ for $x$ and $y$. Then

$$x = \frac{u + v}{3} \qquad y = \frac{u - 2v}{3}$$

The Jacobian is

$$\frac{\partial(x, y)}{\partial(u, v)} = \begin{vmatrix} \dfrac{\partial x}{\partial u} & \dfrac{\partial x}{\partial v} \\[6pt] \dfrac{\partial y}{\partial u} & \dfrac{\partial y}{\partial v} \end{vmatrix} = \begin{vmatrix} \dfrac{1}{3} & \dfrac{1}{3} \\[6pt] \dfrac{1}{3} & -\dfrac{2}{3} \end{vmatrix} = -\frac{2}{9} - \frac{1}{9} = -\frac{1}{3}$$

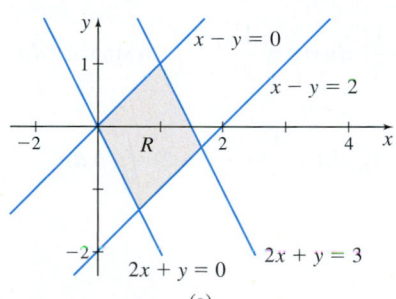

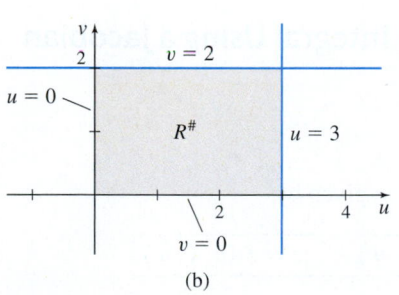

**Figure 63**

Using this change of variables, the integral $\iint\limits_{R} y\, dA$ becomes

$$\iint\limits_{R} y\, dA = \iint\limits_{R^{\#}} \frac{u - 2v}{3} \cdot \left| -\frac{1}{3} \right| du\, dv = \int_0^2 \int_0^3 \frac{1}{9}(u - 2v)\, du\, dv = \frac{1}{9}\int_0^2 \left[ \frac{u^2}{2} - 2uv \right]_0^3 dv$$

$$= \frac{1}{9}\int_0^2 \left( \frac{9}{2} - 6v \right) dv = \frac{1}{9}\left[ \frac{9v}{2} - 3v^2 \right]_0^2 = \frac{1}{9}(9 - 12) = -\frac{1}{3} \qquad \blacksquare$$

**NOW WORK** Problem 21.

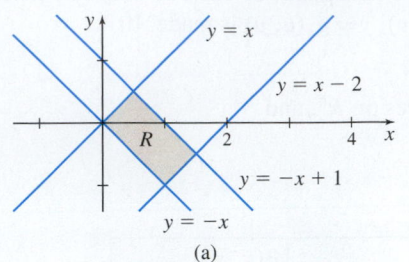

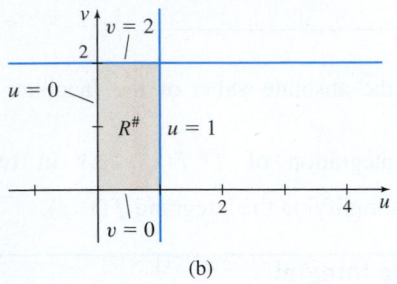

**Figure 64**

EXAMPLE 3   Using a Jacobian in a Double Integral

Find

$$\iint_R (x+y)\sin(x-y)\,dA$$

where $R$ is the region enclosed by $y=x$, $y=x-2$, $y=-x$, and $y=-x+1$.

**Solution** The integrand $(x+y)\sin(x-y)$ suggests changing the variables to $u=x+y$ and $v=x-y$. Then the lines $x-y=0$, $x-y=2$, $x+y=0$, and $x+y=1$ in the $xy$-plane define the region $R$, and the lines $v=0$, $v=2$, $u=0$, and $u=1$ in the $uv$-plane define the region $R^{\#}$. See Figure 64. We solve the system of equations

$$\begin{cases} u=x+y \\ v=x-y \end{cases} \text{ for } x \text{ and } y, \text{ and obtain}$$

$$x=\frac{u+v}{2} \qquad y=\frac{u-v}{2}$$

Using these equations, the Jacobian is

$$\frac{\partial(x,y)}{\partial(u,v)}=\begin{vmatrix} \dfrac{1}{2} & \dfrac{1}{2} \\[2mm] \dfrac{1}{2} & -\dfrac{1}{2} \end{vmatrix}=-\frac{1}{4}-\frac{1}{4}=-\frac{1}{2}$$

The integral under this change of variables becomes

$$\iint_R (x+y)\sin(x-y)\,dA=\iint_{R^{\#}} u\sin v\cdot\left|-\frac{1}{2}\right|\,du\,dv=\frac{1}{2}\int_0^1\int_0^2 u\sin v\,dv\,du$$

$$=\frac{1}{2}\int_0^1 \left[u(-\cos v)\right]_0^2\,du=\frac{1}{2}(1-\cos 2)\int_0^1 u\,du$$

$$=\frac{1}{4}(1-\cos 2) \qquad \blacksquare$$

NOW WORK Problem 23.

## 3 Change the Variables of a Triple Integral Using a Jacobian

In space, the Jacobian is defined as follows.

**DEFINITION** Jacobian: Three Variables

For a change of variables from $x$, $y$, $z$ to $u$, $v$, $w$ given by

$$x=g_1(u,v,w) \qquad y=g_2(u,v,w) \qquad z=g_3(u,v,w)$$

the **Jacobian of $x$, $y$, and $z$ with respect to $u$, $v$, and $w$** is given by

$$\frac{\partial(x,y,z)}{\partial(u,v,w)}=\begin{vmatrix} \dfrac{\partial x}{\partial u} & \dfrac{\partial x}{\partial v} & \dfrac{\partial x}{\partial w} \\[2mm] \dfrac{\partial y}{\partial u} & \dfrac{\partial y}{\partial v} & \dfrac{\partial y}{\partial w} \\[2mm] \dfrac{\partial z}{\partial u} & \dfrac{\partial z}{\partial v} & \dfrac{\partial z}{\partial w} \end{vmatrix}$$

NOW WORK Problem 13.

In changing the variables of a triple integral, we use the following result:

---

**THEOREM  Change in the Variables in a Triple Integral**

Suppose $E$ is a closed, bounded solid of $xyz$-space and $E^\#$ is a closed, bounded solid of $uvw$-space, where each point in $E$ corresponds to one and only one point in $E^\#$ when the change of variables $x = g_1(u, v, w)$, $y = g_2(u, v, w)$, and $z = g_3(u, v, w)$ is made. If:

- $f$ is continuous on $E$,
- $g_1$, $g_2$, and $g_3$ have continuous partial derivatives on $E^\#$, and
- the Jacobian $\dfrac{\partial(x, y, z)}{\partial(u, v, w)} \neq 0$ on $E^\#$, then

$$\iiint\limits_{E} f(x, y, z)\, dV = \iiint\limits_{E^\#} f(g_1(u, v, w),\, g_2(u, v, w),\, g_3(u, v, w)) \left| \frac{\partial(x, y, z)}{\partial(u, v, w)} \right| du\, dv\, dw$$

---

Again notice that in the integral on the right, the absolute value of the Jacobian is introduced as a factor.

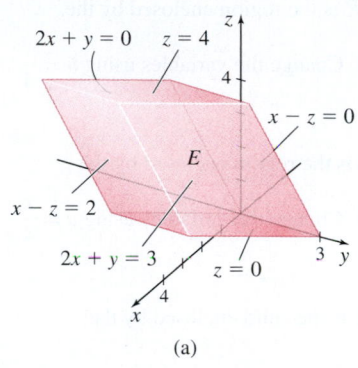

2x + y = 0   z = 4

x − z = 0

E

x − z = 2

2x + y = 3   z = 0

(a)

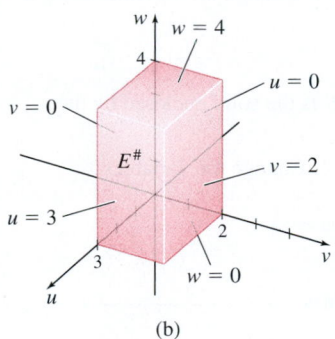

w = 4

v = 0

E^#

v = 2

u = 3

w = 0

u

(b)

**Figure 65**

---

**EXAMPLE 4  Using a Jacobian in a Triple Integral**

Find $\iiint\limits_{E} y\, dV$, where $E$ is the solid enclosed by the planes $2x + y = 0$, $2x + y = 3$, $x - z = 0$, $x - z = 2$, $z = 0$, and $z = 4$.

**Solution**  The solid $E$ is the slanted parallelepiped shown in Figure 65(a). The change of variables $u = 2x + y$, $v = x - z$, and $w = z$ transforms $E$ into the rectangular parallelepiped $E^\#$ shown in Figure 65(b). $E^\#$ is the solid enclosed by the planes $u = 0$, $u = 3$, $v = 0$, $v = 2$, and $w = 0$, $w = 4$.

We solve the system of equations $\begin{cases} u = 2x + y \\ v = x - z \\ w = z \end{cases}$ for $x$, $y$, and $z$ and obtain

$$x = v + w \qquad y = u - 2x = u - 2(v + w) = u - 2v - 2w \qquad z = w$$

The Jacobian is

$$\frac{\partial(x, y, z)}{\partial(u, v, w)} = \begin{vmatrix} \dfrac{\partial x}{\partial u} & \dfrac{\partial x}{\partial v} & \dfrac{\partial x}{\partial w} \\[2mm] \dfrac{\partial y}{\partial u} & \dfrac{\partial y}{\partial v} & \dfrac{\partial y}{\partial w} \\[2mm] \dfrac{\partial z}{\partial u} & \dfrac{\partial z}{\partial v} & \dfrac{\partial z}{\partial w} \end{vmatrix} = \begin{vmatrix} 0 & 1 & 1 \\ 1 & -2 & -2 \\ 0 & 0 & 1 \end{vmatrix} = -1$$

The integral $\iiint\limits_{E} y\, dV$ under this change of variables becomes

$$\iiint\limits_{E} y\, dV = \iiint\limits_{E^\#} (u - 2v - 2w) \left| \frac{\partial(x, y, z)}{\partial(u, v, w)} \right| du\, dv\, dw = \int_0^4 \int_0^2 \int_0^3 (u - 2v - 2w)\, du\, dv\, dw$$

$$= \int_0^4 \int_0^2 \left[ \frac{u^2}{2} - 2vu - 2wu \right]_0^3 dv\, dw = \int_0^4 \int_0^2 \left( \frac{9}{2} - 6v - 6w \right) dv\, dw$$

$$= \int_0^4 \left[ \frac{9}{2}v - 3v^2 - 6wv \right]_0^2 dw = \int_0^4 (9 - 12 - 12w)\, dw = \int_0^4 (-3 - 12w)\, dw$$

$$= \left[ -3w - 6w^2 \right]_0^4 = -12 - 96 = -108 \qquad \blacksquare$$

**NOW WORK**  Problem **25.**

## 14.9 Assess Your Understanding

### Concepts and Vocabulary

1. For a change of variables from $x$ and $y$ to $u$ and $v$ given by $x = g_1(u, v)$ and $y = g_2(u, v)$, the Jacobian is the determinant _____.

2. *True or False*  When introducing a Jacobian into an integral because of a change of variables, the absolute value of the Jacobian is used.

### Skill Building

*In Problems 3–10, find the Jacobian $\dfrac{\partial(x, y)}{\partial(u, v)}$ for each change of variables from $(x, y)$ to $(u, v)$.*

3. $x = u + v$,   $y = u - v$

4. $x = u + 5$,   $y = v - 7$

5. $x = 4u - v$,   $y = 3u + \dfrac{1}{2}v$

6. $x = u^2 - v^2$,   $y = u^2 + v^2$

7. $x = \dfrac{u}{v}$,   $y = u + v$

8. $x = e^u \cos v$,   $y = e^u \sin v$

9. $x = ve^u$,   $y = ue^{-v}$

10. $x = u \cos v$,   $y = v \sin u$

*In Problems 11–16, find the Jacobian $\dfrac{\partial(x, y, z)}{\partial(u, v, w)}$ for each change of variables from $(x, y, z)$ to $(u, v, w)$.*

11. $x = u + v + w$,   $y = u + v - w$,   $z = u - v - w$

12. $x = 2u + v + w$,   $y = u + v - w$,   $z = u - v$

13. $x = u + v$,   $y = v - w$,   $z = u - w$

14. $x = u + v + w$,   $y = u + v$,   $z = w$

15. $x = u$,   $y = v^2$,   $z = w^3$

16. $x = 3(u + v)$,   $y = u - w$,   $z = u^2 - w^2$

17. **Area**  Find the area of the region $R$ enclosed by $y = 4\sqrt{1 - \dfrac{x^2}{9}}$ and $y = 0$, using the change of variables $u = \dfrac{x}{3}$ and $v = \dfrac{y}{2}$.

18. **Area**  Find the area of the region $R$ enclosed by $xy = 1$, $xy = 3$, $y = x$, and $y = 3x$ using the change of variables $u = x$ and $v = xy$. See the figure.

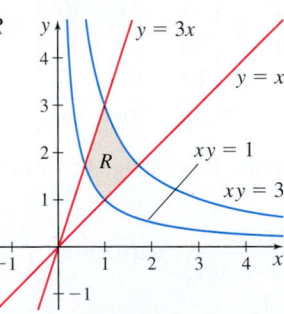

*In Problems 19–22:*
(a) Graph the region $R$ in the $xy$-plane.
(b) Graph the region $R^{\#}$ in the $uv$-plane.
(c) Find the Jacobian $\dfrac{\partial(x, y)}{\partial(u, v)}$.
(d) Change the variables and find the integral.

19. $\displaystyle\iint_R x \, dA$, where $R$ is the region enclosed by $2x - y = 0$, $2x - y = 4$, $x + y = 0$, and $x + y = 3$. Use the change of variables $u = 2x - y$ and $v = x + y$.

20. $\displaystyle\iint_R (x^2 + y^2) \, dA$, where $R$ is the region enclosed by $2x - y = 1$, $2x - y = 3$, $x + y = 1$, and $x + y = 2$. Use the change of variables $u = 2x - y$ and $v = x + y$.

21. $\displaystyle\iint_R xy \, dA$, where $R$ is the triangular region whose vertices are $(-1, 1)$, $(1, 1)$, and $(0, 0)$. Use the change of variables $u = x + y$ and $v = x - y$.

22. $\displaystyle\iint_R (x + y) \sin(x - y) \, dA$, where $R$ is the triangular region whose vertices are $(-1, 1)$, $(1, 1)$, and $(0, 0)$. Use the change of variables $u = x + y$ and $v = x - y$.

*In Problems 23–26, find each integral.*

23. $\displaystyle\iint_R (3x - 2y) \, dA$, where $R$ is the region enclosed by the ellipse $9x^2 + 16y^2 = 144$. Change the variables using $u = \dfrac{x}{4}$ and $v = \dfrac{y}{3}$.

24. $\displaystyle\iint_R (x^2 - y) \, dA$, where $R$ is the region enclosed by the ellipse $9x^2 + 16y^2 = 144$. Change the variables using $u = \dfrac{x}{4}$ and $v = \dfrac{y}{3}$.

25. $\displaystyle\iiint_E (x + 1) \, dV$, where $E$ is the solid enclosed by the ellipsoid $\dfrac{x^2}{4} + \dfrac{y^2}{4} + \dfrac{z^2}{9} = 1$. Change the variables using $u = \dfrac{x}{2}$, $v = \dfrac{y}{2}$, and $w = \dfrac{z}{3}$.

26. $\displaystyle\iiint_E (x + 2y) \, dV$, where $E$ is the solid enclosed by the ellipsoid $\dfrac{x^2}{9} + \dfrac{y^2}{4} + z^2 = 1$. Change the variables using $u = \dfrac{x}{3}$, $v = \dfrac{y}{2}$, and $w = z$.

### Applications and Extensions

*In Problems 27–30, for each integral:*
(a) Graph the region $R$.
(b) Choose a change of variables from $(x, y)$ to $(u, v)$. Graph the new region in the $uv$-plane.
(c) Write each integral using the new variables.
(d) Find each integral.

27. $\displaystyle\iint_R (3x + 9y) \, dA$, where $R$ is the region enclosed by the lines $x + y = 2$, $x + y = 0$, $y = 2x$, and $y = 2x + 4$

28. $\displaystyle\iint_R x \, dA$, where $R$ is the region enclosed by the ellipse $\dfrac{x^2}{9} + \dfrac{y^2}{16} = 1$

**29.** $\iint\limits_{R} x \cos(xy)\, dA$, where $R$ is the region enclosed
by $xy = 1$, $xy = 3$, $x = 1$, and $x = 3$

**30.** $\iint\limits_{R} ye^{xy}\, dA$, where $R$ is the region enclosed
by $xy = 1$, $xy = 3$, $x = 1$, and $x = 3$

*In Problems 31 and 32, for each integral:*

**(a)** *Graph the solid E.*

**(b)** *Choose a change of variables from $(x, y, z)$ to $(u, v, w)$.
Graph the new region in the uvw-space.*

**(c)** *Write each integral using the new variables.*

**(d)** *Find each integral.*

**31.** $\iiint\limits_{E} x\, dV$, where $E$ is the solid enclosed by the planes
$2x - y = 0$, $2x - y = 4$, $x + z = 0$, $x + z = 3$, $z = 0$, and $z = 5$

**32.** $\iiint\limits_{E} yz\, dV$, where $E$ is the solid enclosed by the planes $x - y = 0$,
$x - y = 3$, $z - x = 0$, $z - x = 4$, $z = 0$, and $z = 2$

**33. Area**   Use a change of variables to find the area of the region $R$
enclosed by the lines $y = 3x$, $y - 3x = 2$, $2y + x = 0$,
and $y = -\dfrac{1}{2}x + 5$.

**34. Area**   Use a change of variables to find the area of the region $R$
enclosed by the lines $x + y = 1$, $x + y = -3$, $y = 2x$,
and $y = 2x + 4$.

**35. Volume**   Use a change of variables to find the volume $V$ of the
solid $E$ enclosed by the planes $x + 2y = 0$, $x + 2y = 3$, $y - z = 0$,
$y - z = 2$, $z = 0$, and $z = 6$.

**36. Volume**   Use a change of variables to find the volume $V$ of the
solid $E$ enclosed by the planes $x - y = 0$, $x - y = 3$, $z - 2x = 0$,
$z - 2x = 4$, $z = 1$, and $z = 5$.

**37. Volume**   Find the volume of
the solid $E$ enclosed by the
paraboloid $z = 9 - x^2 - 9y^2$
and the plane $z = 1$, given
the change of variables
$x = 3u \cos v$, $y = u \sin v$,
and $z = w$. See the figure.

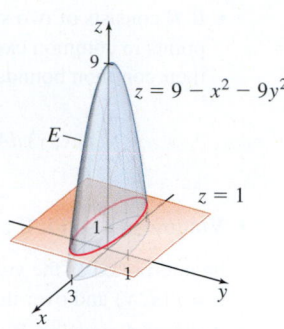

**38. Volume**   Find the volume of the solid $E$ enclosed by the
ellipsoid $\dfrac{x^2}{4} + \dfrac{y^2}{9} + z^2 = 1$ above the $xy$-plane, given the change
of variables $u = \dfrac{x}{2}$, $v = \dfrac{y}{3}$, and $w = z$.

**39.** Show that in changing from rectangular coordinates $(x, y, z)$ to
spherical coordinates $(\rho, \theta, \phi)$, the triple integral of $f$ over $E$
becomes

$$\iiint\limits_{E} f(x, y, z)\, dV$$

$$= \iiint\limits_{E^{\#}} f(\rho \sin \phi \cos \theta,\, \rho \sin \phi \sin \theta,\, \rho \cos \phi)\rho^2 \sin \phi\, d\rho\, d\theta\, d\phi$$

**Challenge Problem** _____

**40.** A region $R$ is enclosed by the graphs of $xy = 2$, $xy = 5$, $y = \sqrt{x}$,
and $y = 3\sqrt{x}$.

**(a)** Graph the region $R$.

**(b)** Use a change of variables that transforms $R$ into a
rectangular region $R^{\#}$.

**(c)** Find the area of $R$.

# Chapter Review

## THINGS TO KNOW

### 14.1 The Double Integral over a Rectangular Region

- **Partial integration** (p. 954)
- **Iterated integrals:** Integrals of the form

$$\int_{a}^{b} \left[ \int_{c}^{d} f(x, y)\, dy \right] dx \quad \text{or} \quad \int_{c}^{d} \left[ \int_{a}^{b} f(x, y)\, dx \right] dy$$

(p. 955)

- **Fubini's Theorem:** If a function $z = f(x, y)$ is continuous
on the closed rectangular region $R$ given by $a \le x \le b$
and $c \le y \le d$, then

$$\iint\limits_{R} f(x, y)\, dA = \int_{a}^{b} \int_{c}^{d} f(x, y)\, dy\, dx$$

$$= \int_{c}^{d} \int_{a}^{b} f(x, y)\, dx\, dy. \quad \text{(p. 955)}$$

### 14.2 The Double Integral over Nonrectangular Regions

- **Theorem:** If a function $z = f(x, y)$ is continuous on a closed,
  bounded region $R$, whose boundary is a piecewise-smooth
  curve, then the double integral $\iint\limits_{R} f(x, y)\, dA$ exists. (p. 961)
- **x-simple region** (p. 961)
- **y-simple region** (p. 963)
- **Fubini's Theorem** for an x-simple region (p. 962)
- **Fubini's Theorem** for a y-simple region (p. 963)
- **Properties of double integrals:** (p. 966)

  - $$\iint\limits_{R} [f(x, y) \pm g(x, y)]\, dA$$

    $$= \iint\limits_{R} f(x, y)\, dA \pm \iint\limits_{R} g(x, y)\, dA$$

  - $$\iint\limits_{R} cf(x, y)\, dA = c \iint\limits_{R} f(x, y)\, dA, \text{ where } c \text{ is a constant}$$

- If $R$ consists of two subregions, $R_1$ and $R_2$, that have no points in common except for points lying on portions of their common boundary, then

$$\iint\limits_{R} f(x, y)\, dA = \iint\limits_{R_1} f(x, y)\, dA + \iint\limits_{R_2} f(x, y)\, dA$$

- **Volume of a solid:** If $z = f(x, y) \geq 0$, then $\iint\limits_{R} f(x, y)\, dA$ can be interpreted as the volume of the solid under the surface $z = f(x, y)$ and over the region $R$. (p. 961)
- **Area $A$ of a region $R$:** $A = \iint\limits_{R} dA$ (p. 967)

## 14.3 Double Integrals Using Polar Coordinates
- **Differential of area in polar coordinates $(r, \theta)$:** (p. 972)
$dA = r\, dr\, d\theta$
- **Iterated double integral in polar coordinates:** (p. 972)

$$\iint\limits_{R} f(x, y)\, dA = \iint\limits_{R} f(r\cos\theta, r\sin\theta)\, r\, dr\, d\theta$$

$$= \int_{\alpha}^{\beta} \int_{r_1(\theta)}^{r_2(\theta)} f(r\cos\theta, r\sin\theta)\, r\, dr\, d\theta$$

- **Area $A$ of a region $R$ in polar coordinates:** (p. 975)

$$A = \iint\limits_{R} r\, dr\, d\theta$$

## 14.4 Center of Mass; Moment of Inertia
- **Mass density:** $\rho = \rho(x, y)$ (p. 978)
- **Mass of a lamina:** $M = \iint\limits_{R} \rho(x, y)\, dA$, where $\rho = \rho(x, y)$ is the mass density (p. 979)
- **Moment of mass $M_x$ about the $x$-axis** (p. 979)
- **Moment of mass $M_y$ about the $y$-axis** (p. 979)
- **Center of mass of a lamina $(\bar{x}, \bar{y})$:** $\bar{x} = \dfrac{M_y}{M}$; $\bar{y} = \dfrac{M_x}{M}$ (p. 979)
- **Moments of inertia $I$:** (pp. 981–982)
$I_x = \iint\limits_{R} y^2 \rho(x, y)\, dA \qquad I_y = \iint\limits_{R} x^2 \rho(x, y)\, dA$
$I_O = \iint\limits_{R} (x^2 + y^2)\rho(x, y)\, dA \qquad I_x + I_y = I_O$
- **Radius of gyration $k$:** $k = \sqrt{\dfrac{I}{M}}$ (p. 982)

## 14.5 Surface Area
- **Surface area $S$ over a region $R$:** (p. 987)

$$S = \iint\limits_{R} \sqrt{[f_x(x, y)]^2 + [f_y(x, y)]^2 + 1}\, dA$$

## 14.6 The Triple Integral
- **Fubini's Theorem for triple integrals** (p. 991)
- **$xy$-simple solid $E$** (p. 992)
- **Triple integral over an $xy$-simple solid:** (p. 993)

$$\iiint\limits_{E} f(x, y, z)\, dV = \iint\limits_{R} \left[ \int_{z_1(x,y)}^{z_2(x,y)} f(x, y, z)\, dz \right] dA$$

- **Volume $V$ of a solid $E$:** (p. 994) $V = \iiint\limits_{E} dV$

## 14.7 Triple Integrals Using Cylindrical Coordinates
- **Cylindrical coordinates $(r, \theta, z)$ of a point $P$:** (p. 1000)
$x = r\cos\theta,\ y = r\sin\theta,\ z = z$
- **Differential of volume in cylindrical coordinates $(r, \theta, z)$:** (p. 1001)

$$dV = r\, dr\, d\theta\, dz$$

## 14.8 Triple Integrals Using Spherical Coordinates
- **Spherical coordinates $(\rho, \theta, \phi)$ of a point $P$:** (p. 1006)
$x = \rho\sin\phi\cos\theta,\ y = \rho\sin\phi\sin\theta,\ z = \rho\cos\phi$
- **Differential of volume in spherical coordinates $(\rho, \theta, \phi)$:** (p. 1008)

$$dV = \rho^2 \sin\phi\, d\rho\, d\theta\, d\phi$$

## 14.9 Change of Variables Using Jacobians
- Jacobian in two variables: (p. 1014)

$$\frac{\partial(x, y)}{\partial(u, v)} = \begin{vmatrix} \dfrac{\partial x}{\partial u} & \dfrac{\partial x}{\partial v} \\ \dfrac{\partial y}{\partial u} & \dfrac{\partial y}{\partial v} \end{vmatrix} = \frac{\partial x}{\partial u}\frac{\partial y}{\partial v} - \frac{\partial x}{\partial v}\frac{\partial y}{\partial u}$$

- Jacobian in three variables: (p. 1016)

$$\frac{\partial(x, y, z)}{\partial(u, v, w)} = \begin{vmatrix} \dfrac{\partial x}{\partial u} & \dfrac{\partial x}{\partial v} & \dfrac{\partial x}{\partial w} \\ \dfrac{\partial y}{\partial u} & \dfrac{\partial y}{\partial v} & \dfrac{\partial y}{\partial w} \\ \dfrac{\partial z}{\partial u} & \dfrac{\partial z}{\partial v} & \dfrac{\partial z}{\partial w} \end{vmatrix}$$

## OBJECTIVES

| Section | You should be able to … | Example | Review Exercises |
|---------|-------------------------|---------|------------------|
| 14.1 | 1 Find double Riemann sums of $z = f(x, y)$ over a closed rectangular region (p. 952) | 1 | 7(a), 7(b) |
| | 2 Find the value of a double integral defined on a closed rectangular region (p. 954) | 2, 3, 4 | 1, 2, 7(c), 8, 9 |
| | 3 Find the volume under a surface and over a rectangular region (p. 956) | 5 | 10, 11 |
| 14.2 | 1 Use Fubini's Theorem for an $x$-simple region (p. 962) | 1 | 3–6, 13 |
| | 2 Use Fubini's Theorem for a $y$-simple region (p. 963) | 2–5 | 3–6, 13, 14, 54 |
| | 3 Use properties of double integrals (p. 966) | 6, 7 | 12 |
| | 4 Use double integrals to find area (p. 967) | 8 | 15–18, 20 |
| 14.3 | 1 Find a double integral using polar coordinates (p. 972) | 1, 2, 3 | 19 |
| | 2 Find area and volume using polar coordinates (p. 974) | 4, 5, 6 | 21, 22 |
| 14.4 | 1 Find the mass and the center of mass of a lamina (p. 978) | 1, 2 | 23–25 |
| | 2 Find moments of inertia (p. 981) | 3 | 26 |

## REVIEW EXERCISES

**1.** Find $\int_0^1 \int_1^e \frac{y}{x}\, dx\, dy$.

**2.** Find $\int_1^3 \int_3^6 (x^2 + y^2)\, dy\, dx$.

**3.** Find $\int_0^1 \int_y^1 ye^{-x^3}\, dx\, dy$.

**4.** Find $\int_0^1 \int_0^{\sqrt{1+x^2}} \frac{1}{x^2 + y^2 + 1}\, dy\, dx$.

**5.** Find $\int_0^{\pi/4} \int_0^{\tan x} \sec x\, dy\, dx$  
**6.** Find $\int_0^1 \int_{-x}^x e^{x+y}\, dy\, dx$

**7.** Let $R$ be the square region $0 \le x \le 4$, $-2 \le y \le 2$ in the $xy$-plane, and suppose that $f(x, y) = x + 2y$.

    **(a)** Partition $R$ into four squares of equal area and evaluate $f$ at the midpoint of each square. Find the corresponding double Riemann sum of $f$ over $R$.

    **(b)** Using the same partition as in (a), but evaluating $f$ at an appropriate point of each square, determine the largest possible double Riemann sum and the smallest. What is the average of these values?

    **(c)** What is the actual value of $\iint_R f(x, y)\, dA$?

**8.** Find $\iint_R xe^y\, dA$ if $R$ is the closed rectangular region defined by $0 \le x \le 2$ and $0 \le y \le 1$.

**9.** Find $\iint_R \sin x \cos y\, dA$ if $R$ is the closed rectangular region defined by $0 \le x \le \frac{\pi}{2}$ and $0 \le y \le \frac{\pi}{3}$.

**10. Volume** Find the volume $V$ under the surface $z = f(x, y) = e^x \cos y$ over the rectangular region $R$ defined by $0 \le x \le 1$ and $0 \le y \le \frac{\pi}{2}$.

**11. Volume** Find the volume $V$ under the surface $z = f(x, y) = 2x^2 + y^2$ over the rectangular region $R$ defined by $0 \le x \le 2$ and $0 \le y \le 1$.

**12.** The functions $f$ and $g$ are continuous on a closed, bounded region $R$, and $\iint_R f(x, y)\, dA = -2$, $\iint_R g(x, y)\, dA = 3$. Use properties of double integrals to find each double integral.

    **(a)** $\iint_R 2[f(x, y) + 3g(x, y)]\, dA$

    **(b)** $\iint_R [g(x, y) - 2f(x, y)]\, dA$

**13.** Find $\iint_R x \sin(y^3)\, dA$, where $R$ is the region enclosed by the triangle with vertices $(0, 0)$, $(0, 2)$, and $(2, 2)$.

**14.** Find $\iint_R (x + y)\, dA$, where $R$ is the region enclosed by $y = x$ and $y^2 = 2 - x$.

**15. Volume** Find the volume of the tetrahedron enclosed by the coordinate planes and the plane $2x + y + 3z = 6$.

**16. Volume** Find the volume of the ellipsoid $4x^2 + y^2 + \frac{z^2}{4} = 1$.

**17. Volume** Find the volume of the solid in the first octant enclosed by the coordinate planes, $4y = 3(4 - x^2)$ and $4z = 4 - x^2$.

**18. Area** Use a double integral to find the area enclosed by the parabola $y^2 = 16x$ and the line $y = 4x - 8$.

**19.** Find $\iint_R \sin\theta\, dA$, where $R$ is the region enclosed by the rays $\theta = 0$ and $\theta = \frac{\pi}{6}$ and the circle $r = 6\sin\theta$.

**20. Area** Use a double integral to find the area of the region in the first quadrant enclosed by $y^2 = x^3$ and $y = x$.

**21. Area** Use a double integral and polar coordinates to find the area of the circle $r = 2\cos\theta$.

**22. Area** Find the area of the region in the first quadrant that lies outside $r = 2a$ and inside $r = 4a\cos\theta$ by using polar coordinates.

**23. Mass** Find the mass of a lamina in the shape of a right triangle of height 6 and base 4 if the mass density is proportional to the square of the distance from the vertex at the right angle.

**24. Center of Mass** Find the center of mass of the homogeneous lamina enclosed by $3x^2 = y$ and $y = 3x$.

**25. Center of Mass** Find the center of mass of the homogeneous lamina in quadrant I, enclosed by $4x = y^2$, $y = 0$, and $x = 4$.

**26. Moments of Inertia** Use double integration to find the moments of inertia for a homogeneous lamina in the shape of the ellipse $2x^2 + 9y^2 = 18$.

**27. Surface Area** Find the area of the first-octant portion of the plane $\frac{x}{2} + y + \frac{z}{3} = 1$.

**28. Surface Area**   Find the surface area of the paraboloid $z = x^2 + y^2$ below the plane $z = 1$.

**29. Surface Area**   Find the surface area of the cone $x^2 + y^2 = 3z^2$ that lies inside the cylinder $x^2 + y^2 = 4y$.

**30.** Find $\iiint\limits_E e^z \cos x \, dV$, where $E$ is the solid given

by $0 \le x \le \dfrac{\pi}{2}, 0 \le y \le 2, 0 \le z \le 1$.

**31.** Find $\int_0^1 \int_0^{2-3x} \int_0^{2y} x^2 \, dz \, dy \, dx$.

**32.** Find $\iiint\limits_E y e^x \, dV$, where $E$ is the solid given

by $0 \le x \le 1, 0 \le y \le 1, y^2 \le z \le y$.

**33.** Find $\iiint\limits_E xyz \, dV$, where $E$ is the solid enclosed by $x = 0$

and $x = 2 - y - z$ whose projection onto the $yz$-plane is the region enclosed by the rectangle $0 \le y \le 1$ and $0 \le z \le 2$.

**34. Center of Mass**   Find the center of mass and the moments of inertia $I_x$ and $I_y$ of the homogeneous hemispherical

shell $0 \le a \le r \le b, 0 \le \phi \le \dfrac{\pi}{2}$.

**35. Volume**   Find the volume in the first octant of the paraboloid $x^2 + y^2 + z = 9$.

**36. Volume**   Find the volume in the first octant of the solid enclosed by $y = 0$, $z = 0$, $x + y = 2$, $x + 2y = 6$, and $y^2 + z^2 = 4$.

*In Problems 37–40:*

*(a) Convert each point given in rectangular coordinates to cylindrical coordinates.*

*(b) Convert each point given in rectangular coordinates to spherical coordinates.*

**37.** $(3, 0, 4)$                                    **38.** $(-2\sqrt{2}, 2\sqrt{2}, 3)$

**39.** $(-1, 1, -2)$                              **40.** $(1, \sqrt{3}, 4)$

**41. Volume**   Use spherical coordinates with triple integrals to find the volume between the spheres $x^2 + y^2 + z^2 = 4$ and $x^2 + y^2 + z^2 = 1$.

**42. Volume**   Find the volume of the solid cut from the ball $\rho = 4$ by

the cone $\phi = \dfrac{\pi}{4}$.

**43.** Use cylindrical coordinates to find

$$\int_0^2 \int_0^{\sqrt{4-x^2}} \int_0^{\sqrt{4-x^2-y^2}} \frac{z}{\sqrt{x^2+y^2}} \, dz \, dy \, dx$$

**44. Volume**   Use cylindrical coordinates to find the volume of the solid in the first octant inside the cylinder $r = 1$ and below the plane $3x + 2y + 6z = 6$.

**45.** Use cylindrical coordinates to find

$$\int_0^a \int_0^{\sqrt{a^2-x^2}} \int_0^{\sqrt{a^2-x^2-y^2}} dz \, dy \, dx$$

**46.** Use spherical coordinates to find the iterated integral given in Problem 45.

*In Problems 47–50, find each integral using either cylindrical or spherical coordinates (whichever is more convenient).*

**47.** $\displaystyle\int_{-2}^2 \int_{-\sqrt{4-y^2}}^{\sqrt{4-y^2}} \int_0^{\sqrt{16-x^2-y^2}} dz \, dx \, dy$

**48.** $\displaystyle\int_0^1 \int_{-\sqrt{1-x^2}}^{\sqrt{1-x^2}} \int_{\sqrt{x^2+y^2}}^{\sqrt{2-x^2-y^2}} dz \, dy \, dx$

**49.** $\displaystyle\int_{-\sqrt{2}}^{\sqrt{2}} \int_{-\sqrt{2-x^2}}^{\sqrt{2-x^2}} \int_0^4 (x^2 + y^2) \, z \, dz \, dy \, dx$

**50.** $\displaystyle\int_0^1 \int_0^1 \int_0^{\sqrt{1-x^2}} dy \, dx \, dz$

*In Problems 51 and 52, find each integral.*

**51.** $\displaystyle\int_{-1}^1 \int_{-\sqrt{1-y^2}}^{\sqrt{1-y^2}} \int_{\sqrt{x^2+y^2}}^{\sqrt{2-x^2-y^2}} dz \, dx \, dy$

**52.** $\displaystyle\int_0^1 \int_y^1 (x^2 + 1)^{2/3} \, dx \, dy$

*In Problems 53–55, find the Jacobian for each change of variables.*

**53.** $x = e^{u+v}, \quad y = e^{v-u}$

**54.** $x = u + 3v, \quad y = 2u - v$

**55.** $x = u^2, \quad y = 3v, \quad \text{and } z = u + w$

**56.** Find $\iint\limits_R y^2 \, dx \, dy$, where the region $R$ is enclosed by $x - 2y = 1$, $x - 2y = 3$, $2x + 3y = 1$, and $2x + 3y = 2$, using the change of variables $u = x - 2y$ and $v = 2x + 3y$.

**57.** Find $\iint\limits_R (x + y)^3 \, dx \, dy$, where the region $R$ is enclosed by the lines $x + y = 2$, $x + y = 5$, $x - 2y = -1$, and $x - 2y = 3$, using the change of variables $u = x + y$ and $v = x - 2y$.

**58.** Find $\iint\limits_R xy \, dx \, dy$, where $R$ is the region enclosed by $2x + y = 0$, $2x + y = 3$, $x - y = 0$, and $x - y = 2$.

**59.** Find $\iiint\limits_E xz^2 \, dx \, dy \, dz$, where $E$ is the solid enclosed by the

ellipsoid $\dfrac{x^2}{4} + \dfrac{y^2}{9} + z^2 = 1$, using the change of

variables $u = \dfrac{x}{2}, v = \dfrac{y}{3}$, and $w = z$.

**60. Mass**   A solid in the first octant enclosed by the surface $z = xy$, the $xy$-plane, and the cylinder whose intersection with the $xy$-plane forms a triangle with vertices $(0, 0, 0)$, $(2, 0, 0)$, and $(0, 1, 0)$ has mass density $\rho = x + y$. Find the mass of the solid.

# CHAPTER 14 PROJECT    The Mass of Stars

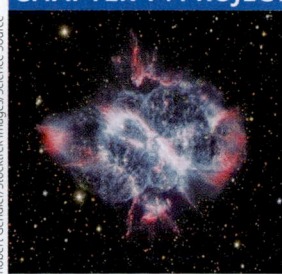

The mass of a star is distributed throughout its volume and can be described by its density $\delta$*   measured in $kg/m^3$. However, while solids often have constant density, the density of a star varies with respect to position within the star's volume. A triple integral can then be used to find the mass $M$ of the star provided its density function is known. We examine several possible density functions for stars. Because stars are almost spherical, we use spherical coordinates and position the sphere so that its center is at the origin.

1. Show that the mass $M$ of a solid enclosed by a sphere of radius $R$ and constant density $\delta$ is given by $M = \dfrac{4}{3}\delta\pi R^3$.

   While we can model the shape of a star as a sphere, its density $\delta$ is not constant. Gravitational forces between the gaseous particles in the star cause the density to be greatest near the center of the star, to become less dense farther from its center, and to eventually become zero at the edge of the star. There are various ways to model the density $\delta$ of a star as a function of the radial distance $\rho$ from the star's center.

   We examine three of these models and use each to determine the mass $M$ of a star. For each these models, $R$ is the radius of the star, $\delta_0$ is a constant, and $\rho$ is the radial distance from the center of the star.

   The **linear density model** is given by

$$\delta(\rho) = \delta_0\left(1 - \frac{\rho}{R}\right)$$

2. Show that the linear density model satisfies the condition that the density at the edge of the star is 0.
3. Show also that in the linear density model the constant $\delta_0$ equals the density at the center of the star.
4. Find the mass $M$ of a star that obeys the linear density model.

   The **quadratic density model** is given by

$$\delta(\rho) = \delta_0\left[1 - \left(\frac{\rho}{R}\right)^2\right]$$

5. Show that the quadratic density model satisfies the condition that the density at the edge of the star is 0.
6. Show also that in the quadratic density model the constant $\delta_0$ equals the density at the center of the star.
7. Find the mass $M$ of a star that obeys the quadratic density model.

   A third model for the density involves an exponential decrease in mass density. In this **exponential density model** the density $\delta = \delta(\rho)$ of a star is given by

$$\delta(\rho) = \delta_0[e^{(1 - \rho/R)} - 1]$$

8. Show that this exponential density model satisfies the condition that the density at the edge of the star is 0.

---

*The widely accepted symbol for density is $\rho$ and the symbol for the radial distance from the center of a star is $r$. Because we use $\rho$ as a spherical coordinate, we use $\delta$ for the density and $\rho$ for the radial distance in this project.

9. Show that the density at the center of a star modeled by the exponential model is $\delta_0(e - 1)\,kg/m^3$.
10. Find the mass $M$ of a star that obeys the exponential model.
11. Using each of the three models, integrate over a volume with radius $\rho < R$ to find an expression for the mass of the star contained within a region of radius $\rho$ about its center. Compare how the mass of a star differs in each of these models at a fixed radial distance $\rho_0$. This is a way to see how the mass within the interior of the star grows as the radial distance from the center increases.

   It should not be a surprise that the structure of a real star is vastly more complicated than the models above indicate.

   The mass of a star at any radial distance $\rho$ is governed by the differential equation of mass conservation,

$$\frac{d}{d\rho}M(\rho) = 4\pi\rho^2\delta(\rho)$$

The balance of gas pressure $P$, which tends to blow the star apart, and gravity, which in the absence of pressure would make the star collapse, gives us the differential equation of hydrostatic equilibrium

$$\frac{dP}{d\rho} = -G\frac{M(\rho)\,\delta}{\rho^2}$$

To these equations we add an equation for the conservation of energy, and an equation that governs the energy transport from the center to the surface. These form a set of four basic coupled differential equations which, for a given stellar mass and composition of elements, can be integrated numerically to give the run of pressure, temperature, mass and energy distribution through the star. As Figure 67 indicates, there are other complicating factors.

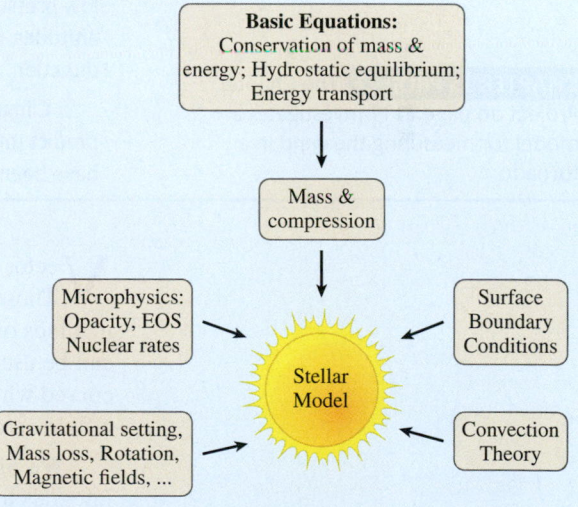

**Figure 67**

*Source*: Brian Chaboyer, Dartmouth College

   Through hundreds of years of careful observation, increasingly sophisticated understanding of the relevant physics, and more and more powerful mathematical techniques, astronomers now have a fairly complete understanding of stellar structure and evolution. It continues as an active area of research and is one of the beautiful triumphs of mathematics.

# 15

# Vector Calculus

Cultura RM Exclusive/Jason Persoff Stormdoctor/Getty Images

**CHAPTER 15 PROJECT** The Chapter Project on page 1111 investigates a model for measuring the wind in a tornado.

## Modeling a Tornado

Most people who live in the wide stretch of land between the Appalachian Mountains in the east and Rockies in the west are familiar with intense supercell thunderstorms and the tornadoes they produce. Although tornadoes are known to form in every state, and on every continent except Antarctica, the unique geographical and meteorological characteristics of Tornado Alley make it an ideal breeding ground for these violent rotating columns of air. Warm moist air from the south is forced upward by fast-moving cold, low pressure systems from the north and west and mixes with cold, drier air at higher altitudes, producing strong shearing winds that begin to rotate in a counterclockwise direction. As the wind increases it sucks up more moist air, forming a funnel-like vortex.

Climatologists and meteorologists are working to develop mathematical models to predict tornadoes and to better understand their behavior. And though several models have been proposed, no single reliable mathematical model exists.

Vector calculus is the culmination of what we have learned throughout the course. Those of you who are studying engineering or science will find the theory and methods of vector calculus rich in engineering applications. For example, line integrals can be used to find the work done by a variable vector force and to find the mass of a curved wire that has variable density. Surface integrals are used to measure the flow of a fluid across a surface.

We also discuss three important theorems that relate line integrals and surface integrals to double integrals and triple integrals.

# 15.1 Vector Fields

**OBJECTIVE** *When you finish this section, you should be able to:*

**1** Describe a vector field (p. 1025)

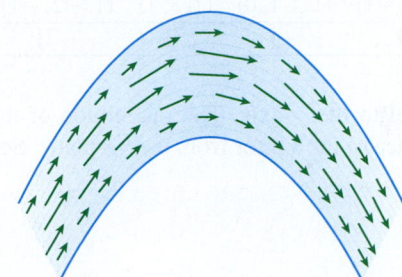

**Figure 1** Velocity field of a river

Suppose you are studying the motion of a fluid such as water or air. The fluid is composed of a multitude of moving particles. At any instant of time $t$, each particle is at a given point and is moving with a particular velocity $\mathbf{v}(t)$. If we assign a velocity vector to each particle at time $t$ and take a picture at that instant, then the snapshot of all the vectors at this time $t$ forms a *vector field*. Figure 1 shows a possible velocity field for the flow of water in a river. The speed of the water is slower near the shore and faster in the middle. On the bend, the water flows faster on the outside bend and slower on the inside bend.

In Chapter 11, we discussed vector functions, $\mathbf{r} = \mathbf{r}(t)$, $a \leq t \leq b$. A vector function associates a vector $\mathbf{r}$ to a real number $t$. So the domain of a vector function is a set of real numbers and its range is a set of vectors. A *vector field* is a function that associates a vector to a point in the plane or to a point in space. That is, a vector field is a function whose domain is a set of points in the plane or in space and whose range is a set of vectors.

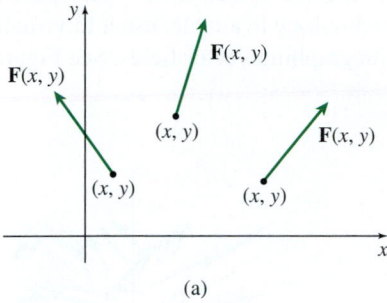

(a)

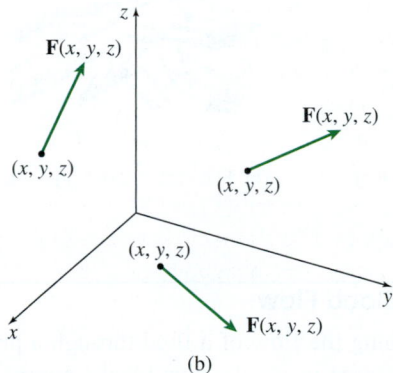

(b)

**Figure 2**

**DEFINITION** Vector Field

Let $P = P(x, y)$ and $Q = Q(x, y)$ be functions of two variables defined on a subset $R$ of the $xy$-plane. A **vector field F over R** is defined as the function

$$\boxed{\mathbf{F} = \mathbf{F}(x, y) = P(x, y)\mathbf{i} + Q(x, y)\mathbf{j}}$$

Let $P = P(x, y, z)$, $Q = Q(x, y, z)$, and $R = R(x, y, z)$ be functions of three variables defined on a subset $E$ of space. A **vector field F over E** is defined as the function

$$\boxed{\mathbf{F} = \mathbf{F}(x, y, z) = P(x, y, z)\mathbf{i} + Q(x, y, z)\mathbf{j} + R(x, y, z)\mathbf{k}}$$

The vector field $\mathbf{F} = \mathbf{F}(x, y)$ consists of vectors in the plane with initial points at $(x, y)$. In general, it is not possible to draw a vector for each point $(x, y)$ in the domain, so a representative set of vectors (arrows) are drawn. See Figure 2(a).

Similarly, the vector field $\mathbf{F} = \mathbf{F}(x, y, z)$ consists of vectors in space with initial points at $(x, y, z)$. See Figure 2(b).

## 1 Describe a Vector Field

**EXAMPLE 1** Describing a Vector Field in the Plane

Describe the vector field $\mathbf{F} = \mathbf{F}(x, y) = -y\mathbf{i} + x\mathbf{j}$ by drawing some of the vectors $\mathbf{F}$.

**Solution** We begin by making a table of vectors. We do this by choosing points $(x, y)$ and finding the values of $\mathbf{F}$.

| $(x, y)$ | $(2, 0)$ | $(0, 2)$ | $(-2, 0)$ | $(0, -2)$ | $(3, 3)$ | $(-3, 3)$ | $(-3, -3)$ | $(3, -3)$ |
|----------|----------|----------|-----------|-----------|----------|-----------|------------|-----------|
| $\mathbf{F}(x, y)$ | $2\mathbf{j}$ | $-2\mathbf{i}$ | $-2\mathbf{j}$ | $2\mathbf{i}$ | $-3\mathbf{i} + 3\mathbf{j}$ | $-3\mathbf{i} - 3\mathbf{j}$ | $3\mathbf{i} - 3\mathbf{j}$ | $3\mathbf{i} + 3\mathbf{j}$ |

Figure 3 illustrates the vectors from the table. Notice that each vector in the field is tangent to a circle centered at the origin, and the direction of the vectors indicates that the field is rotating counterclockwise. This field might represent the motion of a wheel spinning on an axle, with each vector equal to the velocity at a point of the wheel. ∎

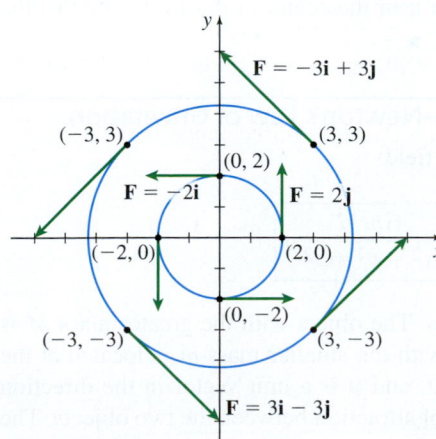

**Figure 3** $\mathbf{F}(x, y) = -y\mathbf{i} + x\mathbf{j}$

**NOW WORK** Problem 7.

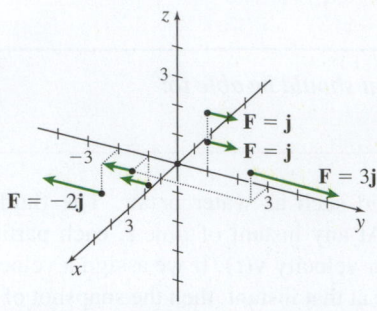

**Figure 4** The vector field $\mathbf{F}(x, y, z) = y\mathbf{j}$

### EXAMPLE 2  Describing a Vector Field in Space

Describe the vector field $\mathbf{F} = \mathbf{F}(x, y, z) = y\mathbf{j}$ by drawing some of the vectors $\mathbf{F}$.

**Solution** Several vectors from the vector field are listed in the table below.

| $(x, y, z)$ | $(0, 0, 0)$ | $(0, 1, 1)$ | $(0, 1, 2)$ | $(0, -1, -1)$ | $(1, -1, 0)$ | $(1, 3, 1)$ | $(1, -2, -1)$ |
|---|---|---|---|---|---|---|---|
| $\mathbf{F}(x, y, z)$ | $\mathbf{0}$ | $\mathbf{j}$ | $\mathbf{j}$ | $-\mathbf{j}$ | $-\mathbf{j}$ | $3\mathbf{j}$ | $-2\mathbf{j}$ |

Each vector in the vector field $\mathbf{F}$ is parallel to the $y$-axis. The magnitude of the vector equals $|y|$ and is proportional to the distance of the vector from the $xz$-plane. See Figure 4. ∎

**NOW WORK** Problem 13.

The vector fields in Examples 1 and 2 were relatively easy to graph by hand. But this is not true for most vector fields, particularly for vector fields in space. There is graphing technology, often found in computer algebra systems, however, with the capability of drawing vector fields. If you have technology available, use it to visualize the vector fields in the text, and to experiment with graphing vector fields. See Figure 5 for some examples.

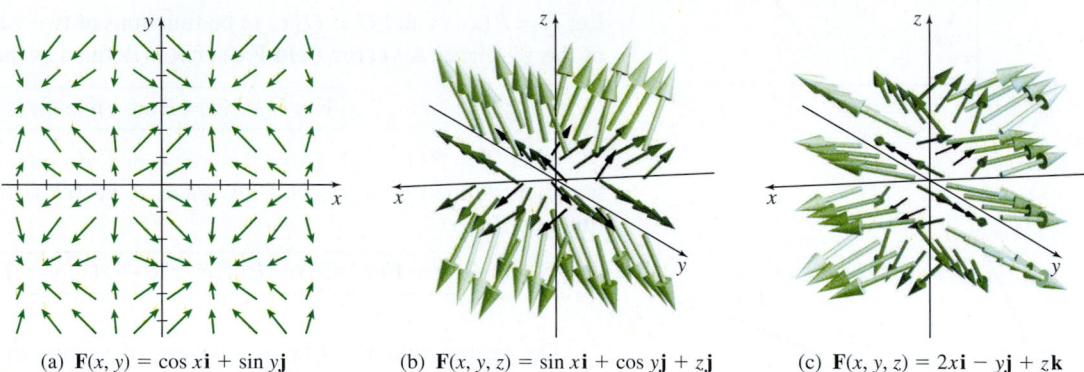

(a) $\mathbf{F}(x, y) = \cos x\mathbf{i} + \sin y\mathbf{j}$     (b) $\mathbf{F}(x, y, z) = \sin x\mathbf{i} + \cos y\mathbf{j} + z\mathbf{j}$     (c) $\mathbf{F}(x, y, z) = 2x\mathbf{i} - y\mathbf{j} + z\mathbf{k}$

**Figure 5**

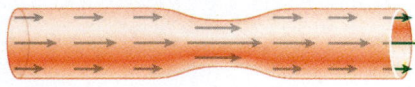

**Figure 6**

### EXAMPLE 3  Describing a Vector Field—Blood Flow

An example of a **velocity field** arises in describing the flow of a fluid through a pipe or the flow of blood through an artery. At each point inside the pipe (or the artery), a vector is defined that represents the velocity of the flow. See Figure 6. The magnitude of each vector represents the speed of the fluid's flow. Notice that the speed of blood flow is slower near the artery wall and is faster near the center of the artery and that the speed increases where the artery is contracted. ∎

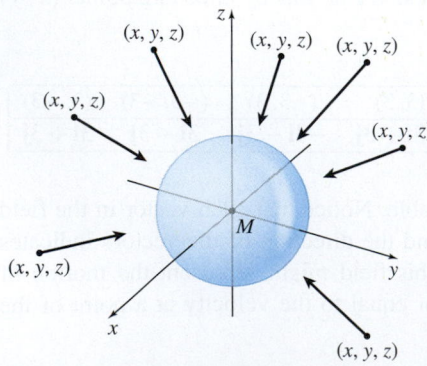

**Figure 7** A gravitational field

### EXAMPLE 4  Describing a Vector Field—Newton's Law of Gravitation

A **gravitational field** is defined by the vector field

$$\mathbf{F} = \mathbf{F}(x, y, z) = -\frac{GmM}{x^2 + y^2 + z^2}\mathbf{u}$$

where $m$ and $M$ are the masses of two objects. The object with the greater mass $M$ is located at the origin $(0, 0, 0)$ and the object with the smaller mass $m$ is located at the point $(x, y, z)$; $G$ is the gravitational constant; and $\mathbf{u}$ is a unit vector in the direction from $(0, 0, 0)$ to $(x, y, z)$. Here $\mathbf{F}$ is the force of attraction between the two objects. The force vectors $\mathbf{F}$ are directed from the object at $(x, y, z)$ toward the object at $(0, 0, 0)$. Figure 7 illustrates a gravitational field. ∎

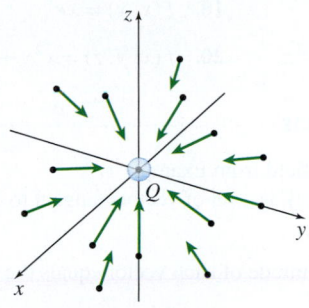

(a) Like charges: $qQ > 0$

(b) Unlike charges: $qQ < 0$

**Figure 8** Electric force fields

| NEED TO REVIEW? Gradients are discussed in Section 13.1, pp. 911–917.

EXAMPLE 5  **Describing a Vector Field—Coulomb's Law**

An **electric force field** is defined by the vector field

$$\mathbf{F} = \mathbf{F}(x, y, z) = \frac{\varepsilon q Q}{x^2 + y^2 + z^2} \mathbf{u}$$

where $\varepsilon$ is a positive constant that depends on the units being used; $q$ and $Q$ are the charges of two objects, one located at the origin $(0, 0, 0)$ and the other at the point $(x, y, z)$; and $\mathbf{u}$ is the unit vector in the direction from $(0, 0, 0)$ to $(x, y, z)$. Here $\mathbf{F}$ is the electric force exerted by the charge at $(0, 0, 0)$ on the charge at $(x, y, z)$. For like charges, we have $qQ > 0$ and the force $\mathbf{F}$ is **repulsive**, as illustrated in Figure 8(a). For unlike charges, we have $qQ < 0$ and the force $\mathbf{F}$ is **attractive**, as illustrated in Figure 8(b). ∎

Recall that the gradient $\nabla f$ of a function $z = f(x, y)$ in the plane is the vector

$$\nabla f(x, y) = f_x(x, y)\mathbf{i} + f_y(x, y)\mathbf{j}$$

and the gradient $\nabla f$ of a function $w = f(x, y, z)$ in space is the vector

$$\nabla f(x, y, z) = f_x(x, y, z)\mathbf{i} + f_y(x, y, z)\mathbf{j} + f_z(x, y, z)\mathbf{k}$$

We see that the gradient $\nabla f$ of a function $f$ is a vector field, called the **gradient vector field.**

CALC
CLIP
EXAMPLE 6  **Describing a Gradient Vector Field**

(a)  Find the gradient vector field of the elliptic paraboloid $z = f(x, y) = x^2 + 4y^2$ and graph it by drawing some of the vectors $\nabla f$.

(b)  Graph the level curves $f(x, y) = c$, for $c = 0, 1, 4$, and $16$.

**Solution** (a) The gradient of $f$ is $\nabla f = \dfrac{\partial f}{\partial x}\mathbf{i} + \dfrac{\partial f}{\partial y}\mathbf{j} = 2x\mathbf{i} + 8y\mathbf{j}$. A graph of the gradient vector field is shown in Figure 9.

(b) Figure 10(a) shows the level curves of the elliptic paraboloid $f(x, y) = x^2 + 4y^2 = c$ for $c = 0, 1, 4$, and $16$. Compare the level curves with the gradient vector field illustrated in Figure 9. Notice that the gradient vectors are orthogonal to the level curves, as shown in Figure 10(b). ∎

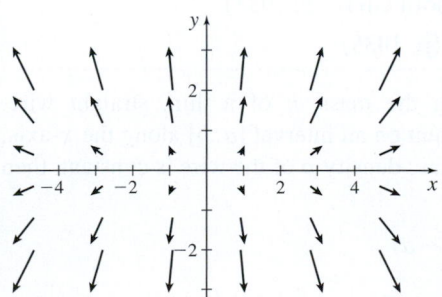

**Figure 9** $\nabla f = 2x\mathbf{i} + 8y\mathbf{j}$

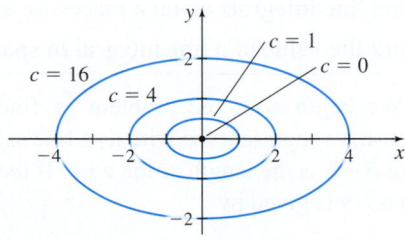

(a)  Level curves are concentric ellipses:
$x^2 + 4y^2 = c$

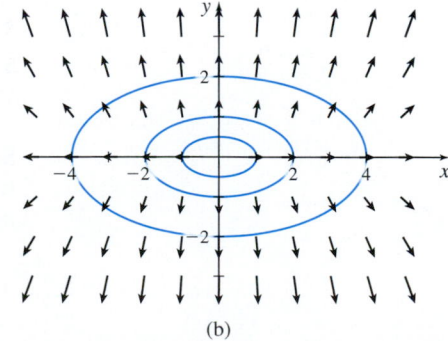

(b)

DF **Figure 10**

NOW WORK Problem **17.**

## 15.1 Assess Your Understanding

### Concepts and Vocabulary

1. A(n) _____ is a function that associates a vector to a point in the plane or to a point in space.

2. *True or False* Let $P = P(x, y, z)$, $Q = Q(x, y, z)$, and $R = R(x, y, z)$ be functions of three variables defined on a subset $E$ of space. A vector field over $E$ is defined as the function $F(x, y, z) = P(x, y, z) + Q(x, y, z) + R(x, y, z)$.

3. The domain of the vector field $\mathbf{F} = \mathbf{F}(x, y)$ is a set of points $(x, y)$ in the plane, and the range of $\mathbf{F}$ is a set of _____ in the plane.

4. *True or False* The gradient of a function $f$ is an example of a vector field.

### Skill Building

*In Problems 5–14, describe each vector field by drawing some of its vectors.*

5. $\mathbf{F} = \mathbf{F}(x, y) = x\mathbf{i} + y\mathbf{j}$

6. $\mathbf{F} = \mathbf{F}(x, y) = x\mathbf{i} - y\mathbf{j}$

7. $\mathbf{F} = \mathbf{F}(x, y) = \mathbf{i} + x\mathbf{j}$

8. $\mathbf{F} = \mathbf{F}(x, y) = y\mathbf{i} - \mathbf{j}$

9. $\mathbf{F} = \mathbf{F}(x, y) = \mathbf{i}$

10. $\mathbf{F} = \mathbf{F}(x, y) = -\mathbf{j}$

11. $\mathbf{F} = \mathbf{F}(x, y) = \mathbf{i} + \mathbf{j}$

12. $\mathbf{F} = \mathbf{F}(x, y) = -\mathbf{i} + \mathbf{j}$

13. $\mathbf{F} = \mathbf{F}(x, y, z) = z\mathbf{k}$

14. $\mathbf{F} = \mathbf{F}(x, y, z) = x\mathbf{i}$

[CAS] *In Problems 15 and 16, use graphing technology to represent each vector field. Then describe the vector field.*

15. $\mathbf{F} = \mathbf{F}(x, y, z) = \dfrac{x\mathbf{i} + y\mathbf{j} + z\mathbf{k}}{\sqrt{x^2 + y^2 + z^2}}$

16. $\mathbf{F} = \mathbf{F}(x, y, z) = -\dfrac{x\mathbf{i} + y\mathbf{j} + z\mathbf{k}}{\sqrt{x^2 + y^2 + z^2}}$

*In Problems 17–20, find the gradient vector field of each function $f$.*

17. $f(x, y) = x \sin y + \cos y$

18. $f(x, y) = xe^y$

19. $f(x, y, z) = x^2y + xy + y^2z$

20. $f(x, y, z) = x^2y + xyz^2$

### Applications and Extensions

21. **(a)** Show that the vector field from Example 1, $\mathbf{F} = \mathbf{F}(x, y) = -y\mathbf{i} + x\mathbf{j}$, is a set of vectors tangent to circles centered at the origin.

    **(b)** Confirm that the magnitude of each vector equals the radius of the circle.

22. **Gravitational Potential Energy** The gravitational field $\mathbf{g}$ due to a very small object of mass $m$ kg that is $r$ meters (m) from a large object is given by $\mathbf{g} = -\dfrac{Gm}{r^3}\mathbf{r}$, where $G = 6.67 \times 10^{-11}$ N m²/kg² and $\mathbf{r} = x\mathbf{i} + y\mathbf{j} + z\mathbf{k}$. Show that the gravitational field is a gradient vector field. That is, show that $\mathbf{g} = -\nabla u$, where $u = -\dfrac{Gm}{r}$. The scalar function $u$ is called the **gravitational potential** due to the mass $m$.

---

# 15.2 Line Integrals of Scalar Functions

**OBJECTIVES** *When you finish this section, you should be able to:*

1. Define a line integral in the plane (p. 1029)
2. Find the value of a line integral along a smooth curve (p. 1030)

   Application: Finding the mass of a wire of variable density (p. 1031)

   Application: Finding the lateral surface area of a cylinder (p. 1032)
3. Find line integrals of the form $\int_C f(x, y)\,dx$ and $\int_C f(x, y)\,dy$ (p. 1032)
4. Find line integrals along a piecewise-smooth curve (p. 1035)
5. Find the value of a line integral in space (p. 1035)

We begin with the problem of finding the mass $m$ of a thin, straight wire. Suppose we represent the wire as a line segment on an interval $[a, b]$ along the $x$-axis, where $b - a$ is the length of the wire. If the mass density $\rho$ of the wire is constant, then its mass $m$ is given by

$$m = \rho(b - a)$$

Now suppose the mass density $\rho$ is not constant and depends on the position $x$, $a \leq x \leq b$. That is, $\rho = \rho(x)$ is a function of position $x$. Following the ideas presented in Chapter 6, we can determine the mass of the wire by partitioning the interval $[a, b]$ into $n$ subintervals, $[x_0, x_1], [x_1, x_2], \ldots, [x_{i-1}, x_i], \ldots, [x_{n-1}, x_n]$, each of length $\Delta x = \dfrac{b - a}{n}$.

**NEED TO REVIEW?** A table with units for mass density in both SI and U.S. units is given in Section 6.7, p. 481.

---

**1.** = NOW WORK problem      [N] = Graphing technology recommended      [CAS] = Computer Algebra System recommended

If $\Delta x$ is small enough, the mass density $\rho$ will not vary much over each subinterval, and we can approximate the mass of the wire by adding up the mass on each subinterval, $\sum_{i=1}^{n} \rho(u_i)\Delta x$, where $u_i$ is a number in the $i$th subinterval $[x_{i-1}, x_1]$, $i = 1, 2, \ldots, n$. If $\rho = \rho(x)$ is continuous on $[a, b]$, the mass $m$ of the wire is given by

$$m = \lim_{\Delta x \to 0} \sum_{i=1}^{n} \rho(u_i)\Delta x = \int_a^b \rho(x)\,dx$$

But what if the wire is twisted into the shape of a smooth curve $x = x(t)$, $y = y(t)$, $a \le t \le b$, where the mass density function $\rho$ depends on the position (point) on the curve? To answer a question like this we introduce a line integral, whose value is determined by the values of a function along a curve.

**NEED TO REVIEW?** Smooth curves are discussed in Section 9.2, pp. 690–691.

## 1 Define a Line Integral in the Plane

Let $C$ be a smooth curve whose parametric equations are given by

$$x = x(t) \quad y = y(t) \qquad a \le t \le b$$

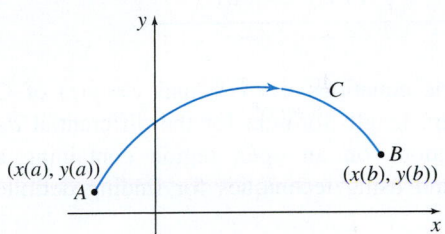

**Figure 11** $x = x(t)$, $y = y(t)$, $a \le t \le b$

See Figure 11. As $t$ increases from $a$ to $b$, the corresponding points $(x(t), y(t))$ trace out the curve $C$ from the point $A = (x(a), y(a))$ to the point $B = (x(b), y(b))$; that is, the orientation of $C$ is from $A$ to $B$. Let $z = f(x, y)$ be a scalar function that is defined on some open region containing the curve $C$.

Suppose we partition the closed interval $[a, b]$ into $n$ subintervals:

$$[a, t_1], \quad [t_1, t_2], \quad \ldots, \quad [t_{i-1}, t_i], \quad \ldots, \quad [t_{n-1}, b]$$

and denote the length of each subinterval by $\Delta t_1, \Delta t_2, \ldots, \Delta t_i, \ldots, \Delta t_n$. Then corresponding to each number $a = t_0, t_1, \ldots, t_i, \ldots, t_n = b$ of the partition, there is a point $P_0, P_1, \ldots, P_i, \ldots, P_n$ on the curve $C$. These points partition the curve $C$ into $n$ subarcs of lengths $\Delta s_1, \Delta s_2, \Delta s_3, \ldots, \Delta s_i, \ldots, \Delta s_n$, as shown in Figure 12. We define the **norm**, $\|\Delta\|$, of the partition to be the length of the subarc with the largest arc length.

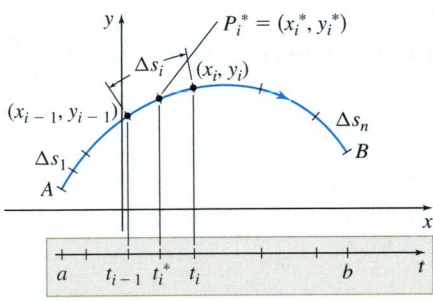

**Figure 12**

Now we pick a number $t_i^*$ in each subinterval $[t_{i-1}, t_i]$, $i = 1, 2, \ldots, n$, and let $P_i^* = (x_i^*, y_i^*)$ be the point in the $i$th subarc corresponding to the number $t_i^*$. Form the sums

$$\sum_{i=1}^{n} f(x_i^*, y_i^*)\Delta s_i$$

**NOTE** The domain of a scalar function is a set of points in the plane or in space and the range is a set of real numbers. From here on we refer to scalar functions simply as functions.

These sums depend on both the choice of the partition of $[a, b]$ and the choice of $P_i^*$. If all such sums can be made as close as we please to a number $L$ by choosing partitions whose norms are sufficiently close to 0, then $L$ is the limit of these sums as $\|\Delta\|$ approaches 0 and we write $L = \lim_{\|\Delta\| \to 0} \sum_{i=1}^{n} f(x_i^*, y_i^*)\Delta s_i$.

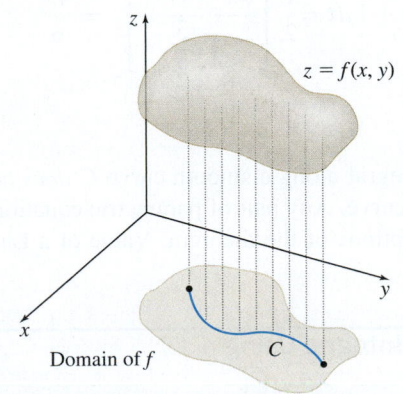

**Figure 13** $C$ lies in the domain of $f$.

> **DEFINITION** Line Integral of $z = f(x, y)$ Along a Smooth Curve $C$
>
> Let $C$ be a smooth curve defined by the parametric equations $x = x(t)$ and $y = y(t)$, $a \le t \le b$. Let $z = f(x, y)$ be a function defined on some open region containing $C$. The **line integral of** $f$ along $C$ from $t = a$ to $t = b$ is
>
> $$\int_C f(x, y)\,ds = \lim_{\|\Delta\| \to 0} \sum_{i=1}^{n} f(x_i^*, y_i^*)\Delta s_i$$
>
> provided this limit exists.

The line integral $\int_C f(x, y)\,ds$ is the sum of the values of $z = f(x, y)$ as $C$ is traversed from $t = a$ to $t = b$. See Figure 13.

## 2 Find the Value of a Line Integral Along a Smooth Curve

We state without proof a method for finding the value of a line integral when the function $f$ is continuous on some open region that contains $C$.

---

**THEOREM** The Value of a Line Integral

Let $C$ be a smooth curve defined by the parametric equations

$$x = x(t) \quad y = y(t) \qquad a \le t \le b$$

Let $z = f(x, y)$ be a function that is continuous on some open region that contains the curve $C$. Then the line integral of $f$ along $C$ from $t = a$ to $t = b$ exists and

$$\int_C f(x, y)\, ds = \int_a^b f(x(t), y(t)) \sqrt{\left(\frac{dx}{dt}\right)^2 + \left(\frac{dy}{dt}\right)^2}\, dt \qquad (1)$$

---

**NOTE** If $f(x, y) = 1$, the line integral of $f$ along $C$ is the arc length of $C$ from $t = a$ to $t = b$.

Formula (1) results when the parametric equations $x = x(t)$ and $y = y(t)$ of $C$ are substituted for $x$ and $y$ in $f$ and the arc length formula for the differential $ds$ is used. Then when a function is continuous on an open region containing a smooth curve, the line integral can be found using techniques for finding definite integrals.

**NEED TO REVIEW?** Arc length is discussed in Section 9.3, pp. 695–697.

**EXAMPLE 1** Finding the Value of a Line Integral Along a Smooth Curve

Find $\int_C y\, ds$ if $C$ is the curve defined by the parametric equations $x(t) = t$ and $y(t) = \sqrt{t}$, $2 \le t \le 6$

**Solution** The curve $C$ is smooth and is part of the graph of $y = \sqrt{x}$, from $(2, \sqrt{2})$ to $(6, \sqrt{6})$, as shown in Figure 14.

The differential $ds$ of arc length along $C$ is given by

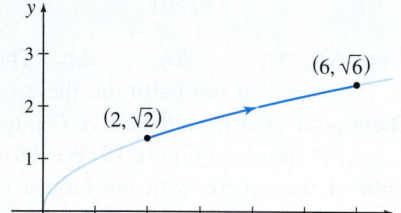

**Figure 14** $x(t) = t$, $y(t) = \sqrt{t}$, $2 \le t \le 6$

$$ds = \sqrt{\left(\frac{dx}{dt}\right)^2 + \left(\frac{dy}{dt}\right)^2}\, dt$$

Since $\dfrac{dx}{dt} = 1$ and $\dfrac{dy}{dt} = \dfrac{d}{dt}\sqrt{t} = \dfrac{1}{2\sqrt{t}}$, we have

$$ds = \sqrt{1^2 + \left(\frac{1}{2\sqrt{t}}\right)^2}\, dt = \sqrt{1 + \frac{1}{4t}}\, dt = \sqrt{\frac{4t+1}{4t}}\, dt = \frac{\sqrt{4t+1}}{2\sqrt{t}}\, dt$$

Now use formula (1).

$$\int_C y\, ds = \int_2^6 \sqrt{t}\, \frac{\sqrt{4t+1}}{2\sqrt{t}}\, dt = \frac{1}{2}\int_2^6 \sqrt{4t+1}\, dt = \frac{1}{2}\left[\frac{(4t+1)^{3/2}}{4 \cdot \dfrac{3}{2}}\right]_2^6 = \frac{49}{6} \qquad \blacksquare$$

$$\uparrow$$
$$y = \sqrt{t}$$

**NOW WORK** Problem 9.

It can be shown that the value of a line integral along a smooth curve $C$ *does not depend on the parametric representation of the curve.* Any pair of parametric equations with the same orientation satisfying the assumptions of the theorem: Value of a Line Integral gives the same value.

**EXAMPLE 2** Finding the Value of a Line Integral Using Different Parametrizations

Find $\int_C (x^2 + y)\, ds$

if $C$ is the line segment from $(0, 0)$ to $(1, 2)$ and $C$ is expressed using the parametric equations:

**(a)** $x(t) = t$ and $y(t) = 2t$,  $0 \le t \le 1$

**(b)** $x(t) = \sin t$ and $y(t) = 2 \sin t$,  $0 \le t \le \dfrac{\pi}{2}$

**Solution**  The two sets of parametric equations in (a) and (b) are just different ways of representing a part of the line $y = 2x$ from $(0, 0)$ to $(1, 2)$, as shown in Figure 15.

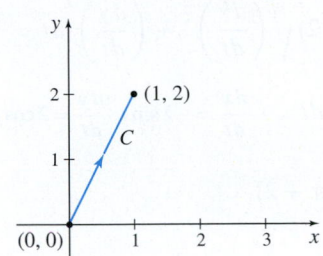

**Figure 15** $y = 2x$ from $(0, 0)$ to $(1, 2)$

**(a)** For $x(t) = t$ and $y(t) = 2t$, we have $\dfrac{dx}{dt} = 1$ and $\dfrac{dy}{dt} = 2$. The differential $ds$ of the arc length $s$ is $ds = \sqrt{1^2 + 2^2}\, dt = \sqrt{5}\, dt$. Then

$$\int_C (x^2 + y)\, ds = \int_0^1 (t^2 + 2t)\sqrt{5}\, dt = \left[\sqrt{5}\left(\frac{t^3}{3} + t^2\right)\right]_0^1 = \frac{4\sqrt{5}}{3}$$

$$\uparrow$$
$$x = t$$
$$y = 2t$$

**(b)** For $x(t) = \sin t$ and $y(t) = 2 \sin t$, we have $\dfrac{dx}{dt} = \cos t$ and $\dfrac{dy}{dt} = 2 \cos t$.

Since $\cos t \ge 0$ on $0 \le t \le \dfrac{\pi}{2}$, the differential $ds$ of the arc length $s$ is

$$ds = \sqrt{\left(\frac{dx}{dt}\right)^2 + \left(\frac{dy}{dt}\right)^2}\, dt = \sqrt{\cos^2 t + 4\cos^2 t}\, dt = \sqrt{5 \cos^2 t}\, dt = \sqrt{5} \cos t\, dt$$

Then

$$\int_C (x^2 + y)\, ds = \int_0^{\pi/2} (\sin^2 t + 2 \sin t)\sqrt{5} \cos t\, dt = \sqrt{5} \int_0^{\pi/2} (\sin^2 t + 2 \sin t) \cos t\, dt$$

$$\uparrow$$
$$x = \sin t$$
$$y = 2 \sin t$$

$$= \sqrt{5}\left[\frac{\sin^3 t}{3} + \sin^2 t\right]_0^{\pi/2} = \frac{4\sqrt{5}}{3}$$

the same value found in (a). ∎

NOW WORK **Problem 21.**

## Application: Finding the Mass of a Wire of Variable Density

The mass of a long thin wire of variable density, whose shape is described by a smooth curve $C$, can be found using a line integral. Suppose $C$ is a smooth curve defined by the parametric equations $x = x(t)$, $y = y(t)$, for $a \le t \le b$. Suppose the variable mass density (mass per unit length) at the point $(x, y)$ on $C$ is $\rho = \rho(x, y)$, where $\rho$ is continuous on some region containing $C$. We partition $[a, b]$ into $n$ subintervals $[t_{i-1}, t_i]$, for $i = 1, 2, \ldots, n$. In each subinterval $[t_{i-1}, t_i]$, we choose a number $t_i^*$, as shown in Figure 16. Then the mass $\Delta M_i$ of the corresponding short piece of wire between $(x_{i-1}, y_{i-1})$ and $(x_i, y_i)$ is $\Delta M_i \approx \rho(x_i^*, y_i^*)\Delta s_i$, where $\Delta s_i$ is the length of the short piece of wire and $(x_i^*, y_i^*)$ is the point corresponding to $t_i^*$. We define the norm, $\|\Delta\|$, of the partition to be largest length $\Delta s_i$. The sums

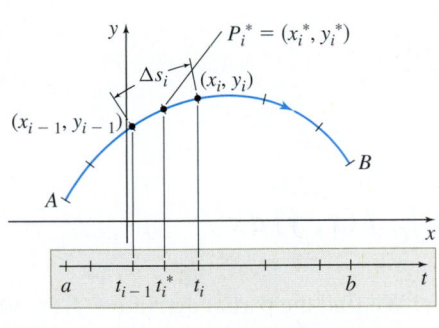

**Figure 16** $\Delta M_i \approx \rho(x_i^*, y_i^*)\Delta s_i$

$$\sum_{i=1}^{n} \Delta M_i = \sum_{i=1}^{n} \rho(x_i^*, y_i^*)\Delta s_i$$

over all the subintervals approximate the mass $M$ of the wire. By taking the limit of these sums as $\|\Delta\|$ approaches 0, we obtain the **mass** $M$ of the wire

$$M = \lim_{\|\Delta\| \to 0} \sum_{i=1}^{n} \rho(x_i^*, y_i^*)\Delta s_i = \int_C \rho(x, y)\, ds \qquad (2)$$

CALC CLIP
**EXAMPLE 3**  Finding the Mass of a Wire with Variable Density

Find the mass $M$ of a thin piece of wire in the shape of a semicircle $x = 2\cos t$, $y = 2\sin t$, $0 \le t \le \pi$, if the variable density of the wire is $\rho(x, y) = y + 2$.

**Solution** Using (2), the mass $M$ of the wire is

$$M = \int_C \rho(x, y)\,ds = \int_C (y + 2)\,ds = \int_0^\pi (y + 2)\sqrt{\left(\frac{dx}{dt}\right)^2 + \left(\frac{dy}{dt}\right)^2}\,dt$$

$$= \int_0^\pi (2\sin t + 2)\sqrt{(-2\sin t)^2 + (2\cos t)^2}\,dt \qquad \frac{dx}{dt} = -2\sin t;\ \frac{dy}{dt} = 2\cos t$$

$$= \int_0^\pi 4(\sin t + 1)\,dt = 4\big[-\cos t + t\big]_0^\pi = 4[\pi + 2] \qquad \blacksquare$$

**NOW WORK** Problem 49.

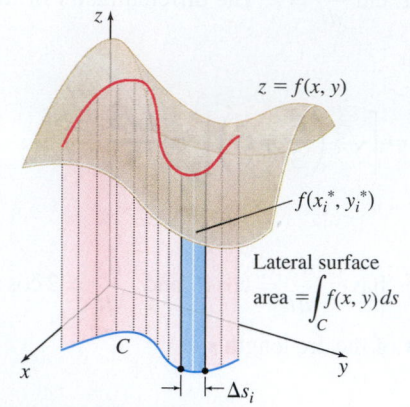

**DF Figure 17** Lateral surface area

### Application: Finding the Lateral Surface Area of a Cylinder

Line integrals also have a geometric application. In Figure 17, the surface $z = f(x, y)$ lies above the $xy$-plane, so $f(x, y) \ge 0$. Suppose $C$ is a smooth curve that lies in the $xy$-plane within the domain of $f$. If we draw lines parallel to the $z$-axis, from $C$ to the surface, the result is a cylinder, sometimes called a **fence**, and its area is called the **lateral surface area of the cylinder**.

**EXAMPLE 4**  Finding the Lateral Surface Area of a Cylinder

Find the lateral surface area $A$ of the cylinder formed by line segments parallel to the $z$-axis that extend from the smooth curve $y = x^2$, $0 \le x \le 2$, to the surface $z = f(x, y) = x$.

**Solution** Figure 18 illustrates the cylinder. Along $C$, $y = x^2$, the differential $ds$ of arc length is

$$ds = \sqrt{1 + \left(\frac{dy}{dx}\right)^2}\,dx = \sqrt{1 + 4x^2}\,dx$$

The lateral surface area $A$ of the cylinder is

$$A = \int_C f(x, y)\,ds = \int_C x\,ds = \int_0^2 x\sqrt{1 + 4x^2}\,dx \qquad \text{Let } u = 1 + 4x^2.$$

$$= \int_1^{17} \frac{1}{8}\sqrt{u}\,du = \left[\frac{1}{4}\frac{u^{3/2}}{3}\right]_1^{17} = \frac{17\sqrt{17} - 1}{12} \qquad \blacksquare$$

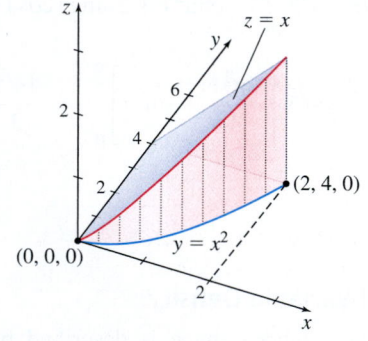

**Figure 18**

**NOW WORK** Problem 51.

## 3 Find Line Integrals of the Form $\int_C f(x, y)\,dx$ and $\int_C f(x, y)\,dy$

Consider a function $f$ of two variables $x$ and $y$. We find the partial derivative $f_x(x, y)$ of a function $z = f(x, y)$ by treating $y$ as a constant ($\Delta y = 0$) and differentiating $z = f(x, y)$ with respect to $x$. Similarly, $f_y(x, y)$ is found by holding $x$ fixed ($\Delta x = 0$) and differentiating $f$ with respect to $y$.

For a line integral $\int_C f(x, y)\,ds$, the differential $ds$ of arc length along the smooth curve $C$ is $(ds)^2 = (dx)^2 + (dy)^2$. If we treat $y$ as a constant, then $dy = 0$, $ds = dx$, and we have the **line integral along $C$ with respect to $x$**, $\int_C f(x, y)\,dx$. Similarly, if $x$ is treated as a constant, then $dx = 0$, $ds = dy$, and we have the **line integral along $C$ with respect to $y$**, $\int_C f(x, y)\,dy$.

These line integrals can be expressed as definite integrals if $C$ is a smooth curve and if the function $f$ is continuous on an open region that contains $C$.

**THEOREM  The Value of a Line Integral Along $C$ with Respect to $x$ or $y$**

Let $C$ be a smooth curve defined by the parametric equations

$$x = x(t) \quad y = y(t) \qquad a \le t \le b$$

Let $z = f(x, y)$ be a function that is continuous on an open region that contains the curve $C$. Then the **line integral of $f$ along $C$ with respect to $x$** is given by

$$\int_C f(x, y)\, dx = \int_a^b f(x(t), y(t))x'(t)\, dt \tag{3}$$

and the **line integral of $f$ along $C$ with respect to $y$** is given by

$$\int_C f(x, y)\, dy = \int_a^b f(x(t), y(t))y'(t)\, dt \tag{4}$$

In general, each of these line integrals has a different value.

**EXAMPLE 5  Finding a Line Integral Along $C$ with Respect to $x$ and with Respect to $y$**

Find

$$\int_C (x - 3y)\, dx \qquad \text{and} \qquad \int_C (x - 3y)\, dy$$

if $C$ is the part of the parabola $x = y^2$ that joins the points $(1, 1)$ and $(4, 2)$.

**Solution** The function $f(x, y) = x - 3y$ is continuous, and $C$ is a smooth curve everywhere in the plane. So, we can use the formulas (3) and (4). Using the parametric equations of the curve $C$, $x = t^2$ and $y = t$, $1 \le t \le 2$, we find $dx = 2t\, dt$ and $dy = dt$. Then

$$\int_C (x - 3y)\, dx = \int_1^2 (t^2 - 3t)2t\, dt = 2\int_1^2 (t^3 - 3t^2)dt = -\frac{13}{2}$$

$$\int_C (x - 3y)\, dy = \int_1^2 (t^2 - 3t)\, dt = -\frac{13}{6}$$    ∎

It is not always necessary to use parametric equations for $C$ when finding the value of the line integral of a function $f$. For example, if a smooth curve $C$ is given by the rectangular equation $y = g(x)$, $a \le x \le b$, then parametric equations of $C$ are

$$x = t \quad y = g(t) \qquad a \le t \le b$$

Since $dx = dt$ and $dy = g'(t)\, dt$, we have

$$\int_C f(x, y)\, dx = \int_a^b f(t, g(t))\, dt = \int_a^b f(x, g(x))\, dx$$

$$\int_C f(x, y)\, dy = \int_a^b f(t, g(t))g'(t)\, dt = \int_a^b f(x, g(x))g'(x)\, dx$$

For example, consider the curve $C$ in Example 5, which has the rectangular equation $y = \sqrt{x}$, $1 \le x \le 4$. Then

$$\underset{\underset{y = \sqrt{x}}{\uparrow}}{\int_C (x - 3y)\, dx} = \int_1^4 (x - 3\sqrt{x})\, dx = -\frac{13}{2}$$

$$\underset{\underset{x = y^2}{\uparrow}}{\int_C (x - 3y)\, dy} = \int_1^2 (y^2 - 3y)\, dy = -\frac{13}{6}$$

**NOW WORK** Problem **25.**

In many applications, the two line integrals $\int_C P(x, y)\, dx$ and $\int_C Q(x, y)\, dy$ are combined and take the form $\int_C (P\, dx + Q\, dy)$.

---

**DEFINITION** Line Integral $\int_C (P\, dx + Q\, dy)$ Along a Smooth Curve $C$

Let $C$ denote a smooth curve, and let $P = P(x, y)$ and $Q = Q(x, y)$ be functions of two variables that are continuous on some open region containing $C$. The **line integral of $P\, dx + Q\, dy$ along** $C$ is defined as

$$\int_C (P\, dx + Q\, dy) = \int_C P(x, y)\, dx + \int_C Q(x, y)\, dy$$

---

**EXAMPLE 6**  Finding a Line Integral of the Form $\displaystyle\int_C (P\, dx + Q\, dy)$

Find the line integral $\int_C (y^2\, dx - x^2\, dy)$ along

**(a)** $C_1$: The parabola $y = x^2$ joining the two points $(0, 0)$ and $(2, 4)$

**(b)** $C_2$: The line $y = 2x$ joining the two points $(0, 0)$ and $(2, 4)$

**Solution** The functions $P(x, y) = y^2$ and $Q(x, y) = -x^2$ are continuous in the $xy$-plane.

**(a)** The curve $C_1$ is smooth, and parametric equations for the parabola are $x(t) = t$, $y(t) = t^2$, $0 \le t \le 2$. Then $dx = dt$ and $dy = 2t\, dt$, so

$$\int_{C_1} (y^2\, dx - x^2\, dy) = \int_{C_1} y^2 dx - \int_{C_1} x^2 dy = \int_0^2 (t^2)^2 dt - \int_0^2 t^2 (2t\, dt)$$

$$= \int_0^2 t^4\, dt - 2\int_0^2 t^3\, dt = \left[\frac{t^5}{5}\right]_0^2 - \left[\frac{t^4}{2}\right]_0^2 = -\frac{8}{5}$$

**(b)** The curve $C_2$ is smooth, and parametric equations for the line segment are $x(t) = t$, $y(t) = 2t$, $0 \le t \le 2$. Then $dx = dt$ and $dy = 2\, dt$, so

$$\int_{C_2} (y^2\, dx - x^2\, dy) = \int_{C_2} y^2 dx - \int_{C_2} x^2 dy = \int_0^2 (2t)^2 dt - \int_0^2 t^2 (2\, dt)$$

$$= \left[\frac{4t^3}{3}\right]_0^2 - \left[\frac{2t^3}{3}\right]_0^2 = \frac{16}{3}$$  ∎

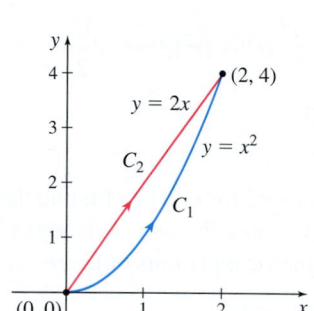

**Figure 19**

Figure 19 shows the curves $C_1$ and $C_2$ joining the points $(0, 0)$ and $(2, 4)$.

Observe that the value of the line integral in Example 6 depends on the curve $C$ over which the integration takes place. In Section 15.4, we investigate conditions under which the value of the integral is independent of the curve taken, that is, conditions under which the value of the integral depends only on the endpoints of the curve.

The orientation of the curve $C$ also plays a role in the value of a line integral over $C$. If $C$ is a smooth curve, then $-C$ is the same curve but with the reverse orientation. Then

$$\int_C (P\, dx + Q\, dy) = -\int_{-C} (P\, dx + Q\, dy)$$

That is, reversing the orientation of the curve $C$ changes the value of the line integral by a factor of $-1$. See Figure 20.

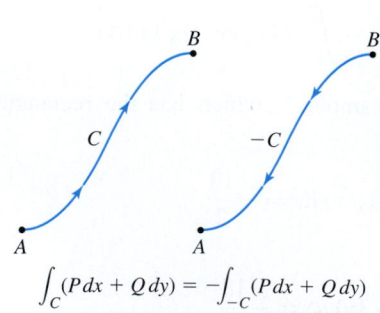

$$\int_C (P\, dx + Q\, dy) = -\int_{-C} (P\, dx + Q\, dy)$$

**DF Figure 20**

NOW WORK Problem 29.

## 4 Find Line Integrals Along a Piecewise-Smooth Curve

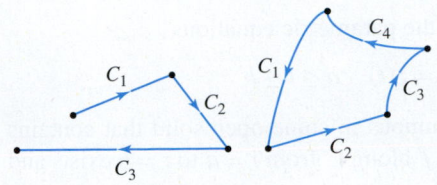

**Figure 21** Two piecewise-smooth curves

The definition of a line integral of $f$ along $C$ assumes that $C$ is a smooth curve. That is, the parametric equations $x(t)$ and $y(t)$ defining $C$ have derivatives $\dfrac{dx}{dt}$ and $\dfrac{dy}{dt}$ that are continuous and are never simultaneously 0 on $a \le t \le b$. Recall if a curve $C$ is not smooth but consists of a finite number of smooth curves, say, $C_1, C_2, \ldots, C_n$, $n \ge 2$, that are joined together end to end, then $C$ is a piecewise-smooth curve. Figure 21 shows two piecewise-smooth curves.

> **DEFINITION** Line Integral $\int_C (P\,dx + Q\,dy)$ Along a Piecewise-Smooth Curve $C$
>
> Let $C$ be a piecewise-smooth curve consisting of $n \ge 2$ smooth curves $C_1, C_2, \ldots, C_n$. Let $P = P(x, y)$ and $Q = Q(x, y)$ be functions of two variables that are each continuous on an open region containing $C$. The **line integral** $\int_C (P\,dx + Q\,dy)$ **along** $C$ is defined as
>
> $$\int_C (P\,dx + Q\,dy) = \int_{C_1}(P\,dx + Q\,dy) + \int_{C_2}(P\,dx + Q\,dy) + \cdots + \int_{C_n}(P\,dx + Q\,dy)$$

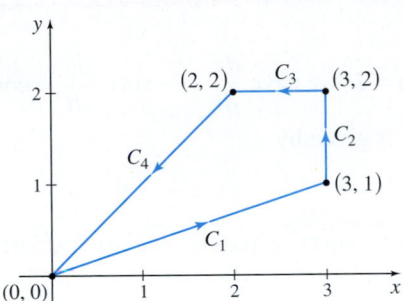

**Figure 22**

**EXAMPLE 7   Finding a Line Integral Along a Piecewise-Smooth Curve**

Find $\int_C (xy\,dx + x^2\,dy)$ along the piecewise-smooth curve $C$ illustrated in Figure 22.

**Solution** The curve $C$ consists of four curves $C_1, C_2, C_3$, and $C_4$. Notice that the orientation of $C$ requires using $C_1$ from $(0, 0)$ to $(3, 1)$, followed by $C_2$ from $(3, 1)$ to $(3, 2)$, then $C_3$ from $(3, 2)$ to $(2, 2)$, and finally $C_4$ from $(2, 2)$ to $(0, 0)$. The values of the line integral along each of the smooth curves $C_1, C_2, C_3$, and $C_4$ are

$$C_1: \quad y = \frac{1}{3}x, \quad dy = \frac{1}{3}dx; \qquad 0 \le x \le 3$$

$$\int_{C_1}(xy\,dx + x^2\,dy) = \int_0^3 \left[ x\left(\frac{1}{3}x\right)dx + x^2\frac{1}{3}dx \right] = 6$$

$$C_2: \quad x = 3, \quad dx = 0; \qquad 1 \le y \le 2$$

$$\int_{C_2}(xy\,dx + x^2\,dy) = \int_1^2 9\,dy = 9$$

$$C_3: \quad y = 2, \quad dy = 0; \qquad \text{Watch the orientation here: } x \text{ varies from 3 to 2.}$$

$$\int_{C_3}(xy\,dx + x^2\,dy) = \int_3^2 2x\,dx = \left[x^2\right]_3^2 = -5$$

$$C_4: \quad y = x, \quad dy = dx; \qquad \text{Watch the orientation here: } x \text{ varies from 2 to 0.}$$

$$\int_{C_4}(xy\,dx + x^2\,dy) = \int_2^0 (x^2\,dx + x^2\,dx) = 2\int_2^0 x^2 dx = -\frac{16}{3}$$

Then

$$\int_C (xy\,dx + x^2\,dy) = 6 + 9 - 5 - \frac{16}{3} = \frac{14}{3} \qquad \blacksquare$$

**NOW WORK** Problem 39.

## 5 Find the Value of a Line Integral in Space

The definition of a line integral in space is similar to that of a line integral in the plane. We state without proof the main results needed to find the value of line integrals in space.

**THEOREM The Value of a Line Integral in Space**

Let $C$ be a smooth curve in space defined by the parametric equations

$$x = x(t) \quad y = y(t) \quad z = z(t) \quad a \leq t \leq b$$

Let $w = f(x, y, z)$ be a function that is continuous in some open solid that contains the smooth curve $C$. Then the line integral of $f$ along $C$ from $t = a$ to $t = b$ exists and

$$\int_C f(x, y, z)\, ds = \int_a^b f(x(t), y(t), z(t)) \sqrt{\left(\frac{dx}{dt}\right)^2 + \left(\frac{dy}{dt}\right)^2 + \left(\frac{dz}{dt}\right)^2}\, dt$$

Observe that if $w = f(x, y, z) = 1$, then

$$\int_C f(x, y, z)\, ds = \int_C ds = \text{arc length of } C \text{ from } a \text{ to } b$$

**EXAMPLE 8  Finding a Line Integral in Space**

Find $\int_C xy\, ds$ along the smooth space curve $C: x(t) = \cos t$, $y(t) = \sin t$, and $z(t) = 2t$, $0 \leq t \leq \dfrac{\pi}{2}$, a part of a helix.

**Solution** For $x(t) = \cos t$, $y(t) = \sin t$, and $z(t) = 2t$ we have $\dfrac{dx}{dt} = -\sin t$, $\dfrac{dy}{dt} = \cos t$, and $\dfrac{dz}{dt} = 2$. The differential of the arc length $s$ is given by

$$ds = \sqrt{\left(\frac{dx}{dt}\right)^2 + \left(\frac{dy}{dt}\right)^2 + \left(\frac{dz}{dt}\right)^2}\, dt = \sqrt{(-\sin t)^2 + (\cos t)^2 + 2^2}\, dt = \sqrt{5}\, dt$$

Then

$$\int_C xy\, ds = \int_0^{\pi/2} \cos t \sin t\, (\sqrt{5}\, dt) = \sqrt{5} \int_0^1 u\, du = \sqrt{5}\left[\frac{u^2}{2}\right]_0^1 = \frac{\sqrt{5}}{2} \qquad \blacksquare$$

$$\underset{\text{Let } u = \sin t.}{\uparrow}$$

**NOW WORK** Problem 59.

Suppose $C$ is a smooth curve in space defined by the parametric equations

$$x = x(t) \quad y = y(t) \quad z = z(t) \qquad a \leq t \leq b$$

In space, we have line integrals along $C$ with respect to $x$, with respect to $y$, and with respect to $z$. If $P(x, y, z)$, $Q(x, y, z)$, and $R(x, y, z)$ are functions that are continuous in some open solid in space containing $C$, then

$$\int_C P(x, y, z)\, dx = \int_a^b P(x(t), y(t), z(t))\, x'(t)\, dt$$

$$\int_C Q(x, y, z)\, dy = \int_a^b Q(x(t), y(t), z(t))\, y'(t)\, dt$$

$$\int_C R(x, y, z)\, dz = \int_a^b R(x(t), y(t), z(t))\, z'(t)\, dt$$

$$\int_C (P\, dx + Q\, dy + R\, dz) = \int_C P\, dx + \int_C Q\, dy + \int_C R\, dz$$

A curve $C$ in space is piecewise-smooth if it consists of a finite number of smooth curves $C_1, C_2, \ldots, C_n$, $n \geq 2$, that are joined together end to end.

If $x = x(t)$, $y = y(t)$, and $z = z(t)$ are parametric equations of a piecewise-smooth curve $C$ and if $P = P(x, y, z)$, $Q = Q(x, y, z)$, and $R = R(x, y, z)$ are functions that are continuous in some open solid containing $C$, then

$$\int_C (P\,dx + Q\,dy + R\,dz) = \int_{C_1} (P\,dx + Q\,dy + R\,dz) + \int_{C_2} (P\,dx + Q\,dy + R\,dz) + \cdots + \int_{C_n} (P\,dx + Q\,dy + R\,dz)$$

EXAMPLE 9  **Finding a Line Integral in Space**

Find $\displaystyle\int_C (xyz\,dx + x^2 y\,dy + z\,dz)$

where the curve $C$ is defined by the parametric equations

$$x = t \qquad y = t^2 \qquad z = t^3 \qquad 0 \le t \le 1$$

**Solution**

$$\frac{dx}{dt} = 1 \qquad \frac{dy}{dt} = 2t \qquad \frac{dz}{dt} = 3t^2$$

and

$$\int_C (xyz\,dx + x^2 y\,dy + z\,dz) = \int_0^1 (t^6 + 2t^5 + 3t^5)dt = \left[\frac{t^7}{7} + 5\frac{t^6}{6}\right]_0^1 = \frac{41}{42} \qquad ■$$

NOW WORK **Problem 55.**

---

# 15.2 Assess Your Understanding

## Concepts and Vocabulary

1. *True or False*  A line integral $\int_C f(x, y)\,ds$ is equal to a definite integral if $C$ is a smooth curve defined on $[a, b]$, and if the function $f$ is continuous on some open region that contains the curve $C$.

2. If $C$ is a smooth curve defined by the parametric equations $x = x(t)$ and $y = y(t)$, $a \le t \le b$, and $f$ is a function that is continuous on some open region that contains the curve $C$, then the line integral of $f$ along $C$ from $t = a$ to $t = b$ exists and is given by $\int_C f(x, y)\,ds = \int_a^b \underline{\qquad}\ dt$.

3. *True or False*  The value of $\int_C f(x, y)\,ds$ depends on the choice of the parametric equations used to represent the smooth curve $C$.

4. If the variable mass density of a wire at a point $(x, y)$ is $\rho(x, y)$, then the mass $M$ of the wire can be found using the line integral $\underline{\qquad}$, where $C$ is the curve representing the shape of the wire.

5. *True or False*  $\int_{-C} (P\,dx + Q\,dy) = \int_C (P\,dx - Q\,dy)$

6. *True or False*  A piecewise-smooth curve $C$ consists of a finite number of smooth curves that are joined together end to end.

7. If a piecewise-smooth curve $C$ consists of smooth curves $C_1, C_2, \ldots, C_n$, $n \ge 2$, then $\int_C (P\,dx + Q\,dy) = \underline{\qquad}$.

8. *True or False*  Suppose $C$ is a smooth curve defined by the parametric equations $x = x(t)$, $y = y(t)$, and $z = z(t)$, $a \le t \le b$, and $P(x, y, z)$, $Q(x, y, z)$, and $R(x, y, z)$ are functions that are continuous in some open solid in space containing $C$, then $\int_C P(x, y, z)\,dx = \int_a^b P(x(t), y(t), z(t))x'(t)\,dt$.

## Skill Building

*In Problems 9–14, find each line integral for the given smooth curve $C$.*

9. $\displaystyle\int_C x^2 y\,ds; \quad C\colon x = \sin t, y = \cos t, 0 \le t \le \frac{\pi}{2}$

10. $\displaystyle\int_C (y - x^2)\,ds; \quad C\colon x = t, y = 3t, 0 \le t \le 1$

11. $\displaystyle\int_C xy\,ds; \quad C\colon x = t^2, y = 4t, 0 \le t \le 2$

12. $\displaystyle\int_C x\,ds; \quad C\colon x = t, y = 3t^2, 0 \le t \le 1$

13. $\displaystyle\int_C \frac{y}{2x^2 - y^2}\,ds; \quad C\colon x = t, y = t, 1 \le t \le 5$

14. $\displaystyle\int_C \frac{x^2}{x^3 + y^3}\,ds; \quad C\colon x = 3t, y = 4t, 1 \le t \le 2$

15. Find $\int_C 2y\,ds$, where $C$ is the curve $y = \frac{1}{2}x^3$ joining $(0, 0)$ to $(2, 4)$.

16. Find $\int_C x\,ds$, where $C$ is the curve $x = 5y^3$ joining $(0, 0)$ to $(5, 1)$.

17. Find $\int_C (xy + 1)\,ds$, where $C$ is the curve $\mathbf{r}(t) = \sin t\,\mathbf{i} + \cos t\,\mathbf{j}$, $0 \le t \le \pi$.

18. Find $\int_C xy^2\,ds$, where $C$ is the curve $\mathbf{r}(t) = \cos t\,\mathbf{i} + \sin t\,\mathbf{j}$, $0 \le t \le \frac{\pi}{2}$.

---

**1.** = NOW WORK problem       ⎿∿⏌ = Graphing technology recommended       CAS = Computer Algebra System recommended

*In Problems 19–22, find each line integral along the two given curves $C_1$ and $C_2$.*

**19.** $\int_C (x + y^2)\, ds$

(a) $C_1$: $x = t$, $y = t$, $0 \leq t \leq 1$

(b) $C_2$: $x = \sin t$, $y = \sin t$, $0 \leq t \leq \dfrac{\pi}{2}$

**20.** $\int_C (x + y)\, ds$

(a) $C_1$: $x = t$, $y = 3t$, $0 \leq t \leq 1$

(b) $C_2$: $x = \sin(2t)$, $y = 3\sin(2t)$, $0 \leq t \leq \dfrac{\pi}{2}$

**21.** $\int_C (x + e^y)\, ds$

(a) $C_1$: $x = 1 - t$, $y = t$, $0 \leq t \leq 1$

(b) $C_2$: $x = \cos t$, $y = 1 - \cos t$, $0 \leq t \leq \dfrac{\pi}{2}$

**22.** $\int_C x e^{x^2}\, ds$

(a) $C_1$: $x = t + 1$, $y = t$, $0 \leq t \leq 1$

(b) $C_2$: $x = 1 + \sin t$, $y = \sin t$, $0 \leq t \leq \dfrac{\pi}{2}$

*In Problems 23–26, find the line integral with respect to $x$ and with respect to $y$ for each function $f(x, y)$ along the given curve $C$.*

**23.** $f(x, y) = 2x + y^2$ along the line $y = 3 - x$ from $(0, 3)$ to $(3, 0)$.

**24.** $f(x, y) = xe^y$ along the line $2x - y = 1$ from $(0, -1)$ to $(1, 1)$.

**25.** $f(x, y) = xe^y$ along the curve $C$ traced out by $x = e^{2y}$ from $(1, 0)$ to $(e^2, 1)$.

**26.** $f(x, y) = x^2 y$ along the parabola $x = 2y^2$ from $(0, 0)$ to $(2\pi, \sqrt{\pi})$.

**27.** Find $\int_C (x^2 y\, dx + xy\, dy)$ along the curve $C$: $x^2 + y^2 = 1$ from $(1, 0)$ to $(0, 1)$ using the following parametric equations:

(a) $x = \cos t$, $y = \sin t$, $0 \leq t \leq \dfrac{\pi}{2}$

(b) $x = \sqrt{1 - y^2}$, $0 \leq y \leq 1$

**28.** Find $\int_C (xy\, dx + xy^2\, dy)$ along the curve $C$: $x^2 + y^2 = 9$ from $(3, 0)$ to $(-3, 0)$ traced out by the parametric equations:

(a) $x = 3\cos t$, $y = 3\sin t$, $0 \leq t \leq \pi$

(b) $y = \sqrt{9 - x^2}$, $-3 \leq x \leq 3$

**29.** Find $\int_C [y^2\, dx + (xy - x^2)\, dy]$ along each of the given curves from $(0, 0)$ to $(1, 3)$:

(a) $C$ is the line $y = 3x$.

(b) $C$ is the parabola $y^2 = 9x$.

**30.** Find $\int_C [(x^2 + y^2)\, dx + 3x^2 y\, dy]$ along each of the given curves from $(-2, 4)$ to $(2, 4)$:

(a) $C$ is the parabola $y = x^2$.

(b) $C$ is the line $y = 4$.

*In Problems 31–34, find $\int_C [(x + 2y)\, dx + (2x + y)\, dy]$ along each curve $C$.*

**31.** $C$ is the curve $y = x^2$ from $(0, 0)$ to $(1, 1)$.

**32.** $C$ is the curve $y = x^3$ from $(0, 0)$ to $(1, 1)$.

**33.** $C$ is the curve $x = \cos t$, $y = \sin t$, $0 \leq t \leq \dfrac{\pi}{2}$.

**34.** $C$ is the curve $x = 1 - t$, $y = t$, $0 \leq t \leq 1$.

*In Problems 35–38, find $\int_C x\, ds$ along each piecewise-smooth curve $C$.*

**35.**

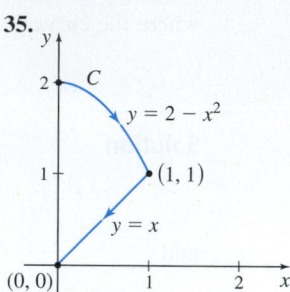

**36.**

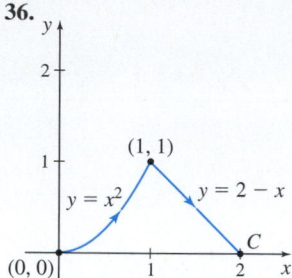

**37.**

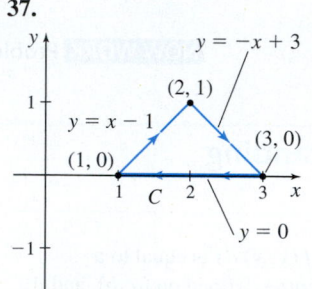

**38.**

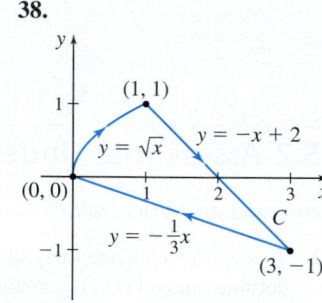

**39.** Find $\int_C [(x \cos y)\, dx - (y \sin x)\, dy]$, where $C$ consists of line segments connecting the points $(0, 0)$, $(1, 0)$, $(1, 1)$, and $(0, 1)$, in that order.

**40.** Find $\int_C [(x \sin y)\, dx - (y \cos x)\, dy]$, where $C$ consists of line segments connecting the points $(0, 0)$, $(0, 1)$, $(1, 1)$, and $(1, 0)$, in that order.

**41.** Find $\int_C (yz\, dx + xz\, dy + xy\, dz)$, where $C$ consists of line segments connecting the points $(0, 0, 0)$, $(1, 0, 0)$, $(1, 1, 0)$, and $(1, 1, 1)$, in that order.

**42.** Find $\int_C [yz\, dx + (y + zx)\, dy + xz\, dz]$, where $C$ consists of line segments connecting the points $(0, 0, 0)$, $(1, 0, 0)$, $(1, 2, 0)$, and $(1, 2, 1)$, in that order.

*In Problems 43–46, find $\int_C [y\, dx + (x - 16y)\, dy]$ along each of the given curves from $(2, 0)$ to $(0, 4)$.*

**43.** The line segment joining the two points

**44.** The parabola $y = 4 - x^2$

**45.** The line segment from $(2, 0)$ to $(2, 2)$, followed by the line segment from $(2, 2)$ to $(0, 4)$.

**46.** The parabola $y = 4 - x^2$ from $(2, 0)$ to $(1, 3)$, followed by the line segment from $(1, 3)$ to $(0, 4)$.

**Mass of a Wire**    *In Problems 47–50, find the mass M of a thin wire in the shape of the curve C whose variable mass density is $\rho$.*

**47.** $C: y = x^2, 1 \le x \le 2;$    $\rho = \rho(x, y) = 4x$

**48.** $C: x = 3 \sin t, y = 3 \cos t, 0 \le t \le \dfrac{\pi}{2};$    $\rho = \rho(x, y) = x + y + 1$

**49.** $C: x = 2t^2, y = t, -2 \le t \le 3;$    $\rho(x, y) = 3x$

**50.** $C: x = t, y = t^2 - 1, 0 \le t \le 3;$    $\rho(x, y) = x(y + 1)$

**Lateral Surface Area**    *In Problems 51–54, find the lateral surface area A of the cylinder that lies above the xy-plane and below the surface $z = f(x, y)$ and is formed by line segments parallel to the z-axis from the smooth curve C to the surface.*

**51.** $z = f(x, y) = 1 - x^2;$    $C: x = \sin t, y = \cos t, 0 \le t \le 2\pi$

**52.** $z = f(x, y) = 1 - y^2;$    $C: x = \sin t, y = \cos t, 0 \le t \le 2\pi$

**53.** $z = f(x, y) = 2xy;$    $C: y = \sqrt{1 - x^2}, 0 \le x \le 1$

**54.** $z = f(x, y) = x + y;$    $C: y = \sqrt{1 - x^2}, 0 \le x \le 1$

*In Problems 55–58, find*

$$\int_C [P(x, y, z)\, dx + Q(x, y, z)\, dy + R(x, y, z)dz]$$

*along each smooth space curve C.*

**55.** $\int_C [xy\, dx + x^2\, dy + xyz\, dz]$, where $C$ is defined by the parametric equations: $x = e^t, y = e^{-t}, z = t^2, 0 \le t \le 1$.

**56.** $\int_C [y\, dx + x^2 y\, dy + z\, dz]$, where $C$ is defined by the parametric equations: $x = e^{2t}, y = e^t, z = t, 0 \le t \le 1$.

**57.** $\int_C [xy\, dx - y\, dy + z\, dz]$, where $C$ is the line segment from $(0, 0, 0)$ to $(1, 3, 2)$.

**58.** $\int_C [xy\, dx - y\, dy + z\, dz]$, where $C$ is the line segment from $(0, 0, 0)$ to $(1, 2, 3)$.

**59.** Find $\int_C (x + y + z)\, ds$ along the smooth space curve $C$ defined by the parametric equations

$$x(t) = \cos t \quad y(t) = t \quad z(t) = \sin t \quad 0 \le t \le \frac{\pi}{4}$$

**60.** Find $\int_C (x^2 + y^2 + z^2)\, ds$ along the smooth space curve $C$ defined by the parametric equations

$$x(t) = \sin t \quad y(t) = \cos t \quad z(t) = t \quad 0 \le t \le \frac{\pi}{2}$$

**Applications and Extensions**

**61.** Find $\int_C \dfrac{ds}{x^2 + y^2 + z^2}$, where $C$ is the helix

$$x = a \cos t, \quad y = a \sin t, \quad z = bt, \quad 0 \le t \le 1$$

**62.** Find $\int_C (y^n\, dx + x^n\, dy)$, where $C$ is the ellipse

$$x = a \sin t, \quad y = b \cos t, \quad 0 \le t \le 2\pi, n \ge 1 \text{ is an integer}$$

**63.** Find $\int_C \sqrt{y}\, ds$, where $C$ is the first arch of the cycloid

$$x = a(t - \sin t), \quad y = a(1 - \cos t)$$

**64.** Find $\int_C \sqrt{x^2 + y^2}\, ds$, where $C$ is the curve

$$x = a(\cos t + t \sin t), \quad y = a(\sin t - t \cos t), \quad 0 \le t \le 2\pi, a > 0$$

**65.** Find $\int_C (z\, dx + x\, dy + y\, dz)$, where $C$ is the circular helix

$$x = a \cos t, \quad y = a \sin t, \quad z = t, 0 \le t \le 2\pi$$

**66.** Find $\int_C [xz\, dx + (y + z)\, dy + x\, dz]$, where $C$ is the curve

$$x = e^t, \quad y = e^{-t}, \quad z = e^{2t} \quad \text{from } t = 0 \text{ to } t = 1$$

**67.** **Mass of a Wire**    A spring is made of a thin wire twisted into the shape of a circular helix $x = 2 \cos t, y = 2 \sin t, z = t$. Find the mass of two turns of the spring if the wire has constant mass density $\rho$.

**68.** **Mass of a Wire**    A spring is made of a thin wire twisted into the shape of a circular helix $x = 3 \sin t, y = 3 \cos t, z = t$. Find the mass of two turns of the spring if the wire has variable mass density $\rho(x, y, z) = z + 2$.

**69.** **Mass of a Wire in Space**    Write a formula similar to $M = \int_C \rho(x, y)\, ds$ for the mass of a thin wire in the shape of a smooth space curve $C$ whose variable mass density is $\rho = \rho(x, y, z)$.

*In Problems 70–72, use the formula developed in Problem 69 to find the mass of each wire.*

**70.** **Mass of a Wire in Space**    The thin wire is in the shape of the helix $C: x = 2 \cos t, y = 2 \sin t, z = 5t, 0 \le t \le 2\pi$, whose mass density is $\rho(x, y, z) = k$, a constant.

**71.** **Mass of a Wire in Space**    The thin wire is in the shape of a helix $C: x = 3 \cos t, y = 3 \sin t, z = 4t, 0 \le t \le 2\pi$, whose variable mass density is $\rho(x, y, z) = 2 + z$.

**72.** **Mass of a Wire in Space**    The thin wire is in the shape of a helix $C: x = 2 \cos t, y = 2 \sin t, z = 5t, 0 \le t \le 2\pi$, if the mass density is proportional to the square of the distance from the origin.

**73.** **Lateral Surface Area**    Derive the formula $A = 2\pi Rh$ for the surface area of a right circular cylinder using the line integral $\int_C f(x, y)\, ds$, where $z = f(x, y) = h, h > 0$, and $C$ is the circle $x^2 + y^2 = R^2$.

**74.** Let $C_1$ be the curve $x = \cos \theta, y = \sin \theta, 0 \le \theta \le 4\pi$, and let $C_2$ be the curve $x = \cos t^2, y = \sin t^2, 0 \le t \le 2\sqrt{\pi}$. For $P(x, y) = x^3 y$ and $Q(x, y) = y^2 x$, show that $\int_{C_1} (P\, dx + Q\, dy) = \int_{C_2} (P\, dx + Q\, dy)$. Explain why this is so.

**Challenge Problems**

**75.** Find $\int_C (x + y)\, ds$, where $C$ is the curve $x = t, y = \dfrac{3t^2}{\sqrt{2}}, 0 \le t \le 1$.

**76.** **Center of Mass**    Find the center of mass of a homogeneous wire in the shape of the hypocycloid

(a) $x = a \cos^3 t, y = a \sin^3 t, 0 \le t \le \dfrac{\pi}{2}$

(b) $x = a \cos^3 t, y = a \sin^3 t, 0 \le t \le 2\pi$

**77.** Suppose the line integral of $f(x, y)$ along the curve $C$ exists.

(a) Use the Mean Value Theorem to derive the formula

$$\int_C f(x, y)\, ds = \int_a^b f(x(t), y(t)) \sqrt{\left(\frac{dx}{dt}\right)^2 + \left(\frac{dy}{dt}\right)^2}\, dt$$

(b) Apply the formula in (a) to the quantity $S(t) = \int_a^t \sqrt{(x')^2 + (y')^2}\, du$, where $S'(t_0)$ is the length of the portion of the curve traced out as $t$ moves from $a$ to $t_0$.

**78.** It follows from Problem 77 that the value of a line integral of a function along a smooth curve, given the hypotheses of the theorem, is independent of the parametrization. Why? In particular, if $x$ and $y$ are differentiable functions of the arc length as one moves along the curve, describe the form of the integral

$$\int_C f(x, y)\, ds = \int_a^b f(x(t), y(t))\sqrt{\left(\frac{dx}{dt}\right)^2 + \left(\frac{dy}{dt}\right)^2}\, dt$$

when $s$ is used as the parameter.

**79.** We have seen two kinds of line integrals: $\int_C (P\, dx + Q\, dy)$ and $\int_C f\, ds$. These are closely connected.

**(a)** If $C$ is expressed using arc length as the parameter, show that $\int_C (P dx + Q dy)$ can be written in the form $\int_C f\, ds$.

**(b)** If the arc length $s$ along $C$ can be expressed as a function of $x$ and $y$, show that $\int_C f\, ds$ can be written in the form $\int_C (P dx + Q dy)$. For example,

$$\int_C (x + y^2)\, ds = \int_C \left[ \frac{x^2 + xy^2}{\sqrt{x^2 + y^2}}\, dx + \frac{xy + y^3}{\sqrt{x^2 + y^2}}\, dy \right]$$

where $C$ is the curve $x = t$, $y = t$, for $0 \le t \le 1$.

---

# 15.3 Line Integrals of Vector Fields; Work

**OBJECTIVES** *When you finish this section, you should be able to:*

1 Find the line integral of a vector field (p. 1040)
2 Compute work (p. 1043)

---

## 1 Find the Line Integral of a Vector Field

The line integral $\int_C (P\, dx + Q\, dy)$ can be written compactly using vectors. Suppose

$$\mathbf{F} = \mathbf{F}(x, y) = P(x, y)\,\mathbf{i} + Q(x, y)\,\mathbf{j}$$

is a vector field that is continuous on some open region containing the smooth curve $C$ traced out by $\mathbf{r} = \mathbf{r}(t) = x(t)\,\mathbf{i} + y(t)\,\mathbf{j}$, $a \le t \le b$. Then $d\mathbf{r} = dx\,\mathbf{i} + dy\,\mathbf{j}$ and the dot product $\mathbf{F} \cdot d\mathbf{r}$ is

**NEED TO REVIEW?** The dot product is discussed in Section 10.4, pp. 754–755.

$$\mathbf{F} \cdot d\mathbf{r} = [P(x, y)\,\mathbf{i} + Q(x, y)\,\mathbf{j}] \cdot [dx\,\mathbf{i} + dy\,\mathbf{j}] = P(x, y)\, dx + Q(x, y)\, dy$$

So

$$\boxed{\int_C \mathbf{F} \cdot d\mathbf{r} = \int_C (P\, dx + Q\, dy)}$$

---

**EXAMPLE 1** **Finding the Line Integral of a Vector Field**

Find

$$\int_C \mathbf{F} \cdot d\mathbf{r}$$

if $\mathbf{F}(x, y) = x\mathbf{i} + xy\mathbf{j}$ and the curve $C$ is traced out by the vector function $\mathbf{r}(t) = t\,\mathbf{i} + t^2\mathbf{j}$, $0 \le t \le 2$.

**Solution** The vector field $\mathbf{F}$ is continuous at every point in the $xy$-plane. Parametric equations of the curve $C$ are

$$x(t) = t \qquad y(t) = t^2 \qquad 0 \le t \le 2$$

So,

$$\mathbf{F} = x\mathbf{i} + xy\mathbf{j} = t\mathbf{i} + t \cdot t^2\mathbf{j} = t\mathbf{i} + t^3\mathbf{j}$$

and

$$d\mathbf{r} = \frac{d\mathbf{r}}{dt}\, dt = \frac{d}{dt}(t\,\mathbf{i} + t^2\mathbf{j})\, dt = (\mathbf{i} + 2t\mathbf{j})\, dt$$

Then

$$\mathbf{F} \cdot d\mathbf{r} = (t\mathbf{i} + t^3\,\mathbf{j}) \cdot (\mathbf{i} + 2t\,\mathbf{j})\,dt = (t + 2t^4)\,dt$$

so that

$$\int_C \mathbf{F} \cdot d\mathbf{r} = \int_0^2 (t + 2t^4)\,dt = \frac{74}{5}$$  ∎

**NOW WORK** Problem 3.

Let $C$ be a piecewise-smooth curve consisting of $n \geq 2$ smooth curves $C_1, C_2, \ldots, C_n$. Let $P = P(x, y)$ and $Q = Q(x, y)$ be functions of two variables that are each continuous on some open region containing $C$.

If $\mathbf{F} = P\mathbf{i} + Q\mathbf{j}$ is a vector field and $C$ is traced out by $\mathbf{r} = \mathbf{r}(t) = x(t)\mathbf{i} + y(t)\mathbf{j}$, $a \leq t \leq b$, then

$$\boxed{\int_C \mathbf{F} \cdot d\mathbf{r} = \int_{C_1} \mathbf{F} \cdot d\mathbf{r} + \int_{C_2} \mathbf{F} \cdot d\mathbf{r} + \cdots + \int_{C_n} \mathbf{F} \cdot d\mathbf{r}}$$

**EXAMPLE 2  Finding the Line Integral of a vector Field Along a Piecewise-Smooth Curve**

Find $\int_C \mathbf{F} \cdot d\mathbf{r}$ if $\mathbf{F}(x, y) = x \sin y\mathbf{i} + \cos y\mathbf{j}$ is a vector field and $C$ is the piecewise-smooth curve traced out by the vector functions $\mathbf{r}_1(t) = t\mathbf{i} + t^2\mathbf{j}$, $0 \leq t \leq 1$, and $\mathbf{r}_2(t) = t\mathbf{i} + t\mathbf{j}$, $1 \leq t \leq \pi$.

**Solution** The vector field $\mathbf{F}$ is continuous at every point in the $xy$-plane. Parametric equations of the vector function $\mathbf{r}_1$ are

$$C_1: \qquad x(t) = t \qquad y(t) = t^2 \qquad 0 \leq t \leq 1$$

So, $\mathbf{F}(x, y) = x \sin y\,\mathbf{i} + \cos y\,\mathbf{j} = t \sin t^2\,\mathbf{i} + \cos t^2\,\mathbf{j}$ and $d\mathbf{r}_1 = \dfrac{d\mathbf{r}_1}{dt}\,dt = (\mathbf{i} + 2t\mathbf{j})\,dt$.

Then $\mathbf{F} \cdot d\mathbf{r}_1 = (t \sin t^2\,\mathbf{i} + \cos t^2\,\mathbf{j}) \cdot (\mathbf{i} + 2t\,\mathbf{j})\,dt = (t \sin t^2 + 2t \cos t^2)\,dt$, and

$$\int_{C_1} \mathbf{F} \cdot d\mathbf{r} = \int_0^1 (t \sin t^2 + 2t \cos t^2)\,dt \underset{\substack{\uparrow \\ u = t^2;\, du = 2t\,dt}}{=} \int_0^1 \left(\frac{1}{2} \sin u + \cos u\right)\,du$$

$$= \left[-\frac{1}{2}\cos u + \sin u\right]_0^1 = -\frac{1}{2}\cos 1 + \sin 1 + \frac{1}{2}$$

Parametric equations of the vector function $\mathbf{r}_2$ are

$$C_2: \qquad x(t) = t \qquad y(t) = t \qquad 1 \leq t \leq \pi$$

So, $\mathbf{F}(x, y) = x \sin y\,\mathbf{i} + \cos y\,\mathbf{j} = t \sin t\,\mathbf{i} + \cos t\,\mathbf{j}$ and $d\mathbf{r}_2 = \dfrac{d\mathbf{r}_2}{dt}\,dt = (\mathbf{i} + \mathbf{j})\,dt$.

Then $\mathbf{F} \cdot d\mathbf{r}_2 = (t \sin t\,\mathbf{i} + \cos t\,\mathbf{j}) \cdot (\mathbf{i} + \mathbf{j})\,dt = (t \sin t + \cos t)\,dt$, and

$$\int_{C_2} \mathbf{F} \cdot d\mathbf{r} = \int_1^\pi (t \sin t + \cos t)\,dt = \int_t^\pi t \sin t\,dt + \int_1^\pi \cos t\,dt$$

$$= \int_1^\pi t \sin t\,dt + [\sin t]_1^\pi = \int_1^\pi t \sin t\,dt - \sin 1$$

Use integration by parts to find $\int_1^\pi t \sin t\,dt$. Let $u = t$ and $dv = \sin t\,dt$. Then $du = dt$, $v = -\cos t$, and

$$\int_1^\pi t \sin t\,dt = [-t \cos t]_1^\pi + \int_1^\pi \cos t\,dt \qquad \int_a^b u\,dv = [uv]_a^b - \int_a^b v\,du$$

$$= [-t \cos t]_1^\pi + [\sin t]_1^\pi$$

$$= \pi + \cos 1 - \sin 1$$

So,

$$\int_{C_2} \mathbf{F} \cdot d\mathbf{r} = \int_1^\pi t \sin t \, dt - \sin 1 = \pi + \cos 1 - 2 \sin 1$$

Then

$$\int_C \mathbf{F} \cdot d\mathbf{r} = \int_{C_1} \mathbf{F} \cdot d\mathbf{r}_1 + \int_{C_2} \mathbf{F} \cdot d\mathbf{r}_2 = \left( -\frac{1}{2} \cos 1 + \sin 1 + \frac{1}{2} \right) + (\pi + \cos 1 - 2 \sin 1)$$

$$= \pi + \frac{1}{2} + \frac{1}{2} \cos 1 - \sin 1 \qquad\blacksquare$$

**NOW WORK** Problem 21.

If $\mathbf{F} = \mathbf{F}(x, y, z) = P(x, y, z)\mathbf{i} + Q(x, y, z)\mathbf{j} + R(x, y, z)\mathbf{k}$ is a vector field in space that is continuous in some open solid containing the smooth curve

$$C : \mathbf{r} = \mathbf{r}(t) = x(t)\,\mathbf{i} + y(t)\,\mathbf{j} + z(t)\,\mathbf{k} \qquad a \le t \le b$$

then

$$\boxed{\int_C \mathbf{F} \cdot d\mathbf{r} = \int_C (P\,dx + Q\,dy + R\,dz)}$$

A curve $C$ in space is piecewise-smooth if it consists of a finite number of smooth curves $C_1, C_2, \ldots, C_n$, $n \ge 2$, that are joined together end to end. If $C: \mathbf{r} = \mathbf{r}(t)$ is piecewise-smooth and if $\mathbf{F} = \mathbf{F}(x, y, z)$ is a vector field that is continuous in some open solid containing $C$, then

$$\boxed{\int_C \mathbf{F} \cdot d\mathbf{r} = \int_{C_1} \mathbf{F} \cdot d\mathbf{r} + \int_{C_2} \mathbf{F} \cdot d\mathbf{r} + \cdots + \int_{C_n} \mathbf{F} \cdot d\mathbf{r}}$$

**EXAMPLE 3** Finding a Line Integral in Space

For each curve $C$, find $\displaystyle\int_C \mathbf{F} \cdot d\mathbf{r}$ if

$$\mathbf{F}(x, y, z) = xy^2\,\mathbf{i} + x^2z\,\mathbf{j} - (y - x)\,\mathbf{k}$$

**(a)** $C_1$: $\mathbf{r}(t) = t\mathbf{i} + t\mathbf{j} + t\mathbf{k}$, $0 \le t \le 1$
**(b)** $C_2$: $\mathbf{r}(t) = t\mathbf{i} + t^2\mathbf{j} + t^3\,\mathbf{k}$, $0 \le t \le 1$

**Solution** The vector field $\mathbf{F}$ is continuous at every point $(x, y, z)$ in space.
**(a)** Parametric equations of the curve $C_1$ are

$$x = t \qquad y = t \qquad z = t \qquad 0 \le t \le 1$$

So,

$$\mathbf{F} = xy^2\mathbf{i} + x^2z\mathbf{j} - (y - x)\mathbf{k} = t^3\mathbf{i} + t^3\mathbf{j}$$

and

$$d\mathbf{r} = \frac{d\mathbf{r}}{dt}dt = (\mathbf{i} + \mathbf{j} + \mathbf{k})dt$$

Then

$$\mathbf{F} \cdot d\mathbf{r} = (t^3 + t^3)dt = 2t^3 dt$$

$$\int_{C_1} \mathbf{F} \cdot d\mathbf{r} = \int_0^1 2t^3 \, dt = \frac{1}{2}$$

**(b)** Parametric equations of the curve $C_2$ are

$$x = t \qquad y = t^2 \qquad z = t^3 \qquad 0 \le t \le 1$$

So,

$$\mathbf{F} = xy^2\,\mathbf{i} + x^2z\,\mathbf{j} - (y - x)\,\mathbf{k} = t^5\mathbf{i} + t^5\mathbf{j} - (t^2 - t)\,\mathbf{k}$$

and

$$d\mathbf{r} = \frac{d\mathbf{r}}{dt}\, dt = (\mathbf{i} + 2t\mathbf{j} + 3t^2\mathbf{k})\, dt$$

Then

$$\mathbf{F} \cdot d\mathbf{r} = [t^5\mathbf{i} + t^5\mathbf{j} - (t^2 - t)\,\mathbf{k}] \cdot (\mathbf{i} + 2t\,\mathbf{j} + 3t^2\,\mathbf{k})\, dt = [t^5 + 2t^6 - 3t^2(t^2 - t)]\, dt$$

$$= (2t^6 + t^5 - 3t^4 + 3t^3)\, dt$$

so that

$$\int_C \mathbf{F} \cdot d\mathbf{r} = \int_0^1 (2t^6 + t^5 - 3t^4 + 3t^3)\, dt = \frac{253}{420}$$    ■

**NOW WORK** Problem 7.

## 2 Compute Work

An important application of a line integral is to work. Recall that **work** is the energy transferred to or from an object by a force acting on the object. The standard units of work are listed in Table 1.

**TABLE 1**

|        | Work Units         | Force Units   | Distance Units |
|--------|--------------------|---------------|----------------|
| SI     | joule (J)          | newton (N)    | meter (m)      |
| U.S.   | foot-pound (ft-lb) | pound (lb)    | foot (ft)      |

Let's review what we already know about work:

- **Constant force:**   The work $W$ done by a *constant* force $F$ in moving an object a distance $s$ along a straight line in the direction of $F$ is defined to be

$$\boxed{W = Fs}$$

When the force $F$ acts in the same direction as the motion, the work done is positive; if the force $F$ acts in a direction opposite to the motion, the work is negative.

- **Variable force:**   In Chapter 6, the definition of work was generalized to the case where the force $F$ is variable. The work $W$ done by a variable force $F = F(x)$ acting in the direction of the motion of an object as that object moves along a straight line from $x = a$ to $x = b$ is

$$\boxed{W = \int_a^b F(x)\,dx}$$

**NEED TO REVIEW?** The work done by a constant force vector in moving an object along a line from $A$ to $B$ is discussed in Section 10.4, pp. 759–760.

- **Constant force vector:**   In Chapter 10, we investigated the work $W$ done by a constant force vector $\mathbf{F}$ acting on an object as that object moves in the direction of a vector $\mathbf{r}$ for a distance $\|\mathbf{r}\|$ and found that

$$\boxed{W = \mathbf{F} \cdot \mathbf{r}}$$

- **Variable force vector:**   Now we investigate the work done by a vector field $\mathbf{F}$, called a **force field**, acting on an object as that object moves along a smooth curve $C$ from the point $A$ on $C$ to the point $B$ on $C$.

Suppose

$$\mathbf{F} = \mathbf{F}(x, y) = P(x, y)\mathbf{i} + Q(x, y)\mathbf{j}$$

is a force field acting on an object at the point $(x, y)$ in some open region $R$. We assume the functions $P$ and $Q$ have continuous, first-order partial derivatives at each point in $R$ and that $C$ is a smooth curve lying entirely in $R$. Finally, suppose $C$ is defined by the vector function

$$\mathbf{r} = \mathbf{r}(t) = x(t)\mathbf{i} + y(t)\mathbf{j} \qquad a \le t \le b$$

Partition the closed interval $[a, b]$ into $n$ subintervals

$$[a, t_1], \quad [t_1, t_2], \quad \ldots, \quad [t_{i-1}, t_i], \quad \ldots, \quad [t_{n-1}, b]$$

and denote the length of each subinterval by $\Delta t_1, \Delta t_2, \ldots, \Delta t_n$. Corresponding to each number $a = t_0, t_1, t_2, \ldots, t_n = b$ of the partition, there is a sequence of points $A = p_0, p_1, \ldots, p_n = B$ on the curve $C$. These points subdivide the curve into $n$ subarcs of lengths $\Delta s_1, \Delta s_2, \ldots, \Delta s_n$, as shown in Figure 23.

The norm $\|\Delta\|$ of the partition is the largest length of the subarcs $\Delta s_i$, $i = 1, 2, \ldots, n$. Let $p_i^* = (x_i^*, y_i^*)$ be a point on the $i$th subarc $p_{i-1} p_i$ of $C$. If the norm $\|\Delta\|$ is small enough, then the work $W_i$ done by the force field $\mathbf{F}$ in moving an object along the arc $p_{i-1} p_i$ can be approximated by the work done by the constant force

$$\mathbf{F}_i^* = \mathbf{F}(x_i^*, y_i^*)$$

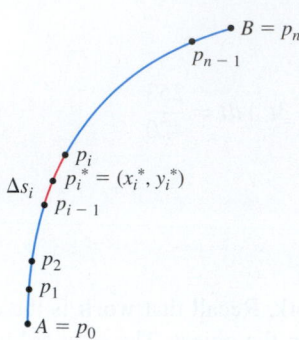

**Figure 23**

in moving the object along the directed line segment $\Delta \mathbf{r}_i = \overrightarrow{p_{i-1} p_i}$. Then

$$W_i \approx \mathbf{F}_i^* \cdot \Delta \mathbf{r}_i$$

The total work $W$ done by $\mathbf{F}$ along $C$ from the point $A : \mathbf{r} = \mathbf{r}(a)$ to the point $B : \mathbf{r} = \mathbf{r}(b)$ is $W = \lim\limits_{\|\Delta\| \to 0} \sum\limits_{i=1}^{n} W_i = \lim\limits_{\|\Delta\| \to 0} \sum\limits_{i=1}^{n} \mathbf{F}_i^* \cdot \Delta \mathbf{r}_i$. This leads to the following definition:

---

**DEFINITION Work**

Let $\mathbf{F} = \mathbf{F}(x, y) = P(x, y)\mathbf{i} + Q(x, y)\mathbf{j}$ be a force field, where the functions $P$ and $Q$ are continuous with continuous partial derivatives at each point of an open region $R$. Let $\mathbf{r} = \mathbf{r}(t)$, $a \le t \le b$, trace out a piecewise-smooth curve $C$ lying entirely in $R$. The work $W$ done by the force field $\mathbf{F}$ in moving an object along $C$ is defined as

$$\boxed{W = \int_C \mathbf{F} \cdot d\mathbf{r} = \int_C (P\,dx + Q\,dy)}$$

---

**CALC CLIP**
**EXAMPLE 4   Computing Work**

Find the work done by the force field $\mathbf{F}(x, y) = -y\mathbf{i} + x\mathbf{j}$, measured in newtons, in moving an object along the half-circle $C$, whose arc length is measured in meters, traced out by

$$\mathbf{r}(t) = \cos t\, \mathbf{i} + \sin t\, \mathbf{j} \qquad 0 \le t \le \pi$$

**Solution** On $C$, $x(t) = \cos t$ and $y(t) = \sin t$, $0 \le t \le \pi$. Then

$$\mathbf{F}(x(t), y(t)) = -y(t)\mathbf{i} + x(t)\mathbf{j} = -\sin t\, \mathbf{i} + \cos t\, \mathbf{j}$$

$$d\mathbf{r}(t) = (-\sin t\, \mathbf{i} + \cos t\, \mathbf{j})\, dt$$

$$\mathbf{F} \cdot d\mathbf{r} = (\sin^2 t + \cos^2 t)\, dt = dt$$

The work $W$ done by $\mathbf{F}$ is

$$W = \int_C \mathbf{F} \cdot d\mathbf{r} = \int_0^\pi dt = \pi \text{ joules} \qquad \blacksquare$$

See Figure 24 for the force field $\mathbf{F}(x, y) = -y\mathbf{i} + x\mathbf{j}$ and the curve $C$.

**NOW WORK** Problem 17.

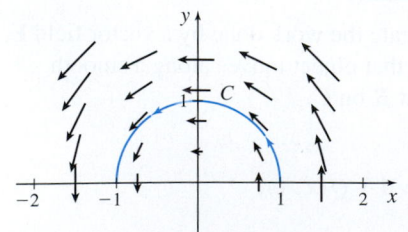

**Figure 24** $\mathbf{F}(x, y) = -y\mathbf{i} + x\mathbf{j}$

The previous discussion for finding the work done by a force field in the plane extends to force fields in space.

Suppose $\mathbf{F} = \mathbf{F}(x, y, z) = P(x, y, z)\mathbf{i} + Q(x, y, z)\mathbf{j} + R(x, y, z)\mathbf{k}$ is a force field in space, and the functions $P$, $Q$, and $R$ are continuous and have continuous first-order partial derivatives at each point in an open solid in space. If $C$ is a smooth curve $\mathbf{r} = \mathbf{r}(t) = x(t)\mathbf{i} + y(t)\mathbf{j} + z(t)\mathbf{k}$, $a \leq t \leq b$, that lies entirely within the solid, then the work $W$ done by the force field $\mathbf{F}$ in moving an object along $C$ is

$$W = \int_C \mathbf{F} \cdot d\mathbf{r} = \int_C P(x, y, z)dx + Q(x, y, z)dy + R(x, y, z)dz$$

The method for finding a line integral $\int_C \mathbf{F} \cdot d\mathbf{r}$ in space is the same as that for the plane.

**NOW WORK** Problem 25.

## 15.3 Assess Your Understanding

### Concepts and Vocabulary

1. Work is the _____ transferred to or from an object by a force acting on the object.

2. *True or False* An interpretation of the line integral $\int_C \mathbf{F} \cdot d\mathbf{r}$ is the work done by the force field $\mathbf{F}$ in moving an object along a piecewise-smooth curve $C$.

### Skill Building

*In Problems 3–10, for each vector field $\mathbf{F}$ and curve $C$: $\mathbf{r} = \mathbf{r}(t)$, find $\int_C \mathbf{F} \cdot d\mathbf{r}$.*

3. $\mathbf{F}(x, y) = (x + 2y)\mathbf{i} + (2x + y)\mathbf{j}$
   $C$ is the curve $\mathbf{r}(t) = t\mathbf{i} + t^2\mathbf{j}$, $0 \leq t \leq 1$.

4. $\mathbf{F}(x, y) = x^2 y\mathbf{i} + xy^2\mathbf{j}$
   $C$ is the curve $\mathbf{r}(t) = 2t\mathbf{i} + t^2\mathbf{j}$, $1 \leq t \leq 3$.

5. $\mathbf{F}(x, y) = xy\mathbf{i} + (x^2 + y^2)\mathbf{j}$
   $C$ is the curve $\mathbf{r}(t) = \sin t\mathbf{i} + \cos t\mathbf{j}$, $0 \leq t \leq \pi$.

6. $\mathbf{F}(x, y) = x^2\mathbf{i} + xy\mathbf{j}$
   $C$ is the curve $\mathbf{r}(t) = \sin t\mathbf{i} + \cos t\mathbf{j}$, $0 \leq t \leq \pi$.

7. $\mathbf{F}(x, y, z) = y\mathbf{i} + x^2\mathbf{j} + xyz\mathbf{k}$
   $C$ is the curve $\mathbf{r}(t) = e^t\mathbf{i} + e^{-t}\mathbf{j} + t^3\mathbf{k}$, $0 \leq t \leq 2$.

8. $\mathbf{F}(x, y, z) = y\mathbf{i} + x^2 y\mathbf{j} + z\mathbf{k}$
   $C$ is the curve $\mathbf{r}(t) = e^{2t}\mathbf{i} + e^t\mathbf{j} + t\mathbf{k}$, $0 \leq t \leq 1$.

9. $\mathbf{F}(x, y, z) = e^x\mathbf{i} + e^y\mathbf{j} + e^z\mathbf{k}$
   $C$ is the curve $\mathbf{r}(t) = t\mathbf{i} + 2t^2\mathbf{j} + t^3\mathbf{k}$, $0 \leq t \leq 2$.

10. $\mathbf{F}(x, y, z) = e^x\mathbf{i} + e^{-y}\mathbf{j} + e^{2z}\mathbf{k}$
    $C$ is the curve $\mathbf{r}(t) = 3t\mathbf{i} + t^2\mathbf{j} + t^3\mathbf{k}$, $0 \leq t \leq 1$.

*In Problems 11–20, find the work done by the force field $\mathbf{F}$ in moving an object along each curve between the indicated points.*

11. $\mathbf{F} = y\mathbf{i} + x\mathbf{j}$ along $\mathbf{r}(t) = t\mathbf{i} + t^2\mathbf{j}$ from $t = 0$ to $t = 1$

12. $\mathbf{F} = xy\mathbf{i} + y^2\mathbf{j}$ along $\mathbf{r}(t) = t\mathbf{i} + t^2\mathbf{j}$ from $t = 0$ to $t = 1$

13. $\mathbf{F} = (x - 2y)\mathbf{i} + xy\mathbf{j}$ along $\mathbf{r}(t) = 3\cos t\mathbf{i} + 2\sin t\mathbf{j}$ from $t = 0$ to $t = \dfrac{\pi}{2}$

14. $\mathbf{F} = x\mathbf{i} - y\mathbf{j}$ along $\mathbf{r}(t) = \cos t\mathbf{i} + \sin t\mathbf{j}$ from $t = 0$ to $t = 2\pi$

15. $\mathbf{F} = x^2 y\mathbf{i} + (x^2 - y^2)\mathbf{j}$ along $y = 2x^2$ from $(0, 0)$ to $(1, 2)$

16. $\mathbf{F} = y\sin x\mathbf{i} - x\cos y\mathbf{j}$ along $y = x$ from $(0, 0)$ to $(1, 1)$

17. $\mathbf{F} = (y - x^2)\mathbf{i} + x\mathbf{j}$ along the upper half of the circle $x^2 + y^2 = 1$ from $(1, 0)$ to $(-1, 0)$

18. $\mathbf{F} = (y - x^2)\mathbf{i} + x\mathbf{j}$ along the line segments joining $(1, 0)$ to $(1, 1)$ to $(-1, 1)$ to $(-1, 0)$

19. $\mathbf{F} = y^3\mathbf{i} + x^3\mathbf{j}$ along the ellipse $\mathbf{r}(t) = a\cos t\mathbf{i} + b\sin t\mathbf{j}$ from $t = 0$ to $t = 2\pi$

20. $\mathbf{F} = -y^2\mathbf{i} + x^2\mathbf{j}$ along the upper half of the ellipse $\dfrac{x^2}{a^2} + \dfrac{y^2}{b^2} = 1$ from $(a, 0)$ to $(-a, 0)$

*In Problems 21–24, find the work done by each force field $\mathbf{F}$ in moving an object along the curve pictured in each graph.*

21. $\mathbf{F} = 3x\mathbf{i} + 2y\mathbf{j}$

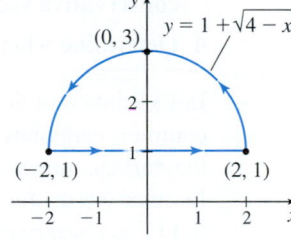

22. $\mathbf{F} = 2xy\mathbf{i} + xy^2\mathbf{j}$

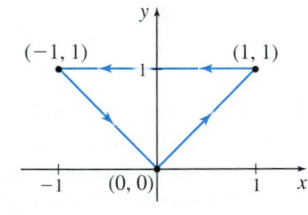

23. $\mathbf{F} = e^x y\mathbf{i} + e^x\mathbf{j}$

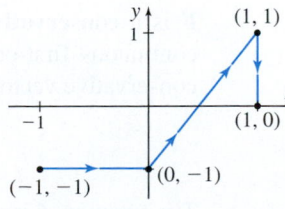

24. $\mathbf{F} = e^x\mathbf{i} + 5y\mathbf{j}$

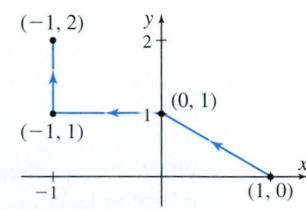

*In Problems 25–28, find the work done by the force field $\mathbf{F}(x, y, z) = xy\mathbf{i} - y\mathbf{j} + z\mathbf{k}$ in moving an object along $C$ between the indicated points.*

25. $C$ is the line segment from $(0, 0, 0)$ to $(1, 2, 3)$.

26. $C$ is the line segment from $(1, 0, 0)$ to $(1, 2, 3)$.

---

**1.** = NOW WORK problem      〰 = Graphing technology recommended     CAS = Computer Algebra System recommended

**27.** $C$ is the line segment from $(0, 2, 0)$ to $(1, 2, 3)$.

**28.** $C$ is the line segment from $(0, 1, 2)$ to $(1, 2, 3)$.

## Applications and Extensions

**29. Work** Find the work done by the force field
$\mathbf{F}(x, y) = \cos x\,\mathbf{i} + \sin y\,\mathbf{j}$ to move an object along the curve $C$
traced out by the parametric equations $x = t$, $y = \sqrt{t}$, $0 \le t \le \pi$.

**30. Work** Find the work done by a force field $\mathbf{F}(x, y) = x^2\,\mathbf{i} + y\,\mathbf{j}$
to move an object along $C$ traced out by the parametric equations
$x = t$, $y = t^2$, $0 \le t \le 3$.

**31. Work** An object moves along the curve
$\mathbf{r}(t) = 64\sqrt{3}t\,\mathbf{i} + (64t - 16t^2)\mathbf{j}$ from $t = 0$ to $t = 4$ and is
acted upon by a force field $\mathbf{F}$ whose magnitude is directly
proportional to the speed of the object and whose direction
is opposite to that of the velocity. Find the work $W$ done by
this force.

**32. Work** An object moves in a clockwise direction from the
origin to the point $(2a, 0)$ along the upper half of the
circle $(x - a)^2 + y^2 = a^2$ and is acted upon by a force field $\mathbf{F}$ with
constant magnitude 3 and with direction $\mathbf{i} + \mathbf{j}$. Find the work $W$
done by this force.

**33. Molecular Work** The repelling force between a charged
particle $P$ at the origin and an oppositely charged particle $Q$
at $(x, y)$ is

$$\mathbf{F}(x, y) = \frac{x}{(x^2 + y^2)^{3/2}}\mathbf{i} + \frac{y}{(x^2 + y^2)^{3/2}}\mathbf{j}$$

Find the work done by $\mathbf{F}$ as $Q$ moves along the line segment from
$(1, 0)$ to $(-1, 2)$.

**34. Minimizing Work** An object moves from the point $(0, 0)$ to the
point $(1, 0)$ along the curve $y = ax(1 - x)$ and is acted upon by a
force given by $\mathbf{F}(x, y) = (y^2 + 1)\mathbf{i} + (x + y)\mathbf{j}$. Find $a$ so that the
work $W$ done is a minimum.

**35. Work** Find the work $W$ done by the force field
$\mathbf{F}(x, y, z) = x^2\mathbf{i} + y^2\mathbf{j} + z^2\mathbf{k}$ in moving an object along the
helix $\mathbf{r} = \mathbf{r}(t) = 2\cos t\,\mathbf{i} + 2\sin t\,\mathbf{j} + 2t\mathbf{k}$, $0 \le t \le \pi$.

**36. Work** Find the work $W$ done by the force field
$\mathbf{F}(x, y, z) = x\mathbf{i} + y\mathbf{j} + \mathbf{k}$ in moving an object along the
helix $\mathbf{r} = \mathbf{r}(t) = \sin t\,\mathbf{i} + \cos t\,\mathbf{j} + t\mathbf{k}$, $0 \le t \le \pi$.

**37. Work** Suppose sea level is the $xy$-plane and the $z$-axis points
upward. Then the gravitational force on an object of mass $m$
near sea level may be taken to be $\mathbf{F} = mg\mathbf{k}$, where $g$ is the
acceleration due to gravity. Show that the work done by gravity
on an object moving from $(x_1, y_1, z_1)$ to $(x_2, y_2, z_2)$ along any
path is $W = mg(z_2 - z_1)$.

# 15.4 Fundamental Theorem of Line Integrals

**OBJECTIVES** *When you finish this section, you should be able to:*

**1** Identify a conservative vector field and its potential function (p. 1047)

**2** Use the Fundamental Theorem of Line Integrals (p. 1047)

**3** Reconstruct a function from its gradient: Finding the potential function for a
conservative vector field (p. 1051)

**4** Determine whether a vector field is conservative (p. 1053)

**NOTE** In Sections 15.4 and 15.5, we
only discuss vector fields in two
dimensions. Vector fields in space are
discussed in Sections 15.8 and 15.9.

In Example 6 of Section 15.2, we found a line integral along two different curves with
common endpoints and obtained two different values. But there are situations where
the curve connecting the two endpoints does not affect the value of the line integral.
In other words, the line integral is *independent of the path*. As it turns out, if a vector
field $\mathbf{F}$ is *conservative*, then $\int_C \mathbf{F} \cdot d\mathbf{r}$ is independent of the path.

**DEFINITION** Conservative Vector Field; Potential Function

$\mathbf{F}$ is a **conservative vector field** if $\mathbf{F}$ is the gradient of some function $f$ that has
continuous first-order partial derivatives on some open region $R$. That is, $\mathbf{F}$ is a
conservative vector field if

$$\boxed{\mathbf{F} = \nabla f}$$

**NEED TO REVIEW?** The gradient
is discussed in Section 13.1,
pp. 911–917.

The function $f$ is called a **potential function** for $\mathbf{F}$.

In Chapter 4 we called a function $f$ an antiderivative of a function $F$ if the
derivative of $f$ equals $F$. In the definition of a conservative vector field we call $f$ a
potential function of the vector field $\mathbf{F}$ if the gradient of $f$ equals $\mathbf{F}$.

## 1 Identify a Conservative Vector Field and Its Potential Function

> **EXAMPLE 1** **Identifying a Conservative Vector Field and Its Potential Function**
>
> $\mathbf{F}(x, y) = 2xy\mathbf{i} + x^2\mathbf{j}$ is a conservative vector field, since $\mathbf{F}$ is the gradient of the function $f(x, y) = x^2 y$. That is,
>
> $$\nabla f(x, y) = f_x(x, y)\mathbf{i} + f_y(x, y)\mathbf{j} = 2xy\mathbf{i} + x^2\mathbf{j} = \mathbf{F}(x, y)$$
>
> The function $f(x, y) = x^2 y$ is the potential function for $\mathbf{F}$. ∎

The next result provides a direct way to find the line integral $\int_C \mathbf{F} \cdot d\mathbf{r}$ when $\mathbf{F}$ is a conservative vector field.

## 2 Use the Fundamental Theorem of Line Integrals

> **THEOREM Fundamental Theorem of Line Integrals**
>
> Let $\mathbf{F} = \mathbf{F}(x, y) = P(x, y)\mathbf{i} + Q(x, y)\mathbf{j}$ be a vector field, where the functions $P(x, y)$ and $Q(x, y)$ are continuous on some open region $R$ containing the points $(x_0, y_0)$ and $(x_1, y_1)$. Let $C$ be a piecewise-smooth curve beginning at the point $(x_0, y_0)$ and ending at the point $(x_1, y_1)$ that lies entirely in $R$. If $\mathbf{F}$ is a conservative vector field on $R$, that is, if $\mathbf{F}$ is the gradient of some function $f$ so that
>
> $$\boxed{\mathbf{F}(x, y) = \nabla f(x, y)}$$
>
> throughout $R$, then
>
> $$\boxed{\int_C \mathbf{F} \cdot d\mathbf{r} = \int_C \nabla f \cdot d\mathbf{r} = \int_{(x_0, y_0)}^{(x_1, y_1)} \nabla f \cdot d\mathbf{r} = \left[ f(x, y) \right]_{(x_0, y_0)}^{(x_1, y_1)} = f(x_1, y_1) - f(x_0, y_0)}$$
>
> $$\boxed{\int_C (P\, dx + Q\, dy) = \int_{(x_0, y_0)}^{(x_1, y_1)} (P\, dx + Q\, dy) = \left[ f(x, y) \right]_{(x_0, y_0)}^{(x_1, y_1)} = f(x_1, y_1) - f(x_0, y_0)}$$

**NOTE** In Chapter 5, the Fundamental Theorem of Calculus states that if $F$ is an antiderivative of $f$, then $\int_a^b f(x)\, dx = F(b) - F(a)$. In other words the value of the integral depends only on the limits of integration. In the Fundamental Theorem of Line Integrals, the potential function $f$ of a vector field $\mathbf{F}$ plays the role of an antiderivative.

**Proof** This proof is for a smooth curve $C$. (The proof for a piecewise-smooth curve is obtained by considering one piece of the curve at a time.) Let $x = x(t)$ and $y = y(t)$, $a \le t \le b$, be parametric equations of a smooth curve $C$. The initial point and endpoint of $C$ can then be written as

$$(x_0, y_0) = (x(a), y(a)) \qquad \text{and} \qquad (x_1, y_1) = (x(b), y(b))$$

Since $\mathbf{F} = P(x, y)\mathbf{i} + Q(x, y)\mathbf{j}$ is a conservative vector field, then $\mathbf{F} = \nabla f = \dfrac{\partial f}{\partial x}\mathbf{i} + \dfrac{\partial f}{\partial y}\mathbf{j}$ for some function $f$. Since $\mathbf{F} = P(x, y)\mathbf{i} + Q(x, y)\mathbf{j}$, it follows that $P(x, y) = \dfrac{\partial f}{\partial x}$ and $Q(x, y) = \dfrac{\partial f}{\partial y}$. Then

$$\int_C (P\, dx + Q\, dy) = \int_C \left( \frac{\partial f}{\partial x}\, dx + \frac{\partial f}{\partial y}\, dy \right) = \int_a^b \left( \frac{\partial f}{\partial x} \frac{dx}{dt} + \frac{\partial f}{\partial y} \frac{dy}{dt} \right) dt$$

$$= \int_a^b \frac{d}{dt}[f(x(t), y(t))]\, dt \quad \text{Use the Chain Rule I: } \frac{d}{dt} f(x(t), y(t)) = \frac{\partial f}{\partial x} \frac{dx}{dt} + \frac{\partial f}{\partial y} \frac{dy}{dt}.$$

$$= \left[ f(x(t), y(t)) \right]_{t=a}^{t=b} \qquad \text{Use the Fundamental Theorem of Calculus.}$$

$$= f(x(b), y(b)) - f(x(a), y(a)) = f(x_1, y_1) - f(x_0, y_0)$$

∎

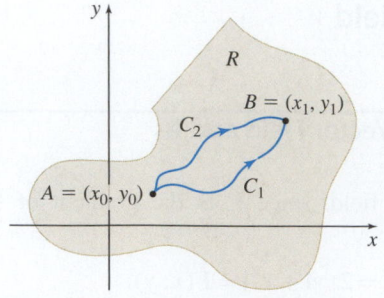

**Figure 25**

In the Fundamental Theorem of Line Integrals, the value of the line integral *depends only on the endpoints* $(x_0, y_0)$ *and* $(x_1, y_1)$ *of the curve* $C$. This means that the value of the line integral does not change if $C$ is replaced by any other piecewise-smooth curve in $R$, provided the new curve connects $(x_0, y_0)$ to $(x_1, y_1)$. Figure 25 illustrates this for two curves $C_1$ and $C_2$ joining $A = (x_0, y_0)$ and $B = (x_1, y_1)$. As a result, line integrals $\int_C \mathbf{F} \cdot d\mathbf{r}$ for which the conditions of the Fundamental Theorem of Line Integral are met are said to be **independent of the path**. In other words, for a conservative vector field $\mathbf{F}$, the line integral $\int_C \mathbf{F} \cdot d\mathbf{r}$ has the same value no matter what path $C$ is taken from $A = (x_0, y_0)$ to $B = (x_1, y_1)$. To summarize:

If $\mathbf{F} = P\mathbf{i} + Q\mathbf{j} = \nabla f$, that is, if $\mathbf{F}$ is conservative, then:

- $\int_C (P\,dx + Q\,dy) = f(x_1, y_1) - f(x_0, y_0)$ for any piecewise-smooth curve $C$ joining $A = (x_0, y_0)$ and $B = (x_1, y_1)$.
- $\int_{C_1} (P\,dx + Q\,dy) = \int_{C_2} (P\,dx + Q\,dy)$ for any piecewise-smooth curves $C_1$ and $C_2$ joining $A$ and $B$.

### EXAMPLE 2  Using the Fundamental Theorem of Line Integrals

$\mathbf{F} = \mathbf{F}(x, y) = (2xy + 24x)\mathbf{i} + (x^2 + 16)\mathbf{j}$ is a conservative vector field, since $\mathbf{F}$ is the gradient of $f(x, y) = x^2 y + 12x^2 + 16y$. Use this fact to find

$$\int_C \left[(2xy + 24x)\,dx + (x^2 + 16)\,dy\right]$$

where $C$ is any piecewise-smooth curve joining the points $(1, 1)$ and $(2, 4)$.

**Solution**  The vector field $\mathbf{F}$ is continuous at every point $(x, y)$ in the plane. We use two methods to find $\int_C [(2xy + 24x)\,dx + (x^2 + 16)\,dy]$.

- **Method I**  Use the potential function $f(x, y) = x^2 y + 12x^2 + 16y$ whose gradient is

$$\nabla f = (2xy + 24x)\mathbf{i} + (x^2 + 16)\mathbf{j} = \mathbf{F}(x, y)$$

Then by the Fundamental Theorem of Line Integrals.

$$\int_C [(2xy + 24x)\,dx + (x^2 + 16)\,dy] = \left[f(x, y)\right]_{(1,1)}^{(2,4)} = f(2, 4) - f(1, 1)$$

$$= (2^2 \cdot 4 + 12 \cdot 2^2 + 16 \cdot 4) - (1^2 \cdot 1 + 12 \cdot 1^2 + 16 \cdot 1) = 99$$

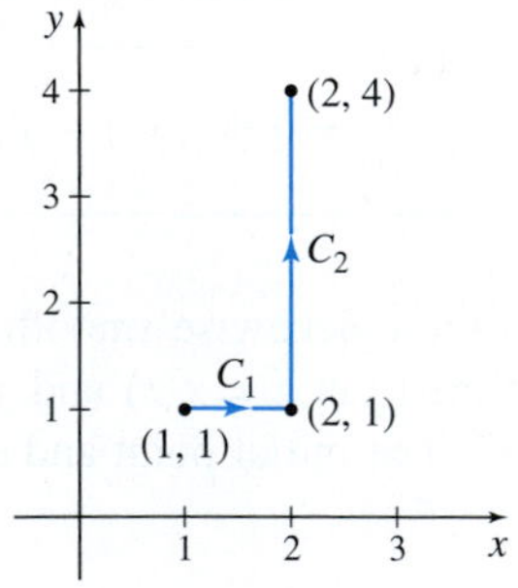

- **Method II**  Use the fact that the given line integral is independent of the path, so it can be integrated along *any* piecewise-smooth curve joining $(1, 1)$ and $(2, 4)$. We choose the path shown in Figure 26 since it makes the integration easy. Along $C_1$, $y = 1$, $dy = 0$, and $1 \leq x \leq 2$; along $C_2$, $x = 2$, $dx = 0$, and $1 \leq y \leq 4$. Then

$$\int_C [(2xy + 24x)\,dx + (x^2 + 16)\,dy] = \int_{C_1} [(2xy + 24x)\,dx + (x^2 + 16)\,dy] \qquad y = 1, dy = 0, 1 \leq x \leq 2$$

$$+ \int_{C_2} [(2xy + 24x)\,dx + (x^2 + 16)\,dy] \qquad x = 2, dx = 0, 1 \leq y \leq 4$$

$$= \int_1^2 (2x + 24x)\,dx + \int_1^4 (4 + 16)\,dy$$

$$= 39 + 60 = 99 \qquad \blacksquare$$

$C_1: 1 \leq x \leq 2, y = 1$
$C_2: x = 2, 1 \leq y \leq 4$

**Figure 26**

Method I is easier to use than Method II, but this is because the potential function $f$ of $\mathbf{F}$ is given. If the potential function $f$ had not been given, then we would have had to find it or use Method II. We discuss how to find potential functions shortly.

**NOW WORK** Problems **15** and **19**.

So, to find the line integral $\int_C \mathbf{F} \cdot d\mathbf{r} = \int_C (P\,dx + Q\,dy)$, where $\mathbf{F} = P\mathbf{i} + Q\mathbf{j}$ is a conservative vector field, either of the methods outlined in Example 2 can be used.

**Finding $\int_C \mathbf{F} \cdot d\mathbf{r} = \int_C (P\,dx + Q\,dy)$ for a Conservative Vector Field F**

**Method I:** Use the Fundamental Theorem of Line Integrals directly. That is, if $\mathbf{F} = \nabla f$,

$$\int_C \mathbf{F} \cdot d\mathbf{r} = \int_C (P\,dx + Q\,dy) = f(x_1, y_1) - f(x_0, y_0)$$

**Method II:** Use the fact that $\int_C (P\,dx + Q\,dy)$ is independent of the path and select some suitable path joining the endpoints of $C$ that makes finding the line integral easy.

Care must be taken when using Method II. For example, suppose we are asked to find

$$\int_C \frac{-y\,dx + x\,dy}{x^2}$$

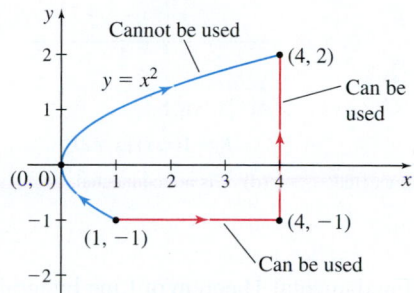

**Figure 27**

along the curve $C$ joining the points $(1, -1)$ and $(4, 2)$. $\mathbf{F} = \mathbf{F}(x, y) = \dfrac{-y\mathbf{i} + x\mathbf{j}}{x^2}$ is a conservative vector field since $\mathbf{F}$ is the gradient of the function $f(x, y) = \dfrac{y}{x}$.

To use Method II to find the line integral, we cannot choose a path (such as $x = y^2$) that intersects the $y$-axis, because $P = -\dfrac{y}{x^2}$ and $Q = \dfrac{1}{x}$ are not defined at $x = 0$. But any path that lies entirely in the first and fourth quadrants can be used. See Figure 27.

A special case of the Fundamental Theorem of Line Integrals occurs when $C$ is a **closed curve,** that is, when the initial point $(x_0, y_0)$ equals the terminal point $(x_1, y_1)$. Then $f(x_1, y_1) = f(x_0, y_0)$ and the value of the line integral is

$$\int_C (P\,dx + Q\,dy) = f(x_1, y_1) - f(x_0, y_0) = 0$$

**COROLLARY Value of a Line Integral over a Closed Curve**

Suppose $\mathbf{F}$ is a conservative vector field on some open region $R$, and $C$ is a closed, piecewise-smooth curve that lies entirely in $R$. Then

$$\boxed{\int_C \mathbf{F} \cdot d\mathbf{r} = \int_C (P\,dx + Q\,dy) = 0}$$

**EXAMPLE 3  Finding a Line Integral over a Closed Curve**

$\mathbf{F} = \mathbf{F}(x, y) = \dfrac{-y\mathbf{i} + x\mathbf{j}}{x^2 + y^2}$ is the gradient of $f(x, y) = \tan^{-1}\dfrac{y}{x}$, $x \neq 0$, since

$$\nabla f = \frac{\partial}{\partial x} \tan^{-1}\frac{y}{x}\mathbf{i} + \frac{\partial}{\partial y} \tan^{-1}\frac{y}{x}\mathbf{j} = \frac{-\dfrac{y}{x^2}}{1 + \dfrac{y^2}{x^2}}\mathbf{i} + \frac{\dfrac{1}{x}}{1 + \dfrac{y^2}{x^2}}\mathbf{j} = \frac{-y\mathbf{i} + x\mathbf{j}}{x^2 + y^2}$$

So, $\mathbf{F}$ is a conservative vector field on any region $R$ that contains no points on the $y$-axis ($x = 0$). Then by the corollary

$$\int_C \mathbf{F} \cdot d\mathbf{r} = 0$$

along any closed, piecewise-smooth curve $C$ that does not cross or touch the $y$-axis. ∎

**NOW WORK** Problem 27.

## The Converse of the Fundamental Theorem of Line Integrals

Simply stated, the Fundamental Theorem of Line Integrals states that if a vector field $\mathbf{F}(x, y) = P(x, y)\mathbf{i} + Q(x, y)\mathbf{j}$ is conservative, then $\int_C \mathbf{F} \cdot d\mathbf{r} = \int_C (P\,dx + Q\,dy)$ is independent of the path.

**NOTE** The converse of the conditional statement "If $p$, then $q$" is "If $q$, then $p$."

To state the converse, we need the following definition.

An open region $R$ is **connected** if any two points in $R$ can be joined by a piecewise-smooth curve $C$ that lies entirely in $R$. Figure 28(a) and (b) illustrate the idea of a connected region. In these two figures, any two points of $R$ can be joined by a piecewise-smooth curve that lies entirely in $R$, so $R$ is connected. In Figure 28(c) and (d), $R$ is not connected, since the points $A$ and $B$ cannot be joined by a curve that lies entirely in $R$.

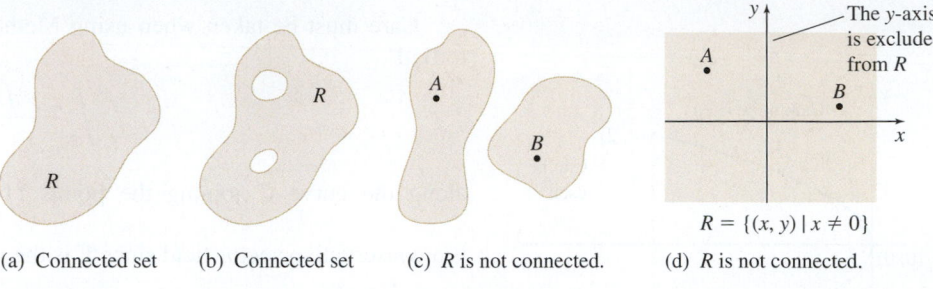

(a) Connected set    (b) Connected set    (c) $R$ is not connected.    (d) $R$ is not connected.

**Figure 28**

If $R$ is connected, then the converse of the Fundamental Theorem of Line Integrals is also true.

**THEOREM The Converse of the Fundamental Theorem of Line Integrals**

Let $\mathbf{F} = \mathbf{F}(x, y) = P(x, y)\mathbf{i} + Q(x, y)\mathbf{j}$ be a vector field, where the functions $P(x, y)$ and $Q(x, y)$ are continuous on an open, connected region $R$ containing the point $(x_0, y_0)$. Suppose for every piecewise-smooth curve $C$ in $R$, the line integral $\int_C \mathbf{F} \cdot d\mathbf{r} = \int_C (P\,dx + Q\,dy)$ is independent of the path. Then $\int_C \mathbf{F} \cdot d\mathbf{r}$ defines a function $f(x, y)$ and $\nabla f = \mathbf{F}$. In other words, $\mathbf{F}$ is a conservative vector field on $R$.

**Proof** Suppose $\int_C (P\,dx + Q\,dy)$ has the same value for every piecewise-smooth curve $C$ in $R$ that joins $(x_0, y_0)$ to $(x, y)$. That is, suppose $\int_C (P\,dx + Q\,dy)$ is independent of the path. If $C$ is any piecewise-smooth curve in $R$ joining a fixed point $(x_0, y_0)$ to an arbitrary point $(x, y)$, the line integral $\int_C (P\,dx + Q\,dy)$ defines a function $f$ that depends only on $(x, y)$ and not on $C$. And, we can define

$$f(x, y) = \int_{(x_0, y_0)}^{(x, y)} (P\,dx + Q\,dy)$$

Since any piecewise-smooth curve $C$ that lies entirely in $R$ is allowed, we choose two paths, one in which a horizontal line segment is used, and another in which a vertical line segment is used. As Figure 29 illustrates:

$C_1$: Consists of a piecewise-smooth curve joining $(x_0, y_0)$ to $(x_1, y)$ plus a horizontal line segment joining $(x_1, y)$ to $(x, y)$.

$C_2$: Consists of a piecewise-smooth curve joining $(x_0, y_0)$ to $(x, y_1)$ plus a vertical line segment joining $(x, y_1)$ to $(x, y)$.

For path $C_1$,

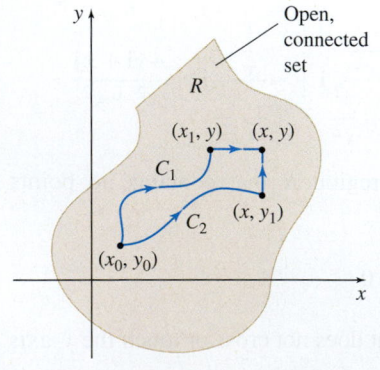

**Figure 29**

$$f(x, y) = \int_{(x_0, y_0)}^{(x, y)} (P\,dx + Q\,dy) = \int_{(x_0, y_0)}^{(x_1, y)} (P\,dx + Q\,dy) + \int_{(x_1, y)}^{(x, y)} (P\,dx + Q\,dy)$$

The first integral on the right is a function of $y$ alone, since $x_0$ and $x_1$ are constants. So, its partial derivative with respect to $x$ equals 0. That is,

$$\frac{\partial}{\partial x} \int_{(x_0, y_0)}^{(x_1, y)} (P\,dx + Q\,dy) = 0$$

In the second integral on the right, the value of $y$ is the same in both the lower and upper limits of integration, so it follows that $dy = 0$. That is,

$$\frac{\partial}{\partial x} \int_{(x_1, y)}^{(x, y)} (P\,dx + Q\,dy) = \frac{\partial}{\partial x} \int_{x_1}^{x} [P\,dx + 0] = P(x, y)$$

Combining these two results, we find

$$\frac{\partial}{\partial x} f(x, y) = \frac{\partial}{\partial x} \int_{(x_0, y_0)}^{(x_1, y)} (P\,dx + Q\,dy) + \frac{\partial}{\partial x} \int_{(x_1, y)}^{(x, y)} (P\,dx + Q\,dy) = P(x, y)$$

For path $C_2$,

$$f(x, y) = \int_{(x_0, y_0)}^{(x, y_1)} (P\,dx + Q\,dy) + \int_{(x, y_1)}^{(x, y)} (P\,dx + Q\,dy)$$

The first integral on the right is a function of $x$ alone, since $y_0$ and $y_1$ are constants. So, the partial derivative with respect to $y$ equals 0. That is,

$$\frac{\partial}{\partial y} \int_{(x_0, y_0)}^{(x, y_1)} (P\,dx + Q\,dy) = 0$$

In the second integral on the right, the value of $x$ is the same in both the lower and upper limits of integration, so it follows that $dx = 0$. That is,

$$\frac{\partial}{\partial y} \int_{(x, y_1)}^{(x, y)} (P\,dx + Q\,dy) = \frac{\partial}{\partial y} \int_{y_1}^{y} [0 + Q(x, y)\,dy] = Q(x, y)$$

Combining these two results, we find

$$\frac{\partial}{\partial y} f(x, y) = \frac{\partial}{\partial y} \int_{(x_0, y_0)}^{(x, y_1)} (P\,dx + Q\,dy) + \frac{\partial}{\partial y} \int_{(x, y_1)}^{(x, y)} (P\,dx + Q\,dy) = Q(x, y)$$

Finally, since $\dfrac{\partial}{\partial x} f(x, y) = P(x, y)$ and $\dfrac{\partial}{\partial y} f(x, y) = Q(x, y)$, we have

$$\nabla f = \frac{\partial f}{\partial x}\mathbf{i} + \frac{\partial f}{\partial y}\mathbf{j} = P(x, y)\mathbf{i} + Q(x, y)\mathbf{j} = \mathbf{F}$$

That is, the vector field $\mathbf{F}$ is conservative. ∎

## 3 Reconstruct a Function from Its Gradient: Finding the Potential Function for a Conservative Vector Field

To use the Fundamental Theorem of Calculus to find a definite integral $\int_a^b f(x)\,dx$, we need to find an antiderivative $F$ of $f$. So, earlier in calculus we developed techniques for finding antiderivatives. Here we show a way to find the potential function of a conservative vector field so we can use the Fundamental Theorem on Line Integrals. Suppose $\mathbf{F} = \mathbf{F}(x, y)$ is a conservative vector field. The proof of the converse of the Fundamental Theorem of Line Integrals provides a way of finding the potential function $f$ when its gradient $\nabla f = \mathbf{F}$ is known. This is called **reconstructing a function from its gradient**.

CALC
● EXAMPLE 4  **Finding a Potential Function for a Conservative Vector Field**
CLIP
If it is known that

$$\mathbf{F} = \mathbf{F}(x, y) = (6xy + y^3)\mathbf{i} + (3x^2 + 3xy^2)\mathbf{j}$$

is a conservative vector field, find a potential function $f$ for $\mathbf{F}$.

**Solution** We seek a function $f(x, y)$ for which

$$\nabla f = \mathbf{F} = (6xy + y^3)\mathbf{i} + (3x^2 + 3xy^2)\mathbf{j}$$

Let

$$P(x, y) = 6xy + y^3 \qquad \text{and} \qquad Q(x, y) = 3x^2 + 3xy^2$$

Since

$$\nabla f = \frac{\partial f}{\partial x}\mathbf{i} + \frac{\partial f}{\partial y}\mathbf{j} = P(x, y)\mathbf{i} + Q(x, y)\mathbf{j}$$

we have

$$\frac{\partial f}{\partial x} = 6xy + y^3 \qquad \text{and} \qquad \frac{\partial f}{\partial y} = 3x^2 + 3xy^2 \qquad (1)$$

Now integrate the left equation partially with respect to $x$.

$$f(x, y) = \int (6xy + y^3)\, dx = 3x^2 y + xy^3 + h(y)$$

where the "constant of integration," $h(y)$, is a function of $y$.

Next differentiate $f$ with respect to $y$:

$$\frac{\partial f}{\partial y} = 3x^2 + 3xy^2 + h'(y)$$

From (1), we have

$$\frac{\partial f}{\partial y} = 3x^2 + 3xy^2$$

So, $h'(y) = 0$ or $h(y) = K$, where $K$ is a constant. A potential function $f$ for $\mathbf{F}$ is

$$f(x, y) = 3x^2 y + xy^3 + K \qquad \blacksquare$$

You can verify that $f$ is a potential function for $\mathbf{F}$ by finding the gradient of $f$:

$$\nabla f = \frac{\partial f}{\partial x}\mathbf{i} + \frac{\partial f}{\partial y}\mathbf{j} = (6xy + y^3)\mathbf{i} + (3x^2 + 3xy^2)\mathbf{j} = \mathbf{F}(x, y)$$

If we had chosen to integrate $\dfrac{\partial f}{\partial y} = 3x^2 + 3xy^2$ partially with respect to $y$, we would have obtained

$$f(x, y) = 3x^2 y + xy^3 + k(x)$$

where the "constant of integration," $k(x)$, is a function of $x$. Then

$$\frac{\partial f}{\partial x} = 6xy + y^3 + k'(x) \qquad \text{and} \qquad \frac{\partial f}{\partial x} = 6xy + y^3 \qquad \text{[from (1)]}$$

So, $k'(x) = 0$ or $k(x) = K$, a constant, and

$$f(x, y) = 3x^2 y + xy^3 + K$$

as before.

NOW WORK **Problem 29.**

## 4 Determine Whether a Vector Field Is Conservative

As the solution to Example 4 illustrates, if **F** is known to be a conservative vector field, then we can find its potential function $f$. But how do we determine if **F** is a conservative vector field? Before giving the theorem, we need some definitions.

> **DEFINITION  Closed Curve; Simple Curve**
>
> Let $\mathbf{r} = \mathbf{r}(t)$, $a \leq t \leq b$, trace out a piecewise-smooth curve $C$.
>
> - $C$ is **closed** if $\mathbf{r}(a) = \mathbf{r}(b)$.
> - $C$ is **simple** if it does not intersect itself for $a < t < b$.

**IN WORDS**  A curve $C$ is closed if its initial point and its terminal point coincide.

See Figure 30.

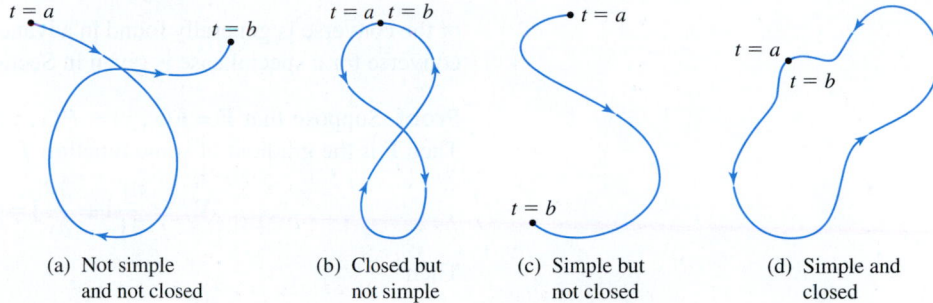

(a) Not simple and not closed    (b) Closed but not simple    (c) Simple but not closed    (d) Simple and closed

**Figure 30**

> **DEFINITION  Simply Connected**
>
> Let $R$ denote a set of points in the plane. A region $R$ is called **simply connected** if its boundary consists of a single, simple closed curve, and the interior of any simple closed curve $C$ in $R$ contains only points of $R$.

Intuitively, a set $R$ is not simply connected if it has "holes" or if it consists of two or more separate pieces.

Figure 31(a) shows a region $R$ that has a "hole." Its boundary consists of two simple closed curves $C_1$ and $C_2$. The curve $C_3$ is a simple closed curve contained completely in the region, but the interior of $C_3$ contains points that are not in the region $R$, so the region $R$ is not simply connected.

Figure 31(b) shows a region $R$ whose boundary consists of one simple closed curve $C$. Since the boundary $C$ of $R$ and the interior of any simple closed curve $C$ in $R$ contains only points in $R$, the region is simply connected.

Figure 31(c) shows a region $R$ whose boundary consists of two simple closed curves $C_1$ and $C_2$, that separate $R$ into two pieces. So, $R$ is not a simply connected region.

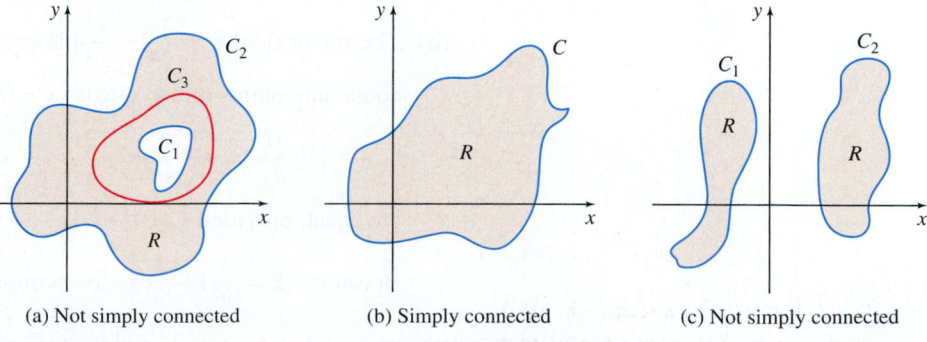

(a) Not simply connected    (b) Simply connected    (c) Not simply connected

**Figure 31**

Now we are ready for the test to determine whether or not a vector field **F** is conservative.

**THEOREM Conservative Vector Field Test**

Let $\mathbf{F} = \mathbf{F}(x, y) = P(x, y)\mathbf{i} + Q(x, y)\mathbf{j}$ be a vector field, where the functions $P$ and $Q$ are continuous on some simply connected, open region $R$. Suppose $\dfrac{\partial P}{\partial y}$ and $\dfrac{\partial Q}{\partial x}$ are also continuous on $R$. Then $\mathbf{F}(x, y) = P(x, y)\mathbf{i} + Q(x, y)\mathbf{j}$ is a conservative vector field on $R$ if and only if

$$\frac{\partial P}{\partial y} = \frac{\partial Q}{\partial x}$$

throughout $R$.

We only prove that if $\mathbf{F}$ is a conservative vector field, then $\dfrac{\partial P}{\partial y} = \dfrac{\partial Q}{\partial x}$. The proof of the converse is generally found in advanced calculus texts. However, the proof of the converse for a special case is given in Section 15.5. See p. 1067.

**Proof** Suppose that $\mathbf{F} = \mathbf{F}(x, y) = P(x, y)\mathbf{i} + Q(x, y)\mathbf{j}$ is a conservative vector field. Then $\mathbf{F}$ is the gradient of some function $f$. That is,

$$\nabla f = \frac{\partial f}{\partial x}\mathbf{i} + \frac{\partial f}{\partial y}\mathbf{j} = P(x, y)\mathbf{i} + Q(x, y)\mathbf{j}$$

Then

$$\frac{\partial f}{\partial x} = P(x, y) \qquad \text{and} \qquad \frac{\partial f}{\partial y} = Q(x, y)$$

so that

$$\frac{\partial^2 f}{\partial y \partial x} = \frac{\partial P}{\partial y} \qquad \text{and} \qquad \frac{\partial^2 f}{\partial x \partial y} = \frac{\partial Q}{\partial x}$$

Since it is assumed that $\dfrac{\partial P}{\partial y}$ and $\dfrac{\partial Q}{\partial x}$ are continuous in $R$, we can use the Equality of Mixed Partials Theorem (p. 878), which states that if the first-order partial derivatives and the second-order mixed partial derivatives exist at each point in some disk centered at $(x_0, y_0)$ and if the mixed partial derivatives are continuous at $(x_0, y_0)$, then $f_{xy}(x_0, y_0) = f_{yx}(x_0, y_0)$. As a result,

$$\frac{\partial P}{\partial y} = \frac{\partial Q}{\partial x}$$

∎

**EXAMPLE 5 Determining Whether F Is a Conservative Vector Field**

(a) $\mathbf{F} = 2xy\mathbf{i} + (x^2 + 1)\mathbf{j}$ is a conservative vector field on the entire $xy$-plane since

$$\frac{\partial P}{\partial y} = \frac{\partial}{\partial y}(2xy) = 2x \qquad \frac{\partial Q}{\partial x} = \frac{\partial}{\partial x}(x^2 + 1) = 2x$$

are equal for any choice of $(x, y)$.

(b) The vector field $\mathbf{F} = \dfrac{x}{y^2}\mathbf{i} - \dfrac{x^2}{y^3}\mathbf{j}$ is conservative on any connected region not containing points on the $x$-axis ($y = 0$) since

$$\frac{\partial P}{\partial y} = \frac{\partial}{\partial y}\frac{x}{y^2} = -\frac{2x}{y^3} \qquad \text{and} \qquad \frac{\partial Q}{\partial x} = \frac{\partial}{\partial x}\left(-\frac{x^2}{y^3}\right) = -\frac{2x}{y^3}$$

are equal, provided $y \neq 0$.

∎

Because $\mathbf{F} = \dfrac{x}{y^2}\mathbf{i} - \dfrac{x^2}{y^3}\mathbf{j}$ is conservative for $y \neq 0$, the line integral

$$\int_C \mathbf{F} \cdot d\mathbf{r} = \int_C \left[\frac{x}{y^2}dx - \frac{x^2}{y^3}dy\right]$$ is independent of the path in any simply connected region not containing points on the $x$-axis.

**NOW WORK** Problem 35.

---

**EXAMPLE 6** **Analyzing a Vector Field**

**(a)** Determine whether the vector field $\mathbf{F} = \mathbf{F}(x, y) = (2xy + 24x)\mathbf{i} + (x^2 + 16)\mathbf{j}$ is conservative in the $xy$-plane.

**(b)** Determine whether the line integral

$$\int_C \mathbf{F} \cdot d\mathbf{r} = \int_C [(2xy + 24x)\, dx + (x^2 + 16)\, dy]$$

is independent of the path in the $xy$-plane.

**(c)** Find a potential function $f$ of the vector field $\mathbf{F}$.

**(d)** Find $\int_C [(2xy + 24x)\, dx + (x^2 + 16)\, dy]$, where $C$ is any piecewise-smooth curve joining $(0, 1)$ to $(1, 2)$.

**Solution**  **(a)** Let $P(x, y) = 2xy + 24x$ and $Q(x, y) = x^2 + 16$. Then $\dfrac{\partial P}{\partial y} = 2x$

and $\dfrac{\partial Q}{\partial x} = 2x$. Since $P$, $Q$, $\dfrac{\partial P}{\partial y}$, and $\dfrac{\partial Q}{\partial x}$ are continuous everywhere in the $xy$-plane

and since $\dfrac{\partial P}{\partial y} = \dfrac{\partial Q}{\partial x}$, then $\mathbf{F} = \mathbf{F}(x, y) = (2xy + 24x)\mathbf{i} + (x^2 + 16)\mathbf{j}$ is a conservative

vector field in the $xy$-plane.

**(b)** Since $\mathbf{F}$ is a conservative vector field, by the Fundamental Theorem of Line Integrals, $\int_C \mathbf{F} \cdot d\mathbf{r}$ is independent of the path in the $xy$-plane.

**(c)** Since $\mathbf{F} = \mathbf{F}(x, y) = (2xy + 24x)\mathbf{i} + (x^2 + 16)\mathbf{j}$ is a conservative vector field, there is a potential function $f$ for $\mathbf{F}$ for which

$$\nabla f = \mathbf{F} = (2xy + 24x)\mathbf{i} + (x^2 + 16)\mathbf{j}$$

Since $\nabla f = \dfrac{\partial f}{\partial x}\mathbf{i} + \dfrac{\partial f}{\partial y}\mathbf{j}$, then

$$\frac{\partial f}{\partial x} = 2xy + 24x \qquad \text{and} \qquad \frac{\partial f}{\partial y} = x^2 + 16$$

Integrate the first of these partial derivatives partially with respect to $x$.

$$f(x, y) = \int (2xy + 24x)\, dx = x^2 y + 12x^2 + h(y)$$

where the "constant of integration," $h(y)$, is a function of $y$. Now differentiate $f$ with respect to $y$, obtaining

$$\frac{\partial f}{\partial y} = \frac{\partial}{\partial y}[x^2 y + 12x^2 + h(y)] = x^2 + h'(y)$$

But $\dfrac{\partial f}{\partial y} = x^2 + 16$, so $h'(y) = 16$. Now we integrate $h'$ with respect to $y$; then

$$h(y) = 16y + K$$

where $K$ is a constant. So,

$$f(x, y) = x^2 y + 12x^2 + 16y + K$$

$K$ a constant, is a potential function for $\mathbf{F}$.

**(d)** Since $\mathbf{F}$ is a conservative vector field, we can use the Fundamental Theorem of Line Integrals. Since $f(x, y) = x^2 y + 12x^2 + 16y + K$ and $\mathbf{F} = \nabla f$,

$$\int_C \mathbf{F} \cdot d\mathbf{r} = \int_C \nabla f \cdot d\mathbf{r} = \left[ f(x, y) \right]_{(0,1)}^{(1,2)} = f(1, 2) - f(0, 1) = 46 + K - 16 - K = 30$$

for any piecewise-smooth curve $C$ connecting $(0, 1)$ to $(1, 2)$.  ∎

**NOW WORK** Problem **37**.

EXAMPLE 7  Analyzing a Vector Field

(a) Determine whether the vector field

$$\mathbf{F} = \mathbf{F}(x, y) = (y \cos x + 2xe^y)\mathbf{i} + (\sin x + x^2 e^y + 4)\mathbf{j}$$

is conservative in the $xy$-plane.

(b) Determine whether the line integral

$$\int_C \mathbf{F} \cdot d\mathbf{r} = \int_C [(y \cos x + 2xe^y)\, dx + (\sin x + x^2 e^y + 4)\, dy]$$

is independent of the path in the $xy$-plane.

(c) Find a potential function $f$ of the vector field $\mathbf{F}$.

(d) Find $\int_C \mathbf{F} \cdot d\mathbf{r}$, where $C$ is any piecewise-smooth curve from $(0, 0)$ to $\left(\dfrac{\pi}{2}, 1\right)$.

**Solution (a)** Let $P(x, y) = y \cos x + 2xe^y$ and $Q(x, y) = \sin x + x^2 e^y + 4$. Then

$$\frac{\partial P}{\partial y} = \cos x + 2xe^y \qquad \text{and} \qquad \frac{\partial Q}{\partial x} = \cos x + 2xe^y$$

Since $P$, $Q$, $\dfrac{\partial P}{\partial y}$, and $\dfrac{\partial Q}{\partial x}$ are continuous everywhere in the $xy$-plane, and $\dfrac{\partial P}{\partial y} = \dfrac{\partial Q}{\partial x}$, the vector field $\mathbf{F} = \mathbf{F}(x, y) = (y \cos x + 2xe^y)\mathbf{i} + (\sin x + x^2 e^y + 4)\mathbf{j}$ is conservative in the $xy$-plane.

**(b)** Since the vector field $\mathbf{F}$ is conservative, by the Fundamental Theorem on Line Integrals, $\int_C \mathbf{F} \cdot d\mathbf{r}$ is independent of the path in the $xy$-plane.

**(c)** Since $\mathbf{F}$ is a conservative vector field, there is a potential function $f$ for $\mathbf{F}$ for which

$$\nabla f = \mathbf{F} = (y \cos x + 2xe^y)\mathbf{i} + (\sin x + x^2 e^y + 4)\mathbf{j}$$

Since $\nabla f = \dfrac{\partial f}{\partial x}\mathbf{i} + \dfrac{\partial f}{\partial y}\mathbf{j}$, we have $\dfrac{\partial f}{\partial x} = y \cos x + 2xe^y$ and $\dfrac{\partial f}{\partial y} = \sin x + x^2 e^y + 4$.

Integrate the function on the right partially with respect to $y$. Then

$$f(x, y) = \int (\sin x + x^2 e^y + 4)\, dy = y \sin x + x^2 e^y + 4y + k(x)$$

in which the "constant of integration," $k(x)$, is a function of $x$. Now differentiate $f$ with respect to $x$ to get

$$\frac{\partial f}{\partial x} = y \cos x + 2xe^y + k'(x)$$

But $\dfrac{\partial f}{\partial x} = y \cos x + 2xe^y$. So, $k'(x) = 0$.

Then $k(x) = K$, where $K$ is a constant. So,

$$f(x, y) = y \sin x + x^2 e^y + 4y + K$$

$K$ a constant, is a potential function for $\mathbf{F}$.

**(d)** Using the potential function $f$ of the vector field $\mathbf{F}$ and the Fundamental Theorem of Line Integrals,

$$\int_C \mathbf{F} \cdot d\mathbf{r} = [f(x, y)]_{(0,0)}^{(\pi/2, 1)} = [y \sin x + x^2 e^y + 4y + K]_{(0,0)}^{(\pi/2, 1)}$$

$$= \left(1 + \frac{\pi^2}{4} e + 4 + K\right) - K = \frac{\pi^2 e}{4} + 5 \qquad \blacksquare$$

NOW WORK Problem 49.

**EXAMPLE 8**  **Determining Whether a Line Integral Is Independent of the Path**

Determine whether $\int_C (x^2 y \, dx + xy^2 \, dy)$ is independent of the path anywhere in the plane.

**Solution** Let $P = x^2 y$ and $Q = xy^2$. Then $\dfrac{\partial P}{\partial y} = x^2$ and $\dfrac{\partial Q}{\partial x} = y^2$. Since these two functions are not equal (except on the graph of the equation $x^2 = y^2$, which is not an open set), the line integral is not independent of the path anywhere in the $xy$-plane.  ∎

## Work and Kinetic Energy

The work $W$ done by a force field $\mathbf{F}$ in moving an object along a piecewise-smooth curve $C$ traced out by $\mathbf{r} = \mathbf{r}(t)$, $a \le t \le b$, is

$$W = \int_C \mathbf{F} \cdot d\mathbf{r} = \int_a^b \mathbf{F} \cdot \mathbf{r}'(t) \, dt$$

where $\mathbf{r} = \mathbf{r}(t)$ is the position vector of the object and $\mathbf{r}' = \mathbf{r}'(t)$ is the velocity of the object at time $t$. Using Newton's Second Law of Motion, $\mathbf{F} = m\mathbf{r}''(t)$, where $\mathbf{r}''(t)$ is the acceleration of the object, we find that the work $W$ done can be expressed as

$$
\begin{aligned}
W &= \int_a^b \mathbf{F} \cdot \mathbf{r}'(t) \, dt = \int_a^b m\mathbf{r}''(t) \cdot \mathbf{r}'(t) \, dt && \mathbf{F} = m\mathbf{r}''(t) \\
&= \frac{m}{2} \int_a^b \frac{d}{dt}[\mathbf{r}'(t) \cdot \mathbf{r}'(t)] \, dt && \frac{d}{dt}[\mathbf{r}'(t) \cdot \mathbf{r}'(t)] = 2[\mathbf{r}''(t) \cdot \mathbf{r}'(t)] \\
&= \frac{m}{2} \int_a^b \frac{d}{dt}\|\mathbf{r}'(t)\|^2 \, dt && \mathbf{u} \cdot \mathbf{u} = \|\mathbf{u}\|^2 \\
&= \frac{m}{2} \left[\|\mathbf{r}'(t)\|^2\right]_a^b && \int_a^b f'(t) \, dt = [f(t)]_a^b \qquad (1)
\end{aligned}
$$

Kinetic energy $K$ is the energy of motion. The kinetic energy $K$ of an object of mass $m$ that moves along a curve with speed $\|\mathbf{v}\|$ is given by

**RECALL** Speed is the magnitude of the velocity $\mathbf{v} = \mathbf{r}'(t)$. That is, speed $= \|\mathbf{v}\| = \|\mathbf{r}'(t)\|$.

$$K = \frac{1}{2}m\|\mathbf{v}\|^2 = \frac{m}{2}\|\mathbf{r}'(t)\|^2 \qquad (2)$$
$$\uparrow$$
$$\mathbf{v} = \mathbf{r}'(t)$$

By combining (1) and (2), we obtain

$$W = K(b) - K(a)$$

which relates work $W$ to kinetic energy $K$. In other words,

$$\left[\begin{array}{l}\text{Work } W \text{ done by a force field } \mathbf{F} \\ \text{in moving an object from } a \text{ to } b\end{array}\right] = \left[\text{Change in the kinetic energy } K \text{ from } a \text{ to } b\right]$$

If the curve $C$ is closed and the force field $\mathbf{F}$ is conservative, then

$$\text{Work} = \int_C \mathbf{F} \cdot d\mathbf{r} = 0$$

### THEOREM  Work on a Closed Path

In a conservative force field $\mathbf{F}$, the work $W$ done by $\mathbf{F}$ in moving an object along a closed path is 0. That is, the object returns to its original position with the same kinetic energy it started with.

## Work and Potential Energy

**NOTE** Potential energy $U$ is the energy of position in a system.

Suppose an object of mass $m$ moves along a smooth curve $C$ traced out by $\mathbf{r} = \mathbf{r}(t)$, $a \leq t \leq b$, in a conservative force field $\mathbf{F} = \mathbf{F}(x, y)$. The **potential energy** $U = U(x, y)$ of the object due to $\mathbf{F}$ is the function $U$ for which

$$\boxed{\nabla U = -\mathbf{F}}$$

That is, $U$ is the negative of the potential function for $\mathbf{F}$. The work $W$ done by $\mathbf{F}$ in moving an object along $C$ from $a$ to $b$ is

$$W = \int_C \mathbf{F} \cdot d\mathbf{r} = -\int_C \nabla U \cdot d\mathbf{r} = -\left[U(x, y)\right]_a^b = -[U(\mathbf{r}(b)) - U(\mathbf{r}(a))] = U(\mathbf{r}(a)) - U(\mathbf{r}(b))$$

This equation relates work $W$ to potential energy $U$, as stated below.

$$\left[\begin{array}{l}\text{The work } W \text{ done by a conservative force field } \mathbf{F} \\ \text{in moving an object from } a \text{ to } b\end{array}\right] = \left[\text{The change in potential energy } U \text{ from } b \text{ to } a\right]$$

Now we state and prove the *Law of Conservation of Energy*.

### Law of Conservation of Energy

In a conservative force field $\mathbf{F}$, the sum of the potential and kinetic energies of an object is constant.

**Proof** The potential energy $U = U(x, y)$ of an object moving alonga curve $C$ in a conservative field of force $\mathbf{F}$ obeys

$$\nabla U = -\mathbf{F}(x, y)$$

If an object of mass $m$ moves along the curve $C$ traced out by $\mathbf{r} = \mathbf{r}(t)$, $a \leq t \leq b$, its kinetic energy $K$ is

$$K = \frac{1}{2}m \left\|\mathbf{r}'(t)\right\|^2 = \frac{1}{2}m[\mathbf{r}'(t) \cdot \mathbf{r}'(t)]$$

Let $E = E(t)$ equal the sum of the potential and kinetic energies of the object at time $t$. Then

$$E(t) = U + K = U(x, y) + \frac{m}{2}[\mathbf{r}'(t) \cdot \mathbf{r}'(t)]$$

We show that $E$ is constant by showing that $E'(t) = 0$:

$$E'(t) = \underset{\uparrow}{\frac{\partial U}{\partial x}\frac{dx}{dt}} + \frac{\partial U}{\partial y}\frac{dy}{dt} + m[\mathbf{r}'(t) \cdot \mathbf{r}''(t)]$$

**Use Chain Rule I with $U$.**

But

$$\mathbf{r}'(t) = \frac{dx}{dt}\mathbf{i} + \frac{dy}{dt}\mathbf{j} \quad \text{and} \quad \frac{\partial U}{\partial x}\mathbf{i} + \frac{\partial U}{\partial y}\mathbf{j} = \nabla U = -\mathbf{F} = \underset{\underset{\mathbf{F} = m\mathbf{a}}{\uparrow}}{-m\mathbf{r}''(t)}$$

So,

$$\frac{\partial U}{\partial x}\frac{dx}{dt} + \frac{\partial U}{\partial y}\frac{dy}{dt} = \nabla U \cdot \mathbf{r}'(t) = -m\,\mathbf{r}''(t) \cdot \mathbf{r}'(t)$$

Then

$$E'(t) = -m[\mathbf{r}''(t) \cdot \mathbf{r}'(t)] + m[\mathbf{r}'(t) \cdot \mathbf{r}''(t)] = 0$$

That is, the energy $E$ of the object is constant. ∎

The summary on the next page lists the important results obtained so far involving line integrals, independence of path, and conservative vector fields.

## Summary

Let $\mathbf{F} = \mathbf{F}(x, y) = P(x, y)\mathbf{i} + Q(x, y)\mathbf{j}$ be a vector field, where $P$ and $Q$ are two functions that are continuous on an open, connected region $R$ containing a piecewise-smooth curve $C$.

- $\mathbf{F}$ is a conservative vector field if $\mathbf{F} = \nabla f$ for a function $f$ that has continuous first-order partial derivatives on $R$.

- If $\mathbf{F} = P\mathbf{i} + Q\mathbf{j}$ is a conservative vector field, then $\int_C \mathbf{F} \cdot d\mathbf{r} = \int_C (P\,dx + Q\,dy)$ is independent of the path.

- If $\int_C (P\,dx + Q\,dy)$ is independent of the path and if $C$ is a closed curve in $R$, then $\int_C \mathbf{F} \cdot d\mathbf{r} = \int_C (P\,dx + Q\,dy) = 0$.

- If $\int_C (P\,dx + Q\,dy)$ is independent of the path in $R$, then $\mathbf{F} = P\mathbf{i} + Q\mathbf{j}$ is a conservative vector field in $R$.

- If $\dfrac{\partial P}{\partial y}$ and $\dfrac{\partial Q}{\partial x}$ are continuous in $R$ and if $R$ is open and simply connected, then $\mathbf{F} = P\mathbf{i} + Q\mathbf{j}$ is a conservative vector field if and only if $\dfrac{\partial P}{\partial y} = \dfrac{\partial Q}{\partial x}$ throughout $R$.

## 15.4 Assess Your Understanding

### Concepts and Vocabulary

1. *True or False* If $\nabla f = \mathbf{F}$ for some function $f$ that has continuous first-order partial derivatives on an open region $R$, then $\mathbf{F}$ is a potential function for $f$.

2. *Multiple Choice* If $\int_{C_1}(P\,dx + Q\,dy) = \int_{C_2}(P\,dx + Q\,dy)$ for any two piecewise-smooth curves $C_1$ and $C_2$ with the same end points lying entirely in an open region $R$, then $\int_C(P\,dx + Q\,dy)$ is said to be [(a) one-to-one (b) singular (c) independent of $P$ and $Q$ (d) independent of the path].

3. *True or False* The line integral $\int_C \mathbf{F} \cdot d\mathbf{r}$, where $C$ is a piecewise-smooth curve and $\mathbf{F}$ is a conservative vector field on an open region $R$, is independent of the path in $R$.

4. *Multiple Choice* Suppose $\mathbf{F} = \mathbf{F}(x, y)$ is a conservative vector field and $\nabla f = \mathbf{F}$. The function $f$ is called a(n) [(a) gradient function (b) independent function (c) potential function (d) gravity function] for $\mathbf{F}$.

5. A piecewise-smooth curve $C$ is _____ if its initial point $(x_0, y_0)$ and its terminal point $(x_1, y_1)$ are the same.

6. If $\mathbf{F}$ is a conservative vector field on some open region $R$, and $C$ is a closed, piecewise-smooth curve that lies entirely in $R$, then $\int_C \mathbf{F} \cdot d\mathbf{r} = $ _____.

7. *True or False* An open region $R$ is connected if there are two points $(x_1, y_1)$ and $(x_2, y_2)$ in $R$ that can be joined by a piecewise-smooth curve $C$ that lies entirely in $R$.

8. *True or False* The converse of the Fundamental Theorem of Line Integrals is true if the open region $R$ is connected.

9. *True or False* Let $\mathbf{r} = \mathbf{r}(t)$, $a \le t \le b$, trace out a piecewise-smooth curve $C$. Then $C$ is closed if there is a least one point where the curve intersects itself.

10. *True or False* A parabola is a simple curve.

11. *True or False* A region $R$ is called simply connected if it is closed.

12. Let $\mathbf{F} = \mathbf{F}(x, y) = P(x, y)\mathbf{i} + Q(x, y)\mathbf{j}$ be a vector field, where the functions $P$ and $Q$ are continuous on some simply connected, open region $R$. Suppose $\dfrac{\partial P}{\partial y}$ and $\dfrac{\partial Q}{\partial x}$ are also continuous on $R$. Then $\mathbf{F}(x, y) = P(x, y)\mathbf{i} + Q(x, y)\mathbf{j}$ is a conservative vector field on $R$ if and only if _____.

13. *Multiple Choice* In a conservative force field $\mathbf{F}$, the work $W$ done by $\mathbf{F}$ in moving an object along a closed path $C$ is [(a) a maximum (b) a minimum (c) $\|\mathbf{F}\|$ (d) 0].

14. Kinetic energy $K$ is the energy of _____; potential energy $U$ is the energy of _____.

### Skill Building

In Problems 15–20, $\mathbf{F} = \mathbf{F}(x, y) = P(x, y)\mathbf{i} + Q(x, y)\mathbf{j}$ is a conservative vector field since $\mathbf{F}$ is the gradient of $f(x, y)$. Use this fact to find $\int_C[P(x, y)dx + Q(x, y)\,dy]$, where $C$ is any piecewise-smooth curve joining the two given points.

15. $\int_C[(2x - y)\,dx + (2y - x)\,dy]$ from $(3, -1)$ to $(5, 4)$; $f(x, y) = x^2 - xy + y^2$

16. $\int_C[(y^2 + 2x)\,dx + 2xy\,dy]$ from $(2, 2)$ to $(-3, 8)$; $f(x, y) = xy^2 + x^2$

17. $\int_C(3x^2y\,dx + x^3\,dy)$ from $(0, -1)$ to $(3, 4)$; $f(x, y) = x^3 y$

18. $\int_C\left(2xe^{x^2+y^2}\,dx + 2ye^{x^2+y^2}\,dy\right)$ from $(0, 0)$ to $(1, 1)$; $f(x, y) = e^{x^2+y^2}$

19. $\int_C\left(\dfrac{x}{x^2+y^2}\,dx + \dfrac{y}{x^2+y^2}\,dy\right)$ from $(3, 4)$ to $(5, 12)$; $f(x, y) = \ln\sqrt{x^2 + y^2}$

20. $\int_C\left(\dfrac{x}{\sqrt{x^2+y^2}}\,dx + \dfrac{y}{\sqrt{x^2+y^2}}\,dy\right)$ from $(0, -4)$ to $(3, 4)$; $f(x, y) = \sqrt{x^2 + y^2}$

In Problems 21–28, $\mathbf{F} = \mathbf{F}(x, y) = P(x, y)\mathbf{i} + Q(x, y)\mathbf{j}$ is a conservative vector field. Use this fact to find $\int_C[P(x, y)\,dx + Q(x, y)\,dy]$, where $C$ is any piecewise-smooth curve joining the two given points.

21. $\int_C(x^3\,dx + y^3\,dy)$ from $(1, 1)$ to $(2, 4)$

22. $\int_C(2x^2\,dx - 4y\,dy)$ from $(-1, 2)$ to $(2, 3)$

23. $\int_C[(x^2 + y^2)\,dx + (2xy + \sin y)\,dy]$ from $(0, 0)$ to $(2, 0)$

24. $\int_C[(x - \cos y)\,dx + x\sin y\,dy]$ from $\left(1, \dfrac{\pi}{2}\right)$ to $(2, \pi)$

25. $\int_C(y^2e^x\,dx + 2ye^x\,dy)$ from $(0, 0)$ to $(\ln 2, 2)$

---

**1.** = NOW WORK problem   📐 = Graphing technology recommended   CAS = Computer Algebra System recommended

**26.** $\int_C \left[ xe^y dx + \dfrac{1}{2}(x^2 - 4)e^y dy \right]$ from $(0, 0)$ to $(2, 1)$

**27.** $\int_C (ye^x dx + e^x dy)$ on a closed curve $C$

**28.** $\int_C (y \cos x\, dx + \sin x\, dy)$ on a closed curve $C$

*In Problems 29–32, $\mathbf{F} = \mathbf{F}(x, y) = P(x, y)\mathbf{i} + Q(x, y)\mathbf{j}$ is a conservative vector field. Find a potential function $f$ for $\mathbf{F}$.*

**29.** $\mathbf{F}(x, y) = (5x - 2y + 1)\mathbf{i} + (4 - 2x)\mathbf{j}$

**30.** $\mathbf{F}(x, y) = (4xy)\mathbf{i} + (2x^2 + y)\mathbf{j}$

**31.** $\mathbf{F}(x, y) = (\ln y + 2x)\mathbf{i} + \left( \dfrac{x}{y} \right)\mathbf{j}$

**32.** $\mathbf{F}(x, y) = (ye^x)\mathbf{i} + e^x\mathbf{j}$

*In Problems 33–36, show that each vector field $\mathbf{F}$ is a conservative vector field. Find a potential function of $\mathbf{F}$.*

**33.** $\mathbf{F}(x, y) = x^2\mathbf{i} + y^2\mathbf{j}$      **34.** $\mathbf{F}(x, y) = xy\mathbf{i} + \dfrac{1}{2}x^2\mathbf{j}$

**35.** $\mathbf{F}(x, y) = xe^y\mathbf{i} + \dfrac{1}{2}x^2 e^y\mathbf{j}$

**36.** $\mathbf{F}(x, y) = (x^2 + y^2)\mathbf{i} + (2xy - \sin y)\mathbf{j}$

*In Problems 37–42:*
*(a) Show that the line integral $\int_C (P\, dx + Q\, dy)$ is independent of the path.*
*(b) Find a function $f$ so that $\nabla f = P(x, y)\mathbf{i} + Q(x, y)\mathbf{j}$.*
*(c) Find $\int_C (P\, dx + Q\, dy)$ for the given points.*

**37.** $\int_C (x\, dx + y\, dy)$, where $C$ is a piecewise-smooth curve joining the points $(1, 3)$ and $(2, 5)$

**38.** $\int_C [2xy\, dx + (x^2 + 1)\, dy]$, where $C$ is a piecewise-smooth curve joining the points $(1, -4)$ and $(-2, 3)$

**39.** $\int_C [(x^2 + 3y)\, dx + 3x\, dy]$, where $C$ is a piecewise-smooth curve joining the points $(1, 2)$ and $(-3, 5)$

**40.** $\int_C [(2x + y + 1)\, dx + (x + 3y + 2)\, dy]$, where $C$ is a piecewise-smooth curve joining the points $(0, 0)$ and $(1, 2)$

**41.** $\int_C [(4x^3 + 20xy^3 - 3y^4)\, dx + (30x^2 y^2 - 12xy^3 + 5y^4)\, dy]$, where $C$ is a piecewise-smooth curve joining the points $(0, 0)$ and $(1, 1)$

**42.** $\int_C (2yx^{-1}\, dx + \ln x^2\, dy)$, where $C$ is a piecewise-smooth curve in the first quadrant joining the points $(1, 1)$ and $(5, 5)$

*In Problems 43–54:*
*(a) Show that each line integral $\int_C \mathbf{F} \cdot d\mathbf{r} = \int_C (P\, dx + Q\, dy)$ is independent of the path in the $xy$-plane.*
*(b) Find a potential function $f$ for $\mathbf{F}$.*

**43.** $\int_C (3x^2 y^2\, dx + 2x^3 y\, dy)$    **44.** $\int_C [(2x + y)\, dx + (x - 2y)\, dy]$

**45.** $\int_C [(x + 3y)\, dx + 3x\, dy]$    **46.** $\int_C [(2x + y)\, dx + (2y + x)\, dy]$

**47.** $\int_C [(2xy - y^2)\, dx + (x^2 - 2xy)\, dy]$

**48.** $\int_C [y^2\, dx + (2yx - e^y)\, dy]$

**49.** $\int_C [(x^2 - x + y^2)\, dx - (ye^y - 2xy)\, dy]$

**50.** $\int_C [(3x^2 y + xy^2 + e^x)\, dx + (x^3 + x^2 y + \sin y)\, dy]$

**51.** $\int_C [(y \cos x - 2 \sin y)\, dx - (2x \cos y - \sin x)\, dy]$

**52.** $\int_C [(e^x \sin y + 2y \sin x)\, dx + (e^x \cos y - 2 \cos x)\, dy]$

**53.** $\int_C [(2x + y \cos x)\, dx + \sin x\, dy]$

**54.** $\int_C [(\cos y - \cos x)\, dx + (e^y - x \sin y)\, dy]$

**55.** Find the line integral $\int_C (e^x \cos y\, dx - e^x \sin y\, dy)$ for each curve $C$:

(a) $C$ consists of the line segments from $\left( 0, \dfrac{\pi}{3} \right)$ to $\left( 1, \dfrac{\pi}{3} \right)$ and from $\left( 1, \dfrac{\pi}{3} \right)$ to $(1, \pi)$.

(b) $C$ consists of the line segment from $\left( 0, \dfrac{\pi}{3} \right)$ to $(1, \pi)$.

(c) $C$ consists of the line segments from $\left( 0, \dfrac{\pi}{3} \right)$ to $\left( 0, \dfrac{\pi}{2} \right)$ and from $\left( 0, \dfrac{\pi}{2} \right)$ to $(1, \pi)$.

**56.** Find the line integral $\int_C (e^y \cos x\, dx + e^y \sin x\, dy)$ for each curve $C$:

(a) $C$ consists of the line segment from $(0, 0)$ to $(\pi, 1)$.

(b) $C$ consists of the line segments from $(0, 0)$ to $\left( \dfrac{\pi}{2}, \dfrac{1}{2} \right)$ and from $\left( \dfrac{\pi}{2}, \dfrac{1}{2} \right)$ to $(\pi, 1)$.

(c) $C$ consists of the line segments from $(0, 0)$ to $(0, 1)$ and from $(0, 1)$ to $(\pi, 1)$.

## Applications and Extensions

**57.** Show that $\displaystyle\int_C \left( -\dfrac{y}{x^2 + y^2}\, dx + \dfrac{x}{x^2 + y^2}\, dy \right)$ is independent of the path in the rectangle $R$ whose vertices are $\left( \dfrac{1}{a}, -a \right)$, $(a, -a)$, $\left( \dfrac{1}{a}, -\dfrac{1}{a} \right)$, and $\left( a, -\dfrac{1}{a} \right)$, $a > 1$. Find a potential function $f$ for $\mathbf{F}$.

**58.** Given the constant vector $\mathbf{c}$, show that $\nabla(\mathbf{c} \cdot \mathbf{r}) = \mathbf{c}$, if $\mathbf{r} = x\mathbf{i} + y\mathbf{j} + z\mathbf{k}$. Use the result to prove that $\int_C \mathbf{c} \cdot d\mathbf{r} = 0$, where $C$ is any closed piecewise smooth curve in space.

**59.** Suppose that $\mathbf{F}(x, y)$ is a force field directed toward the origin with magnitude inversely proportional to the square of the distance from the origin. (Such "inverse square law" forces are common in nature. See Examples 4 and 5 in Section 15.1, pp. 1026–1027.)

(a) Show that $\mathbf{F}$ is a conservative vector field.

(b) Find a potential function $f$ for $\mathbf{F}$.

**60.** Suppose $f$ and $g$ are differentiable functions of one variable. Show that $\int_C [f(x)\, dx + g(y)\, dy] = 0$, where $C$ is any circle in the $xy$-plane.

**61.** Verify that the force field $\mathbf{F} = y\mathbf{i} - x\mathbf{j}$ is not a conservative vector field. Show that the integral $\int_C \mathbf{F} \cdot d\mathbf{r}$ is dependent on the path of integration by using two paths in which the starting point is the origin $(0, 0)$ and the endpoint is $(1, 1)$. For one path, use the line $y = x$. For the other path, move along the $x$-axis from $(0, 0)$ to the point $(1, 0)$ and then move along the line $x = 1$ up to the point $(1, 1)$.

**62. Work**   Show that if the work done by a force $\mathbf{F} = \mathbf{F}(x, y)$ in moving an object along any piecewise-smooth closed curve in an open set $S$ in the plane is 0, then the work done by $\mathbf{F}$ in moving an object from a point $A$ to a point $B$ in $S$ is independent of the piecewise smooth path chosen.

**Challenge Problems**

**63.** Let $f$ and $g$ have continuous partial derivatives in a plane region $R$, and let $C$ be a piecewise-smooth curve in $R$ going from $A$ to $B$. Show that

$$\int_C f\, \nabla g \cdot d\mathbf{r} = f(B)g(B) - f(A)g(A) - \int_C g\, \nabla f \cdot d\mathbf{r}$$

**64. Escape Speed**

(a) Show that the minimum speed needed to escape from the surface of a planet of mass $M$ in kilograms and radius $R$ in meters is $v_{\mathrm{esc}} = \sqrt{\dfrac{2GM}{R}}$, where $G = 6.67 \times 10^{-11}\,\mathrm{N\,m^2/\,kg^2}$.

This speed is called the **escape speed** for the planet. *Hint:* Use the relationship between kinetic energy $K$ and the work $W$ done by a force $\mathbf{F}$.

(b) Find the escape speed for Earth (in kilometers per second) using $M_{\mathrm{Earth}} = 5.97 \times 10^{24}\,\mathrm{kg}$ and $R_{\mathrm{Earth}} = 6.38 \times 10^6\,\mathrm{m}$.

(c) Find the escape speed for Mars (in kilometers per second) using $M_{\mathrm{Mars}} = 6.42 \times 10^{23}\,\mathrm{kg}$ and $R_{\mathrm{Mars}} = 3.40 \times 10^6\,\mathrm{m}$.

(d) Find the escape speed for the Sun (in kilometers per second) using $M_{\mathrm{Sun}} = 1.99 \times 10^{30}\,\mathrm{kg}$ and $R_{\mathrm{Sun}} = 6.96 \times 10^8\,\mathrm{m}$.

# 15.5 Green's Theorem

**OBJECTIVES**   *When you finish this section, you should be able to:*

**1** Use Green's Theorem to find a line integral (p. 1062)

**2** Use Green's Theorem to find area (p. 1064)

**3** Use Green's Theorem with Multiply-Connected Regions (p. 1065)

**ORIGINS**   George Green (1793–1841) was a miller and self-taught mathematician who also worked with applications in electricity, magnetism, and fluid flow.

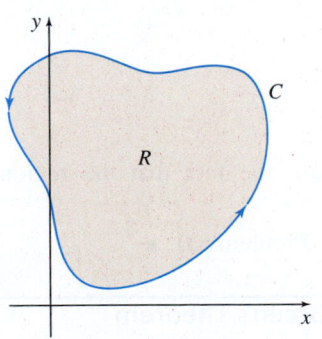

**Figure 32** $R$ lies to the left of $C$ as the curve is traversed.

**RECALL**   A region $R$ is closed if it contains all its boundary points.

In this section, we investigate an amazing theorem, known as *Green's Theorem*, that relates the value of a line integral along a simple closed curve $C$ in the plane to a double integral over the region $R$ enclosed by $C$.

Green's Theorem may be stated in several ways, but the most common form is

$$\oint_C (P\, dx + Q\, dy) = \iint_R \left( \frac{\partial Q}{\partial x} - \frac{\partial P}{\partial y} \right) dx\, dy$$

The plane curve $C$ is the boundary of the region $R$, and the symbol $\oint$ indicates that the simple closed curve $C$ is to be traversed in a counterclockwise direction. That is, the region $R$ always lies to the left as one moves along the curve, as shown in Figure 32.

The equality $\oint_C (P\, dx + Q\, dy) = \iint_R \left( \frac{\partial Q}{\partial x} - \frac{\partial P}{\partial y} \right) dx\, dy$ also requires the following two assumptions:

- The functions $P$ and $Q$ have continuous first-order partial derivatives at each point in the region $R$.
- The curve $C$ is a piecewise-smooth, simple closed curve, and the region $R$ is both simply connected and closed.

The advantage of Green's Theorem is that it can be used to transform line integrals into double integrals and vice versa, allowing us the opportunity to use the easier of the two.

## 1 Use Green's Theorem to Find a Line Integral

### THEOREM  Green's Theorem

Let $C$ be a piecewise-smooth, simple closed curve with a counterclockwise orientation and let $R$ be the simply connected, closed region consisting of $C$ and its interior. Let the functions $P$ and $Q$ have continuous first-order partial derivatives on some open region that contains $R$. Then

$$\oint_C (P\,dx + Q\,dy) = \iint_R \left( \frac{\partial Q}{\partial x} - \frac{\partial P}{\partial y} \right) dx\,dy$$

where the line integral is taken around $C$ in the counterclockwise direction.

**NEED TO REVIEW?** $x$-simple and $y$-simple regions are discussed in Section 14.2, pp. 961 and 963.

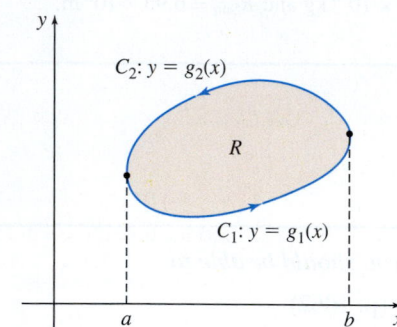

**Figure 33**

We prove Green's Theorem only for regions that are both $x$-simple and $y$-simple.

**Proof**  It is sufficient to show that

$$\oint_C P(x, y)\,dx = -\iint_R \frac{\partial P}{\partial y}\,dx\,dy \qquad \text{and} \qquad \oint_C Q(x, y)\,dy = \iint_R \frac{\partial Q}{\partial x}\,dx\,dy$$

since the result follows from adding these two equations.

Suppose the region $R$ is both $x$-simple and $y$-simple. Then since it is $x$-simple as shown in Figure 33, the boundary of $R$ consists of two smooth curves, $C_1: y = g_1(x)$ and $C_2: y = g_2(x)$, where $g_1(x) \le g_2(x)$, $a \le x \le b$.

To show $\oint_C P(x, y)\,dx = -\iint_R \frac{\partial P}{\partial y}\,dx\,dy$, we begin with the double integral.

Since $R$ is $x$-simple, by Fubini's Theorem, we have

$$-\iint_R \frac{\partial P}{\partial y}\,dx\,dy = -\int_a^b \left[ \int_{g_1(x)}^{g_2(x)} \frac{\partial P}{\partial y}\,dy \right] dx = -\int_a^b \left[ P(x, y) \right]_{g_1(x)}^{g_2(x)} dx$$

$$= -\int_a^b \left[ P(x, g_2(x)) - P(x, g_1(x)) \right] dx = \int_a^b P(x, g_1(x))\,dx + \int_b^a P(x, g_2(x))\,dx$$

$$= \int_{C_1} P(x, y)\,dx + \int_{C_2} P(x, y)\,dx = \oint_C P(x, y)\,dx$$

**NEED TO REVIEW?** Fubini's Theorems are discussed in Section 14.2, pp. 962–964.

To show $\oint_C Q(x, y)\,dy = \iint_R \frac{\partial Q}{\partial x}\,dx\,dy$, use the fact that the region $R$ is also $y$-simple. The derivation is left as an exercise (Problem 53). ∎

### EXAMPLE 1  Finding a Line Integral Using Green's Theorem

Use Green's Theorem to find the line integral

$$\oint_C [(-2xy + y^2)\,dx + x^2\,dy]$$

where $C$ is the boundary of the region $R$ enclosed by $y = 4x$ and $y = 2x^2$.

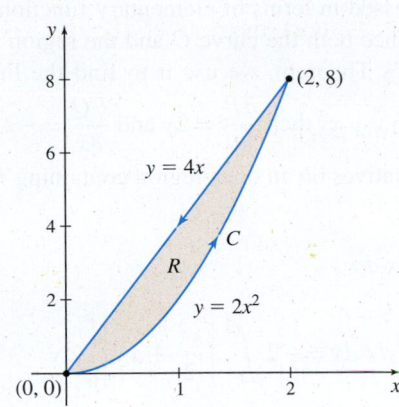

**Figure 34**

**Solution** Figure 34 illustrates the curve $C$ and the region $R$. $C$ is a piecewise-smooth closed curve and $R$ is both simply connected and closed. We let

$$P(x, y) = -2xy + y^2 \qquad \text{and} \qquad Q(x, y) = x^2$$

$$\frac{\partial P}{\partial y} = -2x + 2y \qquad\qquad \frac{\partial Q}{\partial x} = 2x$$

Since $P$ and $Q$ have continuous first-order partial derivatives on an open region containing $R$, we use Green's Theorem. Then

$$\oint_C (P\,dx + Q\,dy) = \iint_R \left( \frac{\partial Q}{\partial x} - \frac{\partial P}{\partial y} \right) dx\,dy$$

$$= \iint_R (4x - 2y)\,dx\,dy \underset{\substack{\uparrow \\ R \text{ is } x\text{-simple.}}}{=} \int_0^2 \int_{2x^2}^{4x} (4x - 2y)\,dy\,dx$$

$$= \int_0^2 \left[ 4xy - y^2 \right]_{2x^2}^{4x} dx = \int_0^2 [(16x^2 - 16x^2) - (8x^3 - 4x^4)]\,dx$$

$$= \left[ -2x^4 + \frac{4}{5}x^5 \right]_0^2 = -32 + \frac{128}{5} = -\frac{32}{5} \qquad \blacksquare$$

**NOW WORK** Problem 9.

---

**EXAMPLE 2**  **Verifying Green's Theorem**

Verify the result from Example 1 by finding the line integral directly.

**Solution** Let $C_1$ be the curve traced out by $y = 2x^2$, $0 \le x \le 2$, and let $C_2$ be the curve traced out by $y = 4x$, from $x = 2$ to $x = 0$. Then

$$\oint_C [(-2xy + y^2)\,dx + x^2\,dy] = \int_{C_1} [(-2xy + y^2)\,dx + x^2\,dy] + \int_{C_2} [(-2xy + y^2)\,dx + x^2\,dy]$$

$$= \int_0^2 [(-4t^3 + 4t^4)\,dt + 4t^3\,dt] + \int_2^0 [(-8t^2 + 16t^2)\,dt + 4t^2\,dt]$$

$$C_1 : x = t,\ y = 2t^2,\ 0 \le t \le 2 \qquad\qquad C_2 : x = t,\ y = 4t;\ \text{from } t = 2 \text{ to } t = 0$$

$$= \left[ 4\frac{t^5}{5} \right]_0^2 + 12 \left[ \frac{t^3}{3} \right]_2^0 = \frac{128}{5} - 32 = -\frac{32}{5} \qquad \blacksquare$$

**NOW WORK** Problem 11.

The next example shows that Green's Theorem can be helpful in finding line integrals that are difficult or impossible to find using the techniques of Section 15.2.

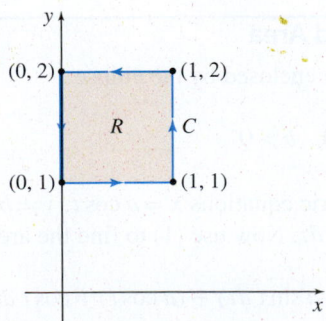

**Figure 35**

**EXAMPLE 3**  **Finding a Line Integral Using Green's Theorem**

Use Green's Theorem to find the line integral

$$\oint_C [(e^{-x^2} + y^2)\,dx + (\ln y - x^2)\,dy]$$

where $C$ is the square illustrated in Figure 35.

**Solution** Recall that $\int e^{-x^2} dx$ cannot be expressed in terms of elementary functions, so finding the integral directly is impossible. Since both the curve $C$ and the region $R$ enclosed by $C$ satisfy the conditions of Green's Theorem, we use it to find the line integral. If $P(x, y) = e^{-x^2} + y^2$ and $Q(x, y) = \ln y - x^2$ then $\dfrac{\partial P}{\partial y} = 2y$ and $\dfrac{\partial Q}{\partial x} = -2x$.

Since $P$ and $Q$ have continuous first-order derivatives on an open region containing $R$, by Green's Theorem, we have

$$\oint_C [(e^{-x^2} + y^2) \, dx + (\ln y - x^2)] \, dy = \iint_R (-2x - 2y) \, dx \, dy$$

$$= -2 \int_1^2 \int_0^1 (x + y) \, dx \, dy = -2 \int_1^2 \left[ \frac{x^2}{2} + xy \right]_0^1 dy$$

$$\underset{\substack{\uparrow \\ R \text{ is } y\text{-simple.}}}{}$$

$$= -2 \int_1^2 \left( \frac{1}{2} + y \right) dy = -2 \left[ \frac{1}{2} y + \frac{y^2}{2} \right]_1^2 = -4 \qquad \blacksquare$$

**NOW WORK** Problem 15.

In Example 3, Green's Theorem was used to convert a line integral into a double integral because the line integral could not be found. Green's Theorem is also used to convert double integrals to line integrals when the line integral is easier to integrate. This is often true when finding area.

## 2 Use Green's Theorem to Find Area

Green's Theorem can be used to express the area $A$ of a region $R$ enclosed by a piecewise-smooth, simple closed curve $C$ as a line integral.

**THEOREM Green's Theorem for Area**

If $C$ is a piecewise-smooth, simple closed curve with a counterclockwise orientation, then the area $A$ of the simply connected closed region $R$ enclosed by $C$ is given by

$$A = \frac{1}{2} \oint_C (-y \, dx + x \, dy) \tag{1}$$

**Proof** In Green's Theorem, let $P = P(x, y) = -y$ and $Q = Q(x, y) = x$. Then

$$\oint_C (-y \, dx + x \, dy) = \iint_R \left[ \frac{\partial}{\partial x} x - \frac{\partial}{\partial y} (-y) \right] dx \, dy = \iint_R 2 \, dx \, dy = 2A \qquad \blacksquare$$

**RECALL** $\iint_R dx \, dy$ equals the area of the region $R$.

In Problem 54, you are asked to show that the area $A$ can also be expressed by the formulas

$$A = \oint_C x \, dy \qquad \text{and} \qquad A = \oint_C (-y) \, dx$$

**EXAMPLE 4 Using Green's Theorem to Find Area**

Use Green's Theorem to find the area of the region enclosed by the ellipse

$$\frac{x^2}{a^2} + \frac{y^2}{b^2} = 1 \qquad a > 0 \quad b > 0$$

**Solution** We express the ellipse using the parametric equations $x = a \cos t$, $y = b \sin t$, $0 \le t \le 2\pi$. Then $dx = -a \sin t \, dt$ and $dy = b \cos t \, dt$. Now use (1) to find the area.

$$A = \frac{1}{2} \oint_C (-y \, dx + x \, dy) = \frac{1}{2} \int_0^{2\pi} [(-b \sin t)(-a \sin t \, dt) + (a \cos t)(b \cos t \, dt)]$$

$$= \frac{1}{2} \int_0^{2\pi} ab(\sin^2 t + \cos^2 t) \, dt = \frac{1}{2} ab \int_0^{2\pi} dt = \frac{1}{2} ab \cdot 2\pi = \pi ab \qquad \blacksquare$$

**NOW WORK** Problem 29.

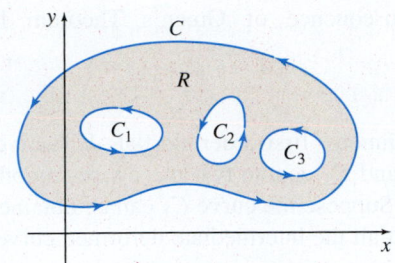

**Figure 36** $R$ is not simply connected, but $R$ is multiply-connected.

## 3 Use Green's Theorem with Multiply-Connected Regions

For Green's Theorem to hold, the double integral is found over a simply connected, closed region $R$ formed by a piecewise-smooth, simple closed curve $C$ and its interior. We now consider Green's Theorem for *multiply-connected regions*.

A **multiply-connected region** is any connected region that is not simply connected. We can think of a multiply-connected region as a connected region with holes. See Figure 36.

**THEOREM  Green's Theorem for Multiply-Connected Regions**

Let $C_1, C_2, \ldots, C_n$ be $n$ piecewise-smooth, simple closed curves for which:

- No two curves $C_i$ and $C_j$, $i \neq j$, $i, j = 1, 2, \ldots, n$, intersect.
- Each of the curves $C_i$, $i = 1, 2, \ldots, n$, lies in the interior of a piecewise-smooth, simple closed curve $C$.
- Any curve $C_i$ is exterior to every curve $C_j$, $i \neq j$, and $i, j = 1, 2, \ldots, n$.

Let $R$ denote the region consisting of the interior of $C$ and its boundary, less the interior of each of the curves $C_1, C_2, \ldots, C_n$. Suppose the functions $P$ and $Q$ have continuous first-order partial derivatives throughout $R$. Then

$$\iint_R \left( \frac{\partial Q}{\partial x} - \frac{\partial P}{\partial y} \right) dx\, dy = \oint_C (P\, dx + Q\, dy) - \sum_{i=1}^{n} \oint_{C_i} (P\, dx + Q\, dy)$$

We give the idea of the proof for $n = 1$. As shown in Figure 37, the curve $C_1$ is a circle contained within the larger circle $C$.

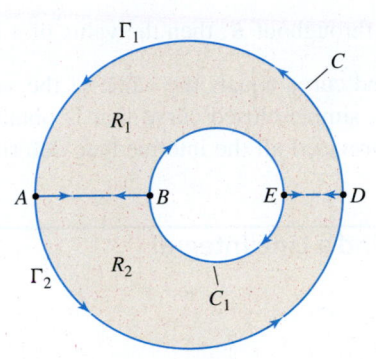

**Figure 37**

**Proof** We construct line segments $\overline{AB}$ and $\overline{ED}$ that join the curves $C$ and $C_1$. This decomposes $R$ into two regions $R_1$ and $R_2$, with boundaries $\Gamma_1$ and $\Gamma_2$, respectively. Since $R_1$ and $R_2$ each satisfy the conditions of Green's Theorem, we have

$$\iint_R \left( \frac{\partial Q}{\partial x} - \frac{\partial P}{\partial y} \right) dx\, dy = \iint_{R_1} \left( \frac{\partial Q}{\partial x} - \frac{\partial P}{\partial y} \right) dx\, dy + \iint_{R_2} \left( \frac{\partial Q}{\partial x} - \frac{\partial P}{\partial y} \right) dx\, dy$$

$$= \oint_{\Gamma_1} (P\, dx + Q\, dy) + \oint_{\Gamma_2} (P\, dx + Q\, dy)$$

$$= \left[ \int_{DA} (P\, dx + Q\, dy) + \int_{\overline{AB}} (P\, dx + Q\, dy) + \int_{BE} (P\, dx + Q\, dy) + \int_{\overline{ED}} (P\, dx + Q\, dy) \right]$$

$$+ \left[ \int_{\overline{DE}} (P\, dx + Q\, dy) + \int_{EB} (P\, dx + Q\, dy) + \int_{\overline{BA}} (P\, dx + Q\, dy) + \int_{AD} (P\, dx + Q\, dy) \right]$$

Along the line segments $\overline{AB}$ and $\overline{ED}$,

$$\int_{\overline{AB}} (P\, dx + Q\, dy) = - \int_{\overline{BA}} (P\, dx + Q\, dy) \quad \text{and} \quad \int_{\overline{ED}} (P\, dx + Q\, dy) = - \int_{\overline{DE}} (P\, dx + Q\, dy)$$

So,

$$\iint_R \left( \frac{\partial Q}{\partial x} - \frac{\partial P}{\partial y} \right) dx\, dy = \oint_C (P\, dx + Q\, dy) + \oint_{-C_1} (P\, dx + Q\, dy)$$

$$= \oint_C (P\, dx + Q\, dy) - \oint_{C_1} (P\, dx + Q\, dy) \qquad \blacksquare$$

The next theorem is an important consequence of Green's Theorem for Multiply-Connected Regions.

**THEOREM**

Let $P$ and $Q$ be functions that have continuous first-order partial derivatives throughout an open, connected set $R$. Let $C_1$ and $C_2$ denote two piecewise-smooth, simple closed curves, each lying entirely in $R$. Suppose the curve $C_1$ can be obtained by continuously deforming the curve $C_2$ so that all the intermediate deformed curves lie entirely in $R$. If $\dfrac{\partial P}{\partial y} = \dfrac{\partial Q}{\partial x}$ throughout $R$, then

$$\int_{C_1} (P\,dx + Q\,dy) = \int_{C_2} (P\,dx + Q\,dy)$$

The significance of this theorem is that it allows us, under the conditions of the theorem, to replace a line integral over a simple closed curve $C_1$ that is difficult to find with an easier line integral over the path $C_2$.

Figure 38 illustrates a set $R$ satisfying the conditions of the theorem, which can be restated informally as follows: If $\dfrac{\partial P}{\partial y} = \dfrac{\partial Q}{\partial x}$ throughout $R$, then the value of a line integral along a piecewise-smooth, simple closed curve equals the value of the same line integral along any other piecewise-smooth, simple closed curve that is obtained by continuously deforming the original curve, provided all the intermediate deformed curves lie entirely in $R$.

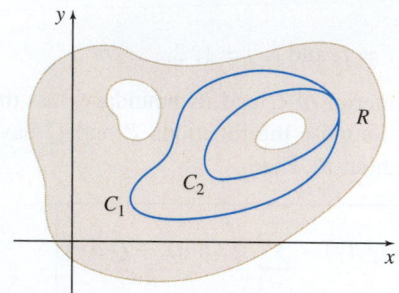

**Figure 38**

---

**EXAMPLE 5** Using Green's Theorem to Find a Line Integral

Find $\displaystyle\oint_C \left( \frac{-y}{x^2+y^2}\,dx + \frac{x}{x^2+y^2}\,dy \right)$ if:

**(a)** $C$ is the curve $x^{2/3} + y^{2/3} = 1$, whose orientation is counterclockwise.

**(b)** $C$ is any piecewise-smooth, simple closed curve not containing the origin in its interior.

**Solution** For $P = \dfrac{-y}{x^2+y^2}$ and $Q = \dfrac{x}{x^2+y^2}$, we have

$$\frac{\partial P}{\partial y} = \frac{y^2 - x^2}{(x^2+y^2)^2} \qquad \text{and} \qquad \frac{\partial Q}{\partial x} = \frac{y^2 - x^2}{(x^2+y^2)^2}$$

Notice that, except at the point $(0,0)$, the partial derivatives $\dfrac{\partial P}{\partial y}$ and $\dfrac{\partial Q}{\partial x}$ are continuous.

**(a)** It is difficult to find the line integral along $C$. Moreover, because $x^{2/3} + y^{2/3} = 1$ contains the point $(0,0)$ in its interior, we cannot use Green's Theorem as we did in Example 2 to change the line integral to a double integral. However, since $\dfrac{\partial P}{\partial y} = \dfrac{\partial Q}{\partial x}$ throughout $R$, we can use a different curve that leads to an easier line integral. For example, suppose we replace $x^{2/3} + y^{2/3} = 1$ by the unit circle $C_1 : x^2 + y^2 = 1$, noting that the conditions required for $\int_C (P\,dx + Q\,dy)$ to equal $\int_{C_1} (P\,dx + Q\,dy)$ are met. See Figure 39.

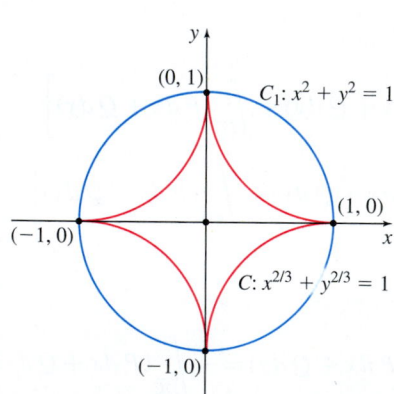

**Figure 39**

On $C_1$, $x = \cos\theta$, $y = \sin\theta$, $0 \le \theta \le 2\pi$. Then $dx = -\sin\theta\,d\theta$, $dy = \cos\theta\,d\theta$, and

$$\oint_C \left( \frac{-y}{x^2+y^2}\,dx + \frac{x}{x^2+y^2}\,dy \right) = \oint_{C_1} \left( \frac{-y}{x^2+y^2}\,dx + \frac{x}{x^2+y^2}\,dy \right)$$

$$= \int_0^{2\pi} (\sin^2\theta + \cos^2\theta)\,d\theta = 2\pi$$

**(b)** In this case, the interior of the curve $C$ does not contain the origin. Since

$$\frac{\partial P}{\partial y} = \frac{\partial Q}{\partial x}$$

in the simply connected region $R$ enclosed by $C$, it follows that $\mathbf{F} = P(x, y)\mathbf{i} + Q(x, y)\mathbf{j}$ is a conservative vector field. Since $C$ is closed,

$$\oint_C \left( \frac{-y}{x^2 + y^2}\, dx + \frac{x}{x^2 + y^2}\, dy \right) = \oint_C \mathbf{F} \cdot d\mathbf{r} = 0 \qquad \blacksquare$$

NOW WORK Problem **27.**

## Application of Green's Theorem to Conservative Vector Fields

In Section 15.4, we proved the following result: If $\mathbf{F} = \mathbf{F}(x, y) = P(x, y)\mathbf{i} + Q(x, y)\mathbf{j}$ is a conservative vector field and the components $P$ and $Q$ have continuous first-order partial derivatives on an open, connected region $R$, then $\dfrac{\partial P}{\partial y} = \dfrac{\partial Q}{\partial x}$ throughout $R$.

We now use Green's Theorem to prove that if $R$ is a simply connected region, then the converse is also true.

> **THEOREM   Converse of the Conservative Vector Field Theorem**
>
> Let $\mathbf{F} = P(x, y)\mathbf{i} + Q(x, y)\mathbf{j}$ be a vector field whose components $P$ and $Q$ have continuous first-order partial derivatives throughout an open, simply connected region $R$. If $\dfrac{\partial P}{\partial y} = \dfrac{\partial Q}{\partial x}$ throughout $R$, then $\mathbf{F}$ is a conservative vector field.

To show that $\mathbf{F}$ is a conservative vector field, we only need to show that for any curve $C$ in $R$, $\int_C \mathbf{F} \cdot d\mathbf{r} = \int_C (P\, dx + Q\, dy)$ is independent of the path. That is, $\int_C \mathbf{F} \cdot d\mathbf{r}$ has the same value for any two piecewise-smooth curves joining a fixed point $A$ in $R$ to an arbitrary point $(x, y)$ in $R$.

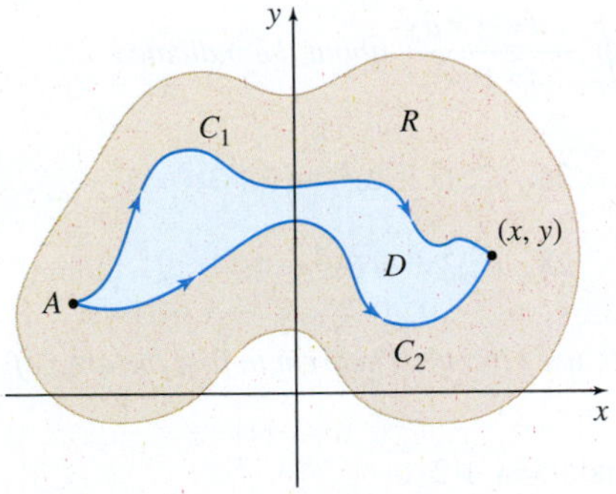

**Proof**   Let $C_1$ and $C_2$ be two piecewise-smooth curves in $R$ joining $A$ to $(x, y)$. Assume that these curves do not intersect or coincide except at $A$ and $(x, y)$.* The curves $C_1$ and $C_2$ form the boundary of a region $D$ that is a subset of $R$, since $R$ is simply connected. See Figure 40.

Since the conditions for Green's Theorem are satisfied on the region $D$, we have

$$\iint_D \left( \frac{\partial Q}{\partial x} - \frac{\partial P}{\partial y} \right) dx\, dy = \int_{C_2 + (-C_1)} (P\, dx + Q\, dy)$$

$$= \int_{C_2} (P\, dx + Q\, dy) + \int_{-C_1} (P\, dx + Q\, dy)$$

$$= \int_{C_2} (P\, dx + Q\, dy) - \int_{C_1} (P\, dx + Q\, dy)$$

Since $\dfrac{\partial P}{\partial y} = \dfrac{\partial Q}{\partial x}$ throughout $R$, then $\displaystyle\iint_D \left( \frac{\partial Q}{\partial x} - \frac{\partial P}{\partial y} \right) dx\, dy = 0$. Therefore,

$$\int_{C_1} (P\, dx + Q\, dy) = \int_{C_2} (P\, dx + Q\, dy)$$

showing that the line integral is independent of the path, and $\mathbf{F}$ is conservative. ∎

**Figure 40**

---

*This assumption may be relaxed so that the curves $C_1$ and $C_2$ may sometimes interesect or even coincide in $R$.

## 15.5 Assess Your Understanding

### Concepts and Vocabulary

1. Green's Theorem relates the value of a(n) —————— integral along a simple closed piecewise-smooth curve $C$ to the value of a(n) —————— integral over the simply connected closed region $R$ enclosed by $C$.

2. If the assumptions of Green's Theorem are met, then
$$\oint_C [P(x, y)\, dx + Q(x, y)\, dy] = \iint_R \underline{\hspace{1cm}} dx\, dy.$$

3. *True or False* The symbol $\oint_C$ indicates that a simple closed curve $C$ is to be traversed in a clockwise direction.

4. If $C$ is a piecewise-smooth, simple closed curve,
then $\dfrac{1}{2} \oint_C (-y\, dx + x\, dy)$ equals the —————— of the closed simply connected region $R$ enclosed by $C$.

### Skill Building

*In Problems 5–10, use Green's Theorem to find each line integral. In each case, assume that $C$ is the perimeter of the rectangle with vertices $(0, 0)$, $(4, 0)$, $(4, 3)$, $(0, 3)$, and that the orientation of $C$ is counterclockwise.*

5. $\oint_C y\, dx$

6. $\oint_C x\, dy$

7. $\oint_C [xy\, dx + (x + y)\, dy]$

8. $\oint_C (3y\, dx - 2x\, dy)$

9. $\oint_C (xy^2\, dx + x^2 y\, dy)$

10. $\oint_C [y\sin(xy)\, dx + x\sin(xy)\, dy]$

11. Use a double integral and Green's Theorem to find $\oint_C xy\, dx + y^2 dy$, where $C$ is the triangle formed by the line segments from $(0, 0)$ to $(1, 0)$, from $(1, 0)$ to $(0, 2)$, and from $(0, 2)$ to $(0, 0)$. Then verify Green's Theorem by evaluating the line integral directly.

12. Use a double integral and Green's Theorem to find $\oint_C xy\, dx + x^2 dy$, where $C$ is the triangle formed by the line segments from $(0, 0)$ to $(1, 0)$, from $(1, 0)$ to $(0, 2)$, and from $(0, 2)$ to $(0, 0)$. Then verify Green's Theorem by evaluating the line integral directly.

13. Use a double integral and Green's Theorem to find $\oint_C y\, dx + (x + y)\, dy$, where $C$ is the closed curve formed by $y = 2x$ and $y = x^2$. Then verify Green's Theorem by evaluating the line integral directly.

14. Use a double integral and Green's Theorem to find $\oint_C (x - y)\, dx + xy\, dy$, where $C$ is the closed curve in the first quadrant formed by the graphs of $y = 2x^2$, $y = 3 - x^2$, and the $y$-axis. Then verify Green's Theorem by evaluating the line integral directly.

*In Problems 15–18, use Green's Theorem to find each line integral. In each problem, assume that $C$ is the unit circle $x^2 + y^2 = 1$, and that the orientation of $C$ is counterclockwise.*

15. $\oint_C (-x^2 y\, dx + y^2 x\, dy)$

16. $\oint_C [y(x^2 + y^2)\, dx - x(x^2 + y^2)\, dy]$

17. $\oint_C [(x^2 - y^3)\, dx + (x^2 + y^2)\, dy]$

18. $\oint_C [(xy^3 + \sin x)\, dx + (x^2 y^2 + 4x)\, dy]$

*In Problems 19–24, use Green's Theorem to find each line integral. In each problem, $C$ is to be traversed in the counterclockwise direction.*

19. $\oint_C [(x^2 + y)\, dx + (x - y^2)\, dy]$, where $C$ is the boundary of the region $R$ bounded by $y = x^{3/2}$, the $x$-axis, and $x = 1$

20. $\oint_C [(4x^2 - 8y^2)\, dx + (y - 6xy)\, dy]$, where $C$ is the boundary of the region $R$ bounded by $y = \sqrt{x}$ and $y = x^2$

21. $\oint_C [(x^3 - x^2 y)\, dx + xy^2\, dy]$, where $C$ is the boundary of the region $R$ bounded by $y = x^2$ and $x = y^2$

22. $\oint_C [(x^2 - y^2)\, dx + xy\, dy]$, where $C$ is the boundary of the region $R$ bounded by $x = y^2$ and $x = 1$

23. $\oint_C \left[ \dfrac{1}{y}\, dx + \dfrac{1}{x}\, dy \right]$, where $C$ is the boundary of the region $R$ bounded by $y = 1$, $x = 9$, and $y = \sqrt{x}$

24. $\oint_C (x^2 y\, dx - y^2 x\, dy)$, where $C$ is the boundary of the region $R$ bounded by $y = \sqrt{a^2 - x^2}$ and $y = 0$

*In Problems 25–28, find $\oint_C \dfrac{y\, dx - x\, dy}{x^2 + y^2}$ about the indicated curve $C$.*

25. $C: x^2 + y^2 = 4$

26. $C: (x - 2)^2 + (y + 3)^2 = 1$

27. $C: x^2 + 4y^2 = 4$

28. $C: 3x^2 + 6y^2 = 4$

**Area** *In Problems 29–36, use Green's Theorem to find the area of each region.*

29. Bounded by $y = x^2$ and $y = x + 2$

30. Bounded by $y = x^2 - 1$ and $y = 0$

31. Bounded by $y = x^3$ and $y = x^2$

32. Bounded by $y = 2\sqrt{x}$ and $y = \dfrac{x^2}{4}$

33. Bounded by $y = \dfrac{1}{\sqrt{x - 1}}$, $y = 0$, $x = 2$, and $x = 5$

34. Bounded by $y = x^{3/2}$ and $y = x$

35. Bounded by the line $x + y = 3$ and the hyperbola $xy = 2$

36. Bounded by the hyperbola $xy = \sqrt{3}$ and the circle $x^2 + y^2 = 4$, in the first quadrant only

## Applications and Extensions

**37. Area**   Find the area under one arch of the cycloid
$\mathbf{r}(t) = [2\pi t - \sin(2\pi t)]\,\mathbf{i} + [1 - \cos(2\pi t)]\,\mathbf{j},\ 0 \le t \le 1.$

**38. Area**   Find the area enclosed by the hypocycloid $x^{2/3} + y^{2/3} = 1$.
*Hint:* Use the parametric equations $x = \cos^3 t$, $y = \sin^3 t$.

**39.** A **folium of Descartes** is given by the parametric
equations $x(t) = \dfrac{6t}{1 + t^3}$, $y(t) = \dfrac{6t^2}{1 + t^3}$. Use Green's Theorem
to show that the area $A$ enclosed by the loop of this folium of
Descartes is 6.

*In Problems 40–44, use the formulas from Problem 56 to find the
center of mass of each region.*

**40.** The region bounded by the $x$-axis and the semi-circle
$y = \sqrt{4 - x^2}$

**41.** The region bounded by the parabola $y = x^2$ and the line $y = 1$

**42.** The region bounded by the triangle with vertices at $(0, 0)$, $(1, 0)$,
and $(0, 1)$

**43.** The region bounded by the trapezoid with vertices at $(0, 0)$,
$(1, 0)$, $(1, 2)$, and $(0, 1)$

**44.** The region bounded by the coordinate axes and the hypocycloid
$x = a \cos^3 t$, $y = a \sin^3 t$, $0 \le t \le \dfrac{\pi}{2}$

**45. Work**   Use Green's Theorem to find the work done by the
variable force $\mathbf{F}(x, y) = x\mathbf{i} + y^2\mathbf{j}$ in moving an object along the
curve $x(t) = t$, $y(t) = t^2$, $0 \le t \le 4$, and then returning to the
point $(0, 0)$ along the line $x(t) = t$, $y(t) = 4t$.

**46. Work**   Use Green's Theorem to find the work done by the
variable force $\mathbf{F}(x, y) = (e^x + 2y)\,\mathbf{i} + (e^y - x)\,\mathbf{j}$ in moving an
object along the closed path $x^2 + y^2 = 9$.

**47.** Find
$$\oint_C \left\{ [\ln(1 + x^2) - x^2 \cos y]\,dx + \left( \frac{x^3}{3} \cos y + e^{\cos y} \right) dy \right\},$$
where $C$ is the boundary of the region enclosed by $y = x^3$, $x = 1$,
and the $x$-axis.

**48.** Find $\oint_C [y^3\,dx - x^3\,dy]$, where $C$ is the boundary of the region
enclosed by $x^2 + y^2 = 9$ and $x^2 + y^2 = 1$.

*In Problems 49–52, explain why each line integral is equal to 0.*

**49.** $\oint_C \left( x e^{x^2 + y^2}\,dx + y e^{x^2 + y^2}\,dy \right)$, where $C$ is any
piecewise-smooth, simple closed curve in the plane

**50.** $\oint_C [e^{xy}(xy + 1)\,dx + x^2 e^{xy}\,dy]$, where $C$ is any
piecewise-smooth, simple closed curve in the plane

**51.** $\oint_C (e^x \sin y\,dx + e^x \cos y\,dy)$, where $C$ is any piecewise-smooth,
simple closed curve in the plane

**52.** $\oint_C \left[ \dfrac{x\,dx}{(x^2 + y^2)^{1/2}} + \dfrac{y\,dy}{(x^2 + y^2)^{1/2}} \right]$, where $C$ is any
piecewise-smooth, simple closed curve not containing the origin
in its interior

**53. Green's Theorem**   Complete the proof of Green's Theorem
for $x$-simple and $y$-simple regions by showing
$$\oint_C Q(x, y)\,dy = \iint_R \frac{\partial Q}{\partial x}\,dx\,dy$$

**54.** Use Green's Theorem for finding the area $A$ of the region $R$
enclosed by simple closed piecewise-smooth curve $C$ to show
that $A = \oint_C x\,dy$ and $A = \oint_C (-y)\,dx$.

**55.** Consider the integral $\displaystyle\oint_C \frac{y\,dx - x\,dy}{(x^2 + y^2)^{3/2}}$. For which simple closed
piecewise-smooth curves $C$ can Green's Theorem be applied to
transform this line integral to a double integral?

## Challenge Problems

**56. Center of Mass**   A homogeneous lamina occupies a region in
the $xy$-plane with boundary $C$ (oriented counterclockwise). Show
that its center of mass has the coordinates given below, where $A$
is the area of the region.
$$\bar{x} = \frac{1}{2A} \oint_C x^2\,dy \qquad \bar{y} = -\frac{1}{2A} \oint_C y^2\,dx$$

**57. Moment of Inertia**   Use Problem 56 to show that the moment of
inertia about the origin is
$$I_o = \frac{1}{3} \oint_C (x^3\,dy - y^3\,dx)$$
Assume that the mass density of the lamina is 1.

**58.** Let $R$ be a simply connected closed region in the $xy$-plane, and
let $R'$ be a simply connected closed region in the $st$-plane.
Suppose $x = x(s, t)$ and $y = y(s, t)$ are differentiable and that,
as $(s, t)$ is allowed to vary throughout $R'$, $(x, y)$ varies
throughout $R$ in a one-to-one correspondence. Suppose further
that $R$ has boundary $C$, which is a piecewise-smooth closed
curve, and $R'$ has boundary $C'$, which is a piecewise-smooth
closed curve. Finally, suppose that as $(s, t)$ moves
along $C'$, $(x(s, t), y(s, t))$ moves along $C$. Show that
$$\iint_R dx\,dy = \iint_{R'} |(x_s y_t - x_t y_s)|\,ds\,dt$$

*Hint:* Using $\displaystyle\iint_R dx\,dy = \frac{1}{2} \oint_C (x\,dy - y\,dx)$, write the integral
on the right in terms of $s$ and $t$. Then apply Green's Theorem to
the resulting integral, which will have the form $\oint_{C'} (P\,ds + Q\,dt)$.

**59. Transformation of a Double Integral**   Let $\phi(x, y)$
be continuous and let $P(x, y) = -\int \phi(x, y)\,dy$
and $Q(x, y) = \int \phi(x, y)\,dx$. Generalize the result of
Problem 54 using $P$ and $Q$ to show that
$$\iint_R \phi(x, y)\,dx\,dy = \iint_{R'} \phi(x(s, t), y(s, t))\,|x_s y_t - x_t y_s|\,ds\,dt$$
Go through a similar process as in Problem 58, using Green's
Theorem on the integral $\oint_C P\,dx + Q\,dy$.

# 15.6 Parametric Surfaces

**OBJECTIVES** *When you finish this section, you should be able to:*

1 Describe surfaces defined parametrically (p. 1071)
2 Find a parametric representation of a surface (p. 1072)
3 Find equations for a tangent plane and a normal line (p. 1074)
4 Find the surface area of a smooth surface (p. 1076)

In Chapter P, we observed that the graph of an explicitly defined function $y = f(x)$ satisfies the Vertical-line Test. We also saw that some graphs, such as circles, do not pass the Vertical-line Test. The equations of such graphs may be defined implicitly as $F(x, y) = 0$. Then in Chapters 9 and 12, we found that parametric equations $x = x(t)$, $y = y(t)$ and vector-valued functions $\mathbf{r} = \mathbf{r}(t) = x(t)\mathbf{i} + y(t)\mathbf{j}$, where $a \leq t \leq b$, allow us to represent a curve in a plane that is not necessarily the graph of function. Similarly, parametric equations $x = x(t)$, $y = y(t)$, $z = z(t)$ and vector functions $\mathbf{r} = \mathbf{r}(t) = x(t)\mathbf{i} + y(t)\mathbf{j} + z(t)\mathbf{k}$, where $a \leq t \leq b$, represent curves in space.

In this section, we define parametric equations and vector-valued functions that represent surfaces in space. Since a surface in space requires two independent variables, the parametric representation of a surface involves two parameters, $u$ and $v$.

**DEFINITION Parametric Surface**

Suppose

$$x = x(u, v) \qquad y = y(u, v) \qquad z = z(u, v)$$

are functions of two variables that are defined on a region $R$ in the $uv$-plane. Also suppose that the functions are continuous and one-to-one* on the interior of $R$. Then $u$ and $v$ are **parameters**; $x = x(u, v)$, $y = y(u, v)$, $z = z(u, v)$ are **parametric equations**; and the set of points defined by the parametric equations is a **parametric surface**.

The vector-valued function

$$\mathbf{r}(u, v) = x(u, v)\mathbf{i} + y(u, v)\mathbf{j} + z(u, v)\mathbf{k}$$

is called a **parametrization** of the surface. The region $R$ is the **parameter domain** of $\mathbf{r}$, and the range of $\mathbf{r}$ is the surface traced out by $\mathbf{r}$.

Just as we can graph a function $z = f(x, y)$ using contour lines or graph a quadric surface using traces, we can graph parametric surfaces using *coordinate curves*.

**DEFINITION Coordinate Curves**

Suppose $S$ is a surface with parametrization $\mathbf{r}(u, v) = x(u, v)\mathbf{i} + y(u, v)\mathbf{j} + z(u, v)\mathbf{k}$ and parameter domain $R$. Then the curve traced out by $\mathbf{r} = \mathbf{r}(u, v)$ as $(u, v)$ ranges over $R$, while holding $v = v_0$ constant,

$$\mathbf{r}(u, v_0) = x(u, v_0)\mathbf{i} + y(u, v_0)\mathbf{j} + z(u, v_0)\mathbf{k}$$

is called a $u$-**coordinate curve**.

Similarly, the curve traced out as $(u, v)$ ranges over $R$, while holding $u = u_0$ constant,

$$\mathbf{r}(u_0, v) = x(u_0, v)\mathbf{i} + y(u_0, v)\mathbf{j} + z(u_0, v)\mathbf{k}$$

is called a $v$-**coordinate curve**.

---

*We require the parametrization to be one-to-one on the interior of $R$ so that the surface does not fold over on itself.

Suppose the surface $S$ is parametrized by $\mathbf{r} = \mathbf{r}(u, v)$ and parameter domain $R$. Then if we hold $u = u_0$ constant in the $uv$-plane, we obtain the vertical line $u = u_0$. Similarly, if we hold $v = v_0$ constant, we obtain a horizontal line $v = v_0$. The portions of these lines that are in the interior of the parameter domain $R$ map to coordinate curves on the surface $S$. By sketching several coordinate curves, we obtain a **wireframe** plot of the surface $S$, as shown in Figure 41.

## 1 Describe Surfaces Defined Parametrically

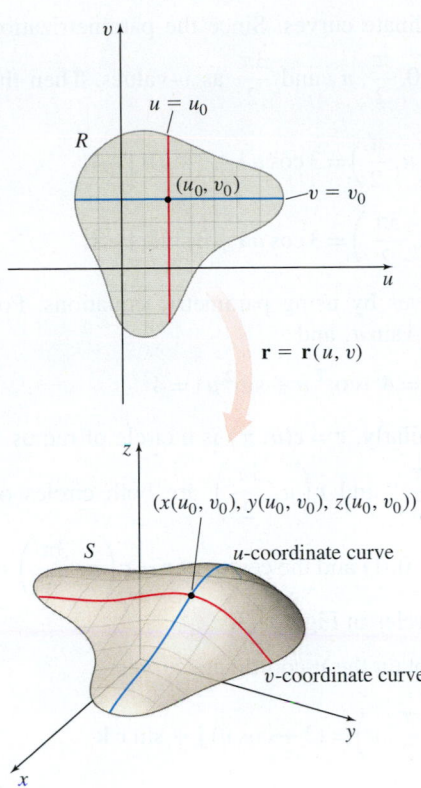

**EXAMPLE 1** Describing a Parametric Surface

**(a)** Describe the parametric surface $S$ given by
$$\mathbf{r}(u, v) = (1 + 2u + v)\mathbf{i} + (u - v)\mathbf{j} + (3 + v)\mathbf{k}.$$
**(b)** Find a rectangular equation for $S$.

**Solution (a)** We begin by finding several coordinate curves. We choose to use $v = 0$, $v = 1$, $u = 0$, and $u = 1$, but any numbers will do.

When $v = 0$, the $u$-coordinate curve is $\mathbf{r}(u, 0) = (1 + 2u)\mathbf{i} + u\mathbf{j} + 3\mathbf{k}$.

When $v = 1$, the $u$-coordinate curve is $\mathbf{r}(u, 1) = (2 + 2u)\mathbf{i} + (u - 1)\mathbf{j} + 4\mathbf{k}$.

When $u = 0$, the $v$-coordinate curve is $\mathbf{r}(0, v) = (1 + v)\mathbf{i} - v\mathbf{j} + (3 + v)\mathbf{k}$.

When $u = 1$, the $v$-coordinate curve is $\mathbf{r}(1, v) = (3 + v)\mathbf{i} + (1 - v)\mathbf{j} + (3 + v)\mathbf{k}$.

The graph of the parametric surface is a plane as shown in Figure 42.

**(b)** To find a rectangular equation for the surface, write the components of $\mathbf{r} = \mathbf{r}(u, v)$ as a system of equations. Since $\mathbf{r} = x\mathbf{i} + y\mathbf{j} + z\mathbf{k} = x(u, v)\mathbf{i} + y(u, v)\mathbf{j} + z(u, v)\mathbf{k}$, we have

$$\begin{cases} x = 2u + v + 1 & (1) \\ y = u - v & (2) \\ z = v + 3 & (3) \end{cases}$$

If we eliminate $u$ and $v$ from this system, we will have an expression involving $x$, $y$, and $z$. We eliminate $v$ by adding equations (1) and (2) and equations (2) and (3).

$$x + y = 3u + 1 \qquad \text{(4)} \quad \text{Add (1) and (2).}$$
$$y + z = u + 3 \qquad \text{(5)} \quad \text{Add (2) and (3).}$$

Now eliminate $u$:

$$x + y = 3u + 1 \qquad \text{(4)}$$
$$\underline{-3y - 3z = -3u - 9} \qquad \qquad \text{Multiply (5) by } -3.$$
$$x - 2y - 3z = -8 \qquad \qquad \text{Add.}$$

This is the equation of a plane whose normal vector is $\mathbf{i} - 2\mathbf{j} - 3\mathbf{k}$. ∎

In Example 1, there were no restrictions on the parameters $u$ and $v$, so the parametric surface $S$ is the entire plane. By restricting the parameter domain, we can define a half-plane or a parallelogram. For some surfaces, such as the one in Example 2, the parameter domain is restricted by the nature of the surface.

**Figure 41** A wireframe plot of $S$.

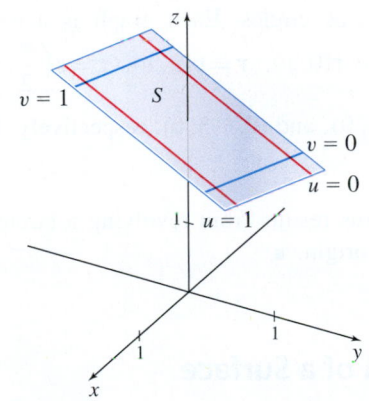

**Figure 42**

**EXAMPLE 2** Describing a Parametric Surface

Describe the parametric surface $S$ parametrized by

$$\mathbf{r}(u, v) = (3 + \cos v) \cos u\, \mathbf{i} + (3 + \cos v) \sin u\, \mathbf{j} + \sin v\, \mathbf{k}$$

where $0 \leq u \leq 2\pi$ and $0 \leq v \leq 2\pi$.

**Solution** We begin by finding some $u$-coordinate curves. Since the parametrization involves trigonometric functions, we choose $0$, $\dfrac{\pi}{2}$, $\pi$, and $\dfrac{3\pi}{2}$ as $v$-values. Then the $u$-coordinate curves are

$$\mathbf{r}(u, 0) = 4\cos u\,\mathbf{i} + 4\sin u\,\mathbf{j} \qquad \mathbf{r}\left(u, \frac{\pi}{2}\right) = 3\cos u\,\mathbf{i} + 3\sin u\,\mathbf{j} + \mathbf{k}$$

$$\mathbf{r}(u, \pi) = 2\cos u\,\mathbf{i} + 2\sin u\,\mathbf{j} \qquad \mathbf{r}\left(u, \frac{3\pi}{2}\right) = 3\cos u\,\mathbf{i} + 3\sin u\,\mathbf{j} - \mathbf{k}$$

We identify the graphs of each of these curves by using parametric equations. For example, for $\mathbf{r}(u, 0)$, we have $x = 4\cos u$, $y = 4\sin u$, and

$$x^2 + y^2 = (4\cos u)^2 + (4\sin u)^2 = 4^2(\cos^2 u + \sin^2 u) = 4^2$$

a circle of radius 4 centered at the origin. Similarly, $\mathbf{r} = \mathbf{r}(u, \pi)$ is a circle of radius 2 centered at the origin. The graphs of $\mathbf{r}\left(u, \dfrac{\pi}{2}\right)$ and $\mathbf{r}\left(u, \dfrac{3\pi}{2}\right)$ are both circles of radius 3, but the center of $\mathbf{r} = \mathbf{r}\left(u, \dfrac{\pi}{2}\right)$ is at $(0, 0, 1)$ and the center of $\mathbf{r} = \mathbf{r}\left(u, \dfrac{3\pi}{2}\right)$ is at $(0, 0, -1)$. Do you see why? See the blue circles in Figure 43.

Using $0$, $\dfrac{\pi}{2}$, $\pi$, and $\dfrac{3\pi}{2}$ as $u$-values, we obtain the $v$-coordinate curves:

$$\mathbf{r}(0, v) = (3 + \cos v)\mathbf{i} + \sin v\,\mathbf{k} \qquad \mathbf{r}\left(\frac{\pi}{2}, v\right) = (3 + \cos v)\mathbf{j} + \sin v\,\mathbf{k}$$

$$\mathbf{r}(\pi, v) = -(3 + \cos v)\mathbf{i} + \sin v\,\mathbf{k} \qquad \mathbf{r}\left(\frac{3\pi}{2}, v\right) = -(3 + \cos v)\mathbf{j} + \sin v\,\mathbf{k}$$

Again, we identify the graphs of these curves as circles. Each graph is a circle of radius 1, but the centers of the circles $\mathbf{r} = \mathbf{r}(0, v)$, $\mathbf{r} = \mathbf{r}(\pi, v)$, $\mathbf{r} = \mathbf{r}\left(\dfrac{\pi}{2}, v\right)$, and $\mathbf{r} = \mathbf{r}\left(\dfrac{3\pi}{2}, v\right)$ are $(3, 0, 0)$, $(-3, 0, 0)$, $(0, 3, 0)$, and $(0, -3, 0)$, respectively. See the red circles in Figure 43.

The surface $S$ is a torus. This particular torus results from revolving a circle of radius 1 about a circle of radius 3 centered at the origin. ∎

**NOW WORK** Problem 5.

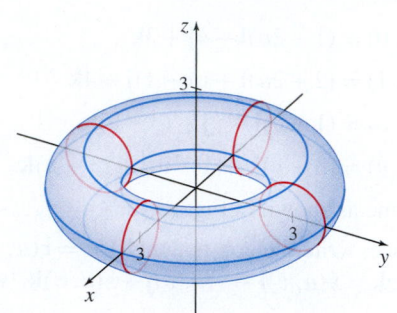

**Figure 43** A torus centered at the origin.

**NEED TO REVIEW?** A torus is discussed in Section 6.8, p. 492.

**NEED TO REVIEW?** Finding parametric representations of functions of one variable is discussed in Section 9.1, pp. 684–685.

## 2 Find a Parametric Representation of a Surface

Suppose we know the implicit equation $F(x, y, z) = 0$ of a surface. If this equation can be solved explicitly for one of the variables, then the function has a natural parametrization. For example, if $z = f(x, y)$, then $x = u$, $y = v$, and $z = f(u, v)$ parametrizes the surface. So, if a surface can be described by an equation explicitly solved for one coordinate in terms of the other two, we can use these two coordinates as parameters and obtain a parametrization.

**EXAMPLE 3** Representing a Surface Parametrically

Find a parametrization of the surface $S$ given by

$$x + y^2 + z = 5 \qquad 0 \le x \le 1 \qquad -1 \le y \le 1$$

**Solution** We begin by expressing $S$ as an explicit function of $x$ and $y$.

$$z = f(x, y) = 5 - x - y^2$$

Since $z$ is a function of $x$ and $y$, we let $x = u$ and $y = v$. Then the parametric equations of $S$ are

$$x = u \qquad y = v \qquad z = 5 - u - v^2$$

Using these parametric equations, we obtain

$$\mathbf{r}(u, v) = u\mathbf{i} + v\mathbf{j} + (5 - u - v^2)\mathbf{k} \qquad 0 \le u \le 1 \qquad -1 \le v \le 1 \qquad ∎$$

The graph of the surface $\mathbf{r}(u, v) = u\mathbf{i} + v\mathbf{j} + (5 - u - v^2)\mathbf{k}$, $0 \le u \le 1$ and $-1 \le v \le 1$, is shown in Figure 44.

Alternatively, in Example 3, we could solve for $x$ and get $x = 5 - z - y^2$. Then using $u = y$ and $v = z$, we obtain

$$\mathbf{r}(u, v) = (5 - v - u^2)\mathbf{i} + u\mathbf{j} + v\mathbf{k}$$

However, in this case, finding the bounds for $u$ and $v$ is not as straightforward. Setting $x = 0$ and $x = 1$ yields $z = 5 - y^2$ and $z = 4 - y^2$, and so $4 - u^2 \le v \le 5 - u^2$ and $-1 \le u \le 1$.

**NOW WORK** Problem 9.

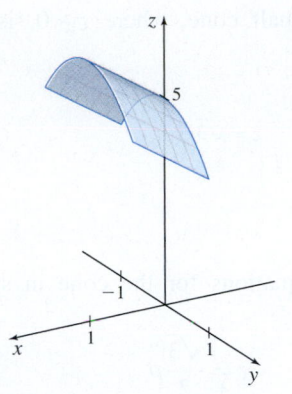

**Figure 44** $\mathbf{r}(u, v) = u\mathbf{i} + v\mathbf{j} + (5 - u - v^2)\mathbf{k}$, $0 \le u \le 1$, $-1 \le v \le 1$

**NEED TO REVIEW?** Spherical coordinates are discussed in Section 14.8, pp. 1006–1011.

### EXAMPLE 4  Parametrizing a Sphere

Find a parametrization for a sphere $S$ of radius 2 centered at the origin.

**Solution** The equation of the sphere, $x^2 + y^2 + z^2 = 4$, in rectangular coordinates cannot be expressed explicitly as a function of two of its variables. However, in spherical coordinates, the sphere has the explicit equation $\rho = 2$. So if we use the spherical coordinates $\theta$ and $\phi$ as parameters, then parametric equations of the sphere are given by

$$x = 2\cos\theta \sin\phi \qquad y = 2\sin\theta \sin\phi \qquad z = 2\cos\phi$$
$$\rho = 2, \; x = \rho\cos\theta \sin\phi \qquad y = \rho\sin\theta \sin\phi \qquad z = \rho\cos\phi$$

and a parametrization of $S$ is

$$\mathbf{r}(\theta, \phi) = 2\cos\theta \sin\phi\,\mathbf{i} + 2\sin\theta \sin\phi\,\mathbf{j} + 2\cos\phi\,\mathbf{k}$$

where $0 \le \theta \le 2\pi$ and $0 \le \phi \le \pi$. ∎

The sphere $\rho = 2$ is shown in Figure 45. Notice the coordinate curves (latitude and longitude lines) on the sphere.

**NOW WORK** Problems 11 and 27.

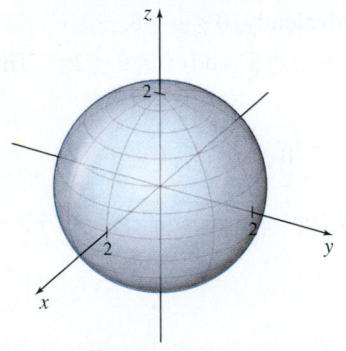

**Figure 45** The sphere of radius 2 centered at the origin.

Different parametrizations of the same surface each produce the same graph but may have different coordinate curves.

### EXAMPLE 5  Parametrizing a Cone Using Different Coordinate Systems

Find a parametrization of the surface $S$ defined by the cone

$$z = \sqrt{3x^2 + 3y^2} \qquad x^2 + y^2 \le 16$$

**(a)** Using rectangular coordinates.
**(b)** Using cylindrical coordinates.
**(c)** Using spherical coordinates.

**Solution** **(a)** The equation is given in rectangular coordinates and is written explicitly as a function $z = f(x, y)$. So, we define $x = u$, $y = v$, $z = \sqrt{3u^2 + 3v^2}$. Then a parametrization of the cone is

$$\mathbf{r}(u, v) = u\mathbf{i} + v\mathbf{j} + \sqrt{3u^2 + 3v^2}\,\mathbf{k} \qquad u^2 + v^2 \le 16$$

**NEED TO REVIEW?** Cylindrical coordinates are discussed in Section 14.7, pp. 1000–1001.

**(b)** In cylindrical coordinates, the cone is given by

$$z = \sqrt{3x^2 + 3y^2} = \sqrt{3}\sqrt{(r\cos\theta)^2 + (r\sin\theta)^2} = \sqrt{3}\sqrt{r^2} = \sqrt{3}\,r$$

where $z$ is expressed explicitly in terms of $r$ and $\theta$. So if we use $r$ and $\theta$ as parameters, then a parametrization of the cone is

$$\mathbf{r}(r, \theta) = r\cos\theta\,\mathbf{i} + r\sin\theta\,\mathbf{j} + \sqrt{3}\,r\,\mathbf{k} \qquad 0 \le r \le 4 \quad 0 \le \theta \le 2\pi$$
$$x = r\cos\theta, \; y = r\sin\theta, \; z = \sqrt{3}r$$

**NEED TO REVIEW?** Spherical coordinates are discussed in Section 14.8, pp. 1006–1008.

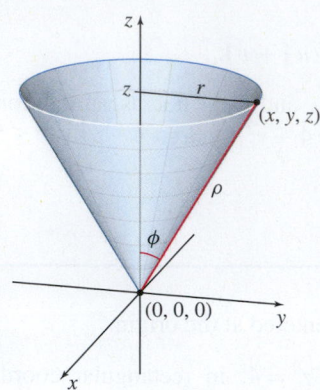

**Figure 46**

**(c)** In spherical coordinates, the equation of a half cone, where $z \geq 0$, is $\phi = a$, $0 < a < \dfrac{\pi}{2}$. We use Figure 46 to find $a$.

$$\tan \phi = \frac{r}{z} = \frac{r}{\sqrt{3}r} = \frac{1}{\sqrt{3}} \qquad z = \sqrt{3}r$$

$$\phi = \tan^{-1} \frac{1}{\sqrt{3}} = \frac{\pi}{6} = a$$

Using $\rho$ and $\theta$ as parameters, parametric equations for the cone in spherical coordinates are

$$x = \underset{\substack{\uparrow \\ \sin \frac{\pi}{6} = \frac{1}{2}}}{\frac{\rho}{2} \cos \theta} \qquad y = \underset{\substack{\uparrow \\ \sin \frac{\pi}{6} = \frac{1}{2}}}{\frac{\rho}{2} \sin \theta} \qquad z = \underset{\substack{\uparrow \\ \cos \frac{\pi}{6} = \frac{\sqrt{3}}{2}}}{\frac{\sqrt{3}}{2} \rho}$$

Now we find the parameter domain. Since

$$z^2 = 3x^2 + 3y^2 \quad \text{and} \quad \rho^2 = x^2 + y^2 + z^2 = x^2 + y^2 + \underbrace{(3x^2 + 3y^2)}_{z^2} = 4x^2 + 4y^2$$

we have $0 \leq \rho^2 = 4(x^2 + y^2) \leq 4 \cdot 16 = 64$, or equivalently, $0 \leq \rho \leq 8$.

The parameter domain of the cone is $0 \leq \rho \leq 8$ and $0 \leq \theta \leq 2\pi$. Then a parametrization of the cone is

$$\mathbf{r}(\rho, \theta) = \frac{\rho}{2} \cos \theta \, \mathbf{i} + \frac{\rho}{2} \sin \theta \, \mathbf{j} + \frac{\sqrt{3}}{2} \rho \, \mathbf{k} \qquad 0 \leq \rho \leq 8 \quad 0 \leq \theta \leq 2\pi \qquad \blacksquare$$

The graphs of the three parametrizations of the cone $z = \sqrt{3x^2 + 3y^2}$, $x^2 + y^2 \leq 16$ are shown in Figure 47.

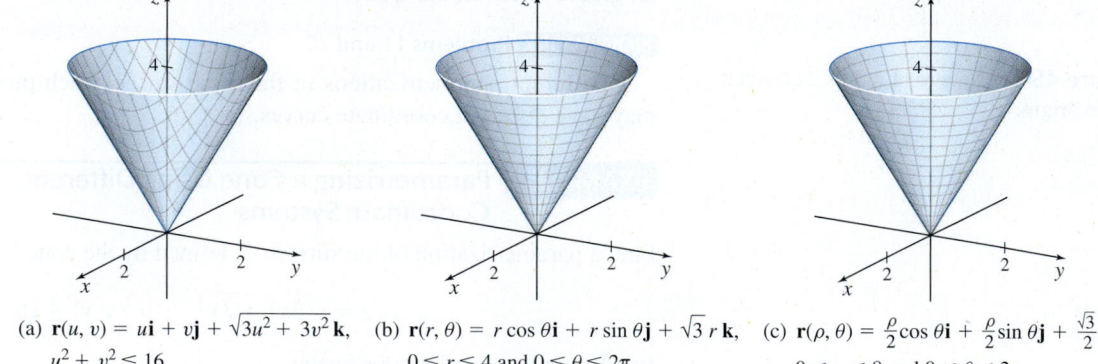

(a) $\mathbf{r}(u, v) = u\mathbf{i} + v\mathbf{j} + \sqrt{3u^2 + 3v^2}\,\mathbf{k}$, $u^2 + v^2 \leq 16$

(b) $\mathbf{r}(r, \theta) = r \cos \theta \, \mathbf{i} + r \sin \theta \, \mathbf{j} + \sqrt{3}\,r\,\mathbf{k}$, $0 \leq r \leq 4$ and $0 \leq \theta \leq 2\pi$

(c) $\mathbf{r}(\rho, \theta) = \frac{\rho}{2}\cos \theta \, \mathbf{i} + \frac{\rho}{2}\sin \theta \, \mathbf{j} + \frac{\sqrt{3}}{2}\rho\,\mathbf{k}$, $0 \leq \rho \leq 8$ and $0 \leq \theta \leq 2\pi$

**Figure 47**

Notice that although the three graphs are identical, the parametrizations are not. The cylindrical and spherical parametrizations are different, but both result in coordinate curves that are rays and circles. The rectangular parametrization results in coordinate curves that are hyperbolas. Notice also that the parameter domains of the cylindrical and spherical parametrizations are rectangular regions, but the parameter domain of the rectangular parametrization is a disk. Because of this, the cylindrical and spherical parametrizations are easier to work with.

**NOW WORK** Problem 33.

## 3 Find Equations for a Tangent Plane and a Normal Line

Now that we have seen how to parametrize a surface $S$, we return to a basic problem of calculus—the tangent problem. If a surface $S$ is parametrized by the vector function $\mathbf{r} = \mathbf{r}(u, v)$, consider the tangent plane to $S$ at a point $P_0 = \mathbf{r}(u_0, v_0)$ on the surface.

The coordinate curves $\mathbf{r} = \mathbf{r}(u, v_0)$ and $\mathbf{r} = \mathbf{r}(u_0, v)$ both lie on the surface $S$ and intersect at the point $P_0 = \mathbf{r}(u_0, v_0)$. Then the partial derivative

$$\mathbf{r}_u(u_0, v_0) = \frac{\partial}{\partial u} x(u_0, v_0)\mathbf{i} + \frac{\partial}{\partial u} y(u_0, v_0)\mathbf{j} + \frac{\partial}{\partial u} z(u_0, v_0)\mathbf{k}$$

provided it exists, is a tangent vector to the $u$-coordinate curve at $P_0$.

Similarly, the partial derivative

$$\mathbf{r}_v(u_0, v_0) = \frac{\partial}{\partial v} x(u_0, v_0)\mathbf{i} + \frac{\partial}{\partial v} y(u_0, v_0)\mathbf{j} + \frac{\partial}{\partial v} z(u_0, v_0)\mathbf{k}$$

provided it exists, is a tangent vector to the $v$-coordinate curve at $P_0$.

To obtain a formula for the tangent plane to a surface $S$, we require $\mathbf{r} = \mathbf{r}(u, v)$ to have certain properties.

**NEED TO REVIEW?** Tangent vectors to a curve are discussed in 11.2, pp. 809–811.

> **DEFINITION  Smooth Parametrization**
>
> A parametrization $\mathbf{r} = \mathbf{r}(u, v)$ is called a **smooth parametrization** if
>
> - $\mathbf{r}$ is continuous and has continuous partial derivatives in the parameter domain $R$, and
> - $\mathbf{r}_u \times \mathbf{r}_v \neq \mathbf{0}$ in $R$.
>
> A surface $S$ that can be given a smooth parametrization is called a **smooth surface**.

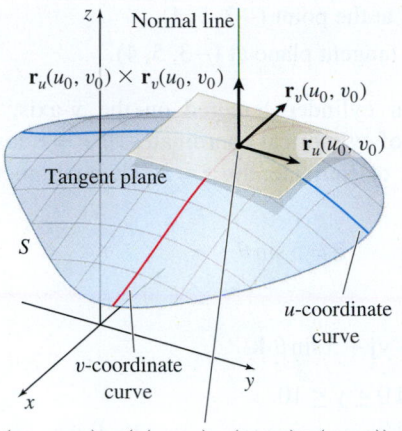

$\mathbf{r}_u(u_0, v_0) \times \mathbf{r}_v(u_0, v_0)$

$(x_0, y_0, z_0) = (x(u_0, v_0), y(u_0, v_0), z(u_0, v_0))$

**DF Figure 48**

A smooth surface has a tangent plane at every point. If $S$ is a smooth surface, then the tangent plane to $S$ at a point $P_0 = \mathbf{r}(u_0, v_0)$ is the plane containing the tangent vectors $\mathbf{r}_u(u_0, v_0)$ and $\mathbf{r}_v(u_0, v_0)$. Also, the vector $\mathbf{r}_u \times \mathbf{r}_v$ is normal to the tangent plane. See Figure 48.

**CALC CLIP**

**EXAMPLE 6  Finding a Tangent Plane and a Normal Line**

(a) Find an equation of the tangent plane to the surface $S$ parametrized by $\mathbf{r} = \mathbf{r}(u, v) = 3u\mathbf{i} + (6 - u^2 - v^2)\mathbf{j} + 2v\mathbf{k}$ at the point $(-3, 1, 4)$.

(b) Find an equation of the normal line to the tangent plane at $(-3, 1, 4)$.

**Solution** (a) We begin by finding the values of the parameters at the point $(-3, 1, 4)$. That is, we solve the system of equations

$$\begin{cases} 3u = -3 & (1) \\ 6 - u^2 - v^2 = 1 & (2) \\ 2v = 4 & (3) \end{cases}$$

From (1) $u = -1$, and from (3) $v = 2$. [Checking, we find these values also satisfy (2).] Now find the tangent vectors $\mathbf{r}_u$ and $\mathbf{r}_v$.

$$\mathbf{r}_u = \frac{\partial}{\partial u}(3u)\mathbf{i} + \frac{\partial}{\partial u}(6 - u^2 - v^2)\mathbf{j} + \frac{\partial}{\partial u}(2v)\mathbf{k} = 3\mathbf{i} - 2u\mathbf{j} \qquad \mathbf{r}_u(-1, 2) = 3\mathbf{i} + 2\mathbf{j}$$

$$\mathbf{r}_v = \frac{\partial}{\partial v}(3u)\mathbf{i} + \frac{\partial}{\partial v}(6 - u^2 - v^2)\mathbf{j} + \frac{\partial}{\partial v}(2v)\mathbf{k} = -2v\mathbf{j} + 2\mathbf{k} \qquad \mathbf{r}_v(-1, 2) = -4\mathbf{j} + 2\mathbf{k}$$

The normal vector $\mathbf{n}$ to the tangent plane at $(-3, 1, 4)$ is

$$\mathbf{n} = \mathbf{r}_u(-1, 2) \times \mathbf{r}_v(-1, 2) = \begin{vmatrix} \mathbf{i} & \mathbf{j} & \mathbf{k} \\ 3 & 2 & 0 \\ 0 & -4 & 2 \end{vmatrix} = 4\mathbf{i} - 6\mathbf{j} - 12\mathbf{k}$$

An equation of the tangent plane at $(-3, 1, 4)$ is

$$4(x + 3) - 6(y - 1) - 12(z - 4) = 0 \qquad \text{or equivalently} \qquad 4x - 6y - 12z = -66$$

See Figure 49.

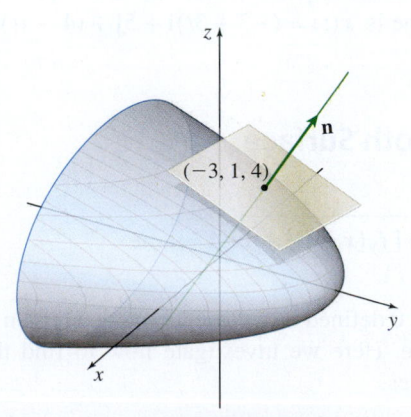

**Figure 49** The tangent plane to $S$ at $(-3, 1, 4)$.

**(b)** The normal line to the tangent plane contains the point $(-3, 1, 4)$ and is parallel to the vector $\mathbf{n} = 4\mathbf{i} - 6\mathbf{j} - 12\mathbf{k}$. A parametrization of the normal line is

$$\mathbf{r}(t) = (-3 + 4t)\mathbf{i} + (1 - 6t)\mathbf{j} + (4 - 12t)\mathbf{k} \qquad \blacksquare$$

**NOW WORK** Problem 13.

**EXAMPLE 7** Finding a Tangent Plane and a Normal Line

A surface $S$ is defined by $x^2 + z^2 = 25$ and $0 \le y \le 10$.

**(a)** Find a smooth parametrization for $S$.

**(b)** Find an equation of the tangent plane to $S$ at the point $(-3, 5, 4)$.

**(c)** Find an equation of the normal line to the tangent plane at $(-3, 5, 4)$.

**Solution (a)** Since $S$ is a part of a circular cylinder centered on the $y$-axis, we choose a parametrization that uses a variation of cylindrical coordinates. If $x = r \cos\theta$, $y = y$, and $z = r \sin\theta$, then $x^2 + z^2 = r^2 = 25$, or equivalently, $r = 5$. Then parametric equations for the cylinder are

$$x = 5 \cos\theta \qquad y = y \qquad z = 5 \sin\theta$$

and a parametrization of $S$ is

$$\mathbf{r}(\theta, y) = 5 \cos\theta\, \mathbf{i} + y\mathbf{j} + 5 \sin\theta\, \mathbf{k}$$

where the parameter domain is $0 \le \theta \le 2\pi$ and $0 \le y \le 10$.

To determine whether this is a smooth parametrization, we find the partial derivatives of $\mathbf{r} = \mathbf{r}(\theta, y)$.

$$\mathbf{r}_\theta(\theta, y) = -5 \sin\theta\, \mathbf{i} + 5 \cos\theta\, \mathbf{k} \qquad \mathbf{r}_y(\theta, y) = \mathbf{j}$$

Each of these is continuous on the parameter domain. Now

$$\mathbf{r}_\theta \times \mathbf{r}_y = \begin{vmatrix} \mathbf{i} & \mathbf{j} & \mathbf{k} \\ -5 \sin\theta & 0 & 5 \cos\theta \\ 0 & 1 & 0 \end{vmatrix} = -5 \cos\theta\, \mathbf{i} - 5 \sin\theta\, \mathbf{k}$$

Since $\|\mathbf{r}_\theta \times \mathbf{r}_y\| = \sqrt{25 \cos^2\theta + 25 \sin^2\theta} = 5$, then $\mathbf{r}_\theta \times \mathbf{r}_y \ne \mathbf{0}$. So, $\mathbf{r} = \mathbf{r}(\theta, y)$ is a smooth parametrization, and $S$ is a smooth surface.

**(b)** At the point $(-3, 5, 4)$, we have $5 \cos\theta = -3$, $y = 5$, and $5 \sin\theta = 4$. Then the normal vector $\mathbf{n}$ to the plane at $(-3, 5, 4)$ is $\mathbf{r}_\theta \times \mathbf{r}_y = -5\left(-\dfrac{3}{5}\right)\mathbf{i} - 5\left(\dfrac{4}{5}\right)\mathbf{k} = 3\mathbf{i} - 4\mathbf{k}$. An equation of the tangent plane at $(-3, 5, 4)$ is

$$3(x + 3) - 4(z - 4) = 0 \qquad \text{or equivalently} \qquad 3x - 4z = -25$$

**(c)** The normal line to the tangent plane contains the point $(-3, 5, 4)$ and is parallel to $3\mathbf{i} - 4\mathbf{k}$. A parametrization of the normal line is $\mathbf{r}(t) = (-3 + 3t)\mathbf{i} + 5\mathbf{j} + (4 - 4t)\mathbf{k}$. See Figure 50. $\blacksquare$

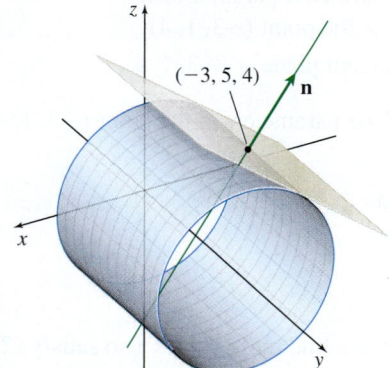

**DF Figure 50** The tangent plane to $\mathbf{r}(\theta, y) = 5 \cos\theta\, \mathbf{i} + y\mathbf{j} + 5 \sin\theta\, \mathbf{k}$ at the point $(-3, 5, 4)$.

## 4 Find the Surface Area of a Smooth Surface

In Section 14.5, we developed the formula

$$S = \iint\limits_R \sqrt{[f_x(x, y)]^2 + [f_y(x, y)]^2 + 1}\; dA$$

for the surface area of a function $z = f(x, y) \ge 0$ defined on a closed, bounded region $R$ whose boundary is a piecewise-smooth curve. Here we investigate how to find the surface area of more general parametric surfaces.

Suppose the smooth surface $S$ has the parametrization

$$\mathbf{r}(u, v) = x(u, v)\mathbf{i} + y(u, v)\mathbf{j} + z(u, v)\mathbf{k}$$

and its parameter domain $R$ is a closed, rectangular region of the $uv$-plane defined by $a \le u \le b$ and $c \le v \le d$. Then as we did in Section 14.5, partition the interval $[a, b]$ into $n$ subintervals of length $\Delta u_i, i = 1, 2, \ldots, n$, and the interval $[c, d]$ into $m$ subintervals of length $\Delta v_j, j = 1, 2, \ldots, m$. The result is $mn$ subrectangles $R_{ij}$ of widths $\Delta u_i$, heights $\Delta v_j$, and areas $\Delta A_{ij} = \Delta u_i \Delta v_j$. The norm $||\Delta||$ of the partition is defined as the length of the largest diagonal of the subrectangles.

The portions of the lines forming the edges of the subrectangle $R_{ij}$ map to portions of their corresponding coordinate curves on the surface $S$. That is, each subrectangle $R_{ij}$ maps into a closed, curved **patch** $S_{ij}^*$ on $S$. See Figure 51.

We now choose a point $(u_{ij}^*, v_{ij}^*)$ in each subrectangle. In a small enough region, a smooth surface can be approximated by its tangent planes. So, the area of $S_{ij}^*$ can be approximated using part of the tangent plane at $\mathbf{r} = \mathbf{r}(u_{ij}^*, v_{ij}^*)$, or more specifically, $S_{ij}^*$ can be approximated by the parallelogram formed by the vectors $\mathbf{r}_u(u_{ij}^*, v_{ij}^*) \Delta u_i$ and $\mathbf{r}_v(u_{ij}^*, v_{ij}^*) \Delta v_j$ (tangents to the coordinate curves). See Figure 52. The area of this parallelogram is

$$\Delta S_{ij}^* = \left\| \mathbf{r}_u(u_{ij}^*, v_{ij}^*) \times \mathbf{r}_v(u_{ij}^*, v_{ij}^*) \right\| \Delta A_{ij} = \left\| \mathbf{r}_u(u_{ij}^*, v_{ij}^*) \times \mathbf{r}_v(u_{ij}^*, v_{ij}^*) \right\| \Delta u_i \Delta v_j$$

Then the **surface area differential** $dS$ is

$$\boxed{dS = \| \mathbf{r}_u(u, v) \times \mathbf{r}_v(u, v) \| \, dA = \| \mathbf{r}_u(u, v) \times \mathbf{r}_v(u, v) \| \, du \, dv}$$

Since, $\Delta S_{ij}^*$ approximates the surface area of $S_{ij}^*$, then the sums

$$\sum_{j=1}^{m} \sum_{i=1}^{n} \Delta S_{ij}^* = \sum_{j=1}^{m} \sum_{i=1}^{n} \left\| \mathbf{r}_u(u_{ij}^*, v_{ij}^*) \times \mathbf{r}_v(u_{ij}^*, v_{ij}^*) \right\| \Delta u_i \Delta v_j$$

approximate the surface area of $S$. Since $\mathbf{r} = \mathbf{r}(u, v)$ is a smooth parametrization, $\displaystyle\lim_{\|\Delta\| \to 0} \sum_{j=1}^{m} \sum_{i=1}^{n} \Delta S_{ij}^*$ exists and the surface area of $S$ is given by a double integral. That is,

$$\boxed{\text{Surface area of } S = \iint_R dS = \iint_R \| \mathbf{r}_u \times \mathbf{r}_v \| \, du \, dv} \tag{1}$$

The next theorem extends (1) to regions that are closed and bounded and whose boundary is a piece-wise smooth curve.

### THEOREM Surface Area of a Smooth Surface

Suppose that $\mathbf{r} = \mathbf{r}(u, v)$ is a parametrization of a smooth surface $S$ whose parameter domain is a closed, bounded region $R$ whose boundary is a piece-wise smooth curve. The surface area of $S$ that lies above $R$ is given by

$$\boxed{S = \iint_R \| \mathbf{r}_u \times \mathbf{r}_v \| \, du \, dv}$$

In the special case that $S$ is the graph of a function $z = f(x, y)$ and we use the parametrization, $x = x$, $y = y$, and $z = f(x, y)$, we obtain

$$dS = \sqrt{[f_x(x, y)]^2 + [f_y(x, y)]^2 + 1} \, dx \, dy$$

and the formula for surface area

$$S = \iint_R \sqrt{[f_x(x, y)]^2 + [f_y(x, y)]^2 + 1} \, dx \, dy$$

from Section 14.5.

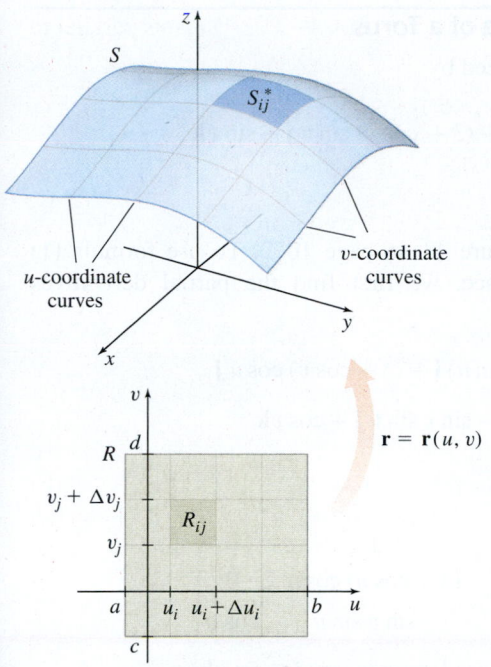

**Figure 51** The subrectangle $R_{ij}$ maps onto patch $S_{ij}^*$.

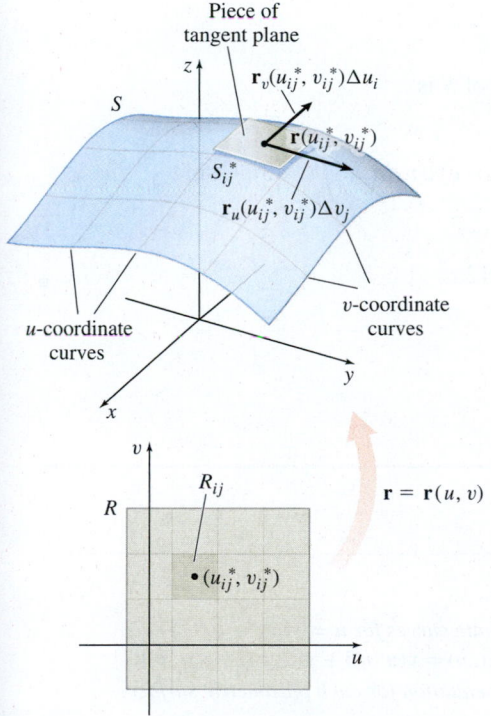

**Figure 52**

---

**EXAMPLE 8**  **Finding the Surface Area of a Torus**

Find the surface area of the torus parametrized by

$$\mathbf{r}(u, v) = (3 + \cos v) \cos u \, \mathbf{i} + (3 + \cos v) \sin u \, \mathbf{j} + \sin v \mathbf{k}$$

where $0 \le u \le 2\pi$ and $0 \le v \le 2\pi$.

**Solution** This is the torus graphed in Figure 43 on page 1072. To use formula (1) to find the area of a parametrized surface, we first find the partial derivatives of $\mathbf{r} = \mathbf{r}(u, v)$.

$$\mathbf{r}_u(u, v) = (3 + \cos v)(-\sin u)\,\mathbf{i} + (3 + \cos v) \cos u \, \mathbf{j}$$

$$\mathbf{r}_v(u, v) = (-\sin v) \cos u\,\mathbf{i} - \sin v \sin u \, \mathbf{j} + \cos v \mathbf{k}$$

Next we find the cross product:

$$\mathbf{r}_u \times \mathbf{r}_v = \begin{vmatrix} \mathbf{i} & \mathbf{j} & \mathbf{k} \\ -(3 + \cos v) \sin u & (3 + \cos v) \cos u & 0 \\ -\sin v \cos u & -\sin v \sin u & \cos v \end{vmatrix}$$

$$= (3 + \cos v)(\cos u \cos v\mathbf{i} + \sin u \cos v\mathbf{j} + \sin v\mathbf{k})$$

Then

$$\|\mathbf{r}_u \times \mathbf{r}_v\| = \sqrt{(3 + \cos v)^2(\cos^2 u \cos^2 v + \sin^2 u \cos^2 v + \sin^2 v)}$$

$$= (3 + \cos v) \sqrt{\cos^2 v(\cos^2 u + \sin^2 u) + \sin^2 v} = (3 + \cos v) \sqrt{\cos^2 v + \sin^2 v}$$

$$= 3 + \cos v$$

Now using formula (1), the surface area of $S$ is

$$\iint_R (3 + \cos v) \, du \, dv = \int_0^{2\pi}\int_0^{2\pi} (3 + \cos v) \, dv \, du = \int_0^{2\pi} \left[3v + \sin v\right]_0^{2\pi} du$$

$$= \int_0^{2\pi} 6\pi \, du = 12\pi^2$$  ∎

**NOW WORK** Problem 17.

---

## 15.6 Assess Your Understanding

### Concepts and Vocabulary

1. *Multiple Choice*  A smooth surface has a [(a) normal plane (b) tangent plane (c) coordinate plane] at each point.

2. *True or False*  The parametrization of a surface is unique.

3. *Multiple Choice*  The parametric equations $x = 5 \cos u \sin v$, $y = 5 \sin u \sin v$, $z = 5 \cos v$ define a [(a) plane (b) circle (c) sphere (d) cylinder (e) paraboloid].

4. *Multiple Choice*  The parametrization $\mathbf{r}(u, v) = (1 - u + 2v)\mathbf{i} + 5u\,\mathbf{j} + (2u + 3v - 7)\,\mathbf{k}$ parametrizes a [(a) plane (b) sphere (c) line (d) cylinder (e) hyperboloid].

### Skill Building

*In Problems 5–8:*

(a) *Identify the coordinate curves for $u = 0$ and $v = 0$ of each parametrization $\mathbf{r}(u, v) = x(u, v)\mathbf{i} + y(u, v)\mathbf{j} + z(u, v)\mathbf{k}$.*

(b) *Find a rectangular equation for each parametric surface.*

5. $x(u, v) = u - 5v$, $y(u, v) = 2u$, $z(u, v) = -u + v + 1$

6. $x(u, v) = u$, $y(u, v) = v$, $z(u, v) = 9 - u^2 - v^2$

7. $x(u, v) = u \cos v$, $y(u, v) = u \sin v$, $z(u, v) = u$; $0 \le u \le 2$, $0 \le v \le \pi$

8. $x(u, v) = \cos u \sin v$, $y(u, v) = \sin u \sin v$, $z(u, v) = \cos v$; $0 \le u \le 2\pi$, $0 \le v \le \dfrac{\pi}{2}$

---

1. = NOW WORK problem      📈 = Graphing technology recommended      CAS = Computer Algebra System recommended

*In Problems 9–12, find a parametrization for each surface.*

**9.** The part of the plane $z = 4 - x - 2y$ that lies in the first octant

**10.** The part of the surface $z = e^{-x^2 + y^2}$ that lies inside the cylinder $x^2 + y^2 = 4$

**11.** The part of the surface $z = \sin(x^2 y)$ that lies above the region bounded by the graphs of $y = x + 2$ and $y = x^2$

**12.** The part of the surface $y + e^{xz} = 5$ that lies inside the cylinder $x^2 + z^2 = 1$

*In Problems 13–16:*

*(a) Find an equation of the tangent plane to each surface at the given point.*

*(b) Find an equation of the normal line to the tangent plane at the point.*

**13.** $\mathbf{r}(u, v) = (3u + 2v)\mathbf{i} + 5u^3\mathbf{j} + v^2\mathbf{k}$ at $(7, 5, 4)$

**14.** $\mathbf{r}(u, v) = (3u - v)\mathbf{i} + (2 - u - v)\mathbf{j} + (1 + 3v)\mathbf{k}$ at $(11, 1, -5)$

**15.** $\mathbf{r}(u, v) = u\,\mathbf{i} + u\cos v\,\mathbf{j} + u\sin v\,\mathbf{k}$ at $\left(5, \dfrac{5\sqrt{2}}{2}, \dfrac{5\sqrt{2}}{2}\right)$

**16.** $\mathbf{r}(u, v) = (3\cos u \sin v + 1)\,\mathbf{i} + (2\sin u \sin v - 1)\,\mathbf{j} + \cos v\,\mathbf{k}$ at $\left(1, 0, \dfrac{\sqrt{3}}{2}\right)$

*In Problems 17–22, find the surface area of each surface S.*

**17.** $S$ is parameterized by $\mathbf{r}(u, v) = u\cos v\,\mathbf{i} + u^3\mathbf{j} + u\sin v\,\mathbf{k}$, $0 \le u \le 1$, $-\pi \le v \le \pi$.

**18.** $S$ is parameterized by $\mathbf{r}(u, v) = (3u^2 + v)\mathbf{i} + u^2\mathbf{j} + (v - u^2)\mathbf{k}$, $0 \le u \le 2$, $-1 \le v \le 1$.

**19.** $S$ is the part of the plane $2x - y + 4z = 3$ that lies inside the cylinder $(x - 2)^2 + y^2 = 4$.

**20.** $S$ is the part of the paraboloid $z = 4 - x^2 - y^2$ that lies above the $xy$-plane.

**21.** $S$ is the part of the sphere $x^2 + y^2 + z^2 = 16$ that lies above the plane $z = 2$.

**22.** $S$ is the frustum of the cone $z = 3\sqrt{x^2 + y^2}$ that lies between $z = 3$ and $z = 6$.

*In Problems 23–28, match each parametrization with its parametric surface, as shown in (A)–(F), top right.*

**23.** $\mathbf{r}(u, v) = u\cos v\,\mathbf{i} + u\sin v\,\mathbf{j} - u\,\mathbf{k}$, $0 \le u \le 4$, $0 \le v \le 2\pi$

**24.** $\mathbf{r}(u, v) = u^3\,\mathbf{i} + u\sin v\,\mathbf{j} + u\cos v\,\mathbf{k}$, $0 \le u \le 4$, $0 \le v \le 2\pi$

**25.** $\mathbf{r}(u, v) = u\sin v\,\mathbf{i} + u\cos v\,\mathbf{j} + u\,\mathbf{k}$, $0 \le u \le 4$, $0 \le v \le 2\pi$

**26.** $\mathbf{r}(u, v) = u\,\mathbf{i} + 4\sin v\,\mathbf{j} + 4\cos v\,\mathbf{k}$, $0 \le u \le 4$, $0 \le v \le 2\pi$

**27.** $\mathbf{r}(u, v) = u\cos v\,\mathbf{i} + u\sin v\,\mathbf{j} - u^3\,\mathbf{k}$, $0 \le u \le 4$, $0 \le v \le 2\pi$

**28.** $\mathbf{r}(u, v) = v\,\mathbf{i} + v^3\,\mathbf{j} + u\,\mathbf{k}$, $0 \le u \le 4$, $-4 \le v \le 4$

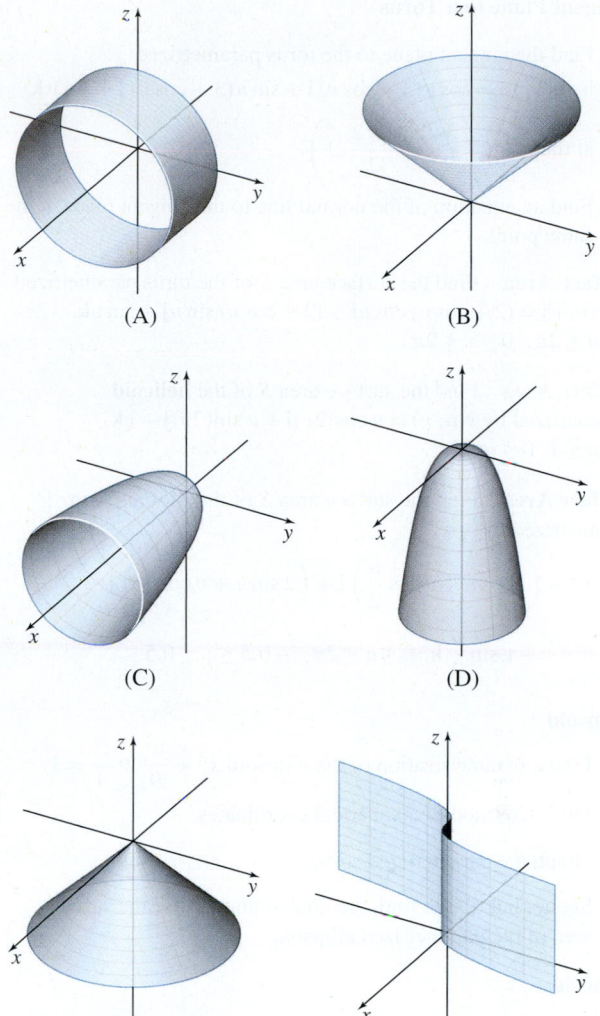

(A)  (B)

(C)  (D)

(E)  (F)

## Applications and Extensions

**29.** Parametrize the part of the cylinder $x^2 + y^2 = 16$ that lies above the $xy$-plane and below $z = 3$.

**30.** Parametrize the sphere $x^2 + y^2 + z^2 = 25$.

**31.** Part of the paraboloid $z = 9 - x^2 - y^2$ lies inside the cylinder $(x - 1)^2 + y^2 = 1$.

    **(a)** Parametrize the surface using rectangular coordinates.

    **(b)** Parametrize the surface using cylindrical coordinates.

**32.** Parametrize the **lumpy sphere**

$$x^2 + y^2 + z^2 = 3\sqrt{x^2 + y^2 + z^2} + z$$

    *Hint:* Use spherical coordinates.

*In Problems 33 and 34, parametrize each surface:*
*(a) Using rectangular coordinates.*
*(b) Using cylindrical coordinates.*
*(c) Using spherical coordinates.*
*Give bounds for the parameters, if necessary.*

**33.** The part of the sphere $x^2 + y^2 + z^2 = 4$ lying in the first octant

**34.** The plane $x - \sqrt{3}y = 0$, where $x \ge 0$, $y \ge 0$

**35. Tangent Plane to a Torus**

(a) Find the tangent plane to the torus parametrized
by $\mathbf{r}(u, v) = \cos u (3 + \cos v) \mathbf{i} + \sin u (3 + \cos v) \mathbf{j} + \sin v \mathbf{k}$
at the point $\left( \dfrac{3\sqrt{2}}{2}, \dfrac{3\sqrt{2}}{2}, 1 \right)$.

(b) Find an equation of the normal line to the tangent plane at the same point.

**36. Surface Area** Find the surface area $S$ of the torus parametrized
by $\mathbf{r}(u, v) = (2 + \cos v) \cos u \mathbf{i} + (2 + \cos v) \sin u \mathbf{j} + \sin v \mathbf{k}$,
$0 \le u \le 2\pi$, $0 \le v \le 2\pi$.

**37. Surface Area** Find the surface area $S$ of the **helicoid**
parametrized by $\mathbf{r}(u, v) = u \cos(2v) \mathbf{i} + u \sin(2v) \mathbf{j} + v \mathbf{k}$,
$0 \le u \le 1$, $0 \le v \le 2\pi$.

**CAS** **38. Surface Area** Find the surface area $S$ of the Möbius strip
parametrized by

$$\mathbf{r}(u, v) = \left( 2 \cos u + v \cos \frac{u}{2} \right) \mathbf{i} + \left( 2 \sin u + v \cos \frac{u}{2} \right) \mathbf{j}$$

$$- v \sin \frac{u}{2} \mathbf{k}, \ 0 \le u \le 2\pi, \ -0.5 \le v \le 0.5$$

**39. Ellipsoid**

(a) Find a parametrization of the ellipsoid $x^2 + \dfrac{y^2}{9} + \dfrac{z^2}{4} = 1$.

*Hint:* Use modified spherical coordinates.

**CAS** (b) Graph the parametrized surface.

(c) Set up, but do not find, the double integral for the surface area of the parametrized ellipsoid.

**CAS** **40. Helicoid**

(a) Graph the helicoid defined by the parametric equations
$x = u \cos(2v)$, $y = u \sin(2v)$, $z = v$, $0 \le u \le 3$ and
$0 \le v \le 2\pi$.

(b) Parametrize the helicoid.

(c) Find an equation of the tangent plane to the helicoid at the point $\left( \dfrac{1}{3}, \pi \right)$.

(d) Find an equation of the normal line to the tangent plane at the point $\left( \dfrac{1}{3}, \pi \right)$.

**CAS** **41. Dini's Surface**

(a) Graph **Dini's surface** defined by the parametric equations
$$x = 6 \cos u \sin v \qquad y = 6 \sin u \sin v,$$
$$z = 6 \left[ \cos v + \ln \left( \tan \frac{v}{2} \right) \right] + u$$
$0.01 \le v \le 1$ and $0 \le u \le 6\pi$.

(b) Find an equation of the tangent plane to Dini's surface at the point $\left( \dfrac{\pi}{3}, \dfrac{\pi}{4} \right)$.

(c) Find an equation of the normal line to the tangent plane at the point $\left( \dfrac{\pi}{3}, \dfrac{\pi}{4} \right)$.

**Challenge Problems**

**42.** Suppose $\triangle ABC$ is a triangle with vertices $A = (a_1, a_2, a_3)$, $B = (b_1, b_2, b_3)$, and $C = (c_1, c_2, c_3)$. Parametrize $\triangle ABC$.

**43. Torus** A torus is formed by rotating a circle about another circle.

(a) Parametrize the torus obtained by rotating a circle of radius $a > 0$ about a circle of radius $b > 0$.

(b) Find an implicit rectangular equation for the torus.

**44. Surface Area** Find the surface area of the part of the cylinder $(x - 1)^2 + y^2 = 1$ that is outside the sphere $x^2 + y^2 + z^2 = 4$, above the $xy$-plane, and below $z = 1$.

**45. An Archimedean Ratio** Suppose $C$ is a right circular cone, $S$ is an upper hemisphere, and $L$ is a right circular cylinder. Suppose that all three surfaces have radii $r$ and the heights $h$ of the cone and the cylinder are $r$. Archimedes was able to show that the ratio of the surface areas among these surfaces is $\sqrt{2} : 2 : 2$. Show this by parametrizing each of these surfaces and deriving the formulas for their surface areas.

**46. Hyperboloid of One Sheet**

(a) Use hyperbolic functions to parametrize the hyperboloid of one sheet: $x^2 + y^2 - z^2 = c$, $c > 0$.

**CAS** (b) Graph the parametric surface for $c = 1$.

**47. Hyperboloid of Two Sheets**

(a) Use hyperbolic functions to parametrize the sheet where
$$x > 0 \text{ in the hyperboloid of two sheets: } \frac{x^2}{a^2} - \frac{y^2}{b^2} - \frac{z^2}{c^2} = 1.$$

**CAS** (b) Graph the parametric surface for $a = 1$, $b = 2$, $c = 3$.

# 15.7 Surface and Flux Integrals

**OBJECTIVES** *When you finish this section, you should be able to:*

1 Find a surface integral using a double integral (p. 1081)

2 Determine the orientation of a surface (p. 1084)

3 Find the flux of a vector field across a surface (p. 1087)

Application: Electric Flux (p. 1089)

Surface integrals are similar to line integrals, but the integration takes place on a surface instead of along a curve.

Suppose the smooth surface $S$ has the parametrization

$$\mathbf{r}(u, v) = x(u, v)\mathbf{i} + y(u, v)\mathbf{j} + z(u, v)\mathbf{k}$$

and its parameter domain $R$ is a closed, rectangular region of the $uv$-plane defined by $a \leq u \leq b$ and $c \leq v \leq d$. Then, as we did in Section 15.6, partition the interval $[a, b]$ into $n$ subintervals of length $\Delta u_i$, $i = 1, 2, \ldots, n$, and the interval $[c, d]$ into $m$ subintervals of length $\Delta v_j$, $j = 1, 2, \ldots, m$. The result is $mn$ subrectangles $R_{ij}$.

The $ij$th subrectangle $R_{ij}$ is formed by the lines $u = u_i$, $u = u_i + \Delta u_i$, $v = v_j$, and $v = v_j + \Delta v_j$. The portions of the lines forming the edges of the subrectangle $R_{ij}$ map to portions of their corresponding coordinate curves on the surface $S$. That is, each subrectangle $R_{ij}$ maps into a closed curved patch $S_{ij}^*$ on $S$ whose area is $\Delta S_{ij}^*$. We define the norm $\|\Delta\|$ to be the largest area $\Delta S_{ij}^*$. We now choose a point $(u_{ij}^*, v_{ij}^*)$ in each subrectangle.

Suppose $w = F(x, y, z)$ is a function that is defined in a solid in space containing the surface $S$. We evelute $F$ at the point $(x(u_{ij}^*, v_{ij}^*), y(u_{ij}^*, v_{ij}^*), z(u_{ij}^*, v_{ij}^*))$ on $S_{ij}^*$ and form the product $F(x(u_{ij}^*, v_{ij}^*), y(u_{ij}^*, v_{ij}^*), z(u_{ij}^*, v_{ij}^*))\Delta S_{ij}^*$. Now we add obtaining the sums

$$\sum_{j=1}^{m}\sum_{i=1}^{n} F(x(u_{ij}^*, v_{ij}^*), y(u_{ij}^*, v_{ij}^*), z(u_{ij}^*, v_{ij}^*))\,\Delta S_{ij}^*$$

The **surface integral of $F$ over $S$** is defined as

$$\iint_S F(x, y, z)\,dS = \lim_{\|\Delta\| \to 0}\sum_{j=1}^{m}\sum_{i=1}^{n} F(x(u_{ij}^*, v_{ij}^*), y(u_{ij}^*, v_{ij}^*), z(u_{ij}^*, v_{ij}^*))\Delta S_{ij}^* \tag{1}$$

provided this limit exists.

This definition can be extended to any smooth surface $S$ whose parameter domain is a closed, bounded region whose boundary is a piecewise-smooth curve.

## 1 Find a Surface Integral Using a Double Integral

It can be shown that the limit in (1) exists if the smooth surface $S$ is defined on a close, bounded region $R$ whose boundary is a piecewise-smooth curve and if the function $w = f(x, y, z)$ is continuous in some solid containing $S$. Further, under these conditions, the surface integral of $F$ over $S$ can be expressed as a double integral.

> **THEOREM   A Surface Integral Expressed as a Double Integral**
>
> Let $S$ be a surface defined on a closed, bounded region $R$ with a smooth parametrization $\mathbf{r}(u, v) = x(u, v)\mathbf{i} + y(u, v)\mathbf{j} + z(u, v)\mathbf{k}$. Also, suppose that $w = F(x, y, z)$ is continuous in a solid containing the surface $S$. Then the surface integral of $F$ over $S$ is given by
>
> $$\iint_S F(x, y, z)\,dS = \iint_R F(x(u, v), y(u, v), z(u, v))\,\|\mathbf{r}_u \times \mathbf{r}_v\|\,du\,dv$$

It is important to note that just as a line integral does not depend on the parametrization used, surface integrals do not depend on the smooth parametrization used in the integration. Also, there are occasions when we want to find a surface integral over a *piecewise-smooth surface*. A **piecewise-smooth surface** is a finite collection of smooth surfaces that intersect only along piecewise-smooth curves. In these cases, just as with line integrals, we integrate over each smooth surface separately and then sum the results.

## EXAMPLE 1  Finding a Surface Integral

Find $\iint\limits_S ye^{x^2+z^2}\,dS$, where $S$ is the part of the surface of the cylinder $x^2+z^2=9$ that lies between $y=0$ and $y=2$.

**Solution** The surface $S$ is shown in Figure 53. Since the surface is a piece of a cylinder, we use a variation of cylindrical coordinates to parametrize the surface. Here we use $x=r\cos\theta$, $y=y$, and $z=r\sin\theta$. (Do you see why?) Then $r^2=x^2+z^2=9$, so $r=3$. We parametrize the surface $S$ using

$$\mathbf{r}(\theta,\,y)=3\cos\theta\,\mathbf{i}+y\mathbf{j}+3\sin\theta\,\mathbf{k}$$

where the parameter domain $R$ is $0\le\theta\le 2\pi$ and $0\le y\le 2$.

The partial derivatives of $\mathbf{r}$ are

$$\mathbf{r}_\theta(\theta,\,y)=-3\sin\theta\,\mathbf{i}+3\cos\theta\,\mathbf{k}\qquad\text{and}\qquad\mathbf{r}_y(\theta,\,y)=\mathbf{j}$$

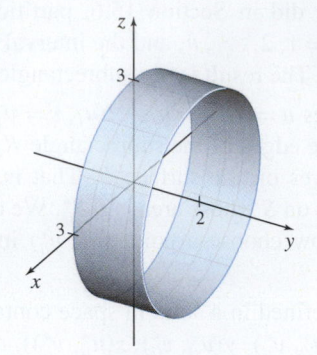

**Figure 53** The surface $x^2+z^2=9,\,0\le y\le 2$.

and their cross product is

$$\mathbf{r}_\theta\times\mathbf{r}_y=\begin{vmatrix}\mathbf{i}&\mathbf{j}&\mathbf{k}\\-3\sin\theta&0&3\cos\theta\\0&1&0\end{vmatrix}=-3\cos\theta\,\mathbf{i}-3\sin\theta\,\mathbf{k}$$

Then $\lVert\mathbf{r}_\theta\times\mathbf{r}_y\rVert=\sqrt{9\cos^2\theta+9\sin^2\theta}=3$. So, we have

$$\iint\limits_S ye^{x^2+z^2}\,dS=\iint\limits_R ye^9\lVert\mathbf{r}_\theta\times\mathbf{r}_y\rVert\,d\theta\,dy=\int_0^2\int_0^{2\pi}ye^9\cdot 3\,d\theta\,dy$$

$$=3e^9\int_0^2 y\,[\theta]_0^{2\pi}\,dy=3e^9\int_0^2 2\pi y\,dy=6\pi e^9\left[\frac{y^2}{2}\right]_0^2=12\pi e^9\quad\blacksquare$$

NOW WORK  Problem 3.

**RECALL** A lamina is a thin sheet of material.

**NEED TO REVIEW?** The center of mass of a lamina with variable density is discussed in Section 14.4, pp. 978–981.

The formulas for finding the center of mass of a lamina can be generalized. The following formulas for the center of mass of a general lamina with variable mass density $\rho=\rho(x,\,y,\,z)$ in the shape of a surface $S$ result from a derivation similar to that found in Section 14.4 for a flat sheet.

| | |
|---|---|
| Mass of $S$: | $M=\iint\limits_S \rho(x,\,y,\,z)\,dS$ |
| Moment about the $yz$-plane: | $M_{yz}=\iint\limits_S x\rho(x,\,y,\,z)\,dS$ |
| Moment about the $xz$-plane: | $M_{xz}=\iint\limits_S y\rho(x,\,y,\,z)\,dS$ |
| Moment about the $xy$-plane: | $M_{xy}=\iint\limits_S z\rho(x,\,y,\,z)\,dS$ |
| Center of mass: | $(\overline{x},\,\overline{y},\,\overline{z})=\left(\dfrac{M_{yz}}{M},\,\dfrac{M_{xz}}{M},\,\dfrac{M_{xy}}{M}\right)$ |

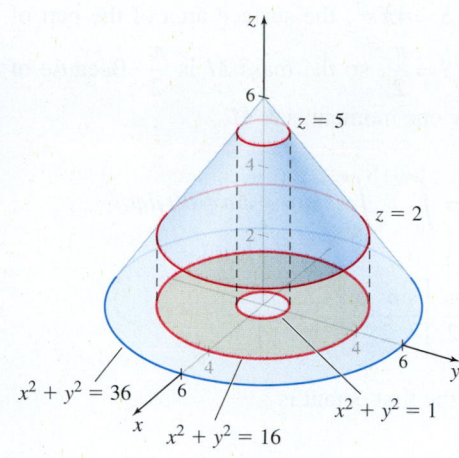

**Figure 54**

**RECALL** The centroid is the geometric center of a lamina, and it is numerically equivalent to the center of mass of a homogeneous lamina that has mass density equal to 1.

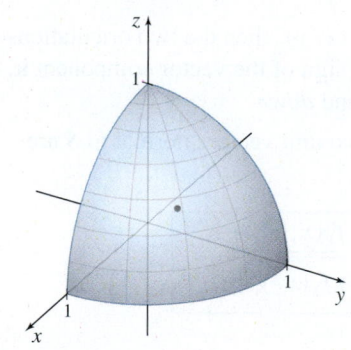

**Figure 55**

**EXAMPLE 2**  Finding the Mass of a Lamina

A lamina in the shape of the cone $z = 6 - \sqrt{x^2 + y^2}$ lies between the planes $z = 2$ and $z = 5$. If the mass density of the lamina is $\rho(x, y, z) = \sqrt{x^2 + y^2}$, find the mass $M$ of the lamina.

**Solution**  Figure 54 illustrates the lamina. Using $x = r\cos\theta$, $y = r\sin\theta$, and $z = 6 - \sqrt{x^2 + y^2} = 6 - r$, we parametrize the surface as

$$\mathbf{r} = \mathbf{r}(r, \theta) = r\cos\theta\,\mathbf{i} + r\sin\theta\,\mathbf{j} + (6 - r)\,\mathbf{k} \qquad 0 \le \theta \le 2\pi \quad 1 \le r \le 4$$

First we find $\mathbf{r}_r \times \mathbf{r}_\theta$.

$$\mathbf{r}_r \times \mathbf{r}_\theta = \begin{vmatrix} \mathbf{i} & \mathbf{j} & \mathbf{k} \\ \cos\theta & \sin\theta & -1 \\ -r\sin\theta & r\cos\theta & 0 \end{vmatrix} = r\cos\theta\,\mathbf{i} + r\sin\theta\,\mathbf{j} + (r\cos^2\theta + r\sin^2\theta)\,\mathbf{k}$$

$$= r\cos\theta\,\mathbf{i} + r\sin\theta\,\mathbf{j} + r\,\mathbf{k}$$

$$\|\mathbf{r}_r \times \mathbf{r}_\theta\| = \sqrt{r^2\cos^2\theta + r^2\sin^2\theta + r^2} = r\sqrt{\cos^2\theta + \sin^2\theta + 1} = r\sqrt{1+1} = \sqrt{2}\,r$$

Since the mass density of the lamina is $\rho(x, y, z) = \sqrt{x^2 + y^2} = r$ and $dS = \|\mathbf{r}_r \times \mathbf{r}_\theta\|\,dr\,d\theta = \sqrt{2}r\,dr\,d\theta$, the mass $M$ of the lamina is given by

$$M = \iint_S \rho(x, y, z)\,dS = \iint_R \rho\|\mathbf{r}_r \times \mathbf{r}_\theta\|\,dr\,d\theta = \iint_R r(\sqrt{2}r)\,dr\,d\theta$$

$$= \sqrt{2}\int_0^{2\pi}\int_1^4 r^2\,dr\,d\theta = \sqrt{2}\int_0^{2\pi}\left[\frac{r^3}{3}\right]_1^4 d\theta = 21\sqrt{2}\int_0^{2\pi} d\theta = 42\sqrt{2}\,\pi \qquad ■$$

**NOW WORK**  Problem 33.

**EXAMPLE 3**  Finding the Centroid of a Surface

Find the centroid of the part of the unit sphere that lies in the first octant. See Figure 55.

**Solution**  In spherical coordinates, the equation of the unit sphere $S$: $x^2 + y^2 + z^2 = 1$ is $\rho = 1$. So using $x = \cos\theta\sin\phi$, $y = \sin\theta\sin\phi$, and $z = \cos\phi$, we obtain the parametrization

$$\mathbf{r} = \mathbf{r}(\theta, \phi) = \cos\theta\sin\phi\,\mathbf{i} + \sin\theta\sin\phi\,\mathbf{j} + \cos\phi\,\mathbf{k}; \qquad 0 \le \theta \le \frac{\pi}{2}, \quad 0 \le \phi \le \frac{\pi}{2}$$

We seek $\|\mathbf{r}_\theta \times \mathbf{r}_\phi\|$ and $dS$. The partial derivatives of $\mathbf{r} = \mathbf{r}(\theta, \phi)$ are

$$\mathbf{r}_\theta = -\sin\theta\sin\phi\,\mathbf{i} + \cos\theta\sin\phi\,\mathbf{j} \qquad \mathbf{r}_\phi = \cos\theta\cos\phi\,\mathbf{i} + \sin\theta\cos\phi\,\mathbf{j} - \sin\phi\,\mathbf{k}$$

$$\mathbf{r}_\theta \times \mathbf{r}_\phi = \begin{vmatrix} \mathbf{i} & \mathbf{j} & \mathbf{k} \\ -\sin\theta\sin\phi & \cos\theta\sin\phi & 0 \\ \cos\theta\cos\phi & \sin\theta\cos\phi & -\sin\phi \end{vmatrix}$$

$$= -\cos\theta\sin^2\phi\,\mathbf{i} - \sin\theta\sin^2\phi\,\mathbf{j} - \sin\phi\cos\phi\,\mathbf{k}$$

$$\|\mathbf{r}_\theta \times \mathbf{r}_\phi\| = \sqrt{\cos^2\theta\sin^4\phi + \sin^2\theta\sin^4\phi + \sin^2\phi\cos^2\phi}$$

$$= \sqrt{(\cos^2\theta + \sin^2\theta)\sin^4\phi + \sin^2\phi\cos^2\phi}$$

$$= \sqrt{\sin^2\phi(\sin^2\phi + \cos^2\phi)} = \sin\phi$$

$$dS = \|\mathbf{r}_\theta \times \mathbf{r}_\phi\|\,d\theta\,d\phi = \sin\phi\,d\theta\,d\phi$$

Since the surface area $S$ of a sphere is $S = 4\pi r^2$, the surface area of the part of the unit sphere $r = 1$ in the first octant is $\frac{1}{8}S = \frac{\pi}{2}$, so the mass $M$ is $\frac{\pi}{2}$. Because of symmetry, $\bar{x} = \bar{y} = \bar{z}$, so we need to find only one moment, say $M_{xy}$.

$$M_{xy} = \iint_S z\, dS = \iint_S \cos\phi\, dS = \int_0^{\pi/2}\int_0^{\pi/2} \cos\phi \sin\phi\, d\theta\, d\phi$$

$$= \int_0^{\pi/2} \frac{\pi}{2}\cos\phi\sin\phi\, d\phi = \frac{\pi}{2}\left[\frac{\sin^2\phi}{2}\right]_0^{\pi/2} = \frac{\pi}{4}$$

The centroid of the part of the unit sphere in the first octant is

$$(\bar{x}, \bar{y}, \bar{z}) = \left(\frac{M_{yz}}{M}, \frac{M_{xz}}{M}, \frac{M_{xy}}{M}\right) = \left(\frac{1}{2}, \frac{1}{2}, \frac{1}{2}\right)$$

∎

## 2 Determine the Orientation of a Surface

Suppose the surface $S$ has a tangent plane at each point, except possibly at its boundary points. Then there are two unit normal vectors, $\mathbf{n}_1$ and $\mathbf{n}_2$, where $\mathbf{n}_1 = -\mathbf{n}_2$, that can be drawn at each point of $S$. See Figure 56.

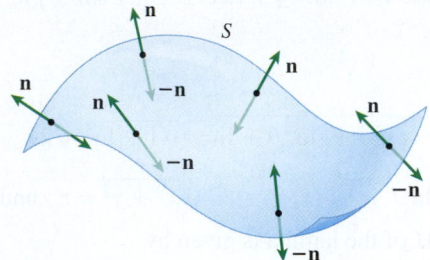

Suppose it is possible to choose a unit normal vector $\mathbf{n}$ at each point of the surface $S$ so that $\mathbf{n}$ varies continuously over the surface, and as $\mathbf{n}$ moves around any closed curve on $S$, $\mathbf{n}$ returns to its original direction, as shown in Figure 57. Such a surface is said to be **orientable**, with the vector $\mathbf{n}$ providing its **orientation**. As a result, a surface $S$ that is orientable has two possible orientations, one for $\mathbf{n}$ and one for $-\mathbf{n}$.

Suppose $S$ has a smooth parametrization $\mathbf{r} = \mathbf{r}(u, v)$, and $(u_0, v_0)$ is in the parameter domain $R$. Then the vector $\mathbf{r}_u(u_0, v_0) \times \mathbf{r}_v(u_0, v_0)$ is normal to the tangent plane at $\mathbf{r}(u_0, v_0)$, and the two unit vectors normal to the surface $S$ are

$$\mathbf{n} = \frac{\mathbf{r}_u \times \mathbf{r}_v}{\|\mathbf{r}_u \times \mathbf{r}_v\|} \qquad \text{and} \qquad -\mathbf{n} = -\frac{\mathbf{r}_u \times \mathbf{r}_v}{\|\mathbf{r}_u \times \mathbf{r}_v\|}$$

When $S$ is defined by the graph of a function $z = f(x, y)$, then the two orientations can be called *upward* and *downward* depending on the sign of the vector component $\mathbf{k}$. For general surfaces, there is no natural choice for *up* and *down*.

Suppose $F(x, y, z) = z - f(x, y)$. Then the two unit vectors normal to $S$ are

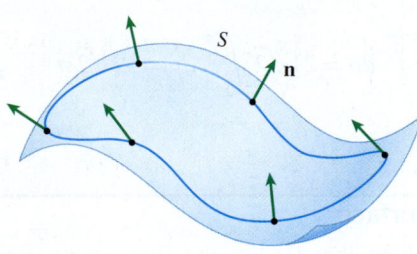

**Figure 56**

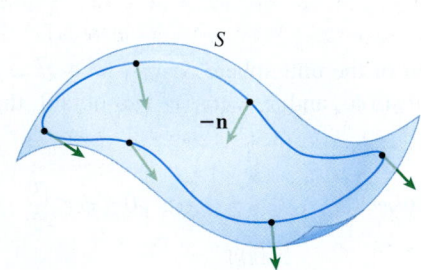

**Figure 57** Two orientations for a surface $S$

$$\mathbf{n} = \frac{\nabla F}{\|\nabla F\|} = \frac{-f_x(x, y)\mathbf{i} - f_y(x, y)\mathbf{j} + \mathbf{k}}{\sqrt{[f_x(x, y)]^2 + [f_y(x, y)]^2 + 1}} \tag{2}$$

and

$$-\mathbf{n} = \frac{-\nabla F}{\|\nabla F\|} = \frac{f_x(x, y)\mathbf{i} + f_y(x, y)\mathbf{j} - \mathbf{k}}{\sqrt{[f_x(x, y)]^2 + [f_y(x, y)]^2 + 1}} \tag{3}$$

Since the $\mathbf{k}$ component of $\mathbf{n}$ is positive, it points in the upward direction. For this reason, $\mathbf{n}$ is called the **upward-pointing unit normal vector** of $S$. The $\mathbf{k}$ component of $-\mathbf{n}$ is negative. So, $-\mathbf{n}$ is called the **downward-pointing unit normal vector** of $S$. Using the two unit vectors, we define the **positive side** or **outside** of $S$ as the side associated with the upward-pointing unit vector, and the **negative side** or **inside** of $S$ as the side associated with the downward-pointing unit vector. See Figure 58.

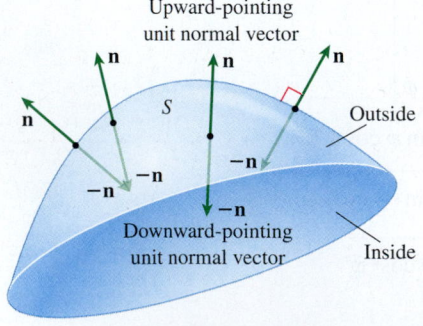

**Figure 58** The outside of the open surface is the side on which the upward-pointing unit normal vectors $\mathbf{n}$ originate.

### EXAMPLE 4  Finding the Orientations of a Surface

Find both orientations for the paraboloid defined by $z = 9 - x^2 - y^2$, $z \geq 0$.

**(a)** Use rectangular coordinates to find the unit normal vectors.

**(b)** Parametrize the surface, and then find the unit normal vectors.

**Solution (a)** Since the paraboloid is given by $f(x, y) = 9 - x^2 - y^2$, $z \geq 0$, we can use the equations (1) and (2) to find the unit normal vectors. We begin by finding the partial derivatives

$$f_x(x, y) = -2x \qquad f_y(x, y) = -2y$$

Then

$$\mathbf{n} = \frac{2x\,\mathbf{i} + 2y\,\mathbf{j} + \mathbf{k}}{\sqrt{[-2x]^2 + [-2y]^2 + 1}} = \frac{2x\,\mathbf{i} + 2y\,\mathbf{j} + \mathbf{k}}{\sqrt{4x^2 + 4y^2 + 1}}$$

$$-\mathbf{n} = \frac{-2x\,\mathbf{i} - 2y\,\mathbf{j} - \mathbf{k}}{\sqrt{[-2x]^2 + [-2y]^2 + 1}} = \frac{-2x\,\mathbf{i} - 2y\,\mathbf{j} - \mathbf{k}}{\sqrt{4x^2 + 4y^2 + 1}}$$

Notice that the $\mathbf{k}$ component of $\mathbf{n}$ is positive, indicating $\mathbf{n}$ is the upward-pointing unit normal vector.

**(b)** We parametrize $z = 9 - x^2 - y^2$, $z \geq 0$, using cylindrical coordinates, by defining the parametric equations

$$x = r \cos\theta \qquad y = r \sin\theta \qquad z = 9 - (x^2 + y^2) = 9 - r^2$$

Then the parametrization is

$$\mathbf{r} = \mathbf{r}(r, \theta) = r \cos\theta\,\mathbf{i} + r \sin\theta\,\mathbf{j} + (9 - r^2)\mathbf{k}$$

We seek $\|\mathbf{r}_r \times \mathbf{r}_\theta\|$. The partial derivatives of $\mathbf{r}$ are

$$\mathbf{r}_r = \cos\theta\,\mathbf{i} + \sin\theta\,\mathbf{j} - 2r\mathbf{k} \qquad \mathbf{r}_\theta = -r \sin\theta\,\mathbf{i} + r \cos\theta\,\mathbf{j}$$

$$\mathbf{r}_r \times \mathbf{r}_\theta = \begin{vmatrix} \mathbf{i} & \mathbf{j} & \mathbf{k} \\ \cos\theta & \sin\theta & -2r \\ -r \sin\theta & r \cos\theta & 0 \end{vmatrix} = 2r^2 \cos\theta\,\mathbf{i} + 2r^2 \sin\theta\,\mathbf{j} + r\mathbf{k}$$

$$\|\mathbf{r}_r \times \mathbf{r}_\theta\| = \sqrt{4r^4 \cos^2\theta + 4r^4 \sin^2\theta + r^2} = \sqrt{4r^4 + r^2} = r\sqrt{4r^2 + 1}$$

The unit normal vectors are

$$\mathbf{n} = \frac{\mathbf{r}_r \times \mathbf{r}_\theta}{\|\mathbf{r}_r \times \mathbf{r}_\theta\|} = \frac{2r^2 \cos\theta\,\mathbf{i} + 2r^2 \sin\theta\,\mathbf{j} + r\mathbf{k}}{r\sqrt{4r^2 + 1}} = \frac{2r \cos\theta\,\mathbf{i} + 2r \sin\theta\,\mathbf{j} + \mathbf{k}}{\sqrt{4r^2 + 1}}$$

$$-\mathbf{n} = \frac{-2r \cos\theta\,\mathbf{i} - 2r \sin\theta\,\mathbf{j} - \mathbf{k}}{\sqrt{4r^2 + 1}}$$

Notice that the unit vectors found in (a) and (b) are equivalent. ∎

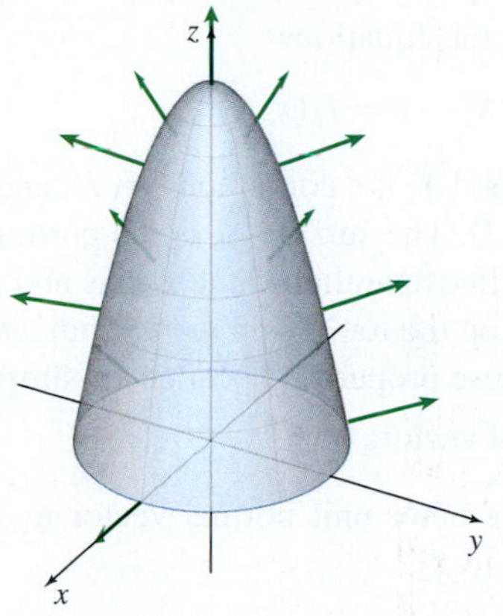

**Figure 59** The outer unit normal vectors of $z = 9 - x^2 - y^2$, $z \geq 0$.

Figure 59 shows the surface and some of the outer (upward-pointing) unit normal vectors of the paraboloid.

**NOW WORK** Problem 17.

For a **closed surface** $S$, that is, a surface that forms the boundary of a closed, bounded solid $E$ in space, such as a sphere, we agree that the **positive orientation** of $S$ has the unit normal vectors $\mathbf{n}$ to $S$ pointing away from $E$ and that the **negative orientation** of $S$ has the unit normal vectors $-\mathbf{n}$ to $S$ pointing inward. For closed surfaces $S$ with a positive orientation, the unit normal vectors are called **outer unit normal vectors**; for closed surfaces $S$ with a negative orientation, the unit normal vectors are called **inner unit normal vectors**.* See Figure 60 on page 1086.

---

*Although at first these definitions seem easy, they contain some hidden difficulties. The discussion here and in the next section is limited to simple types of oriented solids $E$. A discussion of more general solids is found in metric topology.

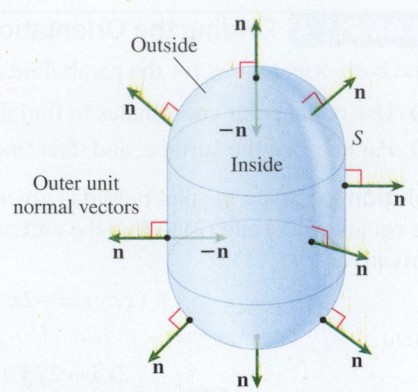

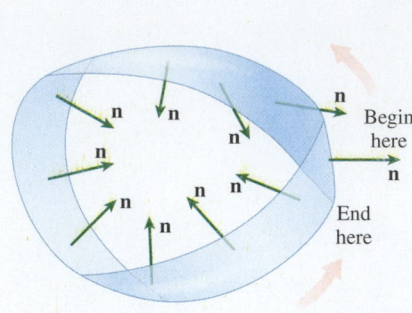

**NOTE** A Möbius strip (Figure 61) is formed by taking a long rectangular strip of paper, giving one end a half-twist, and attaching the two ends. The Möbius strip is named after the German mathematician August Möbius (1790–1868).

**Figure 60** $S$ is a closed surface. The outer unit normal vectors point away from the surface.

**DF Figure 61** Möbius strip: a nonorientable surface.

All curves are orientable, but not all surfaces are orientable. A famous example of a surface that is not orientable is a Möbius strip. See Figure 61. Although it is not immediately apparent, a Möbius strip has only one side and one boundary component. If you were to begin tracing a line on one side of a Möbius strip and continue to follow it once around, you would end on the other side of the strip, that is, the unit normal vector $\mathbf{n}$ now points in the opposite direction.

**EXAMPLE 5** Finding the k Component of the Outer Unit Normal Vectors of an $xy$-Simple Surface

Consider the surface $S$ with positive orientation that forms the boundary of the closed, bounded solid $E$ shown in Figure 62. Notice that the surface $S$ can be decomposed into three surfaces, $S_1$, $S_2$, and $S_3$, and that each of the outer unit normal vectors $\mathbf{n}_1$, $\mathbf{n}_2$, and $\mathbf{n}_3$ of the three surfaces points away from the solid $E$ enclosed by $S$.

Assume the surfaces $S_1$ and $S_2$ are defined by the equations

$$S_1: \quad z = f_1(x, y) \qquad \text{and} \qquad S_2: \quad z = f_2(x, y)$$

where $f_1(x, y) < f_2(x, y)$ on $D$, the functions $f_1$ and $f_2$ are continuous on $D$, and both $f_1$ and $f_2$ have continuous partial derivatives on $D$. The surface $S_3$ is the portion of a cylindrical surface between $S_1$ and $S_2$ formed by lines parallel to the $z$-axis and along the boundary of $D$. The solid $E$ is enclosed by $S_2$ on the top, $S_1$ on the bottom, and the lateral surface $S_3$ on the sides. A surface $S$ with these properties is called **xy-simple**.

Find the **k** component of the outer unit normal vectors of $S$.

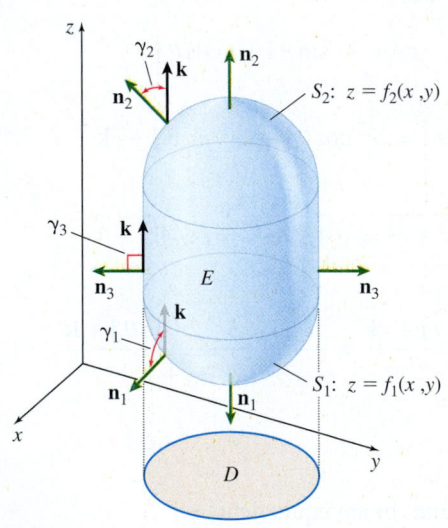

**Figure 62**

**NEED TO REVIEW?** Direction cosines of a vector in space are discussed in Section 10.4, pp. 757–758.

**Solution** For the top surface $S_2$: $z = f_2(x, y)$, the outer unit normal vector $\mathbf{n}_2$ is the same as the upward-pointing unit normal vector $\mathbf{n}$ to $S_2$.

$$\mathbf{n}_2 = \mathbf{n} = \frac{-(f_2)_x(x, y)\mathbf{i} - (f_2)_y(x, y)\mathbf{j} + \mathbf{k}}{\sqrt{[(f_2)_x(x, y)]^2 + [(f_2)_y(x, y)]^2 + 1}}$$

The **k** component of $\mathbf{n}_2$ is given by the direction cosine $\cos \gamma_2$, where

$$\cos \gamma_2 = \mathbf{n}_2 \cdot \mathbf{k} = \frac{1}{\sqrt{[(f_2)_x(x, y)]^2 + [(f_2)_y(x, y)]^2 + 1}}$$

For the bottom surface $S_1$: $z = f_1(x, y)$, the outer unit normal vector $\mathbf{n}_1$ equals the downward-pointing unit normal vector $-\mathbf{n}$ to $S_1$.

$$\mathbf{n}_1 = -\mathbf{n} = \frac{(f_1)_x(x, y)\mathbf{i} + (f_1)_y(x, y)\mathbf{j} - \mathbf{k}}{\sqrt{[(f_1)_x(x, y)]^2 + [(f_1)_y(x, y)]^2 + 1}}$$

The **k** component of $\mathbf{n}_1$ is given by the direction cosine $\cos \gamma_1$, where

$$\cos \gamma_1 = \mathbf{n}_1 \cdot \mathbf{k} = \frac{-1}{\sqrt{[(f_1)_x(x, y)]^2 + [(f_1)_y(x, y)]^2 + 1}}$$

For the lateral surface $S_3$, the outer unit normal vector $\mathbf{n}_3$ is orthogonal to **k** at each point on $S_3$. The **k** component of $\mathbf{n}_3$, given by the direction cosine, $\cos \gamma_3$, is therefore 0. ∎

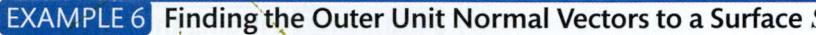

**EXAMPLE 6  Finding the Outer Unit Normal Vectors to a Surface $S$**

Find the outer unit normal vectors to the solid $E$ enclosed by

$$z = f(x, y) = \sqrt{R^2 - x^2 - y^2} \quad \text{and} \quad z = 0 \qquad 0 \le x^2 + y^2 \le R^2$$

**Solution** The solid $E$ is the interior of a hemisphere with center at $(0, 0, 0)$ and radius $R$ as shown in Figure 63. The surface $S$ consists of two surfaces, $S_1$ and $S_2$. The bottom surface $S_1$ and top surface $S_2$ are defined by

$$S_1: \quad z = 0 \quad \text{and} \quad S_2: \quad z = f(x, y) = \sqrt{R^2 - x^2 - y^2} \qquad 0 \le x^2 + y^2 \le R^2$$

The outer unit normal vector $\mathbf{n}_1$ of $S_1$ is $-\mathbf{k}$.

To find the outer unit normal vector $\mathbf{n}_2$ of $S_2$, we find $f_x(x, y)$ and $f_y(x, y)$.

$$f_x(x, y) = \frac{-x}{\sqrt{R^2 - x^2 - y^2}} = \frac{-x}{z} \quad \text{and} \quad f_y(x, y) = \frac{-y}{\sqrt{R^2 - x^2 - y^2}} = \frac{-y}{z}$$

Then

$$\mathbf{n}_2 = \frac{-f_x(x, y)\mathbf{i} - f_y(x, y)\mathbf{j} + \mathbf{k}}{\sqrt{[f_x(x, y)]^2 + [f_y(x, y)]^2 + 1}} = \frac{\dfrac{x}{z}\mathbf{i} + \dfrac{y}{z}\mathbf{j} + \mathbf{k}}{\sqrt{\dfrac{x^2}{z^2} + \dfrac{y^2}{z^2} + 1}}$$

$$= \frac{x\mathbf{i} + y\mathbf{j} + z\mathbf{k}}{\sqrt{x^2 + y^2 + z^2}} = \frac{x\mathbf{i} + y\mathbf{i} + z\mathbf{k}}{R} \qquad ∎$$

**Figure 63**

**NOW WORK** Problem 13.

## 3  Find the Flux of a Vector Field Across a Surface

An important application of a surface integral measures the rate of flow of a fluid across a surface $S$, also called the *flux across $S$*. Consider solar winds (protons and electrons flowing out of the Sun) as they strike Earth (a spherical surface). The rate of flow of the solar winds across the surface of Earth is an example of flux.

Suppose $\mathbf{F} = \mathbf{F}(x, y, z)$ is a vector field that represents the velocity of a fluid, such as air or water, and suppose $\rho = \rho(x, y, z)$ denotes the variable mass density of the fluid. Then $\rho\mathbf{F}$ equals the rate of flow (mass per unit time) per unit area of the fluid. Suppose $S$ is an orientable surface immersed in the fluid and that $S$ does not impede the flow of the fluid. If **n** is a unit normal vector of $S$, then $\rho\mathbf{F} \cdot \mathbf{n}$ equals the rate of flow of fluid in the direction of **n**. See Figure 64.

We seek a formula for the flux across $S$, that is, the mass of fluid that flows across $S$ in a unit of time.

Following the same process used to define surface area in Section 15.6, it can be shown that the mass of fluid per unit time crossing a patch $S^*$ in the direction of the unit normal **n** is approximately

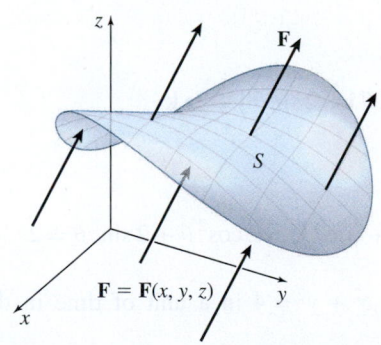

**Figure 64**

$$\boxed{\rho\mathbf{F} \cdot \mathbf{n}\, \Delta S^*}$$

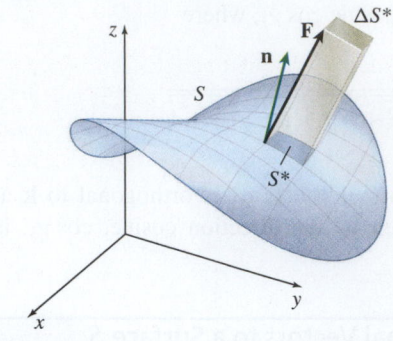

**Figure 65**

where $\mathbf{F}$ and $\mathbf{n}$ are calculated at some point on the patch $S^*$ and $\Delta S^*$ is the area of the patch. See Figure 65.

The **flux of F across** $S$, that is, the mass of fluid flowing across the surface $S$ in a unit of time in the direction of $\mathbf{n}$, is

$$\frac{\text{Mass of fluid}}{\text{Unit time}} = \iint\limits_{S} \rho \mathbf{F} \cdot \mathbf{n} \, dS$$

**CALC**
**CLIP**
▶ **EXAMPLE 7** Finding the Flux of F Across a Cylinder

A fluid has a constant mass density $\rho$. Find the mass of fluid flowing across the cylinder $x^2 + y^2 = 4$, where $0 \le z \le 3$, in a unit of time in the direction outward from the $z$-axis, if the velocity of the fluid at any point on the cylinder is $\mathbf{F} = \mathbf{F}(x, y, z) = x\mathbf{i} + y\mathbf{j} + 2z\mathbf{k}$. That is, find the flux of $\mathbf{F}$ across the cylinder.

**Solution** In cylindrical coordinates $x^2 + y^2 = r^2 = 4$, or equivalently, $r = 2$, which leads to the parametrization $\mathbf{r}(\theta, z) = 2\cos\theta\,\mathbf{i} + 2\sin\theta\,\mathbf{j} + z\mathbf{k}$, $0 \le \theta \le 2\pi$, $0 \le z \le 3$.

We seek $\|\mathbf{r}_\theta \times \mathbf{r}_z\|$ and $dS$. The partial derivatives of $\mathbf{r}$ are

$$\mathbf{r}_\theta = -2\sin\theta\,\mathbf{i} + 2\cos\theta\,\mathbf{j} \qquad \mathbf{r}_z = \mathbf{k}$$

$$\mathbf{r}_\theta \times \mathbf{r}_z = \begin{vmatrix} \mathbf{i} & \mathbf{j} & \mathbf{k} \\ -2\sin\theta & 2\cos\theta & 0 \\ 0 & 0 & 1 \end{vmatrix} = 2\cos\theta\,\mathbf{i} + 2\sin\theta\,\mathbf{j}$$

$$\|\mathbf{r}_\theta \times \mathbf{r}_z\| = \sqrt{4\cos^2\theta + 4\sin^2\theta} = 2$$

$$dS = \|\mathbf{r}_\theta \times \mathbf{r}_z\| \, d\theta \, dz = 2 \, d\theta \, dz$$

Then the unit normal vector $\mathbf{n}$ is

$$\mathbf{n} = \frac{\mathbf{r}_\theta \times \mathbf{r}_z}{\|\mathbf{r}_\theta \times \mathbf{r}_z\|} = \frac{2\cos\theta\,\mathbf{i} + 2\sin\theta\,\mathbf{j}}{2} = \cos\theta\,\mathbf{i} + \sin\theta\,\mathbf{j}$$

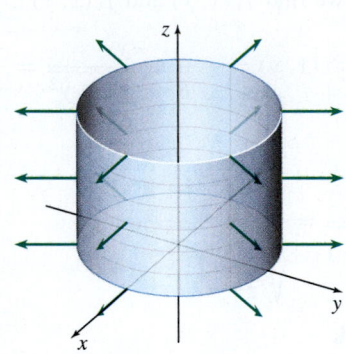

**Figure 66** $x^2 + y^2 = 4$, $0 \le z \le 3$, and $\mathbf{n} = \cos\theta\,\mathbf{i} + \sin\theta\,\mathbf{j}$

These vectors point outward from the surface of the cylinder, so we have the correct orientation. Figure 66 shows the cylinder $x^2 + y^2 = 4$, where $0 \le z \le 3$, and the outer unit normal vectors to the cylinder.

We now write $\mathbf{F}$ using the parametrization.

$$\mathbf{F} = \mathbf{F}(x, y, z) = \mathbf{F}(\theta, z) = 2\cos\theta\,\mathbf{i} + 2\sin\theta\,\mathbf{j} + 2z\mathbf{k}$$

The velocity of the fluid in the direction of $\mathbf{n}$ is

$$\mathbf{F} \cdot \mathbf{n} = (2\cos\theta\,\mathbf{i} + 2\sin\theta\,\mathbf{j} + 2z\mathbf{k}) \cdot (\cos\theta\,\mathbf{i} + \sin\theta\,\mathbf{j}) = 2\cos^2\theta + 2\sin^2\theta = 2$$

The mass of fluid flowing across the cylinder $x^2 + y^2 = 4$ in a unit of time in the direction of $\mathbf{n}$ is given by

$$\iint\limits_{S} \rho\,\mathbf{F} \cdot \mathbf{n}\,dS = \int_0^3 \int_0^{2\pi} 2\rho \cdot 2\,d\theta\,dz = 4\rho \int_0^3 2\pi\,dz = 8\rho\pi \cdot 3 = 24\rho\pi$$

The flux of $\mathbf{F}$ across $S$ is $24\rho\pi$. ∎

**NOW WORK** Problem 39.

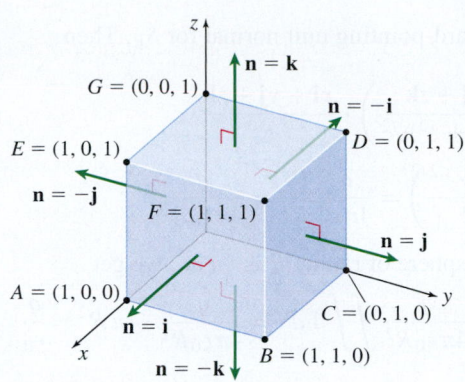

**Figure 67**

EXAMPLE 8  **Finding the Flux of F Across a Cube**

A fluid has constant mass density $\rho$. Find the mass per unit time of fluid flowing across the cube enclosed by the planes $x = 0$, $x = 1$, $y = 0$, $y = 1$, $z = 0$, and $z = 1$ in the direction of the outer unit normal vectors if the velocity of the fluid at any point on the cube is $\mathbf{F} = \mathbf{F}(x, y, z) = 4xz\mathbf{i} - y^2\mathbf{j} + yz\mathbf{k}$. That is, find the flux of $\mathbf{F}$ across the cube.

**Solution** Figure 67 shows the cube and its six faces. The mass per unit time of fluid flowing across the cube in the direction of the outer unit normal vector $\mathbf{n}$ is

$$\iint_S \rho \mathbf{F} \cdot \mathbf{n} \, dS = \rho \iint_S \mathbf{F} \cdot \mathbf{n} \, dS$$

We decompose $S$ into its six faces and find the surface integral over each one, as shown in Table 2.

**TABLE 2**

| Face | | $\mathbf{n}$ | $\mathbf{F}$ | $\mathbf{F} \cdot \mathbf{n}$ | $\rho \iint_S \mathbf{F} \cdot \mathbf{n} \, dS$ |
|---|---|---|---|---|---|
| $S_1 = ABCO$ | $z = 0$ | $-\mathbf{k}$ | $-y^2\mathbf{j}$ | $0$ | $0$ |
| $S_2 = OAEG$ | $y = 0$ | $-\mathbf{j}$ | $4xz\mathbf{i}$ | $0$ | $0$ |
| $S_3 = OCDG$ | $x = 0$ | $-\mathbf{i}$ | $-y^2\mathbf{j} + yz\mathbf{k}$ | $0$ | $0$ |
| $S_4 = ABFE$ | $x = 1$ | $\mathbf{i}$ | $4z\mathbf{i} - y^2\mathbf{j} + yz\mathbf{k}$ | $4z$ | $\rho \int_0^1 \int_0^1 4z \, dy \, dz = 2\rho$ |
| $S_5 = BCDF$ | $y = 1$ | $\mathbf{j}$ | $4xz\mathbf{i} - \mathbf{j} + z\mathbf{k}$ | $-1$ | $-\rho \int_0^1 \int_0^1 dx \, dz = -\rho$ |
| $S_6 = DFEG$ | $z = 1$ | $\mathbf{k}$ | $4x\mathbf{i} - y^2\mathbf{j} + y\mathbf{k}$ | $y$ | $\rho \int_0^1 \int_0^1 y \, dx \, dy = \dfrac{1}{2}\rho$ |

The mass per unit time of fluid flowing across the cube is

$$0 + 0 + 0 + 2\rho - \rho + \frac{1}{2}\rho = \frac{3}{2}\rho \qquad \blacksquare$$

NOW WORK  Problem **23.**

## Application: Electric Flux

The idea of flux, the rate of flow of a fluid across a surface $S$, describes many different phenomena. As an example, we consider the flow of electricity across a surface $S$.

Suppose $\mathbf{E}$ is an electric field and $S$ is a piecewise-smooth surface with orientation $\mathbf{n}$. Then the **electric flux** $\Phi_E$, the rate at which the electric field passes through $S$, is given by

$$\Phi_E = \iint_S \mathbf{E} \cdot \mathbf{n} \, dS$$

Further, suppose we place a point charge of $q$ coulombs (C) at the origin. Then the corresponding electric field is

$$\mathbf{E} = \left(\frac{q}{4\pi \epsilon_0}\right)\left(\frac{x\mathbf{i} + y\mathbf{j} + z\mathbf{k}}{(x^2 + y^2 + z^2)^{3/2}}\right)$$

The field $\mathbf{E}$ is measured in volts per meter (V/m) and $\epsilon_0 \approx 8.854 \times 10^{-12} \, \text{C}^2/\text{Nm}^2$ is the **permittivity of free space**.

Now suppose $S_R$ is the sphere $x^2+y^2+z^2=R^2$, $R>0$. If $F(x, y, z)=x^2+y^2+z^2$,

then $\mathbf{n}=\dfrac{\nabla \mathbf{F}}{\|\nabla \mathbf{F}\|}=\dfrac{x\mathbf{i}+y\mathbf{j}+z\mathbf{k}}{R}$ is the outward-pointing unit normal for $S_R$. Then

$$\mathbf{E}\cdot\mathbf{n}=\left(\frac{q}{4\pi\epsilon_0}\right)\left(\frac{x\mathbf{i}+y\mathbf{j}+z\mathbf{k}}{(x^2+y^2+z^2)^{3/2}}\right)\cdot\frac{x\mathbf{i}+y\mathbf{j}+z\mathbf{k}}{R}$$

$$=\left(\frac{q}{4\pi\epsilon_0}\right)\left(\frac{x^2+y^2+z^2}{R^3\cdot R}\right)=\frac{q}{4\pi\epsilon_0}\frac{R^2}{R^4}=\frac{q}{4\pi\epsilon_0 R^2}$$

Using the fact that the surface area $S_R$ of a sphere of radius $R$ is $4\pi R^2$, we get

$$\Phi_E=\iint_{S_R}\mathbf{E}\cdot\mathbf{n}\,dS=\iint_{S_R}\frac{q}{4\pi\epsilon_0 R^2}\,dS=\frac{q}{4\pi\epsilon_0 R^2}\iint_{S_R}1\,dS=\frac{q}{4\pi\epsilon_0 R^2}\cdot 4\pi R^2=\frac{q}{\epsilon_0}$$

This means that the electric flux of a point charge $q$ originating at the center of a sphere is independent of the size of the sphere.

## 15.7 Assess Your Understanding

### Concepts and Vocabulary

1. *Multiple Choice* If a surface $S$ has a unit normal vector $\mathbf{n}$ at each point of the surface that can be chosen so $\mathbf{n}$ varies continuously over the surface, and as $\mathbf{n}$ moves around any closed curve on the surface, $\mathbf{n}$ returns to its original direction, then the surface is said to be [(a) a Möbius strip (b) integrable (c) orientable].

2. *Multiple Choice* If $\mathbf{F}=\mathbf{F}(x, y, z)$ is a vector field that represents the velocity of a fluid, and $\rho=\rho(x, y, z)$ is the variable mass density of the fluid, then the surface integral $\iint_S \rho\mathbf{F}\cdot\mathbf{n}\,dS$ is called the [(a) speed (b) flux (c) orientation (d) acceleration] of $\mathbf{F}$ across the surface $S$.

### Skill Building

*In Problems 3–12, find each surface integral.*

3. $\displaystyle\iint_S (x^2+y^2)\,dS$;

   $S: z=f(x, y)=2x+3y+5, \ 0\leq x\leq 1, 0\leq y\leq 1$

4. $\displaystyle\iint_S x^2y^2\,dS$;

   $S: z=f(x, y)=3x+4y+8, 0\leq x\leq 1, 0\leq y\leq 1$

5. $\displaystyle\iint_S y\,dS$;   $S: z=f(x, y)=x+5y^2, 0\leq x\leq 2, 0\leq y\leq 1$

6. $\displaystyle\iint_S 4x\,dS$;   $S: z=f(x, y)=2x^2+y, 0\leq x\leq 1, 0\leq y\leq 2$

7. $\iint_S(x+y+z)\,dS$, where $S$ is the portion of the plane $z=x-3y$ above the region enclosed by $y=0$, $x=1$, and $x=3y$ in the $xy$-plane

8. $\iint_S x\,dS$, where $S$ is the portion of the plane $x+y+2z=4$ above the region enclosed by $x=1$, $y=1$, $x=0$, and $y=0$

9. $\iint_S y\,dS$, where $S$ is the portion of the plane $x+y+z=2$ inside the cylinder $x^2+y^2=1$

10. $\iint_S yz\,dS$, where $S$ is the portion of the plane $x+2y+3z=6$ in the first octant

11. $\iint_S x^2z\,dS$, where $S$ is the surface $x^2+y^2=1, 0\leq z\leq 1$

12. $\iint_S(x+y)z\,dS$, where $S$ is the surface $x^2+y^2=9, 1\leq z\leq 3$

*In Problems 13–16, find the outer unit normal vector to the surface $S$ defined by $z=f(x, y)$.*

13. $z=f(x, y)=\sqrt{16-x^2-y^2}, \ 0\leq x^2+y^2\leq 16$

14. $z=f(x, y)=\sqrt{1-x^2-y^2}, \ 0\leq x^2+y^2\leq 1$

15. $z=f(x, y)=\sqrt{36-9x^2-4y^2}, \ 0\leq 9x^2+4y^2\leq 36$

16. $z=f(x, y)=\sqrt{4-x^2-4y^2}, \ 0\leq x^2+4y^2\leq 4$

*In Problems 17 and 18, find both orientations for each surface.*

17. $S: \mathbf{r}(u, v)=(u+v)\mathbf{i}+u^2\mathbf{j}+v^2\mathbf{k}$

18. $S: \mathbf{r}(u, v)=uv\mathbf{i}+vj+e^u\mathbf{k}$

**Flux Across a Surface** *In Problems 19–26, a fluid with constant mass density $\rho$ flows across the cube shown in Figure 67. If the velocity of the fluid at any point on the cube is given by $\mathbf{F}$, find the flux of $\mathbf{F}$ across the cube in the direction of the outer unit normal vectors.*

19. $\mathbf{F}=x\mathbf{i}$          20. $\mathbf{F}=y\mathbf{i}$          21. $\mathbf{F}=z\mathbf{i}$

22. $\mathbf{F}=x\mathbf{i}+y\mathbf{j}$          23. $\mathbf{F}=x\mathbf{i}+y\mathbf{j}+z\mathbf{k}$          24. $\mathbf{F}=z^2\mathbf{i}$

25. $\mathbf{F}=x^2\mathbf{i}+y^2\mathbf{j}+z^2\mathbf{k}$          26. $\mathbf{F}=x^2\mathbf{i}+y^2\mathbf{j}$

### Applications and Extensions

**Flux Across a Surface** *In Problems 27–30, find $\iint_S \mathbf{F}\cdot\mathbf{n}\,dS$, where $\mathbf{n}$ is the outer unit normal of S.*

27. $\mathbf{F}=x\mathbf{i}+y\mathbf{j}+z\mathbf{k}$, and $S$ is the surface $x^2+y^2+z^2=1, z\geq 0$

28. $\mathbf{F}=-y\mathbf{i}+x\mathbf{j}+z\mathbf{k}$, and $S$ is the surface $x^2+y^2+z^2=1, z\geq 0$

---

**1.** = NOW WORK problem          = Graphing technology recommended          CAS = Computer Algebra System recommended

29. $\mathbf{F} = (x + y)\mathbf{i} + (2x - z)\mathbf{j} + y\mathbf{k}$, and $S$ is the tetrahedron formed by the coordinate planes and the plane $z + 2x + 2y = 8$

30. $\mathbf{F} = 2x\mathbf{i} - x^2\mathbf{j} + (z - 2x + 2y)\mathbf{k}$, and $S$ is the tetrahedron formed by the coordinate planes and the plane $2x + 2y + z = 6$

*In Problems 31 and 32, find the flux integral $\iint\limits_S \mathbf{F} \cdot \mathbf{n}\, dS$, given the oriented surface S and vector field **F**.*

31. The surface $S$ is parametrized by $\mathbf{r}(u, v) = uv\mathbf{i} + u^2\mathbf{j} + v\,\mathbf{k}$, $1 \le u \le 2, 0 \le v \le 3$ and the vector field is $\mathbf{F} = \mathbf{F}(x, y, z) = -z\mathbf{j} + \mathbf{k}$.

32. The surface $S$ is parametrized by $\mathbf{r}(u, v) = ve^u\mathbf{i} + v^3\mathbf{j} + u\mathbf{k}$, $0 \le u \le 1, -1 \le v \le 1$ and the vector field is $\mathbf{F} = \mathbf{F}(x, y, z) = xy\mathbf{i} + 2z\mathbf{j}$.

33. **Mass of a Lamina**   Find the mass of a lamina in the shape of a cone $z = \sqrt{x^2 + y^2}$, $2 \le z \le 4$, if the mass density of the lamina is $\rho = \rho(x, y, z) = 8 - z$.

34. **Mass of a Lamina**   Find the mass of a lamina in the shape of a cone $z = \sqrt{x^2 + y^2}$, $2 \le z \le 4$, if the mass density of the lamina is $\rho = \rho(x, y, z) = \sqrt{x^2 + y^2}$.

35. **Center of Mass of a Lamina**   Find the center of mass of a lamina in the shape of the part of the plane $2x + 3y + z = 4$ that lies inside $x^2 + z^2 = 9$, if the mass density of the lamina is $\rho = \rho(x, y, z) = \sqrt{x^2 + z^2}$.

36. **Center of Mass of a Lamina**   Find the center of mass of a lamina in the shape of the unit sphere $x^2 + y^2 + z^2 = 1$, if the mass density of the lamina is $\rho = \rho(x, y, z) = z + 1$.

*In Problems 37 and 38, find each surface integral.*

37. $\iint\limits_S x^2 z\, dS$;   $S$: the portion of the cylinder $x^2 + y^2 = 9$ between $z = 0$ and $z = 5$

38. $\iint\limits_S (x + 2y)\, dS$;   $S$: $\mathbf{r}(u, v) = u\cos v\,\mathbf{i} + (u^2 + 1)\mathbf{j} + u\cos v\,\mathbf{k}$, $0 \le u \le 1$ and $0 \le v \le \dfrac{\pi}{2}$

39. **Flux Across a Surface**   Find the mass of a fluid with constant mass density flowing across the paraboloid $z = x^2 + y^2$, $z \le 5$, in a unit of time in the direction of the outer unit normal, if the velocity of the fluid at any point on the paraboloid is $\mathbf{F} = \mathbf{F}(x, y, z) = -x\mathbf{i} - y\mathbf{j} - z\mathbf{k}$.

40. **Flux Across a Surface**   Find the mass of a fluid with constant mass density flowing across the paraboloid $z = 9 - x^2 - y^2$, $z \ge 0$, in a unit of time in the direction of the outer unit normal, if the velocity of the fluid at any point on the paraboloid is $\mathbf{F} = \mathbf{F}(x, y, z) = x\mathbf{i} + y\mathbf{j} + 3\mathbf{k}$.

41. If $\mathbf{n}$ is the upward-pointing unit normal to a surface $z = f(x, y)$ and $\mathbf{F} = P\mathbf{i} + Q\mathbf{j} + R\mathbf{k}$, show that

$$\iint\limits_S \mathbf{F} \cdot \mathbf{n}\, dS = \iint\limits_D \left( -P\frac{\partial f}{\partial x} - Q\frac{\partial f}{\partial y} + R \right) dx\, dy$$

42. **Electric Flux**   Find the electric flux across the part of the cylinder $x^2 + y^2 = 9$, $0 \le z \le 2$, in the direction of the outer unit normal vectors when the electric field is $\mathbf{E}(x, y, z) = x\mathbf{i} + 2y\mathbf{j} + 3z\mathbf{k}$.

43. CAS **Helicoid**   Find $\iint\limits_S (x^2 + z^2)dS$, where $S$ is the helicoid $\mathbf{r}(u, v) = u\cos(2v)\mathbf{i} + u\sin(2v)\mathbf{j} + v\mathbf{k}$, $0 \le u \le 3, 0 \le v \le 2\pi$.

**Challenge Problems**

44. **Center of Mass**   Find the center of mass of the upper half of the sphere $x^2 + y^2 + z^2 = a^2$ covered by a thin material with mass density at each point proportional to the distance from the $xy$-plane.

45. Find $\iint\limits_S \mathbf{F} \cdot \mathbf{n}\, dS$, where $S$ is a level surface of a function $w = f(x, y, z)$ with outer unit normal vector $\mathbf{n}$, $\mathbf{F} = \nabla f$, and $f$ satisfies the equation $\left(\dfrac{\partial f}{\partial x}\right)^2 + \left(\dfrac{\partial f}{\partial y}\right)^2 + \left(\dfrac{\partial f}{\partial z}\right)^2 = 1$.

*Hint: The answer should depend on S.*

# 15.8 The Divergence Theorem

**OBJECTIVES**   *When you finish this section, you should be able to:*

1  Find the divergence of a vector field (p. 1092)

2  Use the Divergence Theorem (p. 1092)

3  Interpret the divergence of **F** (p. 1096)

   Application: Electric force fields (p. 1097)

Green's Theorem expresses a relationship between a certain double integral over a plane region and a line integral taken around its boundary. There are two ways to generalize this result to space. One of these, known as the *Divergence Theorem*, is the subject of this section, and the other, known as *Stokes' Theorem*, is the subject of the next section.

  The Divergence Theorem expresses a certain relationship between a triple integral involving the *divergence of a vector field* and a surface integral, the flux of a vector field. The theorem is used in applications involving fluid dynamics, electrostatics, and magnetism. We begin with the definition of divergence of a vector field.

**DEFINITION**  Divergence of a Vector Field

Let $\mathbf{F} = \mathbf{F}(x, y, z) = P(x, y, z)\mathbf{i} + Q(x, y, z)\mathbf{j} + R(x, y, z)\mathbf{k}$ be a vector field, where $\dfrac{\partial P}{\partial x}$, $\dfrac{\partial Q}{\partial y}$, and $\dfrac{\partial R}{\partial z}$ exist. The **divergence of F**, denoted by div $\mathbf{F}$, is defined as the function*

$$\operatorname{div} \mathbf{F} = \frac{\partial P}{\partial x} + \frac{\partial Q}{\partial y} + \frac{\partial R}{\partial z} \tag{1}$$

## 1  Find the Divergence of a Vector Field

**EXAMPLE 1**  Finding the Divergence of a Vector Field

The divergence of the vector field

$$\mathbf{F}(x, y, z) = x^2 y z^2 \mathbf{i} + (2xz + y^3)\mathbf{j} + x^2 y^3 z \mathbf{k}$$

is

$$\operatorname{div} \mathbf{F} = \frac{\partial}{\partial x}(x^2 y z^2) + \frac{\partial}{\partial y}(2xz + y^3) + \frac{\partial}{\partial z}(x^2 y^3 z) = 2xyz^2 + 3y^2 + x^2 y^3 \quad \blacksquare$$

Notice that the divergence of a vector field is a scalar.

**NOW WORK** Problem 5.

## 2  Use the Divergence Theorem

The Divergence Theorem (or Gauss' Theorem) states that if certain conditions are satisfied, a triple integral is equal to the flux of a fluid with constant mass density $\rho = 1$ across a surface; that is, a triple integral can be expressed as a surface integral.

**THEOREM**  Divergence Theorem

Let $S$ be a positively oriented surface with outer unit normal $\mathbf{n}$ that encloses a closed, bounded solid $E$ in space. Let $\mathbf{F} = \mathbf{F}(x, y, z) = P(x, y, z)\mathbf{i} + Q(x, y, z)\mathbf{j} + R(x, y, z)\mathbf{k}$ be a vector field for which the functions $P$, $Q$, and $R$ have continuous first-order partial derivatives on an open set containing $E$. Then

$$\iiint_E \operatorname{div} \mathbf{F} \, dV = \iint_S \mathbf{F} \cdot \mathbf{n} \, dS$$

If $\mathbf{n} = \cos\alpha\,\mathbf{i} + \cos\beta\,\mathbf{j} + \cos\gamma\,\mathbf{k}$, then $\mathbf{F} \cdot \mathbf{n} = P\cos\alpha + Q\cos\beta + R\cos\gamma$. Using (1),

$$\iiint_E \left( \frac{\partial P}{\partial x} + \frac{\partial Q}{\partial y} + \frac{\partial R}{\partial z} \right) dV = \iint_S (P\cos\alpha + Q\cos\beta + R\cos\gamma) \, dS$$

The proof of the Divergence Theorem for the general conditions on the solid $E$ can be found in Advanced Calculus texts. The idea behind the proof is to verify each of the three equations

$$\iiint_E \frac{\partial P}{\partial x} \, dV = \iint_S P\cos\alpha \, dS \tag{2}$$

Bettmann/Getty Images

**ORIGINS**  Although Carl Friedrich Gauss (1777–1855) was neither the first to discover nor the first to prove the Divergence Theorem, the theorem is named in honor of him. The Divergence Theorem was proved by the Russian mathematician Mikhail Ostrogradsky (1801–1862), whose name is also associated with the theorem. Gauss is considered among the greatest mathematicians of all time. He was the mentor of many other notable mathematicians including Riemann, Bessel, Dedekind, and Möbius.

---

*Some books define div $\mathbf{F}$ as the dot product $\nabla \cdot \mathbf{F}$, where $\nabla$ is the partial differential operator $\nabla = \dfrac{\partial}{\partial x}\mathbf{i} + \dfrac{\partial}{\partial y}\mathbf{j} + \dfrac{\partial}{\partial z}\mathbf{k}$ and $\mathbf{F} = P\mathbf{i} + Q\mathbf{j} + R\mathbf{k}$.

$$\iiint_E \frac{\partial Q}{\partial y}\, dV = \iint_S Q \cos \beta \, dS \qquad (3)$$

$$\iiint_E \frac{\partial R}{\partial z}\, dV = \iint_S R \cos \gamma \, dS \qquad (4)$$

Once they are verified, the Divergence Theorem follows by adding the three equations.

We prove (4) for a special case of the solid $E$ as described in Example 5 from Section 15.7. As Figure 68 shows, the surface $S$ that encloses $E$ is a positively oriented, $xy$-simple surface consisting of three surfaces, $S_1$, $S_2$, and $S_3$, with the outer unit normals $\mathbf{n}_1$, $\mathbf{n}_2$, and $\mathbf{n}_3$, respectively.

**Proof** We assume that surfaces $S_1$ and $S_2$ are defined by the equations

$$S_1: \quad z = f_1(x, y) \qquad \text{and} \qquad S_2: \quad z = f_2(x, y)$$

where $f_1(x, y) < f_2(x, y)$ on $D$, $f_1$ and $f_2$ are continuous on $D$, and $f_1$ and $f_2$ have continuous partial derivatives on $D$. The surface $S_3$ is the portion of a cylindrical surface between $S_1$ and $S_2$ formed by lines parallel to the $z$-axis and along the boundary of the region $D$ in the $xy$-plane. The solid $E$ is enclosed on the bottom by $S_1$, on the top by $S_2$, and on the sides by the lateral surface $S_3$.

We now show that

$$\iiint_E \frac{\partial R}{\partial z}\, dV = \iint_S R \cos \gamma \, dS \qquad (5)$$

We start with the right side of (5). Then using the fact that $S$ consists of the three surfaces $S_1$, $S_2$, and $S_3$, we have

$$\iint_S R \cos \gamma \, dS = \iint_{S_1} R \cos \gamma \, dS + \iint_{S_2} R \cos \gamma \, dS + \iint_{S_3} R \cos \gamma \, dS$$

For the bottom surface $S_1: z = f_1(x, y)$, the outer unit normal $\mathbf{n}_1$ equals the downward-pointing unit normal vector $-\mathbf{n}$ to $S_1$. So, $\cos \gamma$ is given by

$$\cos \gamma = \mathbf{n}_1 \cdot \mathbf{k} = \frac{-1}{\sqrt{\left[\dfrac{\partial}{\partial x} f_1(x, y)\right]^2 + \left[\dfrac{\partial}{\partial y} f_1(x, y)\right]^2 + 1}} \qquad \text{See Example 5 (p. 1086).}$$

Then

$$\iint_{S_1} R \cos \gamma \, dS = \iint_{S_1} R(x, y, f_1(x, y)) \frac{-1}{\sqrt{\left[\dfrac{\partial}{\partial x} f_1(x, y)\right]^2 + \left[\dfrac{\partial}{\partial y} f_1(x, y)\right]^2 + 1}} \, dS$$

We can express the surface integral as a double integral

$$\iint_{S_1} R \cos \gamma \, dS = -\iint_D R(x, y, f_1(x, y)) \frac{\sqrt{\left[\dfrac{\partial}{\partial x} f_1(x, y)\right]^2 + \left[\dfrac{\partial}{\partial y} f_1(x, y)\right]^2 + 1}}{\sqrt{\left[\dfrac{\partial}{\partial x} f_1(x, y)\right]^2 + \left[\dfrac{\partial}{\partial y} f_1(x, y)\right]^2 + 1}} \, dx\, dy$$

$$= -\iint_D R(x, y, f_1(x, y)) \, dx\, dy \qquad (6)$$

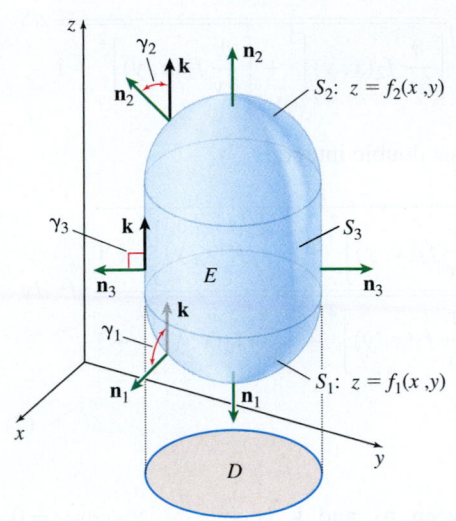

**Figure 68**

For the top surface $S_2$: $z = f_2(x, y)$, the outer unit normal $\mathbf{n}_2$ equals the upward-pointing unit normal vector $\mathbf{n}$ to $S_2$. So, $\cos \gamma$ is given by

$$\cos \gamma = \mathbf{n}_2 \cdot \mathbf{k} = \frac{1}{\sqrt{\left[\dfrac{\partial}{\partial x} f_2(x, y)\right]^2 + \left[\dfrac{\partial}{\partial y} f_2(x, y)\right]^2 + 1}} \qquad \text{See Example 5 (p. 1086).}$$

and

$$\iint_{S_2} R \cos \gamma \, dS = \iint_{S_2} R(x, y, f_2(x, y)) \frac{1}{\sqrt{\left[\dfrac{\partial}{\partial x} f_2(x, y)\right]^2 + \left[\dfrac{\partial}{\partial y} f_2(x, y)\right]^2 + 1}} \, dS$$

Again, we can express the surface integral as a double integral.

$$\iint_{S_2} R \cos \gamma \, dS = \iint_D R(x, y, f_2(x, y)) \frac{\sqrt{\left[\dfrac{\partial}{\partial x} f_2(x, y)\right]^2 + \left[\dfrac{\partial}{\partial y} f_2(x, y)\right]^2 + 1}}{\sqrt{\left[\dfrac{\partial}{\partial x} f_2(x, y)\right]^2 + \left[\dfrac{\partial}{\partial y} f_2(x, y)\right]^2 + 1}} \, dx \, dy$$

$$= \iint_D R(x, y, f_2(x, y)) \, dx \, dy \tag{7}$$

On the lateral surface $S_3$, the angle between $\mathbf{n}_3$ and $\mathbf{k}$ is $\gamma = \dfrac{\pi}{2}$, so $\cos \gamma = 0$. Therefore,

$$\iint_{S_3} R \cos \gamma \, dS = 0 \tag{8}$$

We now obtain $\iint_S R \cos \gamma \, dS$ by summing (6), (7), and (8):

$$\iint_S R \cos \gamma \, dS = \iint_{S_1} R \cos \gamma \, dS + \iint_{S_2} R \cos \gamma \, dS + \iint_{S_3} R \cos \gamma \, dS$$

$$= -\iint_D R(x, y, f_1(x, y)) \, dx \, dy + \iint_D R(x, y, f_2(x, y)) \, dx \, dy + 0$$

$$= \iint_D [R(x, y, f_2(x, y)) - R(x, y, f_1(x, y))] \, dx \, dy$$

$$= \iint_D \left[\int_{f_1(x,y)}^{f_2(x,y)} \frac{\partial R}{\partial z} \, dz\right] dx \, dy = \iiint_E \frac{\partial R}{\partial z} \, dV$$

which proves (4). ∎

### EXAMPLE 2   Using the Divergence Theorem to Find Flux

Find the mass of a fluid of constant mass density $\rho$ flowing across the cube enclosed by the planes $x = 0$, $x = 1$, $y = 0$, $y = 1$, $z = 0$, and $z = 1$ in a unit of time, in the direction of the outer unit normal vectors if the velocity of the fluid at any point on the cube is $\mathbf{F} = \mathbf{F}(x, y, z) = 4xz\mathbf{i} - y^2\mathbf{j} + yz\mathbf{k}$. (This is Example 8 on page 1089.)

**Solution** Suppose $E$ is the solid cube and $S$ is its surface. The mass of the fluid is $\iint_S \rho \mathbf{F} \cdot \mathbf{n}\, dS$, where $\mathbf{n}$ is the outer unit normal vector of $S$. Since the velocity of the fluid is $\mathbf{F} = 4xz\mathbf{i} - y^2\mathbf{j} + yz\mathbf{k}$, we have div $\mathbf{F} = 4z - 2y + y = 4z - y$. Now we use the Divergence Theorem.

$$\iint_S \rho \mathbf{F} \cdot \mathbf{n}\, dS = \rho \iint_S \mathbf{F} \cdot \mathbf{n}\, dS = \rho \iiint_E \operatorname{div} \mathbf{F}\, dV = \rho \int_0^1 \int_0^1 \int_0^1 (4z - y)\, dz\, dy\, dx$$

$$= \rho \int_0^1 \int_0^1 (2 - y)\, dy\, dx = \rho \int_0^1 \frac{3}{2}\, dx = \frac{3}{2}\rho$$

which is the same as the result found in Example 8, Section 15.7. ∎

**NOW WORK** Problem 9.

___

**EXAMPLE 3** **Using the Divergence Theorem to Find a Surface Integral**

Let $S$ be the surface of a cylindrical solid $E$ whose boundary is $x^2 + y^2 = 4$, $z = 0$, and $z = 1$. Let $\mathbf{F} = x^3\mathbf{i} + y^3\mathbf{j} + z^2\mathbf{k}$ and let $\mathbf{n}$ be the outer unit normal to $S$. Use the Divergence Theorem to find $\iint_S \mathbf{F} \cdot \mathbf{n}\, dS$.

**Solution** Figure 69 illustrates the cylindrical surface $S$. We begin by finding div $\mathbf{F} = 3x^2 + 3y^2 + 2z$. Then we use the Divergence Theorem, obtaining

$$\iint_S \mathbf{F} \cdot \mathbf{n}\, dS = \iiint_E (3x^2 + 3y^2 + 2z)\, dV$$

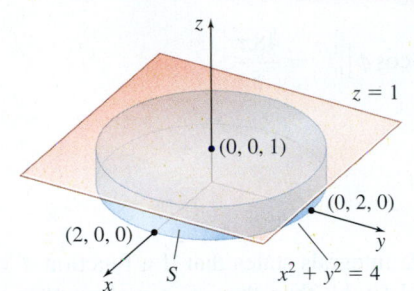

**Figure 69**

**NEED TO REVIEW?** Triple integrals using cylindrical coordinates are discussed in Section 14.7, pp. 1001–1004.

Because $E$ is a cylindrical solid, we use cylindrical coordinates:

$$3x^2 + 3y^2 + 2z = 3r^2 \cos^2 \theta + 3r^2 \sin^2 \theta + 2z = 3r^2(\cos^2 \theta + \sin^2 \theta) + 2z = 3r^2 + 2z$$

The solid $E$ is given by $0 \le r \le 2$, $0 \le \theta \le 2\pi$, $0 \le z \le 1$. Then

$$\iint_S \mathbf{F} \cdot \mathbf{n}\, ds = \iiint_E (3x^2 + 3y^2 + 2z)\, dV = \underset{\underset{dV = r\, dr\, d\theta\, dz}{\uparrow}}{\iiint_E (3r^2 + 2z)\, r\, dr\, d\theta\, dz}$$

$$= \int_0^{2\pi} \int_0^2 \int_0^1 (3r^3 + 2rz)\, dz\, dr\, d\theta$$

$$= \int_0^{2\pi} \int_0^2 \left[3r^3 z + rz^2\right]_0^1\, dr\, d\theta = \int_0^{2\pi} \int_0^2 (3r^3 + r)\, dr\, d\theta$$

$$= \int_0^{2\pi} \left[\frac{3r^4}{4} + \frac{r^2}{2}\right]_0^2\, d\theta = \int_0^{2\pi} 14\, d\theta = 28\pi$$ ∎

**NOW WORK** Problem 13.

___

**CALC CLIP** **EXAMPLE 4** **Using the Divergence Theorem to Find a Surface Integral**

Let $\mathbf{F}(x, y, z) = x^3\mathbf{i} + y^3\mathbf{j} + z^3\mathbf{k}$. If $S$ is the surface of the spherical solid $E$ in the first octant enclosed by $z = \sqrt{4 - x^2 - y^2}$ and $z = 0$, and $\mathbf{n}$ is the outer unit normal to $S$, use the Divergence Theorem to find $\iint_S \mathbf{F} \cdot \mathbf{n}\, dS$.

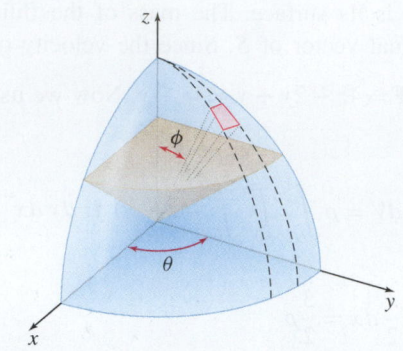

**Figure 70** $dV = \rho^2 \sin\phi \, d\rho \, d\theta \, d\phi$

**NEED TO REVIEW?** Triple integrals using spherical coordinates are discussed in Section 14.8, pp. 1008–1011.

**Solution** See Figure 70. Since div $\mathbf{F} = 3x^2 + 3y^2 + 3z^2 = 3(x^2 + y^2 + z^2)$, we use the Divergence Theorem, obtaining

$$\iint_S \mathbf{F} \cdot \mathbf{n} \, dS = 3 \iiint_E (x^2 + y^2 + z^2) \, dV$$

We use spherical coordinates to find the triple integral. Then $x^2 + y^2 + z^2 = \rho^2$ and $dV = \rho^2 \sin\phi \, d\rho \, d\theta \, d\phi$. The solid $E$ is given by $0 \le \rho \le 2$, $0 \le \theta \le \dfrac{\pi}{2}$, $0 \le \phi \le \dfrac{\pi}{2}$. Then

$$\iint_S \mathbf{F} \cdot \mathbf{n} \, dS = 3 \iiint_E \rho^2 (\rho^2 \sin\phi \, d\rho \, d\theta \, d\phi) = 3 \int_0^{\pi/2} \int_0^{\pi/2} \int_0^2 \rho^4 \sin\phi \, d\rho \, d\theta \, d\phi$$

$$= 3 \int_0^{\pi/2} \int_0^{\pi/2} \left[\frac{\rho^5}{5}\right]_0^2 \sin\phi \, d\theta \, d\phi = 3 \int_0^{\pi/2} \int_0^{\pi/2} \frac{32}{5} \sin\phi \, d\theta \, d\phi$$

$$= \frac{96}{5} \int_0^{\pi/2} [\theta]_0^{\pi/2} \sin\phi \, d\phi$$

$$= \frac{48\pi}{5} \int_0^{\pi/2} \sin\phi \, d\phi = \frac{48\pi}{5} [-\cos\phi]_0^{\pi/2} = \frac{48\pi}{5} \qquad \blacksquare$$

**NOW WORK** Problem 15.

## 3 Interpret the Divergence of F

**NEED TO REVIEW?** The Mean Value Theorem for integrals is discussed in Section 5.4, pp. 390–391.

Recall that the Mean Value Theorem for single integrals states that if a function $f$ of one variable is continuous on a closed interval $[a, b]$, then there is a real number $u$, $a \le u \le b$, for which

$$\boxed{\int_a^b f(x) \, dx = f(u)(b - a)}$$

Similarly, there is a Mean Value Theorem for triple integrals asserting that if a function $f$ of three variables is continuous in a simply connected, closed, bounded solid $E$, then there is a point $(x^*, y^*, z^*)$ in $E$ for which

$$\boxed{\iiint_E f(x, y, z) \, dV = f(x^*, y^*, z^*) V} \qquad (9)$$

where $V$ is the volume of $E$.

Let $\mathbf{F} = \mathbf{F}(x, y, z)$ be a vector field defined and continuous on and within a spherical solid $E_a$ of radius $a$ with its center at $(x_1, y_1, z_1)$. Based on (9), there is a point $(x^*, y^*, z^*)$ in $E_a$ for which

$$\iiint_{E_a} \text{div } \mathbf{F} \, dV_a = \text{div } \mathbf{F}(x^*, y^*, z^*) V_a$$

where $V_a$ is the volume of the sphere of radius $a$. Now use the Divergence Theorem to obtain

$$\iiint_{E_a} \text{div } \mathbf{F} \, dV_a = \iint_{S_a} \mathbf{F} \cdot \mathbf{n} \, dS_a$$

where $S_a$ is the sphere of radius $a$ with its center at $(x_1, y_1, z_1)$. Combining these last two equations, we get

$$\text{div } \mathbf{F}(x^*, y^*, z^*) = \frac{\iint\limits_{S_a} \mathbf{F} \cdot \mathbf{n} \, dS_a}{V_a}$$

The ratio on the right side is the flux of $\mathbf{F}$ per unit volume across $S_a$. If we let the radius $a \to 0$, then $(x^*, y^*, z^*) \to (x_1, y_1, z_1)$, so that

$$\boxed{\text{div } \mathbf{F}(x_1, y_1, z_1) = \lim_{a \to 0} \left( \frac{\iint\limits_{S_a} \mathbf{F} \cdot \mathbf{n} \, dS_a}{V_a} \right)}$$

In other words,

$$\boxed{\text{div } \mathbf{F}(x_1, y_1, z_1) = \text{Limiting value of the flux of } \mathbf{F} \text{ per unit volume across } S_a}$$

If $\mathbf{F}$ is the velocity of a steady fluid flow and if $\text{div } \mathbf{F}(x_1, y_1, z_1) > 0$, then the net flow is out of $S_a$ and the point $(x_1, y_1, z_1)$ is called a **source**. If $\text{div } \mathbf{F}(x_1, y_1, z_1) < 0$, then the net flow is into $S_a$ and the point $(x_1, y_1, z_1)$ is called a **sink**. In terms of this vocabulary, the Divergence Theorem states that the net flow through $S_a$ equals the sum of the sources and sinks across $S_a$.

If there are no sources or sinks across $S$, then $\text{div } \mathbf{F} = 0$, and we call $\mathbf{F}$ a **solenoidal** vector field. For such fields,

$$\text{div } \mathbf{F} = \frac{\partial P}{\partial x} + \frac{\partial Q}{\partial y} + \frac{\partial R}{\partial z} = 0$$

is referred to as the **equation of continuity**.

## Application: Electric Force Fields

We close this section with an example of **Coulomb's law of electrostatic attraction,** an important law in the study of electrostatic fields.

---

**EXAMPLE 5** Applying the Divergence Theorem to an Electric Force Field

Let $\mathbf{F} = \mathbf{F}(x, y, z)$ be an electric force field

$$\mathbf{F} = \mathbf{F}(x, y, z) = \frac{\varepsilon q Q}{x^2 + y^2 + z^2} \mathbf{u}$$

where $\varepsilon$ is a constant that depends on the units used; $q$ and $Q$ are the charges of two objects, one located at the point $(0, 0, 0)$ and the other at a point $(x, y, z)$; and $\mathbf{u}$ is the unit vector in the direction from $(0, 0, 0)$ to $(x, y, z)$. Let $S$ be a closed surface with positive orientation that encloses a solid $E$ in space.

Show that the following statements are true under the assumptions of the Divergence Theorem:

**(a)** If neither $S$ nor its interior contains the point $(0, 0, 0)$, then $\iint\limits_{S} \mathbf{F} \cdot \mathbf{n} \, dS = 0$. That is, the flux of $\mathbf{F}$ across $S$ equals 0.

**(b)** If the interior of $S$ contains the point $(0, 0, 0)$, then $\iint\limits_{S} \mathbf{F} \cdot \mathbf{n} \, dS = 4\pi \varepsilon q Q$. That is, the flux of $\mathbf{F}$ across $S$ equals $4\pi \varepsilon q Q$.

**Solution** The unit vector $\mathbf{u} = \dfrac{x\mathbf{i} + y\mathbf{j} + z\mathbf{k}}{\sqrt{x^2 + y^2 + z^2}}$, so that

$$\mathbf{F} = \mathbf{F}(x, y, z) = \varepsilon q Q \frac{x\mathbf{i} + y\mathbf{j} + z\mathbf{k}}{(x^2 + y^2 + z^2)^{3/2}}$$

The electric force field $\mathbf{F}$ is continuous everywhere except at the point $(0, 0, 0)$.

**(a)** The surface $S$ that forms the boundary of the solid $E$ satisfies the assumptions of the Divergence Theorem. We find div $\mathbf{F}$.

$$\text{div } \mathbf{F} = \varepsilon q \, Q \left\{ \frac{\partial}{\partial x} \left[ \frac{x}{(x^2 + y^2 + z^2)^{3/2}} \right] + \frac{\partial}{\partial y} \left[ \frac{y}{(x^2 + y^2 + z^2)^{3/2}} \right] + \frac{\partial}{\partial z} \left[ \frac{z}{(x^2 + y^2 + z^2)^{3/2}} \right] \right\}$$

$$= \varepsilon q \, Q \left\{ \frac{(x^2 + y^2 + z^2)^{3/2} - 3x^2 \sqrt{x^2 + y^2 + z^2}}{(x^2 + y^2 + z^2)^3} + \frac{(x^2 + y^2 + z^2)^{3/2} - 3y^2 \sqrt{x^2 + y^2 + z^2}}{(x^2 + y^2 + z^2)^3} \right.$$

$$\left. + \frac{(x^2 + y^2 + z^2)^{3/2} - 3z^2 \sqrt{x^2 + y^2 + z^2}}{(x^2 + y^2 + z^2)^3} \right\}$$

$$= \varepsilon q \, Q \, \frac{3(x^2 + y^2 + z^2)^{3/2} - 3\sqrt{x^2 + y^2 + z^2}(x^2 + y^2 + z^2)}{(x^2 + y^2 + z^2)^3} = 0$$

Therefore, by the Divergence Theorem,

$$\iint\limits_{S} \mathbf{F} \cdot \mathbf{n} \, dS = \iiint\limits_{E} \text{div } \mathbf{F} \, dV = 0$$

**(b)** Since the interior of $S$ contains the origin, $\mathbf{F}$ is not continuous at the origin, so we cannot use the Divergence Theorem. We use the following argument to prove **(b)**. Let $E$ be the closed solid enclosed by two separate surfaces: the surface $S$ and a sphere $S_a$ of radius $a$, with its center at $(0, 0, 0)$, as shown in Figure 71. The outer surface is $S$ and the inner surface is $S_a$. Now $\mathbf{F}$ is continuous throughout $E$, so the Divergence Theorem can be used.

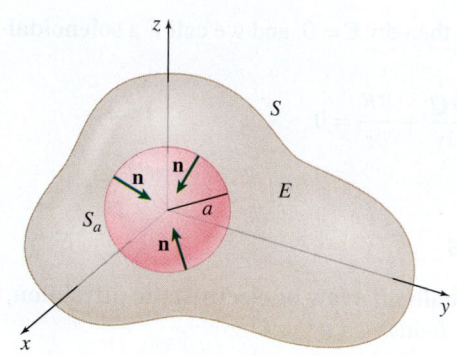

**Figure 71**

$$\iiint\limits_{E} \text{div } \mathbf{F} \, dV = \iint\limits_{S} \mathbf{F} \cdot \mathbf{n} \, dS + \iint\limits_{S_a} \mathbf{F} \cdot \mathbf{n} \, dS$$

From **(a)**, we know that $\iiint\limits_{E} \text{div } \mathbf{F} \, dV = 0$. So,

$$\iint\limits_{S} \mathbf{F} \cdot \mathbf{n} \, dS = - \iint\limits_{S_a} \mathbf{F} \cdot \mathbf{n} \, dS$$

On the inner surface $S_a$, a sphere of radius $a$, the outer unit normal is

$$\mathbf{n} = -\frac{x\mathbf{i} + y\mathbf{j} + z\mathbf{k}}{a} = -\frac{x\mathbf{i} + y\mathbf{j} + z\mathbf{k}}{\sqrt{x^2 + y^2 + z^2}}$$

Then

$$\mathbf{F} \cdot \mathbf{n} = \frac{\varepsilon q \, Q}{x^2 + y^2 + z^2} \mathbf{u} \cdot \left( -\frac{x\mathbf{i} + y\mathbf{j} + z\mathbf{k}}{\sqrt{x^2 + y^2 + z^2}} \right)$$

$$= -\frac{\varepsilon q \, Q}{x^2 + y^2 + z^2} \frac{x\mathbf{i} + y\mathbf{j} + z\mathbf{k}}{\sqrt{x^2 + y^2 + z^2}} \cdot \frac{x\mathbf{i} + y\mathbf{j} + z\mathbf{k}}{\sqrt{x^2 + y^2 + z^2}}$$

$$= -\varepsilon q \, Q \frac{x^2 + y^2 + z^2}{a^4} = -\frac{\varepsilon q \, Q}{a^2}$$

So,

$$\iint\limits_{S} \mathbf{F} \cdot \mathbf{n} \, dS = - \iint\limits_{S_a} \left( -\frac{\varepsilon q \, Q}{a^2} \right) dS = \frac{\varepsilon q \, Q}{a^2} \iint\limits_{S_a} dS = \frac{\varepsilon q \, Q}{a^2} (4\pi a^2) = 4\pi \varepsilon q \, Q$$

Surface area of the sphere $S_a$ is $4\pi a^2$.

# 15.8 Assess Your Understanding

## Concepts and Vocabulary

**1.** *True or False*   The divergence of a vector field **F** is a vector.

**2.** If $\mathbf{F}(x, y, z) = P(x, y, z)\mathbf{i} + Q(x, y, z)\mathbf{j} + R(x, y, z)\mathbf{k}$

is a vector field, where $\dfrac{\partial P}{\partial x}, \dfrac{\partial Q}{\partial y}$, and $\dfrac{\partial R}{\partial z}$ each exist,

then div **F** = _____.

## Skill Building

*In Problems 3–8, find the divergence of* **F**.

**3.** $\mathbf{F}(x, y, z) = x^2\mathbf{i} + y^2\mathbf{j} + z^2\mathbf{k}$

**4.** $\mathbf{F}(x, y, z) = x\mathbf{i} + xy\mathbf{j} + xyz\mathbf{k}$

**5.** $\mathbf{F}(x, y, z) = (x + \cos x)\mathbf{i} + (y + y\sin x)\mathbf{j} + 2z\mathbf{k}$

**6.** $\mathbf{F}(x, y, z) = xye^z\mathbf{i} + x^2e^z\mathbf{j} + x^2ye^z\mathbf{k}$

**7.** $\mathbf{F}(x, y, z) = \sqrt{x^2 + y^2}\mathbf{i} + \sqrt{x^2 + y^2}\mathbf{j} + z\mathbf{k}$

**8.** $\mathbf{F}(x, y, z) = xy^2\mathbf{i} + x^2y\mathbf{j} + xyz\mathbf{k}$

*In Problems 9–18, use the Divergence Theorem to find $\iint\limits_{S} \mathbf{F} \cdot \mathbf{n}\, dS$,*

*where* **n** *is the outer unit normal vector to S.*

**9.** $\mathbf{F} = (2xy + 2z)\mathbf{i} + (y^2 + 1)\mathbf{j} - (x + y)\mathbf{k}$; S is the surface of the solid enclosed by $x + y + z = 4$, $x = 0$, $y = 0$, and $z = 0$.

**10.** $\mathbf{F} = (2xy + z)\mathbf{i} + y^2\mathbf{j} - (x + 4y)\mathbf{k}$; S is the surface of the solid enclosed by $2x + 2y + z = 6$, $x = 0$, $y = 0$, and $z = 0$.

**11.** $\mathbf{F} = x^2\mathbf{i} + y^2\mathbf{j} + z^2\mathbf{k}$; S is the surface of the solid enclosed by $x = 0$, $x = 1$, $y = 0$, $y = 1$, $z = 0$, and $z = 1$.

**12.** $\mathbf{F} = (x - y)\mathbf{i} + (y - z)\mathbf{j} + (x - y)\mathbf{k}$; S is the surface of the solid cube with its center at the origin and faces in the planes $x = \pm 1$, $y = \pm 1$, and $z = \pm 1$.

**13.** $\mathbf{F} = x^2\mathbf{i} + 2y\mathbf{j} + 4z^2\mathbf{k}$; S is the surface of the solid cylinder $x^2 + y^2 \le 4$, $0 \le z \le 2$.

**14.** $\mathbf{F} = x\mathbf{i} + 2y^2\mathbf{j} + 3z^2\mathbf{k}$; S is the surface of the solid cylinder $x^2 + y^2 \le 9$, $0 \le z \le 1$.

**15.** $\mathbf{F} = x\mathbf{i} + y\mathbf{j} + z\mathbf{k}$; S is the sphere $x^2 + y^2 + z^2 = 1$.

**16.** $\mathbf{F} = 2x\mathbf{i} + 2y\mathbf{j} + 2z\mathbf{k}$; S is the sphere $x^2 + y^2 + z^2 = 2$.

**17.** $\mathbf{F} = (x + \cos x)\mathbf{i} + (y + y\sin x)\mathbf{j} + 2z\mathbf{k}$; S is the surface of the solid enclosed by the tetrahedron with vertices $(0, 0, 0)$, $(1, 0, 0)$, $(0, 1, 0)$, and $(0, 0, 1)$.

**18.** $\mathbf{F} = yz\mathbf{i} + xz\mathbf{j} + xy\mathbf{k}$; S is the surface of the solid enclosed by the tetrahedron with vertices $(0, 0, 0)$, $(1, 0, 0)$, $(0, 1, 0)$, and $(0, 0, 1)$.

## Applications and Extensions

**19. Volume**   Let $\mathbf{F} = x\mathbf{i} + y\mathbf{j} + z\mathbf{k}$; let S be the surface of a solid E satisfying the Divergence Theorem; and let **n** be the outer unit normal vector to S. Show that the volume V of E is given by the

formula $V = \dfrac{1}{3} \iint\limits_{S} \mathbf{F} \cdot \mathbf{n}\, dS$.

**Volume**   *In Problems 20–22, use the formula given in Problem 19 to find each volume.*

**20.** A rectangular parallelepiped with sides of length $a$, $b$, and $c$

**21.** A right circular cone with height $h$ and base radius $R$
*Hint:* The calculation is simplified with the cone oriented as shown in the figure.

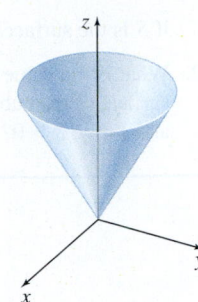

**22.** A sphere of radius $R$

*In Problems 23 and 24, find $\iint\limits_{S} \mathbf{F} \cdot \mathbf{n}\, dS$ and $\iiint\limits_{E} \text{div } \mathbf{F}\, dV$ separately and verify the Divergence Theorem.*

**23.** $\mathbf{F}(x, y, z) = x\mathbf{i} + y\mathbf{j} + z\mathbf{k}$; S is the surface of the sphere $x^2 + y^2 + z^2 = 100$.

**24.** $\mathbf{F}(x, y, z) = x\mathbf{i} + y\mathbf{j} + z\mathbf{k}$; S is the closed cylindrical surface $x^2 + y^2 = 1$ between $z = 0$ and $z = 2$.

**25.** Show that if **F** is a constant vector field and S is the surface of a solid E satisfying the assumptions of the Divergence Theorem, then $\iint\limits_{S} \mathbf{F} \cdot \mathbf{n}\, dS = 0$.

**26.** Let $\mathbf{F} = 3x\mathbf{i} + 4y\mathbf{j} + (7z + 2x)\mathbf{k}$ and $\mathbf{G} = 2x\mathbf{i} + 3y\mathbf{j} + (9z + 6y)\mathbf{k}$. Let S be the surface of a solid E satisfying the assumptions of the Divergence Theorem. Show that $\iint\limits_{S} \mathbf{F} \cdot \mathbf{n}\, dS = \iint\limits_{S} \mathbf{G} \cdot \mathbf{n}\, dS$.

**27. Flux**   Suppose the velocity of a fluid flow in space is constant.

(a) Show that the flux through any closed surface is 0.

(b) In your own words, give a physical interpretation of (a).

**28.** Let $\mathbf{F}(x, y, z) = \dfrac{1}{\rho}(x\mathbf{i} + y\mathbf{j} + z\mathbf{k})$, where $\rho = \sqrt{x^2 + y^2 + z^2}$.

Show that div $\mathbf{F} = \dfrac{2}{\rho}$.

**29.** (a) Use Problem 28 and the Divergence Theorem to find

$$\iiint\limits_{E} \frac{dV}{\sqrt{x^2 + y^2 + z^2}}$$

where E is the solid inside the sphere $x^2 + y^2 + z^2 = 1$.

(b) Verify the solution to (a) by finding the triple integral directly.

**30.** Assume that the hypotheses of the Divergence Theorem hold for S and E. Show that if $f$ satisfies the Laplace equation ($f_{xx} + f_{yy} + f_{zz} = 0$) in a closed, bounded solid E with boundary S with outer unit normal **n**, then $\iint\limits_{S} \nabla f \cdot \mathbf{n}\, dS = 0$.

---

## Challenge Problems

**31.** Let $\mathbf{F} = x^2\mathbf{i} + y^2\mathbf{j} + z^2\mathbf{k}$ and let $\mathbf{n}$ be the outer unit normal to the surface $S$. Use the Divergence Theorem to show that

$$\iint\limits_{S} \mathbf{F} \cdot \mathbf{n}\, dS = \frac{8\pi q^4}{3}$$

if $S$ is the surface $x^2 + y^2 + z^2 = 2qz$, $q > 0$.

**32.** What is the value of the integral in Problem 31 if $S$ is the surface of the cube $x = 0$, $x = q$, $y = 0$, $y = q$, $z = 0$, and $z = q$, $q > 0$?

**33.** Let $f$ and $g$ be two scalar functions and let $S$ be the surface of a solid $E$ satisfying the assumptions of the Divergence Theorem. Show that

$$\iint\limits_{S} f(\nabla g) \cdot \mathbf{n}\, dS = \iiint\limits_{E} [f(\nabla^2 g) + \nabla f \cdot \nabla g]\, dV$$

*Hint:* Let $\mathbf{F} = f(\nabla g)$ in the Divergence Theorem and use $\nabla^2 g = \nabla \cdot \nabla g = g_{xx} + g_{yy} + g_{zz}$.

# 15.9 Stokes' Theorem

**OBJECTIVES** *When you finish this section, you should be able to:*

**1** Find the curl of $\mathbf{F}$ (p. 1101)

**2** Verify Stokes' Theorem (p. 1102)

**3** Use Stokes' Theorem to find an integral (p. 1103)

**4** Use Stokes' Theorem with conservative vector fields (p. 1103)

**5** Interpret the curl of $\mathbf{F}$ (p. 1105)

*Stokes' Theorem* is a second generalization of Green's Theorem to space. Stokes' Theorem expresses a certain relationship between a line integral of a vector field $\mathbf{F}$ around a simple closed curve $C$ in space to a surface integral for which $C$ forms the boundary. The theorem is used in applications of electricity and magnetism, most notably in Ampere's Law and in the Maxwell–Faraday Law.

To get started, we need the definition of the *curl of* $\mathbf{F}$.

**ORIGINS** Stokes' Theorem is named after the Irish mathematician George Gabriel Stokes (1819–1903). Stokes, along with Maxwell and Kelvin, contributed to the prominence of the Cambridge School of Mathematical Physics.

**DEFINITION Curl of F**

Let $\mathbf{F} = \mathbf{F}(x, y, z) = P(x, y, z)\mathbf{i} + Q(x, y, z)\mathbf{j} + R(x, y, z)\mathbf{k}$ be a vector field, where the functions $P$, $Q$, and $R$ have first-order partial derivatives. The **curl of F**, denoted by curl $\mathbf{F}$, is defined as*

$$\text{curl } \mathbf{F} = \left(\frac{\partial R}{\partial y} - \frac{\partial Q}{\partial z}\right)\mathbf{i} - \left(\frac{\partial R}{\partial x} - \frac{\partial P}{\partial z}\right)\mathbf{j} + \left(\frac{\partial Q}{\partial x} - \frac{\partial P}{\partial y}\right)\mathbf{k}$$

Rather than memorize this definition, you can write the curl in the form of the symbolic determinant:

$$\text{curl } \mathbf{F} = \begin{vmatrix} \mathbf{i} & \mathbf{j} & \mathbf{k} \\ \dfrac{\partial}{\partial x} & \dfrac{\partial}{\partial y} & \dfrac{\partial}{\partial z} \\ P & Q & R \end{vmatrix}$$

---

*Some books define curl $\mathbf{F}$ as the cross product $\nabla \times \mathbf{F}$, where $\nabla$ is the partial differential operator $\nabla = \dfrac{\partial}{\partial x}\mathbf{i} + \dfrac{\partial}{\partial y}\mathbf{j} + \dfrac{\partial}{\partial z}\mathbf{k}$ and $\mathbf{F} = P\mathbf{i} + Q\mathbf{j} + R\mathbf{k}$.

## 1 Find the Curl of F

EXAMPLE 1 **Finding the Curl of F**

Find curl $\mathbf{F}$ if $\mathbf{F} = x^2 y \mathbf{i} - 2xz \mathbf{j} + 2yz \mathbf{k}$.

**Solution**

$$
\operatorname{curl} \mathbf{F} = \begin{vmatrix} \mathbf{i} & \mathbf{j} & \mathbf{k} \\ \dfrac{\partial}{\partial x} & \dfrac{\partial}{\partial y} & \dfrac{\partial}{\partial z} \\ x^2 y & -2xz & 2yz \end{vmatrix}
$$

$$
= \left[ \frac{\partial}{\partial y}(2yz) - \frac{\partial}{\partial z}(-2xz) \right] \mathbf{i} - \left[ \frac{\partial}{\partial x}(2yz) - \frac{\partial}{\partial z}(x^2 y) \right] \mathbf{j} + \left[ \frac{\partial}{\partial x}(-2xz) - \frac{\partial}{\partial y}(x^2 y) \right] \mathbf{k}
$$

$$
= (2z + 2x)\mathbf{i} - (0 - 0)\mathbf{j} + (-2z - x^2)\mathbf{k} = (2z + 2x)\mathbf{i} + (-2z - x^2)\mathbf{k} \qquad \blacksquare
$$

NOW WORK **Problem 11.**

*Stokes' Theorem* can be stated in vector form as

$$
\boxed{\oint_C \mathbf{F} \cdot d\mathbf{r} = \iint_S \operatorname{curl} \mathbf{F} \cdot \mathbf{n} \, dS} \qquad (1)
$$

under certain conditions on the surface $S$ and the curve $C$.

The conditions require the surface $S$ to be:

- *Smooth.* That is, for any parametrization $\mathbf{r} = \mathbf{r}(u, v)$ of $S$, $\mathbf{r}$ must have continuous first-order partial derivatives in the parameter domain $R$ and $\mathbf{r}_u \times \mathbf{r}_v \neq \mathbf{0}$.
- *Simply connected.* That is, there can be no "holes" in $S$.
- *Orientable.* That is, it must be possible to choose a unit normal vector $\mathbf{n}$ at each point of $S$, which we take as the positive direction of $\mathbf{n}$. The positively oriented vector $\mathbf{n}$ must vary continuously, and as it moves around any closed curve on the surface, $\mathbf{n}$ must return to its original direction.
- The curve $C$ of the line integral in (1) is a piecewise-smooth, simple closed curve that forms the boundary of the surface $S$. The positively oriented surface $S$ induces a *positive orientation* on $C$ in the sense that if you walk around $C$ in the positive direction with your head pointing in the same direction as the unit normal $\mathbf{n}$ to $S$, then the surface will always be to your left, as shown in Figure 72.

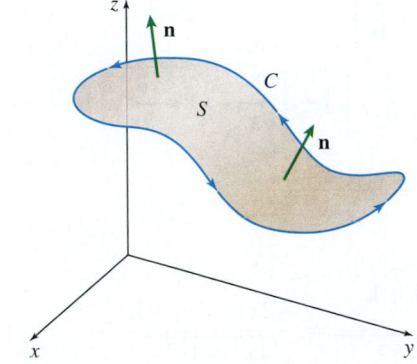

**Figure 72**

---

**THEOREM** Stokes' Theorem

Let $S$ be a smooth, simply connected, orientable surface bounded by a piecewise-smooth, simple closed curve $C$. Let $\mathbf{F} = \mathbf{F}(x, y, z) = P(x, y, z)\mathbf{i} + Q(x, y, z)\mathbf{j} + R(x, y, z)\mathbf{k}$ be a vector field, where $P$, $Q$, and $R$ have continuous first-order partial derivatives throughout a solid $E$ containing $S$ and $C$. Let $\mathbf{n}$ denote the positive unit normal to $S$, and let $C$ be positively oriented, as described earlier. Then

$$
\boxed{\oint_C \mathbf{F} \cdot d\mathbf{r} = \iint_S \operatorname{curl} \mathbf{F} \cdot \mathbf{n} \, dS}
$$

In terms of the components of each vector, Stokes' Theorem takes the form

$$
\boxed{\oint_C (P \, dx + Q \, dy + R \, dz) = \iint_S \left[ \left( \frac{\partial R}{\partial y} - \frac{\partial Q}{\partial z} \right) dy \, dz - \left( \frac{\partial R}{\partial x} - \frac{\partial P}{\partial z} \right) dz \, dx + \left( \frac{\partial Q}{\partial x} - \frac{\partial P}{\partial y} \right) dx \, dy \right]}
$$

The proof of Stokes' Theorem is given in advanced calculus books.

## 2 Verify Stokes' Theorem

**EXAMPLE 2** **Verifying Stokes' Theorem**

Verify Stokes' Theorem for $\mathbf{F} = y\mathbf{i} - x\mathbf{j}$, where the surface $S$ is the paraboloid $z = x^2 + y^2$, with $x^2 + y^2 = 1$, $z = 1$, as its boundary $C$.

**Solution** Figure 73 shows the paraboloid and the circle $C$ in the plane $z = 1$. Stokes' Theorem states the following:

$$\oint_C \mathbf{F} \cdot d\mathbf{r} = \iint_S \operatorname{curl} \mathbf{F} \cdot \mathbf{n} \, dS$$

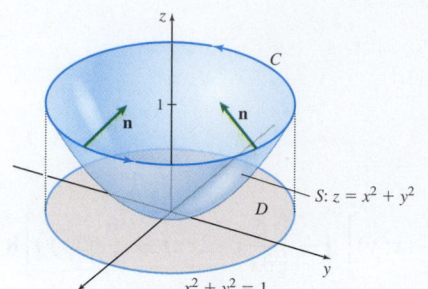

**Figure 73** $C: x^2 + y^2 = 1; z = 1$

To find the line integral $\oint_C \mathbf{F} \cdot d\mathbf{r}$, we use parametric equations for $C$, namely, $x = \cos t$, $y = \sin t$, $z = 1$, $0 \le t \le 2\pi$. Then with $\mathbf{F} = y\mathbf{i} - x\mathbf{j}$, we have

$$\oint_C \mathbf{F} \cdot d\mathbf{r} = \oint_C (y \, dx - x \, dy) = \int_0^{2\pi} [\sin t \,(-\sin t \, dt) - \cos t \cos t \, dt]$$

$$= -\int_0^{2\pi} [\sin^2 t + \cos^2 t] \, dt = -\int_0^{2\pi} dt = -2\pi$$

To find the surface integral, $\iint_S \operatorname{curl} \mathbf{F} \cdot \mathbf{n} \, dS$, we find $\operatorname{curl} \mathbf{F}$ and $\mathbf{n}$:

$$\operatorname{curl} \mathbf{F} = \begin{vmatrix} \mathbf{i} & \mathbf{j} & \mathbf{k} \\ \dfrac{\partial}{\partial x} & \dfrac{\partial}{\partial y} & \dfrac{\partial}{\partial z} \\ P & Q & R \end{vmatrix} = \begin{vmatrix} \mathbf{i} & \mathbf{j} & \mathbf{k} \\ \dfrac{\partial}{\partial x} & \dfrac{\partial}{\partial y} & \dfrac{\partial}{\partial z} \\ y & -x & 0 \end{vmatrix}$$

$$= \left[ \frac{\partial}{\partial y} 0 - \frac{\partial}{\partial z}(-x) \right] \mathbf{i} - \left[ \frac{\partial}{\partial x} 0 - \frac{\partial}{\partial z} y \right] \mathbf{j} + \left[ \frac{\partial}{\partial x}(-x) - \frac{\partial}{\partial y} y \right] \mathbf{k}$$

$$= 0\mathbf{i} - 0\mathbf{j} - 2\mathbf{k} = -2\mathbf{k}$$

For $z = f(x, y) = x^2 + y^2$, we have

$$\mathbf{n} = \frac{-\dfrac{\partial}{\partial x} f(x, y)\mathbf{i} - \dfrac{\partial}{\partial y} f(x, y)\mathbf{j} + \mathbf{k}}{\sqrt{\left[ \dfrac{\partial}{\partial x} f(x, y) \right]^2 + \left[ \dfrac{\partial}{\partial y} f(x, y) \right]^2 + 1}} = \frac{-2x\mathbf{i} - 2y\mathbf{j} + \mathbf{k}}{\sqrt{4x^2 + 4y^2 + 1}}$$

Then

$$\operatorname{curl} \mathbf{F} \cdot \mathbf{n} = -2\mathbf{k} \cdot \frac{1}{\sqrt{4x^2 + 4y^2 + 1}} (-2x\mathbf{i} - 2y\mathbf{j} + \mathbf{k}) = \frac{-2}{\sqrt{4x^2 + 4y^2 + 1}}$$

So,

$$\iint_S \operatorname{curl} \mathbf{F} \cdot \mathbf{n} \, dS = \iint_S \frac{-2}{\sqrt{4x^2 + 4y^2 + 1}} \, dS = \iint_D \frac{-2}{\sqrt{4x^2 + 4y^2 + 1}} \sqrt{4x^2 + 4y^2 + 1} \, dx \, dy$$

$$\underset{\uparrow}{\phantom{=}}$$
$$dS = \sqrt{4x^2 + 4y^2 + 1} \, dx \, dy$$

$$= -2 \iint_D dx \, dy = -2\pi$$

where $D$ is the interior of the circle $x^2 + y^2 = 1$. ∎

**NOW WORK** Problem 19.

## 3 Use Stokes' Theorem to Find an Integral

**EXAMPLE 3   Using Stokes' Theorem to Find a Surface Integral**

Use Stokes' Theorem to find the surface integral $\iint_S \operatorname{curl} \mathbf{F} \cdot \mathbf{n}\, dS$, where

$\mathbf{F} = \mathbf{F}(x, y, z) = y\mathbf{i} + x\mathbf{j} + z\mathbf{k}$ and $S$ is the surface $z = 5 - (x^2 + y^2)$, $z \geq 1$.

**Solution** Figure 74 shows the surface $S$. The boundary curve $C$ is the circle $x^2 + y^2 = 4$ that lies in the plane $z = 1$. The vector form of $C$ is $\mathbf{r} = \mathbf{r}(t) = 2\cos t\, \mathbf{i} + 2\sin t\, \mathbf{j} + \mathbf{k}$, $0 \leq t \leq 2\pi$. Then

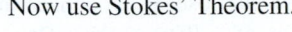

$$d\mathbf{r} = (-2\sin t\, \mathbf{i} + 2\cos t\, \mathbf{j})\, dt$$

$$\mathbf{F} = y\mathbf{i} + x\mathbf{j} + z\mathbf{k} = 2\sin t\, \mathbf{i} + 2\cos t\, \mathbf{j} + \mathbf{k}$$

$$\mathbf{F} \cdot d\mathbf{r} = (-4\sin^2 t + 4\cos^2 t)\, dt = 4\cos(2t)\, dt$$

Now use Stokes' Theorem.

$$\iint_S \operatorname{curl} \mathbf{F} \cdot \mathbf{n}\, dS = \oint_C \mathbf{F} \cdot d\mathbf{r} = \int_0^{2\pi} 4\cos(2t)\, dt = 2\big[\sin(2t)\big]_0^{2\pi} = 0 \qquad \blacksquare$$

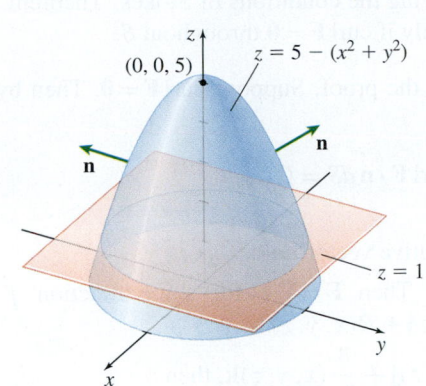

(0, 0, 5)   $z = 5 - (x^2 + y^2)$

$z = 1$

**Figure 74**

**NOW WORK** Problem 25.

In Example 3, we found the surface integral $\iint_S \operatorname{curl} \mathbf{F} \cdot \mathbf{n}\, dS$ using the line integral of $\mathbf{F}$ on the boundary $C$ of the surface $S$. This means that for any surface $S_1$ with the same orientation and the same boundary curve as $S$, the surface integral $\iint_{S_1} \operatorname{curl} \mathbf{F} \cdot \mathbf{n}\, dS_1$ will have the same value as $\iint_S \operatorname{curl} \mathbf{F} \cdot \mathbf{n}\, dS$.

We can use Stokes' Theorem to find a line integral, and avoid having to find a surface integral. However, it is more frequently used in the opposite direction. In cases where it is not easy to find the line integral $\oint_C \mathbf{F} \cdot d\mathbf{r}$, but the quantity $\operatorname{curl} \mathbf{F} \cdot \mathbf{n}$ is simple in form on an open surface with boundary $C$, then it can be easier to find $\iint_S \operatorname{curl} \mathbf{F} \cdot \mathbf{n}\, dS$.

**CALC**
**CLIP**

**EXAMPLE 4   Using Stokes' Theorem to Find a Line Integral**

Find $I = \oint_C [(e^{-x^2/2} - yz)\, dx + (e^{-y^2/2} + xz + 2x)\, dy + (e^{-z^2/2} + 5)\, dz]$, where $C$ is the circle $x = \cos t$, $y = \sin t$, $z = 2$, $0 \leq t \leq 2\pi$.

**Solution** Let $\mathbf{F} = P\mathbf{i} + Q\mathbf{j} + R\mathbf{k}$, with $P = e^{-x^2/2} - yz$, $Q = e^{-y^2/2} + xz + 2x$, and $R = e^{-z^2/2} + 5$. It can be verified that

$$\operatorname{curl} \mathbf{F} = -x\mathbf{i} - y\mathbf{j} + (2 + 2z)\mathbf{k}$$

To use Stokes' Theorem, take $S$ to be a plane region enclosed by the circle $C$ in the plane $z = 2$ so that $\mathbf{n} = \mathbf{k}$. Then $\operatorname{curl} \mathbf{F} \cdot \mathbf{n} = 6$ on $S$, and

$$I = \iint_S \operatorname{curl} \mathbf{F} \cdot \mathbf{n}\, dS = 6 \iint_S dx\, dy = 6\pi \qquad \blacksquare$$

**NOW WORK** Problem 27.

## 4 Use Stokes' Theorem with Conservative Vector Fields

A vector field $\mathbf{F} = \mathbf{F}(x, y, z) = P(x, y, z)\mathbf{i} + Q(x, y, z)\mathbf{j} + R(x, y, z)\mathbf{k}$ is conservative if $\mathbf{F}$ is the gradient of some function $w = f(x, y, z)$. An extension of earlier proofs shows that $\mathbf{F}$ is a conservative vector field if and only if $\int_C \mathbf{F} \cdot d\mathbf{r}$ is independent of the path, where $C$ is a piecewise-smooth curve in space and $\mathbf{F} = \mathbf{F}(x, y, z)$ is defined on some open, connected solid in space. In particular, if $C$ is simple and closed, then $\mathbf{F}$ is a conservative vector field if and only if $\oint_C \mathbf{F} \cdot d\mathbf{r} = 0$ for every piecewise-smooth, simple closed curve $C$.

Stokes' Theorem provides an easy way to determine whether a vector field $\mathbf{F}$ in space is a conservative vector field.

### THEOREM   Conservative Vector Fields in Space

Let $\mathbf{F} = P(x, y, z)\,\mathbf{i} + Q(x, y, z)\,\mathbf{j} + R(x, y, z)\,\mathbf{k}$ be a vector field whose components $P$, $Q$, and $R$ have continuous first-order partial derivatives throughout a solid $E$ whose boundary is a surface $S$ satisfying the conditions of Stokes' Theorem. Then $\mathbf{F}$ is a conservative vector field if and only if curl $\mathbf{F} = \mathbf{0}$ throughout $S$.

**Partial Proof**   We provide only an outline of the proof. Suppose curl $\mathbf{F} = \mathbf{0}$. Then by Stokes' Theorem,

$$\oint_C \mathbf{F} \cdot d\mathbf{r} = \iint_S \operatorname{curl} \mathbf{F} \cdot \mathbf{n}\, dS = 0$$

for any closed curve $C$. That is, $\mathbf{F}$ is a conservative vector field.

Conversely, suppose $\mathbf{F}$ is conservative. Then $\mathbf{F} = \nabla f$ for some function $f$. Since $\mathbf{F} = \mathbf{F}(x, y, z) = P(x, y, z)\mathbf{i} + Q(x, y, z)\mathbf{j} + R(x, y, z)\,\mathbf{k}$,

and $\nabla f(x, y, z) = \dfrac{\partial}{\partial x} f(x, y, z)\mathbf{i} + \dfrac{\partial}{\partial y} f(x, y, z)\mathbf{j} + \dfrac{\partial}{\partial z}(x, y, z)\mathbf{k}$, then

$$\operatorname{curl} \mathbf{F} = \begin{vmatrix} \mathbf{i} & \mathbf{j} & \mathbf{k} \\ \dfrac{\partial}{\partial x} & \dfrac{\partial}{\partial y} & \dfrac{\partial}{\partial z} \\ P & Q & R \end{vmatrix} = \begin{vmatrix} \mathbf{i} & \mathbf{j} & \mathbf{k} \\ \dfrac{\partial}{\partial x} & \dfrac{\partial}{\partial y} & \dfrac{\partial}{\partial z} \\ \dfrac{\partial}{\partial x} f(x, y, z) & \dfrac{\partial}{\partial y} f(x, y, z) & \dfrac{\partial}{\partial z} f(x, y, z) \end{vmatrix}$$

$$= \left( \frac{\partial^2 f}{\partial y \partial z} - \frac{\partial^2 f}{\partial z \partial y} \right)\mathbf{i} - \left( \frac{\partial^2 f}{\partial x \partial z} - \frac{\partial^2 f}{\partial z \partial x} \right)\mathbf{j} + \left( \frac{\partial^2 f}{\partial x \partial y} - \frac{\partial^2 f}{\partial y \partial x} \right)\mathbf{k}$$

$$= 0\mathbf{i} + 0\mathbf{j} + 0\mathbf{k} = \mathbf{0} \qquad \blacksquare$$

**NEED TO REVIEW?**  The Equality of Mixed Partials Theorem is discussed in Section 12.3, p. 878.

The following statements are similar to those listed in the Summary at the end of Section 15.4 for a vector field in the plane.

## Summary

Suppose $\mathbf{F} = P(x, y, z)\,\mathbf{i} + Q(x, y, z)\,\mathbf{j} + R(x, y, z)\,\mathbf{k}$ is a vector field whose components $P$, $Q$, and $R$ have continuous first-order partial derivatives throughout a solid $E$ whose boundary is a smooth, simply connected, oriented surface $S$ bounded by a piecewise-smooth, simple closed curve $C$. Then the following are equivalent statements for a vector field $\mathbf{F} = P\mathbf{i} + Q\mathbf{j} + R\mathbf{k}$ in space:

- $\mathbf{F}$ is a conservative vector field.

- $\mathbf{F}$ is the gradient of some function.
- The work done by $\mathbf{F}$ in moving an object of mass $m$ from a point $A$ to a point $B$ in $S$ is independent of the path from $A$ to $B$.
- The work done by $\mathbf{F}$ in moving an object of mass $m$ along any piecewise-smooth, closed curve $C$ in $S$ is 0.
- curl $\mathbf{F} = \mathbf{0}$

Of all these equivalent statements, the easiest to establish is curl $\mathbf{F} = \mathbf{0}$.

### EXAMPLE 5   Showing That F Is a Conservative Vector Field

Show that $\mathbf{F} = \left( \dfrac{3}{5} y^5 + 2z^2 \right)\mathbf{i} + 3xy^4\,\mathbf{j} + 4xz\,\mathbf{k}$ is a conservative vector field in space.

**Solution**  We find curl $\mathbf{F}$.

$$\operatorname{curl} \mathbf{F} = \begin{vmatrix} \mathbf{i} & \mathbf{j} & \mathbf{k} \\ \dfrac{\partial}{\partial x} & \dfrac{\partial}{\partial y} & \dfrac{\partial}{\partial z} \\ \dfrac{3}{5} y^5 + 2z^2 & 3xy^4 & 4xz \end{vmatrix} = (0 - 0)\mathbf{i} - (4z - 4z)\mathbf{j} + (3y^4 - 3y^4)\mathbf{k} = \mathbf{0}$$

Since curl $\mathbf{F} = \mathbf{0}$, $\mathbf{F}$ is conservative.  ∎

**NOW WORK**  Problem 31.

## 5 Interpret the Curl of F

Stokes' Theorem provides an interpretation for the curl of a vector.

Suppose $Q = (x_1, y_1, z_1)$ is the center of a circular disk $S_\rho$ of radius $\rho$ and $C_\rho$ is the boundary of $S_\rho$, as shown in Figure 75. There is a Mean Value Theorem for double integrals that, when combined with Stokes' Theorem, gives us

$$\oint_{C_\rho} \mathbf{F} \cdot d\mathbf{r} = \iint_{S_\rho} \text{curl}\,\mathbf{F} \cdot \mathbf{n}\, dS = (\text{curl}\,\mathbf{F} \cdot \mathbf{n})_{Q^*} (\pi\rho^2)$$

where $(\text{curl}\,\mathbf{F} \cdot \mathbf{n})_{Q^*}$ is the value of $\text{curl}\,\mathbf{F} \cdot \mathbf{n}$ evaluated at a suitably chosen point $Q^* = (x^*, y^*, z^*)$ in $S_\rho$, and where $\pi\rho^2$ is the area of $S_\rho$. Then

$$(\text{curl}\,\mathbf{F} \cdot \mathbf{n})_{Q^*} = \frac{1}{\pi\rho^2} \oint_{C_\rho} \mathbf{F} \cdot d\mathbf{r}$$

Now if $\rho \to 0$, then $Q^* \to Q$, and

$$(\text{curl}\,\mathbf{F} \cdot \mathbf{n})_Q = \lim_{\rho \to 0} \frac{1}{\pi\rho^2} \oint_{C_\rho} \mathbf{F} \cdot d\mathbf{r} \tag{2}$$

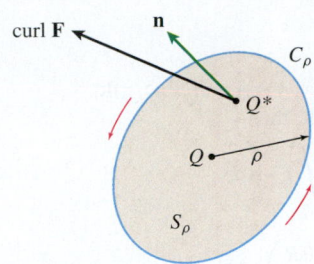

**Figure 75**

In the case where $\mathbf{F}$ is the velocity of a fluid, the integral $\oint_{C_\rho} \mathbf{F} \cdot d\mathbf{r}$ in (2) is referred to as the **circulation**, or **whirling tendency** of the fluid around $C_\rho$. It measures the extent to which the fluid rotates around the circle $C_\rho$ in the direction of the orientation of $C_\rho$. Equation (2) therefore states that the component of $\text{curl}\,\mathbf{F}$ at $(x_1, y_1, z_1)$ in the direction of $\mathbf{n}$ is the limiting ratio of circulation to area for a circle about $(x_1, y_1, z_1)$ with $\mathbf{n}$ as a normal. That is,

Circulation per unit of area at $(x_1, y_1, z_1) = (\text{curl}\,\mathbf{F} \cdot \mathbf{n})_Q$

The expression $(\text{curl}\,\mathbf{F} \cdot \mathbf{n})_Q$ is a maximum at $Q$ when $\mathbf{n}$ has the same direction as $\text{curl}\,\mathbf{F}$.

Suppose that a small paddle wheel of radius $\rho$ is introduced into the fluid at $Q$ with its axle directed along $\mathbf{n}$. The rate of spin of the paddle wheel is affected by the circulation of the fluid around $C_\rho$. The wheel will spin fastest when the circulation integral is maximized; that is, it will spin fastest when the axle of the paddle wheel is in the direction of $\text{curl}\,\mathbf{F}$. See Figure 76.

Another interpretation of the curl is based on motion.

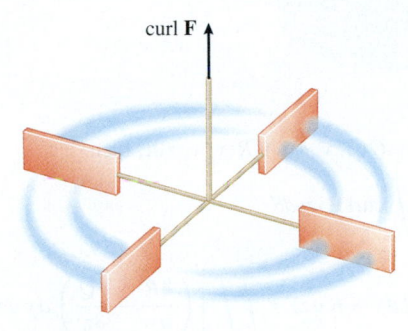

**DF** **Figure 76**

### THEOREM   Curl and Angular Velocity

Let $\mathbf{F}$ be the velocity field of a fluid rotating about a fixed axis and let $\boldsymbol{\omega}$ be the constant angular velocity. Then

$$\text{curl}\,\mathbf{F} = 2\boldsymbol{\omega}$$

**NEED TO REVIEW?**  Angular velocity is discussed in Section 10.5, p. 769.

**Proof**  Since the motion of a fluid is simply a rotation about a given fixed axis in space, the angular velocity $\boldsymbol{\omega}$ can be represented by a constant vector:

$$\boldsymbol{\omega} = \omega_1 \mathbf{i} + \omega_2 \mathbf{j} + \omega_3 \mathbf{k}$$

where the magnitude of $\boldsymbol{\omega}$ is the angular speed and the direction of $\boldsymbol{\omega}$ is along the direction of the axis of rotation, in accordance with the right-hand rule. If the origin of a rectangular coordinate system is on the fixed axis and $\mathbf{r} = x\mathbf{i} + y\mathbf{j} + z\mathbf{k}$, then the velocity vector $\mathbf{F}$ is

$$\mathbf{F} = \boldsymbol{\omega} \times \mathbf{r}$$

**IN WORDS**  If a fluid is rotating, the curl of the velocity vector is a constant vector that equals twice the angular velocity vector.

and

$$\text{curl}\,\mathbf{F} = \begin{vmatrix} \mathbf{i} & \mathbf{j} & \mathbf{k} \\ \dfrac{\partial}{\partial x} & \dfrac{\partial}{\partial y} & \dfrac{\partial}{\partial z} \\ \omega_2 z - \omega_3 y & \omega_3 x - \omega_1 z & \omega_1 y - \omega_2 x \end{vmatrix}$$

Since $\boldsymbol{\omega}$ is constant, then

$$\text{curl}\,\mathbf{F} = 2\omega_1 \mathbf{i} + 2\omega_2 \mathbf{j} + 2\omega_3 \mathbf{k} = 2\boldsymbol{\omega} \qquad \blacksquare$$

## Summary

We opened the chapter by stating that "vector calculus is the culmination of what we have learned throughout the course." Now we end the chapter with a list of several of the formulas we have learned. The theorem formulas given below do not contain the important hypotheses necessary for the formulas to hold. Rather, this summary is meant to "pull everything together" and provide "the big picture of calculus."

### Fundamental Theorem of Calculus
If $F'(x) = f(x)$, then
$$\int_a^b f(x)\,dx = \int_a^b F'(x)\,dx = F(b) - F(a)$$

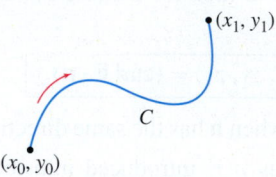

### Fundamental Theorem of Line Integrals
If $\nabla f(x, y) = \mathbf{F}(x, y)$, then
$$\int_C \mathbf{F} \cdot d\mathbf{r} = \int_C \nabla f \cdot d\mathbf{r} = \int_{(x_0, y_0)}^{(x_1, y_1)} \mathbf{F} \cdot d\mathbf{r} = f(x_1, y_1) - f(x_0, y_0)$$

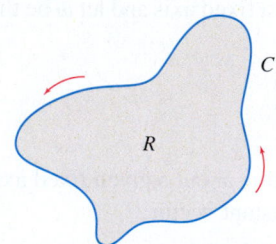

### Green's Theorem
If $\mathbf{F}(x, y) = P(x, y)\mathbf{i} + Q(x, y)\mathbf{j}$, then
$$\oint_C \mathbf{F} \cdot d\mathbf{r} = \oint_C (P\,dx + Q\,dy) = \iint_R \left(\frac{\partial Q}{\partial x} - \frac{\partial P}{\partial y}\right) dx\,dy$$

### Divergence Theorem
If $\mathbf{F}(x, y, z) = P(x, y, z)\mathbf{i} + Q(x, y, z)\mathbf{j} + R(x, y, z)\mathbf{k}$, and $\mathbf{n} = \cos\alpha\,\mathbf{i} + \cos\beta\,\mathbf{j} + \cos\gamma\,\mathbf{k}$, then
$$\iiint_E \operatorname{div}\mathbf{F}\,dV = \iint_S \mathbf{F} \cdot \mathbf{n}\,dS$$
$$\iiint_E \left(\frac{\partial P}{\partial x} + \frac{\partial Q}{\partial y} + \frac{\partial R}{\partial z}\right) dV$$
$$= \iint_S (P\cos\alpha + Q\cos\beta + R\cos\gamma)\,dS$$

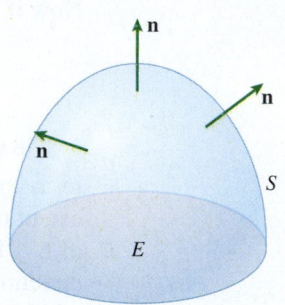

### Stokes' Theorem
If $\mathbf{F}(x, y, z) = P(x, y, z)\mathbf{i} + Q(x, y, z)\mathbf{j} + R(x, y, z)\mathbf{k}$, then
$$\oint_C \mathbf{F} \cdot d\mathbf{r} = \iint_S \operatorname{curl}\mathbf{F} \cdot \mathbf{n}\,dS$$
$$\oint_C (P\,dx + Q\,dy + R\,dz) = \iint_S \left[\left(\frac{\partial R}{\partial y} - \frac{\partial Q}{\partial z}\right) dy\,dz\right.$$
$$\left. - \left(\frac{\partial R}{\partial x} - \frac{\partial P}{\partial z}\right) dz\,dx + \left(\frac{\partial Q}{\partial x} - \frac{\partial P}{\partial y}\right) dx\,dy\right]$$

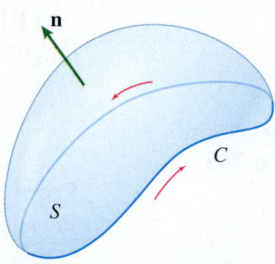

---

## 15.9 Assess Your Understanding

### Concepts and Vocabulary

1. *True or False* The curl of a vector field $\mathbf{F}$ is a scalar.

2. If the vector field
$$\mathbf{F} = (e^{-x^2/2} - yz)\mathbf{i} + (e^{-y^2/2} + xz + 2x)\mathbf{j} + (e^{-z^2/2} + 5)\mathbf{k},$$
then curl $\mathbf{F} = $ _____.

3. *Multiple Choice* A vector field $\mathbf{F}$ is conservative if and only if curl $\mathbf{F} = $ [(a) $\mathbf{F}$ (b) div $\mathbf{F}$ (c) $\mathbf{0}$ (d) $\mathbf{n}$].

4. *True or False* An interpretation of the curl of $\mathbf{F}$ is circulation per unit area of a fluid at a given point on a surface.

5. Suppose $\mathbf{F}$ is the velocity vector of a fluid rotating about a fixed axis and that $\omega$ is a constant angular velocity. Then curl $\mathbf{F} = $ _____.

6. *True or False* If $\mathbf{F}$ is a conservative vector field, then the work done by $\mathbf{F}$ in moving an object of mass $m$ from point $A$ to point $B$ depends on the path taken from $A$ to $B$.

---

**1.** = NOW WORK problem    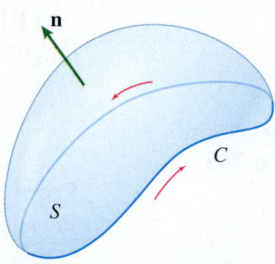 = Graphing technology recommended    **CAS** = Computer Algebra System recommended

## Skill Building

*In Problems 7–18, find* curl **F**.

**7.** $\mathbf{F}(x, y, z) = x\mathbf{i} + y\mathbf{j}$

**8.** $\mathbf{F}(x, y, z) = y\mathbf{i} + x\mathbf{j}$

**9.** $\mathbf{F}(x, y, z) = xyz\mathbf{i} + xz\mathbf{j} + z\mathbf{k}$

**10.** $\mathbf{F}(x, y, z) = 4x\mathbf{i} - y\mathbf{j} - 2z\mathbf{k}$

**11.** $\mathbf{F}(x, y, z) = 3xyz^2\mathbf{i} + y^2\sin z\,\mathbf{j} + xe^{2z}\mathbf{k}$

**12.** $\mathbf{F}(x, y, z) = yz + z^2x\mathbf{j} + yz\mathbf{k}$

**13.** $\mathbf{F}(x, y, z) = \dfrac{x\mathbf{i}}{x^2 + y^2 + z^2} + \dfrac{y\mathbf{j}}{x^2 + y^2 + z^2} + \dfrac{\mathbf{k}}{x^2 + y^2 + z^2}$

**14.** $\mathbf{F}(x, y, z) = e^x\mathbf{i} + x^2y\mathbf{j} + e^z\mathbf{k}$

**15.** $\mathbf{F}(x, y, z) = \cos x\mathbf{i} + \sin y\mathbf{j} + e^{xz}\mathbf{k}$

**16.** $\mathbf{F}(x, y, z) = \sin(xy)\mathbf{i} + \cos(xy^2)\mathbf{j} + x\mathbf{k}$

**17.** $\mathbf{F}(x, y, z) = (x + y)\mathbf{i} + (y + z)\mathbf{j} + (z + x)\mathbf{k}$

**18.** $\mathbf{F}(x, y, z) = (y + z)\mathbf{i} + (z + x)\mathbf{j} + (z + y + x)\mathbf{k}$

*In Problems 19–24, verify Stokes' Theorem for each vector field* **F** *and surface S.*

**19.** $\mathbf{F} = (z - y)\mathbf{i} + (z + x)\mathbf{j} - (x + y)\mathbf{k}$; *S is the portion of the paraboloid* $z = 1 - x^2 - y^2$, $z \geq 0$.

**20.** $\mathbf{F} = y\mathbf{i} + z\mathbf{j} + x\mathbf{k}$; *S is the portion of the paraboloid* $z = 1 - x^2 - y^2$, $z \geq 0$.

**21.** $\mathbf{F} = y\mathbf{i} - x\mathbf{j}$; *S is the hemisphere* $z = \sqrt{1 - x^2 - y^2}$.

**22.** $\mathbf{F} = z\mathbf{i} + x\mathbf{j} + y\mathbf{k}$; *S is the hemisphere* $z = \sqrt{1 - x^2 - y^2}$.

**23.** $\mathbf{F} = y^2\mathbf{i} + x\mathbf{j} - xz\mathbf{k}$; *S is the surface* $z = 1 - x^2 - y^2$, $z \geq 0$.

**24.** $\mathbf{F}(x, y, z) = 4x\mathbf{i} - y\mathbf{j} + 2z\mathbf{k}$; *S is the surface* $z = 1 + x^2 + y^2$, $z \leq 5$.

**25.** Use Stokes' Theorem to find the surface integral $\iint\limits_S \operatorname{curl}\mathbf{F} \cdot \mathbf{n}\,dS$, where $\mathbf{F} = \mathbf{F}(x, y, z) = y\mathbf{i} + x\mathbf{j} + x^2\mathbf{k}$ and *S is the surface enclosed by the paraboloid* $z = 9 - x^2 - y^2$, $z \geq 0$.

**26.** Use Stokes' Theorem to find the surface integral $\iint\limits_S \operatorname{curl}\mathbf{F} \cdot \mathbf{n}\,dS$, where $\mathbf{F} = \mathbf{F}(x, y, z) = 4z\mathbf{i} + 3x\mathbf{j} + 3y\mathbf{k}$ and *S is the surface enclosed by the paraboloid* $z = 10 - x^2 - y^2$, $z \geq 4$.

*In Problems 27–30, use Stokes' Theorem to find each line integral. Verify your answer by a direct calculation of the line integral. Assume the orientation of C is counterclockwise.*

**27.** $\oint_C [(y + z)\,dx + (z + x)\,dy + (x + y)\,dz]$; *C is the curve of intersection of* $x^2 + y^2 + z^2 = 1$, *and* $x + y + z = 0$.

**28.** $\oint_C [(y - z)\,dx + (z - x)\,dy + (x - y)\,dz]$; *C is the curve of intersection of* $x^2 + y^2 = 1$, *and* $x + z = 1$.

**29.** $\oint_C [x\,dx + (x + y)\,dy + (x + y + z)\,dz]$; *C is the curve* $x = 2\cos t$, $y = 2\sin t$, $z = 2$, $0 \leq t \leq 2\pi$.

**30.** $\oint_C (y^2\,dx + z^2\,dy + x^2\,dz)$; *C is the triangle with vertices* $(1, 0, 0)$, $(0, 1, 0)$, *and* $(0, 0, 1)$.

*In Problems 31–34, determine whether the force* **F** *is a conservative vector field.*

**31.** $\mathbf{F} = x\mathbf{i} + y\mathbf{j}$

**32.** $\mathbf{F} = y\mathbf{i} + x\mathbf{j}$

**33.** $\mathbf{F} = xy\mathbf{i} + yz\mathbf{j} + zx\mathbf{k}$

**34.** $\mathbf{F} = yz\mathbf{i} + zx\mathbf{j} + xy\mathbf{k}$

## Applications and Extensions

**35.** Find the value of the constant $c$ so that $\mathbf{F} = xy\mathbf{i} + cx^2\mathbf{j}$ in space is a conservative vector field.

**36.** Find the value of the constant $c$ so that $\mathbf{F} = \dfrac{z}{y}\mathbf{i} + c\dfrac{xz}{y^2}\mathbf{j} + \dfrac{x}{y}\mathbf{k}$, $y \neq 0$, is a conservative vector field.

**37.** Show that $\operatorname{curl}(\mathbf{F} + \mathbf{G}) = \operatorname{curl}\mathbf{F} + \operatorname{curl}\mathbf{G}$.

**38.** Show that $\operatorname{curl}(c\mathbf{F}) = c(\operatorname{curl}\mathbf{F})$, where $c$ is a constant.

**39.** If $\mathbf{F}(x, y, z) = z\mathbf{i} + x\mathbf{j} + y\mathbf{k}$, find $\iint\limits_S \mathbf{F} \cdot \mathbf{n}\,dS$, where *S is the hemisphere* $z = \sqrt{1 - x^2 - y^2}$.

**40.** Rework Problem 39, where *S is the circular region* $x^2 + y^2 \leq 1$, $z = 0$.

**41.** Show that $\mathbf{F} = y\mathbf{i} - x\mathbf{j} + z\mathbf{k}$ is not a conservative vector field. Nevertheless, there are certain paths *C* for which $\oint_C \mathbf{F} \cdot d\mathbf{r} = 0$. Find one.

**42.** Show that $\operatorname{div}(\operatorname{curl}\mathbf{F}) = \mathbf{0}$, where $\mathbf{F} = P(x, y, z)\mathbf{i} + Q(x, y, z)\mathbf{j} + R(x, y, z)\mathbf{k}$ and *P, Q, and R are twice differentiable and the partial derivatives are continuous.*

**43. Work** An object is moved from the origin to the point $(a, b, c)$ in the field of force $\mathbf{F} = (x + y)\mathbf{i} + (x - z)\mathbf{j} + (z - y)\mathbf{k}$. Show that the work done depends on only $a$, $b$, and $c$, and find this value.

**44.** Show that no twice differentiable vector function exists whose curl is $x\mathbf{i} + y\mathbf{j} + z\mathbf{k}$.

## Challenge Problems

**45.** Suppose $\mathbf{F}(x, y, z) = P(x, y, z)\mathbf{i} + Q(x, y, z)\mathbf{j} + R(x, y, z)\mathbf{k}$ is a vector field with continuous and differentiable components in a simply connected region of space.

(a) Use Stokes' Theorem to show that **F** is a conservative vector field if $Q_x = P_y$, $R_y = Q_z$, and $P_z = R_x$.

(b) Why does it follow that $\int_C \mathbf{F} \cdot d\mathbf{r}$ is independent of the path when these conditions hold?

**46.** Use Problem 45 to show that
$$\int_C [(yz - y - z)\,dx + (xz - x - z)\,dy + (xy - x - y)\,dz]$$
is independent of the path.

**47.** Let $\mathbf{F}(x, y, z)$ be a vector field with continuous and differentiable components, and let *S* be a sphere with outer unit normal **n**. Use Stokes' Theorem to show that $\iint\limits_S \operatorname{curl}\mathbf{F} \cdot \mathbf{n}\,dS = 0$.

**48.** Assume that the hypotheses of the Divergence Theorem hold for *S* and *E*. Show that for any vector field **F** with continuous first-order partial derivatives in a closed, bounded solid *E* with orientable boundary *S* with outer unit normal **n**, $\iint\limits_S \operatorname{curl}\mathbf{F} \cdot \mathbf{n}\,dS = 0$.

**49.** Let *C* be a smooth, simple, closed curve lying on an orientable surface *S*, and let *f* and *g* have continuous partial derivatives on *S*. Show that $\int_C (f\nabla g) \cdot d\mathbf{r} = \iint\limits_S (\nabla f \times \nabla g) \cdot \mathbf{n}\,dS$. Assume that the portion of *S* bounded by *C* is smooth and simply connected.

# Chapter Review

## THINGS TO KNOW

### 15.1 Vector Fields

- A vector field $\mathbf{F}$ in the plane:
  $\mathbf{F} = \mathbf{F}(x, y) = P(x, y)\mathbf{i} + Q(x, y)\mathbf{j}$ (p. 1025)

- A vector field $\mathbf{F}$ in space:
  $\mathbf{F} = \mathbf{F}(x, y, z) = P(x, y, z)\mathbf{i} + Q(x, y, z)\mathbf{j} + R(x, y, z)\mathbf{k}$
  (p. 1025)

### 15.2 Line Integrals of Scalar Functions

- Line integral of $f$ along a smooth curve $C$ from $t = a$ to $t = b$
  in the plane (p. 1030):

$$\int_C f(x, y)\, ds = \int_a^b f(x(t), y(t)) \sqrt{\left(\frac{dx}{dt}\right)^2 + \left(\frac{dy}{dt}\right)^2}\, dt$$

  in space (p. 1036):

$$\int_C f(x, y, z)\, ds = \int_a^b f(x(t), y(t), z(t)) \sqrt{\left(\frac{dx}{dt}\right)^2 + \left(\frac{dy}{dt}\right)^2 + \left(\frac{dz}{dt}\right)^2}\, dt$$

- Mass $M$ of a wire of variable density $\rho = \rho(x, y)$ described by
  a smooth curve $C$, $M = \int_C \rho(x, y) ds$ (p. 1031)

- Lateral surface area $A$ of a cylinder $z = f(x, y) \geq 0$:
  $A = \int_C f(x, y) ds$ (p. 1032)

- $\int_C (P\, dx + Q\, dy) = -\int_{-C}(P\, dx + Q\, dy)$ (p. 1034)

- The line integral of $P\, dx + Q\, dy$ along a smooth curve $C$ in
  the plane:
  $\int_C (P\, dx + Q\, dy) = \int_C P(x, y) dx + \int_C Q(x, y) dy$ (p. 1034)

- The line integral of $w = f(x, y, z)$ along a smooth curve $C$ in
  space. (p. 1036)

### 15.3 Line Integrals of Vector Fields; Work

- If $\mathbf{F} = \mathbf{F}(x, y) = P(x, y)\mathbf{i} + Q(x, y)\mathbf{j}$ is a vector field that is
  continuous on some open region containing the smooth curve
  $C$ traced out by $\mathbf{r} = \mathbf{r}(t) = x(t)\mathbf{i} + y(t)\mathbf{j}$, $a \leq t \leq b$, then
  $\int_C \mathbf{F} \cdot d\mathbf{r} = \int_C (P\, dx + Q\, dy)$. (p. 1040)

- Work: $W = \int_C \mathbf{F} \cdot d\mathbf{r} = \int_C (P\, dx + Q\, dy)$. (p. 1044)

### 15.4 Fundamental Theorem of Line Integrals

- $\mathbf{F}$ is a conservative vector field if $\mathbf{F} = \nabla f$; the function $f$ is
  called the potential function for $\mathbf{F}$. (p. 1046)

- Fundamental Theorem of Line Integrals: If $\mathbf{F}$ is a
  conservative vector field, where $\nabla f = \mathbf{F}$, then
  $\int_C \mathbf{F} \cdot d\mathbf{r} = \int_{(x_0, y_0)}^{(x_1, y_1)} \mathbf{F} \cdot d\mathbf{r} = f(x_1, y_1) - f(x_0, y_0)$. (p. 1047)

- If $\mathbf{F}$ is a conservative vector field on an open region $R$
  and $C$ is a closed, piecewise-smooth curve in $R$,
  then $\int_C \mathbf{F} \cdot d\mathbf{r} = \int_C (P\, dx + Q\, dy) = 0$. (p. 1049)

- A piecewise-smooth curve $C$ is closed if its initial point and
  terminal point coincide. The curve $C$ is simple if it does not
  intersect itself. (p. 1053)

- Simply connected region (p. 1053)

- A vector field $\mathbf{F}$ is conservative if and only if $\dfrac{\partial P}{\partial y} = \dfrac{\partial Q}{\partial x}$.
  (p. 1054)

- If $\mathbf{F}$ is a conservative force field, the work $W$ done by $\mathbf{F}$ in
  moving an object along a closed path is 0. (p. 1058)

- Law of Conservation of Energy (p. 1058)

### 15.5 Green's Theorem

- Green's Theorem:

$$\oint_C (P\, dx + Q\, dy) = \iint_R \left(\frac{\partial Q}{\partial x} - \frac{\partial P}{\partial y}\right) dx\, dy \text{ (p. 1062)}$$

- Green's Theorem for Area: $A = \dfrac{1}{2} \oint_C (-y\, dx + x\, dy)$
  (p. 1064)

### 15.6 Parametric Surfaces

- Parametrization; parameter domain (p. 1070)

- Coordinate curves (p. 1070)

- Smooth parametrization; smooth surface (p. 1075)

- Surface area of a smooth surface:

$$S = \iint_R dS = \iint_R \|\mathbf{r}_u \times \mathbf{r}_v\|\, du\, dv \text{ (p. 1077)}$$

### 15.7 Surface and Flux Integrals

- A surface integral expressed as a double integral: (p. 1081)
  $\iint_S F(x, y, z)\, dS$
  $$= \iint_R F(x(u, v), y(u, v), z(u, v)) \|\mathbf{r}_u \times \mathbf{r}_v\|\, du\, dv$$

- Orientable surface (p. 1084)

- The flux of $\mathbf{F}$ across $S$: $\dfrac{\text{Mass of fluid}}{\text{Unit time}} = \iint_S \rho \mathbf{F} \cdot \mathbf{n}\, dS$,

  where $\rho$ is the mass density of the fluid, the vector field $\mathbf{F}$ is
  the velocity of the fluid, and $\mathbf{n}$ is a unit normal vector of $S$.
  (p. 1088)

### 15.8 The Divergence Theorem

$\mathbf{F}(x, y, z) = P(x, y, z)\mathbf{i} + Q(x, y, z)\mathbf{j} + R(x, y, z)\mathbf{k}$

- $\text{div } \mathbf{F} = \dfrac{\partial P}{\partial x} + \dfrac{\partial Q}{\partial y} + \dfrac{\partial R}{\partial z}$ (p. 1092)

- Divergence Theorem: $\iiint_E \text{div } \mathbf{F}\, dV = \iint_S \mathbf{F} \cdot \mathbf{n}\, dS$ (p. 1092)

### 15.9 Stokes' Theorem

$\mathbf{F}(x, y, z) = P(x, y, z)\mathbf{i} + Q(x, y, z)\mathbf{j} + R(x, y, z)\mathbf{k}$

- $\text{curl } \mathbf{F} =$
  $$\left(\frac{\partial R}{\partial y} - \frac{\partial Q}{\partial z}\right)\mathbf{i} - \left(\frac{\partial R}{\partial x} - \frac{\partial P}{\partial z}\right)\mathbf{j} + \left(\frac{\partial Q}{\partial x} - \frac{\partial P}{\partial y}\right)\mathbf{k}$$
  (p. 1100)

- Stokes' Theorem: $\oint_C \mathbf{F} \cdot d\mathbf{r} = \iint_S \text{curl } \mathbf{F} \cdot \mathbf{n}\, dS$ (p. 1101)

- $\mathbf{F}$ is a conservative vector field in space if and only if
  $\text{curl } \mathbf{F} = 0$. (p. 1104)

## OBJECTIVES

## REVIEW EXERCISES

1. Describe the vector field $\mathbf{F}(x, y) = 3\mathbf{i} + \mathbf{j}$ by drawing some of its vectors.

2. Describe the vector field $\mathbf{F}(x, y) = \sin x\mathbf{i} + \mathbf{j}$ by drawing some of its vectors.

3. Find $\int_C \left( \dfrac{dx}{y} + \dfrac{dy}{x} \right)$, where $C$ is the arc of the parabola $y = x^2$ from $(1, 1)$ to $(2, 4)$.

4. Find $\int_C [y \cos(xy)\, dx + x \cos(xy)\, dy]$, where $C$ is the curve $x = t^2$, $y = t^3$, $0 \le t \le 1$.

5. (a) Confirm that the line integral $\int_C [y \cos(xy)\, dx + x \cos(xy)\, dy]$ is independent of the path.

   (b) Find $\int_C [y \cos(xy)\, dx + x \cos(xy)\, dy]$ from $(0, 0)$ to $(1, 1)$ by following the right-angle path from $(0, 0)$ to $(1, 0)$ to $(1, 1)$.

6. (a) Find a function $f$ whose gradient is $y \cos(xy)\mathbf{i} + x \cos(xy)\mathbf{j}$.

   (b) Explain why
   $\int_C [y \cos(xy)\, dx + x \cos(xy)\, dy] = f(1, 1) - f(0, 0)$,
   where $C$ is a smooth curve joining $(0, 0)$ to $(1, 1)$.

7. Find $\int_C [(yz - y - z)\, dx + (xz - x - z)\, dy + (xy - x - y)\, dz]$, where $C$ is the twisted cubic $x = t$, $y = t^2$, $z = t^3$, $0 \le t \le 1$.

8. Find $\int_C \mathbf{F} \cdot d\mathbf{r}$ if $\mathbf{F}(x, y, z) = xy\mathbf{i} + xz\mathbf{j} + (y - x)\mathbf{k}$ and $C$ is the line segment from $(0, 0, 0)$ to $(1, -2, 3)$.

9. (a) Show that $f(x, y) = 3x^2 + y - 2$ is a potential function of $\mathbf{F}(x, y) = 6x\mathbf{i} + \mathbf{j}$.

   (b) Use the Fundamental Theorem of Line Integrals to find $\int_C [6x\, dx + dy]$, where $C$ is any curve joining the two points $(4, 1)$ to $(6, 3)$.

10. Find $\int_C [e^x \sin y\, dx + e^x \cos y\, dy]$, where $C$ is the arc of the parabola $y = x^2$ from $(0, 0)$ to $(1, 1)$.
    *Hint:* $\nabla (e^x \sin y) = e^x \sin y\mathbf{i} + e^x \cos y\mathbf{j}$.

11. **Mass of a Wire**   Find the mass of a thin piece of wire in the shape of a circular arc $x = \cos t$, $y = \sin t$, $0 \le t \le \dfrac{3\pi}{4}$, if the variable mass density of the wire is $\rho(x, y) = x + 1$.

12. Find the line integral $\int_C (x^3 dx + y^3 dy)$

    (a) if $C$ consists of the line segment from $(1, 1)$ to $(2, 4)$.

    (b) if $C$ consists of line segments from $(1, 1)$ to $(1, 5)$, from $(1, 5)$ to $(2, 3)$ and from $(2, 3)$ to $(2, 4)$.

    (c) if $C$ is a part of the parabola $x = t$, $y = t^2$, $1 \le t \le 2$.

13. **Mass of a Lamina**   A lamina is in the shape of the cone $z = \sqrt{x^2 + y^2}$, $1 \le z \le 4$. If the mass density of the lamina is $\rho(x, y, z) = x^2 + y^2$, find the mass $M$ of the lamina.

**14.** Find $\int_C [x^2 y^3\, dx - xy^4\, dy]$, where $C$ is the arc of the parabola $y^2 = x$ from $(0, 0)$ to $(1, 1)$.

**15.** Find $\int_C x^2 y^2\, ds$; $C$: $x = \cos t$, $y = \sin t$; $0 \le t \le \pi$.

**16.** **Work** If $\mathbf{F}(x, y) = \dfrac{y}{(x^2 + y^2)^{3/2}}\mathbf{i} - \dfrac{x}{(x^2 + y^2)^{3/2}}\mathbf{j}$, find the work done by going around the unit circle against $\mathbf{F}$ from $t = 0$ to $t = 2\pi$.

**17.** **Work** Find the work done by the force $\mathbf{F} = y \sin x\mathbf{i} + \sin x\mathbf{j}$ in moving an object along the curve $y = \sin x$ from $x = 0$ to $x = 2\pi$.

**18.** **Work** Find the work done by the force

$$\mathbf{F} = \frac{x}{x^2 + y^2}\mathbf{i} + \frac{y}{x^2 + y^2}\mathbf{j}$$

in moving an object along the curve $\mathbf{r}(t) = t \cos t\mathbf{i} + t \sin t\mathbf{j}$ from $(-\pi, 0)$ to $(2\pi, 0)$.

**19.** **Area** Use Green's Theorem to find the area of the region enclosed by the curves $C_1$: $x(t) = t$, $y(t) = t^2 + 3$ and $C_2$: $x(t) = t$, $y(t) = 30 - 2t^2$.

**20.** Use Green's Theorem to find the line integral

$$\oint_C \left[ \ln(1 + y)\, dx + \frac{xy}{1 + y}\, dy \right]$$ where $C$ is the parallelogram

with vertices $(0, 0)$, $(2, 1)$, $(2, 6)$, and $(0, 5)$ traversed counterclockwise.

**21.** Find $\oint_C (y^2\, dx - x^2\, dy)$, where $C$ is the square with vertices $(0, 0)$, $(1, 0)$, $(1, 1)$, and $(0, 1)$ traversed counterclockwise. Do not use Green's Theorem.

**22.** Rework Problem 21 using Green's Theorem.

**23.** Find $\oint_C [(x - y)\, dx + (x + y)\, dy]$, where $C$ is the ellipse $x = 2 \cos t$, $y = 3 \sin t$, $0 \le t \le 2\pi$, without using Green's Theorem.

**24.** Rework Problem 23 using Green's Theorem.

**25.** **Area** Use Green's Theorem to find the area of the multiply-connected region enclosed by the ellipse $\dfrac{x^2}{4} + \dfrac{y^2}{16} = 1$ which has a small hole $C_1$ given by $x^2 + (y - 1)^2 = 1$ punched out of its interior.

**26.** **(a)** Identify the coordinate curves of the surface parametrized by $\mathbf{r}(u, v) = 3u \cos v\mathbf{i} + 2u \sin v\mathbf{j} + u^2\mathbf{k}$, $0 \le u \le 1$, $0 \le v \le 2\pi$.

**(b)** Find a rectangular equation for the surface.

**27.** Find a parametrization of the part of the cylinder $4x^2 + 25y^2 = 100$ that lies above the plane $z = 1$ and below the plane $z = 6$.

**28.** **(a)** Find an equation of the tangent plane to the surface $\mathbf{r}(u, v) = u \sin v\mathbf{i} + u^2\mathbf{j} + u \cos v\mathbf{k}$, at the point $(\sqrt{3}, 4, 1)$.

**(b)** Find an equation of the normal line to the tangent plane at the point $(\sqrt{3}, 4, 1)$.

**29.** Find the surface area of the part of the paraboloid $\mathbf{r}(u, v) = u \sin v\mathbf{i} + u^2\mathbf{j} + u \cos v\mathbf{k}$, $0 \le u \le 4$, $0 \le v \le 2\pi$.

**30.** Find $\iint_S x\, dS$, where $S$ is the surface parametrized by

$\mathbf{r}(u, v) = \cos v\mathbf{i} + 3 \sin u \sin v\mathbf{j} + 3 \cos u \sin v\mathbf{k}$, $0 \le u \le 2\pi$ and $0 \le v \le \pi/2$.

**CAS** **31.** Find $\iint_S z^2\, dS$, where $S$ is the sphere $x^2 + y^2 + z^2 = 4$.

**32.** Find the outer unit normal vectors to the surface $S$ that forms the boundary of the solid $z = f(x, y) = \sqrt{25 - x^2 - y^2}$, $0 \le x^2 + y^2 \le 25$.

**33.** Find $\iint_S x\, dS$, where $S$ is the surface $x^2 + y^2 = 9$, $-1 \le z \le 1$.

**34.** Find $\iint_S z\, dS$, where $S$ is the surface $z = 9 - x - y$, $x^2 + y^2 \le 9$.

**35.** Find $\iint_S \cos x\, dS$, where $S$ is the portion of the plane $x = y + z$, $x \le \pi$, $y \ge 0$, $z \ge 0$.

**36.** A fluid has a constant mass density $\rho$. Find the mass of fluid flowing across the surface $x^2 + y^2 = 1$, $0 \le z \le 1$ in a unit of time, in the direction outward from the $z$-axis if the velocity of the fluid at any point on the surface is $\mathbf{F} = x^2\mathbf{i} + y\mathbf{j} - z\mathbf{k}$.

*In Problems 37–40, for each vector field* $\mathbf{F}$, *find*

**(a)** div $\mathbf{F}$

**(b)** curl $\mathbf{F}$

**(c)** Verify Stokes' Theorem where $S$ is the paraboloid $z = x^2 + y^2$, with the circle $x^2 + y^2 = 1$ and $z = 1$ as its boundary.

**37.** $\mathbf{F} = z \cos x\mathbf{i} + \sin y\mathbf{j} + e^x\mathbf{k}$

**38.** $\mathbf{F} = x^2\mathbf{i} - 3y\mathbf{j} + 4z^2\mathbf{k}$

**39.** $\mathbf{F} = x\mathbf{i} + y\mathbf{j} + z\mathbf{k}$

**40.** $\mathbf{F} = xe^y\mathbf{i} - ye^z\mathbf{j} + ze^x\mathbf{k}$

**41.** Use the Divergence Theorem to find $\iint_S \mathbf{F} \cdot \mathbf{n}\, dS$, where $S$ is the surface bounded by $x^2 + y^2 = 1$, and $0 \le z \le 1$, and $\mathbf{F} = x^2\mathbf{i} + y\mathbf{j} - z\mathbf{k}$.

**42.** **(a)** Find a function $f(x, y, z)$ whose gradient is $(yz - y - z)\mathbf{i} + (xz - x - z)\mathbf{j} + (xy - x - y)\mathbf{k}$.

**(b)** Use Stokes' Theorem to confirm the answer to $\int_C [(yz - y - z)\, dx + (xz - x - z)\, dy + (xy - x - y)\, dz]$, where $C$ is the twisted cubic $x = t$, $y = t^2$, $z = t^3$, $0 \le t \le 1$.

**43.** Find $\iint_S (xz \cos \alpha + yz \cos \beta + x^2 \cos \gamma)\, dS$, where $S$ is the upper half of the unit sphere together with the plane $z = 0$, and $\cos \alpha$, $\cos \beta$, and $\cos \gamma$ are the direction cosines for the outer unit normal to $S$.

**44.** Find $\iiint_E \text{div } \mathbf{F}\, dV$, where $E$ is the unit ball, $\|\mathbf{r}\| \le 1$, and $\mathbf{F} = \|\mathbf{r}\|^2\mathbf{r}$.

**45.** Use Stokes' Theorem to find $\int_C [(x - y)\, dx + (y - z)\, dy + (z - x)\, dz]$, where $C$ is the boundary of the portion of the plane $x + y + z = 1$, $x \ge 0$, $y \ge 0$, $z \ge 0$ (traversed counterclockwise when viewed from above).

**46.** Let $\mathbf{F}(x, y, z) = x^3\mathbf{i} + y^3\mathbf{j} + z^3\mathbf{k}$ be the velocity of a fluid flow in space, where the mass density of the fluid is 1.

**(a)** Find the flux across the sphere $x^2 + y^2 + z^2 = 1$.

**(b)** Find the circulation around the circle $x^2 + y^2 = 1$ in the $xy$-plane.

**47.** Determine if $\mathbf{F}(x, y, z) = yz\mathbf{i} + xy\mathbf{j} + xy\mathbf{k}$ is a conservative vector field.

**48.** Let $\mathbf{F}(x, y, z) = 2xy^2z\mathbf{i} + 2x^2yz\mathbf{j} + (x^2y^2 - 2z)\mathbf{k}$. Show that $\mathbf{F}$ is a conservative vector field.

## CHAPTER 15 PROJECT    Modeling a Tornado

A tornado is a three-dimensional wind field that can be modeled as a velocity field. In this project we look at a two-dimensional slice of a vortex and use an Oseen vortex velocity field to model a tornado.

In rectangular coordinates, an Oseen velocity field is given by

$$\mathbf{v} = \left( \frac{V_{max}\, y}{\sqrt{x^2 + y^2}} \right) \mathbf{i} - \left( \frac{V_{max}\, x}{\sqrt{x^2 + y^2}} \right) \mathbf{j}$$

We place the center of the tornado at origin, where $\mathbf{v}$ is undefined. Then $\mathbf{v}$ gives the velocity of the wind at any other location within the vortex.

1. Show that in the Oseen model, the speed (magnitude of the velocity) of the wind at any location is $V_{max}$.

2. Suppose $V_{max} = 80\,\text{m/s}$. Describe the velocity field at the following locations

   (a) $(0, -10)$,  (b) $(-10, 0)$,  (c) $(0, 10)$,  (d) $(10, 0)$,  (e) $(10, 10)$

3. Add a few more vectors in the velocity field. What can you conclude about the direction of the wind?

   A fluid, like the air in a tornado, is considered to be **incompressible** if the divergence of its velocity field, div $\mathbf{v}$, equals 0. When a fluid is incompressible, it is possible to find a function, $\psi(x, y)$, called a **stream function**. The level curves of a stream function are called **stream lines**. These are the curves along which particles of fluid would flow. (You may have seen examples of stream lines such as air moving around a car in a wind tunnel.) The stream function, $\psi(x, y)$, for a velocity field, $\mathbf{v} = h(x, y)\,\mathbf{i} + g(x, y)\,\mathbf{j}$, where $h$ and $g$ are functions of $x$ and $y$, satisfies

   $$\frac{\partial \psi}{\partial y} = h(x, y) \quad \text{and} \quad \frac{\partial \psi}{\partial x} = -g(x, y)$$

4. Show that in the Oseen model, the air in a tornado is incompressible.

5. Find a general form for the stream function of a stream line $\psi(x, y) = C$ for an Oseen velocity field.

6. Suppose the speed $V_{max}$ of the wind in a tornado is 80 m/s and the value of the constant $C$ is $3200\,\text{m}^3/\text{s}$ (the units of $C$ represent a volume flow rate).

   (a) Find the stream function and graph the stream line $\psi(x, y) = 3200$.

   (b) Graph the stream lines for $C = 1000, 2000, 3000,$ and $4000$.

   The air in the tornado is turning, and the twisting behavior of the velocity field can be quantified by a vector $\mathbf{w}$, called the *vorticity*. The **vorticity** at a point in space is a vector that is normal to the plane of rotation, and describes the amount of rotation about that point. The vorticity of a vector field is given by the curl of the vector field, curl $\mathbf{v}$. That is,

   $$\mathbf{w} = \text{curl } \mathbf{v}$$

7. Find the vorticity $\mathbf{w}$ of the Oseen velocity field. Interpret the result.

8. In this project, we assumed the wind speed in the vortex was 80 m/s. Research the **Enhanced Fujita scale**, or **EF-scale**, that is used to classify tornados based on the damage they cause. Rework Problems 2, 3, 6, and 7 for a tornado with a wind speed of 110 m/s (an EF-5 tornado). Compare the results to a tornado with a wind speed of 80 m/s.

# 16 Differential Equations

KEENPRESS/Getty Images

## The Melting Arctic Ice Cap

The Arctic ice cap is melting. "The average thickness of the Arctic ice cover is declining because it is rapidly losing its thick component, the *multi-year ice*. At the same time, the surface temperature of the Arctic is going up, which results in a shorter ice-forming season," said senior research scientist Josefino Comiso.

Each year during the Arctic winter, new sea ice, called **seasonal ice**, forms. Most of the seasonal ice melts during the summer melt season, but some of it survives. Scientists use this pattern to differentiate among three types of Arctic ice: seasonal ice forms and melts each year, **perennial ice** is defined as ice that has remained through at least one summer melt season, and **multi-year ice** is perennial ice that has survived at least three summer melt seasons. In a recent study, NASA scientists have discovered that the thick, old multi-year ice of the Arctic Ocean is melting at a faster rate than that of the thin seasonal ice and the one- and two-year perennial ice.

**CHAPTER 16 PROJECT** The Chapter Project on page 1150 investigates the growth of sea ice.

This chapter continues our investigation into methods for finding solutions to differential equations and includes some important applications such as learning curves, falling objects with air resistance, flow rates in mixtures, and sales growth. The analyses and solutions of differential equations are a very useful branch of mathematics, but differential equations are often difficult or impossible to solve algebraically. In practical applications, differential equations are often solved using numerical approximations.

# 16.1 Classification of Ordinary Differential Equations

**OBJECTIVES** *When you finish this section, you should be able to:*

**1** Classify ordinary differential equations (p. 1113)

**2** Verify the solution of an ordinary differential equation (p. 1113)

## 1 Classify Ordinary Differential Equations

An **ordinary differential equation** is an equation involving an independent variable $x$, a dependent variable $y$, and derivatives of $y$ with respect to $x$. Examples of ordinary differential equations are:

**(a)** $y' + y + 5 = 0$    **(b)** $\dfrac{x - y}{x + y} + \dfrac{y + x}{x - y}\dfrac{dy}{dx} = 0$    **(c)** $7y'' - 5y' + 4x^3 y = 0$

**(d)** $y' + y''' = 1$    **(e)** $\left(\dfrac{dy}{dx}\right)^3 = 8\dfrac{d^2 y}{dx^2}$    **(f)** $5x^2 = 6y^3 \left(\dfrac{dy}{dx}\right)^2 + 18$

The **order** of a differential equation is the order of the highest-order derivative of $y$ appearing in the equation. For the equations above, (a), (b), and (f) are first-order, (c) and (e) are second-order, and (d) is a third-order differential equation. The exponent of the highest-order derivative in a differential equation is called the **degree** of the equation. For example, equations (a)–(e) above are of degree 1, and equation (f) is of degree 2. Being able to recognize the order and degree of a differential equation are important steps in finding its solution.

**NOTE** Other variables besides $x$ and $y$ are often used. If $s = f(t)$, then $\dfrac{d^2 s}{dt^2} + t = 0$ is a second-order differential equation.

An important type of differential equation is a *linear differential equation*. If $P_0, P_1, \ldots, P_n$ and $Q$ are functions of the variable $x$, then an equation of the form

$$P_0(x)\frac{d^n y}{dx^n} + P_1(x)\frac{d^{n-1} y}{dx^{n-1}} + \cdots + P_{n-1}(x)\frac{dy}{dx} + P_n(x)y = Q(x)$$

is a **linear differential equation of order** $n$, provided $P_0 \neq 0$. The term *linear* is used to describe these differential equations because

- each derivative including $y$ (which we treat as a derivative of order zero) is of degree 1
- no term contains more than one derivative
- no term contains a product of $y$ and any derivative of $y$ up to the $n$th derivative.

For example,

$$x^2 \frac{d^2 y}{dx^2} + 8\frac{dy}{dx} - 5x^3 y = x + 10$$

is a second-order linear differential equation.

---

**EXAMPLE 1** **Classifying a Differential Equation**

What are the order and degree of the differential equation $xe^x \dfrac{d^2 y}{dx^2} - 5e^x \dfrac{dy}{dx} + 4x^3 y = 0$? Is it linear?

**Solution** This is a second-order differential equation of degree 1. Since $P_0(x) = xe^x$, $P_1(x) = -5e^x$, and $P_2(x) = 4x^3$ are all functions of $x$ alone and no term contains more than one derivative, the differential equation is linear. ∎

**NOW WORK** Problem 9.

## 2 Verify the Solution of an Ordinary Differential Equation

A function $y = f(x)$ is called a **solution** to a differential equation on an interval $I$ if, when $y$ and its derivatives are substituted into the equation, the equation is satisfied for all $x$ in the interval $I$.

The **general solution** of a differential equation represents all the solutions to the differential equation and is in the form of an equation with arbitrary constants. The number of arbitrary constants in the general solution agrees with the order of the differential equation.

As an example, consider the second-order differential equation

$$\frac{d^2y}{dx^2} + y = 0$$

It can be shown that the general solution has the form $y = C_1 \sin x + C_2 \cos x$, where $C_1$ and $C_2$ are constants.

Any solution of a differential equation obtained from the general solution by assigning values to the arbitrary constants is called a **particular solution**. So, for this example, choosing $C_1 = 6$ and $C_2 = 9$ yields the particular solution $y = 6 \sin x + 9 \cos x$.

CALC
CLIP

**EXAMPLE 2** Verifying Solutions to a Differential Equation

(a) Show that $y = f(x) = x^4 - 5x + 1$ is a solution of the differential equation $y'' - 12x^2 = 0$.

(b) Show that $y = g(x) = x^4 + C_1 x + C_2$, where $C_1$ and $C_2$ are constants, is a solution of the differential equation $y'' - 12x^2 = 0$.

**Solution** (a) For $y = f(x) = x^4 - 5x + 1$, we have

$$y' = 4x^3 - 5 \qquad y'' = 12x^2$$

The function $y = f(x)$ satisfies the differential equation $y'' - 12x^2 = 0$, so $y = f(x)$ is a solution.

(b) For $y = g(x) = x^4 + C_1 x + C_2$, we have

$$y' = 4x^3 + C_1 \qquad y'' = 12x^2$$

The function $y = g(x)$ satisfies the differential equation $y'' - 12x^2 = 0$, so $y = g(x)$ is a solution. ∎

Notice in Example 2 that if $C_1 = -5$ and $C_2 = 1$, the solution to (b) results in the solution to (a). It can be shown that $y = x^4 + C_1 x + C_2$ is the general solution of the differential equation $y'' - 12x^2 = 0$. So, $y = x^4 - 5x + 1$ is a particular solution.

**NOW WORK** Problem 21.

**EXAMPLE 3** Verifying a Solution to a Differential Equation

Show that $x^2 - x^3 y + 3y^4 = C$, where $C$ is a constant, is a solution* to the differential equation $\dfrac{dy}{dx} = \dfrac{3x^2 y - 2x}{12y^3 - x^3}$.

**NEED TO REVIEW?** Implicit differentiation is discussed in Section 3.2, pp. 219–222.

**Solution** Differentiate $x^2 - x^3 y + 3y^4 = C$ implicitly with respect to $x$ to find $\dfrac{dy}{dx}$.

$$2x - x^3 \frac{dy}{dx} - 3x^2 y + 12y^3 \frac{dy}{dx} = 0$$

$$(12y^3 - x^3)\frac{dy}{dx} = 3x^2 y - 2x$$

$$\frac{dy}{dx} = \frac{3x^2 y - 2x}{12y^3 - x^3}$$

The function $y = f(x)$ defined by the equation $x^2 - x^3 y + 3y^4 = C$ satisfies the first-order differential equation and so is a solution. ∎

**NOW WORK** Problem 23.

---

*It can be shown that this is the general solution of the differential equation.

## 16.1 Assess Your Understanding

### Concepts and Vocabulary

**1.** The differential equation $\dfrac{d^2y}{dx^2} + 5\left(\dfrac{dy}{dx}\right)^3 - y = 0$ is of order _____ and degree _____.

**2.** *True or False*   The differential equation $(x^2 + 2)\dfrac{dy}{dx} - \dfrac{1}{x}y = 0$ is linear.

**3.** *True or False*   $y = -4x^2 + 100$ is a solution of the differential equation $\dfrac{dy}{dx} = -8x$.

**4.** *True or False*   $y = 8e^{5x}$ is a solution of the differential equation $\dfrac{dy}{dx} + 3 = 8e^{5x}$.

### Skill Building

*In Problems 5–14, state the order and degree of each equation. State whether the equation is linear or nonlinear.*

**5.** $\dfrac{dy}{dx} + x^2 y = xe^x$

**6.** $\dfrac{d^3y}{dx^3} + 4\dfrac{d^2y}{dx^2} - 5\dfrac{dy}{dx} + 3y = \sin x$

**7.** $\dfrac{d^4y}{dx^4} + 3\dfrac{d^2y}{dx^2} + 5y = 0$

**8.** $\dfrac{d^2y}{dx^2} + y\sin x = 0$

**9.** $\dfrac{d^2y}{dx^2} + x\sin y = 0$

**10.** $\dfrac{d^6x}{dt^6} + \dfrac{d^4x}{dt^4} + \dfrac{d^3x}{dt^3} + x = t$

**11.** $\left(\dfrac{dr}{ds}\right)^2 = \left(\dfrac{d^2r}{ds^2}\right)^3 + 1$

**12.** $x(y'')^3 + (y')^4 - y = 0$

**13.** $\dfrac{d^2y}{ds^2} + 3s\dfrac{dy}{ds} = y$

**14.** $\dfrac{dy}{dx} = 1 - xy + y^2$

*In Problems 15–22, show that the function y is a solution of the differential equation.*

**15.** $y = e^x + 3e^{-x}$,   $\dfrac{d^2y}{dx^2} - y = 0$

**16.** $y = 5\sin x + 2\cos x$,   $\dfrac{d^2y}{dx^2} + y = 0$

**17.** $y = \sin(2x)$,   $\dfrac{d^2y}{dx^2} + 4y = 0$

**18.** $y = \cos(2x)$,   $\dfrac{d^2y}{dx^2} + 4y = 0$

**19.** $y = \dfrac{3}{1 - x^3}$,   $\dfrac{dy}{dx} = x^2 y^2$

**20.** $y = 1 + e^{-x^2/2}$,   $\dfrac{dy}{dx} + xy = x$

**21.** $y = e^{-2x}$,   $y'' + 3y' + 2y = 0$

**22.** $y = e^x$,   $2y'' + y' - 3y = 0$

*In Problems 23–30, use implicit differentiation to show that the given equation is a solution of the differential equation. C is a constant.*

**23.** $y^2 + 2xy - x^2 = C$,   $\dfrac{dy}{dx} = \dfrac{x - y}{x + y}$

**24.** $x^2 - y^2 = Cx^2 y^2$,   $\dfrac{dy}{dx} = \dfrac{y^3}{x^3}$   *Hint:* Solve for C first.

**25.** $x^2 - y^2 = Cx$,   $\dfrac{dy}{dx} = \dfrac{x^2 + y^2}{2xy}$   *Hint:* Solve for C first.

**26.** $x^2 y + 12x^2 + 16y = C$,   $\dfrac{dy}{dx} = -\dfrac{2xy + 24x}{x^2 + 16}$

**27.** $2(x - 1)e^x + y^2 = C$,   $\dfrac{dy}{dx} = -\dfrac{xe^x}{y}$

**28.** $x^2 - y^2 + 2(e^{-x} - e^y) = C$,   $\dfrac{dy}{dx} = \dfrac{x - e^{-x}}{y + e^y}$

**29.** $\tan^{-1} y = x + \dfrac{x^2}{2} + C$,   $\dfrac{dy}{dx} = xy^2 + y^2 + x + 1$

**30.** $1 + 2y^2\sqrt{1 + x^2} = Cy^2$,   $\dfrac{dy}{dx} = \dfrac{xy^3}{\sqrt{1 + x^2}}$

*Hint:* Solve for C first.

### Applications and Extensions

*In Problems 31–34, for each differential equation, show that y is a particular solution that satisfies the boundary condition.*

**31.** $\dfrac{dy}{dx} = \dfrac{y}{x}$;   $y = 3x$   boundary condition: $y = 3$ when $x = 1$

**32.** $\dfrac{dy}{dx} = 3y$;   $y = 2e^{3x}$   boundary condition: $y = 2$ when $x = 0$

**33.** $\dfrac{dy}{dx} = y^2$;   $y = \dfrac{1}{1 - x}$   boundary condition: $y = 1$ when $x = 0$

**34.** $x\dfrac{dy}{dx} + y = x^2$;   $y = \dfrac{x^2}{3} + \dfrac{3}{x}$
boundary condition: $y = 4$ when $x = 3$

*In Problems 35–38, show that y is a solution of the differential equation.*

**35.** $y = C_1 e^{ax} + C_2 e^{-ax}$,   $\dfrac{d^2y}{dx^2} - a^2 y = 0$,   $C_1, C_2$ are constants

**36.** $y = \sinh x$,   $\dfrac{d^2y}{dx^2} = y$

**37.** $y = \dfrac{a^2 kt}{1 + akt}$,   $\dfrac{dy}{dt} = k(a - y)^2$,   $a > 0$ is a constant

**38.** $y = \ln(C - e^{-x})$,   $\dfrac{dy}{dx} = e^{-(x+y)}$,   $C$ is a constant

**39.** Find the values of $n$ so that $y = e^{nx}$ is a solution of $y'' + y' - 6y = 0$

**40.** Find a first-order differential equation that has $y = e^x + e^{-x}$ as a solution.

**41. Schrödinger Equation**   In quantum mechanics, the time-independent **Schrödinger equation** in one dimension can be written as $-\dfrac{h^2}{2m}\dfrac{d^2Y(x)}{dx^2} + U(x)Y(x) = EY(x)$. What are the degree and order of this differential equation?

---

**1.** = NOW WORK problem     📉 = Graphing technology recommended     CAS = Computer Algebra System recommended

# 16.2 Separable and Homogeneous First-Order Differential Equations; Slope Fields; Euler's Method

**OBJECTIVES** *When you finish this section, you should be able to:*

1 Solve a separable first-order differential equation (p. 1116)
2 Identify a homogeneous function of degree $k$ (p. 1117)
3 Use a change of variables to solve a homogeneous first-order differential equation (p. 1118)
4 Find orthogonal trajectories (p. 1120)
5 Use a slope field to represent the solution of a first-order differential equation (p. 1122)
6 Use Euler's method to approximate a particular solution of a first-order differential equation (p. 1123)

In this section, we investigate first-order differential equations.

A first-order differential equation is usually written in the form

$$\frac{dy}{dx} = f(x, y)$$

where $f$ is continuous on its domain. If we treat $dy$ and $dx$ as differentials, we can write $\frac{dy}{dx} = f(x, y)$ in **differential form** as

$$M(x, y)\, dx + N(x, y)\, dy = 0$$

We begin our study of first-order differential equations by reviewing separable differential equations, which were discussed in Section 5.6, pp. 413–414.

## 1 Solve a Separable First-Order Differential Equation

**DEFINITION** Separable First-Order Differential Equation

A first-order differential equation is said to be **separable** if it can be written in the form

$$M(x)\, dx + N(y)\, dy = 0$$

where $M$ is a function of $x$ alone, $N$ is a function of $y$ alone, and both $M$ and $N$ are continuous on their domains.

The following steps are used to solve separable first-order differential equations:

**Steps for Solving a Separable First-Order Differential Equation**

**Step 1** Express the given equation in the differential form

$$M(x)\, dx + N(y)\, dy = 0$$

**Step 2** Integrate to obtain the general solution

$$\int M(x)\, dx + \int N(y)\, dy = C$$

where $C$ is the constant of integration.

### EXAMPLE 1　Solving a Separable First-Order Differential Equation

Solve $\dfrac{dy}{dx} = \dfrac{2e^x}{y^2}$

**Solution** Use the steps for solving a separable differential equation.

***Step 1*** Express $\dfrac{dy}{dx} = \dfrac{2e^x}{y^2}$ in the differential form:

$$2e^x\,dx - y^2\,dy = 0$$

***Step 2*** Integrate to obtain the general solution.

$$\int 2e^x\,dx - \int y^2\,dy = C \quad \text{so that} \quad 2e^x - \frac{y^3}{3} = C$$

where $C$ is a constant. This solution is expressed implicitly. To obtain the explicit form, solve for $y$.

$$y^3 = 6e^x + 3C \quad \text{so that} \quad y = \sqrt[3]{6e^x + 3C}$$

where $C$ is a constant. ∎

NOW WORK　Problems **7** and **17**.

Some first-order differential equations are not separable, but can be made separable by a change of variable. This is true for differential equations of the form $\dfrac{dy}{dx} = f(x, y)$, where $f$ is a *homogeneous function*.

## 2　Identify a Homogeneous Function of Degree $k$

**DEFINITION　Homogeneous Function of Degree $k$**

A function $f(x, y)$ is said to be a **homogeneous function of degree $k$ in $x$ and $y$** if, and only if, for $t > 0$,

$$\boxed{f(tx, ty) = t^k f(x, y)}$$

for some real number $k$.

### EXAMPLE 2　Identifying a Homogeneous Function of Degree $k$

**(a)** $f(x, y) = 3x^2 - xy + y^2$ is a homogeneous function of degree 2, since

$$f(tx, ty) = 3(tx)^2 - (tx)(ty) + (ty)^2 = t^2 3x^2 - t^2 xy + t^2 y^2$$
$$= t^2(3x^2 - xy + y^2) = t^2 f(x, y) \quad \text{for all } t > 0$$

**(b)** $f(x, y) = \sqrt{x + 4y}$ is a homogeneous function of degree $\dfrac{1}{2}$, since

$$f(tx, ty) = \sqrt{tx + 4(ty)} = \sqrt{t(x + 4y)} = \sqrt{t}\sqrt{x + 4y}$$
$$= t^{1/2} f(x, y) \quad \text{for all } t > 0$$

**(c)** $f(x, y) = \dfrac{x}{\sqrt{x^2 - y^2}}$ is a homogeneous function of degree 0, since

$$f(tx, ty) = \frac{tx}{\sqrt{(tx)^2 - (ty)^2}} = \frac{tx}{\sqrt{t^2(x^2 - y^2)}} = \frac{tx}{t\sqrt{x^2 - y^2}}$$
$$= t^0 f(x, y) \quad \text{for all } t > 0$$

**(d)** $f(x, y) = x - y^2$ is not a homogeneous function, since

$$f(tx, ty) = tx - (ty)^2 = t(x - ty^2) \neq t^k(x - y^2)$$

for all $t > 0$ and some $k$. ∎

NOW WORK　Problem **23**.

## 3 Use a Change of Variables to Solve a Homogeneous First-Order Differential Equation

**DEFINITION** Homogeneous First-Order Differential Equation

A first-order differential equation of the form

$$M(x, y)\, dx + N(x, y)\, dy = 0$$

is said to be **homogeneous** if $M$ and $N$ are homogeneous functions of the *same* degree.

**NOTE** The word "homogeneous" is used in two different ways: to describe a property of a function and to identify a certain type of first-order differential equation. Context will make it clear which meaning applies.

To solve first-order homogeneous differential equations, we use the following change of variables theorem.

**THEOREM** Change of Variables for Homogeneous First-Order Differential Equations

If $M(x, y)\, dx + N(x, y)\, dy = 0$ is a homogeneous first-order differential equation, then it can be transformed into a first-order differential equation whose variables are separable by using the substitution

$$y = xv(x)$$

where $v = v(x)$ is a differentiable function of $x$.

**NOTE** For simplicity of notation, we usually write $v$ instead of $v(x)$.

**Proof** If $M(x, y)\, dx + N(x, y)\, dy = 0$ is a homogeneous first-order differential equation, use the substitution $y = xv$, where $v = v(x)$ is a differentiable function of $x$. Then $dy = x\, dv + v\, dx$ and

$$M(x, y)\, dx + N(x, y)\, dy = M(x, xv)\, dx + N(x, xv)(x\, dv + v\, dx) = 0$$

Since $M$ and $N$ are each homogeneous functions of degree $k$, it follows that for some number $k$

$$x^k M(1, v)\, dx + x^k N(1, v)(x\, dv + v\, dx) = 0 \qquad M(x, xv) = x^k M(1, v),$$
$$N(x, xv) = x^k N(1, v)$$

$$x^k [M(1, v)\, dx + N(1, v)(x\, dv + v\, dx)] = 0$$

$$M(1, v)\, dx + N(1, v)x\, dv + N(1, v)v\, dx = 0 \qquad \text{Divide out } x^k.$$

$$[M(1, v) + vN(1, v)]\, dx + N(1, v)x\, dv = 0$$

$$\frac{dx}{x} + \frac{N(1, v)}{M(1, v) + vN(1, v)}\, dv = 0$$

provided neither denominator is 0. This first-order differential equation is separable. ∎

**Steps for Solving a Homogeneous First-Order Differential Equation $M(x, y)\, dx + N(x, y)\, dy = 0$**

*Step 1* Confirm that the functions $M$ and $N$ are homogeneous functions of the same degree.

*Step 2* Let $y = xv$. Substitute for $y$ and $dy = x\, dv + v\, dx$.

*Step 3* Express the new equation in differential form. The variables will be separable.

*Step 4* Integrate to obtain the general solution and replace $v$ by $\dfrac{y}{x}$.

**EXAMPLE 3** **Solving a Homogeneous First-Order Differential Equation**

Solve the differential equation $(x^2 - 3y^2)\,dx + 2xy\,dy = 0$.

**Solution** Follow the steps for solving a homogeneous first-order differential equation.

**Step 1** Both $x^2 - 3y^2$ and $2xy$ are homogeneous functions of degree 2. Do you see why?

**Step 2** Let $y = xv$. Then $dy = x\,dv + v\,dx$. Substitute these into the differential equation:

$$(x^2 - 3y^2)\,dx + 2xy\,dy = 0$$

$$[x^2 - 3(xv)^2]\,dx + 2x(xv)(x\,dv + v\,dx) = 0 \qquad \boldsymbol{dy = x\,dv + v\,dx}$$

$$(x^2 - x^2v^2)\,dx + 2x^3v\,dv = 0$$

$$x^2(1 - v^2)\,dx + 2x^3v\,dv = 0$$

**Step 3** Then for $x \neq 0$ and $v \neq \pm 1$, we can separate the variables by dividing by $x^3(1 - v^2)$.

$$\frac{x^2(1 - v^2)\,dx}{x^3(1 - v^2)} + \frac{2x^3v\,dv}{x^3(1 - v^2)} = 0$$

$$\frac{dx}{x} + \frac{2v\,dv}{1 - v^2} = 0$$

**Step 4** Integrate.

$$\int \frac{dx}{x} - \int \frac{2v\,dv}{v^2 - 1} = C_1$$

$$\ln|x| - \ln|v^2 - 1| = C_1$$

$$\ln\left|\frac{x}{v^2 - 1}\right| = C_1$$

Since $C_1$ is a constant, we can write $C_1 = \ln C_2$, where $C_2$ is a constant. Then

$$\ln\left|\frac{x}{v^2 - 1}\right| = \ln C_2$$

$$\left|\frac{x}{v^2 - 1}\right| = C_2$$

$$\left|\frac{x}{\dfrac{y^2}{x^2} - 1}\right| = C_2 \qquad v = \frac{y}{x}$$

$$\left|\frac{x^3}{y^2 - x^2}\right| = C_2$$

$$x^3 = \pm C_2(y^2 - x^2)$$

Since $\pm C_2$ may be either positive or negative, the general solution can be written as $x^3 = C(y^2 - x^2)$, where $C \neq 0$ is a constant. ∎

NOW WORK Problem **39**.

EXAMPLE 4 **Solving a Homogeneous First-Order Differential Equation**

Solve the differential equation $x \, dy + (2xe^{y/x} - y) \, dx = 0$ if $y = 0$ when $x = 1$.

**Solution** Follow the steps for solving a homogeneous first-order differential equation:

**Step 1** Both $x$ and $2xe^{y/x} - y$ are homogeneous of degree 1.

**Step 2** Let $y = xv$. Then $dy = x \, dv + v \, dx$. Substitute these into the differential equation.

$$x \, dy + (2xe^{y/x} - y) \, dx = 0$$
$$x(x \, dv + v \, dx) + (2xe^v - xv) \, dx = 0$$
$$x^2 dv + xv \, dx + 2xe^v dx - xv \, dx = 0$$
$$x^2 dv + 2xe^v dx = 0$$

**Step 3** To separate the variables, divide by $x^2 e^v$. Then for $x \neq 0$,

$$\frac{2 \, dx}{x} + \frac{dv}{e^v} = 0$$

**Step 4** Integrate.

$$\int \frac{2dx}{x} + \int \frac{dv}{e^v} = C$$
$$2 \int \frac{dx}{x} + \int e^{-v} dv = C$$
$$2 \ln |x| - e^{-v} = C$$
$$2 \ln |x| - e^{-y/x} = C \qquad v = \frac{y}{x}$$

This is the general solution to the differential equation. To find the particular solution, substitute $x = 1$ and $y = 0$ to find $C$.

$$2 \ln 1 - e^0 = C$$
$$C = -1$$

The particular solution is

$$2 \ln |x| - e^{-y/x} = -1 \qquad \blacksquare$$

NOW WORK **Problem 43.**

## 4 Find Orthogonal Trajectories

Consider the one-parameter family of circles

$$(x - 1)^2 + (y - 2)^2 = C \qquad C > 0$$

with center at the point $(1, 2)$ and radius $\sqrt{C}$. Figure 1 shows some members of this family of circles. If the equation describing the circles is differentiated with respect to $x$, we obtain

$$2(x - 1) + 2(y - 2)y' = 0$$
$$y' = -\frac{x - 1}{y - 2} \qquad (1)$$

which represents the differential equation of the family of circles.

For a specific point $(x, y)$, $y \neq 2$, on any one of the circles, the differential equation $y' = -\dfrac{x - 1}{y - 2}$ gives the slope of the tangent line to the circle at the point $(x, y)$.

Now consider another example. The family of nonvertical lines passing through the point $(1, 2)$ satisfies the equation

$$y - 2 = m(x - 1)$$

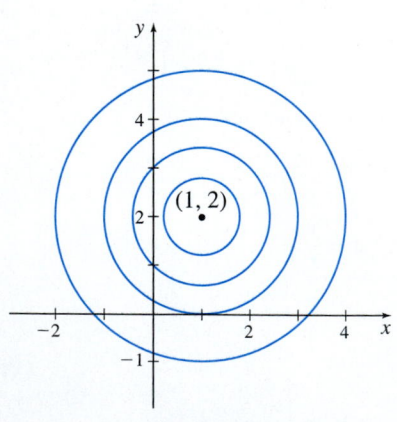

DF **Figure 1** $(x - 1)^2 + (y - 2)^2 = C, C > 0$

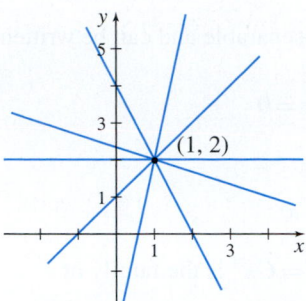

**Figure 2** $y - 2 = m(x - 1)$

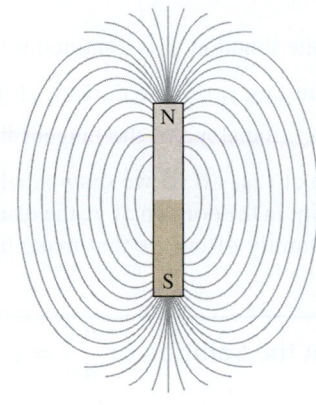

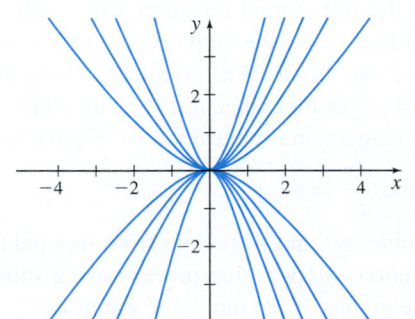

**Figure 3** $y^2 = Cx^3$

where $m$ is the slope of a particular member of the family of lines. The differential equation for this family of lines is

$$y' = m = \frac{y - 2}{x - 1} \qquad (2)$$

Figure 2 shows some members of this family of lines.

If we compare the differential equations (1) and (2), we see that the right side of (1) is the negative reciprocal of the right side of (2). From this, we conclude that if $(x, y)$ is a point of intersection of one of the circles and one of the lines from each family, then the line and the circle are perpendicular (orthogonal) to each other at the point of intersection. Each line in the family $y - 2 = m(x - 1)$ is an *orthogonal trajectory* of the family of circles, and conversely, each circle in the family $(x - 1)^2 + (y - 2)^2 = C$ is an orthogonal trajectory of the family of lines.

In general, if $F(x, y, C) = 0$ and $G(x, y, K) = 0$ are one-parameter families of curves, in which each member of one family intersects the members of the other family at a right angle, then the two families are said to be **orthogonal trajectories** of each other.

Orthogonal trajectories occur naturally. For example, iron filings sprinkled on a pane of glass over a bar magnet arrange themselves along a family of curved lines, indicating the direction of magnetic force. A family of orthogonal trajectories can be determined by locating all the points in the plane of the glass with the same potential energy.

To find a family $G(x, y, K) = 0$ of orthogonal trajectories for a given family $F(x, y, C) = 0$, we use the steps below:

### Steps for Finding a Family of Orthogonal Trajectories

**Step 1** Find the differential equation for the given family.

**Step 2** Replace $y'$ in this differential equation by $-\dfrac{1}{y'}$; the resulting equation is the differential equation for the family of orthogonal trajectories.

**Step 3** Find the solution of the differential equation for the family of orthogonal trajectories obtained in Step 2.

### EXAMPLE 5  Finding a Family of Orthogonal Trajectories

Find the family of orthogonal trajectories for the one-parameter family $y^2 = Cx^3$. See Figure 3 for several graphs in the family.

#### Solution

**Step 1** Differentiate $y^2 = Cx^3$ with respect to $x$.

$$2yy' = 3Cx^2$$

Then eliminate $C$ using this equation and the equation $y^2 = Cx^3$ of the family. Since $C = \dfrac{y^2}{x^3}$, we find

$$2yy' = 3\frac{y^2}{x^3}x^2$$

$$y' = \frac{3y}{2x}$$

**Step 2** Now replace $y'$ with $-\dfrac{1}{y'}$, to obtain the differential equation for the family of orthogonal trajectories.

$$-\frac{1}{y'} = \frac{3y}{2x}$$

$$y' = -\frac{2x}{3y}$$

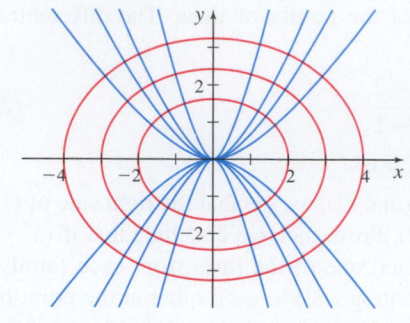

**Figure 4** $y^2 = Cx^3$; $x^2 + \dfrac{3}{2}y^2 = K$

**Step 3** The differential equation $y' = \dfrac{dy}{dx} = -\dfrac{2x}{3y}$ is separable and can be written as

$$2x\,dx + 3y\,dy = 0$$

The general solution is

$$x^2 + \frac{3}{2}y^2 = K$$

The orthogonal trajectories for the family $y^2 = Cx^3$ is the family of ellipses $x^2 + \dfrac{3}{2}y^2 = K$, as shown in Figure 4. ∎

**NOW WORK** Problems 57 and 63.

## 5 Use a Slope Field to Represent the Solution of a First-Order Differential Equation

In a first-order differential equation $\dfrac{dy}{dx} = f(x, y)$, the function $f$ is continuous on its domain, a set of points in the $xy$-plane. At each point $(x_0, y_0)$ in the domain of $f$, we can evaluate $f(x_0, y_0)$ and determine $\dfrac{dy}{dx} = f(x_0, y_0)$, the slope of the tangent line to the graph of the solution to the differential equation at $(x_0, y_0)$. If we draw small line segments representing these tangent lines at every point in the domain of $f$, the resulting display, called a **slope field**, gives a picture of the family of curves that make up the solution of the differential equation.

**EXAMPLE 6** Using a Slope Field to Represent the Solution of $\dfrac{dy}{dx} = y$

(a) Use a slope field to represent the solution of the differential equation $\dfrac{dy}{dx} = y$.

(b) Use the slope field representation to draw some graphs of the family of solutions of the differential equation.

**Solution** (a) Since $\dfrac{dy}{dx} = y$ does not depend on $x$, the derivatives $\dfrac{dy}{dx}$ at all points of the form $(x, y_0)$ will be equal. For example, at all points $(x, 2)$, the slopes of the tangent lines to the graphs of the solution to the differential equation will equal 2. Since the points $(x, 2)$ all lie on the horizontal line $y = 2$, the slope field along $y = 2$ will consist of parallel tangent lines with slope 2. See Figure 5(a). Similarly, along the horizontal line $y = 1$, the slope field will consist of parallel tangent lines with slope 1; along $y = -1$, the slope field consists of parallel tangent lines with slope $-1$. Figure 5(b) illustrates a slope field of the differential equation $\dfrac{dy}{dx} = y$.

(b) Pick a point in the $xy$-plane, say $(0, 2)$. Going left and then right from this point, follow the tangent lines and insert a smooth curve. Repeat this process using other points. See Figure 5(c), which shows four of the graphs of the family of solutions.

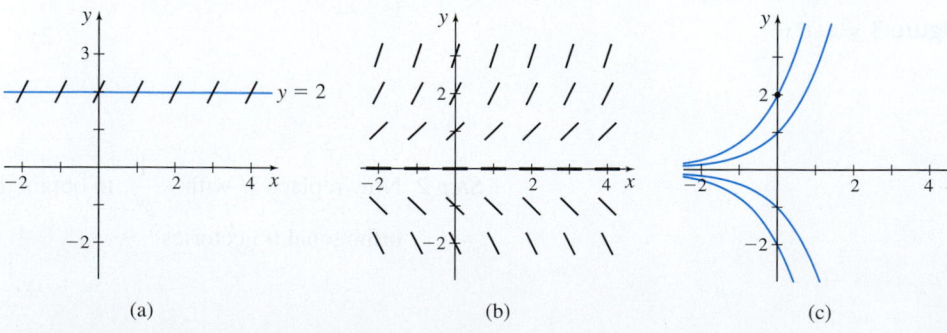

(a)  (b)  (c)

**Figure 5**

**NOW WORK** Problem 45.

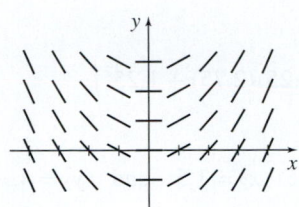

**Figure 6**

**EXAMPLE 7** **Analyzing a Slope Field**

Which of the following differential equations could have the slope field shown in Figure 6?

**(a)** $\dfrac{dy}{dx} = -x$    **(b)** $\dfrac{dy}{dx} = x + y$    **(c)** $\dfrac{dy}{dx} = x$    **(d)** $\dfrac{dy}{dx} = x - y$

**Solution** The tangent lines that lie on a vertical line all have the same slope. That is, for a fixed number $x$, the slopes $\dfrac{dy}{dx}$ of the tangent lines to the graphs of the solutions of the differential equation are the same. This means $\dfrac{dy}{dx}$ does not depend on $y$, so the differential equation does not contain $y$ and choices **(b)** and **(d)** can be eliminated. Points in quadrants II and III give rise to negative slopes and points in quadrants I and IV give rise to positive slopes. The differential equation in **(c)** is the only choice that satisfies these conditions. ∎

**NOW WORK** Problem 51.

## 6 Use Euler's Method to Approximate a Particular Solution of a First-Order Differential Equation

**Euler's method** is a numerical algorithm for approximating a particular solution $y = y(x)$ of a first-order differential equation $\dfrac{dy}{dx} = f(x, y)$ with the boundary condition $y = y_0$ when $x = x_0$. It begins with the boundary condition $(x_0, y_0)$ and a small arbitrary increment $h$ in $x$. The first approximation to $y = y(x)$ is given by

$$y_1 = y_0 + hf(x_0, y_0)$$

The second approximation uses $y_1$ from the first approximation and $x_1 = x_0 + h$. Then

$$y_2 = y_1 + hf(x_1, y_1) \qquad x_1 = x_0 + h$$

Each successive approximation follows the same pattern. The third approximation uses $y_2$ and $x_2 = x_0 + 2h$, and so on.

$$y_3 = y_2 + hf(x_2, y_2) \qquad\qquad x_2 = x_1 + h = x_0 + 2h$$
$$y_4 = y_3 + hf(x_3, y_3) \qquad\qquad x_3 = x_2 + h = x_0 + 3h$$
$$\vdots$$
$$y_n = y_{n-1} + hf(x_{n-1}, y_{n-1}) \qquad x_{n-1} = x_0 + (n-1)h$$

Each successive approximation follows the tangent line to the graph of the solution to the differential equation. That is, it follows the slope field. See Figure 7.

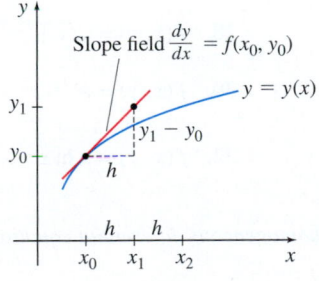

**Figure 7** $y_1 = y_0 + hf(x_0, y_0)$

**EXAMPLE 8** **Using Euler's Method**

Suppose $y = y(x)$ is the solution to the differential equation

$$\frac{dy}{dx} = xy + x^2$$

with the boundary condition $y = 2$ when $x = 1$. Use Euler's method to approximate $y(1.75)$ using $h = 0.25$ as the increment.

**Solution** Begin the first approximation with the boundary condition $x_0 = 1$, and $y_0 = 2$ and the increment $h = 0.25$. Then $x_1 = x_0 + h = 1.25$ and

$$y(x_1) = y(1.25) \approx y_1 = y_0 + hf(x_0, y_0)$$
$$= 2 + 0.25(1 \cdot 2 + 1^2) \qquad f(x, y) = xy + x^2$$
$$= 2.75$$

For the second approximation, we use $x_1 = 1.25$ and $y_1 = 2.75$.

Then $x_2 = x_1 + 0.25 = 1.5$, and

$$y(x_2) = y(1.5) \approx y_2 = y_1 + hf(x_1, y_1)$$
$$= 2.75 + 0.25[(1.25)(2.75) + 1.25^2]$$
$$= 4$$

For the third approximation $y_3$, we use $x_2 = 1.5$ and $y_2 = 4$. Then $x_3 = x_2 + 0.25 = 1.75$, and

$$y(x_3) = y(1.75) \approx y_3 = y_2 + hf(x_2, y_2)$$
$$= 4 + 0.25[(1.5)(4) + 1.5^2]$$
$$= 6.0625$$            ∎

NOW WORK Problem 53.

## 16.2 Assess Your Understanding

### Concepts and Vocabulary

1. The general solution of the differential equation $\dfrac{dy}{dx} = xy$ is _____.

2. *True or False* If $f(tx, ty) = t^k f(x, y)$, for all $t > 0$, where $k$ is a real number, then the function $f$ is said to be homogeneous of degree $k$ in $x$ and $y$.

3. If $F(x, y, C) = 0$ and $G(x, y, K) = 0$ are one-parameter families of curves in which each member of one family intersects the members of the other family orthogonally, then the two families are said to be orthogonal _____ of each other.

4. If $M(x, y)\,dx + N(x, y)\,dy = 0$ is a homogeneous differential equation, then it can be transformed into an equation whose variables are separable by using the substitution _____, where _____ is a differentiable function of $x$.

5. A(n) _____ gives a picture of the family of curves that make up the solution to a first-order differential equation.

6. *True or False* Euler's method is used to approximate a particular solution of a first-order differential equation.

18. $\cos y \dfrac{dy}{dx} = \dfrac{2}{x}$,   $y = \dfrac{\pi}{3}$ when $x = -1$

19. $\dfrac{dy}{dx} = e^{y-x}$;   $y = 0$ when $x = 0$

20. $x\dfrac{dy}{dx} + 2y = 5$;   $y = 1$ when $x = -1$

21. $\dfrac{dy}{dx} = \cos x \dfrac{dy}{dx} - y \sin x$;   $y = \dfrac{1}{2}$ when $x = \pi$

22. $\dfrac{dy}{dx} = x + y\dfrac{dy}{dx}$,   $y = 1$ when $x = 2$

*In Problems 23–32, determine if each function is homogeneous. If it is, find its degree.*

23. $f(x, y) = 2x^2 - 3xy - y^2$          24. $f(x, y) = x^3 - xy^2 + y^3$

25. $f(x, y) = x^3 - xy + y^3$          26. $f(x, y) = x^2 - xy^2 + y^2$

27. $f(x, y) = 2x + \sqrt{x^2 + y^2}$          28. $f(x, y) = \sqrt{x + y}$

29. $f(x, y) = \tan \dfrac{3x}{y}$          30. $f(x, y) = e^{x/y}$

31. $f(x, y) = \ln \dfrac{x}{y}$          32. $f(x, y) = x \ln x - x \ln y$

### Skill Building

*In Problems 7–16, solve each differential equation.*

7. $\dfrac{dy}{dx} = x \sec y$          8. $\cos y \dfrac{dy}{dx} = \dfrac{2}{x}$

9. $\dfrac{dy}{dx} = e^{y-x}$          10. $x\dfrac{dy}{dx} + 2y = 5$

11. $\dfrac{dy}{dx} = \cos x \dfrac{dy}{dx} - y \sin x$          12. $\dfrac{dy}{dx} = x + y\dfrac{dy}{dx}$

13. $\dfrac{dy}{dx} + xy = x$          14. $(3x + 1)\,dx + e^{x+y}dy = 0$

15. $\ln x \dfrac{dx}{dy} = \dfrac{x}{y}$          16. $\dfrac{dy}{dx} = \dfrac{x+2}{2-y}$

*In Problems 17–22, obtain a particular solution of each differential equation. (Use the general solutions obtained for Problems 7–12.)*

17. $\dfrac{dy}{dx} = x \sec y$,   $y = \dfrac{\pi}{4}$ when $x = 1$

*In Problems 33–44, solve each homogeneous differential equation. Follow the steps on p. 1118.*

33. $(x - y)\,dx + x\,dy = 0$          34. $(x + y)\,dx + x\,dy = 0$

35. $(x^2 + y^2)\,dx + (x^2 - xy)\,dy = 0$          36. $xy\,dx + (x^2 + y^2)\,dy = 0$

37. $\dfrac{dy}{dx} = \dfrac{y - x}{y + x}$          38. $\dfrac{dy}{dx} = \dfrac{x + 2y}{2x + y}$

39. $x\dfrac{dy}{dx} = x + y$

40. $x(x^2 - y^2)\dfrac{dy}{dx} - y(x^2 + y^2) = 0$          41. $\dfrac{dy}{dx} = \dfrac{2xy}{x^2 + y^2}$

42. $\dfrac{dy}{dx} = \dfrac{x + 2y}{2x - y}$

43. $x^2\dfrac{dy}{dx} = x^2 + xy - 4y^2$;   $y = 1$ when $x = -1$

44. $x(x^2 - y^2)\dfrac{dy}{dx} - y(x^2 + y^2) = 0$;   $y = -2$ when $x = -1$

---

**1.** = NOW WORK problem          ⟋⟍ = Graphing technology recommended          CAS = Computer Algebra System recommended

*In Problems 45–50,*

*(a) Use a slope field to represent the solution of each differential equation.*

*(b) Use the slope field representation to draw some graphs of the family of solutions to the differential equation.*

**45.** $\dfrac{dy}{dx} = -2y$   **46.** $\dfrac{dy}{dx} = 2x$   **47.** $\dfrac{dy}{dx} = x^2$

**48.** $\dfrac{dy}{dx} = y^2$   **49.** $\dfrac{dy}{dx} = x + y$   **50.** $\dfrac{dy}{dx} = x - y$

**51.** Which of the following differential equations could have the slope field shown below?

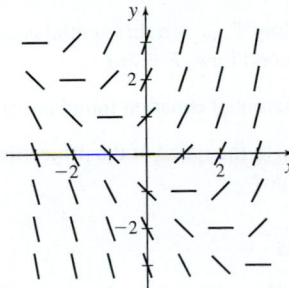

(a) $\dfrac{dy}{dx} = -x$   (b) $\dfrac{dy}{dx} = x + y$   (c) $\dfrac{dy}{dx} = x$   (d) $\dfrac{dy}{dx} = x - y$

**52.** Which of the following differential equations could have the slope field shown below?

(a) $\dfrac{dy}{dx} = -x$   (b) $\dfrac{dy}{dx} = x^2$   (c) $\dfrac{dy}{dx} = 2x + 1$   (d) $\dfrac{dy}{dx} = -x^2$

*In Problems 53–56, use Euler's method to approximate the particular solution to the differential equation with the given boundary condition. Assume $y = y(x)$ is a solution of each differential equation.*

**53.** Approximate $y(0.4)$ if $\dfrac{dy}{dx} = x^2 - y$ and $y = 1$ when $x = 0$.

Use $h = 0.2$ as the increment.

**54.** Approximate $y(0.2)$ if $\dfrac{dy}{dx} = y^2 - x$ and $y = 1$ when $x = 0$.

Use $h = 0.1$ as the increment.

**55.** Approximate $y(1.4)$ if $\dfrac{dy}{dx} = xy^2 - x$ and $y = 2$ when $x = 1$.

Use $h = 0.2$ as the increment.

**56.** Approximate $y(1.2)$ if $\dfrac{dy}{dx} = xy^2 - x$ and $y = 2$ when $x = 1$.

Use $h = 0.1$ as the increment.

*In Problems 57–60, find the orthogonal trajectories of each family of curves. Graph the members of each family that contain the point (2, 1).*

**57.** $xy = C$   **58.** $y^2 = 2(C - x)$

**59.** $y = Cx^2$   **60.** $x^2 = Cy^3$

## Applications and Extensions

*In Problems 61 and 62, solve each differential equation by letting $x = u + h$ and $y = v + k$, where $h$ and $k$ are constants chosen to eliminate the constant terms on the right side.*

**61.** $\dfrac{dy}{dx} = \dfrac{y - x - 3}{y + x + 4}$   **62.** $\dfrac{dy}{dx} = \dfrac{x + 2y - 3}{2x - y + 1}$;   $y = 4$ when $x = 3$

**63. Orthogonal Families**   Find the orthogonal trajectories of the family of parabolas $y^2 = 4Cx$.

**64. Orthogonal Trajectories**   Find the orthogonal trajectories of all circles tangent to the $x$-axis at $(3, 0)$.

**65. Sales Growth**   The annual sales of a new company are expected to grow at a rate that is proportional to the difference between the sales and an upper limit of $20 million. Sales are $0 initially and are $3 million for the second year.

(a) Determine the annual sales at any time $t$.

(b) According to the model, what will sales be during the eighth year of operations?

(c) How long will it take for sales to reach $12 million?

(d) Graph the annual sales of the company for its first 20 years of operation.

**66. Bacteria Growth**   A culture of bacteria is growing in a medium that can support a maximum of 1,000,000 bacteria. The rate of growth of the population at time $t$ is proportional to the difference between 1,000,000 and the number present at time $t$. The culture contains 1000 bacteria initially, and after 1 h there are 1500 bacteria.

(a) How many bacteria will there be after 5 h?

(b) When will the culture contain 9950 bacteria?

**67. Learning Curve**   The number of words per minute $W$ that a person can text increases with practice. Assume that the rate of change of $W$ is proportional to the difference between an upper limit of 150 words per minute and $W$. That is, the rate of change is proportional to $150 - W$. Suppose a beginner cannot text at all, that is, $W = 0$ when $t = 0$, and can text 30 words per minute after 10 h of practice.

(a) How many words per minute can he text after 25 h of practice?

(b) How many hours of practice will be required in order for him to text 100 words per minute?

(c) Graph the number of words per minute $W$ a beginner can text as a function of hours of practice.

**68. Drug Concentration**    A drug is injected into a patient's bloodstream at a constant rate of $r$ mg/s. Simultaneously, the drug is removed from the bloodstream at a rate proportional to the amount $y(t)$ present at time $t$.

(a) Find a differential equation that models this problem.

(b) Find a solution of the differential equation in (a).

(c) Find the particular solution that satisfies the initial condition $y(0) = 0$.

**69. Friction**    The frictional force $f$ on an object sliding over a surface depends on the speed $v$ of the object and can be modeled by the function $f = -Av$, where $A$ is a positive constant that depends on the two surfaces. *Note:* The force is negative because the direction of the friction is opposite to the direction of the velocity.

(a) Express the frictional force $f$ as a differential equation by using Newton's Second Law, $F = ma$.

(b) Solve the differential equation found in (a) for $v(t)$. Assume the object slides over the surface with an initial speed $v_0$.

(c) Suppose that the frictional force on a 25-kg sled is 150 N when the speed of the sled is 10 m/s. Determine the value of the constant $A$.

(d) How long will it take the sled to slow down to 10% of its initial speed?

(e) How far will the sled travel while slowing down to 1.0 m/s? Assume an initial speed of 10 m/s.

**70. Friction**    The frictional force $f$ on an object sliding over a surface depends on the speed $v$ of the object and can be modeled by the function $f = -B\sqrt{v}$, where $B$ is a positive constant that depends on the two surfaces. *Note:* The force is negative because the direction of the friction is opposite to the direction of the velocity.

(a) Express the frictional force $f$ as a differential equation by using Newton's Second Law, $F = ma$.

(b) Solve the differential equation found in (a) for $v(t)$. Assume the object slides over the surface with an initial speed $v_0$.

(c) Suppose that the frictional force on a 25-kg sled is 150 N when the speed of the sled is 10 m/s. Determine the value of the constant $B$.

(d) How long will it take the sled to slow down to 10% of its initial speed?

(e) How far will the sled travel while slowing down to 1.0 m/s? Assume an initial speed of 10 m/s.

**71. Electronics: Discharging a Capacitor**    In a circuit, a capacitor $C$ carries an initial charge $q_0$ and discharges through a resistor $R$. Using Kirchhoff's rules with this circuit, we obtain the differential equation

$$\frac{dq}{dt} + \frac{q}{RC} = 0$$

where $q$ is the charge, in Coulombs, on the capacitor at any time $t$, and $R$ and $C$ are constants.

(a) Solve the differential equation for the charge $q$ as a function of time $t$.

(b) In a typical laboratory circuit, $R = 150\,\Omega$ (ohms) and $C = 5.0\,\mu\text{F}$ (microfarads). How long will it take the discharging capacitor to lose 99% of its initial charge? *Hint:* The product $(1\,\Omega)(1\,\text{F}) = 1$ second.

**72. Air Resistance**    At high speeds, air resistance (**drag**) can be modeled by $F_{\text{drag}} = -Av^2$, where $A$ is a positive constant that depends on the shape of the object, its surface texture, and the conditions of the air. *Note:* The force is negative because the direction of the drag is opposite to the direction of the velocity. Suppose, in a test facility, a projectile of mass $m$ is launched with initial horizontal speed $v_0$ on a frictionless horizontal track. The only force opposing the motion of the projectile is the drag.

(a) Express the drag $F_{\text{drag}}$ as a differential equation by using Newton's Second Law, $F = ma$.

(b) Solve the differential equation found in (a) for $v(t)$.

(c) What happens to the speed of the projectile as $t$ becomes arbitrarily large?

## Challenge Problems

**73. Orthogonal Families**    Find the orthogonal trajectories of the family of hyperbolas $\dfrac{x^2}{C^2} - \dfrac{y^2}{4 - C^2} = 1$, where $0 < C < 2$.

**74. Orthogonal Trajectories**    Find the orthogonal trajectories of the family of parabolas with its vertex at $(1, 2)$ having a vertical axis. Graph the member of each family that contains the point $(2, 3)$.

**75. Velocity of a Projectile**    Near Earth's surface the attraction due to gravity is practically constant, but according to Newton's law, the force of attraction exerted by Earth on a given body is $\dfrac{k}{x^2}$, where $x$ is the distance of the body from the center of the Earth.

(a) Show that if a projectile is shot upward from Earth's surface with initial velocity $v_0$, and if air resistance is neglected, then

$$v^2 = v_0^2 - 2Rg + \frac{2R^2g}{x}$$

where $R = 3960$ mi is the radius of Earth. *Hint:* $\dfrac{d^2x}{dt^2} = \dfrac{dv}{dt} = \dfrac{dv}{dx}\dfrac{dx}{dt} = v\dfrac{dv}{dx}$

(b) Find the smallest value of $v_0$ necessary to keep the projectile going indefinitely (neglecting the attractive forces due to celestial bodies other than Earth). This is called the **escape velocity** for Earth.

**76. Rate of Flow**    From physics, it is known that the rate of flow of water from a tank through an orifice of area $A$ square feet is approximately $\dfrac{dV}{dt} = kA\sqrt{2gh}$, where $k = 0.6$, $g = 32$ ft/s$^2$, and $h$ is the height of the water level, in feet, above the orifice. Suppose a cylindrical tank with height 6 ft and diameter 3 ft is filled with water. At time $t = 0$, a valve in the bottom of the tank is opened and the water begins draining through a circular hole of diameter 1 in.

**(a)** Express the height of the water in the tank as a function of time $t$.

**(b)** How much time is required to empty the tank?

**77. Rate of Flow** If the water in a tank leaks out through a small hole in the bottom, then the rate of flow is proportional to the square root of the height of the water in the tank. Prove that if the

tank is a cylinder with a vertical axis, then the time required for three-fourths of the water to leak out is equal to the time required for the remaining one-fourth of the water to leak out.

**78. Water Level** A cylindrical tank has a leak in the bottom, and water flows out at a rate proportional to the pressure at the bottom. If the tank loses 2% of its water in 24 h, when will it be half empty?

# 16.3 Exact Differential Equations

**OBJECTIVE** *When you finish this section, you should be able to:*

**1** Identify and solve an exact differential equation (p. 1128)

When a first-order differential equation is separable or can be made separable, a solution can be found by integrating. Under certain conditions, this idea can be extended to differential equations of the form

$$M(x, y)\, dx + N(x, y)\, dy = 0$$

in which separation of variables is not possible.

**DEFINITION  Exact Differential Equation**

The differential equation

$$M(x, y)\, dx + N(x, y)\, dy = 0$$

is called an **exact differential equation** in a simply connected, open set $R$ if there is a function $z = f(x, y)$ of two variables with continuous partial derivatives on $R$ for which

$$dz = M(x, y)\, dx + N(x, y)\, dy$$

for all $(x, y)$ in $R$.

**NEED TO REVIEW?** The differential of $z = f(x, y)$ is discussed in Section 12.4, p. 884.

If the differential equation $M(x, y)\, dx + N(x, y)\, dy = 0$ is exact, then the differential of $z = f(x, y)$ is given by

$$dz = \frac{\partial f}{\partial x} dx + \frac{\partial f}{\partial y} dy = M(x, y)\, dx + N(x, y)\, dy$$

In other words, $M(x, y)\, dx + N(x, y)\, dy = 0$ is an exact differential equation if there is a potential function $z = f(x, y)$ whose gradient is $\nabla f = M(x, y)\mathbf{i} + N(x, y)\mathbf{j}$. This happens when the force field given by $\mathbf{F} = M(x, y)\mathbf{i} + N(x, y)\mathbf{j}$ is conservative.

**NEED TO REVIEW?** Gradients are discussed in Section 13.1, pp. 911–917. The Conservative Vector Field Test is discussed in Section 15.4, p. 1046.

**THEOREM  Test for Exactness**

A differential equation of the form

$$M(x, y)\, dx + N(x, y)\, dy = 0$$

is an exact differential equation at every point $(x, y)$ in a simply connected open set $R$ if and only if

$$\frac{\partial M}{\partial y} = \frac{\partial N}{\partial x}$$

for all $(x, y)$ in $R$.

So, if $\dfrac{\partial M}{\partial y} = \dfrac{\partial N}{\partial x}$, then the differential equation $M(x, y)\, dx + N(x, y)\, dy = 0$ is

exact, and it can be written in the equivalent form $\dfrac{\partial f}{\partial x} dx + \dfrac{\partial f}{\partial y} dy = 0$, where $z = f(x, y)$

is a function of two variables with the property that $\dfrac{\partial f}{\partial x} = M$ and $\dfrac{\partial f}{\partial y} = N$.

But $\dfrac{\partial f}{\partial x} dx + \dfrac{\partial f}{\partial y} dy$ is the total differential of $z = f(x, y)$, and so $f_x\, dx + f_y\, dy = 0$

is equivalent to $dz = 0$, whose solution is $f(x, y) = C$, $C$ a constant.

## 1 Identify and Solve an Exact Differential Equation

### EXAMPLE 1  Identifying and Solving an Exact Differential Equation

**(a)** Show that the differential equation

$$(2x + 3x^2 y)\, dx + (x^3 + 2y - 3y^2)\, dy = 0 \tag{1}$$

is an exact differential equation.

**(b)** Find the general solution.

**Solution (a)** Let $M = 2x + 3x^2 y$ and $N = x^3 + 2y - 3y^2$. Then

$$\frac{\partial M}{\partial y} = 3x^2 \qquad \text{and} \qquad \frac{\partial N}{\partial x} = 3x^2$$

Since $\dfrac{\partial M}{\partial y} = \dfrac{\partial N}{\partial x}$ for all $x$ and $y$, the differential equation (1) is exact over

the $xy$-plane.

**NEED TO REVIEW?** Finding a potential function $f$ for a conservative vector field $F$ is discussed in Section 15.4, pp. 1051–1052.

**(b)** Since the differential equation $(2x + 3x^2 y)\, dx + (x^3 + 2y - 3y^2)\, dy = 0$ is exact, there is a function $z = f(x, y)$ so that

$$\frac{\partial f}{\partial x} = M = 2x + 3x^2 y \qquad \text{and} \qquad \frac{\partial f}{\partial y} = N = x^3 + 2y - 3y^2$$

We begin by integrating $\dfrac{\partial f}{\partial x} = M = 2x + 3x^2 y$ partially with respect to $x$ (holding $y$

constant). Then

$$f(x, y) = \int (2x + 3x^2 y)\, dx = x^2 + x^3 y + B(y) \tag{2}$$

where the constant of integration is a function $B$ of $y$ alone, which is as yet unknown.

To determine $B(y)$, we use the fact that $f$ must also satisfy $\dfrac{\partial f}{\partial y} = N = x^3 + 2y - 3y^2$.

Then from (2),

$$\frac{\partial f}{\partial y} = \frac{\partial}{\partial y}[x^2 + x^3 y + B(y)] = x^3 + \frac{\partial}{\partial y} B(y) \qquad \text{and} \qquad \frac{\partial f}{\partial y} = x^3 + 2y - 3y^2$$

Equating the two expressions, we have

$$x^3 + \frac{\partial}{\partial y} B(y) = x^3 + 2y - 3y^2$$

$$\frac{dB}{dy} = 2y - 3y^2 \qquad \textcolor{blue}{B \text{ is a function of } y \text{ alone; } \frac{\partial B}{\partial y} = \frac{dB}{dy}.}$$

Now integrate $\dfrac{dB}{dy} = 2y - 3y^2$ with respect to $y$.

$$B(y) = \int (2y - 3y^2)\, dy = y^2 - y^3 + C$$

Substituting for $B(y)$ in (2), we obtain

$$f(x, y) = x^2 + x^3 y + y^2 - y^3 + C$$

The general solution of the differential equation is

$$x^2 + x^3 y + y^2 - y^3 + C = 0$$

where $C$ is a constant. ∎

**NOW WORK** Problem 9.

CALC
CLIP

**EXAMPLE 2** **Identifying and Solving an Exact Differential Equation**

**(a)** Show that

$$(\cos y - \cos x)\, dx + (e^y - x \sin y)\, dy = 0$$

is an exact differential equation.

**(b)** Find the particular solution that satisfies the boundary condition $y(\pi) = 0$.

**Solution (a)** Let $M = \cos y - \cos x$ and $N = e^y - x \sin y$. Then

$$\frac{\partial M}{\partial y} = -\sin y \qquad \text{and} \qquad \frac{\partial N}{\partial x} = -\sin y$$

Since $\dfrac{\partial M}{\partial y} = \dfrac{\partial N}{\partial x}$ for all $x$ and $y$, the differential equation is exact over the $xy$-plane.

**(b)** Since the differential equation is exact, there is a function $z = f(x, y)$ so that

$$\frac{\partial f}{\partial x} = M = \cos y - \cos x \qquad \text{and} \qquad \frac{\partial f}{\partial y} = N = e^y - x \sin y$$

Integrate $\dfrac{\partial f}{\partial x} = M = \cos y - \cos x$ partially with respect to $x$ (holding $y$ constant). Then

$$f(x, y) = \int (\cos y - \cos x)\, dx = x \cos y - \sin x + B(y) \tag{3}$$

where $B$, which is yet unknown, is a function of $y$ alone. To find $B$, differentiate $f$ with respect to $y$.

$$\frac{\partial f}{\partial y} = \frac{\partial}{\partial y}[x \cos y - \sin x + B(y)] = -x \sin y - 0 + \frac{\partial}{\partial y}B(y) = -x \sin y + \frac{dB}{dy}$$

Since $\dfrac{\partial f}{\partial y} = N = e^y - x \sin y$, we have

$$-x \sin y + \frac{dB}{dy} = e^y - x \sin y$$

$$\frac{dB}{dy} = e^y$$

$$B(y) = e^y + C$$

where $C$ is a constant. Now substitute $B(y)$ into (3) to obtain the general solution

$$f(x, y) = x \cos y - \sin x + e^y + C = 0$$

To find the particular solution that satisfies the boundary condition $y(\pi) = 0$, we need to find $C$ so that $y = 0$ when $x = \pi$. Then

$$\pi \cos 0 - \sin \pi + e^0 + C = 0$$

$$\pi + 1 + C = 0$$

$$C = -(1 + \pi)$$

The particular solution is $x \cos y - \sin x + e^y = \pi + 1$. ∎

**NOW WORK** Problem 25.

## Integrating Factors

There are differential equations that are not exact, but can be made exact by multiplying the equation by an appropriate expression called an *integrating factor*.

**EXAMPLE 3  Transforming a Differential Equation into an Exact Differential Equation**

The equation

$$-y\,dx + x\,dy = 0$$

is not an exact differential equation because

$$\frac{\partial}{\partial y}(-y) = -1 \qquad \text{and} \qquad \frac{\partial}{\partial x}(x) = 1$$

are not equal. The differential equation, however, can be converted into an exact differential equation by multiplying by $\dfrac{1}{x^2}$, an integrating factor.* Then

$$-\frac{y}{x^2}\,dx + \frac{1}{x}\,dy = 0$$

is an exact differential equation because

$$\frac{\partial}{\partial y}\left(-\frac{y}{x^2}\right) = -\frac{1}{x^2} \qquad \text{and} \qquad \frac{\partial}{\partial x}\left(\frac{1}{x}\right) = -\frac{1}{x^2} \qquad ■$$

Problems 27 and 28 discuss two situations for which an integrating factor can be found for a first-order differential equation.

---

*We do not discuss how to find integrating factors in this book. Consult texts on differential equations for more details.

## 16.3 Assess Your Understanding

### Concepts and Vocabulary

**1.** *True or False*   The differential equation $xy\,dx + \dfrac{x^2}{2}dy = 0$ is exact.

**2.** *True or False*   The differential equation $y\cos x\,dx - \sin x\,dy = 0$ is exact.

**3.** *True or False*   If $\dfrac{\partial M}{\partial x} = \dfrac{\partial N}{\partial y}$, then

$M(x, y)\,dx + N(x, y)\,dy = 0$ is an exact differential equation.

**4.** If the differential equation $M(x, y)\,dx + N(x, y)\,dy = 0$ is not exact, but $a(x, y)M(x, y)\,dx + a(x, y)N(x, y)\,dy = 0$ is an exact differential equation, then the expression $a(x, y)$ is called a(n) _____.

### Skill Building

*In Problems 5–22:*
*(a)  Show that each equation is an exact differential equation.*
*(b)  Find the general solution.*

**5.** $(4x - 2y + 5)\,dx + (2y - 2x)\,dy = 0$

**6.** $(3x^2 + 3xy^2)\,dx + (3x^2y - 3y^2 + 2y)\,dy = 0$

**7.** $(a^2 - 2xy - y^2)\,dx - (x + y)^2dy = 0$, $a$ is constant

**8.** $(2ax + by + g)\,dx + (2e^y + bx + h)\,dy = 0$, $a, b, g, h$ are constants

**9.** $\dfrac{1}{y}\,dx - \dfrac{x}{y^2}\,dy = 0$

**10.** $\dfrac{y\,dx - x\,dy}{x^2} = 0$

**11.** $(x - 1)^{-1}y\,dx + [\ln(2x - 2) + y^{-1}]\,dy = 0$

**12.** $2xy^{-1}dy + (2\ln(5y) + x^{-1})\,dx = 0$

**13.** $(x + 3)^{-1}\cos y\,dx - [\sin y\ln(5x + 15) - y - 1]\,dy = 0$

**14.** $p^2\sec(2\theta)\tan(2\theta)\,d\theta + p[\sec(2\theta) + 2]\,dp = 0$

**15.** $\cos(x + y^2)\,dx + 2y\cos(x + y^2)\,dy = 0$

**16.** $[\sin(2\theta) - 2p\cos(2\theta)]\,dp + [2p\cos(2\theta) + 2p^2\sin(2\theta)]\,d\theta = 0$

**17.** $e^{2x}(dy + 2y\,dx) = x^2dx$     **18.** $e^{x^2}(dy + 2xy\,dx) = 3x^2dx$

**19.** $\left[\dfrac{1}{x + y} + y^2\right]dx + \left[\dfrac{1}{x + y} + 2xy\right]dy = 0$

**20.** $\dfrac{y^2 - 2x^2}{xy^2 - x^3}\,dx + \dfrac{2y^2 - x^2}{y^3 - x^2y}\,dy = 0$

**21.** $2y^3\sin(2x)\,dx - 3y^2\cos(2x)\,dy = 0$

**22.** $\dfrac{3y^2}{x^2 + 3x}\,dx + \left(2y\ln\dfrac{5x}{x + 3} + 3\sin y\right)dy = 0$

*In Problems 23–26, each differential equation is exact. Find the particular solution that satisfies the given boundary condition.*

**23.** $(1 + y^2 + xy^2)\,dx + (x^2y + y + 2xy)\,dy = 0$;   $y(1) = 1$

**24.** $(3x^2y^{-1} + 2x)\,dx + (y^2 - x^3y^{-2})\,dy = 0$;   $y(3) = 3$

---

**1.** = NOW WORK problem       ◪ = Graphing technology recommended       [CAS] = Computer Algebra System recommended

**25.** $(2xy - \sin x)\,dx = (2y - x^2)\,dy; \quad y(0) = 1$

**26.** $y[y + \sin x]\,dx - \left[\cos x - 2xy + \dfrac{1}{1 + y^2}\right]dy = 0; \quad y(0) = 1$

## Applications and Extensions

**27. Integrating Factors**   Suppose the equation

$M(x, y)\,dx + N(x, y)\,dy = 0$ has the property that

$\dfrac{\dfrac{\partial N}{\partial x} - \dfrac{\partial M}{\partial y}}{M}$ is a function of $y$ only. If

$$u(y) = e^{\left[\int \frac{\frac{\partial N}{\partial x} - \frac{\partial M}{\partial y}}{M}\,dy\right]} = \exp\left[\int \frac{\frac{\partial N}{\partial x} - \frac{\partial M}{\partial y}}{M}\,dy\right]$$

show that $u(y)\,[M(x, y)\,dx + N(x, y)\,dy] = 0$ is an exact differential equation.

**28. Integrating Factors**   Suppose the equation
$M(x, y)\,dx + N(x, y)\,dy = 0$ has the property that

$\dfrac{\dfrac{\partial M}{\partial y} - \dfrac{\partial N}{\partial x}}{N}$ is a function of $x$ only. If

$$u(x) = e^{\left[\int \frac{\frac{\partial M}{\partial y} - \frac{\partial N}{\partial x}}{N}\,dx\right]} = \exp\left[\int \frac{\frac{\partial M}{\partial y} - \frac{\partial N}{\partial x}}{N}\,dx\right]$$

show that $u(x)[M(x, y)\,dx + N(x, y)\,dy] = 0$ is an exact differential equation.

*In Problems 29–33,*
*(a) Change each equation to an exact differential equation by using one of the integrating factors discussed in Problems 27 and 28.*
*(b) Solve the resulting differential equation.*
*(c) Verify that the solution also satisfies the original equation.*

**29.** $(4x^2 + y^2 + 1)\,dx + (x^2 - 2xy)\,dy = 0$

**30.** $4x^2 y\,dx + (x^3 + y)\,dy = 0$     **31.** $y\,dx + (x^2 y - x)\,dy = 0$

**32.** $(\cos y + x)\,dx + x \sin y\,dy = 0$

**33.** $(x^2 - x \sin y)\,dx + x^2 \cos y\,dy = 0$

## Challenge Problem

**34. Measuring the Effect of Pollution**   A crash in the Gulf of Mexico resulted in an oil spill at the point $A$ shown in the figure below. Several months after the spill, measurements are taken to determine whether the oil is still affecting the marine environment of the Gulf. The contour curves shown in the figure are curves of constant oil concentration. These curves are modeled by the differential equation

$$\frac{2\dfrac{\beta - y}{100} - 2\cos\dfrac{\beta - y}{100}}{2\dfrac{\alpha - x}{100} + 4\cos\dfrac{\alpha - x}{100}}\,y'(x) = 1$$

where $\alpha$ and $\beta$ are specified constants.

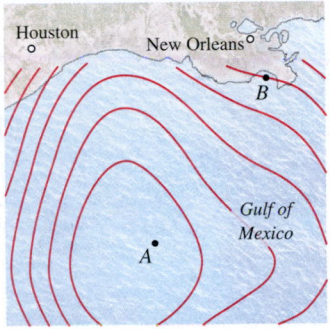

**(a)** Find the solution of the differential equation.

**(b)** The solution from (b) can be interpreted as the oil concentration in pounds of oil per million gallons at a point $(x, y)$. Suppose the point $A$ is at the origin and the point $B$ is (162.14, 250.64). Use the initial condition $f(0, 0) = 74.1$ to find the oil concentration at $B$, given $\alpha = 93.43$ and $\beta = 59.12$.

# 16.4 First-Order Linear Differential Equations; Bernoulli Differential Equations

**OBJECTIVES**   *When you finish this section, you should be able to:*

**1** Solve a first-order linear differential equation (p. 1132)
   Application: Free fall with air resistance (p. 1133)
   Application: Flow rate in mixtures (p. 1135)

**2** Find the general solution of a Bernoulli equation (p. 1136)
   Application: Logistic functions (p. 1137)

In this section, we solve first-order linear differential equations and Bernoulli differential equations. These differential equations occur in the study of free fall with air resistance, flow rates in mixtures, and logistic functions.

## 1 Solve a First-Order Linear Differential Equation

We use integrating factors to solve first-order linear differential equations. Recall that a first-order differential equation is linear if it can be written in the form

$$\frac{dy}{dx} + P(x)\, y = Q(x) \tag{1}$$

where the functions $P$ and $Q$ are continuous on their domains.

For example, the differential equation

$$\frac{dy}{dx} - \frac{3}{x}y = x^2$$

is a first-order linear differential equation.

In the differential equation (1), either $Q(x) = 0$ or $Q(x) \neq 0$. If $Q(x) = 0$, then

$$\frac{dy}{y} + P(x)\, dx = 0$$

and the differential equation is separable. The general solution of this differential equation is

$$\ln|y| + \int P(x)\, dx = C$$

**NOW WORK** Problem 5.

If $Q(x) \neq 0$, we solve (1) by multiplying the first-order differential equation by the integrating factor

$$e^{\int P(x)\, dx} = \exp\left[\int P(x)\, dx\right]$$

$$\frac{dy}{dx} + P(x)\, y = Q(x) \qquad\qquad \textcolor{blue}{\text{Equation (1), } Q(x) \neq 0.}$$

$$\frac{dy}{dx} e^{\int P(x)\, dx} + P(x) y e^{\int P(x)\, dx} = Q(x) e^{\int P(x)\, dx} \qquad \textcolor{blue}{\text{Multiply (1) by } \exp\left[\int P(x)\, dx\right].}$$

The left side of this equation equals $\dfrac{d}{dx}\left(y \cdot e^{\int P(x)\, dx}\right)$, so

$$\frac{d}{dx}\left(y \cdot e^{\int P(x)\, dx}\right) = Q(x) e^{\int P(x)\, dx}$$

Now integrate to obtain

$$y \cdot e^{\int P(x)\, dx} = \int Q(x) e^{\int P(x)\, dx}\, dx + C \tag{2}$$

The general solution of the differential equation $\dfrac{dy}{dx} + P(x)\, y = Q(x)$, $Q(x) \neq 0$, is found by finding the integrals in (2) and solving for $y$.

### EXAMPLE 1  Solving a First-Order Linear Differential Equation

Find the general solution of the first-order linear differential equation

$$\frac{dy}{dx} + \frac{x}{x^2 + 1}y = \frac{x^3}{x^2 + 1}$$

**Solution** Compare the differential equation to (1). Since $P(x) = \dfrac{x}{x^2 + 1}$, the integrating factor is

$$e^{\int P(x)\, dx} = \exp\left[\int \frac{x}{x^2 + 1}\, dx\right] = \exp\left[\frac{1}{2}\ln(x^2 + 1)\right] = \exp\left[\ln(x^2 + 1)^{1/2}\right] = \sqrt{x^2 + 1}$$

Now multiply the differential equation by $\sqrt{x^2+1}$ and obtain

$$\sqrt{x^2+1}\,\frac{dy}{dx} + \frac{x}{\sqrt{x^2+1}}\,y = \frac{x^3}{\sqrt{x^2+1}}$$

The left side of the above equation is the derivative of $\sqrt{x^2+1}\,y$, so

$$\frac{d}{dx}\left(\sqrt{x^2+1}\,y\right) = \frac{x^3}{\sqrt{x^2+1}}$$

Integrating both sides, we find

$$\sqrt{x^2+1}\,y = \int \frac{x^3}{\underset{\uparrow}{\sqrt{x^2+1}}}\,dx = \frac{1}{2}\int \frac{(u-1)}{\sqrt{u}}\,du = \frac{1}{2}\int \left(u^{1/2} - u^{-1/2}\right)du$$

$$\text{Let } u = x^2+1, \; du = 2x\,dx.$$

$$\sqrt{x^2+1}\,y = \frac{1}{2}\left(\frac{u^{3/2}}{\frac{3}{2}} - \frac{u^{1/2}}{\frac{1}{2}}\right) + C = \frac{(x^2+1)^{3/2}}{3} - (x^2+1)^{1/2} + C$$

$$y = \frac{x^2+1}{3} - 1 + \frac{C}{\sqrt{x^2+1}} \qquad \text{Solve for } y.$$

and we have the general solution to the differential equation. ∎

**NOW WORK** Problem 9.

## Application: Free Fall with Air Resistance

In Chapter 4, we saw that if air resistance is ignored, then the position $s = s(t)$ at time $t$ of an object with mass $m$ in free fall obeys the equation

$$ma = -mg$$

where $a = \dfrac{d^2s}{dt^2}$ and $g$ is the acceleration due to gravity.

A more realistic model of the motion of an object in free fall accounts for the effect of air resistance. A common assumption is that air resistance exerts a force proportional to the speed $v$ of the object and opposite to the motion of the object. With this assumption, the motion of an object of mass $m$ in free fall is described by the differential equation

$$\boxed{m\frac{d^2s}{dt^2} = -mg - kv}$$

where $k > 0$ is the constant of proportionality and depends on the particular object and certain air properties. For example, a sheet of paper has a much larger constant of proportionality than a solid lead ball has. Since $v = \dfrac{ds}{dt}$, the second-order differential equation becomes

$$m\frac{dv}{dt} = -mg - kv$$

$$\boxed{\frac{dv}{dt} + \frac{k}{m}v = -g}$$

This is a first-order linear differential equation. To solve the differential equation, multiply both sides by the integrating factor $\exp\left[\int \dfrac{k}{m}\,dt\right] = \exp\left[\dfrac{k}{m}t\right] = e^{kt/m}$.

$$e^{kt/m}\frac{dv}{dt} + e^{kt/m}\frac{k}{m}v = -ge^{kt/m}$$

$$\frac{d}{dt}\left[ve^{kt/m}\right] = -ge^{kt/m}$$

$$ve^{kt/m} = -g\int e^{kt/m}\,dt = -\frac{mg}{k}e^{kt/m} + C \qquad \text{Integrate both sides.}$$

$$v = \frac{-\dfrac{mg}{k}e^{kt/m} + C}{e^{kt/m}} = -\frac{mg}{k} + Ce^{-kt/m} \qquad \text{Solve for } v.$$

The general solution is

$$v(t) = -\frac{mg}{k} + Ce^{-kt/m}$$

If the object has initial velocity $v_0$, then

$$v_0 = v(0) = -\frac{mg}{k} + C, \qquad \text{so} \qquad C = v_0 + \frac{mg}{k}$$

Then

$$\boxed{v(t) = -\frac{mg}{k} + \left(v_0 + \frac{mg}{k}\right)e^{-kt/m}} \tag{3}$$

Notice that $\lim\limits_{t\to\infty} v(t) = -\dfrac{mg}{k}$. That is, the limiting velocity of a freely falling object depends only on the mass of the body and the constant of proportionality $k$.

### EXAMPLE 2 Solving a Free Fall Problem

A parachuter and his parachute together weigh 192 lb. At the instant the parachute opens, he is falling at 150 feet per second (ft/s). Suppose the air resistance exerts a force of 200 lb when the parachuter's speed is 20 ft/s.

**(a)** How fast is he falling 3 s after the parachute opens?
**(b)** What is the parachuter's limiting velocity?

**Solution (a)** In the U.S. customary system of units, length is measured in feet, weight in pounds, time in seconds, and $g \approx 32\,\text{ft/s}^2$. The mass $m$ of the parachuter and his parachute obeys $mg = \text{weight} = 192\,\text{lb}$.

Since the air resistance is 200 lb when the speed is 20 ft/s,

$$200 = k \cdot 20 \qquad F = kv; \text{ force is proportional to speed.}$$
$$k = 10$$

From (3), the velocity of the parachuter at time $t$ after the parachute opens is

$$v(t) = -\frac{mg}{k} + \left(v_0 + \frac{mg}{k}\right)e^{-kt/m}$$

$$v(t) = -\frac{192}{10} + \left(-150 + \frac{192}{10}\right)e^{-10t/6} \qquad v_0 = -150; \ mg = 192; \ g = 32; \ k = 10$$

$$v(t) = -19.2 - 130.8e^{-5t/3}$$

**NOTE** Remember that with falling objects, the positive direction is up, so $v_0 = v(0) = -150\,\text{ft/s}$.

After 3 seconds, the parachuter's velocity is

$$v(3) = -19.2 - 130.8e^{-5} \approx -20.081 \text{ ft/s } (13.692 \text{ mi/h})$$

**(b)** The limiting velocity is $\lim\limits_{t \to \infty} v(t) = -\dfrac{mg}{k} = -19.2 \text{ ft/s } (13.091 \text{ mi/h}).$ ∎

**NOW WORK** Problem **47.**

### Application: Flow Rate in Mixtures

**EXAMPLE 3** **Solving a Flow Rate Problem**

A large tank contains 81 gal of brine in which 20 lb of salt are dissolved. Brine containing 3 lb of dissolved salt per gallon runs into the tank at the rate of 5 gal/min. The mixture, kept uniform by stirring, drains from the tank at the rate of 2 gal/min. How much salt is in the tank at the end of 37 min? See Figure 8.

**Solution** Let $y(t)$ be the amount of salt in the tank at time $t$. The rate of change $\dfrac{dy}{dt}$ of salt in the tank at time $t$ is the rate at which salt enters the tank minus the rate at which salt leaves the tank. Salt enters the tank at the rate of $(3 \text{ lb/gal}) (5 \text{ gal/min}) = 15 \text{ lb/min}$.

To find the rate at which the salt exits the tank, we first need to find the concentration of salt in pounds per gallon at time $t$.

$$\text{Concentration} = \frac{\text{amount of salt}}{\text{number of gallons}} = \frac{y(t)}{81 + (5-2)t} = \frac{y(t)}{81 + 3t} \text{ lb/gal}$$

Then the rate at which the salt exits the tank is

$$\left( \frac{y(t)}{81 + 3t} \text{ lb/gal} \right) (2 \text{ gal/min}) = \frac{2y(t)}{81 + 3t} \text{ lb/min}$$

The mixture problem is modeled by the differential equation

$$\frac{dy}{dt} = \text{rate in} - \text{rate out} = 15 - \frac{2y(t)}{81 + 3t}$$

$$\frac{dy}{dt} + \frac{2}{3(27 + t)} y = 15$$

This is a first-order linear differential equation. We use the integrating factor:

$$e^{\int P(t)\, dt} = \exp\left[ \int \frac{2}{3(27 + t)} dt \right] = \exp\left[ \frac{2}{3} \ln(27 + t) \right]$$

$$= \exp[\ln(27 + t)^{2/3}] = (27 + t)^{2/3}$$

Multiply the differential equation by the integrating factor to obtain

$$(27 + t)^{2/3} \frac{dy}{dt} + \frac{2y}{3(27 + t)^{1/3}} = 15(27 + t)^{2/3}$$

$$\frac{d}{dt}[y(t)(27 + t)^{2/3}] = 15(27 + t)^{2/3}$$

Now integrate both sides with respect to $t$.

$$y(t)(27 + t)^{2/3} = \int 15(27 + t)^{2/3} dt \underset{\underset{\text{Let } u = 27 + t,\, du = dt.}{\uparrow}}{=} 15 \int u^{2/3} du = 15\frac{u^{5/3}}{\frac{5}{3}} + C$$

$$= 9(27 + t)^{5/3} + C$$

$$y(t) = \frac{9(27 + t)^{5/3} + C}{(27 + t)^{2/3}} = 9(27 + t) + C(27 + t)^{-2/3}$$

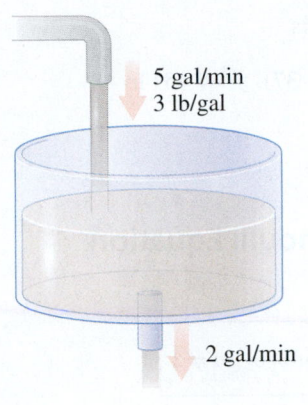

**5 gal/min**
**3 lb/gal**

**2 gal/min**

**DF** **Figure 8**

Next use the initial condition that $y(0) = 20$ lb. Then

$$y(0) = 20 = 9 \cdot 27 + C \cdot 27^{-2/3}$$

$$-223 = \frac{C}{9}$$

$$C = -2007$$

The amount of salt $y(t)$ in the brine at time $t$ min is

$$y(t) = 9(27 + t) - 2007(27 + t)^{-2/3}$$

The amount of salt in the tank at the end of 37 min is

$$y(37) = 9(27 + 37) - 2007(27 + 37)^{-2/3} \approx 451 \text{ lb}$$

∎

NOW WORK **Problem 49.**

**ORIGINS** Jakob (or James) Bernoulli (1654–1705) was a Swiss mathematician. Jakob was one of eight family members, in three generations of Bernoullis, who made major contributions to mathematics and physics.

## 2 Find the General Solution of a Bernoulli Equation

The first-order differential equation

$$\boxed{\frac{dy}{dx} + P(x)y = Q(x)y^n \qquad n \neq 0 \quad n \neq 1} \tag{4}$$

where $P$ and $Q$ are functions of $x$ only, is called a **Bernoulli equation**, named for Jakob Bernoulli, who studied it in 1695.

If we multiply both sides of (4) by $y^{-n}$ and make a judicious substitution, we obtain a first-order linear differential equation. First, multiply (4) by $y^{-n}$:

$$y^{-n}\frac{dy}{dx} + P(x)y^{-n+1} = Q(x) \tag{5}$$

Now use the substitution $v = v(x) = y^{-n+1}$. Then $\dfrac{dv}{dx} = (1-n)y^{-n}\dfrac{dy}{dx}$. Since $n \neq 0$ and $n \neq 1$, we can write $y^{-n}\dfrac{dy}{dx} = \dfrac{1}{1-n}\dfrac{dv}{dx}$. Then (5) becomes

$$\frac{1}{1-n}\frac{dv}{dx} + P(x)v = Q(x)$$

$$\frac{dv}{dx} + (1-n)P(x)v = (1-n)Q(x) \qquad \text{\color{blue}Multiply by } 1-n.$$

This is a first-order linear differential equation in $v = v(x)$.

Use the following steps to solve a Bernoulli equation:

---

**Steps for Solving a Bernoulli Equation**

$$\frac{dy}{dx} + P(x)y = Q(x)y^n \qquad n \neq 0 \quad n \neq 1$$

**Step 1** Multiply both sides of the equation by $y^{-n}$. [This isolates $Q(x)$ on the right.]

**Step 2** Let $v = v(x) = y^{-n+1}$. Then $\dfrac{dv}{dx} = (1-n)y^{-n}\dfrac{dy}{dx}$.

**Step 3** Substitute $v$ and $\dfrac{1}{1-n}\dfrac{dv}{dx}$ into $y^{-n}\dfrac{dy}{dx} + P(x)y^{-n+1} = Q(x)$.

**Step 4** Find the general solution of the resulting first-order linear differential equation.

**Step 5** Rewrite the general solution in terms of the original variables $x$ and $y$.

**CALC CLIP**

**EXAMPLE 4  Solving a Bernoulli Equation**

Find the general solution of the differential equation

$$\frac{dy}{dx} = 4y + 2e^x y^{1/2}$$

**Solution**  Begin by writing the equation in the form of (4):

$$\frac{dy}{dx} - 4y = 2e^x y^{1/2}$$

This is a Bernoulli differential equation with $P(x) = -4$, $Q(x) = 2e^x$, and $n = \dfrac{1}{2}$. Then follow the steps for solving a Bernoulli equation.

**Step 1**  Multiply the differential equation by $y^{-1/2}$ to obtain

$$y^{-1/2}\frac{dy}{dx} - 4y^{1/2} = 2e^x$$

**Step 2**  Let $v = v(x) = y^{1/2}$. Then $\dfrac{dv}{dx} = \dfrac{1}{2}y^{-1/2}\dfrac{dy}{dx}$.

**Step 3**  Substitute $v$ and $\dfrac{dv}{dx}$ into $y^{-1/2}\dfrac{dy}{dx} - 4y^{1/2} = 2e^x$.

$$2\frac{dv}{dx} - 4v = 2e^x$$

$$\frac{dv}{dx} - 2v = e^x$$

This is a first-order linear differential equation in $x$ and $v$.

**Step 4**  Multiply the differential equation by the integrating factor $e^{\int -2\,dx} = e^{-2x}$.

$$e^{-2x}\frac{dv}{dx} - 2e^{-2x}v = e^{-x}$$

$$\frac{d}{dx}(e^{-2x}v) = e^{-x}$$

| NOTE  Since $e^{-2x}v + e^{-x} > 0$, then $C > 0$.

$$e^{-2x}v = \int e^{-x}\,dx = -e^{-x} + C \quad C > 0$$

$$v = Ce^{2x} - e^x$$

**Step 5**  Using $v = y^{1/2}$, write the solution in terms of $x$ and $y$.

$$y^{1/2} = Ce^{2x} - e^x$$

$$y = (Ce^{2x} - e^x)^2 \qquad ■$$

**NOW WORK** Problem 25.

## Application: Logistic Functions

In the mid-nineteenth century, the Belgian mathematical biologist Pierre F. Verhulst (1804–1849) used the differential equation

$$\boxed{\frac{dy}{dt} = ky(M - y)}$$

where $k$ and $M$ are positive constants, to predict the human population of various countries. Verhulst referred to the model as **logistic growth**. Because of this, the equation is known as the **logistic differential equation** and its solutions are called **logistic functions**.

Rewrite the logistic differential equation as

$$\frac{dy}{dt} = kMy - ky^2$$

where the first term on the right side, $kMy$, is a **growth** term and the second term, $-ky^2$, is an "**inhibition**" or "**competition**" term that impedes growth.

The logistic differential equation $\frac{dy}{dt} = kMy - ky^2$ is a Bernoulli differential equation:

$$\frac{dy}{dt} - kMy = -ky^2$$

where $P(t) = -kM$, $Q(t) = -k$, and $n = 2$. To solve it, follow the steps for solving a Bernoulli equation.

**Step 1**  Multiply by $y^{-n} = y^{-2}$.

$$y^{-2}\frac{dy}{dt} - kMy^{-1} = -k$$

**Step 2**  Let $v = y^{-1}$. Then $\frac{dv}{dt} = -y^{-2}\frac{dy}{dt}$.

**Step 3**  Substitute $v$ and $\frac{dv}{dt}$ into $y^{-2}\frac{dy}{dt} - kMy^{-1} = -k$.

$$-\frac{dv}{dt} - kMv = -k$$

$$\frac{dv}{dt} + kMv = k$$

**Step 4**  Multiply by the integrating factor $e^{\int P(t)\,dt} = e^{\int kM\,dt} = e^{kMt}$.

$$e^{kMt}\frac{dv}{dt} + e^{kMt}kMv = ke^{kMt}$$

$$\frac{d}{dt}[e^{kMt}v] = ke^{kMt}$$

$$e^{kMt}v = k\int e^{kMt}\,dt = \frac{1}{M}e^{kMt} + C$$

**Step 5**  Using $v = y^{-1}$, we rewrite the solution in terms of $y$.

$$e^{kMt}\frac{1}{y} = \frac{1}{M}e^{kMt} + C \qquad v = y^{-1} = \frac{1}{y}$$

$$y(t) = \frac{e^{kMt}}{\frac{1}{M}e^{kMt} + C} = \frac{Me^{kMt}}{e^{kMt} + MC} = \frac{M}{1 + MCe^{-kMt}}$$

If we assume that the initial condition is $y(0) = R$, $0 \le R < M$, then $R = \dfrac{M}{1 + MC}$ or $C = \dfrac{M - R}{RM}$ and

$$\boxed{y(t) = \frac{RM}{R + (M - R)e^{-kMt}}} \tag{6}$$

Equation (6) is called a **logistic function**.

The graph of a logistic equation has a distinct "s" shape. See Figure 9. Since $\lim\limits_{t\to\infty} y(t) = M$, $M$ is called the **carrying capacity** of the logistic function.

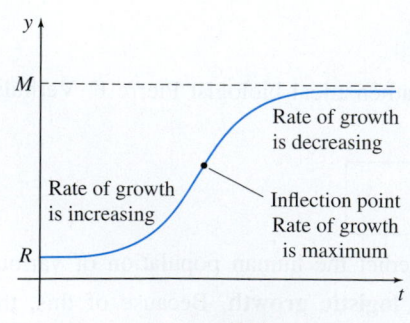

**Figure 9**  $y(t) = \dfrac{RM}{R + (M - R)e^{-kMt}}$

Logistic functions have various applications:

- They are used to predict the growth of certain bacteria in a closed system where the scarcity of space or nutrients limits growth.

- They provide a reasonable model for describing the spread of an epidemic caused by the introduction of an infected individual into a population. In this model, if $M$ is the size of the population and $y = y(t)$ is the number of infected individuals at time $t$, then the logistic equation states that the spread (rate of growth) of the disease is proportional to the product of the number of infected individuals and the number of individuals who have not been infected. At the start of the epidemic, the disease spreads with uninhibited growth, and is modeled by the differential equation $\dfrac{dy}{dt} = kMy$. See Section 5.6 for examples of uninhibited growth. Then since the number of individuals who are exposed is limited, at some point in time the rate of infection begins to decrease.

- In economics, the logistic function is used to predict the growth of sales, the effects of advertising, and company growth. The logistic equation provides a reasonable model for describing the spread of a new technology. For example, after the introduction of the latest version of the iPhone, the rate of sales increases rapidly for a time, until it reaches a maximum. Then although the sales continue to increase, the rate of sales begins to decrease.

- Social scientists use logistic functions to study cultural phenomena in a society. The 2011 Occupy Wall Street movement is one such example. On September 17, 2011, a small group of protestors gathered at Zuccotti Park in New York City to protest the high salaries of Wall Street executives. As the movement spread throughout the city and the United States, the number of protestors increased rapidly for a time, until reaching a maximum rate of increase. Then although more protestors joined the movement, the rate of increase began to decrease.

Science and Society/Science and Society

**ORIGINS** Daniel Bernoulli (1700–1782) was the nephew of Jakob Bernoulli and son of mathematician Daniel Bernoulli. Daniel, the son, was a medical doctor and a mathematician who did much work with Leonhard Euler (1707–1783). In 1760, Daniel Bernoulli used what is now called the logistic equation to study the spread of smallpox.

### EXAMPLE 5 Solving a Logistic Differential Equation

An influenza epidemic is spreading throughout a population of 50,000 people at a rate that is proportional to the product of the number of infected people and the number of noninfected people. Suppose that 100 people were infected initially and 1000 were infected after 10 days.

**(a)** Find the number of infected people at any time $t$. How long will it take for half the population to be infected?

**(b)** Find the time $t$ at which the rate of spread of infection stops increasing and begins to decrease.

**Solution (a)** If $y = y(t)$ is the number of people infected at time $t$, then

$$\frac{dy}{dt} = ky(50{,}000 - y) \tag{7}$$

where $k$ is the constant of proportionality. As given in equation (6), the solution is

$$y(t) = \frac{100(50{,}000)}{100 + (50{,}000 - 100)e^{-50{,}000kt}} \qquad R = 100, \quad M = 50{,}000$$

$$= \frac{100(50{,}000)}{100 + 49{,}900e^{-50{,}000kt}} = \frac{50{,}000}{1 + 499e^{-50{,}000kt}}$$

We find $k$ from the boundary condition $y(10) = 1000$. Then

$$1000 = \frac{50{,}000}{1 + 499e^{-500{,}000k}}$$

$$499e^{-500{,}000k} = 49$$

$$-500{,}000k = \ln\left(\frac{49}{499}\right)$$

$$k \approx 0.00000464 = 4.64 \times 10^{-6}$$

So,

$$y(t) = \frac{50{,}000}{1 + 499e^{-0.232t}}$$

Half the population is infected when $y(t) = 25{,}000$.

$$25{,}000 = \frac{50{,}000}{1 + 499e^{-0.232t}}$$

$$1 + 499e^{-0.232t} = 2$$

$$e^{-0.232t} = \frac{1}{499}$$

$$t = \frac{\ln\dfrac{1}{499}}{-0.232} \approx 27 \text{ days}$$

Half the population is infected in approximately 27 days.

**(b)** The time $t$ we seek is the inflection point of $y = y(t)$. We need to find $t$ so that $y''(t) = 0$.

$$y'(t) = ky(50{,}000 - y) = k(50{,}000y - y^2) \qquad \text{Use (7).}$$

$$y'' = k(50{,}000y' - 2yy') = 0$$

$$2y = 50{,}000$$

$$y = 25{,}000$$

From (a), $y = 25{,}000$ when $t = 27$. That is, on Day 27, the rate of infection stops increasing and begins to decrease. ∎

NOW WORK **Problem 51.**

## 16.4 Assess Your Understanding

### Concepts and Vocabulary

**1.** *True or False*  First-order linear differential equations have the form $\dfrac{dy}{dx} + P(x)y = Q(x)$, where the functions $P$ and $Q$ are continuous on their domains.

**2.** *True or False*  The first-order linear differential equation $\dfrac{dy}{dx} + P(x)y = 0$ is separable.

**3.** *True or False*  Multiplying a first-order linear differential equation $\dfrac{dy}{dx} + P(x)y = Q(x)$, where $Q(x) \neq 0$, by the integrating factor $e^{\int Q(x)\,dx}$ results in a separable differential equation.

**4.** *True or False*  To solve a Bernoulli equation, $\dfrac{dy}{dx} + P(x)y = Q(x)y^n$, $n \neq 0$, $n \neq 1$, the first step is to multiply by $y^n$.

### Skill Building

*In Problems 5–20, find the general solution of each first-order linear differential equation.*

**5.** $\dfrac{dy}{dx} + 2xy = 0$

**6.** $\dfrac{dy}{dx} + \dfrac{1}{x}y = 0$

**7.** $\dfrac{dy}{dx} + \dfrac{1}{x}y = 3x$

**8.** $\dfrac{dr}{d\theta} + \dfrac{4r}{\theta} = \theta$

**9.** $\dfrac{dy}{dx} + \dfrac{y}{x} = x^2$

**10.** $\dfrac{dy}{dx} - \dfrac{2y}{x+1} = 3(x+1)^2$

**11.** $\dfrac{dy}{dx} - 2y = e^{-x}$

**12.** $\dfrac{dy}{dx} - \dfrac{y}{x} = x^{3/2}$

**13.** $\dfrac{dy}{dx} + \dfrac{2y}{x} = x^2 + 1$

**14.** $\dfrac{dy}{dx} + 2xy = 2x$

**15.** $\dfrac{dy}{dx} + e^x y = e^x$

**16.** $\dfrac{dy}{dx} + e^{-x}y = e^{-x}$

**17.** $\dfrac{dy}{dx} + y\tan x = \cos x$

**18.** $\dfrac{dy}{dx} - y\csc x = \sin(2x)$

**19.** $\dfrac{dy}{dx} + y\cot x = \csc^2 x$

**20.** $\dfrac{dy}{dx} + y\tan x = \cos^2 x$

**1.** = NOW WORK problem    ◪ = Graphing technology recommended    CAS = Computer Algebra System recommended

*In Problems 21–26, solve each Bernoulli equation.*

**21.** $\dfrac{dy}{dx} + x^{-1}y = 3x^2y^3$

**22.** $\dfrac{dy}{dx} + x^{-1}y = \dfrac{2}{3}x^2y^4$

**23.** $\dfrac{dy}{dx} + 2xy = xy^4$

**24.** $\dfrac{dy}{dx} + \dfrac{1}{x}y = 3xy^3$

**25.** $\dfrac{dy}{dx} + \dfrac{1}{x}y = 3xy^{1/3}$

**26.** $\dfrac{dy}{dx} + \dfrac{4y}{x} = xy^{1/2}$

## Applications and Extensions

*In Problems 27–38, find the general solution of each differential equation.*

**27.** $x\dfrac{dy}{dx} - y = x^2e^x$

**28.** $\dfrac{dy}{dx} = \dfrac{y}{x - y^2}$

**29.** $dx + (2x - y^2)\,dy = 0$

**30.** $(2x + y)\,dx - x\,dy = 0$

**31.** $\cos x\dfrac{dy}{dx} + y = \sec x$

**32.** $(1 + x^2)\dfrac{dy}{dx} + xy = x^3$

**33.** $dy + y\,dx = 2xy^2e^x\,dx$

**34.** $dx + 2xy^{-1}\,dy = 2x^2y^2\,dy$

**35.** $2\dfrac{dy}{dx} - yx^{-1} = 5x^3y^3$

**36.** $dx - 2xy\,dy = 6x^3y^2e^{-2y^2}\,dy$

**37.** $(x^2 + 2y^2)\dfrac{dx}{dy} + xy = 0$

**38.** $(3x^2y^2 - e^yx^4y)\dfrac{dy}{dx} = 2xy^3$

*In Problems 39–42, find a particular solution of each differential equation that satisfies the given boundary condition.*

**39.** $\dfrac{dy}{dx} + y = e^{-x}$;   $y = 5$ when $x = 0$

**40.** $\dfrac{dy}{dx} + \dfrac{2y}{x} = \dfrac{4}{x}$;   $y = 6$ when $x = 1$

**41.** $\dfrac{dy}{dx} + \dfrac{y}{x} = e^x$;   $y = e^{-1}$ when $x = -1$

**42.** $\dfrac{dy}{dx} + y\cot x = 2\cos x$;   $y = 3$ when $x = \dfrac{\pi}{2}$

[S] *In Problems 43 and 44,*
**(a)** *Use a CAS to find the general solution of each differential equation.*
**(b)** *Find the particular solution that satisfies the given boundary condition.*
**(c)** *Graph the particular solution of each differential equation over the interval $\left[\dfrac{\pi}{2}, \dfrac{5\pi}{2}\right]$.*

**43.** $\dfrac{dy}{dx} + \dfrac{3y}{x} = \sin x$;   $y(\pi) = 0$

**44.** $x\dfrac{dy}{dx} + 2y = \dfrac{5\sin x}{x}$;   $y(\pi) = 0$

**45. Free Fall**  Using $v(t) = -\dfrac{mg}{k} + \left(v_0 + \dfrac{mg}{k}\right)e^{-(k/m)t}$, find the position $s(t)$ of a freely falling object at any time $t$. Assume that its initial position is $s(0) = s_0$.

**46. Free Fall**  An object of mass $m = 2$ kg is dropped from rest from a height of 2000 m. As it falls, the air resistance is equal to $\dfrac{1}{2}v$, where $v$ is the velocity measured in meters per second.

  **(a)** Find the velocity $v$ and the distance $s$ the object has fallen at time $t$.

**(b)** What is the limiting velocity of the object?

**(c)** With what velocity does the object strike Earth?

**47. Velocity of a Skydiver**  A skydiver and her parachute together weigh 160 lb. She free falls from rest from a height of 10,000 ft for 5 s. Assume that there is no air resistance during the fall. After her parachute opens, the air resistance is four times her velocity $v$.

  **(a)** How fast will the skydiver be falling 4 s after the parachute opens?

  **(b)** How long will it take for the skydiver to land on the ground?

  **(c)** What is her velocity $v$ when she lands?

  *Hint:* There are two distinct differential equations that govern the velocity and position of the skydiver: one for the free fall period and the other for the period after the parachute opens.

**48. Spread of a Rumor**  A rumor spreads through a population of 5000 people at a rate that is proportional to the product of the number of people who have heard the rumor and the number who have not heard it. Suppose that 100 people initiated the rumor and 500 have heard it after 3 days.

  **(a)** Write a differential equation that models the rate $\dfrac{dy}{dt}$ at which the rumor spreads, where $y = y(t)$, is the number of people who have heard the rumor after $t$ days.

  **(b)** Solve the differential equation from (a).

  **(c)** How many people will have heard the rumor after 8 days?

  **(d)** How long will it take for half the people to hear the rumor?

**49. Mixtures**  A tank initially contains 100 liters (L) of pure water. Starting at time $t = 0$, brine containing 3 kg of salt per liter flows into the tank at the rate of 8 L/min. The mixture is kept uniform by stirring, and the well-stirred mixture flows out of the tank at the same rate as it flows in. How much salt is in the tank after 5 min? How much salt is in the tank after a very long time?

**50. Flu Epidemic**  A flu virus is spreading through a college campus of 10,000 students at a rate that is proportional to the product of the number of infected students and the number of noninfected students. Assume 10 students were infected initially and 200 students are infected after 10 days.

  **(a)** Write a differential equation that models the rate $\dfrac{dy}{dt}$ of infection with respect to time, where $y = y(t)$ is the number of students infected at time $t$ in days.

  **(b)** Solve the differential equation from (a).

  **(c)** How many students will be infected after 5 days?

  **(d)** How long will it take for 75% of the students to be infected?

  **(e)** Graph $y = y(t)$ for the first 40 days of the flu epidemic.

**51. Urban Planning**  The developers of a planned community assume that the population $P(t)$ of the community will be governed by the logistic equation

$$\frac{dP}{dt} = P(10^{-2} - 10^{-6}P)$$

where $t$ is measured in months.

**(a)** Solve the differential equation if the initial population is estimated to be $P(0) = 1500$.

**(b)** What is the limiting size of the population?

**(c)** When will the population equal one-half of the limiting value?

📈 **(d)** Graph the predicted population for $0 \le t \le 60$ (5 years).

**52. Kirchhoff's Law**    The basic equation governing the amount of current $I$ (in amperes) in a simple $RL$ circuit consisting of a resistance $R$ (in ohms), an inductance $L$ (in henrys), and an electromotive force $E$ (in volts) is

$$\frac{dI}{dt} + \frac{R}{L}I = \frac{E}{L}$$

where $t$ is the time in seconds. Solve the differential equation, assuming that $E$, $R$, and $L$ are constants and $I = 0$ when $t = 0$.

**53. Electrical Charge**    The equation governing the amount of electrical charge $q$ (in coulombs) of an $RC$ circuit consisting of a resistance $R$ (in ohms), a capacitance $C$ (in farads), an electromotive force $E$, and no inductance is

$$\frac{dq}{dt} + \frac{1}{RC}q = \frac{E}{R}$$

where $t$ is the time in seconds. Solve the differential equation, assuming $E$, $R$, and $C$ are constants and $q = 0$ when $t = 0$.

**54. Population Growth**    Consider modeling the growth of a population of fish in a pond as a case of inhibited growth, where the rate of growth of the number of fish is given by the logistic equation $\dfrac{dN}{dt} = kN(N_{max} - N)$, where $t$ is the time in years, $N$ is the number of fish at time $t$, $N_{max}$ is the maximum number of fish the pond can support, and $k$ is a positive constant.

**(a)** Solve the differential equation.

**(b)** If the population doubles after 5 years, the initial population is 20 fish, and $N_{max} = 400$ fish, express $N$ as a function of $t$.

**(c)** What will the population be after 43 years?

**(d)** Show that the population is growing the fastest when $N = \dfrac{N_{max}}{2}$.

📈 **(e)** Graph $N = N(t)$.

**55. Population Growth**    A second model for inhibited growth, called the **Gompertz equation,** is given by the solution to the differential equation $\dfrac{dN}{dt} = pN \ln\left(\dfrac{N_{max}}{N}\right)$, where $N$ is the population at time $t$ in years, $N_{max}$ is the maximum population the environment can support, and $p$ is a positive constant. Suppose the population of fish in the pond from Problem 54 follows a Gompertz equation.

**(a)** Solve the differential equation.

**(b)** If the population doubles after 5 years, the initial population is 20 fish, and $N_{max} = 400$ fish, express $N$ as a function of $t$.

**(c)** What will the population be after 43 years?

**(d)** Show that the population is growing the fastest when $N = \dfrac{N_{max}}{e}$.

📈 **(e)** Graph $N = N(t)$, $0 \le t \le 100$.

**56. Rate of Growth**    If all members of a population are in contact with every other member, the rate of growth of a fad at any time $t$ among the population is proportional to the product $xy$, where $x$ is the number who have adopted the fad and $y$ is the number who have not adopted the fad at time $t$. Suppose that on a certain day $(t = 0)$, two members from a club of 30 members begin wearing a new style of clothing. Each day thereafter, one more member adopts the clothing style.

**(a)** Express the number of members $x$ in the club who have adopted the style as a function of $t$ days.

**(b)** In approximately how many days will half of the club members have adopted the style?

**57.** Consider the logistic differential equation

$$\frac{dy}{dt} = ky(M - y)$$

Show that $\dfrac{dy}{dt}$ is increasing if $y < \dfrac{M}{2}$ and is decreasing if $y > \dfrac{M}{2}$. From this, it follows that the growth rate is a maximum when $y = \dfrac{M}{2}$.

**58.** Solve the logistic differential equation $\dfrac{dy}{dt} = ky(M - y)$ by separating the variables and integrating the resulting rational function using partial fractions. Compare the answer obtained with the "Bernoulli solution."

## Challenge Problems

**59. Air Quality**    A room 150 ft by 50 ft by 20 ft receives fresh air at the rate of 5000 ft³/min. If the fresh air contains 0.04% carbon dioxide and the air in the room initially contained 0.3% carbon dioxide, find the percentage of carbon dioxide after 1 h. What is the percentage after 2 h? Assume the mixed air leaves the room at the rate of 5000 ft³/min.

**60. Renewing Currency**    A nation's federal bank has $3 billion of paper currency in circulation. Each day about $10 million comes into the bank and the same amount is paid out. The federal reserve decides to issue new currency, and whenever the old-style currency comes into the bank, it is destroyed and replaced by the new currency. How long will it take for the currency in circulation to become 95% new?

**61. Kirchhoff's Law**    The equation governing the amount of current $I$ (in amperes) in an $RL$ circuit consisting of a resistance $R$ (in ohms), an inductance $L$ (in henrys), and an electromotive force $E_0 \sin(\omega t)$ volts is given by Kirchhoff's Second Law:

$$L\frac{dI}{dt} + RI = E_0 \sin(\omega t) \qquad \omega > 0$$

Find $I$ as a function of $t$ if $I = I_0$ when $t = 0$. Here $R$, $I$, $E_0$, and $\omega$ are constants.

# 16.5 Power Series Methods

**OBJECTIVE** *When you finish this section, you should be able to:*

**1** Use power series to solve a linear differential equation (p. 1143)

Up to now, we have concentrated on first-order differential equations for which exact solutions could be found. But many first-order differential equations lead to integrals that cannot be expressed in terms of elementary functions. Moreover, most higher-order differential equations cannot be solved exactly. For these reasons, considerable emphasis is placed on methods for approximating solutions of differential equations. One such method, *power series*, is introduced here.

## 1 Use Power Series to Solve a Linear Differential Equation

**Power series methods** assume that a solution $y = y(x)$ of a given differential equation has a power series expansion of the form

**NEED TO REVIEW?** Power series are discussed in Section 8.8, pp. 640–650.

$$y = y(x) = \sum_{k=0}^{\infty} a_k x^k$$

A power series method consists of finding power series for the terms in the differential equation, as well as the power series for $y$, $y'$, $y''$, and so on. These power series expansions are substituted into the differential equation to obtain relationships among the coefficients.

We begin with a simple example to outline the basic idea.

**EXAMPLE 1** **Using Power Series to Solve a Linear Differential Equation**

Use power series to solve the differential equation $y' = y$.

**Solution** Assume that the solution of the differential equation can be expressed as the power series

$$y(x) = \sum_{k=0}^{\infty} a_k x^k \qquad (1)$$

Then

$$y'(x) = \sum_{k=1}^{\infty} k a_k x^{k-1}$$

Since $y' = y$, this leads to

$$\sum_{k=1}^{\infty} k a_k x^{k-1} = \sum_{k=0}^{\infty} a_k x^k$$

To obtain relationships among the coefficients, we write out the terms.

$$a_1 x^0 + 2a_2 x + 3a_3 x^2 + 4a_4 x^3 + \cdots = a_0 x^0 + a_1 x + a_2 x^2 + a_3 x^3 + \cdots$$

Because the coefficients of corresponding powers of $x$ are equal, we have

$$a_1 = a_0 \qquad 2a_2 = a_1 \qquad 3a_3 = a_2 \qquad 4a_4 = a_3 \quad \ldots \quad n a_n = a_{n-1} \quad \ldots$$

Now express these relationships recursively.

$$a_1 = a_0 \qquad a_2 = \frac{1}{2} a_1 = \frac{1}{2!} a_0 \qquad a_3 = \frac{1}{3} a_2 = \frac{1}{3 \cdot 2} a_0 = \frac{1}{3!} a_0$$

$$u_4 = \frac{1}{4} a_3 = \frac{1}{4!} a_0 \quad \ldots \quad a_n = \frac{1}{n} a_{n-1} = \frac{1}{n!} a_0 \quad \cdots$$

The power series (1) takes the form

$$y(x) = \sum_{k=0}^{\infty} a_k x^k = \sum_{k=0}^{\infty} \frac{1}{k!} a_0 x^k = a_0 \sum_{k=0}^{\infty} \frac{x^k}{k!}$$

which is $y(x) = a_0 e^x$. ∎

**RECALL** As we learned in Section 8.9,

$$e^x = \sum_{k=0}^{\infty} \frac{x^k}{k!}.$$

**NOW WORK** Problem 1.

Although this example led to a power series solution we recognized, most solutions will not. The next examples illustrate typical power series solutions.

**EXAMPLE 2** **Using Power Series to Solve a Linear Differential Equation**

Use power series to solve the differential equation $y' - 2y = e^{-x}$.

**Solution** Assume that a solution $y = y(x)$ of the differential equation can be expressed as the power series $y(x) = \sum_{k=0}^{\infty} a_k x^k$. Then $y'(x) = \sum_{k=1}^{\infty} k a_k x^{k-1}$.

Since $e^{-x} = \sum_{k=0}^{\infty} \frac{(-1)^k}{k!} x^k$, the differential equation can be written as

$$\underbrace{\sum_{k=1}^{\infty} k a_k x^{k-1}}_{y'} - 2 \underbrace{\sum_{k=0}^{\infty} a_k x^k}_{y} = \underbrace{\sum_{k=0}^{\infty} \frac{(-1)^k}{k!} x^k}_{e^{-x}} \tag{2}$$

To obtain relationships among the coefficients, write out the terms of the power series.

$$\left(a_1 + 2a_2 x + 3a_3 x^2 + 4a_4 x^3 + 5a_5 x^4 + \cdots\right) - 2\left(a_0 + a_1 x + a_2 x^2 + a_3 x^3 + a_4 x^4 + \cdots\right)$$

$$= 1 - x + \frac{1}{2!} x^2 - \frac{1}{3!} x^3 + \frac{1}{4!} x^4 + \cdots$$

Now combine the coefficients of corresponding powers of $x$ to get

$$\left(a_1 - 2a_0\right) + \left(2a_2 - 2a_1\right)x + \left(3a_3 - 2a_2\right)x^2 + \left(4a_4 - 2a_3\right)x^3 + \cdots$$

$$= 1 - x + \frac{1}{2!} x^2 - \frac{1}{3!} x^3 + \frac{1}{4!} x^4 - \cdots$$

Since the coefficients of corresponding powers of $x$ are equal, we have

$a_1 - 2a_0 = 1$     or equivalently,     $a_1 = 2a_0 + 1$

$2a_2 - 2a_1 = -1$     or equivalently,     $a_2 = \dfrac{2a_1 - 1}{2} = \dfrac{2\left(2a_0 + 1\right) - 1}{2} = 2a_0 + \dfrac{1}{2}$

$3a_3 - 2a_2 = \dfrac{1}{2!}$     or equivalently,     $a_3 = \dfrac{2a_2 + \dfrac{1}{2!}}{3} = \dfrac{2}{3}\left(2a_0 + \dfrac{1}{2}\right) + \dfrac{1}{3 \cdot 2!}$

$\qquad\qquad\qquad\qquad\qquad\qquad = \dfrac{4}{3}a_0 + \dfrac{3}{3!}$

$4a_4 - 2a_3 = -\dfrac{1}{3!}$     or equivalently,     $a_4 = \dfrac{2a_3 - \dfrac{1}{3!}}{4} = \dfrac{1}{2}a_3 - \dfrac{1}{4!}$

$\qquad\qquad\qquad\qquad\qquad\qquad = \dfrac{1}{2}\left(\dfrac{4}{3}a_0 + \dfrac{3}{3!}\right) - \dfrac{1}{4!} = \dfrac{2}{3}a_0 + \dfrac{5}{4!}$

$5a_5 - 2a_4 = \dfrac{1}{4!}$     or equivalently,     $a_5 = \dfrac{2a_4 + \dfrac{1}{4!}}{5} = \dfrac{2}{5}a_4 + \dfrac{1}{5!}$

$\qquad\qquad\qquad\qquad\qquad\qquad = \dfrac{2}{5}\left(\dfrac{2}{3}a_0 + \dfrac{5}{4!}\right) + \dfrac{1}{5!} = \dfrac{4}{15}a_0 + \dfrac{11}{5!}$

This recursive formula $(n+1)a_{n+1} = 2a_n + (-1)^n \dfrac{1}{(n)!}$ can be used to find $a_n$ in terms of $a_0$, as we did for $a_1$, $a_2$, $a_3$, $a_4$, and $a_5$. The power series representation of the general solution of the equation is

$$y(x) = \sum_{k=0}^{\infty} a_k x^k = a_0 + (2a_0 + 1)x + \left(2a_0 + \frac{1}{2!}\right)x^2 + \left(\frac{4}{3}a_0 + \frac{3}{3!}\right)x^3$$

$$+ \left(\frac{2}{3}a_0 + \frac{5}{4!}\right)x^4 + \left(\frac{4}{15}a_0 + \frac{11}{5!}\right)x^5 + \cdots$$

$$= a_0\left(1 + 2x + 2x^2 + \frac{4}{3}x^3 + \frac{2}{3}x^4 + \frac{4}{15}x^5 + \cdots\right)$$

$$+ x + \frac{1}{2!}x^2 + \frac{3}{3!}x^3 + \frac{5}{4!}x^4 + \frac{11}{5!}x^5 + \cdots$$

where $a_0$ is arbitrary. ∎

There is a more direct method for obtaining the recursion formula in Example 2. If we change the index of summation from $k$ to $k+1$ in the first power series on the left side of Equation (2) and distribute the $-2$ inside the second power series, we have

$$\sum_{k=0}^{\infty}(k+1)a_{k+1}x^k + \sum_{k=0}^{\infty}(-2)a_k x^k = \sum_{k=0}^{\infty}\frac{(-1)^k}{k!}x^k$$

Now add the two series on the left.

$$\sum_{k=0}^{\infty}[(k+1)a_{k+1} - 2a_k]x^k = \sum_{k=0}^{n}\frac{(-1)^k}{k!}x^k$$

Then $(n+1)a_{n+1} - 2a_n = \dfrac{(-1)^n}{n!}$, as before.

---

**EXAMPLE 3**  **Using Power Series to Solve a Linear Differential Equation**

Use power series to find the general solution of the linear differential equation

$$y'' + xy' + 2y = 0$$

**Solution** If $y(x) = \sum_{k=0}^{\infty} a_k x^k$ is a solution to the differential equation, then

$$y'(x) = \sum_{k=1}^{\infty} ka_k x^{k-1} \qquad \text{and} \qquad y''(x) = \sum_{k=2}^{\infty} k(k-1)a_k x^{k-2}$$

Now substitute these power series into the differential equation.

$$\sum_{k=2}^{\infty} k(k-1)a_k x^{k-2} + x\sum_{k=1}^{\infty} ka_k x^{k-1} + 2\sum_{k=0}^{\infty} a_k x^k = 0 \quad y'' + xy' + 2y = 0$$

$$\sum_{k=2}^{\infty} k(k-1)a_k x^{k-2} + \sum_{k=1}^{\infty} ka_k x^k + \sum_{k=0}^{\infty} 2a_k x^k = 0 \quad \text{Move } x \text{ and 2 into the summation.}$$

Next adjust the indexes of summation so that $x^k$ appears in each series. Here, only the first series needs modification. If we replace $k$ with $k+2$ in the first series, we obtain

$$\underbrace{\sum_{k=0}^{\infty}(k+2)(k+1)a_{k+2}x^k}_{\text{Replace } k \text{ with } k+2.} + \sum_{k=1}^{\infty} ka_k x^k + \sum_{k=0}^{\infty} 2a_k x^k = 0$$

The index in the first and third series begins at $k = 0$, and the index in the second series begins at $k = 1$. By writing the $k = 0$ term of the first and third series separately, each summation starts at $k = 1$.

$$\underbrace{2 \cdot 1 \cdot a_2 x^0}_{k=0} + \sum_{k=1}^{\infty} (k+2)(k+1)a_{k+2} x^k + \sum_{k=1}^{\infty} k a_k x^k + \underbrace{2a_0 x^0}_{k=0} + \sum_{k=1}^{\infty} 2a_k x^k = 0$$

$$2a_2 + 2a_0 + \sum_{k=1}^{\infty} \left[ (k+2)(k+1)a_{k+2} + (k+2)a_k \right] x^k = 0$$

Equating the coefficients of corresponding powers of $x$ (the coefficients on the right side are all 0), we have

$$2a_2 + 2a_0 = 0 \qquad (n+2)(n+1)a_{n+2} + (n+2)a_n = 0$$

$$a_2 = -a_0 \qquad\qquad a_{n+2} = -\frac{a_n}{(n+1)} \qquad n = 1, 2, 3, \ldots$$

If $n = 1$, we find $a_3 = -\dfrac{a_1}{2}$. Then we use the recursion formula on the left to obtain all of the coefficients in terms of $a_0$ and $a_1$. That is,

$$a_2 = -a_0 \qquad a_4 = -\frac{a_2}{3} = \frac{a_0}{3} \qquad a_6 = -\frac{a_4}{5} = -\frac{a_0}{3 \cdot 5} \qquad a_8 = -\frac{a_6}{7} = \frac{a_0}{3 \cdot 5 \cdot 7}$$

$$a_3 = -\frac{a_1}{2} \qquad a_5 = -\frac{a_3}{4} = \frac{a_1}{2 \cdot 4} \qquad a_7 = -\frac{a_5}{6} = -\frac{a_1}{2 \cdot 4 \cdot 6}$$

and so on. Since $a_0$ and $a_1$ can be chosen arbitrarily, the power series representation for the general solution is

$$y(x) = a_0 \left( 1 - x^2 + \frac{x^4}{3} - \frac{x^6}{3 \cdot 5} + \frac{x^8}{3 \cdot 5 \cdot 7} - \cdots \right)$$

$$+ a_1 \left( x - \frac{x^3}{2} + \frac{x^5}{2 \cdot 4} - \frac{x^7}{2 \cdot 4 \cdot 6} + \cdots \right) \qquad\blacksquare$$

**NOW WORK** Problem 7.

**NEED TO REVIEW?** Maclaurin series are discussed in Section 8.9, pp. 654–660.

A second type of series solution method involves a differential equation with initial conditions and makes use of a Maclaurin series.

CALC
CLIP
**EXAMPLE 4    Using a Maclaurin Series to Solve a Linear Differential Equation**

(a) Use a Maclaurin series to find the solution of

$$y'' = x^2 y + e^x y' \qquad (3)$$

given the initial conditions $y(0) = 1$ and $y'(0) = 1$.

(b) Use the first five terms of the series to approximate values of $y = y(x)$ for $0 \le x \le 1$.

CAS (c) Use a numeric differential equation solver with 10,000 equally spaced numbers in the interval $[0, 1]$ to solve the differential equation in (a). Then construct a table showing the values of $y$ for $x = 0, 0.1, 0.2, \ldots, 0.9, 1$.

**Solution** (a) We assume that the solution of the differential equation is given by the Maclaurin series

$$y(x) = \sum_{k=0}^{\infty} \frac{y^{(k)}(0)}{k!} x^n = y(0) + y'(0)x + \frac{y''(0)}{2!}x^2 + \frac{y'''(0)}{3!}x^3 + \cdots$$

Substitute the initial conditions, $y(0) = 1$ and $y'(0) = 1$, into (3). Then

$$y''(0) = 0^2 \cdot 1 + e^0 \cdot 1 = 1 \qquad x = 0;\ y(0) = 1;\ y'(0) = 1;\ y''(x) = x^2 y + e^x y'$$

Now differentiate $y'' = x^2 y + e^x y'$ with respect to $x$ to find $y'''(0)$.

$$y'''(x) = (x^2 y' + 2xy) + (e^x y'' + e^x y')$$
$$y'''(0) = (0^2 \cdot 1 + 2 \cdot 0 \cdot 1) + (e^0 \cdot 1 + e^0 \cdot 1) = 2$$

Continue differentiating and evaluating the derivative at $x = 0$.

$$y^{(4)}(x) = x^2 y'' + 4xy' + 2y + e^x y''' + 2e^x y'' + e^x y'$$
$$y^{(4)}(0) = 0^2 \cdot 1 + 4 \cdot 0 \cdot 1 + 2 \cdot 1 + e^0 \cdot 2 + 2 \cdot e^0 \cdot 1 + e^0 \cdot 1 = 7$$

and so on. The Maclaurin series then becomes

$$y(x) = 1 + x + \frac{1}{2!}x^2 + \frac{2}{3!}x^3 + \frac{7}{4!}x^4 + \cdots$$

**(b)** We construct Table 1 that uses the first five terms of the series to approximate select values of $y$ in the interval $0 \le x \le 1$.

**TABLE 1** Maclaurin Series Approximation of $y$ Using Five Terms of the Series

| x | 0.0 | 0.1 | 0.2 | 0.3 | 0.4 | 0.5 | 0.6 | 0.7 | 0.8 | 0.9 | 1.0 |
|---|---|---|---|---|---|---|---|---|---|---|---|
| y | 1.0 | 1.1054 | 1.2231 | 1.3564 | 1.5088 | 1.6849 | 1.8898 | 2.1294 | 2.4101 | 2.7394 | 3.125 |

CAS **(c)** Using every thousandth term from the numeric solution, we construct Table 2.

**TABLE 2** Select Values of $y$ Using a Numeric Equation Solver

| x | 0.0 | 0.1 | 0.2 | 0.3 | 0.4 | 0.5 | 0.6 | 0.7 | 0.8 | 0.9 | 1.0 |
|---|---|---|---|---|---|---|---|---|---|---|---|
| y | 1.0 | 1.1054 | 1.2232 | 1.3569 | 1.5114 | 1.6933 | 1.9127 | 2.1836 | 2.5271 | 2.9747 | 3.5752 |

From the results of Tables 1 and 2, we can see how the five-term approximation to the series solution of the differential equation deteriorates as we move away from the center of convergence 0. ∎

NOW WORK Problem 15.

# 16.5 Assess Your Understanding

## Skill Building

*In Problems 1–12, use power series to solve each differential equation.*

1. $y' + 3xy = 0$

2. $y' - x + 3xy = 0$

3. $y'' + y = 0$

4. $y'' + xy = 0$

5. $y'' + x^2 y = 0$

6. $y'' - 2xy = 0$

7. $y'' + x^2 y' + xy = 0$

8. $y'' + 3xy' + 3y = 0$

9. $y''' + y = 0$

10. $y''' - xy = 0$

11. $(1 + x^2)y'' - 4xy' + 6y = 0$

12. $(x^2 + 2)y'' - 3xy' + 4y = 0$

*In Problems 13–22:*

*(a) Use a Maclaurin series to find the solution of each differential equation using the given initial conditions.*

*(b) Use the first five nonzero terms of the series to approximate values of $y$ for $0 \le x \le 1$. Use Table 1 as a guide.*

13. $y'' + xy' + y = 0;$    $y(0) = 1,$   $y'(0) = 0$

14. $y'' - 2xy' + y = 0;$    $y(0) = 2,$   $y'(0) = 1$

15. $y'' - (\sin x)y = 0;$    $y(0) = 0,$   $y'(0) = 1$

16. $y'' + (\cos x)y = 0;$    $y(0) = 0,$   $y'(0) = 1$

17. $y'' + y' + e^x y = 0;$    $y(0) = 2,$   $y'(0) = 1$

---

**1.** = NOW WORK problem     〰 = Graphing technology recommended     CAS = Computer Algebra System recommended

**18.** $y'' + (3+x)y = 0;$   $y(0) = 1,$   $y'(0) = 0$

**19.** $y'' + x^2 y = 0;$   $y(0) = 0,$   $y'(0) = 2$

**20.** $y'' - 3x^2 y' + 2xy = 0;$   $y(0) = 1,$   $y'(0) = 1,$

**21.** $y^{(4)} - \ln(1+x)y = 0;$   $y(0) = 1, y'(0) = 1,$   $y''(0) = 0,$
$y'''(0) = 0$

**22.** $y''' + 4y'' + 2y' - x^3 y = 0;$   $y(0) = 1,$   $y'(0) = 1,$   $y''(0) = 0$

### Applications and Extensions

**CAS** **23. Exact and Series Solutions**

  (a) Use the first six terms of a Maclaurin series to approximate the solution of the differential equation $y'' + 4y' + 4y = 0$ with the initial conditions $y(0) = 1$ and $y'(0) = 2$.

  (b) Solve the differential equation from (a) using a CAS.

  (c) Graph both the series solution from (a) and the exact solution from (b) on the interval $[-1, 2]$.

  (d) Comment on the graphs.

  (e) Repeat (a) using the first eight terms of a Maclaurin series.

  (f) Add the solution from (e) to the graph in (c). What is happening?

**CAS** **24. Exact and Series Solutions**

  (a) Use the first six nonzero terms of a Maclaurin series to approximate the solution of the differential equation $y' = xy(1 - y^2)$ with the initial condition $y(0) = 2$.

  (b) Solve the differential equation from (a) using a CAS.

  (c) Graph the exact solution from (b) and the series solutions using two, four, and six nonzero terms on the interval $[-2, 2]$.

### Challenge Problem

**25. Age of the Earth's Crust.** Uranium has a half-life of $4.5 \times 10^9$ years. The decomposition sequence is very complicated, producing a very large number of intermediate radioactive products, but the final product is an isotope of lead with an atomic weight of 206, called uranium lead.

  (a) Assuming that the change from uranium to lead is direct, show that $u = u_0 e^{-kt}$, $l = u_0(1 - e^{-kt})$, where $u$ and $l$ denote the number of uranium and uranium lead atoms, respectively, present at time $t$. That is, assume $\dfrac{du}{dt} = -ku$, where $k > 0$ is a constant, and $l = u_0 - u$.

  We can measure the ratio $r = \dfrac{l}{u}$ in a rock, and if it is assumed that all the uranium lead came from decomposition of the uranium originally present in the rock, we can obtain a lower bound for the age of Earth's crust. Currently, $r \approx 0.054$.

  (b) Show that this lower bound is given by

$$t = \frac{1}{k}\ln(1+r) = \frac{1}{k}\left(r - \frac{r^2}{2} + \frac{r^3}{3} - \cdots\right) > \frac{r}{k}$$

  (c) What is this lower bound?

---

# Chapter Review

## THINGS TO KNOW

### 16.1 Classification of Ordinary Differential Equations
- Ordinary differential equation  (p. 1113)

- Order of a differential equation  (p. 1113)

- Degree of a differential equation  (p. 1113)

- Linear differential equation of order $n$ (p. 1113)

- General solution; particular solution (p. 1114)

### 16.2 Separable and Homogeneous First-Order Differential Equations; Slope Fields; Euler's Method
- Separable first-order differential equations (p. 1116)

- Steps for Solving a Separable Differential Equation (p. 1116)

- Homogeneous function of degree $k$ in $x$ and $y$, where $k$ is some real number.  (p. 1117)

- Homogeneous first-order differential equations (p. 1118)

- Steps for Solving a Homogeneous First-Order Differential Equation (p. 1118)

- Orthogonal trajectories  (p.1121)

- Steps for finding orthogonal trajectories  (p. 1121)

- Slope field  (p. 1122)

- Euler's method (p. 1123)

### 16.3 Exact Differential Equations
- A differential equation of the form $M(x, y)\,dx + N(x, y)\,dy = 0$ is an exact differential equation at every point $(x, y)$ in a simply connected open set $R$ if and only if $\dfrac{\partial M}{\partial y} = \dfrac{\partial N}{\partial x}$ for all $(x, y)$ in $R$. (p. 1127)

- Integrating factor  (p. 1130)

### 16.4 First-Order Linear Differential Equations; Bernoulli Differential Equations
- First-order linear differential equation: $\dfrac{dy}{dx} + P(x)y = Q(x)$, where $P$ and $Q$ are continuous on their domains (p. 1132)

- If $Q(x) = 0$, the first-order linear differential equation is separable. (p. 1132)

- If $Q(x) \neq 0$, multiply by the integrating factor $e^{\int P(x)\,dx}$. (p. 1132)

- Bernoulli equation: $\dfrac{dy}{dx} + P(x)y = Q(x)y^n, n \neq 0, n \neq 1,$ and $P$ and $Q$ are functions of $x$. (p. 1136)

- Steps for Solving a Bernoulli Equation  (p. 1136)

### 16.5 Power Series Methods
- Assumes a solution $y = y(x)$ of a differential equation has a power series expansion of the form $y = y(x) = \displaystyle\sum_{k=0}^{\infty} a_k x^k$. (p. 1143)

- A Maclaurin series can be used to find a particular solution to a differential equation.  (p. 1146)

## OBJECTIVES

## REVIEW EXERCISES

*In Problems 1–4, state the order and degree of each differential equation. Determine whether the equation is linear or nonlinear.*

**1.** $\dfrac{dy}{dx} + x^2y = \sin x$

**2.** $\dfrac{d^2y}{dx^2} - 5\dfrac{dy}{dx} + 3y = xe^x$

**3.** $\dfrac{d^3y}{dx^3} + 4x\dfrac{d^2y}{dx^2} + \sin^2 x\dfrac{dy}{dx} + 3y = \sin x$

**4.** $\left(\dfrac{dr}{ds}\right)^3 - r = 1$

*In Problems 5–8, verify that the given function is a solution of the differential equation.*

**5.** $y = e^x$,   $\dfrac{d^2y}{dx^2} - y = 0$

**6.** $y = \dfrac{\ln x}{x^2}$,   $\dfrac{d^2y}{dx^2} + \dfrac{5}{x}\dfrac{dy}{dx} + \dfrac{4}{x^2}y = 0$

**7.** $y = e^{-3x}$,   $\dfrac{d^2y}{dx^2} + 2\dfrac{dy}{dx} - 3y = 0$

**8.** $y = x^2 + 3x$,   $x\dfrac{dy}{dx} - y = x^2$

*In Problems 9–12, determine if each function is homogeneous. If it is, state the degree.*

**9.** $f(x, y) = x^3 + 2x^2y + 2xy^2 + y^3$

**10.** $f(x, y) = x^2 - x^2y^2 + y^2$

**11.** $f(x, y) = xe^{y/x}$

**12.** $f(x, y) = \ln\dfrac{y}{x}$

*In Problems 13–24:*

**(a)** Identify each differential equation as separable, homogeneous, exact, linear, or Bernoulli.

**(b)** Solve each differential equation.

**13.** $x(y^2 + 1)\,dx + y(x^2 - x)\,dy = 0$

**14.** $(2y + e^{2x})\,dx + (2x + e^{2y})\,dy = 0$

**15.** $x\dfrac{dy}{dx} + y = x^6$

**16.** $\dfrac{dy}{dx} - 2y = e^{3x}$

**17.** $(2xy^3 + y^2\cos x - 2x)\,dx + (3x^2y^2 + 2y\sin x)\,dy = 0$

**18.** $\dfrac{dy}{dx} + x^2y - y = x^2 - 1$

**19.** $y\sin x\,dx - \cos x\,dy = 0$

**20.** $(2x\sin y - \ln y)\,dx + \left(x^2\cos y - \dfrac{x}{y} + 3y^2\right)dy = 0$

**21.** $\left(x - y\tan\dfrac{y}{x}\right)dx + x\tan\dfrac{y}{x}\,dy = 0$

**22.** $x\dfrac{dy}{dx} - 2y = x + 1$

**23.** $\dfrac{dy}{dx} + \dfrac{2}{x}y = \dfrac{y^3}{x^2}$

**24.** $\dfrac{dy}{dx} - 5y + 2y^2 = 0$

*In Problems 25–34, find the particular solution of each differential equation.*

**25.** $x\dfrac{dy}{dx} + 2y = 6$,   $y(2) = 8$

**26.** $xy\,dy = \left(y^2 + x\sqrt{x^2 + y^2}\right)dx$,   $y(1) = 1$

**27.** $\dfrac{dy}{dx} + \dfrac{y}{x} = 2$,   $y = 0$ when $x = 1$

**28.** $(x^2 + xy + y^2)\,dx - x^2dy = 0$,   $y = 1$ when $x = 1$

**29.** $\dfrac{dy}{dx} = \dfrac{y - 1}{x + 3}$,   $y(-1) = 0$

**30.** $(1 - \cos x)\dfrac{dy}{dx} + y\sin x = 0$,   $y\left(\dfrac{\pi}{4}\right) = 1$

**31.** $\dfrac{dy}{dx} - \dfrac{2y}{x} = x^2\cos x$,   $y\left(\dfrac{\pi}{2}\right) = 3$

**32.** $\dfrac{2x - 1}{y}\,dx + x\cdot\dfrac{1 - x}{y^2}\,dy = 0$,   $y(-1) = 3$

**33.** $(x^2 + y^2)\,dx - xy\,dy = 0$,   $y = 2$ when $x = 1$

**34.** $y^2\sin x\,dx + \left(\dfrac{1}{x} - \dfrac{y}{x}\right)dy = 0$,   $y = 1$ when $x = \pi$

*In Problems 35 and 36, find a power series solution for each differential equation.*

**35.** $(x-2)y' + y = 0$

**36.** $y'' + y' - 2y = 0$

*In Problems 37 and 38, (a) Use a Maclaurin series to find a solution for each differential equation.*
*(b) Use the first five terms of the series to approximate values of $y = y(x)$ for $0 \leq x \leq 1$.*

**37.** $y'' + (\sin x)y = 0$, $\quad y(0) = 1$; $\quad y'(0) = 1$

**38.** $y'' + e^x y' + y = 0$, $\quad y(0) = 1$; $\quad y'(0) = 1$

**39.** (a) Draw a slope field to represent the solution of the differential equation $\dfrac{dy}{dx} = x^2 + y^2$.

(b) Use the slope field representation to graph the solution that satisfies the initial condition that $y = 1$ when $x = 0$.

**40. Flow in Mixtures** A large tank contains 1000 L of water in which 500 mg of chlorine are dissolved. Water containing 2 mg of dissolved chlorine per liter flows into the tank at the rate of 100 L/min. The mixture, kept uniform by stirring, runs out of the tank at the rate of 80 L/min. How much chlorine is in the tank after an hour?

**41. Orthogonal Trajectories** Find the orthogonal trajectories of the family of hyperbolas $xy = c$. Graph a member of each family that contains the point $(2, 3)$.

**42. Spread of Rumors** After a large chip manufacturer posts poor quarterly earnings, the rumor of widespread layoffs spreads among the employees. The rumor spreads throughout the 100,000 employees at a rate that is proportional to the product of the number of employees who have heard the rumor and those who have not. If 10 employees started the rumor, and it has spread to 100 employees after 1 h, how long will it take for 25% of the employees to have heard the rumor?

**43.** Use Euler's method with $h = 0.1$ to approximate the solution $y = y(1.3)$ to the differential equation $\dfrac{dy}{dx} = y + xy^2$ with the boundary condition that $y = 2$ when $x = 1$.

---

**CHAPTER 16 PROJECT** **The Melting Arctic Ice Cap**

In this project, we study the changes in the thickness of sea ice that forms in the Arctic Ocean. There are many factors that affect the behavior of sea ice: abiotic factors such as temperature, both that of the underlying water and air; wind, sea currents, and the motion of the Earth, and biotic factors such as plant and animal life. We examine only one factor: temperature.

We begin with a principle of thermodynamics: The direction of heat transfer is from a warmer to a cooler medium. **Fourier's Law of Conduction** states that the rate of heat flow $\dfrac{dQ}{dt}$ through a homogeneous solid is directly proportional to the area $A$ of the solid orthogonal to the direction of the heat flow, and to the change $\Delta T$ in the temperature $T$, and is inversely proportional to the thickness $x$ of the solid. See Figure 10. The constant of proportionality $k$ depends on the homogeneous solid and is called the **thermal conductivity** of the solid.

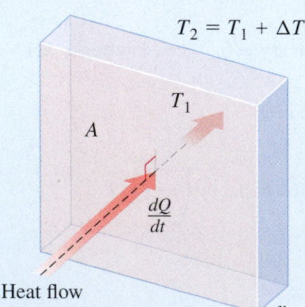

**Figure 10** Fourier's Law of Conduction

**1.** Write Fourier's Law of Conduction as a differential equation.

A simple model depicting sea ice consists of a layer of air, a layer of sea ice, and a layer of sea water. A closer look at this model actually shows another layer directly below the ice. See Figure 11. It is this thin layer of super-cold, almost freezing water that is of interest in this project.

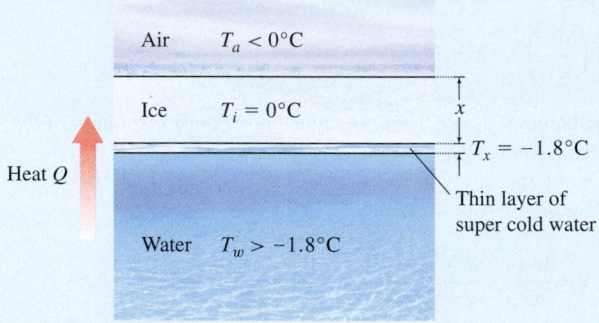

**Figure 11** Model of sea ice

In this model heat flows from the warmer water beneath the ice, through the ice, to the colder air above the ice.

**2.** Using $k = 2.1$ W/m K (the thermal conductivity for sea ice), $A = 1$ m$^2$ as the surface area of the ice, $x$ m as the thickness of the ice, $T_w = -1.8\,°C$ as the near freezing temperature of the layer of sea water directly beneath the ice, and $T_a = -20\,°C$ as the (average winter) temperature of Arctic air, write Fourier's Law of Conduction.

Then as the thin layer of sea water directly below the ice loses heat $Q$, it begins to freeze. The amount of water that freezes satisfies the equation:

$$Q = 334m$$

where 334 J/g is the heat needed to freeze 1 g of water, and $m$ is the mass of the thin layer of water that freezes.

**3.** Find an expression for the mass $m$ of the layer of water directly below the ice that freezes. (The mass density $\rho$ of freezing sea water is 1.025 g/ml.) Recall that the surface area of the ice is $A = 1$ m$^2$.

**4.** Find the rate of change in heat $Q$ with respect to the thickness $x$ of the ice.

**5.** Combine the result from Problem 4 with Fourier's Law of Conductivity (Problem 2) to find the rate of change in thickness of the ice with respect to time.

**6.** Solve the differential equation found in Problem 5. Then use the initial condition that at time $t = 0$, the thickness of the ice is $x = x_0$.

**7.** Interpret the solution of the differential equation in Problem 6. Is the rate of change of the thickness of the ice increasing or decreasing over time?

**8.** Discuss whether this simple model offers a possible explanation for NASA's finding that the multi-year ice is melting faster than seasonal and other perennial ice.

**9.** Investigate some of the other factors that affect the behavior of sea ice and write a position paper on the impact of at least one these other factors on the depth of ice in the Arctic.

To read more and to see how the Arctic Ice cap has changed since 1980, visit http://www.nasa.gov/topics/earth/features/thick-melt.html

*Sources*: http://www.nasa.gov/topics/earth/features/thick-melt.html
http://www.nasa.gov/topics/earth/features/arctic-antarctic-ice.html
http://www.windows2universe.org/earth/Water/density.html
http://plus.maths.org/content/maths-and-climate-change-melting-arctic

# Appendix A  Precalculus Used in Calculus

The topics reviewed here are not exhaustive of the precalculus used in calculus. However, they do represent a large body of the material you will see in calculus. If you encounter difficulty with any of this material, consult a textbook in precalculus for more detail and explanation.

## A.1 Algebra Used in Calculus

**OBJECTIVES** *When you finish this section, you should be able to:*

1 Factor and simplify algebraic expressions (p. A-1)
2 Complete the square (p. A-2)
3 Solve equations (p. A-3)
4 Solve inequalities (p. A-5)
5 Work with exponents (p. A-8)
6 Work with logarithms (p. A-10)

### 1 Factor and Simplify Algebraic Expressions

**EXAMPLE 1**  **Factoring Algebraic Expressions**

Factor each expression completely:

**(a)** $2(x+3)(x-2)^3 + (x+3)^2(3)(x-2)^2$

**(b)** $\dfrac{4}{3}x^{1/3}(2x+1) + 2x^{4/3}$

**Solution (a)** In expression (a), $(x+3)$ and $(x-2)^2$ are **common factors**, factors found in each term. Factor them out.

$$2(x+3)(x-2)^3 + (x+3)^2(3)(x-2)^2$$
$$= (x+3)(x-2)^2[2(x-2)+3(x+3)] \qquad \text{Factor out } (x+3)(x-2)^2.$$
$$= (x+3)(x-2)^2(5x+5) \qquad \text{Simplify.}$$
$$= 5(x+3)(x-2)^2(x+1) \qquad \text{Factor out 5.}$$

**(b)** Begin by writing the term $2x^{4/3}$ as a fraction with a denominator of 3.

$$\frac{4}{3}x^{1/3}(2x+1) + 2x^{4/3} = \frac{4x^{1/3}(2x+1)}{3} + \frac{6x^{4/3}}{3}$$
$$= \frac{4x^{1/3}(2x+1) + 6x^{4/3}}{3} \qquad \text{Add the two fractions.}$$
$$= \frac{2x^{1/3}[2(2x+1) + 3x]}{3} \qquad \text{Factor out the common factors 2 and } x^{1/3}.$$
$$= \frac{2x^{1/3}(7x+2)}{3} \qquad \text{Simplify.} \qquad \blacksquare$$

EXAMPLE 2   **Simplifying Algebraic Expressions**

(a) Simplify $\dfrac{(x^2+1)(3)-(3x+4)(2x)}{(x^2+1)^2}$.

(b) Write the expression $(x^2+1)^{1/2}+x\cdot\dfrac{1}{2}(x^2+1)^{-1/2}\cdot 2x$ as a single quotient in which only positive exponents appear.

**Solution**

(a) $\dfrac{(x^2+1)(3)-(3x+4)(2x)}{(x^2+1)^2}=\dfrac{3x^2+3-(6x^2+8x)}{(x^2+1)^2}=\dfrac{3x^2+3-6x^2-8x}{(x^2+1)^2}$

$=\dfrac{-3x^2-8x+3}{(x^2+1)^2}\underset{\substack{\uparrow\\ \text{Factor}}}{=}\dfrac{-(3x-1)(x+3)}{(x^2+1)^2}$

(b) $(x^2+1)^{1/2}+x\cdot\dfrac{1}{2}(x^2+1)^{-1/2}\cdot 2x=(x^2+1)^{1/2}+\dfrac{x^2}{(x^2+1)^{1/2}}$

$=\dfrac{(x^2+1)^{1/2}(x^2+1)^{1/2}}{(x^2+1)^{1/2}}+\dfrac{x^2}{(x^2+1)^{1/2}}$

$=\dfrac{(x^2+1)+x^2}{(x^2+1)^{1/2}}=\dfrac{2x^2+1}{(x^2+1)^{1/2}}$   ∎

## 2 Complete the Square

We complete the square in one variable by modifying an expression of the form $x^2+bx$ to make it a perfect square. Perfect squares are trinomials of the form

$$x^2+2ax+a^2=(x+a)^2\quad\text{or}\quad x^2-2ax+a^2=(x-a)^2$$

For example, $x^2+6x+9$ is a perfect square because $x^2+6x+9=(x+3)^2$. And $p^2-12p+36$ is a perfect square because $p^2-12p+36=(p-6)^2$.

To make $x^2+6x$ a perfect square, we must add 9. The number to be added is chosen by dividing the coefficient of the first-degree term, which is 6, by 2 and squaring the result $\left[\left(\dfrac{6}{2}\right)^2=9\right]$.

EXAMPLE 3   **Completing the Square**

Determine the number that must be added to each expression to complete the square. Then factor.

| Start | Add | Result | Factored Form |
|---|---|---|---|
| $y^2+8y$ | $\left(\dfrac{1}{2}\cdot 8\right)^2=16$ | $y^2+8y+16$ | $(y+4)^2$ |
| $a^2-20a$ | $\left(\dfrac{1}{2}\cdot(-20)\right)^2=100$ | $a^2-20a+100$ | $(a-10)^2$ |
| $p^2-5p$ | $\left(\dfrac{1}{2}\cdot(-5)\right)^2=\dfrac{25}{4}$ | $p^2-5p+\dfrac{25}{4}$ | $\left(p-\dfrac{5}{2}\right)^2$ |
| $2x^2+6x=2(x^2+3x)$ | $\left(\dfrac{1}{2}\cdot 3\right)^2=\dfrac{9}{4}$ | $2\left(x^2+3x+\dfrac{9}{4}\right)$ | $2\left(x+\dfrac{3}{2}\right)^2$ |

∎

**CAUTION** The original expression $x^2 + bx$ and the perfect square $x^2 + bx + \left(\dfrac{b}{2}\right)^2$ are not equal. So when completing the square within an equation or an inequality, we must not only add $\left(\dfrac{b}{2}\right)^2$, we must also subtract it. That is,

$$x^2 + bx = x^2 + bx + \underbrace{\left(\frac{b}{2}\right)^2 - \left(\frac{b}{2}\right)^2}_{=0} = \left(x + \frac{b}{2}\right)^2 - \left(\frac{b}{2}\right)^2$$

## 3 Solve Equations

To solve a **quadratic equation** $ax^2 + bx + c = 0$, $a \neq 0$, the *quadratic formula* can be used.

> **THEOREM** Quadratic Formula
>
> Consider the quadratic equation
>
> $$ax^2 + bx + c = 0 \quad a \neq 0$$
>
> - If $b^2 - 4ac < 0$, the equation has no real solution.
> - If $b^2 - 4ac \geq 0$, the real solution(s) of the equation is (are) given by the **quadratic formula**:
>
> $$\boxed{x = \frac{-b \pm \sqrt{b^2 - 4ac}}{2a}}$$

The expression $b^2 - 4ac$ is called the **discriminant** of the quadratic equation.

### EXAMPLE 4  Solving Quadratic Equations

Solve the equations:  **(a)** $3x^2 - 5x + 1 = 0$     **(b)** $x^2 + x = -1$

**Solution**  **(a)** The discriminant is $b^2 - 4ac = 25 - 12 = 13$. We use the quadratic formula.

$$x = \frac{-b \pm \sqrt{b^2 - 4ac}}{2a} = \frac{5 \pm \sqrt{13}}{6} \qquad a = 3, b = -5, c = 1$$

The solutions are $x = \dfrac{5 - \sqrt{13}}{6} \approx 0.232$ and $x = \dfrac{5 + \sqrt{13}}{6} \approx 1.434$

**NOTE** Remember to put a quadratic equation in standard form before attempting to solve it. That is, write the quadratic equation in the form $ax^2 + bx + c = 0$.

**(b)** The equation in standard form is $x^2 + x + 1 = 0$. Its discriminant is

$$b^2 - 4ac = 1 - 4 = -3$$

The equation has no real solution. ∎

### EXAMPLE 5  Solving Equations

Solve the equations:

**(a)** $\dfrac{3}{x-2} - \dfrac{1}{x-1} + \dfrac{7}{(x-1)(x-2)}$     **(b)** $x^3 - x^2 - 4x + 4 = 0$

**(c)** $\sqrt{x-1} = x - 7$     **(d)** $|1 - x| = 2$

**NOTE** The set of real numbers that a variable can assume is called the **domain of the variable.**

**Solution** **(a)** First, notice that the domain of the variable is $\{x \mid x \neq 1, x \neq 2\}$. Now clear the equation of rational expressions by multiplying both sides by $(x-1)(x-2)$.

$$\frac{3}{x-2} = \frac{1}{x-1} + \frac{7}{(x-1)(x-2)}$$

$$(x-1)(x-2)\frac{3}{x-2} = (x-1)(x-2)\left[\frac{1}{x-1} + \frac{7}{(x-1)(x-2)}\right] \qquad \text{Multiply both sides by } (x-1)(x-2).$$

$$3x - 3 = (x-1)(x-2)\frac{1}{x-1} + (x-1)(x-2)\frac{7}{(x-1)(x-2)} \qquad \text{Distribute on both sides.}$$

$$3x - 3 = (x-2) + 7 \qquad \text{Simplify.}$$

$$3x - 3 = x + 5$$

$$2x = 8$$

$$x = 4$$

Since 4 is in the domain of the variable, the solution is 4.

**(b)** Group the terms of $x^3 - x^2 - 4x + 4 = 0$ and factor by grouping.

$$x^3 - x^2 - 4x + 4 = 0$$

$$(x^3 - x^2) - (4x - 4) = 0 \qquad \text{Group the terms.}$$

$$x^2(x-1) - 4(x-1) = 0 \qquad \text{Factor out the common factor from each group.}$$

$$(x^2 - 4)(x-1) = 0 \qquad \text{Factor out the common factor } (x-1).$$

$$(x-2)(x+2)(x-1) = 0 \qquad x^2 - 4 = (x-2)(x+2)$$

$$x - 2 = 0 \ \text{ or } \ x + 2 = 0 \ \text{ or } \ x - 1 = 0 \qquad \text{Set each factor equal to 0.}$$

$$x = 2 \qquad\qquad x = -2 \qquad\quad x = 1 \qquad \text{Solve.}$$

The solutions are $-2$, 1, and 2.

**CAUTION** Squaring both sides of an equation may lead to extraneous solutions. Check all apparent solutions.

**(c)** Square both sides of the equation since the index of a square root is 2.

$$\sqrt{x-1} = x - 7$$

$$(\sqrt{x-1})^2 = (x-7)^2 \qquad \text{Square both sides.}$$

$$x - 1 = x^2 - 14x + 49$$

$$x^2 - 15x + 50 = 0 \qquad \text{Put the equation in standard form.}$$

$$(x-10)(x-5) = 0 \qquad \text{Factor.}$$

$$x = 10 \ \text{ or } \ x = 5 \qquad \text{Set each factor equal to 0 and solve.}$$

$$\textit{Check: } x = 10: \ \sqrt{x-1} = \sqrt{10-1} = \sqrt{9} = 3 \text{ and } x - 7 = 10 - 7 = 3$$

$$x = 5: \ \sqrt{x-1} = \sqrt{5-1} = \sqrt{4} = 2 \text{ and } x - 7 = 5 - 7 = -2$$

The apparent solution 5 is extraneous; the only solution of the equation is 10.

**RECALL** $|a| = a$ if $a \geq 0$
$|a| = -a$ if $a < 0$
If $|x| = b$, $b \geq 0$, then $x = b$ or $x = -b$.

**(d)** $|1 - x| = 2$

$$1 - x = 2 \ \text{ or } \ 1 - x = -2 \qquad \text{The expression inside the absolute value bars equals 2 or } -2.$$

$$-x = 1 \qquad\qquad -x = -3 \qquad \text{Simplify.}$$

$$x = -1 \qquad\qquad x = 3 \qquad \text{Simplify.}$$

The solutions are $-1$ and 3. ∎

## 4 Solve Inequalities

In expressing the solution to an inequality, *interval notation* is often used.

**NOTE** Every interval of real numbers contains both rational numbers, numbers that can be expressed as the quotient of two integers, and irrational numbers, numbers that are not rational.

> **DEFINITION  Interval Notation**
>
> Let $a$ and $b$ represent two real numbers with $a < b$.
>
> A **closed interval**, denoted by $[a, b]$, consists of all real numbers $x$ for which $a \leq x \leq b$.
> An **open interval**, denoted by $(a, b)$, consists of all real numbers $x$ for which $a < x < b$.
> The **half-open**, or **half-closed**, **intervals** are:
>
> - $(a, b]$, consisting of all real numbers $x$ for which $a < x \leq b$, and
> - $[a, b)$, consisting of all real numbers $x$ for which $a \leq x < b$.

In each of these definitions, $a$ is called the **left endpoint** and $b$ the **right endpoint** of the interval.

The symbol $\infty$ (read "infinity") is *not* a real number but a notational device used to indicate unboundedness in the positive direction. The symbol $-\infty$ (read "negative infinity") also is not a real number but a notational device used to indicate unboundedness in the negative direction. Using the symbols $\infty$ and $-\infty$, we define five other kinds of intervals:

- $[a, \infty)$, consisting of all real numbers $x$ for which $x \geq a$
- $(a, \infty)$, consisting of all real numbers $x$ for which $x > a$
- $(-\infty, a]$, consisting of all real numbers $x$ for which $x \leq a$
- $(-\infty, a)$, consisting of all real numbers $x$ for which $x < a$
- $(-\infty, \infty)$, consisting of all real numbers $x$

Notice that the endpoints $\infty$ and $-\infty$ are never included in an interval, since neither is a real number.

Table 1 summarizes interval notation, corresponding inequality notation, and their graphs.

**TABLE 1**

| Interval | Inequality | Graph |
|---|---|---|
| The open interval $(a, b)$ | $a < x < b$ | |
| The closed interval $[a, b]$ | $a \leq x \leq b$ | |
| The half-open interval $[a, b)$ | $a \leq x < b$ | |
| The half-open interval $(a, b]$ | $a < x \leq b$ | |
| The interval $[a, \infty)$ | $a \leq x < \infty$ | |
| The interval $(a, \infty)$ | $a < x < \infty$ | |
| The interval $(-\infty, a]$ | $-\infty < x \leq a$ | |
| The interval $(-\infty, a)$ | $-\infty < x < a$ | |

EXAMPLE 6  **Solving Inequalities**

Solve each inequality and graph the solution:

(a) $4x + 7 \geq 2x - 3$    (b) $x^2 - 4x + 3 > 0$    (c) $x^2 + x + 1 < 0$    (d) $\dfrac{1+x}{1-x} > 0$

**Solution** (a)

$$4x + 7 \geq 2x - 3$$

$$4x \geq 2x - 10 \qquad \text{Subtract 7 from both sides.}$$

$$2x \geq -10 \qquad \text{Subtract } 2x \text{ from both sides.}$$

$$x \geq -5 \qquad \text{Divide both sides by 2.}$$

$$\text{(The direction of the inequality symbol is unchanged.)}$$

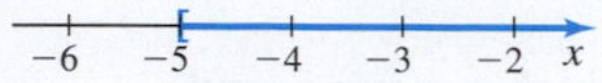

**Figure 1** $x \geq -5$

The solution using interval notation is $[-5, \infty)$. See Figure 1 for the graph of the solution.

(b) This is a quadratic inequality. The related quadratic equation $x^2 - 4x + 3 = (x-1)(x-3) = 0$ has two solutions, 1 and 3. We use these numbers to partition the number line into three intervals. Now select a test number in each interval, and determine the value of $x^2 - 4x + 3$ at the test number. See Table 2.

**NOTE** The test number can be any real number in the interval, but it cannot be an endpoint.

**TABLE 2**

| Interval | Test Number | Value of $x^2 - 4x + 3$ | Sign of $x^2 - 4x + 3$ |
|---|---|---|---|
| $(-\infty, 1)$ | 0 | 3 | Positive |
| $(1, 3)$ | 2 | $-1$ | Negative |
| $(3, \infty)$ | 4 | 3 | Positive |

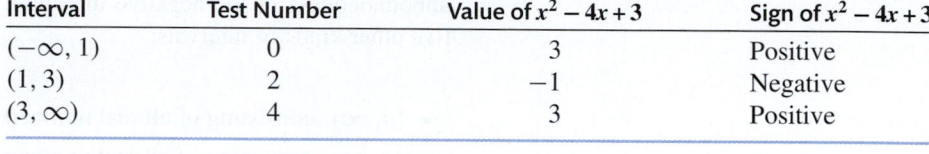

We conclude that $x^2 - 4x + 3 > 0$ on the set $(-\infty, 1) \cup (3, \infty)$. See Figure 2 for the graph of the solution.

**Figure 2** $x < 1$ or $x > 3$

(c) The quadratic equation $x^2 + x + 1 = 0$ has no real solution, since its discriminant is negative. See Example 4(b). When this happens, the quadratic inequality is either positive for all real numbers or negative for all real numbers. To see which is true, evaluate $x^2 + x + 1$ at some number, say, 0. At 0, $x^2 + x + 1 = 1$, which is positive. So, $x^2 + x + 1 > 0$ for all real numbers $x$. The inequality $x^2 + x + 1 < 0$ has no solution.

(d) The only solution of the rational equation $\dfrac{1+x}{1-x} = 0$ is $x = -1$; also, the expression $\dfrac{1+x}{1-x}$ is not defined for $x = 1$. We use the solution $-1$ and the value 1, at which the expression is undefined, to partition the real number line into three intervals. Now select a test number in each interval, and evaluate the rational expression $\dfrac{1+x}{1-x}$ at each test number. See Table 3.

**TABLE 3**

| Interval | Test Number | Value of $\dfrac{1+x}{1-x}$ | Sign of $\dfrac{1+x}{1-x}$ |
|---|---|---|---|
| $(-\infty, -1)$ | $-2$ | $-\dfrac{1}{3}$ | Negative |
| $(-1, 1)$ | 0 | 1 | Positive |
| $(1, \infty)$ | 2 | $-3$ | Negative |

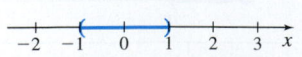

**Figure 3** $-1 < x < 1$

We conclude that $\dfrac{1+x}{1-x} > 0$ on the interval $(-1, 1)$. See Figure 3 for the graph of the solution.  ∎

Below we state an important relationship involving the absolute value of a sum.

**THEOREM Triangle Inequality**

If $x$ and $y$ are real numbers, then

$$\boxed{|x + y| \le |x| + |y|}$$

**NOTE** If $x > 0$ and $y > 0$, the Triangle Inequality is an algebraic statement for the fact that the length of any side of a triangle is less than the sum of the lengths of the other two sides.

Use the following theorem as a guide to solving inequalities involving absolute values.

**THEOREM Inequalities Involving Absolute Value**

If $a$ is a positive real number and if $u$ is an algebraic expression, then

$$\boxed{\begin{array}{l} |u| < a \text{ is equivalent to } -a < u < a \\ |u| > a \text{ is equivalent to } u < -a \quad \text{or} \quad u > a \end{array}}$$

$$(1)$$
$$(2)$$

Similar relationships hold for the nonstrict inequalities $|u| \le a$ and $|u| \ge a$.

---

**EXAMPLE 7** **Solving Inequalities Involving Absolute Value**

Solve each inequality and graph the solution:

(a) $|3 - 4x| < 11$   (b) $|2x + 4| - 1 \le 9$   (c) $\left| \dfrac{4x + 1}{2} - \dfrac{3}{5} \right| > 1$

**Solution (a)** The absolute value is less than the number 11, so statement (1) applies.

$$|3 - 4x| < 11$$

| | |
|---|---|
| $-11 < 3 - 4x < 11$ | Apply statement (1). |
| $-14 < -4x < 8$ | Subtract 3 from each part. |
| $\dfrac{-14}{-4} > x > \dfrac{8}{-4}$ | Divide each part by $-4$, which reverses the inequality signs. |
| $-2 < x < \dfrac{7}{2}$ | Simplify and rearrange the ordering. |

**RECALL** Multiplying (or dividing) an inequality by a negative quantity reverses the direction of the inequality sign.

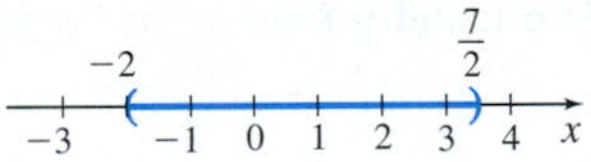

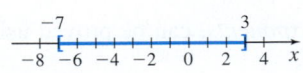

**Figure 4** $-2 < x < \dfrac{7}{2}$

The solutions are all the real numbers in the open interval $\left( -2, \dfrac{7}{2} \right)$. See Figure 4 for the graph of the solution.

**(b)** We begin by putting $|2x + 4| - 1 \le 9$ into the form $|u| \le a$.

| | |
|---|---|
| $|2x + 4| - 1 \le 9$ | |
| $|2x + 4| \le 10$ | Add 1 to each side. |
| $-10 \le 2x + 4 \le 10$ | Apply statement (1) but use $\le$. |
| $-14 \le 2x \le 6$ | |
| $-7 \le x \le 3$ | |

The solutions are all the real numbers in the closed interval $[-7, 3]$. See Figure 5 for the graph of the solution.

**Figure 5** $-7 \le x \le 3$

**(c)** $\left| \dfrac{4x + 1}{2} - \dfrac{3}{5} \right| > 1$ is in the form of statement (2). We begin by simplifying the expression inside the absolute value.

$$\left| \frac{4x + 1}{2} - \frac{3}{5} \right| = \left| \frac{5(4x + 1)}{10} - \frac{2(3)}{10} \right| = \left| \frac{20x + 5 - 6}{10} \right| = \left| \frac{20x - 1}{10} \right|$$

The original inequality is equivalent to the inequality below.

$$\left|\frac{20x - 1}{10}\right| > 1$$

$$\frac{20x - 1}{10} < -1 \quad \text{or} \quad \frac{20x - 1}{10} > 1 \qquad \textbf{Apply statement (2).}$$

$$20x - 1 < -10 \quad \text{or} \quad 20x - 1 > 10$$

$$20x < -9 \quad \text{or} \quad 20x > 11$$

$$x < -\frac{9}{20} \quad \text{or} \quad x > \frac{11}{20}$$

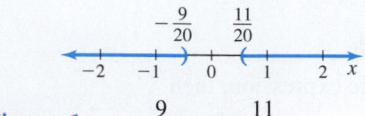

**Figure 6** $x < -\dfrac{9}{20}$ or $x > \dfrac{11}{20}$

The solutions are all the real numbers in the set $\left(-\infty, -\dfrac{9}{20}\right) \cup \left(\dfrac{11}{20}, \infty\right)$. See Figure 6 for the graph of the solution. ∎

## 5 Work with Exponents

Integer exponents provide a shorthand notation for repeated multiplication. For example,

$$2^3 = 2 \cdot 2 \cdot 2 = 8 \qquad \left(\frac{1}{3}\right)^4 = \frac{1}{3} \cdot \frac{1}{3} \cdot \frac{1}{3} \cdot \frac{1}{3} = \frac{1}{81}$$

**DEFINITION**

If $a$ is a real number and $n$ is a positive integer, then the symbol $a^n$ represents the product of $n$ factors of $a$. That is,

$$a^n = \underbrace{a \cdot a \cdot \ldots \cdot a}_{n \text{ factors}}$$

Here, it is understood that $a^1 = a$.

Then $a^2 = a \cdot a$ and $a^3 = a \cdot a \cdot a$, and so on. In the expression $a^n$, the number $a$ is called the **base**, and the number $n$ is called the **exponent** or **power**. We read $a^n$ as "$a$ raised to the power $n$" or as "$a$ to the $n$th power." We usually read $a^2$ as "$a$ squared" and $a^3$ as "$a$ cubed."

**DEFINITION**

If $a \neq 0$, then

$$a^0 = 1$$

If $n$ is a positive integer, then

$$a^{-n} = \frac{1}{a^n} \qquad a \neq 0$$

The following properties, called the *Laws of Exponents*, can be proved using the preceding definitions.

**THEOREM** Laws of Exponents

In each of these properties $a$ and $b$ are real numbers, and $u$ and $v$ are integers.

$$a^u a^v = a^{u+v} \qquad\qquad (a^u)^v = a^{uv} \qquad\qquad (ab)^u = a^u b^u$$

$$\frac{a^u}{a^v} = a^{u-v} = \frac{1}{a^{v-u}} \quad \text{if } a \neq 0 \qquad\qquad \left(\frac{a}{b}\right)^u = \frac{a^u}{b^u} \quad \text{if } b \neq 0$$

**DEFINITION** Principal *n*th Root

The **principal *n*th root of a real number** $a$, where $n \geq 2$ is an integer, symbolized by $\sqrt[n]{a}$, is defined as the solution of the equation $b^n = a$.

$$\sqrt[n]{a} = b \text{ is equivalent to } a = b^n$$

If $n$ is even, then both $a \geq 0$ and $b \geq 0$, and if $n$ is odd, then $a$ and $b$ are any real numbers and both have the same sign.

**IN WORDS** The symbol $\sqrt[n]{a}$ means "the number that, when raised to the *n*th power, equals $a$."

If $a$ is negative and $n$ is even, then $\sqrt[n]{a}$ is not defined. When $\sqrt[n]{a}$ is defined, the principal *n*th root of a number is unique.

The symbol $\sqrt[n]{a}$ for the principal *n*th root of $a$ is called a **radical**; the integer $n$ is called the **index**, and $a$ is called the **radicand**. If the index of a radical is 2, we call $\sqrt[2]{a}$ the **square root** of $a$ and omit the index 2 by writing $\sqrt{a}$. If the index is 3, we call $\sqrt[3]{a}$ the **cube root** of $a$.

Radicals are used to define rational exponents.

**DEFINITION**

If $a$ is a real number and $n \geq 2$ is an integer, then

$$a^{1/n} = \sqrt[n]{a}$$

provided that $a^{1/n} = \sqrt[n]{a}$ exists.

**DEFINITION**

If $a$ is a real number and $m$ and $n$ are integers with $n \geq 2$, then

$$a^{m/n} = (a^{1/n})^m$$

provided that $a^{1/n} = \sqrt[n]{a}$ exists.

From the two definitions, we have

$$a^{m/n} = \sqrt[n]{a^m} = (\sqrt[n]{a})^m$$

In simplifying the rational expression $a^{m/n}$, either $\sqrt[n]{a^m}$ or $(\sqrt[n]{a})^m$ can be used. The choice depends on which is easier to simplify. Generally, taking the root first, as in $(\sqrt[n]{a})^m$, is easier.

But does $a^x$ have meaning, where the base $a$ is a positive real number and the exponent $x$ is an irrational number? The answer is yes, and although a rigorous definition requires methods discussed in calculus, the basis for the definition is easy to follow: Select a rational number $r$ that is formed by truncating (removing) all but a finite number of digits from the irrational number $x$. Then it is reasonable to expect that

$$a^x \approx a^r$$

For example, take the irrational number $\pi = 3.14159\ldots$. Then an approximation to $a^\pi$ is

$$a^\pi \approx a^{3.14}$$

where the digits after the hundredths position have been truncated from the value for $\pi$. A better approximation would be

$$a^\pi \approx a^{3.14159}$$

where the digits after the hundred-thousandths position have been truncated. Continuing in this way, we can obtain approximations to $a^\pi$ to any desired degree of accuracy.

It can be shown that the Laws of Exponents hold for real number exponents $u$ and $v$.

## 6 Work with Logarithms

The definition of a *logarithm* is based on an exponential relationship.

**DEFINITION**

Suppose $y = a^x$, $a > 0$, $a \neq 1$, and $x$ is a real number. The **logarithm with base $a$ of $y$**, symbolized by $\log_a y$, is the exponent to which $a$ must be raised to obtain $y$. That is,

$$\log_a y = x \text{ is equivalent to } y = a^x$$

As this definition states, a logarithm is a name for a certain exponent.

**EXAMPLE 8   Working with Logarithms**

(a) If $x = \log_3 y$, then $y = 3^x$. For example, $4 = \log_3 81$ is equivalent to $81 = 3^4$.

(b) If $x = \log_5 y$, then $y = 5^x$. For example,

$$-1 = \log_5\left(\frac{1}{5}\right) \text{ is equivalent to } \frac{1}{5} = 5^{-1}$$

∎

**THEOREM   Properties of Logarithms**

In the properties given next, $u$ and $a$ are positive real numbers, $a \neq 1$, and $r$ is any real number. The number $\log_a u$ is the exponent to which $a$ must be raised to obtain $u$. That is,

$$a^{\log_a u} = u$$

The logarithm with base $a$ of $a$ raised to a power equals that power. That is,

$$\log_a a^r = r$$

**EXAMPLE 9   Using Properties of Logarithms**

(a) $2^{\log_2 \pi} = \pi$    (b) $\log_{0.2} 0.2^{(-\sqrt{2})} = -\sqrt{2}$    (c) $\log_{1/5}\left(\frac{1}{5}\right)^{kt} = kt$    ∎

**THEOREM   Properties of Logarithms**

In the following properties, $u$, $v$, and $a$ are positive real numbers, $a \neq 1$, and $r$ is any real number:

- **The Log of a Product Equals the Sum of the Logs**

$$\log_a(uv) = \log_a u + \log_a v \tag{3}$$

- **The Log of a Quotient Equals the Difference of the Logs**

$$\log_a\left(\frac{u}{v}\right) = \log_a u - \log_a v \tag{4}$$

- **The Log of a Power Equals the Product of the Power and the Log**

$$\log_a u^r = r \log_a u \tag{5}$$

**EXAMPLE 10   Using Properties (3), (4), and (5) of Logarithms**

(a) $\log_a\left(x\sqrt{x^2+1}\right) = \log_a x + \log_a \sqrt{x^2+1}$     $\log_a(uv) = \log_a u + \log_a v$

$$= \log_a x + \log_a(x^2+1)^{1/2}$$

$$= \log_a x + \frac{1}{2}\log_a(x^2+1) \qquad \log_a u^r = r\log_a u$$

**CAUTION** In using properties (3) through (5), pay attention to the domain of the variable. For example, the domain of the variable for $\log_a x$ is $x > 0$, and for $\log_a(x-1)$ it is $x > 1$. That is, the equality
$$\log_a x + \log_a(x-1) = \log_a[x(x-1)]$$
is true only for $x > 1$.

**(b)** $\log_a \dfrac{x^2}{(x-1)^3} \underset{\underset{\log_a\left(\frac{u}{v}\right)=\log_a u - \log_a v}{\uparrow}}{=} \log_a x^2 - \log_a(x-1)^3 \underset{\underset{\log_a u^r = r\log_a u}{\uparrow}}{=} 2\log_a x - 3\log_a(x-1)$

**(c)** $\log_a x + \log_a 9 + \log_a(x^2+1) - \log_a 5 = \log_a(9x) + \log_a(x^2+1) - \log_a 5$

$$= \log_a[9x(x^2+1)] - \log_a 5$$

$$= \log_a\left[\frac{9x(x^2+1)}{5}\right] \qquad ∎$$

**CAUTION** Common errors made by some students include:

- Expressing the sum of logarithms as the logarithm of a sum

$$\log_a u + \log_a v \quad \text{is } not \text{ equal to} \quad \log_a(u+v)$$

Correct statement: $\qquad \log_a u + \log_a v = \log_a(uv)$ **Property (3)**

- Expressing the difference of logarithms as the quotient of logarithms

$$\log_a u - \log_a v \quad \text{is } not \text{ equal to} \quad \frac{\log_a u}{\log_a v}$$

Correct statement: $\qquad \log_a u - \log_a v = \log_a\left(\frac{u}{v}\right)$ **Property (4)**

- Expressing a logarithm raised to a power as the product of the power and the logarithm

$$(\log_a u)^r \quad \text{is } not \text{ equal to} \quad r\log_a u$$

Correct statement: $\qquad \log_a u^r = r\log_a u$ **Property (5)**

Since most calculators can calculate only logarithms with base 10, called **common logarithms** and abbreviated log, and logarithms with base $e \approx 2.718$, called **natural logarithms** and abbreviated ln, it is often useful to be able to change the bases of logarithms.

**THEOREM  Change-of-Base Formula**

If $a \neq 1$, $b \neq 1$, and $u$ are positive real numbers, then

$$\boxed{\log_a u = \frac{\log_b u}{\log_b a}}$$

For example, to approximate $\log_2 15$, we use the Change-of-Base Formula. Then we use a calculator.

$$\log_2 15 = \underset{\underset{\text{Change-of-Base Formula}}{\uparrow}}{\frac{\log 15}{\log 2}} \approx 3.907$$

# A.2 Geometry Used in Calculus

**OBJECTIVES** *When you finish this section, you should be able to:*

**1** Use properties of triangles and the Pythagorean Theorem (p. A-11)

**2** Work with congruent triangles and similar triangles (p. A-12)

**3** Use geometry formulas (p. A-15)

## 1 Use Properties of Triangles and the Pythagorean Theorem

A **triangle** is a three-sided polygon. The lengths of the sides are labeled with lowercase letters, such as $a$, $b$, and $c$, and the angles are labeled with uppercase letters, such as $A$, $B$, and $C$, with angle $A$ opposite side $a$, angle $B$ opposite side $b$, and angle $C$ opposite side $c$, as shown in Figure 7.

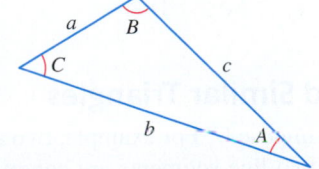

**Figure 7**

Refer to the triangle in Figure 7. The sum of the lengths of any two sides of a triangle is always greater than the length of the remaining side. The sum of the measures of the angles of a triangle, when measured in degrees, equals 180°. That is,

$$a+b>c \qquad a+c>b \qquad b+c>a \qquad A+B+C=180°$$

An **isosceles triangle** is a triangle with two equal sides. In an isosceles triangle, the angles opposite the two equal sides are equal.

An **equilateral triangle** is a triangle with three equal sides. In an equilateral triangle, each angle measures 60°.

The *Pythagorean Theorem* is a statement about *right triangles*. A **right triangle** contains a **right angle**, that is, an angle measuring 90°. The side of the triangle opposite the 90° angle is called the **hypotenuse**; the remaining two sides are called **legs**. In Figure 8, $c$ represents the length of the hypotenuse, and $a$ and $b$ represent the lengths of the legs.

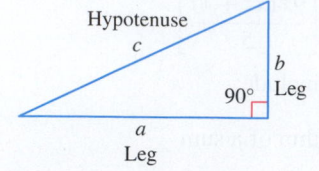

**Figure 8** A right triangle

### THEOREM Pythagorean Theorem

In a right triangle, the square of the length of the hypotenuse is equal to the sum of the squares of the lengths of the legs. That is, in the right triangle shown in Figure 8,

$$\boxed{c^2 = a^2 + b^2}$$

### EXAMPLE 1 Finding the Hypotenuse of a Right Triangle

In a right triangle, one leg has length 4 and the other has length 3. What is the length of the hypotenuse?

**Solution** Since the triangle is a right triangle, we use the Pythagorean Theorem with $a=4$ and $b=3$ to find the length $c$ of the hypotenuse.

$$c^2 = a^2 + b^2$$
$$c^2 = 4^2 + 3^2 = 16 + 9 = 25$$
$$c = \sqrt{25} = 5 \qquad \blacksquare$$

The converse of the Pythagorean Theorem is also true.

### THEOREM Converse of the Pythagorean Theorem

In a triangle, if the square of the length of one side equals the sum of the squares of the lengths of the other two sides, the triangle is a right triangle. The 90° angle is opposite the longest side.

### EXAMPLE 2 Using the Converse of the Pythagorean Theorem

Show that a triangle whose sides have lengths 5, 12, and 13 is a right triangle. Identify the hypotenuse.

**Solution** We square the lengths of the sides.

$$5^2 = 25 \qquad 12^2 = 144 \qquad 13^2 = 169$$

Notice that the sum of the first two squares (25 and 144) equals the third square (169). So, the triangle is a right triangle. The longest side, 13, is the hypotenuse. $\blacksquare$

See Figure 9.

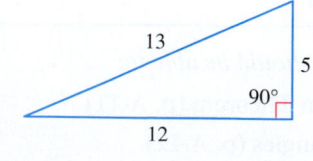

**Figure 9**

## 2 Work with Congruent Triangles and Similar Triangles

The word *congruent* means "coinciding when superimposed." For example, two angles are congruent if they have the same measure, and two line segments are congruent if they have the same length.

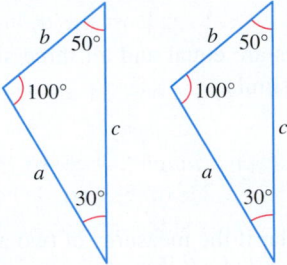

**Figure 10** Congruent triangles.

**DEFINITION Congruent Triangles**

Two triangles are **congruent** if in each triangle, the corresponding angles have the same measure, and the corresponding sides have the same length.

In Figure 10, corresponding angles are equal, and the lengths of the corresponding sides are equal. So, these triangles are congruent.

It is not necessary to verify that all three angles and all three sides have the same measure to determine whether two triangles are congruent.

## Determining Congruent Triangles

- **Angle-Side-Angle Case (ASA)** Two triangles are congruent if the measures of two angles from each triangle are equal and the lengths of the corresponding sides between the two equal angles are equal.

For example, in Figure 11, the two triangles are congruent because both triangles have an angle measuring $40°$, an angle measuring $80°$, and the sides between these angles are both 10 units in length.

- **Side-Side-Side Case (SSS)** Two triangles are congruent if the lengths of the sides of one triangle are equal to the lengths of the sides of the other triangle.

For example, in Figure 12, the two triangles are congruent because both triangles have sides with lengths 8, 15, and 20 units.

- **Side-Angle-Side Case (SAS)** Two triangles are congruent if the lengths of two corresponding sides of the triangles are equal and the angles between the two sides have the same measure.

For example, in Figure 13, the two triangles are congruent because both triangles have sides of length 7 units and 8 units, and the angle between the two congruent sides in each triangle measures $40°$.

- **Angle-Angle-Side Case (AAS)** Two triangles are congruent if the measures of two angles and the length of a nonincluded side of one triangle are equal to the corresponding parts of the other triangle.

For example, in Figure 14, the two triangles are congruent because both triangles have angles measuring $35°$ and $96°$, and the nonincluded side adjacent to the $35°$ angle has length 10 units.

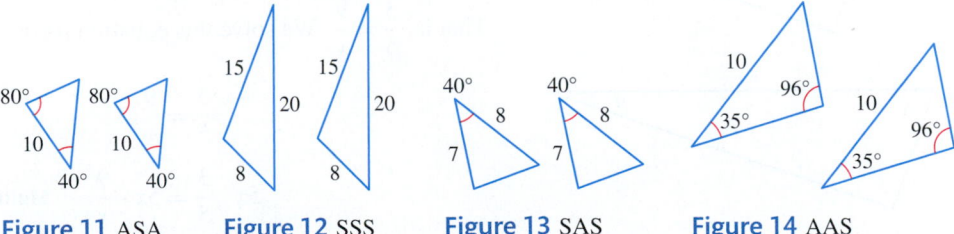

**Figure 11** ASA      **Figure 12** SSS      **Figure 13** SAS      **Figure 14** AAS

We contrast congruent triangles with *similar* triangles.

**DEFINITION Similar Triangles**

Two triangles are **similar** if in each triangle, corresponding angles have the same measure and corresponding sides are proportional in length, that is, the ratio of the lengths of the corresponding sides of each triangle equals the same constant.

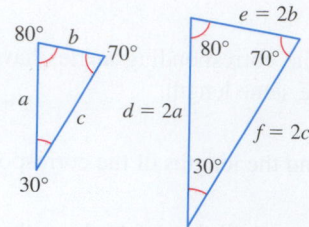

**Figure 15**

For example, the triangles in Figure 15 are similar because each of the corresponding angles of the two triangles has the same measure. Also, the lengths of the corresponding sides are proportional: each side of the triangle on the right is twice as long as the corresponding side of the triangle on the left. That is, the ratio of the corresponding sides is a constant: $\dfrac{d}{a} = \dfrac{e}{b} = \dfrac{f}{c} = 2$.

It is not necessary to verify that all three angles are equal and all three sides are proportional to determine whether two triangles are similar.

## Determining Similar Triangles

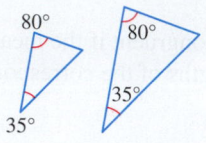

**Figure 16** AA

- **Angle-Angle Case (AA)** Two triangles are similar if the measures of two angles from each triangle are equal.

  For example, the two triangles in Figure 16 are similar because each triangle has an angle measuring 35° and an angle measuring 80°.

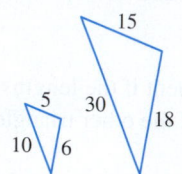

**Figure 17** SSS

- **Side-Side-Side Case (SSS)** Two triangles are similar if the lengths of the sides of one triangle are proportional to the lengths of the sides of the second triangle.

  For example, Figure 17 shows two triangles with sides of lengths 5, 6, 10 and 15, 18, 30. These two triangles are similar because

  $$\frac{5}{15} = \frac{6}{18} = \frac{10}{30} = \frac{1}{3}$$

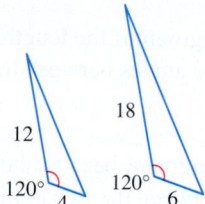

**Figure 18** SAS

- **Side-Angle-Side Case (SAS)** Two triangles are similar if the lengths of two sides of one triangle are proportional to the lengths of two sides of the second triangle, and the angles between the corresponding two sides have equal measure.

  For example, in Figure 18, the two triangles are similar because $\dfrac{4}{6} = \dfrac{12}{18} = \dfrac{2}{3}$ and the angle between sides 4 and 12 and the angle between sides 6 and 18 each measures 120°.

### EXAMPLE 3   Using Similar Triangles

Given that the triangles in Figure 19 are similar, find the missing length $x$ and angles $A$, $B$, and $C$.

**Solution** Because the triangles are similar, corresponding angles have the same measure. So, $A = 71°$, $B = 19°$, and $C = 90°$. Also corresponding sides are proportional. That is, $\dfrac{3}{5} = \dfrac{9}{x}$. We solve this equation for $x$.

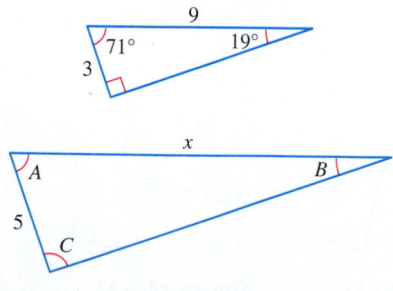

**Figure 19**

$$\frac{3}{5} = \frac{9}{x}$$

$$5x \cdot \frac{3}{5} = 5x \cdot \frac{9}{x} \qquad \text{Multiply both sides by } 5x.$$

$$3x = 45 \qquad \text{Simplify.}$$

$$x = 15 \qquad \text{Divide both sides by 3.}$$

The missing length is 15 units. ∎

## 3 Use Geometry Formulas

Certain formulas from geometry are useful in solving calculus problems. Some of these are listed here.

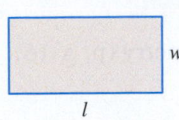

For a rectangle of length $l$ and width $w$,

$$\text{Area} = lw \qquad \text{Perimeter} = 2l + 2w$$

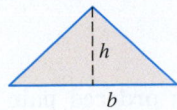

For a triangle with base $b$ and height $h$,

$$\text{Area} = \frac{1}{2}bh$$

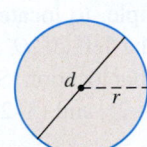

For a circle of radius $r$ (diameter $d = 2r$),

$$\text{Area} = \pi r^2 \qquad \text{Circumference} = 2\pi r = \pi d$$

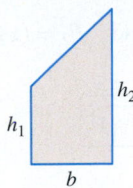

For a trapezoid with base $b$ and parallel heights $h_1$ and $h_2$,

$$\text{Area} = \frac{1}{2}b(h_1 + h_2)$$

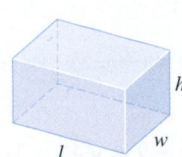

For a closed rectangular box of length $l$, width $w$, and height $h$,

$$\text{Volume} = lwh \qquad \text{Surface area} = 2lh + 2wh + 2lw$$

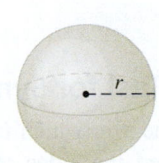

For a sphere of radius $r$,

$$\text{Volume} = \frac{4}{3}\pi r^3 \qquad \text{Surface area} = 4\pi r^2$$

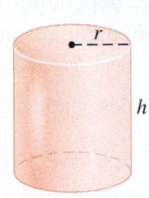

For a closed, right circular cylinder of height $h$ and radius $r$,

$$\text{Volume} = \pi r^2 h \qquad \text{Surface area} = 2\pi r^2 + 2\pi rh$$

If the cylinder has no top or bottom, the surface area is $2\pi rh$.

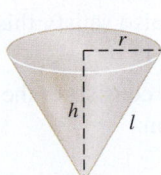

For a right circular cone of height $h$, radius $r$, and slant height $l = \sqrt{h^2 + r^2}$ that is open on top,

$$\text{Volume} = \frac{1}{3}\pi r^2 h \quad \text{Lateral surface area} = \pi rl = \pi r\sqrt{h^2 + r^2}$$

If the cone is closed on top, the surface area is $\pi r^2 + \pi rl$.

# A.3 Analytic Geometry Used in Calculus

**OBJECTIVES** *When you finish this section, you should be able to:*

1 Use the distance formula (p. A-16)
2 Graph equations, find intercepts, and test for symmetry (p. A-16)
3 Work with equations of a line (p. A-18)
4 Work with the equation of a circle (p. A-21)
5 Graph parabolas, ellipses, and hyperbolas (p. A-22)

## 1 Use the Distance Formula

Any point $P$ in the $xy$-plane can be located using an **ordered pair** $(x, y)$ of real numbers. The ordered pair $(x, y)$, called the **coordinates** of $P$, gives enough information to locate the point $P$ in the plane. For example, to locate the point with coordinates $(-3, 1)$, move 3 units along the $x$-axis to the left of $O$ and then move straight up 1 unit. Then **plot** the point by placing a dot at this location. See Figure 20 in which the points with coordinates $(-3, 1)$, $(-2, -3)$, $(3, -2)$, and $(3, 2)$ are plotted.

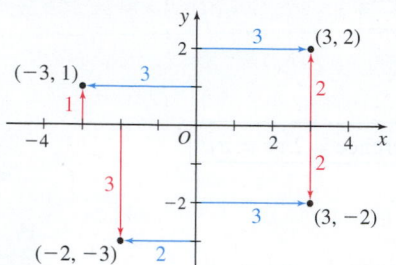

**Figure 20**

**NOTE** If $(x, y)$ are the coordinates of a point $P$, then $x$ is called the $x$-**coordinate** of $P$, and $y$ is called the $y$-**coordinate** of $P$.

> **THEOREM  Distance Formula**
>
> The distance between two points $P_1 = (x_1, y_1)$ and $P_2 = (x_2, y_2)$, denoted by $d(P_1, P_2)$, is
>
> $$d(P_1, P_2) = \sqrt{(x_2 - x_1)^2 + (y_2 - y_1)^2}$$

**EXAMPLE 1  Using the Distance Formula**

Find the distance $d$ between the points $(-3, 5)$ and $(3, 2)$.

**Solution** We use the distance formula with $P_1 = (x_1, y_1) = (-3, 5)$ and $P_2 = (x_2, y_2) = (3, 2)$.

$$d = \sqrt{[3 - (-3)]^2 + (2 - 5)^2} = \sqrt{6^2 + (-3)^2} = \sqrt{36 + 9} = \sqrt{45} = 3\sqrt{5} \approx 6.708$$

## 2 Graph Equations, Find Intercepts, and Test for Symmetry

An **equation in two variables**, say, $x$ and $y$, is a statement in which two expressions involving $x$ and $y$ are equal. The expressions are called the **sides** of the equation. Since an equation is a statement, it may be true or false, depending on the value of the variables. Any values of $x$ and $y$ that result in a true statement are said to **satisfy** the equation.

For example, the following are equations in two variables $x$ and $y$:

$$x^2 + y^2 = 5 \quad 2x - y = 6 \quad y = 2x + 5 \quad x^2 = y$$

The equation $x^2 + y^2 = 5$ is satisfied for $x = 1$ and $y = 2$, since $1^2 + 2^2 = 1 + 4 = 5$. Other choices of $x$ and $y$, such as $x = -1$ and $y = -2$, also satisfy this equation. It is not satisfied for $x = 2$ and $y = 3$, since $2^2 + 3^2 = 4 + 9 = 13 \neq 5$.

The **graph of an equation in two variables** $x$ and $y$ consists of the set of points in the $xy$-plane whose coordinates $(x, y)$ satisfy the equation.

**EXAMPLE 2  Graphing an Equation by Plotting Points**

Graph the equation $y = x^2$.

**Solution** Table 4 lists several points on the graph. In Figure 21(a) the points are plotted and then they are connected with a smooth curve to obtain the graph (a *parabola*). Figure 21(b) shows the graph using graphing technology.

**TABLE 4**

| $x$ | $y = x^2$ | $(x, y)$ |
|-----|-----------|----------|
| $-4$ | 16 | $(-4, 16)$ |
| $-3$ | 9 | $(-3, 9)$ |
| $-2$ | 4 | $(-2, 4)$ |
| $-1$ | 1 | $(-1, 1)$ |
| 0 | 0 | $(0, 0)$ |
| 1 | 1 | $(1, 1)$ |
| 2 | 4 | $(2, 4)$ |
| 3 | 9 | $(3, 9)$ |
| 4 | 16 | $(4, 16)$ |

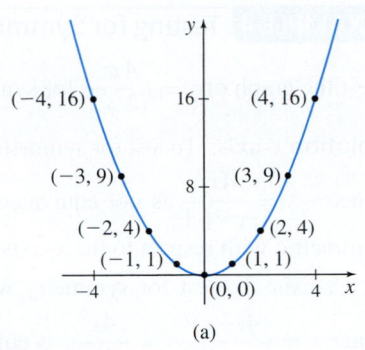

(a)

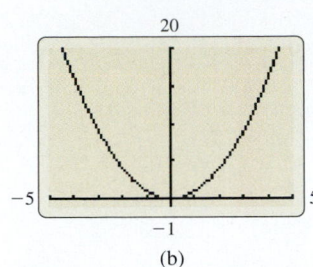

(b)

**Figure 21** $y = x^2$

The graphs in Figure 21 do not show all the points whose coordinates $(x, y)$ satisfy the equation $y = x^2$. For example, the point $(6, 36)$ satisfies the equation $y = x^2$, but it is not shown. It is important when graphing to present enough of the graph so that any viewer will "see" the rest of the graph as an obvious continuation of what is actually there.

The points, if any, at which a graph crosses or touches a coordinate axis are called the **intercepts** of the graph. See Figure 22. The $x$-coordinate of a point where a graph crosses or touches the $x$-axis is an **$x$-intercept**. At an $x$-intercept, the $y$-coordinate equals 0. The $y$-coordinate of a point where a graph crosses or touches the $y$-axis is a **$y$-intercept**. At a $y$-intercept, the $x$-coordinate equals 0. When graphing an equation, all its intercepts should be displayed or be easily inferred from the part of the graph that is displayed.

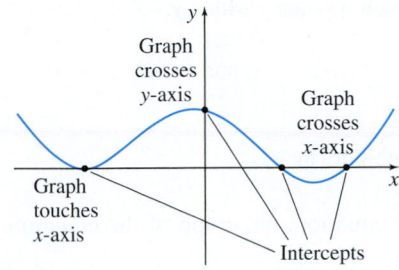

**Figure 22**

### EXAMPLE 3  Finding the Intercepts of a Graph

Find the $x$-intercept(s) and the $y$-intercept(s) of the graph of $y = x^2 - 4$.

**Solution** To find the $x$-intercept(s), we let $y = 0$ and solve the equation

$$x^2 - 4 = 0$$
$$(x + 2)(x - 2) = 0 \qquad \text{Factor.}$$
$$x = -2 \quad \text{or} \quad x = 2 \qquad \text{Set each factor equal to 0 and solve.}$$

The equation has two solutions, $-2$ and $2$. The $x$-intercepts are $-2$ and $2$.

To find the $y$-intercept(s), we let $x = 0$ in the equation.

$$y = 0^2 - 4 = -4$$

The $y$-intercept is $-4$. ∎

### DEFINITION  Symmetry

- A graph is **symmetric with respect to the $x$-axis** if, for every point $(x, y)$ on the graph, the point $(x, -y)$ is also on the graph. See Figure 23.
- A graph is **symmetric with respect to the $y$-axis** if, for every point $(x, y)$ on the graph, the point $(-x, y)$ is also on the graph. See Figure 24.
- A graph is **symmetric with respect to the origin** if, for every point $(x, y)$ on the graph, the point $(-x, -y)$ is also on the graph. See Figure 25.

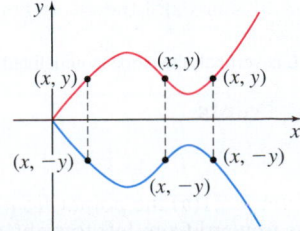

**Figure 23** Symmetry with respect to the $x$-axis.

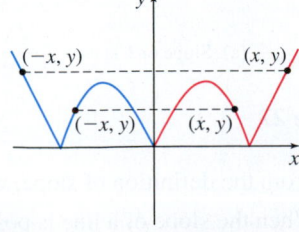

**Figure 24** Symmetry with respect to the $y$-axis.

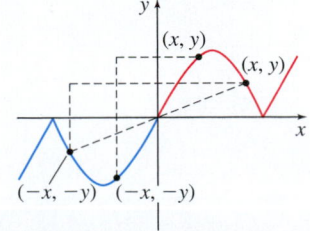

**Figure 25** Symmetry with respect to the origin.

## EXAMPLE 4  Testing for Symmetry

Test the graph of $y = \dfrac{4x^2}{x^2 + 1}$ for symmetry.

**Solution** *x*-axis: To test for symmetry with respect to the *x*-axis, we replace *y* with $-y$.
Since $-y = \dfrac{4x^2}{x^2 + 1}$ is not equivalent to $y = \dfrac{4x^2}{x^2 + 1}$, the graph of the equation is not symmetric with respect to the *x*-axis.

*y*-axis: To test for symmetry with respect to the *y*-axis, we replace *x* with $-x$.
Since $y = \dfrac{4(-x)^2}{(-x)^2 + 1} = \dfrac{4x^2}{x^2 + 1}$ is equivalent to $y = \dfrac{4x^2}{x^2 + 1}$, the graph of the equation is symmetric with respect to the *y*-axis.

Origin:  To test for symmetry with respect to the origin, we replace *x* with $-x$ and *y* with $-y$.

$$-y = \frac{4(-x)^2}{(-x)^2 + 1} \qquad \text{Replace } x \text{ with } -x \text{ and } y \text{ with } -y.$$

$$-y = \frac{4x^2}{x^2 + 1} \qquad \text{Simplify.}$$

$$y = -\frac{4x^2}{x^2 + 1} \qquad \text{Multiply both sides by } -1.$$

Since the result is not equivalent to the original equation, the graph of the equation is not symmetric with respect to the origin. ∎

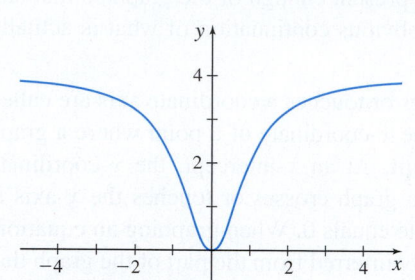

**Figure 26** $y = \dfrac{4x^2}{x^2 + 1}$

Figure 26 shows the graph of $y = \dfrac{4x^2}{x^2 + 1}$ and confirms the symmetry with respect to the *y*-axis.

## 3  Work with Equations of a Line

**DEFINITION  Slope**

Let $P = (x_1, y_1)$ and $Q = (x_2, y_2)$ be two distinct points. If $x_1 \neq x_2$, the **slope** *m* of the nonvertical line *L* containing *P* and *Q* is defined by the formula

$$\boxed{m = \frac{y_2 - y_1}{x_2 - x_1} \qquad x_1 \neq x_2}$$

If $x_1 = x_2$, then *L* is a **vertical line**, and the slope *m* of *L* is **undefined**.

Figure 27(a) provides an illustration of the slope of a nonvertical line; Figure 27(b) illustrates a vertical line.

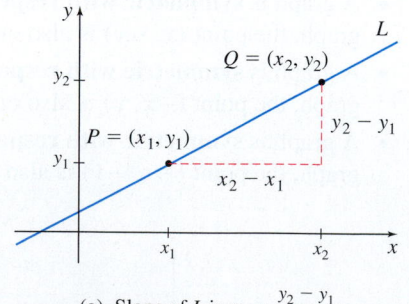

(a)  Slope of *L* is $m = \dfrac{y_2 - y_1}{x_2 - x_1}$       (b)  *L* is vertical; the slope is undefined.

**Figure 27**

From the definition of slope, we conclude:

- When the slope of a line is positive, the line slants upward from left to right.
- When the slope of a line is negative, the line slants downward from left to right.

- When the slope of a line is 0, the line is horizontal.
- When the slope of a line is undefined, the line is vertical.

### THEOREM  Equation of a Vertical Line

A vertical line is given by an equation of the form

$$x = a$$

where $a$ is the $x$-intercept.

### THEOREM  Point-Slope Form of an Equation of a Line

An equation of a nonvertical line with slope $m$ that contains the point $(x_1, y_1)$ is

$$y - y_1 = m\,(x - x_1)$$

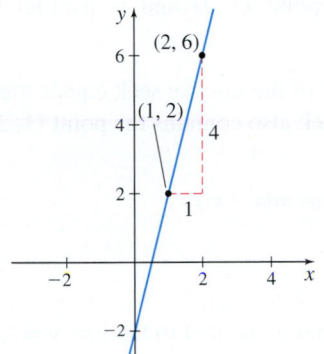

**Figure 28** $y = 4x - 2$

| EXAMPLE 5 | **Using the Point-Slope Form of a Line** |

An equation of the line with slope 4 and containing the point $(1, 2)$ can be found by using the point-slope form with $m = 4$, $x_1 = 1$, and $y_1 = 2$.

$$
\begin{aligned}
y - y_1 &= m(x - x_1) && \text{Point-slope form of a line} \\
y - 2 &= 4(x - 1) && m = 4,\ x_1 = 1,\ y_1 = 2 \\
y &= 4x - 2 && \text{Solve for } y.
\end{aligned}
$$

See Figure 28 for the graph of $y$. ∎

Another useful equation of a line is obtained when the slope $m$ and the $y$-intercept $b$ are known. The point-slope form is used to obtain the following equation of a line with slope $m$, containing the point $(0, b)$:

$$y - b = m(x - 0) \qquad \text{or equivalently} \qquad y = mx + b$$

### THEOREM  Slope-Intercept Form of an Equation of a Line

An equation of a line with slope $m$ and $y$-intercept $b$ is

$$y = mx + b$$

For a horizontal line, the slope $m$ is 0.

### THEOREM  Equation of a Horizontal Line

A horizontal line is given by an equation of the form

$$y = b$$

where $b$ is the $y$-intercept.

| EXAMPLE 6 | **Finding the Slope and $y$-Intercept from an Equation of a Line** |

Find the slope $m$ and the $y$-intercept of the equation $2x + 4y = 8$. Graph the equation.

**Solution** To obtain the slope and $y$-intercept, we write the equation in slope-intercept form by solving for $y$.

$$
\begin{aligned}
2x + 4y &= 8 \\
4y &= -2x + 8 \\
y &= -\frac{1}{2}x + 2 \qquad y = mx + b
\end{aligned}
$$

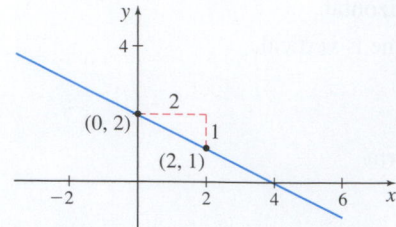

**Figure 29** $2x + 4y = 8$

The coefficient of $x$ is the slope. The slope is $-\dfrac{1}{2}$. The constant 2 is the $y$-intercept, so the point $(0, 2)$ is on the graph. We can use the slope $-\dfrac{1}{2}$ to obtain a second point on the line. Starting at the point $(0, 2)$, move 2 units to the right and then 1 unit down to the point $(2, 1)$. Plot this point and draw a line through the two points. See Figure 29. ∎

When two lines in the plane do not intersect, they are **parallel**.

**THEOREM  Criterion for Parallel Lines**

Two nonvertical lines are parallel if and only if their slopes are equal and their $y$-intercepts are not equal.

**EXAMPLE 7**  **Finding an Equation of a Line Parallel to a Given Line**

Find an equation of the line that contains the point $(1, 2)$ and is parallel to the line $y = 5x$.

**Solution** Since the two lines are parallel, the slope of the line we seek equals the slope of the line $y = 5x$, which is 5. Since the line we seek also contains the point $(1, 2)$, use the point-slope form to obtain the equation.

$$y - 2 = 5(x - 1) \qquad \textcolor{blue}{y - y_1 = m(x - x_1)}$$
$$y - 2 = 5x - 5$$
$$y = 5x - 3$$

The line $y = 5x - 3$ contains the point $(1, 2)$ and is parallel to the line $y = 5x$. See Figure 30. ∎

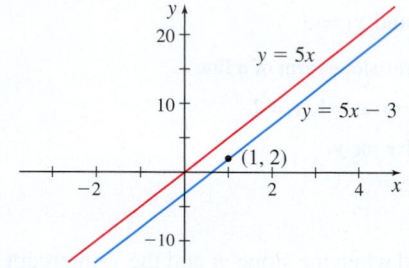

**Figure 30** Parallel lines.

When two lines intersect at a right angle ($90°$), the lines are **perpendicular**. See Figure 31.

**THEOREM  Criterion for Perpendicular Lines**

Two nonvertical lines are perpendicular if and only if the product of their slopes is $-1$.

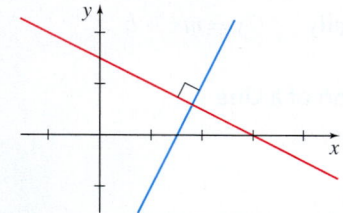

**Figure 31** Perpendicular lines.

**EXAMPLE 8**  **Finding an Equation of a Line Perpendicular to a Given Line**

Find an equation of the line that contains the point $(-1, 3)$ and is perpendicular to the line $4x + y = -1$.

**Solution** Begin by writing the equation of the given line in slope-intercept form to find its slope.

$$4x + y = -1$$
$$y = -4x - 1$$

This line has a slope of $-4$. Any line perpendicular to this line will have slope $\dfrac{1}{4}$. Because the point $(-1, 3)$ is on this line, we use the point-slope form of a line.

$$y - 3 = \frac{1}{4}[x - (-1)] \qquad \textcolor{blue}{y - y_1 = m(x - x_1)}$$
$$y - 3 = \frac{1}{4}x + \frac{1}{4}$$
$$y = \frac{1}{4}x + \frac{13}{4}$$

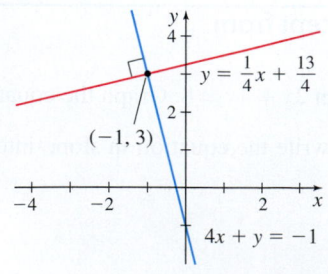

**Figure 32**

Figure 32 shows the graphs of $4x + y = -1$ and $y = \dfrac{1}{4}x + \dfrac{13}{4}$. ∎

## 4 Work with the Equation of a Circle

An advantage of a coordinate system is that it enables a geometric statement to be translated into an algebraic statement, and vice versa. Consider, for example, the following geometric statement that defines a circle.

### DEFINITION Circle

A **circle** is the set of points in the $xy$-plane that is a fixed distance $r$ from a fixed point $(h, k)$. The fixed distance $r$ is called the **radius**, and the fixed point $(h, k)$ is called the **center** of the circle.

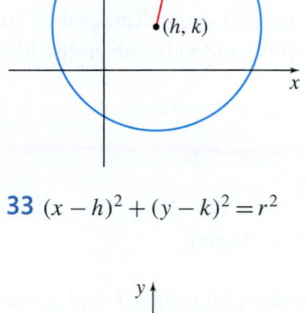

Figure 33 shows the graph of a circle. To find an equation that describes a circle, let $(x, y)$ represent the coordinates of any point on the circle with radius $r$ and center $(h, k)$. Then the distance between the point $(x, y)$ and $(h, k)$ equals $r$. That is, by the distance formula,

$$\sqrt{(x - h)^2 + (y - k)^2} = r$$

Now square both sides to obtain

$$(x - h)^2 + (y - k)^2 = r^2$$

**Figure 33** $(x - h)^2 + (y - k)^2 = r^2$

This form of the equation of a circle is called the **standard form**.

### THEOREM

The standard form of a circle with radius $r$ and center at the origin $(0, 0)$ is

$$x^2 + y^2 = r^2$$

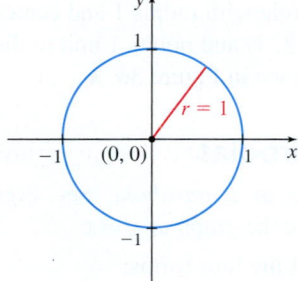

### DEFINITION Unit Circle

The circle with radius $r = 1$ and center at the origin is called the **unit circle** and has the equation

$$x^2 + y^2 = 1$$

**Figure 34** The unit circle: $x^2 + y^2 = 1$

See Figure 34.

### EXAMPLE 9  Graphing a Circle

Graph the equation $(x + 3)^2 + (y - 2)^2 = 16$.

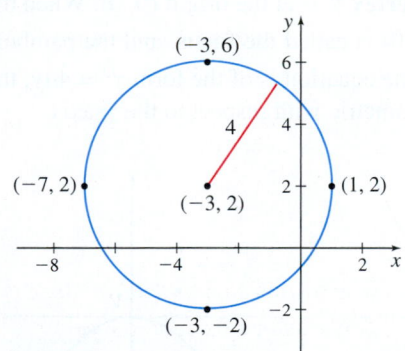

**Solution** This is the standard form of an equation of a circle. To graph the circle, first identify the center and the radius of the circle.

$$(x + 3)^2 + (y - 2)^2 = 16 \qquad (x - h)^2 + (y - k)^2 = r^2$$
$$(x - (-3))^2 + (y - 2)^2 = 4^2$$
$$\quad\uparrow\qquad\qquad\uparrow\quad\ \uparrow$$
$$\quad h\qquad\qquad\ \ k\quad\ r^2$$

The circle has its center at the point $(-3, 2)$; its radius is 4 units. To graph the circle, plot the center $(-3, 2)$. Then locate the four points on the circle that are 4 units to the left, to the right, above, and below the center. These four points are used as guides to obtain the graph. See Figure 35. ∎

**Figure 35** $(x + 3)^2 + (y - 2)^2 = 16$

### DEFINITION General Form of the Equation of a Circle

The equation

$$x^2 + y^2 + ax + bx + c = 0$$

is called the **general form of the equation of a circle.**

**NEED TO REVIEW?** Completing the square is discussed in Section A.1, pp. A-2 and A-3.

If the equation of a circle is given in the general form, we use the method of completing the square to put the equation in standard form to identify the center and radius.

---

**EXAMPLE 10** Graphing a Circle Whose Equation Is in General Form

Graph the equation

$$x^2 + y^2 + 4x - 6y + 12 = 0$$

**Solution** We complete the square in both $x$ and $y$ to put the equation in standard form. First, group the terms involving $x$, group the terms involving $y$, and put the constant on the right side of the equation. The result is

$$(x^2 + 4x) + (y^2 - 6y) = -12$$

Next, complete the square of each expression in parentheses. Remember that any number added to the left side of an equation must also be added to the right side.

$$(x^2 + 4x + 4) + (y^2 - 6y + 9) = -12 + 4 + 9 = 1$$

$$\text{Add } \left(\frac{4}{2}\right)^2 = 4 \quad \text{Add } \left(\frac{-6}{2}\right)^2 = 9$$

$$(x + 2)^2 + (y - 3)^2 = 1 \qquad \text{Factor.}$$

This is the standard form of the equation of a circle with radius 1 and center at the point $(-2, 3)$. To graph the circle, plot the center $(-2, 3)$ and points 1 unit to the right, to the left, above, and below the point $(-2, 3)$, as shown in Figure 36. ∎

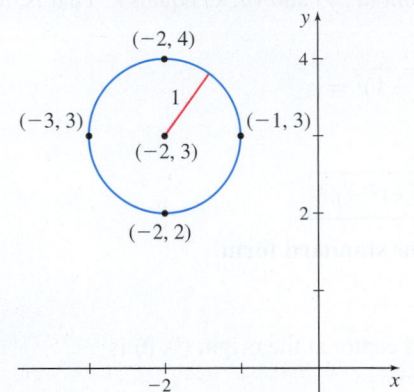

**Figure 36** $x^2 + y^2 + 4x - 6y + 12 = 0$

## 5 Graph Parabolas, Ellipses, and Hyperbolas

The graph of the equation $y = x^2$ is an example of a *parabola*. See Figure 21 on page A-17. In fact, every parabola has a shape like the graph of $y = x^2$.

A **parabola** is the graph of an equation in one of the four forms:

$$\boxed{y^2 = 4ax \qquad y^2 = -4ax \qquad x^2 = 4ay \qquad x^2 = -4ay}$$

where $a > 0$.

In each of the parabolas in Figure 37, the **vertex** $V$ is at the origin $(0, 0)$. When the equation is of the form $y^2 = 4ax$, the point $(a, 0)$ is called the **focus**, and the parabola is symmetric with respect to the $x$-axis. When the equation is of the form $x^2 = 4ay$, the point $(0, a)$ is the focus, and the parabola is symmetric with respect to the $y$-axis.

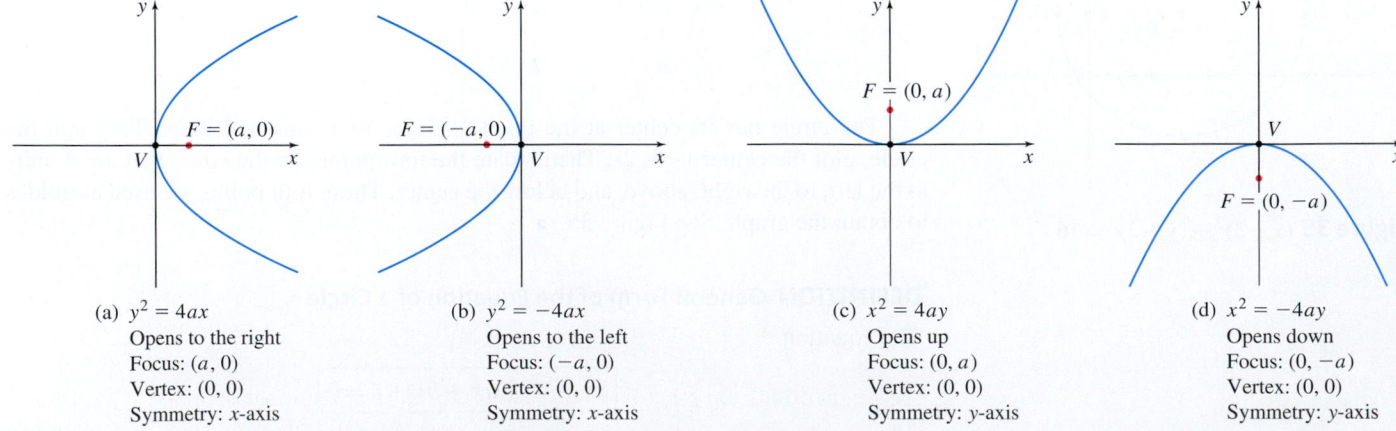

(a) $y^2 = 4ax$
Opens to the right
Focus: $(a, 0)$
Vertex: $(0, 0)$
Symmetry: $x$-axis

(b) $y^2 = -4ax$
Opens to the left
Focus: $(-a, 0)$
Vertex: $(0, 0)$
Symmetry: $x$-axis

(c) $x^2 = 4ay$
Opens up
Focus: $(0, a)$
Vertex: $(0, 0)$
Symmetry: $y$-axis

(d) $x^2 = -4ay$
Opens down
Focus: $(0, -a)$
Vertex: $(0, 0)$
Symmetry: $y$-axis

**Figure 37**

**EXAMPLE 11** **Graphing a Parabola with Vertex at the Origin**

Graph the equation $x^2 = 16y$.

**Solution** The graph of $x^2 = 16y$ is a parabola of the form

$$x^2 = 4ay$$

where $a = 4$. Look at Figure 37(c). The graph will open up, with focus at $(0, 4)$. The vertex is at $(0, 0)$. To graph the parabola, we let $y = a$; that is, let $y = 4$. (Any other positive number will also work.)

$$x^2 = 16y = 16(4) = 64$$
$$\uparrow$$
$$y = a = 4$$
$$x = -8 \quad \text{or} \quad x = 8$$

The points $(-8, 4)$ and $(8, 4)$ are on the parabola and establish its opening. See Figure 38. ∎

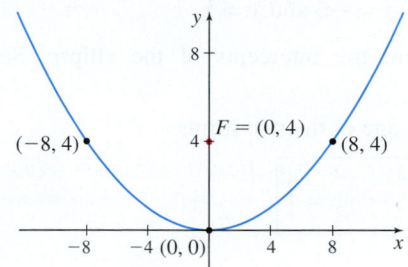

**Figure 38** $x^2 = 16y$

An **ellipse** is the graph of an equation in one of the two forms:

$$\frac{x^2}{a^2} + \frac{y^2}{b^2} = 1 \qquad \frac{x^2}{b^2} + \frac{y^2}{a^2} = 1$$

where $a \geq b > 0$.

**NOTE** If $a = b$, then $\dfrac{x^2}{a^2} + \dfrac{y^2}{a^2} = 1$ is the equation of a circle with radius $a$ and center at the origin.

An ellipse is oval-shaped. The line segment dividing the ellipse in half the long way is the **major axis**; its length is $2a$. The two points of intersection of the ellipse and the major axis are the **vertices** of the ellipse. The midpoint of the vertices is the **center** of the ellipse. The line segment through the center of the ellipse and perpendicular to the major axis is the **minor axis**. Along the major axis $c$ units from the center of the ellipse, where $c^2 = a^2 - b^2$, $c > 0$, are two points, called **foci**, that determine the shape of the ellipse. See Figure 39.

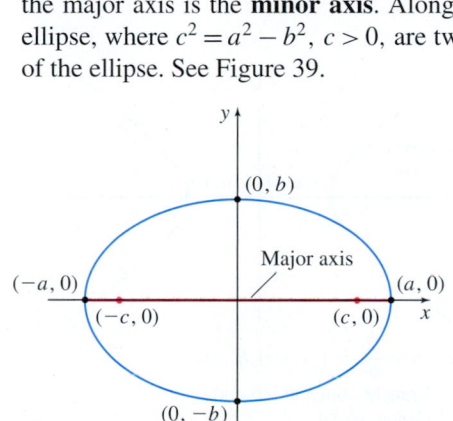

(a) $\dfrac{x^2}{a^2} + \dfrac{y^2}{b^2} = 1,\, a \geq b > 0$
Major axis: along the $x$-axis
Center: $(0, 0)$
Vertices: $(-a, 0), (a, 0)$
Foci: $(-c, 0), (c, 0)$, where $c^2 = a^2 - b^2, c > 0$
Symmetry: $x$-axis, $y$-axis, origin

(b) $\dfrac{x^2}{b^2} + \dfrac{y^2}{a^2} = 1,\, a \geq b > 0$
Major axis: along the $y$-axis
Center: $(0, 0)$
Vertices: $(0, -a), (0, a)$
Foci: $(0, -c), (0, c)$, where $c^2 = a^2 - b^2, c > 0$
Symmetry: $x$-axis, $y$-axis, origin

**Figure 39**

**EXAMPLE 12** **Graphing an Ellipse with Center at the Origin**

Graph the equation $9x^2 + 4y^2 = 36$.

**Solution** To put the equation in standard form, we divide each side by 36.

$$\frac{x^2}{4} + \frac{y^2}{9} = 1$$

The graph of this equation is an ellipse. Since the larger number is under $y^2$, the major axis is along the $y$-axis. The center is at the origin $(0, 0)$. It is easiest to graph an ellipse by finding its intercepts:

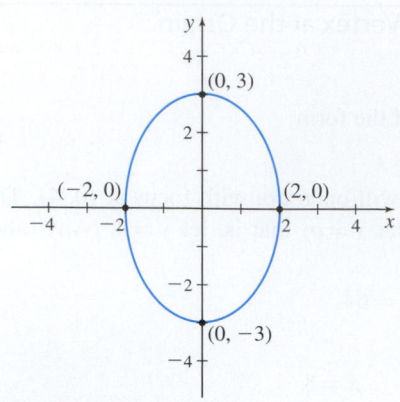

**Figure 40** $9x^2 + 4y^2 = 36$

$x$-intercepts: Let $y = 0$

$$\frac{x^2}{4} + \frac{0^2}{9} = 1$$

$$x^2 = 4$$

$$x = -2 \text{ and } x = 2$$

$y$-intercepts: Let $x = 0$

$$0^2 + \frac{y^2}{9} = 1$$

$$y^2 = 9$$

$$y = -3 \text{ and } y = 3$$

The points $(-2, 0)$, $(2, 0)$, $(0, -3)$, $(0, 3)$ are the intercepts of the ellipse. See Figure 40. ∎

A **hyperbola** is the graph of an equation in one of the two forms:

$$\frac{x^2}{a^2} - \frac{y^2}{b^2} = 1 \qquad \frac{y^2}{a^2} - \frac{x^2}{b^2} = 1$$

where $a > 0$ and $b > 0$.

Two points, called **foci**, determine the shape of the hyperbola. The midpoint of the line segment containing the foci is called the **center** of the hyperbola. The foci are located $c$ units from the center, where $c^2 = a^2 + b^2$, $c > 0$. The line containing the foci is called the **transverse axis**. The **vertices** of a hyperbola are its intercepts.

See Figure 41. The graph of a hyperbola consists of two branches.

- For the hyperbola $\dfrac{x^2}{a^2} - \dfrac{y^2}{b^2} = 1$, the branches of the graph open left and right.

- For the hyperbola $\dfrac{y^2}{a^2} - \dfrac{x^2}{b^2} = 1$, the branches of the graph open up and down.

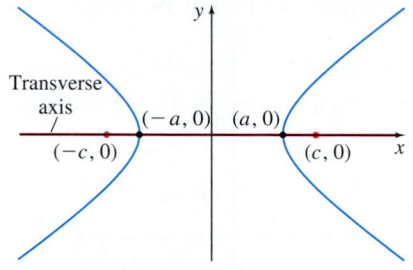

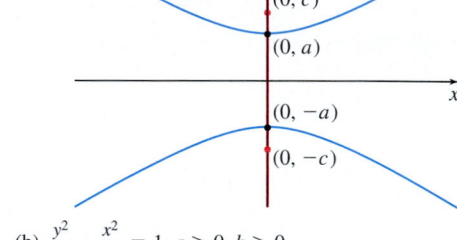

(a) $\dfrac{x^2}{a^2} - \dfrac{y^2}{b^2} = 1$, $a > 0$, $b > 0$
Branches open to left and right
Center: $(0, 0)$
Vertices: $(-a, 0)$, $(a, 0)$
Foci: $(-c, 0)$, $(c, 0)$, where $c^2 = a^2 + b^2$, $c > 0$
Symmetry: $x$-axis, $y$-axis, origin

(b) $\dfrac{y^2}{a^2} - \dfrac{x^2}{b^2} = 1$, $a > 0$, $b > 0$
Branches open up and down
Center: $(0, 0)$
Vertices: $(0, -a)$, $(0, a)$
Foci: $(0, -c)$, $(0, c)$, where $c^2 = a^2 + b^2$, $c > 0$
Symmetry: $x$-axis, $y$-axis, origin

**Figure 41**

---

**EXAMPLE 13** **Graphing a Hyperbola with Center at the Origin**

Graph the equation $\dfrac{y^2}{4} - \dfrac{x^2}{5} = 1$.

**Solution** The graph of $\dfrac{y^2}{4} - \dfrac{x^2}{5} = 1$ is a hyperbola. The hyperbola consists of two branches, one opening up, the other opening down, like the graph in Figure 41(b). The hyperbola has no $x$-intercepts. To find the $y$-intercepts, we let $x = 0$ and solve for $y$.

$$\frac{y^2}{4} = 1$$

$$y^2 = 4$$

$$y = -2 \qquad \text{or} \qquad y = 2$$

The $y$-intercepts are $-2$ and $2$, so the vertices are $(0, -2)$ and $(0, 2)$. The transverse axis is the vertical line $x = 0$. To graph the hyperbola, let $y = \pm 3$ (or any numbers $\geq 2$ or $\leq -2$).

Then

$$\frac{y^2}{4} - \frac{x^2}{5} = 1$$

$$\frac{9}{4} - \frac{x^2}{5} = 1 \qquad y = \pm 3$$

$$\frac{x^2}{5} = \frac{5}{4}$$

$$x^2 = \frac{25}{4}$$

$$x = -\frac{5}{2} \text{ or } x = \frac{5}{2}$$

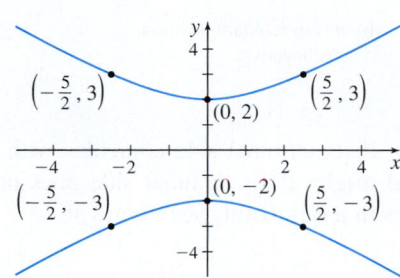

**Figure 42** $\dfrac{y^2}{4} - \dfrac{x^2}{5} = 1$

The points $\left(-\dfrac{5}{2}, 3\right)$, $\left(-\dfrac{5}{2}, -3\right)$, $\left(\dfrac{5}{2}, 3\right)$, and $\left(\dfrac{5}{2}, -3\right)$ are on the hyperbola. See Figure 42 for the graph. ∎

# A.4 Trigonometry Used in Calculus

**OBJECTIVES** *When you finish this section, you should be able to:*

1 Work with angles, arc length of a circle, and circular motion (p. A-25)
2 Define and evaluate trigonometric functions (p. A-27)
3 Determine the domain and the range of the trigonometric functions (p. A-31)
4 Use basic trigonometry identities (p. A-32)
5 Use sum and difference, double-angle and half-angle, and sum-to-product and product-to-sum formulas (p. A-34)
6 Solve triangles using the Law of Sines and the Law of Cosines (p. A-35)

## 1 Work with Angles, Arc Length of a Circle, and Circular Motion

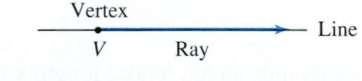

**Figure 43**

A **ray**, or **half-line**, is the portion of a line that starts at a point $V$ on the line and extends indefinitely in one direction. The point $V$ of a ray is called its **vertex**. See Figure 43.

If two rays are drawn with a common vertex, they form an **angle**. We call one ray of an angle the **initial side** and the other ray the **terminal side**. The angle formed is identified by showing the direction and amount of rotation from the initial side to the terminal side. If the rotation is in the counterclockwise direction, the angle is **positive**; if the rotation is clockwise, the angle is **negative**. See Figure 44.

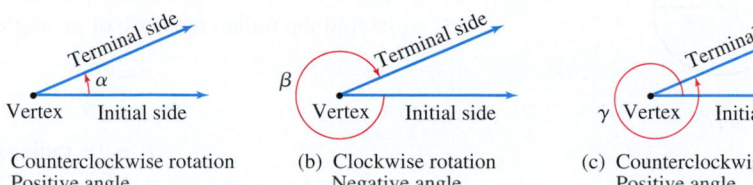

(a) Counterclockwise rotation
    Positive angle

(b) Clockwise rotation
    Negative angle

(c) Counterclockwise rotation
    Positive angle

**Figure 44**

An angle $\theta$ is in **standard position** if its vertex is at the origin of a rectangular coordinate system and its initial side is on the positive $x$-axis. See Figure 45 on page A-26.

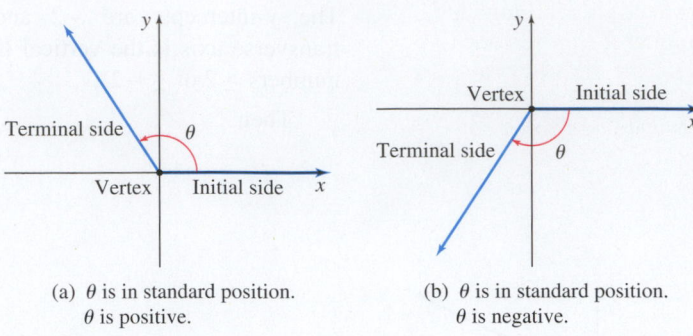

(a) $\theta$ is in standard position.
$\theta$ is positive.

(b) $\theta$ is in standard position.
$\theta$ is negative.

**Figure 45**

Suppose an angle $\theta$ is in standard position. If its terminal side coincides with a coordinate axis, we say that $\theta$ is a **quadrantal angle**. If its terminal side does not coincide with a coordinate axis, we say that $\theta$ **lies in a quadrant**. See Figure 46.

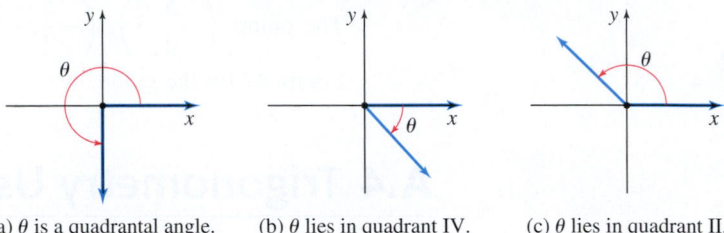

(a) $\theta$ is a quadrantal angle.    (b) $\theta$ lies in quadrant IV.    (c) $\theta$ lies in quadrant II.

**Figure 46**

Angles are measured by determining the amount of rotation needed for the initial side to coincide with the terminal side. The two commonly used measures for angles are *degrees* and *radians*.

The angle formed by rotating the initial side exactly once in the counterclockwise direction until it coincides with itself (one revolution) measures 360 degrees, abbreviated 360°. See Figure 47.

A **central angle** is a positive angle whose vertex is at the center of a circle. The rays of a central angle subtend (intersect) an arc on the circle. If the radius of the circle is $r$ and the arc subtended by the central angle is also of length $r$, then the measure of the angle is **1 radian**. See Figure 48.

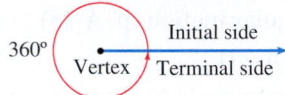

**Figure 47** 1 revolution counterclockwise is 360°.

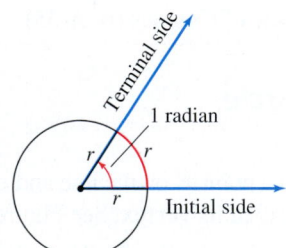

**Figure 48** 1 Radian

**THEOREM Arc Length**

For a circle of radius $r$, a central angle of $\theta$ radians subtends an arc whose length $s$ is

$$s = r\theta$$

See Figure 49.

Consider a circle of radius $r$. A central angle of one revolution will subtend an arc equal to the circumference of the circle, as shown in Figure 50. Because the circumference of a circle equals $2\pi r$, we use $s = 2\pi r$ in the formula for arc length to find the radian measure of an angle of one revolution.

$$s = r\theta$$
$$2\pi r = r\theta \qquad \text{For one revolution: } s = 2\pi r$$
$$\theta = 2\pi \text{ radians} \qquad \text{Solve for } \theta.$$

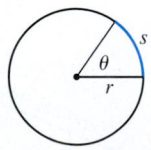

**Figure 49** $s = r\theta$

Since one revolution is equivalent to 360°, we have 360° = $2\pi$ radians so that

$$180° = \pi \text{ radians}$$

In calculus, radians are generally used to measure angles, unless degrees are specifically mentioned. Table 5 lists the radian and degree measures for some common angles.

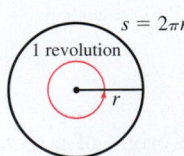

**Figure 50** 1 revolution = $2\pi$ radians

**NOTE** If the measure of an angle is given as a number, it is understood to mean radians. If the measure of an angle is given in degrees, then it is marked either with the symbol ° or the word "degrees."

**TABLE 5**

| Radians | 0 | $\frac{\pi}{6}$ | $\frac{\pi}{4}$ | $\frac{\pi}{3}$ | $\frac{\pi}{2}$ | $\frac{2\pi}{3}$ | $\frac{3\pi}{4}$ | $\frac{5\pi}{6}$ | $\pi$ | $\frac{3\pi}{2}$ | $2\pi$ |
|---|---|---|---|---|---|---|---|---|---|---|---|
| Degrees | 0° | 30° | 45° | 60° | 90° | 120° | 135° | 150° | 180° | 270° | 360° |

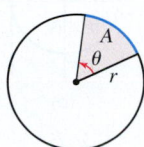

**Figure 51** $A = \frac{1}{2}r^2\theta$

**THEOREM  Area of a Sector**

The area $A$ of the sector of a circle of radius $r$ formed by a central angle of $\theta$ radians is

$$A = \frac{1}{2}r^2\theta$$

See Figure 51.

**DEFINITION  Linear Speed, Angular Speed**

Suppose an object moves around a circle of radius $r$ at a constant speed. If $s$ is the distance traveled along the circle in time $t$, then the **linear speed** $v$ of the object is

$$v = \frac{s}{t}$$

The **angular speed** $\omega$ (the Greek letter omega) of an object moving at a constant speed around a circle of radius $r$ is

$$\omega = \frac{\theta}{t}$$

where $\theta$ is the angle (measured in radians) swept out in time $t$.

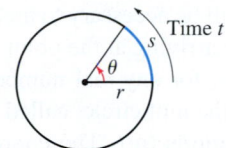

**Figure 52** $v = \frac{s}{t}$; $\omega = \frac{\theta}{t}$

See Figure 52.

Angular speed is used to describe the turning rate of an engine. For example, an engine idling at 900 rpm (revolutions per minute) rotates at an angular speed of

$$900 \ \frac{\text{revolutions}}{\text{minute}} = 900 \ \frac{\text{revolutions}}{\text{minute}} \cdot 2\pi \ \frac{\text{radians}}{\text{revolution}} = 1800\pi \ \frac{\text{radians}}{\text{minute}}$$

**NOTE** In the formula $s = r\theta$ for the arc length of a circle, the angle $\theta$ must be measured in radians.

There is an important relationship between linear speed and angular speed:

$$\text{linear speed} = v = \underset{\underset{s = r\theta}{\uparrow}}{\frac{s}{t}} = \frac{r\theta}{t} = r\left(\frac{\theta}{t}\right) = \underset{\underset{\omega = \frac{\theta}{t}}{\uparrow}}{r\omega}$$

So,

$$v = r\omega$$

where $\omega$ is measured in radians per unit time.

When using the equation $v = r\omega$, remember that $v = \frac{s}{t}$ has the dimensions of length per unit of time (such as meters per second or miles per hour), $r$ has the same length dimension as $s$, and $\omega$ has the dimensions of radians per unit of time. If the angular speed is given in *revolutions* per unit of time (as is often the case), be sure to convert it to *radians* per unit of time, using the fact that one revolution $= 2\pi$ radians, before using the formula $v = r\omega$.

## 2 Define and Evaluate Trigonometric Functions

There are two common approaches to trigonometry: One uses right triangles, and the other uses the unit circle. We suggest you first review the approach you are most familiar with and then read the other approach. The two approaches are given side-by-side on pages A-28 and A-29.

## Right Triangle Approach

Suppose $\theta$ is an **acute angle**; that is, $0 < \theta < \dfrac{\pi}{2}$, as shown in Figure 53(a). Using the angle $\theta$, we form a right triangle, like the one illustrated in Figure 53(b), with hypotenuse of length $c$ and legs of lengths $a$ and $b$. The three sides of the right triangle can be used to form exactly six ratios:

$$\frac{b}{c} \quad \frac{a}{c} \quad \frac{b}{a} \quad \frac{c}{b} \quad \frac{c}{a} \quad \frac{a}{b}$$

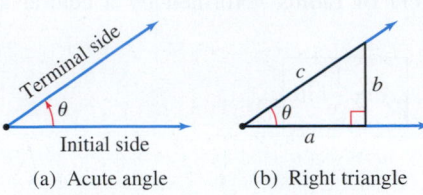

(a) Acute angle       (b) Right triangle

**Figure 53**

Figure 54 shows the relationship of the sides to the angle $\theta$.

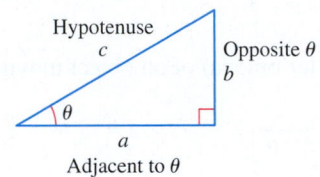

**Figure 54**

Because the ratios depend only on the angle $\theta$ and not on the triangle itself, each ratio is given a name that involves $\theta$.

### DEFINITION Trigonometric Functions
The six ratios of a right triangle are called the **trigonometric functions of an acute angle** $\theta$ and are defined as follows:

| Function Name | Abbreviation | Value | Relation Using Words |
|---|---|---|---|
| sine of $\theta$ | $\sin\theta$ | $\dfrac{b}{c}$ | $\dfrac{\text{opposite}}{\text{hypotenuse}}$ |
| cosine of $\theta$ | $\cos\theta$ | $\dfrac{a}{c}$ | $\dfrac{\text{adjacent}}{\text{hypotenuse}}$ |
| tangent of $\theta$ | $\tan\theta$ | $\dfrac{b}{a}$ | $\dfrac{\text{opposite}}{\text{adjacent}}$ |
| cosecant of $\theta$ | $\csc\theta$ | $\dfrac{c}{b}$ | $\dfrac{\text{hypotenuse}}{\text{opposite}}$ |
| secant of $\theta$ | $\sec\theta$ | $\dfrac{c}{a}$ | $\dfrac{\text{hypotenuse}}{\text{adjacent}}$ |
| cotangent of $\theta$ | $\cot\theta$ | $\dfrac{a}{b}$ | $\dfrac{\text{adjacent}}{\text{opposite}}$ |

To extend the definition of the trigonometric functions to include angles that are not acute, the angle is placed in standard position in a rectangular coordinate system, as shown in Figure 55.

## Unit Circle Approach

Let $t$ be any real number. We position a $t$-axis perpendicular to the $x$-axis so that $t = 0$ coincides with the point $(1, 0)$ of the $xy$-plane and positive values of $t$ are above the $x$-axis. See Figure 56(a).

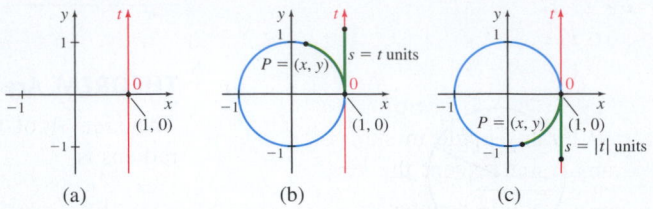

(a)       (b)       (c)

**Figure 56**

Now look at the unit circle in Figure 56(b). If $t > 0$, we begin at the point $(1, 0)$ on the unit circle and travel $s = t$ units in the counterclockwise direction along the circle to arrive at the point $P = (x, y)$. In this sense, the length $s = t$ units on the $t$-axis is being **wrapped** around the unit circle.

If $t < 0$, we begin at the point $(1, 0)$ on the unit circle and travel $s = |t|$ units in the clockwise direction along the circle to the point $P = (x, y)$, as shown in Figure 56(c).

If $t > 2\pi$ or if $t < -2\pi$, it will be necessary to travel around the circle more than once before arriving at the point $P$.

This discussion tells us that, for any real number $t$, there is a unique point $P = (x, y)$ on the unit circle, called **the point on the unit circle that corresponds to $t$**. The coordinates of the point $P = (x, y)$ are used to define the *six trigonometric functions of $t$*.

### DEFINITION Trigonometric Functions of a Number $t$
Let $t$ be a real number and let $P = (x, y)$ be the point on the unit circle that corresponds to $t$.

- The **sine function** is defined as

$$\sin t = y$$

- The **cosine function** is defined as

$$\cos t = x$$

- If $x \neq 0$, the **tangent function** is defined as

$$\tan t = \frac{y}{x} \qquad x \neq 0$$

- If $y \neq 0$, the **cosecant function** is defined as

$$\csc t = \frac{1}{y} \qquad y \neq 0$$

- If $x \neq 0$, the **secant function** is defined as

$$\sec t = \frac{1}{x} \qquad x \neq 0$$

- If $y \neq 0$, the **cotangent function** is defined as

$$\cot t = \frac{x}{y} \qquad y \neq 0$$

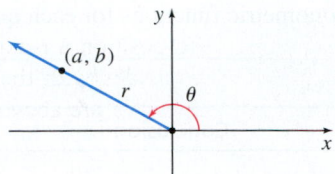

**Figure 55**

Notice in these definitions that if $x=0$, that is, if the point $P$ is on the $y$-axis, then the tangent function and the secant function are not defined. Also, if $y=0$, that is, if the point $P$ is on the $x$-axis, then the cosecant function and the cotangent function are not defined.

Because the unit circle is used in these definitions, the trigonometric functions are sometimes called **circular functions**.

Suppose $\theta$ is an angle in standard position, whose terminal side is the ray from the origin through the point $P$. Since the unit circle has radius $r=1$ unit, then, using the formula for arc length, we find that

$$s = r\theta = \theta$$

So, if $s=|t|$ units, then $\theta=t$ radians, and the trigonometric functions of the angle $\theta$ are defined as

| | | |
|---|---|---|
| $\sin\theta = \sin t$ | $\cos\theta = \cos t$ | $\tan\theta = \tan t$ |
| $\csc\theta = \csc t$ | $\sec\theta = \sec t$ | $\cot\theta = \cot t$ |

**DEFINITION   Trigonometric Functions of $\theta$**

Let $\theta$ be any angle in standard position and let $P=(a,b)$ be any point, except the origin, on the terminal side of $\theta$. If $r=\sqrt{a^2+b^2}$ denotes the distance of the point $P$ from the origin, then the **six trigonometric functions of $\theta$** are defined as the ratios

$$\sin\theta = \frac{b}{r} \qquad \cos\theta = \frac{a}{r} \qquad \tan\theta = \frac{b}{a}$$

$$\csc\theta = \frac{r}{b} \qquad \sec\theta = \frac{r}{a} \qquad \cot\theta = \frac{a}{b}$$

provided no denominator equals 0. If a denominator equals 0, that trigonometric function of the angle $\theta$ is not defined.

## Evaluating Trigonometric Functions

Using properties of right triangles, we can find the values of the six trigonometric functions of $\dfrac{\pi}{6}=30°$, $\dfrac{\pi}{4}=45°$, and $\dfrac{\pi}{3}=60°$ given in Table 6.

**TABLE 6**

| $\theta$ (Radians) | $\theta$ (Degrees) | $\sin\theta$ | $\cos\theta$ | $\tan\theta$ | $\csc\theta$ | $\sec\theta$ | $\cot\theta$ |
|---|---|---|---|---|---|---|---|
| $\dfrac{\pi}{6}$ | $30°$ | $\dfrac{1}{2}$ | $\dfrac{\sqrt{3}}{2}$ | $\dfrac{\sqrt{3}}{3}$ | $2$ | $\dfrac{2\sqrt{3}}{3}$ | $\sqrt{3}$ |
| $\dfrac{\pi}{4}$ | $45°$ | $\dfrac{\sqrt{2}}{2}$ | $\dfrac{\sqrt{2}}{2}$ | $1$ | $\sqrt{2}$ | $\sqrt{2}$ | $1$ |
| $\dfrac{\pi}{3}$ | $60°$ | $\dfrac{\sqrt{3}}{2}$ | $\dfrac{1}{2}$ | $\sqrt{3}$ | $\dfrac{2\sqrt{3}}{3}$ | $2$ | $\dfrac{\sqrt{3}}{3}$ |

The values of the trigonometric functions at the quadrantal angles, $\left(0=0°,\right.$ $\dfrac{\pi}{2}=90°$, $\dfrac{3\pi}{2}=270°$, and $2\pi=360°\left.\right)$, are given in Table 7.

**TABLE 7**

| Radians | Degrees | $\sin\theta$ | $\cos\theta$ | $\tan\theta$ | $\csc\theta$ | $\sec\theta$ | $\cot\theta$ |
|---|---|---|---|---|---|---|---|
| $0$ | $0°$ | $0$ | $1$ | $0$ | Not defined | $1$ | Not defined |
| $\dfrac{\pi}{2}$ | $90°$ | $1$ | $0$ | Not defined | $1$ | Not defined | $0$ |
| $\pi$ | $180°$ | $0$ | $-1$ | $0$ | Not defined | $-1$ | Not defined |
| $\dfrac{3\pi}{2}$ | $270°$ | $-1$ | $0$ | Not defined | $-1$ | Not defined | $0$ |
| $2\pi$ | $360°$ | $0$ | $1$ | $0$ | Not defined | $1$ | Not defined |

Table 8 lists the signs of the six trigonometric functions for each quadrant.

**TABLE 8**

| Quadrant | $\sin \theta, \csc \theta$ | $\cos \theta, \sec \theta$ | $\tan \theta, \cot \theta$ | Conclusion |
|---|---|---|---|---|
| I | Positive | Positive | Positive | All trigonometric functions are positive. |
| II | Positive | Negative | Negative | Only sine and its reciprocal, cosecant, are positive. |
| III | Negative | Negative | Positive | Only tangent and its reciprocal, cotangent, are positive. |
| IV | Negative | Positive | Negative | Only cosine and its reciprocal, secant, are positive. |

For nonacute angles $\theta$ that lie in a quadrant, the acute angle formed by the terminal side of $\theta$ and the $x$-axis, called a **reference angle**, is often used.

Figure 57 illustrates the reference angle for some general angles $\theta$. Note that a reference angle is always an acute angle.

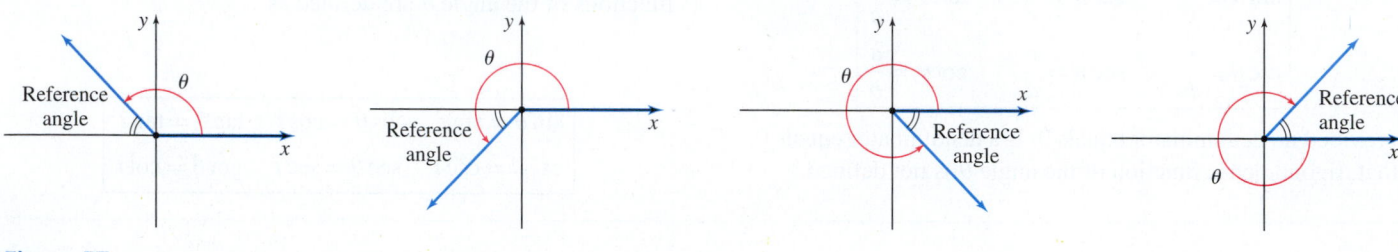

**Figure 57**

Although there are formulas for calculating reference angles, usually it is easier to find the reference angle for a given angle by making a quick sketch of the angle. The advantage of using reference angles is that, except for the correct sign, the values of the trigonometric functions of a general angle $\theta$ equal the values of the trigonometric functions of its reference angle.

**THEOREM  Reference Angles**

If $\theta$ is an angle that lies in a quadrant and if $A$ is its reference angle, then

$$\begin{array}{ccc} \sin \theta = \pm\sin A & \cos \theta = \pm\cos A & \tan \theta = \pm\tan A \\ \csc \theta = \pm\csc A & \sec \theta = \pm\sec A & \cot \theta = \pm\cot A \end{array}$$

where the $+$ or $-$ sign depends on the quadrant in which $\theta$ lies.

**EXAMPLE 1  Using Reference Angles**

Find the exact value of **(a)** $\cos \dfrac{17\pi}{6}$    **(b)** $\tan\left(-\dfrac{\pi}{3}\right)$

**Solution** **(a)** Refer to Figure 58(a). The angle $\dfrac{17\pi}{6}$ is in quadrant II, where the cosine function is negative. The reference angle for $\dfrac{17\pi}{6}$ is $\dfrac{\pi}{6}$. Since $\cos \dfrac{\pi}{6} = \dfrac{\sqrt{3}}{2}$,

$$\cos \dfrac{17\pi}{6} = -\cos \dfrac{\pi}{6} = -\dfrac{\sqrt{3}}{2}$$

**(b)** Refer to Figure 58(b). The angle $-\dfrac{\pi}{3}$ is in quadrant IV, where the tangent function is negative. The reference angle for $-\dfrac{\pi}{3}$ is $\dfrac{\pi}{3}$. Since $\tan \dfrac{\pi}{3} = \sqrt{3}$,

$$\tan\left(-\dfrac{\pi}{3}\right) = -\tan \dfrac{\pi}{3} = -\sqrt{3}$$

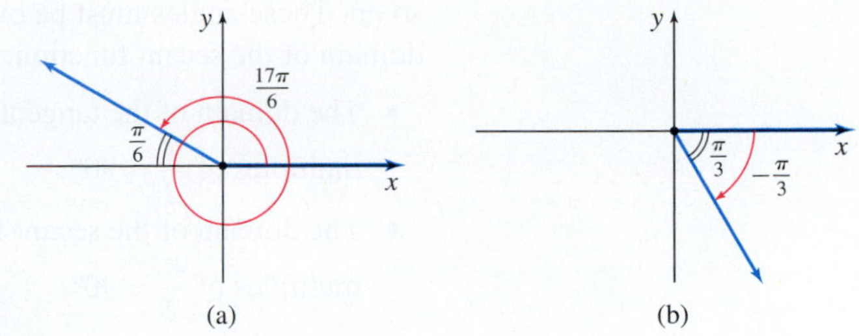

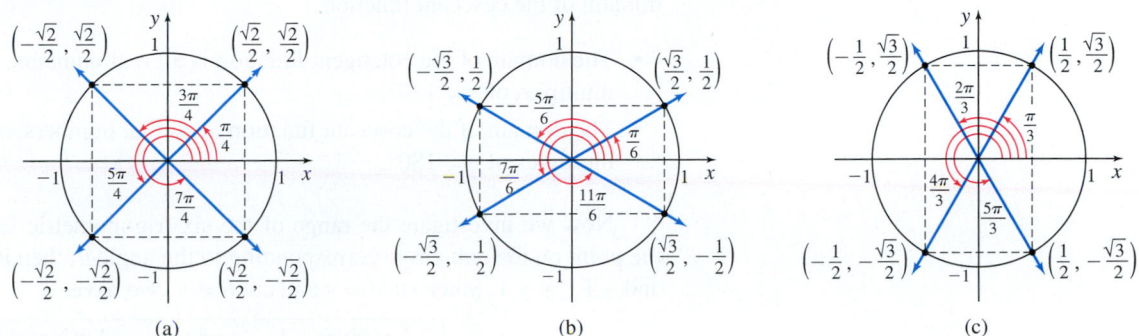

**Figure 58**

The use of symmetry provides information about integer multiples of the angles $\dfrac{\pi}{4} = 45°$, $\dfrac{\pi}{6} = 30°$, and $\dfrac{\pi}{3} = 60°$. See the unit circles in Figure 59(a)–(c).

**Figure 59**

| EXAMPLE 2 | Finding the Exact Values of a Trigonometry Function |

Use reference angles or the symmetry shown in Figure 59 to obtain:

(a) $\sin \dfrac{7\pi}{4} = -\dfrac{\sqrt{2}}{2}$

(b) $\cos \dfrac{7\pi}{6} = -\dfrac{\sqrt{3}}{2}$

(c) $\tan \dfrac{2\pi}{3} = \dfrac{\dfrac{\sqrt{3}}{2}}{-\dfrac{1}{2}} = -\sqrt{3}$

(d) $\cos \left(-\dfrac{3\pi}{4}\right) = -\dfrac{\sqrt{2}}{2}$

(e) $\sin \left(-\dfrac{\pi}{6}\right) = -\dfrac{1}{2}$

(f) $\sin \dfrac{7\pi}{3} = \dfrac{\sqrt{3}}{2}$

∎

## 3 Determine the Domain and the Range of the Trigonometric Functions

Suppose $\theta$ is an angle in standard position and $P = (x, y)$ is the point on the unit circle corresponding to $\theta$. Then the six trigonometric functions are

| | | |
|---|---|---|
| $\sin \theta = y$ | $\cos \theta = x$ | $\tan \theta = \dfrac{y}{x}$ |
| $\csc \theta = \dfrac{1}{y}$ | $\sec \theta = \dfrac{1}{x}$ | $\cot \theta = \dfrac{x}{y}$ |

For $\sin \theta$ and $\cos \theta$, there is no concern about dividing by 0, so $\theta$ can be any angle. It follows that the domain of the sine and cosine functions is all real numbers.

- The domain of the sine function is all real numbers.
- The domain of the cosine function is all real numbers.

For the tangent and secant functions, $x$ cannot be 0 since this results in division by 0. On the unit circle, there are two such points, $(0, 1)$ and $(0, -1)$. These two points correspond to $\dfrac{\pi}{2} = 90°$ and $\dfrac{3\pi}{2} = 270°$ or, more generally, to any angle that is an odd integer multiple of $\dfrac{\pi}{2}$, such as $\pm\dfrac{\pi}{2} = \pm90°$, $\pm\dfrac{3\pi}{2} = \pm270°$, $\pm\dfrac{5\pi}{2} = \pm450°$, and so on. These angles must be excluded from the domain of the tangent function and the domain of the secant function.

- The domain of the tangent function is all real numbers, except odd integer multiples of $\dfrac{\pi}{2} = 90°$.
- The domain of the secant function is all real numbers, except odd integer multiples of $\dfrac{\pi}{2} = 90°$.

For the cotangent and cosecant functions, $y$ cannot be $0$ since this results in division by $0$. On the unit circle, there are two such points, $(1, 0)$ and $(-1, 0)$. These two points correspond to $0 = 0°$ and $\pi = 180°$ or, more generally, to any angle that is an integer multiple of $\pi$, such as $0 = 0°$, $\pm\pi = \pm180°$, $\pm2\pi = \pm360°$, $\pm3\pi = \pm540°$, and so on. These angles must be excluded from the domain of the cotangent function and the domain of the cosecant function.

- The domain of the cotangent function is all real numbers, except integer multiples of $\pi = 180°$.
- The domain of the cosecant function is all real numbers, except integer multiples of $\pi = 180°$.

Now we investigate the range of the six trigonometric functions. If $P = (x, y)$ is the point on the unit circle corresponding to the angle $\theta$, then it follows that $-1 \leq x \leq 1$ and $-1 \leq y \leq 1$. Since $\sin\theta = y$ and $\cos\theta = x$, we have

$$\boxed{-1 \leq \sin\theta \leq 1 \qquad \text{and} \qquad -1 \leq \cos\theta \leq 1}$$

The range of both the sine function and the cosine function is the closed interval $[-1, 1]$. Using absolute value notation, we have $|\sin\theta| \leq 1$ and $|\cos\theta| \leq 1$.

If $\theta$ is not an integer multiple of $\pi = 180°$, then $\csc\theta = \dfrac{1}{y}$. Since $y = \sin\theta$ and $|y| = |\sin\theta| \leq 1$, it follows that $|\csc\theta| = \dfrac{1}{|\sin\theta|} = \dfrac{1}{|y|} \geq 1$. That is, $\dfrac{1}{y} \leq -1$ or $\dfrac{1}{y} \geq 1$. Since $\csc\theta = \dfrac{1}{y}$, the range of the cosecant functions consists of all real numbers in the set $(-\infty, -1] \cup [1, \infty)$. That is,

$$\boxed{\csc\theta \leq -1 \qquad \text{or} \qquad \csc\theta \geq 1}$$

If $\theta$ is not an odd integer multiple of $\dfrac{\pi}{2} = 90°$, then $\sec\theta = \dfrac{1}{x}$. Since $x = \cos\theta$ and $|x| = |\cos\theta| \leq 1$, it follows that $|\sec\theta| = \dfrac{1}{|\cos\theta|} = \dfrac{1}{|x|} \geq 1$. That is, $\dfrac{1}{x} \leq -1$ or $\dfrac{1}{x} \geq 1$. Since $\sec\theta = \dfrac{1}{x}$, the range of the secant function consists of all real numbers in the set $(-\infty, -1] \cup [1, \infty)$. That is,

$$\boxed{\sec\theta \leq -1 \qquad \text{or} \qquad \sec\theta \geq 1}$$

The range of both the tangent function and the cotangent function is all real numbers:

$$\boxed{-\infty < \tan\theta < \infty \qquad \text{and} \qquad -\infty < \cot\theta < \infty}$$

## 4 Use Basic Trigonometry Identities

### DEFINITION Identity

Two expressions $a$ and $b$ involving a variable $x$ are **identically equal** if

$$a = b$$

for every value of $x$ for which both expressions are defined. Such an equation is referred to as an **identity**. An equation that is not an identity is called a **conditional equation**.

For example, the equation $2x + 3 = x$ is a conditional equation because it is true only for $x = -3$. However, the equation $x^2 + 2x = (x + 1)^2 - 1$ is true for any value of $x$, so it is an identity.

The basic trigonometry identities listed below are consequences of the definition of the six trigonometric functions.

## Basic Trigonometry Identities

- **Quotient Identities**

$$\tan\theta = \frac{\sin\theta}{\cos\theta} \qquad \cot\theta = \frac{\cos\theta}{\sin\theta}$$

- **Reciprocal Identities**

$$\csc\theta = \frac{1}{\sin\theta} \qquad \sec\theta = \frac{1}{\cos\theta} \qquad \cot\theta = \frac{1}{\tan\theta}$$

- **Pythagorean Identities**

$$\sin^2\theta + \cos^2\theta = 1 \qquad \tan^2\theta + 1 = \sec^2\theta \qquad \cot^2\theta + 1 = \csc^2\theta$$

- **Even/Odd Identities**

$$\cos(-\theta) = \cos\theta \qquad \sin(-\theta) = -\sin\theta \qquad \tan(-\theta) = -\tan\theta$$

---

**EXAMPLE 3** **Using Basic Trigonometry Identities**

(a) $\displaystyle \tan\frac{\pi}{9} - \frac{\sin\dfrac{\pi}{9}}{\cos\dfrac{\pi}{9}} = \tan\frac{\pi}{9} - \tan\frac{\pi}{9} = 0$

$$\underset{\dfrac{\sin\theta}{\cos\theta} = \tan\theta}{\uparrow}$$

(b) $\displaystyle \sin^2\frac{\pi}{12} + \frac{1}{\sec^2\dfrac{\pi}{12}} = \sin^2\frac{\pi}{12} + \cos^2\frac{\pi}{12} = 1$

$$\underset{\cos\theta = \dfrac{1}{\sec\theta}}{\uparrow} \qquad \underset{\sin^2\theta + \cos^2\theta = 1}{\uparrow}$$

∎

---

**EXAMPLE 4** **Using Basic Trigonometry Identities**

Given that $\sin\theta = \dfrac{1}{3}$ and $\cos\theta < 0$, find the exact value of each of the remaining five trigonometric functions.

**Solution** We begin by solving the identity $\sin^2\theta + \cos^2\theta = 1$ for $\cos\theta$.

$$\sin^2\theta + \cos^2\theta = 1$$
$$\cos^2\theta = 1 - \sin^2\theta$$
$$\cos\theta = \pm\sqrt{1 - \sin^2\theta}$$

Because $\cos\theta < 0$, we choose the negative value of the radical and use the fact that $\sin\theta = \dfrac{1}{3}$.

$$\cos\theta = -\sqrt{1 - \sin^2\theta} = -\sqrt{1 - \frac{1}{9}} = -\sqrt{\frac{8}{9}} = -\frac{2\sqrt{2}}{3}$$

$$\underset{\sin\theta = \dfrac{1}{3}}{\uparrow}$$

The values of $\sin\theta$ and $\cos\theta$ are now known, so we use the quotient and reciprocal identities to obtain

$$\tan\theta = \frac{\sin\theta}{\cos\theta} = \frac{\frac{1}{3}}{-\frac{2\sqrt{2}}{3}} = -\frac{1}{2\sqrt{2}} = -\frac{\sqrt{2}}{4} \qquad \cot\theta = \frac{1}{\tan\theta} = -2\sqrt{2}$$

$$\sec\theta = \frac{1}{\cos\theta} = \frac{1}{-\frac{2\sqrt{2}}{3}} = -\frac{3}{2\sqrt{2}} = -\frac{3\sqrt{2}}{4} \qquad \csc\theta = \frac{1}{\sin\theta} = \frac{1}{\frac{1}{3}} = 3$$

■

## 5 Use Sum and Difference, Double-Angle and Half-Angle, and Sum-to-Product and Product-to-Sum Formulas

**THEOREM** Sum and Difference Formulas for Sine, Cosine, and Tangent

- $\sin(A + B) = \sin A \cos B + \cos A \sin B$
- $\sin(A - B) = \sin A \cos B - \cos A \sin B$
- $\cos(A + B) = \cos A \cos B - \sin A \sin B$
- $\cos(A - B) = \cos A \cos B + \sin A \sin B$
- $\tan(A + B) = \dfrac{\tan A + \tan B}{1 - \tan A \tan B}$
- $\tan(A - B) = \dfrac{\tan A - \tan B}{1 + \tan A \tan B}$

### EXAMPLE 5   Using Trigonometry Identities

If $\sin A = \dfrac{4}{5}$, $\dfrac{\pi}{2} < A < \pi$, and $\sin B = -\dfrac{2}{\sqrt{5}} = -\dfrac{2\sqrt{5}}{5}$, $\pi < B < \dfrac{3\pi}{2}$, find the exact value of:

**(a)** $\cos A$      **(b)** $\cos B$      **(c)** $\cos(A + B)$      **(d)** $\sin(A + B)$

**Solution (a)** We use a Pythagorean identity to find $\cos A$.

$$\cos A = -\sqrt{1 - \sin^2 A} = -\sqrt{1 - \left(\frac{4}{5}\right)^2} = -\sqrt{1 - \frac{16}{25}} = -\sqrt{\frac{9}{25}} = -\frac{3}{5}$$

$\uparrow$
A is in quadrant II
$\cos A < 0$

**(b)** We also find $\cos B$ using a Pythagorean identity.

$$\cos B = -\sqrt{1 - \sin^2 B} = -\sqrt{1 - \frac{4}{5}} = -\sqrt{\frac{1}{5}} = -\frac{\sqrt{5}}{5}$$

**(c)** We use the results from (a) and (b) and a sum formula to find $\cos(A + B)$.

$$\cos(A + B) = \cos A \cos B - \sin A \sin B$$

$$= \left(-\frac{3}{5}\right)\left(-\frac{\sqrt{5}}{5}\right) - \left(\frac{4}{5}\right)\left(-\frac{2\sqrt{5}}{5}\right) = \frac{11\sqrt{5}}{25}$$

**(d)** We use a sum formula to find $\sin(A + B)$.

$$\sin(A + B) = \sin A \cos B + \cos A \sin B = \left(\frac{4}{5}\right)\left(-\frac{\sqrt{5}}{5}\right) + \left(-\frac{3}{5}\right)\left(-\frac{2\sqrt{5}}{5}\right) = \frac{2\sqrt{5}}{25}$$

■

**THEOREM** Double-Angle Formulas

- $\sin(2\theta) = 2\sin\theta\cos\theta$
- $\cos(2\theta) = \cos^2\theta - \sin^2\theta$
- $\tan(2\theta) = \dfrac{2\tan\theta}{1 - \tan^2\theta}$
- $\cos(2\theta) = 1 - 2\sin^2\theta$
- $\cos(2\theta) = 2\cos^2\theta - 1$

The double-angle formulas in the first row are derived directly from the sum formulas by letting $\theta = A = B$. A Pythagorean identity is used to obtain those in the second row.

By rearranging the double-angle formulas, other identities can be obtained that are used in calculus.

- $\sin^2\theta = \dfrac{1 - \cos(2\theta)}{2}$
- $\cos^2\theta = \dfrac{1 + \cos(2\theta)}{2}$

If, in the above identities we replace $\theta$ with $\dfrac{1}{2}\theta$, we obtain these identities.

- $\sin^2\dfrac{\theta}{2} = \dfrac{1 - \cos\theta}{2}$
- $\cos^2\dfrac{\theta}{2} = \dfrac{1 + \cos\theta}{2}$
- $\tan^2\dfrac{\theta}{2} = \dfrac{1 - \cos\theta}{1 + \cos\theta}$

Taking the square root produces the **half-angle formulas**.

- $\sin\dfrac{\theta}{2} = \pm\sqrt{\dfrac{1 - \cos\theta}{2}}$
- $\cos\dfrac{\theta}{2} = \pm\sqrt{\dfrac{1 + \cos\theta}{2}}$
- $\tan\dfrac{\theta}{2} = \pm\sqrt{\dfrac{1 - \cos\theta}{1 + \cos\theta}}$

where the $+$ or $-$ sign is determined by the quadrant of the angle $\dfrac{\theta}{2}$.

The **product-to-sum** and **sum-to-product identities** are also used in calculus. They are a result of combining the sum and difference formulas.

- $\sin A\ \cos B = \dfrac{1}{2}[\sin(A + B) + \sin(A - B)]$
- $\cos A\ \cos B = \dfrac{1}{2}[\cos(A + B) + \cos(A - B)]$
- $\sin B\ \sin B = \dfrac{1}{2}[\cos(A - B) - \cos(A + B)]$

## 6 Solve Triangles Using the Law of Sines and the Law of Cosines

If no angle of a triangle is a right angle, the triangle is called **oblique**. An oblique triangle either has three acute angles or two acute angles and one obtuse angle (an angle between $\dfrac{\pi}{2}$ and $\pi$). See Figure 60.

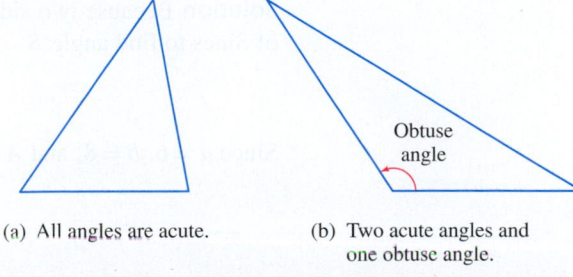

(a) All angles are acute.　　(b) Two acute angles and one obtuse angle.

**Figure 60** Two oblique triangles.

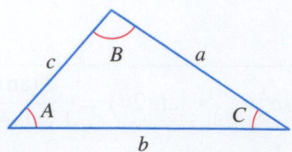

**Figure 61**

**NOTE** To solve an oblique triangle, you must know the length of at least one side. Knowing only angles results in a family of similar triangles.

In the discussion that follows, we label an oblique triangle so that side $a$ is opposite angle $A$, side $b$ is opposite angle $B$, and side $c$ is opposite angle $C$, as shown in Figure 61.

To **solve an oblique triangle** means to find the lengths of its sides and the measures of its angles. To do this, we need to know the length of one side along with (i) two angles; (ii) one angle and one other side; or (iii) the other two sides. There are four possibilities to consider:

*Case 1:* One side and two angles are known (ASA or SAA).
*Case 2:* Two sides and the angle opposite one of them are known (SSA).
*Case 3:* Two sides and the included angle are known (SAS).
*Case 4:* Three sides are known (SSS).

Figure 62 illustrates the four cases.

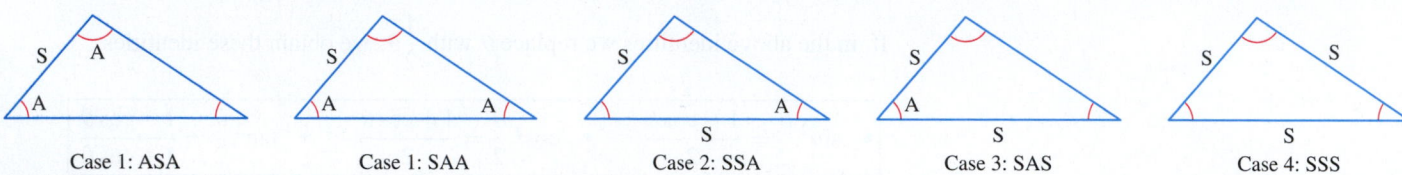

| Case 1: ASA | Case 1: SAA | Case 2: SSA | Case 3: SAS | Case 4: SSS |

**Figure 62**

The *Law of Sines* is used to solve triangles for which Case 1 or 2 holds.

**THEOREM  Law of Sines**

For a triangle with sides $a$, $b$, $c$ and opposite angles $A$, $B$, $C$, respectively,

$$\frac{\sin A}{a} = \frac{\sin B}{b} = \frac{\sin C}{c} \tag{1}$$

The Law of Sines actually consists of three equalities:

$$\frac{\sin A}{a} = \frac{\sin B}{b} \qquad \frac{\sin A}{a} = \frac{\sin C}{c} \qquad \frac{\sin B}{b} = \frac{\sin C}{c}$$

Formula (1) is a compact way to write these three equations.

For Case 1 (ASA or SAA), two angles are given. The third angle can be found using the fact that the sum of the angles in a triangle measures $180°$. Then use the Law of Sines (twice) to find the two missing sides.

Case 2 (SSA), which applies to triangles for which two sides and the angle opposite one of them are known, is referred to as the **ambiguous case** because the information can result in one triangle, two triangles, or no triangle at all.

**EXAMPLE 6  Using the Law of Sines to Solve a Triangle**

Solve the triangle with side $a = 6$, side $b = 8$, and angle $A = 35°$, which is opposite side $a$.

**Solution** Because two sides and an opposite angle are known (SSA), we use the Law of Sines to find angle $B$.

$$\frac{\sin A}{a} = \frac{\sin B}{b}$$

Since $a = 6$, $b = 8$, and $A = 35°$, we have

$$\frac{\sin 35°}{6} = \frac{\sin B}{8}$$

$$\sin B = \frac{8 \sin 35°}{6} \approx 0.765$$

This equation has two solutions

$$B_1 \approx 49.9° \quad \text{and} \quad B_2 \approx 180° - 49.9° = 130.1°$$

For both choices of $B$, we have $A + B < 180°$. So, there are two triangles, one containing the angle $B_1 \approx 49.9°$ and the other containing the angle $B_2 \approx 130.1°$. The third angle $C$ is either

$$C_1 = 180° - A - B_1 \approx 95.1° \qquad \text{or} \qquad C_2 = 180° - A - B \approx 14.9°$$
$$\uparrow \qquad\qquad\qquad\qquad\qquad \uparrow$$
$$A = 35° \qquad\qquad\qquad\qquad\qquad A = 35°$$
$$B_1 = 49.9° \qquad\qquad\qquad\qquad\qquad B_2 = 130.1°$$

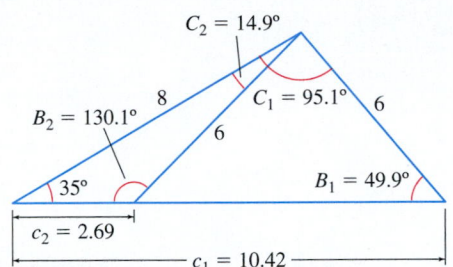

The third side $c$ satisfies the Law of Sines, so

$$\frac{\sin A}{a} = \frac{\sin C_1}{c_1} \qquad\qquad \frac{\sin A}{a} = \frac{\sin C_2}{c_2}$$

$$\frac{\sin 35°}{6} = \frac{\sin 95.1°}{c_1} \qquad\qquad \frac{\sin 35°}{6} = \frac{\sin 14.9°}{c_2}$$

$$c_1 = \frac{6 \sin 95.1°}{\sin 35°} \approx 10.42 \qquad\qquad c_2 = \frac{6 \sin 14.9°}{\sin 35°} \approx 2.69$$

**Figure 63**

The two solved triangles are illustrated in Figure 63. ∎

Cases 3 and 4 are solved using the *Law of Cosines*.

**THEOREM  Law of Cosines**

For a triangle with sides $a$, $b$, $c$ and opposite angles $A$, $B$, $C$, respectively,

- $c^2 = a^2 + b^2 - 2ab \cos C$
- $b^2 = a^2 + c^2 - 2ac \cos B$
- $a^2 = b^2 + c^2 - 2bc \cos A$

**NOTE** If the triangle is a right triangle, the Law of Cosines reduces to the Pythagorean Theorem.

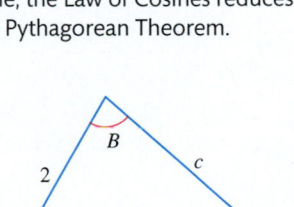

**Figure 64**

**EXAMPLE 7**  **Using the Law of Cosines to Solve a Triangle**

Solve the triangle: $a = 2$, $b = 3$, $C = 60°$ shown in Figure 64.

**Solution** Because two sides $a$ and $b$ and the included angle $C = 60°$ (SAS) are known, we use the Law of Cosines to find the third side $c$.

$$c^2 = a^2 + b^2 - 2ab \cos C = 2^2 + 3^2 - 2 \cdot 2 \cdot 3 \cdot \cos 60° = 13 - \left( 12 \cdot \frac{1}{2} \right) = 7$$

$$c = \sqrt{7}$$

Side $c$ is of length $\sqrt{7}$. To find either the angle $A$ or $B$, use the Law of Cosines. For $A$,

$$a^2 = b^2 + c^2 - 2bc \cos A$$

$$2bc \cos A = b^2 + c^2 - a^2$$

$$\cos A = \frac{b^2 + c^2 - a^2}{2bc} = \frac{9 + 7 - 4}{2 \cdot 3\sqrt{7}} = \frac{12}{6\sqrt{7}} = \frac{2\sqrt{7}}{7}$$

$$A \approx 40.9°$$

Then to find the third angle, use the fact that the sum of the angles of a triangle, when measured in degrees, equals $180°$. That is,

$$40.9° + B + 60° = 180°$$

$$B = 79.1°$$

∎

# Appendix B  Theorems and Proofs

## B.1  Limit Theorems and Proofs

### Uniqueness of a Limit

The limit of a function $f$, if it exists, is unique; that is, a function can have only one limit.

### THEOREM  A Limit Is Unique

If a function $f$ is defined on an open interval containing the number $c$, except possibly at $c$ itself, and if $\lim_{x \to c} f(x) = L_1$ and $\lim_{x \to c} f(x) = L_2$, then $L_1 = L_2$.

**NOTE** An indirect proof (a proof by contradiction) begins by assuming the conclusion is false. Then we show that this assumption leads to a contradiction.

**Proof** Assume that $L_1 \neq L_2$. We will show that this assumption leads to a contradiction. By the definition of the limit of a function $f$, $\lim_{x \to c} f(x) = L_1$ if, for any given number $\varepsilon > 0$, there is a number $\delta_1 > 0$ so that

$$\text{whenever } 0 < |x - c| < \delta_1 \quad \text{then } |f(x) - L_1| < \varepsilon \tag{1}$$

Similarly, $\lim_{x \to c} f(x) = L_2$ if, for any given number $\varepsilon > 0$, there is a number $\delta_2 > 0$ so that

$$\text{whenever } 0 < |x - c| < \delta_2 \quad \text{then } |f(x) - L_2| < \varepsilon \tag{2}$$

Now

$$L_1 - L_2 = L_1 - f(x) + f(x) - L_2$$

**RECALL** The Triangle Inequality: If $x$ and $y$ are real numbers, then

$$|x + y| \leq |x| + |y|$$

See Appendix A.1, p. A-7.

so, by applying the Triangle Inequality, we get

$$|L_1 - L_2| = |L_1 - f(x) + f(x) - L_2| \leq |L_1 - f(x)| + |f(x) - L_2| \tag{3}$$

For any given number $\varepsilon > 0$, let $\delta$ be the smaller of $\delta_1$ and $\delta_2$. Then from (1)–(3), we can conclude that whenever $0 < |x - c| < \delta \leq \delta_1$ and $0 < |x - c| < \delta \leq \delta_2$, we have

$$|L_1 - L_2| < \varepsilon + \varepsilon = 2\varepsilon \tag{4}$$

In particular, (4) is true for $\varepsilon = \dfrac{1}{2}|L_1 - L_2| > 0$. (Remember, $L_1 \neq L_2$.) Then from (4),

$$|L_1 - L_2| < 2\varepsilon = |L_1 - L_2|$$

which is a contradiction. Therefore, $L_1 = L_2$, and the limit, if it exists, is unique. ∎

### Algebra of Limits

### THEOREM  Limit of a Sum

If $f$ and $g$ are functions for which $\lim_{x \to c} f(x)$ and $\lim_{x \to c} g(x)$ both exist, then $\lim_{x \to c} [f(x) + g(x)]$ exists and

$$\lim_{x \to c} [f(x) + g(x)] = \lim_{x \to c} f(x) + \lim_{x \to c} g(x)$$

**Proof** Suppose $\lim\limits_{x \to c} f(x) = L$ and $\lim\limits_{x \to c} g(x) = M$. We need to show that for any number $\varepsilon > 0$, there is a number $\delta > 0$ so that

$$\text{whenever } 0 < |x - c| < \delta \quad \text{then } |[f(x) + g(x)] - [L + M]| < \varepsilon$$

Since $\lim\limits_{x \to c} f(x) = L$, by the definition of a limit, given the number $\dfrac{\varepsilon}{2} > 0$, there is a number $\delta_1 > 0$ so that

$$\text{whenever } 0 < |x - c| < \delta_1 \quad \text{then } |f(x) - L| < \frac{\varepsilon}{2}$$

Since $\lim\limits_{x \to c} g(x) = M$, for this same number $\dfrac{\varepsilon}{2}$, there is a number $\delta_2 > 0$ so that

$$\text{whenever } 0 < |x - c| < \delta_2 \quad \text{then } |g(x) - M| < \frac{\varepsilon}{2}$$

Let $\delta$ be the smaller of $\delta_1$ and $\delta_2$. Then $\delta \leq \delta_1$ and $\delta \leq \delta_2$. Using this $\delta$,

$$\text{whenever } 0 < |x - c| < \delta \quad \text{then } |f(x) - L| < \frac{\varepsilon}{2}$$

$$\text{whenever } 0 < |x - c| < \delta \quad \text{then } |g(x) - M| < \frac{\varepsilon}{2}$$

That is, whenever $0 < |x - c| < \delta$,

$$|[f(x) + g(x)] - [L + M]| = |[f(x) - L] + [g(x) - M]|$$
$$\leq |f(x) - L| + |g(x) - M| \quad \text{Use the Triangle Inequality.}$$
$$< \frac{\varepsilon}{2} + \frac{\varepsilon}{2} = \varepsilon$$

So, $\lim\limits_{x \to c} [f(x) + g(x)] = L + M$. ∎

---

**THEOREM  Limit of a Product**

If $f$ and $g$ are functions for which $\lim\limits_{x \to c} f(x)$ and $\lim\limits_{x \to c} g(x)$ both exist, then $\lim\limits_{x \to c} [f(x) \cdot g(x)]$ exists and

$$\lim_{x \to c} [f(x) \cdot g(x)] = \lim_{x \to c} f(x) \cdot \lim_{x \to c} g(x)$$

---

**Proof** Suppose $\lim\limits_{x \to c} f(x) = L$ and $\lim\limits_{x \to c} g(x) = M$. We need to show that for any number $\varepsilon > 0$, there is a number $\delta > 0$ so that

$$\text{whenever } 0 < |x - c| < \delta \quad \text{then } |f(x) \cdot g(x) - L \cdot M| < \varepsilon$$

Subtracting and adding $f(x) \cdot M$ in the expression $f(x) \cdot g(x) - L \cdot M$ result in terms involving $g(x) - M$ and $f(x) - L$:

$$|f(x) \cdot g(x) - L \cdot M| = |f(x) \cdot g(x) - f(x) \cdot M + f(x) \cdot M - L \cdot M|$$
$$= |f(x) \cdot [g(x) - M] + [f(x) - L] \cdot M|$$
$$\leq |f(x)| \cdot |g(x) - M| + |f(x) - L| \cdot |M| \qquad (5)$$
$$\uparrow$$
$$\text{Use the Triangle Inequality.}$$

Since $\lim\limits_{x \to c} f(x) = L$, there is a number $\delta_1 > 0$ so that whenever $0 < |x - c| < \delta_1$, then

$$|f(x) - L| < 1 \qquad \text{from which} \qquad |f(x)| < 1 + |L| \qquad (6)$$

Also since $\lim\limits_{x \to c} g(x) = M$, given a number $\varepsilon > 0$, there is a number $\delta_2$ so that whenever $0 < |x - c| < \delta_2$, then

$$|g(x) - M| < \frac{\varepsilon}{1 + |L| + |M|} \qquad (7)$$

Given a number $\varepsilon > 0$, there is a number $\delta_3$ so that whenever $0 < |x - c| < \delta_3$, then

$$|f(x) - L| < \frac{\varepsilon}{1 + |L| + |M|} \tag{8}$$

Choose $\delta$ to be the minimum of $\delta_1$, $\delta_2$, and $\delta_3$ and combine (5)–(8). Then for any given $\varepsilon > 0$, there is a $\delta > 0$ so that whenever $0 < |x - c| < \delta$, we have

$$|f(x) \cdot g(x) - L \cdot M| < [1 + |L|]\frac{\varepsilon}{1 + |L| + |M|} + |M|\frac{\varepsilon}{1 + |L| + |M|}$$

$$< [1 + |L| + |M|]\frac{\varepsilon}{1 + |L| + |M|} = \varepsilon$$

That is, $\lim\limits_{x \to c} [f(x) \cdot g(x)] = \lim\limits_{x \to c} f(x) \cdot \lim\limits_{x \to c} g(x)$. ∎

---

**THEOREM** Squeeze Theorem

If the functions $f$, $g$, and $h$ have the property that for all $x$ in an open interval containing $c$, except possibly at $c$ itself,

$$f(x) \leq g(x) \leq h(x)$$

and if

$$\lim_{x \to c} f(x) = \lim_{x \to c} h(x) = L$$

then

$$\lim_{x \to c} g(x) = L$$

---

**Proof** Since $\lim\limits_{x \to c} f(x) = \lim\limits_{x \to c} h(x) = L$, then for any number $\varepsilon > 0$, there are positive numbers $\delta_1$ and $\delta_2$ so that

$$\text{whenever } 0 < |x - c| < \delta_1 \quad \text{then } |f(x) - L| < \varepsilon$$
$$\text{whenever } 0 < |x - c| < \delta_2 \quad \text{then } |h(x) - L| < \varepsilon$$

Choose $\delta$ to be the smaller of the numbers $\delta_1$ and $\delta_2$. Then $0 < |x - c| < \delta$ implies that both $|f(x) - L| < \varepsilon$ and $|h(x) - L| < \varepsilon$. In other words, $0 < |x - c| < \delta$ implies that both

$$L - \varepsilon < f(x) < L + \varepsilon \quad \text{and} \quad L - \varepsilon < h(x) < L + \varepsilon$$

Since $f(x) \leq g(x) \leq h(x)$ for all $x \neq c$ in the open interval, it follows that whenever $0 < |x - c| < \delta$ and $x$ is in the open interval, we have

$$L - \varepsilon < f(x) \leq g(x) \leq h(x) < L + \varepsilon$$

Then for any given number $\varepsilon > 0$, there is a positive number $\delta$ so that whenever $0 < |x - c| < \delta$, then $L - \varepsilon < g(x) < L + \varepsilon$, or equivalently, $|g(x) - L| < \varepsilon$. That is, $\lim\limits_{x \to c} g(x) = L$. ∎

# B.2 Theorems and Proofs Involving Inverse Functions

---

**THEOREM** Continuity of the Inverse Function

If $f$ is a one-to-one function that is continuous on its domain, then its inverse function $f^{-1}$ is also continuous on its domain.

---

**Proof** Let $(a, b)$ be the largest open interval included in the domain of $f$. If $f$ is continuous on its domain, then it is continuous on $(a, b)$. Since $f$ is one-to-one, then $f$ is either increasing on $(a, b)$ or decreasing on $(a, b)$.

Suppose $f$ is increasing on $(a, b)$. Then $f^{-1}$ is also increasing. Let $y_0 = f(x_0)$. We need to show that $f^{-1}$ is continuous at $y_0$, given that $f$ is continuous at $x_0$. The number $f^{-1}(y_0) = x_0$ is in the open interval $(a, b)$. Choose $\varepsilon > 0$ sufficiently small so that $f^{-1}(y_0) - \varepsilon$ and $f^{-1}(y_0) + \varepsilon$ are also in $(a, b)$. Then choose $\delta$ so that

$$f[f^{-1}(y_0) - \varepsilon] < y_0 - \delta \qquad \text{and} \qquad y_0 + \delta < f[f^{-1}(y_0) + \varepsilon]$$

Then whenever $y_0 - \delta < y < y_0 + \delta$, we have

$$f[f^{-1}(y_0) - \varepsilon] < y < f[f^{-1}(y_0) + \varepsilon]$$

This means whenever $0 < |y - y_0| < \delta$, then

$$f^{-1}(y_0) - \varepsilon < f^{-1}(y) < f^{-1}(y_0) + \varepsilon \quad \text{or equivalently} \quad |f^{-1}(y) - f^{-1}(y_0)| < \varepsilon$$

That is, $\lim_{y \to y_0} f^{-1}(y) = f^{-1}(y_0)$, so $f^{-1}$ is continuous at $y_0$.

The case where $f$ is decreasing on $(a, b)$ is proved in a similar way. ∎

---

**THEOREM**  Derivative of the Inverse Function

Let $y = f(x)$ and $x = g(y)$ be inverse functions. If $f$ is differentiable on an open interval containing $x_0$ and if $f'(x_0) \neq 0$, then $g$ is differentiable at $y_0 = f(x_0)$ and

$$\frac{d}{dy} g(y_0) = \frac{1}{f'(x_0)}$$

where the notation $f'(x_0)$ means the value of $f'(x)$ at $x_0$, and the notation $\dfrac{d}{dy} g(y_0)$ means the value of $\dfrac{d}{dy} g(y)$ at $y_0$.

---

**Proof**  Since $f$ and $g$ are inverses of one another, then $f(x) = y$ if and only if $x = g(y)$. So, we have the following identity, where $g(y_0) = x_0$:

$$\frac{g(y) - g(y_0)}{y - y_0} = \frac{x - x_0}{f(x) - f(x_0)} = \frac{1}{\dfrac{f(x) - f(x_0)}{x - x_0}}$$

By the continuity of an inverse function, the continuity of $f$ at $x_0$ implies the continuity of $g$ at $y_0$, and $y \to y_0$ as $x \to x_0$.

Now take the limits of both sides of the above identity. Since $f'(x_0) \neq 0$, we have

$$g'(y_0) = \lim_{y \to y_0} \frac{g(y) - g(y_0)}{y - y_0} = \frac{1}{\displaystyle\lim_{x \to x_0} \frac{f(x) - f(x_0)}{x - x_0}} = \frac{1}{f'(x_0)}$$ ∎

# B.3 Derivative Theorems and Proofs

A proof of the Chain Rule when $\Delta u$ is never 0 appears in Chapter 3. Here, we consider the case when $\Delta u$ may be 0.

---

**THEOREM**  Chain Rule

If a function $g$ is differentiable at $x_0$ and a function $f$ is differentiable at $g(x_0)$, then the composite function $f \circ g$ is differentiable at $x_0$ and

$$(f \circ g)'(x_0) = f'(g(x_0)) \cdot g'(x_0)$$

Using Leibniz notation, if $y = f(u)$ and $u = g(x)$, then

$$\frac{dy}{dx} = \frac{dy}{du} \cdot \frac{du}{dx}$$

where $\dfrac{dy}{du}$ is evaluated at $u_0 = g(x_0)$ and $\dfrac{du}{dx}$ is evaluated at $x_0$.

**Proof** For the fixed number $x_0$, let $\Delta u = g(x_0 + \Delta x) - g(x_0)$. Since the function $u = g(x)$ is differentiable at $x_0$, it is also continuous at $x_0$, and therefore $\Delta u \to 0$ as $\Delta x \to 0$.

For the fixed number $u_0 = g(x_0)$, let $\Delta y = f(u_0 + \Delta u) - f(u_0)$. Since the function $y = f(u)$ is differentiable at the number $u_0$, we can write $\lim\limits_{\Delta u \to 0} \dfrac{\Delta y}{\Delta u} = \dfrac{dy}{du}(u_0)$. This implies that for any $\Delta u \neq 0$:

$$\frac{\Delta y}{\Delta u} = \frac{dy}{du}(u_0) + \alpha \tag{1}$$

where $\alpha = \alpha(\Delta u)$ is a function of $\Delta u$ such that $\alpha \to 0$ as $\Delta u \to 0$. Now we define $\alpha = 0$ when $\Delta u = 0$ so that $\alpha = \alpha(\Delta u)$ is continuous at 0. Multiplying (1) by $\Delta u \neq 0$, we obtain

$$\Delta y = \frac{dy}{du}(u_0) \cdot \Delta u + \alpha(\Delta u) \cdot \Delta u \tag{2}$$

Notice that the equation (2) is true for all $\Delta u$:

- If $\Delta u$ is not equal to 0, then (2) is a consequence of (1).
- If $\Delta u$ equals 0, then the left-hand side of (2) is

$$\Delta y = f(u_0 + \Delta u) - f(u_0) = f(u_0) - f(u_0) = 0$$

and the right-hand side of 2 is also 0.

Now divide (2) by $\Delta x \neq 0$:

$$\frac{\Delta y}{\Delta x} = \frac{dy}{du}(u_0) \cdot \frac{\Delta u}{\Delta x} + \alpha(\Delta u) \cdot \frac{\Delta u}{\Delta x} \tag{3}$$

Since the function $u = g(x)$ is differentiable at $x_0$, $\lim\limits_{\Delta x \to 0} \dfrac{\Delta u}{\Delta x} = \dfrac{du}{dx}(x_0)$. Also, since $\Delta u \to 0$ when $\Delta x \to 0$ and $\alpha(\Delta u)$ is continuous at $\Delta u = 0$, we conclude that $\alpha(\Delta u) \to 0$ as $\Delta x \to 0$. So we can take the limit of (3) as $\Delta x \to 0$, which proves that the derivative $\dfrac{dy}{dx}(x_0)$ exists and is equal to

$$\begin{aligned}
\frac{dy}{dx}(x_0) &= \lim_{\Delta x \to 0} \frac{\Delta y}{\Delta x} = \lim_{\Delta x \to 0}\left[\frac{dy}{du}(u_0) \cdot \frac{\Delta u}{\Delta x} + \alpha(\Delta u) \cdot \frac{\Delta u}{\Delta x}\right] \\
&= \frac{dy}{du}(u_0) \cdot \left[\lim_{\Delta x \to 0}\frac{\Delta u}{\Delta x}\right] + \left[\lim_{\Delta x \to 0}\alpha(\Delta u)\right] \cdot \left[\lim_{\Delta x \to 0}\frac{\Delta u}{\Delta x}\right] \\
&= \frac{dy}{du}(u_0) \cdot \frac{du}{dx}(x_0) + 0 \cdot \frac{du}{dx}(x_0) = \frac{dy}{du}(u_0) \cdot \frac{du}{dx}(x_0) \quad\blacksquare
\end{aligned}$$

### Partial Proof of L'Hôpital's Rule

To prove L'Hôpital's Rule requires an extension of the Mean Value Theorem called *Cauchy's Mean Value Theorem*.

> **THEOREM** Cauchy's Mean Value Theorem
>
> If the functions $f$ and $g$ are continuous on the closed interval $[a, b]$ and differentiable on the open interval $(a, b)$, and if $g'(x) \neq 0$ on $(a, b)$, then there is a number $c$ in $(a, b)$ for which
> $$\frac{f'(c)}{g'(c)} = \frac{f(b) - f(a)}{g(b) - g(a)}$$

**ORIGINS** The theorem was named after the French mathematician Augustin Cauchy (1789–1857).

Notice that under the conditions of Cauchy's Mean Value Theorem, $g(b) \neq g(a)$ because otherwise, by Rolle's Theorem, $g'(c) = 0$ for some $c$ in the interval $(a, b)$.

**Proof** Define the function $h$ as

$$h(x) = [g(b) - g(a)][f(x) - f(a)] - [g(x) - g(a)][f(b) - f(a)] \qquad a \le x \le b$$

Then $h$ is continuous on $[a, b]$ and differentiable on $(a, b)$, and $h(a) = h(b) = 0$. So by Rolle's Theorem, there is a number $c$ in the interval $(a, b)$ for which $h'(c) = 0$. That is,

**NOTE** A special case of Cauchy's Mean Value Theorem is the Mean Value Theorem. To get the Mean Value Theorem, let $g(x) = x$. Then $g'(x) = 1$, $g(b) = b$, and $g(a) = a$, giving the Mean Value Theorem.

$$h'(c) = [g(b) - g(a)]f'(c) - g'(c)[f(b) - f(a)] = 0$$

$$[g(b) - g(a)]f'(c) = g'(c)[f(b) - f(a)]$$

$$\frac{f'(c)}{g'(c)} = \frac{f(b) - f(a)}{g(b) - g(a)}$$

∎

---

**THEOREM** L'Hôpital's Rule

Suppose the functions $f$ and $g$ are differentiable on an open interval $I$ containing the number $c$, except possibly at $c$, and $g'(x) \ne 0$ for all $x \ne c$ in $I$. Let $L$ denote either a real number or $\pm\infty$, and suppose $\dfrac{f(x)}{g(x)}$ is an indeterminate form at $c$ of the type $\dfrac{0}{0}$

or $\dfrac{\infty}{\infty}$. If $\displaystyle\lim_{x \to c} \frac{f'(x)}{g'(x)} = L$, then $\displaystyle\lim_{x \to c} \frac{f(x)}{g(x)} = L$.

---

**Partial Proof** Suppose $\dfrac{f(x)}{g(x)}$ is an indeterminate form at $c$ of the type $\dfrac{0}{0}$, and

suppose $\displaystyle\lim_{x \to c} \frac{f'(x)}{g'(x)} = L$, where $L$ is a real number. We need to prove $\displaystyle\lim_{x \to c} \frac{f(x)}{g(x)} = L$.

First define the functions $F$ and $G$ as follows:

$$F(x) = \begin{cases} f(x) & \text{if} \quad x \ne c \\ 0 & \text{if} \quad x = c \end{cases} \qquad G(x) = \begin{cases} g(x) & \text{if} \quad x \ne c \\ 0 & \text{if} \quad x = c \end{cases}$$

Both $F$ and $G$ are continuous at $c$, since $\displaystyle\lim_{x \to c} F(x) = \lim_{x \to c} f(x) = 0 = F(c)$ and $\displaystyle\lim_{x \to c} G(x) = \lim_{x \to c} g(x) = 0 = G(c)$. Also,

$$F'(x) = f'(x) \qquad \text{and} \qquad G'(x) = g'(x)$$

for all $x$ in the interval $I$, except possibly at $c$. Since the conditions for Cauchy's Mean Value Theorem are met by $F$ and $G$ in either $[x, c]$ or $[c, x]$, there is a number $u$ between $c$ and $x$ for which

$$\frac{F(x) - F(c)}{G(x) - G(c)} = \frac{F'(u)}{G'(u)} = \frac{f'(u)}{g'(u)}$$

Since $F(c) = 0$ and $G(c) = 0$, this simplifies to $\dfrac{f(x)}{g(x)} = \dfrac{f'(u)}{g'(u)}$.

Since $u$ is between $c$ and $x$, it follows that

$$\lim_{x \to c} \frac{f(x)}{g(x)} = \lim_{u \to c} \frac{f'(u)}{g'(u)} = L$$

A similar argument is used if $L$ is infinite. The proof when $\dfrac{f(x)}{g(x)}$ is an indeterminate form at $\infty$ of the type $\dfrac{\infty}{\infty}$ is omitted here, but it may be found in books on advanced calculus. ∎

The use of L'Hôpital's Rule when $c = \infty$ for an indeterminate form of the type $\dfrac{0}{0}$ is justified by the following argument. In $\displaystyle\lim_{x \to \infty} \frac{f(x)}{g(x)}$, let $x = \dfrac{1}{u}$. Then as $x \to \infty$, $u \to 0^+$, and

$$\lim_{x \to \infty} \frac{f(x)}{g(x)} = \lim_{u \to 0^+} \frac{f\left(\dfrac{1}{u}\right)}{g\left(\dfrac{1}{u}\right)} = \lim_{u \to 0^+} \frac{\dfrac{d}{du} f\left(\dfrac{1}{u}\right)}{\dfrac{d}{du} g\left(\dfrac{1}{u}\right)} = \lim_{u \to 0^+} \frac{-\dfrac{1}{u^2} f'\left(\dfrac{1}{u}\right)}{-\dfrac{1}{u^2} g'\left(\dfrac{1}{u}\right)} = \lim_{x \to \infty} \frac{f'(x)}{g'(x)} = L$$

**Chain Rule**            $x = \dfrac{1}{u}$

## Proof That Continuous Partial Derivatives Are Sufficient for Differentiability

**THEOREM**

Let $z = f(x, y)$ be a function of two variables whose domain is $D$. Let $(x_0, y_0)$ be an interior point of $D$. If the partial derivatives $f_x$ and $f_y$ exist at each point of some disk centered at $(x_0, y_0)$, and if $f_x$ and $f_y$ are each continuous at $(x_0, y_0)$, then $f$ is differentiable at $(x_0, y_0)$.

**Proof**  The proof depends on the Mean Value Theorem for derivatives. Let $\Delta x$ and $\Delta y$ be changes, not both 0, in $x$ and in $y$, respectively, so that the point $(x_0 + \Delta x, y_0 + \Delta y)$ lies in some disk centered at $(x_0, y_0)$. The change in $z$ is

$$\Delta z = f(x_0 + \Delta x, \ y_0 + \Delta y) - f(x_0, y_0)$$

Adding and subtracting $f(x_0, y_0 + \Delta y)$ on the right-hand side, we obtain

$$\Delta z = f(x_0 + \Delta x, \ y_0 + \Delta y) - f(x_0, y_0 + \Delta y) + f(x_0, y_0 + \Delta y) - f(x_0, y_0) \qquad (4)$$

The expression $f(x, y_0 + \Delta y)$ is a function of $x$ alone, and its partial derivative $f_x(x, y_0 + \Delta y)$ exists in the disk centered at $(x_0, y_0)$. Then by the Mean Value Theorem, there is a real number $u$ between $x_0$ and $x_0 + \Delta x$ for which

$$f(x_0 + \Delta x, \ y_0 + \Delta y) - f(x_0, y_0 + \Delta y) = f_x(u, \ y_0 + \Delta y)\Delta x \qquad (5)$$

Similarly, the expression $f(x_0, y)$ is a function of $y$ alone, and the partial derivative $f_y(x_0, y)$ exists in the disk centered at $(x_0, y_0)$. Again, by the Mean Value Theorem, there is a real number $v$ between $y_0$ and $y_0 + \Delta y$ for which

$$f(x_0, y_0 + \Delta y) - f(x_0, y_0) = f_y(x_0, v)\Delta y \qquad (6)$$

Substitute (5) and (6) back into equation (4) for $\Delta z$ to obtain

$$\Delta z = f_x(u, \ y_0 + \Delta y)\Delta x + f_y(x_0, v)\Delta y$$

Now introduce the functions $\eta_1$ and $\eta_2$ defined by

$$\eta_1 = f_x(u, \ y_0 + \Delta y) - f_x(x_0, y_0) \qquad \text{and} \qquad \eta_2 = f_y(x_0, v) - f_y(x_0, y_0)$$

As $(\Delta x, \Delta y) \to (0, 0)$, then $u \to x_0$ and $v \to y_0$. Since $f_x$ and $f_y$ are continuous at $(x_0, y_0)$, $\eta_1$ and $\eta_2$ have the desired property that

$$\lim_{(\Delta x, \Delta y) \to (0,0)} \eta_1 = \lim_{(\Delta x, \Delta y) \to (0,0)} [f_x(u, \ y_0 + \Delta y) - f_x(x_0, y_0)]$$

$$= f_x(x_0, y_0) - f_x(x_0, y_0) = 0$$

and

$$\lim_{(\Delta x, \Delta y) \to (0,0)} \eta_2 = \lim_{(\Delta x, \Delta y) \to (0,0)} [f_y(x_0, v) - f_y(x_0, y_0)]$$

$$= f_y(x_0, y_0) - f_y(x_0, y_0) = 0$$

As a result, $\Delta z$ can be written as

$$\Delta z = f_x(u, y_0 + \Delta y)\Delta x + f_y(x_0, v)\Delta y$$
$$= [\eta_1 + f_x(x_0, y_0)]\,\Delta x + [\eta_2 + f_y(x_0, y_0)]\,\Delta y$$
$$= f_x(x_0, y_0)\Delta x + f_y(x_0, y_0)\Delta y + \eta_1\Delta x + \eta_2\Delta y$$

proving that $f$ is differentiable at $(x_0, y_0)$. ■

## B.4 Integral Theorems and Proofs

**THEOREM**

If a function $f$ is continuous on an interval containing the numbers $a$, $b$, and $c$, then

$$\int_a^b f(x)\,dx = \int_a^c f(x)\,dx + \int_c^b f(x)\,dx$$

**Proof** Since $f$ is continuous on an interval containing $a$, $b$, and $c$, the three integrals above exist.

**Part 1** Assume $a < b < c$. Since $f$ is continuous on $[a, b]$ and on $[b, c]$, given any $\varepsilon > 0$, there is a number $\delta_1 > 0$ so that

$$\left| \sum_{i=1}^{k} f(u_i)\Delta x_i - \int_a^b f(x)\,dx \right| < \frac{\varepsilon}{2} \tag{1}$$

for every Riemann sum $\sum_{i=1}^{k} f(u_i)\Delta x_i$ for $f$ on $[a, b]$, where $x_{i-1} \le u_i \le x_i$, $i = 1, 2, \ldots, k$, and whose partition $P_1$ of $[a, b]$ has norm $\|P_1\| < \delta_1$. There is also a number $\delta_2 > 0$ for which

$$\left| \sum_{i=k+1}^{n} f(u_i)\Delta x_i - \int_b^c f(x)\,dx \right| < \frac{\varepsilon}{2} \tag{2}$$

for every Riemann sum $\sum_{i=k+1}^{n} f(u_i)\Delta x_i$ for $f$ on $[b, c]$, where $x_{i-1} \le u_i \le x_i$, $i = k+1, k+2, \ldots, n$, and whose partition $P_2$ of $[b, c]$ has norm $\|P_2\| < \delta_2$.

**NOTE** The partition $P_1$ is $x_0 = a, \ldots, x_k = b$. The partition $P_2$ is $x_k = b, \ldots, x_n = c$.

Let $\delta$ be the smaller of $\delta_1$ and $\delta_2$. Then (1) and (2) hold, with $\delta$ replacing $\delta_1$ and $\delta_2$. If (1) and (2) are added and if $\|P_1\| < \delta$ and $\|P_2\| < \delta$, then

$$\left| \sum_{i=1}^{k} f(u_i)\Delta x_i - \int_a^b f(x)\,dx \right| + \left| \sum_{i=k+1}^{n} f(u_i)\Delta x_i - \int_b^c f(x)\,dx \right| < \frac{\varepsilon}{2} + \frac{\varepsilon}{2} = \varepsilon$$

Using the Triangle Inequality, this result implies that for $\|P_1\| < \delta$ and $\|P_2\| < \delta$,

$$\left| \sum_{i=1}^{k} f(u_i)\Delta x_i - \int_a^b f(x)\,dx + \sum_{i=k+1}^{n} f(u_i)\Delta x_i - \int_b^c f(x)\,dx \right| < \varepsilon \tag{3}$$

Denote $P_1 \cup P_2$ by $P^*$. Then $P^*$ is a partition of $[a, c]$ having the number $b = x_k$ as an endpoint of the $k$th subinterval. So,

$$\sum_{i=1}^{k} f(u_i)\Delta x_i + \sum_{i=k+1}^{n} f(u_i)\Delta x_i = \sum_{i=1}^{n} f(u_i)\Delta x_i$$

are Riemann sums for $f$ on $P^*$. Since $\|P^*\| < \delta$ implies that $\|P_1\| < \delta$ and $\|P_2\| < \delta$, it follows from (3) that

$$\left| \sum_{i=1}^{n} f(u_i)\Delta x_i - \left[ \int_a^b f(x)\,dx + \int_b^c f(x)\,dx \right] \right| < \varepsilon$$

for every Riemann sum $\sum\limits_{i=1}^{n} f(u_i)\Delta x_i$ for $f$ on $[a, c]$ whose partition $P^*$ of $[a, c]$ has $b$ as an endpoint of a subinterval of the partition and has norm $\|P^*\| < \delta$. Therefore,

$$\int_a^c f(x)\, dx = \int_a^b f(x)\, dx + \int_b^c f(x)\, dx$$

**Part 2** There are six possible orderings (permutations) of the numbers $a$, $b$, and $c$:

$$a < b < c \quad a < c < b \quad b < a < c \quad b < c < a \quad c < a < b \quad c < b < a$$

In Part 1, we showed that the theorem is true for the order $a < b < c$. Now consider any other order, say, $b < c < a$. From Part 1,

$$\int_b^c f(x)\, dx + \int_c^a f(x)\, dx = \int_b^a f(x)\, dx \tag{4}$$

But

$$\int_c^a f(x)\, dx = -\int_a^c f(x)\, dx \quad \text{and} \quad \int_b^a f(x)\, dx = -\int_a^b f(x)\, dx$$

Now we substitute this into (4).

$$\int_b^c f(x)\, dx - \int_a^c f(x)\, dx = -\int_a^b f(x)\, dx$$

$$\int_a^b f(x)\, dx + \int_b^c f(x)\, dx = \int_a^c f(x)\, dx$$

proving the theorem for $b < c < a$.

The proofs for the remaining four permutations of $a$, $b$, and $c$ are similar. ∎

---

**THEOREM** Fundamental Theorem of Calculus, Part 1

Let $f$ be a function that is continuous on a closed interval $[a, b]$. The function $I$ defined by

$$I(x) = \int_a^x f(t)\, dt$$

has the property that it is continuous on $[a, b]$ and differentiable on $(a, b)$. Moreover,

$$I'(x) = \frac{d}{dx}\left[\int_a^x f(t)\, dt\right] = f(x)$$

for all $x$ in $(a, b)$.

---

**Proof** Let $x$ and $x + h$, $h \neq 0$, be in the interval $(a, b)$. Then

$$I(x) = \int_a^x f(t)\, dt \qquad I(x+h) = \int_a^{x+h} f(t)\, dt$$

and

$$I(x+h) - I(x) = \int_a^{x+h} f(t)\, dt + \int_x^a f(t)\, dt \qquad \int_x^a f(t)\, dt = -\int_a^x f(t)\, dt$$

$$= \int_x^a f(t)\, dt + \int_a^{x+h} f(t)\, dt = \int_x^{x+h} f(t)\, dt$$

Dividing both sides by $h \neq 0$, we get

$$\frac{I(x+h) - I(x)}{h} = \frac{1}{h}\int_x^{x+h} f(t)\, dt \tag{5}$$

**NEED TO REVIEW?** The Mean Value Theorem for Integrals is discussed in Section 5.4.

Now we use the Mean Value Theorem for Integrals in the integral on the right. There are two possibilities: either $h > 0$ or $h < 0$.

If $h > 0$, there is a number $u$, where $x \leq u \leq x + h$, for which

$$\int_x^{x+h} f(t)\, dt = f(u)h$$

$$\frac{1}{h}\int_x^{x+h} f(t)\, dt = f(u)$$

$$\frac{I(x+h) - I(x)}{h} = f(u) \qquad \text{From (5)}$$

Since $x \leq u \leq x + h$, as $h \to 0^+$, $u$ approaches $x^+$, so

$$\lim_{h \to 0^+} \frac{I(x+h) - I(x)}{h} = \lim_{h \to 0^+} f(u) = \lim_{u \to x^+} f(u) = f(x)$$

$f$ is continuous

Using a similar argument for $h < 0$, we obtain

$$\lim_{h \to 0^-} \frac{I(x+h) - I(x)}{h} = f(x)$$

Since the two one-sided limits are equal,

$$\lim_{h \to 0} \frac{I(x+h) - I(x)}{h} = f(x)$$

The limit is the derivative of the function $I$, meaning $I'(x) = f(x)$ for all $x$ in $(a, b)$. ∎

**THEOREM  Bounds on an Integral**

If a function $f$ is continuous on a closed interval $[a, b]$ and if $m$ and $M$ denote the absolute minimum and absolute maximum values of $f$ on $[a, b]$, respectively, then

$$m(b - a) \leq \int_a^b f(x)\, dx \leq M(b - a)$$

The Bounds on an Integral Theorem is proved by contradiction.

**Proof  Part 1** $m(b - a) \leq \displaystyle\int_a^b f(x)\, dx$

Assume

$$m(b - a) > \int_a^b f(x)\, dx \qquad (6)$$

Since $f$ is continuous on $[a, b]$,

$$\lim_{\|P\| \to 0} \sum_{i=1}^n f(u_i)\Delta x_i = \int_a^b f(x)\, dx$$

By (6), $m(b - a) - \displaystyle\int_a^b f(x)\, dx > 0$. We choose $\varepsilon$ so that

$$\varepsilon = m(b - a) - \int_a^b f(x)\, dx > 0 \qquad (7)$$

Then there is a number $\delta > 0$, so that for all partitions $P$ of $[a, b]$ with norm $\|P\| < \delta$, we have

$$\left| \sum_{i=1}^n f(u_i)\Delta x_i - \int_a^b f(x)\, dx \right| < \varepsilon$$

which is equivalent to

$$\int_a^b f(x)\,dx - \varepsilon < \sum_{i=1}^{n} f(u_i)\Delta x_i < \int_a^b f(x)\,dx + \varepsilon$$

By (7), the right inequality can be expressed as

$$\sum_{i=1}^{n} f(u_i)\Delta x_i < \int_a^b f(x)\,dx + \varepsilon = \int_a^b f(x)\,dx + \left[ m(b-a) - \int_a^b f(x)\,dx \right]$$

$$= m(b-a)$$

Consequently,

$$\sum_{i=1}^{n} f(u_i)\Delta x_i < m(b-a) = \sum_{i=1}^{n} m\Delta x_i$$

implying that for every partition $P$ of $[a, b]$ with $\|P\| < \delta$,

$$f(u_i) < m$$

for some $u_i$ in $[a, b]$. But this is impossible because $m$ is the absolute minimum of $f$ on $[a, b]$. Therefore, the assumption $m(b-a) > \int_a^b f(x)\,dx$ is false. That is,

$$m(b-a) \leq \int_a^b f(x)\,dx$$

**Part 2** To prove $\int_a^b f(x)\,dx \leq M(b-a)$, use a similar argument. ∎

# B.5 A Bounded Monotonic Sequence Converges

### THEOREM

An increasing (or nondecreasing) sequence $\{s_n\}$ that is bounded from above converges. A decreasing (or nonincreasing) sequence $\{s_n\}$ that is bounded from below converges.

To prove this theorem, we need the following property of real numbers. The set of real numbers is defined by a collection of axioms. One of these axioms is the Completeness Axiom.

### Completeness Axiom of Real Numbers

If $S$ is a nonempty set of real numbers that has an upper bound, then it has a least upper bound. Similarly, if $S$ has a lower bound, then it has a greatest lower bound.

As an example, consider the set $S$: $\{x \mid x^2 < 2, \ x > 0\}$. The set of upper bounds to $S$ is the set $\{x \mid x^2 \geq 2, \ x > 0\}$.

**(a)** If our universe is the set of rational numbers, the set of upper bounds has no minimum (since $\sqrt{2}$ is not rational).

**(b)** If our universe is the set of real numbers, then by the Completeness Axiom, the set of upper bounds has a minimum ($\sqrt{2}$). That is, this axiom completes the set of real numbers by incorporating the set of irrational numbers with the set of rational numbers to form the set of real numbers.

We prove the theorem for a nondecreasing sequence $\{s_n\}$. The proofs for the other three cases are similar.

**Proof** Suppose $\{s_n\}$ is a nondecreasing sequence that is bounded from above. Since $\{s_n\}$ is bounded from above, there is a positive number $K$ (an upper bound) so that $s_n \leq K$ for every $n$. From the Completeness Axiom, the set $\{s_n\}$ has a least upper bound $L$. That is, $s_n \leq L$ for every $n$.

Then for any $\varepsilon > 0$, $L - \varepsilon$ is not an upper bound of $\{s_n\}$. That is, $L - \varepsilon < s_N$ for some integer $N$. Since $\{s_n\}$ is nondecreasing, $s_N \leq s_n$ for all $n > N$. Then for all $n > N$,

$$L - \varepsilon < s_n \leq L < L + \varepsilon$$

That is, $|s_n - L| < \varepsilon$ for all $n > N$, so the sequence $\{s_n\}$ converges to $L$. ∎

# B.6 Taylor's Formula with Remainder

**THEOREM** Taylor's Formula with Remainder

Let $f$ be a function whose first $n + 1$ derivatives are continuous on an open interval $I$ containing the number $c$. Then for every $x$ in the interval, there is a number $u$ between $x$ and $c$ for which

$$f(x) = f(c) + f'(c)(x - c) + \frac{f''(c)}{2!}(x - c)^2 + \cdots + \frac{f^{(n)}(c)}{n!}(x - c)^n + R_n(x)$$

where

$$R_n(x) = \frac{f^{(n+1)}(u)}{(n+1)!}(x - c)^{n+1}$$

**Proof** For a fixed number $x \neq c$ in the open interval $I$, there is a number $L$ (depending on $x$) for which

$$f(x) = f(c) + \frac{f'(c)}{1!}(x - c) + \frac{f''(c)}{2!}(x - c)^2 + \cdots + \frac{f^{(n)}(c)}{n!}(x - c)^n + \frac{L}{(n+1)!}(x - c)^{n+1} \quad (1)$$

Define the function $F$ to be

$$F(t) = f(x) - f(t) - \frac{f'(t)}{1!}(x - t) - \frac{f''(t)}{2!}(x - t)^2 - \cdots - \frac{f^{(n)}(t)}{n!}(x - t)^n - \frac{L}{(n+1)!}(x - t)^{n+1} \quad (2)$$

The domain of $F$ is $c \leq t \leq x$ if $x > c$ and $x \leq t \leq c$ if $x < c$. Since $f(t)$, $f'(t)$, $f''(t)$, ..., $f^{(n)}(t)$ are each continuous, then $F$ is continuous on its domain. Furthermore, $F$ is differentiable and

$$\frac{dF}{dt} = F'(t) = -f'(t) + \left[ f'(t) - \frac{f''(t)}{1!}(x - t) \right] + \left[ \frac{f''(t)}{1!}(x - t) - \frac{f'''(t)}{2!}(x - t)^2 \right]$$

$$+ \cdots + \left[ \frac{f^{(n)}(t)}{(n-1)!}(x - t)^{n-1} - \frac{f^{(n+1)}(t)}{n!}(x - t)^n \right] + \frac{L}{n!}(x - t)^n$$

$$= -\frac{f^{(n+1)}(t)}{n!}(x - t)^n + \frac{L}{n!}(x - t)^n$$

for all $t$ between $x$ and $c$. From (1) and (2), we have

$$F(c) = f(x) - f(c) - \frac{f'(c)}{1!}(x - c) - \frac{f''(c)}{2!}(x - c)^2 - \cdots - \frac{f^{(n)}(c)}{n!}(x - c)^n - \frac{L}{(n+1)!}(x - c)^{n+1} = 0$$

Then

$$F(x) = f(x) - f(x) - \frac{f'(x)}{1!}(x - x) - \cdots - \frac{f^{(n)}(x)}{n!}(x - x)^n - \frac{L}{(n+1)!}(x - x)^{n+1} = 0$$

Now apply Rolle's Theorem to $F$. Then there is a number $u$ between $c$ and $x$ for which

$$F'(u) = -\frac{f^{(n+1)}(u)}{n!}(x-u)^n + \frac{L}{n!}(x-u)^n = 0$$

Solving for $L$, we find $L = f^{(n+1)}(u)$. Now let $t = c$ and $L = f^{(n+1)}(u)$ in (2) and solve for $f(x)$. Then

$$f(x) = f(c) + f'(c)(x-c) + \frac{f''(c)}{2!}(x-c)^2 + \cdots + \frac{f^{(n)}(c)}{n!}(x-c)^n + R_n(x)$$

where

$$R_n(x) = \frac{f^{(n+1)}(u)}{(n+1)!}(x-c)^{n+1}$$

# Appendix C — Technology Used in Calculus

This text, as you see it, would not have been possible without technology. All the figures in the text were produced using technology—some on a graphing calculator, most with an interactive graphic system, others in Adobe Illustrator®. The equations and symbols were created and spaced by a computer using the LaTeX equation editor; the page numbering and printing were done electronically.

Since the 1960s, portable computation devices have been available. Their introduction eliminated the need to perform long, tedious arithmetic calculations by hand. Many calculations that were previously impossible can now be done quickly and accurately.

Many calculators today, certainly those you are using, are not so much calculators but small, handheld computers. They *numerically* manipulate data and mathematical expressions.

## C.1 Graphing Calculators

Most graphing calculators have the ability to:

- Graph and compare functions of one variable.
- Graph functions of one variable in several formats: rectangular, parametric, and polar.
- Locate the intercepts, local maxima, and local minima of a graph.
- Graph sequences and explore convergence.
- Solve equations numerically.
- Numerically solve a system of equations.
- Find a function of best fit.
- Find the derivative of a function at a particular number.
- Numerically approximate a definite integral.

As you read the text, you will see examples of graphs generated by a graphing calculator. You will also find problems marked with ⟨⟩ , which alerts you that graphing technology is recommended.

## C.2 Computer Algebra Systems (CAS)

In contrast to a calculator, a CAS *symbolically* manipulates mathematical expressions. This symbolic manipulation usually allows for exact mathematical solutions, as well as for numeric approximations. Most CAS systems are packaged with interactive graphing technology that can produce and manipulate two- and three-dimensional graphs.

At appropriate places in the text, you will see problems marked CAS , which alerts you that a computer algebra system is recommended. If your instructor does not require a CAS, these problems can be omitted. However, they provide insight into calculus and will enrich your calculus experience.

There are many computer algebra systems available. They vary in versatility, ease of use, and price. Here, we list several commonly used systems.

- Maple 2016
- Mathematica 10.4, 1
- Wolfram Alpha
- MATLAB (R2016b)
- TI-Nspire CAS

# Answers

## Chapter P

### Section P.1    (SSM = See Student Solutions Manual; AWV = Answers will vary)

**1.** Independent, dependent   **2.** True   **3.** False   **4.** False   **5.** False   **6.** Vertical   **7.** $5, -3$   **8.** $-2$   **9.** (a)   **10.** (a), (b)   **11.** False
**12.** 8   **13.** (a) $-4$   (b) $3x^2 - 2x - 4$   (c) $-3x^2 - 2x + 4$   (d) $3x^2 + 8x + 1$   (e) $3x^2 + 6xh + 3h^2 + 2x + 2h - 4$

**15.** (a) 4   (b) $|x| + 4$   (c) $-|x| - 4$   (d) $|x + 1| + 4$   (e) $|x + h| + 4$   **17.** $(-\infty, \infty)$   **19.** $(-\infty, -3] \cup [3, \infty)$   **21.** $\{x \,|\, x \neq -2, 0, 2\}$

**23.** $-3$   **25.** $\dfrac{1}{\sqrt{x + h + 7} + \sqrt{x + 7}}$   **27.** $2x + h + 2$   **29.** It is not a function.   **31.** (a) Domain: $[-\pi, \pi]$, range: $[-1, 1]$

  (b) Intercepts: $\left(-\dfrac{\pi}{2}, 0\right), \left(\dfrac{\pi}{2}, 0\right), (0, 1)$   (c) Symmetric with respect to the $y$-axis only

**33.** (a) $f(-1) = 2$,   (b) 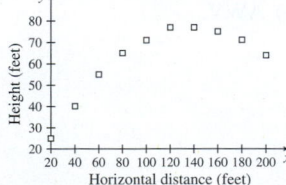   (c) Domain: $[-2, \infty)$,   **35.** (a) $f(-1) = 0$,   (b)   (c) Domain: $(-\infty, \infty)$,
$f(0) = 3$,    range: $(-\infty, 4) \cup \{5\}$,    $f(0) = 0$,    range: $(-\infty, \infty)$,
$f(1) = 5$,    intercepts: $(0, 3), (2, 0)$    $f(1) = 1$,    intercepts: $(-1, 0), (0, 0)$
$f(8) = -6$      $f(8) = 64$

**37.** $f(0) = 3$, $f(-6) = -3$   **39.** Negative   **41.** $(-3, 6) \cup (10, 11]$   **43.** $[-3, 3]$   **45.** 3   **47.** Once   **49.** $-5, 8$   **51.** $[4, 8]$
**53.** $[0, 8]$   **55.** $\{x \,|\, x \neq 6\}$   **57.** $-3, (4, -3)$   **59.** $-2$   **61.** Odd; symmetric with respect to the origin only
**63.** Neither; not symmetric to the $x$-axis, $y$-axis, or the origin   **65.** (a) $-6$   (b) $-8$   (c) $-10$   (d) $-2(x + 1), x \neq 1$

**67.** $f(x) = \begin{cases} -x & \text{if} \quad -1 \le x < 0 \\ \dfrac{1}{2}x & \text{if} \quad 0 \le x \le 2 \end{cases}$ or $f(x) = \begin{cases} -x & \text{if} \quad -1 \le x \le 0 \\ \dfrac{1}{2}x & \text{if} \quad 0 < x \le 2 \end{cases}$ domain: $[-1, 2]$, range: $[0, 1]$

**69.** $f(x) = \begin{cases} -1 & \text{if} \quad x < 1 \\ 0 & \text{if} \quad x = 1 \\ 2 - x & \text{if} \quad 1 < x \le 2 \end{cases}$ domain: $(-\infty, 2]$, range: $\{-1\} \cup [0, 1)$

**71.** $f(x) = \begin{cases} -x^3 & \text{if} \quad -2 < x < 1 \\ 0 & \text{if} \quad x = 1 \\ x^2 & \text{if} \quad 1 < x \le 3 \end{cases}$ domain: $(-2, 3]$, range: $(-1, 9]$   **73.** (a) \$250.60   (b) \$2655.40   (c) AWV.
**75.** (a) The independent variable is $a$; the dependent variable is $P$.   (b) $P(20) = 231.427$ million Americans. AWV.
   (c) $P(0) = 327.287$ million Americans. AWV.

### Section P.2

**1.** (a)   **2.** True   **3.** True   **4.** False   **5.** False   **6.** Zero   **7.** (b)   **8.** True   **9.** True   **10.** False   **11.** D   **13.** F   **15.** C   **17.** G
**19.** (a) 2   (b) 3   (c) $-4$   **21.** (a) 7 with multiplicity 1; $-4$ with multiplicity 3   (b) $x$-intercepts: $7, -4$, $y$-intercept: $-1344$
   (c) Crosses at 7 and at $-4$   (d) The graph of $f$ will resemble the graph of $y = 3x^4$.   **23.** (c), (e), (f)   **25.** Domain: $\{x \,|\, x \neq -3\}$, intercept: $(0, 0)$

**27.** Domain: $(-\infty, \infty)$, intercepts: $(0, 0), \left(\dfrac{1}{3}, 0\right)$   **29.** (a) $A(x) = 16x - x^3$   (b) $[0, 4]$

**31.** (a) The pattern in the scatter plot suggests a quadratic relationship.   (b) 283.8 feet
   (c) 15.63 feet

### Section P.3    (SSM = See Student Solutions Manual; AWV = Answers will vary)

**1.** $\{x \,|\, 0 \le x \le 5\}$   **2.** False   **3.** True   **4.** False   **5.** True   **6.** False   **7.** True   **8.** Horizontal, right   **9.** $y$   **10.** $-5, -2, 2$
**11.** (a) $(f + g)(x) = 5x + 1$, domain: $(-\infty, \infty)$   (b) $(f - g)(x) = x + 7$, domain: $(-\infty, \infty)$   (c) $(f \cdot g)(x) = 6x^2 - x - 12$,

   domain: $(-\infty, \infty)$   (d) $\left(\dfrac{f}{g}\right)(x) = \dfrac{3x + 4}{2x - 3}$, domain: $\left\{x \,\Big|\, x \neq \dfrac{3}{2}\right\}$   **13.** (a) $(f + g)(x) = \sqrt{x + 1} + \dfrac{2}{x}$, domain: $\{x \,|\, x \ge -1, x \neq 0\}$

   (b) $(f - g)(x) = \sqrt{x + 1} - \dfrac{2}{x}$, domain: $\{x \,|\, x \ge -1, x \neq 0\}$   (c) $(f \cdot g)(x) = \dfrac{2\sqrt{x + 1}}{x}$, domain: $\{x \,|\, x \ge -1, x \neq 0\}$

**(d)** $\left(\dfrac{f}{g}\right)(x) = \dfrac{1}{2}x\sqrt{x+1}$, domain: $\{x \mid x \ge -1, x \ne 0\}$   **15. (a)** $(f \circ g)(4) = 98$   **(b)** $(g \circ f)(2) = 49$   **(c)** $(f \circ f)(1) = 4$

**(d)** $(g \circ g)(0) = 4$   **17. (a)** $(f \circ g)(1) = -1$   **(b)** $(f \circ g)(-1) = -1$   **(c)** $(g \circ f)(-1) = 8$   **(d)** $(g \circ f)(1) = 8$   **(e)** $(g \circ g)(-2) = 8$
**(f)** $(f \circ f)(-1) = -7$   **19. (a)** $(g \circ f)(-1) = 4$   **(b)** $(g \circ f)(6) = 2$   **(c)** $(f \circ g)(6) = 1$   **(d)** $(f \circ g)(4) = -2$

**21. (a)** $(f \circ g)(x) = 24x + 1$, domain: $(-\infty, \infty)$   **(b)** $(g \circ f)(x) = 24x + 8$, domain: $(-\infty, \infty)$   **(c)** $(f \circ f)(x) = 9x + 4$, domain: $(-\infty, \infty)$
**(d)** $(g \circ g)(x) = 64x$, domain: $(-\infty, \infty)$   **23. (a)** $(f \circ g)(x) = x$, domain: $\{x \mid x \ge 1\}$   **(b)** $(g \circ f)(x) = |x|$, domain: $(-\infty, \infty)$
**(c)** $(f \circ f)(x) = x^4 + 2x^2 + 2$, domain: $(-\infty, \infty)$   **(d)** $(g \circ g)(x) = \sqrt{\sqrt{x-1}-1}$, domain: $\{x \mid x \ge 2\}$   **25. (a)** $(f \circ g)(x) = \dfrac{2}{2-x}$,

domain: $\{x \mid x \ne 0, x \ne 2\}$   **(b)** $(g \circ f)(x) = \dfrac{2(x-1)}{x}$, domain: $\{x \mid x \ne 0, x \ne 1\}$   **(c)** $(f \circ f)(x) = x$, domain: $\{x \mid x \ne 1\}$

**(d)** $(g \circ g)(x) = x$, domain: $\{x \mid x \ne 0\}$   **27.** $f(x) = x^4$, $g(x) = 2x + 3$   **29.** $f(x) = \sqrt{x}$, $g(x) = x^2 + 1$   **31.** $f(x) = |x|$, $g(x) = 2x + 1$

**33.**   **35.**   **37.**   **39.**

**41.**   **43.**   **45.**

**47. (a)**   **(b)**   **(c)**   **(d)**

**(e)**   **(f)**   **(g)**

**49. (a)**   **(b)**   **(c)** AWV.   **(d)**   **(e)** AWV.

## Section P.4   (SSM = See Student Solutions Manual; AWV = Answers will vary)

**1.** False   **2.** $[4, \infty)$   **3.** False   **4.** True   **5.** False   **6.** False   **7.** AWV.   **8.** AWV. **9.** One-to-one   **11.** Not one-to-one   **13.** One-to-one
**15.** SSM.   **17.** SSM.   **19. (a)** One-to-one   **b)** $\{(5, -3), (9, -2), (2, -1), (11, 0), (-5, 1)\}$,
   **(c)** Domain of $f$: $\{-3, -2, -1, 0, 1\}$, range of $f$: $\{-5, 2, 5, 9, 11\}$, domain of $f^{-1}$: $\{-5, 2, 5, 9, 11\}$, range of $f^{-1}$: $\{-3, -2, -1, 0, 1\}$
**21. (a)** Not one-to-one   **(b)** Does not apply   **(c)** Domain of $f$: $\{-10, -3, -2, 1, 2\}$, range of $f$: $\{0, 1, 2, 9\}$

**23.** 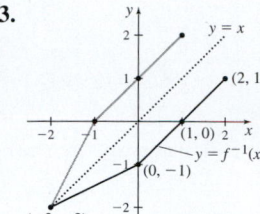 **25.** 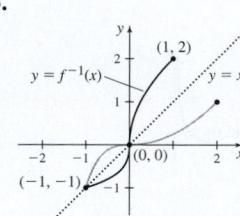 **27.**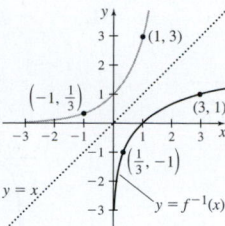

**29. (a)** $f^{-1}(x) = \dfrac{x-2}{4}$ **(b)** Both the domain and range of $f$ are $(-\infty, \infty)$. Both the domain and range of $f^{-1}$ are $(-\infty, \infty)$.

**31. (a)** $f^{-1}(x) = x^3 - 10$ **(b)** Both the domain and range of $f$ are $(-\infty, \infty)$. Both the domain and range of $f^{-1}$ are $(-\infty, \infty)$.

**33. (a)** $f^{-1}(x) = 2 + \dfrac{1}{x}$ **(b)** Domain of $f$: $\{x \mid x \neq 2\}$, range of $f$: $\{y \mid y \neq 0\}$, domain of $f^{-1}$: $\{x \mid x \neq 0\}$, range of $f^{-1}$: $\{y \mid y \neq 2\}$

**35. (a)** $f^{-1}(x) = \dfrac{3-2x}{x-2}$ **(b)** Domain of $f$: $\{x \mid x \neq -2\}$, range of $f$: $\{y \mid y \neq 2\}$, domain of $f^{-1}$: $\{x \mid x \neq 2\}$, range of $f^{-1}$: $\{y \mid y \neq -2\}$

**37. (a)** $f^{-1}(x) = \sqrt{x-4}$ **(b)** Domain of $f$: $\{x \mid x \geq 0\}$, range of $f$: $\{y \mid y \geq 4\}$, domain of $f^{-1}$: $\{x \mid x \geq 4\}$, range of $f^{-1}$: $\{y \mid y \geq 0\}$

## Section P.5 (SSM = See Student Solutions Manual; AWV = Answers will vary)

**1.** $\left(-1, \dfrac{1}{a}\right), (0, 1), (1, a)$ **2.** False **3.** 4 **4.** 3 **5.** False **6.** False **7.** 1 **8.** $\{x \mid x > 0\}$ **9.** $\left(\dfrac{1}{a}, -1\right), (1, 0), (a, 1)$ **10. (a)** **11.** False

**12.** True **13.** True **14.** 1 **15.** AWV. **16.** 0 **17. (a)** $g(-1) = \dfrac{9}{4}, \left(-1, \dfrac{9}{4}\right)$ **(b)** $x = 3, (3, 66)$ **19. (a)** **21. (c)** **23. (b)**

**25.** Domain: $(-\infty, \infty)$, range: $(0, \infty)$ **27.** Domain: $(-\infty, \infty)$, range: $(0, \infty)$ **29.** Domain: $(-\infty, \infty)$, range: $(0, \infty)$

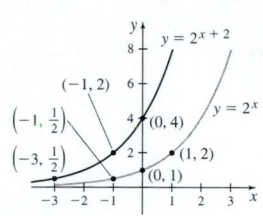 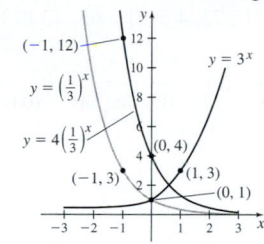 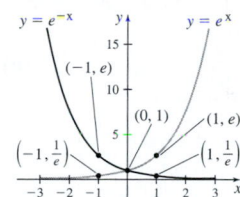

**31.** $\{x \mid x \neq 0\}$ **33.** $\{x \mid x > 1\}$ **35. (b)** **37. (f)** **39. (c)**

**41. (a)** $\{x \mid x > -4\}$ **(b and f)** **43. (a)** $(-\infty, \infty)$ **(b and f)**
**(c)** $(-\infty, \infty)$
**(d)** $f^{-1}(x) = e^x - 4$
**(e)** $(-\infty, \infty)$

**(c)** $\{y \mid y > 2\}$
**(d)** $f^{-1}(x) = \ln\left(\dfrac{x-2}{3}\right)$
**(e)** $\{y \mid y > 2\}$

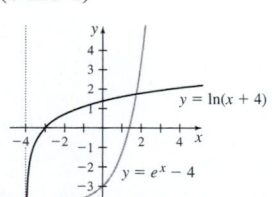 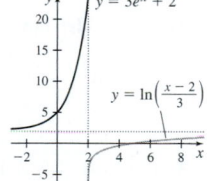

**45.** It shifts the $x$-intercept $c$ units to the left. **47.** $x = 0, x = 2$ **49.** $x = \dfrac{1}{2}$ **51.** $x = \dfrac{1 - \ln 4}{2}$ **53.** $x = \dfrac{\ln\left(\dfrac{9}{5}\right)}{3\ln 2}$

**55.** $x = \dfrac{\ln 3}{\ln 36}$ **57.** $x = \dfrac{7}{2}$ **59.** $x = \dfrac{1}{2}$ **61.** $x = 3$ **63.** $x \approx 2.787$ **65.** $x \approx 1.315$

**67. (a)** 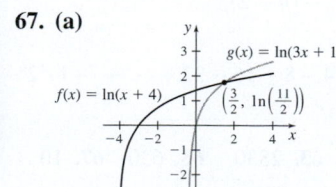 **(b)** $\left(\dfrac{3}{2}, \ln\dfrac{11}{2}\right)$ **(c)** $\left\{x \mid -\dfrac{1}{3} < x < \dfrac{3}{2}\right\}$

## Section P.6 (SSM = See Student Solutions Manual; AWV = Answers will vary)

**1.** $2\pi, \pi$ **2.** All real numbers except odd multiples of $\dfrac{\pi}{2}$ **3.** $\{y \mid -1 \leq y \leq 1\}$ **4.** The period of $\tan x$ is $\pi$. **5.** False **6.** $3, \dfrac{\pi}{3}$ **7.** True

**8.** False **9.** True **10.** Origin **11.** $y$-axis **12.** AWV. **13.** $-1$ **15.** $-\dfrac{2\sqrt{3}}{3}$ **17.** 0 **19.** $-1$ **21. (f)** **23. (a)** **25. (d)**

**27.**

**29.**

**31.**

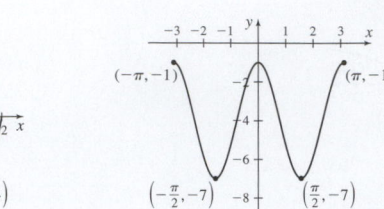

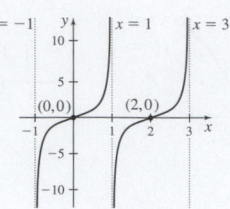

**33.** Amplitude: $\dfrac{1}{2}$, period: 2  **35.** Amplitude: 3, period: $2\pi$  **37.** $f(x)=2\sin(2x)$  **39.** $f(x)=\dfrac{1}{2}\cos(2x)$  **41.** $f(x)=-\sin\left(\dfrac{3}{2}x\right)$

**43.** $f(x)=1-\cos\left(\dfrac{4\pi}{3}x\right)$  **45.** $f(x)=\cot x$  **47.** $f(x)=\tan\left(x-\dfrac{\pi}{2}\right)$

## Section P.7  (SSM = See Student Solutions Manual; AWV = Answers will vary)

**1.** $\sin y$  **2.** False  **3.** False  **4.** True  **5.** False  **6.** True  **7.** True  **8.** True  **9.** 0  **11.** $\pi$  **13.** $-\dfrac{\pi}{4}$  **15.** $\dfrac{\pi}{4}$  **17.** $\dfrac{\pi}{3}$  **19.** $\dfrac{5\pi}{4}$  **21.** $\dfrac{\pi}{5}$

**23.** $-\dfrac{3\pi}{8}$  **25.** $-\dfrac{\pi}{8}$  **27.** $-\dfrac{\pi}{5}$  **29.** $\dfrac{1}{4}$  **31.** 4  **33.** It is not defined.  **35.** $\pi$  **37.** $\sqrt{0.91}$  **39.** $\sqrt{3}$  **41.** $\dfrac{1+4\sqrt{6}}{5\sqrt{5}}$  **43.** $\dfrac{4\sqrt{21}}{17}$

**45.** $\left\{\dfrac{5\pi}{6},\dfrac{11\pi}{6}\right\}$  **47.** $\left\{\dfrac{7\pi}{6},\dfrac{11\pi}{6}\right\}$  **49.** $\left\{\dfrac{\pi}{2},\dfrac{7\pi}{6},\dfrac{11\pi}{6}\right\}$  **51.** $\left\{\dfrac{\pi}{3},\dfrac{2\pi}{3},\dfrac{4\pi}{3},\dfrac{5\pi}{3}\right\}$  **53.** $\left\{\dfrac{\pi}{2}\right\}$  **55.** $\left\{\dfrac{\pi}{6},\dfrac{5\pi}{6},\dfrac{3\pi}{2}\right\}$  **57.** $\left\{\dfrac{3\pi}{4},\dfrac{7\pi}{4}\right\}$

**59.** $\left\{\dfrac{\pi}{3},\dfrac{2\pi}{3},\dfrac{4\pi}{3},\dfrac{5\pi}{3}\right\}$  **61.** $\left\{0,\dfrac{\pi}{3},\dfrac{\pi}{2},\dfrac{2\pi}{3},\pi,\dfrac{4\pi}{3},\dfrac{3\pi}{2},\dfrac{5\pi}{3}\right\}$  **63.** $\{1.373,4.515\}$  **65.** $\{3.871,5.553\}$  **67.** $\cos(\sin^{-1}u)=\sqrt{1-u^2},\ |u|\le 1$

**69.** SSM.  **71. (a and c)**

**(b)** $\left(\dfrac{\pi}{12},\dfrac{7}{2}\right),\ \left(\dfrac{5\pi}{12},\dfrac{7}{2}\right)$  **(c)** See **(a)**.  **(d)** $\left\{x\ \bigg|\ \dfrac{\pi}{12}<x<\dfrac{5\pi}{12}\right\}$

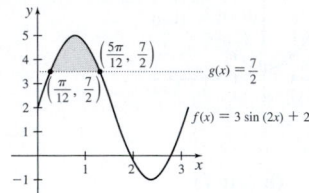

**73. (a)** $\tan^{-1}\sqrt{3}-\tan^{-1}0=\dfrac{\pi}{3}$  **(b)** $\tan^{-1}1-\tan^{-1}\left(-\dfrac{\sqrt{3}}{3}\right)=\dfrac{5\pi}{12}$

## Section P.8

**1.** sequence  **2.** True  **3.** $1\cdot2\cdot3\cdot\dots\cdot(n-1)\cdot n$  **4.** recursively defined  **5.** summation (or sigma)  **6.** True  **7.** 16  **8.** $1;n$  **9.** False

**10.** Binomial Theorem  **11.** 2,4,6,8,10  **13.** $\dfrac{1}{2},\dfrac{2}{3},\dfrac{3}{4},\dfrac{4}{5},\dfrac{5}{6}$  **15.** $-1,4,-9,16,-25$  **17.** $\dfrac{1}{2},-\dfrac{2}{5},\dfrac{2}{7},-\dfrac{8}{41},\dfrac{8}{61}$  **19.** $a_n=\dfrac{n+1}{n}$

**21.** $a_n=\dfrac{1}{n^2}$  **23.** $a_n=(-1)^{n+1}(2n)$  **25.** $a_n=(-1)^{n+1}(n+1)$  **27.** 3,5,7,9,11  **29.** 2,4,7,11,16  **31.** 1,3,9,27,81

**33.** $2,1,\dfrac{1}{3},\dfrac{1}{12},\dfrac{1}{60}$  **35.** $1,-1,-1,1,-1$  **37.** $A,A+d,A+2d,A+3d,A+4d$  **39.** $-1+0+1+2+\dots+(n-2)$

**41.** $\dfrac{1}{2}+\dfrac{4}{3}+\dfrac{9}{4}+\dfrac{16}{5}+\dots+\dfrac{n^2}{n+1}$  **43.** $1+\dfrac{1}{2}+\dfrac{1}{4}+\dfrac{1}{8}+\dots+\dfrac{1}{2^n}$  **45.** $\dfrac{1}{2}+\dfrac{1}{2}+\dfrac{3}{8}+\dfrac{1}{4}+\dots+\dfrac{n+1}{2^{n+1}}$  **47.** $4-8+16-32+\dots+(-1)^n2^n$

**49.** $\displaystyle\sum_{k=1}^{50}2k$  **51.** $\displaystyle\sum_{k=1}^{35}\dfrac{k+1}{k}$  **53.** $\displaystyle\sum_{k=0}^{10}(-1)^k\left(\dfrac{3}{10^k}\right)$  **55.** $\displaystyle\sum_{k=1}^{n}(-1)^{k-1}\dfrac{k}{e^k}$  **57.** 20  **59.** 1395  **61.** 10,000  **63.** 2830  **65.** 630  **67.** 10

**69.** 84  **71.** 50  **73.** $x^4+4x^3+6x^2+4x+1$  **75.** $x^5-15x^4+90x^3-270x^2+405x-243$  **77.** $16x^4+96x^3+216x^2+216x+81$

**79.** $x^6-3x^4y^2+3x^2y^4-y^6$  **81.** After the first payment, John's balance is $2930.

# Chapter 1

## Section 1.1    (SSM = See Student Solutions Manual; AWV = Answers will vary)

**1.** (c)    **2.** True    **3.** False    **4.** False    **5.** False    **6.** False

**7.** $\lim\limits_{x \to 1} 2x = 2$

| | x approaches 1 from the left | | | | x approaches 1 from the right | | |
|---|---|---|---|---|---|---|---|
| x | 0.9 | 0.99 | 0.999 | → 1 ← | 1.001 | 1.01 | 1.1 |
| $f(x) = 2x$ | 1.8 | 1.98 | 1.998 | $f(x)$ approaches 2 | 2.002 | 2.02 | 2.2 |

**9.** $\lim\limits_{x \to 0} (x^2 + 2) = 2$

| | x approaches 0 from the left | | | | x approaches 0 from the right | | |
|---|---|---|---|---|---|---|---|
| x | −0.1 | −0.01 | −0.001 | → 0 ← | 0.001 | 0.01 | 0.1 |
| $f(x) = x^2 + 2$ | 2.01 | 2.0001 | 2.000001 | $f(x)$ approaches 2 | 2.000001 | 2.0001 | 2.01 |

**11.** $\lim\limits_{x \to -3} \dfrac{x^2 - 9}{x + 3} = -6$

| | x approaches −3 from the left | | | | x approaches −3 from the right | | |
|---|---|---|---|---|---|---|---|
| x | −3.5 | −3.1 | −3.01 | → −3 ← | −2.99 | −2.9 | −2.5 |
| $f(x) = \dfrac{x^2 - 9}{x + 3}$ | −6.5 | −6.1 | −6.01 | $f(x)$ approaches −6 | −5.99 | −5.9 | −5.5 |

**13.** $\lim\limits_{x \to 0} \dfrac{2 - 2e^x}{x} = -2$

| | x approaches 0 from the left | | | | x approaches 0 from the right | | |
|---|---|---|---|---|---|---|---|
| x | −0.2 | −0.1 | −0.01 | → 0 ← | 0.01 | 0.1 | 0.2 |
| $f(x) = \dfrac{2 - 2e^x}{x}$ | −1.8127 | −1.9033 | −1.9900 | $f(x)$ approaches −2 | −2.0100 | −2.1034 | −2.2140 |

**15.** $\lim\limits_{x \to 0} \dfrac{1 - \cos x}{x} = 0$

| | x approaches 0 from the left | | | | x approaches 0 from the right | | |
|---|---|---|---|---|---|---|---|
| x | −0.2 | −0.1 | −0.01 | → 0 ← | 0.01 | 0.1 | 0.2 |
| $f(x) = \dfrac{1 - \cos x}{x}$ | −0.09967 | −0.04996 | −0.00500 | $f(x)$ approaches 0 | 0.00500 | 0.04996 | 0.09967 |

**17.** (a) 2    (b) 2    (c) 2    **19.** (a) 3    (b) 6    (c) The limit does not exist.    **21.** 1    **23.** 1    **25.** The limit does not exist because the two one-sided limits are not equal.    **27.** The limit does not exist because the two one-sided limits are not equal.
**29.** 9    **31.** The limit does not exist.    **33.** 2    **35.** The limit does not exist.    **37.** AWV.    **39.** AWV.    **41.** 1
**43.** 0    **45.** 1    **47.** 0    **49.** 0    **51.** (a) $m_{\sec} = 15$    (b) $m_{\sec} = 3(x + 2)$    (c) $\lim\limits_{x \to 2} m_{\sec} = 12$    (d)

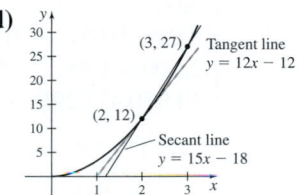

**53.** (a) $m_{\sec} = 2 + \dfrac{1}{2}h$ for $h \neq 0$

(b)

| h | −0.5 | −0.1 | −0.001 | 0.001 | 0.1 | 0.5 |
|---|---|---|---|---|---|---|
| $m_{\sec}$ | 1.75 | 1.95 | 1.9995 | 2.0005 | 2.05 | 2.25 |

(c) $\lim\limits_{h \to 0} m_{\sec} = 2$    (d) $m_{\tan} = 2$

(e)

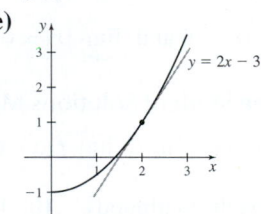

**55.** (a) These values suggest that $\lim\limits_{x \to 0} \cos \dfrac{\pi}{x}$ may be 1.    (b) These values suggest that $\lim\limits_{x \to 0} \cos \dfrac{\pi}{x}$ may be −1.

(c) $\lim\limits_{x \to 0} \cos \dfrac{\pi}{x}$ does not exist. SSM.    (d) SSM.

**57.** (a) $\lim\limits_{x \to 2} \dfrac{x - 8}{2} = -3$    (b) $1.8 \leq x \leq 2.2$    (c) $1.98 \leq x \leq 2.02$

| | x approaches 2 from the left | | | | x approaches 2 from the right | | |
|---|---|---|---|---|---|---|---|
| x | 1.9 | 1.99 | 1.999 | → 2 ← | 2.001 | 2.01 | 2.1 |
| $f(x) = \dfrac{x - 8}{2}$ | −3.05 | −3.005 | −3.0005 | $f(x)$ approaches −3 | −2.9995 | −2.995 | −2.95 |

**59. (a)** $C(w) = \begin{cases} 0.47 & \text{if } 0 < w \le 1 \\ 0.68 & \text{if } 1 < w \le 2 \\ 0.89 & \text{if } 2 < w \le 3 \\ 1.10 & \text{if } 3 < w \le 3.5 \end{cases}$   **(b)** $\{w | 0 < w \le 3.5\}$   **(c)** See the graph below.   **(d)** $\lim\limits_{w \to 2^-} C(w) = 0.68$

$\lim\limits_{w \to 2^+} C(w) = 0.89$

$\lim\limits_{w \to 2} C(w)$ does not exist.

**(e)** $\lim\limits_{w \to 0^+} C(w) = 0.47$

**(f)** $\lim\limits_{w \to 3.5^-} C(w) = 1.10$

**61. (a)** $\lim\limits_{t \to 7} S(t) = 100$   **(b)** Given any $\varepsilon > 0$, there is a number $\delta > 0$ so that whenever $0 < |t - 7| < \delta$ then $|S(t) - 100| < \varepsilon$.

**63.** No, the value of the function at $x = 2$ has no bearing on $\lim\limits_{x \to 2} f(x)$.   **65. (a)** The graph of $f$ is the horizontal line $y = -1$

excluding the point $(3, -1)$.   **(b)** $\lim\limits_{x \to 3^-} f(x) = \lim\limits_{x \to 3^+} f(x) = -1$   **(c)** The graph suggests $\lim\limits_{x \to 3} f(x) = -1$.   **67.** 0   **69.** 0

## Section 1.2   (SSM = See Student Solutions Manual; AWV = Answers will vary)

**1. (a)** $-3$   **(b)** $\pi$   **2.** 243   **3.** 4   **4. (a)** $-1$   **(b)** $e$   **5. (a)** $-2$   **(b)** $\dfrac{7}{2}$   **6. (a)** $-6$   **(b)** 0   **7.** True   **8.** 2   **9.** False   **10.** True

**11.** 14   **13.** 0   **15.** 1   **17.** 6   **19.** $\sqrt{11}$   **21.** $2\sqrt{78}$   **23.** 4   **25.** 0   **27.** 5   **29.** $-\dfrac{31}{8}$   **31.** 10   **33.** $\dfrac{13}{4}$   **35.** 4   **37.** 2   **39.** 2

**41.** $\dfrac{\sqrt{2}}{4}$   **43.** $\dfrac{1}{30}$   **45.** 5   **47.** 6   **49.** 9   **51.** $-1$   **53.** 8   **55.** 0   **57.** $\dfrac{1}{4}$   **59. (a)** 6   **(b)** $-16$   **(c)** $-16$   **(d)** 0   **(e)** $-\dfrac{1}{4}$   **(f)** 0

**61.** 6   **63.** $-4$   **65.** $\dfrac{1}{2}$   **67.** 4   **69.** $6x + 4$   **71.** $-\dfrac{2}{x^2}$   **73.** $\lim\limits_{x \to 1^-} f(x) = -1$, $\lim\limits_{x \to 1^+} f(x) = 2$, $\lim\limits_{x \to 1} f(x)$ does not exist

**75.** $\lim\limits_{x \to 1^-} f(x) = 2$, $\lim\limits_{x \to 1^+} f(x) = 2$, $\lim\limits_{x \to 1} f(x) = 2$   **77.** $\lim\limits_{x \to 1^-} f(x) = 0$, $\lim\limits_{x \to 1^+} f(x) = 0$, $\lim\limits_{x \to 1} f(x) = 0$

**79.** $\lim\limits_{x \to 3^-} f(x) = 6$, $\lim\limits_{x \to 3^+} f(x) = 6$, $\lim\limits_{x \to 3} f(x) = 6$   **81.** The limit does not exist.   **83.** $2x$   **85.** $-\dfrac{1}{x^2}$   **87.** $-\dfrac{1}{16}$   **89.** 6   **91.** 0

**93. (a)** $C(x) = \begin{cases} 9.00 & \text{if } 0 \le x \le 10 \\ 9.00 + 0.95(x - 10) & \text{if } 10 < x \le 30 \\ 28.00 + 1.65(x - 30) & \text{if } 30 < x \le 100 \\ 143.50 + 2.20(x - 100) & \text{if } x > 100 \end{cases}$   **(d)** 9.00   **(e)**

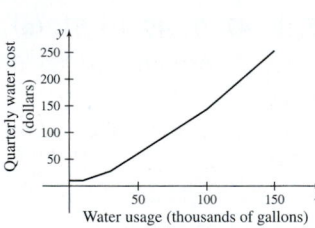

**(b)** $\{x | x \ge 0\}$   **(c)** $\lim\limits_{x \to 5} C(x) = 9.00$, $\lim\limits_{x \to 10} C(x) = 9.00$,

$\lim\limits_{x \to 30} C(x) = 28.00$, $\lim\limits_{x \to 100} C(x) = 143.50$

**95. (a)** 0   **(b)** As the temperature of a gas approaches zero, the molecules in the gas stop moving.

**97.** $\lim\limits_{x \to 0} |x| = 0$ since $\lim\limits_{x \to 0^-} |x| = 0$ and $\lim\limits_{x \to 0^+} |x| = 0$   **99.** AWV.   **101.** AWV.   **103.** SSM.   **105.** $na^{n-1}$   **107.** $\dfrac{m}{n}$   **109.** $\dfrac{a + b}{2}$   **111.** 0

## Section 1.3   (SSM = See Student Solutions Manual; AWV = Answers will vary)

**1.** True   **2.** False   **3.** $f(c)$ is defined, $\lim\limits_{x \to c} f(x)$ exists, $\lim\limits_{x \to c} f(x) = f(c)$   **4.** True   **5.** False   **6.** False   **7.** False

**8.** True   **9.** The function is discontinuous.   **10.** The function is continuous.   **11.** True   **12.** False

**13. (a)** The function is not continuous at $c = -3$.   **(b)** $\lim\limits_{x \to -3} f(x) \ne f(-3)$   **(c)** The discontinuity is removable.   **(d)** $f(-3) = -2$

**15. (a)** The function is not continuous at $c = 2$.   **(b)** $\lim\limits_{x \to 2} f(x)$ does not exist.   **(c)** The discontinuity is not removable.

**17. (a)** The function is continuous at $c = 4$.   **19.** The function is continuous at $c = -1$.   **21.** The function is continuous at $c = -2$.

**23.** The function is continuous at $c = 2$.   **25.** The function is not continuous at $c = 1$.   **27.** The function is not continuous at $c = 1$.

**29.** The function is continuous at $c = 0$.   **31.** The function is not continuous at $c = 0$.   **33.** $f(2) = 4$   **35.** $f(1) = 2$

**37.** The function is not continuous on $[-3, 3]$; it is continuous on the interval $[-3, 3)$.

**39.** The function is not continuous at any number in the interval $[-3, 3]$.   **41.** The function is continuous on $\{x | x \ne 0\}$.

**43.** The function is continuous on the set of all real numbers.   **45.** The function is continuous on $\{x | x \ge 0, x \ne 9\}$.

**47.** The function is continuous on the set $\{x | x < 2\}$.   **49.** The function is continuous on the set of all real numbers.

**51.** $f$ is continuous at 0 because $\lim\limits_{x \to 0} f(x) = f(0)$.   **53.** $f$ is not continuous at 3 because $\lim\limits_{x \to 3} f(x)$ does not exist.

**55.** $f$ is continuous at 1 because $\lim\limits_{x \to 1} f(x) = f(1)$.

**57. (a)** 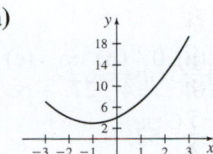   **(b)** Based on the graph, it appears that the function is continuous for all real numbers.
**(c)** $f$ is actually continuous at all real numbers except $x = 2$.
**(d)** AWV.

**59.** Yes. A zero exists on the given interval.   **61.** The IVT gives no information.   **63.** The IVT gives no information.
**65.** 1.154   **67.** 0.211   **69.** 3.134   **71.** 1.157   **73. (a)** Since $f$ is continuous on $[0, 1]$, $f(0) < 0$, and $f(1) > 0$, the
IVT guarantees that $f$ has a zero on the interval $(0, 1)$.   **(b)** 0.828
**75. (a)** The function $f$ is continuous from the left only at $-1$.   **(b)** The function $f$ is continuous from the right only at 1.
**77. (a)** The function $f$ is continuous from the left at $-1$, but not at 5.
**(b)** The function $f$ is continuous from the right at 5, but not at $-1$.

**79. (a)** $C(w) = \begin{cases} 0.47 & \text{if } 0 < w \le 1 \\ 0.68 & \text{if } 1 < w \le 2 \\ 0.89 & \text{if } 2 < w \le 3 \\ 1.10 & \text{if } 3 < w \le 3.5 \end{cases}$   **(b)** $\{w \mid 0 < w \le 3.5\}$

**(c)** The function is continuous on the intervals $(0, 1]$, $(1, 2]$, $(2, 3]$, and $(3, 3.5]$.

**(d)** At $w = 1$, $w = 2$, and $w = 3$, the function has a jump discontinuity.   **(e)** AWV.

**81. (a)** $C(x) = \begin{cases} 7.87 + 0.02173x & \text{if } 0 \le x \le 1000 \\ -2.13 + 0.03173x & \text{if } x > 1000 \end{cases}$   **(b)** $\{x \mid x \ge 0\}$   **(c)** $C$ is continuous on its domain.

**(d)** See (c).   **(e)** AWV.   **83. (a)** $\dfrac{Gm}{R^2}$   **(b)** 1.3 m/s²   **(c)** The gravity on Europa is less than the gravity on Earth.

**85.** $A = 5, B = 7$

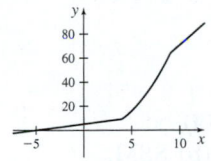

**87. (a)** $\{x \mid x < 2\} \cup \{x \mid x > 2, x \ne 8\}$   **(b)** $f$ is discontinuous at $x = 8$.   **(c)** The discontinuity at $x = 8$ is removable.
**89. (a)** Since $f$ is continuous on $[0, 2]$, $f(0) < 0$ and $f(2) > 0$, the IVT guarantees that $f$ must have a zero on the
interval $(0, 2)$.   **(b)** $x \approx 0.443$   **91.** 1.732   **93.** 1.561   **95.** This does not contradict the IVT. $f(-1)$ and $f(2)$ are both
positive, so there is no guarantee that the function has a zero in the interval $(-1, 2)$.
**97.** This does not contradict the IVT. The IVT guarantees that $f$ must have at least one zero on the interval $(-5, 0)$.
**99. (a)** Since $f(-2)$ and $f(2)$ are both positive, there is no guarantee   **(b)** The graph below indicates that $f$ does not have a
that the function has a zero in the interval $(-2, 2)$.   zero in the interval $(-2, 2)$.

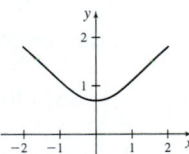

**101.** SSM.   **103. (a)** $x = -2, x = 2, x = 3$   **(b)** $p = 5$   **(c)** $h$ is an even function.   **105.** AWV.   **107.** $(1.125, 1.25)$
**109.** $(0.125, 0.25)$   **111.** $(3.125, 3.25)$   **113.** $(1.125, 1.25)$   **115.** Since $f$ is continuous on $[0, 1]$, $f(0) < 0$, and $f(1) > 0$,
the IVT guarantees that $f$ has a zero on the interval $(0, 1)$. Correct to one decimal place, the zero is $x = 0.8$
**117.** SSM.   **119.** SSM.   **121.** SSM.   **123.** SSM.

## Section 1.4   (SSM = See Student Solutions Manual; AWV = Answers will vary)

**1.** 0   **2.** False   **3.** $L$   **4.** False   **5.** 1   **7.** 1   **9.** 0   **11.** $\dfrac{\sqrt{3}}{2} + \dfrac{1}{2}$   **13.** 1   **15.** $\dfrac{3}{2}$   **17.** 0   **19.** 1   **21.** $\dfrac{1}{2}$   **23.** 7   **25.** 2

**27.** $\dfrac{1}{2}$   **29.** 5   **31.** 0   **33.** 1   **35.** $f$ is continuous at $c = 0$.   **37.** $f$ is continuous at $c = \dfrac{\pi}{4}$.   **39.** $f$ is continuous on $\{x \mid x \ne 4\}$.

**41.** $f$ is continuous on $\left\{ x \mid x \ne \dfrac{3\pi}{2} + 2k\pi \right\}$, $k$ an integer.   **43.** $f$ is continuous on $\{x \mid x > 0, x \ne 3\}$.

**45.** $f$ is continuous on the set of all real numbers.   **47.** 0   **49.** 0   **51.** SSM.   **53.** SSM.   **55.** SSM.   **57.** SSM.
**59.** $f(0) = \pi, f(1) = \pi$   **61.** Yes   **63.** SSM.   **65.** AWV.   **67.** SSM.   **69.** SSM.

## Section 1.5 (SSM = See Student Solutions Manual; AWV = Answers will vary)

**1.** False **2. (a)** $-\infty$ **(b)** $\infty$ **(c)** $-\infty$ **3.** False **4.** Vertical **5. (a)** 0 **(b)** 0 **(c)** $\infty$ **6.** False **7. (a)** 0 **(b)** $\infty$ **(c)** 0
**8.** True **9.** 2 **11.** $\infty$ **13.** $\infty$ **15.** $x = -1, x = 3$ **17.** $-3$ **19.** $\infty$ **21.** 0 **23.** $\infty$ **25.** $x = -3, x = 0, x = 4$ **27.** $-\infty$
**29.** $\infty$ **31.** $\infty$ **33.** $\infty$ **35.** $-\infty$ **37.** $-\infty$ **39.** $-\infty$ **41.** $-\infty$ **43.** 0 **45.** $\dfrac{2}{5}$ **47.** $\infty$ **49.** 0 **51.** $\dfrac{5}{4}$ **53.** 0

**55.** 0 **57.** $\sqrt{3}$ **59.** $-\infty$ **61.** $x = 0$ is a vertical asymptote. $y = 3$ is a horizontal asymptote. **63.** $x = -1$ and $x = 1$
are vertical asymptotes. $y = 1$ is a horizontal asymptote. **65.** $x = \dfrac{3}{2}$ is a vertical asymptote. $y = \dfrac{\sqrt{2}}{2}$ and $y = -\dfrac{\sqrt{2}}{2}$
are horizontal asymptotes. **67. (a)** $\{x \mid x \neq -2, x \neq 0\}$ **(b)** $y = 0$ is a horizontal asymptote. **(c)** $x = 0$ and $x = -2$
are vertical asymptotes. **(d)** SSM. **69. (a)** $\left\{ x \mid x \neq \dfrac{3}{2}, x \neq 2 \right\}$

**(b)** $y = \dfrac{1}{2}$ is a horizontal asymptote. **(c)** $x = \dfrac{3}{2}$ is a vertical asymptote. **(d)** SSM.

**71. (a)** $\{x \mid x \neq 0, x \neq 1\}$ **(b)** There are no horizontal asymptotes. **(c)** $x = 0$ is a vertical asymptote. **(d)** SSM. **73.** AWV.
**75. (a)** $T$ **(b)** $u_0$ **77.** $\infty$ **79. (a)** 500 **(b)** **(c)** AWV. **81. (a)** The graph suggests **(b)** 0 **(c)** AWV.
that $\displaystyle\lim_{t \to \infty} x(t) = 0$.

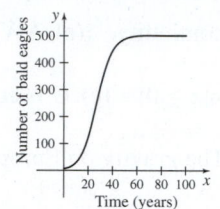

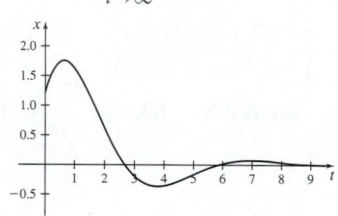

**83. (a)** 0.26 moles **(b)** 395 min **(c)** 0 moles **(d)** SSM. **85.** SSM. **87.** SSM. **89.** SSM.

**91. (a)**

| $x$ | 100 | 10,000 | 1,000,000 | 1,000,000,000 | $\to \infty$ |
|---|---|---|---|---|---|
| $f(x) = \left(1 + \dfrac{1}{x}\right)^x$ | 2.70481 | 2.718146 | 2.718280 | 2.718282 | 2.718282 |

**(b)** $e \approx 2.718281828$ **93. (a)** $\infty$
**(c)** SSM. **(b)** SSM.

## Section 1.6 (SSM = See Student Solutions Manual; AWV = Answers will vary)

**1.** False **2.** True **3.** True **4.** False **5.** True **6.** False **7.** $\delta = 0.005$ **9.** $\delta = \dfrac{1}{12}$ **11.** $\delta = 0.02$

**13. (a)** $\delta \leq 0.025$ **(b)** $\delta \leq 0.0025$ **(c)** $\delta \leq 0.00025$ **(d)** $\delta \leq \dfrac{\varepsilon}{4}$ **15. (a)** $\delta \leq 0.1$ **(b)** $\delta \leq 0.01$ **(c)** $\delta \leq \varepsilon$

**17.** Given any $\varepsilon > 0$, let $\delta \leq \dfrac{\varepsilon}{3}$. SSM for the complete proof. **19.** Given any $\varepsilon > 0$, let $\delta \leq \dfrac{\varepsilon}{2}$. SSM for the complete proof.

**21.** Given any $\varepsilon > 0$, let $\delta \leq \dfrac{\varepsilon}{5}$. SSM for the complete proof.

**23.** Given any $\varepsilon > 0$, let $\delta \leq \min\left\{1, \dfrac{\varepsilon}{3}\right\}$. SSM for the complete proof.

**25.** Given any $\varepsilon > 0$, let $\delta \leq \min\left\{1, \dfrac{2\varepsilon}{7}\right\}$. SSM for the complete proof.

**27.** Given any $\varepsilon > 0$, let $\delta \leq \varepsilon^3$. SSM for the complete proof.

**29.** Given any $\varepsilon > 0$, let $\delta \leq \min\left\{1, \dfrac{\varepsilon}{3}\right\}$. SSM for the complete proof.

**31.** Given any $\varepsilon > 0$, let $\delta \leq \min\{1, 6\varepsilon\}$. SSM for the complete proof. **33.** SSM.

**35.** Given any $\varepsilon > 0$, let $\delta \leq \min\left\{1, \dfrac{234}{7}\varepsilon\right\}$. SSM for the complete proof.

**37.** Given any $\varepsilon > 0$, let $\delta \leq \min\{1, 26\varepsilon\}$. SSM for the complete proof.

**39.** Given any $\varepsilon > 0$, let $\delta \leq \dfrac{\varepsilon}{1 + |m|}$. SSM for the complete proof. **41.** $x$ must be within 0.05 of 3. **43.** SSM. **45.** SSM.

**47.** SSM. **49.** SSM. **51.** SSM. **53.** Given any $\varepsilon > 0$, let $\delta \leq \min\left\{1, \dfrac{\varepsilon}{47}\right\}$. SSM for the complete proof. **55.** $M = 101$ **57.** SSM.

## Review Exercises   (SSM = See Student Solutions Manual; AWV = Answers will vary)

**1.** $\lim\limits_{x\to 0}\dfrac{1-\cos x}{1+\cos x}=0.$

| $x$ | $-0.1$ | $-0.01$ | $-0.001$ | $\to 0 \leftarrow$ | $0.001$ | $0.01$ | $0.1$ |
|---|---|---|---|---|---|---|---|
| | | *x* approaches 0 from the left | | | | *x* approaches 0 from the right | |
| $f(x)=\dfrac{1-\cos x}{1+\cos x}$ | 0.002504 | 0.000025 | 0.00000025 | $f(x)$ approaches 0 | 0.00000025 | 0.000025 | 0.002504 |

**3.** $\lim\limits_{x\to 2} f(x)$ does not exist.   **5.** $-\dfrac{3}{x^2}$   **7.** 1   **9.** $-\pi$   **11.** 0   **13.** 27   **15.** 4   **17.** $\dfrac{2}{3}$   **19.** $-\dfrac{1}{6}$   **21.** 6   **23.** 1   **25.** $-1$   **27.** 8

**29.** $\lim\limits_{x\to 2^-} f(x)=7,\ \lim\limits_{x\to 2^+} f(x)=7,\ \lim\limits_{x\to 2} f(x)=7$   **31.** $f$ is not continuous at $c=1$.   **33.** $f$ is continuous at $c=0$.

**35.** $f$ is not continuous at $c=\dfrac{1}{2}$.   **37. (a)** $2x-3$   **(b)** $-1$   **39.** $f$ is continuous on the set $\{x\,|\,x\neq 3\}$.

**41.** $f$ is continuous on the set $\{x\,|\,x\neq -2,\ x\neq 0\}$.   **43.** $f$ is continuous on the set of all real numbers.   **45.** 0.215

**47.** $\lim\limits_{x\to 0^+}\dfrac{|x|}{x}(1-x)=1,\ \lim\limits_{x\to 0^-}\dfrac{|x|}{x}(1-x)=-1,\ \lim\limits_{x\to 0}\dfrac{|x|}{x}(1-x)$ does not exist.   **49.** $\dfrac{1}{2\sqrt{x}}$   **51.** 1   **53.** $\dfrac{3}{4}$   **55.** 0   **57.** $-\infty$

**59.** 3   **61.** $x=-3$ is a vertical asymptote. $y=4$ is a horizontal asymptote.   **63.** Yes   **65.** $f(0)=-\pi$

**67. (a)** AWV.   **(b)** One possibility is: $f(x)=\dfrac{2x+2}{x-4}$   **69.** SSM.

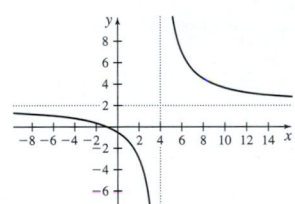

## Chapter 2

## Section 2.1   (SSM = See Student Solutions Manual; AWV = Answers will vary)

**1.** True   **2.** True   **3.** Prime, slope, $(c, f(c))$   **4.** True   **5.** 6   **6.** derivative

**7. (a)** $y=-12x-12$   **9. (a)** $y=12x+16$   **11. (a)** $y=-x+2$   **13. (a)** $y=-\dfrac{1}{36}x+\dfrac{7}{36}$   **15. (a)** $y=-\dfrac{1}{2}x+\dfrac{3}{2}$

**(b)** $y=\dfrac{1}{12}x+\dfrac{73}{6}$   **(b)** $y=-\dfrac{1}{12}x-\dfrac{49}{6}$   **(b)** $y=x$   **(b)** $y=36x-\dfrac{215}{6}$   **(b)** $y=2x-1$

**(c)** 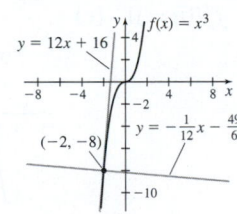   **(c)** 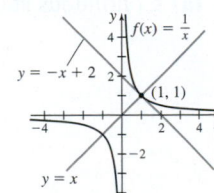   **(c)** 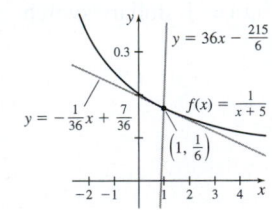   **(c)**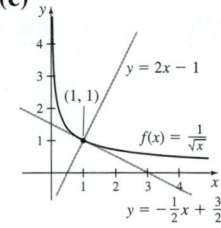

**17. (a)** 5   **(b)** 5   **19. (a)** 0   **(b)** $\dfrac{7}{16}$   **21.** $f'(1)=2$   **23.** $f'(0)=0$   **25.** $f'(-1)=-5$   **27.** $f'(4)=\dfrac{1}{4}$   **29.** $f'(0)=-7$

**31.**

| Time interval | $\Delta t$ | $\dfrac{\Delta s}{\Delta t}$ |
|---|---|---|
| [3, 3.1] | 0.1 | 61 |
| [3, 3.01] | 0.01 | 60.1 |
| [3, 3.001] | 0.001 | 60.01 |

The velocity appears to approach 60 cm/s.

**33.** $v(t)=4$ m/s at $t=0$, $v(t)=16$ m/s at $t=2$, $v(t)=6t_0+4$ m/s at any $t_0$

**35.** $v(1)=7$ cm/s, $v(4)=\dfrac{385}{16}$ cm/s

**37. (a)** $7\leq t\leq 10$ and $11\leq t\leq 13$   **(b)** $0\leq t\leq 4$   **(c)** $4\leq t\leq 7$ and $10\leq t\leq 11$   **(d)** Average speed $\approx 0.675$ mi/min
**(e)** 0.415 mi/min   **39.** $-3$   **41.** No   **43.** $R'(50)-0.59$   **45. (a)** 250 sales per day   **(b)** 200 sales per day

**47. (a)** 32 ft/s **(b)** $t \approx 7.914$ s **(c)** Average velocity $\approx 126.618$ ft/s **(d)** $v(7.914) \approx 253.235$ ft/s
**49. (a)** 9.8 m/s **(b)** $t \approx 4.243$ s **(c)** Average velocity $\approx 20.789$ m/s **(d)** $v(4.243) \approx 41.578$ m/s **51. (a)** AWV. **(b)** AWV.
**53.** SSM. **55. (a)** $d'(t)$ is the rate of change of diameter (in centimeters) with respect to time (in days).
**(b)** $d'(1) > d'(20)$ **(c)** $d'(1)$ is the instantaneous rate of change of the peach's diameter on day 1 and $d'(20)$ is the instantaneous
rate of change of the peach's diameter on day 20. **57. (a)** $\dfrac{\Delta V}{\Delta x} \approx 12.060$ cm³/cm **(b)** $V'(2) = 12$ cm³/cm

## Section 2.2 (SSM = See Student Solutions Manual; AWV = Answers will vary)

**1.** False **2.** False **3. (b)** Vertical **4.** derivative **5.** 0 **7.** 2 **9.** $-2c$ **11.** $f'(x) = 0$, all real numbers
**13.** $f'(x) = 6x + 1$, all real numbers **15.** $f'(x) = \dfrac{5}{2\sqrt{x-1}}, x > 1$

**17.** $f'(x) = \dfrac{1}{3}$ **19.** $f'(x) = 4x - 5$ **21.** $f'(x) = 3x^2 - 8$

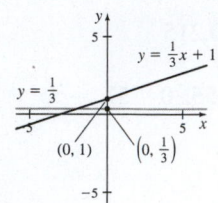 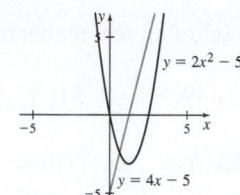 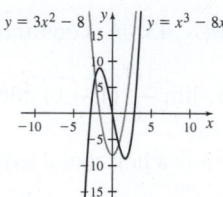

**23.** Not a graph of $f$ and $f'$ **25.** Graph of $f$ and $f'$. The blue curve is the graph of $f$; the green curve is the graph of $f'$.
**27.** **29.**

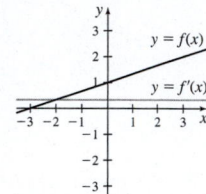 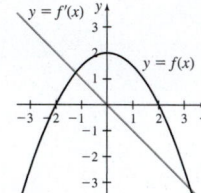

**31.** (B) **33.** (A) **35.** $f'(-8) = -\dfrac{1}{3}$ **37.** $f'(2)$ does not exist. **39.** $f'(1) = 2$ **41.** $f'\left(\dfrac{1}{2}\right)$ does not exist.

**43.** $f'(-1)$ does not exist. **45. (a)** Yes, at $x = 0$ and $x = 4$ **(b)** Yes, a cusp at $x = 2$ **(c)** Yes, at $x = -2$
**47. (a)** Yes, at $x = -2$, $x = 1$, $x = 2$ **(b)** Yes, at $x = -1$, not a cusp **(c)** Yes, at $x = 0$

**49. (a)** $-2$ and 4 **(b)** 0, 2, 6 **51.** $f'(x) = m$ **53.** $f'(x) = -\dfrac{2}{x^3}$ **55.** $f(x) = x^2, c = 2$ **57.** $f(x) = x^2, c = 1$

**59.** $f(x) = \sqrt{x}, c = 9$ **61.** $f(x) = \sin x, c = \dfrac{\pi}{6}$ **63.** $f(x) = 2(x+2)^2 - (x+2), c = 0$ **65.** $f(x) = x^2 + 2x, c = 3$

**67.** $V'(t) = 4$ ft³/s **69.** $C'(x) = 3$; dollars/switch **71. (a)** Continuous at 0 **(b)** $f'(0) = 0$ **(c)**

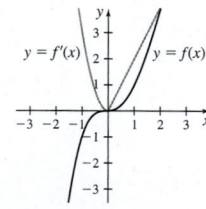

**73. (a)** $s'(4.99) = 74.7003$ ft/s; $s'(5.01) = 0$ ft/s **(b)** Not continuous **(c)** AWV. **75.** $p'(x) = -kp, k > 0$
**77.** $(\sqrt{2}, 4)$ and $(-\sqrt{2}, 4)$ **79. (a)** $2\pi r(\Delta r) + \pi(\Delta r)^2$ **(b)** $2\pi \Delta r$ **(c)** $2\pi r + \pi \Delta r$ **(d)** $2\pi$ **(e)** $2\pi$
**81.** SSM. **83.** SSM. **85. (a)** Parallel **(b)** Perpendicular **87. (a)** $k = 4$ **(b)** $f'(3) = 12$ **(c)** $f'(x) = 4x$

## Section 2.3 (SSM = See Student Solutions Manual; AWV = Answers will vary)

**1.** $0; 3x^2$ **2.** $nx^{n-1}$ **3.** True **4.** $k\left[\dfrac{d}{dx}f(x)\right]$ **5.** $e^x$ **6.** True **7.** $f'(x) = 3$ **9.** $f'(x) = 2x + 3$ **11.** $f'(u) = 40u^4 - 5$

**13.** $f'(s) = 3as^2 + 3s$ **15.** $f'(t) = \dfrac{5}{3}t^4$ **17.** $f'(t) = \dfrac{3}{5}t^2$ **19.** $f'(x) = \dfrac{3x^2 + 2}{7}$ **21.** $f'(x) = 2ax + b$ **23.** $f'(x) = 4e^x$
**25.** $f'(u) = 10u - 2e^u$ **27.** $\sqrt{3}$ **29.** $2\pi R$ **31.** $4\pi r^2$

**33. (a)** 3  **(b)** $y = 3x - 1$  **(c)** $y = -\frac{1}{3}x - 1$

**(d)**
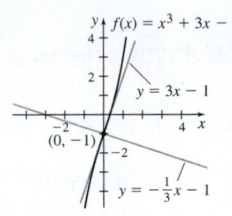

**35. (a)** 6  **(b)** $y = 6x + 1$  **(c)** $y = -\frac{1}{6}x + 1$

**(d)**
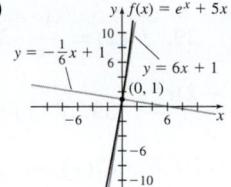

**37. (a)** $(2, -8)$  **(e)**
**(b)** $y = -8$
**(c)** $x > 2$
**(d)** $x < 2$

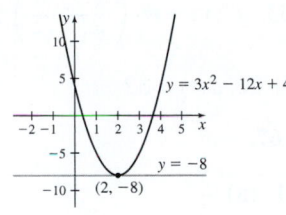

**(f)** $f$ is increasing when $x > 2$ and decreasing when $x < 2$.

**39. (a)** None  **(e)**
**(b)** None
**(c)** All real numbers
**(d)** None

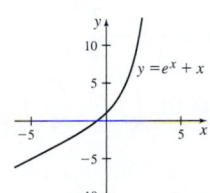

**(f)** $f$ is increasing for all $x$.

**41. (a)** $(1, 0), (-1, 4)$  **(e)**
**(b)** $y = 0, y = 4$
**(c)** $x < -1$ or $x > 1$
**(d)** $-1 < x < 1$

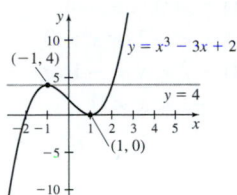

**(f)** $f$ is increasing when $x < -1$ or $x > 1$ and decreasing when $-1 < x < 1$.

**43.** $v(0) = -1$ m/s and $v(5) = 74$ m/s  **45. (a)** $v(t) = 2t - 5$  **(b)** $t = \frac{5}{2}$

**47. (a)** $\frac{4}{5}$  **(b)** $-\frac{7}{20}$  **(c)** $-\frac{9}{5}$  **(d)** $\frac{23}{20}$  **(e)** $-\frac{21}{20}$  **(f)** $-\frac{23}{10}$  **49. (a)** $f'(x) = 24x^2 - 24x + 6$  **(b)** $f'(x) = 6(2x - 1)^2$  **(c)** SSM.

**51.** $\frac{5}{16}$  **53.** $20,480\sqrt{3}$  **55.** $3ax^2$  **57.** $y = 5x - 3$  **59.** $y = x + 1$  **61.** $y = 3x + \frac{5}{3}, y = 3x - 9$

**63. (a)** $y = 45x - 65$

**(b)** $\left( \frac{\sqrt{3}}{3}, -\frac{5\sqrt{3}}{9} - 1 \right), \left( -\frac{\sqrt{3}}{3}, \frac{5\sqrt{3}}{9} - 1 \right)$

**(c)** At $x = \frac{\sqrt{3}}{3}, y = x - \frac{8\sqrt{3}}{9} - 1$ and at $x = -\frac{\sqrt{3}}{3}, y = x + \frac{8\sqrt{3}}{9} - 1$

**(d)**
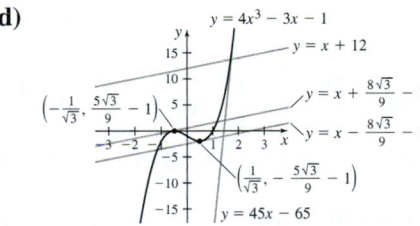

**65.** SSM.  **67.** SSM.  **69.** $a = 3, b = 2, c = 0$  **71.** $(2, 4)$  **73. (a)** 0  **(b)** $-kR$  **(c)** $-2kR$

**75. (a)** $\frac{dL}{dT} = 4\sigma AT^3$  **(b)** $2.694 \times 10^{23} \frac{W}{K}$  **(c)** $2.694 \times 10^{23}$ W  **77.** $F'(x) = mx$  **79.** SSM.  **81.** $Q = \left( \frac{9}{4}, \frac{81}{16} \right)$

**83.** $y = -\frac{7}{64}x^4 + \frac{7}{16}x^3 + \frac{21}{16}x^2 - 4x$  **85. (a)** $c = 1$  **(b)** $y = 12x - 16$ and $y = 12x + 1$

## Section 2.4  (SSM = See Student Solutions Manual; AWV = Answers will vary)

**1.** False  **2.** $f(x)g'(x) + f'(x)g(x)$  **3.** False  **4.** $\dfrac{\left[ \dfrac{d}{dx}f(x) \right] g(x) - f(x) \left[ \dfrac{d}{dx}g(x) \right]}{[g(x)]^2}$  **5.** True  **6.** $-\dfrac{\dfrac{d}{dx}g(x)}{[g(x)]^2}$  **7.** 0  **8.** $\dfrac{d^2s}{dt^2}$

**9.** $f'(x) = e^x(x + 1)$  **11.** $f'(x) = 5x^4 - 2x$  **13.** $f'(x) = 18x^2 + 6x - 10$  **15.** $s'(t) = 16t^7 - 24t^5 + 10t^4 - 4t^3 + 4t - 1$

**17.** $f'(x) = x^3 e^x + 3x^2 e^x + e^x + 3x^2$   **19.** $g'(s) = \dfrac{2}{(s+1)^2}$   **21.** $G'(u) = -\dfrac{4}{(1+2u)^2}$   **23.** $f'(x) = \dfrac{2(6x^2 + 16x + 3)}{(3x+4)^2}$

**25.** $f'(w) = -\dfrac{3w^2}{(w^3-1)^2}$   **27.** $s'(t) = -\dfrac{3}{t^4}$   **29.** $f'(x) = \dfrac{4}{e^x}$   **31.** $f'(x) = -\dfrac{40}{x^5} - \dfrac{6}{x^3}$   **33.** $f'(x) = 9x^2 + \dfrac{2}{3x^3}$

**35.** $s'(t) = -\dfrac{1}{t^2} + \dfrac{2}{t^3} - \dfrac{3}{t^4}$   **37.** $f'(x) = \dfrac{(x-2)e^x}{x^3}$   **39.** $f'(x) = -\dfrac{(x^3 - x^2 + x + 1)}{x^2 e^x}$   **41.** $f'(x) = 6x + 1,\ f''(x) = 6$

**43.** $f'(x) = e^x,\ f''(x) = e^x$   **45.** $f'(x) = e^x(x+6),\ f''(x) = e^x(x+7)$   **47.** $f'(x) = 8x^3 + 3x^2 + 10,\ f''(x) = 24x^2 + 6x$

**49.** $f'(x) = 1 - \dfrac{1}{x^2},\ f''(x) = \dfrac{2}{x^3}$   **51.** $f'(t) = 1 + \dfrac{1}{t^2},\ f''(t) = -\dfrac{2}{t^3}$   **53.** $f'(x) = e^x\left(\dfrac{1}{x} - \dfrac{1}{x^2}\right);\ f''(x) = e^x\left(\dfrac{2}{x^3} - \dfrac{2}{x^2} + \dfrac{1}{x}\right)$

**55. (a)** $y' = -\dfrac{1}{x^2},\ y'' = \dfrac{2}{x^3}$   **(b)** $y' = \dfrac{5}{x^2},\ y'' = -\dfrac{10}{x^3}$   **57.** $v(t) = 32t + 20,\ a(t) = 32$

**59.** $v(t) = 9.8t + 4,\ a(t) = 9.8$   **61.** $f^{(4)}(x) = 0$   **63.** 5040   **65.** $e^u$   **67.** $-e^x$

**69. (a)** $\dfrac{3}{4}$   **(d)**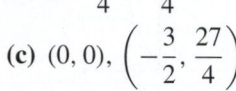

**(b)** $y = \dfrac{3}{4}x + \dfrac{1}{4}$

**(c)** $(0,0),\ (2,4)$

**71. (a)** $\dfrac{5}{4}$   **(d)**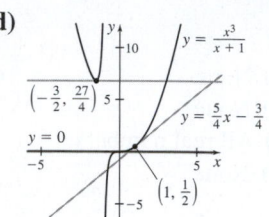

**(b)** $y = \dfrac{5}{4}x - \dfrac{3}{4}$

**(c)** $(0,0),\ \left(-\dfrac{3}{2}, \dfrac{27}{4}\right)$

**73. (a)** $(-2, 5),\ (2, -27)$   **(e)**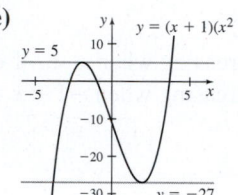
**(b)** $y = 5,\ y = -27$
**(c)** $x < -2$ or $x > 2$
**(d)** $-2 < x < 2$

**(f)** AWV.

**75. (a)** $(0, 0),\ (-2, -4)$   **(e)**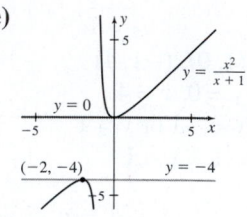
**(b)** $y = 0,\ y = -4$
**(c)** $x < -2$ or $x > 0$
**(d)** $-2 < x < -1$ or
$\quad -1 < x < 0$

**(f)** AWV.

**77. (a)** $(-1,\ -e^{-1})$   **(e)** 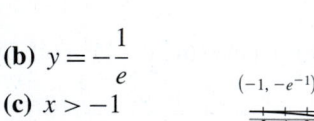   **(f)** AWV.

**(b)** $y = -\dfrac{1}{e}$

**(c)** $x > -1$

**(d)** $x < -1$

**79. (a)** $(-1, -2e),\ \left(3, \dfrac{6}{e^3}\right)$   **(e)** 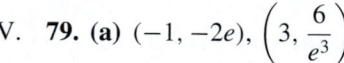 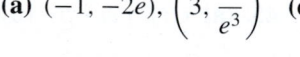   **(f)** AWV.

**(b)** $y = -2e,\ y = \dfrac{6}{e^3}$

**(c)** $-1 < x < 3$

**(d)** $x < -1$ or $x > 3$

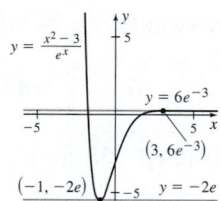

**81. (a)** $\dfrac{8}{5}$   **(b)** $-\dfrac{29}{10}$   **(c)** $\dfrac{1}{9}$   **(d)** $-\dfrac{19}{160}$   **(e)** $\dfrac{1}{9}$   **(f)** $\dfrac{12}{25}$   **83. (a)** $v(t) = -9.8t + 39.2$   **(b)** $t = 4$ seconds   **(c)** 78.4 m
**(d)** $-9.8$ m/s$^2$   **(e)** $t = 8$ s   **(f)** $v = -39.2$ m/s   **(g)** 156.8 m

**85. (a)** $0 \le x \le 100$   **(b)** 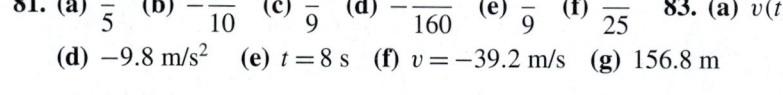   **(c)** \$13,333.33   **(d)** $C'(x) = \dfrac{550}{(110-x)^2}$

**(e)** $C'(40) = 0.112 = \$112/\%,\ C'(60) = 0.220 = \$220/\%,$
$\quad C'(80) = 0.611 = \$611/\%,\ C'(90) = 1.375 = \$1,375/\%$
**(f)** AWV.

**87. (a)** $f'(t) = \dfrac{0.4 - 0.8t^2}{(2t^2 + 1)^2}$ mg/L/h   **(d)**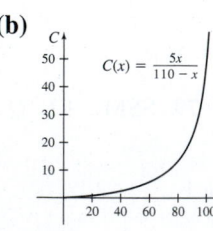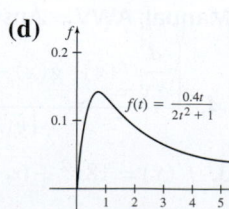

**(e)** Concentration is highest at approximately
45 min and is about 0.14 mg/L.

**(b)** $f'\left(\dfrac{1}{6}\right) \approx 0.339$ mg/L/h, $f'\left(\dfrac{1}{2}\right) \approx 0.089$ mg/L/h,

$\quad f'(1) \approx -0.044$ mg/L/h

**(c)** AWV.

**89. (a)** $-(100,000)\dfrac{2p+10}{(p^2+10p+50)^2}$ books/dollar  **(b)** $D'(5)=-128$, $D'(10)=-48$, $D'(15)=-22.145$  **(c)** AWV.

**91. (a)** $\dfrac{dp}{dr}=-\dfrac{9nRT}{4\pi r^4}$  **(b)** As the radius increases, pressure within the container decreases.  **(c)** $p'\left(\dfrac{1}{4}\right)\approx-415945.358$ Pa/m

**93. (a)** $v(t)=3t^2-1$, $a(t)=6t$, $J(t)=6$, $S(t)=0$  **(b)** $t=\pm\dfrac{\sqrt{3}}{3}$ s  **(c)** $a(2)=12$ m/s², $a(5)=30$ m/s²  **(d)** No

**(e)** AWV.  **95. (a)** $a(t)=1.6+1.998t$ m/s²  **(b)** $J(t)=1.998$ m/s³

**97. (a)** $\dfrac{dJ}{dr}=-\dfrac{2I}{\pi r^3}$  **(b)** As radius increases, current density decreases.  **(c)** $-1.273\times10^{10}$ amps/m³

**99.** SSM.  **101.** $y'=6x^5-5x^4+20x^3-18x^2+2x-5$  **103.** $y'=9x^2(x^3+1)^2$  **105.** $y'=\dfrac{1}{x^2}+\dfrac{2}{x^3}-\dfrac{4}{x^5}-\dfrac{5}{x^6}+\dfrac{6}{x^7}$

**107.** SSM.  **109.** $f'(x)=\dfrac{6x^5-4x^3+2x}{x^4+1}-4x^3\left[\dfrac{x^6-x^4+x^2}{(x^4+1)^2}\right]$

**111. (a)** $f'(x)=\dfrac{2}{x+1}-\dfrac{2x}{(x+1)^2}$  **(b)** $f'(x)=\dfrac{2}{(x+1)^2}$  **(c)** $f^{(5)}(x)=\dfrac{240}{(x+1)^6}$

**113. (a)** $[f_1'(x)\cdot f_2(x)\cdots f_n(x)]+[f_1(x)\cdot f_2'(x)\cdot f_3(x)\cdots f_n(x)]+\cdots+[f_1(x)\cdots f_{n-1}(x)\cdot f_n'(x)]$

**(b)** $-\dfrac{1}{f_1(x)\cdots f_n(x)}\left[\dfrac{f_1'(x)}{f_1(x)}+\cdots+\dfrac{f_n'(x)}{f_n(x)}\right]$

**115. (a)** $f_1(x)=\dfrac{x}{x-1}$, $f_2(x)=\dfrac{2x-1}{x}$, $f_3(x)=\dfrac{3x-1}{2x-1}$, $f_4(x)=\dfrac{5x-2}{3x-1}$, $f_5(x)=\dfrac{8x-3}{5x-2}$

**(b)** $a_0=1$, $a_1=1$, $a_2=2$, $a_3=3$, $a_4=5$, $a_5=8$, …  or $\{1, 1, 2, 3, 5, 8, \ldots\}$  **(c)** $a_{n+2}=a_{n+1}+a_n$

**(d)** $f_0'(x)=1$, $f_1'(x)=-\dfrac{1}{(x-1)^2}$, $f_2'(x)=\dfrac{1}{x^2}$, $f_3'(x)=-\dfrac{1}{(2x-1)^2}$, $f_4'(x)=\dfrac{1}{(3x-1)^2}$, $f_5'(x)=-\dfrac{1}{(5x-2)^2}$

## Section 2.5  (SSM = See Student Solutions Manual; AWV = Answers will vary)

**1.** False  **2.** False  **3.** True  **4.** False  **5.** $y'=1-\cos x$  **7.** $y'=\sec^2 x-\sin x$  **9.** $y'=3\cos\theta+2\sin\theta$

**11.** $y'=\cos^2 x-\sin^2 x=\cos(2x)$  **13.** $y'=\cos t-t\sin t$  **15.** $y'=e^x(\sec^2 x+\tan x)$  **17.** $y'=\pi\sec u(\sec^2 u+\tan^2 u)$

**19.** $y'=-\dfrac{x\csc^2 x+\cot x}{x^2}$  **21.** $y'=x(x\cos x+2\sin x)$  **23.** $y'=t\sec^2 t+\tan t-\sqrt{3}\sec t\tan t$  **25.** $y'=-\dfrac{1}{1-\cos\theta}$

**27.** $y'=\dfrac{\cos t+t\cos t-\sin t}{(1+t)^2}$  **29.** $y'=\dfrac{\cos x-\sin x}{e^x}$  **31.** $y'=-\dfrac{2}{(\sin\theta-\cos\theta)^2}$  **33.** $y'=\dfrac{\sec t\tan t+t\tan^2 t-t-\tan t}{(1+t\sin t)^2}$

**35.** $y'=-\csc\theta(\csc^2\theta+\cot^2\theta)$  **37.** $y'=\dfrac{2\sec^2 x}{(1-\tan x)^2}$  **39.** $y''=-\sin x$  **41.** $y''=2\tan\theta\sec^2\theta$  **43.** $y''=2\cos t-t\sin t$

**45.** $y''=2e^x\cos x$  **47.** $y''=3\cos u-2\sin u$  **49.** $y''=-(a\sin x+b\cos x)$

**51. (a)** $y=x$  **(b)**

**53. (a)** $y=x$  **(b)**

**55. (a)** $y=\sqrt{2}$  **(b)**

**57. (a)** $\left(\tan^{-1}2+n\pi,\,(-1)^n\sqrt{5}\right)$ where $n$ is an integer  **(b)**

**59. (a)** $(n\pi,\,(-1)^n)$ where $n$ is an integer  **(b)**

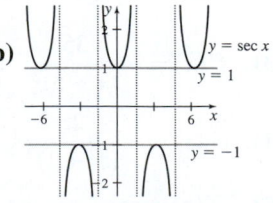

**61.** $f^{(n)}(x)=\begin{cases}(-1)^{n/2}\sin x & \text{if }n\text{ is even}\\ (-1)^{(n-1)/2}\cos x & \text{if }n\text{ is odd}\end{cases}$

**63.** $-1$   **65. (a)** $v(t) = -\dfrac{1}{8}\sin t$ m/s

**(b)** $\dfrac{\pi}{2} + n\pi$, where $n$ is an integer

**(c)** $a(t) = -\dfrac{1}{8}\cos t$ m/s$^2$

**(d)** $\dfrac{\pi}{2} + n\pi$, where $n$ is an integer

**(e)**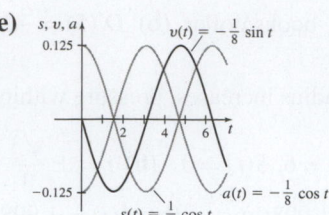

**67. (a)** $\dfrac{dy}{d\theta} = -8\sin\theta$   **(b)** $\dfrac{dy}{d\theta} = -4\sqrt{2}$ ft/radian

**69. (a)**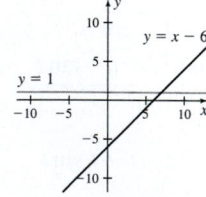

**(b)** Max $= 4$, Min $= \dfrac{4}{3}$

**(c)** $x = \pi, 2\pi, 3\pi$

**(d)** $w'(\pi - 0.1) \approx 0.395$, $w'(2\pi - 0.1) \approx -0.045$

**(e)** A symmetric wave would give slopes having equal magnitudes but opposite signs.

**71. (a)** $10,000, $25,093, $15,411, $37,570

**(b)** $R'(t) = \cos t + 0.3$

**(c)** $\dfrac{\$10,539}{2}$ per month

**(d)**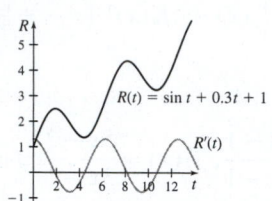

**(e)** AWV.

**73.** $n = 4$   **75.** SSM.   **77.** SSM.   **79.** SSM.   **81.** SSM.

## Review Exercises

**1. (a)** $\dfrac{1}{2}$   **(b)** $\dfrac{1}{4}$   **(c)** $\dfrac{1}{2\sqrt{c}}$   **3.** 2   **5.** 5   **7.** 2

**9.** $f'(x) = 1$

**11.** $f'(x) = -\dfrac{3}{2x^4}$

**13.** $f$ does not have a derivative at $c = 1$. AWV.

**15.** Not the graph of a function and its derivative.

**17.**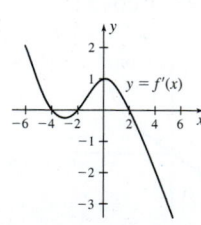

**19.** $f'(x) = 5x^4$   **21.** $f'(x) = x^3$   **23.** $f'(x) = 4x - 3$   **25.** $F'(x) = 14x$   **27.** $f'(x) = 15(x^2 - 6x + 6)$   **29.** $f'(x) = 2 + \dfrac{3}{x^2}$

**31.** $f'(x) = -\dfrac{35}{(x - 5)^2}$   **33.** $f'(x) = 4x + \dfrac{10}{x^3}$   **35.** $f'(x) = -\dfrac{a}{x^2} + \dfrac{3b}{x^4}$   **37.** $f'(x) = -\dfrac{6(2x - 3)}{(x^2 - 3x)^3}$   **39.** $s'(t) = \dfrac{2t^2(t - 3)}{(t - 2)^2}$

**41.** $F'(z) = -\dfrac{2z}{(z^2 + 1)^2}$   **43.** $g'(z) = -\dfrac{2z - 1}{(1 - z + z^2)^2}$   **45.** $s'(t) = -e^t$   **47.** $f'(x) = -\dfrac{x}{e^x}$   **49.** $f'(x) = x\cos x + \sin x$

**51.** $G'(u) = \sec u(\sec u + \tan u)$   **53.** $f'(x) = e^x(\cos x + \sin x)$   **55.** $f'(x) = 2(\cos^2 x - \sin^2 x) = 2\cos(2x)$

**57.** $f'(x) = 2\cos x \sin x = \sin(2x)$   **59.** $f'(\theta) = -\dfrac{\sin\theta + \cos\theta}{2e^\theta}$   **61.** $f'(x) = 50x + 30$, $f''(x) = 50$

**63.** $g'(u) = \dfrac{1}{(2u+1)^2}$, $g''(u) = -\dfrac{4}{(2u+1)^3}$  **65.** $f'(u) = -\dfrac{\sin u + \cos u}{e^u}$, $f''(u) = \dfrac{2\sin u}{e^u}$

**67.** **(a)** $y = -7x + 5$  **(c)**

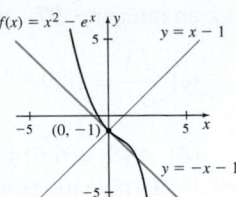

**(b)** $y = \dfrac{1}{7}x + \dfrac{85}{7}$

**69.** **(a)** $y = -x - 1$  **(c)**

**(b)** $y = x - 1$

**71.** **(a)** $-1$ m/s **(b)** $v(0) = -6$ m/s, $v(5) = 4$ m/s, $v(t) = 2t - 6$ m/s  **(c)** $a(t) = 2$ m/s$^2$

**73.** **(a)** $\dfrac{dp}{dx} = -\dfrac{50{,}000}{(5x+100)^2}$  **(b)** $R(x) = \dfrac{10{,}000x}{5x+100} - 5x$  **(c)** $R'(10) = \dfrac{355}{9} \approx \$39.44$/lb, $R'(40) = \dfrac{55}{9} \approx \$6.11$/lb

**75.** $f'(3)$ does not exist.  **77.** $8x - 16$, $f(x) = (4-2x)^2$

# Chapter 3

## Section 3.1  (SSM = See Student Solutions Manual; AWV = Answers will vary)

**1.** Chain  **2.** True  **3.** False  **4.** $\tan u$; $1 + \cos x$  **5.** $100(x^3 + 4x + 1)^{99}(3x^2 + 4)$  **6.** $6xe^{3x^2+5}$  **7.** True  **8.** $2x\cos(x^2)$

**9.** $y = (x^3+1)^5$; $\dfrac{dy}{dx} = 15x^2(x^3+1)^4$  **11.** $y = \dfrac{x^2+1}{x^2+2}$; $\dfrac{dy}{dx} = \dfrac{2x}{(x^2+2)^2}$  **13.** $y = \left(\dfrac{1}{x}+1\right)^2$; $\dfrac{dy}{dx} = -2\left(\dfrac{1}{x}+1\right)\left(\dfrac{1}{x^2}\right)$

**15.** $f'(x) = 6(3x+5)$  **17.** $f'(x) = -\dfrac{18}{(6x-5)^4}$  **19.** $g'(x) = 8x(x^2+5)^3$  **21.** $f'(u) = 3\left(u - \dfrac{1}{u}\right)^2\left(1 + \dfrac{1}{u^2}\right)$

**23.** $g'(x) = 3(4x + e^x)^2(4 + e^x)$  **25.** $f'(x) = 2\tan x \sec^2 x$  **27.** $f'(z) = 2(\tan z + \cos z)(\sec^2 z - \sin z)$

**29.** $y' = 4x(x^2+4)(2x^3-1)^3 + 18x^2(x^2+4)^2(2x^3-1)^2 = 2x(x^2+4)(2x^3-1)^2(13x^3 + 36x - 2)$

**31.** $y' = \dfrac{2\sin x(x\cos x - \sin x)}{x^3}$  **33.** $y' = 4\cos(4x)$  **35.** $y' = 4(x+1)\cos(x^2+2x-1)$  **37.** $y' = -\dfrac{1}{x^2}\cos\dfrac{1}{x}$

**39.** $y' = 4\sec(4x)\tan(4x)$  **41.** $y' = -\dfrac{1}{x^2}e^{1/x}$  **43.** $y' = -\dfrac{4x^3 - 2}{(x^4 - 2x + 1)^2}$  **45.** $y' = \dfrac{9900e^{-x}}{(1+99e^{-x})^2}$

**47.** $y' = (\ln 2)2^{\sin x}\cos x$  **49.** $y' = (\ln 6)6^{\sec x}\sec x \tan x$  **51.** $y' = 5e^{3x} + 15xe^{3x}$

**53.** $y' = 2x\sin(4x) + 4x^2\cos(4x)$  **55.** $y' = -ae^{-ax}\sin(bx) + be^{-ax}\cos(bx)$  **57.** $y' = \dfrac{2ae^{ax}}{(e^{ax}+1)^2}$

**59.** $y = \left(\dfrac{48}{x^4}+1\right)^3$; $\dfrac{dy}{dx} = -\dfrac{576}{x^5}\left(\dfrac{48}{x^4}+1\right)^2$  **61.** $y = \dfrac{16}{x^4}+1$; $\dfrac{dy}{dx} = -\dfrac{64}{x^5}$  **63.** $y' = -2e^{-2x}\cos(3x) - 3e^{-2x}\sin(3x)$

**65.** $y' = -2xe^{x^2}\sin(e^{x^2})$  **67.** $y' = -4\sin(4x)e^{\cos(4x)}$  **69.** $y' = 24\sin(3x)\cos(3x)$

**71.** **(a)** $y' = 6x^2(x^3+1)$  **(b)** $y' = 6x^2(x^3+1)$  **(c)** $y' = 6x^5 + 6x^2$  **(d)** SSM.

**73.** **(a)** $y = 0$

**75.** **(a)** $y = -\dfrac{7}{27}x + \dfrac{16}{27}$

**77.** **(a)** $y = 2x + 1$

**(b)** $x = 1$

**(b)** $y - \dfrac{2}{27} = \dfrac{27}{7}(x-2)$

**(b)** $y = -\dfrac{1}{2}x + 1$

**(c)**

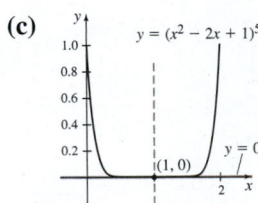

**(c)**

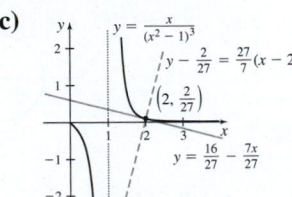

**(c)**

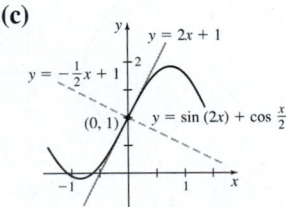

**79.** $\dfrac{d^2y}{dx^2} = -25x^8\cos(x^5) - 20x^3\sin(x^5)$  **81.** $h'(1) = -12$  **83.** $h'(0) = 0$  **85.** 78  **87.** $\dfrac{df}{dx} = f'(x^2+1)(2x)$

**89.** $\dfrac{df}{dx} = f'\left(\dfrac{x+1}{x-1}\right)\left(\dfrac{-2}{(x-1)^2}\right)$  **91.** $\dfrac{df}{dx} = f'(\sin x)\cos x$  **93.** $\dfrac{d^2f}{dx^2} = f''(\cos x)\sin^2 x - f'(\cos x)\cos x$

**95. (a)** $v(t) = -A\omega \sin(\omega t + \phi)$ **(b)** $t = \dfrac{k\pi - \phi}{\omega}$, $k$ an integer **(c)** $a(t) = -A\omega^2 \cos(\omega t + \phi)$

**(d)** $t = \dfrac{(2k-1)\dfrac{\pi}{2} - \phi}{\omega}$, $k$ an integer **97.** $a(t) = \dfrac{-20\pi}{9} \sin\dfrac{\pi t}{6}$ **99.** $\dfrac{dR}{dT} = -3.597 \times 10^{-6}$ ohms/K

**101. (a)** 102 **(c)** $\dfrac{dA}{dt} = 18.9e^{-0.21t}$ **(e)** Explanations will vary.

**(b)**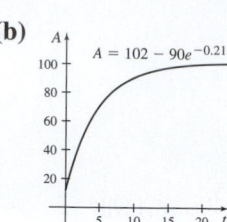

**(d)** $A'(5) = 6.614$; $A'(10) = 2.314$
Explanations will vary.

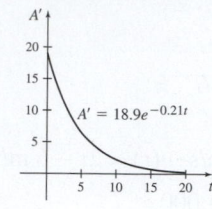

**103. (a)** $a(t) = ge^{-\frac{kt}{m}}$ m/s$^2$ **(b)** $\dfrac{mg}{k}$ m/s; explanations will vary. **(c)** 0 m/s$^2$; explanations will vary.

**105.** SSM. **107. (a)** $\omega(t) = -\dfrac{\pi}{6}\sqrt{\dfrac{2k}{5}} \sin\left(\dfrac{1}{2}\sqrt{\dfrac{2k}{5}}t\right)$ **(b)** $\omega(3) = -\dfrac{\pi}{6}\sqrt{\dfrac{2k}{5}} \sin\left(\dfrac{3}{2}\sqrt{\dfrac{2k}{5}}\right)$

**109.** $F'(1) = 2$ **111.** SSM. **113.** SSM. **115.** SSM. **117.** SSM. **119.** SSM.

**121. (a)** $y^{(n)}(x) = a^n e^{ax}$ **(b)** $y^{(n)}(x) = (-1)^n a^n e^{-ax}$ **123. (a)** $\dfrac{d^{11}\cos(ax)}{dx^{11}} = a^{11}\sin(ax)$ **(b)** $\dfrac{d^{12}\cos(ax)}{dx^{12}} = a^{12}\cos(ax)$

**(c)** $f^{(n)}(x) = -a^n \sin(ax)$, $n = 1 + 4k$, $k = 0, 1, 2, 3, \ldots$; $f^{(n)}(x) = -a^n \cos(ax)$, $n = 2 + 4k$, $k = 0, 1, 2, 3, \ldots$;
$f^{(n)}(x) = a^n \sin(ax)$, $n = 3 + 4k$, $k = 0, 1, 2, 3, \ldots$; $f^{(n)}(x) = a^n \cos(ax)$, $n = 4 + 4k$, $k = 0, 1, 2, 3, \ldots$
**125.** SSM. **127.** SSM. **129.** SSM. **131.** SSM. **133. (a)** $-72,000\pi^2$cm/s$^2$ **(b)** $-72,000\pi^2$kg $\cdot$ cm/s$^2$

## Section 3.2 (SSM = See Student Solutions Manual; AWV = Answers will vary)

**1.** True **2.** False **3.** $\dfrac{1}{x^{2/3}}$ **4.** $3x(x^2+1)^{1/2}$ **5.** $y' = -\dfrac{3}{2}$ **7.** $y' = -\dfrac{x}{y}$ **9.** $y' = \dfrac{\cos x}{e^y}$ **11.** $y' = \dfrac{-e^{x+y}}{e^{x+y}-1}$ **13.** $y' = -\dfrac{2y}{x}$

**15.** $y' = \dfrac{2x-y}{2y+x}$ **17.** $y' = -\dfrac{y^2}{x^2}$ **19.** $y' = \dfrac{x^3+y}{x(1-xy)}$ **21.** $y' = \dfrac{e^y \sin x - e^x \sin y}{e^x \cos y + e^y \cos x}$ **23.** $y' = \dfrac{-6x(x^2+y)^2}{3(x^2+y)^2-1}$

**25.** $y' = \dfrac{\sec^2(x-y)}{1+\sec^2(x-y)}$ **27.** $y' = \dfrac{\sin y}{1 - x\cos y}$ **29.** $y' = \dfrac{ye^{xy}-2xy}{x^2-xe^{xy}}$ **31.** $y' = \dfrac{2}{3x^{1/3}}$ **33.** $y' = \dfrac{2}{3x^{1/3}}$ **35.** $y' = \dfrac{1}{3x^{2/3}} + \dfrac{1}{3x^{4/3}}$

**37.** $y' = \dfrac{3x^2}{2(x^3-1)^{1/2}}$ **39.** $y' = (x^2-1)^{1/2} + \dfrac{x^2}{(x^2-1)^{1/2}}$ **41.** $y' = \dfrac{xe^{(x^2-9)^{1/2}}}{(x^2-9)^{1/2}}$ **43.** $y' = \dfrac{3}{2}(x^2\cos x)^{1/2}(2x\cos x - x^2\sin x)$

**45.** $y' = 3x(x^2-3)^{1/2}(6x+1)^{5/3} + 10(x^2-3)^{3/2}(6x+1)^{2/3} = (6x+1)^{2/3}(x^2-3)^{1/2}(28x^2 + 3x - 30)$

**47.** $y' = -\dfrac{x}{y}$, $y'' = -\dfrac{x^2+y^2}{y^3}$ **49.** $y' = \dfrac{2x-5}{2y}$, $y'' = \dfrac{4y^2-(2x-5)^2}{4y^3}$ **51.** $y' = \dfrac{x}{(x^2+1)^{1/2}}$, $y'' = \dfrac{1}{(x^2+1)^{3/2}}$

**53. (a)** $-1/2$ **(b)** $y = -\dfrac{x}{2} + \dfrac{5}{2}$ **55. (a)** 3 **(b)** $y = 3x - 8$ **57. (a)** $1/2$ **(b)** $y = \dfrac{x}{2} + 2$

**(c)**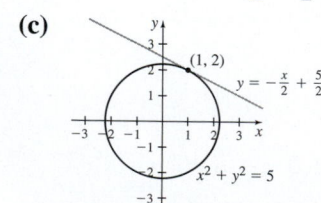

**(c)**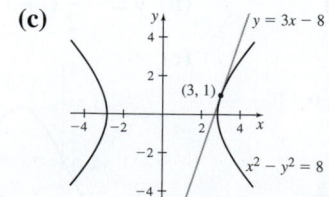

**(c)**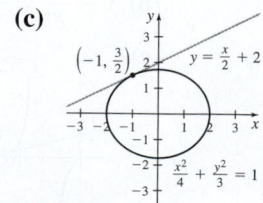

**59.** $y' = -\dfrac{4}{9}$; $y'' = -\dfrac{100}{243}$ **61.** $y' = \dfrac{3x}{|x|}$ **63.** $y' = \dfrac{2(2x-1)}{|2x-1|}$ **65.** $y' = \dfrac{-\sin x \cos x}{|\cos x|}$ **67.** $y' = \cos|x| \cdot \dfrac{|x|}{x}$ **69.** $y = -2x + 10$

**71. (a)** $y' = \dfrac{-y - 1}{x + 4y}$  **(b)** $y = -\dfrac{1}{3}x + \dfrac{5}{3}$  **(c)** $(6, -3), (2, 1)$  **(d)**  **73.** SSM.  **75.** SSM.

**77. (a)** $y' = \dfrac{3x^2 - 2y}{2x - 3y^2}$  **(b)** $y = -x + 2$  **(c)** $\left(\dfrac{2^{4/3}}{3}, \dfrac{2^{5/3}}{3}\right)$  **(d)**  Explanations will vary.

**79.** $\left(\dfrac{-5}{\sqrt{2}}, \dfrac{5}{\sqrt{2}}\right)$  **81.** SSM.  **83.** SSM.  **85.** SSM.

**87. (a)** $\dfrac{dy}{dx} = -\dfrac{y}{x}$ for the first and $\dfrac{dy}{dx} = \dfrac{x}{y}$ for the second.  **(b)**  **89.** SSM.

## Section 3.3   (SSM = See Student Solutions Manual; AWV = Answers will vary)

**1.** True  **2.** True  **3.** False  **4.** $\dfrac{1}{\sqrt{1 - x^2}}$  **5.** $g'(4) = -\dfrac{1}{2}$  **7.** $f'(-2) = 2$  **9.** $g'(32) = \dfrac{1}{80}$  **11.** $g'(6) = \dfrac{1}{4}$  **13.** $g'(2) = 12$

**15.** $g'(49) = \dfrac{1}{8}$  **17.** $f'(x) = \dfrac{4}{\sqrt{1 - 16x^2}}$  **19.** $g'(x) = \dfrac{1}{x\sqrt{9x^2 - 1}}$  **21.** $s'(t) = \dfrac{2}{4 + t^2}$  **23.** $f'(x) = -\dfrac{2x}{2x^4 - 2x^2 + 1}$

**25.** $f'(x) = \dfrac{2x}{(x^2 + 2)\sqrt{(x^2 + 2)^2 - 1}}$  **27.** $F'(x) = \dfrac{e^x}{\sqrt{1 - e^{2x}}}$  **29.** $g'(x) = -\dfrac{1}{x^2 + 1}$  **31.** $g'(x) = \dfrac{x}{\sqrt{1 - x^2}} + \sin^{-1} x$

**33.** $s'(t) = \dfrac{3t}{\sqrt{t^6 - 1}} + 2t\sec^{-1} t^3$  **35.** $f'(x) = \dfrac{\cos x}{1 + \sin^2 x}$  **37.** $G'(x) = \dfrac{\cos(\tan^{-1} x)}{1 + x^2}$  **39.** $f'(x) = \dfrac{3e^{\tan^{-1}(3x)}}{1 + 9x^2}$

**41.** $g'(x) = \dfrac{1}{1 - x^2} + \dfrac{x\sin^{-1} x}{(1 - x^2)^{3/2}}$  **43.** $\dfrac{dy}{dx} = \dfrac{2\sqrt{1 - y^2}}{2y\sqrt{1 - y^2} + 1}$  **45.** $\dfrac{dy}{dx} = \dfrac{3\pi x^2 y(1 + y^4)}{80y - \pi x^3(1 + y^4)}$  **47.** $g'(0) = \dfrac{1}{2}$, $g'(3) = \dfrac{1}{5}$

**49. (a)** $y = \dfrac{1}{2}x$  **51.** $y - 2 = \dfrac{1}{23}(x - 14)$  **53.** $5x + 3y = 18$  **55. (a)** $-\dfrac{\sqrt{3}}{6}$ m/s  **(b)** $\dfrac{7\sqrt{3}}{36}$ m/s²  **57.** SSM.  **59.** SSM.

**(b)**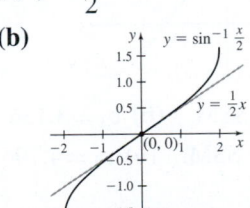

## Section 3.4   (SSM = See Student Solutions Manual; AWV = Answers will vary)

**1.** $\dfrac{1}{x}$  **2.** True  **3.** False  **4.** False  **5.** $\dfrac{1}{x}$  **6.** $e$  **7.** $f'(x) = \dfrac{5}{x}$  **9.** $s'(t) = \dfrac{1}{t \ln 2}$  **11.** $g'(x) = \dfrac{1}{x}\cos x - \sin x \ln x$  **13.** $F'(x) = \dfrac{1}{x}$

**15.** $s'(t) = \dfrac{e^t + e^{-t}}{e^t - e^{-t}}$  **17.** $f'(x) = \dfrac{2x^2}{x^2 + 4} + \ln(x^2 + 4)$  **19.** $f'(x) = \dfrac{3x^2}{(3x + 5)\ln 5} + 2x\log_5(3x + 5)$  **21.** $f'(x) = \dfrac{1}{1 - x^2}$

**23.** $f'(x) = \dfrac{1}{x \ln x}$  **25.** $g'(x) = \dfrac{1}{x(x^2 + 1)}$  **27.** $f'(x) = \dfrac{4x}{x^2 + 1} - \dfrac{1}{x} - \dfrac{x}{x^2 - 1}$  **29.** $F'(\theta) = \cot \theta$  **31.** $g'(x) = \dfrac{1}{\sqrt{x^2 + 4}}$

**33.** $f'(x) = \dfrac{2x}{(1+x^2)\ln 2}$  **35.** $f'(x) = \dfrac{1}{x[1+(\ln x)^2]}$  **37.** $s'(t) = \dfrac{1}{\tan^{-1}t} \cdot \dfrac{1}{1+t^2}$  **39.** $G'(x) = \dfrac{1}{2x(\ln x)^{1/2}}$

**41.** $f'(\theta) = \cos(\ln \theta) \cdot \dfrac{1}{\theta}$  **43.** $g'(x) = \dfrac{3(\log_3 x)^{1/2}}{2x \ln 3}$  **45.** $\dfrac{dy}{dx} = -\dfrac{y^2 + xy \ln y}{x^2 + xy \ln x}$  **47.** $\dfrac{dy}{dx} = \dfrac{x^2 + y^2 - 2x}{2y - x^2 - y^2}$  **49.** $\dfrac{dy}{dx} = \dfrac{y}{x(1-y)}$

**51.** $y' = (x^2+1)^2(2x^3-1)^4 \left( \dfrac{4x}{x^2+1} + \dfrac{24x^2}{2x^3-1} \right)$  **53.** $y' = \dfrac{x^2(x^3+1)}{\sqrt{x^2+1}} \left( \dfrac{2}{x} + \dfrac{3x^2}{x^3+1} - \dfrac{x}{x^2+1} \right)$

**55.** $y' = \dfrac{x \cos x}{(x^2+1)^3 \sin x} \left( \dfrac{1}{x} - \tan x - \dfrac{6x}{x^2+1} - \cot x \right)$  **57.** $y' = (3x)^x[\ln(3x)+1]$  **59.** $y' = 2x^{\ln x}\dfrac{\ln x}{x}$  **61.** $y' = x^{x^2}(2x \ln x + x)$

**63.** $y' = x^{e^x} \left( e^x \ln x + \dfrac{e^x}{x} \right)$  **65.** $y' = x^{\sin x} \left( \cos x \ln x + \dfrac{\sin x}{x} \right)$  **67.** $y' = (\sin x)^x (\ln(\sin x) + x \cot x)$

**69.** $y' = (\sin x)^{\cos x}(-\sin x \ln(\sin x) + \cos x \cot x)$  **71.** $y' = \dfrac{-y}{x \ln x}$  **73.** $y = 5x - 1$  **75.** $y = -5x + 18$

**77.** $e^2$  **79.** $e^{1/3}$  **81.** $\dfrac{d^{10}y}{dx^{10}} = \dfrac{362880}{x}$  **83.** $\dfrac{dy}{dx} = 2$  **85.** $y' = x^x(\ln x + 1)$  **87.** $y' = \tan^{-1}\dfrac{x}{a}$

**89. (a)** SSM.  **(b)** \$6107.01  **(c)** 28.881 years  **(d)** SSM.  **91. (a)** $-\dfrac{10}{\ln 10}$ dB/m  **(b)** AWV.  **93.** SSM.  **95.** SSM.  **97.** SSM.

## Section 3.5  (SSM = See Student Solutions Manual; AWV = Answers will vary)

**1. (d)**  **2.** $f(x_0) + f'(x_0)(x - x_0)$  **3.** True  **4.** $\dfrac{|\Delta Q|}{Q(x_0)}$  **5.** True  **6.** True  **7.** $dy = (3x^2 - 2)\,dx$  **9.** $dy = 12x(x^2+1)^{1/2}\,dx$

**11.** $dy = (6\cos(2x)+1)\,dx$  **13.** $dy = -e^{-x}\,dx$  **15.** $dy = (e^x + xe^x)\,dx$  **17.** $dy = \dfrac{2\,dx}{\sqrt{1-4x^2}}$

**19. (a)** $dy = e^x dx$  **(b) (i)** $dy = \dfrac{1}{2}e \approx 1.3591, \Delta y = e^{1.5} - e \approx 1.7634$  **(ii)** $dy = \dfrac{1}{10}e \approx 0.2718, \Delta y = e^{1.1} - e \approx 0.2859$

    **(iii)** $dy = \dfrac{1}{100}e \approx 0.0272, \Delta y = e^{1.01} - e \approx 0.0273$;  **(c) (i)** 0.4043  **(ii)** 0.0141  **(iii)** 0.000136

**21. (a)** $dy = \dfrac{2}{3\sqrt[3]{x}}\,dx$  **(b) (i)** $dy \approx 0.2646, \Delta y \approx 0.2546$  **(ii)** $dy \approx 0.0529, \Delta y \approx 0.0525$  **(iii)** $dy \approx 0.0053, \Delta y \approx 0.0053$;

    **(c) (i)** 0.009952,  **(ii)** 0.000431,  **(iii)** $4 \times 10^{-6}$  **23. (a)** $dy = -\sin x\,dx$  **(b) (i)** $dy = 0, \Delta y \approx 0.1224$

    **(ii)** $dy = 0, \Delta y \approx 0.005$  **(iii)** $dy = 0, \Delta y \approx 0.00005$;  **(c) (i)** 0.1224,  **(ii)** 0.005  **(iii)** 0.00005

**25. (a)** $L(x) = 405x - 567$  **27. (a)** $L(x) = \dfrac{x}{4} + 1$  **29. (a)** $L(x) = x - 1$  **31. (a)** $L(x) = -\dfrac{\sqrt{3}x}{2} + \dfrac{1}{2} + \dfrac{\sqrt{3}\pi}{6}$

**(b)**
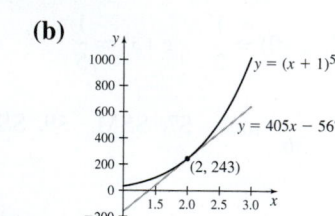
**(b)**
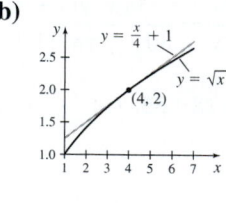
**(b)**
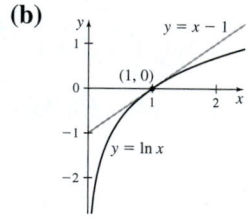
**(b)**
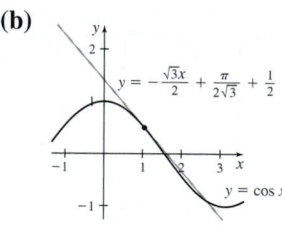

**33. (a)** 0.006  **(b)** 0.00125  **35. (a)** SSM.  **(b)** $c_3 = 1.155$  **37. (a)** SSM.  **(b)** $c_3 = 0.214$  **39. (a)** SSM.  **(b)** $c_3 = 3.136$

**41. (a)** SSM.  **(b)** AWV.  **43. (a)** SSM.  **(b)** $c_5 = -0.567$  **45. (a)** SSM.  **(b)** $c_5 = -0.791$  **47. (a)** SSM.  **(b)** $c_5 = 4.796$

**49.** $2\pi$  **51.** $3.6\pi$  **53. (a)** $1.6\pi$  **(b)** Underestimate  **(c)** SSM.  **55.** 0.9%  **57.** 72 min

**59.** $-0.0019$ kg  **61. (a)** The relative error in the radius is no more than $\dfrac{1}{300}$.  **(b)** $\dfrac{1}{3}$%  **(c)** AWV.

**63. (a)** The relative error in the pressure is no more than 0.01.  **(b)** 1%  **(c)** AWV.  **65.** 6%  **67.** 1.310

**69.** $-0.703$  **71.** 2.718  **73.** AWV.  **75.** SSM.  **77.** SSM.  **79.** SSM.  **81.** 9.871

## Section 3.6  (SSM = See Student Solutions Manual; AWV = Answers will vary)

**1.** False  **2.** $\dfrac{\sinh x}{\cosh x}$  **3.** False  **4. (a)**  **5.** False  **6.** False  **7.** True  **8.** False  **9.** $\dfrac{3}{4}$  **11.** 1  **13.** 0

**15.** SSM.  **17.** SSM.  **19.** SSM.  **21.** SSM.  **23.** SSM.  **25.** $f'(x) = 3\cosh(3x)$

**27.** $f'(x) = 2x \sinh(x^2 + 1)$ **29.** $g'(x) = \dfrac{\operatorname{csch}^2\left(\frac{1}{x}\right)}{x^2}$ **31.** $F'(x) = 4 \sinh x \sinh(4x) + \cosh x \cosh(4x)$

**33.** $s'(t) = 2 \sinh t \cosh t$ **35.** $G'(x) = e^x(\sinh x + \cosh x)$ **37.** $f'(x) = 2x \operatorname{sech} x - x^2 \tanh x \operatorname{sech} x$ **39.** $v'(t) = \dfrac{4}{\sqrt{16t^2 - 1}}$

**41.** $f'(x) = \dfrac{2}{2x - x^3}$ **43.** $g'(x) = \dfrac{x}{\sqrt{x^2 + 1}} + \sinh^{-1} x$ **45.** $s'(t) = \dfrac{\sec^2 t}{1 - \tan^2 t}$ **47.** $f'(x) = \dfrac{x}{\sqrt{x^2 - 1}\,\sqrt{x^2 - 2}}$

**49.** $y' = \dfrac{1 - \cosh x}{\sinh y - 1}$ **51.** $y' = \dfrac{1 - xe^x \cosh y}{xe^x \sinh y}$ **53.** $y'' = 2(\sinh^2 x + \cosh^2 x)$ **55.** $y'' = 3x(2 \sinh x^3 + 3x^3 \cosh x^3)$

**57.** $y = 2x$ **59.** $y = x + 1$ **61.** $\theta \approx 0.031$ radians $\approx 1.776°$

**63.** (a) 625.092 (b) 580.744 (c) $\pm 0.384$ (d) $\pm 2.766$ (e) $82.026°$ (f)

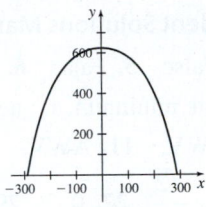

**65.** (a) SSM. (b) $x > 1$ **67.** SSM. **69.** SSM. **71.** SSM. **73.** SSM. **75.** 0 **77.** SSM.

## Review Exercises (SSM = See Student Solutions Manual; AWV = Answers will vary)

**1.** $an(ax + b)^{n-1}$ **3.** $(1 - x)^{1/2} - \dfrac{x}{2(1 - x)^{1/2}}$ **5.** $3x(x^2 + 4)^{1/2}$ **7.** $-\dfrac{a}{x\sqrt{2ax - x^2}}$ **9.** $(e^x - x)^{5x}\left(5 \ln(e^x - x) + \dfrac{5x(e^x - 1)}{e^x - x}\right)$

**11.** $-\dfrac{2x}{(x - 1)^3}$ **13.** $\sec(2x) + 2x \sec(2x) \tan(2x)$ **15.** $\dfrac{a \cos \frac{x}{a}}{2\left(a^2 \sin \frac{x}{a}\right)^{1/2}}$ **17.** $\cos v - \sin^2 v \cos v = \cos^3 v$ **19.** $1.05^x \ln 1.05$

**21.** $-\dfrac{1}{\sqrt{u^2 + 25}}$ **23.** $2 \cot(2x)$ **25.** $\dfrac{2x - 2}{x^2 - 2x}$ **27.** $e^{-x}\left(-\ln x + \dfrac{1}{x}\right)$ **29.** $\dfrac{12}{x(144 - x^2)}$ **31.** $\dfrac{e^x(x^2 + 4)}{(x - 2)}\left[1 + \dfrac{2x}{x^2 + 4} - \dfrac{1}{x - 2}\right]$

**33.** $\dfrac{\sqrt{x}}{1 + x}$ **35.** $\dfrac{1}{\sqrt{x - x^2}}$ **37.** $\tan^{-1}(x)$ **39.** $\dfrac{1}{2}\operatorname{sech}^2\left(\dfrac{x}{2}\right) + \dfrac{8 - 2x^2}{(4 + x^2)^2}$ **41.** $\dfrac{\cosh x}{2(\sinh x)^{1/2}}$ **43.** $\dfrac{1}{5y^4 + 1}$ **45.** $\dfrac{y(x \cos y - 1)}{x(1 + xy \sin y)}$

**47.** $\dfrac{1 + y \cos(xy)}{1 - x \cos(xy)}$ **49.** $y' = \dfrac{10 - y}{6y + x}$; $y'' = \dfrac{2(y - 10)(3y + x + 30)}{(6y + x)^3}$ **51.** $y' = \dfrac{8x - e^y}{xe^y}$; $y'' = \dfrac{-64x^2 + 8xe^y + e^{2y}}{x^2 e^{2y}}$

**53.** $\dfrac{1}{2}$ **55.** $e^{2/5}$ **57.** 0 **59.** SSM. **61.** $x \neq (2k - 1)\dfrac{\pi}{2}$, $k$ an integer

**63.** (a) Domain: $\left\{x \mid x \geq -\dfrac{1}{6}\right\}$; Range: $\{y \mid y \geq 0\}$ (b) $\dfrac{3}{5}$ (c) $\left(0, \dfrac{13}{5}\right)$ (d) $\left(\dfrac{4}{3}, 3\right)$ **65.** $dy = \dfrac{2xy - 3x^2}{4y - x^2}\,dx$

**67.** $L(x) = 2x - 1$ **69.** (a) $dy = dx$; $\Delta y = \tan(\Delta x)$ (b) (i) $dy = 0.5$, $\Delta y \approx 0.546$; (ii) $dy = 0.1$, $\Delta y \approx 0.100$;
(iii) $dy = 0.01$, $\Delta y \approx 0.010$ **71.** $-\dfrac{1}{2}(1 + \ln 8)$ **73.** $-2$ **75.** (a) SSM. (b) 2.568 **77.** $y = 6x - 3$

## Chapter 4

## Section 4.1 (SSM = See Student Solutions Manual; AWV = Answers will vary)

**1.** $10 \text{ m}^3/\text{min}$ **2.** $0.5 \text{ m/min}$ **3.** $-\dfrac{8}{3}$ **4.** $-1$ **5.** $5\pi^2$ **7.** $40$ **9.** $900 \text{ cm}^3/\text{s}$ **11.** $-0.000625 \text{ cm/h}$

**13.** $-\dfrac{8}{\sqrt{2009}} \approx -0.178 \text{ cm/min}$ **15.** $\dfrac{2\sqrt{3}\pi}{45} \approx 0.242 \text{ cm}^2/\text{min}$ **17.** shrinking at a rate of $0.75 \text{ m}^2/\text{min}$

**19.** decreasing at a rate of $\dfrac{1}{6} \text{ rad/s}$ **21.** $0.2 \text{ m/min}$ **23.** $\dfrac{4}{\pi} \approx 1.273 \text{ m/min}$

**25.** (a) $\dfrac{16}{3\pi} \approx 1.698 \text{ m/min}$ (b) $\dfrac{96 - 9\pi}{32} \approx 2.116 \text{ m}^3/\text{min}$ **27.** $\dfrac{18}{\sqrt{17}} \approx 4.366 \text{ units/s}$

**29.** $50\sqrt{2}$ ft/s   **31.** 1.75 kg/cm²/min   **33.** $100.8\pi \approx 316.673$ ft²/min

**35. (a)** $\dfrac{1.5}{\sqrt{55}} \approx 0.202$ m/s   **(b)** $\dfrac{0.5}{\sqrt{3}} \approx 0.289$ m/s   **(c)** $\dfrac{1.5}{\sqrt{7}} \approx 0.567$ m/s   **37.** $\dfrac{24\pi}{5} \approx 15.08$ km/s   **39.** 4 m/min

**41.** He is rising at $\dfrac{25\sqrt{3}\pi}{2} \approx 68.017$ ft/min. He is moving horizontally at $-\dfrac{25\pi}{2} \approx -39.270$ ft/min.

**43.** $-\dfrac{9\sqrt{29}}{29} \approx -1.671$ m/s   **45. (a)** \$150/day   **(b)** \$200/day   **(c)** \$50/day   **47.** $\approx -4.864$ lb/s

**49.** $-100\pi \approx -314.159$ ft/s   **51.** $-316.8$ rad/h   **53. (a)** AWV.   **(b)** AWV.   **55.** $\approx 32.071$ ft/s; AWV.   **57.** $\approx 0.165$ in²/min

## Section 4.2   (SSM = See Student Solutions Manual; AWV = Answers will vary)

**1.** False   **2.** (b)   **3.** False   **4.** False   **5.** False   **6.** False   **7.** $x_1$: neither, $x_2$: local maximum,
$x_3$: local minimum and absolute minimum, $x_4$: neither, $x_5$: local maximum, $x_6$: neither, $x_7$: local minimum,
$x_8$: absolute maximum   **9.** AWV.   **11.** AWV.   **13.** 4   **15.** 0, 2   **17.** $-1, 0, 1$   **19.** 0

**21.** 0   **23.** $\pi$   **25.** $-\dfrac{\sqrt{2}}{2}, -1, 1, \dfrac{\sqrt{2}}{2}$   **27.** 0, 2   **29.** $-3, 0, 1$   **31.** 3, 4   **33.** $-3\sqrt{3}, -3, 3, 3\sqrt{3}$   **35.** 1

**37.** Absolute maximum 20 at $x = 10$, absolute minimum $-16$ at $x = 4$   **39.** Absolute maximum 16 at $x = 4$,
absolute minimum $-4$ at $x = 2$   **41.** Absolute maximum 9 at $x = 2$, absolute minimum 0 at $x = 1$

**43.** Absolute maximum 1 at $x = -1, 1$, absolute minimum 0 at $x = 0$   **45.** Absolute maximum 4 at $x = 4$,
absolute minimum 2 at $x = 1$   **47.** Absolute maximum $\pi$ at $x = \pi$, absolute minimum 0 at $x = 0$

**49.** Absolute maximum $\dfrac{1}{2}$ at $x = \dfrac{\sqrt{2}}{2}$, absolute minimum $-\dfrac{1}{2}$ at $x = -\dfrac{\sqrt{2}}{2}$   **51.** Absolute maximum 0 at $x = 0$,
absolute minimum $-\dfrac{1}{2}$ at $x = -1, \dfrac{1}{2}$   **53.** Absolute maximum $128\sqrt[3]{2}$ at $x = 5$, absolute minimum 0 at $x = -3, 1$

**55.** Absolute maximum $\dfrac{1}{3}$ at $x = 4$, absolute minimum $-1$ at $x = 2$   **57.** Absolute maximum $\dfrac{\sqrt[3]{18}}{3\sqrt{3}}$ at $x = 3\sqrt{3}$,
absolute minimum 0 at $x = 3$   **59.** Absolute maximum 1 at $x = 0$,   absolute minimum $e - 3$ at $x = 1$
**61.** Absolute maximum 9 at $x = 3$, absolute minimum 1 at $x = 0$   **63.** Absolute maximum 8 at $x = 2$, absolute minimum 0 at $x = 0$

**65. (a)** $f'(x) = 12x^3 - 6x^2 - 42x + 36$   **(b)** $-2, 1, \dfrac{3}{2}$   **(c)**

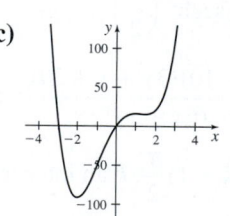

Absolute minimum at $x = -2$,
local maximum at $x = 1$,
local minimum at $x = \dfrac{3}{2}$

**67. (a)** $f'(x) = \dfrac{3x^4 + 5x^3 + 8x^2 - 10x - 96}{2\sqrt{x+5}(x^2+2)^{3/2}}$   **(b)** $-5, \approx -2.364, \approx 1.977$   **(c)**

Local maximum at $x \approx -2.364$, absolute minimum at $x \approx 1.977$, neither a local minimum nor a local maximum at $x = -5$

**69. (a)** $f'(x) = 4x^3 - 37.2x^2 + 98.48x - 68.64$   **(b)** Absolute maximum 0 at $x = 0$, absolute minimum $-37.2$ at $x = 5$
**(c)**

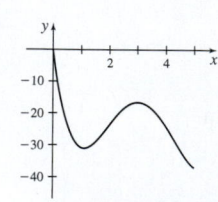

**71. (a)** 50 mi/h **(b)**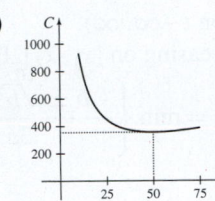

**73. (a)** $67.5°$, $0.0183\, v_0^2$ ft **(b)** 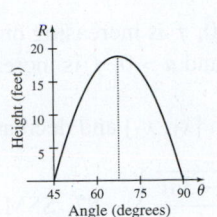 **(c)** AWV.

**75. (a)** SSM. **(b)** $\approx 961.1$ m

**77.** A tax rate of $\dfrac{3\sqrt{6}}{2} \approx 3.674$, maximizes revenue. The maximum revenue generated is $\dfrac{9\sqrt{3}}{2} \approx 7.794$. **79.** 5 m

**81. (a)** $0.04, 0.07$ **(b)** Absolute maximum $-3.719$ at $x = 1$, absolute minimum $-4.391$ at $x = -1$ **(c)**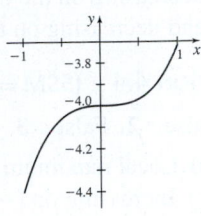

**83.** Absolute maximum $\sqrt{10} + 1$ at $x = 3$, absolute minimum $\sqrt{5}$ at $x = 2$ **85. (a)** $[-4, -3] \cup [3, 4]$ **(b)** $\dfrac{7}{2}$ **87. (c) (e)**

**89.** SSM. **91.** SSM. **93. (a)** SSM. **(b)** AWV.

## Section 4.3 (SSM = See Student Solutions Manual; AWV = Answers will vary)

**1.** False **2.** AWV. **3.** True **4.** False **5.** $\dfrac{3}{2}$ **7.** 1 **9.** $-\dfrac{\sqrt{3}}{3}$ **11.** $\pm\dfrac{\sqrt{3}}{3}$ **13.** $\pm 1, 0$ **15.** $\dfrac{\pi}{4}, \dfrac{3\pi}{4}$

**17.** $f(-2) \neq f(1)$ **19.** $f$ is not differentiable at $x = 0$, so $f$ is not differentiable on $(-1, 1)$. **21. (a)** SSM. **(b)** 1 **(c)** AWV.

**23. (a)** SSM. **(b)** $e - 1$ **(c)** AWV. **25. (a)** SSM. **(b)** $\dfrac{7}{3}$ **(c)** AWV. **27. (a)** SSM. **(b)** $\sqrt{3}$ **(c)** AWV.

**29. (a)** SSM. **(b)** $\dfrac{2744}{729}$ **(c)** AWV. **31.** $f$ is increasing on $(-\infty, \infty)$.

**33.** $f$ is increasing on $(-\infty, -1] \cup [1, \infty)$ and decreasing on $[-1, 1]$.

**35.** $f$ is increasing on $[-\sqrt{2}, 0] \cup [\sqrt{2}, \infty)$ and decreasing on $(-\infty, -\sqrt{2}] \cup [0, \sqrt{2}]$.

**37.** $f$ is increasing on $[-1, 0] \cup [1, \infty)$ and decreasing on $(-\infty, -1] \cup [0, 1]$.

**39.** $f$ is increasing on $[-\sqrt[3]{3}, \infty)$ and decreasing on $(-\infty, -\sqrt[3]{3}]$.

**41.** $f$ is increasing on $\left[0, \dfrac{\pi}{2}\right] \cup \left[\dfrac{3\pi}{2}, 2\pi\right]$ and decreasing on $\left[\dfrac{\pi}{2}, \dfrac{3\pi}{2}\right]$.

**43.** $f$ is increasing on $[-1, \infty)$ and decreasing on $(-\infty, -1]$.

**45.** $f$ is increasing on $\left[0, \dfrac{3\pi}{4}\right] \cup \left[\dfrac{7\pi}{4}, 2\pi\right]$ and decreasing on $\left[\dfrac{3\pi}{4}, \dfrac{7\pi}{4}\right]$.

**47. (a)** $\{x \mid x \neq -2, x \neq 2\}$ **(b)** $-2, 0, 2, 4$ **(c)** 0 and 4 **(d)** 2 **(e)** $-2$ **(f)** $(-\infty, 0], [2, 4]$ **(g)** $[0, 2], [4, \infty)$

**49. (a)** $\{x \mid x \neq -1, x \neq 0\}$ **(b)** $-2, -1, 0, 1, 2$ **(c)** $-2, 1, 2$ **(d)** $-1$ **(e)** 0 **(f)** $(-\infty, -1], [2, \infty)$ **(g)** $[1, 2]$

**51.** SSM. **53.** SSM. **55.** AWV. **57.** SSM. **59.** SSM.

**61. (a)** SSM. **(b)** $d(0) = d(5) = 0$ **(c)** $\approx 2.892$ ft; $d(2.892) \approx -0.423$ ft. **(d)**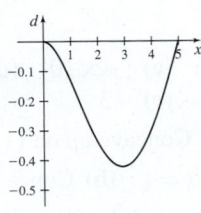

**63.** SSM. **65.** SSM. **67.** AWV. **69.** $\sqrt{1 - \dfrac{4}{\pi^2}}$

**71. (a)** The function $f$ is increasing on the interval $(0, \infty)$. **(b)** Yes. **73. (b)** **75.** SSM. **77.** SSM. **79.** SSM. **81.** SSM.

**83.** SSM.  **85.** For $b^2 - 3ac \le 0$, $f$ is increasing on $(-\infty, \infty)$ for $a > 0$ and decreasing on $(-\infty, \infty)$ for $a < 0$. For $b^2 - 3ac > 0$ and $a > 0$, $f$ is increasing on $(-\infty, x_1] \cup [x_2, \infty)$ and decreasing on $[x_1, x_2]$. For $b^2 - 3ac > 0$ and $a < 0$, $f$ is increasing on $[x_1, x_2]$ and decreasing on $(-\infty, x_1] \cup [x_2, \infty)$, where $x_1 = \min\left\{ \dfrac{-b \pm \sqrt{b^2 - 3ac}}{3a} \right\}$ and $x_2 = \max\left\{ \dfrac{-b \pm \sqrt{b^2 - 3ac}}{3a} \right\}$.  **87.** SSM.  **89.** SSM.

**91.** 0 is the only critical number. If $n$ is odd and $ad - bc > 0$, then $f$ is increasing on the domain. If $n$ is odd and $ad - bc < 0$, then $f$ is decreasing on the domain. If $n$ is even and $ad - bc > 0$, then $f$ is increasing on the domain where $x > 0$ and decreasing on the domain where $x < 0$. If $n$ is even and $ad - bc < 0$, then $f$ is increasing on the domain where $x < 0$ and decreasing on the domain where $x > 0$.  **93.** SSM.  **95.** SSM.

## Section 4.4  (SSM = See Student Solutions Manual; AWV = Answers will vary)

**1.** False  **2.** False  **3.** (a)  **4.** (b)  **5.** (a)  **6.** (c)  **7.** True  **8.** False

**9. (a)** Local maximum at $(-1, 0)$, local minima at $(-2.5, -4)$ and $(0.5, -4)$, inflection points $(-1.8, -2)$ and $(-0.2, -2)$
**(b)** Increasing on $[-2.5, -1] \cup [0.5, \infty)$, decreasing on $(-\infty, -2.5] \cup [-1, 0.5]$, concave up on $(-\infty, -1.8) \cup (-0.2, \infty)$, concave down on $(-1.8, -0.2)$

**11. (a)** Local maxima at $(-2, 3)$ and $(12, 10)$, local minimum at $(0, 0)$, no inflection point  **(b)** Increasing on $(-\infty, -2] \cup [0, 12]$, decreasing on $[-2, 0] \cup [12, \infty)$, never concave up, concave down on $(-\infty, 0) \cup (0, \infty)$

**13. (a)** $0, 4$  **(b)** Local maximum 2 at 0, local minimum $-30$ at 4

**15. (a)** $0, 1$  **(b)** Local minimum $-1$ at 1  **17. (a)** $\dfrac{3}{2}$  **(b)** Local maximum $2e^{3/2}$ at $\dfrac{3}{2}$

**19. (a)** $-\dfrac{1}{8}, 0$  **(b)** Local minimum $-\dfrac{1}{4}$ at $-\dfrac{1}{8}$  **21. (a)** $-1, 0, 1$  **(b)** Local maximum 0 at 0, local minimum $-3$ at $-1$ and 1

**23. (a)** $\sqrt[3]{e}$  **(b)** Local maximum $\dfrac{1}{3e}$ at $\sqrt[3]{e}$  **25. (a)** $k\pi - \tan^{-1}\dfrac{1}{2}$, $k$ an integer  **(b)** Local maximum $\sqrt{5}$ at $(2k+1)\pi - \tan^{-1}\dfrac{1}{2}$, local minimum $-\sqrt{5}$ at $2k\pi - \tan^{-1}\dfrac{1}{2}$, $k$ an integer

**27. (a)** The object moves right on $(1, \infty)$, left on $(0, 1)$  **(d)**   **(e)**
**(b)** $t = 1$  **(c)** The velocity is increasing on $(0, \infty)$

**29. (a)** The object moves right on $(1, \infty)$, left on $(0, 1)$  **(d)**   **(e)**
**(b)** $t = 1$  **(c)** The velocity is increasing on $(0, \infty)$

**31. (a)** The object moves right on $(0, \infty)$  **(d)**   **(e)**
**(b)** No direction reverse  **(c)** The velocity is decreasing on $(0, \infty)$

**33. (a)** The object moves right on $\left(0, \dfrac{\pi}{6}\right)$ and $\left(\dfrac{\pi}{2}, \dfrac{2\pi}{3}\right)$, left on $\left(\dfrac{\pi}{6}, \dfrac{\pi}{2}\right)$  **(b)** $t = \dfrac{\pi}{6}$ and $t = \dfrac{\pi}{2}$

**(c)** The velocity is increasing on $\left(\dfrac{\pi}{3}, \dfrac{2\pi}{3}\right)$, decreasing on $\left(0, \dfrac{\pi}{3}\right)$  **(d)**   **(e)**

**35. (a)** $0, 1$  **(b)** $[0, 1], [1, \infty)$  **(c)** $(-\infty, 0]$  **(d)** 0  **(e)** none  **37. (a)** $-3, 0, 1$  **(b)** $(-\infty, -3], [1, \infty)$
**(c)** $[-3, 0], [0, 1]$  **(d)** 1  **(e)** $-3$

**39. (a)** No local extrema  **(b)** Concave up on $(1, \infty)$, concave down on $(-\infty, 1)$  **(c)** $(1, -1)$

**41. (a)** Local minimum $-3$ at $x = 1$  **(b)** Concave up on $(-\infty, 0) \cup (0, \infty)$  **(c)** No inflection point

**43. (a)** Local maximum 256 at $x = 4$, local minimum 0 at $x = 0$
**(b)** Concave up on $(-\infty, 0) \cup (0, 3)$, concave down on $(3, \infty)$  **(c)** $(3, 162)$

**45. (a)** Local maximum 64 at $x = -2$, local minimum $-64$ at $x = 2$
**(b)** Concave up on $(-\sqrt{2}, 0) \cup (\sqrt{2}, \infty)$, concave down on $(-\infty, -\sqrt{2}) \cup (0, \sqrt{2})$  **(c)** $(0, 0), (\sqrt{2}, -28\sqrt{2}), (-\sqrt{2}, 28\sqrt{2})$

**47. (a)** Local maximum $4e^{-2}$ at $x = -2$, local minimum $0$ at $x = 0$  **(b)** Concave up on $(-\infty, -2 - \sqrt{2}) \cup (-2 + \sqrt{2}, \infty)$,
concave down on $(-2 - \sqrt{2}, -2 + \sqrt{2})$  **(c)** $(-2 - \sqrt{2}, (6 + 4\sqrt{2})e^{-2 - \sqrt{2}})$, $(-2 + \sqrt{2}, (6 - 4\sqrt{2})e^{-2 + \sqrt{2}})$

**49. (a)** Local minimum $-\dfrac{9}{8}$ at $x = \dfrac{1}{8}$  **(b)** Concave up on $\left(-\infty, -\dfrac{1}{4}\right) \cup (0, \infty)$, concave down on $\left(-\dfrac{1}{4}, 0\right)$  **(c)** $\left(-\dfrac{1}{4}, \dfrac{9\sqrt[3]{2}}{4}\right)$, $(0, 0)$

**51. (a)** Local maximum $0$ at $x = 0$, local minimum $-6\sqrt[3]{2}$ at $x = -\sqrt{2}$ and $\sqrt{2}$  **(b)** Concave up on $(-\infty, 0) \cup (0, \infty)$
**(c)** No inflection point

**53. (a)** Local minimum $\dfrac{1}{2} + \dfrac{1}{2}\ln 2$ at $x = \dfrac{\sqrt{2}}{2}$  **(b)** Concave up on $(0, \infty)$  **(c)** No inflection point

**55. (a)** Local maximum $\dfrac{16\sqrt{5}}{125}$ at $x = \dfrac{1}{2}$, local minimum $-\dfrac{16\sqrt{5}}{125}$ at $x = -\dfrac{1}{2}$  **(b)** Concave up on $\left(-\dfrac{\sqrt{3}}{2}, 0\right) \cup \left(\dfrac{\sqrt{3}}{2}, \infty\right)$,
concave down on $\left(-\infty, -\dfrac{\sqrt{3}}{2}\right) \cup \left(0, \dfrac{\sqrt{3}}{2}\right)$  **(c)** $\left(-\dfrac{\sqrt{3}}{2}, -\dfrac{16\sqrt{3}}{7^{5/2}}\right)$, $\left(\dfrac{\sqrt{3}}{2}, \dfrac{16\sqrt{3}}{7^{5/2}}\right)$, $(0, 0)$

**57. (a)** Local maximum $\dfrac{2\sqrt{3}}{9}$ at $x = \pm\dfrac{\sqrt{6}}{3}$, local minimum $0$ at $x = 0$  **(b)** Concave up on $\left(-\dfrac{1}{6}\sqrt{27 - 3\sqrt{33}}, \dfrac{1}{6}\sqrt{27 - 3\sqrt{33}}\right)$,
concave down on $\left(-1, -\dfrac{1}{6}\sqrt{27 - 3\sqrt{33}}\right) \cup \left(\dfrac{1}{6}\sqrt{27 - 3\sqrt{33}}, 1\right)$  **(c)** $\left(\pm\dfrac{1}{6}\sqrt{27 - 3\sqrt{33}}, \dfrac{9 - \sqrt{33}}{12}\sqrt{\dfrac{3 + \sqrt{33}}{12}}\right)$

**59. (a)** Local maximum $\dfrac{5\pi}{3} + \sqrt{3}$ at $x = \dfrac{5\pi}{3}$, local minimum $\dfrac{\pi}{3} - \sqrt{3}$ at $x = \dfrac{\pi}{3}$
**(b)** Concave up on $(0, \pi)$, concave down on $(\pi, 2\pi)$  **(c)** $(\pi, \pi)$

**61. (a)** Local minimum $1$ at $x = 0$  **(b)** Concave up on $(-\infty, \infty)$  **(c)** No inflection point.

**63. (a)** $-\sqrt{3}, 0, \sqrt{3}$  **(b)** $[-\sqrt{3}, 0], [\sqrt{3}, \infty)$  **(c)** $(-\infty, -\sqrt{3}], [0, \sqrt{3}]$  **(d)** $-\sqrt{3}, \sqrt{3}$  **(e)** $0$
**(f)** $(-\infty, -1), (1, \infty)$  **(g)** $(-1, 1)$  **(h)** $(-1, f(-1)), (1, f(1))$

**65. (a)** $0, 1$  **(b)** $(-\infty, 0], [1, \infty)$  **(c)** $[0, 1]$  **(d)** $1$  **(e)** $0$  **(f)** $\left(\sqrt[3]{\dfrac{1}{4}}, \infty\right)$  **(g)** $\left(-\infty, \sqrt[3]{\dfrac{1}{4}}\right)$  **(h)** $\left(\sqrt[3]{\dfrac{1}{4}}, f\left(\sqrt[3]{\dfrac{1}{4}}\right)\right)$

**67. (a)** and **(b)** Local maximum $-20$ at $x = 3$, local minimum $-21$ at $x = 2$  **(c)** AWV

**69. (a)** and **(b)** Local minima $-21$ at $x = -2$ and $x = 2$; local maximum of $-5$ at $x = 0$  **(c)** AWV.

**71. (a)** Local minimum of $1$ at $x = 0$; local maximum of $\dfrac{337}{81}$ at $x = -\dfrac{4}{3}$

**(b)** No information about $x = 0$; local maximum of $\dfrac{337}{81}$ at $x = -\dfrac{4}{3}$  **(c)** AWV.

**73. (a)** and **(b)** Local maximum $4e$ at $x = 1$, local minimum $0$ at $x = 3$  **(c)** AWV.

**75.** AWV.
One possible answer below.

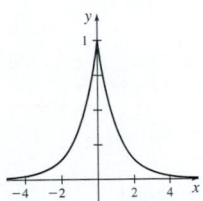

**77.** AWV.
One possible answer below.

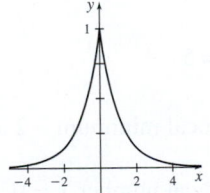

**79.** AWV.
One possible answer below.

**81.** AWV.
One possible answer below.

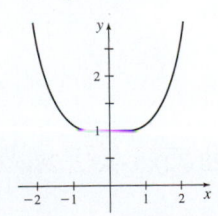

**83.** AWV.
One possible answer below.

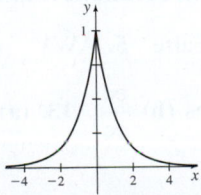

**85.** AWV.
One possible answer below.

**87. (a)** Concave up on $\left(-\infty, 2-\dfrac{\sqrt{2}}{2}\right) \cup \left(2+\dfrac{\sqrt{2}}{2}, \infty\right)$, concave down on $\left(2-\dfrac{\sqrt{2}}{2}, 2+\dfrac{\sqrt{2}}{2}\right)$

**(b)** $\left(2-\dfrac{\sqrt{2}}{2}, \dfrac{1}{\sqrt{e}}\right), \left(2+\dfrac{\sqrt{2}}{2}, \dfrac{1}{\sqrt{e}}\right)$   **(c)**

**89. (a)** Concave up on $(-\infty, 0.135) \cup (0.721, 5.144)$, concave down on $(0.135, 0.721) \cup (5.144, \infty)$
**(b)** $(0.135, 2.433), (0.721, 2.139), (5.144, -0.072)$
**(c)**

**91.** $a = -3, b = 9$   **93. (a)** $N'(t) = \dfrac{99,990,000e^{-t}}{(1+9999e^{-t})^2}$   **(b)** $N'(t)$ is increasing on the interval $(0, \ln 9999)$ and decreasing on the interval $(\ln 9999, \infty)$.   **(c)** $\ln 9999$   **(d)** $(\ln 9999, 5000)$   **(e)** AWV.

**95. (a)**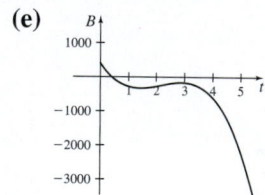

Population (millions)

1880 1900 1920 1940 1960 1980 2000 2020  $t$

**(b)** AWV.   **(c)** $P'(t) = \dfrac{107,952,118.9e^{-0.0162t}}{(1+8.743e^{-0.0162t})^2}$

**(d)** $P'(t)$ is increasing on the interval $[0, 133.843]$ and decreasing on the interval $[133.843, \infty)$.

**(e)** The rate of change in population is maximum when $t = 133.843$.

**(f)** $(133.843, 381088358.9)$

**(g)** AWV.

**97. (a)** Local minimum $B(2.80) \approx -328.53$ billion dollars, local maximum $B(5.71) \approx -170.80$ billion dollars.
**(b)** Both represent a budget deficit.   **(c)** Concave up on $(0, 4.26)$, concave down on $(4.26, 9)$; $(4.26, -249.27)$ is a point of inflection.
**(d)** To the left of the inflection point, the budget is increasing at an increasing rate. To the immediate right of the inflection point, the budget is increasing at a decreasing rate.
**(e)**

**99.** $a = -\dfrac{7}{8}, b = \dfrac{21}{4}, c = 0$, and $d = 5$

**101.** Local maximum 2 at $x = \dfrac{\pi}{3}$, local minimum $-2$ at $x = \dfrac{4\pi}{3}$. Inflection points $\left(\dfrac{5\pi}{6}, 0\right), \left(\dfrac{11\pi}{6}, 0\right)$.

**103.** $f''(x)$ does not exist at the critical number $x = 0$.   **105.** (e)   **107.** SSM.   **109.** SSM.   **111.** SSM.   **113.** SSM.
**115.** SSM.   **117.** SSM.   **119.** SSM.   **121.** SSM.   **123.** SSM.

## Section 4.5   (SSM = See Student Solutions Manual; AWV = Answers will vary)

**1.** False   **2.** False   **3.** False   **4.** False   **5.** AWV.   **6.** AWV.   **7. (a)** Yes   **(b)** $\dfrac{0}{0}$

**9. (a)** No   **(b)** AWV.   **11. (a)** Yes **(b)** $\dfrac{\infty}{\infty}$   **13. (a)** No   **(b)** AWV.   **15. (a)** Yes   **(b)** $\dfrac{0}{0}$

**17. (a)** Yes **(b)** $\dfrac{0}{0}$ **19. (a)** Yes **(b)** $0 \cdot \infty$ **21. (a)** Yes **(b)** $\infty - \infty$ **23. (a)** Yes **(b)** $\infty^0$

**25. (a)** No **(b)** AWV. **27.** $\dfrac{0}{0}, 5$ **29.** $\dfrac{0}{0}, \dfrac{1}{2}$ **31.** $\dfrac{0}{0}, 2$ **33.** $\dfrac{0}{0}, -\pi$ **35.** $\dfrac{\infty}{\infty}, 0$

**37.** $\dfrac{\infty}{\infty}, 0$ **39.** $\dfrac{0}{0}, 1$ **41.** $\dfrac{0}{0}, -\dfrac{1}{6}$ **43.** $0 \cdot \infty, 0$ **45.** $0 \cdot \infty, 1$ **47.** $\infty - \infty, 0$ **49.** $\infty - \infty, -1$ **51.** $0^0, 1$

**53.** $\infty^0, 1$ **55.** $\infty^0, 1$ **57.** $1^\infty, 1$ **59.** 2 **61.** 0 **63.** 0 **65.** 2 **67.** 1 **69.** $\dfrac{1}{4}$ **71.** 0 **73.** 0 **75.** 1 **77.** $\dfrac{4a^2}{\pi}$ **79.** $\dfrac{1}{2}$ **81.** $-1$

**83.** 1 **85.** 1 **87.** 1 **89.** 1 **91. (a)** 252 wolves **(b)** AWV. **(c)**

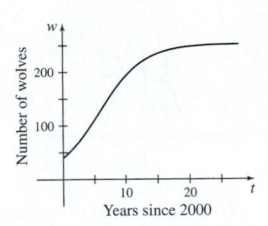

**93. (a)** $\dfrac{E}{R}, \dfrac{Et}{L}$ **(b)** AWV.

**95.** SSM. **97.** SSM. **99.** 0 **101.** SSM. **103.** SSM. **105. (a)** SSM. $f'(0) = 0$ **(b)**

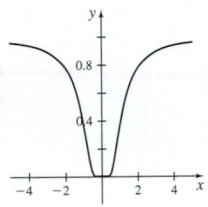

**107.** SSM. **109. (a)** 0 **(b)** $a$ **(c)** $\infty$ **(d)** Does not exist **(e)** 0 **(f)** $a$ **(g)** 1 **(h)** $a$ **(i)** Does not exist **(j)** $e^{-1}$

**111. (a)** $f''(x)$ **(b)** $f'''(x)$ **(c)** $\displaystyle\lim_{h \to 0} \dfrac{\sum_{k=0}^{n} (-1)^k \dfrac{n!}{k!\,(n-k)!} f(x + (n-k)h)}{h^n} = f^{(n)}(x)$ **113.** SSM.

## Section 4.6   (SSM = See Student Solutions Manual; AWV = Answers will vary)

**1.** 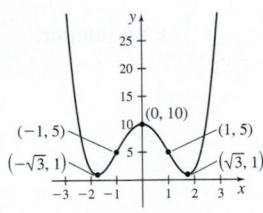 **3.** 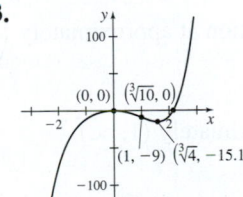 **5.** 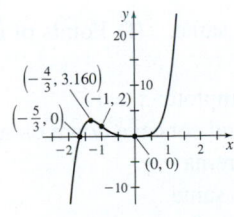 **7.** 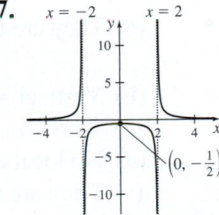 **9.**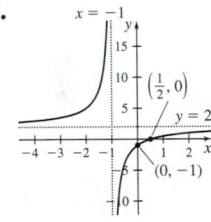

**11.** 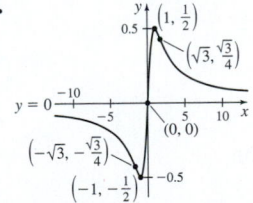 **13.** 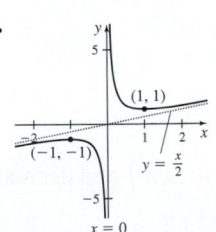 **15.** 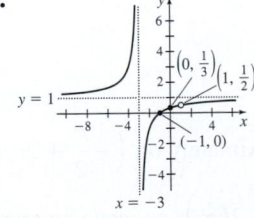 **17.** 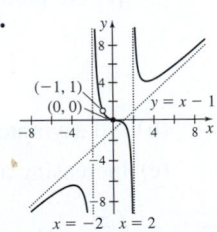 **19.**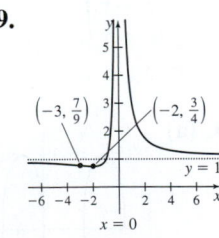

**21.** 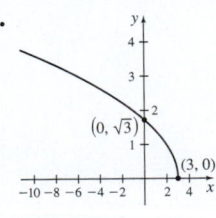 **23.** 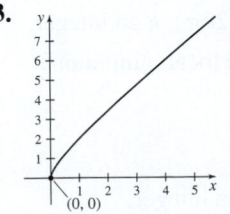 **25.** 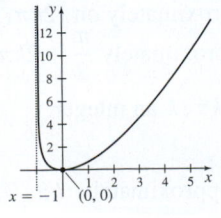 **27.** 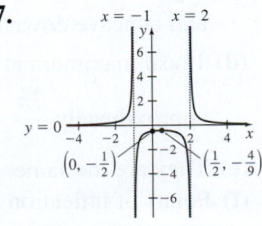 **29.**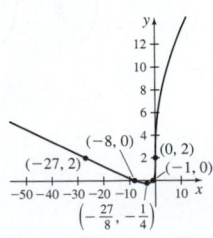

**31.**

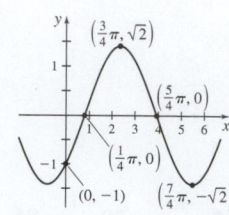

**33.**

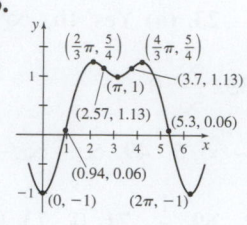

**35.**

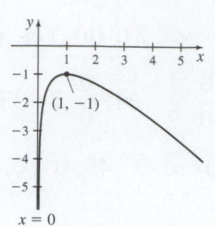

**37.**

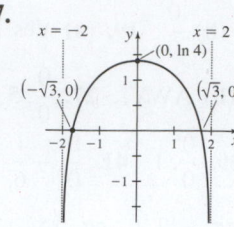

**39.**

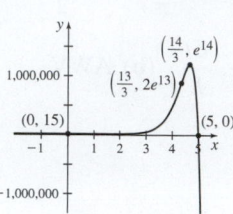

**41.**

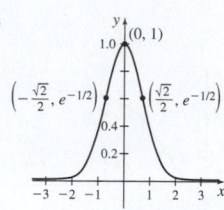

**43. (a)**

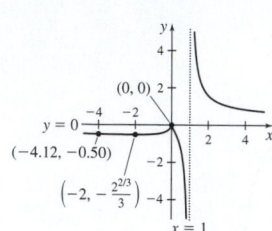

(b) Vertical asymptote: $x = 1$, horizontal asymptote: $y = 0$

(c) Decreasing on approximately $(-\infty, -2) \cup (0, 1) \cup (1, \infty)$ and increasing on approximately $(-2, 0)$, concave down on approximately $(-\infty, -4) \cup (0.1, 1)$ and concave up on approximately $(-4, 0) \cup (0, 0.1) \cup (1, \infty)$

(d) Local minimum at approximately $-2$ and local maximum at approximately $0$

(e) They are the same.

(f) Points of inflection are approximately $\left(-4, -\dfrac{2\sqrt[3]{2}}{5}\right)$ and $(0.1, -0.239)$

**45. (a)**

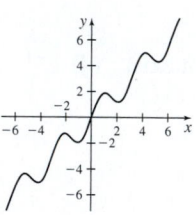

(b) No asymptotes

(c) Increasing on $\left(-\dfrac{\pi}{3} + k\pi, \dfrac{\pi}{3} + k\pi\right)$ and decreasing on $\left(\dfrac{\pi}{3} + k\pi, \dfrac{2\pi}{3} + k\pi\right)$, $k$ an integer,

concave up on $\left(-\dfrac{\pi}{2} + k\pi, k\pi\right)$ and concave down on $\left(k\pi, \dfrac{\pi}{2} + k\pi\right)$, $k$ an integer

(d) Local maximum at approximately $\dfrac{\pi}{3} + k\pi$ and local minimum at approximately

$\dfrac{2\pi}{3} + k\pi$, $k$ an integer

(e) They are the same.   (f) Points of inflection at approximately $\left(\dfrac{k\pi}{2}, \dfrac{k\pi}{2}\right)$, $k$ an integer.

**47. (a)**

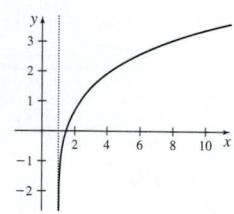

(b) Vertical asymptote: $x = 1$

(c) Increasing and concave down on approximately $(1, \infty)$

(d) No local extrema

(e) They are the same.

(f) No points of inflection

**49. (a)**

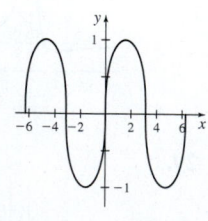

(b) No asymptotes

(c) Increasing approximately on $\left(-\dfrac{\pi}{2} + 2k\pi, \dfrac{\pi}{2} + 2k\pi\right)$ and decreasing on

$\left(\dfrac{\pi}{2} + 2k\pi, \dfrac{3\pi}{2} + 2k\pi\right)$, concave up approximately on $(-\pi + 2k\pi, 2k\pi)$

and concave down approximately on $(2k\pi, \pi + 2k\pi)$, $k$ an integer.

(d) Local maximum at approximately $\dfrac{\pi}{2} + 2k\pi$ and local minimum at

approximately $\dfrac{3\pi}{2} + 2k\pi$, $k$ an integer.

(e) They are the same.

(f) Points of inflection at approximately $(k\pi, 0)$, $k$ an integer.

**51. (a)**

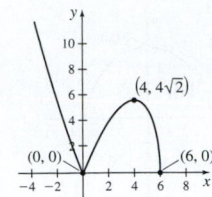

**(b)** There are no asymptotes.

**(c)** The function is increasing on $[0, 4]$, it is decreasing on $(-\infty, 0]$ and $[4, 6]$. It is concave up on $(-\infty, 0)$ and is concave down on $(0, 6)$.

**(d)** There is a local minimum at $(0, 0)$ and a local maximum at $(4, 4\sqrt{2})$.

**(e)** They are the same.

**(f)** $(0, 0)$ is an inflection point.

**53.** AWV.  **55.** AWV.  **57.** AWV.

**59. (a)** Absolute maximum at $x = \dfrac{1}{e}$ and absolute minimum at $x = 1$

**(b)** Concave up on $\left(\dfrac{1}{e}, 2\right)$

**(c)**

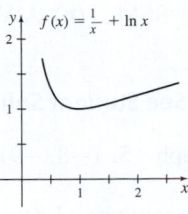

**61.**

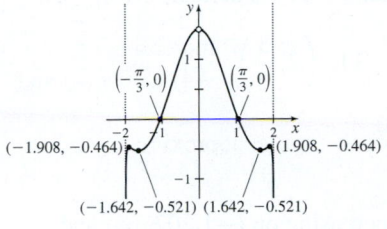

$x = \pm 2$ are two vertical asymptotes.

**63.**

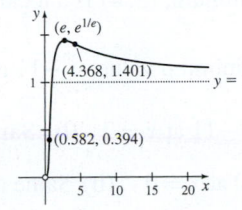

$y = 1$ is a horizontal asymptote.

## Section 4.7   (SSM = See Student Solutions Manual; AWV = Answers will vary)

**1.** $1{,}125{,}000 \text{ m}^2$   **3.** $\dfrac{L}{4} \text{ m} \times \dfrac{L}{4} \text{ m}$   **5.** $\dfrac{5000}{3} \text{ m}^2$   **7.** $8 \text{ cm} \times 8 \text{ cm} \times 2 \text{ cm}$

**9.** $10\sqrt[3]{4} \times 10\sqrt[3]{4} \times 5\sqrt[3]{4} \approx 15.874 \text{ cm} \times 15.874 \text{ cm} \times 7.937 \text{ cm}$

**11.** Radius $\sqrt[3]{\dfrac{15}{4\pi}} \approx 1.061 \text{ m}$, height $\dfrac{8}{3}\sqrt[3]{\dfrac{15}{4\pi}} \approx 2.829 \text{ m}$   **13.** \$21 per day   **15.** $(1, 1)$   **17.** $(1.784, 0.817)$

**19. (a)** $40 \text{ mi/h}$  **(b)** $\approx 46.6 \text{ mi/h}$  **(c)** $\approx 51.0 \text{ mi/h}$   **21. (a) and (b)** $\sqrt{p^2 + (q + r)^2}$   **23.** $\approx 9.582 \text{ m}, \approx 36.383°$

**25.** Width $2\sqrt{3}$, depth $\dfrac{4\sqrt{15}}{3}$   **27.** 594 cases, \$456 per case   **29.** $\dfrac{\pi}{2}$   **31.** $\dfrac{2}{\pi + 1}$   **33.** 2 m from the weak light source.

**35.** minimum length: 28.359 cm, maximum length: 42.758 cm

**37. (a)** Cut 19.603 cm for a square and 15.397 cm for a circle.

**(b)** Cut 0 cm for a square and 35 cm for a circle.

**(c)**

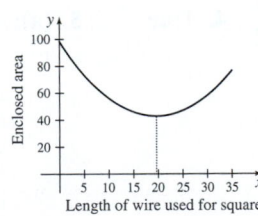

**39.** SSM.   **41.** $6\sqrt{3}$   **43.** $y = -x\sqrt[3]{\dfrac{b}{a}} + b + a\sqrt[3]{\dfrac{b}{a}}$   **45.** SSM.

**47.** Observer should stand approximately 4.583 m from the wall.   **49.** SSM.   **51.** 20 ft   **53.** $(-3.758, 6.547)$

## Section 4.8   (SSM = See Student Solutions Manual; AWV = Answers will vary)

**1.** Antiderivative   **2.** True   **3.** $\ln|x| + C$   **4.** True   **5.** False   **6.** True   **7.** True   **8.** True   **9.** $2x + C$   **11.** $\dfrac{2}{3}x^6 + C$

**13.** $2x^{5/2} + C$  **15.** $-\dfrac{2}{x} + C$   **17.** $\dfrac{2}{3}x^{3/2} + C$   **19.** $x^4 - x^3 + x + C$   **21.** $3x^3 - 6x^2 + 4x + C$   **23.** $3x - 2\ln|x| + C$

**25.** $6x^{1/2} - 4\ln|x| + C$   **27.** $x^2 - 3\sin x + C$   **29.** $4e^x + \dfrac{1}{2}x^2 + C$   **31.** $7\tan^{-1} x + C$   **33.** $e^x + \sec^{-1} x + C$

**35.** $3\cosh x + C$   **37.** $y = x^3 - x^2 + x - 5$   **39.** $y = \dfrac{3}{4}x^{4/3} + \dfrac{2}{5}x^{5/2} - 2x + \dfrac{57}{20}$   **41.** $s = \dfrac{1}{4}t^4 - \dfrac{1}{t} + \dfrac{11}{4}$

**43.** $f(x) = \dfrac{1}{2}x^2 + 2\cos x + 2 - \dfrac{1}{2}\pi^2$   **45.** $y = e^x - x + 1$   **47.** $s(t) = -16t^2 + 128t$   **49.** $s(t) = \dfrac{1}{2}t^3 + 18t + 2$

**51.** $s(t) = 6t - \sin t$   **53.** $\dfrac{1}{6}u^2 + \dfrac{7}{3}u + C$   **55.** $f(t) = \dfrac{1}{4}t^4 + 3t - \ln|t| - 3$   **57.** $F(x) = x\cos x + \sin x + 1$

**59.** $v = 10\sqrt{3}$ m/s $\approx 38.74$ mi/h   **61.** 220 ft   **63.** 13.133 m/s   **65.** 8 N   **67.** $\approx 16.759$ m/s   **69.** 15.00625 m

**71.** $x^2 + 3\tan^{-1}x + C$   **73.** (a) SSM.   (b) AWV.   (c) $I(x) = I_0 e^{-kx}$   (d) $-\dfrac{1}{2}\ln 0.1$ cm$^{-1}$

## Review Exercises   (SSM = See Student Solutions Manual; AWV = Answers will vary)

**1.** $-\dfrac{4}{5}$ cm$^2$/min   **3.** 318.953 mph   **5.** $(-8, -9)$ is the absolute minimum, $(-5, 0)$ is neither, $(-2, 9)$ is a local

maximum and the absolute maximum, $(1, 0)$ is a local minimum, $(3, 4)$ is a local maximum, $(5, 0)$ is neither   **7.** $0, \dfrac{\pi}{2}, \pi$

**9.** Absolute maximum value of 21 at $x = -2$ and absolute minimum value of $-11$ at $x = 2$   **11.** $\left(2, \dfrac{3}{2}\right)$

**13.** (a) Local maximum $\dfrac{203}{27}$ at $x = -\dfrac{4}{3}$ and local minimum $-11$ at $x = 2$   (b) Same as (a)

**15.** (a) Local maximum $16e^{-4}$ at $x = 2$ and local minimum $0$ at $x = 0$   (b) Same as (a)

**17.** 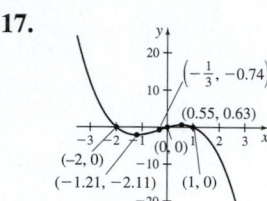   **19.** 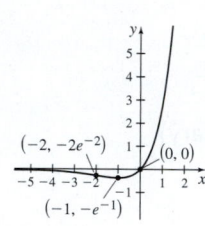   **21.** 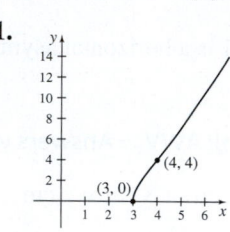   **23.** (a) Increasing on $(-1.207, \infty)$ and
decreasing on $(-\infty, -1.207)$
(b) Concave up on $(-\infty, \infty)$
(c) No point of inflection

**25.** (B)   **27.** AWV.   **29.** $10 - 2\sqrt{7}$ in.   **31.** $F(x) = C$   **33.** $F(x) = \sin x + C$   **35.** $F(x) = 2\ln|x| + C$

**37.** $F(x) = x^4 - 3x^3 + 5x^2 - 3x + C$   **39.** $\dfrac{31}{5}$ cm/s   **41.** Yes, $\dfrac{0}{0}$   **43.** Yes, $\infty - \infty$   **45.** 9   **47.** 2   **49.** 1   **51.** $\dfrac{1}{6}$   **53.** 1

**55.** $\dfrac{8}{9}$   **57.** $y = e^x + 1$   **59.** $y = 2\ln|x| + 4$   **61.** 1075 items   **63.** $\dfrac{1}{e}$

## Chapter 5

### Section 5.1   (SSM = See Student Solutions Manual; AWV = Answers will vary)

**1.** AWV.   **2.** False   **3.** $\dfrac{1}{2}$   **4.** True   **5.** (a) Area is $\dfrac{35}{4}$.   (b) Area is $\dfrac{37}{4}$.   (c) $\dfrac{35}{4} < 9 < \dfrac{37}{4}$

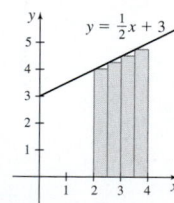   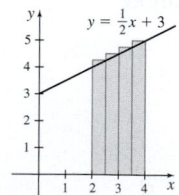

**7.** (a) 3   (b) 6   **9.** $[1, 2], [2, 3], [3, 4]$

**11.** $\left[-1, -\dfrac{1}{2}\right], \left[-\dfrac{1}{2}, 0\right], \left[0, \dfrac{1}{2}\right], \left[\dfrac{1}{2}, 1\right], \left[1, \dfrac{3}{2}\right], \left[\dfrac{3}{2}, 2\right], \left[2, \dfrac{5}{2}\right], \left[\dfrac{5}{2}, 3\right], \left[3, \dfrac{7}{2}\right], \left[\dfrac{7}{2}, 4\right]$   **13.** (a) 14   (b) 48

**15.** (a) [graph with $y = x$]   (b) $\left[0, \dfrac{3}{n}\right], \left[\dfrac{3}{n}, 2 \cdot \dfrac{3}{n}\right], \ldots, \left[(n-1) \cdot \dfrac{3}{n}, 3\right]$   (c) SSM.

(d) SSM.   (e) SSM.

**Answers; Section 5.1–5.2 AN-29**

**17. (a)** $s_4 = 40$, $s_8 = 44$  **(b)** $S_4 = 56$, $S_8 = 52$  **19. (a)** $s_4 = 34$, $s_8 = \dfrac{77}{2}$  **(b)** $S_4 = 50$, $S_8 = \dfrac{93}{2}$

**21. (a)** $s_4 = \dfrac{\sqrt{2}}{4}\pi \approx 1.111$, $s_8 \approx 1.582$  **(b)** $S_4 = \dfrac{\sqrt{2}+2}{4}\pi \approx 2.682$, $S_8 \approx 2.367$

**23.** $s_n = \displaystyle\sum_{i=1}^{n}\left(3(i-1)\dfrac{10}{n}\right)\dfrac{10}{n} = 150 - \dfrac{150}{n}$; $\displaystyle\lim_{n\to\infty} s_n = 150$

**25. (a)** $A = \displaystyle\lim_{n\to\infty} s_n = \lim_{n\to\infty}\left(20 - \dfrac{16}{n}\right) = 20$  **(b)** $A = \displaystyle\lim_{n\to\infty} S_n = \lim_{n\to\infty}\left(20 + \dfrac{16}{n}\right) = 20$  **(c)** AWV.

**27. (a)** $A = \displaystyle\lim_{n\to\infty} s_n = \lim_{n\to\infty}\left(24 - \dfrac{24}{n}\right) = 24$  **(b)** $A = \displaystyle\lim_{n\to\infty} S_n = \lim_{n\to\infty}\left(24 + \dfrac{24}{n}\right) = 24$  **(c)** AWV.

**29. (a)** $A = \displaystyle\lim_{n\to\infty} s_n = \lim_{n\to\infty}\left(\dfrac{32}{3} - \dfrac{16}{n} + \dfrac{16}{3n^2}\right) = \dfrac{32}{3}$  **(b)** $A = \displaystyle\lim_{n\to\infty} S_n = \lim_{n\to\infty}\left(\dfrac{32}{3} + \dfrac{16}{n} + \dfrac{16}{3n^2}\right) = \dfrac{32}{3}$  **(c)** AWV.

**31. (a)** $A = \displaystyle\lim_{n\to\infty} s_n = \lim_{n\to\infty}\left(\dfrac{16}{3} - \dfrac{4}{n} - \dfrac{4}{3n^2}\right) = \dfrac{16}{3}$  **(b)** $A = \displaystyle\lim_{n\to\infty} S_n = \lim_{n\to\infty}\left(\dfrac{16}{3} + \dfrac{4}{n} - \dfrac{4}{3n^2}\right) = \dfrac{16}{3}$  **(c)** AWV.

**33.** 10  **35.** 18  **37.** $\dfrac{58}{3}$  **39.** $A \approx 25{,}994$  **41.** $A \approx 1.693$

**43. (a)**

**(b)** $\left[1, 1+\dfrac{3}{n}\right]$, $\left[1+\dfrac{3}{n}, 1+2\cdot\dfrac{3}{n}\right]$, ..., $\left[1+(n-1)\cdot\dfrac{3}{n}, 4\right]$.

**(c)** SSM.
**(d)** SSM.
**(e)**

| $n$ | 5 | 10 | 50 | 100 |
|---|---|---|---|---|
| $s_n$ | 4.754 | 5.123 | 5.456 | 5.500 |
| $S_n$ | 6.554 | 6.023 | 5.636 | 5.590 |

**(f)** $5.500 \le A \le 5.590$

**45.** SSM.  **47.** SSM.

## Section 5.2  (SSM = See Student Solutions Manual; AWV = Answers will vary)

**1.** False  **2.** (c) 2  **3.** True  **4.** Lower limit of integration; upper limit of integration; integral sign; integrand  **5.** 0

**6.** False  **7.** True  **8.** (b) 10  **9.** $\dfrac{13}{8}$  **11.** 80  **13.** $7e$  **15.** $3\pi$  **17.** 0  **19.** $\int_0^2 (e^x + 2)\, dx$  **21.** $\int_0^{2\pi} \cos x\, dx$

**23.** $\displaystyle\int_1^4 \dfrac{2}{x^2}\, dx$  **25.** $\int_1^e x\ln x\, dx$  **27.** $\int_2^6 \left(2 + \sqrt{4-(x-4)^2}\right) dx$  **29.** $\int_{-2}^4 [\sin(1.5x) + 3]\, dx$

**31. (a)** $[-4, -1]$, $[-1, 0]$, $[0, 1]$, $[1, 3]$, $[3, 5]$, $[5, 6]$  **(b)** 0  **(c)** 13  **33.** $\displaystyle\lim_{\max \Delta x_i \to 0} \sum_{i=1}^{n} \sin u_i \Delta x_i$ on $[0, \pi]$

**35.** $\displaystyle\lim_{\max \Delta x_i \to 0} \sum_{i=1}^{n} (u_i - 2)^{1/3} \Delta x_i$ on $[1, 4]$  **37.** $\displaystyle\lim_{\max \Delta x_i \to 0} \sum_{i=1}^{n} (|u_i| - 2)\Delta x_i$ on $[1, 4]$  **39. (a)** 80  **(b)** 172  **(c)** $\approx 138.667$

**41. (a)** 56  **(b)** 40  **(c)** $\approx 42.667$  **43. (a)** $\pi$  **(b)** $\pi$  **(c)** $\approx 3.142$  **45. (a)** $\approx 932.832$  **(b)** $\approx 4714.224$  **(c)** $\approx 2979.958$

**47. (a)**

| $n$ | 10 | 50 | 100 |
|---|---|---|---|
| Left | 14.536 | 14.737 | 14.762 |
| Right | 15.030 | 14.836 | 14.812 |
| Mid | 14.789 | 14.787 | 14.787 |

**49. (a)**

| $n$ | 10 | 50 | 100 |
|---|---|---|---|
| Left | 4.702 | 4.712 | 4.712 |
| Right | 4.702 | 4.712 | 4.712 |
| Mid | 4.717 | 4.713 | 4.712 |

**(b)** $\approx 14.787$

**(b)** $\dfrac{3\pi}{2} \approx 4.712$

**51. (a)** $-\dfrac{7}{2}$  **(b)** $\displaystyle\int_0^1 (x-4)\, dx$ does not represent area.  **53. (a)** 6  **(b)** $\displaystyle\int_0^2 (2x+1)\, dx$ could represent area.

**55. (a)** 72  **(b)** 56  **(c)** 64  **(d)** $\int_{-3}^5 (2x+6)\, dx$; 64

**57. (a)** 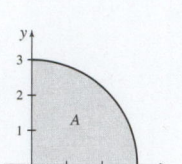　**(b)** $A = \int_0^3 \sqrt{9 - x^2}\, dx$　**(c)** $A \approx 7.069$　**(d)** $A = \dfrac{9\pi}{4} \approx 7.069$

**59. (a)** 　**(b)** $A = \int_0^6 (3 - \sqrt{6x - x^2})\, dx$　**(c)** $A \approx 3.863$　**(d)** $A = 18 - \dfrac{9\pi}{2} \approx 3.863$

**61.** $\dfrac{4448}{6435} \approx 0.691$　**63. (a)** Using a Left Riemann sum: 22　**(b)** meters/second　**(c)** AWV.

**65. (a)** Using a Right Riemann sum: 46　**(b)** millions of dollars　**(c)** AWV.　**67.** newton-meters　**69.** meters

**71.** $\int_0^1 x\, dx$　**73.** $\int_0^3 \sqrt{x}\, dx$　**75.** SSM.　**77.** SSM.

## Section 5.3　(SSM = See Student Solutions Manual; AWV = Answers will vary)

**1.** $f(x)$　**2.** False　**3.** False　**4.** False　**5.** $\sqrt{x^2 + 1}$　**7.** $(3 + t^2)^{3/2}$　**9.** $\ln x$　**11.** $6x^2 \sqrt{4x^6 + 1}$

**13.** $5x^4 \sec(x^5)$　**15.** $-\sin(x^2)$　**17.** $-10x(30x^2)^{2/3}$　**19.** 5　**21.** $\dfrac{15}{4}$　**23.** $\dfrac{2}{3}$　**25.** $\sqrt{3}$

**27.** $\sqrt{2} - 1$　**29.** $\dfrac{e - 1}{e}$　**31.** 1　**33.** $\dfrac{\pi}{4}$　**35.** $\dfrac{99}{5}$　**37.** $-\dfrac{\sqrt{2}}{2}$　**39.** $\dfrac{15}{4}$　**41.** $e^2 - 1$

**43. (a)** $F(x) = \int_0^x \cos t\, dt$　**(b)** $F\left(\dfrac{\pi}{6}\right) = \dfrac{1}{2}; F\left(\dfrac{\pi}{2}\right) = 1; F\left(\dfrac{4\pi}{3}\right) = -\dfrac{\sqrt{3}}{2}$　**(c)**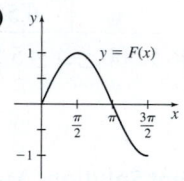

**45. (a)** $F(x) = \int_{-2}^x t^3\, dt$　**(b)** $F(-1) = -\dfrac{15}{4}; F(0) = -4; F(3) = \dfrac{65}{4}$　**(c)**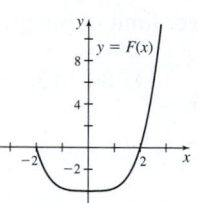

**47. (a)** $F(x) = \int_{-1}^x (4 - t^2)\, dt$　**(b)** $F(0) = \dfrac{11}{3}; F(2) = 9; F(3) = \dfrac{20}{3}$　**(c)**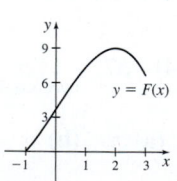

**49. (a)** $F(x) = \int_0^x (\sqrt{t} - 2)\, dt$　**(b)** $F(1) = -\dfrac{4}{3}; F(4) = -\dfrac{8}{3}; F(9) = 0$　**(c)**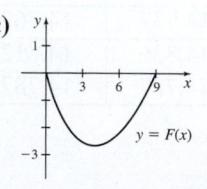

**51.** 160　**53.** $\dfrac{\pi}{6}$　**55.** $\dfrac{\pi}{3}$　**57.** $2\sqrt{r} - 2$, which goes to infinity as $r \to \infty$　**59.** The corporation has \$23 million in sales over two years.　**61. (a)** $\int_a^b H(t)\, dt$　**(b)** $cm^3$　**(c)** 100 $cm^3$ of helium leaked out in the first 5 h.

**63. (a)** 44.740 g  **(b)** AWV.  **(c)** 59.740 g  **65. (a)** AWV.  **(b)** $\dfrac{3}{4}[50^{4/3}-40^{4/3}]\approx 35.553$ hundreds

of dollars in additional revenue  **67. (a)** $\int_0^{5.2} 9.8t\,dt$  **(b)** 132.5 m  **69.** $P(x)=3x^2+6x+6$

**71. (a)** $\dfrac{64}{3}$ and $\dfrac{128}{3}$  **(b)** Even  **(c)** AWV.  **73. (a)** 1 and 2  **(b)** Even  **(c)** AWV.

**75.** $\dfrac{1}{\sqrt[3]{2}}$  **77.** $F'(c)$  **79.** $e^4-1$  **81.** 4  **83.** SSM.

**85. (a)** SSM.  **(b)** SSM.  **(c)** AWV.  **87.** $a=\dfrac{\sqrt{5}-1}{2}$

**89. (a)** $\dfrac{\cos x}{\sqrt{1-\sin^2 x}}=\dfrac{\cos x}{|\cos x|}$  **(b)** Yes  **(c)** Yes

## Section 5.4   (SSM = See Student Solutions Manual; AWV = Answers will vary)

**1.** True  **2.** False  **3.** True  **4.** −2  **5.** Average value  **6.** False  **7.** 7  **9.** 31  **11.** 17  **13.** 45  **15.** 139  **17.** $-\dfrac{1}{15}$

**19.** 4  **21.** $\dfrac{\pi}{2}$  **23.** $-\dfrac{76}{3}$  **25.** $\dfrac{5}{3}$  **27.** −3  **29.** $e+\dfrac{1}{e}-2$  **31.** $\dfrac{1}{2}+2\ln\dfrac{3}{2}$  **33.** $\cosh 2-\cosh 1+1$  **35.** $\dfrac{\pi+15}{6}$  **37.** $\dfrac{10}{3}$

**39.** $-\dfrac{2}{3}$  **41.** $\dfrac{1}{24}(\pi^3-3\pi^2+3\pi\sqrt{2}+24-12\sqrt{2})$  **43.** $\dfrac{679}{32}$  **45.** $\dfrac{131}{4}$  **47.** SSM.

**49.** SSM.  **51.** 12 to 32  **53.** $\dfrac{\sqrt{2}\,\pi}{8}$ to $\dfrac{\pi}{4}$  **55.** 1 to $\sqrt{2}$  **57.** 1 to $e$  **59.** $u=\sqrt{3}$  **61.** $u=\dfrac{4\sqrt{3}}{3}$

**63.** $u=\dfrac{\pi}{2},\dfrac{3\pi}{2}$  **65.** $e-1$  **67.** $\dfrac{3}{5}$  **69.** $\dfrac{2}{\pi}$  **71.** $\dfrac{2}{3}$  **73.** $\dfrac{2}{\pi}(e^{\pi/2}-2)$  **75. (a)** 12  **(b)** 4  **(c)** AWV.

**77. (a)** $e^2+3-\dfrac{1}{e}$  **(b)** $\dfrac{1}{3}\left(e^2+3-\dfrac{1}{e}\right)$  **(c)** AWV.  **79. (a)** 54 m  **(b)** 74 m  **81. (a)** $\dfrac{235}{12}$ m  **(b)** $\dfrac{395}{12}$ m

**83.** $8\sqrt{4x^2+1}-\sqrt{x^2+4}$  **85.** $(9x^2-4x)\ln x$  **87.** 9  **89.** $\dfrac{13}{3}$  **91.** 37.5 °C  **93. (a)** $\dfrac{1875}{16}=117.1875$ N  **(b)** 4 m

**95.** $\approx 1000.345$ kg/m³  **97.** $\dfrac{13\pi}{3}$ m²  **99. (a)** 2  **(b)** $\dfrac{1}{2}bh=2$  **101.** 12 m/s  **103. (a)** $\dfrac{f(b)-f(a)}{b-a}$  **(b)** AWV.

**105.** SSM.  **107.** $k=\dfrac{\pi}{6}$  **109.** $\approx 8.296$  **111.** Only III need not be true.  **113.** $\dfrac{1}{6}$  **115. (a)** $\dfrac{3}{2}g$  **(b)** $2g$

**117. (a)** $(-\infty,\infty)$  **(b)** $[0,\infty)$  **(c)** $(-\infty,\infty)$  **(d)** All $x\neq 1$  **(e)** $\dfrac{1}{6}$  **119.** $F(x)=\dfrac{x^2}{4}$  **121.** $\dfrac{1}{2}$

**123.** SSM.  **125. (a)** SSM.  **(b)** AWV.

## Section 5.5   (SSM = See Student Solutions Manual; AWV = Answers will vary)

**1.** $f(x)$  **2.** False  **3.** $\dfrac{x^{a+1}}{a+1}+C$  **4.** True  **5.** $\dfrac{1}{2}\sin u$  **6.** False  **7.** (c)  **8.** True  **9.** $\dfrac{3}{5}x^{5/3}+C$  **11.** $\sin^{-1}x+C$

**13.** $\dfrac{5}{2}x^2+2e^x-\ln|x|+C$  **15.** $\sec x+C$  **17.** $\dfrac{2}{5}\sin^{-1}x+C$  **19.** $\cosh t+C$  **21.** $\dfrac{1}{3}e^{3x+1}+C$  **23.** $-\dfrac{1}{14}(1-t^2)^7+C$

**25.** $\dfrac{1}{3}\sin^{-1}x^3+C$  **27.** $-\dfrac{1}{3}\cos(3x)+C$  **29.** $-\dfrac{1}{3}\cos^3 x+C$  **31.** $-e^{1/x}+C$  **33.** $\dfrac{1}{2}\ln|x^2-1|+C$  **35.** $2\sqrt{1+e^x}+C$

**37.** $-\dfrac{2}{3(\sqrt{x}+1)^3}+C$  **39.** $4(e^x-1)^{3/4}+C$  **41.** $\dfrac{1}{2}\ln|2\sin x-1|+C$  **43.** $\dfrac{1}{5}\ln|\sec(5x)+\tan(5x)|+C$  **45.** $\dfrac{2}{3}(\tan x)^{3/2}+C$

**47.** $\sec x+C$  **49.** $-e^{\cos x}+C$  **51.** $\dfrac{2}{5}(x+3)^{5/2}-2(x+3)^{3/2}+C$  **53.** $\dfrac{1}{3}\sin(3x)-\cos x+C$  **55.** $\dfrac{1}{5}\tan^{-1}\dfrac{x}{5}+C$

**57.** $\sin^{-1}\dfrac{x}{3}+C$  **59.** $\dfrac{1}{2}\cosh^2 x+C$  **61. (a) (b)** $-\dfrac{2}{21}$  **(c)** AWV.  **63. (a) (b)** $\dfrac{1}{3}(e^2-e)$  **(c)** AWV.

**65. (a) (b)** $\dfrac{232}{5}$  **(c)** AWV.  **67. (a) (b)** $\dfrac{1640}{\ln 3}$  **(c)** AWV.  **69.** $\left(\dfrac{2}{3}\right)^{5/2}$  **71.** $\dfrac{1}{2}\ln\dfrac{e^4+1}{2}$  **73.** $\ln\left(\dfrac{\ln 3}{\ln 2}\right)$

**75.** $\sin e^{\pi} - \sin 1$   **77.** $\dfrac{\pi}{8}$   **79.** $-\dfrac{32}{3}$   **81.** 0   **83.** $\dfrac{3\pi}{2}$   **85.** 50   **87.** 2   **89.** 4   **91. (a)** $\dfrac{1}{3}(x+1)^3 + C$   **(b)** $\dfrac{1}{3}x^3 + x^2 + x + C$

**93. (a)** 0   **(b)** 0   **(c)** 0   **(d)** SSM.   **95.** $\dfrac{124}{15}$   **97.** $\dfrac{\sqrt{3}\pi}{9}$   **99.** $a^2 \sinh 1 + a(b-a)$   **101.** $\dfrac{2\ln 2}{\pi}$

**103.** 8   **105.** 4   **107.** $(\ln 10)\ln|\ln x| + C$   **109.** $(\ln 10)\ln 2$   **111.** $(\ln 2)\ln 3$   **113.** $b = \sqrt[3]{2}$   **115.** $\dfrac{1}{2}\ln(x^2+1) + \tan^{-1}x + C$

**117.** $\dfrac{1}{14}(2\sqrt{x^2+3} - \dfrac{4}{x} + 9)^7 + C$   **119.** $\dfrac{2}{3}x^{3/2} + \dfrac{8}{7}x^{7/2} + C$   **121.** $\dfrac{4}{9}(4 + t^{3/2})^{3/2} + C$   **123.** $\dfrac{3^{2x+1}}{2\ln 3} + C$

**125.** $-\sin^{-1}\dfrac{\cos x}{2} + C$   **127.** $\dfrac{1}{6}\ln\dfrac{187}{27}$   **129.** SSM.   **131.** SSM.

**133. (a)** 5   **(b)** 5   **(c)** 5   **(d)** $a=2, b=4$   **(e)** $a=k, b=k+2$   **135.** Hint: Split the integral at 0; SSM.

**137.** $c=1$   **139.** SSM.   **141.** $\dfrac{n}{b(n+1)}(a+bx)^{(n+1)/n} + C$   **143. (a)** 2   **(b)** $\dfrac{\pi^2}{4}$   **145.** SSM.   **147.** $V(z) = \dfrac{Q}{\sqrt{R^2 + z^2}} + C$

**149. (a)** $v(t) = \dfrac{mg}{k}(1 - e^{-kt/m})$   **(b)** $s(t) = \dfrac{mg}{k}\left[\dfrac{m}{k}(e^{-kt/m} - 1) + t\right]$   **(c)** As $t \to \infty$, $a \to 0$, $v \to \dfrac{mg}{k}$, $s \to \infty$

**(d)** SSM.   **151.** $\dfrac{x^3}{3} + x - \dfrac{1}{2(x^2+1)} + C$   **153.** $6\sqrt{x+1} + 2\ln|\sqrt{x+1} - 1| - 2\ln|\sqrt{x+1} + 1| + C$

**155. (a)** $y' = \sec x$   **(b)** SSM.   **(c)** SSM.   **157. (a)** $y' = \sqrt{a^2 - x^2}$   **(b)** SSM.   **159. (a)** $y = \tan^{-1}(\sinh x)$   **(b)–(d)** SSM.

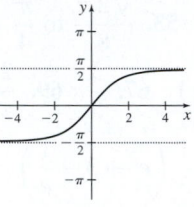

## Section 5.6

**1.** True   **2.** False   **3.** $\ln|y| = \dfrac{x^2}{2} + C$   **5.** $\ln|y| = \dfrac{2x^{3/2}}{3} + C$   **7.** $e^{-y} = -\dfrac{x^2}{2} + C$   **9.** $\ln|\csc(2y) - \cot(2y)| = 2e^x + C$

**11.** $e^y = \dfrac{5}{2}e^{2x} + C$   **13.** $\tan^{-1}y = \dfrac{x^3}{3} + C$   **15.** $y = \dfrac{x^4}{16}$   **17.** $|y-1| = |x-1|$   **19.** $\sin y = \dfrac{x^2}{2} - 2$   **21.** $y^2 = 8e^x - 4$

**23.** $\approx 6944$ mosquitoes   **25. (a)** $\dfrac{dP}{dt} = kP$   **(b)** $P = P_0 e^{kt}$   **(c)** $P = 4000e^{\frac{\ln 2}{18}t}$   **(d)** $\approx 25{,}398$ people

**27. (a)** $\dfrac{dA}{dt} = kA$   **(b)** $A = A_0 e^{kt}$   **(c)** $A = 8e^{-\frac{\ln 2}{1690}t}$   **(d)** $\approx 7.679$ g

**29. (a)** $\dfrac{dP}{dt} = 0.00289P$   **(b)** $P = P(t) = P_0 e^{0.00289t}$   **(c)** $P(t) = (1.251 \times 10^9)e^{0.00289t}$, where $t$ is the number of years since 2015.

**(d)** If the population continues to follow this model, the projected population in 2025 will be approximately $1.288 \times 10^9$ people.

**31.** $\approx 52.67\%$ remains   **33.** $\approx 0.262$ mol and $\approx 394.7$ min   **35.** $V = 20e^{-\frac{\ln 2}{2}t}$ and $t = 4$ s

**37. (a)** $\dfrac{du}{dt} = k[u(t) - 30], k > 0$   **(b)** $u = -26e^{kt} + 30$   **(c)** $\approx 16.507\,°\text{C}$

**39. (a)** $\dfrac{du}{dt} = -0.159[u(t) - 12]$   **(b)** $u = (u_0 - 12)e^{-0.159t} + 12$   **(c)** $u = 25e^{-0.159t} + 12$   **(d)** $\approx 8{:}50$ a.m.

**(e)** It takes about 13 h, 20 min to cool to $15\,°\text{C}$.

## Review Exercises   (SSM = See Student Solutions Manual; AWV = Answers will vary)

**1.** $s_4 = 16$, $S_4 = 24$, $s_8 = 18$, $S_8 = 22$   **3.** $A = \lim\limits_{n \to \infty} s_n = 18$

**5. (a)** 14   **(b)** $\lim\limits_{n \to \infty} \sum\limits_{i=1}^{n} \dfrac{4}{n}\left[\left(-1 + \dfrac{4i}{n}\right)^2 - 3\left(-1 + \dfrac{4i}{n}\right) + 3\right] = \displaystyle\int_{-1}^{3}(x^2 - 3x + 3)\,dx$   **(c)** $\dfrac{28}{3}$   **(d)** $\dfrac{28}{3}$   **7.** $x^{2/3}\sin x$

**9.** $-2x \tan x^2$   **11.** $1 - \dfrac{\sqrt{2}}{2}$   **13.** $\dfrac{\pi}{4}$   **15.** 4   **17.** $\dfrac{14}{\ln 2}$   **19.** $2e^x + \ln|x| + C$   **21.** The train traveled a distance of 460 km in 16 h.

**23.** $\dfrac{22}{3}$   **25.** 0   **27.** $2 \le \int_0^2 e^{x^2}\, dx \le 2e^4$   **29.** $\sin^{-1}\dfrac{2}{\pi}$ and $\pi - \sin^{-1}\dfrac{2}{\pi}$   **31.** 0   **33.** $\dfrac{e^2-1}{2e}$   **35.** $\sqrt{\dfrac{1}{1+4x^2}}$

**37.** $-\dfrac{1}{y-2} - \dfrac{1}{(y-2)^2} + C$   **39.** $-\dfrac{2}{3}\left(\dfrac{1+x}{x}\right)^{3/2} + C$   **41.** $\dfrac{1}{12}\left(\dfrac{7}{8}\right)^4$   **43.** $-\ln|1 - 2\sqrt{x}| + C$   **45.** $\dfrac{99}{10\ln 10}$   **47.** $\dfrac{\pi}{8}$

**49.** $\dfrac{1}{4}(x^3 + 3\cos x)^{4/3} + C$   **51.** $\dfrac{2}{3}$   **53.** 5   **55.** (a) Using a Left Riemann sum: 91.5   (b) square meters   (c) AWV.

**57.** 15,552 gallons   **59.** (a) $\dfrac{dA}{dt} = kA, k < 0$   (b) $A = A_0 e^{kt}$   (c) $A = 10 e^{-\frac{\ln 2}{1690}t}$   (d) $\approx 8.842$ g   **61.** $y = 4 e^{3x^2/2}$

**63.** 18   **65.** AWV.   **67.** $\displaystyle\int_0^3 x^2 dx$

## Chapter 6

### Section 6.1   (SSM = See Student Solutions Manual; AWV = Answers will vary)

**1.** $\int_0^1 \left(\sqrt{x} - x^2\right) dx$   **2.** $\int_{-1}^1 \left(1 - y^2\right) dy$   **3.** $\dfrac{1}{2}$   **5.** $\dfrac{1}{6}$   **7.** $\dfrac{1}{2}$   **9.** $\dfrac{4}{15}$   **11.** $\dfrac{\sqrt{3}}{2} - \dfrac{\pi}{6}$   **13.** $\dfrac{9}{2}$   **15.** $\dfrac{32}{3}$   **17.** $\dfrac{9}{2}$

**19.** $e^2 - 3$   **21.** $2\sqrt{2}$   **23.** $\dfrac{32}{3}$   **25.** (a) $\int_0^1 \left(\sqrt{x} - x^3\right) dx = \dfrac{5}{12}$   (b) $\int_0^1 \left(\sqrt[3]{y} - y^2\right) dy = \dfrac{5}{12}$

**27.** (a) $\int_0^1 \left((x+1) - (x^2+1)\right) dx = \dfrac{1}{6}$   (b) $\int_1^2 \left(\sqrt{y-1} - (y-1)\right) dy = \dfrac{1}{6}$

**29.** (a) $\int_0^3 \left(\sqrt{9-x} - \sqrt{9-3x}\right) dx + \int_3^9 \sqrt{9-x}\, dx = 12$   (b) $\int_0^3 \left((9 - y^2) - \left(3 - \dfrac{y^2}{3}\right)\right) dy = 12$

**31.** (a) $\int_2^3 \sqrt{x-2}\, dx + \int_3^4 \left(\sqrt{x-2} - \sqrt{2x-6}\right) dx = \dfrac{2\sqrt{2}}{3}$   (b) $\int_0^{\sqrt{2}} \left(\left(\dfrac{y^2}{2} + 3\right) - (y^2 + 2)\right) dy = \dfrac{2\sqrt{2}}{3}$

**33.** $\dfrac{16\sqrt{2}}{3}$   **35.** $\dfrac{64}{3}$   **37.** 2   **39.** $\dfrac{\sqrt{3}}{2} - \dfrac{\pi}{6}$   **41.** $\dfrac{e^2-3}{2}$   **43.** $\dfrac{125}{24}$   **45.** $1 - \dfrac{\pi}{4}$   **47.** SSM.

**49.** (a) $\left(-\dfrac{\pi}{4}, \dfrac{1}{2}\right), \left(\dfrac{\pi}{4}, \dfrac{1}{2}\right)$   (b) $\dfrac{1}{2}$   **51.** $\dfrac{\sqrt{3}}{2} + \dfrac{\pi}{12} - \dfrac{9}{8}$

**53.** (a)   (b) $(0, 0), \approx (0.489, 0.511), (1, 0)$   (c) $A \approx 0.253$   **55.** (a)   (b) $(0, 1), (1, 1)$ and $\approx (1.317, 0.366)$   (c) $A \approx 0.295$

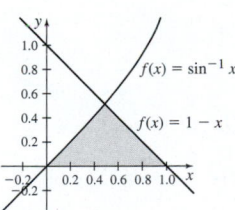

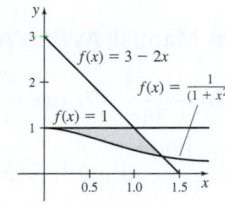

**57.** (a)   (b) $(0, 0), (1, 0)$ and $\approx (0.599, 0.642)$   (c) $A \approx 0.325$

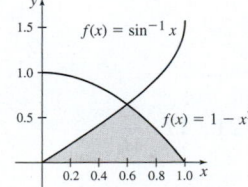

**59.** (a) $A(k) = k\tan^{-1} k - \ln\sec(\tan^{-1} k)$   (b) $A(1) = \dfrac{\pi}{4} - \dfrac{\ln 2}{2}$   (c) $\dfrac{\pi}{40}$   **61.** $\dfrac{4}{3}$

### Section 6.2   (SSM = See Student Solutions Manual; AWV = Answers will vary)

**1.** $\pi \int_a^b [f(x)]^2\, dx$   **2.** False   **3.** False   **4.** False   **5.** $V = 30\pi$   **7.** $\dfrac{3\pi}{4}$   **9.** $2\pi(\tan 1 - 1)$   **11.** $\dfrac{4\pi}{5}$   **13.** $\dfrac{1}{2}\pi\left(1 - \dfrac{1}{e^4}\right)$

**15.** $\dfrac{15\pi}{2}$   **17.** $\dfrac{128\pi}{5}$   **19.** $32\pi$   **21.** $\dfrac{2\pi}{5}$   **23.** $\dfrac{\pi}{2}$   **25.** $\dfrac{59{,}049\pi}{5}$   **27.** $\dfrac{129\pi}{7}$   **29.** $\dfrac{117\pi}{2}$   **31.** $\dfrac{64\pi}{45}$   **33.** $\dfrac{512\pi}{21}$

**35.** $\left(\dfrac{e^2}{2} - \dfrac{5}{6}\right)\pi$  **37.** $\pi$  **39.** $\left(\dfrac{e^4}{2} + 2e^2 - \dfrac{5}{2}\right)\pi$  **41.** $\dfrac{\pi}{6}$  **43.** $\dfrac{1024\pi}{15}$  **45.** $\dfrac{363\pi}{64}$

**47.** (a) $\dfrac{243}{5}\pi$  (b) $\dfrac{333}{5}\pi$  (c) $\dfrac{657}{5}\pi$  (d) $\dfrac{90a - 243}{5}\pi$  **49.** (a) $8\pi$  (b) $\dfrac{184}{3}\pi$  (c) $\dfrac{136}{3}\pi$  (d) $\dfrac{32b - 24}{3}\pi$

**51.** (a) 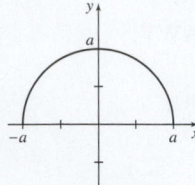  (b) SSM.  **53.** (a) $\pi \displaystyle\int_0^1 \dfrac{1}{(x^2 + 4)^2}\,dx$  (b) $\approx 0.170$

**55.** (a) 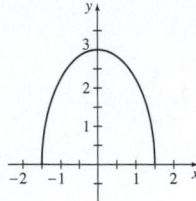  (b) $18\pi$  **57.** (a) $\dfrac{k^5\pi}{30}$  (b) $\dfrac{k^3}{6}$  **59.** $\left(2 - \dfrac{\pi}{4}\right)\pi$

**61.** (a) $k = \dfrac{e^{-b^2}}{b^2}$, $P(x) = \dfrac{e^{-b^2}}{b^2}x^2$  (b) $V = \dfrac{b^2 e^{-b^2}\pi}{2}$  (c) $b = 1$; AWV.

## Section 6.3  (SSM = See Student Solutions Manual; AWV = Answers will vary)

**1.** False  **2.** False  **3.** False  **4.** True  **5.** $\dfrac{3\pi}{2}$  **7.** $\dfrac{3\pi}{10}$  **9.** $\dfrac{768\pi}{7}$  **11.** $\dfrac{4\pi}{5}$  **13.** $\dfrac{2\pi}{15}$  **15.** $\pi\left(1 - \dfrac{1}{e^4}\right)$  **17.** $16\pi$  **19.** $\dfrac{17\pi}{6}$

**21.** $\dfrac{\pi}{2}$  **23.** (a) $\dfrac{128\pi}{105}$  (b) $\dfrac{16\pi}{5}$  (c) AWV.  **25.** (a) $\dfrac{206\pi}{15}$  (b) $12\pi$  (c) AWV.  **27.** $\dfrac{64\pi}{5}$

**29.** $\dfrac{4\pi}{15}$  **31.** $\dfrac{308\pi}{3}$  **33.** $\dfrac{80\pi}{3}$  **35.** $\dfrac{\pi}{12}$  **37.** $\dfrac{1177\pi}{10}$  **39.** $\dfrac{\pi}{2}$  **41.** (a) $\dfrac{16}{15}\pi$  (b) $\dfrac{8\pi}{3}$  (c) $\dfrac{16\pi}{3}$  (d) $\dfrac{8}{5}\pi$  **43.** $f(x) = \dfrac{x^2 + 2x}{\sqrt{\pi}}$

**45.** SSM.  **47.** (a) $2\pi \int_0^{\pi/2} x \cos x\,dx$ or $\pi \int_0^1 (\cos^{-1} y)^2\,dy$  (b) $\pi(\pi - 2) \approx 3.586$  **49.** SSM.  **51.** SSM.

## Section 6.4  (SSM = See Student Solutions Manual; AWV = Answers will vary)

**1.** AWV.  **2.** False  **3.** $\dfrac{128}{3}$  **5.** (a) $\dfrac{9}{35}\pi$  (b) $\dfrac{18\sqrt{3}}{35}$  **7.** (a) $\dfrac{32}{5}$  (b) $\dfrac{4}{5}\pi$  (c) $\dfrac{8\sqrt{3}}{5}$

**9.** (a) $\dfrac{729}{70}$  (b) $\dfrac{729}{560}\pi$  (c) $\dfrac{729\sqrt{3}}{280}$  **11.** $\dfrac{256{,}000}{3}$ m$^3$  **13.** SSM.  **15.** $\dfrac{37}{1320}\pi$  **17.** $\approx 3.758$

**19.** $\dfrac{2}{3}hr^2$, where $h$ is the height of the glass and $r$ is its radius  **21.** $\left(\dfrac{500}{3} - 28\sqrt{21}\right)\pi$  **23.** $\dfrac{\pi}{3}bah$

## Section 6.5  (SSM = See Student Solutions Manual; AWV = Answers will vary)

**1.** False  **2.** True  **3.** $2\sqrt{10}$  **5.** $\sqrt{13}$  **7.** $\dfrac{80\sqrt{10} - 13\sqrt{13}}{27}$  **9.** $\dfrac{80\sqrt{10} - 8}{27}$  **11.** $\dfrac{4\sqrt{2} - 2}{3}$  **13.** 45  **15.** 21  **17.** $\dfrac{33}{16}$

**19.** $\ln\dfrac{\sqrt{3}}{3} - \ln(2 - \sqrt{3})$  **21.** $\dfrac{79}{2}$  **23.** $\dfrac{13\sqrt{13} - 8}{27}$  **25.** $\dfrac{20\sqrt{10} - 2}{27}$  **27.** (a) $\displaystyle\int_0^2 \sqrt{1 + (2x)^2}\,dx$  (b) $s \approx 4.647$

**29.** (a) $\displaystyle\int_0^4 \dfrac{5}{\sqrt{25 - x^2}}\,dx$  (b) $s \approx 4.636$  **31.** (a) $\displaystyle\int_0^{\pi/2} \sqrt{1 + \cos^2 x}\,dx$  (b) $s \approx 1.910$

**33.** $32\sqrt{10}\,\pi$  **35.** $\dfrac{\sqrt{3}}{6}\pi(11\sqrt{11} - 7\sqrt{7})$  **37.** $\dfrac{8}{3}\pi(5\sqrt{5} - 2\sqrt{2})$  **39.** $\dfrac{\pi}{27}(145^{3/2} - 1)$  **41.** $4\pi$

**43. (a)** $S = 2\pi \int_1^3 x^2\sqrt{1+4x^2}\, dx$ **(b)** $S \approx 257.508$ **45. (a)** $S = 2\pi \int_1^8 x^{2/3}\sqrt{1+\dfrac{4}{9x^{2/3}}}\, dx$ **(b)** $S \approx 126.220$

**47. (a)** $S = 2\pi \int_0^3 e^x \sqrt{1+e^{2x}}\, dx$ **(b)** $S \approx 1273.371$ **49. (a)** $S = 2\pi \int_0^{\pi/2} \sin x \sqrt{1+\cos^2 x}\, dx$ **(b)** $S \approx 7.212$ **51.** 8

**53.** $2(e^6 - 1)$ **55.** $6a$ **57.** $\sqrt{2} + \dfrac{1}{27}(13\sqrt{13}-8)$ **59.** $\dfrac{17}{12}$ **61.** $-\ln(\sqrt{2}-1)$

**63. (a)** $s = \displaystyle\int_0^{2\sqrt{5}/5} \sqrt{\dfrac{1+3y^2}{1-y^2}}\, dy$ **(b)** $\approx 1.519$ **(c)** $P \approx 9.688$ **65.** $\dfrac{\pi}{6}(2\sqrt{2}-1)\,\mathrm{m}^2$

**67. (a)** $b + \dfrac{5}{2\sinh^{-1}\left(\frac{3}{4}\right)}$ **(b)** $\dfrac{15}{\sinh^{-1}\left(\frac{3}{4}\right)}$ **69.** $\dfrac{\pi}{2}[\sinh{(2b)} - \sinh{(2a)}] + \pi(b-a)$

**71. (a)** SSM. **(b)** SSM. **(c)** AWV. **73.** $f(x) = \cosh x$ **75.** AWV.

## Section 6.6

**1.** True **2.** $W = Fx$ **3.** joule; foot-pound **4.** $W = \int_a^b F(x)\, dx$ **5.** Equilibrium **6.** True **7.** True **8.** True

**9.** $\dfrac{825}{2}$ J **11.** 23,520 J **13.** $k = 12$ N/m **15.** $-0.9$ J **17. (a)** $(62.42)(1152)\pi \approx 2.259 \times 10^5$ ft lb

**(b)** $(62.42)(4032)\pi \approx 7.907 \times 10^5$ ft lb **19. (a)** $\dfrac{512}{75}\rho g \approx 1197.534$ J **(b)** $\dfrac{256}{15}\rho g \approx 2993.835$ J **21.** $6.074 \times 10^6$ ft lb

**23.** 8330 J **25.** $-\dfrac{3}{4}$ J **27.** 10 ft **29.** $\dfrac{64}{3}\pi\rho g \approx 6.568 \times 10^5$ J **31.** $36\pi\rho g \approx 1.108 \times 10^6$ J **33. (a)** $18\pi\rho g \approx 5.542 \times 10^5$ J

**(b)** $54\pi\rho g \approx 1.663 \times 10^6$ J **35.** $t \approx 8.5$ m **37.** $6.965 \times 10^9$ J **39.** 475 ft lb **41.** $\dfrac{112}{3}\pi\rho g$ **43.** $2.092 \times 10^5$ in lb

## Section 6.7

**1.** Force **2.** True **3. (b)** **4.** True **5.** $\dfrac{75}{2}\rho g = 367,500$ N **7.** $\dfrac{5}{3}\rho g = \dfrac{49,000}{3}$ N **9.** $36\rho g = 352,800$ N **11.** 22,500 lb

**13.** $\dfrac{5}{6}\rho g = \dfrac{24,500}{3}$ N **15.** $\dfrac{16}{3}\rho g = \dfrac{156,800}{3}$ N **17.** $300,000\,\rho g = 2.940 \times 10^9$ N **19.** $5\pi\rho g \approx 1.578 \times 10^5$ N

**21.** $\dfrac{1}{96}\rho g = \dfrac{1127}{16}$ N

## Section 6.8 (SSM = See Student Solutions Manual; AWV = Answers will vary)

**1. (c)** **2. (c)** **3.** True **4. (a)** **5.** False **6. (a)** **7.** $\bar{x} = \dfrac{58}{7}$ **9.** $\bar{x} = \dfrac{29}{15}$ **11.** $M_y = 20,\ M_x = 24,\ (\bar{x}, \bar{y}) = \left(\dfrac{20}{13}, \dfrac{24}{13}\right)$

**13.** $M_y = 29,\ M_x = 62,\ (\bar{x}, \bar{y}) = \left(\dfrac{29}{15}, \dfrac{62}{15}\right)$ **15.** $\left(\dfrac{7}{8}, \dfrac{19}{8}\right)$ **17.** $\left(\dfrac{9}{4}, \dfrac{27}{10}\right)$ **19.** $\left(2, \dfrac{8}{5}\right)$ **21.** $\left(\dfrac{12}{5}, \dfrac{3}{4}\right)$ **23.** $72\pi^2$

**25.** $\left(1, \dfrac{4+3\pi}{4+\pi}\right)$ **27.** $\left(\dfrac{a}{3}, \dfrac{b}{3}\right)$ **29.** $\left(\dfrac{3a}{4}, \dfrac{3h}{10}\right)$ **31. (a)** $M = \dfrac{kL^2}{2}$ **(b)** $k = \dfrac{2M}{L^2}$ **(c)** $\bar{x} = \dfrac{2}{3}L$

**(d)** AWV. **(e)** AWV. **33.** $\dfrac{22\pi}{3}$ **35.** $\dfrac{28\pi}{3}$ **37.** SSM.

**39.** $\left(0, \dfrac{3}{7}\right)$ **41.** $\left(0, \dfrac{38}{35}\right)$ **43.** $\left(-\dfrac{1}{2}, \dfrac{12}{5}\right)$ **45.** SSM. **47.** $\left(\dfrac{836}{217}, \dfrac{80}{31}\right)$ **49.** $\left(\dfrac{227}{74}, \dfrac{607}{148}\right)$ **51.** $\left(\dfrac{113}{38}, \dfrac{113}{76}\right)$

## Review Exercises

**1.** $4\ln 4 - 3$ **3.** $\dfrac{8}{3}$ **5.** $\dfrac{125}{6}$ **7.** $\dfrac{5\pi}{6}$ **9.** $\dfrac{\pi}{2}$ **11.** $\dfrac{512\pi}{15}$ **13.** $\dfrac{32\pi}{5}$ **15.** $\dfrac{3456\pi}{35}$ **17.** $\dfrac{209}{6}$ **19.** 9 **21.** 4 **23.** $8\pi$ **25.** $\dfrac{10\pi}{3}$

**27.** 345,000 ft lb **29.** $256\pi\rho g \approx 7.882 \times 10^6$ J **31.** 0.3 m **33.** $\dfrac{250}{3}\rho g \approx 6.019 \times 10^5$ N **35.** $\left(\dfrac{27}{5}, \dfrac{9}{8}\right)$ **37.** $72\pi$

**39.** $\dfrac{16\sqrt{2}}{3}\pi(12^{3/2} - 27)$

## Chapter 7

### Section 7.1  (SSM = See Student Solutions Manual; AWV = Answers will vary)

**1.** True   **2.** $uv - \int v\,du$   **3.** $\dfrac{1}{2}xe^{2x} - \dfrac{1}{4}e^{2x} + C$   **5.** $x\sin x + \cos x + C$   **7.** $\dfrac{2}{3}x^{3/2}\ln x - \dfrac{4}{9}x^{3/2} + C$

**9.** $x\cot^{-1}x + \dfrac{1}{2}\ln(x^2+1) + C$   **11.** $x(\ln x)^2 - 2x\ln x + 2x + C$   **13.** $-x^2\cos x + 2x\sin x + 2\cos x + C$

**15.** $\dfrac{1}{2}x\cos x\sin x + \dfrac{1}{4}x^2 - \dfrac{1}{4}\sin^2 x + C$   **17.** $x\cosh x - \sinh x + C$   **19.** $x\cosh^{-1}x - \sqrt{x^2-1} + C$

**21.** $\dfrac{x}{2}[\sin(\ln x) - \cos(\ln x)] + C$   **23.** $x(\ln x)^3 - 3x(\ln x)^2 + 6x\ln x - 6x + C$   **25.** $\dfrac{x^3}{27}[9(\ln x)^2 - 6\ln x + 2] + C$

**27.** $\dfrac{1}{3}x^3\tan^{-1}x - \dfrac{1}{6}x^2 + \dfrac{1}{6}\ln(1+x^2) + C$   **29.** $7^x\left[\dfrac{x}{\ln 7} - \dfrac{1}{(\ln 7)^2}\right] + C$   **31.** $\dfrac{e^{-x}}{5}[2\sin(2x) - \cos(2x)] + C$

**33.** $\dfrac{e^{2x}}{5}(2\sin x - \cos x) + C$   **35.** $-\dfrac{e^\pi + 1}{2}$   **37.** $\dfrac{2}{27}\left(1 - \dfrac{25}{e^6}\right)$   **39.** $\dfrac{\sqrt{2}}{4}\pi - \ln(\sqrt{2}+1)$

**41.** $9\ln 3 - 4$   **43.** $e - 2$   **45.** $\dfrac{9}{2}\ln 3 - 4$   **47.** $\dfrac{e^\pi + 1}{2}$   **49.** $1 - \dfrac{2}{e}$

**51. (a)** $v(t) = -\dfrac{1}{5}e^{-2t}(\cos t + 2\sin t) + \dfrac{41}{5}$   **(b)** $s(t) = \dfrac{1}{25}e^{-2t}(3\sin t + 4\cos t) + \dfrac{41}{5}t - \dfrac{4}{25}$   **53.** $2\pi$

**55.** $e\pi - 2\pi$   **57.** $\dfrac{\pi}{9}(5e^6 + 1)$   **59.** $-37$   **61. (a)** 1   **(b)** $\dfrac{\pi}{2}(e^2+1)$   **63.** $2\sin\sqrt{x} - 2\sqrt{x}\cos\sqrt{x} + C$

**65.** $(\sin x)\ln(\sin x) - \sin x + C$   **67.** $\dfrac{1}{2}(e^{2x}\sin e^{2x} + \cos e^{2x}) + C$   **69.** $\dfrac{1}{2}e^{x^2}(x^2-1) + C$

**71.** $\dfrac{1}{2}e^x(x\sin x + x\cos x - \sin x) + C$   **73.** SSM.   **75.** SSM.

**77. (a)** $\dfrac{1}{5}e^{5x}\left(x^2 - \dfrac{2}{5}x + \dfrac{2}{25}\right) + C$   **(b)** $\int x^n e^{kx}\,dx = \dfrac{1}{k}\left(x^n e^{kx} - n\int x^{n-1}e^{kx}\,dx\right)$

**79. (a)** $f(x) = \dfrac{1}{2}\sin^2 x$   **(b)** $g(x) = -\dfrac{1}{2}\cos^2 x$   **(c)** $h(x) = -\dfrac{1}{4}\cos(2x)$   **(d)** $C_2 = \dfrac{1}{2} + C_1$   **(e)** $C_3 = \dfrac{1}{4} + C_1$

**81. (a)**   **(b)** $\approx 1.079$   **(c)** $\approx 1.890$

**83.** SSM.   **85. (a)** $\dfrac{5\pi}{32}$   **(b)** $\dfrac{8}{15}$   **(c)** $\dfrac{35\pi}{256}$   **(d)** $\dfrac{5\pi}{32}$   **87.** SSM.   **89.** SSM.   **91.** SSM.

### Section 7.2  (SSM = See Student Solutions Manual; AWV = Answers will vary)

**1.** True   **2.** True   **3.** $\dfrac{\sin^5 x}{5} - \dfrac{2\sin^3 x}{3} + \sin x + C$   **5.** $\dfrac{5}{16}x - \dfrac{1}{4}\sin(2x) + \dfrac{3}{64}\sin(4x) + \dfrac{1}{48}\sin^3(2x) + C$

**7.** $\dfrac{x}{2} - \dfrac{\sin(2\pi x)}{4\pi} + C$   **9.** 0   **11.** $\dfrac{\cos^5 x}{5} - \dfrac{\cos^3 x}{3} + C$   **13.** $\dfrac{8}{45}$   **15.** $\dfrac{3}{10}(\cos x)^{10/3} - \dfrac{3}{4}(\cos x)^{4/3} + C$

**17.** $\dfrac{2}{3}\sin^3\dfrac{x}{2} - \dfrac{2}{5}\sin^5\dfrac{x}{2} + C$   **19.** $\dfrac{\sec^5 x}{5} - \dfrac{\sec^3 x}{3} + C$   **21.** $\dfrac{2}{5}\tan^{5/2}x + \dfrac{2}{9}\tan^{9/2}x + C$   **23.** $\dfrac{2}{7}(\sec x)^{7/2} - \dfrac{2}{3}(\sec x)^{3/2} + C$

**25.** $\csc x - \dfrac{\csc^3 x}{3} + C$   **27.** $-\dfrac{1}{8}\cos(4x) - \dfrac{1}{4}\cos(2x) + C$   **29.** $\dfrac{1}{8}\sin(4x) + \dfrac{1}{4}\sin(2x) + C$   **31.** $\dfrac{1}{4}\sin(2x) - \dfrac{1}{12}\sin(6x) + C$

**33.** $\dfrac{2}{3}$   **35.** $\dfrac{8}{105}$   **37.** $\sec x + \cos x + C$   **39.** 0   **41.** 0   **43.** $\dfrac{1}{2}\tan^2 x - \ln|\sec x| + C$   **45.** $-\dfrac{1}{2}\cot^2 x + 2\ln|\tan x| + \dfrac{1}{2}\tan^2 x + C$

**47.** $\dfrac{\csc^4 x}{4} - \dfrac{\csc^6 x}{6} + C$ or $-\dfrac{\cot^4 x}{4} - \dfrac{\cot^6 x}{6} + C$   **49.** $-\dfrac{1}{8}\csc^4(2x) + C$   **51.** $\dfrac{7\sqrt{2}}{48} + \dfrac{\ln(1+\sqrt{2})}{16}$   **53.** 0

**55.** $\dfrac{1}{2}\sin x - \dfrac{1}{4}\sin(2x) + C$ **57.** $\dfrac{\pi^2}{2}$ **59.** $\dfrac{2}{15}\pi$ **61. (a)** $\bar{y} = \dfrac{4}{3\pi}$ **(b)** AWV. **(c)**

**63.** $\dfrac{\pi}{4}$

**65. (a)** $\dfrac{\pi}{4} - \ln\sqrt{2}$ **(b)** $\dfrac{\pi^2}{2} - \pi$ **67. (a)** $\dfrac{3x}{8} - \dfrac{1}{4}\sin(2x) + \dfrac{1}{32}\sin(4x) + C$ **(b)** $-\dfrac{1}{4}\sin^3 x\cos x - \dfrac{3}{8}\sin x\cos x + \dfrac{3}{8}x + C$

**(c)** SSM. **(d)** Same as (a). **69.** $\dfrac{1}{2(m-n)}\sin[(m-n)x] - \dfrac{1}{2(m+n)}\sin[(m+n)x] + C$

**71.** $\dfrac{1}{2(m-n)}\sin[(m-n)x] + \dfrac{1}{2(m+n)}\sin[(m+n)x] + C$ **73.** $\dfrac{\pi}{4} - \dfrac{1}{2}$

## Section 7.3 (SSM = See Student Solutions Manual; AWV = Answers will vary)

**1.** True **2.** (d) **3.** (c) **4.** (b) **5.** $\dfrac{x}{2}\sqrt{4-x^2} + 2\sin^{-1}\dfrac{x}{2} + C$ **7.** $-\dfrac{x}{2}\sqrt{16-x^2} + 8\sin^{-1}\dfrac{x}{4} + C$

**9.** $-\dfrac{1}{x}\sqrt{4-x^2} - \sin^{-1}\dfrac{x}{2} + C$ **11.** $\dfrac{x(x^2-2)}{4}\sqrt{4-x^2} + 2\sin^{-1}\dfrac{x}{2} + C$ **13.** $\dfrac{1}{4}\dfrac{x}{\sqrt{4-x^2}} + C$

**15.** $\dfrac{1}{2}x\sqrt{4+x^2} + 2\ln\left|\dfrac{\sqrt{4+x^2}+x}{2}\right| + C$ **17.** $\ln\left|\dfrac{\sqrt{x^2+16}+x}{4}\right| + C$ **19.** $\dfrac{x}{2}\sqrt{1+9x^2} + \dfrac{1}{6}\ln\left|\sqrt{1+9x^2}+3x\right| + C$

**21.** $\dfrac{1}{18}x\sqrt{4+9x^2} - \dfrac{2}{27}\ln\left|\dfrac{\sqrt{4+9x^2}+3x}{2}\right| + C$ **23.** $-\dfrac{\sqrt{x^2+4}}{4x} + C$ **25.** $\dfrac{x}{4\sqrt{x^2+4}} + C$

**27.** $\dfrac{x}{2}\sqrt{x^2-25} + \dfrac{25}{2}\ln\left|\dfrac{x+\sqrt{x^2-25}}{5}\right| + C$ **29.** $\sqrt{x^2-1} - \sec^{-1}x + C$ **31.** $\dfrac{\sqrt{x^2-36}}{36x} + C$

**33.** $\dfrac{1}{2}\ln\left|\dfrac{2x+\sqrt{4x^2-9}}{3}\right| + C$ **35.** $-\dfrac{1}{9}\dfrac{x}{\sqrt{x^2-9}} + C$ **37.** $-\dfrac{x}{\sqrt{x^2-9}} + \ln\left|\dfrac{x+\sqrt{x^2-9}}{3}\right| + C$ **39.** $x - 4\tan^{-1}\dfrac{x}{4} + C$

**41.** $\dfrac{x}{2}\sqrt{4-25x^2} + \dfrac{2}{5}\sin^{-1}\dfrac{5x}{2} + C$ **43.** $\dfrac{x}{4\sqrt{4-25x^2}} + C$ **45.** $\dfrac{1}{2}x\sqrt{4+25x^2} + \dfrac{2}{5}\ln\left|\dfrac{\sqrt{4+25x^2}+5x}{2}\right| + C$

**47.** $\dfrac{1}{128}\sec^{-1}\dfrac{x}{4} + \dfrac{\sqrt{x^2-16}}{32x^2} + C$ **49.** $\dfrac{\pi}{4}$ **51.** $\dfrac{\sqrt{2}+\ln(1+\sqrt{2})}{2}$ **53.** $10 - 2\sqrt{7} + 9\ln 3 - \dfrac{9}{2}\ln(4+\sqrt{7})$ **55.** $\dfrac{\sqrt{3}}{3} - \dfrac{\pi}{6}$

**57.** $3 - \dfrac{3\pi}{4}$ **59.** $\dfrac{\pi}{40} + \dfrac{\pi}{16}\tan^{-1}\dfrac{1}{2}$ **61.** $\bar{y} = \sin^{-1}\dfrac{1}{3} \approx 0.340$ **63.** $\sin^{-1}\dfrac{2}{3}$ **65.** $\dfrac{1}{2}\ln(2\sqrt{6}+5) - \dfrac{1}{2}\ln(2\sqrt{2}+3) - 3\sqrt{2} + 5\sqrt{6}$

**67. (a)** $\dfrac{4\sqrt{7}}{3} - 3\ln\dfrac{4+\sqrt{7}}{3}$ **(b)** $\dfrac{14\sqrt{10}}{3} - 3\ln\dfrac{7+2\sqrt{10}}{3}$ **69.** $\dfrac{5}{2}\sqrt{26} + \dfrac{1}{4}\ln(\sqrt{26}+5) - \dfrac{1}{4}\ln(\sqrt{26}-5)$

**71. (a)** $\pi - \left[4\sin^{-1}\dfrac{\sqrt{15}}{8} + \sin^{-1}\dfrac{\sqrt{15}}{4} - \dfrac{\sqrt{15}}{2}\right]$ **(b)** $4\pi - \left[4\sin^{-1}\dfrac{\sqrt{15}}{8} + \sin^{-1}\dfrac{\sqrt{15}}{4} - \dfrac{\sqrt{15}}{2}\right]$ **73.** $\sin^{-1}(x-2) + C$

**75.** $\dfrac{1}{2}\ln\left|\dfrac{2x-1}{2} + \dfrac{\sqrt{4x^2-4x-3}}{2}\right| + C$ **77.** $\dfrac{e^x}{2}\sqrt{25-e^{2x}} + \dfrac{25}{2}\sin^{-1}\dfrac{e^x}{5} + C$ **79.** $\dfrac{1}{2}x^2\sin^{-1}x - \dfrac{1}{4}\sin^{-1}x + \dfrac{1}{4}x\sqrt{1-x^2} + C$

**81. (a)** $\dfrac{x}{2}\sqrt{x^2+a^2} + \dfrac{a^2}{2}\ln\left|\dfrac{x+\sqrt{x^2+a^2}}{a}\right| + C$ **(b)** $\dfrac{x}{2}\sqrt{x^2+a^2} + \dfrac{a^2}{2}\sinh^{-1}\dfrac{x}{a} + C$ **83.** SSM.

**85.** SSM. **87.** SSM. **89.** SSM. **91.** $\ln\left|\tan x - 3 + \sqrt{\tan^2 x - 6\tan x + 8}\right| + C$

## Section 7.4 (SSM = See Student Solutions Manual; AWV = Answers will vary)

**1.** $\tan^{-1}(x+2)+C$  **3.** $\dfrac{1}{2}\tan^{-1}\dfrac{x+2}{2}+C$  **5.** $\dfrac{2\sqrt{5}}{5}\tan^{-1}\dfrac{2x+1}{\sqrt{5}}+C$  **7.** $\dfrac{1}{4}\ln(2x^2+2x+3)-\dfrac{\sqrt{5}}{10}\tan^{-1}\dfrac{2x+1}{\sqrt{5}}+C$

**9.** $\sin^{-1}\dfrac{x-1}{3}+C$  **11.** $\sin^{-1}\dfrac{x-2}{2}+C$  **13.** $\ln\left|\dfrac{\sqrt{x^2+2x+2}-1}{x+1}\right|+C$  **15.** $\sin^{-1}\dfrac{x+1}{5}+C$

**17.** $\sqrt{x^2-2x+5}-4\ln\left|\dfrac{x-1+\sqrt{x^2-2x+5}}{2}\right|+C$  **19.** $\ln(\sqrt{2}+1)$  **21.** $\ln\left(e^x+\dfrac{1}{2}+\sqrt{e^{2x}+e^x+1}\right)+C$

**23.** $-2\sqrt{4x-x^2-3}+\sin^{-1}(x-2)+C$  **25.** $\dfrac{x-1}{9\sqrt{x^2-2x+10}}+C$  **27.** $\ln\left|x+1+\sqrt{x^2+2x-3}\right|+C$

**29.** $\sqrt{5+4x-x^2}-3\ln\left|3+\sqrt{5+4x-x^2}\right|+3\ln|x-2|+C$  **31.** $\sqrt{x^2+2x-3}-\ln\left|x+1+\sqrt{x^2+2x-3}\right|+C$

**33.** SSM.  **35.** $a\sin^{-1}\dfrac{x}{a}-\sqrt{a^2-x^2}+C$

## Section 7.5 (SSM = See Student Solutions Manual; AWV = Answers will vary)

**1.** (a)  **2.** True  **3.** True  **4.** False  **5.** $x-1+\dfrac{2}{x+1}$  **7.** $x^2+2x+7+\dfrac{10}{x-2}$  **9.** $2x+3+\dfrac{x}{x^2-9}$

**11.** $x^2-4x+16-\dfrac{64x+1}{x^2+4x}$  **13.** $2x^2-7+\dfrac{26}{x^2+4}$  **15.** $\dfrac{x^2}{2}-x+2\ln|x+1|+C$  **17.** $\dfrac{x^3}{3}+x^2+7x+10\ln|x-2|+C$

**19.** $x^2+3x+\dfrac{1}{2}\ln|x^2-9|+C$  **21.** $\dfrac{1}{3}\ln|x-2|-\dfrac{1}{3}\ln|x+1|+C$  **23.** $-\ln|x-1|+2\ln|x-2|+C$

**25.** $\dfrac{2}{21}\ln|3x-2|+\dfrac{1}{14}\ln|2x+1|+C$  **27.** $-5\ln|x+2|+5\ln|x+1|+\dfrac{4}{x+1}+C$

**29.** $\dfrac{1}{4}\ln|x+1|+\dfrac{3}{4}\ln|x-1|-\dfrac{1}{2(x-1)}+C$  **31.** $\ln|x|-\dfrac{1}{2}\ln(x^2+1)+C$  **33.** $\dfrac{1}{6}\ln(x^2+2x+4)+\dfrac{2}{3}\ln|x+1|+C$

**35.** $\dfrac{x-32}{32(x^2+16)}+\dfrac{1}{128}\tan^{-1}\dfrac{x}{4}+C$  **37.** $-\dfrac{1}{2(x^2+16)}+\dfrac{4}{(x^2+16)^2}+C$  **39.** $\dfrac{1}{4}\ln|x-1|+\dfrac{3}{4}\ln|x+3|+C$

**41.** $\dfrac{14}{3}\ln|x-1|-\dfrac{4}{x-1}-\dfrac{7}{3}\ln(x^2+2)+\dfrac{2\sqrt{2}}{3}\tan^{-1}\dfrac{\sqrt{2}x}{2}+C$  **43.** $-\ln|x|-\ln|x+1|+2\ln|x-3|+C$

**45.** $4\ln|x-2|-3\ln|x-1|+\dfrac{1}{x-1}+C$  **47.** $\ln|x-1|-\dfrac{1}{2}\ln|x^2+x+1|+\dfrac{\sqrt{3}}{3}\tan^{-1}\dfrac{\sqrt{3}(2x+1)}{3}+C$  **49.** $-\dfrac{1}{6}\ln 2$

**51.** $\dfrac{1}{8}\ln 21$  **53.** $\dfrac{1}{5}\ln(2-\sin\theta)-\dfrac{1}{5}\ln(\sin\theta+3)+C$  **55.** $-\ln|\cos\theta|+\dfrac{1}{2}\ln(\cos^2\theta+1)+C$

**57.** $\dfrac{1}{3}\ln|e^t-1|-\dfrac{1}{3}\ln(e^t+2)+C$  **59.** $\dfrac{1}{2}\ln|e^x-1|-\dfrac{1}{2}\ln(e^x+1)+C$  **61.** $t-\dfrac{1}{2}\ln(e^{2t}+1)+C$

**63.** $\ln|\sin x-1|-\dfrac{1}{\sin x-1}+C$  **65.** $\dfrac{1}{54}\tan^{-1}\dfrac{\sin x}{3}+\dfrac{1}{18}\dfrac{\sin x}{\sin^2 x+9}+C$  **67.** $\ln\dfrac{15}{7}$

**69.** $\dfrac{4\pi\sqrt{3}}{3}+\dfrac{4}{3}\ln 3$  **71.** $\sqrt{e^2+1}-\sqrt{2}+\dfrac{1}{2}\ln\left|\dfrac{\sqrt{e^2+1}-1}{\sqrt{e^2+1}+1}\right|-\dfrac{1}{2}\ln\left(\dfrac{\sqrt{2}-1}{\sqrt{2}+1}\right)$

**73.** (a) $\dfrac{dP}{dt}=0.15P\left(1-\dfrac{P}{50}\right)$, $P(0)=1$  (b) $P(t)=\dfrac{50}{1+49e^{-0.15t}}$  (c) $t=\dfrac{-\ln 49}{-0.15}\approx 25.945$. By the 26th day half the

population is infected.  (d) $t=\dfrac{-\ln 196}{-0.15}\approx 35.187$. On the 35th day 80% of the population is infected.

**75.** (a) $\dfrac{dP}{dt}=0.20P\left(1-\dfrac{P}{600,000}\right)$, $P(0)=100$  (b) $P(t)=\dfrac{600,000}{1+5999e^{-0.20t}}$

(c) The population exceeds 100,000 on the 35th day.

**77.** (a) $-4,-2,3$  (b) $(x+4)(x+2)(x-3)$  (c) $\dfrac{13}{10}\ln|x+2|+\dfrac{2}{35}\ln|x-3|-\dfrac{19}{14}\ln|x+4|+C$  **79.** $2\sqrt{x}-4\ln(\sqrt{x}+2)+C$

**81.** $\dfrac{3}{5}x^{5/3}+\dfrac{3}{4}x^{4/3}+\dfrac{3}{2}x^{2/3}+x+3x^{1/3}+3\ln|x^{1/3}-1|+C$  **83.** $\dfrac{2}{3}x^{1/2}+\dfrac{1}{3}x^{1/3}+\dfrac{2}{9}x^{1/6}+\dfrac{2}{27}\ln|3x^{1/6}-1|+C$

**85.** $\dfrac{2}{3}(2x+1)^{3/4}+C$  **87.** $3(1+x)^{1/3}+C$  **89.** $-\dfrac{3x^{1/6}}{x^{1/3}+1}+3\tan^{-1}x^{1/6}+C$

**91.** $-\dfrac{2}{1+\tan\dfrac{x}{2}}+C$    **93.** $\dfrac{2\sqrt{5}}{5}\tan^{-1}\left(\dfrac{\sqrt{5}}{5}\tan\dfrac{x}{2}\right)+C$    **95.** $-\ln\left|1-\tan\dfrac{x}{2}\right|+C$

**97.** $\dfrac{1}{2}\ln\left|\tan^2\dfrac{x}{2}+2\tan\dfrac{x}{2}-1\right|-\dfrac{1}{2}\ln\left(\tan^2\dfrac{x}{2}+1\right)-\dfrac{x}{2}+C$    **99.** $\dfrac{\sqrt{5}}{5}\ln\left|\tan\dfrac{x}{2}-\dfrac{\sqrt{5}}{2}+\dfrac{1}{2}\right|-\dfrac{\sqrt{5}}{5}\ln\left|\tan\dfrac{x}{2}+\dfrac{\sqrt{5}}{2}+\dfrac{1}{2}\right|+C$

**101.** $-\dfrac{1}{2}\ln\left|\tan\dfrac{x}{2}-1\right|+\dfrac{1}{2}\ln\left|\tan\dfrac{x}{2}+1\right|+\dfrac{1}{\tan\dfrac{x}{2}+1}-\dfrac{1}{\left(\tan\dfrac{x}{2}+1\right)^2}+C$

**103.** $\dfrac{1}{3}\ln\dfrac{\sqrt{3}}{3}-\dfrac{1}{3}\ln\left(\sqrt{2}-1\right)-\dfrac{4}{15}\ln\left(\sqrt{3}+1\right)-\dfrac{4}{15}\ln\left(4-\sqrt{2}\right)+\dfrac{4}{15}\ln(3\sqrt{2}-2)+\dfrac{4}{15}\ln\left(3-\dfrac{\sqrt{3}}{3}\right)$    **105.** $\dfrac{\pi}{2}+\ln 2$

**107.** SSM.    **109.** SSM.

## Section 7.6    (SSM = See Student Solutions Manual; AWV = Answers will vary)

**1.** True    **2.** True    **3.** 22 and 20    **5.** 26 and $\dfrac{58}{3}$    **7. (a)** 0.483    **(b)** 0.007    **(c)** 26    **9. (a)** 0.743    **(b)** 0.010    **(c)** 41

**11. (a)** 0.907    **(b)** 0.005    **(c)** 29    **13. (a)** 0. 480    **(b)** 0.004    **(c)** 19    **15. (a)** 0. 749    **(b)** 0.005    **(c)** 29

**17. (a)** 0. 911    **(b)** 0.003    **(c)** 21    **19. (a)** 3.059    **(b)** $5.309\times10^{-4}$    **(c)** 8    **21. (a)** 0.747    **(b)** $2.604\times10^{-4}$    **(c)** 6

**23. (a)** 0.910    **(b)** $3.255\times10^{-4}$    **(c)** 6    **25. (a)** SSM.    **(b)** 0.696    **(c)** 0.692    **(d)** 0.693

**27. (a)** 1.910    **(b)** 1.910    **(c)** 1.910    **29.** $\approx62.983$ in lb    **31.** 16,787.5 m³    **33.** T: 131, 787.5 m³, S: 132, 625 m³

**35.** $\dfrac{105}{2}\pi\approx164.934$    **37.** 9.5 m    **39. (a)** Using the disk method: $V\approx1.6095$. Using the shell method: $V\approx1.471$.

   **(b)** Using the disk method: $V\approx1.971$. Using the shell method: $V\approx1.323$.    **(c)** Using the disk method: $V\approx1.333$.
   Using the shell method: $V\approx1.542$.    **41. (a)** 1.845    **(b)** 1.856    **(c)** 1.852    **43.** SSM.

## Section 7.7    (SSM = See Student Solutions Manual; AWV = Answers will vary)

**1. (c)**    **2. (b)**    **3.** False    **4.** True    **5.** False    **6.** $\displaystyle\lim_{t\to b^-}\int_a^t f(x)\,dx$    **7.** Yes, $\infty$ is a limit of integration.    **9.** No.

**11.** Yes, the function is undefined at 0.    **13.** Yes, the function is undefined at 1.    **15.** Converges to $\dfrac{1}{2}$.    **17.** Diverges

**19.** Diverges    **21.** Converges to $\dfrac{1}{24}$    **23.** Converges to $\dfrac{\pi}{2}$    **25.** Diverges    **27.** Diverges    **29.** Converges to 4    **31.** Converges to 0

**33.** Diverges    **35.** Converges to 1    **37.** Diverges    **39.** Diverges    **41.** Converges to 1    **43.** Diverges    **45.** Converges to $\dfrac{4}{3}$

**47.** Converges to $\pi$    **49.** Diverges    **51.** Converges to $\sqrt[3]{486}$    **53.** Diverges    **55.** Converges to 0    **57.** Converges to 4

**59.** Diverges    **61.** Converges to $\dfrac{1}{2}$    **63.** Diverges    **65.** Diverges    **67. (a)** Converges    **(b)** $\approx0.673$    **69. (a)** Converges    **(b)** $\dfrac{\pi}{2}$

**71.** $\ln 2$    **73.** $\dfrac{\pi}{2}$    **75.** $2\pi a^2$    **77. (a)** \$1250    **(b)** \$30,612    **79.** $\dfrac{\pi NI}{5r}\left(1-\dfrac{x}{\sqrt{r^2+x^2}}\right)$    **81.** $GmM$    **83.** 1    **85.** $\dfrac{1}{2}$    **87.** $\dfrac{1}{4}$

**89.** $\dfrac{\pi}{2}-\sec^{-1}2=\dfrac{\pi}{6}$    **91.** SSM.    **93.** SSM.    **95.** SSM.    **97.** SSM.    **99.** SSM.    **101.** $\dfrac{1}{s^2}$    **103.** $\dfrac{1}{1+s^2}$

**105.** $\dfrac{1}{s-a}$    **107.** 2    **109.** 1    **111.** SSM.    **113.** $\dfrac{a+b}{2}$    **115.** $\sigma^2=\dfrac{(b-a)^2}{12}$, $\sigma=\dfrac{b-a}{2\sqrt{3}}$

## Section 7.8    (SSM = See Student Solutions Manual; AWV = Answers will vary)

**1.** $\dfrac{e^{2x}}{5}(\sin x+2\cos x)+C$    **3.** $\dfrac{(2+3x)^5}{9}\left(\dfrac{2+3x}{6}-\dfrac{2}{5}\right)+C$    **5.** $\dfrac{2}{15}(32-8x+3x^2)\sqrt{6+3x}+C$

**7.** $\sqrt{x^2+4}-2\ln\left|\dfrac{2+\sqrt{x^2+4}}{x}\right|+C$    **9.** $-\dfrac{x}{4\sqrt{x^2-4}}+C$    **11.** $\dfrac{x^4(\ln x)^2}{4}-\dfrac{x^4}{8}\left(\ln x-\dfrac{1}{4}\right)+C$

**13.** $\dfrac{x}{2}\sqrt{x^2-16}-8\ln\left|x+\sqrt{x^2-16}\right|+C$    **15.** $\dfrac{x}{4}(6-x^2)^{3/2}+\dfrac{9}{4}x\sqrt{6-x^2}+\dfrac{27}{2}\sin^{-1}\dfrac{\sqrt{6}x}{6}+C$

**17.** $\dfrac{x-5}{2}\sqrt{10x-x^2}+\dfrac{25}{2}\cos^{-1}\dfrac{5-x}{5}+C$  **19.** $\dfrac{\sin(11x)}{22}+\dfrac{\sin(5x)}{10}+C$  **21.** $\dfrac{x^2+1}{2}\tan^{-1}x-\dfrac{x}{2}+C$

**23.** $\dfrac{x^5}{5}\left(\ln x-\dfrac{1}{5}\right)+C$  **25.** $\dfrac{\sinh(2x)}{4}-\dfrac{x}{2}+C$  **27.** $-\dfrac{\sqrt{8x-x^2}}{12x^2}-\dfrac{\sqrt{8x-x^2}}{48x}+C$  **29.** $\dfrac{1}{8}\left[\dfrac{x}{4+x^2}+\dfrac{1}{2}\tan^{-1}\dfrac{x}{2}\right]+C$

**31.** $x\cosh x-\sinh x+C$  **33.** $\dfrac{1}{60}(6x+5)(4x+5)^{3/2}+C$  **35.** $\dfrac{135}{8}\sin^{-1}\dfrac{1}{3}-\dfrac{9\sqrt{2}}{4}$

**37.** $\dfrac{e^{2x}}{5}(\sin x+2\cos x)$  **39.** $\dfrac{27}{2}x^6+\dfrac{216}{5}x^5+54x^4+32x^3+8x^2$  **41.** $\dfrac{2}{45}\sqrt{3}\sqrt{x+2}\,(3x^2-8x+32)$

**43.** $\sqrt{x^2+4}-2\operatorname{arctanh}\dfrac{2}{\sqrt{x^2+4}}$  **45.** $-\dfrac{1}{4}\dfrac{x}{\sqrt{x^2-4}}$  **47.** $\dfrac{1}{32}x^4(8\ln^2 x-4\ln x+1)$

**49.** $\dfrac{1}{2}x\sqrt{x^2-16}-8\ln(x+\sqrt{x^2-16})$  **51.** $\dfrac{9}{4}x\sqrt{6-x^2}-\dfrac{27}{2}i\ln\left(ix+\sqrt{6-x^2}\right)+\dfrac{1}{4}x(6-x^2)^{3/2}$

**53.** $\dfrac{1}{2}x\sqrt{-x(x-10)}-\dfrac{25}{2}i\ln(ix+\sqrt{-x(x-10)}-5i)-\dfrac{5}{2}\sqrt{-x(x-10)}$  **55.** $\dfrac{1}{10}\sin 5x+\dfrac{1}{22}\sin 11x$

**57.** $\dfrac{1}{2}\arctan x-\dfrac{1}{2}x-\dfrac{1}{4}\pi+\dfrac{1}{2}x^2\arctan x$  **59.** $\dfrac{1}{5}x^5\ln x-\dfrac{1}{25}x^5$  **61.** $-\dfrac{1}{8e^{2x}}(-e^{2(2x)}+4xe^{2x}+1)$  **63.** $-\dfrac{(x+4)\sqrt{-(x-8)x}}{48x^2}$

**65.** $\dfrac{1}{32(x^2+4)}\left(4x-4\pi+8\arctan\dfrac{1}{2}x+2x^2\arctan\dfrac{1}{2}x-\pi x^2\right)$  **67.** $\dfrac{1}{2e^{-x}}(x+e^{2(-x)}+xe^{2(-x)}-1)$  **69.** $\dfrac{1}{60}(4x+5)^{3/2}(6x+5)$

**71.** $-\dfrac{9\sqrt{2}}{4}+\dfrac{135}{8}\cos^{-1}\dfrac{2\sqrt{2}}{3}$  **73.** AWV.  **75.** AWV.  **77.** AWV.

## Section 7.9

**1.** $\dfrac{1}{12}[2x^3-\sin(2x^3)]+C$  **3.** $(5x-1)\sin x+5\cos x+C$  **5.** $\dfrac{1}{2}\ln\left|\dfrac{\sqrt{4x^2+25}}{5}+\dfrac{2x}{5}\right|+C$  **7.** $4(8\ln 2-5\ln 3)$

**9.** $\dfrac{1}{3}\sec^3 x+C$  **11.** $\dfrac{1}{5}\tan^5 x+\dfrac{1}{3}\tan^3 x+C$  **13.** $\dfrac{1}{4}\sec^{-1}\dfrac{x}{4}+C$  **15.** $\dfrac{3}{10}(\tan x)^{10/3}+\dfrac{3}{4}(\tan x)^{4/3}+C$

**17.** $-\dfrac{1}{2}\cos(x^4+2x)+C$  **19.** $\dfrac{1}{3}x^3\sin^{-1}x-\dfrac{1}{9}(1-x^2)^{3/2}+\dfrac{1}{3}(1-x^2)^{1/2}+C$  **21.** $\dfrac{4\sqrt{2}}{15}$

**23.** $18\sin^{-1}\left(\dfrac{x}{6}\right)+\dfrac{x\sqrt{36-x^2}}{2}+C$  **25.** $\dfrac{1}{2}\ln(x^2+2x+5)-\dfrac{1}{2}\tan^{-1}\left(\dfrac{x+1}{2}\right)+C$  **27.** Converges; $\dfrac{\pi}{2}$

**29.** $\ln|x-1|+\tan^{-1}x-\dfrac{1}{2}\ln(x^2+1)+C$  **31.** $\dfrac{3}{10}(\cos x)^{10/3}-\dfrac{3}{4}(\cos x)^{4/3}+C$

## Review Exercises  (SSM = See Student Solutions Manual; AWV = Answers will vary)

**1.** $\dfrac{1}{4}\tan^{-1}\dfrac{x+2}{4}+C$  **3.** $\dfrac{1}{3}\sec^3\phi+C$  **5.** $\dfrac{1}{3}\cos^3\phi-\cos\phi+C$  **7.** $\ln\left|x+2+\sqrt{(x+2)^2-1}\right|+C$

**9.** $-v\cot v+\ln|\sin v|+C$  **11.** $6\sin^{-1}\dfrac{x}{2}+2x\sqrt{4-x^2}+\dfrac{1}{4}x\sqrt{4-x^2}\,(2-x^2)+C$  **13.** $e^t+2\ln\left|e^t-2\right|+C$

**15.** $\dfrac{1}{16}\ln\left|\dfrac{x^2-4}{x^2+4}\right|+C$  **17.** $\ln|y+1|+\dfrac{2}{y+1}-\dfrac{1}{2(y+1)^2}+C$  **19.** $x\tan x-\ln|\sec x|+C$

**21.** $y\ln|1-y|-y-\ln|1-y|+C$  **23.** $-2\ln|x|+\dfrac{2}{x}+5\ln|x-1|+C$  **25.** $\dfrac{1}{3}x^3\sin^{-1}x+\dfrac{1}{3}\sqrt{1-x^2}-\dfrac{1}{9}\left(1-x^2\right)^{3/2}+C$

**27.** $\dfrac{1}{2}\ln\left|\dfrac{x}{x+2}\right|+C$  **29.** $\dfrac{1}{2}\ln|1-w|-\dfrac{3}{2}\ln|w+1|+C$  **31.** $\sqrt{x}+\sin\sqrt{x}\cos\sqrt{x}+C$  **33.** $\dfrac{1}{2}\cos x-\dfrac{1}{6}\cos(3x)+C$

**35.** 1  **37.** SSM.  **39.** Converges to $\dfrac{2}{e}$.  **41.** Converges to 1.  **43.** SSM.  **45.** Diverges  **47.** $f(x)=x^2\sin x$

**49. (a)** 1.910  **(b)** 1.910  **(c)** 1.910  **51.** 3  **53. (a)** $M=100$  **(b)** 0.24 or 24%  **(c)** 50

# Chapter 8

## Section 8.1 (SSM = See Student Solutions Manual; AWV = Answers will vary)

**1.** False **2.** False **3.** True **4.** (b) **5.** False **6.** True **7.** False **8.** False **9.** False **10.** (b) **11.** True **12.** False

**13.** $2, \dfrac{3}{2}, \dfrac{4}{3}, \dfrac{5}{4}$ **15.** $0, \ln 2, \ln 3, \ln 4$ **17.** $\dfrac{1}{3}, -\dfrac{1}{5}, \dfrac{1}{7}, -\dfrac{1}{9}$ **19.** $1, -1, 1, -1$ **21.** $\dfrac{1}{2}, \dfrac{1}{2}, \dfrac{3}{4}, \dfrac{3}{2}$ **23.** $a_n = 2n$ **25.** $a_n = 2^n$

**27.** $a_n = \dfrac{(-1)^{n+1}}{n+1}$ **29.** $a_n = \dfrac{n}{n+1}$ **31.** $a_n = (n-1)!$ **33.** 0 **35.** 1 **37.** 4 **39.** 0 **41.** 0 **43.** 1 **45.** $\ln \dfrac{1}{3}$ **47.** $e^{-2}$

**49.** 0 **51.** 1 **53.** $-\dfrac{1}{2}$ **55.** 1 **57.** 0 **59.** 0 **61.** 0 **63.** Diverges **65.** Diverges **67.** Converges **69.** Diverges

**71.** Converges **73.** Bounded from above and from below **75.** Bounded from below **77.** Bounded from below

**79.** Bounded from above and from below **81.** Nonmonotonic **83.** Nonmonotonic **85.** Monotonic decreasing

**87.** Nonmonotonic **89.** Decreasing, bounded from below **91.** Increasing, bounded from above

**93.** Increasing, bounded from above **95.** Converges; 6 **97.** Converges; $\ln \dfrac{1}{3}$ **99.** Diverges **101.** Converges; 0

**103.** Converges; 0 **105.** Converges; 0 **107.** Converges; 1 **109.** Converges; 1 **111.** Converges; 1 **113.** Converges; 1

**115.** 0 **117.** Converges **119.** Converges **121.** Converges **123.** Diverges **125.** Converges **127.** Converges

**129.** $p_n = 3000 r^n + h\dfrac{r^n - 1}{r-1}$; Converges; $\lim\limits_{n \to \infty} p_n = \dfrac{h}{1-r}$ **131.** (a) $I_n = 0.95^n I_0$ (b) 77 **133.** $e^2$ **135.** $e^3$

**137.** SSM. **139.** SSM. **141.** SSM. **143.** SSM. **145.** AWV. **147.** SSM. **149.** Diverges

**151.** SSM. **153.** SSM. **155.** SSM. **157.** SSM. **159.** SSM.

## Section 8.2 (SSM = See Student Solutions Manual; AWV = Answers will vary)

**1.** (b) **2.** (d) **3.** True **4.** False **5.** $\dfrac{a}{1-r}$ **6.** False **7.** $\dfrac{175}{64}$ **9.** 10 **11.** $\dfrac{1}{3}$ **13.** $-\dfrac{1}{3}$ **15.** $\dfrac{1}{2}$ **17.** Diverges

**19.** Converges; 6 **21.** Converges; $\dfrac{21}{2}$ **23.** Converges; $\dfrac{50}{69}$ **25.** Converges; 6 **27.** Converges; $\dfrac{1}{3}$ **29.** Diverges

**31.** Diverges **33.** Converges; $\dfrac{3}{2}$ **35.** Diverges **37.** Converges; $\dfrac{1}{42}$ **39.** Diverges **41.** Converges; $\dfrac{1}{99}$ **43.** Diverges

**45.** Converges; $\dfrac{2}{3}$ **47.** Diverges **49.** Diverges **51.** Diverges **53.** Converges; $-\dfrac{1}{4}$ **55.** Diverges **57.** Converges; $\sin 1$

**59.** $\dfrac{5}{9}$ **61.** $\dfrac{3857}{900}$ **63.** 90 ft **65.** (a) $\dfrac{h}{1-r}$ (b) $h = 2000$ **67.** (a) $P\dfrac{1+r}{r-i}$ (b) \$6.67 **69.** SSM.

**71.** (a) $T(n) = p\displaystyle\sum_{k=1}^{n}(1-e)^{k-1}$ (b) $\lim\limits_{n \to \infty} T(n) = \dfrac{p}{e}$ (c) $e_{\min} = \dfrac{p}{L}$ (d) $e_{\min} = \dfrac{2}{3}$; $T(365) = 150$ kg

**73.** SSM. **75.** $n = 11$ **77.** SSM. **79.** SSM. **81.** SSM. **83.** SSM. **85.** SSM. **87.** SSM.

## Section 8.3 (SSM = See Student Solutions Manual; AWV = Answers will vary)

**1.** (a) **2.** False **3.** True **4.** True **5.** False **6.** True **7.** True **8.** False **9.** $p > 1; 0 < p \leq 1$ **10.** Converges

**11.** Diverges **12.** False **13.** Diverges **15.** Diverges **17.** Diverges **19.** Converges **21.** Diverges **23.** Converges

**25.** Converges **27.** Diverges **29.** Converges **31.** Diverges **33.** Converges **35.** Converges **37.** Converges

**39.** Diverges **41.** Diverges **43.** Diverges **45.** Diverges **47.** Diverges **49.** Diverges **51.** Diverges **53.** Diverges

**55.** $1 < \displaystyle\sum_{k=1}^{\infty} \dfrac{1}{k^2} < 2$ **57.** $\dfrac{1}{e-1} < \displaystyle\sum_{k=1}^{\infty} \dfrac{1}{k^e} < \dfrac{e}{e-1}$ **59.** $2 < 1 + \dfrac{1}{2\sqrt{2}} + \dfrac{1}{3\sqrt{3}} + \dfrac{1}{4\sqrt{4}} + \cdots < 3$ **61.** $\dfrac{1}{2e} < \displaystyle\sum_{k=1}^{\infty} ke^{-k^2} < \dfrac{3}{2e}$

**63.** $\dfrac{1}{3}\left(\dfrac{\pi}{2}-\tan^{-1}\dfrac{1}{3}\right) < \displaystyle\sum_{k=1}^{\infty}\dfrac{1}{k^2+9} < \dfrac{1}{10}+\dfrac{1}{3}\left(\dfrac{\pi}{2}-\tan^{-1}\dfrac{1}{3}\right)$ **65.** SSM. **67.** SSM. **69.** $\pi^2\approx 9.8099$ **71.** Converges

**73.** Converges **75.** SSM. **77.** SSM. **79.** SSM. **81.** SSM. **83.** $(-\infty,\,-1)$

**85. (a)** 1.1975 **(b)** Upper is 1.5, lower is 0.5. **(c)** SSM.

## Section 8.4 (SSM = See Student Solutions Manual; AWV = Answers will vary)

**1. (b)** **2.** False **3.** False **4.** False **5.** Converges **7.** Converges **9.** Diverges **11.** Converges **13.** Converges

**15.** Converges **17.** Diverges **19.** Converges **21.** Converges **23.** Diverges **25.** Converges **27.** Converges **29.** Converges

**31.** Converges **33.** Converges **35.** Converges **37.** Diverges **39.** Converges **41.** Diverges **43.** Converges **45.** Diverges

**47.** Converges **49.** Converges **51.** Diverges **53.** Diverges **55.** Converges **57.** SSM. **59.** SSM. **61.** Diverges

**63.** Converges **65.** SSM. **67.** Diverges **69.** Converges **71.** SSM. **73.** SSM. **75.** Diverges **77.** SSM.

## Section 8.5 (SSM = See Student Solutions Manual; AWV = Answers will vary)

**1.** False **2.** False **3.** False **4.** False **5.** False **6.** True **7.** Converges **9.** Diverges **11.** Diverges **13.** Converges

**15.** Diverges **17.** Diverges **19.** Converges **21.** Diverges **23. (a)** SSM. **(b)** 3 terms **(c)** $\dfrac{7565}{7776}\approx 0.973$

**25. (a)** SSM. **(b)** 6 terms **(c)** $\dfrac{91}{144}\approx 0.632$ **27. (a)** SSM. **(b)** 9 terms **(c)** $\dfrac{171}{512}\approx 0.334$

**29.** 0.8611; upper estimate to the error is 0.0625 **31.** 0.9498; upper estimate to the error is 0.0039

**33.** 0.7222; upper estimate to the error is 0.00617 **35.** 0.3218; upper estimate to the error is $6.53\times 10^{-5}$

**37.** SSM. **39.** SSM. **41.** Absolutely convergent **43.** Diverges **45.** Absolutely convergent

**47.** Absolutely convergent **49.** Conditionally convergent **51.** Absolutely convergent **53.** Conditionally convergent

**55.** Absolutely convergent **57.** Diverges **59.** Absolutely convergent **61. (a)** SSM. **(b)** SSM. **(c)** $\dfrac{1}{2}$

**63.** SSM. **65.** SSM. **67.** SSM. **69.** Absolutely convergent for $|r|<1$; divergent if $|r|\geq 1$ **71.** SSM. **73.** SSM.

**75.** SSM. **77.** Absolutely convergent **79.** Absolutely convergent if $2<p<3$; conditionally convergent if $p\geq 3$

**81.** Absolutely convergent **83.** Converges **85.** SSM.

## Section 8.6 (SSM = See Student Solutions Manual; AWV = Answers will vary)

**1.** False **2.** False **3.** False **4.** False **5.** Converges **7.** Converges **9.** Converges **11.** Converges **13.** Diverges

**15.** Converges **17.** Converges **19.** Converges **21.** Diverges **23.** Diverges **25.** Converges **27.** Diverges

**29.** Converges **31.** Diverges **33.** Converges **35.** Diverges **37.** The Root Test provides no information.

**39.** Diverges **41.** Converges **43.** Converges **45.** Converges **47.** Converges **49.** Converges **51.** Converges

**53.** Converges **55.** Converges **57.** SSM. **59.** AWV. **61. (a)** SSM. **(b)** $\dfrac{1-e^3}{e^3}$ **63.** SSM.

**65.** Converges if $|x|\leq 1$; diverges if $|x|>1$. **67.** SSM. **69.** SSM. **71.** SSM.

## Section 8.7 (SSM = See Student Solutions Manual; AWV = Answers will vary)

**1.** False **2.** False **3.** True **4.** False **5.** False **6.** $a_k, b_k$ **7.** Diverges by the Test for Divergence

**9.** Absolutely convergent by the Geometric Series Test **11.** Absolutely convergent by the Limit Comparison Test

**13.** Diverges by the Comparison Test for Divergence **15.** Diverges by the Ratio Test

**17.** Diverges by the Test for Divergence **19.** Conditionally convergent by the Alternating Series Test

**21.** Diverges by the Ratio Test **23.** Absolutely convergent by the Ratio Test **25.** Diverges by the Limit Comparison Test

**27.** Absolutely convergent by the Comparison Test for Convergence **29.** Diverges by the Test for Divergence

**31.** Absolutely convergent by the Root Test **33.** Absolutely convergent by the Root Test

**35.** Absolutely convergent by the Integral Test **37.** Diverges by the Root Test

**39.** Absolutely convergent by the Comparison Test for Convergence **41.** Converges; $\dfrac{11}{6}$ **43.** Converges by the Ratio Test.

**45. (a)** SSM. **(b)** $\dfrac{14}{5}$ **47.** Converges

## Section 8.8   (SSM = See Student Solutions Manual; AWV = Answers will vary)

**1.** True   **2.** True   **3.** True   **4.** False   **5.** True   **6.** False   **7.** False   **8.** True   **9.** True   **10.** False   **11.** False

**12.** False   **13.** $-1 < x < 1$   **15.** $-4 < x < 2$   **17. (a, b)** $R = 2; -2 \le x < 2$   **(c)** AWV.

**19. (a, b)** $R = 1; -1 \le x < 1$   **(c)** AWV.   **21. (a, b)** $R = 3; -3 < x < 3$   **(c)** AWV.

**23. (a, b)** $R = 1; -1 < x < 1$   **(c)** AWV.   **25. (a, b)** $R = 1; 2 < x < 4$   **(c)** AWV.

**27.** $R = 1; -1 \le x \le 1$   **29.** $R = 1; 1 \le x \le 3$   **31.** $R = \infty; -\infty < x < \infty$   **33.** $R = 1; -1 < x < 1$   **35.** $R = 4; -4 < x < 4$

**37.** $R = 3; 0 < x < 6$   **39.** $R = \infty; -\infty < x < \infty$   **41.** $R = \infty; -\infty < x < \infty$   **43.** $R = \dfrac{1}{e}; -\dfrac{1}{e} < x < \dfrac{1}{e}$

**45. (a)** $-3 < x < 3$   **(b)** $f(2) = 3, f(-1) = \dfrac{3}{4}$   **(c)** $f(x) = \dfrac{3}{3-x}$   **47. (a)** $0 < x < 4$   **(b)** $f(1) = \dfrac{2}{3}, f(2) = 1$   **(c)** $f(x) = \dfrac{2}{4-x}$

**49.** Converges at $x = 2$; no information about $x = 5$   **51. (a)** True   **(b)** False   **(c)** False   **(d)** False   **(e)** True   **(f)** True

**53. (a)** $f(x) = \displaystyle\sum_{k=0}^{\infty} (-1)^k x^{3k}$   **(b)** $R = 1; -1 < x < 1$   **55. (a)** $f(x) = \dfrac{1}{6}\displaystyle\sum_{k=0}^{\infty} \left(\dfrac{x}{3}\right)^k$   **(b)** $R = 3; -3 < x < 3$

**57. (a)** $f(x) = \displaystyle\sum_{k=0}^{\infty} (-1)^k x^{3k+1}$   **(b)** $R = 1; -1 < x < 1$

**59. (a)** $f'(x) = \displaystyle\sum_{k=0}^{\infty} \dfrac{(-1)^k x^{2k}}{(2k)!}$   **(b)** $\displaystyle\int_0^x f(t)\, dt = \sum_{k=0}^{\infty} \dfrac{(-1)^k x^{2k+2}}{(2k+2)!}$

**61. (a)** $f'(x) = \displaystyle\sum_{k=1}^{\infty} \dfrac{x^{k-1}}{(k-1)!}$   **(b)** $\displaystyle\int_0^x f(t)\, dt = \sum_{k=0}^{\infty} \dfrac{x^{k+1}}{(k+1)!}$

**63.** $f(x) = \displaystyle\sum_{k=1}^{\infty} (-1)^{k-1} k x^{k-1}, -1 < x < 1$   **65.** $f(x) = \dfrac{2}{3}\displaystyle\sum_{k=1}^{\infty} k x^{k-1}, -1 < x < 1$   **67.** $f(x) = \displaystyle\sum_{k=0}^{\infty} \dfrac{(-1)^{k+1} x^{k+1}}{k+1}, -1 < x \le 1$

**69.** $f(x) = -\displaystyle\sum_{k=0}^{\infty} \dfrac{x^{2k+2}}{k+1}, -1 < x < 1$   **71.** $-1 \le x < 1$   **73.** $-1 \le x < 1$   **75.** $-1 < x < 1$   **77.** $-3 < x < 5$

**79. (a)** $\displaystyle\sum_{k=0}^{\infty} x^{2k}$   **(b)** $-1 < x < 1$   **81.** 0.693   **83.** SSM.   **85.** SSM.

**87.** SSM.   **89.** $-1 \le x < 5$   **91.** $\displaystyle\sum_{k=0}^{\infty} \dfrac{(-1)^k x^{2k+1}}{(2k+1)!}; R = \infty$

## Section 8.9   (SSM = See Student Solutions Manual; AWV = Answers will vary)

**1.** Taylor series   **2.** Maclaurin   **3.** $2 - 4x + \dfrac{3}{2}x^2 - \dfrac{1}{3}x^3$   **4.** $3 + 2(x-5) - \dfrac{1}{2}(x-5)^2 - \dfrac{2}{3}(x-5)^3$

**5.** $f(x) = -6 + 5x + 2x^2 + 3x^3$   **7.** $f(x) = 4 + 18(x-1) + 11(x-1)^2 + 3(x-1)^3$

**9.** $\displaystyle\sum_{k=0}^{\infty} \dfrac{x^{2k}}{k!}$; interval: $(-\infty, \infty)$   **11.** $\displaystyle\sum_{k=0}^{\infty} \dfrac{x^{k+1}}{k!}$; interval: $(-\infty, \infty)$

**13.** $\displaystyle\sum_{k=0}^{\infty} \dfrac{(-1)^k x^{2k+3}}{(2k+1)!}$; interval: $(-\infty, \infty)$   **15.** $\displaystyle\sum_{k=0}^{\infty} x^{2k}$; interval: $(-1, 1)$

**17.** $\displaystyle\sum_{k=0}^{\infty} \dfrac{(3x)^k}{k}$; interval: $\left[-\dfrac{1}{3}, \dfrac{1}{3}\right)$   **19.** $\displaystyle\sum_{k=0}^{\infty} (-1)^k \dfrac{x^{2k+2}}{k+1}$; interval: $[-1, 1]$

**21.** $\displaystyle\sum_{k=0}^{\infty} \binom{-3}{k} x^k = 1 - 3x + 6x^2 - 10x^3 + \cdots$; interval: $(-1, 1)$

**23.** $\displaystyle\sum_{k=0}^{\infty} \binom{1/2}{k} x^{2k} = 1 + \dfrac{x^2}{2} - \dfrac{x^4}{8} + \dfrac{x^6}{16} + \cdots$; interval: $[-1, 1]$   **25.** $\displaystyle\sum_{k=0}^{\infty} \binom{1/5}{k} x^k = 1 + \dfrac{x}{5} - \dfrac{2x^2}{25} + \dfrac{6x^3}{125} + \cdots$; interval: $(-1, 1]$

**27.** $\displaystyle\sum_{k=0}^{\infty} \binom{-1/2}{k} x^{2k} = 1 - \dfrac{x^2}{2} + \dfrac{3x^4}{8} - \dfrac{5x^6}{16} + \cdots$; interval: $[-1, 1]$

**29.** $2\sum_{k=0}^{\infty}\binom{-1/2}{k}(-1)^k x^{k+1} = 2x + x^2 + \frac{3}{4}x^3 + \frac{5}{8}x^4 + \cdots$; interval: $[-1, 1)$

**31.** $2 - x - \frac{x^3}{3!} + \frac{2x^4}{4!} - \frac{x^5}{5!}$  **33.** $1 + 2x + \frac{5}{2}x^2 + \frac{8}{3}x^3 + \frac{65}{24}x^4$  **35.** $1 + x + \frac{x^2}{2} + \frac{x^3}{2} + \frac{13}{24}x^4$

**37.** $x - x^2 + \frac{x^3}{3} - \frac{x^5}{30} + \frac{x^6}{90}$  **39.** $f(x) = \sqrt{x} = 1 + \frac{1}{2}(x-1) - \frac{1}{8}(x-1)^2 + \frac{1}{16}(x-1)^3 - \cdots$

**41.** $f(x) = \ln x = (x-1) - \frac{(x-1)^2}{2} + \frac{(x-1)^3}{3} - \frac{(x-1)^4}{4} + \cdots$  **43.** $f(x) = \sum_{k=0}^{\infty}(-1)^k(x-1)^k$

**45.** $f(x) = \sum_{k=0}^{\infty}\frac{\sin\left(\frac{1}{6}(\pi+3k\pi)\right)\left(x-\frac{\pi}{6}\right)^k}{k!}$  **47.** $e^2\sum_{k=0}^{\infty}\frac{2^k(x-1)^k}{k!}$  **49.** $x + \frac{x^3}{3} + \frac{x^5}{10} + \frac{x^7}{42} + \frac{x^9}{216} + \cdots$

**51.** $\frac{x^3}{3} - \frac{x^7}{42} + \frac{x^{11}}{1320} - \frac{x^{15}}{15\cdot 7!} + \cdots$  **53.** $\sec x = 1 + \frac{x^2}{2} + \frac{5x^4}{24} + \frac{61x^6}{720} + \frac{277x^8}{8064}$  **55.** SSM.

**57.** SSM.  **59.** SSM.  **61.** $\frac{\pi}{4} - \frac{1}{2}$  **63.** SSM.  **65.** SSM.  **67.** $1 + x^2 - \frac{x^3}{2} + \frac{5x^4}{6} + \cdots$

## Section 8.10  (SSM = See Student Solutions Manual; AWV = Answers will vary)

**1.** $P_5(x) = 2(x-1) - (x-1)^2 + \frac{2}{3}(x-1)^3 - \frac{1}{2}(x-1)^4 + \frac{2}{5}(x-1)^5$

**3.** $P_5(x) = 1 - (x-1) + (x-1)^2 - (x-1)^3 + (x-1)^4 - (x-1)^5$

**5.** $P_5(x) = -\left(x-\frac{\pi}{2}\right) + \frac{\left(x-\frac{\pi}{2}\right)^3}{3!} - \frac{\left(x-\frac{\pi}{2}\right)^5}{5!}$  **7.** $P_5(x) = 1 + 2x + 2x^2 + \frac{4}{3}x^3 + \frac{2}{3}x^4 + \frac{4}{15}x^5$

**9.** $P_5(x) = 1 + 2x + 4x^2 + 8x^3 + 16x^4 + 32x^5$  **11.** $P_5(x) = (x-1) + \frac{1}{2}(x-1)^2 - \frac{1}{6}(x-1)^3 + \frac{1}{12}(x-1)^4 - \frac{1}{20}(x-1)^5$

**13. (a)** $P_6(x) = 1 - \frac{x^2}{2!} + \frac{x^4}{4!} - \frac{x^6}{6!}$  **(b)** 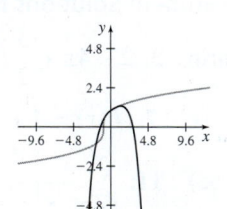  **(c)** $\cos\frac{\pi}{90} \approx 0.999$  **(d)** $|E| \le 5.47 \times 10^{-17}$  **(e)** $|x| \le 1.190$

**15. (a)** $\sum_{k=0}^{\infty}\binom{1/3}{k}x^k$  **(b)** $[-1, 1]$  **(c)** $P_4(x) = 1 + \frac{x}{3} - \frac{x^2}{9} + \frac{5x^3}{81} - \frac{10x^4}{243}$  **(d)** 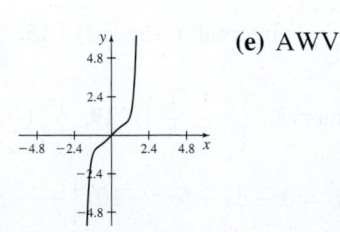  **(e)** AWV.  **(f)** $\approx 0.965$; the error is $\le 3.018 \times 10^{-7}$

**17. (a)** $\sum_{k=0}^{\infty}\frac{(-1)^k x^{1+2k}}{1+2k}$  **(b)** $[-1, 1]$  **(c)** $P_9(x) = x - \frac{x^3}{3} + \frac{x^5}{5} - \frac{x^7}{7} + \frac{x^9}{9}$  **(d)**  **(e)** AWV.

**19.** 0.310  **21.** 0.111  **23.** 0.200  **25.** 0.487  **27.** SSM.

**29. (a)** $P_4(x) = 0.34 - 0.34\left(\frac{\ln 2}{5600}\right)x + 0.17\left(\frac{\ln 2}{5600}\right)^2 x^2 - \frac{0.17}{3}\left(\frac{\ln 2}{5600}\right)^3 x^3 + \frac{0.17}{12}\left(\frac{\ln 2}{5600}\right)^4 x^4$  **(b)**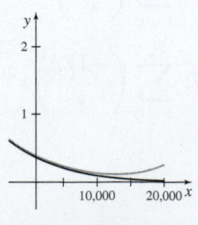

**31. (a)** $P_3(x) = 3.205 + 0.620x + 0.058x^2 + 0.0034x^3$ **(b)**

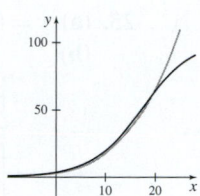

**33. (a)** 0.0698 **(b)** 0.9659 **(c)** 0.04996 **35.** $\dfrac{1538665}{489888} \approx 3.14085$ **37.** $P_2(x) = -(1+\lambda)x + \pi$; SSM.

**39. (a)** $(-1)^{k+1}\dfrac{2}{k}$ **(b)** SSM. **(c)** SSM. **41. (a)** SSM. **(b)** 1.571 **(c)** 0.524

## Review Exercises  (SSM = See Student Solutions Manual; AWV = Answers will vary)

**1.** $1, -\dfrac{1}{16}, \dfrac{1}{81}, -\dfrac{1}{256}, \dfrac{1}{625}$ **3.** $a_n = 2\left(-\dfrac{3}{4}\right)^{n-1}$ **5.** 0 **7.** 0 **9.** Diverges **11.** Converges; 0 **13.** SSM. **15.** $\dfrac{4}{5}$

**17.** Converges; $\dfrac{\ln 2}{\ln 2 - 1}$ **19.** Diverges **21.** SSM. **23.** Converges **25.** Converges **27.** Converges **29.** Converges

**31.** Converges; 1.079 **33.** Converges; −1.715 **35.** Conditionally convergent **37.** Conditionally convergent

**39.** Converges **41.** Diverges **43.** Diverges **45.** Converges **47.** Diverges **49.** Converges **51.** Converges

**53.** $6.244 < \displaystyle\sum_{k=2}^{\infty} \dfrac{6}{k(\ln k)^3} < 15.252$ **55. (a)** $R=1$ **(b)** $2 \le x \le 4$ **57. (a)** $R=\infty$ **(b)** all $x$

**59. (a)** $R=1$ **(b)** $0 \le x < 2$ **61.** $f(x) = \dfrac{2}{3}\displaystyle\sum_{k=0}^{\infty}(-1)^k\left(\dfrac{x}{3}\right)^k, -3 < x < 3$ **63. (a)** $\displaystyle\sum_{k=0}^{\infty}\dfrac{3^k}{2k+1}x^{2k+1}$ **(b)** 0.758

**65.** $\displaystyle\sum_{k=0}^{\infty}\dfrac{\sqrt{e}(x-1)^k}{2^k k!}$ **67.** $1 + 2\left(x - \dfrac{\pi}{4}\right) + 2\left(x - \dfrac{\pi}{4}\right)^2 + \dfrac{8}{3}\left(x - \dfrac{\pi}{4}\right)^3 + \dfrac{10}{3}\left(x - \dfrac{\pi}{4}\right)^4 + \dfrac{64}{15}\left(x - \dfrac{\pi}{4}\right)^5 + \cdots$

**69.** $\displaystyle\sum_{k=0}^{\infty}\binom{-4}{k}x^k = 1 - 4x + 10x^2 - 20x^3 + \dots; -1 < x < 1$

**71.** $\displaystyle\sum_{k=0}^{\infty}\binom{-1/2}{k}(-1)^k x^k = 1 + \dfrac{1}{2}x + \dfrac{3}{8}x^2 + \dfrac{5}{16}x^3 + \dots; -1 < x \le 1$ **73.** 1.349 **75.** 0.545

# Chapter 9
## Section 9.1  (SSM = See Student Solutions Manual; AWV = Answers will vary)

**1.** Plane curve; parameter **2.** (c) **3.** (d) **4.** Cycloid **5.** False **6.** True

**7. (a)** $y = \dfrac{1}{2}x + \dfrac{3}{2}$ **9. (a)** $x = 2y - 3, 1 \le x \le 5$ **11. (a)** $x = e^y$ **13. (a)** $x^2 + y^2 = 1$

**(b)**
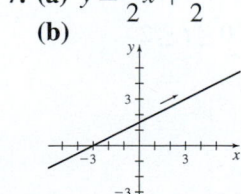
**(b)**
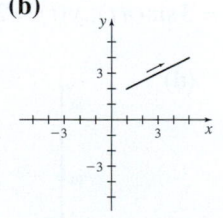
**(b)**
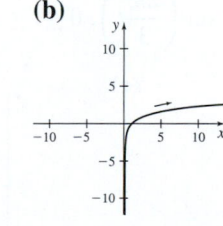
**(b)**
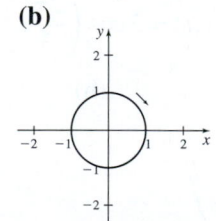

**15. (a)** $\dfrac{x^2}{4} + \dfrac{y^2}{9} = 1$ **17. (a)** $\dfrac{(x+3)^2}{4} + \dfrac{(y-1)^2}{4} = 1$ **19. (a)** $x = \dfrac{y^2}{4}$ **21. (a)** $y = \dfrac{x^2}{4} + 4$ **(c)** $x > 0, y > 4$

**(b)**
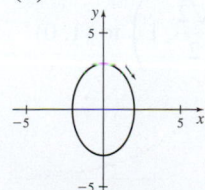
**(b)**
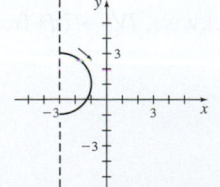
**(b)**
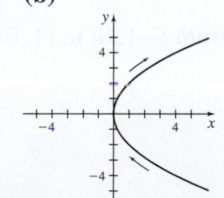
**(b)**
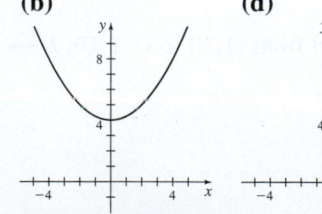
**(d)**

**23. (a)** $x = y^2 + 5$    **(c)** $x \geq 5, y \geq 0$      **25. (a)** $y = (x-1)^3$    **(c)** $x \geq 2, y \geq 1$

**(b)**      **(d)**        **(b)**       **(d)**

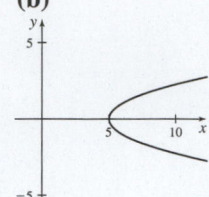 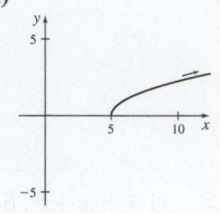 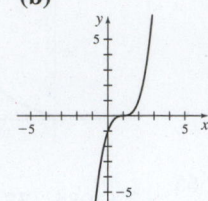 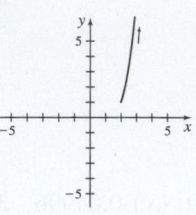

**27. (a)** $x = \sec \tan^{-1} y$    **(c)** $x \geq 1$      **29. (a)** $x = y^2$    **(c)** $x \geq 0, y \geq 0$

**(b)**      **(d)**        **(b)**       **(d)**

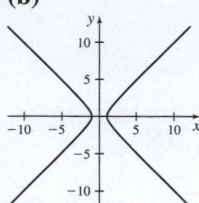 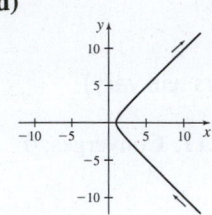  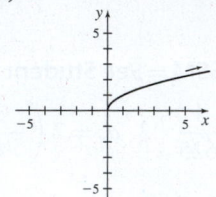

**31. (a)** $x = \left(\dfrac{y+1}{2}\right)^2$    **(c)** $x \geq 0$      **33. (a)** $x = \left(\dfrac{y+1}{2}\right)^2$    **(c)** $x > 0, y \neq -1$

**(b)**      **(d)**        **(b)**       **(d)**

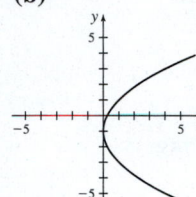 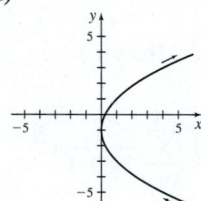 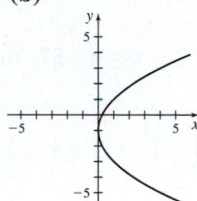 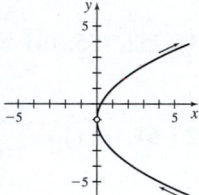

**35. (a)** $\dfrac{x+2}{3} + \dfrac{y^2}{4} = 1$    **(c)** $-2 \leq x \leq 1, -2 \leq y \leq 2$

**(b)**      **(d)**

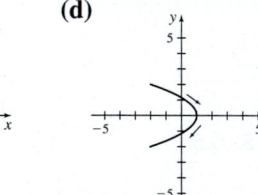

**37.** $y = \dfrac{2x}{1+x}, x \geq 0, 0 \leq y < 2$; AWV.    **39.** $y = 2\sqrt{x^2 - 4}, x \geq 2, y \geq 0$; AWV.

**41.** $(x+2)^2 + \left(\dfrac{4-y}{2}\right)^2 = 1, -3 \leq x \leq -1, 2 \leq y \leq 6$; AWV.    **43.** AWV.    **45.** AWV.    **47.** AWV.    **49.** AWV.    **51.** AWV.

**53.** AWV.    **55.** $x(t) = 3\cos\left(\dfrac{2\pi}{3}t\right), y(t) = 2\sin\left(\dfrac{2\pi}{3}t\right), 0 \leq t \leq 3$    **57.** $x(t) = 3\sin(\pi t), y(t) = 2\cos(\pi t), 0 \leq t \leq 2$

**59. (a)**      **(b)**      **(c)**      **(d)**

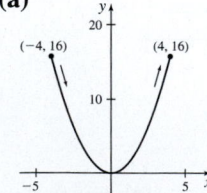 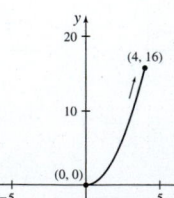 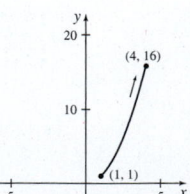 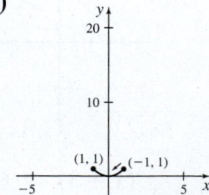

**61.** $I \to$ (d) counterclockwise, $II \to$ (a) counterclockwise, $III \to$ (b) counterclockwise, $IV \to$ (c) counterclockwise

**63.** $I \to$ (c) from $(1, 0)$ to $(-1, 0), II \to$ (b) from $(-1, 0)$ to $(1, 0), III \to$ (a) clockwise, $IV \to$ (d) from $\left(-\dfrac{\sqrt{2}}{2}, 1\right)$ to $(1, 0)$

**65. (a)** 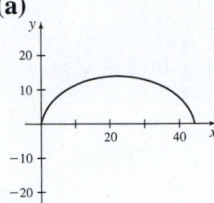   **(b)** $x \approx 8.66$, $y \approx 10.53$   **67.** AWV.

**69. (a)** $x(t) = (125 \cos 40°)t$, $y(t) = -16t^2 + (125 \sin 40°)t + 3$   **(b)** $y \approx 99.7$ ft
   **(c)** $x \approx 191.5$ ft   **(d)** $t \approx 3.13$ s   **(e)** $y \approx 97.7$ ft   **(f)** $t \approx 5.06$ s   **(g)** $x \approx 484.5$ ft
**71. (a)** $x(t) = (80 \cos 35°)t$, $y(t) = -16t^2 + (80 \sin 35°)t + 6$   **(b)** $y \approx 35.9$ ft   **(c)** $x \approx 65.5$ ft   **(d)** $t \approx 1.83$ s   **(e)** $y \approx 36.4$ ft
**73.** The first circle is counterclockwise from $(2, 0)$; the second is clockwise from $(0, 2)$. AWV.

**75.** $x = \dfrac{R(1 - m^2)}{1 + m^2}$, $y = \dfrac{2Rm}{1 + m^2}$   **77.** SSM.   **79.** $x(t) = (a + b) \cos t - b \cos \left( \dfrac{a + b}{b} t \right)$, $y(t) = (a + b) \sin t - b \sin \left( \dfrac{a + b}{b} t \right)$

**Section 9.2**   (SSM = See Student Solutions Manual; AWV = Answers will vary)

**1. (a)**   **2.** True   **3.** Horizontal; vertical   **4.** False   **5.** $\dfrac{dy}{dx} = \dfrac{\sin t + \cos t}{\cos t - \sin t}$   **7.** $\dfrac{dy}{dx} = \dfrac{1}{1 - \dfrac{1}{t^2}}$   **9.** $\dfrac{dy}{dx} = \tan t$   **11.** $\dfrac{dy}{dx} = \dfrac{1}{2 \cot t}$

**13. (a)** $y = \dfrac{1}{8}x + 1$   **15. (a)** $y = \dfrac{4}{3}x - 3$   **17. (a)** $y = -\dfrac{1}{4}x + \dfrac{3}{4}$   **19. (a)** $y = -2x + 2$

**(b)** 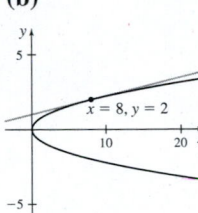   **(b)** 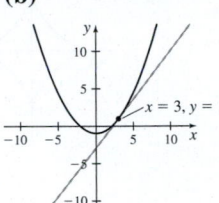   **(b)** 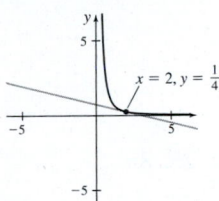   **(b)**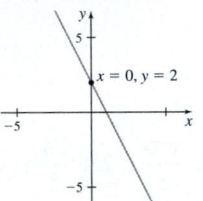

**21. (a)** $y = -x + 2$   **23. (a)** $y = -x + \sqrt{2}$   **25. (a)** $y = -\dfrac{3\sqrt{3}}{4}x + 6$

**(b)** 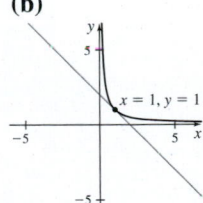   **(b)** 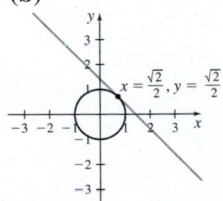   **(b)**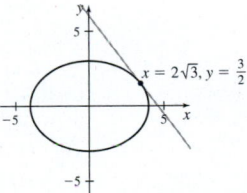

**27.** Horizontal at $\left( \dfrac{4}{3}, -\dfrac{16\sqrt{3}}{9} \right)$, $\left( \dfrac{4}{3}, \dfrac{16\sqrt{3}}{9} \right)$; vertical at $(0, 0)$   **29.** Horizontal at $(1, 0)$, $(1, 2)$; vertical at $(0, 1)$, $(2, 1)$

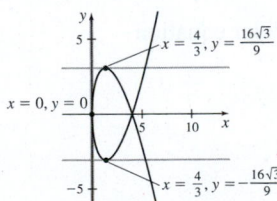   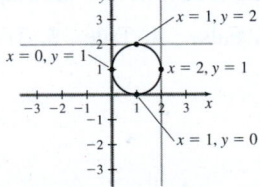

**31. (a)** Horizontal at $\left( \dfrac{10}{3}, -\dfrac{16\sqrt{3}}{9} \right)$, $\left( \dfrac{10}{3}, \dfrac{16\sqrt{3}}{9} \right)$; vertical at $(2, 0)$   **(d)**

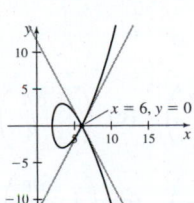

**(b)** $t = 2$, $t = -2$; SSM.
**(c)** $y = 2x - 12$, $y = -2x + 12$

**33.** SSM.   **35.** $\dfrac{1}{3a \cos^4 \theta \sin \theta}$

## Section 9.3

**1.** False   **2.** $S = 2\pi \displaystyle\int_a^b x(t)\sqrt{\left(\dfrac{dx}{dt}\right)^2 + \left(\dfrac{dy}{dt}\right)^2}\,dt$   **3.** False   **4.** False   **5.** $\dfrac{8}{27}\left(10\sqrt{10} - 1\right)$   **7.** $\sqrt{5} + \dfrac{1}{2}\ln(2 + \sqrt{5})$

**9.** $4\pi$   **11.** $4\pi$   **13. (a)** $s = \int_0^{2\pi}\sqrt{[-4\sin(2t)]^2 + (2t)^2}\,dt$   **(b)** $s \approx 44.528$   **(c)**

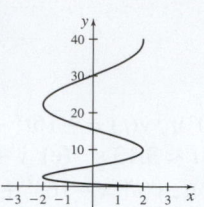

**15. (a)** $s = \displaystyle\int_{-2}^2 \sqrt{(2t)^2 + \left(\dfrac{1}{2\sqrt{t+2}}\right)^2}\,dt$   **(b)** $s \approx 8.429$   **(c)**

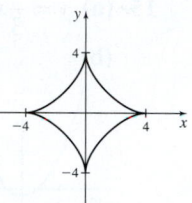

**17. (a)** $s = 4\int_0^{\pi/2}\sqrt{[-3\sin t - 3\sin(3t)]^2 + [3\cos t - 3\cos(3t)]^2}\,dt$   **(b)** $s = 24$   **(c)**

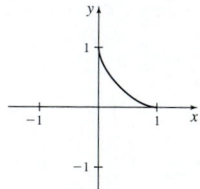

**19.** $24\pi(2\sqrt{2} - 1)$   **21.** $\dfrac{6\pi}{5}$   **23.** $\dfrac{24}{5}\pi(\sqrt{2} + 1)$   **25.** $8\pi$

**27. (a)** $s = \dfrac{3b}{2}$   **(b)** graphed using $b = 1$   **29.** $s = 5 + \dfrac{9}{4}\ln 3$   **31.** $s = \dfrac{3\sqrt{11} - \sqrt{3}}{2} + \ln\left(\dfrac{3 + \sqrt{11}}{1 + \sqrt{3}}\right)$

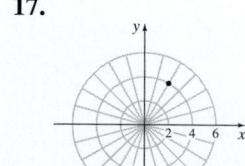

**33.** $s = \sqrt{5}$   **35.** $768\pi$   **37.** $0.734$   **39.** $\sqrt{e^{0.4a} - 2e^{0.2a} + e^{0.4b} - 2e^{0.2b} + 2}$   **41.** $\left(\dfrac{68}{3}, -8\right)$ and $\left(\dfrac{4}{3}, -24\right)$

## Section 9.4   (SSM = See Student Solutions Manual; AWV = Answers will vary)

**1.** Pole, polar axis   **2.** False   **3.** False   **4.** True   **5.** True   **6.** True   **7.** $x = r\cos\theta,\ y = r\sin\theta$   **8.** Polar equation
**9.** A   **11.** C   **13.** B   **15.** A

**17.**

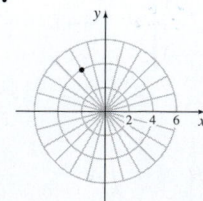

**19.**

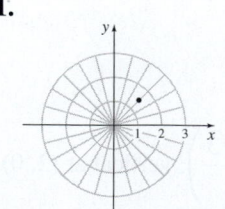

**21.**

**23.**

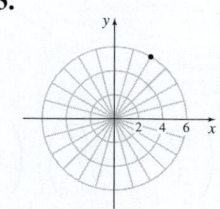

**25. (a)** $\left(5, -\dfrac{4\pi}{3}\right)$   **(b)** $\left(-5, \dfrac{5\pi}{3}\right)$   **(c)** $\left(5, \dfrac{8\pi}{3}\right)$   **27. (a)** $(2, -2\pi)$   **(b)** $(-2, \pi)$   **(c)** $(2, 2\pi)$

**29. (a)** $\left(1, -\dfrac{3\pi}{2}\right)$   **(b)** $\left(-1, \dfrac{3\pi}{2}\right)$   **(c)** $\left(1, \dfrac{5\pi}{2}\right)$   **31. (a)** $\left(3, -\dfrac{5\pi}{4}\right)$   **(b)** $\left(-3, \dfrac{7\pi}{4}\right)$   **(c)** $\left(3, \dfrac{11\pi}{4}\right)$

**33.** $(3\sqrt{3}, 3)$   **35.** $(-3\sqrt{3}, 3)$   **37.** $(0, 5)$   **39.** $(2, -2)$

**41.**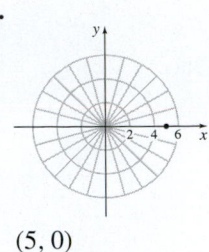

$(5, 0)$

**43.**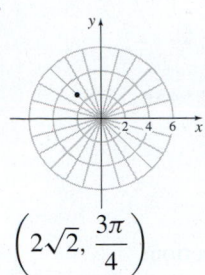

$\left(2\sqrt{2}, \dfrac{3\pi}{4}\right)$

**45.**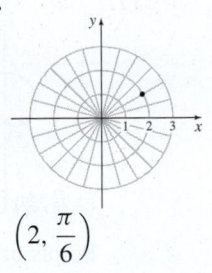

$\left(2, \dfrac{\pi}{6}\right)$

**47.**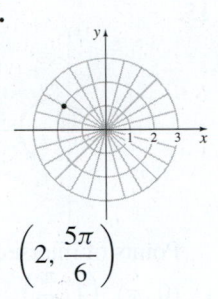

$\left(2, \dfrac{5\pi}{6}\right)$

**49.**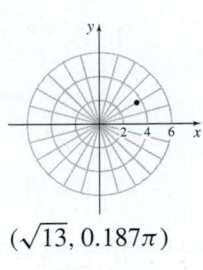

$(\sqrt{13}, 0.187\pi)$

**51.** E **53.** F **55.** H **57.** D

**59.** A circle centered at $(0, 0)$ of radius 4.

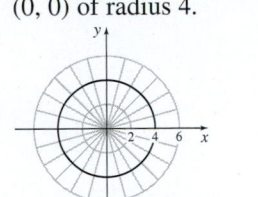

**61.** A line through the origin.

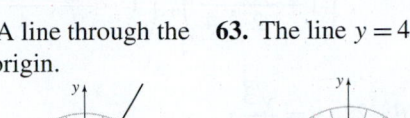

**63.** The line $y = 4$.

**65.** The line $x = -2$.

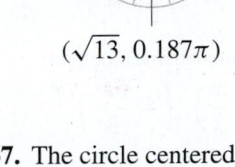
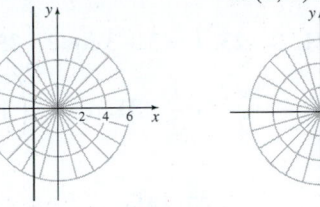

**67.** The circle centered at $(1, 0)$ of radius 1.

**69.** The circle centered at $(0, -2)$ of radius 2.

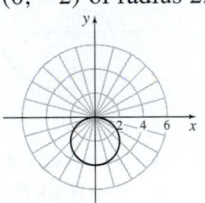

**71.** The circle, excluding the pole, centered at $(2, 0)$ of radius 2.

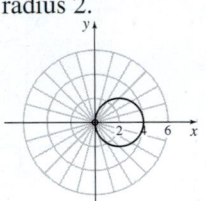

**73.** The circle, excluding the pole, centered at $(0, -1)$ of radius 1.

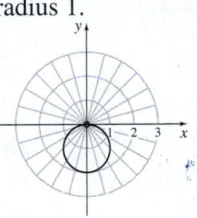

**75.** $r = 6\dfrac{\sqrt{9\cos^2\theta + 4\sin^2\theta}}{9\cos^2\theta + 4\sin^2\theta}$ **77.** $r = 4\cos\theta$ **79.** $r^2\cos^2\theta + 4r\sin\theta - 1 = 0$ **81.** $r = \dfrac{\sqrt{\cos\theta\sin\theta}}{\cos\theta\sin\theta}$ **83.** $\left(x - \dfrac{1}{2}\right)^2 + y^2 = \dfrac{1}{4}$

**85.** $y = (x^2 + y^2)^{3/2}$ **87.** $\sqrt{x^2 + y^2} - x = 4$ **89.** $y = x\tan(x^2 + y^2)$ **91.** $y = \sqrt{4 - x^2}$ **93.** $y = 4x$

**95.** (a) $x = -10, y = 36$ (b) $(37.363, 1.842)$ (c) $x = -3, y = -35$ (d) $(35.128, 4.627)$ **97.** SSM. **99.** SSM. **101.** SSM.

**103.** (a) (b) AWV. (c) (d) AWV. **105.** SSM. **107.** SSM.

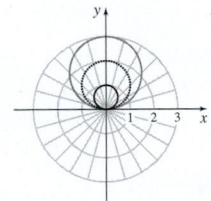
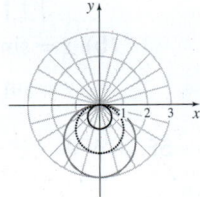

## Section 9.5 (SSM = See Student Solutions Manual; AWV = Answers will vary)

**1.** True **2.** (a), (c), (b) **3.** True **4.** 3

**5.** (a)

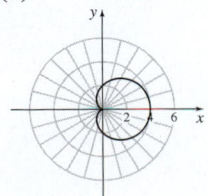

(b) $x = 2(1 + \cos\theta)\cos\theta,$
$y = 2(1 + \cos\theta)\sin\theta$

**7.** (a)

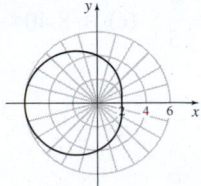

(b) $x = 2(2 - \cos\theta)\cos\theta,$
$y = 2(2 - \cos\theta)\sin\theta$

**9.** (a)

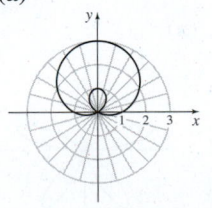

(b) $x = (1 + 2\sin\theta)\cos\theta,$
$y = (1 + 2\sin\theta)\sin\theta$

**11.** (a)

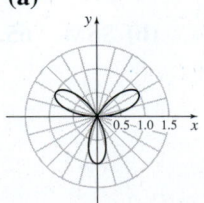

(b) $x = \sin(3\theta)\cos\theta,$
$y = \sin(3\theta)\sin\theta$

**13.**

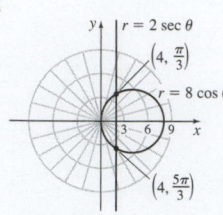

Points of intersection:

$$\left(4, \frac{\pi}{3}\right), \left(4, \frac{5\pi}{3}\right)$$

**15.**

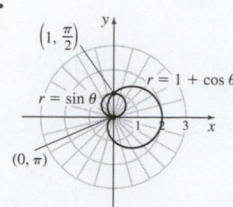

Points of intersection:

$$(0, \pi), \left(1, \frac{\pi}{2}\right)$$

**17.**

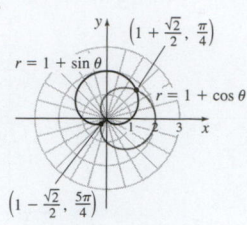

Points of intersection:

$$\left(1 + \frac{\sqrt{2}}{2}, \frac{\pi}{4}\right), \left(1 - \frac{\sqrt{2}}{2}, \frac{5\pi}{4}\right), (0, \theta) \text{ (the pole)}$$

**19.** $\sqrt{5}(e - 1)$   **21.** 2   **23.** $r = 3 + 3\cos\theta$   **25.** $r = 4 + \sin\theta$   **27.** $y = \frac{\sqrt{3}}{3}x$

**29.** $y = (\sqrt{2} - 2)x + \frac{5\sqrt{2}}{2} - \frac{1}{2}$   **31.** $y = -\left(1 + \frac{5\sqrt{2}}{4}\right)x + \frac{57\sqrt{2}}{8} + 10$

**33. (a)**

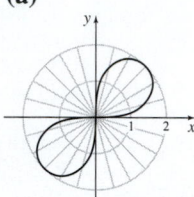

**(b)** $x = 2\sqrt{\sin(2\theta)}\cos\theta$,
$y = 2\sqrt{\sin(2\theta)}\sin\theta$

**35. (a)**

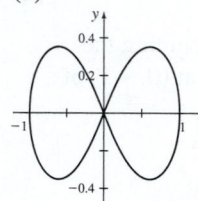

**(b)** $x = \sqrt{\cos(2\theta)}\cos\theta$,
$y = \sqrt{\cos(2\theta)}\sin\theta$

**37. (a)**

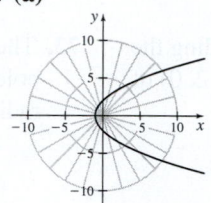

**(b)** $x = \frac{2\cos\theta}{1 - \cos\theta}$,

$y = \frac{2\sin\theta}{1 - \cos\theta}$

**39. (a)**

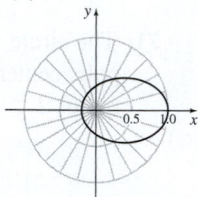

**(b)** $x = \frac{\cos\theta}{3 - 2\cos\theta}$,

$y = \frac{\sin\theta}{3 - 2\cos\theta}$

**41. (a)**

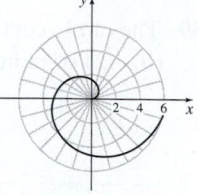

**(b)** $x = \theta\cos\theta$,
$y = \theta\sin\theta$

**43. (a)**

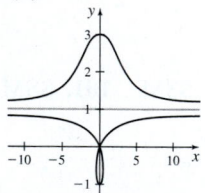

**(b)** $x = (\csc\theta - 2)\cos\theta$,
$y = (\csc\theta - 2)\sin\theta$

**45. (a)**

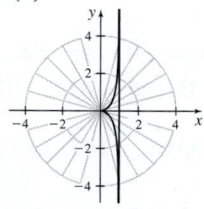

**(b)** $x = \sin^2\theta$,
$y = \sin^2\theta\tan\theta$

**47. (a)**

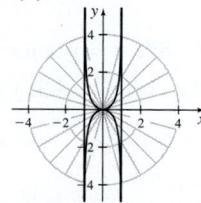

**(b)** $x = \sin\theta$,
$y = \sin\theta\tan\theta$

**49.** SSM.   **51.** $\pi\sqrt{1 + 4\pi^2} + \frac{1}{2}\ln\left(2\pi + \sqrt{4\pi^2 + 1}\right)$   **53.** 8   **55.** AWV.

**57.** Horizontal: $y = \frac{9\sqrt{3}}{4}, y = -\frac{9\sqrt{3}}{4}$; vertical: $x = -\frac{3}{4}, x = 6$   **59.** Horizontal: $y = \frac{4}{3\sqrt{6}}, y = -\frac{4}{3\sqrt{6}}, y = 2, y = -2$;

vertical: $x = \frac{4}{3\sqrt{6}}, x = -\frac{4}{3\sqrt{6}}, x = 2, x = -2$   **61.** Horizontal: $y = \frac{3^{3/4}}{\sqrt{2}}, y = -\frac{3^{3/4}}{\sqrt{2}}$; vertical: $x = \frac{3^{3/4}}{\sqrt{2}}, x = -\frac{3^{3/4}}{\sqrt{2}}$

**63. (a)** AWV.   **(b)** SSM.   **65.** SSM.   **67. (a)**

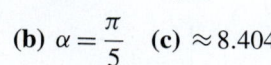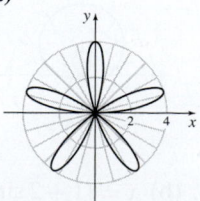

**(b)** $\alpha = \frac{\pi}{5}$   **(c)** $\approx 8.404$

## Section 9.6 (SSM = See Student Solutions Manual; AWV = Answers will vary)

**1.** $\dfrac{\theta r^2}{2}$   **2.** True   **3.** False   **4.** True   **5.** $\dfrac{\pi}{4}$   **7.** $8 + 3\pi$   **9.** $\dfrac{3}{16}\left(3\sqrt{3} + 4\pi\right)$   **11.** $\dfrac{4\pi^3 a^2}{3}$   **13.** $\dfrac{3\pi}{2}$   **15.** $\dfrac{19\pi}{2}$   **17.** $16\pi$

**19.** $2\pi$   **21.** 2   **23.** $\dfrac{\sqrt{3}}{2} + \dfrac{\pi}{3}$   **25.** $1 - \dfrac{\pi}{4}$   **27.** $\dfrac{\pi^2}{2}$   **29.** $\dfrac{2}{5}\sqrt{2}(1 + e^{2\pi})\pi$   **31.** $\pi - \dfrac{3\sqrt{3}}{2}$   **33.** $\sqrt{3} + \dfrac{4\pi}{3} - 4\ln\left(2 + \sqrt{3}\right)$

**35.** $\dfrac{3\pi}{2}$   **37.** $4\sqrt{3} + \dfrac{32\pi}{3}$   **39.** $\dfrac{9\sqrt{3}}{2} - \pi$   **41.** $\dfrac{7\pi}{12} - \sqrt{3}$   **43.** $2\sqrt{2}$   **45.** $\dfrac{1 - e^{-2}}{4}$   **47.** $\dfrac{\pi - 1}{2\pi}$   **49.** $\sqrt{3} + \dfrac{4\pi}{3} - 4\ln\left(2 + \sqrt{3}\right)$

**51.** $4\pi R\sqrt{R^2 - a^2}$   **53.** $2 - \dfrac{\pi}{2}$   **55.** SSM.   **57.** $2\sqrt{5} - \sin^{-1}\left(\dfrac{2}{3}\right)$

## Section 9.7 (SSM = See Student Solutions Manual; AWV = Answers will vary)

**1.** (a)   **2.** AWV.   **3.** They are both parabolas. AWV.   **4.** False

**5.** A parabola, $e = 1$, directrix perpendicular to the polar axis $p = 1$ unit to the right of the pole

**7.** A hyperbola, $e = \dfrac{3}{2}$, directrix parallel to the polar axis $p = \dfrac{4}{3}$ units below the pole

**9.** An ellipse, $e = \dfrac{1}{2}$, directrix perpendicular to the polar axis $p = \dfrac{3}{2}$ units to the left of the pole

**11.** A parabola, $e = 1$, directrix parallel to the polar axis $p = \dfrac{4}{3}$ units above the pole

**13.** (a) Ellipse   (b) $7\left(y + \dfrac{24}{7}\right)^2 + 16x^2 = \dfrac{1024}{7}$   **15.** (a) Hyperbola   (b) $3(x + 2)^2 - y^2 = 3$

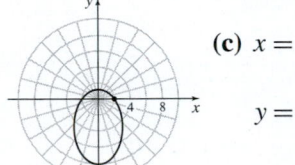

(c) $x = \dfrac{8\cos\theta}{4 + 3\sin\theta}$,
$y = \dfrac{8\sin\theta}{4 + 3\sin\theta}$

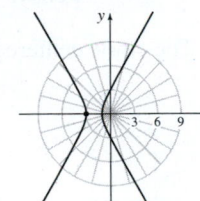

(c) $x = \dfrac{9\cos\theta}{3 - 6\cos\theta}$,
$y = \dfrac{9\sin\theta}{3 - 6\cos\theta}$

**17.** (a) Ellipse   (b) $9x^2 + 5\left(y - \dfrac{12}{5}\right)^2 = \dfrac{324}{5}$   **19.** (a) Ellipse   (b) $4y^2 + 3(x - 2)^2 = 48$

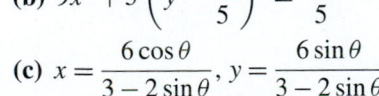

(c) $x = \dfrac{6\cos\theta}{3 - 2\sin\theta}$, $y = \dfrac{6\sin\theta}{3 - 2\sin\theta}$

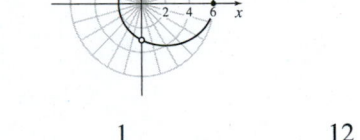

(c) $x = \dfrac{6\cos\theta}{2 - \cos\theta}$, $y = \dfrac{6\sin\theta}{2 - \cos\theta}$

**21.** undefined   **23.** 0   **25.** undefined   **27.** $r = \dfrac{12}{5 - 4\cos\theta}$   **29.** $r = \dfrac{1}{1 + \sin\theta}$   **31.** $r = \dfrac{12}{1 - 6\sin\theta}$

**33.** (a) 0.967   (b) 0.587 AU   (c) 35 AU   (d)

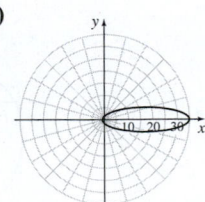

**35.** (a) SSM.   (b) and (c) AWV.   **37.** SSM.   **39.** SSM.

## Review Exercises   (SSM = See Student Solutions Manual; AWV = Answers will vary)

**1. (a)** $x = -4y + 2$
**(b)**

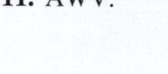

**(c)** No restrictions

**3. (a)** $y = \dfrac{1}{x}$
**(b)**

**(c)** $x > 0$, $y > 0$

**5. (a)** $y + 1 = x$
**(b)**

**(c)** $1 \le x \le 2$,
$0 \le y \le 1$

**7. (a)** $y = \dfrac{1}{2}x + \dfrac{5}{2}$
**(b)**

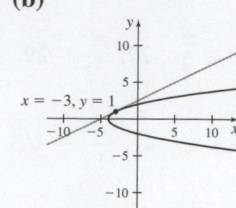

**9. (a)** $y = -\dfrac{81}{2\sqrt{10}}x + \dfrac{9}{2\sqrt{10}} + \sqrt{10}$
**(b)**

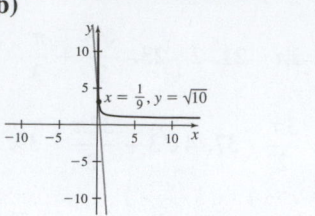

**11.** AWV.

**13.** $\dfrac{x^2}{4} + y^2 = 1$

**15.** $(1, \sqrt{3})$

**17.** $(-2\sqrt{2}, 2\sqrt{2})$

**19.** $(5, 0.295\pi)$, $(-5, 1.295\pi)$   **21.** $\left(3\sqrt{2}, \dfrac{3\pi}{4}\right)$, $\left(-3\sqrt{2}, \dfrac{7\pi}{4}\right)$   **23.** $y = x \tan \ln(x^2 + y^2)$   **25.** $x^2 + y^2 = a\sqrt{x^2 + y^2} - y$

**27.** $y = x \tan \sqrt{x^2 + y^2}$   **29.** $\cos^2 \theta - \sin^2 \theta = r^2$   **31.** $r = \dfrac{6\sqrt{9\cos^2 \theta + 4\sin^2 \theta}}{9\cos^2 \theta + 4\sin^2 \theta}$

**33.** The circle centered at $(1, 0)$ of radius 1.

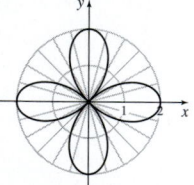

**35.** The circle centered at $\left(\dfrac{-5}{2}, 0\right)$ of radius $\dfrac{5}{2}$.

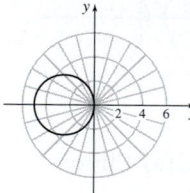

**37. (a)**

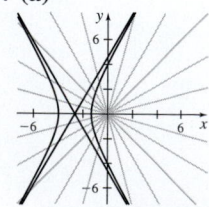

**(b)** $x = 4\cos(2\theta)\cos\theta$, $y = 4\cos(2\theta)\sin\theta$

**39. (a)**

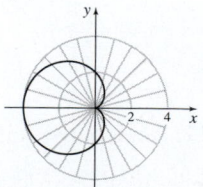

**(b)** $x = \dfrac{4\cos\theta}{1 - 2\cos\theta}$, $y = \dfrac{4\sin\theta}{1 - 2\cos\theta}$

**41. (a)**

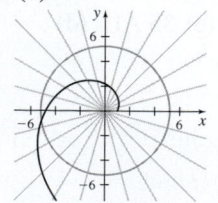

**(b)** $x = (2 - 2\cos\theta)\cos\theta$, $y = (2 - 2\cos\theta)\sin\theta$

**43. (a)**

**(b)** $x = e^{0.5\theta}\cos\theta$, $y = e^{0.5\theta}\sin\theta$

**45. (a)**

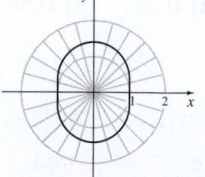

**(b)** $x = \sqrt{1 + \sin^2\theta}\cos\theta$, $y = \sqrt{1 + \sin^2\theta}\sin\theta$

**47. (a)**  **(b)** $35\left(x - \dfrac{6}{35}\right)^2 + 36y^2 = \dfrac{1296}{35}$  **(c)** $x = \dfrac{6\cos\theta}{6 - \cos\theta}, y = \dfrac{6\sin\theta}{6 - \cos\theta}$

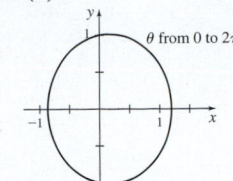
$\theta$ from 0 to $2\pi$

**49.** $x = 4\sin\left(\dfrac{2\pi t}{5}\right)$, $y = 3\cos\left(\dfrac{2\pi t}{5}\right)$, $0 \le t \le 5$  **51.** Vertical: $(0, 2)$, $(2, 2)$; horizontal: $(1, 5)$, $(1, -1)$

**53.** $\dfrac{\sqrt{13}}{6} + \dfrac{3}{4}\ln\dfrac{\sqrt{13}+2}{3}$  **55.** $\dfrac{3}{2} + \dfrac{\ln 2}{4}$  **57.** $\sqrt{2}(1 - e^{-2\pi})$  **59.** $2(4 - 2\sqrt{2})$  **61.** $4\sqrt{3} - \dfrac{2\pi}{3}$

**63.** $\pi[\sqrt{2} + \ln(1 + \sqrt{2})]$  **65.** $16\pi$

# Chapter 10
## Section 10.1

**1.** True  **2.** False  **3.** True  **4.** True  **5.** Sphere  **6.** $(0, 0, 0)$
**7.** $(1, 1, 1)$  **9.** $(0, 2, 5)$  **11.** $(-3, 1, 0)$  **13.** $(2, 0, 0)$, $(0, 1, 0)$, $(0, 0, 3)$, $(2, 1, 0)$, $(2, 0, 3)$, $(0, 1, 3)$

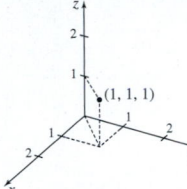  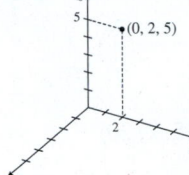  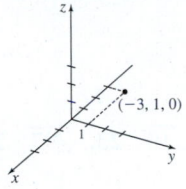

**15.** $(1, 4, 3)$, $(3, 4, 3)$, $(3, 2, 3)$, $(1, 2, 5)$, $(1, 4, 5)$, $(3, 2, 5)$  **17.** $(-1, 0, 5)$, $(4, 2, 2)$, $(-1, 2, 2)$, $(4, 0, 5)$, $(4, 0, 2)$, $(-1, 2, 5)$

**19.** A plane parallel to the $xz$ plane passing through the point $(0, -2, 0)$

**21.** A plane in space, also known as the $yz$-plane  **23.** A line parallel to the $z$-axis through the point $(1, 0, 0)$

**25.** The region of space where $x > -2$ and the $y$ and $z$ values are any real numbers

**27.** All points whose distance from the origin is at most 1.

**29.** All points $(x, y, z)$ that lie either on the plane $x = 0$ or on the plane $y = 0$.

**31.** All points $(x, y, z)$ that lie either on the plane $x = 1$ or on the plane $z = -3$.  **33.** 5  **35.** $\sqrt{57}$

**37.** $\sqrt{26}$  **39.** $(x - 3)^2 + (y - 1)^2 + (z - 1)^2 = 1$  **41.** $(x + 1)^2 + (y - 1)^2 + (z - 2)^2 = 9$

**43.** Radius 2; center $(-1, 1, 0)$  **45.** Radius 3; center $(-2, 2, -1)$  **47.** Radius $\dfrac{9}{4}$; center $\left(2, 0, -\dfrac{5}{4}\right)$

**49.** $x^2 + (y - 3)^2 + (z - 6)^2 = 17$  **51.** $(x + 3)^2 + (y - 2)^2 + (z - 1)^2 = 62$  **53.** $(x - 2)^2 + (y - 1)^2 + (z + 2)^2 = 4$

**55. (a)** $(0, 0, 0)$, $(0.287, 0, 0)$, $(0.287, 0.287, 0)$, $(0, 0.287, 0)$, $(0, 0, 0.287)$, $(0.287, 0, 0.287)$, $(0, 0.287, 0.287)$,
$(0.287, 0.287, 0.287)$, $(0.1435, 0.1435, 0.1435)$  **(b)** $(0.287, 0.287, 0.287)$  **(c)** $(0.1435, 0.1435, 0.1435)$  **(d)** $\approx 0.249$ nm

**57. (a)** $(x - 4)^2 + y^2 + (z + 2)^2 = 25$  **(b)** The circle $(x - 4)^2 + y^2 = 21$; center $(4, 0)$, radius $\sqrt{21}$

**59.** 1  **61.** $(x - 1)^2 + (y + 2)^2 + (z + 1)^2 = 9$

## Section 10.2  (SSM = See Student Solutions Manual; AWV = Answers will vary)

**1.** True  **2.** Magnitude  **3.** Scalar multiple  **4.** (a); (d)  **5.** Scalars: (a), (b), (d), (e), (f), (g); Vectors: (c), (h), (i)
**7.** $-2\mathbf{v}$  **9.** $\mathbf{v} - \mathbf{w}$  **11.** $\mathbf{v} - 2\mathbf{w}$  **13.** $\mathbf{v} + (\mathbf{w} + 3\mathbf{u})$

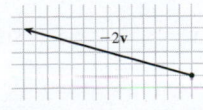  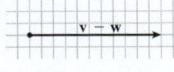  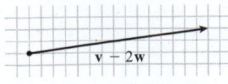  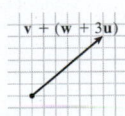

**15.** $x = A$  **17.** $C = -D - E - F$  **19.** $E = G + H - D$  **21.** $0$  **23.**

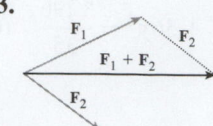

**25. (a)**

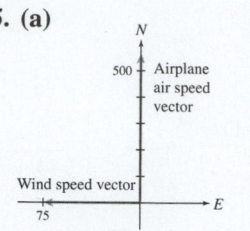

**(b)**

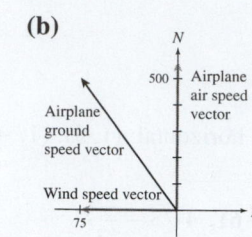

**(c)** AWV.  **27.** $u = \dfrac{1}{2}(v + w)$

## Section 10.3  (SSM = See Student Solutions Manual; AWV = Answers will vary)

**1.** $\langle 1, 0, 0 \rangle$; $\langle 0, 1, 0 \rangle$; $\langle 0, 0, 1 \rangle$  **2.** Unit  **3.** Components  **4.** True  **5.** False  **6.** True  **7.** $\dfrac{1}{\sqrt{3}}i + \dfrac{1}{\sqrt{3}}j + \dfrac{1}{\sqrt{3}}k$  **8.** False

**9. (a)** $\langle 4, -5 \rangle$  **(b)** $4i - 5j$    **11. (a)** $\langle -1, 1 \rangle$  **(b)** $-i + j$    **13. (a)** $\langle -3, -2, 1 \rangle$  **(b)** $-3i - 2j + k$

**(c)**

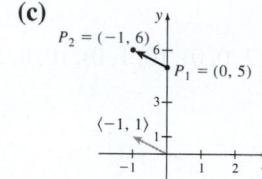

**(c)**

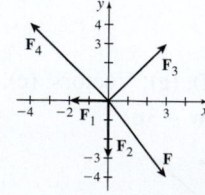

**15. (a)** $\langle 3, 0, -1 \rangle$  **(b)** $3i - k$  **17.** $\langle 9, -4 \rangle$  **19.** $\left\langle -\dfrac{1}{2}, \dfrac{2}{3} \right\rangle$  **21.** $\sqrt{10}$  **23.** $\sqrt{97}$  **25.** $\langle 15, 2, 20 \rangle$

**27.** $\langle -19, -2, -28 \rangle$  **29.** $\sqrt{38}$  **31.** $\sqrt{1313}$  **33.** $5$  **35.** $\sqrt{2}$  **37.** $\sqrt{21}$  **39.** $\sqrt{2}$  **41.** $1$  **43.** $\left\langle \dfrac{5}{13}, \dfrac{12}{13} \right\rangle, \left\langle -\dfrac{5}{13}, -\dfrac{12}{13} \right\rangle$

**45.** $\dfrac{2}{\sqrt{5}}i + \dfrac{1}{\sqrt{5}}j;\ -\dfrac{2}{\sqrt{5}}i - \dfrac{1}{\sqrt{5}}j$  **47.** $\dfrac{1}{2}i + \dfrac{\sqrt{3}}{2}j;\ -\dfrac{1}{2}i - \dfrac{\sqrt{3}}{2}j$  **49.** $\left\langle \dfrac{1}{\sqrt{3}}, \dfrac{1}{\sqrt{3}}, \dfrac{1}{\sqrt{3}} \right\rangle; \left\langle -\dfrac{1}{\sqrt{3}}, -\dfrac{1}{\sqrt{3}}, -\dfrac{1}{\sqrt{3}} \right\rangle$

**51.** $\dfrac{\cos\theta}{3}i + \dfrac{\sin\theta}{3}j + \dfrac{2\sqrt{2}}{3}k;\ -\dfrac{\cos\theta}{3}i - \dfrac{\sin\theta}{3}j - \dfrac{2\sqrt{2}}{3}k$  **53.** $\dfrac{3}{13}i - \dfrac{4}{13}j + \dfrac{12}{13}k;\ -\dfrac{3}{13}i + \dfrac{4}{13}j - \dfrac{12}{13}k$

**55.** $3i - 8j$  **57.** $\dfrac{7}{6}i$  **59.** $\sqrt{26}$  **61.** $-3i + j - 5k$  **63.** $17i - 3j + 13k$  **65.** $\sqrt{13}$  **67.** $\sqrt{14} + \sqrt{11} + \sqrt{38}$

**69.** $\dfrac{4}{\sqrt{38}}i - \dfrac{1}{\sqrt{38}}j + \dfrac{2}{\sqrt{38}}k$  **71.** Parallel; same  **73.** Not parallel  **75.** Parallel; opposite

**77.** $\dfrac{8}{\sqrt{29}}i - \dfrac{12}{\sqrt{29}}j + \dfrac{16}{\sqrt{29}}k$  **79.** $a = \pm 1$  **81.** $x = -5, 1$  **83.** $2\sqrt{3}i + 2j$  **85.** $2\sqrt{5}i + \sqrt{5}j, -2\sqrt{5}i - \sqrt{5}j$

**87.** $-i + j$  **89.** $\sqrt{\dfrac{19}{2}}$ N/kg  **91. (a)** $(0, -2), (2, -4), (6, -1), (4, 1)$

**(b)** $\left( -\dfrac{3}{2}, -1 \right), \left( \dfrac{1}{2}, -3 \right), \left( \dfrac{9}{2}, 0 \right), \left( \dfrac{5}{2}, 2 \right)$

**93.** $-275i + 75\sqrt{3}j$  **95.** $F = 3i - 4j$    **97.** Left: $\approx 1000$ lb; right: $\approx 845.237$ lb.

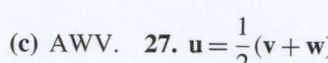

**99.** (a) $\mathbf{A} = \langle 1100, 0 \rangle$, $\mathbf{B} = \langle -525, 525\sqrt{3} \rangle$, $\mathbf{C} = \langle -500, -500\sqrt{3} \rangle$  (b) $75\mathbf{i} + 25\sqrt{3}\mathbf{j}$  **101.** $\mathbf{F} = \langle 20\sqrt{3}, 20 \rangle$

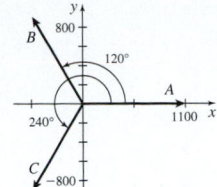
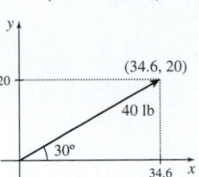

**103.** (a) $(0, 0, 0)$, $(0.408, 0, 0)$, $(0, 0.408, 0)$, $(0, 0, 0.408)$, $(0.408, 0.408, 0)$, $(0.408, 0, 0.408)$, $(0, 0.408, 0.408)$,

$(0.408, 0.408, 0.408)$, $(0.204, 0, 0.204)$, $(0, 0.204, 0.204)$, $(0.204, 0.204, 0)$, $(0.204, 0.408, 0.204)$, $(0.204, 0.204, 0.408)$,

$(0.408, 0.204, 0.204)$  (b) (i) $0.408\mathbf{i} + 0.408\mathbf{j} + 0.408\mathbf{k}$  (ii) $0.204\mathbf{i} + 0.408\mathbf{j} + 0.204\mathbf{k}$  (iii) $0.204\mathbf{i} + 0.204\mathbf{j} + 0.408\mathbf{k}$

(c) $\approx 0.707$ nm; $\approx 0.500$ nm; $\approx 0.500$ nm

**105.** $8\sqrt{2}$ mi/h from northwest  **107.** SSM.  **109.** SSM.

**111.** $\overrightarrow{AC} = \mathbf{b} + \mathbf{d}$; $\overrightarrow{AF} = \mathbf{b} + \mathbf{e}$; $\overrightarrow{AG} = \mathbf{b} + \mathbf{d} + \mathbf{e}$; $\overrightarrow{FG} = \mathbf{d}$; $\overrightarrow{EG} = \mathbf{b} + \mathbf{d}$  **113.** SSM.

## Section 10.4  (SSM = See Student Solutions Manual; AWV = Answers will vary)

**1.** False  **2.** False  **3.** $\dfrac{\mathbf{v} \cdot \mathbf{w}}{||\mathbf{v}|| \; ||\mathbf{w}||}$  **4.** True  **5.** False  **6.** True  **7.** (a) 6  (b) $\approx 0.388$ radians  **9.** (a) $-1$  (b) $\dfrac{2\pi}{3}$

**11.** (a) $-8$  (b) $\approx 2.967$ radians  **13.** (a) 0  (b) $\dfrac{\pi}{2}$  **15.** $\mathbf{v}$ and $\mathbf{w}$ are orthogonal.  **17.** $\mathbf{v}$ and $\mathbf{w}$ are orthogonal.

**19.** $\mathbf{v}$ and $\mathbf{w}$ are not orthogonal, the angle between them is $\cos^{-1}\dfrac{7\sqrt{6}}{30}$  **21.** $a = 1$  **23.** $a = -2$

**25.** (a) $||\mathbf{v}|| = 7$; $\cos\alpha = \dfrac{3}{7}$; $\cos\beta = -\dfrac{6}{7}$; $\cos\gamma = -\dfrac{2}{7}$  (b) $\mathbf{v} = 7\left(\dfrac{3}{7}\mathbf{i} - \dfrac{6}{7}\mathbf{j} - \dfrac{2}{7}\mathbf{k}\right)$

**27.** (a) $||\mathbf{v}|| = \sqrt{3}$; $\cos\alpha = \dfrac{1}{\sqrt{3}}$; $\cos\beta = \dfrac{1}{\sqrt{3}}$; $\cos\gamma = \dfrac{1}{\sqrt{3}}$  (b) $\mathbf{v} = \sqrt{3}\left(\dfrac{1}{\sqrt{3}}\mathbf{i} + \dfrac{1}{\sqrt{3}}\mathbf{j} + \dfrac{1}{\sqrt{3}}\mathbf{k}\right)$

**29.** (a) $||\mathbf{v}|| = \sqrt{2}$; $\cos\alpha = \dfrac{1}{\sqrt{2}}$; $\cos\beta = 0$; $\cos\gamma = -\dfrac{1}{\sqrt{2}}$  (b) $\mathbf{v} = \sqrt{2}\left(\dfrac{1}{\sqrt{2}}\mathbf{i} - \dfrac{1}{\sqrt{2}}\mathbf{k}\right)$

**31.** (a) $||\mathbf{v}|| = \sqrt{38}$; $\cos\alpha = \dfrac{3}{\sqrt{38}}$; $\cos\beta = \dfrac{-5}{\sqrt{38}}$; $\cos\gamma = \dfrac{2}{\sqrt{38}}$  (b) $\mathbf{v} = \sqrt{38}\left(\dfrac{3}{\sqrt{38}}\mathbf{i} - \dfrac{5}{\sqrt{38}}\mathbf{j} + \dfrac{2}{\sqrt{38}}\mathbf{k}\right)$

**33.** (a) $\text{proj}_{\mathbf{w}}\mathbf{v} = 2\mathbf{i} - 2\mathbf{j} + 2\mathbf{k}$  (b) $\mathbf{v}_1 = 2\mathbf{i} - 2\mathbf{j} + 2\mathbf{k}$; $\mathbf{v}_2 = -\mathbf{j} - \mathbf{k}$

**35.** (a) $\text{proj}_{\mathbf{w}}\mathbf{v} = -\dfrac{1}{2}\mathbf{j} - \dfrac{1}{2}\mathbf{k}$  (b) $\mathbf{v}_1 = -\dfrac{1}{2}\mathbf{j} - \dfrac{1}{2}\mathbf{k}$; $\mathbf{v}_2 = \mathbf{i} - \dfrac{1}{2}\mathbf{j} + \dfrac{1}{2}\mathbf{k}$

**37.** (a) $\text{proj}_{\mathbf{w}}\mathbf{v} = \dfrac{8}{3}\mathbf{i} + \dfrac{4}{3}\mathbf{j} - \dfrac{4}{3}\mathbf{k}$  (b) $\mathbf{v}_1 = \dfrac{8}{3}\mathbf{i} + \dfrac{4}{3}\mathbf{j} - \dfrac{4}{3}\mathbf{k}$; $\mathbf{v}_2 = \dfrac{1}{3}\mathbf{i} - \dfrac{1}{3}\mathbf{j} + \dfrac{1}{3}\mathbf{k}$  **39.** $a = 0$, $a = 4$

**41.** Speed $\approx 459.536$ km/h, direction $\approx 84.703°$.  **43.** The boat should be steered at an angle of about $70.529°$ to the shore.

**45.** 738 lb, 5248 lb  **47.** $1000\sqrt{3}$ ft-lb  **49.** 2 J  **51.** (a) 7361.216 J; 6010.408 J  (b) AWV.

**53.** SSM.  **55.** SSM.  **57.** $\dfrac{\pi}{3}$  **59.** $\sqrt{\dfrac{5}{4} - \cos\theta}$  **61.** $2\sqrt{13}$; $2\sqrt{7}$  **63.** SSM.  **65.** SSM.  **67.** SSM.  **69.** $\gamma = \dfrac{\pi}{4}$

**71.** $\mathbf{v} = \sqrt{2}\mathbf{i} - \mathbf{j} + \mathbf{k}$  **73.** $\mathbf{v} = \dfrac{1}{8}\mathbf{i} + \dfrac{1}{8}\mathbf{j} + \dfrac{\sqrt{14}}{8}\mathbf{k}$  **75.** $\mathbf{v} = 0$  **77.** SSM.  **79.** SSM.  **81.** SSM.

**83.** $c_1 = \dfrac{(\mathbf{v} \cdot \mathbf{u}_1)(\mathbf{u}_2 \cdot \mathbf{u}_2) - (\mathbf{v} \cdot \mathbf{u}_2)(\mathbf{u}_1 \cdot \mathbf{u}_2)}{(\mathbf{u}_1 \cdot \mathbf{u}_1)(\mathbf{u}_2 \cdot \mathbf{u}_2) - (\mathbf{u}_1 \cdot \mathbf{u}_2)^2}$, $c_2 = \dfrac{(\mathbf{v} \cdot \mathbf{u}_2)(\mathbf{u}_1 \cdot \mathbf{u}_1) - (\mathbf{v} \cdot \mathbf{u}_1)(\mathbf{u}_1 \cdot \mathbf{u}_2)}{(\mathbf{u}_1 \cdot \mathbf{u}_1)(\mathbf{u}_2 \cdot \mathbf{u}_2) - (\mathbf{u}_1 \cdot \mathbf{u}_2)^2}$; $c_1 = \mathbf{v} \cdot \mathbf{u}_1$, $c_2 = \mathbf{v} \cdot \mathbf{u}_2$

**85.** SSM.  **87.** SSM.  **89.** SSM.  **91.** SSM.  **93.** SSM.

**95.** $\mathbf{w}_1 = -\dfrac{1}{\sqrt{2}}\mathbf{i} + \dfrac{1}{\sqrt{2}}\mathbf{j}$; $\mathbf{w}_2 = \dfrac{1}{\sqrt{3}}\mathbf{i} + \dfrac{1}{\sqrt{3}}\mathbf{j} + \dfrac{1}{\sqrt{3}}\mathbf{k}$; $\mathbf{w}_3 = -\dfrac{1}{\sqrt{6}}\mathbf{i} - \dfrac{1}{\sqrt{6}}\mathbf{j} + \dfrac{2}{\sqrt{6}}\mathbf{k}$

**97.** $\|\mathbf{u}\|^2 - \dfrac{(\mathbf{u} \cdot \mathbf{v})^2}{\|\mathbf{v}\|^2}$; SSM.

**99.** $x + y \geq 0$

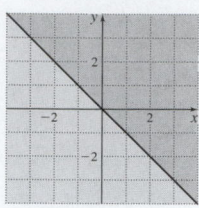

## Section 10.5 (SSM = See Student Solutions Manual; AWV = Answers will vary)

**1.** True  **2.** False  **3.** (b)  **4.** False  **5.** (d)  **6.** True  **7.** (c)  **8.** False  **9.** 2  **11.** 5

**13.** (a) $-3\mathbf{j} - 3\mathbf{k}$  (b) SSM.  **15.** (a) $-\mathbf{i} + \mathbf{j} - \mathbf{k}$  (b) SSM.

**17.** (a) $-\mathbf{i} + \mathbf{j} + 5\mathbf{k}$  (b) SSM.  **19.** (a) $-6\mathbf{i} + 21\mathbf{j} - 58\mathbf{k}$  (b) SSM.

**21.** (a) $-8\mathbf{i} + 12\mathbf{j} + 5\mathbf{k}$  (b) SSM.  **23.** 0  **25.** $\mathbf{i} - 13\mathbf{j} - 4\mathbf{k}$  **27.** $3\mathbf{i} - 39\mathbf{j} - 12\mathbf{k}$

**29.** $\mathbf{i} - 13\mathbf{j} - 4\mathbf{k}$; AWV.  **31.** $2\mathbf{i} - 2\mathbf{j}$; AWV.

**33.** $-\dfrac{8}{3\sqrt{65}}\mathbf{i} + \dfrac{20}{3\sqrt{65}}\mathbf{j} + \dfrac{11}{3\sqrt{65}}\mathbf{k}$ (the opposite of the given vector is also correct)

**35.** $\dfrac{3}{7}\mathbf{i} - \dfrac{2}{7}\mathbf{j} + \dfrac{6}{7}\mathbf{k}$ (the opposite of the given vector is also correct)  **37.** $2\sqrt{3}$  **39.** $\sqrt{158}$

**41.** $\sqrt{1013}$  **43.** 58  **45.** $46\sqrt{2}$  **47.** $5\sqrt{3}$  **49.** 58  **51.** $10\sqrt{78}$ m/s  **53.** SSM.

**55.** (a) AWV.  (b) $\tau_1 = \dfrac{125}{2}\mathbf{k}$ Nm; $\tau_2 = \dfrac{125\sqrt{2}}{4}\mathbf{k}$ Nm; $\tau_3 = 0$ Nm  (c) AWV.

**57.** $(0.0138125\mathbf{i} - 0.015625\mathbf{j} - 0.0036125\mathbf{k})$ N  **59.** SSM.  **61.** SSM.

**63.** AWV.  **65.** 249  **67.** SSM.  **69.** SSM.  **71.** SSM.  **73.** SSM.  **75.** SSM.  **77.** SSM.  **79.** SSM.

**81.** (a) $\|\mathbf{v}\| \approx 8.4896 \times 10^5$ m/s; The velocity $\mathbf{v}$ is parallel to $-\mathbf{k}$.  (b) AWV.

**83.** (a) 0 N, south to north  (b) 0.05 N, east to west  (c) $30°$

## Section 10.6 (SSM = See Student Solutions Manual; AWV = Answers will vary)

**1.** False  **2.** False  **3.** True  **4.** False  **5.** False  **6.** AWV.  **7.** AWV.  **8.** Skew

**9.** (a) $\mathbf{r}(t) = (1 + 2t)\mathbf{i} + (2 - t)\mathbf{j} + (3 + t)\mathbf{k}$  (b) $x = 1 + 2t$, $y = 2 - t$, $z = 3 + t$  (c) $\dfrac{x - 1}{2} = \dfrac{y - 2}{-1} = \dfrac{z - 3}{1}$

**11.** (a) $\mathbf{r}(t) = (1 + 3t)\mathbf{i} + (-1 + 3t)\mathbf{j} + (3 - 2t)\mathbf{k}$  (b) $x = 1 + 3t$; $y = -1 + 3t$; $z = 3 - 2t$  (c) $\dfrac{x - 1}{3} = \dfrac{y + 1}{3} = \dfrac{z - 3}{-2}$

**13.** $x = -1 + 5t$, $y = 5 + 4t$, $z = 6 - 3t$  **15.** AWV.  **17.** $y = 2$, $\dfrac{x - 4}{2} = \dfrac{z - 1}{1}$

**19.** $\dfrac{x}{8} = \dfrac{y}{-10} = \dfrac{z}{-7}$ or $\dfrac{x}{-8} = \dfrac{y}{10} = \dfrac{z}{7}$  **21.** $x = 1$, $\dfrac{y - 2}{4} = \dfrac{z + 1}{8}$

**23.** (a) Parallel  **25.** (a) Intersect when $t = \dfrac{1}{4}$ at the point $\left(\dfrac{17}{4}, \dfrac{11}{4}, \dfrac{3}{2}\right)$  **27.** (a) Parallel  **29.** (a) Skew

(b)  (b)  (b)  (b)

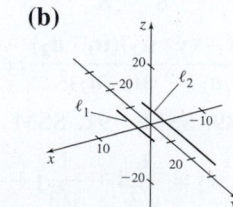

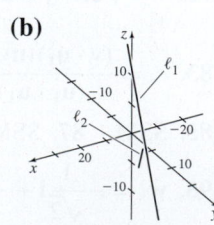

**31.** $2x - y + z = 5$   **33.** $x + 2y - z = 12$   **35.** $2x + 5y - 2z = 21$   **37.** $z = 4$   **39.** $y = -2$

**41. (a)** $x + y + 3z = 0$   **43. (a)** $3x + 7y - 6z = 11$   **45. (a)** $2x + 6y + 29z = 2$
**(b)**   **(b)**   **(b)**

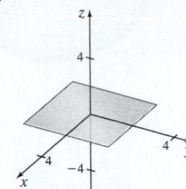   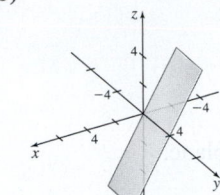   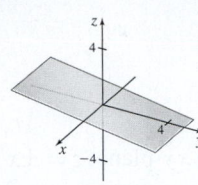

**47.** $\dfrac{x + 14}{-1} = \dfrac{y - 9}{2} = \dfrac{z}{1}$   **49.** $\dfrac{x - 4}{1} = \dfrac{y - 2}{1} = \dfrac{z}{1}$   **51.** $\dfrac{\sqrt{6}}{3}$   **53.** $\dfrac{12\sqrt{11}}{11}$   **55.** $\dfrac{4}{3}$   **57.** $\dfrac{4\sqrt{5}}{5}$   **59.** $(3, 1, 2)$   **61.** $(4, -2, 3)$

**63.** $\dfrac{\sqrt{41}}{3}$   **65.** $\dfrac{\sqrt{6}}{2}$   **67.** $(1, 6, -4); \theta = \sin^{-1}\sqrt{\dfrac{5}{11}}$   **69.** $(-2, -3, -5); \theta = \sin^{-1}\dfrac{10}{\sqrt{198}}$

**71. (a)** $x = 1 + 2t, y = 2 - t, z = -1 + t$   **(b)**

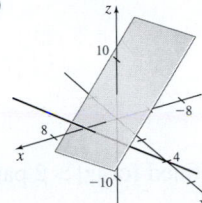

**73.** $\ell_1$ and $\ell_2$ are orthogonal.   **75.** $\cos^{-1}\dfrac{\sqrt{2}}{3}$   **77.** $\cos^{-1}\dfrac{17}{\sqrt{406}}$   **79.** $\cos^{-1}\dfrac{5\sqrt{318}}{106}$

**81. (a)** Each path is a straight line.   **(b)** $\cos^{-1}\dfrac{5\sqrt{7}}{21}$   **(c)** $(-2, 2, 3)$   **(d)** No, AWV.

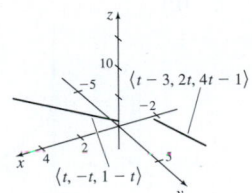

**83.** $x = 1 + 3t, y = 3t, z = -1$; AWV.   **85.** $\dfrac{x - \dfrac{10}{3}}{2} = \dfrac{y + \dfrac{2}{3}}{-7} = \dfrac{z}{-3}$   **87.** $(4, 0, -4)$   **89.** $x + y - 2z = 1$

**91.** $\dfrac{x}{-12} = \dfrac{y + 5}{5} = \dfrac{z + 3}{1}$   **93. (a)** SSM.   **(b)** SSM.   **(c)** AWV.   **95.** SSM.

**97.** Minimum distance: $2\sqrt{42}$. The point on $l_1 = (5, 6, -2)$, the point on $l_2 = (-3, 4, 8)$.

**Section 10.7**   (SSM = See Student Solutions Manual; AWV = Answers will vary)

**1. (c)**   **2.** $(4, 0, 0), (-4, 0, 0), (0, 0, -4)$   **3.** True   **4. (b)**   **5.** Hyperbolic cylinder   **6.** Saddle point

**7.** C   **9.** B   **11.** E   **13.** L   **15.** K   **17.** J

**19. (a)** Elliptic paraboloid
**(b)** Intercept: $(0, 0, 0)$; traces: $(0, 0, 0)$ in the $xy$-plane, $z = x^2$ in the $xz$-plane, $z = y^2$ in the $yz$-plane

**(c)**

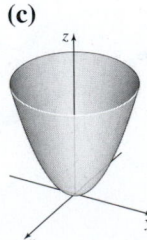

**21. (a)** Ellipsoid

(c)

**(b)** Intercepts: $(1, 0, 0)$, $(-1, 0, 0)$, $(0, 2, 0)$, $(0, -2, 0)$, $(0, 0, 1)$ and $(0, 0, -1)$; traces: $4x^2 + y^2 = 4$ in the $xy$-plane, $x^2 + z^2 = 1$ in the $xz$-plane, $y^2 + 4z^2 = 4$ in the $yz$-plane

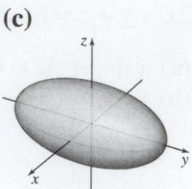

**23. (a)** Elliptic cone

(c)

**(b)** Intercept: $(0, 0, 0)$; traces: $(0, 0, 0)$ in the $xy$-plane, $z = \pm x$ in the $xz$-plane, $z = \pm\sqrt{2}y$ in the $yz$-plane.

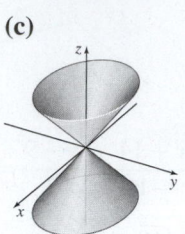

**25. (a)** Parabolic cylinder

(c)

**(b)** The intercepts are $(0, y, 0)$ for all real numbers $y$. Traces: $x = 4z^2$ in the $xz$-plane, $x = 0$ in the $xy$-plane, $z = 0$ in the $yz$-plane

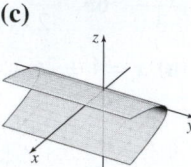

**27. (a)** Hyperboloid of two sheets

(c)

**(b)** Intercepts: $(0, 0, 2)$ and $(0, 0, -2)$; traces: ellipses defined for $|z| > 2$ parallel to the $xy$-plane, $\dfrac{z^2}{4} - \dfrac{y^2}{2} = 1$ in the $yz$-plane, $\dfrac{z^2}{4} - \dfrac{x^2}{4} = 1$ in the $xz$-plane

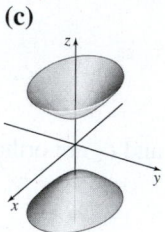

**29. (a)** Parabolic cylinder   **(b)** The intercepts are $(0, 0, z)$ for all real numbers $z$; trace $2x = y^2$ in the $xy$-plane

(c)

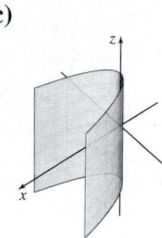

**31.** AWV.     **33. (a)** 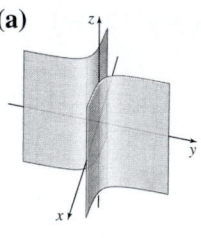     **(b)** 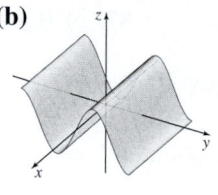     **(c)** 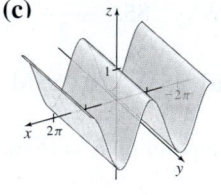     **(d)**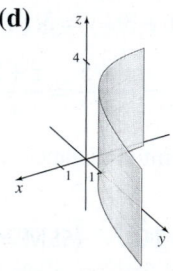

**35.** Cone          **37.** Circular cylinder     **39.** Paraboloid of revolution     **41.** Half of a hyperboloid of two sheets

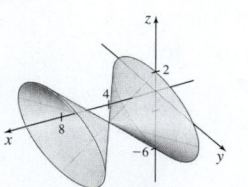          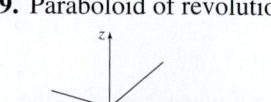     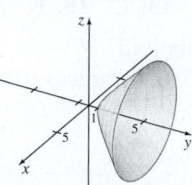

**43.** $(x - 1)^2 + (y + 4)^2 + z^2 = 21$; sphere     **45.** $x = \dfrac{(y - 1)^2}{4} + \dfrac{z^2}{4}$; paraboloid of revolution

**47.** $4(x-1)^2 + (y+2)^2 - z^2 = 6$; hyperboloid of one sheet   **49.** $x^2 + (y-3)^2 + z^2 = 9$   **51.**

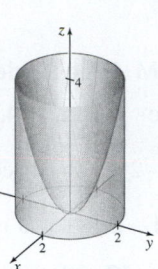

## Review Exercises

**1.** $(1, 0, 4)$, $(2, 3, 2)$, $(2, 0, 2)$, $(2, 0, 4)$, $(1, 3, 4)$, $(1, 3, 2)$   **3. (a)** $\mathbf{v} = \langle 2, 1, 5 \rangle$   **(b)** $\mathbf{v} = 2\mathbf{i} + \mathbf{j} + 5\mathbf{k}$

**5.** 7   **7.** Center $(2, -4, 0)$; radius 5   **9.** $\sqrt{6}$   **11.** $-\dfrac{25}{2}\mathbf{i} + \dfrac{25\sqrt{3}}{2}\mathbf{j}$

**13. (a)** $\|\mathbf{v}\| = \sqrt{93}$; $\cos\alpha = \dfrac{5}{\sqrt{93}}$, $\cos\beta = \dfrac{8}{\sqrt{93}}$, $\cos\gamma = -\dfrac{2}{\sqrt{93}}$   **(b)** $\mathbf{v} = \sqrt{93}\left(\dfrac{5}{\sqrt{93}}\mathbf{i} + \dfrac{8}{\sqrt{93}}\mathbf{j} - \dfrac{2}{\sqrt{93}}\mathbf{k}\right)$

**15.** 9 J   **17.** $\cos^{-1}\left(-\dfrac{1}{2}\right) = \dfrac{2\pi}{3}$ radians   **19.** $\dfrac{2\sqrt{3}}{3}$   **21.** $x = 1 + 3t$, $y = 3t$, $z = -1$

**23.** $a = 12$   **25.** $x = \dfrac{7}{4} - t$, $y = -\dfrac{5}{4} + 3t$, $z = 4t$

**27. (a)** $(2+3)\mathbf{v}$   **(b)** $2(\mathbf{u} - \mathbf{w})$   **(c)** $(\mathbf{u} + \mathbf{v}) + \mathbf{w}$   **(d)** $\mathbf{v} + (-\mathbf{v})$   **(e)** $\mathbf{u} - 3\mathbf{v}$

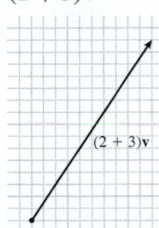

**29.** $2x + 3y + z = 5$   **31.** $\dfrac{x}{-12} = \dfrac{y+5}{5} = \dfrac{z+3}{1}$

**33. (a)** Hyperbolic cylinder

   **(b)** Intercepts: $(2, 0, 0)$ and $(-2, 0, 0)$; trace in the $xy$-plane is $\dfrac{x^2}{4} - \dfrac{y^2}{9} = 1$.

   The traces in the $xz$-plane are the lines $x = -2$ and $x = 2$.

**(c)**

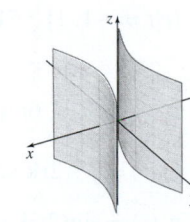

**35. (a)** Hyperbolic paraboloid.

   **(b)** Intercept: $(0, 0, 0)$; traces: $xy$-plane, the pair of lines $\dfrac{y}{3} = \pm\dfrac{x}{2}$, which intersect at the origin;

   The trace in the $xz$-plane is the parabola $z = \dfrac{x^2}{4}$, and the trace in the $yz$-plane is the

   parabola $z = -\dfrac{y^2}{9}$.

**(c)**

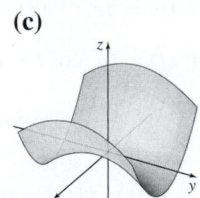

**37. (a)** Ellipsoid

   **(b)** Intercepts: $(2, 0, 0)$, $(-2, 0, 0)$, $(0, 3, 0)$, $(0, -3, 0)$, $(0, 0, 1)$, $(0, 0, -1)$;

   traces: $xy$-plane, $\dfrac{x^2}{4} + \dfrac{y^2}{9} = 1$; $yz$-plane, $\dfrac{y^2}{9} + z^2 = 1$; $xz$-plane, $\dfrac{x^2}{4} + z^2 = 1$

**(c)**

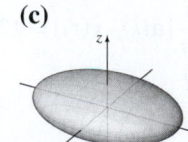

## Chapter 11

### Section 11.1 (SSM = See Student Solutions Manual; AWV = Answers will vary)

**1.** $\{t\,|\,t \le 4\}$  **2.** False  **3.** True  **4.** True  **5.** False  **6.** $\mathbf{u}' \times \mathbf{v} + \mathbf{u} \times \mathbf{v}'$

**7. (a)** $\mathbf{i} - 2\mathbf{j}$  **(b)** $\mathbf{i} - 2\mathbf{j}$  **(c)** $\mathbf{r}$ is continuous at $t_0 = 1$  **9. (a)** $\dfrac{\sqrt{2}}{2}\mathbf{i} - \dfrac{\sqrt{2}}{2}\mathbf{j}$  **(b)** $\dfrac{\sqrt{2}}{2}\mathbf{i} - \dfrac{\sqrt{2}}{2}\mathbf{j}$  **(c)** $\mathbf{v}$ is continuous at $t_0 = \dfrac{\pi}{4}$

**11. (a)** $\mathbf{i} - 2\mathbf{k}$  **(b)** $\mathbf{i} - 2\mathbf{k}$  **(c)** $\mathbf{u}$ is continuous at $t_0 = 0$  **13. (a)** $4\mathbf{i} + \mathbf{j} + 8\mathbf{k}$  **(b)** $4\mathbf{i} + \mathbf{j} + 8\mathbf{k}$  **(c)** $\mathbf{g}$ is continuous at $t_0 = 4$

**15.** All real numbers  **17.** $\{t\,|\,t \ge 0\}$  **19.** $\{t\,|\,t \ne 0\}$  **21.** $\{t\,|\,t > 0\}$  **23.** $\{t\,|\,1 < t \le 4\}$

**25.**

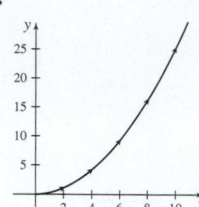

**27.**

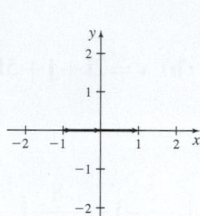

**29.**

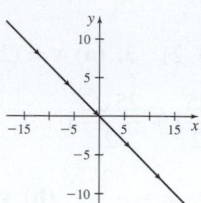

**31.**

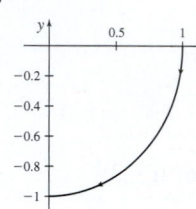

**33.**

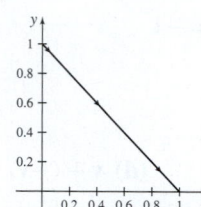

**35.**

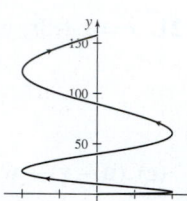

**37.** D  **39.** C

**41.**

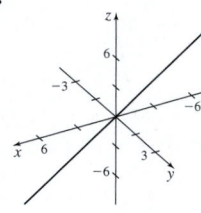

**43.**

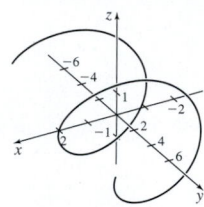

**45.**

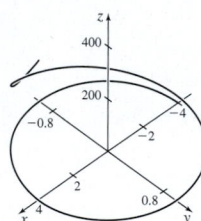

**47.**

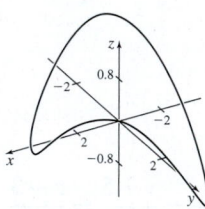

**49.** $\{t\,|\,t \ne -1, 1\}$  **51.** $\left\{t\,\Big|\,t \ne \dfrac{(2k+1)\pi}{2}\right\}$, $k$ an integer  **53.** $\{t\,|\,t \ne -1\}$  **55.** $\left\{t\,\Big|\,t \ne \dfrac{(2k+1)\pi}{2}\right\}$, $k$ an integer

**57.** $\mathbf{r}'(t) = 8t\mathbf{i} - 6t^2\mathbf{j}$ and $\mathbf{r}''(t) = 8\mathbf{i} - 12t\mathbf{j}$  **59.** $\mathbf{r}'(t) = \dfrac{2\sqrt{t}}{t}\mathbf{i} + 2e^t\mathbf{j}$ and $\mathbf{r}''(t) = -\dfrac{1}{t^{3/2}}\mathbf{i} + 2e^t\mathbf{j}$

**61.** $\mathbf{r}'(t) = \mathbf{j} + 2t\mathbf{k}$ and $\mathbf{r}''(t) = 2\mathbf{k}$  **63.** $\mathbf{r}'(t) = 2t\mathbf{i} + 3t^2\mathbf{j} - \mathbf{k}$ and $\mathbf{r}''(t) = 2\mathbf{i} + 6t\mathbf{j}$

**65.** $\mathbf{r}'(t) = \sin(2t)\mathbf{i} + \sin(2t)\mathbf{j}$ and $\mathbf{r}''(t) = 2\cos(2t)\mathbf{i} + 2\cos(2t)\mathbf{j}$

**67.** $\mathbf{r}'(t) = 5e^{t^2+t}(2t+1)\mathbf{i} - (2t+1)\mathbf{j}$ and $\mathbf{r}''(t) = 5e^{t^2+t}(4t^2+4t+3)\mathbf{i} - 2\mathbf{j}$

**69.** $\mathbf{r}'(t) = (e^t\cos t - e^t\sin t)\mathbf{i} + (e^t\sin t + e^t\cos t)\mathbf{j} + \mathbf{k}$ and $\mathbf{r}''(t) = -2e^t\sin t\,\mathbf{i} + 2e^t\cos t\,\mathbf{j}$

**71.** $\mathbf{r}'(t) = (1 - 3t^2)\mathbf{i} + (1 + 3t^2)\mathbf{j} - \mathbf{k}$ and $\mathbf{r}''(t) = -6t\mathbf{i} + 6t\mathbf{j}$  **73.** $\dfrac{d}{dt}[\mathbf{u}(t) \cdot \mathbf{v}(t)] = 0$

**75.** $\dfrac{d}{dt}[\mathbf{u}(t) \cdot \mathbf{v}(t)] = te^t + e^t + t^2e^{-t} - 2te^{-t}$  **77.** $\dfrac{d}{dt}[\mathbf{u}(t) \cdot \mathbf{v}(t)] = \omega\cos(\omega t) - \omega\sin(\omega t)$

**79.** $\dfrac{d}{dt}[\mathbf{u}(t) \cdot \mathbf{v}(t)] = 12t^2$ and $\dfrac{d}{dt}[\mathbf{u}(t) \times \mathbf{v}(t)] = (2t+10)\mathbf{i} - (10t+2)\mathbf{j} + (8t - 4t^3)\mathbf{k}$

**81.** $\dfrac{d}{dt}[\mathbf{u}(t) \cdot \mathbf{v}(t)] = \sin t\cos(2t) - \sin(2t)\cos t$

$\dfrac{d}{dt}[\mathbf{u}(t) \times \mathbf{v}(t)] = [2\cos(2t) - \cos t]\mathbf{i} + [2\sin(2t) - \sin t]\mathbf{j} - [\sin(2t)\sin t + \cos(2t)\cos t]\mathbf{k}$

**83.** $\dfrac{d\mathbf{u}}{dt} = -\omega\sin(\omega t)\mathbf{i} + \omega\cos(\omega t)\mathbf{j}$ and $\left\|\dfrac{d\mathbf{u}}{dt}\right\| = |\omega|$    **85.** SSM.

**87.**                   **89.**                   **91.**                   **93.**

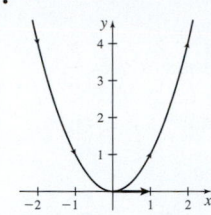   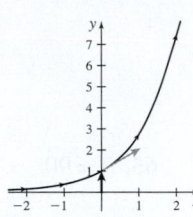   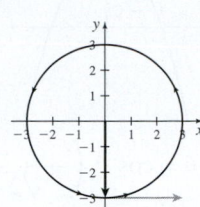   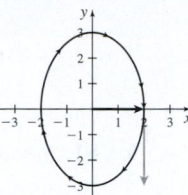

**95. (a)** $y = 1 - x$    **(b)**    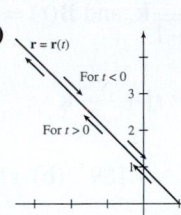    **97. (a)** $y = \sqrt{\dfrac{x}{1-x}}$ for $0 < x < 1$    **(b)**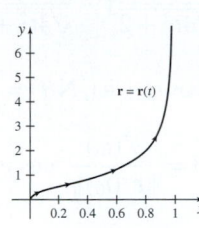

**99.** SSM.   **101.** SSM.   **103.** SSM.   **105.** SSM.   **107.** SSM.

## Section 11.2   (SSM = See Student Solutions Manual; AWV = Answers will vary)

**1.** False   **2.** False   **3.** False   **4.** False   **5.** True   **6.** True

**7. (a)** $\mathbf{r}'(1) = \mathbf{i} - 2\mathbf{j}$    **9. (a)** $\mathbf{r}'(1) = 2\mathbf{i} - \mathbf{j}$    **11. (a)** $\mathbf{r}'(1) = 4\mathbf{i} - \dfrac{1}{2}\mathbf{j}$    **13. (a)** $\mathbf{r}'(0) = \mathbf{i} - \mathbf{j}$
**(b)** and **(c)**        **(b)** and **(c)**        **(b)** and **(c)**        **(b)** and **(c)**

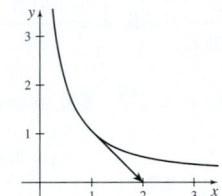

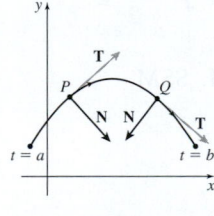   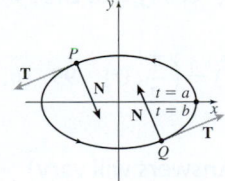

**15.** $\mathbf{r}'(0) = -3\mathbf{i} + 2\mathbf{j} - \mathbf{k}$   **17.** $\mathbf{r}'\left(\dfrac{\pi}{4}\right) = -2\mathbf{i}$   **19.** $\mathbf{r}'\left(\dfrac{\pi}{6}\right) = -\mathbf{i} + \sqrt{3}\mathbf{k}$   **21.** $\mathbf{r}'(0) = \mathbf{i} + \mathbf{j} + \mathbf{k}$

**23.** Figure below not drawn to scale.   **25.** Figure below not drawn to scale.

**27.** $\mathbf{T}(1) = \dfrac{\sqrt{5}}{5}\mathbf{i} + \dfrac{2\sqrt{5}}{5}\mathbf{j}$ and $\mathbf{N}(1) = -\dfrac{2\sqrt{5}}{5}\mathbf{i} + \dfrac{\sqrt{5}}{5}\mathbf{j}$   **29.** $\mathbf{T}(1) = \dfrac{2\sqrt{5}}{5}\mathbf{i} - \dfrac{\sqrt{5}}{5}\mathbf{j}$ and $\mathbf{N}(1) = \dfrac{\sqrt{5}}{5}\mathbf{i} + \dfrac{2\sqrt{5}}{5}\mathbf{j}$

**31.** $\mathbf{T}(1) = \dfrac{2\sqrt{13}}{13}\mathbf{i} - \dfrac{3\sqrt{13}}{13}\mathbf{j}$ and $\mathbf{N}(1) = -\dfrac{3\sqrt{13}}{13}\mathbf{i} - \dfrac{2\sqrt{13}}{13}\mathbf{j}$   **33.** $\mathbf{T}(0) = \dfrac{\sqrt{2}}{2}\mathbf{i} - \dfrac{\sqrt{2}}{2}\mathbf{j}$ and $\mathbf{N}(0) = \dfrac{\sqrt{2}}{2}\mathbf{i} + \dfrac{\sqrt{2}}{2}\mathbf{j}$

**35.** $\mathbf{T}(0) = -\dfrac{3\sqrt{14}}{14}\mathbf{i} + \dfrac{2\sqrt{14}}{14}\mathbf{j} - \dfrac{\sqrt{14}}{14}\mathbf{k}$ and $\mathbf{N}(0)$ is undefined   **37.** $\mathbf{T}\left(\dfrac{\pi}{6}\right) = -\dfrac{1}{2}\mathbf{i} + \dfrac{\sqrt{3}}{2}\mathbf{k}$ and $\mathbf{N}\left(\dfrac{\pi}{6}\right) = -\dfrac{\sqrt{3}}{2}\mathbf{i} - \dfrac{1}{2}\mathbf{k}$

**39.** $\mathbf{T}\left(\dfrac{\pi}{4}\right) = -\mathbf{i}$ and $\mathbf{N}\left(\dfrac{\pi}{4}\right) = -\mathbf{j}$   **41.** $\mathbf{T}(0) = \dfrac{\sqrt{3}}{3}\mathbf{i} + \dfrac{\sqrt{3}}{3}\mathbf{j} + \dfrac{\sqrt{3}}{3}\mathbf{k}$ and $\mathbf{N}(0) = -\dfrac{\sqrt{2}}{2}\mathbf{i} + \dfrac{\sqrt{2}}{2}\mathbf{j}$   **43.** $s = \dfrac{1}{27}\left(76^{3/2} - 13^{3/2}\right)$

**45.** $s = \dfrac{33}{2}$   **47.** $s = \sqrt{6}$   **49.** $s = \sqrt{5}\pi$   **51.** $s = e - \dfrac{1}{e}$   **53.** $s = \sqrt{3}(e^{2\pi} - 1)$

**55. (a)** 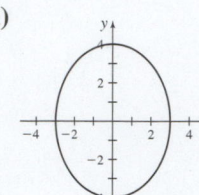    **(b)** $s \approx 11.052$    **57. (a)** 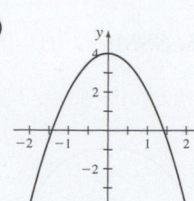    **(b)** $s \approx 33.637$

**59.** $\dfrac{1}{4}\mathbf{i} - \dfrac{3}{4}\mathbf{j} + \dfrac{1}{2}\mathbf{k}, -\mathbf{j}, \dfrac{1}{4}\mathbf{i} - \dfrac{3}{4}\mathbf{j} - \dfrac{1}{2}\mathbf{k}$   **61.** $\theta \approx 37°$   **63.** $\theta = \cos^{-1}\left(-\dfrac{\pi}{\sqrt{2+2\pi^2}}\right)$    **65.** $\theta = 90°$

**67.** $\mathbf{T}(t) = \dfrac{2t}{\sqrt{4t^2+2}}\mathbf{i} - \dfrac{1}{\sqrt{4t^2+2}}\mathbf{j} - \dfrac{1}{\sqrt{4t^2+2}}\mathbf{k}, \mathbf{N}(t) = \dfrac{1}{\sqrt{2t^2+1}}\mathbf{i} + \dfrac{t}{\sqrt{2t^2+1}}\mathbf{j} + \dfrac{t}{\sqrt{2t^2+1}}\mathbf{k}$, and $\mathbf{B}(t) = -\dfrac{\sqrt{2}}{2}\mathbf{j} + \dfrac{\sqrt{2}}{2}\mathbf{k}$

**69.** $\mathbf{T}(t) = \dfrac{\sqrt{2}}{2}(-\sin t\,\mathbf{i} + \cos t\,\mathbf{j} + \mathbf{k}), \mathbf{N}(t) = -\cos t\,\mathbf{i} - \sin t\,\mathbf{j}$, and $\mathbf{B}(t) = \dfrac{\sqrt{2}}{2}\sin t\,\mathbf{i} - \dfrac{\sqrt{2}}{2}\cos t\,\mathbf{j} + \dfrac{\sqrt{2}}{2}\mathbf{k}$

**71.** $\cos \alpha = \dfrac{x'(t_0)}{\|\mathbf{r}'(t_0)\|}, \cos \beta = \dfrac{y'(t_0)}{\|\mathbf{r}'(t_0)\|}, \cos \gamma = \dfrac{z'(t_0)}{\|\mathbf{r}'(t_0)\|}$   **73.** $s \approx 10.516$   **75. (a)** $s = \dfrac{4\pi}{3} \approx 4.189$   **(b)** $s \approx 33.510$   **(c)** SSM.

## Section 11.3   (SSM = See Student Solutions Manual; AWV = Answers will vary)

**1.** (d)   **2.** False   **3.** False   **4.** True   **5.** False   **6.** $\dfrac{|f''(x)|}{(1+[f'(x)]^2)^{3/2}}$   **7.** $\dfrac{1}{3}$   **8.** False   **9.** No   **11.** No   **13.** No   **15.** Yes

**17.** No   **19.** $P\,R\,Q$   **21.** $\kappa = \dfrac{3t^4}{2(t^6+1)^{3/2}}$   **23.** $\kappa = \dfrac{1}{2}$   **25.** $\kappa = \dfrac{2}{3(t^2+1)^2}$   **27.** $\kappa = 0$   **29.** $\kappa = \dfrac{4}{5}$   **31.** $\kappa = \dfrac{\sqrt{2}e^{2t}}{(e^{2t}+1)^2}$

**33.** $\kappa = \dfrac{1}{|3\sin t\cos t|}$   **35.** $\kappa = \dfrac{2\sqrt{5}}{25}$   **37.** $\kappa = \sqrt{2}$   **39.** $\kappa = \dfrac{2}{27}$   **41.** $\kappa = \dfrac{2}{9}$   **43.** $\kappa = \dfrac{\sqrt{2}}{4}$   **45.** $\rho = \dfrac{5\sqrt{10}}{3}$   **47.** $\rho = 1$

**49.** $\rho = 3\sqrt{2}$   **51.** $\rho = \sqrt{2}$   **53.** $\rho = 6$   **55.** $\rho = \dfrac{27}{4}$   **57.** $\rho = \dfrac{5}{4}$   **59.** $\rho = 2\sqrt{2}$   **61.** $\rho = 12a$   **63.** SSM.

**65.** $\left(\dfrac{\sqrt{2}}{2}, \ln\dfrac{\sqrt{2}}{2}\right)$   **67.** $\left(\pm\left(\dfrac{1}{5}\right)^{1/4}, \pm\dfrac{1}{3}\left(\dfrac{1}{5}\right)^{3/4}\right)$   **69.** $\alpha = 1$   **71.** SSM.   **73.** $\kappa = \dfrac{1}{4}$

**75.** $\kappa = \dfrac{e^t\sqrt{2}}{2}$. AWV.   **77.** SSM.   **79.** $\mathbf{r}(s) = \dfrac{2s}{3}\mathbf{i} + \dfrac{2s-3}{3}\mathbf{j} + \dfrac{s}{3}\mathbf{k}, 0 \le s \le 6$   **81.** SSM.   **83.** SSM.   **85.** SSM.

**87.** $\kappa = \dfrac{\sqrt{2}}{2}, \mathbf{T}(s) = \dfrac{\sqrt{2}}{2}\cos s\,\mathbf{i} - \dfrac{\sqrt{2}}{2}\sin s\,\mathbf{j} + \dfrac{\sqrt{2}}{2}\mathbf{k}, \mathbf{N}(s) = -\sin s\,\mathbf{i} - \cos s\,\mathbf{j}$, and $\mathbf{B}(s) = \dfrac{\sqrt{2}}{2}\cos s\,\mathbf{i} - \dfrac{\sqrt{2}}{2}\sin s\,\mathbf{j} - \dfrac{\sqrt{2}}{2}\mathbf{k}$

**89.** $\kappa = \dfrac{23}{7^{3/2}}$   **91.** $\kappa = \dfrac{3\sqrt{2}}{4a}$   **93.** $\kappa = \dfrac{8\sqrt{7}}{49}$   **95.** SSM.   **97.** $(h, k) = \left(\dfrac{\pi}{2}, 0\right)$   **99.** $(h, k) = \left(\dfrac{7}{4}, \dfrac{7}{4}\right)$   **101.** SSM.

## Section 11.4   (SSM = See Student Solutions Manual; AWV = Answers will vary)

**1.** True   **2.** False   **3.** True   **4.** Tangential, normal

**5. (a)** $\mathbf{v}(t) = \mathbf{i} + 2t\mathbf{j}, \mathbf{a}(t) = 2\mathbf{j}$, and $v(t) = \sqrt{1+4t^2}$    **7. (a)** $\mathbf{v}(t) = 2t\mathbf{i} - \mathbf{j}, \mathbf{a}(t) = 2\mathbf{i}$, and $v(t) = \sqrt{4t^2+1}$    **9. (a)** $\mathbf{v}(t) = 4\mathbf{i} - 3t^2\mathbf{j}, \mathbf{a}(t) = -6t\mathbf{j}$, and $v(t) = \sqrt{16+9t^4}$

**(b)** 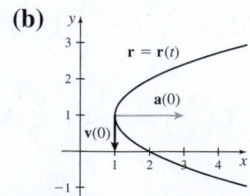      **(b)** 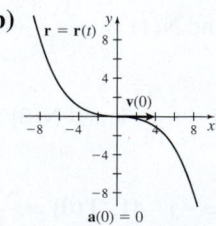      **(b)**

**11.** $\mathbf{v}(t) = 2t\mathbf{i} + \mathbf{j} - 9t^2\mathbf{k}, \mathbf{a}(t) = 2\mathbf{i} - 18t\mathbf{k}$, and $v(t) = \sqrt{81t^4+4t^2+1}$

**13.** $\mathbf{v}(t) = -\dfrac{t}{\sqrt{16-t^2}}\mathbf{i} + 2t\mathbf{j} + \mathbf{k}, \mathbf{a}(t) = -\dfrac{16}{(16-t^2)^{3/2}}\mathbf{i} + 2\mathbf{j}$, and $v(t) = \sqrt{\dfrac{t^2}{16-t^2} + 4t^2 + 1}$

**15.** $\mathbf{v}(t) = -2\sin t\,\mathbf{i} + \cos t\,\mathbf{j} + \mathbf{k}$, $\mathbf{a}(t) = -2\cos t\,\mathbf{i} - \sin t\,\mathbf{j}$, and $v(t) = \sqrt{3\sin^2 t + 2}$

**17.** (a) $\mathbf{v}(t) = \cos t\,\mathbf{i} - \sin t\,\mathbf{j} + 2\cos(2t)\mathbf{k}$, $\mathbf{a}(t) = -\sin t\,\mathbf{i} - \cos t\,\mathbf{j} - 4\sin(2t)\mathbf{k}$, and $v(t) = \sqrt{1 + 4\cos^2(2t)}$    **(c)**

   **(b)** $\mathbf{v}\left(\dfrac{\pi}{2}\right) = -\mathbf{j} - 2\mathbf{k}$, $\mathbf{a}\left(\dfrac{\pi}{2}\right) = -\mathbf{i}$, and $v\left(\dfrac{\pi}{2}\right) = \sqrt{5}$

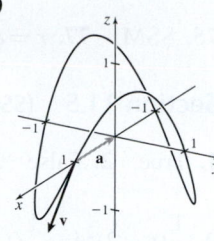

**19.** (a) $\mathbf{F}(t) = m\mathbf{r}(t)$   (b) $\|\mathbf{a}(t)\| = \sqrt{e^{2t} + e^{-2t}}$    **21.** (a) $\mathbf{F}(t) = m\,e^t\mathbf{j}$   (b) $\|\mathbf{a}(t)\| = e^t$    **23.** (a) $\mathbf{F}(t) = -4m\mathbf{r(t)}$   (b) $\|\mathbf{a}(t)\| = 12$

**25.** (a) $\mathbf{F}(t) = -m\mathbf{r}(t)$   (b) $\|\mathbf{a}(t)\| = \sqrt{5\sin^2 t + 4}$    **27.** (a) $\mathbf{v}(t) = 2\mathbf{i} + \mathbf{j}$, $\mathbf{a}(t) = 0$, and $v(t) = \sqrt{5}$    (b) $a_T = 0$, $a_N = 0$

**29.** (a) $\mathbf{v}(t) = e^t\mathbf{i} + 2e^{2t}\mathbf{j}$, $\mathbf{a}(t) = e^t\mathbf{i} + 4e^{2t}\mathbf{j}$, and $v(t) = e^t\sqrt{1 + 4e^{2t}}$   (b) $a_T = \dfrac{e^t(1 + 8e^{2t})}{\sqrt{1 + 4e^{2t}}}$, $a_N = \dfrac{2e^{2t}}{\sqrt{1 + 4e^{2t}}}$

**31.** (a) $\mathbf{v}(t) = 2\cos t\,\mathbf{i} - \sin t\,\mathbf{j}$, $\mathbf{a}(t) = -2\sin t\,\mathbf{i} - \cos t\,\mathbf{j}$, and $v(t) = \sqrt{3\cos^2 t + 1}$   (b) $a_T = -\dfrac{3\sin t\cos t}{\sqrt{3\cos^2 t + 1}}$, $a_N = \dfrac{2}{\sqrt{3\cos^2 t + 1}}$

**33.** (a) $\mathbf{v}(t) = -3\mathbf{i} + 2\mathbf{j} - \mathbf{k}$, $\mathbf{a}(t) = 0$, and $v(t) = \sqrt{14}$   (b) $a_T = 0$, $a_N = 0$

**35.** (a) $\mathbf{v}(t) = \mathbf{i} + 2t\mathbf{j} + 3t^2\mathbf{k}$, $\mathbf{a}(t) = 2\mathbf{j} + 6t\mathbf{k}$, and $v(t) = \sqrt{1 + 4t^2 + 9t^4}$   (b) $a_T = \dfrac{18t^3 + 4t}{\sqrt{1 + 4t^2 + 9t^4}}$, $a_N = \dfrac{2\sqrt{1 + 9t^2 + 9t^4}}{\sqrt{1 + 4t^2 + 9t^4}}$

**37.** (a) $\mathbf{v}(t) = -\sin t\,\mathbf{j} + \cos t\,\mathbf{k}$, $\mathbf{a}(t) = -\cos t\,\mathbf{j} - \sin t\,\mathbf{k}$, and $v(t) = 1$   (b) $a_T = 0$, $a_N = 1$

**39.** (a) $\mathbf{v}(t) = \dfrac{1}{t}\mathbf{i} + \dfrac{1}{2\sqrt{t}}\mathbf{j} + \dfrac{3\sqrt{t}}{2}\mathbf{k}$, $\mathbf{a}(t) = -\dfrac{1}{t^2}\mathbf{i} - \dfrac{1}{4t^{3/2}}\mathbf{j} + \dfrac{3}{4\sqrt{t}}\mathbf{k}$, and $v(t) = \dfrac{\sqrt{9t^3 + t + 4}}{2t}$

   **(b)** $a_T = \dfrac{9t^3 - t - 8}{4t^2\sqrt{9t^3 + t + 4}}$, $a_N = \dfrac{\sqrt{9t^3 + 81t^2 + 1}}{2t^{3/2}\sqrt{9t^3 + t + 4}}$

**41.** (a) $\mathbf{v}(t) = (e^t\cos t - e^t\sin t)\mathbf{i} + (e^t\sin t + e^t\cos t)\mathbf{j} + e^t\mathbf{k}$, $\mathbf{a}(t) = -2e^t\sin t\,\mathbf{i} + 2e^t\cos t\,\mathbf{j} + e^t\mathbf{k}$, and $v(t) = \sqrt{3}e^t$

   **(b)** $a_T = \sqrt{3}e^t$, $a_N = \sqrt{2}e^t$

**43.** (a) $\mathbf{v}(t) = -a\sin t\,\mathbf{i} + b\cos t\,\mathbf{j} + c\mathbf{k}$, $\mathbf{a}(t) = -a\cos t\,\mathbf{i} - b\sin t\,\mathbf{j}$, and $v(t) = \sqrt{a^2\sin^2 t + b^2\cos^2 t + c^2}$

   **(b)** $a_T = \dfrac{(a^2 - b^2)(\sin t\cos t)}{\sqrt{a^2\sin^2 t + b^2\cos^2 t + c^2}}$, $a_N = \sqrt{\dfrac{a^2c^2\cos^2 t + b^2(a^2 + c^2\sin^2 t)}{a^2\sin^2 t + b^2\cos^2 t + c^2}}$

**45.** $\mathbf{v}(t) = [\pi - \pi\cos(\pi t)]\mathbf{i} + \pi\sin(\pi t)\mathbf{j}$, $\mathbf{a}(t) = \pi^2\sin(\pi t)\mathbf{i} + \pi^2\cos(\pi t)\mathbf{j}$    **47.** $\mathbf{a}(0) = \dfrac{2}{3}\mathbf{j}$ and $\mathbf{a}(6) = -\dfrac{2}{3}\mathbf{j}$

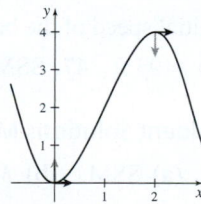

**49.** SSM.    **51.** $a_T = 0$ and $a_N = \dfrac{4}{(1 + x^2)^{3/2}}$    **53.** (a) 350 N   (b) 35 km/h slower

**55.** $2.118 \times 10^{13}$ $m$ N    **57.** $4\sqrt{5}$ $m$    **59.** (a) $\mathbf{v}(t) = -\dfrac{4t}{(t^2 + 1)^2}\mathbf{i} - \dfrac{2(t^2 - 1)}{(t^2 + 1)^2}\mathbf{j}$   (b) No. AWV.   (c) $(-1, 0)$

**61.** (a) SSM.   (b)     (c) $\mathbf{a}(n\pi) = \pm 4\mathbf{i} + 8\mathbf{j}$; $n$ an integer

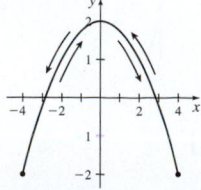

**63.** $v \approx 3750$ m/s   **65.** SSM.   **67.** SSM.   **69.** AWV.   **71.** SSM.

**73. (a)** $\mathbf{v}(t) = 2t\cos(\omega t)\mathbf{i} + 2t\sin(\omega t)\mathbf{j} + t^2\mathbf{v_d}$ and $\mathbf{a}(t) = 2\cos(\omega t)\mathbf{i} + 2\sin(\omega t)\mathbf{j} + 4t\mathbf{v_d} + t^2\mathbf{a_d}$   **(b)** $2\cos(\omega t)\mathbf{i} + 2\sin(\omega t)\mathbf{j} + 4t\mathbf{v_d}$

**75.** SSM.   **77.** $r = e^{2t}$, where $\theta = t$; $\mathbf{v}(t) = 2r\mathbf{u}_r + r\mathbf{u}_\theta$; $\mathbf{a}(t) = 3r\mathbf{u}_r + 4r\mathbf{u}_\theta$

## Section 11.5   (SSM = See Student Solutions Manual; AWV = Answers will vary)

**1.** True   **2.** False   **3.** $-\cos t\mathbf{i} - \sin t\mathbf{j} + \frac{1}{2}t^2\mathbf{k} + \mathbf{c}$   **5.** $\frac{1}{3}t^3\mathbf{i} - \frac{1}{2}t^2\mathbf{j} + e^t\mathbf{k} + \mathbf{c}$   **7.** $(t\ln t - t)\mathbf{i} + \left(\frac{t^2}{4} - \frac{t^2}{2}\ln t\right)\mathbf{j} - 2t\mathbf{k} + \mathbf{c}$

**9.** $\frac{1}{2}(t-2)^2\mathbf{i} - \frac{1}{3}(t-2)^3\mathbf{j} + t\mathbf{k} + \mathbf{c}$   **11.** $\mathbf{i} + 27\mathbf{j}$   **13.** $-\frac{1}{3}\mathbf{i} + \mathbf{j}$   **15.** $20\mathbf{i} + 68\mathbf{j} + 12\mathbf{k}$   **17.** $\frac{1}{2}(e-1)\mathbf{i} + (e-1)\mathbf{j} + (e^2-1)\mathbf{k}$

**19.** $\mathbf{r}(t) = (e^t - e)\mathbf{i} + (t - t\ln t)\mathbf{j} + t^2\mathbf{k}$   **21.** $\mathbf{r}(t) = (3 - 2\cos t)\mathbf{i} + (\sin t - 1)\mathbf{j} + t\mathbf{k}$

**23.** $\mathbf{r}(t) = (\ln t + 1)\mathbf{i} + \frac{1}{2}(t^2 + 1)\mathbf{j} + \frac{1}{3}(t^3 + 2)\mathbf{k}$   **25.** $\mathbf{v}(t) = -32t\mathbf{k}$, $v(t) = 32t$, and $\mathbf{r}(t) = -16t^2\mathbf{k}$

**27.** $\mathbf{v}(t) = (\sin t + 1)\mathbf{i} + (1 - \cos t)\mathbf{j}$, $v(t) = \sqrt{3 + 2\sin t - 2\cos t}$, and $\mathbf{r}(t) = (t - \cos t + 1)\mathbf{i} + (t - \sin t + 1)\mathbf{j}$

**29.** $\mathbf{v}(t) = \mathbf{i} - 9.8t\mathbf{k}$, $v(t) = \sqrt{1 + 96.04t^2}$, and $\mathbf{r}(t) = t\mathbf{i} + (5 - 4.9t^2)\mathbf{k}$

**31.** $\mathbf{v}(t) = (2 - e^{-t})\mathbf{i} + (t + 1)\mathbf{j}$, $v(t) = \sqrt{e^{-2t} - 4e^{-t} + t^2 + 2t + 5}$, and $\mathbf{r}(t) = (2t + e^{-t})\mathbf{i} + \left(\frac{1}{2}t^2 + t - 1\right)\mathbf{j}$

**33.** The range $\approx 23{,}895$ m, the time of flight $\approx 53$ s, and the greatest height reached $\approx 3449$ m.

**35. (a)** $x = \dfrac{1200t}{13}$ and $y = -4.9t^2 + \dfrac{500t}{13}$   **(b)** Range $\approx 724.6$ m.   **(c)** Time of flight $\approx 7.8$ s.   **(d)**

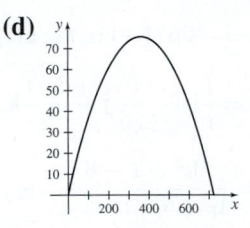

**37. (a)** The projectile travels approximately 152.5 ft up the hill.   **(b)** The projectile is in the air for approximately 1.9 s.

**39. (a)** $\mathbf{v}(t) = 5t\mathbf{i} + (-9.8t + 3)\mathbf{j}$ and $v(t) = \sqrt{121.04t^2 - 58.8t + 9}$   **(b)** $\mathbf{r}(t) = \frac{5}{2}t^2\mathbf{i} + (-4.9t^2 + 3t)\mathbf{j}$
**(c)** $\approx 0.612$ s   **(d)**

**41.** $v_0 \approx 54.521$ ft/s   **43.** The initial speed of the ball is approximately 114 ft/s. It took the ball approximately 5 s to reach the vines.

**45. (a)** Range $\approx 114.342$ ft   **(b)** $\approx 93$ ft   **47.** SSM.   **49.** SSM.   **51.** SSM.   **53.** SSM.

## Section 11.6   (SSM = See Student Solutions Manual; AWV = Answers will vary)

**1.** $\approx 79.479$ days   **3.** AWV.   **5. (a)** SSM.   **(b)** $M \approx 2.0 \times 10^{30}$ kg   **7.** SSM.

## Chapter Review

**1. (a)** All real numbers.   **(b)**
**(c)** $\mathbf{r}'(2) = 4\mathbf{i} + 3\mathbf{j}$

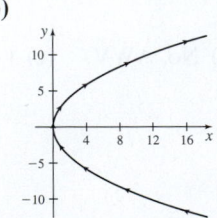

**3. (a)** All real numbers.   **(b)**
**(c)** $\mathbf{r}'(0) = \mathbf{i} + 2\mathbf{k}$

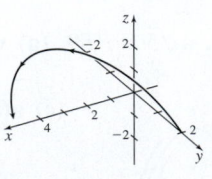

**5.** $\frac{1}{2}\mathbf{j} - \frac{4}{3}\mathbf{k}$   **7.** Continuous   **9.** $\mathbf{r}'(t) = -2\sin t\mathbf{i} - 3\sin t\mathbf{j} + \mathbf{k}$ and $\mathbf{r}''(t) = -2\cos t\mathbf{i} - 3\cos t\mathbf{j}$

**11.** $[\mathbf{f}(t) \cdot \mathbf{g}(t)]' = 4t - 2\sin(2t)\cos t - \cos(2t)\sin t - 5\cos t$

$[\mathbf{f}(t) \times \mathbf{g}(t)]' = [-2\sin(2t)\sin t + \cos(2t)\cos t - 5\sin t]\mathbf{i} + (-10 - \sin t - t\cos t)\mathbf{j} + [\cos t - t\sin t + 4t\sin(2t) - 2\cos(2t)]\mathbf{k}$

**13.** (a) $\mathbf{r}'(0) = \mathbf{i} + 2\mathbf{j}$ (b) $\mathbf{T}(0) = \dfrac{\sqrt{5}}{5}\mathbf{i} + \dfrac{2\sqrt{5}}{5}\mathbf{j}$ (c) $\mathbf{N}(0) = -\mathbf{k}$ **15.** $\theta = \cos^{-1}\dfrac{3\sqrt{10}}{10}$ **17.** $s = \sin^{-1}\dfrac{\sqrt{5}}{4}$ **19.** No **21.** $\dfrac{1}{4}$

**23.** $\kappa = \dfrac{2\sqrt{4e^{2t} + 100e^{-2t} + 25}}{(e^{2t} + 25e^{-2t} + 16)^{3/2}}$ **25.** $\kappa = \dfrac{16}{65^{3/2}}$ **27.** $\rho = \infty$ **29.** $\kappa = \dfrac{6x^{1/3}}{(9x^{4/3} + 1)^{3/2}}$

**31.** (a) $\mathbf{v}(t) = -2\sin t\,\mathbf{i} + \cos t\,\mathbf{j}$, $\mathbf{a}(t) = -2\cos t\,\mathbf{i} - \sin t\,\mathbf{j}$, and $v(t) = \sqrt{3\sin^2 t + 1}$ (b) $a_\mathbf{T} = \dfrac{3\sin t\cos t}{\sqrt{3\sin^2 t + 1}}, a_\mathbf{N} = \dfrac{2}{\sqrt{3\sin^2 t + 1}}$

**33.** (a) $\mathbf{v}(t) = e^t\mathbf{i} - e^{-t}\mathbf{j}$, $\mathbf{a}(t) = e^t\mathbf{i} + e^{-t}\mathbf{j}$, and $v(t) = \sqrt{e^{2t} + e^{-2t}}$ (b) $a_\mathbf{T} = \dfrac{e^{2t} - e^{-2t}}{\sqrt{e^{2t} + e^{-2t}}}, a_\mathbf{N} = \dfrac{2}{\sqrt{e^{2t} + e^{-2t}}}$

**35.** $\mathbf{v}(t) = [\pi - \pi\cos(\pi t)]\mathbf{i} + \pi\sin(\pi t)\mathbf{j}$ and $\mathbf{a}(t) = \pi^2\sin(\pi t)\mathbf{i} + \pi^2\cos(\pi t)\mathbf{j}$ **37.** $\left(\dfrac{1}{3}t^3 - 2t\right)\mathbf{i} - \dfrac{(t-2)^3}{3}\mathbf{j} + \dfrac{1}{2}e^{2t}\mathbf{k} + \mathbf{c}$

**39.** $\mathbf{r}(t) = \left(\dfrac{1}{2}e^{2t} - \dfrac{1}{2}e^2\right)\mathbf{i} + (t\ln t - t + 2)\mathbf{j} + (2e^t - 2e + 1)\mathbf{k}$ **41.** $\mathbf{r}(t) = (\sin t + 1)\mathbf{i} + (\cos t - 1)\mathbf{j} + \dfrac{t^2}{4}\mathbf{k}$ **43.** $\approx 25{,}540$ m away

# Chapter 12

## Section 12.1 (SSM = See Student Solutions Manual; AWV = Answers will vary)

**1.** $\sqrt{3}$ **2.** True **3.** $x, y, z; w$ **4.** True **5.** (a) $2$ (b) $10$ (c) $3x + 3(\Delta x) + 2y + xy + (\Delta x)y$

(d) $3x + 2y + 2(\Delta y) + xy + x(\Delta y)$ **7.** (a) $0$ (b) $at + a^2$ (c) $\sqrt{xy + (\Delta x)y} + x + \Delta x$ (d) $\sqrt{xy + x\Delta y} + x$

**9.** (a) $1$ (b) $0$ (c) $3\sin t\cos t$ **11.** (a) $0$ (b) $2$ (c) $92$

**13.** $\{(x, y) \mid x \geq 0, y > 0\}$ **15.** $\{(x, y) \mid x \geq 0, y \geq 0$ **17.** $\{(x, y) \mid x \neq 0, y \neq k\pi,$ **19.** $\{(x, y) \mid x^2 - y^2 > 0\}$

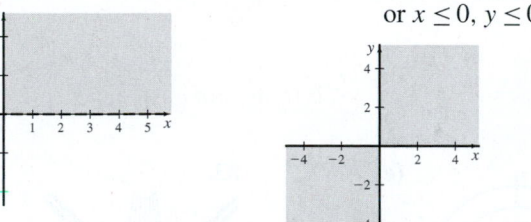  or $x \leq 0, y \leq 0\}$  $k$ an integer$\}$

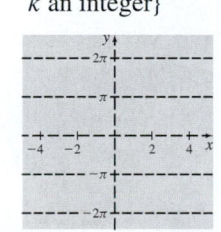 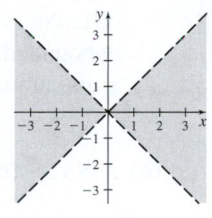

**21.** $\{(x, y) \mid x > 0, y > 0,$ **23.** $\{(x, y) \mid x^2 + y^2 < 9\}$ **25.** $\{(x, y) \mid y \geq x, y \geq -x$ or $y \leq x, y \leq -x, y \neq 0\}$

$y \neq 1\}$

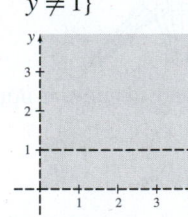 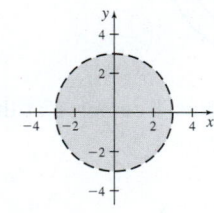 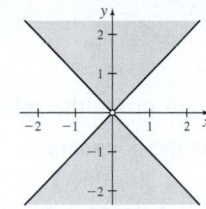

**27.** $\{(x, y, z) \mid z \neq 0\}$ **29.** $\left\{(x, y, z) \mid y \neq \dfrac{(2k+1)\pi}{2} \text{ for } k \text{ an integer}\right\}$

**31.** 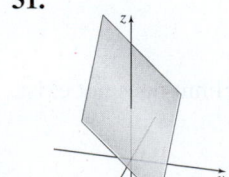 **33.** 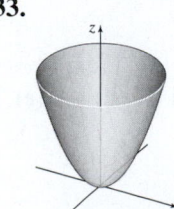 **35.** 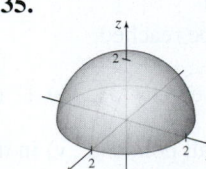 **37.** 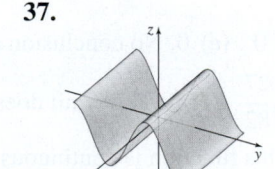 **39.**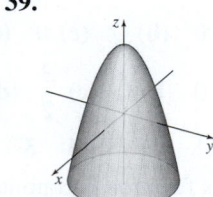

**41. (a)** 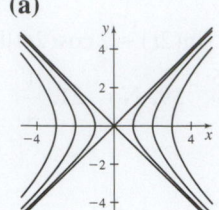 **(b)** 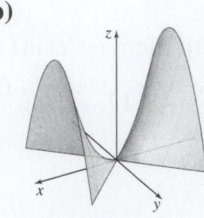 **43. (a)** 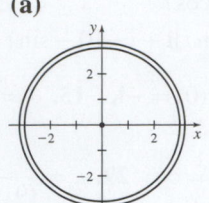 **(b)**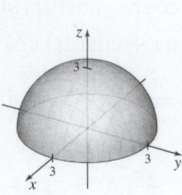

**45. (a)** 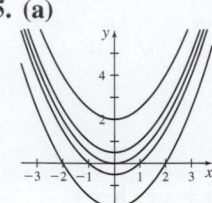 **(b)** 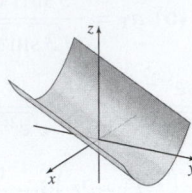 **47. (a)** 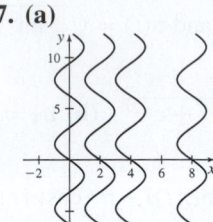 **(b)**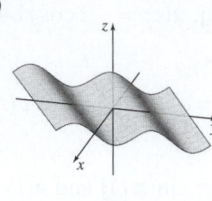

**49. (a)** AWV. **(b)** AWV. **(c)** AWV. **51. (a)** AWV. **(b)** AWV. **(c)** AWV.

**53. (a)** AWV. **(b)** AWV. **(c)** AWV. **55.** F **57.** B **59.** A

**61. (a)** 47°F **(b)** As you move south, the temperature change is increasing. **(c)** If you wanted to move as quickly as possible

to the next warmer temperature zone, you would move toward the southwest. Your path would be perpendicular
to the isotherm. **(d)** AWV.

**63. (a)** $\approx 83$ **(b)** 120 **(c)** $\approx 14$ **65. (a)** 6.75 **(b)** 18 **(c)** 2 **(d)** 1.5 **67. (a)** $\approx 104.715°F$ **(b)** $\approx 43.830\%$

**69.** $S(x, y, z) = xy + 2xz + 2yz$ **71.** $C(x, y, z) = 4xy + 6xz + 4yz$

**73.** 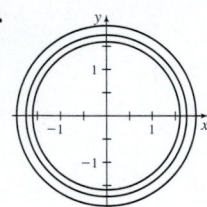 **75.** Level surfaces for $E(x, y, z)$ are all cylinders with different radii unbounded in the $x$ direction. 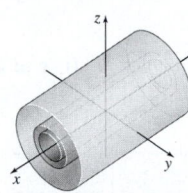 **77. (a)** AWV. **(b)** AWV. **(c)** AWV. **79. (a)** AWV. **(b)** AWV.

**81. (a)** 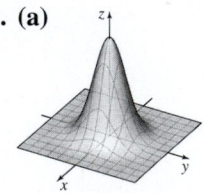 **(b)** AWV. **(c)** 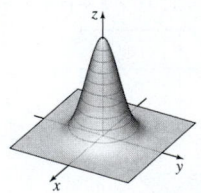 **(d)** 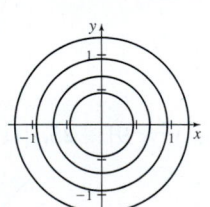 **(e)** AWV. **83.**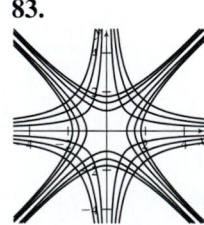

**85.** These inequalities describe the set of points that are inside the sphere of radius 1 centered at the origin, and inside the upper
half of the elliptic cone $x^2 + y^2 = z^2$, not including the boundary.

## Section 12.2 (SSM = See Student Solutions Manual; AWV = Answers will vary)

**1. (c)** **2.** 16 **3.** False **4.** True **5. (c)** **6. (a)** **7.** False **8.** 1 **9.** 7 **11.** $-1$ **13.** $-1$ **15.** $-\dfrac{3}{5}$ **17.** $-\dfrac{1}{\pi}$

**19.** $-3$ **21.** 0 **23.** $-\dfrac{1}{4}$ **25. (a)** 0 **(b)** 0 **(c)** 1 **(d)** $\dfrac{9}{11}$ **(e)** 0. This limit does not exist.

**27. (a)** 0 **(b)** 0 **(c)** 0 **(d)** 0 **(e)** 0. No conclusion can be reached.

**29. (a)** 0 **(b)** 0 **(c)** $\dfrac{3}{2}$ **(d)** $\dfrac{27}{82}$ **(e)** 0. This limit does not exist. **31. (a)** 1 **(b)** 0 **(c)** 1 **(d)** $\dfrac{2}{5}$ **(e)** 1. This limit does not exist.

**33.** SSM. **35.** SSM. **37.** This function is continuous for all points $(x, y)$ in the plane.

**39.** This function is continuous for all points $(x, y)$ in the plane except on the line $y = x$.

**41.** This function is continuous for all points $(x, y)$ in the plane. **43.** This function is continuous for all points $(x, y)$ in the plane.

**45.** This function is continuous for all points $(x, y)$ in the plane. **47.** This function is continuous for all points $(x, y)$ in the plane.

**49.** 1  **51.** 1  **53.** 0  **55.** $e^{\pi^2}\cos\pi^2$  **57.** $\dfrac{\pi}{4}$  **59.** 4  **61.** $\dfrac{1}{2}$  **63.** 1  **65.** $\dfrac{2}{3}$  **67.** SSM.

**69. (a)** The function is discontinuous at $(0,0)$.  **(b)** Yes, it is possible to define $f(x,y)$ at $(0,0)$.  **(c)** 0

**71. (a)** The function is discontinuous at $(0,0)$.  **(b)** No, it is not possible to define $f(x,y)$ at $(0,0)$.

**73. (a)** The function is discontinuous at $(0,0)$.  **(b)** No, it is not possible to define $f(x,y)$ at $(0,0)$.

**75. (a)** The function is continuous at $(0,0)$.  **77.** 0  **79.** This limit does not exist.  **81.** SSM.  **83.** 5

## Section 12.3  (SSM = See Student Solutions Manual; AWV = Answers will vary)

**1.** False  **2.** (b)  **3.** True  **4.** Mixed  **5.** False  **6.** $x$; $z$; $y$  **7.** $f_x(x,y)=2x+3y$; $f_y(x,y)=3x-1$

**9.** $f_x(x,y)=-2xy+3$; $f_y(x,y)=6-x^2$  **11.** $f_x(x,y)=y^2+\cos x$; $f_y(x,y)=2xy$

**13.** $f_x(x,y)=2xy$; $f_y(x,y)=x^2+12y$  **15.** $f_x(x,y)=\dfrac{2y}{(x+y)^2}$; $f_y(x,y)=-\dfrac{2x}{(x+y)^2}$

**17.** $f_x(x,y)=-e^y\sin x+e^x\sin y$; $f_y(x,y)=e^y\cos x+e^x\cos y$  **19.** $f_x(x,y)=2xe^{xy}+x^2ye^{xy}$; $f_y(x,y)=x^3e^{xy}$

**21.** $f_x(x,y)=\dfrac{y}{y^2+x^2}$; $f_y(x,y)=-\dfrac{x}{y^2+x^2}$  **23.** $f_x(x,y)=\ln(x^2+y^2)+\dfrac{2x^2}{x^2+y^2}$; $f_y(x,y)=\dfrac{2xy}{x^2+y^2}$

**25.** $\dfrac{\partial z}{\partial x}=2xy$; $\dfrac{\partial z}{\partial y}=x^2-\sin y$  **27.** $\dfrac{\partial z}{\partial x}=-y\sin(xy+y^2)$; $\dfrac{\partial z}{\partial y}=-(x+2y)\sin(xy+y^2)$

**29.** $\dfrac{\partial z}{\partial x}=\dfrac{y^2}{1+x^2y^4}$; $\dfrac{\partial z}{\partial y}=\dfrac{2xy}{1+x^2y^4}$  **31.** $\dfrac{\partial z}{\partial x}=\dfrac{xe^{\sqrt{x^2+y^2}}}{\sqrt{x^2+y^2}}$; $\dfrac{\partial z}{\partial y}=\dfrac{ye^{\sqrt{x^2+y^2}}}{\sqrt{x^2+y^2}}$

**33.** $\dfrac{\partial z}{\partial x}=2xye^{x^2y}\cos(e^{x^2y})$; $\dfrac{\partial z}{\partial y}=x^2e^{x^2y}\cos(e^{x^2y})$  **35.** $f_{xx}=12$; $f_{xy}=-8$; $f_{yx}=-8$; $f_{yy}=18$

**37.** $f_{xx}=12$; $f_{xy}=5$; $f_{yx}=5$; $f_{yy}=-12$  **39.** $f_{xx}=\dfrac{-3x(x^3-2y^2)}{(x^3+y^2)^2}$; $f_{xy}=\dfrac{-6x^2y}{(x^3+y^2)^2}$; $f_{yx}=\dfrac{-6x^2y}{(x^3+y^2)^2}$; $f_{yy}=\dfrac{2(x^3-y^2)}{(x^3+y^2)^2}$

**41.** $f_{xx}=-4x^2y^6\cos(x^2y^3)-2y^3\sin(x^2y^3)$; $f_{xy}=-6x^3y^5\cos(x^2y^3)-6xy^2\sin(x^2y^3)$;
$f_{yx}=-6x^3y^5\cos(x^2y^3)-6xy^2\sin(x^2y^3)$; $f_{yy}=-9x^4y^4\cos(x^2y^3)-6x^2y\sin(x^2y^3)$

**43.** $f_x(x,y,z)=y+z$; $f_y(x,y,z)=x+z$; $f_z(x,y,z)=x+y$

**45.** $f_x(x,y,z)=y\sin z-yz\cos x$; $f_y(x,y,z)=x\sin z-z\sin x$; $f_z(x,y,z)=xy\cos z-y\sin x$

**47.** $f_x(x,y,z)=-\dfrac{yz}{x^2+y^2}$; $f_y(x,y,z)=\dfrac{xz}{x^2+y^2}$; $f_z(x,y,z)=\tan^{-1}\dfrac{y}{x}$

**49.** $f_x(x,y,z)=\dfrac{2x\cos[\ln(x^2+y^2+z^2)]}{x^2+y^2+z^2}$; $f_y(x,y,z)=\dfrac{2y\cos[\ln(x^2+y^2+z^2)]}{x^2+y^2+z^2}$; $f_z(x,y,z)=\dfrac{2z\cos[\ln(x^2+y^2+z^2)]}{x^2+y^2+z^2}$

**51.** $f_x(0,0)=1$; $f_y(0,0)=1$  **53.** $z-5=2(x-1)$, $y=2$  **55.** $z-\dfrac{\sqrt{3}}{2}=-\dfrac{\sqrt{3}}{3}\left(y-\dfrac{1}{2}\right)$, $x=0$

**57.** $z-2\sqrt{5}=\dfrac{\sqrt{5}}{5}(y-2)$, $x=4$  **59.** $z-\dfrac{\sqrt{3}}{2}=\dfrac{1}{2}\left(y-\dfrac{\pi}{3}\right)$, $x=0$  **61. (a)** $\dfrac{3}{25}$  **(b)** $\dfrac{4}{25}$

**63. (a)** SSM.  **(b)** $\dfrac{40}{\ln 2}\approx 57.708$; $\dfrac{80}{\ln 2}\approx 115.416$. AWV.  **(c)** 0; $\dfrac{100}{\ln 2}\approx 144.270$. AWV.  **65.** SSM.

**67. (a)** $\dfrac{\partial P}{\partial K}\approx 0.249$; $\dfrac{\partial P}{\partial L}\approx 0.761$  **(b)** $\dfrac{\partial P}{\partial K}\approx 0.144$; $\dfrac{\partial P}{\partial L}\approx 0.913$  **(c)** AWV.

**69.** $\dfrac{\partial w}{\partial x}=\dfrac{y^{1/3}z^{1/6}}{x^{1/2}}$; $\dfrac{\partial w}{\partial y}=\dfrac{2x^{1/2}z^{1/6}}{3y^{2/3}}$; $\dfrac{\partial w}{\partial z}=\dfrac{x^{1/2}y^{1/3}}{3z^{5/6}}$  **71.** SSM.

**73.** $\dfrac{\partial x}{\partial r}=\cos\theta$; $\dfrac{\partial x}{\partial\theta}=-r\sin\theta$; $\dfrac{\partial y}{\partial r}=\sin\theta$; $\dfrac{\partial y}{\partial\theta}=r\cos\theta$  **75. (a)** 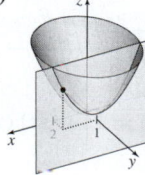  **(b)** $z-\dfrac{7}{3}=2(x-2)$, $y=1$

**77.** SSM. **79.** SSM. **81.** SSM. **83.** SSM. **85.** SSM. **87.** SSM. **89.** SSM. **91.** SSM.

**93. (a)** $f_x(x, y) = \begin{cases} \dfrac{-y^3(x^2 - y^2)}{(x^2 + y^2)^2} & (x, y) \neq (0, 0) \\ 0 & (x, y) = (0, 0) \end{cases}$ $f_y(x, y) = \begin{cases} \dfrac{xy^2(3x^2 + y^2)}{(x^2 + y^2)^2} & (x, y) \neq (0, 0) \\ 0 & (x, y) = (0, 0) \end{cases}$ **(b)** SSM.

**95.** No; AWV. **97.** SSM. **99.** SSM. **101.** SSM.

**103.** $\dfrac{\partial a}{\partial b} = \dfrac{b - c \cos A}{\sqrt{b^2 + c^2 - 2bc \cos A}}$; $\dfrac{\partial a}{\partial c} = \dfrac{c - b \cos A}{\sqrt{b^2 + c^2 - 2bc \cos A}}$; $\dfrac{\partial a}{\partial A} = \dfrac{bc \sin A}{\sqrt{b^2 + c^2 - 2bc \cos A}}$

**105.** $f_x = x^{y-1}y$; $f_y = x^y \ln x$ **107.** $f_x = \dfrac{(y + z)x^{y+z}}{x}$; $f_y = x^{y+z} \ln x$; $f_z = x^{y+z} \ln x$

**109.** $f_x = (x + y)^{z-1}z$; $f_y = (x + y)^{z-1}z$; $f_z = (x + y)^z \ln(x + y)$ **111.** $f_x(0, 0) = 0$; $f_y(0, 0) = 0$ **113.** SSM. **115.** SSM.

**117. (a)** $f_x(1, 6) = 6$; $f_y(1, 6) = 432$ **(b)** $\bar{f}_{x'}(a, b) = 219\sqrt{2}$; $\bar{f}_{y'}(a, b) = 213\sqrt{2}$

## Section 12.4 (SSM = See Student Solutions Manual; AWV = Answers will vary)

**1.** True **2.** True **3.** True **4.** True **5.** True **6.** False **7.** $\Delta z = 1.45$ **9.** $\Delta z \approx -0.318$ **11.** $dz = (2x)\,dx + (2y)\,dy$

**13.** $dz = (\sin y + y \cos x)\,dx + (x \cos y + \sin x)\,dy$ **15.** $dz = -\dfrac{y}{x^2 + y^2}\,dx + \dfrac{x}{x^2 + y^2}\,dy$ **17.** $dz = -\dfrac{1}{x}\,dx + \dfrac{1}{y}\,dy$

**19.** $dz = ye^{xy}\,dx + xe^{xy}\,dy$ **21.** $dz = 2xy\,dx + (x^2 + 2ye^{y^2})\,dy$

**23.** $dw = (e^{yz} + yze^{xz} + yze^{xy})\,dx + (xze^{yz} + e^{xz} + xze^{xy})\,dy + (xye^{yz} + xye^{xz} + e^{xy})\,dz$

**25.** $dw = \dfrac{2x}{x^2 + y^2 + z^2}\,dx + \dfrac{2y}{x^2 + y^2 + z^2}\,dy + \dfrac{2z}{x^2 + y^2 + z^2}\,dz$ **27.** $dw = (yz)\,dx + (xz)\,dy + (xy)\,dz$

**29. (a)** $\Delta z = (y^2 - 2y)\Delta x + (2xy - 2x)\Delta y + (2y\Delta y - 2\Delta y)\Delta x + (x\Delta y + \Delta x\Delta y)\Delta y$

  **(b)** $\eta_1 = 2y\Delta y - 2\Delta y$; $\eta_2 = x\Delta y + \Delta x\Delta y$ **(c)** SSM.

**31. (a)** $\Delta z = -\dfrac{y}{x^2}\Delta x + \dfrac{1}{x}\Delta y + \left(\dfrac{y}{x^2} - \dfrac{y}{x(x + \Delta x)}\right)\Delta x + \left(\dfrac{-1}{x} + \dfrac{1}{x + \Delta x}\right)\Delta y$

  **(b)** $\eta_1 = \dfrac{y}{x^2} - \dfrac{y}{x(x + \Delta x)}$; $\eta_2 = -\dfrac{1}{x} + \dfrac{1}{x + \Delta x}$ **(c)** SSM.

**33.** $f(2.3, -2.1) \approx 1.6$ **35.** $f(2.03, 0.1) \approx 1.4$ **37.** The area of the triangle increases by $\approx 0.225 \text{ cm}^2$

**39. (a)** $\approx 0.016 \text{ cm}^3$ **(b)** $\approx 0.126 \text{ cm}^2$ **41.** $\approx 15\%$ **43.** $\approx 3\%$ **45. (a)** $2.173 \text{ g/cm}^3$ **(b)** $16\%$

**47. (a)** $f_x(x, y) = \dfrac{x - 1}{\sqrt{x^2 + y^2 - 2x - 6y + 10}}$; **(b)** The partial derivatives are continuous for all points $(x, y)$ provided $(x, y) \neq (1, 3)$. **(c)** 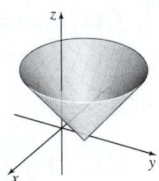 **49.** SSM. **51.** SSM. **53. (a)** $\approx 0.578\Omega$ **(b)** $\approx 15.516\%$ **(c)** The increase in the resistivity

$f_y(x, y) = \dfrac{y - 3}{\sqrt{x^2 + y^2 - 2x - 6y + 10}}$

  **(d)** AWV.

## Section 12.5 (SSM = See Student Solutions Manual; AWV = Answers will vary)

**1.** False **2.** False **3.** $\dfrac{dz}{dt} = (2x)(\cos t) - 4y \sin(2t) = 2 \sin t \cos t - 4 \sin(2t) \cos(2t)$

**5.** $\dfrac{dz}{dt} = (2x)(te^t + e^t) + (2y)(-te^{-t} + e^{-t}) = 2t^2 e^{2t} + 2te^{2t} - 2t^2 e^{-2t} + 2te^{-2t}$

**7.** $\dfrac{dz}{dt} = \dfrac{e^u \sin v}{2\sqrt{t}} + (e^u \cos v)(\pi) = \dfrac{e^{\sqrt{t}} \sin(\pi t)}{2\sqrt{t}} + \pi e^{\sqrt{t}} \cos(\pi t)$

**9.** $\dfrac{dz}{dt} = \dfrac{e^{u/v}(te^t + e^t)}{v} - \dfrac{ue^{u/v}(2te^{t^2})}{v^2} = \dfrac{te^{t + te^t - t^2} + e^{t + te^t - t^2}}{e^{t^2}} - \dfrac{2t^2 e^{t + t^2 + te^t - t^2}}{e^{2t^2}}$

**11.** $\dfrac{dz}{dt} = (2xe^{x^2 + y^2})(2\cos(2t)) - (2ye^{x^2 + y^2})(\sin t) = 4 \sin(2t) \cos(2t)e^{\sin^2(2t) + \cos^2 t} - 2 \sin t \cos t e^{\sin^2(2t) + \cos^2 t}$

**13.** $\dfrac{dz}{dt} = \dfrac{y(-x^2 + y^2)}{(x^2 + y^2)^2} \cdot \cos t + \dfrac{x(x^2 - y^2)}{(x^2 + y^2)^2} \cdot (-\sin t) = \cos^2 t - \sin^2 t$

**15.** $\dfrac{dp}{dt} = 2x(te^t + e^t) + 2y(-te^{-t} + e^{-t}) - 2z \cdot 2e^{2t} = 2t^2 e^{2t} + 2te^{2t} - 2t^2 e^{-2t} + 2te^{-2t} - 4e^{4t}$

**17.** $\dfrac{dp}{dt} = (e^x \sin y \cos z) \cdot \dfrac{1}{2\sqrt{t}} + (e^x \cos y \cos z) \cdot \pi - (e^x \sin y \sin z) \cdot \dfrac{1}{2}$

$$= \dfrac{e^{\sqrt{t}} \sin(\pi t) \cos \dfrac{t}{2}}{2\sqrt{t}} + \pi e^{\sqrt{t}} \cos(\pi t) \cos \dfrac{t}{2} - \dfrac{e^{\sqrt{t}} \sin(\pi t) \sin \dfrac{t}{2}}{2}$$

**19.** $\dfrac{dp}{dt} = \dfrac{w}{u} \cdot (te^t + e^t) - \dfrac{w}{v} \cdot (2te^{t^2}) + \left(\ln \dfrac{u}{v}\right)(2e^{2t}) = e^{2t} + \dfrac{e^{2t}}{t} + 2e^{2t} \ln t - 2t^2 e^{2t}$

**21.** $\dfrac{dp}{dt} = (2uvw)(\cos t) - (u^2 w)(\sin t) + (u^2 v)(e^t) = 2e^t \sin t \cos^2 t - e^t \sin^3 t + e^t \sin^2 t \cos t$

**23.** $\dfrac{\partial z}{\partial u} = (2x)(e^v) + (2y)(ve^u) = 2ue^{2v} + 2v^2 e^{2u}; \quad \dfrac{\partial z}{\partial v} = (2x)(ue^v) + (2y)(e^u) = 2u^2 e^{2v} + 2ve^{2u}$

**25.** $\dfrac{\partial z}{\partial u} = (e^x \sin y)(2uv) + (e^x \cos y)\left(\dfrac{1}{u}\right) = 2uve^{u^2 v} \sin(\ln(uv)) + \dfrac{e^{u^2 v} \cos(\ln(uv))}{u};$

$\dfrac{\partial z}{\partial v} = (e^x \sin y)(u^2) + (e^x \cos y)\left(\dfrac{1}{v}\right) = u^2 e^{u^2 v} \sin(\ln(uv)) + \dfrac{e^{u^2 v} \cos(\ln(uv))}{v}$

**27.** $\dfrac{\partial z}{\partial u} = \left(\dfrac{2x}{x^2 + y^2}\right)\left(-\dfrac{v^2}{u^2}\right) + \left(\dfrac{2y}{x^2 + y^2}\right)\left(\dfrac{1}{v^2}\right) = \dfrac{-2v^8}{u(u^4 + v^8)} + \dfrac{2u^3}{u^4 + v^8}$

$\dfrac{\partial z}{\partial v} = \left(\dfrac{2x}{x^2 + y^2}\right)\left(\dfrac{2v}{u}\right) + \left(\dfrac{2y}{x^2 + y^2}\right)\left(-\dfrac{2u}{v^3}\right) = \dfrac{4v^7}{u^4 + v^8} - \dfrac{4u^4}{v(u^4 + v^8)}$

**29.** $\dfrac{\partial z}{\partial u} = (2x)(\cos(u - v)) - (2y)(\sin(u + v)) = 2\sin(u - v)\cos(u - v) - 2\sin(u + v)\cos(u + v)$

$\dfrac{\partial z}{\partial v} = -(2x)(\cos(u - v)) - (2y)(\sin(u + v)) = -2\sin(u - v)\cos(u - v) - 2\sin(u + v)\cos(u + v)$

**31.** $\dfrac{\partial z}{\partial u} = (e^r)\left(\dfrac{-v}{u^2}\right) + (se^r)(2u) = -\dfrac{ve^{u^2 + v^2}}{u^2} + 2ve^{u^2 + v^2}$

$\dfrac{\partial z}{\partial v} = (e^r)\left(\dfrac{1}{u}\right) + (se^r)(2v) = \dfrac{e^{u^2 + v^2}}{u} + \dfrac{2v^2 e^{u^2 + v^2}}{u}$

**33.** $\dfrac{\partial z}{\partial u} = 2y^2 w^3 + 10xyw^3 + 6xy^2 w^2$

$= 2(5u - 3v)^2(2u + 3v)^3 + 10(2u + v)(5u - 3v)(2u + 3v)^3 + 6(2u + v)(5u - 3v)^2(2u + 3v)^2$

$\dfrac{\partial z}{\partial v} = y^2 w^3 - 6xyw^3 + 9xy^2 w^2$

$= (5u - 3v)^2(2u + 3v)^3 - 6(2u + v)(5u - 3v)(2u + 3v)^3 + 9(2u + v)(5u - 3v)^2(2u + 3v)^2$

**35.** $\dfrac{\partial f}{\partial u} = (2x)(v) + (2y)(e^{u + 2v + 3w}) = 2uv^2 + 2e^{2u + 4v + 6w}$

$\dfrac{\partial f}{\partial v} = (2x)(u) + (2y)(2e^{u + 2v + 3w}) + (2z)(2) = 2u^2 v + 4e^{2u + 4v + 6w} + 4(2v + 3w)$

$\dfrac{\partial f}{\partial w} = (2y)(3e^{u + 2v + 3w}) + (2z)(3) = 6e^{2u + 4v + 6w} + 6(2v + 3w)$

**37.** $\dfrac{\partial f}{\partial u} = [\cos(u^2 + v^2 + w^2) + w\sin(uvw) + 2uvw^2(u^2 + v^2 + w^2)](vw) + [-uvw\sin(u^2 + v^2 + w^2) + u^2 v^2 w^3](2u)$

$\dfrac{\partial f}{\partial v} = [\cos(u^2 + v^2 + w^2) + w\sin(uvw) + 2uvw^2(u^2 + v^2 + w^2)](uw) + [-uvw\sin(u^2 + v^2 + w^2) + u^2 v^2 w^3](2v)$

$\dfrac{\partial f}{\partial w} = [\cos(u^2 + v^2 + w^2) + w\sin(uvw) + 2uvw^2(u^2 + v^2 + w^2)](uv) + [-uvw\sin(u^2 + v^2 + w^2) + u^2 v^2 w^3](2w)$

$\quad + [-\cos(uvw) + u^2 v^2 w^2(u^2 + v^2 + w^2)]$

**39.** $\dfrac{\partial f}{\partial u} = t + 4ye^{u+2v+3w+4t} - 2z = t + 4e^{2u+4v+6w+8t} - 2\left(u + \dfrac{v}{2} + 4t\right)$

$\dfrac{\partial f}{\partial v} = 8ye^{u+2v+3w+4t} - z = 8e^{2u+4v+6w+8t} - u - \dfrac{v}{2} - 4t$

$\dfrac{\partial f}{\partial w} = 12ye^{u+2v+3w+4t} = 12e^{2u+4v+6w+8t}$

$\dfrac{\partial f}{\partial t} = u + 16ye^{u+2v+3w+4t} - 8z = u + 16e^{2u+4v+6w+8t} - 8\left(u + \dfrac{v}{2} + 4t\right)$

**41.** $-\pi$   **43.** $-\dfrac{\sqrt{42}}{14}$   **45.** 5   **47.** $\dfrac{dy}{dx} = -\dfrac{2xy - y^2 + y}{x^2 - 2xy + x}$   **49.** $\dfrac{dy}{dx} = -\dfrac{\sin y + y\cos x}{x\cos y + \sin x}$

**51.** $\dfrac{dy}{dx} = -\dfrac{y^{2/3}}{x^{2/3}}$   **53.** $\dfrac{\partial z}{\partial x} = -\dfrac{z + 2xy^3}{x + 6yz - 5}$; $\dfrac{\partial z}{\partial y} = -\dfrac{3z^2 + 3x^2y^2}{x + 6yz - 5}$

**55.** $\dfrac{\partial z}{\partial x} = -\dfrac{yz}{\cos z - y\sin z + xy}$; $\dfrac{\partial z}{\partial y} = -\dfrac{\cos z + xz}{\cos z - y\sin z + xy}$   **57.** $\dfrac{\partial z}{\partial x} = -\dfrac{e^{yz} + yze^{xz} + yz}{xye^{yz} + xye^{xz} + xy}$; $\dfrac{\partial z}{\partial y} = -\dfrac{xze^{yz} + e^{xz} + xz}{xye^{yz} + xye^{xz} + xy}$

**59.** $\dfrac{\partial w}{\partial x} = 8z(2x + 3y)^{4z-1}$; $\dfrac{\partial w}{\partial y} = 12z(2x + 3y)^{4z-1}$; $\dfrac{\partial w}{\partial z} = 4(2x + 3y)^{4z}\ln(2x + 3y)$   **61.** 151.2 m³/s

**63.** (a) SSM.   (b) SSM.   (c) SSM.   (d) SSM.   **65.** $\dfrac{\partial y}{\partial s}(1, 3) = 22$; $\dfrac{\partial y}{\partial t}(1, 3) = -3$

**67.** $\dfrac{\partial^2 z}{\partial u^2} = \dfrac{\partial z}{\partial x}\dfrac{\partial^2 x}{\partial u^2} + \dfrac{\partial z}{\partial y}\dfrac{\partial^2 y}{\partial u^2} + \dfrac{\partial^2 z}{\partial x^2}\left(\dfrac{\partial x}{\partial u}\right)^2 + 2\dfrac{\partial^2 z}{\partial x\partial y}\left(\dfrac{\partial x}{\partial u}\right)\left(\dfrac{\partial y}{\partial u}\right) + \dfrac{\partial^2 z}{\partial y^2}\left(\dfrac{\partial y}{\partial u}\right)^2$

$\dfrac{\partial^2 z}{\partial u\partial v} = \dfrac{\partial z}{\partial x}\dfrac{\partial^2 x}{\partial u\partial v} + \dfrac{\partial z}{\partial y}\dfrac{\partial^2 y}{\partial u\partial v} + \dfrac{\partial^2 z}{\partial x^2}\dfrac{\partial x}{\partial u}\dfrac{\partial x}{\partial v} + \dfrac{\partial^2 z}{\partial y^2}\dfrac{\partial y}{\partial u}\dfrac{\partial y}{\partial v} + \dfrac{\partial^2 z}{\partial x\partial y}\left(\dfrac{\partial y}{\partial u}\dfrac{\partial x}{\partial v} + \dfrac{\partial y}{\partial v}\dfrac{\partial x}{\partial u}\right)$

$\dfrac{\partial^2 z}{\partial v^2} = \dfrac{\partial z}{\partial x}\dfrac{\partial^2 x}{\partial v^2} + \dfrac{\partial z}{\partial y}\dfrac{\partial^2 y}{\partial v^2} + \dfrac{\partial^2 z}{\partial x^2}\left(\dfrac{\partial x}{\partial v}\right)^2 + 2\dfrac{\partial^2 z}{\partial x\partial y}\left(\dfrac{\partial x}{\partial v}\right)\left(\dfrac{\partial y}{\partial v}\right) + \dfrac{\partial^2 z}{\partial y^2}\left(\dfrac{\partial y}{\partial v}\right)^2$

**69.** SSM.   **71.** SSM.   **73.** SSM.   **75.** SSM.   **77.** (a) SSM.   (b) $\dfrac{d^2y}{dx^2} = \dfrac{-6x(3x - 3y^2)^2 + 6(3x^2 + 3y)(3x - 3y^2) + 6y(3x^2 + 3y)^2}{(3x - 3y^2)^3}$

## Chapter Review   (SSM = See Student Solutions Manual; AWV = Answers will vary)

**1.** (a) 0   (b) $e^{x+\Delta x}\ln y$   (c) $e^x \ln(y + \Delta y)$   **3.** (a) $\dfrac{3\pi}{2}$   (b) $\dfrac{e\pi}{6}$   (c) 0   **5.** $\{(x, y) \mid x^2 + 4y^2 \le 9\}$   **7.** $\left\{(x, y) \mid y \ne -\dfrac{1}{2}x\right\}$

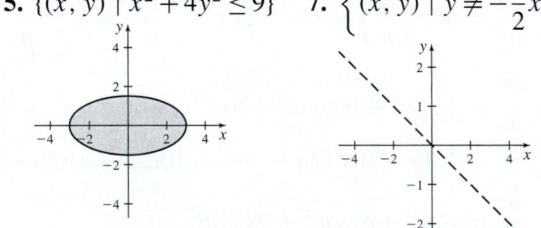

**9.** $\{(x, y, z) \mid z > 0\}$   **11.**   **13.**   **15.**   **17.** The level surfaces associated with the function $w = 4x^2 + y^2 + z^2$ are ellipsoids and the point $(0, 0, 0)$.

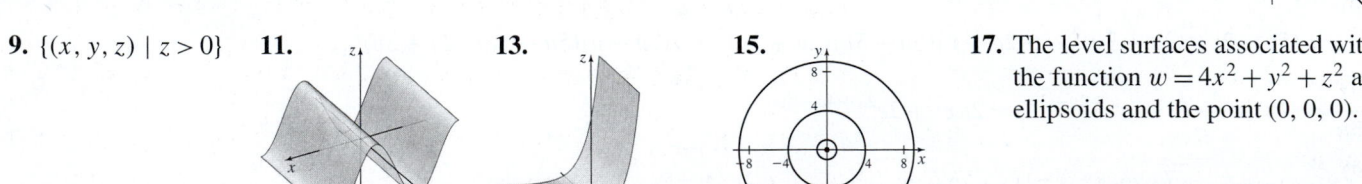

**19.** $\dfrac{2}{\pi}$   **21.** (a) SSM.   (b) $\dfrac{3}{2}$   (c) This limit does not exist.   **23.** The function is continuous at all points $(x, y)$ in the plane.

**25.** (a) This function is continuous at all points $(x, y)$ in the plane except the origin, $(0, 0)$.   (b) $\dfrac{\pi}{4}$

**27.** $f_x(x, y) = \dfrac{1}{y}$; $f_y(x, y) = -\dfrac{x}{y^2}$   **29.** $\dfrac{\partial z}{\partial x} = e^x \ln(5x + 2y) + \dfrac{5e^x}{5x + 2y}$; $\dfrac{\partial z}{\partial y} = \dfrac{2e^x}{5x + 2y}$

**31.** $F_x(0, \pi/6) = \dfrac{1}{2}$; $F_y(0, \pi/6) = \dfrac{\sqrt{3}}{2}$   **33.** (a) $z - 7 = 8(x - 1)$, $y = -2$   (b) $x = 1$, $z - 7 = 4(y + 2)$

**35.** $f_{xx}(x, y) = 6e^{3x} + 9(x + y^2)e^{3x}$; $f_{xy}(x, y) = 6ye^{3x}$; $f_{yx}(x, y) = 6ye^{3x}$; $f_{yy}(x, y) = 2e^{3x}$

**37.** $f_x = yze^{xyz}$; $f_y = xze^{xyz}$; $f_z = xye^{xyz}$  **39.** $f_x = e^x \sin y$; $f_y = e^x \cos y + e^y \sin z$; $f_z = e^y \cos z$  **41.** SSM.

**43. (a)** $\Delta z = y\Delta x + (x - 10y)\Delta y + \Delta y\Delta x - 5\Delta y\Delta y$  **(b)** $\eta_1 = \Delta y$; $\eta_2 = -5\Delta y$  **(c)** SSM.

**45.** $dz = \dfrac{1}{\sqrt{y^2 - x^2}}\,dx - \dfrac{x}{y\sqrt{y^2 - x^2}}\,dy$  **47.** $dw = \dfrac{1}{x}dx + \dfrac{1}{y}dy + \dfrac{1}{z}dz$  **49.** $\Delta z \approx 0.1$

**51.** The approximate maximum percentage error is 5%.

**53.** $\dfrac{dw}{dt} = \left(\dfrac{1}{y} - \dfrac{z}{x^2}\right)\left(\dfrac{-1}{t^2}\right) + \left(\dfrac{1}{z} - \dfrac{x}{y^2}\right)\left(\dfrac{-2}{t^3}\right) + \left(\dfrac{1}{x} - \dfrac{y}{z^2}\right)\left(\dfrac{-3}{t^4}\right) = 2 - \dfrac{2}{t^3}$

**55.** $\dfrac{\partial u}{\partial r} = \dfrac{x}{\sqrt{x^2 + y^2 + z^2}}(\cos s) + \dfrac{y}{\sqrt{x^2 + y^2 + z^2}}(\sin s) + \dfrac{z}{\sqrt{x^2 + y^2 + z^2}}\left(\dfrac{r}{\sqrt{r^2 + s^2}}\right) = \dfrac{2r}{\sqrt{2r^2 + s^2}}$

$\dfrac{\partial u}{\partial s} = \dfrac{x}{\sqrt{x^2 + y^2 + z^2}}(-r\sin s) + \dfrac{y}{\sqrt{x^2 + y^2 + z^2}}(r\cos s) + \dfrac{z}{\sqrt{x^2 + y^2 + z^2}}\left(\dfrac{s}{\sqrt{r^2 + s^2}}\right) = \dfrac{s}{\sqrt{2r^2 + s^2}}$

**57.** $\dfrac{\partial z}{\partial x} = \dfrac{-\sin y - y\cos x - yz}{xy}$; $\dfrac{\partial z}{\partial y} = \dfrac{-x\cos y - \sin x - xz}{xy}$

# Chapter 13

## Section 13.1  (SSM = See Student Solutions Manual; AWV = Answers will vary)

**1.** True  **2.** True  **3.** False  **4.** True  **5.** False  **6.** $\dfrac{1}{\|\nabla f(x_0, y_0)\|}\nabla f(x_0, y_0)$  **7.** $-\nabla f(x_0, y_0)$  **8.** $\|\nabla f(x_0, y_0)\|$

**9. (a)** $D_{\mathbf{u}}f(-1, 2) = 1 - 2\sqrt{3} \approx -2.46$  **(b)** AWV.  **(c)** AWV.

**11. (a)** $D_{\mathbf{u}}f(-1, 3) = -3 - 4\sqrt{3} \approx -9.93$  **(b)** AWV.  **(c)** AWV.

**13. (a)** $D_{\mathbf{u}}f(0, 0) = \dfrac{1 + \sqrt{3}}{2} \approx 1.37$  **(b)** AWV.  **(c)** AWV.

**15. (a)** $D_{\mathbf{u}}f(1, 1) = -\dfrac{7}{10} = -0.7$  **(b)** AWV.  **(c)** AWV.

**17. (a)** $\nabla f(1, 2) = 6\mathbf{i} + 4\mathbf{j}$  **(b)** $D_{\mathbf{u}}f(1, 2) = \dfrac{14\sqrt{5}}{5}$  **19. (a)** $\nabla f(0, 3) = 6\mathbf{i}$  **(b)** $D_{\mathbf{u}}f(0, 3) = \dfrac{12\sqrt{5}}{5}$

**21. (a)** $\nabla f(0, 1) = 2\mathbf{i}$  **(b)** $D_{\mathbf{u}}f(0, 1) = \dfrac{2\pi}{\sqrt{\pi^2 + 1}}$  **23. (a)** $\nabla f(1, 0) = \mathbf{j}$  **(b)** $D_{\mathbf{u}}f(1, 0) = \dfrac{\pi}{\sqrt{\pi^2 + 9}}$

**25. (a)** $\nabla f(2, 0) = 4\mathbf{i} + 4\mathbf{j}$  **(b)** $D_{\mathbf{u}}f(2, 0) = 4$  **27. (a)** $\nabla f(0, 1, 2) = -4\mathbf{i}$  **(b)** $D_{\mathbf{u}}f(0, 1, 2) = -\dfrac{4\sqrt{11}}{11}$

**29. (a)** $\nabla f(1, 1, 3) = -\dfrac{3}{2}\mathbf{i} + \dfrac{3}{2}\mathbf{j} + \dfrac{\pi}{4}\mathbf{k}$  **(b)** $D_{\mathbf{u}}f(1, 1, 3) = -\dfrac{(12 + 5\pi)\sqrt{3}}{36}$

**31. (a)** $\nabla f(3, 4, 0) = \dfrac{3}{5}\mathbf{i} + \dfrac{4}{5}\mathbf{j}$  **(b)** $D_{\mathbf{u}}f(3, 4, 0) = -\dfrac{13\sqrt{30}}{75}$  **33. (a)** $\mathbf{u} = \dfrac{\sqrt{5}}{5}\mathbf{i} - \dfrac{2\sqrt{5}}{5}\mathbf{j}$  **(b)** $\|\nabla f(-1, 2)\| = 2\sqrt{5}$

**35. (a)** $\mathbf{u} = \dfrac{\sqrt{2}}{2}\mathbf{i} + \dfrac{\sqrt{2}}{2}\mathbf{j}$  **(b)** $\|\nabla f(0, 0)\| = \sqrt{2}$  **37. (a)** $\mathbf{u} = \dfrac{3}{5}\mathbf{i} - \dfrac{4}{5}\mathbf{j}$  **(b)** $\|\nabla f(1, 2)\| = \dfrac{1}{5}$

**39. (a)** $\mathbf{u} = -\dfrac{6}{\sqrt{\pi^2 + 72}}\mathbf{i} + \dfrac{6}{\sqrt{\pi^2 + 72}}\mathbf{j} + \dfrac{\pi}{\sqrt{\pi^2 + 72}}\mathbf{k}$  **(b)** $\|\nabla f(1, 1, 3)\| = \sqrt{\dfrac{9}{2} + \dfrac{\pi^2}{16}}$

**41.**

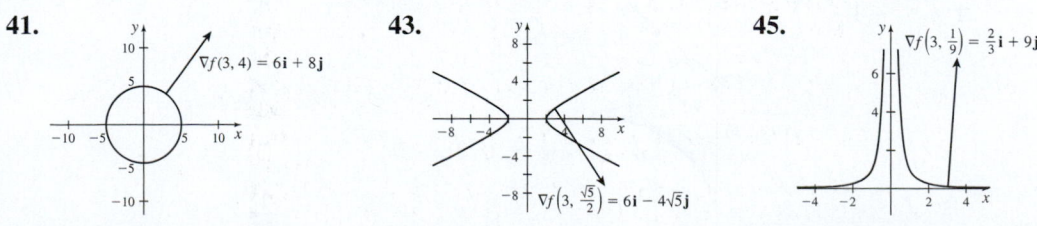

**47. (a)** $D_{\mathbf{u}}T(0, 0) = -\dfrac{1}{5}$. AWV.  **(b)** The temperature is increasing most rapidly in the direction $\nabla T(0, 0) = \mathbf{i} + \mathbf{j}$.

**(c)** The temperature is decreasing most rapidly in the direction $-\nabla T(0, 0) = -\mathbf{i} - \mathbf{j}$.

**(d)** The rate of change of temperature is 0 in the directions $-\mathbf{i} + \mathbf{j}$ and $\mathbf{i} - \mathbf{j}$.

**49.** Water will flow in the direction $\mathbf{u} = \dfrac{\sqrt{2}}{2}\mathbf{i} + \dfrac{\sqrt{2}}{2}\mathbf{j}$.

**51.** The rate of change of temperature is $\dfrac{\sqrt{3}}{2}$°C per unit of distance in the $\mathbf{u}$-direction.

**53. (a)** The direction is $\mathbf{u} = \dfrac{8\sqrt{145}}{145}\mathbf{i} + \dfrac{9\sqrt{145}}{145}\mathbf{j}$. **(b)** SSM. **(c)** The maximum rate of change is $2\sqrt{145}$.

**55.** Vector $\mathbf{u} = -\dfrac{4\sqrt{17}}{17}\mathbf{i} + \dfrac{\sqrt{17}}{17}\mathbf{j}$ is normal to the level curve of $f$ at $P$.

**57. (a)** The level curves are circles of the form
$$x^2 + y^2 = \left(\dfrac{250}{T}\right)^2$$
**(b)** 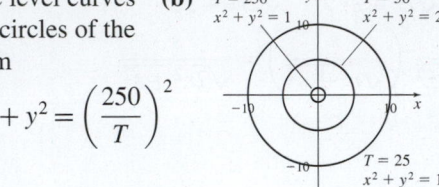 **(c)** The level curve is $x^2 + y^2 = 25$. **(d)** 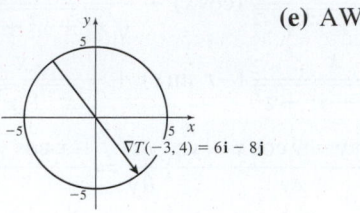 **(e)** AWV.

**59.** The charge moves in the direction $\mathbf{u} = \dfrac{2}{3}\mathbf{i} - \dfrac{2}{3}\mathbf{j} + \dfrac{1}{3}\mathbf{k}$. The electrical potential changes at a rate of $-3e^{-2}$ volts per unit change in the $\mathbf{u}$-direction. **61.** $-\dfrac{x}{\sqrt{4 - x^2}}$ **63.** SSM. **65.** SSM. **67.** SSM. **69.** SSM.

**71.** The function $f$ is not necessarily differentiable. SSM. **73.** SSM. **75.** SSM. **77.** SSM.

## Section 13.2 (SSM = See Student Solutions Manual; AWV = Answers will vary)

**1.** True **2.** True **3.** $4y + z = 14$ **5.** $8x + 18y - z = 13$ **7. (a)** $x - 2y + 3z = 14$ **(b)** $\dfrac{x - 1}{1} = \dfrac{y + 2}{-2} = \dfrac{z - 3}{3}$

**9. (a)** $4x + y + 2z = 3$ **(b)** $\dfrac{x - 1}{4} = \dfrac{y - 1}{1} = \dfrac{z + 1}{2}$ **11. (a)** $4x - 2y + z = -5$ **(b)** $\dfrac{x + 2}{-4} = \dfrac{y - 1}{2} = \dfrac{z - 5}{-1}$

**13. (a)** $2x - z = 0$ **(b)** $\dfrac{x - 1}{2} = \dfrac{z - 2}{-1}; \; y = 0$ **15. (a)** $3x - 3y - 4z = 2$ **(b)** $\dfrac{x - 3}{3} = \dfrac{y - 1}{-3} = \dfrac{z - 1}{-4}$

**17. (a)** $x\sqrt{\dfrac{\pi}{2}} + y\sqrt{\dfrac{\pi}{2}} - z = \dfrac{\pi}{2} - \ln\sqrt{2}$ **(b)** $\dfrac{x - \sqrt{\dfrac{\pi}{8}}}{\sqrt{\dfrac{\pi}{2}}} = \dfrac{y - \sqrt{\dfrac{\pi}{8}}}{\sqrt{\dfrac{\pi}{2}}} = \dfrac{z - \ln\sqrt{2}}{-1}$

**19. (a)** $y + z = \dfrac{\pi}{2}$ **(b)** $\dfrac{y - \dfrac{\pi}{2}}{-1} = \dfrac{z}{-1}; \; x = 0$ **(c)**

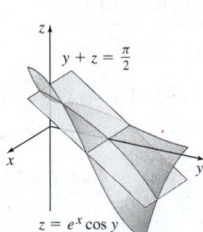

**21. (a)** $2x + y - z = 18$ **(b)** $\dfrac{x - 1}{2} = \dfrac{y - 8}{1} = \dfrac{z + 8}{-1}$ **(c)**

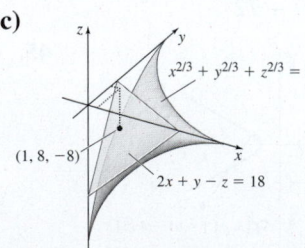

**23.** SSM. **25. (a)** $(3, -1, 11)$ **(b)** $z = 11$ **27. (a)** $(0, 0, 0)$ and $(0, 2, 4e^{-2})$ **(b)** $z = 0$ and $z = 4e^{-2}$

**29. (a)** $(-1, 0, -1)$ **(b)** $z = -1$ **31. (a)** $(0, -1, 1)$, $\left(\dfrac{1}{2}, -1, \dfrac{7}{8}\right)$, and $\left(-\dfrac{1}{2}, -1, \dfrac{7}{8}\right)$ **(b)** $z = 1$ and $z = \dfrac{7}{8}$.

**33.** The engineer should drill in the direction $-2\mathbf{i} - 2\mathbf{j} - \mathbf{k}$.

**35. (a)** SSM.      **(b)**

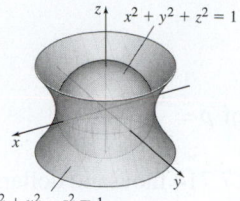

**37.** SSM.   **39.** The equation of the tangent plane is $y + z = \pi$.

The symmetric equations for the normal line are $\dfrac{y - \dfrac{\pi}{2}}{-1} = \dfrac{z - \dfrac{\pi}{2}}{-1}, x = 0$.   **41.** SSM.

**43.** SSM.   **45.** An equation of the tangent plane is $(\ln 2)x + 2y + (1 + \ln 2)z = 4(1 + \ln 2)$.

Symmetric equations of the normal line are $\dfrac{x - 2}{\ln 2} = \dfrac{y - 1}{2} = \dfrac{z - 2}{1 + \ln 2}$.   **47.** No. SSM.

## Section 13.3   (SSM = See Student Solutions Manual; AWV = Answers will vary)

**1. (a)**   **2.** False   **3.** False   **4. (c)**   **5.** True   **6.** boundary   **7.** The critical points are $(0, 0)$, $(1, 0)$, and $(-1, 0)$.

**9.** The critical points are $(0, 0)$, $(1, 1)$, and $(-1, -1)$.   **11.** There are no critical points.

**13.** There is a local minimum at $(1, -2)$; the local minimum value is $z = -3$.

**15.** There is a local minimum at $(2, -1)$; the local minimum value is $z = -9$.

**17.** The point $(0, 0, 0)$ is a saddle point of $z$. There is a local minimum at $(2, 2)$; the local minimum value is $z = -8$.

**19.** The point $(0, 2, 4)$ is a saddle point of $z$.

**21.** The point $(0, 0, 0)$ is a saddle point of $z$. There is a local maximum at $(-1, -1)$; the local maximum value is $z = 1$.

**23.** The points $\left(\dfrac{1}{2}, 1, \dfrac{2}{3}\right)$ and $\left(-\dfrac{1}{2}, -1, -\dfrac{2}{3}\right)$ are saddle points of $z$.   **25.** There are no critical points.

**27.** There is a local minimum at $(1, 1)$; the local minimum value is $z = 3$.

**29.** The function has critical points at $(x, y) = (m\pi, n\pi)$ for integers $m$ and $n$. For odd values of $m$ and $n$, the function has a local minimum; the local minimum value is $z = -2$. For even values of $m$ and $n$, the function has a local maximum; the local maximum value is $z = 2$. For odd $m$ and even $n$ OR even $m$ and odd $n$, $(m\pi, n\pi, 0)$ is a saddle point of $z$.

**31.** The local minima are at $(n\pi, 3)$ for $n$ an even integer and at $(n\pi, -3)$ for $n$ an odd integer. The local minimum value is $z = -3$.

For $n$, an integer, the points $\left(\dfrac{\pi}{2} + \pi n, 0, 6\right)$ are saddle points of $z$.

**33.** The point $(0, 0, 1)$ is a saddle point of $z$.   **35.** There is a local maximum at $\left(\dfrac{\pi}{2}, 0\right)$; the local maximum value is $z = 1$.

There is a local minimum at $\left(\dfrac{3\pi}{2}, 0\right)$; the local minimum value is $z = -1$.

**37.** The absolute maximum value of $f$ is 4; the absolute minimum value of $f$ is 0.

**39.** The absolute maximum value of $f$ is $4 + 4\sqrt{2}$; the absolute minimum value of $f$ is $-2$.

**41.** The absolute maximum value of $f$ is 33; the absolute minimum value of $f$ is $-7$.   **43. (a)** $(2, 1, 1)$ **(b)** $\sqrt{6}$

**45. (a)** $(0, 0, 2)$, $(0, 0, -2)$ **(b)** 2   **47.** At its highest peak, the mountain is 14 km above sea level.

**49.** The volume is a maximum when the length and width are $\sqrt{\dfrac{S}{3}}$ and the depth is $\dfrac{1}{2}\sqrt{\dfrac{S}{3}}$.

**51.** The dimensions of the closed box of volume $V = 500$ cm³ that uses the least amount of material

occur when the length, width, and depth are all $\sqrt[3]{500}$ cm.

**53. (a)** The rectangular box with maximum volume has a cross section of dimension 18 in. by 18 in. and a length of 36 in.

**(b)** The cylindrical box with maximum volume has a radius of $\dfrac{36}{\pi}$ in. and a length of 36 in.

**55. (a)** For a fixed amount $x$ of Drug A, $y = \dfrac{2b}{3}$ units of Drug B produces the maximum reaction.

**(b)** For a fixed amount $y$ of Drug B, $x = \dfrac{2a}{3}$ units of Drug A produces the maximum reaction.

**(c)** Reaction is a maximum when $x = \dfrac{2a}{3}$ units of Drug A and $y = \dfrac{2b}{3}$ units of Drug B are used.

**57.** The location $\left(\dfrac{1}{3}, \dfrac{1}{2}\right)$ will minimize the cost of connecting the plants to $D$.

**59.** The location with the highest reading is $(x, y) = \left(\dfrac{1}{2}, \dfrac{1}{8}\right)$.

**61.** Production is maximized when $x = \dfrac{20}{7} \approx 2.857$ thousand units are sold at a price of $p = \dfrac{64}{7} \approx 9.143$

thousand dollars and $y = \dfrac{2}{7} \approx 0.286$ thousand units are sold at a price of $q = \dfrac{54}{7} \approx 7.714$ thousand dollars.

The maximum profit is \$18,286. **63.** SSM.

**65. (a)** SSM. **(b)** The equation of the least-squares estimate is $y = 0.7x + 1.2$.

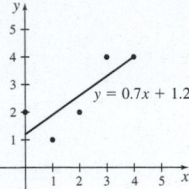

## Section 13.4 (SSM = See Student Solutions Manual; AWV = Answers will vary)

**1.** (c) **2.** True **3.** False **4.** False **5.** The maximum value of $f$ is $4\sqrt{6}$. The minimum value of $f$ is $-4\sqrt{6}$.

**7.** The maximum value of $f$ is $4 + \dfrac{1}{2}\sqrt{37}$. The minimum value of $f$ is $4 - \dfrac{1}{2}\sqrt{37}$.

**9.** The maximum value of $f$ is 1. The minimum value of $f$ is $-\dfrac{17}{8}$.

**11.** The maximum value of $f$ is 2. The minimum value of $f$ is $-2$.

**13.** The maximum value of $f$ is 20. The minimum value of $f$ is 0. **15.** The minimum value of $f$ is $-\dfrac{75}{4}$.

**17.** The absolute minimum value of $f$ is $-2$. The absolute maximum value of $f$ is 6.

**19.** The absolute minimum value of $f$ is $6 - 4\sqrt{2}$. The absolute maximum value of $f$ is $6 + 4\sqrt{2}$.

**21.** The point $\left(\dfrac{3}{5}, -\dfrac{9}{5}\right)$ is closest to the origin. **23.** The maximum product is $\dfrac{441}{8}$.

**25. (a)** The points $(0, 3, \sqrt{7})$ and $(0, 3, -\sqrt{7})$ are farthest from the origin.

    **(b)** The points $\left(\dfrac{\sqrt{7}}{2}, 3, 0\right)$ and $\left(-\dfrac{\sqrt{7}}{2}, 3, 0\right)$ are closest to the origin.

**27.** The product is a maximum at $\left(1, \dfrac{\sqrt{2}}{2}\right)$ and $\left(-1, -\dfrac{\sqrt{2}}{2}\right)$.

**29.** The optimal dimensions are obtained when $r = \sqrt[3]{\dfrac{V}{2\pi}}$ and $h = 2r = 2 \cdot \sqrt[3]{\dfrac{V}{2\pi}}$.

**31.** An open-topped box with a square base of $2\sqrt[3]{3}$ m by $2\sqrt[3]{3}$ m and a height of $\sqrt[3]{3}$ m will minimize the cost.

**33.** A cubical suitcase with each side measuring 15 inches will maximize the volume.

**35. (a)** Width $x$ must be no less than 20 m and no more than 50 m.

    **(b)** The greatest number of square meters that can be enclosed by the two fields is 3400 m².

**37.** Production levels of $x = 1800$ thousand units and $y = 3600$ thousand units will minimize production cost.

**39. (a)** Invest 250 units of capital and 1050 units of labor to maximize production. **(b)** The maximum production is 740.8 units.

**41.** Closest: $(0, 2, 0)$ and $\left(\dfrac{8}{5}, -\dfrac{6}{5}, 0\right)$; furthest: $\left(-\dfrac{4\sqrt{5}}{5}, -\dfrac{2\sqrt{5}}{5}, 2 + 2\sqrt{5}\right)$. **43.** SSM.

**45.** On the union of the two curves, $f$ attains a maximum value at $(\pm 1, 0)$ and $(0, \pm 1)$ and a minimum value at $\left(\pm\dfrac{\sqrt{2}}{2}, \pm\dfrac{\sqrt{2}}{2}\right)$.

**47.** $\dfrac{D^4}{(A^{4/3} + B^{4/3} + C^{4/3})^3}$ **49.** $\left(\dfrac{1}{2}, \dfrac{1}{2}, \dfrac{3}{2}\right)$

## Chapter Review   (SSM = See Student Solutions Manual; AWV = Answers will vary)

**1.** $\nabla f\left(\sqrt{\frac{\pi}{8}}, \sqrt{\frac{\pi}{8}}\right) = \sqrt{\frac{\pi}{2}}\mathbf{i} + \sqrt{\frac{\pi}{2}}\mathbf{j}$; $D_\mathbf{u} f\left(\sqrt{\frac{\pi}{8}}, \sqrt{\frac{\pi}{8}}\right) = \sqrt{\pi}$

**3.** $\nabla f(0, 1) = \mathbf{i}$; $D_\mathbf{u} f(0, 1) = \dfrac{3\sqrt{10}}{10}$   **5.** $\nabla f(0, 0, 0) = \mathbf{j}$; $D_\mathbf{u} f(0, 0, 0) = \dfrac{1}{3}$

**7. (a)** $\mathbf{u} = \dfrac{\sqrt{2}}{2}\mathbf{i} + \dfrac{\sqrt{2}}{2}\mathbf{j}$

    **(b)** $\left\| \nabla f\left(\sqrt{\frac{\pi}{8}}, \sqrt{\frac{\pi}{8}}\right) \right\| = \sqrt{2\pi}$ is the maximum value; $-\left\| \nabla f\left(\sqrt{\frac{\pi}{8}}, \sqrt{\frac{\pi}{8}}\right) \right\| = -\sqrt{2\pi}$ is the minimum value.

**9. (a)** $\mathbf{u} = \mathbf{i}$   **(b)** $\|\nabla f(0, 1)\| = 1$ is the maximum value; $-\|\nabla f(0, 1)\| = -1$ is the minimum value.

**11. (a)** $4x - z = 2$   **(b)** $\dfrac{x - 1}{4} = \dfrac{z - 2}{-1}$; $y = 0$   **(c)** $(0, 0, 0)$   **13.** The critical point is $(-4, 2)$.   **15.** The critical point is $(0, 0)$.

**17.** For integers $m$ and $n$, there is a local maximum value at $\left(\dfrac{\pi}{2} + 2\pi m, \dfrac{\pi}{2} + 2\pi n\right)$; the local maximum value is $z = 2$.

    For integers $m$ and $n$, there is a local minimum value at $\left(-\dfrac{\pi}{2} + 2\pi m, -\dfrac{\pi}{2} + 2\pi n\right)$; the local minimum value is $z = -2$.

    For integers $m$ and $n$, $\left(-\dfrac{\pi}{2} + 2\pi m, \dfrac{\pi}{2} + 2\pi n, 0\right)$ and $\left(\dfrac{\pi}{2} + 2\pi m, -\dfrac{\pi}{2} + 2\pi n, 0\right)$ are saddle points of $z$.

**19.** The absolute maximum value is 9; the absolute minimum value is $-7$.

**21.** There is a minimum value at $\left(\dfrac{130}{19}, -\dfrac{120}{19}\right)$; the minimum value is $z = \dfrac{5900}{19}$.

**23.** There is a minimum value at $\left(\dfrac{8}{5}, \dfrac{4}{5}\right)$; the minimum value is $z = \dfrac{16}{5}$.

**25. (a)** $\dfrac{7}{5}$   AWV.   **(b)** The temperature is increasing most rapidly in the direction $\nabla T(0, 0) = \mathbf{i} + \mathbf{j}$.

    **(c)** The temperature is decreasing most rapidly in the direction $-\nabla T(0, 0) = -\mathbf{i} - \mathbf{j}$.

    **(d)** The rate of change of temperature is 0 in the directions $\mathbf{i} - \mathbf{j}$ and $-\mathbf{i} + \mathbf{j}$.

**27. (a)** Invest 306 units of capital and 357 units of labor to maximize production.   **(b)** The maximum production is 3408.7 units.

**29.** The points $(1, 1, 1)$, $(-1, 1, -1)$, $(-1, -1, 1)$, and $(1, -1, -1)$ are all closest to the origin.

**31.** A box with a square base with each side measuring $\sqrt[3]{\dfrac{V}{3}}$ and a height of $3\sqrt[3]{\dfrac{V}{3}}$ will minimize the total cost of the box.

## Chapter 14

## Section 14.1   (SSM = See Student Solutions Manual; AWV = Answers will vary)

**1.** True   **2.** False   **3.** (d)   **4.** True   **5. (a)** 36   **(b)** 9   **7. (a)** 60   **(b)** 164   **9.** 320   **11.** $x$   **13.** $\dfrac{\pi^2}{8}\sin y$   **15.** $e - 1$

**17.** $\dfrac{2}{3}$   **19.** 27   **21.** $\dfrac{2}{3}$   **23.** 0   **25.** $\sqrt{2}$   **27.** $\dfrac{63}{2}$   **29.** $e - 1$   **31.** 0   **33.** $1 + e^2$   **35.** $\dfrac{(\pi - 2)\ln 2}{2}$   **37.** 30   **39.** 28

**41.** $5(1 - e)$   **43.** $\dfrac{9}{2}$   **45.** $\ln\dfrac{5}{3}$   **47.** $\dfrac{\pi - 2}{2}$   **49.** $\dfrac{1}{2} - \dfrac{1}{2}\cos\left(\dfrac{\pi^2}{4}\right)$   **51.** $\dfrac{111}{76} - \dfrac{\sqrt{2}}{8}\tan^{-1}\left(3\sqrt{2}\right)$   **53.** $\dfrac{1}{3}(e^8 - 9)$

**55.** 5 cubic units   **57.** $\dfrac{10}{3}$ cubic units   **59.** 1 cubic unit   **61.** $3\ln 2$ cubic units   **63.** $(e - 1)^2(e + 1)$ cubic units

**65.** $\dfrac{8}{3}$ cubic units   **67. (a)** 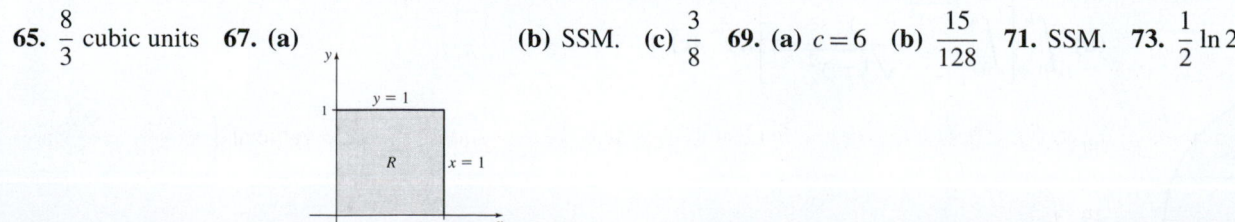   **(b)** SSM.   **(c)** $\dfrac{3}{8}$   **69. (a)** $c = 6$   **(b)** $\dfrac{15}{128}$   **71.** SSM.   **73.** $\dfrac{1}{2}\ln 2$

## Section 14.2 (SSM = See Student Solutions Manual; AWV = Answers will vary)

**1.** $x$-simple region  **2.** True  **3.** True  **4.** AWV.  **5. (a)** $\int_1^3 \left[ \int_0^{9-y^2} xy^2 \, dx \right] dy$  **(b)** $\int_0^8 \left[ \int_1^{\sqrt{9-x}} xy^2 \, dy \right] dx$  **(c)** $\dfrac{2504}{35}$

**7. (a)** $\int_0^3 \left[ \int_{3-y}^{\sqrt{9-y^2}} xy^2 \, dx \right] dy$  **(b)** $\int_0^3 \left[ \int_{3-x}^{\sqrt{9-x^2}} xy^2 \, dy \right] dx$  **(c)** $\dfrac{243}{20}$

**9.** $\dfrac{1}{3}$  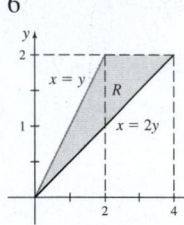   **11.** $\dfrac{9}{2}$  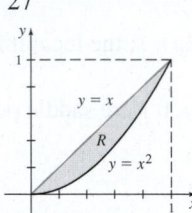   **13.** 1  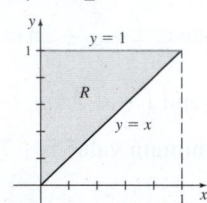   **15.** $\dfrac{1}{3}$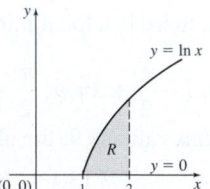

**17.** 6  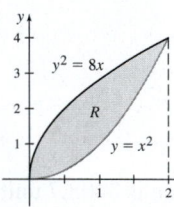   **19.** $\dfrac{2}{27}$  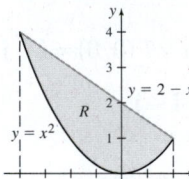   **21.** $\dfrac{\pi}{4} - \dfrac{\ln 2}{2}$  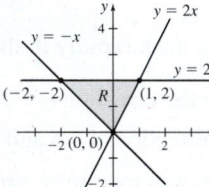   **23.** $\dfrac{5}{6}$

**25.** $\dfrac{36}{5}$  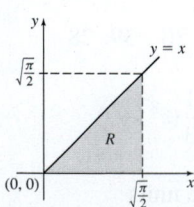   **27.** $\dfrac{423}{28}$  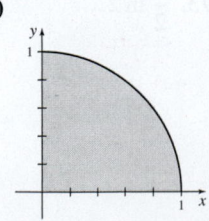   **29.** $\dfrac{12}{5}$

**31.** $A = \dfrac{1}{12}$ square units  **33.** $A = 72$ square units  **35.** $A = 2$ square units  **37.** $A = \dfrac{3}{2} - 2\ln 2$ square units

**39.** $\int_0^1 \left[ \int_y^1 f(x, y) \, dx \right] dy$  **41.** $\int_0^a \left[ \int_0^{\sqrt{a^2-x^2}} f(x, y) \, dy \right] dx$  **43.** $\int_0^2 \left[ \int_{x^4}^{8x} f(x, y) \, dy \right] dx$  **45.** $\int_0^{\ln 2} \left[ \int_{e^x}^2 f(x, y) \, dy \right] dx$

**47.** 14  **49.** 0  **51.** $A = \dfrac{56}{3}$ square units  **53.** $V = \int_{-1}^4 \int_{x^2-2x}^{x+4} e^{x^2+y^2} \, dy \, dx$  **55.** $V = \dfrac{2}{3}$ cubic units

**57.** $V = \dfrac{1}{6}$ cubic units  **59.** $V = 8\pi$ cubic units  **61.** $V = \dfrac{\pi}{2}$ cubic units  **63.** $V = \dfrac{124}{15}$ cubic units

**65. (a)**   **(b)** $\int_0^{\sqrt{\pi/2}} \left[ \int_0^x \sin x^2 \, dy \right] dx$  **(c)** $\dfrac{1}{2}$

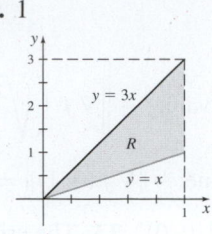

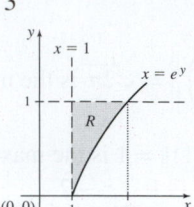

**67. (a)**   **(b)** $\int_0^1 \left[ \int_0^{\sqrt{1-y^2}} \dfrac{1}{\sqrt{1-y^2}} \, dx \right] dy$  **(c)** 1

**69. (a)**

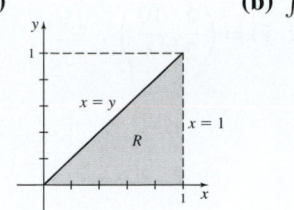

**(b)** $\int_0^1 \left[ \int_0^x \sqrt{2+x^2}\, dy \right] dx$   **(c)** $\sqrt{3} - \dfrac{2\sqrt{2}}{3}$

**71. (a)**

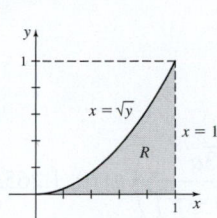

**(b)** $\int_0^1 \left[ \int_0^{x^2} e^{y/x}\, dy \right] dx$   **(c)** $\dfrac{1}{2}$   **73.** $\iint\limits_R f(x, y)\, dA = -2$   **75.** $\iint\limits_R f(x, y)\, dA = 4$

**77.** The solid is the volume under the hemisphere $z = \sqrt{1 - x^2 - y^2}$ in the first octant.

**79.** The solid is a rectangular box with height $z = 3$ units, with a rectangular base of dimensions $x = 4$ units and $y = 1$ unit.

**81. (a)** 0   **(b)** $\dfrac{4}{3}$   **83. (a)** $\dfrac{1}{2}$   **(b)** $\int_0^1 \left[ \int_0^{\cos^{-1} y} \sin x\, dx \right] dy = \dfrac{1}{2}$

**85.** $V = 4 \left[ (a-1)\sqrt{2a - a^2} + \sin^{-1} \sqrt{2a - a^2} \right]$ m³   **87. (a)** $\dfrac{1}{3}$   **(b)** $\dfrac{5}{9}$   **89. (a)** $c = 8$   **(b)** $\dfrac{9}{16}$

**91.** AWV.   **93.** 4   **95.** $12 - 4e$   **97.** SSM.

## Section 14.3   (SSM = See Student Solutions Manual; AWV = Answers will vary)

**1.** $f(r, \theta) = 2r$   **2.** $r\, dr\, d\theta$   **3.** False   **4.** True   **5.** $\int_0^{2\pi} \int_0^2 f(r\cos\theta, r\sin\theta) r\, dr\, d\theta$   **7.** $\int_0^{2\pi} \int_1^3 f(r\cos\theta, r\sin\theta) r\, dr\, d\theta$

**9.** $\int_0^{\pi/2} \int_0^3 r^2(\cos\theta + \sin\theta)\, dr\, d\theta$   **11.** $\int_0^{\pi/4} \int_0^2 r^2 \cos\theta\, dr\, d\theta$   **13.** $\int_0^{2\pi} \int_1^2 (r^2 \cos\theta + r^3 \sin^2\theta)\, dr\, d\theta$   **15.** $\dfrac{2\sqrt{2}}{3}$   **17.** $\dfrac{\pi^3}{24}$

**19.** $\dfrac{\pi + 2}{16}$   **21.** 0   **23.** 6   **25.** $\dfrac{2}{3}$   **27.** 0

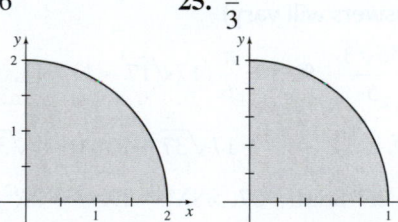

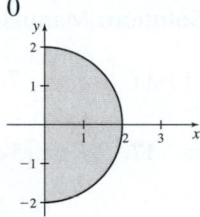

**29.** $\dfrac{\pi}{4}$   **31.** $4\pi$   **33.** $\dfrac{\pi}{4} \sin(1)$   **35.** $\dfrac{\pi(e^3 - 1)}{4e^4}$   **37.** $\dfrac{\pi}{8}$   **39.** $21\pi$   **41.** $\dfrac{64\pi}{3}$   **43.** $V = \dfrac{16\pi}{3}$ cubic units

**45.** $V = \dfrac{64\pi}{3} - \dfrac{256}{9}$ cubic units   **47.** $A = \dfrac{9}{2}$ square units   **49.** $A = 6 + \dfrac{3\pi}{4}$ square units

**51.** $A = \dfrac{5\pi}{6} + \dfrac{7\sqrt{3}}{8}$ square units   **53.** $\int_0^{1/2} \int_{\sqrt{y - y^2}}^{\sqrt{1/2 - y^2}} dx\, dy$   **55.** $V = \pi a \left( 1 - \dfrac{a^2}{3} \right)$ m³

**57. (a)**

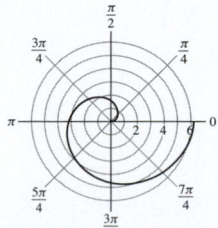

**(b)** $A = \dfrac{4\pi^3}{3}$ square units   **59.** $A = \dfrac{\pi}{12}$ square units.

**61. (a)** SSM.   **(b)** SSM.   **(c)** $J_a = \dfrac{\pi}{4}\left( 1 - e^{-a^2} \right)$   **(d)** SSM.   **(c)** SSM.

## Section 14.4 (SSM = See Student Solutions Manual; AWV = Answers will vary)

**1.** Mass **2.** False **3.** $M_y = \iint\limits_{R} x\rho(x, y)\, dA$ **4.** False **5.** True **6.** False **7.** $M = \dfrac{486}{5}$ and $(\bar{x}, \bar{y}) = \left(\dfrac{5}{2}, \dfrac{10}{3}\right)$

**9.** $M = 64$ and $(\bar{x}, \bar{y}) = \left(\dfrac{3}{2}, \dfrac{8}{3}\right)$ **11.** $M = \dfrac{31}{20}$ and $(\bar{x}, \bar{y}) = \left(\dfrac{150}{217}, \dfrac{44}{93}\right)$

**13.** $M = \dfrac{k}{15}$, where $k$ is the constant of proportionality, and $(\bar{x}, \bar{y}) = \left(\dfrac{15}{28}, \dfrac{5}{8}\right)$

**15.** $M = 5k$, where $k$ is the constant of proportionality, and $(\bar{x}, \bar{y}) = \left(\dfrac{6}{5}, \dfrac{7}{10}\right)$

**17.** $M = \dfrac{5k\pi}{3}$, where $k$ is the constant of proportionality, and $(\bar{x}, \bar{y}) = \left(0, \dfrac{21}{20}\right)$

**19.** $M = 2ka\left(\sqrt{3} - \dfrac{\pi}{3}\right) \approx 1.370ka$, where $k$ is the constant of proportionality, and $(\bar{x}, \bar{y}) = \left(0, \dfrac{3\sqrt{3}a}{2(3\sqrt{3} - \pi)}\right) \approx (0, 1.265a)$

**21.** $M = 3\rho,\ I_x = 2\rho,\ k = \dfrac{\sqrt{6}}{3}$ **23.** $M = ab\rho,\ I_y = \dfrac{a^3 b\rho}{3},\ k = \dfrac{\sqrt{3}a}{3}$ **25.** $M = \dfrac{8\rho}{3},\ I_y = \dfrac{8\rho}{15},\ k = \dfrac{\sqrt{5}}{5}$

**27.** $M = 2\pi k \ln 2$, where $k$ is the constant of proportionality

**29.** $M = ka(\pi - 2) \approx 1.142ka$, where $k$ is the constant of proportionality, and $(\bar{x}, \bar{y}) = \left(\dfrac{a\pi}{2(\pi - 2)}, 0\right) \approx (1.376a, 0)$

**31.** $M = \dfrac{1}{27}$ **33.** $M = \dfrac{423}{28}$ and $(\bar{x}, \bar{y}) = \left(-\dfrac{574}{235}, -\dfrac{469}{470}\right)$ **35.** $(\bar{x}, \bar{y}) = \left(\dfrac{a}{2}, \dfrac{2b}{5}\right)$ **37.** (a) $\dfrac{\rho s^4}{3}$ (b) $\dfrac{\rho s^4}{12}$ (c) $\dfrac{\rho s^4}{6}$

**39.** (a) $M = 2\pi k \ln\left(\dfrac{b}{a}\right)$, where $k$ is the constant of proportionality. As $a \to 0^+$, $M \to \infty$.

(b) $I = k\pi(b^2 - a^2)$, where $k$ is the constant of proportionality. As $a \to 0^+$, $I \to k\pi b^2$. **41.** SSM.

**43.** SSM. **45.** (a) SSM. (b) $I = \dfrac{mR^2}{4}$ (c) $I = \dfrac{5mR^2}{4}$

## Section 14.5 (SSM = See Student Solutions Manual; AWV = Answers will vary)

**1.** True **2.** False **3.** $S = 3\pi$ **5.** $S = 15\sin^{-1}\left(\dfrac{2}{3}\right)$ **7.** $S = \dfrac{52}{3} - \dfrac{36\sqrt{3}}{5}$ **9.** $S = \dfrac{\pi}{6}(17\sqrt{17} - 1)$ **11.** $S = \sqrt{2}\pi$

**13.** $S = \dfrac{\pi}{6}[(a^2 + 1)^{3/2} - 1]$ **15.** $S = 8\pi$ **17.** $S = 6\pi(5\sqrt{5} - 1)$ **19.** $S = \dfrac{1}{12}\left(1 + 17\sqrt{37} + 9\ln\left(6 + \sqrt{37}\right)\right)$

**21.** $S = 16\pi\left(2 - \sqrt{3}\right)$ **23.** (a) 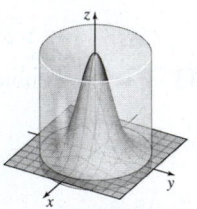 (b) $S \approx 55.436$ **25.** SSM. **27.** SSM. **29.** $S = 2R^2(\pi - 2)$ square units

## Section 14.6 (SSM = See Student Solutions Manual; AWV = Answers will vary)

**1.** True **2.** False **3.** True **4.** False

**5.** (a) $\int_0^4 \int_0^2 \int_0^1 xy^2\, dx\, dy\, dz$; $\int_0^2 \int_0^4 \int_0^1 xy^2\, dx\, dz\, dy$;

$\int_0^4 \int_0^1 \int_0^2 xy^2\, dy\, dx\, dz$; $\int_0^1 \int_0^4 \int_0^2 xy^2\, dy\, dz\, dx$;

$\int_0^2 \int_0^1 \int_0^4 xy^2\, dz\, dx\, dy$; $\int_0^1 \int_0^2 \int_0^4 xy^2\, dz\, dy\, dx$ (b) $\dfrac{16}{3}$

**7.** (a) $\int_0^3 \int_0^2 \int_0^1 \left(x^2 + y^2 + z^2\right) dx\, dy\, dz$; $\int_0^2 \int_0^3 \int_0^1 \left(x^2 + y^2 + z^2\right) dx\, dz\, dy$;

$\int_0^3 \int_0^1 \int_0^2 \left(x^2 + y^2 + z^2\right) dy\, dx\, dz$; $\int_0^1 \int_0^3 \int_0^2 \left(x^2 + y^2 + z^2\right) dy\, dz\, dx$;

$\int_0^2 \int_0^1 \int_0^3 \left(x^2 + y^2 + z^2\right) dz\, dx\, dy$; $\int_0^1 \int_0^2 \int_0^3 \left(x^2 + y^2 + z^2\right) dz\, dy\, dx$ (b) 28

**9. (a)** $\int_0^1 \int_0^{\pi/2} \int_0^{\pi/4} e^z \cos x \sin y\, dx\, dy\, dz$;  $\int_0^{\pi/2} \int_0^1 \int_0^{\pi/4} e^z \cos x \sin y\, dx\, dz\, dy$;

$\int_0^1 \int_0^{\pi/4} \int_0^{\pi/2} e^z \cos x \sin y\, dy\, dx\, dz$;  $\int_0^{\pi/4} \int_0^1 \int_0^{\pi/2} e^z \cos x \sin y\, dy\, dz\, dx$;

$\int_0^{\pi/2} \int_0^{\pi/4} \int_0^1 e^z \cos x \sin y\, dz\, dx\, dy$;  $\int_0^{\pi/4} \int_0^{\pi/2} \int_0^1 e^z \cos x \sin y\, dz\, dy\, dx$  **(b)** $(e-1)\left(1 - \frac{\sqrt{2}}{2}\right)$

**11.** $-\dfrac{2}{3}$  **13.** 81  **15.** $2(e-2)$  **17.** $1 - \dfrac{\pi^2}{8} + \dfrac{\pi^3}{48}$  **19.** $\dfrac{21}{4}$  **21.** $\dfrac{1}{180}$  **23.** $-\dfrac{2}{15}$  **25.** $-0.4142$  **27.** $\dfrac{1}{8}$  **29.** $\dfrac{648}{5}$

**31. (a)** The semi-cylinder is given by $x^2 + y^2 = 4$  **(b)**
and the planes $x=0$, $z=0$, and $z=1$.

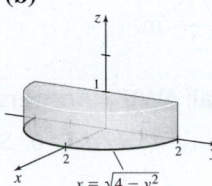

$x = \sqrt{4 - y^2}$

**33. (a)** The solid is bounded by $y = x^2$ and  **(b)**
the planes $z=0$, $x=1$, and $z=y$.

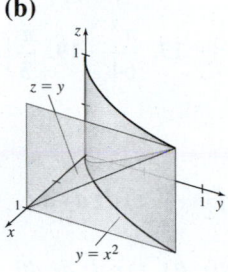

$z = y$

$y = x^2$

**35. (a)** $\int_0^6 \int_0^{3-x/2} \int_0^{2-x/3-2y/3} dz\, dy\, dx$;  $\int_0^3 \int_0^{6-2y} \int_0^{2-x/3-2y/3} dz\, dx\, dy$;  $\int_0^2 \int_0^{3-3z/2} \int_0^{6-2y-3z} dx\, dy\, dz$;

$\int_0^3 \int_0^{2-2y/3} \int_0^{6-2y-3z} dx\, dz\, dy$;  $\int_0^6 \int_0^{2-x/3} \int_0^{3-x/2-3z/2} dy\, dz\, dx$;  $\int_0^2 \int_0^{6-3z} \int_0^{3-x/2-3z/2} dy\, dx\, dz$

**(b)** $V = 6$ cubic units

**37.** $V = \dfrac{1}{60}$ cubic units  **39.** $V = \int_{-1}^1 \int_{-\sqrt{1-x^2}}^{\sqrt{1-x^2}} \int_{1-x-y}^{3-x-y} xy\, dz\, dy\, dx$ cubic units

**41.** $V = 4\int_0^{2\sqrt{2}} \int_0^{\sqrt{8-x^2}} \int_{x^2+y^2}^{16-x^2-y^2} dz\, dy\, dx$ cubic units  **43.** $V = \int_{-2}^1 \int_{-\sqrt{\frac{9}{4}-(x+\frac{1}{2})^2}}^{\sqrt{\frac{9}{4}-(x+\frac{1}{2})^2}} \int_{x^2+y^2}^{2-x} dz\, dy\, dx$

**45.** $V = 4\int_0^2 \int_0^{\sqrt{1-(x-1)^2}} \int_0^{2\sqrt{x}} dz\, dy\, dx$ cubic units

**47.** $M = \int_0^1 \int_0^{1-x} \int_0^{1-x-y} k(xyz)\, dz\, dy\, dx$, where $k$ is the constant of proportionality

**49.** $M = ka^5$, where $k$ is the constant of proportionality

**51.** $I_x = \int_{-3}^3 \int_{-\sqrt{9-x^2}}^{\sqrt{9-x^2}} \int_0^{\sqrt{9-x^2-y^2}} k(y^2+z^2)z\, dz\, dy\, dx$

$I_y = \int_{-3}^3 \int_{-\sqrt{9-x^2}}^{\sqrt{9-x^2}} \int_0^{\sqrt{9-x^2-y^2}} k(x^2+z^2)z\, dz\, dy\, dx$, where $k$ is the constant of proportionality

**53.** SSM. When $a=b=c$, the formula for the volume of an ellipsoid reduces to the formula
for the volume of a sphere.  **55.** $V = 2$ cubic units  **57.** SSM.

## Section 14.7

**1.** $x = r\cos\theta$, $y = r\sin\theta$, and $z = z$  **2.** $r = 2z$  **3.** $dV = r\, dr\, d\theta\, dz$  **4.** True  **5.** $\left(2, \dfrac{7\pi}{6}, -5\right)$

**7.** $\left(\sqrt{2}, \dfrac{\pi}{4}, \sqrt{2}\right)$  **9.** $(2, 0, 4)$  **11.** $\left(3, \dfrac{\pi}{2}, 4\right)$  **13.** $\left(\sqrt{3}, 1, -5\right)$  **15.** $(1, 0, 8)$  **17.** $(0, 2, 0)$

**19.** The triple integral represents the volume of a hemisphere of radius 1 to the right of the $xz$-plane, that is $y \geq 0$.

**21.** The triple integral represents the volume inside the vertical cylinder $x^2 + y^2 = 4$, bounded below by the $xy$-plane
and bounded above by the surface of the cone $z = \sqrt{x^2 + y^2}$.  **23.** $9\sqrt{3}$  **25.** $\dfrac{\pi}{6e}$  **27.** $24\pi$  **29.** 0  **31.** 0

**33.** $V = 4\pi \left(9 - \dfrac{7\sqrt{7}}{3}\right)$ cubic units   **35.** $V = 64\pi$ cubic units   **37.** $V = \dfrac{128\sqrt{2}}{15}$ cubic units

**39.** $M = \dfrac{4k\pi a^5}{5}$, where $k$ is the constant of proportionality   **41.** $I_x = I_y = \dfrac{243}{8}k\pi$, where $k$ is the constant of proportionality

**43.** $(\bar{x}, \bar{y}, \bar{z}) = \left(0, 0, \dfrac{4}{3}\right)$   **45.** $(\bar{x}, \bar{y}, \bar{z}) = \left(0, 0, \dfrac{8}{3}\right)$   **47.** $M = \dfrac{16}{3}\rho$

**49. (a)** $r = R\sqrt{\dfrac{2^{2/3} - 1}{2^{2/3}}}$   **(b)** $r \approx 6.083$ cm   **51.** $r = \dfrac{\sqrt[3]{15}}{2}$ m

## Section 14.8   (SSM = See Student Solutions Manual; AWV = Answers will vary)

**1.** Origin   **2.** $z$-axis   **3.** False   **4.** True   **5.** $dV = \rho^2 \sin\phi\, d\rho\, d\theta\, d\phi$   **6.** Sphere

**7. (a)** $(0, 0, -4)$   **(b)** $(-1, \sqrt{3}, 0)$   **9.** $\left(4, \dfrac{5\pi}{4}, \dfrac{\pi}{6}\right)$   **11.** $\left(2, \dfrac{\pi}{4}, \dfrac{\pi}{4}\right)$

**13.** $\left(\sqrt{14}, \tan^{-1}(2), \cos^{-1}\left(\dfrac{3}{\sqrt{14}}\right)\right)$   **15.** $\left(6, \dfrac{\pi}{2}, \dfrac{\pi}{3}\right)$   **17.** $\dfrac{\pi^2}{64}$   **19.** $\dfrac{\pi}{3}\left[\sqrt{2} + \ln\left(\sqrt{2} - 1\right)\right]$

**21.** 0   **23.** $\dfrac{\pi}{24}(2 - \sqrt{2})$   **25.** $4\pi(2 + \sqrt{3})$   **27.** $\dfrac{8}{3}\pi$

**29.** $\iiint\limits_E f(x, y, z)\, dV = \int_{-a}^{a} \int_{-\sqrt{a^2-x^2}}^{\sqrt{a^2-x^2}} \int_{-\sqrt{a^2-x^2-y^2}}^{\sqrt{a^2-x^2-y^2}} f(x, y, z)\, dz\, dy\, dx$

$$= \int_0^{2\pi} \int_0^a \int_{-\sqrt{a^2-r^2}}^{\sqrt{a^2-r^2}} f(x(r, \theta), y(r, \theta), z)\, r\, dz\, dr\, d\theta$$

$$= \int_0^{2\pi} \int_0^{\pi} \int_0^a f(x(\rho, \theta, \phi), y(\rho, \theta, \phi), z(\rho, \theta, \phi))\, \rho^2 \sin\phi\, d\rho\, d\phi\, d\theta$$

**31.** $\int_{-2\sqrt{2}}^{2\sqrt{2}} \int_0^{\sqrt{8-x^2}} \int_{-\sqrt{16-x^2-y^2}}^{-\sqrt{x^2+y^2}} (x^2 + y^2 + z^2)z\, dz\, dy\, dx$   **33.** $18\pi$ and $\dfrac{196}{3}\pi$   **35.** $\dfrac{2\pi}{3}a^3(1 - \cos\alpha)$

**37.** $M = \dfrac{4\pi}{5}ka^5$, where $k$ is the constant of proportionality   **39.** $(\bar{x}, \bar{y}, \bar{z}) = \left(0, 0, \dfrac{2a}{5}\right)$

**41.** $(\bar{x}, \bar{y}, \bar{z}) = \left(0, 0, \dfrac{9}{16}(2 + \sqrt{2})\right)$   **43.** $I = \dfrac{256}{15}\pi\delta$, where $\delta$ is the mass density   **45.** $\left(0, 0, \dfrac{3a}{8}\right)$

**47. (a)** $k = 1.238 \times 10^{-6}$ m$^{-1}$ and $D_0 = 2500$ kg/m$^3$   **(b)** The calculated mass of Dione is $1.10 \times 10^{21}$ kg, which differs from the measured mass by only 4.76%. The density model for the asteroid is reasonable.   **49.** SSM.

## Section 14.9   (SSM = See Student Solutions Manual; AWV = Answers will vary)

**1.** $\begin{vmatrix} \dfrac{\partial x}{\partial u} & \dfrac{\partial x}{\partial v} \\[2mm] \dfrac{\partial y}{\partial u} & \dfrac{\partial y}{\partial v} \end{vmatrix}$   **2.** True   **3.** $\dfrac{\partial(x, y)}{\partial(u, v)} = -2$   **5.** $\dfrac{\partial(x, y)}{\partial(u, v)} = 5$   **7.** $\dfrac{\partial(x, y)}{\partial(u, v)} = \dfrac{u + v}{v^2}$   **9.** $\dfrac{\partial(x, y)}{\partial(u, v)} = -e^{u-v}(1 + uv)$

**11.** $\dfrac{\partial(x, y, z)}{\partial(u, v, w)} = -4$   **13.** $\dfrac{\partial(x, y, z)}{\partial(u, v, w)} = -2$   **15.** $\dfrac{\partial(x, y, z)}{\partial(u, v, w)} = 6vw^2$   **17.** $A = 6\pi$ square units

**19. (a)**       **(b)**       **(c)** $\dfrac{\partial(x, y)}{\partial(u, v)} = \dfrac{1}{3}$   **(d)** $\dfrac{14}{3}$

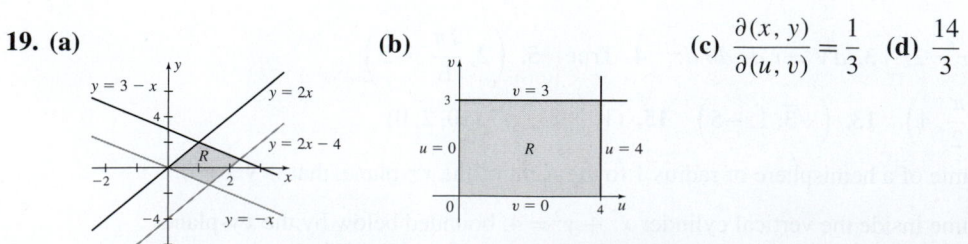

**21. (a)** 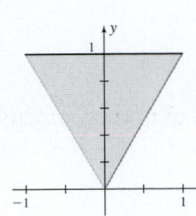 **(b)** 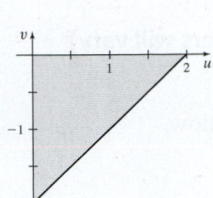 **(c)** $\dfrac{\partial(x, y)}{\partial(u, v)} = -\dfrac{1}{2}$ **(d)** 0 **23.** 0 **25.** $16\pi$

**27. (a)** 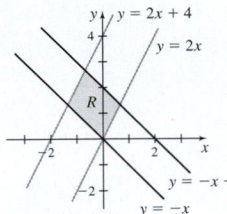 **(b)** 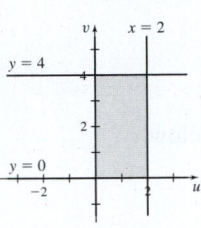 **(c)** $\dfrac{1}{3}\displaystyle\int_0^4 \int_0^2 (7u + 2v)\,du\,dv$ **(d)** $\dfrac{88}{3}$

**29. (a)** 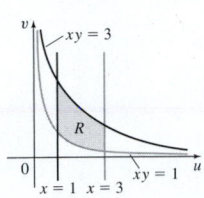 **(b)** 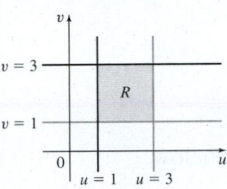 **(c)** $\int_1^3 \int_1^3 \cos(v)\,dv\,du$ **(d)** $2(\sin 3 - \sin 1)$

**31. (a)** 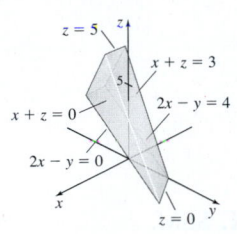 **(b)** 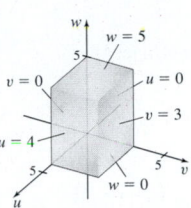 **(c)** $\int_0^5 \int_0^3 \int_0^4 (v - w)\,du\,dv\,dw$ **(d)** $-60$

**33.** $A = \dfrac{20}{7}$ square units  **35.** $V = 36$ cubic units  **37.** $V = \dfrac{21\pi}{2}$ cubic units  **39.** SSM.

## Review Exercises

**1.** $\dfrac{1}{2}$  **3.** $\dfrac{e - 1}{6e}$  **5.** $\sqrt{2} - 1$

**7. (a)** The midpoint Riemann sum is 32.

**(b)** The largest Riemann sum is 80, the lowest Riemann sum is $-16$, and their average is 32.  **(c)** 32

**9.** $\dfrac{\sqrt{3}}{2}$  **11.** $V = 6$ cubic units  **13.** $\dfrac{1}{6}(1 - \cos 8)$  **15.** $V = 6$ cubic units  **17.** $V = \dfrac{16}{5}$ cubic units  **19.** $12 - \dfrac{27\sqrt{3}}{4}$

**21.** $A = \pi$ square units  **23.** $M = 104k$, where $k$ is the constant of proportionality  **25.** $(\bar{x}, \bar{y}) = \left(\dfrac{12}{5}, \dfrac{3}{2}\right)$  **27.** $S = \dfrac{7}{2}$

**29.** $S = \dfrac{16\sqrt{3}}{3}\pi$  **31.** $\dfrac{2}{15}$  **33.** $\dfrac{5}{36}$  **35.** $V = \dfrac{81}{8}\pi$ cubic units  **37. (a)** $(r, \theta, z) = (3, 0, 4)$  **(b)** $(\rho, \theta, \phi) = \left(5, 0, \cos^{-1}\left(\dfrac{4}{5}\right)\right)$

**39. (a)** $(r, \theta, z) = \left(\sqrt{2}, \dfrac{3\pi}{4}, -2\right)$  **(b)** $(\rho, \theta, \phi) = \left(\sqrt{6}, \dfrac{3\pi}{4}, \cos^{-1}\left(\dfrac{-2}{\sqrt{6}}\right)\right)$  **41.** $V = \dfrac{28}{3}\pi$ cubic units

**43.** $\dfrac{4\pi}{3}$  **45.** $\dfrac{\pi}{6}a^3$  **47.** $\pi\left(\dfrac{128}{3} - 16\sqrt{3}\right)$  **49.** $16\pi$  **51.** $\dfrac{4\pi}{3}(\sqrt{2} - 1)$

**53.** $\dfrac{\partial(x, y)}{\partial(u, v)} = 2e^{2v}$  **55.** $\dfrac{\partial(x, y, z)}{\partial(u, v, w)} = 6u$  **57.** 203  **59.** 0

## Chapter 15

### Section 15.1   (SSM = See Student Solutions Manual; AWV = Answers will vary)

**1.** Vector field   **2.** False   **3.** Vectors   **4.** True

**5.** The vector field is pictured below.   **7.** The vector field is pictured below.
   SSM for a description.   SSM for a description.

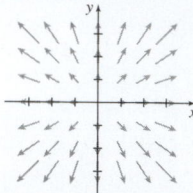   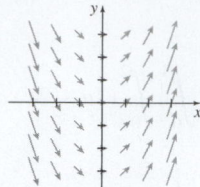

**9.** The vector field is pictured below.   **11.** The vector field is pictured below.
   SSM for a description.   SSM for a description.

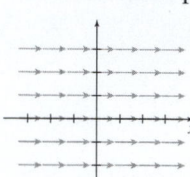   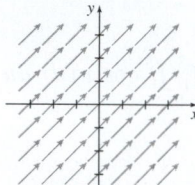

**13.** The vector field is pictured below.   **15.** The vector field is pictured below.
   SSM for a description.   SSM for a description.

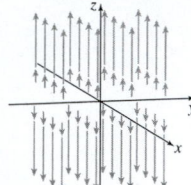   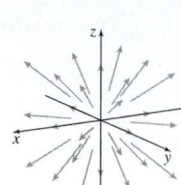

**17.** $\nabla f(x, y) = \sin y\mathbf{i} + (x \cos y - \sin y)\mathbf{j}$   **19.** $\nabla f(x, y, z) = (2xy + y)\mathbf{i} + (x^2 + x + 2yz)\mathbf{j} + y^2\mathbf{k}$   **21.** SSM.

### Section 15.2   (SSM = See Student Solutions Manual; AWV = Answers will vary)

**1.** True   **2.** $f(x(t), y(t))\sqrt{\left(\dfrac{dx}{dt}\right)^2 + \left(\dfrac{dy}{dt}\right)^2}$   **3.** False   **4.** $\int_C \rho(x, y)\, ds$   **5.** False   **6.** True

**7.** $\int_{C_1} (P\,dx + Q\,dy) + \int_{C_2} (P\,dx + Q\,dy) + \cdots + \int_{C_n} (P\,dx + Q\,dy)$   **8.** False   **9.** $\dfrac{1}{3}$   **11.** $\dfrac{512}{15}(1 + \sqrt{2})$   **13.** $\sqrt{2}\ln 5$

**15.** $\dfrac{74}{27}\sqrt{37} - \dfrac{2}{27}$   **17.** $\pi$   **19.** (a) $\dfrac{5}{6}\sqrt{2}$ (b) $\dfrac{5}{6}\sqrt{2}$   **21.** (a) $\sqrt{2}e - \dfrac{1}{2}\sqrt{2}$ (b) $\sqrt{2}e - \dfrac{1}{2}\sqrt{2}$

**23.** $\int_C f(x, y)dx = 18;\ \int_C f(x, y)dy = -18$   **25.** $\int_C f(x, y)dx = \dfrac{2}{5}(e^5 - 1);\ \int_C f(x, y)dy = \dfrac{1}{3}(e^3 - 1)$

**27.** (a) $\dfrac{1}{3} - \dfrac{\pi}{16}$ (b) $\dfrac{1}{3} - \dfrac{\pi}{16}$   **29.** (a) 5 (b) 6.15   **31.** 3   **33.** 0   **35.** $\dfrac{5}{12}\sqrt{5} - \dfrac{1}{12} - \dfrac{1}{2}\sqrt{2}$   **37.** $4\sqrt{2} - 4$

**39.** $\dfrac{1}{2}(1 - \sin 1 - \cos 1)$   **41.** 1   **43.** $-128$   **45.** $-128$   **47.** $\dfrac{1}{3}(17\sqrt{17} - 5\sqrt{5})$

**49.** $\dfrac{2601}{64}\sqrt{145} + \dfrac{387}{32}\sqrt{65} - \dfrac{3}{256}[\ln(\sqrt{145} + 12) - \ln(\sqrt{65} - 8)]$   **51.** $\pi$   **53.** 1   **55.** $\dfrac{1}{2}$   **57.** $-\dfrac{3}{2}$

**59.** $\sqrt{2}\left(\dfrac{\pi^2}{32} + 1\right)$   **61.** $\dfrac{\sqrt{a^2 + b^2}}{ab}\tan^{-1}\left(\dfrac{b}{a}\right)$   **63.** $(2a)^{3/2}\pi$   **65.** $2\pi a + a^2\pi$   **67.** $M = 4\pi\rho\sqrt{5}$   **69.** $M = \int_C \rho(x, y, z)\, ds$

**71.** $M = 20\pi + 40\pi^2$   **73.** SSM.   **75.** $\dfrac{1}{54}(19\sqrt{19} - 1) + \dfrac{37\sqrt{38}}{96} - \dfrac{1}{288}\ln(\sqrt{19} + 3\sqrt{2})$

**77.** SSM.   **79.** (a and b) SSM.

## Section 15.3 (SSM = See Student Solutions Manual; AWV = Answers will vary)

**1.** Energy **2.** True **3.** 3 **5.** $-\dfrac{4}{3}$ **7.** $35 - e^2$ **9.** $2e^8 + e^2 - 3$ **11.** 1 **13.** $3\pi - \dfrac{1}{2}$ **15.** $-\dfrac{19}{15}$ **17.** $\dfrac{2}{3}$ **19.** $\dfrac{3}{4}ab\pi(a^2 - b^2)$

**21.** 0 **23.** $\dfrac{1}{e}$ **25.** $\dfrac{19}{6}$ **27.** $\dfrac{11}{2}$ **29.** $1 - \cos\sqrt{\pi}$ **31.** The work done is $-\dfrac{163{,}840}{3}k$, where $k$ is a constant of proportionality.

**33.** $1 - \dfrac{\sqrt{5}}{5}$ **35.** $\dfrac{8}{3}\pi^3 - \dfrac{16}{3}$ **37.** SSM.

## Section 15.4 (SSM = See Student Solutions Manual; AWV = Answers will vary)

**1.** False **2.** (d) **3.** True **4.** (c) **5.** Closed **6.** 0 **7.** False **8.** True **9.** False **10.** True **11.** False **12.** $\dfrac{\partial P}{\partial y} = \dfrac{\partial Q}{\partial x}$

**13.** (d) **14.** motion; position **15.** 8 **17.** 108 **19.** $\ln 13 - \ln 5$ **21.** $\dfrac{135}{2}$ **23.** $\dfrac{8}{3}$ **25.** 8 **27.** 0

**29.** $f(x, y) = \dfrac{5}{2}x^2 - 2xy + x + 4y + K$ **31.** $f(x, y) = x \ln y + x^2 + K$

**33.** SSM. $f(x, y) = \dfrac{x^3}{3} + \dfrac{y^3}{3} + K$ **35.** SSM. $f(x, y) = \dfrac{x^2}{2}e^y + K$

**37.** (a) $P, Q, \dfrac{\partial P}{\partial y}$, and $\dfrac{\partial Q}{\partial x}$ are continuous everywhere in the $xy$-plane, and $\dfrac{\partial P}{\partial y} = \dfrac{\partial Q}{\partial x} = 0$. (b) $f(x, y) = \dfrac{1}{2}x^2 + \dfrac{1}{2}y^2 + K$ (c) $\dfrac{19}{2}$

**39.** (a) $P, Q, \dfrac{\partial P}{\partial y}$, and $\dfrac{\partial Q}{\partial x}$ are continuous everywhere in the $xy$-plane, and $\dfrac{\partial P}{\partial y} = \dfrac{\partial Q}{\partial x} = 3$.

(b) $f(x, y) = \dfrac{1}{3}x^3 + 3xy + K$ (c) $-\dfrac{181}{3}$

**41.** (a) $P, Q, \dfrac{\partial P}{\partial y}$, and $\dfrac{\partial Q}{\partial x}$ are continuous everywhere in the $xy$-plane, and $\dfrac{\partial P}{\partial y} = \dfrac{\partial Q}{\partial x} = 60xy^2 - 12y^3$.

(b) $f(x, y) = x^4 + 10x^2y^3 - 3xy^4 + y^5 + K$ (c) 9

**43.** (a) $P, Q, \dfrac{\partial P}{\partial y}$, and $\dfrac{\partial Q}{\partial x}$ are continuous everywhere in the $xy$-plane, and $\dfrac{\partial P}{\partial y} = \dfrac{\partial Q}{\partial x} = 6x^2y$. (b) $f(x, y) = x^3y^2 + K$

**45.** (a) $P, Q, \dfrac{\partial P}{\partial y}$, and $\dfrac{\partial Q}{\partial x}$ are continuous everywhere in the $xy$-plane, and $\dfrac{\partial P}{\partial y} = \dfrac{\partial Q}{\partial x} = 3$. (b) $f(x, y) = \dfrac{1}{2}x^2 + 3xy + K$

**47.** (a) $P, Q, \dfrac{\partial P}{\partial y}$, and $\dfrac{\partial Q}{\partial x}$ are continuous everywhere in the $xy$-plane, and $\dfrac{\partial P}{\partial y} = \dfrac{\partial Q}{\partial x} = 2x - 2y$. (b) $f(x, y) = x^2y - xy^2 + K$

**49.** (a) $P, Q, \dfrac{\partial P}{\partial y}$, and $\dfrac{\partial Q}{\partial x}$ are continuous everywhere in the $xy$-plane, and $\dfrac{\partial P}{\partial y} = \dfrac{\partial Q}{\partial x} = 2y$.

(b) $f(x, y) = \dfrac{1}{3}x^3 - \dfrac{1}{2}x^2 + xy^2 + e^y - ye^y + K$

**51.** (a) $P, Q, \dfrac{\partial P}{\partial y}$, and $\dfrac{\partial Q}{\partial x}$ are continuous everywhere in the $xy$-plane, and $\dfrac{\partial P}{\partial y} = \dfrac{\partial Q}{\partial x} = \cos x - 2\cos y$.

(b) $f(x, y) = y \sin x - 2x \sin y + K$

**53.** (a) $P, Q, \dfrac{\partial P}{\partial y}$, and $\dfrac{\partial Q}{\partial x}$ are continuous everywhere in the $xy$-plane, and $\dfrac{\partial P}{\partial y} = \dfrac{\partial Q}{\partial x} = \cos x$. (b) $f(x, y) = x^2 + y \sin x + K$

**55.** (a) $-\dfrac{1}{2} - e$ (b) $-\dfrac{1}{2} - e$ (c) $-\dfrac{1}{2} - e$ **57.** $P, Q, \dfrac{\partial P}{\partial y}$, and $\dfrac{\partial Q}{\partial x}$ are continuous everywhere in the rectangle,

and $\dfrac{\partial P}{\partial y} = \dfrac{\partial Q}{\partial x} = \dfrac{y^2 - x^2}{(x^2 + y^2)^2}$; $f(x, y) = \tan^{-1}\left(\dfrac{y}{x}\right) + K$. **59.** (a) SSM.

(b) $f(x, y) = \dfrac{k}{\sqrt{x^2 + y^2}} + K$ **61.** SSM. **63.** SSM.

## Section 15.5 (SSM = See Student Solutions Manual; AWV = Answers will vary)

**1.** line; double **2.** $\dfrac{\partial Q}{\partial x} - \dfrac{\partial P}{\partial y}$ **3.** False **4.** Area **5.** $-12$ **7.** $-12$ **9.** 0 **11.** $-\dfrac{1}{3}$; SSM.

**13.** 0; SSM. **15.** $\dfrac{\pi}{2}$ **17.** $\dfrac{3}{4}\pi$ **19.** 0 **21.** $\dfrac{6}{35}$ **23.** $\dfrac{32}{9}$ **25.** $-2\pi$ **27.** $-2\pi$ **29.** $\dfrac{9}{2}$ **31.** $\dfrac{1}{12}$

**33.** 2 **35.** $\dfrac{3}{2} - 2\ln 2$ **37.** $3\pi$ **39.** SSM. **41.** $(\bar{x}, \bar{y}) = \left(0, \dfrac{3}{5}\right)$ **43.** $(\bar{x}, \bar{y}) = \left(\dfrac{5}{9}, \dfrac{7}{9}\right)$ **45.** 0

**47.** $\dfrac{1}{3}(\sin 1 - \cos 1)$ **49.** AWV. **51.** AWV. **53.** SSM. **55.** Any piecewise smooth, simple closed curve not passing through the origin and not containing the origin in its interior **57.** SSM. **59.** SSM.

## Section 15.6 (SSM = See Student Solutions Manual; AWV = Answers will vary)

**1.** (b) **2.** False **3.** (c) **4.** (a) **5.** (a) For $u = 0$: the line $\mathbf{r}(0, v) = -5v\mathbf{i} + (1 + v)\mathbf{k}$; for $v = 0$: the line $\mathbf{r}(u, 0) = u\mathbf{i} + 2u\mathbf{j} + (1 - u)\mathbf{k}$.
(b) $x + 2y + 5z = 5$ **7.** (a) For $u = 0$: the point $(0, 0, 0)$; for $v = 0$: the line $\mathbf{r}(u, 0) = u\mathbf{i} + u\mathbf{k}$ (b) $x^2 + y^2 = z^2$, $0 \le z \le 2$

**9.** $\mathbf{r}(u, v) = u\mathbf{i} + v\mathbf{j} + (4 - u - 2v)\mathbf{k}$; $0 \le u \le 4$, $0 \le v \le 2 - \dfrac{1}{2}u$ **11.** $\mathbf{r}(u, v) = u\mathbf{i} + v\mathbf{j} + \sin(u^2 v)\mathbf{k}$; $-1 \le u \le 2$, $u^2 \le v \le u + 2$

**13.** (a) $10x - 2y - 5z = 40$ (b) $\mathbf{r}(t) = (7 + 10t)\mathbf{i} + (5 - 2t)\mathbf{j} + (4 - 5t)\mathbf{k}$

**15.** (a) $\sqrt{2}x - y - z = 0$ (b) $\mathbf{r}(t) = (5 + \sqrt{2}t)\mathbf{i} + \left(\dfrac{5\sqrt{2}}{2} - t\right)\mathbf{j} + \left(\dfrac{5\sqrt{2}}{2} - t\right)\mathbf{k}$ **17.** $\dfrac{1}{6}\pi(3\sqrt{10} + \ln(3 + \sqrt{10}))$ **19.** $\sqrt{21}\pi$

**21.** $16\pi$ **23.** (E) **25.** (B) **27.** (D) **29.** $\mathbf{r}(\theta, z) = 4\cos\theta\mathbf{i} + 4\sin\theta\mathbf{j} + z\mathbf{k}$; $0 \le \theta \le 2\pi$, $0 \le z \le 3$

**31.** (a) $\mathbf{r}(u, v) = u\mathbf{i} + v\mathbf{j} + (9 - u^2 - v^2)\mathbf{k}$, $-\sqrt{1 - (u - 1)^2} \le v \le \sqrt{1 - (u - 1)^2}$, $0 \le u \le 2$
(b) $\mathbf{r}(r, \theta) = (1 + r\cos\theta)\mathbf{i} + r\sin\theta\mathbf{j} + (8 - r^2 - 2r\cos\theta)\mathbf{k}$, $0 \le r \le 1$, $0 \le \theta \le 2\pi$

**33.** (a) $\mathbf{r}(u, v) = u\mathbf{i} + v\mathbf{j} + \sqrt{4 - u^2 - v^2}\,\mathbf{k}$; $0 \le u \le 2$, $0 \le v \le \sqrt{4 - u^2}$
(b) $\mathbf{r}(r, \theta) = r\cos\theta\mathbf{i} + r\sin\theta\mathbf{j} + \sqrt{4 - r^2}\,\mathbf{k}$; $0 \le r \le 2$, $0 \le \theta \le \dfrac{\pi}{2}$
(c) $\mathbf{r}(\theta, \phi) = 2\cos(\theta)\sin(\phi)\mathbf{i} + 2\sin(\theta)\sin(\phi)\mathbf{j} + 2\cos(\phi)\mathbf{k}$; $0 \le \theta \le \dfrac{\pi}{2}$, $0 \le \phi \le \dfrac{\pi}{2}$

**35.** (a) $z = 1$ (b) $\mathbf{r}(t) = \dfrac{3\sqrt{2}}{2}\mathbf{i} + \dfrac{3\sqrt{2}}{2}\mathbf{j} + (1 + 3t)\mathbf{k}$ **37.** $\sqrt{5}\pi + \dfrac{1}{2}\pi\ln(2 + \sqrt{5})$

**39.** (a) $\mathbf{r}(\theta, \phi) = \cos(\theta)\sin(\phi)\mathbf{i} + 3\sin(\theta)\sin(\phi)\mathbf{j} + 2\cos(\phi)\mathbf{k}$; $0 \le \theta \le 2\pi$, $0 \le \phi \le \pi$ (b) SSM.
(c) $S = \int_0^{2\pi} \int_0^{\pi} \sqrt{36\cos^2\theta\sin^4\phi + 4\sin^2\theta\sin^4\phi + 9\sin^2\phi\cos^2\phi}\,d\phi\,d\theta$

**41.** (a) (b) $\left(9 - \dfrac{3\sqrt{6}}{2}\right)\left(x - \dfrac{3\sqrt{2}}{2}\right) + \left(9\sqrt{3} + \dfrac{3\sqrt{2}}{2}\right)\left(y - \dfrac{3\sqrt{6}}{2}\right)$

$$-18\left\{z - \left(\dfrac{\pi}{3} + 6\left[\dfrac{\sqrt{2}}{2} + \ln\left(\tan\dfrac{\pi}{8}\right)\right]\right)\right\} = 0$$

(c) $\mathbf{r}(t) = \left[\dfrac{3}{2}\sqrt{2} + \left(9 - \dfrac{3}{2}\sqrt{6}\right)t\right]\mathbf{i} + \left[\dfrac{3}{2}\sqrt{6} + \left(9\sqrt{3} + \dfrac{3}{2}\sqrt{2}\right)t\right]\mathbf{j} + \left(3\sqrt{2} + 6\ln\left(\tan\dfrac{\pi}{8}\right)\right) - 18t\,\mathbf{k}$

**43.** (a) $\mathbf{r}(u, v) = (b + a\cos u)\cos v\mathbf{i} + (b + a\cos u)\sin v\mathbf{j} + a\sin u\mathbf{k}$ (b) $\left(b - \sqrt{x^2 + y^2}\right)^2 + z^2 = a^2$
**45.** SSM. **47.** (a) $\mathbf{r}(u, v) = a\cosh u\mathbf{i} + (b\sinh u\cos v)\mathbf{j} + (c\sinh u\sin v)\mathbf{k}$ (b)

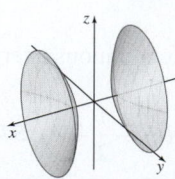

## Section 15.7 (SSM = See Student Solutions Manual; AWV = Answers will vary)

**1.** (c) **2.** (b) **3.** $\dfrac{2}{3}\sqrt{14}$ **5.** $\dfrac{17}{25}\sqrt{102} - \dfrac{1}{75}\sqrt{2}$ **7.** $\dfrac{5}{27}\sqrt{11}$ **9.** 0 **11.** $\dfrac{\pi}{2}$ **13.** $\dfrac{x}{4}\mathbf{i} + \dfrac{y}{4}\mathbf{j} + \dfrac{\sqrt{16 - x^2 - y^2}}{4}\mathbf{k}$

**15.** $\dfrac{9x}{2\sqrt{9 + 18x^2 + 3y^2}}\mathbf{i} + \dfrac{4y}{2\sqrt{9 + 18x^2 + 3y^2}}\mathbf{j} + \dfrac{\sqrt{36 - 9x^2 - 4y^2}}{2\sqrt{9 + 18x^2 + 3y^2}}\mathbf{k}$

**17.** $\mathbf{n} = \dfrac{2uv}{\sqrt{4u^2v^2 + v^2 + u^2}}\mathbf{i} + \dfrac{-v}{\sqrt{4u^2v^2 + v^2 + u^2}}\mathbf{j} + \dfrac{-u}{\sqrt{4u^2v^2 + v^2 + u^2}}\mathbf{k}$

$-\mathbf{n} = \dfrac{-2uv}{\sqrt{4u^2v^2 + v^2 + u^2}}\mathbf{i} + \dfrac{v}{\sqrt{4u^2v^2 + v^2 + u^2}}\mathbf{j} + \dfrac{u}{\sqrt{4u^2v^2 + v^2 + u^2}}\mathbf{k}$

**19.** $\rho$  **21.** 0  **23.** $3\rho$  **25.** $3\rho$  **27.** $2\pi$  **29.** $\dfrac{160}{3}$  **31.** $-5$  **33.** $\dfrac{176}{3}\sqrt{2}\pi$  **35.** $(\bar{x}, \bar{y}, \bar{z}) = \left(0, \dfrac{4}{3}, 0\right)$

**37.** $\dfrac{675}{2}\pi$  **39.** $\dfrac{25}{2}\pi\rho$  **41.** SSM.

**43.** $\approx 936.58$  **45.** $\iint\limits_{S} \mathbf{F} \cdot \mathbf{n}\,dS$ is equal to the surface area of $S$.

## Section 15.8  (SSM = See Student Solutions Manual; AWV = Answers will vary)

**1.** False  **2.** $\dfrac{\partial P}{\partial x} + \dfrac{\partial Q}{\partial y} + \dfrac{\partial R}{\partial z}$  **3.** div $\mathbf{F} = 2x + 2y + 2z$  **5.** div $\mathbf{F} = 4$  **7.** div $\mathbf{F} = \dfrac{x+y}{\sqrt{x^2+y^2}} + 1$  **9.** $\dfrac{128}{3}$  **11.** 3  **13.** $80\pi$

**15.** $4\pi$  **17.** $\dfrac{2}{3}$  **19.** SSM.  **21.** $V = \dfrac{1}{3}\pi R^2 h$  **23.** $4000\pi$  **25.** SSM.  **27.** SSM.  **29.** $2\pi$  **31.** SSM.  **33.** SSM.

## Section 15.9  (SSM = See Student Solutions Manual; AWV = Answers will vary)

**1.** False  **2.** $-x\mathbf{i} - y\mathbf{j} + (2z+2)\mathbf{k}$  **3.** (c)  **4.** False  **5.** $2\omega$  **6.** False  **7.** curl $\mathbf{F} = \mathbf{0}$  **9.** curl $\mathbf{F} = -x\mathbf{i} + xy\mathbf{j} + (z - xz)\mathbf{k}$

**11.** curl $\mathbf{F} = -y^2\cos z\,\mathbf{i} + (6xyz - e^{2z})\mathbf{j} - 3xz^2\mathbf{k}$  **13.** curl $\mathbf{F} = \dfrac{2yz - 2y}{(x^2+y^2+z^2)^2}\mathbf{i} + \dfrac{2x - 2xz}{(x^2+y^2+z^2)^2}\mathbf{j}$  **15.** curl $\mathbf{F} = -ze^{xz}\mathbf{j}$

**17.** curl $\mathbf{F} = -\mathbf{i} - \mathbf{j} - \mathbf{k}$  **19.** $2\pi$  **21.** $-2\pi$  **23.** $\pi$  **25.** 0  **27.** 0  **29.** $4\pi$  **31.** curl $\mathbf{F} = \mathbf{0}$. Force $\mathbf{F}$ is a conservative vector field.

**33.** curl $\mathbf{F} \neq \mathbf{0}$. Force $\mathbf{F}$ is not a conservative vector field.  **35.** $c = \dfrac{1}{2}$  **37.** SSM.  **39.** 0

**41.** curl $\mathbf{F} = -2\,\mathbf{k}$. Answers for paths may vary.  **43.** Force $\mathbf{F}$ is a conservative vector field. The work done is $\dfrac{a^2 + 2ab - 2bc + c^2}{2}$.

**45.** SSM.  **47.** SSM.  **49.** SSM.

## Review Exercises

**1.**

**3.** $\dfrac{5}{2}$  **5.** (a) $\dfrac{\partial P}{\partial y} = \dfrac{\partial Q}{\partial x} = \cos(xy) - xy\sin(xy)$  (b) $\sin 1$  **7.** $-2$  **9.** (a) $\nabla f(x, y) = \mathbf{F}(x, y)$  (b) 62  **11.** $\dfrac{3}{4}\pi + \dfrac{\sqrt{2}}{2}$

**13.** $\dfrac{255}{2}\sqrt{2}\pi$  **15.** $\dfrac{\pi}{8}$  **17.** $\pi$  **19.** 108  **21.** $-2$  **23.** $12\pi$  **25.** $7\pi$  **27.** $\mathbf{r}(\theta, z) = (5\cos\theta)\mathbf{i} + (2\sin\theta)\mathbf{j} + z\mathbf{k}; 0 \le \theta \le 2\pi, 1 \le z \le 6$

**29.** $\dfrac{65\sqrt{65}}{6}\pi - \dfrac{\pi}{6}$  **31.** $\dfrac{64}{3}\pi$  **33.** 0  **35.** $-2\sqrt{3}$  **37.** (a) div $\mathbf{F} = -z\sin x + \cos y$  (b) curl $\mathbf{F} = (\cos x - e^x)\mathbf{j}$  (c) 0

**39.** (a) div $\mathbf{F} = 3$  (b) curl $\mathbf{F} = \mathbf{0}$  (c) 0  **41.** 0  **43.** $\dfrac{\pi}{2}$  **45.** $\dfrac{3}{2}$  **47.** $\mathbf{F}$ is not a conservative vector field because curl $\mathbf{F} \neq \mathbf{0}$.

## Chapter 16

## Section 16.1  (SSM = See Student Solutions Manual; AWV = Answers will vary)

**1.** 2; 1  **2.** True  **3.** True  **4.** False  **5.** This is a first-order differential equation of degree 1. The differential equation is linear.

**7.** This is a fourth-order differential equation of degree 1. The differential equation is linear.

**9.** This is a second-order differential equation of degree 1. The differential equation is nonlinear.

**11.** This is a second-order differential equation of degree 3. The differential equation is nonlinear.

**13.** This is a second-order differential equation of degree 1. The differential equation is linear.

**15.** SSM.  **17.** SSM.  **19.** SSM.  **21.** SSM.  **23.** SSM.  **25.** SSM.  **27.** SSM.  **29.** SSM.  **31.** SSM.  **33.** SSM.

**35.** SSM.  **37.** SSM.  **39.** $n = -3, 2$  **41.** The Schrödinger equation is a second-order differential equation of degree 1.

## Section 16.2    (SSM = See Student Solutions Manual; AWV = Answers will vary)

**1.** $y = Ce^{0.5x^2}$  **2.** True  **3.** Trajectories  **4.** $y = xv(x); v = v(x)$  **5.** slope field  **6.** True  **7.** $\sin y = \frac{1}{2}x^2 + C$

**9.** $y = -\ln(e^{-x} + C)$  **11.** $y = \dfrac{C}{1 - \cos x}$  **13.** $y = 1 - Ce^{-0.5x^2}$  **15.** $y = C\left(\sqrt{x}\right)^{\ln x}$ where $C \neq 0$ is a constant.

**17.** $y = \sin^{-1}\left(\dfrac{1}{2}x^2 + \dfrac{\sqrt{2}-1}{2}\right)$  **19.** $y = x$  **21.** $y = \dfrac{1}{1 - \cos x}$  **23.** Function $f$ is a homogeneous function of degree 2.

**25.** Function $f$ is not a homogeneous function.  **27.** Function $f$ is a homogeneous function of degree 1.

**29.** Function $f$ is a homogeneous function of degree 0.  **31.** Function $f$ is a homogeneous function of degree 0.

**33.** $y = Cx - x \ln|x|$  **35.** $\dfrac{y}{x} - 2\ln|x+y| + \ln|x| + C = 0$  **37.** $\dfrac{1}{2}\ln(x^2 + y^2) + \tan^{-1}\left(\dfrac{y}{x}\right) = C$

**39.** $y = x(C + \ln|x|)$  **41.** $\dfrac{y^2 - x^2}{y} = C$  **43.** $\dfrac{x^4(2y - x)}{2y + x} = 3$

**45. (a)**    **(b)**

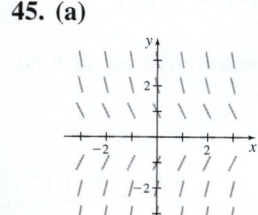    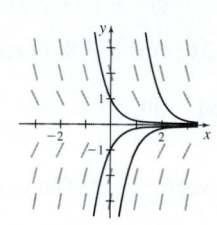

**47. (a)**    **(b)**

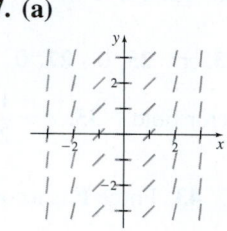    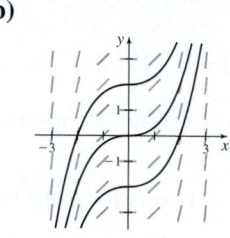

**49. (a)**    **(b)**

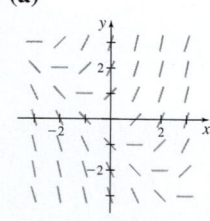    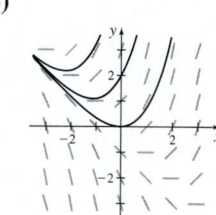

**51.** (b)    **53.** $y(0.4) \approx 0.648$    **55.** $y(1.4) \approx 3.982$

**57.** The orthogonal trajectories for the family $xy = C$ is the family $y^2 = x^2 + K$. The graph of $xy = 2$ and $y^2 = x^2 - 3$ is shown.

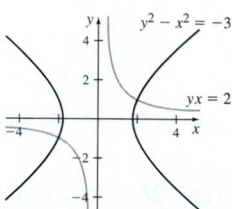

**59.** The orthogonal trajectories for the family $y = Cx^2$ is the family $x^2 + 2y^2 = K$. The graph of $y = \dfrac{1}{4}x^2$ and $x^2 + 2y^2 = 6$ is shown.

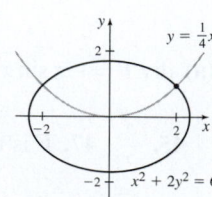

**61.** $\dfrac{1}{2}\ln\left((2y+1)^2 + (2x+7)^2\right) + \tan^{-1}\left(\dfrac{2y+1}{2x+7}\right) = C$  **63.** $2x^2 + y^2 = K$

**65.** (a) $y(t) = 20(1 - e^{(0.5 \ln 0.85)t})$

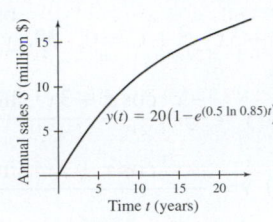

   (b) During the eighth year of operation, the sales will be approximately $9.56 million dollars.
   (c) Sales will reach $12 million dollars in approximately 11.28 years.
   (d) The graph of the annual sales of the company for its first 20 years of operation is shown.

**67.** (a) After 25 hours of practice, the person can text approximately 64.1 words per minute.

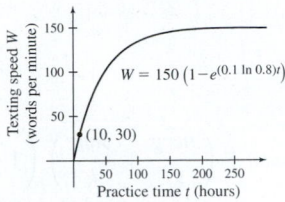

   (b) The person will need approximately 49.2 hours of practice in order to
     text 100 words per minute.
   (c) The graph of texting speed as a function of hours of practice is shown.

**69.** (a) $\dfrac{dv}{dt} = \dfrac{-A}{m} v$  (b) $v = v_0 e^{\frac{-A}{m}t}$  (c) $A = 15$ kg/s  (d) The sled will slow down to 10% of its initial speed
in approximately 3.84 seconds.  (e) While slowing down to 1.0 m/s, the sled will travel 15 meters.

**71.** (a) $q = q_0 \cdot e^{\frac{-1}{RC}t}$  (b) The discharging capacitor will lose 99% of its initial charge in approximately 0.00345 seconds.

**73.** $yx^{\frac{c^2}{4-c^2}} = K$, where $C^2 = \dfrac{1}{2}\left[x^2 + y^2 + 4 - \sqrt{(x^2+y^2+4)^2 - 16x^2}\right]$  **75.** (a) SSM.  (b) $v_0 = \sqrt{2Rg}$  **77.** SSM.

## Section 16.3   (SSM = See Student Solutions Manual; AWV = Answers will vary)

**1.** True  **2.** False  **3.** False  **4.** Integrating factor  **5.** (a) $\dfrac{\partial M}{\partial y} = \dfrac{\partial N}{\partial x} = -2$  (b) $2x^2 - 2xy + y^2 + 5x + C = 0$

**7.** (a) $\dfrac{\partial M}{\partial y} = \dfrac{\partial N}{\partial x} = -2x - 2y$  (b) $a^2x - x^2y - xy^2 - \dfrac{1}{3}y^3 + C = 0$  **9.** (a) $\dfrac{\partial M}{\partial y} = \dfrac{\partial N}{\partial x} = \dfrac{-1}{y^2}$  (b) $\dfrac{x}{y} + C = 0$

**11.** (a) $\dfrac{\partial M}{\partial y} = \dfrac{\partial N}{\partial x} = \dfrac{1}{x-1}$  (b) $y \ln(x-1) + y \ln(2) + \ln|y| + C = 0$

**13.** (a) $\dfrac{\partial M}{\partial y} = \dfrac{\partial N}{\partial x} = \dfrac{-1}{x+3}\sin y$  (b) $\ln(x+3)\cos y + \ln(5)\cos y + \dfrac{1}{2}y^2 + y + C = 0$

**15.** (a) $\dfrac{\partial M}{\partial y} = \dfrac{\partial N}{\partial x} = -2y\sin(x+y^2)$  (b) $\sin(x+y^2) + C = 0$  **17.** (a) $\dfrac{\partial M}{\partial y} = \dfrac{\partial N}{\partial x} = 2e^{2x}$  (b) $ye^{2x} - \dfrac{1}{3}x^3 + C = 0$

**19.** (a) $\dfrac{\partial M}{\partial y} = \dfrac{\partial N}{\partial x} = 2y - \dfrac{1}{(x+y)^2}$  (b) $\ln|x+y| + xy^2 + C = 0$  **21.** (a) $\dfrac{\partial M}{\partial y} = \dfrac{\partial N}{\partial x} = 6y^2\sin(2x)$  (b) $-y^3\cos(2x) + C = 0$

**23.** $x + xy^2 + \dfrac{1}{2}x^2y^2 + \dfrac{1}{2}y^2 - 3 = 0$  **25.** $x^2y + \cos x - y^2 = 0$  **27.** SSM.

**29.** (a) $u(x) = \dfrac{1}{x^2}$  (b) $4x - \dfrac{y^2}{x} - \dfrac{1}{x} + y + C = 0$  (c) SSM.  **31.** (a) $u(x) = \dfrac{1}{x^2}$  (b) $-\dfrac{y}{x} + \dfrac{1}{2}y^2 + C = 0$  (c) SSM.

**33.** (a) $u(x) = \dfrac{1}{x^3}$  (b) $\ln|x| + \dfrac{\sin y}{x} + C = 0$  (c) SSM.

## Section 16.4   (SSM = See Student Solutions Manual; AWV = Answers will vary)

**1.** True  **2.** True  **3.** False  **4.** False  **5.** $y = Ce^{-x^2}$  **7.** $y = x^2 + \dfrac{C}{x}$  **9.** $y = \dfrac{1}{4}x^3 + \dfrac{C}{x}$  **11.** $y = -\dfrac{1}{3}e^{-x} + Ce^{2x}$

**13.** $y = \dfrac{1}{5}x^3 + \dfrac{1}{3}x + \dfrac{C}{x^2}$  **15.** $y = 1 + Ce^{-e^x}$  **17.** $y = x\cos x + C\cos x$  **19.** $y = \dfrac{\ln|\csc(x) - \cot(x)| + C}{\sin(x)}$

**21.** $y = \dfrac{\pm 1}{x\sqrt{C - 6x}}$  **23.** $y = \sqrt[3]{\dfrac{2}{1 + Ce^{3x^2}}}$  **25.** $y^{\frac{2}{3}} = \dfrac{3}{4}x^2 + \dfrac{C}{x^{\frac{2}{3}}}$  **27.** $y = x(e^x + C)$  **29.** $x - \dfrac{1}{4} + \dfrac{1}{2}y - \dfrac{1}{2}y^2 - Ce^{-2y} = 0$

**31.** $y = \dfrac{\sec x \tan x + \tan^2 x + \ln|\sec x + \tan x| + C}{2(\sec x + \tan x)}$  **33.** $y = \dfrac{1}{e^x(C - x^2)}$  **35.** $y = \pm\sqrt{\dfrac{x}{C - x^5}}$

**37.** $x^6 + 3x^4y^2 + C = 0$   **39.** $y = \dfrac{x+5}{e^x}$   **41.** $y = e^x - \dfrac{e^x}{x} + \dfrac{e^{-1}}{x}$

**43. (a)** $y = \dfrac{-x^3\cos x + 3x^2\sin x - 6\sin x + 6x\cos x + C}{x^3}$

**(b)** $y = \dfrac{-x^3\cos x + 3x^2\sin x - 6\sin x + 6x\cos x + 6\pi - \pi^3}{x^3}$

**(c)** The graph of the particular solution over the interval [0, 10] is shown below.

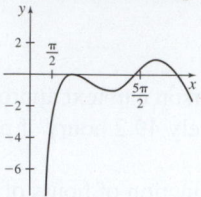

**45.** $s(t) = s_0 + \left(\dfrac{m^2g}{k^2} + \dfrac{v_0 m}{k}\right)\left(1 - e^{\frac{-k}{m}t}\right) - \dfrac{mg}{k}t$

**47. (a)** The skydiver will be falling at a velocity of 44.891 ft/s.

**(b)** The skydiver will land on the ground in approximately 241.250 seconds.

**(c)** The skydiver will land with a velocity of 40 ft/s.

**49.** After 5 minutes, there are approximately 98.9 kg of salt in the tank. After a long time, there will be 300 kg of salt in the tank.

**51. (a)** The solution of the differential equation is $P(t) = \dfrac{30{,}000}{3 + 17e^{-0.01t}}$

**(b)** The limiting size of the population is 10,000.

**(c)** The population will be equal to 5000 (i.e. one-half of the limiting value) in 173.46 months.

**(d)** The graph of the predicted population for $0 \le t \le 600$ months is shown.

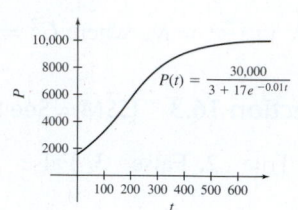

**53.** $q(t) = EC\left(1 - e^{\frac{-t}{RC}}\right)$   **55. (a)** $N(t) = N_{\max}\exp(-e^{-p(t+C)})$

**(b)** $N(t) = 400\exp(-e^{-0.0526(t - 20.847)})$

**(c)** After 43 years, there will be approximately 293 fish.

**(d)** SSM.

**(e)** The graph of $N = N(t)$ is shown.

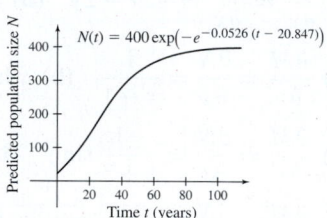

**57.** SSM.   **59.** After 1 hour, the percentage of carbon dioxide is approximately 0.0752%. After 2 hours, the percentage of carbon dioxide is approximately 0.0448%.   **61.** $I = \dfrac{E_0(R\sin(\omega t) - \omega L\cos(\omega t) + \omega L e^{-Rt/L})}{R^2 + \omega^2 L^2} + I_0 e^{-Rt/L}$

## Section 16.5   (SSM = See Student Solutions Manual; AWV = Answers will vary)

**1.** $y(x) = a_0 \displaystyle\sum_{k=0}^{\infty}\left(-\dfrac{3}{2}\right)^k \dfrac{x^{2k}}{k!}$   **3.** $y(x) = a_0 \displaystyle\sum_{k=0}^{\infty}(-1)^k\dfrac{x^{2k}}{(2k)!} + a_1\displaystyle\sum_{k=0}^{\infty}(-1)^k\dfrac{x^{2k+1}}{(2k+1)!}$

**5.** $y(x) = a_0\left(1 - \dfrac{x^4}{3\cdot4} + \dfrac{x^8}{3\cdot4\cdot7\cdot8} - \dfrac{x^{12}}{3\cdot4\cdot7\cdot8\cdot11\cdot12} + \dfrac{x^{16}}{3\cdot4\cdot7\cdot8\cdot11\cdot12\cdot15\cdot16} - \cdots\right)$

$+ a_1\left(x - \dfrac{x^5}{4\cdot5} + \dfrac{x^9}{4\cdot5\cdot8\cdot9} - \dfrac{x^{13}}{4\cdot5\cdot8\cdot9\cdot12\cdot13} + \dfrac{x^{17}}{4\cdot5\cdot8\cdot9\cdot12\cdot13\cdot16\cdot17} - \cdots\right)$

**7.** $y(x) = a_0\left(1 - \dfrac{1\cdot x^3}{2\cdot3} + \dfrac{1\cdot4\cdot x^6}{2\cdot3\cdot5\cdot6} - \dfrac{1\cdot4\cdot7\cdot x^9}{2\cdot3\cdot5\cdot6\cdot8\cdot9} + \dfrac{1\cdot4\cdot7\cdot10\cdot x^{12}}{2\cdot3\cdot5\cdot6\cdot8\cdot9\cdot11\cdot12} - \cdots\right)$

$+ a_1\left(x - \dfrac{2\cdot x^4}{3\cdot4} + \dfrac{2\cdot5\cdot x^7}{3\cdot4\cdot6\cdot7} - \dfrac{2\cdot5\cdot8\cdot x^{10}}{3\cdot4\cdot6\cdot7\cdot9\cdot10} + \dfrac{2\cdot5\cdot8\cdot11\cdot x^{13}}{3\cdot4\cdot6\cdot7\cdot9\cdot10\cdot12\cdot13} - \cdots\right)$

**9.** $y(x) = a_0\left(1 - \dfrac{x^3}{3!} + \dfrac{x^6}{6!} - \dfrac{x^9}{9!} + \dfrac{x^{12}}{12!} - \cdots\right) + a_1\left(x - \dfrac{x^4}{4!} + \dfrac{x^7}{7!} - \dfrac{x^{10}}{10!} + \dfrac{x^{13}}{13!} - \cdots\right)$

**11.** $y(x) = a_0\left(1 - 3x^2\right) + a_1\left(x - \dfrac{1}{3}x^3\right) + a_2\left(x^2 - \dfrac{2\cdot x^5}{5!} + \dfrac{2\cdot x^8}{8!} - \dfrac{2\cdot x^{11}}{11!} + \dfrac{2\cdot x^{14}}{14!} - \cdots\right)$

**13. (a)** $y(x) = 1 - \dfrac{1}{2!}x^2 + \dfrac{3}{4!}x^4 - \dfrac{15}{6!}x^6 + \dfrac{105}{8!}x^8 - \cdots$

**(b)**

| $x$ | 0.0 | 0.1 | 0.2 | 0.3 | 0.4 | 0.5 | 0.6 | 0.7 | 0.8 | 0.9 | 1.0 |
|---|---|---|---|---|---|---|---|---|---|---|---|
| $y(x)$ | 1.0000 | 0.9950 | 0.9802 | 0.9560 | 0.9231 | 0.8825 | 0.8353 | 0.7827 | 0.7262 | 0.6671 | 0.6068 |

**15. (a)** $y(x) = x + \dfrac{2}{4!}x^4 - \dfrac{4}{6!}x^6 + \dfrac{10}{7!}x^7 + \dfrac{6}{8!}x^8 + \cdots$

**(b)**

| $x$ | 0 | 0.1 | 0.2 | 0.3 | 0.4 | 0.5 | 0.6 | 0.7 | 0.8 | 0.9 | 1.0 |
|---|---|---|---|---|---|---|---|---|---|---|---|
| $y(x)$ | 0 | 0.100008 | 0.2001 | 0.3007 | 0.4021 | 0.5051 | 0.6106 | 0.7195 | 0.8331 | 0.9527 | 1.0799 |

**17. (a)** $y(x) = 2 + x - \dfrac{3}{2!}x^2 - \dfrac{1}{4!}x^4 + \dfrac{5}{5!}x^5 + \cdots$

**(b)**

| $x$ | 0.0 | 0.1 | 0.2 | 0.3 | 0.4 | 0.5 | 0.6 | 0.7 | 0.8 | 0.9 | 1.0 |
|---|---|---|---|---|---|---|---|---|---|---|---|
| $y(x)$ | 2.0000 | 2.0850 | 2.1399 | 2.1648 | 2.1594 | 2.1237 | 2.0578 | 1.9620 | 1.8366 | 1.6823 | 1.5000 |

**19. (a)** $y(x) = 2x - \dfrac{12}{5!}x^5 + \dfrac{504}{9!}x^9 - \dfrac{55{,}440}{13!}x^{13} + \dfrac{11{,}642{,}400}{17!}x^{17} - \cdots$

**(b)**

| $x$ | 0.0 | 0.1 | 0.2 | 0.3 | 0.4 | 0.5 | 0.6 | 0.7 | 0.8 | 0.9 | 1.0 |
|---|---|---|---|---|---|---|---|---|---|---|---|
| $y(x)$ | 0.0000 | 0.2000 | 0.4000 | 0.5998 | 0.7990 | 0.9969 | 1.1922 | 1.3832 | 1.5674 | 1.7415 | 1.9014 |

**21. (a)** $y(x) = 1 + x + \dfrac{1}{5!}x^5 + \dfrac{1}{6!}x^6 - \dfrac{1}{7!}x^7 + \cdots$

**(b)**

| $x$ | 0.0 | 0.1 | 0.2 | 0.3 | 0.4 | 0.5 | 0.6 | 0.7 | 0.8 | 0.9 | 1.0 |
|---|---|---|---|---|---|---|---|---|---|---|---|
| $y(x)$ | 1.0000 | 1.1000 | 1.2000 | 1.3000 | 1.4001 | 1.5003 | 1.6007 | 1.7015 | 1.8031 | 1.9056 | 2.0095 |

**23. (a)** $y(x) \approx 1 + 2x - 6x^2 + \dfrac{20}{3}x^3 - \dfrac{14}{3}x^4 + \dfrac{12}{5}x^5$ **(b)** $y(x) = e^{-2x}(1 + 4x)$

**(d)** SSM.

**(e)** $y(x) \approx 1 + 2x - 6x^2 + \dfrac{20}{3}x^3 - \dfrac{14}{3}x^4 + \dfrac{12}{5}x^5 - \dfrac{44}{45}x^6 + \dfrac{104}{315}x^7$

**(c and f)** The graphs are shown. SSM for explanation.

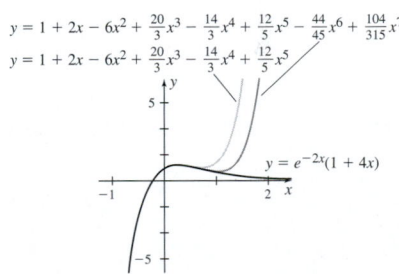

$y = 1 + 2x - 6x^2 + \frac{20}{3}x^3 - \frac{14}{3}x^4 + \frac{12}{5}x^5 - \frac{44}{45}x^6 + \frac{104}{315}x^7$

$y = 1 + 2x - 6x^2 + \frac{20}{3}x^3 - \frac{14}{3}x^4 + \frac{12}{5}x^5$

$y = e^{-2x}(1 + 4x)$

**25. (a)** SSM.  **(b)** SSM.  **(c)** The lower bound is $3.506 \times 10^8$ years.

## Review Exercises  (SSM = See Student Solutions Manual; AWV = Answers will vary)

**1.** This is a first-order differential equation of degree 1. The differential equation is linear.

**3.** This is a third-order differential equation of degree 1. The differential equation is linear.  **5.** SSM.  **7.** SSM.

**9.** Function $f$ is a homogeneous function of degree 3.  **11.** Function $f$ is a homogeneous function of degree 1.

**13. (a)** The differential equation is separable.  **(b)** $y = \pm\sqrt{\dfrac{C}{(1-x)^2} - 1}$

**15. (a)** The differential equation is first-order linear and exact.  **(b)** $y = \dfrac{1}{7}x^6 + \dfrac{C}{x}$

**17. (a)** The differential equation is exact. **(b)** $x^2y^3 + y^2\sin x - x^2 + C = 0$

**19. (a)** The differential equation is separable and exact. **(b)** $y = \dfrac{C}{\cos x}$

**21. (a)** The differential equation is homogeneous of degree 1. **(b)** $y = x\cos^{-1}(Cx)$

**23. (a)** The differential equation is Bernoulli. **(b)** $y = \pm\sqrt{\dfrac{5x}{2 + Cx^5}}$ **25.** $y = 3 + \dfrac{20}{x^2}$

**27.** $y = x - \dfrac{1}{x}$ **29.** $y = \dfrac{-1}{2} - \dfrac{1}{2}x$ **31.** $y = x^2\left(\sin x + \dfrac{12}{\pi^2} - 1\right)$ **33.** $y = x\sqrt{4 + 2\ln x}$ **35.** $y(x) = a_0\displaystyle\sum_{k=0}^{\infty}\left(\dfrac{x}{2}\right)^k$

**37. (a)** $y(x) = 1 + x - \dfrac{1}{6}x^3 - \dfrac{1}{12}x^4 + \dfrac{1}{120}x^5 + \dots$

**(b)**

| $x$ | 0 | 0.1 | 0.2 | 0.3 | 0.4 | 0.5 | 0.6 | 0.7 | 0.8 | 0.9 | 1.0 |
|---|---|---|---|---|---|---|---|---|---|---|---|
| $y(x)$ | 1.00000 | 1.09983 | 1.19854 | 1.29485 | 1.38729 | 1.47422 | 1.55385 | 1.62423 | 1.68326 | 1.72875 | 1.75833 |

**39. (a)** 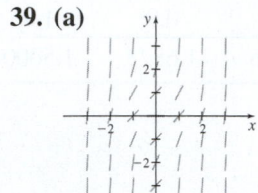 **(b)**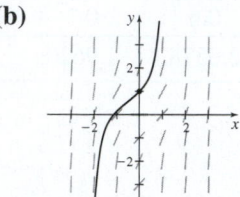

**41.** The orthogonal trajectories for the family $xy = c$ is the family $y^2 - x^2 = K$. The graph of $xy = 6$ and $y^2 - x^2 = 5$ is shown.

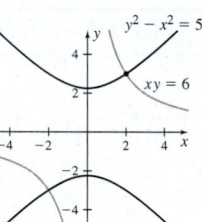

**43.** $y(1.3) \approx 5.522$

# Index

Note: **Boldface** indicates a definition, *italics* indicates a figure, and *t* indicates a table.

# TABLE OF DERIVATIVES

1. $\dfrac{d}{dx}A = 0$,   $A$ is a constant

2. $\dfrac{d}{dx}[ku(x)] = k\dfrac{d}{dx}u(x)$, $k$ a constant

3. $\dfrac{d}{dx}[u(x)+v(x)] = \dfrac{d}{dx}u(x) + \dfrac{d}{dx}v(x)$

4. $\dfrac{d}{dx}[u(x)-v(x)] = \dfrac{d}{dx}u(x) - \dfrac{d}{dx}v(x)$

5. $\dfrac{d}{dx}[u(x)v(x)] = u(x)\left[\dfrac{d}{dx}v(x)\right] + \left[\dfrac{d}{dx}u(x)\right]v(x)$

6. $\dfrac{dy}{dx}\left[\dfrac{u(x)}{v(x)}\right] = \dfrac{\left[\dfrac{d}{dx}u(x)\right]v(x) - u(x)\left[\dfrac{d}{dx}v(x)\right]}{[v(x)]^2}$

7. $\dfrac{dy}{dx} = \dfrac{dy}{du}\dfrac{du}{dx}$,   $y = f(u)$,   $u = g(x)$

8. $\dfrac{d}{dx}[u(x)]^a = a[u(x)]^{a-1}u'(x)$

9. $\dfrac{d}{dx}e^{u(x)} = e^{u(x)}u'(x)$

10. $\dfrac{d}{dx}\ln u(x) = \dfrac{u'(x)}{u(x)}$

11. $\dfrac{d}{dx}a^{u(x)} = a^{u(x)}u'(x)\ln a$

12. $\dfrac{d}{dx}\log_a u(x) = \dfrac{u'(x)}{u(x)\ln a}$, $a > 0, a \ne 1$

13. $\dfrac{d}{dx}\sin u(x) = \cos u(x)\, u'(x)$

14. $\dfrac{d}{dx}\cos u(x) = -\sin u(x)\, u'(x)$

15. $\dfrac{d}{dx}\tan u(x) = \sec^2 u(x)\, u'(x)$

16. $\dfrac{d}{dx}\cot u(x) = -\csc^2 u(x)\, u'(x)$

17. $\dfrac{d}{dx}\sec u(x) = \sec u(x)\tan u(x)\, u'(x)$

18. $\dfrac{d}{dx}\csc u(x) = -\csc u(x)\cot u(x)\, u'(x)$

19. $\dfrac{d}{dx}\sin^{-1} u(x) = \dfrac{u'(x)}{\sqrt{1-[u(x)]^2}}$

20. $\dfrac{d}{dx}\tan^{-1} u(x) = \dfrac{u'(x)}{1+[u(x)]^2}$

21. $\dfrac{d}{dx}\sec^{-1} u(x) = \dfrac{u'(x)}{u(x)\sqrt{[u(x)]^2-1}}$

22. $\dfrac{d}{dx}\cos^{-1} u(x) = -\dfrac{u'(x)}{\sqrt{1-[u(x)]^2}}$

23. $\dfrac{d}{dx}\cot^{-1} u(x) = -\dfrac{u'(x)}{1+[u(x)]^2}$

24. $\dfrac{d}{dx}\csc^{-1} u(x) = -\dfrac{u'(x)}{u(x)\sqrt{[u(x)]^2-1}}$

25. $\dfrac{d}{dx}\sinh u(x) = \cosh u(x)\, u'(x)$

26. $\dfrac{d}{dx}\cosh u(x) = \sinh u(x)\, u'(x)$

27. $\dfrac{d}{dx}\tanh u(x) = \operatorname{sech}^2 u(x)\, u'(x)$

28. $\dfrac{d}{dx}\coth u(x) = -\operatorname{csch}^2 u(x)\, u'(x)$

29. $\dfrac{d}{dx}\operatorname{sech} u(x) = -\operatorname{sech} u(x)\tanh u(x)\, u'(x)$

30. $\dfrac{d}{dx}\operatorname{csch} u(x) = -\operatorname{csch} u(x)\coth u(x)\, u'(x)$

31. $\dfrac{d}{dx}\sinh^{-1} u(x) = \dfrac{u'(x)}{\sqrt{1+[u(x)]^2}}$

32. $\dfrac{d}{dx}\cosh^{-1} u(x) = \dfrac{u'(x)}{\sqrt{[u(x)]^2-1}}$

33. $\dfrac{d}{dx}\tanh^{-1} u(x) = \dfrac{u'(x)}{1-[u'(x)]^2}$

34. $\dfrac{d}{dx}\coth^{-1} u(x) = \dfrac{u'(x)}{1-[u'(x)]^2}$

35. $\dfrac{d}{dx}\operatorname{sech}^{-1} u(x) = -\dfrac{u'(x)}{u(x)\sqrt{[u(x)]^2-1}}$

36. $\dfrac{d}{dx}\operatorname{csch}^{-1} u(x) = -\dfrac{u'(x)}{|u(x)|\sqrt{[u(x)]^2+1}}$

# TABLE OF INTEGRALS
## General Formulas

1. $\displaystyle\int [f(x)+g(x)]\,dx = \int f(x)\,dx + \int g(x)\,dx$

2. $\displaystyle\int [f(x)-g(x)]\,dx = \int f(x)\,dx - \int g(x)\,dx$

3. $\displaystyle\int kf(x)\,dx = k\int f(x)\,dx$,   $k$ a constant

4. **Substitution:** $\displaystyle\int f'(g(x))\,g'(x)\,dx = \int f'(u)\,du$

5. **Integration by Parts Formula:** $\displaystyle\int u\,dv = uv - \int v\,du$

## Basic Integrals

1. $\displaystyle\int x^a\,dx = \dfrac{x^{a+1}}{a+1} + C$,   $a \ne -1$

2. $\displaystyle\int \dfrac{1}{x}\,dx = \ln|x| + C$

3. $\displaystyle\int e^x\,dx = e^x + C$

4. $\displaystyle\int a^x\,dx = \dfrac{a^x}{\ln a} + C$, $a > 0, a \ne 1$

5. $\displaystyle\int \ln x\,dx = x\ln x - x + C$

6. $\displaystyle\int \sin x\,dx = -\cos x + C$

7. $\displaystyle\int \cos x\,dx = \sin x + C$

8. $\displaystyle\int \sec^2 x\,dx = \tan x + C$

9. $\displaystyle\int \sec x\tan x\,dx = \sec x + C$

10. $\displaystyle\int \csc x\cot x\,dx = -\csc x + C$

11. $\displaystyle\int \csc^2 x\,dx = -\cot x + C$

12. $\displaystyle\int \tan x\,dx = \ln|\sec x| + C$

13. $\displaystyle\int \sec x\,dx = \ln|\sec x + \tan x| + C$

14. $\displaystyle\int \cot x\,dx = \ln|\sin x| + C$

15. $\displaystyle\int \csc x\,dx = \ln|\csc x - \cot x| + C$

**16.** $\displaystyle\int \frac{dx}{\sqrt{a^2 - x^2}} = \sin^{-1}\frac{x}{a} + C, \quad a > 0$

**17.** $\displaystyle\int \frac{dx}{a^2 + x^2} = \frac{1}{a}\tan^{-1}\frac{x}{a} + C, \quad a > 0$

**18.** $\displaystyle\int \frac{dx}{x\sqrt{x^2 - a^2}} = \frac{1}{a}\sec^{-1}\frac{x}{a} + C, \quad a > 0$

**19.** $\displaystyle\int \sinh x \, dx = \cosh x + C$

**20.** $\displaystyle\int \cosh x \, dx = \sinh x + C$

**21.** $\displaystyle\int \operatorname{sech}^2 x \, dx = \tanh x + C$

**22.** $\displaystyle\int \operatorname{csch}^2 x \, dx = -\coth x + C$

**23.** $\displaystyle\int \operatorname{sech} x \, \tanh x \, dx = -\operatorname{sech} x + C$

**24.** $\displaystyle\int \operatorname{csch} x \, \coth x \, dx = -\operatorname{csch} x + C$

## Integrals Involving $a + bx$ $\quad a \neq 0, b \neq 0$

**25.** $\displaystyle\int \frac{dx}{a + bx} = \frac{1}{b}\ln|a + bx| + C$

**26.** $\displaystyle\int \frac{x \, dx}{a + bx} = \frac{1}{b^2}(a + bx - a\ln|a + bx|) + C$

**27.** $\displaystyle\int \frac{x \, dx}{(a + bx)^2} = \frac{a}{b^2(a + bx)} + \frac{1}{b^2}\ln|a + bx| + C$

**28.** $\displaystyle\int \frac{x^2 \, dx}{(a + bx)^2} = \frac{1}{b^3}\left(a + bx - \frac{a^2}{a + bx} - 2a\ln|a + bx|\right) + C$

**29.** $\displaystyle\int \frac{dx}{x(a + bx)^2} = \frac{1}{a(a + bx)} - \frac{1}{a^2}\ln\left|\frac{a + bx}{x}\right| + C$

**30.** $\displaystyle\int \frac{dx}{x^2(a + bx)} = -\frac{1}{ax} + \frac{b}{a^2}\ln\left|\frac{a + bx}{x}\right| + C$

**31.** $\displaystyle\int x(a + bx)^n \, dx = \frac{(a + bx)^{n+1}}{b^2}\left(\frac{a + bx}{n + 2} - \frac{a}{n + 1}\right) + C, \quad n \neq -1, -2$

**32.** $\displaystyle\int \frac{x \, dx}{(a + bx)(c + dx)} = \frac{1}{bc - ad}\left(-\frac{a}{b}\ln|a + bx| + \frac{c}{d}\ln|c + dx|\right) + C, \quad bc - ad \neq 0$

**33.** $\displaystyle\int \frac{x \, dx}{(a + bx)^2(c + dx)} = \frac{1}{bc - ad}\left[\frac{a}{b(a + bx)} + \frac{c}{bc - ad}\ln\left|\frac{a + bx}{c + dx}\right|\right] + C, \quad bc - ad \neq 0$

## Integrals Involving $a^2 - x^2$, $a^2 + x^2$, $x^2 - a^2$ $\quad a > 0$

**34.** $\displaystyle\int \frac{dx}{a^2 - x^2} = \frac{1}{2a}\ln\left|\frac{x + a}{x - a}\right| + C$

**35.** $\displaystyle\int \frac{dx}{x^2 - a^2} = \frac{1}{2a}\ln\left|\frac{x - a}{x + a}\right| + C$

**36.** $\displaystyle\int \frac{dx}{(a^2 \pm x^2)^n} = \frac{1}{2(n - 1)a^2}\left[\frac{x}{(a^2 \pm x^2)^{n-1}} + (2n - 3)\int \frac{dx}{(a^2 \pm x^2)^{n-1}}\right], \quad n \neq 1$

**37.** $\displaystyle\int \frac{dx}{(x^2 - a^2)^n} = \frac{1}{2(n - 1)a^2}\left[-\frac{x}{(x^2 - a^2)^{n-1}} - (2n - 3)\int \frac{dx}{(x^2 - a^2)^{n-1}}\right], \quad n \neq 1$

## Integrals Containing $\sqrt{a+bx}$   $a \neq 0, b \neq 0$

**38.** $\displaystyle\int x\sqrt{a+bx}\,dx = \frac{2}{15b^2}(3bx-2a)(a+bx)^{3/2}+C$

**39.** $\displaystyle\int x^n\sqrt{a+bx}\,dx = \frac{2}{b(2n+3)}[x^n(a+bx)^{3/2}-na\int x^{n-1}\sqrt{a+bx}\,dx]$

**40.** $\displaystyle\int \frac{x\,dx}{\sqrt{a+bx}} = \frac{2}{3b^2}(bx-2a)\sqrt{a+bx}+C$

**41.** $\displaystyle\int \frac{x^2\,dx}{\sqrt{a+bx}} = \frac{2}{15b^3}(8a^2-4abx+3b^2x^2)\sqrt{a+bx}+C$

**42.** $\displaystyle\int \frac{x^n\,dx}{\sqrt{a+bx}} = \frac{2x^n\sqrt{a+bx}}{b(2n+1)} - \frac{2na}{b(2n+1)}\int \frac{x^{n-1}\,dx}{\sqrt{a+bx}}$

**43.** $\displaystyle\int \frac{dx}{x\sqrt{a+bx}} = \begin{cases} \dfrac{1}{\sqrt{a}}\ln\left|\dfrac{\sqrt{a+bx}-\sqrt{a}}{\sqrt{a+bx}+\sqrt{a}}\right|+C, & a>0 \\[3mm] \dfrac{2}{\sqrt{-a}}\tan^{-1}\sqrt{\dfrac{a+bx}{-a}}+C, & a<0 \end{cases}$

**44.** $\displaystyle\int \frac{dx}{x^n\sqrt{a+bx}} = -\frac{\sqrt{a+bx}}{a(n-1)x^{n-1}} - \frac{b(2n-3)}{2a(n-1)}\int \frac{dx}{x^{n-1}\sqrt{a+bx}}$

**45.** $\displaystyle\int \frac{\sqrt{a+bx}}{x}\,dx = 2\sqrt{a+bx} + a\int \frac{dx}{x\sqrt{a+bx}}$

**46.** $\displaystyle\int \frac{\sqrt{a+bx}}{x^2}\,dx = -\frac{\sqrt{a+bx}}{x} + \frac{b}{2}\int \frac{dx}{x\sqrt{a+bx}}$

## Integrals Containing $\sqrt{x^2 \pm a^2}$   $a > 0$

**47.** $\displaystyle\int \sqrt{x^2\pm a^2}\,dx = \frac{x}{2}\sqrt{x^2\pm a^2}\pm\frac{a^2}{2}\ln\left|x+\sqrt{x^2\pm a^2}\right|+C$

**48.** $\displaystyle\int x\sqrt{x^2\pm a^2}\,dx = \frac{1}{3}(x^2\pm a^2)^{3/2}+C$

**49.** $\displaystyle\int x^2\sqrt{x^2\pm a^2}\,dx = \frac{x}{8}(2x^2\pm a^2)\sqrt{x^2\pm a^2} - \frac{a^4}{8}\ln\left|x+\sqrt{x^2\pm a^2}\right|+C$

**50.** $\displaystyle\int \frac{\sqrt{x^2+a^2}}{x}\,dx = \sqrt{x^2+a^2} - a\ln\left|\frac{a+\sqrt{x^2+a^2}}{x}\right|+C$

**51.** $\displaystyle\int \frac{\sqrt{x^2-a^2}}{x}\,dx = \sqrt{x^2-a^2} - a\sec^{-1}\frac{x}{a}+C$

**52.** $\displaystyle\int \frac{\sqrt{x^2\pm a^2}}{x^2}\,dx = -\frac{\sqrt{x^2\pm a^2}}{x} + \ln\left|x+\sqrt{x^2\pm a^2}\right|+C$

**53.** $\displaystyle\int \frac{dx}{\sqrt{x^2\pm a^2}} = \ln\left|x+\sqrt{x^2\pm a^2}\right|+C$

**54.** $\displaystyle\int \frac{x^2\,dx}{\sqrt{x^2\pm a^2}} = \frac{x}{2}\sqrt{x^2\pm a^2} \mp \frac{a^2}{2}\ln\left|x+\sqrt{x^2\pm a^2}\right|+C$

**55.** $\displaystyle\int \frac{dx}{x\sqrt{x^2+a^2}} = -\frac{1}{a}\ln\left|\frac{a+\sqrt{x^2+a^2}}{x}\right|+C$

**56.** $\displaystyle\int \frac{dx}{x\sqrt{x^2-a^2}} = \frac{1}{a}\sec^{-1}\frac{x}{a}+C$

**57.** $\displaystyle\int \frac{dx}{x^2\sqrt{x^2\pm a^2}} = \mp\frac{\sqrt{x^2\pm a^2}}{a^2x}+C$

**58.** $\displaystyle\int (x^2\pm a^2)^{3/2}\,dx = \frac{x}{8}(2x^2\pm 5a^2)\sqrt{x^2\pm a^2} + \frac{3a^4}{8}\ln\left|x+\sqrt{x^2\pm a^2}\right|+C$

**59.** $\displaystyle\int \frac{dx}{(x^2\pm a^2)^{3/2}} = \pm\frac{x}{a^2\sqrt{x^2\pm a^2}}+C$

## Integrals Containing $\sqrt{a^2 - x^2}$    $a > 0$

**60.** $\displaystyle\int \sqrt{a^2 - x^2}\,dx = \frac{x}{2}\sqrt{a^2 - x^2} + \frac{a^2}{2}\sin^{-1}\frac{x}{a} + C$

**61.** $\displaystyle\int x^2\sqrt{a^2 - x^2}\,dx = \frac{x}{8}(2x^2 - a^2)\sqrt{a^2 - x^2} + \frac{a^4}{8}\sin^{-1}\frac{x}{a} + C$

**62.** $\displaystyle\int \frac{\sqrt{a^2 - x^2}}{x}\,dx = \sqrt{a^2 - x^2} - a\ln\left|\frac{a + \sqrt{a^2 - x^2}}{x}\right| + C$

**63.** $\displaystyle\int \frac{\sqrt{a^2 - x^2}}{x^2}\,dx = -\frac{\sqrt{a^2 - x^2}}{x} - \sin^{-1}\frac{x}{a} + C$

**64.** $\displaystyle\int \frac{x^2}{\sqrt{a^2 - x^2}}\,dx = -\frac{x}{2}\sqrt{a^2 - x^2} + \frac{a^2}{2}\sin^{-1}\frac{x}{a} + C$

**65.** $\displaystyle\int \frac{dx}{x\sqrt{a^2 - x^2}} = -\frac{1}{a}\ln\left|\frac{a + \sqrt{a^2 - x^2}}{x}\right| + C$

**66.** $\displaystyle\int \frac{dx}{x^2\sqrt{a^2 - x^2}} = -\frac{\sqrt{a^2 - x^2}}{a^2 x} + C$

**67.** $\displaystyle\int (a^2 - x^2)^{3/2}\,dx = \frac{x}{4}(a^2 - x^2)^{3/2} + \frac{3a^2 x}{8}\sqrt{a^2 - x^2} + \frac{3a^4}{8}\sin^{-1}\frac{x}{a} + C$

**68.** $\displaystyle\int \frac{dx}{(a^2 - x^2)^{3/2}} = \frac{x}{a^2\sqrt{a^2 - x^2}} + C$

## Integrals Containing $\sqrt{2ax - x^2}$    $a > 0$

**69.** $\displaystyle\int \sqrt{2ax - x^2}\,dx = \frac{x - a}{2}\sqrt{2ax - x^2} + \frac{a^2}{2}\cos^{-1}\left(\frac{a - x}{a}\right) + C$

**70.** $\displaystyle\int x\sqrt{2ax - x^2}\,dx = \frac{2x^2 - ax - 3a^2}{6}\sqrt{2ax - x^2} + \frac{a^3}{2}\cos^{-1}\left(\frac{a - x}{a}\right) + C$

**71.** $\displaystyle\int \frac{\sqrt{2ax - x^2}}{x}\,dx = \sqrt{2ax - x^2} + a\cos^{-1}\left(\frac{a - x}{a}\right) + C$

**72.** $\displaystyle\int \frac{\sqrt{2ax - x^2}}{x^2}\,dx = -\frac{2\sqrt{2ax - x^2}}{x} - \cos^{-1}\left(\frac{a - x}{a}\right) + C$

**73.** $\displaystyle\int \frac{dx}{\sqrt{2ax - x^2}} = \cos^{-1}\left(\frac{a - x}{a}\right) + C$

**74.** $\displaystyle\int \frac{x\,dx}{\sqrt{2ax - x^2}} = -\sqrt{2ax - x^2} + a\cos^{-1}\left(\frac{a - x}{a}\right) + C$

**75.** $\displaystyle\int \frac{x^2\,dx}{\sqrt{2ax - x^2}} = -\frac{x + 3a}{2}\sqrt{2ax - x^2} + \frac{3a^2}{2}\cos^{-1}\left(\frac{a - x}{a}\right) + C$

**76.** $\displaystyle\int \frac{dx}{x\sqrt{2ax - x^2}} = -\frac{\sqrt{2ax - x^2}}{ax} + C$

**77.** $\displaystyle\int \frac{\sqrt{2ax - x^2}}{x^n}\,dx = \frac{(2ax - x^2)^{3/2}}{(3 - 2n)ax^n} + \frac{n - 3}{(2n - 3)a}\int \frac{\sqrt{2ax - x^2}}{x^{n-1}}\,dx, \quad n \neq \frac{3}{2}$

**78.** $\displaystyle\int \frac{x^n\,dx}{\sqrt{2ax - x^2}} = -\frac{x^{n-1}\sqrt{2ax - x^2}}{n} + \frac{a(2n - 1)}{n}\int \frac{x^{n-1}}{\sqrt{2ax - x^2}}\,dx$

**79.** $\displaystyle\int \frac{dx}{x^n\sqrt{2ax - x^2}} = \frac{\sqrt{2ax - x^2}}{a(1 - 2n)x^n} + \frac{n - 1}{(2n - 1)a}\int \frac{dx}{x^{n-1}\sqrt{2ax - x^2}}$

**80.** $\displaystyle\int \frac{dx}{(2ax - x^2)^{3/2}} = \frac{x - a}{a^2\sqrt{2ax - x^2}} + C$

**81.** $\displaystyle\int \frac{x\,dx}{(2ax - x^2)^{3/2}} = \frac{x}{a\sqrt{2ax - x^2}} + C$

# Integrals Containing Trigonometric Functions

**82.** $\displaystyle\int \sin^2 x \, dx = \frac{x}{2} - \frac{\sin(2x)}{4} + C$

**83.** $\displaystyle\int \cos^2 x \, dx = \frac{x}{2} + \frac{\sin(2x)}{4} + C$

**84.** $\displaystyle\int \tan^2 x \, dx = \tan x - x + C$

**85.** $\displaystyle\int \cot^2 x \, dx = -\cot x - x + C$

**86.** $\displaystyle\int \sec^3 x \, dx = \frac{1}{2}\sec x \tan x + \frac{1}{2}\ln|\sec x + \tan x| + C$

**87.** $\displaystyle\int \csc^3 x \, dx = -\frac{1}{2}\csc x \cot x + \frac{1}{2}\ln|\csc x - \cot x| + C$

**88.** $\displaystyle\int \sin^n x \, dx = -\frac{1}{n}\sin^{n-1} x \cos x + \frac{n-1}{n}\int \sin^{n-2} x \, dx$

**89.** $\displaystyle\int \cos^n x \, dx = \frac{1}{n}\cos^{n-1} x \sin x + \frac{n-1}{n}\int \cos^{n-2} x \, dx$

**90.** $\displaystyle\int \tan^n x \, dx = \frac{1}{n-1}\tan^{n-1} x - \int \tan^{n-2} x \, dx$

**91.** $\displaystyle\int \cot^n x \, dx = \frac{-1}{n-1}\cot^{n-1} x - \int \cot^{n-2} x \, dx$

**92.** $\displaystyle\int \sec^n x \, dx = \frac{1}{n-1}\tan x \sec^{n-2} x + \frac{n-2}{n-1}\int \sec^{n-2} x \, dx$

**93.** $\displaystyle\int \csc^n x \, dx = \frac{-1}{n-1}\cot x \csc^{n-2} x + \frac{n-2}{n-1}\int \csc^{n-2} x \, dx$

**94.** $\displaystyle\int \sin(mx)\sin(nx) \, dx = -\frac{\sin[(m+n)x]}{2(m+n)} + \frac{\sin[(m-n)x]}{2(m-n)} + C, \ m^2 \neq n^2$

**95.** $\displaystyle\int \cos(mx)\cos(nx) \, dx = \frac{\sin[(m+n)x]}{2(m+n)} + \frac{\sin[(m-n)x]}{2(m-n)} + C, \ m^2 \neq n^2$

**96.** $\displaystyle\int \sin(mx)\cos(nx) \, dx = -\frac{\cos[(m+n)x]}{2(m+n)} - \frac{\cos[(m-n)x]}{2(m-n)} + C, \ m^2 \neq n^2$

**97.** $\displaystyle\int x \sin x \, dx = \sin x - x \cos x + C$

**98.** $\displaystyle\int x \cos x \, dx = \cos x + x \sin x + C$

**99.** $\displaystyle\int x^2 \sin x \, dx = 2x \sin x + (2 - x^2)\cos x + C$

**100.** $\displaystyle\int x^2 \cos x \, dx = 2x \cos x + (x^2 - 2)\sin x + C$

**101.** $\displaystyle\int x^n \sin x \, dx = -x^n \cos x + n \int x^{n-1} \cos x \, dx$

**102.** $\displaystyle\int x^n \cos x \, dx = x^n \sin x - n \int x^{n-1} \sin x \, dx$

**103.** **(a)** $\displaystyle\int \sin^m x \cos^n x \, dx = -\frac{\sin^{m-1} x \cos^{n+1} x}{m+n} + \frac{m-1}{m+n}\int \sin^{m-2} x \cos^n x \, dx$

**(b)** $\displaystyle\int \sin^m x \cos^n x \, dx = \frac{\sin^{m+1} x \cos^{n-1} x}{m+n} + \frac{n-1}{m+n}\int \sin^m x \cos^{n-2} x \, dx$

(if $m = -n$ use formula 90 or 91.)

## Integrals Containing Inverse Trigonometric Functions

**104.** $\displaystyle\int \sin^{-1}x \, dx = x\sin^{-1}x + \sqrt{1-x^2} + C$

**105.** $\displaystyle\int \cos^{-1}x \, dx = x\cos^{-1}x - \sqrt{1-x^2} + C$

**106.** $\displaystyle\int \tan^{-1}x \, dx = x\tan^{-1}x - \frac{1}{2}\ln(1+x^2) + C$

**107.** $\displaystyle\int x\sin^{-1}x \, dx = \frac{2x^2-1}{4}\sin^{-1}x + \frac{x\sqrt{1-x^2}}{4} + C$

**108.** $\displaystyle\int x\cos^{-1}x \, dx = \frac{2x^2-1}{4}\cos^{-1}x - \frac{x\sqrt{1-x^2}}{4} + C$

**109.** $\displaystyle\int x\tan^{-1}x \, dx = \frac{x^2+1}{2}\tan^{-1}x - \frac{x}{2} + C$

**110.** $\displaystyle\int x^n \sin^{-1}x \, dx = \frac{1}{n+1}\left( x^{n+1}\sin^{-1}x - \int \frac{x^{n+1}dx}{\sqrt{1-x^2}} \right), \quad n \neq -1$

**111.** $\displaystyle\int x^n \cos^{-1}x \, dx = \frac{1}{n+1}\left( x^{n+1}\cos^{-1}x + \int \frac{x^{n+1}dx}{\sqrt{1-x^2}} \right), \quad n \neq -1$

**112.** $\displaystyle\int x^n \tan^{-1}x \, dx = \frac{1}{n+1}\left( x^{n+1}\tan^{-1}x - \int \frac{x^{n+1}dx}{1+x^2} \right), \quad n \neq -1$

## Integrals Containing Exponential and Logarithmic Functions

**113.** $\displaystyle\int xe^{ax} \, dx = \frac{1}{a^2}(ax-1)e^{ax} + C$

**114.** $\displaystyle\int x^n e^{ax} \, dx = \frac{1}{a}x^n e^{ax} - \frac{n}{a}\int x^{n-1}e^{ax} \, dx$

**115.** $\displaystyle\int \frac{e^x}{x^n} \, dx = -\frac{e^x}{(n-1)x^{n-1}} + \frac{1}{n-1}\int \frac{e^x}{x^{n-1}} \, dx$

**116.** $\displaystyle\int (\ln x)^n \, dx = x(\ln x)^n - n\int (\ln x)^{n-1} \, dx$

**117.** $\displaystyle\int x^n \ln x \, dx = \left( \frac{x^{n+1}}{n+1} \right)\left( \ln x - \frac{1}{n+1} \right) + C$

**118.** $\displaystyle\int \frac{(\ln x)^n}{x} \, dx = \frac{(\ln x)^{n+1}}{n+1} + C, \quad n \neq -1$

**119.** $\displaystyle\int \frac{dx}{x\ln x} = \ln|\ln x| + C$

**120.** $\displaystyle\int x^m (\ln x)^n \, dx = \frac{x^{m+1}(\ln x)^n}{m+1} - \frac{n}{m+1}\int x^m (\ln x)^{n-1} \, dx$

**121.** $\displaystyle\int \frac{x^m}{(\ln x)^n} \, dx = -\frac{x^{m+1}}{(n-1)(\ln x)^{n-1}} + \frac{m+1}{n-1}\int \frac{x^m}{(\ln x)^{n-1}} \, dx$

**122.** $\displaystyle\int e^{ax}\sin(bx) \, dx = \frac{e^{ax}}{a^2+b^2}[a\sin(bx) - b\cos(bx)] + C$

**123.** $\displaystyle\int e^{ax}\cos(bx) \, dx = \frac{e^{ax}}{a^2+b^2}[a\cos(bx) + b\sin(bx)] + C$

## Integrals Containing Hyperbolic Functions

**124.** $\displaystyle\int \tanh x \, dx = \ln\cosh x + C$

**125.** $\displaystyle\int \coth x \, dx = \ln|\sinh x| + C$

**126.** $\displaystyle\int \operatorname{sech} x \, dx = \tan^{-1}(\sinh x) + C$

**127.** $\displaystyle\int \operatorname{csch} x \, dx = \ln\left|\tanh \frac{x}{2}\right| + C$

**128.** $\displaystyle\int \sinh^2 x \, dx = \frac{\sinh(2x)}{4} - \frac{x}{2} + C$

**129.** $\displaystyle\int \cosh^2 x \, dx = \frac{\sinh(2x)}{4} + \frac{x}{2} + C$

**130.** $\displaystyle\int \tanh^2 x \, dx = x - \tanh x + C$

**131.** $\displaystyle\int \coth^2 x \, dx = x - \coth x + C$

**132.** $\displaystyle\int x\sinh x \, dx = x\cosh x - \sinh x + C$

**133.** $\displaystyle\int x\cosh x \, dx = x\sinh x - \cosh x + C$

**134.** $\displaystyle\int x^n \sinh x \, dx = x^n \cosh x - n\int x^{n-1}\cosh x \, dx$

**135.** $\displaystyle\int x^n \cosh x \, dx = x^n \sinh x - n\int x^{n-1}\sinh x \, dx$